Calculus with Analytic Geometry

Richard A. Silverman

Prentice-Hall, Inc., Englewood Cliffs, New Jersey 07632

Library of Congress Cataloging in Publication Data

Silverman, Richard A.
  Calculus with analytic geometry.

  Includes index.
  1. Calculus. 2. Geometry, Analytic. I. Title.
QA303.S55429 1985   515′.15   84-17813
ISBN 0-13-111634-7

Editorial/production supervision: Zita de Schauensee
Interior design: Caliber
Cover design: Maureen Eide
Illustrator: Eric G. Hieber
Manufacturing buyer: John Hall
Cover illustration: ICON-Communications/FPG

© 1985 by Prentice-Hall, Inc., Englewood Cliffs, New Jersey 07632

All rights reserved. No part of this book may be
reproduced, in any form or by any means,
without permission in writing from the publisher.

Printed in the United States of America

10 9 8 7 6 5 4 3 2 1

ISBN 0-13-111634-7 01

Prentice-Hall International, Inc., *London*
Prentice-Hall of Australia Pty. Limited, *Sydney*
Editora Prentice-Hall do Brasil, Ltda., *Rio de Janeiro*
Prentice-Hall Canada Inc., *Toronto*
Prentice-Hall of India Private Limited, *New Delhi*
Prentice-Hall of Japan, Inc., *Tokyo*
Prentice-Hall of Southeast Asia Pte. Ltd., *Singapore*
Whitehall Books Limited, *Wellington, New Zealand*

Contents

Preface *xi*

To the Student *xvi*

0 Precalculus *1*

0.1 Sets and Numbers 1
0.2 Inequalities and Laws of Exponents 7
0.3 Absolute Values and Intervals 14
0.4 Coordinates in the Plane 21
0.5 Straight Lines and Their Equations 26
Supplementary Problems 36

1 Functions and Limits *39*

1.1 The Function Concept 39
1.2 Operations on Functions; Graphs of Functions 44
1.3 Trigonometric Functions 52
1.4 More About Trigonometric Functions 61
1.5 The Limit Concept and Continuity 65
1.6 A Closer Look at Limits 71
1.7 Algebraic Operations on Limits 80
1.8 The Limit of a Composite Function 87
1.9 One-Sided Limits and Continuity 91
1.10 Properties of Continuous Functions 96
Supplementary Problems 102

2 Derivatives *105*

2.1 Velocity and Rates of Change; The Derivative Concept 105
2.2 The Tangent Line to a Curve 115
2.3 The Tangent Line Approximation and Differentials 122

v

  2.4 The Product and Quotient Rules 127
  2.5 The Chain Rule 133
  2.6 Higher Derivatives 140
  2.7 Implicit Differentiation 145
  2.8 Related Rates 150
  Supplementary Problems 157

3 Further Applications of Differentiation *159*

  3.1 The Mean Value Theorem 159
  3.2 Local Extrema and Monotonic Functions 166
  3.3 Concavity and Inflection Points 174
  3.4 Limits Involving Infinity; Indeterminate Forms 181
  3.5 Asymptotes and Vertical Tangents 190
  3.6 L'Hospital's Rule 196
  3.7 Optimization Problems 203
  3.8 Applications to Business and Economics (Optional) 214
  Supplementary Problems 223

4 Integrals *225*

  4.1 Sigma Notation 225
  4.2 Area under a Curve and the Definite Integral 229
  4.3 More About Area; The Mean Value Theorem for Integrals 238
  4.4 Antiderivatives and the Indefinite Integral 248
  4.5 The Fundamental Theorem of Calculus 256
  4.6 Integration of the Velocity Function; Differential Equations 261
  4.7 Newtonian Mechanics; Kinetic Energy and Work 266
  Supplementary Problems 275

5 Inverse Functions *278*

  5.1 The Inverse of a One-to-One Function 278
  5.2 The Derivative of an Inverse Function 286
  5.3 Inverse Trigonometric Functions 290
  Supplementary Problems 298

6 Logarithms and Exponentials *300*

  6.1 The Natural Logarithm 300
  6.2 Some Integrals Leading to Logarithms 307
  6.3 The Exponential Function; Exponentials to Any Base 312
  6.4 Logarithms to Any Base 319
  6.5 The General Power Function; More About Indeterminate Forms 324
  6.6 Separable Differential Equations; Exponential Growth and Decay 333
  6.7 Further Applications of Exponentials 341

6.8 Hyperbolic Functions 349
6.9 Inverse Hyperbolic Functions 357
Supplementary Problems 362

7 Methods of Integration 366

7.1 Integration by Substitution 367
7.2 Integration by Parts 373
7.3 Reduction Formulas 379
7.4 Trigonometric Integrals 382
7.5 Trigonometric and Hyperbolic Substitutions 388
7.6 Integration of Rational Functions; Partial Fractions 397
7.7 Rationalizing Substitutions (Optional) 407
7.8 Approximate Integration and Simpson's Rule 413
7.9 Improper Integrals 424
Supplementary Problems 433

8 Further Applications of Integration 436

8.1 Volume by the Method of Cross Sections 436
8.2 Volume of a Solid of Revolution 442
8.3 Curves in Parametric Form 450
8.4 Length of a Plane Curve 459
8.5 Area of a Surface of Revolution 466
8.6 More About Work 471
8.7 Fluid Pressure (Optional) 477
Supplementary Problems 483

9 Sequences and Series 487

9.1 Infinite Sequences 487
9.2 Infinite Series 497
9.3 Nonnegative Series; Comparison Tests and the Integral Test 508
9.4 Absolute and Conditional Convergence 517
9.5 The Ratio and Root Tests 525
9.6 Power Series 529
9.7 Differentiation and Integration of Power Series 537
9.8 Taylor's Theorem and Its Uses 548
9.9 Taylor and Maclaurin Series 557
9.10 Newton's Method 563
Supplementary Problems 571

10 Plane Analytic Geometry and Polar Coordinates 575

10.1 Translation and Rotation of Axes 576
10.2 Parabolas 581
10.3 Ellipses 587

Contents vii

- 10.4 Hyperbolas 595
- 10.5 Conic Sections 604
- 10.6 Second Degree Curves (Optional) 613
- 10.7 Polar Coordinates 620
- 10.8 Conics in Polar Coordinates 628
- 10.9 The Tangent Line to a Polar Curve 631
- 10.10 Area in Polar Coordinates; Length of a Polar Curve 635
  Supplementary Problems 642

11 Vectors in the Plane *647*

- 11.1 The Vector Concept; Operations on Vectors 647
- 11.2 The Dot Product 655
- 11.3 Vector Functions; Velocity and the Unit Tangent Vector 661
- 11.4 The Unit Normal Vector; Curvature and Acceleration 671
- 11.5 Applications to Mechanics 682
  Supplementary Problems 691

12 Vectors in Space and Solid Analytic Geometry *694*

- 12.1 Rectangular Coordinates in Space 694
- 12.2 From Vectors in the Plane to Vectors in Space 701
- 12.3 The Cross Product 707
- 12.4 Lines and Planes in Space 715
- 12.5 Space Curves and Orbital Motion 722
- 12.6 Quadric Surfaces 737
  Supplementary Problems 744

13 Partial Differentiation *747*

- 13.1 Functions of Several Variables 747
- 13.2 Limits and Continuity 754
- 13.3 Partial Derivatives 759
- 13.4 Differentiability and Differentials 765
- 13.5 The Chain Rule and Implicit Differentiation 771
- 13.6 The Tangent Plane to a Surface 779
- 13.7 The Directional Derivative and Gradient 783
- 13.8 Extrema of Functions of Several Variables 789
- 13.9 Lagrange Multipliers 798
  Supplementary Problems 806

14 Multiple Integration *810*

- 14.1 Double Integrals 810
- 14.2 Triple Integrals 824
- 14.3 The Center of Mass and Centroids 833

14.4 More About Centroids; The Theorems of Pappus   842
14.5 Moments of Inertia   850
14.6 Double Integrals in Polar Coordinates   855
14.7 Triple Integrals in Cylindrical Coordinates   862
14.8 Triple Integrals in Spherical Coordinates   867
Supplementary Problems   874

15 Line and Surface Integrals   *877*

15.1 Line Integrals   877
15.2 Path Independence and Gradient Fields   886
15.3 Surface Area and Surface Integrals   897
15.4 Green's Theorem; Change of Variables in Multiple Integrals   909
15.5 The Divergence Theorem   921
15.6 Stokes' Theorem   929
Supplementary Problems   938

16 Elementary Differential Equations   *941*

16.1 Exact Equations and Integrating Factors   942
16.2 First Order Linear Equations   946
16.3 Second Order Linear Equations with Constant Coefficients   951
16.4 Simple Harmonic Motion; Damped and Forced Oscillations   960
Supplementary Problems   968

Appendix   *A-1*

Mathematical Induction   A-1
Areas and Volumes   A-5
The Greek Alphabet   A-5
Numerical Tables   A-7

Answers and Hints to Odd-Numbered Problems   *A-16*

Index   *I-1*

Preface

A mainstream calculus book such as this one must in the first instance have sufficient richness of content to encompass not only the hypothetical "standard course," but also any number of variants that may be desired by different users. This requirement inevitably leads to a substantial book, containing much optional material that can be included or not, depending on the needs of the individual class or syllabus. The attribute just described can be summarized as *versatility*, and a perusal of the present book will quickly reveal that it is very versatile indeed.

But versatility by itself is not enough to carry the day, for once some subset of a calculus book has been selected as the course to be taught, whether over three semesters or several quarters, the book's usefulness then rests on the equally important attribute of *accessibility*. In writing an accessible book, that is, one which the average student finds intelligible, the author must face up to the fact that calculus is generally regarded as a difficult and demanding subject, at least the first time around. Thus students of calculus need all the help they can get, and it is the book's duty to give it to them. The minimalistic style of the advanced treatise has no place whatsoever in the teaching of elementary calculus. What is called for instead is a compassionate approach, in which precedence is given to the helpful hint, insightful comment and amplifying remark, rather than to the terse definition or flashy proof. Accordingly, I have spared no pains to enhance the accessibility of this text by adopting a tone of *patient explanation* throughout, especially at those points where experience shows that misunderstandings are most likely to occur.

So much for the book's general outlook and pedagogical philosophy. Now for a description of some of its important special features.

- The book contains 142 problem sets, one at the end of each section and an extra set at the end of each chapter. There are over 6000 problems in all. They are carefully graded, and range from routine drill exercises to problems that will challenge even the ablest students. A great many of the problems are of an applied character. Classical topics in physics and engineering are well represented, but there are also numerous problems originating in fields of more contemporary interest such as astrophysics, biomedical science and information theory. Some idea of the extent of these problems can be obtained by using the detailed index to look up subjects such as neutron star, critical mass, nutrient uptake and DNA, or names such as Einstein, Shannon and Keynes.

• In keeping with the abundance of problem material, the book is equipped with a very large number (843) of illustrative examples, most of which are in the problem-solution format. It is mainly by assimilating solved examples that the student prepares for the problems assigned as homework, and thus it is a disservice to the student to make all the examples easier than the harder problems. Therefore, in compiling the examples, I have seen to it that some of them (like Examples 4 and 5 of Section 3.7, to cite just two) would be regarded as challenging problems if their solutions were omitted.

• The book begins with an introductory Chapter 0, containing a review of those topics from college algebra which constitute the mathematical underpinnings of calculus. Because of its preliminary character this material, usually called precalculus, should be covered as rapidly as the degree of preparation of the class permits. The number 0 serves as a reminder not to linger over this prefatory chapter.

• Some books give an extended treatment of limits before mentioning continuity at all, but in my opinion it is better to discuss continuity early. Thus both limits and continuity are introduced in the same Section 1.5. Students have a tendency to confuse the limit L of a function f at a point a with the value of f at a, and a good way to forestall this confusion is to tell them that $L = f(a)$ if and only if f belongs to a special class of functions, said to be continuous at a.

• The rules governing algebraic operations on limits are proved in a way that avoids artifice and leads to relatively easy proofs (see Theorem 3 of Section 1.6 and Theorems 4 to 6 of Section 1.7). There is no immediate need for limits involving infinity, and hence this topic is deferred until Sections 3.4 and 3.5 where time allows it to be discussed in detail. The basic properties of continuous functions are presented in Section 1.10, which contains such extras as the bisection method and the interval mapping theorem.

• Chapters 2 and 3 are devoted to the derivative and its applications. The tangent line approximation and differentials are given a higher priority than usual (see Section 2.3). Every result concerning monotonicity and local extrema has an exact analogue for concavity and inflection points, and this parallelism is developed fully in Sections 3.2 and 3.3.

• L'Hospital's rule puts in an early appearance in Section 3.6, as an application of differentiation, and is used freely thereafter. This powerful tool for evaluating limits is simply too practical to be kept under wraps any longer than is absolutely necessary!

• The optional Section 3.8 caters to those with a special interest in the applications of calculus to business and economics. It is shown by concrete examples that the output level maximizing profit is in general different from that maximizing revenue or minimizing cost.

• The study of integration begins in Chapter 4, after a preliminary section on sigma notation. In presenting the definite integral I have adopted a Riemann sum approach from the very outset, in preference to the method of upper and lower sums favored by some authors. Antiderivatives and the indefinite integral are introduced in Section 4.4. The fundamental theorem of calculus is then proved with the help of the mean value theorem for integrals.

- Integration is put to immediate use in Sections 4.6 and 4.7 as a tool for solving differential equations. In particular, the key ideas of one-dimensional mechanics, including kinetic energy and work, emerge in the course of integrating Newton's second law of motion. Mechanical problems in the plane and in space are considered in Section 11.5 and in the part of Section 12.5 that deals with Kepler's laws. Some elementary topics in the mechanics of fluids are treated in Sections 8.6 and 8.7.

- The stockpile of functions at the student's disposal is brought up to size in Chapter 5 on inverse functions and in Chapter 6 on logarithms and exponentials. Separable differential equations are introduced in Section 6.6, and then used to solve a host of applied problems involving growth and decay. The last two sections of Chapter 6 are devoted to hyperbolic functions and their inverses.

- Chapter 7 is devoted to methods of integration, and includes sections on approximate integration and improper integrals. These methods encompass the ad hoc methods of Chapters 4 and 6; for example, the useful rule (iv) of Section 4.4 is just the form taken by the method of integration by substitution if the substitution is linear. Armed with the techniques of Chapter 7 and the concept of a parametric curve, presented in Section 8.3, the student is now in a position to calculate volumes of solids, lengths of curves, and areas of surfaces of revolution. Some books present this material earlier, at the expense of drastically curtailing the class of problems solvable with the means at hand.

- Both Chapter 9 on sequences and series and Chapter 10 on analytic geometry and polar coordinates are written on a very generous scale, allowing the instructor many options in the choice of topics to be covered. My preference is to treat sequences and series before analytic geometry and vectors, but some may prefer to reverse this order. Newton's method with a quadratic error estimate is given as Theorem 19 of Section 9.10 (whose proof is optional), but if the notion of the limit of a sequence is regarded as intuitively clear, this important topic can be treated much earlier as an application of the derivative. By the same token, almost all the material on conics in Sections 10.1 to 10.6 is "calculus-free" and could just as well be discussed at any stage of the course.

- For pedagogical reasons I have split the discussion of vectors into two parts, Chapter 11 on vectors in the plane and Chapter 12 on vectors in space. However those who prefer a full three-dimensional treatment from the outset can achieve this objective by judiciously interweaving Sections 11.1 to 11.5 and 12.1 to 12.6, beginning of course with Section 12.1 on three-dimensional rectangular coordinates.

- Chapters 13 to 15 are devoted to the differential and integral calculus of functions of several variables. Section 13.9 contains a detailed exposition of the important topic of Lagrange multipliers. The treatment of multiple integration is unified by introducing double integrals and triple integrals in consecutive sections. These integrals are defined without recourse to "inner partitions," in a way that eliminates the need for a preliminary treatment of rectangular regions of integration. The subject of surface area is deferred until Section 15.3 on surface integrals, as seems only natural. The coverage of centroids (of both regions and curves), the theorems of Pappus, and moments of inertia is particularly thorough, and can be abbreviated at the instructor's discretion. Chapter 15 on line and surface integrals culminates in a careful presentation of Green's theorem, the divergence theorem and Stokes' theorem, material often called vector calculus or the calculus of vector fields.

• Chapter 16 returns to the important subject of differential equations. After Section 16.1 on exact equations and integrating factors, the rest of the chapter is devoted to the key topic of linear equations. Section 16.4 shows how second order linear equations are used to investigate diverse oscillatory phenomena. A brief discussion of mathematical induction, with an accompanying problem set, is given in the Appendix.

A few words about format are now in order. Explicit page references are given throughout, this being by far the most convenient scheme of cross-referencing. Figures and theorems are numbered consecutively in each chapter, while formulas and examples are numbered consecutively in each section. Every theorem has a name. The end of a solution to a solved example is denoted by the symbol □, and the end of the proof of a theorem is denoted by ■. At the end of the book there is a short answer or hint for every odd-numbered problem without exception, including a graph whenever a graph is requested in the problem statement. There is a list of key terms and topics at the end of each chapter and sometimes other review lists as well. Two separate solutions manuals are available from the publisher: a student's manual containing detailed solutions to all the odd-numbered problems and an instructor's manual containing detailed solutions to all the even-numbered problems. Both manuals are profusely illustrated and unusually complete.

Acknowledgments

The author expresses his gratitude to the following reviewers and colleagues for a profusion of helpful suggestions, both general and specific, which played a key role in shaping this book:

Neil E. Berger, University of Illinois, Chicago
Peter Colwell, Iowa State University
Arthur Crummer, University of Florida
Milos Dostal, Stevens Institute of Technology
Daniel Drucker, Wayne State University
Leon Gerber, St. John's University
Richard Grassl, University of New Mexico
Gary Grimes, Mt. Hood Community College
Nathaniel Grossman, University of California, Los Angeles
Lance L. Littlejohn, University of Texas, San Antonio
Robert Martin, Tarrant County Junior College
David Minda, University of Cincinnati
Donald J. Newman, Temple University
Peter Nyikos, University of South Carolina
Chester Palmer, Auburn University
James V. Peters, C. W. Post Center of Long Island University
Richard Pollack, New York University
Harold B. Reiter, University of Maryland
Raymond C. Roan, Hughes Aircraft Company, Los Angeles
Frederick Solomon, State University of New York, Purchase

John Spellman, Southwest Texas State University
Grace Marmor Spruch, Rutgers University
Larry Spruch, New York University
Richard B. Thompson, University of Arizona
Peter A. Tomas, University of Texas
John D. Vargas, Mercy College
Kathy Vargas, Mercy College
Frank Wattenberg, University of Massachusetts

The above list is perforce incomplete, due to the fact that I do not know the identity of a number of other reviewers who saw parts of the manuscript at early stages of its development.

The book is much the better for having been subjected twice to the close scrutiny and searching analysis of Leon Gerber, once in manuscript and then in page proof. The book also benefited from Milos Dostal's mathematically alert reading of the galley proof.

I am especially grateful to Norton Starr, Amherst College, for providing the computer-generated graphics in Section 13.1. The illustrations for the book and the accompanying manuals were produced in close collaboration with the ever-obliging Eric Hieber.

No pains have been spared to guarantee the accuracy of the book, but stray errors have a way of getting through any filter, however fine. Needless to say, I am eager to eliminate residual misprints, and would be very happy to hear from readers who find any.

The assistance of the staff of Prentice-Hall in guiding this text through successive revisions and final production has been of vital importance. I am particularly indebted to Robert Sickles, executive editor, who served as both devil's advocate and sounding board; to his assistant Jo Marie Jacobs, who provided unstinting logistic support; to Eleanor Henshaw Hiatt, production manager, who cut one Gordian knot after another; and to Zita de Schauensee, production editor, who exhibited care and patience far beyond the call of duty.

No project of this magnitude can see the light of day without a source of constant encouragement. In my case that source has been and is Joan L. Silverman, truest friend.

New York, N.Y.   *Richard A. Silverman*

To the Student

Before beginning, a bit of advice: In working through this book, or any other mathematics text, you should always have a paper and pencil at hand. Be prepared to make a little calculation or sketch a rough figure at a moment's notice. Calculus cannot be learned without solving a lot of problems. In fact, this is really a workshop course, where instead of writing short stories, painting pictures or baking pies, *you must solve problems*. Do not be discouraged if you have to read the same passage several times before you grasp its full meaning. This is par for the course. Bear in mind that many of the key ideas of calculus are novel, and were introduced, somewhat reluctantly, only after it gradually dawned on mathematicians that they were actually indispensable. For example, one of the central concepts of calculus, namely the notion of a *limit*, has been fully understood for only a century or so, after having eluded philosophers and mathematicians for thousands of years. So don't be hard on yourself if these same concepts give you a little trouble the first time around. This is where the exercises come in, for it is only by solving them that you can master the concepts they illustrate.

    Another point: In studying calculus you will have to get used to a certain degree of generality, especially in connection with the function concept. Don't fight this generality. It is a blessing in disguise, often allowing the same method to be used in solving a great variety of problems. For example, calculus deals with *rates of change* in general, and not just special kinds of rates of change like the slope of a curve, the velocity of a particle, the current in an electric circuit, or the marginal cost of a commodity, to mention only a few. From the standpoint of calculus, there are often deep similarities between things that are seemingly unrelated. Thus a certain breadth of outlook is called for from the very outset.

    Calculus is the brainchild of two men of genius, Sir Isaac Newton (1642–1727) and Gottfried Wilhelm Leibniz (1646–1716), who appear to have discovered it independently. However, the historical roots of the subject are very deep, and can be traced back as far as the work of Archimedes of Syracuse (287–212 B.C.). In fact, modern scholarship has made it clear that by the second half of the seventeenth century, calculus was an idea whose time had come. Newton put it best when he wrote, "If I have seen farther than Descartes, it is because I have stood on the shoulders of giants."

0 Precalculus

This chapter deals with a number of topics in algebra and geometry which are prerequisites for the study of calculus, and thus are gathered together under the heading of "precalculus." Don't be lulled into a false sense of security by the fact that some of these topics may already be familiar to you, in one guise or another, from earlier courses. Instead go from familiarity to mastery, adding to your knowledge where necessary, so that your subsequent study of calculus itself will not be undermined by inadequate preparation.

0.1 Sets and Numbers

The language of sets is often helpful in simplifying mathematical discussions. However, guard against the tendency to overdo this language, which should be used sparingly and only when the occasion really calls for it.

A collection of objects of any kind is called a *set*, and the objects themselves are called *elements* or *members* of the set. Sets are often denoted by capital letters and their elements by lowercase letters. If x is an element of a set A, we write $x \in A$, where the symbol $\in$ is read "is an element of." Other ways of reading $x \in A$ are "x is a member of A," "x belongs to A," and "A contains x."

Example 1 The English alphabet is a set containing 26 elements, namely all the letters from a to z.

If every element of a set A is also an element of a set B, we write $A \subset B$, which reads "A is a *subset* of B." If A is a subset of B, but B is not a subset of A, we say that A is a *proper subset* of B. This means that B not only contains all the elements of A, but also contains one or more extra elements.

Example 2 Two proper subsets of the English alphabet are the set of vowels and the set of consonants.

One way of describing a set is to write its elements between curly brackets. Thus the set $\{a, b, c\}$ is made up of the elements a, b and c. Changing the order of the elements does *not* change the set. For example, the set $\{b, c, a\}$ is the same as $\{a, b, c\}$. Repeating an element does not change a set. Thus the set $\{a, a, b, c, c\}$ is the same as $\{a, b, c\}$.

Example 3 The set of all possible days of the month which are divisible by 7 is $\{7, 14, 21, 28\}$.

We can also describe a set by giving properties that uniquely determine its elements. Thus

$$\{1, -1\} = \{\text{all } x \text{ such that } x^2 = 1\} = \{x : x^2 = 1\},$$

where in the last expression the colon stands for the phrase "such that" and we omit the superfluous word "all."

Example 4 The set $\{x : x = x^2\}$ is the set of all numbers equal to their own squares. You can easily verify that this set contains only the two elements 0 and 1.

If a set has no elements at all, it is said to be *empty* and is denoted by the symbol $\varnothing$. For example, the set of pink elephants in the Bronx Zoo is empty, and so is the set of months with more than five Sundays. By convention, an empty set is considered to be a subset of every set.

We say that two sets A and B are *equal*, and we write $A = B$, if A and B have the same elements. Thus $\{x : x = x^2\} = \{0, 1\}$, as already noted in Example 4. If A is empty, we write $A = \varnothing$.

Example 5 Every number x is equal to itself, and hence $\{x : x \neq x\} = \varnothing$.

Notice that if $A \subset B$ and $B \subset A$ hold simultaneously, then every element of A is an element of B and every element of B is an element of A, so that $A = B$. In particular, all empty sets are equal, for if $\varnothing$ and $\varnothing'$ are both empty, then $\varnothing \subset \varnothing'$ and $\varnothing' \subset \varnothing$ (why?), so that $\varnothing = \varnothing'$. Thus we talk about "the" empty set.

Two sets A and B with no elements in common are said to be *disjoint*. This is not to be confused with the idea of *distinct* sets, where "distinct" is just another word for "unequal." Thus the sets $A = \{1, 2\}$ and $B = \{2, 3\}$ are distinct, but not disjoint since they share the element 2.

Numbers and Their Representation

Next we discuss numbers of various kinds, beginning with integers and rational numbers, and moving on to irrational numbers and real numbers. The set of all real numbers is the number system which is indispensable for the study of calculus.

Suppose we construct a horizontal straight line L through a point O, and imagine that L extends indefinitely in both directions. Selecting any unit of measurement, we make marks on L, both to the right and to the left of O, at distances of 1 unit, 2 units, 3 units, and so on. As in Figure 1, the marks to the right of O represent the *positive integers* $1, 2, 3, \ldots$ and those to the left of O represent the *negative integers* $-1, -2, -3, \ldots$ (Here the dots $\ldots$ mean "and so on forever.") The line L is called a *number line*; the point O is called the *origin* (of L), and corresponds to 0, the number *zero*, which is regarded as an integer which is neither positive nor negative. The direction

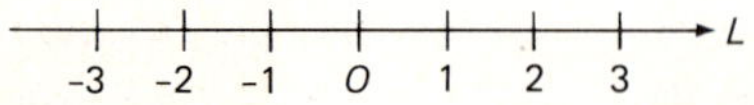

Figure 1

from negative to positive numbers along L is called the *positive* direction, and is indicated by an arrowhead, as in the figure.

If we add or multiply two positive integers, we always get another positive integer. This fact is summarized by saying that the set of positive integers is *closed* under the operations of addition and multiplication. For example, $2 + 3 = 5$ and $2 \cdot 3 = 6$, where 5 and 6 are positive integers. However, the set of positive integers is *not* closed under subtraction. For example, $2 - 3 = -1$, where -1 is not a positive integer, but a negative integer.

Following mathematical tradition, we use the letter Z to denote the set of all integers, whether positive, negative or zero. The set Z, unlike the set of positive integers (denoted by Z^+), is closed under subtraction, as well as under addition and multiplication. For example, $4 - 2 = 2$, $3 - 3 = 0$ and $6 - 11 = -5$, where the numbers 2, 0 and -5 are all integers.

An integer n is said to be an *even number* if $n = 2k$, where k is itself an integer, that is, if n is divisible by 2. On the other hand, an integer n is said to be an *odd* number if $n = 2k + 1$, where k is an integer. It is clear that every integer is either even or odd. For example, $-22 = 2(-11)$ is even, while $7 = 3(2) + 1$ is odd. Also $0 = 2(0)$ is even, while $-1 = 2(-1) + 1$ is odd (in these two cases the integers n and k happen to be the same).

Example 6 Show that the square of an even number is even, while the square of an odd number is odd.

Solution If n is even, then $n = 2k$ where k is an integer, and hence

$$n^2 = (2k)^2 = 4k^2 = 2(2k^2),$$

which is an even number, since it is of the form $2m$, where $m = 2k^2$ is an integer. On the other hand, if n is odd, then $n = 2k + 1$ (k an integer), and hence

$$n^2 = (2k + 1)^2 = 4k^2 + 4k + 1 = 2(2k^2 + 2k) + 1,$$

which is an odd number, since it is of the form $2m + 1$, where this time the integer m is $2k^2 + 2k$. □

Rational Numbers

The set Z of all integers is not closed under division. This means that the quotient of two integers is not always another integer. For example, $2 \div 3 = \frac{2}{3}$ and $-5 \div 4 = -\frac{5}{4}$, where $\frac{2}{3}$ and $-\frac{5}{4}$ are fractions, not integers. Of course, the quotient of two integers is *sometimes* an integer; thus $12 \div 3 = 4$ and $10 \div (-5) = -2$. However, to make division possible in general, we need a bigger set of numbers than Z. Thus we introduce *rational numbers*, that is, fractions of the form m/n, where m and n are both integers and the denominator n is not zero. Notice that every integer m, including zero, is a rational number, since $m/1 = m$. A rational number is said to be *in lowest terms* if its numerator and denominator have no integer factors in common. (The numbers 1 and -1 do not count as common factors here.) For example, $\frac{8}{12}$ is not in lowest terms, but dividing both the numerator and denominator by 4, we get the rational number $\frac{2}{3}$, which is in lowest terms.

Let Q denote the set of all rational numbers. Then the set Q is closed under the four basic arithmetical operations of addition, subtraction, multiplication and division, provided that we never divide by zero. To see why the operation of division by zero is excluded, we first observe that if $a/b = c$, then certainly $a = bc$. Suppose $b = 0$, which corresponds to dividing a by zero. Then

$$a = 0 \cdot c, \qquad (1)$$

which is impossible unless $a = 0$, since the expression on the right equals zero. But if $a = 0$, formula (1) becomes

$$0 = 0 \cdot c, \qquad (1')$$

which is true for *every* number c. Hence there is either no number c satisfying $a/0 = c$ if $a \neq 0$, or else any number c at all will do if $a = 0$. (For this reason, the expression $0/0$ is often called an *indeterminate form*.) Accordingly, division by zero is either impossible or indeterminate, and thus in any event meaningless.

Irrational Numbers

When plotted on a number line, the rational numbers occupy the points corresponding to the integers and many but *not all* of the points in between. In other words, there are points of the number line which do *not* correspond to rational numbers. To see this, suppose we construct a right triangle PAO with sides PA and AO of length 1, as in Figure 2(a). Then, by the familiar Pythagorean theorem, the side OP is of length $\sqrt{1^2 + 1^2} = \sqrt{2}$. We now place the side OP on the number line, as in Figure 2(b), with the point O coinciding with the origin of the line. Then the point P corresponds to the number $\sqrt{2}$. But, as discovered long ago, the number $\sqrt{2}$ cannot be rational, and therefore P is a point of the number line which does not correspond to a rational number.

By an *irrational number* we simply mean a number, like $\sqrt{2}$, which is not rational. The fact that $\sqrt{2}$ is irrational is proved indirectly, by showing that the assumption that $\sqrt{2}$ is rational leads to a contradiction.

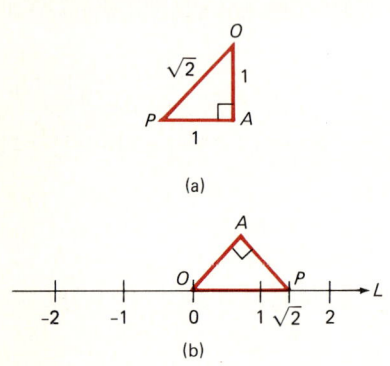

Figure 2

Optional
Thus suppose $\sqrt{2}$ is a rational number. Then

$$\sqrt{2} = \frac{m}{n},$$

where m and n are positive integers and it can be assumed that the fraction m/n has *already* been reduced to lowest terms. Squaring both sides of this equation, we get $2 = m^2/n^2$, or equivalently

$$m^2 = 2n^2.$$

Thus m^2 is divisible by 2, and hence is an even number. But then m itself must be even, since if m were odd, m^2 would also be odd, as shown in Example 6. Since m is even, we can write m in the form $m = 2k$, where k is a positive integer. Therefore $m^2 = 4k^2$, and when this formula is compared with $m^2 = 2n^2$, we find that $4k^2 = 2n^2$, or equivalently

$$n^2 = 2k^2.$$

Therefore n^2 is an even number, and hence so is n, for the reason just given in connection with m^2 and m.

Thus we have managed to show that *m* and *n* are both even numbers, that is, both *m* and *n* are exactly divisible by 2. But this contradicts the original assumption that the fraction *m/n* has been reduced to lowest terms. Since we run into a contradiction if we assume that $\sqrt{2}$ is a rational number, we must conclude that $\sqrt{2}$ is an *irrational* number. This fact was known to the ancient Greeks, who proved it in just the same way.

There are many other irrational numbers. For example, the square roots $\sqrt{3}$, $\sqrt{5}$ and $\sqrt{7}$ are all irrational, and so is π (lowercase Greek pi), the ratio of the circumference of any circle to its diameter.

Real Numbers

We now define a real number as the number, rational or irrational, corresponding to any point whatsoever of some number line. The set of all real numbers is called the *real number system,* and the number line is also known as the *real line.* From now on, when we use the word "number" without further qualification, we will always mean a real number. Figure 3 shows the approximate positions of several rational and irrational numbers on the real line.

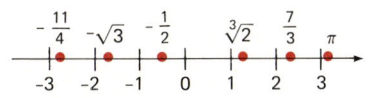

Figure 3

The connection between real numbers and decimals is noteworthy, and will give you further insight into the difference between rational and irrational numbers. If a rational number is expressed in decimal form, the decimal either terminates, as in

$$\frac{5}{8} = 0.625, \qquad (2)$$

or else has a block of digits that repeats itself endlessly, like the block 037 in

$$\frac{28}{27} = 1.037037037\ldots$$

We can abbreviate the last decimal by writing

$$\frac{28}{27} = 1.\overline{037},$$

where the bar covers the repeating block, in this case 037. However, if an irrational number is expressed in decimal form, the decimal neither terminates nor has an endlessly repeating block of digits. For example,

$$\sqrt{2} = 1.414213562373\ldots,$$

with no pattern at all in the unending string of digits. Actually we can also write the rational number given by (2) as

$$\frac{5}{8} = 0.625000\ldots = 0.625\overline{0}.$$

This shows that any terminating decimal can be regarded as a repeating decimal with an endless string of zeros after a certain decimal place. So when you get right down to it, there are only two kinds of decimals, repeating decimals which correspond to rational numbers, and nonrepeating decimals which correspond to irrational numbers. It seems intuitively clear that there are

Section 0.1 Sets and Numbers **5**

many more of the latter than of the former, and in fact this is true in a sense that can be made precise.

Example 7 Other decimal representations of rational numbers are

$$\frac{7}{16} = 0.4375, \qquad -\frac{93}{32} = -2.90625, \qquad \frac{1}{7} = 0.\overline{142857}$$

(the first two are terminating decimals), and other decimal representations of irrational numbers are

$$\sqrt{3} = 1.732050807568\ldots, \qquad \sqrt[3]{2} = 1.259921049894\ldots,$$
$$\pi = 3.141592653589\ldots$$

Remark It can be shown that the correspondence between decimals and real numbers is *one-to-one*, meaning that for each decimal there is a unique real number and for each real number there is a unique decimal. For this statement to be true, we must identify decimals having an endless string of nines after a certain decimal place with the "next highest" terminating decimal. Thus, for example,

$$0.14999\ldots = 0.14\bar{9} = 0.15.$$

The basic rules governing arithmetical operations on (real) numbers are given in the following list, where a, b and c denote arbitrary numbers.

$a + b = b + a,$	$ab = ba$	(*Commutative laws*)
$(a + b) + c = a + (b + c),$	$(ab)c = a(bc)$	(*Associative laws*)
$(a + b)c = ac + bc,$	$a(b + c) = ab + ac$	(*Distributive laws*)
$a + 0 = a,$	$a + (-a) = 0,$	
$1 \cdot a = a,$	$a\left(\dfrac{1}{a}\right) = 1 \quad (a \neq 0)$	
$0 \cdot a = 0,$	$(-1)a = -a,$	
$(-a)b = a(-b) = -ab,$	$(-a)(-b) = ab.$	

Some of these rules can be derived from others, but that is a technical matter. The *negative* of a, denoted by $-a$, is the number such that $a + (-a) = 0$. In particular, this implies $-(-a) = a$. Notice that $0 = -0$, since $0 + 0 = 0$, but no number other than 0 is its own negative. The *reciprocal* of a, denoted by $1/a$, is the number such that $a(1/a) = 1$; here we must insist that $a \neq 0$ to avoid dividing by zero. Subtraction of b is equivalent to addition of the negative of b:

$$a - b = a + (-b).$$

Similarly, division by b is equivalent to multiplication by the reciprocal of b:

$$\frac{a}{b} = a\left(\frac{1}{b}\right) \qquad (b \neq 0).$$

Example 8 With the help of the rules listed above, prove that if $ab = 0$, then at least one of the factors a and b is zero.

Solution If $a = 0$, the proof is complete. If $a \neq 0$, we multiply both sides of the equality $ab = 0$ by $1/a$. This gives

$$(ab)\left(\frac{1}{a}\right) = 0 \cdot \frac{1}{a} = 0.$$

But we also have

$$(ab)\left(\frac{1}{a}\right) = \left[a\left(\frac{1}{a}\right)\right]b = 1 \cdot b = b,$$

and hence $b = 0$. □

Problems

Write each of the following sets in another way, by listing its elements.

1. $\{x: x = -x\}$
2. $\{x: x^2 = 9\}$
3. $\{x: x + 7 = 13\}$
4. $\{x: x^2 - 5x + 6 = 0\}$
5. $\{x: x = x^3\}$
6. $\{x: x = x^4\}$

Let A be the set $\{1, 2, \{3\}, \{4, 5\}\}$. (Notice that two elements of A are sets themselves!) State whether the given formula is true or false, and explain your answer.

7. $1 \in A$
8. $3 \in A$
9. $\{2\} \in A$
10. $\{2\} \subset A$
11. $\{1, 2\} \in A$
12. $\{1, \{3\}\} \subset A$
13. $\{4, 5\} \subset A$
14. $A = \{1, 2, 3, 4, 5\}$
15. $\emptyset \subset \{A\}$

16. How many subsets does the set $A = \{a, b, c\}$ have? List them.

Express the given rational number as a terminating decimal.

17. $\dfrac{1}{4}$
18. $-\dfrac{1}{5}$
19. $\dfrac{1}{8}$
20. $\dfrac{1}{25}$
21. $\dfrac{1}{125}$
22. $\dfrac{1}{64}$

Express the given rational number as a repeating decimal.

23. $\dfrac{1}{3}$
24. $\dfrac{1}{6}$
25. $\dfrac{1}{9}$
26. $\dfrac{1}{11}$
27. $-\dfrac{1}{22}$
28. $\dfrac{1}{33}$

Reduce the given rational number to lowest terms.

29. $\dfrac{57}{133}$
30. $-\dfrac{161}{99}$
31. $\dfrac{81}{363}$
32. $\dfrac{91}{169}$

33. Show that if two numbers are rational, then so are their sum and product.
34. Without using decimals, show that the number $1 - \sqrt{2}$ is irrational.
35. Give an example of two irrational numbers with a rational sum.
36. Give an example of two (unequal) irrational numbers with a rational product.
37. How many digits long is the repeating block in the decimal representation of $\tfrac{1}{17}$?

0.2 Inequalities and Laws of Exponents

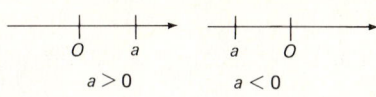

Figure 4

Let a be any *nonzero* real number. Then either a is *positive*, written $a > 0$, or a is *negative*, written $a < 0$. If we regard a as a point of a real line L directed from left to right, then $a > 0$ means that a lies to the right of the origin O, and $a < 0$ means that a lies to the left of O, as illustrated in Figure 4.

The Symbols > and <

Now let a and b be any pair of *unequal* numbers. Then either $a - b > 0$ or $a - b < 0$. In the first case we say that *a is greater than b*, written $a > b$, and in the second case we say that *a is less than b*, written $a < b$. Geometrically,

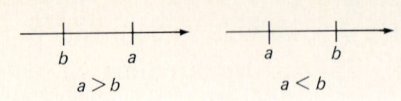

Figure 5

$a > b$ means that a lies to the right of b, and $a < b$ means that a lies to the left of b, as illustrated in Figure 5. Any formula of the type $a > b$ or $a < b$ is called an *inequality*.

Example 1 Some typical inequalities are
$$\sqrt{2} > 0, \quad -3 < 0, \quad -2 > -3, \quad 3 < \pi, \quad 4 > \pi,$$
$$-1 > -1000, \quad -\frac{1}{7} < -\frac{1}{8}, \quad -1 < -0.001, \quad 3.2999 > 3.2998.$$

The following facts about positive and negative numbers will be regarded as known and used freely.

(i) a is positive if and only if $-a$ is negative, that is, a and $-a$ have opposite signs.
(ii) If a is positive, so is the reciprocal $1/a$.
(iii) If a and b are positive, so are the sum $a + b$ and the product ab.
(iv) The product ab is positive if and only if a and b have the same sign, and negative if and only if a and b have opposite signs.

In particular, choosing $b = a$ in rule (iv), we find that $a^2 > 0$ for all nonzero a. This, together with the formula $0^2 = 0$, shows that *the square of any real number is always nonnegative*. (A nonnegative number is one that is either positive or zero.)

Inequalities are often combined. Thus $a < b < c$ has the same meaning as the pair of inequalities $a < b$ and $b < c$. Similarly, $a > b > c$ means the same as $a > b$ and $b > c$. For example, $3 < \pi < 4$ and $2 > \sqrt{2} > 1$.

Next we establish some easy theorems which give us the tools for manipulating inequalities algebraically.

Theorem 1 *(Addition rule for inequalities).* If $a < b$, then
$$a + c < b + c \tag{1}$$
for every number c.

Proof If $a < b$, or equivalently $b - a > 0$, then
$$(b + c) - (a + c) = (b - a) + (c - c) = b - a > 0,$$
which is equivalent to (1). ∎

Theorem 2 *(Multiplication rule for inequalities).* If $a < b$, then
$$ac < bc \tag{2}$$
for positive c, but
$$ac > bc \tag{2'}$$
for negative c.

Proof Let $a < b$, so that $b - a$ is positive. Then $(b - a)c$ has the same sign as c. Therefore $bc - ac > 0$ for positive c, which is equivalent to (2), while $bc - ac < 0$ for negative c, which is equivalent to (2'). ∎

8 Chapter 0 Precalculus

Notice the important respect in which Theorem 1 differs from Theorem 2. The former says that if the same number, positive or negative, is added to both sides of an inequality, then the result is another correct inequality. On the other hand, according to Theorem 2, the inequality resulting from multiplication of both sides of an inequality by the same nonzero number will be correct only if the number is positive, and in fact the direction of the inequality will be *reversed* if the number is negative.

> **Theorem 3** *(Transitivity of inequalities).* If $a < b$ and $b < c$, then $a < c$.

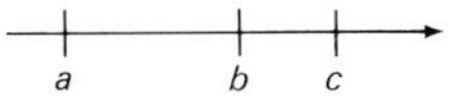

Proof Think of a, b and c as points of the real line. Then a lies to the left of b, and b lies to the left of c (see Figure 6). It follows that a lies to the left of c. ∎

Figure 6

It is easy to see that Theorems 1–3 remain true if we reverse the direction of all the inequalities, replacing every occurrence of the symbol $<$ by the opposite symbol $>$.

> **Theorem 4** *(Reciprocals rule for inequalities).* If $0 < a < b$ or $a < b < 0$, then
> $$\frac{1}{a} > \frac{1}{b}. \tag{3}$$

Proof If $0 < a < b$ or $a < b < 0$, then a and b have the same sign and $a < b$, so that $ab > 0$ and $b - a > 0$. Since the quotient of two positive numbers is positive, it follows that
$$\frac{1}{a} - \frac{1}{b} = \frac{b-a}{ab} > 0,$$
which is equivalent to (3). ∎

The following examples illustrate the use of these theorems. Notice that an inequality is preserved if both sides are *divided* by the same positive number, since division by a positive number c amounts to multiplication by the reciprocal $1/c$, which is also positive.

Example 2 Solve the inequality
$$3x - 5 < \pi, \tag{4}$$
that is, find all numbers x for which (4) is satisfied.

Solution By Theorem 1, $(3x - 5) + 5 < \pi + 5$ and hence
$$3x < \pi + 5.$$
Dividing both sides of this inequality by 3, we get
$$x < \frac{\pi + 5}{3}. \quad \square \tag{5}$$

Example 3 Solve the inequality
$$x^2 - x - 6 > 0.$$

Section 0.2 Inequalities and Laws of Exponents

Solution Factoring the expression on the left, we write (5) in the form
$$(x + 2)(x - 3) > 0.$$

We then make a table showing how the signs of both factors $x + 2$ and $x - 3$, and of their product $(x + 2)(x - 3)$, depend on x:

Condition on x	Sign of $x + 2$	Sign of $x - 3$	Sign of $(x + 2)(x - 3)$
$x < -2$	−	−	+
$-2 < x < 3$	+	−	−
$x > 3$	+	+	+

Here we are guided by the fact that an expression of the form $x - a$ is negative for $x < a$ and positive for $x > a$. It is immediately apparent from the table that the inequality $(x + 2)(x - 3) > 0$ holds if and only if $x < -2$ or $x > 3$, and hence the same is true of the original inequality (5). It is incorrect to combine the last two inequalities into the single formula $-2 > x > 3$. This is because there is no number x satisfying both inequalities $x < -2$ and $x > 3$ *simultaneously*. □

Example 4 Solve the inequality
$$\frac{x + 1}{2 - x} > 1. \tag{6}$$

Solution Subtracting 1 from both sides of (6), we get
$$\frac{x + 1}{2 - x} - 1 > 0,$$
or equivalently
$$\frac{(x + 1) - (2 - x)}{2 - x} = \frac{2x - 1}{2 - x} > 0.$$

We then make a table showing how the signs of the numerator $2x - 1$ and denominator $2 - x$, and of their quotient $(2x - 1)/(2 - x)$, depend on x:

Condition on x	Sign of $2x - 1$	Sign of $2 - x$	Sign of $\frac{2x - 1}{2 - x}$
$x < \frac{1}{2}$	−	+	−
$\frac{1}{2} < x < 2$	+	+	+
$x > 2$	+	−	−

It is clear from the table that the inequality $(2x - 1)/(2 - x) > 0$ holds if and only if $\frac{1}{2} < x < 2$. Hence the same is true of the original inequality (6). □

The Symbols ≥ and ≤

Given any two numbers a and b (not necessarily unequal), by $a \geq b$ we mean that *a is greater than or equal to b*. In other words, if $a \geq b$, then either $a > b$ or $a = b$. Similarly, $a \leq b$ means that *a is less than or equal to b*, that is, either $a < b$ or $a = b$. For example

$$\sqrt{3} \geq \sqrt{2} \geq 1^3 \geq 1, \qquad -3 \leq 0 \leq \frac{1-1}{2} \leq 1. \tag{7}$$

The symbols $\geq$ and $\leq$ give "weaker" estimates than the symbols $>$, $<$ and $=$, and in fact the "sharper" versions of (7) are

$$\sqrt{3} > \sqrt{2} > 1^3 = 1, \qquad -3 < 0 = \frac{1-1}{2} < 1.$$

Inequalities involving the symbols $>$ and $<$ are sometimes called *strict* inequalities, as opposed to those involving the symbols $\geq$ and $\leq$. If the two inequalities $a \leq b$ and $a \geq b$ hold simultaneously, then $a = b$. In fact, $a \leq b$ implies $a < b$ or $a = b$, but $a < b$ is incompatible with $a \geq b$.

The Symbols max and min

Given any n numbers $a_1, a_2, \ldots, a_n$, at least one of them, call it M, is greater than or equal to all the others. Similarly, at least one of the numbers, call it m, is less than or equal to all the others. The numbers M and m are known as the *maximum* and *minimum* of $a_1, a_2, \ldots, a_n$, denoted by $\max\{a_1, a_2, \ldots, a_n\}$ and $\min\{a_1, a_2, \ldots, a_n\}$, respectively. Thus, for example

$$\max\{-1, 2, 2\} = 2, \qquad \min\{-1, 1, -3\} = -3.$$

Clearly $M \geq m$, where $M = m$ if and only if $a_1 = a_2 = \cdots = a_n$.

Powers and Roots

If a is any number and n any positive integer, the product

$$\underbrace{a \cdot a \cdots a}_{n \text{ factors}}$$

is called the *n*th *power* of a, written a^n. One then defines

$$a^{-n} = \frac{1}{a^n}, \qquad a^0 = 1,$$

where in these formulas it is assumed that $a \neq 0$. Thus, for example,

$$4^2 = 16, \quad (-2)^2 = 4, \quad (-1)^3 = -1, \quad 3^3 = 27, \quad 10^4 = 10000,$$

$$2^6 = 64, \quad 2^{-2} = \frac{1}{4}, \quad 10^{-3} = \frac{1}{1000}, \quad \left(-\frac{1}{2}\right)^{-3} = -8, \quad \pi^0 = 1.$$

Integer powers of real numbers obey the well-known *laws of exponents*

$$a^m a^n = a^{m+n}, \qquad \frac{a^m}{a^n} = a^{m-n}, \qquad (a^m)^n = a^{mn}, \qquad (ab)^n = a^n b^n. \tag{8}$$

Here m and n are arbitrary integers, positive, negative or zero, while a and b are arbitrary numbers, except that $a \neq 0$ in the second formula. We must also insist that the number 0 should not be raised to a nonpositive power.

If $a \geq 0$, there is exactly one *nonnegative* number whose square equals a. This number, denoted by $\sqrt{a}$, is called the *square root* of a. Thus $\sqrt{0} = 0$ and $\sqrt{4} = 2$, but $\sqrt{4} = -2$ is wrong even though $(-2)^2 = 4$, since $\sqrt{4}$ is nonnegative by definition. The square root of a negative number cannot be a real number, because the square of any real number is always nonnegative.

Thus, for example, $\sqrt{-4}$ is not a real number. By introducing numbers of a more general kind, called *complex numbers*, it does become possible to take square roots of negative numbers. However, complex numbers are never used in this book, and hence the square root of a negative number will be regarded as undefined.

More generally, if $a \geq 0$ and n is any positive integer, there is exactly one *nonnegative* number whose nth power equals a, called the *nth root* of a and written $\sqrt[n]{a}$. For example,

$$\sqrt[3]{27} = 3, \qquad \sqrt[4]{0} = 0, \qquad \sqrt[5]{\frac{1}{32}} = \frac{1}{2}, \qquad \sqrt[6]{64} = 2.$$

In particular, $\sqrt[2]{a} = \sqrt{a}$, but one always writes just $\sqrt{a}$. With this definition of the nth root, we have the familiar rules

$$\sqrt[n]{a^n} = a, \qquad \sqrt[n]{ab} = \sqrt[n]{a}\sqrt[n]{b}, \qquad \sqrt[n]{\frac{a}{b}} = \frac{\sqrt[n]{a}}{\sqrt[n]{b}}, \qquad \sqrt[m]{\sqrt[n]{a}} = \sqrt[mn]{a}$$

Here m and n are arbitrary positive integers, while a and b are arbitrary nonnegative numbers, except that $b \neq 0$ in the third formula.

If n is odd, we can also define $\sqrt[n]{a}$ for negative values of a, as the unique (negative) number whose nth power is a. For example, $\sqrt[3]{-1} = -1$ and $\sqrt[5]{-32} = -2$. However, if n is even, $\sqrt[n]{a}$ is undefined for negative a, just as in the case of the square root. This is because if n is even, then $n = 2k$ where k is an integer, so that $a^n = (a^k)^2 \geq 0$.

Example 5 Show that if $0 < a < b$, then $\sqrt{a} < \sqrt{b}$.

Solution We have $b - a > 0$, or equivalently $B^2 - A^2 > 0$ in terms of the square roots $A = \sqrt{a}$, $B = \sqrt{b}$. Factoring $B^2 - A^2$, we find that

$$B^2 - A^2 = (B - A)(B + A) > 0. \qquad (9)$$

The factor $B + A$ is positive, since it is the sum of two positive numbers. Therefore (9) implies $B - A > 0$, that is, $A < B$ which is equivalent to $\sqrt{a} < \sqrt{b}$. □

Example 6 More generally, show that if $0 < a < b$ and n is any positive integer, then $\sqrt[n]{a} < \sqrt[n]{b}$.

Solution Again $b - a > 0$, or equivalently $B^n - A^n > 0$ in terms of the nth roots $A = \sqrt[n]{a}$, $B = \sqrt[n]{b}$. Using a formula from college algebra to factor $B^n - A^n$, we find that

$$B^n - A^n = (B - A)(B^{n-1} + B^{n-2}A + \cdots + BA^{n-2} + A^{n-1}) > 0. \qquad (9')$$

The factor $B^{n-1} + B^{n-2}A + \cdots + BA^{n-2} + A^{n-1}$ is positive, since it is the sum of n positive numbers. Therefore (9') implies $B - A > 0$, that is, $A < B$ or equivalently $\sqrt[n]{a} < \sqrt[n]{b}$. □

Finally we discuss the meaning of the expression a^r, where r is a rational number, that is, a number of the form m/n where m and n are integers (and $n \neq 0$). There is no loss of generality in assuming that n is positive, since if r is negative, we can always attribute this to m being negative. For simplicity, we will also assume that a is positive, so that the nth root of a is always

defined, and that m/n is in lowest terms. The definition of $a^{m/n}$ is then
$$a^{m/n} = \sqrt[n]{a^m},$$
or equivalently
$$a^{m/n} = (\sqrt[n]{a})^m$$
(notice that $a^{1/n} = \sqrt[n]{a}$). For example,
$$8^{2/3} = \sqrt[3]{8^2} = \sqrt[3]{64} = 4, \qquad 32^{-3/5} = (\sqrt[5]{32})^{-3} = 2^{-3} = \frac{1}{8},$$
and $0^r = 0$ if r is positive. Since
$$a^{-m/n} = \sqrt[n]{a^{-m}} = \sqrt[n]{\frac{1}{a^m}} = \frac{1}{\sqrt[n]{a^m}} = \frac{1}{a^{m/n}},$$
we have $a^{-r} = 1/a^r$, and it is not much harder to show that
$$a^r a^s = a^{r+s}, \qquad \frac{a^r}{a^s} = a^{r-s}, \qquad (a^r)^s = a^{rs}, \qquad (ab)^r = a^r b^r \qquad (10)$$
(the details are given in any college algebra text). Here r and s are arbitrary rational numbers, while a and b are arbitrary positive real numbers. If r and s are integers, and if we drop the requirement that a and b be positive, the *laws of exponents* (10) reduce to the corresponding laws for integer powers.

Example 7 Simplify $2^{1/6} \cdot 8^{1/9}$.

Solution Using two of the formulas (10), we find that
$$2^{1/6} \cdot 8^{1/9} = 2^{1/6} \cdot (2^3)^{1/9} = 2^{1/6} \cdot 2^{3/9}$$
$$= 2^{1/6} \cdot 2^{1/3} = 2^{(1/6)+(1/3)} = 2^{1/2} = \sqrt{2}. \qquad \square$$

Problems

Show that
1. If $a < b$, then $-a > -b$
2. If $a < b$, then $c - b < c - a$
3. If $a < b$ and $c < d$, then $a + c < b + d$
4. If $0 < a < b$ and $0 < c < d$, then $0 < ac < bd$
5. If $0 < a < 1$, then $\sqrt[n]{a} > a$ for every integer $n \geq 2$

Without making numerical calculations, determine which number is larger.

6. $\dfrac{1}{\pi}$ or $\dfrac{1}{3}$
7. $\dfrac{1}{\sqrt{2}}$ or $\dfrac{1}{2}$
8. $94 \cdot 98$ or 96^2
9. $23 \cdot 24 \cdot 26 \cdot 27$ or 25^4
10. $\sqrt{3} + \sqrt{5}$ or $\sqrt{2} + \sqrt{6}$
11. $\sqrt{7} - \sqrt{3}$ or $\sqrt{6} - 2$

12. Let $p = m/n$ and $p' = m'/n'$ be two different rational numbers, written with positive denominators (this is always possible). Show that $p < p'$ is equivalent to $mn' < m'n$, while $p > p'$ is equivalent to $mn' > m'n$.

Without dividing, use the preceding problem to determine which number is larger.

13. $\dfrac{33}{10}$ or $\dfrac{10}{3}$
14. $\dfrac{46}{25}$ or $\dfrac{11}{6}$
15. $\dfrac{7}{19}$ or $\dfrac{18}{49}$
16. $-\dfrac{10}{3}$ or $-\dfrac{167}{50}$

17. What does $a^2 + b^2 = 0$ imply about a and b?
18. Verify that
$$\frac{4}{7} < \frac{1}{4} + \frac{1}{5} + \frac{1}{6} + \frac{1}{7} < 1.$$

19. The *geometric mean* of two positive numbers x and y is defined by $g = \sqrt{xy}$ and the *arithmetic mean* (or *average*) by $a = \frac{1}{2}(x+y)$. Show that $g < a$ unless $x = y$, in which case $g = a$.

20. Use the preceding problem to show that of all rectangles with a given perimeter p, the square has the largest area. What is this area?

21. If $2 \le a \le 4$ and $4 \le b \le 6$, what can be said about the size of $\dfrac{1}{a} + \dfrac{1}{b}$?

Find all x for which
22. $2 < 4x - 5 < 7$
23. $(x-1)(x+1) \le (x+1)(x+2)$
24. $x^2 < 5x - 6$
25. $\dfrac{x}{x+2} > 0$
26. $\dfrac{x+1}{x+2} < 1$
27. $(x+1)(x+2)(x+3) > 0$
28. $(x-1)^2 x(x+1) > 0$
29. $(x-1)(x+1)(x^2+1) < 0$

Find
30. $\max\{-2, (-2)^2, (-2)^3\}$
31. $\min\{-1, (-1)^2, (-1)^3\}$
32. $\min\{-\sqrt{2}, -\sqrt{3}, -2\}$
33. $\max\{1, 2, \tfrac{4}{2}, (\sqrt{2})^2\}$

Write the given expression without negative exponents.
34. $\left(\dfrac{x^2}{yz}\right)^{-3}$
35. $\dfrac{a^3 b^2 c}{a^{-1} b^{-2} c^{-3}}$
36. $\dfrac{a^{-1} b^{-2} c^{-3}}{a^3 b^2 c}$

Write the given expression as a power of 2.
37. $\left(\dfrac{1}{4}\right)^7$
38. $\left(\dfrac{1}{8}\right)^{-5}$
39. $2 \cdot 4 \cdot 8 \cdot 16$

Simplify the given expression.
40. $\sqrt{4 \cdot 16 \cdot 36}$
41. $\sqrt[3]{64 a^9}$
42. $\sqrt[3]{1.728 \times 10^6}$
43. $\sqrt[4]{81 a^4}$
44. $\sqrt[5]{-0.00001}$
45. $\sqrt{4 - 2\sqrt{3}}$
46. $\sqrt{\sqrt{12} - \sqrt{3}}$
47. $\dfrac{\sqrt{2}+1}{\sqrt{2}-1}$
48. $(8a^6)^{4/3}$
49. $(125 a^3)^{-2/3}$
50. $(32)^{-4/5}(16)^{5/4}$
51. $2^{1/6} \cdot 2^{1/3} \cdot 2^{1/2}$
52. $(0.0001)^{3/2}$
53. $\left(\dfrac{81}{625}\right)^{-3/4}$

54. Music students will recognize the sequence C, C♯, D, D♯, E, F, F♯, G, G♯, A, A♯, B, C as a chromatic scale of C, consisting of 12 consecutive semitones. The second C is an octave higher than the first C, and has a frequency (pitch) twice as great. Suppose the frequency of each note of the scale is a fixed number r times the frequency of the preceding note, as in the system of *equal temperament* used to tune keyboard instruments. Find r. What is the ratio of the frequencies of two consecutive whole tones (like C and D)?

55. Show that if $p < q$ and $r = \tfrac{1}{2}(p+q)$, then $p < r < q$, where r is rational if p and q are.

56. Use the preceding problem to show that there is no largest rational (or real) number less than 1, and no smallest rational (or real) number greater than 0.

0.3 Absolute Value and Intervals

The Symbol | |

By the *absolute value* of the number a, written $|a|$ with a pair of vertical bars, we mean the number equal to a itself if a is nonnegative and to $-a$ if a is negative. In other words,

$$|a| = \begin{cases} a & \text{if } a \ge 0, \\ -a & \text{if } a < 0, \end{cases} \tag{1}$$

where the brace is used to combine two formulas into one. Thus, for example,

$$|0| = 0, \quad |-1.45| = -(-1.45) = 1.45,$$
$$|(-3)^2| = |9| = 9, \quad |(-3)^3| = |-27| = -(-27) = 27,$$
$$|x-2| = -(x-2) = 2 - x \text{ if } x < 2.$$

It is easy to see that

$$|a| = \sqrt{a^2}. \tag{2}$$

In fact, if $a \ge 0$, then $|a| = a$ and $\sqrt{a^2} = a$, while if $a < 0$, then $|a| = -a$ and $\sqrt{a^2} = \sqrt{(-a)(-a)} = -a$. (Why is $\sqrt{a^2} = a$ wrong if a is negative?) Squaring

14   Chapter 0   Precalculus

both sides of (2), we get

$$|a|^2 = a^2. \qquad (2')$$

Thus the square of the absolute value of a number is the same as the square of the number itself. It follows from (2) that

$$|ab| = \sqrt{(ab)^2} = \sqrt{a^2 b^2} = \sqrt{a^2}\sqrt{b^2},$$

or equivalently

$$|ab| = |a|\,|b|. \qquad (3)$$

This important formula is valid for arbitrary numbers a and b. A similar argument shows that

$$\left|\frac{a}{b}\right| = \frac{|a|}{|b|} \qquad (b \neq 0),$$

and is left as an exercise. Choosing $b = -1$ in (3), we find that

$$|-a| = |a|.$$

It is an immediate consequence of (1) that

$$a = \begin{cases} |a| & \text{if } a \geq 0, \\ -|a| & \text{if } a < 0. \end{cases} \qquad (1')$$

Therefore

$$-|a| \leq a \leq |a|$$

for every number a.

The Triangle Inequality

The next theorem is of great importance in calculus.

> **Theorem 5** *(Triangle inequality)*. *The inequality*
> $$|a + b| \leq |a| + |b| \qquad (4)$$
> *holds for arbitrary numbers a and b.*

Proof It follows from the inequality

$$ab \leq |ab| = |a|\,|b|$$

that

$$|a + b|^2 = (a + b)^2 = a^2 + 2ab + b^2 \leq a^2 + 2|a|\,|b| + b^2 = |a|^2 + 2|a|\,|b| + |b|^2.$$

Therefore

$$|a + b|^2 \leq (|a| + |b|)^2,$$

and taking square roots of both sides (with the help of Example 5, page 12), we obtain (4). ∎

The triangle inequality (4) reduces to the equality $|a + b| = |a| + |b|$ if a and b have the same sign, or if at least one of the numbers a and b is zero, since these are the conditions under which $ab = |ab|$. However, if a and b have opposite signs, the triangle inequality becomes the strict inequality

$|a + b| < |a| + |b|$, since then $ab < |ab|$. For example,
$$4 = |3 - 7| < |3| + |-7| = 10.$$

The important point is that (4) always holds, regardless of the signs of a and b.

Replacing a by $a - b$ in (4) gives
$$|a| = |(a - b) + b| \leq |a - b| + |b|,$$
which implies
$$|a - b| \geq |a| - |b|.$$

Similarly, replacing b by $b - a$ in (4) gives
$$|b| = |a + (b - a)| \leq |a| + |b - a| = |a| + |a - b|,$$
which implies
$$|a - b| \geq |b| - |a|.$$

But one of the two numbers $|a| - |b|$ and $|b| - |a|$ is the absolute value of $|a| - |b|$, and therefore
$$|a - b| \geq ||a| - |b||. \tag{5}$$

We will have occasional need of this inequality.

Coordinates and Distance Between Points

By the *coordinate* of a point P on the real line we mean the real number corresponding to P. Let P_1 and P_2 be two points on the real line, with coordinates x_1 and x_2, and let $|P_1 P_2|$ denote the distance between P_1 and P_2, or equivalently the length of the line segment $P_1 P_2$. Then $|P_1 P_2|$ is just the difference between the larger and the smaller of the two numbers x_1 and x_2, as illustrated in Figure 7. (The points P_1 and P_2 coincide if $x_1 = x_2$.) More exactly,

$$|P_1 P_2| = \begin{cases} x_1 - x_2 & \text{if } x_1 \geq x_2, \\ x_2 - x_1 & \text{if } x_1 < x_2, \end{cases}$$

which is equivalent to

$$|P_1 P_2| = \begin{cases} x_1 - x_2 & \text{if } x_1 - x_2 \geq 0, \\ x_2 - x_1 & \text{if } x_1 - x_2 < 0. \end{cases}$$

This looks complicated, but in terms of the absolute value it reduces to
$$|P_1 P_2| = |x_1 - x_2|.$$

It was in anticipation of this result that we used vertical bars in our notation for the distance between P_1 and P_2. Notice that $|P_2 P_1| = |x_2 - x_1| = |x_1 - x_2|$ by a property of the absolute value, and therefore $|P_1 P_2| = |P_2 P_1|$, as is to be expected on geometric grounds.

Since $|x| = |x - 0|$, the absolute value of x is just the distance between the point x and the origin of the real line. By "the point x" we mean of course the point with coordinate x. Similarly, $|x - 3|$ is the distance between the point x and the point 3, while $|x + 3| = |x - (-3)|$ is the distance between x

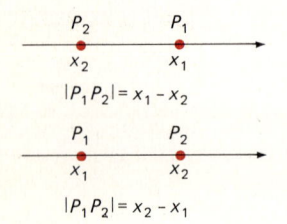

Figure 7

16 Chapter 0 Precalculus

and the point -3. Make sure you understand why $+3$ must be thought of as $-(-3)$ in order to interpret $|x + 3|$ as a distance.

The inequality
$$|x| < a \quad (a > 0)$$
says that the distance between the point x and the origin is less than a. The set of all x for which this is true is just the set of all x such that
$$-a < x < a.$$
Similarly, $|x| \le a$ if and only if $-a \le x \le a$. As an exercise, show that $|x| > a$ ($a > 0$) if and only if $x > a$ or $x < -a$, and $|x| \ge a$ if and only if $x \ge a$ or $x \le -a$.

Example 1 Find all x for which $|x^2 - 2| \le 7$.

Solution This double inequality is equivalent to $-7 \le x^2 - 2 \le 7$, which in turn holds if and only if
$$-5 \le x^2 \le 9. \tag{6}$$
The first inequality in (6) is automatically satisfied, since x^2 is nonnegative for all x, while the second can be written as $|x|^2 \le 9$ or equivalently $|x| \le 3$. Therefore x satisfies the given inequality $|x^2 - 2| \le 7$ if and only if
$$-3 \le x \le 3. \quad \square$$

Example 2 Solve the equation
$$|x| + |x - 1| = 2. \tag{7}$$

Solution A geometric approach works well here. Equation (7) says that the distance between the point x and the origin plus the distance between the point x and the point 1 is equal to 2. Imagine a string of length 2 with its ends tied to the points 0 and 1. Then the loop of string will become taut if we pull it to the point $\frac{3}{2}$, one half unit to the right of 1 [see Figure 8(a)], or to the point $-\frac{1}{2}$, one half unit to the left of 0 [see Figure 8(b)]. In other words, equation (7) has the two solutions $x = \frac{3}{2}$ and $x = -\frac{1}{2}$. $\quad \square$

Example 3 Solve equation (7) *algebraically*.

Solution If
$$x \ge 1,$$
then x and $x - 1$ are both nonnegative, so that (7) reduces to the equation $x + (x - 1) = 2$ or $2x = 3$, with solution $x = \frac{3}{2}$. If
$$0 \le x < 1,$$
then x is nonnegative and $x - 1$ is negative. In this case (7) becomes $x - (x - 1) = 2$ or equivalently $1 = 2$, which is patently false; therefore (7) has no solutions for $0 \le x < 1$. Finally, if
$$x < 0,$$
then x and $x - 1$ are both negative and (7) reduces to the equation $-x - (x - 1) = 2$ or $-2x = 1$, with solution $x = -\frac{1}{2}$. Having considered all the possibilities, we conclude that equation (7) has just the two solutions $x = \frac{3}{2}$ and $x = -\frac{1}{2}$. $\quad \square$

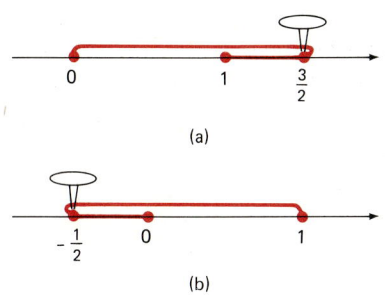

Figure 8

Section 0.3 Absolute Value and Intervals

Types of Intervals

Let a and b be two points of the real line such that $a < b$. Then the set of all points between a and b is called an *interval*. Here we are temporarily vague about whether or not the *endpoints* a and b themselves belong to the interval. Actually there are four possibilities:

(i) The set $\{x\colon a \le x \le b\}$ of all points x between a and b *including* the endpoints a and b is called a *closed interval* and is denoted by $[a, b]$, where we use two *square brackets*.

(ii) The set $\{x\colon a < x < b\}$ of all points x between a and b *excluding* the endpoints a and b is called an *open interval* and is denoted by (a, b), where we use two *parentheses*.

(iii) The set $\{x\colon a \le x < b\}$ of all points x between a and b *including* the left endpoint a and *excluding* the right endpoint b is called a *half-open interval* and is denoted by $[a, b)$, where we use a square bracket on the left and a parenthesis on the right.

(iv) The set $\{x\colon a < x \le b\}$ of all points between a and b *excluding* the left endpoint a and *including* the right endpoint b is also called a *half-open interval*, this time denoted by $(a, b]$, where we use a parenthesis on the left and a square bracket on the right.

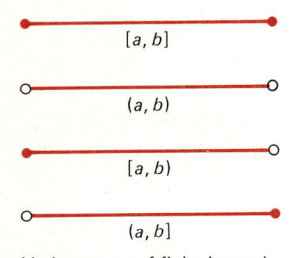

All four intervals $[a, b]$, (a, b), $[a, b)$ and $(a, b]$ have the same endpoints a and b, and hence are assigned the same length $b - a$ (to see why, think of a point as having zero length). The geometric meaning of the various kinds of intervals is shown in Figure 9, where included endpoints are indicated by solid dots and excluded endpoints by hollow dots. By an *interior point* of any of these intervals we mean a point other than an endpoint, that is, a point of the *open* interval (a, b).

Throughout the scientific literature you will find phrases like "the interval $1 \le x \le 4$" or "the interval $-2 < y < 5$," where what is really meant is the interval $\{x\colon 1 \le x \le 4\}$ or the interval $\{y\colon -2 < y < 5\}$. After all, an interval is a set, not a double inequality. However, once this distinction has been made, there is no harm done in using the "bare inequality" notation for intervals, and we will do so whenever it suits our convenience.

Example 4 Show that any interval with endpoints a and b has the midpoint $\frac{1}{2}(a + b)$.

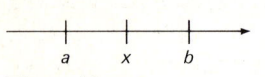

Solution Let x be the midpoint of the interval, as in Figure 10, and suppose $a < b$. Then $a < x < b$ and
$$x - a = b - x,$$
since x is equidistant from a and b. Solving this equation for x, we find that $x = \frac{1}{2}(a + b)$. The same result is obtained if $b < a$ (why?). □

Example 5 Find the set I of all points whose distance from the point 4 is less than 0.1.

Solution Clearly
$$I = \{x\colon |x - 4| < 0.1\} = \{x\colon -0.1 < x - 4 < 0.1\},$$
so that
$$I = \{x\colon 4 - 0.1 < x < 4 + 0.1\} = \{x\colon 3.9 < x < 4.1\}.$$

Therefore I is the open interval $(3.9, 4.1)$. □

[a, b]

(a, b)

[a, b)

(a, b]

Various types of finite intervals
Figure 9

Figure 10

18   Chapter 0   Precalculus

Neighborhoods

Generalizing Example 5, let I be the set of all points whose distance from a fixed point a is less than a given number $\delta > 0$. (The use of δ, lowercase Greek delta, is traditional in this context.) Then

$$I = \{x: |x - a| < \delta\} = \{x: -\delta < x - a < \delta\},$$

so that

$$I = \{x: a - \delta < x < a + \delta\}.$$

Therefore I is the open interval $(a - \delta, a + \delta)$ of length 2δ with midpoint a, as indicated schematically in Figure 11(a). An interval of this kind is called a *neighborhood* of a, or more exactly, the *δ-neighborhood* of a.

Since the inequality $|x - a| < \delta$ defines the δ-neighborhood of a, the double inequality

$$0 < |x - a| < \delta \tag{8}$$

defines the same neighborhood with a single missing point, namely the midpoint a itself [see Figure 11(b)]. In fact, (8) holds if and only if

$$a - \delta < x < a \quad \text{or} \quad a < x < a + \delta,$$

that is, the points x satisfying (8) make up the *pair* of open intervals

$$(a - \delta, a) \quad \text{and} \quad (a, a + \delta).$$

The technical name for the set of points defined by (8) is the *deleted δ-neighborhood* of a. Thus, for example, the deleted 0.01-neighborhood of the point 2 is the open interval (1.99, 2.01) with the point 2 itself missing, or equivalently the pair of open intervals (1.99, 2) and (2, 2.01). We could just as well have written (8) in the form

$$0 \neq |x - a| < \delta, \tag{8'}$$

which makes the fact that a is deleted even clearer, but (8) looks nicer and is standard.

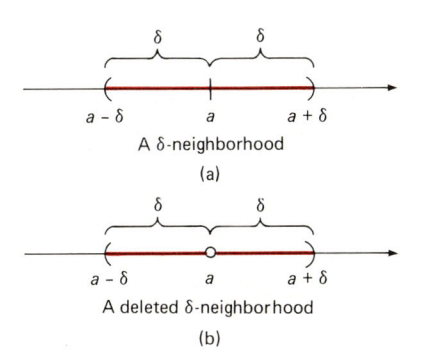

Figure 11

A δ-neighborhood (a)

A deleted δ-neighborhood (b)

Infinite Intervals

Each of the intervals considered so far is *finite* or *bounded*, which means that it has a definite length. We can also consider *infinite* or *unbounded* intervals, that is, intervals which "go on forever" in one or both directions along the real line, and therefore cannot be assigned any length at all. (You might say, though, that such intervals are infinitely long.) To describe infinite intervals, we introduce two new symbols. These are ∞, called *(plus) infinity*, and $-\infty$, called *minus infinity*. The symbols ∞ and $-\infty$ must not be thought of as numbers, even though they can appear in inequalities. Using ∞ and $-\infty$, we now introduce the following kinds of infinite intervals, where c is any fixed number.

(i) The set $\{x: x \geq c\}$, denoted by $[c, \infty)$ or $c \leq x < \infty$;
(ii) The set $\{x: x \leq c\}$, denoted by $(-\infty, c]$ or $-\infty < x \leq c$;
(iii) The set $\{x: x > c\}$, denoted by (c, ∞) or $c < x < \infty$;
(iv) The set $\{x: x < c\}$, denoted by $(-\infty, c)$ or $-\infty < x < c$;
(v) The whole real line, denoted by $(-\infty, \infty)$ or $-\infty < x < \infty$.

The intervals (i) and (ii) are regarded as closed, while the intervals (iii), (iv) and (v) are regarded as open. The geometric meaning of the various kinds of infinite intervals is shown in Figure 12, where included endpoints are

Various types of infinite intervals

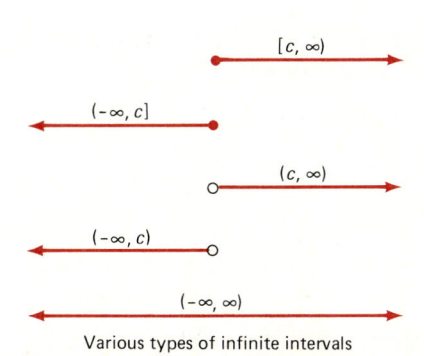

Figure 12

Section 0.3 Absolute Value and Intervals 19

indicated by heavy dots, excluded endpoints by hollow dots, and infinite endpoints by arrowheads suggestively pointing "out to infinity." Since ∞ and $-\infty$ are not numbers, we cannot allow either $x = \infty$ or $x = -\infty$. Therefore it is wrong to write $x \leq \infty$ or $-\infty \leq x$, or to put a square bracket next to the symbol ∞ or $-\infty$.

Example 6 According to Example 3, page 9, x satisfies the quadratic inequality $x^2 - x - 6 > 0$ if and only if x belongs to one of the infinite open intervals $(3, \infty)$ or $(-\infty, -2)$.

Remark We have just introduced various types of intervals, finite and infinite. Each of these intervals is "connected" in the sense that if it contains two distinct points x and y, then it contains every point between x and y. Conversely, every connected set of points on the real line must be an interval of one of the above types. This is intuitively clear, but a rigorous proof involves certain technicalities that will not be given here.

Problems

Calculate the given absolute value.
1. $|6 - 15|$
2. $|1 - \sqrt{3}|$
3. $|\sqrt{2} - 1|$
4. $|-|-3||$
5. $||-1| - |-2||$
6. $|1 - (\frac{1}{2})^{-2}|$
7. $|\pi - \sqrt{11}|$
8. $|(-1)^n|$ (n any integer)

Let $|a - 10| < 2$ and $|b - 6| < 1$. What can be said about the size of the given quantity?
9. $a + b$
10. $a - b$
11. $a^2 - b^2$
12. $a^2 + ab + 1$

Express the given statement in terms of absolute values.
13. x is closer to the origin than to the point -2
14. The distance between x and the point 2 does not exceed π
15. x is twice as far from the point 1 as from the point -1

Find all points x which are
16. Equidistant from the points -3 and 7
17. One fourth as far from the point -1 as from the point 4
18. At distance 2 from the point -1
19. Closer to the point 1 than to the origin
20. Farther from the point 2 than from the point -3
21. Such that the sum of the distances from x to the points 1 and -1 is less than 4

Find all x satisfying the given equality.
22. $|1 - x| = 2$
23. $|x - 1| = |3 - x|$
24. $|x + 1| = |3 + x|$
25. $|2x| = |x - 2|$
26. $|x + 2| + |x - 1| = 2$
27. $|x + 2| + |x - 1| = 4$

28. How many *positive* integers have absolute value less than 5? How many integers?

Find all x satisfying the given inequality.
29. $|x + 1| \geq 2$
30. $|2x + 1| < |3x + 4|$
31. $|x^2 - 1| > 3$
32. $|x| - |x - 1| \geq 0$
33. $|x| - |x - 1| < 1$
34. $|x| - |x - 1| > 1$
35. $\left|\dfrac{x - 1}{x + 1}\right| > 0$
36. $\left|\dfrac{x - 1}{x + 1}\right| < 1$
37. $\left|\dfrac{x - 1}{x + 1}\right| \geq 1$

38. Give an alternative proof of the inequality (5), starting from the fact that $|a - b|^2 = (a - b)^2 = a^2 - 2ab + b^2$.

Find the midpoint of any interval with the given endpoints.
39. 3, 6
40. $-1, 8$
41. $-7, -1$
42. $-\sqrt{2}, \sqrt[4]{4}$

Write each interval in another way.
43. $-2 < x \leq 3$
44. $(2, \pi)$
45. $[-1, 1)$
46. $-\infty < x < 3$
47. $5 < x < 13$
48. $2 \leq x < \infty$
49. $(3, \infty)$
50. $(-\infty, -1]$
51. $-4 \leq x < -2$
52. $(-3, 4]$
53. $[-2, -1]$
54. $3 \leq x \leq 9$
55. $-\infty < x \leq -5$
56. $-1 < x < \infty$
57. $(-\infty, 5)$
58. $[-\pi, \infty)$
59. $|x - 3| \leq 2$
60. $|x| < \sqrt{2}$
61. $|x + 2| < 1$
62. $|x - \frac{1}{2}| \leq \frac{3}{2}$

20 Chapter 0 Precalculus

Write each neighborhood with the help of the absolute value.

63. The 2-neighborhood of 1

64. The $\sqrt{3}$-neighborhood of -2

65. The deleted 1-neighborhood of -1

66. The deleted $\frac{1}{2}$-neighborhood of π

0.4 Coordinates in the Plane

So far we only have coordinates on a line. To introduce coordinates in a given plane (the plane of this page, say), we draw two directed lines which intersect at right angles, and think of them both as number lines. The point of intersection of the lines serves as a common *origin* O from which distance is measured along both lines with the same unit of length. The two lines are called the *coordinate axes*. One line, called the *x-axis*, is usually drawn horizontally pointing to the right, and the other line, called the *y-axis*, is usually drawn vertically pointing upward, as in Figure 13. The plane determined by the *x*- and *y*-axes is called the *xy-plane*.

Ordered Pairs and Rectangular Coordinates

Now let (a, b) be an *ordered pair* of real numbers, where the notation indicates that a comes first and b second.† Plotting the first element a of the pair (a, b) as a point of the *x*-axis and the second element b as a point of the *y*-axis, we erect a perpendicular to the *x*-axis at a and a perpendicular to the *y*-axis at b. As in Figure 13, these perpendiculars intersect in a point P, which we regard as representing the ordered pair (a, b). The point P is said to have (*rectangular*) *coordinates* a and b, specifically *x-coordinate* a and *y-coordinate* b. By reversing this construction, that is, by drawing perpendiculars from P to the coordinate axes, we can find the coordinates and hence the ordered pair corresponding to any given point P.

Thus there is a unique point in the plane corresponding to any given ordered pair, and conversely a unique ordered pair corresponding to any given point in the plane. Because of this one-to-one correspondence, we will usually make little or no distinction between ordered pairs and the points representing them. In particular, $P = (a, b)$ means that P is the point with *x*-coordinate a and *y*-coordinate b. (Some authors write $P(a, b)$ for this point.) Notice that the origin O is the point $(0, 0)$. Equality of two ordered pairs (a, b) and (c, d) means of course that the two pairs have the same first element and the same second element, so that $a = c$ and $b = d$. Thus $(\sqrt{4}, 1) = (2, 1)$, but $(2, \sqrt{3}) \neq (2, 1)$.

We have just set up a *system of rectangular coordinates* in the plane. More generally, the axes may be labeled with letters other than x and y, and the units of measurement along the two axes may be different. For example, in meteorology one might label the horizontal axis with t for time, measured in seconds, and the vertical axis with p for atmospheric pressure, measured in millibars. By the *abscissa* of a point we mean its coordinate along the horizontal axis, and by the *ordinate* we mean its coordinate along

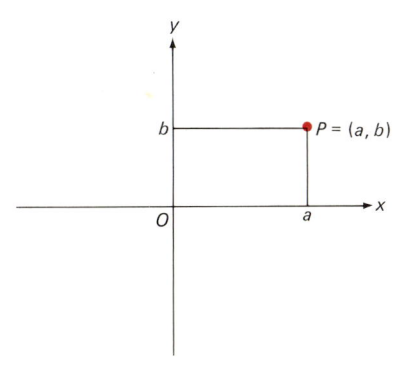

Figure 13

---

† The same kind of two-parenthesis notation is used to denote both ordered pairs and open intervals, but the context will always prevent any confusion between these two concepts.

the vertical axis. These terms are useful because they stop short of naming the symbols used to denote the coordinates. Having made these observations, we now return to the *xy*-plane, where the abscissa is *x* and the ordinate is *y*, and the units of measurement are the same along both axes.

Example 1 A square has one side along the *x*-axis and its diagonals intersect in the point $(0, \frac{1}{2})$. Where are the vertices of the square?

Solution The answer is apparent from Figure 14. Notice that the side length of the square is 1. □

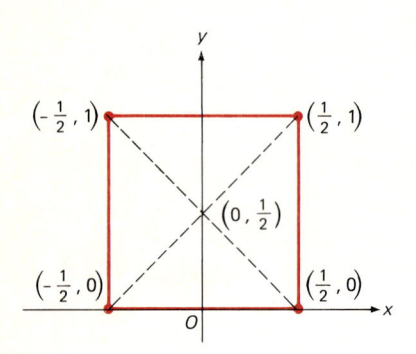

Figure 14

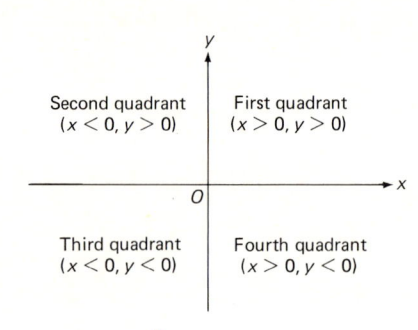

Figure 15 The four quadrants of the *xy*-plane

A point (x, y) lies on the *x*-axis if and only if $y = 0$ and on the *y*-axis if and only if $x = 0$. The *positive x*-axis consists of the points $(x, 0)$ with $x > 0$ and the *negative x*-axis of those with $x < 0$. Similarly, the *positive y*-axis consists of the points $(0, y)$ with $y > 0$ and the *negative y*-axis of those with $y < 0$. The coordinate axes divide the *xy*-plane into four parts, called *quadrants*, as shown in Figure 15. The first quadrant consists of all points (x, y) for which $x > 0$, $y > 0$, and the other three quadrants are defined by the conditions on *x* and *y* given in the figure.

The Distance Formula

The distance between two points P_1 and P_2 in the plane is denoted by the same notation $|P_1P_2|$ as for points on the line. The next theorem shows how to calculate this distance.

> **Theorem 6** (*Distance between two points in the plane*). *The distance between two points $P_1 = (x_1, y_1)$ and $P_2 = (x_2, y_2)$ in the plane is given by the formula*
>
> $$|P_1P_2| = \sqrt{(x_1 - x_2)^2 + (y_1 - y_2)^2}. \tag{1}$$

Proof Drawing perpendiculars from P_1 and P_2 to the *x*- and *y*-axes, we find that P_1P_2 is the hypotenuse of the right triangle P_1QP_2 shown in Figure 16, where Q is the point (x_2, y_1). It is clear that $|P_1Q| = |AB|$ and $|QP_2| = |CD|$, where *A* and *B* have coordinates x_1 and x_2 regarded as points of the *x*-axis, while *C* and *D* have coordinates y_1 and y_2 regarded as points of the *y*-axis. Therefore, by the Pythagorean theorem and the formula for the distance between two points on the line (see page 16), we have

$$|P_1P_2|^2 = |P_1Q|^2 + |QP_2|^2 = |AB|^2 + |CD|^2 = |x_1 - x_2|^2 + |y_1 - y_2|^2,$$

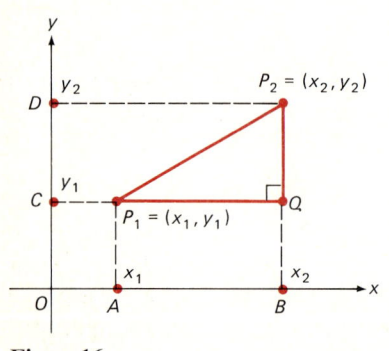

Figure 16

22   Chapter 0   Precalculus

or equivalently
$$|P_1P_2|^2 = (x_1 - x_2)^2 + (y_1 - y_2)^2. \tag{2}$$

Taking square roots of both sides of (2), we get the desired formula (1). ∎

Figure 16 is drawn under the assumption that $x_1 < x_2$ and $y_1 < y_2$, but it is easy to see that the same distance formula (1) is obtained if the direction of either (or both) of these inequalities is reversed.

Example 2 Find the distance between the points $P_1 = (1, 7)$ and $P_2 = (13, 2)$.

Solution According to (1),
$$|P_1P_2| = \sqrt{(1-13)^2 + (7-2)^2} = \sqrt{(-12)^2 + 5^2}$$
$$= \sqrt{144 + 25} = \sqrt{169} = 13.$$
□

Example 3 Is the triangle ABC with vertices $A = (-1, -4)$, $B = (2, -1)$, $C = (-2, 3)$ a right triangle?

Solution Using formula (2) to calculate the squares of the side lengths of ABC, we obtain
$$|AB|^2 = (-3)^2 + (-3)^2 = 18, \qquad |BC|^2 = 4^2 + (-4)^2 = 32,$$
$$|AC|^2 = 1^2 + (-7)^2 = 50.$$

Thus the side lengths of ABC satisfy the Pythagorean formula
$$|AC|^2 = |AB|^2 + |BC|^2.$$

Therefore ABC is a right triangle, with the side AC as its hypotenuse (see Figure 17). Here we actually use the converse of the Pythagorean theorem.
□

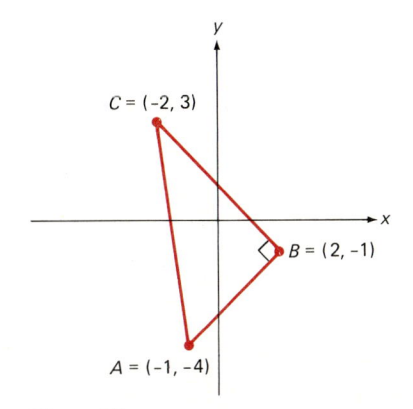

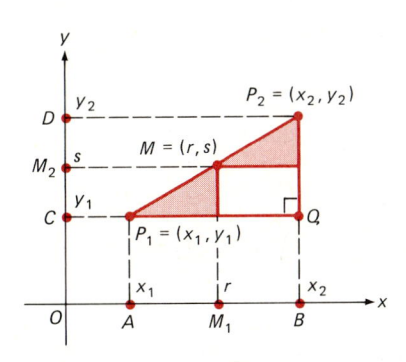

Figure 17

Example 4 Find the midpoint M of the segment P_1P_2 joining the points $P_1 = (x_1, y_1)$ and $P_2 = (x_2, y_2)$.

Solution Examining Figure 18, which is a modification of Figure 16, we see that the shaded right triangles are congruent (why?). Therefore M_1 and M_2, the feet of the perpendiculars drawn from M to the x- and y-axes, are the midpoints of the segments AB and CD. Thus if $M = (r, s)$, then
$$r = \frac{x_1 + x_2}{2}, \qquad s = \frac{y_1 + y_2}{2},$$

by Example 4, page 18, and we have the *midpoint formula*
$$M = \left(\frac{x_1 + x_2}{2}, \frac{y_1 + y_2}{2} \right). \quad □$$

M is the midpoint of the segment P_1P_2.

Figure 18

Graphs of Equations and Inequalities

By the *graph* of an equation or inequality in two variables x and y we mean the set of points (x, y) in the xy-plane whose coordinates satisfy the given equation or inequality. For example, the graph of the equation
$$xy = 0$$
consists of the two coordinates axes, since $xy = 0$ if and only if $x = 0$ or $y = 0$ (or both). One of the variables x and y may be absent. For instance, the graph of $x = a$ is the vertical line through the point $(a, 0)$, while the graph of $y = b$ is the horizontal line through the point $(0, b)$. The word "graph" will also be used as a verb, meaning "sketch the graph of."

Example 5 Graph the equation

$$x^2 + y^2 = 1. \tag{3}$$

Solution Since $x^2 + y^2$ is the square of the distance between the point (x, y) and the origin $O = (0, 0)$, the point (x, y) belongs to the graph of (3) if and only if the distance between (x, y) and O is equal to 1. Thus the graph of equation (3) is the *unit circle*, that is, the circle of radius 1 with its center at O (see Figure 19). ◻

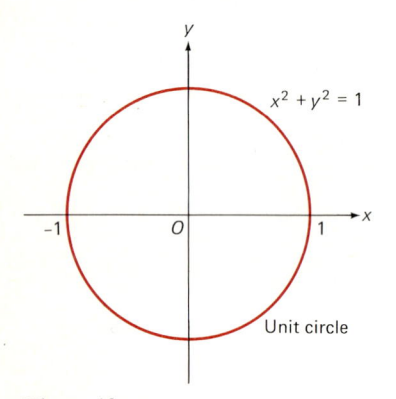

Figure 19

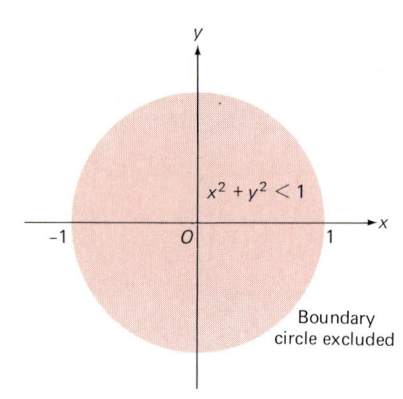

Figure 20

Example 6 Graph the inequality

$$x^2 + y^2 < 1. \tag{4}$$

Solution According to (4), the square of the distance between the point (x, y) and the origin O is less than 1, and hence the same is true of the distance itself. Thus the graph of (4) is the region *inside* the unit circle (the shaded region in Figure 20). Similarly, the graph of the inequality $x^2 + y^2 > 1$ is the region *outside* the unit circle. ◻

Circles and Completing the Square

Generalizing Example 5, we find that the coordinates of the point (x, y) satisfy the equation

$$(x - a)^2 + (y - b)^2 = r^2 \quad (r > 0) \tag{5}$$

if and only if the square of the distance between (x, y) and the fixed point (a, b) equals r^2, or equivalently if and only if the distance between (x, y) and (a, b) equals r. Thus the coordinates of (x, y) satisfy (5) if and only if (x, y) lies on the circle of radius r centered at (a, b). Notice that (5) reduces to equation (3) of the unit circle if we choose $a = b = 0$ and $r = 1$.

Now suppose we are given an equation of the form

$$x^2 + y^2 + Ax + By + C = 0, \tag{6}$$

where A, B and C are constants (fixed numbers). Is this the equation of a circle? To answer the question, we try to convert (6) into a form resembling (5). First we *complete the square* in each of the expressions $x^2 + Ax$ and $y^2 + By$, that is, we write

$$x^2 + Ax = \left(x + \frac{A}{2}\right)^2 - \frac{A^2}{4}, \quad y^2 + By = \left(y + \frac{B}{2}\right)^2 - \frac{B^2}{4}.$$

24 Chapter 0 Precalculus

Then, substituting these expressions into equation (6), we get the equivalent equation

$$\left(x + \frac{A}{2}\right)^2 + \left(y + \frac{B}{2}\right)^2 = D, \tag{6'}$$

where

$$D = \frac{A^2}{4} + \frac{B^2}{4} - C.$$

Everything now depends on the sign of the quantity D. If $D \geq 0$, the graph of (6'), and therefore of (6), is the circle of radius $\sqrt{D}$ with its center at the point $(-A/2, -B/2)$, where the circle "degenerates" to the single point $(-A/2, -B/2)$ if $D = 0$, since the radius of the circle is then zero. On the other hand, if $D < 0$, there are no points (x, y) whose coordinates satisfy equation (6), since the left side of the equivalent equation (6') is always non-negative. In this case we say that (6) has no graph, or more formally that the graph of (6) is the empty set.

Example 7 Graph the equation

$$x^2 + y^2 + 6x - 4y + 9 = 0. \tag{7}$$

Solution This is of the form (6) or (6') with $A = 6$, $B = -4$, $C = 9$,

$$D = \frac{6^2}{4} + \frac{(-4)^2}{4} - 9 = 9 + 4 - 9 = 4 > 0.$$

Hence the graph of (7) is the circle of radius $\sqrt{D} = 2$ with its center at the point $(-A/2, -B/2) = (-3, 2)$. Note that the x-axis is tangent to the circle at the point $(-3, 0)$, as shown in Figure 21. An even better way of solving the problem is to complete the squares in (7) *directly*, without using the general equations (6) and (6'). In fact, substituting

$$x^2 + 6x = (x + 3)^2 - 9, \qquad y^2 - 4y = (y - 2)^2 - 4$$

into (7), we get

$$(x + 3)^2 - 9 + (y - 2)^2 - 4 + 9 = 0,$$

or equivalently

$$(x + 3)^2 + (y - 2)^2 = 4, \tag{8}$$

which you will recognize at once as an equation of the circle of radius 2 centered at the point $(-3, 2)$. Hence the original equation (7) also has this circle as its graph. ∎

Observe that in the next-to-last sentence of Example 7 we say "an" equation rather than "the" equation. This is because there are any number of equations with the same graph. In fact, if both sides of an equation are multiplied by a nonzero number, if terms of an equation are transposed from one side to another, or if certain algebraic operations (like calculating squares) are actually carried out explicitly, the new equation has the same graph as the old one. For example, the equations

$$\frac{1}{4}(x + 3)^2 + \frac{1}{4}(y - 2)^2 = 1$$

and

$$(x + 3)^2 = 4 - (y - 2)^2$$

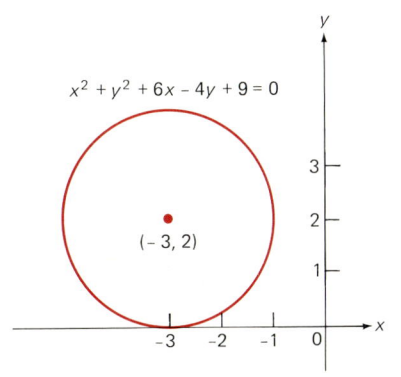

Figure 21

have the same graph as (8), and so does the original equation (7). Thus each of these equations is also *an* equation of the circle of radius 2 centered at the point $(-3, 2)$.

Problems

1. Plot the points $A = (2, 0)$, $B = (0, 3)$, $C = (-2, 0)$, $D = (-2, 2)$, $E = (0, -1)$, $F = (2, 2)$ on ordinary graph paper. Then join A to B to C to A and also D to E to F to D. What is the resulting figure?

2. Suppose the figure in the preceding problem is shifted one unit to the right and two units upward. Then A, B, C, D, E, F go into new points A', B', C', D', E', F'. What are these new points?

3. Which of the points $(0, 2)$, $(-2, 0)$, $(2, -1)$, $(-2, 1)$, $(0, -2)$, $(2, 0)$ lie in the fourth quadrant?

4. Show that if the points $P_1 = (x_1, y_1)$ and $P_2 = (x_2, y_2)$ lie on the same horizontal line, then $|P_1P_2| = |x_1 - x_2|$, while if they lie on the same vertical line, then $|P_1P_2| = |y_1 - y_2|$.

Find the distance between the given pair of points.

5. $(1, 3), (5, 7)$
6. $(-2, -3), (1, 1)$
7. $(1, 3), (1, 4)$
8. $(6, 2), (4, 2)$
9. $(1, -1), (-1, 1)$
10. $(0, 1), (1, 0)$
11. $(-1, 1), (3, 3)$
12. $(3, 5), (-2, -4)$
13. $(2, -1), (-1, 3)$
14. $(7, 11), (3, 9)$
15. $(\pi, \sqrt{2}), (\sqrt{2}, -\pi)$
16. $(2\sqrt{2}, 4), (2, -\sqrt{2})$

Is the triangle with the given vertices a right triangle?

17. $(2, 3), (-3, 3), (1, 1)$
18. $(7, -4), (5, -3), (7, 1)$
19. $(3, 1), (2, 0), (0, 1)$
20. $(\frac{3}{2}, -\frac{1}{2}), (4, 2), (-1, 2)$

Find the midpoint of the segment joining the given pair of points.

21. $(-1, 3), (11, 5)$
22. $(3, -2), (-4, 3)$
23. $(100, -50), (-100, 50)$
24. $(-2, -5), (18, 3)$

25. Plot the points $A = (4, 0)$, $B = (3, 4)$, $C = (-1, 3)$, $D = (0, -1)$. Show that the figure $ABCD$ is a square. What is the side length of the square?

26. Find the midpoints of the sides of the square $ABCD$ in the preceding problem.

Find an equation of the circle with the given radius and center.

27. $1, (-1, 1)$
28. $\sqrt{2}, (0, 1)$
29. $3, (4, -5)$
30. $\frac{3}{4}, (-1, 0)$

Describe the graph of the given equation.

31. $(x + 2)^2 + y^2 = 64$
32. $(x - 4)^2 + (y + 3)^2 = 36$
33. $(x + 5)^2 + (y - 2)^2 = 0$
34. $x^2 + (y - 5)^2 = 5$
35. $x^2 + y^2 - 2x + 4y - 20 = 0$
36. $x^2 + y^2 + 4x - 2y + 5 = 0$

Find an equation of the circle

37. With $(3, 2)$ and $(-1, 6)$ as the endpoints of a diameter
38. Of radius 1 with the positive coordinate axes as tangents
39. Going through the points $(1, 1), (2, 2), (3, 1)$

Determine whether the point $(1, -2)$ lies inside, outside or on the circle which is the graph of the given equation.

40. $x^2 + y^2 = 1$
41. $x^2 + y^2 = 5$
42. $x^2 + y^2 = 9$
43. $x^2 + y^2 - 10x + 8y = 0$
44. $x^2 + y^2 - 8x - 4y - 5 = 0$

Graph the region of the xy-plane determined by the given pair of simultaneous inequalities.

45. $x \leq 2, y \leq -1$
46. $|x| \geq 1, |y| \geq 2$
47. $xy > 0, |y| < 2$
48. $xy < 0, x^2 + y^2 < 4$

49. By completing the square, show that the quadratic equation $ax^2 + bx + c = 0$ $(a \neq 0)$ has two solutions

$$x = \frac{-b \pm \sqrt{b^2 - 4ac}}{2a}$$

if $b^2 > 4ac$, only one solution $x = -b/2a$ if $b^2 = 4ac$, and no (real) solutions if $b^2 < 4ac$.

0.5 Straight Lines and Their Equations

Slope of a Line

Let L be any *nonvertical* straight line in the xy-plane, and let $P_1 = (x_1, y_1)$ and $P_2 = (x_2, y_2)$ be any two distinct points of L. Then by the *slope* of L we mean the ratio

$$m = \frac{y_2 - y_1}{x_2 - x_1}. \tag{1}$$

The assumption that L is nonvertical guarantees that $x_1 \neq x_2$, so that the denominator of this ratio is nonzero and the slope m is defined. The slope of a vertical line is not defined, since $x_1 = x_2$ for such a line.

To interpret the slope geometrically, we assume that the point P_1 comes before the point P_2 on the line L as it goes from left to right. Actually there is no loss of generality in making this assumption, since interchanging the labels 1 and 2 on the points P_1 and P_2 and their coordinates does not change the value of m given by formula (1). Let Q be the point of intersection of the line through P_1 parallel to the x-axis and the line through P_2 parallel to the y-axis, so that P_1QP_2 is a right triangle with legs P_1Q and P_2Q. Then the slope is equal to

$$m = \frac{y_2 - y_1}{x_2 - x_1} = \frac{|P_2Q|}{|P_1Q|} > 0 \tag{2}$$

if the line L rises in going from left to right, as in Figure 22(a). The quantity $|P_2Q|/|P_1Q|$ is the ratio of the length of the vertical segment P_2Q to the length of the horizontal segment P_1Q. In civil engineering this is called the *rise-to-run ratio*, and measures the rate of ascent of an uphill road. If the line L *falls* in going from left to right, then, instead of (2), we have

$$m = \frac{y_2 - y_1}{x_2 - x_1} = -\frac{y_1 - y_2}{x_2 - x_1} = -\frac{|P_2Q|}{|P_1Q|} < 0, \tag{2'}$$

as in Figure 22(b). From an engineering standpoint, a road of negative slope m goes *downhill*, and $|m|$ measures its rate of *descent*.

Alternatively, as a point moves along the line L from P_1 to P_2, its x-coordinate changes from x_1 to x_2, while its y-coordinate changes from y_1 to y_2. Thus

$$\text{Slope} = m = \frac{y_2 - y_1}{x_2 - x_1} = \frac{\text{Change in } y}{\text{Change in } x} = \frac{\Delta y}{\Delta x},$$

where in writing $\Delta y/\Delta x$ we anticipate the increment notation to be introduced in Chapter 2 (Δ is capital Greek delta).

It should be noted that the slope of a line L does not depend on the particular choice of the points P_1 and P_2 used to compute the slope. This is

A line of positive slope
(a)

A line of negative slope
(b)

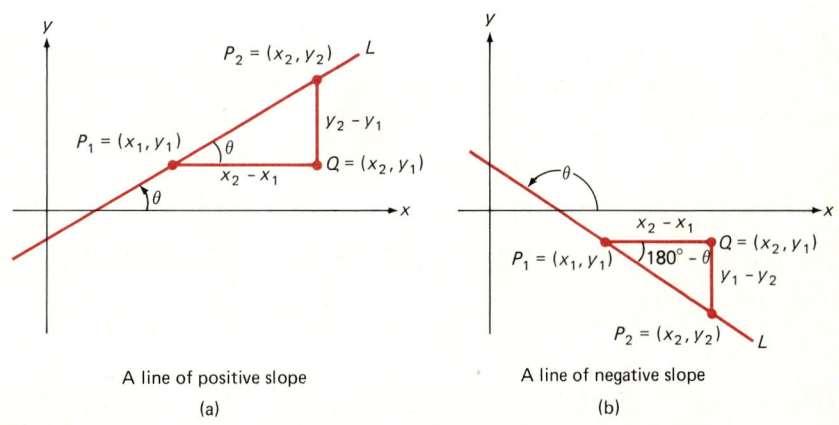

Figure 22

Section 0.5 Straight Lines and Their Equations

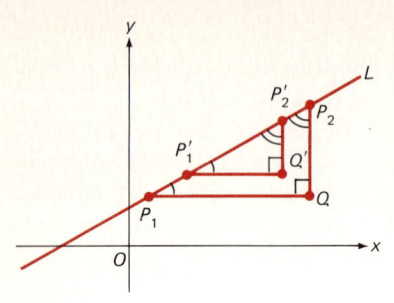

Figure 23

simply because any two right triangles with hypotenuses along the line L and the other sides parallel to the coordinate axes are similar, so that the ratios of corresponding sides are equal. For example, in Figure 23 the slope is

$$m = \frac{y_2 - y_1}{x_2 - x_1} = \frac{|P_2 Q|}{|P_1 Q|}$$

if computed by using the points $P_1 = (x_1, y_1)$, $P_2 = (x_2, y_2)$, $Q = (x_2, y_1)$, and

$$m' = \frac{y'_2 - y'_1}{x'_2 - x'_1} = \frac{|P'_2 Q'|}{|P'_1 Q'|}$$

if computed by using the points $P'_1 = (x'_1, y'_1)$, $P'_2 = (x'_2, y'_2)$, $Q' = (x'_2, y'_1)$. But

$$\frac{|P_2 Q|}{|P_1 Q|} = \frac{|P'_2 Q'|}{|P'_1 Q'|},$$

by the similarity of the right triangles $P_1 Q P_2$ and $P'_1 Q' P'_2$, so that $m = m'$.

Inclination of a Line

By the *angle of inclination*, or simply the *inclination*, of a straight line L in the xy-plane we mean the smallest angle θ between the positive x-axis and L, measured from the x-axis to L in the counterclockwise direction (θ is lowercase Greek theta). Any line parallel to the x-axis is assigned inclination zero. It is clear from Figure 24 that the inclination θ of any straight line must satisfy the condition $0° \leq \theta < 180°$. Thus, for example, the inclination of the line L in the figure is $150°$, not $330°$ or $-30°$.

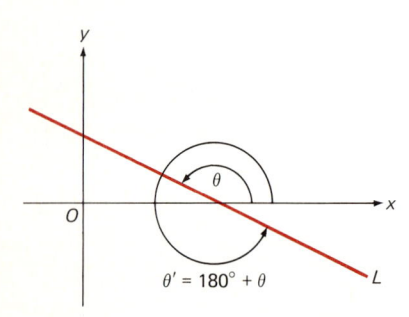

With the help of a little trigonometry† it is easy to establish a formula relating the slope m of a line L to its inclination θ. Suppose L rises in going from left to right. Then, as in Figure 22(a),

$$m = \frac{y_2 - y_1}{x_2 - x_1} = \tan \theta,$$

The inclination of L is θ, not θ'.
Figure 24

where $\tan \theta$ is the *tangent* of the angle θ, that is, the length of the side opposite θ in the right triangle $P_1 Q P_2$ divided by the length of the side adjacent to θ. If the inclination θ lies between $90°$ and $180°$, the line L *falls* in going from left to right, but the formula

$$m = \tan \theta \tag{3}$$

still holds. In fact, in this case

$$m = -\frac{y_1 - y_2}{x_2 - x_1} = -\tan(180° - \theta),$$

as in Figure 22(b), where

$$\tan(180° - \theta) = -\tan \theta$$

by a familiar formula of trigonometry, so that

$$m = -\tan(180° - \theta) = \tan \theta,$$

as before. Figure 25 shows various lines, together with their inclinations and

---

† Trigonometry will be reviewed in Section 1.3. For now we only need the definition of the tangent and the formulas $\tan(180° - \theta) = -\tan \theta$ and $\tan(90° + \theta) = -1/\tan \theta$, established in Example 3, page 59.

28 Chapter 0 Precalculus

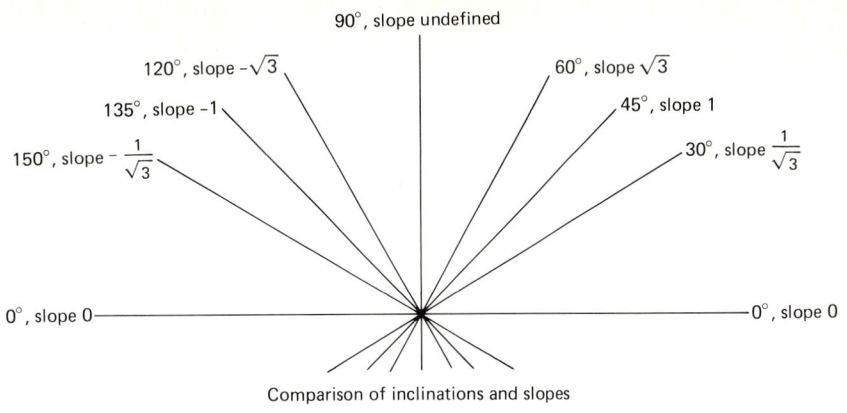

Figure 25

slopes, related by formula (3). Notice that a vertical line has inclination 90°, even though its slope is undefined.

Example 1 Find the slope and inclination of the line L through the points $P_1 = (1, -1)$ and $P_2 = (4, 2)$.

Solution Here

$$m = \frac{y_2 - y_1}{x_2 - x_1} = \frac{2 - (-1)}{4 - 1} = \frac{3}{3} = 1.$$

Therefore, by (3) and the definition of inclination, θ is the smallest positive angle whose tangent equals 1, namely 45° (see Figure 26). □

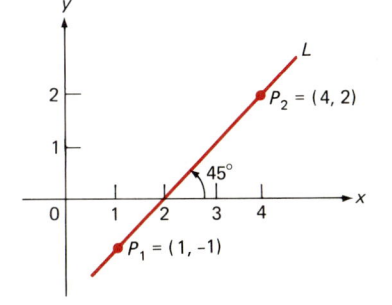

Figure 26

Example 2 Find the slope m of the line whose inclination is 15°.

Solution Here

$$m = \tan 15° = 0.26795$$

to five decimal places, where we consult a table of tangents or use a scientific calculator. □

Intercepts of a Line

Consulting Figure 27(a), we see that if a straight line L is *oblique*, that is, neither horizontal nor vertical, then L intersects the x-axis in a point $(a, 0)$ and the y-axis in a point $(0, b)$. We call (the number) a the *x-intercept* of L and

Two intercepts
(a)

x-intercept only
(b)

y-intercept only
(c)

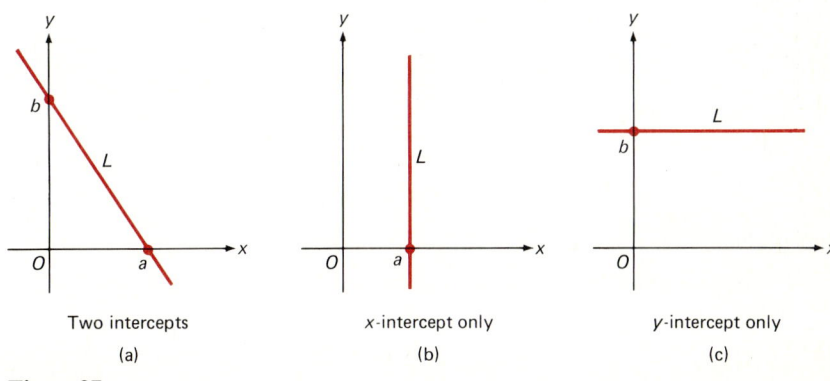

Figure 27

Section 0.5 Straight Lines and Their Equations

b the *y-intercept* of L, using the term *intercept* to mean either an *x*-intercept or a *y*-intercept. A nonoblique line has only one intercept. Thus if L is vertical as in Figure 27(b), L has an *x*-intercept a but no *y*-intercept, while if L is horizontal as in Figure 27(c), L has a *y*-intercept b but no *x*-intercept. This applies to the coordinate axes as well, that is, the *y*-axis is regarded as having only an *x*-intercept (equal to 0), and the *x*-axis is regarded as having only a *y*-intercept (also equal to 0).

> **Theorem 7** *(Point-slope and slope-intercept equations of a line). An equation of the line with slope m going through a given point $P_1 = (x_1, y_1)$ is*
>
> $$y - y_1 = m(x - x_1). \qquad (4)$$
>
> *An equation of the line with slope m and y-intercept b is*
>
> $$y = mx + b. \qquad (5)$$

Proof Let $P = (x, y)$ be a variable point of the line. Then if $x \neq x_1$, the line has slope $(y - y_1)/(x - x_1)$, since it goes through P_1 and P. Equating this expression to the given slope m, we get

$$\frac{y - y_1}{x - x_1} = m,$$

which is equivalent to (4). Equation (4) also holds for the excluded value $x = x_1$ (why?). If the line has *y*-intercept b, the point $(0, b)$ lies on the line, and setting $x_1 = 0$, $y_1 = b$, we get $y - b = mx$ or equivalently (5). ∎

Example 3 By (4), an equation of the line with slope -2 going through the point $(-1, 5)$ is

$$y = -2(x + 1) + 5 = -2x + 3.$$

By (5), an equation of the line with slope 9 and *y*-intercept -7 is

$$y = 9x - 7.$$

Example 4 Graph the equation

$$3x - 4y - 2 = 0. \qquad (6)$$

Solution We can write (6) in the form

$$y = \frac{3}{4}x - \frac{1}{2}. \qquad (6')$$

But (6') can be recognized at once as an equation of the line with slope $\frac{3}{4}$ and *y*-intercept $-\frac{1}{2}$ (see Figure 28). This line has *x*-intercept $\frac{2}{3}$, as we see by setting $y = 0$ in (6) and solving for x. □

More generally, as you can easily verify, the graph of any equation of the form

$$Ax + By + C = 0, \qquad (7)$$

where A, B and C are constants, is a straight line, and conversely every straight line is the graph of an equation of this form; here of course it is assumed that A and B are not both zero. Therefore an equation of the form (7) is said to be *linear*.

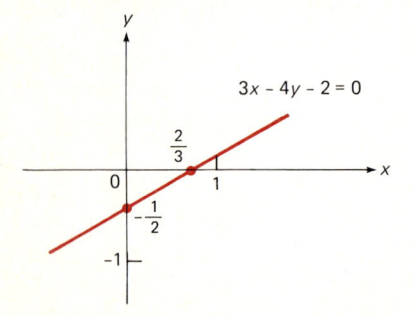

Figure 28

> **Theorem 8** *(Two-point and two-intercept equations of a line). An equation of the line going through two given points $P_1 = (x_1, y_1)$ and $P_2 = (x_2, y_2)$ is*
>
> $$y - y_1 = \frac{y_2 - y_1}{x_2 - x_1}(x - x_1), \qquad (8)$$
>
> *if $x_1 \neq x_2$. An equation of the line with x-intercept a and y-intercept b is*
>
> $$\frac{x}{a} + \frac{y}{b} = 1 \qquad (9)$$
>
> *if $a \neq 0$, $b \neq 0$.*

Proof Substituting $m = (y_2 - y_1)/(x_2 - x_1)$ into (4), we get (8). If the line has x-intercept a and y-intercept b, then both points $(a, 0)$ and $(0, b)$ lie on the line. Thus for this line we can choose $x_1 = a$, $y_1 = 0$ and $x_2 = 0$, $y_2 = b$ in (8), obtaining

$$y = \frac{b}{-a}(x - a),$$

or equivalently

$$bx + ay = ab,$$

which reduces to (9) after dividing through by ab. ∎

Example 5 By (8), an equation of the line going through the points $(-2, 3)$ and $(4, -1)$ is

$$y - 3 = \frac{-1 - 3}{4 - (-2)}(x + 2) = -\frac{2}{3}(x + 2),$$

or equivalently

$$2x + 3y - 5 = 0$$

in the form (7). By (9), an equation of the line with x-intercept -2 and y-intercept 5 is

$$\frac{x}{-2} + \frac{y}{5} = 1$$

or equivalently

$$5x - 2y + 10 = 0.$$

We now turn our attention to pairs of straight lines.

Example 6 Find the point of intersection P of the two lines with equations

$$x + y - 3 = 0, \qquad x - 2y + 2 = 0. \qquad (10)$$

Solution The point P lies on both lines, so that the coordinates of P must satisfy both equations (10) simultaneously. Thus, finding P is algebraically equivalent to solving the system (10) of two linear equations in two unknowns x and y. To do this, we add twice the first equation to the second:

$$2(x + y - 3) + (x - 2y + 2) = 0.$$

The terms involving y then cancel out, leaving just the equation

$$3x - 4 = 0,$$

with solution $x = \frac{4}{3}$. We then substitute $x = \frac{4}{3}$ back into either of the original equations, obtaining $y = \frac{5}{3}$. Therefore $P = (\frac{4}{3}, \frac{5}{3})$ is the point of intersection of the two given lines, as shown in Figure 29. □

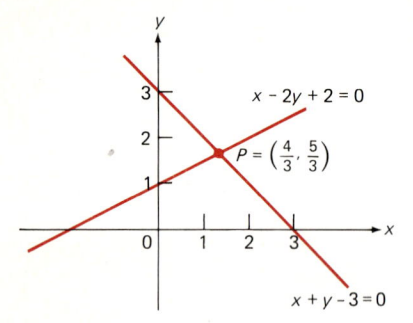

Figure 29

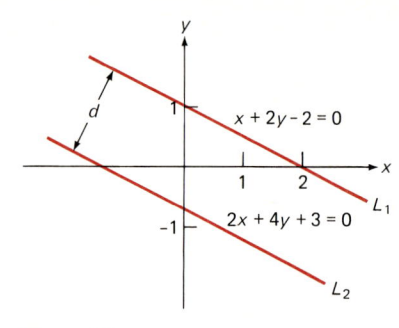

Figure 30

Parallel Lines

Clearly two lines in the xy-plane are parallel if and only if they have the same inclination. Therefore two nonvertical lines are parallel if and only if they have the same slope. Of course, all vertical lines are parallel.

Example 7 The two lines L_1 and L_2 with equations

$$x + 2y - 2 = 0, \qquad 2x + 4y + 3 = 0 \tag{11}$$

have the same slope $-\frac{1}{2}$, and hence are parallel (see Figure 30). If you try to solve the system of equations (11), you will quickly run into trouble. In fact, subtracting twice the first equation from the second gives $7 = 0$, which is false. It follows that the system (11) has no solution. Geometrically, this corresponds to the fact that distinct parallel lines do not intersect.

Perpendicular Lines

Next we establish a condition for two lines to be perpendicular, that is, to intersect at right angles.

> **Theorem 9** (*Perpendicularity condition*). *Two oblique lines are perpendicular if and only if the product of their slopes is* -1.

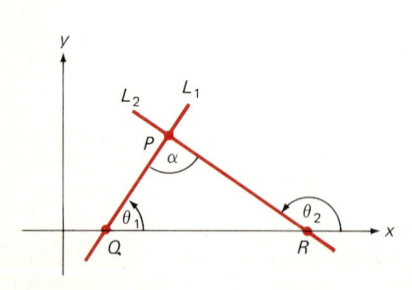

Figure 31

Proof One line, call it L_1, has positive slope m_1 and angle of inclination θ_1 between $0°$ and $90°$, and the other line, call it L_2, has negative slope and angle of inclination θ_2 between $90°$ and $180°$. It can be assumed that the x-axis lies below the point of intersection P of the lines L_1 and L_2 (otherwise we can choose a new x-axis parallel to the old one, with the point of intersection above it, since this does not change the slopes of the lines). Thus we have the situation shown in Figure 31, where θ_1 and α (lowercase Greek alpha) are interior angles of the triangle PQR and θ_2 is an exterior angle of PQR. Any exterior angle of a triangle equals the sum of the remote interior angles. Therefore $\theta_2 = \alpha + \theta_1$, or equivalently $\alpha = \theta_2 - \theta_1$. If L_1 and L_2 are

32   Chapter 0   Precalculus

perpendicular, then $\alpha = 90°$ and

$$m_2 = \tan \theta_2 = \tan(90° + \theta_1) = -\frac{1}{\tan \theta_1},$$

by a familiar formula of trigonometry, so that

$$m_1 m_2 = \tan \theta_1 \left(-\frac{1}{\tan \theta_1}\right) = -1.$$

Conversely, if $m_1 m_2 = \tan \theta_1 \tan \theta_2 = -1$, then

$$\tan \theta_2 = -\frac{1}{\tan \theta_1} = \tan(90° + \theta_1),$$

which implies $\theta_2 = 90° + \theta_1$ or $\alpha = 90°$, so that the lines L_1 and L_2 are perpendicular. ∎

Thus two oblique lines with slopes m_1 and m_2 are perpendicular if and only if

$$m_1 m_2 = -1,$$

that is, if and only if the slope of each line is the negative reciprocal of the slope of the other line.

Remark In concluding that $\tan \theta_2 = \tan(90° + \theta_1)$ implies $\theta_2 = 90° + \theta_1$, we tacitly relied on the fact that θ_2 and $90° + \theta_1$ both lie in the interval $90° < \theta < 180°$. This guarantees that both tangents are defined and that the tangents of the two angles θ_2 and $90° + \theta_1$ are equal only if the angles themselves are equal.

Example 8 Verify that the two lines with equations $2x + 5y - 7 = 0$ and $15x - 6y + 4 = 0$ are perpendicular.

Solution The first line has slope $-\frac{2}{5}$ and the second has slope $\frac{15}{6} = \frac{5}{2}$. Since $-\frac{2}{5} \cdot \frac{5}{2} = -1$, the lines are perpendicular. □

Example 9 Find the perpendicular bisector of the line segment with endpoints $P_1 = (-4, 3)$ and $P_2 = (2, -1)$.

Solution The midpoint of the segment $P_1 P_2$ is the point

$$\left(\frac{-4 + 2}{2}, \frac{3 - 1}{2}\right) = (-1, 1)$$

(see Example 4, page 18), and $P_1 P_2$ has slope

$$\frac{-1 - 3}{2 - (-4)} = -\frac{4}{6} = -\frac{2}{3}.$$

Hence the perpendicular bisector of $P_1 P_2$ is the line through $(-1, 1)$ with slope $\frac{3}{2}$, equal to the negative reciprocal of $-\frac{2}{3}$. By Theorem 7, an equation of this line is

$$y = \frac{3}{2}(x + 1) + 1 = \frac{3}{2}x + \frac{5}{2}$$

(see Figure 32). □

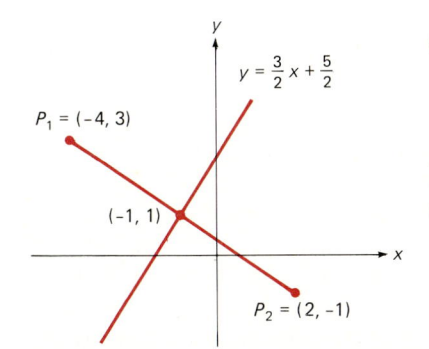

Figure 32

Distance Between a Point and a Line

Finally, we prove a theorem which recapitulates the material in this section and is of considerable importance in its own right.

> **Theorem 10** *(Distance between a point and a line).* *The distance d between the point $P_1 = (x_1, y_1)$ and the line L with equation*
>
> $$Ax + By + C = 0 \qquad (12)$$
>
> *is given by*
>
> $$d = \frac{|Ax_1 + By_1 + C|}{\sqrt{A^2 + B^2}}. \qquad (13)$$

Proof As in Figure 33, let $P_2 = (x_2, y_2)$ be the foot of the perpendicular drawn from P_1 to L. Then the distance d between P_1 and L is defined as the length of the line segment P_1P_2. Since the slope of L equals $-A/B$, as can be seen by solving (12) for y, the slope of the line L' through P_1 perpendicular to L equals B/A, the negative reciprocal of $-A/B$. Thus an equation for L' is

$$y - y_1 = \frac{B}{A}(x - x_1).$$

But P_2 lies on L'. Therefore

$$y_2 - y_1 = \frac{B}{A}(x_2 - x_1),$$

or equivalently

$$\frac{x_2 - x_1}{A} = \frac{y_2 - y_1}{B}. \qquad (14)$$

Denoting the ratio (14) by q, we find that

$$x_2 - x_1 = Aq, \qquad y_2 - y_1 = Bq,$$

and hence

$$d = |P_1P_2| = \sqrt{(x_1 - x_2)^2 + (y_1 - y_2)^2} = \sqrt{A^2q^2 + B^2q^2} = \sqrt{A^2 + B^2}\,|q|. \qquad (15)$$

But P_2 also lies on L, so that x_2 and y_2 satisfy equation (12). Therefore

$$Ax_2 + By_2 + C = A(Aq + x_1) + B(Bq + y_1) + C = 0.$$

Solving for q, we get

$$q = -\frac{Ax_1 + By_1 + C}{A^2 + B^2}. \qquad (16)$$

Finally we substitute (16) into (15), obtaining (13). ∎

In proving Theorem 10, we have tacitly assumed that both A and B are nonzero. You should verify that formula (13) continues to work if either A or B is zero.

The distance from P_1 to L is d.

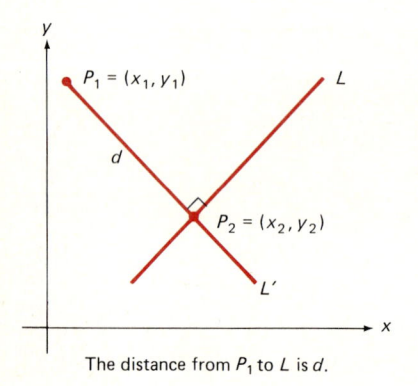

Figure 33

Example 10 Find the distance between the point (3, 1) and the line $3x + 4y - 3 = 0$.

Solution By the line $3x + 4y - 3 = 0$ we mean of course the line *with equation* $3x + 4y - 3 = 0$ (this kind of abbreviated language is customary). With the help of (13), we find that

$$d = \frac{|3(3) + 4(1) - 3|}{\sqrt{3^2 + 4^2}} = \frac{10}{\sqrt{25}} = \frac{10}{5} = 2. \quad \square$$

Example 11 Find the distance d between the parallel lines in Example 7.

Solution Clearly d equals the distance between L_1 and any point of L_2, or between L_2 and any point of L_1 (see Figure 30). The point (0, 1) lies on L_1 and therefore

$$d = \frac{|2(0) + 4(1) + 3|}{\sqrt{2^2 + 4^2}} = \frac{7}{\sqrt{20}}. \quad \square$$

Problems

Use a table of tangents or a scientific calculator to find the slope (to three decimal places) of the line with the given inclination.
1. 20°
2. 100°
3. 50°
4. 165°
5. 140°
6. 89°

Find the slope of the line through the given pair of points.
7. (2, −3), (4, 2)
8. (3, −5), (−3, −2)
9. (3, 8), ($\sqrt{2}$, 8)
10. ($\frac{1}{3}$, $\frac{1}{6}$), ($\frac{1}{2}$, $\frac{1}{2}$)
11. (π, 7), (π, −1)

Find the inclination of the line through the given pair of points.
12. (2, 4), (4, 6)
13. (2, 3), (2, 5)
14. (2, −4), (4, −6)
15. (0, −1), (−$\sqrt{3}$, 0)
16. (2, $\sqrt{2}$), (−2, $\sqrt{2}$)

17. Show that $y = mx$ is an equation of the line with slope m through the origin.

Find an equation of the line with slope m through the given point P.
18. $m = 2$, $P = (1, 2)$
19. $m = -1$, $P = (2, -1)$
20. $m = \frac{1}{2}$, $P = (3, 1)$
21. $m = -2$, $P = (-1, -2)$
22. $m = 1$, $P = (\frac{1}{4}, \frac{1}{2})$

Find an equation of the line with the given slope m and y-intercept b.
23. $m = \frac{2}{3}$, $b = 3$
24. $m = 3$, $b = 0$
25. $m = 0$, $b = -2$
26. $m = -\frac{3}{4}$, $b = 1$
27. $m = -7$, $b = -3$

28. Verify that if a line has both an x-intercept a and a y-intercept b, then either $a = b = 0$ or a and b are both nonzero.

Find the slope m, x-intercept a and y-intercept b of the given line.
29. $5x - y + 3 = 0$
30. $2x + 3y - 5 = 0$
31. $5x + 2y + 2 = 0$
32. $3x + 2y = 0$
33. $2y - 4 = 0$

Find an equation of the line through the given pair of points.
34. (2, −5), (3, 2)
35. (−$\frac{1}{2}$, 0), (0, $\frac{1}{4}$)
36. (−3, 1), (7, 8)
37. (5, 3), (−1, 6)
38. (−3, −7), (−4, −5)

Find an equation of the line with the given x-intercept a and y-intercept b.
39. $a = -1$, $b = 2$
40. $a = -\frac{1}{3}$, $b = -1$
41. $a = 4$, $b = -\frac{1}{2}$
42. $a = \frac{1}{4}$, $b = \frac{1}{8}$
43. $a = 5$, $b = \frac{1}{5}$

If the inclination of a line has the given value, how are the x-intercept a and y-intercept b related?
44. 45°
45. 60°
46. 135°

Find the point of intersection of the given pair of lines.
47. $5x + y - 2 = 0$, $2x - 2y + 1 = 0$
48. $3x - 2y + 4 = 0$, $3x + y - 5 = 0$
49. $2x + 6y - 1 = 0$, $x + 3y + 4 = 0$
50. $-4x + 5y + 1 = 0$, $3x + 4y + 7 = 0$

Find an equation of the line through the point P parallel to the given line.
51. $P = (0, 0)$, $x + y + 1 = 0$
52. $P = (2, -3)$, $3x - 7y + 3 = 0$
53. $P = (1, 2)$, $x + 9y - 11 = 0$
54. $P = (-4, 1)$, $16x - 24y - 7 = 0$

Find an equation of the line through the point P perpendicular to the given line.
55. $P = (0, 0)$, $3x - y + 2 = 0$
56. $P = (2, 3)$, $4x + 3y + 5 = 0$
57. $P = (-1, 4)$, $x - 2y - 7 = 0$
58. $P = (0, 5)$, $2x - 5y + 6 = 0$

Find an equation of the perpendicular bisector of the line segment joining the given points.
59. $(2, 1), (1, 2)$ **60.** $(7, 4), (-3, 5)$
61. $(3, 3), (0, -1)$ **62.** $(-5, -2), (6, -4)$

Find the distance between the point P and the given line.
63. $P = (2, -1)$, $4x + 3y + 10 = 0$
64. $P = (0, 3)$, $5x - 12y - 29 = 0$
65. $P = (-2, 3)$, $2x - y - 3 = 0$
66. $P = (1, -2)$, $x - 2y - 5 = 0$

Find the distance between the given pair of parallel lines.
67. $3x - 4y - 10 = 0$, $6x - 8y + 5 = 0$
68. $5x - 12y + 26 = 0$, $5x - 12y - 13 = 0$
69. $4x - 3y + 15 = 0$, $8x - 6y + 25 = 0$
70. $24x - 10y + 39 = 0$, $12x - 5y - 26 = 0$

Key Terms and Topics
Sets and numbers, the empty set
Rational and irrational numbers, real numbers and the real line
Algebraic manipulation of inequalities
Maximum and minimum of a set of n numbers
Powers and roots, laws of exponents
Absolute value and the triangle inequality
Closed, open and half-open intervals; infinite intervals
Neighborhoods and deleted neighborhoods
Ordered pairs and rectangular coordinates
Distance between two points in the plane
Graphs of equations and inequalities
Equations of circles and completing the square
Slope, inclination and intercepts of a line
Point-slope and slope-intercept equations of a line
The perpendicularity condition
Distance between a point and a line

Supplementary Problems

The set of all elements belonging to at least one of two given sets A and B is called the *union* of A and B, denoted by $A \cup B$, and the set of all elements belonging to both A and B is called the *intersection* of A and B, denoted by $A \cap B$. Find $A \cup B$ and $A \cap B$ if
1. $A = \{1, 2, 3, 4\}$, $B = \{0, 1, 2, 3\}$
2. $A = \{x: x^2 = 4\}$, $B = \{x: 2x = 4\}$
3. $A = \{x: x \geq 1\}$, $B = \{x: |x| > 1\}$
4. $A = \{x: x^2 - 2x + 1 = 0\}$, $B = \{x: x^2 + 1 = 0\}$

Given two sets A and B, by the *difference* between A and B, denoted by $A - B$, is meant the set of all elements belonging to A but not to B. Let $A = \{1, 2, 3\}$. Find $A - B$ if
5. $B = \{1, 2\}$ **6.** $B = \{4, 5\}$
7. $B = \varnothing$ **8.** $B = \{1, 2, 3\}$

9. Express $2^4 \cdot 4^3 \cdot 8^2 \cdot 16$ as a power of 2, and as a power of 4.

10. Show that a^n is nonnegative for all a if n is even, while a^n has the same sign as a ($\neq 0$) if n is odd.

36   Chapter 0   Precalculus

11. Insert another rational number between $\frac{31}{100}$ and $\frac{3111}{10000}$.
12. Let $r = m/n$ be a rational number in lowest terms, with positive n. Show that a^r is defined for negative a if n is odd, but not if n is even.

Show that if $a > 0$, then

13. $a + \dfrac{1}{a} \geq 2$
14. $\sqrt{a+1} - \sqrt{a} < \dfrac{1}{\sqrt{a+1}}$

Verify that

15. $|a - b| = \max\{a, b\} - \min\{a, b\}$
16. $\max\{a, b\} = \dfrac{(a+b) + |a-b|}{2}$,
    $\min\{a, b\} = \dfrac{(a+b) - |a-b|}{2}$

17. What happens to the point $(1-x)a + xb$ on the real line as x varies from 0 to 1? (Assume that $a \neq b$.)
18. When does the point x^2 lie to the right of x? To the left of x? Coincide with x?
19. Without making numerical calculations, show that

$$\frac{1}{2}\frac{3}{4}\frac{5}{6} \cdots \frac{99}{100} < \frac{1}{10}.$$

20. Show that $ab + bc + ac \leq a^2 + b^2 + c^2$ for arbitrary numbers a, b and c.

Find $M = \max\{a, a^2, \ldots, a^n\}$ and $m = \min\{a, a^2, \ldots, a^n\}$ for $n > 1$ if

21. $0 < a < 1$
22. $a > 1$
23. $a = 0, \pm 1$
24. $-1 < a < 0$
25. $a < -1$

Write each of the given sets, which is a union or intersection of two intervals (see the preamble to Problems 1–4), as a single interval.

26. $(-\infty, 1) \cup (0, \infty)$
27. $[-1, 2) \cup [2, 4)$
28. $[-2, 3] \cap [0, 4]$
29. $(-\infty, 1] \cap (-2, \infty)$

Describe the graph of the given equation.
30. $x^2 + y^2 - 2x + 4y + 14 = 0$
31. $x^2 + y^2 + x = 0$
32. $x^2 + y^2 + y = 0$
33. $x^2 + y^2 + 6x - 4y + 12 = 0$
34. Graph the equation $x^2 - y^2 = 0$.
35. Graph the inequality $x^2 - y^2 > 0$.
36. Graph the simultaneous inequalities $x^2 + y^2 < 1$, $x^2 - y^2 > 0$.
37. Find all points of the xy-plane which are equidistant from the x-axis, the y-axis and the point $(3, 6)$.
38. Find all points of the circle $x^2 + y^2 = 1$ which are equidistant from the points $(1, 3)$ and $(-2, 2)$.

39. How many points (m, n), where m and n are both integers, lie inside the circle of radius $\frac{5}{2}$ with its center at the origin?
40. Find the path traversed by a point $P = (x, y)$ which moves in such a way that the difference between the squares of its distances from the points $(-1, 1)$ and $(1, -1)$ is always equal to 4.
41. What is the area of the triangular region bounded by the coordinate axes and the line $2x + 5y - 20 = 0$?
42. Of all the lines through the point $(2, 3)$, two have equal intercepts. Find these lines.
43. Find the line through the point $(1, 2)$ which is perpendicular to the line through the points $(2, 4)$ and $(3, 5)$. What is the point of intersection of these two lines?

Find the line joining the origin to the point of intersection of the given pair of lines.
44. $x + 2y - 3 = 0$, $x - 3y + 7 = 0$
45. $2x + 3y + 4 = 0$, $x - 2y - 3 = 0$

46. Verify that the quadrilateral with vertices $(2, -2)$, $(5, 1)$, $(3, 6)$ and $(0, 3)$ is a parallelogram. Find the equations of its diagonals. What is the point of intersection of the diagonals?

Given two intersecting lines L_1 and L_2, there are two other lines, the dashed lines in Figure 34, which are the *bisectors* of the angles between L_1 and L_2 (why are the bisectors always perpendicular?). Find the bisectors of the angles between the given pair of lines, and in each case verify that they are perpendicular.
47. $x - y = 0$, $x + y = 0$
48. $2x + 3y - 4 = 0$, $3x + 2y + 1 = 0$
49. $6x + 2y + 1 = 0$, $x - 3y - 2 = 0$
50. $x + y + 2 = 0$, $2x - 2y - 3 = 0$

Hint. The points of each bisector are equidistant from L_1 and L_2.

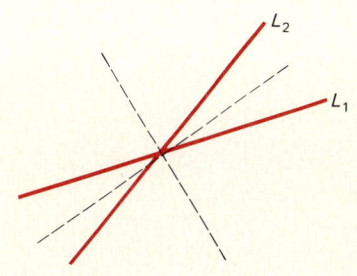

Figure 34

Graph the given linear inequality.
51. $x + 2y - 2 < 0$
52. $x + y - 3 > 0$
53. $3x - 4y + 6 \geq 0$
54. $2x - 3y - 3 \leq 0$

Supplementary Problems 37

Problems 55–62 illustrate how straight lines are used to solve problems arising in business and economics. Let q_d be the quantity of a commodity demanded at price p, and let q_s be the quantity supplied at price p. Then it is often a reasonable approximation to assume that q_d and q_s depend linearly on p, which means that

(i) $$q_d = a + bp,$$
(ii) $$q_s = c + dp,$$

where a, b, c and d are constants. Under normal market conditions, a price increase leads to a decrease in q_d and an increase in q_s. Also q_d, q_s and p are intrinsically nonnegative, because of their economic meaning, and $q_s = 0$ if p is less than a certain number (there is no supply at too low a price). It follows that the coefficients a and d are positive, while b and c are negative (see Figure 35). Note that b is the slope of the *demand curve* (i), while d is the slope of the *supply curve* (ii).

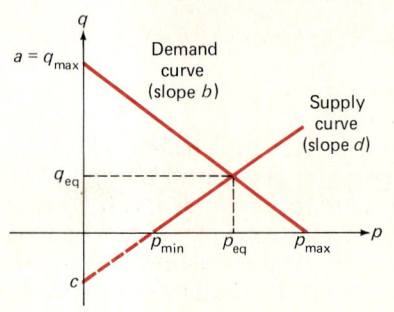

Figure 35

55. Suppose the maximum demand for a certain commodity at any price is q_{max}, and the price at which demand just ceases is p_{max}. Find the demand curve (i).

56. Suppose the price at which a given commodity just begins to be supplied is p_{min}, while the supply goes up d units for every unit increase in the price. Find the supply curve (ii).

57. The market for a commodity will be in equilibrium when the quantity demanded equals the quantity supplied. Determine the corresponding *equilibrium price* p_{eq} and *equilibrium demand* (or *equilibrium supply*) q_{eq} for the linear market model (i), (ii).

58. Suppose the demand for a commodity increases by the same amount at every price, as might occur in the sugar market, say, after the government enforces a ban on certain sugar substitutes suspected of being carcinogenic. Show that this has the effect of increasing both the equilibrium price and the equilibrium demand.

Find the equilibrium price p_{eq} and equilibrium demand q_{eq} for the market with the given demand and supply curves.
59. $q_d = 450 - 3p$, $q_s = -100 + 2p$
60. $q_d = 1000 - 40p$, $q_s = -50 + 10p$
61. $q_d = 3000 - 12p$, $q_s = -2000 + 38p$
62. $q_d = 1600 - 5p$, $q_s = 75p$

1 Functions and Limits

In the first four sections of this chapter we investigate functions and their graphs, with special emphasis on the important *trigonometric functions*. The function concept stands at the border between precalculus mathematics and calculus, and it is only in Section 1.5, when we arrive at the idea of a *limit* and the closely related idea of *continuity*, that we start doing calculus as such. In fact, calculus is often defined as that part of mathematics in which limits play a predominant role.

One of the main concerns of this book is a special kind of limit, known as the *derivative*, which is of supreme importance in the applications. So bear in mind that the limit techniques presented here, despite their theoretical flavor, are really being stockpiled for later use when we study derivatives.

1.1 The Function Concept

Functions and Variables

By a *function* we mean a correspondence between two sets of numbers with the following key property: To each number in the first set, called the *domain* (*of definition*) of the function, there corresponds *one and only one* number in the second set. It is traditional to regard the numbers in the first set as the values of an *independent variable* and the numbers in the second set that correspond to numbers in the first set as the values of a *dependent variable*; the word "variable" means a symbol used to represent an unspecified member of some set. We then say that the dependent variable *is a function of* the independent variable, whatever these variables may be in any particular problem. Note that in this language the domain of the function is the set of all values taken by the independent variable. The set of all values taken by the *dependent* variable is called the *range* of the function.

Example 1 The area of a square is a function of its side length, since if s is the side length of the square, its area A is given by the formula

$$A = s^2.$$

Here s is the independent variable, and A is the dependent variable. The domain and range of the function are the same set, namely the set of all positive numbers. The area of a square is also a function of its perimeter. In fact, a square of side length s has perimeter $p = 4s$. Therefore $s = \frac{1}{4}p$ and $A = s^2 = (\frac{1}{4}p)^2$, so that

$$A = \frac{1}{16}p^2.$$

Here the dependent variable is the area A as before, but the independent variable is now the perimeter p.

Example 2 Is the area of a rectangle a function of its perimeter?

Solution No, since knowledge of the perimeter of a general rectangle (as opposed to a square) does not uniquely determine its area. Thus, for example, a rectangle of length 15 and width 3 has perimeter $15 + 3 + 15 + 3 = 36$ and area $15 \cdot 3 = 45$, while a square of side 9 has the same perimeter $9 + 9 + 9 + 9 = 36$, but a different area $9^2 = 81$ (see Figure 1). □

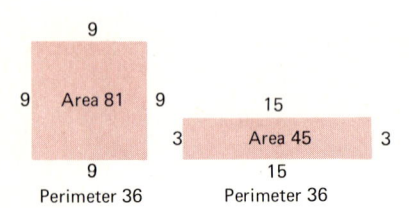

Figure 1

Function Notation

To be explicit, let us denote the independent variable by x, the dependent variable by y, and the function by f. The idea of representing a *function* by a symbol like f is a major intellectual breakthrough, due to the great Swiss mathematician Leonhard Euler (1707–1783). After all, the values of x and y are numbers, but f is something more abstract. In fact, f can be thought of as a rule or procedure establishing a correspondence between the values of x and those of y, and assigning a *unique* value of y to any given value of x. We express this symbolically by writing

$$y = f(x),$$

which is read as "y equals f of x." If x is assigned any particular value c, the corresponding value of y is denoted by $f(c)$ and is called the *value of f at c*. However, it is simpler to use the same letter to denote both the independent variable and its values, and with this understanding we refer to $f(x)$ as the value of f at x. We say that f is *defined on* (or *in*) a given set if every point of the set belongs to the domain of f.

Example 3 Let f be the function which takes the square root of a given number x. Then

$$f(x) = \sqrt{x}. \tag{1}$$

There is admittedly a distinction between the function f (the "square root taker") and its value at x, but it would be pedantic to burden our language with this distinction. Thus we do not say "the function f such that $f(x) = \sqrt{x}$," although this is logically correct. Rather we simply say "the function $f(x) = \sqrt{x}$," or "the function $y = \sqrt{x}$" if y is the dependent variable, or even more concisely "the function $\sqrt{x}$." Notice that the largest set on which the function (1) can be defined is the set of all nonnegative numbers,

40   Chapter 1   Functions and Limits

since we cannot take square roots of negative numbers. Some typical values of the function (1) are

$$f(0) = \sqrt{0} = 0, \qquad f(9) = \sqrt{9} = 3, \qquad f(50) = \sqrt{50} = 5\sqrt{2}. \qquad \square$$

Whenever a function is given by an explicit formula like (1), with no information about the values of the independent variable, it is understood that the domain of the function is the *largest* set of values for which the formula makes sense. This set is called the *natural domain* of the function. For example, the natural domain of the function (1) is the interval $0 \leq x < \infty$. Similarly, if

$$y = f(x) = \sqrt{1 - x^2}, \tag{2}$$

the natural domain of f is the interval $-1 \leq x \leq 1$, since $1 - x^2$ is negative if x lies outside this interval. Any set smaller than the natural domain of a function f can also serve as the domain of f, but in such cases we will always explicitly indicate the domain, as in the formula

$$y = f(x) = \sqrt{1 - x^2} \qquad (0 < x < \tfrac{1}{2}), \tag{2'}$$

where the domain is now the interval $0 < x < \tfrac{1}{2}$, instead of $-1 \leq x \leq 1$.

Example 4 Let

$$y = f(x) = \frac{1}{x^2 - 4}. \tag{3}$$

Find $f(0)$, $f(1)$, $f(2)$ and $f(-3)$.

Solution To find $f(0)$, we substitute $x = 0$ into formula (3). This gives

$$f(0) = \frac{1}{0^2 - 4} = -\frac{1}{4},$$

and similarly

$$f(1) = \frac{1}{1^2 - 4} = -\frac{1}{3}, \qquad f(-3) = \frac{1}{(-3)^2 - 4} = \frac{1}{5}.$$

On the other hand, the quantity

$$f(2) = \frac{1}{2^2 - 4} = \frac{1}{4 - 4}$$

is not defined, since it involves division by zero! Notice that the natural domain of f consists of all numbers x except 2 and -2. $\quad\square$

There is nothing obligatory about the use of the letters x, y and f for the independent variable, dependent variable and function, although these are frequent choices. Thus in Example 1 we can write the function leading from the perimeter p of a square to its area A as

$$A = \phi(p) = \frac{1}{16} p^2,$$

choosing the symbol ϕ (lowercase Greek phi) for the function. As in this case, the variables are often denoted by letters that suggest the geometric or physical quantities under discussion. For example, A is habitually used for area, V for volume, t for time, and so on. The term *argument* is a traditional synonym for the independent variable. Thus x is the argument of the function $f(x) = x^2$, u is the argument of the function $g(u) = u^3 - 1$, and so on.

Our functions will usually be defined with the help of formulas, as in the examples considered above, but *there is no reason whatsoever for this to be the case in general*, and the next example gives a function for which there is no formula relating the independent and dependent variables. In fact, in the last analysis, a function is just a collection of distinct ordered pairs of real numbers, no two of which have the same first element (see Problem 58). Also the functions considered here are *real functions of a single real variable*, that is, both the independent and dependent variables are real numbers or equivalently points on the real line. Later in the book we will generalize the concept of a function by allowing first the dependent variable and then the independent variable to be a point in the plane or in a space of three (or more) dimensions.

Example 5 Let p be the closing price of U.S. Steel on the New York Stock Exchange, and let d be any past date on which the Exchange was open. (You can regard d as an eight-digit number, with the first two digits giving the month, the next two the day, and the last four the year, so that 10241929 is October 24, 1929, 07041976 is July 4, 1976, and so on.) Then p is a function of d. Although there is no explicit formula relating the variables p and d, we can always find the value of p corresponding to a given value of d by *looking up* the one and only value of p listed in the financial section of an evening newspaper published on day d. The fact that p is unique guarantees that p is a function of d. We can express this symbolically by writing $p = h(d)$, this time choosing the symbol h for the function.

Example 6 Suppose we drop a stone into a deep dry well. Let s be the distance in feet fallen by the stone, and let t be the elapsed time in seconds after dropping the stone. Then, as shown in physics, the formula

$$s = 16t^2 \tag{4}$$

expresses s as a function of t, to a good approximation. However, formula (4) is valid only for a limited time, since the stone eventually hits the bottom of the well. If the well is 64 ft deep, the stone hits the bottom of the well after an elapsed time of 2 sec and is subsequently motionless (assuming that it does not bounce). In this case, formula (4) makes sense only for $0 \leq t \leq 2$, that is, the domain of the function (4) is the interval $0 \leq t \leq 2$. We can make this explicit by writing

$$s = 16t^2 \quad (0 \leq t \leq 2), \tag{4'}$$

instead of (4). The subsequent behavior of the stone is described by the formula $s = 64$, or more exactly by

$$s = 64 \quad (t > 2).$$

Incidentally, here we see the desirability of having *constant* functions at our disposal, that is, functions that take only one value.

The last two formulas can be combined into a single formula

$$s = \begin{cases} 16t^2 & \text{if } 0 \leq t \leq 2, \\ 64 & \text{if } t > 2. \end{cases} \tag{4''}$$

The domain of this new function is the infinite interval $0 \leq t < \infty$. Notice that the two functions (4') and (4''), although different, have the same range, namely the interval $0 \leq s \leq 64$.

Problems

Let $f(x) = x^2 + 3x + 5$. Find
1. $f(0)$
2. $f(-1)$
3. $f(1)$
4. $f(2)$
5. $f(-7)$
6. $f(\sqrt{3})$

Let $g(x) = x^3 - 6x^2 + 11x - 6$. Find
7. $g(-1)$
8. $g(0)$
9. $g(1)$
10. $g(2)$
11. $g(3)$
12. $g(4)$

Let $h(x) = |x|/x$. Find
13. $h(0)$
14. $h(1)$
15. $h(-1)$
16. $h(\pi)$
17. $h(100)$
18. $h(-99)$

Let $F(s) = (s-1)/(s+1)$. Find
19. $F(1)$
20. $F(0)$
21. $F(\frac{9}{10})$
22. $F(-1)$
23. $F(\sqrt{2})$
24. $F(1+a)$

Let $G(t) = \sqrt{4-3t}$. Find
25. $G(1)$
26. $G(0)$
27. $G(1.33)$
28. $G(1.34)$
29. $G(4)$
30. $G(-4)$

Let $\phi(u) = 2u^2 - |u|$. Find
31. $\phi(3)$
32. $\phi(-\frac{1}{2})$
33. $\phi(-\sqrt{5})$
34. $\phi(\sqrt[4]{2})$
35. $\phi(1-\pi)$
36. $\phi(\sqrt{3}-2)$

37. Is the area of a circle a function of its circumference?
38. Is the area of a triangle a function of its perimeter?
39. Let c be the number of commas on page p of this book. Is c a function of p? Is p a function of c?
40. Is the weight of a first-class letter a function of its postage?
41. Let $f(n) = a_n$, where

$$3.a_1 a_2 \ldots a_n \ldots \qquad (0 \leq a_n \leq 9)$$

is the decimal representation of the number π. Which is larger, $f(4)$ or $f(5)$?

42. The function some of whose values are given by the table

x	0	20	—	60	80	100
y	32	68	104	140	—	212

is familiar from everyday life. What is it? Find a formula expressing y as a function of x and one expressing x as a function of y. Fill in the two missing entries in the table. What is y if $x = -40$?

43. Calculate

$$\frac{f(1+a) - f(1)}{a} \qquad (a \neq 0),$$

where $f(x) = x^2$. (Quotients of this type arise in the study of derivatives.)

44. Calculate

$$\frac{f(-2+a) - f(-2)}{a} \qquad (a \neq 0),$$

where $f(x) = x^3$.

Find the (natural) domain of the given function.
45. $y = \dfrac{x(x+1)}{x}$
46. $y = \dfrac{1}{2x-3}$
47. $y = \dfrac{1}{x+|x|}$
48. $y = \sqrt{x+2}$
49. $y = \sqrt[3]{x+1}$
50. $y = \sqrt{x^2-9}$
51. $y = \sqrt{16-x^2}$
52. $y = \sqrt{4x-x^2}$
53. $y = \sqrt{\dfrac{x-1}{2-x}}$

Find a function $y = f(x)$ with the given interval as its natural domain.
54. The closed interval $[0, 1]$
55. The open interval $(0, 1)$
56. The half-open interval $[0, 1)$
57. The half-open interval $(0, 1]$

58. Verify that the following definitions agree in all essentials with those given in this section. Let f be a set of distinct ordered pairs of real numbers such that no two pairs (x, y) in f have the same first element, and let $D = \{x : (x, y) \in f\}$, that is, let D be the set of all first elements of pairs in f. Then f is said to be a *function defined on D*, and D is called the *domain* of f. If (x, y) is an ordered pair in f, then y, the second element of the pair, is called the *value of f at x*, written $f(x)$. The set $\{y : (x, y) \in f\}$, that is, the set of all second elements of pairs in f, is called the *range* of f.

Hint. Regard x as the independent variable and y as the dependent variable. Then knowledge of x uniquely determines y.

1.2 Operations on Functions; Graphs of Functions

Given two functions f and g, let D be the largest set on which f and g are both defined (D is assumed to be nonempty). Then by the *sum* $f + g$ we mean the function whose value at every point x in D is the sum of the value of f at x and the value of g at x. More concisely,

$$(f + g)(x) = f(x) + g(x)$$

for every x in D. Other algebraic operations on f and g are defined similarly, that is,

$$(cf)(x) = cf(x)$$
$$(f - g)(x) = f(x) - g(x),$$
$$(fg)(x) = f(x)g(x)$$
$$f^n(x) = \underbrace{f(x)f(x) \cdots f(x)}_{n \text{ factors}},$$

$$\left(\frac{f}{g}\right)(x) = \frac{f(x)}{g(x)},$$

where in the last formula it is of course assumed that $g(x) \neq 0$ for all x in D.

Example 1 Let

$$f(x) = \sqrt{x - 1}, \qquad g(x) = \frac{1}{x + 1}.$$

Find the values of $f + g$ and fg at the point $x = 5$.

Solution Since

$$f(5) = \sqrt{5 - 1} = 2, \qquad g(5) = \frac{1}{5 + 1} = \frac{1}{6},$$

we have

$$(f + g)(5) = f(5) + g(5) = 2 + \frac{1}{6} = \frac{13}{6},$$

$$(fg)(5) = f(5)g(5) = 2\left(\frac{1}{6}\right) = \frac{1}{3}.$$

The natural domain of the functions $f + g$ and fg is the interval $[1, \infty)$, since this is the largest set on which f and g are *both* defined. ☐

The Symbol ≡

By $f(x) \equiv g(x)$ we mean that the functions f and g have the same domain D and $f(x) = g(x)$ for every x in D. The formula $f(x) \equiv g(x)$, called an *identity*, is read "$f(x)$ is identically equal to $g(x)$." Equality of functions means identical equality, that is, $f = g$ means $f(x) \equiv g(x)$.

Example 2 The functions $f(x) = |x|$ and $g(x) = \sqrt{x^2}$ are (identically) equal. Their common domain of definition is the whole real line $-\infty < x < \infty$.

In cases where the context makes it clear that a formula is an identity, the equality sign $=$ is often used instead of the identity sign $\equiv$. For example, the familiar factorization formula $x^2 - 1 = (x + 1)(x - 1)$ is an identity, since it is valid for all x.

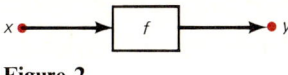

Figure 2

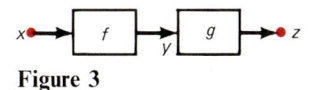

Figure 2 shows a useful way of thinking of a function f as an input-output device. A number x is fed into the device f, and out comes the number $y = f(x)$, the value of f at x. This is also described by saying that *f maps x into y*, or that *y is the image of x under f*. The fact that f is a function means that a given input x always produces the same output y. In general, we may not know how the device works, and engineers call a device a "black box" when its contents are unknown or ignored.

Composite Functions

It is now a natural question to ask what happens if the output of one device serves as the input of another device, as in Figure 3, where x is fed into a device f, resulting in the output y, which is then fed in turn into a second device g, resulting in the final output z. Since $z = g(y)$ and $y = f(x)$, it is clear that $z = g(f(x))$. A function like $g(f(x))$ is called a *composite* function, and such functions occur throughout calculus. The operation of combining functions in this way is called *composition*. Of course, in Figure 3 we must insist that the "intermediate" output y be an acceptable input for the second device g; this is because $g(f(x))$ is defined only for values of x such that $f(x)$ is in the domain of g. In the same way, we can define other composite functions, like $f(g(x))$, $f(f(x))$, and so on.

Figure 3

Example 3 Let

$$f(x) = x + 1, \qquad g(x) = x^2. \tag{1}$$

Then straightforward substitution gives

$$g(f(x)) = g(x + 1) = (x + 1)^2 = x^2 + 2x + 1 \tag{2}$$

and

$$f(g(x)) = f(x^2) = x^2 + 1. \tag{3}$$

In the same way,

$$f(f(x)) = f(x + 1) = (x + 1) + 1 = x + 2$$

and

$$g(g(x)) = g(x^2) = (x^2)^2 = x^4. \quad \square$$

The Symbol ∘

The function leading from x to $g(f(x))$ is denoted by $g \circ f$. Thus

$$(g \circ f)(x) = g(f(x)),$$

and similarly $(f \circ g)(x) = f(g(x))$, $(f \circ f)(x) = f(f(x))$, etc. With this notation (2) becomes

$$(g \circ f)(x) = x^2 + 2x + 1, \tag{2'}$$

(3) becomes

$$(f \circ g)(x) = x^2 + 1, \tag{3'}$$

and so on. The meaning of $g \circ f$ as an input-output device is illustrated in Figure 4(a). It is the single device which has the same effect as the two devices f and g connected "in series," with f first and g second. If f and g are connected in the opposite order, with g first and f second, the combined effect of the two devices is the same as that of the single device $f \circ g$ [see Figure 4(b)].

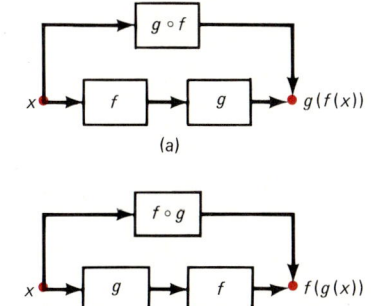

Figure 4

Section 1.2 Operations on Functions; Graphs of Functions

Never make the mistake of confusing the composite function $f \circ g$ with the product function fg (the symbol $\circ$ calls for composition, not multiplication). For example, the product of the functions (1) is

$$(fg)(x) = (x+1)x^2 = x^3 + x^2,$$

which is entirely different from the composite functions (2') and (3'). Composition of functions is *noncommutative*, that is, in general $f \circ g \neq g \circ f$, as in the case of the functions just considered, whereas multiplication of functions is of course always commutative ($fg = gf$).

Example 4 Let $f(x) = \sqrt{x}$ and $g(x) = -x^2$. Then

$$(f \circ g)(x) = f(g(x)) = \sqrt{-x^2}$$

is undefined for every value of x except $x = 0$, and $(f \circ g)(0) = 0$, while

$$(g \circ f)(x) = g(f(x)) = -(\sqrt{x})^2 = -x$$

is defined for every nonnegative value of x.

The operation of composition can involve more than two functions. It is then performed in stages, moving from right to left in the case of the $\circ$ notation, and from the inside out in the case of parentheses.

Example 5 Let

$$f(x) = \frac{1}{x}, \qquad g(x) = x^2, \qquad h(x) = \sqrt{x}.$$

Then

$$(f \circ g \circ h)(x) = f(g(h(x))) = f(g(\sqrt{x})) = f((\sqrt{x})^2) = f(x) = \frac{1}{x},$$

$$(h \circ g \circ f)(x) = h(g(f(x))) = h\left(g\left(\frac{1}{x}\right)\right) = h\left(\left(\frac{1}{x}\right)^2\right) = h\left(\frac{1}{x^2}\right)$$

$$= \sqrt{\frac{1}{x^2}} = \frac{1}{|x|},$$

and so on. Notice that the domain of $f \circ g \circ h$ is the set of all $x > 0$, while the domain of $h \circ g \circ f$ is the set of all $x \neq 0$. As an exercise, show that composition of functions is *associative*, meaning that $(f \circ g) \circ h = f \circ (g \circ h)$, as assumed in writing $f \circ g \circ h$ without parentheses.

The Vertical Line Property

We turn now to the pictorial representation of functions. By the *graph* of a function

$$y = f(x) \qquad (4)$$

we mean the geometric figure in the xy-plane obtained by plotting all the points (x, y) whose coordinates satisfy formula (4), regarded as an equation in the two variables x and y. The graph of a function $y = f(x)$, as opposed to the graph of a more general equation in the variables x and y, has the following distinguishing property: *No vertical line, that is, no line parallel to the y-axis, can intersect the graph of $y = f(x)$ in more than one point.* For if some vertical line $x = c$, say, were to intersect the graph of $y = f(x)$ in two or more points, then two or more values of y, namely the ordinates of these

46 Chapter 1 Functions and Limits

points, would correspond to the same value of x, namely c, and this would contradict the very definition of a function.

Example 6 No vertical line intersects the curve G in Figure 5(a) in more than one point. Therefore G is the graph of a function. On the other hand, there are vertical lines intersecting the curve G' in Figure 5(b) in more than one point; for example, the line $x = c$ intersects G' in three points. Therefore G' is *not* the graph of a function.

G is the graph of a function.
(a)

G' is not the graph of a function.
(b)

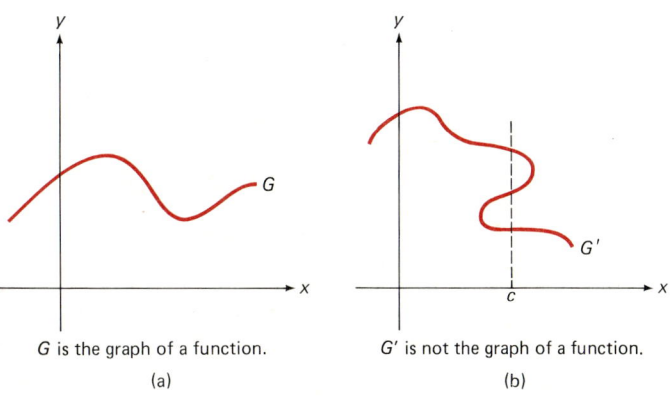

Figure 5

Example 7 Graph the absolute value function

$$y = |x|. \tag{5}$$

Solution If $x \geq 0$, then $|x| = x$ and (5) reduces to the straight line $y = x$ of slope 1 going through the origin, while if $x < 0$, then $|x| = -x$ and (2) reduces to the straight line $y = -x$ of slope -1 going through the origin. Thus the graph of the function (5), shown in Figure 6, is made up of pieces of the lines $y = x$ and $y = -x$. Notice that the graph has a sharp corner at the origin, where the lines $y = x$ and $y = -x$ intersect. □

A function like $|x|$, whose graph is made up of pieces of straight lines, is said to be *piecewise linear* (if there is just one line, the function is said to be *linear*).

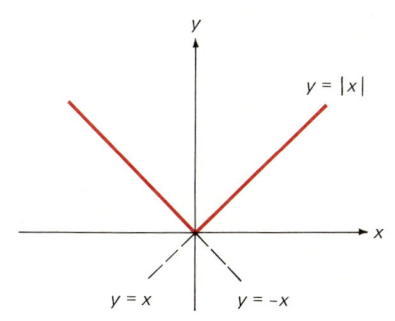

Figure 6

Example 8 Graph the absolute value function

$$y = x^2. \tag{6}$$

Solution We begin by making a little table of some values of x and the corresponding values of y given by formula (6):

x	0	$\frac{1}{2}$	$-\frac{1}{2}$	1	-1	$\frac{3}{2}$	$-\frac{3}{2}$
y	0	$\frac{1}{4}$	$\frac{1}{4}$	1	1	$\frac{9}{4}$	$\frac{9}{4}$

Then we plot the pairs (x, y) as points in the xy-plane and connect them with a smooth curve. This gives the curve shown in Figure 7, an example of a

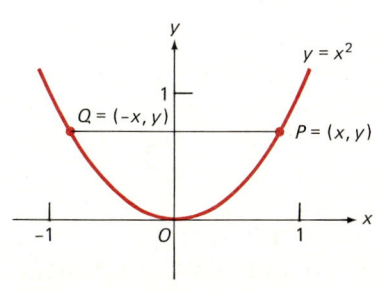

Figure 7

Section 1.2 Operations on Functions; Graphs of Functions 47

parabola. Since the domain and range of the function (6) are both infinite intervals (which ones?), the figure is actually only the part of the graph near the origin. If the point $P = (x, y)$ belongs to the graph of (6), then so does the point $Q = (-x, y)$, since $(-x)^2 = x^2$. Also P and Q are equidistant from the y-axis and lie on the same horizontal line. Thus for every point P of the graph on one side of the y-axis, there is a point Q of the graph on the other side such that the y-axis is the perpendicular bisector of the segment PQ. This is summarized by saying that the curve $y = x^2$ is *symmetric about the y-axis*. ☐

Remark The fact that we lose no essential features of the graph of (6) by connecting a few "typical" points in this way seems entirely plausible, and can be justified with the help of the calculus methods for curve sketching developed in Chapter 3 or the geometric characterization of the parabola given in Chapter 10. More generally, the graph G of any function of the form $y = ax^2 + bx + c$, where a, b and c are constants (with $a \neq 0$) is a parabola, and so is any figure obtained by shifting or rotating G in the xy-plane.

Even Functions

A function f is said to be *even* if

$$f(-x) \equiv f(x). \tag{7}$$

We have just shown that the function $f(x) = x^2$ is even and that its graph is symmetric about the y-axis. The graph of every other even function has the same symmetry property. For example, the function $f(x) = |x|$ is even, since $|-x| \equiv |x|$, and hence the graph of $y = |x|$ is symmetric about the y-axis, as is apparent from Figure 6.

Example 9 Graph the function

$$y = x^3. \tag{8}$$

Solution Plotting some typical points (x, y) with y given by (8) and connecting them by a smooth curve, we get the graph shown in Figure 8. If the point $P = (x, y)$ belongs to the graph of (8), then so does the point $Q = (-x, -y)$, since $(-x)^3 = -x^3$. Also by Example 4, page 23, the midpoint of the segment PQ is the point

$$\left(\frac{x + (-x)}{2}, \frac{y + (-y)}{2}\right) = (0, 0),$$

that is, the origin O. Thus for every point P of the graph on one side of the y-axis, there is a point Q of the graph on the other side of the y-axis such that the origin O is the midpoint of the segment PQ. This is summarized by saying that the curve $y = x^3$ is *symmetric about the origin*. ☐

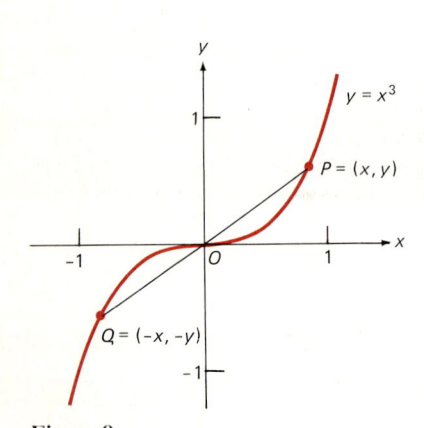

Figure 8

Odd Functions

A function f is said to be *odd* if

$$f(-x) \equiv -f(x). \tag{7'}$$

We have just shown that the function $f(x) = x^3$ is odd and that its graph is symmetric about the origin. The graph of every other odd function has the same symmetry property. For example, every line $y = mx$ through the origin has this property.

The property of evenness or oddness of functions plays an important role in applied mathematics and physics. Given a function f, the question "What is the *parity* of f?" simply means "Is f even or odd?" Of course, most functions are neither even nor odd. This is true, for example, of the function $f(x) = x^2 + x^3$, which does not satisfy either (7) or (7′).

Increasing and Decreasing Functions

Next suppose the graph of a function $y = f(x)$ *rises* as a variable point P on the graph moves from left to right, with its abscissa x in some interval I. Then f is said to be *increasing* on I. Similarly, if the graph of f *falls* as P moves from left to right, with its abscissa x in some interval I, we say that $f(x)$ is *decreasing* on I.

Example 10 The functions $|x|$ and x^2 are both increasing on the interval $0 \leq x < \infty$, as we see at once from Figures 6 and 7. It is also apparent from these figures that $|x|$ and x^2 are decreasing on $-\infty < x \leq 0$.

Example 11 The function x^3 is increasing on the whole real line, that is, on the interval $-\infty < x < \infty$, as we see from Figure 8.

Example 12 The graph of the constant function $f(x) \equiv c$ is the horizontal line $y = c$, which neither rises nor falls. Therefore a constant function is neither increasing nor decreasing (on any interval).

It is easy to give an algebraic definition of an increasing or decreasing function. A function f defined on a set X, usually an interval, is said to be *increasing on* X if $f(s) < f(t)$ whenever $s < t$ (here s and t both lie in X). Similarly, f is said to be *decreasing on* X if the opposite inequality $f(s) > f(t)$ holds whenever $s < t$. The graphical meaning of these definitions is illustrated in Figure 9(a) for an increasing function and in Figure 9(b) for a decreasing function.

Increasing f
(a)

Decreasing f
(b)

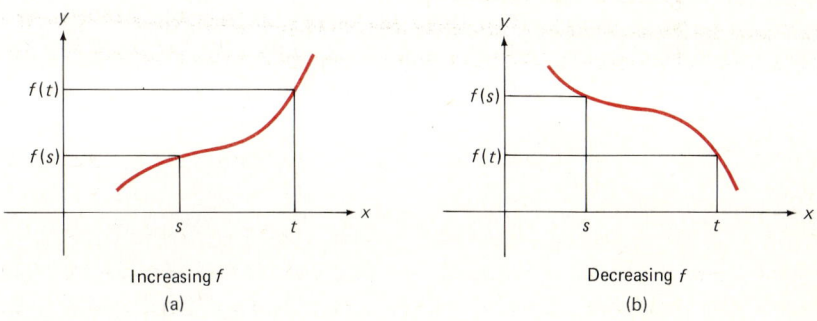

Figure 9

Example 13 Show algebraically that $f(x) = x^2$ is increasing on $[0, \infty)$ and decreasing on $(-\infty, 0]$.

Solution If $0 \leq s < t < \infty$, then $t + s > 0$ and $t - s > 0$, so that

$$f(t) - f(s) = t^2 - s^2 = (t + s)(t - s) > 0,$$

and therefore $f(s) < f(t)$. Similarly, if $-\infty < s < t \leq 0$, then $t + s < 0$ and $t - s > 0$, so that

$$f(t) - f(s) = (t + s)(t - s) < 0,$$

which implies the opposite inequality $f(s) > f(t)$. □

Shifting Graphs

The following theorem shows how to shift the graph of a function in either the horizontal or the vertical direction.

Theorem 1 *(Shifting the graph of a function).* *Given a function*

$$y = f(x), \tag{9}$$

with graph G, the graph of the function

$$y = f(x - c) \tag{10}$$

*is the result of subjecting G to a **horizontal** shift of $|c|$ units, to the right if $c > 0$ and to the left if $c < 0$. Similarly, the graph of the function*

$$y = f(x) + c \tag{10'}$$

*is the result of subjecting G to a **vertical** shift of $|c|$ units, upward if $c > 0$ and downward if $c < 0$.*

Proof The point (x, y) satisfies (9) if and only if the horizontally shifted point $(x + c, y)$ satisfies (10). Similarly, (x, y) satisfies (9) if and only if the vertically shifted point $(x, y + c)$ satisfies (10'). ∎

The fact that there is a minus sign in (10) and a plus sign in (10') seems puzzling until we realize that the exact "y-analogue" of (10) is

$$y - c = f(x),$$

rather than the equivalent equation (10') in which the constant c has been transposed to the right side, as is customary.

Example 14 Let G be the graph of the function $y = |x|$. Then the graph of $y = |x - 1|$ is obtained by shifting G one unit to the right, and the graph of $y = |x + 1|$ is obtained by shifting G one unit to the left [see Figure 10(a)].

Horizontal shifts of a graph
(a)

Vertical shifts of a graph
(b)

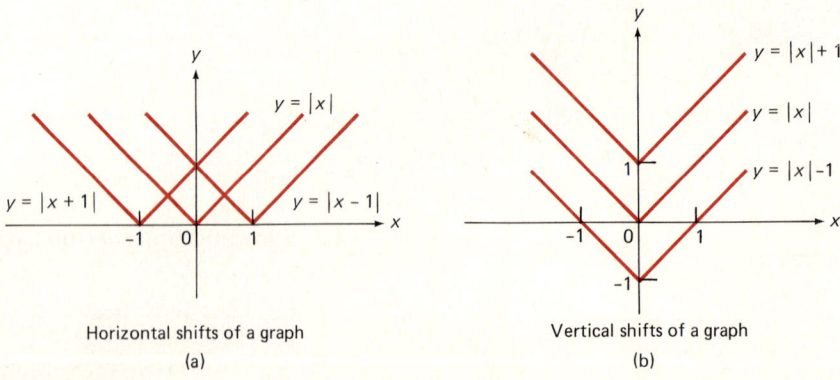

Figure 10

50   Chapter 1   Functions and Limits

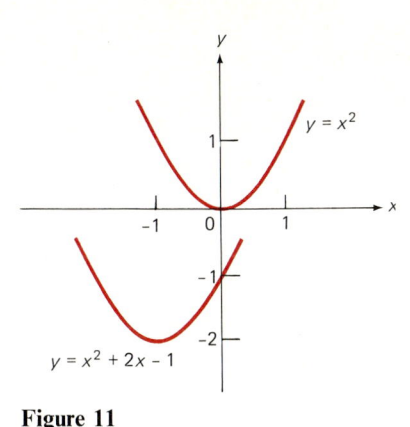

Similarly, the graph of $y = |x| + 1$ is obtained by shifting G one unit upward, and the graph of $y = |x| - 1$ is obtained by shifting G one unit downward [see Figure 10(b)].

Example 15 Graph the function
$$y = x^2 + 2x - 1. \tag{11}$$

Solution Completing the square in (11), we get
$$y = (x + 1)^2 - 2. \tag{11'}$$

By Theorem 1, the graph of (11') is the graph of $y = (x + 1)^2$ shifted 2 units downward, and the graph of $y = (x + 1)^2$ is in turn the graph of $y = x^2$ shifted 1 unit to the left. Thus the graph of (11'), or equivalently of (11), is the graph of the parabola $y = x^2$ shifted 1 unit to the left and 2 units downward, as shown in Figure 11. ☐

Figure 11

Problems

Let $f(x) = \sqrt{x + 1}$ and $g(x) = 1/x$. Evaluate

1. $(f + g)(3)$
2. $(f - g)(8)$
3. $(fg)(-2)$
4. $(2f + 3g)(15)$
5. $(f^2)(24)$
6. $(f/g)(35)$

Let f and g be the given functions. Is it true that $f(x) \equiv g(x)$?

7. $f(x) = x$, $g(x) = (\sqrt{x})^2$
8. $f(x) = x/x$, $g(x) \equiv 1$
9. $f(x) = |x|/x$, $g(x) = x/|x|$
10. $f(x) = (1 + x)^2 - (1 - x)^2$, $g(x) = 4x$

Let $f(x) = x^2 + 1$ and $g(x) = x^2 - 1$. Evaluate

11. $f(f(2))$
12. $f(g(1))$
13. $g(f(-1))$
14. $g(g(0))$
15. $g(f(g(1)))$
16. $f(f(f(0)))$

Let $f(x) = \dfrac{1-x}{1+x}$ and $g(x) = \dfrac{1+x}{1-x}$. Find the specified composite function.

17. $(f \circ f)(x)$
18. $(f \circ g)(x)$
19. $(g \circ f)(x)$
20. $(g \circ g)(x)$

Let $f(x) = x^3$, $g(x) = x - 1$ and $h(x) = \sqrt{x}$. Find the specified composite function.

21. $(f \circ g \circ h)(x)$
22. $g(g(f(x)))$
23. $f(f(g(x)))$
24. $(h \circ g \circ f)(x)$
25. $(g \circ h \circ f)(x)$
26. $f(h(g(x)))$

Find a linear function $f(x) = ax + b$ satisfying the given identity.

27. $f(x + 1) \equiv 2x$
28. $f(2x + 3) \equiv 3x - 2$
29. $f(1 - x) \equiv 5x + 1$
30. $f(f(x)) \equiv 4x + 3$

Graph the given function.

31. $y = x^2 + x + 1$
32. $y = 1 - x^2$
33. $y = \sqrt{x^2 + 4}$
34. $y = \sqrt{9 - x^2}$
35. $y = \dfrac{3}{x^2 + 1}$
36. $y = 3 - \sqrt{16 - x^2}$

37. $y = |x + 1| + |x - 1|$
38. $y = |x| + |x + 1| + |x + 2|$

Identify the given function as even or odd (or neither).

39. $f(x) \equiv 10$
40. $f(x) = \dfrac{1}{x}$
41. $f(x) = x^3 - x + 1$
42. $f(x) = x^4 + x^2 - 7$
43. $f(x) = x^5 - x^3 + x$
44. $f(x) = x^6 + x^3$
45. $f(x) = \dfrac{x^2}{x^4 + 1}$
46. $f(x) = \dfrac{x}{x^2 - 1}$
47. $f(x) = \dfrac{x^3}{\sqrt{x^3 + 2}}$
48. $f(x) = \dfrac{1}{\sqrt[3]{x^3 + x}}$

49. Let f be any function whose domain contains $-x$ whenever it contains x. Show that the identity
$$f(x) = \frac{f(x) + f(-x)}{2} + \frac{f(x) - f(-x)}{2}$$
represents f as the sum of an even function and an odd function.

Write the given function as the sum of an even function and an odd function.

50. $f(x) = \dfrac{2}{x^2 - x + 1}$
51. $f(x) = (1 + x)^{100}$

52. Where is the function $y = x^2 - 4x + 3$ increasing? Decreasing?
53. Where is the function $y = |x + 1| + |x - 1|$ increasing? Decreasing? Constant? (See Problem 37.)

Section 1.2 Operations on Functions; Graphs of Functions

54. A piece of string of length 12 is used to make a rectangular contour, one of whose sides is of length x. Find the largest interval $a \leq x \leq b$ on which the area enclosed by the contour is an increasing function of x.

Find the function $y = g(x)$ whose graph is obtained by subjecting the graph of $y = f(x) = x^2 + x + 1$ to the given pair of shifts.

55. 5 units to the right and 6 units upward
56. 3 units to the right and 4 units downward
57. 2 units to the left and 7 units downward
58. 10 units to the left and 1000 units upward

59. What pair of shifts transforms the graph of $y = f(x) = x^2 + x$ into the graph of $y = g(x) = x^2 + 2x$?

60. Suppose that $f(x + 1) - f(x) \equiv 8x + 3$, where $f(x) = ax^2 + bx + 5$. Find a and b.

61. Show that a function f is increasing on an interval I if and only if its negative $-f$ is decreasing on I.

62. Show that a function f of fixed sign (that is, a function whose values are either all positive or all negative) is increasing on an interval I if and only if its reciprocal $1/f$ is decreasing on I.

1.3 Trigonometric Functions

Let OAB be a right triangle with hypotenuse OA, and let θ (lowercase Greek theta) be the acute angle at the vertex O, as in Figure 12. Then, regarding θ as the independent variable, we introduce the following *trigonometric functions*. First we define the *sine* and *cosine* of θ as the ratios

$$\sin\theta = \frac{\text{side opposite }\theta}{\text{hypotenuse}} = \frac{|AB|}{|OA|}, \qquad \cos\theta = \frac{\text{side adjacent to }\theta}{\text{hypotenuse}} = \frac{|OB|}{|OA|}.$$

We then define the *tangent* and *cotangent* of θ as

$$\tan\theta = \frac{\sin\theta}{\cos\theta}, \qquad \cot\theta = \frac{\cos\theta}{\sin\theta} = \frac{1}{\tan\theta}, \qquad (1)$$

and the *secant* and *cosecant* of θ as

$$\sec\theta = \frac{1}{\cos\theta}, \qquad \csc\theta = \frac{1}{\sin\theta}. \qquad (2)$$

In terms of the triangle OAB,

$$\tan\theta = \frac{\text{side opposite }\theta}{\text{side adjacent to }\theta} = \frac{|AB|}{|OB|},$$

and

$$\cot\theta = \frac{|OB|}{|AB|}, \qquad \sec\theta = \frac{|OA|}{|OB|}, \qquad \csc\theta = \frac{|OA|}{|AB|}.$$

The suitability of these definitions of the six trigonometric functions rests on the fact that all right triangles with the same acute angle θ are similar (why?).

Example 1 In the right triangle shown in Figure 13 we have

$$\sin\theta = \frac{2}{\sqrt{5}}, \qquad \cos\theta = \frac{1}{\sqrt{5}}, \qquad \tan\theta = 2,$$

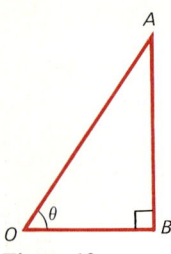

$$\cot\theta = \frac{1}{2}, \qquad \sec\theta = \sqrt{5}, \qquad \csc\theta = \frac{\sqrt{5}}{2}.$$

Figure 12

Figure 13

Notice that once the legs of the triangle are known to be of lengths 1 and 2, the fact that the hypotenuse is of length $\sqrt{5} = \sqrt{1^2 + 2^2}$ follows from the Pythagorean theorem.

Suppose we replace OAB by a similar right triangle OPQ with hypotenuse of length 1. Then OPQ can be placed in the unit circle $x^2 + y^2 = 1$ in such a way that O is the center of the circle, OP is a radius of the circle, and OQ lies along the positive x-axis, as shown in Figure 14. Let the point P have abscissa x and ordinate y. Then

$$\cos\theta = \frac{|OQ|}{|OP|} = \frac{x}{1}, \qquad \sin\theta = \frac{|PQ|}{|OP|} = \frac{y}{1},$$

so that

$$\cos\theta = x, \qquad \sin\theta = y. \qquad (3)$$

To define the cosine and sine for arbitrary θ, and not just for the case where the angle θ is acute (that is, between 0° and 90°), we simply continue to use the formulas (3). By the same token, formulas (1) and (2) are taken to be the definitions of the tangent, cotangent, secant and cosecant for arbitrary values of θ. As always, angles are regarded as positive when measured in the counterclockwise direction, and as negative when measured in the clockwise direction.

As an immediate consequence of these definitions, we obtain the important formulas

$$\cos^2\theta + \sin^2\theta = 1, \qquad (4)$$

$$|\cos\theta| \le 1, \qquad |\sin\theta| \le 1, \qquad (5)$$

valid for all θ. In fact, according to (3), $\cos\theta$ and $\sin\theta$ are the x- and y-coordinates of a variable point P of the unit circle $x^2 + y^2 = 1$, and this immediately implies formula (4), in which it is customary to write $\cos^2\theta$ and $\sin^2\theta$ instead of $(\cos\theta)^2$ and $(\sin\theta)^2$. Also it is clear that the coordinates of any point $P = (x, y)$ of the unit circle satisfy the inequalities $|x| \le 1$, $|y| \le 1$, equivalent to (5). Other trigonometric formulas can be deduced from (4). For example,

$$1 + \tan^2\theta = \sec^2\theta, \qquad (6)$$

since

$$1 + \tan^2\theta = 1 + \frac{\sin^2\theta}{\cos^2\theta} = \frac{\cos^2\theta + \sin^2\theta}{\cos^2\theta} = \frac{1}{\cos^2\theta} = \sec^2\theta,$$

with the help of (1), (2) and (4). Similarly,

$$1 + \cot^2\theta = \csc^2\theta, \qquad (6')$$

as you can easily verify.

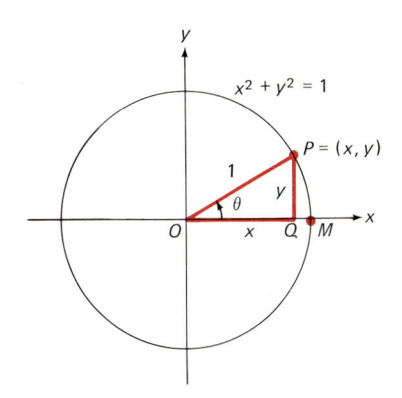

Figure 14

Radian Measure

In Figure 14 let $M = (1, 0)$ be the point of the unit circle lying on the positive x-axis. Then by the radian measure of the angle θ we mean the length of the arc $\overset{\frown}{MP}$ regarded as a "dimensionless" number, that is, without the

Section 1.3 Trigonometric Functions 53

physical dimension of length (as measured in inches or meters, say). Since the circumference of the unit circle is 2π, it follows that 360 degrees = 2π radians, or equivalently 180 degrees = π radians. Therefore

$$1 \text{ degree} = \frac{\pi}{180} \text{ radians} \approx 0.01745 \text{ radian},$$

$$1 \text{ radian} = \frac{180}{\pi} \text{ degrees} \approx 59.29578 \text{ degrees},$$

where the symbol $\approx$ denotes approximate equality. To avoid confusion between degree measure and radian measure, we adopt the convention that *an angle is regarded as measured in radians unless it is accompanied by the word "degrees" or by the degree symbol °*. Thus, with this understanding,

$$\frac{\pi}{6} = \frac{\pi}{6}\frac{180}{\pi} \text{ degrees} = 30°, \quad \frac{\pi}{4} = \frac{\pi}{4}\frac{180}{\pi} \text{ degrees} = 45°,$$

$$60° = 60\left(\frac{\pi}{180}\right) = \frac{\pi}{3}, \quad 90° = 90\left(\frac{\pi}{180}\right) = \frac{\pi}{2},$$

and so on.

More generally, let the circle in Figure 14 be of radius r, so that $|OP| = r$, $M = (r, 0)$ instead of $|OP| = 1$, $M = (1, 0)$, and let L be the length of the arc $\overset{\frown}{MP}$. Then the radian measure of θ is defined as the ratio L/r (if $r = 1$, this reduces to the previous definition). The unit of length cancels out in forming the ratio L/r, and this is why radian measure is "dimensionless." We can now write a formula for the length L of a circular arc of radius r, subtending an angle of θ radians at its center, as in Figure 15. In fact, since $\theta = L/r$, it follows at once that

$$L = r\theta. \tag{7}$$

There is also a simple formula for the area A of a circular sector of radius r whose central angle is θ radians (see Figure 15). Clearly A is directly proportional to θ and equals zero when $\theta = 0$. Therefore $A = k\theta$, where k is a positive constant of proportionality. To determine k, we observe that $A = \pi r^2$ when $\theta = 2\pi$, since the area of a full circular disk of radius r is πr^2. Thus $\pi r^2 = 2\pi k$, or equivalently $k = \frac{1}{2}r^2$, and the formula $A = k\theta$ becomes

$$A = \frac{1}{2}r^2\theta. \tag{8}$$

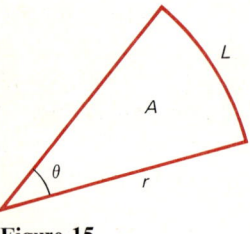

Figure 15

Remark The formulas to be derived later for calculating limits, derivatives and integrals of trigonometric functions will all be based on the tacit assumption that angles are measured in radians, and they take a different, more complicated form if angles are measured in degrees. Therefore make sure to use radian measure in solving calculus problems. Failure to do so is a common mistake. However, it is often desirable to express *answers* to calculus problems in degrees, since degree measure is more usual than radian measure in everyday life.

Let n be any integer. Then adding $2n\pi$ radians to the angle θ in Figure 14 causes the radius OP to make $|n|$ complete revolutions about the origin, in

54 Chapter 1 Functions and Limits

the counterclockwise direction if $n > 0$ and in the clockwise direction if $n < 0$ (OP does not move at all if $n = 0$). This has no effect on the final position of OP, and hence no effect on the coordinates of P, that is, on the values of the cosine and sine. Therefore

$$\cos(\theta + 2n\pi) = \cos\theta, \qquad \sin(\theta + 2n\pi) = \sin\theta$$

for every integer n. Also the radius OP is vertical if and only if $\theta = (n + \tfrac{1}{2})\pi$ and horizontal if and only if $\theta = n\pi$, where n is any integer. Since $P = (\cos\theta, \sin\theta)$, it follows that $\cos\theta = 0$ if and only if $\theta = (n + \tfrac{1}{2})\pi$, while $\sin\theta = 0$ if and only if $\theta = n\pi$. In particular,

$$\cos\frac{\pi}{2} = \cos\frac{3\pi}{2} = 0$$

and

$$\sin 0 = \sin \pi = 0.$$

The radius OP lies along the positive x-axis if $\theta = 0$ and along the negative x-axis if $\theta = \pi$, so that

$$\cos 0 = 1, \qquad \cos \pi = -1,$$

while OP lies along the positive y-axis if $\theta = \pi/2$ and along the negative y-axis if $\theta = 3\pi/2$, so that

$$\sin\frac{\pi}{2} = 1, \qquad \sin\frac{3\pi}{2} = -1.$$

Also, changing θ to $-\theta$ causes the point $P = (x, y) = (\cos\theta, \sin\theta)$ to go into the point $P' = (x, -y) = (\cos\theta, -\sin\theta)$, its reflection in the x-axis (see Figure 16). But $P' = (\cos(-\theta), \sin(-\theta))$, and therefore, by equality of ordered pairs,

$$\cos(-\theta) = \cos\theta, \qquad \sin(-\theta) = -\sin\theta. \tag{9}$$

The first of these important formulas tells us that $\cos\theta$ is an *even* function, and the second that $\sin\theta$ is an *odd* function.

There are some other special values of θ, besides integer multiples of $\pi/2$, for which the values of the trigonometric functions can be found without resorting to numerical tables or to a scientific calculator. Figure 17(a) shows an isosceles right triangle with legs of length 1, from which we can read off

$$\sin\frac{\pi}{4} = \frac{|AB|}{|OA|} = \frac{1}{\sqrt{2}}, \qquad \cos\frac{\pi}{4} = \frac{|OB|}{|OA|} = \frac{1}{\sqrt{2}}, \qquad \tan\frac{\pi}{4} = \frac{|AB|}{|OB|} = 1,$$

and Figure 17(b) shows a bisected equilateral triangle with sides of length 2, from which we can read off

$$\sin\frac{\pi}{6} = \frac{|AB|}{|OA|} = \frac{1}{2}, \qquad \cos\frac{\pi}{6} = \frac{|OB|}{|OA|} = \frac{\sqrt{3}}{2}, \qquad \tan\frac{\pi}{6} = \frac{|AB|}{|OB|} = \frac{1}{\sqrt{3}},$$

$$\sin\frac{\pi}{3} = \frac{|OB|}{|OA|} = \frac{\sqrt{3}}{2}, \qquad \cos\frac{\pi}{3} = \frac{|AB|}{|OA|} = \frac{1}{2}, \qquad \tan\frac{\pi}{3} = \frac{|OB|}{|AB|} = \sqrt{3}.$$

These values of the sine, cosine and tangent functions should be memorized, because they are encountered so frequently.

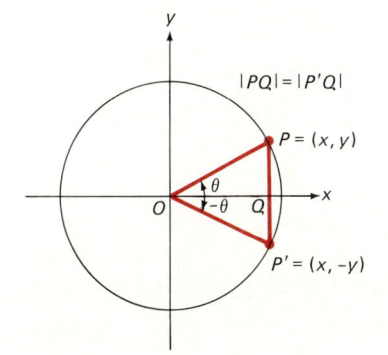

Figure 16

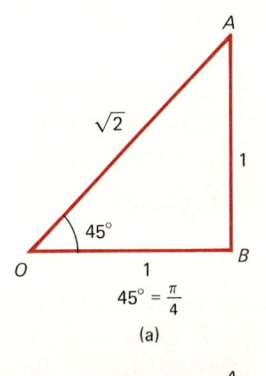

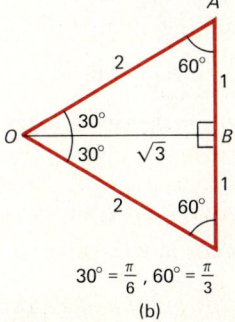

Figure 17

Section 1.3 Trigonometric Functions

The Law of Cosines

The following generalization of the Pythagorean theorem is a particularly useful tool for solving trigonometry problems.

> **Theorem 2** *(Law of cosines).* Let OAB be any triangle (not necessarily a right triangle), and let θ be the angle at the vertex O. Then
> $$c^2 = a^2 + b^2 - 2ab\cos\theta, \tag{10}$$
> where $a = |OA|$, $b = |OB|$ and $c = |AB|$.

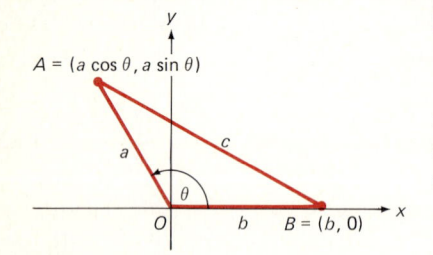

Proof Place the triangle OAB with the vertex O at the origin of the xy-plane and the side OB along the positive x-axis. Then $A = (a\cos\theta, a\sin\theta)$ and $B = (b, 0)$, as in Figure 18, and with the help of formula (2), page 23, for the square of the distance between two points, we find that

$$c^2 = |AB|^2 = (a\cos\theta - b)^2 + (a\sin\theta)^2$$
$$= a^2(\cos^2\theta + \sin^2\theta) + b^2 - 2ab\cos\theta,$$

which, because of (4), is equivalent to (10). ■

Notice that if $\theta = \pi/2$, the triangle OAB becomes a right triangle, and the law of cosines reduces to the Pythagorean theorem $c^2 = a^2 + b^2$.

Example 2 The angle between two sides of a triangle is 60°. If these sides have lengths 2 and 3, what is the length of the third side?

Solution Let a and b be the lengths of the given sides, and let c be the length of the third side. We are told that $a = 2$, $b = 3$, and also that $\theta = \pi/3$. Therefore, by formula (10),

$$c^2 = 2^2 + 3^2 - 2(2)(3)\left(\cos\frac{\pi}{3}\right) = 4 + 9 - 12\left(\frac{1}{2}\right) = 7,$$

so that $c = \sqrt{7}$. □

With the help of Theorem 2, we can prove the key formula

$$\cos(\alpha - \beta) = \cos\alpha\cos\beta + \sin\alpha\sin\beta \tag{11}$$

for the cosine of the difference between two angles α and β (lowercase Greek alpha and beta). Consider the triangle OAB in Figure 19, where O is the origin, $A = (\cos\alpha, \sin\alpha)$ and $B = (\cos\beta, \sin\beta)$. The angle opposite the side AB is $\theta = \alpha - \beta$, and $|OA| = |OB| = 1$. On the one hand,

$$|AB|^2 = |OA|^2 + |OB|^2 - 2|OA||OB|\cos\theta = 2 - 2\cos(\alpha - \beta),$$

by the law of cosines applied to the triangle OAB, and on the other hand,

$$|AB|^2 = (\cos\alpha - \cos\beta)^2 + (\sin\alpha - \sin\beta)^2$$
$$= \cos^2\alpha - 2\cos\alpha\cos\beta + \cos^2\beta + \sin^2\alpha - 2\sin\alpha\sin\beta + \sin^2\beta$$
$$= (\cos^2\alpha + \sin^2\alpha) + (\cos^2\beta + \sin^2\beta) - 2(\cos\alpha\cos\beta + \sin\alpha\sin\beta)$$
$$= 2 - 2(\cos\alpha\cos\beta + \sin\alpha\sin\beta),$$

by the formula for the square of the distance between the points A and B.

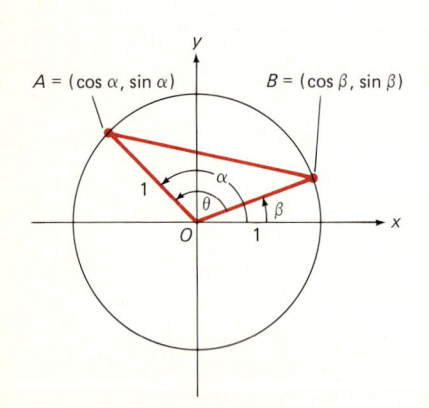

56   Chapter 1   Functions and Limits

It follows that
$$2 - 2\cos(\alpha - \beta) = 2 - 2(\cos\alpha\cos\beta + \sin\alpha\sin\beta),$$
which is equivalent to (11).

Figure 19 is drawn under the tacit assumption that $\alpha > \beta > 0$. You should consider the other possibilities and convince yourself that the same formula (11) is obtained in every case.

Changing β to $-\beta$ in (11) and using the formulas (9), we get the companion formula
$$\cos(\alpha + \beta) = \cos\alpha\cos\beta - \sin\alpha\sin\beta. \tag{11'}$$

To obtain analogous formulas for $\sin(\alpha + \beta)$ and $\sin(\alpha - \beta)$, we first choose $\alpha = (\pi/2) + \theta$, $\beta = \pi/2$ in (11). This gives
$$\cos\theta = \cos\left(\frac{\pi}{2} + \theta\right)\cos\frac{\pi}{2} + \sin\left(\frac{\pi}{2} + \theta\right)\sin\frac{\pi}{2},$$
or equivalently
$$\sin\left(\frac{\pi}{2} + \theta\right) = \cos\theta, \tag{12}$$
since $\cos(\pi/2) = 0$ and $\sin(\pi/2) = 1$. Next we choose $\alpha = \pi/2$, $\beta = \theta$ in (11'), which gives
$$\cos\left(\frac{\pi}{2} + \theta\right) = \cos\frac{\pi}{2}\cos\theta - \sin\frac{\pi}{2}\sin\theta,$$
or equivalently
$$\cos\left(\frac{\pi}{2} + \theta\right) = -\sin\theta. \tag{12'}$$

We now replace α by $(\pi/2) + \alpha$ in (11'), obtaining
$$\cos\left(\frac{\pi}{2} + \alpha + \beta\right) = \cos\left(\frac{\pi}{2} + \alpha\right)\cos\beta - \sin\left(\frac{\pi}{2} + \alpha\right)\sin\beta,$$
which becomes
$$\sin(\alpha + \beta) = \sin\alpha\cos\beta + \cos\alpha\sin\beta, \tag{13}$$
after applying (12) and (12') and multiplying by -1. Finally, changing β to $-\beta$ in (13), we get the companion formula
$$\sin(\alpha - \beta) = \sin\alpha\cos\beta - \cos\alpha\sin\beta. \tag{13'}$$

Addition Laws

Formulas (11), (11'), (13) and (13'), which we now group together in a form more suitable for memorization, are called the *addition laws* for the sine and cosine functions.

$$\sin(\alpha + \beta) = \sin\alpha\cos\beta + \cos\alpha\sin\beta,$$
$$\sin(\alpha - \beta) = \sin\alpha\cos\beta - \cos\alpha\sin\beta,$$
$$\cos(\alpha + \beta) = \cos\alpha\cos\beta - \sin\alpha\sin\beta,$$
$$\cos(\alpha - \beta) = \cos\alpha\cos\beta + \sin\alpha\sin\beta.$$

The following table gives some other useful trigonometric formulas, and an indication of how each is proved.

Formula	Proof
$\sin\left(\frac{\pi}{2} - \theta\right) = \cos\theta$	Change θ to $-\theta$ in the formula for $\sin\left(\frac{\pi}{2} + \theta\right)$.
$\cos\left(\frac{\pi}{2} - \theta\right) = \sin\theta$	Change θ to $-\theta$ in the formula for $\cos\left(\frac{\pi}{2} + \theta\right)$.
$\sin(\pi + \theta) = -\sin\theta$	Choose $\alpha = \pi$, $\beta = \theta$ in the formula for $\sin(\alpha + \beta)$.
$\sin(\pi - \theta) = \sin\theta$	Change θ to $-\theta$ in the above formula.
$\cos(\pi + \theta) = -\cos\theta$	Choose $\alpha = \pi$, $\beta = \theta$ in the formula for $\cos(\alpha + \beta)$.
$\cos(\pi - \theta) = -\cos\theta$	Change θ to $-\theta$ in the above formula.

Double-Angle Formulas

Setting $\alpha = \beta = \theta$ in the formulas for $\sin(\alpha + \beta)$ and $\cos(\alpha + \beta)$, we get the important *double-angle formulas*

$$\sin 2\theta = 2 \sin\theta \cos\theta, \tag{14}$$

$$\cos 2\theta = \cos^2\theta - \sin^2\theta. \tag{14'}$$

Then the observation that

$$1 + \cos 2\theta = 1 + (\cos^2\theta - \sin^2\theta) = (1 - \sin^2\theta) + \cos^2\theta = 2\cos^2\theta,$$
$$1 - \cos 2\theta = 1 - (\cos^2\theta - \sin^2\theta) = (1 - \cos^2\theta) + \sin^2\theta = 2\sin^2\theta$$

leads to another pair of useful formulas:

$$\cos^2\theta = \frac{1 + \cos 2\theta}{2}, \qquad \sin^2\theta = \frac{1 - \cos 2\theta}{2}.$$

We already have a formula for the sine of the difference between two angles. We can also derive a formula for the difference between two sines. In fact,

$$\sin\alpha - \sin\beta = \sin\left(\frac{\alpha+\beta}{2} + \frac{\alpha-\beta}{2}\right) - \sin\left(\frac{\alpha+\beta}{2} - \frac{\alpha-\beta}{2}\right)$$

$$= \sin\frac{\alpha+\beta}{2}\cos\frac{\alpha-\beta}{2} + \cos\frac{\alpha+\beta}{2}\sin\frac{\alpha-\beta}{2}$$

$$- \sin\frac{\alpha+\beta}{2}\cos\frac{\alpha-\beta}{2} + \cos\frac{\alpha+\beta}{2}\sin\frac{\alpha-\beta}{2},$$

and therefore

$$\sin\alpha - \sin\beta = 2\cos\frac{\alpha+\beta}{2}\sin\frac{\alpha-\beta}{2}, \tag{15}$$

after canceling a pair of terms. This formula will be found useful later.

The above formulas, which involve sines and cosines, can be used to establish analogous formulas involving the other trigonometric functions.

58 Chapter 1 Functions and Limits

Example 3 Show that

$$\tan(\pi - \theta) = -\tan\theta, \qquad \tan\left(\frac{\pi}{2} + \theta\right) = -\cot\theta,$$

and

$$\tan(\alpha - \beta) = \frac{\tan\alpha - \tan\beta}{1 + \tan\alpha\tan\beta}. \tag{16}$$

Solution Using the corresponding formulas for the sine and cosine, we have

$$\tan(\pi - \theta) = \frac{\sin(\pi - \theta)}{\cos(\pi - \theta)} = \frac{\sin\theta}{-\cos\theta} = -\tan\theta,$$

$$\tan\left(\frac{\pi}{2} + \theta\right) = \frac{\sin\left(\frac{\pi}{2} + \theta\right)}{\cos\left(\frac{\pi}{2} + \theta\right)} = \frac{\cos\theta}{-\sin\theta} = -\cot\theta.$$

Also

$$\tan(\alpha - \beta) = \frac{\sin(\alpha - \beta)}{\cos(\alpha - \beta)} = \frac{\sin\alpha\cos\beta - \cos\alpha\sin\beta}{\cos\alpha\cos\beta + \sin\alpha\sin\beta},$$

which reduces to (16) after dividing the numerator and denominator by the product $\cos\alpha\cos\beta$. □

The Angle Between Two Lines

As an application of formula (16), we now discuss the angle between two lines. Let L_1 and L_2 be two lines which intersect in a point P, which can be assumed to be above the x-axis (as in the proof of Theorem 9, page 32). Then the smallest angle α through which L_1 can be rotated about P in the counterclockwise direction to be brought into coincidence with L_2 is called the *angle between L_1 and L_2*, or more exactly *the angle from L_1 to L_2*. (The angle between two parallel or coincident lines is regarded as 0.) It is clear from Figure 20 that α always lies in the interval $0 \leq \alpha < \pi$. Let θ_1 be the inclination of L_1, and θ_2 the inclination of L_2. If $\theta_1 < \theta_2$ as in Figure 20(a), then $\theta_2 = \alpha + \theta_1$ or $\alpha = \theta_2 - \theta_1$ (see the proof of the cited theorem), while if $\theta_1 > \theta_2$ as in Figure 20(b), then $\theta_1 = (\pi - \alpha) + \theta_2$ or $\alpha = \pi + (\theta_2 - \theta_1)$. Suppose L_1 and L_2 are nonvertical and nonperpendicular with slopes m_1 and m_2, so that $m_1 = \tan\theta_1$ and $m_2 = \tan\theta_2$. If $\theta_1 < \theta_2$, then

$$\tan\alpha = \tan(\theta_2 - \theta_1) = \frac{\tan\theta_2 - \tan\theta_1}{1 + \tan\theta_1\tan\theta_2} = \frac{m_2 - m_1}{1 + m_1 m_2},$$

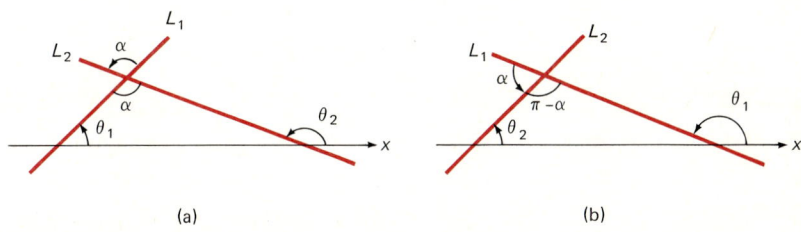

Figure 20

Section 1.3 Trigonometric Functions

with the help of (16), while if $\theta_1 > \theta_2$, then

$$\tan \alpha = \tan[\pi + (\theta_2 - \theta_1)] = \frac{\sin[\pi + (\theta_2 - \theta_1)]}{\cos[\pi + (\theta_2 - \theta_1)]}$$

$$= \frac{-\sin(\theta_2 - \theta_1)}{-\cos(\theta_2 - \theta_1)} = \tan(\theta_2 - \theta_1),$$

so that

$$\tan \alpha = \frac{m_2 - m_1}{1 + m_1 m_2}, \qquad (17)$$

as before. Notice that $m_1 m_2 \neq -1$ since L_1 and L_2 are nonperpendicular, so that the denominator $1 + m_1 m_2$ in (17) is nonzero.

Example 4 Find the angle α between the lines $y = 3x - 7$ and $y = -2x + 1$.

Solution The first line has slope $m_1 = 3$, and the second line has slope $m_2 = -2$. Substituting these values into formula (17), we get

$$\tan \alpha = \frac{-2 - 3}{1 \quad 6} = 1.$$

Since $0° \le \alpha < 180°$, we deduce that the angle α is 45°, measured from the first line to the second line (see Figure 21). □

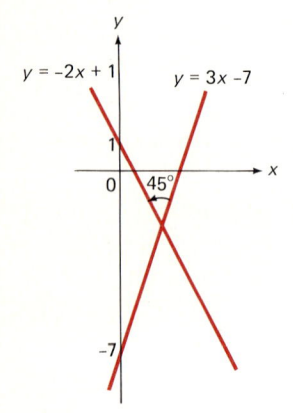

Figure 21

Problems

Convert the given angle from degrees to radians.
1. 15°
2. 150°
3. 1500°
4. $-72°$
5. $-220°$
6. 423°

Convert the given angle from radians to degrees.
7. $\pi/15$
8. $\pi/45$
9. $-\pi/12$
10. -5
11. π^2
12. 60

Find the length L of the circular arc and the area A of the circular sector with the given radius r and central angle θ.
13. $r = 2, \theta = \pi/4$
14. $r = 3, \theta = \pi/6$
15. $r = 5, \theta = 120°$
16. $r = 10, \theta = 5\pi/6$
17. $r = 50, \theta = 1°$
18. $r = 4, \theta = 36°$

Given a triangle with sides a, b, c and angle θ opposite the side c, find
19. c if $a = 3, b = 5, \theta = \pi/6$
20. c if $a = 4, b = 6, \theta = \pi/4$
21. b if $a = 5, c = 7, \theta = 2\pi/3$
22. a if $b = 6, c = 9, \theta = \pi/3$
23. a if $b = 10, c = 15, \theta = 3\pi/4$
24. θ if $a = 5, b = 12, c = 13$

As is customary, the same symbols are used to denote the sides and their lengths.

In Problems 25–48 evaluate the given expression, without the help of tables or a calculator.

25. $\tan 135°$
26. $\sec 45°$
27. $\csc 60°$
28. $\cot 30°$
29. $\csc 30°$
30. $\cot 45°$
31. $\sec \dfrac{\pi}{6}$
32. $\tan \dfrac{2\pi}{3}$
33. $\sec \dfrac{\pi}{3}$
34. $\tan \dfrac{5\pi}{6}$
35. $\cot \dfrac{\pi}{3}$
36. $\csc \dfrac{\pi}{4}$
37. $\cos(-120°)$
38. $\sin 765°$
39. $\sec 150°$
40. $\cot 210°$
41. $\csc 135°$
42. $\tan 300°$
43. $\cot\left(-\dfrac{35\pi}{4}\right)$
44. $\sin\left(\dfrac{17\pi}{3}\right)$
45. $\tan \dfrac{19\pi}{6}$
46. $\csc\left(-\dfrac{15\pi}{4}\right)$
47. $\sec \dfrac{11\pi}{3}$
48. $\cos \dfrac{99\pi}{4}$

Is the given number positive, negative or zero?
49. $\sin 200\pi$
50. $\sec 5$
51. $\cos 899°$
52. $\sin 2^4$

53. Given a triangle with sides a, b, c and angles A, B, C opposite these sides, verify the *law of sines*

$$\frac{a}{\sin A} = \frac{b}{\sin B} = \frac{c}{\sin C}.$$

60 Chapter 1 Functions and Limits

54. Show that the area of the triangle in the preceding problem is equal to

$$\frac{1}{2} bc \sin A = \frac{1}{2} ac \sin B = \frac{1}{2} ab \sin C.$$

55. What is the area of an equilateral triangle with side length s?

56. What is the area of an isosceles right triangle with hypotenuse of length h?

Verify the given trigonometric identity.
57. $\sin 3\theta = 3 \sin \theta - 4 \sin^3 \theta$
58. $\cos 3\theta = 4 \cos^3 \theta - 3 \cos \theta$
59. $\cos 4\theta = 8 \cos^4 \theta - 8 \cos^2 \theta + 1$
60. $\sin 4\theta = 8 \sin \theta \cos^3 \theta - 4 \sin \theta \cos \theta$
61. $\cos \alpha + \cos \beta = 2 \cos \frac{\alpha + \beta}{2} \cos \frac{\alpha - \beta}{2}$
62. $\cos \alpha - \cos \beta = -2 \sin \frac{\alpha + \beta}{2} \sin \frac{\alpha - \beta}{2}$
63. $\tan \theta = \frac{\sin 2\theta}{1 + \cos 2\theta}$
64. $\tan \theta = \frac{1 - \cos 2\theta}{\sin 2\theta}$
65. $\tan 2\theta = \frac{2 \tan \theta}{1 - \tan^2 \theta}$
66. $\tan 2\theta = \frac{2}{\cot \theta - \tan \theta}$

Find the angle between the given pair of lines, measured from the first line to the second line.
67. $5x - y + 7 = 0$, $3x + 2y = 0$
68. $3x - 2y + 7 = 0$, $2x + 3y - 3 = 0$
69. $2x - y - 4 = 0$, $x - 3y + 5 = 0$
70. $\sqrt{3}x - y + 1 = 0$, $\sqrt{3}x + y - 1 = 0$

71. The approximation $\pi \approx \frac{355}{113}$ is surprisingly accurate. How accurate?

72. Show that the length of the spiral groove in a 12-inch phonograph record is about a quarter of a mile. Assume that the record lasts 20 minutes and ends 2 inches from the center, turning at the rate of $33\frac{1}{3}$ rpm (revolutions per minute).

1.4 More About Trigonometric Functions

We now study the graphs of the trigonometric functions, denoting the independent variable by x instead of θ, and assuming that x is measured in radians. We begin with the sine and cosine. The graphs of $\sin x$ and $\cos x$ are shown in Figure 22 in the same system of rectangular coordinates, and can be obtained either directly from the geometric meaning of the functions, or indirectly with the help of a scientific calculator or tables of the trigonometric functions. Each of the graphs is a curve which oscillates up and down in a repetitive fashion. Of course, no figure can show more than a piece of the graph of a function like $\sin x$ which is defined on the whole real line, but it is clear what is happening: The part of the graph of $\sin x$ over every interval

$$2n\pi \leq x \leq 2(n+1)\pi \quad (n = \pm 1, \pm 2, \ldots)$$

is an exact copy of the part of the graph over the interval $0 \leq x \leq 2\pi$, and the same is true of the graph of $\cos x$.

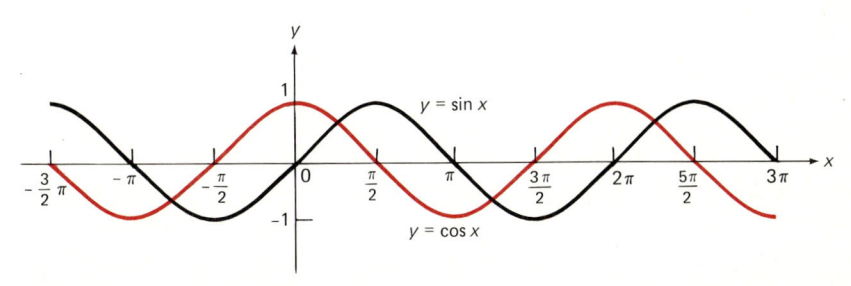

Figure 22

Periodic Functions

This repetitive property of the graphs of the sine and cosine (and of many other functions, including the other trigonometric functions) is called *periodicity*. More exactly, a function f is said to be *periodic*, with *period* p ($p \neq 0$), if

$$f(x + p) \equiv f(x),$$

that is, if adding p to the argument of f does not change the value of f; here it is assumed that the domain of f contains $x \pm p$ whenever it contains x. In this language the functions $\sin x$ and $\cos x$ are both periodic, with period 2π. They are also periodic with period $2n\pi$, where n is any positive or negative integer. The smallest positive period of a periodic function is called its *fundamental period*. It is clear from Figure 22 that the fundamental period of $\sin x$ and $\cos x$ is 2π.

Example 1 Figure 23 shows that the fundamental period of $\sin^2 x$ is π, half the fundamental period of the function $\sin x$ itself.

Optional
This can also be seen algebraically as follows. The number π is a period of $\sin^2 x$, since $\sin^2 (x + \pi) \equiv (-\sin x)^2 \equiv \sin^2 x$. Suppose $\sin^2 x$ has a positive period p smaller than π. Then

$$\sin^2 (x + p) \equiv \sin^2 x,$$

where $0 < p < \pi$. Choosing $x = 0$ in the last formula, we get

$$\sin^2 p = \sin^2 0 = 0,$$

or equivalently $\sin p = 0$. But this is impossible, since $\sin p$ is nonzero for $0 < p < \pi$. Therefore π is the *smallest* positive period of $\sin^2 x$.

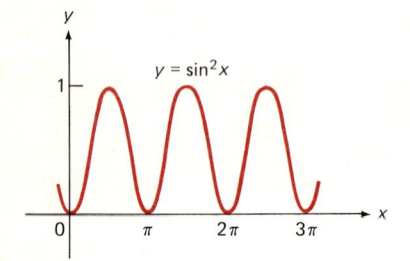

Figure 23

Bounded and Unbounded Functions

The graphs of the functions $y = \sin x$ and $y = \cos x$ are both entirely confined to the horizontal strip $-1 \leq y \leq 1$. This corresponds to the fact that $|\sin x| \leq 1$ and $|\cos x| \leq 1$ for all x (see page 53). If the graph of a function $y = f(x)$ is entirely confined to some horizontal strip $-C \leq y \leq C$, or equivalently if $|f(x)| \leq C$ for all x, we say that f is *bounded*, but otherwise f is said to be *unbounded*. More generally, a function $y = f(x)$ is said to be *bounded on an interval* I if $-C \leq f(x) \leq C$, or equivalently $|f(x)| \leq C$, for all values of x in I and some $C > 0$, but otherwise we say that f is *unbounded on* I.

Example 2 The linear function

$$f(x) = mx + b \quad (m \neq 0)$$

is bounded on $[-1, 1]$ and unbounded on $[0, \infty)$. More generally, f is bounded on every finite interval and unbounded on every infinite interval.

Because of Theorem 1, page 50, and the formulas

$$\cos x = \sin\left(x + \frac{\pi}{2}\right), \quad \sin x = \cos\left(x - \frac{\pi}{2}\right),$$

shifting the graph of $\sin x$ to the left by $\pi/2$ gives the graph of $\cos x$, and equivalently, shifting the graph of $\cos x$ to the right by $\pi/2$ gives the graph

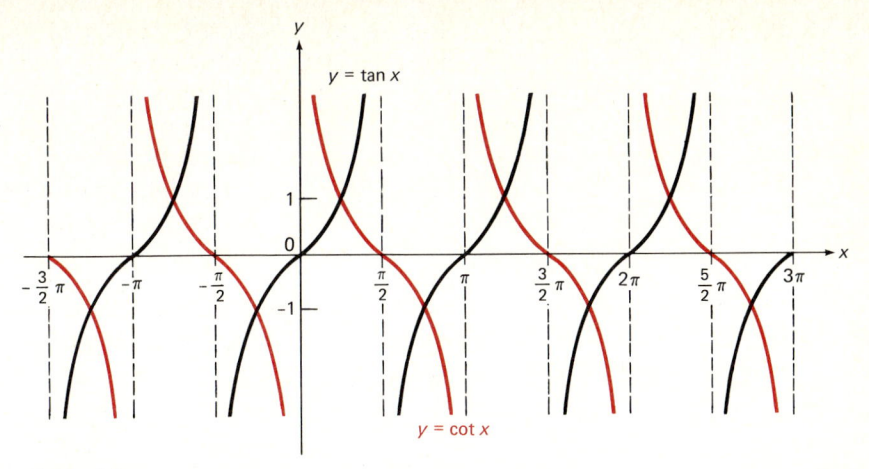

Figure 24

of sin x (see Figure 22). In the language of engineering, where the sine and cosine functions are used to describe oscillatory phenomena and x is the time, cos x *leads* ("precedes") sin x by $\pi/2$ (or 90°), while sin x *lags* ("follows") cos x by $\pi/2$; by the same token, the two "waveforms" sin x and cos x are said to be 90° *out of phase*. Notice that the graph of cos x is symmetric about the y-axis, as is to be expected since cos x is an even function, and the graph of sin x is symmetric about the origin, since sin x is an odd function. Also observe that cos x is increasing on $[-\pi, 0]$ and decreasing on $[0, \pi]$, while sin x is increasing on $[-\pi/2, \pi/2]$ and decreasing on $[\pi/2, 3\pi/2]$.

Next we consider the functions tan x and cot x. Figure 24 shows the graphs of these two functions in the same system of rectangular coordinates. The function tan x is undefined for $x = (n + \frac{1}{2})\pi$, where n is any integer. Hence we can think of tan x as made up of infinitely many† separate functions

$$y = \tan x \qquad ((n - \tfrac{1}{2})\pi < x < (n + \tfrac{1}{2})\pi),$$

called the *branches* of tan x, each with its own restricted domain of definition and its own graph (also called a branch), lying over an open interval of length π. Note that the branches of tan x are separated by the vertical lines $x = (n + \frac{1}{2})\pi$. Similarly, the function cot x is undefined for $x = n\pi$, and therefore it also consists of infinitely many branches

$$y = \cot x \qquad (n\pi < x < (n + 1)\pi),$$

separated by the vertical lines $x = n\pi$.

It is clear from Figure 24 that each branch of tan x is an unbounded increasing function, while each branch of cot x is an unbounded decreasing function. The figure also shows that both functions tan x and cot x are periodic, with fundamental period π (a periodic function need not be defined for all values of x). Of course, 2π is also a period of these functions, but it is not the *fundamental* period, that is, the *smallest* positive period. Note that tan $x = 0$ if and only if $x = n\pi$, while cot $x = 0$ if and only if $x = (n + \frac{1}{2})\pi$.

---

† If a set contains n elements, where n is some nonnegative integer, we say that the set is *finite*; otherwise we say that the set is *infinite*, or that it contains *infinitely many* elements. For example, the set of all integers is infinite, sin $x = 0$ for infinitely many values of x, and so on.

Section 1.4   More about Trigonometric Functions   **63**

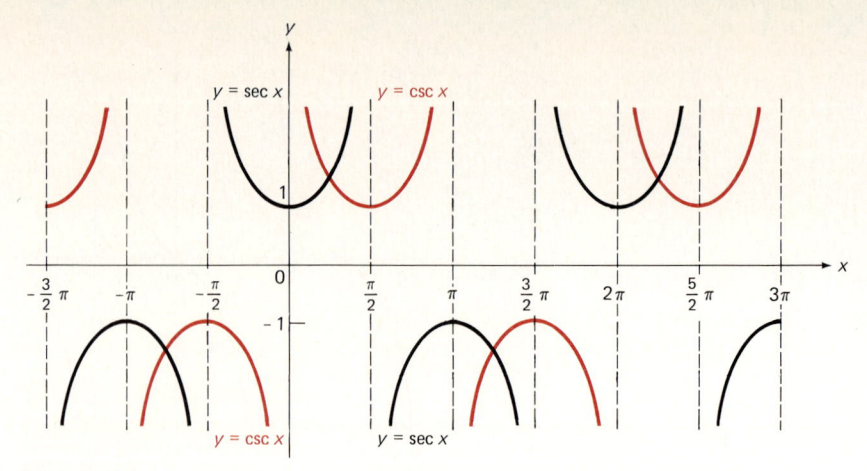

Figure 25

The graphs of the remaining trigonometric functions sec x and csc x are shown in Figure 25. It is apparent from the figure that each of these functions is unbounded and periodic, with fundamental period 2π, and has infinitely many branches. Observe that sec x leads csc x by $\pi/2$, while csc x lags sec x by $\pi/2$. This property is "inherited" from the functions sin x and cos x, of which csc x and sec x are the reciprocals.

Example 3 Find all values of x for which sec $x = 2$.

Solution The function sec x is equal to 2 whenever $\cos x = 1/\sec x$ is equal to $\frac{1}{2}$. This occurs at the points $x = \pm \pi/3$, but at no other points of the interval $[-\pi, \pi]$, since cos x is increasing on $[-\pi, 0]$ and decreasing on $[0, \pi]$. Also $\sec(x + 2n\pi) \equiv \sec x$ for every integer n. Therefore sec $x = 2$ if and only if $x = (2n \pm \frac{1}{3})\pi$, where n is any integer. □

Some Trigonometric Inequalities

Finally, we derive some useful inequalities satisfied by the sine and cosine functions. Consider the construction shown in Figure 26, where the circle is of unit radius and the angle x lies in the interval $0 < x < \pi/2$. Since

Area of triangle AOB < Area of sector AOB < Area of triangle AOC,

it follows that

$$\frac{1}{2} \sin x < \frac{1}{2} x < \frac{1}{2} \tan x, \tag{1}$$

where we use formula (8), page 54, and the fact that the area of a triangle is half the product of its base and its altitude. Dividing (1) by the positive quantity $\frac{1}{2} \sin x$, we get

$$1 < \frac{x}{\sin x} < \frac{1}{\cos x},$$

or equivalently

$$\cos x < \frac{\sin x}{x} < 1 \tag{2}$$

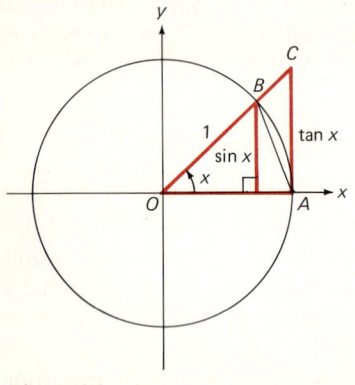

Figure 26

(see Theorem 4, page 9). This double inequality holds for $-\pi/2 < x < 0$, as well as for $0 < x < \pi/2$, and hence is true for $0 < |x| < \pi/2$, since changing x to $-x$ does not change the value of the even function $\cos x$ or of the function $(\sin x)/x$, which is also even:

$$\frac{\sin(-x)}{-x} \equiv \frac{-\sin x}{-x} \equiv \frac{\sin x}{x} \qquad (x \neq 0).$$

Since $(\sin x)/x$ is positive if $0 < |x| < \pi/2$, the second inequality in (2) can be written as

$$\left|\frac{\sin x}{x}\right| < 1,$$

or equivalently

$$|\sin x| < |x|. \tag{3}$$

This inequality holds for $|x| \geq \pi/2$, as well as for $0 < |x| < \pi/2$, since $|\sin x| \leq 1$ and $\pi/2 \approx 1.57 > 1$. Therefore (3) holds for all $x \neq 0$, and in fact

$$|\sin x| \leq |x| \tag{3'}$$

holds for all x, including $x = 0$, since $|\sin 0| = |0|$.

Problems

Let n be any integer. Show that
1. $\sin(x + n\pi) = (-1)^n \sin x$
2. $\cos(x + n\pi) = (-1)^n \cos x$

Let the function f be periodic with period p, so that $f(x + p) \equiv f(x)$. Show that
3. $f(x - p) \equiv f(x)$
4. $f(x + np) \equiv f(x)$ for every integer n

Show that
5. $\tan x$ and $\cot x$ are both odd functions
6. $\sec x$ is an even function, while $\csc x$ is an odd function

Identify the given function as even or odd (or neither).
7. $f(x) = \sin x + \cos x$
8. $f(x) = x + \sin x$
9. $f(x) = \sin^2 x - \cos x$
10. $f(x) = \dfrac{\sin x}{2 + \cos x}$

Find all values of x for which
11. $\cos x = 1$
12. $\sin x = 1$
13. $\sin x = -1$
14. $\cos x = -1$
15. $\sin x = \frac{1}{2}$
16. $\cos x = -\frac{1}{2}$
17. $\tan x = 1$
18. $\cot x = -1$
19. $\csc x = -2$
20. $\sec x = \sqrt{2}$
21. $\sec x$ is undefined
22. $\csc x$ is undefined

Find all intervals of length π on which
23. $\sin x$ is increasing
24. $\sin x$ is decreasing
25. $\cos x$ is decreasing
26. $\cos x$ is increasing

Find the fundamental period of the given function.
27. $\sqrt{\sin x}$
28. $\cos^3 x$
29. $|\cos x|$
30. $\cos x + \sin x$

31. Draw the graph of the periodic function $y = f(x)$, with period 2, whose values on the interval $-1 \leq x \leq 1$ are given by the formula $y = 1 - |x|$.

32. Verify that

$$\tan\left(\frac{\pi}{4} - x\right) = \cot\left(\frac{\pi}{4} + x\right).$$

What does this imply about the connection between the graphs of $\tan x$ and $\cot x$?

1.5 The Limit Concept and Continuity

We now pose a question central to the subject of calculus: How do the values of a function $f(x)$ behave when its argument x gets closer and closer to a given value a? In some cases the answer is intuitively clear. For example, if $f(x) = x^2 + 1$ and if x gets closer and closer to the number 2, it seems evident that $f(x)$ gets closer and closer to the number $f(2) = 2^2 + 1 = 5$, the value of the function at $x = 2$. This situation is summarized by saying that the *limit* of $f(x)$ as x approaches 2 is equal to 5. However, in other cases we cannot find the limit so simply, by making an easy substitution. In fact, it may well happen that the function is not even *defined* at the point where the limit is to be evaluated! The following examples show how this can come about. In the next chapter you will see that these examples, far from being oddities, are instances of what happens *every time* we try to evaluate a particularly important kind of limit, known as the *derivative*. Thus it is a good idea to learn about limits now, so that the story of the derivative can be told without interruption later.

Example 1 The function

$$f(x) = \frac{x^2 - 1}{x - 1} \tag{1}$$

is defined for all $x \neq 1$, but not for $x = 1$, since the denominator $x - 1$ equals zero if $x = 1$. In fact, the right side of (1) becomes the indeterminate form $0/0$ if $x = 1$. Since

$$f(x) = \frac{(x + 1)(x - 1)}{x - 1} = x + 1 \quad \text{if} \quad x \neq 1, \tag{1'}$$

but $f(1)$ is not defined, the graph of (1) is the straight line $y = x + 1$ with a single missing point, namely the point $(1, 2)$. This graph is shown in Figure 27, where the missing point is indicated, as usual, by a hollow dot. It is clear from the figure that although $f(x)$ is not defined for $x = 1$, $f(x)$ is very close to 2 if x is very close to 1, a fact which is also apparent algebraically from formula (1'), since if x is almost (but not quite) equal to 1, $x + 1$ must be almost equal to 2. We summarize all this by saying that *the limit of $f(x)$ as x approaches 1 is equal to 2*, or symbolically

$$\lim_{x \to 1} f(x) = 2. \tag{2}$$

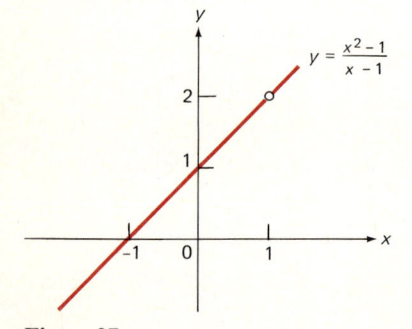

Figure 27

Alternatively, we can say that $f(x)$ *approaches* 2 *as x approaches* 1, or briefly

$$f(x) \to 2 \quad \text{as} \quad x \to 1. \quad \square \tag{2'}$$

We have yet to formalize the limit concept and make it more precise, and this will be done in the next section. But in the meantime we mention that we have just shown informally that

$$\lim_{x \to 1} \frac{x^2 - 1}{x - 1} = 2. \tag{3}$$

We mention for the record that the limit in (3) is an example of a derivative, namely the derivative of the function x^2 at the point 1. More generally, the derivative of x^2 at the point a is the limit

$$\lim_{x \to a} \frac{x^2 - a^2}{x - a},$$

which turns out to be $2a$, by virtually the same argument as just used to establish formula (3), corresponding to the case $a = 1$. Much more will be said about this later.

To evaluate the limit (3), we needed only a little algebra. Some limits are not evaluated so easily, as the next example shows.

Example 2 Evaluate

$$\lim_{x \to 0} \frac{\sin x}{x}. \qquad (4)$$

Solution The function $(\sin x)/x$ is not defined at $x = 0$, which is the very point at which we want to evaluate its limit. In this respect, the situation resembles Example 1, suggesting that we try to divide the numerator by the denominator for $x \neq 0$. But now there is no way of carrying out the division algebraically. To evaluate the limit (4), we will eventually resort to another method. For the time being, though, let's see if we can manage to guess the value of the limit.

To this end, we use a scientific calculator to compute the values of $(\sin x)/x$ for typical values of x that get closer and closer to 0, bearing in mind that $(\sin x)/x$ is an even function and therefore takes the same value at both points x and $-x$. The results of our calculations are displayed in the following table, where x is measured in radians and the sign $\pm$ in parentheses serves as a reminder that each x entry can be either the listed number or its negative.

x	$\dfrac{\sin x}{x}$	x	$\dfrac{\sin x}{x}$
($\pm$) 1.2	0.77670	($\pm$) 0.10	0.99833
1.0	0.84147	0.08	0.99893
0.8	0.89670	0.06	0.99940
0.6	0.94107	0.04	0.99973
0.4	0.97355	0.02	0.99993
0.2	0.99335	0	Undefined

The table strongly suggests that as x gets closer and closer to 0, the values of the function $(\sin x)/x$ get closer and closer to 1, in the way exhibited in Figure 28, which shows the graph of $(\sin x)/x$. The limiting behavior of the function $(\sin x)/x$ at $x = 0$, that is, the behavior of the function as $x \to 0$, has been indicated by drawing two arrowheads at the point $(0, 1)$ which is approached by the graph as $x \to 0$. The point $(0, 1)$ itself does not belong to the graph, as indicated by the hollow dot at $(0, 1)$; this is simply because the function $(\sin x)/x$ is not defined at $x = 0$. Thus it certainly seems as if

$$\lim_{x \to 0} \frac{\sin x}{x} = 1,$$

and we will eventually give a rigorous proof of this fact in Example 7, page 85. □

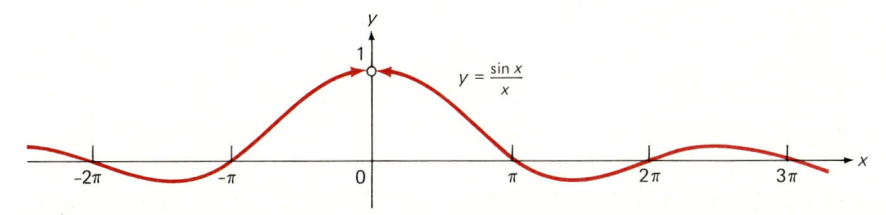

Figure 28

Section 1.5   The Limit Concept and Continuity

Informal Definition of a Limit

Next we put Examples 1 and 2 in a more general setting by talking about *any* function $f(x)$, *any* limit L, and *any* fixed point a approached by the independent variable or argument x. Thus in Example 1

$$f(x) = \frac{x^2 - 1}{x - 1}, \qquad a = 1, \qquad L = 2,$$

and in Example 2

$$f(x) = \frac{\sin x}{x}, \qquad a = 0, \qquad L = 1.$$

We now say that $f(x)$ *approaches the limit L as $x \to a$*, or that *$f(x)$ has the limit L at a*, if $f(x)$ gets closer and closer to L as x gets closer and closer to a (without ever being equal to a). This is expressed by writing

$$\lim_{x \to a} f(x) = L \tag{5}$$

or

$$f(x) \to L \quad \text{as} \quad x \to a. \tag{5'}$$

When we say that $f(x)$ gets closer and closer to L as $x \to a$, what we really mean, as illustrated by the above examples, is that $f(x)$ *achieves and maintains* any preselected degree of closeness to L as $x \to a$. In other words, it must be possible to make $f(x)$ as close as we please to L for *all* x sufficiently close to a, whether x lies to the left or to the right of a (this "two-sidedness" is essential). The smaller the quantity $|f(x) - L|$, the closer $f(x)$ is to L, that is, smallness of $|f(x) - L|$ means closeness of $f(x)$ to L, and similarly, smallness of $|x - a|$ means closeness of x to a. Thus, in terms of absolute values, (5) says that $|f(x) - L|$ can be made "arbitrarily small" for all "sufficiently small" (but nonzero) values of $|x - a|$. These ideas will be made more precise in the next section.

It is often convenient to talk about a function having a limit without specifying what the limit is. Thus a function $f(x)$ is said to *have a limit at a* if there is some number L such that (5) holds, and in this case we say that the limit in the left side of (5) *exists*. If there is no such number L, we say that the limit *does not exist*, or that $f(x)$ *has no limit at a*.

Example 3 Show that the function

$$f(x) = \frac{|x|}{x}$$

has no limit at $x = 0$.

Solution It is easy to see that the function $|x|/x$ is defined for all $x \neq 0$ and is exactly the same as the function

$$f(x) = \begin{cases} 1 & \text{if } x > 0, \\ -1 & \text{if } x < 0, \end{cases}$$

graphed in Figure 29. If $f(x) \to L$ as $x \to 0$, then $f(x)$ must stay arbitrarily close to L, within distance $\frac{1}{2}$ say, for all x sufficiently close to 0. But this is impossible, since every deleted neighborhood of $x = 0$, no matter how small,

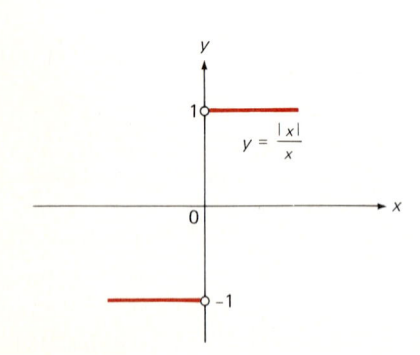

Figure 29

68 Chapter 1 Functions and Limits

contains points $x > 0$ for which $f(x) = 1$ and also points $x < 0$ for which $f(x) = -1$, and there is no number L within distance $\frac{1}{2}$ of both 1 and -1. It follows that $f(x)$ has no limit at $x = 0$. ☐

Example 4 Evaluate $\lim_{x \to 0} \cos x$.

Solution As in Example 2, we use a calculator to compute values of the function $\cos x$ for typical values of x approaching 0. The results are given in the following table (remember that $\cos x$ is an even function).

x	$\cos x$	x	$\cos x$
($\pm$) 1.2	0.3624	($\pm$) 0.10	0.9950
1.0	0.5403	0.08	0.9968
0.8	0.6967	0.06	0.9982
0.6	0.8253	0.04	0.9992
0.4	0.9211	0.02	0.9998
0.2	0.9801	0	Defined and equal to 1

Inspection of these numbers strongly suggests that

$$\lim_{x \to 0} \cos x = 1,$$

and in fact it can be shown that this is true. But there is a crucial difference between this table and the one in Example 2, for $\cos x$ *is defined* for $x = 0$, whereas the function $(\sin x)/x$ is *not*. Furthermore, the limit of $\cos x$ at $x = 0$ is the same as the *value* of $\cos x$ at $x = 0$, namely $\cos 0 = 1$. This key fact is expressed by saying that the function $\cos x$ is *continuous at* $x = 0$. ☐

Continuity and Reasons for Discontinuity

More generally, let f be a function defined at the point a, and suppose the limit of f at $x = a$ exists and equals $f(a)$, so that

$$\lim_{x \to a} f(x) = f(a). \tag{6}$$

Then f is said to be *continuous at* a. A function f can be *discontinuous at* a (that is, fail to be continuous at a) for one of three reasons:

(i) The limit of f at a may fail to exist, as in Example 3.
(ii) The limit of f at a may exist, but f may not be defined at a, as in Examples 1 and 2.
(iii) The limit of f at a may exist and f may be defined at a, but the limit may not equal $f(a)$, as in Example 7 below.

If a function f is discontinuous at a, the discontinuity will always be revealed by some abnormality of the graph of f along the line $x = a$. Roughly speaking, the graph of a function continuous at a cannot have a "break" at a, and in particular cannot have a "hole" or a "jump" at a.

Example 5 The function

$$f(x) = \frac{|x|}{x}$$

graphed in Figure 29 is discontinuous at $x = 0$ for reason (i), as revealed by the jump that the graph makes in going from the line $y = -1$ to the line $y = 1$.

Section 1.5 The Limit Concept and Continuity **69**

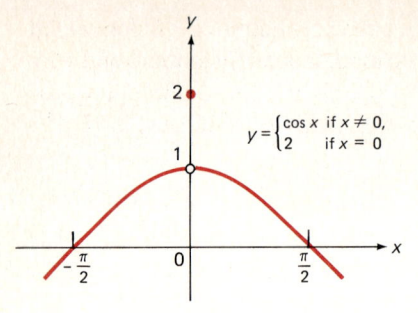

Figure 30

Example 6 In Figure 27 the graph of the function

$$f(x) = \frac{x^2 - 1}{x - 1}$$

comes arbitrarily close to the point (1, 2), but the point itself is missing from the graph, and the presence of the corresponding hole tells us at once that the function is discontinuous at $x = 1$ for reason (ii).

Example 7 The function

$$f(x) = \begin{cases} \cos x & \text{if } x \neq 0, \\ 2 & \text{if } x = 0, \end{cases}$$

graphed in Figure 30, is discontinuous at $x = 0$ for reason (iii). In fact, $f(x) \to 1$ as $x \to 0$ for the same reason as in Example 4, but $f(0) = 2$, so that

$$\lim_{x \to 0} f(x) = 1 \neq f(0),$$

thereby violating the condition for continuity of f at $x = 0$. The discontinuity of f at $x = 0$ is again revealed by the presence of a hole in the graph, this time at the point (0, 1). The graph also has an "isolated" solid dot at the point (0, 2), which expresses the fact that $f(0) = 2$.

It will be shown in Theorem 8, page 84, that any *polynomial function* (or simply *polynomial*), that is, any function of the form

$$P(x) = a_0 + a_1 x + a_2 x^2 + \cdots + a_n x^n \qquad (a_n \neq 0),$$

is continuous for all x. Here n is a nonnegative integer, called the *degree* of the polynomial, and $a_0, a_1, a_2, \ldots, a_n$ are constants, called the *coefficients* of the polynomial. Moreover, it will be shown in Theorem 9, page 84, that each of the six trigonometric functions $\sin x$, $\cos x$, $\tan x$, $\cot x$, $\sec x$ and $\csc x$ is continuous at every point where the function is defined.

Example 8 By the continuity of the polynomial $x^2 + 1$,

$$\lim_{x \to 2} (x^2 + 1) = 2^2 + 1 = 5,$$

as in the first paragraph of this section. Similarly, by the continuity of the polynomial $x^3 - 3x + 2$,

$$\lim_{x \to -2} (x^3 - 3x + 2) = (-2)^3 - 3(-2) + 2 = 0.$$

Example 9 By the continuity of $\sin x$,

$$\lim_{x \to 0} \sin x = \sin 0 = 0.$$

In our still rather informal definition of the limit of a function f at a point a, the independent variable x must be able to take values arbitrarily close to a, but on the other hand, the function f does not have to be defined at the point a itself, as in the case of the function $(\sin x)/x$ which is not defined at $x = 0$. Hence the limit of f at a is influenced only by the values of f in the immediate vicinity of a. Thus, in order to talk about the limit of a function f at a point a, we need only assume that f is defined in some *deleted* neighborhood of *a*. Similarly, to talk about the continuity of a function *f* at a point *a*, we need only assume that *f* is defined in some neighborhood of *a*, but it must now be an ordinary *nondeleted* neighborhood containing the point *a* itself. This is because formula (6) defining the continuity of *f* at *a* involves the *value* of *f* at *a*.

Problems

Use the same kind of reasoning as in Example 1 to evaluate

1. $\lim\limits_{x \to 4} \dfrac{x^2 - 2x - 8}{x - 4}$
2. $\lim\limits_{x \to -2} \dfrac{2x^2 + 3x - 2}{x + 2}$
3. $\lim\limits_{x \to 3} \dfrac{x^2 - 6x + 9}{x - 3}$
4. $\lim\limits_{x \to 5} \dfrac{x^3 - 125}{x - 5}$
5. $\lim\limits_{x \to -1} \dfrac{x^3 + 1}{x + 1}$
6. $\lim\limits_{x \to 2} \dfrac{\frac{1}{x} - \frac{1}{2}}{x - 2}$

7. By making a table of values of the function $f(x) = (x - 1)/(\sqrt{x} - 1)$ near $x = 1$, convince yourself that the function approaches a limit as $x \to 1$. What is this limit? Prove the same result algebraically. Graph the function near $x = 1$.

8. By making a table of values of the function $(\tan x)/x$ near $x = 0$, convince yourself that the function approaches a limit as $x \to 0$. What is this limit? Graph the function near $x = 0$.

Use an informal argument to evaluate

9. $\lim\limits_{x \to 0} |x|$
10. $\lim\limits_{x \to 2} \dfrac{x}{x}$
11. $\lim\limits_{x \to 0} \dfrac{x}{x}$
12. $\lim\limits_{x \to -3} \dfrac{x}{x}$
13. $\lim\limits_{x \to 0.1} \dfrac{|x|}{x}$
14. $\lim\limits_{x \to -0.2} \dfrac{|x|}{x}$
15. $\lim\limits_{x \to 0} \dfrac{x^2}{x}$
16. $\lim\limits_{x \to 2} \dfrac{x^2}{x}$
17. $\lim\limits_{x \to 0} \dfrac{\sqrt{x^2}}{x}$
18. $\lim\limits_{x \to -2} \dfrac{\sqrt{x^2}}{x}$
19. $\lim\limits_{x \to 0} \dfrac{|\sin x|}{x}$
20. $\lim\limits_{x \to 0} \dfrac{\sin |x|}{x}$

Use the continuity of polynomials to evaluate

21. $\lim\limits_{x \to -3} (x^2 + 3x + 4)$
22. $\lim\limits_{x \to 4} (x^3 + 1)$
23. $\lim\limits_{x \to 3} (x^4 - 9x - 6)$
24. $\lim\limits_{x \to -2} (x^5 + 5x^2)$
25. $\lim\limits_{x \to 1} (2x^6 - 3x^5 + x^4)$
26. $\lim\limits_{x \to -1} (x^{11} - x^7 + 2)$

Use the continuity of trigonometric functions to evaluate

27. $\lim\limits_{x \to 1} \sin x$
28. $\lim\limits_{x \to \pi} \cos x$
29. $\lim\limits_{x \to -\pi/4} \tan x$
30. $\lim\limits_{x \to \pi/2} \cot x$
31. $\lim\limits_{x \to 0} \sec x$
32. $\lim\limits_{x \to \pi/4} \csc x$

Figure 31 shows the graph of a function *f*. Find

33. $\lim\limits_{x \to -1} f(x)$
34. $\lim\limits_{x \to 1} f(x)$
35. $\lim\limits_{x \to 2} f(x)$
36. $\lim\limits_{x \to 3} f(x)$

37. At which of the points $x = \pm 1, 2, 3$ is the function graphed in Figure 31 continuous?

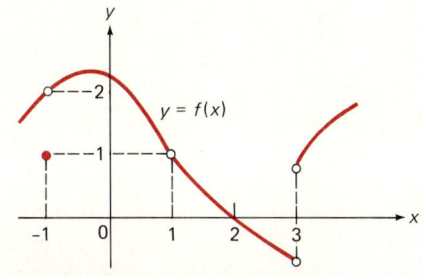

Figure 31

1.6 A Closer Look at Limits

You will recall that we said that $f(x) \to L$ as $x \to a$ means that $|f(x) - L|$ can be made "arbitrarily small" for all "sufficiently small" but nonzero values of $|x - a|$. Can this intuitive definition of a limit be made completely rigorous? Yes it can, by resorting to a method introduced by the German mathematicians Weierstrass and Heine about a hundred years ago, some two hundred years after the invention of calculus by Newton and Leibniz. This is the "ε, δ method," so called because it involves numbers traditionally denoted by the Greek letters ε and δ (lowercase Greek epsilon and delta). What does it mean to say that $|f(x) - L|$ is arbitrarily small for all sufficiently small $|x - a|$? Just this: Suppose someone we call the "challenger" presents us with any positive number ε that he pleases. Then we must be able to find another positive number δ such that $|f(x) - L| < \varepsilon$ for all $x \neq a$ satisfying the inequality $|x - a| < \delta$. At this point, you may well ask what all this has to do with the numbers $|f(x) - L|$ and $|x - a|$ being small. The answer is simply that we allow our challenger to present us with *any positive number ε whatsoever*, in particular with a number ε which is as small as the challenger pleases, that is, arbitrarily small. We must then find a corresponding positive number δ, which in general is subject to certain size restrictions and hence must be sufficiently small, such that $|f(x) - L| < \varepsilon$ whenever $|x - a| < \delta$ and $x \neq a$.

Formal Definition of a Limit

To recapitulate, in more formal language, let $f(x)$ be defined in a deleted neighborhood of the point a. Then

$$\lim_{x \to a} f(x) = L \tag{1}$$

means that given any $\varepsilon > 0$ *(no matter how small)*, we can find a *(sufficiently small)* $\delta > 0$ such that

$$|f(x) - L| < \varepsilon$$

whenever

$$0 < |x - a| < \delta. \tag{2}$$

Here the double inequality (2) is just a nice way of writing $|x - a| < \delta$ and $x \neq a$ at the same time, and has already been encountered on page 19. The phrases in parentheses can be omitted once you get used to the definition. Also this is the time to look at Problem 31, which is relevant to the present discussion.

> **Remark** It is important to realize that if the inequality $|f(x) - L| < \varepsilon$ is satisfied whenever (2) holds, then it continues to be satisfied if we replace δ by *any smaller positive number*. In particular, δ is not a function of ε, even though δ depends on ε, since δ is not *uniquely* determined by ε (any smaller δ will work just as well).

What happens if we replace the double inequality (2) by the single inequality

$$|x - a| < \delta, \tag{2'}$$

and at the same time require that $f(x)$ be defined in an ordinary, nondeleted neighborhood of a, so that in particular $f(a)$ is defined? In other words, what does it mean to say that given any $\varepsilon > 0$, we can find a $\delta > 0$ such that $|f(x) - L| < \varepsilon$ whenever (2') holds? Some thought shows that this must mean that $f(x)$ is *continuous* at a. In fact, $f(x)$ still has the limit L

at a, since if $|f(x) - L| < \varepsilon$ whenever (2′) holds, it is certainly true that $|f(x) - L| < \varepsilon$ whenever the more restrictive condition (2) holds. Moreover, since $x = a$ is now allowed, and since (2′) is automatically satisfied for $x = a$, regardless of the value of δ, we must have $|f(a) - L| < \varepsilon$ for every $\varepsilon > 0$, which is possible only if $f(a) = L$. In other words, instead of (1) we now have

$$\lim_{x \to a} f(x) = L = f(a), \tag{1′}$$

and this is exactly what is meant by saying that $f(x)$ is continuous at a.

We now illustrate the implementation of the ε, δ method.

Example 1 Show that

$$\lim_{x \to a} x = a \tag{3}$$

for every a.

Solution Given any $\varepsilon > 0$, we need only choose $\delta = \varepsilon$, as in Figure 32. Then it is evident that $|x - a| < \varepsilon$ whenever $0 < |x - a| < \delta$, or for that matter, whenever $|x - a| < \delta$. Despite its seeming triviality, (3) expresses an important fact, namely that the function $f(x) = x$ is continuous for all x. □

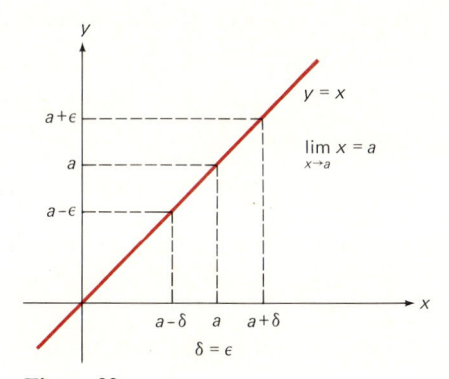

Figure 32

Example 2 Let c be any constant. Show that

$$\lim_{x \to a} c = c \tag{4}$$

for every a. More exactly, show that if $f(x) \equiv c$, then

$$\lim_{x \to a} f(x) = c \tag{4′}$$

for every a.

Solution We observe that in this case

$$|f(x) - c| \equiv |c - c| = 0.$$

Therefore, given any $\varepsilon > 0$, $|f(x) - c| < \varepsilon$ is automatically true whenever $0 < |x - a| < \delta$, or for that matter, whenever $|x - a| < \delta$, regardless of the

Figure 33
Any δ > 0 works.

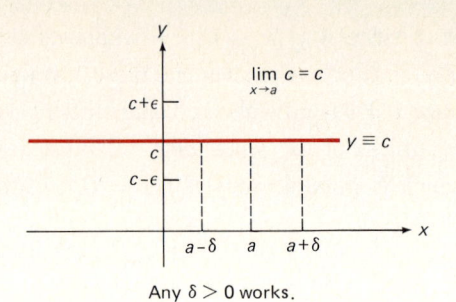

choice of δ (see Figure 33). Formula (4) says that the limit of a constant is the constant itself, or equivalently that a constant function is continuous everywhere (that is, for all x). □

Example 3 Show that
$$\lim_{x \to 1} (3x + 2) = 5. \tag{5}$$

Solution We have already solved problems of this kind by using the continuity of polynomials (see Example 8, page 70), but it is instructive to see how an ε, δ proof goes. Given any ε > 0, we want a δ > 0 such that
$$|(3x + 2) - 5| = |3(x - 1)| = 3|x - 1| < \varepsilon$$
whenever $0 < |x - 1| < \delta$, for then (5) will be proved. Since $3|x - 1| < \varepsilon$ is equivalent to $|x - 1| < \varepsilon/3$, an appropriate choice is $\delta = \varepsilon/3$, as illustrated in Figure 34. □

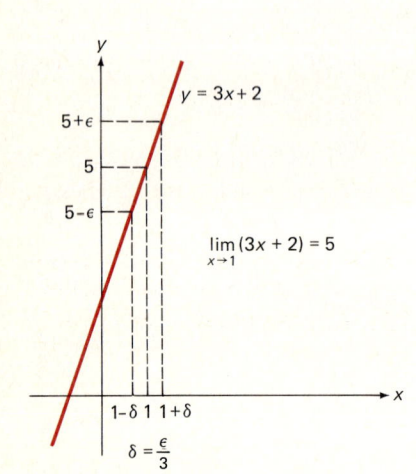

Figure 34

Example 4 Show that
$$\lim_{x \to a} |x| = |a| \tag{6}$$
for every a.

Solution Applying inequality (5), page 16, we have
$$||x| - |a|| < |x - a|.$$
Thus, given any ε > 0, we need only choose $\delta = \varepsilon$ to make $||x| - |a|| < \varepsilon$ whenever $0 < |x - a| < \delta$ (or $|x - a| < \delta$). This is illustrated in Figure 35 for the case of negative a. Formula (6) says that $|x|$, the absolute value function, is continuous everywhere. □

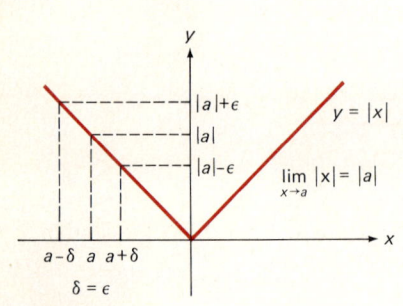

Figure 35

Example 5 Give a rigorous proof of the limit
$$\lim_{x \to 1} \frac{x^2 - 1}{x - 1} = 2,$$
evaluated informally in Example 1, page 66.

Solution Given any ε > 0, we want a δ > 0 such that
$$\left| \frac{x^2 - 1}{x - 1} - 2 \right| = \left| \frac{(x + 1)(x - 1)}{x - 1} - 2 \right| = |(x + 1) - 2| = |x - 1| < \varepsilon$$

74   Chapter 1   Functions and Limits

whenever $0 < |x - 1| < \delta$. An appropriate choice is of course $\delta = \varepsilon$. Note that the first part of the double inequality $0 < |x - 1| < \delta$ is now indispensable, since the function $(x^2 - 1)/(x - 1)$ is not defined for $x = 1$. ☐

Example 6 Show that
$$\lim_{x \to 0} \frac{x^3 + x}{x} = 1. \tag{7}$$

Solution Given any $\varepsilon > 0$, we look for a $\delta > 0$ such that
$$\left| \frac{x^3 + x}{x} - 1 \right| = |(x^2 + 1) - 1| = |x^2| = |x|^2 < \varepsilon$$
whenever $0 < |x| < \delta$. Since $|x|^2 < \varepsilon$ is equivalent to $|x| < \sqrt{\varepsilon}$, an appropriate choice is $\delta = \sqrt{\varepsilon}$. Alternatively, dividing the numerator $x^3 + x$ by the denominator x and using the continuity of the resulting polynomial $x^2 + 1$, we find that
$$\lim_{x \to 0} \frac{x^3 + x}{x} = \lim_{x \to 0} (x^2 + 1) = 0^2 + 1 = 1.$$
(Problem 9 makes an important point in this regard.) The validity of (7) can also be seen more informally by inspecting the graph of the function $(x^3 + x)/x$, which is a parabola with the single point $(0, 1)$ missing (see Figure 36). ☐

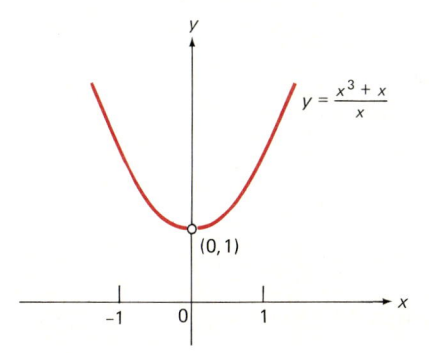

Figure 36

Geometric Meaning of a Limit

Suppose $f(x) \to L$ as $x \to a$. Then, given any $\varepsilon > 0$, there is a $\delta > 0$ such that $|f(x) - L| < \varepsilon$ whenever $0 < |x - a| < \delta$. But the inequality $|f(x) - L| < \varepsilon$ is equivalent to $-\varepsilon < f(x) - L < \varepsilon$ or
$$L - \varepsilon < f(x) < L + \varepsilon. \tag{8}$$
Thus the ε, δ definition of a limit has a simple geometric interpretation: *The part of the graph of the function $y = f(x)$ corresponding to values of x in the deleted neighborhood $0 < |x - a| < \delta$ is entirely confined to the horizontal strip $L - \varepsilon < y < L + \varepsilon$ of width 2ε parallel to the x-axis.* This fact can be used to give a graphical construction of the largest deleted neighborhood of a for which $|f(x) - L| < \varepsilon$. The two parts of Figure 37 illustrate this construction for the same function f and two different values of ε. Make sure you understand why the largest value of δ that works is determined by the behavior of f to the *left* of a in Figure 37(a), but by the behavior of f to the

Section 1.6 A Closer Look at Limits 75

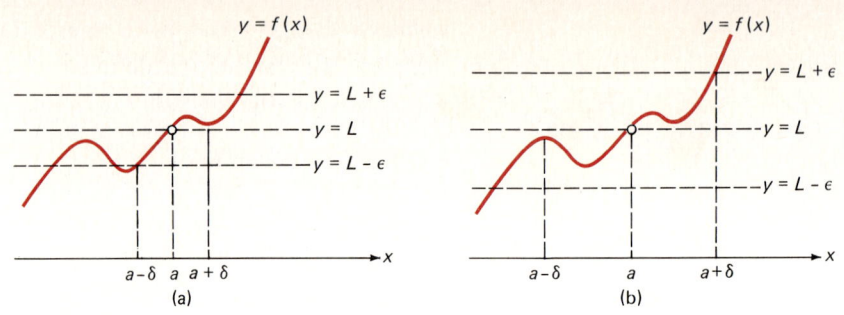

Figure 37 How to find the largest δ for a given ε

right of *a* in Figure 37(b). We put a hollow dot at the point (*a*, *L*) to indicate that the function *f* may not be defined at *a*, or that its value at *a* may not coincide with *L*. The hollow dot becomes a solid dot if *f* is continuous at *a*.

Basic Limit Rules

With the help of this geometric interpretation of the ε, δ procedure, we can establish a number of basic rules obeyed by limits. In each case we start from the double inequality (8), valid for all x in some deleted δ-neighborhood $0 < |x - a| < \delta$.

(i) *If $f(x)$ has a limit L at a, then $f(x)$ is bounded in a deleted neighborhood of a*, that is, there are positive numbers C and δ such that $|f(x)| \leq C$, or equivalently $-C \leq f(x) \leq C$, whenever $0 < |x - a| < \delta$. In fact, we need only pick C large enough to make the strip $-C \leq y \leq C$ contain the strip $L - \varepsilon < y < L + \varepsilon$, which can always be done.

(ii) *If $f(x)$ has a nonzero limit L at a, there is a deleted neighborhood of a in which $f(x)$ is nonzero and has the same sign as L*. To see this, we first choose ε small enough to make $L - \varepsilon > 0$ if $L > 0$ or $L + \varepsilon < 0$ if $L < 0$. Then, as shown in Figure 38(a) for $L > 0$ and in Figure 38(b) for $L < 0$, the function $f(x)$ has the same sign as L for all x in a suitable neighborhood of *a*.

(iii) *If $f(x)$ is nonnegative in a deleted neighborhood of a, then $f(x)$ cannot have a negative limit at a*. The same is true with the words *nonnegative* and *negative* replaced by *nonpositive* and *positive*. This is just another way of stating rule (ii).

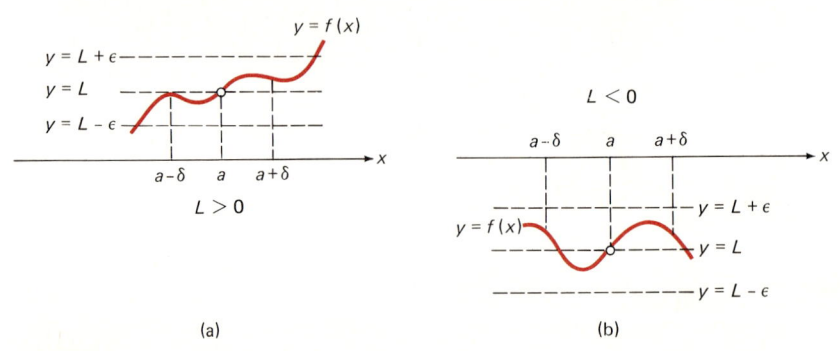

Figure 38

76   Chapter 1   Functions and Limits

The last rule is more technical, and is included only because it will be used in the proof of Theorem 6, page 82.

Optional

(iv) *If $f(x)$ has a nonzero limit L at a, then the reciprocal function $1/f(x)$ is bounded in a deleted neighborhood of a.* To verify this rule, we again choose ε small enough to make $L - \varepsilon > 0$ if $L > 0$ and $L + \varepsilon < 0$ if $L < 0$ (see Figure 38). Then $L - \varepsilon < f(x) < L + \varepsilon$ implies

$$\frac{1}{L+\varepsilon} < \frac{1}{f(x)} < \frac{1}{L-\varepsilon},$$

with the help of Theorem 4, page 9. Thus the part of the graph of the function $y = 1/f(x)$ corresponding to values of x in the deleted neighborhood $0 < |x - a| < \delta$ is entirely confined to the horizontal strip $1/(L + \varepsilon) < y < 1/(L - \varepsilon)$, and to complete the proof, we choose C large enough to make this strip lie inside the strip $-C \le y \le C$.

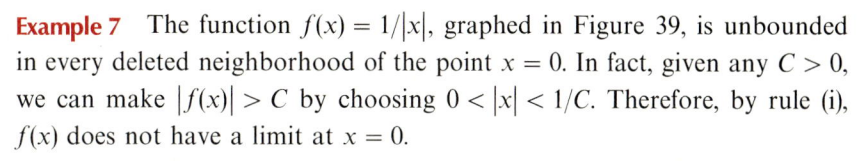

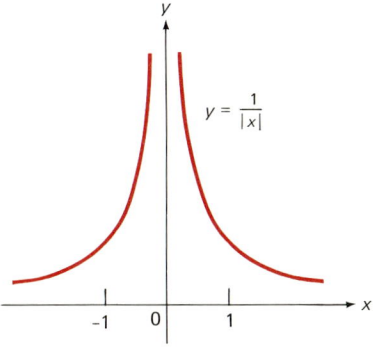

Example 7 The function $f(x) = 1/|x|$, graphed in Figure 39, is unbounded in every deleted neighborhood of the point $x = 0$. In fact, given any $C > 0$, we can make $|f(x)| > C$ by choosing $0 < |x| < 1/C$. Therefore, by rule (i), $f(x)$ does not have a limit at $x = 0$.

Figure 39

We now begin a study, to be continued in the next two sections, of limits involving two or more functions. Our first result of this sort says that if a bounded function is multiplied by a function approaching zero, then the product also approaches zero.

Theorem 3 *(Preservation of approach to zero).* If $f(x)$ is bounded in a deleted neighborhood of a and if $g(x) \to 0$ as $x \to a$, then $f(x)g(x) \to 0$ as $x \to a$.

Proof (Optional) Since $f(x)$ is bounded in a deleted neighborhood of a, there are numbers $C > 0$ and $\delta_f > 0$ such that $|f(x)| \le C$ whenever $0 < |x - a| < \delta_f$. Also, since $g(x) \to 0$ as $x \to a$, for any given $\varepsilon > 0$ there is a number $\delta_g > 0$ such that $|g(x) - 0| = |g(x)| < \varepsilon/C$ whenever $0 < |x - a| < \delta_g$. (The use of ε/C instead of ε anticipates the last step of the proof.) We now choose δ to be the smaller of the two numbers δ_f and δ_g, that is, we choose $\delta = \min\{\delta_f, \delta_g\}$. This choice of δ is dictated by our need to have a deleted neighborhood of a in which both inequalities $|f(x)| \le C$ and $|g(x)| < \varepsilon/C$ hold *simultaneously*. We then have

$$|f(x)g(x) - 0| = |f(x)||g(x)| < C \cdot \frac{\varepsilon}{C} = \varepsilon$$

whenever $0 < |x - a| < \delta$. But in ε, δ language this means that $f(x)g(x) \to 0$ as $x \to a$. ■

Corollary *If $f(x)$ has a limit at a and if $g(x) \to 0$ as $x \to a$, then $f(x)g(x) \to 0$ as $x \to a$.*

Proof Apply Theorem 3, observing that if $f(x)$ has a limit at a, then $f(x)$ is bounded in a deleted neighborhood of a, because of rule (i). ■

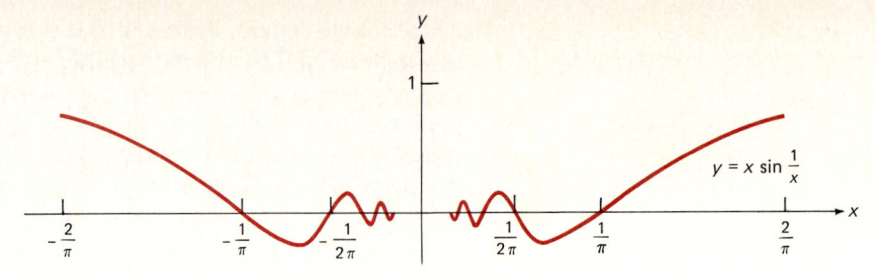

The limit at x = 0 exists and is equal to 0.

Figure 40

Example 8 Consider the function $x \sin (1/x)$, graphed in Figure 40, which makes more and more oscillations as $x \to 0$ (the function is not defined for $x = 0$). Since $|\sin (1/x)| \leq 1$ for all $x \neq 0$ and since $\lim_{x \to 0} x = 0$, it follows from Theorem 3 that

$$\lim_{x \to 0} x \sin \frac{1}{x} = 0.$$

Example 9 Let $f(x) = 1/|x|$ and $g(x) = |x|$. Then

$$\lim_{x \to 0} g(x) = \lim_{x \to 0} |x| = 0$$

(see Example 4), while

$$\lim_{x \to 0} f(x)g(x) = \lim_{x \to 0} 1 = 1.$$

This does not contradict Theorem 3, since $f(x)$ is not bounded in a deleted neighborhood of $x = 0$, as already noted in Example 7.

Example 10 In Example 8 the oscillations of $x \sin (1/x)$ get progressively weaker as $x \to 0$, because of the factor x. We now remove this factor and consider the behavior of the function $\sin (1/x)$ itself. As $x \to 0$, $\sin (1/x)$ makes more and more oscillations between the values -1 and 1, and at the same time the distance between successive crossings of the x-axis by the graph of the function becomes smaller and smaller (see Figure 41). It is hard to imagine that $\sin (1/x)$ stays close to any number at all near $x = 0$, since there

The limit at x = 0 does not exist.

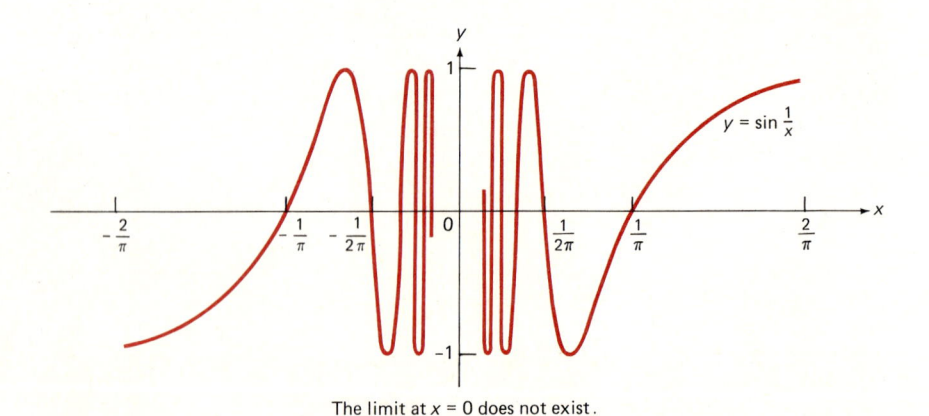

Figure 41

78 Chapter 1 Functions and Limits

is no deleted neighborhood of $x = 0$ in which the function fails to make a complete oscillation (in fact, infinitely many oscillations!). Thus intuition strongly suggests that $\sin(1/x)$ has no limit at $x = 0$.

Optional

To *prove* that this is actually true, we use the ε, δ method to show that the assumption that $\sin(1/x)$ has a limit L at $x = 0$ leads to a contradiction. Suppose $f(x) \to L$ as $x \to 0$. Then, choosing $\varepsilon = \frac{1}{2}$, we can find a number $\delta > 0$ such that

$$\left|\sin\frac{1}{x} - L\right| < \frac{1}{2} \tag{9}$$

whenever $0 < |x| < \delta$. Let n be any integer whose absolute value is so large that both points $x_1 = 1/(2n + \frac{1}{2})\pi$ and $x_2 = 1/(2n - \frac{1}{2})\pi$ belong to the deleted neighborhood $0 < |x| < \delta$. Then $\sin(1/x_1) = \sin(2n + \frac{1}{2})\pi = \sin\frac{1}{2}\pi = 1$, while $\sin(1/x_2) = \sin(2n - \frac{1}{2})\pi = \sin(-\frac{1}{2}\pi) = -1$. It follows from (9) that

$$\left|\sin\frac{1}{x_1} - L\right| = |1 - L| < \frac{1}{2}, \qquad \left|\sin\frac{1}{x_2} - L\right| = |-1 - L| < \frac{1}{2}$$

These two inequalities are incompatible, since the first implies that $L > \frac{1}{2}$ and the second implies that $L < -\frac{1}{2}$. Thus the assumption that $\sin(1/x)$ has a limit at $x = 0$ leads to a contradiction. Therefore $\sin(1/x)$ does *not* have a limit at $x = 0$.

Problems

First find the limit L of the given function f at the point a. Then, for the given number ε, find a number $\delta > 0$ such that $|f(x) - L| < \varepsilon$ whenever $0 < |x - a| < \delta$, and explain your choice. Try to make δ as large as possible (some experimentation with a calculator is called for in Problems 5–8).

1. $f(x) = 5x$, any a, any $\varepsilon > 0$
2. $f(x) = mx + b$ ($m \neq 0$), any a, any $\varepsilon > 0$
3. $f(x) = \dfrac{3x^2 + 8x - 3}{x + 3}$, $a = -3$, $\varepsilon = 0.15$
4. $f(x) = \dfrac{2x^2 - 7x - 4}{x - 4}$, $a = 4$, $\varepsilon = 0.25$
5. $f(x) = x^2$, $a = 2$, $\varepsilon = 1$
6. $f(x) = x^2 \sin x$, $a = 0$, $\varepsilon = 0.5$
7. $f(x) = x^3$, $a = -1$, $\varepsilon = 0.1$
8. $f(x) = \dfrac{x^3 - 1}{x - 1}$, $a = 1$, $\varepsilon = 0.05$

9. Suppose the function f has a limit L at a but is undefined at a. Suppose further that $f(x) = g(x)$ for all x in some deleted neighborhood of a, where g is continuous at a. Show that $L = g(a)$.
10. Suppose the function f has a limit L at a, but either is undefined at a or has a value different from L at a. Then f is discontinuous at a, as we know from page 69. Show that this discontinuity is *removable* in the sense that it can be eliminated by suitably defining (or redefining) f at a. More exactly, show that if the function f is extended (or modified) by defining its value at a to be L, then the new function obtained in this way is continuous at a.
11. Let f be discontinuous at a point a. When is the discontinuity *nonremovable*?
12. Is the discontinuity of the function $(\sin x)/x$ at $x = 0$ removable? How about the discontinuity of $|x|/x$? Of $\sin(1/x)$?
13. Is the discontinuity of the function $(x^2 - 1)/(x - 1)$ at $x = 1$ removable?
14. Give an example of a function which is bounded in every deleted neighborhood of a point a, but has no limit at a.

Evaluate the given limit (if it exists).

15. $\lim\limits_{x \to 0} x \cos x$
16. $\lim\limits_{x \to 0} \dfrac{\cos x}{x}$
17. $\lim\limits_{x \to \pi/2} \dfrac{\cos x}{x}$
18. $\lim\limits_{x \to 0} x \cos \dfrac{1}{x}$
19. $\lim\limits_{x \to 0} \cos \dfrac{1}{x}$
20. $\lim\limits_{x \to 0} \sin x \cos x$

21. $\lim\limits_{x \to 0} \sin x \sin \dfrac{1}{x}$

22. $\lim\limits_{x \to \pi/2} \tan x$

23. $\lim\limits_{x \to 0} \sin x \tan x$

24. $\lim\limits_{x \to 0} \sin^n x$ (n any positive integer)

Show that

25. If $\lim\limits_{x \to a} f(x) = L$, then $\lim\limits_{x \to a} |f(x)| = |L|$

26. If $\lim\limits_{x \to a} |f(x)| = 0$, then $\lim\limits_{x \to a} f(x) = 0$

27. If $\lim\limits_{x \to a} |f(x)|$ exists and is nonzero, then $\lim\limits_{x \to a} f(x)$ may fail to exist

28. Show that if a function f is continuous at a, then so is its absolute value $|f|$. Is the converse true?

Let $\lim\limits_{x \to a} f(x) = L$ and $\lim\limits_{x \to a} g(x) = M$. Show that

29. If $L < M$, then $f(x) < g(x)$ near a (that is, in some deleted neighborhood of a)

30. If $f(x) \leq g(x)$ near a, then $L \leq M$, where $L = M$ is a possibility even if $f(x) < g(x)$

31. Use the ε, δ method to show that if
$$\lim\limits_{x \to a} f(x) = L \quad \text{and} \quad \lim\limits_{x \to a} f(x) = M,$$
then $L = M$. In other words, verify that if the limit of the function f at the point a exists, then this limit is *unique*, in the sense of having only one possible value, as we have been tacitly assuming all along.

1.7 Algebraic Operations on Limits

In our subsequent work we must be able to evaluate limits of algebraic expressions involving two or more functions. These calculations will be greatly facilitated by the fact, which we now proceed to prove (in Theorems 4–6), that the limit of a sum of two functions equals the sum of the limits of the separate functions, and similarly with the word *sum* replaced in both places by *difference*, *product* and *quotient*. The proofs are simple, but technical, and thus have been made optional. If you skip the proofs, be sure that you understand the statements of the theorems themselves, which will be used freely. The same warning also applies to Theorems 9 and 10, whose proofs will be found at the end of the section.

> **Theorem 4** *(Limit of the sum or difference of two functions).* If $f(x) \to L$ and $g(x) \to M$ as $x \to a$, then $f(x) + g(x) \to L + M$ and $f(x) - g(x) \to L - M$ as $x \to a$.

Proof (Optional) The argument closely resembles the one used to prove Theorem 3, page 77. Since $f(x) \to L$ as $x \to a$, for any given $\varepsilon > 0$ there is a number $\delta_f > 0$ such that $|f(x) - L| < \varepsilon/2$ whenever $0 < |x - a| < \delta_f$. (The use of $\varepsilon/2$ instead of ε anticipates the last step of the proof.) Also, since $g(x) \to M$ as $x \to a$, there is a number $\delta_g > 0$ such that $|g(x) - M| < \varepsilon/2$ whenever $0 < |x - a| < \delta_g$. Therefore, by the triangle inequality (Theorem 5, page 15),

$$|f(x) + g(x) - (L + M)| = |[f(x) - L] + [g(x) - M]|$$
$$\leq |f(x) - L| + |g(x) - M| < \frac{\varepsilon}{2} + \frac{\varepsilon}{2} = \varepsilon$$

whenever $0 < |x - a| < \delta = \min\{\delta_f, \delta_g\}$. But in ε, δ language this means that $f(x) + g(x) \to L + M$ as $x \to a$. The proof that $f(x) - g(x) \to L - M$ as $x \to a$ is virtually the same. ∎

The result extends at once to the case of more than two functions.

Corollary 1 *If $f_1(x) \to L_1, f_2(x) \to L_2, \ldots, f_n(x) \to L_n$ as $x \to a$, then*

$$f_1(x) \pm f_2(x) \pm \cdots \pm f_n(x) \to L_1 \pm L_2 \pm \cdots \pm L_n \tag{1}$$

as $x \to a$.

Proof Here we can pick either $+$ or $-$ in each occurrence of the symbol $\pm$, with the understanding that the *same* choice is made in corresponding places in both sides of formula (1). The proof is by repeated application of Theorem 4. ∎

Corollary 2 *The function $f(x) \to L$ as $x \to a$ if and only if $f(x) = L + e(x)$, where $e(x) \to 0$ as $x \to a$.*

Proof Suppose $f(x) \to L$ as $x \to a$, and let $e(x) = f(x) - L$. Then

$$\lim_{x \to a} e(x) = \lim_{x \to a} [f(x) - L] = \lim_{x \to a} f(x) - \lim_{x \to a} L = L - L = 0,$$

since the limit of a constant is the constant itself. Conversely, suppose $f(x) = L + e(x)$, where $e(x) \to 0$ as $x \to a$. Then

$$\lim_{x \to a} f(x) = \lim_{x \to a} [L + e(x)] = \lim_{x \to a} L + \lim_{x \to a} e(x) = L + 0 = L. \quad ∎$$

We can think of $e(x) = f(x) - L$ as the "error" of the approximation $f(x) \approx L$ near a. Corollary 2 says that $f(x) \to L$ as $x \to a$ if and only if this error approaches 0 as $x \to a$.

Theorem 5 *(Limit of the product of two functions).* *If $f(x) \to L$ and $g(x) \to M$ as $x \to a$, then $f(x)g(x) \to LM$ as $x \to a$.*

Proof (Optional) Let $e_f(x) = f(x) - L$ and $e_g(x) = g(x) - M$ be the errors of the two approximations $f(x) \approx L$ and $g(x) \approx M$ near a. Then $e_f(x) \to 0$ and $e_g(x) \to 0$ as $x \to a$, by Corollary 2. Moreover,

$$\begin{aligned}f(x)g(x) - LM &= [L + e_f(x)][M + e_g(x)] - LM \\ &= Me_f(x) + Le_g(x) + e_f(x)e_g(x),\end{aligned}$$

by easy algebra. But by Theorem 3, page 77, and its corollary, each of the three terms on the right approaches zero as $x \to a$ (every constant function is bounded), and hence so does the whole expression on the right, by Corollary 1 above. Therefore, as $x \to a$, $f(x)g(x) - LM \to 0$ or equivalently $f(x)g(x) \to LM$. ∎

Corollary 1 *If $f_1(x) \to L_1, f_2(x) \to L_2, \ldots, f_n(x) \to L_n$ as $x \to a$, then*

$$f_1(x)f_2(x) \cdots f_n(x) \to L_1 L_2 \cdots L_n$$

as $x \to a$.

Proof Apply Theorem 5 repeatedly. ∎

Corollary 2 *If c is any constant and $f(x) \to L$ as $x \to a$, then $cf(x) \to cL$ as $x \to a$.*

Proof Choose $g(x) \equiv c$ in Theorem 5. ■

> **Theorem 6** *(Limit of the quotient of two functions). If $f(x) \to L$ and $g(x) \to M \neq 0$ as $x \to a$, then $f(x)/g(x) \to L/M$ as $x \to a$.*

Proof (Optional) Writing
$$\frac{1}{g(x)} - \frac{1}{M} = \frac{M - g(x)}{Mg(x)}, \tag{2}$$
we observe that
$$\lim_{x \to a} [M - g(x)] = \lim_{x \to a} M - \lim_{x \to a} g(x) = M - M = 0,$$
by Theorem 4. Moreover
$$\lim_{x \to a} Mg(x) = M^2 \neq 0,$$
by Corollary 2 of Theorem 5, so that $1/Mg(x)$ is bounded in a deleted neighborhood of a, by rule (iv), page 77. Therefore, by Theorem 3, page 77, the right side of (2) approaches zero as $x \to a$, or equivalently $1/g(x) \to 1/M$ as $x \to a$. But $f(x)/g(x)$ is the product of $f(x)$ and $1/g(x)$, the reciprocal of $g(x)$, and hence by Theorem 5, $f(x)/g(x) = f(x)[1/g(x)] \to L(1/M) = L/M$ as $x \to a$. ■

To summarize Theorems 4–6, they say that *if*
$$\lim_{x \to a} f(x) = L, \qquad \lim_{x \to a} g(x) = M, \tag{3}$$
then
$$\lim_{x \to a} [f(x) \pm g(x)] = L \pm M, \tag{4}$$
$$\lim_{x \to a} f(x)g(x) = LM, \tag{5}$$
$$\lim_{x \to a} \frac{f(x)}{g(x)} = \frac{L}{M} \qquad (M \neq 0). \tag{6}$$

We now illustrate the use of these formulas.

Example 1 Use (4) and (5) to show that
$$\lim_{x \to -4} \left(x^2 + \frac{3}{2}x \right) = 10.$$

Solution We already know that
$$\lim_{x \to -4} x = -4, \qquad \lim_{x \to -4} \frac{3}{2} = \frac{3}{2}.$$
Therefore
$$\lim_{x \to -4} x^2 = \lim_{x \to -4} x \cdot \lim_{x \to -4} x = (-4)(-4) = 16,$$
$$\lim_{x \to -4} \frac{3}{2}x = \lim_{x \to -4} \frac{3}{2} \cdot \lim_{x \to -4} x = \frac{3}{2}(-4) = -6,$$

by two applications of formula (5). An application of (4) then gives

$$\lim_{x \to -4} \left(x^2 + \frac{3}{2}x \right) = \lim_{x \to -4} x^2 + \lim_{x \to -4} \frac{3}{2}x = 16 + (-6) = 10. \quad \square$$

Example 2 Evaluate

$$L = \lim_{x \to 3} \frac{x^2 - 5x + 6}{x^2 - 8x + 15}.$$

Solution The denominator approaches zero as $x \to 3$, but this difficulty can be avoided by canceling a common factor of $x - 3$ from the numerator and denominator:

$$\frac{x^2 - 5x + 6}{x^2 - 8x + 15} = \frac{(x-3)(x-2)}{(x-3)(x-5)} = \frac{x-2}{x-5} \quad (x \neq 3, 5).$$

The rest of the calculation is straightforward. Using (6) and (4), we have

$$L = \lim_{x \to 3} \frac{x-2}{x-5} = \frac{\lim_{x \to 3}(x-2)}{\lim_{x \to 3}(x-5)} = \frac{\lim_{x \to 3} x - \lim_{x \to 3} 2}{\lim_{x \to 3} x - \lim_{x \to 3} 5} = \frac{3-2}{3-5} = -\frac{1}{2}. \quad \square$$

A quotient of two polynomials, like $(x^2 - 5x + 6)/(x^2 - 8x + 15)$, is called a *rational function*.

Algebraic Operations on Continuous Functions

Next let $f(x)$ and $g(x)$ be two functions, both of which are continuous at a. Then instead of (3), we have

$$\lim_{x \to a} f(x) = f(a), \quad \lim_{x \to a} g(x) = g(a), \qquad (3')$$

so that formulas (4)–(6) become

$$\lim_{x \to a} [f(x) \pm g(x)] = f(a) \pm g(a), \qquad (4')$$

$$\lim_{x \to a} f(x)g(x) = f(a)g(a), \qquad (5')$$

$$\lim_{x \to a} \frac{f(x)}{g(x)} = \frac{f(a)}{g(a)} \quad (g(a) \neq 0). \qquad (6')$$

But this is exactly what is meant by saying that the functions $f(x) \pm g(x)$, $f(x)g(x)$ and $f(x)/g(x)$ are continuous at a. Thus we have just proved the following basic result.

> **Theorem 7** *(Continuity of combinations of functions).* If the functions $f(x)$ and $g(x)$ are both continuous at a, then so are the sum $f(x) + g(x)$, the difference $f(x) - g(x)$, the product $f(x)g(x)$, and the quotient $f(x)/g(x)$, provided that $g(a) \neq 0$ in the last case.

Corollary *If the functions $f_1(x), f_2(x), \ldots, f_n(x)$ are all continuous at a, then so are the algebraic sum $f_1(x) \pm f_2(x) \pm \cdots \pm f_n(x)$ and the product $f_1(x)f_2(x) \cdots f_n(x)$.*

Proof By definition, an algebraic sum is a sum each of whose terms can have either sign. To prove the corollary, apply Theorem 7 repeatedly. ∎

Section 1.7   Algebraic Operations on Limits

Continuity of Polynomials, Rational Functions and Trigonometric Functions

The next two theorems were anticipated on page 70, and have already been used unofficially to solve limit problems.

> **Theorem 8** (*Continuity of polynomials and rational functions*). *Any polynomial*
> $$P(x) = a_0 + a_1 x + a_2 x^2 + \cdots + a_n x^n \qquad (a_n \neq 0)$$
> *is continuous for all x. Any rational function*
> $$R(x) = \frac{a_0 + a_1 x + a_2 x^2 + \cdots + a_n x^n}{b_0 + b_1 x + b_2 x^2 + \cdots + b_N x^N} \qquad (a_n \neq 0, b_N \neq 0)$$
> *is continuous at every point of its domain of definition, that is, at every point where its denominator is nonzero.*

Proof Each term $a_k x^k$ ($k = 0, 1, \ldots, n$) of the polynomial $P(x)$ is continuous, since it is the product of $k + 1$ continuous functions, namely the constant a_k and k factors of x. But then $P(x)$, which is the sum of $n + 1$ such terms, is also continuous. Since polynomials are continuous, so is any quotient $R(x) = P(x)/Q(x)$ of two polynomials, except at the points (if any) at which the denominator $Q(x)$ equals zero. ∎

Once it is known that polynomials are continuous, the calculations in Example 1 can be abbreviated to

$$\lim_{x \to -4} \left(x^2 + \frac{3}{2} x \right) = (-4)^2 + \frac{3}{2}(-4) = 16 - 6 = 10.$$

The continuity of rational functions can also be used to bypass certain steps in Example 2 (which ones?).

Example 3 The rational function $1/(1 + x^2)$ is continuous for all x, since its denominator is never zero.

Example 4 The rational function $x/(1 - x^2)$ is continuous everywhere except at the two "points of discontinuity" $x = 1$ and $x = -1$ at which its denominator takes the value 0.

> **Theorem 9** (*Continuity of trigonometric functions*). *Each of the trigonometric functions* $\sin x$, $\cos x$, $\tan x$, $\cot x$, $\sec x$ *and* $\csc x$ *is continuous at every point of its domain of definition.*

The graphs of the trigonometric functions strongly suggest that they are continuous, since the graph of each branch is an "unbroken" curve, but a formal proof is still called for. It is a bit technical, and hence is deferred to the end of the section.

Example 5 By the continuity of $\cos x$,

$$\lim_{x \to \pi} (\cos^2 x + \cos x + 1) = \cos^2 \pi + \cos \pi + 1 = (-1)^2 + (-1) + 1 = 1.$$

The Sandwich Theorem

The next theorem gives another useful tool for evaluating limits.

> **Theorem 10** *(Sandwich theorem). If*
> $$f(x) \leq g(x) \leq h(x)$$
> *in some deleted neighborhood of a, and if*
> $$\lim_{x \to a} f(x) = \lim_{x \to a} h(x) = L,$$
> *then*
> $$\lim_{x \to a} g(x) = L.$$

In other words, a function sandwiched or squeezed between two functions approaching the same limit L must also approach L. The validity of this theorem is strongly suggested by Figure 42 (where else can $g(x)$ go as $x \to a$ if not to L?), but you will find a rigorous proof below, following the proof of Theorem 9.

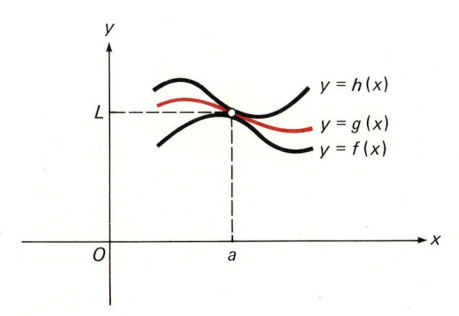

Figure 42

Example 6 Given any positive integer n, show that
$$\lim_{x \to 0} \sqrt[n]{1 + x} = 1. \tag{7}$$
Since $\sqrt[n]{1 + 0} = 1$, this says that the function $\sqrt[n]{1 + x}$ is continuous at $x = 0$.

Solution If $|x| < 1$, then
$$1 - |x| \leq \sqrt[n]{1 + x} \leq 1 + |x|.$$
(The details are left as an exercise, based on the meaning of $|x|$ and Example 6, page 12.) Since
$$\lim_{x \to 0} (1 \pm |x|) = \lim_{x \to 0} 1 \pm \lim_{x \to 0} |x| = 1 \pm 0 = 1,$$
formula (8) now follows at once from the sandwich theorem. □

Example 7 Give a rigorous proof of the limit
$$\lim_{x \to 0} \frac{\sin x}{x} = 1, \tag{8}$$
evaluated informally in Example 2, page 67.

Section 1.7 Algebraic Operations on Limits

Solution By formula (2), page 64,

$$\cos x < \frac{\sin x}{x} < 1 \qquad (0 < |x| < \pi/2),$$

where $\cos x$ and the constant 1 both approach 1 as $x \to 0$. Applying the sandwich theorem, we immediately obtain (8). ∎

Proof of Theorem 9 (Optional) Setting $\alpha = x$, $\beta = a$ in formula (15), page 58, we get

$$\sin x - \sin a = 2 \cos \frac{x+a}{2} \sin \frac{x-a}{2},$$

which implies the inequality

$$|\sin x - \sin a| = 2\left|\cos \frac{x+a}{2}\right|\left|\sin \frac{x-a}{2}\right| \le 2\left|\sin \frac{x-a}{2}\right| \le 2\left|\frac{x-a}{2}\right|,$$

or equivalently

$$|\sin x - \sin a| \le |x - a|, \tag{9}$$

valid for all x and a. (Here we use the fact that $|\cos (x + a)/2| \le 1$, together with inequality (3'), page 65.) Also, because of (9),

$$|\cos x - \cos a| = \left|\sin\left(x + \frac{\pi}{2}\right) - \sin\left(a + \frac{\pi}{2}\right)\right|$$

$$\le \left|\left(x + \frac{\pi}{2}\right) - \left(a + \frac{\pi}{2}\right)\right| = |x - a|,$$

and we have the companion inequality

$$|\cos x - \cos a| \le |x - a|, \tag{9'}$$

also valid for all x and a. From (9) and (9') it follows at once that $|\sin x - \sin a|$ and $|\cos x - \cos a|$ will both be less than any given $\varepsilon > 0$ whenever x is within distance $\delta = \varepsilon$ of a. Therefore

$$\lim_{x \to a} \sin x = \sin a, \qquad \lim_{x \to a} \cos x = \cos a,$$

and the continuity of $\sin x$ and $\cos x$ is proved. Since the functions $\sin x$ and $\cos x$ are both continuous for all x, the reciprocals $\sec x = 1/\cos x$, $\csc x = 1/\sin x$ and quotients $\tan x = \sin x/\cos x$, $\cot x = \cos x/\sin x$ are continuous wherever they are defined, that is, wherever their denominators are nonzero. ∎

Proof of Theorem 10 Given any $\varepsilon > 0$, we can find a number $\delta_f > 0$ such that $|f(x) - L| < \varepsilon$ or $-\varepsilon < f(x) - L < \varepsilon$ whenever $0 < |x - a| < \delta_f$, and a number $\delta_h > 0$ such that $|h(x) - L| < \varepsilon$ or $-\varepsilon < h(x) - L < \varepsilon$ whenever $0 < |x - a| < \delta_h$. Together with $f(x) \le g(x) \le h(x)$, these inequalities imply that $-\varepsilon < f(x) - L \le g(x) - L \le h(x) - L < \varepsilon$ whenever $0 < |x - a| < \delta = \min\{\delta_f, \delta_g\}$. Therefore $-\varepsilon < g(x) - L < \varepsilon$ or $|g(x) - L| < \varepsilon$ whenever $0 < |x - a| < \delta$, so that $g(x) \to L$ as $x \to a$. ∎

Problems

Evaluate

1. $\lim_{x \to 1.6} \dfrac{25x^2 - 64}{5x - 8}$

2. $\lim_{x \to -0.3} \dfrac{100x^2 - 9}{10x + 3}$

3. $\lim_{x \to 0} \dfrac{x^2 - 2x + 3}{x^2 + 2x - 1}$

4. $\lim_{x \to 5} \dfrac{2x^2 - 11x + 5}{3x^2 - 14x - 5}$

5. $\lim_{x \to -4} \dfrac{x^3 + 64}{x + 4}$

6. $\lim_{x \to 3} \dfrac{x^4 - 81}{x - 3}$

7. $\lim_{x \to 2} \dfrac{x^3 - 2x^2 - 4x + 8}{x^4 - 8x^2 + 16}$

8. $\lim_{x \to 0} \dfrac{(1 + x)(1 + 2x) - 1}{x}$

9. $\lim_{x \to 0} \dfrac{(1 + x)(1 + 2x)(1 + 3x) - 1}{x}$

10. $\lim_{x \to -1} \dfrac{x^{100} - 10x^{10} + 999}{x^{50} - 5x^5 + 99}$

11. $\lim_{x \to 0} \dfrac{(1 + x)^5 - (1 + 5x)}{x^2 + x^5}$

12. $\lim_{x \to 0} \dfrac{(1 + 3x)^4 - (1 + 4x)^3}{x^2}$

13. Find the limit L of $f(x) = (x + 1)/(x - 2)$ at $x = 5$. Then, given any $\varepsilon > 0$, find a $\delta > 0$ such that $0 < |x - 5| < \delta$ implies $|f(x) - L| < \varepsilon$.

Find the points, if any, at which the given function is discontinuous.

14. $(x - 1)^2$

15. $101x^{11} - 1001x$

16. $\dfrac{x^2}{x^2 - 49}$

17. $\dfrac{2x + 3}{x^2 + x + 1}$

18. $\dfrac{1}{2x^2 + x - 1}$

19. $\dfrac{x - 2}{x^2 - 4}$

20. $\dfrac{3x - 5}{x^2 - 3x + 2}$

21. $\dfrac{x^2 - x + 1}{x^3 - 6x^2 + 11x - 6}$

22. $\dfrac{x^3 + 10}{x^5 - 2x^3 + x}$

Evaluate

23. $\lim_{x \to 1} (\cos^2 x - \sin^2 x)$

24. $\lim_{x \to -1} (\cos^2 x + \sin^2 x)$

25. $\lim_{x \to \pi} (\cos^3 x + \sin^3 x)$

26. $\lim_{x \to -\pi/2} (\cos^4 x - \sin^5 x)$

27. $\lim_{x \to 0} \dfrac{\sin^2 x}{x}$

28. $\lim_{x \to 0} \dfrac{\sin^2 x}{x^2}$

29. $\lim_{x \to \pi/6} \dfrac{1}{1 + \tan^2 x}$

30. $\lim_{x \to \pi/4} (2 \sec^4 x - 1)$

31. $\lim_{x \to \pi/3} (\cot x + \csc x)$

32. It was shown in Example 8, page 78, that
$$\lim_{x \to 0} x \sin \dfrac{1}{x} = 0.$$
Use the sandwich theorem to give another proof of this fact.

1.8 The Limit of a Composite Function

To evaluate limits easily, we must still learn how to deal with limits of composite functions, like $\sqrt{2x + 3}$ and $\sin (\cos x)$. The next theorem shows how this is done.

Theorem 11 *(Limit of a composite function). If*
$$\lim_{x \to a} f(x) = L$$
and g is continuous at L, then
$$\lim_{x \to a} g(f(x)) = g(L). \tag{1}$$

Proof (Optional) Let $y = f(x)$, and choose any $\varepsilon > 0$. Since g is continuous at L, we can find a $\delta_g > 0$ such that $|g(y) - g(L)| < \varepsilon$ whenever $|y - L| < \delta_g$. Moreover, since f has the limit L at a, we can also find a $\delta_f > 0$

such that $|f(x) - L| = |y - L| < \delta_g$ whenever $0 < |x - a| < \delta_f$. Therefore $0 < |x - a| < \delta_f$ implies $|y - L| < \delta_g$, which in turn implies $|g(y) - g(L)| = |g(f(x)) - g(L)| < \varepsilon$. Thus $|g(f(x)) - g(L)| < \varepsilon$ whenever $0 < |x - a| < \delta_f$, and (1) is proved. ∎

Corollary *(Continuity of a composite function). If f is continuous at a and g is continuous at $f(a)$, then $g(f(x))$ is continuous at a.*

Proof Use the theorem with $L = f(a)$. ∎

Informally, the corollary states that a continuous function of a continuous function is itself continuous.

Example 1 Given any positive integer n, show that

$$\lim_{x \to 1} \sqrt[n]{x} = 1. \tag{2}$$

Since $\sqrt[n]{1} = 1$, this says that the nth root function $\sqrt[n]{x}$ is continuous at the point $x = 1$.

Solution Choosing $f(x) = x - 1$, $g(x) = \sqrt[n]{1 + x}$, we have $g(f(x)) = \sqrt[n]{x}$. Moreover

$$\lim_{x \to 1} f(x) = 0,$$

and $g(x)$ is continuous at $x = 0$, by Example 6, page 85. It follows from Theorem 11 that

$$\lim_{x \to 1} g(f(x)) = \lim_{x \to 1} \sqrt[n]{x} = \sqrt[n]{1 + 0} = 1. \quad \square$$

Example 2 Show that the function $\sqrt[n]{x}$ is continuous at every point $a > 0$.

Solution We have

$$\lim_{x \to a} \sqrt[n]{x} = \lim_{x \to a} \sqrt[n]{a \frac{x}{a}} = \lim_{x \to a} \sqrt[n]{a} \cdot \sqrt[n]{\frac{x}{a}} = \sqrt[n]{a} \lim_{x \to a} \sqrt[n]{\frac{x}{a}},$$

since $\sqrt[n]{a}$ is a constant. Introducing a new variable $t = x/a$, and observing that $x \to a$ implies $t \to 1$, we get

$$\lim_{x \to a} \sqrt[n]{x} = \sqrt[n]{a} \lim_{t \to 1} \sqrt[n]{t} = \sqrt[n]{a} \cdot 1 = \sqrt[n]{a}, \tag{3}$$

after using formula (2) with x replaced with t. But (3) says that $\sqrt[n]{x}$ is continuous at $x = a$. (Note that introducing the new variable $t = x/a$ is equivalent to using Theorem 11 with $f(x) = x/a$ and $g(x) = \sqrt[n]{ax}$.)

The reason for the condition $a > 0$ is that $\sqrt[n]{x}$ is not defined for negative x if n is even, but the condition can be dropped if n is odd. In fact, if n is odd, $\sqrt[n]{x}$ is defined for all x and the argument just given shows that $\sqrt[n]{x} \to \sqrt[n]{a}$ as $x \to a \neq 0$, while $\sqrt[n]{x} \to \sqrt[n]{0} = 0$ as $x \to 0$ by an argument given in Example 1 of the next section. $\square$

Example 3 Show that $\lim_{x \to 11} \sqrt{2x + 3} = 5$.

Solution The functions $f(x) = 2x + 3$ and $g(x) = \sqrt{x}$ are both continuous, $f(x)$ because it is a polynomial and $g(x)$ by the preceding example. Hence the

function $g(f(x)) = \sqrt{2x + 3}$ is also continuous, by the corollary. But then
$$\lim_{x \to 11} \sqrt{2x + 3} = \sqrt{2(11) + 3} = \sqrt{25} = 5. \quad \square$$

Example 4 Find the limit of cos (sin x) at the point $x = 0$.

Solution Since sin x and cos x are continuous for all x, so is the composite function cos (sin x). Therefore
$$\lim_{x \to 0} \cos (\sin x) = \cos (\sin 0) = \cos 0 = 1. \quad \square$$

Resolving an Indeterminacy

In the following examples we make full use of all the techniques for evaluating limits that have been developed in this chapter. In each case we will be evaluating the limit at a of an expression which reduces to the indeterminate form $0/0$ if we set $x = a$. The process of evaluating such limits is often called "resolving an indeterminacy of the form $0/0$."

Example 5 Evaluate $L = \lim_{x \to 0} \dfrac{1 - \sqrt{x + 1}}{x}$.

Solution The function $(1 - \sqrt{x + 1})/x$ is not defined for $x = 0$; in fact, the substitution $x = 0$ reduces it to the indeterminate form $0/0$. We try to eliminate the troublesome x in the denominator by multiplying both numerator and denominator by the factor $1 + \sqrt{x + 1}$. This device works beautifully, because it leads to an x in the numerator that cancels the x already in the denominator. In detail,

$$L = \lim_{x \to 0} \frac{1 - \sqrt{x + 1}}{x} \frac{1 + \sqrt{x + 1}}{1 + \sqrt{x + 1}} = \lim_{x \to 0} \frac{1 - (x + 1)}{x(1 + \sqrt{x + 1})}$$

$$= \lim_{x \to 0} \frac{-x}{x(1 + \sqrt{x + 1})} = \lim_{x \to 0} \frac{-1}{1 + \sqrt{x + 1}}.$$

It follows that

$$L = \frac{-1}{\lim\limits_{x \to 0} (1 + \sqrt{x + 1})} = \frac{-1}{1 + \lim\limits_{x \to 0} \sqrt{x + 1}} = \frac{-1}{1 + 1} = -\frac{1}{2},$$

where we use the continuity of $\sqrt{x + 1}$ at $x = 0$. $\quad \square$

Example 6 Evaluate $L = \lim_{\theta \to 0} \dfrac{\sin c\theta}{\theta}$, where c is any constant.

Solution We first write
$$L = \lim_{\theta \to 0} c \, \frac{\sin c\theta}{c\theta} = c \lim_{\theta \to 0} \frac{\sin c\theta}{c\theta}.$$

Then, making the substitution $t = c\theta$ and noting that $t \to 0$ as $\theta \to 0$, we get
$$L = c \lim_{t \to 0} \frac{\sin t}{t} = c \cdot 1 = c. \quad \square$$

Example 7 Evaluate $L = \lim\limits_{x \to \pi} \dfrac{\sin^2 x}{1 + \cos^3 x}$.

Section 1.8 The Limit of a Composite Function

Solution Here we are dealing with the indeterminate form 0/0, since $\sin \pi = 0$, $\cos \pi = -1$. Replacing $\sin^2 x$ by $1 - \cos^2 x$, we factor the numerator and denominator. Then after canceling the common factor $1 + \cos x$, we get

$$L = \lim_{x \to \pi} \frac{1 - \cos^2 x}{1 + \cos^3 x} = \lim_{x \to \pi} \frac{(1 + \cos x)(1 - \cos x)}{(1 + \cos x)(1 - \cos x + \cos^2 x)}$$

$$= \lim_{x \to \pi} \frac{1 - \cos x}{1 - \cos x + \cos^2 x}.$$

It was the presence of the factor $1 + \cos x$, which equals zero at $x = \pi$, that caused the original indeterminacy, but now that it is gone, we can use the continuity of $\cos x$ to evaluate L. In other words, we are now entitled to make the substitution $x = \pi$, obtaining

$$L = \frac{1 - (-1)}{1 - (-1) + (-1)^2} = \frac{1 + 1}{1 + 1 + 1} = \frac{2}{3}. \quad \square$$

Example 8 Evaluate $L = \lim_{x \to 0} \dfrac{1 - \cos x}{x}$.

Solution Here we are again dealing with the indeterminate form 0/0. Multiplying both numerator and denominator by $1 + \cos x$, we get

$$L = \lim_{x \to 0} \frac{1 - \cos x}{x} \frac{1 + \cos x}{1 + \cos x} = \lim_{x \to 0} \frac{1 - \cos^2 x}{x(1 + \cos x)} = \lim_{x \to 0} \frac{\sin^2 x}{x(1 + \cos x)},$$

where, unlike the preceding example, there is no trouble with the factor $1 + \cos x$, since it is nonzero at $x = 0$. There is still an x in the denominator which gives us pause. However, the last limit is easily evaluated by writing it as a product of two limits, one of which is already known, while the other can be found by continuity. In detail,

$$L = \lim_{x \to 0} \frac{\sin x}{x} \frac{\sin x}{1 + \cos x} = \lim_{x \to 0} \frac{\sin x}{x} \lim_{x \to 0} \frac{\sin x}{1 + \cos x} = 1 \cdot \frac{0}{1 + 1} = 0. \quad \square$$

Problems

Evaluate

1. $\lim\limits_{x \to 0} \sin(\cos x)$

2. $\lim\limits_{x \to 0} \cos(\tan x)$

3. $\lim\limits_{x \to 0} \sin(\cos x) \sin x$

4. $\lim\limits_{x \to 0} \cos(\sin x) \cos x$

5. $\lim\limits_{x \to \pi/6} \tan(\sin x)$

6. $\lim\limits_{x \to 2\pi/3} \sin^2(\cos x)$

7. $\lim\limits_{x \to 0} \sin\left(x \sin \dfrac{1}{x}\right)$

8. $\lim\limits_{x \to -1} \sqrt{\dfrac{3 - x}{8 - x}}$

9. $\lim\limits_{x \to 8} \dfrac{1 + \sqrt{x + 1}}{x}$

10. $\lim\limits_{x \to 2} \sqrt{\dfrac{x^2 + 1}{x^2 - 3}}$

11. $\lim\limits_{x \to 4} \sqrt[3]{\dfrac{x^3 - 16}{x + 2}}$

12. $\lim\limits_{x \to 3} \sqrt[5]{2x^3 - 3x^2 + x + 2}$

13. $\lim\limits_{x \to -2} \sqrt{x^4 - 2x^3 + 4x^2 - 8x + 16}$

14. $\lim\limits_{x \to \pi} \sqrt{\cos^2 x - \cos x + 1}$

15. $\lim\limits_{x \to \sqrt{3}} \tan\sqrt{1 + x^2}$

16. $\lim\limits_{x \to -\pi} \sin\sqrt{10 + \cos x}$

17. $\lim\limits_{x \to 0} \dfrac{\sin \pi x}{x}$

18. $\lim\limits_{z \to 0} \dfrac{z}{\sin \frac{1}{2}z}$

19. $\lim\limits_{\theta \to 0} \dfrac{\sin 2\theta}{\sin 3\theta}$

20. $\lim\limits_{\alpha \to 0} \dfrac{\tan 2\alpha}{\alpha}$

21. $\lim_{\beta \to 0} \dfrac{\tan \beta}{\tan 3\beta}$

22. $\lim_{x \to -4} \sqrt[5]{4x^3 + x^2 - 3}$

31. $\lim_{x \to 0} \dfrac{\tan x}{\sqrt{1 + \tan x} - 1}$

32. $\lim_{x \to 0} \dfrac{\tan x - \sin x}{\sin^3 x}$

23. $\lim_{x \to 2} \sqrt{\dfrac{4 - x^2}{2 - x}}$

24. $\lim_{x \to 1} \dfrac{1 - \sqrt{x}}{1 - x^2}$

33. $\lim_{x \to 1} \dfrac{1}{1 - x} \cos \dfrac{\pi x}{2}$

34. $\lim_{x \to 0} \sin(\cot x) \sin x$

25. $\lim_{x \to -2} \sqrt[3]{\dfrac{x + 1}{x^2 + 4}}$

26. $\lim_{x \to 16} \dfrac{\sqrt[4]{x} - 2}{\sqrt{x} - 4}$

27. $\lim_{x \to 0} \dfrac{\sin(\sin x)}{x}$

28. $\lim_{x \to 0} \dfrac{1 - \cos(\cos x)}{\cos x}$

35. It is tempting to make the following generalization of Theorem 11: If $f(x) \to L$ as $x \to a$ and $g(x) \to M$ as $x \to L$, then $g(f(x)) \to M$ as $x \to a$. Give an example showing that this assertion is false as it stands. Give a supplementary condition on the behavior of the function f near a that makes the assertion true.

29. $\lim_{x \to 0} \dfrac{1 - \cos(\tan x)}{\tan x}$

30. $\lim_{x \to 0} \dfrac{\sin(\sin(\sin x))}{\sin(\sin x)}$

1.9 One-Sided Limits and Continuity

The graph of the function

$$f(x) = \dfrac{|x - a|}{x - a} = \begin{cases} 1 & \text{if } x > a, \\ -1 & \text{if } x < a, \end{cases} \qquad (1)$$

shown in Figure 43, jumps from -1 to 1 as its argument x goes from the left of a to the right of a, and therefore f has no limit at a. However, f would approach a limiting value at a if the approach of x to a were *from one side only*. For then we could ignore the behavior of f on the other side of a, regarding f as either the constant function $f(x) \equiv 1$ to the right of a, or the constant function $f(x) \equiv -1$ to the left of a. The limiting behavior of f as x approaches a from one side or the other is indicated by the two arrowheads in the figure. One arrowhead corresponds to the limit from the right and touches the line $x = a$ at the point with y-coordinate 1, and the other corresponds to the limit from the left and touches the line $x = a$ at the point with y-coordinate -1. (We represent these points by hollow dots, since neither of them belongs to the graph of f; in fact, f is not defined at a.) Thus the function f has *one-sided limits* at a, even though the ordinary limit of f at a does not exist.

If the variable point x approaches the fixed point a from the right, taking only values larger than a ($x > a$), we write $x \to a^+$, while if x approaches a from the left, taking only values smaller than a ($x < a$), we write

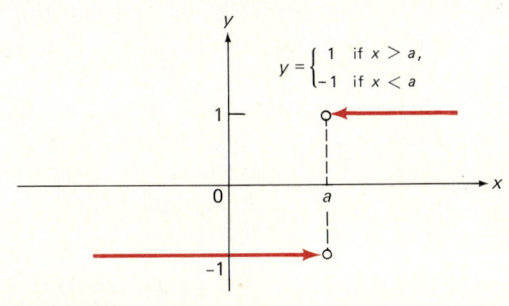

Figure 43

$x \to a^-$. Thus we have just observed that in the case of the function (1), $f(x) \to 1$ as $x \to a^+$ and $f(x) \to -1$ as $x \to a^-$, or equivalently

$$\lim_{x \to a^+} f(x) = 1, \qquad \lim_{x \to a^-} f(x) = -1.$$

The first of these one-sided limits is called the *right-hand limit* of f at a, and the second is called the *left-hand limit* of f at a.

It is easy to give an ε, δ definition of one-sided limits. Suppose f has an ordinary (two-sided) limit L at the point a. In ε, δ language this means that given any $\varepsilon > 0$, we can find a $\delta > 0$ such that $|f(x) - L| < \varepsilon$ whenever $0 < |x - a| < \delta$, that is, whenever either

$$a < x < a + \delta \qquad (2)$$

or

$$a - \delta < x < a. \qquad (2')$$

Of course, the points x satisfying (2) lie immediately to the right of a, while those satisfying (2') lie immediately to the left of a. Therefore, to go over to one-sided limits, we need only keep (2) and drop (2') or keep (2') and drop (2). More exactly, $f(x) \to L$ as $x \to a^+$ means that given any $\varepsilon > 0$, we can find a $\delta > 0$ such that $|f(x) - L| < \varepsilon$ whenever $a < x < a + \delta$, while $f(x) \to L$ as $x \to a^-$ means that $|f(x) - L| < \varepsilon$ whenever $a - \delta < x < a$.

One-Sided vs. Ordinary Limits

The following table spells out the similarities and differences between an ordinary limit and one-sided limits.

	Ordinary Limit	**One-Sided Limits**	
Let f be defined in	a deleted neighborhood of a	an open interval with left endpoint a	an open interval with right endpoint a
Then f is said to have	the limit L at a	the right-hand limit L at a	the left-hand limit L at a
and we write	$\lim_{x \to a} f(x) = L$	$\lim_{x \to a^+} f(x) = L$	$\lim_{x \to a^-} f(x) = L$
if for any $\varepsilon > 0$, we can find a $\delta > 0$ such that $|f(x) - L| < \varepsilon$ whenever	$0 < |x - a| < \delta$	$a < x < a + \delta$	$a - \delta < x < a$

Here we choose the first and second columns of the table to get the definition of an ordinary limit, the first and third columns to get the definition of a right-hand limit, and the first and last columns to get the definition of a left-hand limit. It is apparent from the parallelism of the three definitions that every proposition stated for an ordinary limit has an exact analogue for one-sided limits. In particular, this applies to Theorems 4–6 of Section 1.7, so that algebraic operations on one-sided limits obey the same rules as algebraic operations on ordinary limits. Thus, for example, if the functions f and g both have right-hand limits at a, then

$$\lim_{x \to a^+} [f(x) + g(x)] = \lim_{x \to a^+} f(x) + \lim_{x \to a^+} g(x),$$

if f and g both have left-hand limits at a, then

$$\lim_{x \to a^-} f(x)g(x) = \lim_{x \to a^-} f(x) \cdot \lim_{x \to a^-} g(x),$$

and so on.

The following theorem is almost self-evident, but it is important enough to merit a formal statement.

Theorem 12 *(Conditions for existence of a limit).* *The limit*

$$\lim_{x \to a} f(x)$$

exists if and only if the one-sided limits

$$\lim_{x \to a^+} f(x) \quad \text{and} \quad \lim_{x \to a^-} f(x)$$

both exist and are equal. If this is the case, then

$$\lim_{x \to a} f(x) = \lim_{x \to a^+} f(x) = \lim_{x \to a^-} f(x),$$

that is, the limit of f at a equals the common value of the right- and left-hand limits of f at a.

Proof Left as an exercise. Consult the table. ∎

Example 1 Show that

$$\lim_{x \to 0^+} \sqrt[n]{x} = 0.$$

Solution Given any $\varepsilon > 0$, we must find a $\delta > 0$ such that

$$|\sqrt[n]{x} - 0| = |\sqrt[n]{x}| = \sqrt[n]{x} < \varepsilon$$

whenever $0 < x < \delta$. An appropriate choice is $\delta = \varepsilon^n$, since

$$0 < x < \varepsilon^n$$

implies

$$0 < \sqrt[n]{x} < \sqrt[n]{\varepsilon^n} = \varepsilon$$

(see Example 6, page 12). If n is even, $\sqrt[n]{x}$ is not defined for negative x, and it makes no sense to talk about the limit of $\sqrt[n]{x}$ as $x \to 0^-$. However, $\sqrt[n]{x}$ is defined for negative x if n is odd, and then virtually the same argument shows that

$$\lim_{x \to 0^-} \sqrt[n]{x} = 0.$$

Thus, if n is odd, it follows from Theorem 12 that

$$\lim_{x \to 0} \sqrt[n]{x} = 0. \quad \square$$

Section 1.9 One-Sided Limits and Continuity

Example 2 Given any number x, the largest integer not exceeding x is denoted by $[\![x]\!]$. In other words, $[\![x]\!]$ is the unique integer n such that $n \leq x < n + 1$. Therefore

$$[\![\tfrac{1}{2}]\!] = 0, \qquad [\![0]\!] = 0, \qquad [\![\sqrt{2}]\!] = 1, \qquad [\![-\tfrac{3}{2}]\!] = -2,$$

and so on. In Figure 44 we graph the function $y = [\![x]\!]$, called the *greatest integer function* ($[\![x]\!]$ is also called the *integer part* of x). As usual, the points indicated by solid dots belong to the graph, but not those indicated by hollow dots. It is clear from the graph that $[\![x]\!]$ has an ordinary two-sided limit everywhere except at the integer points

$$x = n \qquad (n = 0, \pm 1, \pm 2, \ldots).$$

At each such point $[\![x]\!]$ has a right-hand limit

$$\lim_{x \to n^+} [\![x]\!] = n \tag{3}$$

and a left-hand limit

$$\lim_{x \to n^-} [\![x]\!] = n - 1, \tag{3'}$$

but not an ordinary limit, since these one-sided limits are unequal.

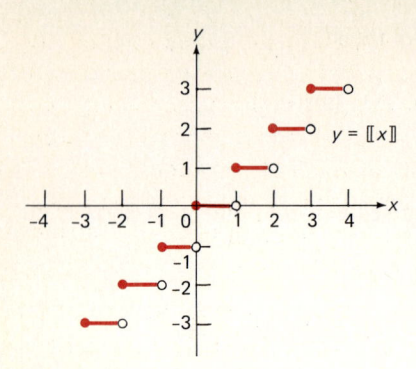

Figure 44

One-Sided Continuity

There is also a kind of continuity involving one-sided limits. If

$$\lim_{x \to a^+} f(x) = f(a),$$

we say that the function f is *continuous from the right at a*, while if

$$\lim_{x \to a^-} f(x) = f(a),$$

we say that f is *continuous from the left at a* (here it must be assumed that f is defined at a). It is an immediate consequence of Theorem 12 that a function f is continuous at the point a if and only if it is continuous both from the right and from the left at a. In fact, according to the theorem,

$$\lim_{x \to a} f(x) = f(a)$$

if and only if

$$\lim_{x \to a^+} f(x) = \lim_{x \to a^-} f(x) = f(a).$$

Example 3 If n is an integer, then $[\![n]\!] = n$. It follows from formulas (3) and (3') that the greatest integer function $[\![x]\!]$ is continuous from the right, but not from the left, at each integer point $x = n$. This is also apparent from inspection of Figure 44. The function $[\![x]\!]$ is continuous in the ordinary sense at every other point.

Continuity on an Interval

A function f is said to be *continuous on an interval I* if it is continuous at every point of I, with the understanding that one-sided continuity is enough at an endpoint of I (that belongs to I). Thus continuity of f on the *open* interval (a, b) means that f is continuous at every point of (a, b), while continuity of f on the *closed* interval $[a, b]$ means that f is continuous at every point of (a, b), continuous from the right at the left endpoint a, and continuous from the left at the right endpoint b. This makes sense, since x cannot approach a from the left or b from the right without going outside the interval $[a, b]$, and we want to be able to define continuity for a function whose entire domain of definition is a given closed interval. Similarly, f is continuous on the half-open interval $[a, b)$ if f is continuous on (a, b) and continuous from the right at a, f is continuous on $(-\infty, b]$ if f is continuous on $(-\infty, b)$ and continuous from the left at b, and so on.

Example 4 Let n be any integer. Then the greatest integer function $[\![x]\!]$ is continuous on the interval $[n, n + 1)$, but not on the interval $(n, n + 1]$. Reflecting the graph of $[\![x]\!]$ in the y-axis (see Figure 45), we find that the function $[\![-x]\!]$ is continuous on $(n, n + 1]$, but not on $[n, n + 1)$.

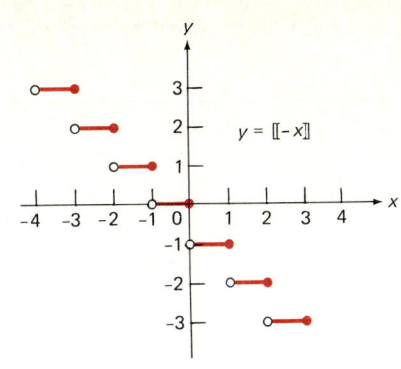

Figure 45

Problems

Find
1. $[\![-\pi]\!]$
2. $[\![\pi^2]\!]$
3. $[\![\sqrt{2} - \sqrt{3}]\!]$
4. $[\![(\frac{2}{3})^4]\!]$
5. $[\![(1.1)^{10}]\!]$
6. $[\![(-\frac{7}{11})^{13}]\!]$

Evaluate the given limit (if it exists).

7. $\lim_{x \to 1^-} \sqrt{x - 1}$
8. $\lim_{x \to 1^+} \sqrt{x - 1}$
9. $\lim_{x \to 1^-} \sqrt[3]{x - 1}$
10. $\lim_{x \to 0^+} x\sqrt{1 + \frac{4}{x^2}}$
11. $\lim_{x \to 0^-} x\sqrt{\frac{1}{x^2} - 1}$
12. $\lim_{x \to 0^+} \frac{|x|}{\sqrt{x}}$
13. $\lim_{x \to 1^+} \frac{[\![x]\!]}{x}$
14. $\lim_{x \to 1^-} \frac{[\![x]\!]}{x}$
15. $\lim_{x \to 0^+} \frac{[\![x]\!]}{x}$
16. $\lim_{x \to 0^-} \frac{[\![x]\!]}{x}$
17. $\lim_{x \to -1^+} \frac{[\![x]\!]}{x}$
18. $\lim_{x \to -1^-} \frac{[\![x]\!]}{x}$

19. Find the one-sided limits at $x = 0$ of the function
$$f(x) = \frac{x + x^2}{|x|} \quad (x \neq 0).$$

20. Find the one-sided limits at $x = 2$ of the function
$$f(x) = \begin{cases} x & \text{if } x < 2, \\ 1.1x & \text{if } x \geq 2. \end{cases}$$

21. Given that
$$f(x) = \begin{cases} 3x & \text{if } 0 \leq x < 1, \\ 2 - x & \text{if } 1 \leq x \leq 2, \end{cases}$$

find the two intervals with endpoints 0 and 1 on which the function f is continuous. Show that f is continuous on every interval with endpoints 1 and 2.

22. Let
$$f(x) = \begin{cases} -|x + 1| & \text{if } x < 0, \\ |x - 1| & \text{if } x > 0, \end{cases}$$

where $f(0)$ is undefined. How should $f(0)$ be defined to make f continuous from the right at $x = 0$? Continuous from the left at $x = 0$?

Given a function with both right- and left-hand limits at the point a, suppose these limits are unequal. Then the quantity
$$J = \lim_{x \to a^+} f(x) - \lim_{x \to a^-} f(x)$$
is called the *jump* of f at a, and f is said to have a *jump discontinuity* at a. Thus, for example, every discontinuity of the greatest integer function $[\![x]\!]$, graphed in Figure 44, is a jump discontinuity, and the jump at every point of discontinuity $x = 0, \pm 1, \pm 2, \ldots$ has the same value 1.

23. What is the jump of the function
$$f(x) = \frac{J}{2} \frac{|x - a|}{x - a},$$
and at what point does it occur?

24. Show that a jump discontinuity cannot be removable (recall Problem 10, page 79).

25. Find the discontinuities of the function $f(x) = [\![x]\!] + [\![-x]\!]$. Are they jump discontinuities? Are they removable?

26. Find the jumps of the function
$$f(x) = \begin{cases} 3x & \text{if } x < 1, \\ 0 & \text{if } x = 1, \\ x & \text{if } 1 < x < 3, \\ 4 & \text{if } x \geq 3. \end{cases}$$

Section 1.9 One-Sided Limits and Continuity

27. Given that

$$f(x) = \begin{cases} -2\sin x & \text{if } x < -\pi/2, \\ a\sin x + b & \text{if } -\pi/2 \leq x \leq \pi/2, \\ \cos x & \text{if } x > \pi/2, \end{cases}$$

what choice of the numbers a and b makes the function f continuous everywhere?

28. Show that the smallest integer not less than x is equal to $-[\![-x]\!]$.

1.10 Properties of Continuous Functions

We now study the properties that a function f has simply as a consequence of being continuous on some interval.

The Intermediate Value Theorem

Theorem 13 *(Intermediate value theorem).* *If f is continuous on an interval I and if f takes different values $f(s)$ and $f(t)$ at two points s and t of I, then f takes every intermediate value k, that is, every value k between $f(s)$ and $f(t)$, at some point r between s and t.*

The proof is omitted, since it involves a concept (the *completeness* of the real number system), which is beyond the scope of a first course in calculus. However, the geometric meaning of the theorem is quite clear: The graph of a continuous function f cannot cross from one side of the horizontal line $y = k$ to the other without actually intersecting the line, as illustrated in Figure 46(a). The continuity of f is essential for the truth of the theorem, since the graph of a discontinuous function can simply jump over the line $y = k$ without intersecting it, as illustrated in Figure 46(b).

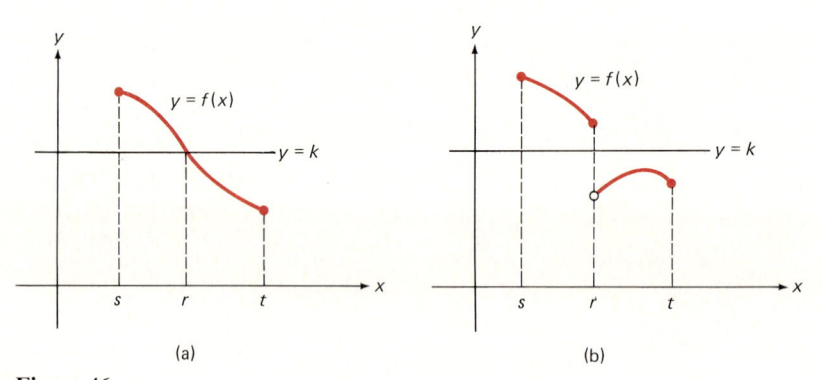

Figure 46

The Bisection Method

Our first example shows how the intermediate value theorem can be used to solve a difficult equation to any desired accuracy.

Example 1 Show that the equation

$$2x^5 + 2x^2 + x - 3 = 0 \tag{1}$$

has a root between 0 and 1. Find this root to within $\frac{1}{16}$.

Solution By a *root* (or *solution*) of equation (1) we mean any value of x satisfying the equation. Introducing the function

$$f(x) = 2x^5 + 2x^2 + x - 3,$$

96   Chapter 1   Functions and Limits

which is continuous (since it is a polynomial), we observe that $f(0) = -3$ and $f(1) = 2$. Since $-3 < 0 < 2$, it follows from the intermediate value theorem that $f(x) = 0$ for some x between 0 and 1. Therefore equation (1) has a root r between 0 and 1, that is, in the interval (0, 1).† To locate r more accurately, we apply the following procedure, known as the *bisection method*. First we calculate the value of f at the midpoint $\frac{1}{2}$ of the interval (0, 1):

$$f\left(\frac{1}{2}\right) = 2\left(\frac{1}{2}\right)^5 + 2\left(\frac{1}{2}\right)^2 + \frac{1}{2} - 3 = -\frac{31}{16} < 0.$$

But $f(1) > 0$, and hence by the intermediate value theorem again, $f(x)$ must equal 0 for some x between $\frac{1}{2}$ and 1. Therefore $\frac{1}{2} < r < 1$, and the search for r has been narrowed down to the interval $(\frac{1}{2}, 1)$ of length $\frac{1}{2}$. Making another bisection, we calculate the value of f at the midpoint $\frac{3}{4}$ of the interval $(\frac{1}{2}, 1)$:

$$f\left(\frac{3}{4}\right) = 2\left(\frac{3}{4}\right)^5 + 2\left(\frac{3}{4}\right)^2 + \frac{3}{4} - 3 = -\frac{333}{512} < 0.$$

Since $f(\frac{3}{4}) < 0$ and $f(1) > 0$, the function f takes values with opposite signs at the points $\frac{3}{4}$ and 1, and therefore it is now known that $f(x) = 0$ for some x between $\frac{3}{4}$ and 1, so that r lies in the interval $(\frac{3}{4}, 1)$ of length $\frac{1}{4}$. This is still not enough to determine r to the desired accuracy, so we make a third bisection and find that $f(\frac{7}{8}) > 0$ (the calculation is left as an exercise). Since $f(\frac{3}{4}) < 0$ and $f(\frac{7}{8}) > 0$, we can now be sure that r lies in the interval $(\frac{3}{4}, \frac{7}{8})$ of length $\frac{1}{8}$. But the midpoint $\frac{1}{2}(\frac{3}{4} + \frac{7}{8}) = \frac{13}{16}$ of this interval is within distance $\frac{1}{16}$ of any of its interior points, and hence is within distance $\frac{1}{16}$ of the root r, as required.

Thus the approximation $r \approx \frac{13}{16} = 0.8125$ is accurate to within $\frac{1}{16}$. A more exact calculation shows that $r = 0.8305$ to four decimal places, so that the error of our approximation to r, based on three bisections, is about $0.8305 - 0.8125 = 0.0180$. Since $\frac{1}{16} = 0.0625$, this error is actually smaller than we have a right to expect (see Problem 3). The graphical meaning of the three consecutive bisections is indicated schematically by the three horizontal line segments in Figure 47, where the shortest segment connects the points $\frac{3}{4}$ and $\frac{7}{8}$ of the x-axis. □

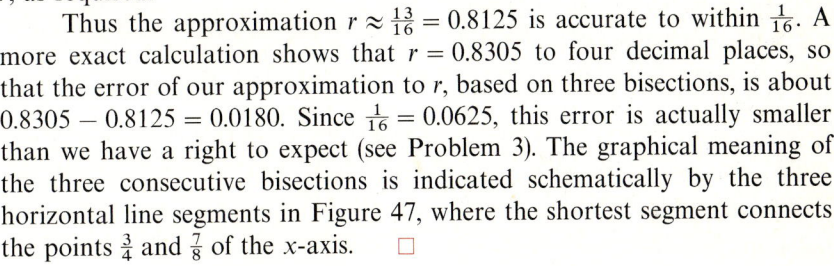

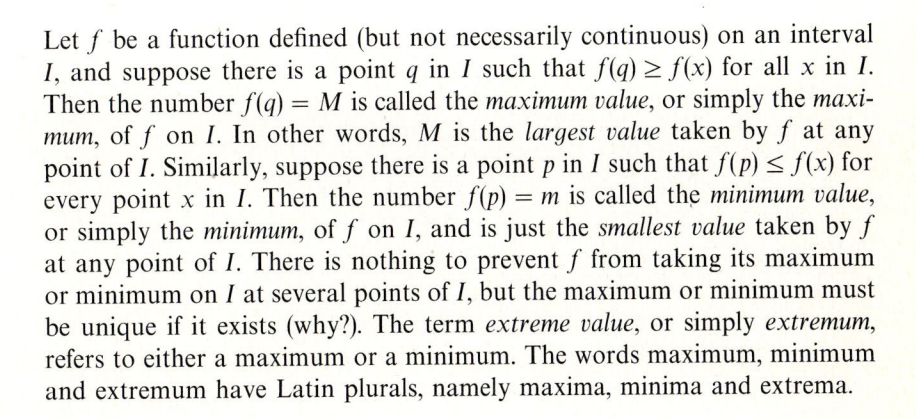

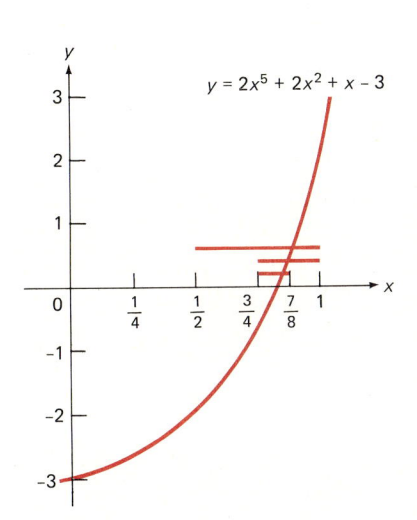

Figure 47

Extreme Values

Let f be a function defined (but not necessarily continuous) on an interval I, and suppose there is a point q in I such that $f(q) \geq f(x)$ for all x in I. Then the number $f(q) = M$ is called the *maximum value*, or simply the *maximum*, of f on I. In other words, M is the *largest value* taken by f at any point of I. Similarly, suppose there is a point p in I such that $f(p) \leq f(x)$ for every point x in I. Then the number $f(p) = m$ is called the *minimum value*, or simply the *minimum*, of f on I, and is just the *smallest value* taken by f at any point of I. There is nothing to prevent f from taking its maximum or minimum on I at several points of I, but the maximum or minimum must be unique if it exists (why?). The term *extreme value*, or simply *extremum*, refers to either a maximum or a minimum. The words maximum, minimum and extremum have Latin plurals, namely maxima, minima and extrema.

---

† Actually, it can be shown with the help of a later test that f is increasing on (0, 1), so that r is the *only* root of equation (1) between 0 and 1 (see Example 4, page 170).

Section 1.10 Properties of Continuous Functions

Example 2 Let $f(x) \equiv k$ be a constant function defined on any interval I. Then f has a maximum and a minimum, both equal to k, at every point of I. Conversely, if f has a maximum M and a minimum m on some interval I, and if M and m are equal, with common value k, then $f(x) \equiv k$ on I.

Example 3 Let $f(x) = x^2$ and let I be the half-open interval $[-1, 1)$, as in Figure 48(a). Then it is apparent that f has both a maximum $M = 1$ and a minimum $m = 0$ on I. In fact, f takes its maximum at $x = -1$, the abscissa of the highest point of the graph of f, and its minimum at $x = 0$, the abscissa of the lowest point of the graph. If I is the *closed* interval $[-1, 1]$, then f has the same extrema, but the maximum M is now taken at $x = 1$ as well as at $x = -1$ [see Figure 48(b)].

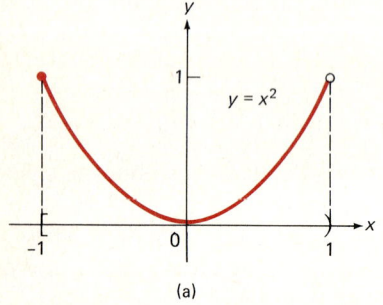

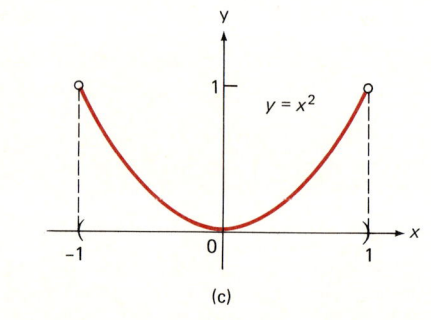

Figure 48

(a) (b) (c)

Example 4 Again let $f(x) = x^2$, but this time let I be the *open* interval $(-1, 1)$. Then it is clear from Figure 48(c) that f still has a minimum $m = 0$ at $x = 0$, as in the preceding example. However, f no longer has a maximum on I, since f takes values arbitrarily close to 1 without ever taking the value 1 itself (there is no largest number less than 1).

Example 5 Let I be the closed interval $[0, 2]$, and let f be the *discontinuous* function

$$f(x) = \begin{cases} -x & \text{if } 0 \leq x < 1, \\ 2 - x & \text{if } 1 \leq x \leq 2, \end{cases}$$

graphed in Figure 49. Then it is apparent that f has a maximum $M = 1$ on I, taken at $x = 1$ (the abscissa of the highest point of the graph), but no

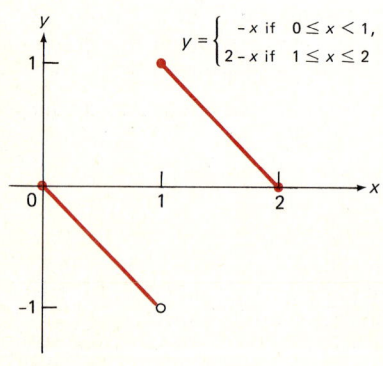

Figure 49

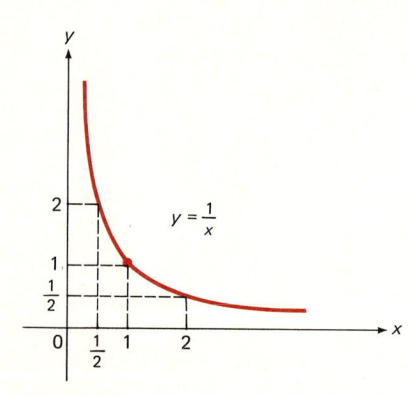

Figure 50

98 Chapter 1 Functions and Limits

minimum on I, since f takes values arbitrarily close to -1 without ever taking the value -1 (there is no smallest number greater than -1).

Example 6 Let I be the unbounded closed interval $[1, \infty)$, and let $f(x) = 1/x$. Then it is apparent from Figure 50 that f has a maximum $M = 1$ on I, taken at $x = 1$, but no minimum, since f takes values arbitrarily close to 0 without ever taking the value 0 (there is no smallest positive number).

The Extreme Value Theorem

As the preceding examples make clear, there are no general conclusions that can be drawn about the existence of extreme values of an arbitrary function f on an arbitrary interval I. However, in the case where f is *continuous* on I, and I is both *bounded* and *closed*, the existence of extreme values can be guaranteed:

> **Theorem 14** *(Extreme value theorem)*. *If f is continuous on a bounded closed interval $I = [a, b]$, then f is bounded on I, and f has both a maximum M and a minimum m on I, that is, there are points p and q in I such that*
> $$m = f(p) \leq f(x) \leq f(q) = M$$
> *for all x in I.*

The proof is omitted, since it involves the same kind of technicalities as the proof of the intermediate value theorem. However, the fact that the truth of the extreme value theorem depends on f being continuous is shown by Example 5, and the fact that it also depends on I being both closed and bounded is shown by Examples 4 and 6. The geometric meaning of this theorem is that the graph of a function continuous on a closed bounded interval I must have both a highest point $Q = (q, M)$ and a lowest point $P = (p, m)$. The point q can be either an interior point or an endpoint of I, and the same is true for p. Some of the possibilities are illustrated in Figure 51.

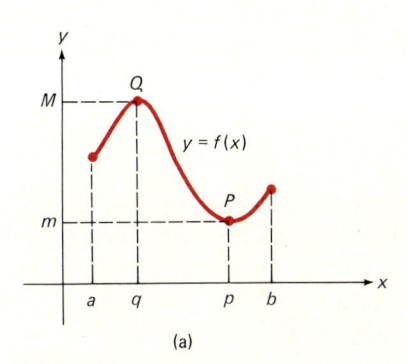

Figure 51

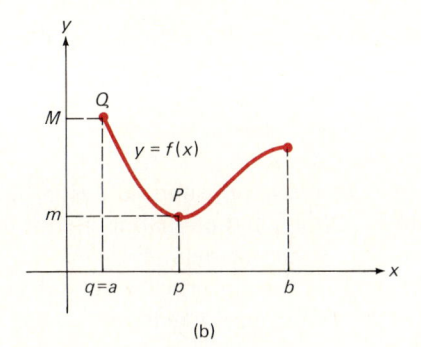

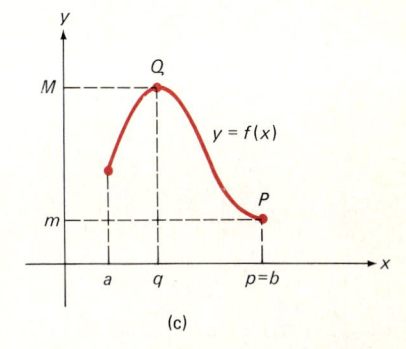

Example 7 The function $f(x) = 1/\sqrt{x}$ is continuous on the closed bounded interval $I = [1, 4]$, and in keeping with Theorem 14, f is bounded on $[1, 4]$, with maximum $M = f(1) = 1$ and minimum $m = f(4) = \frac{1}{2}$ (notice that f is decreasing). The function f is also continuous on the interval $(0, 1)$, but as

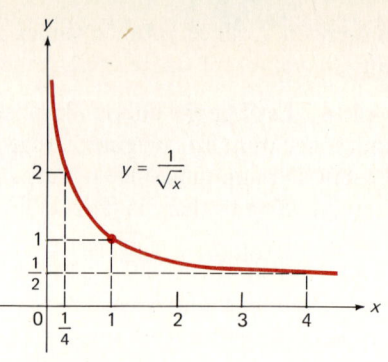

Figure 52

is apparent from Figure 52, f has no maximum on $(0, 1)$, and is in fact unbounded on $(0, 1)$. This is compatible with Theorem 14, since the interval $(0, 1)$ is *open*.

Behavior of Intervals Under Continuous Mappings

We conclude our discussion of the properties of continuous functions with a useful theorem on the behavior of intervals under continuous mappings. Given a function f defined on a set X, let Y be the set $\{y : y = f(x), x \in X\}$, that is, the set of all values taken by f as the independent variable x varies over the set X. Then f is said to *map X onto Y*, and the set Y is called the *image of X under f*, just as the number y is called the image of x under f.

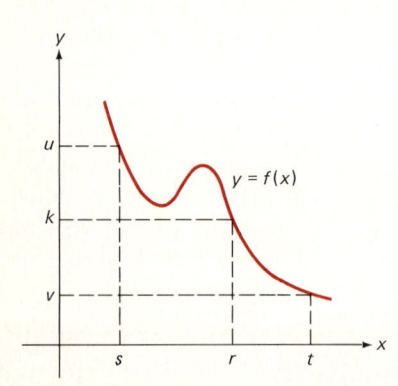

> **Theorem 15** *(Interval mapping theorem).* Let f be continuous on an interval I of any kind, and let J be the image of I under f. Then
>
> (i) J is also an interval;
> (ii) If I is a bounded closed interval, so is J.

Proof (Optional) To prove (i), let u and v be distinct points in J. Then there are distinct points s and t in I such that $f(s) = u$ and $f(t) = v$ (see Figure 53). Let k be any point between u and v. Then, by the intermediate value theorem, there is a point r between s and t such that $f(r) = k$. Therefore k belongs to J. Thus whenever the set J contains two distinct points u and v, it also contains every point k between u and v. Therefore J is an interval, by the remark on page 20.

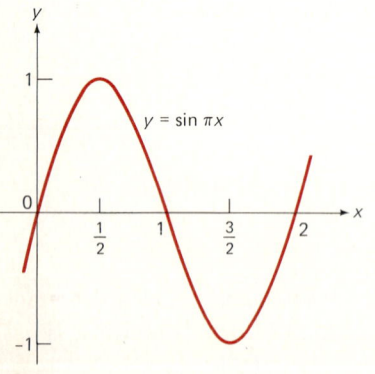

To prove (ii), suppose I is bounded and closed, and let M and m be the maximum and minimum of f on I, whose existence is guaranteed by the extreme value theorem. Then the set J, which is known to be an interval because of (i), can only be the closed bounded interval $[m, M]$. If f is a constant function, J reduces to a set containing only one point. To cover this case, we will henceforth consider a set containing only one point, say k, to be a closed interval of the form $[k, k]$, with coincident endpoints. ∎

In concise language Theorem 15 says that a continuous function maps intervals onto intervals and bounded closed intervals onto bounded closed intervals. The continuous image of an interval that is not bounded and closed can be any kind of interval at all. For instance, for the function considered

Figure 53

Figure 54

100   Chapter 1   Functions and Limits

in Example 7, the image of the bounded interval (0, 1) is the unbounded interval (1, ∞).

Example 8 Let $f(x) = \sin \pi x$. Then inspection of Figure 54 shows that the continuous function f maps the open interval $(\frac{1}{2} - \delta, \frac{3}{2} + \delta)$ onto the closed interval $[-1, 1]$ if $\delta > 0$.

Problems

1. If $f(x) = 1/x$, then $f(-1) = -1$ and $f(1) = 1$, but f does not take the value zero between $x = -1$ and $x = 1$. Why doesn't this contradict the intermediate value theorem?

2. Show that the equation $8x^3 - 12x^2 - 2x + 3 = 0$ has a root r_1 between -1 and 0, a root r_2 between 0 and 1, and a root r_3 between 1 and 2. By making the substitution $x = t/2$, find the exact values of these roots. Are there any other roots?

3. In Example 1 three bisections give the approximation $r \approx \frac{13}{16}$, which happens to have an error of about 0.0180. How many more bisections are necessary to *guarantee* an error this small?

4. Show that the equation $x^6 + 2x - 2 = 0$ has a root r_1 between -2 and -1, and another root r_2 between 0 and 1. Show that there are no other roots. Use the bisection method to approximate r_1 and r_2 to within $\frac{1}{16}$.
Hint. The line $y = 2 - 2x$ intersects the curve $y = x^6$ in exactly two points.

5. Show that the equation $\cos \pi x = x$ has one and only one root r between 0 and $\frac{1}{2}$. Use the bisection method to approximate r to within $\frac{1}{32}$.

6. Show that the equation $\sin \pi x = x$ has a root r_1 between $\frac{1}{2}$ and $\frac{3}{4}$, and another root r_2 between $-\frac{3}{4}$ and $-\frac{1}{2}$. What is the relation between r_1 and r_2? Does the equation have any additional roots other than $x = 0$? Use the bisection method to approximate r_1 and r_2 to within $\frac{1}{64}$.

7. Let f be continuous on the interval $[a, b]$, and suppose $a \leq f(x) \leq b$ for all x in $[a, b]$. Show that there is a point c in $[a, b]$, called a *fixed point* of f, such that $f(c) = c$.

8. An elevator goes up a tall building in 5 minutes, stopping at various floors to take on and discharge passengers. It then descends in 3 minutes. Show that regardless of the details of the trip, there is some position (in general between floors) which is visited by the elevator at two instants of time exactly 4 minutes apart.

For the given function f and interval I, find the maximum M and minimum m of f on I, and the points of I at which M and m are taken (specify any extrema that fail to exist). Also find the interval J which is the image of I under f.

9. $f(x) = x^2 - 2x - 2$, $I = (0, 3]$
10. $f(x) = 1 + x - x^2$, $I = [-1, 2]$
11. $f(x) = x^3 + 1$, $I = [-1, 1]$
12. $f(x) = x^2 + 2x + 3$, $I = [-2, 1)$
13. $f(x) = 1/(x - 1)$, $I = (1, \infty)$
14. $f(x) = |x| + |x + 1|$, $I = [-3, 1]$
15. $f(x) = 1/(x^2 + 1)$, $I = (-\infty, 0]$
16. $f(x) = \sqrt{1 - x^2}$, $I = (-1, 1)$
17. $f(x) = \sqrt[3]{1 - x}$, $I = [1, 9]$
18. $f(x) = \sqrt[4]{1 + x}$, $I = [0, 15]$
19. $f(x) = \cos^2 x - \sin^2 x$, $I = [0, \pi]$
20. $f(x) = |\sin x|$, $I = [0, 2\pi)$

21. Suppose the function f has a maximum M and a minimum m on an interval I. Show that $-f$ also has extreme values on I. What are they, and where are they taken?

22. Show that an increasing function f always has both a maximum M and a minimum m on a bounded closed interval $I = [a, b]$, even if f is discontinuous at points of I. Where does f take its extreme values? Discuss the case of decreasing f.

Key Terms and Topics
Definition of a function, independent and dependent variables
Domain and range of a function, the natural domain
Argument and values of a function
Algebraic operations on functions, equality of functions
Identical equality and identities
Composite functions and the operation of composition
The graph of a function and the vertical line property

Even and odd functions, increasing and decreasing functions
Shifting the graph of a function
Trigonometric functions and their graphs
Radian vs. degree measure
Length of a circular arc, area of a circular sector
Law of cosines, addition laws for the sine and cosine
The angle between two lines
Periodic functions, the fundamental period
Bounded and unbounded functions
The limit of a function $f(x)$ as $x \to a$
Continuity and reasons for discontinuity
The ε, δ definition of a limit and its geometric meaning
Algebraic operations on limits and continuous functions
The sandwich theorem
Right- and left-hand limits
Continuity from the right and from the left, continuity on an interval
The intermediate value theorem, the bisection method
Extreme values of a function, the extreme value theorem
The interval mapping theorem

Frequently Used Values of the Trigonometric Functions

Function	θ — 0° or 0 radians	30° or $\frac{\pi}{6}$ radians	45° or $\frac{\pi}{4}$ radians	60° or $\frac{\pi}{3}$ radians	90° or $\frac{\pi}{2}$ radians
$\sin \theta$	0	$\frac{1}{2}$	$\frac{1}{\sqrt{2}}$	$\frac{\sqrt{3}}{2}$	1
$\cos \theta$	1	$\frac{\sqrt{3}}{2}$	$\frac{1}{\sqrt{2}}$	$\frac{1}{2}$	0
$\tan \theta$	0	$\frac{1}{\sqrt{3}}$	1	$\sqrt{3}$	—

Supplementary Problems

Let $f(x) = (2x + 1)/(3x^2 - 1)$. Find
1. $f(-1)$
2. $f(0)$
3. $f(\frac{1}{2})$
4. $f(1/\sqrt{2})$
5. $f(1/\sqrt{3})$
6. $f(-3)$

Let $f(x) = x^2 - 2x + 3$. Find all solutions of the given equation.
7. $f(x) = 6$
8. $f(x) = 0$
9. $f(x) = 2$
10. $f(x) = 11$

11. Is the perimeter of an equilateral triangle a function of its area?

12. Let d be the number of positive integers which are (exact) divisors of a positive integer n; for example, $d = 6$ if $n = 12$, since 12 has the divisors 1, 2, 3, 4, 6 and 12. Is n a function of d?

Let $f(t) = 2^t$ and $g(t) = t^2$. Evaluate
13. $(2f - g)(0)$
14. $(f + \frac{1}{2}g)(1)$
15. $(fg)(-1)$
16. $(1 + g^2)(2)$
17. $(f/g)(-2)$
18. $(f^2 - g^2)(4)$

If
$$f(x) = \begin{cases} 1 & \text{if } x > 0, \\ 0 & \text{if } x \leq 0, \end{cases} \quad g(x) = \begin{cases} 0 & \text{if } x > 0, \\ 1 & \text{if } x \leq 0, \end{cases}$$
find the specified composite function.
19. $f(f(x))$
20. $f(g(x))$
21. $g(f(x))$
22. $g(g(x))$

23. Find a nonconstant function f such that $f \circ f = f$.

24. Show that $fg \equiv 0$ is compatible with $f \neq 0$, $g \neq 0$. In other words, show that a product of two nonzero functions can be zero. Contrast this with the situation for numbers, as opposed to functions.

Find a simple formula involving absolute values for the function f with the graph shown in

25. Figure 55   **26.** Figure 56

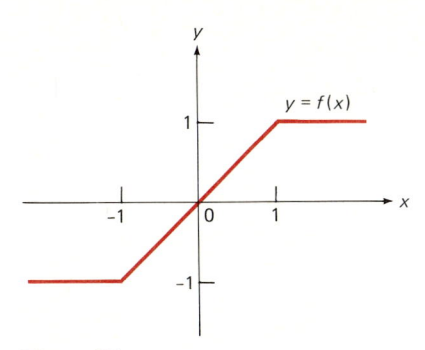

Figure 55

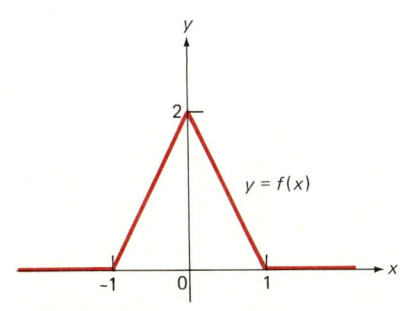

Figure 56

Identify the given function as even or odd (or neither).
27. $f(x) = \sin(\sin x)$ **28.** $f(x) = \cos(\sin x)$
29. $f(x) = \tan(\sec x)$ **30.** $f(x) = [\![x]\!]$

31. What function f is both even and odd?
32. Show that the graph of a nonzero function f cannot be symmetric about the x-axis.
33. Find the two lines through the origin making an angle of $45°$ with the line $3x - 2y + 6 = 0$.
34. Show that each of the equations $\sin x = \cos x$, $\tan x = \cot x$ and $\sec x = \csc x$ has infinitely many solutions.
35. Determine the fundamental period of the function $|\sin x| + |\cos x|$.
36. Show that the function $f(x) = x - [\![x]\!]$ is periodic, with fundamental period 1, and draw its graph. Find the discontinuities of f.
37. If $f + g$ and f have limits at a, must the same be true of g? If $f + g$ has a limit at a, must the same be true of f and g?
38. If fg and f have limits at a, must the same be true of g? If fg has a limit at a, must the same be true of f and g?

What choice or alteration of $f(0)$ makes the given function continuous at $x = 0$?

39. $f(x) = \dfrac{x^2}{x}$ **40.** $f(x) = \dfrac{5x^2 - 3x}{2x}$

41. $f(x) = \begin{cases} x^2 + 1 & \text{if } x \neq 0, \\ 2 & \text{if } x = 0 \end{cases}$

42. $f(x) = \begin{cases} \sqrt{x + 2} & \text{if } x \neq 0, \\ -1 & \text{if } x = 0 \end{cases}$

Find
43. $[\![n - \tfrac{1}{2}]\!]$ (n an integer)   **44.** $[\![(2.05)^4]\!]$
45. $[\![\pi - \sqrt{10}]\!]$ **46.** $[\![(-0.9)^{90}]\!]$

47. Is it true that $[\![|x|]\!] \equiv |[\![x]\!]|$?

Given that
$$f(x) = \begin{cases} -1 & \text{if } x < 0, \\ 0 & \text{if } x = 0, \\ 1 & \text{if } x > 0, \end{cases}$$
find all the discontinuities of the composite functions $f \circ g$ and $g \circ f$ if
48. $g(x) = 1 + x^2$
49. $g(x) = x(1 - x^2)$
50. $g(x) = x - [\![x]\!]$

Evaluate
51. $\lim\limits_{x \to 1} (x^{100} + x^{50} + 1)$ **52.** $\lim\limits_{x \to -1} (x^{99} + x^{49} + 1)$

53. $\lim\limits_{x \to -4} \dfrac{x^3 + 4x + 12}{x^4 - 3x + 4}$ **54.** $\lim\limits_{x \to 3} \sqrt{\dfrac{x^2 + 10x - 39}{x - 3}}$

55. $\lim\limits_{x \to 1} \dfrac{(x^2 - 1)^{10}}{(x^2 - 2x + 1)^5}$ **56.** $\lim\limits_{x \to 2} \dfrac{(x^2 - x - 2)^{20}}{(x^3 - 12x + 16)^{10}}$

57. $\lim\limits_{x \to 1} \dfrac{x^3 - 3x + 2}{x^4 - 4x + 3}$ **58.** $\lim\limits_{x \to -1} \dfrac{x^3 - 2x - 1}{x^5 - 2x - 1}$

59. $\lim\limits_{x \to 1} \dfrac{1 - \sqrt{x}}{1 - \sqrt[3]{x}}$

60. $\lim\limits_{x \to 1} \dfrac{x^m - 1}{x^n - 1}$ (m, n positive integers)

Hint. According to the *factor theorem* (proved on page 399), a polynomial $Q(x)$ is divisible by the linear factor $x - c$ if and only if $Q(c) = 0$. This observation is helpful in solving Problems 56–58.

61. Show that if the functions f and g are continuous everywhere, then so are the functions $M(x) = \max\{f(x), g(x)\}$ and $m(x) = \min\{f(x), g(x)\}$. [Figures 57(b) and 57(c) show $M(x)$ and $m(x)$ for the functions f and g graphed in Figure 57(a).]

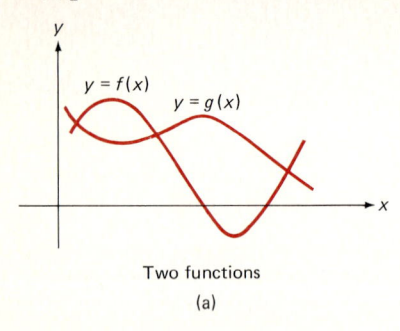

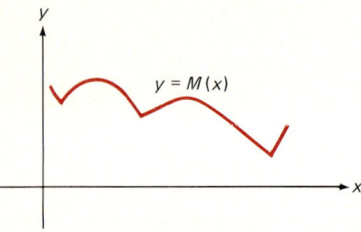

Two functions
(a)

Their maximum
(b)

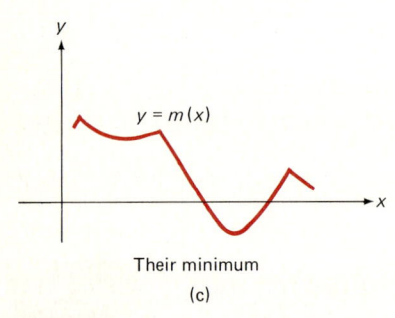

Their minimum
(c)

Figure 57

62. Use the preceding problem to show that if the function f is continuous everywhere, then so is the "clipped" function

$$f_c(x) = \begin{cases} c & \text{if } f(x) > c, \\ f(x) & \text{if } |f(x)| \leq c, \\ -c & \text{if } f(x) < -c \end{cases}$$

for every $c > 0$. [Figure 58(b) shows a clipped version of the function f graphed in Figure 58(a).]

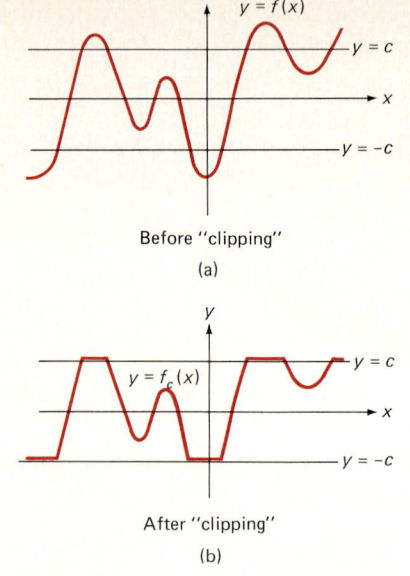

Before "clipping"
(a)

After "clipping"
(b)

Figure 58

63. Show that between any two distinct real numbers there can always be found both a rational number and an irrational number, and in fact infinitely many of each.

64. Use the preceding problem to show that the function

$$f(x) = \begin{cases} 1 & \text{if } x \text{ is rational,} \\ 0 & \text{if } x \text{ is irrational} \end{cases}$$

fails to have a limit at every point a, and hence is discontinuous everywhere.

Let $f(x)$ be the same as in Problem 64. Discuss the limiting behavior (at every point) of the function

65. $xf(x)$ **66.** $|x|f(x)$

67. $\dfrac{|x|}{x} f(x)$ **68.** $f(x) \sin x$

69. Given arbitrary numbers a, b and c, where $a \neq 0$, show that the equation $ax + b \sin x = c$ always has a solution.

70. Show that the equation $\cot \pi x = x$ has exactly one solution r_n in every interval $I_n = (n, n+1)$, where n is any integer. Use the bisection method to estimate r_0, r_1 and r_2 to within $\frac{1}{32}$.

2 Derivatives

The notion of the *rate of change* of one variable with respect to another permeates both the natural sciences and the quantitative social sciences. The mathematical concept corresponding to a rate of change is a special kind of limit known as the *derivative*. We now undertake a detailed study of derivatives, interweaving the basic theory with a variety of applications. Of particular importance are the use of the derivative to define the instantaneous velocity of a particle undergoing rectilinear motion (see Section 2.1), and the use of the derivative to find the tangent line to a curve of general shape (see Section 2.2).

2.1 Velocity and Rates of Change; The Derivative Concept

By a *particle* we mean an object whose actual size can be neglected in a given problem, and which can therefore be idealized as a point. Thus, depending on circumstances, an electron, an automobile or a planet might be regarded as a particle. Consider the motion of a particle P along a straight line L, and let s be the particle's position, that is, its coordinate on L. (It is assumed that the line L, which we can regard as the s-axis, has been assigned an origin O, a positive direction and a unit of length.) Let the particle's position at time t be given by the *position function*

$$s = f(t)$$

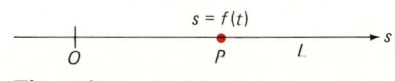

Figure 1

(see Figure 1). It seems intuitively clear that at each instant of time t the particle has a certain velocity $v = v(t)$, which is itself a function of t. But how should this velocity be defined?

Average Velocity

In an attempt to answer this question, we first define the *average velocity* v_{av} of the particle between two times t and u as the quotient

$$v_{av} = \frac{f(u) - f(t)}{u - t}. \tag{1}$$

The denominator of (1) is just the change in the time, the independent variable, in going from the "old" value t to the "new" value u, and the numerator $f(u) - f(t)$ is the corresponding change in the particle's position, the dependent variable, in going from the old value $f(t)$ to the new value $f(u)$. It is convenient to introduce a special notation for these changes or differences. Thus we write

$$\Delta t = u - t, \quad \Delta s = f(u) - f(t),$$

where the expressions Δt and Δs, pronounced "delta t" and "delta s," should be thought of as *single* entities and not as products of the separate symbols Δ and t, or Δ and s. We also call Δt the *increment of* t and Δs the *increment of* s. Notice that

$$u = t + \Delta t, \quad f(u) = f(t + \Delta t),$$

and therefore

$$\Delta s = f(t + \Delta t) - f(t).$$

In terms of the increments Δt and Δs, the expression (1) for the average velocity takes the form

$$v_{av} = \frac{\Delta s}{\Delta t},$$

or equivalently

$$v_{av} = \frac{f(t + \Delta t) - f(t)}{\Delta t}. \tag{2}$$

Instantaneous Velocity

So far, so good. But we must still contend with the fact that the expression (2) depends on Δt, and hence differs from our intuitive idea of what is meant by the *instantaneous* velocity at time t, the quantity we are trying to define precisely. What intuition suggests is that the instantaneous velocity $v(t)$ is the result of calculating the average velocity v_{av} with an arbitrarily small "averaging time" $|\Delta t|$. We cannot actually set Δt equal to 0 in formula (2) for the average velocity, since (2) will then reduce to the indeterminate form 0/0. But we *can* let Δt approach 0. Thus for a particle with position function $s = f(t)$, we are led to define the *instantaneous velocity* (or simply the *velocity*) at time t as the limit

$$v(t) = \lim_{\Delta t \to 0} v_{av} = \lim_{\Delta t \to 0} \frac{\Delta s}{\Delta t},$$

that is,

$$v(t) = \lim_{\Delta t \to 0} \frac{f(t + \Delta t) - f(t)}{\Delta t}. \tag{3}$$

106 Chapter 2 Derivatives

We can also write (3) in the equivalent form

$$v(t) = \lim_{u \to t} \frac{f(u) - f(t)}{u - t}. \tag{3'}$$

Each of the limits in (3) and (3') is a way of expressing the *rate of change* of the position function f with respect to t. Thus this whole discussion can be summarized by saying that the velocity of a moving particle is the rate of change of the particle's position with respect to time.

> **Remark** Velocity is a quantity with a *sign*, and its absolute value is often called the *speed*. According to an apocryphal story, a driver going the wrong way on a one-way street was once arrested for a "velocity violation." Explain why.

Example 1 What is the velocity of a stone dropped into a deep dry well t seconds after the stone is released? One and a half seconds after release?

Solution As in Example 6, page 42, the stone's position at time t is

$$s = 16t^2,$$

where t is measured in seconds and s is measured in feet vertically downward from the top of the well (see Figure 2). Therefore, by (3) with $f(t) = 16t^2$, the instantaneous velocity of the stone at time t is

$$v(t) = \lim_{\Delta t \to 0} \frac{16(t + \Delta t)^2 - 16t^2}{\Delta t} = 16 \lim_{\Delta t \to 0} \frac{t^2 + 2t\,\Delta t + (\Delta t)^2 - t^2}{\Delta t}$$

$$= 16 \lim_{\Delta t \to 0} \frac{2t\,\Delta t + (\Delta t)^2}{\Delta t} = 16 \lim_{\Delta t \to 0} (2t + \Delta t) = 16(2t) = 32t.$$

In particular, 1.5 seconds after release the stone's velocity is $32(1.5) = 48$ feet per second (more concisely 48 ft/sec). Alternatively, by (3'),

$$v(t) = \lim_{u \to t} \frac{16u^2 - 16t^2}{u - t} = 16 \lim_{u \to t} \frac{(u + t)(u - t)}{u - t}$$

$$= 16 \lim_{u \to t} (u + t) = 16(2t) = 32t. \quad \square$$

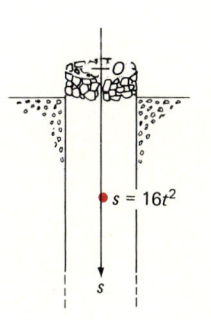

Figure 2

Acceleration

Suppose we calculate the rate of change with respect to time of the velocity function $v(t)$, which is itself a rate of change. This gives us a new function

$$a(t) = \lim_{\Delta t \to 0} \frac{v(t + \Delta t) - v(t)}{\Delta t}, \tag{4}$$

called the *acceleration*. Negative acceleration is often called *deceleration*.

Example 2 It follows from formula (4) and the preceding example that the acceleration of a falling stone is just

$$a(t) = \lim_{\Delta t \to 0} \frac{32(t + \Delta t) - 32t}{\Delta t} = \lim_{\Delta t \to 0} \frac{32\,\Delta t}{\Delta t} = \lim_{\Delta t \to 0} 32 = 32,$$

that is, in this case the acceleration has the constant value of 32 feet per second per second (more concisely 32 ft/sec^2). The acceleration of a falling object, denoted by g, is due to the earth's gravitational attraction. Actually g varies somewhat with geographical location, but $g \approx 32$ ft/sec^2 is a good approximation. In the metric system $g \approx 9.8$ meters/sec^2.

Average and Exact Density

Here is another quite different physical problem in which the idea of a rate of change arises, this time of mass with respect to distance rather than of position or velocity with respect to time. Consider a thin *nonhomogeneous* metal rod AB, which we take to lie along the positive x-axis with the origin at A (see Figure 3). Let the mass m of the piece of the rod between A and the point x be given by some *mass function*

$$m = f(x).$$

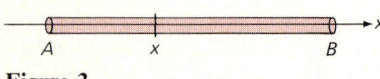

Figure 3

It seems intuitively clear that at each point x the rod has a certain density $d(x)$, which is itself a function of x. But how should this density be defined?

To answer this question, we first define the *average density* d_{av} of the piece of rod with endpoints x and $x + \Delta x$ as the mass Δm of this piece of rod divided by its length:

$$d_{av} = \frac{\Delta m}{\Delta x} = \frac{f(x + \Delta x) - f(x)}{\Delta x}.$$

(Why does this formula still work if $\Delta x < 0$?) Our intuition suggests that the *exact* density $d(x)$ is the result of calculating the average density d_{av} with an arbitrarily small "averaging length" $|\Delta x|$. Although we cannot set $\Delta x = 0$ in the formula for d_{av}, since this leads to the indeterminate form $0/0$, we *can* let Δx approach 0. Thus we are led to define the *exact density*, or simply the *density*, of the rod at the point x as the limit

$$d(x) = \lim_{\Delta x \to 0} d_{av} = \lim_{\Delta x \to 0} \frac{\Delta m}{\Delta x},$$

that is,

$$d(x) = \lim_{\Delta x \to 0} \frac{f(x + \Delta x) - f(x)}{\Delta x}.$$

Substituting $u = x + \Delta x$, we also write $d(x)$ in the equivalent form

$$d(x) = \lim_{u \to x} \frac{f(u) - f(x)}{u - x}.$$

The analogy between these formulas for the density $d(x)$ and the formulas (3) and (3') for the velocity $v(t)$ is complete, although velocity and density are entirely different concepts from a physical point of view. What they have in common is that they both lead to the same kind of limit, known in calculus as the *derivative*. This motivates us to make a special investigation of the derivative, and we now proceed to do so.

Definition of the Derivative

Let f be a function defined in a neighborhood of a point x. Then by the *derivative of f at x*, denoted by $f'(x)$, we mean the limit

$$f'(x) = \lim_{\Delta x \to 0} \frac{f(x + \Delta x) - f(x)}{\Delta x}, \tag{5}$$

provided that the limit exists, or equivalently

$$f'(x) = \lim_{u \to x} \frac{f(u) - f(x)}{u - x} \tag{5'}$$

(let $u = x + \Delta x$). If f has a derivative at x, we also say that f is *differentiable at x.*

It is important to bear in mind that in calculating the limit (5) or (5'), we regard x as *fixed*, so that only Δx varies in (5) and only u varies in (5'). Thus for each fixed value of x the derivative is a *number*, which we denote by $f'(x)$. However, as the notation itself suggests, it is natural to regard the number $f'(x)$ as the *value*, at the point x, of a new *function f'*, called just the derivative of f, rather than the derivative of f at x. It will always be clear from the context in which sense we are thinking of the derivative, whether as a number with fixed x or as a function with variable x.

The Differentiation Operation

The operation leading from a function f to its derivative f' is called *differentiation*, with respect to the independent variable. We denote this operation by the symbol D written before the function being differentiated, attaching the symbol for the independent variable to D as a subscript. Thus D_x denotes differentiation with respect to x, D_t denotes differentiation with respect to t, and so on. The expression

$$\frac{f(x + \Delta x) - f(x)}{\Delta x}, \tag{6}$$

or equivalently

$$\frac{f(u) - f(x)}{u - x}, \tag{6'}$$

of which the derivative $f'(x)$ is the limit as $\Delta x \to 0$ or as $u \to x$, is called the *difference quotient*. The numerator of (6) or (6') is known as the *increment of the function f at the point x*, denoted by $\Delta f(x)$. Thus

$$\Delta f(x) = f(x + \Delta x) - f(x) = f(u) - f(x), \tag{7}$$

where, as always with increments, Δf is regarded as a single entity (pronounced "delta f") and not as a product of separate factors Δ and f. Naturally $\Delta f(x)$ depends not only on the point x, but also on Δx, the increment of the independent variable, or on u, the new value of the independent variable, but we do not indicate this explicitly.

You will observe that the difference quotient (6) or (6') reduces to the indeterminate form $0/0$ if $\Delta x = 0$ or $u = x$. Thus the problem of resolving indeterminacies of the form $0/0$, far from being a mere technicality, lies at the very heart of differential calculus!

It is often convenient to introduce a dependent variable. If $y = f(x)$, say, we can abbreviate the derivative $D_x f(x)$ or $f'(x)$ to $D_x y$ or simply y'. In this notation

$$y' = \lim_{\Delta x \to 0} \frac{\Delta y}{\Delta x},$$

where Δy, the increment of y, is another name for the quantity (7). The geometric meaning of $\Delta y = f(x + \Delta x) - f(x)$ is shown in Figure 4.

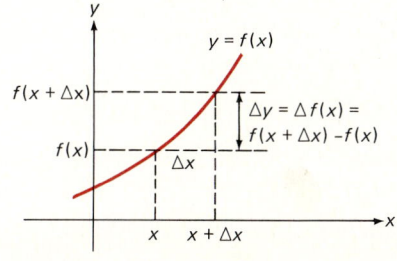

Figure 4

Section 2.1 Velocity and Rates of Change; The Derivative Concept

Example 3 Find the increment Δy and the different quotient $\Delta y/\Delta x$ for the function $y = f(x) = x^5$ if $x = 2$ and $\Delta x = 0.1$.

Solution For the given values of x and Δx we have

$$\Delta y = f(x + \Delta x) - f(x) = f(2.1) - f(2)$$
$$= (2.1)^5 - 2^5 = 40.84101 - 32 = 8.84101,$$

$$\frac{\Delta y}{\Delta x} = \frac{8.84101}{0.1} = 88.4101. \quad \square$$

Differentiation Rules

We now establish a few simple rules for evaluating derivatives. This is just to get us started, and additional rules will be given in subsequent sections.

(i) *The result of differentiating any constant function $f(x) \equiv c$ is zero.* In fact

$$D_x f(x) = D_x c = \lim_{\Delta x \to 0} \frac{c - c}{\Delta x} = \lim_{\Delta x \to 0} \frac{0}{\Delta x} = 0,$$

or alternatively

$$D_x c = \lim_{u \to x} \frac{c - c}{u - x} = \lim_{u \to x} \frac{0}{u - x} = 0,$$

so that

$$D_x c = 0.$$

(ii) *If c is any constant, then*

$$D_x cf(x) = cD_x f(x).$$

In other words, the derivative of a constant times a function is the constant times the derivative of the function. The proof is immediate:

$$D_x cf(x) = \lim_{\Delta x \to 0} \frac{cf(x + \Delta x) - cf(x)}{\Delta x}$$
$$= c \lim_{\Delta x \to 0} \frac{f(x + \Delta x) - f(x)}{\Delta x} = cD_x f(x).$$

Alternatively,

$$D_x cf(x) = \lim_{u \to x} \frac{cf(u) - cf(x)}{u - x} = c \lim_{u \to x} \frac{f(u) - f(x)}{u - x} = cD_x f(x).$$

(iii) *The derivative of x is identically equal to 1.* In fact,

$$D_x x = \lim_{\Delta x \to 0} \frac{(x + \Delta x) - x}{\Delta x} = \lim_{\Delta x \to 0} \frac{\Delta x}{\Delta x} = \lim_{\Delta x \to 0} 1 = 1,$$

or alternatively

$$D_x x = \lim_{u \to x} \frac{u - x}{u - x} = \lim_{u \to x} 1 = 1,$$

so that

$$D_x x = 1.$$

(iv) *The functions x^2 and x^3 have derivatives $2x$ and $3x^2$, respectively.* To verify this, we make the following straightforward calculations:

$$D_x x^2 = \lim_{u \to x} \frac{u^2 - x^2}{u - x} = \lim_{u \to x} (u + x) = x + x = 2x,$$

$$D_x x^2 = \lim_{u \to x} \frac{u^3 - x^3}{u - x} = \lim_{u \to x} (u^2 + ux + x^2) = x^2 + x^2 + x^2 = 3x^2.$$

(Remember that x is regarded as fixed in evaluating these limits.) Thus

$$D_x x^2 = 2x,$$
$$D_x x^3 = 3x^2.$$

(v) More generally, *the derivative of the function x^n, where n is any positive integer, is given by the formula*

$$D_x x^n = nx^{n-1}. \tag{8}$$

Notice that (8) reduces to rules (iii) and (iv) if we set $n = 1$, 2 and 3 in turn. To prove formula (8) for an arbitrary positive integer n, we observe that

$$D_x x^n = \lim_{\Delta x \to 0} \frac{(x + \Delta x)^n - x^n}{\Delta x}$$

$$= \lim_{\Delta x \to 0} \frac{1}{\Delta x} \left[x^n + nx^{n-1} \Delta x + \frac{n(n-1)}{2} x^{n-2} (\Delta x)^2 + \cdots + (\Delta x)^n - x^n \right]$$

$$= \lim_{\Delta x \to 0} \left[nx^{n-1} + \frac{n(n-1)}{2} x^{n-2} \Delta x + \cdots + (\Delta x)^{n-1} \right],$$

with the help of the binomial theorem (see Example 5, page 227, or any college algebra text). Therefore

$$D_x x^n = nx^{n-1} + \lim_{\Delta x \to 0} \left[\frac{n(n-1)}{2} x^{n-2} \Delta x + \cdots + (\Delta x)^{n-1} \right],$$

where the missing terms indicated by the dots all contain Δx raised to a power greater than 1. Hence the last limit is equal to 0, and (8) is proved. Alternatively, if you remember how to divide $u^n - x^n$ by $u - x$,

$$D_x x^n = \lim_{u \to x} \frac{u^n - x^n}{u - x} = \lim_{u \to x} \underbrace{(u^{n-1} + u^{n-2}x + \cdots + ux^{n-2} + x^{n-1})}_{n \text{ terms}},$$

which immediately implies (8), since each of the indicated n terms approaches the same limit x^{n-1} as $u \to x$.

(vi) *The derivative of the function $\sqrt{x}$ is given by the formula*

$$D_x \sqrt{x} = \frac{1}{2\sqrt{x}}.$$

This follows from the calculation

$$D_x \sqrt{x} = \lim_{u \to x} \frac{\sqrt{u} - \sqrt{x}}{u - x} = \lim_{u \to x} \frac{\sqrt{u} - \sqrt{x}}{(\sqrt{u} + \sqrt{x})(\sqrt{u} - \sqrt{x})} = \lim_{u \to x} \frac{1}{\sqrt{u} + \sqrt{x}} = \frac{1}{2\sqrt{x}},$$

where we use the continuity of $\sqrt{x}$, established in Example 2, page 88.

(vii) *The derivatives of the functions* sin x *and* cos x *are given by the formulas*

$$D_x \sin x = \cos x \qquad (9)$$

and

$$D_x \cos x = -\sin x. \qquad (9')$$

To prove (9), we set $\alpha = u$, $\beta = x$ in formula (15), page 58, obtaining

$$\sin u - \sin x = 2 \cos \frac{u + x}{2} \sin \frac{u - x}{2}. \qquad (10)$$

Therefore

$$D_x \sin x = \lim_{u \to x} \frac{\sin u - \sin x}{u - x} = \lim_{u \to x} \cos \frac{u + x}{2} \lim_{u \to x} \frac{\sin \frac{u - x}{2}}{\frac{u - x}{2}}$$

$$= \lim_{u \to x} \cos \frac{u + x}{2} \lim_{t \to 0} \frac{\sin t}{t},$$

with $t = (u - x)/2$. It follows that

$$D_x \sin x = \left(\cos \frac{x + x}{2} \right) \cdot 1 = \cos x,$$

where we use the continuity of the cosine function and the fact that $(\sin t)/t \to 1$ as $t \to 0$. To prove (9′), we replace u and x by $u + (\pi/2)$ and $x + (\pi/2)$ in formula (10), obtaining

$$\sin \left(u + \frac{\pi}{2} \right) - \sin \left(x + \frac{\pi}{2} \right) = 2 \cos \left(\frac{u + x}{2} + \frac{\pi}{2} \right) \sin \frac{u - x}{2},$$

or equivalently

$$\cos u - \cos x = -2 \sin \frac{u + x}{2} \sin \frac{u - x}{2},$$

so that

$$D_x \cos x = \lim_{u \to x} \frac{\cos u - \cos x}{u - x} = -\lim_{u \to x} \sin \frac{u + x}{2} \lim_{u \to x} \frac{\sin \frac{u - x}{2}}{\frac{u - x}{2}} = -\sin x.$$

You should learn all these rules and also the following theorem, which says that the derivative of the sum of two functions can be calculated by differentiating the sum *term by term*.

> **Theorem 1** *(Derivative of the sum of two functions)*. If f and g are differentiable at x, then so is the sum $f + g$, and
> $$D_x(f + g) = D_x f + D_x g, \qquad (11)$$

Proof To keep the notation simple, we omit the arguments of f and g in formula (11). The proof is an immediate consequence of the fact that the limit of a sum is the sum of the limits of the separate terms:

$$D_x(f + g) = \lim_{u \to x} \frac{f(u) + g(u) - [f(x) + g(x)]}{u - x}$$

$$= \lim_{u \to x} \frac{f(u) - f(x)}{u - x} + \lim_{u \to x} \frac{g(u) - g(x)}{u - x} = D_x f + D_x g. \blacksquare$$

Corollary If $f_1, f_2, \ldots, f_n$ are all differentiable at x, then so is the algebraic sum $f_1 \pm f_2 \pm \cdots \pm f_n$, and

$$D_x(f_1 \pm f_2 \pm \cdots \pm f_n) = D_x f_1 \pm D_x f_2 \pm \cdots \pm D_x f_n.$$

Proof Apply the theorem repeatedly. To handle minus signs, use the fact that $D_x(-f) = -D_x f$, which follows from rule (ii) with $c = -1$. $\blacksquare$

The following examples demonstrate the use of Theorem 1 and of rules (i)–(vii).

Example 4 Differentiate $7x^3 + 3x^2 - 5x + 8$.

Solution By the corollary to Theorem 1 and rules (i)–(iv), we have

$$D_x(7x^3 + 3x^2 - 5x + 8) = D_x(7x^3) + D_x(3x^2) - D_x(5x) + D_x 8$$
$$= 7D_x x^3 + 3D_x x^2 - 5D_x x + 0$$
$$= 7(3x^2) + 3(2x) - 5(1) = 21x^2 + 6x - 5.$$

After you have mastered the differentiation rules, you will find it unnecessary to write down all the intermediate steps in a calculation of this kind. □

Example 5 Differentiate $2\sqrt{x} - \frac{1}{3}\sin x$.

Solution By rules (vi) and (vii),

$$D_x\left(2\sqrt{x} - \frac{1}{3}\sin x\right) = 2D_x\sqrt{x} - \frac{1}{3}D_x \sin x = \frac{1}{\sqrt{x}} - \frac{1}{3}\cos x.$$

Which other rules have been used? □

Example 6 Differentiate the general polynomial function

$$P(x) = a_0 + a_1 x + a_2 x^2 + a_3 x^3 + \cdots + a_n x^n,$$

where $a_0, a_1, a_2, a_3, \ldots, a_n$ are constants.

Solution We assume that $a_n \neq 0$, so that $P(x)$ is of degree n. If $n = 0$, $P(x)$ is the constant a_0 and the derivative of $P(x)$ is identically zero. If $n \geq 1$, then by repeated application of formula (8) of rule (v),

$$P'(x) = D_x P(x) = D_x a_0 + a_1 D_x x + a_2 D_x x^2 + a_3 D_x x^3 + \cdots + a_n D_x x^n$$
$$= 0 + a_1(1) + a_2(2x) + a_3(3x^2) + \cdots + a_n(nx^{n-1})$$
$$= a_1 + 2a_2 x + 3a_3 x^2 + \cdots + na_n x^{n-1},$$

which is a polynomial of degree $n - 1$, since $na_n \neq 0$. ☐

Example 7 The position at time $t \geq 0$ of a particle moving along a straight line is given by $s = \frac{1}{3}t^3 - 2t^2 + 3t$. Find the particle's velocity and acceleration at time t. When does the direction of motion of the particle change? When does the particle return to its initial position?

Solution The particle's velocity is

$$v = D_t s = D_t\left(\frac{1}{3}t^3 - 2t^2 + 3t\right) = \frac{1}{3}D_t t^3 - 2D_t t^2 + 3D_t t$$

$$= \frac{1}{3}(3t^2) - 2(2t) + 3 = t^2 - 4t + 3,$$

and its acceleration is

$$a = D_t v = D_t(t^2 - 4t + 3) = D_t t^2 - 4D_t t + D_t 3 = 2t - 4.$$

The direction of motion of the particle changes when its velocity v changes sign, and this happens when the velocity equals zero, that is, when $t = 1$ or $t = 3$. In fact, since $v = t^2 - 4t + 3 = (t - 1)(t - 3)$, v is positive for $0 \leq t < 1$, zero for $t = 1$, negative for $1 < t < 3$, zero for $t = 3$, and positive again for $t > 3$. The particle's initial position is $s = 0$, its position at time $t = 0$; since

$$s = \frac{1}{3}t^3 - 2t^2 + 3t = \frac{1}{3}t(t^2 - 6t + 9) = \frac{1}{3}t(t - 3)^2,$$

the particle returns to its initial position when $t = 3$. Figure 5 shows the particle's position, velocity and acceleration as functions of time, and gives a detailed picture of the particle's motion. ☐

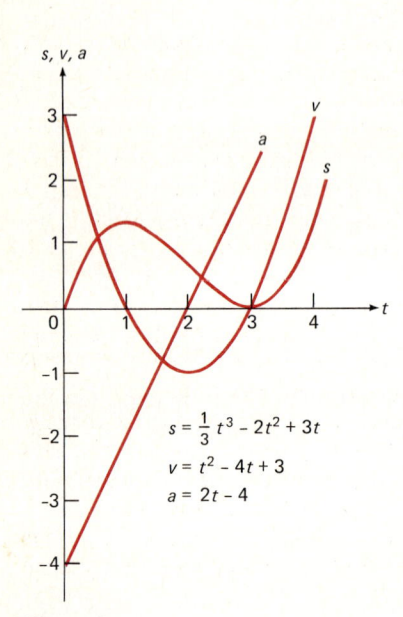

Figure 5

Example 8 The rate of change of electric charge with respect to time is called *current*. Suppose $q = 2t^2 - 3t + 1$ coulombs of charge flow through a conducting wire in t seconds. Find the resulting current i in amperes (coulombs per second) after 3 sec. When does the current reverse itself? If there is a 15-amp fuse in the line, how long will it last?

Solution Since

$$i = D_t q = D_t(2t^2 - 3t + 1) = 4t - 3,$$

the current after 3 sec equals $4(3) - 3 = 9$ amp. The current reverses itself when i changes sign, and this happens when $i = 0$, that is, when $t = \frac{3}{4} = 0.75$ sec. In fact, since $i = 4(t - \frac{3}{4})$, i is negative for $0 \leq t < \frac{3}{4}$, zero for $t = \frac{3}{4}$, and positive for $t > \frac{3}{4}$. The fuse burns out when $i = 4t - 3 = 15$, that is, when $t = \frac{18}{4} = 4.5$ sec. ☐

Problems

1. The position at time t of a particle moving along a straight line is given by $s = 10t + 5t^2$, where s is measured in feet and t in seconds. Let $v_{av}(\Delta t)$ be the average velocity of the particle between the times $t = 20$ and $t = 20 + \Delta t$. Calculate $v_{av}(\Delta t)$ for $\Delta t = 1, 0.1, 0.01, 0.001$. What is the particle's instantaneous velocity v at the time $t = 20$?

2. In solving Examples 1 and 2, two derivatives have in effect been calculated. Which ones?

Find the increment $\Delta y = f(x + \Delta x) - f(x)$ and the difference quotient $\Delta y / \Delta x$ for the given function $y = f(x)$, point x and increment Δx.

3. $y = x^2$, $x = 3$, $\Delta x = 0.1$
4. $y = x^3$, $x = -1$, $\Delta x = -0.1$
5. $y = x^4$, $x = 0$, $\Delta x = -0.2$
6. $y = 1/x$, $x = 2$, $\Delta x = 2$
7. $y = 1/x^2$, $x = 1$, $\Delta x = -3$
8. $y = 1/x^3$, $x = -2$, $\Delta x = 1$
9. $y = \sqrt{x}$, $x = 16$, $\Delta x = -7$
10. $y = \sin x$, $x = 0$, $\Delta x = \pi/6$
11. $y = \cos x$, $x = \pi/2$, $\Delta x = \pi/4$
12. $y = \tan x$, $x = \pi/4$, $\Delta x = -\pi/2$

Differentiate

13. $x^3 + x^2 + x + 1$
14. $2s^3 - 4s^2 + 8s - 16$
15. $-t^3 + 9t^2 + 5t$
16. $4u^3 - 3u^2 - u + \pi$
17. $(x + 1)(x^2 - x + 1)$
18. $(x - 2)(x^2 + 2x + 4)$
19. $(x^2 - 4)(x + 4)$
20. $\cos^2 x + \sin^2 x$
21. $x^4 + 3x^2 - 6$
22. $x^5 - 2\sqrt{x}$
23. $x^9 + x^6 + x^3 + 1$
24. $2x^{10} - 3x^7 + 5x^3 + 20$
25. $x^{100} - 2x^{50} + 25x$
26. $3 \sin x - 4 \cos x$
27. $s^4 - s^3 + 3s^2 + 8$
28. $10t^5 - 100t^4 + 1000t$
29. $5u^7 - 7u^6 + 9u^5 - 11$
30. $8w^3 - 4\sqrt{w} + \cos w$
31. $(x^2 + 3)(x^2 - 3)$
32. $(x - 1)(x + 1)(x^2 + 1)(x^4 + 1)$

33. Interpret
$$\lim_{x \to -2} \frac{x^3 + 8}{x + 2}$$
as a derivative, and evaluate the limit.

34. The position at time $t \geq 0$ of a particle moving along a straight line is given by $s = t^4 - 4t^3 + 4t^2$. Find the particle's velocity v and acceleration a at time t. When does the direction of motion of the particle change? When does the particle return to its initial position?

35. The height above its initial position of a stone thrown vertically upward with an initial velocity of v_0 ft/sec is given by $s = v_0 t - 16t^2$, where s is measured in feet and t in seconds (this is shown in Example 3, page 268. Let $v_0 = 96$ ft/sec. When does the stone stop rising and begin to fall? What is the maximum height reached by the stone? What is the stone's acceleration?

36. A stone is thrown vertically upward with an initial velocity of 32 ft/sec by a person standing at the edge of a roof 48 ft above the ground. When does the stone hit the ground if it misses the roof on the way down? What is the stone's speed when it hits the ground?

37. In the preceding problem, how high should the roof be if the stone is to hit the ground four seconds later? Five seconds later?

38. The position t seconds later of a car starting from rest is $s = \frac{1}{2} kt^2$, where s is measured in feet and t in seconds. Interpret the constant k, and find its value if the car reaches a speed of 60 mph (miles per hour) in 10 seconds flat. Can the formula $s = \frac{1}{2} kt^2$ be expected to describe the car's motion for large t?

39. Suppose $q = \frac{1}{3} t^3 - t^2 + t$ coulombs of charge flow through a conducting wire in t seconds. Find the resulting current i after 4 sec. Does the current ever reverse itself? If there is a 25-amp fuse in the line, how long will it last?

40. A thin nonhomogeneous metal rod AB is 6 meters long. The mass of the first 2 meters of the rod starting from A is 1 kilogram. Suppose the mass of the piece AP of the rod is proportional to the cube of the distance from A to P. What is the density of the rod at its midpoint? What is the average density of the whole rod? What is the average density of the middle third of the rod?

41. Show that
$$\lim_{u \to x} \frac{uf(x) - xf(u)}{u - x} = f(x) - xf'(x).$$

42. Let $f(x) = (x - a)g(x)$, where the function g is continuous at a. Show that the derivative of f at a is equal to $g(a)$.

Find the derivative of the given function by direct evaluation of the limit defining the derivative.

43. $f(x) = 1/x$
44. $f(x) = 1/x^2$
45. $f(x) = 1/(x^2 + 1)$

2.2 The Tangent Line to a Curve

We now use the derivative to solve the important geometric problem of finding the *tangent line*, or simply the *tangent*,† to a curve of general shape. The problem of finding the tangent to a curve is inseparable from that of

---
† In calculus the words *tangent* and *secant* can refer either to lines, as in this section, or to functions, as in Sections 1.3 and 1.4. The context will always make it clear which meaning is intended.

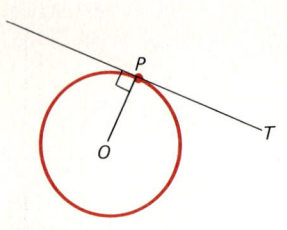

Figure 6

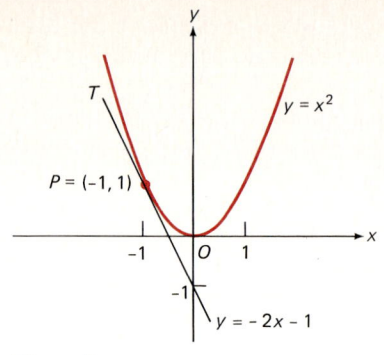

Figure 7

defining the tangent in the first place. Suppose the curve C is a circle, and let P be any point of C. Then, by elementary geometry, the tangent to C at P is either

(i) The line intersecting C in the point P and in no other points, or alternatively,

(ii) The line through P perpendicular to the radius of C drawn to P.

This is the line T shown in Figure 6. Definition (ii), involving the radius, has no apparent counterpart for an arbitrary curve, and definition (i) is also not susceptible to generalization. For example, in the case of the curve $y = x^2$ graphed in Figure 7 (a parabola), there are *two* lines, namely the x-axis and the y-axis, intersecting the curve in the origin O and in no other point. But common sense rejects the idea of the y-axis being the tangent to the curve at O, and at the same time accepts the x-axis as a plausible candidate for the tangent at O. These considerations suggest that the key property of the tangent is that it "clings" to the curve very closely near the point of tangency. For example, this seems to be true of the line T in Figure 7, which, as will be shown in Example 2 is the tangent to the parabola $y = x^2$ at the point $P = (-1, 1)$.

The Tangent Line as the Limiting Position of the Secant Line

To give precise mathematical meaning to this intuitive idea, we reason as follows. Let $P = (a, b)$ be a fixed point and $Q = (x, y)$ a variable point of a given curve $y = f(x)$, where it is assumed that f is continuous on some open interval containing a and x. Let S be the straight line going through the points P and Q; such a line is called a *secant line*, or simply a *secant*, of the curve. If S is nonvertical, the slope of S is given by

$$m_S = \frac{y - b}{x - a} = \frac{f(x) - f(a)}{x - a}, \qquad (1)$$

as illustrated in Figure 8. We now vary the point Q along the curve, making Q move progressively closer to the fixed point P (the successive positions of Q need not all be on the same side of P). Then x approaches a, and at the same time the slope of the secant line through P and Q varies. Suppose the limit

$$m = \lim_{x \to a} m_S \qquad (2)$$

Figure 8

Chapter 2 Derivatives

exists. Then the *tangent* to the curve $y = f(x)$ at the point P is defined as the straight line through P with slope m. We might say that the tangent at P has the limiting slope of the secant S through P and Q as the variable point Q approaches the fixed point P. This behavior is illustrated in Figure 9, where as the variable point takes successive positions $Q_1, Q_2, Q_3, Q_4, \ldots$, getting closer and closer to P, the secant takes successive positions $S_1, S_2, S_3, S_4, \ldots$, getting closer and closer to the position of the tangent line T. The figure also shows that unlike the case of a circle, the tangent to a general curve C may well intersect C in points other than the point of tangency (here T intersects C in the point R as well as in P).

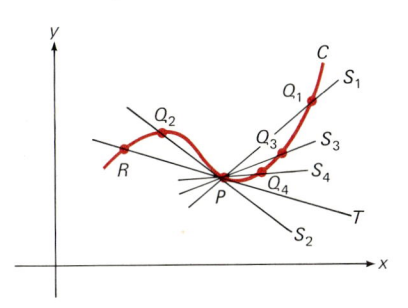

Figure 9

Definition and Equation of the Tangent Line

All that remains now is to substitute (1) into (2). This gives

$$m = \lim_{x \to a} \frac{f(x) - f(a)}{x - a},$$

which you will immediately recognize as the limit defining $f'(a)$, the *derivative of f at a*. It follows that *the curve $y = f(x)$ has a nonvertical tangent line T at the point $P = (a, f(a))$ if and only if f is differentiable at a, and the slope of T is then equal to $f'(a)$*. This slope is also called the *slope of the curve* itself at P. Thus the slope of the curve at P can be thought of as the rate of change of the y-coordinate of the curve $y = f(x)$ with respect to its x-coordinate, evaluated at $x = a$. Since an equation of the straight line with slope m going through the point (a, b) is $y = m(x - a) + b$, an equation of the tangent line to the curve $y = f(x)$ at the point $(a, f(a))$ is

$$y = f'(a)(x - a) + f(a). \tag{3}$$

Notice that if $f'(a) = 0$, the tangent is the *horizontal* line $y = f(a)$.

Example 1 Show that the tangent to the line $y = f(x) = mx + b$ at every point of the line is just the line itself, as is to be expected.

Solution Since $f'(x) = m$, $f'(a) = m$ and $f(a) = ma + b$, in this case equation (3) reduces to $y = m(x - a) + (ma + b) = mx + b$. □

Example 2 Find an equation of the tangent T to the parabola $y = x^2$ at the point $P = (-1, 1)$.

Solution Here $f(x) = x^2$, so that $f'(x) = 2x$. Substituting $a = -1$, $f(-1) = (-1)^2 = 1$ and $f'(-1) = 2(-1) = -2$ into equation (3), we find that an equation of the tangent T is

$$y = -2(x + 1) + 1 = -2x - 1$$

(T is shown in Figure 7). □

Example 3 Where is the tangent to the curve $y = f(x) = x^3 + x^2 - x + 1$ horizontal?

Solution The slope of the tangent at the point $(x, f(x))$ is

$$f'(x) = D_x(x^3 + x^2 - x + 1) = 3x^2 + 2x - 1 = (3x - 1)(x + 1).$$

Since $f'(x) = 0$ if and only if $x = \frac{1}{3}$ or $x = -1$, the tangent is horizontal at just two points of the curve, namely $(\frac{1}{3}, f(\frac{1}{3})) = (\frac{1}{3}, \frac{22}{27})$ and $(-1, f(-1)) = (-1, 2)$, as shown in Figure 10. □

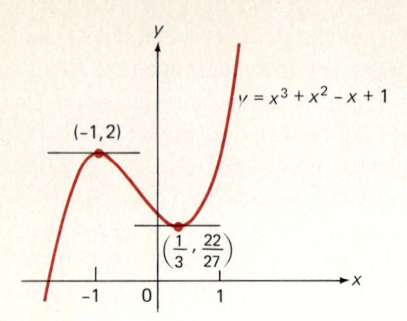

Figure 10

Example 4 Show that the curve $y = f(x) = x^4$ has two tangents going through the point $(\frac{3}{4}, 0)$ of the x-axis.

Solution The point $(\frac{3}{4}, 0)$ does not lie on the curve $y = x^4$ itself (see Figure 11). However, by formula (3) with $f(x) = x^4$ and $f'(x) = 4x^3$, an equation of the tangent to the curve $y = x^4$ at the point (a, a^4) is

$$y = 4a^3(x - a) + a^4 = 4a^3 x - 3a^4, \qquad (4)$$

and (4) must have the solution $x = \frac{3}{4}$, $y = 0$ if the tangent is to go through the point $(\frac{3}{4}, 0)$. Setting $x = \frac{3}{4}$, $y = 0$ in (4), we obtain

$$4a^3 \left(\frac{3}{4}\right) - 3a^4 = 0,$$

or equivalently $a^3 - a^4 = a^3(1 - a) = 0$. This equation has two solutions $a = 0$ and $a = 1$. Substituting these values of a into (4), we find that there are two tangents satisfying the conditions of the problem, one with equation $y = 0$ (the x-axis), the other with equation $y = 4x - 3$, as shown in the figure. □

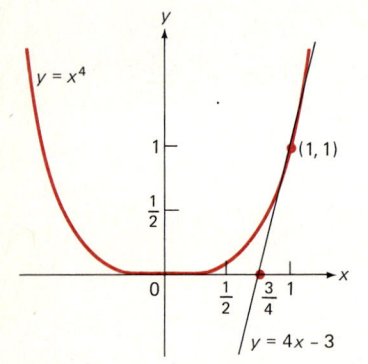

Figure 11

Example 5 The position at time t of a particle moving along a straight line is given by the function $s = f(t)$. Interpret the particle's instantaneous velocity v as the slope of a curve.

Solution Suppose we graph the curve $s = f(t)$ with t as abscissa and s as ordinate. Then the velocity v is just the slope of the curve $s = f(t)$, since $v = D_t s = f'(t)$. In particular, v equals zero whenever the curve has a horizontal tangent. For the position function graphed in Figure 5, page 114, this happens when $t = 1$ and $t = 3$, and correspondingly the velocity v equals zero at precisely these times. □

Example 6 Show that the graph of $y = |x|$ has no tangent at the origin.

Solution Let $f(x) = |x|$. Then, by Example 3, page 68, the limit

$$f'(0) = \lim_{x \to 0} \frac{|x| - |0|}{x - 0} = \lim_{x \to 0} \frac{|x|}{x}$$

defining the derivative of f at $x = 0$ does not exist. Hence the graph of $y = |x|$ has no tangent at the origin. □

On the other hand, the graph of $y = |x|$ does have a tangent at every point other than the origin. In fact, if P lies to the right of the origin, the

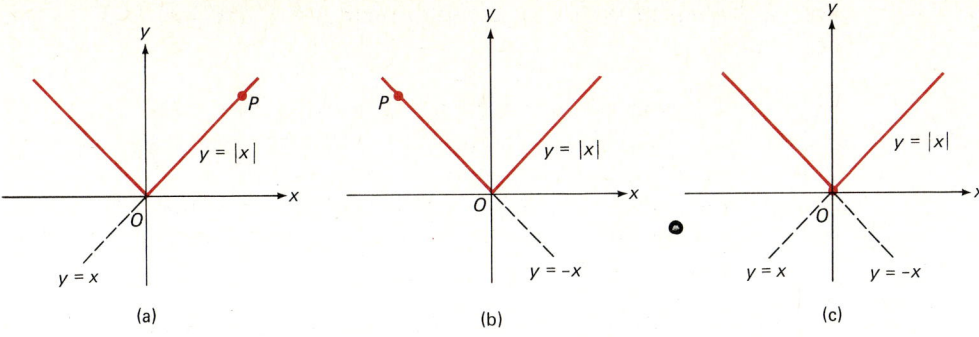

Figure 12

tangent at P is just the line $y = x$, shown in Figure 12(a), while if P lies to the left of the origin, the tangent at P is just the line $y = -x$, as shown in Figure 12(b). (In this regard, see Example 1.)

In the case of the function $f(x) = |x|$, both the "right-hand" derivative

$$f'_+(0) = \lim_{x \to 0^+} \frac{|x| - |0|}{x - 0} = \lim_{x \to 0^+} \frac{|x|}{x} = 1,$$

and the "left-hand" derivative

$$f'_-(0) = \lim_{x \to 0^-} \frac{|x|}{x} = -1$$

exist at $x = 0$, but they are unequal, giving rise to a sharp corner at the origin. Since the one-sided derivatives $f'_+(0)$ and $f'_-(0)$ are unequal, the ordinary (two-sided) derivative $f'(0)$ does not exist (see Theorem 12, page 93), and thus there can be no ordinary (two-sided) tangent at the origin O. However, we can define a right-hand tangent at O, with slope $f'_+(0)$, and a left-hand tangent at O, with slope $f'_-(0)$. Of course, these tangents turn out to be the lines $y = x$ and $y = -x$, and the fact that these lines are different shows again that there is no tangent in the ordinary sense at O [see Figure 12(c)].

More generally, the limit

$$f'_+(a) = \lim_{x \to a^+} \frac{f(x) - f(a)}{x - a}$$

is called the *right-hand derivative* of the function f at the point a, and the limit

$$f'_-(a) = \lim_{x \to a^-} \frac{f(x) - f(a)}{x - a}$$

is called the *left-hand derivative* of f at a. It follows from Theorem 12, page 12, that the ordinary derivative $f'(a)$ exists if and only if $f'_+(a)$ and $f'_-(a)$ both exist and are equal, in which case $f'(a) = f'_+(a) = f'_-(a)$. If $f'_+(a)$ and $f'_-(a)$ both exist but are unequal, the curve $y = f(x)$ is said to have a *corner* at the point $P = (a, f(a))$. In this case there is a *right-hand tangent* T_+ at P, with equation

$$y = f'_+(a)(x - a) + f(a),$$

Section 2.2 The Tangent Line to a Curve

A curve with two corners

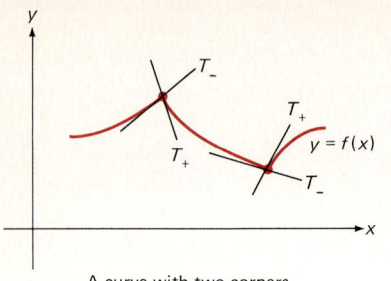

Figure 13

and a *left-hand tangent* T_- at P, with equation

$$y = f'_-(a)(x - a) + f(a),$$

but no ordinary tangent at P. Figure 13 shows a curve with two corners, together with the corresponding right- and left-hand tangents T_+ and T_-. There are cases where the graph of a continuous function not only fails to have a tangent at some point P, but also fails to have one-sided tangents at P (see Problem 23).

Although the function $y = |x|$ has no derivative at $x = 0$, as we saw in Example 6, it is continuous at $x = 0$. This shows, in concise language, that continuity does *not* imply differentiability. On the other hand, as we now prove, differentiability *does* imply continuity, that is, a function must be continuous at any point where it has a derivative.

Theorem 2 *(Differentiability implies continuity). If f is differentiable at a, then f is continuous at a.*

Proof Suppose f is differentiable at a. Then the limit

$$f'(a) = \lim_{x \to a} \frac{f(x) - f(a)}{x - a}$$

exists. It follows that

$$\lim_{x \to a} [f(x) - f(a)] = \lim_{x \to a} \frac{f(x) - f(a)}{x - a} (x - a)$$

$$= \lim_{x \to a} \frac{f(x) - f(a)}{x - a} \lim_{x \to a} (x - a) = f'(a) \cdot 0 = 0,$$

or equivalently

$$\lim_{x \to a} f(x) = f(a),$$

that is, f is continuous at a. ∎

The Normal Line to a Curve

Suppose a curve $y = f(x)$ has a tangent T at the point $P = (a, f(a))$. Then the line N through P perpendicular to T is called the *normal line*, or simply the *normal*, to the curve at the point P. Since the slope of T is $f'(a)$, it follows from the perpendicularity condition (Theorem 9, page 32) that the slope of N is $-1/f'(a)$, provided that $f'(a) \neq 0$. But if $f'(a) = 0$, the tangent T is the horizontal line $y = f(a)$ through P, and the normal is then the vertical line $x = a$ through P. Thus an equation of the normal to the curve $y = f(x)$ at the point $P = (a, f(a))$ is

$$y = -\frac{1}{f'(a)}(x - a) + f(a) \tag{5}$$

if $f'(a) \neq 0$ and

$$x = a \tag{5'}$$

if $f'(a) = 0$. The case where the tangent is vertical and the normal is horizontal involves the notion of an infinite derivative, and will be considered in Section 3.5.

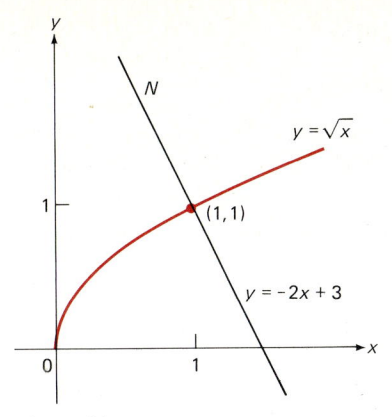

Figure 14

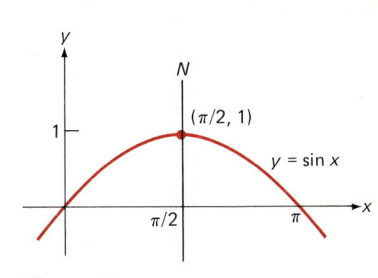

Figure 15

Example 7 Find the normal N to the curve $y = \sqrt{x}$ at the point $(1, 1)$.

Solution Here $f(x) = \sqrt{x}$, $a = 1$, so that

$$f'(x) = \frac{1}{2\sqrt{x}}, \quad f'(a) = \frac{1}{2\sqrt{1}} = \frac{1}{2}.$$

Thus, by (5), the normal N is the line $y = -2(x - 1) + 1$ or $y = -2x + 3$, as shown in Figure 14. □

Example 8 Find the normal N to the curve $y = \sin x$ at the point $(\pi/2, 1)$.

Solution This time $f(x) = \sin x$, $a = \pi/2$, so that $f'(x) = \cos x$, $f'(a) = \cos(\pi/2) = 0$. It follows that N is the vertical line $x = \pi/2$, as shown in Figure 15. □

Problems

Find an equation of the tangent to the curve
1. $y = x^2 + 2x + 3$ at $(-1, 2)$
2. $y = 2 - x - x^2$ at $(2, -4)$
3. $y = \frac{1}{3}x^3 - 1$ at $(-1, -\frac{4}{3})$
4. $y = 2x^4$ at $(\frac{1}{2}, \frac{1}{8})$
5. $y = \sqrt{x}$ at $(4, 2)$
6. $y = \sin x$ at $(\pi, 0)$
7. $y = \cos x$ at $(\pi/2, 0)$
8. $y = \sin x + \cos x$ at $(\pi/4, \sqrt{2})$

Find the tangent (or tangents) to the given curve going through the point P. (In each case P does *not* lie on the curve.)
9. $y = x^2$, $P = (3, 8)$
10. $y = x^3$, $P = (0, 2)$
11. $y = x^4$, $P = (0, -3)$

Does the given curve have two distinct parallel tangents?
12. $y = x^2$
13. $y = x^3$
14. $y = \sin x$

Does the given curve have two perpendicular tangents?
15. $y = x^3$
16. $y = x^2$
17. $y = \cos x$

18. At what point of the curve $y = x^2$ is the tangent parallel to the secant going through the points of the curve with x-coordinates 1 and 3?
19. At what point of the curve $y = x^2 - 2x + 5$ is the tangent perpendicular to the line $y = x$?
20. Let T be the tangent to the parabola $y = x^2$ at any point $P = (a, a^2)$ except the origin O. Then, as shown in Figure 16,

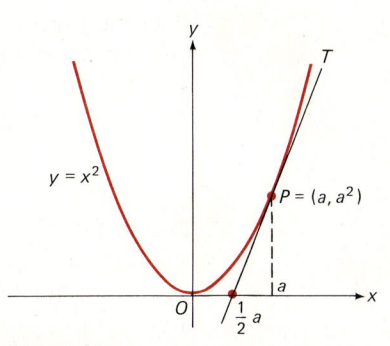

Figure 16

Section 2.2 The Tangent Line to a Curve

T is the line going through the point P and the point $(\frac{1}{2}a, 0)$ of the x-axis. Why is this true?

For the given function f evaluate the one-sided derivatives $f'_+(0)$ and $f'_-(0)$, if they exist.

21. $f(x) = |x|$

22. $f(x) = |x - x^2|$

23. $f(x) = \begin{cases} x \sin \dfrac{1}{x} & \text{if } x \neq 0, \\ 0 & \text{if } x = 0 \end{cases}$

24. $f(x) = \begin{cases} \sin x & \text{if } x \leq 0, \\ \cos x & \text{if } x > 0 \end{cases}$

25. For what values of m and b is the function
$$f(x) = \begin{cases} mx + b & \text{if } x < a, \\ x^2 & \text{if } x \geq a \end{cases}$$
differentiable at a?

26. Graph the function $y = |x| + |x - 1|$. Find the one-sided tangents at the corners of the graph.

27. For what values of the positive integer n does the curve
$$y = f(x) = \begin{cases} b & \text{if } x < a, \\ b + (x - a)^n & \text{if } x \geq a \end{cases}$$
have a corner at the point (a, b)?

28. Show that if the curve $y = f(x)$ has a corner at $(a, f(a))$, then f is continuous at a.

Find an equation of the normal to the curve

29. $y = x^2 + 1$ at $(0, 1)$

30. $y = 1 - x^3$ at $(-1, 2)$

31. $y = x^4 - 6x^2 + 2$ at $(2, -6)$

32. $y = 2 \sin x + 3 \cos x$ at $(\pi/2, 2)$

Find the normal (or normals) to the curve $y = x^2$ going through the given point P, which does not lie on the curve.

33. $P = (3, 0)$ **34.** $P = (0, \frac{9}{2})$

As in Figure 17, the angle α ($0 \leq \alpha < \pi$) between two intersecting curves C_1 and C_2 is defined as the angle between their tangents T_1 and T_2 at the point of intersection P, measured from T_1 to T_2 in the counterclockwise direction. With the help of formula (17), page 60, find the angle α between the given pair of curves, measured from the first curve to the second curve.

35. $y = x^2$ and $y = x^3$ at $(0, 0)$ and $(1, 1)$

36. $y = \sqrt{x}$ and $y = x^2$ at $(1, 1)$

37. $y = \frac{1}{2}x^2$ and $y = 1 - \frac{1}{2}x^2$ at $(\pm 1, \frac{1}{2})$

38. $y = \sin x$ and $y = \cos x$ at $(\pi/4, 1/\sqrt{2})$

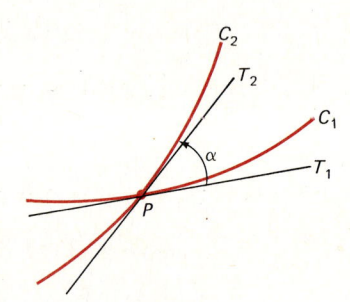

The angle between the curves C_1 and C_2 at P is α.

Figure 17

39. What choice of the constant c makes the curves $y = 1/x$ and $y = cx^3$ *orthogonal*, that is, makes them intersect at right angles at both points of intersection?

40. What choice of c makes the curves $y = x^2$ and $y = 1 - cx^2$ orthogonal?

2.3 The Tangent Line Approximation and Differentials

There is a method of approximating functions which is intimately related to the concept of the tangent to a curve. Suppose a given curve $y = f(x)$ has a tangent line T at the point $P = (a, f(a))$. Then, according to formula (3), page 117, T is the graph of the function $y = t(x)$, where

$$t(x) = f(a) + f'(a)(x - a). \tag{1}$$

Notice that $t(a) = f(a)$, since the point P belongs to both the curve $y = f(x)$ and the line $y = t(x)$, as shown in Figure 18.

The figure suggests that the tangent line $y = t(x)$ is a good approximation to the curve $y = f(x)$, at least in the vicinity of the point P, and this is quite generally true (see Problem 29). Thus, if $|x - a|$ is small but nonzero,

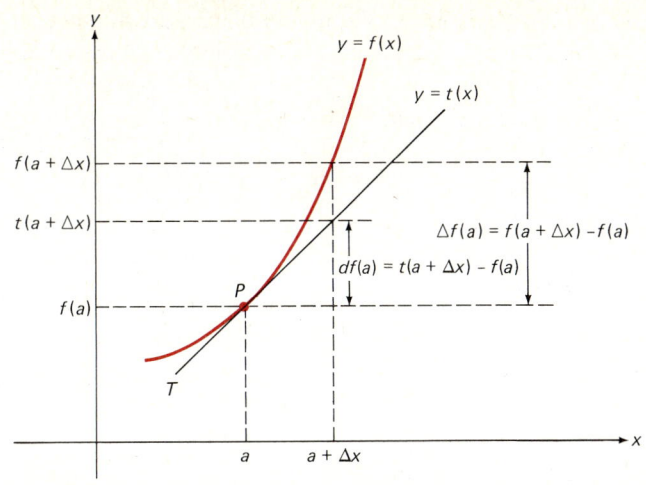

Geometric meaning of the tangent line approximation and the differential

Figure 18

we have $f(x) \approx t(x)$, that is,

$$f(x) \approx f(a) + f'(a)(x - a).$$

This approximation, called the *tangent line approximation*, takes the form

$$f(a + \Delta x) \approx f(a) + f'(a)\,\Delta x \qquad (2)$$

in terms of the increment $\Delta x = x - a$. It can also be written very concisely as

$$\Delta f(a) \approx f'(a)\,\Delta x \qquad (2')$$

where

$$\Delta f(a) = f(a + \Delta x) - f(a) \qquad (3)$$

is the increment of the function f at a.

Definition of the Differential

Formulas (2) and (2') involve the expression $f'(a)\,\Delta x$, which we denote by $df(a)$ and call the *differential of the function f at a*. Here df is regarded as a single entity, and not as a product of separate factors d and f. Thus, by definition,

$$df(a) = f'(a)\,\Delta x.$$

Substituting $\Delta x = x - a$ in formula (1), we find that

$$t(a + \Delta x) = f(a) + f'(a)\,\Delta x.$$

Hence the differential of f at a is given by the formula

$$df(a) = t(a + \Delta x) - f(a),$$

which is to be compared with formula (3) for the increment $\Delta f(a)$ of f at a (it is understood that $df(a)$ and $\Delta f(a)$ both depend on Δx, as well as on a). Thus the differential $df(a)$ is the change in the ordinate (*y*-coordinate) of the tangent line $y = t(x)$ as x changes from a to $a + \Delta x$, while the increment $\Delta f(a)$ is the corresponding change in the ordinate of the *curve* $y = f(x)$ itself (see Figure 18).

We now replace a by x, since x is no longer needed to denote the x-coordinate of a variable point of the curve $y = f(x)$ or of the tangent line $y = t(x)$. As a result, the formula $df(a) = f'(a)\,\Delta x$ becomes

$$df(x) = f'(x)\,\Delta x. \qquad (4)$$

Suppose we make the special choice $f(x) = x$, for which $df(x) = dx$ and $f'(x) = 1$. Then (4) reduces to just

$$dx = \Delta x,$$

that is, *the differential and the increment of the independent variable are equal.* This fact enables us to write

$$df(x) = f'(x)\,dx, \qquad (4')$$

instead of (4). Dividing both sides of (4') by dx, we get the formula

$$\frac{df(x)}{dx} = f'(x).$$

Thus we can interpret the derivative $f'(x)$ as the quotient of the differentials $df(x)$ and dx. For example, in this notation the formula for the derivative of the power x^n becomes

$$\frac{d(x^n)}{dx} = nx^{n-1}, \qquad (5)$$

the rule for differentiating the sum of two functions f and g becomes

$$\frac{d(f+g)}{dx} = \frac{df}{dx} + \frac{dg}{dx} \qquad (6)$$

(if we omit arguments), and so on. As suggested by these formulas, we can also use the "bare" expression d/dx to denote the operation of differentiation with respect to x, denoted so far by D_x. For example, the formula

$$D_x(x^4 + \sin x) = 4x^3 + \cos x$$

can also be written as

$$\frac{d}{dx}(x^4 + \sin x) = 4x^3 + \cos x.$$

The Leibniz Notation

Because of its simplicity and flexibility, we will tend to favor the use of the notation $df(x)/dx$, introduced by Leibniz in the year 1684, although the other ways of writing the derivative of f at x, namely $f'(x)$ and $D_x f(x)$, are also valuable and will be retained. You should become thoroughly familiar with all these equivalent notations. Despite the interpretation of $df(x)/dx$ as a quotient of differentials, it is usually best to think of $df(x)/dx$ as just an alternative notation for the derivative of f at x.

Since

$$df(x) = f'(x)\,dx = \frac{df(x)}{dx}\,dx,$$

every formula for evaluating a derivative gives rise to an analogous formula for evaluating a differential. Thus, multiplying both sides of (5) by dx, we

obtain the formula
$$d(x^n) = nx^{n-1}\,dx$$
for the differential of x^n. Similarly, multiplying both sides of (6) by dx gives
$$d(f + g) = df + dg,$$
which is the analogue for differentials of the rule for differentiating a sum. Similarly, if c is any constant,
$$dc = 0, \qquad d(cf) = c\,df$$
(these are the analogues for differentials of rules (i) and (ii), page 110).

Example 1 Evaluate $d(x^5 + 3\cos x)$.

Solution Since
$$\frac{d}{dx}(x^5 + 3\cos x) = \frac{d}{dx}x^5 + 3\frac{d}{dx}\cos x = 5x^4 - 3\sin x,$$
we have
$$d(x^5 + 3\cos x) = (5x^4 - 3\sin x)\,dx. \quad \square$$

Example 2 Evaluate $d(9 - 2\sqrt{x})$.

Solution This time we make direct use of some of the rules governing differentials:
$$d(9 - 2\sqrt{x}) = d9 - 2d\sqrt{x} = 0 - 2 \cdot \frac{1}{2\sqrt{x}}\,dx = -\frac{1}{\sqrt{x}}\,dx = -\frac{dx}{\sqrt{x}}.$$

Notice how bringing dx into the numerator allows the answer to be written somewhat more concisely. $\quad \square$

Example 3 Evaluate $d(t^3 + t)$.

Solution The independent variable is now t, and accordingly
$$d(t^3 + t) = \frac{d(t^3 + t)}{dt}\,dt = (3t^2 + 1)\,dt. \quad \square$$

As already noted, the approximation (2) amounts to replacing the curve $y = f(x)$ in the vicinity of the point $P = (a, f(a))$ by its tangent line at P. An equivalent approximation consists in writing $\Delta f(a) \approx df(a)$, that is, in approximating the increment $\Delta f(a)$ by the differential $df(a)$. As illustrated by the following examples, these approximations are quite accurate for small $|\Delta x|$.

Example 4 Estimate $\sqrt{98}$.

Solution Observing that 98 is close to the number 100, whose square root is exactly 10, we choose $f(x) = \sqrt{x}$, $a = 100$, $\Delta x = -2$ in formula (2). Since $f'(x) = 1/2\sqrt{x}$, this gives
$$\sqrt{a + \Delta x} \approx \sqrt{a} + \frac{1}{2\sqrt{a}}\Delta x,$$
$$\sqrt{98} = \sqrt{100 - 2} \approx \sqrt{100} + \frac{1}{2\sqrt{100}}(-2) = 10 - \frac{1}{10} = 9.9,$$

as compared with the exact value of 9.8995 to four decimal places. The error is gratifyingly small. ∎

Example 5 Estimate sin 46°.

Solution This time we use the fact that 46° is close to the angle 45°, whose sine and cosine we know to be $1/\sqrt{2}$. Thus we choose $f(x) = \sin x$, $a = 45° = \pi/4$, $\Delta x = 1° = \pi/180$ in formula (2). Since $f'(x) = \cos x$ (if x is measured in radians), the result is

$$\sin(a + \Delta x) \approx \sin a + \Delta x \cos a,$$

$$\sin 46° = \sin\left(\frac{\pi}{4} + \frac{\pi}{180}\right) \approx \sin\frac{\pi}{4} + \frac{\pi}{180}\cos\frac{\pi}{4} = \frac{1}{\sqrt{2}}\left(1 + \frac{\pi}{180}\right) = 0.71945$$

to five decimal places. This is in close agreement with the exact value of 0.71934, to the same number of decimal places. ∎

Remark The utility of the tangent line approximation will be enhanced in Section 9.8, when we develop a way of estimating the accuracy of the approximation *in advance*.

Example 6 About how much would the earth's surface area increase if its radius were increased by 1 foot?

Solution The surface area of a sphere of radius R is $S = 4\pi R^2$, and the earth's radius is approximately 3960 miles. Estimating the actual increment ΔS of the surface area by the differential dS evaluated at $R = 3960$, we get

$$\Delta S \approx dS = d(4\pi R^2) = 8\pi R \,\Delta R = 8\pi(3960) \cdot \frac{1}{5280} \text{ square miles,}$$

since there are 5280 feet in a mile. Doing the arithmetic, we find that $\Delta S \approx 18.85$ sq mi. For comparison, the area of Manhattan Island is approximately 22 sq mi. ∎

Actually, in Example 6 we can readily determine the accuracy of the approximation $\Delta S \approx dS$, since it is so easy to calculate ΔS exactly. In fact,

$$\Delta S = 4\pi(R + \Delta R)^2 - 4\pi R^2 = 8\pi R\,\Delta R + 4\pi(\Delta R)^2,$$

so that

$$\Delta S - dS = 4\pi(\Delta R)^2 = 4\pi\left(\frac{1}{5280}\right)^2 < \frac{1}{2} \times 10^{-6} \text{ sq mi.}$$

Example 7 How much would the earth's volume decrease if its radius were to shrink by 1 inch?

Solution The volume of a spherical ball of radius R is $V = \frac{4}{3}\pi R^3$. Estimating the actual increment ΔV of the volume by the differential dV, we find that

$$\Delta V \approx dV = d\left(\frac{4}{3}\pi R^3\right) = 4\pi R^2 \,\Delta R$$

$$= -4\pi(3960)^2 \cdot \frac{1}{12(5280)} \approx -3110 \text{ cubic miles}$$

(1 inch = $\frac{1}{12}$ ft). As large as this volume loss is in absolute terms, it represents only about one millionth of a percent of the earth's total volume, since

$$\frac{4\pi R^2 |\Delta R|}{\frac{4}{3}\pi R^3} = \frac{3|\Delta R|}{R} = \frac{3}{3960(12)(5280)} \approx 1.2 \times 10^{-8}. \quad \square$$

Problems

Find the differential of
1. $7x - 11$
2. $2x^2 - 3x + 5$
3. $x^5 + 6x^3 + x$
4. $t^4 - 3t^3 + 2t^2 - 10$
5. $5u^6 - \cos u$
6. $4\sqrt{v} + 5 \sin v$
7. $\cos x - \sin x$
8. $10x^9 - 9x^{10}$

Find the differential $df(a) = f'(a)\Delta x$ of the given function for the indicated values of a and Δx. In each case compare $df(a)$ with the increment $\Delta f(a) = f(a + \Delta x) - f(a)$ to a suitable number of decimal places.

9. $f(x) = 2x + 1$, $a = 3$, $\Delta x = 0.001$
10. $f(x) = x^2$, $a = -\frac{1}{4}$, $\Delta x = 1$
11. $f(x) = x^3 - 1$, $a = -1$, $\Delta x = 0.01$
12. $f(x) = x^4 + 3x^2$, $a = 2$, $\Delta x = -0.02$
13. $f(x) = \sqrt{x}$, $a = 4$, $\Delta x = -0.1$
14. $f(x) = \sin x + \cos x$, $a = \Delta x = \pi/6$

Use the tangent line approximation (2) to estimate the given quantity, and compare the result with the exact answer to a suitable number of decimal places.

15. $(3.1)^2 + (3.1)^3 + (3.1)^4$
16. $\sqrt{35}$
17. $\sqrt{49.6}$
18. $\sqrt{611}$
19. $\sin 28°$
20. $\cos 121°$

Justify the given approximation for small $|x|$.
21. $\sin x \approx x$
22. $\cos x \approx 1$

23. Estimate the percentage increase in the area of a circular disk if its radius is increased by 5%. If its circumference is increased by 3%.
24. Imagine that the earth's circumference is increased by 1 foot. Estimate the distance between the surface of the "new" earth and the surface of the "old" earth.

Estimate the amount by which the radius of the earth must be increased to produce additional surface area the size of
25. Rhode Island ($\approx$ 1215 sq. mi)
26. Ohio ($\approx$ 41,220 sq. mi)
27. Texas ($\approx$ 267,340 sq. mi)
28. Antarctica ($\approx$ 5,500,000 sq. mi)

29. Let $e(\Delta x)$ be the error of the tangent line approximation (2), that is, let $e(\Delta x) = f(a + \Delta x) - f(a) - f'(a)\Delta x$. Show that

$$\lim_{\Delta x \to 0} e(\Delta x) = 0, \quad \lim_{\Delta x \to 0} \frac{e(\Delta x)}{\Delta x} = 0.$$

Thus, as $\Delta x \to 0$, the error $e(\Delta x)$ not only approaches zero, but in fact approaches zero "faster" than Δx.

30. Show that the tangent line approximation (2) is the *best linear approximation* to $f(a + \Delta x)$ near a in the following sense: If $e(\Delta x)$ is the error of the tangent line approximation, and if $E(\Delta x)$ is the error made in approximating $f(a + \Delta x)$ near a by any line L through the point $(a, f(a))$ other than the tangent line, then $|e(\Delta x)| < |E(\Delta x)|$ for all $\Delta x \neq 0$ of sufficiently small absolute value, that is, for all Δx in some deleted neighborhood of zero.

31. Let $\tilde{q}$ be an approximation to a quantity whose exact value is q. Then the error $q - \tilde{q}$ of the approximation $q \approx \tilde{q}$ is often called the *absolute error* to distinguish it from the *relative error*, defined as $|(q - \tilde{q})/q|$. (The relative error is often expressed as a percentage; thus a relative error of 0.05 is also called a relative error of 5%.) According to Problem 29, the absolute error of the tangent line approximation approaches zero faster than Δx as $\Delta x \to 0$. Give an example where the relative error of the tangent line approximation is 100%, regardless of the size of Δx.

2.4 The Product and Quotient Rules

We now resume our study of the technique of differentiation, learning how to calculate the derivatives of products, reciprocals and quotients of functions. We begin with products. By a misguided analogy with the case of limits, one might be tempted to write the derivative of the product of two functions as the product of the derivatives of the factors, but this must be wrong, as shown at once by the fact that $D_x(1 \cdot x) = D_x x = 1$ is *not* equal to $(D_x 1)(D_x x) = 0 \cdot 1 = 0$. The next theorem gives the correct way to calculate the derivative of a product.

> **Theorem 3** (*Product rule*). *If the functions f and g are differentiable at x, then so is the product fg, and*
>
> $$\frac{d}{dx}(fg) = \frac{df}{dx}g + f\frac{dg}{dx} = f'g + fg'. \tag{1}$$

Proof Here the prime denotes differentiation with respect to x, and in the interest of brevity we sometimes omit arguments of functions (this is a common practice). By definition,

$$\frac{d}{dx}(fg) = \lim_{u \to x} \frac{f(u)g(u) - f(x)g(x)}{u - x},$$

where x is regarded as fixed. The numerator $f(u)g(u) - f(x)g(x)$ can be written in the form

$$f(u)g(u) - f(x)g(u) + f(x)g(u) - f(x)g(x),$$

where we subtract out the term $f(x)g(u)$ and then add it right back in again! This gambit allows us to write

$$\frac{d}{dx}(fg) = \lim_{u \to x}\left[\frac{f(u)g(u) - f(x)g(u)}{u - x} + \frac{f(x)g(u) - f(x)g(x)}{u - x}\right]$$

$$= \lim_{u \to x}\frac{f(u) - f(x)}{u - x}g(u) + \lim_{u \to x} f(x)\frac{g(u) - g(x)}{u - x},$$

and leads to the appearance of the *difference quotients* of f and g at the last step, just as we would like. Continuing the calculation, we get

$$\frac{d}{dx}(fg) = \lim_{u \to x}\frac{f(u) - f(x)}{u - x}\lim_{u \to x} g(u) + \lim_{u \to x} f(x) \lim_{u \to x}\frac{g(u) - g(x)}{u - x} \tag{2}$$

$$= f'(x)\lim_{u \to x} g(u) + f(x)g'(x),$$

where we use the definitions of the derivatives $f'(x)$ and $g'(x)$, and the fact that $f(x)$ is a constant. But g is differentiable at x, by hypothesis, and therefore continuous at x, by Theorem 2, page 120, so that

$$\lim_{u \to x} g(u) = g(x). \tag{3}$$

Substituting (3) into (2), we obtain the desired result (1). ∎

Thus the derivative of the product of two differentiable functions is the derivative of the first function times the second function plus the first function times the derivative of the second function.

Corollary *If the functions $f_1, f_2, \ldots, f_n$ are all differentiable at x, then so is the product $f_1 f_2 \cdots f_n$, and*

$$\frac{d}{dx}(f_1 f_2 \cdots f_n) = \frac{df_1}{dx}f_2 \cdots f_n + f_1 \frac{df_2}{dx}f_3 \cdots f_n + \cdots + f_1 f_2 \cdots f_{n-1}\frac{df_n}{dx}$$

$$= f'_1 f_2 \cdots f_n + f_1 f'_2 f_3 \cdots f_n + \cdots + f_1 f_2 \cdots f_{n-1} f'_n.$$

Proof Apply the theorem repeatedly. For example,

$$\frac{d}{dx}(fgh) = \frac{d(fg)}{dx}h + fg\frac{dh}{dx} = \left(\frac{df}{dx}g + f\frac{dg}{dx}\right)h + fg\frac{dh}{dx}$$

$$= \frac{df}{dx}gh + f\frac{dg}{dx}h + fg\frac{dh}{dx},$$

or equivalently

$$(fgh)' = (fg)'h + (fg)h' = (f'g + fg')h + fgh' = f'gh + fg'h + fgh'. \quad \blacksquare$$

Example 1 Differentiate $(2x - 1)(x^2 - 6x + 3)$.

Solution By the product rule,

$$\frac{d}{dx}[(2x-1)(x^2-6x+3)] = \frac{d(2x-1)}{dx}(x^2-6x+3) + (2x-1)\frac{d(x^2-6x+3)}{dx}$$

$$= 2(x^2-6x+3) + (2x-1)(2x-6) = 6x^2 - 26x + 12.$$

The alternative is to multiply first and then differentiate:

$$\frac{d}{dx}[(2x-1)(x^2-6x+3)] = \frac{d}{dx}(2x^3 - 13x^2 + 12x - 3) = 6x^2 - 26x + 12. \quad \square$$

Example 2 Differentiate $x^2 \sin x$.

Solution There is now no way to avoid the use of the product rule:

$$\frac{d}{dx}(x^2 \sin x) = \frac{d(x^2)}{dx}\sin x + x^2\frac{d \sin x}{dx} = 2x \sin x + x^2 \cos x. \quad \square$$

Example 3 Differentiate $\sqrt{x} \sin x \cos x$.

Solution Using the corollary, we have

$$\frac{d}{dx}(\sqrt{x} \sin x \cos x) = \frac{d\sqrt{x}}{dx}\sin x \cos x + \sqrt{x}\frac{d \sin x}{dx}\cos x + \sqrt{x} \sin x \frac{d \cos x}{dx}$$

$$= \frac{1}{2\sqrt{x}}\sin x \cos x + \sqrt{x} \cos^2 x - \sqrt{x} \sin^2 x.$$

Notice that this can be written more simply as

$$\frac{d}{dx}(\sqrt{x} \sin x \cos x) = \frac{\sin 2x}{4\sqrt{x}} + \sqrt{x} \cos 2x,$$

with the help of formulas (14) and (14'), page 58. $\quad \square$

Theorem 4 (*Derivative of the reciprocal of a function*). *If the function g is differentiable at x, then so is the reciprocal of g, and*

$$\frac{d}{dx}\frac{1}{g} = -\frac{1}{g^2}\frac{dg}{dx} = -\frac{g'}{g^2},$$

provided that $g(x) \neq 0$.

Section 2.4 The Product and Quotient Rules 129

Proof We have

$$\frac{d}{dx}\frac{1}{g} = \lim_{u \to x} \frac{\frac{1}{g(u)} - \frac{1}{g(x)}}{u - x} = \lim_{u \to x} \frac{g(x) - g(u)}{g(u)g(x)(u - x)}$$

$$= -\lim_{u \to x} \frac{1}{g(u)g(x)} \frac{g(u) - g(x)}{u - x}$$

$$= -\frac{1}{g(x)} \lim_{u \to x} \frac{1}{g(u)} \lim_{u \to x} \frac{g(u) - g(x)}{u - x} = -\frac{1}{g^2(x)} g'(x),$$

with the help of (3). Since $g(x) \neq 0$, by hypothesis, there is no need to worry about $g(u)$ being equal to zero in any of the denominators (recall rule (ii), page 76). ■

Example 4 Differentiate $1/(x^2 + 1)$.

Solution By Theorem 4,

$$\frac{d}{dx}\frac{1}{x^2 + 1} = -\frac{1}{(x^2 + 1)^2}\frac{d}{dx}(x^2 + 1) = -\frac{2x}{(x^2 + 1)^2}. \quad □$$

Example 5 Differentiate $1/(\sin x + \cos x)$.

Solution By the same theorem,

$$\frac{d}{dx}\frac{1}{\sin x + \cos x} = -\frac{1}{(\sin x + \cos x)^2}\frac{d}{dx}(\sin x + \cos x)$$

$$= -\frac{\cos x - \sin x}{(\sin x + \cos x)^2} = \frac{\sin x - \cos x}{(\sin x + \cos x)^2}. \quad □$$

Differentiation of Negative Powers

We are now able to differentiate the reciprocal of x^n, where n is any positive integer. By Theorem 4 again,

$$\frac{d}{dx}\frac{1}{x^n} = -\frac{1}{x^{2n}}\frac{d}{dx}x^n = -\frac{1}{x^{2n}}nx^{n-1} = -\frac{n}{x^{n+1}},$$

provided that $x \neq 0$. We can write this in the form

$$\frac{d}{dx}x^{-n} = -nx^{-n-1}. \tag{4}$$

But (4) is exactly what is obtained by replacing n by $-n$ in the formula

$$\frac{d}{dx}x^n = nx^{n-1}, \tag{5}$$

which is already part of our repertory. In fact, we just used (5) for $n > 0$ to prove (4)! Thus we see that the basic differentiation formula (5) continues to hold for *negative* integers. The formula also works for $n = 0$, at least in a formal sense, since if $x \neq 0$, then $x^0 \equiv 1$, so that $D_x x^0 = D_x 1 = 0$, whereas (5) gives $D_x x^0 = 0x^{-1} = 0$. Therefore (5) is valid for an *arbitrary* integer, positive, negative or zero.

Example 6 By two applications of formula (5),

$$\frac{d}{dx}\left(\frac{1}{x} + \frac{1}{x^3}\right) = \frac{d}{dx}(x^{-1} + x^{-3}) = -1x^{-2} - 3x^{-4} = -\frac{1}{x^2} - \frac{3}{x^4}.$$

> **Theorem 5** (*Quotient rule*). *If the functions f and g are differentiable at x, then so is the quotient f/g, and*
>
> $$\frac{d}{dx}\frac{f}{g} = \frac{\frac{df}{dx}g - f\frac{dg}{dx}}{g^2} = \frac{f'g - fg'}{g^2},$$
>
> *provided that* $g(x) \neq 0$.

Proof Using first the product rule and then the rule for differentiating the reciprocal of a function, we have

$$\frac{d}{dx}\frac{f}{g} = \frac{d}{dx}\left(f \cdot \frac{1}{g}\right) = f' \cdot \frac{1}{g} + f \cdot \frac{-g'}{g^2} = \frac{f'}{g} - \frac{fg'}{g^2} = \frac{f'g - fg'}{g^2}. \quad \blacksquare$$

Example 7 Differentiate $2x/(1 - x^2)$.

Solution By the quotient rule,

$$\frac{d}{dx}\frac{2x}{1-x^2} = \frac{\frac{d(2x)}{dx}(1-x^2) - 2x\frac{d(1-x^2)}{dx}}{(1-x^2)^2}$$

$$= \frac{2(1-x^2) - 2x(-2x)}{(1-x^2)^2} = \frac{2(1+x^2)}{(1-x^2)^2},$$

provided that $x \neq \pm 1$. □

Example 8 Differentiate $(x^3 + 1)/(x^2 - x - 2)$.

Solution Before differentiating, think! Perhaps the given function can be simplified. It can, by recognizing that the numerator and denominator have the common factor $x + 1$. In fact

$$\frac{x^3 + 1}{x^2 - x - 2} = \frac{(x + 1)(x^2 - x + 1)}{(x + 1)(x - 2)} = \frac{x^2 - x + 1}{x - 2} = x + 1 + \frac{3}{x - 2},$$

where in the last step we use long division to divide $x^2 - x + 1$ by $x - 2$ (give the details). We now differentiate, obtaining

$$\frac{d}{dx}\frac{x^3 + 1}{x^2 - x - 2} = \frac{d}{dx}(x + 1) + \frac{d}{dx}\frac{3}{x - 2} = 1 - \frac{3}{(x - 2)^2} = \frac{x^2 - 4x + 1}{(x - 2)^2}.$$

Here it is assumed that $x \neq -1, 2$. Failure to make the preliminary cancellation leads to a much more complicated answer, with $(x + 1)^2$ as an unrecognizable common factor of the numerator and denominator. □

Differentiation of Trigonometric Functions

It was shown in rule (vii), page 112, that

$$\frac{d}{dx}\sin x = \cos x, \qquad \frac{d}{dx}\cos x = -\sin x. \tag{6}$$

We now establish formulas for the derivatives of the other four trigonometric functions. The derivative of tan x is found by using the quotient rule, and the derivatives of cot x, sec x and csc x are found by using the rule for differentiating the reciprocal of a function (Theorem 4). The calculations are straightforward, and go as follows:

$$\frac{d}{dx}\tan x = \frac{d}{dx}\frac{\sin x}{\cos x} = \frac{\cos x \frac{d\sin x}{dx} - \sin x \frac{d\cos x}{dx}}{\cos^2 x}$$

$$= \frac{\cos x \cos x - \sin x(-\sin x)}{\cos^2 x} = \frac{\cos^2 x + \sin^2 x}{\cos^2 x}$$

$$= \frac{1}{\cos^2 x} = \sec^2 x,$$

$$\frac{d}{dx}\cot x = \frac{d}{dx}\frac{1}{\tan x} = -\frac{1}{\tan^2 x}\frac{d}{dx}\tan x = -\frac{1}{\tan^2 x}\sec^2 x$$

$$= -\frac{\cos^2 x}{\sin^2 x}\frac{1}{\cos^2 x} = -\frac{1}{\sin^2 x} = -\csc^2 x,$$

$$\frac{d}{dx}\sec x = \frac{d}{dx}\frac{1}{\cos x} = -\frac{1}{\cos^2 x}\frac{d}{dx}\cos x = -\frac{1}{\cos^2 x}(-\sin x)$$

$$= \frac{1}{\cos x}\frac{\sin x}{\cos x} = \sec x \tan x,$$

$$\frac{d}{dx}\csc x = \frac{d}{dx}\frac{1}{\sin x} = -\frac{1}{\sin^2 x}\frac{d}{dx}\sin x = -\frac{1}{\sin^2 x}\cos x$$

$$= -\frac{1}{\sin x}\frac{\cos x}{\sin x} = -\csc x \cot x.$$

Thus we have

$$\frac{d}{dx}\tan x = \sec^2 x, \qquad \frac{d}{dx}\cot x = -\csc^2 x, \qquad (7)$$

and

$$\frac{d}{dx}\sec x = \sec x \tan x, \qquad \frac{d}{dx}\csc x = -\csc x \cot x. \qquad (8)$$

Formulas (7) and (8) are somewhat more complicated than the formulas (6), but they too should be memorized.

Example 9 Differentiate $(\tan x)/x$.

Solution By the quotient rule and (7),

$$\frac{d}{dx}\frac{\tan x}{x} = \frac{\frac{d\tan x}{dx}x - \tan x \frac{dx}{dx}}{x^2} = \frac{x\sec^2 x - \tan x}{x^2}. \quad \square$$

Example 10 Differentiate $\sec^2 x$.

Solution By the product rule and (8),
$$\frac{d}{dx}\sec^2 x = \frac{d}{dx}(\sec x \sec x) = \frac{d\sec x}{dx}\sec x + \sec x \frac{d\sec x}{dx}$$
$$= 2\sec x \frac{d\sec x}{dx} = 2\sec^2 x \tan x. \quad \square$$

Problems

Differentiate

1. $3x^{-2} - 5x^{-3}$
2. $6x^{-4} + 5x^{-5} + 4x^{-6}$
3. $\dfrac{1}{x} + \dfrac{2}{x^2} + \dfrac{3}{x^3}$
4. $\dfrac{1}{x^{10}} + \dfrac{1}{x^5} + 100$
5. $(1 + x^2)(2 - x^2)$
6. $(1 + 4x^2)(1 + 2x^2)$
7. $(1 - x^2)(1 + x^3)$
8. $(x - 1)(x^2 + x + 1)$
9. $(x + 1)(x^2 - x + 1)$
10. $(1 - x^{99})(1 + x^{99})$
11. $(2 - x^3)(1 + x^{-2})$
12. $(2 + x^2)(1 - x^{-3})$
13. $(1 - x^{-5})^2$
14. $(1 + x + x^{-1})(1 + x - x^{-1})$
15. $(1 + x^2)(1 + x^3)(1 + x^4)$
16. $(1 - x)(1 + x + x^2)(1 + x^3)$
17. $\dfrac{1}{1 + x}$
18. $\dfrac{1 - x}{1 + x}$
19. $\dfrac{x^2}{1 + x^2}$
20. $\dfrac{x^3 + 1}{x^2 - 1}$
21. $\dfrac{x^3 - 1}{x^3 + 1}$
22. $\dfrac{x^2 + 5x}{x^3 - 3}$
23. $\dfrac{1 + x - x^2}{1 - x + x^2}$
24. $\dfrac{x^3 - 1}{x^2 + 2x - 3}$
25. $\dfrac{t^4 - 1}{t^2 - 3t + 2}$
26. $\dfrac{\sqrt{t}}{1 + \sqrt{t}}$
27. $\dfrac{1 - \sqrt{u}}{1 + \sqrt{u}}$
28. $u^2 \cos u$
29. $\dfrac{\sin v}{v}$
30. $\dfrac{\tan v}{\sqrt{v}}$
31. $\sqrt{w} \cot w$
32. $w^3 \sec w$
33. $2x \csc x$
34. $x^2 \sin^2 x$
35. $x \sin x \tan x$
36. $\dfrac{1}{1 + \sin x}$
37. $\dfrac{1}{1 - \cos x}$
38. $\dfrac{1 - \sin x}{1 + \cos x}$
39. $\dfrac{1 + \tan x}{1 - \tan x}$
40. $\dfrac{\sin^2 x}{\cos x}$
41. $\dfrac{\cos x}{\sin^2 x}$
42. $\dfrac{\sin x - x \cos x}{\cos x + x \sin x}$

Find the differential of

43. $x^{-2} + x^{-1} + 7$
44. $\dfrac{1}{1 - x}$
45. $\dfrac{x}{1 + x}$
46. $x \tan x$
47. $\sqrt{x} \sin x$
48. $\dfrac{x}{\sin x}$
49. $\dfrac{x}{\cos x}$
50. $\csc^2 x$

51. With as little effort as possible, show that
$$\frac{d}{dx}\frac{x^4 - 5x^2 + 4}{x^3 + 2x^2 - x - 2} = 1.$$

52. Prove the following analogues for differentials of the product and quotient rules for derivatives:
$$d(fg) = g\,df + f\,dg, \qquad d\left(\frac{f}{g}\right) = \frac{g\,df - f\,dg}{g^2}.$$

2.5 The Chain Rule

None of the differentiation rules proved so far tells us how to differentiate composite functions like $\sqrt{x^2 + 1}$ or $\sin(\sin x)$. Thus we now establish a further rule expressing the derivative of a composite function in terms of the derivatives of its component functions. The rule itself, known as the *chain rule*, is simple enough, although its proof is a little tricky. Suppose f and g are two functions such that f is differentiable at x and g is differentiable at $f(x)$, the *value* taken by f at x. Then, as will be shown in a moment, the composite function $g(f(x))$ is also differentiable at x, and has the derivative

$$\frac{d}{dx}g(f(x)) = g'(f(x))f'(x), \qquad (1)$$

where the prime denotes differentiation with respect to the argument of the function to which it is attached. For example, if $f(x) = x^2 + 1$ and $g(x) = \sqrt{x}$, then $g(f(x)) = \sqrt{x^2 + 1}$. Moreover, f is differentiable for all x, while g is differentiable for all $x > 0$ and hence for all values of f, since $x^2 + 1 \geq 1$. In this case $g'(x) = 1/2\sqrt{x}$ and $f'(x) = 2x$, so that the chain rule (1) takes the form

$$\frac{d}{dx}\sqrt{x^2 + 1} = \left(\frac{1}{2\sqrt{x^2 + 1}}\right) 2x = \frac{x}{\sqrt{x^2 + 1}}, \tag{2}$$

and is valid for all x.

The structure of formula (1) becomes clearer if we introduce two dependent variables $y = f(x)$ and $z = g(y) = g(f(x))$, for then (1) can be written in the particularly simple form

$$\frac{dz}{dx} = \frac{dz}{dy}\frac{dy}{dx}. \tag{3}$$

For example, if $y = x^2 + 1$ and $z = \sqrt{y}$, then $z = \sqrt{y} = \sqrt{x^2 + 1}$ and (3) says that

$$\frac{d}{dx}\sqrt{x^2 + 1} = \frac{d\sqrt{y}}{dy}\frac{d(x^2 + 1)}{dx} = \left(\frac{1}{2\sqrt{y}}\right) 2x = \frac{x}{\sqrt{y}} = \frac{x}{\sqrt{x^2 + 1}},$$

in agreement with (2). Despite its suggestive form, (3) is still a theorem of differential calculus requiring *proof*, and not a trivial algebraic identity. In other words, the fact that the dy's on the right in (3) can actually be canceled is a *consequence* of the following theorem, rather than a legitimate way of proving it.

> **Theorem 6** *(Chain rule).* Let f be differentiable at x, and let g be differentiable at $f(x)$. Then the composite function $g \circ f$ is differentiable at x, and its derivative is given by
>
> $$\frac{d}{dx}(g \circ f)(x) = g'(f(x))f'(x). \tag{1'}$$

Proof (Optional) Since g is differentiable at $y = f(x)$, we have

$$\lim_{\Delta y \to 0} \frac{g(y + \Delta y) - g(y)}{\Delta y} = g'(y),$$

or equivalently

$$\frac{g(y + \Delta y) - g(y)}{\Delta y} = g'(y) + \varepsilon(\Delta y), \tag{4}$$

where

$$\lim_{\Delta y \to 0} \varepsilon(\Delta y) = 0$$

(recall Corollary 2, page 81). Multiplying (4) by Δy, we find that

$$g(y + \Delta y) - g(y) = [g'(y) + \varepsilon(\Delta y)]\Delta y. \tag{4'}$$

It is important to realize that although (4) is written under the assumption that $\Delta y \neq 0$, the formula (4') obtained from (4) for $\Delta y \neq 0$ remains true even if $\Delta y = 0$, since the increment $g(y + \Delta y) - g(y)$ is indeed zero for $\Delta y = 0$, regardless of the choice of $\varepsilon(0)$. So far $\varepsilon(\Delta y)$ has been defined only for $\Delta y \neq 0$, but we now enlarge the domain of $\varepsilon(\Delta y)$ by setting $\varepsilon(0) = 0$, thereby making $\varepsilon(\Delta y)$ *continuous* at $\Delta y = 0$.

We now divide (4') by Δx, the increment of the independent variable, and take the limit as $\Delta x \to 0$. This gives

$$\lim_{\Delta x \to 0} \frac{g(y + \Delta y) - g(y)}{\Delta x} = \lim_{\Delta x \to 0} [g'(y) + \varepsilon(\Delta y)] \lim_{\Delta x \to 0} \frac{\Delta y}{\Delta x}, \tag{5}$$

where

$$\Delta y = f(x + \Delta x) - f(x) \tag{6}$$

is the increment of the dependent variable y. Moreover, since f is differentiable at x, and hence continuous at x (recall Theorem 2, page 120), $\Delta x \to 0$ implies $\Delta y \to 0$, so that

$$\lim_{\Delta x \to 0} \varepsilon(\Delta y) = \lim_{\Delta y \to 0} \varepsilon(\Delta y) = 0. \tag{7}$$

The fact that $\varepsilon(0) = 0$, and consequently that $\varepsilon(\Delta y)$ is continuous at $\Delta y = 0$, is crucial in drawing this conclusion, since we cannot guarantee that $\Delta y \neq 0$ (remember that Δy is not arbitrary, but is determined by the value of Δx). It follows from (5)–(7) that

$$\lim_{\Delta x \to 0} \frac{g(y + \Delta y) - g(y)}{\Delta x} = g'(y)f'(x) = g'(f(x))f'(x),$$

or equivalently

$$\lim_{\Delta x \to 0} \frac{g(f(x + \Delta x)) - g(f(x))}{\Delta x} = g'(f(x))f'(x), \tag{8}$$

since $y = f(x)$ and $y + \Delta y = f(x + \Delta x)$. The proof of the chain rule (1) or (1') is now complete, since the limit in (8) is just the derivative of the composite function $g(f(x)) = (g \circ f)(x)$. ∎

Informal Treatment of the Chain Rule

There is only one reason for the complexity of the proof of the chain rule: We must get around the fact that the increment $\Delta y = f(x + \Delta x) - f(x)$ may equal zero even if $\Delta x \neq 0$, giving rise to troublesome zero denominators. Were it not for this "$\Delta y = 0$ difficulty," the proof of the chain rule would be quite simple. In fact, we could then write

$$\Delta z = g(y + \Delta y) - g(y)$$

and argue as follows. The function f is differentiable at x, and hence continuous at x. Therefore $\Delta x \to 0$ implies $\Delta y \to 0$, so that

$$\frac{dz}{dx} = \lim_{\Delta x \to 0} \frac{\Delta z}{\Delta x} = \lim_{\Delta x \to 0} \frac{\Delta z}{\Delta y} \frac{\Delta y}{\Delta x} = \lim_{\Delta x \to 0} \frac{\Delta z}{\Delta y} \lim_{\Delta x \to 0} \frac{\Delta y}{\Delta x}$$

$$= \lim_{\Delta y \to 0} \frac{\Delta z}{\Delta y} \lim_{\Delta x \to 0} \frac{\Delta y}{\Delta x} = \frac{dz}{dy} \frac{dy}{dx}, \tag{9}$$

which is the chain rule in the form (3). This reasoning gives some insight into the proof of Theorem 6, but it does not come to grips with the "$\Delta y = 0$ difficulty."

Remark Although technically defective, the quick proof (9) of the chain rule is good enough for all practical purposes. The point is that a function bizarre enough to exhibit the "$\Delta y = 0$ difficulty" (an example is given in Problem 65) can hardly be expected to arise in the applications.

Example 1 Suppose a metal rod of length L expands by 2 millimeters for every degree Celsius that its temperature T is raised, so that

$$\frac{dL}{dT} = 2 \text{ mm/°C}.$$

The rod is placed in an oven and heated in such a way that its temperature increases steadily at the rate of 3° per minute, that is,

$$\frac{dT}{dt} = 3°/\text{min},$$

where t is the time. How fast is the rod's length increasing?

Solution Common sense tells us that since T increases by 3° every minute, and since L increases by 2mm for every degree by which T increases, then L must increase at the rate of $3 \cdot 2 = 6$ mm per minute. The chain rule says the same thing more concisely:

$$\frac{dL}{dt} = \frac{dL}{dT}\frac{dT}{dt} = 2 \cdot 3 = 6 \text{ mm/min}. \quad \square$$

Let c be any constant. Then, by the chain rule,

$$\frac{d}{dx}f(cx) = f'(cx)\frac{d}{dx}cx,$$

which at once yields the important differentiation formula

$$\frac{d}{dx}f(cx) = cf'(cx). \tag{10}$$

Be careful not to confuse (10) with the formula

$$\frac{d}{dx}cf(x) = cf'(x),$$

in which the function rather than its argument is multiplied by c.

Example 2 Differentiate $\sin 2x$.

Solution By formula (10), with $c = 2$, $f(x) = \sin x$ and $f'(x) = \cos x$,

$$\frac{d}{dx}\sin 2x = 2\cos 2x.$$

Equivalently, you can just use the chain rule directly, with $2x$ as the "inside" function. $\quad \square$

Example 3 Differentiate $(1 + 5x)^{10}$.

Solution Writing $y = 1 + 5x$ and $z = y^{10}$, we use the chain rule in the form (3). This gives

$$\frac{d}{dx}(1+5x)^{10} = \frac{dz}{dx} = \frac{dz}{dy}\frac{dy}{dx} = 10y^9 \cdot 5 = 50(1+5x)^9.$$

The idea of raising $1 + 5x$ to the tenth power and differentiating the resulting polynomial should never be entertained! □

Example 4 Differentiate $1/(x^2 + 2)^3$.

Solution This time we choose $y = x^2 + 2$ and $z = y^{-3}$. Then

$$\frac{d}{dx}\frac{1}{(x^2+2)^3} = \frac{dz}{dx} = \frac{dz}{dy}\frac{dy}{dx} = (-3y^{-4})(2x) = -\frac{6x}{(x^2+2)^4}.$$ □

Example 5 Differentiate $(x + \sqrt{x})^2$.

Solution By the chain rule,

$$\frac{d}{dx}(x+\sqrt{x})^2 = 2(x+\sqrt{x})\frac{d}{dx}(x+\sqrt{x}) = 2(x+\sqrt{x})\left(1+\frac{1}{2\sqrt{x}}\right)$$
$$= 2x + 3\sqrt{x} + 1,$$

where we evaluate the derivative of the "outside" function $(\cdots)^2$ at $x + \sqrt{x}$, and then multiply the result by the derivative of the "inside" function $x + \sqrt{x}$. Of course, "intermediate" variables (like y and z in the preceding two examples) could have been used, but once you get the hang of the chain rule, this becomes superfluous, or in any event can be done mentally. □

Example 6 Differentiate $\cos(2x^3 - 1)$.

Solution Applying the chain rule, we get

$$\frac{d}{dx}\cos(2x^3-1) = -\sin(2x^3-1)\frac{d}{dx}(2x^3-1) = -6x^2\sin(2x^3-1). \quad \Box$$

Example 7 Differentiate $\sin(\sin x)$.

Solution An application of the chain rule gives

$$\frac{d}{dx}\sin(\sin x) = \cos(\sin x)\frac{d}{dx}\sin x = \cos(\sin x)\cos x. \quad \Box$$

Let $h(g(f(x))) = (h \circ g \circ f)(x)$ be the composite of *three* functions f, g and h. Then assuming the necessary differentiability and applying the chain rule twice in succession, we have

$$\frac{d}{dx}h(g(f(x))) = h'(g(f(x)))\frac{d}{dx}g(f(x)) = h'(g(f(x)))g'(f(x))f'(x).$$

Thus, roughly speaking, we "peel off" the layers of parentheses one at a time, from the outside in, differentiating each function in its turn.

Example 8 Differentiate $(1 + \tan\sqrt{x})^2$.

Solution The function $(1 + \tan\sqrt{x})^2$ is of the form $h(g(f(x)))$ with $f(x) = \sqrt{x}$, $g(x) = 1 + \tan x$ and $h(x) = x^2$. Two consecutive applications of the chain rule give

$$\frac{d}{dx}(1 + \tan\sqrt{x})^2 = 2(1 + \tan\sqrt{x})\frac{d}{dx}(1 + \tan\sqrt{x})$$

$$= 2(1 + \tan\sqrt{x})\sec^2\sqrt{x}\,\frac{d}{dx}\sqrt{x}$$

$$= 2(1 + \tan\sqrt{x})\sec^2\sqrt{x}\,\frac{1}{2\sqrt{x}} = \frac{(1 + \tan\sqrt{x})\sec^2\sqrt{x}}{\sqrt{x}}.$$

∎

Example 9 Given a function $y = f(x)$, find the derivative of y^n, where n is any integer.

Solution By the chain rule,

$$\frac{d(y^n)}{dx} = \frac{d(y^n)}{dy}\frac{dy}{dx} = ny^{n-1}\frac{dy}{dx},$$

or equivalently

$$\frac{d}{dx}y^n = ny^{n-1}y',$$

where y' is the derivative of y with respect to x. To make sure that you understand just what this very concise formula means, we go to the other extreme and write it out in full:

$$\frac{d}{dx}[f(x)]^n = n[f(x)]^{n-1}f'(x). \quad \square$$

Differentiation of Rational Powers

It has already been shown that the formula

$$\frac{d}{dx}x^n = nx^{n-1} \tag{11}$$

is valid for every integer n. With the help of the chain rule, we now show that (11) continues to hold if n is an arbitrary *rational number*. We begin by calculating the derivative of $\sqrt[n]{x}$, where n is any positive integer, directly from its definition:

$$\frac{d}{dx}\sqrt[n]{x} = \lim_{u \to x}\frac{\sqrt[n]{u} - \sqrt[n]{x}}{u - x}.$$

This limit is easily evaluated by setting $y = \sqrt[n]{x}$ and $v = \sqrt[n]{u}$. Then $u \to x$ implies $v \to y$ by the continuity of $\sqrt[n]{x}$, and hence

$$\frac{d}{dx}\sqrt[n]{x} = \lim_{v \to y}\frac{v - y}{v^n - y^n} = \lim_{v \to y}\frac{1}{\frac{v^n - y^n}{v - y}} = \frac{1}{\lim_{v \to y}\frac{v^n - y^n}{v - y}}.$$

But the limit in the denominator is just the derivative of y^n with respect to

138 Chapter 2 Derivatives

y. Therefore

$$\frac{dy}{dx} = \frac{d}{dx}\sqrt[n]{x} = \frac{1}{\dfrac{d}{dy}y^n} = \frac{1}{ny^{n-1}} = \frac{y}{ny^n}, \tag{12}$$

or equivalently

$$\frac{d}{dx}\sqrt[n]{x} = \frac{\sqrt[n]{x}}{nx}. \tag{12'}$$

If n is even, formula (12') is valid only for $x > 0$, but if n is odd, it is valid for all $x \neq 0$ (why?).

Now let r be any rational number, so that $r = m/n$, where m and n are integers. It can be assumed that $n > 0$ and that m/n is in lowest terms. Writing

$$z = x^r = x^{m/n} = (\sqrt[n]{x})^m = y^m$$

and applying the chain rule, we obtain

$$\frac{d}{dx}x^r = \frac{dz}{dx} = \frac{dz}{dy}\frac{dy}{dx} = my^{m-1}\frac{y}{ny^n} = \frac{m}{n}\frac{y^m}{y^n},$$

with the help of (12). Therefore

$$\frac{d}{dx}x^r = \frac{m}{n}\frac{(\sqrt[n]{x})^m}{(\sqrt[n]{x})^n} = \frac{m}{n}\frac{x^{m/n}}{x} = \frac{m}{n}x^{(m/n)-1},$$

so that finally

$$\frac{d}{dx}x^r = rx^{r-1}. \tag{13}$$

This is the desired extension of formula (11) to the case of an arbitrary rational exponent. If $x > 0$, formula (13) is always valid, but for some values of r it also holds for $x < 0$ or $x \leq 0$ (explain further).

Example 10 Differentiate $x^{2/3} + x^{3/4}$.

Solution By formula (13),

$$\frac{d}{dx}(x^{2/3} + x^{3/4}) = \frac{2}{3}x^{(2/3)-1} + \frac{3}{4}x^{(3/4)-1} = \frac{2}{3}x^{-1/3} + \frac{3}{4}x^{-1/4}. \quad \square$$

Example 11 Differentiate $(1 - x^2)^{-1/5}$.

Solution By the chain rule and formula (13),

$$\frac{d}{dx}(1-x^2)^{-1/5} = -\frac{1}{5}(1-x^2)^{-6/5}\frac{d}{dx}(1-x^2) = \frac{2}{5}x(1-x^2)^{-6/5}. \quad \square$$

Problems

Differentiate

1. $(x^2 - 2x + 1)^2$
2. $(x^2 + 1)^3$
3. $(2x + 3)^4$
4. $(x^3 + 1)^5$
5. $(1 - 3x^2)^6$
6. $(x^2 + x - 1)^7$
7. $(x + 1)^3(x - 1)^4$
8. $(1 - x)(1 - x^2)^2$
9. $(x + 1)(x + 2)^2(x + 3)^3$
10. $\dfrac{x^2}{(x + 1)^2}$

11. $\dfrac{(1-x)^2}{(1+x)^3}$

12. $\left(\dfrac{1-2x}{1+2x}\right)^2$

13. $\dfrac{(x^2+1)^2}{(x-1)^2}$

14. $\dfrac{x}{(1-x)^2(1+x)^2}$

15. $\left(\dfrac{1+x}{1-x}\right)^{10}$

16. $x\sqrt{1+x^2}$

17. $\sqrt{\dfrac{1-x}{1+x}}$

18. $\sqrt{x+\sqrt{x}}$

19. $x^{3/2} - x^{4/3} + 1$

20. $9x^{10/9} - 10x^{9/10}$

21. $x^{0.99} + x^{-0.99}$

22. $(s^2 + s + 2)^{7/11}$

23. $\sqrt[3]{t} + \sqrt[4]{t} + \sqrt[5]{t}$

24. $\dfrac{1}{u} + \dfrac{1}{\sqrt{u}} + \dfrac{1}{\sqrt[3]{u}}$

25. $\sqrt[3]{v^2} - \dfrac{3}{\sqrt{v}}$

26. $\sqrt[3]{1+\sqrt{x}}$

27. $\sqrt{1+\sqrt[3]{x}}$

28. $\sqrt[3]{1+\sqrt[3]{x}}$

29. $\sin^2 x + \tan^2 x$

30. $\sin^3 x + \cos^3 x$

31. $(1+\sin 2x)^3$

32. $\tfrac{1}{3}\tan^3 x + \cot 3x$

33. $3\cos 2x + 2\sec 3x$

34. $(2-\cos x)^{-2}$

35. $x \sin(1/x)$

36. $x^2 \cos(1/x)$

37. $\sqrt{1+\sin u}$

38. $\sqrt{\dfrac{1+\cos v}{1-\sin v}}$

39. $\cos w^3$

40. $(1+\tan x)^{2/3}$

41. $\cos(\cos x)$

42. $\sin(\tan x)$

43. $\sin(\sin(\sin x))$

44. $\sin(\cos x^2)$

Find the differential of

45. $(x^4 - 2)^3$

46. $\sqrt{1-x^2}$

47. $x^{3/5} + x^{5/3}$

48. $\sin(1/x^2)$

49. $\cos\sqrt{x}$

50. $\sin(\cos x)$

51. $\tan u^2$

52. $\tan(\cot v)$

53. $(\sec w)^{2/3}$

54. Find $f'(0)$ if $f(x) = (2x+3)^{100}$.
55. Find $f'(1)$ if $f(x) = (1+x^{-2})^{50}$.
56. Deduce each of the double-angle formulas $\sin 2x = 2\sin x \cos x$, $\cos 2x = \cos^2 x - \sin^2 x$ from the other by differentiation.
57. Use the chain rule to prove the formula $D_x(1/g) = -g'/g^2$ for the derivative of the reciprocal of the function g (Theorem 4, page 129).
58. Show that differentiation changes parity, which means that the derivative of an even function is odd, while the derivative of an odd function is even.
59. Show that differentiation preserves periodicity, which means that the derivative of a periodic function is periodic with the same period.

Find the tangent to the curve

60. $y = \sqrt{x^2+1}$ at $(1, \sqrt{2})$
61. $y = \sqrt[3]{x+1}$ at $(7, 2)$
62. $y = \cos(\sin x)$ at $((n+\tfrac{1}{2})\pi, \cos 1)$, where n is any integer

Find the normal to the curve

63. $y = \sqrt{x^3+1}$ at $(0, 1)$
64. $y = x^{3/2}$ at $(4, 8)$

65. Let

$$f(x) = \begin{cases} x^2 \sin \dfrac{1}{x} & \text{if } x \neq 0, \\ 0 & \text{if } x = 0. \end{cases}$$

Verify that the derivative $f'(0)$ exists. Show that f exhibits the "$\Delta y = 0$ difficulty" at $x = 0$. In other words, show that every neighborhood of $x = 0$ contains points $x \neq 0$ for which the increment $\Delta y = f(x) - f(0)$ is zero.

66. Show that the expression for the differential of a function is the same whether or not its argument is the independent variable. More exactly, show that if $z = g(y)$ where $y = f(x)$, then $dz = g'(y)\,dy$ even though y is a *dependent* variable.

2.6 Higher Derivatives

A function $f(x)$ is said to be *differentiable on an interval I* if it has a derivative $f'(x)$ at every point of I. Let $f(x)$ be differentiable on an open interval I, and suppose the derivative $f'(x)$ is itself differentiable on I, with derivative

$$D_x f'(x).$$

Then the function $D_x f'(x)$ is called the *second derivative* of $f(x)$, and is denoted by $f''(x)$, read as "f double prime of x," or by $f^{(2)}(x)$. Similarly, if the second derivative $f''(x)$ is in turn differentiable on I, with derivative

$$D_x f''(x),$$

we call $D_x f''(x)$ the *third derivative* of $f(x)$, denoted by $f'''(x)$ or by $f^{(3)}(x)$. After n consecutive differentiations of the original function $f(x)$, we get the *derivative of order n* of $f(x)$, or simply the *nth derivative* of $f(x)$, denoted by $f^{(n)}(x)$, and defined recursively by

$$f^{(n)}(x) = D_x f^{(n-1)}(x) \qquad (n = 1, 2, \ldots), \tag{1}$$

where it is assumed that the derivative $f^{(n-1)}(x)$ of order $n - 1$ exists and is differentiable on I. To make the formula (1) work for $n = 1$, we set

$$f^{(0)}(x) = f(x),$$

by definition, that is, the "zeroth derivative" of a function is the function itself. It is not customary to use more than three primes in writing higher derivatives.

Example 1 The function

$$f(x) = \frac{1}{x}$$

is differentiable on any open interval I that does not contain the point $x = 0$. The derivative

$$f'(x) = -\frac{1}{x^2}$$

is also differentiable on I, and hence $f(x)$ has a second derivative

$$f''(x) = D_x f'(x) = D_x\left(-\frac{1}{x^2}\right) = \frac{2}{x^3}$$

on I. Since $f''(x)$ is in turn differentiable on I, $f(x)$ has a third derivative

$$f'''(x) = D_x f''(x) = D_x\left(\frac{2}{x^3}\right) = -\frac{6}{x^4}$$

on I, and so on (see Problem 10). ☐

In terms of the d notation, $f^{(n)}(x)$ is written as

$$f^{(n)}(x) = \frac{d^n f(x)}{dx^n} = \frac{d^n}{dx^n} f(x).$$

In the numerator the exponent n is attached to the symbol d, but in the denominator it is attached to the independent variable x instead. The expression d^n/dx^n should be thought of as a single entity calling for n-fold differentiation (that is, differentiation n times) of any function written after it. Similarly, $d^n f(x)/dx^n$ should be regarded as just another way of writing the nth derivative $f^{(n)}(x)$, without attempting to ascribe separate meaning to the different symbols making up the expression. Higher derivatives of the dependent variable are defined in the natural way. Thus, if $y = f(x)$, we have

$$y' = \frac{dy}{dx} = f'(x), \quad y'' = \frac{d^2 y}{dx^2} = f''(x), \ldots, \quad y^{(n)} = \frac{d^n y}{dx^n} = f^{(n)}(x).$$

The symbol D_x^n is sometimes used as an alternative to d^n/dx^n.

Section 2.6 Higher Derivatives   **141**

Example 2 If $y = x^4$, then

$$\frac{dy}{dx} = 4x^3, \quad \frac{d^2y}{dx^2} = \frac{d}{dx} 4x^3 = 12x^2, \quad \frac{d^3y}{dx^3} = \frac{d}{dx} 12x^2 = 24x,$$

$$\frac{d^4y}{dx^4} = \frac{d}{dx} 24x = 24, \quad \frac{d^5y}{dx^5} = \frac{d}{dx} 24 = 0,$$

and all the derivatives of order $n > 5$ are also zero. Notice that the fourth derivative of x^4 equals $4 \cdot 3 \cdot 2 \cdot 1 = 24$. More generally, let $y = x^n$. Then

$$\frac{dy}{dx} = nx^{n-1}, \quad \frac{d^2y}{dx^2} = n(n-1)x^{n-2}, \ldots,$$

$$\frac{d^k y}{dx^k} = n(n-1)\cdots(n-k+1)x^{n-k}, \ldots, \quad \frac{d^n y}{dx^n} = n(n-1)\cdots 2 \cdot 1 = n!,$$

where $n!$ denotes n *factorial*, the product of the first n positive integers ($1! = 1$, $2! = 2 \cdot 1 = 2$, $3! = 3 \cdot 2 \cdot 1 = 6$, $4! = 4 \cdot 3 \cdot 2 \cdot 1 = 24, \ldots$). Thus each of the first n differentiations lowers the degree of x^n by 1, until the constant $n!$ is obtained, and all derivatives of order higher than n are identically equal to zero.

Example 3 It follows from the preceding example that the nth derivative of the polynomial

$$P(x) = a_0 + a_1 x + a_2 x^2 + \cdots + a_n x^n \qquad (a_n \neq 0)$$

of degree n equals

$$P^{(n)}(x) = n! a_n,$$

and $P^{(n+1)}(x) \equiv P^{(n+2)}(x) \equiv \cdots \equiv 0$.

Example 4 Find the second and third derivatives of $f(x)g(x)$.

Solution By the product rule,

$$(fg)' = f'g + fg',$$

where we omit arguments for simplicity. Therefore, by two more applications of the product rule,

$$(fg)'' = \frac{d}{dx}(f'g + fg') = \frac{d}{dx}f'g + \frac{d}{dx}fg'$$

$$= (f''g + f'g') + (f'g' + fg'') = f''g + 2f'g' + fg''. \qquad (2)$$

In the same way,

$$(fg)''' = \frac{d}{dx}(f''g + 2f'g' + fg'') = \frac{d}{dx}f''g + 2\frac{d}{dx}f'g' + \frac{d}{dx}fg''$$

$$= (f'''g + f''g') + 2(f''g' + f'g'') + (f'g'' + fg''')$$

$$= f'''g + 3f''g' + 3f'g'' + fg'''. \qquad □ \qquad (3)$$

Remark The general formula for the nth derivative of the product $f(x)g(x)$ is given in Problem 35, page 229. It explains why the coefficients 1, 2, 1 in formula (2) and 1, 3, 3, 1 in formula (3) are the same as in the familiar *algebraic* formulas $(a + b)^2 = a^2 + 2ab + b^2$ and $(a + b)^3 = a^3 + 3a^2 b + 3ab^2 + b^3$.

Newton's Second Law of Motion

Consider a particle of mass m moving along a straight line, and let $s = s(t)$ be the particle's position at time t, as measured from a suitable origin. Here we use one and the same symbol s to denote both the function and the dependent variable, a common practice which prevents proliferation of superfluous notation. Suppose the particle is acted on by a force F which in general is variable. Then *Newton's second law of motion* states that

$$F = ma, \tag{4}$$

where $a = a(t)$ is the particle's acceleration. As on page 107, a is defined as the derivative with respect to time of the particle's velocity $v = v(t)$:

$$a = \frac{dv}{dt}.$$

But v is in turn defined as the derivative with respect to time of the particle's position $s = s(t)$:

$$v = \frac{ds}{dt}.$$

Therefore

$$a = \frac{dv}{dt} = \frac{d^2s}{dt^2}, \tag{5}$$

that is, the acceleration is the second derivative with respect to time of the particle's position. Substituting (5) into (4), we get

$$m\frac{d^2s}{dt^2} = F. \tag{6}$$

Differential Equations

An equation like (6), involving at least one derivative of a function and possibly the function itself, is called a *differential equation*. By the *order* of a differential equation we mean the order of the highest derivative appearing in the equation. Thus Newton's second law $F = ma$ is actually a second order differential equation with far-reaching physical consequences. Of course, the precise form of the equation depends on the nature of the force F. For example, $F = 0$ if the particle is "free," meaning that it is acted on by no forces, $F = -ks$ ($k > 0$) if the particle is acted on by an attractive force proportional to its distance from the origin, and so on. This important topic will be pursued in Section 4.7.

Example 5 Verify that the function $y = \sin x$ satisfies the second order differential equation

$$y'' + y = 0. \tag{7}$$

Solution The validity of equation (7) for $y = \sin x$ follows at once from the fact that $y' = \cos x$ and $y'' = -\sin x$. As an exercise, show that (7) is also satisfied by $y = \cos x$ and by any function of the form $y = a \sin x + b \cos x$, where a and b are arbitrary constants. ☐

Example 6 Verify that the function $y = (x - 1)/(x + 1)$ satisfies the second order differential equation $2y'^2 = (y - 1)y''$.

Solution Dividing the numerator $x - 1$ by the denominator $x + 1$, we get

$$y = 1 - \frac{2}{x + 1}.$$

We then differentiate y twice with the help of the chain rule, obtaining

$$y' = \frac{(-2)(-1)}{(x + 1)^2} \frac{d}{dx}(x + 1) = \frac{2}{(x + 1)^2},$$

$$y'' = \frac{2(-2)}{(x + 1)^3} \frac{d}{dx}(x + 1) = -\frac{4}{(x + 1)^3}$$

(it is assumed that $x \neq -1$). Therefore

$$(y - 1)y'' = \left[-\frac{2}{x + 1}\right]\left[-\frac{4}{(x + 1)^3}\right] = \frac{8}{(x + 1)^4}$$

$$= 2\left[\frac{2}{(x + 1)^2}\right]^2 = 2y'^2,$$

as required. □

Problems

Find the second and third derivatives y'' and y''' if
1. $y = 5x^{10} + 10x^5 + 1$
2. $y = (1 + x^2)^3$
3. $y = x/(1 + x)$
4. $y = (1 + x^{1/2})^2$
5. $y = x \sin x$
6. $y = x^2 \cos x$
7. $y = \sin x^2$
8. $y = \tan x$
9. $y = \sec x$

10. Show that

$$\frac{d^n}{dx^n} \frac{1}{x} = (-1)^n \frac{n!}{x^{n+1}} \quad (x \neq 0)$$

for all $n = 1, 2, \ldots$

Let $y = x(2x - 1)^2(x + 3)^3$. With as little effort as possible, find
11. $y^{(6)}$
12. $y^{(7)}$

Show that
13. $\dfrac{d^8}{dx^8} \dfrac{x^2}{x - 1} = \dfrac{8!}{(x - 1)^9}$
14. $\dfrac{d^6}{dx^6} \sin^2 x = 32 \cos 2x$

15. $\dfrac{d^n}{dx^n} \dfrac{1}{x(1 - x)} = n!\left[\dfrac{(-1)^n}{x^{n+1}} + \dfrac{1}{(1 - x)^{n+1}}\right] \quad (n = 1, 2, \ldots)$

16. Let $y = \sin x$. Show that
$$y^{(4n)} = \sin x, \qquad y^{(4n+1)} = \cos x,$$
$$y^{(4n+2)} = -\sin x, \qquad y^{(4n+3)} = -\cos x$$
for all $n = 0, 1, 2, \ldots$

17. For what choice of the constants a, b and c does the function

$$f(x) = \begin{cases} x^3 & \text{if } x \leq 1, \\ ax^2 + bx + c & \text{if } x > 1 \end{cases}$$

have a second derivative at $x = 1$?

18. Let

$$f(x) = \begin{cases} 0 & \text{if } x \leq 0, \\ x^4 & \text{if } x > 0. \end{cases}$$

Find $f'(x)$, $f''(x)$ and $f'''(x)$. Show that $f^{(4)}(0)$ does not exist.

19. Show that the function

$$f(x) = \begin{cases} x^2 \sin \dfrac{1}{x} & \text{if } x \neq 0, \\ 0 & \text{if } x = 0 \end{cases}$$

is differentiable on $(-\infty, \infty)$, but has no second derivative at $x = 0$.

20. Verify that the function $y = \sqrt{1 - x^2}$ satisfies the differential equation $yy' + x = 0$.

21. Verify that the function $y = \sqrt{2x - x^2}$ satisfies the differential equation $y^3 y'' + 1 = 0$.

22. Let $y = (x + \sqrt{x^2 + 1})^n$, where n is an arbitrary integer. Verify that y satisfies the differential equation

$$(x^2 + 1)y'' + xy' - n^2 y = 0.$$

2.7 Implicit Differentiation

Of the great variety of equations involving two variables x and y, some are easily solved for y in terms of the variable x. For example, if

$$x^2 + y^2 = 1, \tag{1}$$

we can find two solutions expressing y as a continuous function of x, namely

$$y = \sqrt{1 - x^2} \tag{2}$$

and

$$y = -\sqrt{1 - x^2}. \tag{2'}$$

However, in the case of the equation

$$x^2 - xy + y^3 = 1,$$

it is not so easy to find an explicit formula expressing y in terms of x (we would have to solve a cubic equation in y), and in the case of the equation

$$xy + \sin y + x^2 = 1,$$

it is impossible to do so. Nevertheless, for each of the last two equations there is some function $y = f(x)$ which satisfies the given equation, at least for certain values of x (this can be shown with the help of the intermediate value theorem), although we cannot or prefer not to write an explicit formula for the function.

Implicit Functions

A function $y = f(x)$ defined in this way, by an equation in the two variables x and y, is called an *implicit function*. Conditions for the existence of implicit functions are given at the end of Section 13.5. Here we will simply take it for granted that an equation in x and y can very often be used to *define* y as a function of x, even when the equation cannot be solved for y as an explicit expression in x. Surprisingly, it is always possible to calculate the derivative $y' = dy/dx$ implicitly, that is, without ever solving for y in terms of x (however, see the comments following Example 1). For if we differentiate the given equation with respect to x, simply thinking of y as a function of x, it always turns out that the resulting equation can be readily solved for y'. This way of finding y' is called *implicit differentiation*. Bear in mind that differentiating a function of y with respect to x will always lead to a factor of y', because of the chain rule. Thus

$$\frac{d}{dx} y^n = \frac{d(y^n)}{dy} \frac{dy}{dx} = n y^{n-1} y', \tag{3}$$

as in Example 9, page 138,

$$\frac{d}{dx} \sin y = \frac{d \sin y}{dy} \frac{dy}{dx} = \cos y \frac{dy}{dx} = y' \cos y,$$

and so on.

Example 1 Show that implicit differentiation of equation (1) leads to the same expression for the derivative y' as ordinary (explicit) differentiation of formula (2) or (2′).

Solution Differentiating both sides of (1), we get

$$\frac{d}{dx}(x^2 + y^2) = \frac{d}{dx}1,$$

or

$$2x + 2yy' = 0,$$

with the help of (3). This equation is easily solved for y' and gives

$$y' = -\frac{x}{y}. \tag{4}$$

We can stop here, or we can substitute (2) and (2′) into (4), obtaining

$$y' = -\frac{x}{\sqrt{1-x^2}}, \tag{5}$$

and

$$y' = \frac{x}{\sqrt{1-x^2}}. \tag{5′}$$

On the other hand, differentiating (2) and (2′) explicitly with the help of the chain rule, we get

$$\frac{d}{dx}\sqrt{1-x^2} = \frac{1}{2\sqrt{1-x^2}}\frac{d}{dx}(1-x^2) = -\frac{x}{\sqrt{1-x^2}},$$

which agrees with (5), and

$$\frac{d}{dx}(-\sqrt{1-x^2}) = -\frac{1}{2\sqrt{1-x^2}}\frac{d}{dx}(1-x^2) = \frac{x}{\sqrt{1-x^2}},$$

which agrees with (5′). ☐

The method of implicit differentiation cannot be used blindly, since it gives a formal answer for y' even in cases where y does not exist as a function of x. For example, regardless of the value of x, there is no value of y satisfying the equation $x^2 + y^2 = -1$, and yet implicit differentiation of this equation gives the same answer (4) as in the case of the equation $x^2 + y^2 = 1$. Notice also that when we solve for y', after implicit differentiation of some equation in x and y, the answer will in general be an expression involving both variables x and y.

Example 2 Evaluate the derivative (4) at the point $x = \frac{1}{2}$.

Solution Substituting $x = \frac{1}{2}$ into (1), we get the equation

$$y^2 = 1 - \frac{1}{4} = \frac{3}{4},$$

which has two solutions $y = \pm\frac{1}{2}\sqrt{3}$. Using (4), we find that the corresponding values of y' are

$$y'\big|_{x=\frac{1}{2},\, y=\frac{1}{2}\sqrt{3}} = -\frac{\frac{1}{2}}{\frac{1}{2}\sqrt{3}} = -\frac{1}{\sqrt{3}}$$

146   Chapter 2   Derivatives

and
$$y'\big|_{x=\frac{1}{2},\, y=-\frac{1}{2}\sqrt{3}} = -\frac{\frac{1}{2}}{-\frac{1}{2}\sqrt{3}} = \frac{1}{\sqrt{3}},$$

Here $y'\big|_{x=x_0,\, y=y_0}$ stands for the value of y' corresponding to $x = x_0$, $y = y_0$ (we will often find this kind of single vertical bar notation useful). Why have we obtained *two* different values of y' here? Simply because the graph of equation (1), namely the circle of radius 1 with its center at the origin, has *two* different points with the same x-coordinate $\frac{1}{2}$, and the slopes of the tangents to the circle at these points are different, as shown in Figure 19. In other words, one value of the slope y' goes with the upper semicircle (2), whose graph is shown in Figure 20(a), while the other value of y' goes with the lower semicircle (2'), whose graph is shown in Figure 20(b). Naturally, we get the same two values of y' if we set $x = \frac{1}{2}$ in (5) and (5'). □

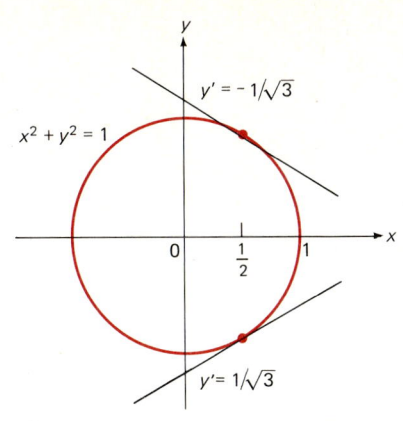

Figure 19

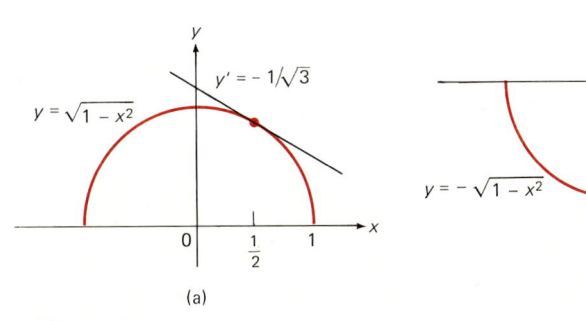

Figure 20

Example 3 Given that
$$x^2 - xy + y^3 = 1, \tag{6}$$
evaluate the derivative y' at $x = 1$.

Solution By implicit differentiation, we have
$$\frac{d}{dx}(x^2 - xy + y^3) = \frac{d}{dx} 1,$$
or
$$2x - y - xy' + 3y^2 y' = 0,$$
with the help of (3). Notice that $(xy)' = y + xy'$, by the product rule. Solving for y', we find that
$$y' = \frac{2x - y}{x - 3y^2}. \tag{7}$$

Substituting $x = 1$ into (6), we get the equation
$$1 - y + y^3 = 1,$$
or
$$y^3 = y,$$

Section 2.7 Implicit Differentiation

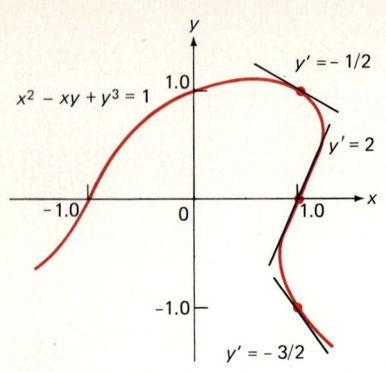

which has three solutions $y = 0$ and $y = \pm 1$. The corresponding values of y' obtained from (7) are

$$y'\big|_{x=1, y=0} = \frac{2(1) - 0}{1 - 3(0)^2} = 2,$$

$$y'\big|_{x=1, y=1} = \frac{2(1) - 1}{1 - 3(1)^2} = -\frac{1}{2},$$

$$y'\big|_{x=1, y=-1} = \frac{2(1) + 1}{1 - 3(-1)^2} = -\frac{3}{2}.$$

Figure 21 shows the geometric meaning of these three values of y'. ☐

Example 4 Use implicit differentiation to show that

$$\frac{d}{dx} x^r = rx^{r-1}, \qquad (8)$$

where r is any rational number.

Solution We have already proved (8) by another method in Section 2.6. Let $y = x^{m/n}$. Then

$$y^n = (x^{m/n})^n = x^m,$$

and differentiating with respect to x, we get

$$ny^{n-1} y' = mx^{m-1},$$

with the help of (3). It follows that

$$y' = \frac{m}{n} \frac{x^{m-1}}{y^{n-1}} = \frac{m}{n} \frac{x^{m-1}}{(x^{m/n})^{n-1}} = \frac{m}{n} \frac{x^{m-1}}{x^{m-(m/n)}} = \frac{m}{n} x^{(m/n)-1},$$

which is equivalent to (8). The method of implicit differentiation shows to great advantage here. ☐

Implicit differentiation often simplifies the task of finding higher derivatives, as illustrated by the next example.

Example 5 Given that

$$x^2 + y^2 = 25, \qquad (9)$$

find the values of y'' and y''' at the point $(3, 4)$.

Solution Differentiating (9) implicitly with respect to x, we obtain

$$2x + 2yy' = 0, \qquad (10)$$

and therefore

$$y' = -\frac{x}{y},$$

just as in Example 1, so that in particular

$$y'\big|_{x=3, y=4} = -\frac{3}{4}. \qquad (11)$$

148 Chapter 2 Derivatives

Dividing (10) by 2 and differentiating again, we get

$$\frac{d}{dx}(x + yy') = 1 + y'^2 + yy'' = 0, \tag{12}$$

or

$$y'' = -\frac{1 + y'^2}{y}.$$

Thus, with the help of (11),

$$y''|_{x=3, y=4} = -\frac{1 + \frac{9}{16}}{4} = -\frac{25}{64}. \tag{13}$$

Another implicit differentiation, this time of equation (12), gives

$$\frac{d}{dx}(1 + y'^2 + yy'') = 2y'y'' + y'y'' + yy''' = 3y'y'' + yy''' = 0,$$

or

$$y''' = -\frac{3y'y''}{y}.$$

Using (11) and (13) to evaluate y''' at the point (3, 4), we find that

$$y'''|_{x=3, y=4} = -\frac{3(-\frac{3}{4})(-\frac{25}{64})}{4} = -\frac{225}{1024}.$$

Try calculating these higher derivatives by explicit differentiation of $y = \sqrt{25 - x^2}$, and see how much extra work it is. □

Problems

Use implicit differentiation to find y' if x and y satisfy the given equation.

1. $\dfrac{x^2}{4} + \dfrac{y^2}{9} = 1$
2. $\dfrac{x^2}{9} - \dfrac{y^2}{4} = -1$
3. $\dfrac{1}{x^2} + \dfrac{1}{y^2} = 1$
4. $x^3 + y^3 = 3xy$
5. $(xy)^3 = 3(x + y)$
6. $x^3 + x^2y + xy^2 + y^3 = 1$
7. $x^{2/3} + y^{-2/3} = 8$
8. $y^2 = \dfrac{1}{x + y}$
9. $\dfrac{1}{x + y} - \dfrac{1}{x - y} = 2$
10. $\dfrac{1 + xy}{x + y} = 10$
11. $\sin x + \cos y = 0$
12. $y = \sin(x + y)$
13. $\cos xy = x$
14. $y \tan y = x^2$

15. In the equation $x^2 + y^2 = r^2$ let x be regarded as a function of y. Use implicit differentiation to calculate dx/dy (not dy/dx). Then show that the tangent to the circle $x^2 + y^2 = r^2$ is *vertical* at the points $(\pm r, 0)$.

16. Use calculus to show that the tangent at any point P of the circle $x^2 + y^2 = r^2$ is perpendicular to the radius joining the origin to P.
17. Graph the equation $\sqrt{x} + \sqrt{y} = 4$. Find y' by implicit differentiation. Solve for y as a function of x, and then find y' by ordinary differentiation. Evaluate $y'|_{x=4}$ by both methods.

Find the tangent to the graph of the equation
18. $y^2 + xy - 5 = 0$ at $(4, -5)$
19. $y^3 - y^2 - 4x + x^2 = 0$ at $(2, 2)$
20. $y^4 - 2x^2y^3 - 27 = 0$ at $(-1, 3)$

Let $x^2 - xy + y^2 = 1$. Use implicit differentiation to evaluate the first three derivatives y', y'', y''' at
21. $x = 0$
22. $x = 1$
23. $x = -1$

Find the points of the graph of the equation

$$x^2 - xy + y^2 = 1$$

at which the tangent is a line with the given inclination.
24. 0°
25. 90°
26. 45°
27. 135°

Section 2.7 Implicit Differentiation

28. Changing x to $-x$ in equation (6) yields the equation $x^2 + xy + y^3 = 1$, whose graph is the reflection in the y-axis of the graph shown in Figure 21. Use implicit differentiation to evaluate the first four derivatives y', y'', y''', $y^{(4)}$ at $x = 1$.

29. Blind application of implicit differentiation to the equation $x^4 + y^4 = x^2 y^2$ leads to the formula

$$y' = \frac{2x^3 - xy^2}{x^2 y - 2y^3}.$$

Why is this result meaningless?

2.8 Related Rates

There is an important class of problems in which we are given the rate of change of one quantity, usually with respect to time, and are asked to find the rate of change of another related quantity. The method of solving such problems involving *related rates* closely resembles the technique of implicit differentiation, but there are now *two* dependent variables.

Example 1 A large spherical balloon is losing air at the rate of one tenth of a cubic foot per second (more concisely, 0.1 ft³/sec). How fast is the radius of the balloon decreasing when its diameter is 6 ft?

Solution Let R be the radius and V the volume of the balloon. Then

$$V = \frac{4}{3}\pi R^3. \tag{1}$$

Since the size of the balloon is changing, both V and R are functions of the time t. We could express this fact by writing $V = V(t)$, $R = R(t)$, but it is better to just remember that V and R depend on time. Differentiating both sides of (1) with respect to t, we get

$$\frac{dV}{dt} = \frac{4}{3}\pi \left(3R^2 \frac{dR}{dt}\right) = 4\pi R^2 \frac{dR}{dt}.$$

We then solve this equation for dR/dt, obtaining

$$\frac{dR}{dt} = \frac{1}{4\pi R^2} \frac{dV}{dt}. \tag{2}$$

This is the key step, since we have now expressed dR/dt, the unknown rate of change of the radius of the balloon, in terms of dV/dt, the *given* rate of change of its volume. In fact, we are told that the balloon's volume is decreasing (it is *losing* air) at the rate of 0.1 ft³/sec, which means that the derivative dV/dt has the constant value of -0.1 ft³/sec. Substituting this value of dV/dt into (2), we find that

$$\frac{dR}{dt} = -\frac{0.1}{4\pi R^2}. \tag{3}$$

When the balloon's diameter is 6 ft, its radius is 3 ft. At this instant, (3) gives

$$\frac{dR}{dt} = -\frac{0.1}{4\pi(3)^2} = -\frac{1}{360\pi} \approx -0.00088 \text{ ft/sec}.$$

Chapter 2 Derivatives

Since this number is so small, we convert it to inches per minute, obtaining

$$\frac{dR}{dt} = -\frac{1}{360\pi} \frac{\text{ft}}{\text{sec}} \cdot \frac{12 \text{ in}}{\text{ft}} \cdot \frac{60 \text{ sec}}{\text{min}} = -\frac{12(60)}{360\pi} \frac{\text{in}}{\text{min}},$$

where we cancel units of measurement in the way taught in elementary physics. Therefore

$$\frac{dR}{dt} = -\frac{2}{\pi} \approx -0.64 \text{ in/min},$$

that is, the radius R is decreasing at the rate of about 0.64 in/min, a rather slow leak for such a large balloon. Note that dR/dt is itself a function of the radius. In fact, the smaller the balloon, the larger $|dR/dt|$, as shown by (3). □

Example 2 In the preceding example, how fast is the balloon's surface area S decreasing when its radius is 4 ft?

Solution A sphere of radius R has surface area

$$S = 4\pi R^2.$$

Therefore

$$\frac{dS}{dt} = 4\pi \left(2R \frac{dR}{dt}\right) = 8\pi R \frac{dR}{dt}, \qquad (4)$$

after differentiating with respect to time. This expresses the rate of change of the balloon's surface area in terms of the rate of change of its radius, but it is not quite what we want. However, the desired relation between dS/dt and the given quantity $dV/dt = -0.1 \text{ ft}^3/\text{sec}$ is easily found by substituting (2) into (4):

$$\frac{dS}{dt} = \frac{8\pi R}{4\pi R^2} \frac{dV}{dt} = \frac{2}{R} \frac{dV}{dt} = -\frac{0.2}{R}.$$

When the balloon's radius is 4 ft, this gives

$$\frac{dS}{dt} = -\frac{0.2}{4} = -0.05 \text{ ft}^2/\text{sec},$$

or equivalently

$$\frac{dS}{dt} = -0.05(12)^2 = -7.2 \text{ in}^2/\text{sec},$$

that is, the balloon's surface area is decreasing at the rate of 7.2 square inches per second. □

Example 3 A ladder 20 ft long is leaning against a wall. Suppose the top of the ladder slides down the wall at the constant rate of 1.5 ft/sec. How fast is the bottom of the ladder moving away from the wall when the top of the ladder is 16 ft from the ground?

Solution Idealizing the ladder as a straight line segment, we introduce rectangular coordinates, as shown in Figure 22, where x is the distance between the wall and the bottom of the ladder, and y is the height above ground of

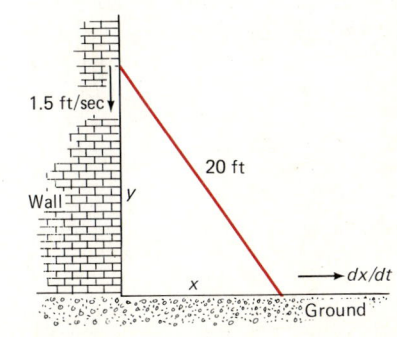

Figure 22

Section 2.8 Related Rates

the top of the ladder. By the Pythagorean theorem,

$$x^2 + y^2 = 20^2 = 400. \tag{5}$$

Since the position of the ladder is changing, both x and y are functions of the time t, a fact that could be emphasized by writing $x = x(t)$, $y = y(t)$. To find dx/dt, we differentiate both sides of (5) with respect to t, obtaining the equation

$$x \frac{dx}{dt} + y \frac{dy}{dt} = 0,$$

which we then solve for dx/dt. The result is

$$\frac{dx}{dt} = -\frac{y}{x} \frac{dy}{dt}. \tag{6}$$

According to the statement of the problem, the top of the ladder is sliding down the wall at a constant rate of 1.5 ft/sec. Therefore the coordinate y is decreasing at the rate of 1.5 ft/sec, which means that the derivative dy/dt equals -1.5 ft/sec. Substituting this value of dy/dt into (6), we get

$$\frac{dx}{dt} = \frac{1.5 y}{x}. \tag{7}$$

Solving (5) for x in terms of y, we find that $x = \sqrt{400 - y^2}$. Therefore, when $y = 16$, that is, when the top of the ladder is 16 ft from the ground, $x = \sqrt{400 - 16^2} = \sqrt{400 - 256} = \sqrt{144} = 12$, so that the bottom of the ladder is 12 ft from the wall. Substituting these values of x and y into (7), we find that

$$\frac{dx}{dt} = \frac{(1.5)(16)}{12} = 2 \text{ ft/sec}.$$

Thus at the exact moment when the top of the ladder is 16 ft from the ground, the bottom of the ladder is moving away from the wall at the rate of 2 ft/sec. □

Example 4 A man 6 ft tall walks at a speed of 4 ft/sec toward a street lamp which is 14 ft above the ground. How fast is the length of the man's shadow decreasing?

Solution Let x be the man's distance from the base of the lamppost and s the length of his shadow, as shown in Figure 23 (the angle θ will play a role in the next example). The triangles ABB' and $A'B'C$ have the same angles, and hence are similar. Hence the ratios of corresponding sides are equal. In particular,

$$\frac{|AB|}{|AB'|} = \frac{|A'B'|}{|A'C|},$$

that is,

$$\frac{s}{6} = \frac{x}{8},$$

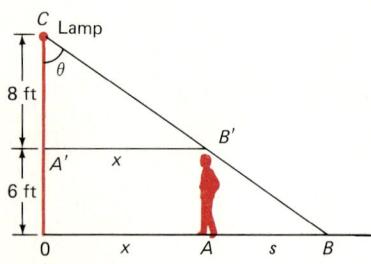

Figure 23

152 Chapter 2 Derivatives

or equivalently

$$s = \frac{3}{4}x. \tag{8}$$

Equation (8) expresses the length of the man's shadow in terms of his distance from the post, which is just what we want.

We now differentiate both sides of (8) with respect to the time t, obtaining

$$\frac{ds}{dt} = \frac{3}{4}\frac{dx}{dt}. \tag{9}$$

According to the statement of the problem, the man is walking toward the post at a speed of 4 ft/sec, so that $dx/dt = -4$ ft/sec. Substituting this value into (9), we get

$$\frac{ds}{dt} = \frac{3}{4}(-4) = -3 \text{ ft/sec.}$$

Thus the length of the man's shadow is decreasing at the constant rate of 3 ft/sec, regardless of his distance from the post. □

Example 5 In the preceding problem, let θ be the angle subtended by the man's shadow at his head; this is also the angle between the lamppost and a light ray from the lamp to his head. How fast is θ changing when the man is 12 ft from the post?

Solution Taking another look at Figure 23, we see that

$$x = 8 \tan \theta.$$

Differentiating both sides of this equation with respect to the time t, we obtain

$$\frac{dx}{dt} = 8\frac{d\tan\theta}{d\theta}\frac{d\theta}{dt} = 8\sec^2\theta\frac{d\theta}{dt}. \tag{10}$$

Solving (10) for $d\theta/dt$, the quantity we are trying to express in terms of the data of the problem, we get

$$\frac{d\theta}{dt} = \frac{\cos^2\theta}{8}\frac{dx}{dt}.$$

Moreover, it is clear from the figure that

$$\cos\theta = \frac{|A'C|}{|B'C|} = \frac{8}{\sqrt{8^2 + x^2}}.$$

Combining the last two equations, we obtain

$$\frac{d\theta}{dt} = \frac{8}{64 + x^2}\frac{dx}{dt}, \tag{11}$$

so that we have managed to express $d\theta/dt$ entirely in terms of the given data, namely $dx/dt = -4$ ft/sec and $x = 12$ ft. Substituting these values into (11), we find that

$$\frac{d\theta}{dt} = -\frac{8(4)}{64 + 12^2} = -\frac{2}{13} \approx -0.154 \text{ rad/sec.}$$

The answer comes out in radians per second, because in using the formula $D_\theta \tan \theta = \sec^2 \theta$ to differentiate $\tan \theta$ we have tacitly assumed that θ is measured in radians. Of course, once we have *found* the answer, we can convert it to degree measure. In fact,

$$\frac{d\theta}{dt} = -\frac{2}{13} \frac{180}{\pi} \approx -8.8°/\text{sec},$$

that is, the angle subtended by the man's shadow at his head is decreasing at about 8.8 degrees per second. It is apparent from formula (11) that in this problem, unlike the preceding one, the answer does depend on the man's distance from the lamppost. □

How to Solve Related Rates Problems

The following step-by-step procedure will help you solve related rates problems.

1. Label the related variables, choosing letters that call to mind the actual meaning of the variables. Initial letters are a good choice, like V for volume, L for length, and T for temperature (t is reserved for time).

2. Write an equation involving the related variables. This step is very often facilitated by drawing a suitable figure, and you should become adept at drawing "good enough" figures, that is, figures that show the essential features of a problem without being too fancy.

3. Differentiate both sides of the equation with respect to the *independent* variable. This variable, which is typically (but not always) the time t, is not one of the related variables, and therefore you can expect the chain rule to be needed.

4. Solve the differentiated equation for the required rate of change, and then (but not before!) complete the solution by substituting any numerical data supplied in the statement of the problem, including the known rate of change and given or calculated values of the variables.

Problems

1. A point $P = (x, y)$ in the first quadrant moves away from the origin along the curve $y = x^3/48$ in such a way that dx/dt is constant. Which coordinate, x or y, is increasing faster? (The answer depends on the size of x.)

2. The radius of the circular ripple produced by tossing a stone into a pond grows at the rate of 3 m/sec (meters per second). How fast is the area enclosed by the ripple increasing when the ripple is 10 m in diameter?

3. The length of a rectangle decreases at 3 cm/sec (centimeters per second), while its width increases at 2 cm/sec. At a certain moment the rectangle is 50 cm long and 20 cm wide. Is the area of the rectangle increasing or decreasing, and how fast?

4. Air is being pumped into a large spherical balloon at the rate of 10 ft³/min. How fast is the balloon's radius increasing when its diameter is 4 ft? How fast is its surface area increasing at the same moment?

5. A balloon's volume is increasing at 15 ft³/sec at the same moment as its surface area is increasing at 5 ft²/sec. What is its radius? How fast is the radius increasing at this moment?

6. The volume of an evaporating mothball decreases at a rate proportional to its surface area. Show that this implies that the radius of the mothball decreases at a constant rate.

7. Two ships A and B sail away from a point P along perpendicular routes. Ship A is going at a speed of 15 knots and ship B at a speed of 20 knots. Suppose that at a certain time, A is 5 nautical miles from P, and B is 10 nautical miles from P. How fast are the ships moving apart 1 hour later? (One knot = one nautical mile per hour.)

8. A solvent is being poured at the rate of 8 cm³/sec into a cylindrical vessel of radius 2 cm. How fast is the level of the solvent rising?

9. One end of a rope is attached to a boat 3 ft below a dock. The other end, attached to a windlass at the dock's edge, is being reeled in at the rate of 1 ft/sec. How fast is the boat approaching the dock when it is 4 ft away? How fast is the angle between the rope and the surface of the water changing at this moment?

10. A car doing a steady 60 mph along a straight road passes directly under a balloon rising vertically upward at 10 mph. Suppose the balloon is 1 mile up when the car passes under it. How fast is the distance between the car and the balloon increasing 1 minute later? How fast is the angle between the road and the line joining the car and the balloon changing at this moment?

11. Water is being pumped at the rate of 720 in³/min into a storage tank shaped like an inverted right circular cone of height 6 ft and radius 2 ft at the top (see Figure 24, where the tank is filled to depth h and the liquid surface is of radius r). How fast is the water level rising when the tank is one eighth full? One half full? How long does it take to fill the tank? Is calculus needed to answer the last question? (The volume of a right circular cone of height h and base radius r is $V = \frac{1}{3}\pi r^2 h$.)

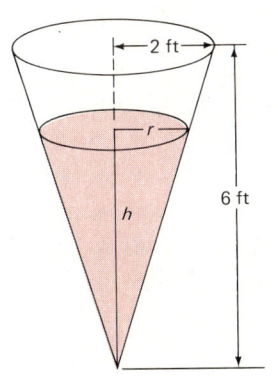

Figure 24

12. A swimming pool 12 ft wide, 24 ft long, 2 ft deep at the shallow end, and 8 ft deep at the deep end, is being filled at the rate of 16 ft³/min (see Figure 25, where the pool is filled to depth h). How fast is the water level rising when the water is 1 ft deep at the shallow end? When the water is 2 ft deep at the deep end? One half hour after starting to fill the pool?

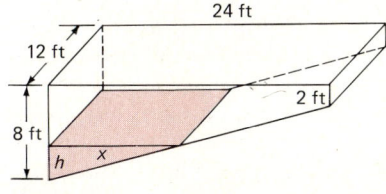

Figure 25

13. A racing car is doing a steady 150 km/hr (kilometers per hour) around a circular track. Suppose there is a light source at the center O of the track and a fence tangent to the track at a point P (see Figure 26, where the track is of radius r, and the car, at the point Q, is moving in the counterclockwise direction). How fast is the car's shadow (the point S in the figure) moving along the fence when the car has gone one eighth of a lap from P?

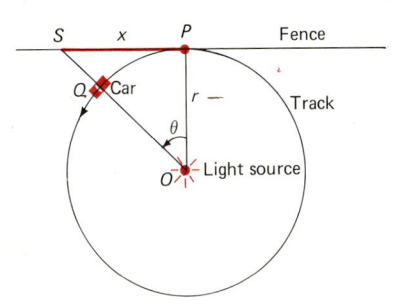

Figure 26

14. A cameraman is filming a race. Suppose his perpendicular distance from the track at the finish line is 40 ft, and suppose the camera must be turned at the rate of 18 degrees per second to stay trained on the lead runner at the moment the runner is 30 ft from the finish line. How fast is the runner going at that moment? (Assume that the track is straight.)

15. A baseball player runs from first to second base in 3.6 seconds flat (see Figure 27, where the player is at distance x from first base). How fast is the angle between the first base line and the line from home plate to the player changing when he is halfway between first and second base? (Assume that the player's speed is constant.)

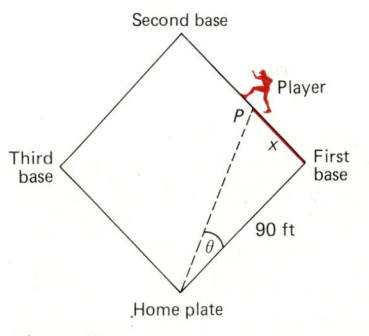

Figure 27

16. The top of a ladder leaning against a wall slides down the wall at the rate of 3 in/sec. When the bottom of the ladder is 6 ft from the wall, it is moving away from the wall at the rate of 4 in/sec. How long is the ladder?

Section 2.8 Related Rates   155

In the ladder problem of Example 3,

17. How fast is the bottom of the ladder moving away from the wall when it is 10 ft from the wall?

18. How fast is the angle between the ladder and the ground changing when the bottom of the ladder is 8 ft from the wall?

19. What is the acceleration of the bottom of the ladder at the moment when the top of the ladder is 12 ft from the ground? When the top of the ladder strikes the ground?

20. The same substance is made into both a rod of length L and a cube of volume V. The *coefficient of linear expansion* of the substance is α, which means that when the rod is heated, its length changes in accordance with the formula $\alpha = (1/L)(dL/dT)$, where T is the temperature. The *coefficient of volume expansion* of the substance is β, which means that when the cube is heated, its volume changes in accordance with the formula $\beta = (1/V)(dV/dT)$. Show that $\beta = 3\alpha$.

21. A straight line parallel to the y-axis moves from the position $x = -1$ to the position $x = 1$ at constant velocity v, intersecting the unit circle $x^2 + y^2 = 1$ and dividing it into a left-hand segment of area A and a right-hand segment of area $\pi - A$ (see Figure 28). How fast is A increasing when the line is at the position $x = \tfrac{1}{2}$?

Hint. Express A in terms of the angle θ shown in the figure.

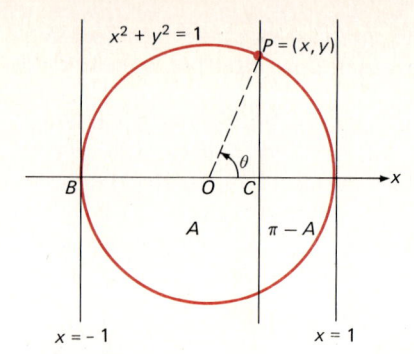

Figure 28

Key Terms and Topics
Average and instantaneous velocity, acceleration
Average and exact density
Definition of the derivative
Differentiation and differentiability
The difference quotient and increments
The tangent and normal lines to a curve
One-sided derivatives and tangents
Continuity of a differentiable function
The tangent line approximation and differentials
The Leibniz notation for derivatives
The product and quotient rules
The chain rule
Higher derivatives
Implicit differentiation
Related rates

Basic Differentiation Rules and Formulas

Function	Derivative	
f	$f'(x) = \lim\limits_{\Delta x \to 0} \dfrac{f(x + \Delta x) - f(x)}{\Delta x}$	(Definition)
f	$f'(x) = \lim\limits_{u \to x} \dfrac{f(u) - f(x)}{u - x}$	(Equivalent definition)
c (constant c)	0	
cf	cf'	
x	1	
x^2	$2x$	

156   Chapter 2   Derivatives

Function	Derivative
x^3	$3x^2$
x^n (integer n)	nx^{n-1}
$\sqrt{x}$	$\dfrac{1}{2\sqrt{x}}$
x^r (rational r)	rx^{r-1} (This rule contains the above five.)
$f+g$	$f'+g'$
$f-g$	$f'-g'$
fg	$f'g+fg'$ (Product rule)
$\dfrac{f}{g}$	$\dfrac{f'g-fg'}{g^2}$ (Quotient rule)
$z=z(y)$, where $y=y(x)$	$\dfrac{dz}{dx}=\dfrac{dz}{dy}\dfrac{dy}{dx}$ (Chain rule)
$\sin x$	$\cos x$
$\cos x$	$-\sin x$
$\tan x$	$\sec^2 x$
$\cot x$	$-\csc^2 x$
$\sec x$	$\sec x \tan x$
$\csc x$	$-\csc x \cot x$

Supplementary Problems

1. A person standing on the edge of a cliff draws a pistol and fires a bullet vertically downward with muzzle velocity $v_0 = 480$ mph. The bullet hits the ground exactly one half second later. How high is the cliff?

2. A car is going v_0 mph when its brakes are suddenly applied. Suppose its subsequent motion is described by the function $s = v_0 t - \tfrac{1}{2}kt^2$ ($k > 0$). Interpret the constant k, and find its value if $v_0 = 60$ mph and the car brakes to a complete stop in 5.5 sec. How far does the car go before stopping? Show that the distance traveled by the car after braking is proportional to the square of its speed v_0.

Let a, b, c and d be arbitrary constants. Differentiate

3. $ax + \dfrac{b}{x}$ **4.** $(x-a)(x-b)$

5. $x(x-a)(x-b)$ **6.** $(x-a)(x-b)(x-c)$

7. $\dfrac{x-a}{x+a}$ **8.** $\dfrac{x-a}{x+b}$

9. $\dfrac{x+a}{x-b}$ **10.** $\dfrac{ax+b}{cx+d}$

11. $\dfrac{x^2-a}{x^2-b}$ **12.** $\dfrac{x^2+ax+b}{x^2+cx+d}$

13. $\sqrt{x^2+a^2}$ **14.** $\dfrac{x}{\sqrt{a^2-x^2}}$

15. $\sqrt{\dfrac{a+x}{a-x}}$ **16.** $\sin^2(ax+b)$

17. $\cos(ax^2+bx+c)$

Two "calculus-free" (and hence inadequate) definitions of the tangent line to a circle at a point P were given on page 116.

18. Find a curve other than a circle for which definition (ii) continues to work although definition (i) fails.

19. Find a curve other than a circle for which definition (i) continues to work.

Find an equation of the tangent to the curve

20. $y = 1/x$ at $(2, \tfrac{1}{2})$ **21.** $y = 1/x$ at $(-1, -1)$

22. $y = 1/x^2$ at $(-\tfrac{1}{2}, 4)$ **23.** $y = \tan x$ at $(0, 0)$

24. $y = \csc x$ at $(\pi/4, \sqrt{2})$ **25.** $y = 8/(x^2+4)$ at $(2, 1)$

Find an equation of the normal to the curve

26. $y = 1/x$ at $(-\tfrac{1}{2}, -2)$ **27.** $y = 1/x^2$ at $(2, \tfrac{1}{4})$

28. $y = \cot x$ at $(\pi/2, 0)$ **29.** $y = \sec x$ at $(-\pi/4, \sqrt{2})$

30. The curve $y = 1/x$ has a tangent going through the point $(0, 1)$. Find it.
31. Find the angle between the curves $y = 1/x^2$ and $y = 1/x$ at the point $(1, 1)$.
32. Show that the segment of any tangent to the curve $y = 1/x$ cut off by the coordinate axes is bisected by the point of tangency.
33. Graph the curve $y = |x - 1| - |x| + |x + 1|$. Find the one-sided tangents at the corners of the curve.
34. Is the function $|x|^2$ differentiable at $x = 0$?
35. Is the derivative of a rational function always a rational function?

Find the differential $df(a) = f'(a)\Delta x$ of the given function for the indicated values of a and Δx. In each case compare $df(a)$ with the increment $\Delta f(a) = f(a + \Delta x) - f(a)$ to a suitable number of decimal places.

36. $f(x) = 1/x$, $a = 5$, $\Delta x = -0.1$
37. $f(x) = (1 + x)/(1 - x)$, $a = 0$, $\Delta x = 0.1$
38. $f(x) = 1/\sqrt{x}$, $a = 9$, $\Delta x = 0.5$
39. $f(x) = \sec x$, $a = \pi/3$, $\Delta x = \pi/60$

40. Use the tangent line approximation to show that
$$(1 + x)^n \approx 1 + nx \quad (n = \pm 1, \pm 2, \ldots),$$
$$\sqrt[n]{1 + x} \approx 1 + \frac{x}{n} \quad (n = 1, 2, \ldots)$$
for small $|x|$.

Use the tangent line approximation to estimate the given quantity, and compare the result with the exact answer to a suitable number of decimal places.

41. $(2.9)^{-2}$
42. $\sqrt[3]{26}$
43. $(8.6)^{2/3}$
44. $\csc 32°$
45. $\tan 63°$
46. $1/\sqrt{4.1}$
47. $\sec 1°$
48. $(82)^{-1/4}$
49. $\sqrt[4]{17}$
50. $\cot 43°$
51. $(7.8)^{-1/3}$
52. $(0.9)^{0.9}$

53. Estimate $f(x) = (x - 4)(x - 3)^2(x - 2)^3$ for $x = 4.001$.
54. Use the product rule to give another proof of the formula $D_x x^n = nx^{n-1}$ ($n = 2, 3, \ldots$).

Find a function $y = f(x)$ such that
55. $y' = (x + 1)^3$
56. $y' = x(x^2 + 1)^3$
57. $y' = x^2(x^3 + 1)^2$
58. $y' = x^3(x^4 + 1)^3$
Hint. Remember the chain rule.

Find a function $y = f(x)$ such that
59. $y'' = x$
60. $y'' = x^2$
61. $y'' = x^3$
62. $y''' = 1$
63. $y''' = x$
64. $y''' = x^2$
Hint. Bear in mind that each differentiation of a power of x lowers the exponent by 1.

Differentiate

65. $\dfrac{1}{\sqrt{1 + x^2}(x + \sqrt{1 + x^2})}$
66. $\tan(\sec x^2)$
67. $\sqrt{\sin(\tan\sqrt{x})}$
68. $\sqrt{x + \sqrt{x + \sqrt{x}}}$
69. $\sin(\cos(\sin x^2))$
70. $\sin(\cos(\tan(\cot x)))$

71. Let $p(x) = f(x)g(x)$ and suppose $f'(x)g'(x) = c$, where c is a constant and the third derivatives f''' and g''' exist. Show that
$$\frac{p'''(x)}{p(x)} = \frac{f'''(x)}{f(x)} + \frac{g'''(x)}{g(x)}.$$

72. Find the tangent and the normal to the graph of the equation $y^6 + y^5 - xy + 2 = 0$ at the point $(4, 1)$.

A man 6 ft tall walks at a speed of 2 ft/sec toward a building, keeping his eyes on a window 4 ft high and 10 ft above the ground. Let θ be the angle subtended at the man's eyes by the window (see Figure 29, neglecting the size of the man's head).

73. How fast is θ changing when the man's distance from the building is 16 ft? 8 ft? 4 ft?
74. At what distance from the building does the angle θ, which is initially increasing, begin to decrease?

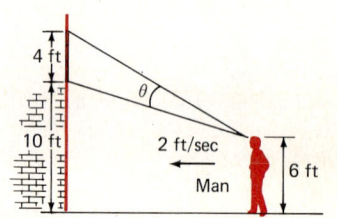

Figure 29

75. Show that the curve $y = \sin x$ has infinitely many distinct tangents going through the origin. Show that the points of tangency have the same x-coordinates as the points in which the line $y = x$ intersects the graph of $y = \tan x$.

3 Further Applications of Differentiation

In the first three sections of this chapter we present some of the more powerful techniques of differential calculus, which will enable you to make a detailed analysis of the behavior of functions and sketch their graphs with confidence. In the next three sections we discuss limits involving "approach to infinity" and introduce an effective method, *L'Hospital's rule*, for evaluating limits with the help of differentiation. In Section 3.7 calculus is used to solve a great variety of practical problems of *optimization*, and in Section 3.8, which is optional, we demonstrate the power of calculus as applied to problems of business and economics.

3.1 The Mean Value Theorem

Let f be a function continuous on a closed interval $I = [a, b]$, and consider the difference $f(b) - f(a)$ between the values of f at the endpoints of I. If the derivative $f'(a)$ exists, we can use $f'(a)$ to estimate $f(b) - f(a)$ by writing

$$f(b) - f(a) \approx f'(a)(b - a), \tag{1}$$

where the approximation is good if $b - a$ is small. In fact, this is just the tangent line approximation (2), page 123, with Δx replaced by $b - a$. Actually, as we will show in a moment, the approximation (1) can be replaced by the *exact* formula

$$f(b) - f(a) = f'(c)(b - a), \tag{1'}$$

where the derivative f' is evaluated at a suitable point c between a and b, rather than at the endpoint a. (Here we assume that f is differentiable at every point between a and b, and the choice of c will depend on the particular function f.) This result, known as the *mean value theorem*, has many applications in calculus.

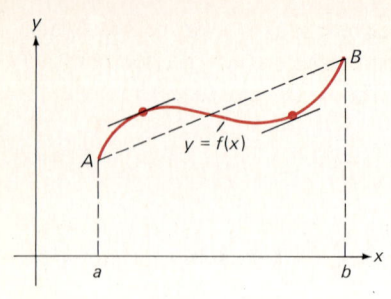

Geometric meaning of the mean value theorem

Figure 1

To interpret the mean value theorem geometrically, we write equation (1′) in the form

$$f'(c) = \frac{f(b) - f(a)}{b - a}, \qquad (2)$$

and observe that the right side of (2) is the slope of the chord joining the endpoints $A = (a, f(a))$ and $B = (b, f(b))$ of the curve

$$y = f(x) \qquad (a \le x \le b). \qquad (3)$$

Thus the mean value theorem says that the tangent to the curve (3) at some point of the curve other than its endpoints is parallel to the chord AB. This is illustrated in Figure 1, where the curve has two tangents parallel to AB.

Rolle's Theorem

If $f(a) = f(b)$, the endpoints of the curve (3) have the same y-coordinate, and the chord AB joining the endpoints is *horizontal*. In this case the mean value theorem reduces to *Rolle's theorem*,† which states that the tangent to the curve at some point of the curve other than its endpoints is horizontal, or equivalently that the derivative f' equals 0 at some point c between a and b. This somewhat simpler situation is illustrated in Figure 2, where the curve has a horizontal tangent at four points. The fact that the y-coordinates of two of these points are the maximum M and the minimum m of the function f on the interval $[a, b]$ is no coincidence (see the proof of Theorem 1).

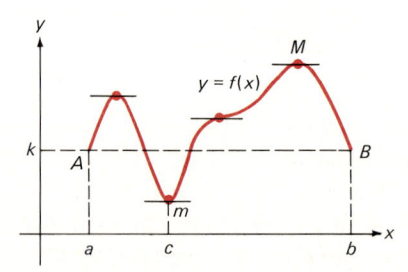

Geometric meaning of Rolle's theorem

Figure 2

First we prove Rolle's theorem, and then use it to prove the mean value theorem.

> **Theorem 1** (*Rolle's theorem*). *Let the function f be continuous on the closed interval $I = [a, b]$ and differentiable, with derivative f', on the open interval (a, b), that is, at every interior point of I. Suppose further that $f(a) = f(b) = k$. Then there is a point c in (a, b) such that $f'(c) = 0$.*

Proof By the extreme value theorem (see page 99), f has both a maximum M and a minimum m on I. Clearly $m \le k \le M$, since k is a value taken by f on I (at the points a and b). If $m = k = M$, then f reduces to the constant function $f(x) \equiv k$, whose derivative equals 0 at *every* point of (a, b), and the

---

† Named after the French academician Michel Rolle (1652–1719), an early opponent of calculus, who was eventually persuaded of its validity by one of his colleagues.

theorem is proved. Otherwise we have either $m < k$ or $k < M$ (or both), but in any event f takes at least one of its extreme values at an interior point of I, that is, at some point c in (a, b). By hypothesis, the derivative

$$f'(c) = \lim_{x \to c} \frac{f(x) - f(c)}{x - c}$$

exists at c. Also, this limit has the same value $f'(c)$ both as $x \to c^+$ and as $x \to c^-$ (recall Theorem 12, page 93). Suppose, to be explicit, that $f(c) = m$, so that $f(c) \leq f(x)$, or equivalently $f(x) - f(c) \geq 0$, for all x in I. Then the difference quotient

$$\frac{f(x) - f(c)}{x - c} \qquad (4)$$

is nonnegative for $x > c$ ($x - c > 0$) and nonpositive for $x < c$ ($x - c < 0$). Hence the limit of the quotient (4) as $x \to c^+$ cannot be *negative*, by rule (iii), page 76 (which applies equally well to one-sided limits), and similarly the limit of (4) as $x \to c^-$ cannot be *positive*. It follows that $f'(c)$ can be neither positive nor negative. The only remaining possibility is that $f'(c) = 0$. The case where $f(c) = M$ is treated in virtually the same way. ∎

Remark Clearly, the conclusion that $f'(c)$ can be neither positive nor negative, implying that $f'(c) = 0$, does not depend on the validity of the inequality $f(c) \leq f(x)$ for *all* x in the interval I, but only on its validity for all x in some neighborhood of the point c.

Example 1 The function

$$f(x) = \sqrt{x(4 - x)},$$

graphed in Figure 3, is defined only on the closed interval $[0, 4]$, since the expression under the radical sign becomes negative outside this interval. Since f is not defined in a full neighborhood of the endpoint $x = 0$ or $x = 4$, it is not differentiable at either point. In fact, even the one-sided derivatives $f'_+(0)$ and $f'_-(4)$ fail to exist (why?). However, f is continuous from the right at $x = 0$ and from the left at $x = 4$, and has the derivative

$$f'(x) = \frac{1}{2\sqrt{x(4-x)}} \frac{d}{dx}[x(4-x)] = \frac{2-x}{\sqrt{x(4-x)}} \qquad (5)$$

at every interior point of $[0, 4]$, so that f is continuous on the closed interval $[0, 4]$ and differentiable on the open interval $(0, 4)$. Moreover, f takes the same value 0 at the endpoints $x = 0$ and $x = 4$. Therefore, by Rolle's theorem, the derivative f' must equal 0 at a point between $x = 0$ and $x = 4$. It does, at $x = 2$, as is apparent both from (5) and the graph of f, which is actually a semicircle of radius 2 with its center at $(2, 0)$.

Example 2 The function

$$f(x) = |x - 2| \qquad (1 \leq x \leq 3)$$

graphed in Figure 4, is continuous on $[1, 3]$, and takes the same value 1 at both $x = 1$ and $x = 3$. However, there is no point c in $(1, 3)$ such that $f'(c) = 0$. This does not contradict Rolle's theorem, since one of the conditions of the theorem is not satisfied. In fact, f is not differentiable at $x = 2$, since the graph of f has a corner at the point $(2, 0)$.

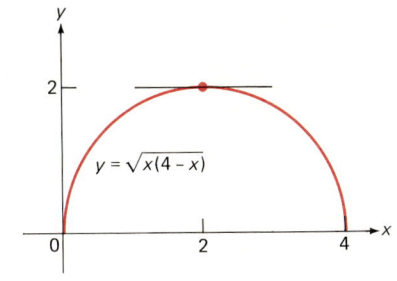

Figure 3

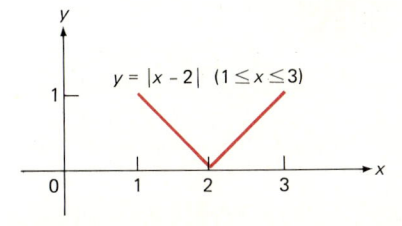

Figure 4

Section 3.1 The Mean Value Theorem   **161**

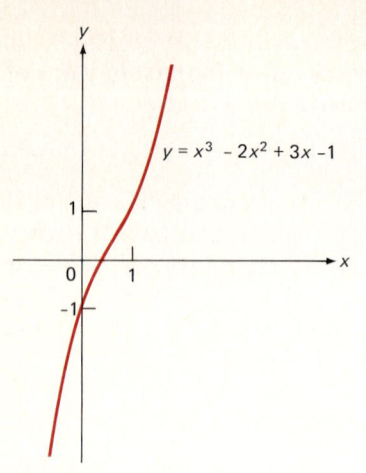

Figure 5

Example 3 Show that the cubic equation

$$x^3 - 2x^2 + 3x - 1 = 0 \qquad (6)$$

has a real root, but no more than one.

Solution Writing $f(x) = x^3 - 2x^2 + 3x - 1$, we have $f(0) = -1$, $f(1) = 1$. Therefore, by the intermediate value theorem (see page 96), f takes the value 0 at a point r_1 between 0 and 1 (actually $r_1 \approx 0.43$). This is one root of equation (6). If there were another root r_2, then $f(r_1) = f(r_2) = 0$, so that the derivative f' would take the value 0 at some point between r_1 and r_2, because of Rolle's theorem. But this is impossible, since

$$f'(x) = 3x^2 - 4x + 3 = 3\left(x - \frac{2}{3}\right)^2 + \frac{5}{3} \geq \frac{5}{3}$$

for all x, so that f' never equals 0. Therefore r_1 is the only root of equation (6). Graphically, this means that the curve $y = f(x)$, shown in Figure 5, has only one x-intercept. In fact, f is increasing on $(-\infty, \infty)$, as can be proved by using a test (Theorem 7, page 169) established in the next section. □

Proof of the Mean Value Theorem

Now that Rolle's theorem is available, the mean value theorem can be proved quite easily.

> **Theorem 2** *(Mean value theorem).* Let the function f be continuous on the closed interval $[a, b]$ and differentiable, with derivative f', on the open interval (a, b). Then there is a point c in (a, b) such that
>
> $$f(b) - f(a) = f'(c)(b - a). \qquad (7)$$

Proof Introducing a new function

$$g(x) = f(x) - kx,$$

where k is a constant, we choose k so as to make g take the same value at both endpoints of the interval $[a, b]$. In other words, we require that k satisfy the equation

$$g(a) = f(a) - ka = f(b) - kb = g(b).$$

Solving for k, we get

$$k = \frac{f(b) - f(a)}{b - a}.$$

With this choice of k, the function g satisfies all the conditions of Rolle's theorem, and hence there is a point c in (a, b) such that

$$g'(c) = f'(c) - k = 0.$$

It follows that

$$f'(c) = k = \frac{f(b) - f(a)}{b - a},$$

which is equivalent to (7). ∎

162   Chapter 3   Further Applications of Differentiation

Suppose $b < a$ instead of $a < b$, as tacitly assumed in Theorem 2 and formula (7), and let f be continuous on $[b, a]$ and differentiable on (b, a). Then we have

$$f(a) - f(b) = f'(c)(a - b) \qquad (7')$$

instead of (7), where c is now a point of (b, a). Multiplying both sides of (7') by -1, we get back (7). Thus the mean value theorem can always be used in the form (7) if we say that c *lies between a and b*, for this way of expressing the condition on c works in both cases $a < b$ and $b < a$.

To illustrate the geometric meaning of the function $g(x)$ introduced in the proof of the mean value theorem, suppose the curve

$$y = f(x) \qquad (a \leq x \leq b)$$

lies above the graph of

$$y = kx = \frac{f(b) - f(a)}{b - a} x,$$

which is the line through the origin parallel to the chord AB joining the endpoints of the curve. Then, if kx is positive, every point $(x, g(x))$ of the graph of g lies a distance kx below the corresponding point $(x, f(x))$ of the graph of f. For example, the top curve in Figure 6 is the graph of the same function f as in Figure 1, and the bottom curve is the graph of the function g corresponding to this function f. Notice that whenever the tangent to the graph of f is parallel to AB, the tangent to the graph of g at the point with the same x-coordinate is horizontal.

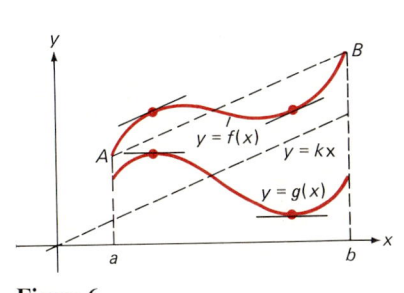

Figure 6

Example 4 Find a point c satisfying the mean value theorem for the function $f(x) = x^2$.

Solution Here $f'(x) = 2x$, and formula (7) becomes

$$b^2 - a^2 = 2c(b - a).$$

Solving for c, we get

$$c = \frac{b^2 - a^2}{2(b - a)} = \frac{b + a}{2}.$$

Thus in this special case c is the midpoint of the interval with endpoints a and b, regardless of the choice of a and b. ☐

Example 5 Find a point c satisfying the mean value theorem for the function $f(x) = 1/x$ if $a = 4$ and $b = 9$, and also if $a = -1$ and $b = -4$.

Solution This time $f'(x) = -1/x^2$, and formula (7) becomes

$$\frac{1}{b} - \frac{1}{a} = -\frac{b - a}{c^2}.$$

Solving for c, we get $c^2 = ab$, or equivalently $c = \pm\sqrt{ab}$, where to make c lie between a and b, we choose the plus sign if a and b are both positive and the minus sign if a and b are both negative. Thus $c = \sqrt{4(9)} = 6$ if $a = 4$ and $b = 9$ but $c = -\sqrt{(-1)(-4)} = -2$ if $a = -1$ and $b = -4$. Notice that the mean value theorem is not applicable here if a and b have opposite signs, since then the point $x = 0$, at which $f(x) = 1/x$ is not defined, lies between a and b. □

We already know that the derivative of a constant function is identically equal to zero, that is, if $f(x) \equiv$ constant, then $f'(x) \equiv 0$. As an example of the application of the mean value theorem, we now prove the converse, namely that $f'(x) \equiv 0$ implies $f(x) \equiv$ constant if the domain of f is an interval.

> **Theorem 3** *(Constancy of a function with a zero derivative).* Let the function f be differentiable on an interval I, with derivative f', and suppose f' equals zero at every point of I. Then f is constant on I, that is, f takes the same value at every point of I.

Proof We fix some point a of I, and let x be any other point of I. Then $[a, x]$ or $[x, a]$, depending on whether $x > a$ or $x < a$, is a subinterval of I, and f is differentiable (and hence continuous) on this subinterval. Thus, by the mean value theorem,

$$f(x) - f(a) = f'(c)(x - a), \tag{8}$$

where c lies between a and x. But $f'(c) = 0$, since c belongs to I, and therefore $f(x) - f(a) = 0$, or equivalently $f(x) = f(a)$. Since x is an arbitrary point of I, it follows that $f(x) = f(a)$ for every x in I, that is, f is constant on I. ∎

Example 6 The function

$$f(x) = \frac{|x|}{x} = \begin{cases} 1 & \text{if } x > 0, \\ -1 & \text{if } x < 0 \end{cases}$$

is not a constant, but its derivative equals zero wherever f is defined. This does not contradict Theorem 3, since the domain of f is not an interval, but rather the pair of nonoverlapping intervals $(0, \infty)$ and $(-\infty, 0)$. In this case we are justified in using formula (8) only when the points a and x belong to the *same* interval $(0, \infty)$ or $(-\infty, 0)$. We then correctly conclude that f is constant on each interval separately, although the value 1 taken by f on $(0, \infty)$ differs from the value -1 taken by f on $(-\infty, 0)$.

Cauchy's Mean Value Theorem (Optional)

Finally we go a step further and prove a useful mean value theorem involving *two* functions, which is due to the great French mathematician Augustin Cauchy (1789–1857). For applications of Cauchy's mean value theorem, see Sections 3.6 and 9.8.

> **Theorem 4** *(Cauchy's mean value theorem).* Let the functions f and g be continuous on the closed interval $[a, b]$ and differentiable, with derivatives f' and g', on the open interval (a, b). Suppose further that g' is nonzero at every point of (a, b). Then there is a point c in (a, b) such that
>
> $$\frac{f(b) - f(a)}{g(b) - g(a)} = \frac{f'(c)}{g'(c)}. \tag{9}$$

Proof If $g(a) = g(b)$, then by Rolle's theorem, $g'(c) = 0$ for some c in (a, b), contrary to hypothesis. Therefore $g(a) \neq g(b)$, so that the left side of

(9) is defined. The rest of the proof parallels that of the ordinary mean value theorem, to which Theorem 4 reduces if $g(x) = x$. Introducing a new function
$$h(x) = f(x) - kg(x),$$
where k is a constant, we choose k so as to make h take the same value at both endpoints of the interval $[a, b]$. In other words, we require that k satisfy the equation
$$h(a) = f(a) - kg(a) = f(b) - kg(b) = h(b).$$
Solving for k, we get
$$k = \frac{f(b) - f(a)}{g(b) - g(a)}.$$
With this choice of k, the function h satisfies all the conditions of Rolle's theorem, and hence there is a point c in (a, b) such that
$$h'(c) = f'(c) - kg'(c) = 0.$$
It follows that
$$\frac{f'(c)}{g'(c)} = k = \frac{f(b) - f(a)}{g(b) - g(a)},$$
which is equivalent to (9). ∎

Example 7 Find a point c satisfying Cauchy's mean value theorem for the functions $f(x) = \cos x$ and $g(x) = \sin x$ if $a = 0$ and $b = \pi/2$.

Solution Here $f'(x) = -\sin x$ and $g'(x) = \cos x$, so that in particular g' is nonzero at every point of $(0, \pi/2)$. The left side of (9) is
$$\frac{\cos(\pi/2) - \cos 0}{\sin(\pi/2) - \sin 0} = \frac{0 - 1}{1 - 0} = -1,$$
and the right side is
$$\frac{-\sin c}{\cos c} = -\tan c.$$
Therefore formula (9) becomes $\tan c = 1$. Solving for c subject to the condition $0 < c < \pi/2$, we find at once that $c = \pi/4$. □

Problems

Find a point c satisfying Rolle's theorem (so that $f'(c) = 0$) if

1. $f(x) = x^2 - 3x + 5$, $a = 1$, $b = 2$
2. $f(x) = 2x^3 - x^2 - 2x + 5$, $a = -1$, $b = 1$
3. $f(x) = 1/(x^2 + 1)$, $a = -2$, $b = 2$
4. $f(x) = x^{1/2} + (1 - x)^{1/2}$, $a = 0$, $b = 1$
5. $f(x) = x^{1/2}(2 - x)^{1/3}$, $a = 0$, $b = 2$
6. $f(x) = (3 - x)^{4/3}$, $a = 2$, $b = 4$
7. $f(x) = \sin 2x$, $a = \pi$, $b = 3\pi/2$
8. $f(x) = \sec x$, $a = -1.5$, $b = 1.5$

9. The function $f(x) = 1 - x^{2/3}$ is continuous on $[-1, 1]$ and equals 0 at $x = \pm 1$. Show that there is no point c in $(-1, 1)$ such that $f'(c) = 0$. Why doesn't this contradict Rolle's theorem? Graph the function.

10. Check the validity of Rolle's theorem for the function $f(x) = (x - 1)(x - 2)(x - 3)$. In other words, show that the derivative f' equals 0 at a point in the interval $(1, 2)$ and at a point in the interval $(2, 3)$.

Use Rolle's theorem to show that

11. The equation $x^3 + x^2 + x + 1 = 0$ has a real root, but no more than one
12. The equation $x^8 + x - 1 = 0$ has two real roots, but no more than two
13. The equation $x^5 - 5x + 1 = 0$ has three real roots, but no more than three

Section 3.1 The Mean Value Theorem 165

Find a point c satisfying the mean value theorem (7) if

14. $f(x) = x^2 + x + 1, a = 1, b = 2$
15. $f(x) = x^3 - 1, a = 0, b = 3$
16. $f(x) = \dfrac{1}{2-x}, a = 1, b = 0$
17. $f(x) = \dfrac{x-1}{x+1}, a = 2, b = 3$
18. $f(x) = \sqrt{x}, a = 9, b = 25$
19. $f(x) = x^{1/3}, a = 8, b = 0$
20. $f(x) = x^{4/3}, a = -8, b = -1$
21. $f(x) = \sin x, a = 0, b = \pi/2$

22. Justify the following "kinematic" interpretation of the mean value theorem: If a train goes the distance between two stations at an average velocity v_{av}, then there is a moment when the train's instantaneous velocity is exactly equal to v_{av}.

23. Show that the inequality $|\sin a - \sin b| \leq |a - b|$ holds for arbitrary numbers a and b.

24. Show that the inequality $|\tan a - \tan b| \leq 4|a - b|$ holds for arbitrary numbers a and b in the interval $[-\pi/3, \pi/3]$.

25. Give an example of a function f with infinitely many different values, whose derivative is identically equal to zero wherever f is defined.

26. Find the flaw in the following "proof" of Cauchy's mean value theorem: By two applications of the ordinary mean value theorem,

$$f(b) - f(a) = f'(c)(b-a), \quad g(b) - g(a) = g'(c)(b-a).$$

Therefore, dividing the first of these inequalities by the second, we get

$$\frac{f(b) - f(a)}{g(b) - g(a)} = \frac{f'(c)}{g'(c)}.$$

Find a point c satisfying Cauchy's mean value theorem (9) if

27. $f(x) = x^3, g(x) = x^2 + 2x, a = -1, b = 2$
28. $f(x) = x^2, g(x) = 1/x, a = -2, b = -1$
29. $f(x) = 1/x, g(x) = 1/x^2, a = 1, b = 2$
30. $f(x) = \sin x, g(x) = \tan x, a = 0, b = \pi/4$

3.2 Local Extrema and Monotonic Functions

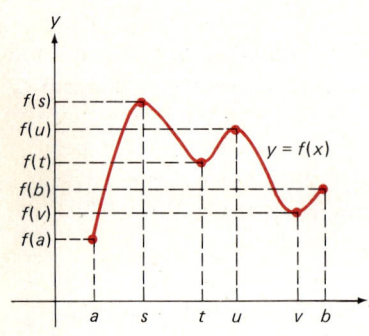

Figure 7

Figure 7 shows a curve which is the graph of a function continuous on a closed interval $[a, b]$. It is clear that f takes its maximum on $[a, b]$, equal to $f(s)$, at the highest point of the curve, namely $(s, f(s))$, and its minimum on $[a, b]$, equal to $f(a)$, at the lowest point of the curve, namely $(a, f(a))$, which here happens to be an endpoint of the curve. But there is also something special about the behavior of the curve at the points $(t, f(t))$, $(u, f(u))$ and $(v, f(v))$. In fact, $(u, f(u))$ is higher than all "nearby" points of the curve, although it is not as high as $(s, f(s))$. Moreover $(t, f(t))$ is lower than all nearby points of the curve, although it is not as low as $(a, f(a))$, and the same is true of the point $(v, f(v))$. Thus the curve in Figure 7 has six special points, the "peaks" $(s, f(s))$ and $(u, f(u))$, the "valleys" $(t, f(t))$ and $(v, f(v))$, and the endpoints $(a, f(a))$ and $(b, f(b))$. (Endpoints are always worthy of special consideration.)

Definition of Local Extrema

We now make these qualitative notions more precise. Given a function f and a number c, which can be regarded as a point on the real line, suppose $f(c) \geq f(x)$ for all x sufficiently close to c, that is, for all x in some neighborhood of c (it is assumed that f is defined at every point of this neighborhood). Then f is said to have a *local maximum* at c, equal to the number $f(c)$. This local maximum is said to be *strict* if $f(c) > f(x)$, with $>$ instead of $\geq$, for all x sufficiently close to c but not equal to c, that is, for all x in some *deleted* neighborhood of c. Similarly, if $f(c) \leq f(x)$ for all x in some neighborhood of c, then f is said to have a *local minimum at c*,

166   Chapter 3   Further Applications of Differentiation

equal to the number $f(c)$, and the local minimum is said to be *strict* if $f(c) < f(x)$, with $<$ instead of $\leq$, for all x in some deleted neighborhood of c. The term *local extremum* refers to either a local maximum or a local minimum. In some books local extrema are called "relative" extrema.

Example 1 The function f with the graph shown in Figure 7 has four local extrema, strict local maxima $f(s)$ and $f(u)$ at the points s and u, and strict local minima $f(t)$ and $f(v)$ at the points t and v. It does *not* have local extrema at the endpoints a and b of its interval of definition $I = [a, b]$, simply because the definition of a local extremum involves comparison of the value of f at the given point c with the values of f *on both sides of* c, and such a comparison is only possible at an interior point c of I.

Example 2 The piecewise linear function f with the graph shown in Figure 8 has a local maximum at $x = 1$, a local minimum at $x = 2$, and both a local maximum and a local minimum at every point of the interval $(1, 2)$ on which f is constant. All of these local extrema are equal to 2, and none of them is strict.

Figure 8

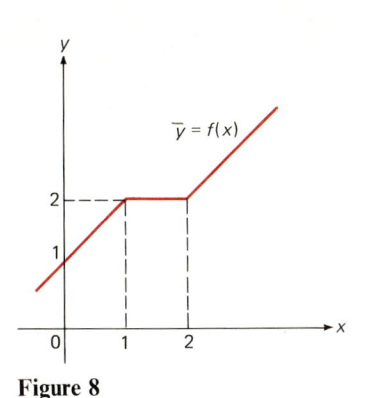

Local vs. Absolute Extrema

Make sure you understand the crucial distinction between *local* extrema, as just defined, and the extrema of a function *on an interval*, defined on page 97. The latter will henceforth be called *absolute* extrema, whenever there is any possibility of confusion. For example, to qualify as the absolute minimum of a function f on an interval I, taken at a point c in I, the number $m = f(c)$ must be less than or equal to the values of f taken at *all* points of I, whereas to be a local minimum, m need only be less than or equal to the values taken by f at all points in some neighborhood of c, no matter how small.

If f is a function whose domain is an interval I, then an absolute extremum of f taken at an *interior* point of I is automatically a local extremum of f. (We have already seen that f cannot have a local extremum at an endpoint of I.) For example, suppose f has an absolute maximum on I at an interior point c. Then $f(c) \geq f(x)$ for all x in I, and therefore it is certainly true that $f(c) \geq f(x)$ for all x in any neighborhood of c small enough to be contained in I.

According to the extreme value theorem (see page 99), a function f continuous on a bounded closed interval always has absolute extrema. We now give a useful rule for finding these extrema.

Theorem 5 *(Test for absolute extrema)*. *Let f be continuous on a bounded closed interval $[a, b]$ and suppose f has local extrema at the points c_1, $c_2, \ldots, c_n$ of the open interval (a, b) and only at these points. Then the largest of the numbers*

$$f(a), f(c_1), f(c_2), \ldots, f(c_n), f(b)$$

is the absolute maximum of f on $[a, b]$, and the smallest of these numbers is the absolute minimum of f on $[a, b]$.

Proof If an absolute extremum of f occurs at an interior point of $[a, b]$, it will be found among the local extrema $f(c_1), f(c_2), \ldots, f(c_n)$. But there is also the possibility that it occurs at an endpoint of $[a, b]$, and hence these n values of f must be compared with $f(a)$ and $f(b)$. □

Example 3 The function f with the graph shown in Figure 7 has local extrema at the points s, t, u and v, as already noted in Example 1. It is clear from the figure that the largest of the numbers

$$f(a), f(s), f(t), f(u), f(v), f(b)$$

is $f(s)$ and the smallest is $f(a)$. Hence the (absolute) maximum of f on $[a, b]$ is $M = f(s)$, taken at the interior point s, and the (absolute) minimum of f on $[a, b]$ is $m = f(a)$, taken at the endpoint a. Since we say "on $[a, b]$," the word *absolute* is somewhat redundant here, and hence has been put in parentheses.

We now need a systematic way of finding all the local extrema of a given function f. We will assume that f is continuous and differentiable on some interval I, with the possible exception of certain special points at which f fails to be differentiable but is still continuous. If f has a local extremum at c, then as we now show, this is reflected in the behavior of the derivative of f at c.

> **Theorem 6** *(Necessary condition for a local extremum).* If f has a local extremum at a point c, then the derivative $f'(c)$ either fails to exist, or it exists and is equal to zero.

Proof To be explicit, suppose f has a local minimum at c, so that $f(c) \leq f(x)$ for all x in some neighborhood of c, that is,

$$f(x) - f(c) \geq 0 \qquad (|x - c| < \delta) \tag{1}$$

for some $\delta > 0$. Either $f'(c)$ fails to exist, in which case there is nothing more to prove, or else

$$f'(c) = \lim_{x \to c} \frac{f(x) - f(c)}{x - c} \tag{2}$$

does exist. But, because of (1), the difference quotient

$$\frac{f(x) - f(c)}{x - c}$$

is nonnegative if $c < x < c + \delta$ and nonpositive if $c - \delta < x < c$. Thus the limit in (2), that is, the derivative $f'(c)$, can be neither negative nor positive, by the same argument as in the proof of Rolle's theorem (see the remark on page 161. The only remaining possibility is that $f'(c) = 0$. The case where f has a local maximum at c is treated in virtually the same way. ∎

Interpreted geometrically, Theorem 6 says that if a function f has a local extremum at a point c, then either the graph of f has no tangent at the point $P = (c, f(c))$, if $f'(c)$ fails to exist, or the graph of f has a *horizontal* tangent at P, if $f'(c) = 0$. These two possibilities are illustrated in Figures 9(a) and 9(b), which show two functions, each with a (strict) local extremum at c.

Local minimum:
$f'(c)$ does not exist

(a)

Local maximum:
$f'(c) = 0$

(b)

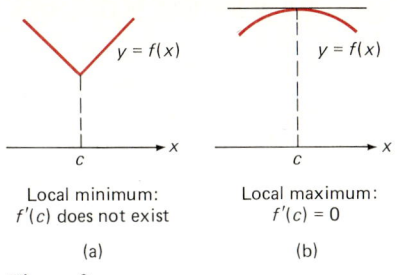

Figure 9

No extremum:
$f'(c)$ does not exist

(a)

No extremum:
$f'(c) = 0$

(b)

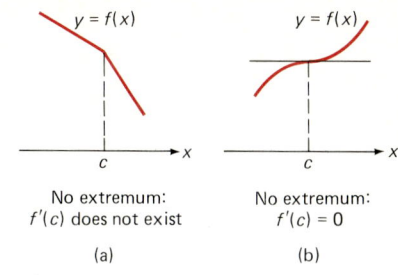

Figure 10

Critical Points

A point c in the domain of a function f is called a *critical point* of f if the derivative $f'(c)$ either fails to exist or is equal to zero. According to Theorem 6, if f has a local extremum at c, then c is a critical point of f. On the other hand, and this is essential to an understanding of the theory of extrema, if c is a critical point of f, the function f may well *fail* to have a local extremum at c. This is illustrated in Figures 10(a) and 10(b), which show two functions, each with a critical point at c, but neither with a local extremum at c.

The Monotonicity Test

Thus what we really want are conditions on a function f which *compel* f to have a local extremum at a given point c. In the language of logic, these are *sufficient* conditions for a local extremum, as opposed to the *necessary* condition given in Theorem 6. Such conditions will be presented in the form of two tests for a local extremum (Theorems 8 and 9 below). As a tool for proving these tests, we first establish the following theorem, of great importance in its own right, which gives conditions for a function to be monotonic. A function f is said to be *monotonic* on an interval I if f is either increasing on I or decreasing on I, and *monotonicity* is the property of being monotonic.

Theorem 7 *(Monotonicity test).* Let f be continuous on an interval I, and suppose the derivative f' exists and has the same sign at every interior point of I. Then

(i) f is increasing on I if f' is positive at every interior point of I;
(ii) f is decreasing on I if f' is negative at every interior point of I.

Proof Suppose f' is positive at every interior point of I, and let a and b be any two points of I such that $a < b$. Then, by the mean value theorem,

$$f(b) - f(a) = f'(c)(b - a)$$

for some point c between a and b. Clearly c is an interior point of I, so that $f'(c) > 0$. Therefore $f'(c)(b - a)$ is positive, being the product of two positive numbers. But then $f(b) - f(a)$ is also positive. In other words, $f(a) < f(b)$ for every pair of points a and b in I such that $a < b$. Thus f is increasing on I, and part (i) is proved. The proof of (ii) is virtually the same, and is left as an exercise. ∎

The conclusions of Theorem 7 are insensitive to the behavior of f' at the endpoints of I (if the interval I contains one or both of its endpoints), and in fact f' may well take the value 0, or even fail to exist, at an endpoint of I. Also, if I is open, every point of I is an interior point, and the word "interior" can be dropped in three places in the statement of the theorem.

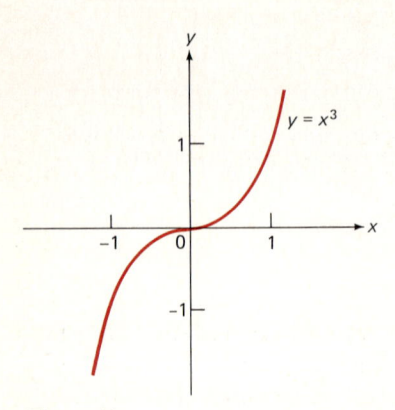

Example 4 It was claimed in Example 1, page 97, that the function

$$f(x) = 2x^5 + 2x^2 + x - 3$$

is increasing on $(0, 1)$. Using the monotonicity test, we now see that this follows at once from the fact that the derivative

$$f'(x) = 10x^4 + 4x + 1$$

is positive on $(0, 1)$.

Example 5 For the function $f(x) = x^3$ we have $f'(x) = 3x^2 \geq 0$, with equality only if $x = 0$. Therefore f' is positive at every interior point of the intervals $(-\infty, 0]$ and $[0, \infty)$. It follows from the monotonicity test that f is increasing on both of these intervals, and hence that f is increasing on the whole real line $(-\infty, \infty)$, as is apparent from the graph of f, shown in Figure 11.

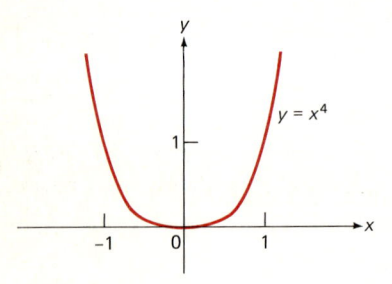

Example 6 For the function $f(x) = x^4$ we have $f'(x) = 4x^3$, so that $f'(x) > 0$ if $x > 0$ and $f'(x) < 0$ if $x < 0$, while $f'(0) = 0$. Therefore f' is positive at every interior point of the interval $[0, \infty)$ and negative at every interior point of the interval $(-\infty, 0]$. This time the monotonicity test tells us that f is increasing on $[0, \infty)$ and decreasing on $(-\infty, 0]$, as is apparent from the graph of f, shown in Figure 12.

The First and Second Derivative Tests

We are now in a position to prove the promised tests for a local extremum.

> **Theorem 8** *(First derivative test for a local extremum). Let c be a critical point of f, and suppose the derivative f' changes sign at c, that is, suppose f' has one sign on an interval (a, c) to the left of c and the opposite sign on an interval (c, b) to the right of c. Then f has a strict local extremum at c. The extremum is a minimum if f' changes sign from minus to plus, and a maximum if f' changes sign from plus to minus.*

Proof Nothing is said about the differentiability of f at the point c itself, and the derivative $f'(c)$ may fail to exist (although it is assumed, as always, that f is continuous at c). Suppose f' changes sign from minus to plus at c. Then f' is negative on some interval (a, c) and positive on some interval (c, b). It follows from the monotonicity test that f is decreasing on $(a, c]$ and increasing on $[c, b)$, where we can now include the endpoint c in both intervals. But then f has a strict local minimum at c, since $f(c) < f(x)$ for every point in some deleted neighborhood of c [see Figure 13(a)]. Similarly, if f' changes sign from plus to minus at c, then f is increasing on $(a, c]$ and decreasing on $[c, b)$, so that f has a strict local maximum at c [see Figure 13(b)]. We have drawn corners at the point $(c, f(c))$ in both figures, to stress again that the derivative $f'(c)$ may not exist. ∎

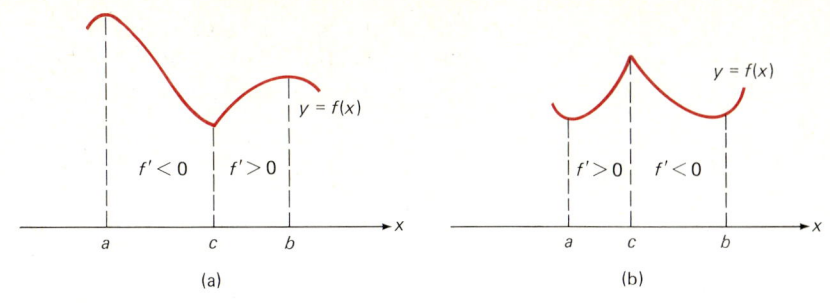

Figure 13

Theorem 9 *(Second derivative test for a local extremum).* Let c be a critical point of f, and suppose the second derivative $f''(c)$ exists and is nonzero. Then f has a strict local extremum at c. The extremum is a minimum if $f''(c) > 0$ and a maximum if $f''(c) < 0$.

Proof By definition,

$$f''(c) = \lim_{x \to c} \frac{f'(x) - f'(c)}{x - c}, \tag{3}$$

so that $f''(c)$ can exist only if the *first* derivative $f'(x)$ exists in some neighborhood of c. In particular, $f'(c)$ exists and equals 0, since c is a critical point. Thus (3) reduces to

$$f''(c) = \lim_{x \to c} \frac{f'(x)}{x - c}.$$

If $f''(c) > 0$, there is a deleted neighborhood of c in which $f'(x)/(x - c)$ is positive, and in this deleted neighborhood the sign of $f'(x)$ is the same as that of $x - c$. Therefore f' changes sign from minus to plus at c, so that f has a strict local minimum at c, by the first derivative test (Theorem 8). The case $f''(c) < 0$ is treated similarly, and leads to a strict local maximum. ∎

If $f''(c) = 0$ at a critical point c, the second derivative test is inconclusive. In fact, let

$$f(x) = x^3, \quad g(x) = x^4.$$

Then, calculating second derivatives, we get

$$f''(x) = 6x, \quad g''(x) = 12x^2,$$

so that

$$f''(0) = 0, \quad g''(0) = 0.$$

But f has no extremum at $x = 0$, since f is increasing on $(-\infty, \infty)$, as shown in Example 5, while g has a strict local (and absolute) minimum at $x = 0$, since $g(0) = 0$ and $g(x) = x^4 > 0$ for all $x \neq 0$. Thus the fact that the second derivative of a function equals zero at a critical point c does not allow us to decide whether or not it has a local extremum at c.

Example 7 Find the local extrema of the function

$$f(x) = \frac{1}{2}x^3 - \frac{3}{2}x^2 + 5. \tag{4}$$

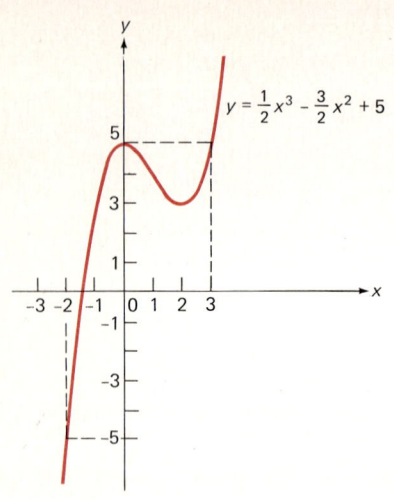

Figure 14

$y = \frac{1}{2}x^3 - \frac{3}{2}x^2 + 5$

Solution Here f is differentiable for all x, and the only critical points of f are the points at which the derivative

$$f'(x) = \frac{3}{2}x^2 - 3x = \frac{3}{2}x(x-2) \qquad (4')$$

equals zero, namely the points $x = 0$ and $x = 2$. Another differentiation gives $f''(x) = 3x - 3$, so that $f''(0) = -3 < 0$ and $f''(2) = 3 > 0$. Therefore, by the second derivative test, f has a strict local maximum at $x = 0$, equal to $f(0) = 5$, and a strict local minimum at $x = 2$, equal to $f(2) = 3$. The first derivative test leads to the same conclusion, since (4') shows that f' changes sign from plus to minus at $x = 0$ and from minus to plus at $x = 2$. Also, by the monotonicity test, f is increasing on the intervals $(-\infty, 0]$ and $[2, \infty)$, and decreasing on the interval $[0, 2]$. This information can now be used, with the help of a calculator, to graph the function f (see Figure 14). $\square$

Example 8 Find the absolute extrema of the function (4) on the interval $I = [-2, 3]$.

Solution By the test for absolute extrema (Theorem 5), it is enough to compare the numbers

$$f(-2) = -5, \quad f(0) = 5, \quad f(2) = 3, \quad f(3) = 5,$$

that is, the local extrema of f at interior points of I and the values of f at the endpoints of I. The largest of these numbers, namely 5, is the absolute maximum of f on I, taken at both the interior point $x = 0$ and the right endpoint $x = 3$, while the smallest of the numbers, namely -5, is the absolute minimum, taken at the left endpoint $x = -2$ (reexamine Figure 14). $\square$

Example 9 Find the local extrema of the function $f(x) = \sin^4 x + \cos^4 x$.

Solution Again f is differentiable for all x, and the only critical points of f are the points at which the derivative

$$f'(x) = 4\sin^3 x \cos x - 4\cos^3 x \sin x$$
$$= 4(\sin^2 x - \cos^2 x)\sin x \cos x = -4\cos 2x \sin x \cos x$$

equals zero. These are the points $x = 0, \pm\pi, \pm 2\pi, \ldots$ at which $\sin x = 0$, the points $x = \pm\pi/2, \pm 3\pi/2, \ldots$ at which $\cos x = 0$, and the points $x = \pm\pi/4, \pm 3\pi/4, \ldots$ at which $\cos 2x = 0$. Another differentiation gives

$$f''(x) = 12\sin^2 x \cos^2 x - 4\sin^4 x + 12\cos^2 x \sin^2 x - 4\cos^4 x$$
$$= 24\sin^2 x \cos^2 x - 4\sin^4 x - 4\cos^4 x,$$

so that

$$f''(x) = 24(0) - 4(0) - 4(1) = -4 \quad \text{if} \quad x = 0, \pm\pi, \pm 2\pi, \ldots,$$
$$f''(x) = 24(0) - 4(1) - 4(0) = -4 \quad \text{if} \quad x = \pm\pi/2, \pm 3\pi/2, \ldots,$$

and

$$f''(x) = 24\left(\frac{1}{4}\right) - 4\left(\frac{1}{4}\right) - 4\left(\frac{1}{4}\right) = 4 \quad \text{if} \quad x = \pm\pi/4, \pm 3\pi/4, \ldots$$

Therefore, by the second derivative test, f has strict local maxima at the points $x = 0, \pm\pi/2, \pm\pi, \pm 3\pi/2, \pm 2\pi, \ldots$ and strict local minima at the points $x = \pm\pi/4, \pm 3\pi/4, \ldots$ Verify that the maxima are all equal to 1, and

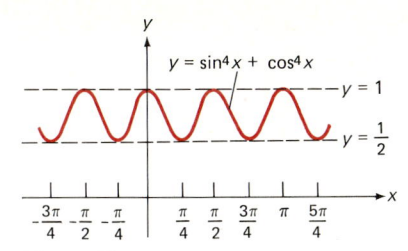

Figure 15

the minima are all equal to $\frac{1}{2}$. The graph of f, part of which is shown in Figure 15, is periodic with fundamental period $\pi/2$. Notice that there are infinitely many extrema, as we would expect because of the periodicity. □

Example 10 Find the local extrema of the function $f(x) = |x^2 - 1|$.

Solution Graphing the function $x^2 - 1$ inside the absolute value sign, we get the parabola shown in Figure 16(a). Taking the absolute value of $x^2 - 1$ gives the function f, and causes the part of the parabola below the x-axis [the dashed curve in Figure 16(b)] to be reflected in the x-axis. Thus the graph of f is the solid curve in Figure 16(b). Because of the reflection, this curve has corners at the points $(1, 0)$ and $(-1, 0)$. Correspondingly, f fails to be differentiable at $x = 1$ and $x = -1$, and hence has critical points at $x = 1$ and $x = -1$. Since f is differentiable for all $x \neq \pm 1$, other critical points of f can occur only at values of x for which the derivative f' takes the value 0. It is easy to see that this happens only at $x = 0$. In fact,

$$f(x) = \begin{cases} x^2 - 1 & \text{if } x \leq -1 \text{ or } x \geq 1, \\ 1 - x^2 & \text{if } -1 < x < 1, \end{cases}$$

$$f'(x) = \begin{cases} 2x & \text{if } x < -1 \text{ or } x > 1, \\ -2x & \text{if } -1 < x < 1, \end{cases}$$

so that $f'(x) = 0$ if and only if $x = 0$. Moreover, f' changes sign from plus to minus at $x = 0$, since $f'(x) > 0$ if $-1 < x < 0$ and $f'(x) < 0$ if $0 < x < 1$. Therefore, by the first derivative test, f has a strict local maximum at $x = 0$, equal to $f(0) = 1$. The second derivative test leads to the same result, since $f''(0) = -2$. At the other critical points $x = 1$ and $x = -1$, f has strict local minima, equal to $f(\pm 1) = 0$, as follows at once from the fact that $f(\pm 1) = 0$, while $f(x) > 0$ if $x \neq \pm 1$. As an exercise, show that these minima can also be found by the first derivative test. Does the function f have absolute extrema on $(-\infty, \infty)$, and if so, where? □

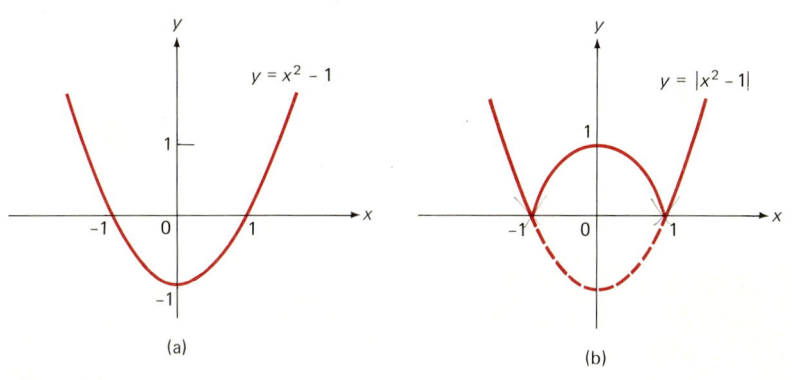

Figure 16

Problems

By investigating all critical points, find the local extrema of the given function. On which intervals is the function increasing? Decreasing? Graph the function.

1. $f(x) = |x + 1| - 1$
2. $f(x) = x^2 - 2x$
3. $f(x) = 3x - x^3$
4. $f(x) = 2x^3 - 9x^2 + 12x$
5. $f(x) = x^4 - 4x^2 + 4$
6. $f(x) = x + \dfrac{1}{x}$
7. $f(x) = \dfrac{8}{x^2 + x + 2}$
8. $f(x) = \dfrac{6x}{x^2 + 1}$

Section 3.2 Local Extrema and Monotonic Functions 173

9. $f(x) = \dfrac{x^2}{x+1}$

10. $f(x) = \sqrt{x^2 + 2x}$

11. $f(x) = \cos x + \sin x$

12. $f(x) = x + \sin x$

Find the absolute extrema of the given function on the specified interval.

13. $f(x) = x^2 - 4x + 6$ on $[-3, 10]$
14. $f(x) = x^3 - 2x + 1$ on $[0, 1]$
15. $f(x) = |x^2 - 3x + 2|$ on $[-10, 10]$
16. $f(x) = \dfrac{x}{x^2 + 1}$ on $[0, 2]$
17. $f(x) = x + \dfrac{1}{x}$ on $[0.01, 100]$
18. $f(x) = \sqrt{5 - 4x}$ on $[-1, 1]$
19. $f(x) = \sqrt{x^2 + x + 1}$ on $[-1, 0]$
20. $f(x) = \sqrt{2 - x - x^2}$ on $[-2, 1]$

21. $f(x) = 4\sin x + 2\cos 2x$ on $[0, \pi]$
22. $f(x) = \cos^2 x + 2\sin^2 x$ on $[\pi/4, 3\pi/4]$
23. $f(x) = \sin x^2$ on $[1, 2]$
24. $f(x) = \sin(\sin x)$ on $[0, 2\pi]$

25. Show that if $f(x)$ is an even function with a local maximum (minimum) at $x = c$, then $f(x)$ also has a local maximum (minimum) at $x = -c$. To which of Problems 1–12 is this result applicable?

26. Show that if $f(x)$ is an odd function with a local maximum (minimum) at $x = c$, then $f(x)$ has a local minimum (maximum) at the point $x = -c$. To which of Problems 1–12 is this result applicable?

27. What value of c minimizes the maximum of the function $f(x) = |x^2 + c|$ on the interval $[-1, 1]$?

28. Let f be continuous on an interval I, and suppose f has strict local maxima at two points s and t of I. Show that f must have a local minimum (not necessarily strict) at a point between s and t.

3.3 Concavity and Inflection Points

Concave Functions

Let f be a function which is differentiable on an interval I, and suppose the *derivative* f' is monotonic on I. Then we say that f is *concave upward* on I if the derivative f' is increasing on I, and *concave downward* on I if f' is decreasing on I. The concept of concavity has a simple geometric interpretation. Since $f'(x)$ is the slope of the curve $y = f(x)$ at a variable point $P = (x, f(x))$, that is, the slope of the tangent line to the curve at P, the part of the curve over I bends upward [see Figure 17(a)] if f' is increasing on I, and bends downward [see Figure 17(b)] if f' is decreasing on I. Thus you might say that the part of the curve over I "holds water" if f is concave upward, but "spills water" if f is concave downward on I. Also, the figures suggest that the part of the curve over I lies above each of its tangent lines if f is concave upward on I, and below each of its tangent lines if f is concave downward on I, and it is not hard to show that this is true (see Problem 25).

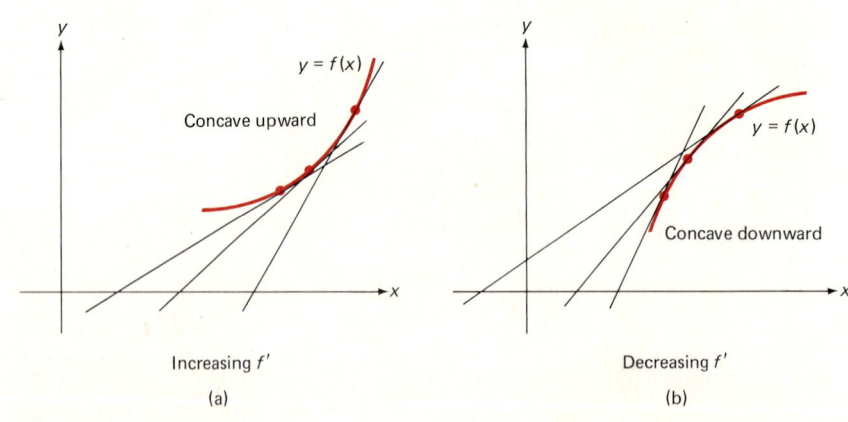

Figure 17

By a simple reworking of the monotonicity test (Theorem 7), we obtain conditions for a function to be concave upward or downward.

> **Theorem 10** *(Concavity test).* Let f have a continuous derivative f' on an interval I, and suppose the second derivative f'' exists and has the same sign at every interior point of I. Then
>
> (i) f is concave upward on I if f'' is positive at every interior point of I;
> (ii) f is concave downward on I if f'' is negative at every interior point of I.

Proof Apply Theorem 7 to the *derivative* f' rather than to the function f itself, bearing in mind the definition of upward and downward concavity. ∎

The conclusions of Theorem 10 are insensitive to the behavior of f'' at the endpoints of I (if the interval I contains one or both of its endpoints), and in fact f'' may well take the value 0, or even fail to exist at an endpoint of I. Also, if I is open, every point of I is an interior point of I, and the word "interior" can be dropped in three places in the statement of the theorem, just as in the case of Theorem 7.

Definition of an Inflection Point

A function f is said to have an *inflection point* at c if f changes its concavity at c. This means that there is an interval $L = (a, c]$ to the left of c and an interval $R = [c, b)$ to the right of c such that f is concave upward on L and concave downward on R [see Figure 18(a)], or concave downward on L and concave upward on R [see Figure 18(b)]. Here it is assumed that $f'(c)$ exists. There is also the possibility that $f'(c)$ does not exist, as in Example 3 below. In this case we take L and R to be the open intervals (a, c) and (c, b), with the endpoint c deleted, assuming as always that f is continuous at c. Thus f has an inflection point at c if and only if the derivative f' is increasing on L and decreasing on R, or decreasing on L and increasing on R. It follows at once that *if f has an inflection point at c and if $f'(c)$ exists, then the derivative f' has a strict local extremum at c*. In fact, f' has a strict local maximum at c if f' is increasing on L and decreasing on R, and a strict local minimum at c if f' is decreasing on L and increasing on R.

If a function f has an inflection point at c, the *curve* $y = f(x)$ is also said to have an inflection point, at $P = (c, f(c))$. If the curve has a tangent at

Each curve has an inflection point at P.

(a)   (b)

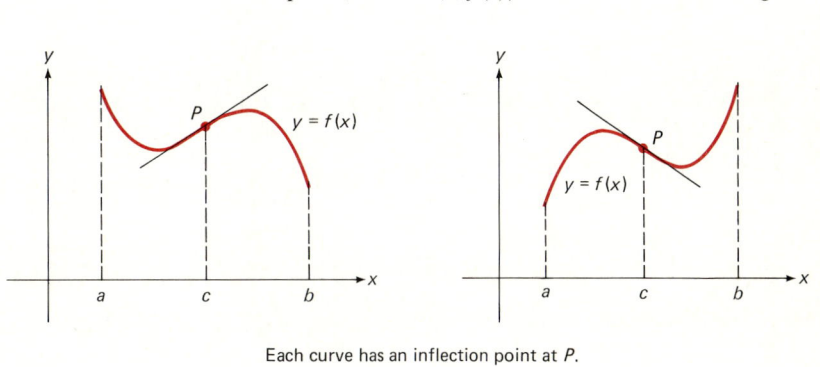

Figure 18

P, then as shown in Figures 18(a) and 18(b), the curve crosses from one side of the tangent to the other at P (see Problem 25). A tangent at an inflection point is sometimes called an *inflectional tangent*.

Example 1 For the function $f(x) = x^3$, graphed in Figure 11, page 170, $f'(x) = 3x^2$ and $f''(x) = 6x$. Therefore $f''(x) < 0$ if $x < 0$ and $f''(x) > 0$ if $x > 0$. Thus, by the concavity test, f is concave upward on $[0, \infty)$ and concave downward on $(-\infty, 0]$, with an inflection point at $x = 0$.

Example 2 For the function $f(x) = x^4$, graphed in Figure 12, page 170, $f'(x) = 4x^3$ and $f''(x) = 12x^2$. Therefore $f''(x) \geq 0$ for all x, with equality if and only if $x = 0$. This time the concavity test tells us that f is concave upward on both intervals $(-\infty, 0]$ and $[0, \infty)$, and hence on the whole real line $(-\infty, \infty)$. Since the concavity of f never changes, f has no inflection points.

Example 3 If $f(x) = x^{1/3}$, then $f'(x) = \frac{1}{3}x^{-2/3}$ and $f''(x) = -\frac{2}{9}x^{-5/3}$ for all $x \neq 0$, but both $f'(0)$ and $f''(0)$ fail to exist. In fact, the limit defining $f'(0)$, namely

$$\lim_{x \to 0} \frac{f(x) - f(0)}{x - 0} = \lim_{x \to 0} \frac{x^{1/3} - 0^{1/3}}{x} = \lim_{x \to 0} x^{-2/3}$$

does not exist, since $x^{-2/3}$ is unbounded in every deleted neighborhood of the point $x = 0$, and the nonexistence of the first derivative $f'(0)$ clearly implies the nonexistence of the second derivative

$$f''(0) = \lim_{x \to 0} \frac{f'(x) - f'(0)}{x - 0}.$$

Investigating the sign of $f''(x)$ for $x \neq 0$, we find that $f''(x) > 0$ if $x < 0$ and $f''(x) < 0$ if $x > 0$. Here we use the fact that

$$x^{-5/3} = \frac{1}{(x^{1/3})^5} \qquad (x \neq 0)$$

has the same sign as $x^{1/3}$, which in turn has the same sign as x. It follows from the concavity test that f is concave upward on $(-\infty, 0)$ and concave downward on $(0, \infty)$, with an inflection point at $x = 0$, as is apparent from the graph of f, shown in Figure 19. The graph seems to have a *vertical tangent* at the origin, and on page 194 it will be explained why this is true. Notice that we cannot include the endpoint $x = 0$ in the intervals of concavity $(-\infty, 0)$ and $(0, \infty)$, since $f'(0)$ does not exist.

Example 4 If $f(x) = x^{4/3}$, then $f'(x) = \frac{4}{3}x^{1/3}$ for all x and $f''(x) = \frac{4}{9}x^{-2/3}$ for all $x \neq 0$, but $f''(0)$ does not exist, since here the limit

$$f''(0) = \lim_{x \to 0} \frac{f'(x) - f'(0)}{x - 0} = \lim_{x \to 0} \frac{4}{3} \frac{x^{1/3} - 0^{1/3}}{x} = \lim_{x \to 0} \frac{4}{3} x^{-2/3}$$

defining $f''(0)$ is unbounded in every deleted neighborhood of $x = 0$. Investigating the sign of the second derivative, we find that $f''(x) > 0$ for all $x \neq 0$. Therefore, by the concavity test, f is concave upward on both intervals $(-\infty, 0]$ and $[0, \infty)$, and hence on the whole real line $(-\infty, \infty)$, as shown in Figure 20. Despite the nonexistence of $f''(0)$, why can we now include the endpoint $x = 0$ in the intervals of concavity $(-\infty, 0]$ and $[0, \infty)$?

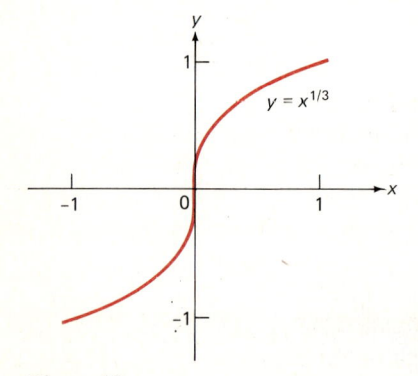

Figure 19

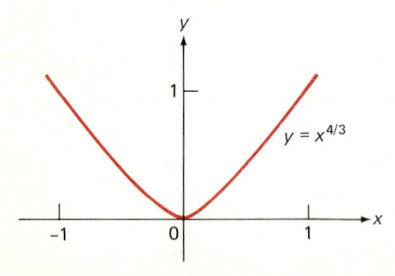

Figure 20

176 Chapter 3 Further Applications of Differentiation

Example 5 If $f(x) = x - \cos x$, then $f'(x) = 1 + \sin x \geq 0$ for all x, where $f'(x) = 0$ if and only if

$$x = (2n - \tfrac{1}{2})\pi \qquad (n = 0, \pm 1, \pm 2, \ldots).$$

Therefore, by the monotonicity test, f is increasing on every interval $[(2n - \tfrac{1}{2})\pi, (2n + \tfrac{3}{2})\pi]$, and hence on the whole real line $(-\infty, \infty)$; in particular, f has no local extrema. Investigating the second derivative $f''(x) = \cos x$, we find that f'' has the value zero at the points

$$x = (n + \tfrac{1}{2})\pi \qquad (n = 0, \pm 1, \pm 2, \ldots), \tag{1}$$

and changes sign at each of these points, from plus to minus if n is even and from minus to plus if n is odd. It follows from the concavity test that f is concave upward on every interval $[(n - \tfrac{1}{2})\pi, (n + \tfrac{1}{2})\pi]$ with even n and concave downward on every such interval with odd n, with every point (1) as an inflection point. All this is apparent from the graph of f, given in Figure 21.

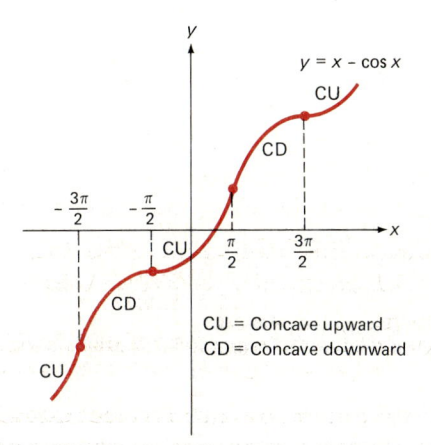

Figure 21

Tests for Inflection Points

As shown by the following rules, the theory of concave functions and inflection points is completely analogous to the theory of monotonic functions and local extrema.

(i) *If f has an inflection point at c, then the second derivative $f''(c)$ either fails to exist, or it exists and equals zero,* that is, c is a critical point of the derivative f'. This *necessary condition for an inflection point* is an immediate consequence of the necessary condition for a local extremum (Theorem 6, page 168), of which it is the exact analogue. In fact, if f has an inflection point at c and $f'(c)$ exists, then f' has a local extremum at c (see page 175), so that either $f''(c)$ fails to exist or $f''(c) = 0$, while if $f'(c)$ does not exist, then neither does $f''(c)$, since the limit

$$\lim_{x \to c} \frac{f'(x) - f'(c)}{x - c}$$

defining $f''(c)$ involves $f'(c)$.

(ii) *If c is a critical point of the derivative f' and if the second derivative f'' changes sign at c, then f has an inflection point at c.* Here we assume the existence of f'' at least in a deleted neighborhood of c. This *second derivative test for an inflection point* is the exact analogue of the first derivative test for a local extremum (Theorem 8, page 170), and follows at once from the concavity test. In fact, we have already tacitly used this second derivative test in solving Examples 1, 3 and 5.

(iii) *If $f''(c) = 0$ and if the third derivative $f'''(c)$ exists and is nonzero, then f has an inflection point at c.* This *third derivative test for an inflection point* is the exact analogue of the second derivative test for a local extremum (Theorem 9, page 171), and is proved in virtually the same way, as follows.

Optional

Since the third derivative

$$f'''(c) = \lim_{x \to c} \frac{f''(x) - f''(c)}{x - c} = \lim_{x \to c} \frac{f''(x)}{x - c}$$

exists, the second derivative $f''(x)$ must exist in some neighborhood of c. If $f'''(c) > 0$, there is a deleted neighborhood of c in which the sign of $f''(x)$ is the same as that of $x - c$, while if $f'''(c) < 0$, there is a deleted neighborhood of c in which the sign of $f''(x)$ is the opposite of that of $x - c$. In either case f'' changes sign at c, and therefore f has an inflection point at c, by the second derivative test (ii).

It should be emphasized that the critical points of the derivative f' are merely *candidates* for the inflection points of f, just as the critical points of the function f itself are merely candidates for the local extrema of f, and it may well turn out that f fails to have an inflection point at a critical point of f'.

Example 6 In Example 1 the second derivative $f''(x) = 6x$ is zero if and only if $x = 0$. Therefore, by rule (i), $x = 0$ is the only candidate for an inflection point of f. Since $f'''(x) = 6 \neq 0$, it follows from the third derivative test (iii) that $x = 0$ actually is an inflection point of f, as already shown by an argument equivalent to the second derivative test (ii).

Example 7 In Example 3 the second derivative of f is nonzero except at $x = 0$, where it fails to exist. Therefore, by rule (i), $x = 0$ is the only candidate for an inflection point of f. The fact that $x = 0$ actually is an inflection point of f is a consequence of the second derivative test, since as already noted, f'' changes sign at $x = 0$.

Example 8 In Example 5 the second derivative $f''(x) = \cos x$ is zero at the points $x = (n + \frac{1}{2})\pi$ (n any integer), and is nonzero everywhere else. Hence, by rule (i) again, these points are the only candidates for inflection points of f, and in fact the third derivative test shows that they really are inflection points, since

$$f'''((n + \tfrac{1}{2})\pi) = -\sin((n + \tfrac{1}{2})\pi) = (-1)^{n+1} \neq 0$$

for every integer n. Of course, we have already arrived at the same conclusion by an argument that amounts to the repeated use of the second derivative test.

Curve Sketching

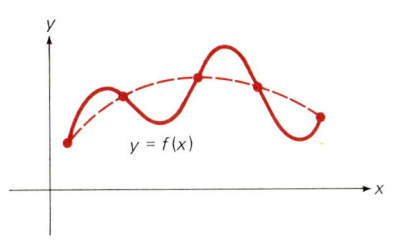

The danger of "connecting the dots"

Figure 22

By now it should be apparent that information about the first few derivatives of a given function f is of great value in constructing the graph of f, that is, in sketching the curve $y = f(x)$. Certainly no clear idea of the behavior of a function can be formed without (at the very least) locating all its extrema and inflection points. Figure 22 shows what can go wrong if we try to graph a function f without doing this first. The solid curve is the actual graph of f, and the dashed curve is the quite misleading result of drawing a smooth curve through five "badly chosen" points of the graph of f.

The following table lists all the techniques for *curve sketching* that are now at our disposal. The first entry in each row gives a property of a continuous function f or its derivatives, and the second entry gives a consequence of the property.

If:	Then:
f has a local extremum at c	f has a critical point at c, that is, either $f'(c)$ does not exist or $f'(c) = 0$ (*Necessary condition for a local extremum*)
f is continuous on an interval I, and f' is positive (negative) at every interior point of I	f is increasing (decreasing) on I (*Monotonicity test*)
f has a critical point at c, and either f' changes sign from minus to plus at c or $f''(c) > 0$	f has a strict local minimum at c
f has a critical point at c, and either f' changes sign from plus to minus at c or $f''(c) < 0$	f has a strict local maximum at c

First and second derivative tests for a local extremum

If:	Then:
f' is increasing (decreasing) on an interval I	f is concave upward (downward) on I (*Definition*)
f' is continuous on an interval I, and f'' is positive (negative) at every interior point of I	f is concave upward (downward) on I (*Concavity test*)
f changes its concavity at c	f has an inflection point at c (*Definition*)
f has an inflection point at c	f' has a critical point at c, that is, either $f''(c)$ does not exist or $f''(c) = 0$ (*Necessary condition for an inflection point at c*)
f' has a critical point at c, and either f'' changes sign at c or $f'''(c) \neq 0$	f has an inflection point at c (*Second and third derivative tests for an inflection point*)

In sketching a curve $y = f(x)$, be on the lookout for possible symmetries (there may be none). If f is an even function, the curve will be symmetric about the *y*-axis, while if f is an odd function, the curve will be symmetric about the origin, but there are other possibilities, as illustrated by the next example. Also try to find the *x-intercepts* (if any) of the curve, that is, the *x*-coordinates of the points in which it intersects the *x*-axis. Exact determination of the *x*-intercepts may be difficult, since it requires solving the equation $f(x) = 0$, and an approximation technique like the bisection method (see page 97) may be called for.

Section 3.3 Concavity and Inflection Points   **179**

Example 9 Use calculus to investigate the function

$$f(x) = 2x^4 - 8x^3 + 8x^2 \tag{2}$$

and sketch its graph.

Solution Factoring (2), we get

$$f(x) = 2x^4 - 8x^3 + 8x^2 = 2x^2(x^2 - 4x + 4) = 2x^2(x-2)^2,$$

and this shows that the curve $y = f(x)$ has two x-intercepts 0 and 2. The function f is neither even nor odd, so that the curve has no symmetry about the y-axis or the origin, but it has another kind of symmetry which will be apparent in a moment. Differentiating f three times, we obtain

$$f'(x) = 8x^3 - 24x^2 + 16x = 8x(x^2 - 3x + 2) = 8x(x-1)(x-2),$$

$$f''(x) = 24x^2 - 48x + 16 = 24\left(x^2 - 2x + \frac{2}{3}\right),$$

$$f'''(x) = 48x - 48.$$

Setting $f'(x)$ equal to zero, we find that f has three critical points $x = 0, 1, 2$. Evaluating the second derivative f'' at these points, we find that $f''(0) = 16 > 0$, $f''(1) = -8 < 0$, $f''(2) = 16 > 0$. Therefore f has strict local minima at $x = 0$ and $x = 2$, equal to $f(0) = f(2) = 0$ (the two minima happen to be equal), and a strict local maximum at $x = 1$, equal to $f(1) = 2$. Setting $f''(x)$ equal to zero, we get the quadratic equation

$$x^2 - 2x + \frac{2}{3} = (x-1)^2 - \frac{1}{3} = 0,$$

with solutions

$$r_1 = 1 - \frac{1}{\sqrt{3}} \approx 0.42, \qquad r_2 = 1 + \frac{1}{\sqrt{3}} \approx 1.58.$$

The points r_1 and r_2 are candidates for inflection points of f, and in fact they actually are inflection points, since $f'''(r_1) = -16\sqrt{3} \neq 0$ and $f'''(r_2) = 16\sqrt{3} \neq 0$.

There is still more to be learned about f by taking a closer look at the first and second derivatives f' and f''. It follows from the formula $f'(x) = 8x(x-1)(x-2)$ that $f'(x) < 0$ if $x < 0$, $f'(x) > 0$ if $0 < x < 1$, $f'(x) < 0$ if $1 < x < 2$, $f'(x) > 0$ if $x > 2$. Therefore f is increasing on the intervals $[0, 1]$ and $[2, \infty)$, and decreasing on the intervals $(-\infty, 0]$ and $[1, 2]$. Also, writing $f''(x) = 24(x - r_1)(x - r_2)$, we find that $f''(x) > 0$ if $x < r_1$, $f''(x) < 0$ if $r_1 < x < r_2$, $f''(x) > 0$ if $x > r_2$. Thus f is concave upward on the intervals $(-\infty, r_1]$ and $[r_2, \infty)$, and concave downward on the interval $[r_1, r_2]$.

Using all this information, we can now graph the function (2). The resulting curve is shown in Figure 23. A calculator was used to plot about a dozen points on the curve, but it is only with the help of calculus that we can be certain that the points have been properly connected and that no important features of the curve $y = f(x)$ have been overlooked. In particular, this is because we have been sure to plot the points of the curve corresponding to the extrema and inflection points of the function f. The graph reveals

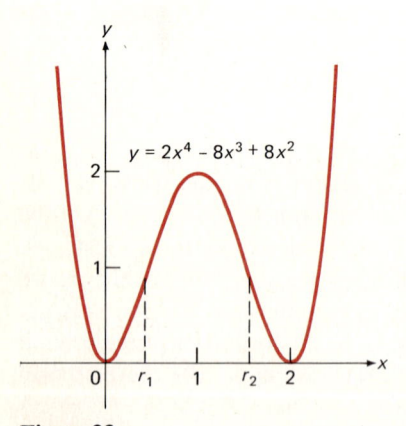

Figure 23

that the curve is symmetric about the vertical line $x = 1$. This could have been anticipated if we had noticed that the function $f(x) = 2x^2(2 - x)^2$ satisfies the easily verified identity $f(1 - x) \equiv f(1 + x)$. □

Problems

Find all inflection points of the given function.

1. $f(x) = \sin x$
2. $f(x) = \cos x$
3. $f(x) = \cot x$
4. $f(x) = \tan x$
5. $f(x) = \sec x$
6. $f(x) = \csc x$
7. $f(x) = \tan x + \cot x$
8. $f(x) = x + \sin x$

9. Show that the third derivative test for an inflection point is inconclusive if the third derivative is equal to zero.

Find all local extrema and inflection points of the given function. On which intervals is the function increasing? Decreasing? Concave upward? Concave downward? Graph the function.

10. $f(x) = x^{2/3}$
11. $f(x) = x^{5/3}$
12. $f(x) = 2 + x - x^2$
13. $f(x) = x^3 - 3x^2 + 4$
14. $f(x) = \frac{1}{8}x^3 - \frac{3}{4}x^2 + \frac{3}{2}x + 1$
15. $f(x) = 4x^2 - 2x^4$
16. $f(x) = 3x^5 - 5x^3$
17. $f(x) = \dfrac{3x^2}{x^2 + 3}$
18. $f(x) = \dfrac{x + 1}{x^2 + 1}$

19. Show that the function $f(x) = x^2 + ax + b$ has no inflection points, regardless of the values of a and b. Is the same true of the function $g(x) = x^4 + ax + b$?

20. Show that the function $f(x) = x^3 + ax^2 + bx + c$ always has an inflection point, regardless of the values of a, b, and c. For what value of a does f have an inflection point at $x = 1$?

21. Show that the function $f(x) = x^4 + ax^3 + bx^2 + cx + d$ has no inflection points if $3a^2 \leq 8b$ and two inflection points if $3a^2 > 8b$, regardless of the values of c and d.

22. Where is the inflection point of the function $f(x) = (x - a)(x - b)(x - c)$?

23. For what values of a and b is the point $(1, 3)$ an inflection point of the curve $y = ax^3 + bx^2$?

24. Show that the three inflection points of the graph of the function in Problem 18 lie on the same line. What is this line?

25. Let f be differentiable on an open interval $I = (a, b)$, and let c be any point of I. Then the tangent to the curve $y = f(x)$ at $P = (c, f(c))$ is the line $y = t(x) = f'(c)(x - c) + f(c)$, and the difference between the y-coordinate of the curve $y = f(x)$ and that of the line $y = t(x)$, regarded as a function of x, is given by

$$g(x) = f(x) - t(x) = f(x) - f(c) - f'(c)(x - c).$$

Show that if f is concave upward (downward) on I, then g is positive (negative) on I except at the point c where g has the value 0, so that the curve $y = f(x)$ lies above (below) its tangent on both sides of the point P. Show that if the curve has an inflectional tangent at P, then g changes sign at c, so that the curve crosses from one side of its tangent to the other at P.

3.4 Limits Involving Infinity; Indeterminate Forms

The graph of the function

$$y = f(x) = \frac{1}{x} \qquad (x \neq 0)$$

is shown in Figure 24. Examining this graph, we see that f has some interesting limit properties of a type not yet discussed:

(i) As x takes smaller and smaller values of either sign (positive or negative), y takes larger and larger values of the same sign.

(ii) As x takes larger and larger values of either sign, y takes smaller and smaller values (of the same sign).

Here by a small or large negative number we mean, of course, a negative number with small or large absolute value.

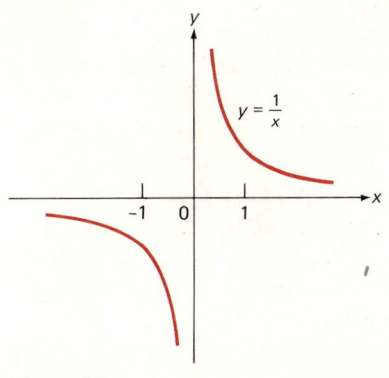

Figure 24

These properties of f express a kind of limiting behavior in which largeness plays a role, as well as smallness. How do we modify the language of limits to include situations of this type? Very simply. If a variable, say x, takes larger and larger positive values, we say that x *approaches* (*plus*) *infinity* and write $x \to \infty$, while if x takes larger and larger negative values, we say that x *approaches minus infinity* and write $x \to -\infty$. This is in keeping with the use of the symbols ∞ and $-\infty$ in writing infinite intervals. Once again we emphasize that ∞ and $-\infty$ are *not* numbers.

Infinite Limits vs. Limits at Infinity

We can now express properties (i) and (ii) of the function $y = 1/x$ more concisely, writing (i) as

$$\lim_{x \to 0^+} \frac{1}{x} = \infty, \qquad \lim_{x \to 0^-} \frac{1}{x} = -\infty, \tag{1}$$

and (ii) as

$$\lim_{x \to \infty} \frac{1}{x} = 0, \qquad \lim_{x \to -\infty} \frac{1}{x} = 0, \tag{2}$$

or even more concisely as

$$\lim_{x \to \pm\infty} \frac{1}{x} = 0. \tag{2'}$$

In (1) we have *infinite limits* and in (2) *limits at infinity*. This is in contrast to the limits considered so far, which are all *finite*, that is, of the type

$$\lim_{x \to a} f(x) = L, \qquad \lim_{x \to a^+} f(x) = L, \qquad \lim_{x \to a^-} f(x) = L,$$

where a and L are *numbers*, and not one of the symbols ∞ and $-\infty$.

The limits (1) are one-sided, but infinite limits can also be two-sided. Thus Figure 25 shows that

$$\lim_{x \to 0} \frac{1}{x^2} = \infty.$$

Clearly $f(x) \to \infty$ as $x \to a$ if and only if $f(x) \to \infty$ both as $x \to a^+$ and as $x \to a^-$, and the same is true with ∞ replaced by $-\infty$. We can also have infinite limits at infinity. For example, it is apparent from Figures 19 and 20, page 176, that

$$\lim_{x \to \infty} x^{1/3} = \infty, \qquad \lim_{x \to -\infty} x^{1/3} = -\infty$$

and

$$\lim_{x \to \pm\infty} x^{4/3} = \infty.$$

These considerations can be made rigorous by a modification of ε, δ language in which symbols other than ε and δ, namely C and A, are used for numbers that are typically *large*; this avoids the suggestion of *smallness* associated with ε and δ. For example, $f(x) \to \infty$ as $x \to a^+$ means that given any $C > 0$ (no matter how large), we can find a (sufficiently small) $\delta > 0$ such that $f(x) > C$ whenever $a < x < a + \delta$. Similarly, $f(x) \to -\infty$ as $x \to \infty$ means that given any $C > 0$ (no matter how large), we can find a (sufficiently

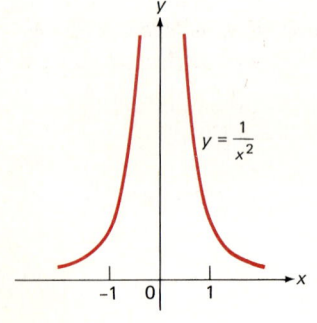

Figure 25

182 Chapter 3 Further Applications of Differentiation

large) $A > 0$ such that $f(x) < -C$ whenever $x > A$; here it is assumed that f is defined on some infinite interval of the type (c, ∞). Or, to give another example, $f(x) \to L$ as $x \to -\infty$ means that given any $\varepsilon > 0$ (no matter how small), we can find a (sufficiently large) $A > 0$ such that $|f(x) - L| < \varepsilon$ whenever $x < -A$, where it is now assumed that f is defined on some infinite interval of the type $(-\infty, c)$. The phrases in parentheses can be omitted once you get used to these definitions.

Example 1 Give a rigorous proof of the limit formulas (1).

Solution Given any $C > 0$, let $\delta = 1/C$. If $0 < x < \delta$, we have

$$\frac{1}{x} > \frac{1}{\delta} = C,$$

while if $-\delta < x < 0$, then

$$\frac{1}{x} < \frac{1}{-\delta} = -C,$$

by the reciprocals rule for inequalities (Theorem 4, page 9). But this is equivalent to (1) in "C, δ language." □

Example 2 Let n be any positive integer. Show that

$$\lim_{x \to 0^+} \frac{1}{x^n} = \infty, \quad \lim_{x \to 0^-} \frac{1}{x^n} = \begin{cases} \infty & \text{if } n \text{ is even,} \\ -\infty & \text{if } n \text{ is odd.} \end{cases} \quad (3)$$

Solution Given any $C > 0$, let $\delta = \sqrt[n]{1/C}$, and suppose first that $0 < x < \delta$. Then $0 < x^n < \delta^n$, and therefore

$$\frac{1}{x^n} > \frac{1}{\delta^n} = C. \quad (4)$$

Suppose next that $-\delta < x < 0$. Then $0 < x^n < \delta^n$ if n is even, and we again get the inequality (4), while $-\delta^n < x^n < 0$ if n is odd, and we get

$$\frac{1}{x^n} < \frac{1}{-\delta^n} = -C \quad (4')$$

instead, thereby completing the proof of (3) in C, δ language. Observe that (3) reduces to (1) if $n = 1$. □

Example 3 Show that

$$\lim_{x \to \pm\infty} \frac{1}{x^n} = 0, \quad (5)$$

where n is any positive integer.

Solution Given any $\varepsilon > 0$, let $A = \sqrt[n]{1/\varepsilon}$. Then if $x > A$ or $x < -A$, that is, if $|x| > A$, we have

$$\left|\frac{1}{x^n} - 0\right| = \frac{1}{|x|^n} < \frac{1}{A^n} = \varepsilon,$$

and we have proved (5), this time in "ε, A language." Observe that (5) reduces to (2') if $n = 1$. □

Example 4 Let m and n be any positive integers. Show that

$$\lim_{x \to \infty} x^{m/n} = \infty. \tag{6}$$

Solution To make $x^{m/n}$ exceed any given $C > 0$, we choose $x > A = C^{n/m}$, since then $x^{m/n} > A^{m/n} = C$. This proves (6) in "C, A language." □

Example 5 Let m and n be positive integers such that n is odd and m/n is in lowest terms. Show that

$$\lim_{x \to -\infty} x^{m/n} = \begin{cases} \infty & \text{if } m \text{ is even,} \\ -\infty & \text{if } m \text{ is odd.} \end{cases} \tag{6'}$$

Solution An informal proof will suffice. Since n is odd, $x^{m/n} = (\sqrt[n]{x})^m$ is defined for negative x. If x takes arbitrarily large negative values, so does $\sqrt[n]{x}$. Therefore $x^{m/n} = (\sqrt[n]{x})^m$ takes arbitrarily large positive values if m is even and arbitrarily large negative values if m is odd. □

For instance,

$$\lim_{x \to \infty} x^{5/6} = \infty, \quad \lim_{x \to -\infty} x^{3/7} = -\infty, \quad \lim_{x \to -\infty} x^{4/7} = \infty,$$

with the help of formulas (6) and (6'), but

$$\lim_{x \to -\infty} x^{5/6}$$

is meaningless.

Operations on Limits at Infinity

The theorems governing algebraic operations on ordinary limits, summarized on page 82, remain in force for limits at infinity. Thus, if

$$\lim_{x \to \infty} f(x) = L, \quad \lim_{x \to \infty} g(x) = M, \tag{7}$$

then

$$\lim_{x \to \infty} [f(x) \pm g(x)] = L \pm M, \tag{8}$$

$$\lim_{x \to \infty} f(x)g(x) = LM, \tag{9}$$

$$\lim_{x \to \infty} \frac{f(x)}{g(x)} = \frac{L}{M} \quad (M \neq 0), \tag{10}$$

and the same formulas hold if the symbol ∞ is replaced by $-\infty$ everywhere. The proofs are virtually the same as for ordinary limits.

Optional

For example, the proof of (8) is the exact analogue of that of Theorem 4, page 80, and goes as follows. Because of (7), given any $\varepsilon > 0$, we can find a number $A_f > 0$ such that $|f(x) - L| < \varepsilon/2$ whenever $x > A_f$ and a number $A_g > 0$ such that $|g(x) - M| < \varepsilon/2$ whenever $x > A_g$. Therefore, by the triangle inequality,

$$|f(x) \pm g(x) - (L \pm M)| \leq |f(x) - L| + |g(x) - M| < \frac{\varepsilon}{2} + \frac{\varepsilon}{2} = \varepsilon$$

Chapter 3   Further Applications of Differentiation

whenever $x > A = \max\{A_f, A_g\}$, that is, whenever x exceeds both A_f and A_g. But this is exactly what (8) means in ε, A language.

Example 6 Evaluate $\lim\limits_{x \to \infty} \dfrac{x^3}{(x+4)(2x^2+1)}$.

Solution First we divide the numerator and denominator by x^3:

$$\frac{x^3}{(x+4)(2x^2+1)} = \frac{\dfrac{x^3}{x^3}}{\dfrac{x+4}{x}\dfrac{2x^2+1}{x^2}} = \frac{1}{\left(1+\dfrac{4}{x}\right)\left(2+\dfrac{1}{x^2}\right)}.$$

This is a great help, since the expression on the right involves the functions $1/x$ and $1/x^2$, which, as we already know from Example 3, both approach 0 as $x \to \infty$. Using formulas (8)–(10), we now take the limit as $x \to \infty$. The result is

$$\lim_{x \to \infty} \frac{x^3}{(x+4)(2x^2+1)} = \frac{1}{\lim\limits_{x \to \infty}\left[\left(1+\dfrac{4}{x}\right)\left(2+\dfrac{1}{x^2}\right)\right]}$$

$$= \frac{1}{\lim\limits_{x \to \infty}\left(1+\dfrac{4}{x}\right)\lim\limits_{x \to \infty}\left(2+\dfrac{1}{x^2}\right)} = \frac{1}{1 \cdot 2} = \frac{1}{2},$$

since

$$\lim_{x \to \infty}\left(1+\frac{4}{x}\right) = \lim_{x \to \infty} 1 + 4 \lim_{x \to \infty} \frac{1}{x} = 1 + 4 \cdot 0 = 1,$$

$$\lim_{x \to \infty}\left(2+\frac{1}{x^2}\right) = \lim_{x \to \infty} 2 + \lim_{x \to \infty} \frac{1}{x^2} = 2 + 0 = 2. \quad \square$$

We have just shown that

$$\lim_{x \to \infty} \frac{x^3}{(x+4)(2x^2+1)} = \lim_{x \to \infty} \frac{x^3}{2x^3 + 8x^2 + x + 4} = \frac{1}{2}.$$

The same result can be obtained by the following more informal argument: Of the four terms in the denominator $2x^3 + 8x^2 + x + 4$, the term $2x^3$ makes by far the largest contribution if x is large. We can therefore write

$$\frac{x^3}{2x^3 + 8x^2 + x + 4} \approx \frac{x^3}{2x^3} = \frac{1}{2},$$

where the approximation gets better and better as x gets larger and larger.

Operations on Infinite Limits

The rules for manipulating infinite limits, as opposed to limits at infinity, are of a different type. For example, if

$$\lim_{x \to a} f(x) = L, \quad \lim_{x \to a} g(x) = \infty,$$

then

$$\lim_{x \to a}[f(x) + g(x)] = \infty. \tag{11}$$

Section 3.4 Limits Involving Infinity; Indeterminate Forms

Informally, suppose $f(x)$ approaches a limit as $x \to a$. Then if $g(x)$ takes arbitrarily large positive values as $x \to a$, so does $f(x) + g(x)$. This is certainly true, since the sum of a very large positive number and a number which is near a given number L must also be a very large positive number.

Optional

If a rigorous proof of this fact is desired, here's how it goes. Since $f(x) \to L$ as $x \to a$, there is a number k (not necessarily positive) and a positive number δ_f such that $f(x) > k$ whenever $0 < |x - a| < \delta_f$. Also, since $g(x) \to \infty$ as $x \to a$, then, given any $C > 0$, there is a positive number δ_g such that $g(x) > C - k$ whenever $0 < |x - a| < \delta_g$. But then $f(x) + g(x) > k + (C - k) = C$ whenever $0 < |x - a| < \delta = \min\{\delta_f, \delta_g\}$, and (11) is proved in C, δ language.

In an abbreviated notation, we have just shown that if $f(x) \to L$ and $g(x) \to \infty$, then $f(x) + g(x) \to \infty$ (the phrase "as $x \to a$" is omitted in three places). Similarly, it is easy to see that if $f(x) \to L$ and $g(x) \to -\infty$, then $f(x) + g(x) \to -\infty$. In fact, we can combine this and the preceding rule in a single rule:

(i) If $f(x) \to L$ and $g(x) \to \pm\infty$, then $f(x) + g(x) \to \pm\infty$, with the understanding that the same sign, plus or minus, must be chosen in both occurrences of the symbol $\pm$ (more generally, if the symbol $\pm$ or $\mp$ occurs in two or more places, we agree to pick the top sign in all the places or the bottom sign in all the places). A related rule is

(i') If $f(x) \to \pm\infty$ and $g(x) \to \pm\infty$, then $f(x) + g(x) \to \pm\infty$, which can be proved informally by observing that the sum of two arbitrarily large positive numbers is an arbitrarily large positive number, while the sum of two arbitrarily large negative numbers is an arbitrarily large negative number.

Continuing in this manner, we now give further rules for dealing with infinite limits. In rules (iv) and (iv') the notation $g(x) \to 0^+$ means that $g(x) \to 0$ as $x \to a$ and $g(x)$ is *positive* for all x in some deleted neighborhood of a, while $g(x) \to 0^-$ means that $g(x) \to 0$ as $x \to a$ and $g(x)$ is *negative* for all x in some deleted neighborhood of a.

(ii) If $f(x) \to L \neq 0$ and $g(x) \to \pm\infty$, then $f(x)g(x) \to \pm\infty$ if $L > 0$, while $f(x)g(x) \to \mp\infty$ if $L < 0$;

(ii') If $f(x) \to \pm\infty$ and $g(x) \to \pm\infty$, then $f(x)g(x) \to \pm\infty$ if $f(x) \to \infty$, while $f(x)g(x) \to \mp\infty$ if $f(x) \to -\infty$;

(iii) If $f(x) \to L$ and $g(x) \to \pm\infty$, then $f(x)/g(x) \to 0$;

(iv) If $f(x) \to L \neq 0$ and $g(x) \to 0^\pm$, then $f(x)/g(x) \to \pm\infty$ if $L > 0$, while $f(x)/g(x) \to \mp\infty$ if $L < 0$;

(iv') If $f(x) \to \pm\infty$ and $g(x) \to 0^\pm$, then $f(x)/g(x) \to \pm\infty$ if $f(x) \to \infty$, while $f(x)/g(x) \to \mp\infty$ if $f(x) \to -\infty$.

For example, the gist of rule (iv) is that the result of dividing any nonzero number by a tiny number of the same sign is a huge positive number, while the result of dividing any nonzero number by a tiny number of the opposite sign is a huge negative number. Make sure that you have a similar intuitive understanding of what each of these rules says.

In the above rules it is understood that the functions $f(x)$ and $g(x)$ approach their limits, finite or infinite, as $x \to a$, but all the rules remain valid if $x \to a^+$ or $x \to a^-$ instead, or even if $x \to \infty$ or $x \to -\infty$. For a limit as $x \to a^+$ the notation $g(x) \to 0^+$ means that $g(x) \to 0$ and $g(x)$ is positive for all x in some interval $(a, a + \delta)$, for a limit as $x \to \infty$ the notation $g(x) \to 0^-$ means that $g(x) \to 0$ and $g(x)$ is negative in some interval (c, ∞), and so on.

Example 7 Since $\cos x \to 1$ and $1/x \to \infty$ as $x \to 0^+$, application of rule (i) gives

$$\lim_{x \to 0^+} \left(\cos x + \frac{1}{x} \right) = \infty.$$

Example 8 Since $x^{2/3} \to \infty$ and $x^{5/3} \to \infty$ as $x \to \infty$, by Example 4, it follows from rule (i') that

$$\lim_{x \to \infty} (x^{2/3} + x^{5/3}) = \infty,$$

and then from rule (iii) that

$$\lim_{x \to \infty} \frac{1}{x^{2/3} + x^{5/3}} = 0.$$

Example 9 Since $(\sin x)/x \to 1$ and $1/x \to -\infty$ as $x \to 0^-$, application of rule (ii) gives

$$\lim_{x \to 0^-} \frac{\sin x}{x^2} = -\infty.$$

How does rule (iv) immediately lead to the same result?

The Indeterminate Forms 0/0, 0 · ∞, ∞/∞ and ∞ − ∞

Although rules (i)–(iv) tell us a great deal about the behavior of infinite limits, there are still a few cases which they do not cover. Specifically, rule (i') does not include $f(x) \to \infty$ and $g(x) \to -\infty$ (or $f(x) \to -\infty$ and $g(x) \to \infty$), rule (ii) does not include $f(x) \to 0$ and $g(x) \to \pm\infty$, rule (iii) does not include $f(x) \to \infty$ and $g(x) \to \pm\infty$ (or $f(x) \to -\infty$ and $g(x) \to \pm\infty$), and rule (iv) does not include $f(x) \to 0$ and $g(x) \to 0^{\pm}$. These four cases are denoted by the *indeterminate forms* $\infty - \infty$, $0 \cdot \infty$, ∞/∞ and $0/0$, respectively.† Each indeterminate form is shorthand for a limit that can take any value at all (including ∞ or $-\infty$), or even fail to exist. The word *indeterminacy* has the same meaning as indeterminate form, and by *resolving* an indeterminacy we mean finding the limit (if any) corresponding to the indeterminacy.

The indeterminate form 0/0 is shorthand for

$$\lim_{x \to a} \frac{f(x)}{g(x)},$$

where $f(x) \to 0$ and $g(x) \to 0$ as $x \to a$ (instead of $x \to a$, there might be $x \to a^+$, $x \to a^-$, $x \to \infty$ or $x \to -\infty$). We have already investigated many limits of this type, and in particular the evaluation of every derivative

$$f'(a) = \lim_{x \to a} \frac{f(x) - f(a)}{x - a}$$

† The expressions 0^0, ∞^0 and 1^∞ are also indeterminate forms, and will be considered in Section 6.5.

Section 3.4 Limits Involving Infinity; Indeterminate Forms 187 amounts to resolving the indeterminacy 0/0, since the numerator and denominator of the difference quotient on the right both approach 0 as $x \to a$, and in fact equal 0 for $x = a$. Choosing $f(x) = Lx$ and $g(x) = x$, we have

$$\lim_{x \to 0} \frac{f(x)}{g(x)} = \lim_{x \to 0} L = L,$$

while the choice $f(x) = \pm x$ and $g(x) = x^3$ gives

$$\lim_{x \to 0} \frac{f(x)}{g(x)} = \lim_{x \to 0} \frac{\pm 1}{x^2} = \pm \infty.$$

Moreover, if $f(x) = x \sin(1/x)$ and $g(x) = x$, then

$$\lim_{x \to 0} \frac{f(x)}{g(x)} = \lim_{x \to 0} \sin \frac{1}{x}$$

does not exist, by Example 10, page 78. But in all three cases $f(x) \to 0$ and $g(x) \to 0$ as $x \to a$. Thus the limit corresponding to the indeterminacy 0/0 can take any value at all, including ∞ or $-\infty$, or even fail to exist. The same is true of the indeterminacies $0 \cdot \infty$ and ∞/∞, since an expression $f(x)/g(x)$, reducing to 0/0 as $x \to a$, can also be written in the form

$$f(x) \frac{1}{g(x)}$$

reducing to $0 \cdot \infty$, or in the form

$$\frac{1/g(x)}{1/f(x)}$$

reducing to ∞/∞ (by rule (iv), if a function approaches zero, its reciprocal approaches infinity). For example, it follows from

$$\lim_{x \to 0^+} \frac{\pi x}{x} = \lim_{x \to 0^+} \pi x \frac{1}{x} = \lim_{x \to 0^+} \frac{1/x}{1/\pi x} = \pi$$

that π is a possible value of the limit corresponding to each of the indeterminate forms 0/0, $0 \cdot \infty$ and ∞/∞.

The indeterminate form $\infty - \infty$ is shorthand for

$$\lim_{x \to a} [f(x) - g(x)],$$

where $f(x) \to \infty$ and $g(x) \to \infty$ as $x \to a$. Choosing $f(x) = L + (1/x^2)$ and $g(x) = 1/x^2$, we have

$$\lim_{x \to 0} [f(x) - g(x)] = \lim_{x \to 0} L = L,$$

while the choice $f(x) = 2/x^2$ and $g(x) = 1/x^2$ gives

$$\lim_{x \to 0} [f(x) - g(x)] = \lim_{x \to 0} \frac{1}{x^2} = \infty$$

(similarly, $f(x) - g(x) \to -\infty$ as $x \to 0$ if $f(x) = 1/x^2$ and $g(x) = 2/x^2$). Moreover, if $f(x) = \sin(1/x) + 1/x^2$ and $g(x) = 1/x^2$, then

$$\lim_{x \to 0} [f(x) - g(x)] = \lim_{x \to 0} \sin \frac{1}{x}$$

188 Chapter 3 Further Applications of Differentiation

does not exist. But in all three cases $f(x) \to \infty$ and $g(x) \to \infty$ as $x \to 0$. Thus the limit corresponding to the indeterminacy $\infty - \infty$ can take any value at all, including ∞ or $-\infty$, or even fail to exist.

Example 10 Evaluate $\lim_{x \to \infty} (\sqrt{x^2 + 2x} - \sqrt{x^2 - 2x})$.

Solution As you can easily verify, $\sqrt{x^2 + 2x} \to \infty$ and $\sqrt{x^2 - 2x} \to \infty$ as $x \to \infty$. Thus, in evaluating this limit, we are resolving an indeterminacy of the form $\infty - \infty$. To get rid of the difference between the square roots, we multiply and divide the given expression by the sum of the square roots. In detail,

$$\sqrt{x^2 + 2x} - \sqrt{x^2 - 2x} = \frac{(\sqrt{x^2 + 2x} - \sqrt{x^2 - 2x})(\sqrt{x^2 + 2x} + \sqrt{x^2 - 2x})}{\sqrt{x^2 + 2x} + \sqrt{x^2 - 2x}}$$

$$= \frac{x^2 + 2x - (x^2 - 2x)}{\sqrt{x^2 + 2x} + \sqrt{x^2 - 2x}}$$

$$= \frac{4x}{\sqrt{x^2\left(1 + \frac{2}{x}\right)} + \sqrt{x^2\left(1 - \frac{2}{x}\right)}}$$

$$= \frac{4x}{x\sqrt{1 + \frac{2}{x}} + x\sqrt{1 - \frac{2}{x}}} = \frac{4}{\sqrt{1 + \frac{2}{x}} + \sqrt{1 - \frac{2}{x}}},$$

where in the next to the last step it can be assumed that x is positive (since $x \to \infty$), so that $\sqrt{x^2} = x$. Therefore

$$\lim_{x \to \infty} (\sqrt{x^2 + 2x} - \sqrt{x^2 - 2x}) = \lim_{x \to \infty} \frac{4}{\sqrt{1 + \frac{2}{x}} + \sqrt{1 - \frac{2}{x}}} = \frac{4}{1 + 1} = 2.$$

As an exercise, use the fact that $\sqrt{x^2} = -x$ if $x < 0$ to show that the limit is -2 instead of 2 if $x \to -\infty$ instead of $x \to \infty$. □

Problems

Evaluate the given limit (∞ or $-\infty$ is a possible value).

1. $\lim_{x \to \infty} x^{5/3}$
2. $\lim_{x \to 0^+} x^{-3/4}$
3. $\lim_{x \to -\infty} x^{11/7}$
4. $\lim_{x \to 0^-} x^{-7/11}$
5. $\lim_{x \to 2^+} \frac{2}{x^2 - 4}$
6. $\lim_{x \to 3^-} \frac{x + 3}{x^2 - 9}$
7. $\lim_{x \to 3^+} \frac{x^2 + 9}{x^2 - 9}$
8. $\lim_{x \to -1^+} \frac{6}{x^3 + 1}$
9. $\lim_{x \to 1^-} \frac{2x + 1}{x^3 - 1}$
10. $\lim_{x \to 0^+} \frac{x - 4}{\sqrt{x}}$
11. $\lim_{x \to -\infty} \frac{10}{x^3 + 2}$
12. $\lim_{x \to 0^-} \frac{x^2 + 1}{x^{2/3}}$
13. $\lim_{x \to 0^+} \frac{\sin x}{x^2}$
14. $\lim_{x \to 0} \frac{\cos x}{x^2}$
15. $\lim_{x \to 0^+} \frac{\tan x}{x^{1.1}}$

Evaluate the given limit, each an indeterminacy of the form ∞/∞.

16. $\lim_{x \to \infty} \frac{3 - x}{2 + x}$
17. $\lim_{x \to \infty} \frac{1 + x}{1 - x^2}$
18. $\lim_{x \to -\infty} \frac{1 - x^2}{1 + x}$
19. $\lim_{x \to -\infty} \frac{2x^2 + 1}{x^2 - 2}$
20. $\lim_{x \to \infty} \frac{\sqrt{x^2 + 1}}{x + 2}$
21. $\lim_{x \to -\infty} \frac{\sqrt{x^2 + 3}}{2x + 1}$

22. If $f(x) = (x - 1)/(x + 2)$, then $f(x) \to 1$ as $x \to \pm \infty$. Find all x such that $|f(x) - 1| < 0.01$.

Section 3.4 Limits Involving Infinity; Indeterminate Forms 189

23. If $f(x) = x/(x-3)$, then $f(x) \to \infty$ as $x \to 3^+$ and also $f(x) \to -\infty$ as $x \to 3^-$. Find all x such that $f(x) > 1000$ and also all x such that $f(x) < -1000$.

Evaluate the given limit, each an indeterminacy of the form $0 \cdot \infty$.

24. $\lim\limits_{x \to \infty} x^{-1/2}(x^{1/4} - 1)$

25. $\lim\limits_{x \to \infty} x^{-2/3}(1 - x^{2/3})$

26. $\lim\limits_{x \to 0^+} x^{1/2}(1 + x^{-3/4})$

27. $\lim\limits_{x \to 0^-} x^{1/3}(4 + x^{-2/3})$

28. $\lim\limits_{x \to 0} x \cot x$

29. $\lim\limits_{x \to 0^+} \sqrt{x} \csc x$

30. $\lim\limits_{x \to 0^+} x\sqrt{\csc x}$

31. A function bounded on an interval of the type (c, ∞) is said to be *bounded near* ∞, and a function bounded on an interval of the type $(-\infty, c)$ is said to be *bounded near* $-\infty$. Show that if $f(x)$ is bounded near ∞ and $g(x) \to 0$ as $x \to \infty$, then $f(x)g(x) \to 0$ as $x \to \infty$, and that the same is true with ∞ replaced by $-\infty$.

With the help of Problem 31, evaluate the given limit (if it exists).

32. $\lim\limits_{x \to \pm\infty} \sin x$

33. $\lim\limits_{x \to \pm\infty} \dfrac{\sin x}{x}$

34. $\lim\limits_{x \to \infty} \dfrac{\cos x}{\sqrt{x}}$

35. $\lim\limits_{x \to -\infty} \dfrac{\sin x + \cos x}{x^{1/3}}$

36. $\lim\limits_{x \to \infty} \dfrac{\sin x^2}{x}$

37. $\lim\limits_{x \to -\infty} \dfrac{1}{x} \sin \dfrac{1}{x}$

38. $\lim\limits_{x \to \infty} \dfrac{1}{\sqrt{x}} \cos \dfrac{1}{x}$

39. $\lim\limits_{x \to \infty} \dfrac{\sqrt{\cos x}}{x}$

40. $\lim\limits_{x \to -\infty} \dfrac{\tan x}{x}$

Evaluate the given limit, each an indeterminacy of the form $\infty - \infty$.

41. $\lim\limits_{x \to \infty} (x - \sqrt{x^2 - x + 1})$

42. $\lim\limits_{x \to \infty} (\sqrt{x^2 + 3x} - x)$

43. $\lim\limits_{x \to -\infty} (\sqrt{x^2 + 1} - \sqrt{x^2 - 4x})$

44. $\lim\limits_{x \to -\infty} (\sqrt{x^2 + 1} - \sqrt{x^2 - 1})$

45. $\lim\limits_{x \to \infty} (x\sqrt{x^2 + 1} - x^2)$

46. $\lim\limits_{x \to \infty} (\sqrt{(x + a)(x + b)} - x)$

47. $\lim\limits_{x \to \infty} (\sqrt{x^2 + x + 1} - \sqrt{x^2 - x})$

48. $\lim\limits_{x \to -\infty} (\sqrt{x^2 + x - 1} - \sqrt{x^2 - x + 1})$

3.5 Asymptotes and Vertical Tangents

Horizontal Asymptotes

Let f be a continuous function such that

$$\lim_{x \to \infty} f(x) = b, \tag{1}$$

where b is finite. Writing (1) in the equivalent form

$$\lim_{x \to \infty} |f(x) - b| = 0,$$

we recognize that $|f(x) - b|$ is the distance between the horizontal line $y = b$ and a variable point $P = (x, f(x))$ on the graph of the function f (see Figure 26). Therefore, as $x \to \infty$, the point P gets closer and closer to the line $y = b$. More generally, if $P = (x, f(x))$ gets closer and closer to a given straight line L as the distance between P and the origin approaches infinity, we say that L is an *asymptote* of f, or that the graph of f approaches L *asymptotically*. (For this definition to apply, the graph of f must "go to infinity" in at least one direction, which is of course not always true.) Thus we have just shown that if (1) holds, then f has the line $y = b$ as a *horizontal asymptote*.

Virtually the same argument shows that if

$$\lim_{x \to -\infty} f(x) = c, \tag{1'}$$

where c is finite, then f has the line $y = c$ as a horizontal asymptote. Notice that in the case where (1) and (1') both hold, f has two distinct horizontal

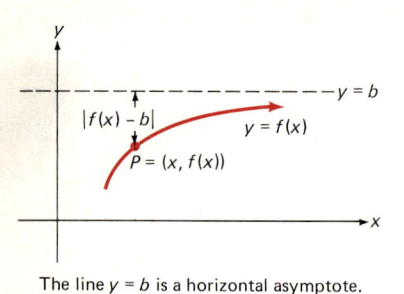

The line $y = b$ is a horizontal asymptote.
Figure 26

asymptotes if $b \neq c$, but only one if $b = c$. A function f cannot have more than two horizontal asymptotes, since as the point $P = (x, f(x))$ moves farther and farther away from the origin, it cannot get arbitrarily close to more than two horizontal lines, one as it moves to the right $(x \to \infty)$ and another as it moves to the left $(x \to -\infty)$. Also if f has a horizontal asymptote, at least one of the formulas (1) and (1') must hold. Thus, in looking for horizontal asymptotes, if any, of a function f, we need only examine the limiting behavior of f as $x \to \pm\infty$.

Example 1   If

$$f(x) = \frac{x}{1 + |x|},$$

then

$$\lim_{x \to \infty} f(x) = \lim_{x \to \infty} \frac{x}{1 + |x|} = \lim_{x \to \infty} \frac{x}{1 + x} = \lim_{x \to \infty} \frac{1}{\frac{1}{x} + 1} = 1,$$

since $|x| = x$ if $x > 0$, while

$$\lim_{x \to -\infty} f(x) = \lim_{x \to -\infty} \frac{x}{1 + |x|} = \lim_{x \to -\infty} \frac{x}{1 - x} = \lim_{x \to -\infty} \frac{1}{\frac{1}{x} - 1} = -1,$$

since $|x| = -x$ if $x < 0$. Therefore f has both lines $y = 1$ and $y = -1$ as horizontal asymptotes. This, together with the easily verified fact that f is odd and increasing on $(-\infty, \infty)$, gives f the "S-shaped" graph shown in Figure 27.

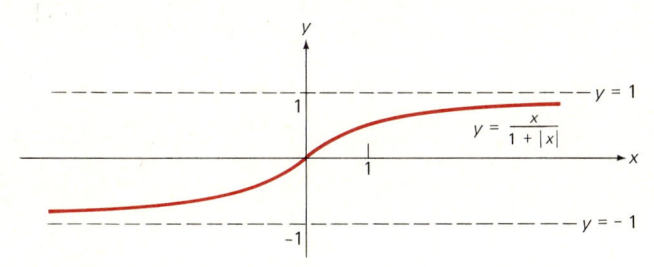

Figure 27

Example 2   If

$$f(x) = \frac{x^2 + 2}{x^2 + 1},$$

then

$$\lim_{x \to \pm\infty} f(x) = \lim_{x \to \pm\infty} \frac{x^2 + 2}{x^2 + 1} = \lim_{x \to \pm\infty} \frac{1 + \frac{2}{x^2}}{1 + \frac{1}{x^2}} = 1.$$

Hence in this case f has only one horizontal asymptote, namely the line $y = 1$, as shown in Figure 28.

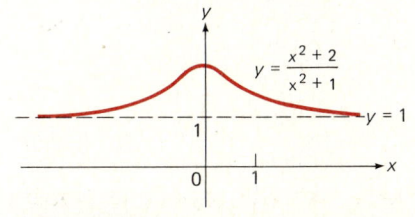

Figure 28

Section 3.5   Asymptotes and Vertical Tangents   **191**

Vertical Asymptotes

Next suppose f is a continuous function such that

$$\lim_{x \to a^+} f(x) = \infty \quad (\text{or } -\infty), \tag{2}$$

Writing (2) in the equivalent form

$$\lim_{x-a \to 0^+} f(x) = \infty \quad (\text{or } -\infty),$$

we recognize that $x - a$ is the distance between the vertical line $x = a$ and a variable point $P = (x, f(x))$ on the graph of f (see Figure 29). Hence, as $x \to a^+$, the point P gets closer and closer to the line $x = a$, while moving farther and farther away from the origin, either in the upward direction or in the downward direction. Thus, if (2) holds, the function f has the line $x = a$ as a *vertical asymptote*, and the graph of f approaches this asymptote from the *right*. Virtually the same argument shows that if

$$\lim_{x \to a^-} f(x) = \infty \quad (\text{or } -\infty), \tag{2'}$$

then f again has the line $x = a$ as a vertical asymptote, but the graph of f now approaches the asymptote from the *left*. Although a function f can have at most two horizontal asymptotes, it can have any number of vertical asymptotes (see Example 4 and Problem 1). Also if f has a vertical asymptote with x-intercept a, then at least one of the formulas (2) and (2') must hold, and if both hold, the graph of f approaches the asymptote from *both sides*. Thus, in looking for vertical asymptotes, we can confine our attention to the points, if any, at which f approaches infinity (∞ or $-\infty$).

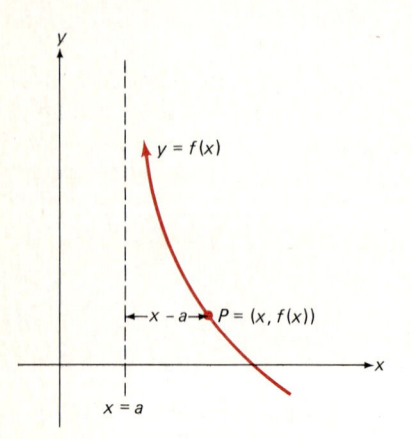

The line $x = a$ is a vertical asymptote.
Figure 29

Example 3 If

$$f(x) = \frac{1}{x^2 - 1},$$

then

$$\lim_{x \to 1^{\pm}} f(x) = \lim_{x \to 1^{\pm}} \frac{1}{x^2 - 1} = \lim_{x \to 1^{\pm}} \frac{1}{x + 1} \cdot \frac{1}{x - 1}.$$

But

$$\lim_{x \to 1^{\pm}} \frac{1}{x + 1} = \frac{1}{2},$$

and

$$\lim_{x \to 1^{\pm}} \frac{1}{x - 1} = \lim_{t \to 0^{\pm}} \frac{1}{t} = \pm \infty,$$

with the help of the substitution $t = x - 1$. Therefore, by rule (ii), page 186,

$$\lim_{x \to 1^+} \frac{1}{x^2 - 1} = \infty, \quad \lim_{x \to 1^-} \frac{1}{x^2 - 1} = -\infty,$$

and in virtually the same way it can be shown that

$$\lim_{x \to -1^+} \frac{1}{x^2 - 1} = -\infty, \quad \lim_{x \to -1^-} \frac{1}{x^2 - 1} = \infty$$

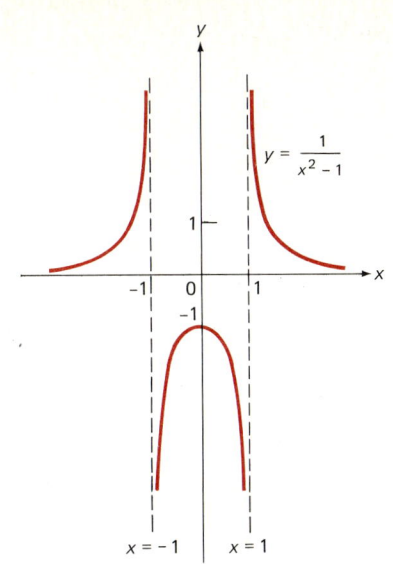

Figure 30

(give the details). Thus f has the lines $x = 1$ and $x = -1$ as vertical asymptotes, and the graph of f approaches these asymptotes from both sides, as shown in Figure 30. There are no other vertical asymptotes, since 1 and -1 are the only points at which f approaches infinity. However, the line $y = 0$ (the x-axis) is a horizontal asymptote of f. This follows at once from the fact that

$$\lim_{x \to \pm\infty} \frac{1}{x^2 - 1} = 0. \quad \square$$

As illustrated by the last example, a rational function approaches infinity (in absolute value) at precisely those points, if any, where its denominator equals zero. In saying this, we assume that the function is *in lowest terms*, which means that the numerator and denominator have no common factors. Thus, for example, the function $(x + 1)/(x^2 - 1)$ approaches ∞ as $x \to 1^+$ and $-\infty$ as $x \to 1^-$, but it has a finite limit as $x \to -1^+$ and as $x \to -1^-$, since

$$\lim_{x \to -1^\pm} \frac{x+1}{x^2 - 1} = \lim_{x \to -1^\pm} \frac{x+1}{(x+1)(x-1)} = \lim_{x \to -1^\pm} \frac{1}{x-1} = -\frac{1}{2}.$$

Example 4 Examining the graphs of the functions $\tan x$ and $\cot x$ (see Figure 24, page 63), we find that

$$\lim_{x \to 0^+} \cot x = \infty, \qquad \lim_{x \to 0^-} \cot x = -\infty,$$

$$\lim_{x \to \frac{1}{2}\pi^+} \tan x = -\infty, \qquad \lim_{x \to \frac{1}{2}\pi^-} \tan x = \infty.$$

More generally,

$$\lim_{x \to n\pi^+} \cot x = \infty, \qquad \lim_{x \to n\pi^-} \cot x = -\infty,$$

$$\lim_{x \to (n+\frac{1}{2})\pi^+} \tan x = -\infty, \qquad \lim_{x \to (n+\frac{1}{2})\pi^-} \tan x = \infty$$

for every integer $n = 0, \pm 1, \pm 2, \ldots$ Therefore every line $x = n\pi$ is a vertical asymptote of $\cot x$, while every line $x = (n + \frac{1}{2})\pi$ is a vertical asymptote of $\tan x$. Thus each of the functions $\tan x$ and $\cot x$ has infinitely many vertical asymptotes!

It is also possible for a function to have an *oblique* asymptote, that is, an asymptote which is neither horizontal nor vertical.

Example 5 The function

$$f(x) = \left| x + \frac{1}{x} \right|,$$

graphed in Figure 31, has both lines $y = \pm x$ as oblique asymptotes. This is apparent from the fact that the difference $f(x) - |x|$ becomes smaller and smaller as $x \to \pm\infty$, since $1/x \to 0$ as $x \to \pm\infty$, but it can also be proved more formally (see Problem 13). Notice that f also has the y-axis as a vertical asymptote, since $f(x) \to \infty$ as $x \to 0$.

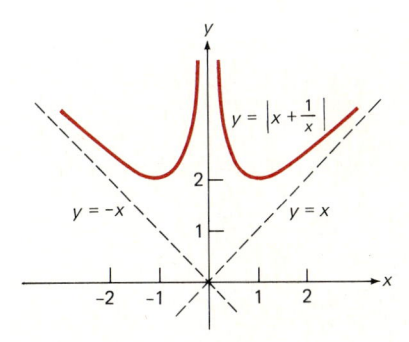

The lines $y = \pm x$ are oblique asymptotes.
Figure 31

Section 3.5   Asymptotes and Vertical Tangents   **193**

Infinite Derivatives and Vertical Tangents

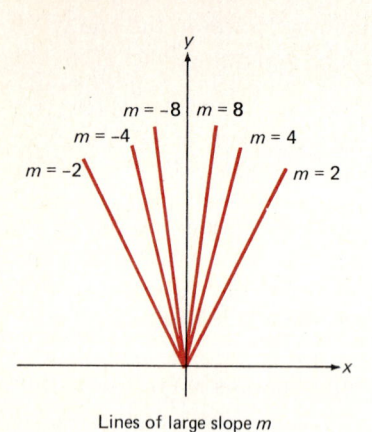

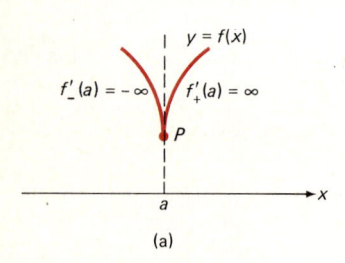

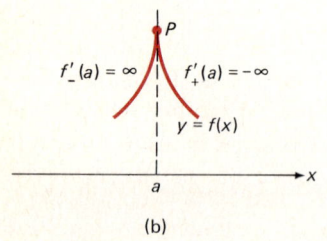

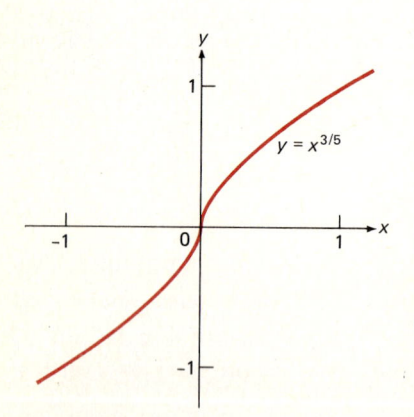

Finally, suppose the derivative of a function f turns out to be infinite at a, which means that

$$f'(a) = \lim_{x \to a} \frac{f(x) - f(a)}{x - a} = \infty$$

or

$$f'(a) = \lim_{x \to a} \frac{f(x) - f(a)}{x - a} = -\infty.$$

Then the function

$$m(x) = \frac{f(x) - f(a)}{x - a}$$

approaches ∞ or $-\infty$ as $x \to a$. But $m(x)$ is the slope of the secant line through the fixed point $P = (a, f(a))$ and the variable point $Q = (x, f(x))$ of the curve $y = f(x)$, and as the slope of a line takes larger and larger positive or negative values, the line gets closer and closer to being vertical (see Figure 32). Guided by these considerations, we define the *vertical line* $x = a$ to be the tangent (line) to the curve $y = f(x)$ at the point P in the case where $f'(a) = \infty$ or $f'(a) = -\infty$. In writing the last two formulas, we are not saying that ∞ and $-\infty$ are numbers, which they are not, but only that $m(x)$ approaches (plus or minus) infinity as $x \to a$.

In terms of one-sided derivatives, $f'(a) = \infty$ is equivalent to $f'_+(a) = f'_-(a) = \infty$, and $f'(a) = -\infty$ is equivalent to $f'_+(a) = f'_-(a) = -\infty$. There is also the possibility that the one-sided derivatives are infinite but unequal, in the sense that

$$f'_+(a) = \lim_{x \to a^+} m(x) = \infty, \qquad f'_-(a) = \lim_{x \to a^-} m(x) = -\infty, \qquad (3)$$

or

$$f'_+(a) = -\infty, \qquad f'_-(a) = \infty. \qquad (3')$$

Then we still define the tangent at P to be the vertical line $x = a$, but now the curve $y = f(x)$ has a sharp point or *cusp* at P, as illustrated in Figure 33(a) for the case (3) and in Figure 33(b) for the case (3').

Example 6 Investigate the behavior of the curve $y = f(x) = x^{3/5}$ at the origin.

Solution Using rule (iv), page 186, and the fact that $x^{2/5} \to 0^+$ as $x \to 0$, we calculate the derivative of f at $x = 0$:

$$f'(0) = \lim_{x \to 0} \frac{x^{3/5} - 0^{3/5}}{x - 0} = \lim_{x \to 0} \frac{1}{x^{2/5}} = \infty.$$

Therefore $f'_+(0) = f'_-(0) = \infty$, and the curve $y = x^{3/5}$ has a vertical tangent, but no cusp, at the origin (see Figure 34). □

Example 7 Investigate the behavior of the curve $y = f(x) = x^{2/5}$ at the origin.

Solution Using the same rule and the fact that $x^{3/5} \to 0^{\pm}$ as $x \to 0^{\pm}$, we calculate the one-sided derivatives of f at $x = 0$:

$$f'_+(0) = \lim_{x \to 0^+} \frac{x^{2/5} - 0^{2/5}}{x - 0} = \lim_{x \to 0^+} \frac{1}{x^{3/5}} = \infty, \qquad f'_-(0) = \lim_{x \to 0^-} \frac{1}{x^{3/5}} = -\infty.$$

Figure 32 Lines of large slope m

Figure 33
(a) $f'_-(a) = -\infty$, $f'_+(a) = \infty$, $y = f(x)$
(b) $f'_-(a) = \infty$, $f'_+(a) = -\infty$, $y = f(x)$

Figure 34 $y = x^{3/5}$

Since $f'_+(0)$ and $f'_-(0)$ are infinite but unequal, the curve $y = x^{2/5}$ has a vertical tangent and a cusp at the origin (see Figure 35). ☐

There is a close analogy between cusps and corners, the latter corresponding to the case where $f'_+(a)$ and $f'_-(a)$ are finite but unequal (see page 119), or where one of the derivatives $f'_+(a)$ and $f'_-(a)$ is infinite and the other is finite. However, although a curve has a vertical tangent at a cusp, a curve has no tangent at a corner. How do you explain this difference?

If a curve $y = f(x)$ has a vertical tangent T at the point $P = (a, f(a))$, then the normal N to the curve at P, defined as the line through P perpendicular to T, is the horizontal line $y = f(a)$.

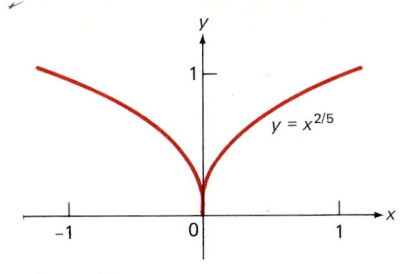

Figure 35

Example 8 Find the tangent T and normal N to the curve

$$y = f(x) = (x - 1)^{1/3} + 2$$

at the point $P = (1, 2)$.

Solution Calculating the derivative of f at $x = 1$, we get

$$f'(1) = \lim_{x \to 1} \frac{[(x-1)^{1/3} + 2] - 2}{x - 1} = \lim_{x \to 1} \frac{(x-1)^{1/3}}{x - 1},$$

so that

$$f'(1) = \lim_{t \to 0} \frac{t^{1/3}}{t} = \lim_{t \to 0} \frac{1}{t^{2/3}} = \infty$$

after substituting $t = x - 1$ and observing that $t^{2/3} \to 0^+$ as $t \to 0$. Hence the tangent T to the curve $y = f(x)$ at the point P is the vertical line $x = 1$ (there is no cusp), and the normal N at P is the horizontal line $y = f(1) = 2$, as shown in Figure 36. ☐

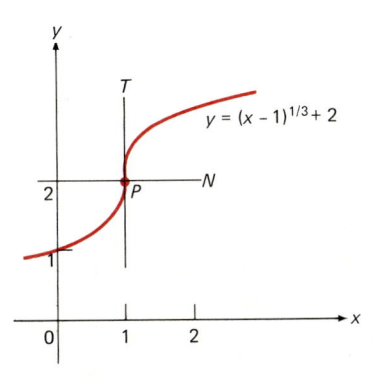

Figure 36

You will recall that a function f is said to be *differentiable* at a point x if f has a derivative $f'(x)$ at x. This definition was made on page 109, before the notion of an infinite derivative had entered the picture. We now stress that the derivative $f'(x)$ figuring in the definition of differentiability must be *finite*. Thus infinite derivatives do not "exist" in the previously understood sense of the word. In particular, if a function f has an infinite derivative at c, then c is a critical point of c (see page 169).

Problems

★ **1.** Give an example of a function f with exactly n vertical asymptotes.

Find all asymptotes (horizontal and vertical) of the given function, and sketch its graph.

2. $f(x) = \dfrac{2}{x - 1}$

3. $f(x) = \dfrac{x - 4}{2x + 4}$

4. $f(x) = \dfrac{x}{x^2 - 1}$

5. $f(x) = \dfrac{x^2}{x^2 - 4}$

6. $f(x) = \dfrac{1 + |x|}{x}$

7. $f(x) = \dfrac{4 - x^2}{4 + x^2}$

8. $f(x) = \dfrac{x}{\sqrt{1 - x^2}}$

9. $f(x) = \dfrac{\sqrt{x} + 1}{\sqrt{x} - 1}$

10. $f(x) = \dfrac{3 \cos x}{x + 1}$

11. The function

$$f(x) = \dfrac{1}{x^7 + 128}$$

has exactly two asymptotes. What are they?

Section 3.5 Asymptotes and Vertical Tangents

12. The function

$$f(x) = \frac{1}{x^8 - 256}$$

has exactly three asymptotes. What are they?

13. Show that the line

$$y = mx + b \quad (m \neq 0)$$

is an oblique asymptote of the function $y = f(x)$ if and only if at least one of the conditions

$$\lim_{x \to \infty} [f(x) - mx - b] = 0 \quad \text{(i)}$$

or

$$\lim_{x \to -\infty} [f(x) - mx - b] = 0 \quad \text{(ii)}$$

holds. Show that (i) is equivalent to

$$m = \lim_{x \to \infty} \frac{f(x)}{x}, \quad b = \lim_{x \to \infty} [f(x) - mx], \quad \text{(i')}$$

while (ii) is equivalent to

$$m = \lim_{x \to -\infty} \frac{f(x)}{x}, \quad b = \lim_{x \to -\infty} [f(x) - mx]. \quad \text{(ii')}$$

What is the largest number of oblique asymptotes that a function can have?

With the help of Problem 13, find all asymptotes (oblique as well as horizontal and vertical) of the given function, and sketch its graph.

14. $f(x) = \dfrac{1 + x^2}{1 + x}$

15. $f(x) = \dfrac{4 - x^3}{x^2}$

16. $f(x) = \dfrac{x^3}{4 - x^2}$

17. $f(x) = \dfrac{2x^2}{1 + |x|}$

18. $f(x) = \sqrt{x^2 + x + 1}$

19. $f(x) = \sqrt[3]{x^2 + 1}$

Show that

20. A polynomial $P(x)$ of degree greater than 1 has no asymptotes

21. The function $f(x) = x + \sin x$ has no asymptotes

22. If r is a positive rational number, the function $f(x) = x^r$ has no asymptotes unless $r = 1$

Find the tangent T and normal N to the given curve at the point P. Is there a cusp at P?

23. $y = (x - 1)^{3/5}, P = (1, 0)$

24. $y = (x + 1)^{2/3} + 1, P = (-1, 1)$

25. $y = \sqrt{|x - 3|} - 1, P = (3, -1)$

26. $y = \sqrt[3]{x^2 - 2x} + 3, P = (2, 3)$

27. At which points does the curve $y = \sqrt[3]{\cos x}$ have vertical tangents? Are these points cusps?

3.6 L'Hospital's Rule

We now use differentiation to develop a powerful technique for resolving indeterminacies, which will very often allow us to substitute routine computation for manipulative ingenuity. We begin with indeterminacies of the form $0/0$, but afterwards allow infinite limits, leading to the indeterminate forms ∞/∞, $0 \cdot \infty$ and $\infty - \infty$ as well.

Theorem 11 (*L'Hospital's rule for 0/0*).† If

(i) f and g are differentiable on (a, b), with derivatives f' and g',

(ii) g' is nonzero at every point of (a, b),

(iii) $\lim_{x \to a^+} f(x) = \lim_{x \to a^+} g(x) = 0$,

and if

$$\lim_{x \to a^+} \frac{f'(x)}{g'(x)} = L,$$

then

$$\lim_{x \to a^+} \frac{f(x)}{g(x)} = L.$$

---

† Actually discovered by John Bernoulli (1667–1748) and given to the Marquis de l'Hospital (1661–1704), the author of the first calculus book, in return for salary. The letter *s* in L'Hospital is silent, and the alternative spelling L'Hôpital is often encountered.

Proof (Optional) If f and g are already continuous from the right at a, then $f(a) = g(a) = 0$, because of (iii), but otherwise we set $f(a) = g(a) = 0$ by definition. Then f and g are continuous on every interval $[a, x]$ such that $a < x < b$ (why?). By Theorem 4, page 164 (Cauchy's mean value theorem),

$$\frac{f(x)}{g(x)} = \frac{f(x) - f(a)}{g(x) - g(a)} = \frac{f'(c)}{g'(c)},$$

where $a < c < x$. But $x \to a^+$ implies $c \to a^+$, and hence

$$\lim_{x \to a^+} \frac{f(x)}{g(x)} = \lim_{c \to a^+} \frac{f'(c)}{g'(c)} = L. \quad \blacksquare$$

For simplicity, we proved L'Hospital's rule for the case $x \to a^+$, but virtually the same argument shows that it continues to hold for $x \to a^-$, with the interval (a, b) replaced by (b, a), $b < a$. L'Hospital's rule also holds for $x \to a$, with appropriate small changes in the hypotheses. Specifically, if

(i') f and g are differentiable, with derivatives f' and g', in a deleted neighborhood D of the point a,
(ii') g' is nonzero at every point of D,
(iii') $\lim_{x \to a} f(x) = \lim_{x \to a} g(x) = 0$,

and if

$$\lim_{x \to a} \frac{f'(x)}{g'(x)} = L,$$

then

$$\lim_{x \to a} \frac{f(x)}{g(x)} = L.$$

This version of L'Hospital's rule is, of course, an immediate consequence of Theorem 11 and the companion theorem for $x \to a^-$.

Example 1 Evaluate $\lim_{x \to 0^+} \dfrac{\tan x}{\sqrt{x}}$.

Solution By L'Hospital's rule, we have

$$\lim_{x \to 0^+} \frac{\tan x}{\sqrt{x}} = \lim_{x \to 0^+} \frac{\dfrac{d}{dx} \tan x}{\dfrac{d}{dx} \sqrt{x}} = \lim_{x \to 0^+} \frac{\sec^2 x}{1/(2\sqrt{x})}$$

$$= \lim_{x \to 0^+} \frac{2\sqrt{x}}{\cos^2 x} = \frac{2\sqrt{0}}{\cos^2 0} = 0. \quad \square$$

Example 2 Show that

$$\lim_{x \to 0} \frac{1 - \cos x}{x^2} = \frac{1}{2}. \tag{1}$$

Solution By L'Hospital's rule,

$$\lim_{x \to 0} \frac{1 - \cos x}{x^2} = \lim_{x \to 0} \frac{\frac{d}{dx}(1 - \cos x)}{\frac{d}{dx} x^2} = \lim_{x \to 0} \frac{\sin x}{2x} = \frac{1}{2} \lim_{x \to 0} \frac{\sin x}{x} = \frac{1}{2}.$$ ☐

Remark You might be tempted to evaluate the limit

$$\lim_{x \to 0} \frac{\sin x}{x} \qquad (2)$$

itself with the help of L'Hospital's rule, instead of regarding it as known, observing that

$$\lim_{x \to 0} \frac{\sin x}{x} = \lim_{x \to 0} \frac{\frac{d}{dx} \sin x}{\frac{d}{dx} x} = \lim_{x \to 0} \frac{\cos x}{1} = 1,$$

by the continuity of $\cos x$. However, this is circular reasoning, since if you look back, you will see that formula (2) was used to prove the differentiation formula $D_x \sin x = \cos x$ in the first place!

Example 3 Evaluate $\lim_{x \to 0} \dfrac{x - \sin x}{x^3}$.

Solution By L'Hospital's rule,

$$\lim_{x \to 0} \frac{x - \sin x}{x^3} = \lim_{x \to 0} \frac{\frac{d}{dx}(x - \sin x)}{\frac{d}{dx} x^3} = \lim_{x \to 0} \frac{1 - \cos x}{3x^2} = \frac{1}{6},$$

where in the last step we use formula (1), which was itself established by applying L'Hospital's rule. This shows that several consecutive applications of L'Hospital's rule may be needed to evaluate a limit.   ☐

Example 4 There is no guarantee that L'Hospital's rule will help resolve a given indeterminacy. In fact, it may turn out that the limit

$$\lim_{x \to a} \frac{f'(x)}{g'(x)}$$

fails to exist, even though the limit

$$\lim_{x \to a} \frac{f(x)}{g(x)}$$

exists and can be found easily by some other method. For example, if

$$f(x) = x^2 \sin \frac{1}{x}, \qquad g(x) = x,$$

then

$$\lim_{x \to 0} \frac{f(x)}{g(x)} = \lim_{x \to 0} x \sin \frac{1}{x} = 0$$

(see Example 8, page 78), whereas the limit

$$\lim_{x \to 0} \frac{f'(x)}{g'(x)} = \lim_{x \to 0} \left(2x \sin \frac{1}{x} - \cos \frac{1}{x} \right)$$

fails to exist (why?).

There is a rule similar to Theorem 11 for using differentiation to resolve indeterminacies of the form ∞/∞. For simplicity, we state the rule for the case $x \to a^+$, but it is also true for $x \to a^-$ and $x \to a$, with the same slight changes of hypotheses already described in connection with Theorem 11. The proof is quite technical, and hence is omitted; it can be found in any advanced calculus book.

Theorem 11' (*L'Hospital's rule for ∞/∞*). If

(i) f and g are differentiable on (a, b), with derivatives f' and g',
(ii) g' is nonzero at every point of (a, b),
(iii) $\lim_{x \to a^+} f(x) = \lim_{x \to a^+} g(x) = \infty$,

and if

$$\lim_{x \to a^+} \frac{f'(x)}{g'(x)} = L,$$

then

$$\lim_{x \to a^+} \frac{f(x)}{g(x)} = L.$$

It can be shown that Theorem 11' remains true if condition (iii) is replaced by

$$\lim_{x \to a^+} |f(x)| = \lim_{x \to a^+} |g(x)| = \infty,$$

and that both Theorems 11 and 11' remain true if L is replaced by ∞ or $-\infty$. Also, Theorems 11 and 11' can be extended to the case $x \to \infty$ by making the substitution $x = 1/t$ and observing that

$$\lim_{x \to \infty} \frac{f(x)}{g(x)} = \lim_{t \to 0^+} \frac{f(1/t)}{g(1/t)} = \lim_{t \to 0^+} \frac{(-1/t^2)f'(1/t)}{(-1/t^2)g'(1/t)}$$

$$= \lim_{t \to 0^+} \frac{f'(1/t)}{g'(1/t)} = \lim_{x \to \infty} \frac{f'(x)}{g'(x)},$$

and a similar result holds for $x \to -\infty$. Here we assume that L'Hospital's rule can be applied at the second step of the calculation, and that conditions (i) and (ii) of Theorems 11 and 11' hold on an infinite interval of the type (c, ∞) or $(-\infty, c)$, as the case may be.

Example 5 Evaluate $L = \lim_{x \to \infty} \frac{3x^2 + 4x - 2}{6x^2 - 5x + 8}$.

Solution The limit L is an indeterminacy of the form ∞/∞. By two applications of Theorem 11', that is, by two consecutive differentiations of the numerator and denominator, we have

$$L = \lim_{x \to \infty} \frac{\dfrac{d}{dx}(3x^2 + 4x - 2)}{\dfrac{d}{dx}(6x^2 - 5x + 8)} = \lim_{x \to \infty} \frac{6x + 4}{12x - 5}$$

$$= \lim_{x \to \infty} \frac{\dfrac{d}{dx}(6x + 4)}{\dfrac{d}{dx}(12x - 5)} = \lim_{x \to \infty} \frac{6}{12} = \frac{1}{2}.$$

As an exercise, evaluate L by the method of Section 3.4. ∎

Example 6 Evaluate $L = \lim\limits_{x \to 1} (1 - x) \tan \dfrac{\pi x}{2}$.

Solution As it stands, the limit corresponds to an indeterminacy of the type $0 \cdot \infty$, but it can be rewritten as

$$L = \lim_{x \to 1} \frac{1 - x}{\cot \dfrac{\pi x}{2}},$$

corresponding to the indeterminate form $0/0$. The old way of evaluating L is to make the substitution $t = 1 - x$. Then

$$L = \lim_{t \to 0} \frac{t}{\cot\left(\dfrac{\pi}{2} - \dfrac{\pi t}{2}\right)} = \lim_{t \to 0} \frac{t}{\tan \dfrac{\pi t}{2}} = \lim_{t \to 0} \frac{t \cos \dfrac{\pi t}{2}}{\sin \dfrac{\pi t}{2}}$$

$$= \frac{2}{\pi} \lim_{t \to 0} \frac{\dfrac{\pi t}{2}}{\sin \dfrac{\pi t}{2}} \cdot \lim_{t \to 0} \cos \dfrac{\pi t}{2} = \frac{2}{\pi} \cdot 1 \cdot \cos 0 = \frac{2}{\pi},$$

after considerable effort. The new way, based on L'Hospital's rule, is much easier:

$$L = \lim_{x \to 1} \frac{\dfrac{d}{dx}(1 - x)}{\dfrac{d}{dx} \cot \dfrac{\pi x}{2}} = \lim_{x \to 1} \frac{-1}{-\dfrac{\pi}{2} \csc^2 \dfrac{\pi x}{2}}$$

$$= \frac{2}{\pi} \lim_{x \to 1} \sin^2 \frac{\pi x}{2} = \frac{2}{\pi} \sin^2 \frac{\pi}{2} = \frac{2}{\pi}.$$

Moreover, the preliminary substitution is no longer necessary, and would just be wasted effort. ∎

Example 7 Evaluate $L = \lim\limits_{x \to \pi/2} (\tan x - \sec x)$.

Solution This is an indeterminacy of the form $\infty - \infty$, which can be converted into an indeterminacy of the form 0/0 by observing that

$$L = \lim_{x \to \pi/2} \frac{\sin x - 1}{\cos x}.$$

With the help of L'Hospital's rule, we immediately get

$$L = \lim_{x \to \pi/2} \frac{\frac{d}{dx}(\sin x - 1)}{\frac{d}{dx} \cos x} = \lim_{x \to \pi/2} \frac{\cos x}{-\sin x} = \frac{0}{-1} = 0.$$

To find L without using L'Hospital's rule requires much more work, as well as considerable ingenuity:

$$L = \lim_{x \to \pi/2} \left(\frac{\sin x - 1}{\cos x} \cdot \frac{\sin x + 1}{\sin x + 1} \right) = \lim_{x \to \pi/2} \frac{\sin^2 x - 1}{(\sin x + 1)\cos x}$$

$$= \lim_{x \to \pi/2} \frac{-\cos^2 x}{(\sin x + 1)\cos x} = \lim_{x \to \pi/2} \frac{-\cos x}{\sin x + 1} = \frac{-0}{1 + 1} = 0. \quad \square$$

When *Not* To Use L'Hospital's Rule

L'Hospital's rule is certainly a very powerful tool for evaluating limits. However, as illustrated by the following examples, there are many situations in which L'Hospital's rule is not applicable, for one reason or another.

Example 8 Evaluate $L = \lim\limits_{x \to 0} \dfrac{\cos x}{x + 2}$.

Solution By continuity, we see at once that

$$L = \frac{\cos 0}{0 + 2} = \frac{1}{2}.$$

Blind application of L'Hospital's rule gives

$$L = \lim_{x \to 0} \frac{\frac{d}{dx} \cos x}{\frac{d}{dx}(x + 2)} = \lim_{x \to 0} \frac{-\sin x}{1} = -\sin 0 = 0,$$

which is *false*. But L'Hospital's rule is not applicable here. In fact,

$$\lim_{x \to 0} \cos x = 1, \quad \lim_{x \to 0} (x + 2) = 2,$$

so that we are not even dealing with an indeterminate form! $\quad \square$

Example 9 Evaluate $L = \lim\limits_{x \to \infty} \dfrac{x - \sin x}{x}$.

Solution This time we have an indeterminacy of the form ∞/∞, and the limit can be evaluated at once:

$$L = \lim_{x \to \infty} \left(1 - \frac{\sin x}{x} \right) = 1 - \lim_{x \to \infty} \frac{\sin x}{x} = 1$$

Section 3.6 L'Hospital's Rule

(see Problem 31, page 190). Attempting to evaluate L by L'Hospital's rule, we get

$$L = \lim_{x \to \infty} \frac{\dfrac{d}{dx}(x - \sin x)}{\dfrac{d}{dx}x} = \lim_{x \to \infty} (1 - \cos x).$$

But the limit on the right does not exist (why not?), and hence L'Hospital's rule is not applicable here either. □

Example 10 Evaluate

$$L = \lim_{x \to \pi/2} \frac{\tan x}{\sec x}. \tag{3}$$

Solution Here we have another indeterminacy of the form ∞/∞, which can be evaluated effortlessly, since

$$L = \lim_{x \to \pi/2} \frac{\sin x}{\cos x} \cos x = \lim_{x \to \pi/2} \sin x = 1. \tag{4}$$

Attempting to evaluate L by L'Hospital's rule, we get

$$L = \lim_{x \to \pi/2} \frac{\dfrac{d}{dx}\tan x}{\dfrac{d}{dx}\sec x} = \lim_{x \to \pi/2} \frac{\sec^2 x}{\sec x \tan x} = \lim_{x \to \pi/2} \frac{\sec x}{\tan x},$$

which is again of the form ∞/∞. Another application of L'Hospital's rule gives

$$L = \lim_{x \to \pi/2} \frac{\dfrac{d}{dx}\sec x}{\dfrac{d}{dx}\tan x} = \lim_{x \to \pi/2} \frac{\sec x \tan x}{\sec^2 x} = \lim_{x \to \pi/2} \frac{\tan x}{\sec x},$$

and we are back where we started from, completely stymied! If we write (3) in the form

$$L = \lim_{x \to \pi/2} \frac{\cos x}{\cot x}, \tag{3'}$$

which is now an indeterminacy of the form 0/0, then one application of L'Hospital's rule gives the answer, since

$$L = \lim_{x \to \pi/2} \frac{\dfrac{d}{dx}\cos x}{\dfrac{d}{dx}\cot x} = \lim_{x \to \pi/2} \frac{-\sin x}{-\csc^2 x} = \lim_{x \to \pi/2} \sin^3 x = 1.$$

However, in view of (4), this is still doing things the hard way. □

As a continuing exercise, go back and use L'Hospital's rule to evaluate as many limits as you can from among those appearing in earlier examples and problems (starting with Section 1.5). You will discover that use of the rule very often converts a difficult calculation into a routine exercise.

Problems

Describe the indeterminacy corresponding to the given limit. Then use L'Hospital's rule to evaluate the limit.

1. $\lim\limits_{x \to 7} \dfrac{x^2 - 15x + 56}{x^2 - 3x - 28}$

2. $\lim\limits_{x \to -1} \dfrac{3x^3 + 3x^2 + x + 1}{x^2 - 1}$

3. $\lim\limits_{x \to 1} \dfrac{x^{1/3} - x^{1/2}}{x^{1/4} - x^{1/5}}$

4. $\lim\limits_{x \to 4} \dfrac{2 - \sqrt{x}}{3 - \sqrt{2x + 1}}$

5. $\lim\limits_{x \to -8} \dfrac{\sqrt{1-x} - 3}{2 + \sqrt[3]{x}}$

6. $\lim\limits_{x \to \infty} \dfrac{x + \sqrt{x}}{2x + 1}$

7. $\lim\limits_{x \to \infty} x \sin \dfrac{1}{x}$

8. $\lim\limits_{x \to 0^+} \dfrac{\sin \sqrt{x}}{\sin x}$

9. $\lim\limits_{x \to 0} \dfrac{\cos x - \cos 3x}{x^2}$

10. $\lim\limits_{x \to 0^-} \dfrac{\sin x}{1 - \cos x}$

11. $\lim\limits_{x \to 0} x \cot 2x$

12. $\lim\limits_{x \to 0} \dfrac{x - \sin x}{x - \tan x}$

13. $\lim\limits_{x \to 0^-} \dfrac{\sec x - 1}{x^3}$

14. $\lim\limits_{x \to 0} \left(\cot x - \dfrac{1}{x} \right)$

15. $\lim\limits_{x \to 0} \left(\csc x - \dfrac{1}{x} \right)$

16. $\lim\limits_{x \to \pi/3} \dfrac{\sin\left(x - \dfrac{\pi}{3}\right)}{1 - 2\cos x}$

17. $\lim\limits_{x \to \pi/4} \dfrac{\sqrt{2}\cos x - 1}{1 - \tan^2 x}$

18. $\lim\limits_{x \to \pi/4} \left(\dfrac{\pi}{4} - x\right) \csc\left(\dfrac{3\pi}{4} + x\right)$

19. Let F be differentiable in a neighborhood of a, with derivative F', and suppose $\lim\limits_{x \to a} F'(x)$ exists and is finite. Use L'Hospital's rule to show that $F'(a) = \lim\limits_{x \to a} F'(x)$.

20. Without using Cauchy's mean value theorem, prove the following simplified version of L'Hospital's rule for 0/0, which is often effective. Suppose

    (i) f and g are differentiable at a, with $g'(a) \neq 0$,
    (ii) $\lim\limits_{x \to a} f(x) = \lim\limits_{x \to a} g(x) = 0$.

    Then
    $$\lim\limits_{x \to a} \dfrac{f(x)}{g(x)} = \dfrac{f'(a)}{g'(a)}.$$

    Give an example of an indeterminacy of the form 0/0 which cannot be resolved by this method.

    Evaluate

21. $\lim\limits_{x \to 0} \left(\cot^2 x - \dfrac{1}{x^2} \right)$

22. $\lim\limits_{x \to \pi/4} \dfrac{\tan 2x}{\cot\left(\dfrac{\pi}{4} - x\right)}$

23. Problem 19 suggests (erroneously!) that $F'(a)$ is always equal to $\lim\limits_{x \to a} F'(x)$. Give an example of a function F for which the derivative F' is defined in a neighborhood of a, but $F'(a) \neq \lim\limits_{x \to a} F'(x)$. Reconcile this with Problem 19.

24. Assuming that the limit in Example 3 exists, evaluate it without using L'Hospital's rule.
    Hint. Use the formula $\sin 3x = 3 \sin x - 4 \sin^3 x$.

3.7 Optimization Problems

A great variety of practical problems involve the determination of *largest* size, *least* cost, *shortest* time, *greatest* revenue, and so on. Problems of this type ask for the "best" value of some variable quantity, and hence are called *optimization problems*. Many of them can be solved with the help of the tools developed in the last few sections, but others require more advanced techniques (some of which will be introduced in Sections 13.8 and 13.9). The problems considered in this section have the following property in common: The quantity to be optimized can be expressed as a function of a *single* variable defined on some interval I, and the optimum value of the quantity can then be identified with an extreme value of the function on I. As in all problems stated in narrative form, great care must be taken in translating English into mathematics, and the penalty for mistranslation is that all ensuing calculations will be a complete waste of time. In other words, "garbage in, garbage out," to borrow a slogan from computer science.

The following examples will give you a good idea of how to go about solving optimization problems. Most of them are of a geometric nature or pertain to the natural sciences. Optimization in business and economics is a topic in its own right, which we take up in the next section.

Example 1 A farmer has 800 feet of fencing with which to enclose a rectangular field. What is the largest area he can enclose?

Solution Let x, y and A be the length, width and area of the field, as in Figure 37, and let L be the length of the fence. Then $L = 2x + 2y$ and $A = xy$, and the fact that there are 800 feet of available fencing is expressed by the condition

$$L = 2(x + y) = 800.$$

Solving for y in terms of x, we get $y = 400 - x$. This allows us to express the area of the field as a function of the variable x alone. In fact,

$$A = xy = x(400 - x) = 400x - x^2. \qquad (1)$$

Since area is nonnegative, the admissible values of x vary from 0 to 400. Thus our problem is to determine the value of x for which the area function (1) takes its (absolute) maximum on the interval $I = [0, 400]$. This maximum, whose existence is guaranteed by the extreme value theorem (see page 99), must be achieved at an interior point of I, since $A > 0$ if $0 < x < 400$ and $A|_{x=0} = A|_{x=400} = 0$. (The notation $A|_{x=a}$ is shorthand for the value of the function $A = A(x)$ at $x = a$.) Hence the maximum is a local maximum, and can only be achieved at a critical point of A. Since A is differentiable for all x, with derivative

$$\frac{dA}{dx} = 400 - 2x,$$

the only critical point of A is the point $x = 200$ at which dA/dx equals zero, and the maximum of the area A on the interval I must be achieved at this point. Moreover,

$$y|_{x=200} = (400 - x)|_{x=200} = 200.$$

Thus the rectangular field of largest area enclosed by a fence 800 feet long is a *square* field 200 feet on a side, and its area is

$$A|_{x=200} = (200)^2 = 40{,}000 \text{ square feet.}$$

As an exercise, use either the first or the second derivative test to verify that A has a strict local maximum at $x = 200$. □

Example 2 A square box with no top is made by cutting little squares out of the four corners of a square sheet of metal 12 inches on a side, as in Figure 38(a), and then folding up the resulting flaps, as in Figure 38(b). What is the largest volume of a box which can be made in this way?

Solution Let x be the side length of each little square. Then, as is evident from the figure, the volume of the box is

$$V = x(12 - 2x)^2 = 4x(6 - x)^2. \qquad (2)$$

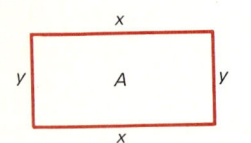

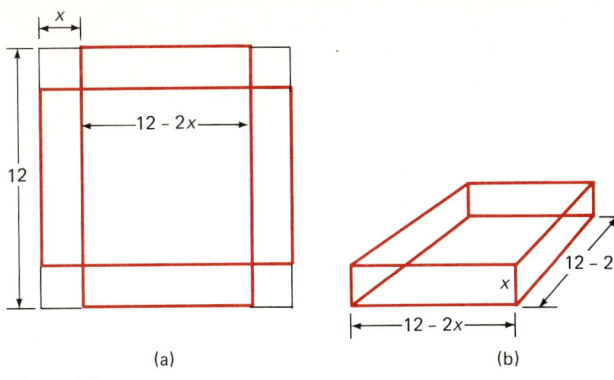

Figure 38

The admissible values of x vary from 0 to 6, since volume is nonnegative and it is impossible to cut away overlapping squares. Thus our problem is to find the value of x for which the volume function (2) takes its (absolute) maximum on the interval $I = [0, 6]$. This maximum must be achieved at an interior point of I, since $V > 0$ if $0 < x < 6$ and $V|_{x=0} = V|_{x=6} = 0$. Hence the maximum is a local maximum, and can only be achieved at a critical point of V. Since V is differentiable for all x, the critical points of V are the points $x = 2$ and $x = 6$ at which the derivative

$$\frac{dV}{dx} = 4(6-x)^2 - 8x(6-x) = 12(6-x)(2-x)$$

equals zero. Of these two points only $x = 2$ is an interior point of I, so that the maximum of the volume V on the interval I must be achieved at $x = 2$. Thus the box of largest volume is obtained by cutting squares 2 inches on a side out of the metal sheet, and its volume is

$$V|_{x=2} = 4(2)(6-2)^2 = 128 \text{ cubic inches.}$$

As an exercise, use either the first or the second derivative test to verify that V has a strict local maximum at $x = 2$. □

Example 3 The Public Works Department wants to construct a new road from Town A to Town B. Town A lies along an abandoned east-west road, while Town B lies 20 miles north of this road and 40 miles east of Town A (see Figure 39). It is proposed that the new road consist of a restored section of the old road, together with an entirely new section that leaves the old road at a point to be determined and heads cross-country straight for Town B. If the cost of building the first section is \$300,000 per mile, while the cost of building the second section is \$600,000 per mile, how much of the old road should be restored in order to minimize the cost of the new two-section road?

Figure 39

Solution As in the figure, let P be the point up to which the old road is restored, let Q be the point of the old road due south of Town B, and let $|PQ| = x$. Then the cost of constructing the new road in millions of dollars is given by

$$C(x) = 0.3|AP| + 0.6|PB| = 0.3(40-x) + 0.6\sqrt{x^2 + 400}.$$

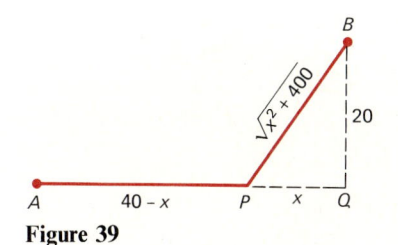

We want the (absolute) minimum of the cost function $C(x)$ on the interval $I = [0, 40]$. The only critical point of $C(x)$ is the point x at which

$$\frac{dC(x)}{dx} = -0.3 + 0.6 \frac{x}{\sqrt{x^2 + 400}} = 0,$$

or equivalently

$$\sqrt{x^2 + 400} = 2x,$$

that is, $x = 20/\sqrt{3} \approx 11.55$. Comparing $C(20/\sqrt{3})$ with the values of the cost function at the endpoints of the interval I, we get

$$C(0) = 24, \quad C(20/\sqrt{3}) = 12 + 6\sqrt{3} \approx 22.39, \quad C(40) = 12\sqrt{5} \approx 26.83.$$

The smallest of these numbers, namely $C(20/\sqrt{3})$, is the minimum of $C(x)$ on I (recall Theorem 5, page 167). Thus we see that $40 - 11.55 \approx 28.45$ miles of the old road should be restored before cutting cross-country to Town B. In this way, the Public Works Department will save about 1.61 million dollars over the L-shaped route AQB, and about 4.44 million dollars over the direct route AB. □

Example 4 Suppose one corner of a long rectangular strip of paper 3 inches wide is folded over until it just reaches the opposite edge, thereby creating a triangle EFG of area A as in Figure 40. What is the largest possible value of A?

Solution Let x and y be the lengths of the legs of the right triangle EFG, as in the figure. Then, by elementary geometry,

$$A = \frac{1}{2} xy, \tag{3}$$

and we must somehow use the special nature of the problem, that is, the fact that EFG is not an arbitrary triangle, but rather one obtained by folding a strip of paper in the way described, to express one of the two variables x and y in terms of the other. This is the hard part of the problem, so try to avoid a feeling of frustration if the "gimmick" eludes you for a while. Don't give up, for it is intuitively clear that a knowledge of x uniquely determines y (fold a piece of paper and see). The key observation is that when the triangle DEG is folded over, the vacated space is a congruent triangle DEG'. Moreover, the side EG' of the latter triangle, which is easily seen to be of length $3 - y$ (remember that the strip is of width 3), becomes the hypotenuse EG of the triangle EFG after the fold is made. Therefore, by the Pythagorean theorem,

$$x^2 + y^2 = |EG|^2 = (3 - y)^2 = 9 - 6y + y^2.$$

Canceling y^2 and solving for y, we get the formula

$$y = \frac{9 - x^2}{6}, \tag{4}$$

which expresses y as a function of x.

The rest is straightforward. Substituting (4) into (3), we get

$$A = \frac{9x - x^3}{12},$$

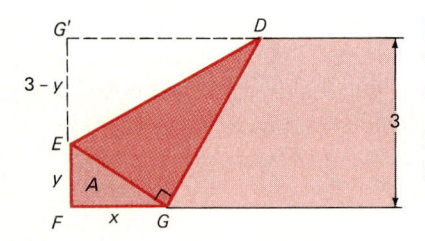

Figure 40

206 Chapter 3 Further Applications of Differentiation

and the problem reduces to finding the (absolute) maximum of the area A on the interval $I = [0, 3]$. (Why this interval?) The maximum must be taken at an interior point of I, since $A > 0$ if $0 < x < 3$ and $A|_{x=0} = A|_{x=3} = 0$. Hence the value of x maximizing A is a critical point of A. Since A is differentiable for all x, the critical points of A are the points $x = \sqrt{3}$ and $x = -\sqrt{3}$ at which the derivative

$$\frac{dA}{dx} = \frac{9 - 3x^2}{12} = \frac{3 - x^2}{4}$$

equals zero, but only $x = \sqrt{3}$ is an interior point of I. Thus the maximum value of A is

$$A|_{x=\sqrt{3}} = \left.\frac{9x - x^3}{12}\right|_{x=\sqrt{3}} = \frac{9\sqrt{3} - 3\sqrt{3}}{12} = \frac{\sqrt{3}}{2} \text{ square inches.}$$

As an exercise, show that if $x = \sqrt{3}$, both the folded-over flap DEG and the triangle EFG are right triangles with acute angles of 30° and 60°. □

A Useful Tool for Optimization

Before giving further examples, we establish a theorem which is often useful in solving optimization problems.

Theorem 12 *(Extreme value theorem for functions approaching infinity).* Let f be a function continuous on an open interval $I = (a, b)$, where the cases $a = -\infty$ and $b = \infty$ are allowed. Then f has a minimum on I if

$$\lim_{x \to a^+} f(x) = \infty, \qquad \lim_{x \to b^-} f(x) = \infty, \tag{5}$$

and a maximum on I if

$$\lim_{x \to a^+} f(x) = -\infty, \qquad \lim_{x \to b^-} f(x) = -\infty \tag{5'}$$

(change $x \to a^+$ to $x \to -\infty$ if $a = -\infty$ and $x \to b^-$ to $x \to \infty$ if $b = \infty$).

Proof (Optional) We give the proof for the case (5) only, since virtually the same proof works in the other cases. Let c be any point in (a, b). Then, because of (5), there are points a_1 and b_1 such that $a < a_1 < c < b_1 < b$ and $f(x) > f(c)$ whenever $a < x < a_1$ and $b_1 < x < b$ (see Figure 41). Let m be the minimum of f on the bounded closed interval $[a_1, b_1]$; the existence of m is guaranteed by Theorem 15, page 99 (the ordinary extreme value theorem). Then m is also the minimum of f on (a, b). In fact, $f(x) \geq m$ for all x in $[a_1, b_1]$, by the meaning of m, while $f(x) > f(c) \geq m$ for all x in (a, a_1) and (b_1, b), by the choice of a_1 and a_2. ∎

Naturally, the function f has no maximum if (5) holds and no minimum if (5') holds. Theorem 12 plays a role in the solution of the following examples.

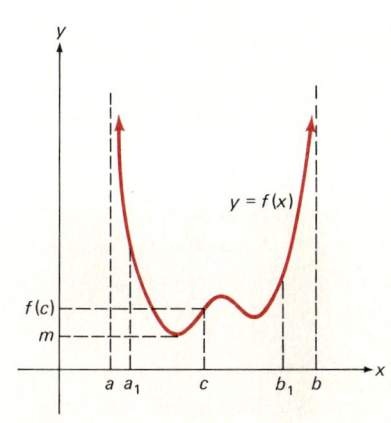

Figure 41

Example 5 Find the smallest volume of a right circular cone circumscribed about a sphere of radius R.

Solution A right circular cone of base radius r and height h has volume $V = \frac{1}{3}\pi r^2 h$, and we must first express one of the two variables r and h in terms of the other. The fact that the cone is circumscribed about the sphere leads to the configuration shown in Figure 42, where the right triangles ADE and ACB share the angle θ equal to half the vertex angle of the cone. It follows that

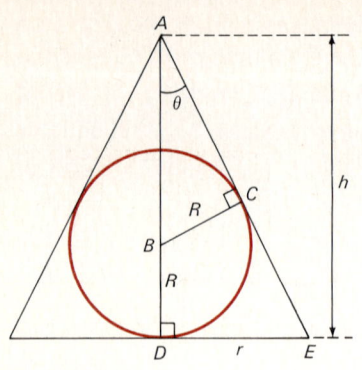

$$\tan\theta = \frac{|DE|}{|AD|} = \frac{r}{h}, \qquad \sin\theta = \frac{|BC|}{|AB|} = \frac{R}{h-R}. \tag{6}$$

But

$$\tan^2\theta = \frac{\sin^2\theta}{\cos^2\theta} = \frac{\sin^2\theta}{1-\sin^2\theta}, \tag{7}$$

and combining (6) and (7), we obtain

$$\frac{r^2}{h^2} = \frac{R^2/(h-R)^2}{1-[R^2/(h-R)^2]} = \frac{R^2}{(h-R)^2 - R^2} = \frac{R^2}{h^2 - 2Rh},$$

or equivalently

$$r^2 = \frac{R^2 h^2}{h^2 - 2Rh} = \frac{R^2 h}{h - 2R},$$

which expresses r^2 as a function of h. Therefore

$$V = V(h) = \frac{1}{3}\pi r^2 h = \frac{1}{3}\pi R^2 \frac{h^2}{h-2R}, \tag{8}$$

where $2R < h < \infty$ (why this interval?).

The problem now reduces to finding the (absolute) minimum of V on the open infinite interval $I = (2R, \infty)$. It follows from (8) that

$$\lim_{h \to 2R^+} V(h) = \infty, \qquad \lim_{h \to \infty} V(h) = \infty.$$

Hence the minimum exists, by Theorem 12, and is achieved at a critical point of V, since it is a local minimum (every point of the open interval I is an interior point). Since V is differentiable on I, with derivative

$$\frac{dV}{dh} = \frac{1}{3}\pi R^2 \frac{2h(h-2R) - h^2}{(h-2R)^2} = \frac{1}{3}\pi R^2 \frac{h(h-4R)}{(h-2R)^2},$$

the only critical point of V is the point $h = 4R$ at which $dV/dh = 0$, so that the minimum of the volume V on the interval I must be achieved at $h = 4R$. Thus the smallest volume of a cone that can be circumscribed about a sphere of radius R is

$$V\big|_{h=4R} = \frac{1}{3}\pi R^2 \frac{16R^2}{4R - 2R} = \frac{8}{3}\pi R^3.$$

Surprisingly, this is just twice the volume of the inscribed sphere. □

Example 6 Let P_1 and P_2 be two points on the same side of a plane mirror, and consider all paths P_1QP_2 made up of two line segments P_1Q and QP_2, where Q is an arbitrary point of the mirror. Then, according to a law of optics known as *Fermat's principle*, the actual path taken by a ray of light emanating from P_1 and reflected by the mirror back to P_2 is the path which is traversed in the *least time*. Find this path.

208 Chapter 3 Further Applications of Differentiation

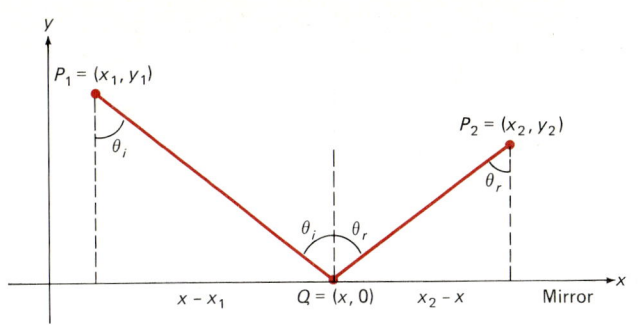

Figure 43

Solution Introducing a rectangular coordinate system with the x-axis along the mirror, we assign the points P_1, Q and P_2 the coordinates shown in Figure 43. Let v be the velocity of light in air. Then the time it takes a ray of light to traverse the path P_1QP_2 is just $T = L/v$, where L is the total length of P_1QP_2. Thus Fermat's principle calls for minimization of the function

$$T = T(x) = \frac{1}{v}(|P_1Q| + |QP_2|) = \frac{1}{v}[\sqrt{(x-x_1)^2 + y_1^2} + \sqrt{(x-x_2)^2 + y_2^2}], \qquad (9)$$

where x is the abscissa of the point Q. Differentiating T with respect to x and setting the result equal to zero, we obtain

$$\frac{dT}{dx} = \frac{x-x_1}{v\sqrt{(x-x_1)^2 + y_1^2}} + \frac{x-x_2}{v\sqrt{(x-x_2)^2 + y_2^2}} = 0,$$

which implies

$$\frac{x-x_1}{|P_1Q|} = \frac{x_2-x}{|QP_2|}. \qquad (10)$$

The fact that T actually achieves its minimum on $(-\infty, \infty)$ at the point x determined by (10) is a consequence of Theorem 12, since T has only one critical point and $T \to \infty$ as $x \to \pm\infty$. Notice that minimizing the time T is equivalent to minimizing the length L, because of the constancy of the velocity v.

Now let θ_i be the *angle of incidence*, that is, the angle between the incident ray P_1Q and the perpendicular to the mirror, and let θ_r be the *angle of reflection*, that is, the angle between the reflected ray QP_2 and the perpendicular to the mirror. Then it is apparent from the figure that (10) is equivalent to

$$\sin \theta_i = \sin \theta_r,$$

which in turn implies

$$\theta_i = \theta_r,$$

a result known as the *law of reflection*. In words, the angle of incidence equals the angle of reflection. □

Example 7 Let P_1 and P_2 be two points on opposite sides of a plane interface between two media, for instance air and water. Use Fermat's principle to find the path taken by a ray of light emanating from P_1 and refracted at the interface to P_2.

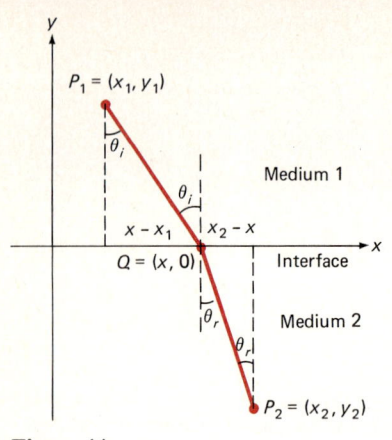

Solution We parallel the solution of Example 6, choosing the coordinates shown in Figure 44. Suppose the velocity of light is v_1 in the first medium and v_2 in the second medium. Then, instead of (9), we now have

$$T = T(x) = \frac{|P_1Q|}{v_1} + \frac{|QP_2|}{v_2} = \frac{\sqrt{(x-x_1)^2 + y_1^2}}{v_1} + \frac{\sqrt{(x-x_2)^2 + y_2^2}}{v_2}. \qquad (9')$$

Setting the derivative dT/dx equal to zero, we are led at once to the condition

$$\frac{x - x_1}{v_1|P_1Q|} = \frac{x_2 - x}{v_2|QP_2|}, \qquad (10')$$

which differs from (10) by the presence of the factors v_1 and v_2 in the denominators. Consulting the figure, we see that (10') is equivalent to

$$\frac{\sin \theta_i}{v_1} = \frac{\sin \theta_r}{v_2}, \qquad (11)$$

where θ_i is still the angle of incidence, but θ_r is now the angle of *refraction*, that is, the angle between the refracted ray QP_2 and the perpendicular to the interface. Formula (11) is the celebrated *law of refraction*, also known as *Snell's law*, of great importance in optics. It is often written in the form

$$\frac{\sin \theta_i}{\sin \theta_r} = \frac{v_1}{v_2} = \text{constant.} \quad \square$$

Remark In the preceding two examples, it was assumed tacitly that the points P_1, Q and P_2 lie in the same plane, and explicitly that the partial paths P_1Q and QP_2 are rectilinear. These assumptions are justified, since otherwise the light takes *longer* to go from P_1 to P_2 (why?).

Example 8 Find the shortest distance between the point $P = (2, 0)$ and the curve $y = \sqrt{x}$.

Solution Let $L = |PQ|$ be the distance between P and a variable point $Q = (x, \sqrt{x})$ of the curve $y = \sqrt{x}$. Then $L = L(x)$ is a function of x, and

$$L^2 = |PQ|^2 = (x - 2)^2 + (\sqrt{x} - 0)^2 = x^2 - 3x + 4.$$

Since L is positive, L has a minimum at a point c if and only if L^2 has a minimum at c (explain further). The only critical point of L^2 is at the point $x = \frac{3}{2}$ for which

$$\frac{d}{dx} L^2 = \frac{d}{dx}(x^2 - 3x + 4) = 2x - 3 = 0.$$

Also, since $d^2(L^2)/dx^2 \equiv 2 > 0$, it follows from the second derivative test that L^2 has a strict local minimum at $x = \frac{3}{2}$. Therefore L also has a strict local minimum at $x = \frac{3}{2}$, equal to

$$L\left(\frac{3}{2}\right) = \sqrt{\left(\frac{3}{2}\right)^2 - 3\left(\frac{3}{2}\right) + 4} = \sqrt{\frac{7}{4}} = \frac{1}{2}\sqrt{7} \approx 1.32.$$

We are looking for the (absolute) minimum of L on $I = [0, \infty)$, the interval on which the curve $y = \sqrt{x}$ is defined. Since $L(0) = 2 > L(\frac{3}{2})$ and $L \to \infty$ as $x \to \infty$, this minimum exists (by an argument like that used to prove Theorem 12) and is taken at an interior point of I. Hence the minimum of L on

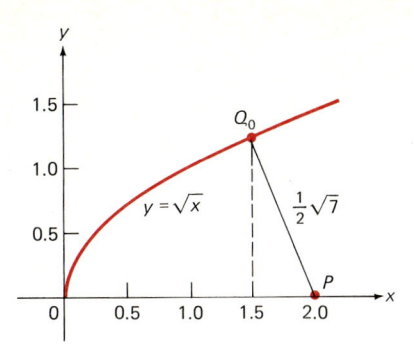

Figure 45

I coincides with the local minimum of L at $x = \frac{3}{2}$, and the shortest distance between P and the curve $y = \sqrt{x}$ is $L(\frac{3}{2}) = \frac{1}{2}\sqrt{7}$. This is the distance between P and Q_0, the point of the curve with x-coordinate $\frac{3}{2}$ (see Figure 45). As an exercise, show that the segment PQ_0 is perpendicular to the tangent to the curve at Q_0. □

How to Solve Optimization Problems

The following step-by-step procedure will help you solve optimization problems.

1. Identify the quantity which is to be maximized or minimized, and label it with a suitable letter, preferably one that reminds you of its concrete meaning. This is the *dependent* variable, and your aim is to eventually express it as a function of a single independent variable.

2. Identify the other quantities which play a role in the problem, and label them with suitable letters too. Finding these quantities involves a certain amount of experimentation, and you can expect to make a few false starts. Everybody does.

3. Look for formulas relating the auxiliary quantities chosen in Step 2 to each other and to the dependent variable chosen in Step 1. The discovery of such formulas is usually facilitated by drawing a figure or two.

4. After due deliberation, select one of the auxiliary quantities as the *independent* variable, and eliminate any others. You should now have a formula expressing the quantity to be optimized as a function f of a single independent variable defined on some interval I. The interval I may or may not include endpoints, and I may be smaller than the natural domain of f, because of limitations inherent in the concrete meaning of the problem.

5. The hard part of the problem has now been accomplished, and the rest is relatively easy. You are looking for the maximum or minimum of a function f on an interval I, and this absolute extremum can be found with the help of the theory developed in Section 3.2. In particular, use differentiation to locate the critical points of f, namely the points at which the derivative f' equals zero or fails to exist, since f may have a local extremum at any such point. You may want to use the first or second derivative test to verify that f actually *has* an extremum at a critical point, and to decide whether the extremum is a maximum or a minimum. Comparison of the values of the function f at its critical points with the values of f at endpoints of I may be called for, in the way described in

Theorem 5, page 167. If f approaches infinity at endpoints of I, use of Theorem 12, page 207, may be helpful.

6. Don't forget to translate your final answer back from mathematics to English, responding to the specific questions posed in the statement of the problem.

Problems

1. As in Example 1, a farmer has 800 ft of fencing with which to enclose a rectangular field, but this time one side of the field lies along a straight river bank. What is the largest area the farmer can enclose, assuming that no fence is needed next to the river?

2. A farmer has 600 ft of fencing with which to enclose five equal rectangular lots, arranged as in Figure 46. For what dimensions of the lots is the total enclosed area the largest? How is the answer changed if one boundary line, shared by all the lots, can be located along a straight river bank, with no fence needed next to the river?

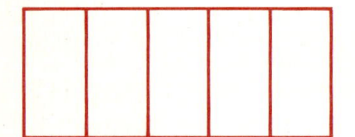

Figure 46

3. Find the rectangle of area A with the smallest perimeter.

4. What is the maximum value of the sum of two numbers (not necessarily positive) whose product is a given number c? What is the minimum value?

5. What is the maximum value of the product of two numbers whose sum is a given number c? What is the minimum value?

6. What is the maximum value of the product of two numbers whose *difference* is a given number c? What is the minimum value?

7. A menu of total area 100 sq in. is printed with 2-in. margins at the top and bottom and 1-in. margins at the sides. For what dimensions of the menu is the printed area the largest?

8. Find the largest volume of a rectangular box with no top made by cutting little squares out of the four corners of an 8 in. × 15 in. rectangular sheet of metal, and then folding up the resulting flaps.

Find the largest area of a rectangle inscribed in

9. A circle of radius R

10. A semicircle of radius R, with one side of the rectangle on the diameter

11. An isosceles triangle of base b and common side length s, with one side of the rectangle on the base of the triangle

12. The strength of a beam of rectangular cross section is proportional to the product of its width and the square of its height. Suppose a wooden beam is cut out of a circular log. For what ratio of height to width is its strength the greatest?

13. What is the largest area of a trapezoid with three nonparallel sides of equal length s? What is the length of the fourth side when the area is the largest?

14. Of all right triangles for which the sum of the lengths of the hypotenuse and one leg is a given constant c, find the one with the largest area.

15. Let BPC be a triangle inscribed in a circle, and let the side BC be parallel to the tangent to the circle at P, the vertex opposite BC. For what choice of BC is the area of BPC the largest?

16. For what ratio of height to radius of a cylindrical oil drum of given volume is the total surface area of the drum the smallest?

17. What is the maximum volume of a cup of given surface area S, in the form of a right circular cylinder with no top? What is the minimum surface area for a given volume V? Find a connection between these two problems.

18. Find the largest volume of a right circular cone of slant height s.

19. Of all right circular cylinders generated by revolving a rectangle of perimeter p about one of its sides, find the cylinder of largest volume.

20. Which isosceles triangle of perimeter p leads to the solid of largest volume when the triangle is revolved about its base?

21. A conical cup is made by cutting a circular sector of central angle θ out of a paper disk (see Figure 47(a), where the disk is of radius R), and then taping the straight edges of the sector together (see Figure 47(b), where the cup is of

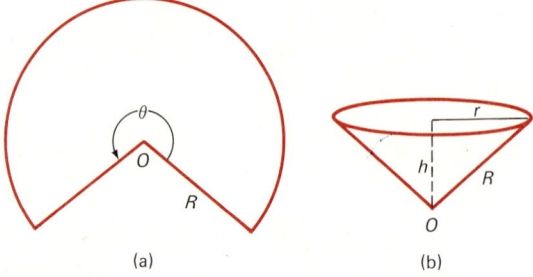

Figure 47

height h and its top is of radius r). What value of θ leads to the cup of largest volume?

22. Find the largest volume of a right circular cylinder inscribed in a sphere of radius R (see Figure 48).

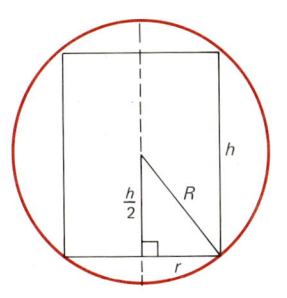

Figure 48

23. Find the smallest volume of a right circular cone circumscribed about a hemisphere of radius R (see Figure 49).

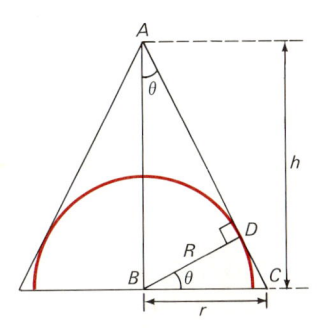

Figure 49

24. Find the shortest distance between the point $(4, 1)$ and the parabola $y = \frac{1}{2}x^2$.

25. Find the largest vertical distance between the curves $y = \sqrt{x}$ and $y = \sqrt[3]{x}$ on the interval $0 \le x \le 1$.

26. Let the function f be differentiable on $(-\infty, \infty)$. Show that the shortest distance to the curve $y = f(x)$ from a fixed point $P = (a, b)$ not on the curve is along a normal line to the curve. Check this result as applied to Problem 24.

27. In Example 3 suppose the cost of building the second section of road is $500,000 per mile. How much of the old road should now be restored?

28. In Example 3 how low a cost of building entirely new road allows the Public Works Department to drop the plan of building a two-section road and opt for the direct route AB? Does it ever make sense to build the L-shaped route AQB?

29. Two ships A and B sail toward a point P along perpendicular routes. Ship A has a speed of 12 knots and is initially 25 nautical miles from P, while ship B has a speed of 16 knots and is initially 20 nautical miles from P. When are the ships closest together? What is the distance of closest approach?

30. Given two points $P_1 = (0, 3)$ and $P_2 = (4, 5)$, find the point Q on the x-axis for which the sum of the distances $|P_1Q|$ and $|QP_2|$ is the smallest. Relate this to Example 6.

An island lies 4 miles offshore from a straight beach. Down the beach 5 miles from the point of the beach nearest the island there is a general store. An islander visits the store regularly, using a rowboat to make part of the trip and jogging the rest of the way to the store. The islander can jog at a speed of 5 mph and row at a speed of 3 mph in average seas.

31. How far from the store should he beach the boat in order to reach the store in the least time?

32. Suppose the sea is particularly calm, so that he is able to row at a speed of 4 mph. How does this change his itinerary?

33. Suppose his objective is to minimize the energy he expends in making the trip to the store, instead of the time required to reach the store, and suppose he expends twice as much energy in rowing a given distance as in jogging the same distance. How far from the store should he now beach the boat?

34. Find the tangent to the curve $y = x^2 - 1$ which cuts off the triangle of minimum area from the fourth quadrant. What is the minimum area?

35. Given a point $P = (a, b)$ in the first quadrant, find the line through P which cuts off the triangle of minimum area from the first quadrant. What is the minimum area?

36. The illumination produced by a light source is directly proportional to the strength of the source and inversely proportional to the square of the distance between the source and the illuminated object. Consider the illumination at a variable point of the line connecting two light sources a distance d apart, one k times stronger than the other. Show that at the point where the illumination is the weakest, the stronger source provides $\sqrt[3]{k}$ times more illumination than the weaker one.

37. A wire of length L is cut into two pieces. One piece is then bent into a circle of radius r, while the other is bent into a square of side length s. What is the smallest total area enclosed by the two pieces? How is it achieved?

38. A window is shaped like a rectangle of base x and height y, surmounted by a semicircle of diameter x (see Figure 50). If the window is of perimeter p, find the dimensions of the window admitting the most light.

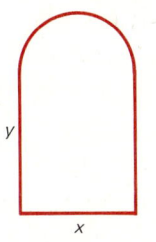

Figure 50

Section 3.7 Optimization Problems

39. A rigid beam of length L is moved on rollers from a corridor of width a into another corridor of width b which is perpendicular to the first corridor (see Figure 51). Find the length of the longest beam that can make the right angle turn without getting stuck. Neglect the width of the beam, and assume that it is always kept horizontal.
Hint. Let x and h be the distances shown in the figure. Express h as a function of x, and show that the maximum of h is $(L^{2/3} - b^{2/3})^{3/2}$.

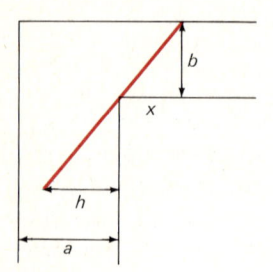

Figure 51

40. In Problem 39 show that the longest beam capable of making the turn is also the shortest beam which touches both outside walls and the common point of the inside walls of the corridors (in the way shown in Figure 52). By regarding this fact as known, give an alternative solution of the problem.

41. Justify the following construction for finding the reflected ray QP_2 and the point of reflection Q in Example 6. Let P'_2 be the image of the point P_2 in the mirror, that is, the reflection of P_2 in the x-axis, and draw the line segment $P_1P'_2$. Then Q is the point in which $P_1P'_2$ intersects the x-axis, and QP_2 is the reflection of the segment QP'_2 in the x-axis.

42. Suppose a ray of light parallel to a plane mirror M and at distance h from M enters a wedge with vertex angle $32°$, formed by M and another plane mirror M' sharing an edge with M (see Figure 53). Show that h is also the distance of closest approach of the multiply reflected ray to the vertex V of the wedge, that is, to the common edge of the mirrors. Does this result depend on the size of the vertex angle? How many reflections has the ray experienced at the time of closest approach to V? What is the total number of reflections experienced by the multiply reflected ray before leaving the wedge?
Hint. Make repeated use of the law of reflection.

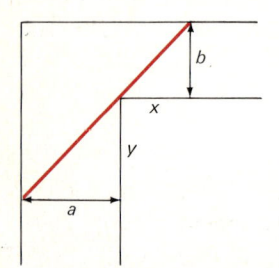

Figure 52

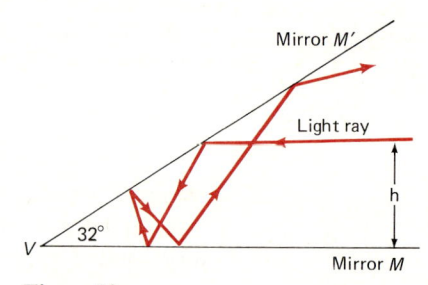

Figure 53

3.8 Applications to Business and Economics (Optional)

The Concept of Marginality

The word "marginal" is encountered repeatedly in business and economics, in expressions like marginal cost, marginal revenue, marginal profit, etc. The second word in each expression is always some function, and the word *marginal* calls for taking the rate of change of the given function with respect to its argument. Thus to find a marginal quantity, we must perform the mathematical operation of differentiation. The word "average," as used in economic theory, also calls for a mathematical operation, namely, dividing whatever function follows the word by the independent variable, but this can be accomplished without a knowledge of calculus.

For example, the total cost to a firm of producing a quantity q of a given commodity is some function of q, called the (*total*) *cost function* and denoted by $C(q)$. The derivative of this function, namely

$$C'(q) = \frac{dC(q)}{dq}, \tag{1}$$

is called the *marginal cost*, denoted by MC(q). Here we follow the convention, standard in economic theory, of denoting certain functions by consecutive capital letters like MC for marginal cost, AR for average revenue, and so on. Do not think of these pairs of letters as products! In this notation, (1) takes the form

$$\text{MC}(q) = \frac{dC(q)}{dq}. \tag{1'}$$

Similarly, the average cost of producing a quantity q of the commodity in question is given by

$$\text{AC}(q) = \frac{C(q)}{q}.$$

In writing $C(q)$ we tacitly assume that $C(q)$ is defined for all real numbers $q \geq 0$ (the output is inherently nonnegative), and not just for the integers $q = 0, 1, 2, \ldots$ This assumption is certainly appropriate for oil or salt, but it also makes sense for commodities that "come one at a time" if the output is large and if we are not too literal-minded. Thus if the answer to a production problem is "Make 31.5 TV sets a day," we can either make 63 sets in 2 days, or else settle for making 31 or 32 sets a day.

Under these conditions, it is a good approximation to regard the marginal cost, at a given output level q, as the extra cost of producing one more unit, the marginal revenue as the extra revenue received from the sale of one more unit, and so on. Thus, for example, approximating the derivative in (1') by the difference quotient, we have

$$\text{MC}(q) = C'(q) \approx \frac{C(q + \Delta q) - C(q)}{\Delta q}, \tag{2}$$

where the approximation is good if Δq is "suitably small." As is customary in this subject, we assume that $\Delta q = 1$ meets the requirement of being suitably small. Then, setting $\Delta q = 1$ in (2), we get

$$\text{MC}(q) = C(q + 1) - C(q), \tag{2'}$$

to a good approximation. In other words, the marginal cost at each level of production is the extra cost of producing the next unit of output.

Example 1 The Rainy Day Co. finds that it costs a total of

$$C(q) = 9720 + 500q - 1.5q^2 + 0.005q^3 \tag{3}$$

dollars to produce q lots of its latest board game "Cats 'n Dogs," each lot consisting of 100 games. Find the corresponding marginal and average costs. Interpret the constant term in the expression for $C(q)$. Analyze the behavior of the marginal cost as a function of the output q.

Solution Figure 54 shows the graph of the cost function $C(q)$. For the marginal cost we have

$$\text{MC}(q) = \frac{dC(q)}{dq} = 500 - 3q + 0.015q^2, \tag{4}$$

and for the average cost

$$\text{AC}(q) = \frac{C(q)}{q} = \frac{9720}{q} + 500 - 1.5q + 0.005q^2. \tag{5}$$

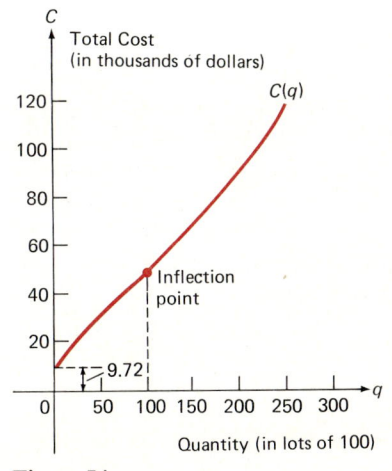

Figure 54

MC, AC
Marginal Cost, Average Cost (in dollars)

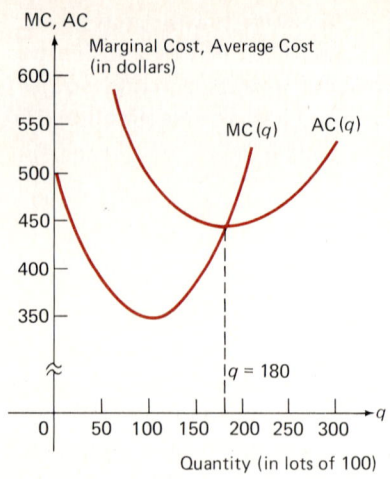

Figure 55

Quantity (in lots of 100), $q = 180$

These two functions are graphed together in Figure 55. The constant term 9720 in formula (3) for the total cost represents the *overhead*, that is, the fixed costs of production (plant, equipment, insurance, etc.). These costs are incurred even in the absence of any output, and thus the overhead equals $C(0)$. Since the derivative of a constant equals zero, *the marginal cost is independent of the overhead*. This is not true of the average cost, as can be seen at once from formula (5).

Rewriting the expression (4) for the marginal cost in the form

$$\text{MC}(q) = 0.015(q - 100)^2 + 350, \tag{4'}$$

we find that $\text{MC}(q)$ is positive for all $q \geq 0$, and in fact takes its minimum value of 350 at $q = 100$. Had it not turned out that $\text{MC}(q) > 0$, there would be something wrong with our model, since the economic meaning of the marginal cost makes it intrinsically positive. We also observe that

$$\frac{d}{dq}\text{MC}(q) = 0.03(q - 100),$$

so that $D_q \text{MC}(q) < 0$ if $q < 100$, while $D_q \text{MC}(q) > 0$ if $q > 100$. It follows that $\text{MC}(q)$ is a decreasing function of the output for moderate levels of production ($q < 100$), but eventually becomes an increasing function of output for higher levels of production ($q > 100$). This is typical, since large-scale production initially achieves certain savings ("economies of scale"), which are eventually outweighed by other factors (for example, inadequate plant capacity) as production is pushed to higher levels. The fact that $\text{MC}(q) > 0$ makes the total cost $C(q)$ an increasing function, while the fact that $D_q \text{MC}(q)$ changes sign at $q = 100$ leads to an inflection point of $C(q)$ at $q = 100$ (see Figure 54). ☐

Example 2 Find the minimum average cost of producing the board game in the preceding example. Show that the curves of marginal and average cost intersect at the point corresponding to minimum average cost. Is this a coincidence?

Solution Setting the derivative of the function (5) equal to zero, we get

$$\frac{d}{dq}\text{AC}(q) = -\frac{9720}{q^2} - 1.5 + 0.01q = 0,$$

or equivalently

$$q^3 - 150q^2 = (q - 150)q^2 = 972{,}000 = 30(180)^2. \tag{6}$$

It follows that $q = 180$ is a root of the cubic equation (6). In fact, $q = 180$ is the only real root, since (6) is equivalent to

$$q^3 - 150q^2 - 972{,}000 = (q - 180)(q^2 + 30q + 5400) = 0,$$

and the quadratic factor is always positive (why?). Calculating the second derivative of $\text{AC}(q)$, we find that

$$\frac{d^2}{dq^2}\text{AC}(q) = \frac{19{,}440}{q^3} + 0.01 > 0 \qquad (q > 0).$$

Chapter 3 Further Applications of Differentiation

Therefore AC(q) has a strict local minimum at $q = 180$, and moreover the curve of average cost is concave upward. Since AC(q) $\to \infty$ as $q \to 0^+$ and as $q \to \infty$, this local minimum is also the absolute minimum of AC(q) on $(0, \infty)$. Substitution of $q = 180$ into (4) and (5) gives

$$MC(180) = 500 - 3(180) + 0.015(180)^2$$
$$= 500 - 540 + 486 = 446,$$

$$AC(180) = \frac{9720}{180} + 500 - 1.5(180) + 0.005(180)^2$$
$$= 54 + 500 - 270 + 162 = 446.$$

Thus the minimum average cost of producing the board game is $446 per lot of 100 games, or $4.46 per game. The fact that MC(180) = AC(180), making the curves of marginal and average cost intersect at the point corresponding to minimum average cost (see Figure 55), is not a coincidence. In fact,

$$\frac{d}{dq} AC(q) = \frac{d}{dq} \frac{C(q)}{q} = \frac{C'(q)q - C(q)}{q^2},$$

and since the minimum average cost is a critical point of AC(q), it must be a root of the equation $C'(q)q - C(q) = 0$, which is equivalent to

$$MC(q) = C'(q) = \frac{C(q)}{q} = AC(q). \quad \square$$

In the preceding two examples, we analyzed the economic activity of the Rainy Day Co. from the standpoint of the *cost* of producing its new board game "Cats 'n Dogs." In the following examples we bring the consumers (buyers) of the product into the picture.

Example 3 A market analysis done by the Rainy Day Co. reveals that the number of lots of "Cats 'n Dogs" ordered by the wholesalers when the game is offered at a price of p dollars per lot is given by the formula

$$q = 600 - 0.4p. \tag{7}$$

Find the company's total, marginal and average revenues. What is the maximum total revenue, and at what output level and price is it achieved?

Solution The economic meaning of the *linear demand function* (7) is that no games will be ordered at a price of $p = \$1500$ per lot ($15 per game), but for every $100 reduction in the price of a lot, 40 more lots will be ordered. The company's total revenue $R(q)$ is simply

$$R(q) = pq,$$

where p is the price and q the quantity sold at that price. To express $R(q)$ as a function of q alone, we first solve (7) for p in terms of q, obtaining

$$p = 1500 - 2.5q. \tag{7'}$$

We then have

$$R(q) = pq = (1500 - 2.5q)q = 1500q - 2.5q^2. \tag{8}$$

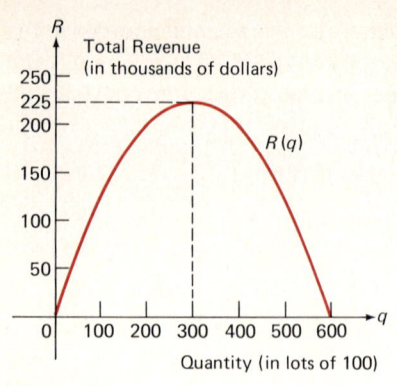

Figure 56

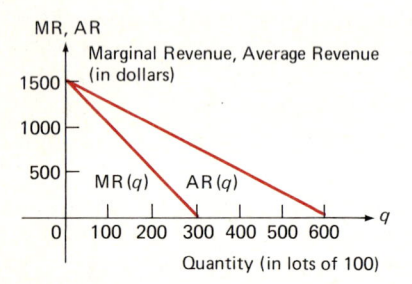

Figure 57

The graph of this function is shown in Figure 56. Note that $R(q)$ is nonnegative, and hence economically meaningful only on the interval $[0, 600]$. For the marginal revenue we have

$$\text{MR}(q) = \frac{dR(q)}{dq} = 1500 - 5q,$$

and for the average revenue

$$\text{AR}(q) = \frac{R(q)}{q} = 1500 - 2.5q.$$

These two functions are graphed in Figure 57. Notice that the average revenue curve (a straight line) is identical with the curve (7'). This is because $R(q) = pq$ is equivalent to $R(q)/q = p$. The marginal revenue curve is also a straight line, with twice the negative slope of the average revenue curve and the same point of intersection $(0, 1500)$ with the vertical axis. Completing the square in (8), we find that

$$R(q) = -2.5(q - 300)^2 + 225{,}000,$$

from which it follows at once that the maximum of $R(q)$ on the interval $[0, 600]$ is at the point $q = 300$ (see Figure 56). The corresponding maximum total revenue is $R(300) = \$225{,}000$, and is obtained by selling 300 lots of "Cats 'n Dogs" at a price of $1500 - 2.5(300) = \$750$ per lot (\$7.50 per game). □

To continue the analysis begun in Examples 1–3, we now take the essential step of assuming that the Rainy Day Co., like any good business, operates in such a way as to maximize its (total) profit. The company's profit $P(q)$ is of course just the difference between its total revenue and its total costs, that is,

$$P(q) = R(q) - C(q). \tag{9}$$

Example 4 What is the maximum profit that the Rainy Day Co. can make from the sale of the game "Cats 'n Dogs," and at what output level is it achieved?

Solution Substituting (3) and (8) into (9), we obtain the function

$$\begin{aligned} P(q) &= (1500q - 2.5q^2) - (9720 + 500q - 1.5q^2 + 0.005q^3) \\ &= -9720 + 1000q - q^2 - 0.005q^3, \end{aligned} \tag{10}$$

with derivative

$$\frac{dP(q)}{dq} = 1000 - 2q - 0.015q^2.$$

Setting $dP(q)/dq$ equal to zero, we find that the critical points of $P(q)$ satisfy the quadratic equation

$$0.015q^2 + 2q - 1000 = 0.$$

The roots of this equation are

$$q = \frac{-2 \pm \sqrt{4 + 4(0.015)(1000)}}{2(0.015)} = \frac{-1 \pm \sqrt{1 + 15}}{0.015}$$

218   Chapter 3   Further Applications of Differentiation

(see Problem 49, page 26). Since the negative root has no economic meaning, we discard it, obtaining

$$q = \frac{-1 + 4}{0.015} = \frac{3}{0.015} = 200.$$

The profit corresponding to this value of q is

$$P(200) = -9720 + 1000(200) - (200)^2 - 0.005(200)^3$$
$$= -9720 + 200{,}000 - 40{,}000 - 40{,}000 = \$110{,}280.$$

This must be the absolute maximum of $P(q)$ on the interval $[0, \infty)$, since $P(0) = -9720 < 0$ and $P(q) \to -\infty$ as $q \to \infty$ (explain further). Thus the company's profit-maximizing policy is to produce 200 lots (20,000) games of "Cats 'n Dogs," selling each lot for $p = 1500 - 2.5(200) = \$1000$ (\$10 per game). Observe that the output that maximizes profit ($q = 200$) is different from that which minimizes average cost ($q = 180$) or that which maximizes total revenue ($q = 300$). In fact, the company would make a somewhat smaller profit (\$108,720) at the output level minimizing average cost, and a much smaller profit (\$65,280) at the level maximizing total revenue.

Thus, to summarize, at the level maximizing profit ($q = 200$) the Rainy Day Co. spends $C(200) = \$89{,}720$ to produce 20,000 games of "Cats 'n Dogs," which it sells at \$10 each for a total revenue of \$200,000, making a handsome profit of \$110,280. The whole "profit picture," at any output level, is contained in the graph of the function $P(q)$, shown in Figure 58. □

The output level q_0 maximizing total profit is a critical point of $P(q)$, and hence the *marginal profit*

$$\text{MP}(q) = \frac{dP(q)}{dq} = \frac{dR(q)}{dq} - \frac{dC(q)}{dq} = \text{MR}(q) - \text{MC}(q) \quad (11)$$

equals 0 at $q = q_0$. This is reasonable, since there is no point in producing more of the commodity if the extra profit to be made from the sale of another unit of output is zero. Since $\text{MP}(q_0) = 0$, it follows from (11) that $\text{MR}(q_0) = \text{MC}(q_0)$. This is also reasonable, since there is no further possibility of making a profit if the sale of another unit of output brings in no more money than it costs to produce it! However, the condition $\text{MR}(q_0) = \text{MC}(q_0)$ is by itself not sufficient to guarantee that the output q_0 maximizes profit (see Problem 19).

Example 5 For the total profit (10) we have the marginal profit

$$\text{MP}(q) = 1000 - 2q - 0.015q^2,$$

as already calculated in the course of solving Example 4. Figure 58 shows the graph of this function, along with the total profit function $P(q)$ of which it is the derivative. The maximum of $P(q)$ is at $q = 200$, and correspondingly $\text{MP}(200) = 0$, as is evident from the figure. □

The Concept of Elasticity

Finally, we introduce the concept of *elasticity*, which is of importance in business and economics. By the derivative with respect to x of a function $y = f(x)$ we mean, of course, the quantity

$$\frac{dy}{dx} = \lim_{\Delta x \to 0} \frac{\text{Change in } y}{\text{Change in } x} = \lim_{\Delta x \to 0} \frac{f(x + \Delta x) - f(x)}{\Delta x} = \lim_{\Delta x \to 0} \frac{\Delta y}{\Delta x}.$$

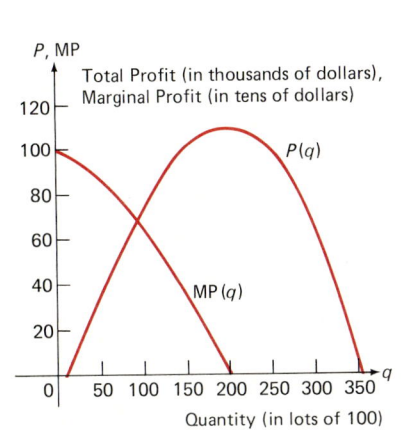

Figure 58

Suppose we replace the changes Δx and Δy by the *proportional* changes $\Delta x/x$ and $\Delta y/y$. We then get an analogous quantity

$$e_{yx} = \lim_{\Delta x \to 0} \frac{\text{Proportional change in } y}{\text{Proportional change in } x} = \lim_{\Delta x \to 0} \frac{\frac{\Delta y}{y}}{\frac{\Delta x}{x}} = \frac{x}{y} \lim_{\Delta x \to 0} \frac{\Delta y}{\Delta x} = \frac{x}{y} \frac{dy}{dx},$$

(12)

called the *elasticity* of the function $y = f(x)$, at the point x. As a measure of the "change in y due to a change in x," the elasticity has the merit of being independent of the units of measurement of x and y. In fact, the units cancel out in forming the ratios $\Delta x/x$ and $\Delta y/y$, thereby making elasticity a "dimensionless" quantity (unlike the derivative itself). This is a great convenience in certain business problems, where, for example, y might be the quantity of a given commodity demanded at the price x. Then changing the units of measurement of x from dollars to pesos, say, or the units of measurement of y from bushels to carloads, would leave the value of the elasticity e_{yx} unchanged.

Let the demand for a commodity produced by a monopolistic firm[†] be described by a function $q = q(p)$, where q is the quantity demanded at the price p. For example, according to formula (7), the demand for the game "Cats 'n Dogs" produced by the Rainy Day Co. (Examples 1–5) is the linear function $q = 600 - 0.4p$. Then the quantity

$$e_D = -\frac{p}{q} \frac{dq}{dp} \qquad (13)$$

is called the *elasticity of demand*, at the price p. Don't be bothered by the extra minus sign in (13). Its sole purpose is to make the elasticity of demand come out positive, in keeping with economic convention and in anticipation of the fact that a demand curve typically has negative slope (the greater the price, the less the demand). The demand is said to be *elastic* if $e_D > 1$ and *inelastic* if $e_D < 1$.

Example 6 Let the demand function be $q = 600 - 0.4p$, as in Example 3. Find e_D. At what price is the demand elastic? Inelastic?

Solution Using (13) to calculate the elasticity of demand, we get

$$e_D = -\frac{p}{q} \frac{dq}{dp} = \frac{0.4p}{600 - 0.4p} = \frac{p}{1500 - p}.$$

Therefore $e_D > 1$ and the demand is elastic if $p > 750$, while $e_D < 1$ and the demand is inelastic if $p < 750$. (The conditions $p < 1500$ and $p \geq 0$ must also be satisfied.) We already know from Example 3 that the total revenue is maximized for $p = 750$. Thus we see that to maximize revenue, the price should be lowered if the demand is elastic and raised if the demand is inelastic. Moreover, the revenue is a maximum at the price $p = 750$ for which the

---

[†] In the presence of competition, a firm will not be able to manipulate demand by adjusting the price, for it must bring its product to market at roughly the prevailing price.

elasticity is exactly equal to 1. These conclusions are valid for every decreasing demand function, linear or not (see Problem 23). □

The economic meaning of all this is clear: If the demand is elastic at a given price, then a price decrease by a certain proportion leads to a proportionately larger increase in sales, so that the total revenue, which is the product of price and quantity, is increased. On the other hand, if the demand is inelastic, then a price increase by a certain proportion leads to a proportionately smaller decrease in sales, so that the revenue is again increased.

Problems

1. Let the cost function be linear, so that $C(q) = a + bq$, where a and b are positive constants. Find the corresponding marginal and average costs. Is there a minimum average cost?

2. Can the cost function $C(q) = 1000 + 100q - 0.5q^2$ be valid for $q = 50$? For $q = 150$? Explain your answer.

3. Suppose the total cost to a firm of producing a quantity q of some commodity is $C(q) = 490 + 20q + 0.1q^2$ dollars. Find the corresponding marginal and average costs. Find the minimum average cost and the output level at which it is achieved. Verify that $MC(q) = AC(q)$ at this output level. Show that a 10¢ error is made at every output level in writing $C(q + 1) - C(q) = MC(q)$. Why is this error unimportant?

4. Let the total cost to a firm of producing a quantity q of some commodity be $C(q) = 3380 + 18q - 0.06q^2 + 0.001q^3$ dollars. Find the corresponding marginal and average costs. Find the minimum average cost and the output level at which it is achieved. Compare the marginal cost with the cost of producing the next unit of output at this level.
Hint. Note that

$$q^3 - 30q^2 - 1{,}690{,}000 = (q - 130)(q^2 + 100q + 13{,}000).$$

5. The *variable cost* $VC(q)$ is defined as the total cost $C(q)$ minus the fixed cost (overhead). Find the average variable cost $AVC(q)$ for the cost function in the preceding problem. Find the minimum average variable cost and the output level at which it is achieved. Verify that $MC(q) = AVC(q)$ at this output level. Why must this be true in general?

6. Find the minimum average variable cost and the output level at which it is achieved for the cost function (3) of the Rainy Day Co.

7. Suppose a firm has a cubic total cost function $C(q) = a + bq + cq^2 + dq^3$, where a, b, c, d are constants, and suppose that as output increases, the marginal cost first decreases and then increases, as in Example 1. Show that $a > 0$, $b > 0$, $c < 0$, $d > 0$, $c^2 < 3bd$. Show that $C(q)$ has an inflection point at $q_0 = -c/3d$.

8. If a straight line is drawn from the origin tangent to the total cost curve, the abscissa of the point of tangency is the output level minimizing the average cost. Why?

9. In Example 3 find the price and quantity minimizing the total revenue by first expressing the revenue as a function of price (rather than of quantity).

10. A firm with overhead C_0 produces quantity q of a commodity, which it sells at a fixed unit price of p dollars. Suppose it costs k dollars to produce each additional unit of output, where $k < p$. At what output level does the firm break even, and what is the graphical interpretation of this *break-even point*?

11. Suppose it costs \$3.50 to produce a prerecorded cassette that sells for \$6.00 (with no middleman). How many cassettes must be sold to break even if the overhead is \$10,000?

12. The fare for a midnight river cruise run by the Zephyr Boat Co. is \$12.50, but to stimulate business, the company offers to refund each passenger 5¢ for every passenger over 100 taking the cruise. The boat can hold 200 passengers. What is the maximum revenue the company can expect under these terms? For what number of passengers is it achieved? The company should withdraw its offer if the average number of persons who might take the cruise at full fare exceeds a certain number. What is this number?

13. A street vendor finds that he can sell 200 soft drinks a day at a price of 50¢ a drink, but that he can sell 50 more for every 5¢ that he lowers the price of a drink. What price maximizes his total revenue? Suppose his supplier charges him 20¢ a drink. What price maximizes his total profit? How much money will he lose if he makes the mistake of maximizing his revenue rather than his profit?

Find the price that maximizes the total revenue if the demand function is

14. $q = 1200 - 1.5p$
15. $q = 1600 - 5p$
16. $q = 675 - p^2$
17. $q = \sqrt{1800 - p^2}$
18. $q = 2500 - p^{3/2}$

In each case find the interval of elastic demand ($e_D > 1$) and the interval of inelastic demand ($e_D < 1$).

19. Why is the condition $MR(q_0) = MC(q_0)$ by itself not sufficient to guarantee that the output level q_0 maximizes profit? Show that q_0 does maximize profit if marginal revenue is increasing more slowly than marginal cost at this level.

20. Show that the definition

$$e_{yx} = \lim_{\Delta x \to 0} \frac{\text{Percentage change in } y}{\text{Percentage change in } x}$$

of the elasticity of a function $y = f(x)$ is equivalent to (12).

21. Show that if $y = f(x)$ has elasticity e_{yx}, then $xf(x)$ has elasticity $1 + e_{yx}$.

22. Verify that $e_{zx} = e_{zy}e_{yx}$ (the chain rule for elasticities).

23. Given any demand curve $q = q(p)$, where $q(p)$ is a decreasing function, show that the total revenue $R(p) = pq$ is maximized at the price for which the elasticity of demand e_D is equal to 1. Show that if the demand is elastic ($e_D > 1$) at a given price, lowering the price will increase revenue, while if the demand is inelastic ($e_D < 1$), raising the price will increase revenue.

Find the maximum profit that the Rainy Day Co. studied in Examples 1–6 can make from the sale of the game "Cats 'n Dogs," and the output level at which it is achieved, if the conditions of the problem are modified as follows:

24. The cost function is $C(q) = 9720 + 500q$ instead of (3)
25. The cost function is $C(q) = 9720 + 500q + 1.5q^2$ instead of (3)
26. The demand function is $q = 775 - 0.4p$ instead of (7)
27. The demand function is $q = 900 - p$ instead of (7)
28. Competition forces the company to sell the games for $6.50 apiece

29. Each week of operation a distillery produces q gallons of whiskey at a cost of $C(q) = 2000 + 5q + 0.001q^2$ dollars, selling its output at $15 a gallon. The government decides to levy a tax of r dollars a gallon on the distillery's output, knowing that the distillers will add the tax to their costs and adjust the output to maximize profit after taxes. What is the government's maximum tax revenue $T = rq$, and at what tax rate r is it achieved? What is the distillery's maximum profit after taxes, and at what output level is it achieved?

30. In Problem 29 suppose the distillers convince the government that the rate of taxation maximizing tax revenue is punitive, and the government agrees to lower the tax rate by $1.50 a gallon. Show that this has the effect of lowering the tax revenue by less than 10%, while at the same time more than doubling the distillery's profit.

31. In Problem 29 suppose the distillers persuade the government to grant them the privilege of operating the distillery without paying taxes proportional to output, in return for payment of a $750,000 a year licensing fee. Show that this has the effect of increasing the government's revenue by $100,000 a year, while at the same time more than doubling the distillery's profit.

Key Terms and Topics

The mean value theorem and Rolle's theorem
Cauchy's mean value theorem
Local maxima and minima of a function
Local vs. absolute extrema
The test for absolute extrema
The necessary condition for a local extremum
Critical points
Monotonic functions and the monotonicity test
First and second derivative tests for a local extremum
Concave functions and the concavity test
Inflection points
Second and third derivative tests for an inflection point
Curve sketching techniques
Infinite limits and limits at infinity
The indeterminate forms $0/0$, $0 \cdot \infty$, ∞/∞ and $\infty - \infty$
Horizontal, vertical and oblique asymptotes
Infinite derivatives and vertical tangents
L'Hospital's rule and its variants
Optimization problems
Marginality and elasticity in business and economics

Supplementary Problems

1. Show that if the function f is differentiable on an interval I containing the points a and $a + \Delta x$, then the increment of f at a can be written in the form

$$\Delta f(a) = f(a + \Delta x) - f(a) = f'(a + t\,\Delta x)\,\Delta x, \quad \text{(i)}$$

where $0 < t < 1$.

Find a point t satisfying formula (i) if
2. $f(x) = x^2 + x + 1$, $a = 2$, $\Delta x = 0.1$
3. $f(x) = 1/x$, $a = 1$, $\Delta x = -0.1$
4. $f(x) = \sqrt{x}$, $a = 9$, $\Delta x = -5$
5. $f(x) = \sin x$, $a = 0$, $\Delta x = 1$
6. Given that

$$f(x) = \begin{cases} 3 - x^2 & \text{if } x \leq 1, \\ \dfrac{2}{x} & \text{if } x > 1, \end{cases}$$

find two points t satisfying formula (i) if $a = 0$, $\Delta x = 2$.

7. Show that Theorem 4, page 164 (Cauchy's mean value theorem), which says that

$$\frac{f(b) - f(a)}{g(b) - g(a)} = \frac{f'(c)}{g'(c)} \quad (a < c < b), \quad \text{(ii)}$$

remains true if we replace the condition that g' be nonzero at every point of (a, b) by the condition that f' and g' are not simultaneously zero at any point of (a, b) and that $g(a) \neq g(b)$.

Find a point c satisfying formula (ii) if
8. $f(x) = 2x^3$, $g(x) = x^2 + 2x$, $a = -2$, $b = 2$
9. $f(x) = \frac{1}{2}x^2$, $g(x) = \frac{1}{3}x^3 - x$, $a = \frac{1}{2}$, $b = \frac{3}{2}$
10. $f(x) = \cos x$, $g(x) = \sin x$, $a = 0$, $b = 3\pi/2$
Notice that Problem 7 plays a role here, since in each case g' is zero at a point in (a, b).

*11. Prove the following *generalized Rolle's theorem*: Let f be continuous on the closed interval $[a, b]$, and let f have a finite nth derivative $f^{(n)}(x)$ at every point of the open interval (a, b). Suppose further that there are $n - 1$ points $x_1, x_2, \ldots, x_{n-1}$ such that $a < x_1 < x_2 < \cdots < x_{n-1} < b$ and $f(a) = f(x_1) = f(x_2) = \cdots = f(x_{n-1}) = f(b)$. Then there is a point c in (a, b) such that $f^{(n)}(c) = 0$.

Find a point c satisfying the generalized Rolle's theorem if
12. $f(x) = x^3 - 9x^2 + 26x - 14$, $n = 2$, $a = 2$, $x_1 = 3$, $b = 4$
13. $f(x) = \sin x + \cos x$, $n = 3$, $a = -5\pi/4$, $x_1 = -\pi/4$, $x_2 = 3\pi/4$, $b = 7\pi/4$
14. $f(x) = x^5 - 10x^4 + 35x^3 - 50x^2 + 24x - 5$, $n = 4$, $a = 0$, $x_1 = 1$, $x_2 = 2$, $x_3 = 3$, $b = 4$
15. $f(x) = x^6 - 14x^4 + 49x^2 - 29$, $n = 5$, $a = -3$, $x_1 = -2$, $x_2 = -1$, $x_3 = 1$, $x_4 = 2$, $b = 3$

16. Find the absolute extrema of the cubic function $f(x) = 2x^3 - 9x^2 + 12x - 6$ on the interval $[-1, 1]$. On $[0, 2]$. On $[1, 3]$.

17. For what choice of the constants a and b does the function

$$f(x) = \frac{ax + b}{(x - 1)(x - 4)}$$

have a strict local maximum, equal to -1, at the point $x = 2$?

18. Use extrema to show that $|3x - x^3| \leq 2$ if $|x| \leq 2$.
19. Let $f(x) = x^r + (1 - x)^r$. Use extrema to show that $1 \leq f(x) \leq 2^{1-r}$ for $0 < r < 1$, while $2^{1-r} \leq f(x) \leq 1$ for $r \geq 1$.
20. Find the local extrema of the function $f(x) = x^m(1 - x)^n$, where m and n are arbitrary positive integers.
21. What is the smallest value of the function $f(x) = \max\{2|x|, |1 + x|\}$?
22. The function $f(x) = \sin^3 x + \cos^3 x$ has six inflection points in the interval $[0, 2\pi]$. Find them.

Find all local extrema and inflection points of the given function. On which intervals is the function increasing? Decreasing? Concave upward? Concave downward? Find all global extrema, asymptotes, vertical tangents and cusps. Graph the function.
23. $f(x) = 2(x - 3)x^{1/2}$
24. $f(x) = 3x(x - 2)^{2/3}$
25. $f(x) = (1 - x)x^{2/3}$
26. $f(x) = 4(x - 1)x^{4/3}$
27. $f(x) = x^{1/2}(2 - x)^{3/2}$
28. $f(x) = x^{1/3}(1 - x)^{2/3}$

Evaluate
29. $\displaystyle\lim_{x \to \infty} \frac{(x - 1)(x - 2)(x - 3)(x - 4)(x - 5)}{(5x - 1)^5}$
30. $\displaystyle\lim_{x \to -\infty} \frac{(2x - 3)^{20}(3x + 2)^{30}}{(2x + 1)^{50}}$
31. $\displaystyle\lim_{x \to 0} \left(\frac{1}{\sin^2 x} - \frac{1}{4 \sin^2 (x/2)} \right)$
32. $\displaystyle\lim_{x \to \infty} (\sqrt{x + \sqrt{x}} - \sqrt{x})$
33. $\displaystyle\lim_{x \to \infty} (\sqrt{x + \sqrt{x + \sqrt{x}}} - \sqrt{x})$
34. $\displaystyle\lim_{x \to \infty} (\sqrt{x + \sqrt{x + \sqrt{x}}} - \sqrt{x + \sqrt{x}})$
35. $\displaystyle\lim_{x \to \infty} \frac{\sqrt{x + \sqrt{x + \sqrt{x}}}}{\sqrt{x + 100}}$
36. $\displaystyle\lim_{x \to \infty} (\sin \sqrt{x + 1} - \sin \sqrt{x})$

Evaluate the given limit, where m and n are arbitrary positive integers.

37. $\lim\limits_{x \to 0} \dfrac{\sqrt[n]{1+x} - 1}{x}$

38. $\lim\limits_{x \to 1} \dfrac{x + x^2 + \cdots + x^n - n}{x - 1}$

39. $\lim\limits_{x \to 1} \dfrac{x^{n+1} - (n+1)x + n}{(x-1)^2}$

40. $\lim\limits_{x \to 0} \dfrac{(1+mx)^n - (1+nx)^m}{x^2}$

41. $\lim\limits_{x \to 1} \left(\dfrac{m}{1 - x^m} - \dfrac{n}{1 - x^n} \right)$

42. $\lim\limits_{x \to -1} \dfrac{x^m + 1}{x^n + 1}$

Hint. Use L'Hospital's rule where applicable.

43. Find the largest area of a rectangle of perimeter p.
44. Find the largest area of an isosceles triangle with equal sides of length s.
45. What fraction of the volume enclosed by a sphere is occupied by the right circular cone of largest volume inscribed in the sphere?
46. The results of n measurements of an unknown quantity x are $x_1, x_2, \ldots, x_n$. What value of x is the *least squares estimate* of x, in the sense of minimizing the expression $(x - x_1)^2 + (x - x_2)^2 + \cdots + (x - x_n)^2$?

What ratio of height to radius minimizes the cost of manufacturing a cylindrical can of given volume if the side of the can is made of material which is
47. Three times as expensive as the material used to make the top and bottom?
48. Half as expensive as the material used to make the top and bottom?

Find the largest area of a rectangle inscribed in
49. A right triangle of leg lengths a and b, with two sides of the rectangle on the legs of the triangle
50. The region bounded by the x-axis and the parabola $y = 3 - x^2$, with one side of the rectangle on the x-axis

Assume that a package can be sent by parcel post only if the sum of its length and its girth (the perimeter of a cross section) does not exceed 96 inches. Find the volume of the largest mailable package whose cross section is
51. A square
52. A rectangle, with sides in the ratio $3:2$
53. A regular hexagon   54. A circle

55. In Problems 51–54 the length of the largest mailable package is the same in each case (and hence so is the girth). How do you explain this?
56. Given a point P inside an acute angle, let L be the straight line through P cutting off the triangle of least area from the angle. Show that P is the midpoint of the part of L inside the angle. Show that this property also characterizes the point P in Problem 35, page 213.
57. The bottom of a line segment of length s lies at distance h from a straight line L which is perpendicular to the direction of the segment (see Figure 59). For what point P of the line is the angle θ subtended by the segment at P the largest?

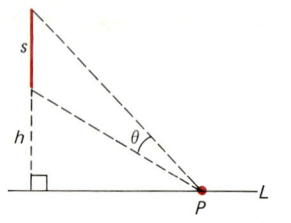

Figure 59

58. Use the preceding problem to give an alternative solution of Problem 74, page 158.
59. Parallel to the side of a tall office building there is a security fence. Suppose the fence is 27 ft high and 64 ft from the building (see Figure 60). Firemen from a hook and ladder company want to reach the building with a ladder resting on top of the fence. How long is the shortest ladder that will reach the building?

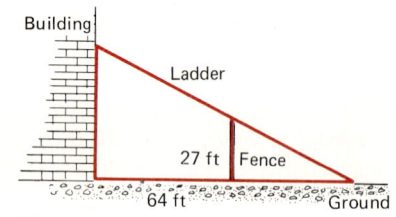

Figure 60

60. If $C = C(q)$ is a firm's total cost function, the quantity

$$e_C = \dfrac{q}{C} \dfrac{dC}{dq}$$

is called the *elasticity of cost*, at the output q. Show that if $e_C > 1$ at a given output, then the marginal cost (MC) is greater than the average cost (AC), and AC increases as output increases, while if $e_C < 1$, then MC is less than AC, and AC decreases as output increases.

4 Integrals

In giving precise meaning to the intuitive idea of a rate of change we were led to the concept of the derivative of a function, and the last two chapters have been devoted to *differential calculus*, the study of derivatives and their applications. Another basic concept of calculus is the *integral* of a function, which arises in an attempt to give precise meaning to the intuitive idea of the area of a region with one or more curved sides. The study of integrals and their applications, to which we now turn our attention, is known as *integral calculus*. The key result of this chapter is the *fundamental theorem of calculus* (see Section 4.5), which reveals a deep connection between differential and integral calculus. In fact, it turns out that the *integral* of a function f can be calculated from two values of another function F which has f as its *derivative*. In Sections 4.6 and 4.7 we use integrals to solve a variety of physical problems involving motion along a straight line.

4.1 Sigma Notation

Our study of integration will be greatly facilitated by the introduction of a concise notation for sums. Let f be a function whose domain includes the positive integers. Then

$$\sum_{i=1}^{n} f(i) \tag{1}$$

is shorthand for

$$f(1) + f(2) + \cdots + f(n). \tag{2}$$

The operation leading from the function f to the sum (1) or (2) is called *summation*, and the symbol $\sum$ (capital Greek sigma) is called the *summation sign*. The numbers 1 and n are called the *lower and upper limits of summation*,

225

respectively. In this context, the word "limit" has its colloquial meaning of "boundary," rather than the technical meaning that it has had so far. The three dots in (2) mean "and so on up to," but if $n = 1, 2, 3$, we have simply

$$\sum_{i=1}^{1} f(i) = f(1), \quad \sum_{i=1}^{2} f(i) = f(1) + f(2), \quad \sum_{i=1}^{3} f(i) = f(1) + f(2) + f(3).$$

The symbol i appearing in (1) is called the *index of summation*, and is a "dummy index," in the sense that any other symbol would do just as well. Thus

$$\sum_{i=1}^{n} f(i) = \sum_{j=1}^{n} f(j) = \sum_{k=1}^{n} f(k) = \sum_{p=1}^{n} f(p),$$

and so on.

Example 1 Calculate $\sum_{i=1}^{5} \frac{1}{i}$.

Solution Expanding the sum, that is, writing it out in full, we have

$$\sum_{i=1}^{5} \frac{1}{i} = \frac{1}{1} + \frac{1}{2} + \frac{1}{3} + \frac{1}{4} + \frac{1}{5} = \frac{137}{60}. \quad \square$$

Sometimes a sum begins with a "zeroth term." Thus

$$\sum_{i=0}^{n} f(i)$$

is shorthand for

$$f(0) + f(1) + f(2) + \cdots + f(n)$$

(the domain of f is now assumed to include 0 as well as the positive integers). Similarly, if m is less than n,

$$\sum_{i=m}^{n} f(i) \qquad (3)$$

means

$$f(m) + f(m + 1) + \cdots + f(n),$$

where the total number of terms equals $n - m + 1$. The sum (3) reduces to the single term $f(m)$ if $m = n$, and is not defined if $m > n$.

Example 2 Calculate $\sum_{i=0}^{5} 2^i$.

Solution Expanding the sum, we get

$$\sum_{i=0}^{5} 2^i = 2^0 + 2^1 + 2^2 + 2^3 + 2^4 + 2^5 = 1 + 2 + 4 + 8 + 16 + 32 = 63. \quad \square$$

Example 3 Calculate $\sum_{n=3}^{6} n!$.

Solution As on page 142, the symbol $n!$ (n factorial) denotes the product $n(n - 1) \cdots 2 \cdot 1$ of the first n positive integers. Therefore

$$\sum_{n=3}^{6} n! = 3! + 4! + 5! + 6! = 6 + 24 + 120 + 720 = 870. \quad \square$$

Let

$$\sum_{i=m}^{n} g(i) \qquad (3')$$

be another sum of the form (3), involving another function g. Then it is easy to see that

$$\sum_{i=m}^{n} [af(i) + bg(i)] = a \sum_{i=m}^{n} f(i) + b \sum_{i=m}^{n} g(i), \qquad (4)$$

where a and b are arbitrary constants. The detailed verification of (4) is left as an exercise (expand the sums, carry out the multiplications, and rearrange the terms).

Example 4 If $S = \sum_{i=1}^{7} 3^i$ and $T = \sum_{i=1}^{7} (6 - 3^{i+1})$, calculate $S + \frac{1}{3} T$.

Solution With the help of formula (4), read from right to left, we obtain

$$S + \frac{1}{3} T = \sum_{i=1}^{7} \left[3^i + \frac{1}{3}(6 - 3^{i+1}) \right] = \sum_{i=1}^{7} [3^i + (2 - 3^i)] = \sum_{i=1}^{7} 2 = 14.$$

The last step follows from the fact that if c is a constant, then

$$\sum_{i=1}^{n} c = \underbrace{c + c + \cdots + c}_{n \text{ terms}} = nc. \qquad \square$$

The Binomial Theorem

Our last example illustrates the great brevity that can be achieved by using sigma notation.

Example 5 A celebrated proposition of algebra known as the *binomial theorem* states that if n is a positive integer, then the nth power of $a + b$ has the expansion

$$(a + b)^n = a^n + \binom{n}{1} a^{n-1} b + \binom{n}{2} a^{n-2} b^2 + \cdots + \binom{n}{n-1} ab^{n-1} + b^n$$

$$= a^n + na^{n-1} b + \frac{n(n-1)}{2} a^{n-2} b^2 + \cdots + nab^{n-1} + b^n. \qquad (5)$$

Here the numbers $\binom{n}{k}$, known as the *binomial coefficients*, are defined by

$$\binom{n}{k} = \frac{n!}{k!(n-k)!} \qquad (k = 1, \ldots, n-1). \qquad (6)$$

Using sigma notation, we can write (5) very concisely as

$$(a + b)^n = \sum_{k=0}^{n} \binom{n}{k} a^{n-k} b^k, \qquad (5')$$

with the understanding that

$$\binom{n}{0} = \binom{n}{n} = 1.$$

Section 4.1 Sigma Notation   **227**

This corresponds to using the formula (6) for $k = 0, 1, \ldots, n$ after defining $0! = 1$. The binomial theorem can be proved with the help of a method known as *mathematical induction*. The details are given in Example 3 of the Appendix (see page A-3). □

To calculate binomial coefficients, we observe that

$$\binom{n}{k} = \binom{n}{n-k}, \quad \binom{n}{1} = \binom{n}{n-1} = n,$$

and

$$\binom{n}{k} = \frac{n(n-1)\cdots(n-k+1)(n-k)(n-k-1)\cdots 2 \cdot 1}{k!(n-k)(n-k-1)\cdots 2 \cdot 1}$$

$$= \frac{n(n-1)\cdots(n-k+1)}{k!}.$$

Thus, for instance,

$$\binom{8}{3} = \binom{8}{5} = \frac{8 \cdot 7 \cdot 6}{3 \cdot 2 \cdot 1} = 56,$$

$$\binom{100}{98} = \binom{100}{2} = \frac{100 \cdot 99}{2 \cdot 1} = 50 \cdot 99 = 4950.$$

Problems

Calculate the given sum.

1. $\sum_{i=3}^{7} i^2$

2. $\sum_{j=1}^{6} 2^{-j}$

3. $\sum_{k=0}^{5} (-1)^k 3^k$

4. $\sum_{m=0}^{8} (m^2 - 1)$

5. $\sum_{n=2}^{5} n^{n-1}$

6. $\sum_{i=0}^{10} (2i - 1)$

7. $\sum_{j=0}^{6} \frac{j-1}{j+2}$

8. $\sum_{k=2}^{7} \frac{k+1}{k-1}$

9. $\sum_{m=0}^{5} \sin \frac{m\pi}{3}$

10. $\sum_{n=1}^{9} \cos \frac{n\pi}{4}$

11. Show that

$$\sum_{i=m}^{n} [f(i+1) - f(i)] = f(n+1) - f(m).$$

A sum like this, most of whose terms cancel each other out, is said to be *telescoping*.

Use the preceding problem to calculate the given sum.

12. $\sum_{i=0}^{9} (2^{i+1} - 2^i)$

13. $\sum_{j=2}^{15} \left(\frac{1}{j} - \frac{1}{j+1}\right)$

14. $\sum_{k=1}^{35} (\sqrt{k+1} - \sqrt{k})$

Use sigma notation to write the given sum.

15. $2 + 5 + 8 + \cdots + 32$

16. $5 - 9 + 13 - 17 + \cdots + 45$

17. $3 + 2 + \frac{5}{3} + \frac{6}{4} + \frac{7}{5} + \frac{8}{6} + \frac{9}{7} + \frac{10}{8}$

18. $\frac{1}{3} - \frac{3}{5} + \frac{5}{7} - \cdots - \frac{51}{53}$

Is the given formula true or false?

19. $\sum_{n=3}^{7} n = \sum_{m=4}^{8} (m-1)$

20. $\sum_{i=1}^{50} 1 = 49$

21. $\sum_{k=1}^{n} (1 + 2k) = 1 + 2\sum_{k=1}^{n} k$

22. $\sum_{j=1}^{11} (-1)^{j-1} = 1$

Calculate the given binomial coefficient.

23. $\binom{12}{6}$

24. $\binom{14}{5}$

25. $\binom{19}{4}$

26. $\binom{20}{11}$

27. $\binom{31}{29}$

28. $\binom{101}{3}$

Use the binomial theorem to expand

29. $(1 + \sqrt{2})^5$

30. $(a + b)^6$

31. $(a - b)^7$

32. $(1 + x)^8 - (1 - x)^8$

228 Chapter 4 Integrals

Show that

33. $\binom{n}{0} + \binom{n}{1} + \binom{n}{2} + \cdots + \binom{n}{n} = 2^n$

34. $\binom{n}{0} - \binom{n}{1} + \binom{n}{2} - \cdots + (-1)^n \binom{n}{n} = 0$

35. Use the binomial theorem to establish *Leibniz's rule*

$$\frac{d^n}{dx^n}[f(x)g(x)] = \sum_{k=0}^{n} \binom{n}{k} f^{(n-k)}(x) g^{(k)}(x)$$

for the *n*th derivative of the product of two functions.

Use Leibniz's rule to calculate

36. The eighth derivative of $x^3 \cos x$

37. The tenth derivative of $x^4 \sin x$

38. Given any numbers $a_0, a_1, \ldots, a_n$ and $b_0, b_1, \ldots, b_n$, show that

$$\sum_{i=1}^{n}(a_i - a_{i-1})b_{i-1} = (a_n b_n - a_0 b_0) - \sum_{i=1}^{n} a_i(b_i - b_{i-1}).$$

This result is known as the formula for *summation by parts*.

4.2 Area Under a Curve and the Definite Integral

The kind of limit leading to the concept of the *definite integral* comes up time and again in problems involving the "summation of a very large number of individually small terms." The prototype of all such problems is that of finding the area under a curve, to which we now turn our attention.

Let f be a function which is continuous and nonnegative on a bounded closed interval $[a, b]$. Then by the *area under the curve* $y = f(x)$, from $x = a$ to $x = b$, we mean the area A of the plane region R bounded by the vertical lines $x = a$ and $x = b$, the x-axis and the curve $y = f(x)$, as shown in Figure 1(a). The region R has at least one straight side (three in the case where f is positive), but its top side is curved. Regions of this kind are not considered in elementary geometry. Thus, in the course of calculating the area A, we must decide what is *meant* by A in the first place. This is the same philosophy as in the problem of finding, after first *defining*, the tangent to a general curve (see page 115), and just as the tangent problem led to the concept of the derivative, the area problem will now lead to the concept of the integral.

With a view to defining the area A, we divide the interval $[a, b]$ into a large number of smaller subintervals, by introducing *points of subdivision*

$$a = x_0, x_1, x_2, \ldots, x_{n-1}, x_n = b, \tag{1}$$

(a)

(b)

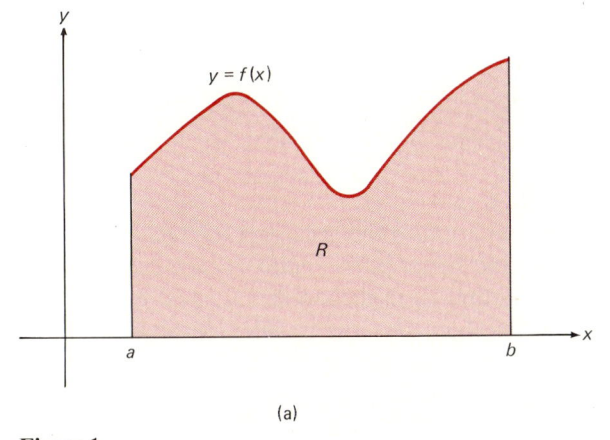

Figure 1

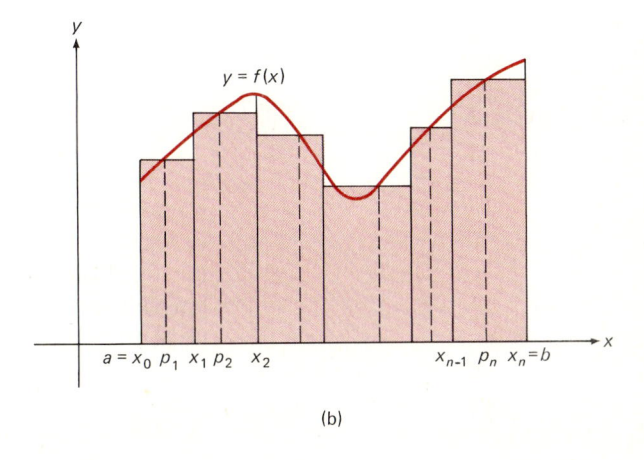

satisfying the inequalities
$$a = x_0 < x_1 < x_2 < \cdots < x_{n-1} < x_n = b.$$

Here, in the interest of a uniform notation, the endpoints a and b of the original interval $[a, b]$ are regarded as points of subdivision and assigned alternative symbols x_0 and x_n. The points (1), which are said to form a *partition* of the interval $[a, b]$, divide $[a, b]$ into exactly n subintervals
$$[x_0, x_1], [x_1, x_2], \ldots, [x_{n-1}, x_n].$$

Let
$$\Delta x_i = x_i - x_{i-1} \qquad (i = 1, 2, \ldots, n)$$
be the length of the ith subinterval $[x_{i-1}, x_i]$, and let μ (lowercase Greek mu) be the maximum length of all the subintervals, so that
$$\mu = \max\{x_1 - x_0, x_2 - x_1, \ldots, x_n - x_{n-1}\} = \max\{\Delta x_1, \Delta x_2, \ldots, \Delta x_n\}.$$

The number μ is called the *mesh size* (or *norm*) of the partition (1); it measures the "fineness" of the partition, in the sense that the smaller the mesh size μ, the larger the number of points of subdivision and the closer together they are (see Problem 25). Thus smallness of μ guarantees largeness of n, but the converse is not true. For example, the points
$$\frac{1}{2^{n-1}}, \ldots, \frac{1}{8}, \frac{1}{4}, \frac{1}{2}$$
divide the unit interval $[0, 1]$ into n subintervals, but the mesh size μ is always equal to $\frac{1}{2}$, the length of the longest subinterval $[\frac{1}{2}, 1]$, regardless of how large n is.

Next we draw the vertical lines $x = x_0, x = x_1, x = x_2, \ldots, x = x_{n-1}, x = x_n$. These lines divide the region R into n narrow strips, as shown in Figure 1(b). The function f is continuous, and hence its value changes only slightly on the subinterval $[x_{i-1}, x_i]$, at least if Δx_i is small enough. Thus it seems a good approximation to regard f as having the constant value $f(p_i)$ on $[x_{i-1}, x_i]$, where p_i is an *arbitrary* point of $[x_{i-1}, x_i]$. The arbitrariness of p_i may give you pause, since it seems as if some choices of p_i are better than others, but the difference between various choices of p_i will soon "disappear in the limit," since we are going to make μ approach zero! Replacing $f(x)$ by $f(p_i)$ on each subinterval $[x_{i-1}, x_i]$ is equivalent to replacing the strips with curved tops by the shaded rectangles shown in the figure. The sum of the areas of these rectangles is given by

$$\sum_{i=1}^{n} f(p_i)(x_i - x_{i-1}) = \sum_{i=1}^{n} f(p_i) \Delta x_i, \qquad (2)$$

where we use the sigma notation introduced in Section 4.1. It is reasonable to regard (2) as a good approximation to the area A of the region R, where the approximation gets better and better as the width of *each and every* rectangle gets smaller and smaller, that is, as the number μ equal to the maximum width of the rectangles gets smaller and smaller. Motivated by this argument, we now *define* A as the limit

$$A = \lim_{\mu \to 0} \sum_{i=1}^{n} f(p_i)(x_i - x_{i-1}) = \lim_{\mu \to 0} \sum_{i=1}^{n} f(p_i) \Delta x_i. \qquad (3)$$

Definition of the Integral
$\int_a^b f(x)\,dx$

These considerations lead naturally to the following definition. Given a function f defined on a bounded closed interval $[a, b]$, we introduce an arbitrary partition of $[a, b]$, that is, any set of points $a = x_0, x_1, x_2, \ldots, x_{n-1}, x_n = b$ satisfying the inequalities $a = x_0 < x_1 < x_2 < \cdots < x_{n-1} < x_n = b$. In each subinterval $[x_{i-1}, x_i]$ of length $\Delta x_i = x_i - x_{i-1}$, we choose an arbitrary point p_i and form the sum

$$S = \sum_{i=1}^n f(p_i)\,\Delta x_i. \tag{4}$$

Suppose that as the mesh size $\mu = \max\{\Delta x_1, \Delta x_2, \ldots, \Delta x_n\}$ approaches zero, the sum S approaches a finite limit I, *regardless of the choice of the points of subdivision x_i and the "intermediate" points p_i*. (Actually μ approaches zero from the right only, since μ is inherently positive.) Then the limit I is called the *definite integral*, or simply the *integral*, of f from a to b, written

$$\int_a^b f(x)\,dx,$$

and the function f is said to be *integrable* on $[a, b]$, or over $[a, b]$. The sum (4) is known as a *Riemann sum*, in honor of the illustrious German mathematician G. F. Bernhard Riemann (1826–1866). The definite integral is often called the *Riemann integral*, to distinguish it from other kinds of integrals encountered in higher mathematics.

Area Under a Curve as an Integral

In terms of the integral, formula (3) for the area under the curve $y = f(x)$ from $x = a$ to $x = b$ can be written more concisely as

$$A = \int_a^b f(x)\,dx. \tag{5}$$

Although the definition of the integral was motivated by the area problem, it will become apparent in the course of this and subsequent chapters that the integral has far-reaching applications to a great variety of problems that have nothing to do with area.

The integral $\int_a^b f(x)\,dx$ is a *number*, and the operation leading from the function f to this number is called *integration*. The symbol $\int$, introduced three centuries ago by Leibniz, is called the *integral sign*; historically, it is a stylized letter S for sum, suggesting that integration has something to do with summation, at least in the limit. The numbers a and b are called the *lower and upper limits of integration*, respectively. Here, as in the case of limits of summation, the word "limit" has its colloquial meaning of "boundary," rather than its usual technical meaning. The interval $[a, b]$, which has the numbers a and b as its endpoints, is called the *interval of integration*. The function f is called the *integrand* of the integral $\int_a^b f(x)\,dx$. The argument of f, in this case x, is called the *variable of integration*, and is a "dummy variable," in the sense that any other symbol would do just as well. Thus

$$\int_a^b f(x)\,dx = \int_a^b f(s)\,ds = \int_a^b f(t)\,dt = \int_a^b f(z)\,dz,$$

and so on. The situation is exactly the same as for a dummy index of summation (see page 226).

The expression $\int_a^b f(x)\,dx$ for the integral of the function f from a to b contains the differential dx of the variable of integration. The differential dx "matches" the increment Δx_i in the Riemann sum $\sum_{i=1}^{n} f(p_i)\Delta x_i$, of which the integral is the limit as the mesh size μ approaches zero. In fact, as $\mu \to 0$, the summation sign $\sum$ and the increment Δx_i are replaced by the integral sign $\int$ and the differential dx, and the limits of summation are replaced by the limits of integration. Although this is very suggestive, we could just as well write the integral simply as $\int_a^b f$, omitting both the variable of integration and its differential, and we will occasionally do so. After all, integration is an operation leading from a function to a number called its integral over a given interval, and thus it is enough to specify just the function and the endpoints of the interval. However, we will ordinarily use the full notation $\int_a^b f(x)\,dx$ (in the case where x is the name of the variable of integration). The advantage of this notation will become evident later (see page 367).

Remark Both the Riemann sum $S = \sum_{i=1}^{n} f(p_i)\Delta x_i$ and the mesh size $\mu = \max\{\Delta x_1, \Delta x_2, \ldots, \Delta x_n\}$ figuring in the definition of the integral $I = \int_a^b f(x)\,dx$ depend on the choice of the partition of the interval of integration $[a, b]$. We can make this explicit by writing $S = S(X)$ and $\mu = \mu(X)$, where X denotes the set of points of subdivision $a = x_0, x_1, x_2, \ldots, x_{n-1}, x_n = b$. Bear in mind that S also depends on the choice of the "intermediate" points $p_1, p_2, \ldots, p_n$, one in each of the subintervals $[a, x_1], [x_1, x_2], \ldots, [x_{n-1}, b]$. Thus to say that S approaches the limit I as $\mu \to 0$ means that the difference $S(X) - I$ is arbitrarily near zero for all X such that $\mu(X)$ is sufficiently small, *regardless of the choice of the points p_i associated with X*. Or in ε, δ language, it means that given any $\varepsilon > 0$, we can find a $\delta > 0$ such that

$$\left| S(X) - I \right| = \left| S(X) - \int_a^b f(x)\,dx \right| < \varepsilon$$

whenever $0 < \mu(X) < \delta$, for every choice of the p_i associated with X.

Integrability of Continuous Functions

In defining the integral $\int_a^b f(x)\,dx$, we said nothing about the continuity of f, and a discontinuous function may or may not be integrable (see Problems 29 and 31). On the other hand, *every continuous function is integrable*. In particular, this justifies the approximations that were made in defining the area under the graph of a continuous function, and guarantees the existence of the area.

> **Theorem 1** *(Continuity implies integrability)*. *If the function f is continuous on an interval $[a, b]$, then f is integrable on $[a, b]$.*

Theorem 1 is a further testimonial, if one be needed, to the importance of continuity in calculus. The proof of the theorem involves a concept (uniform continuity) that lies beyond the scope of this book. The interested reader is referred to any text on advanced calculus.

We now evaluate a number of simple integrals, making direct use of the definition of the integral as the limit of a Riemann sum.

Example 1 Evaluate the integer $\int_a^b 1\, dx$, usually written as $\int_a^b dx$.

Solution Here we are integrating the constant function $f(x) \equiv 1$. Therefore

$$\int_a^b dx = \lim_{\mu \to 0} \sum_{i=1}^n f(p_i)\Delta x_i = \lim_{\mu \to 0} \sum_{i=1}^n \Delta x_i,$$

since $f(p_i) = 1$ for every p_i. Expanding the sum, we get

$$\sum_{i=1}^n \Delta x_i = \sum_{i=1}^n (x_i - x_{i-1})$$

$$= (x_1 - x_0) + (x_2 - x_1) + \cdots + (x_{n-1} - x_{n-2}) + (x_n - x_{n-1})$$

$$= -x_0 + (x_1 - x_1) + (x_2 - x_2) + \cdots + (x_{n-1} - x_{n-1}) + x_n.$$

Since all the terms in the sum on the right equal 0 except the first and the last terms, the sum "telescopes" and reduces simply to

$$\sum_{i=1}^n \Delta x_i = x_n - x_0 = b - a.$$

This could have been anticipated, because the sum of the lengths of the subintervals $[x_{i-1}, x_i]$ must equal the length of the interval $[a, b]$. Thus $\int_a^b dx$ is the limit as $\mu \to 0$ of the constant $b - a$, so that

$$\int_a^b dx = b - a. \tag{6}$$

Geometrically, the integral (6) is the area under the line $y = 1$ from $x = a$ to $x = b$, that is, the area of the shaded rectangle in Figure 2, of length $b - a$ and height 1. The area of this rectangle, or equivalently the length of the interval $[a, b]$, is of course $b - a$. as given by (6). □

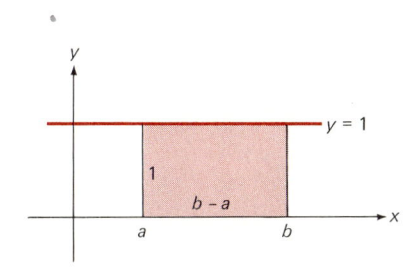

Figure 2

While evaluating the integral in Example 1, we incidentally proved its *existence*. The existence of the integral also follows from Theorem 1 and the continuity of the integrand $f(x) \equiv 1$. In the next two examples, we first use the continuity of the integrand to deduce the existence of the given integral, and then use the existence of the integral and the consequent arbitrariness of the points p_i appearing in the sum (4) to evaluate the integral.

Example 2 Evaluate the integral $\int_a^b x\, dx$.

Solution The integrand $f(x) = x$ is continuous, and hence integrable, by Theorem 1. Thus

$$\int_a^b x\, dx = \lim_{\mu \to 0} \sum_{i=1}^n f(p_i)\Delta x_i = \lim_{\mu \to 0} \sum_{i=1}^n p_i(x_i - x_{i-1}),$$

where the limit exists and is the same for every choice of the points p_i in the subintervals $[x_{i-1}, x_i]$, $i = 1, \ldots, n$. Suppose we choose p_i to be the midpoint of $[x_{i-1}, x_i]$, so that $p_i = \frac{1}{2}(x_{i-1} + x_i)$. Then

$$\int_a^b x\, dx = \lim_{\mu \to 0} \frac{1}{2} \sum_{i=1}^n (x_i + x_{i-1})(x_i - x_{i-1}) = \frac{1}{2} \lim_{\mu \to 0} \sum_{i=1}^n (x_i^2 - x_{i-1}^2)$$

$$= \frac{1}{2} \lim_{\mu \to 0} (x_n^2 - x_0^2) = \frac{1}{2} \lim_{\mu \to 0} (b^2 - a^2),$$

Section 4.2   Area Under a Curve and the Definite Integral   **233** since our choice of p_i has led to another telescoping sum, and therefore

$$\int_a^b x\,dx = \frac{1}{2}(b^2 - a^2). \tag{7}$$

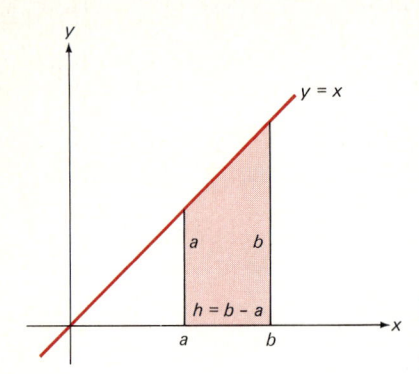

Geometrically, the integral (7) is the area under the line $y = x$ from $x = a$ to $x = b$, that is, the area of the shaded trapezoid in Figure 3. Since the area of a trapezoid with parallel sides of lengths a and b, a distance h apart, is $\frac{1}{2}(a + b)h$, the trapezoid in the figure is of area

$$\frac{1}{2}(b + a)(b - a) = \frac{1}{2}(b^2 - a^2),$$

in keeping with formula (7). □

Example 3 Evaluate the integral $\int_a^b x^2\,dx$.

Solution Again the integrand is continuous, and hence integrable. Thus

$$\int_a^b x^2\,dx = \lim_{\mu \to 0} \sum_{i=1}^n f(p_i)\Delta x_i = \lim_{\mu \to 0} \sum_{i=1}^n p_i^2(x_i - x_{i-1}),$$

where the limit exists and is the same for every choice of the points p_i in the subintervals $[x_{i-1}, x_i]$, $i = 1, \ldots, n$. To facilitate our calculations, we assume that $a \geq 0$ (it can be shown that the answer does not depend on this assumption). Then $0 \leq x_{i-1} < x_i$ for all i, and therefore

$$x_{i-1}^2 < \frac{1}{3}(x_{i-1}^2 + x_{i-1}x_i + x_i^2) < x_i^2,$$

so that the point

$$p_i = \sqrt{\frac{1}{3}(x_{i-1}^2 + x_{i-1}x_i + x_i^2)}$$

belongs to the subinterval $[x_{i-1}, x_i]$. This choice of p_i is "premeditated," for with it the evaluation of the integral leads to yet another telescoping sum. In detail

$$\int_a^b x^2\,dx = \lim_{\mu \to 0} \frac{1}{3} \sum_{i=1}^n (x_i^2 + x_i x_{i-1} + x_{i-1}^2)(x_i - x_{i-1})$$

$$= \frac{1}{3} \lim_{\mu \to 0} \sum_{i=1}^n (x_i^3 - x_{i-1}^3) = \frac{1}{3} \lim_{\mu \to 0} (x_n^3 - x_0^3) = \frac{1}{3} \lim_{\mu \to 0} (b^3 - a^3),$$

with the help of the algebraic identity $(u^2 + uv + v^2)(u - v) = u^3 - v^3$. It follows that

$$\int_a^b x^2\,dx = \frac{1}{3}(b^3 - a^3). \tag{8}$$

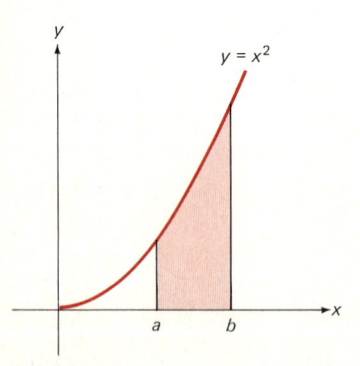

This time the integral is the area of a region of a kind not familiar from elementary geometry, namely the shaded region in Figure 4, bounded by the vertical lines $x = a$ and $x = b$, the x-axis and the parabola $y = x^2$. □

In Section 4.5 we will establish a general method for evaluating definite integrals that completely avoids the explicit calculation of Riemann sums and their limits. We will then no longer require special tricks of the kind used to solve the preceding two examples. This is indeed fortunate, for otherwise, we would have been at our wit's end to evaluate even relatively simple integrals! In fact, it turns out that formulas (6)–(8) are all special cases of the general formula

$$\int_a^b x^n \, dx = \frac{1}{n+1}(b^{n+1} - a^{n+1}) \qquad (n = 0, 1, 2, \ldots), \tag{9}$$

which will be proved on page 257. As of now, we will regard formula (9) as known, and use it freely.

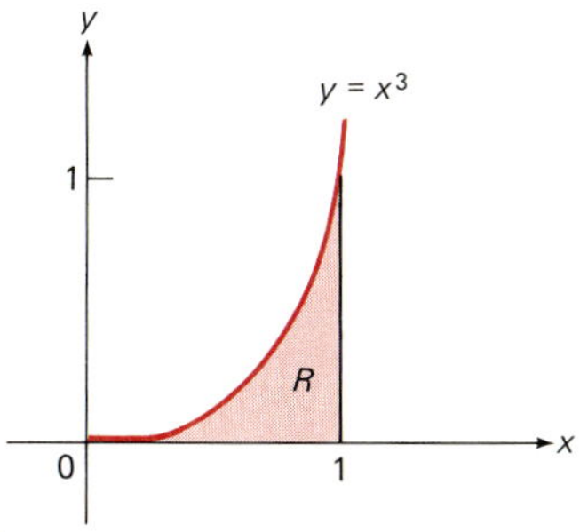

Example 4 Find the area A of the region R bounded by the curve $y = x^3$, the x-axis and the line $x = 1$ (see Figure 5).

Solution Since

$$A = \int_0^1 x^3 \, dx,$$

we find with the help of formula (9) with $n = 3$ that

$$A = \frac{1}{4}(1^4 - 0^4) = \frac{1}{4}. \qquad \square$$

Figure 5

Integration Rules

Next, we establish a number of important rules obeyed by integrals.

(i) *If f is integrable on $[a, b]$ and c is any constant, then cf is also integrable on $[a, b]$, and*

$$\int_a^b cf(x) \, dx = c \int_a^b f(x) \, dx. \tag{10}$$

In fact, if $S = \sum_{i=1}^n f(p_i) \, \Delta x_i$ is a Riemann sum for f, then

$$cS = c \sum_{i=1}^n f(p_i) \, \Delta x_i = \sum_{i=1}^n cf(p_i) \, \Delta x_i$$

is a Riemann sum for cf. But

$$\lim_{\mu \to 0} cS = c \lim_{\mu \to 0} S$$

(by the analogue for Riemann sums of Corollary 2, page 82), and hence

$$\lim_{\mu \to 0} \sum_{i=1}^n cf(p_i) \, \Delta x_i = c \lim_{\mu \to 0} \sum_{i=1}^n f(p_i) \, \Delta x_i,$$

which is equivalent to (10). Thus in a definite integral any constant multiplying the integrand can be factored out and placed in front of the integral sign.

(ii) *If f and g are both integrable on $[a, b]$, then the sum $f + g$ is also integrable on $[a, b]$, and*

$$\int_a^b [f(x) + g(x)] \, dx = \int_a^b f(x) \, dx + \int_a^b g(x) \, dx. \tag{11}$$

To prove this, we observe that if $S_h = \sum_{i=1}^n h(p_i) \, \Delta x_i$ is any Riemann sum

Section 4.2   Area Under a Curve and the Definite Integral   **235** for $h = f + g$, then $S_f = \sum_{i=1}^{n} f(p_i) \Delta x_i$ and $S_g = \sum_{i=1}^{n} g(p_i) \Delta x_i$ are Riemann sums for f and g (based on the same points p_i and x_i). But

$$S_h = \sum_{i=1}^{n} h(p_i) \Delta x_i = \sum_{i=1}^{n} [f(p_i) + g(p_i)] \Delta x_i = \sum_{i=1}^{n} f(p_i) \Delta x_i + \sum_{i=1}^{n} g(p_i) \Delta x_i,$$

and therefore, by the analogue for Riemann sums of Theorem 4, page 80,

$$\lim_{\mu \to 0} S_h = \lim_{\mu \to 0} (S_f + S_g) = \lim_{\mu \to 0} S_f + \lim_{\mu \to 0} S_g,$$

which is equivalent to (11). Thus the integral of the sum of two functions is the sum of the integrals of the separate functions.

(iii) If $f_1, f_2, \ldots, f_n$ are integrable on $[a, b]$ and $c_1, c_2, \ldots, c_n$ are any constants, then the "linear combination" $c_1 f_1 + c_2 f_2 + \cdots + c_n f_n$ is also integrable on $[a, b]$, and

$$\int_a^b (c_1 f_1 + c_2 f_2 + \cdots + c_n f_n) = c_1 \int_a^b f_1 + c_2 \int_a^b f_2 + \cdots + c_n \int_a^b f_n \tag{12}$$

(here we use the abbreviated notation discussed on page 232). The proof of (12) is by repeated application of formulas (10) and (11). Thus the integral of a linear combination of functions is a linear combination, with the same coefficients, of the integrals of the separate functions.

(iv) If f is integrable on $[a, b]$ and $f(x) \geq 0$, then

$$\int_a^b f(x) \, dx \geq 0.$$

In fact, the Riemann sum $S = \sum_{i=1}^{n} f(p_i) \Delta x_i$ is always nonnegative, since $f(p_i) \geq 0$ for every choice of the points p_i. Therefore

$$\int_a^b f(x) \, dx = \lim_{\mu \to 0} S \geq 0,$$

since if $\lim_{\mu \to 0} S < 0$, we could find a partition of $[a, b]$ and points p_i giving rise to a negative Riemann sum, which is impossible. Thus the integral of a nonnegative function is itself nonnegative.

(v) If f and g are integrable on $[a, b]$ and $f(x) \geq g(x)$, then

$$\int_a^b f(x) \, dx \geq \int_a^b g(x) \, dx. \tag{13}$$

To see this, we note that $f(x) - g(x) \geq 0$, and hence, by rules (iii) and (iv),

$$\int_a^b f(x) \, dx - \int_a^b g(x) \, dx = \int_a^b [f(x) - g(x)] \, dx \geq 0,$$

which is equivalent to (13).

(vi) If f is integrable on $[a, b]$ and $c \leq f(x) \leq C$, where c and C are constants, then

$$c(b - a) \leq \int_a^b f(x) \, dx \leq C(b - a). \tag{14}$$

In fact, every constant is integrable, and hence, by rule (v),

$$\int_a^b c\,dx \le \int_a^b f(x)\,dx \le \int_a^b C\,dx,$$

which is equivalent to (14), since

$$\int_a^b c\,dx = c\int_a^b dx = c(b-a), \qquad \int_a^b C\,dx = C\int_a^b dx = C(b-a).$$

From now on, we will make free use of all these rules.

Example 5 Evaluate the integral $\int_2^4 \left(\frac{3}{8}x^2 + 5x - 6\right)dx$.

Solution With the help of rule (iii) and formulas (6)–(8),

$$\int_2^4 \left(\frac{3}{8}x^2 + 5x - 6\right)dx = \frac{3}{8}\int_2^4 x^2\,dx + 5\int_2^4 x\,dx - 6\int_2^4 dx$$

$$= \frac{3}{8}\left(\frac{1}{3}\right)(4^3 - 2^3) + 5\left(\frac{1}{2}\right)(4^2 - 2^2) - 6(4-2)$$

$$= \frac{1}{8}(56) + \frac{5}{2}(12) - 6(2) = 25. \quad \square$$

Example 6 Find the area between the curve $y = 4 - x^2$ and the x-axis.

Solution We are looking for the area A of the shaded region R in Figure 6, which is the area under the curve $y = 4 - x^2$ (a parabola) from $x = -2$ to $x = 2$, the x-intercepts of the curve. Therefore

$$A = \int_{-2}^2 (4 - x^2)\,dx = 4\int_{-2}^2 dx - \int_{-2}^2 x^2\,dx$$

$$= 4[2 - (-2)] - \frac{1}{3}[2^3 - (-2)^3]$$

$$= 16 - \frac{16}{3} = \frac{32}{3}. \quad \square$$

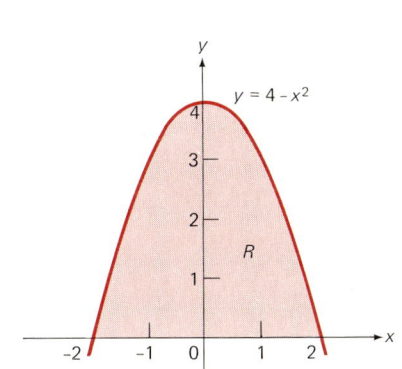

Figure 6

Problems

1. Why is the function $\sin x$ integrable on every closed interval $[a, b]$?

2. Why is the function $\cot x$ integrable on every closed interval $[a, b]$ that does not contain any of the points $x = 0$, $\pm\pi$, $\pm 2\pi$, ...?

3. Find the area A under the line $y = (b/a)x$ from $x = 0$ to $x = a$, and explain the result geometrically.

With the help of formulas (9) and (12), evaluate

4. $\int_1^2 2x^3\,dx$

5. $\int_{-1}^0 \frac{1}{2}x^4\,dx$

6. $\int_0^1 8x^7\,dx$

7. $\int_2^3 |x^2 - 4x|\,dx$

8. $\int_{-3}^0 |x^2 - 4x|\,dx$

9. $\int_0^{\sqrt{2}} (t^2 + \frac{3}{4}t - \frac{2}{3})\,dt$

10. $\int_{-2}^1 (t^3 - t^2 + t - 1)\,dt$

11. $\int_{-1}^1 (3u^5 - 5u^3)\,du$

12. $\int_0^1 (2u^{99} - u^{49} + \pi)\,du$

13. $\int_0^{\sqrt{3}} (\frac{1}{9}v - \frac{1}{27}v^3)\,dv$

14. $\int_{1/2}^1 (\frac{1}{4} - \frac{1}{3}v + \frac{1}{7}v^2 - \frac{1}{5}v^3)\,dv$

15. $\int_4^5 (\sqrt{x} + 1)(\sqrt{x} - 1)\,dx$

16. $\int_{-4}^6 (x-1)(x^2 + x + 1)\,dx$

17. $\int_2^8 \frac{x^3 - 1}{x - 1}\,dx$

18. $\int_{-1}^2 \frac{x^3 + 8}{x + 2}\,dx$

Find the area A of the region R
19. Under the curve $y = x^2 - 1$ from $x = 1.2$ to $x = 1.8$
20. Under the curve $y = x^2 + x + 1$ from $x = -1$ to $x = 1$
21. Between the curve $y = 2 + x - x^2$ and the x-axis
22. Between the curve $y = 2x - x^2$ and the x-axis
23. Under the curve $y = 2x^3 - 2x$ from $x = -1$ to $x = 0$
24. Between the curve $y = x^4 - 4x^3 + 4x^2$ and the x-axis
In each case sketch the region R.

25. Let the interval $[a, b]$ be divided into n subintervals $[a, x_1], [x_1, x_2], \ldots, [x_{n-1}, b]$ by a partition $a = x_0, x_1, x_2, \ldots, x_{n-1}, x_n = b$ of mesh size μ. What is the smallest possible value of n for a given mesh size μ? Show that as $\mu \to 0$, this smallest value approaches infinity.
26. The interval $[0, 10]$ is divided into n subintervals by a partition of mesh size $\sqrt{2}$. What is the smallest possible value of n?
27. The interval $[0, 10]$ is divided into 100 subintervals by a partition $0 = x_0, x_1, x_2, \ldots, x_{99}, x_{100} = 10$. What is the smallest possible value of the mesh size μ? The largest possible value of μ?
28. Show that if the integral as defined on page 232 exists, then it is unique, in the sense of having only one possible value.
29. The function
$$f(x) = \begin{cases} 0 & \text{if } x \neq 0, \\ 1 & \text{if } x = 0 \end{cases}$$

is discontinuous at $x = 0$, and hence fails to be continuous on any interval $[a, b]$ containing the point $x = 0$. Show that nevertheless f is integrable on $[a, b]$, with integral equal to 0.
30. How do we know that
$$0 \leq \int_0^1 x^{10}\sqrt{1 + x^2}\, dx \leq \sqrt{2}$$
without being able to evaluate the integral?
31. Let
$$f(x) = \begin{cases} 1 & \text{if } x \text{ is rational,} \\ 0 & \text{if } x \text{ is irrational.} \end{cases}$$
Show that f is nonintegrable on every interval $[a, b]$.
32. Give an example of a function which is bounded but not integrable on an interval $[a, b]$.
33. Show that if f is integrable on $[a, b]$, then f must be bounded on $[a, b]$.
34. Let
$$f(x) = \begin{cases} 1/x & \text{if } x \neq 0, \\ 0 & \text{if } x = 0. \end{cases}$$
Is f integrable on $[-1, 1]$?
35. Show that if f is continuous on $[a, b]$, then
$$\left| \int_a^b f(x)\, dx \right| \leq \int_a^b |f(x)|\, dx.$$
This is the analogue for integrals of the *triangle inequality* for sums.

4.3 More About Area; The Mean Value Theorem for Integrals

So far, in writing the integral
$$\int_a^b f(x)\, dx,$$
it has been assumed that $a < b$. We now allow the case $a \geq b$, setting
$$\int_a^b f(x)\, dx = -\int_b^a f(x)\, dx, \tag{1}$$
by definition. Suppose $a = b$ in (1). Then
$$\int_a^a f(x)\, dx = -\int_a^a f(x)\, dx,$$
which leads us to define
$$\int_a^a f(x)\, dx = 0. \tag{2}$$
This makes sense, since we expect the area of a "region of zero width" to be zero.

Example 1 Evaluate the integral $\int_2^1 (x^2 - x)\,dx$.

Solution With the help of (1),

$$\int_2^1 (x^2 - x)\,dx = -\int_1^2 (x^2 - x)\,dx = \int_1^2 (x - x^2)\,dx = \int_1^2 x\,dx - \int_1^2 x^2\,dx$$

$$= \frac{1}{2}(2^2 - 1^2) - \frac{1}{3}(2^3 - 1^3) = \frac{3}{2} - \frac{7}{3} = -\frac{5}{6}. \quad \square$$

Integration on Adjacent Intervals

Next we consider what happens when the interval of integration is "split up."

> **Theorem 2** *(Additivity of the integral on adjacent intervals).* If f is continuous on $[a, b]$ and if c is an interior point of $[a, b]$, then
>
> $$\int_a^b f(x)\,dx = \int_a^c f(x)\,dx + \int_c^b f(x)\,dx \qquad (a < c < b). \tag{3}$$

Proof As in the definition of the integral, we divide the interval $[a, b]$ into a large number of small subintervals by introducing points of subdivision, but this time we insist that one of the points of subdivision be the fixed point c. In other words, we now choose points of subdivision x_i ($i = 0, 1, \ldots, n$) such that

$$a = x_0 < x_1 < \cdots < x_{m-1} < x_m = c < x_{m+1} < \cdots < x_{n-1} < x_n = b,$$

where the subscript m depends, of course, on the number of points x_i to the left of c. The partition of $[a, b]$ formed by the points $x_0, x_1, \ldots, x_n$ then automatically gives rise to a partition of $[a, c]$ formed by the points $x_0, x_1, \ldots, x_m$, and also a partition of $[c, b]$ formed by the points $x_m, x_{m+1}, \ldots, x_n$. Correspondingly, the Riemann sum

$$S = \sum_{i=1}^n f(p_i)\,\Delta x_i,$$

used to define the integral of f from a to b, splits up into the sum

$$S = S' + S'',$$

where

$$S' = \sum_{i=1}^m f(p_i)\,\Delta x_i, \qquad S'' = \sum_{i=m+1}^n f(p_i)\,\Delta x_i$$

(Δx_i and p_i have the same meaning as on page 230). You will recognize S' and S'' as the Riemann sums used to define the integral of f from a to c and the integral of f from c to b. Let μ, μ' and μ'' be the mesh sizes of the partitions of $[a, b]$, $[a, c]$ and $[c, b]$, respectively, that is, let

$$\mu = \max\{\Delta x_1, \ldots, \Delta x_n\}, \qquad \mu' = \max\{\Delta x_1, \ldots, \Delta x_m\},$$
$$\mu'' = \max\{\Delta x_{m+1}, \ldots, \Delta x_n\}.$$

Then clearly $\mu \to 0$ implies $\mu' \to 0$ and $\mu'' \to 0$, so that

$$\int_a^b f(x)\,dx = \lim_{\mu \to 0} S = \lim_{\mu \to 0}(S' + S'') = \lim_{\mu \to 0} S' + \lim_{\mu \to 0} S''$$

$$= \lim_{\mu' \to 0} S' + \lim_{\mu'' \to 0} S'' = \int_a^c f(x)\,dx + \int_c^b f(x)\,dx,$$

where the existence of all three integrals follows from the assumption that f is continuous on $[a, b]$, and hence on $[a, c]$ and $[c, b]$ as well. ∎

As we now show, the points a, b and c in formula (3) need not satisfy the condition $a < c < b$, and in fact can be *arbitrary*.

Corollary *If f is continuous on an interval containing the points a, b, and c, then*

$$\int_a^b f(x)\,dx = \int_a^c f(x)\,dx + \int_c^b f(x)\,dx \qquad (a, b, c \text{ arbitrary}). \tag{4}$$

Proof Formula (4) follows at once from (1) and (2) if two or three of the points a, b and c coincide. Moreover, (4) reduces to (3) if $a < c < b$. The other cases can be dealt with by using (1), together with (3). For example, if $c < b < a$, then, by (3), with c, b, a instead of a, c, b,

$$\int_c^a f(x)\,dx = \int_c^b f(x)\,dx + \int_b^a f(x)\,dx.$$

Hence, by two applications of (1),

$$-\int_a^c f(x)\,dx = \int_c^b f(x)\,dx - \int_a^b f(x)\,dx,$$

which implies

$$\int_a^b f(x)\,dx = \int_a^c f(x)\,dx + \int_c^b f(x)\,dx.$$

The remaining cases $a < b < c$, $b < a < c$, $b < c < a$, $c < a < b$ are treated similarly. ∎

Thus the validity of the interval additivity property (4) does not depend on the order of the points a, b and c. This fact testifies to the aptness of the definition (1), which plays a key role in the proof of the corollary.

Theorem 2 has a simple geometric interpretation in the case where f is nonnegative on $[a, b]$. Then $\int_a^b f(x)\,dx$ is the area under the curve $y = f(x)$ from $x = a$ to $x = b$, while $\int_a^c f(x)\,dx$ is the area under the curve from $x = a$ to $x = c$ and $\int_c^b f(x)\,dx$ is the area under the curve from $x = c$ to $x = b$. These are the areas of the regions $abBA$, $acCA$ and $cbBC$ in Figure 7, and equation (3) says that

$$\text{Area of } abBA = (\text{Area of } acCA) + (\text{Area of } cbBC),$$

a fact which is geometrically evident, since the regions $acCA$ and $cbBC$ share no points other than the line segment cC, their common boundary. The figure is drawn for the case where f is positive on $[a, b]$, but it is easy to see that this additivity of nonoverlapping areas is still valid if f equals zero at one or more points of $[a, b]$.

Example 2 Evaluate the integral $\int_0^2 |1 - x|\,dx$.

Solution Evaluation of the integral is equivalent to finding the area under the curve $y = |1 - x|$ from $x = 0$ to $x = 2$. It is apparent from Figure 8 that this area equals 1, since each shaded triangle is an isosceles right triangle with legs

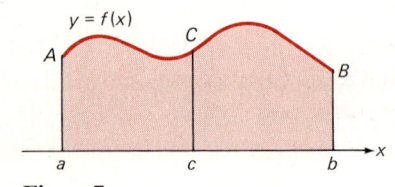

Figure 7

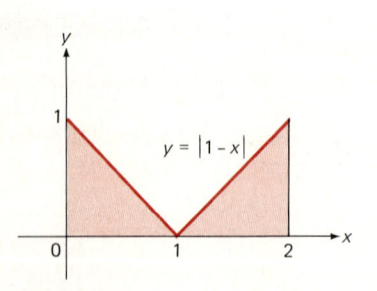

Figure 8

of length 1 and area $\frac{1}{2}$. Alternatively, we can evaluate the integral with the help of Theorem 2:

$$\int_0^2 |1-x|\,dx = \int_0^1 |1-x|\,dx + \int_1^2 |1-x|\,dx$$
$$= \int_0^1 (1-x)\,dx + \int_1^2 (x-1)\,dx$$

(this way of splitting up the interval $[0, 2]$ is motivated by the observation that $1 - x$ changes sign at $x = 1$). Therefore

$$\int_0^2 |1-x|\,dx = \int_0^1 dx - \int_0^1 x\,dx + \int_1^2 x\,dx - \int_1^2 dx$$
$$= 1 - \frac{1}{2}(1^2 - 0^2) + \frac{1}{2}(2^2 - 1^2) - 1 = 1 - \frac{1}{2} + \frac{3}{2} - 1 = 1,$$

as already shown by the geometric argument. Notice the care that must be taken in dealing with integrands involving absolute values. □

The Area Between Two Curves

We have already shown the appropriateness of using the integral $\int_a^b f(x)\,dx$ as the definition of the area under the curve $y = f(x)$ from $x = a$ to $x = b$, that is, the area of the plane region bounded by the vertical lines $x = a$ and $x = b$, the x-axis and the curve $y = f(x)$, where $f(x) \geq 0$. We now consider the more general problem of finding the *area between two curves* $y = f(x)$ and $y = g(x)$, from $x = a$ to $x = b$, where the functions f and g are both continuous and $f(x) \geq g(x)$. This is the area of the plane region R shown in Figure 9(a), bounded by the vertical lines $x = a$ and $x = b$ and by the curves $y = f(x)$ and $y = g(x)$, with the curve $y = f(x)$ as its upper boundary and the curve $y = g(x)$ as its lower boundary (we have chosen $f(x) > g(x) > 0$ for simplicity).

The region R will in general have two curved sides (CD and EF in the figure), and such regions are not considered in elementary geometry. Thus, in the process of calculating the area A of the region R, we must again concern ourselves with the proper definition of A. To this end, we parallel

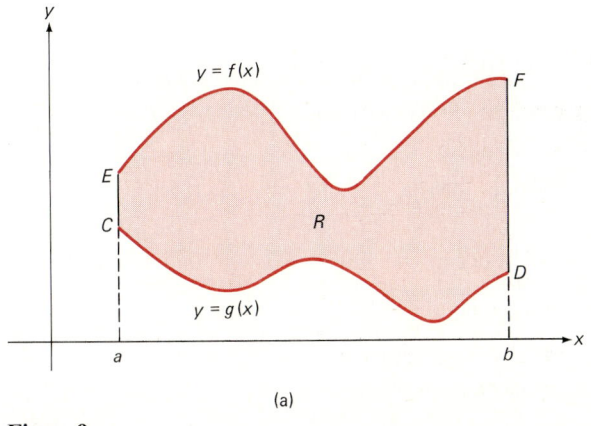

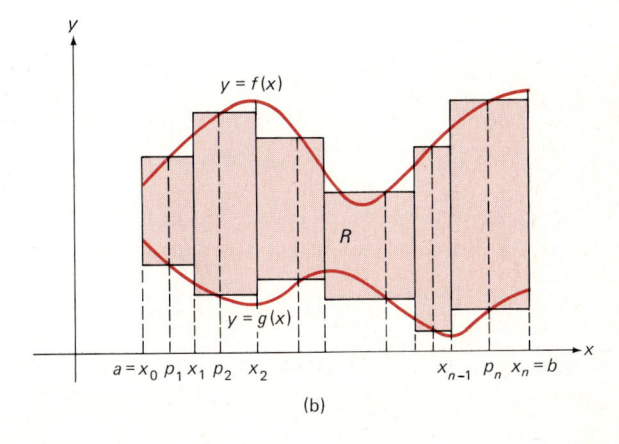

Figure 9

Section 4.3 More About Area; The Mean Value Theorem for Integrals 241 the construction on pages 229–231, approximating A by a sum of the form

$$\sum_{i=1}^{n} [f(p_i) - g(p_i)] \Delta x_i, \tag{5}$$

based on a partition of the interval $[a, b]$ into n subintervals $[x_0, x_1]$, $[x_1, x_2], \ldots, [x_{n-1}, x_n]$, of lengths $\Delta x_1, \Delta x_2, \ldots, \Delta x_n$, respectively, with p_i an arbitrary point in $[x_{i-1}, x_i]$. This approximation corresponds to replacing the strips with curved tops and bottoms in Figure 9(b) by the indicated shaded rectangles. We then *define* A as the limit of the sum (5) as the mesh size $\mu = \max \{\Delta x_1, \Delta x_2, \ldots, \Delta x_n\}$ approaches zero:

$$A = \lim_{\mu \to 0} \sum_{i=1}^{n} [f(p_i) - g(p_i)] \Delta x_i.$$

As we now know, the limit in question is the integral

$$A = \int_{a}^{b} [f(x) - g(x)] \, dx, \tag{6}$$

whose existence follows from the continuity of the function $f - g$. This is the desired formula for the area A between the two curves $y = f(x)$ and $y = g(x)$. The nonnegativity of A, required by its geometric meaning, is a consequence of the inequality $f(x) - g(x) \geq 0$ and rule (iv), page 236. It is intuitively clear that $A > 0$ unless the curves $y = f(x)$ and $y = g(x)$ coincide, and this follows from the fact that the integral of a continuous nonnegative function is positive unless the function is identically zero (see Example 8). If $g(x) \equiv 0$, (6) reduces to the formula

$$A = \int_{a}^{b} f(x) \, dx$$

for the area under the curve $y = f(x)$, as one would expect.

Alternatively, formula (6) can be derived by the following argument. It can be assumed without loss of generality that f and g are both nonnegative, since otherwise this can be achieved by subjecting the curves $y = f(x)$ and $y = g(x)$ to the same upward shift, which does not change the function $f - g$ or the area A between the curves. Then, as in Figure 9(a),

$$A = \text{Area of } R = (\text{Area of } abFE) - (\text{Area of } abDC),$$

since the regions $abFE$ and $abDC$ share no points other than their common boundary CD. But

$$\text{Area of } abFE = \int_{a}^{b} f(x) \, dx, \quad \text{Area of } abDC = \int_{a}^{b} g(x) \, dx,$$

and hence

$$A = \int_{a}^{b} f(x) \, dx - \int_{a}^{b} g(x) \, dx = \int_{a}^{b} [f(x) - g(x)] \, dx.$$

Example 3 Find the area A between the line $y = 2 - x$ and the parabola $y = x^2$.

Solution To find the x-coordinates of the points in which the line and the parabola intersect, we solve the quadratic equation $2 - x = x^2$, obtaining the two roots $x = -2$ and $x = 1$. On the interval $[-2, 1]$ the line is the upper curve and the parabola is the lower curve (see Figure 10). Therefore,

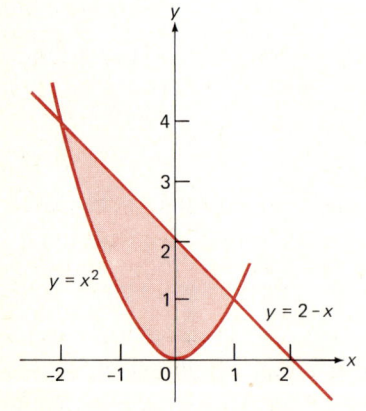

Figure 10

242   Chapter 4   Integrals

by (6),
$$A = \int_{-2}^{1} (2 - x - x^2)\,dx = 2\int_{-2}^{1} dx - \int_{-2}^{1} x\,dx - \int_{-2}^{1} x^2\,dx$$
$$= 2[1 - (-2)] - \frac{1}{2}[1^2 - (-2)^2] - \frac{1}{3}[1^3 - (-2)^3]$$
$$= 2(3) - \frac{1}{2}(-3) - \frac{1}{3}(9) = \frac{9}{2}. \quad \square$$

Let $y = f(x)$ and $y = g(x)$ be the two intertwining curves shown in Figure 11. Then $y = f(x)$ is the upper curve and $y = g(x)$ the lower curve on the intervals $[a, c_1]$, $[c_2, c_3]$ and $[c_4, b]$, but the roles of the two curves are reversed on the intervals $[c_1, c_2]$ and $[c_3, c_4]$, with $y = f(x)$ becoming the lower curve and $y = g(x)$ the upper curve. Correspondingly, the contributions to the integral in (6) from $[a, c_1]$, $[c_2, c_3]$ and $[c_4, b]$ are *positive*, but those from $[c_1, c_2]$ and $[c_3, c_4]$ are *negative*, as indicated by the signs in the figure Specifically, with the help of Theorem 2,
$$\int_a^b [f(x) - g(x)]\,dx = \sum_{n=1}^{5} I_n,$$
where
$$I_n = \int_{c_{n-1}}^{c_n} [f(x) - g(x)]\,dx \quad (c_0 = a, c_5 = b),$$
and I_1, I_3 and I_5 are positive, while I_2 and I_4 are negative. Thus (6) does not give the area between the two curves in question, but rather the sum of the areas of the three regions marked with a plus sign less the sum of the areas of the two regions marked with a minus sign.

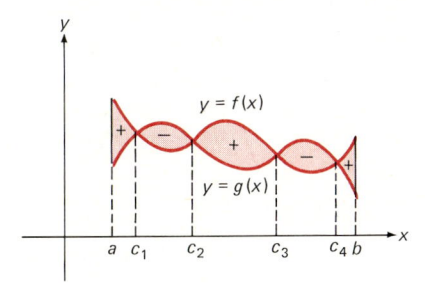

Figure 11

The appropriate definition for the area A between two intertwining curves $y = f(x)$ and $y = g(x)$ is
$$A = \int_a^b |f(x) - g(x)|\,dx, \tag{6'}$$
involving the *absolute value* of the difference $f(x) - g(x)$. With this definition, the area between the two curves in Figure 11 is the sum of the areas of all five shaded regions without regard for sign. Notice that formula (6') reduces to the previous formula (6) if $f(x) \geq g(x)$.

Suppose $f(x)$ can take both positive and negative values. Then choosing $g(x) \equiv 0$ in (6'), we find that $\int_a^b |f(x)|\,dx$ is the area A between the curve $y = f(x)$ and the x-axis. It is easy to see that $\int_a^b f(x)\,dx = A_+ - A_-$, where A_+ is the part of A above the x-axis and A_- is the part below the x-axis.

Example 4 Find the area A between the curve $y = x^3 - x^2 - 2x$ and the x-axis. This is the sum of the areas of the two shaded regions shown in Figure 12.

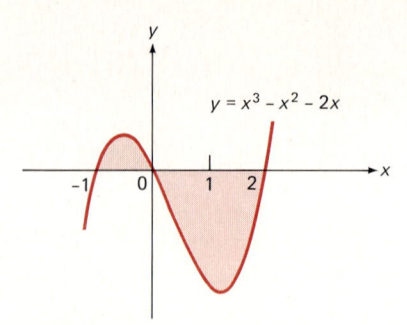

Figure 12

Solution Solving the cubic equation $x^3 - x^2 - 2x = x(x + 1)(x - 2) = 0$, we find that the curve $y = x^3 - x^2 - 2x$ has three x-intercepts, $x = -1$, $x = 0$ and $x = 2$. Moreover, the curve lies above the x-axis between $x = -1$ and $x = 0$, and below the x-axis between $x = 0$ and $x = 2$, as shown in the figure. Therefore

$$A = \int_{-1}^{2} |x^3 - x^2 - 2x| \, dx = \int_{-1}^{0} (x^3 - x^2 - 2x) \, dx - \int_{0}^{2} (x^3 - x^2 - 2x) \, dx$$

$$= \int_{-1}^{0} x^3 \, dx - \int_{-1}^{0} x^2 \, dx - 2\int_{-1}^{0} x \, dx - \int_{0}^{2} x^3 \, dx + \int_{0}^{2} x^2 \, dx + 2\int_{0}^{2} x \, dx$$

$$= \frac{1}{4}[0^4 - (-1)^4] - \frac{1}{3}[0^3 - (-1)^3] - 2\left(\frac{1}{2}\right)[0^2 - (-1)^2]$$

$$- \frac{1}{4}(2^4 - 0^4) + \frac{1}{3}(2^3 - 0^3) + 2\left(\frac{1}{2}\right)(2^2 - 0^2)$$

$$= -\frac{1}{4} - \frac{1}{3} + 1 - 4 + \frac{8}{3} + 4 = \frac{7}{3} + \frac{3}{4} = \frac{37}{12}. \quad \square$$

The Mean Value of a Function

Let f be a function integrable on an interval $[a, b]$. Then the number

$$\frac{1}{b-a} \int_{a}^{b} f(x) \, dx$$

is called the *mean value* (or *average*) of f on $[a, b]$, or over $[a, b]$. If, in addition, f is continuous on $[a, b]$, then, as we will show in a moment (see Theorem 3), there is always at least one point c in $[a, b]$ such that $f(c)$ equals the mean value of f on $[a, b]$.

Example 5 The integral of the function $f(x) = x^2 + 1$ on the interval $[-2, 1]$ is

$$\int_{-2}^{1} (x^2 + 1) \, dx = \int_{-2}^{1} x^2 \, dx + \int_{-2}^{1} dx$$

$$= \frac{1}{3}[1^3 - (-2)^3] + [1 - (-2)] = 3 + 3 = 6.$$

244 Chapter 4 Integrals

Hence the mean value of f on $[-2, 1]$ is

$$\frac{1}{1-(-2)}\int_{-2}^{1}(x^2+1)\,dx = \frac{1}{3}(6) = 2.$$

Notice that f takes this value at the two points 1 and -1, both of which belong to the interval $[-2, 1]$.

Example 6 The mean value of the function $f(x) = 1 - x^3$ on the interval $[0, 4]$ is

$$\frac{1}{4-0}\int_0^4 (1-x^3)\,dx = \frac{1}{4}\int_0^4 dx - \frac{1}{4}\int_0^4 x^3\,dx$$

$$= \frac{1}{4}(4-0) - \frac{1}{4}\left(\frac{1}{4}\right)(4^4 - 0^4) = 1 - 16 = -15,$$

and f takes this value at the point $\sqrt[3]{16} \approx 2.52$, which belongs to $[0, 4]$.

> **Theorem 3** *(Mean value theorem for integrals).* If f is continuous on $[a, b]$, there is a point c in $[a, b]$ such that
>
> $$\frac{1}{b-a}\int_a^b f(x)\,dx = f(c),$$
>
> or equivalently
>
> $$\int_a^b f(x)\,dx = (b-a)f(c). \tag{7}$$

Proof By the extreme value theorem (see page 99), f has both a minimum m and a maximum M on $[a, b]$, taken at points p and q of $[a, b]$. Since $m \leq f(x) \leq M$, we have

$$m(b-a) \leq \int_a^b f(x)\,dx \leq M(b-a),$$

with the help of rule (vi), page 236, or equivalently

$$m \leq \frac{1}{b-a}\int_a^b f(x)\,dx \leq M.$$

Thus the mean value of f on $[a, b]$, which we denote by h, belongs to the interval $[m, M]$. If $h = m$ or $h = M$, then $h = f(p)$ or $h = f(q)$, and the theorem is proved with $c = p$ or $c = q$. Otherwise $m < h < M$, and it follows from the intermediate value theorem (see page 96) that there is a point c between p and q, and hence certainly in $[a, b]$, such that $h = f(c)$. ∎

Theorem 3 has a simple geometric interpretation. Suppose f is continuous and nonnegative on $[a, b]$. Then formula (7) says that there is a rectangle of length $b - a$, and of height equal to the value of f at a point c in $[a, b]$, whose area is equal to the area under the curve $y = f(x)$ from $x = a$ to $x = b$. Alternatively, writing (7) in the equivalent form

$$\int_a^b [f(x) - f(c)]\,dx = 0, \tag{7'}$$

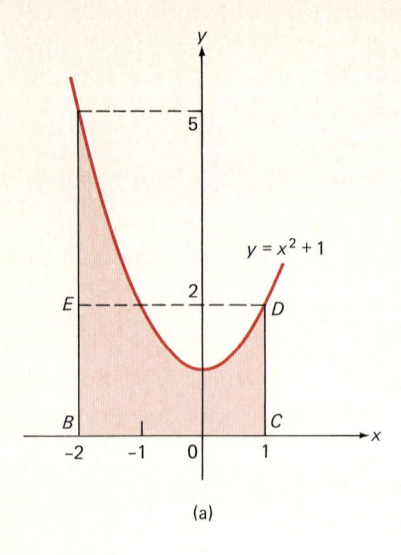

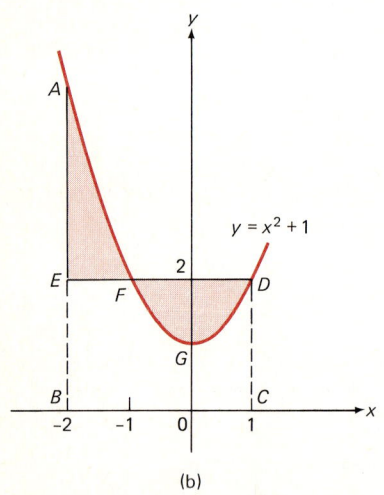

we see that there is a horizontal line $y = f(c)$, where c is a point in $[a, b]$, such that the area under the curve $y = f(x)$ and above the line, from $x = a$ to $x = b$, is exactly equal to the area under the line and above the curve (here it is unnecessary to assume that f is nonnegative).

Example 7 Interpret Theorem 3 geometrically, as applied to the function $y = f(x) = x^2 + 1$ on the interval $[-2, 1]$.

Solution As shown in Example 5, the mean value of f on $[-2, 1]$ is 2. Correspondingly, the area under the parabola $y = x^2 + 1$ from $x = -2$ to $x = 1$, that is, the area of the shaded region in Figure 13(a), is equal to the area of the rectangle $BCDE$, of length 3 and height 2 (both areas are equal to 6), and the height of the rectangle is equal to the value of f at the two points 1 and -1 of the interval $[-2, 1]$. Alternatively, the area of the shaded region AEF in Figure 13(b) under the parabola $y = x^2 + 1$ and above the line $y = 2$ is equal to the area of the shaded region FGD under the line and above the parabola. In fact, both of these areas are equal to $\frac{4}{3}$ (check this). □

Example 8 With the help of Theorem 3, show that if f is continuous and nonnegative on $[a, b]$ and if $f(c) \neq 0$ for at least one point c in $[a, b]$, then

$$\int_a^b f(x)\,dx > 0. \qquad (8)$$

Solution Clearly $f(c) > 0$, since f is nonnegative on $[a, b]$. If $a < c < b$, then by rule (ii), page 76, there is a subinterval $[c - \delta, c + \delta]$ of the interval $[a, b]$ on which f is positive, since f has a positive limit $f(c)$ at c. Also

$$\int_a^b f(x)\,dx = \int_a^{c-\delta} f(x)\,dx + \int_{c-\delta}^{c+\delta} f(x)\,dx + \int_{c+\delta}^b f(x)\,dx,$$

by Theorem 2, where the first and third integrals in the sum are nonnegative (why?). But, by Theorem 3, the second integral in the sum is equal to $2\delta f(p)$ for some point p in $[c - \delta, c + \delta]$, and hence it is positive, since $f(p) > 0$. Thus the inequality (8) holds. The proof is virtually the same if $c = a$ or $c = b$ (this time split $[a, b]$ into just two subintervals). The details are left as an exercise. □

In particular, it follows from Example 8 that if f is continuous and nonnegative on $[a, b]$ and if

$$\int_a^b f(x)\,dx = 0,$$

then f is identically equal to zero on $[a, b]$.

Remark If $b < a$ instead of $a < b$, as tacitly assumed in Theorem 3 and formula (7), and if f is continuous on $[b, a]$, we have

$$\int_b^a f(x)\,dx = (a - b)f(c),$$

instead of (7), where c is now a point in $[b, a]$. Multiplying both sides of this formula by -1, we get back (7). Thus the mean value theorem for integrals can always be used in the form (7), where c is a point in the interval with endpoints a and b. Actually, it turns out (see Problem 39) that we can always choose c to be a point *between* a and b, just as in the case of the mean value theorem for derivatives.

Problems

1. According to formula (9), page 235,
$$\int_a^b x^n \, dx = \frac{1}{n+1}(b^{n+1} - a^{n+1}) \quad (n = 1, 2, \ldots),$$
where it was assumed tacitly that $a < b$. Show that the formula remains valid for $a > b$.

Evaluate

2. $\int_{11}^{7} x \, dx$

3. $\int_{1}^{-1}(x^2 - x) \, dx$

4. $\int_{-2}^{2} |x + 1| \, dx$

5. $\int_{-1}^{3} |x^2 - 2x| \, dx$

6. $\int_0^1 t^{10} \, dt + \int_1^0 t^{10} \, dt$

7. $\int_0^1 u^{10} \, du - \int_1^0 u^{10} \, du$

8. $\int_0^1 v^2 \, dv + \int_1^3 (v^2 - 1) \, dv + \int_3^2 (v^2 + 1) \, dv$

Let
$$f(x) = \begin{cases} 2 - x & \text{if } x < 1, \\ x^3 & \text{if } x \geq 1. \end{cases}$$

Evaluate

9. $\int_{-2}^{3} f(x) \, dx$

10. $\int_{2}^{-1} f(x) \, dx$

11. $\int_0^2 f(x+1) \, dx$

12. $\int_1^2 [f(x) - f(x-1)] \, dx$

Find the area A of the region R between the curves
13. $y = x^2 - 1$ and $y = 1 - x^2$
14. $y = x^2 - 4$ and $y = 2x - x^2$
15. $y = x^2$ and $y = x^5$
16. $y = x + 1$ and $y = |x| + |x - 1|$
17. $y = |x|$ and $y = 2 - x^2$
18. $y = |2x - 1|$ and $y = 4 - x^2$

In each case sketch the region R.

19. Find the area A between the curve $y = x^3 - x^2 - 2x$ (see Figure 12) and the line $y = 4x$.
20. Let R be a region bounded by the vertical lines $x = a$ and $x = b$, an upper curve $y = f(x)$ and a lower curve $y = g(x) \leq f(x)$. Then R is said to be of *width* $w(x) = f(x) - g(x)$, where $w(x)$ is a function defined on $[a, b]$. Prove Cavalieri's principle for area, which says that two regions of this type with the same width $w(x)$, like the two shaded regions in Figure 14, have the same area A, regardless of the choice of the upper and lower curves.
21. What is the geometric meaning of the average of the function x over the interval $[a, b]$?

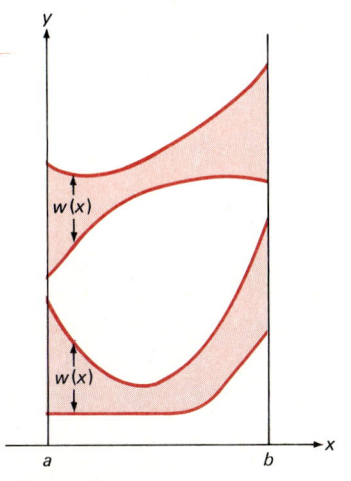

Figure 14

22. Show that the average of the function x^2 over the interval $[a, b]$ is equal to $\frac{1}{3}(a^2 + ab + b^2)$.

For the given function f and points a and b find the mean value of f on $[a, b]$.
23. $f(x) = 1 - x - x^2$, $a = 0$, $b = 4$
24. $f(x) = x^3 - 2x + 1$, $a = -2$, $b = 3$
25. $f(x) = |1 - x|$, $a = -1$, $b = 2$
26. $f(x) = x^4 + 5x^2 - 10$, $a = -3$, $b = -1$

Find a point c satisfying the mean value formula (7) if
27. $f(x) = x$, $a = 1$, $b = 7$
28. $f(x) = 2x + 3$, $a = -1$, $b = 3$
29. $f(x) = x^2$, $a = 2$, $b = 0$
30. $f(x) = 3x^2 + 1$, $a = 4$, $b = 1$
31. $f(x) = |x^2 - 1|$, $a = -2$, $b = 2$
32. $f(x) = \begin{cases} 2x^2 & \text{if } x < 0, \\ x & \text{if } x \geq 0, \end{cases}$ $a = -1$, $b = 2$

33. Let f and g be continuous on $[a, b]$, and suppose $f(x) \geq g(x)$ with $f(c) \neq g(c)$ for at least one point c in $[a, b]$. Show that
$$\int_a^b f(x) \, dx > \int_a^b g(x) \, dx.$$

34. Verify that
$$\frac{1}{6} < \int_0^2 \frac{1}{10 + x} \, dx < \frac{1}{5}.$$

Section 4.3 More About Area; The Mean Value Theorem for Integrals

Without attempting to evaluate any integrals, determine which integral is larger.

35. $\int_0^1 x\,dx$ or $\int_0^1 x^2\,dx$ 36. $\int_1^2 x\,dx$ or $\int_1^2 x^2\,dx$

37. $\int_0^{\pi/2} \sin x\,dx$ or $\int_0^{\pi/2} x\,dx$

38. $\int_0^{\pi/2} \sin^2 x\,dx$ or $\int_0^{\pi/2} \sin^{10} x\,dx$

39. Show that the point c in Theorem 3 can always be chosen to be an *interior* point of the interval $[a, b]$.

40. Suppose f is continuous on $[1, 4]$ and $f(3) \neq 0$. Which number is larger,

$$I_1 = \int_1^4 f(x)\,dx + \int_4^2 f(x)\,dx + \int_2^3 f(x)\,dx + \int_3^1 f(x)\,dx$$

or

$$I_2 = \int_{\sqrt{5}}^\pi f^2(x)\,dx?$$

4.4 Antiderivatives and the Indefinite Integral

The concepts introduced in this section will play a key role in our further study of integral calculus. With their help we will soon prove the fundamental theorem of calculus (Theorem 6, page 256), which will enable us to evaluate definite integrals without resorting to the explicit calculation of Riemann sums.

Definition of an Antiderivative

Let $f(x)$ be a function defined on an interval I, and let $F(x)$ be another function defined on I whose derivative is $f(x)$, so that

$$\frac{dF(x)}{dx} = F'(x) = f(x)$$

for every x in I. Then $F(x)$ is said to be an *antiderivative* of $f(x)$, on the interval I. Here we use the letter x for the independent variable, but another letter would do just as well.

Example 1 The function $\frac{1}{2}x^2$ is an antiderivative of x on $(-\infty, \infty)$, since

$$\frac{d}{dx}\frac{1}{2}x^2 = \frac{1}{2}(2x) = x$$

for every x.

Example 2 The function $\frac{2}{3}t^{3/2}$ is an antiderivative of $t^{1/2}$ on $(0, \infty)$, since

$$\frac{d}{dt}\frac{2}{3}t^{3/2} = \frac{2}{3}\left(\frac{3}{2}t^{1/2}\right) = t^{1/2}$$

for every positive t.

Example 3 The function $\tan u$ is an antiderivative of $\sec^2 u$ on any interval that does not contain any of the points $u = \pm\frac{1}{2}\pi, \pm\frac{3}{2}\pi, \ldots$. In fact,

$$\frac{d}{du}\tan u = \sec^2 u$$

except at these points, where both $\tan u$ and $\sec u$ are undefined.

The General Antiderivative

If $F(x)$ is an antiderivative of $f(x)$ on an interval I, then so is $G(x) = F(x) + C$, where C is an arbitrary constant, since

$$\frac{dG(x)}{dx} = \frac{d}{dx}[F(x) + C] = \frac{dF(x)}{dx} + \frac{dC}{dx} = F'(x) = f(x).$$

As we now show, $G(x)$ is the general antiderivative of $f(x)$ on I, in the sense that *every* antiderivative of $f(x)$ on I is of the form $G(x)$.

> **Theorem 4** *(Form of the general antiderivative).* Let $F(x)$ be any antiderivative of $f(x)$ on an interval I. Then every other antiderivative of $f(x)$ on I is of the form $F(x) + C$, where C is a constant.

Proof Let $G(x)$ be any other antiderivative of $f(x)$ on I, and let $H(x) = G(x) - F(x)$. Then

$$H'(x) = G'(x) - F'(x) = f(x) - f(x) = 0$$

for every x in I, that is, the derivative $H'(x)$ equals zero at every point of I. It follows from Theorem 3, page 164, that $H(x)$ has the same value, call it C, at every point of I. Therefore

$$H(x) = G(x) - F(x) \equiv C,$$

or equivalently $G(x) \equiv F(x) + C$. ∎

Thus two functions with the same derivative on an interval can differ only by a constant. Geometrically, this means that if two curves over the same interval have the same slope at every pair of points with the same abscissa, then each curve can be obtained from the other by making a suitable vertical shift, as illustrated by the two curves $y = F(x)$ and $y = G(x)$ in Figure 15.

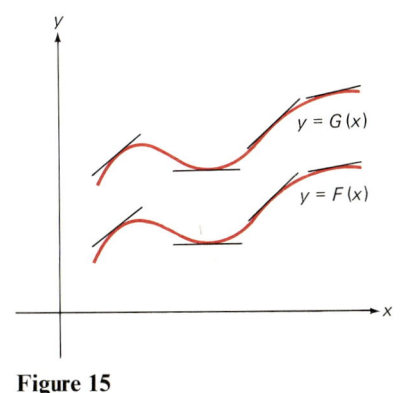

Figure 15

Example 4 It follows from

$$\frac{d}{dx}(4 - \cos x) = \sin x, \qquad \frac{d}{dx}\left(2 \sin^2 \frac{x}{2}\right) = 2 \sin \frac{x}{2} \cos \frac{x}{2} = \sin x$$

that

$$4 - \cos x \equiv 2 \sin^2 \frac{x}{2} + C,$$

and choosing $x = 0$, we find that $C = 3$. As an exercise, verify this trigonometric identity.

Definition of the Indefinite Integral

We have just shown that if $F(x)$ is an antiderivative of $f(x)$ on I, then the general antiderivative of $f(x)$ on I is given by $F(x) + C$, where C is an arbitrary constant. The expression $F(x) + C$ is also called the *indefinite integral* of $f(x)$, denoted by $\int f(x)\,dx$. Thus, by definition

$$\int f(x)\,dx = F(x) + C, \tag{1}$$

so that the indefinite integral is defined only to within an arbitrary "additive constant." Here the notation is the same as for the definite integral, *except that there are no limits of integration.* The absence of limits of integration tells us that the indefinite integral $\int f(x)\,dx$ is a *function* (plus an arbitrary constant), as opposed to the definite integral $\int_a^b f(x)\,dx$, which is a *number*. As before, the function $f(x)$ is called the *integrand*, its argument (in this case x) is called the *variable of integration*, and the operation leading from $f(x)$ to the expression (1) is called (indefinite) *integration*. The constant C in (1) is called the *constant of integration*.

In writing (1), it is tacitly assumed that the formula is an identity for all x in some underlying interval I on which f and F are both defined; however, I is usually left unspecified. Differentiating (1), we get

$$\frac{d}{dx}\int f(x)\,dx = \frac{d}{dx}[F(x) + C] = F'(x),$$

so that

$$\frac{d}{dx}\int f(x)\,dx = f(x). \qquad (2)$$

Since it is an antiderivative, the indefinite integral must have the same argument as the integrand. For example,

$$\int x\,dx = \frac{1}{2}x^2 + C, \qquad \int t\,dt = \frac{1}{2}t^2 + C,$$

and in this sense

$$\int x\,dx \neq \int t\,dt.$$

Here, unlike the case of the definite integral, the variable of integration is not a dummy variable.

Since any differentiable function $f(x)$ is an antiderivative of its own derivative $f'(x)$, we have

$$\int f'(x)\,dx = f(x) + C. \qquad (3)$$

This formula can be used to derive an integration formula from every differentiation formula. For example, if r is a rational number not equal to -1, then

$$\frac{d}{dx}\frac{x^{r+1}}{r+1} = \frac{(r+1)x^r}{r+1} = x^r,$$

and application of (3) gives

$$\int x^r\,dx = \frac{x^{r+1}}{r+1} + C \qquad (r \neq -1). \qquad (4)$$

Choosing $r = 1, \frac{1}{2}, \frac{1}{3}, 0$ and $-\frac{1}{2}$ in turn, we get

$$\int x\,dx = \frac{1}{2}x^2 + C,$$

$$\int x^{1/2}\,dx = \frac{2}{3}x^{3/2} + C,$$

$$\int x^{1/3} \, dx = \frac{3}{4} x^{4/3} + C,$$

$$\int dx = x + C,$$

$$\int \frac{dx}{\sqrt{x}} = 2\sqrt{x} + C.$$

The first two formulas were anticipated in Examples 1 and 2. In the last formula we follow the common practice of writing

$$\int \frac{1}{f(x)} \, dx \quad \text{as} \quad \int \frac{dx}{f(x)}.$$

In the same way, the formulas for the derivatives of the trigonometric functions (see page 132) lead to the following integration formulas:

$$\int \cos x \, dx = \sin x + C,$$

$$\int \sin x \, dx = -\cos x + C,$$

$$\int \sec^2 x \, dx = \tan x + C,$$

$$\int \csc^2 x \, dx = -\cot x + C,$$

$$\int \sec x \tan x \, dx = \sec x + C,$$

$$\int \csc x \cot x \, dx = -\csc x + C.$$

Rules for Indefinite Integration

Next we establish some simple rules obeyed by indefinite integrals.

(i) *If f has an indefinite integral (antiderivative) and c is any constant, then*

$$\int cf(x) \, dx = c \int f(x) \, dx. \tag{5}$$

In fact, the expression on the right is an antiderivative of $cf(x)$, since

$$\frac{d}{dx}\left(c \int f(x) \, dx\right) = c \frac{d}{dx} \int f(x) \, dx = cf(x),$$

with the help of formula (2). Moreover, $\int f(x) \, dx$ is defined only to within an arbitrary additive constant, and hence the same is true of the product $c \int f(x) \, dx$. Thus in an indefinite integral any constant multiplying the integrand can be factored out and placed in front of the integral sign, just as in the case of the definite integral.

(ii) *If f and g have indefinite integrals (antiderivatives) on the same interval, then*

$$\int [f(x) + g(x)] \, dx = \int f(x) \, dx + \int g(x) \, dx. \tag{6}$$

Section 4.4   Antiderivatives and the Indefinite Integral   **251**

To prove this, we observe that the sum of integrals on the right is an antiderivative of $f + g$, since

$$\frac{d}{dx}\left(\int f(x)\,dx + \int g(x)\,dx\right) = \frac{d}{dx}\int f(x)\,dx + \frac{d}{dx}\int g(x)\,dx = f(x) + g(x).$$

Moreover, each of the integrals $\int f(x)\,dx$ and $\int g(x)\,dx$ is defined only to within an arbitrary additive constant, and hence the same is true of their sum. Thus the indefinite integral of the sum of two functions is the sum of the indefinite integrals of the separate functions. This is the analogue for indefinite integrals of rule (ii), page 235.

(iii) If $f_1, f_2, \ldots, f_n$ have indefinite integrals on the same interval and $c_1, c_2, \ldots, c_n$ are any constants, then

$$\int [c_1 f_1(x) + c_2 f_2(x) + \cdots + c_n f_n(x)]\,dx$$

$$= c_1 \int f_1(x)\,dx + c_2 \int f_2(x)\,dx + \cdots + c_n \int f_n(x)\,dx, \qquad (7)$$

just as in the case of definite integrals (see page 236). Formula (7) is proved by repeated application of formulas (5) and (6). Thus the indefinite integral of a linear combination of functions is a linear combination, with the same coefficients, of the indefinite integrals of the separate functions.

(iv) If f has an antiderivative F, so that $\int f(x)\,dx = F(x) + C$, then

$$\int f(ax + b)\,dx = \frac{F(ax + b)}{a} + C \qquad (8)$$

for arbitrary constants $a \neq 0$ and b. In fact, since $F'(x) = f(x)$, we have

$$\frac{d}{dx}\frac{F(ax + b)}{a} = \frac{1}{a}\frac{d}{dx}F(ax + b) = \frac{1}{a}F'(ax + b)\frac{d}{dx}(ax + b)$$

$$= \frac{a}{a}F'(ax + b) = f(ax + b),$$

with the help of the chain rule. Thus $(1/a)F(ax + b)$ is an antiderivative of $f(ax + b)$, which proves (8).

We are now in a position to evaluate a number of indefinite integrals. Further techniques of integration will be developed as we pursue our study of integral calculus. In particular, rule (iv) is an important special case of a general method called *integration by substitution* (see Section 7.1).

Example 5 Evaluate $\int \left(5x^4 - 6x^2 + \dfrac{2}{x^2}\right) dx$.

Solution By rule (iii), we have

$$\int \left(5x^4 - 6x^2 + \frac{2}{x^2}\right) dx = 5\int x^4\,dx - 6\int x^2\,dx + 2\int x^{-2}\,dx$$

$$= 5\left(\frac{x^5}{5}\right) - 6\left(\frac{x^3}{3}\right) + 2\left(\frac{x^{-1}}{-1}\right) + C$$

$$= x^5 - 2x^3 - \frac{2}{x} + C,$$

after applying formula (4) three times (with $r = 4, 2$ and -2). Notice that the arbitrary constants of integration contributed by each of the three integrals separately have been combined into a single constant of integration C. □

Example 6 Evaluate the indefinite integral of an arbitrary polynomial

$$P(x) = a_0 + a_1 x + a_2 x^2 + \cdots + a_n x^n,$$

where $a_0, a_1, a_2, \ldots, a_n$ are constants.

Solution We assume that $a_n \neq 0$, so that $P(x)$ is of degree n. By rule (iii) and repeated application of formula (4), we have

$$\int P(x) \, dx = a_0 \int dx + a_1 \int x \, dx + a_2 \int x^2 \, dx + \cdots + a_n \int x^n \, dx$$

$$= a_0 x + \frac{a_1}{2} x^2 + \frac{a_2}{3} x^3 + \cdots + \frac{a_n}{n+1} x^{n+1} + C,$$

which is another polynomial, in fact a polynomial of degree $n + 1$, since the coefficient of x^{n+1} is nonzero. It should be noted that differentiation of this new polynomial immediately gives back the original polynomial $P(x)$, and shows that our calculations have been done correctly. □

Example 7 Evaluate $\int \cos 2x \, dx$.

Solution By rule (iv), with $a = 2$, $b = 0$, $f(x) = \cos x$ and $F(x) = \sin x$,

$$\int \cos 2x \, dx = \frac{\sin 2x}{2} + C. \quad \square$$

Example 8 Evaluate $\int \cos^2 x \, dx$.

Solution Since

$$1 + \cos 2x = 1 + \cos^2 x - \sin^2 x = 2 \cos^2 x,$$

we have

$$\int \cos^2 x \, dx = \frac{1}{2} \int (1 + \cos 2x) \, dx$$

$$= \frac{1}{2} \int dx + \frac{1}{2} \int \cos 2x \, dx = \frac{1}{2} x + \frac{1}{4} \sin 2x + C,$$

with the help of Example 7. We can think of C as one half the sum of the arbitrary constants of integration contributed by the integrals $\int dx$ and $\int \cos 2x \, dx$, or we can simply supply an arbitrary constant of integration at the very end of the calculation. □

Example 9 Evaluate $\int (1 - u)(1 + u + u^2) \, du$.

Solution Here the variable of integration is u, instead of x. Carrying out the multiplication in the integrand, we find that

$$\int (1 - u)(1 + u + u^2) \, du = \int (1 - u^3) \, du = u - \frac{1}{4} u^4 + C,$$

either by rule (iii) and formula (4), or simply by recognizing that $u - \frac{1}{4} u^4$ is an antiderivative of $1 - u^3$. □

Existence of Antiderivatives of Continuous Functions

We already know that every continuous function has a definite integral (see Theorem 1, page 232). We now show that every continuous function has an antiderivative, and hence an indefinite integral.

> **Theorem 5** *(Continuity implies existence of an antiderivative).* Let f be continuous on an interval I, and let
> $$F(x) = \int_a^x f(t)\,dt, \tag{9}$$
> where a is any fixed point of I and x is a variable point of I. Then F is an antiderivative of f on I, that is, $F'(x) = f(x)$ for every x in I.

Proof Before starting the proof, we point out that the definite integral (9), whose existence is guaranteed by the continuity of f, is a *function* of its variable upper limit of integration x. In fact, the presence of x in the upper limit forces us to use another letter (here chosen to be t) for the variable of integration.

Now suppose x and $x + \Delta x$ both belong to the interval I. Then
$$F(x + \Delta x) = \int_a^{x+\Delta x} f(t)\,dt = \int_a^x f(t)\,dt + \int_x^{x+\Delta x} f(t)\,dt,$$

by the corollary on page 240, or equivalently
$$F(x + \Delta x) - F(x) = \int_x^{x+\Delta x} f(t)\,dt.$$

Applying the mean value theorem for integrals to the integral on the right, which is independent of the fixed point a, we get
$$F(x + \Delta x) - F(x) = (x + \Delta x - x)f(c) = f(c)\,\Delta x,$$

where $x \leq c \leq x + \Delta x$ or $x + \Delta x \leq c \leq x$, depending on whether Δx is positive or negative. The point c depends on Δx, and in particular $c \to x$ as $\Delta x \to 0$. Therefore $f(c) \to f(x)$ as $\Delta x \to 0$, by the continuity of f. Thus the derivative of F at every point x of I is equal to †

$$F'(x) = \lim_{\Delta x \to 0} \frac{F(x + \Delta x) - F(x)}{\Delta x} = \lim_{\Delta x \to 0} \frac{f(c)\,\Delta x}{\Delta x} = \lim_{\Delta x \to 0} f(c) = f(x),$$

so that F is an antiderivative of f on I. ∎

Differentiation of an Integral with a Variable Upper Limit

The function F is of course continuous on I, since it is differentiable on I. Theorem 5 can be written concisely as
$$\frac{d}{dx}\int_a^x f(t)\,dt = f(x),$$

and has a simple geometric interpretation. Let $I = [a, b]$ and $f(t) \geq 0$. Then, as in Figure 16(a), $F(x)$ is the area of the shaded region R under the curve $y = f(t)$ from $t = a$ to $t = x$, and the theorem says that as x increases, $F(x)$ increases at a rate equal to the height $f(x)$ of the region R at its upper right-hand corner. This makes sense, since increasing x to $x + \Delta x$ increases the

---

† If x is the left or right endpoint of I, we let $\Delta x \to 0^+$ or $\Delta x \to 0^-$ instead, interpreting $F'(x)$ as a right-hand or left-hand derivative. It is then unnecessary to assume that f is defined outside I. This issue does not arise if the interval I is open.

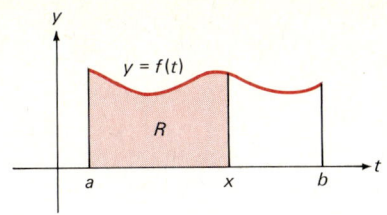

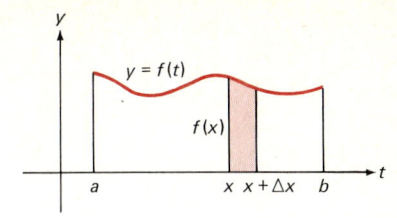

The shaded area is $F(x) = \int_a^x f(t)\,dt$.

(a)

The shaded area is exactly $F(x + \Delta x) - F(x)$ and approximately $f(x)\Delta x$.

(b)

Figure 16

area of R by an amount $\Delta F(x) = F(x + \Delta x) - F(x)$ equal to the area of the narrow, almost rectangular strip under the curve $y = f(t)$ from $t = x$ to $t = x + \Delta x$, and the area of this strip, shown in Figure 16(b), is approximately $f(x)\Delta x$ with a relative error that goes to zero as $\Delta x \to 0$.

It is an immediate consequence of Theorem 5 that *if f is continuous on an interval I, then f has an indefinite integral on I.* In fact, since the function (9) is an antiderivative of f on I, the indefinite integral of f is given by

$$\int f(x)\,dx = \int_a^x f(t)\,dt + C,$$

where C is an arbitrary constant.

Example 10 We cannot use the basic integration formula

$$\int x^r\,dx = \frac{x^{r+1}}{r+1} + C$$

if $r = -1$. In fact, setting $r = -1$ leads to a zero denominator on the right. On the other hand, the function $1/x$ is continuous on every interval that does not contain the point $x = 0$, and therefore, by Theorem 5, it has an antiderivative on every such interval. In other words, the indefinite integral

$$\int \frac{dx}{x}$$

exists, although we do not yet know the name of this function. In Section 6.1 we will study the function in question, known as the *natural logarithm* of x and denoted by $\ln x$.

Problems

Find the general antiderivative of the given function.
1. $x^2 + x + 2$
2. $x(x - 1)(x - 2)$
3. $x^{49} - 5x^{24} + 20x^9 - 10$
4. $\frac{2}{3}x^{3/2} + \frac{3}{2}x^{2/3}$
5. $x^{-3/4} - x^{-4/3}$
6. $(1 + x + x^2)/x^4$
7. $2\sin x - 3\cos x$
8. $5\sec^2 x + 4\csc^2 x$
9. $(1 - t)(1 + t)(1 + t^2)$
10. $(3 + 2u)(9 - 6u + 4u^2)$
11. $(2 - 3v)(4 + 6v + 9v^2)$
12. $\frac{1}{4}w^4 - \frac{1}{2}w^2 + 8\sec w \tan w$

13. Show that if $F(x)$ is an antiderivative of $f(x)$, then $-F(-x)$ is an antiderivative of $f(-x)$.

14. Use differentiation to show that $\sin^2 x = C - \frac{1}{2}\cos 2x$, where C is a constant, and then find C.

15. What can be said about a function $f(x)$ whose nth derivative $f^{(n)}(x)$ is identically equal to zero?

Evaluate

16. $\int (x + 5)(x - 6)\,dx$

17. $\int (x^4 - 3x^2 + 2x - 4)\,dx$

18. $\int x(1 + x)(1 - x)\,dx$

Section 4.4 Antiderivatives and the Indefinite Integral

19. $\displaystyle\int \left(x^3 - x + \frac{1}{x^2} - \sin 3x\right) dx$

20. $\displaystyle\int t^2(5-t)^4 \, dt$

21. $\displaystyle\int (1-u)(1-2u)(1-3u) \, du$

22. $\displaystyle\int \frac{v+1}{\sqrt{v}} \, dv$

23. $\displaystyle\int \sin^2 x \, dx$

24. $\displaystyle\int \tan^2 x \, dx$

25. $\displaystyle\int \cot^2 x \, dx$

26. $\displaystyle\int \sin x \cos x \, dx$

27. $\displaystyle\int \frac{\sin x}{\cos^2 x} \, dx$

28. $\displaystyle\int \frac{dx}{\sin^2 x \cos^2 x}$

29. $\displaystyle\int \frac{\sin 2x}{\sin x} \, dx$

30. $\displaystyle\int \frac{\cos x}{\sin^2 x} \, dx$

31. $\displaystyle\int \frac{\cos 3u}{\cos u} \, du$

32. $\displaystyle\int \frac{\sin 3v}{\sin v} \, dv$

33. $\displaystyle\int \frac{w^4 - 1}{w - 1} \, dw$

34. $\displaystyle\int \frac{z^4 - 16}{z + 2} \, dz$

35. $\displaystyle\int \frac{1 - \sin^3 x}{1 - \sin x} \, dx$

36. $\displaystyle\int \frac{1 + \cos^3 x}{1 + \cos x} \, dx$

37. $\displaystyle\frac{d}{dx} \int_a^b f(x) \, dx$

38. $\displaystyle\frac{d}{da} \int_a^b f(x) \, dx$

39. $\displaystyle\frac{d}{db} \int_a^b f(x) \, dx$

40. $\displaystyle\frac{d}{dx} \int_0^x t^{50}(1-t)^{50} \, dt$

41. $\displaystyle\frac{d}{dt} \int_1^t (1 + \sin x)^{25} \, dx$

42. $\displaystyle\frac{d}{dt} \int_0^1 (2 + \tan t)^{99} \, dt$

43. Show that Theorem 5 can be proved without using the mean value theorem for integrals. Then show that the mean value theorem for integrals can be deduced from the mean value theorem for derivatives (Theorem 2, page 162).

4.5 The Fundamental Theorem of Calculus

The following basic theorem reveals the intimate connection between differential and integral calculus. At the same time, it gives us a powerful tool for evaluating definite integrals.

Theorem 6 *(Fundamental theorem of calculus)*. *If f is continuous on $[a, b]$, then*

$$\int_a^b f(x) \, dx = F(b) - F(a), \tag{1}$$

where F is any antiderivative of f on $[a, b]$.

Proof By Theorem 5, page 254,

$$F_0(x) = \int_a^x f(t) \, dt$$

is an antiderivative of f on $[a, b]$. Here we supply F with the subscript zero to emphasize that it is a particular antiderivative of f, rather than an arbitrary antiderivative of f. Let F be any other antiderivative of f on $[a, b]$. Then, by Theorem 4, page 249,

$$F_0(x) = F(x) + C, \tag{2}$$

where C is a constant. To determine C, we observe that

$$F(a) + C = F_0(a) = \int_a^a f(t) \, dt = 0,$$

which implies $C = -F(a)$. Substituting this value of C into (2), we get

$$F_0(x) = F(x) - F(a).$$

256 Chapter 4 Integrals

Finally, setting $x = b$ and changing the dummy variable of integration from t to x, we obtain

$$F_0(b) = \int_a^b f(x)\,dx = F(b) - F(a),$$

and the proof is complete. ∎

In some treatments of the subject you will find Theorems 5 and 6 combined into a single two-part theorem, called the fundamental theorem of calculus. There is also a proof of Theorem 6 based on the mean value theorem for *derivatives*, rather than on Theorem 5 (see Problem 41).

The fact that the right side of formula (1) does not depend on the choice of the antiderivative of f is easily checked by direct calculation: Let G be any other antiderivative of f on $[a, b]$. Then $G = F + C$, where C is a constant, and therefore

$$G(b) - G(a) = [F(b) + C] - [F(a) + C] = F(b) - F(a),$$

that is, the constant C cancels out in forming the difference between the values of the antiderivative at a and b. It should also be noted that formula (1) remains true for $b < a$, provided that f is continuous on $[b, a]$, since we then have

$$\int_a^b f(x)\,dx = -\int_b^a f(x)\,dx = -[F(a) - F(b)] = F(b) - F(a).$$

A little extra notation is useful here. Given any function $F(x)$ defined for $x = a$ and $x = b$, let

$$\left[F(x) \right]_a^b \quad \text{or} \quad \left. F(x) \right|_a^b$$

denote the *difference* $F(b) - F(a)$. With this notation, we can write (1) more compactly as

$$\int_a^b f(x)\,dx = \left[F(x) \right]_a^b. \tag{1'}$$

Moreover, since

$$\left[F(x) \right]_a^b = \left[F(x) + C \right]_a^b = \left[\int f(x)\,dx \right]_a^b,$$

we can in turn write (1') as

$$\int_a^b f(x)\,dx = \left[\int f(x)\,dx \right]_a^b.$$

This last version of the fundamental theorem of calculus shows the connection between the definite and indefinite integrals of f very explicitly.

It follows from (1') and formula (4), page 250, that

$$\int_a^b x^r\,dx = \left[\frac{x^{r+1}}{r+1} \right]_a^b = \frac{b^{r+1} - a^{r+1}}{r+1}$$

if r is a rational number not equal to -1. Choosing r to be a positive integer n, we immediately get formula (9), page 235, which we have been using freely for some time now. The interval $[a, b]$ (or $[b, a]$ if $b < a$) must not contain

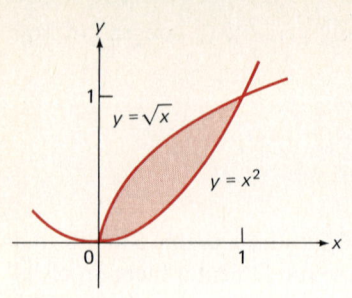

Figure 17

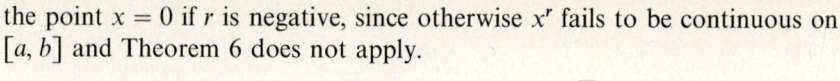

Example 1 Find the area between the curves $y = \sqrt{x}$ and $y = x^2$.

Solution We are looking for the area A of the shaded region in Figure 17. To find the x-coordinates of the points in which the curves intersect, we solve the equation $\sqrt{x} = x^2$, obtaining two roots $x = 0$ and $x = 1$. Therefore

$$A = \int_0^1 (\sqrt{x} - x^2)\,dx,$$

since $y = \sqrt{x}$ is the upper curve and $y = x^2$ the lower curve on the interval $[0, 1]$. Using Theorem 6 to evaluate the integral, we get

$$A = \left[\frac{2}{3}x^{3/2} - \frac{1}{3}x^3\right]_0^1 = \frac{2}{3} - \frac{1}{3} = \frac{1}{3}. \quad \square$$

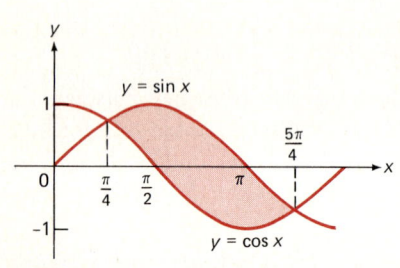

Example 2 Find the area between the curves $y = \sin x$ and $y = \cos x$ from $x = \pi/4$ to $x = 5\pi/4$.

Solution This time we want the area A of the shaded region in Figure 18. Since $y = \sin x$ is the upper curve and $y = \cos x$ is the lower curve on the interval $[\pi/4, 5\pi/4]$, it follows that

$$A = \int_{\pi/4}^{5\pi/4} (\sin x - \cos x)\,dx = \left[-\cos x - \sin x\right]_{\pi/4}^{5\pi/4}$$

$$= -\cos\frac{5\pi}{4} - \sin\frac{5\pi}{4} + \cos\frac{\pi}{4} + \sin\frac{\pi}{4} = \frac{4}{\sqrt{2}} = 2\sqrt{2}. \quad \square$$

Figure 18

More About the Area Between Two Curves

In specifying curves, it is often convenient to choose the ordinate y as the independent variable and the abscissa x as the dependent variable; this is just the reverse of what has been done so far. Let $f(y)$ and $g(y)$ be two functions continuous on some interval $a \leq y \leq b$, and suppose $f(y) \geq g(y)$. Then the *horizontal* lines $y = a$ and $y = b$ and the curves $x = f(y)$ and $x = g(y)$ form the boundary of a region R, as shown in Figure 19(a). By virtually the same argument as in Section 4.3, but applied to horizontal strips instead of

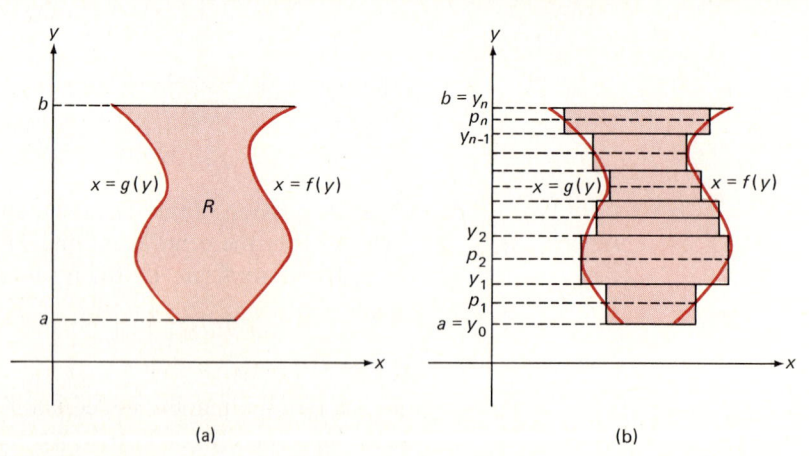

(a)   (b)

Figure 19

vertical strips, we find that the area A of the region R is given by the formula

$$A = \int_a^b [f(y) - g(y)] \, dy. \tag{3}$$

In detail, we approximate A by a sum of the form

$$\sum_{i=1}^n [f(p_i) - g(p_i)] \Delta y_i, \tag{4}$$

based on a partition of the interval $[a, b]$ into n subintervals $[y_{i-1}, y_i]$, $i = 1, 2, \ldots, n$, with $y_0 = a$, $y_n = b$, where $[y_{i-1}, y_i]$ is of length $\Delta y_i = y_i - y_{i-1}$ and p_i is an arbitrary point in $[y_{i-1}, y_i]$. This approximation corresponds to replacing the strips with curved sides in Figure 19(b) by the indicated shaded rectangles. We then define A as the limit of the sum (4) as the mesh size $\mu = \max \{\Delta y_1, \Delta y_2, \ldots, \Delta y_n\}$ approaches zero:

$$A = \lim_{\mu \to 0} \sum_{i=1}^n [f(p_i) - g(p_i)] \Delta y_i = \int_a^b [f(y) - g(y)] \, dy.$$

Formula (3) is of course completely analogous to the formula

$$A = \int_a^b [f(x) - g(x)] \, dx \tag{3'}$$

for the area of the region bounded by the *vertical* lines $x = a$ and $x = b$ and the curves $y = f(x)$ and $y = g(x)$, where $f(x) \geq g(x)$, and to get one formula from the other, we need only replace x by y or y by x.

Example 3 Find the area between the curves $x = y^2$ and $x = \frac{1}{2}y^2 + 2$.

Solution We are looking for the area A of the shaded region in Figure 20, bounded by the given curves, which are parabolas symmetric about the x-axis. The y-coordinates of the points of intersection of the parabolas are the roots $y = 2$ and $y = -2$ of the equation $y^2 = \frac{1}{2}y^2 + 2$. Therefore, by formula (3),

$$A = \int_{-2}^2 \left(\frac{1}{2}y^2 + 2 - y^2\right) dy = \int_{-2}^2 \left(2 - \frac{1}{2}y^2\right) dy = \left[2y - \frac{1}{6}y^3\right]_{-2}^2$$

$$= 2(2) - \frac{1}{6}(2)^3 - 2(-2) + \frac{1}{6}(-2)^3 = \frac{16}{3}.$$

Since the shaded region OBD is symmetric about the x-axis, the two subregions OBC and ODC have the same area. Thus the calculation could have been simplified somewhat by writing

$$A = 2\int_0^2 \left(2 - \frac{1}{2}y^2\right) dy = 2\left[2y - \frac{1}{6}y^3\right]_0^2$$

$$= 2\left[2(2) - \frac{1}{6}(2)^3\right] = \frac{16}{3} \tag{5}$$

from the very beginning.

If we insist on regarding x as the independent variable and y as the dependent variable, the calculation becomes needlessly complicated. We must now distinguish between *four* functions, namely $y = \pm\sqrt{x}$, obtained

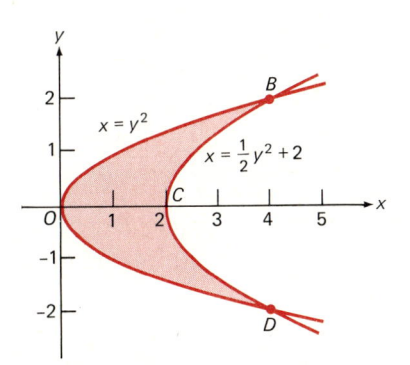

Figure 20

by solving $x = y^2$ for y, and $y = \pm\sqrt{2x - 4}$, obtained by solving $x = \frac{1}{2}y^2 + 2$ for y. We can again avoid some work by calculating the area of the subregion OBC and doubling the answer. But there is now a new complication, for although $y = \sqrt{x}$ is the upper curve on the whole interval $0 \le x \le 4$, the x-axis is the lower curve on the subinterval $0 \le x \le 2$, while $y = \sqrt{2x - 4}$ is the lower curve on the subinterval $2 \le x \le 4$. Taking all this into account, we have

$$A = 2\int_0^2 \sqrt{x}\, dx + 2\int_2^4 (\sqrt{x} - \sqrt{2x-4})\, dx$$

$$= 2\left[\frac{2}{3}x^{3/2}\right]_0^2 + 2\left[\frac{2}{3}x^{3/2} - \frac{2}{3}\cdot\frac{1}{2}(2x-4)^{3/2}\right]_2^4$$

$$= \frac{4}{3}(2)^{3/2} + \frac{4}{3}(4)^{3/2} - \frac{2}{3}(4)^{3/2} - \frac{4}{3}(2)^{3/2} = \frac{2}{3}(4)^{3/2} = \frac{2}{3}(8) = \frac{16}{3}.$$

The factor of $\frac{1}{2}$ before $(2x - 4)^{3/2}$ comes from the application of formula (8), page 252. Naturally we get the same answer for A as before, but this is certainly a long and circuitous calculation as compared with (5)! □

Problems

1. Use the fundamental theorem of calculus and Problem 13, page 255, to show that $\int_{-a}^{a} f(x)\, dx = 2\int_0^a f(x)\, dx$ if f is even, which has the geometric meaning shown in Figure 21(a). Also show that $\int_{-a}^{a} f(x)\, dx = 0$ if f is odd, which has the geometric meaning shown in Figure 21(b). Assume that f is continuous on $[-a, a]$.

2. Express $[F(x) + G(x)]_a^b$ and $[F(x)G(x)]_a^b$ in terms of $[F(x)]_a^b$ and $[G(x)]_a^b$.

Evaluate

3. $\int_0^2 (x^3 - 2x^2 + 3x - 4)\, dx$

4. $\int_{-1}^{1} (x^9 + 5x^8 + 10x^7)\, dx$

5. $\int_2^1 \left(\frac{1}{x^2} - \frac{1}{x^3} + \frac{1}{x^4}\right) dx$

6. $\int_1^9 (1 + \sqrt{s})\, ds$

7. $\int_1^2 \left(3s^3 - \frac{5}{s^4}\right) ds$

8. $\int_0^3 \frac{dt}{(2t + 1)^2}$

9. $\int_4^{16} \frac{1 - t}{\sqrt{t}}\, dt$

10. $\int_{-1}^{1} \frac{du}{(2 - u)^3}$

11. $\int_1^{27} u^{-2/3}\, du$

12. $\int_{-8}^{1} (1 + v^{2/3})\, dv$

13. $\int_1^4 (v^{3/2} - v^{1/2})\, dv$

14. $\int_0^\pi (2 + 3 \sin x)\, dx$

15. $\int_{-\pi/2}^{\pi/2} (2 \cos x - 1)\, dx$

16. $\int_0^{\pi/2} \sin^2 x\, dx$

17. $\int_{-\pi}^{\pi} \cos^2 x\, dx$

18. $\int_0^{2\pi} \sin x \cos x\, dx$

19. $\int_0^{\pi/3} \frac{dt}{\cos^2 t}$

$\int_{-a}^{a} f(x)\, dx = A + A = 2A = 2\int_0^a f(x)\, dx$.
(a)

$\int_{-a}^{a} f(x)\, dx = A_+ - A_- = 0$, since $A_+ = A_-$.
(b)

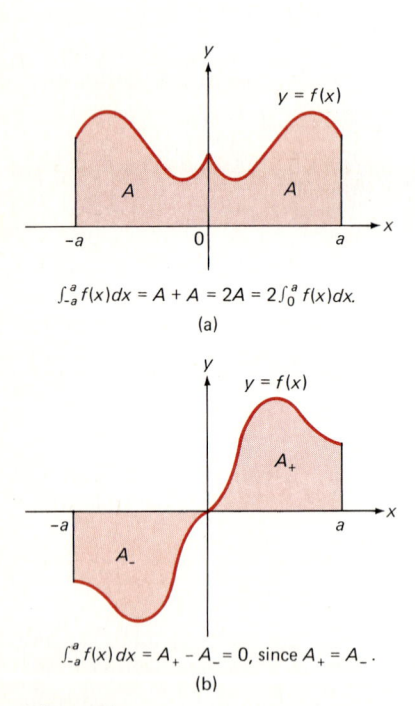

Figure 21

20. $\int_{\pi/4}^{\pi/2} \dfrac{du}{\sin^2 u}$
21. $\int_{\pi/6}^{\pi/3} \dfrac{dv}{\sin^2 v \cos^2 v}$
22. $\int_{-\pi}^{\pi} \sin^5 x \, dx$
23. $\int_{-1}^{1} \sin^4 x \tan x \, dx$
24. $\int_{\pi/4}^{3\pi/4} \dfrac{\cot x}{\sin x} \, dx$
25. $\int_{0}^{\pi/4} \dfrac{\tan x}{\cos x} \, dx$

Find the area A of the region R between the curves

26. $y = \sqrt{x}$ and $y = \sqrt[3]{x}$
27. $y = x^3$ and $y = \sqrt[3]{x}$ $(x \geq 0)$
28. $y = x^{2/3}$ and $y = x^{3/2}$
29. $x = 3y + 2$ and $x = 2y^2$
30. $x = -y^2$ and $x = 4 - 2y^2$
31. $x = 4 - y^2$ and $x = y^2 + 2y$

In each case sketch the region R, and solve the problem in two ways, that is, by integration with respect to x and integration with respect to y. (In Problems 29–31 the integrations with respect to y present no difficulty, but the integrations with respect to x are tricky.)

32. Four regions are formed by the intersection of the curves $y = \sin x$ and $y = \sin 2x$ on the interval $[0, 2\pi]$. Graph the curves and label each region with its area A.

33. Three regions are formed by the intersection of the curves $y = \cos x$ and $y = \cos 2x$ on the interval $[0, 2\pi]$. Graph the curves and label each region with its area A.

Find the mean value of the function

34. $f(x) = \sqrt{x}$ on $[0, 4]$
35. $f(x) = 1/x^2$ on $[-3, -1]$
36. $f(x) = \sin x$ on $[0, \pi]$
37. $f(x) = \cos x$ on $[0, 2\pi]$
38. $f(x) = \sin^2 x$ on $[0, 2\pi]$
39. $f(x) = \sec^2 x$ on $[-\pi/4, \pi/4]$
40. $f(x) = \sec x \tan x$ on $[0, \pi/3]$

41. Let F be any antiderivative of f on $[a, b]$, and let $a = x_0 < x_1 < x_2 < \cdots < x_{n-1} < x_n = b$. Then

$$\sum_{i=1}^{n} [F(x_i) - F(x_{i-1})] = F(b) - F(a), \qquad (i)$$

since the sum on the left is telescoping. Use formula (i) and the mean value theorem for derivatives (Theorem 2, page 162) to give a direct proof of the fundamental theorem of calculus.

4.6 Integration of the Velocity Function; Differential Equations

Let $s = s(t)$ be the position at time t of a particle P moving along a straight line L (see Figure 22), where as always we assume that L has been assigned an origin O, a positive direction and a unit of length. Then the particle's (instantaneous) velocity $v = v(t)$ at time t is given by the derivative

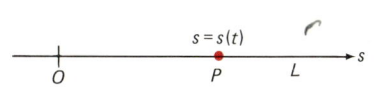

$$v(t) = \dfrac{ds(t)}{dt}. \qquad (1)$$

Figure 22

Thus the particle's velocity function is found by *differentiating* its position function. Conversely, as illustrated by the following examples, the particle's position function can be found by *integrating* its velocity function.

Example 1 The velocity at time t of a particle moving along a straight line is

$$v = v(t) = 3t^2 - 2t + 4.$$

Find the distance between the positions of the particle at the times $t = 2$ and $t = 5$. What is the average velocity of the particle between these two times?

Solution Integrating both sides of equation (1) over the interval $[2, 5]$, we find that

$$\int_{2}^{5} v(t) \, dt = \int_{2}^{5} \dfrac{ds(t)}{dt} \, dt = \left[s(t) \right]_{2}^{5} = s(5) - s(2),$$

where the fundamental theorem of calculus is used at the second step. Therefore $s(5) - s(2)$, the distance between the positions of the particle at the times $t = 2$ and $t = 5$, is equal to

$$\int_2^5 v(t)\,dt = \int_2^5 (3t^2 - 2t + 4)\,dt = \left[t^3 - t^2 + 4t\right]_2^5$$
$$= (125 - 25 + 20) - (8 - 4 + 8) = 120 - 12 = 108.$$

The average velocity of the particle between these two times is

$$\frac{s(5) - s(2)}{5 - 2} = \frac{108}{3} = 36. \quad \square$$

Example 2   Find the position function $s = s(t)$ of a particle with the same velocity function as in Example 1 if $s(0) = 6$, that is, if the particle's position coordinate is 6 at the time $t = 0$.

Solution   It follows from (1) that $s(t)$ is an antiderivative of $v(t)$, and therefore,

$$s(t) = \int v(t)\,dt = \int (3t^2 - 2t + 4)\,dt.$$

Evaluating the integral, we get

$$s(t) = t^3 - t^2 + 4t + C, \tag{2}$$

where further information is needed to determine the constant of integration C. This information is given by the condition $s(0) = 6$. In fact, setting $t = 0$ in (2) gives $s(0) = C$, so that $C = 6$. With this choice of C, the particle's position function (2) becomes

$$s(t) = t^3 - t^2 + 4t + 6. \quad \square$$

More generally, it is clear that we can always choose the constant C in such a way as to satisfy any condition $s(t_0) = s_0$. Writing (1) in the form

$$\frac{ds}{dt} = v(t), \tag{3}$$

we can look upon the problem of determining the position function from a knowledge of the velocity function as the problem of finding the function $s = s(t)$ satisfying equation (3) and the *initial condition*

$$s(t_0) = s_0. \tag{3'}$$

The use of the word "initial" stems from the fact that t_0 is usually, but not always, the time at which the motion begins.

First Order Differential Equations

Equation (3) is a *first order differential equation* (recall page 143). It is of a particularly simple type, in which the right-hand side is a function of the independent variable, but not of the dependent variable. To solve equation (3), that is, to find the function $s = s(t)$ for which it reduces to an identity, we integrate both sides. This gives

$$s = \int \frac{ds}{dt}\,dt = \int v(t)\,dt = V(t) + C, \tag{4}$$

where $V(t)$ is an antiderivative of $v(t)$ and C is an arbitrary constant of integration. We call (4) the *general solution* of the differential equation (3), since every solution is of the form (4) for some choice of the constant C; this follows from the fact that $V(t) + C$ is the general antiderivative of $v(t)$. We can also write the general solution in the form

$$s = \int v(t)\,dt + C,$$

with the understanding that here $\int v(t)\,dt$ denotes any *fixed* antiderivative of $v(t)$. We will favor this convention in problems involving differential equations. The solutions of the differential equation (3) obtained by assigning C various values are called *particular solutions*. Typically, C is chosen in such a way as to make s satisfy an initial condition (3').

The same terminology and method of solution is used regardless of the concrete scientific meaning of the independent and dependent variables, or in the absence of any such meaning. Thus consider the differential equation

$$y' = \frac{dy}{dx} = f(x), \tag{5}$$

involving an unknown function $y = y(x)$ and a given function $f(x)$. To solve (5), subject to the condition

$$y(x_0) = y_0$$

(which is still called an *initial condition*), we proceed in exactly the same way. Integrating (5), we get

$$y = \int \frac{dy}{dx}\,dx = \int f(x)\,dx + C = F(x) + C, \tag{6}$$

where $F(x) = \int f(x)\,dx$ is any fixed antiderivative of $f(x)$ and C is an arbitrary constant of integration. This is the general solution of (5), and we want the *particular* solution satisfying (5'). It is easy to see that this solution is obtained by choosing $C = y_0 - F(x_0)$.

Example 3 Find the particular solution of the differential equation

$$y' = \frac{dy}{dx} = x, \tag{7}$$

satisfying the initial condition

$$y(1) = 2. \tag{7'}$$

Solution First we use integration to find the *general* solution of (7):

$$y = \int \frac{dy}{dx}\,dx = \int x\,dx + C = \frac{1}{2}x^2 + C.$$

To determine the constant of integration C, we now impose the condition (7'), setting $x = 1$, $y = 2$ in the general solution $y = \frac{1}{2}x^2 + C$. This gives

$$2 = \frac{1}{2} + C,$$

Section 4.6 Integration of the Velocity Function; Differential Equations so that $C = \frac{3}{2}$. Choosing this value of C in the general solution, we get the desired particular solution of (7):

$$y = \frac{1}{2}x^2 + \frac{3}{2}.$$

The fact that this solution satisfies both (7) and (7′) is easily verified by direct calculation. □

Second Order Differential Equations

The simplest *second order* differential equation is

$$y'' = \frac{d^2y}{dx^2} = f(x). \tag{8}$$

Integrating (8), we obtain

$$y' = \frac{dy}{dx} = \int \frac{d^2y}{dx^2}\,dx = \int f(x)\,dx + C_1 = F(x) + C_1, \tag{9}$$

where $F(x) = \int f(x)\,dx$ is any fixed antiderivative of $f(x)$ and C_1 is an arbitrary constant of integration. You will observe that (9) is now a *first order* differential equation, and is in fact of the form (5). Integrating (9) in turn, we get

$$y = \int \frac{dy}{dx}\,dx = \int [F(x) + C_1]\,dx + C_2 = \int F(x)\,dx + C_1 x + C_2,$$

where C_2 is another constant of integration. Thus the *general* solution of the differential equation (8) involves *two* arbitrary constants, and this is a characteristic feature of the general solution of a *second order* differential equation. Hence, to single out a *particular* solution of (8), we must now impose *two* initial conditions, since this will give us two algebraic equations that can be solved for the constants C_1 and C_2.

Example 4 Find the particular solution of the differential equation

$$y'' = \frac{d^2y}{dx^2} = x \tag{10}$$

satisfying the initial conditions

$$y(1) = \frac{1}{2}, \qquad y'(1) = -\frac{1}{2}. \tag{10′}$$

Solution Notice that one initial condition involves the function $y = y(x)$, while the other involves its derivative y'. Integrating (10) twice with respect to x, we get first

$$y' = \frac{dy}{dx} = \int x\,dx + C_1 = \frac{1}{2}x^2 + C_1, \tag{11}$$

and then

$$y = \int \frac{1}{2}x^2\,dx + \int C_1\,dx + C_2 = \frac{1}{6}x^3 + C_1 x + C_2. \tag{12}$$

Imposing the initial conditions (10′), that is, setting $x = 1$, $y' = -\frac{1}{2}$ in (11) and $x = 1$, $y = \frac{1}{2}$ in (12), we find that

$$\frac{1}{2} + C_1 = -\frac{1}{2},$$

$$\frac{1}{6} + C_1 + C_2 = \frac{1}{2}.$$

Solving for C_1 and C_2, we then obtain

$$C_1 = -1, \quad C_2 = \frac{4}{3}.$$

Substituting these values of C_1 and C_2 into (12), we finally get the desired particular solution of (10) satisfying the initial conditions (10′):

$$y = \frac{1}{6}x^3 - x + \frac{4}{3}$$

(check this by direct calculation). □

The problem of finding a solution of a given differential equation satisfying specified initial conditions, as in Examples 2–4, is called an *initial value problem*.

Example 5 Find the particular solution of the differential equation (10) satisfying the conditions

$$y(0) = 1, \quad y(1) = 2, \tag{10″}$$

instead of the initial conditions (10′).

Solution Note that instead of imposing one condition on the function y and the other on its derivative y' at the same point, we are now imposing two conditions on y at two different points. Correspondingly, the conditions (10″) are called *boundary conditions*, and we are now solving a *boundary value problem* instead of an initial value problem. To impose the boundary conditions (10″), we first set $x = 0$, $y = 1$ in the general solution (12) and then set $x = 1$, $y = 2$. This gives two equations satisfied by the constants of integration C_1 and C_2:

$$C_2 = 1,$$

$$\frac{1}{6} + C_1 + C_2 = 2.$$

Solving for C_1 and C_2, we find that

$$C_1 = \frac{5}{6}, \quad C_2 = 1,$$

and then substituting these values of C_1 and C_2 into (12), we get the desired particular solution of (10) satisfying the boundary conditions (10″):

$$y = \frac{1}{6}x^3 + \frac{5}{6}x + 1. \quad □$$

Problems

1. Let $s = s(t)$ be the position function and $v = v(t)$ the velocity function of a particle moving along a straight line. Then the average velocity over the interval $[a, b]$ is

$$v_{av} = \frac{1}{b-a} \int_a^b v(t)\,dt$$

if we use the definition of average on page 244, and

$$v_{av} = \frac{s(b) - s(a)}{b-a}$$

if we use the earlier definition on page 106. Show that these two definitions are equivalent.

Starting at time $t = 0$, a particle moves along a straight line with velocity

$$v(t) = 20\left(1 - \frac{1}{\sqrt{t+1}}\right) \text{ cm/sec.}$$

2. How far does the particle go during the first 15 sec, and what is its average velocity over this interval?
3. How far does the particle go during the first 2 min, and what is its average velocity over this interval?
4. Let $v_{av}(T)$ be the average velocity of the particle over the interval $[0, T]$. Why is the approximation $v_{av}(T) \approx v(T)$ good if T is large? Compare $v_{av}(T)$ with $v(T)$ for $T = 15$ min.

A car starts from rest and accelerates to a velocity of

$$v(t) = 75\left(1 - \frac{100}{(t+10)^2}\right) \text{ mph}$$

in t seconds.

5. How far does the car go during the first 30 sec, and what is its average velocity over this interval?
6. How far does the car go during the first 90 sec, and what is its average velocity over this interval?
7. Let $v_{av}(T)$ be the average velocity of the car over the interval $[0, T]$. Why is the approximation $v_{av}(T) \approx v(T)$ good if T is large? Compare $v_{av}(T)$ with $v(T)$ for $T = 10$ min.
8. Show that the particular solution of the differential equation $y' = f(x)$ satisfying the initial condition $y(x_0) = y_0$ can be written in the form

$$y = \int_{x_0}^x f(t)\,dt + y_0.$$

Find the particular solution of the given first order differential equation satisfying the specified initial condition.

9. $y' = 2x$, $y(2) = 1$
10. $x^2 y' = 1$, $y(1) = 2$
11. $y' = x(x-1)$, $y(3) = \frac{1}{2}$
12. $y' = \sqrt{x}$, $y(0) = 3$
13. $y' = 3x^2 + \cos x$, $y(\pi) = 2$
14. $y' = 2\cos^2 x + \sin 2x$, $y(0) = -2$

Find the particular solution of the given second order differential equation satisfying the specified initial conditions or boundary conditions.

15. $y'' = x(x+1)$, $y(1) = 0$, $y'(1) = 1$
16. $(x+1)^3 y'' = 1$, $y(0) = 1$, $y'(0) = -1$
17. $\sqrt{x} y'' = 1$, $y(4) = 2$, $y'(4) = 0$
18. $y'' = \sin x$, $y(0) = -1$, $y'(0) = 3$
19. $x^3 y'' = 3$, $y(2) = -1$, $y(3) = 1$
20. $y'' = x^2 + x + 1$, $y(-1) = 0$, $y(1) = 3$

21. Find the particular solution of the differential equation

$$y' + \frac{y}{x} = \frac{\sin x}{x}$$

satisfying the initial condition $y(\pi/2) = 0$.

4.7 Newtonian Mechanics; Kinetic Energy and Work

With the help of integration, we now use the physical ideas of Sir Isaac Newton to make a detailed study of motion along a straight line (rectilinear motion). As in the preceding section, let $s = s(t)$ be the position at time t of a particle of mass m moving along a line L. As we know, the particle's velocity $v = v(t)$ and acceleration $a = a(t)$ are given by the derivatives

$$v = \frac{ds}{dt}, \qquad a = \frac{dv}{dt} = \frac{d^2s}{dt^2}.$$

Suppose the particle is subject to a force F, acting along the line L. Then *Newton's second law of motion*, discussed in a preliminary way on page 143, tells us that

$$m\frac{d^2s}{dt^2} = F \qquad (1)$$

(mass times acceleration equals the applied force). Thus once F is known, the particle's position as a function of time can be determined by solving the second order differential equation (1), subject to appropriate initial conditions. The following examples illustrate how this is done.

Example 1 Find the free motion of a particle, that is, the motion in the absence of any external force.

Solution In this case there is no force, so that $F \equiv 0$ in (1). Therefore

$$a = \frac{d^2s}{dt^2} = 0,$$

after canceling out the mass, which plays no role here. Integrating this differential equation twice, we get first

$$v = \int \frac{dv}{dt}\,dt = \int a\,dt = \int 0\,dt = C_1, \qquad (2)$$

and then

$$s = \int \frac{ds}{dt}\,dt = \int v\,dt = \int C_1\,dt + C_2 = C_1 t + C_2. \qquad (2')$$

To determine the constants of integration C_1 and C_2, we impose the initial conditions

$$s(0) = s_0, \qquad v(0) = v_0,$$

where s_0 and v_0 are the position and the velocity of the particle at the initial time, conveniently chosen to be $t = 0$. Setting $t = 0$, $v = v_0$ in (2) and $t = 0$, $s = s_0$ in (2'), we find that $C_1 = v_0$, $C_2 = s_0$. Hence (2) and (2') become

$$v = v_0$$

and

$$s = v_0 t + s_0,$$

where you will observe that s has the constant value s_0 if $v_0 = 0$. Thus, unless acted on by an external force, a body at rest ($v_0 = 0$) remains at rest, and a body in motion ($v_0 \neq 0$) continues to move with constant velocity. This is the one-dimensional version of *Newton's first law of motion*, discussed further in Problem 31, page 736. □

Example 2 Find the motion of a stone of mass m dropped from a high tower or dropped into a deep dry well.

Solution We regard the stone as a particle, neglecting its size. Let $s = s(t)$ be the stone's position, as measured along a vertical axis with the positive direction pointing downward and the origin O at the initial position of the stone (see Figure 23). As shown in physics, the force acting on the stone is its *weight*, equal to

$$F = mg,$$

Figure 23

Section 4.7 Newtonian Mechanics; Kinetic Energy and Work

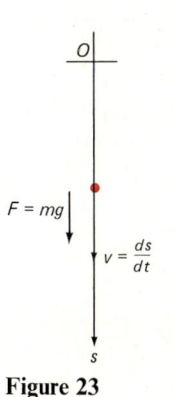

where g is the acceleration due to gravity (≈ 32 ft/sec^2 or 9.8 m/sec^2),† and we neglect the effect of air resistance. With this choice of the force F, Newton's second law (1) becomes

$$m\frac{d^2s}{dt^2} = mg,$$

or

$$a = \frac{d^2s}{dt^2} = g, \tag{3}$$

after dividing through by m. The differential equation (3) says that the acceleration a has the constant value g. Integrating (3) twice, we get first

$$v = \int \frac{dv}{dt}\,dt = \int a\,dt = \int g\,dt + C_1 = gt + C_1, \tag{4}$$

and then

$$s = \int \frac{ds}{dt}\,dt = \int v\,dt = \int (gt + C_1)\,dt + C_2 = \frac{1}{2}gt^2 + C_1 t + C_2. \tag{4'}$$

This time the initial conditions are

$$s(0) = 0, \qquad v(0) = 0,$$

since the stone is dropped (that is, released with initial velocity zero) from the point chosen as the origin. Setting $t = 0$, $v = 0$ in (4) and $t = 0$, $s = 0$ in (4'), we find at once that $C_1 = C_2 = 0$. Therefore

$$v = gt \tag{5}$$

and

$$s = \frac{1}{2}gt^2, \tag{5'}$$

at least until the stone hits the ground or the bottom of the well. If s is measured in feet and t in seconds, then $s \approx 16t^2$, as anticipated in Example 6, page 42. ∎

Example 3 Find the motion of a stone thrown vertically upward with initial velocity v_0.

Solution It is now more natural to measure the stone's position s along a vertical axis with the positive direction pointing *upward* (and the origin at the initial position of the stone), for then s will again be nonnegative, as in the preceding example. This choice of the positive direction has the effect of changing g to $-g$ in (4) and (4'), since the force of gravity points *downward* (see Figure 24). The initial conditions are now

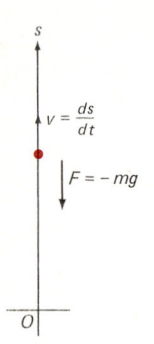

$$s(0) = 0, \qquad v(0) = v_0.$$

Figure 24

---

† Thus, to get the mass of a body from its weight, we must divide the weight by g. The English unit of mass is not the pound, which is a force, but the *slug*, defined as the mass of a body whose acceleration is 1 ft/sec^2 when acted on by a force of 1 pound. Therefore a weight of 1 pound has a mass of about $\frac{1}{32}$ slug, and a mass of 1 slug has a weight of about 32 pounds.

268   Chapter 4   Integrals

Setting $t = 0$, $v = v_0$ in (4) and $t = 0$, $s = 0$ in (4′), we find that $C_1 = v_0$, $C_2 = 0$. Thus in this case (4) and (4′) reduce to

$$v = v_0 - gt \tag{6}$$

and

$$s = v_0 t - \frac{1}{2} gt^2, \tag{6′}$$

after changing g to $-g$. □

Definition of Kinetic Energy and Work

Next we show how the concepts of kinetic energy and work arise in Newtonian mechanics. Suppose a particle of mass m, moving along a straight line, is acted on by a (net) force $F = F(s)$ which is a continuous function of its position s. Then, according to Newton's second law,

$$ma = m \frac{dv}{dt} = F(s),$$

or, by the chain rule,

$$m \frac{dv}{ds} \frac{ds}{dt} = mv \frac{dv}{ds} = F(s), \tag{7}$$

if we think of the velocity v as a function of s rather than of t (here we assume that v is of fixed sign, so that the particle moves in one direction only). Let

$$v_0 = v(s_0), \qquad v_1 = v(s_1)$$

be the particle's velocity at two different positions s_0 and s_1. Then, integrating (7) with respect to s from s_0 to s_1, we get

$$\int_{s_0}^{s_1} mv \frac{dv}{ds} ds = \int_{s_0}^{s_1} \frac{d}{ds}\left(\frac{1}{2} mv^2\right) ds = \int_{s_0}^{s_1} F(s)\, ds,$$

and therefore

$$\left[\frac{1}{2} mv^2\right]_{s_0}^{s_1} = \frac{1}{2} mv_1^2 - \frac{1}{2} mv_0^2 = \int_{s_0}^{s_1} F(s)\, ds. \tag{8}$$

In other words, as a result of the action of the force, the quantity

$$K = \frac{1}{2} mv^2,$$

which is called the *kinetic energy* of the moving particle, changes by an amount

$$W = \int_{s_0}^{s_1} F(s)\, ds. \tag{9}$$

The quantity (9) is called the *work* done by the force on the particle in moving it from the position $s = s_0$ to the position $s = s_1$. It is clear that W is just the area under the curve $F = F(s)$ from $s = s_0$ to $s = s_1$ (see Figure 25).

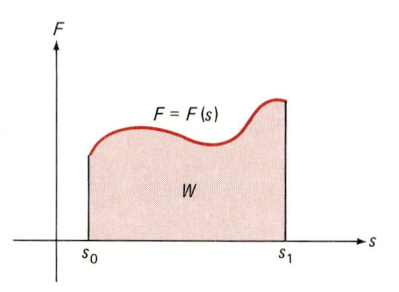

Figure 25

Example 4 In the absence of any force, we have $F \equiv 0$ and the work (9) is equal to zero. Then (8) reduces to

$$\frac{1}{2}mv_1^2 = \frac{1}{2}mv_0^2, \tag{10}$$

so that the kinetic energy remains unchanged, or, in the language of physics, is *conserved*. This is hardly surprising, since we already know from Example 1 that $v_1 = v_0$ if $F \equiv 0$.

Example 5 If $F(s)$ has the constant value F, formula (9) becomes

$$W = \int_{s_0}^{s_1} F(s)\,ds = F\int_{s_0}^{s_1} ds = F \cdot (s_1 - s_0). \tag{11}$$

Thus, in this case, the work equals the product of the force F and the "displacement" $s_1 - s_0$. It is sometimes *assumed* that the work done by a constant force equals the product of the force and the displacement. Then the natural *definition* of the work done by a variable force $F(s)$ turns out to be the integral (9). There is no need to give the argument, which is virtually the same as the one given in Section 4.2 to define the area under a curve.

Example 6 If $F = mg$, $s_0 = 0$, $v_0 = 0$, we have the problem of the falling stone, as in Example 2. Then (11) gives $W = mgs_1$ and (8) reduces to

$$\frac{1}{2}mv^2 = mgs,$$

after dropping the subscript 1 twice. Solving this equation for v, we get

$$v = \sqrt{2gs}. \tag{12}$$

The same result can be obtained by eliminating t from formulas (5) and (5'), but here we have used the concepts of work and kinetic energy to find the connection between the stone's velocity and its position without having to find either as a function of time. For instance, neglecting the effect of air resistance and choosing $g = 32$ ft/sec^2, we deduce from (12) that the velocity of a stone that has fallen 400 ft is

$$v = \sqrt{2(32)(400)} = 160 \text{ ft/sec.}$$

Hooke's Law

Suppose a spring of "natural" (that is, unstretched) length l is fastened at one end. Then, according to *Hooke's law*, to increase the length of the spring from l to $l + s$, we must apply a force $F = ks$ to its free end. Here k is a positive constant called the *stiffness* or *spring constant*. Hooke's law works for negative s as well, corresponding to compression rather than elongation of the spring. However, the accuracy of Hooke's law deteriorates if $|s|$ is too large. Let P be the free end of the spring, and choose the origin O and the positive direction of the s-axis in the way shown in Figure 26, so that P is at O if the spring is unstretched. This is a logical choice, since it makes the elongation s equal to the coordinate of P. Then an "external" force $F = ks$ is required to hold P at the point s, and by (9), the work that this force does on P in moving P from $s = s_0$ to $s = s_1$ is given by

$$W = \int_{s_0}^{s_1} F\,ds = \int_{s_0}^{s_1} ks\,ds = \frac{1}{2}k(s_1^2 - s_0^2) \tag{13}$$

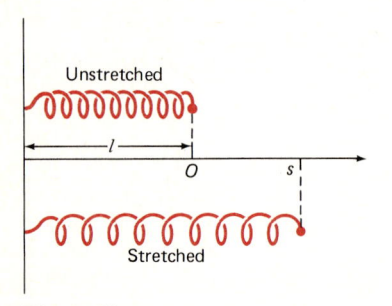

Figure 26

(think of P as a particle). At the same time, application of the external force F creates an equal and opposite *elastic restoring force* $-F = -ks$ in the spring, and the work that this force does on P is just the negative of (13).

Example 7 Suppose it takes a force of 10 lb to stretch a spring 6 in. beyond its natural length of 2 ft. Find the work W done in stretching the spring from a length of 3 ft to a length of 4 ft, assuming that Hooke's law is valid for elongations of this size.

Solution Since a force of 10 lb elongates the spring 0.5 ft beyond its natural length, we have $0.5k = 10$ and therefore $k = 20$ lb/ft. The elongation is $3 - 2 = 1$ ft when the spring is 3 ft long and $4 - 2 = 2$ ft when the spring is 4 ft long, so that $s_0 = 1$, $s_1 = 2$. It follows from (13) that

$$W = \frac{1}{2}k(s_1^2 - s_0^2) = \frac{1}{2}(20)(2^2 - 1^2) = 30 \text{ foot-pounds.} \quad \square$$

Gravitation and Escape Velocity

In the next example, we use the concepts of work and kinetic energy to solve an important problem of space flight.

Example 8 With what velocity v_0 must a rocket be fired vertically upward in order to completely escape the earth's gravitational attraction?

Solution According to *Newton's law of gravitation*, the force attracting the rocket back to earth is given by the *inverse square law*

$$F = F(s) = -\frac{GMm}{s^2}, \qquad (14)$$

where G is a positive constant, called the *universal gravitational constant*, M is the mass of the earth, m is the mass of the rocket, and s is the distance between the rocket (regarded as a particle) and the center of the earth. The choice of the s-axis, with the origin O at the center of the earth, is shown in Figure 27; of course, "vertically upward" could just as well mean away from the earth along any other line through the earth's center. The minus sign in (14) expresses the fact that the force of gravitation is *attractive*, pulling the rocket back to earth.

The work done on the rocket by the earth's gravitational pull as the rocket leaves the surface of the earth and goes off to a distant point is given by the integral

$$W = \int_{s_0}^{s_1} F(s)\,ds = -\int_{s_0}^{s_1} \frac{GMm}{s^2}\,ds = \left[\frac{GMm}{s}\right]_{s_0}^{s_1} = \frac{GMm}{s_1} - \frac{GMm}{s_0},$$

where s_0 equals R, the radius of the earth, and s_1 is a very large number. Therefore

$$W = -\frac{GMm}{R}, \qquad (15)$$

after dropping the negligibly small number GMm/s_1. The work W equals the change

$$\frac{1}{2}mv_1^2 - \frac{1}{2}mv_0^2 \qquad (16)$$

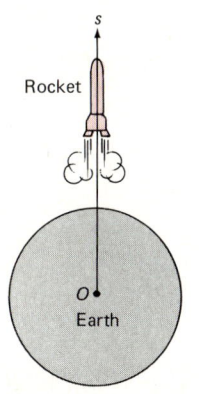

Figure 27

Section 4.7 Newtonian Mechanics; Kinetic Energy and Work   **271**

in the rocket's kinetic energy in going from the earth's surface to the distant point (v_0 is the initial velocity and v_1 the final velocity of the rocket). Since we are looking for the smallest value of v_0 that will allow the rocket to escape the earth's gravitational pull, we choose v_1 equal to zero, so that the rocket will arrive at the distant point with its initial kinetic energy completely "spent." Equating (15) and (16), with $v_1 = 0$, we get

$$\frac{1}{2}mv_0^2 = \frac{GMm}{R}.$$

Thus v_0 is given by the formula

$$v_0 = \sqrt{\frac{2GM}{R}}, \tag{17}$$

and is independent of the rocket's mass.

To calculate (17), we observe that the force acting on the rocket at the earth's surface is $-GMm/R^2$ by (14) and $-mg$ in terms of the constant g, the acceleration due to gravity, which plays a role in problems involving gravitation at or near the earth's surface. Therefore $-GMm/R^2 = -mg$, or equivalently

$$\frac{GM}{R} = gR. \tag{18}$$

With the help of this formula, we can calculate v_0 without knowing the mass of either the earth or the rocket. In fact, substituting (18) into (17), we obtain

$$v_0 = \sqrt{2gR}. \tag{19}$$

Since, to a good approximation, $R = 3960$ miles and $g = 32$ ft/sec^2, we finally have

$$v_0 = \sqrt{\frac{2(32)(3960)}{5280}} \approx 6.9 \text{ mi/sec}$$

(1 mi = 5280 ft). The quantity v_0 is usually called the *escape velocity* for the earth, although to be more precise, it is actually a *speed* of escape from the earth's surface.† A rocket fired upward with a speed less than v_0 must eventually fall back to earth, unless it is captured by the gravitational attraction of some other celestial body. ☐

---

† This lack of precision in distinguishing between velocity and speed (the absolute value of velocity) is widespread in the study of motion along a line, where the two quantities differ by at most a sign. The distinction must be carefully preserved in treating motion in two or three dimensions (see Chapters 11 and 12).

Problems

1. Find the motion of a particle of mass m acted on by a constant force F, given that the particle is initially at rest at the point $s = 0$.

2. A particle of mass m moves under the action of a constant force F. Suppose the particle's position at time $t = t_0$ is $s = s_0$. What velocity v_0 must the particle have at time $t = t_0$ in order to arrive at the point $s = s_1$ at time $t = t_1$?

3. Find the motion of a particle of mass m acted on by a force $F = kt$, proportional to the elapsed time since the onset of motion, given that the particle starts from the point $s = 0$ with initial velocity v_0.

4. Which has more kinetic energy, a 1-ounce bullet going 500 mph or a 10-ton truck going 1 mph? What is the answer if the bullet's velocity is 600 mph?

5. The same amount of work done on two particles starting from rest causes one to go twice as fast as the other. What can be said about the masses of the particles?

6. Who does more work against gravity, a woman holding a 2.5-pound weight at arm's length for 1 minute or a man climbing up a flight of stairs?

7. A particle of mass m initially at rest is acted on by a force $F = 5 \cos 2t$. What is the maximum kinetic energy of the particle?

8. Suppose it takes a force of 15 lb to stretch a spring 2 in. beyond its natural length of 8 in. Assuming the validity of Hooke's law, how much work is done in doubling the natural length of the spring? In stretching it from a length of 10 in. to a length of 14 in? In compressing it from its natural length to a length of 6 in?

9. What is the maximum height h reached by a stone thrown vertically upward with initial velocity v_0? What is the effect on h of doubling v_0? Find v_0 if $h = 100$ ft.

10. Suppose a stone thrown vertically upward returns to the ground 8 sec later. What is the stone's initial velocity? Its final velocity? What is the maximum height reached by the stone? Show that the stone's height at time $8 - t$ is the same as its height at time t $(0 \leq t \leq 8)$. Show that the stone's velocity at time $8 - t$ is the negative of its velocity at time t.

11. An object A falls out of a window 260 ft up, and exactly 1 sec later an object B falls out of a window 200 ft up. Does A ever catch up with B, and if so, when?

12. A particle is attracted to each of two fixed points A and B with a force proportional to the distance between the particle and the point. How much work is done in moving the particle from A to B along the segment AB? Assume that the constant of proportionality k is the same for both A and B.

13. A rocket fired vertically upward from the earth's surface reaches a maximum height of h. Show that the rocket's initial velocity is

$$v_0 = \sqrt{\frac{2gRh}{R+h}}, \qquad \text{(i)}$$

where g is the acceleration due to gravity and R is the radius of the earth. How can the escape velocity for the earth be deduced from this formula?

14. A rocket fired vertically upward from the earth's surface reaches a maximum height equal to the radius of the earth. What is the rocket's initial velocity?

15. How high will a rocket rise if it is fired vertically upward from the earth's surface with a velocity of 1 mi/sec? Of 2 mi/sec?

16. Estimate the escape velocity for the moon, given that the moon has approximately $\frac{3}{11}$ the radius and $\frac{1}{81}$ the mass of the earth.

17. Suppose an astronaut in his space suit can jump 2.5 ft high on the earth. How high can he jump on the moon? Use the data given in the preceding problem.

18. The mass of the sun is approximately 2×10^{30} kg, while its radius is approximately 7×10^5 km. Estimate the escape velocity for the sun, given that the universal gravitational constant G is approximately 6.67×10^{-20} km^3/kg sec^2.

19. Estimate the escape velocity for a *white dwarf*, a type of star in which a mass about equal to that of the sun is compressed into a volume about equal to that of the earth. (The radius of the earth is approximately 6400 km.)

20. Estimate the escape velocity for a *neutron star*, a type of star in which a mass about equal to that of the sun is compressed into a sphere of radius 10 km or so.

21. Suppose a mass M is compressed into a sphere whose radius R is shrinking, as in the case of a "dying" star. Then for some value of R, called the *gravitational radius* and denoted by R_0, the velocity of escape from the contracting mass will be exactly equal to the velocity of light $c \approx 300{,}000$ km/sec. According to this oversimplified model of the behavior of light in a strong gravitational field, when $R = R_0$ light can no longer emerge from the sphere, which then becomes an invisible *black hole*. Show that

$$R_0 = \frac{2GM}{c^2}. \qquad \text{(ii)}$$

Comment. Since light particles or photons have no mass, the escape velocity model is not really applicable here. However, the problem can be solved rigorously with the help of Einstein's *general theory of relativity*, which remarkably enough also predicts the existence of black holes and leads to the same value of R_0.

Find the gravitational radius of

22. The sun   23. The earth   24. The moon

(The mass of the earth is approximately 6×10^{24} kg.)

25. According to the currently accepted cosmological model, the universe came into being some 15–20 billion years ago in an explosion called the "big bang." Since that time the universe has been expanding in such a manner that the velocity v of a galaxy at distance R from our galaxy (the Milky Way) is given by *Hubble's law* $v = HR$, where H is *Hubble's constant*, approximately equal to 15 km/sec per million light years. (A light year is the distance traveled by light in one year, about 9.5×10^{12} km.) It is not known whether or not this expansion of the universe will continue indefinitely. If the universe contains enough matter, the gravitational forces exerted by this matter on itself will eventually bring the expansion to an end, followed by a period of contraction culminating in a complete gravitational collapse, called the "big crunch," in which the universe as we know it is destroyed. Show that the universe will continue to expand forever if the density (mass per unit volume) of matter in the universe at the present time does

not exceed the *critical density*†

$$\rho_c = \frac{3H^2}{8\pi G}.$$

Estimate ρ_c.

26. The muzzle velocity of a $\frac{1}{4}$-ounce bullet fired from a recoilless rifle is 1200 mph. The rifle's barrel is 2 ft long. What is the force acting on the bullet while it is in the barrel, assuming that the force is constant? How long does the bullet stay in the barrel?

27. By the *momentum* of a particle is meant the product $p = mv$ of its mass m and its velocity $v = v(t)$. Suppose the particle is acted on by a force $F = F(t)$ which is a continuous function of the time t. Let Δp be the change in the particle's momentum during the time interval $[t_0, t_1]$. Then

$$\Delta p = mv\Big|_{t_0}^{t_1} = mv_1 - mv_0,$$

where $v_0 = v(t_0)$ and $v_1 = v(t_1)$. Show that

$$\Delta p = \int_{t_0}^{t_1} F(t)\,dt,$$

where the integral on the right is called the *impulse* of the force during the interval $[t_0, t_1]$.

28. A baseball pitcher throws a 90-mph fastball to a batter, who hits a low line drive right back at the pitcher, who fortunately ducks in time! Suppose the ball leaves the bat with a speed of 102 mph. What is the average force on the bat if the time of contact between the bat and the ball is 0.01 sec? (A regulation baseball weighs 5 oz.)

Hint. First find the change in the ball's momentum as a result of being struck by the bat.

29. Suppose a particle of mass m and velocity $v = v(t)$ is acted on by a time-dependent force $F = F(t)$, as in Problem 27. Then the particle's kinetic energy changes during the time interval $[t_0, t]$ by an amount

$$\frac{1}{2}mv^2\Big|_{t_0}^{t} = \frac{1}{2}mv^2 - \frac{1}{2}mv_0^2,$$

where $v_0 = v(t_0)$. By a slight modification of the language on page 269, this change in kinetic energy is called the *work* done on the particle by the force during the interval $[t_0, t]$, denoted by $W = W(t)$. Moreover, the derivative of W with respect to time is called the *power*, denoted by $P = P(t)$. Show that

$$W = \int_{t_0}^{t} F(u)v(u)\,du, \qquad P = \frac{dW}{dt} = Fv. \qquad \text{(iii)}$$

30. To visit a friend living in a four-story walk-up, a man weighing 165 lb climbs 50 ft in 1 min. Find the average horsepower expended by the man in making the climb. What is the equivalent power in watts? (1 horsepower = 550 ft-lb/sec = 746 watts.)

† ρ is lowercase Greek rho, pronounced "roe."

Key Terms and Topics
Summation, limits of summation, index of summation
The binomial theorem, binomial coefficients
Partitions of an interval, mesh size of a partition
The definite integral and integrability, Riemann sums
The area under a curve as a definite integral
Integration, limits of integration, variable of integration
Integrability of continuous functions
Integration on adjacent intervals, area between two curves
The mean value theorem for integrals
Antiderivatives, form of the general antiderivative
The indefinite integral
Existence of antiderivatives of continuous functions
Differentiation of an integral with a variable upper limit
The fundamental theorem of calculus
The position function as the integral of the velocity function
The differential equations $y' = f(x)$ and $y'' = f(x)$
General solutions and particular solutions
Initial conditions and boundary conditions
Initial value and boundary value problems
Rectilinear motion and Newton's second law
Kinetic energy and work, Hooke's law
Gravitation and escape velocity

Supplementary Problems

Calculate

1. $\sum_{n=2}^{6} |1 - 3n|$

2. $\sum_{n=3}^{8} (n^2 - 2n)$

3. $\sum_{n=0}^{4} \frac{1 - n^2}{1 + n^2}$

4. $\sum_{n=1}^{7} \tan \frac{(2n - 1)\pi}{4}$

Calculate the given sum with as little work as possible.

5. $\sum_{n=10}^{99} \frac{1}{n(n + 1)}$

6. $\sum_{n=1}^{49} (3n^2 + 3n + 1)$

Hint. Notice that
$$\frac{1}{n(n + 1)} = \frac{1}{n} - \frac{1}{n + 1}, \quad 3n^2 + 3n + 1 = (n + 1)^3 - n^3.$$

7. The two functions
$$F(x) = \begin{cases} x^2 & \text{if } x > 0, \\ x^2 + 1 & \text{if } x < 0, \end{cases} \quad G(x) = \begin{cases} x^2 + 1 & \text{if } x > 0 \\ x^2 & \text{if } x < 0 \end{cases}$$
have the same derivative $2x$, but do *not* differ by a constant. Why doesn't this contradict Theorem 4, page 249?

8. Use differentiation to show that $\tan^2 x = \sec^2 x + C$ on every interval not containing any of the points $x = \pm\frac{1}{2}\pi$, $\pm\frac{3}{2}\pi, \ldots$ Then find the constant C.

Evaluate

9. $\int \left(3x^2 - \frac{1}{x^2} + \frac{1}{x^3}\right) dx$

10. $\int \frac{(x + 2)^2}{x^4} dx$

11. $\int \cos(\pi x + \sqrt{2}) \, dx$

12. $\int (1 + \sec^2 3t) \, dt$

13. $\int \frac{\cos 2u}{\cos^2 u \sin^2 u} \, du$

14. $\int \frac{(\sqrt{v} + 1)^3}{\sqrt{v}} \, dv$

15. $\int_{1}^{2} \left(x - \frac{4}{x}\right)^2 dx$

16. $\int_{-1}^{1} (1 - \sqrt[3]{x})^3 \, dx$

17. $\int_{-\pi/4}^{\pi/4} \tan x \, dx$

18. $\int_{-\pi/4}^{\pi/4} \tan^2 t \, dt$

19. $\int_{\pi/4}^{3\pi/4} \cot^2 u \, du$

20. $\int_{-1}^{1} \cos^2 v \tan v \, dv$

21. Given any integer $n \geq 0$, show that
$$\frac{n + 1}{2} \int_{-1}^{1} x^n \, dx = \begin{cases} 1 & \text{if } n \text{ is even,} \\ 0 & \text{if } n \text{ is odd.} \end{cases}$$

Find the mean value of the function
22. $f(x) = (1 - x)^3$ on $[-2, 2]$
23. $f(x) = (x + 1)^{2/3}$ on $[-1, 7]$
24. $f(x) = 2\cos x - 3\sin x$ on $[0, \pi/2]$
25. $f(x) = (\cos x - x)^2$ on $[-\pi, \pi]$

26. Deduce from the formula
$$\int_{0}^{1} (x^2 - x + 1) \, dx = \frac{5}{6}$$
that the quadratic equation $6x^2 - 6x + 1$ has a root between 0 and 1. Verify this directly.
Hint. Use the mean value theorem for integrals.

27. Prove the following *generalized mean value theorem for integrals*: Let f and g be continuous on $[a, b]$, and suppose g does not change sign on $[a, b]$. Then there is a point c in $[a, b]$ such that
$$\int_{a}^{b} f(x)g(x) \, dx = f(c) \int_{a}^{b} g(x) \, dx. \qquad (i)$$
(If $g(x) \equiv 1$, this reduces to the ordinary mean value theorem for integrals.)

Find a point c satisfying formula (i) if
28. $f(x) = x^2$, $g(x) = x$, $a = 0$, $b = 2$
29. $f(x) = x$, $g(x) = x^2 - 1$, $a = -1$, $b = 0$
30. $f(x) = \cos x$, $g(x) = \sin x$, $a = \pi$, $b = 3\pi/2$
31. $f(x) = \sin x$, $g(x) = \cos x$, $a = 0$, $b = \pi/2$

Find the area A of the region R
32. Between the curves $y = |x|$ and $y = \sqrt{2} \cos(\pi x/4)$
33. Between the curves $y = 1/x^2$ and $y = \frac{15}{8}x - \frac{7}{8}x^2$
34. Bounded by the x-axis, the parabola $x = y^2 + 1$ and the tangent to the parabola at the point $(2, 1)$
35. Inside the loop of the curve $y^2 = x(x - 1)^2$
In each case sketch the region R.

36. Show that
$$\int_{0}^{1} x^r \, dx + \int_{0}^{1} x^{1/r} \, dx = 1$$
for every positive rational number r. Interpret this geometrically.

37. If $0 < a < 1$, find $\int_{0}^{1} f(x) \, dx$, where
$$f(x) = \begin{cases} x & \text{if } 0 \leq x \leq a, \\ \dfrac{1 - x}{1 - a} & \text{if } a < x \leq 1. \end{cases}$$

38. Show that if the second derivative f'' is continuous on $[a, b]$, then
$$\int_{a}^{b} xf''(x) \, dx = bf'(b) - af'(a) + f(a) - f(b).$$

Find the particular solution of the given first order differential equation satisfying the indicated initial condition.
39. $y' = x^5 - x^3$, $y(0) = 4$ **40.** $y' = \tan^2 x$, $y(\pi/4) = 0$
41. $y' \cos^2 x = \sin x$, $y(0) = 7$
42. $y' \sin^2 x = \cos x$, $y(\pi/2) = -1$

Find the particular solution of the given second order differential equation satisfying the indicated initial conditions or boundary conditions.

43. $y'' = \sqrt{x}$, $y(1) = 2$, $y'(1) = 1$
44. $y'' = \sin^2 x$, $y(0) = -1$, $y'(0) = 1$
45. $y'' = 7x^{1/3}$, $y(-1) = -2$, $y(1) = 0$
46. $y'' = \cos^2 x$, $y(0) = 1$, $y(\pi) = 1$

47. The marginal cost to a firm of producing a certain commodity is $MC(q) = 3q^2 - 90q + 1200$ dollars at output level q. Find the total cost function $C(q)$ if the overhead is $7200. Express the cost (exclusive of overhead) of producing the second dozen units of the commodity as an integral, and evaluate it. Find the average cost per unit (inclusive of overhead) of producing the first three dozen units.

48. A car is going 90 mph when its brakes are suddenly applied. How many seconds elapse before the car comes to a stop if the braking resists the car's forward motion with a force equal to one half its weight? How far does the car go after the brakes are applied? Assume that $g = 32$ ft/sec^2.

49. A car weighing 2700 lb accelerates uniformly from rest to a speed of 75 mph in 15 seconds flat. Find the instantaneous power delivered to the car by its engine as a function of time. What is the average power delivered to the car over the interval of acceleration?

50. A particle is repelled by a force inversely proportional to the square of its distance from a fixed point O. Suppose the force on the particle is 100 dynes when it is 1 cm away from O. How much work must be done on the particle to bring it from far away to a point 0.001 cm from O? (A force of 1 dyne imparts an acceleration of 1 cm/sec^2 to a mass of 1 gram, and 1 dyne-cm is called an *erg*.)

51. A worker drops a wrench into an open elevator shaft and hears the wrench hit the bottom of the shaft exactly 3 sec later. How deep is the shaft? (Take g to be 32 ft/sec^2 and the speed of sound, in warm air, to be 1156 ft/sec.)

52. How far from the moon is the point where the gravitational attraction of the earth is just balanced by that of the moon? (The earth's mass is about 81 times that of the moon, and the earth–moon distance is about 240,000 mi.)

53. Newton's second law of motion $m(dv/dt) = F$ takes the form $d(mv)/dt = dp/dt = F$ in terms of the momentum $p = mv$ of a particle of mass m and velocity v, and remains true in this form even if m is not a constant, as assumed so far. Suppose a spherical raindrop falls through an atmosphere saturated with water vapor. As a result of condensation, the mass of the raindrop increases at a rate proportional to its surface area, with constant of proportionality k. Show that the raindrop falls with constant acceleration $a = \frac{1}{4}g \approx 8$ ft/sec^2, assuming that its initial size is negligible and that its initial velocity is zero.

54. According to Einstein's *special theory of relativity*, the Newtonian law of motion $d(mv)/dt = F$ should be replaced by

$$\frac{d}{dt}\frac{mv}{\sqrt{1-(v/c)^2}} = F, \qquad \text{(ii)}$$

where c is the velocity of light ($\approx 300{,}000$ km/sec). If v is tiny compared to c (as is true in all problems of ordinary engineering mechanics), Newton's law is a superb approximation to Einstein's law. However, if v is an appreciable fraction of c (as in the case of subatomic particles of very high energy), Einstein's law of motion (ii) leads to "relativistic effects" that are not predicted by Newton's law. Show, for example, that according to Einstein's law, a particle of nonzero mass acted on by a constant force F can never achieve the velocity of light, no matter how long the force acts. What does Newton's law have to say about this?

Hint. Integration of (ii) gives $v/\sqrt{1-(v/c)^2} = Ft/m$ if the particle starts from rest.

Problems 55–64 treat integration of functions continuous only on parts of their intervals of integration. Specifically, a function f is said to be *piecewise continuous* on an interval $[a, b]$ if there is a partition $a = c_0, c_1, c_2, \ldots, c_{n-1}, c_n = b$ ($c_0 < c_1 < c_2 < \cdots < c_{n-1} < c_n$) of $[a, b]$ such that (1) f is discontinuous at $c_1, c_2, \ldots, c_{n-1}$, (2) f is continuous on the *open* intervals $(c_0, c_1), (c_1, c_2), \ldots, (c_{n-1}, c_n)$, and (3) the one-sided limits

$$f(c_0^+), f(c_1^-), f(c_1^+), \ldots, f(c_{n-1}^-), f(c_{n-1}^+), f(c_n^-)$$

all exist and are finite, where for brevity we write

$$f(c_i^+) = \lim_{x \to c_i^+} f(x), \qquad f(c_i^-) = \lim_{x \to c_i^-} f(x).$$

The values of f at the points c_i can be arbitrary (or even undefined). Note that if $n = 1$ and $f(c_0^+) = f(a)$, $f(c_1^-) = f(b)$, then f is continuous on $[a, b]$, so that the class of piecewise continuous functions includes all continuous functions. Also note that the discontinuities of f at the points $c_1, c_2, \ldots, c_{n-1}$ (there are no such points if $n = 1$) must be jump discontinuities, as defined on page 95. It can be shown by a technical argument, which we omit, that if f is piecewise continuous on $[a, b]$, then f is integrable on $[a, b]$, with integral

$$\int_a^b f(x)\,dx = \int_a^{c_1} f_1(x)\,dx + \int_{c_1}^{c_2} f_2(x)\,dx + \cdots + \int_{c_{n-1}}^b f_n(x)\,dx, \qquad \text{(iii)}$$

where f_i is the *continuous* function on $[c_{i-1}, c_i]$ which coincides with f on (c_{i-1}, c_i), that is,

$$f_i(x) = \begin{cases} f(c_{i-1}^+) & \text{if } x = c_{i-1}, \\ f(x) & \text{if } c_{i-1} < x < c_i, \\ f(c_i^-) & \text{if } x = c_i. \end{cases}$$

This is entirely plausible in terms of the area interpretation of the integral. For example, the function f graphed in Figure 28(a) is piecewise continuous on $[a, b]$, with discontinuities at c_1 and c_2, and its integral is the sum of the areas of the three shaded regions. A prime example of a piecewise continuous function is the greatest integer function $[\![x]\!]$, which is piecewise continuous on every interval $[a, b]$. On the other hand, the function f graphed in Figure 28(b) is not piecewise continuous, since $f(c_2^-) = \lim_{x \to c_2^-} f(x) = \infty$, and in fact f is unbounded and therefore nonintegrable on $[a, b]$ (see Problem 33, page 238).

Use formula (iii) to evaluate the given integral (in each case the integrand is piecewise continuous).

55. $\int_{-1}^{2} [\![x]\!]\, dx$ **56.** $\int_{-2}^{1} [\![x]\!]\, dx$

57. $\int_{\sqrt{2}}^{\pi} [\![x]\!]\, dx$ **58.** $\int_{0}^{2} [\![2x + 1]\!]\, dx$

59. $\int_{0}^{5} f(x)\, dx$, where $f(x) = \begin{cases} 2 - x & \text{if } 0 \leq x \leq 2, \\ -1 & \text{if } 2 < x < 3, \\ 10 & \text{if } x = 3, \\ x^2 & \text{if } 3 < x \leq 5 \end{cases}$

60. $\int_{0}^{2\pi} f(x)\, dx$, where $f(x) = \begin{cases} \sin x & \text{if } 0 \leq x < \pi/2, \\ \cos x & \text{if } \pi/2 \leq x < \pi, \\ \sin x & \text{if } \pi \leq x < 3\pi/2, \\ \cos x & \text{if } 3\pi/2 \leq x \leq 2\pi \end{cases}$

61. $\int_{-2}^{8} f(x)\, dx$, where $f(x) = \begin{cases} 6 & \text{if } -2 < x < 1, \\ -4 & \text{if } 1 < x < 3, \\ 5 & \text{if } 3 < x < 8 \end{cases}$

62. Let
$$f(x) = \begin{cases} a_1 & \text{if } 0 \leq x < 1, \\ a_2 & \text{if } 1 \leq x < 2, \\ \cdots \\ a_n & \text{if } n - 1 \leq x \leq n. \end{cases}$$

Show that the mean value or average of the function f on the interval $[0, n]$ is equal to $(x_1 + x_2 + \cdots + x_n)/n$, the arithmetic mean or average of the n numbers $a_1, a_2, \ldots, a_n$.

63. Find the mean value of $[\![x]\!]$ on the interval $[0, n]$, where n is any positive integer.

64. Find the mean value of $[\![x]\!]$ on the interval $[-a, a]$ for arbitrary $a > 0$.

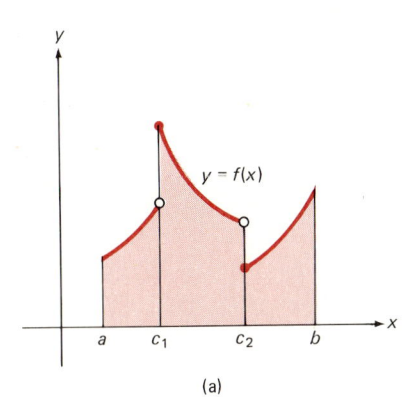

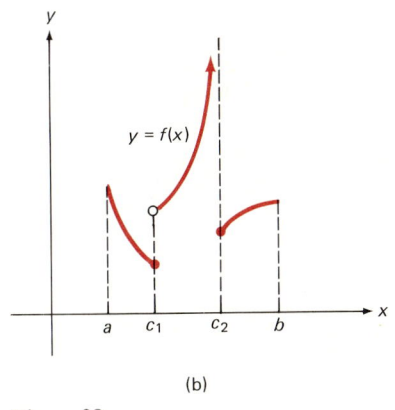

Figure 28

Supplementary Problems

5 Inverse Functions

Every function $y = f(x)$ has the property that the value of the dependent variable y is uniquely determined by the value of the independent variable x, but some functions have the additional property that the value of x is uniquely determined by the value of y. Functions of this special kind are called *one-to-one functions*. With every such function $y = f(x)$ we can associate an *inverse function* $x = f^{-1}(y)$, which maps the values of y back into the values of x. Moreover, as shown in Sections 5.1 and 5.2, if f is continuous, so is f^{-1}, and if f has a nonzero derivative, so does f^{-1}. It sometimes turns out that the inverse of a familiar one-to-one function is a new function worthy of study in its own right. This is true of the important and useful *inverse trigonometric functions*, introduced in Section 5.3, and of a number of other functions to be considered in the next chapter.

5.1 The Inverse of a One-to-One Function

Let f be a function defined on a set X, and let Y be the image of X under f, that is, the set of all values $y = f(x)$ taken by f as x varies over the set X; in the notation of set theory, $Y = \{y : y = f(x), x \in X\}$. Then knowledge of x uniquely determines y, in the sense that for every point x in X there is one and only one point y in Y such that $y = f(x)$. In fact, this uniqueness is the essence of the function concept. Of course, f may well map more than one point of X into the same point of Y, and in general this will happen. For example, if $y = f(x) = |x|$, then f maps both points $x = 3$ and $x = -3$ into the same point $y = 3$.

Suppose, however, that f does *not* map more than one point of X into the same point of Y. Then for every point y in Y there is one and only one point x in X such that $f(x) = y$, and we say that f is a *one-to-one function*, or that f is *one-to-one on X*. In other words, if f is one-to-one on X and if x and x' are points of X, the equality $f(x) = f(x')$ is possible only if $x = x'$. Put concisely, a one-to-one function f maps distinct points of X into distinct points of Y.

Inverse Functions

In the case of a one-to-one function, and *only* in this case, we can introduce another function, denoted by f^{-1} and called the *inverse function* (or simply the *inverse*) of the original function f. The inverse function is defined on Y and maps any given y in Y into the unique point x in X from which it "originated," that is, into the unique point x in X for which $f(x) = y$. It is clear that X is the image of Y under f^{-1}, just as Y is the image of X under f. If X is the largest set on which f is defined, then X is the domain of f and the range of f^{-1}, while Y is the range of f and the domain of f^{-1}. It is easy to see that f^{-1} is itself a one-to-one function, with f as its inverse, so that $(f^{-1})^{-1} = f$.

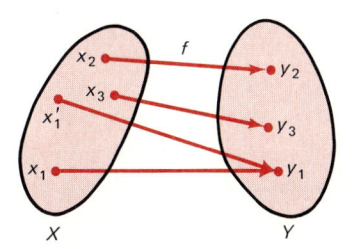

Figure 1

Example 1 Consider the function f specified by the "mapping diagram" in Figure 1. The patches labeled X and Y represent the domain and range of f, and the values of the independent variable x and dependent variable y are represented by the points in the patches. Each value of x is connected by an arrow to the value of y into which it is mapped by f, and the direction of the arrows indicates the direction of the mapping (from X to Y). The function f is not one-to-one on the whole domain X, since two arrows terminate on the same point y_1. However, as is apparent from Figure 2(a), f is one-to-one on the set X^* obtained by deleting the point x'_1 from X, and the corresponding inverse function f^{-1}, depicted in Figure 2(b), is obtained by reversing the directions of the arrows, making them go from the points of Y to the points of X^*. Note that f is one-to-one, and hence has an inverse, on any subset of X that does not contain both of the points x_1 and x'_1.

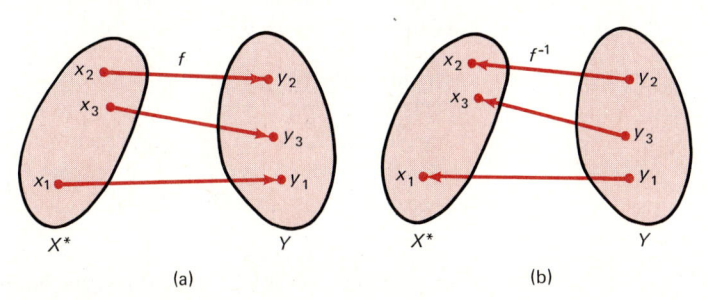

Figure 2

Section 5.1 The Inverse of a One-to-One Function

Example 2 Show that the function

$$y = \frac{1+x}{1-x} \quad (x \neq 1) \tag{1}$$

is one-to-one, and find its inverse.

Solution It follows from (1) that $1 + x = y - xy$, or equivalently $x + xy = y - 1$. Therefore

$$x = \frac{y-1}{y+1} \quad (y \neq -1), \tag{2}$$

and we have managed very easily to solve (1) for x in terms of y. Moreover, formula (2) certainly assigns a unique value of x to any given $y \neq -1$. Hence (1) is a one-to-one function $y = f(x)$, and (2) is its inverse function $x = f^{-1}(y)$. Let X be the set of all $x \neq 1$ and Y the set of all $y \neq -1$. Then f has domain X and range Y, while f^{-1} has domain Y and range X. The graphs of these functions are shown in Figures 3(a) and 3(b). Note that in Figure 3(b) the *ordinate* is x and the *abscissa* is y. ☐

Example 3 The function $y = f(x) = x^2$ is not one-to-one on the interval $(-\infty, \infty)$, since given any $y > 0$, there are two points in $(-\infty, \infty)$, namely $x = \sqrt{y}$ and $x = -\sqrt{y}$, for which $f(x) = y$. (As always, $\sqrt{y}$ denotes the *positive* square root of y.) However, f is one-to-one on the interval $[0, \infty)$, since there is precisely one point in $[0, \infty)$, namely $x = \sqrt{y}$, for which $f(x) = y$. In other words, the function

$$y = f(x) = x^2 \quad (0 \leq x < \infty)$$

is a one-to-one function, with inverse

$$x = f^{-1}(y) = \sqrt{y} \quad (0 \leq y < \infty). \tag{3}$$

Moreover, it is easy to see that f is one-to-one on any interval where x is of fixed sign, but not on any interval where x changes sign.

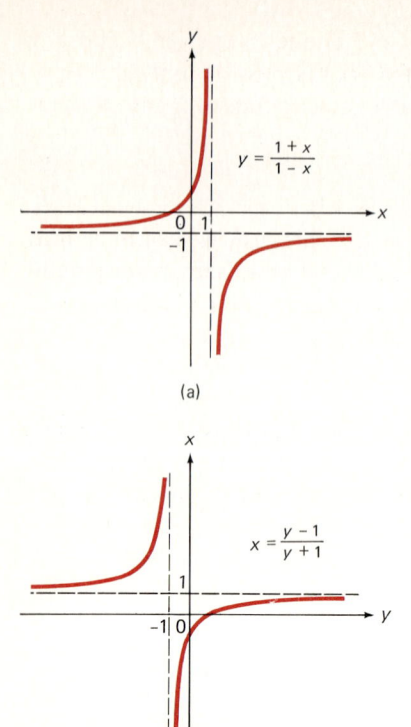

Figure 3

The Horizontal Line Property

It will be recalled from page 46 that the graph of any function $y = f(x)$ has the property that no vertical line, that is, no line parallel to the y-axis, can intersect the graph in more than one point. The graph of a *one-to-one* function has the additional property that no *horizontal* line, that is, no line parallel to the x-axis, can intersect the graph in more than one point. For if some horizontal line $y = c$, say, were to intersect the graph of $y = f(x)$ in two or more points, then two or more values of x, namely the abscissas of these points, would correspond to the same value of y, namely c, and this would contradict the very definition of a one-to-one function.

Example 4 Inspection of Figure 3(a) reveals that no horizontal line intersects the graph of the function (1) in more than one point. Therefore (1) is a one-to-one function, as we already know. On the other hand, the function $y = f(x)$ with the graph G shown in Figure 4 cannot be one-to-one, since there are horizontal lines intersecting G in more than one point (such a line is shown in the figure).

Let $y = f(x)$ be a one-to-one function with inverse $x = f^{-1}(y)$. Substituting $y = f(x)$ into $x = f^{-1}(y)$, and also substituting $x = f^{-1}(y)$ into

Figure 4

280   Chapter 5   Inverse Functions

$y = f(x)$, we get the important pair of identities

$$f^{-1}(f(x)) \equiv x, \qquad f(f^{-1}(y)) \equiv y,$$

or equivalently

$$f^{-1}(f(x)) \equiv x, \qquad f(f^{-1}(x)) \equiv x, \tag{4}$$

where in the interest of a uniform notation we have changed y to x in the second identity. According to (4), each of the functions f and f^{-1} nullifies the action of the other. In other words, the result of successively applying the functions f and f^{-1} in either order is to leave the value of x unchanged. The formulas (4) can be written more concisely as

$$f^{-1} \circ f = f \circ f^{-1} = I, \tag{5}$$

where I is the *identity function*, that is, the function mapping every number x into itself.† Equation (5) shows that f^{-1} is the reciprocal of f with respect to the operation of *composition*. Do not confuse f^{-1} with the reciprocal of f with respect to the operation of *multiplication*, which is denoted by $1/f$. The function $1/f$ is defined by the formula

$$\left(\frac{1}{f}\right)(x) = \frac{1}{f(x)},$$

provided that $f(x) \neq 0$.

Example 5 Use the formula $f(f^{-1}(x)) \equiv x$ to find the inverse of the function (1) considered in Example 2.

Solution Writing $y = f(x)$ in (1), we have

$$f(x) = \frac{1+x}{1-x} \qquad (x \neq 1). \tag{1'}$$

Application of the indicated formula then gives

$$f(f^{-1}(x)) = \frac{1 + f^{-1}(x)}{1 - f^{-1}(x)} = x.$$

Therefore

$$1 + f^{-1}(x) = x - xf^{-1}(x),$$

and solving for $f^{-1}(x)$, we get

$$f^{-1}(x) = \frac{x-1}{x+1} \qquad (x \neq -1). \tag{2'}$$

Superficial differences notwithstanding, formulas (2) and (2') are actually two equivalent ways of writing the same function. In fact, to convert (2) into (2'), we need only interchange the variables x and y, and then set $y = f^{-1}(x)$. How does one convert (2') into (2)? □

Relation Between the Graphs of f and f^{-1}

Let $y = f(x)$ be a one-to-one function with inverse function f^{-1}. Then *the graph of f^{-1} is the reflection of the graph of f in the line $y = x$* (the line of slope 1 through the origin). To see this, let a be any point in the domain of

---

† More exactly, $f^{-1} \circ f = I$ for all x in the domain of f, while $f \circ f^{-1} = I$ for all x in the range of f (the domain of f^{-1}).

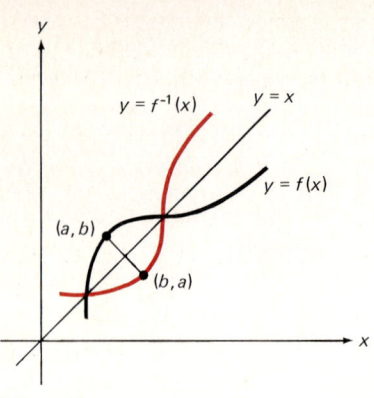

Figure 5

f. If $b = f(a)$, then $a = f^{-1}(b)$. Therefore if the point (a, b) belongs to the graph of f, the point (b, a) belongs to the graph of f^{-1} (see Figure 5). But these two points are *symmetric about the line $y = x$*, that is, the line $y = x$ is the perpendicular bisector of the segment joining the points (a, b) and (b, a). In fact, the line through the points (a, b) and (b, a) has slope

$$\frac{a - b}{b - a} = -1$$

and hence is perpendicular to the line $y = x$, while the midpoint of the segment joining the points (a, b) and (b, a) is

$$\left(\frac{a + b}{2}, \frac{a + b}{2}\right)$$

and hence lies on the line $y = x$. Note that to graph a function f and its inverse f^{-1} in the same system of rectangular coordinates, we must use the same symbol for the arguments of both functions, as in Figure 5.

Example 6 In Figure 6 we graph the function

$$y = f(x) = x^2 \qquad (0 \le x < \infty)$$

and its inverse

$$y = f^{-1}(x) = \sqrt{x} \qquad (0 \le x < \infty)$$

in the same system of rectangular coordinates. (The expression for $f^{-1}(x)$ can be obtained from formula (3) by interchanging the variables x and y.) Note that each graph is half of a parabola, and that reflection in the line $y = x$ transforms either one of the graphs into the other. How can one tell at a glance that the functions f and f^{-1} are one-to-one?

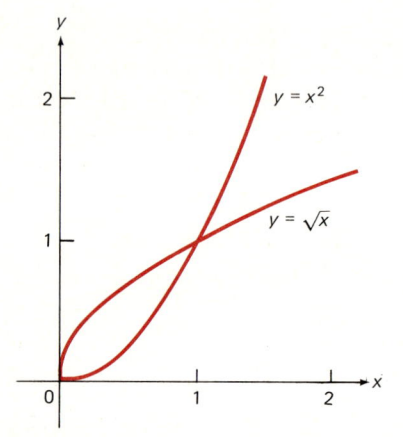

Figure 6

The Inverse of a Monotonic Function

As you may have already guessed, every monotonic function (that is, every increasing or decreasing function) is automatically one-to-one, and hence has an inverse. After all, if the graph of a function f rises steadily, as in Figure 7(a), or falls steadily, as in Figure 7(b), then no horizontal line can intersect the graph of f in more than one point. Moreover, the inverse of a monotonic function is itself a monotonic function, as we now show.

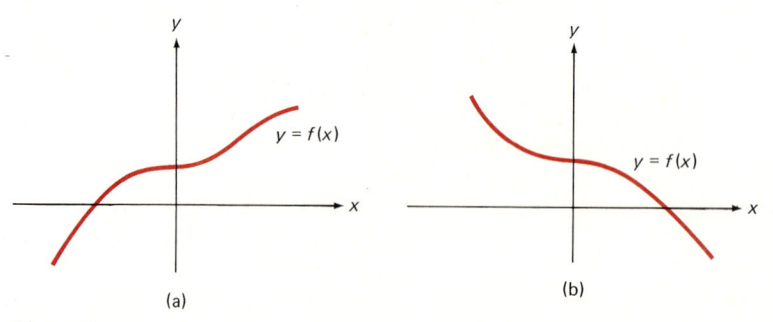

Figure 7

> **Theorem 1** (*Inverse of a monotonic function*). *Every increasing function is one-to-one with an increasing inverse. Every decreasing function is one-to-one with a decreasing inverse.*

Proof Let f be increasing (on some set), and suppose $x \neq x'$. Then either $x < x'$ or $x' < x$. In the first case $y < y'$, where $y = f(x)$, $y' = f(x')$, and in the second case $y' < y$, but in either case $y \neq y'$. Therefore f takes different values at different points, that is, f is one-to-one, and hence has an inverse function f^{-1}. Suppose $y < y'$, and let $x = f^{-1}(y)$, $x' = f^{-1}(y')$. Then $x < x'$, since $x = x'$ implies $y = y'$ (f is a function), while $x > x'$ implies $y > y'$ (f is increasing). Therefore f^{-1} is also increasing (on Y, the image of X under f). The proof for decreasing f is virtually the same. ∎

Example 7 Let

$$y = f(x) = x^n \quad (0 \leq x < \infty),$$

where n is any positive integer. By the monotonicity test (Theorem 7, page 169), f is increasing on its domain $[0, \infty)$, since the derivative $f'(x) = nx^{n-1}$ is positive on $(0, \infty)$. It follows from Theorem 1 that f is one-to-one, with inverse function

$$x = f^{-1}(y) = y^{1/n},$$

or

$$y = f^{-1}(x) = x^{1/n}, \tag{6}$$

after interchanging the variables x and y, and that f^{-1} is increasing on J, the image of I under f. The set J must be an interval, since f is continuous on I (recall Theorem 15, page 100). In fact, J is the interval $[0, \infty)$, since $f(0) = 0$, f is nonnegative on $[0, \infty)$, and

$$\lim_{x \to \infty} f(x) = \lim_{x \to \infty} x^n = \infty.$$

Since $J = [0, \infty)$ is the domain of the inverse function (6), this *proves* that the nth root $\sqrt[n]{x} = x^{1/n}$ exists and is unique for every nonnegative x, as we have been tacitly assuming all along.

Example 8 Show that the function

$$f(x) = x^{11} + 3x^7 + 2x + \sin 2x - 13 \tag{7}$$

is one-to-one.

Solution Differentiating (7), we get the function

$$f'(x) = 11x^{10} + 21x^6 + 2(1 + \cos 2x),$$

which is positive for all x (why?). It follows from the monotonicity test that f is increasing on $(-\infty, \infty)$. Therefore, by Theorem 1, f is one-to-one on its domain $(-\infty, \infty)$. Moreover, although it is impossible to find an explicit formula for the inverse function f^{-1}, the theorem tells us that f^{-1} is an increasing function. As an exercise, show that the range of f, and hence the domain of f^{-1}, is again the interval $(-\infty, \infty)$. □

Continuous One-to-One Functions

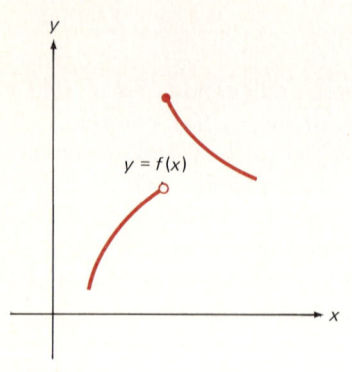

Figure 8

According to Theorem 1, every monotonic function must be one-to-one. The converse is not true, that is, there are one-to-one functions which are not monotonic. The graph of such a function f is shown in Figure 8. It is no accident that this function is discontinuous, for as we now show, every *continuous* one-to-one function must be monotonic.

Theorem 2 *(Continuous one-to-one functions are monotonic).* If f is continuous and one-to-one on an interval I, then f is either increasing on I or decreasing on I.

Proof It can be assumed that f is nonconstant, since a constant function is certainly not one-to-one. Suppose f is continuous and one-to-one on I, but neither increasing on I nor decreasing on I. Then there exist three points a, b, c in I such that $a < b < c$ and

$$f(a) < f(b), \qquad f(c) < f(b), \tag{8}$$

or

$$f(a) > f(b), \qquad f(c) > f(b). \tag{8'}$$

Suppose (8) holds, and choose a number k satisfying the inequalities

$$f(a) < k < f(b), \qquad f(c) < k < f(b)$$

[see Figure 9(a)]. By the intermediate value theorem, there is a point p in the interval (a, b) such that $f(p) = k$ and a point q in the interval (b, c) such that $f(q) = k$. But then $f(p) = f(q)$, contrary to the assumption that f is one-to-one on I. A similar contradiction can be deduced from (8') by choosing

$$f(a) > k > f(b), \qquad f(c) > k > f(b)$$

[see Figure 9(b)]. ∎

It seems intuitively evident that the inverse of a continuous function is itself continuous, since if the graph of f is unbroken, it must remain so after reflection in the line $y = x$ (the operation yielding the graph of f^{-1}). This suggests the following theorem, whose proof (given at the end of the section) can be skipped, because of its technical nature.

Theorem 3 *(Continuity of the inverse function).* Let f be continuous and one-to-one on an interval I, and let J be the image of I under f (J is an interval, by the continuity of f). Then f^{-1} is continuous on J.

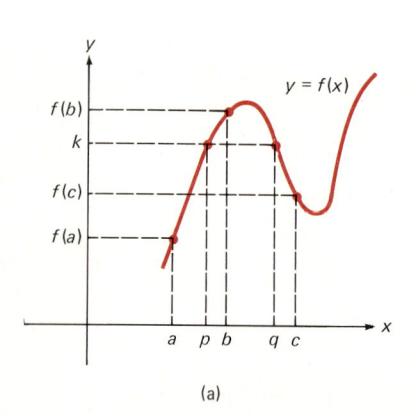

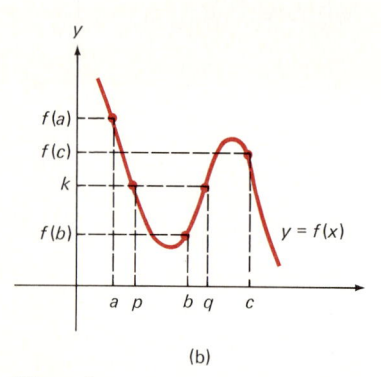

Figure 9

Example 9 The function $f(x) = x^n$ (n a positive integer) is continuous on $(-\infty, \infty)$, and in particular on $[0, \infty)$. Moreover f is increasing, and hence one-to-one, on $[0, \infty)$, with inverse $f^{-1}(x) = x^{1/n} = \sqrt[n]{x}$ (see Example 7). Since the image of the interval $[0, \infty)$ under f is again $[0, \infty)$, it follows from Theorem 3 that $\sqrt[n]{x}$ is continuous on $[0, \infty)$. The continuity of $\sqrt[n]{x}$ was proved by an entirely different method in Sections 1.8 and 1.9.

Example 10 The one-to-one function

$$f(x) = \frac{1 + x}{1 - x}$$

considered in Examples 2 and 5, is continuous on $I_1 = (-\infty, 1)$ and $I_2 = (1, \infty)$. Therefore its inverse

$$f^{-1}(x) = \frac{x-1}{x+1}$$

is continuous on the images of these two intervals under f, namely the intervals $J_1 = (-1, \infty)$ and $J_2 = (-\infty, -1)$.

Proof of Theorem 3 (Optional) According to Theorem 2, f is either increasing or decreasing on I. Suppose f is increasing on I. Then f^{-1} is increasing on J, by Theorem 1. Let k be any interior point of J, and let $c = f^{-1}(k)$. Then c is an interior point of I (why?). To show that f^{-1} is continuous at k, we argue as follows. Let ε be any positive number small enough to make both points $c \pm \varepsilon$ belong to I, and let $k_1 = f(c - \varepsilon)$, $k_2 = f(c + \varepsilon)$, which implies $c - \varepsilon = f^{-1}(k_1)$, $c + \varepsilon = f^{-1}(k_2)$, as shown schematically in Figure 10(a). Then $k_1 < k < k_2$, since f is increasing. If y belongs to the interval (k_1, k_2), that is, if $k_1 < y < k_2$, then $f^{-1}(k_1) < f^{-1}(y) < f^{-1}(k_2)$, since f^{-1} is increasing, or equivalently $c - \varepsilon < f^{-1}(y) < c + \varepsilon$. In other words, we can make $f^{-1}(y)$ as close as we please to $c = f^{-1}(k)$, that is, within distance ε of c, by choosing y sufficiently close to k, that is, within the interval (k_1, k_2). Therefore $f^{-1}(y) \to f^{-1}(k)$ as $y \to k$, so that f^{-1} is continuous at k, where k is an arbitrary interior point of J. If k is an endpoint of J belonging to J, a slight modification (and simplification!) of the above argument, illustrated in Figure 10(b), establishes that f^{-1} is continuous from the right (left) at k if k is the left (right) endpoint of J. The last two assertions taken together show that f^{-1} is continuous on the interval J. The proof for decreasing f is the same as that for increasing f if we reverse the direction of increasing y, taking account of the fact that then $k_2 < k < k_1$ in Figure 10(a) or $k_2 < k$ in Figure 10(b). ∎

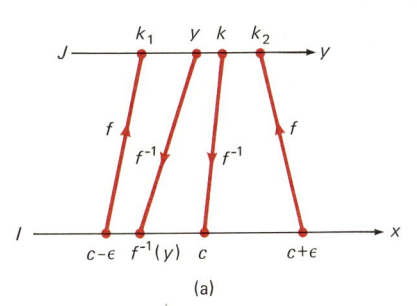

Figure 10

Problems

1. What condition on the constant a makes the linear function $f(x) = ax + b$ one-to-one? Find the corresponding inverse function f^{-1}. What extra condition on a makes the graph of f intersect the graph of f^{-1} in a single point? Find this point.

2. Show that an even function f cannot be one-to-one on any interval (or set) which is symmetric about the origin.

3. Show that if f is one-to-one and odd, then f^{-1} is also odd.

4. Show that if f and g are one-to-one, then $f \circ g$ is also one-to-one and $(f \circ g)^{-1} = g^{-1} \circ f^{-1}$.

Is the given function one-to-one on the specified interval? Explain your answer in each case.

5. $f(x) = x^3 - 6x^2 + 9x$ on $(-\infty, 1]$
6. $f(x) = x^3 - 6x^2 + 9x$ on $[2, 4]$
7. $f(x) = x^4 - 2x^2 + 1$ on $[-1, 1]$
8. $f(x) = x^4 - 2x^2 + 1$ on $[1, \infty)$

9. $f(x) = \sqrt{x(4-x)}$ on $[1, 3]$
10. $f(x) = \sqrt{x(4-x)}$ on $[0, 2]$
11. $f(x) = x^{3/5}$ on $(-\infty, \infty)$
12. $f(x) = x^{8/3}$ on $(-\infty, \infty)$
13. $f(x) = x^{-2/5}$ on $(-\infty, 0)$
14. $f(x) = x^{-5/4}$ on $(0, \infty)$
15. $f(x) = \sin x$ on $[\pi/4, 3\pi/4]$
16. $f(x) = \cos x$ on $[0, \pi]$
17. $f(x) = \tan x$ on $(-\pi/2, \pi/2)$
18. $f(x) = \sec x$ on $(-\pi/2, \pi/2)$
19. $f(x) = \cos x + \sin x$ on $[0, \pi]$
20. $f(x) = x + \cos x$ on $(-\infty, \infty)$

Find all intervals of length π on which
21. $\sin x$ is one-to-one
22. $\cos x$ is one-to-one

Find the inverse of the given one-to-one function by solving for x as a function of y.
23. $y = -x$
24. $y = 2x + 1$
25. $y = \dfrac{1}{x}$
26. $y = \dfrac{1}{1-x}$
27. $y = \dfrac{x}{x+1}$
28. $y = \dfrac{3x-1}{3x+1}$
29. $y = x^3 - 2$
30. $y = \sqrt[3]{x-1}$
31. $y = \sqrt{x(8-x)}$ $(0 \le x \le 4)$
32. $y = \sqrt{x(8-x)}$ $(4 \le x \le 8)$

Find the inverse of the given one-to-one function by using the formula $f(f^{-1}(x)) \equiv x$, as in Example 5.
33. $f(x) = 1 - 3x$
34. $f(x) = \dfrac{1}{x} - 1$

35. $f(x) = \dfrac{1}{x^2+1}$ $(0 \le x < \infty)$
36. $f(x) = \dfrac{1}{x^2+1}$ $(-\infty < x \le 0)$
37. $f(x) = 2 - \sqrt{x-3}$
38. $f(x) = (x^3 + 1)^{1/3}$
39. $f(x) = x^2 - 2x + 5$ $(-\infty < x \le 1)$
40. $f(x) = x^2 - 2x + 5$ $(1 \le x < \infty)$

41. Characterize the graph of a function $y = f(x)$ which is its own inverse.

Show that each of the following functions is its own inverse.
42. $f(x) = 2 - x$
43. $f(x) = \dfrac{1-x}{1+x}$
44. $f(x) = \dfrac{x-2}{x-1}$
45. $f(x) = \dfrac{3x+5}{4x-3}$
46. $f(x) = \sqrt{9-x^2}$ $(0 \le x \le 3)$

47. Use Theorem 3 to show that $\sqrt[n]{x}$ is continuous on $(-\infty, \infty)$ if n is odd.
48. In the ordered pair definition of a function (see Problem 58, page 43), there can be no two ordered pairs in f with the same first element. What extra condition makes f one-to-one? If f is one-to-one, how is the inverse function f^{-1} obtained?
49. Show that the inverse of the function (7) is continuous on $(-\infty, \infty)$.

5.2 The Derivative of an Inverse Function

As we now show, there is a simple relationship between the derivative of f^{-1}, the inverse of a one-to-one function f, and the derivative of f itself.

> **Theorem 4** *(Derivative of an inverse function).* Let f be continuous and one-to-one in a neighborhood of the point x, and suppose f has a finite nonzero derivative at x, equal to $f'(x)$. Then f^{-1} has a derivative at the point $y = f(x)$, equal to
> $$(f^{-1})'(y) = \dfrac{1}{f'(x)}, \tag{1}$$
> where $x = f^{-1}(y)$.

Chapter 5 Inverse Functions

Proof Let I be the neighborhood of x, and let J be the image of I under f. Then f^{-1} is continuous on J, in particular at the point $y = f(x)$, which is an interior point of J (these assertions follow from Theorems 2 and 3, page 284). Let v be a point of J other than y, and let $u = f^{-1}(v)$. Then

$$(f^{-1})'(y) = \lim_{v \to y} \frac{f^{-1}(v) - f^{-1}(y)}{v - y}$$

$$= \lim_{v \to y} \frac{u - x}{f(u) - f(x)} = \lim_{v \to y} \frac{1}{\frac{f(u) - f(x)}{u - x}}$$

(why are there no zero denominators?). But f^{-1} is continuous at y, and hence $v \to y$ implies $f^{-1}(v) \to f^{-1}(y)$, or equivalently $u \to x$. It follows that

$$(f^{-1})'(y) = \lim_{u \to x} \frac{1}{\frac{f(u) - f(x)}{u - x}} = \frac{1}{\lim_{u \to x} \frac{f(u) - f(x)}{u - x}} = \frac{1}{f'(x)}. \blacksquare$$

This method of proof was anticipated on page 138 in showing that $D_x x^r = r x^{r-1}$ for an arbitrary rational number r. In the d notation (1) takes the form

$$\frac{df^{-1}(y)}{dy} = \frac{1}{\frac{df(x)}{dx}}.$$

More concisely, we have

$$\frac{dx}{dy} = \frac{1}{\frac{dy}{dx}},$$

or equivalently

$$\frac{dy}{dx} = \frac{1}{\frac{dx}{dy}}, \qquad \frac{dy}{dx}\frac{dx}{dy} = 1.$$

The last three formulas resemble algebraic identities, but they do not, of course, constitute a proof of the theorem. They do, however, bear further testimony to the aptness of the d notation.

Example 1 The function

$$y = f(x) = \frac{1 + x}{1 - x} \qquad (x \neq 1) \tag{2}$$

is continuous, differentiable and one-to-one, with inverse

$$x = f^{-1}(y) = \frac{y - 1}{y + 1} \qquad (y \neq -1) \tag{3}$$

(see Example 2, page 280). Hence, by Theorem 4,

$$\frac{df^{-1}(y)}{dy} = \frac{1}{f'(x)} \qquad (y \neq -1),$$

Section 5.2 The Derivative of an Inverse Function **287**

provided that $f'(x) \neq 0$. Differentiating (2) with respect to x, we get

$$f'(x) = \frac{2}{(1-x)^2} \quad (x \neq 1),$$

which is never zero. Therefore

$$\frac{df^{-1}(y)}{dy} = \frac{1}{\dfrac{2}{(1-x)^2}} = \frac{(1-x)^2}{2}$$

for all $x \neq 1$, $y \neq -1$. Substituting (3) into this formula, we obtain

$$\frac{df^{-1}(y)}{dy} = \frac{\left(1 - \dfrac{y-1}{y+1}\right)^2}{2} = \frac{2}{(y+1)^2}, \tag{4}$$

or equivalently

$$\frac{df^{-1}(x)}{dx} = \frac{2}{(x+1)^2},$$

if the symbol x is chosen for the independent variable. As an exercise, check the validity of (4) by direct differentiation of (3) with respect to y.

Example 2 If $y = \sqrt{x}$, then $x = y^2$. Therefore

$$\frac{d\sqrt{x}}{dx} = \frac{dy}{dx} = \frac{1}{\dfrac{dx}{dy}} = \frac{1}{2y} = \frac{1}{2\sqrt{x}} \quad (x > 0),$$

as we already know.

Example 3 Given that

$$f(x) = x^8 + 3x^4 + 2x^2 + 1 \quad (0 \leq x < \infty),$$

find $(f^{-1})'(7)$.

Solution Since $f'(x) = 8x^7 + 12x^3 + 4x > 0$ if $x > 0$, we know that f is increasing on $[0, \infty)$. Therefore f is one-to-one on $[0, \infty)$, with inverse function f^{-1}. It is easy to see that $f(1) = 7$, $f'(1) = 24$. Hence, by Theorem 4,

$$(f^{-1})'(7) = \frac{1}{f'(1)} = \frac{1}{24}. \quad \square$$

Notice that Theorem 4 enables us to calculate the value of the derivative of f^{-1} even in cases where it is impossible to find an explicit formula for f^{-1}. The situation is very reminiscent of the technique of implicit differentiation, and this is not a coincidence (see Problem 1).

Example 4 Given that

$$f(x) = 2x + \sin^3 x,$$

find $(f^{-1})'(0)$.

Solution We have

$$f'(x) = 2 + 3 \sin^2 x \cos x = 2 + \frac{3}{2} \sin x \sin 2x,$$

where

$$|\sin x \sin 2x| = |\sin x| |\sin 2x| \leq 1.$$

Therefore $f'(x) \geq \frac{1}{2} > 0$ for all x, and f is increasing on $(-\infty, \infty)$. It follows that f is one-to-one on $(-\infty, \infty)$, with inverse function f^{-1}. Moreover $f(0) = 0$, $f'(0) = 2$, and hence, by Theorem 4,

$$(f^{-1})'(0) = \frac{1}{f'(0)} = \frac{1}{2}.$$

This is another case where it is impossible to find an explicit formula for the inverse f^{-1}. □

Problems

1. Establish formula (1) with the help of implicit differentiation, assuming that f and f^{-1} are differentiable.
2. Let $y = f(x) = x^2 + 2x + 1$, where $x \geq -1$. Calculate $(f^{-1})'(9)$ by using Theorem 4. Then check the answer by first solving for x as a function of y.

Use Theorem 4 to evaluate $(f^{-1})'(c)$ if
3. $f(x) = 4x^2 - 5x + 1$, $c = 10$
4. $f(x) = x^3 - 3x^2 + 4$, $c = -16$
5. $f(x) = \frac{1}{8}x^5 + \frac{1}{2}x^3 - 1$, $c = 7$
6. $f(x) = \dfrac{x+1}{x-1}$, $c = 5$
7. $f(x) = \dfrac{2x-1}{x+2}$, $c = -3$
8. $f(x) = \dfrac{x^3}{x^2+1}$, $c = \dfrac{1}{2}$
9. $f(x) = \sqrt{25 - x^2}$, $c = 4$
10. $f(x) = x^{21} + 2x^{11} + 5x^7$, $c = -8$
11. $f(x) = x + \cos x$, $c = \pi - 1$
12. $f(x) = x^3 + x + \sin x$, $c = 0$
13. $f(x) = \tan x$, $c = -\sqrt{3}$
14. $f(x) = \cot^3 x$, $c = 1$

In each case verify that f is one-to-one on a suitable interval.

In Problems 15–20 each function f is one-to-one, with an inverse f^{-1}. Find the tangent to the curve $y = f^{-1}(x)$ at the given point P.

15. $f(x) = \dfrac{x}{x-4}$, $P = (-3, 3)$

16. $f(x) = \dfrac{x+1}{x-5}$, $P = (2, 11)$

17. $f(x) = x^3 + x$, $P = (-10, -2)$

18. $f(x) = x + \sin x$, $P = \left(\dfrac{\pi}{2} + 1, \dfrac{\pi}{2}\right)$

19. $f(x) = \sqrt{169 - x^2}$ $(0 \leq x \leq 13)$, $P = (12, 5)$
20. $f(x) = \sqrt{169 - x^2}$ $(-13 \leq x \leq 0)$, $P = (5, -12)$

21. Let $f(x) = \int_1^x \sqrt{2 + \sin^{11} t}\, dt$. Show that f is one-to-one on $(-\infty, \infty)$. Let $a = f(\pi)$, $b = f(3\pi/2)$, although we are unable to calculate these numbers. Find $(f^{-1})'(a)$ and $(f^{-1})'(b)$. Also find $(f^{-1})'(0)$.

Let $f(x) = \int_0^x \sqrt{1 + u^6}\, du$. Evaluate $(f^{-1})'(c)$ if
22. $c = 0$ 23. $c = f(\frac{1}{2})$ 24. $c = f(\sqrt{2})$

25. Let f satisfy the same conditions as in Theorem 4, and suppose that in addition f has a finite second derivative at the point x, equal to $f''(x)$. Show that f^{-1} has a second derivative at the point $y = f(x)$, equal to

$$(f^{-1})''(y) = -\frac{f''(x)}{[f'(x)]^3}. \qquad (i)$$

In Problems 26–31 each function is one-to-one, with an inverse f^{-1}. Use formula (i) to evaluate $(f^{-1})''(c)$ if
26. $f(x) = x^{3/2}$, $c = 8$

27. $f(x) = \dfrac{3x+1}{3x-1}$, $c = 2$

28. $f(x) = \dfrac{1}{x^3+2}$, $c = 1$

29. $f(x) = x + \sin x$, $c = 0$
30. $f(x) = \int_0^x \sqrt{1 + v^2}\, dv$, $c = f(1)$
31. $f(x) = \tan^3 x$ $(-\pi/2 < x < \pi/2)$, $c = -1$

32. Let f be differentiable, with derivative f', on an interval I, and suppose f' is nonzero on I. Then f is monotonic and one-to-one on I, with an inverse f^{-1}. Let the interval J be

the image of I under f. Show that if f is concave upward (downward) on I, then f^{-1} is concave upward (downward) on J in the case of decreasing f and concave downward (upward) on J in the case of increasing f. Show that if f has an inflection point at an interior point c of I, then f^{-1} has an inflection point at $f(c)$.

5.3 Inverse Trigonometric Functions

We now turn our attention to the problem of defining appropriate inverse functions for the trigonometric functions sin x, cos x, tan x, cot x, sec x and csc x. Each of these six functions is periodic, and hence cannot be one-to-one on its natural domain of definition. For example, sin x is periodic on $(-\infty, \infty)$, with fundamental period 2π, and the periodicity condition $\sin(x + 2\pi) \equiv \sin x$ itself prevents sin x from being one-to-one, since it says that the sine takes the same value at the distinct points x and $x + 2\pi$. To make the function sin x one-to-one, so that it can have an inverse function, we must suitably restrict its domain X, without choosing X to be unnecessarily small. The standard choice of X is the interval $[-\pi/2, \pi/2]$, on which sin x is increasing and hence one-to-one. (It is easy to see that sin x fails to be one-to-one on any interval of length greater than π.) The range of sin x is then the interval $Y = [-1, 1]$, which is the same as the range of sin x when defined on the whole interval $(-\infty, \infty)$.

The Inverse Sine

Having restricted the domain of sin x to the interval $X = [-\pi/2, \pi/2]$, we can now take the inverse of sin x, obtaining a function called the *inverse sine*. Specifically, the inverse sine of y, denoted by arcsin y, is the unique number x in the interval $-\pi/2 \le x \le \pi/2$ such that $y = \sin x$. An alternative notation is $\sin^{-1} y$, but in using this notation be careful to avoid confusion with the reciprocal of sin y, which is denoted by $1/\sin y$ or $(\sin y)^{-1}$; a similar warning applies to the other inverse trigonometric functions. Both notations are common in the literature. The "arc notation" is less ambiguous, but the "-1 notation" takes up less space, and hence is preferable for engraving the keys of pocket scientific calculators. Notice that the function arcsin y has domain $Y = [-1, 1]$ and range $X = [-\pi/2, \pi/2]$.

Before graphing the inverse sine function, we replace arcsin y by arcsin x, to make x the independent variable. The graph of arcsin x is shown in Figure 11, and is the reflection in the line $y = x$ of part of the graph of sin x, namely the part between the lines $x = \pm \pi/2$. You will observe that arcsin x is increasing on $[-1, 1]$, in keeping with Theorem 1, and continuous on $[-1, 1]$, in keeping with Theorem 3. Also note that arcsin x is odd on $[-1, 1]$, as is to be expected (see Problem 3, page 285).

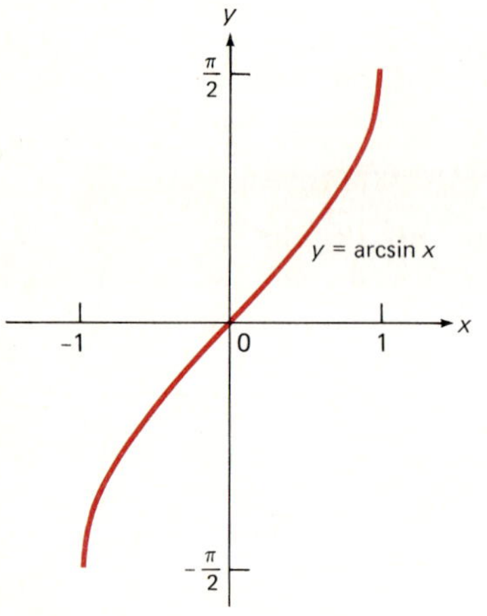

Example 1 Evaluate sin (arcsin 1) and arcsin (sin π).

Figure 11

Solution The sine of the angle whose sine is 1 must clearly be 1, and hence sin (arcsin 1) = 1. This is a special case of the formula $f(f^{-1}(x)) = x$, valid for any one-to-one function f and its inverse f^{-1}. In view of the matching formula $f^{-1}(f(x)) = x$, we might be tempted to write arcsin (sin π) = π, but this is wrong! In fact, π is not a possible value of the inverse sine function, whose range is $[-\pi/2, \pi/2]$. On the other hand, sin $\pi = 0$ and there is a unique angle in $[-\pi/2, \pi/2]$ whose sine equals 0, namely the angle 0. It follows that arcsin (sin π) = 0. ☐

Example 2 Evaluate $\tan(\arcsin \frac{2}{5})$.

Solution By $\arcsin \frac{2}{5}$ is meant the angle θ between $-\pi/2$ and $\pi/2$ whose sine is $\frac{2}{5}$. Figure 12 shows a right triangle with θ as an acute angle, 2 as the length of the side opposite θ, and 5 as the length of the hypotenuse. By the Pythagorean theorem, the length of the other side is $\sqrt{25 - 4} = \sqrt{21}$. Hence $\tan \theta$, the ratio of the length of the side opposite θ to that of the side adjacent to θ, is $2/\sqrt{21}$, and accordingly $\tan(\arcsin \frac{2}{5}) = 2/\sqrt{21}$. □

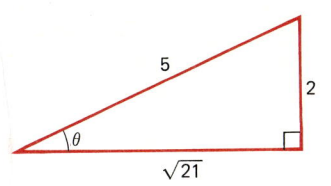

Figure 12

To differentiate the inverse sine, we set $y = \arcsin x$, $x = \sin y$ and apply Theorem 4, page 286. As a result, we obtain

$$\frac{d}{dx} \arcsin x = \frac{dy}{dx} = \frac{1}{\frac{dx}{dy}} = \frac{1}{\frac{d}{dy}(\sin y)} = \frac{1}{\cos y},$$

provided that $\cos y \neq 0$. Observe that this formula holds only on the open interval $-\pi/2 < y < \pi/2$, where $\cos y > 0$ and

$$\cos y = \sqrt{1 - \sin^2 y} = \sqrt{1 - x^2}.$$

Combining the last two formulas, we find that

$$\frac{d}{dx} \arcsin x = \frac{1}{\sqrt{1 - x^2}}, \tag{1}$$

provided that $-1 < x < 1$. It is an immediate consequence of (1) that

$$\int \frac{dx}{\sqrt{1 - x^2}} = \arcsin x + C \qquad (|x| < 1). \tag{2}$$

Moreover, using the chain rule to differentiate $\arcsin(x/a)$, where $a > 0$, we get

$$\frac{d}{dx} \arcsin \frac{x}{a} = \frac{1/a}{\sqrt{1 - (x/a)^2}} = \frac{1}{\sqrt{a^2 - x^2}}.$$

It follows that

$$\int \frac{dx}{\sqrt{a^2 - x^2}} = \arcsin \frac{x}{a} + C \qquad (a > 0, |x| < a), \tag{2'}$$

which generalizes (2).

Example 3 Evaluate $I = \int_0^{\sqrt{2}/3} \frac{dx}{\sqrt{4 - 9x^2}}$.

Solution Since

$$I = \int_0^{\sqrt{2}/3} \frac{dx}{\sqrt{9(\frac{4}{9} - x^2)}} = \frac{1}{3} \int_0^{\sqrt{2}/3} \frac{dx}{\sqrt{\frac{4}{9} - x^2}},$$

it follows from formula (2') with $a = \frac{2}{3}$ that

$$I = \frac{1}{3}\left[\arcsin \frac{3x}{2}\right]_0^{\sqrt{2}/3} = \frac{1}{3}\left(\arcsin \frac{1}{\sqrt{2}} - \arcsin 0\right) = \frac{1}{3}\left(\frac{\pi}{4}\right) = \frac{\pi}{12}.$$

Alternatively, since $\sqrt{4 - 9x^2} = \sqrt{4 - (3x)^2}$, the same answer can be obtained by using rule (iv), page 252, and formula (2') with $a = 2$. □

Section 5.3 Inverse Trigonometric Functions

The Inverse Cosine

To define the *inverse cosine*, denoted by arccos x or $\cos^{-1} x$, we proceed similarly. From the outset, we reverse the roles of x and y, writing $x = \cos y$, so that when we form the inverse function, x will be the independent variable and y the dependent variable. Then we restrict the domain of $\cos y$ to the interval $[0, \pi]$. The function $\cos y$ is decreasing and hence one-to-one on this interval, which it maps onto the interval $[-1, 1]$. The corresponding inverse function $y = \arccos x$ is decreasing and continuous on $[-1, 1]$, and has the graph shown in Figure 13. Comparison of Figures 11 and 13 shows that the graph of arccos x can be obtained from that of arcsin x by reflection in the y-axis, followed by an upward shift of $\pi/2$ units. Therefore

$$\arccos x = \arcsin(-x) + \frac{\pi}{2},$$

or equivalently

$$\arccos x = \frac{\pi}{2} - \arcsin x, \tag{3}$$

since arcsin x is odd. Differentiating (3), we find that

$$\frac{d}{dx} \arccos x = -\frac{d}{dx} \arcsin x = -\frac{1}{\sqrt{1-x^2}}, \tag{4}$$

provided that $-1 < x < 1$.

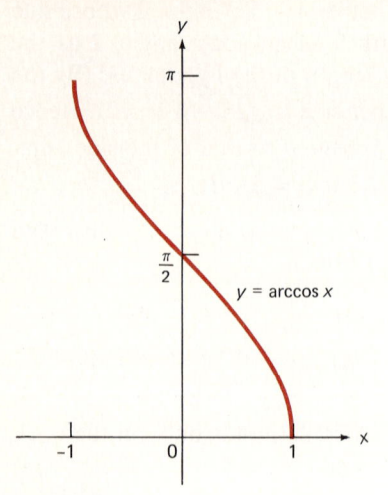

Figure 13

The Inverse Tangent

Next we turn to the *inverse tangent*, denoted by arctan x or $\tan^{-1} x$. Writing $x = \tan y$, we select one of the infinitely many branches of $\tan y$ by restricting the domain of $\tan y$ to the open interval $(-\pi/2, \pi/2)$. The function $\tan y$ is increasing and hence one-to-one on this interval, which it maps onto the interval $(-\infty, \infty)$. The corresponding inverse function $y = \arctan x$ is increasing and continuous on the interval $(-\infty, \infty)$, which it maps onto the interval $(-\pi/2, \pi/2)$, and has the graph shown in Figure 14. The graph of arctan x is the reflection in the line $y = x$ of the specified branch of $\tan x$, and under the reflection the vertical asymptotes $x = \pm \pi/2$ of $\tan x$ become the horizontal asymptotes $y = \pm \pi/2$ of arctan x. Note that arctan x is an odd function, like $\tan x$.

To differentiate the inverse tangent, we set $y = \arctan x$, $x = \tan y$ and apply Theorem 4, obtaining

$$\frac{d}{dx} \arctan x = \frac{dy}{dx} = \frac{1}{\dfrac{dx}{dy}} = \frac{1}{\dfrac{d}{dy} \tan y} = \frac{1}{\sec^2 y}.$$

But

$$\sec^2 y = \tan^2 y + 1 = x^2 + 1$$

(see formula (6), page 53), and therefore

$$\frac{d}{dx} \arctan x = \frac{1}{x^2 + 1}. \tag{5}$$

It is an immediate consequence of (5) that

$$\int \frac{dx}{x^2 + 1} = \arctan x + C. \tag{6}$$

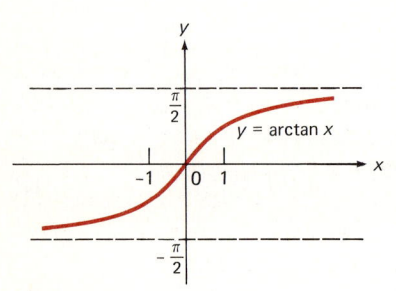

Figure 14

Moreover, using the chain rule to differentiate arctan (x/a), we find that

$$\frac{d}{dx} \arctan \frac{x}{a} = \frac{1/a}{(x/a)^2 + 1} = \frac{a}{x^2 + a^2}.$$

It follows that

$$\int \frac{dx}{x^2 + a^2} = \frac{1}{a} \arctan \frac{x}{a} + C \quad (a \neq 0), \tag{6'}$$

which generalizes (6). Actually, there is no loss of generality in assuming that $a > 0$.

Example 4 Find the area between the curves $y = 1/(x^2 + 1)$ and $y = \frac{1}{2}x^2$.

Solution We are looking for the area A of the shaded region in Figure 15. To find the x-coordinates of the points in which the two curves intersect, we solve the equation $1/(x^2 + 1) = \frac{1}{2}x^2$, which is equivalent to

$$x^4 + x^2 - 2 = (x^2 + 2)(x^2 - 1) = 0. \tag{7}$$

Since the factor $x^2 + 2$ is never zero, equation (7) has only two real roots, $x = 1$ and $x = -1$. On the interval $[-1, 1]$ the "bell-shaped" curve $y = 1/(x^2 + 1)$ is the upper curve and the parabola $y = \frac{1}{2}x^2$ is the lower curve. It follows with the help of (6) that

$$A = \int_{-1}^{1} \left(\frac{1}{x^2 + 1} - \frac{1}{2}x^2 \right) dx = 2 \int_{0}^{1} \left(\frac{1}{x^2 + 1} - \frac{1}{2}x^2 \right) dx$$

$$= 2 \left[\arctan x - \frac{1}{6}x^3 \right]_0^1 = 2 \left(\arctan 1 - \frac{1}{6} \right) = \frac{\pi}{2} - \frac{1}{3}.$$

The evenness of the integrand was used in the second step (recall Problem 1, page 260). □

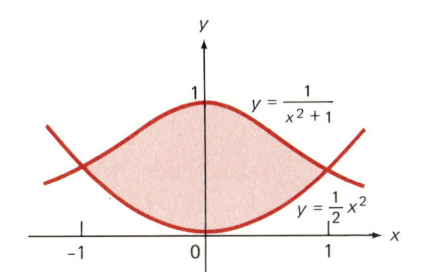

Figure 15

The Inverse Cotangent

To define the *inverse cotangent*, denoted by arccot x or $\cot^{-1} x$, we write $x = \cot y$ and restrict the domain of $\cot y$ to the interval $(0, \pi)$. The function $\cot y$ is decreasing and hence one-to-one on this interval, which it maps onto the interval $(-\infty, \infty)$. The corresponding inverse function $y = \text{arccot } x$ is decreasing and continuous on the interval $(-\infty, \infty)$, which it maps onto the interval $(0, \pi)$, and has the graph shown in Figure 16. Comparison of Figures 14 and 16 shows that the graph of arccot x can be obtained from that of arctan x by reflection in the x-axis, followed by an upward shift of $\pi/2$ units. Therefore

$$\text{arccot } x = \frac{\pi}{2} - \arctan x. \tag{8}$$

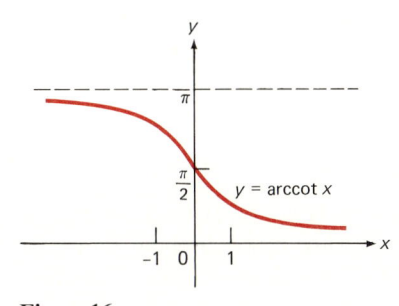

Figure 16

Differentiating (8), we find that

$$\frac{d}{dx} \text{arccot } x = -\frac{d}{dx} \arctan x = -\frac{1}{x^2 + 1}. \tag{9}$$

The Inverse Secant and Cosecant

The remaining two inverse trigonometric functions are the *inverse secant*, denoted by arcsec x or $\sec^{-1} x$, and the *inverse cosecant*, denoted by arccsc x or $\csc^{-1} x$. We begin with arccsc x, since it is actually the simpler of the two

Section 5.3 Inverse Trigonometric Functions 293

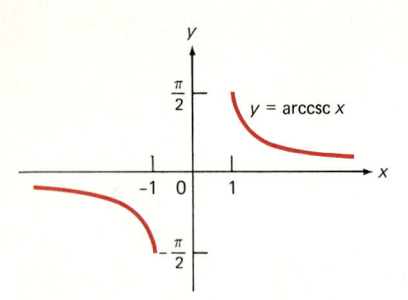

Figure 17

functions, being odd. Writing $x = \csc y$, we choose the domain of $\csc y$ to be the *pair* of intervals $[-\pi/2, 0)$ and $(0, \pi/2]$; the point 0 must be excluded, since $\csc 0$ is undefined. The function $\csc y$ is decreasing and hence one-to-one on each of these two intervals. Moreover, it is continuous on each of the intervals $[-\pi/2, 0)$ and $(0, \pi/2]$, which it maps onto the intervals $(-\infty, -1]$ and $[1, \infty)$, respectively. All this information can be inferred from Figure 25, page 64. The corresponding inverse function $y = \text{arccsc } x$ is decreasing and continuous on each of the intervals $(-\infty, -1]$ and $[1, \infty)$, which it maps onto the intervals $[-\pi/2, 0)$ and $(0, \pi/2]$, respectively, but it is not defined on the interval $(-1, 1)$. The graph of arccsc x is shown in Figure 17, from which it is apparent that arccsc x is an odd function, like $\csc x$.

To differentiate the inverse cosecant, we proceed in the usual way, writing $x = \csc y$, $y = \text{arccsc } x$ and applying Theorem 4. As a result, we get

$$\frac{d}{dx} \text{arccsc } x = \frac{dy}{dx} = \frac{1}{\frac{dx}{dy}} = \frac{1}{\frac{d}{dy} \csc y} = \frac{1}{-\csc y \cot y}.$$

But

$$\cot^2 y = \csc^2 y - 1 = x^2 - 1$$

(see formula (6'), page 53), and hence

$$\cot y = \pm\sqrt{x^2 - 1}. \tag{10}$$

We must be very careful about the choice of sign in (10). In fact, since $\cot y$ is positive when $0 < y < \pi/2$ and negative when $-\pi/2 < y < 0$, we must choose the plus sign when $0 < y < \pi/2$, that is, when $x = \csc y > 1$, and the minus sign when $-\pi/2 < y < 0$, that is, when $x = \csc y < -1$. Therefore

$$\frac{d}{dx} \text{arccsc } x = \begin{cases} -\dfrac{1}{x\sqrt{x^2 - 1}} & \text{if } x > 1, \\ \dfrac{1}{x\sqrt{x^2 - 1}} & \text{if } x < -1, \end{cases}$$

or, more simply,

$$\frac{d}{dx} \text{arccsc } x = -\frac{1}{|x|\sqrt{x^2 - 1}}, \tag{11}$$

provided that $|x| > 1$.

Finally, to define arcsec x, we write $x = \sec y$ and choose the domain of $\sec y$ to be the *pair* of intervals $[0, \pi/2)$ and $(\pi/2, \pi]$; the point $\pi/2$ must be excluded, since $\sec(\pi/2)$ is undefined. The function $\sec y$ is increasing and hence one-to-one on each of these two intervals. Moreover, it is continuous on each of the intervals $[0, \pi/2)$ and $(\pi/2, \pi]$, which it maps onto the intervals $[1, \infty)$ and $(-\infty, -1]$, respectively. All this information can again be inferred from Figure 25, page 64. The corresponding inverse function arcsec x is increasing and continuous on each of the intervals $[1, \infty)$ and $(-\infty, -1]$, which it maps onto the intervals $[0, \pi/2)$ and $(\pi/2, \pi]$, respectively, but it is not defined on the interval $(-1, 1)$. The graph of arcsec x is shown in Figure 18. Comparison of Figures 17 and 18 shows that the graph of arcsec x can be obtained from that of arccsc x by reflection in the x-axis, followed by an

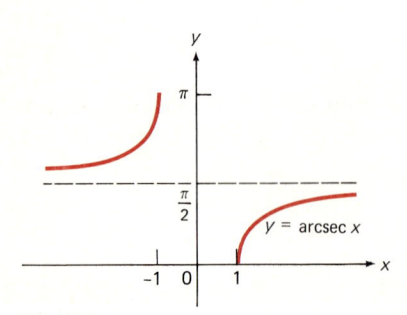

Figure 18

294   Chapter 5   Inverse Functions

upward shift of $\pi/2$ units. Therefore

$$\operatorname{arcsec} x = \frac{\pi}{2} - \operatorname{arccsc} x. \tag{12}$$

Differentiating (12), we find that

$$\frac{d}{dx}\operatorname{arcsec} x = -\frac{d}{dx}\operatorname{arccsc} x = \frac{1}{|x|\sqrt{x^2-1}}, \tag{13}$$

provided that $|x| > 1$.

Example 5 Show that

$$\operatorname{arcsec} x = \operatorname{arccos}\frac{1}{x}, \tag{14}$$

$$\operatorname{arccsc} x = \operatorname{arcsin}\frac{1}{x}, \tag{14'}$$

provided that $|x| \geq 1$.

Solution If $y = \operatorname{arcsec} x$, then $\sec y = x$, or equivalently $\cos y = 1/x$. It follows that $y = \operatorname{arccos}(1/x)$, since y is a number in the interval $0 \leq y \leq \pi$. Comparing the two expressions for y, we get (14). A similar argument, left as an exercise, establishes formula (14'). □

Collecting formulas (1), (4), (5), (9), (11) and (13) in one place, we have

$$\frac{d}{dx}\operatorname{arcsin} x = \frac{1}{\sqrt{1-x^2}}, \qquad \frac{d}{dx}\operatorname{arccos} x = -\frac{1}{\sqrt{1-x^2}},$$

$$\frac{d}{dx}\operatorname{arctan} x = \frac{1}{x^2+1}, \qquad \frac{d}{dx}\operatorname{arccot} x = -\frac{1}{x^2+1},$$

$$\frac{d}{dx}\operatorname{arcsec} x = \frac{1}{|x|\sqrt{x^2-1}}, \qquad \frac{d}{dx}\operatorname{arccsc} x = -\frac{1}{|x|\sqrt{x^2-1}}.$$

Example 6 Using one of these formulas and the chain rule to differentiate

$$f(x) = \operatorname{arctan}\frac{1}{x},$$

we obtain

$$f'(x) = \frac{1}{(1/x)^2+1}\frac{d}{dx}\frac{1}{x} = \frac{-1/x^2}{(1/x)^2+1} = -\frac{1}{x^2+1},$$

where it must be assumed that $x \neq 0$ since $f(0)$, and hence $f'(0)$, is undefined. Therefore $f(x)$ has the same derivative as $-\operatorname{arctan} x$ on each of the intervals $(-\infty, 0)$ and $(0, \infty)$. It follows that

$$\operatorname{arctan}\frac{1}{x} = -\operatorname{arctan} x + C_1 \qquad (0 < x < \infty), \tag{15}$$

$$\operatorname{arctan}\frac{1}{x} = -\operatorname{arctan} x + C_2 \qquad (-\infty < x < 0), \tag{15'}$$

where C_1 and C_2 are two constants, which need not be the same. In fact, C_1 and C_2 are unequal, for setting $x = 1$ in (15) gives

$$\arctan 1 = -\arctan 1 + C_1, \qquad \frac{\pi}{4} = -\frac{\pi}{4} + C_1,$$

so that $C_1 = \pi/2$, while setting $x = -1$ in (15′) gives

$$\arctan(-1) = -\arctan(-1) + C_2, \qquad -\frac{\pi}{4} = \frac{\pi}{4} + C_2,$$

so that $C_2 = -\pi/2$. With these values of C_1 and C_2, we can combine (15) and (15′) into the single formula

$$\arctan x + \arctan \frac{1}{x} = \begin{cases} \dfrac{\pi}{2} & \text{if } x > 0, \\ -\dfrac{\pi}{2} & \text{if } x < 0. \end{cases} \qquad (16)$$

Problems

Evaluate without using tables or a calculator.
1. $\arcsin \frac{1}{2}$
2. $\operatorname{arccot}(-1)$
3. $\operatorname{arcsec} \sqrt{2}$
4. $\operatorname{arccsc}(2/\sqrt{3})$
5. $\arctan(-1/\sqrt{3})$
6. $\arccos 1$
7. $\operatorname{arccot}(-\sqrt{3})$
8. $\operatorname{arcsec} 2$
9. $\operatorname{arccsc}(-\sqrt{2})$
10. $\arctan \sqrt{3}$
11. $\arccos(-\frac{1}{2})$
12. $\arcsin(1/\sqrt{2})$
13. $\arcsin(\sin(3\pi/2))$
14. $\sin(\arccos(-1/\sqrt{2}))$
15. $\cos(\arcsin \frac{1}{3})$
16. $\tan(\arccos \frac{1}{4})$
17. $\operatorname{arcsec}(\sec(5\pi/4))$
18. $\operatorname{arccot}(\tan(4\pi/3))$

19. What is the inverse of the function $\sin x$ if its domain is restricted to the interval $[\pi/2, 3\pi/2]$?

Express without using trigonometric or inverse trigonometric functions.
20. $\sin(\operatorname{arcsec} x)$
21. $\cos(\arctan x)$
22. $\tan(\arcsin x)$
23. $\sin(2 \arccos x)$
24. $\cos(2 \arccos x)$
25. $\cos(2 \arcsin x)$

Differentiate
26. $(\arccos x)^2$
27. $\operatorname{arcsec}(2x + 1)$
28. $\arctan \dfrac{1-x}{1+x}$
29. $\operatorname{arccot} \dfrac{2t}{1-t^2}$
30. $\arcsin t^2$
31. $\operatorname{arccsc} \dfrac{1}{t}$
32. $\arcsin \sqrt{1 - u^2}$
33. $\arccos \dfrac{1-u}{\sqrt{2}}$

34. $\operatorname{arcsec} \sqrt{u^2 - 1}$
35. $\dfrac{1}{\sqrt{3}} \operatorname{arccot} \dfrac{\sqrt{3}}{v}$
36. $\operatorname{arccsc} \sqrt{v}$
37. $\arctan \dfrac{3 \sin v}{4 + 5 \cos v}$

38. Starting from formulas (14) and (14′), deduce the formulas for the derivatives of $\operatorname{arcsec} x$ and $\operatorname{arccsc} x$.

39. Show that

$$\operatorname{arccot} x + \operatorname{arccot} \frac{1}{x} = \begin{cases} \dfrac{\pi}{2} & \text{if } x > 0, \\ \dfrac{3\pi}{2} & \text{if } x < 0. \end{cases}$$

40. Show that

$$\arctan x + \arctan \frac{1-x}{1+x} = \begin{cases} \dfrac{\pi}{4} & \text{if } x > -1, \\ -\dfrac{3\pi}{4} & \text{if } x < -1. \end{cases}$$

41. Verify the integration formula

$$\int \frac{dx}{x\sqrt{x^2 - 1}} = \operatorname{arcsec} |x| + C \qquad (|x| > 1), \qquad \text{(i)}$$

and more generally

$$\int \frac{dx}{x\sqrt{x^2 - a^2}} = \frac{1}{a} \operatorname{arcsec} \frac{|x|}{a} + C \qquad (|x| > a > 0). \qquad \text{(i′)}$$

Evaluate
42. $\displaystyle\int \frac{dx}{\sqrt{25 - x^2}}$
43. $\displaystyle\int \frac{dx}{x^2 + 121}$
44. $\displaystyle\int \frac{dt}{t\sqrt{t^2 - 49}}$

45. $\int \dfrac{dt}{\sqrt{16-4t^2}}$
46. $\int \dfrac{du}{64u^2+36}$
47. $\int \dfrac{dv}{v\sqrt{121v^2-144}}$
48. $\int_{-1/2}^{1/2} \dfrac{dx}{\sqrt{1-x^2}}$
49. $\int_{1/\sqrt{3}}^{\sqrt{3}} \dfrac{dx}{x^2+1}$
50. $\int_{\sqrt{2}}^{2} \dfrac{dx}{x\sqrt{x^2-1}}$

51. A picture 5 ft high is hung on a wall with its bottom edge 6 ft from the floor. A child whose eye level is 3.5 ft above the floor wants to get the best view of the picture. Assuming that the best view is obtained when the angle θ subtended by the picture at the child's eyes (see Figure 19) is the largest, how far from the wall should the child stand?

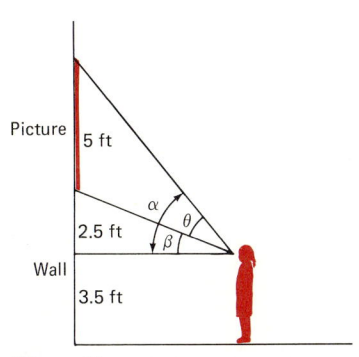

Figure 19

52. A space vehicle is 20,000 km from the surface of the moon, which it approaches at a velocity of 2 km/sec. How fast is the angle subtended by the moon at the vehicle's position P increasing? (This is the angle θ in Figure 20.) The moon's radius is 1738 km.

Find the area A of the region R bounded by
53. The x-axis, the line $x = 1$ and the curve $y = \arcsin x$
54. The y-axis, the line $y = \pi/2$ and the curve $y = \arcsin x$
55. The coordinate axes and the curve $y = \arccos x$
In each case sketch the region R.
Hint. Integrate with respect to y.

56. Four of the six inverse trigonometric functions have inflection points. Which are they, and where are the inflection points?

57. Verify the formulas

$$\arcsin x + \arcsin y = \arcsin(x\sqrt{1-y^2} + y\sqrt{1-x^2}), \quad \text{(ii)}$$

$$\arctan x + \arctan y = \arctan \dfrac{x+y}{1-xy}, \quad \text{(iii)}$$

where in (ii) it is assumed that $|\arcsin x + \arcsin y| \le \pi/2$ and in (iii) that $|\arctan x + \arctan y| < \pi/2$.

Show that

58. $\arcsin \dfrac{4}{5} - \arcsin \dfrac{3}{5} = \arcsin \dfrac{7}{25}$

59. $\arctan \dfrac{1}{2} + \arctan \dfrac{1}{3} = \dfrac{\pi}{4}$

60. $\arctan \dfrac{1}{4} + \arctan \dfrac{3}{5} = \dfrac{\pi}{4}$

61. $\arctan \dfrac{1}{3} + \arctan \dfrac{1}{4} + \arctan \dfrac{2}{9} = \dfrac{\pi}{4}$

62. Verify the formula

$$\pi = 16 \arctan \dfrac{1}{5} - 4 \arctan \dfrac{1}{239}.$$

In 1706 John Machin used this formula to calculate the first hundred decimal places of π. More than eight million decimal places of π are now known!

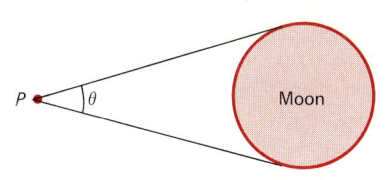

Figure 20

Key Terms and Topics
One-to-one functions
The inverse of a one-to-one function
The graph of a one-to-one function and the horizontal line property
Relation between the graph of f^{-1} and the graph of f
Properties of continuous one-to-one functions
The derivative of an inverse function
The inverse sine and cosine
The inverse tangent and cotangent
The inverse secant and cosecant
The derivatives of the inverse trigonometric functions

Supplementary Problems

1. Show that if f is even on $[-a, a]$ and one-to-one on $[0, a]$, then f is one-to-one on $[-a, 0]$. Is f one-to-one on $[-a, a]$?
2. Show that if f is odd on $[-a, a]$ and one-to-one on $[0, a]$, then f is one-to-one on $[-a, 0]$. Is f one-to-one on $[-a, a]$?

Is the given function one-to-one on the specified interval?

3. $f(x) = \sqrt{x^2 + x + 1}$ on $(-\infty, 0]$
4. $f(x) = \dfrac{x}{1 - x^2}$ on $(-1, 1)$
5. $f(x) = \dfrac{x}{\sqrt{x^2 - 1}}$ on $(1, \infty)$
6. $f(x) = \dfrac{x}{1 + x^2}$ on $[0, \infty)$

Find the inverse of the given one-to-one function.

7. $f(s) = s^2 + s + 1 \quad (-\tfrac{1}{2} \le s < \infty)$
8. $g(t) = \dfrac{t^2}{t^2 + 1} \quad (0 \le t < \infty)$
9. $h(u) = (u^3 + 1)^{1/5}$
10. $k(v) = (10 + v^{1/3})^5$

11. Give an example of a nonconstant function f that fails to be one-to-one on every interval, no matter how small.
12. Let n be a positive integer. Show that $f(x) = x^n$ is one-to-one on $(-\infty, 0]$. What is the inverse of f on $(-\infty, 0]$?

Show that each of the following functions is its own inverse.

13. $f(x) = (a^{2/3} - x^{2/3})^{3/2} \quad (0 \le x \le a)$
14. $f(x) = \dfrac{9x + 11}{13x - 9}$
15. $f(x) = \sqrt[3]{27 - x^3}$
16. $f(x) = \sqrt[4]{16 - x^4} \quad (0 \le x \le 2)$

Evaluate $(f^{-1})'(c)$ if

17. $f(x) = x^7 + x^3 + 2x, \; c = 4$
18. $f(x) = \dfrac{x - 4}{3 - x}, \; c = -2$
19. $f(x) = \dfrac{3x^2 - 1}{x^2 + 1} \; (x \ge 0), \; c = 1$
20. $f(x) = \tan^3 x, \; c = 3\sqrt{3}$

In each case verify that f is one-to-one on a suitable interval.

Evaluate $(f^{-1})''(c)$ if

21. $f(x) = \dfrac{x + 5}{x - 6}, \; c = 0$
22. $f(x) = \dfrac{1}{x^3 + 1}, \; c = \tfrac{1}{2}$
23. $f(x) = 2x + \cos x, \; c = 1$
24. $f(x) = \cot^5 x, \; c = -1$

25. Let f be continuous and increasing on $[0, \infty)$, with $f(0) = 0$. Then f is one-to-one, with an inverse f^{-1}. Use Figure 21 to prove *Young's inequality*

$$\int_0^a f(x)\,dx + \int_0^b f^{-1}(x)\,dx \ge ab,$$

where a and b are arbitrary positive numbers. When does the inequality become an equality?

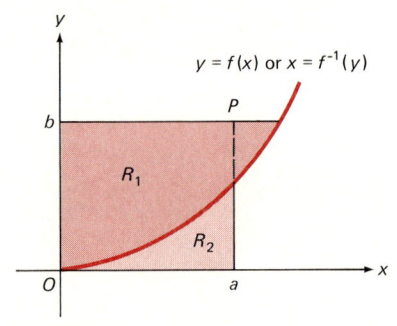

Figure 21

26. Let a, b, c and d be constants such that $c^2 + d^2 \ne 0$ and $ad - bc \ne 0$. Show that the *fractional linear transformation*

$$f(x) = \dfrac{ax + b}{cx + d}$$

(many cases of which have already been considered) is one-to-one on its domain. What is this domain? What happens if $ad - bc = 0$? When is the function f its own inverse? Calculate $(f^{-1})'(0)$ and $(f^{-1})''(0)$.

Evaluate without using tables or a calculator.

27. $\operatorname{arcsec}(-2/\sqrt{3})$
28. $\arcsin(-\sqrt{3}/2)$
29. $\operatorname{arccot} 1$
30. $\arccos(\sqrt{3}/2)$
31. $\operatorname{arccsc}(-2)$
32. $\arctan(-1)$
33. $\sec(\arctan 2)$
34. $\cot(\operatorname{arcsec}(-3))$
35. $\arccos(\tan \pi)$
36. $\operatorname{arccsc}(\sec \pi)$
37. $\csc(\operatorname{arccot} \tfrac{2}{3})$
38. $\arctan(-\tan(5\pi/4))$
39. $\sin(\arcsin \tfrac{1}{2} - \arcsin \tfrac{1}{3})$
40. $\cos(\arccos \tfrac{2}{3} + \arcsin \tfrac{3}{4})$
41. $\tan(\arctan 5 - \arctan 4)$
42. $\cot(\arctan \sqrt{3} + \operatorname{arccot} 1)$

Differentiate

43. $\dfrac{1}{2}x\sqrt{a^2 - x^2} + \dfrac{1}{2}a^2 \arcsin \dfrac{x}{a}$ $(a > 0)$

44. $x(\arcsin x)^2 + 2\sqrt{1 - x^2} \arcsin x - 2x$

45. $\arctan\left(\dfrac{x}{1 + \sqrt{1 - x^2}}\right)$

46. $\operatorname{arccot}\left(\dfrac{\sin x + \cos x}{\sin x - \cos x}\right)$

Integrate

47. $\displaystyle\int_{-3/7}^{0} \dfrac{dx}{\sqrt{36 - 49x^2}}$

48. $\displaystyle\int_{-1}^{2} \dfrac{dx}{x^2 + 2}$

49. $\displaystyle\int_{5\sqrt{2}/3}^{10/3} \dfrac{dx}{x\sqrt{9x^2 - 25}}$

50. Graph the function $f(x) = \arcsin(\sin x)$, and show that it is periodic, with fundamental period 2π.

Use L'Hospital's rule to evaluate

51. $\displaystyle\lim_{x \to 0} \dfrac{\arcsin x}{x}$

52. $\displaystyle\lim_{x \to 0} \dfrac{\arctan x}{x}$

53. $\displaystyle\lim_{x \to 0} \dfrac{\arcsin x - x}{x^3}$

54. $\displaystyle\lim_{x \to 0} \dfrac{x - \arctan x}{x - \arcsin x}$

55. The curves $y = \arcsin x$ and $y = \arccos x$ divide the rectangle bounded by the lines $x = 0$, $x = 1$, $y = 0$ and $y = \pi/2$ into four regions. Graph the curves and the rectangle, and label each region with its area A.

56. Show that $|\arctan a - \arctan b| \leq |a - b|$ for arbitrary numbers a and b.

57. Find the inverse of the function

$$f(x) = 4 \arcsin \sqrt{1 - x^2} \quad (0 \leq x \leq 1).$$

Supplementary Problems

6 Logarithms and Exponentials

It would be hard to exaggerate the importance of the logarithm and exponential functions introduced in this chapter. They are indispensable in both pure and applied mathematics, and turn up time and again in the physical, biological and social sciences. We begin our treatment of these key functions by defining the (natural) *logarithm* as a certain integral with a variable upper limit of integration. After deducing the properties of the logarithm from this definition, and in particular establishing that it is a one-to-one function, we define the *exponential* as the inverse function of the logarithm. The high point of our study of these key functions occurs in Sections 6.6 and 6.7, where exponentials and logarithms are used to solve a variety of practical problems involving growth and decay. In the last two sections we investigate some important functions which are closely related to the exponential and logarithm, namely the *hyperbolic* and *inverse hyperbolic functions*.

6.1 The Natural Logarithm

Suppose we differentiate all nonnegative integer powers

$$x^0 = 1, x, x^2, x^3, \ldots \tag{1}$$

and all negative integer powers

$$x^{-1}, x^{-2}, x^{-3}, \ldots \tag{2}$$

of an independent variable x. We then find that the derivatives of the powers (1) are

$$0, 1, 2x, 3x^2, \ldots, \tag{1'}$$

while those of the powers (2) are

$$-x^{-2}, -2x^{-3}, -3x^{-4}, \ldots \tag{2'}$$

Examining the derivatives (1′) and (2′), we discover a curious fact: *Every integer power of x appears except* x^{-1}, *the reciprocal of x*. But surely there is a function whose derivative is x^{-1} on any interval not containing the point $x = 0$. Indeed there is, as we now show, and it turns out that this function is a close relative of the common logarithm of high school mathematics!

Definition of the Logarithm as an Integral

Guided by these considerations, we now introduce a new function ln x, called the *natural logarithm*, or simply the *logarithm*, and defined for all positive x by the formula

$$\ln x = \int_1^x \frac{1}{t}\, dt,$$

which can be written more concisely as

$$\ln x = \int_1^x \frac{dt}{t}. \tag{3}$$

This is a definite integral with a *variable* upper limit of integration. It is an immediate consequence of the definition (3) and Theorem 5, page 254, that ln x is an antiderivative of the function $1/x = x^{-1}$ on the interval $(0, \infty)$, as anticipated in Example 10, page 255. Thus we have the basic formula

$$\frac{d}{dx} \ln x = \frac{1}{x} \qquad (x > 0), \tag{4}$$

valid for all positive x. Since for $x = 1$ the integral defining ln x reduces to

$$\int_1^1 \frac{dt}{t} = 0,$$

we see that ln x is the antiderivative of $1/x$ singled out by the condition

$$\ln 1 = 0. \tag{5}$$

The logarithm is of course continuous on $(0, \infty)$, since it is differentiable on $(0, \infty)$. Also, by the monotonicity test (Theorem 7, page 169), ln x is increasing on $(0, \infty)$, since

$$\frac{d}{dx} \ln x = \frac{1}{x} > 0 \qquad (0 < x < \infty).$$

It follows that ln x is one-to-one on $(0, \infty)$.

Example 1 Differentiate $x \ln x$.

Solution By the product rule and formula (4),

$$\frac{d}{dx}(x \ln x) = \frac{dx}{dx} \ln x + x \frac{d \ln x}{dx} = \ln x + x\left(\frac{1}{x}\right) = \ln x + 1. \quad \square$$

To interpret the logarithm geometrically, we consider the cases $x > 1$ and $0 < x < 1$ separately. If $x > 1$, then ln x is the area of the shaded region in Figure 1(a), that is, the area under the curve $y = 1/t$ from $t = 1$ to $t = x$. On the other hand, if $0 < x < 1$, then, since

$$\int_1^x \frac{dt}{t} = -\int_x^1 \frac{dt}{t},$$

ln x is the *negative* of the area of the shaded region in Figure 1(b), that is,

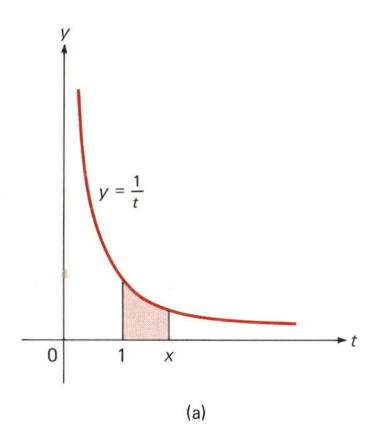

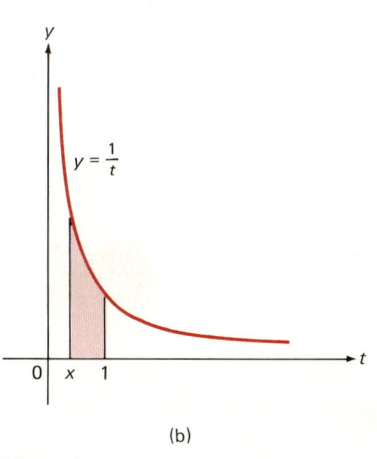

Figure 1

the negative of the area under the curve $y = 1/t$ from $t = x$ to $t = 1$. Thus $\ln x > 0$ if $x > 1$, while $\ln x < 0$ if $0 < x < 1$. The function $\ln x$ is not defined for $x \leq 0$, since the integrand $1/t$ in (1) is unbounded on any interval containing the point $t = 0$ (see Problem 33, page 238).

Example 2 The function $\ln (x^2 + x + 1)$ is defined for all x, since

$$x^2 + x + 1 = \left(x + \frac{1}{2}\right)^2 + \frac{3}{4} > 0$$

for all x. However, the function $\ln (2x^2 - x - 1)$ is defined only for $x > 1$ or $x < -\frac{1}{2}$, since

$$2x^2 - x - 1 = (2x + 1)(x - 1) \leq 0$$

for $-\frac{1}{2} \leq x \leq 1$. To differentiate these functions, we apply formula (4) and the chain rule:

$$\frac{d}{dx} \ln (x^2 + x + 1) = \frac{1}{x^2 + x + 1} \frac{d}{dx} (x^2 + x + 1) = \frac{2x + 1}{x^2 + x + 1},$$

$$\frac{d}{dx} \ln (2x^2 - x - 1) = \frac{1}{2x^2 - x - 1} \frac{d}{dx} (2x^2 - x - 1) = \frac{4x - 1}{2x^2 - x - 1}. \quad \square$$

Formula (4) for the derivative of the logarithm can be extended, giving the more general result

$$\frac{d}{dx} \ln |x| = \frac{1}{x} \qquad (x \neq 0), \tag{4'}$$

valid for all nonzero x. In fact, if x is positive, $|x| = x$ and (4') reduces to (4), while if x is negative, $|x| = -x$ and therefore

$$\frac{d}{dx} \ln |x| = \frac{d}{dx} \ln (-x) = \frac{1}{-x} \frac{d}{dx} (-x) = -\frac{1}{x}(-1) = \frac{1}{x}.$$

It is an immediate consequence of (4') that

$$\int \frac{dx}{x} = \ln |x| + C. \tag{6}$$

You should memorize this important integration formula.

The Logarithm of a Product

We now establish a basic property of the logarithm.

> **Theorem 1** (*Logarithm of a product*). *If a and b are positive, then*
>
> $$\ln ab = \ln a + \ln b. \tag{7}$$

Proof Differentiating the function $\ln ax$, we find that

$$\frac{d}{dx} \ln ax = \frac{1}{ax} \frac{d}{dx} ax = \frac{a}{ax} = \frac{1}{x}.$$

Thus both functions $\ln ax$ and $\ln x$ have the same derivative $1/x$. In other words, $\ln ax$ and $\ln x$ are both antiderivatives of $1/x$ on the interval $(0, \infty)$.

It follows that
$$\ln ax = \ln x + C,$$
where C is a constant. To determine C, we set $x = 1$, obtaining
$$\ln a = \ln 1 + C = C,$$
with the help of (5), so that $C = \ln a$. Therefore
$$\ln ax = \ln a + \ln x,$$
and setting $x = b$ in this formula, we get (7). ∎

According to (7), the logarithm of the product of two factors is the sum of the logarithms of the separate factors. Equivalently, reading (7) from right to left, we see that the sum of two given logarithms is itself a logarithm, whose argument is the product of the arguments of the given logarithms. The common logarithm studied in high school has the same property, and in fact differs from the natural logarithm only by a constant factor, as will be shown on page 320.

Formula (7) leads at once to two other important formulas. Since
$$\ln a + \ln \frac{1}{a} = \ln \left(a \cdot \frac{1}{a} \right) = \ln 1 = 0,$$
we have
$$\ln \frac{1}{a} = -\ln a,$$
and then, since
$$\ln \frac{a}{b} = \ln \left(a \cdot \frac{1}{b} \right) = \ln a + \ln \frac{1}{b},$$
we also have
$$\ln \frac{a}{b} = \ln a - \ln b. \tag{8}$$

The next theorem gives another basic property of the logarithm.

Theorem 2 *(Logarithm of a rational power).* *If $x > 0$, then*
$$\ln x^r = r \ln x \tag{9}$$
for any rational number r.

Proof The method of proof resembles that of Theorem 1. Differentiating the function $\ln x^r$, where $x > 0$ and r is rational, we find that
$$\frac{d}{dx} \ln x^r = \frac{1}{x^r} \frac{d}{dx} x^r = \frac{1}{x^r} r x^{r-1} = \frac{r}{x},$$
which is the same as the derivative of the function $r \ln x$. Therefore $\ln x^r$ and $r \ln x$ are both antiderivatives of r/x on the interval $(0, \infty)$. It follows that
$$\ln x^r = r \ln x + C, \tag{9'}$$
where C is a constant. Setting $x = 1$, we get $\ln 1 = r \ln 1 + C$, so that $C = 0$ and (9') reduces to (9). ∎

In proving Theorem 2, we used the differentiation formula $D_x x^r = rx^{r-1}$, established for rational r on pages 139 and 148. It will be shown in Section 6.5 that this formula, and hence Theorem 2, remains valid for arbitrary real r (not necessarily rational).

Example 3 Express $\ln 72$, $\ln 6^{1/5}$ and $\ln \sqrt{\frac{2}{27}}$ in terms of $\ln 2$ and $\ln 3$.

Solution Making free use of formulas (7)–(9), we have

$$\ln 72 = \ln (2^3 \cdot 3^2) = \ln 2^3 + \ln 3^2 = 3 \ln 2 + 2 \ln 3,$$

$$\ln 6^{1/5} = \frac{1}{5} \ln 6 = \frac{1}{5} \ln (2 \cdot 3) = \frac{1}{5} (\ln 2 + \ln 3),$$

$$\ln \sqrt{\frac{2}{27}} = \ln \left(\frac{2}{3^3}\right)^{1/2} = \ln \frac{2^{1/2}}{3^{3/2}} = \ln 2^{1/2} - \ln 3^{3/2} = \frac{1}{2} \ln 2 - \frac{3}{2} \ln 3. \quad \square$$

Next we examine the behavior of $\ln x$ as $x \to \infty$. Given any positive number C, no matter how large, let n be any integer greater than $C/\ln 2$. Then to make $\ln x$ exceed C, we need only choose $x > 2^n$. In fact, since $\ln x$ is an increasing function, $x > 2^n$ implies

$$\ln x > \ln 2^n = n \ln 2 > \frac{C}{\ln 2} (\ln 2) = C,$$

where at the second step we use formula (9) with $x = 2$ and $r = n$ (note that $\ln 2 > 0$). Therefore

$$\lim_{x \to \infty} \ln x = \infty. \tag{10}$$

Also

$$\lim_{x \to 0^+} \ln x = -\infty, \tag{11}$$

since

$$\lim_{x \to 0^+} \ln x = \lim_{t \to \infty} \ln \frac{1}{t} = -\lim_{t \to \infty} \ln t = -\infty,$$

with the help of the substitution $x = 1/t$. According to (10) and (11), $\ln x$ takes arbitrarily large positive values and arbitrarily large negative values. This fact, together with the intermediate value theorem, implies that $\ln x$ takes every real value. In other words, the range of $\ln x$ is the whole real line $(-\infty, \infty)$.

Figure 2 shows the graph of the function $\ln x$. It is apparent from the figure that $\ln x$ is increasing on $(0, \infty)$, has the range $(-\infty, \infty)$ and satisfies the condition $\ln 1 = 0$. You will also observe that $\ln x$ is concave downward on $(0, \infty)$. This follows at once from the concavity test (Theorem 10, page 175), since

$$\frac{d^2}{dx^2} \ln x = \frac{d}{dx} \frac{1}{x} = -\frac{1}{x^2} < 0 \qquad (0 < x < \infty).$$

Moreover, because of (11), $\ln x$ has the y-axis as an asymptote.

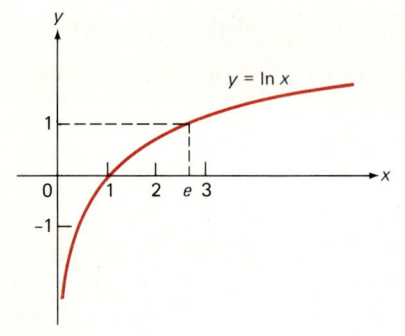

Figure 2

The Number e

Let e be the number such that

$$\ln e = 1,$$

304   Chapter 6   Logarithms and Exponentials

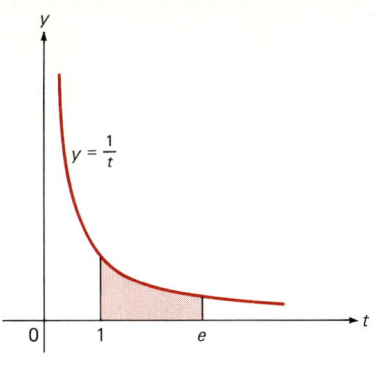

Figure 3

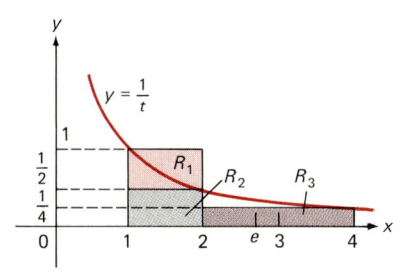

Figure 4

as in Figure 2, or equivalently

$$\int_1^e \frac{dt}{t} = 1,$$

so that the area under the curve $y = 1/t$ from $t = 1$ to $t = e$ (the shaded area in Figure 3) is precisely equal to 1. It is clear from the construction in Figure 4, involving three rectangles R_1, R_2 and R_3, each of area $\frac{1}{2}$, that

$$\ln 2 < (\text{Area of } R_1) + (\text{Area of } R_2) = 1 = (\text{Area of } R_2) + (\text{Area of } R_3) < \ln 4.$$

Therefore e is a number between 2 and 4. This number, called the *base of the natural logarithm*, is a constant of great importance in calculus and its applications. It turns out that e is irrational and

$$e = 2.718281828459045\ldots$$

(the fact that the block of digits 1828 occurs twice in succession is fortuitous, but makes the number e easy to remember†). As we will see in Section 6.5, after deciding what is meant by a^x when x is irrational, the number e is also given by the formula

$$e = \lim_{x \to \infty} \left(1 + \frac{1}{x}\right)^x.$$

Example 4 The function $\ln(\ln(\ln x))$ is defined only when $\ln(\ln x) > 0$, that is, only when $\ln x > 1$, or equivalently $x > e$, and its derivative is

$$\frac{d}{dx} \ln(\ln(\ln x)) = \frac{1}{\ln(\ln x)} \frac{d}{dx} \ln(\ln x)$$

$$= \frac{1}{\ln(\ln x)} \frac{1}{\ln x} \frac{d}{dx} \ln x = \frac{1}{x \ln x \cdot \ln(\ln x)}. \quad \square$$

According to formulas (10) and (11), the logarithm approaches infinity (in absolute value) both as $x \to \infty$ and as $x \to 0^+$, but it does so more slowly than x as $x \to \infty$, and more slowly than $1/x$ as $x \to 0^+$. More exactly,

$$\lim_{x \to \infty} \frac{\ln x}{x} = 0, \tag{12}$$

$$\lim_{x \to 0^+} \frac{\ln x}{1/x} = \lim_{x \to 0^+} x \ln x = 0. \tag{13}$$

Both of these formulas are easily proved with the help of L'Hospital's rule. In detail,

$$\lim_{x \to \infty} \frac{\ln x}{x} = \lim_{x \to \infty} \frac{D_x \ln x}{D_x x} = \lim_{x \to \infty} \frac{1/x}{1} = \lim_{x \to \infty} \frac{1}{x} = 0,$$

$$\lim_{x \to 0^+} \frac{\ln x}{1/x} = \lim_{x \to 0^+} \frac{D_x \ln x}{D_x (1/x)} = \lim_{x \to 0^+} \frac{1/x}{-1/x^2} = \lim_{x \to 0^+} (-x) = 0.$$

We can also deduce (13) from (12) by observing that

$$\lim_{x \to 0^+} x \ln x = \lim_{t \to \infty} \frac{1}{t} \ln \frac{1}{t} = -\lim_{t \to \infty} \frac{\ln t}{t} = 0,$$

after making the substitution $x = 1/t$.

† The next six digits 459045 are also easy to remember, since they call to mind the angles of an isosceles right triangle!

Section 6.1 The Natural Logarithm

Logarithmic Differentiation

The calculation of the derivative of a function $f(x)$ can often be simplified by the technique of *logarithmic differentiation*. This consists in first using formula (4') and the chain rule to calculate the *logarithmic derivative*

$$\frac{d}{dx} \ln |f(x)| = \frac{1}{f(x)} \frac{d}{dx} f(x) = \frac{f'(x)}{f(x)}, \tag{14}$$

and then multiplying (14) by $f(x)$ to get the ordinary derivative

$$f'(x) = f(x) \frac{d}{dx} \ln |f(x)|.$$

Notice that this technique is applicable only at points where $f(x) \neq 0$, since otherwise $\ln |f(x)|$ is not defined.

Example 5 Use logarithmic differentiation to calculate the derivative of the function

$$f(x) = \frac{(6x+1)^{7/3} \cos^9 x}{(x^2-4)^5}.$$

Solution Since

$$|f(x)| = \frac{|6x+1|^{7/3} |\cos x|^9}{|x^2-4|^5},$$

we find with the help of formulas (7)–(9) that

$$\ln |f(x)| = \frac{7}{3} \ln |6x+1| + 9 \ln |\cos x| - 5 \ln |x^2-4|.$$

We now differentiate $\ln |f(x)|$, obtaining the logarithmic derivative

$$\frac{d}{dx} \ln |f(x)| = \frac{7}{3(6x+1)} \frac{d}{dx}(6x+1) + \frac{9}{\cos x} \frac{d}{dx} \cos x - \frac{5}{x^2-4} \frac{d}{dx}(x^2-4)$$

$$= \frac{14}{6x+1} - \frac{9 \sin x}{\cos x} - \frac{10x}{x^2-4}.$$

Multiplication by $f(x)$ then gives the desired derivative

$$f'(x) = \frac{(6x+1)^{7/3} \cos^9 x}{(x^2-4)^5} \left(\frac{14}{6x+1} - 9 \tan x - \frac{10x}{x^2-4} \right).$$

Explain why this formula is not valid if $x = \pm 2$, $-\frac{1}{6}$, or if $x = (n + \frac{1}{2})\pi$, where n is any integer. ∎

Problems

1. Are the functions $\ln x^2$ and $2 \ln x$ identically equal?

Find all x for which the given function is defined.

2. $\ln (x^2 - 9)$

3. $\ln (\sqrt{x-4} + \sqrt{6-x})$

4. $\ln (\sin \pi x)$

5. $\arcsin (\ln x)$

6. $\ln (\ln (1 - x^2))$

7. $\sqrt{\ln (x-2)}$

Express in terms of $\ln 2$, $\ln 3$ and $\ln 5$.

8. $\ln \frac{125}{36}$

9. $\ln \sqrt[3]{\frac{24}{25}}$

10. $\ln (810)^{3/4}$

11. $\ln (0.002)$

12. $\ln \sqrt{0.005}$

13. $\ln (4.5 \times 10^4)$

14. Verify that the function $\ln (x + \sqrt{1 + x^2})$ is odd.

15. Find the average of the function $1/x$ over the interval $[a, b]$, where a and b have the same sign.

Differentiate
16. $\ln(6 - x^2)$
17. $\ln(x^3 - 2x + 5)$
18. $(\ln x)^2$
19. $x^2 \ln x$
20. $\dfrac{\ln x}{x}$
21. $\dfrac{\ln x}{x^2 + 1}$
22. $\ln \tan \dfrac{x}{2}$
23. $x[\sin(\ln x) - \cos(\ln x)]$
24. $\ln(\arcsin x)$
25. $\ln \dfrac{1 + t}{1 - t}$
26. $\ln \sqrt{\dfrac{1 - t}{1 + t}}$
27. $\ln(t + \sqrt{1 + t^2})$

Find
28. The fourth derivative of $x^2 \ln x$
29. The fifth derivative of $\dfrac{\ln x}{x}$

30. Of all the lines tangent to the curve $y = \ln x$ only one goes through the origin. Find this line.

31. Verify that
$$\int \ln x \, dx = x \ln x - x + C.$$

32. Show that the function $\ln x$ has no asymptotes other than the y-axis.

Find the area A of the region R
33. Bounded by the x-axis, the line $x = e$ and the curve $y = \ln x$
34. Under the curve $y = 2/(x + 1)$ from $x = 0$ to $x = 3$
35. Between the curves $y = 2x - x^2$ and $y = 1/x$
36. Between the curves $y = 2/x$ and $y = 10/(x^2 + 4)$
In each case sketch the region R.

Find all local extrema and inflection points of the given function. On which intervals is the function increasing? Decreasing? Concave upward? Concave downward? Find all absolute extrema and asymptotes. Graph the function.
37. $f(x) = x \ln x$
38. $f(x) = (\ln x)^2$
39. $f(x) = \ln(1 + x^2)$
40. $f(x) = \dfrac{1}{\ln x}$

41. Use the mean value theorem to show that
$$\frac{b - a}{b} < \ln \frac{b}{a} < \frac{b - a}{a}$$
if $0 < a < b$.

42. Show that $x - \tfrac{1}{2}x^2 < \ln(1 + x) < x$ for all $x > 0$.

Use logarithmic differentiation to find the derivative of
43. $(2x^2 - 1)^{3/4}(x^3 + 1)^{4/3}$
44. $\dfrac{(x + 1)^2}{(x + 2)^3(x + 3)^4}$
45. $\sqrt[3]{\dfrac{x(x^2 + 1)}{(x - 1)^2}}$
46. $\dfrac{\sqrt{4x + 1}}{(x + 2)^7(\ln x)^3}$
47. $\dfrac{(1 + \sin x)^5}{(1 - \cos x)^6}$
48. $\dfrac{(2 - \cot x)^3}{(3 + \sec x)^2}$

Use L'Hospital's rule to evaluate
49. $\lim\limits_{x \to 1} \dfrac{\ln x}{x - 1}$
50. $\lim\limits_{x \to 0^+} \dfrac{\ln x}{\ln \sin x}$
51. $\lim\limits_{x \to 0} \dfrac{\ln \cos x}{x}$
52. $\lim\limits_{x \to 0^+} \dfrac{\ln \sin 2x}{\ln \sin x}$
53. $\lim\limits_{x \to 1} \left(\dfrac{1}{\ln x} - \dfrac{1}{x - 1} \right)$
54. $\lim\limits_{x \to \frac{1}{2}\pi^+} \left[\cos x \ln \left(x - \dfrac{\pi}{2} \right) \right]$

6.2 Some Integrals Leading to Logarithms

As illustrated by the following examples, introduction of the logarithm function considerably enhances our ability to evaluate integrals.

Example 1 Evaluate $\displaystyle\int \dfrac{dx}{ax + b}$ $(a \neq 0)$.

Solution The function $F(x) = \ln|x|$ is an antiderivative of $1/x$. Therefore, by rule (iv), page 252,

$$\int \dfrac{dx}{ax + b} = \dfrac{1}{a} F(ax + b) + C = \dfrac{1}{a} \ln|ax + b| + C.$$

Section 6.2 Some Integrals Leading to Logarithms

Thus

$$\int \frac{dx}{5x+7} = \frac{1}{5} \ln|5x+7| + C,$$

$$\int \frac{dx}{1-2x} = \frac{1}{-2} \ln|1-2x| + C = -\frac{1}{2} \ln|2x-1| + C,$$

and so on. ☐

Example 2 Evaluate $\int \frac{x+a}{x+b} dx$.

Solution Dividing $x+b$ into $x+a$, we find that

$$\frac{x+a}{x+b} = 1 + \frac{a-b}{x+b},$$

and thus

$$\int \frac{x+a}{x+b} dx = \int \left(1 + \frac{a-b}{x+b}\right) dx = x + (a-b) \ln|x+b| + C. \quad \square \quad (1)$$

Example 3 Evaluate $\int \frac{2x+3}{6x-1} dx$.

Solution Since

$$\int \frac{2x+3}{6x-1} dx = \int \frac{2(x+\frac{3}{2})}{6(x-\frac{1}{6})} dx = \frac{1}{3} \int \frac{x+\frac{3}{2}}{x-\frac{1}{6}} dx,$$

it follows from formula (1), with $a = \frac{3}{2}$, $b = -\frac{1}{6}$, that

$$\int \frac{2x+3}{6x-1} dx = \frac{1}{3}\left[x + \left(\frac{3}{2} + \frac{1}{6}\right) \ln\left|x-\frac{1}{6}\right| + k\right],$$

$$= \frac{1}{3} x + \frac{5}{9} \ln\left|x-\frac{1}{6}\right| + C,$$

where k and $C = \frac{1}{3}k$ are arbitrary constants. Since

$$\frac{5}{9} \ln\left|x-\frac{1}{6}\right| = \frac{5}{9} \ln\left|\frac{6x-1}{6}\right| = \frac{5}{9} \ln|6x-1| - \frac{5}{9} \ln 6,$$

we can also write

$$\int \frac{2x+3}{6x-1} dx = \frac{1}{3} x + \frac{5}{9} \ln|6x-1| + C,$$

after absorbing $-\frac{5}{9} \ln 6$ into the arbitrary constant of integration C. ☐

Integration of a Logarithmic Derivative

Next, let $f(x)$ be a differentiable function that does not take the value zero. Then, as on page 306, $f(x)$ has the logarithmic derivative

$$\frac{d}{dx} \ln|f(x)| = \frac{f'(x)}{f(x)}, \quad (2)$$

308 Chapter 6 Logarithms and Exponentials

where the prime denotes differentiation with respect to x. It follows that

$$\int \frac{f'(x)}{f(x)} dx = \ln |f(x)| + C. \tag{3}$$

This formula is a particularly useful tool for evaluating integrals.

Example 4 Evaluate $\int \frac{2x - 3}{x^2 - 3x + 2} dx$.

Solution The numerator of the integrand is the derivative of the denominator. Therefore, by (3),

$$\int \frac{2x - 3}{x^2 - 3x + 2} dx = \ln |x^2 - 3x + 2| + C$$

on any interval containing neither of the points $x = 1, 2$ at which the denominator of the integrand equals zero. □

Example 5 Evaluate $\int \tan x \, dx$.

Solution Since

$$\tan x = \frac{\sin x}{\cos x} = -\frac{(\cos x)'}{\cos x},$$

we find with the help of (3) that

$$\int \tan x \, dx = -\ln |\cos x| + C. \tag{4}$$

Alternatively,

$$\tan x = \frac{\sec x \tan x}{\sec x} = \frac{(\sec x)'}{\sec x},$$

which implies

$$\int \tan x \, dx = \ln |\sec x| + C. \tag{4'}$$

Formulas (4) and (4') are equivalent, since

$$-\ln |\cos x| = \ln \left| \frac{1}{\cos x} \right| = \ln |\sec x|.$$

Both formulas are valid on any interval containing none of the points at which $\cos x$ equals zero, that is, none of the points $x = (n + \frac{1}{2})\pi$, where n is any integer. □

Suppose we choose

$$f(x) = \frac{x + a}{x + b} \qquad (a \neq b)$$

in formula (2). Then

$$\frac{f'(x)}{f(x)} = \frac{d}{dx} \ln |f(x)| = \frac{d}{dx} (\ln |x + a| - \ln |x + b|)$$

$$= \frac{1}{x + a} - \frac{1}{x + b} = \frac{b - a}{(x + a)(x + b)},$$

Section 6.2 Some Integrals Leading to Logarithms

and (3) becomes

$$\int \frac{b-a}{(x+a)(x+b)} dx = \ln \left| \frac{x+a}{x+b} \right| + C \quad (a \neq b),$$

or equivalently

$$\int \frac{dx}{(x+a)(x+b)} = \frac{1}{b-a} \ln \left| \frac{x+a}{x+b} \right| + C \quad (a \neq b). \tag{5}$$

Assuming that $a \neq 0$ and setting $b = -a$, we find that

$$\int \frac{dx}{x^2 - a^2} = -\frac{1}{2a} \ln \left| \frac{x+a}{x-a} \right| + C \quad (a \neq 0),$$

or equivalently

$$\int \frac{dx}{x^2 - a^2} = \frac{1}{2a} \ln \left| \frac{x-a}{x+a} \right| + C \quad (a \neq 0). \tag{6}$$

This is in contrast to the formula

$$\int \frac{dx}{x^2 + a^2} = \frac{1}{a} \arctan \frac{x}{a} + C \quad (a \neq 0),$$

derived on page 293. In the last three formulas we might just as well choose $a > 0$.

Example 6 Evaluate $\int \frac{dx}{x^2 + x - 6}$.

Solution Since

$$\frac{1}{x^2 + x - 6} = \frac{1}{(x-2)(x+3)},$$

it follows from formula (5), with $a = -2$, $b = 3$, that

$$\int \frac{dx}{x^2 + x - 6} = \frac{1}{3 - (-2)} \ln \left| \frac{x-2}{x+3} \right| + C = \frac{1}{5} \ln \left| \frac{x-2}{x+3} \right| + C.$$

Verify that the same answer is obtained if we choose $a = 3$, $b = -2$ instead. □

Example 7 Evaluate $\int_0^1 \frac{dx}{x^2 + x - 6}$.

Solution By the fundamental theorem of calculus and the preceding example,

$$\int_0^1 \frac{dx}{x^2 + x - 6} = \left[\frac{1}{5} \ln \left| \frac{x-2}{x+3} \right| \right]_0^1 = \frac{1}{5} \left(\ln \left| \frac{1-2}{1+3} \right| - \ln \left| \frac{0-2}{0+3} \right| \right)$$

$$= \frac{1}{5} \left(\ln \frac{1}{4} - \ln \frac{2}{3} \right) = \frac{1}{5} \ln \frac{3}{8} \approx -0.196.$$

Notice that the validity of this calculation depends on the fact that the interval of integration does not contain either of the points $x = 2, -3$ at which the integrand equals zero. □

Example 8 Evaluate $\int \dfrac{dx}{3x^2 + 5x - 2}$.

Solution Since
$$\frac{1}{3x^2 + 5x - 2} = \frac{1}{(3x - 1)(x + 2)} = \frac{1}{3(x - \frac{1}{3})(x + 2)},$$
it follows from formula (5), with $a = -\frac{1}{3}$, $b = 2$, that
$$\int \frac{dx}{3x^2 + 5x - 2} = \frac{1}{3[2 - (-\frac{1}{3})]} \ln\left|\frac{x - \frac{1}{3}}{x + 2}\right| = \frac{1}{7} \ln\left|\frac{x - \frac{1}{3}}{x + 2}\right| + C.$$

As an exercise, show that this can also be written in the form
$$\int \frac{dx}{3x^2 + 5x - 2} = \frac{1}{7} \ln\left|\frac{3x - 1}{x + 2}\right| + C$$
(see the argument at the end of Example 3). □

Example 9 Evaluate $\int \dfrac{dx}{4x^2 - 9}$.

Solution By rule (iv), page 252, and formula (6) with $a = 3$, we have
$$\int \frac{dx}{4x^2 - 9} = \int \frac{dx}{(2x)^2 - 3^2} = \frac{1}{2(2)(3)} \ln\left|\frac{2x - 3}{2x + 3}\right| + C$$
$$= \frac{1}{12} \ln\left|\frac{2x - 3}{2x + 3}\right| + C. \quad \square$$

In all but one of the above examples the integrand is a simple rational function. A technique for evaluating the integral of an *arbitrary* rational function will be given in Section 7.6.

Problems

Evaluate

1. $\int \dfrac{dx}{15x + 5}$

2. $\int \dfrac{dx}{2x - 9}$

3. $\int \dfrac{ds}{11 - 7s}$

4. $\int_1^7 \dfrac{dx}{20x + 10}$

5. $\int_2^4 \dfrac{dt}{8 - 5t}$

6. $\int_{-2}^2 \dfrac{du}{12u + 25}$

Use formula (1) to evaluate

7. $\int \dfrac{x - 2}{4x + 3} dx$

8. $\int \dfrac{2x + 1}{2x - 1} dx$

9. $\int \dfrac{6t + 1}{5 - t} dt$

10. $\int_0^3 \dfrac{1 - 3x}{2 + 4x} dx$

11. $\int_{-2}^4 \dfrac{v}{v + 5} dv$

12. $\int_0^1 \dfrac{8w + 4}{8 - 4w} dw$

Use formula (3) to evaluate

13. $\int \dfrac{x^2}{x^3 - 1} dx$

14. $\int \dfrac{x^3 + x}{x^4 + 2x^2 + 3} dx$

15. $\int \dfrac{dx}{x \ln x}$

16. $\int_4^9 \dfrac{dx}{\sqrt{x}(\sqrt{x} + 1)}$

17. $\int_0^\pi \dfrac{\cos x}{2 + \sin x} dx$

18. $\int_{e^2}^{e^4} \dfrac{dx}{x \ln x \cdot \ln (\ln x)}$

Show that

19. $\int \cot x \, dx = \ln |\sin x| + C$

20. $\int \sec x \csc x \, dx = \int \dfrac{dx}{\cos x \sin x} = \ln |\tan x| + C$

Use formula (5) or (6) to evaluate

21. $\int \dfrac{dx}{x^2 - 4x - 5}$

22. $\int \dfrac{dx}{4x^2 + 4x - 3}$

23. $\int \dfrac{dx}{9x^2 - 25}$

24. $\int_0^1 \dfrac{dx}{x^2 - 5x + 6}$

25. $\int_{-3}^0 \dfrac{dy}{y^2 + 3y - 4}$

26. $\int_{-2}^1 \dfrac{dz}{16 - z^2}$

6.3 The Exponential Function; Exponentials to Any Base

The Exponential as the Inverse of the Logarithm

The logarithm function $f(x) = \ln x$, defined in the preceding section, is increasing and continuous on the interval $(0, \infty)$, which it maps onto the interval $(-\infty, \infty)$. Therefore f has an increasing and continuous inverse function f^{-1} on the interval $(-\infty, \infty)$, and f^{-1} maps $(-\infty, \infty)$ onto the interval $(0, \infty)$. The function f^{-1} is one of the most important functions in mathematics. It is called the *exponential to the base e*, or simply the *exponential*, and is denoted by

$$\exp x.$$

The function $\exp x$ is defined for all x, since its domain is the range of $\ln x$, namely the interval $(-\infty, \infty)$. Moreover, $\exp x$ is *positive* for all x, since its range is the domain of $\ln x$, namely the interval $(0, \infty)$. Since each of the functions $\exp x$ and $\ln x$ is the inverse of the other, we have the identities

$$\exp (\ln x) = x \quad (x > 0), \qquad \ln (\exp x) = x \quad (\text{all } x). \tag{1}$$

In particular, it follows from the first of these identities and the formulas

$$\ln 1 = 0, \qquad \ln e = 1$$

that

$$\exp 0 = 1 \tag{2}$$

and

$$\exp 1 = e. \tag{3}$$

As is true for every one-to-one function and its inverse, the graph of each of the functions $\ln x$ and $\exp x$ is the reflection of the other in the line $y = x$ (see Figure 5). It is apparent from the figure that $\exp x$ is positive and increasing on $(-\infty, \infty)$, and satisfies the conditions (2) and (3). Since $\exp x$ is increasing on $(-\infty, \infty)$ and has the range $(0, \infty)$, we see at once that

$$\lim_{x \to \infty} \exp x = \infty, \tag{4}$$

$$\lim_{x \to -\infty} \exp x = 0, \tag{5}$$

and this behavior is also apparent from the figure. It follows from (5) that $\exp x$ has the x-axis as an asymptote.

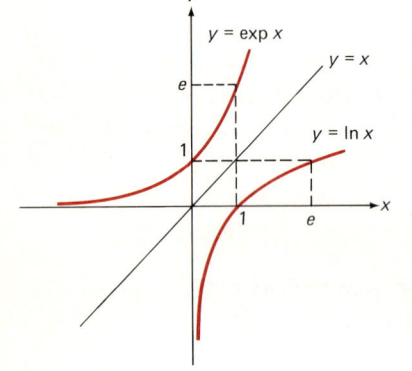

Figure 5

The Exponential of a Sum

Next we establish a key property of the exponential, which it "inherits" from the formula for the logarithm of a product.

312 Chapter 6 Logarithms and Exponentials

Theorem 3 *(Exponential of a sum).* *The formula*

$$\exp(x+y) = (\exp x)(\exp y) \tag{6}$$

holds for arbitrary x and y.

Proof Let $X = \exp x$, $Y = \exp y$, so that $x = \ln X$, $y = \ln Y$. Then

$$x + y = \ln X + \ln Y = \ln XY,$$

by Theorem 1, page 302, and therefore

$$\exp(x+y) = \exp(\ln(XY)) = XY = (\exp x)(\exp y). \quad \blacksquare$$

According to (6), the exponential of a sum of two terms is the product of the exponentials of the separate terms. Equivalently, reading (6) from right to left, we see that the product of two given exponentials is itself an exponential, whose argument is the sum of the arguments of the given exponentials.

Let r be any rational number. Then, choosing $x = e$ in formula (9), page 303, we find that

$$\ln e^r = r \ln e = r,$$

and therefore

$$\exp r = e^r.$$

But $\exp x$ is also defined for *irrational* x, although for such x we have not yet assigned meaning to e^x. We now do so, by the simple expedient of *defining*

$$e^x = \exp x \tag{7}$$

for all x, rational or irrational. Moreover, for the reason given in Problem 41, this is the only possible definition of e^x if the function e^x is to be continuous. Thus we have now managed to assign a unique meaning to irrational powers of e, such as $e^{\sqrt{2}}$ or e^π. To appreciate this accomplishment, consider the problem of interpreting

$$e^{\sqrt{2}} = (2.718281828459\ldots)^{1.414213562373\ldots}$$

without the help of calculus.

The notation e^x will henceforth be preferred to $\exp x$, but the latter notation will occasionally come in handy. In terms of e^x, the identities (1) take the compact form

$$e^{\ln x} = x \quad (x > 0), \qquad \ln e^x = x \quad (\text{all } x).$$

Similarly, formula (6) can be written (from right to left) as

$$e^x e^y = e^{x+y}. \tag{8}$$

The validity of (8) for rational x and y was known before making the definition (7), but we now see that (8) holds for arbitrary real numbers, in particular for irrational numbers. Setting $y = -x$ in (8), we get

$$e^x e^{-x} = e^{x-x} = e^0 = 1,$$

which immediately implies

$$e^{-x} = \frac{1}{e^x}. \tag{9}$$

Section 6.3 The Exponential Function; Exponentials to Any Base

The function e^{-x} is important in its own right. Figure 6 shows the graphs of e^x and e^{-x} in the same system of rectangular coordinates. Notice that each graph is the reflection of the other in the y-axis.

In terms of e^x, formulas (4) and (5) become

$$\lim_{x \to \infty} e^x = \infty \tag{4'}$$

and

$$\lim_{x \to -\infty} e^x = 0. \tag{5'}$$

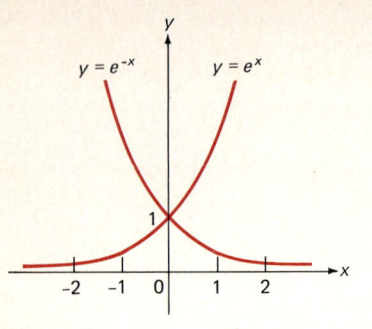

Figure 6

Note that (5') is an immediate consequence of (4'), since

$$\lim_{x \to -\infty} e^x = \lim_{t \to \infty} e^{-t} = \lim_{t \to \infty} \frac{1}{e^t} = 0,$$

with the help of the substitution $x = -t$ and formula (9). It should also be noted that

$$\lim_{x \to \infty} e^{-x} = \lim_{t \to -\infty} e^t = 0, \qquad \lim_{x \to -\infty} e^{-x} = \lim_{t \to \infty} e^t = \infty,$$

as is apparent from the graph of e^{-x}.

The Derivative and Integral of e^x

To differentiate the exponential function, we use Theorem 4, page 286, observing that the conditions of the theorem are satisfied. Writing $y = e^x$ and $x = \ln y$, we have

$$\frac{d}{dx} e^x = \frac{dy}{dx} = \frac{1}{\frac{dx}{dy}} = \frac{1}{\frac{d}{dy} \ln y} = \frac{1}{\frac{1}{y}} = y = e^x,$$

so that

$$\frac{d}{dx} e^x = e^x. \tag{10}$$

As this basic formula shows, the function e^x has the remarkable property of being its own derivative, and therefore of being unaffected by any number of consecutive differentiations. Thus

$$\frac{d^n}{dx^n} e^x = e^x$$

for every positive integer n. It follows from (10) that

$$\int e^x \, dx = e^x + C. \tag{11}$$

Also observe that

$$\frac{d^2}{dx^2} e^x = e^x > 0,$$

so that e^x is concave upward on $(-\infty, \infty)$, by the concavity test.

314 Chapter 6 Logarithms and Exponentials

Example 1 Differentiate xe^x.

Solution By the product rule,
$$\frac{d}{dx}(xe^x) = \frac{dx}{dx}e^x + x\frac{de^x}{dx} = e^x + xe^x. \quad \square$$

Example 2 Differentiate $\sqrt{1+e^x}$.

Solution By the chain rule,
$$\frac{d}{dx}\sqrt{1+e^x} = \frac{1}{2\sqrt{1+e^x}}\frac{d}{dx}(1+e^x) = \frac{e^x}{2\sqrt{1+e^x}}. \quad \square$$

Example 3 Evaluate $\int \frac{e^{3x}+1}{e^x+1}dx$.

Solution Dividing the numerator by the denominator, we get
$$\frac{e^{3x}+1}{e^x+1} = \frac{(e^x)^3+1}{e^x+1} = (e^x)^2 - e^x + 1 = e^{2x} - e^x + 1.$$

Therefore
$$\int \frac{e^{3x}+1}{e^x+1}dx = \int (e^{2x} - e^x + 1)dx = \frac{1}{2}e^{2x} - e^x + x + C,$$

with the help of (11). $\quad \square$

The Exponential to the Base a

Having assigned meaning to arbitrary real powers of the number e, we now wish to do the same for any positive number a. What we need is a continuous function $\exp_a x$ which takes the value a^r when x is a rational number r. The appropriate choice is

$$\exp_a x = \exp(x \ln a) = e^{x \ln a}.$$

In fact, since
$$\ln a^r = r \ln a, \tag{12}$$

for rational r, by Theorem 2, page 303, with $x = a$, it follows that
$$\exp_a r = \exp(r \ln a) = \exp(\ln a^r) = a^r,$$

as desired, and moreover $\exp_a x$ is continuous, since it is a continuous function $\exp x$ of a continuous function $x \ln a$. The function $\exp_a x$ is called the *exponential to the base a*, and reduces to e^x if $a = e$. To assign meaning to a^x for arbitrary real x, in particular for irrational x, we now set

$$a^x = \exp_a x = e^{x \ln a} \quad (a > 0), \tag{13}$$

by *definition*, in complete analogy with (7).

The function a^x, as defined by (13), "inherits" its properties from the corresponding properties of e^x. For example,
$$a^{-x} = e^{-x \ln a} = \frac{1}{e^{x \ln a}},$$

so that
$$a^{-x} = \frac{1}{a^x}.$$

Section 6.3 The Exponential Function; Exponentials to Any Base   **315**

Moreover,
$$a^x a^y = e^{x \ln a} e^{y \ln a} = e^{x \ln a + y \ln a} = e^{(x+y) \ln a},$$
and therefore
$$a^x a^y = a^{x+y}. \tag{14}$$

Taking logarithms of both sides of (13), we find that
$$\ln a^x = \ln (e^{x \ln a}).$$
It follows that
$$\ln a^x = x \ln a,$$
which generalizes formula (12) from the case where x is a rational number r to the case of arbitrary real x.

Another important property of a^x is given by the formula
$$(a^x)^y = a^{xy}, \tag{15}$$
whose validity for rational x and y is already known. To prove (15) for arbitrary real x and y, we first use (13) with a and x replaced by e^x and y to deduce
$$(e^x)^y = e^{y \ln e^x} = e^{yx} = e^{xy}, \tag{15'}$$
which is just the form taken by (15) for $a = e$. But then
$$(a^x)^y = (e^{x \ln a})^y = e^{xy \ln a} = a^{xy},$$
which is (15) for general $a > 0$. The validity of formulas (14) and (15) for arbitrary real x and y testifies further to the aptness of the definition (13). Furthermore
$$\frac{a^x}{a^y} = a^x a^{-y},$$
and therefore
$$\frac{a^x}{a^y} = a^{x-y},$$
while if b is another positive number, then
$$a^x b^x = e^{x \ln a} e^{x \ln b} = e^{x(\ln a + \ln b)} = e^{x \ln ab},$$
which implies
$$a^x b^x = (ab)^x.$$

Thus, to summarize, we have proved the same *laws of exponents*
$$a^x a^y = a^{x+y}, \quad \frac{a^x}{a^y} = a^{x-y}, \quad (a^x)^y = a^{xy}, \quad (ab)^x = a^x b^x$$
as on page 13, but this time for *arbitrary real* exponents x and y, that is, for both rational and irrational exponents x and y.

The behavior of the function a^x depends in an essential way on whether the positive number a is greater than or less than 1 (note that $a^x \equiv 1$ if $a = 1$). Let $t = x \ln a$, so that (13) takes the concise form $a^x = e^t$. If $a > 1$, then

316 Chapter 6 Logarithms and Exponentials

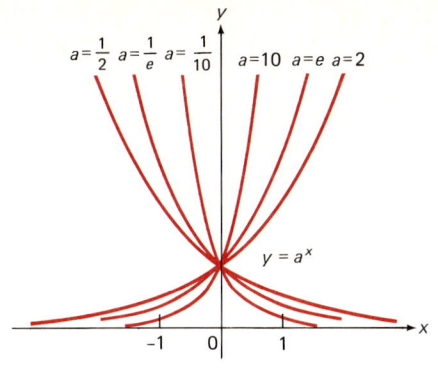

Figure 7

$\ln a > 0$. Therefore t has the same sign as x, and

$$\lim_{x \to \infty} a^x = \lim_{t \to \infty} e^t = \infty, \qquad \lim_{x \to -\infty} a^x = \lim_{t \to -\infty} e^t = 0 \qquad (a > 1), \qquad (16)$$

since $x \to \pm\infty$ implies $t \to \pm\infty$. On the other hand, if $0 < a < 1$, then $\ln a < 0$, so that t has the opposite sign from x, and

$$\lim_{x \to \infty} a^x = \lim_{t \to -\infty} e^t = 0, \qquad \lim_{x \to -\infty} a^x = \lim_{t \to \infty} e^t = \infty \qquad (0 < a < 1), \qquad (16')$$

instead of (16), since now $x \to \pm\infty$ implies $t \to \mp\infty$. This basic difference between the behavior of the function a^x for $a > 1$ and its behavior for $0 < a < 1$ is illustrated in Figure 7, where we graph a^x for $a = 2, e, 10, 1/2, 1/e, 1/10$ in the same system of rectangular coordinates. Notice that each of the pair of curves $y = a^x$ and $y = (1/a)^x$ is a reflection of the other in the y-axis. This is an immediate consequence of the fact that $(1/a)^x = a^{-x}$. Explain why the curves $y = a^x$ all go through the point $(0, 1)$, but have no other points in common.

The Derivative and Integral of a^x

The derivative of the function a^x is easily found. In fact, by (10) and the chain rule

$$\frac{d}{dx} a^x = \frac{d}{dx} e^{x \ln a} = e^{x \ln a} \frac{d}{dx}(x \ln a),$$

and therefore

$$\frac{d}{dx} a^x = a^x \ln a. \qquad (17)$$

It is an immediate consequence of (17) that

$$\int a^x \, dx = \frac{a^x}{\ln a} + C. \qquad (18)$$

Example 4 Differentiate $x\left(\dfrac{1}{3}\right)^x$.

Solution By the product rule and (17),

$$\frac{d}{dx}\left[x\left(\frac{1}{3}\right)^x\right] = \left(\frac{1}{3}\right)^x + x\left(\frac{1}{3}\right)^x \ln \frac{1}{3} = \left(\frac{1}{3}\right)^x (1 - x \ln 3). \qquad \square$$

Example 5 Differentiate $2^{\sin x}$.

Solution By (17) and the chain rule,

$$\frac{d}{dx} 2^{\sin x} = 2^{\sin x} \ln 2 \, \frac{d}{dx} \sin x = 2^{\sin x} \cos x \ln 2. \quad \square$$

Example 6 Evaluate $\int_{-1}^{1} 10^x \, dx$.

Solution By formula (18),

$$\int_{-1}^{1} 10^x \, dx = \frac{10^x}{\ln 10} \bigg|_{-1}^{1} = \frac{10^1 - 10^{-1}}{\ln 10} = \frac{9.9}{\ln 10} \approx 4.3. \quad \square$$

Example 7 Show that $\lim_{x \to 0} \dfrac{a^x - 1}{x} = \ln a$.

Solution This is an indeterminacy of the form 0/0, which can be resolved by using L'Hospital's rule:

$$\lim_{x \to 0} \frac{a^x - 1}{x} = \lim_{x \to 0} \frac{D_x(a^x - 1)}{D_x x} = \lim_{x \to 0} a^x \ln a = \ln a. \quad \square$$

Problems

1. Verify that the graph of the function ce^x ($c > 0$) can be obtained by subjecting the graph of e^x to a horizontal shift.

Differentiate

2. e^{4x+5} **3.** e^{-6x} **4.** xe^{2x}

5. $x^2 e^x$ **6.** $\dfrac{e^x - 1}{e^x + 1}$ **7.** e^{x^2}

8. $e^x \ln x$ **9.** $e^{1/x}$ **10.** $\dfrac{e^x}{x}$

11. $e^{\sqrt{x}}$ **12.** $\cos(e^x)$ **13.** $e^{\tan x}$
14. $\arcsin(e^{x/2})$ **15.** $x \, 10^x$ **16.** 3^{-x}
17. 5^{x^2} **18.** $\pi^{x^2 - x}$ **19.** $\exp_2(4^x)$

20. $\ln |e^x - 1|$ **21.** $\dfrac{10^x - 1}{5^x}$ **22.** $\exp(e^t)$

23. $\exp(\ln u - u)$ **24.** $\ln(\sqrt{e^v})$

Find
25. The third derivative of xe^{x^2}
26. The fourth derivative of $x^2 e^{2x}$
27. The fifth derivative of $e^x \ln x$
Hint. In Problems 26 and 27 it is best to use Leibniz's rule (Problem 35, page 229).

28. Show that the function e^x has no asymptotes other than the x-axis.

29. Use logarithmic differentiation to find the derivative of $e^{x^2 + 2x}/(2^x \ln x)$.
30. Find the average of the function e^x over the interval $[\ln a, \ln b]$, where $0 < a < b$.
31. The function e^x has the property of being its own derivative. Show that every other function $y = f(x)$ with this property is of the form ce^x, where c is a constant.
32. Verify that

$$\int xe^x \, dx = xe^x - e^x + C.$$

Find the area A of the region R
33. Bounded by the line $x = 1$ and by the curves $y = e^x$ and $y = e^{-x}$
34. Bounded by the lines $x = 1$ and $x = 2$ and by the curves $y = \ln x$ and $y = e^{x/2}$
35. Between the curves $y = xe^{1-x}$ and $y = 4x^2 - 3x$
36. Bounded by the line $x = 1$ and by the curves $y = 2^x$ and $y = 4^x$
In each case sketch the region R.

Find all local extrema and inflection points of the given function. On which intervals is the function increasing? Decreasing? Concave upward? Concave downward? Find all absolute extrema and asymptotes. Graph the function.
37. $f(x) = e^{-x^2/2}$ **38.** $f(x) = 2^x + 2^{-x}$
39. $f(x) = 4xe^{-2x}$ **40.** $f(x) = \exp(-e^{-x})$

41. Let h be a continuous function defined on $(-\infty, \infty)$, and suppose $h(x) = 0$ for all rational x. Show that $h(x) = 0$ for all irrational x as well, so that $h(x) \equiv 0$. Use this to show that a continuous function defined on $(-\infty, \infty)$ is uniquely determined by its values for rational x.
Hint. By considering $h = f - g$, show that two continuous functions f and g which coincide for all rational x are identically equal.

42. Solve the equation $2^x - 2x = 0$.

43. Of all the lines tangent to the curve $y = e^x$ only one goes through the origin. Find this line.

Without attempting to evaluate any integrals, determine which integral is larger.

44. $\int_0^1 e^x \, dx$ or $\int_0^1 e^{x^2} \, dx$

45. $\int_0^1 e^{-x} \, dx$ or $\int_0^1 e^{-x^2} \, dx$

46. $\int_{-2}^{-1} (\frac{1}{3})^x \, dx$ or $\int_{-2}^{-1} 3^x \, dx$

47. $\int_1^e \sqrt{\ln x} \sin x \, dx$ or $\int_1^e \ln x \sin x \, dx$

48. Show that
$$2e^{-1/4} < \int_0^2 e^{x^2 - x} \, dx < 2e^2.$$

49. Verify that the function $y = ae^{2x} + be^{3x}$ satisfies the second order differential equation $y'' - 5y' + 6y = 0$ for arbitrary constants a and b.

Evaluate

50. $\int \dfrac{e^x}{e^x + 1} \, dx$

51. $\int x a^x \, dx$

52. $\int \dfrac{e^{4x} - 1}{e^x - 1} \, dx$

53. $\int_0^1 (3^x + 3^{-x}) \, dx$

54. $\int_2^4 \dfrac{dx}{1 - e^{-x}}$

55. $\int_{-1}^3 5^x \, dx$

56. $\int_0^4 2^{-t} 3^t \, dt$

57. $\int_0^1 4^u e^u \, du$

58. $\int_{-1}^1 v \, 2^v \, dv$

Use L'Hospital's rule to evaluate

59. $\lim\limits_{x \to 0} \dfrac{e^{ax} - e^{bx}}{x}$

60. $\lim\limits_{x \to 1} \dfrac{e^x - e}{x - 1}$

61. $\lim\limits_{x \to 0} \dfrac{e^x - e^{-x}}{\sin x}$

62. $\lim\limits_{x \to \infty} x(e^{1/x} - 1)$

63. $\lim\limits_{x \to 0} \left(\dfrac{1}{x} - \dfrac{1}{e^x - 1} \right)$

64. $\lim\limits_{x \to 0} \dfrac{2^x - 1}{\sqrt{1 + x} - 1}$

65. $\lim\limits_{x \to 0} \dfrac{3^{\tan x} - 1}{\sin x}$

66. $\lim\limits_{x \to \pi/2} \dfrac{e^{\sin x} - e}{\cos x}$

67. $\lim\limits_{x \to 0} \dfrac{4^{\sin x} - 1}{8^{\tan x} - 1}$

6.4 Logarithms to Any Base

In the preceding section we found that a^x, the exponential to the base a (where $a > 0$), is given by the formula

$$a^x = e^{x \ln a}. \tag{1}$$

The inverse of the function (1) is called the *logarithm to the base a*, denoted by $\log_a x$. Here a is positive, but we must exclude the case $a = 1$, since the function $1^x \equiv 1$ is not one-to-one, and hence has no inverse. Since the function a^x has domain $(-\infty, \infty)$ and range $(0, \infty)$, its inverse, the function $\log_a x$, has domain $(0, \infty)$ and range $(-\infty, \infty)$. Thus for all $x > 0$ we have

$$a^{\log_a x} = x, \tag{2}$$

or equivalently

$$e^{\log_a x \cdot \ln a} = x.$$

It follows from the last formula that

$$\log_a x \cdot \ln a = \ln x,$$

Section 6.4 Logarithms to Any Base

and therefore

$$\log_a x = \frac{\ln x}{\ln a}. \tag{3}$$

Note that if $a = e$, (3) reduces to

$$\log_e x = \frac{\ln x}{\ln e} = \ln x,$$

that is, the logarithm to the base e is simply the natural logarithm.

It is apparent from (2) that $\log_a x$ is just the power to which a must be raised to obtain x. We also have the companion formula

$$\log_a a^x = x, \tag{4}$$

which is valid for all x. The formulas (2) and (4) are, of course special cases of the general formulas $f(f^{-1}(x)) \equiv x$ and $f^{-1}(f(x)) \equiv x$, which are satisfied by every one-to-one function f and its inverse f^{-1}.

The properties of $\log_a x$ are analogous to those of $\ln x$, and are an immediate consequence of the definition (3). Thus

$$\log_a 1 = \frac{\ln 1}{\ln a} = 0, \qquad \log_a a = \frac{\ln a}{\ln a} = 1,$$

$$\log_a \frac{1}{x} = \frac{\ln(1/x)}{\ln a} = \frac{-\ln x}{\ln a} = -\log_a x,$$

$$\log_a xy = \frac{\ln xy}{\ln a} = \frac{\ln x + \ln y}{\ln a} = \log_a x + \log_a y,$$

and so on. The formula

$$\log_a b^x = \frac{\ln b^x}{\ln a} = \frac{x \ln b}{\ln a} = x \log_a b \qquad (b > 0)$$

should also be noted. For $a = 10$ we get the logarithm to the base 10 or *common logarithm* $\log_{10} x$ of high school mathematics, which is often denoted by $\log x$ without the subscript 10. The connection between the common logarithm and the natural logarithm is given by the formula

$$\log x = \frac{\ln x}{\ln 10} \approx 0.43429 \ln x.$$

Since $\ln a > 0$ if $a > 1$, it follows from formulas (10) and (11), page 304, and the definition of $\log_a x$ that

$$\lim_{x \to \infty} \log_a x = \infty, \qquad \lim_{x \to 0^+} \log_a x = -\infty \qquad (a > 1).$$

On the other hand, since $\ln a < 0$ if $0 < a < 1$, we have

$$\lim_{x \to \infty} \log_a x = -\infty, \qquad \lim_{x \to 0^+} \log_a x = \infty \qquad (0 < a < 1).$$

This basic difference between the behavior of the function $\log_a x$ for $a > 1$ and its behavior for $0 < a < 1$ is illustrated in Figure 8, where we graph $\log_a x$ for $a = 2, e, 10, 1/2, 1/e, 1/10$ in the same system of rectangular coordinates. Notice that each of the pair of curves $y = \log_a x$ and $y = \log_{1/a} x$ is the reflection of the other in the x-axis. This is an immediate consequence of the

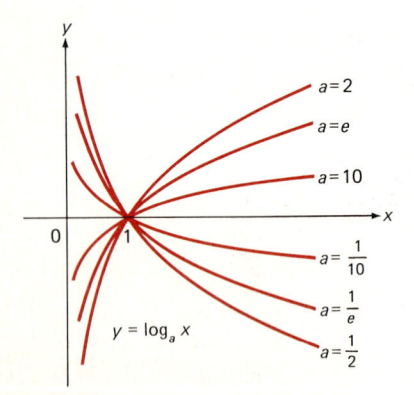

Figure 8

320   Chapter 6   Logarithms and Exponentials

fact that
$$\log_{1/a} x = \frac{\ln x}{\ln (1/a)} = \frac{\ln x}{-\ln a} = -\log_a x.$$

Explain why the curves $y = \log_a x$ all go through the point $(1, 0)$, but have no other points in common.

It follows from formulas (12) and (13), page 305, and the definition of $\log_a x$ that

$$\lim_{x \to \infty} \frac{\log_a x}{x} = 0, \qquad \lim_{x \to 0^+} x \log_a x = 0,$$

both for $a > 1$ and for $0 < a < 1$.

If a and b are two positive numbers different from 1, then

$$\log_a x = \frac{\ln x}{\ln a} = \frac{\ln b}{\ln a} \frac{\ln x}{\ln b},$$

so that

$$\log_a x = \log_a b \cdot \log_b x.$$

In particular, choosing $x = a$, we find that

$$1 = \log_a b \cdot \log_b a,$$

or equivalently

$$\log_a b = \frac{1}{\log_b a}. \tag{5}$$

For $b = e$ this becomes

$$\log_a e = \frac{1}{\log_e a} = \frac{1}{\ln a}. \tag{5'}$$

Example 1 Find $\log_2 64$ and $\log_{64} 2$.

Solution We have

$$\log_2 64 = \log_2 2^6 = 6,$$

with the help of (4). It then follows from (5) that

$$\log_{64} 2 = \frac{1}{\log_2 64} = \frac{1}{6}.$$

Actually, since $\log_a x$ is the power to which the number a must be raised to get the number x, we can calculate $\log_2 64$ and $\log_{64} 2$ at once mentally, by observing that $2^6 = 64$, $64^{1/6} = 2$. □

The derivative of the function $\log_a x$ is easily found. In fact,

$$\frac{d}{dx} \log_a x = \frac{d}{dx} \frac{\ln x}{\ln a} = \frac{1}{x \ln a},$$

or equivalently

$$\frac{d}{dx} \log_a x = \frac{1}{x} \log_a e, \tag{6}$$

with the help of (5').

Example 2 Differentiate $\log_3 (\sin x)$.

Solution By (6) and the chain rule,

$$\frac{d}{dx} \log_3 (\sin x) = \frac{1}{\sin x} \log_3 e \frac{d}{dx} \sin x = \frac{\cos x}{\sin x} \log_3 e = \cot x \log_3 e. \quad \square$$

Example 3 Differentiate $\log_x a$.

Solution This time the independent variable x is the *base* of the logarithm. By (3) and the chain rule,

$$\frac{d}{dx} \log_x a = \frac{d}{dx} \frac{\ln a}{\ln x} = -\frac{\ln a}{(\ln x)^2} \frac{d}{dx} \ln x = -\frac{\ln a}{x(\ln x)^2}. \quad \square$$

Example 4 Show that

$$\lim_{x \to 0} \frac{\log_a (1+x)}{x} = \log_a e. \tag{7}$$

Solution This is an indeterminacy of the form 0/0, which we resolve by using L'Hospital's rule:

$$\lim_{x \to 0} \frac{\log_a (1+x)}{x} = \lim_{x \to 0} \frac{D_x \log_a (1+x)}{D_x x} = \lim_{x \to 0} \frac{\log_a e}{1+x} = \log_a e.$$

If $a = e$, formula (7) becomes

$$\lim_{x \to 0} \frac{\ln (1+x)}{x} = \log_e e = 1. \tag{8}$$

Alternatively, since

$$\lim_{x \to 0} \frac{\ln (1+x)}{x} = \lim_{x \to 0} \frac{\ln (1+x) - \ln 1}{x},$$

we recognize that the limit in (8) is actually the derivative of $\ln x$ at the point $x = 1$, so that

$$\lim_{x \to 0} \frac{\ln (1+x)}{x} = \frac{d \ln x}{dx}\bigg|_{x=1} = \frac{1}{x}\bigg|_{x=1} = 1.$$

As an exercise, prove formula (7) in the same way. $\quad \square$

Problems

Simplify
1. $\log_2 1024$
2. $\log_{10} (0.001)$
3. $\log_3 \frac{1}{81}$
4. $\log_{81} 3$
5. $\log_{1/2} \sqrt{2}$
6. $\log_4 (0.0625)$
7. $\log_2 3 \cdot \log_3 4$
8. $\log_5 (2.5 \times 10^4)$
9. $\log_{0.1} (0.2)$
10. $\log_3 (\log_3 27)$
11. $\log_\pi \pi x^2 \quad (x \neq 0)$
12. $\log_x \pi x^2 \quad (x > 0)$
13. $\log_{\sqrt{2}} \sqrt{3}$
14. $\log_{\sqrt{e}} (\ln e^e)$
15. $\log_2 (\log_2 (\log_2 16))$
16. $\log_2 (\log_4 (\log_8 64))$
17. $\log_2 (\log_3 (\log_4 64))$
18. $\log_2 (3^{\ln 4})$

19. Show that if $1 < a < b$ or $0 < a < b < 1$, then $\log_a x > \log_b x$, provided that $x > 1$. Does this conclusion remain true if $0 < x < 1$?

20. Show that if $\log_a y$ is a nonconstant linear function of x, then y is proportional to an exponential function of x.

322   Chapter 6   Logarithms and Exponentials

Show that the converse is true if the constant of proportionality is positive.

21. Express y as a function of x if $\log_3 y = 1 - 2x$. Express $\log_4 y$ as a function of x if $y = \frac{1}{8}(2^x)$.

Find all x for which the given function is defined.

22. $\log_5 (x + 1) + \log_{0.5} (x + 2)$
23. $\sqrt{\log_a x}$
24. $\arcsin (1 - x) + \log_2 (\log_2 x)$
25. $\log_{10} (1 - \log_{10} (x^2 - 5x + 16))$
26. $\log_2 (\log_3 (\log_4 x))$
27. $\arcsin (\log_{10}(x/10))$

Differentiate

28. $\log_\pi |x|$
29. $x \log_{10} x$
30. $x^2 \log_3 x$
31. $\log_4 (2^{\ln x})$
32. $\dfrac{x}{\log_2 x}$
33. $5^{\log_7 x}$

34. Show that the derivative of $\log_a (\log_b x)$ is independent of the choice of b.

Evaluate

35. $\displaystyle\int \log_2 x \, dx$
36. $\displaystyle\int_1^{10} \log_{10} x \, dx$

Use L'Hospital's rule to evaluate

37. $\displaystyle\lim_{x \to 0} \dfrac{\log_3 (1 + 2x)}{\log_2 (1 + 3x)}$
38. $\displaystyle\lim_{x \to 0} \dfrac{\log_\pi (1 + \sin x)}{\tan x}$
39. $\displaystyle\lim_{x \to 0} \dfrac{\log_{10} (1 + x)}{10^x - 1}$
40. $\displaystyle\lim_{x \to 0} \dfrac{\log_2 (x^2 + 1)}{x \sin x}$
41. $\displaystyle\lim_{x \to e} \dfrac{\log_2 (\ln x)}{x - e}$
42. $\displaystyle\lim_{x \to 0^+} \dfrac{\log_4 (\tan x)}{\log_5 (\sin x)}$

43. A number a is said to be k *orders of magnitude* larger than a number b if $a \approx 10^k b$, and by the same token, b is then said to be k orders of magnitude smaller than a. This language is particularly useful in the physical and biological sciences. If a is k orders of magnitude larger than b, what is the relation between $\log_{10} a$ and $\log_{10} b$?

44. How does the speed of light ($\approx 186{,}000$ miles per second) compare in order of magnitude with the speed of sound (≈ 1150 feet per second)?

45. How does the weight of a mouse (≈ 1 oz) compare in order of magnitude with the weight of a man?

46. The pH of an aqueous solution is defined by the formula $\text{pH} = -\log_{10} [H^+]$, where $[H^+]$ is the concentration of hydrogen ions, measured in moles per liter. The pH of pure distilled water is 7, acidic solutions have pH values less than 7, and alkaline solutions have pH values greater than 7. Litmus paper, which is naturally pink in color, turns red in acidic solutions and blue in alkaline solutions. Is litmus paper red or blue in a solution for which $[H^+] = 4 \times 10^{-9}$? In a solution for which $[H^+] = 0.00002$?

47. The *intensity I* of a sound wave is defined as the rate at which acoustic energy is transported by the wave across a unit area perpendicular to the direction of propagation of the wave. The faintest sound audible to the human ear, corresponding to the *threshold of hearing*, has an intensity of about 10^{-16} watt/cm^2, while the loudest sound tolerable to the ear, corresponding to the *threshold of pain*, has an intensity of about 10^{-4} watt/cm^2. Thus the ear responds to sounds whose intensities can differ by a factor of 10^{12} (one trillion). The subjective sensation of *loudness*, denoted by L, seems to be proportional to the logarithm of the intensity I, and is customarily defined by the formula

$$L = 10 \log_{10} \dfrac{I}{I_0},$$

where I_0 is the intensity of the faintest audible sound. When defined in this way, L is said to be measured in *decibels*, abbreviated as dB. It turns out that a 1-dB change in loudness is about the smallest change which the human ear can ordinarily detect. Show that L increases by exactly 10 dB if I is multiplied by 10. What percentage increase in I leads to a 1-dB increase in L?

48. Show that to an excellent approximation, a 3-dB increase in the loudness of a sound corresponds to a doubling of its intensity.

49. What is the loudness in decibels of a sound 50,000 times more intense than the faintest audible sound? Of a sound 200 times less intense than the loudest tolerable sound?

50. Estimate the intensity of the rustle of leaves, with a loudness of about 10 dB. Of busy street traffic, with a loudness of about 70 dB.

51. In ancient times the stars visible to the naked eye were classified into six groups. The stars were assigned *magnitudes* ranging from 1 to 6, with the brightest being assigned magnitude 1 and the faintest magnitude 6. In modern times this classification has been greatly extended, and the formula

$$m_2 - m_1 = 2.5 \log_{10} \dfrac{I_1}{I_2} \qquad \text{(i)}$$

is now used to relate the magnitudes m_1, m_2 and intensities I_1, I_2 of two stars S_1, S_2. How many times brighter than a star of magnitude 6 is a star of magnitude 1? The faintest stars that can be photographed with the 200-inch telescope at Mount Palomar Observatory in California have a magnitude of about 23.5. How many times more sensitive is the telescope than the naked eye?

52. Show that to an excellent approximation, a decrease of 1 in the magnitude of a star corresponds to a 2.5-fold increase in its intensity.

53. Sirius, the brightest star in the sky, has magnitude -1.4, and Canopus, the second brightest star, has magnitude

−0.7. How many times brighter (in terms of intensity) is Sirius than Canopus?

54. A coin is tossed repeatedly. Suppose the probability of the coin coming up heads is p, where the number p lies in the interval $[0, 1]$, since it is a probability. Then the probability of tails is $q = 1 - p$, and q also lies in $[0, 1]$. In the case of a fair coin, heads and tails are equally likely, that is, $p = q = \frac{1}{2}$. We then have no idea in advance what the outcome of a toss of the coin will be, so that one binary digit or *bit* is required to tell us the outcome (1 for heads, say, and 0 for tails). If the coin is completely biased, for example, if $p = 1$, so that the outcome of every toss is heads, then we know in advance what the outcome of a toss will be. Thus 1 bit of information is needed to resolve our uncertainty about the outcome of a coin toss if $p = \frac{1}{2}$, whereas no information at all is required if $p = 1$ (or $q = 1$). According to Claude Shannon, the founder of *information theory*, in the case of arbitrary p the *uncertainty* or *entropy* of a coin toss should be defined as

$$H(p) = -p \log_2 p - q \log_2 q$$
$$= -p \log_2 p - (1 - p) \log_2 (1 - p) \qquad \text{(ii)}$$

bits, and this is the *amount of information* conveyed by a message giving the outcome of the toss. By convention $0 \log_2 0 = 0$, which makes $H(p)$ continuous on $[0, 1]$.

Show that the entropy function $H(p)$ is nonnegative on $[0, 1]$, taking its maximum, equal to 1, for $p = \frac{1}{2}$, and its minimum, equal to 0, for $p = 0$ or $p = 1$. Show that $H(p)$ is concave downward on $[0, 1]$ and symmetric about the line $p = \frac{1}{2}$. The graph of $H(p)$ is shown in Figure 9, and exhibits all these features.

H (in bits)

$H(p) = -p \log_2 p - q \log_2 q$ $(q = 1 - p)$

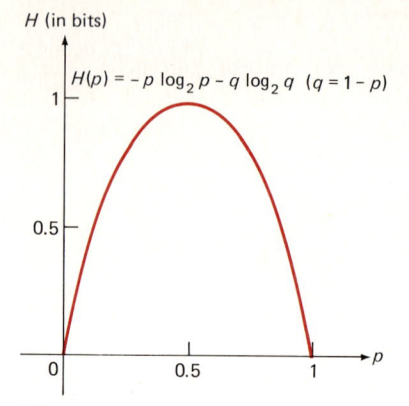

Figure 9

55. Suppose you are given the outcome of one toss of a biased coin which is twice as likely to come up heads as tails. How much information have you been given?

56. In information theory it is shown that there are $\log_2 N$ bits of information in a message telling the outcome of a random experiment with N equally likely outcomes. Suppose you are told the birthday of a complete stranger. How much information have you been given? (Neglect leap years.)

6.5 The General Power Function; More About Indeterminate Forms

We already know what is meant by x^a if a is rational. In fact, if $a = m/n$, where m and $n > 0$ are integers, then x^a means

$$x^{m/n} = \sqrt[n]{x^m}. \qquad (1)$$

We now wish to assign meaning to x^a in the case where a is irrational. The appropriate definition is

$$x^a = e^{a \ln x} \qquad (x > 0), \qquad (2)$$

where it must be assumed that x is positive, so that $\ln x$ is defined. From an algebraic standpoint, formula (2) is just formula (13), page 315, with the roles of a and x interchanged, but the behavior of x^a, regarded as a function of x, is altogether different from that of a^x.

The function (2), called the *general power function*, is continuous on $(0, \infty)$, and takes the value $\sqrt[n]{x^m}$ when a is a rational number m/n. To see this, we substitute $a = m/n$ in (2), obtaining

$$x^{m/n} = e^{(m/n) \ln x} = e^{mn^{-1} \ln x}.$$

It follows that

$$(x^{m/n})^n = (e^{mn^{-1} \ln x})^n = e^{mn^{-1} n \ln x} = e^{m \ln x} = (e^{\ln x})^m,$$

324   Chapter 6   Logarithms and Exponentials

and therefore
$$(x^{m/n})^n = x^m,$$
which is equivalent to formula (1). However, if a is irrational, we cannot represent a as the ratio m/n of two integers, and then formula (2) is the only way of defining x^a.

Remark Let $a = m/n$ be a rational number in lowest terms, with $n > 0$, and suppose n is *odd*. Then formula (1) actually goes further than formula (2), with which it coincides for positive x, since it also defines $x^{m/n}$ for *negative* x, and for $x = 0$ as well if $m > 0$ (in this case $0^{m/n} = 0$). In fact, if n is odd, then $x^{m/n}$ has the same parity as m, that is, $x^{m/n}$ is an even function if m is even and an odd function if m is odd. However, formula (1) does not define $x^{m/n}$ for negative x if n is *even*, since then m is odd (remember that m/n is in lowest terms), and hence if x is negative, so is x^m, and $\sqrt[n]{x^m}$ calls for taking an even root of a negative number, which is impossible.

Thus we have now assigned meaning to irrational powers of x, like $x^{\sqrt{2}}$ and x^e. It follows from (2) that
$$x^{-a} = e^{-a \ln x} = \frac{1}{e^{a \ln x}},$$
and hence
$$x^{-a} = \frac{1}{x^a}. \tag{3}$$

Moreover, if a and b are arbitrary real numbers, then
$$(x^a)^b = (e^{a \ln x})^b = e^{ab \ln x},$$
that is,
$$(x^a)^b = x^{ab}. \tag{4}$$

Similarly,
$$x^a x^b = e^{a \ln x} e^{b \ln x} = e^{a \ln x + b \ln x} = e^{(a+b) \ln x},$$
so that
$$x^a x^b = x^{a+b}. \tag{5}$$

Also, by (3),
$$\frac{x^a}{x^b} = x^a x^{-b},$$
and therefore
$$\frac{x^a}{x^b} = x^{a-b}, \tag{6}$$
with the help of (5). Formulas (4)–(6) are just variants of some of the laws of exponents on page 316.

The behavior of x^a depends in an essential way on the sign of the exponent a (the case $a = 0$ is exceptional, since $x^0 \equiv 1$). Let $t = a \ln x$, so that

Section 6.5 The General Power Function; More About Indeterminate Forms

(2) takes the concise form $x^a = e^t$. If $a > 0$, then t has the same sign as $\ln x$, and

$$\lim_{x \to 0^+} x^a = \lim_{t \to -\infty} e^t = 0, \quad \lim_{x \to \infty} x^a = \lim_{t \to \infty} e^t = \infty \quad (a > 0), \quad (7)$$

since $x \to 0^+$ implies $t \to -\infty$, while $x \to \infty$ implies $t \to \infty$. It is apparent from the first of these formulas that if we set

$$0^a = 0 \quad (a > 0),$$

by definition, then the function x^a, originally defined only on the open interval $(0, \infty)$, will be continuous on the closed interval $[0, \infty)$. Therefore, for positive a, we make the extended definition

$$x^a = \begin{cases} e^{a \ln x} & \text{if } x > 0, \\ 0 & \text{if } x = 0 \end{cases} \quad (a > 0).$$

On the other hand, if $a < 0$, then $t = a \ln x$ has the opposite sign from $\ln x$, and

$$\lim_{x \to 0^+} x^a = \lim_{t \to \infty} e^t = \infty, \quad \lim_{x \to \infty} x^a = \lim_{t \to -\infty} e^t = 0 \quad (a < 0), \quad (7')$$

since now $x \to 0^+$ implies $t \to \infty$, while $x \to \infty$ implies $t \to -\infty$. This basic difference between the behavior of the function x^a for $a > 0$ and its behavior for $a < 0$ is illustrated in Figure 10, where we graph x^a for $a = \pm 1/\sqrt{2}, \pm e$. We have chosen irrational values of a, for which the use of $x^a = e^{a \ln x}$ is indispensable. How do you account for the fact that all the curves $y = x^a$ go through the point $(1, 1)$, but have no other points in common?

To differentiate the function x^a, we use its definition and the chain rule. This gives

$$\frac{d}{dx} x^a = \frac{d}{dx} e^{a \ln x} = e^{a \ln x} \frac{d}{dx}(a \ln x) = x^a \frac{a}{x} = a \frac{x^a}{x}$$

if $x > 0$, which becomes simply

$$\frac{d}{dx} x^a = ax^{a-1}, \qquad (8)$$

after using (6) with $b = 1$. We have been using formula (8) all along for the case of rational a, and we now see that it remains true for irrational a. For example,

$$\frac{d}{dx} x^{\sqrt{2}} = \sqrt{2} x^{\sqrt{2}-1},$$

$$\frac{d}{dx} x^e = ex^{e-1},$$

and so on. It is an immediate consequence of (8) that

$$\int x^a \, dx = \frac{x^{a+1}}{a+1} + C \quad (a \neq -1)$$

for both rational and irrational values of x; for example,

$$\int x^\pi \, dx = \frac{x^{\pi+1}}{\pi+1} + C.$$

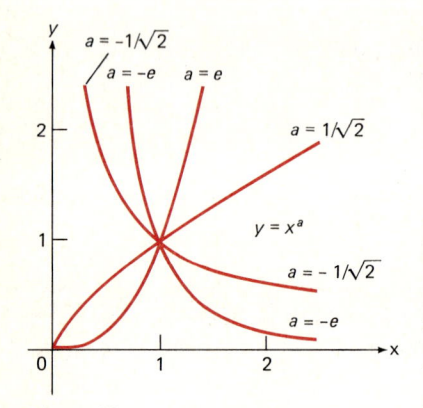

Figure 10

326   Chapter 6   Logarithms and Exponentials

The case $a = -1$ causes no trouble, since

$$\int x^{-1} \, dx = \int \frac{dx}{x} = \ln |x| + C.$$

Thus we are now able to differentiate and integrate an *arbitrary real* power of x.

Example 1 Differentiate x^x.

Solution We have

$$\frac{d}{dx} x^x = \frac{d}{dx} e^{x \ln x} = e^{x \ln x} \frac{d}{dx} (x \ln x) = x^x(\ln x + 1).$$

Equivalently, by logarithmic differentiation,

$$\frac{d}{dx} x^x = x^x \frac{d}{dx} \ln x^x = x^x \frac{d}{dx} (x \ln x) = x^x(\ln x + 1). \quad \square$$

Example 2 Show that

$$\lim_{x \to 1} \frac{x^a - 1}{x - 1} = a \tag{9}$$

for arbitrary a.

Solution This is an indeterminacy of the form $0/0$, which can be resolved by using L'Hospital's rule and formula (8):

$$\lim_{x \to 1} \frac{x^a - 1}{x - 1} = \lim_{x \to 1} \frac{D_x(x^a - 1)}{D_x(x - 1)} = \lim_{x \to 1} ax^{a-1} = a. \quad \square$$

We have already shown on page 305 that

$$\lim_{x \to \infty} \frac{\ln x}{x} = 0, \quad \lim_{x \to 0^+} x \ln x = 0.$$

The next two examples show that these formulas remain true if x is replaced by any positive power of x.

Example 3 Show that

$$\lim_{x \to \infty} \frac{\ln x}{x^a} = 0 \tag{10}$$

for every $a > 0$.

Solution This is an indeterminacy of the form ∞/∞, which we resolve by L'Hospital's rule, using (8) and the second of the formulas (7):

$$\lim_{x \to \infty} \frac{\ln x}{x^a} = \lim_{x \to \infty} \frac{D_x \ln x}{D_x x^a} = \lim_{x \to \infty} \frac{1/x}{ax^{a-1}} = \lim_{x \to \infty} \frac{1}{ax^a} = 0.$$

According to (10), as $x \to \infty$, the logarithm $\ln x$ grows more slowly than any positive power of x, no matter how small. Thus, for example,

$$\lim_{x \to \infty} \frac{\ln x}{x^{0.001}} = 0. \quad \square$$

Section 6.5 The General Power Function; More About Indeterminate Forms

Example 4 Show that

$$\lim_{x \to 0^+} x^a \ln x = 0$$

for every $a > 0$.

Solution This time we have an indeterminacy of the form $0 \cdot \infty$, which we resolve with the help of the preceding example and the substitution $x = 1/t$:

$$\lim_{x \to 0^+} x^a \ln x = \lim_{t \to \infty} \left(\frac{1}{t}\right)^a \ln \frac{1}{t} = -\lim_{t \to \infty} \frac{\ln t}{t^a} = 0. \quad \square$$

The Indeterminate Forms 0^0, ∞^0 and 1^∞

On page 187 it was stated that 0^0, ∞^0 and 1^∞ are indeterminate forms. We are now in a position to investigate these indeterminacies. Let F and G be continuous functions, where F is also positive, and consider the limit as $x \to a$ of the expression $[F(x)]^{G(x)}$. (As usual in dealing with indeterminate forms, we write $x \to a$ only for convenience, and other possibilities are $x \to a^+$, $x \to a^-$, $x \to \infty$ and $x \to -\infty$.) Then

$$\lim_{x \to a} [F(x)]^{G(x)} = \lim_{x \to a} e^{G(x) \ln F(x)} = e^L, \quad (11)$$

where

$$L = \lim_{x \to a} [G(x) \ln F(x)], \quad (12)$$

provided that the limit L exists and is finite. This is an immediate consequence of Theorem 11, page 87, with $f(x) = G(x) \ln F(x)$ and $g(x) = e^x$. (It is easy to see that the limit (11) is ∞ if $L = \infty$ and 0 if $L = -\infty$.) But (12), and therefore (11), is indeterminate if either of the functions $\ln F(x)$ or $G(x)$ approaches zero, while the other approaches infinity. This can happen in three ways, namely

(i) $F(x) \to 0^+$, or equivalently $\ln F(x) \to -\infty$, and $G(x) \to 0$,
(ii) $F(x) \to \infty$, or equivalently $\ln F(x) \to \infty$, and $G(x) \to 0$,
(iii) $F(x) \to 1$, or equivalently $\ln F(x) \to 0$, and $G(x) \to \infty$,

corresponding to the indeterminate forms 0^0, ∞^0 and 1^∞, respectively. Thus evaluation of these forms reduces to resolution of an indeterminacy of the form $0 \cdot \infty$.

Example 5 Evaluate $\lim\limits_{x \to 0^+} x^x$.

Solution To resolve this indeterminacy, of the form 0^0, we observe that

$$\lim_{x \to 0^+} x^x = \lim_{x \to 0^+} e^{x \ln x} = e^L,$$

where

$$L = \lim_{x \to 0^+} x \ln x.$$

But we already know that $L = 0$, and hence

$$\lim_{x \to 0^+} x^x = e^0 = 1. \quad \square$$

328 Chapter 6 Logarithms and Exponentials

Example 6 Evaluate $\lim_{x \to \infty} x^{1/x}$.

Solution This time the indeterminacy is of the form ∞^0. Observing that
$$\lim_{x \to \infty} x^{1/x} = \lim_{x \to \infty} e^{(1/x)\ln x} = e^L,$$
where
$$L = \lim_{x \to \infty} \frac{\ln x}{x},$$
we immediately have
$$\lim_{x \to \infty} x^{1/x} = e^0 = 1,$$
since $L = 0$, as we already know. $\square$

Example 7 Evaluate $\lim_{x \to 0} (1 + x)^{1/x}$.

Solution We now have an indeterminacy of the form 1^∞. Since
$$(1 + x)^{1/x} = e^{(1/x)\ln(1+x)},$$
we see that
$$\lim_{x \to 0} (1 + x)^{1/x} = e^L,$$
where
$$L = \lim_{x \to 0} \frac{\ln(1+x)}{x}.$$
But $L = 1$, by Example 4, page 322, and thus
$$\lim_{x \to 0} (1 + x)^{1/x} = e^1 = e. \quad \square \tag{13}$$

Example 8 Evaluate $\lim_{x \to \infty} \left(1 + \frac{1}{x}\right)^x$.

Solution This is again an indeterminacy of the form 1^∞, and in fact is just a variant of the limit considered in the preceding example. Making the substitution $t = 1/x$, we find that
$$\lim_{x \to \infty} \left(1 + \frac{1}{x}\right)^x = \lim_{t \to 0^+} (1 + t)^{1/t},$$
and therefore
$$\lim_{x \to \infty} \left(1 + \frac{1}{x}\right)^x = e, \tag{14}$$
with the help of (13). Restricting the values of x to positive integers n, we get the important formula
$$\lim_{n \to \infty} \left(1 + \frac{1}{n}\right)^n = e, \tag{14'}$$
expressing e as the limit of an "infinite sequence." In Example 11, page 491, we will say more about the meaning of this formula. $\square$

Example 9 Evaluate $\lim_{x \to \infty} \left(1 + \dfrac{a}{x}\right)^x$, where a is an arbitrary number.

Solution To resolve this indeterminacy of the form 1^∞, we make the substitution $t = a/x$, obtaining

$$\lim_{x \to \infty} \left(1 + \frac{a}{x}\right)^x = \lim_{t \to 0^+} (1 + t)^{a/t} = \lim_{t \to 0^+} e^{(a/t)\ln(1+t)} = e^L,$$

where

$$L = \lim_{t \to 0^+} a\,\frac{\ln(1+t)}{t} = a \cdot 1 = a.$$

It follows that

$$\lim_{x \to \infty} \left(1 + \frac{a}{x}\right)^x = e^a. \tag{15}$$

For $a = 1$ this reduces to formula (14). □

Remark Formula (15) remains true if we replace $x \to \infty$ by $x \to -\infty$. In fact, making the substitution $x = -t$, we have

$$\lim_{x \to -\infty} \left(1 + \frac{a}{x}\right)^x = \lim_{t \to \infty} \left(1 - \frac{a}{t}\right)^{-t} = \lim_{t \to \infty} \frac{1}{\left(1 - \dfrac{a}{t}\right)^t}$$

$$= \frac{1}{\lim_{t \to \infty}\left(1 - \dfrac{a}{t}\right)^t} = \frac{1}{e^{-a}},$$

and therefore

$$\lim_{x \to -\infty} \left(1 + \frac{a}{x}\right)^x = e^a. \tag{15'}$$

Compound Interest

As we now show, formula (15) has an important financial application. Suppose money invested at an annual interest rate r, or equivalently at $100r$ percent, is compounded n times a year. Let A_i be the amount of money in the bank account at the end of the ith interest period, assuming that no money is withdrawn after the initial deposit. Then

$$A_{i+1} = A_i + A_i\,\frac{r}{n} = A_i\left(1 + \frac{r}{n}\right), \tag{16}$$

since the interest is computed on the accrued amount at a rate equal to r, the nominal interest rate, divided by n, the number of compoundings per year. The initial amount A_0 is of course the principal P. The amount of money in the bank account after t years is just the amount in the account after nt interest periods. To calculate this amount, which we denote simply by A, we use formula (16) repeatedly, obtaining

$$A = A_{nt} = A_{nt-1}\left(1 + \frac{r}{n}\right) = A_{nt-2}\left(1 + \frac{r}{n}\right)\left(1 + \frac{r}{n}\right) = A_{nt-2}\left(1 + \frac{r}{n}\right)^2$$

$$= \cdots = A_1\left(1 + \frac{r}{n}\right)^{nt-1} = A_0\left(1 + \frac{r}{n}\right)^{nt}.$$

330 Chapter 6 Logarithms and Exponentials

Therefore

$$A = P\left(1 + \frac{r}{n}\right)^{nt}, \qquad (17)$$

since $A_0 = P$.

Now suppose interest is *compounded continuously*, that is, suppose the number of compoundings per year is made larger and larger, so that the time interval between consecutive compoundings becomes smaller and smaller. Then the amount after t years is given by

$$A = P \lim_{n \to \infty} \left(1 + \frac{r}{n}\right)^{nt},$$

or equivalently, after setting $x = nt$,

$$A = P \lim_{x \to \infty} \left(1 + \frac{rt}{x}\right)^{x}.$$

But the limit on the right is just the limit (15), with $a = rt$. Therefore

$$A = Pe^{rt}. \qquad (18)$$

This allows us to interpret the number e in financial language! Suppose $1 is deposited at 100% interest and compounded continuously. Then $P = \$1$, $r = 1$, and the amount after 1 year is precisely e dollars, that is, $2.72 to the nearest cent.

Example 10 Let $P = \$1000$, $r = 0.06$ (6%) and $t = 1$ year. Compare the values of the amount A given by formula (17) for various values of n (the number of compoundings per year) with the value of A given by formula (18) for continuous compounding.

Solution The results are given in the following table for annual, semiannual, quarterly, monthly, daily, and continuous compounding (the last indicated by ∞):

n	1	2	4	12	365	∞
A	$1060.00	$1060.90	$1061.36	$1061.68	$1061.83	$1061.84

It is clear that the distinction between daily compounding (offered by some banks) and continuous compounding is of little monetary significance. □

Example 11 What amount of money must be deposited now to be worth $10,000 in 4 years if compounded continuously at an annual interest rate of 5%?

Solution Solving equation (18) for P, we find that

$$P = \frac{A}{e^{rt}} = Ae^{-rt}. \qquad (18')$$

Setting $A = \$10{,}000$, $r = 0.05$ and $t = 4$ in this formula, we get

$$P = \$10{,}000 e^{-0.2} = \$8187.31$$

to the nearest cent. In financial language P is called the *present value* (or *discounted value*) of $10,000 due in 4 years at 5% interest compounded continuously. □

Section 6.5 The General Power Function; More About Indeterminate Forms

Problems

Differentiate

1. $x^a \ln x$
2. $x^a e^x$
3. $x^a a^x$
4. $x^\pi 5^x$
5. $x^{1/x}$
6. $x^{\sqrt{x}}$
7. $(\ln x)^x$
8. $x^{\ln x}$
9. $x^{\tan x}$
10. $(\ln x)^{\ln x}$
11. $(\sin x)^{\cos x}$
12. x^{x^2}
13. e^{x^x}
14. x^{e^x}
15. x^{4^x}
16. x^{x^x}

17. Verify that
$$\int x^{a-1} \ln x \, dx = \frac{x^a \ln x}{a} - \frac{x^a}{a^2} + C \quad (a \neq 0).$$

Evaluate

18. $\int \dfrac{\ln x}{\sqrt{x}} \, dx$

19. $\int \sqrt{x} \ln x \, dx$

20. Show that if $\log_a y$ is a nonconstant linear function of $\log_a x$, then y is proportional to a power function of x. Show that the converse is true if the constant of proportionality is positive.

21. Express y as a function of x if $\log_2 y = \pi \log_2 x - 1$. Express $\log_3 y$ as a function of $\log_3 x$ if $y = 9x^{\sqrt{2}}$.

22. Show that the equation $x^a = a^x$ ($a > 0$) has a nontrivial solution, that is, a solution distinct from $x = a$, if and only if $1 < a < e$ or $a > e$. What is the solution for $a = 2$?

23. Let a be any real number. Show that
$$\lim_{x \to \infty} \frac{e^x}{x^a} = \infty, \tag{i}$$
or equivalently
$$\lim_{x \to \infty} \frac{x^a}{e^x} = 0.$$
(Thus e^x grows faster than any power of x, no matter how large.)

24. Show that 0^∞ and ∞^∞ are not indeterminate forms.

Describe the indeterminacy corresponding to the given limit. Then evaluate the limit, with the help of L'Hospital's rule or formula (15).

25. $\lim\limits_{x \to 1} \dfrac{x^e - 1}{x^\pi - 1}$

26. $\lim\limits_{x \to 0} \dfrac{(1+x)^{\sqrt{2}} - 1}{x}$

27. $\lim\limits_{x \to \infty} \dfrac{e^x}{x^{101}}$

28. $\lim\limits_{x \to 0^+} x^{\sqrt{x}}$

29. $\lim\limits_{x \to 0^+} (\sin x)^x$

30. $\lim\limits_{x \to \infty} \dfrac{(\ln x)^a}{x}$ $(a > 0)$

31. $\lim\limits_{x \to 0^-} (2^{\cot x})^x$

32. $\lim\limits_{x \to 0^+} x^{x \ln x}$

33. $\lim\limits_{x \to 0} (1 + \pi x)^{1/x}$

34. $\lim\limits_{x \to -\infty} \left(1 - \dfrac{e}{x}\right)^x$

35. $\lim\limits_{x \to \infty} \left(\dfrac{x+1}{x-1}\right)^x$

36. $\lim\limits_{x \to \infty} \left(1 + \dfrac{\ln x}{x}\right)^x$

37. $\lim\limits_{x \to 1^+} \left(\dfrac{1}{x-1}\right)^{\ln x}$

38. $\lim\limits_{x \to 0} (e^x + x)^{1/x}$

39. $\lim\limits_{x \to 1^-} x^{x/(1-x)}$

40. $\lim\limits_{x \to 0^+} \left(\ln \dfrac{1}{x}\right)^x$

41. $\lim\limits_{x \to 0^+} (e^{1/x})^{\tan x}$

42. $\lim\limits_{x \to 0^+} (\csc x)^x$

43. How large does an initial deposit of $1000 become in 5 years if compounded quarterly at an annual interest rate of 8%? If compounded continuously?

44. A continuously compounded initial deposit doubles in 10 years. What is the annual interest rate?

45. How long does it take $15,000 compounded continuously at an annual interest rate of 7% to grow to $25,000? To $35,000?

46. What is the present value of $60,000 due in 5 years at an annual interest rate of 9% compounded continuously? At the same rate compounded monthly?

47. The value V of a case of Napoleon brandy is increasing with time, in accordance with the formula $V = V_0 e^{\sqrt{t}/5}$ (t in years). When the brandy is eventually sold, the owner plans to deposit the proceeds in a bank account paying interest at an annual rate of r percent compounded continuously. After how many years should she sell the brandy to maximize her subsequent holdings?

48. If n is the number of compoundings per year, justify calling
$$r_E = \left(1 + \frac{r}{n}\right)^n - 1 \tag{ii}$$
the *effective* (annual) interest rate, as opposed to the *nominal* (annual) interest rate r.

Find the effective interest rate corresponding to a nominal rate of

49. 8% compounded semiannually
50. 7.2% compounded monthly
51. 6.5% compounded continuously

52. What nominal interest rate "annualizes" to an effective rate of 8% if compounded quarterly? Monthly? Continuously?

6.6 Separable Differential Equations; Exponential Growth and Decay

To prepare for further study of the exponential function and its applications, we digress briefly to investigate first order differential equations of the form

$$y' = \frac{dy}{dx} = \frac{f(x)}{g(y)} \qquad (g(y) \neq 0). \tag{1}$$

Here $f(x)$ and $g(y)$ are two given continuous functions, and $y = y(x)$ is the unknown function, assumed to be differentiable. The salient feature of equation (1) is that $g(y)$ is a function of the *dependent* variable y. An equation of the form (1) is said to have *separated variables*, and an equation that can be brought into this form is said to be *separable*. For example, the equation $y' = y^3 \sin x$ is separable, as we see at once by choosing $f(x) = \sin x$ and $g(y) = 1/y^3$, but the equation $y' = \sin xy$ is not separable.

To solve (1), we first multiply both sides of the equation by $g(y)$. This gives

$$g(y) \frac{dy}{dx} = f(x). \tag{2}$$

We then observe that the left side of (2) is just the derivative with respect to x of $G(y) = G(y(x))$, where $G(y)$ is any antiderivative of $g(y)$. In fact,

$$\frac{dG(y)}{dx} = \frac{dG(y)}{dy} \frac{dy}{dx} = g(y) \frac{dy}{dx},$$

with the help of the chain rule. Thus (2) can be written in the form

$$\frac{dG(y)}{dx} = f(x). \tag{3}$$

We now integrate both sides of (3) with respect to x, obtaining

$$G(y) = \int f(x)\,dx + C = F(x) + C, \tag{4}$$

where $F(x) = \int f(x)\,dx$ is any fixed antiderivative of $f(x)$, and C is an arbitrary constant of integration. Notice that if $g(y) \equiv 1$, we can choose $G(y) = y$. Then the differential equation (1) reduces to $y' = f(x)$, and (4) reduces to formula (6), page 263, for the general solution of $y' = f(x)$.

Separation of Variables

Alternatively, multiplying both sides of (1) by $g(y)\,dx$ and interpreting dy/dx as a quotient of differentials, we obtain the equation

$$g(y)\,dy = f(x)\,dx, \tag{1'}$$

where the left side involves only the variable y and the right side only the variable x. It is in this sense that the variables are *separated* both in (1') and in the original differential equation (1), and the process leading from (1) to (1') is called *separation of variables*. If we now simply integrate both sides of (1'), we get

$$\int g(y)\,dy = \int f(x)\,dx + C,$$

which is just another way of writing (4). At first glance, it looks as if this argument avoids the use of the chain rule, but it really doesn't, since $y = y(x)$ is the dependent variable and its differential is $dy = y'(x)\,dx$.

Notice that as it stands, equation (4) defines the function $y = y(x)$ implicitly, but in many cases it is easy to solve (4) for y as an explicit function of x. In any event, equation (4) or the equation obtained from (4) by solving for y, is called the *general solution* of the differential equation (1). The general solution contains an arbitrary constant of integration C, and to find the *particular solution* of (1) satisfying a given initial condition

$$y(x_0) = y_0, \tag{5}$$

we must determine the constant C. Substituting $x = x_0$, $y = y_0$ into (4) and solving the resulting equation for C, we find at once that

$$C = G(y_0) - F(x_0).$$

This particular solution is unique. In fact, if y satisfies the differential equation (1) and the initial condition (5), then $G(y) = F(x) + G(y_0) - F(x_0)$, as we see by substituting C into (4). But $dG(y)/dy = g(y)$ is nonzero by assumption, so that $G(y)$ is monotonic and hence one-to-one. Thus there is only one solution of (1) satisfying (5), namely $y = G^{-1}(F(x) + G(y_0) - F(x_0))$, where G^{-1} is the inverse function of G.

Example 1 Find the particular solution of the differential equation

$$\frac{dy}{dx} = 2xy \tag{6}$$

satisfying the initial condition

$$y(0) = 3. \tag{6'}$$

Solution Assuming that y is never zero, we multiply both sides of (6) by dx/y. This gives the equation

$$\frac{dy}{y} = 2x\,dx,$$

in which the variables are separated. We then integrate, obtaining

$$\int \frac{dy}{y} = \int 2x\,dx + k,$$

so that

$$\ln |y| = x^2 + k.$$

Here we denote the arbitrary constant of integration by k, saving the symbol C for later. Taking exponentials of both sides of the last equation, we find that

$$|y| = e^{x^2 + k} = e^k e^{x^2} = Ce^{x^2}, \tag{7}$$

where $C = e^k$ is now an arbitrary *positive* constant (why?). The function $y = y(x)$ is continuous (since it is differentiable) and never zero. Therefore y has the same sign for all x, which must be positive if (6') is to be satisfied.

Thus $|y| = y$ and (7) becomes

$$y = Ce^{x^2}.$$

Substituting $x = 0$ and $y = 3$ into this formula, we immediately obtain $C = 3$. Hence the desired particular solution of equation (6) is $y = 3e^{x^2}$. □

Remark The formula $y = Ce^{x^2}$ gives the general solution of the differential equation (6), if we relax the condition that C be positive and allow C to take any value, positive, negative or zero. In fact, since e^{-x^2} is nonzero and

$$\frac{d}{dx}(ye^{-x^2}) = \left(\frac{dy}{dx} - 2xy\right)e^{-x^2},$$

(6) holds if and only if the derivative of ye^{-x^2} is zero, that is, if and only if $ye^{-x^2} = C$, or equivalently $y = Ce^{x^2}$, where C is an arbitrary constant.

Exponential Growth and Decay

We are now ready to solve problems of exponential growth and decay. Suppose the dependence of one variable, say y, on another variable, say t (typically the time), is given by the formula

$$y = y_0 e^{rt}, \qquad (8)$$

where $y_0 > 0$ and r are constants. Then the rate of change of $y = y(t)$, with respect to t, is the derivative

$$\frac{dy}{dt} = y_0 r e^{rt}.$$

Thus y satisfies the simple differential equation

$$\frac{dy}{dt} = ry, \qquad (9)$$

that is, *the rate of change of the variable y is proportional to the value of y.* If r is *positive*, the function (8) is an *increasing* function of t, since then $dy/dt = y_0 r e^{rt} > 0$ for all t, and we say that y *grows exponentially* with t, or that y is an *exponentially increasing* function of t. On the other hand, if r is *negative*, (8) is a *decreasing* function of t, since then $dy/dt = y_0 r e^{rt} < 0$ for all t, and we say that y *decays* or (*falls off*) *exponentially* with t, or that y is an *exponentially decreasing* function of t. This basic difference between the behavior of exponentially increasing and exponentially decreasing functions is shown in Figure 11, where we graph the function e^{rt} for $r = \pm 0.25, \pm 0.5$ on the interval $0 \leq t < \infty$ in the same system of rectangular coordinates. What happens if $r = 0$?

Setting $t = 0$ in the "exponential law" (8), we find that

$$y(0) = y_0 \qquad (y_0 > 0). \qquad (9')$$

Thus the constant y_0 is just the *initial value* of y, that is, the value of y at $t = 0$, and we see that (8) is the particular solution of the differential equation (9) satisfying the initial condition (9'). This can also be shown directly by separation of variables. (Do so as an exercise, using the same reasoning as in Example 1, but with (9) instead of (6) and $\int r \, dt = rt$ instead of $\int 2x \, dx = x^2$.)

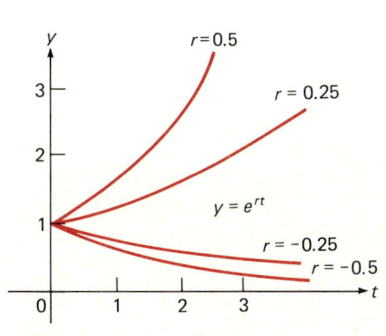

Figure 11

Section 6.6 Separable Differential Equations; Exponential Growth and Decay

It follows from (9) that

$$r = \frac{1}{y}\frac{dy}{dt}, \qquad (10)$$

or equivalently

$$r = \frac{d}{dt}\ln y, \qquad (10')$$

that is, r is the *logarithmic derivative* of y. Thus r is not the rate of change of y, but rather the rate of change of y divided by the "current" value of y. In other words, instead of being the "absolute" growth rate dy/dt, r is the *relative* or *fractional* growth rate (10). Despite this important distinction, r is often simply called the *growth rate* (or just the *rate*) in problems where the context prevents any ambiguity.

Example 2 As we saw on page 331, if an initial deposit of P dollars is compounded continuously at an annual interest rate of r, then the amount of money in the bank account after t years is given by

$$A = Pe^{rt} \qquad (11)$$

dollars. Thus A grows exponentially with time, and the (relative) growth rate is just the interest rate r. In Figure 12 we graph the function (13) over a period of many years for an initial deposit of $1000 and an interest rate of 7.5%.

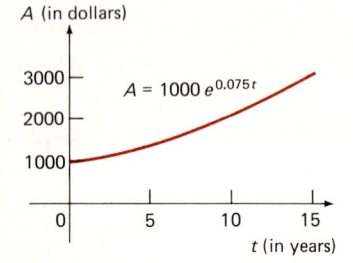

Figure 12

Population Growth

The theory of exponential growth has important applications in the subject of population biology. Let $N = N(t)$ be the size of a population of living organisms (bacteria, insects, people, etc.) at time t. We will treat N as a continuous function, although, in point of fact, the values of N can only be integers. Since N is typically very large, the error introduced by this approximation is entirely negligible. Suppose the population is growing exponentially, with relative growth rate r. Then

$$N = N(t) = N_0 e^{rt}, \qquad (12)$$

where N_0 is the population size at time $t = 0$. The function N is of course just the particular solution of the differential equation

$$\frac{dN}{dt} = rN \qquad (13)$$

satisfying the initial condition

$$N(0) = N_0. \qquad (13')$$

The differential equation (13) says that the rate of change of the population size at any given time is proportional to the size of the population at that time. This is plausible, at least under ordinary circumstances and for limited periods of time. In fact, on the one hand, we have

$$\frac{dN}{dt} = B - D, \qquad (14)$$

336 Chapter 6 Logarithms and Exponentials

where B and D are the (absolute) birth and death rates, respectively. On the other hand, both B and D are often proportional to the population size (there are more maternity wards and more cemeteries in large cities than in small towns), and then $B - D$ is also proportional to N. Comparing (13) and (14), we find that

$$r = \frac{B - D}{N}.$$

In other words, the relative rate of population growth is equal to the per capita excess of the birth rate over the death rate.

The Doubling Time

A population undergoing exponential growth at the rate r doubles its size during every time period of length

$$T = \frac{\ln 2}{r} \tag{15}$$

($\ln 2 \approx 0.6931$), and for this reason T is called the *doubling time* of the population. In fact, it follows from (12) that

$$N(t + T) = N_0 e^{r(t+T)} = N_0 e^{rt} e^{rT} = e^{rT} N(t).$$

Thus $N(t + T) = 2N(t)$ for every $t \geq 0$ if and only if

$$e^{rT} = 2,$$

which is equivalent to (15).

In problems involving annual growth rates, r is usually expressed in percent per year. Formula (15) then leads to the approximation

$$T = \frac{100 \ln 2}{r} \approx \frac{69}{r} \text{ years} \tag{15'}$$

for the doubling time. For example, a country's population will double in about $69/3 = 23$ years if its annual growth rate is a constant 3%, money deposited in a bank at an annual interest rate of 7.5% compounded continuously will double in about $69/7.5 = 9.2$ years (see Figure 12), and so on.

Example 3 The much investigated bacterium *Escherichia coli* (briefly *E. coli*) is a one-celled organism which normally inhabits the human intestinal tract. Under ideal conditions, a cell of *E. coli*, with a mass of approximately 5×10^{-13} grams, reproduces asexually about 20 minutes after its "birth" by undergoing *binary fission*, that is, by dividing into two cells. About how long would it take a single one of these bacteria to generate a culture of mass 3 grams if reproduction were to continue at this rate?

Solution Here it is natural to talk about the mass of the culture, rather than the number of cells in the culture. After t minutes of growth, the mass of the culture in grams is

$$m = m(t) = m_0 e^{rt},$$

where m_0 is its initial mass, equal to 5×10^{-13} g, and r is the growth rate. Let t_1 be the time it takes the culture to achieve a mass of 3 g. Then

$$m_0 e^{rt_1} = 3,$$

or equivalently

$$t_1 = \frac{1}{r} \ln \frac{3}{m_0}.$$

Moreover, since $T = 20$ min, formula (15) implies

$$r = \frac{\ln 2}{20}.$$

Therefore

$$t_1 = \frac{20}{\ln 2} \ln \frac{3}{m_0} = \frac{20}{\ln 2} \ln \frac{3}{5 \times 10^{-13}} = \frac{20}{\ln 2} \ln (6 \times 10^{12})$$

$$= 20 \frac{\ln 6 + 12 \ln 10}{\ln 2} \approx 849 \text{ min} = 14.15 \text{ hr.} \quad \square$$

Radioactive Decay

We now turn to problems of exponential *decay*, of which radioactive decay is a prime example. Let $m = m(t)$ be the mass at time t of a single radioactive substance, like radium. Then, as the substance disintegrates, due to instability of the nuclei of its constituent atoms, the rate of disappearance of its mass is proportional at every instant to the mass of the remaining substance. Thus m satisfies the differential equation

$$\frac{dm}{dt} = rm,$$

where r is a constant. Since r is negative (mass is disappearing), we write $r = -k$, where k is a positive number called the *decay constant*. Therefore, to find the function $m = m(t)$, we must solve the differential equation

$$\frac{dm}{dt} = -km,$$

subject to the initial condition

$$m(0) = m_0,$$

where m_0 is the mass of the substance present at time $t = 0$. By separation of variables (or simply by inspection), we find that

$$m = m(t) = m_0 e^{-kt} \qquad (16)$$

is the solution of this initial value problem. Thus the mass of the radioactive substance decays exponentially at a rate determined by the constant k (the larger k, the faster the decay). In Figure 13 we graph the function (16), measuring t in units of $1/k$.

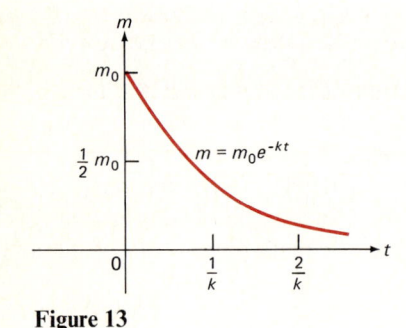

Figure 13

The Half-Life

A substance undergoing radioactive decay, with decay constant k, loses half its mass during every time period of length

$$T = \frac{\ln 2}{k}, \qquad (17)$$

338 Chapter 6 Logarithms and Exponentials

and for this reason T (often written as $T_{1/2}$) is called the *half-life* of the substance. In fact, it follows from (16) that

$$m(t + T) = m_0 e^{-k(t+T)} = m_0 e^{-kt} e^{-kT} = e^{-kT} m(t).$$

Thus $m(t + T) = \frac{1}{2}m(t)$ for every $t \geq 0$ if and only if

$$e^{-kT} = \frac{1}{2},$$

which is equivalent to (17). Notice the complete analogy between formula (17) for the half-life and formula (15) for the doubling time.

Example 4 How long does it take for 99% of a sample of strontium-90 to disappear? The half-life of strontium-90, a particularly dangerous component of radioactive fallout, is 28.1 years.

Solution Disappearance of 99% of the sample means that the initial mass m_0 has decayed to $\frac{1}{100}m_0$. Therefore, if t_1 is the time it takes for 99% of the sample to disappear, we have

$$m_0 e^{-kt_1} = \frac{1}{100} m_0,$$

or equivalently

$$t_1 = \frac{\ln 100}{k}.$$

But

$$k = \frac{\ln 2}{28.1},$$

by formula (17), and hence

$$t_1 = 28.1 \frac{\ln 100}{\ln 2} \approx 186.7 \text{ years.} \quad \square$$

Problems

Use separation of variables to solve the given initial value problem.

1. $x\dfrac{dy}{dx} + y = 0$, $y(2) = 1$

2. $x\dfrac{dy}{dx} = y^2 + 1$, $y(-1) = 1$

3. $2xy\dfrac{dy}{dx} + y^2 + 1 = 0$, $y(1) = 2$

4. $x^3 \dfrac{dy}{dx} = 2y$, $y(\sqrt{\log_2 e}) = 3$

5. $(x^2 + x)\dfrac{dy}{dx} = 2y + 1$, $y(1) = 0$

6. $(e^x + 1)y\dfrac{dy}{dx} = e^x$, $y(0) = 1$

7. $(x^2 + 1)\dfrac{dy}{dx} + y^2 + 1 = 0$, $y(3) = 2$

8. $x\dfrac{dy}{dx} + y \ln y = 0$, $y(1) = e$

9. A curve goes through the point (0, 2) and has the property that the slope of the curve at every point P is three times the ordinate of P. What is the curve?

10. A curve has the property that the normal to the curve at every point goes through a fixed point A. Show that the curve is a circle with A as its center (or an arc of such a circle).

Section 6.6 Separable Differential Equations; Exponential Growth and Decay 339

11. Suppose the radius R of a balloon grows exponentially at the rate of 2.5% per minute. How does the balloon's surface area S behave?

12. An exponentially growing population doubles its size in 50 years. What is its annual growth rate?

13. Show that if T is the doubling time of an exponentially growing population of size N, then $N = N_0 \, 2^{t/T}$, where N_0 is the initial population size.

14. An exponentially growing population increases by 20% in 5 years. What is its doubling time?

15. The world's population, equal to 4.5 billion in 1980, is growing exponentially at the rate of about 1.8% per year. Estimate the world's population in the year 2010 if growth continues at this rate.

16. Suppose total consumption grows exponentially at the rate of r% per year, while population grows exponentially at the rate of s% per year. How does the per capita consumption behave?

17. The number of bacteria in a culture doubles every 15 minutes. How long does it take 500 bacteria to grow to a million?

18. The number of bacteria in a culture at time t is $N = 1500(2^{2.5t})$, where t is measured in hours. What is the time between successive fissions of the bacteria?

19. A culture contains two types of bacteria, type A and type B. Bacteria of type A reproduce (by binary fission) every hour, while bacteria of type B reproduce every 2 hours. After 2 hours the culture contains 3.5 times the initial number of bacteria. Find the initial composition of the culture. How large does the culture become after 4 hours of growth?

20. Find the purchasing power of today's dollar after a decade of sustained inflation at 8% per year. At 12% per year. (Assume exponential inflation.)

21. Bread that cost 10¢ a loaf in 1936 costs $1.35 a loaf in 1986. Estimate the annual inflation rate over this 50-year period.

22. How long does it take a sample of plutonium-239 to lose 90% of its radioactivity? (The half-life of plutonium-239, produced in nuclear reactors of the "breeder" type, is 24,360 years.)

23. One tenth of a radioactive substance disappears in 20 years. What is the half-life of the substance?

24. If 30% of a radioactive substance disappears in 10 days, how long does it take for 60% to disappear?

25. Let $C = C(t)$ be the concentration of an ingested drug in the bloodstream. Then, as the body eliminates the drug, C decreases at a rate proportional to its current value, that is,

$$\frac{dC}{dt} = -kC \qquad (k > 0),$$

where the number k is known as the *elimination constant* of the drug. If the initial concentration is C_0, what is the concentration at time t? How long does it take the body to eliminate 95% of an ingested drug if it eliminates half the drug in 36 hours?

26. Radioactive drugs are often used as "tracers" in medical diagnosis. Suppose a patient is given a dose of a radioactive tracer, with a half-life of 8 days, and suppose half the dose is eliminated metabolically in 2 days (metabolism involves biochemical processes and has nothing to do with radioactivity, which involves nuclear processes). How long does it take for the radioactivity in the patient's body to fall to 1% of its initial value? How long does this take in the absence of metabolic elimination?

27. The average amount of radium in the earth's crust is about 1 atom in 10^{12}. Does it make sense to assume that this is the radium left over from a larger amount present at an earlier time? Explain your answer. (The half-life of radium is 1620 years.)

Radioactive carbon-14 (*radiocarbon*), with a half-life of 5730 years, is being produced continually in the upper atmosphere by the action of cosmic rays on nitrogen. Incorporated in carbon dioxide, the radiocarbon is mixed into the lower atmosphere, and is absorbed first by plants, during photosynthesis, and then by animals eating the plants. As long as they are alive, the plants and animals take in fresh radiocarbon, but when they die, the process ceases and the radiocarbon in their tissues slowly disintegrates, dropping to half the original amount in 5730 years. This fact leads to a method, called *radiocarbon dating*, of great importance in archaeology for estimating the ages of ancient objects. For example, the age of a sliver of a mummy case can be estimated by comparing the amount of radioactivity in the sliver with the amount of radioactivity in a piece of fresh wood of the same kind and size. It is by this method that the Dead Sea scrolls are estimated to be about 2000 years old.

28. Suppose a Geiger counter records m disintegrations from an old carboniferous specimen of unknown age α during the same time period in which it records n disintegrations from a similar contemporary specimen ($n > m$). Show that

$$\alpha = \frac{5730}{\ln 2} \ln \frac{n}{m} \text{ years.} \qquad (i)$$

29. Heartwood from a giant sequoia tree has about 75% of the radioactivity of the younger outer wood. Estimate the age of the tree.

30. Charcoal and animal bones from an excavated prehistoric settlement are found to have about 55% of the radioactivity of similar contemporary samples. Estimate the age of the settlement.

Suppose that disintegration of each atom of a radioactive substance A, with decay constant a, produces an atom of a new radioactive substance B, with decay constant b ($\neq a$). Let $m_A = m_A(t)$ be the mass of substance A and $m_B = m_B(t)$ the mass of substance B present at time t. Then, since disappearance of A leads to appearance of B, the evolution of this decay process is described by the *pair* of differential equations

$$\frac{dm_A}{dt} = -am_A,$$

$$\frac{dm_B}{dt} = am_A - bm_B.$$

31. Solve the first of these equations for m_A, subject to the initial condition $m_A(0) = m_0$, where m_0 is the initial mass of substance A, and eliminate m_A from the second equation. Then multiply the resulting differential equation for m_B by e^{bt}, and solve it, subject to the initial condition $m_B(0) = 0$ (no B initially present).

32. Show that the largest value of m_B is $m_0(b/a)^{b/(a-b)}$, taken at the time

$$t = \frac{1}{a-b} \ln \frac{a}{b}.$$

33. If the half-life of A is 100 years, while that of B is 150 years, and if a sample which was initially all A now contains A and B in equal amounts, how old is the sample? What is the largest value of m_B, and when is it achieved?

34. Show that if the half-life of A is less than that of B, then $m_B/m_A \to \infty$ as $t \to \infty$, so that a sample which was originally all A eventually becomes almost all B. What happens if the half-life of A is greater than that of B?

6.7 Further Applications of Exponentials

Logistic Growth (Optional)

In the last section we saw that the differential equation

$$\frac{dN}{dt} = rN \tag{1}$$

$(r > 0)$, subject to the initial condition

$$N(0) = N_0, \tag{1'}$$

leads to population growth in accordance with the exponential law

$$N = N(t) = N_0 e^{rt}. \tag{2}$$

Formula (2) inevitably leads to a "population explosion," in which any given population level is exceeded in a surprisingly short time. For example, at a 3% annual growth rate a country's population would increase by a factor of 16 after 4 doubling times, that is, in about $4(23) = 92$ years. Eventually, of course, population growth must slow down, due to lack of food, spread of infectious diseases, loss of fertility due to overcrowding, wars fought for dwindling resources, and so on. It turns out that these effects of "overpopulation" are described remarkably well in many cases by introducing an extra term $-sN^2$ in the right side of equation (1), where s (like r) is a positive constant. (For an explanation of how such a term might arise, see Problem 8.) The differential equation governing growth then becomes

$$\frac{dN}{dt} = rN - sN^2, \tag{3}$$

instead of (1), and is subject to the same initial condition (1').

To solve the differential equation (3), we separate variables and integrate. This gives

$$\int \frac{dN}{rN - sN^2} = \int dt + c = t + c, \tag{4}$$

where c is a constant of integration. The integral on the left is not hard to evaluate. Setting

$$N_1 = \frac{r}{s}, \tag{5}$$

we have

$$\int \frac{dN}{rN - sN^2} = -\frac{1}{s}\int \frac{dN}{N^2 - \frac{r}{s}N} = -\frac{1}{s}\int \frac{dN}{N(N - N_1)}.$$

Formula (5), page 310, is then applicable (with $a = 0$, $b = -N_1$), and leads to

$$\int \frac{dN}{rN - sN^2} = \frac{1}{sN_1}\ln\left|\frac{N}{N - N_1}\right| = \frac{1}{r}\ln\left|\frac{N}{N - N_1}\right|.$$

Thus (4) becomes

$$\ln\left|\frac{N}{N - N_1}\right| = rt + k,$$

where $k = rc$, or

$$\left|\frac{N}{N - N_1}\right| = Ce^{rt},$$

where $C = e^k$. Applying the initial condition $N(0) = N_0$, we get

$$C = \left|\frac{N_0}{N_0 - N_1}\right|,$$

so that

$$\left|\frac{N}{N - N_1}\right| = \left|\frac{N_0}{N_0 - N_1}\right|e^{rt}.$$

The absolute value signs can now be dropped, since N and N_0 are both positive, while $N - N_1$ and $N_0 - N_1 = N(0) - N_1$ have the same sign (in evaluating the integral, it was tacitly assumed that $N - N_1 \neq 0$ for all $t \geq 0$). Doing this and solving for N, we find after some straightforward manipulation that

$$N = \frac{N_1 e^{rt}}{e^{rt} + \left(\dfrac{N_1}{N_0} - 1\right)},$$

or equivalently

$$N = \frac{N_1}{1 + \left(\dfrac{N_1}{N_0} - 1\right)e^{-rt}}. \tag{6}$$

Assuming that $N_1 > N_0$ (the other cases are treated in Problems 9 and 10), we see that the denominator of the function $N = N(t)$ defined by (6) has a negative derivative, and hence is a decreasing function. It follows that N is an increasing function. Graphing this function, we get the S-shaped curve shown in Figure 14 for the case $N_1 = 10N_0$. The population growth is now restricted, since as $t \to \infty$, the exponential e^{-rt} approaches zero, so that (6)

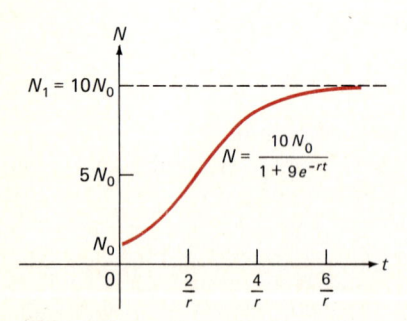

Figure 14

342 Chapter 6 Logarithms and Exponentials

approaches the *limiting* or *stable population size* N_1, given by formula (5). Notice that N_1 is independent of the initial population size N_0. In fact, N_1 is the population size at which the right side of the differential equation (3) equals zero, that is, at which the death rate equals the birth rate. The validity of the *logistic growth law* (6) has been confirmed by many observations, not only of human populations, but also of experimental populations of yeast, protozoa and flies. It has also been used to describe the growth of individual multicellular organisms.

Example 1 A population obeys the logistic growth law (6), with $N_1 > 2N_0$. When is the rate of growth of the population largest?

Solution The rate of population growth is the derivative dN/dt, given by the differential equation (3). Because of (5), we can write (3) in the form

$$\frac{dN}{dt} = s(NN_1 - N^2) = sP(N), \qquad (3')$$

in terms of the quadratic polynomial

$$P(N) = NN_1 - N^2.$$

Since $s > 0$, the growth rate dN/dt is largest when $P(N)$ is largest. Differentiating $P(N)$ with respect to N, we find that

$$P'(N) = \frac{d}{dN}(NN_1 - N^2) = N_1 - 2N.$$

Therefore $P'(N)$ is positive if $N_0 \leq N < \frac{1}{2}N_1$, zero if $N = \frac{1}{2}N_1$ and negative if $\frac{1}{2}N_1 < N < N_1$ (N never reaches the value N_1). It follows from the first derivative test that $P(N)$ has a strict local (and absolute) maximum at $N = \frac{1}{2}N_1$, equal to

$$P\left(\frac{1}{2}N_1\right) = \frac{1}{2}N_1^2 - \frac{1}{4}N_1^2 = \frac{1}{4}N_1^2.$$

But $dN/dt = sP(N)$, and hence dN/dt, regarded as a function of N, also has a maximum at $N = \frac{1}{2}N_1$, equal to

$$\left.\frac{dN}{dt}\right|_{N=\frac{1}{2}N_1} = sP\left(\frac{1}{2}N_1\right) = \frac{1}{4}sN_1^2 = \frac{1}{4}rN_1.$$

This maximum is achieved at the time t_1 for which $N = \frac{1}{2}N_1$. Also, by the monotonicity test, $P(N)$ is increasing on $[N_0, \frac{1}{2}N_1]$ and decreasing on $[\frac{1}{2}N_1, N_1)$, and hence the same is true of dN/dt, regarded as a function of N.

To find the time t_1 at which $N = \frac{1}{2}N_1$, we observe that at time t_1 the denominator in formula (6) for N is equal to 2. Therefore

$$1 + \left(\frac{N_1}{N_0} - 1\right)e^{-rt_1} = 2,$$

or equivalently

$$e^{rt_1} = \frac{N_1}{N_0} - 1. \qquad (7)$$

Solving for t_1, we get

$$t_1 = \frac{1}{r}\ln\left(\frac{N_1}{N_0} - 1\right). \qquad (8)$$

For the logistic function graphed in Figure 14, corresponding to the case $N_1 = 10N_0$, we find that

$$t_1 = \frac{\ln 9}{r} \approx \frac{2.2}{r}.$$

At this time $N = 5N_0$ and dN/dt has its maximum value, which is equal to $2.5rN_0$. $\square$

With the help of (7), we find that (6) can be written in the alternative form

$$N = \frac{N_1}{1 + e^{-r(t-t_1)}}. \tag{6'}$$

The function $N = N(t)$ has an inflection point at $t = t_1$ (provided that $N_1 > 2N_0$). In fact, as shown in Example 1, the derivative dN/dt, regarded as a function of N, is increasing on $[N_0, \frac{1}{2}N_1]$ and decreasing on $[\frac{1}{2}N_1, N_1]$. But N is an increasing function of t on $[0, \infty)$, and $N(t_1) = \frac{1}{2}N_1$. It follows that dN/dt, regarded as a function of t, is increasing on $[0, t_1]$ and decreasing on $[t_1, \infty)$. Therefore N is concave upward on $[0, t_1]$ and concave downward on $[t_1, \infty)$, with an inflection point at $t = t_1$, as is apparent from Figure 14 for the case $N_1 = 10N_0$.

Example 2 Reexamining Example 3, page 337, from a more realistic point of view, suppose the growth of the bacterial culture is logistic, rather than exponential, with a limiting mass of 3 grams. How long does it take the culture to achieve a fraction q $(0 < q < 1)$ of its limiting mass?

Solution Let $m = m(t)$ be the mass of the culture after t minutes of logistic growth. Then

$$m = m(t) = \frac{m_1}{1 + \left(\dfrac{m_1}{m_0} - 1\right) e^{-rt}},$$

by the analogue of formula (6) for mass, where $m_0 = 5 \times 10^{-13}$ g, the mass of a single cell of E. coli, and $m_1 = 3$ g, the limiting mass. We still have

$$r = \frac{\ln 2}{20},$$

since this is the relative growth rate in the absence of any effects of overpopulation ($s = 0$). Let T_q be the time at which the culture achieves mass qm_1. Then

$$m(T_q) = \frac{m_1}{1 + \left(\dfrac{m_1}{m_0} - 1\right) e^{-rT_q}} = qm_1,$$

so that

$$\frac{1}{1 + \dfrac{m_1}{m_0} e^{-rT_q}} = q.$$

344 Chapter 6 Logarithms and Exponentials

Here we have replaced $(m_1/m_0) - 1$ by m_1/m_0 without even calling it an approximation, since m_1/m_0 is so enormous (6×10^{12}). It follows that

$$\frac{m_1}{m_0} e^{-rT_q} = \frac{1}{q} - 1,$$

which implies

$$T_q = \frac{1}{r} \ln \left(\frac{m_1}{m_0} \frac{q}{1-q} \right) = \frac{1}{r} \ln \frac{m_1}{m_0} + \frac{1}{r} \ln \frac{q}{1-q}. \quad (9)$$

The first term on the right is the time t_1 that it takes an *exponentially* growing culture to achieve mass m_1, which on page 338 was found to be about 849 min. A logistically growing culture can never quite reach mass m_1. In fact, at time t_1, the mass of such a culture is only $\frac{1}{2}m_1 = 1.5$ g, as we see by setting $q = \frac{1}{2}$ in (9), which gives $T_{1/2} = t_1$. Typical values of T_q, calculated from (9), are given in the following table. Notice how for small q the mass of the culture doubles about every 20 min, while for large q the mass of the culture changes only slightly in 20 min.

q	T_q	q	T_q
0.0005	629.7	0.25	817.3
0.0010	649.7	0.50	849.0
0.0025	676.2	0.75	880.7
0.005	696.2	0.90	912.4
0.010	716.4	0.95	933.9
0.025	743.3	0.99	981.5
0.05	764.0	0.999	1048.2
0.10	785.6	0.9999	1114.7

Some Physical Applications of Exponentials

Exponential functions arise in a great variety of physical problems other than radioactivity. In particular, they play an important role in the study of electric circuits.

Example 3 A switch is suddenly closed, applying a source of constant voltage V (a battery, say) to the electric circuit shown in Figure 15, consisting of a resistor, of resistance R ohms, connected in series to an inductor, of inductance L henries. Find the resulting current $i = i(t)$ in the circuit. (With these units i will be in amperes.)

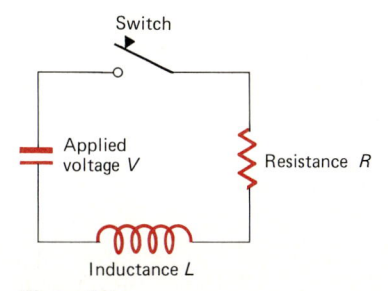

Figure 15

Solution By the theory of electric circuits, the voltage across the resistor is Ri (*Ohm's law*) and the voltage across the inductor is $L\, di/dt$. Moreover, the sum of these two voltages must equal the applied voltage V. Hence the current satisfies the differential equation

$$Ri + L\frac{di}{dt} = V, \quad (10)$$

or equivalently

$$\frac{di}{dt} = \frac{R}{L}\left(\frac{V}{R} - i\right), \quad (11)$$

subject to the initial condition

$$i(0) = 0 \quad (11')$$

(there is no current in the circuit until the switch is closed at time $t = 0$). Separating variables in (11) and integrating, we get

$$\int \frac{di}{(V/R) - i} = \int \frac{R}{L} dt + c,$$

where c is a constant of integration. It follows that

$$-\ln\left|\frac{V}{R} - i\right| = \frac{R}{L} t + c.$$

Since $(V/R) - i > 0$ (why?), we can drop the absolute value sign, obtaining

$$\ln\left(\frac{V}{R} - i\right) = -\frac{R}{L} t - c.$$

Therefore

$$\frac{V}{R} - i = ke^{-Rt/L},$$

where $k = e^{-c}$, that is,

$$i = \frac{V}{R} - ke^{-Rt/L}.$$

Applying the initial condition (11′), we immediately get $k = V/R$. Thus, finally,

$$i = \frac{V}{R}(1 - e^{-Rt/L}).$$

The behavior of i as a function of t is shown in Figure 16. Observe that i is the difference between two terms, a *steady-state* constant current

$$i_0 = \frac{V}{R},$$

which is the solution of (10) in the absence of any inductance, and an exponentially decaying *transient* current

$$i_{tr} = \frac{V}{R} e^{-Rt/L},$$

which dies out rapidly. In fact, i_{tr} drops to $100/e \approx 37\%$ of its initial value in a time $T = L/R$, called the *time constant* of the circuit. For all practical purposes, i reaches its steady-state value i_0 in a time approximately equal to $5T$ (note that $1 - e^{-5} \approx 0.993$). ☐

The following example illustrates how exponentials arise in problems of mechanics.

Example 4 A bullet is fired with initial velocity v_0 into a medium which resists the advance of the bullet with a force proportional to the square of its velocity. Find the bullet's velocity v after it has penetrated a distance s into the medium.

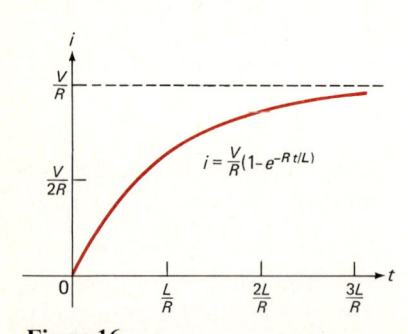

Figure 16

Solution Let m be the mass of the bullet. Then, by Newton's second law of motion,

$$m\frac{dv}{dt} = F,$$

where F is the force with which the medium resists the bullet's advance. Since F is proportional to v^2 and acts in the direction opposite to that of the velocity v, we have $F = -bv^2$, where b is a positive constant. Therefore

$$m\frac{dv}{dt} = -bv^2,$$

so that

$$\frac{dv}{dt} = -\frac{b}{m}v^2 = -kv^2,$$

where $k = b/m > 0$. To express v as a function of s, the distance of penetration of the bullet into the medium, we use the chain rule in the same way as on page 269, writing

$$\frac{dv}{dt} = \frac{dv}{ds}\frac{ds}{dt} = v\frac{dv}{ds}.$$

It follows from the last two equations that

$$v\frac{dv}{ds} = -kv^2,$$

or equivalently

$$\frac{dv}{ds} = -kv. \qquad (12)$$

The appropriate initial condition is

$$v|_{s=0} = v_0, \qquad (12')$$

since the bullet enters the medium with velocity v_0. The solution of (12) satisfying (12') is

$$v = v_0 e^{-ks},$$

found by inspection or by separation of variables. Thus the bullet's velocity falls off exponentially with the distance of penetration s. Notice the complete mathematical analogy between this problem and the physically dissimilar problem of radioactive decay. □

Problems

An insect population is growing logistically with initial size 100 and limiting size 10,100. Suppose the population reaches a size of 5050 after 20 days of growth.

1. What is the population size after 25 days?

2. What is the population size after 30 days?
3. How long does it take the population to reach a size of 10,000?
4. What is the doubling time of the population in the early stages of its growth?

Section 6.7 Further Applications of Exponentials 347

A bacterial culture is growing logistically with initial mass 10^{-6} g and limiting mass m_1. The culture achieves mass $\frac{1}{2}m_1$ in 10 hr and mass $\frac{15}{16}m_1$ in 12 hr.

5. What is the limiting mass m_1?

6. At what time does the culture achieve 99% of its limiting mass?

7. How long does it take a bacterium to undergo binary fission?

8. Suppose each individual in a growing population of N individuals produces a toxic effect by releasing metabolic waste products or other pollutants into the environment. Show that the combined effect of this toxicity might be expected to lower the growth rate dN/dt by an amount proportional to N^2, the *square* of the population size. *Hint.* Justify replacing r by $r - sN$ in equation (1).

9. Suppose the initial size N_0 of a population governed by the differential equation (3) *exceeds* $N_1 = r/s$. Show that the population size is a *decreasing* function, which approaches the limiting value N_1 as $t \to \infty$. In this context, N_1 is called the *carrying capacity* of the environment.

10. Graph the logistic function (6) for each of the four values $N_0 = \frac{1}{4}N_1, \frac{1}{2}N_1, N_1, 2N_1$ in the same coordinate system. How do the resulting four curves differ?

11. The differential equation

$$\frac{dN}{dt} = sN^2 - rN, \qquad \text{(i)}$$

differing from (3) in the sign of the right side, is used as a model for studying the population growth of endangered species. The reasoning is that the birth rate is proportional to the number of encounters between males and females of the species, which is in turn proportional to N^2 if the encounters are random and the male and female population sizes are equal. This accounts for the term sN^2, and the term $-rN$ corresponds to a constant per capita death rate (in the absence of effects of overpopulation). Show that a population whose size N is governed by (i) is destined for extinction if its initial size N_0 is less than the *critical population size* $N_1 = r/s$. What happens if $N_0 > N_1$?

12. Solve equation (i) for the case $N_0 = \frac{1}{2}N_1$, and graph the solution.

13. Solve the initial value problem

$$\frac{dN}{dt} = rN - s, \qquad N(0) = N_0 \qquad \text{(ii)}$$

($r > 0, s > 0$), which corresponds to exponential population growth together with a constant *emigration* rate s. What condition on s leads to a "population explosion"? Maintains the population at a constant size? Leads to a "population drain"?

14. The absorption of light by sea water is described by the exponential law $I = I_0 e^{-kx}$, where I_0 is the intensity of light at the surface of the sea and $I = I(x)$ is its intensity at depth

x. Of what initial value problem is I the solution? Find the coefficient k, called the *absorption coefficient*, if the intensity of light at a depth of 5 meters is one thousandth of its intensity at the surface. At what depth is the intensity of light one hundred thousandth of its intensity at the surface?

15. According to *Newton's law of cooling*, an object at temperature T cools at a rate proportional to the difference between T and the temperature T_1 of the surrounding air, so that

$$\frac{dT}{dt} = -k(T - T_1) \qquad (k > 0).$$

Find the solution of this differential equation satisfying the initial condition $T = T_0$ at time $t = 0$.

16. Suppose the air temperature is $20°$ (Celsius), and a heated object cools from $140°$ to $80°$ in 10 minutes. How much later does the object cool to $35°$?

17. A thermometer is taken from a room where the temperature is $72°$ (Fahrenheit) and carried outdoors. A minute later the thermometer reads $56°$, and after another minute it reads $44°$. How cold is it outside? What does the thermometer read after 5 minutes?

18. A tank of salt water initially contains 50 lb of salt dissolved in 240 gallons of water. To clean out the tank, fresh water is pumped in at the rate of 6 gallons per minute and the solution is pumped out at the same rate, with the contents of the tank being continually stirred to promote uniform mixing. How long does it take to reduce the amount of salt in the tank to 1 oz?

19. A ship slows down under the action of water resistance, which retards the ship's motion with a force proportional to its velocity. Suppose the ship's initial velocity (at $t = 0$) is 12 ft/sec, and that its velocity is 8 ft/sec at $t = 10$ sec. When does the ship's velocity drop to 1 ft/sec?

$+$ **20.** A bullet with velocity v_0 ft/sec pierces a board h feet thick and emerges with velocity v_1 ft/sec. Suppose the board resists the bullet's advance with a force proportional to the square of the bullet's velocity. Show that it takes the bullet

$$T = \frac{h(v_0 - v_1)}{v_0 v_1 \ln(v_0/v_1)}$$

seconds to go through the board. Find T if $v_0 = 600$ ft/sec, $v_1 = 200$ ft/sec and $h = 6$ in.

21. The volume, and hence the mass, of an evaporating mothball decreases at a rate proportional to its surface area. Suppose an 8-gram mothball loses 1 gram of its mass in the first day of evaporation. How many days does it take the mothball to lose half its mass? To shrink to a mass of 1 gram? Does the mothball ever disappear? Does this problem involve exponentials?

22. A 12-volt battery is suddenly connected to a 20-ohm resistor and a 5-henry inductor in series. How long does it take the current to build up to 99% of its steady-state value? Does the answer change as the battery ages and loses its voltage?

23. A switch is suddenly closed, applying a source of constant voltage V to the electric circuit shown in Figure 17, consisting of a resistor, of resistance R ohms, connected in series to a capacitor, of capacitance C farads. Find the resulting charge $q = q(t)$ on the capacitor and current $i = i(t)$ in the circuit. (With these units q will be in coulombs and i in amperes.) Start from the fact that the voltage across the capacitor is q/C.

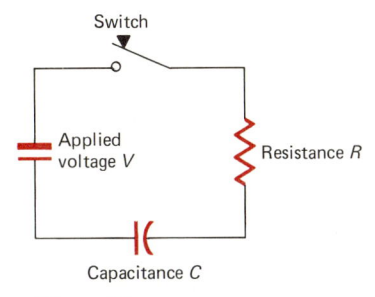

Figure 17

24. A 5-microfarad capacitor is discharged by connecting a 2-megohm resistor across its terminals. How long does it take the charge to drop to 10% of its initial value? Does the answer depend on the initial charge on the capacitor? (1 microfarad = 10^{-6} farad, 1 megohm = 10^6 ohms.)

25. Suppose every member of a population belongs to one or the other of two classes A and B, where members of class A can "infect" members of class B. For example, A may consist of people who have a certain disease and B of those who do not, or A may consist of people who have heard a rumor and B of those who have not. Let N_A be the size of class A, N_B the size of class B, and N the total population size, so that $N_A + N_B = N$. Then N_A will increase as a result of contacts between members of the two classes, and one can expect the rate of change of N_A to be proportional to the number of these contacts, which is in turn proportional to the product $N_A N_B = N_A(N - N_A)$. This leads to the differential equation

$$\frac{dN_A}{dt} = kN_A(N - N_A),$$

where k is a positive constant, or equivalently

$$\frac{dy}{dt} = ky(1 - y),$$

where $y = N_A/N$ is the fraction of the total population belonging to class A. Solve this equation, subject to the initial condition $y = y_0 < 1$ at time $t = 0$, and show that the disease or rumor eventually spreads to the whole population. Find the time T that it takes for half of the population to catch the disease or hear the rumor, assuming that $y_0 < \frac{1}{2}$. What flaws do you see in this model, as applied to the spread of disease, say?

26. The amount of detritus or litter L (dead organic matter) in an ecosystem is governed by the differential equation

$$\frac{dL}{dt} = I - kL,$$

where I is the rate of input to the litter layer and k is a rate constant for decomposition of the litter. Show that even if litter production is low, large amounts of litter can accumulate if k is small enough. (For example, $k \approx 0.02$ in pine forests, where low temperatures inhibit decomposer metabolism, allowing substantial accumulation of pine needles.)

6.8 Hyperbolic Functions

The Hyperbolic Cosine and Sine

Next we consider certain functions related to the exponential which warrant special names and separate study, because of the frequency with which they arise in problems involving the evaluation of integrals and the solution of differential equations.† The two most important of these functions are the *hyperbolic cosine*

$$\cosh x = \frac{1}{2}(e^x + e^{-x}) \qquad (1)$$

and the *hyperbolic sine*

$$\sinh x = \frac{1}{2}(e^x - e^{-x}) \qquad (2)$$

(the symbol sinh is usually pronounced "cinch"). Both functions e^x and e^{-x} are continuous and differentiable on $(-\infty, \infty)$, and hence the same is true

---

† For instance, see Example 7, page 688, where these functions are used to find the shape of a hanging chain with two points of suspension.

of the functions cosh x and sinh x. Figure 18 shows the graphs of cosh x and sinh x, along with the graphs of $\frac{1}{2}e^x$ and $\frac{1}{2}e^{-x}$ for comparison. By first adding and then subtracting the y-coordinates of the curves $y = \frac{1}{2}e^x$ and $y = \frac{1}{2}e^{-x}$, we see that cosh x is positive for all x, while sinh x is positive for $x > 0$ and negative for $x < 0$. Moreover, setting $x = 0$ in formulas (1) and (2) gives

$$\cosh 0 = 1, \qquad \sinh 0 = 0.$$

Notice also that cosh x is even, since

$$\cosh(-x) = \frac{1}{2}(e^{-x} + e^x) = \frac{1}{2}(e^x + e^{-x}) = \cosh x,$$

while sinh x is odd, since

$$\sinh(-x) = \frac{1}{2}(e^{-x} - e^x) = -\frac{1}{2}(e^x - e^{-x}) = -\sinh x.$$

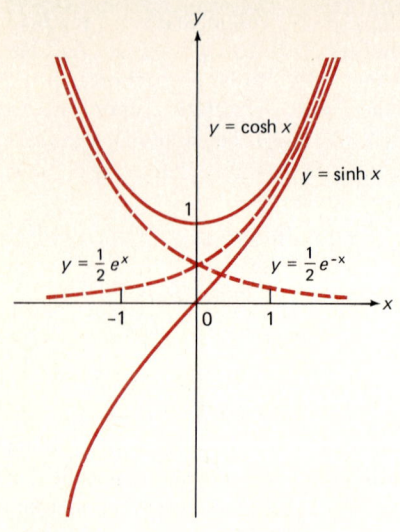

Figure 18

The derivatives of cosh x and sinh x are easily found. In fact,

$$\frac{d}{dx}\cosh x = \frac{1}{2}\frac{d}{dx}(e^x + e^{-x}) = \frac{1}{2}(e^x - e^{-x}),$$

that is,

$$\frac{d}{dx}\cosh x = \sinh x, \qquad (3)$$

while

$$\frac{d}{dx}\sinh x = \frac{1}{2}\frac{d}{dx}(e^x - e^{-x}) = \frac{1}{2}(e^x + e^{-x}),$$

that is

$$\frac{d}{dx}\sinh x = \cosh x. \qquad (4)$$

It follows from (3) and (4) that

$$\int \sinh x \, dx = \cosh x + C \qquad (3')$$

and

$$\int \cosh x \, dx = \sinh x + C. \qquad (4')$$

Since $D_x \cosh x = \sinh x$ is positive for $x > 0$, zero for $x = 0$, and negative for $x < 0$, it follows from the monotonicity test that cosh x is increasing on $[0, \infty)$ and decreasing on $(-\infty, 0]$. Hence the function cosh x has no maximum, and takes its absolute minimum, equal to 1, at $x = 0$. Similarly, since $D_x \sinh x = \cosh x > 0$ for all x, sinh x is increasing on the whole interval $(-\infty, \infty)$, with no extrema. These properties of cosh x and sinh x are apparent from Figure 18, which also shows that

$$\lim_{x \to \pm\infty} \cosh x = \infty,$$

while

$$\lim_{x \to \infty} \sinh x = \infty, \qquad \lim_{x \to -\infty} \sinh x = -\infty.$$

Thus the range of cosh x is $[1, \infty)$, while the range of sinh x is $(-\infty, \infty)$. You should check the validity of these limit formulas by deducing them directly from the definitions (1) and (2).

Example 1 Differentiate sinh (cosh x).

Solution Using (3) and (4), we have

$$\frac{d}{dx} \sinh (\cosh x) = \cosh (\cosh x) \frac{d}{dx} \cosh x = \cosh (\cosh x) \sinh x,$$

with the help of the chain rule. ☐

Example 2 Evaluate $\int_0^{\ln 2} \cosh x \, dx$.

Solution Using (4′), we find that

$$\int_0^{\ln 2} \cosh x \, dx = \sinh (\ln 2) - \sinh 0 = \sinh (\ln 2)$$

$$= \frac{1}{2}(e^{\ln 2} - e^{-\ln 2}) = \frac{1}{2}\left(2 - \frac{1}{2}\right) = \frac{3}{4}. \qquad ☐$$

Example 3 Investigate the concavity of cosh x and sinh x.

Solution Since

$$\frac{d^2}{dx^2} \cosh x = \frac{d}{dx} \sinh x = \cosh x > 0$$

for all x, it follows from the concavity test that cosh x is concave upward on $(-\infty, \infty)$. Similarly, since

$$\frac{d^2}{dx^2} \sinh x = \frac{d}{dx} \cosh x = \sinh x$$

is positive for $x > 0$, zero for $x = 0$, and negative for $x < 0$, we find that sinh x is concave upward on $[0, \infty)$ and concave downward on $(-\infty, 0]$, with an inflection point at $x = 0$ (see Figure 18). ☐

Hyperbolic Identities

The hyperbolic functions satisfy a number of formulas closely resembling those satisfied by the trigonometric functions. The most important of these are

$$\cosh^2 x - \sinh^2 x = 1, \tag{5}$$

and

$$\sinh (x + y) = \sinh x \cosh y + \cosh x \sinh y, \tag{6}$$

$$\cosh (x + y) = \cosh x \cosh y + \sinh x \sinh y. \tag{7}$$

Section 6.8 Hyperbolic Functions **351**

They are all easily proved by returning to the definitions (1) and (2). Thus

$$\cosh^2 x - \sinh^2 x = \frac{1}{4}(e^x + e^{-x})^2 - \frac{1}{4}(e^x - e^{-x})^2$$

$$= \frac{1}{4}(e^{2x} + 2 + e^{-2x} - e^{2x} + 2 - e^{-2x}) = \frac{1}{4}(4) = 1,$$

which proves (5). Similarly

$\sinh x \cosh y + \cosh x \sinh y$

$$= \frac{1}{4}(e^x - e^{-x})(e^y + e^{-y}) + \frac{1}{4}(e^x + e^{-x})(e^y - e^{-y})$$

$$= \frac{1}{4}(e^{x+y} - e^{-x+y} + e^{x-y} - e^{-x-y} + e^{x+y} + e^{-x+y} - e^{x-y} - e^{-x-y})$$

$$= \frac{1}{2}(e^{x+y} - e^{-x-y}) = \sinh(x+y),$$

which proves (6). The proof of (7) is equally straightforward, and is left as an exercise.

Replacing y by $-y$ in (6) and (7), and using the fact that the hyperbolic cosine is even while the hyperbolic sine is odd, we find that

$$\sinh(x - y) = \sinh x \cosh y - \cosh x \sinh y, \qquad (6')$$

$$\cosh(x - y) = \cosh x \cosh y - \sinh x \sinh y. \qquad (7')$$

Also, replacing y by x in (6) and (7), we get the formulas

$$\sinh 2x = 2 \sinh x \cosh x, \qquad \cosh 2x = \cosh^2 x + \sinh^2 x. \qquad (8)$$

The analogy between (6) and the corresponding trigonometric formula

$$\sin(x + y) = \sin x \cos y + \cos x \sin y$$

is complete; to change this formula into (6), we need only change sin to sinh and cos to cosh. On the other hand, to get (5) and (7) from the corresponding trigonometric formulas

$$\cos^2 x + \sin^2 x = 1 \qquad (9)$$

and

$$\cos(x + y) = \cos x \cos y - \sin x \sin y,$$

we must change the signs of the terms involving products of sines, besides replacing sin and cos by sinh and cosh. Similar changes convert the trigonometric double-angle formulas

$$\sin 2x = 2 \sin x \cos x, \qquad \cos 2x = \cos^2 x - \sin^2 x$$

into their hyperbolic analogues (8).

There is another respect in which the hyperbolic functions resemble the trigonometric functions. It follows from formula (9) that the point $P = (\cos \theta, \sin \theta)$ lies on the unit circle

$$x^2 + y^2 = 1$$

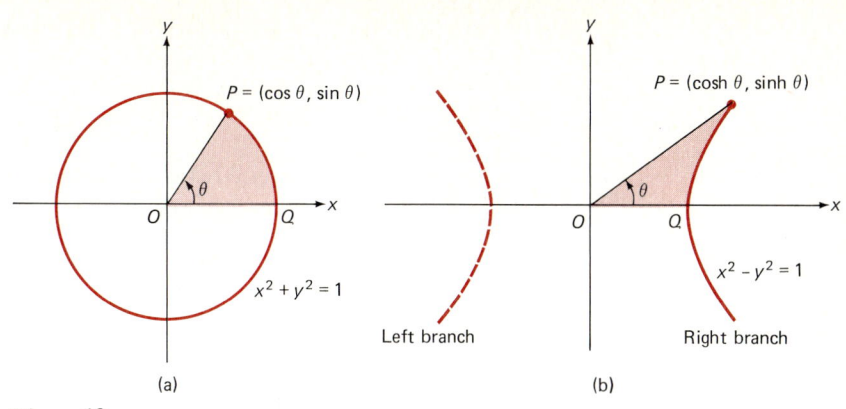

Figure 19

(a) $P = (\cos\theta, \sin\theta)$ on $x^2 + y^2 = 1$

(b) $P = (\cosh\theta, \sinh\theta)$ on $x^2 - y^2 = 1$ (Left branch / Right branch)

[see Figure 19(a)], and in fact θ is the angle between the radius OP and the positive x-axis. If θ is measured in radians and if $0 \leq \theta \leq 2\pi$, then θ is twice the area of the shaded circular sector POQ, bounded by the radius OP, the x-axis and the unit circle (see formula (8), page 54). Similarly, it follows from formula (5) that the point $P = (\cosh\theta, \sinh\theta)$ lies on the curve

$$x^2 - y^2 = 1.$$

This curve, known as the *unit hyperbola*, has two separate parts or *branches* [see Figure 19(b)], but because of the condition $\cosh\theta > 0$, we can confine our attention to the right branch. It is now natural to inquire about the geometric meaning of the variable θ in this case, and remarkably enough, it turns out (see Problem 37, page 397) that θ is just twice the area of the shaded "hyperbolic sector" POQ, bounded by the segment OP, the x-axis and the unit hyperbola. This explains why $\sinh x$, $\cosh x$, etc. are called hyperbolic functions (and, incidentally, why $\sin x$, $\cos x$, etc. are sometimes called *circular functions* instead of trigonometric functions).

> **Remark** Of course, the analogy between trigonometric functions and hyperbolic functions can be carried only so far. For example, $\cosh x$ and $\sinh x$ are neither bounded nor periodic, unlike $\cos x$ and $\sin x$.

Next we introduce the four remaining hyperbolic functions, namely, the *hyperbolic tangent*

$$\tanh x = \frac{\sinh x}{\cosh x} = \frac{e^x - e^{-x}}{e^x + e^{-x}},$$

the *hyperbolic cotangent*

$$\coth x = \frac{\cosh x}{\sinh x} = \frac{1}{\tanh x} = \frac{e^x + e^{-x}}{e^x - e^{-x}},$$

the *hyperbolic secant*

$$\text{sech } x = \frac{1}{\cosh x} = \frac{2}{e^x + e^{-x}},$$

Section 6.8 Hyperbolic Functions

and the *hyperbolic cosecant*

$$\operatorname{csch} x = \frac{1}{\sinh x} = \frac{2}{e^x - e^{-x}}.$$

Notice the analogy between these definitions and the corresponding definitions for the trigonometric functions.

The Hyperbolic Tangent

The functions sinh x and cosh x are both continuous and differentiable on $(-\infty, \infty)$, and cosh x is never equal to zero. Therefore tanh x, the quotient of sinh x and cosh x, is also continuous and differentiable on $(-\infty, \infty)$. Also, as already noted, sinh x is positive for $x > 0$, zero for $x = 0$ and negative for $x < 0$, and hence the same is true of tanh x, since cosh x is always positive. The derivative of tanh x is easily found by using the quotient rule. In fact,

$$\frac{d}{dx} \tanh x = \frac{d}{dx} \frac{\sinh x}{\cosh x} = \frac{\frac{d \sinh x}{dx} \cosh x - \sinh x \frac{d \cosh x}{dx}}{\cosh^2 x}$$

$$= \frac{\cosh^2 x - \sinh^2 x}{\cosh^2 x} = \frac{1}{\cosh^2 x},$$

with the help of (3), (4) and (5), so that

$$\frac{d}{dx} \tanh x = \operatorname{sech}^2 x. \qquad (10)$$

Since $\operatorname{sech}^2 x$ is positive for all x, it follows from (10) that tanh x is increasing on $(-\infty, \infty)$, and in particular has no extrema. Moreover,

$$\tanh(-x) = \frac{\sinh(-x)}{\cosh(-x)} = \frac{-\sinh x}{\cosh x} = -\tanh x,$$

$$\lim_{x \to \infty} \tanh x = \lim_{x \to \infty} \frac{e^x - e^{-x}}{e^x + e^{-x}} = \lim_{x \to \infty} \frac{1 - e^{-2x}}{1 + e^{-2x}} = 1,$$

$$\lim_{x \to -\infty} \tanh x = \lim_{t \to \infty} \tanh(-t) = -\lim_{t \to \infty} \tanh t = -1$$

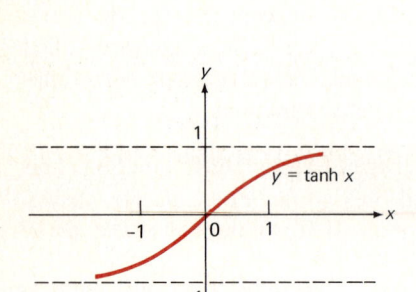

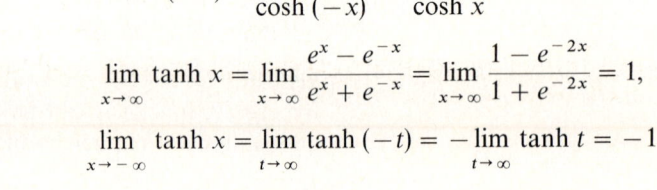

$(e^{-2x} \to 0$ as $x \to \infty)$. Therefore tanh x is an odd function, with range $(-1, 1)$, which has the lines $y = \pm 1$ as horizontal asymptotes. The graph of tanh x is shown in Figure 20, and exhibits all these features.

Figure 20

Example 4 Evaluate $\int \tanh x \, dx$.

Solution Observing that

$$\tanh x = \frac{\sinh x}{\cosh x} = \frac{(\cosh x)'}{\cosh x},$$

we apply formula (3), page 309, and obtain

$$\int \tanh x \, dx = \ln \cosh x + C. \qquad \square$$

354   Chapter 6   Logarithms and Exponentials

Example 5 Evaluate $\int_0^{1/2} \text{sech}^2 x \, dx$.

Solution It follows from (10) that $\tanh x$ is an antiderivative of $\text{sech}^2 x$. Therefore

$$\int_0^{1/2} \text{sech}^2 x \, dx = \tanh \frac{1}{2} - \tanh 0 = \tanh \frac{1}{2}$$

$$= \frac{e^{1/2} - e^{-1/2}}{e^{1/2} + e^{-1/2}} = \frac{e - 1}{e + 1} \approx 0.46. \quad \square$$

The formulas for the derivatives of the hyperbolic functions are

$$\frac{d}{dx} \cosh x = \sinh x,$$

$$\frac{d}{dx} \sinh x = \cosh x,$$

$$\frac{d}{dx} \tanh x = \text{sech}^2 x,$$

$$\frac{d}{dx} \coth x = -\text{csch}^2 x,$$

$$\frac{d}{dx} \text{sech} \, x = -\text{sech} \, x \tanh x,$$

$$\frac{d}{dx} \text{csch} \, x = -\text{csch} \, x \coth x.$$

(Describe the similarities and differences between these formulas and the corresponding formulas for the derivatives of the trigonometric functions.) The first three formulas have already been proved, and to prove the other three, we make repeated use of the rule for differentiating the reciprocal of a function:

$$\frac{d}{dx} \coth x = \frac{d}{dx} \frac{1}{\tanh x} = -\frac{1}{\tanh^2 x} \frac{d}{dx} \tanh x = -\frac{1}{\tanh^2 x} \text{sech}^2 x$$

$$= -\frac{\cosh^2 x}{\sinh^2 x} \frac{1}{\cosh^2 x} = -\frac{1}{\sinh^2 x} = -\text{csch}^2 x,$$

$$\frac{d}{dx} \text{sech} \, x = \frac{d}{dx} \frac{1}{\cosh x} = -\frac{1}{\cosh^2 x} \frac{d}{dx} \cosh x = -\frac{1}{\cosh^2 x} \sinh x$$

$$= -\frac{1}{\cosh x} \frac{\sinh x}{\cosh x} = -\text{sech} \, x \tanh x,$$

$$\frac{d}{dx} \text{csch} \, x = \frac{d}{dx} \frac{1}{\sinh x} = -\frac{1}{\sinh^2 x} \frac{d}{dx} \sinh x = -\frac{1}{\sinh^2 x} \cosh x$$

$$= -\frac{1}{\sinh x} \frac{\cosh x}{\sinh x} = -\text{csch} \, x \coth x.$$

Example 6 Investigate the concavity of $\tanh x$.

Solution The second derivative

$$\frac{d^2}{dx^2}\tanh x = \frac{d}{dx}\operatorname{sech}^2 x = 2\operatorname{sech} x \frac{d}{dx}\operatorname{sech} x = -2\operatorname{sech}^2 x \tanh x$$

is positive for $x < 0$, zero for $x = 0$, and negative for $x > 0$. Hence, by the concavity test, $\tanh x$ is concave upward on $(-\infty, 0]$ and concave downward on $[0, \infty)$, with an inflection point at $x = 0$ (see Figure 20). □

As compared with $\sinh x$, $\cosh x$ and $\tanh x$, the functions $\coth x$, $\operatorname{sech} x$ and $\operatorname{csch} x$ are of minor importance. Thus investigation of these functions is relegated to Problems 30–32.

Problems

Express without using hyperbolic functions
1. $\cosh x + \sinh x$
2. $\cosh x - \sinh x$
3. $\cosh (\ln x)$
4. $\tanh (\ln 2x)$
5. $\sinh (\frac{1}{2} \ln x)$
6. $\cosh^2 (\ln x) + \sinh^2 (\ln x)$

Show that
7. $\cosh^2 x = \dfrac{\cosh 2x + 1}{2}$
8. $\sinh^2 x = \dfrac{\cosh 2x - 1}{2}$
9. $1 - \tanh^2 x = \operatorname{sech}^2 x$
10. $1 - \coth^2 x = -\operatorname{csch}^2 x$
11. $\tanh (x + y) = \dfrac{\tanh x + \tanh y}{1 + \tanh x \tanh y}$

Find the values of the other five hyperbolic functions at the point c if
12. $\cosh c = 2$
13. $\sinh c = -1$
14. $\tanh c = \frac{1}{2}$

Differentiate
15. $\sinh^2 x + \cosh^2 x$
16. $\cosh^3 x$
17. $\sqrt{\cosh 2x}$
18. $\tanh x^2$
19. $\ln (\operatorname{sech} x)$
20. $\coth (\tan x)$
21. $\operatorname{csch} \sqrt{x}$
22. $\sinh e^x$
23. $\tanh (\ln x)$
24. $\log_2 (\cosh x)$
25. $e^{\coth x}$
26. $3^{\sinh x}$

27. Find a simple second order differential equation satisfied by the function $y = a \sinh cx + b \cosh cx$ for arbitrary constants a, b and c. Do the same for the function $y = a \sin cx + b \cos cx$.

28. Find the area A under the curve $y = \cosh x$ from $x = \ln 3$ to $x = \ln 4$.

29. Does either of the functions $\cosh x$ or $\sinh x$ have asymptotes? Explain your answer.

30. Show that the hyperbolic cotangent $\coth x$, graphed in Figure 21, is positive, decreasing and concave upward on $(0, \infty)$, and negative, decreasing and concave downward on $(-\infty, 0)$. Show that $\coth x$ is an odd function, with horizontal asymptotes $y = \pm 1$ and the y-axis as a vertical asymptote. Does $\coth x$ have extrema or inflection points?

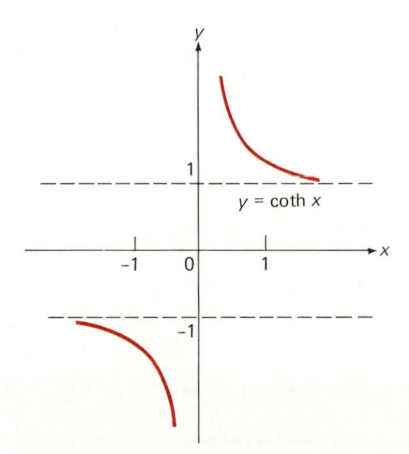

Figure 21

31. Show that the hyperbolic secant $\operatorname{sech} x$, graphed in Figure 22, is a positive even function, with the x-axis as a horizontal asymptote. Show that $\operatorname{sech} x$ is increasing on $(-\infty, 0]$ and decreasing on $[0, \infty)$, with an absolute maximum equal to 1 at $x = 0$, and no minimum. Investigate the concavity of $\operatorname{sech} x$. Where are the inflection points of $\operatorname{sech} x$?

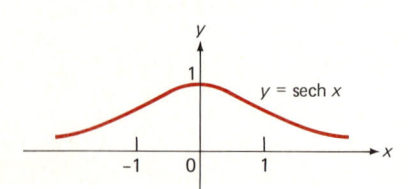

Figure 22

32. Show that the hyperbolic cosecant csch x, graphed in Figure 23, is positive, decreasing and concave upward on $(0, \infty)$, and negative, decreasing and concave downward on $(-\infty, 0)$. Show that csch x is an odd function, with the x-axis as a horizontal asymptote and the y-axis as a vertical asymptote. Does csch x have extrema or inflection points?

Evaluate

33. $\int \cosh^2 x \, dx$

34. $\int \sinh^2 x \, dx$

35. $\int \coth x \, dx$

36. $\int \dfrac{dx}{\sinh x \cosh x}$

37. $\int \text{sech}\,(\ln x) \, dx$

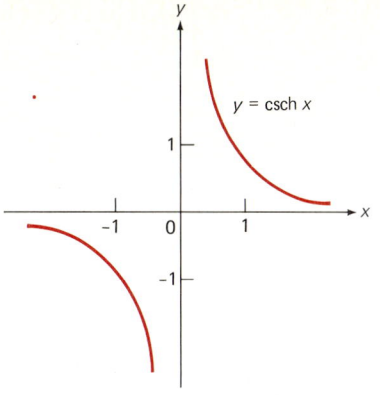

38. $\int \dfrac{\sinh x}{3 \cosh x + 2} \, dx$

Figure 23

6.9 Inverse Hyperbolic Functions

Of the six inverse hyperbolic functions, we now examine the two that are most commonly encountered, namely the inverse hyperbolic sine and the inverse hyperbolic tangent. The other four inverse hyperbolic functions are treated in Problems 11 and 13–15.

The Inverse Hyperbolic Sine

To define the inverse hyperbolic sine, we use a procedure already familiar from the case of the inverse trigonometric functions (see Section 5.3). Let $x = \sinh y$. The continuous function $\sinh y$ is increasing and hence one-to-one on the interval $(-\infty, \infty)$, which it maps onto the same interval $(-\infty, \infty)$. Therefore $x = \sinh y$ has an inverse function $y = \sinh^{-1} x$, the *inverse hyperbolic sine*, which is continuous and increasing on $(-\infty, \infty)$. The graph of this function, shown in Figure 24, can be obtained by reflecting the graph of $\sinh x$ in the line $y = x$.

To differentiate the inverse hyperbolic sine, we write $y = \sinh^{-1} x$, $x = \sinh y$, and apply Theorem 4, page 286, obtaining

$$\frac{d}{dx} \sinh^{-1} x = \frac{dy}{dx} = \frac{1}{\frac{dx}{dy}} = \frac{1}{\cosh y}.$$

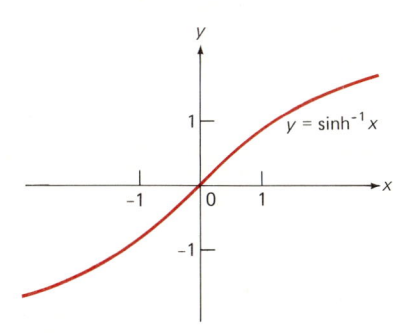

Figure 24

But

$$\cosh y = \pm\sqrt{\sinh^2 y + 1} = \pm\sqrt{x^2 + 1},$$

by formula (5), page 351, where we must choose the plus sign, since $\cosh y$ is positive. Therefore

$$\frac{d}{dx} \sinh^{-1} x = \frac{1}{\sqrt{x^2 + 1}}. \qquad (1)$$

It is an immediate consequence of (1) that

$$\int \frac{dx}{\sqrt{x^2 + 1}} = \sinh^{-1} x + C. \qquad (2)$$

Moreover, using the chain rule to differentiate $\sinh^{-1}(x/a)$, where $a > 0$, we get

$$\frac{d}{dx}\sinh^{-1}\frac{x}{a} = \frac{1/a}{\sqrt{(x/a)^2 + 1}} = \frac{1}{\sqrt{x^2 + a^2}}.$$

It follows that

$$\int \frac{dx}{\sqrt{x^2 + a^2}} = \sinh^{-1}\frac{x}{a} + C, \tag{2'}$$

which generalizes (2).

Example 1 Evaluate $\int \frac{dx}{\sqrt{9x^2 + 4}}$

Solution By formula (2') with $a = \frac{2}{3}$,

$$\int \frac{dx}{\sqrt{9x^2 + 4}} = \frac{1}{3}\int \frac{dx}{\sqrt{x^2 + \frac{4}{9}}} = \frac{1}{3}\sinh^{-1}\frac{3x}{2} + C. \quad \square$$

There is a simple connection between the inverse hyperbolic sine and the logarithm. Let $x = \sinh y$. Then, since

$$e^y = \frac{e^y - e^{-y}}{2} + \frac{e^y + e^{-y}}{2} = \sinh y + \cosh y$$

$$= \sinh y + \sqrt{\sinh^2 y + 1} = x + \sqrt{x^2 + 1},$$

it follows that

$$y = \sinh^{-1} x = \ln(x + \sqrt{x^2 + 1}). \tag{3}$$

Therefore, if $a > 0$,

$$\sinh^{-1}\frac{x}{a} = \ln\frac{x + \sqrt{x^2 + a^2}}{a} = \ln(x + \sqrt{x^2 + a^2}) - \ln a, \tag{3'}$$

and we can now write (2') in the more useful form

$$\int \frac{dx}{\sqrt{x^2 + a^2}} = \ln(x + \sqrt{x^2 + a^2}) + C, \tag{4}$$

after absorbing $-\ln a$ into the constant of integration C. You should check the validity of (4) by differentiating the expression on the right. $\quad \square$

Example 2 Evaluate $\int_0^3 \frac{dx}{\sqrt{x^2 + 2}}$.

Solution By formula (4) with $a = \sqrt{2}$,

$$\int_0^3 \frac{dx}{\sqrt{x^2 + 2}} = \left[\ln(x + \sqrt{x^2 + 2})\right]_0^3 = \ln(3 + \sqrt{11}) - \ln\sqrt{2}$$

$$= \ln\frac{3 + \sqrt{11}}{\sqrt{2}} \approx 1.5 \quad \square$$

358 Chapter 6 Logarithms and Exponentials

The Inverse Hyperbolic Tangent

To define the inverse hyperbolic tangent, let $x = \tanh y$. The continuous function $\tanh y$ is increasing and hence one-to-one on the interval $(-\infty, \infty)$, which it maps onto the interval $(-1, 1)$. Therefore $x = \tanh y$ has an inverse function $y = \tanh^{-1} x$, the *inverse hyperbolic tangent*, which is continuous and increasing on $(-1, 1)$. The graph of this function, shown in Figure 25, can be obtained by reflecting the graph of $\tanh x$ in the line $y = x$.

As in the case of $\sinh^{-1} x$, there is a simple connection between the function $\tanh^{-1} x$ and the logarithm. Writing $x = \tanh y$, we have

$$x = \frac{e^y - e^{-y}}{e^y + e^{-y}}.$$

Therefore

$$(e^y + e^{-y})x = e^y - e^{-y},$$

or equivalently

$$(e^{2y} + 1)x = e^{2y} - 1.$$

This is a linear equation in e^{2y}, with solution

$$e^{2y} = \frac{1 + x}{1 - x}.$$

It follows that

$$y = \tanh^{-1} x = \frac{1}{2} \ln \frac{1 + x}{1 - x} \qquad (-1 < x < 1). \tag{5}$$

In particular,

$$\frac{d}{dx} \tanh^{-1} x = \frac{1}{2} \frac{d}{dx} \ln \frac{1 + x}{1 - x} = \frac{1}{2} \frac{d}{dx} [\ln(1 + x) - \ln(1 - x)]$$

$$= \frac{1}{2} \left(\frac{1}{1 + x} + \frac{1}{1 - x} \right) = \frac{1}{(1 + x)(1 - x)},$$

and hence

$$\frac{d}{dx} \tanh^{-1} x = \frac{1}{1 - x^2} \qquad (-1 < x < 1). \tag{6}$$

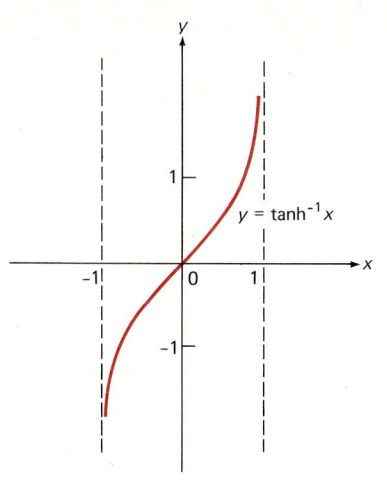

Figure 25

Example 3 Differentiate $\tanh^{-1}(\sin x)$.

Solution Using (6), we find that

$$\frac{d}{dx} \tanh^{-1}(\sin x) = \frac{1}{1 - \sin^2 x} \frac{d}{dx} \sin x = \frac{\cos x}{\cos^2 x} = \frac{1}{\cos x} = \sec x. \quad \square$$

Example 4 Evaluate $\int_0^{\pi/6} \sec x \, dx$.

Solution By the preceding example, $\tanh^{-1}(\sin x)$ is an antiderivative of $\sec x$. Therefore, with the help of (5),

$$\int_0^{\pi/6} \sec x \, dx = \tanh^{-1}\left(\sin \frac{\pi}{6}\right) - \tanh^{-1}(\sin 0)$$

$$= \tanh^{-1} \frac{1}{2} = \frac{1}{2} \ln \frac{1 + \frac{1}{2}}{1 - \frac{1}{2}} = \frac{1}{2} \ln 3 \approx 0.55 \quad \square$$

Section 6.9 Inverse Hyperbolic Functions

Problems

Differentiate
1. $x \sinh^{-1} x$
2. $\sin(\sinh^{-1} x)$
3. $\sinh^{-1}(\cos x)$
4. $x^2 \tanh^{-1} x$
5. $\tanh^{-1}(\ln x)$
6. $\sinh^{-1}(\tanh^{-1} x)$

7. Verify formula (1) with the help of formula (3).
8. Verify formula (6) with the help of Theorem 4, page 286.
9. What is the smallest value taken by the function $\sinh x + 2 \cosh x$?
10. Does the function $2 \sinh x + \cosh x$ have a smallest value?
11. The inverse of the function $x = \cosh y$ $(0 \le y < \infty)$ is called the *inverse hyperbolic cosine*, denoted by $y = \cosh^{-1} x$, and has the graph shown in Figure 26. Show that

$$\cosh^{-1} x = \ln(x + \sqrt{x^2 - 1}) \qquad (x \ge 1),$$

$$\frac{d}{dx} \cosh^{-1} x = \frac{1}{\sqrt{x^2 - 1}} \qquad (x > 1),$$

and

$$\int \frac{dx}{\sqrt{x^2 - a^2}} = \cosh^{-1} \frac{x}{a} + C$$

$$= \ln(x + \sqrt{x^2 - a^2}) + C \qquad (x > a > 0). \quad \text{(i)}$$

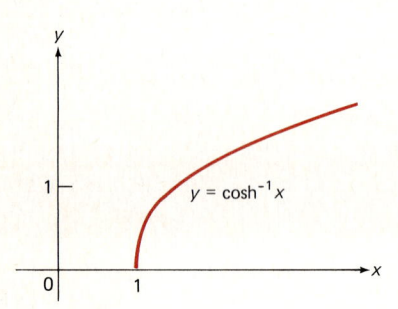

Figure 26

12. Show that

$$\int \frac{dx}{a^2 - x^2} = \frac{1}{a} \tanh^{-1} \frac{x}{a} + C$$

$$= \frac{1}{2a} \ln \frac{a + x}{a - x} + C \qquad (|x| < a), \quad \text{(ii)}$$

which you will recognize as a variant of formula (6), page 310.

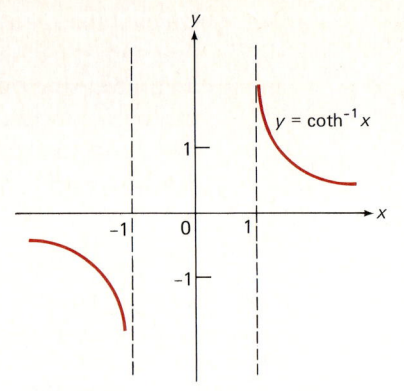

Figure 27

13. The inverse of the function $x = \coth y$ is called the *inverse hyperbolic cotangent*, denoted by $y = \coth^{-1} x$, and has the graph shown in Figure 27. Show that

$$\coth^{-1} x = \frac{1}{2} \ln \frac{x+1}{x-1} = \tanh^{-1} \frac{1}{x} \qquad (|x| > 1),$$

$$\frac{d}{dx} \coth^{-1} x = \frac{1}{1 - x^2} \qquad (|x| > 1),$$

and

$$\int \frac{dx}{a^2 - x^2} = \frac{1}{a} \coth^{-1} \frac{x}{a} + C$$

$$= \frac{1}{2a} \ln \frac{x + a}{x - a} + C \qquad (|x| > a > 0), \quad \text{(iii)}$$

where (iii) will be recognized as another variant of formula (6), page 310.

14. The inverse of the function $x = \text{sech } y$ $(0 \le y < \infty)$ is called the *inverse hyperbolic secant*, denoted by $y = \text{sech}^{-1} x$, and has the graph shown in Figure 28. Show that

$$\text{sech}^{-1} x = \ln \frac{1 + \sqrt{1 - x^2}}{x} = \cosh^{-1} \frac{1}{x} \qquad (0 < x \le 1),$$

$$\frac{d}{dx} \text{sech}^{-1} x = -\frac{1}{x\sqrt{1 - x^2}} \qquad (0 < x < 1),$$

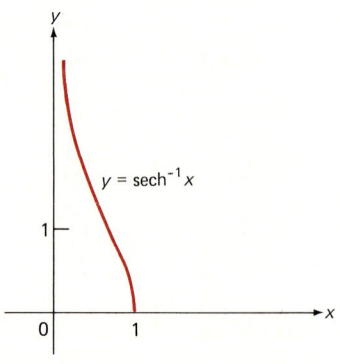

Figure 28

360   Chapter 6   Logarithms and Exponentials

and

$$\int \frac{dx}{x\sqrt{a^2-x^2}} = -\frac{1}{a}\text{sech}^{-1}\frac{|x|}{a} + C$$

$$= -\frac{1}{a}\ln\frac{a+\sqrt{a^2-x^2}}{|x|} + C \quad (0 < |x| < a).$$

(iv)

15. The inverse of the function $x = \text{csch } y$ is called the *inverse hyperbolic cosecant*, denoted by $y = \text{csch}^{-1} x$, and has the graph shown in Figure 29. Show that

$$\text{csch}^{-1} x = \ln\left(\frac{1}{x} + \frac{\sqrt{1+x^2}}{|x|}\right) = \sinh^{-1}\frac{1}{x} \quad (x \neq 0),$$

$$\frac{d}{dx}\text{csch}^{-1} x = -\frac{1}{|x|\sqrt{1+x^2}} \quad (x \neq 0),$$

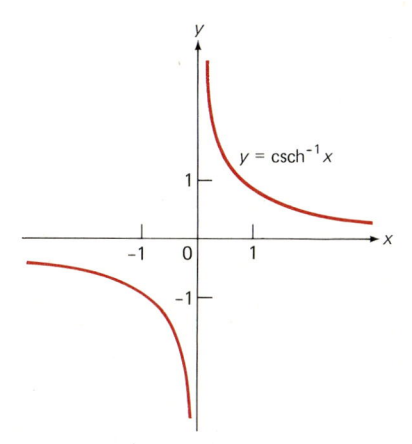

Figure 29

and

$$\int \frac{dx}{x\sqrt{a^2+x^2}} = -\frac{1}{a}\text{csch}^{-1}\frac{|x|}{a} + C$$

$$= -\frac{1}{a}\ln\frac{a+\sqrt{a^2+x^2}}{|x|} + C \quad (a > 0, x \neq 0).$$

(v)

Differentiate

16. $\dfrac{\cosh^{-1} x}{x}$ **17.** $\cosh^{-1}(\cos x)$ **18.** $\ln(\coth^{-1} x)$

19. $\coth^{-1}\sqrt{x}$ **20.** $\text{sech}^{-1}(e^{-x})$ **21.** $\text{csch}^{-1}(\ln x)$

Evaluate

22. $\displaystyle\int_0^2 \frac{dx}{\sqrt{x^2+4}}$ **23.** $\displaystyle\int_{\sinh 1}^{\sinh 2} \frac{dx}{\sqrt{x^2+1}}$

24. $\displaystyle\int_2^4 \frac{dx}{\sqrt{x^2-3}}$ **25.** $\displaystyle\int_{2/3}^1 \frac{dx}{\sqrt{9x^2-1}}$

26. $\displaystyle\int_0^2 \frac{dx}{9-x^2}$ **27.** $\displaystyle\int_4^6 \frac{dx}{9-x^2}$

28. $\displaystyle\int_{1/4}^{1/2} \frac{dx}{x\sqrt{1-x^2}}$ **29.** $\displaystyle\int_2^4 \frac{dx}{x\sqrt{1+x^2}}$

30. $\displaystyle\int_1^2 \frac{dx}{x\sqrt{4+x^2}}$

31. Three of the six inverse hyperbolic functions have inflection points. Which are they, and where are the inflection points?

Key Terms and Topics
Definition of the natural logarithm as an integral
The logarithm of a product and of a power
Definition of the number *e*
Logarithmic differentiation, integration of a logarithmic derivative
Definition of the exponential as the inverse of the logarithm
The exponential of a sum
Laws of exponents for arbitrary real exponents
Exponentials and logarithms to the base *a*
The general power function x^a
The indeterminate forms 0^0, ∞^0 and 1^∞
The mathematics of compound interest, continuous compounding
Separable differential equations, separation of variables
Exponential growth and decay
Population growth, radioactive decay
Doubling time, half-life
Logistic growth
The hyperbolic functions and their derivatives
The inverse hyperbolic functions

Differentiation Formulas Involving Logarithms and Exponentials

Function	Derivative	
$\ln x$	$\dfrac{1}{x}$	$(x > 0)$
$\ln |x|$	$\dfrac{1}{x}$	$(x \neq 0)$
$\ln |f(x)|$	$\dfrac{f'(x)}{f(x)}$	
$\log_a x$	$\dfrac{1}{x} \log_a x$	$(a > 0)$
e^x	e^x	
a^x	$a^x \ln a$	$(a > 0)$
x^a	ax^{a-1}	$(x > 0)$
$\cosh x$	$\sinh x$	
$\sinh x$	$\cosh x$	

Other Key Formulas

$$\ln x = \int_1^x \frac{dt}{t} \quad (x > 0), \quad \ln 1 = 0, \quad \ln e = 1$$

$$e^{\ln x} = x \quad (x > 0), \quad \ln e^x = x \quad (\text{all } x)$$

$$\ln xy = \ln x + \ln y, \quad e^x e^y = e^{x+y}$$

$$\lim_{x \to \infty} \ln x = \infty, \quad \lim_{x \to 0^+} \ln x = -\infty, \quad \lim_{x \to 0^+} x \ln x = 0$$

$$\lim_{x \to \infty} e^x = \infty, \quad \lim_{x \to -\infty} e^x = 0$$

$$\lim_{x \to \infty} \left(1 + \frac{a}{x}\right)^x = e^a, \quad \lim_{n \to \infty} \left(1 + \frac{1}{n}\right)^n = e$$

$$a^x = e^{x \ln a} \quad (a > 0), \quad \log_a x = \frac{\ln x}{\ln a} \quad (a > 0)$$

$$x^a = e^{a \ln x} \quad (x > 0), \quad \ln x^a = a \ln x$$

$$\cosh x = \frac{1}{2}(e^x + e^{-x}), \quad \sinh x = \frac{1}{2}(e^x - e^{-x}), \quad \cosh^2 x - \sinh^2 x = 1$$

Supplementary Problems

Solve for x.

1. $\ln x = \frac{1}{2}(\ln 4 + \ln 9)$
2. $\ln x^3 - \ln x = \ln 32 - \ln 8$
3. $\log_2 x = \log_2 4 + \log_4 8 + \log_{16} 64$
4. $\log_{100} x + \log_{0.1} x = 1$

5. Is the function $\log_a \dfrac{1-x}{1+x}$ even or odd?
6. Without doing any numerical calculations, show that $\pi^e < e^\pi$ and $\sqrt{10}^\pi < \pi^{\sqrt{10}}$.
   Hint. First show that $(\ln x)/x$ is decreasing on $[e, \infty)$.

362   Chapter 6   Logarithms and Exponentials

Evaluate

7. $\int_0^{\pi/4} \tan s \, ds$

8. $\int_{\pi/4}^{3\pi/4} \cot t \, dt$

9. $\int_{\pi/6}^{\pi/3} \dfrac{du}{\sin u \cos u}$

10. Show that
$$\sum_{n=10}^{29} \ln\left(1 + \frac{1}{n}\right) = \ln 3.$$

Differentiate

11. $e^{\cosh x}$ 12. $\pi^{\ln x}$ 13. 2^{3x}
14. $\ln(\tanh^{-1} x)$ 15. $x^{\sinh x}$ 16. $[\ln(\ln x)]^x$

Find the local extrema of the given function.
17. $f(x) = (x+1)^{10} e^{-x}$
18. $f(x) = ae^{cx} + be^{-cx}$ $(a^2 + b^2 \ne 0, c \ne 0)$
19. $f(x) = x^e 2^{-x}$

20. Show that the function $f(x) = e^x + cx^3$ has no inflection point if $-e/6 \le c \le 0$, one inflection point if $c > 0$, and two inflection points if $c < -e/6$.

Does the given function have an inflection point, and if so, where?
21. $f(x) = x^2 + \ln x$ 22. $f(x) = x^4 + x^2 + e^x$
23. $f(x) = x^x$ 24. $f(x) = e^{\arctan x}$
25. $f(x) = e^{x^{1/3}}$

Use L'Hospital's rule to evaluate

26. $\lim\limits_{x \to e} \dfrac{\ln x - 1}{x - e}$ 27. $\lim\limits_{x \to \infty} \dfrac{\pi - 2 \arctan x}{\ln\left(1 + \dfrac{1}{x}\right)}$

28. $\lim\limits_{x \to 0} \dfrac{e^{\sin 2x} - e^{\tan x}}{x}$ 29. $\lim\limits_{x \to 0^+} \dfrac{x \ln x}{\sin x}$

30. $\lim\limits_{x \to 0} \dfrac{e^{x^2} - \cos x}{x^2}$

31. $\lim\limits_{x \to \infty} x[\ln(x + \pi) - \ln x]$

32. $\lim\limits_{x \to -\infty} \left(1 - \dfrac{4}{x}\right)^{x-1}$ 33. $\lim\limits_{x \to (1/4)\pi^-} (\tan x)^{\tan 2x}$

Use separation of variables to solve the given initial value problem.

34. $\dfrac{dy}{dx} \cot x = y \ln y$, $y(0) = e$

35. $\sqrt{x^2 + 1} \dfrac{dy}{dx} = y$, $y(0) = 1$

36. $\sqrt{1 - x^2} \dfrac{dy}{dx} = y^2 + 1$, $y(0) = 0$

37. $\dfrac{dy}{dx} = e^{x+y}$, $y(0) = 0$

38. $(x + \sqrt{x}) \dfrac{dy}{dx} + y = 0$, $y(1) = (\tfrac{1}{4})$

39. $(1 - x^2) \dfrac{dy}{dx} + y = 0$, $y(\tfrac{5}{3}) = 2$

40. By making the preliminary substitution $z = 2x - y - 1$ and then separating variables, solve the initial value problem $dy/dx = 2x - y - 1$, $y(0) = 1$.

41. A curve has the property that the slope of the curve at every point P is n times the slope of the line joining the origin to P. What is the curve?

42. How long does it take money invested at an annual interest rate of 10% to triple if compounded continuously?

43. Find the minimum investment in a municipal bond fund paying an annual interest rate of 9% compounded continuously that will allow the investor to withdraw $10,000 every year in perpetuity (that is, forever).

44. $50,000 is deposited in a bank account paying 6% interest per year compounded monthly. When does the money in the account first exceed $75,000?

45. A continuously compounded initial deposit of $1250 grows to $2000 in 5 years. What is the annual interest rate?

46. What is the present value of $25,000 due in 6 years at an annual interest rate of 7.5% compounded continuously? At the same rate compounded monthly?

47. Every 3 months a saver deposits $125 in a bank account paying an annual interest rate of 8% compounded quarterly. Find the amount of money in the account just before making the twenty-first deposit, that is, after 5 years.

48. The population of a city increases from 125,000 to 180,000 in 15 years. What is the annual rate of growth of the population?

49. An exponentially growing bacterial culture increases from 2×10^5 cells to 8×10^7 cells in 4 hr. Find the time between successive binary fissions (cell divisions).

50. A mutation occurs in a culture of 1024 cells growing exponentially with a doubling time of 1 hr. The mutant cells have a doubling time of only 30 min. When will the population of mutant cells equal that of the cells of the original type? When will there be 16 mutant cells for each cell of the original type?

51. Experiments show that when bacteria are grown in a nutrient medium, the rate of *nutrient uptake*, that is, the rate of change of the concentration C of the nutrient in the bacteria, is proportional to $C_1 - C$, where C_1 is the final concentration of the nutrient (it is found that C_1 is much larger than the concentration in the medium itself). Therefore

$$\frac{dC}{dt} = k(C_1 - C).$$

Solve this differential equation, subject to the initial condition $C = 0$ at time $t = 0$.

52. A bacterial culture is growing logistically, with initial size $N_0 = 4000$ and limiting size N_1. The culture achieves size $\frac{1}{2}N_1$ in 12 hr and size $\frac{9}{10}N_1$ in 15 hr. What is the limiting size N_1?

53. The growth of solid tumors is accurately described by the differential equation

$$\frac{dm}{dt} = rm \ln \frac{m_1}{m}, \qquad (i)$$

where r and m_1 are positive constants, and $m = m(t)$ is the mass of the tumor at time t. Show that the solution of (i) satisfying the initial condition $m(0) = m_0$ ($m_0 < m_1$) is given by the function

$$m = m_1 \exp\left(e^{-rt} \ln \frac{m_0}{m_1}\right) = m_1 \left(\frac{m_0}{m_1}\right)^{e^{-rt}}, \qquad (ii)$$

known as the *Gompertz growth law*. Verify that (ii) is an increasing function of t and that $m \to m_1$ as $t \to \infty$, so that m_1 is the limiting mass of the tumor. Figure 30 shows the Gompertz law (ii) for the case $m_1 = 10m_0$. Note that the curve is S-shaped and closely resembles the corresponding logistic curve in Figure 14, page 342.

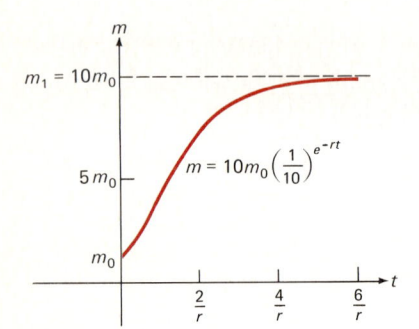

Figure 30

54. Show that if $m_1 > em_0$, the Gompertz law (ii) has an inflection point at $t = t_1$, where t_1 is the time at which m equals m_1/e.

55. To explain how the Gompertz law arises, solve the problem of exponential growth governed by the differential equation $dm/dt = km$ and the initial condition $m(0) = m_0$, where the relative growth rate k, instead of being constant, is an exponentially decreasing function $k_0 e^{-rt}$. Show that

the solution of this problem is

$$m = m_0 \exp\left[\frac{k_0}{r}(1 - e^{-rt})\right], \qquad (ii')$$

where $m \to m_0 e^{k_0/r}$ as $t \to \infty$. Show that (ii) and (ii') coincide if $m_1 = m_0 e^{k_0/r}$.

56. One fourth of a radioactive substance disappears in 15 years. What is the half-life of the substance?

57. Naturally occurring uranium consists of two radioactive isotopes, namely uranium-238 with a half-life of about 4.5×10^9 years, and uranium-235 with a half-life of about 7×10^8 years. In contemporary samples uranium-238 is about 137.8 times more abundant than uranium-235. Assuming that the two isotopes were equally abundant at the time the uranium was created, presumably as the result of a supernova explosion, how old is the uranium?

58. Although stable when bound in the nucleus of an atom, a free neutron will decay, with a half-life of 12.8 min, into a proton, an electron and an antineutrino. Suppose a beam of free neutrons is emitted into space with a velocity of 25 km/sec. How far does the beam travel during the time it takes one tenth of the neutrons to decay?

59. Let N be the number of free neutrons in a solid sphere of uranium-235 of radius R. Then N satisfies the differential equation

$$\frac{dN}{dt} = aN - \frac{bN}{R}, \qquad (iii)$$

with constants $a \approx 2 \times 10^8$/sec and $b \approx 17 \times 10^8$ cm/sec. The term aN describes the "multiplication" of neutrons due to nuclear fission (most of the uranium nuclei hit by neutrons disintegrate, each releasing two or three "fresh" neutrons), while the term $-bN/R$ describes the loss of neutrons escaping through the surface of the sphere (the surface-to-volume ratio of the sphere is proportional to $1/R$). Show that N increases dramatically when the radius R reaches a certain value R_{cr}, called the *critical radius*, or equivalently, when the mass of the sphere reaches a certain value m_{cr}, called the *critical mass*. Find R_{cr} and m_{cr}, given that the density of uranium-235 is 18.7 g/cm^3. This is a simplified model of the "chain reaction" occurring in an atomic bomb.

60. Consider a chemical reaction in which one molecule of a substance A interacts with one molecule of a substance B to produce one molecule of a third substance C. Suppose a and b are the initial concentrations of A and B, and let $y = y(t)$ be the concentration of C at time t. Then the concentrations of A and B at time t are given by $a - y$ and $b - y$, respectively, and the rate at which the reaction proceeds is proportional to the product of these concentrations. This leads to the differential equation

$$\frac{dy}{dt} = k(a - y)(b - y)$$

($k > 0$), known as the *rate law* of the reaction. Find y as a function of t, given that $y(0) = 0$. Show that $y \to \min\{a, b\}$

as $t \to \infty$ if $a \neq b$. What happens if $a = b$? Let T be the time it takes the reaction to be 99% complete. Calculate T for $a = b$ and for $a = \frac{1}{2}b$. Which time is larger, and why?

61. An object of mass m, initially at rest, falls in a medium resisting its motion with a force proportional to the square of its velocity $v = v(t)$. Thus, by Newton's second law,

$$m \frac{dv}{dt} = mg - bv^2,$$

where g is the acceleration due to gravity and b is a positive constant. Show that the object has a *limiting* or *terminal velocity* v_1, which is approached but never exceeded. Show that v_1 is proportional to the square root of the object's mass. Find the object's position s as a function of time.

62. A stone of mass m is thrown vertically upward with initial velocity v_0. Suppose the stone's motion is opposed by air resistance proportional to the square of its velocity, with constant of proportionality b. Show that the (unsigned) velocity of the stone upon returning to its initial position is

$$v_1 = v_0 \sqrt{\frac{mg}{mg + bv_0^2}}.$$

63. An object of mass m, initially at rest, falls in a medium resisting its motion with a force proportional to its velocity $v = v(t)$, rather than to the square of its velocity as in Problem 61. (This occurs for a sufficiently small object like a raindrop or for a sufficiently viscous medium like heavy oil.) Show that just as in the case of square law resistance, the object has a *limiting* or *terminal velocity* v_1, which is approached but never exceeded. Show that v_1 is proportional to the object's mass. Find the object's position s as a function of time.

64. Use Leibniz's rule (Problem 35, page 229) to find the hundredth derivative of $x \sinh x$.

Show that

65. $\cosh x - \cosh y = 2 \sinh \dfrac{x+y}{2} \sinh \dfrac{x-y}{2}$

66. $\sinh x - \sinh y = 2 \cosh \dfrac{x+y}{2} \sinh \dfrac{x-y}{2}$

67. $\sinh^2 x - \sinh^2 y = \sinh(x+y)\sinh(x-y)$
68. $(\cosh x + \sinh x)^a = \cosh ax + \sinh ax$ (a arbitrary)
69. $\sinh 3x = 3 \sinh x + 4 \sinh^3 x$
70. $\cosh 4x = 8 \cosh^4 x - 8 \cosh^2 x + 1$

Use L'Hospital's rule to evaluate

71. $\lim\limits_{x \to 0} \dfrac{\sinh x}{x}$

72. $\lim\limits_{x \to 0} \dfrac{\tanh x}{x}$

73. $\lim\limits_{x \to 0} \dfrac{1 - \cosh x}{x^2}$

74. $\lim\limits_{x \to 0} \dfrac{\tanh x - x}{x^3}$

75. $\lim\limits_{x \to 0} \dfrac{x - \sinh^{-1} x}{x^3}$

76. $\lim\limits_{x \to 0} \dfrac{\tanh^{-1} x}{\sinh^{-1} x}$

77. $\lim\limits_{x \to 0^+} x^{\sinh x}$

78. $\lim\limits_{x \to 0^+} (\tanh x)^{\tan x}$

79. $\lim\limits_{x \to 0} (1 + \sinh x)^{\operatorname{csch} x}$

Suppose that for all x in its domain a function $y = f(x)$ satisfies an equation of the form

$$P_0(x) + P_1(x)y + P_2(x)y^2 + \cdots + P_n(x)y^n = 0, \quad \text{(iv)}$$

where $P_0(x), P_1(x), P_2(x), \ldots, P_n(x) \neq 0$ are all polynomials in x. Then the function f is said to be *algebraic*. Show that
80. Rational functions are algebraic
81. The function $y = \sqrt[3]{x - 2\sqrt{x}}$ is algebraic
82. The inverse of a one-to-one algebraic function is itself algebraic.

Comment. A function which is not algebraic is said to be *transcendental*. It is known that exponentials, trigonometric functions and hyperbolic functions are transcendental. It then follows from Problem 82 that the same is true of logarithms, inverse trigonometric functions and inverse hyperbolic functions.

Evaluate

83. $\displaystyle\int \dfrac{dx}{\sqrt{25x^2 - 16}}$

84. $\displaystyle\int \dfrac{dx}{x\sqrt{1 - 4x^2}}$

85. $\displaystyle\int \dfrac{dx}{x\sqrt{9 + 49x^2}}$

86. The *hyperbolic argument* or *Gudermannian* is the function $y = \operatorname{gd} x$ defined by the formula

$$y = \operatorname{gd} x = \int_0^x \frac{dt}{\cosh t}.$$

Show that

$$\operatorname{gd} x = \arctan(\sinh x) = 2 \arctan(e^x) - \frac{\pi}{2},$$

$$\sinh x = \tan(\operatorname{gd} x), \qquad \cosh x = \sec(\operatorname{gd} x),$$
$$\tanh x = \sin(\operatorname{gd} x).$$

Also show that $y = \operatorname{gd} x$ is a one-to-one function, with inverse

$$x = \operatorname{gd}^{-1} y = \int_0^y \sec t\, dt.$$

Supplementary Problems   **365**

7 Methods of Integration

The stock of integration formulas established in Chapters 4–6, although sizable, is still quite inadequate for the needs of practical calculus. Thus in the present chapter we make a concerted attack on the problem of evaluating integrals, both indefinite and definite. The two most important methods of integration are *integration by substitution* (see Section 7.1) and *integration by parts* (see Section 7.2). These methods are also the main tools in subsequent sections as well, where we consider a variety of integrals involving trigonometric functions and radicals. In addition there is a general technique for integrating an arbitrary rational function, given in Section 7.6.

A surprising feature of the subject is the existence of many integrals, some quite simple in appearance, which cannot be evaluated "in closed form." Thus, for example, none of the integrals

$$\int e^{-x^2}\,dx, \quad \int \frac{\sin x}{x}\,dx, \quad \int \cos x^2\,dx \qquad (1)$$

can be expressed as an explicit formula involving only a finite number of algebraic operations, including composition and extraction of roots, performed on the functions studied in the preceding chapters (a function which can be expressed by such a formula is said to be *elementary*). To deal with integrals like (1), we must resort to the approximation methods of Section 7.8, or to the use of infinite series (see Chapter 9). There is also the possibility of regarding such integrals as defining "brand new" functions, and then making a special study of these functions, just as we did in the case of the integral $\int (1/x)\,dx$ defining the natural logarithm.

The concept of the integral can be enlarged to include *improper integrals*, that is, integrals with unbounded integrands and integrals over unbounded intervals. Section 7.9 is devoted to this important topic.

7.1 Integration by Substitution

The chain rule for differentiating a composite function leads to an important method of integration, known as integration by *substitution* (or integration by *change of variables*), which can be used to evaluate both indefinite and definite integrals. First we consider the case of indefinite integrals.

> **Theorem 1** *(Evaluation of an indefinite integral by substitution). Let $f(u)$ be a continuous function, and let $u = u(x)$ be a function with a continuous derivative $u'(x)$. Then*
>
> $$\int f(u(x))u'(x)\,dx = \left.\int f(u)\,du\right|_{u=u(x)}, \qquad (1)$$
>
> *where the notation on the right means that the substitution $u = u(x)$ is made in the integral $\int f(u)\,du$ after its evaluation.*

Proof To establish the equality of the two integrals in (1), let F be an antiderivative of f, where the existence of F is guaranteed by the continuity of f (see Theorem 5, p. 254). Then, by the chain rule,

$$\frac{d}{dx} F(u(x)) = F'(u(x))u'(x),$$

and hence

$$\int f(u(x))u'(x)\,dx = \int F'(u(x))u'(x)\,dx = F(u(x)) + C, \qquad (2)$$

where C is an arbitrary constant. But, on the other hand,

$$\left.\int f(u)\,du\right|_{u=u(x)} = \left.[F(u) + C]\right|_{u=u(x)} = F(u(x)) + C, \qquad (2')$$

and comparison of (2) and (2') immediately gives (1). ∎

Observe that the extra factor of $u'(x)$ on the left in equation (1) is "concealed" in the differential du on the right, since $du = u'(x)\,dx$. We can now appreciate the merit of using a notation for integrals in which the expression behind the integral sign is written as the product of the function being integrated (the integrand) and the *differential* of the variable of integration. In fact, if the variable of integration is changed, the differential automatically takes account of the action of the chain rule. Manipulation of differentials will play a key role in the examples that follow.

Example 1 Evaluate $\int \sin^3 x \cos x \, dx$.

Solution Let $u = \sin x$. Then $du = (D_x \sin x)\,dx = \cos x\,dx$, and

$$\int \sin^3 x \cos x \, dx = \int u^3 \, du = \frac{1}{4} u^4 + C.$$

Therefore

$$\int \sin^3 x \cos x \, dx = \frac{1}{4} \sin^4 x + C,$$

after replacing u by $\sin x$. Here, as in any problem involving the evaluation of an indefinite integral, it is a good idea to check the work by verifying that the derivative of the alleged answer is actually equal to the original integrand (do so). In this example the functions figuring in Theorem 1 are $f(u) = u^3$ and $u = \sin x$, and equation (1) takes the form

$$\int \sin^3 x \cos x \, dx = \int u^3 \, du \bigg|_{u=\sin x}. \qquad \square$$

Example 2 Evaluate $\int \sin^3 x \, dx$.

Solution In Example 1 the proper choice of the substitution $u = u(x)$ was immediately apparent, but it is now less clear what choice to make. At first we might be tempted to make the same substitution $u = \sin x$, but this choice is inappropriate, since $du = \cos x \, dx$ and there is no factor of $\cos x$ in the given integral. However, suppose we try the substitution $u = \cos x$ instead. Then $du = (D_x \cos x) \, dx = -\sin x \, dx$, and the integrand already contains $\sin x$ as a factor, so that

$$\int \sin^3 x \, dx = \int (-\sin^2 x) \, du.$$

Further progress now depends on expressing $-\sin^2 x$ in terms of the variable u. This is easily accomplished, since

$$-\sin^2 x = \cos^2 x - 1 = u^2 - 1.$$

Therefore

$$\int \sin^3 x \, dx = \int (u^2 - 1) \, du = \frac{1}{3} u^3 - u + C,$$

which gives

$$\int \sin^3 x \, dx = \frac{1}{3} \cos^3 x - \cos x + C,$$

after replacing u by $\cos x$. The functions figuring in Theorem 1 are now $f(u) = u^2 - 1$ and $u = \cos x$, and equation (1) takes the form

$$\int \sin^3 x \, dx = \int (u^2 - 1) \, du \bigg|_{u=\cos x}. \qquad \square$$

Once you have got the idea of how to integrate by substitution, you can omit some of the intermediate steps, even leaving out explicit introduction of the auxiliary variable u. Thus a more concise solution of Example 2 is

$$\int \sin^3 x \, dx = \int \sin^2 x \sin x \, dx = \int (-\sin^2 x) \, d(\cos x)$$

$$= \int (\cos^2 x - 1) \, d(\cos x) = \frac{1}{3} \cos^3 x - \cos x + C,$$

where the whole expression $\cos x$ is treated as a variable of integration.

Example 3 Evaluate $\int x\sqrt{2+3x}\,dx$.

Solution Let $u = 2 + 3x$. Then $du = 3\,dx$, $dx = \frac{1}{3}du$, $x = \frac{1}{3}(u-2)$, and therefore

$$\int x\sqrt{2+3x}\,dx = \frac{1}{9}\int (u-2)\sqrt{u}\,du = \frac{1}{9}\int (u^{3/2} - 2u^{1/2})\,du$$

$$= \frac{2}{45}u^{5/2} - \frac{4}{27}u^{3/2} + C$$

$$= \frac{2}{45}(2+3x)^{5/2} - \frac{4}{27}(2+3x)^{3/2} + C.$$

Check this result by direct differentiation of the expression on the right. □

Example 4 Evaluate $\int \sec x\,dx$.

Solution Guided by the tactic that led to formula (4'), page 309, for the integral of $\tan x$, you might discover that if $\sec x$ is written in the form

$$\sec x = \sec x\,\frac{\sec x + \tan x}{\sec x + \tan x} = \frac{\sec x \tan x + \sec^2 x}{\sec x + \tan x},$$

then the numerator of the last expression is the derivative of its denominator! Thus, choosing $u = \sec x + \tan x$, we have $du = (\sec x \tan x + \sec^2 x)\,dx$, and hence

$$\int \sec x\,dx = \int \frac{\sec x \tan x + \sec^2 x}{\sec x + \tan x}\,dx = \int \frac{du}{u} = \ln|u| + C,$$

so that

$$\int \sec x\,dx = \ln|\sec x + \tan x| + C. \tag{3}$$

Equivalently, (3) can be obtained by setting $f(x) = \sec x + \tan x$ in the formula

$$\int \frac{f'(x)}{f(x)}\,dx = \ln|f(x)| + C, \tag{4}$$

established on page 309 by direct differentiation of $\ln|f(x)|$. As an exercise, verify (4) by making the substitution $u = f(x)$. □

Next we consider the analogue of Theorem 1 for the case of definite integrals.

Theorem 1' (*Evaluation of a definite integral by substitution*). Let $u = u(x)$ and $f(u)$ satisfy the conditions of Theorem 1 on an interval $a \le x \le b$ and on the interval $A \le u \le B$ which is the image of $a \le x \le b$ under the substitution $u = u(x)$. Then

$$\int_a^b f(u(x))u'(x)\,dx = \int_{u(a)}^{u(b)} f(u)\,du.$$

Proof Let F be an antiderivative of f. Then, by formula (2) and two applications of the fundamental theorem of calculus,

$$\int_a^b f(u(x))u'(x)\,dx = F(u(x))\Big|_a^b = F(u(b)) - F(u(a)) = \int_{u(a)}^{u(b)} f(u)\,du. \blacksquare$$

Notice that it was *not* assumed that $u = u(x)$ is monotonic on $[a, b]$, so that there is no necessary connection between $u(a)$, $u(b)$ and A, B. On the other hand, monotonic substitutions are the most common, and then $A = u(a)$ and $B = u(b)$ if u is increasing on $[a, b]$, while $A = u(b)$ and $B = u(a)$ if u is decreasing on $[a, b]$.

Example 5 Evaluate $\int_1^e \dfrac{\ln x}{x}\,dx$.

Solution Making the substitution $u = u(x) = \ln x$, we have $du = dx/x$. Moreover, $u(1) = \ln 1 = 0$, $u(e) = \ln e = 1$. It follows from Theorem 1' that

$$\int_1^e \frac{\ln x}{x}\,dx = \int_0^1 u\,du = \left[\frac{1}{2}u^2\right]_0^1 = \frac{1}{2}. \tag{5}$$

Alternatively, since

$$\int \frac{\ln x}{x}\,dx = \int u\,du = \frac{1}{2}u^2 + C = \frac{1}{2}(\ln x)^2 + C,$$

we could have used the fundamental theorem of calculus to write

$$\int_1^e \frac{\ln x}{x}\,dx = \left[\frac{1}{2}(\ln x)^2\right]_1^e = \frac{1}{2}(\ln e)^2 - \frac{1}{2}(\ln 1)^2 = \frac{1}{2},$$

but this approach is inefficient, since there is actually no need to return from u to the original variable x. In fact, once the second integral in (5) has been evaluated, the first integral is also known, since both are *definite* integrals and therefore *numbers*. $\square$

Example 6 Starting from the definition

$$\ln x = \int_1^x \frac{dt}{t}$$

of the natural logarithm, give another proof of the formula

$$\ln ab = \ln a + \ln b \qquad (a, b \text{ positive}),$$

already established in Theorem 1, page 302.

Solution We have

$$\ln b = \int_1^b \frac{dt}{t} = \int_1^b \frac{a\,dt}{at} = \int_a^{ab} \frac{du}{u},$$

where in the last step we go over to a new variable $u = at$ and make the corresponding changes in the limits of integration. Therefore

$$\ln b = \int_a^{ab} \frac{dt}{t},$$

370 Chapter 7 Methods of Integration

after reverting to the original (dummy) variable of integration. It follows that

$$\ln ab = \int_1^{ab} \frac{dt}{t} = \int_1^a \frac{dt}{t} + \int_a^{ab} \frac{dt}{t} = \ln a + \ln b. \quad \square$$

An Alternative Approach

So far, in applying the method of integration by substitution, we have looked for a substitution $u = u(x)$ and an easily integrated function $f(u)$ such that the function we actually want to integrate, call it $g(x)$, can be recognized as being of the form

$$g(x) = f(u(x))u'(x). \tag{6}$$

For then

$$\int g(x)\,dx = \int f(u)\,du \bigg|_{u=u(x)}, \tag{7}$$

by Theorem 1, and

$$\int_a^b g(x)\,dx = \int_{u(a)}^{u(b)} f(u)\,du, \tag{7'}$$

by Theorem 1'. An alternative approach is to make a substitution of the form $x = x(u)$ *directly* in the integral $\int g(x)\,dx$. If $x(u)$ has a continuous derivative $x'(u)$, so that in particular $dx = x'(u)\,du$, this substitution "transforms" $\int g(x)\,dx$ into

$$\int g(x(u))x'(u)\,du = \int f(u)\,du,$$

where

$$f(u) = g(x(u))x'(u), \tag{8}$$

and if the substitution $x = x(u)$ has been carefully chosen, the function $f(u)$, unlike the original function $g(x)$, can be integrated easily.

Optional
There is a simple connection between these two approaches to the method of integration by substitution. Suppose $x = x(u)$ is a one-to-one function, with inverse $u = u(x)$. Then, by Theorem 4, page 286, on the derivative of an inverse function,

$$u'(x) = \frac{1}{x'(u)}$$

(here we make the additional assumption that $x'(u)$ is nonzero), so that

$$x'(u) = \frac{1}{u'(x)}.$$

But substitution of this expression for $x'(u)$ into formula (8) leads at once to formula (6), and hence to formulas (7) and (7'). Thus the two approaches are essentially the same.

Example 7 Evaluate $\int \dfrac{dx}{\sqrt{x}\,(1 + \sqrt{x})}$ in two ways.

Solution Guided by the fact that $\sqrt{x}$ appears twice in the integrand, we make the substitution $u = \sqrt{x}$. Then

$$du = d\sqrt{x} = \dfrac{dx}{2\sqrt{x}},$$

so that $dx = 2\sqrt{x}\,du = 2u\,du$. Therefore

$$\int \dfrac{dx}{\sqrt{x}\,(1 + \sqrt{x})} = 2 \int \dfrac{u\,du}{u(1 + u)} = 2 \int \dfrac{du}{1 + u} = 2 \ln |1 + u| + C,$$

or

$$\int \dfrac{dx}{\sqrt{x}\,(1 + \sqrt{x})} = 2 \ln (1 + \sqrt{x}) + C,$$

after returning to the variable x. The expression $2 \ln (1 + \sqrt{x})$ can be replaced by $\ln (1 + \sqrt{x})^2$, if you prefer.

Alternatively, the substitution $x = u^2$ (the inverse of $u = \sqrt{x}$) seems a good choice, since it gets rid of the radical. With this substitution, $dx = 2u\,du$, $\sqrt{x} = u$, and

$$\int \dfrac{dx}{\sqrt{x}\,(1 + \sqrt{x})} = \int \dfrac{2u\,du}{u(1 + u)} = 2 \int \dfrac{du}{1 + u},$$

leading to the same result as before. □

Problems

Use integration by substitution to evaluate

1. $\int (1 - 2x)^9\,dx$
2. $\int (1 + x^2)^{49}\,x\,dx$
3. $\int \sqrt{4 + 5x}\,dx$
4. $\int \dfrac{dx}{\sqrt{\pi + 2x}}$
5. $\int \dfrac{x}{\sqrt{3x - 1}}\,dx$
6. $\int (x^2 - 4x + 4)^{-5/3}\,dx$
7. $\int x\sqrt[3]{1 - x}\,dx$
8. $\int \dfrac{x^2}{\sqrt{x^3 + 1}}\,dx$
9. $\int \left(\dfrac{2 + \sqrt{x}}{x}\right)^{1/2}\,dx$
10. $\int \dfrac{dx}{\sqrt{x}\,\sqrt{1 - x}}$
11. $\int \dfrac{dx}{x\sqrt{x - 1}}$
12. $\int \cos^3 x \sin x\,dx$
13. $\int \cos^3 x\,dx$
14. $\int \sin^4 x \cos x\,dx$
15. $\int \sin^5 x\,dx$
16. $\int \dfrac{\cos x}{1 + \sin^2 x}\,dx$
17. $\int \dfrac{\sin x}{\sqrt{1 + 2 \cos x}}\,dx$
18. $\int \cos (\tan x) \sec^2 x\,dx$
19. $\int x \sec x^2 \tan x^2\,dx$
20. $\int \dfrac{\sec^2 x^{1/3}}{x^{2/3}}\,dx$
21. $\int e^{x^2} x\,dx$
22. $\int \dfrac{e^{\sqrt{x}}}{\sqrt{x}}\,dx$
23. $\int \dfrac{dx}{x\,(\ln x)^2}$
24. $\int \dfrac{\sqrt{1 + \ln x}}{x}\,dx$
25. $\int \dfrac{e^{3x}}{e^{3x} - 1}\,dx$
26. $\int \dfrac{dx}{\sqrt{x}\,(1 + e^{-\sqrt{x}})}$
27. $\int \dfrac{e^x}{e^{2x} + 1}\,dx$
28. $\int \dfrac{2^x}{4^x + 1}\,dx$
29. $\int \dfrac{2^x}{\sqrt{4^x + 1}}\,dx$
30. $\int x \cosh x^2\,dx$
31. $\int \dfrac{\sinh \sqrt{x} \cosh \sqrt{x}}{\sqrt{x}}\,dx$
32. $\int \dfrac{\sqrt{\arctan x}}{x^2 + 1}\,dx$

33. $\displaystyle\int \frac{1}{x^2} \tan \frac{1}{x} \, dx$

34. Use integration by substitution to verify rule (iv), page 252, which says that if f has an antiderivative F, then

$$\int f(ax + b) \, dx = \frac{1}{a} F(ax + b) + C \qquad (a \neq 0).$$

Show that

35. $\displaystyle\int \csc x \, dx = -\ln|\csc x + \cot x| + C$
$\qquad = \ln|\csc x - \cot x| + C$

36. $\displaystyle\int \csc x \, dx = \ln\left|\tan \frac{x}{2}\right| + C,$

$\displaystyle\int \sec x \, dx = \ln\left|\tan\left(\frac{x}{2} + \frac{\pi}{4}\right)\right| + C$

37. $\displaystyle\int \sec x \, dx = \tanh^{-1}(\sin x) + C,$

$\displaystyle\int \csc x \, dx = -\tanh^{-1}(\cos x) + C$

38. $\displaystyle\int \sec x \, dx = \frac{1}{2} \ln \frac{1 + \sin x}{1 - \sin x} + C,$

$\displaystyle\int \csc x \, dx = \frac{1}{2} \ln \frac{1 - \cos x}{1 + \cos x} + C$

39. Use integration by substitution to show that

$$\int_{-a}^{a} f(x) \, dx = 2 \int_{0}^{a} f(x) \, dx$$

if f is even, while

$$\int_{-a}^{a} f(x) \, dx = 0$$

if f is odd, as already established by another method in Problem 1, page 260. Assume that f is continuous on $[-a, a]$.

40. Given any function f continuous on $[0, a]$, show that

$$\int_{0}^{a} f(a - x) \, dx = \int_{0}^{a} f(x) \, dx.$$

Evaluate

41. $\displaystyle\int_{-1}^{1} (1 + x^3)^7 x^2 \, dx$

42. $\displaystyle\int_{0}^{2} s\sqrt{1 + 2s^2} \, ds$

43. $\displaystyle\int_{4}^{9} \frac{\sqrt{t}}{\sqrt{t} - 1} \, dt$

44. $\displaystyle\int_{0}^{1} \frac{x}{\sqrt{x} + 1} \, dx$

45. $\displaystyle\int_{1}^{3} \frac{ds}{\sqrt{s}(1 + s)}$

46. $\displaystyle\int_{0}^{\pi} \cos^4 t \sin t \, dt$

47. $\displaystyle\int_{0}^{\pi/2} \cos^5 u \, du$

48. $\displaystyle\int_{1}^{e} \frac{dv}{v[1 + (\ln v)^2]}$

49. $\displaystyle\int_{0}^{2} \frac{3^w}{3^w + 1} \, dw$

50. $\displaystyle\int_{-\pi/3}^{\pi/3} \sec x \, dx$

51. $\displaystyle\int_{\pi/4}^{3\pi/4} \csc x \, dx$

52. $\displaystyle\int_{\pi/6}^{\pi/3} \sec x \csc x \, dx$

53. Suppose f is continuous on $(-\infty, \infty)$ and periodic with period p, so that $f(x + p) \equiv f(x)$. Verify the formula

$$\int_{a}^{a+p} f(x) \, dx = \int_{0}^{p} f(x) \, dx \qquad (a \text{ arbitrary}),$$

which shows that f has the same integral over every interval of length p.

54. Show that

$$\int_{a}^{a+2\pi} \sin^3 x \, dx = 0 \qquad (a \text{ arbitrary}).$$

55. Show that

$$\int_{-1/2}^{1/2} \ln \frac{1 + x}{1 - x} \sin^{10} x \, dx = 0.$$

56. Use Problem 40 to show that if f is continuous on $[0, 1]$, then

$$\int_{0}^{\pi/2} f(\cos x) \, dx = \int_{0}^{\pi/2} f(\sin x) \, dx.$$

7.2 Integration by Parts

Next we consider another important method of integration, which is a consequence of the product rule for differentiation. Let $u = u(x)$ and $v = v(x)$ be two differentiable functions, with continuous derivatives $u'(x)$ and $v'(x)$. Differentiating the product $u(x)v(x)$ with respect to x, we obtain

$$\frac{d}{dx}[u(x)v(x)] = [u(x)v(x)]' = u'(x)v(x) + u(x)v'(x),$$

or equivalently

$$u(x)v'(x) = [u(x)v(x)]' - v(x)u'(x). \tag{1}$$

Section 7.2 Integration by Parts 373

Integration of both sides of (1) then gives

$$\int u(x)v'(x)\,dx = \int [u(x)v(x)]'\,dx - \int v(x)u'(x)\,dx.$$

But

$$\int [u(x)v(x)]'\,dx = u(x)v(x) + C,$$

and therefore

$$\int u(x)v'(x)\,dx = u(x)v(x) - \int v(x)u'(x)\,dx, \qquad (2)$$

where C can be dropped, since there is a constant of integration implicit in each of the two indefinite integrals. Introducing the differentials

$$du = u'(x)\,dx, \qquad dv = v'(x)\,dx,$$

and omitting arguments of functions for simplicity, we can write (2) in the particularly concise form

$$\int u\,dv = uv - \int v\,du. \qquad (3)$$

Equation (3), called the formula for *integration by parts*, is one of the most valuable techniques of integration, often allowing us to express a difficult integral in terms of an easier one, as will be illustrated in the examples given below.

To find a corresponding formula for *definite* integrals, we integrate both sides of (1) with respect to x from a to b. This gives

$$\int_a^b u(x)v'(x)\,dx = \int_a^b [u(x)v(x)]'\,dx - \int_a^b v(x)u'(x)\,dx.$$

But

$$\int_a^b [u(x)v(x)]'\,dx = u(x)v(x)\Big|_a^b,$$

by the fundamental theorem of calculus, and hence

$$\int_a^b u(x)v'(x)\,dx = u(x)v(x)\Big|_a^b - \int_a^b v(x)u'(x)\,dx, \qquad (2')$$

or more concisely

$$\int_a^b u\,dv = uv\Big|_a^b - \int_a^b v\,du, \qquad (3')$$

where $uv\big|_a^b$ is shorthand for

$$u(x)v(x)\Big|_a^b = u(b)v(b) - u(a)v(a).$$

Observe that (3') is obtained from (3) by putting limits of integration on the indefinite integrals $\int u\,dv$ and $\int v\,du$, and replacing the function uv by the difference $uv\big|_a^b$, which is of course a number.

374 Chapter 7 Methods of Integration

Example 1 Evaluate $\int x \sin x \, dx$.

Solution Let $u = x$, $dv = \sin x \, dx$. Then $du = dx$, $v = -\cos x$, and therefore by (3),

$$\int x \sin x \, dx = -x \cos x - \int (-\cos x) \, dx$$

$$= -x \cos x + \int \cos x \, dx = -x \cos x + \sin x + C,$$

where C is an arbitrary constant of integration. □

Proper Choice of Parts

In Example 1 we chose $u = x$ instead of $u = \sin x$, because x, unlike $\sin x$, is simplified by differentiation. The whole point of integration by parts is to make $\int v \, du$ easier to evaluate than $\int u \, dv$, by choosing the "parts" u and dv judiciously. Had we chosen $u = \sin x$, $dv = x \, dx$, then $du = \cos x \, dx$, $v = \frac{1}{2} x^2$, and formula (3) would have led to

$$\int x \sin x \, dx = \frac{1}{2} x^2 \sin x - \frac{1}{2} \int x^2 \cos x \, dx,$$

where the integral on the right is more challenging than the original one on the left!

When integrating by parts, it is unnecessary to introduce an extra constant of integration k in going from dv to v (thus in Example 1 we wrote $v = -\cos x$ rather than $v = -\cos x + k$). In fact, this would only lead to extra terms which cancel out in forming the right sides of (3) and (3'). The details are left as an exercise (replace v by $v + k$ and see what happens).

Example 2 Evaluate $\int \ln x \, dx$.

Solution Here the immediate choice is $u = \ln x$, $dv = dx$, and it works at once. In fact, $du = dx/x$, $v = x$, and (3) gives

$$\int \ln x \, dx = x \ln x - \int dx = x \ln x - x + C$$

(as anticipated in Problem 31, page 307). □

Example 3 Evaluate $\int x \ln x \, dx$.

Solution There are various possibilities here. We can choose $u = x$, $dv = \ln x \, dx$, or $u = \ln x$, $dv = x \, dx$, or even $u = x \ln x$, $dv = dx$. The only good choice is $u = \ln x$, $dv = x \, dx$, since only this choice makes $\int v \, du$ simpler than $\int u \, dv$, by getting rid of the logarithm. We then have $du = dx/x$, $v = \frac{1}{2} x^2$, and therefore by (3),

$$\int x \ln x \, dx = \frac{1}{2} x^2 \ln x - \frac{1}{2} \int x \, dx = \frac{1}{2} x^2 \ln x - \frac{1}{4} x^2 + C. \tag{4}$$

Equivalently, we can avoid explicit introduction of the auxiliary variables u and v by taking differentials of functions. Thus

$$\int x \ln x \, dx = \int \ln x \, d\left(\frac{1}{2}x^2\right) = \frac{1}{2}x^2 \ln x - \frac{1}{2}\int x^2 \, d(\ln x)$$

$$= \frac{1}{2}x^2 \ln x - \frac{1}{2}\int x \, dx = \frac{1}{2}x^2 \ln x - \frac{1}{4}x^2 + C,$$

where it will be noted that $d(\frac{1}{2}x^2) = x \, dx$, $d(\ln x) = dx/x$. □

Example 4 Evaluate $\int_1^e x \ln x \, dx$.

Solution Using (3′) with the same choice of u and dv as in Example 3, we obtain

$$\int_1^e x \ln x \, dx = \frac{1}{2}x^2 \ln x \Big|_1^e - \frac{1}{2}\int_1^e x \, dx = \frac{1}{2}e^2 \ln e - \frac{1}{2}\ln 1 - \frac{1}{4}x^2 \Big|_1^e$$

$$= \frac{1}{2}e^2 - \frac{1}{4}e^2 + \frac{1}{4} = \frac{1}{4}(e^2 + 1). \quad \square$$

Example 5 Evaluate $\int_0^1 \arctan x \, dx$.

Solution Here our only hope is that the choice $u = \arctan x$, $dv = dx$ will be effective, and it is. In fact, we then have $du = dx/(x^2 + 1)$, $v = x$, and (3′) becomes

$$\int_0^1 \arctan x \, dx = x \arctan x \Big|_0^1 - \int_0^1 \frac{x}{x^2 + 1} \, dx$$

$$= \arctan 1 - \frac{1}{2}\ln(x^2 + 1)\Big|_0^1 = \frac{\pi}{4} - \frac{1}{2}\ln 2. \quad \square$$

To evaluate an integral, it is often necessary to integrate by parts more than once.

Example 6 Evaluate $\int x^2 e^x \, dx$.

Solution Let $u = x^2$, $dv = e^x \, dx$. Then $du = 2x \, dx$, $v = e^x$, and integration by parts gives

$$\int x^2 e^x \, dx = x^2 e^x - 2 \int x e^x \, dx. \tag{5}$$

To evaluate the integral on the right, we integrate by parts *again*, this time choosing $u = x$, $dv = e^x \, dx$, $du = dx$, $v = e^x$:

$$\int x e^x \, dx = x e^x - \int e^x \, dx = x e^x - e^x + k \tag{6}$$

(k arbitrary). Substituting (6) into (5), we obtain

$$\int x^2 e^x \, dx = x^2 e^x - 2x e^x + 2 e^x + C$$

($C = -2k$ arbitrary). □

376 Chapter 7 Methods of Integration

The method of integration by parts sometimes leads to an *equation* that can be *solved* for the given integral.

Example 7 Evaluate $\int \sec^3 x \, dx$.

Solution Let $u = \sec x$, $dv = \sec^2 x \, dx$. Then $du = \sec x \tan x \, dx$, $v = \tan x$, and integration by parts gives

$$\int \sec^3 x \, dx = \sec x \tan x - \int \tan^2 x \sec x \, dx.$$

Since $\tan^2 x = \sec^2 x - 1$, this can be written as

$$\int \sec^3 x \, dx = \sec x \tan x - \int (\sec^2 x - 1) \sec x \, dx,$$

that is,

$$\int \sec^3 x \, dx = \sec x \tan x - \int \sec^3 x \, dx + \int \sec x \, dx,$$

where the integral $\int \sec^3 x \, dx$ which we are trying to evaluate has reappeared on the right with the opposite sign and can be solved for! Doing so, we find at once that

$$\int \sec^3 x \, dx = \frac{1}{2} \sec x \tan x + \frac{1}{2} \int \sec x \, dx.$$

Using the formula for $\int \sec x \, dx$ already found in Example 4, page 369, we finally obtain

$$\int \sec^3 x \, dx = \frac{1}{2} \sec x \tan x + \frac{1}{2} \ln |\sec x + \tan x| + C. \quad \square \quad (7)$$

Example 8 Evaluate $\int e^{ax} \cos bx \, dx$ $(ab \neq 0)$.

Solution First we choose

$$u = e^{ax}, \quad dv = \cos bx \, dx, \quad du = ae^{ax} \, dx, \quad v = \frac{\sin bx}{b},$$

and integrate by parts:

$$\int e^{ax} \cos bx \, dx = \frac{e^{ax} \sin bx}{b} - \frac{a}{b} \int e^{ax} \sin bx \, dx. \quad (8)$$

Although the integral on the right is as hard as the one we are trying to evaluate, we integrate it by parts anyway, this time choosing

$$u = e^{ax}, \quad dv = \sin bx \, dx, \quad du = ae^{ax} \, dx, \quad v = -\frac{\cos bx}{b}.$$

As a result, we obtain

$$\int e^{ax} \sin bx \, dx = -\frac{e^{ax} \cos bx}{b} + \frac{a}{b} \int e^{ax} \cos bx \, dx, \quad (8')$$

and substitution of (8') into (8) then gives

$$\int e^{ax} \cos bx \, dx = \frac{e^{ax} \sin bx}{b} + \frac{ae^{ax} \cos bx}{b^2} - \frac{a^2}{b^2} \int e^{ax} \cos bx \, dx.$$

Section 7.2 Integration by Parts

We are now able to solve for the integral $\int e^{ax} \cos bx\, dx$, which has obligingly reappeared on the right, obtaining

$$\left(1 + \frac{a^2}{b^2}\right) \int e^{ax} \cos bx\, dx = e^{ax} \frac{a \cos bx + b \sin bx}{b^2},$$

or equivalently

$$\int e^{ax} \cos bx\, dx = e^{ax} \frac{a \cos bx + b \sin bx}{a^2 + b^2} + C, \tag{9}$$

where C is an arbitrary constant.† Similarly, substituting (8) into (8′) and solving the resulting equation for $\int e^{ax} \sin bx\, dx$, we get the companion formula

$$\int e^{ax} \sin bx\, dx = e^{ax} \frac{a \sin bx - b \cos bx}{a^2 + b^2} + C. \quad \square \tag{9′}$$

† An indefinite integral is defined only to within an arbitrary constant of integration, and it is often convenient to "supply" (that is, add in) this constant at the very end of the calculation.

Problems

Use integration by parts to evaluate

1. $\int x \cos x\, dx$
2. $\int (x-1) \ln x\, dx$
3. $\int x^2 \ln x\, dx$
4. $\int x e^{-2x}\, dx$
5. $\int \arcsin x\, dx$
6. $\int e^{2x} \cos 3x\, dx$
7. $\int e^{-x} \sin 2x\, dx$
8. $\int \sinh^{-1} x\, dx$
9. $\int (\ln x)^2\, dx$
10. $\int x 3^x\, dx$
11. $\int \sqrt{x} \ln x\, dx$
12. $\int x^2 \sin 2x\, dx$
13. $\int \sin x \cos 2x\, dx$
14. $\int x^5 \ln x\, dx$
15. $\int x^3 \cos x\, dx$
16. $\int x \csc^2 x\, dx$
17. $\int x \operatorname{sech}^2 x\, dx$
18. $\int x^2 \sinh x\, dx$
19. $\int x\sqrt{2x+3}\, dx$
20. $\int e^{\sqrt{x}}\, dx$
21. $\int \csc^3 x\, dx$
22. $\int x^3 e^{-x^2}\, dx$
23. $\int \sin(\ln x)\, dx$
24. $\int x(x+1)^9\, dx$
25. $\int_0^1 x e^{-x}\, dx$
26. $\int_0^{e-1} \ln(x+1)\, dx$
27. $\int_0^\pi x^2 \cos x\, dx$
28. $\int_0^{\pi/2} e^{2x} \sin x\, dx$
29. $\int_0^\pi x^3 \sin x\, dx$
30. $\int_1^2 x \log_2 x\, dx$
31. $\int_1^e (\ln x)^3\, dx$
32. $\int_0^{\pi/4} y \sec^2 y\, dy$
33. $\int_{-1}^1 \arccos z\, dz$

34. Let f have a continuous second derivative f'' on $[a, b]$. Show that

$$\int_a^b x f''(x)\, dx = [bf'(b) - f(b)] - [af'(a) - f(a)].$$

Find the area A of the region R between the curves

35. $y = \ln x$ and $y = (\ln x)^2$
36. $y = x \ln x$ and $y = (4 \ln x)/x$

In each case sketch the region R.

37. Let the functions $u = u(x)$ and $v = v(x)$ have continuous derivatives $u', v', u'', v'', \ldots, u^{(n)}, v^{(n)}, u^{(n+1)}, v^{(n+1)}$ of all orders up to $n + 1$ inclusive. Show that

$$\int uv^{(n+1)}\, dx = uv^{(n)} - u'v^{(n-1)} + u''v^{(n-2)} - \cdots + (-1)^n u^{(n)} v$$

$$+ (-1)^{n+1} \int u^{(n+1)} v\, dx. \tag{i}$$

38. Let $u = u(x)$ be a polynomial of degree n, and let the function $v = v(x)$ have continuous derivatives $v', v'', \ldots, v^{(n)}, v^{(n+1)}$ of all orders up to $n + 1$ inclusive. Show that

$$\int uv^{(n+1)}\, dx = uv^{(n)} - u'v^{(n-1)} + u''v^{(n-2)}$$

$$- \cdots + (-1)^n u^{(n)} v + C. \tag{i′}$$

Use this formula to evaluate $\int x^4 e^x\, dx$ and $\int x^5 \sin x\, dx$.

39. Show that

$$\int_0^1 x^m (1-x)^n\, dx = \int_0^1 x^n (1-x)^m\, dx,$$

where m and n are arbitrary positive integers. Then evaluate the integral.

378 Chapter 7 Methods of Integration

7.3 Reduction Formulas

Integration by parts can be used to prove *reduction formulas*, that is, formulas in which integrals involving powers of some expression are expressed in terms of integrals involving lower powers of the expression.

Example 1 Show that

$$\int \sin^n x \, dx = -\frac{1}{n} \sin^{n-1} x \cos x + \frac{n-1}{n} \int \sin^{n-2} x \, dx \qquad (n = 2, 3, \ldots). \tag{1}$$

Solution We integrate by parts, choosing $u = \sin^{n-1} x$, $dv = \sin x \, dx$. Then $du = (n-1) \sin^{n-2} x \cos x \, dx$, $v = -\cos x$, and

$$\int \sin^n x \, dx = -\sin^{n-1} x \cos x + (n-1) \int \sin^{n-2} x \cos^2 x \, dx.$$

But $\cos^2 x = 1 - \sin^2 x$, so that

$$\int \sin^n x \, dx = -\sin^{n-1} x \cos x + (n-1) \int \sin^{n-2} x \, dx - (n-1) \int \sin^n x \, dx,$$

where the last term on the right contains the integral we are trying to evaluate. Transposing this term to the left side of the equation and combining the two terms containing $\int \sin^n x \, dx$, we get the formula

$$n \int \sin^n x \, dx = -\sin^{n-1} x \cos x + (n-1) \int \sin^{n-2} x \, dx,$$

which is equivalent to (1). □

Example 2 Show that

$$\int \cos^n x \, dx = \frac{1}{n} \cos^{n-1} x \sin x + \frac{n-1}{n} \int \cos^{n-2} x \, dx \qquad (n = 2, 3, \ldots). \tag{2}$$

Solution Let $u = \cos^{n-1} x$, $dv = \cos x \, dx$, and integrate by parts. The details are much the same as in Example 1, and are left as an exercise. □

By repeated application of the reduction formulas (1) and (2), we can reduce the evaluation of the integrals $\int \sin^n x \, dx$ and $\int \cos^n x \, dx$ to the evaluation of one of the easy integrals $\int dx$, $\int \sin x \, dx$ and $\int \cos x \, dx$.

Example 3 Evaluate $\int \sin^4 x \, dx$.

Solution Choosing first $n = 4$ and then $n = 2$ in formula (1), we get

$$\int \sin^4 x \, dx = -\frac{1}{4} \sin^3 x \cos x + \frac{3}{4} \int \sin^2 x \, dx$$

and

$$\int \sin^2 x \, dx = -\frac{1}{2} \sin x \cos x + \frac{1}{2} \int dx.$$

Therefore, after substituting for $\int \sin^2 x \, dx$,

$$\int \sin^4 x \, dx = -\frac{1}{4} \sin^3 x \cos x - \frac{3}{8} \sin x \cos x + \frac{3}{8} \int dx$$

$$= -\frac{1}{4} \sin^3 x \cos x - \frac{3}{8} \sin x \cos x + \frac{3}{8} x + C. \quad \square$$

Example 4 Evaluate $\int \cos^5 x \, dx$.

Solution Choosing first $n = 5$ and then $n = 3$ in formula (2), we get

$$\int \cos^5 x \, dx = \frac{1}{5} \cos^4 x \sin x + \frac{4}{5} \int \cos^3 x \, dx.$$

and

$$\int \cos^3 x \, dx = \frac{1}{3} \cos^2 x \sin x + \frac{2}{3} \int \cos x \, dx.$$

Therefore, after substituting for $\int \cos^3 x \, dx$,

$$\int \cos^5 x \, dx = \frac{1}{5} \cos^4 x \sin x + \frac{4}{15} \cos^2 x \sin x + \frac{8}{15} \int \cos x \, dx$$

$$= \frac{1}{5} \cos^4 x \sin x + \frac{4}{15} \cos^2 x \sin x + \frac{8}{15} \sin x + C. \quad \square$$

Example 5 Show that

$$\int \frac{dx}{(x^2 + a^2)^{n+1}} = \frac{1}{2na^2} \frac{x}{(x^2 + a^2)^n} + \frac{2n - 1}{2na^2} \int \frac{dx}{(x^2 + a^2)^n}, \quad (3)$$

where $n = 1, 2, \ldots$ and it can be assumed that $a > 0$.

Solution We use integration by parts to evaluate the integral on the *right*, choosing

$$u = \frac{1}{(x^2 + a^2)^n}, \quad dv = dx, \quad du = -\frac{2nx}{(x^2 + a^2)^{n+1}} dx, \quad v = x.$$

This gives

$$\int \frac{dx}{(x^2 + a^2)^n} = \frac{x}{(x^2 + a^2)^n} + 2n \int \frac{x^2}{(x^2 + a^2)^{n+1}} dx.$$

But $x^2 = (x^2 + a^2) - a^2$, and therefore

$$\int \frac{x^2}{(x^2 + a^2)^{n+1}} dx = \int \frac{(x^2 + a^2) - a^2}{(x^2 + a^2)^{n+1}} dx$$

$$= \int \frac{dx}{(x^2 + a^2)^n} - a^2 \int \frac{dx}{(x^2 + a^2)^{n+1}}.$$

Combining the last two equations, we obtain

$$\int \frac{dx}{(x^2 + a^2)^n} = \frac{x}{(x^2 + a^2)^n} + 2n \int \frac{dx}{(x^2 + a^2)^n} - 2na^2 \int \frac{dx}{(x^2 + a^2)^{n+1}},$$

which is equivalent to (3). $\quad \square$

By repeated application of the reduction formula (3), we can reduce the evaluation of the integral on the left to the evaluation of the known integral

$$\int \frac{dx}{x^2 + a^2} = \frac{1}{a} \arctan \frac{x}{a} + C \qquad (4)$$

(see p. 293). Thus, choosing $n = 1, 2, \ldots$ in formula (3), we have

$$\int \frac{dx}{(x^2 + a^2)^2} = \frac{1}{2a^2} \frac{x}{x^2 + a^2} + \frac{1}{2a^2} \int \frac{dx}{x^2 + a^2}$$

$$= \frac{1}{2a^2} \frac{x}{x^2 + a^2} + \frac{1}{2a^3} \arctan \frac{x}{a} + C, \qquad (5)$$

$$\int \frac{dx}{(x^2 + a^2)^3} = \frac{1}{4a^2} \frac{x}{(x^2 + a^2)^2} + \frac{3}{4a^2} \int \frac{dx}{(x^2 + a^2)^2}$$

$$= \frac{1}{4a^2} \frac{x}{(x^2 + a^2)^2} + \frac{3}{8a^4} \frac{x}{x^2 + a^2} + \frac{3}{8a^5} \arctan \frac{x}{a} + C, \qquad (6)$$

and so on. Actually, the constant C in (5) is $1/2a^2$ times the constant C in (4), and the constant C in (6) is $3/4a^2$ times the constant C in (5), but there is no need to make this explicit, since each is an *arbitrary* constant of integration.

Problems

Verify the given reduction formula, where n is any positive integer.

1. $\int x^n e^x \, dx = x^n e^x - n \int x^{n-1} e^x \, dx$

2. $\int (\ln x)^n \, dx = x (\ln x)^n - n \int (\ln x)^{n-1} \, dx$

3. $\int x^n \sin x \, dx = -x^n \cos x + n \int x^{n-1} \cos x \, dx$

4. $\int x^n \cos x \, dx = x^n \sin x - n \int x^{n-1} \sin x \, dx$

Use a reduction formula to evaluate

5. $\int \cos^4 x \, dx$ 6. $\int \sin^5 x \, dx$ 7. $\int x^5 e^x \, dx$

8. $\int (\ln x)^3 \, dx$ 9. $\int x^4 \sin x \, dx$ 10. $\int x^6 \cos x \, dx$

11. $\int \frac{dx}{(x^2 + 1)^4}$ 12. $\int \frac{e^x}{(e^{2x} + 1)^3} \, dx$

13. Show that

$$\int_0^{\pi/2} \sin^n x \, dx = \frac{1}{2} \frac{3}{4} \cdots \frac{n-1}{n} \frac{\pi}{2} \quad (n = 2, 4, \ldots) \quad (i)$$

for every even integer $n > 0$. Thus, for example,

$$\int_0^{\pi/2} \sin^6 x \, dx = \frac{1}{2} \frac{3}{4} \frac{5}{6} \frac{\pi}{2} = \frac{5\pi}{32}.$$

Also show that

$$\int_0^{\pi/2} \sin^n x \, dx = \frac{2}{3} \frac{4}{5} \cdots \frac{n-1}{n} \quad (n = 3, 5, \ldots), \quad (i')$$

with no factor of $\pi/2$, for every odd integer $n > 1$. Thus, for example,

$$\int_0^{\pi/2} \sin^7 x \, dx = \frac{2}{3} \frac{4}{5} \frac{6}{7} = \frac{16}{35}.$$

14. Use Problem 56, page 373, to show that

$$\int_0^{\pi/2} \cos^n x \, dx = \int_0^{\pi/2} \sin^n x \, dx$$

for every nonnegative integer n.

Evaluate

15. $\int_0^{\pi/2} \sin^{10} x \, dx$ 16. $\int_0^{\pi/2} \cos^{12} x \, dx$

17. $\int_0^{\pi/4} \sin^5 x \cos^5 x \, dx$ 18. $\int_{-\pi/2}^{\pi/2} \sin^{14} t \, dt$

19. $\int_0^{\pi} \sin^{15} u \, du$ 20. $\int_{\pi/2}^{3\pi/2} \cos^9 v \, dv$

Section 7.3 Reduction Formulas **381**

21. Verify the reduction formula

$$\int x^a (\ln x)^n \, dx = \frac{1}{a+1} x^{a+1} (\ln x)^n$$

$$- \frac{n}{a+1} \int x^a (\ln x)^{n-1} \, dx, \quad \text{(ii)}$$

where $a \neq -1$ and n is any positive integer.

Comment. The case $a = -1$ is easily treated, since

$$\int x^{-1} (\ln x)^n \, dx = \int (\ln x)^n \, d(\ln x) = \frac{1}{n+1} (\ln x)^{n+1} + C.$$

Use formula (ii) to evaluate

22. $\int x^3 (\ln x)^2 \, dx$ **23.** $\int \sqrt{x} \, (\ln x)^3 \, dx$

24. $\int_1^e x (\ln x)^3 \, dx$ **25.** $\int_e^3 x^2 (\ln x)^2 \, dx$

26. Verify the reduction formulas

$$\int \tan^n x \, dx = \frac{1}{n-1} \tan^{n-1} x - \int \tan^{n-2} x \, dx,$$

$$\int \sec^n x \, dx = \frac{1}{n-1} \sec^{n-1} x \tan x + \frac{n-2}{n-1} \int \sec^{n-2} x \, dx,$$

where n is any integer greater than 1.

7.4 Trigonometric Integrals

Integrands of the Form $\sin^p x \cos^q x$

Integrals with integrands that are combinations of trigonometric functions are called *trigonometric integrals*, and a number of such integrals have already been evaluated in the preceding sections. We now focus our attention on the case where the integrand is a product of powers of trigonometric functions. We begin with the integral

$$\int \sin^p x \cos^q x \, dx,$$

involving two exponents p and q. It is easy to evaluate this integral if one of the exponents p, q is a positive odd integer, or if both exponents are positive even integers. In the first case, we use one of the identities

$$\sin^2 x = 1 - \cos^2 x, \qquad \cos^2 x = 1 - \sin^2 x \qquad (1)$$

to bring the integrand into the form $f(\sin x) \cos x$ or $f(\cos x) \sin x$, after which the integral can be evaluated by making the substitution $u = \sin x$ or $u = \cos x$. In the second case, we make repeated use of the identities

$$\sin^2 x = \frac{1 - \cos 2x}{2}, \qquad \cos^2 x = \frac{1 + \cos 2x}{2} \qquad (2)$$

to eliminate all powers of $\sin x$ and $\cos x$ higher than 1. The following examples illustrate these techniques.

Example 1 Evaluate $\int \sin^4 x \cos^7 x \, dx$.

Solution Here $p = 4$, $q = 7$, and q is a positive odd integer. We separate a factor of $\cos x$ from $\cos^7 x$, and then use the second of the formulas (1) to express the remaining factor $\cos^6 x$ entirely in terms of $\sin x$. This gives

$$\int \sin^4 x \cos^7 x \, dx = \int \sin^4 x \cos^6 x \cos x \, dx$$

$$= \int \sin^4 x \, (1 - \sin^2 x)^3 \cos x \, dx.$$

The integrand is now of the form $f(\sin x) \cos x$, and hence the substitution $u = \sin x$, $du = \cos x \, dx$ is effective. In detail,

$$\int \sin^4 x \cos^7 x \, dx = \int u^4 (1 - u^2)^3 \, du = \int u^4 (1 - 3u^2 + 3u^4 - u^6) \, du$$

$$= \int (u^4 - 3u^6 + 3u^8 - u^{10}) \, du$$

$$= \frac{1}{5} u^5 - \frac{3}{7} u^7 + \frac{1}{3} u^9 - \frac{1}{11} u^{11} + C$$

$$= \frac{1}{5} \sin^5 x - \frac{3}{7} \sin^7 x + \frac{1}{3} \sin^9 x - \frac{1}{11} \sin^{11} x + C. \quad \square$$

Example 2 Evaluate $\int \dfrac{\sin^5 x}{\sqrt{\cos x}} \, dx$.

Solution Here $p = 5$, $q = -\frac{1}{2}$, and p is a positive odd integer. We separate a factor of $\sin x$ from $\sin^5 x$, and then use the first of the formulas (1) to express the remaining factor $\sin^4 x$ entirely in terms of $\cos x$. As a result, we obtain

$$\int \frac{\sin^5 x}{\sqrt{\cos x}} \, dx = \int \frac{\sin^4 x}{\sqrt{\cos x}} \sin x \, dx = \int \frac{(1 - \cos^2 x)^2}{\sqrt{\cos x}} \sin x \, dx.$$

The integrand is now of the form $f(\cos x) \sin x$, and hence the substitution $u = \cos x$, $du = -\sin x \, dx$ is effective. In detail,

$$\int \frac{\sin^5 x}{\sqrt{\cos x}} \, dx = -\int \frac{(1 - u^2)^2}{\sqrt{u}} \, du = -\int \frac{u^4 - 2u^2 + 1}{\sqrt{u}} \, du$$

$$= -\int (u^{7/2} - 2u^{3/2} + u^{-1/2}) \, du$$

$$= -\frac{2}{9} u^{9/2} + \frac{4}{5} u^{5/2} - 2u^{1/2} + C$$

$$= -\frac{2}{9} (\cos x)^{9/2} + \frac{4}{5} (\cos x)^{5/2} - 2(\cos x)^{1/2} + C. \quad \square$$

Example 3 Evaluate $\int \sin^2 x \cos^4 x \, dx$.

Solution Here $p = 2$, $q = 4$, and both exponents are positive even integers. Therefore we use the formulas (2) to lower the exponents of $\sin^2 x$ and

$\cos^4 x$, as follows:

$$\int \sin^2 x \cos^4 x \, dx = \frac{1}{8} \int (1 - \cos 2x)(1 + \cos 2x)^2 \, dx$$

$$= \frac{1}{8} \int (1 - \cos^2 2x)(1 + \cos 2x) \, dx$$

$$= \frac{1}{8} \int (1 + \cos 2x - \cos^2 2x - \cos^3 2x) \, dx$$

$$= \frac{1}{8} \int dx + \frac{1}{8} \int \cos 2x \, dx - \frac{1}{8} \int \cos^2 2x \, dx - \frac{1}{8} \int \cos^3 2x \, dx$$

$$= \frac{1}{8} x + \frac{1}{16} \sin 2x - \frac{1}{8} \int \cos^2 2x \, dx - \frac{1}{8} \int \cos^3 2x \, dx. \tag{3}$$

The last two integrals require some more work. To evaluate $\int \cos^2 2x \, dx$, we use the second of the formulas (2) again. This gives

$$\int \cos^2 2x \, dx = \frac{1}{2} \int (1 + \cos 4x) \, dx = \frac{1}{2} \int dx + \frac{1}{2} \int \cos 4x \, dx = \frac{1}{2} x + \frac{1}{8} \sin 4x \tag{4}$$

(a constant of integration will be supplied at the end of the calculation). As for the integral $\int \cos^3 2x \, dx$, it involves a positive odd power of $\cos x$, and hence can be evaluated by the method of Example 1. Thus

$$\int \cos^3 2x \, dx = \int \cos^2 2x \cos 2x \, dx = \int (1 - \sin^2 2x) \cos 2x \, dx,$$

and making the substitution $u = \sin 2x$, $du = 2 \cos 2x \, dx$, we get

$$\int \cos^3 2x \, dx = \frac{1}{2} \int (1 - u^2) \, du = \frac{1}{2} u - \frac{1}{6} u^3 = \frac{1}{2} \sin 2x - \frac{1}{6} \sin^3 2x. \tag{5}$$

Substituting (4) and (5) into (3) and supplying a constant of integration C, we find after a little algebraic manipulation that

$$\int \sin^2 x \cos^4 x \, dx = \frac{1}{16} x - \frac{1}{64} \sin 4x + \frac{1}{48} \sin^3 2x + C.$$

As an exercise, show that this integral can also be written in the form

$$\int \sin^2 x \cos^4 x \, dx$$

$$= \frac{1}{16} x - \frac{1}{16} \sin x \cos^3 x + \frac{1}{16} \sin^3 x \cos x + \frac{1}{6} \sin^3 x \cos^3 x + C,$$

where the sines and cosines on the right all have the same argument x. □

Integrands of the Form $\tan^p x \sec^q x$

Next we turn to the integral

$$\int \tan^p x \sec^q x \, dx,$$

involving two exponents p and q. It is easy to evaluate this integral if p is

384 Chapter 7 Methods of Integration

a positive odd integer or if q is a positive even integer. In the first case, we use the identity

$$\tan^2 x = \sec^2 x - 1 \tag{6}$$

to bring the integrand into the form $f(\sec x) \sec x \tan x$, after which the integral can be evaluated by making the substitution $u = \sec x$. In the second case, we use the identity

$$\sec^2 x = \tan^2 x + 1 \tag{7}$$

to bring the integrand into the form $f(\tan x) \sec^2 x$, after which the integral can be evaluated by making the substitution $u = \tan x$.

Example 4 Evaluate $\int \sqrt{\tan x} \sec^4 x \, dx$.

Solution Here $p = \frac{1}{2}$, $q = 4$, and q is a positive even integer. We separate a factor of $\sec^2 x$ from $\sec^4 x$, and use formula (7) to express the remaining factor of $\sec^2 x$ in terms of $\tan x$. This gives

$$\int \sqrt{\tan x} \sec^4 x \, dx = \int (\sqrt{\tan x} \sec^2 x) \sec^2 x \, dx$$

$$= \int \sqrt{\tan x} (\tan^2 x + 1) \sec^2 x \, dx.$$

Making the substitution $u = \tan x$, $du = \sec^2 x \, dx$, we finally get

$$\int \sqrt{\tan x} \sec^4 x \, dx = \int \sqrt{u}(u^2 + 1) \, du = \int (u^{5/2} + u^{1/2}) \, du$$

$$= \frac{2}{7} u^{7/2} + \frac{2}{3} u^{3/2} + C = \frac{2}{7} (\tan x)^{7/2} + \frac{2}{3} (\tan x)^{3/2} + C.$$

□

Example 5 Evaluate $\int \tan^3 x \sec^5 x \, dx$.

Solution Here $p = 3$, $q = 5$, and p is a positive odd integer, but q is not a positive even integer. Thus the appropriate move is to separate a factor of $\sec x \tan x$ from the integrand and use formula (6) to express the remaining factor entirely in terms of $\sec x$. In detail,

$$\int \tan^3 x \sec^5 x \, dx = \int \tan^2 x \sec^4 x \, (\sec x \tan x) \, dx$$

$$= \int (\sec^2 x - 1) \sec^4 x \, (\sec x \tan x) \, dx$$

$$= \int (\sec^6 x - \sec^4 x) \sec x \tan x \, dx.$$

The substitution $u = \sec x$, $du = \sec x \tan x \, dx$ then gives

$$\int \tan^3 x \sec^5 x \, dx = \int (u^6 - u^4) \, du = \frac{1}{7} u^7 - \frac{1}{5} u^5 + C$$

$$= \frac{1}{7} \sec^7 x - \frac{1}{5} \sec^5 x + C. \quad \square$$

Section 7.4 Trigonometric Integrals

Example 6 Evaluate $\int \tan^2 x \sec^3 x \, dx$.

Solution Here the above considerations do not apply, since $p = 2$ is even and $q = 3$ is odd, but we have the tools to evaluate the integral anyway. First we observe that

$$\int \tan^2 x \sec^3 x \, dx = \int (\sec^2 x - 1) \sec^3 x \, dx$$

$$= \int \sec^5 x \, dx - \int \sec^3 x \, dx, \tag{8}$$

with the help of formula (6). To evaluate the integral $\int \sec^5 x \, dx$, we use integration by parts, choosing $u = \sec^3 x$, $dv = \sec^2 x \, dx$. Then

$$du = 3 \sec^3 x \tan x \, dx, \qquad v = \tan x,$$

and

$$\int \sec^5 x \, dx = \sec^3 x \tan x - 3 \int \tan^2 x \sec^3 x \, dx, \tag{9}$$

where the integral we are trying to evaluate shows up again on the right! Substituting (9) into (8) and solving for this integral, we obtain

$$\int \tan^2 x \sec^3 x \, dx = \frac{1}{4} \sec^3 x \tan x - \frac{1}{4} \int \sec^3 x \, dx$$

$$= \frac{1}{4} \sec^3 x \tan x - \frac{1}{8} \sec x \tan x$$

$$- \frac{1}{8} \ln |\sec x + \tan x| + C,$$

where at the last step we use Example 7, page 377. As an exercise, obtain the same answer by applying the reduction formula for $\int \sec^n x \, dx$ given in Problem 26, page 382. □

Because of the parallelism between the properties of the functions $\tan x$, $\sec x$ and those of $\cot x$, $\csc x$, virtually the same techniques can be used to evaluate the integral.

$$\int \cot^p x \csc^q x \, dx,$$

with the help of the formulas

$$\cot^2 x = \csc^2 x - 1, \tag{6'}$$

$$\csc^2 x = \cot^2 x + 1. \tag{7'}$$

Example 7 Evaluate $\int \cot^3 x \csc^3 x \, dx$.

Solution We have

$$\int \cot^3 x \csc^3 x \, dx = \int \cot^2 x \csc^2 x \, (\csc x \cot x) \, dx$$

$$= -\int (\csc^2 - 1) \csc^2 x \, d(\csc x)$$

$$= -\int (\csc^4 x - \csc^2 x) \, d(\csc x)$$

$$= -\frac{1}{5} \csc^5 x + \frac{1}{3} \csc^3 x + C,$$

where we treat the whole expression $\csc x$ as a variable of integration, thereby avoiding the need to make an explicit substitution $u = \csc x$. □

Integration of Products of Sines and Cosines with Different Arguments

The trigonometric integrals

$$\int \sin ax \cos bx \, dx, \quad \int \cos ax \cos bx \, dx, \quad \int \sin ax \sin bx \, dx$$

are frequently encountered in applied problems dealing with oscillatory phenomena, like mechanical vibrations and radio waves. To evaluate integrals of this type, we use the trigonometric identities

$$\sin ax \cos bx = \frac{1}{2}[\sin (a + b)x + \sin (a - b)x],$$

$$\cos ax \cos bx = \frac{1}{2}[\cos (a + b)x + \cos (a - b)x],$$

$$\sin ax \sin bx = \frac{1}{2}[\cos (a - b)x - \cos (a + b)x],$$

which follow at once from the addition laws for the sine and cosine functions (see p. 57).

Example 8 Evaluate $\int \sin 2x \cos 5x \, dx$.

Solution With the help of the first of the above identities, we have

$$\int \sin 2x \cos 5x \, dx = \frac{1}{2} \int [\sin (2 + 5)x + \sin (2 - 5)x] \, dx$$

$$= \frac{1}{2} \int \sin 7x \, dx - \frac{1}{2} \int \sin 3x \, dx$$

$$= -\frac{1}{14} \cos 7x + \frac{1}{6} \cos 3x + C. \quad □$$

Problems

Evaluate

1. $\int \sin^2 x \cos^2 x \, dx$
2. $\int \sin^2 x \cos^3 x \, dx$
3. $\int \sin^3 4x \cos^2 4x \, dx$
4. $\int \sin^5 x \cos^4 x \, dx$
5. $\int \sin^7 x \sqrt{\cos x} \, dx$
6. $\int (\sin x)^{2/3} \cos^5 x \, dx$
7. $\int \frac{\cos^3 x}{\sin^4 x} \, dx$
8. $\int \sin^4 x \cos^2 x \, dx$

9. $\displaystyle\int \frac{\sin^3 x}{\cos^2 x}\,dx$

10. $\displaystyle\int_0^{\pi/6} \sin^3 s \cos^3 s\,ds$

11. $\displaystyle\int_0^{\pi/2} \sin^4 t \cos^4 t\,dt$

12. $\displaystyle\int_{\pi/3}^{2\pi/3} \frac{\cos^5 u}{\sin^4 u}\,du$

13. $\displaystyle\int \tan^3 2x \sec 2x\,dx$

14. $\displaystyle\int \tan^3 x \sec^3 x\,dx$

15. $\displaystyle\int \tan^2 x \sec^4 x\,dx$

16. $\displaystyle\int \tan^3 x\,dx$

17. $\displaystyle\int \cot^2 x \csc^4 x\,dx$

18. $\displaystyle\int \sec^4 x\,dx$

19. $\displaystyle\int \tan^4 3x\,dx$

20. $\displaystyle\int \cot^3 x \csc x\,dx$

21. $\displaystyle\int \cot^3 x\,dx$

22. $\displaystyle\int_0^{\pi/4} (\tan s)^{3/2} \sec^4 s\,ds$

23. $\displaystyle\int_0^{\pi/3} \tan^3 t \sec t\,dt$

24. $\displaystyle\int_{1/6}^{5/6} \cot^2 \pi u \csc^2 \pi u\,du$

25. $\displaystyle\int \sin 5x \cos 3x\,dx$

26. $\displaystyle\int \cos 4x \cos 5x\,dx$

27. $\displaystyle\int \sin 3x \sin 6x\,dx$

28. $\displaystyle\int \sin \pi x \cos 2\pi x\,dx$

29. $\displaystyle\int \cos\frac{2x}{3} \cos\frac{x}{3}\,dx$

30. $\displaystyle\int \sin\frac{5x}{2} \sin\frac{x}{2}\,dx$

31. $\displaystyle\int_0^{\pi} \sin 3x \cos 4x\,dx$

32. $\displaystyle\int_{-\pi/2}^{3\pi/2} \sin 6y \sin 9y\,dy$

33. $\displaystyle\int_{-a}^{a} \cos 2z \cos 4z\,dz$

34. Let m and n be arbitrary nonzero integers. Verify the following formulas, of great importance in applied mathematics:

$$\int_0^{2\pi} \sin mx \cos nx\,dx = 0,$$

$$\int_0^{2\pi} \sin mx \sin nx\,dx = \int_0^{2\pi} \cos mx \cos nx\,dx$$

$$= \begin{cases} 0 & \text{if } m \neq n, \\ \pi & \text{if } m = n. \end{cases}$$

Show that these formulas continue to hold if the limits of integration $0, 2\pi$ are replaced by $-\pi, \pi$.

7.5 Trigonometric and Hyperbolic Substitutions

The following examples serve as an introduction to the type of substitutions considered in this section.

Example 1 Evaluate the indefinite integral

$$\int \sqrt{a^2 - x^2}\,dx \quad (a > 0).$$

Solution Let $x = a \sin u$, where $-\pi/2 \leq u \leq \pi/2$. Then $dx = a \cos u\,du$, and

$$\sqrt{a^2 - x^2} = \sqrt{a^2 - a^2 \sin^2 u} = a\sqrt{1 - \sin^2 u} = a\sqrt{\cos^2 u} = a \cos u,$$

since $\cos u$ is nonnegative for the indicated values of u. Therefore

$$\int \sqrt{a^2 - x^2}\,dx = a^2 \int \cos^2 u\,du.$$

But

$$\int \cos^2 u\,du = \frac{1}{2}\int (\cos 2u + 1)\,du = \frac{1}{4}\sin 2u + \frac{1}{2}u + k$$

$$= \frac{1}{2}\sin u \cos u + \frac{1}{2}u + k,$$

where k is an arbitrary constant of integration, so that

$$\int \sqrt{a^2 - x^2}\,dx = \frac{1}{2}a^2 \sin u \cos u + \frac{1}{2}a^2 u + C, \tag{1}$$

where the constant $C = a^2 k$ is also arbitrary. Moreover

$$\sin u = \frac{x}{a}, \qquad u = \arcsin \frac{x}{a},$$

$$\cos u = \sqrt{1 - \sin^2 u} = \sqrt{1 - \frac{x^2}{a^2}} = \frac{1}{a}\sqrt{a^2 - x^2},$$

and substituting these expressions for $\sin u$, $\cos u$ and u into (1), we finally obtain

$$\int \sqrt{a^2 - x^2}\, dx = \frac{1}{2} x\sqrt{a^2 - x^2} + \frac{1}{2} a^2 \arcsin \frac{x}{a} + C. \qquad (2)$$

It is interesting to differentiate the right side of (2) and observe how the resulting three terms combine to form $\sqrt{a^2 - x^2}$. □

Example 2 Evaluate the definite integral

$$\int_0^a \sqrt{a^2 - x^2}\, dx \qquad (3)$$

Solution It follows from (2) that

$$\int_0^a \sqrt{a^2 - x^2}\, dx = \left[\frac{1}{2} x\sqrt{a^2 - x^2} + \frac{1}{2} a^2 \arcsin \frac{x}{a}\right]_0^a$$

$$= \frac{1}{2} a^2 \arcsin 1 = \frac{1}{2} a^2 \left(\frac{\pi}{2}\right) = \frac{1}{4} \pi a^2$$

(at the second step three terms turn out to be zero). We could also have evaluated (3) directly, without bothering to calculate the indefinite integral (2). In fact, making the same substitution $x = a \sin u$ ($-\pi/2 \le u \le \pi/2$) as in Example 1, and observing that $x = 0$ implies $u = 0$, while $x = a$ implies $u = \pi/2$ (the substitution is one-to-one), we get

$$\int_0^a \sqrt{a^2 - x^2}\, dx = a^2 \left[\frac{1}{4} \sin 2u + \frac{1}{2} u\right]_0^{\pi/2} = \frac{1}{4} \pi a^2.$$

An even simpler way of evaluating the integral (3) is to recognize that (3) is the area under the curve $y = \sqrt{a^2 - x^2}$ in the first quadrant, equal to one fourth the area A enclosed by the circle $x^2 + y^2 = a^2$ of radius a centered at the origin (see Figure 1). But $A = \pi a^2$, and thus the integral is equal to $\frac{1}{4}\pi a^2$. □

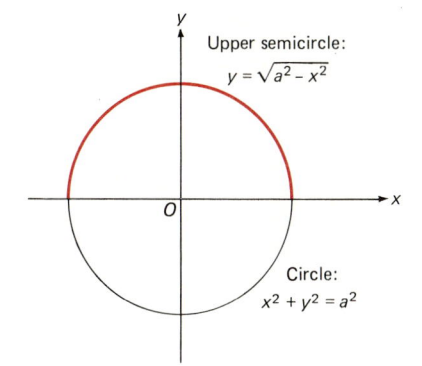

Figure 1

Upper semicircle: $y = \sqrt{a^2 - x^2}$

Circle: $x^2 + y^2 = a^2$

The Substitutions
$x = a \sin u$, $x = \tan u$
and $x = a \sec u$

More generally, an integral containing one of the irrational expressions

$$\sqrt{a^2 - x^2}, \qquad \sqrt{x^2 + a^2}, \qquad \sqrt{x^2 - a^2} \qquad (a > 0) \qquad (4)$$

can often be evaluated by making a *trigonometric substitution*, that is, a substitution $x = x(u)$ involving a trigonometric function. Figure 2 illustrates schematically the substitutions that are appropriate in these three cases. Notice that in each part of the figure x and a are the lengths of two sides of a right triangle, the length of the other side is one of the above expressions, and the new variable u is an angle of the triangle (measured in radians). In each triangle $0 < u < \pi/2$, but the indicated substitutions continue to work for certain other values of u (see below). After the given integral has been

Section 7.5 Trigonometric and Hyperbolic Substitutions **389**

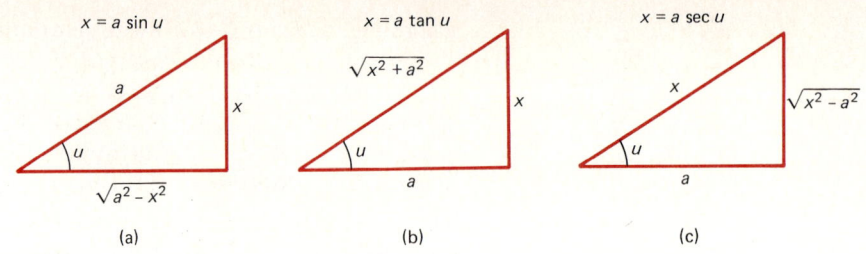

Figure 2

(a) $x = a \sin u$

(b) $x = a \tan u$

(c) $x = a \sec u$

evaluated in terms of the new variable u, the same triangle suggests the substitutions required to return to the original variable x. Here we assume that the substitution $x = x(u)$ is one-to-one, and this will be guaranteed by suitably restricting the values of u.

The triangles in Figure 2 are just a convenient way to remember the following more detailed information.

(i) In an integral involving $\sqrt{a^2 - x^2}$ make the substitution $x = a \sin u$, where $-\pi/2 \leq u \leq \pi/2$. Then $dx = a \cos u\, du$, and

$$\sqrt{a^2 - x^2} = \sqrt{a^2 - a^2 \sin^2 u} = a\sqrt{1 - \sin^2 u} = a\sqrt{\cos^2 u} = a \cos u,$$

since $\cos u \geq 0$ for these values of u. If the expression $\sqrt{a^2 - x^2}$ appears in the denominator of the integrand, we must restrict u to the *open* interval $-\pi/2 < u < \pi/2$, since $\sqrt{a^2 - x^2} = a \cos u$ equals zero for $u = \pm \pi/2$.

(ii) In an integral involving $\sqrt{x^2 + a^2}$ make the substitution $x = a \tan u$, where $-\pi/2 < u < \pi/2$ (note that $\tan u$ is not defined for $u = \pm \pi/2$). Then $dx = a \sec^2 u\, du$, and

$$\sqrt{x^2 + a^2} = \sqrt{a^2 \tan^2 u + a^2} = a\sqrt{\tan^2 u + 1} = a\sqrt{\sec^2 u} = a \sec u,$$

since $\sec u \geq 0$ for these values of u.

(iii) In an integral involving $\sqrt{x^2 - a^2}$ make the substitution $x = a \sec u$, where $0 \leq u < \pi/2$ (note that $\sec u$ is not defined for $u = \pi/2$). Then $dx = a \sec u \tan u\, du$, and

$$\sqrt{x^2 - a^2} = \sqrt{a^2 \sec^2 u - a^2} = a\sqrt{\sec^2 u - 1} = a\sqrt{\tan^2 u} = a \tan u,$$

since $\tan u \geq 0$ for these values of u. We can also allow u to take the values $\pi/2 < u \leq \pi$, but then $\sqrt{x^2 - a^2} = a\sqrt{\tan^2 u} = -a \tan u$, since $\tan u \leq 0$ for these values of u. If the expression $\sqrt{x^2 - a^2}$ appears in the denominator of the integrand, we must exclude the values $u = 0$ and $u = \pi$, since $\sqrt{x^2 - a^2} = a \tan u$ equals zero for these values of u.

It should be noted that in each case we have restricted the values of u in such a way as to make the function $x = x(u)$ one-to-one. Thus in (i) the function $x = a \sin u$ is one-to-one if $-\pi/2 \leq u \leq \pi/2$, with inverse $u = \arcsin(x/a)$, in (ii) the function $x = a \tan u$ is one-to-one if $-\pi/2 < u < \pi/2$, with inverse function $u = \arctan(x/a)$, and in (iii) the function $x = a \sec u$ is one-to-one if $0 \leq u < \pi/2$ or $\pi/2 < u \leq \pi$, with inverse $u = \operatorname{arcsec}(x/a)$.

The Substitutions
$x = a \sinh u$
and $x = a \cosh u$

The first of the expressions (4) can also be simplified by the substitution $x = a \cos u$, the second by $x = a \cot u$, and the third by $x = a \csc u$ (verify this statement). However, these substitutions are redundant, since anything they can do can also be achieved, and more simply at that, by the substitutions $x = a \sin u$, $x = a \tan u$ and $x = a \sec u$. Of greater interest are the *hyperbolic substitutions* $x = a \sinh u$ and $x = a \cosh u$, which are just as effective in simplifying the expressions $\sqrt{x^2 + a^2}$ and $\sqrt{x^2 - a^2}$ $(a > 0)$ as the trigonometric substitutions $x = a \tan u$ and $x = a \sec u$. In fact, if $x = a \sinh u$, where $-\infty < u < \infty$, then $dx = a \cosh u\, du$, and

$$\sqrt{x^2 + a^2} = \sqrt{a^2 \sinh^2 u + a^2} = a\sqrt{\sinh^2 u + 1} = a\sqrt{\cosh^2 u} = a \cosh u,$$

since $\cosh u > 0$, while if $x = a \cosh u$, where $0 \leq u < \infty$, then $dx = a \sinh u\, du$, and

$$\sqrt{x^2 - a^2} = \sqrt{a^2 \cosh^2 u - a^2} = a\sqrt{\cosh^2 u - 1} = a\sqrt{\sinh^2 u} = a \sinh u,$$

since $\sinh u \geq 0$ if u is nonnegative.

Example 3 Use a hyperbolic substitution to evaluate

$$\int \sqrt{x^2 + a^2}\, dx \qquad (a > 0).$$

Solution Let $x = a \sinh u$, so that $dx = a \cosh u$, $\sqrt{x^2 + a^2} = a \cosh u$, and

$$\int \sqrt{x^2 + a^2}\, dx = a^2 \int \cosh^2 u\, du.$$

But

$$\int \cosh^2 u\, du = \frac{1}{2}\int (\cosh 2u + 1)\, du = \frac{1}{4} \sinh 2u + \frac{1}{2} u + C,$$

and therefore

$$\int \sqrt{x^2 + a^2}\, dx = \frac{1}{2} a^2 \sinh u \cosh u + \frac{1}{2} a^2 u + C, \tag{5}$$

since $\sinh 2u = 2 \sinh u \cosh u$. Moreover,

$$\sinh u = \frac{x}{a}, \qquad u = \sinh^{-1} \frac{x}{a},$$

$$\cosh u = \sqrt{\sinh^2 u + 1} = \sqrt{\left(\frac{x}{a}\right)^2 + 1} = \frac{\sqrt{x^2 + a^2}}{a}.$$

Substituting these expressions for $\sinh u$, $\cosh u$ and u into (5), we finally obtain

$$\int \sqrt{x^2 + a^2}\, dx = \frac{1}{2} x\sqrt{x^2 + a^2} + \frac{1}{2} a^2 \sinh^{-1} \frac{x}{a} + C. \tag{6}$$

The resemblance between formulas (2) and (6) is noteworthy. Using formula (3′), page 358, we can write (6) in the equivalent form

$$\int \sqrt{x^2 + a^2}\, dx = \frac{1}{2} x\sqrt{x^2 + a^2} + \frac{1}{2} a^2 \ln(x + \sqrt{x^2 + a^2}) + C, \tag{6′}$$

after absorbing $-\frac{1}{2} a^2 \ln a$ into the arbitrary constant of integration C.

□

Example 4 Solve Example 3 by making a *trigonometric* substitution.

Solution Let $x = a \tan u$, $dx = a \sec^2 u \, du$, so that $\sqrt{x^2 + a^2} = a \sec u$, and

$$\int \sqrt{x^2 + a^2} \, dx = a^2 \int \sec^3 u \, du$$

$$= \frac{1}{2} a^2 \sec u \tan u + \frac{1}{2} a^2 \ln |\sec u + \tan u| + C,$$

with the help of Example 7, page 377. Consulting Figure 2(b), in which $\tan u = x/a$, we find that

$$\sec u = \frac{\sqrt{x^2 + a^2}}{a}.$$

Therefore

$$\int \sqrt{x^2 + a^2} \, dx = \frac{1}{2} x \sqrt{x^2 + a^2} + \frac{1}{2} a^2 \ln \left| \frac{\sqrt{x^2 + a^2}}{a} + \frac{x}{a} \right| + C$$

$$= \frac{1}{2} x \sqrt{x^2 + a^2} + \frac{1}{2} a^2 \ln (x + \sqrt{x^2 + a^2}) + C,$$

and we again have (6'), after absorbing $-\frac{1}{2} a^2 \ln a$ into C. (Why are we justified in dropping the absolute value sign?) □

Example 5 Evaluate $\displaystyle\int \frac{dx}{(9x^2 + 4)^{3/2}}$.

Solution This time let $x = \frac{2}{3} \tan u$, so that $dx = \frac{2}{3} \sec^2 u \, du$, and

$$(9x^2 + 4)^{3/2} = (4 \tan^2 u + 4)^{3/2} = 8 \sec^3 u.$$

Then

$$\int \frac{dx}{(9x^2 + 4)^{3/2}} = \int \frac{\frac{2}{3} \sec^2 u}{8 \sec^3 u} \, du = \frac{1}{12} \int \frac{du}{\sec u}$$

$$= \frac{1}{12} \int \cos u \, du = \frac{1}{12} \sin u + C.$$

Inspection of Figure 3, in which $\tan u = 3x/2$, shows that

$$\sin u = \frac{3x}{\sqrt{9x^2 + 4}}.$$

Therefore

$$\int \frac{dx}{(9x^2 + 4)^{3/2}} = \frac{x}{4\sqrt{9x^2 + 4}} + C.$$

As an exercise, show that the hyperbolic substitution $x = \frac{2}{3} \sinh u$ leads to the same answer. □

Example 6 Evaluate

$$\int \frac{dx}{(16x^2 - 25)^{3/2}}. \tag{7}$$

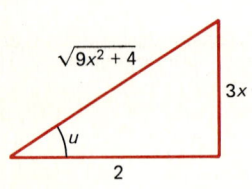

Figure 3

Solution Let $x = \frac{5}{4} \sec u$ $(0 < u < \pi/2)$, so that $dx = \frac{5}{4} \sec u \tan u \, du$, and
$$(16x^2 - 25)^{3/2} = (25 \sec^2 u - 25)^{3/2} = 125 \tan^3 u.$$
Then
$$\int \frac{dx}{(16x^2 - 25)^{3/2}} = \int \frac{\frac{5}{4} \sec u \tan u}{125 \tan^3 u} \, du = \frac{1}{100} \int \frac{du}{\cos u \frac{\sin^2 u}{\cos^2 u}}$$

$$= \frac{1}{100} \int \frac{1}{\sin u} \frac{\cos u}{\sin u} \, du = \frac{1}{100} \int \csc u \cot u \, du$$

$$= -\frac{1}{100} \csc u + C.$$

Consulting Figure 4, in which $\sec u = 4x/5$, we find that
$$\csc u = \frac{4x}{\sqrt{16x^2 - 25}},$$
and therefore
$$\int \frac{dx}{(16x^2 - 25)^{3/2}} = -\frac{1}{25} \frac{x}{\sqrt{16x^2 - 25}} + C,$$
provided that $x > \frac{5}{4}$. As an exercise, show that the same formula holds if $x < -\frac{5}{4}$. □

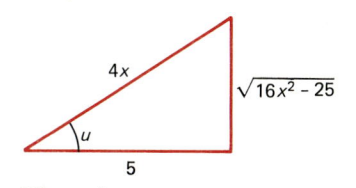

Figure 4

Example 7 Solve Example 6 by making a *hyperbolic* substitution.

Solution Let $x = \frac{5}{4} \cosh u$ $(0 \leq u < \infty)$, so that $dx = \frac{5}{4} \sinh u \, du$, and
$$(16x^2 - 25)^{3/2} = (25 \cosh^2 u - 25)^{3/2} = 125 \sinh^3 u.$$
Then
$$\int \frac{dx}{(16x^2 - 25)^{3/2}} = \int \frac{\frac{5}{4} \sinh u}{125 \sinh^3 u} \, du = \frac{1}{100} \int \frac{du}{\sinh^2 u}$$

$$= \frac{1}{100} \int \text{csch}^2 u \, du = -\frac{1}{100} \coth u + C$$

$$= -\frac{1}{100} \frac{\cosh u}{\sinh u} + C.$$

But
$$\cosh u = \frac{4x}{5}, \qquad \sinh u = \frac{\sqrt{16x^2 - 25}}{5},$$
and therefore
$$\int \frac{dx}{(16x^2 - 25)^{3/2}} = -\frac{1}{100} \frac{4x}{5} \frac{5}{\sqrt{16x^2 - 25}} + C = -\frac{1}{25} \frac{x}{\sqrt{16x^2 - 25}} + C,$$
as before. This takes care of the case $x > \frac{5}{4}$, and the case $x < -\frac{5}{4}$ can be handled by substituting $x = -\frac{5}{4} \cosh u$ (explain further). □

Section 7.5 Trigonometric and Hyperbolic Substitutions

Example 8 Evaluate $\int \dfrac{dx}{\sqrt{x^2 + x + 1}}$.

Solution There is now a first power of x under the radical, but we can get rid of it by first completing the square. In fact,

$$x^2 + x + 1 = \left(x + \frac{1}{2}\right)^2 + \frac{3}{4} = y^2 + \frac{3}{4},$$

where $y = x + \frac{1}{2}$, and therefore

$$\int \frac{dx}{\sqrt{x^2 + x + 1}} = \int \frac{dy}{\sqrt{y^2 + \frac{3}{4}}}$$

($dx = dy$). To evaluate the integral on the right, we substitute $y = (\sqrt{3}/2) \tan u$. Then $dy = (\sqrt{3}/2) \sec^2 u\, du$, $\sqrt{y^2 + \frac{3}{4}} = (\sqrt{3}/2) \sec u$, and

$$\int \frac{dy}{\sqrt{y^2 + \frac{3}{4}}} = \int \frac{\sec^2 u}{\sec u}\, du = \int \sec u\, du = \ln |\sec u + \tan u| + C,$$

with the help of Example 4, page 369. Inspection of Figure 5, in which $\tan u = 2y/\sqrt{3}$, shows that

$$\sec u = \frac{\sqrt{4y^2 + 3}}{\sqrt{3}},$$

and hence

$$\int \frac{dy}{\sqrt{y^2 + \frac{3}{4}}} = \ln \left| \frac{\sqrt{4y^2 + 3}}{\sqrt{3}} + \frac{2y}{\sqrt{3}} \right| + C = \ln (\sqrt{4y^2 + 3} + 2y) + C,$$

after absorbing $-\ln \sqrt{3}$ into the arbitrary constant of integration C (justify dropping the absolute value sign). Returning to the original variable x, we have

$$\int \frac{dx}{\sqrt{x^2 + x + 1}} = \ln \left[\sqrt{4\left(x + \frac{1}{2}\right)^2 + 3} + 2\left(x + \frac{1}{2}\right) \right] + C$$

$$= \ln \left[\sqrt{4x^2 + 4x + 4} + 2\left(x + \frac{1}{2}\right) \right] + C$$

$$= \ln \left(\sqrt{x^2 + x + 1} + x + \frac{1}{2} \right) + \ln 2 + C.$$

Absorbing $\ln 2$ into C, we finally obtain

$$\int \frac{dx}{\sqrt{x^2 + x + 1}} = \ln \left(\sqrt{x^2 + x + 1} + x + \frac{1}{2} \right) + C. \quad \square$$

Example 9 Evaluate $\int_1^3 \dfrac{x}{\sqrt{4x - x^2}}\, dx$.

Solution Completing the square in the expression $4x - x^2$, we get

$$4x - x^2 = 4 - (x - 2)^2 = 4 - y^2,$$

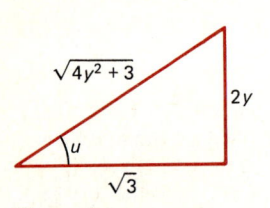

Figure 5

394 Chapter 7 Methods of Integration

where $y = x - 2$. Then $x = 2 + y$, $dx = dy$, and

$$\int_1^3 \frac{x}{\sqrt{4x - x^2}} dx = \int_{-1}^1 \frac{2 + y}{\sqrt{4 - y^2}} dy$$

$$= 2\int_{-1}^1 \frac{dy}{\sqrt{4 - y^2}} + \int_{-1}^1 \frac{y}{\sqrt{4 - y^2}} dy.$$

The second integral in the sum on the right equals zero, since the integrand is odd and the interval of integration is symmetric about the origin. To evaluate the first integral, we use formula (2′), page 291, obtaining

$$\int_{-1}^1 \frac{dy}{\sqrt{4 - y^2}} = \left[\arcsin \frac{y}{2}\right]_{-1}^1 = 2\left(\arcsin \frac{1}{2}\right) = 2\left(\frac{\pi}{3}\right) = \frac{2\pi}{3}.$$

Thus, finally,

$$\int_1^3 \frac{x}{\sqrt{4x - x^2}} dx = \frac{2\pi}{3}. \quad \square$$

In some cases an integral that can be evaluated by making a trigonometric or hyperbolic substitution can also be evaluated by making a simpler algebraic substitution, and you should always be on the lookout for this possibility. For example, making the trigonometric substitution $x = 2 \sin u$, we find that

$$\int x\sqrt{4 - x^2}\, dx = 8 \int \cos^2 u \sin u\, du$$

$$= -\frac{8}{3} \cos^3 u + C = -\frac{1}{3}(4 - x^2)^{3/2} + C,$$

but the algebraic substitution $u = 4 - x^2$ is more straightforward and hence preferable:

$$\int x\sqrt{4 - x^2}\, dx = -\frac{1}{2}\int u^{1/2}\, du$$

$$= -\frac{1}{3} u^{3/2} + C = -\frac{1}{3}(4 - x^2)^{3/2} + C.$$

It should also be noted that a trigonometric or hyperbolic substitution often leads to the rediscovery of a formula that is already known. For example, making the trigonometric substitution $x = \sin u$, we find that

$$\int \frac{dx}{\sqrt{1 - x^2}} = \int \frac{\cos u}{\cos u}\, du = \int du = u + C = \arcsin x + C,$$

as in formula (2), page 291, while the trigonometric substitution $x = \tan u$ gives

$$\int \frac{dx}{x^2 + 1} = \int \frac{\sec^2 u}{\sec^2 u}\, du = \int du = u + C = \arctan x + C,$$

Section 7.5 Trigonometric and Hyperbolic Substitutions

as in formula (6), page 292. Similarly, making the hyperbolic substitution $x = 3 \sinh u$, we get

$$\int \frac{dx}{\sqrt{x^2 + 9}} = \int \frac{3 \cosh u}{3 \cosh u} du = \int du = u + C = \sinh^{-1} \frac{x}{3} + C,$$

which is just a special case of formula (2'), page 358. Thus there will often be different ways of evaluating the same integral, and you should try to find the way that is most efficient. Needless to say, it is always a good idea to use differentiation to check that any given integral has been evaluated correctly.

Problems

Evaluate

1. $\int \sqrt{1 - 9x^2}\, dx$

2. $\int \sqrt{4x^2 + 25}\, dx$

3. $\int \sqrt{16x^2 - 1}\, dx$

4. $\int \frac{dx}{\sqrt{9 - 4x^2}}$

5. $\int \frac{dx}{\sqrt{x^2 + 64}}$

6. $\int \frac{dx}{\sqrt{36x^2 - 49}}$

7. $\int \frac{x^2}{\sqrt{1 - 3x^2}}\, dx$

8. $\int \frac{x^2}{\sqrt{x^2 - 25}}\, dx$

9. $\int \frac{dx}{x\sqrt{x^2 + 4}}$

10. $\int \frac{dx}{x\sqrt{1 - 100x^2}}$

11. $\int \frac{dx}{x^2 \sqrt{x^2 - 2}}$

12. $\int \frac{x^3}{\sqrt{x^2 + 1}}\, dx$

13. $\int \frac{x^3}{\sqrt{16 - x^2}}\, dx$

14. $\int (1 - x^2)^{3/2}\, dx$

15. $\int (x^2 - 1)^{3/2}\, dx$

16. $\int \frac{\sqrt{4 - x^2}}{x}\, dx$

17. $\int \frac{\sqrt{x^2 - 9}}{x^2}\, dx$

18. $\int \frac{dx}{(x^2 + 1)^{5/2}}$

19. $\int \frac{dx}{x^3 \sqrt{x^2 - 4}}$

20. $\int \sqrt{x^2 + 2x + 2}\, dx$

21. $\int \sqrt{2x - x^2}\, dx$

22. $\int \frac{dx}{\sqrt{x^2 + 4x + 6}}$

23. $\int \frac{dx}{\sqrt{x^2 + 3x - 4}}$

24. $\int \frac{dx}{(4x^2 + 8x - 1)^{3/2}}$

25. $\int_{1/2}^{2} \sqrt{x^2 - x + 1}\, dx$

26. $\int_{-1}^{1} \frac{ds}{\sqrt{4 - 2s - s^2}}$

27. $\int_{1}^{3} \frac{t}{\sqrt{t^2 + 2t + 5}}\, dt$

28. $\int_{0}^{2} \frac{x^2}{\sqrt{12 + 4x - x^2}}\, dx$

29. $\int_{1}^{2} \frac{ds}{s^2 \sqrt{9 - s^2}}$

30. $\int_{1/\sqrt{3}}^{1} \frac{dt}{t(t^2 + 1)}$

31. $\int_{\ln 2}^{\ln 3} \frac{e^u}{\sqrt{e^{2u} - 1}}\, du$

32. $\int_{0}^{\pi/2} \frac{\sin v}{\sqrt{\cos^2 v + 1}}\, dv$

33. $\int_{0}^{\pi/4} \frac{\sec^2 w}{\sqrt{4 - \tan^2 w}}\, dw$

34. Find the areas of the two regions into which the parabola $y = \frac{1}{2}x^2$ divides the interior of the circle $x^2 + y^2 = 8$ (see Figure 6).

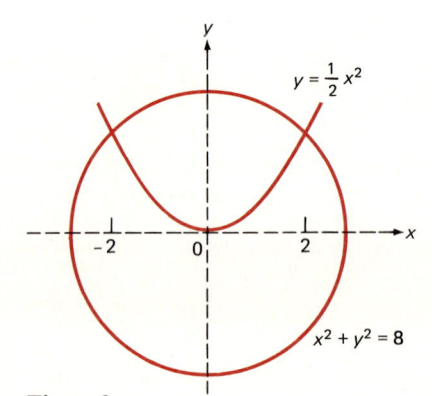

Figure 6

35. The centers of two circular disks of unit radius are a distance $2a$ apart ($0 \le a \le 1$). Find the area A of the region in which the two disks overlap (the lens-shaped region in Figure 7).

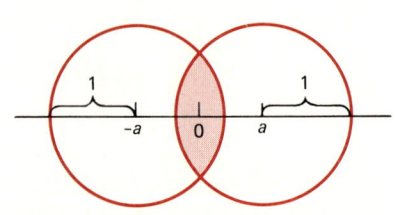

Figure 7

36. Show that

$$\int \sqrt{x^2 - a^2}\, dx = \frac{1}{2} x\sqrt{x^2 - a^2} - \frac{1}{2} a^2 \ln |x + \sqrt{x^2 - a^2}| + C$$

for $|x| \geq a > 0$.

+ **37.** Find the area A of the hyperbolic sector POQ shown in Figure 19(b), page 353.

Hint. Note that $A = \frac{1}{2} \cosh \theta \sinh \theta - \int_1^{\cosh \theta} \sqrt{x^2 - 1}\, dx$.

38. Use a trigonometric substitution to deduce the reduction formula

$$\int \frac{dx}{(x^2 + a^2)^{n+1}} = \frac{1}{2na^2} \frac{x}{(x^2 + a^2)^n} + \frac{2n - 1}{2na^2} \int \frac{dx}{(x^2 + a^2)^n}$$

(see Example 5, page 380) from the reduction formula

$$\int \cos^n x\, dx = \frac{1}{n} \cos^{n-1} x \sin x + \frac{n - 1}{n} \int \cos^{n-2} x\, dx$$

(see Example 2, page 379).

7.6 Integration of Rational Functions; Partial Fractions

Proper and Improper Rational Functions

In this section we show how to integrate rational functions. You will recall that a rational function is defined as the quotient of two polynomials, that is, as a function of the form

$$R(x) = \frac{P(x)}{Q(x)} = \frac{a_0 + a_1 x + a_2 x^2 + \cdots + a_n x^n}{b_0 + b_1 x + b_2 x^2 + \cdots + b_N x^N} \quad (a_n \neq 0, b_N \neq 0). \quad (1)$$

The numerator $P(x)$ is a polynomial of degree n, since $a_n \neq 0$, while the denominator $Q(x)$ is a polynomial of degree N, since $b_N \neq 0$. (It will always be assumed that the numerator and denominator have no common factors, since otherwise the common factors can be canceled at the outset.) If $n < N$, that is, if the degree of the numerator is less than that of the denominator, we say that the rational function $R(x)$ is *proper*. If $R(x)$ is *improper*, that is, if $n \geq N$, we can divide $Q(x)$ into $P(x)$ by ordinary long division, thereby expressing $R(x)$ as the sum of a polynomial and another rational function $R_1(x)$, with the same denominator $Q(x)$, where $R_1(x)$ is now proper. But polynomials are easy to integrate (see Example 6, page 253), and hence the nub of the problem is to integrate $R_1(x)$.

Example 1 Evaluate $\int \frac{x^2 + x + 1}{x - 1}\, dx$.

Solution The integrand is a rational function, but it is improper, since the degree of the numerator exceeds that of the denominator. Therefore we divide the denominator into the numerator, obtaining

$$\begin{array}{r} x + 2 \\ x - 1 \overline{\smash{\big)}\, x^2 + x + 1} \\ \underline{x^2 - x} \\ 2x + 1 \\ \underline{2x - 2} \\ 3 \end{array}$$

where $x + 2$ is the quotient and 3 is the remainder. Thus

$$\frac{x^2 + x + 1}{x - 1} = x + 2 + \frac{3}{x - 1},$$

and we have expressed the integrand as the sum of the polynomial $x + 2$ and the proper rational function $3/(x - 1)$. It follows that

$$\int \frac{x^2 + x + 1}{x - 1} \, dx = \int (x + 2) \, dx + 3 \int \frac{dx}{x - 1}$$

$$= \frac{1}{2} x^2 + 2x + 3 \ln |x - 1| + C,$$

where C is a constant of integration. ∎

Example 2 Evaluate $\int \frac{x^2 - 1}{x^2 + 1} \, dx$.

Solution The integrand is again an improper rational function, since the numerator and denominator are of the same degree. This time

$$\frac{x^2 - 1}{x^2 + 1} = 1 - \frac{2}{x^2 + 1},$$

by long division or simply by observing that the numerator $x^2 - 1$ can be written in the form $(x^2 + 1) - 2$. Therefore

$$\int \frac{x^2 - 1}{x^2 + 1} \, dx = \int dx - 2 \int \frac{dx}{x^2 + 1} = x - 2 \arctan x + C. \quad ∎$$

Irreducible Quadratic Polynomials

In Example 1 the denominator $x - 1$ is a linear polynomial, while in Example 2 the denominator $x^2 + 1$ is an *irreducible* quadratic polynomial, that is, a quadratic polynomial that cannot be factored into a product of linear polynomials. To see that $x^2 + 1$ is irreducible, we observe that if $x^2 + 1$ had a factorization

$$x^2 + 1 = (x - a)(x - b),$$

then $x^2 + 1$ would equal zero for $x = a$ or $x = b$, that is, the quadratic equation $x^2 + 1 = 0$ would have a solution. But this is impossible, since $x^2 + 1 \geq 1 > 0$ for all real x.

More generally, a quadratic polynomial $x^2 + px + q$ is irreducible if and only if the quadratic equation $x^2 + px + q = 0$ has no (real) solutions. By the formula for the roots of a quadratic equation, this is the case if and only if $p^2 - 4q < 0$, or equivalently $p^2 < 4q$. In fact, if $p^2 \geq 4q$, the equation $x^2 + px + q = 0$ has the roots $a = \frac{1}{2}(-p + \sqrt{p^2 - 4q})$ and $b = \frac{1}{2}(-p - \sqrt{p^2 - 4q})$, which coincide if $p^2 = 4q$, and then, as is easily verified,

$$x^2 + px + q = (x - a)(x - b),$$

but if $p^2 < 4q$, then

$$x^2 + px + q = \left(x + \frac{p}{2}\right)^2 + \left(q - \frac{p^2}{4}\right) \geq q - \frac{p^2}{4} > 0$$

for all real x.

Factorization of $Q(x)$

To treat the case of a general rational function of the form (1), the first step is to factor the denominator

$$Q(x) = b_0 + b_1 x + b_2 x^2 + \cdots + b_N x^N \qquad (b_N \neq 0). \qquad (2)$$

Here we rely on a theorem of algebra which states that *any polynomial with real coefficients, like Q(x), can be written as a product of linear and irreducible quadratic factors.* In detail, if N is the degree of $Q(x)$, then

$$Q(x) = a(x - c_1)^{r_1} \cdots (x - c_k)^{r_k}(x^2 + p_1 x + q_1)^{s_1} \cdots (x^2 + p_m x + q_m)^{s_m}, \tag{3}$$

where $k, m, r_1, \ldots, r_k, s_1, \ldots, s_m$ are positive integers such that

$$r_1 + \cdots + r_k + 2s_1 + \cdots + 2s_m = N, \tag{4}$$

and $a, c_1, \ldots, c_k, p_1, q_1, \ldots, p_m, q_m$ are real constants such that each pair p_i, q_i satisfies the irreducibility condition $p_i^2 < 4q_i$ (why does (4) hold?). Also, the factorization (3) is unique, apart from the order of the factors. It is of course assumed that no two linear factors or quadratic factors in (3) are the same.

The following theorem facilitates the search for linear factors of a polynomial.

Theorem 2 *(Factor theorem).* *The polynomial $Q(x)$ is divisible by the linear factor $x - c$ if and only if $Q(c) = 0$.*

Proof (Optional) If $Q(x)$ is divisible by $x - c$, then $Q(x) = (x - c)S(x)$, where $S(x)$ is another polynomial, and hence

$$Q(c) = (c - c)S(c) = 0.$$

Conversely, if $Q(x)$ is given by (2) and $Q(c) = 0$, then

$$Q(x) = Q(x) - Q(c)$$
$$= (b_0 + b_1 x + b_2 x^2 + \cdots + b_N x^N) - (b_0 + b_1 c + b_2 c^2 + \cdots + b_N c^N)$$
$$= b_1(x - c) + b_2(x^2 - c^2) + \cdots + b_N(x^N - c^N). \tag{5}$$

But

$$x^n - c^n = (x - c)(x^{n-1} + x^{n-2}c + \cdots + xc^{n-2} + c^{n-1}) \qquad (n = 1, 2, \ldots, N).$$

Therefore $x - c$ is a factor of every term on the right in (5), and hence is a factor of the polynomial $Q(x)$ as well. ∎

Example 3 Factor $Q(x) = x^4 + x^3 - 2x^2 - 6x - 4$.

Solution After a little experimentation, we discover that $Q(2) = 0$. Therefore $Q(x)$ is divisible by $x - 2$. Carrying out the long division, we find that $Q(x) = (x - 2)S(x)$, where

$$S(x) = x^3 + 3x^2 + 4x + 2.$$

Further testing reveals that $S(-1) = 0$, so that $S(x)$ is divisible by $x + 1$. This time the division gives

$$S(x) = (x + 1)(x^2 + 2x + 2),$$

where the quadratic factor is irreducible (why?). Thus, finally,

$$Q(x) = (x - 2)S(x) = (x - 2)(x + 1)(x^2 + 2x + 2),$$

which is of the form (3), with $k = 2$, $m = 1$, $r_1 = r_2 = s_1 = 1$ ($N = 4$), $a = 1$, $c_1 = 2$, $c_2 = -1$, $p_1 = q_1 = 2$ ($p_1^2 < 4q_1$). □

Partial Fraction Expansions

Once the denominator $Q(x)$ of the given rational function $R(x)$ has been factored, we can prepare $R(x)$ for integration by using the following theorem, whose proof will be omitted, since it is rather technical and is more algebra than calculus. However, the statement and meaning of the theorem are simple enough. The idea is to represent $R(x)$ as a sum of particularly simple rational functions, called *partial fractions*. These functions are of the form

$$\frac{A}{(x - c)^n} \quad (n = 1, 2, \ldots)$$

or

$$\frac{Bx + C}{(x^2 + px + q)^n} \quad (p^2 < 4q, n = 1, 2, \ldots),$$

and as we will see below, their integrals can be evaluated quite readily. The sum of partial fractions representing $R(x)$ is called the *partial fraction expansion* of $R(x)$.

Theorem 3 *(Partial fraction expansion of a rational function).* Let $R(x)$ be a proper rational function with denominator

$$Q(x) = a(x - c_1)^{r_1} \cdots (x - c_k)^{r_k}(x^2 + p_1 x + q_1)^{s_1} \cdots (x^2 + p_m x + q_m)^{s_m},$$

where no two linear or quadratic factors coincide and the quadratic factors are all irreducible. Then $R(x)$ is the sum of k blocks of terms of the form†

$$\frac{A_1}{x - c_i} + \frac{A_2}{(x - c_i)^2} + \cdots + \frac{A_{r_i}}{(x - c_i)^{r_i}}, \quad (6)$$

one for each distinct linear factor $x - c_i$, and m blocks of terms of the form

$$\frac{B_1 x + C_1}{x^2 + p_i x + q_i} + \frac{B_2 x + C_2}{(x^2 + p_i x + q_i)^2} + \cdots + \frac{B_{s_i} x + C_{s_i}}{(x^2 + p_i x + q_i)^{s_i}}, \quad (7)$$

one for each distinct quadratic factor $x^2 + p_i x + q_i$. The coefficients $A_1, \ldots, A_{r_i}, B_1, \ldots, B_{s_i}, C_1, \ldots, C_{s_i}$ are real constants that are uniquely determined by the function $R(x)$.

The block (6) consists of only one term, the first, if $r_i = 1$, and the same is true of the block (7) if $s_i = 1$. The polynomial $Q(x)$ ordinarily has only a few factors, and a simpler notation can be adopted for the coefficients. We will favor the use of the consecutive capital letters $A, B, C, D, E, F, \ldots,$ without subscripts.

---

† To keep the notation from getting out of hand, we do not bother to distinguish coefficients corresponding to different blocks of the form (6) or (7).

Identity of Polynomials

Before giving a number of examples illustrating the application of Theorem 3, we must acquire one last algebraic tool:

> **Theorem 4** (*Identical polynomials have the same coefficients*). *If two polynomials in x are identically equal, then the polynomials are of the same degree, and identical powers of x have the same coefficients. In other words, if*
> $$a_0 + a_1 x + a_2 x^2 + \cdots + a_n x^n \equiv b_0 + b_1 x + b_2 x^2 + \cdots + b_N x^N, \qquad (8)$$
> *where $a_n \neq 0$, $b_N \neq 0$, then $n = N$, $a_0 = b_0$, $a_1 = b_1$, $a_2 = b_2, \ldots, a_n = b_n$.*

Proof (Optional) Suppose that $n \neq N$. Then, if $n > N$, we differentiate the identity (8) n times, obtaining $n! a_n = 0$ (see Example 3, page 142), while if $n < N$, we differentiate (8) N times, obtaining $N! b_N = 0$. Hence, if $n \neq N$, either $a_n = 0$ or $b_N = 0$, contrary to assumption. It follows that $n = N$, so that (8) takes the form

$$a_0 + a_1 x + a_2 x^2 + \cdots + a_n x^n \equiv b_0 + b_1 x + b_2 x^2 + \cdots + b_n x^n. \qquad (8')$$

Setting $x = 0$ in (8'), we immediately get $a_0 = b_0$. Moreover, differentiating (8') n times in succession and setting $x = 0$ in each of the resulting equations (except the last), we get $a_1 = b_1, a_2 = b_2, \ldots, a_n = b_n$. ∎

Example 4 Expand the rational function

$$\frac{x^2 + 2}{(x - 2)(x + 1)^2}$$

in partial fractions, and then find its integral.

Solution Applying Theorem 3, we see that this rational function is the sum of a single term

$$\frac{A}{x - 2},$$

corresponding to the factor $x - 2$ in the denominator, and a block of two terms

$$\frac{B}{x + 1} + \frac{C}{(x + 1)^2},$$

corresponding to the other factor $(x + 1)^2$. Therefore

$$\frac{x^2 + 2}{(x - 2)(x + 1)^2} = \frac{A}{x - 2} + \frac{B}{x + 1} + \frac{C}{(x + 1)^2}, \qquad (9)$$

where, for simplicity, we denote the coefficients by the consecutive letters A, B and C.

To determine the coefficients, we multiply both sides of equation (9) by $(x - 2)(x + 1)^2$. This gives

$$x^2 + 2 = A(x + 1)^2 + B(x - 2)(x + 1) + C(x - 2), \qquad (10)$$

or equivalently

$$x^2 + 2 = (A + B)x^2 + (2A - B + C)x + (A - 2B - 2C). \qquad (10')$$

If (10) or (10′) holds for all x, then certainly (9) holds for all x except for the values $x = 2$ and $x = -1$, which lead to zero denominators. Applying Theorem 4 to the polynomial identity (10′), we find that the coefficients of x^2 in the left and right sides of (10′) must be equal, and that the same must be true of the coefficients of x and the constant terms (the latter can be regarded as the coefficients of x^0). This leads at once to the following system of three linear equations in the three unknowns A, B and C:†

$$A + B = 1,$$
$$2A - B + C = 0, \qquad (11)$$
$$A - 2B - 2C = 2.$$

Solving this system, we find that

$$A = \frac{2}{3}, \quad B = \frac{1}{3}, \quad C = -1.$$

In fact, adding twice the second equation to the third equation, we obtain $5A - 4B = 2$, which together with $A + B = 1$, or $B = 1 - A$, implies $5A - 4(1 - A) = 2$, $9A = 6$, $A = \frac{2}{3}$, $B = \frac{1}{3}$, $C = B - 2A = \frac{1}{3} - \frac{4}{3} = -1$. Substituting these values of the coefficients A, B, C into (9), we find that the given rational function has the partial fraction expansion

$$\frac{x^2 + 2}{(x - 2)(x + 1)^2} = \frac{2}{3}\left(\frac{1}{x - 2}\right) + \frac{1}{3}\left(\frac{1}{x + 1}\right) - \frac{1}{(x + 1)^2}.$$

The rational function is now easily integrated:

$$\int \frac{x^2 + 2}{(x - 2)(x + 1)^2}\,dx = \frac{2}{3}\int \frac{dx}{x - 2} + \frac{1}{3}\int \frac{dx}{x + 1} - \int \frac{dx}{(x + 1)^2}$$
$$= \frac{2}{3}\ln|x - 2| + \frac{1}{3}\ln|x + 1| + \frac{1}{x + 1} + C. \quad \square$$

There is another, more efficient way of determining the coefficients A, B, C in Example 4, which bypasses the need to solve the system (11). The method is based on the observation that if two polynomials in x are identically equal, their values must be the same for every choice of x. But a glance at (10) shows that the expression on the right takes a particularly simple form if we choose $x = 2$ or $x = -1$, since in each case two of the three terms equal zero. Thus, setting $x = 2$ in (10), we immediately get $6 = 9A$, or $A = \frac{2}{3}$, and setting $x = -1$, we get $3 = -3C$, or $C = -1$. It then follows from the first of the equations (11) that $B = \frac{1}{3}$. Alternatively, we can set $A = \frac{2}{3}$, $C = -1$, $x = 0$ in (10), obtaining $2 = \frac{2}{3} - 2B + 2$, which implies $B = \frac{1}{3}$.

Example 5 Expand the rational function

$$\frac{3x^2 + x + 4}{x(x^2 + 2)^2}$$

in partial fractions, and then find its integral.

---

† By a *linear equation* in n variables (or "unknowns") $x_1, x_2, \ldots, x_n$ is meant an equation of the first degree in the variables, that is, an equation of the form $a_1 x_1 + a_2 x_2 + \cdots + a_n x_n = b$, where $a_1, a_2, \ldots, a_n$, b are constants. Some of these constants may be zero, and the linear equation is said to be *homogeneous* if $b = 0$.

Solution Since $x^2 + 2$ is irreducible, it follows from Theorem 3 that the given rational function is the sum of a single term

$$\frac{A}{x},$$

corresponding to the factor x in the denominator, and a block of two terms

$$\frac{Bx + C}{x^2 + 2} + \frac{Dx + E}{(x^2 + 2)^2},$$

corresponding to the other factor $(x^2 + 2)^2$. Therefore

$$\frac{3x^2 + x + 4}{x(x^2 + 2)^2} = \frac{A}{x} + \frac{Bx + C}{x^2 + 2} + \frac{Dx + E}{(x^2 + 2)^2}, \tag{12}$$

where this time the coefficients are denoted by the letters A, B, C, D and E. Multiplying both sides of (12) by $x(x^2 + 2)^2$, we get

$$3x^2 + x + 4 = A(x^2 + 2)^2 + (Bx + C)(x^2 + 2)x + (Dx + E)x,$$

or equivalently

$$3x^2 + x + 4 = (A + B)x^4 + Cx^3 + (4A + 2B + D)x^2 + (2C + E)x + 4A. \tag{13}$$

Applying Theorem 4 to this identity, we obtain a system of five linear equations in the five unknowns A, B, C, D and E:

$$A + B = 0,$$
$$C = 0,$$
$$4A + 2B + D = 3,$$
$$2C + E = 1,$$
$$4A = 4.$$

Despite its complicated appearance, this system of equations can be solved with very little effort. Indeed the fifth and second equations tell us at once that $A = 1$, $C = 0$, and then the other equations immediately yield $B = -A = -1$, $D = 3 - 4A - 2B = 1$, $E = 1$. Thus

$$A = 1, \quad B = -1, \quad C = 0, \quad D = 1, \quad E = 1,$$

and substituting these values of the coefficients into (12), we find that our rational function has the partial fraction expansion

$$\frac{3x^2 + x + 4}{x(x^2 + 2)^2} = \frac{1}{x} - \frac{x}{x^2 + 2} + \frac{x + 1}{(x^2 + 2)^2}$$

$$= \frac{1}{x} - \frac{x}{x^2 + 2} + \frac{x}{(x^2 + 2)^2} + \frac{1}{(x^2 + 2)^2}.$$

It is now an easy matter to integrate the rational function. In fact,

$$\int \frac{3x^2 + x + 4}{x(x^2 + 2)^2} dx = \int \frac{dx}{x} - \int \frac{x}{x^2 + 2} dx + \int \frac{x}{(x^2 + 2)^2} dx + \int \frac{dx}{(x^2 + 2)^2}$$

$$= \ln |x| - \frac{1}{2} \int \frac{d(x^2)}{x^2 + 2} + \frac{1}{2} \int \frac{d(x^2)}{(x^2 + 2)^2} + \int \frac{dx}{(x^2 + 2)^2}$$

$$= \ln |x| - \frac{1}{2} \ln (x^2 + 2) - \frac{1}{2} \left(\frac{1}{x^2 + 2} \right) + \int \frac{dx}{(x^2 + 2)^2}.$$

To evaluate the last integral, we set $a = \sqrt{2}$ in formula (5), page 381, obtaining

$$\int \frac{dx}{(x^2+2)^2} = \frac{1}{4}\left(\frac{x}{x^2+2}\right) + \frac{1}{4\sqrt{2}} \arctan \frac{x}{\sqrt{2}} + C.$$

Therefore, finally,

$$\int \frac{3x^2 + x + 4}{x(x^2+2)^2} dx = \ln|x| - \frac{1}{2}\ln(x^2+2) - \frac{1}{2}\left(\frac{1}{x^2+2}\right) + \frac{1}{4}\left(\frac{x}{x^2+2}\right)$$

$$+ \frac{1}{4\sqrt{2}} \arctan \frac{x}{\sqrt{2}} + C$$

$$= \frac{1}{2} \ln \frac{x^2}{x^2+2} + \frac{x-2}{4(x^2+2)} + \frac{1}{4\sqrt{2}} \arctan \frac{x}{\sqrt{2}} + C. \quad \square$$

Example 6 Evaluate $\int \frac{3x}{x^3 - 1} dx$.

Solution Factoring the denominator of the integrand, we get

$$x^3 - 1 = (x - 1)(x^2 + x + 1),$$

where the quadratic polynomial $x^2 + x + 1$ is irreducible (why?). Hence, by Theorem 3,

$$\frac{3x}{x^3 - 1} = \frac{A}{x-1} + \frac{Bx + C}{x^2 + x + 1}.$$

Multiplying both sides of this equation by $x^3 - 1$, we obtain

$$3x = A(x^2 + x + 1) + (Bx + C)(x - 1). \quad (14)$$

Choosing $x = 1$, we find that $3 = 3A$, or $A = 1$. With this value of A, (14) can be written in the form

$$3x - (x^2 + x + 1) = -x^2 + 2x - 1 = (Bx + C)(x - 1),$$

which implies

$$Bx + C = \frac{-x^2 + 2x - 1}{x - 1} = \frac{(x-1)(-x+1)}{x - 1} = -x + 1.$$

Thus the partial fraction expansion of the integrand is

$$\frac{3x}{x^3 - 1} = \frac{1}{x - 1} + \frac{-x + 1}{x^2 + x + 1}.$$

It follows that

$$\int \frac{3x}{x^3 - 1} dx = \int \frac{dx}{x - 1} + \int \frac{-x+1}{x^2+x+1} dx = \ln|x-1| + \int \frac{-x+1}{(x+\frac{1}{2})^2 + \frac{3}{4}} dx$$

$$= \ln|x-1| + \int \frac{-y + \frac{3}{2}}{y^2 + \frac{3}{4}} dy$$

$$= \ln|x-1| - \frac{1}{2} \int \frac{d(y^2)}{y^2 + \frac{3}{4}} + \frac{3}{2} \int \frac{dy}{y^2 + \frac{3}{4}},$$

where $y = x + \frac{1}{2}$, and consequently

$$\int \frac{3x}{x^3 - 1} dx = \ln|x - 1| - \frac{1}{2}\ln\left(y^2 + \frac{3}{4}\right) + \frac{3}{2}\left(\frac{2}{\sqrt{3}}\arctan\frac{2y}{\sqrt{3}}\right) + C$$

$$= \ln|x - 1| - \frac{1}{2}\ln(x^2 + x + 1) + \sqrt{3}\arctan\frac{2x + 1}{\sqrt{3}} + C,$$

with the help of formula (6'), page 293. □

We conclude this section by showing that the method of partial fractions enables us to evaluate the integral of an *arbitrary* rational function, at least in theory.

Optional
According to Theorem 3, the integral of any partial fraction corresponding to a linear factor in the denominator of a proper rational function is of the form

$$\int \frac{A}{(x - c)^n} dx \qquad (n = 1, 2, \ldots), \qquad (15)$$

while the integral of any partial fraction corresponding to an irreducible quadratic factor is of the form

$$\int \frac{Bx + C}{(x^2 + px + q)^n} dx \qquad (p^2 < 4q, n = 1, 2, \ldots). \qquad (16)$$

We can evaluate (15) at once, with the help of the substitution $u = x - c$, obtaining

$$\int \frac{A}{x - c} dx = A\int \frac{du}{u} = A \ln|u| = A \ln|x - c|$$

if $n = 1$, and

$$\int \frac{A}{(x - c)^n} dx = A\int \frac{du}{u^n} = A\int u^{-n} du = A\left(\frac{u^{-n+1}}{-n + 1}\right) = -\frac{A}{(n - 1)(x - c)^{n-1}}$$

if $n > 1$. For simplicity, we omit constants of integration in writing integrals of partial fractions, with the understanding that a constant of integration will be supplied at the end of the whole calculation.

To evaluate (16), we first complete the square in the denominator. This gives

$$x^2 + px + q = \left(x + \frac{p}{2}\right)^2 + \left(q - \frac{p^2}{4}\right) = u^2 + a^2,$$

where

$$u = x + \frac{p}{2}, \qquad a = \sqrt{q - \frac{p^2}{4}}.$$

Thus if $n = 1$, we have

$$\int \frac{Bx + C}{x^2 + px + q} dx = \int \frac{Bu + [C - (Bp/2)]}{u^2 + a^2} du = \frac{B}{2} \int \frac{2u}{u^2 + a^2} du$$

$$+ \left(C - \frac{Bp}{2}\right) \int \frac{du}{u^2 + a^2}$$

$$= \frac{B}{2} \ln(u^2 + a^2) + \frac{1}{a}\left(C - \frac{Bp}{2}\right) \arctan \frac{u}{a}$$

$$= \frac{B}{2} \ln(x^2 + px + q) + \frac{2C - Bp}{\sqrt{4q - p^2}} \arctan \frac{2x + p}{\sqrt{4q - p^2}}. \tag{17}$$

If $n > 1$, we have

$$\int \frac{Bx + C}{(x^2 + px + q)^n} dx = \int \frac{Bu + [C - (Bp/2)]}{(u^2 + a^2)^n} du$$

$$= \frac{B}{2} \int \frac{2u}{(u^2 + a^2)^n} du + \left(C - \frac{Bp}{2}\right) \int \frac{du}{(u^2 + a^2)^n}.$$

To evaluate the second integral, we use the reduction formula established in Example 5, page 380, while to evaluate the first integral, we make the substitution $v = u^2 + a^2$:

$$\int \frac{2u}{(u^2 + a^2)^n} du = \int \frac{dv}{v^n} = \int v^{-n} dv = \frac{v^{-n+1}}{-n+1} = -\frac{1}{(n-1)(u^2 + a^2)^{n-1}}.$$

The rest of the calculation is just algebra, involving nothing more than expressing u and a in terms of x, p and q, as we did in the case $n = 1$.

Problems

Express the given polynomial as a product of linear and irreducible quadratic factors.
1. $x^4 - x^3 + x^2 - x$
2. $2x^3 - 8x^2 + 2x + 12$
3. $x^4 + x^3 + 2x^2 + x + 1$
4. $x^4 - 6x^2 + 8x - 3$
5. $x^4 - x^3 - 91x^2 + x + 90$
6. $x^5 + x^4 + 4x^3 + 4x^2 + 4x + 4$

7. It was shown on page 310 that
$$\int \frac{dx}{(x+a)(x+b)} = \frac{1}{b-a} \ln\left|\frac{x+a}{x+b}\right| + C \quad (a \neq b).$$
Use partial fractions to derive this formula.

Evaluate

8. $\int \frac{x}{x-2} dx$

9. $\int \frac{8x - 3}{4x + 1} dx$

10. $\int \frac{dx}{x^2 + 4x - 77}$

11. $\int \frac{dx}{(x+2)(3x+4)}$

12. $\int \frac{x^2 + 2}{x^2 - 1} dx$

13. $\int \frac{x}{x^2 - x - 6} dx$

14. $\int \frac{x}{(2x+1)(2x+3)} dx$

15. $\int \left(\frac{x-1}{x+2}\right)^2 dx$

16. $\int \frac{dx}{(x-1)(x-2)(x-3)}$

17. $\int \frac{dx}{x^3 + 1}$

18. $\int \frac{x^2 - x + 1}{x^2 + x + 1} dx$

19. $\int \frac{x^2 - 3x + 2}{x(x^2 + 2x + 1)} dx$

20. $\int \frac{x^3 - 1}{4x^3 - x} dx$

21. $\int \frac{dx}{x^2(x^2 + 1)^2}$

22. $\int \frac{dx}{x^4 - 1}$

23. $\int \frac{1 - 2x - x^2}{(x^2 + 1)^2} dx$

24. $\int \dfrac{x^3}{(x^2+9)^3}\, dx$

25. $\int \dfrac{32x}{(2x-1)(2x-3)(2x-5)}\, dx$

26. $\int \dfrac{x^2+1}{x^3+8}\, dx$

27. $\int \dfrac{x+2}{(x^2+1)(x^2+3)}\, dx$

28. $\int \dfrac{2x^7+3x^4+x-6}{x^3-1}\, dx$

29. $\int \dfrac{x^3}{(x-1)^{100}}\, dx$

30. $\int \dfrac{6-9x-3x^2}{x^4-5x^2+4}\, dx$

31. $\int \dfrac{x^2}{x^4-16}\, dx$

32. $\int \dfrac{x^4+1}{x^6-1}\, dx$

33. $\int \dfrac{x^9}{(x^4-1)^2}\, dx$

34. $\int \dfrac{x^{11}}{(x^8+1)^2}\, dx$

35. $\int_{-1}^{1} \dfrac{s^2+s}{s^3-9s^2+26s-24}\, ds$

36. $\int_{2}^{3} \dfrac{dt}{t^4+2t^3-2t-1}$

37. $\int_{0}^{1} \dfrac{u+2}{(u^2+1)^4}\, du$

38. Show that the integral
$$\int \dfrac{ax^2+bx+c}{x^3(x-1)^2}\, dx$$
is a rational function if and only if $a+2b+3c=0$.

7.7 Rationalizing Substitutions (Optional)

It is often possible to simplify an integral by making a *rationalizing substitution*, that is, a substitution converting the integrand into a rational function of the new variable. The integral can then be evaluated by using the method developed in the preceding section.

Example 1 Evaluate $\int \dfrac{dx}{\sqrt{x}+2\sqrt[3]{x}}$.

Solution Observing that 6 is the least common multiple of 2 and 3, we expect the substitution $x=u^6$ to be effective in getting rid of both the square root and the cube root, and in fact
$$\sqrt{x}=(u^6)^{1/2}=u^3, \qquad \sqrt[3]{x}=(u^6)^{1/3}=u^2.$$
Moreover $dx=6u^5\, du$, and therefore
$$\int \dfrac{dx}{\sqrt{x}+2\sqrt[3]{x}} = 6\int \dfrac{u^5}{u^3+2u^2}\, du = 6\int \dfrac{u^3}{u+2}\, du.$$
Since the rational function $u^3/(u+2)$ is improper, we use long division to express it as the sum of a polynomial and a proper rational function, obtaining
$$\dfrac{u^3}{u+2}=u^2-2u+4-\dfrac{8}{u+2}.$$
The integration is now easily accomplished:
$$\int \dfrac{dx}{\sqrt{x}+2\sqrt[3]{x}} = 6\int \left(u^2-2u+4-\dfrac{8}{u+2}\right) du$$
$$= 2u^3-6u^2+24u-48\ln|u+2|+C$$
$$= 2\sqrt{x}-6\sqrt[3]{x}+24\sqrt[6]{x}-48\ln(\sqrt[6]{x}+2)+C. \quad \square$$

Example 2 Evaluate

$$\int \frac{\sqrt{1+x} - \sqrt{1-x}}{\sqrt{1+x} + \sqrt{1-x}} \, dx. \tag{1}$$

Solution Multiplying both numerator and denominator of the integrand by $\sqrt{1+x} - \sqrt{1-x}$, we get

$$\frac{\sqrt{1+x} - \sqrt{1-x}}{\sqrt{1+x} + \sqrt{1-x}} = \frac{\sqrt{1+x} - \sqrt{1-x}}{\sqrt{1+x} + \sqrt{1-x}} \frac{\sqrt{1+x} - \sqrt{1-x}}{\sqrt{1+x} - \sqrt{1-x}}$$

$$= \frac{(1+x) - 2\sqrt{1-x^2} + (1-x)}{(1+x) - (1-x)} = \frac{1 - \sqrt{1-x^2}}{x}.$$

Therefore (1) can be written in the form

$$\int \frac{\sqrt{1+x} - \sqrt{1-x}}{\sqrt{1+x} + \sqrt{1-x}} \, dx = \int \frac{1 - \sqrt{1-x^2}}{x^2} x \, dx.$$

Now let $u^2 = 1 - x^2$. Then $u \, du = -x \, dx$, and

$$\int \frac{1 - \sqrt{1-x^2}}{x^2} x \, dx = \int \frac{(1-u)(-u)}{1-u^2} \, du = -\int \frac{u}{1+u} \, du$$

$$= -\int \left(1 - \frac{1}{1+u}\right) du = -u + \ln|1+u| + C.$$

Returning to the variable x, we find that

$$\int \frac{\sqrt{1+x} - \sqrt{1-x}}{\sqrt{1+x} + \sqrt{1-x}} \, dx = -\sqrt{1-x^2} + \ln(1 + \sqrt{1-x^2}) + C. \quad \square$$

Example 3 Evaluate $\int \frac{e^x + 1}{e^{2x} - e^x + 2} \, dx.$

Solution Let $u = e^x$. Then $du = e^x \, dx$, $dx = du/u$, so that

$$\int \frac{e^x + 1}{e^{2x} - e^x + 2} \, dx = \int \frac{u+1}{u(u^2 - u + 2)} \, du,$$

and we have reduced the problem to the integration of a rational function of u, with partial fraction expansion

$$\frac{u+1}{u(u^2 - u + 2)} = \frac{1}{2}\left(\frac{1}{u} - \frac{u-3}{u^2 - u + 2}\right).$$

Therefore

$$\int \frac{u+1}{u(u^2 - u + 2)} \, du = \frac{1}{2} \int \frac{du}{u} - \frac{1}{2} \int \frac{u-3}{u^2 - u + 2} \, du$$

$$= \frac{1}{2} \ln|u| - \frac{1}{4} \ln(u^2 - u + 2) + \frac{5}{2\sqrt{7}} \arctan \frac{2u-1}{\sqrt{7}} + C,$$

408   Chapter 7   Methods of Integration

with the help of formula (17), page 406. Returning to the variable x, we find that

$$\int \frac{e^x + 1}{e^{2x} - e^x + 2} \, dx = \frac{1}{2} x - \frac{1}{4} \ln(e^{2x} - e^x + 2) + \frac{5}{2\sqrt{7}} \arctan \frac{2e^x - 1}{\sqrt{7}} + C.$$

∎

Integration of Rational Functions in sin x and cos x

By a *polynomial in two variables* x and y we mean any sum of a finite number of terms of the form $ax^m y^n$, where m and n are nonnegative integers and a is any constant. For example,

$$\sqrt{5} + 7xy^2 + 9x^2 y^3 - \frac{1}{2} y^4$$

is a polynomial in x and y. A quotient

$$R(x, y) = \frac{P(x, y)}{Q(x, y)}$$

of two polynomials $P(x, y)$ and $Q(x, y)$ in x and y is called a *rational function in x and y*.† An example of such a function is

$$\frac{x - y}{1 - 2x^2 + 3xy}. \tag{2}$$

If $R(x, y)$ is a rational function in x and y, then $R(\sin x, \cos x)$ is called a *rational function in* $\sin x$ *and* $\cos x$. Thus, replacing x by $\sin x$ and y by $\cos x$ in (2), we get the following rational function in $\sin x$ and $\cos x$:

$$\frac{\sin x - \cos x}{1 - 2 \sin^2 x + 3 \sin x \cos x}. \tag{2'}$$

As we now show, the integral of any rational function in $\sin x$ and $\cos x$ can always be evaluated with the help of a suitable rationalizing substitution.

Theorem 5 *(Integration of a rational function in* **sin** x *and* **cos** x). Let $R(\sin x, \cos x)$ be a rational function in $\sin x$ and $\cos x$, and let

$$u = \tan \frac{x}{2} \quad (-\pi < x < \pi). \tag{3}$$

Then

$$\int R(\sin x, \cos x) \, dx = \int R_1(u) \, du,$$

where $R_1(u)$ is a rational function of the single variable u. In particular, since the integral on the right can always be evaluated by the method of partial fractions, the same is true of the integral on the left.

Proof First we express $\sin x$ and $\cos x$ in terms of the new variable u. By the double-angle formulas for the sine and cosine, together with the

---

† In writing $P(x, y)$, $Q(x, y)$ and $R(x, y)$, we anticipate the notation for functions of two variables (see Section 13.1).

identity $\cos^2(x/2) + \sin^2(x/2) = 1$, we have

$$\sin x = \sin 2\left(\frac{x}{2}\right) = 2\sin\frac{x}{2}\cos\frac{x}{2} = \frac{2\sin\frac{x}{2}\cos\frac{x}{2}}{\cos^2\frac{x}{2} + \sin^2\frac{x}{2}} = \frac{2\tan\frac{x}{2}}{1 + \tan^2\frac{x}{2}},$$

$$\cos x = \cos 2\left(\frac{x}{2}\right) = \cos^2\frac{x}{2} - \sin^2\frac{x}{2} = \frac{\cos^2\frac{x}{2} - \sin^2\frac{x}{2}}{\cos^2\frac{x}{2} + \sin^2\frac{x}{2}} = \frac{1 - \tan^2\frac{x}{2}}{1 + \tan^2\frac{x}{2}},$$

where at the last step of each calculation we divide both the numerator and the denominator by $\cos^2(x/2)$. After making the substitution (3), known as the *half-angle substitution*, these formulas become

$$\sin x = \frac{2u}{1 + u^2}, \qquad \cos x = \frac{1 - u^2}{1 + u^2}, \tag{4}$$

which shows that $\sin x$ and $\cos x$ are both rational functions of u. The derivative dx/du is also a rational function of u. In fact, (3) is equivalent to

$$x = 2\arctan u, \tag{3'}$$

and differentiation of (3') immediately gives

$$\frac{dx}{du} = \frac{2}{1 + u^2}. \tag{5}$$

With the help of (4) and (5), we now write the integral of $R(\sin x, \cos x)$ in the form

$$\int R(\sin x, \cos x)\,dx = \int R(\sin x, \cos x)\frac{dx}{du}\,du$$

$$= \int R\left(\frac{2u}{1 + u^2}, \frac{1 - u^2}{1 + u^2}\right)\frac{2}{1 + u^2}\,du = \int R_1(u)\,du,$$

where

$$R_1(u) = R\left(\frac{2u}{1 + u^2}, \frac{1 - u^2}{1 + u^2}\right)\frac{2}{1 + u^2}$$

is a rational function of the single variable u. This follows from the fact that the quotient of two polynomials in $2u/(1 + u^2)$ and $(1 - u^2)/(1 + u^2)$ becomes a rational function of u after multiplying both the numerator and denominator by a suitable power of $1 + u^2$ (explain further). Since $R_1(u)$ is a rational function, it can be integrated by the method of partial fractions, leading to an integral $I_1(u)$, which in general is a sum of rational functions, logarithms and inverse tangents. The integral of $R(\sin x, \cos x)$ is then

$$I_1\left(\tan\frac{x}{2}\right),$$

after returning to the original variable x. ∎

Example 4 Evaluate $\displaystyle\int\frac{dx}{3\sin x + 2\cos x + 2}$.

Solution Making the half-angle substitution (3) and using formulas (4) and (5), we find that

$$\int \frac{dx}{3 \sin x + 2 \cos x + 2} = \int \frac{1}{\frac{6u}{1+u^2} + 2\frac{1-u^2}{1+u^2} + 2} \frac{2}{1+u^2} du$$

$$= \int \frac{2}{6u + 2(1-u^2) + 2(1+u^2)} du = \int \frac{du}{3u+2}$$

$$= \frac{1}{3} \ln |3u + 2| + C = \frac{1}{3} \ln \left| 3 \tan \frac{x}{2} + 2 \right| + C. \quad \square$$

Example 5 Evaluate $\int \frac{1 - \sin x}{1 + \cos x} dx.$

Solution This time we have

$$\int \frac{1 - \sin x}{1 + \cos x} dx = \int \frac{1 - \frac{2u}{1+u^2}}{1 + \frac{1-u^2}{1+u^2}} \frac{2}{1+u^2} du = \int \frac{1 + u^2 - 2u}{1+u^2} du$$

$$= \int \left(1 - \frac{2u}{1+u^2} \right) du = u - \ln(1 + u^2) + C$$

$$= \tan \frac{x}{2} - \ln \left(1 + \tan^2 \frac{x}{2} \right) + C = \tan \frac{x}{2} - \ln \left(\sec^2 \frac{x}{2} \right) + C,$$

or equivalently

$$\int \frac{1 - \sin x}{1 + \cos x} dx = \tan \frac{x}{2} + \ln \left(\cos^2 \frac{x}{2} \right) + C. \quad \square$$

According to Theorem 5, the half-angle substitution (3) is *universal*, which means that in principle it can be used to integrate an *arbitrary* rational function in sin x and cos x. However, in practice, other substitutions are often much more appropriate, as illustrated by the following examples.

Example 6 Evaluate $\int \frac{\sin^3 x}{\cos x + 2} dx.$

Solution Observing that

$$\int \frac{\sin^3 x}{\cos x + 2} dx = \int \frac{\sin^2 x}{\cos x + 2} \sin x \, dx = \int \frac{1 - \cos^2 x}{\cos x + 2} \sin x \, dx,$$

we make the substitution $u = \cos x$. Then $du = -\sin x \, dx$, and

$$\int \frac{\sin^3 x}{\cos x + 2} dx = \int \frac{u^2 - 1}{u + 2} du = \int \left(u - 2 + \frac{3}{u+2} \right) du$$

$$= \frac{1}{2} u^2 - 2u + 3 \ln |u + 2| + C$$

$$= \frac{1}{2} \cos^2 x - 2 \cos x + 3 \ln (\cos x + 2) + C$$

Section 7.7 Rationalizing Substitutions 411

(why can we drop the absolute value sign?). Suppose that instead of the substitution $u = \cos x$, we had made the half-angle substitution (3). Then instead of the relatively simple integrand $(u^2 - 1)/(u + 2)$, we would have obtained the complicated integrand

$$\frac{16u^3}{(u^2 + 3)(u^2 + 1)^3},$$

which is much harder to integrate (see Problem 25). ☐

Example 7 Evaluate

$$\int \frac{dx}{a^2 \sin^2 x + b^2 \cos^2 x},$$

where a and b are arbitrary positive constants.

Solution Dividing the numerator and denominator by $a^2 \cos^2 x$, we get

$$\int \frac{dx}{a^2 \sin^2 x + b^2 \cos^2 x} = \int \frac{\frac{1}{a^2 \cos^2 x}}{\frac{\sin^2 x}{\cos^2 x} + \frac{b^2}{a^2}} \, dx = \frac{1}{a^2} \int \frac{\sec^2 x}{\tan^2 x + \frac{b^2}{a^2}} \, dx,$$

from which it is apparent that in this case the appropriate rationalizing substitution is $u = \tan x$, rather than the half-angle substitution $u = \tan(x/2)$. In fact, if $u = \tan x$, then $du = \sec^2 x \, dx$, and

$$\int \frac{dx}{a^2 \sin^2 x + b^2 \cos^2 x} = \frac{1}{a^2} \int \frac{du}{u^2 + \frac{b^2}{a^2}} = \frac{1}{a^2} \left(\frac{a}{b} \arctan \frac{au}{b} \right) + C$$

$$= \frac{1}{ab} \arctan \left(\frac{a}{b} \tan x \right) + C. \quad \square$$

Problems

Evaluate

1. $\int \dfrac{dx}{\sqrt[4]{x} + 1}$

2. $\int \dfrac{dx}{4\sqrt{x} + \sqrt[3]{x}}$

3. $\int \dfrac{\sqrt[3]{x}}{\sqrt{x} + 1} \, dx$

4. $\int \dfrac{\sqrt{x} + 1}{\sqrt{x} - 1} \, dx$

5. $\int \dfrac{\sqrt{x} - 1}{\sqrt[3]{x} + 1} \, dx$

6. $\int \dfrac{dx}{\sqrt[3]{x} - \sqrt[4]{x}}$

7. $\int \sqrt{\dfrac{x + 1}{x - 1}} \, dx$

8. $\int \dfrac{1}{x^2} \sqrt{\dfrac{1 + x}{1 - x}} \, dx$

9. $\int \dfrac{dx}{\sqrt{1 + \sqrt{x}}}$

10. $\int \dfrac{\sqrt[3]{1 + \sqrt[4]{x}}}{\sqrt{x}} \, dx$

11. $\int \dfrac{e^x}{1 - e^{2x}} \, dx$

12. $\int \dfrac{dx}{\sqrt{e^x + 1}}$

13. $\int \dfrac{dx}{2 + \cos x}$

14. $\int \dfrac{dx}{3 - \sin x}$

15. $\int \dfrac{dx}{\sin x + \cos x}$

16. $\int \dfrac{dx}{1 - \sin x + \cos x}$

17. $\int \dfrac{\cos^3 x}{2 \sin x - 1} \, dx$

18. $\int \dfrac{dx}{2 \sin x - \cos x + 5}$

19. $\int \dfrac{\cos x}{1 + \sin x - \cos x} \, dx$

20. $\int \dfrac{\sin^2 x}{1 + \sin^2 x} \, dx$

21. $\int \dfrac{dx}{\cos x \sin x - 2 \sin x}$

22. $\int \dfrac{\sin^2 x \cos x}{\sin x + \cos x} \, dx$

23. $\int \dfrac{dx}{a + b \tan x} \quad (ab \neq 0)$

24. $\int \dfrac{dx}{\sin^4 x \cos^2 x}$

412 Chapter 7 Methods of Integration

25. Solve Example 6 the hard way, by making the substitution $u = \tan(x/2)$.

26. Use the substitution
$$x = a\cos^2 u + b\sin^2 u \quad (0 < u < \pi/2)$$
to show that
$$\int \frac{dx}{\sqrt{(x-a)(b-x)}} = 2\arctan\sqrt{\frac{x-a}{b-x}} + C \quad (a < x < b).$$

Evaluate

27. $\int_0^{\ln 2} \sqrt{e^x - 1}\, dx$

28. $\int_0^{\pi/6} \frac{dx}{9\sin^2 x + \cos^2 x}$

29. $\int_0^{\pi} \frac{x \sin x}{1 + \cos^2 x}\, dx$

30. $\int_0^{\pi/4} \frac{ds}{2 + \tan s}$

31. $\int_0^{\pi/2} \frac{dt}{2 - \sin t + \cos t}$

32. $\int_2^5 \frac{du}{\sqrt{(u-1)(6-u)}}$

7.8 Approximate Integration and Simpson's Rule

Consider the problem of evaluating the definite integral
$$\int_a^b f(x)\, dx \tag{1}$$

of a continuous function f. The easiest way of evaluating I is to use the fundamental theorem of calculus, which says that
$$I = \int_a^b f(x)\, dx = F(b) - F(a),$$

where F is any antiderivative (or equivalently, the indefinite integral) of the integrand f. But by now we know that it may be difficult or even impossible to find an explicit formula for F, even though the *existence* of F is guaranteed by Theorem 5, page 254.

The Midpoint Rule

Nevertheless, in such cases we can still calculate the integral I to any desired accuracy. The idea is to approximate I by a suitable sum. (It is hardly surprising that this can be done, since the integral I was defined in the first place as the limiting value of the Riemann sum of f as the maximum length of the subintervals of the interval of integration $[a, b]$ becomes smaller and smaller.) In fact, choosing any positive even number $N = 2n$, suppose we partition the interval $[a, b]$ by introducing *equally spaced* points of subdivision

$$x_i = a + \frac{b-a}{N} i \quad (i = 0, 1, \ldots, N),$$

a distance

$$h = \frac{b-a}{N} = \frac{b-a}{2n}$$

apart (note that $x_0 = a$, $x_N = b$). Then, of the three approximate or *numerical* methods of integration to be described in this section, the first, known as the *midpoint rule*, is just the approximation of I by a Riemann sum of the form

$$\sum_{i=1}^{N} f\left(\frac{x_{2i-2} + x_{2i}}{2}\right)(x_{2i} - x_{2i-2}), \tag{2}$$

involving the n subintervals $[x_0, x_2], [x_2, x_4], \ldots, [x_{2n-2}, x_{2n}]$, each of the same length $x_{2i} - x_{2i-2}$ equal to

$$2h = \frac{b-a}{n}.$$

The point

$$\frac{x_{2i-2} + x_{2i}}{2} \quad (i = 1, 2, \ldots, n)$$

is the midpoint x_{2i-1} of the interval $[x_{2i-2}, x_{2i}]$, and we can write (2) in the form

$$2h \sum_{i=1}^{n} f(x_{2i-1}) = \frac{b-a}{n} \sum_{i=1}^{n} y_{2i-1}, \tag{2'}$$

where $y_{2i-1} = f(x_{2i-1})$ is the value of the function f at x_{2i-1}. Thus the midpoint rule consists of the approximation

$$\int_a^b f(x)\,dx \approx \frac{b-a}{n}(y_1 + y_3 + \cdots + y_{2n-1}), \tag{3}$$

Geometrically this corresponds to replacing the area under the curve $y = f(x)$ by the sum of the areas of n rectangles, where every rectangle has the same width $2h = (b-a)/n$ and the ith rectangle is of height y_{2i-1} and area $2hy_{2i-1}$, as illustrated in Figure 8 for the case of 6 rectangles ($n = 6$, $N = 12$).

The *error* E_M of the midpoint rule is defined as the number E_M which must be added to the right side of (3) to make the equation exact, that is

$$\int_a^b f(x)\,dx = \frac{b-a}{n}(y_1 + y_3 + \cdots + y_{2n-1}) + E_M.$$

It is clear that E_M is a function of n, the number of subintervals, a fact which we can emphasize by writing $E_M = E_M(n)$. Suppose f has a continuous second derivative f'' on the interval $[a, b]$. Then it can be shown (by an argument which is too technical for this course) that

$$E_M = \frac{(b-a)^3}{24n^2} f''(c) \tag{4}$$

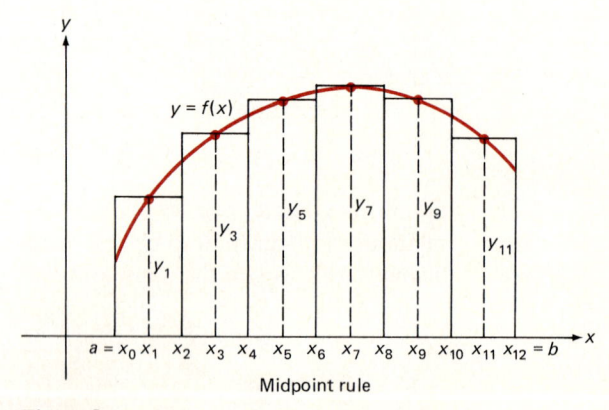

Figure 8

Midpoint rule

Chapter 7 Methods of Integration

for some point c in $[a, b]$. In particular, (4) implies

$$|E_M| \le \frac{(b-a)^3}{24n^2} \max |f''|, \tag{4'}$$

where $\max |f''|$ is the maximum value of $|f''(x)|$ on the interval $[a, b]$. The key feature of formula (4) is that E_M is inversely proportional to the square of n. Hence the error E_M can be made as small as we please by choosing a large enough value of n, that is, a fine enough subdivision of the interval of integration $[a, b]$.

Example 1 Use the midpoint rule with $n = 10$ to approximate the integral

$$I = \int_1^2 \frac{dx}{x}.$$

What is the accuracy of the approximation?

Solution Since we already know that $I = \ln 2 = 0.693147\ldots$, the purpose of this example is just to illustrate the implementation of the midpoint rule. Here $a = 1$, $b = 2$, $f(x) = 1/x$, $N = 2n = 20$, and the subintervals are $[1.0, 1.1], [1.1, 1.2], \ldots, [1.9, 2.0]$, with midpoints $x_1 = 1.05$, $x_3 = 1.15, \ldots$, $x_{19} = 1.95$. Calculating the corresponding ordinates $y_1, y_3, \ldots, y_{19}$ to four decimal places, we get

$$
\begin{array}{ll}
x_1 = 1.05 & y_1 = 0.9524 \\
x_3 = 1.15 & y_3 = 0.8696 \\
x_5 = 1.25 & y_5 = 0.8000 \\
x_7 = 1.35 & y_7 = 0.7407 \\
x_9 = 1.45 & y_9 = 0.6897 \\
x_{11} = 1.55 & y_{11} = 0.6452 \\
x_{13} = 1.65 & y_{13} = 0.6061 \\
x_{15} = 1.75 & y_{15} = 0.5714 \\
x_{17} = 1.85 & y_{17} = 0.5405 \\
x_{19} = 1.95 & y_{19} = 0.5128 \\
\hline
& \text{Sum} = 6.9284
\end{array}
$$

Therefore, by (3),

$$I = \int_1^2 \frac{dx}{x} \approx \frac{1}{10}(y_1 + y_3 + \cdots + y_{19}) = \frac{6.9284}{10} = 0.69284.$$

To determine the accuracy of this approximation, we first observe that

$$f''(x) = \frac{d^2}{dx^2}\frac{1}{x} = \frac{2}{x^3}.$$

Hence formula (4') becomes

$$|E_M| \le \frac{1}{24n^2} \max \left|\frac{2}{x^3}\right|,$$

or

$$|E_M| \le \frac{1}{12n^2}, \tag{5}$$

since the maximum of $|2/x^3|$ on $[1, 2]$ is equal to 2, taken at the point $x = 1$. Actually, using formula (4) and the fact that f'' is positive on $[1, 2]$, we see that E_M is also positive. Therefore (5) can be replaced by

$$0 < E_M \le \frac{1}{12n^2}. \tag{5'}$$

Thus in this case the midpoint rule *underestimates* the value of the integral I. Substituting $n = 10$ into (5'), we find that

$$0 < E_M \le \frac{1}{1200} < 0.00084.$$

Each ordinate y_i was calculated to four decimal places, and therefore has a round-off error of less than 0.00005. But the quantity 6.9284/10 is the average of the 10 ordinates $y_1, y_3, \ldots, y_{19}$, and hence its round-off error is also less than 0.00005 (why?). This fact, together with our estimate of the error E_M, shows that I lies between $0.69284 - 0.00005 = 0.69279$ and $0.69284 + 0.00005 + 0.00084 = 0.69373$. Thus we can certainly conclude that $I = 0.693$ to within 0.001 (or even that $I = 0.69325$ to within 0.0005). ☐

The Trapezoidal Rule

Next we turn to another method of approximate integration, known as the *trapezoidal rule*. The idea behind this method is to approximate the given integral I by a sum of the form

$$\sum_{i=1}^{n} \frac{f(x_{i-1}) + f(x_i)}{2} (x_i - x_{i-1}), \tag{6}$$

involving equally spaced points of subdivision

$$x_i = a + \frac{b-a}{n} i \qquad (i = 0, 1, \ldots, n)$$

and n subintervals $[x_0, x_1], [x_1, x_2], \ldots, [x_{n-1}, x_n]$, each of the same length $x_i - x_{i-1}$ equal to

$$h = \frac{b-a}{n}.$$

(The notation in (6) is simpler than in the case of the midpoint rule (2), since there is no longer any need to have extra points of subdivision to serve as labels for the midpoints of the subintervals.) Observe that each term in the sum (6) contains the average of *two* values of the function f, namely its values at the endpoints of a subinterval, whereas each term in the sum (2) involves only a *single* value of f, namely its value at the midpoint of a subinterval. We can also write (6) in the form

$$\frac{h}{2} \sum_{i=1}^{n} [f(x_{i-1}) + f(x_i)] = \frac{b-a}{2n} \sum_{i=1}^{n} (y_{i-1} + y_i), \tag{6'}$$

where $y_i = f(x_i)$. Thus the trapezoidal rule consists of the approximation

$$\int_a^b f(x)\, dx \approx \frac{b-a}{2n} (y_0 + 2y_1 + 2y_2 + \cdots + 2y_{n-1} + y_n) \tag{7}$$

Trapezoidal rule

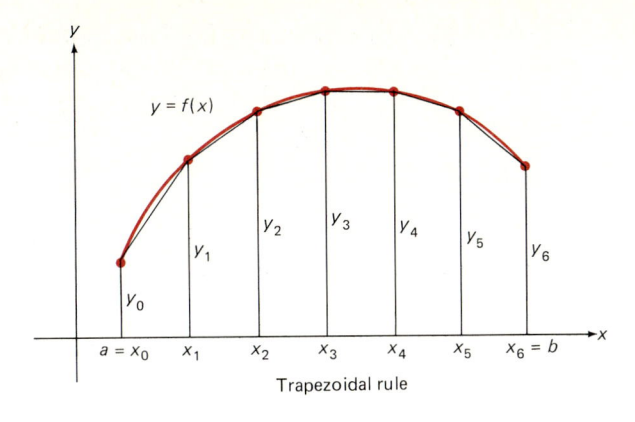

Figure 9

(every ordinate y_i except y_0 and y_n occurs in a pair of consecutive terms $y_{i-1} + y_i$, and therefore has a coefficient of 2 in the sum on the right). Geometrically, (7) corresponds to replacing the area under the curve $y = f(x)$ by the sum of the areas of n trapezoids, where every trapezoid has the same width $h = (b - a)/n$ and the ith trapezoid has parallel sides y_{i-1}, y_i and area $h(y_{i-1} + y_i)/2$, by a familiar formula of elementary geometry. This approximation is illustrated in Figure 9 for the case of six trapezoids ($n = 6$).

The *error* $E_T = E_T(n)$ of the trapezoidal rule is defined as the number E_T which must be added to the right side of (7) to make the equation exact, that is

$$\int_a^b f(x)\,dx = \frac{b-a}{2n}(y_0 + 2y_1 + 2y_2 + \cdots + 2y_{n-1} + y_n) + E_T.$$

Suppose f has a continuous second derivative f'' on the interval $[a, b]$. Then it can be shown that

$$E_T = -\frac{(b-a)^3}{12n^2} f''(c) \tag{8}$$

for some point c in $[a, b]$. In particular, (8) implies

$$|E_T| \leq \frac{(b-a)^3}{12n^2} \max |f''|, \tag{8'}$$

where $\max |f''|$ is the maximum value of $|f''(x)|$ on the interval $[a, b]$. Clearly the larger the number n, the smaller the error E_T, and in fact E_T is inversely proportional to n^2.

Example 2 Use the trapezoidal rule with $n = 10$ to approximate the integral

$$I = \int_1^2 \frac{dx}{x} \quad (=\ln 2).$$

What is the error of the approximation?

Solution Here $a = 1$, $b = 2$, $f(x) = 1/x$, $n = 10$, as in Example 1, and the subintervals are again $[1.0, 1.1]$, $[1.1, 1.2]$, ..., $[1.9, 2.0]$, with endpoints

Section 7.8 Approximate Integration and Simpson's Rule   **417**

$x_0 = 1.0$, $x_1 = 1.1$, $x_2 = 1.2, \ldots, x_9 = 1.9$, $x_{10} = 2.0$. Calculating the corresponding ordinates $y_0, y_1, y_2, \ldots, y_9, y_{10}$ to four decimal places, we get

$x_0 = 1.0$	$y_0 = 1.0000$	$x_1 = 1.1$	$y_1 = 0.9091$
$x_{10} = 2.0$	$y_{10} = 0.5000$	$x_2 = 1.2$	$y_2 = 0.8333$
	Sum = 1.5000	$x_3 = 1.3$	$y_3 = 0.7692$
		$x_4 = 1.4$	$y_4 = 0.7143$
		$x_5 = 1.5$	$y_5 = 0.6667$
		$x_6 = 1.6$	$y_6 = 0.6250$
		$x_7 = 1.7$	$y_7 = 0.5882$
		$x_8 = 1.8$	$y_8 = 0.5556$
		$y_9 = 1.9$	$y_9 = 0.5263$
			Sum = 6.1877

Therefore, by (7),

$$I = \int_1^2 \frac{dx}{x} \approx \frac{1}{2(10)} (y_0 + 2y_1 + 2y_2 + \cdots + 2y_9 + y_{10})$$

$$= \frac{1}{20} [(y_0 + y_{10}) + 2(y_1 + y_2 + \cdots + y_9)]$$

$$= \frac{1}{20} [1.5000 + 2(6.1877)] = \frac{13.8754}{20} = 0.69377.$$

To determine the accuracy of this approximation, we observe that once again $f''(x) = 2/x^3$, max $|f''| = 2$, $n = 10$. Thus it follows from (8) and (8') that $E_T < 0$ and

$$|E_T| \le \frac{1}{6n^2} = \frac{1}{600} < 0.00167.$$

Notice that in this case the trapezoidal rule *overestimates* the value of I. The round-off error of each ordinate is less than 0.00005 in absolute value, and hence the same is true of the quantity 13.8754/20 (observe that the sum $\frac{1}{20}(y_0 + 2y_1 + 2y_2 + \cdots + 2y_9 + y_{10})$ is actually the average of the 20 numbers $y_0, y_1, y_1, y_2, y_2, \ldots, y_9, y_9, y_{10}$, where each of the numbers $y_1, y_2, \ldots, y_9$ appears twice). This fact, together with our estimate of the error E_T, shows that I lies between $0.69377 - 0.00005 - 0.00167 = 0.69205$ and $0.69377 + 0.00005 = 0.69382$. Thus we can certainly conclude that $I = 0.693$ to within 0.001, just as in Example 1. ☐

Simpson's Rule

The trapezoidal rule is based on approximating the integral

$$\int_{x_{i-1}}^{x_i} f(x)\, dx \qquad (i = 1, 2, \ldots, n)$$

over each of the n subintervals $[x_{i-1}, x_i]$, making up the interval of integration $[a, b]$, by the area under the graph of a *linear* function $y = Ax + B$, namely the straight line segment joining the two points (x_{i-1}, y_{i-1}) and (x_i, y_i), where $y_i = f(x_i)$. We now turn to a much more powerful method of

418 Chapter 7 Methods of Integration

numerical integration, known as *Simpson's rule*. Here it is convenient to use the same notation as in the case of the midpoint rule, since we will need to label the midpoints of our subintervals, as well as their endpoints. Thus, choosing any positive even number $N = 2n$, we introduce equally spaced points of subdivision

$$x_i = a + \frac{b-a}{N} i \qquad (i = 0, 1, \ldots, N),$$

a distance

$$h = \frac{b-a}{N} = \frac{b-a}{2n}$$

apart, just as on page 413. The points $a = x_0, x_2, x_4, \ldots, x_{2n-2}, x_{2n} = b$ divide $[a, b]$ into n subintervals, each of length $2h$, where it will be noted that x_{2i-1} is the midpoint of $[x_{2i-2}, x_{2i}]$. We then write the given integral as

$$\int_a^b f(x)\,dx = \sum_{i=1}^n \int_{x_{2i-2}}^{x_{2i}} f(x)\,dx, \tag{9}$$

and approximate each integral in the sum on the right by the area under the graph of a *quadratic* function

$$y = P(x) = A + Bx + Cx^2,$$

where the coefficients A, B and C are such that the curve $y = P(x)$ goes through the three points (x_{2i-2}, y_{2i-2}), (x_{2i-1}, y_{2i-1}) and (x_{2i}, y_{2i}). If these points are noncollinear, then $C \neq 0$, and the curve $y = P(x)$ is a parabola whose axis of symmetry is vertical, as in Figure 10.

Next, setting $r = x_{2i-2}$ and $s = x_{2i}$ for brevity, we find that

$$\int_r^s P(x)\,dx = \int_r^s (A + Bx + Cx^2)\,dx = \left[Ax + \frac{1}{2}Bx^2 + \frac{1}{3}Cx^3\right]_r^s$$

$$= A(s-r) + \frac{1}{2}B(s^2 - r^2) + \frac{1}{3}C(s^3 - r^3)$$

$$= \frac{s-r}{6}\left[6A + 3B(r+s) + 2C(r^2 + rs + s^2)\right].$$

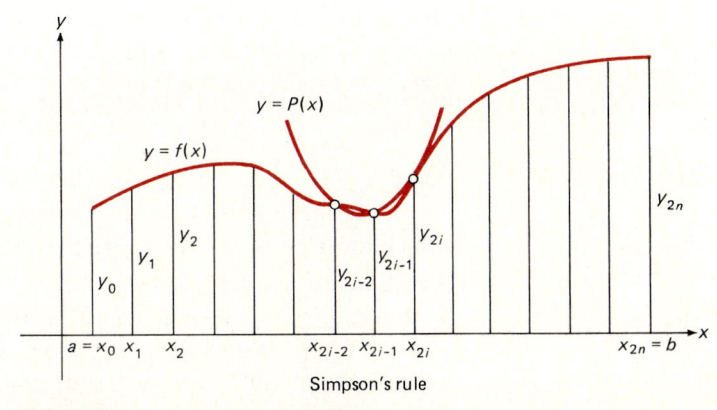

Figure 10

Simpson's rule

Section 7.8 Approximate Integration and Simpson's Rule **419**

Therefore

$$\int_r^s P(x)\,dx$$

$$= \frac{s-r}{6}\left[A + Br + Cr^2 + 4A + 4B\left(\frac{r+s}{2}\right) + 4C\left(\frac{r+s}{2}\right)^2 + A + Bs + Cs^2\right]$$

after some algebraic manipulation, and hence

$$\int_r^s P(x)\,dx = \frac{s-r}{6}\left[P(r) + 4P\left(\frac{r+s}{2}\right) + P(s)\right]. \tag{10}$$

Since $s - r = x_{2i} - x_{2i-2} = 2h$ and $\frac{1}{2}(r+s) = \frac{1}{2}(x_{2i-2} + x_{2i}) = x_{2i-1}$, formula (10) becomes

$$\int_{x_{2i-2}}^{x_{2i}} P(x)\,dx = \frac{h}{3}\left[P(x_{2i-2}) + 4P(x_{2i-1}) + P(x_{2i})\right].$$

in terms of x_{2i-2}, x_{2i-1} and x_{2i}. But $f(x)$ and $P(x)$ coincide for $x = x_{2i-2}$, $x = x_{2i-1}$ and $x = x_{2i}$, since the curve $y = P(x) = A + Bx + Cx^2$ has been chosen to go through the points (x_{2i-2}, y_{2i-2}), (x_{2i-1}, y_{2i-1}) and (x_{2i}, y_{2i}). Thus

$$\int_{x_{2i-2}}^{x_{2i}} P(x)\,dx = \frac{h}{3}\left[f(x_{2i-2}) + 4f(x_{2i-1}) + f(x_{2i})\right]$$

$$= \frac{b-a}{6n}(y_{2i-2} + 4y_{2i-1} + y_{2i})$$

(observe that there is no need to determine the coefficients A, B and C explicitly). Approximating each of the n integrals in the right side of (9) by an expression of this type (the area under a parabola), we finally get Simpson's rule

$$\int_a^b f(x)\,dx \approx \frac{b-a}{6n}\sum_{i=1}^n (y_{2i-2} + 4y_{2i-1} + y_{2i}), \tag{11}$$

or equivalently

$$\int_a^b f(x)\,dx$$

$$\approx \frac{b-a}{6n}\left[(y_0 + y_n) + 2(y_2 + y_4 + \cdots + y_{2n-2}) + 4(y_1 + y_3 + \cdots + y_{2n-1})\right].$$

$$\tag{11'}$$

The *error* $E_S = E_S(n)$ of Simpson's rule is defined as the number E_S which must be added to the right side of (11) to make the equation exact, that is

$$\int_a^b f(x)\,dx = \frac{b-a}{6n}\sum_{i=1}^n (y_{2i-2} + 4y_{2i-1} + y_{2i}) + E_S.$$

Suppose f has a continuous *fourth* derivative $f^{(4)}$ on the interval $[a, b]$.

Then it can be shown by an advanced argument that

$$E_S = -\frac{(b-a)^5}{180(2n)^4} f^{(4)}(c) \tag{12}$$

for some point c in $[a, b]$. In particular, (12) implies

$$|E_S| \leq \frac{(b-a)^5}{180(2n)^4} \max |f^{(4)}|, \tag{12'}$$

where $\max |f^{(4)}|$ is the maximum value of $|f^{(4)}(x)|$ on the interval $[a, b]$. The key feature of formula (12) is that E_S is inversely proportional to the fourth power of n, so that E_S approaches zero very rapidly as n is increased.

Example 3 Use Simpson's rule with $n = 5$ to approximate the integral

$$I = \int_1^2 \frac{dx}{x} \quad (= \ln 2).$$

What is the error of the approximation?

Solution As in Examples 1 and 2, $a = 1$, $b = 2$, $f(x) = 1/x$, but the subintervals are now $[1.0, 1.2]$, $[1.2, 1.4]$, ..., $[1.8, 2.0]$, with endpoints $x_0 = 1.0$, $x_2 = 1.2$, $x_4 = 1.4$, ..., $x_8 = 1.8$, $x_{10} = 2.0$ and midpoints $x_1 = 1.1$, $x_3 = 1.3$, ..., $x_9 = 1.9$. Calculating the corresponding ordinates to five decimal places, we get

$x_0 = 1.0$	$y_0 = 1.00000$	$x_1 = 1.1$	$y_1 = 0.90909$
$x_{10} = 2.0$	$y_{10} = 0.50000$	$x_3 = 1.3$	$y_3 = 0.76923$
	Sum $= 1.50000$	$x_5 = 1.5$	$y_5 = 0.66667$
		$x_7 = 1.7$	$y_7 = 0.58824$
		$x_9 = 1.9$	$y_9 = 0.52632$
			Sum $= 3.45955$
$x_2 = 1.2$	$y_2 = 0.83333$		
$x_4 = 1.4$	$y_4 = 0.71429$		
$x_6 = 1.6$	$y_6 = 0.62500$		
$x_8 = 1.8$	$y_8 = 0.55556$		
	Sum $= 2.72818$		

Therefore, by (11') with $n = 5$,

$$I = \int_1^2 \frac{dx}{x} \approx \frac{1}{6(5)}[(y_0 + y_{10}) + 2(y_2 + y_4 + y_6 + y_8) + 4(y_1 + y_3 + y_5 + y_7 + y_9)]$$

$$= \frac{1}{30}[1.500000 + 2(2.72818) + 4(3.45955)] = \frac{20.79456}{30} = 0.693152.$$

To determine the accuracy of this approximation, we observe that the fourth derivative of the integrand $f(x) = 1/x$ is

$$f^{(4)}(x) = \frac{d^4}{dx^4}\frac{1}{x} = \frac{24}{x^5}.$$

Therefore $E_S < 0$, by (12), so that in this case Simpson's rule *overestimates* the value of I. Moreover, by (12'),

$$|E_S| \leq \frac{24}{180(10)^4} = \frac{1}{75000} < 0.000014,$$

since the maximum of $|24/x^5|$ on the interval $[1, 2]$ is equal to 24, taken at the point $x = 1$. Each ordinate was calculated to five decimal places, and therefore has a round-off error of less than 0.000005. Hence the round-off error of the quantity 20.79456/30 is also less than 0.000005 (interpret 20.79456/30 as the average of 30 numbers). This fact, together with our estimate of the error E_S, shows that I lies between $0.693152 - 0.000005 - 0.000014 = 0.693133$ and $0.693152 + 0.000005 = 0.693157$. Thus we are able to conclude that $I = 0.693145$ to within 0.000012. Actually, as already noted, $I = \ln 2 = 0.693147\ldots$ ☐

A comparison of Examples 1–3 reveals the great power of Simpson's rule. In fact, as applied to the evaluation of the integral $\int_1^2 dx/x$, Simpson's rule with only five subintervals is very much more accurate than the midpoint or trapezoidal rules with twice as many subintervals!

Example 4 Use Simpson's rule to approximate

$$I = \int_0^1 e^{-x^2} dx.$$

Solution As already noted on page 366, the indefinite integral $\int e^{-x^2} dx$ is not an elementary function. Thus we must calculate I by approximate integration. To this end we apply Simpson's rule, choosing $n = 5$. The maximum of the absolute value of the fourth derivative of the integrand e^{-x^2} on the interval $[0, 1]$ is 12 (see Problem 23), and hence, by the estimate (12'),

$$|E_S| \leq \frac{12}{180(10)^4} = \frac{1}{150000} < 0.000007.$$

Using a scientific calculator to compute the ordinates to five decimal places, we obtain

$x_0 = 0.0$ $y_0 = 1.00000$ $x_1 = 0.1$ $y_1 = 0.99005$
$x_{10} = 1.0$ $y_{10} = 0.36788$ $x_3 = 0.3$ $y_3 = 0.91393$
                 Sum = 1.36788 $x_5 = 0.5$ $y_5 = 0.77880$
                                    $x_7 = 0.7$ $y_7 = 0.61263$
                                    $x_9 = 0.9$ $y_9 = 0.44486$
                                                          Sum = 3.74027

$x_2 = 0.2$ $y_2 = 0.96079$
$x_4 = 0.4$ $y_4 = 0.85214$
$x_6 = 0.6$ $y_6 = 0.69768$
$x_8 = 0.8$ $y_8 = 0.52729$
                 Sum = 3.03790

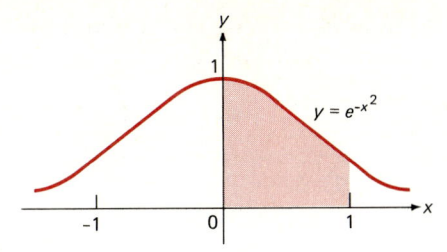

Figure 11

Therefore, by the same version of Simpson's rule as used in Example 3,

$$I = \int_0^1 e^{-x^2}\,dx \approx \frac{1}{30}[1.36788 + 2(3.03790) + 4(3.74027)]$$

$$= \frac{22.40476}{30} = 0.746825.$$

After taking account of the round-off error and our estimate of $|E_S|$, we find that here I is only known to within $0.000007 + 0.000005 = 0.000012$. Hence we can only be sure that the first four decimal places are correct, but a more exact calculation shows that the approximation $I \approx 0.746825$ is actually correct to within 0.000001. In Figure 11 we graph the curve $y = e^{-x^2}$, which is "bell-shaped." The integral I is the area of the shaded region under the curve from $x = 0$ to $x = 1$. (For another way of approximating I, see Example 9, page 545.) □

Remark The nonelementary function

$$\int_0^x e^{-t^2}\,dt \tag{13}$$

is of great importance in mathematics, especially in probability theory and its applications. The function

$$\operatorname{erf} x = \frac{2}{\sqrt{\pi}} \int_0^x e^{-t^2}\,dt, \tag{13'}$$

which differs from (13) by the presence of the factor $2/\sqrt{\pi}$, is called the *error function*. The integral (13) approaches $\sqrt{\pi}/2$ as $x \to \infty$ (see Example 3, page 860), and hence the factor $2/\sqrt{\pi} = 1.128379\ldots$ in (13') makes erf x approach 1 as $x \to \infty$.

Problems

Approximate the given integral by using first the midpoint rule, and then the trapezoidal rule with the specified value of n, the number of subintervals. Calculate ordinate values to four decimal places, and give the answer to three decimal places (without attempting to estimate the error).

1. $\int_0^2 \sqrt{x^4 + 1}\,dx$, $n = 4$

2. $\int_0^1 \frac{dx}{\sqrt{x^3 + 1}}$, $n = 4$

3. $\int_0^{\pi/3} \sqrt{\cos x}\,dx$, $n = 5$

4. $\int_0^\pi \frac{\sin x}{x}\,dx$, $n = 5$

5. $\int_1^2 \frac{e^x}{x}\,dx$, $n = 5$

6. $\int_0^1 e^{-x^2}\,dx$, $n = 10$

In Problem 4 the value of the integrand at $x = 0$ is taken to be 1.

7. Let M_n be the approximation to the integral $I = \int_a^b f(x)\,dx$ based on the midpoint rule with n subintervals, and let T_n be the approximation to I based on the trapezoidal rule with n subintervals. Express T_{2n} in terms of T_n and M_n.

8. Show that the sum (6) figuring in the trapezoidal rule is actually a Riemann sum for the function f on the interval $[a, b]$.

9. Use formula (4') to find the number of subintervals n guaranteeing that the error E_M made in approximating the integral $\int_0^1 e^{-x^2}\,dx$ by the midpoint rule is less than 0.0001 in absolute value. Use formula (8') to do the same for the trapezoidal rule.

10. Show that both the midpoint rule and the trapezoidal rule give the exact value of the integral $I = \int_a^b f(x)\,dx$ if the integrand is a linear function $f(x) = Ax + B$, but not if it is a quadratic function $f(x) = A + Bx + Cx^2$ ($C \neq 0$).

11. Find exact expressions for the errors $E_M = E_M(n)$ and $E_T = E_T(n)$ made in using the midpoint and trapezoidal rules to approximate the integral $I = \int_{-1}^{1} |x|\,dx$.

12. Let $I = \int_a^b f(x)\,dx$, where f has a continuous second derivative f'' on $[a, b]$ which is nonzero at every point of $[a, b]$. Show that if f is concave upward on $[a, b]$, then the midpoint rule underestimates I and the trapezoidal rule overestimates I, while if f is concave downward on $[a, b]$, the midpoint rule overestimates I and the trapezoidal rule underestimates I.

13. Show that the *prismoidal formula*

$$\int_r^s P(x)\,dx = \frac{s-r}{6}\left[P(r) + 4P\left(\frac{r+s}{2}\right) + P(s)\right] \quad \text{(i)}$$

is valid for any polynomial $P(x) = A + Bx + Cx^2 + Dx^3$ of degree 3 or less. (For $D = 0$, (i) has already been proved in the course of establishing Simpson's rule.) Give an example showing the failure of the prismoidal formula for a polynomial of degree 4.

14. Show that Simpson's rule gives the exact value of the integral $I = \int_a^b f(x)\,dx$ if the integrand is a cubic function $f(x) = A + Bx + Cx^2 + Dx^3$, but not if it is a quartic function $f(x) = A + Bx + Cx^2 + Dx^3 + Ex^4$ ($E \neq 0$). *Hint*. Use formula (12).

Use the prismoidal formula (i) to evaluate

15. $\int_0^2 (x^3 + x^2 + x + 1)\,dx$

16. $\int_{1/2}^{3/2} (8x^3 - 4x^2 + 2x - 1)\,dx$

17. Suppose the table of values

x	1.05	1.10	1.15	1.20	1.25	1.30	1.35
$f(x)$	2.36	2.50	2.74	3.04	3.46	3.98	4.60

is all that is known about the function f. Use Simpson's rule to estimate the integral $\int_{1.05}^{1.35} f(x)\,dx$.

18. Find the exact value of the integral

$$I = \int_0^1 \frac{dx}{x^2 + 1},$$

and verify that the approximation to I obtained by using Simpson's rule with only two subintervals is accurate to within 0.00001.

Approximate the given integral by using Simpson's rule with the specified value of n, the number of subintervals. Calculate ordinate values to five decimal places, and give the answer to four decimal places (without attempting to estimate the error).

19. $\int_0^{\pi/2} \sqrt{1 - \tfrac{1}{2}\sin^2 x}\,dx$, $n = 3$

20. $\int_2^3 \frac{dx}{\ln x}$, $n = 4$

21. $\int_1^2 \sqrt{\ln x}\,dx$, $n = 4$

22. $\int_0^2 \sin\frac{\pi x^2}{2}\,dx$, $n = 8$

23. If $f(x) = e^{-x^2}$, show that the fourth derivative $f^{(4)}(x)$ does not exceed 12 in absolute value on the interval $[0, 1]$. Is $f^{(4)}$ of fixed sign on $[0, 1]$?

7.9 Improper Integrals

In introducing the concept of the definite integral $\int_a^b f(x)\,dx$, we assumed from the outset that the interval of integration $[a, b]$ is closed and *bounded*. Moreover, the integrand f must be a *bounded* function on $[a, b]$, for otherwise the limit defining the integral will fail to exist (see Problem 33, page 238). Thus, as of now, the integral

$$\int_1^\infty \frac{dx}{x^2} \tag{1}$$

has no meaning, since the interval of integration is unbounded. The integral

$$\int_0^1 \frac{dx}{\sqrt{x}} \tag{2}$$

is also meaningless, since the integrand $1/\sqrt{x}$ approaches infinity as $x \to 0^+$, and therefore is unbounded on the interval $[0, 1]$.

Remark The function $1/\sqrt{x}$ is not defined at the point $x = 0$, but this is only incidental to the real reason for the nonexistence of (2), namely the unboundedness of $1/\sqrt{x}$ on the interval of integration. For example, the function

$$f(x) = \begin{cases} 1/\sqrt{x} & \text{if } x > 0, \\ 0 & \text{if } x = 0 \end{cases}$$

is also unbounded and hence nonintegrable on $[0, 1]$, although it is defined at every point of $[0, 1]$.

An integral with an unbounded interval of integration or an unbounded integrand (or both) is said to be *improper*, as opposed to the ordinary or *proper* integrals considered so far. As we now show, there is a way of assigning a numerical value to certain improper integrals, in particular to the integrals (1) and (2).

Unbounded Intervals of Integration

First we consider improper integrals like the integral (1), with an unbounded interval of integration. Let f be continuous on the infinite interval $[a, \infty)$, and suppose the limit

$$\lim_{u \to \infty} \int_a^u f(x)\, dx \qquad (u > a) \tag{3}$$

exists and is finite. (Since the function f is continuous on $[a, \infty)$, it is continuous and hence integrable on every subinterval $[a, u]$.) Then we say that the improper integral

$$\int_a^\infty f(x)\, dx \tag{4}$$

is *convergent* (or *converges*), and we assign it the value (3). However, if the limit (3) is infinite or fails to exist, we say that the integral (4) is *divergent* (or *diverges*). Similarly, let f be continuous on the interval $(-\infty, b]$, and suppose the limit

$$\lim_{u \to -\infty} \int_u^b f(x)\, dx \qquad (u < b) \tag{3'}$$

exists and is finite. Then the improper integral

$$\int_{-\infty}^b f(x)\, dx \tag{4'}$$

is said to be convergent and is assigned the value (3'), but we call the integral (4') divergent if the limit (3') is infinite or fails to exist.

There is also the case of improper integrals of the type

$$\int_{-\infty}^\infty f(x)\, dx, \tag{5}$$

Section 7.9 Improper Integrals 425

where *both* limits of integration are infinite and f is continuous on the whole real line $(-\infty, \infty)$. Suppose both (improper) integrals

$$I_1 = \int_{-\infty}^{a} f(x)\,dx \quad \text{and} \quad I_2 = \int_{a}^{\infty} f(x)\,dx$$

are convergent, where a is any real number. Then the integral (5) is said to be *convergent* and is assigned $I_1 + I_2$ as its value, that is

$$\int_{-\infty}^{\infty} f(x)\,dx = \int_{-\infty}^{a} f(x)\,dx + \int_{a}^{\infty} f(x)\,dx, \tag{6}$$

but otherwise we say that (5) is *divergent*. As an exercise, show that the particular choice of the number a does not affect the convergence or divergence of I_1 and I_2, or the value of $I_1 + I_2$ in the case of convergence, so that we could just as well have chosen $a = 0$, say.

Example 1 The improper integral

$$\int_{1}^{\infty} \frac{dx}{x^2},$$

discussed at the beginning of the section, is convergent. In fact,

$$\lim_{u \to \infty} \int_{1}^{u} \frac{dx}{x^2} = \lim_{u \to \infty} \left[-\frac{1}{x}\right]_{1}^{u} = \lim_{u \to \infty} \left(1 - \frac{1}{u}\right) = 1,$$

and therefore

$$\int_{1}^{\infty} \frac{dx}{x^2} = \lim_{u \to \infty} \int_{1}^{u} \frac{dx}{x^2} = 1.$$

By a natural extension of the definition of the area under a curve for the case of a bounded interval of integration, we can regard the value of this improper integral, namely the number 1, as the area under the curve $y = 1/x^2$ from $x = 1$ to $x = \infty$.† Thus it is quite possible for an unbounded region, here the shaded region in Figure 12 "extending out to infinity," to have a finite area. ☐

Technically, a region is said to be *bounded* or *finite* if it lies entirely inside some (sufficiently large) circle with its center at the origin; otherwise the region is said to be *unbounded* or *infinite*. In other words, an unbounded region contains points which are arbitrarily far from the origin.

Example 2 The improper integral

$$\int_{1}^{\infty} \frac{dx}{x}$$

is divergent, since

$$\lim_{u \to \infty} \int_{1}^{u} \frac{dx}{x} = \lim_{u \to \infty} \ln x \Big|_{1}^{u} = \lim_{u \to \infty} \ln u = \infty.$$

Thus the area under the curve $y = 1/x$ from $x = 1$ to $x = \infty$, that is, the area of the unbounded shaded region in Figure 13, must be regarded as infinite.

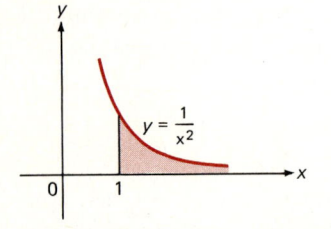

Figure 12

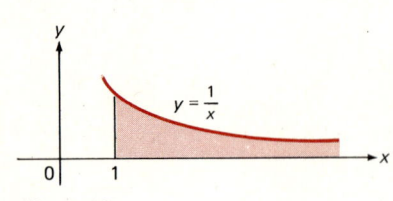

Figure 13

---

† In writing $x = \infty$, we relax an earlier prohibition (against putting the symbol ∞ or $-\infty$ after the equality sign), in the interest of a uniform notation.

426 Chapter 7 Methods of Integration

Example 3 Since the function

$$\int_0^u \sin x \, dx = -\cos x \Big|_0^u = 1 - \cos u$$

oscillates back and forth between the values 0 and 2 as $u \to \infty$, we see at once that the limit

$$\lim_{u \to \infty} \int_0^u \sin x \, dx$$

does not exist. Hence the improper integral

$$\int_0^\infty \sin x \, dx$$

is divergent.

Example 4 The improper integral

$$\int_{-\infty}^\infty e^{-|x|} \, dx$$

converges. In fact,

$$\int_0^\infty e^{-|x|} \, dx = \int_0^\infty e^{-x} \, dx = \lim_{u \to \infty} \int_0^u e^{-x} \, dx$$

$$= \lim_{u \to \infty} (-e^{-x}) \Big|_0^u = \lim_{u \to \infty} (1 - e^{-u}) = 1,$$

since $|x| = x$ for $x \geq 0$ and $e^{-u} \to 0$ as $u \to \infty$, while

$$\int_{-\infty}^0 e^{-|x|} \, dx = \int_{-\infty}^0 e^x \, dx = \lim_{u \to -\infty} \int_u^0 e^x \, dx = \lim_{u \to -\infty} e^x \Big|_u^0 = \lim_{u \to -\infty} (1 - e^u) = 1,$$

since $|x| = -x$ for $x \leq 0$ and $e^u \to 0$ as $u \to -\infty$. Therefore, by (6) with $a = 0$,

$$\int_{-\infty}^\infty e^{-|x|} \, dx = \int_{-\infty}^0 e^{-|x|} \, dx + \int_0^\infty e^{-|x|} \, dx = 1 + 1 = 2.$$

Thus we can regard the number 2 as being the area of the infinite shaded region in Figure 14 under the curve $y = e^{-|x|}$ from $x = -\infty$ to $x = \infty$.

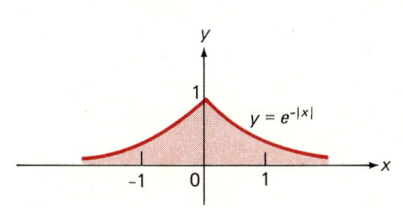

Figure 14

Unbounded Integrands

We now consider improper integrals like the integral (2), with an unbounded integrand. Let $f(x)$ be a function continuous on the half-open interval $[a, b)$ which approaches infinity (∞ or $-\infty$) as $x \to b^-$, and suppose the limit

$$\lim_{u \to b^-} \int_a^u f(x) \, dx \quad (a < u < b) \tag{7}$$

exists and is finite. (Why is f integrable on every subinterval $[a, u]$?) Then we say that the improper integral

$$\int_a^b f(x) \, dx \tag{8}$$

is *convergent* (or *converges*), and we assign it the value (7). However, if the limit (7) is infinite or fails to exist, we say that the integral (8) is *divergent* (or *diverges*). Similarly, let $f(x)$ be a function continuous on $(a, b]$ which approaches infinity as $x \to a^+$, and suppose the limit

$$\lim_{u \to a^+} \int_u^b f(x)\,dx \qquad (a < u < b) \tag{7'}$$

exists and is finite. Then the improper integral (8) is said to be convergent and is assigned the value (7'), but we call the integral (8) divergent if the limit (7') is infinite or fails to exist.

There is also the case where $f(x)$ becomes infinite as x approaches an interior point c of $[a, b]$. More exactly, let $f(x)$ be continuous on both sides of c, that is, on the intervals $[a, c)$ and $(c, b]$, and suppose $f(x)$ approaches infinity as x approaches c from one or both sides. Then if both integrals

$$I_1 = \int_a^c f(x)\,dx, \qquad I_2 = \int_c^b f(x)\,dx$$

converge (or if one is convergent and the other is proper), the integral (8) is said to be convergent and is assigned the value $I_1 + I_2$, that is,

$$\int_a^b f(x)\,dx = \int_a^c f(x)\,dx + \int_c^b f(x)\,dx,$$

but otherwise we say that (8) is divergent.

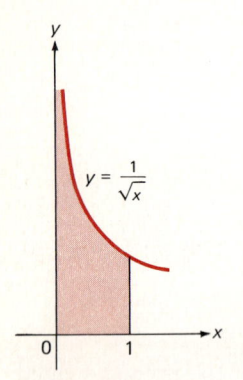

Example 5 The improper integral

$$\int_0^1 \frac{dx}{\sqrt{x}},$$

discussed at the beginning of the section, is convergent. In fact,

$$\lim_{u \to 0^+} \int_u^1 \frac{dx}{\sqrt{x}} = \lim_{u \to 0^+} 2\sqrt{x}\Big|_u^1 = \lim_{u \to 0^+} (2 - 2\sqrt{u}) = 2,$$

and therefore

$$\int_0^1 \frac{dx}{\sqrt{x}} = \lim_{u \to 0^+} \int_u^1 \frac{dx}{\sqrt{x}} = 2.$$

Thus we can regard the number 2 as the area under the curve $y = 1/\sqrt{x}$ from $x = 0$ to $x = 1$ (the area of the unbounded shaded region in Figure 15). This is another example of an infinite region with a finite area.

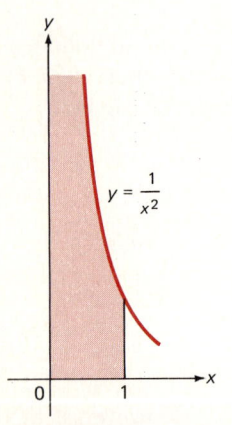

Example 6 The improper integral

$$\int_0^1 \frac{dx}{x^2} \tag{9}$$

diverges, since

$$\lim_{u \to 0^+} \int_u^1 \frac{dx}{x^2} = \lim_{u \to 0^+} \left[-\frac{1}{x}\right]_u^1 = \lim_{u \to 0^+} \left(\frac{1}{u} - 1\right) = \infty.$$

Thus the area under the curve $y = 1/x^2$ from $x = 0$ to $x = 1$ (the area of the unbounded shaded region in Figure 16) must be regarded as infinite.

Figure 15 $y = \dfrac{1}{\sqrt{x}}$

Figure 16 $y = \dfrac{1}{x^2}$

Example 7 Since the integral (9) is divergent, so is the integral

$$\int_{-1}^{1} \frac{dx}{x^2}$$

(see the discussion preceding Example 5). Suppose we make the mistake of calculating this integral formally, ignoring the fact that the integrand approaches infinity at the origin $x = 0$. Then we get the absurd result

$$\int_{-1}^{1} \frac{dx}{x^2} = \left[-\frac{1}{x}\right]_{-1}^{1} = -2,$$

seemingly an example of a positive function with a negative integral!

Example 8 The improper integral

$$I = \int_{0}^{1} \frac{dx}{\sqrt{1-x^2}}$$

converges and is equal to $\pi/2$. In fact,

$$I = \lim_{u \to 1^-} \int_{0}^{u} \frac{dx}{\sqrt{1-x^2}} = \lim_{u \to 1^-} \arcsin x \Big|_{0}^{u}$$

$$= \lim_{u \to 1^-} \arcsin u = \arcsin 1 = \frac{\pi}{2},$$

by the continuity of the inverse sine. Geometrically, I is the area of the unbounded shaded region in Figure 17.

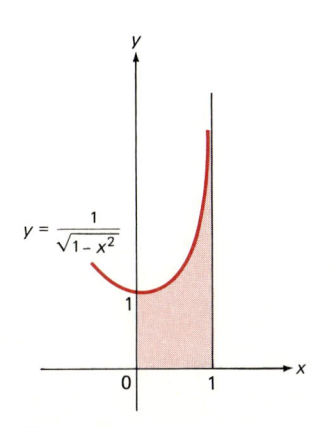

Figure 17

The following theorem provides a powerful tool for investigating improper integrals.

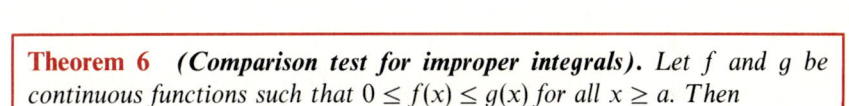

Theorem 6 *(Comparison test for improper integrals)*. *Let f and g be continuous functions such that $0 \le f(x) \le g(x)$ for all $x \ge a$. Then*
(i) *If $\int_{a}^{\infty} g(x)\,dx$ converges, so does $\int_{a}^{\infty} f(x)\,dx$;*
(ii) *If $\int_{a}^{\infty} f(x)\,dx$ diverges, so does $\int_{a}^{\infty} g(x)\,dx$.*

Although the proof of Theorem 6 involves certain technicalities, and hence is omitted, the intuitive content of the theorem is apparent from Figure 18: If the area under the top curve $y = g(x)$ is finite, then so is the area under the bottom curve $y = f(x)$, since the latter cannot exceed the former. On the other hand, if the area under the bottom curve is infinite, then so is the area under the top curve, since the latter cannot be less than the former. Notice that since the functions f and g are nonnegative, oscillations like those in Example 3 cannot occur here, and divergence of the integrals $\int_{a}^{\infty} f(x)\,dx$ and $\int_{a}^{\infty} g(x)\,dx$ can only mean that

$$\lim_{u \to \infty} \int_{a}^{u} f(x)\,dx = \infty \quad \text{and} \quad \lim_{u \to \infty} \int_{a}^{u} g(x)\,dx = \infty.$$

Comparison of two functions on an unbounded interval

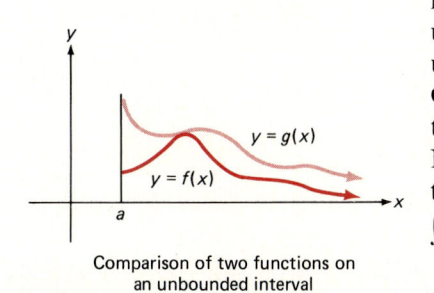

Figure 18

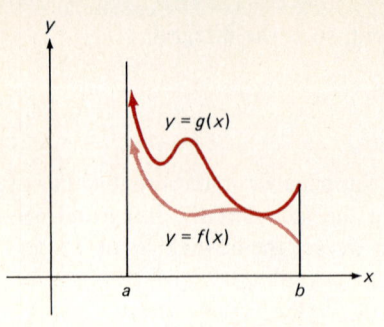

There are analogues of the comparison test for other kinds of improper integrals. For example, let f and g be continuous functions such that

$$\lim_{x \to a^+} f(x) = \lim_{x \to a^+} g(x) = \infty,$$

and let $0 \leq f(x) \leq g(x)$ for all $a < x \leq b$. Then

(i') If $\int_a^b g(x)\, dx$ converges, so does $\int_a^b f(x)\, dx$;
(ii') If $\int_a^b f(x)\, dx$ diverges, so does $\int_a^b g(x)\, dx$.

You should interpret (i') and (ii') geometrically, by referring to Figure 19.

Figure 19 Comparison of two unbounded functions

Example 9 Show that the improper integral

$$I = \int_0^\infty e^{-x^2}\, dx$$

is convergent.

Solution Since I is the sum of the proper integral

$$\int_0^1 e^{-x^2}\, dx$$

and the improper integral

$$\int_1^\infty e^{-x^2}\, dx, \tag{10}$$

it is enough to show that the integral (10) converges. Observing that $0 \leq e^{-x^2} \leq e^{-x}$ for all $x \geq 1$ (but not for $0 < x < 1$), and applying Theorem 6, we can deduce the convergence of the "difficult" integral (10) from that of the much "easier" integral

$$\int_1^\infty e^{-x}\, dx = \lim_{u \to \infty} \int_1^u e^{-x}\, dx = \lim_{u \to \infty} (-e^{-x})\Big|_1^u$$

$$= \lim_{u \to \infty} (e^{-1} - e^{-u}) = e^{-1}.$$

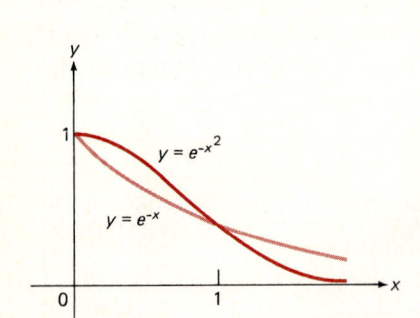

Figure 20 shows the geometric meaning of the comparison test, as applied to this example. As already noted in the remark on page 423, it turns out that $I = \sqrt{\pi}/2$. ☐

Figure 20

Example 10 Show that the improper integral

$$\int_0^{\pi/2} \frac{dx}{x \sin x}$$

is divergent.

Solution There is no need to attempt the evaluation of the integral. We observe first that

$$\frac{1}{x \sin x} \geq \frac{1}{x} > 0$$

for all $0 < x \leq \pi/2$ (why?), and then that the integral

$$\int_0^{\pi/2} \frac{dx}{x}$$

430   Chapter 7   Methods of Integration

diverges, since

$$\lim_{u \to 0^+} \int_u^{\pi/2} \frac{dx}{x} = \lim_{u \to 0^+} \ln x \Big|_u^{\pi/2} = \lim_{u \to 0^+} \left(\ln \frac{\pi}{2} - \ln u \right) = \infty.$$

The divergence of I is now a consequence of the comparison test, in the form (ii′). □

There are also improper integrals of "mixed type," with both an unbounded interval of integration and an unbounded integrand. For example, suppose $f(x)$ is continuous on the interval (a, ∞) and approaches infinity as $x \to a^+$, and suppose both improper integrals

$$I_1 = \int_a^c f(x)\,dx, \quad I_2 = \int_c^\infty f(x)\,dx \quad (a < c < \infty)$$

are convergent. Then the improper integral

$$\int_a^\infty f(x)\,dx \tag{11}$$

is said to be convergent and is assigned the value $I_1 + I_2$, that is,

$$\int_a^\infty f(x)\,dx = \int_a^c f(x)\,dx + \int_c^\infty f(x)\,dx, \tag{12}$$

but otherwise we say that (11) is divergent. Here we rely on the easily verified fact that the convergence or divergence of the integrals I_1 and I_2, and the value of their sum, if they both converge, does not depend on the particular choice of the intermediate point c (show this).

Example 11 The improper integral

$$\int_0^\infty \frac{dx}{x^2}$$

is divergent. In fact, choosing $c = 1$ in (12), we have

$$\int_0^\infty \frac{dx}{x^2} = \int_0^1 \frac{dx}{x^2} + \int_1^\infty \frac{dx}{x^2}.$$

Of the two integrals on the right, the second converges, by Example 1, but the first diverges, by Example 6. Hence the integral on the left also diverges.

Example 12 The improper integral

$$\int_0^\infty \frac{e^{-x}}{\sqrt{x}}\,dx$$

is convergent. To show this, we first write

$$\int_0^\infty \frac{e^{-x}}{\sqrt{x}}\,dx = \int_0^1 \frac{e^{-x}}{\sqrt{x}}\,dx + \int_1^\infty \frac{e^{-x}}{\sqrt{x}}\,dx.$$

We then observe that the first integral on the right converges, by comparison with the convergent integral $\int_0^1 dx/\sqrt{x}$ (see Example 5), while the second integral converges, by comparison with the convergent integral $\int_1^\infty e^{-x}\,dx = e^{-1}$ (the details are left as an exercise). Hence the integral on the left is also convergent, with the value given in Problem 49.

Problems

If the given improper integral is convergent, find its value. Otherwise identify it as divergent.

1. $\int_1^\infty \dfrac{dx}{x^3}$

2. $\int_0^1 \dfrac{dx}{x^4}$

3. $\int_{-\infty}^\infty \dfrac{dx}{x^2 + 1}$

4. $\int_0^\infty \dfrac{dx}{x^2 + 2x + 2}$

5. $\int_{-1}^0 \dfrac{dx}{x^2 - 1}$

6. $\int_0^2 \dfrac{dx}{x^2 - 4x + 3}$

7. $\int_{-1}^{27} \dfrac{dx}{\sqrt[3]{x}}$

8. $\int_0^{16} \dfrac{ds}{\sqrt[4]{s}}$

9. $\int_{-32}^1 \dfrac{dt}{\sqrt[5]{t}}$

10. $\int_1^2 \dfrac{x}{\sqrt{x-1}}\,dx$

11. $\int_3^5 \dfrac{dx}{\sqrt{x^2 - 9}}$

12. $\int_{-4}^0 \dfrac{dx}{\sqrt{16 - x^2}}$

13. $\int_2^\infty \dfrac{dx}{x\sqrt{x^2 - 1}}$

14. $\int_0^1 \dfrac{dx}{\sqrt{x - x^2}}$

15. $\int_0^1 \dfrac{x}{\sqrt{1 - x^2}}\,dx$

16. $\int_{-\infty}^0 \cos x\,dx$

17. $\int_0^{\pi/2} \tan y\,dy$

18. $\int_0^\infty e^{-3z}\,dz$

19. $\int_0^1 \ln x\,dx$

20. $\int_1^\infty \dfrac{\ln x}{x}\,dx$

21. $\int_0^\infty e^{-\sqrt{x}}\,dx$

22. $\int_0^\infty e^{-3x} \cos 2x\,dx$

23. $\int_0^\infty e^{-2x} \sin x\,dx$

24. $\int_1^\infty \dfrac{\ln x}{x^2}\,dx$

25. $\int_1^2 \dfrac{dx}{x \ln x}$

26. $\int_2^\infty \dfrac{dx}{x \ln x}$

27. $\int_0^\infty x e^{-x^2}\,dx$

28. $\int_e^\infty \dfrac{dx}{x(\ln x)^2}$

29. $\int_0^1 \dfrac{\arcsin x}{\sqrt{1 - x^2}}\,dx$

30. $\int_1^\infty \dfrac{\arctan x}{x^2}\,dx$

Given that $a > 0$, show that

31. $\int_0^a x^p\,dx$ is convergent if and only if $p > -1$

32. $\int_a^\infty x^p\,dx$ is convergent if and only if $p < -1$

33. $\int_a^\infty x^p\,dx$ is divergent for all p

Given that $a > 0$, evaluate

34. $\int_0^\infty e^{-ax}\,dx$

35. $\int_0^\infty e^{-ax} \cos bx\,dx$

36. $\int_0^\infty e^{-ax} \sin bx\,dx$

Find the area A of the region R between the curves
37. $y = x^{-1/2}$ and $y = \tfrac{1}{2}x^{-1/3}$ from $x = 0$ to $x = 1$
38. $y = e^{-x}$ and $y = 1/(x^2 + 1)$ from $x = 0$ to $x = \infty$
39. $y = \cosh x$ and $y = \sinh x$ in the first quadrant
In each case sketch the region R.

Use the comparison test to determine whether the given improper integral is convergent or divergent.

40. $\int_0^\infty \dfrac{dx}{\sqrt{x^3 + 1}}$

41. $\int_1^\infty \dfrac{dx}{\sqrt{x^2 + x}}$

42. $\int_0^1 \dfrac{dx}{x \cos x}$

43. $\int_1^\infty \dfrac{dx}{x^2 + \ln x}$

44. $\int_0^1 \dfrac{dx}{\sqrt{1 - x^3}}$

45. $\int_0^{\pi/2} \dfrac{\sin x}{x^{3/2}}\,dx$

46. $\int_0^{\pi/2} \dfrac{\cos x}{x}\,dx$

47. $\int_0^\infty \dfrac{dx}{\sqrt{x^4 + x^2}}$

48. $\int_0^\infty \dfrac{dx}{\sqrt{x^3 + x}}$

Regarding the formula $\int_0^\infty e^{-x^2}\,dx = \dfrac{\sqrt{\pi}}{2}$ as known, show that

49. $\int_0^\infty \dfrac{e^{-x}}{\sqrt{x}}\,dx = \sqrt{\pi}$

50. $\int_0^\infty x^2 e^{-x^2}\,dx = \dfrac{\sqrt{\pi}}{4}$

Find the number x such that

51. $\int_0^x e^{-t}\,dt = \int_x^\infty e^{-t}\,dt$

52. $\int_0^x \dfrac{dt}{\sqrt{t}} = \int_x^1 \dfrac{dt}{\sqrt{t}}$

53. $\int_0^x \dfrac{dt}{t^2 + 1} = \int_x^\infty \dfrac{dt}{t^2 + 1}$

54. Let f be continuous on $(-\infty, \infty)$, and suppose the integral $\int_{-\infty}^\infty f(x)\,dx$ is convergent. Show that

$$\int_{-\infty}^\infty f(x)\,dx = 2\int_0^\infty f(x)\,dx$$

if f is even, while
$$\int_{-\infty}^{\infty} f(x)\,dx = 0$$
if f is odd.

55. Give an example of a function f for which
$$\lim_{u \to \infty} \int_{-u}^{u} f(x)\,dx = 0,$$
even though $\int_{-\infty}^{\infty} f(x)\,dx$ is divergent.

56. The owner of a rent-controlled building, receiving D dollars per year in combined rent from his tenants (paid in monthly installments), puts the building up for sale. Why does he regard D/r dollars as a fair price for the building if he ordinarily invests the rent as it is received at an annual interest rate of $100r$ percent compounded continuously? *Hint.* Use an improper integral to approximate the present value of the "income stream" consisting of all future rent receipts.

Key Terms and Topics
Integration by substitution: $\int f(u(x))u'(x)\,dx = \int f(u)\,du\big|_{u=u(x)}$
Integration by parts: $\int u\,dv = uv - \int v\,du$
Reduction formulas
Integration of products of powers of $\sin x$ and $\cos x$
Integration of products of powers of $\tan x$ and $\sec x$
Integration of products of sines and cosines with different arguments
The substitutions $x = a \sin u$, $x = a \tan u$ and $x = a \sec u$
The substitutions $x = a \sinh u$ and $x = a \cosh u$
Proper and improper rational functions
Factorization of polynomials with real coefficients
The partial fraction expansion of a rational function
Integration of rational functions with the help of partial fractions
Rationalizing substitutions, the half-angle substitution
Integration of rational functions in $\sin x$ and $\cos x$
The midpoint rule and its error
The trapezoidal rule and its error
Simpson's rule and its error
Improper integrals with an unbounded interval of integration
Improper integrals with an unbounded integrand
The comparison test for improper integrals

For a review of basic integration formulas, see the front endpaper of this book, Numbers 1–28.

Supplementary Problems

Evaluate the given integral by any method.

1. $\int x^5 e^{x^2}\,dx$

2. $\int \dfrac{dx}{\sqrt{5-4x}}$

3. $\int \dfrac{x^2}{(x-1)^9}\,dx$

4. $\int \sin 7x \cos 8x\,dx$

5. $\int \dfrac{x^3}{x^3+1}\,dx$

6. $\int \dfrac{ds}{\tan^5 s}$

7. $\int (x^2+1)^{3/2}\,dx$

8. $\int \sin^6 t \cos^5 t\,dt$

9. $\int \dfrac{dx}{\sqrt[4]{x^4+1}}$

10. $\int \sin^2 x \cot^3 x\,dx$

11. $\int \dfrac{dx}{\sin x \cos 2x}$

12. $\int x^2 \sinh x\,dx$

13. $\int \sqrt{x^4+x}\,dx$

14. $\int \dfrac{x^2}{(4-x^2)^{3/2}}\,dx$

15. $\int \dfrac{y}{\sqrt{y+1}+1}\,dy$

16. $\int \dfrac{x}{\cos^2 x}\,dx$

17. $\int \tanh^{-1} x\,dx$

18. $\int \dfrac{dx}{\sqrt{1-x^2}\,(\arccos x)^2}$

19. $\int (1-x^2)^{5/2}\,dx$

20. $\int \dfrac{\tan^2 z + 1}{\tan^2 z - 1}\,dz$

21. $\int \operatorname{sech} x \, dx$

22. $\int \operatorname{csch} x \, dx$

23. $\int \sqrt{1-x^2} \arcsin x \, dx$

24. $\int (\arcsin x)^2 \, dx$

25. $\int x \coth^2 x \, dx$

26. $\int \sec^2 u \csc^2 u \, du$

27. $\int x e^x \sin x \, dx$

28. $\int v^2 \ln \dfrac{v-1}{v} \, dv$

29. $\int \dfrac{3x^2 - 1}{x^4 + x^3 + 2x^2 + x + 1} \, dx$

30. $\int e^{2x} \sqrt{e^x + 2} \, dx$

31. $\int \dfrac{\sin 2x}{\sin^4 x + \cos^4 x} \, dx$

32. $\int \cot^4 w \csc^4 w \, dw$

33. $\int \dfrac{x^2}{\sqrt{x^2 + 2}} \, dx$

34. $\int \dfrac{dx}{\sqrt{x}(1 + \sqrt[4]{x})^3}$

35. $\int \dfrac{dx}{\sin^6 x}$

36. $\int \dfrac{dx}{(x^2 - 4x + 4)(x^2 - 4x + 5)}$

37. $\int (\tan y - \sec y)^2 \, dy$

38. $\int \dfrac{1}{x}\left(\dfrac{x+2}{x-1}\right)^2 dx$

39. $\int x \cos \sqrt{x} \, dx$

40. $\int \sin^3 x \tan x \, dx$

41. $\int \dfrac{dx}{(x-1)^2 (x+1)^3}$

42. $\int \dfrac{\ln z}{\sqrt{z}} \, dz$

43. $\int \sin x \sinh x \, dx$

44. $\int \sinh 2x \cosh x \, dx$

45. $\int \dfrac{\arcsin x}{\sqrt{1+x}} \, dx$

46. $\int \cos \pi v \cos v \, dv$

47. $\int \dfrac{x^2 + x + 1}{x^4 - x^3 - 91x^2 + x + 90} \, dx$

48. $\int \cos^2 x \csc^3 x \, dx$

49. $\int \dfrac{x + \sin x \cos x}{(x \sin x + \cos x)^2} \, dx$

50. $\int \arctan \sqrt{2x - 1} \, dx$

51. $\int \dfrac{\sqrt{x+1} - 1}{\sqrt{x+1} + 1} \, dx$

52. $\int \left(\dfrac{u}{u \cos u - \sin u}\right)^2 du$

53. $\int \dfrac{\tan^2 x}{1 + \cos^2 x} \, dx$

54. $\int \sqrt{1 - \sin x} \, dx$

55. $\int \cos x \tan^4 x \, dx$

56. $\int \dfrac{x^2}{\sqrt{x^2 + x + 1}} \, dx$

57. $\int \dfrac{dx}{x^4 + 1}$

58. $\int \dfrac{x^4 + 1}{x^6 + 1} \, dx$

59. $\int \dfrac{x^4 - 1}{x^6 - 1} \, dx$

60. $\int \tan^3 x \sec^4 x \, dx$

61. $\int \dfrac{ds}{e^{3s} - e^s}$

62. $\int \dfrac{dx}{1 + \sin^2 x}$

63. $\int \dfrac{t^5 + 1}{t^6 + t^4} \, dt$

64. $\int \cot^6 x \, dx$

65. $\int \dfrac{\sqrt{\tan z}}{\sin 2z} \, dz$

66. $\int \cos x \cos 2x \cos 3x \, dx$

67. $\int \dfrac{dx}{\sin(x+1)\sin(x-1)}$

68. $\int \dfrac{du}{\sin^3 u \cos^5 u}$

69. $\int y^4 \ln y \, dy$

70. $\int \dfrac{dx}{\sqrt{\tan x}}$

71. $\int \dfrac{\sin v}{\sin 3v} \, dv$

72. $\int \tanh^4 x \, dx$

73. $\int \operatorname{sech}^3 x \, dx$

74. $\int \dfrac{ds}{s\sqrt{s^3 - 1}}$

75. $\int \dfrac{dt}{2 \sin t + \sin 2t}$

76. $\int \left(\dfrac{x^2 + 1}{x + 1}\right)^2 dx$

77. $\int \dfrac{\sin x}{2 + \sin x + \cos x} \, dx$

78. $\int \dfrac{dx}{1 - \tanh x}$

79. Suppose $Q(x) = (x - c_1)(x - c_2) \cdots (x - c_n)$, where $c_1, c_2, \ldots, c_n$ are distinct constants (that is, $c_i \neq c_j$ if $i \neq j$), and let $P(x)$ be any polynomial of degree less than n. Show that

$$\int \dfrac{P(x)}{Q(x)} dx = \sum_{i=1}^{n} \dfrac{P(c_i)}{Q'(c_i)} \ln|x - c_i| + C. \quad (i)$$

Use formula (i) to evaluate

80. $\int \dfrac{x}{(x-1)(x-2)(x-3)} \, dx$

81. $\int \dfrac{x^2 + 5}{(x-1)(x+2)(x+3)} \, dx$

82. $\int \dfrac{x^3}{(x^2 - 1)(x^2 - 4)} \, dx$

83. Show that $\displaystyle\int_0^1 \dfrac{\ln(x+1)}{x^2 + 1} \, dx = \dfrac{1}{8}\pi \ln 2$.

Use the specified rules to approximate the given integral, where n is the number of subintervals. Calculate ordinate values to four decimal places, and give the answer to three decimal places (without attempting to estimate the error).

84. $\displaystyle\int_1^2 \dfrac{\cos x}{x} \, dx$, midpoint and trapezoidal rules, $n = 5$

85. $\displaystyle\int_0^{\pi/2} \sqrt{\sin x} \, dx$, midpoint and trapezoidal rules, $n = 6$

86. $\displaystyle\int_0^1 e^{x^2} \, dx$, trapezoidal and Simpson's rule, $n = 4$

87. What value of n, the number of subintervals, guarantees that the error E_S made in using Simpson's rule to approximate the integral in Problem 86 is less than 0.00001 in absolute value?

88. With the help of the inequality $x^2 \geq 2ax - a^2$, show that $\int_3^\infty e^{-x^2} dx < 0.00003$. Then use Simpson's rule with $n = 12$ to approximate the *improper* integral $I = \int_0^\infty e^{-x^2} dx$. Calculate ordinate values to five decimal places, and by analyzing the error, find a value of I which is accurate to within 0.0001. Check the answer by comparing it with the exact value of I.

If the given improper integral is convergent, find its value. Otherwise identify it as divergent.

89. $\int_0^\infty \dfrac{dx}{x+1}$

90. $\int_2^\infty \dfrac{dy}{\ln y}$

91. $\int_0^\infty \dfrac{dx}{x^4+1}$

92. $\int_a^b \dfrac{dx}{\sqrt{(x-a)(b-x)}}$ $(a<b)$

93. $\int_1^\infty \dfrac{dx}{x\sqrt{1+x^2}}$

94. $\int_0^\infty \dfrac{x}{x^2-1} dx$

95. $\int_0^1 \dfrac{dz}{z\sqrt{1+z^2}}$

96. $\int_{1/2}^1 \dfrac{dx}{x\sqrt{1-x^2}}$

97. $\int_0^\pi \dfrac{dt}{1+\cos t}$

98. $\int_0^\infty e^{-x} \sin x \cos x\, dx$

99. $\int_0^\infty e^{-2x} \sinh x\, dx$

100. $\int_0^\infty e^{-x} \cosh x\, dx$

101. Show that the improper integral
$$\int_2^\infty \dfrac{dx}{x(\ln x)^p}$$
is convergent if and only if $p > 1$.

102. Use repeated integration by parts to show that
$$\int_0^\infty x^n e^{-x} dx = n!,$$
where n is any positive integer.

103. By making the substitution $x = \tan t$, show that
$$\int_0^\infty \dfrac{dx}{(x^2+1)^n} = \begin{cases} \dfrac{\pi}{2} & \text{if } n = 1, \\ \dfrac{1}{2}\dfrac{3}{4} \cdots \dfrac{2n-3}{2n-2}\dfrac{\pi}{2} & \text{if } n = 2, 3, \ldots \end{cases}$$

104. With the help of Problem 102 and the substitution $x = -(m+1)\ln t$, show that
$$\int_0^1 x^m (\ln x)^n dx = \dfrac{(-1)^n n!}{(m+1)^{n+1}},$$
where m is any nonnegative integer and n is any positive integer.

Evaluate

105. $\int_0^\infty x^6 e^{-x} dx$

106. $\int_0^\infty \dfrac{dx}{(x^2+1)^4}$

107. $\int_0^1 x^2 (\ln x)^3 dx$

108. Show that the improper integral $\int_0^\infty t^{x-1} e^{-t} dt$ is convergent if and only if $x > 0$.

109. For positive x the integral in the preceding problem defines a function of x, known as the *gamma function* and denoted by $\Gamma(x)$. Thus
$$\Gamma(x) = \int_0^\infty t^{x-1} e^{-t} dt \qquad (x > 0),$$
where the symbol Γ is capital Greek gamma (the equivalent of English G). Show that
$$\Gamma(x+1) = x\Gamma(x) \qquad (x > 0),$$
and in particular
$$\Gamma(n+1) = n! \qquad (n = 1, 2, \ldots).$$

110. Show that
$$\int_0^\infty \dfrac{\ln x}{x^2+1} dx = 0.$$

Supplementary Problems **435**

8 Further Applications of Integration

In this chapter we pursue the study of applications of integration begun in Chapter 4. We first show how integration can be used to calculate the volume of a solid, in particular a solid of revolution. We then generalize the notion of a curve in such a way as to free it from the rather unnatural restriction that the curve be the graph of a single function. This leads to the concept of a "parametric curve," and puts us in a position to calculate both the length of a general curve and the area of a general surface of revolution. Finally we turn from problems of geometry to problems of physics and engineering, and investigate a number of topics in the mechanics of fluids.

8.1 Volume by the Method of Cross Sections

In this section and the next we use integration to calculate volumes of solids. As in Figure 1(a), let S be a solid, that is, a three-dimensional region, and let L be a line, which we choose to be the x-axis. Some of the planes perpendicular to L intersect S in two-dimensional regions, called *cross sections* of S. Let $A(x)$ be the area of the cross section of S at x, by which we mean the cross section cut from S by the plane perpendicular to L at the point of L with coordinate x (see the figure). Suppose further that

(i) A plane perpendicular to L intersects S if and only if the plane intersects L in a point of the closed interval $[a, b]$;
(ii) The function $A(x)$ is continuous on $[a, b]$.

Then, as we now show, the appropriate definition of the volume V of the solid S is given by the formula

$$V = \int_a^b A(x)\, dx. \qquad (1)$$

The reasoning closely parallels the construction on pages 229–231, leading to the integral $\int_a^b f(x)\,dx$ as the appropriate definition of the area under the graph of a continuous function $y = f(x)$ from $x = a$ to $x = b$.

Let

$$a = x_0, x_1, x_2, \ldots, x_{n-1}, x_n = b \qquad (2)$$

be points of subdivision of the interval $[a, b]$ satisfying the inequalities $a = x_0 < x_1 < x_2 < \cdots < x_{n-1} < x_n = b$. These points divide $[a, b]$ into n subintervals

$$[x_{i-1}, x_i] \qquad (i = 1, 2, \ldots, n),$$

where $[x_{i-1}, x_i]$ is of length $\Delta x_i = x_i - x_{i-1}$. As usual, let

$$\mu = \max\{\Delta x_1, \Delta x_2, \ldots, \Delta x_n\}$$

be the mesh size of the subdivision (2), that is, the maximum length of all the subintervals. The planes perpendicular to the line L at the points x_i ($i = 0, 1, \ldots, n$) divide the solid S into n thin slices $S_1, S_2, \ldots, S_n$, as shown in Figure 1(b). Consider the ith slice, of thickness Δx_i. The function $A(x)$ is continuous, and hence its value changes only slightly on the subinterval $[x_{i-1}, x_i]$, at least if Δx_i is small enough. Thus, in calculating the volume of S, it seems a good approximation to simultaneously

(i) Regard $A(x)$ as having the constant value $A(p_i)$ on $[x_{i-1}, x_i]$, where p_i is an arbitrary point of $[x_{i-1}, x_i]$;†

(ii) Ignore the fact that the sides of S_i are in general slanted, as in Figure 2, and replace S_i by the indicated *right cylinder* C_i, generated by moving the cross section of S at p_i a distance Δx_i along a line perpendicular to the cross section, that is, parallel to L.

In geometry the volume of a right cylinder of cross-sectional area A and height h is defined as the product of A and h. Therefore, if V_i is the volume of the cylinder C_i, we have

$$V_i = A(p_i)\,\Delta x_i. \qquad (3)$$

Since the volume V of the whole solid S is equal to the sum of the volumes of the n slices $S_1, S_2, \ldots, S_n$, and since the volume of each slice S_i is approximately equal to the volume (3) of the corresponding cylinder C_i, it follows that

$$V \approx \sum_{i=1}^{n} A(p_i)\,\Delta x_i. \qquad (4)$$

You will recognize the right side of (4) as a Riemann sum of the function $A(x)$ on the interval $[a, b]$.

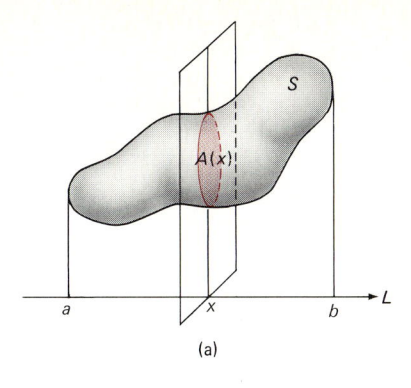

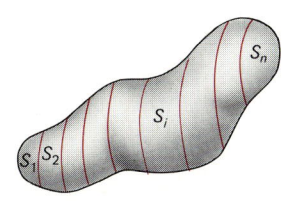

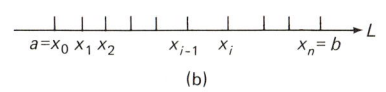

Figure 1

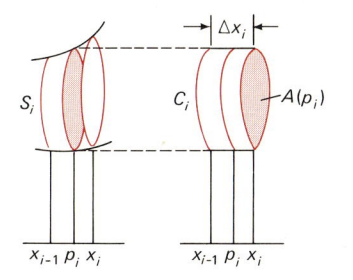

Figure 2

Volume as an Integral

The final step follows a pattern that is by now familiar. Let all the slices S_i, and hence all the approximating cylinders C_i, get thinner and thinner; this is achieved by making the mesh size μ approach zero. Then it is reasonable to expect the approximation (4) to get better and better. Motivated by these

---

† The dependence of the approximation on the particular choice of the points p_i will disappear in the limit when we make $\mu \to 0$.

considerations, we now *define* the volume V of the solid S as the limit

$$V = \lim_{\mu \to 0} \sum_{i=1}^{n} A(p_i) \Delta x_i.$$

This establishes formula (1), since the limit in question is simply the integral $\int_a^b A(x)\,dx$, the existence of which is guaranteed by the assumed continuity of $A(x)$.

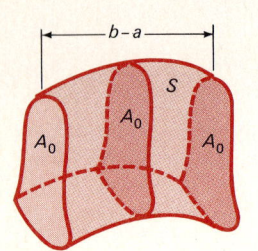

Example 1 Suppose every cross section of the solid S has the same area, as in Figure 3, so that $A(x)$ has the constant value A_0. Then formula (1) implies

$$V = \int_a^b A(x)\,dx = \int_a^b A_0\,dx = A_0(b-a).$$

This is the same as the volume of a right cylinder of base area A_0 and height $b - a$, but of course S can have constant cross-sectional area without being a cylinder, much less a right cylinder.

Example 2 Find the volume V of a right circular cone of height h, whose base is of radius r.

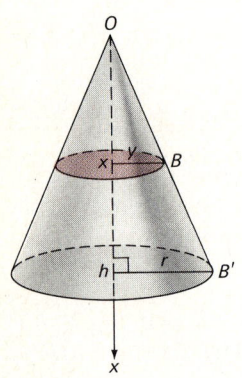

Solution As in Figure 4, we choose the axis of symmetry of the cone as the x-axis (pointing downward) and the vertex of the cone as the origin O. Then $A(x) = \pi y^2$, where $y = y(x)$ is the radius of the cross section at x (a circular disk). The right triangles OBx and $OB'h$ are similar, and hence

$$\frac{y}{r} = \frac{x}{h}.$$

Therefore

$$y = \frac{r}{h}x, \qquad A(x) = \frac{\pi r^2}{h^2}x^2,$$

and formula (1) gives

$$V = \int_0^h A(x)\,dx = \frac{\pi r^2}{h^2}\int_0^h x^2\,dx = \frac{1}{3}\pi r^2 h. \qquad \square \qquad (5)$$

Example 3 Find the volume V of a right pyramid of height h, whose base is a square of side length a.

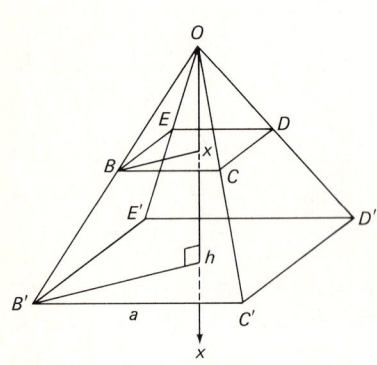

Solution As in Figure 5, we choose the x-axis to be a line pointing downward through the apex of the pyramid, perpendicular to the base of the pyramid. Let the origin O be at the apex, and let h be the coordinate of the point in which the x-axis intersects the base. The area $A(x)$ of the cross section of the pyramid at x is the area of the square $BCDE$ in the figure. We have

$$\frac{|BC|}{|B'C'|} = \frac{|OB|}{|OB'|},$$

by the similarity of the triangles OBC and $OB'C'$. Moreover,

$$\frac{|OB|}{|OB'|} = \frac{|Ox|}{|Oh|} = \frac{x}{h},$$

Figure 3

Figure 4

Figure 5

438   Chapter 8   Further Applications of Integration

by the similarity of the triangles OBx and $OB'h$. It follows that

$$\frac{|BC|}{|B'C'|} = \frac{x}{h},$$

that is,

$$|BC| = \frac{|B'C'|}{h} x = \frac{a}{h} x.$$

Therefore

$$A(x) = |BC|^2 = \frac{a^2}{h^2} x^2,$$

and this time formula (1) gives

$$V = \int_0^h A(x)\, dx = \frac{a^2}{h^2} \int_0^h x^2\, dx = \frac{1}{3} a^2 h. \qquad \square \qquad (6)$$

The base area of the cone in Example 2 is πr^2, while that of the pyramid in Example 3 is a^2. Thus both formulas (5) and (6) take the same form $\frac{1}{3} A_0 h$ in terms of the base area A_0 and the height h of the given solid. Of course, this is not a coincidence (see Problem 19).

Example 4 Find the volume V of the wedge cut from a right circular cylinder of radius r by a plane through a diameter of the base of the cylinder making angle θ with the base.

Solution As in Figure 6, we introduce a rectangular coordinate system in the base, with x-axis along the diameter BB', origin O at the midpoint of BB' (on the axis of the cylinder), and y-axis perpendicular to BB' at O. The base of the wedge is bounded by the line segment BB' and the semicircle $x^2 + y^2 = r^2$, $y \geq 0$, and the cross section of the wedge at x is the right triangle CDE, with area

$$A(x) = \frac{1}{2} |CD| |DE| = \frac{1}{2} |CD|^2 \tan\theta = \frac{1}{2} y^2 \tan\theta = \frac{1}{2} (r^2 - x^2) \tan\theta.$$

Thus, applying formula (1), we find that

$$V = \int_{-r}^{r} A(x)\, dx = \frac{1}{2} \tan\theta \int_{-r}^{r} (r^2 - x^2)\, dx$$

$$= \frac{1}{2} \tan\theta \left[r^2 x - \frac{1}{3} x^3 \right]_{-r}^{r} = \frac{2}{3} r^3 \tan\theta.$$

If h is the height of the wedge, then $h = |FG| = r \tan\theta$ (see the figure), so that the volume of the wedge can also be written in the form $V = \frac{2}{3} r^2 h$. $\quad\square$

Example 5 Two right circular cylinders of radius r intersect at right angles, as in Figure 7. Find the volume V of the solid S common to the two cylinders.

Solution The shaded region in the figure is one eighth of the solid S. Choosing the x-axis as indicated, with origin O at the point of intersection of the axes of the cylinders, we find that the section of the solid S at x is the

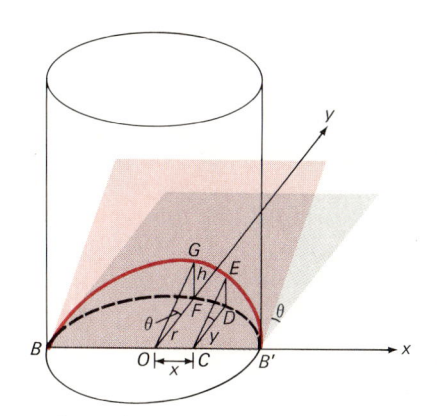

Figure 6

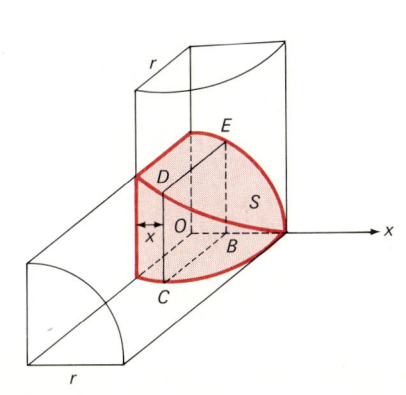

Figure 7

Section 8.1 Volume by the Method of Cross Sections   **439** square $BCDE$, of side length $|BC| = \sqrt{r^2 - x^2}$. Therefore $A(x) = r^2 - x^2$, so that

$$V = 8 \int_0^r (r^2 - x^2)\,dx = \frac{16}{3} r^3. \quad \square$$

Problems

Find the volume of the solid S whose base is the region bounded by the parabola $y = x^2 + 1$, the coordinate axes and the line $x = 1$ if every cross section of S by a plane perpendicular to the x-axis is

1. A square
2. A semicircle with its diameter in the xy-plane

The base of a solid S is the region between the x-axis and the curve $y = \sin x$ from $x = 0$ to $x = \pi$. Find the volume of S if every cross section of S by a plane perpendicular to the x-axis is

3. A rectangle of height 1
4. A square

Find the volume of the solid S whose base is the circular disk $x^2 + y^2 \leq 9$ if every cross section of S by a plane perpendicular to the x-axis is

5. A square
6. A semicircle with its diameter in the xy-plane
7. An equilateral triangle
8. An isosceles right triangle with its hypotenuse in the xy-plane

Find the volume of

9. An oblique circular cylinder of height h and base radius r (see Figure 8)
10. An oblique circular cone of height h and base radius r (see Figure 9)

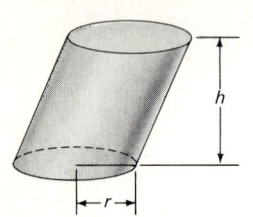

Figure 8

Figure 9

11. The solid shown in Figure 10 is called a *frustum* of a right circular cone. Express the volume of the frustum in terms of its upper radius r_1, base radius r_2 and height h.

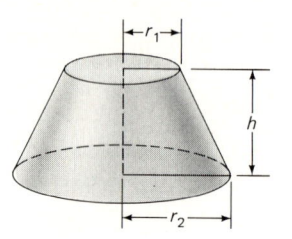

Figure 10

12. Water is poured into a hemispherical basin r feet in radius. What is the volume of water in the basin when the water is h feet deep, as measured from the bottom of the basin? To what percent of capacity is the basin filled when the water is $\frac{1}{2}r$ feet deep? About how deep is the water when the basin is 75% full?

The base of a solid S is the region between the two parabolas $y = x^2 - 1$ and $y = 1 - x^2$. Find the volume of S if every cross section of S by a plane perpendicular to the x-axis is

13. A right triangle with its hypotenuse in the xy-plane and one angle equal to 30°
14. A regular hexagon with one edge in the xy-plane

As in Problems 3 and 4, the base of a solid S is the region between the x-axis and the curve $y = \sin x$ from $x = 0$ to $x = \pi$. Find the volume of S if every cross section of S by a plane perpendicular to the y-axis is

15. A rectangle of height 1
16. A square

17. The base of a solid S is an equilateral triangle T of altitude h. Suppose every cross section of S by a plane perpendicular to a fixed altitude of T is a circular disk with its diameter on T. Find the volume of S in two ways.

18. At a distance of x feet from the tip of a church spire, the cross section of the spire by a plane perpendicular to its axis of symmetry is a square whose diagonal is of length $\frac{1}{25}x^2$ feet. What is the volume of the spire if it is 10 feet tall?

19. Let R be a plane region of area A_0, and let P be a point not in R, at distance h from the plane of R. Then the line segments drawn from P to the points of R sweep out a solid C, called a *general cone* (see Figure 11). Show that the volume of C is equal to $\frac{1}{3}A_0 h$.

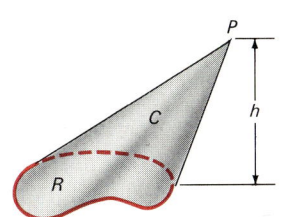

Figure 11

20. Let S be a *prismoid*, that is, a solid bounded by two plane faces and a lateral surface. The plane faces, which need not be polygonal,† are called the *bases* of S. Let the bases be a distance h apart, and suppose the area of a cross section of S by a plane parallel to the bases is a polynomial function $A(x) = a + bx + cx^2 + dx^3$, of degree 3 or less, where the x-axis is perpendicular to the bases, and any point of the x-axis can serve as the origin. With the help of the prismoidal formula (i), page 424, show that the volume of the solid S is given by the formula

$$V = \frac{1}{6}h(B_1 + 4B + B_2),$$

where B_1, B_2 are the areas of the bases, and B is the area of a cross section of S by the plane halfway between the bases. This formula for V is also known as the prismoidal formula (and explains the origin of the term "prismoidal"). Use it to simplify the calculation in Problem 11.

21. Find the volume of a tetrahedron (triangular pyramid) with three mutually perpendicular faces and three mutually perpendicular edges, of lengths a, b and c (see Figure 12).

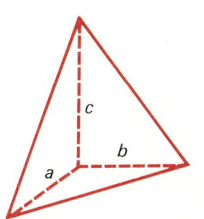

Figure 12

† For example, a conical frustum is a prismoid, and so is a spherical segment (see Problem 27). The prismoid is called a *prism* if its bases are congruent polygons, and if the lateral surface consists of parallelograms joining corresponding vertices of the bases.

22. A tetrahedron is said to be *regular* if its faces are equilateral triangles. Find the height h and volume V of a regular tetrahedron of edge length s (see Figure 13).

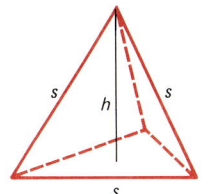

Figure 13

23. Find the volume of the horn-shaped solid in Figure 14, whose cross sections by planes perpendicular to the x-axis are circular disks. The endpoints of a diameter of each disk lie on the curves $y = \sqrt{x}$ and $y = 2\sqrt{x}$, and the solid extends from its tip at the origin of the xy-plane to its cross section at $x = 1$.

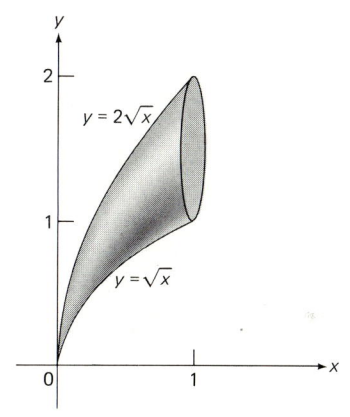

Figure 14

24. Two solids lie between a pair of parallel planes, a distance h apart, and the cross sections of the solids by any plane parallel to the given planes have the same area. Prove *Cavalieri's principle*, which says that the two solids have the same volume. (Cavalieri's principle for area is given in Problem 20, page 247.)

25. Suppose a conical hole of base radius r and depth r is drilled along the axis of symmetry of a circular cylinder of radius r and height r. Show that the resulting object has the same volume as a solid hemisphere of radius r.

26. Calculate the volume of the wedge in Example 4 by using cross sections by planes perpendicular to the y-axis (rather than to the x-axis).

27. The solid S bounded by a sphere and two parallel planes intersecting the sphere and its interior in circular disks is called a *spherical segment* of two bases, with the disks as its bases (see Figure 15, where the sphere is of radius R centered at the origin, and the x-axis is vertical). Find the volume of S if one base is of radius r_1, the other base is of radius r_2, and the distance between the bases is h.

28. Find the volume of the infinite solid whose base is the region bounded by the positive coordinate axes and the curve $y = e^{-x}$, and whose cross sections by planes perpendicular to the x-axis are semicircles with diameters in the xy-plane.

29. Find the volume of the infinite solid whose base is the region bounded by the positive coordinate axes, the line $x = 1$ and the curve $y = x^{-1/3}$, and whose cross sections by planes perpendicular to the x-axis are squares. What is the volume if $y = x^{-1/2}$ instead?

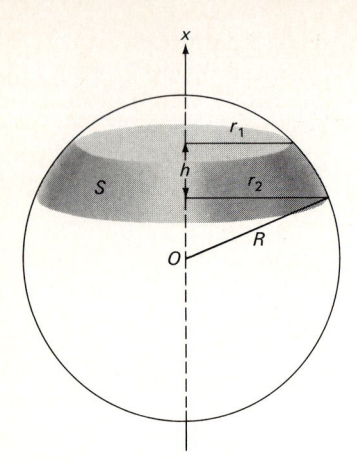

Figure 15

8.2 Volume of a Solid of Revolution

The Method of Disks

If a plane region R is revolved about a line L in the plane of R, the region sweeps out or "generates" a solid S, called a *solid of revolution*, with the line L as the *axis of revolution*. Let R be the region of the xy-plane bounded by the graph of a continuous nonnegative function

$$y = f(x) \quad (a \le x \le b),$$

the x-axis and the vertical lines $x = a$ and $x = b$, as in Figure 16(a), and suppose the x-axis is chosen as the axis of revolution. Then R generates a solid of revolution S of the form shown in Figure 16(b), whose cross section by the plane perpendicular to the x-axis at the point with coordinate x is a circular disk of radius $y = f(x)$ and area $A(x) = \pi y^2 = \pi [f(x)]^2$. Let V be the volume of the solid S. Then $V = \int_a^b A(x)\, dx$, by the method of cross sections, and hence

$$V = \pi \int_a^b y^2\, dx = \pi \int_a^b [f(x)]^2\, dx. \tag{1}$$

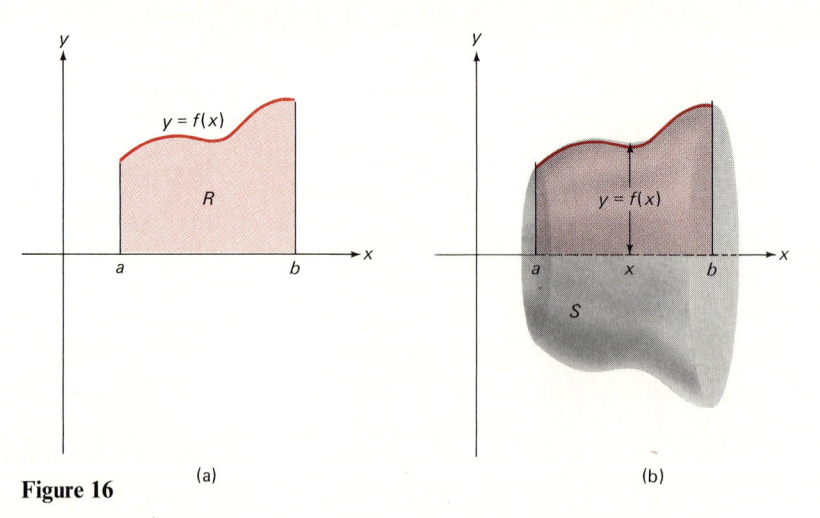

Figure 16

442 Chapter 8 Further Applications of Integration

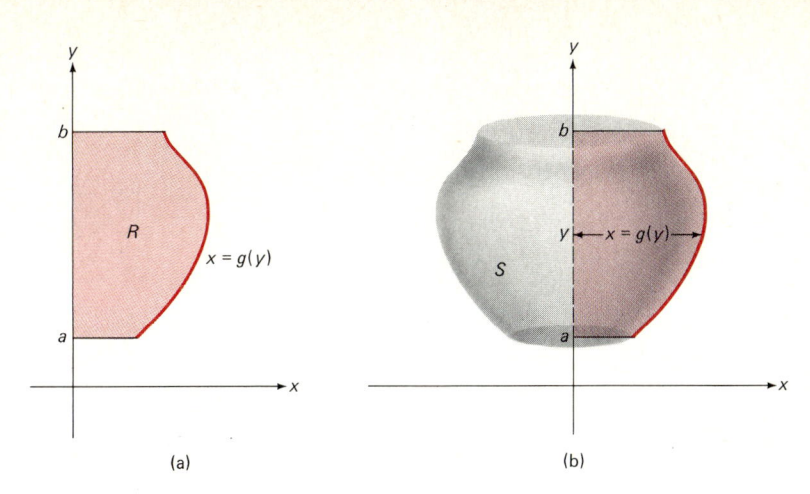

Figure 17

Sometimes the axis of revolution is chosen to be the y-axis, rather than the x-axis. Thus, suppose the region R bounded by the graph of a continuous nonnegative function

$$x = g(y) \quad (a \leq y \leq b),$$

the y-axis and the horizontal lines $y = a$ and $y = b$, as in Figure 17(a), is revolved about the y-axis, generating a solid of revolution S of the form shown in Figure 17(b). The cross section of S by the plane perpendicular to the y-axis at the point with coordinate y is a circular disk of radius $x = g(y)$ and area $\pi x^2 = \pi[g(y)]^2$. Therefore, by the method of cross sections, the volume V of the solid is now given by the formula

$$V = \pi \int_a^b x^2 \, dy = \pi \int_a^b [g(y)]^2 \, dy, \tag{1'}$$

which is analogous to (1). The use of formula (1) or (1') to calculate the volume of a solid of revolution is called the *method of disks*. The method was anticipated in Example 2, page 438, where it was used to calculate the volume of a right circular cone.

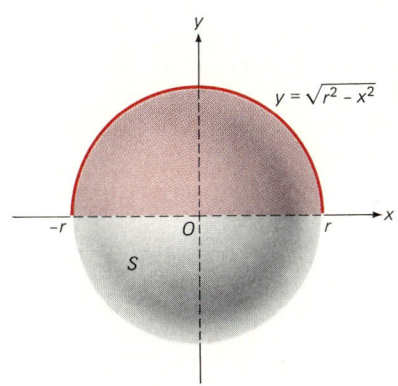

Example 1 Find the volume V of a solid sphere of S radius r.

Solution We can generate S by revolving the region bounded by the x-axis and the graph of the function

$$y = \sqrt{r^2 - x^2} \quad (-r \leq x \leq r)$$

about the x-axis (see Figure 18). Note that in this case the vertical line segments in Figure 16(a) reduce to points, since $y = 0$ for $x = \pm r$. Applying formula (1), we immediately obtain

$$V = \pi \int_{-r}^{r} y^2 \, dx = \pi \int_{-r}^{r} (r^2 - x^2) \, dx = \pi \left[r^2 x - \frac{1}{3} x^3 \right]_{-r}^{r} = \frac{4}{3} \pi r^3,$$

thereby *proving* a formula that we have been using freely all along. □

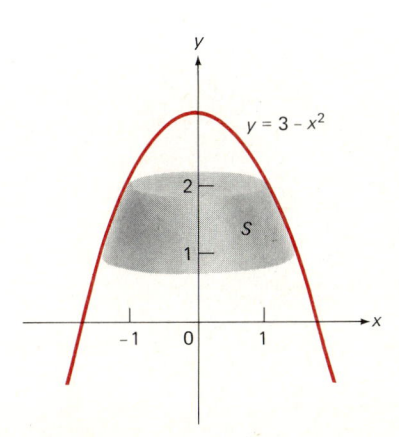

Example 2 Find the volume V of the solid S, shown in Figure 19, generated by revolving the region bounded by the curve $y = 3 - x^2$, the y-axis, and the lines $y = 1$ and $y = 2$ about the y-axis.

Section 8.2 Volume of a Solid of Revolution 443

Solution The solid S is a frustum of the solid whose boundary is the *paraboloid of revolution* generated by revolving the parabola $y = 3 - x^2$ about the y-axis. (A *frustum* is the part of a solid between two parallel planes cutting the solid.) Since $y = 3 - x^2$, we have $x^2 = 3 - y$, and formula (1') gives

$$V = \pi \int_1^2 x^2 \, dy = \pi \int_1^2 (3 - y) \, dy = \pi \left[3y - \frac{1}{2} y^2 \right]_1^2 = \frac{3}{2} \pi. \quad \square$$

The Method of Washers

Next let R be the region in the xy-plane bounded by two curves $y = f(x)$ and $y = g(x)$, with $f(x) \geq g(x) \geq 0$, and two vertical lines $x = a$ and $x = b$ ($a < b$), as in Figure 20(a). Then revolving R about the x-axis generates a solid of revolution S of the form shown in Figure 20(b), whose cross section by the plane perpendicular to the x-axis at the point with coordinate x is an *annulus* or washer-shaped region of outer radius $f(x)$ and inner radius $g(x)$, as in Figure 20(c). The area $A(x)$ of this washer is the difference between the area of a circular disk of radius $f(x)$ and the area of a concentric disk of radius $g(x)$:

$$A(x) = \pi [f(x)]^2 - \pi [g(x)]^2.$$

Let V be the volume of the solid S. Then, by the method of cross sections,

$$V = \int_a^b A(x) \, dx = \pi \int_a^b \{[f(x)]^2 - [g(x)]^2\} \, dx. \tag{2}$$

Of course, this formula can also be obtained without resorting explicitly to washers by observing that the volume of the solid of revolution S in Figure 20(b) is the difference between the volumes of two solids of revolution of a type considered earlier, namely an "outer" solid of volume $\pi \int_a^b [f(x)]^2 \, dx$, generated by revolving the region bounded by the curve $y = f(x)$, the x-axis

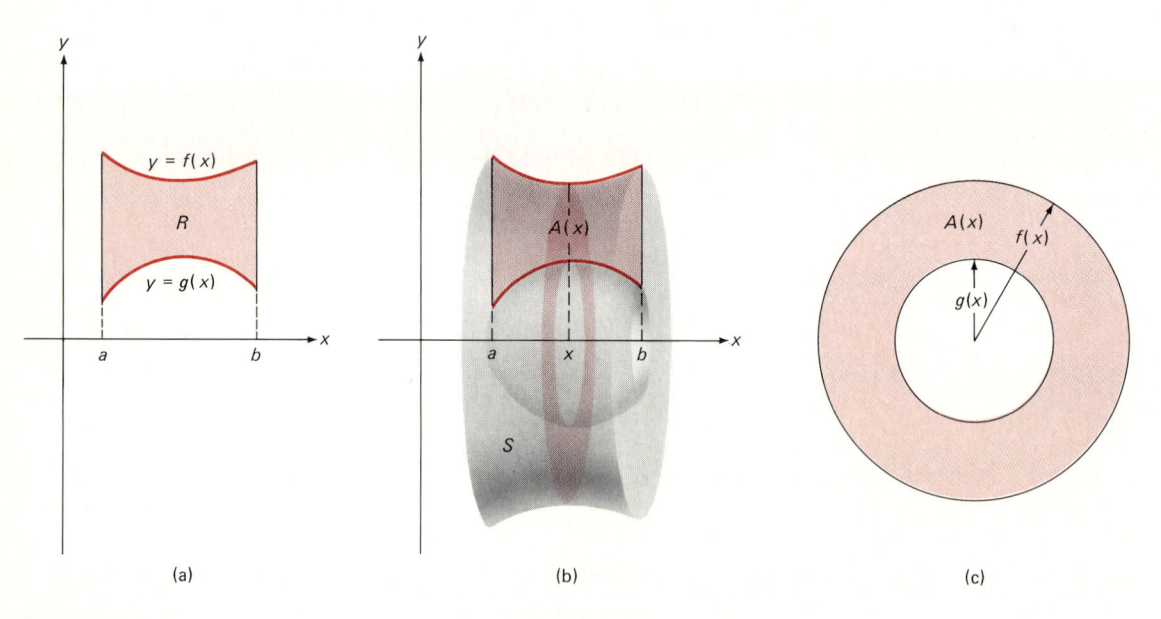

(a)   (b)   (c)

Figure 20

(a)

(b)

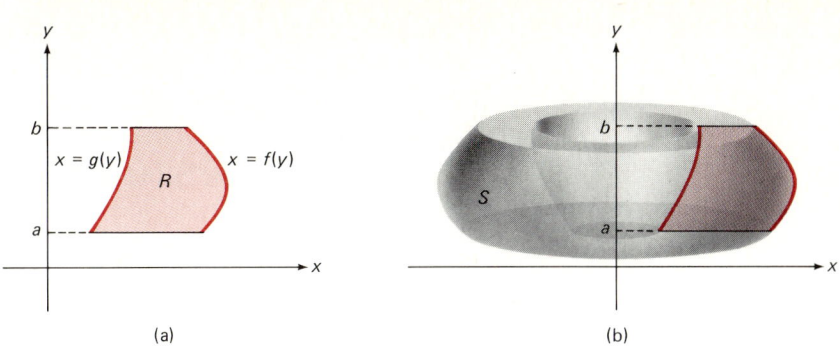

Figure 21

and the lines $x = a$ and $x = b$ about the x-axis, and an "inner" solid of volume $\pi \int_a^b [g(x)]^2 \, dx$, generated by doing the same to the region bounded by the curve $y = g(x)$, the x-axis and the lines $x = a$ and $x = b$.

As before, the axis of revolution is sometimes chosen to be the y-axis, rather than the x-axis. Thus suppose R is the region bounded by two curves $x = f(y)$ and $x = g(y)$, with $f(y) \geq g(y) \geq 0$, and two horizontal lines $y = a$ and $y = b$ ($a < b$), as in Figure 21(a). Then revolving R about the y-axis generates a solid of revolution S of the form shown in Figure 21(b). It is easy to see that the volume of S is given by the formula

$$V = \pi \int_a^b \{[f(y)]^2 - [g(y)]^2\} \, dy, \qquad (2')$$

which is analogous to (2). The use of formula (2) or (2') to calculate the volume of a solid of revolution is called the *method of washers*.

Example 3 A hole of length $2a$ is bored in a solid sphere of radius r along a diameter. What is the volume V of the remaining solid S? Assume that $a < r$.

Solution The solid S, shown in Figure 22, is shaped like a napkin ring, and is generated by revolving the region bounded by the circular arc BCD and the line segment BD about the x-axis. The arc BCD is the graph of the function $f(x) = \sqrt{r^2 - x^2}$ and the segment BD is the graph of the constant function $g(x) = h = \sqrt{r^2 - a^2}$. Therefore, by formula (2),

$$V = \pi \int_{-a}^{a} [(r^2 - x^2) - (r^2 - a^2)] \, dx = \pi \int_{-a}^{a} (a^2 - x^2) \, dx$$

$$= \pi \left[a^2 x - \frac{1}{3} x^3 \right]_{-a}^{a} = \frac{4}{3} \pi a^3.$$

Thus the volume of the solid S is the same as that of a sphere of radius a, and is independent of the size of the sphere in which the hole is bored, provided only that the diameter of the sphere exceeds the length of the hole. This result is less surprising when it is realized that to bore a short hole in a large sphere requires a drill of radius only slightly less than that of the sphere. □

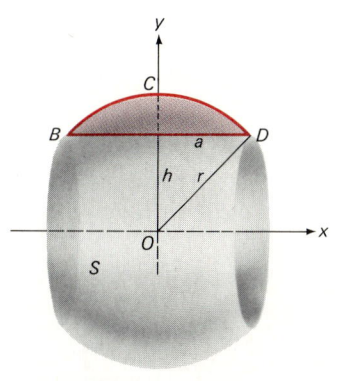

Figure 22

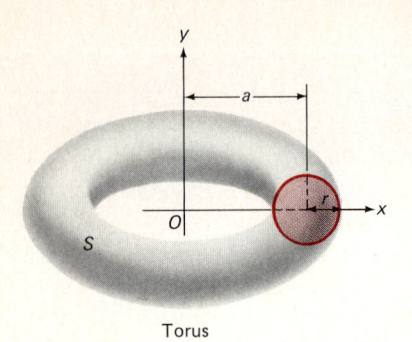

Example 4 Find the volume V of the solid S generated by revolving a circular disk about a line in its plane that does not intersect the disk.

Solution Let the circular disk be the region $(x - a)^2 + y^2 \leq r^2$ in the xy-plane, and let the axis of revolution be the y-axis; r is the radius of the disk, and a is the distance between the center of the disk and the y-axis ($r < a$ since the y-axis does not intersect the disk). Then S is the doughnut-shaped solid shown in Figure 23, known as an *anchor ring* or *torus*. Applying formula (2') with

$$f(y) = a + \sqrt{r^2 - y^2}, \qquad g(y) = a - \sqrt{r^2 - y^2},$$

we find that

$$V = \pi \int_{-r}^{r} \{[a + \sqrt{r^2 - y^2}]^2 - [a - \sqrt{r^2 - y^2}]^2\} \, dy$$

$$= 4\pi a \int_{-r}^{r} \sqrt{r^2 - y^2} \, dy = 8\pi a \int_{0}^{r} \sqrt{r^2 - y^2} \, dy,$$

where the evenness of the integrand is used in the last step. You will recognize the last integral as one fourth the area of a circular disk of radius r, namely $\frac{1}{4}\pi r^2$. It follows that

$$V = 8\pi a \left(\frac{1}{4}\pi r^2\right) = 2\pi^2 a r^2. \quad \square$$

Figure 23 Torus

The Method of Shells

We now develop another way of calculating the volume of a solid of revolution, known as the *method of shells*. As the name implies, the idea of this method is to partition the given solid into cylindrical shells, rather than into circular disks or washers. Consider the region R in Figure 24(a), bounded by the x-axis, the vertical lines $x = a$ and $x = b$, and the graph of a continuous nonnegative function $y = f(x)$. If the region R is revolved about the y-axis, it generates the solid of revolution S in Figure 24(b). Let

$$a = x_0, x_1, x_2, \ldots, x_{n-1}, x_n = b \tag{3}$$

be points of subdivision of the interval $[a, b]$ which satisfy the inequalities $a = x_0 < x_1 < x_2 < \cdots < x_{n-1} < x_n = b$. These points divide $[a, b]$ into n subintervals

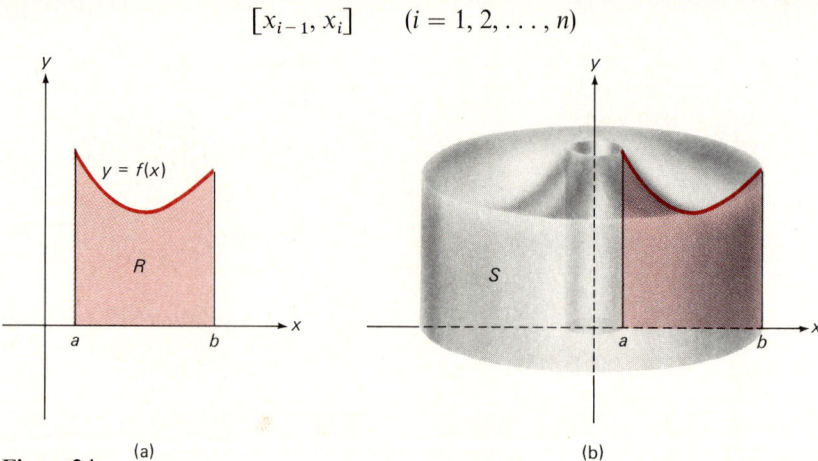

$$[x_{i-1}, x_i] \qquad (i = 1, 2, \ldots, n)$$

Figure 24
(a)   (b)

where $[x_{i-1}, x_i]$ is of length $\Delta x_i = x_i - x_{i-1}$. As usual, let

$$\mu = \max\{\Delta x_1, \Delta x_2, \ldots, \Delta x_n\}$$

be the mesh size of the partition (3), that is, the maximum length of all the subintervals. The vertical lines $x = x_0, x = x_1, x = x_2, \ldots, x = x_{n-1}, x = x_n$ divide the region R into n narrow strips $R_1, R_2, \ldots, R_n$, as in Figure 25(a). Let S_i be the cylindrical shell, with a curved top, generated by revolving the strip R_i about the y-axis, as in Figure 25(b). The function $f(x)$ is continuous, and hence its value changes only slightly on the subinterval $[x_{i-1}, x_i]$, at least if Δx_i is small enough. Thus it seems a good approximation to regard $f(x)$ as having the constant value $f(p_i)$ on $[x_{i-1}, x_i]$, where p_i is an arbitrary point of $[x_{i-1}, x_i]$. For a reason that will be apparent in a moment, we choose p_i to be the midpoint of $[x_{i-1}, x_i]$, so that

$$p_i = \frac{x_{i-1} + x_i}{2}. \tag{4}$$

The result of replacing $f(x)$ by $f(p_i)$ on the interval $[x_{i-1}, x_i]$, and then revolving the resulting rectangular region about the y-axis, is to replace the cylindrical shell S_i in Figure 25(b) by the cylindrical shell C_i in Figure 25(c), with the same inner radius x_{i-1} and outer radius x_i, but with a flat top of height $f(p_i)$, rather than a curved top. Since the volume V of the whole solid of revolution S is equal to the sum of the volumes of the n shells $S_1, S_2, \ldots, S_n$ with curved tops, and since the volume of each shell S_i is approximately equal to the volume V_i of the corresponding shell C_i with a flat top, it follows that

$$V \approx \sum_{i=1}^{n} V_i. \tag{5}$$

But V_i is the difference between the volume of an outer right circular cylinder, of base area πx_i^2 and height $f(p_i)$, and a concentric inner cylinder, of base area πx_{i-1}^2 and the same height. Therefore

$$V_i = \pi f(p_i)(x_i^2 - x_{i-1}^2) = \pi f(p_i)(x_i + x_{i-1})(x_i - x_{i-1}),$$

(a)

(b)

(c)

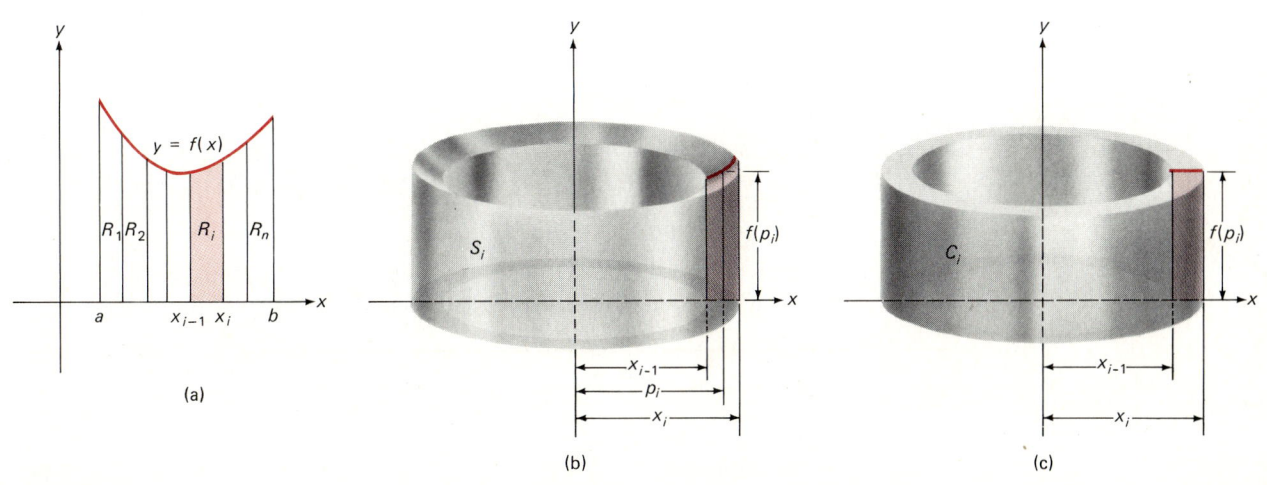

Figure 25

Section 8.2 Volume of a Solid of Revolution

which can be written in the form

$$V_i = 2\pi f(p_i) \frac{x_{i-1} + x_i}{2} \Delta x_i = 2\pi p_i f(p_i) \Delta x_i,$$

where the reason for the choice (4) is now clear. Substituting this expression for V_i into (5), we get

$$V \approx \sum_{i=1}^{n} 2\pi p_i f(p_i) \Delta x_i. \tag{6}$$

You will recognize the right side of (6) as a Riemann sum for the function $2\pi x f(x)$ on the interval $[a, b]$.

Finally, in the familiar fashion, we make the mesh size μ approach zero. Then all the shells S_i with curved tops get thinner and thinner, and so do the approximating shells C_i with flat tops. Thus as $\mu \to 0$, it is reasonable to expect the approximation (6) to get better and better. These considerations suggest that the volume V of the solid S be defined as the limit

$$V = \lim_{\mu \to 0} 2\pi \sum_{i=1}^{n} p_i f(p_i) \Delta x_i,$$

that is,

$$V = 2\pi \int_a^b x f(x)\, dx = 2\pi \int_a^b xy\, dx, \tag{7}$$

where the existence of the integral is guaranteed by the continuity of $f(x)$, and therefore of $xf(x)$. We now adopt (7) as the definition of volume, as calculated by the shell method.

Notice that the product $2\pi x f(x)$ is the area of a cylindrical surface of radius x and height $f(x)$. Therefore (7) can also be written in the form

$$V = \int_a^b A(x)\, dx, \tag{7'}$$

where $A(x)$ is the area of the section of the solid S by the cylindrical surface of radius x with the axis of revolution (here the y-axis) as its axis of symmetry. This is the same as the formula used in the method of cross sections, but in that method $A(x)$ is the area of the section of S by the plane perpendicular to the axis of revolution at the point with coordinate x.

More generally, if R is a region of the type shown in Figure 20(a), bounded by two curves $y = f(x)$ and $y = g(x)$, with $f(x) \geq g(x) \geq 0$, and two vertical lines $x = a$ and $x = b$ ($a < b$), the volume of the solid of revolution S generated by revolving R about the y-axis is given by

$$V = 2\pi \int_a^b x[f(x) - g(x)]\, dx \tag{8}$$

(why?). As an exercise, write the formulas for V in the case where S is generated by revolving a region of the type shown in Figure 17(a) or 21(a) about the x-axis.

Example 5 Use the shell method to find the volume V of a right circular cone of height h and base radius r.

Solution We have already found V by the disk method in Example 2, page 438. To calculate V by the shell method, we observe that the cone can be generated by revolving the triangular region bounded by the positive coordinate axes and the line $y = h[1 - (x/r)]$ about the y-axis (see Figure 26). Therefore, by formula (7),

$$V = 2\pi h \int_0^r x\left(1 - \frac{x}{r}\right) dx = 2\pi h \int_0^r \left(x - \frac{x^2}{r}\right) dx$$

$$= 2\pi h \left[\frac{x^2}{2} - \frac{x^3}{3r}\right]_0^r = \frac{1}{3}\pi r^2 h,$$

as we already know. □

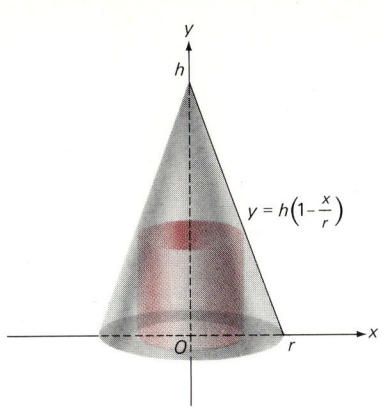

Example 6 Find the volume V of the solid S generated by revolving the region between the curves $y = x^2$ and $y = x^3$ about the y-axis.

Solution Figure 27 shows the solid S, which is bowl-shaped. The two curves intersect when $x^2 = x^3$, that is, when $x = 0$ or $x = 1$. The only finite region between the curves lies over the interval $[0, 1]$, and on this interval $y = x^2$ is the upper curve and $y = x^3$ the lower curve. Thus we choose $a = 0$, $b = 1$, $f(x) = x^2$ and $g(x) = x^3$ in formula (8), obtaining

$$V = 2\pi \int_0^1 x(x^2 - x^3) dx = 2\pi \int_0^1 (x^3 - x^4) dx$$

$$= 2\pi \left[\frac{x^4}{4} - \frac{x^5}{5}\right]_0^1 = \frac{\pi}{10}. \quad \square$$

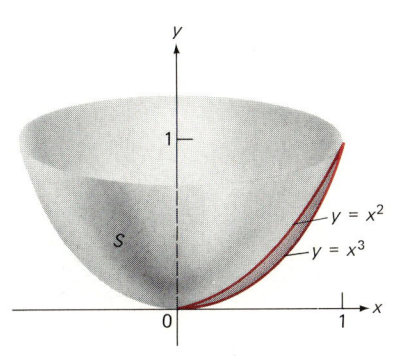

Example 7 The unbounded region in the first quadrant between the x-axis and the curve

$$y = f(x) = \frac{1}{x^4 + 1} \tag{9}$$

is revolved about the y-axis. Find the volume V of the resulting unbounded solid of revolution S.

Solution Figure 28 shows the solid S, which is mound-shaped. By the natural extension of formula (7) for the case of an unbounded interval of integration, we express V as an improper integral:

$$V = 2\pi \int_0^\infty xf(x) dx = 2\pi \int_0^\infty \frac{x}{x^4 + 1} dx.$$

With the help of the substitution $t = x^2$, we then obtain

$$V = 2\pi \lim_{u \to \infty} \int_0^u \frac{x}{x^4 + 1} dx = \pi \lim_{u \to \infty} \int_0^{u^2} \frac{dt}{t^2 + 1}$$

$$= \pi \lim_{u \to \infty} \left[\arctan t\right]_0^{u^2} = \pi \lim_{u \to \infty} \arctan u^2 = \frac{\pi^2}{2}.$$

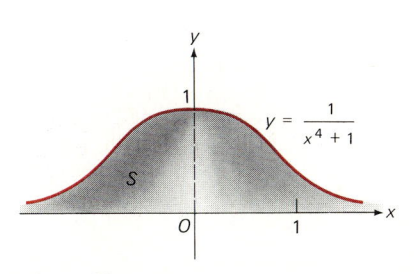

As an exercise, show that V is infinite if the denominator of (9) is $x^2 + 1$ instead of $x^4 + 1$. □

Example 8 The unbounded region under the curve $y = f(x) = x^{-3/2}$ from $x = 0$ to $x = 1$ is revolved about the y-axis. Find the volume V of the resulting unbounded solid of revolution.

Section 8.2 Volume of a Solid of Revolution **449**

Solution This time we have

$$V = 2\pi \int_0^1 xf(x)\,dx = 2\pi \int_0^1 x(x^{-3/2})\,dx = 2\pi \int_0^1 \frac{dx}{\sqrt{x}},$$

where the improper integral on the right is equal to 2, as shown in Example 5, page 428. It follows that

$$V = 2\pi(2) = 4\pi. \quad \square$$

Problems

Use any method (disks, washers or shells) to find the volume of the solid generated by revolving the region under the given curve and over the given interval about the specified axis.

1. $y = x^2$, $-2 \le x \le 1$, x-axis
2. $y = 4 - x^2$, $0 \le x \le 2$, y-axis
3. $y = x^3$, $-1 \le x \le 1$, y-axis
4. $y = |x|$, $-1 \le x \le 3$, x-axis
5. $y = \sqrt{25 - x^2}$, $3 \le x \le 4$, y-axis
6. $y = \cos x$, $0 \le x \le \pi/2$, y-axis
7. $y = \sec x$, $0 \le x \le \pi/3$, x-axis
8. $y = \ln x$, $1 \le x \le e$, x-axis
9. $y = \sinh x$, $-1 \le x \le 1$, x-axis
10. $y = \cosh x$, $0 \le x \le 1$, y-axis
11. $y = e^{-x}$, $0 \le x \le 1$, y-axis
12. $y = \arcsin x$, $0 \le x \le 1$, x-axis
13. $y = x^{-4/3}$, $0 < x \le 1$, y-axis
14. $y = xe^{-x}$, $0 \le x < \infty$, x-axis
15. $y = e^{-x^2}$, $0 \le x < \infty$, y-axis
16. $y = (x^2 + 1)^{-3/2}$, $0 \le x < \infty$, x-axis

17. Use the shell method to find the volume V of a solid sphere of radius r.
18. Solve Example 4 by the shell method.
19. Solve Example 6 by the washer method.
20. Solve Example 7 by the disk method.

Calculate the volume of the solid generated by revolving the given region OAB or OBC in Figure 29 about the specified axis. The curve OB is the graph of the function $y = \sqrt{x}$ over the interval $0 \le x \le 4$. In each case use both the disk or washer method and the shell method.

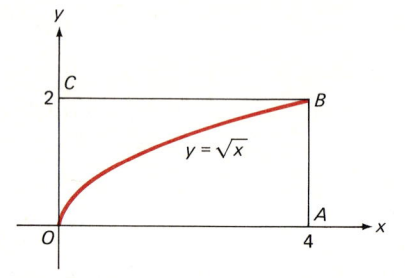

Figure 29

21. OAB about the x-axis
22. OBC about the x-axis
23. OBC about the y-axis
24. OAB about the y-axis
25. OAB about the line $x = 4$
26. OBC about the line $x = 4$
27. OBC about the line $y = 2$
28. OAB about the line $y = 2$

The parabola $y = x^2$ divides the triangle with vertices $(0, 0)$, $(2, 0)$ and $(0, 2)$ into two regions, one above the parabola and the other below the parabola. Find the volume of the solid generated by revolving each region about

29. The x-axis
30. The y-axis

Find the volume of the solid generated by revolving the triangle with vertices $(1, 0)$, $(3, 1)$, $(2, 2)$ about

31. The x-axis
32. The y-axis

33. Find the volume of the solid generated by revolving the region between the curves $y = \sqrt{x}$ and $y = x^2$ about the line $y = x$.

8.3 Curves in Parametric Form

So far the word "curve" has meant the graph of a continuous function $y = f(x)$, with x as the independent variable and y as the dependent variable, and on occasion it has also meant the graph of a continuous function $x = g(y)$, with y as the independent variable and x as the dependent variable. No vertical line can intersect the graph of a curve of the form $y = f(x)$ in more than one point, and similarly no horizontal line can intersect the graph of a curve of the form $x = g(y)$ in more than one point; this is an immediate consequence of the fact that f and g are functions. We now want to free ourselves from this restriction, so that, for example, the drawing labeled C in Figure 30 (which violates both the vertical and the horizontal line properties) is nevertheless a curve, as our intuition demands.

Parametric Equations of a Curve

This is easily accomplished by treating the x-coordinate and the y-coordinate of a variable point of C in exactly the same way, making them *both* functions of a new variable t, called a *parameter*. Thus we henceforth define a *curve* (*in parametric form*) to be the graph of a *pair* of (parametric) equations

$$x = x(t), \qquad y = y(t), \tag{1}$$

that is, the set of all points (x, y) whose coordinates x and y satisfy (1); here $x(t)$ and $y(t)$ are two continuous functions with the same domain of definition, which we will always take to be some interval I. In dealing with parametric equations, we use the same symbols x and y to denote both dependent variables and functions. We also assume that $x(t)$ and $y(t)$ are not both constant functions, since otherwise the curve (1) would reduce to a single point. As the parameter t, which may be thought of as the time, varies over the interval I, the point $P = (x, y)$ takes various positions in the xy-plane, and traces out the (parametric) curve (1). This is suggested in Figure 30, where positions of P for various values of t are indicated. It is now quite possible for a curve to intersect itself, as at the point Q in the figure, for it is only necessary that there be two different values t_1 and t_2 of the parameter such that $x(t_1) = x(t_2)$ and $y(t_1) = y(t_2)$. (In the figure these values are $t_1 = 1$ and $t_2 = 7$.) By an *arc* of the curve (1) we mean any curve with the same parametric equations, but where the domain of $x(t)$ and $y(t)$ is some subinterval of I.

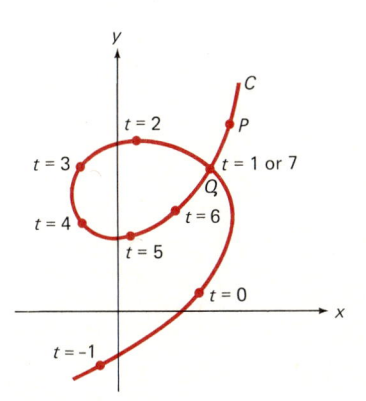

Figure 30

Example 1 The graph of a continuous function

$$y = f(x) \qquad (a \le x \le b)$$

can be represented in parametric form by identifying x with the parameter t and writing

$$x = t, \qquad y = f(t) \qquad (a \le t \le b).$$

(Here we have chosen the domain of f to be a closed interval, but any other kind of interval, finite or infinite, would do just as well.) To represent the graph of a continuous function

$$x = g(y) \qquad (a \le y \le b)$$

in parametric form, we write

$$x = g(t), \qquad y = t \qquad (a \le t \le b).$$

Thus our new definition of a curve has the desirable property of including as a special case whatever was previously called a curve.

Section 8.3 Curves in Parametric Form

Example 2 The curve
$$x = 1 + 2t, \qquad y = -1 + t \qquad (-\infty < t < \infty) \qquad (2)$$
is the straight line through the point $(1, -1)$ with slope $\frac{1}{2}$. In fact, solving the first of the equations (2) for the parameter t and substituting the result in the second, we get the single equation
$$y = -1 + t = -1 + \frac{1}{2}(x - 1), \qquad (2')$$
whose graph is clearly the line of slope $\frac{1}{2}$ going through the point $(1, -1)$. Examining (2), we find at once that the point $(1, -1)$ corresponds to the parameter value $t = 0$. The same line is also represented by the parametric equations
$$x = 1 + 2t^3, \qquad y = -1 + t^3 \qquad (-\infty < t < \infty),$$
since as t increases from $-\infty$ to ∞, so does t^3, or by
$$x = 1 + 2\tan t, \qquad y = -1 + \tan t \qquad (-\pi/2 < t < \pi/2),$$
since $\tan t$ increases from $-\infty$ to ∞ as t increases from $-\pi/2$ to $\pi/2$.

Remark There are infinitely many continuous functions $f(t)$ which increase from $-\infty$ to ∞ as t increases on some interval I. Thus the straight line (2) has infinitely many parametric representations, of the form
$$x = 1 + 2f(t), \qquad y = -1 + f(t) \qquad (t \text{ in } I).$$
A similar argument shows that every curve has infinitely many parametric representations.

Orientation of a Curve

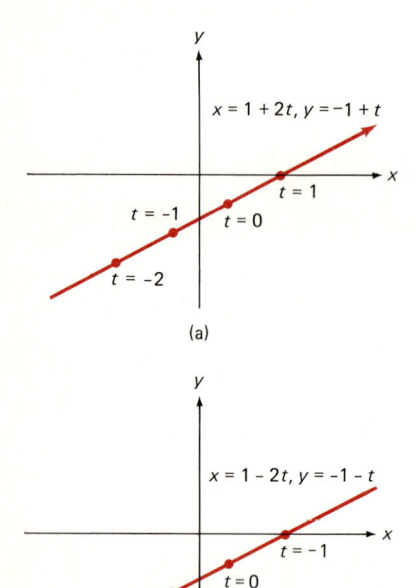

Curves in parametric form have a natural *orientation* or direction of traversal, associated with the direction of increasing t. For example, as t increases from $-\infty$ to ∞, a variable point $P = (x, y)$ on the line (2) traverses the line upward from left to right [see Figure 31(a)]. However, as t increases from $-\infty$ to ∞, the line
$$x = 1 - 2t, \qquad y = -1 - t \qquad (-\infty < t < \infty),$$
which consists of the same *points* as the line (2), is traversed in the opposite direction, that is, downward from right to left [see Figure 31(b)]. As an exercise, show that the equations
$$x = 1 + 2\ln t, \qquad y = -1 + \ln t \qquad (0 < t < \infty)$$
represent the same line L as (2), with the same orientation, while the equations
$$x = 1 + 2\cot t, \qquad y = -1 + 2\cot t \qquad (0 < t < \pi)$$
represent the line L traversed in the opposite direction.

Example 3 Since t^2 is inherently nonnegative, the curve
$$x = 1 + 2t^2, \qquad y = -1 + t^2 \qquad (-\infty < t < \infty) \qquad (3)$$
is not the same as the line (2), but is instead a *ray* or *half-line* consisting of

Figure 31

only part of the line (2), as shown in Figure 32. Let t increase from $-\infty$ to ∞. Then the point $P = (x, y)$ traverses the ray (3) twice, once in the downward direction and once in the upward direction, turning at the point $(1, -1)$, which corresponds to the parameter value $t = 0$. As an exercise, show that the equations

$$x = 1 + 2e^t, \qquad y = -1 + e^t \qquad (-\infty < t < \infty)$$

represent the same ray *without the point* $(1, -1)$, traversed once in the upward direction.

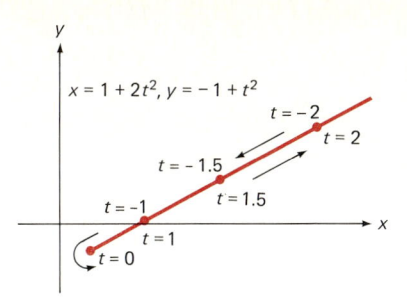

Figure 32

Example 4 More generally, the curve

$$x = x_1 + at, \qquad y = y_1 + bt \qquad (-\infty < t < \infty, a \neq 0) \tag{4}$$

is the straight line through the point (x_1, y_1) with slope b/a. In fact, solving the first equation for t and substituting the result into the second equation, we get

$$y = y_1 + \frac{b}{a}(x - x_1) = \frac{b}{a}x + \left(y_1 - \frac{b}{a}x_1\right), \tag{4'}$$

which you will recognize as the equation of a straight line L of slope b/a. Moreover, the point (x_1, y_1) lies on L, as we see by setting $t = 0$ in (4) or $x = x_1$ in (4'). Notice that the curve (4) is a *vertical* line if $a = 0$ and $b \neq 0$.

Example 5 Consider the curve

$$x = a \cos t, \qquad y = b \sin t \qquad (0 \leq t \leq 2\pi), \tag{5}$$

where $a > 0$, $b > 0$. Since $x/a = \cos t$, $y/b = \sin t$ and $\cos^2 t + \sin^2 t \equiv 1$, we can easily eliminate the parameter t from the two parametric equations (5). This gives the single equation

$$\frac{x^2}{a^2} + \frac{y^2}{b^2} = 1, \tag{5'}$$

called the *cartesian equation* of the curve because the only variables appearing in it are the cartesian or rectangular coordinates x and y. (Rectangular coordinates are also called *cartesian coordinates*, in honor of the French philosopher and mathematician René Descartes (1596–1650), the founder of analytic geometry.) The graph of (5') is a curve known as the *ellipse*, shown in Figure 33 for the case $a > b$. Every point with coordinates satisfying the parametric equations (5) also satisfies the cartesian equation (5'), and therefore lies on the ellipse. Conversely, for each point (x, y) of the ellipse (5'), there is a value of t such that x and y satisfy (5); to see this, check that the variable point $P = (x, y) = (a \cos t, b \sin t)$ actually traces out the whole ellipse once in the counterclockwise direction as t increases from 0 to 2π, starting and ending at the same point $(a, 0)$. Hence the parametric equations (5) and the cartesian equation (5') are *coextensive*, which means that the graphs of (5) and (5') coincide. (This coincidence of graphs is not automatic, as we will see in the next example.) The parametric equations (5) represent the same ellipse even if a and b are allowed to be negative, except that P traces out the ellipse in the *clockwise* direction if $ab < 0$ (why?).

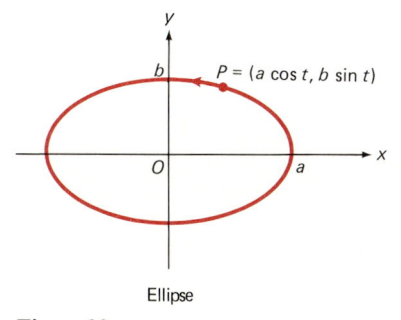

Figure 33 Ellipse

Section 8.3 Curves in Parametric Form 453

If $a = b$, (5') reduces to the equation
$$x^2 + y^2 = a^2$$
of the circle of radius a with its center at the origin, and the corresponding parametric equations are
$$x = a \cos t, \quad y = a \sin t \quad (0 \leq t \leq 2\pi).$$
Here, of course, the parameter t is the central angle of the point $P = (x, y)$, that is, the angle between the positive x-axis and the line joining the origin to P, measured in the counterclockwise direction, but this interpretation of t fails in the case of an ellipse with $a \neq b$ (explain why). If $a = 0$, $b \neq 0$, the curve (5) reduces to the vertical line segment joining the points $(0, -b)$ and $(0, b)$, while if $a \neq 0$, $b = 0$, it reduces to the horizontal line segment joining the points $(-a, 0)$ and $(a, 0)$, where both segments are traversed twice as t increases from 0 to 2π (check this). The case $a = b = 0$ is excluded, since then the curve (5) "degenerates" into the single point $(0, 0)$.

Example 6 Next consider the curve
$$x = a \cosh t, \quad y = b \sinh t \quad (-\infty < t < \infty), \tag{6}$$
where $a > 0$, $b > 0$. Since $x/a = \cosh t$, $y/b = \sinh t$ and $\cosh^2 t - \sinh^2 t \equiv 1$, we can eliminate the parameter t from the equations (6), obtaining the cartesian equation
$$\frac{x^2}{a^2} - \frac{y^2}{b^2} = 1. \tag{6'}$$
The graph of (6'), known as a *hyperbola*, is shown in Figure 34 and consists of two disjoint curves (that is, two curves with no points in common), a *left* branch lying in the left half-plane $x < 0$, and a *right* branch lying in the right half-plane $x > 0$. Since every point of the left branch has a negative x coordinate and since $a \cosh t$ is always positive if $a > 0$, it is clear that (6) cannot represent the left branch of the hyperbola, but only the right branch. Specifically, it follows from (6) and familiar properties of the functions $\cosh t$ and $\sinh t$ that as t increases from $-\infty$ to 0, x decreases from ∞ to a and y increases from $-\infty$ to 0, while as t increases from 0 to ∞, x increases from a to ∞ and y increases from 0 to ∞. Therefore as t increases from $-\infty$ to ∞, the variable point $P = (x, y) = (a \cosh t, b \sinh t)$ moves upward along the right branch of the hyperbola, and the left branch does not appear at all. Thus, unlike Example 5, here the parametric equations and the cartesian equation obtained by elimination of the parameter t are *not* coextensive.

Hyperbola

Figure 34

454   Chapter 8   Further Applications of Integration

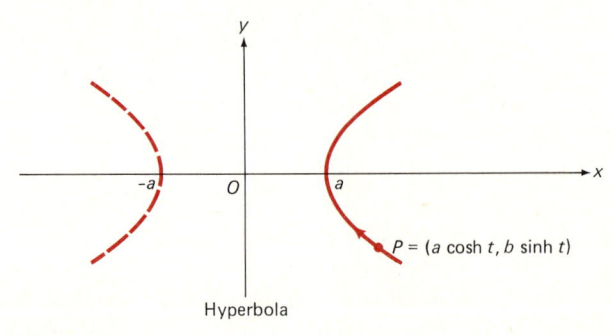

If a is positive, while b is negative, the curve (6) is again the right branch of the hyperbola, but this time traversed in the *downward* direction. To get the left branch of the hyperbola from (6), we must allow the constant a to be negative. In fact, if $a < 0$, the curve (6) becomes the left branch of the hyperbola (the dashed curve in the figure), traversed in the upward direction if $b > 0$, and in the downward direction if $b < 0$.

The discussion of the ellipse and hyperbola in Examples 5 and 6 is just a preview. In Chapter 10 we will make a much more detailed study of these important curves.

Example 7 Suppose a circle of radius a rolls without sliding along a horizontal straight line. Find the curve traced out by a fixed point P of the circumference of the circle. (You might think of P as a pebble stuck in the tread of a tire.)

Solution Let the straight line be the x-axis, and let t be the angle in radians through which the circle has rotated, starting from the position in which P coincides with the origin O. Then the curve traced out by P, called a *cycloid*, has parametric equations

$$x = a(t - \sin t), \qquad y = a(1 - \cos t) \qquad (-\infty < t < \infty). \tag{7}$$

In fact, examining Figure 35(a), we find that P has abscissa

$$x = |OQ| = |OO'| - |QO'| = |OO'| - |PP'|.$$

But $|OO'|$ equals the length of the arc $\overset{\frown}{O'P}$, since the circle rolls without sliding, so that $|OO'| = at$, while $|PP'| = a \sin t$, as can be seen from Figure 35(b), an enlargement of part of Figure 35(a). It follows that

$$x = at - a \sin t = a(t - \sin t).$$

In the same way, we have

$$y = |O'A'| - |P'A'| = a - a \cos t = a(1 - \cos t).$$

Note that this curve is the graph of a function $y = f(x)$, where f is periodic with fundamental period $2\pi a$ and has a vertical tangent (in fact, a cusp) at

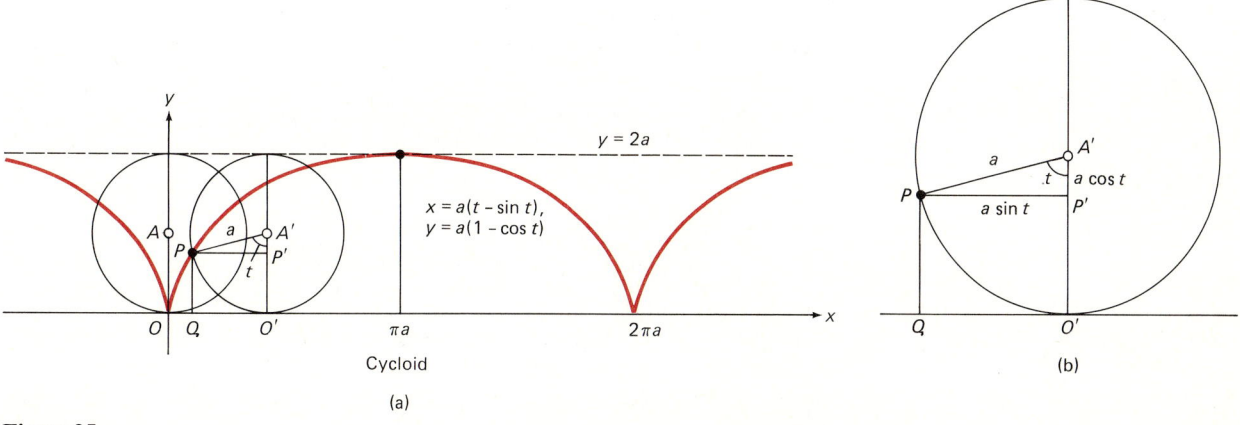

Figure 35

(a) Cycloid

(b)

every point
$$x = 2n\pi a \quad (n = 0, \pm 1, \pm 2, \ldots).$$
Moreover, every arc of the cycloid over an interval of the form
$$(n-1)\pi \le t \le n\pi \quad (n = 0, \pm 1, \pm 2, \ldots),$$
is the graph of a function $x = g(y)$. Explain why, and in particular show that
$$x = a \arccos\left(\frac{a-y}{a}\right) - \sqrt{2ay - y^2}$$
over the interval $0 \le t \le \pi$. □

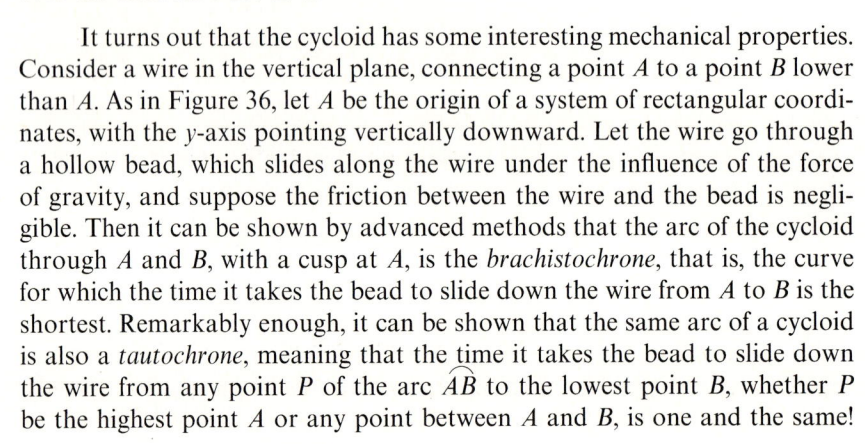

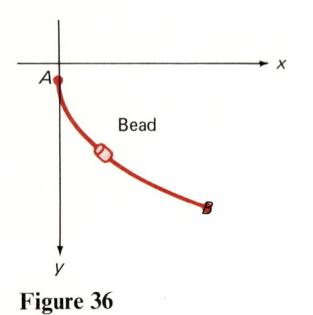

It turns out that the cycloid has some interesting mechanical properties. Consider a wire in the vertical plane, connecting a point A to a point B lower than A. As in Figure 36, let A be the origin of a system of rectangular coordinates, with the y-axis pointing vertically downward. Let the wire go through a hollow bead, which slides along the wire under the influence of the force of gravity, and suppose the friction between the wire and the bead is negligible. Then it can be shown by advanced methods that the arc of the cycloid through A and B, with a cusp at A, is the *brachistochrone*, that is, the curve for which the time it takes the bead to slide down the wire from A to B is the shortest. Remarkably enough, it can be shown that the same arc of a cycloid is also a *tautochrone*, meaning that the time it takes the bead to slide down the wire from any point P of the arc $\widehat{AB}$ to the lowest point B, whether P be the highest point A or any point between A and B, is one and the same!

Figure 36

The Tangent Line to a Parametric Curve

Given a parametric curve, that is, a curve with parametric equations
$$x = x(t), \quad y = y(t) \quad (t \text{ in } I), \tag{8}$$
where I is some interval, let a be an interior point of I, and suppose the derivatives $x'(a)$ and $y'(a)$ both exist and are finite, with $x'(a) \ne 0$. Then the curve has a tangent (line) at the point $P = (x(a), y(a))$, and the slope of the tangent is
$$m = \left.\frac{y'(t)}{x'(t)}\right|_{t=a} = \frac{y'(a)}{x'(a)}. \tag{9}$$

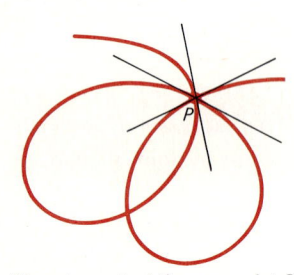

As usual, m is also called the slope of the curve itself at P; note that the slope of the curve at P may well take a value different from m if the curve passes through the same point P again for other parameter values, and at such a point the curve can have two or more tangents (Figure 37 illustrates this possibility). To prove (9), we start from the definition of m given on page 116, observing that
$$m = \lim_{t \to a} \frac{y(t) - y(a)}{x(t) - x(a)}, \tag{10}$$

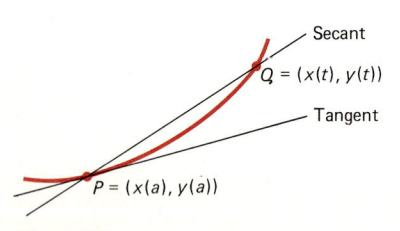

Three tangents at the same point P

Figure 37

since the quotient on the right is the slope of the secant line through the fixed point $P = (x(a), y(a))$ and a variable point $Q = (x(t), y(t))$, as in Figure 38. Here we rely on the fact that $x(t) - x(a) \ne 0$ if t is sufficiently close to a, which follows from the assumption that $x'(a) \ne 0$ (explain further). Dividing both the numerator and the denominator of the difference quotient in (10) by $t - a$,

Figure 38

456 Chapter 8 Further Applications of Integration

we find that

$$m = \lim_{t \to a} \frac{\frac{y(t) - y(a)}{t - a}}{\frac{x(t) - x(a)}{t - a}} = \frac{\lim_{t \to a} \frac{y(t) - y(a)}{t - a}}{\lim_{t \to a} \frac{x(t) - x(a)}{t - a}} = \frac{y'(a)}{x'(a)}, \tag{11}$$

which establishes (9).

We can write the equation of the tangent line to the curve (8) at the point $P = (x(a), y(a))$ either in the cartesian form

$$\frac{y - y(a)}{x - x(a)} = m = \frac{y'(a)}{x'(a)},$$

that is,

$$y'(a)[x - x(a)] - x'(a)[y - y(a)] = 0, \tag{12}$$

or, guided by Example 2, in the parametric form

$$x = x(a) + x'(a)t, \qquad y = y(a) + y'(a)t \qquad (-\infty < t < \infty). \tag{12'}$$

Example 8 Find the tangent to the curve

$$x = t^3, \qquad y = t^2 + 1 \qquad (-\infty < t < \infty)$$

at the point $P = (-1, 2)$ and at the point $Q = (0, 1)$.

Solution Since t^3 is an increasing function, the curve has no points of self-intersection and hence no multiple tangents. The points P and Q correspond to the parameter values $t = -1$ and $t = 0$, respectively. We have $x'(t) = 3t^2$, $y'(t) = 2t$, and hence $x'(-1) = 3$, $y'(-1) = -2$. Therefore, by (9) and (12), the tangent at P has slope $m = -2/3$ and equation

$$2x + 3y - 4 = 0,$$

as in Figure 39. We cannot use formula (11) at Q, since $x'(0) = y'(0) = 0$, and hence we apply formula (10) directly, obtaining

$$m = \lim_{t \to 0} \frac{y(t) - y(0)}{x(t) - x(0)} = \lim_{t \to 0} \frac{(t^2 + 1) - 1}{t^3} = \lim_{t \to 0} \frac{1}{t}.$$

Thus the limiting value of the difference quotient is either ∞ or $-\infty$, depending on whether t approaches 0 from the right or from the left. Therefore, by the same argument as on page 194, the curve has a *vertical tangent* (namely the y-axis) at Q. In fact, the curve has a *cusp* at Q, as is apparent from the figure. □

Example 9 Find the tangents to the curve

$$x = t^2 - 2t + 2, \qquad y = t^3 - 3t^2 + 2t + 3 \qquad (-\infty < t < \infty)$$

at the point $P = (2, 3)$.

Solution The point P corresponds to *two* parameter values, namely $t = 0$ and $t = 2$ (check this). Thus the curve goes through P twice, and there are two tangents at P. Since $x'(t) = 2t - 2$, $y'(t) = 3t^2 - 6t + 2$, we have $x'(0) = -2$, $y'(0) = 2$ and $x'(2) = 2$, $y'(2) = 2$. Therefore, by (9), one tangent has slope -1 and the other has slope 1. By (12), the tangents have the equations $x + y - 5 = 0$ and $x - y + 1 = 0$, as in Figure 40. As an exercise, use (12') to write parametric equations for the tangents. □

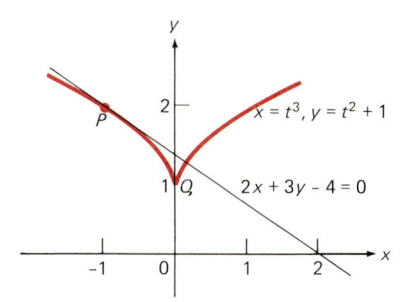

Figure 39

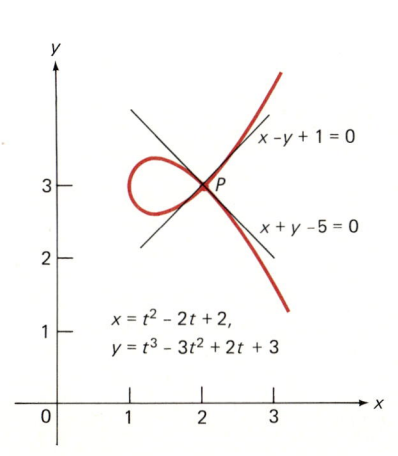

Figure 40

Section 8.3 Curves in Parametric Form 457

Problems

Find a function or a cartesian equation of which the given parametric curve is the graph. Sketch the curve, and indicate its orientation.

1. $x = 1 + 2t$, $y = 4t + 3$ $\quad (-1 \leq t \leq 0)$
2. $x = -1 + 3t$, $y = 2 - 2t$ $\quad (0 \leq t \leq 1)$
3. $x = t^4$, $y = t^2$ $\quad (0 \leq t < \infty)$
4. $x = t^3$, $y = 6 \ln t$ $\quad (0 < t < \infty)$
5. $x = e^{-t}$, $y = e^{2t}$ $\quad (-\infty < t < \infty)$
6. $x = t$, $y = \sqrt{t^2 - 9}$ $\quad (3 \leq t < \infty)$
7. $x = \sqrt{4 - t^2}$, $y = t$ $\quad (-2 \leq t \leq 2)$
8. $x = \cos t$, $y = \cos 2t$ $\quad (0 \leq t \leq \pi)$
9. $x = \sec t$, $y = \cos t$ $\quad (\pi/2 < t \leq \pi)$
10. $x = \tan t$, $y = \sec t$ $\quad (-\pi/4 \leq t \leq \pi/4)$
11. $x = 2 \cos t$, $y = 2 \sin t$ $\quad (0 \leq t \leq 2\pi)$
12. $x = 2 - 3 \sin t$, $y = -1 + 3 \cos t$ $\quad (2\pi \leq t \leq 4\pi)$
13. $x = 5 \cos t$, $y = 4 \sin t$ $\quad (0 \leq t \leq 2\pi)$
14. $x = 2 \sin t$, $y = 3 \cos t$ $\quad (0 \leq t \leq 2\pi)$
15. $x = 3 \cosh t$, $y = 4 \sinh t$ $\quad (-\infty < t < \infty)$
16. $x = -2 \sinh t$, $y = 3 \cosh t$ $\quad (-\infty < t < \infty)$

Write parametric equations for the line
17. With slope -1 and y-intercept 4
18. With slope 2 and x-intercept -3
19. With x-intercept 2 and y-intercept 6
20. Through the point $(3, -4)$ with slope 5
21. Through the points $(-3, 2)$ and $(4, 7)$
22. Through the points $(1, 8)$ and $(9, -2)$

Write parametric equations for
23. The circle of radius 5 with center at the point $(-2, 3)$, traversed in the counterclockwise direction
24. The circle of diameter 6 with center at the point $(4, -8)$, traversed in the clockwise direction
25. The ellipse $9x^2 + 4y^2 = 36$, traversed in the clockwise direction
26. The ellipse $2x^2 + 5y^2 = 4$, traversed in the counterclockwise direction
27. The left branch of the hyperbola $16x^2 - 9y^2 = 144$, traversed in the upward direction
28. The right branch of the hyperbola $3x^2 - 2y^2 = 6$, traversed in the downward direction
29. The curve $x^{2/3} + y^{2/3} = 1$, traversed from the point $(1, 0)$ to the point $(0, 1)$

30. Describe the curve with parametric equations

$$x(t) = \begin{cases} t & \text{if } 0 \leq t \leq 1, \\ 1 & \text{if } 1 \leq t \leq 2, \\ 3-t & \text{if } 2 \leq t \leq 3, \\ 0 & \text{if } 3 \leq t \leq 4, \end{cases} \quad y(t) = \begin{cases} 0 & \text{if } 0 \leq t \leq 1, \\ t-1 & \text{if } 1 \leq t \leq 2, \\ 1 & \text{if } 2 \leq t \leq 3, \\ 4-t & \text{if } 3 \leq t \leq 4. \end{cases}$$

Find the slope m of the given curve at the point corresponding to the specified value of the parameter t.
31. $x = t^2$, $y = 2t$, $t = 4$
32. $x = \sqrt{t}$, $y = t^2 + 1$, $t = 9$
33. $x = t^3$, $y = \ln (\ln t)$, $t = e$
34. $x = \cos t$, $y = t + \sin t$, $t = \pi/3$
35. $x = 2 \cos^2 t$, $y = 3 \sin t$, $t = \pi/6$
36. $x = e^t \sin t$, $y = e^t \cos t$, $t = \pi$

37. Find a parametric representation of the hyperbola (6') which does not involve hyperbolic functions.

Find the tangent(s) to the curve
38. $x = 4 \cos t$, $y = 2\sqrt{3} \sin t$ at $(2, 3)$
39. $x = t - t^4$, $y = t^2 - t^3$ at $(0, 0)$
40. $x = t^3 + 1$, $y = t^2 + t + 1$ at $(0, 1)$
41. $x = \cosh t$, $y = 2 \sinh t$ at $(\frac{5}{4}, \frac{3}{2})$

42. Let
$$x = x(t), \quad y = y(t) \quad (t \text{ in } I)$$
be a parametric curve, and let a be an interior point of the interval I. Suppose $x(t)$, $y(t)$ have finite derivatives $x'(t)$, $y'(t)$ in a neighborhood of a, where $x'(a) \neq 0$, and finite second derivatives $x''(a)$, $y''(a)$ at the point a itself. Use the chain rule and Theorem 4, page 286, to show that

$$\left.\frac{d^2y}{dx^2}\right|_{x=x(a)} = \frac{x'(a)y''(a) - y'(a)x''(a)}{[x'(a)]^3}. \quad \text{(i)}$$

Find d^2y/dx^2 at $x = x(a)$ if
43. $x = 2t^2$, $y = 3t^3$, $a = 1$
44. $x = 5 \cos t$, $y = 4 \sin t$, $a = \pi/6$
45. $x = t \cos t$, $y = t \sin t$, $a = 0$
46. $x = \cos^3 t$, $y = \sin^3 t$, $a = \pi/4$

Where does the given curve have vertical tangents?
47. $x = -1 + 2 \cos t$, $y = 1 - 2 \sin t$ $\quad (0 \leq t \leq 2\pi)$
48. $x = -3 \cos t$, $y = 4 \sin t$ $\quad (0 \leq t \leq 2\pi)$
49. $x = 2 \cosh t$, $y = 3 \sinh t$ $\quad (-\infty < t < \infty)$
50. $x = t^4 - 2t^2$, $y = t^3 + 1$ $\quad (-\infty < t < \infty)$

51. Show that if $x'(a) = y'(a) = 0$, the curve $x = x(t)$, $y = y(t)$ may or may not have a vertical tangent at the point $P = (x(a), y(a))$.

52. The curve
$$x = \frac{3at}{t^3 + 1}, \quad y = \frac{3at^2}{t^3 + 1} \quad (-\infty < t < \infty, a > 0),$$

shown in Figure 41 for the case $a = 1$, is called the *folium of Descartes*. Find a cartesian equation for the folium. What values of the parameter t give the part of the curve in the first quadrant? In the second quadrant? In the fourth quadrant? Show that the folium has both a horizontal and a vertical tangent at the origin.

53. Show that a variable point $P = (x, y)$ of the curve

$$x = \frac{1}{t^2}(1 + \ln t), \qquad y = \frac{1}{t}(3 + 2\ln t) \qquad (0 < t < \infty)$$

satisfies the differential equation

$$y\frac{dy}{dx} = 1 + 2x\left(\frac{dy}{dx}\right)^2.$$

54. Show that the tangent and normal to the cycloid (7) at any point P go through the highest and lowest points of the circle generating the cycloid (with P on its circumference).

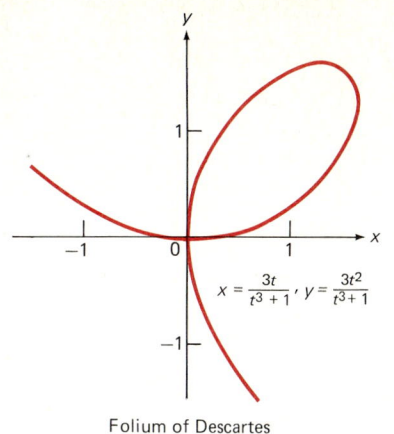

Figure 41

$$x = \frac{3t}{t^3+1}, \quad y = \frac{3t^2}{t^3+1}$$

Folium of Descartes

55. Find the three tangents to the curve

$$x = t^3 - 3t^2 + 2t + 1, \quad y = t^6 - 5t^4 + 4t^2 + 2 \quad (-\infty < t < \infty)$$

at the point (1, 2).

8.4 Length of a Plane Curve

We now consider the problem of finding the length of a plane curve C with parametric equations

$$x = x(t), \qquad y = y(t) \qquad (a \le t \le b),$$

where C has no more than a finite number of self-intersections, which in particular implies that no arc of C is traced out more than once. The point $A = (x(a), y(a))$ is called the *initial point* and the point $B = (x(b), y(b))$ the *final point* of C, since as t increases from a to b, the moving point $P = (x(t), y(t))$ traces out C, starting at A and ending at B. The points A and B are called the *endpoints* of C. If A and B coincide, the curve C is said to be *closed*. A curve with no points of self-intersection, except for the possibility of coincident endpoints, is said to be *simple*. Figure 42(a) shows a simple curve with distinct endpoints, and Figure 42(b) shows a simple curve with coincident endpoints, that is, a *simple closed* curve. The curve shown in Figure 42(c), which has three points of self-intersection, is neither simple nor closed.

Remark According to the celebrated *Jordan curve theorem*, every simple closed curve C divides the plane into two distinct regions with C as their common boundary, where one region, called the *interior of C*, is bounded, and the other region, called the *exterior of C*, is unbounded. Although this result seems geometrically obvious, it is actually quite hard to *prove*! Remember that we are assuming only that the functions $x(t)$ and $y(t)$ are continuous, which allows them very great freedom of behavior.

In elementary geometry, length is defined only for line segments and circular arcs (and curves made up of such segments and arcs). Thus our first task is to *define* what is meant by the length of C, just as we had to define

A simple curve
(a)

A simple closed curve
(b)

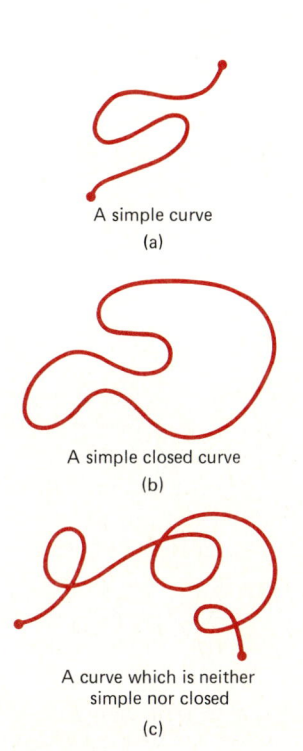

A curve which is neither simple nor closed
(c)

Figure 42

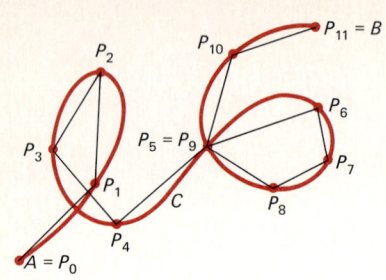

Figure 43
A polygonal path inscribed in a curve

the area of the region between two curves or the volume of a solid. It is natural to try to approximate C by a curve whose length we already know. With this aim, we begin by dividing the interval $[a, b]$ into n subintervals by introducing parameter values $a = t_0, t_1, t_2, \ldots, t_{n-1}, t_n = b$, which satisfy the inequalities $a = t_0 < t_1 < t_2 < \cdots < t_{n-1} < t_n = b$. These values of the parameter t determine $n + 1$ points

$$P_i = (x(t_i), y(t_i)) \qquad (i = 0, 1, \ldots, n)$$

on the curve C, dividing C into n arcs, where P_0 is the initial point A and P_n the final point B (see Figure 43 for the case $n = 11$). Since a finite number of points of self-intersection are allowed, some of the points $P_0, P_1, \ldots, P_n$ may coincide, although they correspond to different values of t. For example, the points P_5 and P_9 coincide for the curve shown in the figure.

Polygonal Paths and Rectifiability

Now suppose we join each of the points $P_0, P_1, P_2, \ldots, P_n$ to the next by a line segment, as shown in the figure. Then C is approximated by a *polygonal path* $P_0P_1P_2 \ldots P_{n-1}P_n$, said to be *inscribed* in C, which consists of the n line segments $P_0P_1, P_1P_2, \ldots, P_{n-1}P_n$ joined end to end. (The points $P_0, P_1, P_2, \ldots, P_n$ are called the *vertices* of $P_0P_1P_2 \ldots P_n$.) The length of this polygonal path is of course just the sum of the lengths of its segments, that is, the path is of length

$$\sum_{i=1}^{n} |P_{i-1}P_i|. \tag{1}$$

Let

$$\mu = \max \{\Delta t_1, \Delta t_2, \ldots, \Delta t_n\},$$

where $\Delta t_i = t_i - t_{i-1}$ is the length of the ith subinterval $[t_{i-1}, t_i]$. Then it seems reasonable to regard (1) as a good approximation to the length of C, where the approximation gets better and better as μ gets smaller and smaller. This suggests *defining* the length L of the curve C as the limit

$$L = \lim_{\mu \to 0} \sum_{i=1}^{n} |P_{i-1}P_i|,$$

provided that the limit exists and is finite, in which case C is said to be *rectifiable*.

Length of a Curve as an Integral

It turns out that continuity of the functions $x(t)$ and $y(t)$ is not enough to guarantee rectifiability of the curve C. In other words, there are curves $x = x(t), y = y(t)$ ($a \le t \le b$), with continuous $x(t)$ and $y(t)$, in which we can inscribe polygonal paths of arbitrarily great length. (An example of such a nonrectifiable curve is given in Problem 23.) However, suppose the functions $x(t)$ and $y(t)$ are *continuously differentiable* on $[a, b]$, which means that the derivatives $x'(t)$ and $y'(t)$ exist and are continuous on $[a, b]$.† Then C is recti-

---

† The derivatives of $x(t)$ and $y(t)$ at the endpoints a and b should be interpreted as the right-hand derivatives $x'_+(a), y'_+(a)$ and left-hand derivatives $x'_-(b), y'_-(b)$. It is then unnecessary to assume that $x(t)$ and $y(t)$ are defined outside the interval $[a, b]$.

fiable, with length L given by

$$L = \int_a^b \sqrt{[x'(t)]^2 + [y'(t)]^2}\, dt = \int_a^b \sqrt{\left(\frac{dx}{dt}\right)^2 + \left(\frac{dy}{dt}\right)^2}\, dt. \qquad (2)$$

To see this, we first use the formula for the distance between two points to express the length of the inscribed polygonal path in the form

$$\sum_{i=1}^n |P_{i-1}P_i| = \sum_{i=1}^n \sqrt{[x(t_i) - x(t_{i-1})]^2 + [y(t_i) - y(t_{i-1})]^2}.$$

We then apply the mean value theorem for derivatives (Theorem 2, page 162) to each of the coordinate differences $x(t_i) - x(t_{i-1})$, $y(t_i) - y(t_{i-1})$, obtaining

$$x(t_i) - x(t_{i-1}) = x'(u_i)\,\Delta t_i \qquad (t_{i-1} < u_i < t_i),$$
$$y(t_i) - y(t_{i-1}) = y'(v_i)\,\Delta t_i \qquad (t_{i-1} < v_i < t_i).$$

This enables us to write

$$\sum_{i=1}^n |P_{i-1}P_i| = \sum_{i=1}^n \sqrt{[x'(u_i)]^2 + [y'(v_i)]^2}\,\Delta t_i. \qquad (3)$$

If u_i and v_i were equal for every i, the expression on the right would be a Riemann sum for the continuous function $\sqrt{[x'(t)]^2 + [y'(t)]^2}$ on the interval $[a, b]$, and we would then have

$$L = \lim_{\mu \to 0} \sum_{i=1}^n \sqrt{[x'(u_i)]^2 + [y'(u_i)]^2}\,\Delta t_i = \int_a^b \sqrt{[x'(t)]^2 + [y'(t)]^2}\, dt,$$

thereby proving formula (2). In general $u_i \neq v_i$, but it seems intuitively clear that as $\mu \to 0$, the right side of (3) must still approach the same limit

$$\int_a^b \sqrt{[x'(t)]^2 + [y'(t)]^2}\, dt = \int_a^b \sqrt{\left(\frac{dx}{dt}\right)^2 + \left(\frac{dy}{dt}\right)^2}\, dt$$

(after all, $\mu \to 0$ does imply $u_i - v_i \to 0$ for every i). This is in fact true, as can be proved rigorously by an argument involving a concept (uniform continuity) that lies beyond the scope of a first course in calculus. Thus we will henceforth regard formula (2) as established.

Example 1 Find the length (called the circumference) of the circle

$$x = a \cos t, \qquad y = a \sin t \qquad (0 \leq t \leq 2\pi)$$

of radius a.

Solution Here

$$\frac{dx}{dt} = -a \sin t, \qquad \frac{dy}{dt} = a \cos t,$$

and formula (2) gives

$$L = \int_0^{2\pi} \sqrt{(-a \sin t)^2 + (a \cos t)^2}\, dt = a \int_0^{2\pi} \sqrt{\sin^2 t + \cos^2 t}\, dt$$

$$= a \int_0^{2\pi} dt = 2\pi a,$$

as we already know from elementary geometry. □

Example 2 Find the length L of one arch of the cycloid shown in Figure 35(a), page 455.

Solution The arch has parametric equations

$$x = a(t - \sin t), \qquad y = a(1 - \cos t) \qquad (0 \leq t \leq 2\pi, a > 0),$$

so that

$$\frac{dx}{dt} = a(1 - \cos t), \qquad \frac{dy}{dt} = a \sin t,$$

and

$$\left(\frac{dx}{dt}\right)^2 + \left(\frac{dy}{dt}\right)^2 = a^2(1 - \cos t)^2 + a^2 \sin^2 t = a^2(2 - 2\cos t)$$

$$= 2a^2\left(1 - \cos^2 \frac{t}{2} + \sin^2 \frac{t}{2}\right) = 4a^2 \sin^2 \frac{t}{2}.$$

Since $a > 0$ and $\sin(t/2) \geq 0$ for $0 \leq t \leq 2\pi$, it follows from (2) that

$$L = \int_0^{2\pi} \sqrt{4a^2 \sin^2 \frac{t}{2}}\, dt = 2a \int_0^{2\pi} \sin \frac{t}{2}\, dt$$

$$= 2a\left[-2\cos \frac{t}{2}\right]_0^{2\pi} = 4a(-\cos \pi + \cos 0) = 8a.$$

This is $8/2\pi \approx 1.27$ times larger than the circumference $2\pi a$ of the circle of radius a figuring in the construction of the cycloid. Thus for every mile traveled by a car, a pebble stuck in the tread of one of its tires travels an extra distance of about $[(4/\pi) - 1](5280) \approx 1443$ ft, regardless of the tire size. $\square$

Suppose C is the graph of a continuously differentiable function

$$y = f(x) \qquad (a \leq x \leq b). \tag{4}$$

Then C has the parametric representation

$$x = t, \qquad y = f(t) \qquad (a \leq t \leq b), \tag{4'}$$

and (2) takes the form

$$L = \int_a^b \sqrt{1 + [f'(t)]^2}\, dt = \int_a^b \sqrt{1 + \left(\frac{dy}{dt}\right)^2}\, dt,$$

or equivalently

$$L = \int_a^b \sqrt{1 + [f'(x)]^2}\, dx = \int_a^b \sqrt{1 + \left(\frac{dy}{dx}\right)^2}\, dx, \tag{5}$$

after changing the variable of integration back from t to x.

Example 3 If C is the semicircle

$$y = \sqrt{a^2 - x^2} \qquad (-a \leq x \leq a),$$

then

$$\frac{dy}{dx} = -\frac{x}{\sqrt{a^2 - x^2}},$$

462   Chapter 8   Further Applications of Integration

and (5) gives

$$L = \int_{-a}^{a} \sqrt{1 + \frac{x^2}{a^2 - x^2}}\, dx = a \int_{-a}^{a} \frac{dx}{\sqrt{a^2 - x^2}},$$

where we are now dealing with an improper integral! Although the integral is easily evaluated and turns out to be π (show this), it would have been much simpler to represent C by the parametric equations

$$x = a \cos t, \quad y = a \sin t \quad (0 \le t \le \pi),$$

for then we immediately get

$$L = a \int_0^{\pi} \sqrt{\sin^2 t + \cos^2 t}\, dt = a \int_0^{\pi} dt = \pi a$$

(compare with Example 1).

Actually, if C is a curve of the form (4), where $f(x)$ is continuously differentiable on $[a, b]$, we can give a direct proof of formula (5) without using formula (2) and the parametric representation (4'). In fact, let

$$a = x_0 < x_1 < x_2 < \cdots < x_{n-1} < x_n = b,$$
$$\mu = \max \{\Delta x_1, \Delta x_2, \ldots, \Delta x_n\} \quad (\Delta x_i = x_i - x_{i-1}),$$

and let $P_i = (x_i, f(x_i))$, so that the polygonal path $P_0 P_1 P_2 \ldots P_{n-1} P_n$ is inscribed in C. Then

$$\sum_{i=1}^{n} |P_{i-1} P_i| = \sum_{i=1}^{n} \sqrt{\Delta x_i^2 + [f(x_i) - f(x_{i-1})]^2},$$

where

$$f(x_i) - f(x_{i-1}) = f'(u_i)\, \Delta x_i \quad (x_{i-1} < u_i < x_i),$$

by the mean value theorem. Therefore

$$\sum_{i=1}^{n} |P_{i-1} P_i| = \sum_{i=1}^{n} \sqrt{1 + [f'(u_i)]^2}\, \Delta x_i$$

is a *true* Riemann sum for the continuous function $\sqrt{1 + [f'(x)]^2}$ on the interval $[a, b]$, and hence the length of the curve C is

$$L = \lim_{\mu \to 0} \sum_{i=1}^{n} |P_{i-1} P_i| = \int_a^b \sqrt{1 + [f'(x)]^2}\, dx = \int_a^b \sqrt{1 + \left(\frac{dy}{dx}\right)^2}\, dx.$$

Thus we have established formula (5) directly and rigorously, avoiding the "u_i, v_i technicality" arising in the proof of formula (2). As an exercise, show both directly and from formula (2) that if the curve C is the graph of a continuously differentiable function

$$x = g(y) \quad (a \le y \le b),$$

then the length of C is given by the formula

$$L = \int_a^b \sqrt{[g'(y)]^2 + 1}\, dy = \int_a^b \sqrt{\left(\frac{dx}{dy}\right)^2 + 1}\, dy, \tag{5'}$$

analogous to (5).

Example 4 Find the length L of the arc of the parabola $y = \frac{1}{2}x^2$ joining the origin to the point $(1, \frac{1}{2})$.

Solution The arc in question is shown in Figure 44. Here $dy/dx = x$, and (5) gives

$$L = \int_0^1 \sqrt{1+x^2}\, dx = \left[\frac{1}{2} x\sqrt{1+x^2} + \frac{1}{2}\ln(x+\sqrt{1+x^2})\right]_0^1$$

$$= \frac{1}{2}[\sqrt{2} + \ln(1+\sqrt{2})] \approx 1.15,$$

with the help of Example 3, page 391. □

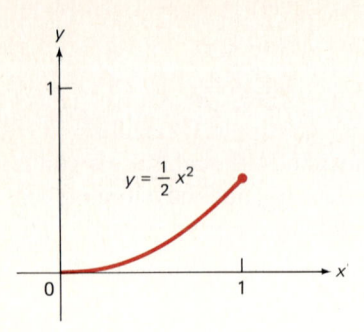

Figure 44

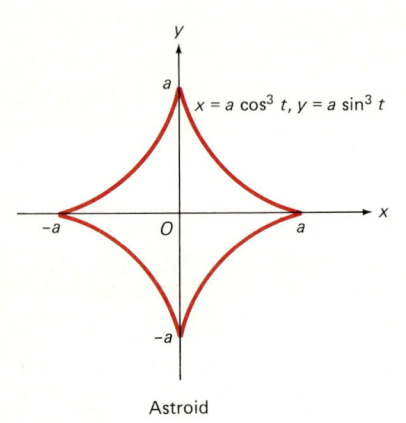

Example 5 The curve

$$x = a \cos^3 t, \qquad y = a \sin^3 t \qquad (0 \le t \le 2\pi, \ a > 0),$$

shown in Figure 45, is known as the *astroid*. Find its length L.

Solution It is apparent from the symmetry of the astroid that

$$L = 4 \int_0^{\pi/2} \sqrt{[x'(t)]^2 + [y'(t)]^2}\, dt.$$

Since

$$\frac{dx}{dt} = -3a\cos^2 t \sin t, \qquad \frac{dy}{dt} = 3a\sin^2 t \cos t,$$

we have

$$\left(\frac{dx}{dt}\right)^2 + \left(\frac{dy}{dt}\right)^2 = 9a^2(\cos^4 t \sin^2 t + \sin^4 t \cos^2 t)$$

$$= 9a^2 \cos^2 t \sin^2 t \ (\cos^2 t + \sin^2 t)$$

$$= 9a^2 \cos^2 t \sin^2 t.$$

Therefore, by (2), since $\cos t \ge 0$, $\sin t \ge 0$ for $0 \le t \le \pi/2$,

$$L = 12a \int_0^{\pi/2} \cos t \sin t \, dt$$

$$= 6a \int_0^{\pi/2} \sin 2t \, dt = -3a \cos 2t \Big|_0^{\pi/2} = 6a. \quad \square$$

Figure 45 Astroid

Problems

Find the length of the given parametric curve.
1. $x = 2t^3, y = 3t^2$ $(-1 \le t \le 1)$
2. $x = t^2, y = t^3$ $(0 \le t \le 2)$
3. $x = \ln t, y = 1/t$ $(1 \le t \le 2)$
4. $x = \cos^2 t, y = \sin^2 t$ $(-\pi/2 \le t \le 0)$
5. $x = e^t, y = e^{2t}$ $(-\infty < t \le 0)$
6. $x = 3\sin^2 t, y = 2\cos^3 t$ $(0 \le t \le \pi/2)$
7. $x = \cos t, y = t + \sin t$ $(-\pi \le t \le \pi)$
8. $x = t \cos t, y = t \sin t$ $(0 \le t \le 1)$
9. $x = e^{-t}\cos t, y = e^{-t}\sin t$ $(a \le t \le b)$
10. $x = \cos t + t\sin t, y = \sin t - t\cos t$ $(0 \le t \le 10)$

11. Verify that formula (2) gives the correct result for the length L of the line segment joining two arbitrary points $P_1 = (x_1, y_1)$ and $P_2 = (x_2, y_2)$.

12. Show that the curve $y = \cosh x$ has the property that the length of the curve over the interval $[a, b]$ equals the area under the curve from $x = a$ to $x = b$.

Find the length of the given curve (the graph of a function of x or y).

13. $x = \frac{1}{6} y^3 + \frac{1}{2y}$ $(1 \le y \le 3)$

464 Chapter 8 Further Applications of Integration

14. $y = \frac{1}{8}x^4 + \frac{1}{4x^2}$ $(1 \le x \le 2)$

15. $y = 2x^{3/2} + 3$ $(0 \le x \le 7)$

16. $x = \frac{1}{3}y^{3/2} - \sqrt{y}$ $(0 \le y \le 1)$

17. $y = \ln x$ $(\sqrt{3} \le x \le \sqrt{8})$

18. $x = \ln \cos y$ $(0 \le y \le \pi/4)$

19. $y = \ln \sin x$ $(\pi/3 \le x \le 2\pi/3)$

20. $y = \frac{1}{8}x^2 - \ln x$ $(1 \le x \le e)$

21. $x = \int_0^y \sqrt{\cosh t}\, dt$ $(0 \le y \le 2)$

22. $y = \int_{-\pi/2}^x \sqrt{\cos t}\, dt$ $(-\pi/2 \le x \le \pi/2)$

23. Verify that the graph of the continuous function

$$y = f(x) = \begin{cases} \sqrt{x} \cos \dfrac{\pi}{x} & \text{if } 0 < x \le 1, \\ 0 & \text{if } x = 0, \end{cases}$$

shown in Figure 46, is a nonrectifiable curve C.

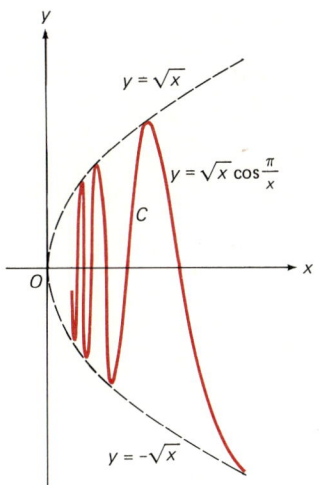

A nonrectifiable curve
Figure 46

24. Show that the astroid with parametric equations $x = a \cos^3 t$, $y = a \sin^3 t$ $(0 \le t \le 2\pi, a > 0)$ has the cartesian equation $x^{2/3} + y^{2/3} = a^{2/3}$. Find the length of the astroid by using formula (5), rather than formula (2) as in Example 5. Verify that the astroid has cusps at the four points $(\pm a, 0)$, $(0, \pm a)$, as is apparent from Figure 45.

25. Suppose a circle of radius b rolls without slipping along the outside of a fixed circle of radius $a > b$. Then a point $P = (x, y)$ on the circumference of the rolling circle traces out a curve known as the *epicycloid*. Let P be initially at the point $(a, 0)$, and let t be the angle between the positive x-axis and the line joining the centers of the two circles, as in Figure 47. Show that the epicycloid has the parametric equations

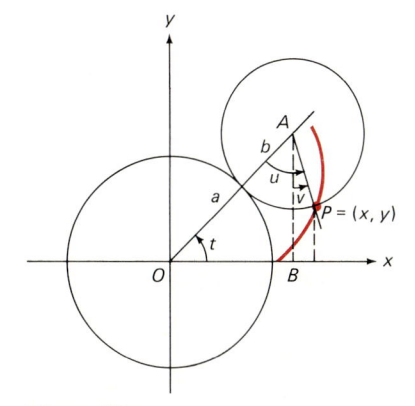

$$x = (a + b) \cos t - b \cos \frac{a + b}{b} t,$$

$$y = (a + b) \sin t - b \sin \frac{a + b}{b} t.$$
(i)

Figure 47

26. In the preceding problem, suppose the circle of radius b rolls without slipping along the *inside* of the circle of radius $a > b$. Then P traces out a curve known as the *hypocycloid*. Show that the hypocycloid has the parametric equations

$$x = (a - b) \cos t + b \cos \frac{a - b}{b} t,$$

$$y = (a - b) \sin t - b \sin \frac{a - b}{b} t.$$
(ii)

Notice that changing the sign of b has the effect of converting the equations (i) into the equations (ii).

27. If $b = a$, the epicycloid becomes the curve shown in Figure 48, which is known as the *cardioid* because it is heart-shaped. Find the length of the cardioid.

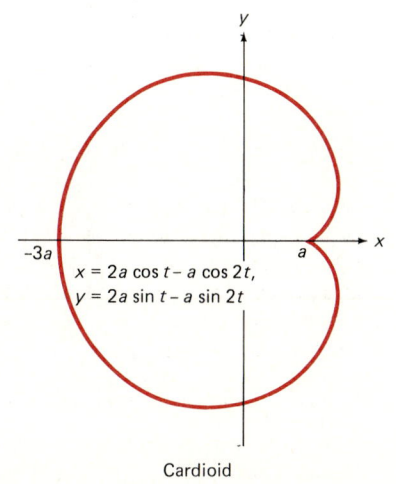

$x = 2a \cos t - a \cos 2t,$
$y = 2a \sin t - a \sin 2t$

Cardioid
Figure 48

Section 8.4 Length of a Plane Curve

28. Show that if $b = \frac{1}{4}a$, the hypocycloid becomes the astroid, investigated in Example 5 and Problem 24. (For this reason, the astroid is also known as the *hypocycloid of four cusps*.) What does the hypocycloid become if $b = \frac{1}{2}a$?

29. If a/b is an integer, what is the length of the epicycloid (i)? Of the hypocycloid (ii)? Check the answers by applying them to the cardioid and the astroid.

30. The integral

$$E(k) = \int_0^{\pi/2} \sqrt{1 - k^2 \sin^2 t}\, dt \quad (0 < k < 1)$$

defines a nonelementary function of k, known as the *complete elliptic integral of the second kind*. Show that the ellipse

$$\frac{x^2}{a^2} + \frac{y^2}{b^2} = 1 \quad (a > b > 0)$$

is of length $L = 4aE(\sqrt{a^2 - b^2}/a)$.

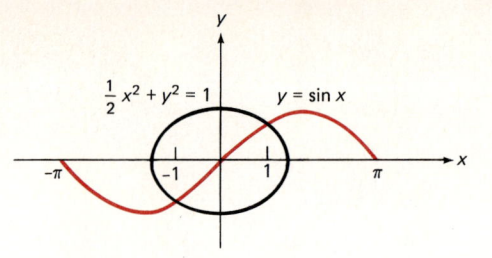

Two curves with the same length
Figure 49

31. Verify that one complete oscillation of the sine curve $y = \sin x$ has the same length L as the ellipse $\frac{1}{2}x^2 + y^2 = 1$ (see Figure 49). Approximate L with the help of Problem 19, page 424.

8.5 Area of a Surface of Revolution

Let

$$x = x(t), \quad y = y(t) \quad (a \le t \le b)$$

be the parametric equations of a simple curve C in the xy-plane, where the functions $x(t)$ and $y(t)$ are continuously differentiable and $y(t)$ is nonnegative. Suppose we revolve C about the x-axis, as in Figure 50(a). This generates a surface S, called a *surface of revolution*, with the x-axis as its axis of revolution. Then, as we now show, the appropriate definition of the area A of the surface S is given by the formula

$$A = 2\pi \int_a^b y(t)\sqrt{[x'(t)]^2 + [y'(t)]^2}\, dt = 2\pi \int_a^b y\sqrt{\left(\frac{dx}{dt}\right)^2 + \left(\frac{dy}{dt}\right)^2}\, dt. \quad (1)$$

To this end, we first inscribe a polygonal path $P_0 P_1 P_2 \ldots P_{n-1} P_n$ in the curve C, just as in the construction on page 460 leading to the def-

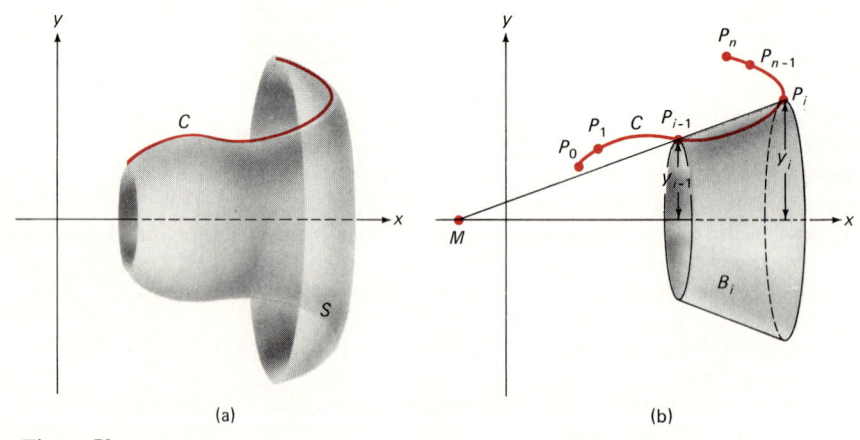

Figure 50

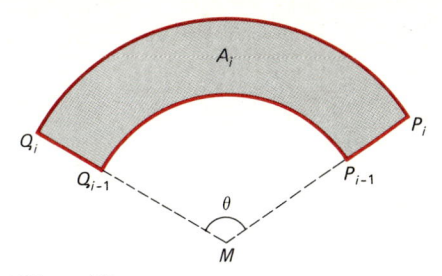

Figure 51

inition of the length of C. As before, the vertices of the polygonal path are the points

$$P_i = (x(t_i), y(t_i)) \qquad (i = 0, 1, \ldots, n),$$

where $a = t_0 < t_1 < t_2 < \cdots < t_{n-1} < t_n = b$. When C is revolved about the x-axis, each line segment $P_{i-1}P_i$ sweeps out a conical band B_i (the lateral surface of a conical frustum), as shown in Figure 50(b). The area of B_i is equal to

$$A_i = \pi(y_{i-1} + y_i)|P_{i-1}P_i|, \tag{2}$$

where $|P_{i-1}P_i|$ is the length of $P_{i-1}P_i$ and

$$y_i = y(t_i) \qquad (i = 0, 1, \ldots, n).$$

To verify (2), we slit the band B_i along the segment $P_{i-1}P_i$ and flatten it out, obtaining the shaded region in Figure 51 (the band can be reconstructed by pasting the edges $P_{i-1}P_i$ and $Q_{i-1}Q_i$ back together). Then A_i is the difference between the areas of two circular sectors with the same central angle θ, the sector MP_iQ_i of radius $R = |MP_i|$ and the sector $MP_{i-1}Q_{i-1}$ of radius $r = |MP_{i-1}|$.† Therefore

$$A_i = \frac{1}{2}R^2\theta - \frac{1}{2}r^2\theta$$

(see page 54). But

$$R\theta = 2\pi y_i, \qquad r\theta = 2\pi y_{i-1}, \qquad R - r = |P_{i-1}P_i|,$$

by construction, and hence

$$A_i = \frac{1}{2}(R+r)(R-r)\theta = \frac{1}{2}(R\theta + r\theta)|P_{i-1}P_i| = \pi(y_{i-1} + y_i)|P_{i-1}P_i|,$$

which establishes (2).

Surface Area as an Integral

It seems reasonable to regard the sum of the areas of all n conical bands B_i as a good approximation to the area of the surface of revolution S, where the approximation gets better and better as the mesh size

$$\mu = \max\{\Delta t_1, \Delta t_2, \ldots, \Delta t_n\} \qquad (\Delta t_i = t_i - t_{i-1})$$

† Here it is assumed that $y_{i-1} < y_i$, as in Figure 50(b). If $y_{i-1} > y_i$, then $R = |MP_{i-1}|$ and $r = |MP_i|$, while if $y_{i-1} = y_i$, then B_i is a cylindrical band of area $2\pi y_i|P_{i-1}P_i|$.

gets smaller and smaller. Thus we now *define* the area of S as the limit

$$A = \lim_{\mu \to 0} \sum_{i=1}^{n} A_i = \lim_{\mu \to 0} \sum_{i=1}^{n} \pi(y_{i-1} + y_i)|P_{i-1}P_i|$$

$$= 2\pi \lim_{\mu \to 0} \sum_{i=1}^{n} \frac{y_{i-1} + y_i}{2} |P_{i-1}P_i|,$$

provided that the limit exists and is finite. Calculating the length $|P_{i-1}P_i|$ in the same way as on page 461, we can write this limit as

$$A = 2\pi \lim_{\mu \to 0} \sum_{i=1}^{n} \frac{y_{i-1} + y_i}{2} \sqrt{[x'(u_i)]^2 + [y'(v_i)]^2} \, \Delta t_i,$$

where $t_{i-1} < u_i < t_i$, $t_{i-1} < v_i < t_i$. Moreover the number $\frac{1}{2}(y_{i-1} + y_i)$ lies between (or possibly coincides with) the numbers y_{i-1}, y_i, and hence by the intermediate value theorem, $y(w_i) = \frac{1}{2}(y_{i-1} + y_i)$ for some w_i in the interval $[t_{i-1}, t_i]$. It follows that

$$A = 2\pi \lim_{\mu \to 0} \sum_{i=1}^{n} y(w_i) \sqrt{[x'(u_i)]^2 + [y'(v_i)]^2} \, \Delta t_i. \qquad (3)$$

If u_i, v_i and w_i were equal for every i, the expression on the right would be a Riemann sum for the continuous function $y(t)\sqrt{[x'(t)]^2 + [y'(t)]^2}$ on the interval $[a, b]$, and we would then have

$$A = 2\pi \lim_{\mu \to 0} \sum_{i=1}^{n} y(u_i) \sqrt{[x'(u_i)]^2 + [y'(u_i)]^2} \, \Delta t_i$$

$$= 2\pi \int_a^b y(t) \sqrt{[x'(t)]^2 + [y'(t)]^2} \, dt,$$

thereby proving formula (1). In general u_i, v_i and w_i do not coincide, but it seems intuitively clear that as $\mu \to 0$, the right side of (3) must still approach the same limit

$$2\pi \int_a^b y(t) \sqrt{[x'(t)]^2 + [y'(t)]^2} \, dt = 2\pi \int_a^b y \sqrt{\left(\frac{dx}{dt}\right)^2 + \left(\frac{dy}{dt}\right)^2} \, dt$$

(after all, $\mu \to 0$ does imply $u_i - v_i \to 0$, $u_i - w_i \to 0$ for every i). This is in fact true, as can be proved rigorously by a technical argument which will not be given here. Thus we will henceforth regard formula (1) as established.

As an exercise, show that if $x(t)$ is nonnegative and the curve C is revolved about the y-axis instead of the x-axis, then the area A of the resulting surface of revolution S is given by the formula

$$A = 2\pi \int_a^b x(t) \sqrt{[x'(t)]^2 + [y'(t)]^2} \, dt = 2\pi \int_a^b x \sqrt{\left(\frac{dx}{dt}\right)^2 + \left(\frac{dy}{dt}\right)^2} \, dt. \qquad (1')$$

Example 1 Find the surface area A of a sphere of radius r.

Solution A sphere of radius r can be generated by revolving the semicircle

$$x = r \cos t, \qquad y = r \sin t \qquad (0 \le t \le \pi)$$

about the x-axis (see Figure 18, page 443, where the same semicircle is specified as the graph of the function $y = \sqrt{r^2 - x^2}$). Applying formula (1), we immediately obtain

$$A = 2\pi \int_0^\pi r \sin t \sqrt{(-r \sin t)^2 + (r \cos t)^2}\, dt$$

$$= 2\pi r^2 \int_0^\pi \sin t\, dt = -2\pi r^2 \cos t \Big|_0^\pi = 4\pi r^2,$$

which *proves* a formula we have already used more than once. □

Example 2 Find the surface area A of the torus in Figure 23, page 446, that is, the area of surface generated by revolving the circle $(x - a)^2 + y^2 = r^2$ about the y-axis.

Solution The circle can be represented parametrically by the equations

$$x = a + r \cos t, \qquad y = r \sin t \qquad (0 \leq t \leq 2\pi).$$

This time we apply formula (1'), obtaining

$$A = 2\pi \int_0^{2\pi} (a + r \cos t)\sqrt{(-r \sin t)^2 + (r \cos t)^2}\, dt$$

$$= 2\pi r \int_0^{2\pi} (a + r \cos t)\, dt = 2\pi r \left[at + r \sin t \right]_0^{2\pi} = 4\pi^2 r a. \qquad □$$

Example 3 Let C be one arch of a cycloid. Find the area A of the surface S generated by revolving C about the x-axis (the football-shaped surface in Figure 52).

Solution As in Example 2, page 462, C has the parametric equations

$$x = a(t - \sin t), \qquad y = a(1 - \cos t) \qquad (0 \leq t \leq 2\pi, a > 0)$$

and

$$\left(\frac{dx}{dt}\right)^2 + \left(\frac{dy}{dt}\right)^2 = 4a^2 \sin^2 \frac{t}{2}.$$

Since $a > 0$ and $\sin(t/2) \geq 0$ for $0 \leq t \leq 2\pi$, we find with the help of (1) that

$$A = 2\pi \int_0^{2\pi} a(1 - \cos t) \sqrt{4a^2 \sin^2 \frac{t}{2}}\, dt = 4\pi a^2 \int_0^{2\pi} (1 - \cos t) \sin \frac{t}{2}\, dt$$

$$= 8\pi a^2 \int_0^{2\pi} \sin^3 \frac{t}{2}\, dt = 16\pi a^2 \int_0^\pi \sin^3 u\, du,$$

after substituting $u = t/2$. But

$$\int_0^\pi \sin^3 u\, du = \int_0^\pi (1 - \cos^2 u) \sin u\, du$$

$$= \left[-\cos u + \frac{\cos^3 u}{3} \right]_0^\pi = 1 - \frac{1}{3} + 1 - \frac{1}{3} = \frac{4}{3},$$

and hence

$$A = \frac{64}{3} \pi a^2. \qquad □$$

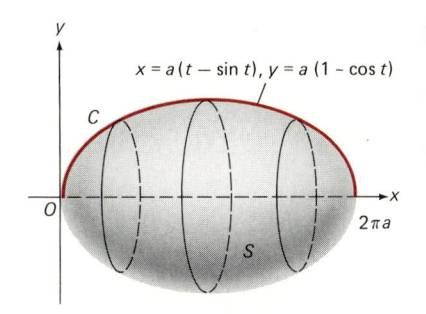

Figure 52 $x = a(t - \sin t), y = a(1 - \cos t)$

Suppose C is the graph of a continuously differentiable function
$$y = f(x) \quad (a \le x \le b).$$
Then C has parametric equations
$$x = t, \quad y = f(t) \quad (a \le t \le b),$$
and formula (1) for the area of the surface generated by revolving C about the x-axis becomes
$$A = 2\pi \int_a^b f(x)\sqrt{1 + [f'(x)]^2}\, dx = 2\pi \int_a^b y\sqrt{1 + \left(\frac{dy}{dx}\right)^2}\, dx, \quad (4)$$
after changing the variable of integration back from t to x. Similarly, in this case formula (1') for the area of the surface generated by revolving C about the y-axis becomes
$$A = 2\pi \int_a^b x\sqrt{1 + [f'(x)]^2}\, dx = 2\pi \int_a^b x\sqrt{1 + \left(\frac{dy}{dx}\right)^2}\, dx. \quad (5)$$
As an exercise, show that if the curve C is the graph of a continuously differentiable function
$$x = g(y) \quad (a \le y \le b),$$
then the area of the surface generated by revolving C about the x-axis is
$$A = 2\pi \int_a^b y\sqrt{[g'(y)]^2 + 1}\, dy = 2\pi \int_a^b y\sqrt{\left(\frac{dx}{dy}\right)^2 + 1}\, dy, \quad (4')$$
while the area of the surface generated by revolving C about the y-axis is
$$A = 2\pi \int_a^b g(y)\sqrt{[g'(y)]^2 + 1}\, dy = 2\pi \int_a^b x\sqrt{\left(\frac{dx}{dy}\right)^2 + 1}\, dy. \quad (5')$$

Example 4 Let C be the same parabolic arc as in Figure 44, page 464, namely
$$y = \frac{1}{2}x^2 \quad (0 \le x \le 1), \quad (6)$$
and let S be the surface generated by revolving C about the x-axis (the horn-shaped surface in Figure 53). Then $dy/dx = x$, and (4) gives
$$A = \pi \int_0^1 x^2 \sqrt{1 + x^2}\, dx.$$
To evaluate the integral, we substitute $x = \tan u$, obtaining
$$\int_0^1 x^2\sqrt{1+x^2}\, dx = \int_0^{\pi/4} \tan^2 u \sqrt{1 + \tan^2 u}\, \sec^2 u\, du = \int_0^{\pi/4} \tan^2 u \sec^3 u\, du$$
$$= \left[\frac{1}{4}\sec^3 u \tan u - \frac{1}{8}\sec u \tan u - \frac{1}{8}\ln|\sec u + \tan u|\right]_0^{\pi/4}$$
$$= \frac{1}{2}\sqrt{2} - \frac{1}{8}\sqrt{2} - \frac{1}{8}\ln(\sqrt{2} + 1),$$

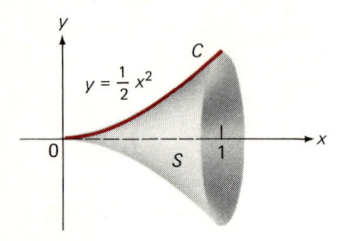

Figure 53

470 Chapter 8 Further Applications of Integration

with the help of Example 6, page 386. It follows that

$$A = \left[\frac{3}{8}\sqrt{2} - \frac{1}{8}\ln(\sqrt{2}+1)\right]\pi \approx 0.42\pi.$$

Example 5 Revolving the parabolic arc (6) about the y-axis, we generate the dish-shaped surface S in Figure 54 (part of a *paraboloid of revolution*). Then, by (5),

$$A = 2\pi \int_0^1 x\sqrt{1+x^2}\,dx = \frac{2\pi}{3}(1+x^2)^{3/2}\Big|_0^1 = \frac{2\pi}{3}(2^{3/2}-1) \approx 1.22\pi.$$

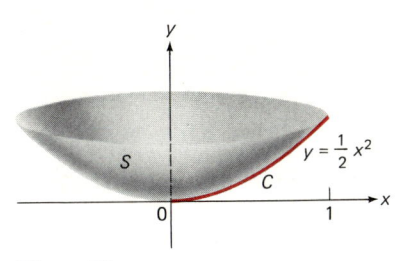

Figure 54

Problems

Find the area of the surface generated by revolving the given parametric curve about the specified axis.

1. $x = t^3$, $y = \frac{3}{2}t^2$ $(0 \le t \le 1)$, x-axis
2. $x = 1/t$, $y = \ln t$ $(\frac{1}{2} \le t \le 1)$, y-axis
3. $x = \sin^2 t$, $y = \cos^2 t$ $(0 \le t \le \pi/2)$, y-axis
4. $x = t + \sin t$, $y = \cos t$ $(0 \le t \le \pi/3)$, x-axis
5. $x = e^t \sin t$, $y = e^t \cos t$ $(0 \le t \le \pi/2)$, x-axis
6. $x = 2e^{-t}$, $y = e^{-2t}$ $(0 \le t < \infty)$, y-axis
7. $x = a\cos^3 t$, $y = a\sin^3 t$ $(0 \le t \le \pi, a > 0)$, y-axis
8. $x = 2\cos t + \cos 2t$, $y = 2\sin t - \sin 2t$ $(0 \le t \le 2\pi/3)$, x-axis

9. Suppose the region under the curve

$$y = a\cosh\frac{x}{a} \quad (0 \le x \le b, a > 0)$$

is revolved about the x-axis. What is the ratio of the volume V of the resulting solid of revolution to its lateral surface area A?

10. Verify that formula (1) gives the correct result for the surface area A of the conical band generated by revolving the line segment joining two arbitrary points $P_1 = (x_1, y_1)$ and $P_2 = (x_2, y_2)$ about the x-axis (assume that $y_1 \ge 0$, $y_2 \ge 0$). As a special case, derive the formula $A = \pi r L$ for the lateral area of a right circular cone of base radius r and slant height L.

Find the area of the surface generated by revolving the given curve (the graph of a function of x or y) about the specified axis.

11. $x = y^{1/3}$ $(0 \le y \le 1)$, y-axis
12. $y = \sin x$ $(0 \le x \le \pi/2)$, the line $x = -1$
13. $x = \sin y$ $(-\pi \le y \le 0)$, the line $x = 1$
14. $y = |1-x| + 1 - x$ $(0 \le x < \infty)$, y-axis
15. $y = \frac{1}{6}x^3 + \frac{1}{2}x^{-1}$ $(1 \le x \le \sqrt{2})$, x-axis
16. $y = e^{-x}$ $(0 \le x < \infty)$, x-axis
17. $y = \frac{1}{4}x^2 - \frac{1}{2}\ln x$ $(1 \le x \le e)$, x-axis
18. $x = \frac{1}{8}y^2 - \ln y$ $(1 \le y \le 4)$, x-axis

19. Let S be a *spherical zone* of altitude h, that is, the surface cut from a sphere of radius R by two parallel planes a distance h apart $(0 < h \le 2R)$, both of which intersect the sphere. Find the area of S, and show that it is independent of the location of the planes (see Figure 15, page 442).

20. Suppose the curve $y = 1/x$ $(x \ge 1)$ is revolved about the x-axis, generating an unbounded solid of revolution S. Show that S has finite volume V, but infinite surface area A. Find V. Can the surface of S (sometimes called "Gabriel's horn") be painted with a finite amount of paint?

21. Let S be the surface generated by revolving the arc of the unit circle $x^2 + y^2 = 1$ in the first quadrant about the line $x + y = 1$. What is the area of S?

8.6 More About Work

Let s be the position coordinate of a particle moving along a straight line, and suppose the particle is acted on by a variable force $F = F(s)$. Then, as on page 269, the work done on the particle by the force is given by

$$W = \int_a^b F(s)\,ds, \tag{1}$$

where a is the initial position and b the final position of the particle. In particular, if $F(s)$ has the constant value F, formula (1) becomes

$$W = F \int_a^b ds = F \cdot (b - a),$$

that is, the work equals the product of the force and the displacement $b - a$ (the distance moved by the particle). We now consider problems in which work is done on a "continuous medium," like a rope or a fluid, which can be regarded as made up of a very large number of particles.

Example 1 A heavy rope of length L feet, weighing c pounds per foot, hangs over the edge of a roof. How much work is required to pull the rope up to the roof? Neglect any work done on the rope after it clears the edge of the roof.

Solution As in Figure 55, let the s-axis point vertically downward, with the origin at the edge of the roof. Then the rope initially occupies the interval $[0, L]$. By introducing points of subdivision s_i which satisfy the inequalities $0 = s_0 < s_1 < s_2 < \cdots < s_{n-1} < s_n = L$, we partition $[0, L]$ into a large number of subintervals $[s_{i-1}, s_i]$, thereby dividing the rope into a large number of small "elements," each of which can be regarded as a particle. The weight of the element initially occupying the subinterval $[s_{i-1}, s_i]$ is $c \, \Delta s_i$ pounds, where $\Delta s_i = s_i - s_{i-1}$, and this is the gravitational force acting on the element that must be overcome in lifting it. Moreover, in being lifted to the roof, the ith element of rope experiences a displacement approximately equal to s_i (or any other number in the interval $[s_{i-1}, s_i]$), and hence the amount of work required to lift the element is approximately $(c \, \Delta s_i) s_i$ foot-pounds. Any reasonable definition of the total amount of work W required to lift the whole rope to the roof must satisfy the condition that W be the sum of the amounts of work required to lift the individual elements of rope. Therefore

$$W \approx \sum_{i=1}^{n} c s_i \, \Delta s_i, \qquad (2)$$

where the approximation improves as the elements of rope all become smaller, that is, as the mesh size $\mu = \max \{\Delta s_1, \Delta s_2, \ldots, \Delta s_n\}$ approaches zero. Recognizing the right side of (2) as a Riemann sum for the function cs on the interval $[0, L]$, we now set

$$W = \lim_{\mu \to 0} \sum_{i=1}^{n} c s_i \, \Delta s_i = \int_0^L cs \, ds,$$

by definition. The value of this integral is of course just

$$W = \frac{1}{2} c s^2 \bigg|_0^L = \frac{1}{2} c L^2.$$

For example, if the rope is 30 ft long and weighs 1.5 lb/ft, the amount of work required to lift it to the roof is $\frac{1}{2}(1.5)(30)^2 = 675$ ft-lb. □

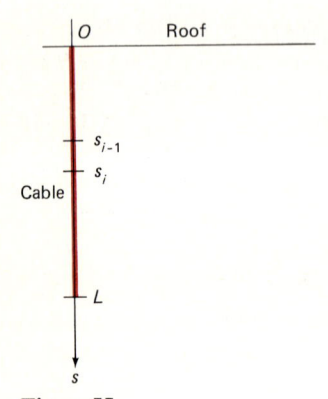

Figure 55

The Work Expended in Pumping a Fluid

In the next example we use the same kind of reasoning to solve a problem of hydraulics.

Example 2 A tank in the form of an inverted right circular cone of base radius r and height h is completely filled with a fluid whose density (mass per unit volume) is ρ. How much work is required to pump all the fluid to the top of the tank, if it is pumped out through a hose kept just below its surface?

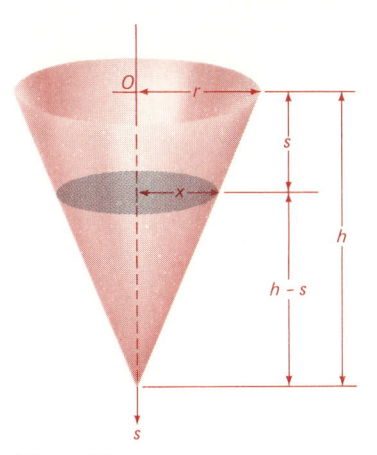

Solution As in Figure 56, we choose the axis of symmetry of the cone as the s-axis, pointing vertically downward, with the origin at the center of the base of the cone. Let the interval $[0, h]$ be divided into a large number of subintervals $[s_{i-1}, s_i]$, by introducing points of subdivision s_i satisfying the inequalities $0 = s_0 < s_1 < s_2 < \cdots < s_{n-1} < s_n = h$. Then the planes perpendicular to the s-axis at the points with coordinates s_i ($i = 0, 1, \ldots, n$) divide the fluid-filled cone into n layers. Let $A(s)$ be the area of the cross section of the cone at s, that is, of the circular disk cut from the cone by the plane perpendicular to the axis of the cone at the point with coordinate s. The ith layer, between the planes perpendicular to the s-axis at the points with coordinates s_{i-1} and s_i, is actually a conical frustum, but its volume is approximately that of a right circular cylinder of base area $A(s_i)$ and height $\Delta s_i = s_i - s_{i-1}$, namely $A(s_i)\Delta s_i$ (the same type of approximation was used throughout Section 8.1, in calculating volumes by the method of cross sections). Thus the weight of the fluid in the ith layer is approximately $\rho g A(s_i) \Delta s_i$, where ρ is the density of the fluid and g is the acceleration due to gravity. Moreover, in being lifted to the top of the tank, the ith layer of fluid experiences a displacement approximately equal to s_i, and hence the amount of work required to lift the layer is approximately $\rho g A(s_i) s_i \Delta s_i$. Let W be the total amount of work required to pump all the fluid to the top of the tank. Then, by the same argument as in the preceding example,

$$W \approx \sum_{i=1}^{n} \rho g A(s_i) s_i \Delta s_i, \qquad (3)$$

where the approximation improves as the layers of fluid all become thinner, that is, as the mesh size $\mu = \max \{\Delta s_1, \Delta s_2, \ldots, \Delta s_n\}$ approaches zero. Therefore, recognizing the right side of (3) as a Riemann sum for the function $\rho g A(s) s$ on the interval $[0, h]$, we set

$$W = \lim_{\mu \to 0} \sum_{i=1}^{n} \rho g A(s_i) s_i \Delta s_i = \int_0^h \rho g A(s) s\, ds,$$

by definition.

To evaluate the integral, let x be the radius of the cross section of the cone at s, as in the figure. Then, by similar triangles,

$$\frac{x}{h-s} = \frac{r}{h},$$

and therefore

$$x = \frac{r}{h}(h-s), \qquad A(s) = \pi x^2 = \pi\left(\frac{r}{h}\right)^2 (h-s)^2.$$

It follows that

$$W = \int_0^h \rho g A(s) s\, ds = \pi \rho g \left(\frac{r}{h}\right)^2 \int_0^h (h^2 - 2hs + s^2) s\, ds$$

$$= \pi \rho g \left(\frac{r}{h}\right)^2 \left[\frac{1}{2} h^2 s^2 - \frac{2}{3} h s^3 + \frac{1}{4} s^4\right]_0^h = \frac{1}{12} \pi \rho g r^2 h^2.$$

For example, suppose the tank is of base radius 2 m and height 6 m, and is filled with glycerine, whose density is 1260 kg/m³. Then, choosing $g = 9.8$ m/sec², we find that the amount of work required to pump out the tank is

$$W = \frac{1}{12}\pi(1260)(9.8)(2^2)(6^2) = 148{,}176\pi \approx 465{,}510 \text{ joules},$$

in the metric system, or about 343,360 ft-lb in the engineering system.† For water, whose density is 1000 kg/m³ (at 4°C), the amount of work required to pump out the tank is about $(1000/1260)(465{,}510) \approx 369{,}450$ joules. □

Power

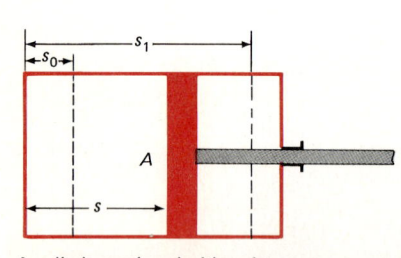

The rate of change of work with respect to time, that is, the derivative dW/dt, is called the *power* (see Problem 29, page 274). The mks unit of power is the *watt*, equal to 1 joule per second. Another important unit of power is the *horsepower*, equal to 746 watts or 550 ft-lb/sec.

Example 3 It takes a pump 30 minutes to empty the tankful of glycerine considered in the preceding example. What is the power of the pump?

Solution The pump does about 465,510 joules of work in 30 minutes. Therefore its power is

$$\frac{465{,}510}{30(60)} \text{ joules/sec} \approx 258.6 \text{ watts},$$

or equivalently

$$\frac{258.6}{746} \approx 0.35 \text{ horsepower}.$$

Here we assume that the pump does work at a constant rate and that it ejects the fluid with negligible velocity, so that all of its power is expended in lifting the fluid and none in imparting excess kinetic energy to the fluid. To keep things simple, it is also assumed (rather unrealistically) that the pump is so efficient that the difference between its input power and its output power is negligible. □

A cylinder equipped with a piston
Figure 57

Example 4 A gas is contained in a circular cylinder of cross-sectional area A, equipped with a movable piston (see Figure 57). Find the work done by the gas on the piston in expanding from initial volume V_0 to final volume V_1.

Solution Let p be the pressure (that is, the force per unit area) exerted by the gas on the face of the piston, and let s be the distance between the piston and the cylinder head. Suppose the expanding gas moves the piston from initial position s_0 to final position s_1. As the gas expands, its pressure changes (in fact, decreases), and hence the same is true of the force pA on the piston. Thus the work W done on the piston by the expanding gas is equal to

$$W = \int_{s_0}^{s_1} pA \, ds.$$

† In the centimeter-gram-second (cgs) system of units, the unit of force is the *dyne*, defined as the force which imparts an acceleration of 1 cm/sec² to a mass of 1 gram, while in the meter-kilogram-second (mks) system, the unit of force is the *newton*, defined as the force which imparts an acceleration of 1 m/sec² to a mass of 1 kg. The corresponding units of work are the *dyne-centimeter* or *erg* and the *newton-meter* or *joule* (1 joule = 10^7 ergs = 0.7376 ft-lb).

474 Chapter 8 Further Applications of Integration

Here the pressure is a function of the coordinate s, but it can just as well be regarded as a function $p = p(V)$ of the volume $V = As$ of the confined gas. Therefore, changing the variable of integration from s to V, we can express W in the equivalent form

$$W = \int_{V_0}^{V_1} p\, dV, \tag{4}$$

where V_0 is the initial volume and V_1 is the final volume. □

Isothermal Expansion

Pursuing Example 4, suppose the gas expands *isothermally*, that is, at constant temperature. Then, to a good approximation, the pressure and volume are related by *Boyle's law*

$$pV = C = \text{constant}, \tag{5}$$

at least if neither the pressure nor the temperature is too low. Notice that $C = p_0 V_0 = p_1 V_1$, where p_0 is the initial pressure and p_1 the final pressure of the gas. It follows from (4) and (5) that

$$W = \int_{V_0}^{V_1} p\, dV = C \int_{V_0}^{V_1} \frac{dV}{V} = C \ln V \Big|_{V_0}^{V_1} = C \ln \frac{V_1}{V_0} = p_0 V_0 \ln \frac{V_1}{V_0}. \tag{6}$$

For example, in expanding isothermally from an initial volume of 2 ft^3 to a final volume of 10 ft^3, a gas initially at a pressure of 50 lb/in^2 does work equal to

$$144(50)(2) \ln \frac{10}{2} = 14{,}400 \ln 5 \approx 23{,}176 \text{ ft-lb}$$

(144 is the conversion factor from lb/in^2 to lb/ft^2).

Adiabatic Expansion (Optional)

On the other hand, suppose the gas expands *adiabatically*, that is, without exchanging heat with its surroundings. Then the pressure and volume of the gas are related by the formula

$$pV^k = C = \text{constant}, \tag{5'}$$

where k is another constant which depends on the nature of the gas; k is approximately equal to 1.67 for monatomic gases and 1.40 for diatomic gases. It follows from (4) and (5') that

$$W = \int_{V_0}^{V_1} p\, dV = C \int_{V_0}^{V_1} V^{-k}\, dV = C \frac{V^{1-k}}{1-k}\Big|_{V_0}^{V_1} = \frac{C}{1-k}(V_1^{1-k} - V_0^{1-k}).$$

But $C = p_0 V_0^k = p_1 V_1^k$, where p_0 is the initial pressure and p_1 the final pressure of the gas, and therefore

$$W = \frac{p_0 V_0 - p_1 V_1}{k - 1}. \tag{7}$$

For expansion $V_1 > V_0$, while for compression $V_1 < V_0$. In the case of adiabatic compression, the work W' done *on the gas* by the piston in compressing the gas is the negative of (7), namely

$$W' = -W = \frac{p_1 V_1 - p_0 V_0}{k - 1}. \tag{7'}$$

This can be written in the form

$$W' = \frac{p_0 V_0}{k-1}\left[\left(\frac{V_0}{V_1}\right)^{k-1} - 1\right], \qquad (8)$$

after substituting the value of p_1 obtained by solving the equation $p_1 V_1^k = p_0 V_0^k$. For example, suppose air initially at a pressure of 25 lb/in² is compressed adiabatically from an initial volume of 900 in³ to a final volume of 60 in³. Then, since $k = 1.4$ for air,

$$W' = \frac{1}{12}\frac{25(900)}{0.4}\left[\left(\frac{900}{60}\right)^{0.4} - 1\right] = 4687.5[(15)^{0.4} - 1] \approx 9160 \text{ ft-lb}$$

($\frac{1}{12}$ is the conversion factor from in-lb to ft-lb). As an exercise, show that only about 55% as much work is needed to compress the same air isothermally.

Problems

1. A heavy cable, of length 20 m and linear density 5 kg/m, is initially lying on the ground. It is then hoisted vertically upward until its free end hangs 4 m above the ground. How much work is done on the cable?

A 1-ton elevator car, suspended by a cable 100 ft long weighing 10 lb/ft, is lifted by winding the cable onto a winch at the top of the elevator shaft.

2. As the car rises 50 ft from its lowest position, how much work is done on the car? On the cable?

3. As the car rises from a height of 50 ft to a height of 75 ft, how much work is done on both the car and the cable?

4. A bucket full of water is lifted vertically upward at the rate of 2.5 ft/sec. The bucket weighs 5 lb and initially contains 45 lb of water, but as it is lifted, water leaks out of it at the rate of 1.25 lb/sec. How much work is required to lift the leaking bucket to a height of 50 ft? Of 100 ft?

5. How long will it take a 2-hp pump to lift 150,000 ft³ of water from the surface of a lake to a height 12 ft above the lake? (The weight density of water is 62.5 lb/ft³, to a good approximation.)

6. Two waterfalls are located one after the other on the same river. One is 45 ft high, and the other is 30 ft high. What is the total power of the two waterfalls if the flow of water in the river is 2250 ft³/sec?

7. A tank of height h is filled to depth d with a fluid of density ρ. Suppose the horizontal plane a distance s below the top of the tank intersects the tank in a region of area $A(s)$. Find the work W required to pump all the fluid to the top of the tank.

8. A well 4 ft in diameter and 30 ft deep is half full of water. Find the work required to pump all the water to the top of the well. How much work is required to pump up 250 gallons of water? (One gallon equals 231 cubic inches.)

9. A bowl full of fluid of density ρ is shaped like a hemisphere of radius r. How much work is required to pump all the fluid to a height h above the top of the bowl?

10. Suppose it requires an amount of work W_1 to fill a tank of height h by pumping fluid in through a hole in its bottom. How is W_1 related to W_2, the amount of work required to empty the tank by pumping the fluid to its top?

11. A tank in the form of an inverted right circular cone of base radius 3 ft and height 5 ft is filled with water to a depth of 4 ft. Use the prismoidal formula (i), page 424, to find the work required to pump all the water to the top of the tank. To a height 2 ft above the top of the tank.

12. A tank is in the form of a right pyramid of height h, whose base is a square of side length a (such a pyramid is shown in Figure 5, page 438). Find the work required to fill the tank with fluid of density ρ by pumping the fluid in through a hole in the bottom of the tank (the square base). Also find the work required to empty the tank by pumping the fluid out through a hole in the top of the tank (the apex of the pyramid).

13. The Great Pyramid of Cheops, at Gizeh near Cairo, was originally of height 147 m, with a square base 230 m on a side. It is made of limestone blocks, with a density of about 2500 kg/m³. Estimate both the amount of work and the number of man-years of labor required to build the pyramid. Assume that the average laborer worked on the pyramid 50 hours a week, year in and year out, lifting about 50 kg a distance of about 1 m in each minute of work.

14. The piston of a steam engine makes 90 strokes per minute, each of length 15 in. Suppose the cross-sectional area of the cylinder is 48 in², and the average pressure on the piston during a stroke is 60 lb/in². What is the average power output of the engine, assuming that it is perfectly efficient?

15. How much work is needed to compress 720 in³ of helium at an initial pressure of 20 lb/in² to a final volume of 40 in³ if the compression is isothermal? If the compression is adiabatic? (Helium is a monatomic gas.)

16. A gas does 1568 ft-lb of work in expanding isothermally from an initial volume of 40 in³ and pressure of 16 atmospheres. Find the final volume and pressure of the gas. (One atmosphere equals 14.7 lb/in².)

17. Let $p = p(h)$ and $\rho = \rho(h)$ be the pressure and density of air at altitude h above sea level. If the temperature is constant, then $p = k\rho$, where k is a constant of proportionality (this is another version of Boyle's law). Neglecting the variation of atmospheric temperature with altitude, derive the *barometric equation* $p = p_0 e^{-gh/k}$, where p_0 is the atmospheric pressure at sea level and g is the acceleration due to gravity.

18. From a mechanical standpoint, the human heart is a pump (see Figure 58), in which blood enters the left ventricle through the mitral valve, and is then forced out through the aortic valve when the heart muscle contracts, thereby decreasing the volume of the heart. During each contraction, the pressure exerted by the heart wall on the blood increases in an approximately linear fashion from a typical diastolic pressure of 80 mm Hg (millimeters of mercury) to a typical systolic pressure of 120 mm Hg, for a heart that is young and healthy. Estimate the work W done by the heart in a single heartbeat, given that the blood volume changes by about 75 cm³ during one contraction. (100 mm Hg $\approx 1.33 \times 10^5$ dynes/cm².)

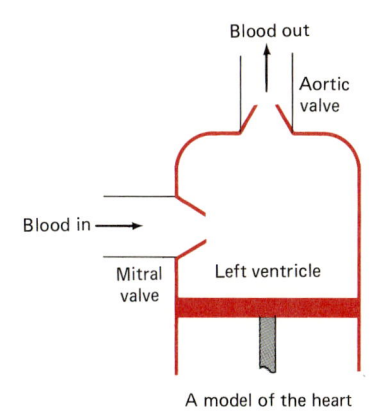

Figure 58
A model of the heart

8.7 Fluid Pressure (Optional)

Consider a thin flat plate in the shape of a plane region R of area A, which is submerged in a fluid at rest. Let the fluid be of mass density ρ, or equivalently of weight density $\delta = \rho g$, where g is the acceleration due to gravity. Suppose R is horizontal and lies at depth h, so that R and the free surface of the fluid are a distance h apart, as in Figure 59. Then the force F exerted by the fluid on R is just the weight of the fluid in a right cylinder of height h and base R (here we are making a tacit assumption that will be examined in the remark following Example 1). Since the volume of the cylinder is Ah, it follows that

$$F = \delta A h \tag{1}$$

Thus the force F is proportional to the depth h. Dividing F by the area A, we find that the *pressure*, that is, the force per unit area, at the submerged plate is also proportional to h:

$$p = \delta h. \tag{2}$$

The force (1) and pressure (2) are called "hydrostatic," even when the fluid is not water.

Figure 59

Pascal's Principle

Actually, these results are more general than is at first apparent, for according to *Pascal's principle*, the pressure at any point of a fluid is *the same in all directions*. Thus a tiny "test surface" of area a, submerged at depth h in a fluid of weight density δ, experiences one and the same force $F = \delta a h$ regardless of the orientation of the surface. In making this statement, we assume that the surface is so small that there is negligible variation of pressure over the surface even when it is vertical (the "worst case"). Otherwise, to get the force on the surface, we must integrate the pressure over the surface, in a way to be described below.

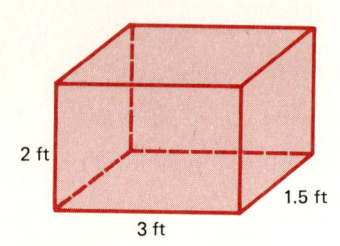

Example 1 An aquarium is shaped like a rectangular parallelepiped of length 3 ft, width 1.5 ft and height 2 ft (see Figure 60). Find the force exerted by the water on the bottom of the aquarium. What is the pressure at the bottom? At a point 6 in. deep?

Solution We assume that the aquarium is filled to the top, and neglect the possible presence of fish, vegetation etc. Choosing the weight density δ of water to be 62.5 lb/ft^3 (actually δ varies somewhat with temperature), we deduce from (1) that the force on the bottom of the aquarium, a rectangle of area 4.5 ft^2, is $(62.5)(4.5)(2) = 562.5$ lb. Moreover, by (2), the pressure at the bottom is $(62.5)(2) = 125$ lb/ft^2, and the pressure at a point 6 in. deep is $(62.5)(0.5) = 31.25$ lb/ft^2. The forces exerted by the water on the vertical sides of the aquarium will be found in Example 2. □

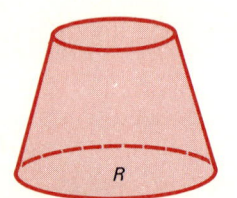

Remark You may have wondered what happens if it is impossible to construct a fluid-filled cylinder over the region R. For example, if R is the bottom of the vessel shown in Figure 61, the slanting sides of the vessel prevent the construction of such a cylinder. Remarkably enough, it turns out that formulas (1) and (2) remain valid even under these circumstances. At first this "hydrostatic paradox" seems inexplicable, until we realize that the forces exerted on the fluid by the walls of the vessel have a *downward* component, which is transmitted by the fluid to the bottom of the vessel. Thus the total force acting on the bottom is actually the sum of this downward component and the weight of the fluid in the vessel. In fact, a detailed analysis shows that *regardless of the shape of the vessel, F is exactly what it would be if the vessel were a right cylinder with the same base R, filled to the same depth!*

The Force on a Submerged Plate

We now consider the problem of finding the force on a submerged plate R which is not horizontal, so that the pressure varies from point to point, confining ourselves to the important case where R is vertical (however, see Problems 13 and 14). As in Figure 62, we introduce rectangular coordinates x and y, with the x-axis pointing vertically downward and the y-axis along the free surface of the fluid. Let R be the region in the xy-plane bounded by the horizontal lines $x = a$, $x = b$ ($a < b$) and the graphs of the functions $y = f(x)$, $y = g(x)$, where f and g are continuous on $[a, b]$ and $f(x) \geq g(x)$. We partition the given interval $[a, b]$ into a large number of subintervals $[x_{i-1}, x_i]$, by introducing points of subdivision x_i which satisfy the inequalities $a = x_0 < x_1 < x_2 < \cdots < x_{n-1} < x_n = b$. Correspondingly, the lines $x = x_i$ divide R into a large number of horizontal strips R_i, where R_i is the region bounded by the lines $x = x_{i-1}$, $x = x_i$ and the curves $y = f(x)$, $y = g(x)$. Each strip R_i is almost rectangular, of area approximately equal to $[f(x_i) - g(x_i)]\Delta x_i$, where $\Delta x_i = x_i - x_{i-1}$ and we might just as well have chosen x_i to be any other point of the interval $[x_{i-1}, x_i]$. Moreover, although there are points of R_i at depths varying from x_{i-1} to x_i, if Δx_i is small this variation of depth is slight and we can regard all the points of R_i as being at approximately the same depth, say x_i. Therefore, by formula (1) and Pascal's principle, the force exerted by the fluid on the strip R_i is approximately $\delta[f(x_i) - g(x_i)]x_i \Delta x_i$.

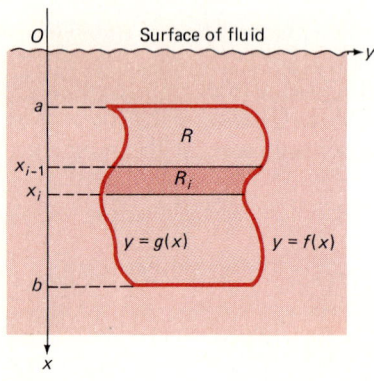

478   Chapter 8   Further Applications of Integration

The rest of the argument follows a familiar pattern. Let F be the total force exerted by the fluid on the whole plate R. Then F is the sum of the forces acting on the individual strips R_i. Therefore

$$F \approx \delta \sum_{i=1}^{n} [f(x_i) - g(x_i)] x_i \Delta x_i, \tag{3}$$

where the approximation improves as the strips R_i all become narrower, that is, as the mesh size $\mu = \max \{\Delta x_1, \Delta x_2, \ldots, \Delta x_n\}$ approaches zero. Recognizing the right side of (3) as a Riemann sum for the function $[f(x) - g(x)]x$ on the interval $[a, b]$, we now set

$$F = \delta \lim_{\mu \to 0} \sum_{i=1}^{n} [f(x_i) - g(x_i)] x_i \Delta x_i = \delta \int_a^b [f(x) - g(x)] x \, dx,$$

by definition. This can be written in the form

$$F = \delta \int_a^b w(x) x \, dx, \tag{4}$$

where $w(x) = f(x) - g(x)$ is the width of the region R at depth x below the surface of the fluid.

Example 2 Find the forces exerted by the water on the vertical sides of the aquarium in Example 1.

Solution Two of the vertical sides are rectangles of width 3 ft and height 2 ft, and the other two are rectangles of width 1.5 ft and height 2 ft. Thus $w(x) \equiv 3$ in the first case, while $w(x) \equiv 1.5$ in the second case. Also $a = 0$ (the aquarium is filled to the top) and $b = 2$. It follows that the force F_1 on each of the larger sides is

$$F_1 = 62.5 \int_0^2 3x \, dx = 187.5 \left. \frac{x^2}{2} \right|_0^2 = 375 \text{ lb},$$

while the force on each of the smaller sides is

$$F_2 = 62.5 \int_0^2 1.5x \, dx = 93.75 \left. \frac{x^2}{2} \right|_0^2 = 187.5 \text{ lb}.$$

Suppose the aquarium is only half full of water. Then we still have $a = 0$ (remember that the origin is at the free surface of the water), but now $b = 1$. As an exercise, show that this makes both forces F_1 and F_2 four times smaller. ☐

Example 3 A distillery tank, filled with ethyl alcohol of weight density $\delta = 7950$ newtons/m^3, has a circular glass window in one of its vertical sides. The window is of radius 0.5 m, and the highest point of the window is 1.5 m below the surface of the alcohol. What is the force exerted by the alcohol on the window?

Solution Choosing coordinates as in Figure 63, we find that the rim of the window is the circle with equation

$$(x - 2)^2 + y^2 = \frac{1}{4}.$$

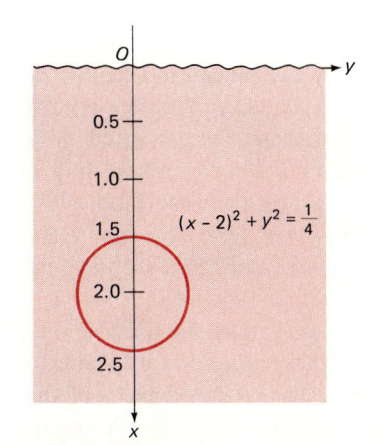

Figure 63

Accordingly, the window is the region between the curves $y = f(x) = \sqrt{\frac{1}{4} - (x-2)^2}$ and $y = g(x) = -\sqrt{\frac{1}{4} - (x-2)^2}$ over the interval $[1.5, 2.5]$, so that

$$w(x) = 2\sqrt{\frac{1}{4} - (x-2)^2} \quad (1.5 \leq x \leq 2.5).$$

It follows from formula (4) that the force F exerted by the alcohol on the window is

$$F = 15{,}900 \int_{1.5}^{2.5} \sqrt{\frac{1}{4} - (x-2)^2}\, x\, dx.$$

To evaluate the integral, we make the substitution $u = x - 2$, obtaining

$$F = 15{,}900 \int_{-0.5}^{0.5} \sqrt{\frac{1}{4} - u^2}\, (2+u)\, du$$

$$= 31{,}800 \int_{-0.5}^{0.5} \sqrt{\frac{1}{4} - u^2}\, du + 15{,}900 \int_{-0.5}^{0.5} u\sqrt{\frac{1}{4} - u^2}\, du$$

$$= 31{,}800 \int_{-0.5}^{0.5} \sqrt{\frac{1}{4} - u^2}\, du,$$

where we exploit the fact that the function $u\sqrt{\frac{1}{4} - u^2}$ is odd (explain further). The last integral is just the area of a semicircular disk of radius $\frac{1}{2}$, and hence is equal to $\frac{1}{8}\pi$. Therefore

$$F = \frac{31{,}800\pi}{8} = 3975\pi \approx 12{,}488 \text{ newtons}.$$

As an exercise verify that the force F is the product of the area of the window and the pressure at its center. ☐

Buoyancy and Archimedes' Principle

Example 4 A block of wood of length a, width b and height c floats partly submerged in a lake. Find the depth h to which the block sinks below the surface of the water [see Figure 64(a)]. Assume that the block has no tendency to topple over.

Solution Let δ be the weight density of water. Then, by formula (1), the water exerts an upward "buoyancy" force $F = \delta abh$ on the bottom of the block. But abh is the volume of the water displaced by the block, and hence *the buoyancy force is equal to the weight of the displaced water*. This is a special case of *Archimedes' principle*, which states that the net force exerted on a floating or submerged object *of arbitrary shape* by the surrounding fluid is an upward buoyancy force equal to the weight of the displaced fluid. Note that in the present problem, the forces exerted by the water on the four vertical sides of the block cancel each other out in pairs, and therefore have no net effect on the block.

To find the depth h to which the floating block sinks, we observe that in equilibrium the buoyancy force F on the block is equal and opposite to the block's weight w. But $w = \delta' abc$, where δ' is the weight density of the wood of which the block is made. Thus the equilibrium condition becomes

$$F = \delta abh = \delta' abc = w. \tag{5}$$

480   Chapter 8   Further Applications of Integration

Solving for h, we get $h = (\delta'/\delta)c$, or equivalently

$$h = \alpha c,$$

where $\alpha = \delta'/\delta$ is the *specific gravity* of the wood, that is, the ratio of its density to the density of water. □

Example 5 Suppose the wooden block in Example 4 is lifted vertically upward by a rope fastened to the midpoint of its upper face. Find the work required to lift the block up until its bottom just clears the surface of the water. (Neglect the weight of the rope and any effects due to surface tension.)

Solution When the bottom of the block is a distance x below the surface of the water, the block experiences an upward buoyancy force δabx, equal to the weight of the water displaced by the submerged part of the block [see Figure 64(b)]. The block is also acted on by a downward force equal to its weight w, and hence the net downward force on the block is $F(x) = w - \delta abx$. But $w = \delta abh$, by formula (5), and thus

$$F(x) = \delta ab(h - x).$$

Therefore, to lift the block out of the water, we must overcome this force by doing an amount of work

$$W = -\int_h^0 F(x)\,dx = \int_0^h F(x)\,dx = \delta ab \int_0^h (h - x)\,dx$$

$$= \delta ab \left[hx - \frac{1}{2}x^2 \right]_0^h = \frac{1}{2}\delta abh^2 = \frac{1}{2}wh. \quad \square$$

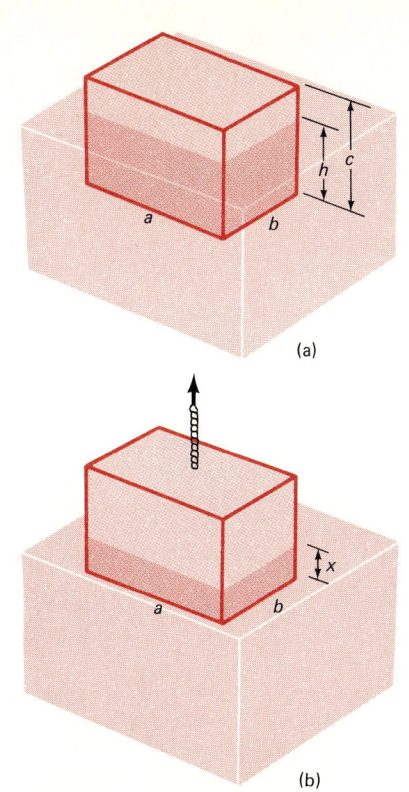

Figure 64

Problems

1. Find the hydrostatic force on a vertical rectangular dam 40 ft wide and 30 ft high when the water level is 4 ft below the top of the dam.

2. Find the hydrostatic forces on all six faces of a submerged cube of edge length 2 ft, whose upper face is 5 ft below the surface of a lake, parallel to the surface.

3. A cup in the form of an inverted conical frustum is of height h, with a bottom of area A (see Figure 65). What is the pressure at the bottom of the cup if it is filled with a fluid of weight density δ? Is the force on the bottom equal to the weight of the fluid in the cup? What forces support the part of the fluid that does not lie directly above the bottom of the cup?

4. A rectangular plate is submerged vertically with one edge parallel to (or along) the surface of a fluid of weight density δ. Show that the force on one side of the plate is equal to the product of its area and the pressure at its center (the point in which the diagonals intersect).

5. Find the hydrostatic force exerted on a square plate of side length a submerged vertically in a fluid of weight density δ, with one vertex of the square at the surface of the fluid and one diagonal parallel to the surface.

6. A trough with triangular ends of area A is completely filled to depth h with a fluid of weight density δ (see Figure 66). Find the force F exerted by the fluid on each end. Does F depend on the shape of the triangle?

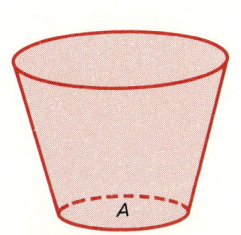

Figure 65

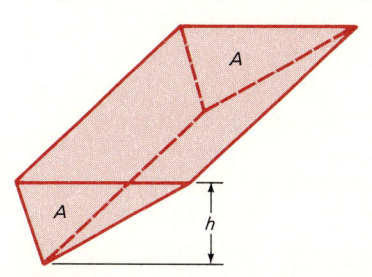

Figure 66

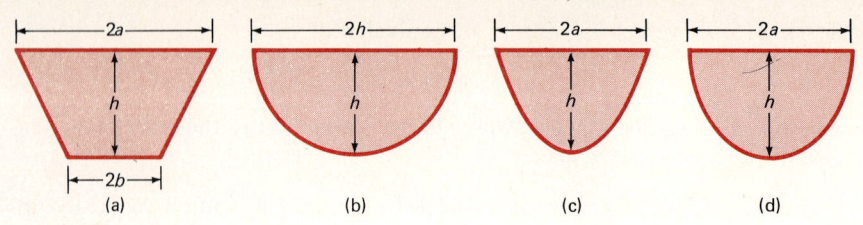

Figure 67

Find the hydrostatic force on one side of a vertical plate submerged in a fluid of weight density δ if the plate is shaped like

7. The isosceles trapezoid in Figure 67(a)
8. The semicircular disk in Figure 67(b)
9. The parabolic segment in Figure 67(c)
10. The semielliptic region in Figure 67(d)

In each case the upper edge of the plate lies along the surface of the fluid.

Given a thin plate in the form of the region bounded by the x-axis and the graph of the function

$$y = b \cos \frac{\pi x}{2a} \quad (-a \leq x \leq a, b > 0),$$

find the hydrostatic force exerted on the plate if it is submerged vertically in fluid of weight density δ with the straight edge of the plate

11. Along the surface of the fluid
12. Perpendicular to the surface of the fluid, with one end at the surface

13. A circular disk 1 ft in radius is submerged in fluid of weight density δ in such a way that its plane makes angle θ with the vertical and its highest point lies 2 ft below the surface of the fluid. What is the force exerted by the fluid on one side of the disk?

14. A swimming pool is 10 ft wide, 24 ft long, 4 ft deep at the shallow end, and 8 ft deep at the deep end. What is the hydrostatic force on the bottom of the pool if the pool is full? If the water is 6 ft deep at the deep end?

15. How much work is required to push down the wooden block in Example 4 until it is just submerged?

16. Given that the specific gravity of ice is 0.92 and that of sea water is 1.03, what fraction of an iceberg lies below the surface of the sea?

A buoy in the shape of an inverted right circular cone of height H and base radius R floats in a lake. The buoy is of weight w and specific gravity α. Find

17. The equilibrium depth h of the bottom of the buoy (the vertex of the cone)
18. The work required to lift the buoy up until it just clears the water
19. The work required to push the buoy down until it is just submerged

20. A metal object, claimed to be made of solid gold, is suspected of having a hollow core. It weighs 25 oz in air and 23 oz in water. Show that the suspicion is justified, and find the volume of the core. (The specific gravity of gold is 19.3.)

Key Terms and Topics
Calculation of volume by the method of cross sections
Calculation of volume by the method of disks or washers
Calculation of volume by the method of shells
Curves in parametric form (parametric curves)
The tangent line to a parametric curve
The length of a plane curve
The area of a surface of revolution
Work done on a continuous medium (ropes, liquids, gases)
Fluid pressure and the force on a submerged plate

Supplementary Problems

1. A right circular cone C of altitude H is cut into two pieces of equal volume by a plane parallel to its base. What is the distance between this plane and the vertex of C? Does the answer depend on the vertex angle of C?

2. Solve the preceding problem if the cone C is to be cut into three pieces of equal volume by planes parallel to its base. Is the answer the same for a general cone?

Find the volume of the solid S whose base is the ellipse
$$\frac{x^2}{a^2} + \frac{y^2}{b^2} = 1$$
($a > 0$, $b > 0$) if every cross section of S by a plane perpendicular to the x-axis is

3. A square

4. An isosceles right triangle with its hypotenuse in the xy-plane

5. A semicircle with its diameter in the xy-plane

6. Find the volume of the infinite solid whose base is the region between the curves $y = \cosh x$ and $y = \sinh x$ in the first quadrant, and whose cross sections by planes perpendicular to the x-axis are squares.

7. Find the volume of the infinite solid whose base is the region bounded by the positive x-axis, the negative y-axis and the curve $y = \ln x$, and whose cross sections by planes perpendicular to the x-axis are semicircles with diameters in the xy-plane.

Use any method (disks, washers or shells) to find the volume of the solid generated by revolving the region under the given curve and over the given interval about the specified axis.

8. $y = \sqrt{\ln x}$, $1 \leq x \leq e^2$, x-axis
9. $y = xe^x$, $1 \leq x \leq 2$, y-axis
10. $y = \arctan x$, $0 \leq x \leq 1$, y-axis
11. $y = \tanh x$, $-1 \leq x \leq 1$, x-axis
12. $y = \sin x^2$, $0 \leq x \leq \sqrt{\pi/3}$, y-axis
13. $y = \csc x$, $\pi/4 \leq x \leq \pi/2$, x-axis
14. $y = e^{-|x|}$, $-\infty < x < \infty$, x-axis
15. $y = e^{-\sqrt{x}}$, $0 \leq x < \infty$, y-axis

Calculate the volume of the solid generated by revolving the given region OAB or OBC in Figure 68 about the specified axis. The curve OB is the graph of the function $y = \sin x$ over the interval $0 \leq x \leq \pi/2$. In each case use both the disk or washer method and the shell method.

16. OAB about the x-axis
17. OBC about the x-axis
18. OBC about the y-axis
19. OAB about the y-axis
20. OAB about the line $x = \pi/2$
21. OBC about the line $x = \pi/2$
22. OBC about the line $y = 1$
23. OAB about the line $y = 1$

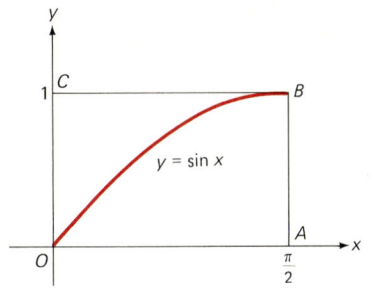

Figure 68

Find the values of the parameter t for which the point P lies on the given parametric curve (defined for all t).

24. $P = (0, 0)$ on $x = t^3 + 2t^2 - t - 2$, $y = t^4 - 5t^2 + 4$
25. $P = (-3, 0)$ on $x = 2 \cos t - \cos 2t$, $y = 2 \sin t - \sin 2t$
26. $P = (1, 2)$ on $x = \tan t$, $y = 2 \sin^2 t + \sin 2t$

Graph the given parametric curve, where t varies over any interval of length $\geq 2\pi$. Each curve is known as a *Lissajous figure*, and will appear on the screen of a cathode ray oscilloscope if voltages equal to $x = x(t)$ and $y = y(t)$ are applied to the horizontal and vertical plates of the oscilloscope.

27. $x = \sin t$, $y = \sin 2t$
28. $x = \sin t$, $y = \sin 3t$
29. $x = \sin 2t$, $y = \sin 3t$
30. $x = \sin 3t$, $y = \sin 4t$

31. Given a parametric curve
$$x = x(t), \quad y = y(t) \quad (a \leq t \leq b),$$
suppose the derivatives $x'(t)$, $y'(t)$ exist and are finite on the open interval (a, b), with $[x'(t)]^2 + [y'(t)]^2 \neq 0$ at every point of (a, b), and suppose the endpoints $A = (x(a), y(a))$, $B = (x(b), y(b))$ of the curve do not coincide. Show that there is at least one point of the curve (not an endpoint) where the tangent to the curve is parallel to the chord joining A and B. [The chord and tangent may be nonvertical as in Figure 69(a), or vertical as in Figure 69(b).]

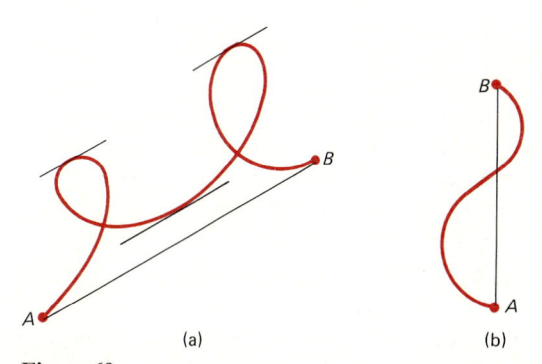

Figure 69

Supplementary Problems **483**

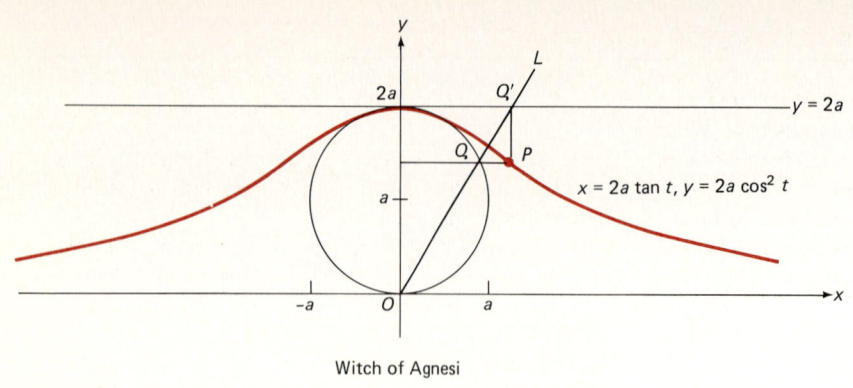

Figure 70

Witch of Agnesi

32. The bell-shaped curve

$$x = 2a \tan t, \quad y = 2a \cos^2 t \quad (-\pi/2 < t < \pi/2, a > 0),$$

shown in Figure 70, is known as the *witch of Agnesi*. Write a cartesian equation for the witch. Show that it is symmetric about the y-axis and has the x-axis as an asymptote. Find its inflection points. Show that if horizontal and vertical lines are drawn through an arbitrary point P of the witch, and if the horizontal line first intersects the circle $x^2 + (y - a)^2 = a^2$ in the point Q, while the vertical line intersects the line $y = 2a$ in the point Q', then the points Q and Q' lie on a line L through the origin O, as shown in the figure. What is the geometric meaning of the parameter t?

33. The curve

$$x = t - a \tanh \frac{t}{a}, \quad y = a \operatorname{sech} \frac{t}{a} \quad (-\infty < t < \infty, a > 0),$$

shown in Figure 71, is known as the *tractrix*. Verify that the tractrix is symmetric about the y-axis and has the x-axis as an asymptote, and also that it is the graph of a function $y = f(x)$ and has a cusp at the point $(0, a)$. Show that the tractrix is *equitangential*, in the sense that if T is the tangent line to the curve at an arbitrary point P and if Q is the x-intercept of T (see the figure), then $|PQ|$ has the constant value a. What is the geometric meaning of the parameter t?

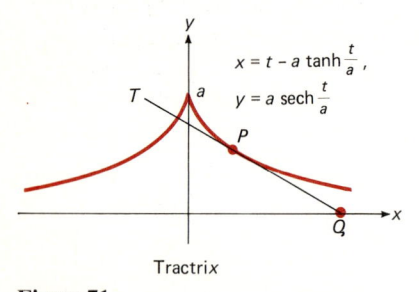

Tractrix

Figure 71

34. A dog initially at the point $(0, a)$ of the positive y-axis chases a fox initially at the origin. The fox runs along the positive x-axis, with the dog in hot pursuit. Suppose the dog always points directly at the fox, while remaining at distance a from the fox. What is the dog's pursuit curve?

35. The curve

$$x = 3a(t^2 - 3), \quad y = at(t^2 - 3) \quad (-\infty < t < \infty, a > 0),$$

shown in Figure 72, is known as *Tschirnhausen's cubic*. Write a cartesian equation for the cubic, and show that it is symmetric about the x-axis. Find the tangents to the cubic at the origin.

Tschirnhausen's cubic

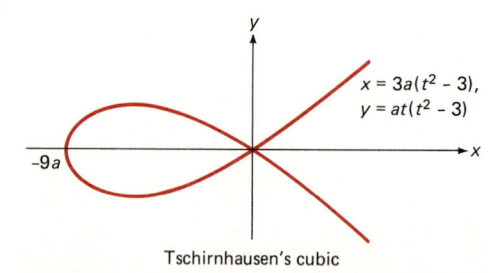

Figure 72

36. If $b = \frac{1}{3}a$, the hypocycloid considered in Problem 26, page 465, becomes a curve known as the *deltoid* or *hypocycloid of three cusps*. Write parametric equations for the deltoid. Sketch the deltoid, and calculate its length.

Find the length of the given curve.

37. $x = \dfrac{1 - t^2}{1 + t^2}, \ y = \dfrac{2t}{1 + t^2} \quad (-\infty < t < \infty)$

38. $x = -\dfrac{1}{t^2} \cos t + \dfrac{1}{t} \sin t, \ y = \dfrac{1}{t^2} \sin t + \dfrac{1}{t} \cos t \quad (1 \le t \le 2)$

39. $x = 3 \cos t - \cos 3t, \ y = 3 \sin t - \sin 3t \quad (-\pi \le t \le \pi)$

40. $y = \dfrac{1}{10} x^5 + \dfrac{1}{6x^3} \quad (1 \le x \le 2)$

41. $x = \ln \csc y \quad (\pi/4 \le y \le 3\pi/4)$

484   Chapter 8   Further Applications of Integration

Find the area of the surface generated by revolving the given curve about the specified axis.

42. $x = 2\cos t + \cos 2t$, $y = 2\sin t - \sin 2t$ ($\frac{2}{3}\pi \leq t \leq \frac{4}{3}\pi$), y-axis
43. $x = 3\cos t - \cos 3t$, $y = 3\sin t - \sin 3t$ ($0 \leq t \leq \pi$), x-axis
44. $y = |x-1| + |x|$ ($-1 \leq x \leq 2$), the line $y = 2$
45. $x = 2y^{1/4} - \frac{2}{7}y^{7/4}$ ($0 \leq y \leq 1$), x-axis

46. Let S be the surface generated by revolving the ellipse

$$\frac{x^2}{a^2} + \frac{y^2}{b^2} = 1 \qquad (a > b > 0)$$

about the x-axis. Then S is called an *ellipsoid of revolution* or *spheroid*. Since $a > b$, the spheroid S is *prolate*, that is, stretched or elongated like a football, or a cigar if very elongated [see Figure 73(a)]. Show that S has surface area

$$A = 2\pi b^2 + \frac{2\pi ab}{e} \arcsin e,$$

where the number

$$e = \frac{\sqrt{a^2 - b^2}}{a}$$

(not to be confused with the base of the natural logarithms!) is called the *eccentricity* of the ellipse. Suppose the ellipse is revolved about the y-axis, instead of the x-axis. Then, since $b < a$, the resulting spheroid S' is *oblate*, that is, compressed or flattened like a beach ball with a child sitting on it, or a frisbee if very flattened [see Figure 73(b)]. Show that S' has surface area

$$A' = 2\pi a^2 + \frac{\pi b^2}{e} \ln \frac{1+e}{1-e},$$

where e is again the eccentricity. Show that A and A' approach the same limit $4\pi a^2$ as $e \to 0^+$ with a held fixed. Why is this to be expected?

47. Let S and S' be the same as in the preceding problem. Find the volume V of the solid prolate spheroid bounded by S, and the volume V' of the solid oblate spheroid bounded by S'.

48. The planet Jupiter is an oblate spheroid with equatorial radius $\approx 71{,}600$ km and polar radius $\approx 67{,}300$ km. Estimate the volume and surface area of Jupiter. (The earth and the sun are also oblate, but only very slightly.)

49. A regular hexagon of side length a is revolved about one of its sides. Find the area of the resulting surface of revolution. What is the volume of the solid bounded by this surface?

50. A wading pool is 8 ft wide, 16 ft long, 2 ft deep at the shallow end, and 4 ft deep at the deep end. How much work is required to pump all the water in the pool to the top if the pool is full? If the water is 6 in. deep at the shallow end? If the water is 1 ft deep at the deep end? How long will it take a 0.125-hp pump to empty the pool if it is 75% full?

51. A trough shaped like half of a right circular cylinder of radius r and length L (see Figure 74) is filled with fluid of density ρ. How much work is required to pump all the fluid to a height h above the top of the trough?

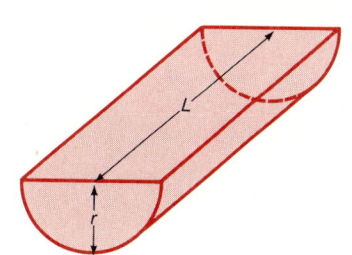

Figure 74

52. Figure 75 shows a tank shaped like part of a paraboloid of revolution, of height h and radius r at the top. Find the work required to fill the tank with fluid of density ρ through a hole in its bottom (the vertex of the paraboloid).

Prolate spheroid
(a)

Oblate spheroid
(b)

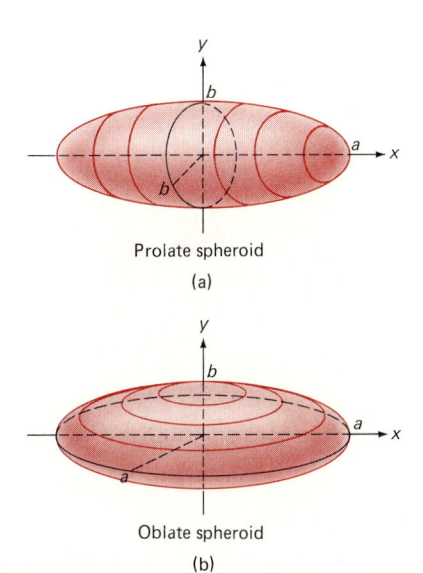

Figure 73

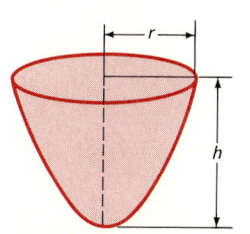

Figure 75

53. Find the hydrostatic force on one side of a vertical plate submerged in a fluid of weight density δ, if the plate is shaped like a circular washer of inner radius r_1 and outer radius r_2 with its center a distance $h \geq r_2$ below the surface of the fluid (see Figure 76).

54. A tank full of fluid of weight density δ has a vertical rectangular wall with its top edge along the surface of the fluid. Suppose the wall is divided into two parts by a diagonal of the rectangle. Show that the force exerted by the fluid on one part of the wall is twice the force on the other part.

A wooden ball 1 ft in radius is floating half submerged in a fluid of weight density δ. Find the work required

55. To lift the ball up until it just clears the fluid

56. To push the ball down until it is just completely submerged

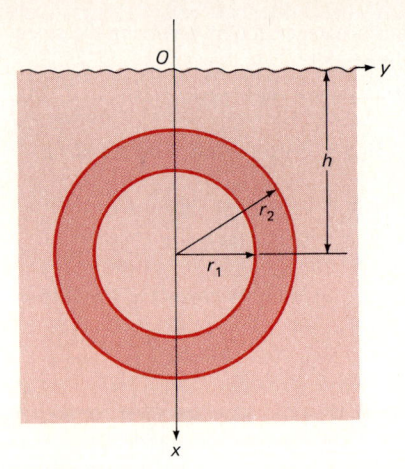

Figure 76

9 Sequences and Series

The techniques developed in this chapter are indispensable for carrying out numerical calculations. With their help, we can approximate numbers like e and π or values of functions like $\ln x$ and $\sin x$ *to any desired accuracy*. This is accomplished by representing the number or the function value as an *infinite series*, that is, as a sum with an infinite number of terms.

As a stepping stone to the study of infinite series, we first consider the allied topic of *infinite sequences* (see Section 9.1); these can be regarded as functions defined only on the set of positive integers. Sections 9.2–9.5 are devoted to a systematic investigation of infinite series whose terms are *numbers*. The study of *power series* is next on the agenda (see Sections 9.6–9.7); these series, whose terms are *functions*, can be regarded as polynomials with infinitely many terms and of arbitrarily large degree. It is sometimes possible to recognize the function which is the sum of a given power series. But of greater importance is the ability to start from a given function f and represent it as the sum of a power series. In Section 9.8 we show that f can be written as the sum of a polynomial and a "remainder term" if f is sufficiently well-behaved, and in Section 9.9 we go a step further and show that f can be written as the sum of a power series, called the *Taylor series* of f, in those commonly encountered cases in which the remainder term approaches zero as the degree of the approximating polynomial is made arbitrarily large.

Of the many powerful computational techniques presented in this chapter, *Newton's method* for solving the equation $f(x) = 0$ (see Section 9.10) deserves special mention not only in its own right, but also as a prime example of the *iterative methods* used throughout applied mathematics.

9.1 Infinite Sequences

Suppose that with each positive integer n we associate a real number a_n. Then the list

$$a_1, a_2, \ldots, a_n, \ldots \tag{1}$$

with the subscripts written in increasing order from left to right, is called an *infinite sequence*, or simply a *sequence*. The numbers in the list are called the *terms* of the sequence. Thus a_1 is the first term of the sequence, a_2 the second term, and so on up to the nth term a_n, after which the sequence continues on indefinitely, as indicated by the second group of three dots. The nth term a_n is also known as the *general term* of the sequence.

It is of course understood that the terms of a sequence are uniquely determined, in the sense that there is one and only one term with a given subscript. Therefore, from a formal standpoint, a sequence is just a special kind of function, namely one whose domain is the set of all positive integers. In less formal language, a sequence is completely determined once we know its "law of formation," that is, the function $a_n = f(n)$ leading from the subscript n (which plays the role of the independent variable) to the general term a_n.

Example 1 Find the second, fifth and seventh terms of the sequence with general term
$$a_n = 2^n.$$

Solution Choosing $n = 2, 5$ and 7, we have
$$a_2 = 2^2 = 4, \quad a_5 = 2^5 = 32, \quad a_7 = 2^7 = 128.$$

This sequence can be written as
$$2, 4, \ldots, 2^n, \ldots,$$
or, in more detail, as
$$2, 4, 8, 16, 32, 64, \ldots, 2^n, \ldots \quad \square$$

Another way of denoting the sequence (1) is to write its general term a_n between braces:†
$$\{a_n\}. \tag{1'}$$

For example, in this notation the sequence $2, 4, \ldots, 2^n, \ldots$ takes the concise form $\{2^n\}$.

Example 2 Write the first four terms of the sequence $\{(-1)^n\}$.

Solution Since $(-1)^1 = -1, (-1)^2 = 1, (-1)^3 = -1, (-1)^4 = 1$, the first four terms are $-1, 1, -1, 1$, with "alternating" signs. The sequence $\{(-1)^n\}$, like every infinite sequence, has infinitely many terms, but they only take one of two values, namely 1 or -1. This is to be contrasted with the sequence $\{2^n\}$, where no two terms have the same value. $\quad \square$

Example 3 Write the first seven terms of the sequence $\{n!\}$, whose general term is n factorial (see page 142). What is the eleventh term?

Solution The first seven terms are easily found to be
$$1, 2, 6, 24, 120, 720, 5040,$$
from which it is apparent that the terms of this sequence rapidly become very large. Calculating the product of the first 11 positive integers, we find that the eleventh term is $11! = 39{,}916{,}800$. $\quad \square$

---

† The context prevents any confusion between the sequence $\{a_n\}$ and the set whose only element is a_n.

Example 4 Find the average of the first five terms of the sequence $\{b_n\}$, where

$$b_n = \begin{cases} n & \text{if } n \text{ is odd,} \\ \dfrac{1}{n} & \text{if } n \text{ is even.} \end{cases}$$

Solution Observe that this time we denote the sequence by $\{b_n\}$ instead of $\{a_n\}$. (The notation for sequences enjoys as much freedom as that for functions.) The average of the first five terms is

$$\frac{1}{5}(b_1 + b_2 + b_3 + b_4 + b_5) = \frac{1}{5}\left(1 + \frac{1}{2} + 3 + \frac{1}{4} + 5\right) = \frac{39}{20}. \quad \square$$

Recursion Formulas

The law of formation of a sequence is often given by an explicit formula for its general term, as in the preceding examples. A sequence may also be defined *recursively*, that is, by giving a formula, called a *recursion formula*, which shows how each term can be obtained from the terms with lower subscripts.

Example 5 Let $\{a_n\}$ be the sequence defined by the recursion formula

$$a_1 = 1, \qquad a_n = a_{n-1} + n \quad \text{if} \quad n \geq 2. \tag{2}$$

Then

$$a_1 = 1, \quad a_2 = a_1 + 2 = 3, \quad a_3 = a_2 + 3 = 6, \quad a_4 = a_3 + 4 = 10, \ldots,$$

and the sequence $\{a_n\}$ begins with the terms $1, 3, 6, 10, \ldots$ As an exercise, verify that $\{a_n\}$ can also be defined explicitly (nonrecursively) as the sequence with general term $a_n = \frac{1}{2}n(n+1)$.

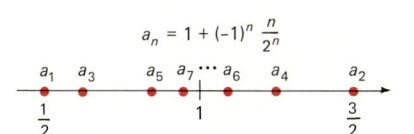

Example 6 Since $n! = n(n-1)!$, the sequence $\{a_n\} = \{n!\}$ is the same as the sequence defined by the recursion formula

$$a_1 = 1, \qquad a_n = na_{n-1} \quad \text{if} \quad n \geq 2. \quad \square$$

Sequences can be represented graphically in two ways, either by plotting the terms $a_1, a_2, \ldots, a_n, \ldots$ as points on a number line, or by plotting the ordered pairs $(1, a_1), (2, a_2), \ldots, (n, a_n), \ldots$, one for each term, as points in a coordinate plane. These two ways of representing the sequence $\{a_n\} = \{1 + (-1)^n(n/2^n)\}$ are shown in Figures 1 and 2. The first way is simpler, but the second has the merit of emphasizing that sequences are functions.

Figure 1

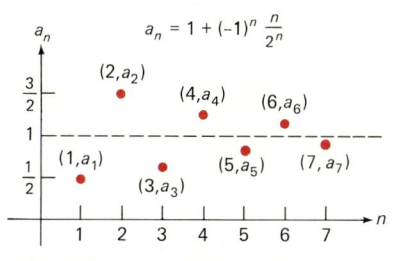

Figure 2

The Limit of a Sequence

Just as a function $f(x)$ can approach a limit as $x \to \infty$, a sequence $\{a_n\}$ can approach a limit as $n \to \infty$. Given any sequence $\{a_n\}$, suppose the general term a_n can be made as close as we please to some number L by choosing a large enough value of n. Then we say that the sequence $\{a_n\}$ *approaches* or *has the limit* L (as n approaches infinity), and we write

$$\lim_{n \to \infty} a_n = L \tag{3}$$

or

$$a_n \to L \qquad (\text{as } n \to \infty). \tag{3'}$$

Section 9.1 Infinite Sequences

The exact meaning of

$$\lim_{x \to \infty} f(x) = L$$

is that given any $\varepsilon > 0$, we can find a number $A > 0$ such that $|f(x) - L| < \varepsilon$ for all $x > A$, and similarly the exact meaning of (3) is that given any $\varepsilon > 0$, we can find an integer $N > 0$ such that $|a_n - L| < \varepsilon$ for all $n > N$, that is, for all $n = N + 1, N + 2, \ldots$ Equivalently, $a_n \to L$ means that given any $\varepsilon > 0$, the interval $(L - \varepsilon, L + \varepsilon)$ contains all the terms of the sequence $\{a_n\}$ with subscripts exceeding some integer N, where of course N depends on the choice of ε. In particular, choosing $\varepsilon = 1$, we see that if $a_n \to L$, then all the terms of the sequence $\{a_n\}$ with subscripts exceeding some integer N lie in the interval $(L - 1, L + 1)$; this fact will be used in the proof of Theorem 1 below.

Convergent and Divergent Sequences

If a sequence has a finite limit, it is said to *converge* (to this limit) or to be *convergent*, but otherwise the sequence is said to *diverge* or to be *divergent*. Two kinds of divergent sequences are worthy of special mention. We say that a sequence $\{a_n\}$ *diverges to* ∞, and write $a_n \to \infty$, if given any $C > 0$, there is an integer $N > 0$ such that $a_n > C$ for all $n > N$; similarly, we say that a sequence $\{a_n\}$ *diverges to* $-\infty$, and write $a_n \to -\infty$, if given any $C > 0$, there is an integer $N > 0$ such that $a_n < -C$ for all $n > N$. Of course, there are other ways in which a sequence can be divergent (see Example 9).

Example 7 The sequence $\{a_n\} = \{1/n\}$ is convergent with limit 0. In fact, given any $\varepsilon > 0$, let N be any integer greater than $1/\varepsilon$. Then

$$|a_n - 0| = \left|\frac{1}{n}\right| = \frac{1}{n} < \varepsilon$$

for all $n > N$, since

$$\frac{1}{n} < \frac{1}{N} < \varepsilon$$

for all such n. As an exercise, use a similar argument to show that the sequence $\{a_n\} = \{(n-1)/n\}$ is convergent with limit 1.

Example 8 Given any $\varepsilon > 0$, it is apparent from Figure 3 (a modification of Figure 2) that the interval $(1 - \varepsilon, 1 + \varepsilon)$ contains all but a finite number of terms of the sequence $\{a_n\} = \{1 + (-1)^n (n/2^n)\}$. In other words, given any interval $(1 - \varepsilon, 1 + \varepsilon)$ there is some integer N such that a_n lies in $(1 - \varepsilon, 1 + \varepsilon)$ for all $n > N$. Thus, without bothering to find the value of N corresponding to any given ε, we are able to conclude (informally) that

$$\lim_{n \to \infty} \left[1 + (-1)^n \frac{n}{2^n}\right] = 1.$$

Admittedly, this conclusion should be confirmed by giving a formal proof (see Problem 53).

Example 9 The sequence $\{a_n\} = \{(-1)^n\}$ is divergent, as shown by the following argument. Take any proposed limit L, and make ε so small that the interval $I = (L - \varepsilon, L + \varepsilon)$ fails to contain at least one of the points 1 and

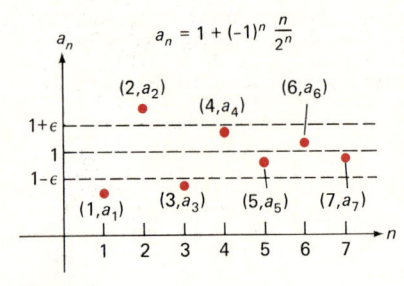

Figure 3

-1. Clearly this can always be done, even if $L = 1$ or $L = -1$. Since $(-1)^n = 1$ if n is even, all the terms a_n with even n lie outside the interval I if it fails to contain the point 1, and since $(-1)^n = -1$ if n is odd, all the terms a_n with odd n lie outside the interval I if it fails to contain the point -1. Thus, in any event, the sequence cannot be convergent.

Example 10 The sequence $\{2^n\}$ diverges to ∞, since 2^n becomes arbitrarily large for sufficiently large n. In fact, given any $C > 0$, we can make $2^n > C$ by choosing $n > \log_2 C$.

Example 11 It was shown in Example 8, page 329, that

$$\lim_{x \to \infty} \left(1 + \frac{1}{x}\right)^x = e,$$

where $e = 2.7182818\ldots$ is the base of the natural logarithms. This means that given any $\varepsilon > 0$, there is a number $A > 0$ such that

$$\left|\left(1 + \frac{1}{x}\right)^x - e\right| < \varepsilon$$

for every real number x exceeding A. But then it is certainly true that

$$\left|\left(1 + \frac{1}{n}\right)^n - e\right| < \varepsilon$$

for every integer n exceeding A, and hence

$$\lim_{n \to \infty} \left(1 + \frac{1}{n}\right)^n = e,$$

as anticipated on page 329. More generally, if f is a function defined on $[1, \infty)$ such that $f(x) \to L$ as $x \to \infty$, virtually the same argument shows that $f(n) \to L$ as $n \to \infty$. In particular, it follows from formula (15), page 330, that

$$\lim_{n \to \infty} \left(1 + \frac{a}{n}\right)^n = e^a,$$

where a is an arbitrary number.

Remark Let $\{a_n\}$ and $\{b_n\}$ be two sequences which differ only in the values of a finite number of terms. Then it is easy to see that either $\{a_n\}$ and $\{b_n\}$ are both convergent, with the same limit, or $\{a_n\}$ and $\{b_n\}$ are both divergent.

Bounded and Unbounded Sequences

A sequence $\{a_n\}$ is said to be *bounded* if there is some number $C > 0$ such that $-C \le a_n \le C$, or equivalently $|a_n| \le C$, for all n, but if no such number exists, we say that $\{a_n\}$ is *unbounded*. (These are the analogues of the corresponding definitions for functions on page 62.) For example, the sequences $\{1/n\}$ and $\{(-1)^n\}$ are both bounded, since $|1/n| \le 1$ and $|(-1)^n| \le 1$ for all n. On the other hand, the sequences $\{n\}$ and $\{n!\}$ are both unbounded, since n and $n!$ exceed any given positive number C if n is large enough (note that $n! > n$ if $n > 2$). As the following theorem shows, the concept of a bounded sequence arises in a very natural way when studying convergent sequences.

Theorem 1 *(Boundedness of a convergent sequence).* Every convergent sequence is bounded.

Proof Let $\{a_n\}$ be a convergent sequence with limit L. Then there is an integer N such that all the terms $a_{N+1}, a_{N+2}, \ldots,$ that is, all the terms of the sequence with subscripts exceeding N, lie in the interval $(L - 1, L + 1)$. By choosing $C > 0$ large enough, we can see to it that the interval $[-C, C]$, with its midpoint at the origin, contains not only the interval $(L - 1, L + 1)$, and with it all the terms $a_{N+1}, a_{N+2}, \ldots,$ but also all the remaining terms $a_1, a_2, \ldots, a_N$ as well. But then $|a_n| \leq C$ for all $n = 1, 2, \ldots,$ so that the sequence $\{a_n\}$ is bounded. ∎

Example 12 The construction used in the proof of Theorem 1 is illustrated in Figure 4 for the sequence with general term

$$a_n = 1 + (-1)^n \frac{5}{n}.$$

Here $L = 1$, $N = 5$, and any $C \geq 4$ works.

Since a convergent sequence must be bounded, *an unbounded sequence must be divergent.* For example, the unbounded sequences $\{n\}$ and $\{n!\}$ are divergent. On the other hand, a bounded sequence need not be convergent. In fact, we have already observed that the sequence $\{(-1)^n\}$ is both bounded and divergent.

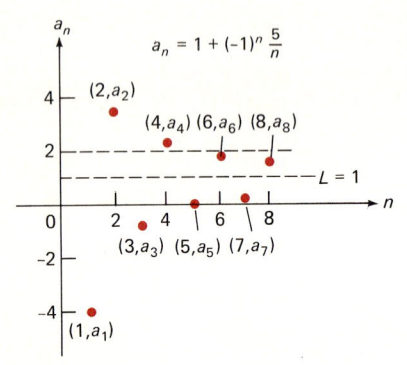

Figure 4

Limit Rules for Sequences

Algebraic operations on sequences are defined in the same way as for functions, that is, the sum, difference, product and quotient of two given sequences $\{a_n\}$ and $\{b_n\}$ are the sequences with general terms $a_n + b_n$, $a_n - b_n$, $a_n b_n$ and a_n/b_n. Let $\{a_n\}$ and $\{b_n\}$ be convergent sequences, and suppose that

$$\lim_{n \to \infty} a_n = L, \quad \lim_{n \to \infty} b_n = M. \tag{4}$$

Then

$$\lim_{n \to \infty} (a_n + b_n) = L + M, \tag{5}$$

$$\lim_{n \to \infty} (a_n - b_n) = L - M, \tag{6}$$

$$\lim_{n \to \infty} a_n b_n = LM, \tag{7}$$

$$\lim_{n \to \infty} \frac{a_n}{b_n} = \frac{L}{M}, \tag{8}$$

provided that $M \neq 0$ in the last formula. (A finite number of terms of the sequence $\{a_n/b_n\}$ may have zero denominators and hence fail to exist, but $b_n \neq 0$ for large enough n, since $b_n \to M \neq 0$ as $n \to \infty$.) In other words, the sum $\{a_n + b_n\}$, difference $\{a_n - b_n\}$, product $\{a_n b_n\}$ or quotient $\{a_n/b_n\}$ of two given convergent sequences $\{a_n\}$ and $\{b_n\}$ is itself a convergent sequence, with a limit equal to the sum, difference, product or quotient of the limits of $\{a_n\}$ and $\{b_n\}$. The rules (5)–(8) are the analogues of the corresponding rules for limits of functions, and they are proved in essentially the same way.

Optional
As an example of how this is done, we show that (4) implies (5), using a slight modification of the argument used to prove Theorem 4, page 80. Suppose (4) holds. Then given any $\varepsilon > 0$, we can find positive integers N_a and N_b such that $|a_n - L| < \varepsilon/2$ for all $N > N_a$ and $|b_n - M| < \varepsilon/2$ for all $n > N_b$. Therefore, by the triangle inequality,

$$|(a_n + b_n) - (L + M)| = |(a_n - L) + (b_n - M)|$$

$$\leq |a_n - L| + |b_n - M| < \frac{\varepsilon}{2} + \frac{\varepsilon}{2} = \varepsilon$$

for all $n > N = \max\{N_a, N_b\}$. But in "ε, N language" this means that (5) holds.

Given any number c, the *constant sequence* $\{c\}$, all of whose terms equal c, is clearly convergent, with limit c, so that

$$\lim_{n \to \infty} c = c.$$

It follows from (7) that if $\{a_n\}$ converges to L, then $\{ca_n\}$ converges to cL.

Example 13 Evaluate $\lim\limits_{n \to \infty} \dfrac{2n}{3n + 1}$.

Solution With the help of Example 7 and some of the above rules, we have

$$\lim_{n \to \infty} \frac{2n}{3n+1} = \lim_{n \to \infty} \frac{2}{3 + \frac{1}{n}} = \frac{\lim_{n \to \infty} 2}{\lim_{n \to \infty}\left(3 + \frac{1}{n}\right)} = \frac{2}{\lim_{n \to \infty} 3 + \lim_{n \to \infty} \frac{1}{n}} = \frac{2}{3 + 0} = \frac{2}{3}.$$

A more informal way of evaluating this limit, which avoids many steps, is to observe that $3n + 1 \approx 3n$ for large n, so that

$$\frac{2n}{3n+1} \approx \frac{2n}{3n} = \frac{2}{3},$$

where the approximation gets better and better as $n \to \infty$. □

Monotonic Sequences

A sequence $\{a_n\}$ is said to be *increasing* if $a_n \leq a_{n+1}$ for all n and *decreasing* if $a_n \geq a_{n+1}$ for all n. By a *monotonic* sequence is meant a sequence which is either increasing or decreasing. Here we write $\leq$ and $\geq$, instead of $<$ and $>$ as in the corresponding definitions for functions (see page 49), calling the sequence $\{a_n\}$ *strictly increasing* if $a_n < a_{n+1}$ for all n and *strictly decreasing* if $a_n > a_{n+1}$ for all n (of course, strictly increasing and strictly decreasing sequences are monotonic). These definitions prepare us for the following basic result.

> **Theorem 2** *(Convergence of a bounded monotonic sequence).* *Every bounded monotonic sequence is convergent.*

Proof (Optional) Suppose $\{a_n\}$ is a bounded increasing sequence. Then, by the boundedness of $\{a_n\}$, there is a number C such that $a_n \leq C$ for all $n = 1, 2, \ldots$ Such a number C is called an *upper bound* of the sequence $\{a_n\}$. There are of course infinitely many upper bounds of $\{a_n\}$, and in fact any number greater than C is also an upper bound, but one of these upper bounds, which we denote by L for *least* upper bound, is the smallest.† Now, given any $\varepsilon > 0$, there must be a term of the sequence $\{a_n\}$, say a_N, such that

$$L - \varepsilon < a_N \leq L,$$

since otherwise the number $L - \varepsilon$, which is smaller than L, would be an upper bound of $\{a_n\}$, contrary to the very definition of L. But then since $\{a_n\}$ is increasing (and L is an upper bound), we have

$$L - \varepsilon < a_N \leq a_n \leq L$$

for all $n > N$. Therefore $|a_n - L| < \varepsilon$ for all $n > N$, that is, $\{a_n\}$ is convergent with limit L.

On the other hand, suppose $\{a_n\}$ is a bounded *decreasing* sequence. Then $\{-a_n\}$ is a bounded increasing sequence. Hence, as just proved, $\{-a_n\}$ is convergent, with a limit which we denote by $-L$. But then

$$\lim_{n \to \infty} a_n = -\lim_{n \to \infty} (-a_n) = -(-L) = L,$$

so that $\{a_n\}$ is also convergent, with limit L. ∎

The intuitive meaning of Theorem 2 is clear. If the terms of a sequence are confined to a finite interval and can never get smaller as n increases, then they must all eventually "cluster" or "accumulate" at some point L, which is the limit of the sequence. Figure 5 illustrates this phenomenon for the bounded increasing sequence $\{a_n\} = \{1 - (1/n)\}$, which has the number $L = 1$ as both its least upper bound and its limit.

Figure 5

Example 14 Investigate the convergence of the sequence $\{r^n\}$, where r is any real number.

Solution Suppose first that $0 < r < 1$. Then the sequence $\{r^n\}$ is bounded, since $0 < r^n < 1$ for all n, and it is also (strictly) decreasing, since $r^{n+1} = r(r^n) < r^n$ for all n. Hence, by Theorem 2, $\{r^n\}$ converges to some limit L. To determine L, we observe that

$$L = \lim_{n \to \infty} r^{n+1} = \lim_{n \to \infty} r(r^n) = r \lim_{n \to \infty} r^n = rL.$$

Since $r \neq 1$, this is possible only if $L = 0$, and therefore

$$\lim_{n \to \infty} r^n = 0 \quad (0 < r < 1). \tag{9}$$

Next let $-1 < r < 0$. Then $0 < |r| < 1$, so that $\{|r|^n\}$ converges to 0, by (9) with r replaced by $|r|$. But then $\{r^n\}$ also converges to 0, since $|r^n - 0| = |r^n| = |r|^n$ becomes as small as we please for sufficiently large n. It follows that

$$\lim_{n \to \infty} r^n = 0 \quad (-1 < r < 0). \tag{9'}$$

Combining this with (9) and the evident fact that the sequence $\{0^n\} = \{0\}$

---

† In assuming the existence of L, we are relying on a fundamental property of the real number system, known as *completeness*, which states that any set of real numbers with an upper bound has a least upper bound. For more on completeness, see any book on advanced calculus.

converges to 0, we find that

$$\lim_{n \to \infty} r^n = 0 \quad (-1 < r < 1).$$

Thus $\{r^n\}$ converges to 0 if $-1 < r < 1$, or equivalently if $|r| < 1$.

If $r = 1$, $\{r^n\}$ is the constant sequence $\{1\}$, which obviously converges to 1, while if $r = -1$, $\{r^n\}$ is the divergent sequence $\{(-1)^n\}$. If $r > 1$, then

$$r^n = [1 + (r-1)]^n \geq 1 + n(r-1) > n(r-1),$$

with the help of the binomial theorem (see page 227), so that $\{r^n\}$ diverges to ∞. In fact, given any $C > 0$, we have $r^n > C$ for all $n > C/(r-1)$; alternatively, if $r > 1$, then $r^x \to \infty$ as $x \to \infty$ (see page 317), which implies $r^n \to \infty$ as $n \to \infty$. If $r < -1$, then $|r| > 1$ and $\{|r|^n\}$ diverges to ∞. But then

$$r^n = (-|r|)^n = (-1)^n |r|^n$$

takes arbitrarily large positive values if n is even and arbitrarily large negative values if n is odd. Therefore $\{r^n\}$ has no limit if $r < -1$. Thus, finally, *the sequence $\{r^n\}$ converges to 0 if $|r| < 1$ and converges to 1 if $r = 1$, while it diverges to ∞ if $r > 1$ and has no limit at all if $r \leq -1$.* ☐

A letter other than n is often chosen for the variable subscript of the general term of a sequence, usually another letter from the middle of the alphabet. Thus $\{2^i\}$, $\{2^j\}$, $\{2^k\}$ are three other ways of denoting the same sequence $\{2^n\}$ as in Example 1.

Given any sequence $a_1, a_2, \ldots, a_n, \ldots$, let

$$a_{n_1}, a_{n_2}, \ldots, a_{n_i}, \ldots \quad (n_1 < n_2 < \cdots < n_i < \cdots)$$

be any "subsequence" extracted from $\{a_n\}$ by omitting a finite or infinite number of terms. Then, if $\{a_n\}$ is convergent with limit L, $\{a_{n_i}\}$ must also converge to L (why?). In particular, the subsequence $\{a_{n+N}\}$ obtained from $\{a_n\}$ by omitting the first N terms must converge to L, the subsequence $\{a_{2k}\}$ consisting of the even-numbered terms must converge to L, and so on. By the same token, $\{a_n\}$ must be divergent if $\{a_n\}$ has two subsequences converging to different limits.

Example 15 We already know from Example 9 that the sequence $\{(-1)^n\}$ is divergent. This also follows from the observation that the even-numbered terms converge to 1, while the odd-numbered terms converge to -1. In fact,

$$\lim_{k \to \infty} a_{2k} = \lim_{k \to \infty} (-1)^{2k} = \lim_{k \to \infty} 1 = 1,$$

while

$$\lim_{k \to \infty} a_{2k+1} = \lim_{k \to \infty} (-1)^{2k+1} = \lim_{k \to \infty} (-1) = -1.$$

Problems

Write the first six terms and find the limit L (if it exists) of the sequence $\{a_n\}$ with the given general term. (The case $L = \infty$ or $L = -\infty$ is allowed.)

1. $a_n = \dfrac{2n-1}{n+1}$

2. $a_n = \dfrac{1}{n^2+1}$

3. $a_n = \dfrac{n^2+n+1}{(n+1)^2}$

4. $a_n = (-1)^{n-1} n^2$

5. $a_n = n^{(-1)^n}$

6. $a_n = \dfrac{2^n-1}{2^n+1}$

Section 9.1 Infinite Sequences 495

7. $a_n = \dfrac{2^n + 1}{3^n}$

8. $a_n = (-1)^n + \dfrac{1}{n}$

9. $a_n = \dfrac{1 + (-1)^n}{2n}$

10. $a_n = 0.\underbrace{333\ldots3}_{n\text{ digits}}$

11. $a_n = \sqrt{2}$ to n decimal places

12. $a_n = \begin{cases} \dfrac{1}{2} - \dfrac{1}{n} & \text{if } n \text{ is odd,} \\ \dfrac{1}{2} + \dfrac{1}{n} & \text{if } n \text{ is even} \end{cases}$

13. $a_n = \begin{cases} \dfrac{1}{2} + \dfrac{1}{n} & \text{if } n \text{ is odd,} \\ 1 - \dfrac{1}{n} & \text{if } n \text{ is even} \end{cases}$

14. $a_n = \begin{cases} \sqrt{n} & \text{if } n \text{ is odd,} \\ n^2 & \text{if } n \text{ is even} \end{cases}$

15. $a_n = \cos\dfrac{n\pi}{3}$

16. $a_n = \dfrac{1}{n}\sin\dfrac{n\pi}{2}$

17. $a_n = \left(\dfrac{n}{n+1}\right)^n$

18. $a_n = \dfrac{n^2}{2^n}$

19. $a_n = \ln\dfrac{1}{n}$

20. $a_n = \dfrac{1}{n}\ln\dfrac{1}{n}$

Write the general term a_n of the given sequence and find the limit L of the sequence (if it exists), assuming that the law of formation suggested by the first few terms of the sequence holds for all the terms.

21. $0, 3, 8, 15, \ldots$

22. $1, \tfrac{1}{3}, \tfrac{1}{5}, \tfrac{1}{7}, \ldots$

23. $-\tfrac{1}{3}, \tfrac{3}{5}, -\tfrac{5}{7}, \tfrac{7}{9}, \ldots$

24. $0, \tfrac{3}{4}, \tfrac{8}{9}, \tfrac{15}{16}, \ldots$

25. $1, \tfrac{1}{6}, \tfrac{1}{15}, \tfrac{1}{28}, \ldots$

26. $\tfrac{1}{2}, -\tfrac{1}{4}, \tfrac{1}{8}, -\tfrac{1}{16}, \ldots$

27. $5, 0, -5, -10, \ldots$

28. $-5, 10, -15, 20, \ldots$

29. For what values of n are the terms of the sequence $\{(-\tfrac{1}{2})^n\}$ within 10^{-6} of its limit?

30. For what values of n are the terms of the sequence $\{(2n-1)/(2-3n)\}$ within 10^{-3} of its limit?

Write the first six terms of the sequence $\{a_n\}$ defined by the recursion formula

31. $a_1 = 0$, $a_n = 3a_{n-1} + 2$ if $n \geq 2$

32. $a_1 = -1$, $a_n = 1 - 4a_{n-1}$ if $n \geq 2$

33. $a_1 = \dfrac{1}{2}$, $a_n = \dfrac{1}{a_{n-1}} + 1$ if $n \geq 2$

34. $a_1 = 3$, $a_n = \dfrac{2}{a_{n-1} + 1}$ if $n \geq 2$

† 35. The sequence $\{a_n\}$, defined by the recursion formula

$$a_1 = 1, \quad a_2 = 1, \quad a_n = a_{n-1} + a_{n-2} \quad \text{if } n \geq 3,$$

is known as the *Fibonacci sequence*. Write the first ten terms of $\{a_n\}$.

36. A child starts a rabbit colony by introducing a pair of newborn rabbits, one male and one female, into a large pen. Suppose it takes a newborn pair of rabbits 1 month to achieve sexual maturity and 1 more month to produce another pair of rabbits. Assuming that no rabbits die and that each litter consists of one male and one female rabbit, born on the first day of the new month, show that the number of pairs of rabbits in the pen after n months is the nth term of the Fibonacci sequence.

37. Starting with any sequence $\{a_n\}$, let $\{s_n\}$ be another sequence defined by the recursion formula

$$s_1 = a_1, \quad s_n = s_{n-1} + a_n \quad \text{if } n \geq 2.$$

Write an expression for s_n which involves only terms of $\{a_n\}$. Write a simple formula for s_n in the case where $a_n = 2n - 1$.

38. Is the least upper bound of the sequence $\{1 + (-1)^n/n\}$ the same as its limit? Explain your answer.

Find a bounded sequence which has

39. A largest but no smallest term

40. A largest and a smallest term

41. Neither a largest nor a smallest term

42. A smallest but no largest term

43. Let $\{a_n\}$ be a bounded sequence (in particular, any convergent sequence), and let $\{b_n\}$ be a sequence converging to zero. Show that the sequence $\{a_n b_n\}$ also converges to zero.

44. Show that the two sequences with general terms a_n and

$$b_n = a_n + (n-1)(n-2)\cdots(n-N)$$

have the same first N terms, but differ in all subsequent terms. (Thus knowledge of any finite number of initial terms can never uniquely determine a sequence.)

† 45. Which terms of the two sequences $\{n^3 - 6n^2\}$ and $\{6 - 11n\}$ coincide?

46. Which of the sequences in Problems 1–28 are increasing? Decreasing?

Hint. If $a_n = f(n)$, the sequence $\{a_n\}$ is (strictly) increasing if $df(x)/dx > 0$ for all $x \geq 1$ and (strictly) decreasing if $df(x)/dx < 0$ for all $x \geq 1$. Explain why.

Let r be any real number. Evaluate

47. $\lim\limits_{n\to\infty} \dfrac{r^n}{1 + r^n}$ $(r \neq -1)$

48. $\lim\limits_{n\to\infty} \dfrac{r^n}{1 + r^{2n}}$

49. $\lim\limits_{n\to\infty} \dfrac{r^n}{n!}$

Find two divergent sequences $\{a_n\}$ and $\{b_n\}$ such that
50. $\{a_n + b_n\}$ converges   **51.** $\{a_n b_n\}$ converges
52. $\{a_n/b_n\}$ converges

53. Let c be any number greater than 1. Show that
$$\lim_{n\to\infty} \frac{n}{c^n} = 0.$$
In particular, use this to verify the limit in Example 8.

54. Let $\{a_n\}$ and $\{b_n\}$ be two convergent sequences with the same limit L, and suppose the sequence $\{c_n\}$ is such that $a_n \leq c_n \leq b_n$ for all n (or for all sufficiently large n). Show that $\{c_n\}$ also converges to L.
Hint. This is the analogue for sequences of Theorem 10, page 85 (the sandwich theorem).

55. Show that if $a_n \to L$ as $n \to \infty$ and if f is a function continuous at L, then $f(a_n) \to f(L)$ as $n \to \infty$.

56. With the help of the preceding problem, show that
$$\lim_{n\to\infty} \sqrt[n]{c} = 1,$$
where c is any positive number. Also show that
$$\lim_{n\to\infty} \sqrt[n]{n} = 1.$$

Use differentiation to find the largest term of the sequence with general term

57. $a_n = \sqrt[n]{n}$ **58.** $a_n = \dfrac{\sqrt{n}}{n + 2500}$

59. $a_n = \dfrac{n^{10}}{2^n}$

60. After verifying that the sequence $\{[1 + (1/n)]^n\}$ is strictly increasing, while $\{[1 + (1/n)]^{n+1}\}$ is strictly decreasing, show that
$$\left(1 + \frac{1}{n}\right)^n < e < \left(1 + \frac{1}{n}\right)^{n+1} \quad (n = 1, 2, \ldots). \quad \text{(i)}$$
What is the disadvantage of using this double inequality to estimate the number e?

61. Suppose k_n congruent circular disks occupying n rows are inscribed in an equilateral triangle in the way illustrated in Figure 6, so that
$$k_1 = 1, \quad k_2 = 1 + 2 = 3, \quad k_3 = 1 + 2 + 3 = 6, \ldots$$
Let A be the area of the triangle and A_n the total area of the k_n disks. Show that
$$\lim_{n\to\infty} \frac{A_n}{A} = \frac{\pi}{2\sqrt{3}}.$$

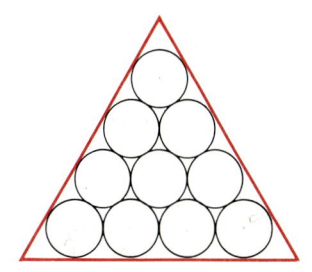

Figure 6

62. Let
$$I_m = \int_0^{\pi/2} \sin^m x \, dx.$$
Then, by the formulas established in Problem 13, page 381,
$$\frac{I_{2n+1}}{I_{2n}} = \frac{\dfrac{2}{3}\dfrac{4}{5}\cdots\dfrac{2n}{2n+1}}{\dfrac{1}{2}\dfrac{3}{4}\cdots\dfrac{2n-1}{2n}\dfrac{\pi}{2}} = \frac{2}{\pi}\frac{2}{1}\frac{2}{3}\frac{4}{3}\frac{4}{5}\cdots\frac{2n}{2n-1}\frac{2n}{2n+1}.$$
Show that $I_{2n+1}/I_{2n} \to 1$ as $n \to \infty$, and hence
$$\pi = 2 \lim_{n\to\infty} \frac{2}{1}\frac{2}{3}\frac{4}{3}\frac{4}{5}\cdots\frac{2n}{2n-1}\frac{2n}{2n+1}. \quad \text{(ii)}$$
This remarkable formula for π was discovered in 1650 by the English mathematician John Wallis.
Hint. Use Problem 54.

9.2 Infinite Series

Let $\{a_n\}$ be an infinite sequence. Then the expression
$$a_1 + a_2 + \cdots + a_n + \cdots, \quad (1)$$
or more concisely
$$\sum_{n=1}^{\infty} a_n, \quad (1')$$

is called an *infinite series*, or simply a *series*. In writing (1'), we use a modification of the sigma notation introduced on page 225, in which the upper limit of summation is the symbol ∞, rather than an integer. The numbers $a_1, a_2, \ldots, a_n, \ldots$ are called the *terms* of the series (as well as of the sequence $\{a_n\}$), and a_n is called the *n*th or *general term*.

Convergent and Divergent Series

At this point, the series (1) or (1') is not a number, but merely a formal expression, since we have not yet decided what is meant by the sum of *infinitely many* terms, a concept that plays no role in elementary mathematics. To attach meaning to such an "infinite sum," we proceed as follows. The sum of the first *n* terms of the series (1) or (1') is the perfectly well-defined number

$$s_n = a_1 + a_2 + \cdots + a_n,$$

called the *n*th *partial sum* of the series. (After all, there is no doubt about what is meant by the sum of a *finite* number of terms.) The consecutive partial sums form a sequence

$$s_1 = a_1, \; s_2 = a_1 + a_2, \ldots, s_n = \sum_{i=1}^{n} a_i, \ldots$$

Suppose this sequence is convergent with the limit

$$S = \lim_{n \to \infty} s_n, \tag{2}$$

as defined in Section 9.1. Then we say that the series (1) or (1') is *convergent* (or *converges*), and we assign it the *sum S*. However, if the sequence of partial sums $\{s_n\}$ is divergent, that is, if $\{s_n\}$ fails to approach a finite limit, we say that the series is *divergent* (or *diverges*), with no sum. If the series is convergent with sum *S*, we write

$$a_1 + a_2 + \cdots + a_n + \cdots = S,$$

or

$$\sum_{n=1}^{\infty} a_n = S$$

in sigma notation.

Just as in the case of a finite sum, the index of summation is a "dummy index," in the sense that any other symbol would do just as well. For example,

$$\sum_{i=1}^{\infty} a_i, \quad \sum_{k=1}^{\infty} a_k, \quad \sum_{n=1}^{\infty} a_n, \quad \sum_{p=1}^{\infty} a_p$$

all denote the same series. This observation allows us to write (2) in the form

$$\sum_{i=1}^{\infty} a_i = \lim_{n \to \infty} \sum_{i=1}^{n} a_i,$$

which is completely analogous to the definition

$$\int_{c}^{\infty} f(x)\,dx = \lim_{u \to \infty} \int_{c}^{u} f(x)\,dx$$

of a convergent improper integral with an unbounded interval of integration.

The process of finding the sum of a convergent series is called *summing the series*, although it actually consists of evaluating the limit of the sequence of partial sums of the series. If *m* is any positive integer (not necessarily 1), the series

$$\sum_{n=m}^{\infty} a_n$$

means

$$a_m + a_{m+1} + \cdots + a_n + \cdots,$$

that is, the series obtained from (1) by omitting the first $m - 1$ terms. Sometimes $m = 0$, so that the infinite series begins with a "zeroth term," as in the case of the important series considered in the following example.

The Geometric Series

Example 1 Investigate the convergence of the *geometric series*

$$\sum_{n=0}^{\infty} ar^n = a + ar + ar^2 + \cdots + ar^n + \cdots, \qquad (3)$$

where r and $a \neq 0$ are arbitrary constants.

Solution Notice that the geometric series begins with the term a, and each successive term is obtained by multiplying the preceding term by the number r, called the *ratio* of the series. The sum of the first n terms of the geometric series, that is, its nth partial sum, is just

$$s_n = a + ar + \cdots + ar^{n-1}.$$

To get a simple formula for s_n, we observe that

$$s_n - rs_n = (a + ar + \cdots + ar^{n-1}) - (ar + ar^2 + \cdots + ar^n),$$

where all but two terms cancel out, leaving

$$s_n - rs_n = a - ar^n.$$

It follows that $s_n(1 - r) = a(1 - r^n)$, or equivalently

$$s_n = \frac{a}{1 - r}(1 - r^n) \qquad (r \neq 1).$$

If $|r| < 1$, then $r^n \to 0$ as $n \to \infty$, by Example 14, page 494, and therefore

$$\lim_{n \to \infty} s_n = \frac{a}{1 - r},$$

which implies

$$\sum_{n=0}^{\infty} ar^n = \frac{a}{1 - r} \qquad (|r| < 1).$$

By the same example, $\{r^n\}$ is divergent if $|r| > 1$, and hence the same is true of the sequence $\{s_n\}$, and the series (3) is divergent. If $r = 1$, the series becomes

$$a + a + a + a + \cdots,$$

while if $r = -1$, it becomes

$$a - a + a - a + \cdots.$$

In the first case $s_n = na$, while in the second case

$$s_n = \begin{cases} a & \text{if } n \text{ is odd,} \\ 0 & \text{if } n \text{ is even,} \end{cases}$$

but in both cases $\{s_n\}$ is again divergent, and hence so is the series (3). Thus, to summarize, *the geometric series (3) is convergent with sum $a/(1-r)$ if $-1 < r < 1$ and divergent otherwise.* ∎

Example 2 Choosing $a = 1$, $r = \frac{1}{2}$ in the geometric series, we obtain

$$\sum_{n=0}^{\infty} \left(\frac{1}{2}\right)^n = 1 + \frac{1}{2} + \frac{1}{4} + \frac{1}{8} + \cdots + \frac{1}{2^n} + \cdots = \frac{1}{1 - \frac{1}{2}} = \frac{1}{\frac{1}{2}} = 2,$$

while the choice $a = 1$, $r = -\frac{1}{2}$ gives

$$\sum_{n=0}^{\infty} \left(-\frac{1}{2}\right)^n = 1 - \frac{1}{2} + \frac{1}{4} - \frac{1}{8} + \cdots + \frac{(-1)^n}{2^n} + \cdots = \frac{1}{1 - (-\frac{1}{2})} = \frac{1}{\frac{3}{2}} = \frac{2}{3}.$$

Example 3 For $a = 3$, $r = \frac{1}{10}$ the geometric series becomes

$$\sum_{n=0}^{\infty} 3\left(\frac{1}{10}\right)^n = 3 + \frac{3}{10} + \frac{3}{100} + \cdots + \frac{3}{10^n} + \cdots = \frac{3}{1 - \frac{1}{10}} = \frac{3}{\frac{9}{10}} = \frac{10}{3},$$

a result which is hardly surprising, since this sum is just the repeating decimal

$$3.333\ldots = 3.\overline{3} = \frac{10}{3}.$$

(The relationship between decimals and infinite series, and in particular between repeating decimals and geometric series, will be pursued in the next section.) On the other hand, for $a = \frac{1}{10}$, $r = 3$, we get the divergent geometric series

$$\sum_{n=0}^{\infty} \frac{3^n}{10} = \frac{1}{10} + \frac{3}{10} + \frac{9}{10} + \cdots + \frac{3^n}{10} + \cdots$$

Example 4 Show that the series

$$\sum_{n=1}^{\infty} \frac{1}{n(n+1)} = \frac{1}{1 \cdot 2} + \frac{1}{2 \cdot 3} + \cdots + \frac{1}{n(n+1)} + \cdots$$

is convergent, and find its sum.

Solution Expanding the general term in partial fractions, we find that

$$\frac{1}{n(n+1)} = \frac{1}{n} - \frac{1}{n+1},$$

and hence

$$\sum_{n=1}^{\infty} \frac{1}{n(n+1)} = \sum_{n=1}^{\infty} \left(\frac{1}{n} - \frac{1}{n+1}\right).$$

Thus the *n*th partial sum of the series is

$$s_n = \left(1 - \frac{1}{2}\right) + \left(\frac{1}{2} - \frac{1}{3}\right) + \left(\frac{1}{3} - \frac{1}{4}\right) + \cdots + \left(\frac{1}{n-1} - \frac{1}{n}\right) + \left(\frac{1}{n} - \frac{1}{n+1}\right)$$

$$= 1 + \left(-\frac{1}{2} + \frac{1}{2}\right) + \left(-\frac{1}{3} + \frac{1}{3}\right) + \cdots + \left(-\frac{1}{n} + \frac{1}{n}\right) - \frac{1}{n+1}.$$

Since all the terms in the sum on the right equal 0 except the first and the last, the sum "telescopes," reducing simply to

$$s_n = 1 - \frac{1}{n+1}.$$

Therefore

$$\lim_{n \to \infty} s_n = \lim_{n \to \infty} \left(1 - \frac{1}{n+1}\right) = 1,$$

from which it follows that the given series is convergent with sum 1. □

Example 5 Show that the series

$$\sum_{n=1}^{\infty} \ln\left(1 + \frac{1}{n}\right)$$

is divergent.

Solution Since

$$\sum_{n=1}^{\infty} \ln\left(1 + \frac{1}{n}\right) = \sum_{n=1}^{\infty} \ln \frac{n+1}{n} = \sum_{n=1}^{\infty} [\ln(n+1) - \ln n],$$

the nth partial sum

$$s_n = (\ln 2 - \ln 1) + (\ln 3 - \ln 2) + \cdots + [\ln(n+1) - \ln n],$$

of the series telescopes and reduces to just

$$s_n = \ln(n+1) - \ln 1 = \ln(n+1).$$

But

$$\lim_{n \to \infty} s_n = \lim_{n \to \infty} \ln(n+1) = \infty,$$

and hence the given series diverges. □

The Harmonic Series

Example 6 Show that the series

$$\sum_{n=1}^{\infty} \frac{1}{n} = 1 + \frac{1}{2} + \frac{1}{3} + \frac{1}{4} + \cdots + \frac{1}{n} + \cdots, \qquad (4)$$

known as the *harmonic series*, is divergent.

Solution Let s_n be the nth partial sum of the harmonic series. Then

$$s_{2^k} = 1 + \frac{1}{2} + \frac{1}{3} + \frac{1}{4} + \cdots + \frac{1}{2^k}$$

$$= 1 + \frac{1}{2} + \left(\frac{1}{3} + \frac{1}{4}\right) + \left(\frac{1}{5} + \frac{1}{6} + \frac{1}{7} + \frac{1}{8}\right)$$

$$+ \cdots + \left(\frac{1}{2^{k-1}+1} + \frac{1}{2^{k-1}+2} + \cdots + \frac{1}{2^k}\right)$$

$$> 1 + \frac{1}{2} + \left(\frac{1}{4} + \frac{1}{4}\right) + \left(\frac{1}{8} + \frac{1}{8} + \frac{1}{8} + \frac{1}{8}\right) + \cdots + \underbrace{\left(\frac{1}{2^k} + \frac{1}{2^k} + \cdots + \frac{1}{2^k}\right)}_{2^{k-1} \text{ terms}}$$

$$= 1 + \frac{1}{2} + \frac{2}{4} + \frac{4}{8} + \cdots + \frac{2^{k-1}}{2^k} = 1 + \underbrace{\frac{1}{2} + \frac{1}{2} + \cdots + \frac{1}{2}}_{k \text{ terms}} = 1 + \frac{k}{2}$$

(use the fact that $2^{k-1} + j < 2^k$ for $j = 1, 2, \ldots, 2^{k-2}$). For example,

$$s_4 = s_{2^2} > 1 + \frac{2}{2} = 2, \qquad s_8 = s_{2^3} > 1 + \frac{3}{2} = \frac{5}{2}, \qquad s_{16} = s_{2^4} > 1 + \frac{4}{2} = 3,$$

and so on. Thus the sequence of partial sums $\{s_n\}$ contains arbitrarily large terms, and in fact, given any $C > 0$, we have $s_{2^k} > C$ for all $k > 2C - 2$. Therefore $\{s_n\}$ is divergent, and hence so is the harmonic series (4). ∎

Remark It can be shown that to an excellent approximation the nth partial sum of the harmonic series is given by the formula

$$s_n \approx C + \ln n,$$

where $C = 0.5772156649\ldots$ is a number known as *Euler's constant* and the error of the approximation rapidly approaches 0 as $n \to \infty$. Using this formula, we find that

$$s_{1000} \approx 7.48, \qquad s_{10,000} \approx 9.79,$$
$$s_{100,000} \approx 12.09, \qquad s_{1,000,000} \approx 14.39.$$

Thus the rate of divergence of the harmonic series is extraordinarily slow.

Necessary Condition for Convergence

Next we establish a simple condition which a series must satisfy if it is to be convergent.

> **Theorem 3** (*Necessary condition for convergence of a series*). If the series
>
> $$\sum_{n=1}^{\infty} a_n = a_1 + a_2 + \cdots + a_n + \cdots \tag{5}$$
>
> is convergent, then
>
> $$\lim_{n \to \infty} a_n = 0. \tag{6}$$
>
> Equivalently, the series is divergent if
>
> $$\lim_{n \to \infty} a_n \neq 0 \tag{6'}$$
>
> (*in this form the theorem is often called the "nth term test for divergence"*).

Proof Clearly

$$a_n = s_{n+1} - s_n,$$

where $s_n = a_1 + a_2 + \cdots + a_n$ is the nth partial sum of the series. Assuming that the series is convergent, let S be its sum. Then

$$\lim_{n \to \infty} a_n = \lim_{n \to \infty} (s_{n+1} - s_n) = \lim_{n \to \infty} s_{n+1} - \lim_{n \to \infty} s_n = S - S = 0. \qquad \blacksquare$$

Example 7 The series

$$\sum_{n=1}^{\infty} \frac{n}{2n+1} = \frac{1}{3} + \frac{2}{5} + \frac{3}{7} + \cdots + \frac{n}{2n+1} + \cdots$$

is divergent, since
$$\lim_{n \to \infty} \frac{n}{2n+1} = \frac{1}{2} \neq 0. \quad \square$$

Checking that every convergent geometric series satisfies the condition (6), we observe that it is indeed true that
$$\lim_{n \to \infty} ar^n = 0 \quad \text{if} \quad -1 < r < 1.$$

Similarly, the convergent series in Example 4 clearly satisfies the condition
$$\lim_{n \to \infty} \frac{1}{n(n+1)} = 0,$$

as required by Theorem 3. However, the converse of Theorem 3 is not true, that is, *the validity of* (6) *does not imply the convergence of the series* (5). For example, the harmonic series is divergent, even though
$$\lim_{n \to \infty} \frac{1}{n} = 0,$$

and the series in Example 5 is divergent, even though
$$\lim_{n \to \infty} \ln\left(1 + \frac{1}{n}\right) = \ln 1 = 0.$$

In the language of logic, (6) is a *necessary* but not a sufficient condition for the convergence of the series (5). Much of this chapter is devoted to the study of "convergence tests." These are *sufficient* conditions for convergence, that is, conditions which actually guarantee the convergence of a given series.

Algebraic Operations on Series

By definition, the product of a series $\sum a_n$ and a number c is the series $\sum ca_n$, and the sum or difference of two series $\sum a_n$ and $\sum b_n$ is the series $\sum (a_n + b_n)$ or $\sum (a_n - b_n)$. (The "bare" symbol $\sum$ is an abbreviation for $\sum_{n=m}^{\infty}$, where the nonnegative integer m is the lower index of summation, here equal to 1). These definitions apply whether or not the series $\sum a_n$ and $\sum b_n$ are convergent, but if they are, with sums S and S', respectively, then

$$\sum_{n=1}^{\infty} ca_n = \lim_{n \to \infty} (ca_1 + \cdots + ca_n) = c \lim_{n \to \infty} (a_1 + \cdots + a_n) = cS,$$

$$\sum_{n=1}^{\infty} (a_n + b_n) = \lim_{n \to \infty} [(a_1 + b_1) + \cdots + (a_n + b_n)]$$
$$= \lim_{n \to \infty} [(a_1 + \cdots + a_n) + (b_1 + \cdots + b_n)]$$
$$= \lim_{n \to \infty} (a_1 + \cdots + a_n) + \lim_{n \to \infty} (b_1 + \cdots + b_n) = S + S',$$

and similarly $\sum (a_n - b_n) = S - S'$. Notice that the word "sum" is being used with two meanings here, both as the sum of a convergent series, which is a number, and as the formal sum $\sum (a_n + b_n)$ of two series $\sum a_n$ and $\sum b_n$, which may not be convergent, but the context always makes it clear which meaning is intended.

Example 8 The two geometric series

$$\sum_{n=0}^{\infty} \left(\frac{1}{2}\right)^n \quad \text{and} \quad \sum_{n=0}^{\infty} \left(-\frac{1}{2}\right)^n$$

are convergent, with sums 2 and $\frac{2}{3}$, respectively (see Example 2). Therefore

$$\sum_{n=0}^{\infty} \left[\left(\frac{1}{2}\right)^n + \left(-\frac{1}{2}\right)^n\right] = 2 + \frac{2}{3} = \frac{8}{3},$$

as can be checked directly by observing that

$$\sum_{n=0}^{\infty} \left[\left(\frac{1}{2}\right)^n + \left(-\frac{1}{2}\right)^n\right] = 2 + 0 + 2\left(\frac{1}{2}\right)^2 + 0 + 2\left(\frac{1}{2}\right)^4 + \cdots$$

$$= 2 + 2\left(\frac{1}{4}\right) + 2\left(\frac{1}{4}\right)^2 + \cdots = \frac{2}{1 - \frac{1}{4}} = \frac{2}{\frac{3}{4}} = \frac{8}{3}.$$

Similarly,

$$\sum_{n=0}^{\infty} \left[\left(\frac{1}{2}\right)^n - \left(-\frac{1}{2}\right)^n\right] = 2 - \frac{2}{3} = \frac{4}{3},$$

and more generally

$$\sum_{n=0}^{\infty} \left[A\left(\frac{1}{2}\right)^n + B\left(-\frac{1}{2}\right)^n\right] = 2A + \frac{2}{3}B. \quad \square$$

Let $\sum a_n$ and $\sum b_n$ be two series, with nth partial sums s_n and t_n, respectively, and suppose the series differ in only a finite number of terms, so that $a_n = b_n$ for all n exceeding some integer N. Then

$$\sum_{n=1}^{\infty} a_n = a_1 + \cdots + a_N + a_{N+1} + \cdots + a_n + \cdots,$$

$$\sum_{n=1}^{\infty} b_n = b_1 + \cdots + b_N + a_{N+1} + \cdots + a_n + \cdots,$$

and therefore, if $n > N$,

$$s_n - t_n = (a_1 + \cdots + a_N) - (b_1 + \cdots + b_N) = s_N - t_N,$$

or equivalently

$$s_n = t_n + c,$$

where $c = s_N - t_N$ is a fixed number. It follows that the two sequences $\{s_n\}$ and $\{t_n\}$ are either both convergent or both divergent, and hence the same is true of the two series $\sum a_n$ and $\sum b_n$. In other words, two series which differ in only a *finite* number of terms are either both convergent or both divergent. As an exercise, show that if a series is altered by omission of a finite number of terms, or by insertion of a finite number of extra terms (in arbitrary positions), then the altered series converges if the original series converges and diverges if the original series diverges. Of course, in the case of convergence, the sum of the altered series is in general different from the sum of the original series.

The Remainder of a Series

In particular, suppose we omit the first n terms of the series

$$\sum_{k=1}^{\infty} a_k = a_1 + a_2 + \cdots + a_k + \cdots, \tag{7}$$

obtaining the new series

$$\sum_{k=1}^{\infty} a_{n+k} = a_{n+1} + a_{n+2} + \cdots + a_{n+k} + \cdots. \tag{7'}$$

Then, if (7) is convergent, so is (7'). We denote the sum of (7') by R_n, and call it the *remainder after n terms* of the original series (7). Let S be the sum of (7). Then

$$S = (a_1 + a_2 + \cdots + a_n) + R_n,$$

and hence

$$\lim_{n \to \infty} R_n = S - \lim_{n \to \infty} (a_1 + a_2 + \cdots + a_n) = S - S = 0.$$

Thus *the remainder after n terms of a convergent series converges to 0 as $n \to \infty$*.

Example 9 The remainder after n terms of the convergent geometric series

$$\sum_{k=0}^{\infty} \frac{1}{2^k} = 1 + \frac{1}{2} + \frac{1}{4} + \cdots + \frac{1}{2^k} + \cdots = 2$$

is

$$R_n = \sum_{k=0}^{\infty} \frac{1}{2^{n+k}} = \frac{1}{2^n} + \frac{1}{2^{n+1}} + \frac{1}{2^{n+2}} + \cdots + \frac{1}{2^{n+k}} + \cdots$$

$$= \frac{1}{2^n}\left(1 + \frac{1}{2} + \frac{1}{4} + \cdots + \frac{1}{2^k} + \cdots\right) = \frac{2}{2^n} = \frac{1}{2^{n-1}},$$

so that

$$\lim_{n \to \infty} R_n = \lim_{n \to \infty} \frac{1}{2^{n-1}} = 0.$$

Changing the Index of Summation

In the last example we could just as well have written the remainder in the form

$$R_n = \sum_{k=n}^{\infty} \frac{1}{2^k},$$

with n as the lower limit of summation. More generally, it is always possible to change the index of the general term of an infinite series by making a corresponding change in the lower limit of summation. Thus

$$\sum_{n=0}^{\infty} ar^n, \quad \sum_{n=3}^{\infty} ar^{n-3}, \quad \sum_{n=-2}^{\infty} ar^{n+2}$$

are three different ways of writing the same geometric series

$$a + ar + ar^2 + \cdots,$$

but the second and third ways are clumsy as compared to the first, which is the natural way of writing the series. Of course, no matter how we write a

Problems

Calculate the first five partial sums of the series

1. $\sum_{n=1}^{\infty} 3$
2. $\sum_{n=1}^{\infty} \dfrac{1}{n^2}$
3. $\sum_{n=1}^{\infty} \dfrac{1}{n!}$
4. $\sum_{n=1}^{\infty} n(n+1)$
5. $\sum_{n=1}^{\infty} \ln n$
6. $\sum_{n=2}^{\infty} \dfrac{1}{n(n-1)}$
7. $\sum_{n=0}^{\infty} \left(\dfrac{1}{2^n} - \dfrac{1}{2^{n+1}} \right)$
8. $\sum_{n=3}^{\infty} \sin \dfrac{n\pi}{2}$

9. Let a_n be the nth term of the series $\sum a_n$, and let s_n be its nth partial sum. Express the sequence $\{a_n\}$ in terms of the sequence $\{s_n\}$.

Write the series whose nth partial sum is

10. $s_n = \dfrac{n+2}{n+1}$
11. $s_n = \dfrac{2^n - 1}{2^n}$
12. $s_n = \dfrac{(-1)^n}{n+1}$

If the given series is convergent, find its sum. Otherwise identify it as divergent. In the case of a series specified by giving a few initial terms, assume that the law of formation suggested by these terms holds for all the terms of the series.

13. $1 - \dfrac{1}{3} - \dfrac{1}{9} - \dfrac{1}{27} - \cdots$
14. $1 + \dfrac{1}{2} - \dfrac{1}{4} - \dfrac{1}{8} + \dfrac{1}{16} + \dfrac{1}{32} - \cdots$
15. $\sum_{n=1}^{\infty} \dfrac{1}{\sqrt{n}}$
16. $1 + \dfrac{1}{2} - 1 - \dfrac{1}{4} + 1 + \dfrac{1}{8} - 1 - \dfrac{1}{16} + \cdots$
17. $\sum_{n=0}^{\infty} 4 \left(-\dfrac{5}{6} \right)^n$
18. $\sum_{n=0}^{\infty} 3^{-n/2}$
19. $\sum_{n=1}^{\infty} \dfrac{1}{n(n+2)}$
20. $\sum_{n=2}^{\infty} \dfrac{n}{\ln n}$
21. $\sum_{n=0}^{\infty} \cos \dfrac{n\pi}{3}$
22. $\sum_{n=1}^{\infty} \dfrac{2 + (-1)^n}{2^n}$
23. $\sum_{n=0}^{\infty} \left(\dfrac{3}{2^n} - \dfrac{2}{3^n} \right)$
24. $\dfrac{1}{1 \cdot 3} + \dfrac{1}{3 \cdot 5} + \dfrac{1}{5 \cdot 7} + \cdots$
25. $\dfrac{3}{1 \cdot 2} - \dfrac{5}{2 \cdot 3} + \dfrac{7}{3 \cdot 4} - \dfrac{9}{4 \cdot 5} + \cdots$
26. $\sum_{n=0}^{\infty} e^n \pi^{1-n}$
27. $\sum_{n=1}^{\infty} \dfrac{1}{n(n+1)(n+2)}$
28. $\sum_{n=1}^{\infty} [\arctan(n+1) - \arctan n]$

Find the remainder R_n after n terms of the given series, and verify that $R_n \to 0$ as $n \to \infty$.

29. $\sum_{n=0}^{\infty} \dfrac{e-1}{e^{n+1}}$
30. $\sum_{n=1}^{\infty} \dfrac{1}{n(n+2)}$
31. $\sum_{n=1}^{\infty} \dfrac{3}{n(n+3)}$

32. Show that if $\sum a_n$ is convergent and $\sum b_n$ is divergent, then the sum $\sum (a_n + b_n)$ of the two series is divergent.

33. Is there any value of the number r for which the series

$$\sum_{n=0}^{\infty} \left(r^n + \dfrac{1}{r^n} \right)$$

is convergent?

34. Let s_n be the nth partial sum of a convergent series. Show that given any $\varepsilon > 0$, there is an integer $N > 0$ such that $|s_m - s_n| < \varepsilon$ for all $m > N$ and $n > N$. Use this to give another proof of the divergence of the harmonic series (4).

35. Achilles runs along the road in pursuit of a tortoise which has a head start of 20 ft, as shown in Figure 7, where Achilles is initially at P_0 and the tortoise is initially at P_1. Suppose Achilles runs at a speed of 20 ft/sec, while the tortoise runs at a speed of 10 ft/sec.† Then it takes Achilles 1 sec to run from P_0 to P_1, but in the meantime the tortoise

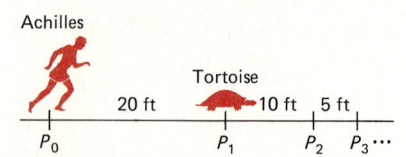

Figure 7

† This is admittedly fast for a tortoise, but it keeps the arithmetic simple.

has moved 10 ft farther down the road to position P_2. To go from P_1 to P_2 takes Achilles another $\frac{1}{2}$ sec, but by that time the tortoise has moved another 5 ft down the road to position P_3, and so on indefinitely. Since Achilles is always in the process of catching up to the tortoise's last position, it seems as if Achilles will never be able to overtake and pass the tortoise, although he runs twice as fast as the tortoise. This paradox is due to the Greek philosopher Zeno of Elea, who lived in the fifth century B.C. Use an elementary argument to show that actually Achilles overtakes the tortoise after running for 2 sec, and then resolve Zeno's paradox by summing an appropriate geometric series.

36. Suppose the tortoise in the preceding problem is able to run at any speed less than 20 ft/sec, and that it chooses to run at $20(\frac{1}{2}) = 10$ ft/sec as Achilles goes from P_0 to P_1, at $20(\frac{2}{3}) = 13\frac{1}{3}$ ft/sec as Achilles goes from P_1 to P_2, at $20(\frac{3}{4}) = 15$ ft/sec as Achilles goes from P_2 to P_3, and in general at $20n/(n+1)$ ft/sec as Achilles goes from P_{n-1} to P_n. Show that in this case Achilles will never overtake the tortoise, although the tortoise never runs as fast as Achilles!

37. A rubber ball bounces up and down after being dropped from a height of 45 ft onto a hard pavement. The ball's elasticity is such that on each rebound it rises to two thirds its previous height. How far has the ball traveled when it reaches the top of its sixth rebound? How far will it travel before finally coming to rest?

38. When money is spent on goods and services, those receiving the money also spend some of it, those receiving the twice-spent money spend some of it in turn, and so on indefinitely. Suppose the original expenditure is D dollars, and each recipient of spent or respent money spends $100c$ percent of it, while saving $100s$ percent. The quantities c and s, known as the *marginal propensity to consume* and the *marginal propensity to save*, are both numbers between 0 and 1; clearly $c + s = 1$, since money is either consumed (spent) or saved. Then the income of the community as a whole (the national income in the case of an entire country) is eventually increased by kD dollars, where the factor k is called the *multiplier*. All these key concepts of macroeconomics are due to the English economist John Maynard Keynes (1883–1940). Show that $k = 1/s > 1$, leading to the "multiplier effect," of basic importance in Keynesian economics. (For example, if $s = 0.2$, then $k = 5$, so that \$1 spent or invested leads to a \$5 increase in national income.)

39. Paul and Steve, who are initially 250 ft apart, run toward one another, each at a speed of 10 ft/sec. At the same time, Bozo the dog runs back and forth between Paul and Steve at a speed of 15 ft/sec. How far has Bozo run when the two boys meet? Solve the problem both by an elementary method and by summing an appropriate series.

40. Solve the preceding problem, again in two ways, assuming that Steve runs *away* from Paul at 5 ft/sec, rather than toward him at 10 ft/sec.

41. The hour and minute hands of a clock coincide at noon. Find the time (to the nearest second) at which the hands first coincide again. Do this in two ways.

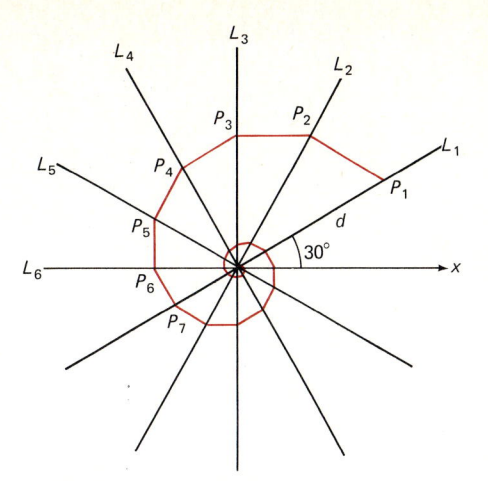

Figure 8

42. As in Figure 8, let $L_1, L_2, \ldots, L_6$ be six lines such that L_1 makes an angle of 30° with the positive x-axis and the angle between each pair of adjacent lines is also 30° (notice that L_3 coincides with the y-axis and L_6 with the x-axis). From a point P_1 on L_1, a distance d from the origin, drop a perpendicular to L_2 intersecting L_2 at the point P_2, from P_2 drop a perpendicular to L_3 intersecting L_3 in the point P_3, and so on (the line following L_6 is again L_1). The consecutive line segments $P_1P_2, P_2P_3, \ldots, P_nP_{n+1}, \ldots$ form a spiral-like polygonal path $P_1P_2 \ldots P_n \ldots$, which winds around the origin and simultaneously shrinks into it. Show that this path is of length $(2 + \sqrt{3})d$.

43. Figure 9 shows the region bounded by two tangent circles of radius 1 and by a line tangent to both circles. A sequence of smaller circles is inscribed in the region in the way shown. It is apparent from the geometry that the lengths of the diameters of these circles are the terms of a series whose sum is 1. What is the series?

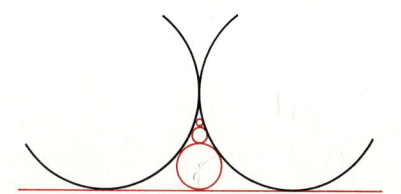

Figure 9

44. Show that the series $\cos \theta + \cos 2\theta + \cdots + \cos n\theta + \cdots$ is divergent for all θ, while the series $\sin \theta + \sin 2\theta + \cdots + \sin n\theta + \cdots$ is divergent unless $\theta = k\pi$, where k is an integer.

Hint. Use the identities

$$\cos(n-1)\theta = \cos n\theta \cos \theta + \sin n\theta \sin \theta,$$
$$\sin(n-1)\theta = \sin n\theta \cos \theta - \cos n\theta \sin \theta,$$
$$\cos^2 n\theta + \sin^2 n\theta = 1.$$

45. Jack has high blood pressure, and his doctor prescribes an extended course of treatment with a certain drug. Jack takes the first dose at time $t = 0$, and subsequent doses at times $t = T, 2T, \ldots, nT, \ldots$ Each dose quickly increases the concentration of the drug in Jack's bloodstream by an amount C_0, but at the same time his body acts to eliminate the drug. Let $C = C(t)$ be the concentration of the drug in Jack's bloodstream at time t. Then, as in Problem 25, page 340, C satisfies the differential equation

$$\frac{dC}{dt} = -kC,$$

where $k > 0$ is the elimination constant. Show that C eventually oscillates between the level

$$R = \frac{C_0}{e^{kT} - 1},$$

called the *residual concentration*, and the level $R + C_0$. Suppose the doctor wants to be sure that the concentration never falls below the level C_e at which the drug is effective, but never rises above the level C_s at which the drug is safe. Show that this can be achieved by choosing

$$C_0 = C_s - C_e, \qquad T = \frac{1}{k} \ln \frac{C_s}{C_e}.$$

9.3 Nonnegative Series; Comparison Tests and the Integral Test

An infinite series

$$\sum_{n=1}^{\infty} a_n = a_1 + a_2 + \cdots + a_n + \cdots$$

is said to be *nonnegative* if all its terms are nonnegative, that is, if $a_n \geq 0$ for all $n = 1, 2, \ldots$ Every partial sum $s_n = a_1 + a_2 + \cdots + a_n$ of a nonnegative series $\sum a_n$ is a sum of finitely many nonnegative numbers, and hence is itself a nonnegative number. Moreover, $\{s_n\}$ is an increasing sequence, since

$$s_n = a_1 + \cdots + a_n \leq a_1 + \cdots + a_n + a_{n+1} = s_{n+1} \qquad (n = 1, 2, \ldots).$$

These observations lead at once to the following proposition, which is basic to the study of nonnegative series.

> **Theorem 4** *(Convergence criterion for nonnegative series).* The nonnegative series $\sum a_n$ is convergent if the sequence $\{s_n\}$ of its partial sums has an upper bound, that is, if there is a number $C > 0$ such that
>
> $$s_n = a_1 + a_2 + \cdots + a_n \leq C$$
>
> for all n. If there is no such number, the series is divergent.

Proof If $\{s_n\}$ has an upper bound, then $\{s_n\}$ is a bounded increasing sequence. But then $\{s_n\}$ is convergent, by Theorem 2, page 493, and hence so is $\sum a_n$. On the other hand, if $\{s_n\}$ has no upper bound, then s_n exceeds any given $C > 0$ for all sufficiently large n, so that $s_n \to \infty$ as $n \to \infty$. In this case $\{s_n\}$ is divergent, and therefore so is $\sum a_n$. ∎

Decimals and Infinite Series

Example 1 The nonterminating decimal $0.c_1 c_2 \ldots c_n \ldots$, where $0 \leq c_n \leq 9$ for all n, is shorthand for the infinite series

$$\sum_{n=1}^{\infty} \frac{c_n}{10^n} = \frac{c_1}{10} + \frac{c_2}{10^2} + \cdots + \frac{c_n}{10^n} + \cdots. \tag{1}$$

Show that every such series converges, as we have been tacitly assuming all along.

Solution The series (1) is nonnegative, and therefore, by Theorem 4, its convergence will be proved if we can show that the sequence $\{s_n\}$ of its partial sums has an upper bound. But this is true, since

$$s_n = \frac{c_1}{10} + \frac{c_2}{10^2} + \cdots + \frac{c_n}{10^n} \le \frac{9}{10} + \frac{9}{10^2} + \cdots + \frac{9}{10^n}$$

$$= \frac{9}{10}\left[1 + \frac{1}{10} + \cdots + \left(\frac{1}{10}\right)^{n-1}\right] = \frac{9}{10}\frac{1 - (\frac{1}{10})^n}{1 - \frac{1}{10}}$$

$$= 1 - \left(\frac{1}{10}\right)^n < 1$$

for all n. □

Example 2 Let $0.a_1 \ldots a_m\overline{b_1 \ldots b_n}$ ($\ne 0$) be a repeating decimal, where the block of digits $b_1 \ldots b_n$ of length n covered by the bar repeats itself endlessly. (For instance, in the decimal $0.517\overline{29} = 0.517292929\ldots$ we have $a_1 = 5$, $a_2 = 1$, $a_3 = 7$, $b_1 = 2$, $b_2 = 9$.) Show that as anticipated on page 5, every such decimal represents a rational number, defined as the quotient

$$\frac{p}{q} \quad (q \ne 0)$$

of two integers p and q.

Solution Let $A = a_1 \ldots a_m$ and $B = b_1 \ldots b_n$, where these are strings of digits, not products! Then A and B are positive integers, and

$$0.a_1 \ldots a_m\overline{b_1 \ldots b_n} = \frac{A}{10^m} + \frac{1}{10^m}\sum_{k=1}^{\infty} B\left(\frac{1}{10^n}\right)^k = \frac{A}{10^m} + \frac{B}{10^{m+n}}\sum_{k=0}^{\infty}\left(\frac{1}{10^n}\right)^k$$

$$= \frac{A}{10^m} + \frac{B}{10^{m+n}}\frac{1}{1 - \frac{1}{10^n}} = \frac{A}{10^m} + \frac{B}{10^m}\frac{1}{10^n - 1},$$

after summing a convergent geometric series. Therefore

$$0.a_1 \ldots a_m\overline{b_1 \ldots b_n} = \frac{A(10^n - 1) + B}{10^m(10^n - 1)}, \tag{2}$$

and we have expressed the given repeating decimal as a quotient of two positive integers, namely $p = A(10^{n-1} - 1) + B$ and $q = 10^m(10^n - 1)$. This quotient is not necessarily in lowest terms. □

Example 3 Find the rational number in lowest terms represented by the repeating decimal $4.3\overline{21}$.

Solution Writing $4.3\overline{21}$ as $4 + 0.3\overline{21}$, we apply the method of the preceding example to the repeating decimal $0.3\overline{21}$. Here $A = 3$, $B = 21$, $m = 1$, $n = 2$, and formula (2) gives

$$0.3\overline{21} = \frac{3(10^2 - 1) + 21}{10(10^2 - 1)} = \frac{3(99) + 21}{10(99)} = \frac{318}{990},$$

Section 9.3 Nonnegative Series; Comparison Tests and the Integral Test where the last fraction is clearly not in lowest terms, since the numerator and denominator are both (exactly) divisible by 2 and also by 3. Therefore

$$0.3\overline{21} = \frac{106}{330} = \frac{53}{165},$$

which is now in lowest terms, since 53 is not divisible by 3, 5 or 11, the prime factors of $165 = 3(5)(11)$. Thus, finally,

$$4.3\overline{21} = 4 + \frac{53}{165} = \frac{713}{165},$$

which is still in lowest terms (why?). □

Despite the success that we had in Section 9.2 in summing certain special series, it is usually difficult or impossible to find the exact sum of a convergent series. Fortunately, if the convergence of a series is "rapid" enough, its sum can be well approximated by adding up a rather small number of initial terms of the series. However, before attempting to find the exact or approximate sum of a series, we must be sure that it converges in the first place! The divergent harmonic series serves as a warning here, since the fact that it diverges is far from obvious.

The Comparison Test

Thus our immediate objective is to develop tests allowing us to decide whether or not a given series converges. We begin by using Theorem 4 to establish a test in which the convergence behavior of one series determines that of another series. This "comparison test" is the exact analogue of Theorem 6, page 429, for improper integrals.

> **Theorem 5** *(Comparison test)*. Let $\sum a_n$ and $\sum b_n$ be two nonnegative series such that $a_n \leq b_n$ for all sufficiently large n. Then
>
> (i) If $\sum b_n$ converges, so does $\sum a_n$;
> (ii) If $\sum a_n$ diverges, so does $\sum b_n$.

Proof We can assume that $a_n \leq b_n$ for *all* n, since as shown on page 504, two series which differ in only a finite number of terms are either both convergent or both divergent. Let the nth partial sums of $\sum a_n$ and $\sum b_n$ be s_n and t_n, respectively. Then

$$s_n = a_1 + \cdots + a_n \leq b_1 + \cdots + b_n = t_n$$

for all n. If $\sum b_n$ is convergent with sum T, then $t_n \leq T$ for all n (why?), and therefore, since $s_n \leq t_n$, we have $s_n \leq T$ for all n, so that the sequence $\{s_n\}$ has an upper bound. It follows from Theorem 4 that $\sum a_n$ is also convergent. By the same token, if $\sum a_n$ diverges, then so does $\sum b_n$, for, as just shown, convergence of $\sum b_n$ would imply convergence of $\sum a_n$. ∎

Given two series $\sum a_n$ and $\sum b_n$, we say that $\sum b_n$ *dominates* $\sum a_n$ if $b_n \geq a_n$ for all sufficiently large n. Thus, according to Theorem 5, *a series dominated by a convergent series is itself convergent, while a series dominating a divergent series is itself divergent.* Here, as elsewhere in this section, the series under consideration are assumed to be nonnegative.

Example 4 The series

$$\sum_{n=0}^{\infty} \frac{1}{n!} = 1 + 1 + \frac{1}{2!} + \frac{1}{3!} + \cdots + \frac{1}{n!} + \cdots \qquad (3)$$

($0! = 1$ by definition) is dominated by the series

$$1 + 1 + \frac{1}{2} + \frac{1}{2^2} + \cdots + \frac{1}{2^n} + \cdots. \qquad (4)$$

To check this, observe that although the inequality $n! > 2^n$ fails for $n = 0$, 1, 2, 3, it holds for all $n \geq 4$, so that every term of the series (3) beginning with the fifth is less than the corresponding term of the series (4). But the series (4) is convergent, since from the second term on it is a geometric series with ratio $\frac{1}{2}$, and in fact its sum is 3. Therefore, by the comparison test, the series (3) is also convergent, and its sum turns out to be e, as will be shown in Example 5, page 542.

Example 5 Let p be any number less than 1. Then, since $1^p = 1$ and n^x is an increasing function of x for $n \geq 2$, we have

$$n^p \leq n \qquad (n = 1, 2, \ldots),$$

or equivalently

$$\frac{1}{n} \leq \frac{1}{n^p} \qquad (n = 1, 2, \ldots).$$

Therefore, if $p < 1$, the series

$$1 + \frac{1}{2^p} + \frac{1}{3^p} + \cdots + \frac{1}{n^p} + \cdots,$$

called the *p-series*, dominates the divergent harmonic series

$$1 + \frac{1}{2} + \frac{1}{3} + \cdots + \frac{1}{n} + \cdots,$$

and hence is itself divergent, by Theorem 5. In fact, the *p*-series diverges if $p \leq 1$, since for $p = 1$ it reduces to the harmonic series. It will be shown in Example 10 that the *p*-series converges if $p > 1$.

The Limit Comparison Test

From Theorem 5 we can deduce another even more useful comparison test:

Theorem 6 *(Limit comparison test).* Let $\sum a_n$ and $\sum b_n$ be two series with positive terms such that

$$\lim_{n \to \infty} \frac{a_n}{b_n} = L, \qquad (5)$$

where the case $L = \infty$ is allowed. Then convergence of $\sum b_n$ implies convergence of $\sum a_n$ if $0 \leq L < \infty$, while divergence of $\sum b_n$ implies divergence of $\sum a_n$ if $L > 0$ or $L = \infty$. In particular, if L is a positive number, the two series $\sum a_n$ and $\sum b_n$ are either both convergent or both divergent.

Proof Since a_n and b_n are positive, the ratio a_n/b_n and its reciprocal b_n/a_n are defined for all n. Suppose (5) holds with $0 \leq L < \infty$ (needless to say, L cannot be negative). Then, given any $\varepsilon > 0$, there is an integer N such that

$$\frac{a_n}{b_n} < L + \varepsilon,$$

or equivalently

$$a_n < (L + \varepsilon)b_n,$$

for all $n > N$. If $\sum b_n$ converges, so does the nonnegative series $\sum (L + \varepsilon)b_n$, obtained by multiplying the terms of $\sum b_n$ by the positive factor $L + \varepsilon$ (see page 503), and then the convergence of $\sum a_n$ follows by the ordinary comparison test (Theorem 5). On the other hand, if $\sum b_n$ diverges and $L > 0$ or $L = \infty$, then, since the reciprocal ratio b_n/a_n approaches a nonnegative limit as $n \to \infty$ (equal to 0 if $L = \infty$), the series $\sum a_n$ also diverges, for otherwise, as just shown, $\sum b_n$ would converge, contrary to assumption. ∎

Example 6 Let

$$a_n = \frac{1}{n^2}, \qquad b_n = \frac{1}{n(n+1)},$$

so that $\sum a_n$ is the p-series for $p = 2$, and $\sum b_n$ is the series studied in Example 4, page 500. Then $\sum b_n$ is convergent (with sum 1), and

$$\lim_{n \to \infty} \frac{a_n}{b_n} = \lim_{n \to \infty} \frac{n(n+1)}{n^2} = \lim_{n \to \infty} \left(1 + \frac{1}{n}\right) = 1.$$

Therefore, by Theorem 6, the series $\sum a_n = \sum (1/n^2)$ is also convergent. It turns out that the sum of this series is $\pi^2/6$.

Example 7 We can use Theorem 6 to give another proof of the divergence of the harmonic series $\sum (1/n)$. Let

$$a_n = \ln\left(1 + \frac{1}{n}\right), \qquad b_n = \frac{1}{n}.$$

Then $\sum a_n$ is divergent, as shown very simply in Example 5, page 501, and

$$\lim_{n \to \infty} \frac{a_n}{b_n} = \lim_{n \to \infty} \frac{\ln\left(1 + \frac{1}{n}\right)}{\frac{1}{n}} = \lim_{x \to \infty} \frac{\ln\left(1 + \frac{1}{x}\right)}{\frac{1}{x}}$$

$$= \lim_{u \to 0^+} \frac{\ln(1 + u)}{u} = \lim_{u \to 0^+} \frac{D_u \ln(1 + u)}{D_u u} = \lim_{u \to 0^+} \frac{1}{1 + u} = 1,$$

with the help of the substitution $u = 1/x$ and L'Hospital's rule. Therefore, by Theorem 6, the series $\sum b_n$, namely the harmonic series, is also divergent.

Example 8 Determine whether the series

$$\sum_{n=1}^{\infty} \frac{n+2}{\sqrt{n^3 + n}} \tag{6}$$

converges or diverges.

Solution The approximation

$$\frac{n+2}{\sqrt{n^3+n}} \approx \frac{n}{\sqrt{n^3}} = \frac{1}{\sqrt{n}}$$

is good for large n, and since $1/\sqrt{n} \geq 1/n$, this suggests that the series (6) dominates the divergent harmonic series and hence is itself divergent. Thus we compare (6) with the harmonic series, choosing

$$a_n = \frac{n+2}{\sqrt{n^3+n}}, \qquad b_n = \frac{1}{n}.$$

Then

$$\lim_{n \to \infty} \frac{a_n}{b_n} = \lim_{n \to \infty} \frac{n^2+2n}{\sqrt{n^3+n}} = \lim_{n \to \infty} \frac{n^2+2n}{\sqrt{n^3\left(1+\frac{1}{n^2}\right)}} = \lim_{n \to \infty} \frac{\sqrt{n}+\frac{2}{\sqrt{n}}}{\sqrt{1+\frac{1}{n^2}}} = \infty,$$

and since $\sum b_n$ is the divergent harmonic series, it follows from Theorem 6 that the series $\sum a_n$, that is, the series (6), is also divergent. □

Example 9 Determine whether the series

$$\sum_{n=1}^{\infty} \frac{2n+1}{\sqrt{n^6+n^2}} \qquad (7)$$

converges or diverges.

Solution The approximation

$$\frac{2n+1}{\sqrt{n^6+n^2}} \approx \frac{2n}{\sqrt{n^6}} = \frac{2}{n^2}$$

is good for large n, suggesting that we compare the series (7) with the series $\sum (1/n^2)$, whose convergence was established in Example 6. Thus we choose

$$a_n = \frac{2n+1}{\sqrt{n^6+n^2}}, \qquad b_n = \frac{1}{n^2}.$$

Then

$$\lim_{n \to \infty} \frac{a_n}{b_n} = \lim_{n \to \infty} \frac{n^2(2n+1)}{\sqrt{n^6+n^2}} = \lim_{n \to \infty} \frac{n^3\left(2+\frac{1}{n}\right)}{\sqrt{n^6\left(1+\frac{1}{n^4}\right)}} = \lim_{n \to \infty} \frac{2+\frac{1}{n}}{\sqrt{1+\frac{1}{n^4}}} = 2,$$

and since $\sum b_n = \sum (1/n^2)$ is convergent, it follows from Theorem 6 that the series $\sum a_n$, that is, the series (7), is also convergent. □

The Integral Test

The analogy between infinite series and improper integrals has already been pointed out (see page 498). The next convergence test will often allow us to deduce the convergence or divergence of an infinite series from that of an associated improper integral.

Theorem 7 *(Integral test).* Let $f(x)$ be a continuous positive function which is decreasing on the interval $1 \leq x < \infty$, and let $a_n = f(n)$ for all $n = 1, 2, \ldots$ Then the series

$$\sum_{n=1}^{\infty} a_n = a_1 + a_2 + \cdots + a_n + \cdots$$

and the improper integral

$$\int_1^{\infty} f(x)\,dx = \lim_{u \to \infty} \int_1^u f(x)\,dx \qquad (8)$$

are either both convergent or both divergent.

Proof In Figure 10(a) the area under the curve $y = f(x)$ from $x = 1$ to $x = n + 1$ is *less* than the total area of the shaded *circumscribed* rectangles, and therefore

$$\int_1^{n+1} f(x)\,dx < a_1 + a_2 + \cdots + a_n = s_n, \qquad (9)$$

where s_n is the nth partial sum of the series $\sum a_n$. In the somewhat different Figure 10(b) the area under the curve $y = f(x)$ from $x = 1$ to $x = n$ is *greater* than the total area of the shaded *inscribed* rectangles, so that this time

$$a_2 + a_3 + \cdots + a_n < \int_1^n f(x)\,dx. \qquad (9')$$

Suppose the improper integral (8) is divergent. Then, since $f(x)$ is positive, this can only mean that the limit in (8) is ∞. Thus the left side of (9) approaches ∞ as $n \to \infty$, and hence so does the right side s_n, that is, $\sum a_n$ is divergent. On the other hand, suppose the integral (8) is convergent with limit L. Then, because of (9'),

$$s_n = a_1 + a_2 + a_3 + \cdots + a_n < a_1 + \int_1^n f(x)\,dx < a_1 + L$$

for all n, so that the partial sums of the nonnegative series $\sum a_n$ have an upper bound. But then $\sum a_n$ is convergent, by Theorem 4. ∎

If $\sum a_n$ is a series for which the general term a_n is given by an explicit formula, the simplest way to find a function $f(x)$ such that $f(n) = a_n$ is to replace n by x in the formula for a_n. Then the integral test is applicable if the resulting function is continuous, positive and decreasing on $[1, \infty)$.

Example 10 Show that the p-series

$$\sum_{n=1}^{\infty} \frac{1}{n^p} = 1 + \frac{1}{2^p} + \frac{1}{3^p} + \cdots + \frac{1}{n^p} + \cdots$$

converges if $p > 1$ and diverges if $p \leq 1$.

Solution Changing x to n in the general term of the series, we get a continuous positive function $1/x^p$ which is decreasing on $[1, \infty)$ if $p > 0$, so that the integral test is applicable. For $p = 1$, $1/x^p$ becomes $1/x$ and

$$\int_1^{\infty} \frac{dx}{x} = \lim_{u \to \infty} \int_1^u \frac{dx}{x} = \lim_{u \to \infty} \ln x \Big|_1^u = \lim_{u \to \infty} \ln u = \infty,$$

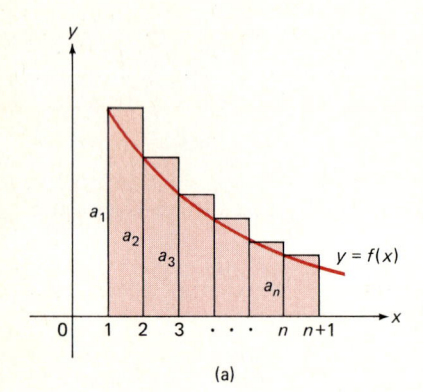

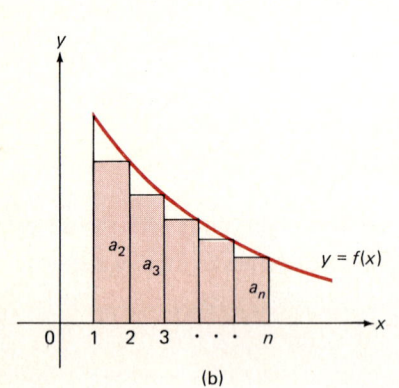

Figure 10

while for $0 < p < 1$ or $p > 1$,

$$\int_1^\infty \frac{dx}{x^p} = \lim_{u \to \infty} \int_1^u \frac{dx}{x^p} = \lim_{u \to \infty} \frac{x^{1-p}}{1-p}\bigg|_1^u = \lim_{u \to \infty} \left(\frac{u^{1-p}}{1-p} - \frac{1}{1-p}\right).$$

But, as is easily verified,

$$\lim_{u \to \infty} \left(\frac{u^{1-p}}{1-p} - \frac{1}{1-p}\right) = \begin{cases} \frac{1}{p-1} & \text{if } p > 1, \\ \infty & \text{if } 0 < p < 1. \end{cases}$$

Therefore, by the integral test, the p-series converges if $p > 1$ and diverges if $0 < p \leq 1$. (Notice that we have incidentally given still another proof of the divergence of the harmonic series.) If $p \leq 0$, the nth term $1/n^p$ does not approach 0 as $n \to \infty$, and the p-series diverges, by Theorem 3, page 502. Combining cases, we finally find that the p-series converges if $p > 1$ and diverges if $p \leq 1$. $\square$

Example 11 The series

$$\sum_{n=2}^\infty \frac{1}{n \ln n}$$

is divergent, by the integral test, since the associated improper integral diverges:

$$\int_2^\infty \frac{dx}{x \ln x} = \lim_{u \to \infty} \int_2^u \frac{dx}{x \ln x} = \lim_{u \to \infty} \ln(\ln x)\bigg|_2^u = \lim_{u \to \infty} \left[\ln(\ln u) - \ln(\ln 2)\right] = \infty.$$

Here, since $1/(n \ln n)$ is not defined for $n = 1$, we choose 2 rather than 1 as both the lower limit of summation and the lower limit of integration.

Example 12 The series

$$\sum_{n=2}^\infty \frac{1}{n(\ln n)^2}$$

is convergent, by the integral test, since the associated improper integral now converges:

$$\int_2^\infty \frac{dx}{x(\ln x)^2} = \lim_{u \to \infty} \int_2^u \frac{dx}{x(\ln x)^2} = \lim_{u \to \infty} \left[-\frac{1}{\ln x}\right]_2^u = \lim_{u \to \infty} \left(\frac{1}{\ln 2} - \frac{1}{\ln u}\right) = \frac{1}{\ln 2}.$$

Problems

Find the rational number in lowest terms represented by the given repeating decimal.

1. $0.4\overline{9}$
2. $3.\overline{79}$
3. $0.\overline{215}$
4. $6.3\overline{63}$
5. $4.000\overline{72}$
6. $0.05\overline{44}$
7. $0.\overline{047619}$
8. $0.\overline{0384615}$
9. $5.10\overline{285714}$

10. Verify that a decimal ending in an infinite run of nines represents the same rational number as the "next highest" terminating decimal. For example, $0.\overline{9} = 1$, $1.325\overline{9} = 1.326$, and so on.

Use either comparison test (Theorem 5 or 6) to determine whether the given series is convergent or divergent.

11. $\sum_{n=1}^\infty \frac{1}{4n-3}$
12. $\sum_{n=1}^\infty \frac{1}{\sqrt{3n(n+1)}}$
13. $\sum_{n=1}^\infty \frac{1}{\sqrt{n(n^2+2)}}$
14. $\sum_{n=1}^\infty \frac{1+\sin n}{n^2}$
15. $\sum_{n=1}^\infty \frac{(n+1)(n+3)}{n(n+2)(n+4)}$
16. $\sum_{n=2}^\infty \frac{1}{\ln n}$

Section 9.3 Nonnegative Series; Comparison Tests and the Integral Test

17. $\sum_{n=1}^{\infty} \dfrac{\ln n}{n^{3/2}}$

18. $\sum_{n=1}^{\infty} \dfrac{3n^4 - n^2 + 1}{5n^6 - n^3 - 2}$

19. $\sum_{n=1}^{\infty} \dfrac{1}{n^n}$

20. $\sum_{n=1}^{\infty} \dfrac{2^n + 1}{3^n - 1}$

21. $\sum_{n=1}^{\infty} \dfrac{1}{\sqrt[3]{n(n+1)(n+2)}}$

22. $\sum_{n=1}^{\infty} \dfrac{1}{\sqrt{n(n+2)(n+4)}}$

23. $\sum_{n=1}^{\infty} \dfrac{1}{nc^n} \quad (c > 1)$

24. $\sum_{n=1}^{\infty} \dfrac{2}{100n - 99}$

25. $\sum_{n=1}^{\infty} \dfrac{n!}{n^n}$

26. $\sum_{n=1}^{\infty} \dfrac{(n!)^2}{(2n)!}$

27. $\sum_{n=1}^{\infty} \sin \dfrac{1}{n}$

28. $\sum_{n=1}^{\infty} 2^n \sin \dfrac{1}{3^n}$

29. $\sum_{n=1}^{\infty} \dfrac{1}{n\sqrt[n]{n}}$

30. $\sum_{n=3}^{\infty} \dfrac{\arctan n}{n^2 - 5}$

31. $\sum_{n=1}^{\infty} \operatorname{sech} n$

32. $\sum_{n=1}^{\infty} \dfrac{\cos^2 n}{(\sqrt{n} + 1)^3}$

33. $\sum_{n=1}^{\infty} \left(1 - \cos \dfrac{1}{n}\right)$

34. $\sum_{n=1}^{\infty} \dfrac{1}{1 + c^n} \quad (c > 0)$

Use the integral test (Theorem 7) to determine whether the given series is convergent or divergent.

35. $\sum_{n=1}^{\infty} \dfrac{1}{2n - 1}$

36. $\sum_{n=1}^{\infty} \dfrac{1}{\sqrt{3n - 2}}$

37. $\sum_{n=1}^{\infty} \dfrac{1}{n^2 + 1}$

38. $\sum_{n=1}^{\infty} \dfrac{\arctan n}{n^2 + 1}$

39. $\sum_{n=3}^{\infty} \dfrac{1}{n^2 + n - 6}$

40. $\sum_{n=2}^{\infty} \dfrac{n}{n^2 - 3}$

41. $\sum_{n=1}^{\infty} \dfrac{1}{\sqrt{n + 1}}$

42. $\sum_{n=2}^{\infty} \dfrac{1}{n(n^2 - 1)}$

43. $\sum_{n=0}^{\infty} e^{-n/2}$

44. $\sum_{n=1}^{\infty} ne^{-n^2}$

45. $\sum_{n=1}^{\infty} n^2 e^{-n}$

46. $\sum_{n=2}^{\infty} \dfrac{1}{n(\ln n)^3}$

47. $\sum_{n=3}^{\infty} \dfrac{1}{n \ln n \cdot \ln (\ln n)}$

48. $\sum_{n=2}^{\infty} \dfrac{1}{n\sqrt{\ln n}}$

49. $\sum_{n=3}^{\infty} \dfrac{1}{n \ln n \, [\ln (\ln n)]^2}$

50. $\sum_{n=1}^{\infty} e^{-2n} \sinh n$

51. $\sum_{n=0}^{\infty} \left(\dfrac{\pi}{2} - \arctan n\right)$

52. $\sum_{n=1}^{\infty} \operatorname{csch}^2 n$

53. Let $\sum a_n$ be a convergent nonnegative series. Show that the series $\sum a_n^2$ is also convergent.

54. Let $\sum a_n$ and $\sum b_n$ be two convergent nonnegative series. Show that the series $\sum a_n b_n$ is also convergent.

55. Let $\sum a_n$ be a convergent nonnegative series. Show that
$$\lim_{n \to \infty} na_n = 0 \qquad (i)$$
if the limit exists. Find a convergent nonnegative series $\sum a_n$ which does not satisfy (i).

56. Show that the condition (i) is satisfied by any convergent nonnegative series for which $\{a_n\}$ is a decreasing sequence.

57. Must a nonnegative series $\sum a_n$ be convergent if the sequence $\{a_n\}$ is decreasing and satisfies the condition (i)?

58. Let $\{a_n\}$ be a decreasing sequence of positive numbers. Show that the series
$$\sum_{n=1}^{\infty} a_n = a_1 + a_2 + a_3 + a_4 + \cdots$$
is convergent if and only if the associated series
$$\sum_{k=0}^{\infty} 2^k a_{2^k} = a_1 + 2a_2 + 4a_4 + 8a_8 + \cdots$$
is convergent. This result is known as *Cauchy's condensation test*.

Hint. Let s_n be the nth partial sum of $\sum a_n$. Verify that $s_n \leq a_1 + 2a_2 + \cdots + 2^{k-1} a_{2^{k-1}}$ if $n < 2^k$, while on the other hand $s_n \geq \tfrac{1}{2}(a_1 + 2a_2 + \cdots + 2^k a_{2^k})$ if $n > 2^k$.

59. Use Cauchy's condensation test to show that the p-series converges if $p > 1$ and diverges if $p \leq 1$, as already proved by the integral test in Example 10.

60. Use Cauchy's condensation test to show that the series
$$\sum_{n=2}^{\infty} \dfrac{1}{n(\ln n)^p}$$
converges if $p > 1$ and diverges if $p \leq 1$ (this generalizes Examples 11 and 12).

61. Show that the series
$$\sum_{n=2}^{\infty} \dfrac{1}{(\ln n)^p}$$
is divergent for every value of p.

62. Show that the series
$$\sum_{n=2}^{\infty} \dfrac{1}{(\ln n)^{\ln n}} \quad \text{and} \quad \sum_{n=3}^{\infty} \dfrac{1}{[\ln (\ln n)]^{\ln n}}$$
are both convergent, while the series
$$\sum_{n=3}^{\infty} \dfrac{1}{(\ln n)^{\ln (\ln n)}}$$
is divergent.

9.4 Absolute and Conditional Convergence

Absolutely Convergent Series

The preceding section was devoted to a special investigation of nonnegative series, that is, series whose terms are all nonnegative numbers. We now return to the study of arbitrary series (as in Section 9.2), allowing the series under consideration to have both positive and negative terms. A series

$$\sum_{n=1}^{\infty} a_n = a_1 + a_2 + \cdots + a_n + \cdots \tag{1}$$

is said to be *absolutely convergent* if the related series

$$\sum_{n=1}^{\infty} |a_n| = |a_1| + |a_2| + \cdots + |a_n| + \cdots, \tag{1'}$$

whose terms are the absolute values of those of (1), is convergent. Notice that this new concept plays no role in the theory of nonnegative series, since the series (1) and (1') coincide if $a_n \geq 0$ for all n. The true significance of absolute convergence arises only in dealing with series which have both positive and negative terms (and in fact infinitely many terms of both signs), for a convergent series of this type may or may not be absolutely convergent. More exactly, as we will see in a moment, although an absolutely convergent series must be convergent, a series can be convergent without being absolutely convergent.

Example 1 The geometric series

$$\sum_{n=0}^{\infty} \left(-\frac{2}{3}\right)^n = 1 - \frac{2}{3} + \frac{4}{9} - \frac{8}{27} + \cdots \tag{2}$$

is convergent with sum

$$\frac{1}{1-(-\frac{2}{3})} = \frac{3}{5}$$

(set $a = 1$, $r = -\frac{2}{3}$ in Example 1, page 499). It is also absolutely convergent, since

$$\sum_{n=0}^{\infty} \left|\left(-\frac{2}{3}\right)^n\right| = \sum_{n=0}^{\infty} \left(\frac{2}{3}\right)^n = 1 + \frac{2}{3} + \frac{4}{9} + \frac{8}{27} + \cdots \tag{2'}$$

is another convergent geometric series, this time nonnegative, with sum

$$\frac{1}{1-\frac{2}{3}} = 3.$$

How do you account for the fact that the second sum is so much larger than the first sum?

Example 2 The series

$$\sum_{n=1}^{\infty} \frac{(-1)^{n-1}}{n} = 1 - \frac{1}{2} + \frac{1}{3} - \frac{1}{4} + \cdots, \tag{3}$$

known as the *alternating harmonic series*, is not absolutely convergent, since

$$\sum_{n=1}^{\infty} \left|\frac{(-1)^{n-1}}{n}\right| = \sum_{n=1}^{\infty} \frac{1}{n} = 1 + \frac{1}{2} + \frac{1}{3} + \frac{1}{4} + \cdots \tag{3'}$$

is the ordinary harmonic series, which is of course divergent. Despite the fact that the series (3) is not absolutely convergent, it *is* convergent. This will be shown in Example 4, with the help of a special convergence test.

Conditionally Convergent Series

As noted at the beginning of the section and as illustrated by the last example, there are series which are convergent without being absolutely convergent. Such a series is said to be *conditionally convergent*. The existence of conditionally convergent series shows that *convergence does not imply absolute convergence*. On the other hand, as shown by the next theorem, *an absolutely convergent series must be convergent*.

> **Theorem 8** (*Absolute convergence implies convergence*). Let $\sum a_n$ be a series such that $\sum |a_n|$ is convergent. Then $\sum a_n$ is also convergent.

Proof Let $\sum |a_n|$ be convergent. The series

$$\sum_{n=1}^{\infty} (a_n + |a_n|) \tag{4}$$

is nonnegative, since

$$a_n + |a_n| = \begin{cases} 2a_n = 2|a_n| & \text{if } a_n \geq 0, \\ 0 & \text{if } a_n < 0, \end{cases}$$

and moreover (4) is dominated by the convergent nonnegative series $\sum 2|a_n|$. It follows from the comparison test (Theorem 5, page 510) that the series (4) is also convergent. But then so is the series

$$\sum_{n=1}^{\infty} a_n = \sum_{n=1}^{\infty} (a_n + |a_n|) - \sum_{n=1}^{\infty} |a_n|,$$

since the difference between two convergent series is convergent (see page 503). ∎

Example 3 Consider the series

$$1 + \frac{1}{2} - \frac{1}{4} - \frac{1}{8} + \frac{1}{16} + \frac{1}{32} - \frac{1}{64} - \frac{1}{128} + \cdots, \tag{5}$$

obtained from the geometric series

$$1 + \frac{1}{2} + \frac{1}{4} + \frac{1}{8} + \frac{1}{16} + \frac{1}{32} + \frac{1}{64} + \frac{1}{128} + \cdots \tag{5'}$$

by changing the signs of every other *pair* of terms, starting with the third and fourth terms. Since the series (5′) is convergent (with sum 2), the series (5) is absolutely convergent, and hence, by Theorem 8, it is also convergent. As an exercise, show that the sum of this modified geometric series is $\frac{6}{5}$.

Alternating Series

An infinite series is said to be *alternating* if its terms are alternately positive and negative, that is, if consecutive terms always have opposite signs. Thus the series (2) and (3) in Examples 1 and 2 are alternating, but the series (5) considered in the preceding example is not, since it contains (infinitely many) pairs of consecutive terms with the same sign. It is clear that every alternating series can be written in one of the two forms

$$\sum_{n=1}^{\infty} (-1)^{n-1} a_n = a_1 - a_2 + a_3 - a_4 + \cdots$$

or

$$\sum_{n=1}^{\infty} (-1)^n a_n = -a_1 + a_2 - a_3 + a_4 - \cdots,$$

where the numbers $a_1, a_2, \ldots, a_n, \ldots$ are all positive. Notice that in the first of these forms, we could just as well have written $(-1)^{n+1} a_n$ instead of $(-1)^{n-1} a_n$.

The Alternating Series Test

For alternating series there is an important special convergence test due to Leibniz:

> **Theorem 9** *(Alternating series test with error estimate).* If $\{a_n\}$ is a strictly decreasing sequence of positive numbers such that
>
> $$\lim_{n \to \infty} a_n = 0, \qquad (6)$$
>
> then the alternating series
>
> $$a_1 - a_2 + a_3 - a_4 + \cdots + (-1)^{n-1} a_n + \cdots \qquad (7)$$
>
> is convergent. Let $R_n = S - s_n$ be the error made in approximating the sum S of the series (7) by its nth partial sum, that is, let
>
> $$R_n = (-1)^n a_{n+1} + (-1)^{n+1} a_{n+2} + (-1)^{n+2} a_{n+3} + \cdots \qquad (8)$$
>
> be the remainder after n terms of the series (7). Then R_n is smaller in absolute value than the first unused term in the approximation $S \approx s_n$ and has the same sign as the first unused term, that is, $|R_n| < a_{n+1}$ and R_n has the same sign as $(-1)^n a_{n+1}$ (this is the sign of $(-1)^n$ since a_{n+1} is positive).

Proof The condition (6) must be imposed, since otherwise the series (7) would be divergent (why?). Let s_n be the nth partial sum of the series. Then the even-numbered partial sums s_{2k} can be written in the form

$$s_{2k} = (a_1 - a_2) + (a_3 - a_4) + \cdots + (a_{2k-1} - a_{2k}),$$

where the right side is a sum of positive numbers, since $a_n > a_{n+1}$ and hence $a_n - a_{n+1} > 0$ for all n. Thus the even-numbered partial sums are all positive, and form a strictly increasing sequence $s_2, s_4, \ldots, s_{2k}, \ldots$ This sequence is also bounded, since

$$s_{2k} = a_1 - (a_2 - a_3) - \cdots - (a_{2k-2} - a_{2k-1}) - a_{2k} < a_1$$

(the subtracted numbers $a_2 - a_3, \ldots, a_{2k-2} - a_{2k-1}, a_{2k}$ are all positive). Therefore, by Theorem 2, page 493, the sequence $s_2, s_4, \ldots, s_{2k}, \ldots$ has a finite limit S, that is,

$$\lim_{k \to \infty} s_{2k} = S,$$

where S is clearly positive. As for the odd-numbered partial sums, we have

$$s_{2k+1} = s_{2k} + a_{2k+1},$$

and hence

$$\lim_{k\to\infty} s_{2k+1} = \lim_{k\to\infty} s_{2k} + \lim_{k\to\infty} a_{2k} = S + 0 = S,$$

because of (6). Since the even-numbered and odd-numbered partial sums approach the same limit S, the series (7) is convergent with sum S, and the series (8) is also convergent, for the reason given on page 504.

To prove the error estimate, we observe that the error or remainder R_n can be written in the form

$$R_n = (-1)^n(a_{n+1} - a_{n+2} + a_{n+3} - \cdots),$$

where the series in parentheses converges to a positive sum, since it is an alternating series of the same type as the original series (7). Therefore R_n has the same sign as $(-1)^n$, and

$$|R_n| = a_{n+1} - a_{n+2} + a_{n+3} - a_{n+4} + a_{n+5} - \cdots$$
$$= a_{n+1} - (a_{n+2} - a_{n+3}) - (a_{n+4} - a_{n+5}) - \cdots < a_{n+1},$$

where the inequality follows from the fact that the subtracted numbers $a_{n+2} - a_{n+3}, a_{n+4} - a_{n+5}, \ldots$ are all positive. ∎

It is easy to see that if $\{a_n\}$ is a strictly decreasing sequence of positive numbers satisfying the condition (6), then the alternating series

$$-a_1 + a_2 - a_3 + a_4 - \cdots + (-1)^n a_n + \cdots, \tag{7'}$$

which begins with $-a_1$ instead of a_1, is also convergent, with sum S' equal to the negative of the sum S of the series (7). This is an immediate consequence of the fact that the series (7') is the result of multiplying the series (7) by -1. As an exercise, show that if R'_n is the remainder after n terms of the series (7'), then $|R'_n| < a_{n+1}$ and R'_n has the same sign as $(-1)^{n+1} a_{n+1}$, that is, the sign of $(-1)^{n+1}$.

The proof of Theorem 9 seems intricate, but its intuitive meaning is quite simple. Suppose we plot the partial sums $s_1, s_2, s_3, s_4, \ldots$ of the alternating series (7) on a number line, as shown in Figure 11. Then s_1 lies to the right of the origin O, since $s_1 = a_1 > 0$, s_2 lies to the left of s_1, since $s_2 - s_1 = -a_2 < 0$, s_3 lies to the right of s_2, since $s_3 - s_2 = a_3 > 0$, s_4 lies to the left of s_3, since $s_4 - s_3 = -a_4 < 0$, and so on. (Why does every partial sum s_n lie to the right of the origin?) Also, as illustrated in the figure, the even-numbered partial sums $s_2, s_4, \ldots$ form an increasing sequence, while the odd-numbered partial sums $s_1, s_3, \ldots$ form a decreasing sequence. Both sequences are bounded and monotonic, and hence convergent; moreover, they must have the same limit S, since the terms of the two sequences get closer and closer together as $n \to \infty$ (note that $|s_{n+1} - s_n| = a_{n+1} \to 0$ as $n \to \infty$). Clearly, s_n lies to the left of S if n is even and to the right of S if n is odd, and correspondingly the remainder $R_n = S - s_n$ is positive if n is even and negative if n is odd. But then R_n has the same sign as $(-1)^n a_{n+1}$, since $(-1)^n a_{n+1}$ is positive if n is even and negative if n is odd. Finally, since S lies between every pair of consecutive partial sums s_n and s_{n+1}, it follows that

$$|R_n| = |S - s_n| < |s_{n+1} - s_n| = a_{n+1}.$$

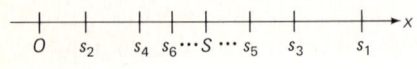

Figure 11

Example 4 If $p > 0$, the sequence $\{1/n^p\}$ is strictly decreasing and converges to 0. Therefore, by Theorem 9, the alternating series

$$1 - \frac{1}{2^p} + \frac{1}{3^p} - \frac{1}{4^p} + \cdots + (-1)^{n-1}\frac{1}{n^p} + \cdots \tag{9}$$

is convergent. If $p > 1$, the series is absolutely convergent, since the series

$$1 + \frac{1}{2^p} + \frac{1}{3^p} + \frac{1}{4^p} + \cdots + \frac{1}{n^p} + \cdots, \tag{9'}$$

whose terms are the absolute values of the terms of the alternating series (9), is a convergent p-series. However, if $0 < p \leq 1$, the series (9') is a divergent p-series, and in this case the series (9) is not absolutely convergent, but only conditionally convergent. For $p = 1$ we get the alternating harmonic series

$$1 - \frac{1}{2} + \frac{1}{3} - \frac{1}{4} + \cdots + (-1)^{n-1}\frac{1}{n} + \cdots,$$

already discussed in Example 2. The sum of this conditionally convergent series turns out to be $\ln 2$ (see Example 7, page 544).

Example 5 The alternating series

$$1 - \frac{1}{3} + \frac{1}{5} - \frac{1}{7} + \cdots + (-1)^{n-1}\frac{1}{2n-1} + \cdots, \tag{10}$$

satisfies the conditions of Theorem 9, and therefore is convergent. However, the series

$$1 + \frac{1}{3} + \frac{1}{5} + \frac{1}{7} + \cdots + \frac{1}{2n-1} + \cdots, \tag{10'}$$

whose terms are the absolute values of the terms of the series (10), is divergent. In fact,

$$\lim_{n \to \infty} \frac{1/(2n-1)}{1/n} = \lim_{n \to \infty} \frac{n}{2n-1} = \frac{1}{2},$$

so that the divergence of (10') follows from that of the harmonic series $\sum (1/n)$, with the help of the limit comparison test (Theorem 6, page 511). Thus the series (10) is conditionally convergent. It will be shown in Example 8, page 544, that the sum of this series is $\pi/4$.

Example 6 The alternating series

$$\sum_{n=0}^{\infty} \frac{(-1)^n}{n!} = 1 - 1 + \frac{1}{2!} - \frac{1}{3!} + \frac{1}{4!} - \frac{1}{5!} + \cdots \tag{11}$$

converges, by Theorem 9, but its convergence also follows from the fact that it is absolutely convergent, since the series

$$\sum_{n=0}^{\infty} \frac{1}{n!} = 1 + 1 + \frac{1}{2!} + \frac{1}{3!} + \frac{1}{4!} + \frac{1}{5!} + \cdots \tag{11'}$$

converges, as shown in Example 4, page 511. With the help of the second part of Theorem 9, we can estimate the error committed in approximating the sum of the series (11) by the sum of its first n terms. For example, let S be the sum of (11), and suppose we retain the first 8 terms, writing

$$S = 1 - 1 + \frac{1}{2!} - \frac{1}{3!} + \frac{1}{4!} - \frac{1}{5!} + \frac{1}{6!} - \frac{1}{7!} + R_8.$$

Then $1/8!$ is the first unused term, so that R_8 is positive and less than $1/8! = 0.0000248\ldots$, while

$$1 - 1 + \frac{1}{2!} - \frac{1}{3!} + \frac{1}{4!} - \frac{1}{5!} + \frac{1}{6!} - \frac{1}{7!} = \frac{1854}{5040} = \frac{103}{280} = 0.367857\ldots.$$

Thus we can conclude that the error of the approximation $S \approx 103/280$ is positive and less than 0.000025. Actually $S = e^{-1} = 0.367879\ldots$, as shown in Example 5, page 542. □

The alternating series test (Theorem 9) continues to work if the sequence $\{a_n\}$ is decreasing, but not *strictly* decreasing, provided that the error estimate is changed from $|R_n| < a_{n+1}$ to $|R_n| \le a_{n+1}$. The proof is a slight modification of the one for strictly decreasing $\{a_n\}$, and is left as an exercise.

Example 7 The sequence

$$1, 1, \frac{1}{2}, \frac{1}{2}, \frac{1}{3}, \frac{1}{3}, \ldots, \frac{1}{n}, \frac{1}{n}, \ldots,$$

is decreasing, but not strictly decreasing. The associated alternating series

$$1 - 1 + \frac{1}{2} - \frac{1}{2} + \frac{1}{3} - \frac{1}{3} + \cdots + \frac{1}{n} - \frac{1}{n} + \cdots$$

is conditionally convergent with sum 0, and the remainder R_n is either 0 if n is even or exactly equal to the first unused term if n is odd.

Example 8 The sum of the first $2n$ terms of the alternating series

$$a_1 - a_2 + a_3 - a_4 + \cdots + a_{2n-1} - a_{2n} + \cdots$$

$$= \frac{1}{\sqrt{2}-1} - \frac{1}{\sqrt{2}+1} + \frac{1}{\sqrt{3}-1} - \frac{1}{\sqrt{3}+1}$$

$$+ \cdots + \frac{1}{\sqrt{n}-1} - \frac{1}{\sqrt{n}+1} + \cdots \tag{12}$$

is equal to twice the nth partial sum of the divergent harmonic series, since

$$a_1 - a_2 + a_3 - a_4 + \cdots + a_{2n-1} - a_{2n} = \sum_{k=2}^{n+1}\left(\frac{1}{\sqrt{k}-1} - \frac{1}{\sqrt{k}+1}\right)$$

$$= \sum_{k=2}^{n+1} \frac{2}{k-1} = 2\sum_{k=1}^{n} \frac{1}{k}.$$

Hence the series (12) is also divergent, although $a_n \to 0$ as $n \to \infty$. This does not contradict the alternating series test, since the sequence $\{a_n\}$ is not decreasing. In fact, as you can easily verify, a_{2k+1} exceeds a_{2k} for $k = 1, 2, \ldots$

Remark You should not get the idea that an alternating series which fails to satisfy the conditions of Theorem 9 is necessarily divergent.

For example, the series

$$2 - \frac{1}{2^2} + \frac{2}{3^2} - \frac{1}{4^2} + \frac{2}{5^2} - \frac{1}{6^2} + \cdots,$$

obtained from the series $\sum (-1)^{n-1}(1/n^2)$ by multiplying all the odd-numbered terms by 2, is absolutely convergent by comparison with the series $\sum (2/n^2)$, but every odd-numbered term starting from the fifth is *larger* in absolute value than the preceding term.

Rearrangement of Series

If a series is absolutely convergent, its terms can be rearranged arbitrarily without changing the sum of the series, while if a series is conditionally convergent, its terms can be rearranged to give a new series with any sum whatsoever! The proof of this assertion is beyond the scope of a first course in calculus, but can be found in any advanced calculus text. Thus, qualitatively speaking, absolute convergence of a series is due to the intrinsic smallness of its terms, so that the series remains convergent even if all its negative terms are replaced by their absolute values, while conditional convergence of a series comes about only because of "advantageous" mutual cancellation of its positive and negative terms, and therefore depends in a critical way on the order in which these terms appear. In fact, one can even find *divergent* rearrangements of a conditionally convergent series.

Example 9 Let S be the sum of the conditionally convergent alternating harmonic series

$$1 - \frac{1}{2} + \frac{1}{3} - \frac{1}{4} + \frac{1}{5} - \frac{1}{6} + \cdots$$

($S = \ln 2 \approx 0.693$, as noted in Example 4). Find a rearrangement of the series with sum $\frac{1}{2}S$.

Solution Since

$$S = 1 - \frac{1}{2} + \frac{1}{3} - \frac{1}{4} + \frac{1}{5} - \frac{1}{6} + \cdots,$$

it follows that

$$\frac{1}{2}S = \frac{1}{2} - \frac{1}{4} + \frac{1}{6} - \frac{1}{8} + \frac{1}{10} - \frac{1}{12} + \cdots.$$

But any number of zero terms can be inserted between the terms of a series, without affecting either its convergence or the value of its sum, and therefore

$$S = 1 + 0 - \frac{1}{2} + \frac{1}{3} + 0 - \frac{1}{4} + \frac{1}{5} + 0 - \frac{1}{6} + \cdots, \qquad (13)$$

$$\frac{1}{2}S = 0 + \frac{1}{2} - \frac{1}{4} + 0 + \frac{1}{6} - \frac{1}{8} + 0 + \frac{1}{10} - \frac{1}{12} + \cdots. \qquad (13')$$

Subtracting the two series term by term, that is, subtracting each term of (13') from the term of (13) just above it, we get the desired rearrangement:

$$\frac{1}{2}S = 1 - \frac{1}{2} - \frac{1}{4} + \frac{1}{3} - \frac{1}{6} - \frac{1}{8} + \frac{1}{5} - \frac{1}{10} - \frac{1}{12} + \cdots. \quad \square$$

Problems

Determine whether the given series (with both positive and negative terms) is absolutely convergent, conditionally convergent or divergent. In the case of a series specified by giving a few initial terms, assume that the law of formation suggested by these terms holds for all the terms of the series.

1. $1 - \dfrac{1}{\sqrt{3}} + \dfrac{1}{\sqrt{5}} - \dfrac{1}{\sqrt{7}} + \cdots$

2. $1 - \dfrac{1}{2!} + \dfrac{1}{4!} - \dfrac{1}{6!} + \cdots$

3. $\sum_{n=1}^{\infty} \dfrac{(-1)^n}{(2n+1)^3}$

4. $\sum_{n=1}^{\infty} \dfrac{(-1)^{n-1}}{3n-1}$

5. $\dfrac{1}{8} - \dfrac{2}{12} + \dfrac{3}{16} - \dfrac{4}{20} + \cdots$

6. $\dfrac{1}{2 \ln 2} - \dfrac{1}{3 \ln 3} + \dfrac{1}{4 \ln 4} - \dfrac{1}{5 \ln 5} + \cdots$

7. $\sum_{n=1}^{\infty} \dfrac{(-1)^{n-1}}{\sqrt[n]{n}}$

8. $\sum_{n=1}^{\infty} \dfrac{(-1)^{n+1}}{\ln(n+1)}$

9. $1 - \dfrac{1}{101} + \dfrac{1}{201} - \dfrac{1}{301} + \cdots$

10. $1 - \dfrac{1}{2} - \dfrac{1}{4} + \dfrac{1}{8} - \dfrac{1}{16} + \dfrac{1}{32} + \dfrac{1}{64} - \dfrac{1}{128} - \dfrac{1}{256} + \cdots$

11. $\sum_{n=1}^{\infty} (-1)^n \dfrac{n}{e^{2n}}$

12. $\sum_{n=2}^{\infty} \dfrac{(-1)^n}{n(\ln n)^2}$

13. $\sum_{n=1}^{\infty} (-1)^n \dfrac{\ln n}{\sqrt{n}}$

14. $\sum_{n=1}^{\infty} (-1)^{n+1} \ln \dfrac{n+1}{n}$

15. $\dfrac{1}{1 \cdot 2} - \dfrac{1}{3 \cdot 4} + \dfrac{1}{5 \cdot 6} - \dfrac{1}{7 \cdot 8} + \cdots$

16. $\dfrac{1}{\sqrt{1 \cdot 2}} - \dfrac{1}{\sqrt{2 \cdot 3}} + \dfrac{1}{\sqrt{3 \cdot 4}} - \dfrac{1}{\sqrt{4 \cdot 5}} + \cdots$

17. $\sum_{n=1}^{\infty} \left(\dfrac{1}{n} - 1 \right)^n$

18. $1 + \dfrac{1}{4} - \dfrac{1}{9} - \dfrac{1}{16} + \dfrac{1}{25} + \dfrac{1}{36} - \dfrac{1}{49} - \dfrac{1}{64} + \cdots$

19. $\sum_{n=1}^{\infty} \dfrac{\sin n}{n^2}$

20. $\sum_{n=1}^{\infty} \dfrac{\cos(n\pi/2)}{n}$

21. $\sum_{n=1}^{\infty} \dfrac{2 \sin(2n\pi/3)}{\sqrt{3} n}$

22. $\sum_{n=1}^{\infty} (-1)^{n-1} \sin \dfrac{a}{n}$ $(a > 0)$

23. $a - \dfrac{1}{2}a^2 + \dfrac{1}{3}a^3 - \dfrac{1}{4}a^4 + \cdots$ $(a > 0)$

24. $\sum_{n=1}^{\infty} \dfrac{(-1)^{n(n+1)/2}}{n(n+1)}$

Each of the following series is convergent, by the alternating series test (Theorem 9). Let R_n be the remainder of the series after n terms, that is, the error made in approximating the sum of the series by its nth partial sum. In Problems 25–29 find the sign of R_n and an upper bound for $|R_n|$ if n has the specified value. In Problems 30–34 find the smallest value of n for which R_n satisfies the specified inequality. (As always, assume that the law of formation suggested by the first few terms holds for all the terms of the series.)

25. $1 - \dfrac{1}{2} + \dfrac{1}{3} - \dfrac{1}{4} + \cdots, n = 99$

26. $-1 + \dfrac{1}{2^2} - \dfrac{1}{3^3} + \dfrac{1}{4^4} - \cdots, n = 5$

27. $1 - \dfrac{1}{2!} + \dfrac{1}{4!} - \dfrac{1}{6!} + \cdots, n = 6$

28. $\dfrac{1}{\ln 2} - \dfrac{1}{\ln 3} + \dfrac{1}{\ln 4} - \dfrac{1}{\ln 5} + \cdots, n = 998$

29. $-1 + \dfrac{1}{3} - \dfrac{1 \cdot 4}{3 \cdot 6} + \dfrac{1 \cdot 4 \cdot 7}{3 \cdot 6 \cdot 9} - \cdots, n = 8$

30. $1 - \dfrac{1}{\sqrt{2}} + \dfrac{1}{\sqrt{3}} - \dfrac{1}{\sqrt{4}} + \cdots, |R_n| < 0.001$

31. $1 - \dfrac{1}{3!} + \dfrac{1}{5!} - \dfrac{1}{7!} + \cdots, |R_n| < 0.0002$

32. $\dfrac{1}{1!2!} - \dfrac{1}{2!3!} + \dfrac{1}{3!4!} - \dfrac{1}{4!5!} + \cdots, |R_n| < 10^{-8}$

33. $1 - \dfrac{1}{2^2} + \dfrac{1}{3^2} - \dfrac{1}{4^2} + \cdots, |R_n| < 0.0005$

34. $-\dfrac{1}{3 \cdot 2^2} + \dfrac{1}{4 \cdot 2^3} - \dfrac{1}{5 \cdot 2^4} + \dfrac{1}{6 \cdot 2^5} - \cdots, |R_n| < 0.0001$

35. Find a divergent rearrangement of the conditionally convergent alternating series

$$\sum_{n=1}^{\infty} \dfrac{(-1)^{n-1}}{\sqrt{n}} = 1 - \dfrac{1}{\sqrt{2}} + \dfrac{1}{\sqrt{3}} - \dfrac{1}{\sqrt{4}} + \dfrac{1}{\sqrt{5}} - \dfrac{1}{\sqrt{6}} + \cdots$$

The sum of the alternating harmonic series is $\ln 2$. Find a rearrangement of the series which has the sum

36. $\tfrac{3}{2} \ln 2$ 37. $2 \ln 2$ 38. 0

39. Let $\{a_n\}$ be a decreasing sequence of positive numbers converging to 0, and let

$$\sum_{n=1}^{\infty} b_n = b_1 + b_2 + \cdots + b_n + \cdots$$

be a series (in general divergent!) whose partial sums $B_n = b_1 + b_2 + \cdots + b_n$ form a bounded sequence $\{B_n\}$. With the help of Problem 38, page 229, show that the series

$$\sum_{n=1}^{\infty} a_n b_n = a_1 b_1 + a_2 b_2 + \cdots + a_n b_n + \cdots$$

is convergent. This convergence test is known as *Dirichlet's test*. Show that it contains the alternating series test as a special case.

Use Dirichlet's test to verify the convergence of the given series.

40. $1 - \dfrac{2}{3} + \dfrac{1}{5} + \dfrac{1}{7} - \dfrac{2}{9} + \dfrac{1}{11} + \dfrac{1}{13} - \dfrac{2}{15} + \dfrac{1}{17} - \cdots$

41. $1 + \dfrac{2}{\sqrt{2}} - \dfrac{3}{\sqrt{3}} + \dfrac{1}{\sqrt{4}} + \dfrac{2}{\sqrt{5}} - \dfrac{3}{\sqrt{6}} + \dfrac{1}{\sqrt{7}} + \dfrac{2}{\sqrt{8}} - \dfrac{3}{\sqrt{9}} + \cdots$

42. $1 - \dfrac{3}{2} - \dfrac{5}{3} + \dfrac{7}{4} + \dfrac{1}{5} - \dfrac{3}{6} - \dfrac{5}{7} + \dfrac{7}{8} + \dfrac{1}{9} - \dfrac{3}{10} - \dfrac{5}{11} + \dfrac{7}{12} + \cdots$

Notice that none of these series is an alternating series.

9.5 The Ratio and Root Tests

We now present two further convergence tests. These tests, of importance in their own right, are indispensable tools for the investigation of power series that begins in the next section.

The Ratio Test

Theorem 10 *(Ratio test)*. *Let* $\sum a_n$ *be a series with nonzero terms such that*

$$\lim_{n \to \infty} \left| \frac{a_{n+1}}{a_n} \right| = L, \qquad (1)$$

where the case $L = \infty$ *is allowed. Then the series is absolutely convergent if* $0 \leq L < 1$ *and divergent if* $L > 1$ *or* $L = \infty$. *The test is inconclusive if* $L = 1$.

Proof Suppose (1) holds with $0 \leq L < 1$ (needless to say, L cannot be negative), and let r be any number in the open interval $(L, 1)$. Then, because of (1),

$$\left| \frac{a_{n+1}}{a_n} \right| < r < 1$$

for all n starting from some integer N. Therefore

$$|a_{N+1}| < |a_N| r,$$
$$|a_{N+2}| < |a_{N+1}| r < |a_N| r^2,$$
$$|a_{N+3}| < |a_{N+2}| r < |a_N| r^3,$$

and in general

$$|a_{N+n}| < |a_N| r^n \qquad (n = 1, 2, \ldots), \qquad (2)$$

so that

$$|a_{N+1}| + |a_{N+2}| + |a_{N+3}| + \cdots < |a_N| r + |a_N| r^2 + |a_N| r^3 + \cdots$$
$$= |a_N| r (1 + r + r^2 + \cdots).$$

The series on the right converges, since it is a convergent geometric series, and hence the series on the left also converges, by the comparison test (see Theorem 5, page 510). But then $\sum |a_n|$ converges, since omission of a finite number of terms of a series has no effect on whether or not the series is convergent. In other words, $\sum a_n$ is absolutely convergent.

Next suppose $L > 1$ or $L = \infty$, and let r be any number in the open interval $(1, L)$ if $L > 1$ or in the interval $(1, \infty)$ if $L = \infty$. Because of (1), there is now an integer N such that

$$\left|\frac{a_{n+1}}{a_n}\right| > r > 1$$

for all n starting from some integer N, and instead of (2) we get

$$|a_{N+n}| > |a_N|r^n \quad (n = 1, 2, \ldots), \tag{2'}$$

by the same argument as before, with all the inequalities reversed. But then

$$|a_{N+n}| > |a_N| > 0 \quad (n = 1, 2, \ldots),$$

which is incompatible with the necessary condition

$$\lim_{n \to \infty} a_n = 0$$

for convergence of $\sum a_n$, so that $\sum a_n$ is divergent.

Finally, to show that the test is inconclusive if $L = 1$, we apply it to the three series

$$\sum_{n=1}^{\infty} \frac{1}{n^2}, \quad \sum_{n=1}^{\infty} \frac{(-1)^{n-1}}{n}, \quad \sum_{n=1}^{\infty} \frac{1}{n}.$$

Then $L = 1$ in each case, as you can easily verify, although the first series is absolutely convergent, the second conditionally convergent and the third divergent. ∎

Of course, if the series $\sum a_n$ is nonnegative, the word "absolutely" in the statement of the theorem is superfluous and can be dropped, and in this case (1) simplifies to

$$\lim_{n \to \infty} \frac{a_{n+1}}{a_n} = L. \tag{1'}$$

The ratio test is particularly effective when the nth term a_n involves factorials, since then many common factors of the numerator and denominator can be cancelled in calculating the ratio $|a_{n+1}/a_n|$.

Example 1 It was shown in Example 4, page 511, that the series

$$\sum_{n=1}^{\infty} \frac{1}{n!}$$

is convergent. This also follows from the ratio test, since

$$\lim_{n \to \infty} \frac{a_{n+1}}{a_n} = \lim_{n \to \infty} \frac{n!}{(n+1)!} = \lim_{n \to \infty} \frac{1}{n+1} = 0$$

(here, and below, a_n denotes the nth term of the series under consideration).

Example 2 The series
$$\sum_{n=1}^{\infty} (-1)^{n-1} \frac{n!}{n^n}$$
is absolutely convergent, since
$$\lim_{n\to\infty} \left|\frac{a_{n+1}}{a_n}\right| = \lim_{n\to\infty} \frac{(n+1)!}{(n+1)^{n+1}} \frac{n^n}{n!} = \lim_{n\to\infty} \frac{n^n}{(n+1)^n}$$
$$= \lim_{n\to\infty} \frac{1}{\left(1+\frac{1}{n}\right)^n} = \frac{1}{e} \approx 0.37 < 1.$$

Example 3 The series
$$\sum_{n=1}^{\infty} \frac{(2n)!}{(n!)^2}$$
is divergent, since
$$\lim_{n\to\infty} \frac{a_{n+1}}{a_n} = \lim_{n\to\infty} \frac{(2n+2)!}{[(n+1)!]^2} \frac{(n!)^2}{(2n)!} = \lim_{n\to\infty} \frac{(2n+2)(2n+1)}{(n+1)(n+1)} = 4 > 1.$$

Example 4 The case $L = \infty$ is illustrated by the series $\sum n!$, which we already know to be divergent (its nth term does not approach 0 as $n \to \infty$), and in fact for this series the ratio test gives
$$\lim_{n\to\infty} \frac{a_{n+1}}{a_n} = \lim_{n\to\infty} \frac{(n+1)!}{n!} = \lim_{n\to\infty} (n+1) = \infty.$$

The Root Test

The proof of the next convergence test resembles that of the ratio test, but is actually somewhat simpler.

> **Theorem 11** *(Root test)*. Let $\sum a_n$ be a series such that
> $$\lim_{n\to\infty} \sqrt[n]{|a_n|} = L, \qquad (3)$$
> where the case $L = \infty$ is allowed. Then the series is absolutely convergent if $0 \leq L < 1$ and divergent if $L > 1$ or $L = \infty$. The test is inconclusive if $L = 1$.

Proof If $0 \leq L < 1$, let r be any number in the open interval $(L, 1)$. Then, because of (3),
$$\sqrt[n]{|a_n|} < r < 1$$
for all n starting from some integer N, or equivalently
$$|a_n| < r^n \qquad (n = N, N+1, \ldots). \qquad (4)$$
Therefore
$$|a_N| + |a_{N+1}| + |a_{N+2}| + \cdots < r^N + r^{N+1} + r^{N+2} + \cdots$$
$$= r^N(1 + r + r^2 + \cdots),$$

Section 9.5 The Ratio and Root Tests

where the series on the right is a convergent geometric series. Hence, by the comparison test, $\sum |a_n|$ is also convergent, that is, $\sum a_n$ is absolutely convergent. If $L > 1$ or $L = \infty$, let r be any number in the interval $(1, L)$. Then

$$|a_n| > r^n > 1 \quad (n = N, N+1, \ldots), \tag{4'}$$

instead of (4), so that the sequence $\{a_n\}$ fails to approach 0 and the series $\sum a_n$ diverges. If $L = 1$, the root test is inconclusive, as you can check by applying it to the same three series as in the last part of the proof of the ratio test. ∎

In applying the root test, it helps to bear in mind that

$$\lim_{n \to \infty} \sqrt[n]{c} = \lim_{n \to \infty} e^{(1/n) \ln c} = e^0 = 1,$$

where $c > 0$, and similarly

$$\lim_{n \to \infty} \sqrt[n]{n} = \lim_{n \to \infty} e^{(1/n) \ln n} = e^0 = 1,$$

as anticipated in Problem 56, page 497.

Example 5 The series

$$\sum_{n=2}^{\infty} \frac{(-1)^{n-1}}{(\ln n)^n}$$

is absolutely convergent, since

$$\lim_{n \to \infty} \sqrt[n]{\left|\frac{(-1)^{n-1}}{(\ln n)^n}\right|} = \lim_{n \to \infty} \frac{1}{\ln n} = 0.$$

Example 6 The series

$$\sum_{n=1}^{\infty} \frac{n}{3^n}$$

is convergent, since

$$\lim_{n \to \infty} \sqrt[n]{\frac{n}{3^n}} = \lim_{n \to \infty} \frac{\sqrt[n]{n}}{3} = \frac{1}{3} < 1.$$

Example 7 The series

$$\sum_{n=1}^{\infty} \frac{2^n}{n^4}$$

is divergent, since

$$\lim_{n \to \infty} \sqrt[n]{\frac{2^n}{n^4}} = \lim_{n \to \infty} \frac{2}{(\sqrt[n]{n})^4} = 2 > 1.$$

Example 8 The case $L = \infty$ is illustrated by the series $\sum n^n$, which we already know to be divergent (why?), and in fact for this series the root test gives

$$\lim_{n \to \infty} \sqrt[n]{n^n} = \lim_{n \to \infty} n = \infty.$$

Problems

Use the ratio test or root test to investigate the convergence of the given series. If both tests are inconclusive, use any other convergence test. In the case of series with both positive and negative terms, distinguish between absolute and conditional convergence.

1. $\sum_{n=1}^{\infty} \dfrac{n^2}{(2n-1)!}$

2. $\sum_{n=1}^{\infty} \dfrac{e^n}{n^{100}}$

3. $\sum_{n=1}^{\infty} (-1)^n \dfrac{2^n}{n^2}$

4. $\sum_{n=2}^{\infty} \dfrac{n}{(\ln n)^n}$

5. $\sum_{n=1}^{\infty} (-1)^{n-1} \dfrac{n^n}{(2n)!}$

6. $\sum_{n=1}^{\infty} \left(\dfrac{1-n}{1+n}\right)^n$

7. $\sum_{n=1}^{\infty} \dfrac{\ln n}{5^n}$

8. $\sum_{n=1}^{\infty} (-1)^n \dfrac{4n^2-1}{(1.01)^n}$

9. $\sum_{n=1}^{\infty} \dfrac{1}{n\sqrt[n]{n}}$

10. $\sum_{n=1}^{\infty} \dfrac{n^{4/9}}{3^n}$

11. $\sum_{n=1}^{\infty} \dfrac{n!}{n^n}\left(\dfrac{19}{7}\right)^n$

12. $\sum_{n=1}^{\infty} \dfrac{(n+1)^n}{n^{n+1}}$

13. $\sum_{n=1}^{\infty} \dfrac{n(n+1)(n+2)}{4^n}$

14. $\sum_{n=1}^{\infty} \dfrac{n!(-3)^n}{(2n+1)!}$

15. $\sum_{n=1}^{\infty} \dfrac{10^{n/2}}{n\pi^n}$

16. $\sum_{n=1}^{\infty} (-1)^n \dfrac{(2n)^n}{(n+1)^{n+1}}$

17. $\sum_{n=1}^{\infty} \dfrac{(n!)^2}{e^{n^2}}$

18. $\sum_{n=1}^{\infty} \dfrac{1 \cdot 3 \cdot 5 \cdots (2n-1)}{3 \cdot 6 \cdot 9 \cdots 3n}$

19. $\sum_{n=1}^{\infty} \dfrac{1 \cdot 4 \cdot 7 \cdots (3n-2)}{2 \cdot 4 \cdot 6 \cdots 2n}$

20. $\sum_{n=1}^{\infty} \dfrac{c(c+1) \cdots (c+n-1)}{n^n}$

21. $\sum_{n=2}^{\infty} \dfrac{(-1)^n}{(\ln n)^2}$

22. $\sum_{n=1}^{\infty} \dfrac{n!(2n)!}{(3n)!}$

23. $\sum_{n=1}^{\infty} (-1)^{n-1} \dfrac{n^p}{n!}$

24. $\sum_{n=1}^{\infty} \dfrac{n!}{1 \cdot 5 \cdot 9 \cdots (4n-3)}$

25. Let $\{a_n\}$ be any sequence with nonzero terms such that
$$\lim_{n \to \infty} \left|\dfrac{a_{n+1}}{a_n}\right| < 1.$$
Show that $\{a_n\}$ converges to 0.

Verify that

26. $\lim\limits_{n \to \infty} \dfrac{c^n}{n!} = 0$ (c arbitrary)

27. $\lim\limits_{n \to \infty} \dfrac{n!}{n^n} = 0$

28. $\sum_{n=1}^{\infty} \left(\dfrac{n!}{n^n}\right)^n$ is convergent

29. It can be shown that if
$$\lim_{n \to \infty} \left|\dfrac{a_{n+1}}{a_n}\right| = L, \tag{i}$$
where the case $L = \infty$ is allowed, then
$$\lim_{n \to \infty} \sqrt[n]{|a_n|} = L. \tag{i'}$$
Give an example of a series whose convergence can be proved by using the root test, but for which the ratio test does not work because (i) fails to exist. (Thus, although the limit (i) is often easier to evaluate than the limit (i') in cases where both exist, the root test is actually more powerful than the ratio test.)

Verify that

30. $\lim\limits_{n \to \infty} \dfrac{n}{\sqrt[n]{n!}} = e$

31. $\lim\limits_{n \to \infty} \dfrac{\sqrt[n]{(n+1)(n+2)\cdots 2n}}{n} = \dfrac{4}{e}$

32. Investigate the convergence of the series
$$a + ab + a^2b + a^2b^2 + \cdots + a^nb^{n-1} + a^nb^n + \cdots,$$
where a and b are distinct positive numbers.

9.6 Power Series

Let x be an independent variable, and let $\{a_n\}$ be any sequence of real numbers. Then an infinite series of the form

$$\sum_{n=0}^{\infty} a_n x^n = a_0 + a_1 x + a_2 x^2 + \cdots + a_n x^n + \cdots \tag{1}$$

is called a *power series* (in x), and the numbers $a_0, a_1, a_2, \ldots, a_n, \ldots$ are called the *coefficients* of the series or of its terms. Here the lower limit of summation is 0 rather than 1, since a power series conventionally begins with a "zeroth term." Notice that the terms of a power series $\sum a_n x^n$ are *functions*, rather than numbers as in the case of the series considered earlier (which are sometimes called *numerical* series to distinguish them from series whose terms are functions). If $a_N \neq 0$ and $a_n = 0$ for all $n > N$, the power series (1) reduces to

$$\sum_{n=0}^{N} a_n x^n = a_0 + a_1 x + a_2 x^2 + \cdots + a_N x^N \qquad (a_N \neq 0),$$

which is a polynomial of degree N. Thus, roughly speaking, power series are polynomials with infinitely many terms. If $x = 0$, the right side of (1) reduces to the single constant term a_0. To make the left side of (1) also reduce to a_0 for $x = 0$, we adopt the convention that here $0^0 = 1$, so that $a_0 x^0 = a_0$ even if $x = 0$.

If c is any constant, we can also consider the more general power series

$$\sum_{n=0}^{\infty} a_n(x-c)^n = a_0 + a_1(x-c) + a_2(x-c)^2 + \cdots + a_n(x-c)^n + \cdots,$$
$$(1')$$

involving powers of $x - c$ instead of powers of x. Naturally, (1') reduces to (1) if $c = 0$. There is really no need to make a separate study of series of the form (1'), since anything we want to know about the series (1') can be learned by analyzing the series (1) with the same coefficients, and then transforming back to the variable $x - c$ (see the discussion on page 534).

Let x take a fixed value r, where r is any real number. Then the power series $\sum a_n x^n$, whose terms are functions, becomes the series $\sum a_n r^n$, whose terms are numbers, and the numerical series $\sum a_n r^n$ can be investigated by using the methods of Sections 9.2–9.5. Every power series $\sum a_n x^n$ converges (trivially) for $x = 0$, since it then reduces to the single constant term a_0. The series may converge only for $x = 0$, or it may converge for every value of x, but in general it will converge for some nonzero values of x and diverge for other values.

Example 1 Let $a_n = n!$, where $0! = 1$. Then the power series $\sum a_n x^n$ converges only for $x = 0$. This follows from the ratio test, since

$$\lim_{n \to \infty} \left| \frac{a_{n+1} x^{n+1}}{a_n x^n} \right| = \lim_{n \to \infty} \left| \frac{(n+1)! x^{n+1}}{n! x^n} \right| = \lim_{n \to \infty} (n+1)|x| = \infty,$$

unless $x = 0$.

Example 2 If $a_n = 1/n^n$, the power series $\sum a_n x^n$ (with lower limit of summation 1) is absolutely convergent for all x. This follows from the root test, since

$$\lim_{n \to \infty} \sqrt[n]{|a_n x^n|} = \lim_{n \to \infty} \sqrt[n]{\frac{|x|^n}{n^n}} = \lim_{n \to \infty} \frac{|x|}{n} = 0$$

for every value of x.

Example 3 One of the simplest power series is the geometric series

$$\sum_{n=0}^{\infty} x^n = 1 + x + x^2 + \cdots + x^n + \cdots,$$

530   Chapter 9   Sequences and Series

which, as we already know from Example 1, page 499, converges for all x in the open interval $(-1, 1)$ and diverges for all other values of x.

Convergence of Power Series

Our investigation of the convergence of power series rests on the following key result.

> **Theorem 12** (*Basic convergence property of power series*). *If a power series $\sum a_n x^n$ is convergent for $x = r$, where $r \neq 0$, then it is absolutely convergent for every x such that $|x| < |r|$. If the series is divergent for $x = s$, then it is also divergent for every x such that $|x| > |s|$.*

Proof If the series $\sum a_n r^n$ is convergent, then the sequence $\{a_n r^n\}$ converges to 0, by Theorem 3, page 502. In particular, $\{a_n r^n\}$ is bounded, by Theorem 1, page 492, which means that there is a constant $C > 0$ such that $|a_n r^n| \leq C$ for all n. But then

$$|a_n x^n| = |a_n r^n| \left|\frac{x}{r}\right|^n \leq C \left|\frac{x}{r}\right|^n,$$

so that the series $\sum |a_n x^n|$ is dominated by the geometric series

$$\sum_{n=0}^{\infty} C \left|\frac{x}{r}\right|^n = C\left(1 + \left|\frac{x}{r}\right| + \left|\frac{x}{r}\right|^2 + \cdots\right).$$

If $|x/r| < 1$, or equivalently $|x| < |r|$, this geometric series is convergent, and then $\sum |a_n x^n|$ is also convergent, by the comparison test. In other words, $\sum a_n x^n$ is absolutely convergent for every x such that $|x| < |r|$.

To prove the second assertion, we observe that if the series $\sum a_n x^n$ diverges for $x = s$, it cannot converge at a point x such that $|x| > |s|$. For then, by the first part of the proof, $\sum a_n s^n$ would also converge, contrary to assumption. ∎

The Interval of Convergence

Let I be the set of points for which a given power series $\sum a_n x^n$ converges. Then, as anticipated by the notation, I always turns out to be an interval, called the *interval of convergence* (possibly the whole real line $(-\infty, \infty)$ or the "degenerate interval" $I = [0, 0]$ containing only the point 0).

> **Theorem 13** (*Interval of convergence of a power series*). *Let I be the set of all points x for which the power series $\sum a_n x^n$ is convergent. Then I is an interval with 0 as its midpoint.*

Proof (Optional) The point 0 always belongs to I, since every power series converges for $x = 0$. If $x = 0$ is the only point in I, then I reduces to the interval $[0, 0]$, but otherwise I contains at least two distinct points u and v. Let r be any point between u and v. Then $|r| < |u|$ or $|r| < |v|$, since r must be closer to the origin than at least one or both of the points u and v. Hence, by Theorem 12, the series $\sum a_n x^n$ is (absolutely) convergent for $x = r$, that is, r belongs to I too. Thus whenever the set I contains two distinct points u and v, it also contains every point r between u and v. Therefore I is an interval (see the remark on page 20). Moreover, if r is an interior point of I, then $-r$ also belongs to I. In fact, in this case we can always find points u and v in I such that r lies between u and v. But then $\sum |a_n r^n|$ converges, by the argument just given, and hence so does $\sum |a_n(-r)^n|$. In other words, $\sum a_n x^n$ is (absolutely) convergent for $x = -r$, that is, $-r$ belongs to I. It follows that I has 0 as its midpoint. ∎

To recapitulate, it is an immediate consequence of Theorem 13 that any given power series $\sum a_n x^n$ behaves in exactly one of the following ways:

(i) The series converges only for $x = 0$, as in Example 1, and then the interval of convergence I reduces to the interval $[0, 0]$, containing the single point 0.
(ii) The series is (absolutely) convergent for all x, as in Example 2, and then I is the whole real line $(-\infty, \infty)$.
(iii) The series converges for some nonzero values of x and diverges for other values. Then I is a finite interval of the form $(-R, R)$, $[-R, R]$, $[-R, R)$ or $(-R, R]$, where $R > 0$, depending on how the series behaves at the points $x = R$ and $x = -R$, which must be investigated separately. Here it is important to observe that the proof of Theorem 13 does not allow us to conclude that the endpoints of I belong to I, and in fact the interval of convergence I may fail to contain one or both of its endpoints, as shown in subsequent examples. In other words, the series may or may not converge for $x = R$ or $x = -R$.

The Radius of Convergence

The number R in case (iii) is called the *radius of convergence* of the power series $\sum a_n x^n$. Case (i) can be regarded as the subcase of case (iii) corresponding to $R = 0$, and case (ii) as the subcase of (iii) corresponding to $R = \infty$.

Example 4 For the geometric series $\sum x^n$ considered in Example 3, the radius of convergence is 1. The interval of convergence is the open interval $(-1, 1)$, since $\sum x^n$ is divergent for $|x| \geq 1$, in particular for $x = \pm 1$. More generally, the geometric series

$$\sum_{n=0}^{\infty} \left(\frac{x}{R}\right)^n = 1 + \frac{x}{R} + \left(\frac{x}{R}\right)^2 + \cdots + \left(\frac{x}{R}\right)^n + \cdots$$

has radius of convergence R and interval of convergence $(-R, R)$.

Example 5 Applying the ratio test to the power series

$$\sum_{n=0}^{\infty} \frac{x^n}{n+1} = 1 + \frac{x}{2} + \frac{x^2}{3} + \frac{x^3}{4} + \cdots, \qquad (2)$$

we find that

$$\lim_{n \to \infty} \left|\frac{x^{n+1}}{n+2} \frac{n+1}{x^n}\right| = \lim_{n \to \infty} \frac{n+1}{n+2} |x| = |x|.$$

Hence the series is absolutely convergent if $|x| < 1$ and divergent if $|x| > 1$, so that the radius of convergence is 1. The interval of convergence is the half-open interval $[-1, 1)$. In fact, for $x = 1$ the series (2) becomes the divergent harmonic series

$$1 + \frac{1}{2} + \frac{1}{3} + \frac{1}{4} + \cdots,$$

while for $x = -1$ it becomes the conditionally convergent alternating harmonic series

$$1 - \frac{1}{2} + \frac{1}{3} - \frac{1}{4} + \cdots.$$

Changing x to $-x$ in (2), we get the series

$$\sum_{n=0}^{\infty} \frac{(-1)^n x^n}{n+1} = 1 - \frac{x}{2} + \frac{x^2}{3} - \frac{x^3}{4} + \cdots,$$

which again has the radius of convergence 1, but is now conditionally convergent for $x = 1$ and divergent for $x = -1$. Therefore this time the interval of convergence is $(-1, 1]$, the other half-open interval with the same endpoints ± 1.

Example 6 The alternating series

$$\sum_{n=0}^{\infty} (-1)^n \frac{x^{2n}}{n+1} = 1 - \frac{x^2}{2} + \frac{x^4}{3} - \frac{x^6}{4} + \cdots \tag{3}$$

is a power series containing only even powers of x. Application of the ratio test now gives

$$\lim_{n \to \infty} \left| \frac{x^{2n+2}}{n+2} \frac{n+1}{x^{2n}} \right| = \lim_{n \to \infty} \frac{n+1}{n+2} |x|^2 = x^2.$$

Hence the series is absolutely convergent if $x^2 < 1$ or equivalently $|x| < 1$, and divergent if $x^2 > 1$ or equivalently $|x| > 1$, so that the radius of convergence is 1. The interval of convergence is the closed interval $[-1, 1]$. In fact, substituting $x = 1$ and $x = -1$ in the series (3), we get the same conditionally convergent series, namely the alternating harmonic series.

Example 7 Applying the ratio test to the power series

$$\sum_{n=0}^{\infty} \frac{x^n}{5^n (n+1)^2} = 1 + \frac{x}{5 \cdot 2^2} + \frac{x^2}{5^2 \cdot 3^2} + \frac{x^3}{5^3 \cdot 4^2} + \cdots, \tag{4}$$

we find that

$$\lim_{n \to \infty} \left| \frac{x^{n+1}}{5^{n+1}(n+2)^2} \frac{5^n (n+1)^2}{x^n} \right| = \lim_{n \to \infty} \frac{(n+1)^2}{(n+2)^2} \frac{|x|}{5} = \frac{|x|}{5}.$$

Hence the series is absolutely convergent if $|x| < 5$ and divergent if $|x| > 5$, so that the radius of convergence is 5. The interval of convergence is the closed interval $[-5, 5]$. In fact, for $x = 5$ the series (4) becomes the convergent series

$$1 + \frac{1}{2^2} + \frac{1}{3^2} + \frac{1}{4^2} + \cdots$$

(the *p*-series with $p = 2$), while for $x = -5$ it becomes the absolutely convergent series

$$1 - \frac{1}{2^2} + \frac{1}{3^2} - \frac{1}{4^2} + \cdots.$$

It is apparent from the preceding examples that a power series may be absolutely convergent, conditionally convergent or divergent at both endpoints of its interval of convergence *I*, or conditionally convergent at one endpoint of *I* and divergent at the other. As an exercise, show that it is impossible for a power series to be absolutely convergent at one endpoint of *I* and divergent or conditionally convergent at the other endpoint.

Of course, a power series need not have a constant term, and in fact may begin with any power of *x* and may lack any number (finite or infinite) of powers of *x*.

Example 8 Applying the root test to the power series

$$\sum_{n=2}^{\infty} \frac{n^2 x^{2n}}{(\ln n)^n} = \frac{2^2 x^4}{(\ln 2)^2} + \frac{3^2 x^6}{(\ln 3)^3} + \frac{4^2 x^8}{(\ln 4)^4} + \cdots,$$

which contains only even powers of *x* beginning with x^4, we find that

$$\lim_{n \to \infty} \sqrt[n]{\left|\frac{n^2 x^{2n}}{(\ln n)^n}\right|} = \lim_{n \to \infty} \frac{(\sqrt[n]{n})^2}{\ln n} |x|^2 = 0$$

for every value of *x* (recall that $\sqrt[n]{n} \to 1$ as $n \to \infty$). Hence this series is absolutely convergent for all *x*, that is, it has radius of convergence ∞ and interval of convergence $(-\infty, \infty)$.

For a power series of the more general form $\sum a_n(x - c)^n$, where *c* is an arbitrary constant, the set of points for which the series converges is again an interval *I*, called the *interval of convergence*, and in fact the three possibilities listed on page 532 become

(i') The series converges only for $x = c$, and then the interval of convergence *I* reduces to the interval $[c, c]$, containing the single point *c*.
(ii') The series is (absolutely) convergent for all *x*, and then *I* is the whole real line $(-\infty, \infty)$.
(iii') The series converges for some values of *x* not equal to *c* and diverges for other values. Then *I* is a finite interval of the form $(c - R, c + R)$, $[c - R, c + R]$, $[c - R, c + R)$ or $(c - R, c + R]$, depending on how the series behaves at the points $x = c + R$ and $x = c - R$, which must be investigated separately.

To get these cases from the previous cases (i), (ii) and (iii) for a power series of the form $\sum a_n x^n$, we need only observe that if the variable of a power series is changed from *x* to $x - c$, then its interval of convergence is shifted *c* units to the right along the real line if $c > 0$, and $|c|$ units to the left if $c < 0$. As before, the number *R* in case (iii') is called the *radius of convergence* of the power series $\sum a_n(x - c)^n$, and cases (i') and (ii') correspond to $R = 0$ and $R = \infty$. Notice that *R* is always one half the length of the interval of convergence, with the length of the intervals $[c, c]$ and $(-\infty, \infty)$ being interpreted as 0 and ∞.

Example 9 The series

$$\sum_{n=0}^{\infty} \frac{(x-6)^n}{n+1} = 1 + \frac{x-6}{2} + \frac{(x-6)^2}{3} + \frac{(x-6)^3}{4} + \cdots \qquad (2')$$

is the power series in $x - 6$ obtained if x is replaced by $x - 6$ in the series (2). Since (2) has radius of convergence 1 and interval of convergence $[-1, 1)$, the series (2') has the same radius of convergence 1 but a different interval of convergence, namely the interval $[6 - 1, 6 + 1) = [5, 7)$ six units to the right.

Alternatively, applying the ratio test *directly* to the series (2'), we get

$$\lim_{n \to \infty} \left| \frac{(x-6)^{n+1}}{n+2} \cdot \frac{n+1}{(x-6)^n} \right| = \lim_{n \to \infty} \frac{n+1}{n+2} |x - 6| = |x - 6|.$$

It follows that the series (2') is absolutely convergent if $|x - 6| < 1$, that is, if $-1 < x - 6 < 1$ or equivalently $5 < x < 7$, and divergent if $|x - 6| > 1$, that is, if $x - 6 > 1$ or $x - 6 < -1$, which is equivalent to $x > 7$ or $x < 5$. Moreover, for $x = 7$ the series becomes the divergent harmonic series, while for $x = 5$ it becomes the (conditionally) convergent alternating harmonic series. Hence the series has radius of convergence 1 and interval of convergence $[5, 7)$, as before.

Our last example gives an explicit way of constructing a whole class of power series with the same radius of convergence, and will be used in the next section.

Example 10 Let $\sum a_n x^n$ be any power series, and let $\{c_n\}$ be any sequence of positive numbers such that

$$\lim_{n \to \infty} \sqrt[n]{c_n} = 1. \qquad (5)$$

Show that the series $\sum c_n a_n x^n$ has the same radius of convergence as $\sum a_n x^n$.

Solution (Optional) Let R be the radius of convergence of $\sum a_n x^n$ and R' the radius of convergence of $\sum c_n a_n x^n$. Suppose that $R \neq 0$, $R \neq \infty$. Given any point $x \neq 0$ in $(-R, R)$, we first choose a point r such that $|x| < |r| < R$, so that $\sum |a_n r^n|$ converges, and afterwards an integer N such that

$$\sqrt[n]{c_n} < \frac{|r|}{|x|},$$

or equivalently

$$c_n |x|^n < |r|^n,$$

for all $n > N$ (this can be done because of (5) and the fact that $|r|/|x| > 1$). Then

$$|c_n a_n x^n| = c_n |a_n| |x|^n < |a_n| |r|^n = |a_n r^n|$$

if $n > N$, so that the series $\sum |c_n a_n x^n|$ is dominated by the convergent series $\sum |a_n r^n|$. Therefore $\sum c_n a_n x^n$ is (absolutely) convergent, by the comparison test. Thus we have shown that x belongs to the interval of convergence of

$\sum c_n a_n x^n$ whenever x belongs to the interval of convergence of $\sum a_n x^n$. It follows that $R' \geq R$ (the convergence of both series for $x = 0$ is trivial). Next let

$$a'_n = c_n a_n, \quad c'_n = \frac{1}{c_n}.$$

Then

$$\lim_{n \to \infty} \sqrt[n]{c'_n} = \lim_{n \to \infty} \frac{1}{\sqrt[n]{c_n}} = 1,$$

and exactly the same argument applied to the series $\sum a'_n x^n = \sum c_n a_n x^n$ with radius of convergence R' and the series $\sum c'_n a'_n x^n = \sum a_n x^n$ with radius of convergence R shows that $R \geq R'$. But the inequalities $R' \geq R$ and $R \geq R'$ taken together imply $R' = R$.

Finally, if $R = 0$, the inequality $R \geq R'$ implies $R' = 0$, while if $R = \infty$, then $\sum c_n a_n x^n$ converges for every x (why?), so that $R' = \infty$. Thus $R' = R$ in every case, that is, the two series $\sum a_n x^n$ and $\sum c_n a_n x^n$ have the same radius of convergence. □

As an application of Example 10, we observe that the two series $\sum a_n x^n$ and $\sum n^p a_n x^n$ have the same radius of convergence for every real number p, since

$$\lim_{n \to \infty} \sqrt[n]{n^p} = \lim_{n \to \infty} (\sqrt[n]{n})^p = 1.$$

Problems

Find the radius of convergence and interval of convergence of the given power series.

1. $\sum_{n=1}^{\infty} n^n (x - \pi)^n$

2. $\sum_{n=0}^{\infty} (-1)^n x^{2n+1}$

3. $\sum_{n=0}^{\infty} \frac{x^n}{\sqrt{n+1}}$

4. $\sum_{n=0}^{\infty} (2n + 1) x^{2n}$

5. $\sum_{n=0}^{\infty} \frac{(x + 7)^n}{n!}$

6. $\sum_{n=0}^{\infty} \frac{(-1)^n x^n}{(n + 1)^{3/2}}$

7. $\sum_{n=0}^{\infty} x^{n!}$

8. $\sum_{n=0}^{\infty} \frac{n + 1}{(n + 2)(n + 3)} (x + 5)^n$

9. $\sum_{n=1}^{\infty} \left(1 + \frac{1}{n}\right)^{n^2} x^n$

10. $\sum_{n=2}^{\infty} n^{\ln n} x^n$

11. $\sum_{n=0}^{\infty} c^{n^2} x^n \quad (0 < c < 1)$

12. $\sum_{n=0}^{\infty} \frac{(n + 2)!}{n!(n + 4)!} x^n$

13. $\sum_{n=0}^{\infty} \frac{1 \cdot 4 \cdot 7 \cdots (3n + 1)}{1 \cdot 5 \cdot 9 \cdots (4n + 1)} x^{2n}$

14. $\sum_{n=0}^{\infty} \frac{2^n}{n!} (x - 3)^n$

15. $\sum_{n=0}^{\infty} (-1)^n e^{n^2} (x + e)^n$

16. $\sum_{n=0}^{\infty} \frac{x^{n^2}}{2^n}$

17. $\sum_{n=0}^{\infty} \frac{x^n}{a^n + b^n} \quad (a > 0, b > 0)$

18. $\sum_{n=2}^{\infty} \frac{(-1)^n \ln n}{n} x^n$

19. $\sum_{n=1}^{\infty} \left(\frac{3^n}{n} + \frac{2^n}{n^2}\right) x^n$

20. $\sum_{n=1}^{\infty} \left(\frac{1}{n} + \frac{2^n}{n^2}\right) x^n$

21. $\sum_{n=0}^{\infty} (1 + 2 + \cdots + 2^n) x^n$

22. $\sum_{n=1}^{\infty} \left(1 + \frac{1}{2} + \cdots + \frac{1}{n}\right) x^n$

23. $\sum_{n=1}^{\infty} \frac{(x - 4)^{n^2}}{n!}$

24. $\sum_{n=2}^{\infty} \frac{(4x)^n}{\ln n}$

25. $\sum_{n=1}^{\infty} \frac{(n!)^3}{(3n)!} x^{3n}$

26. $\sum_{n=0}^{\infty} 10^n \left(\frac{x - 1}{5}\right)^n$

27. $\sum_{n=0}^{\infty} (n + c^n) x^n \quad (0 \leq c < \infty)$

28. $\sum_{n=1}^{\infty} \dfrac{2 \cdot 4 \cdot 6 \cdots 2n}{1 \cdot 3 \cdot 5 \cdots (2n-1)} x^{4n}$

29. $\sum_{n=2}^{\infty} \dfrac{(x+1)^n}{n \ln n}$

30. $\sum_{n=1}^{\infty} \dfrac{n! \, x^n}{3^{n^2}}$

31. Given a power series $\sum a_n x^n$, suppose that

$$\lim_{n \to \infty} \left| \dfrac{a_{n+1}}{a_n} \right| = L, \quad \text{(i)}$$

or

$$\lim_{n \to \infty} \sqrt[n]{|a_n|} = L, \quad \text{(i')}$$

where the case $L = \infty$ is allowed. Show that the radius of convergence R of the series is given by the formula

$$R = \begin{cases} 1/L & \text{if } 0 < L < \infty, \\ 0 & \text{if } L = \infty, \\ \infty & \text{if } L = 0. \end{cases} \quad \text{(ii)}$$

(This problem summarizes in capsule form the use of the ratio and root tests to find the radius of convergence of a power series for which one of the limits (i) and (i') exists. Actually, the existence of (i) implies that of (i'), but not conversely (see Problem 29, page 529). Thus there are power series for which (i') exists, but (i) does not. There are also series, like the one in Problem 37, for which neither (i) nor (i') exists.)

Given a power series $\sum a_n x^n$ with radius of convergence R, where $0 < R < \infty$, find the radius of convergence of the series

32. $\sum_{n=0}^{\infty} \dfrac{a_n}{n!} x^n$

33. $\sum_{n=0}^{\infty} \dfrac{(n!)^2}{(2n)!} a_n x^n$

34. $\sum_{n=0}^{\infty} n^n a_n x^n$

35. $\sum_{n=0}^{\infty} a_n^p x^n$ (p arbitrary)

36. $\sum_{n=0}^{\infty} a_n x^{pn}$ ($p = 2, 3, \ldots$)

Assume the existence of either limit (i) or (i'), as needed.

37. Show that neither of the limits (i) and (i') exists for the power series

$$\sum_{n=0}^{\infty} [3 + (-1)^n]^n x^n.$$

Find the radius of convergence and interval of convergence of this series.

9.7 Differentiation and Integration of Power Series

The Sum of a Power Series

Let $\sum a_n x^n$ be a power series with interval of convergence I, and let f be the function defined on I whose value at any point r in I is just the sum of the series $\sum a_n r^n$. Then f is called the *sum* of the power series $\sum a_n x^n$, and we write

$$f(x) = \sum_{n=0}^{\infty} a_n x^n. \quad (1)$$

Always bear in mind that the sum of a power series $\sum a_n x^n$ is a *function* of the variable x, unlike the sum of the series $\sum a_n r^n$, which is a *number*. This is to be expected, since $\sum a_n x^n$ is a series whose terms are functions of x, while $\sum a_n r^n$ is a *numerical* series, that is, a series whose terms are numbers (r is a particular value of x). Formula (1) is called the *power series expansion of* f (*at* $x = 0$).

Example 1 The power series $\sum x^n$ is a geometric series with interval of convergence $I = (-1, 1)$. Moreover,

$$\sum_{n=0}^{\infty} x^n = 1 + x + x^2 + \cdots + x^n + \cdots = \dfrac{1}{1-x},$$

as we know from Example 1, page 499. Hence the function

$$f(x) = \dfrac{1}{1-x}$$

is the sum of the series $\sum x^n$, but only on the interval I, since the series diverges outside I. Correspondingly,

$$\frac{1}{1-x} = 1 + x + x^2 + \cdots + x^n + \cdots \tag{2}$$

is the power series expansion of this function. □

More generally, let $\sum a_n(x-c)^n$ be a power series in $x-c$ with interval of convergence I, and let f be the function defined on I whose value at any point r in I is just the sum of the numerical series $\sum a_n(r-c)^n$. Then f is called the *sum* of the power series $\sum a_n(x-c)^n$, and we write

$$f(x) = \sum_{n=0}^{\infty} a_n (x-c)^n. \tag{1'}$$

Formula (1') is called the *power series expansion of f at $x = c$*,† and reduces to formula (1) if $c = 0$. It will be shown in Example 4 that the coefficients a_n ($n = 0, 1, \ldots$) in the expansion (1') are uniquely determined by the function f and the constant c.

Example 2 The geometric series

$$\sum_{n=0}^{\infty} (x-1)^n = 1 + (x-1) + (x-1)^2 + \cdots \tag{3}$$

converges if and only if $|x-1| < 1$, while the series

$$\sum_{n=0}^{\infty} \frac{x^n}{2^{n+1}} = \frac{1}{2} \sum_{n=0}^{\infty} \left(\frac{x}{2}\right)^n = \frac{1}{2} + \frac{x}{4} + \frac{x^2}{8} + \cdots, \tag{3'}$$

which is also geometric, converges if and only if $|x/2| < 1$ or equivalently $|x| < 2$. Replacing x in formula (2) first by $x-1$ and then by $x/2$, we find that both series (3) and (3') have the same sum

$$\frac{1}{1-(x-1)} = \frac{1}{2} \frac{1}{1-\frac{x}{2}} = \frac{1}{2-x}.$$

Thus the function

$$f(x) = \frac{1}{2-x}$$

has the power series expansion

$$\frac{1}{2-x} = \frac{1}{2} + \frac{x}{4} + \frac{x^2}{8} + \cdots \qquad (|x| < 2)$$

at $x = 0$ and the expansion

$$\frac{1}{2-x} = 1 + (x-1) + (x-1)^2 + \cdots \qquad (|x-1| < 1)$$

---

† The expansion is "at $x = c$" in the sense that the interval of convergence of the power series has its midpoint at $x = c$. The phrase "about $x = c$" is sometimes used instead, with the same meaning.

538 Chapter 9 Sequences and Series

at $x = 1$. Notice that the interval of convergence of the series (3′) is $(-2, 2)$, while that of (3) is the smaller interval $(0, 2)$. This makes sense, because the interval of convergence of the power series of f cannot contain the point $x = 2$ at which f approaches infinity. □

Given a power series

$$\sum_{n=0}^{\infty} a_n x^n = a_0 + a_1 x + a_2 x^2 + \cdots + a_n x^n + \cdots, \tag{4}$$

we can form two new power series by differentiating and integrating (4) term by term, that is, we can form the series

$$\sum_{n=1}^{\infty} n a_n x^{n-1} = a_1 + 2a_2 x + \cdots + n a_n x^{n-1} + \cdots, \tag{5}$$

whose general term is the derivative of $a_n x^n$, and the series

$$\sum_{n=0}^{\infty} \frac{a_n}{n+1} x^{n+1} = a_0 x + \frac{a_1}{2} x^2 + \cdots + \frac{a_n}{n+1} x^{n+1} + \cdots, \tag{6}$$

whose general term is the integral of $a_n x^n$ (from 0 to x). *All three series (4), (5) and (6) have the same radius of convergence.* In fact, multiplication of a power series by x or division of a power series with no constant term by x has no effect on the radius of convergence (why not?), so that the series (5) has the same radius of convergence as

$$\sum_{n=1}^{\infty} n a_n x^n, \tag{5′}$$

while (6) has the same radius of convergence as

$$\sum_{n=0}^{\infty} \frac{a_n}{n+1} x^n. \tag{6′}$$

But (5′) and (6′) are both of the form $\sum c_n a_n x^n$, where $\sqrt[n]{c_n} \to 1$ as $n \to \infty$, since $\sqrt[n]{n} \to 1$ as $n \to \infty$ (see page 528) and

$$\lim_{n \to \infty} \sqrt[n]{\frac{1}{n+1}} = \lim_{n \to \infty} e^{-(1/n) \ln(n+1)} = e^0 = 1.$$

Therefore, by Example 10, page 535, the two series (5′) and (6′) have the same radius of convergence as the original series (4), and hence so do the differentiated and integrated series (5) and (6). Repeated application of this argument shows that *the result of differentiating or integrating a power series $\sum a_n x^n$ term by term any number of times is another power series with the same radius of convergence.*

Differentiation of Power Series

The following theorem establishes the connection between the function which is the sum of a power series and the function which is the sum of the differentiated series. Despite the simplicity of the statement of the theorem, its proof is technical and hence is omitted. The interested reader is referred to any text on advanced calculus.

Section 9.7   Differentiation and Integration of Power Series   **539**

> **Theorem 14** (*Term by term differentiation of a power series*). *Let $\sum a_n x^n$ be a power series with radius of convergence R. Then the function*
>
> $$f(x) = \sum_{n=0}^{\infty} a_n x^n,$$
>
> *that is, the sum of the power series, is differentiable on the open interval $(-R, R)$, and has the derivative*
>
> $$f'(x) = \sum_{n=1}^{\infty} n a_n x^{n-1}, \tag{7}$$
>
> *equal to the sum of the series obtained by differentiating the given series term by term.*

By repeated application of Theorem 14, we find that f has derivatives of all orders at every point of $(-R, R)$, a fact summarized by saying that f is *infinitely differentiable* on $(-R, R)$. Moreover, the differentiability of f on $(-R, R)$ implies the continuity of f on $(-R, R)$. Thus *the sum of every power series is a continuous function.*

Integration of Power Series

With the help of Theorem 14, it is easy to establish the connection between f and the function which is the sum of the *integrated* series.

> **Theorem 15** (*Term by term integration of a power series*). *Let $\sum a_n x^n$ be a power series with radius of convergence R. Then the function*
>
> $$f(x) = \sum_{n=0}^{\infty} a_n x^n$$
>
> *is integrable on every closed subinterval $[0, x]$ of $(-R, R)$, with integral*
>
> $$\int_0^x f(t)\,dt = \sum_{n=0}^{\infty} \frac{a_n}{n+1} x^{n+1}, \tag{8}$$
>
> *equal to the sum of the series obtained by integrating the given series term by term.*

Proof The integrability of f on $[0, x]$ is an immediate consequence of the continuity of f on $(-R, R)$. It has already been shown that the series

$$F(x) = \sum_{n=0}^{\infty} \frac{a_n}{n+1} x^{n+1}$$

has radius of convergence R, and clearly $F(0) = 0$. Term by term differentiation of this series immediately yields the original series $f(x) = \sum a_n x^n$, and therefore $F'(x) = f(x)$, by Theorem 14. It follows that

$$F(x) = \int_0^x f(t)\,dt$$

(notice how this incorporates the condition $F(0) = 0$). The desired formula (8) is now obtained by equating the two expressions for $F(x)$. ∎

Next we give some examples of term by term differentiation of power series.

Example 3 Application of the ratio or root test shows at once that the power series

$$f(x) = \sum_{n=1}^{\infty} \frac{x^n}{n^2} \tag{9}$$

has radius of convergence 1. Using Theorem 14 to differentiate (9) twice, we get

$$f'(x) = \sum_{n=1}^{\infty} \frac{x^{n-1}}{n} \tag{9'}$$

and

$$f''(x) = \sum_{n=1}^{\infty} \frac{n-1}{n} x^{n-2}. \tag{9''}$$

The three series (9), (9') and (9'') all have the same radius of convergence 1, but it is easily verified (do so as an exercise) that the interval of convergence of (9) is the *closed* interval $[-1, 1]$, that of (9') is the *half-open* interval $[-1, 1)$, and that of (9'') is the *open* interval $(-1, 1)$. Thus, although differentiation of a power series preserves convergence at every interior point of the interval of convergence I, it may well destroy convergence at the endpoints of I.

Example 4 If two power series $\sum a_n x^n$ and $\sum b_n x^n$ have the same sum in a neighborhood of the point $x = 0$, then the two series are identical, that is, identical powers of x have the same coefficients. In fact, setting $x = 0$ in the identity

$$a_0 + a_1 x + a_2 x^2 + \cdots \equiv b_0 + b_1 x + b_2 x^2 + \cdots, \tag{10}$$

we immediately get $a_0 = b_0$. Moreover, because of Theorem 14, the identity (10) remains valid if both sides are differentiated the same number of times. Differentiating (10) n times in succession, we find that

$$n! a_n + (n+1)! a_{n+1} x + \frac{(n+2)!}{2!} a_{n+2} x^2 + \cdots$$

$$= n! b_n + (n+1)! b_{n+1} x + \frac{(n+2)!}{2!} b_{n+2} x^2 + \cdots,$$

which gives $a_n = b_n$ after setting $x = 0$. Therefore $a_n = b_n$ for all $n = 0, 1, 2, \ldots$ You will recognize this as the power series analogue of the argument used to prove Theorem 4, page 401, on the coefficients of identical polynomials. More generally, if $\sum a_n(x-c)^n$ and $\sum b_n(x-c)^n$ are two power series in $x - c$ which have the same sum in a neighborhood of the point $x = c$, then the two series are identical (set $x = c$ in the identity $\sum a_n(x-c)^n \equiv \sum b_n(x-c)^n$ and those obtained from it by repeated differentiation). Thus the coefficients of a power series expansion $f(x) = \sum a_n(x-c)^n$ are *uniquely* determined by the function f, and also of course by the choice of the constant c.

Example 5 It is an immediate consequence of the ratio test that the power series

$$f(x) = \sum_{n=0}^{\infty} \frac{x^n}{n!} = 1 + x + \frac{x^2}{2!} + \frac{x^3}{3!} + \cdots + \frac{x^n}{n!} + \cdots$$

converges for all x. By Theorem 14,

$$f'(x) = \sum_{n=1}^{\infty} \frac{nx^{n-1}}{n!} = \sum_{n=1}^{\infty} \frac{x^{n-1}}{(n-1)!} = \sum_{n=0}^{\infty} \frac{x^n}{n!},$$

so that the differentiated series is exactly the same as the original series! Thus the sum function $y = f(x)$ satisfies the differential equation

$$y' = y,$$

subject to the initial condition

$$y|_{x=0} = 1,$$

obtained by setting $x = 0$ in the series for $f(x)$. As we know from Section 6.6, $y = e^x$ is the unique solution of this initial value problem. Therefore

$$e^x = \sum_{n=0}^{\infty} \frac{x^n}{n!} = 1 + x + \frac{x^2}{2!} + \frac{x^3}{3!} + \cdots + \frac{x^n}{n!} + \cdots, \quad (11)$$

and we have found the power series expansion of e^x at $x = 0$.

Choosing first $x = 1$ and then $x = -1$ in formula (11) enables us to sum two previously encountered numerical series, namely

$$\sum_{n=0}^{\infty} \frac{1}{n!} = 1 + 1 + \frac{1}{2!} + \frac{1}{3!} + \frac{1}{4!} + \frac{1}{5!} + \cdots = e \quad (12)$$

and

$$\sum_{n=0}^{\infty} \frac{(-1)^n}{n!} = 1 - 1 + \frac{1}{2!} - \frac{1}{3!} + \frac{1}{4!} - \frac{1}{5!} + \cdots = e^{-1}. \quad (12')$$

In fact, we can now use (12) to calculate the number e to any desired accuracy. For example, suppose we retain the first eight terms of the series (12) and neglect the remaining terms. Then the error committed is equal to

$$\frac{1}{8!} + \frac{1}{9!} + \frac{1}{10!} + \cdots = \frac{1}{8!} + \frac{1}{8! \cdot 9} + \frac{1}{8! \cdot 9 \cdot 10} + \cdots,$$

which is certainly less than the sum of the geometric series

$$\frac{1}{8!} \left[1 + \frac{1}{9} + \left(\frac{1}{9}\right)^2 + \cdots \right] = \frac{1}{8!} \frac{1}{1 - \frac{1}{9}} = \frac{1}{8!} \frac{9}{8} \approx 2.8 \times 10^{-5}.$$

Therefore we can be sure that the approximation

$$e \approx 1 + 1 + \frac{1}{2!} + \frac{1}{3!} + \frac{1}{4!} + \frac{1}{5!} + \frac{1}{6!} + \frac{1}{7!}$$

$$= 1 + 1 + \frac{1}{2} + \frac{1}{6} + \frac{1}{24} + \frac{1}{120} + \frac{1}{720} + \frac{1}{5040} = \frac{685}{252} \approx 2.7183$$

is accurate to four decimal places (actually $e = 2.7182818\ldots$). Here we use the fact that an approximation is said to be *accurate to n decimal places* if

542   Chapter 9 Sequences and Series

the error of the approximation is less than $0.5 \times 10^{-n} = 5 \times 10^{-n-1}$ in absolute value.

Remark In the preceding example there is a slight distinction between the approximation $e \approx \frac{685}{252}$ and the approximation $e \approx 2.7183$. The first approximation is subject only to the *truncation error* made in dropping all but the first eight terms of the series (12), while the second approximation is also subject to an additional *round-off error* made in representing the fraction $\frac{685}{252}$ as a decimal to four places. If the absolute value of the round-off error were near its upper bound of 5×10^{-5} and if the truncation error had the same sign as the round-off error, the sum of the truncation and round-off errors might exceed 5×10^{-5}; thus the fourth decimal place of the approximation $e \approx 2.7183$ can be regarded as uncertain (but only by ± 1). Further analysis shows that this does not occur, so that the approximation $e \approx 2.7183$ is in fact accurate to four decimal places. To keep matters simple, we will henceforth ignore round-off errors in approximations based on the use of infinite series.

Example 6 We already know from Example 1 that

$$\frac{1}{1-x} = 1 + x + x^2 + x^3 + \cdots + x^n + \cdots \qquad (|x| < 1). \qquad (13)$$

Using Theorem 14 to differentiate this series term by term, we get

$$\frac{1}{(1-x)^2} = 1 + 2x + 3x^2 + \cdots + nx^{n-1} + \cdots \qquad (|x| < 1), \qquad (13')$$

which is the power series expansion of the function $1/(1-x)^2$ at $x = 0$. Notice that (13′) can also be obtained from (13) by formally multiplying the "infinite polynomial" $1 + x + x^2 + x^3 + \cdots$ by itself:

$$\begin{array}{r} 1 + x + x^2 + x^3 + \cdots \\ 1 + x + x^2 + x^3 + \cdots \\ \hline 1 + x + x^2 + x^3 + \cdots \\ x + x^2 + x^3 + \cdots \\ x^2 + x^3 + \cdots \\ x^3 + \cdots \\ \hline 1 + 2x + 3x^2 + 4x^3 + \cdots \end{array}$$

The legitimacy of this procedure is a consequence of the following theorem due to Cauchy, which we cite without proof: Let $f(x) = \sum a_n x^n$ and $g(x) = \sum b_n x^n$ be two power series, with radii of convergence R_a and R_b. Then

$$f(x)g(x) = a_0 b_0 + (a_0 b_1 + a_1 b_0)x + (a_0 b_2 + a_1 b_1 + a_2 b_0)x^2$$
$$+ \cdots + (a_0 b_n + a_1 b_{n-1} + \cdots + a_n b_0)x^n + \cdots$$

if $|x| < \min\{R_a, R_b\}$.

We now turn to some examples of term-by-term integration of power series, with the help of Theorem 15.

Example 7 Integrating the convergent geometric series

$$\sum_{n=0}^{\infty} (-x)^n = 1 - x + x^2 - x^3 + \cdots$$

$$= \frac{1}{1-(-x)} = \frac{1}{1+x} \qquad (|x| < 1) \qquad (14)$$

term by term, we get

$$x - \frac{x^2}{2} + \frac{x^3}{3} - \frac{x^4}{4} + \cdots = \int_0^x \frac{dt}{1+t}.$$

But

$$\int_0^x \frac{dt}{1+t} = \ln(1+x),$$

since $|x| < 1$, and therefore

$$\ln(1+x) = \sum_{n=1}^{\infty} (-1)^{n-1} \frac{x^n}{n} = x - \frac{x^2}{2} + \frac{x^3}{3} - \frac{x^4}{4} + \cdots \qquad (|x| < 1), \qquad (15)$$

which is the power series expansion of the function $\ln(1+x)$ at $x = 0$. Theorem 15 guarantees the validity of this expansion only for $-1 < x < 1$, but we suspect that it may also hold for $x = 1$, since for $x = 1$ the right side of (15) becomes the conditionally convergent alternating harmonic series. In fact, the expansion (15) *does* hold for $x = 1$, allowing us to conclude that

$$1 - \frac{1}{2} + \frac{1}{3} - \frac{1}{4} + \cdots = \ln 2. \qquad (15')$$

This is a consequence of the following theorem due to the Norwegian mathematical prodigy Neils Abel (1802–1829): *If a power series $\sum a_n x^n$ with radius of convergence R is convergent for $x = R$, then its sum is continuous from the left at $x = R$, that is,*

$$\lim_{x \to R^-} \sum_{n=0}^{\infty} a_n x^n = \sum_{n=0}^{\infty} a_n R^n.$$

As applied to the series (15), which converges at $x = 1$, Abel's theorem asserts that

$$\lim_{x \to 1^-} \ln(1+x) = 1 - \frac{1}{2} + \frac{1}{3} - \frac{1}{4} + \cdots,$$

and this validates formula (15′), since the limit is of course just $\ln 2$.

Example 8 Changing x to x^2 in the expansion (14), we find that

$$\sum_{n=0}^{\infty} (-x^2)^n = 1 - x^2 + x^4 - x^6 + \cdots = \frac{1}{1+x^2} \qquad (|x| < 1).$$

Term-by-term integration of this power series gives

$$x - \frac{x^3}{3} + \frac{x^5}{5} - \frac{x^7}{7} + \cdots = \int_0^x \frac{dt}{1+t^2}.$$

But

$$\int_0^x \frac{dt}{1+t^2} = \arctan x,$$

so that the power series expansion of arctan x at $x = 0$ is

$$\arctan x = \sum_{n=0}^{\infty} (-1)^n \frac{x^{2n+1}}{2n+1} = x - \frac{x^3}{3} + \frac{x^5}{5} - \frac{x^7}{7} + \cdots \qquad (|x| < 1). \qquad (16)$$

This result is known as *Gregory's series*, after the Scottish mathematician James Gregory who discovered it in 1671. For $x = 1$ the right side of (16) becomes the series $1 - \frac{1}{3} + \frac{1}{5} - \frac{1}{7} + \cdots$, which is conditionally convergent by the alternating test. Therefore Abel's theorem tells us that

$$\lim_{x \to 1^-} \arctan x = 1 - \frac{1}{3} + \frac{1}{5} - \frac{1}{7} + \cdots.$$

Since the limit is arctan $1 = \pi/4$, it follows that

$$1 - \frac{1}{3} + \frac{1}{5} - \frac{1}{7} + \cdots = \frac{\pi}{4}.$$

This series converges too slowly to be useful for calculating π, but there is another way of using Gregory's series to calculate π to great accuracy (see Problem 84, page 573).

Example 9 In Example 4, page 422, we used Simpson's rule to approximate the integral

$$I = \int_0^1 e^{-x^2} dx$$

to four decimal places. We now give a much simpler way of approximating I, based on the use of power series, in which numerical values of the integrand are not needed. Changing x to $-x^2$ in formula (11), we get the power series expansion

$$e^{-x^2} = 1 - x^2 + \frac{x^4}{2!} - \frac{x^6}{3!} + \cdots + (-1)^n \frac{x^{2n}}{n!} + \cdots,$$

which is valid for all x. Then, integrating this series term by term from 0 to 1, we find that

$$I = \int_0^1 e^{-x^2} dx = \int_0^1 \left(1 - x^2 + \frac{x^4}{2!} - \frac{x^6}{3!} + \cdots + (-1)^n \frac{x^{2n}}{n!} + \cdots \right) dx$$

$$= 1 - \frac{1}{3} + \frac{1}{5 \cdot 2!} - \frac{1}{7 \cdot 3!} + \cdots + (-1)^n \frac{1}{(2n+1)n!} + \cdots.$$

Let R_n be the error made in approximating I by the sum of the first n terms of the numerical series on the right. Then by Theorem 9, page 519, R_n is smaller in absolute value than the first unused term, that is,

$$|R_n| < \frac{1}{(2n+1)n!},$$

Section 9.7 Differentiation and Integration of Power Series

and if $|R_n| < 5 \times 10^{-5}$, we can be sure that the approximation is accurate to four decimal places. Since

$$\frac{1}{13 \cdot 6!} \approx 1.1 \times 10^{-4}, \qquad \frac{1}{15 \cdot 7!} \approx 1.3 \times 10^{-5},$$

the smallest integer n such that $|R_n| < 5 \times 10^{-5}$ is 7. It follows that the approximation

$$I \approx 1 - \frac{1}{3} + \frac{1}{5 \cdot 2!} - \frac{1}{7 \cdot 3!} + \frac{1}{9 \cdot 4!} - \frac{1}{11 \cdot 5!} + \frac{1}{13 \cdot 6!}$$

$$= 1 - \frac{1}{3} + \frac{1}{10} - \frac{1}{42} + \frac{1}{216} - \frac{1}{1320} + \frac{1}{9360} \approx 0.7468$$

is accurate to four decimal places.

The Binomial Series

According to the binomial theorem, if r is any nonnegative integer, then

$$(1 + x)^r = 1 + rx + \frac{r(r-1)}{2} x^2$$

$$+ \cdots + \frac{r(r-1) \cdots (r-n+1)}{n!} x^n + \cdots + x^r \qquad (17)$$

(set $a = 1$, $b = x$, $n = r$ in formula (5), page 227). Our last example generalizes this formula to the case where r is an arbitrary real number.

Example 10 The power series

$$f(x) = 1 + \sum_{n=1}^{\infty} \frac{r(r-1) \cdots (r-n+1)}{n!} x^n$$

$$= 1 + rx + \frac{r(r-1)}{2} x^2 + \cdots + \frac{r(r-1) \cdots (r-n+1)}{n!} x^n + \cdots, \qquad (18)$$

where r is an arbitrary real number, is called the *binomial series*. If r is a nonnegative integer, the binomial series terminates and reduces to the polynomial (17) of degree r, but otherwise there are infinitely many terms and the series has radius of convergence 1 (show this, with the help of the ratio test). Hence the function $f(x)$ is defined on the interval $-1 < x < 1$. Differentiating (18) term by term, we get

$$f'(x) = r + r(r-1)x + \cdots + \frac{r(r-1) \cdots (r-n+1)}{(n-1)!} x^{n-1} + \cdots, \qquad (18')$$

which becomes

$$xf'(x) = rx + r(r-1)x^2 + \cdots + \frac{r(r-1) \cdots (r-n+1)}{(n-1)!} x^n + \cdots$$

after multiplication by x. We now add the last two series, after observing that the coefficient of x^n in (18′) is not the last term written, but the following (unwritten) one, namely

$$\frac{r(r-1) \cdots (r-n+1)(r-n)}{n!} x^n.$$

546 Chapter 9 Sequences and Series

As a result, we obtain

$$f'(x) + xf'(x)$$
$$= r + \sum_{n=1}^{\infty} \left[\frac{r(r-1)\cdots(r-n+1)(r-n)}{n!} + \frac{r(r-1)\cdots(r-n+1)}{(n-1)!} \right] x^n$$
$$= r + \sum_{n=1}^{\infty} \frac{r(r-1)\cdots(r-n+1)}{(n-1)!} \left(\frac{r-n}{n} + 1 \right) x^n$$
$$= r + r \sum_{n=1}^{\infty} \frac{r(r-1)\cdots(r-n+1)}{n!} x^n$$
$$= r \left[1 + \sum_{n=1}^{\infty} \frac{r(r-1)\cdots(r-n+1)}{n!} x^n \right] = rf(x).$$

Thus the sum function $y = f(x)$ satisfies the differential equation

$$(1 + x)y' = ry, \qquad (19)$$

subject to the initial condition

$$y|_{x=0} = 1, \qquad (19')$$

obtained by setting $x = 0$ in (18).

Equation (19) can be solved by separation of variables, but it is simpler to guess the answer $y = (1 + x)^r$, suggested by the observation that while differentiation lowers the degree of $(1 + x)^r$ by 1, this is compensated by the extra factor of $1 + x$ in the left side of (19). We see at once that this solution happens to be the one satisfying the initial condition (19'). Thus we have finally found that the function $f(x)$ in (18) is $(1 + x)^r$, so that (18) can be written in the form

$$(1 + x)^r = 1 + \sum_{n=1}^{\infty} \frac{r(r-1)\cdots(r-n+1)}{n!} x^n \qquad (-1 < x < 1). \qquad (20)$$

This binomial series expansion of $(1 + x)^r$ generalizes the binomial theorem (17) to the case of an arbitrary real exponent r.

Problems

Expand the given rational function in a power series at $x = 0$, and specify the radius of convergence R.

1. $\dfrac{1}{3-x}$
2. $\dfrac{x}{1+x}$
3. $\dfrac{1}{(1+x)^2}$
4. $\dfrac{x^{11}}{1-x}$
5. $\dfrac{x}{2x+1}$
6. $\dfrac{1}{1-x^2}$
7. $\dfrac{x}{(1-x^2)^2}$
8. $\dfrac{x^2}{(1-x)^3}$
9. $\dfrac{1}{x^2 - 3x + 2}$

10. Let $f(x) = \sum a_n x^n$. With the help of Example 4, show that $a_1 = a_3 = a_5 = \cdots = 0$ if f is even, while $a_0 = a_2 = a_4 = \cdots = 0$ if f is odd.

Let $f(x) = 1/(2-x)$, as in Example 2. Find the power series expansion of f

11. At $x = 3$
12. At $x = -1$

Expand the given function in a power series at $x = 0$, and specify the radius of convergence R.

13. $x^2 e^{-x}$
14. xe^{x^2}
15. $\sinh x$
16. $\cosh x$
17. a^x $(a > 0)$
18. $\ln \dfrac{1}{1-x^2}$
19. $\ln \dfrac{1+x}{1-x}$
20. $(1+x)\ln(1+x)$
21. $\ln(1 - x + x^2)$

With the help of formula (20), find the first five terms of the binomial series expansion of the given function.

22. $(1 + x)^{1/2}$ **23.** $(1 + x)^{1/3}$ **24.** $(1 + x)^{-1/2}$
25. $(4 - x)^{3/2}$ **26.** $(8 + x)^{2/3}$ **27.** $(1 - x^2)^{-10}$

Find the power series expansion of
28. $\ln x$ at $x = 1$
29. $\ln(2 + 2x + x^2)$ at $x = -1$

30. The series
$$\sum_{n=0}^{\infty} e^{-nx} = 1 + e^{-x} + e^{-2x} + \cdots + e^{-nx} + \cdots \quad (i)$$
is *not* a power series, since the terms are exponentials, not powers of x. Show that the series (i) converges if and only if x lies in the interval $(0, \infty)$. What function does this series have as its sum on $(0, \infty)$? Is $(0, \infty)$ a possible interval of convergence for a power series?

Use the first three terms of the binomial series (20) to approximate the given root. In each case verify that the approximation is accurate to five decimal places.

31. $\sqrt{1.04}$ **32.** $\sqrt[3]{0.975}$ **33.** $\sqrt[4]{79}$
34. $\sqrt[5]{33}$ **35.** $\sqrt[6]{65}$ **36.** $\sqrt[7]{126}$

37. Use the binomial series to show that
$$\arcsin x = x + \frac{1}{2}\frac{x^3}{3} + \frac{1 \cdot 3}{2 \cdot 4}\frac{x^5}{5} + \frac{1 \cdot 3 \cdot 5}{2 \cdot 4 \cdot 6}\frac{x^7}{7} + \cdots,$$
$$\sinh^{-1} x = x - \frac{1}{2}\frac{x^3}{3} + \frac{1 \cdot 3}{2 \cdot 4}\frac{x^5}{5} - \frac{1 \cdot 3 \cdot 5}{2 \cdot 4 \cdot 6}\frac{x^7}{7} + \cdots,$$
provided that $|x| < 1$.

38. By direct differentiation of the power series expansions of $\cosh x$ and $\sinh x$ (see Problems 15 and 16), verify that $D_x \cosh x = \sinh x$, $D_x \sinh x = \cosh x$.

39. Use Gregory's series (16) to show that
$$1 - \frac{1}{3 \cdot 3} + \frac{1}{5 \cdot 3^2} - \frac{1}{7 \cdot 3^3} + \cdots = \frac{\pi}{2\sqrt{3}}.$$

40. Show that the sum $y = f(x)$ of the power series
$$\sum_{n=0}^{\infty} \frac{x^n}{(n!)^2}$$
satisfies the differential equation $xy'' + y' - y = 0$.

41. The alternating harmonic series $1 - \frac{1}{2} + \frac{1}{3} - \frac{1}{4} + \cdots = \ln 2$ converges much too slowly to be of any practical value in calculating $\ln 2$. Show that $\ln 2$ is also the sum of the much more rapidly converging series
$$\ln 2 = \frac{2}{3}\left(1 + \frac{1}{3 \cdot 9} + \frac{1}{5 \cdot 9^2} + \frac{1}{7 \cdot 9^3} + \cdots\right). \quad (ii)$$

42. Use formula (ii) to approximate $\ln 2$ to five decimal places.

Find the sum of the power series
43. $1 - 4x + 7x^2 - 10x^3 + \cdots$ $(|x| < 1)$
44. $1 - 3x^2 + 5x^4 - 7x^6 + \cdots$ $(|x| < 1)$

Find the power series expansion of the function of x defined by the given integral.

45. $\int_0^x \frac{e^{t^2} - 1}{t^2} dt$ **46.** $\int_0^x \frac{1 - e^{-t}}{t} dt$

47. $\int_0^x \frac{\ln(1 + t)}{t} dt$ **48.** $\int_0^x \cosh t^2 \, dt$

In Problems 45-47 define the integrand $f(x)$ at $x = 0$ by continuity, that is, set $f(0) = \lim_{x \to 0} f(x)$.

Use power series to approximate the given integral to an accuracy of four decimal places.

49. $\int_0^{1/2} \sqrt{1 + x^4} \, dx$ **50.** $\int_0^{1/5} \frac{\sinh x}{x} dx$

51. $\int_0^1 e^{x^2} dx$ **52.** $\int_0^{2/3} \frac{dx}{1 + x^5}$

53. Let
$$f(x) = \sum_{n=0}^{\infty} a_n x^n, \quad g(x) = \frac{f(x)}{1 - x}.$$
Find the power series expansion of g at $x = 0$.

9.8 Taylor's Theorem and Its Uses

In the next section we will be concerned with the problem of finding the power series expansion of a given function. The solution of this problem will require the use of a basic theorem on the approximation of functions by polynomials, due to the English mathematician Brook Taylor (1685–1731). As an introduction to Taylor's theorem, we first consider a special case, known as the *extended mean value theorem*.

The Extended Mean Value Theorem

Let f be a function which is differentiable on an interval I, that is, suppose f has a finite derivative f' at every point of I. Then, writing x and t instead of b and c in the mean value theorem (7), page 162, we have

$$f(x) = f(a) + f'(t)(x-a), \tag{1}$$

where t is a point between a and x. If t is replaced by a, we get the *tangent line approximation*

$$f(x) \approx f(a) + f'(a)(x-a) \qquad (x \neq a). \tag{2}$$

When this approximation was introduced in Section 2.3, we observed that the error of the approximation is small for small $|x-a|$, but we made no attempt to estimate the error for a given value of $x-a$ (apart from Problems 29 and 30, page 127). We now prove an extension of the mean value theorem (1) which will allow us to estimate the error in the case where f has a second derivative f'', as well as a first derivative f'.

Theorem 16 (*Extended mean value theorem*). *Suppose f has a finite second derivative f'' at every point of an interval I,† and let a and x be arbitrary points of I. Then there is a point t between a and x such that*

$$f(x) = f(a) + f'(a)(x-a) + \frac{f''(t)}{2}(x-a)^2. \tag{3}$$

Proof (Optional) Let

$$h(x) = f(x) - f(a) - f'(a)(x-a), \qquad g(x) = (x-a)^2,$$

so that in particular

$$h(a) = g(a) = 0, \qquad h'(x) = f'(x) - f'(a).$$

Then, applying Cauchy's mean value theorem (see Theorem 4, page 164) to the functions h and g on the interval $[a, x]$ if $x > a$, or on $[x, a]$ if $x < a$, we have

$$\frac{h(x)}{g(x)} = \frac{h(x) - h(a)}{g(x) - g(a)} = \frac{h'(t_1)}{g'(t_1)} = \frac{f'(t_1) - f'(a)}{2(t_1 - a)},$$

where t_1 lies between a and x (verify that the hypotheses of Cauchy's theorem are satisfied). But by the ordinary mean value theorem applied to the function f' on the interval $[a, t_1]$ or $[t_1, a]$,

$$f'(t_1) - f'(a) = f''(t)(t_1 - a),$$

where t lies between a and t_1, and hence between a and x. Thus there is a point t between a and x such that

$$\frac{h(x)}{g(x)} = \frac{f''(t)(t_1 - a)}{2(t_1 - a)} = \frac{f''(t)}{2}.$$

---

† The existence of the second derivative f'' on I implies the existence and continuity of the function f and its first derivative f' on I (why?). More generally, the existence of the derivative $f^{(n+1)}$ of order $n+1$ on I implies the existence and continuity of $f, f', \ldots, f^{(n)}$ on I.

It follows that
$$h(x) = \frac{f''(t)}{2} g(x) = \frac{f''(t)}{2} (x-a)^2,$$
which is equivalent to formula (3). In general, the point t depends on both a and x. ∎

The Error of the Tangent Line Approximation

The last term on the right in (3) is the error
$$E_{\text{TL}} = \frac{f''(t)}{2} (x-a)^2$$
of the tangent line approximation (2). In particular, if f'' is continuous on I, it follows that
$$|E_{\text{TL}}| \leq \frac{1}{2}(x-a)^2 \max |f''|, \qquad (4)$$
where $\max |f''|$ is the maximum value of $|f''(t)|$ on the closed interval with endpoints a and x.

Example 1 Estimate $\cos 47°$ by using the tangent line approximation (2) with $a = 45° = \pi/4$. Then use the inequality (4) to show that this approximation is accurate to three decimal places.

Solution If $f(x) = \cos x$, then $f'(x) = -\sin x$, $f''(x) = -\cos x$, and formula (2) becomes
$$\cos x = f(x) \approx f(a) + f'(a)(x-a) = \cos a - (x-a)\sin a.$$
Choosing
$$x = 47° = \frac{\pi}{4} + \frac{\pi}{90},$$
we have
$$\cos 47° = \cos\left(\frac{\pi}{4} + \frac{\pi}{90}\right) \approx \cos \frac{\pi}{4} - \frac{\pi}{90} \sin \frac{\pi}{4} = \frac{1}{\sqrt{2}}\left(1 - \frac{\pi}{90}\right) \approx 0.682.$$

The maximum of $|f''(t)| = |\cos t|$ on the interval $[45°, 47°]$ is $\cos 45° = 1/\sqrt{2}$. Therefore, by (4),
$$|E_{\text{TL}}| \leq \frac{1}{2\sqrt{2}} \left(\frac{\pi}{90}\right)^2 \approx 4.3 \times 10^{-4} < 5 \times 10^{-4},$$
and we can be sure that our approximation of $\cos 47°$ is accurate to three decimal places. □

Taylor's Theorem

Theorem 16 is just the first of a whole series of refinements of the mean value theorem involving derivatives of the function f of progressively higher order, together with progressively higher powers of $x - a$. They are all special cases of the following theorem, whose proof is rather technical, and hence is deferred to the end of the section.

> **Theorem 17** *(Taylor's theorem). Suppose f has a finite derivative $f^{(n+1)}$ of order $n+1$ at every point of an interval I, and let a and x be arbitrary points of I. Then there is a point t between a and x such that*
>
> $$f(x) = \sum_{k=0}^{n} \frac{f^{(k)}(a)}{k!}(x-a)^k + \frac{f^{(n+1)}(t)}{(n+1)!}(x-a)^{n+1}$$
>
> $$= f(a) + f'(a)(x-a) + \frac{f''(a)}{2!}(x-a)^2$$
>
> $$+ \cdots + \frac{f^{(n)}(a)}{n!}(x-a)^n + \frac{f^{(n+1)}(t)}{(n+1)!}(x-a)^{n+1}. \quad (5)$$

Taylor's Formula and Taylor Polynomials

Formula (5) is known as *Taylor's formula* (*with remainder*). It represents the function $f(x)$ near $x = a$ as the sum

$$f(x) = P_n(x) + R_n(x)$$

of a polynomial

$$P_n(x) = \sum_{k=0}^{n} \frac{f^{(k)}(a)}{k!}(x-a)^k \quad (6)$$

of degree n in the variable $x - a$ and a term

$$R_n(x) = \frac{f^{(n+1)}(t)}{(n+1)!}(x-a)^{n+1},$$

called the *remainder*. (Since $P_n(x)$ is in general the sum of $n + 1$ terms, the remainder $R_n(x)$, despite the subscript n, is actually the remainder after $n + 1$ terms.) The polynomial $P_n(x)$ is called the *nth Taylor polynomial of f at $x = a$*, and has the key property that its value and the values of its first n derivatives at the point $x = a$ coincide with the value of the given function f and the values of the first n derivatives of f at $x = a$. In fact, j successive differentiations of formula (6) give

$$P_n^{(j)}(x) = \sum_{k=j}^{n} \frac{f^{(k)}(a)}{(k-j)!}(x-a)^{k-j}$$

$$= f^{(j)}(a) + f^{(j+1)}(a)(x-a) + \cdots + \frac{f^{(n)}(a)}{(n-j)!}(x-a)^{n-j}$$

$$(j = 0, 1, \ldots, n),$$

and then setting $x = a$, we find at once that

$$P_n^{(j)}(a) = f^{(j)}(a) \quad (j = 0, 1, \ldots, n).$$

In view of this matching of derivatives, we would expect $P_n(x)$ to be a good approximation to $f(x)$ near $x = a$, which improves as n is made larger, and this is usually true,† as illustrated by the following example.

Example 2 Approximate the function $f(x) = e^x$ near $x = 0$ by its first four Taylor polynomials.

---
† We hedge here because of the existence of a function for which the approximation $f(x) \approx P_n(x)$ is poor for all n (see Example 7 at the end of the section).

Solution Since $f^{(n)}(x) = D_x^n e^x = e^x$ for all n, the first four Taylor polynomials (with $a = 0$) are

$$P_0(x) = f(0) = 1,$$
$$P_1(x) = f(0) + f'(0)x = 1 + x,$$
$$P_2(x) = f(0) + f'(0)x + \frac{f''(0)}{2!}x^2 = 1 + x + \frac{1}{2}x^2,$$
$$P_3(x) = f(0) + f'(0)x + \frac{f''(0)}{2!}x^2 + \frac{f'''(0)}{3!}x^3 = 1 + x + \frac{1}{2}x^2 + \frac{1}{6}x^3.$$

Figure 12 shows the graphs of e^x and of these polynomials in the same system of rectangular coordinates. The approximation $e^x \approx P_0(x)$ is of course very crude, since $P_0(x)$ has the constant value 1, but $e^x \approx P_1(x)$ is the tangent line approximation, which is quite good for small values of $|x|$. A better approximation is given by $e^x \approx P_2(x)$, in which the graph of e^x is approximated near $x = 0$ by the parabola

$$y = P_2(x) = 1 + x + \frac{1}{2}x^2 = \frac{1}{2}(x + 1)^2 + \frac{1}{2},$$

rather than by the straight line $y = P_1(x) = 1 + x$. But the approximation of e^x by the *cubic* polynomial $P_3(x)$ is better still. In fact, by Taylor's theorem,

$$e^x = P_3(x) + R_3(x),$$

where

$$R_3(x) = \frac{f^{(4)}(t)}{4!}x^4 = \frac{e^t}{24}x^4 \qquad (t \text{ between } 0 \text{ and } x),$$

so that the error of the approximation $e^x \approx P_3(x)$ is positive and less than

$$\frac{e^{1/2}}{24}\left(\frac{1}{2}\right)^4 \approx 0.0043$$

over the whole interval $[-\frac{1}{2}, \frac{1}{2}]$. Here we have used the fact that the maximum value of e^t on this interval is $e^{1/2}$, since e^t is an increasing function. ∎

Example 3 Write Taylor's formula for the function $f(x) = 1/x$ at the point $a = 1$, choosing $n = 3$.

Solution The function and its first four derivatives are

$$f(x) = \frac{1}{x}, \quad f'(x) = -\frac{1}{x^2}, \quad f''(x) = \frac{2}{x^3}, \quad f'''(x) = -\frac{6}{x^4}, \quad f^{(4)}(x) = \frac{24}{x^5}.$$

Setting $a = 1$, $n = 3$ in formula (5) and assuming that $x > 0$ (why?), we find that

$$\frac{1}{x} = f(x) = f(1) + f'(1)(x - 1) + \frac{f''(1)}{2!}(x - 1)^2$$
$$+ \frac{f'''(1)}{3!}(x - 1)^3 + \frac{f^{(4)}(t)}{4!}(x - 1)^4$$
$$= 1 - (x - 1) + (x - 1)^2 - (x - 1)^3 + \frac{(x - 1)^4}{t^5},$$

where t lies between 1 and x. ∎

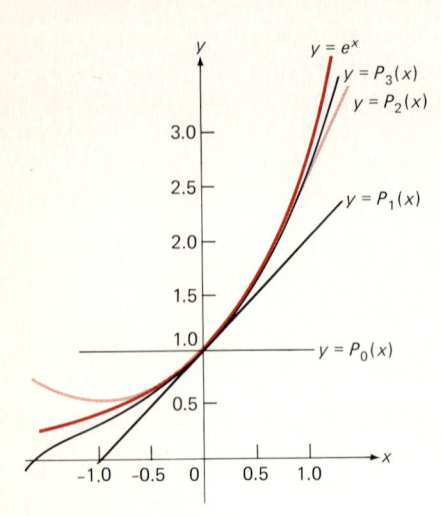

Approximation of e^x by Taylor polynomials
Figure 12

The following examples exhibit some of the many ways in which Taylor's formula can be used to advantage as a computational tool.

Example 4 Use Taylor's formula with $n = 3$ to improve the tangent line approximation of $\cos 47°$ given in Example 1.

Solution Let $f(x) = \cos x$. Then $f'(x) = -\sin x$, $f''(x) = -\cos x$, $f'''(x) = \sin x$, $f^{(4)}(x) = \cos x$, and choosing $n = 3$ in formula (5), we have

$$\cos x = f(x) = f(a) + f'(a)(x - a) + \frac{f''(a)}{2!}(x - a)^2$$

$$+ \frac{f'''(a)}{3!}(x - a)^3 + \frac{f^{(4)}(t)}{4!}(x - a)^4$$

$$= \cos a - (x - a)\sin a - \frac{(x - a)^2}{2}\cos a + \frac{(x - a)^3}{6}\sin a + \frac{(x - a)^4}{24}\cos t,$$

where t lies between a and x. For $a = 45° = \pi/4$, $x = 47°$ this gives

$$\cos 47° = \cos\left(\frac{\pi}{4} + \frac{\pi}{90}\right) = \cos\frac{\pi}{4} - \frac{\pi}{90}\sin\frac{\pi}{4} - \frac{1}{2}\left(\frac{\pi}{90}\right)^2 \cos\frac{\pi}{4}$$

$$+ \frac{1}{6}\left(\frac{\pi}{90}\right)^3 \sin\frac{\pi}{4} + \frac{1}{24}\left(\frac{\pi}{90}\right)^4 \cos t \quad (45° < t < 47°),$$

and hence

$$\cos 47° \approx \frac{1}{\sqrt{2}}\left[1 - \frac{\pi}{90} - \frac{1}{2}\left(\frac{\pi}{90}\right)^2 + \frac{1}{6}\left(\frac{\pi}{90}\right)^3\right] \approx 0.6819983, \quad (7)$$

where the error of this approximation, which differs from the tangent line approximation in Example 1 by the presence of two "higher order corrections" involving the second and third powers of $\pi/90$, is given by the remainder

$$R_3 = \frac{1}{24}\left(\frac{\pi}{90}\right)^4 \cos t \quad (45° < t < 47°).$$

But the maximum of $\cos t$ on the interval $[45°, 47°]$ is $\cos 45° = 1/\sqrt{2}$. Therefore

$$R_3 \leq \frac{1}{24\sqrt{2}}\left(\frac{\pi}{90}\right)^4 \approx 4.4 \times 10^{-8} < 5 \times 10^{-8},$$

and we can be sure that the approximation (7) is accurate to seven decimal places. □

Example 5 Transform the polynomial $Q(x) = 4 - 2x - x^2 + x^3$ into a polynomial in the new variable $x + 1$.

Solution Choosing $f(x) = Q(x)$, $a = -1$, $n = 3$ in Taylor's formula, we have

$$Q(x) = Q(-1) + Q'(-1)(x + 1) + \frac{Q''(-1)}{2!}(x + 1)^2 + \frac{Q'''(-1)}{3!}(x + 1)^3 + R_3(x),$$

where the remainder $R_3(x)$ is zero since $Q^{(4)}(x) \equiv 0$ (explain further). But

$$Q'(x) = -2 - 2x + 3x^2, \quad Q''(x) = -2 + 6x, \quad Q'''(x) = 6,$$

Section 9.8 Taylor's Theorem and Its Uses 553

and therefore
$$Q(-1) = 4, \quad Q'(-1) = 3, \quad Q''(-1) = -8, \quad Q'''(-1) = 6,$$
so that
$$Q(x) = 4 + 3(x+1) - 4(x+1)^2 + (x+1)^3.$$
This is easily checked by verifying algebraically that the expression on the right is identically equal to $4 - 2x - x^2 + x^3$. □

Example 6 Use Taylor's formula to show that
$$\lim_{x \to 0} \frac{x - \sin x}{x^3} = \frac{1}{6}.$$

Solution Choosing $f(x) = \sin x$, $a = 0$, $n = 4$ in Taylor's formula, and observing that $f'(x) = \cos x$, $f''(x) = -\sin x$, $f'''(x) = -\cos x$, $f^{(4)}(x) = \sin x$, we have
$$\sin x = \sin 0 + x \cos 0 - \frac{1}{2} x^2 \sin 0 - \frac{1}{6} x^3 \cos 0 + \frac{1}{24} x^4 \sin t$$
$$= x - \frac{1}{6} x^3 + \frac{1}{24} x^4 \sin t,$$
where t lies between 0 and x. Therefore
$$\lim_{x \to 0} \frac{x - \sin x}{x^3} = \lim_{x \to 0} \frac{\frac{1}{6}x^3 - \frac{1}{24}x^4 \sin t}{x^3} = \lim_{x \to 0} \left(\frac{1}{6} - \frac{1}{24} x \sin t \right) = \frac{1}{6}$$
($t \to 0$ as $x \to 0$). This limit has already been evaluated with the help of L'Hospital's rule in Example 3, page 198. □

Despite the fact that one usually finds that a function is well approximated by its Taylor polynomials of high enough degree, our next example shows that exceptional cases can occur.

Example 7 Let $P_n(x)$ be the nth Taylor polynomial of the function
$$f(x) = \begin{cases} e^{-1/x^2} & \text{if } x \neq 0, \\ 0 & \text{if } x = 0, \end{cases}$$
graphed in Figure 13. Show that $P_n(x) \equiv 0$ for all $n = 0, 1, 2, \ldots$

Solution (Optional) Since
$$\lim_{x \to 0} e^{-1/x^2} = \lim_{t \to \infty} e^{-t} = 0$$
($t = 1/x^2$), the function f is continuous at $x = 0$. Moreover, if $x \neq 0$,
$$f'(x) = \frac{2}{x^3} e^{-1/x^2}, \quad f''(x) = \left(\frac{4}{x^6} - \frac{6}{x^4} \right) e^{-1/x^2}, \ldots, \quad f^{(n)}(x) = Q_{3n}\left(\frac{1}{x}\right) e^{-1/x^2},$$
where $Q_{3n}(x)$ is some polynomial of degree $3n$. But
$$\lim_{x \to 0} \frac{e^{-1/x^2}}{x^n} = \lim_{t \to \infty} \frac{t^{n/2}}{e^t} = \lim_{t \to \infty} \left(\frac{t}{e^{2t/n}} \right)^{n/2} = \left(\lim_{t \to \infty} \frac{t}{e^{2t/n}} \right)^{n/2}$$
$$= \lim_{t \to \infty} \left(\frac{D_t t}{D_t e^{2t/n}} \right)^{n/2} = \left(\lim_{t \to \infty} \frac{1}{(2/n) e^{2t/n}} \right)^{n/2} = 0$$

Graph of $y = f(x) = \begin{cases} e^{-1/x^2} & \text{if } x \neq 0, \\ 0 & \text{if } x = 0 \end{cases}$

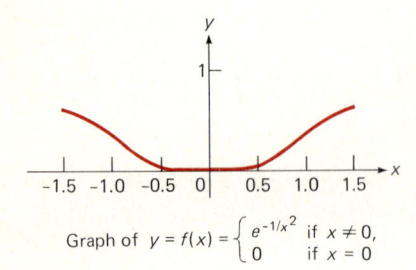

Figure 13

for all $n = 1, 2, \ldots$, with the help of the substitution $t = 1/x^2$ and L'Hospital's rule, so that

$$\lim_{x \to 0} Q\left(\frac{1}{x}\right) e^{-1/x^2} = 0$$

for every polynomial $Q(x)$. Suppose $f^{(k)}(0) = 0$. Then

$$f^{(k+1)}(0) = \lim_{x \to 0} \frac{f^{(k)}(x) - f^{(k)}(0)}{x} = \lim_{x \to 0} \frac{f^{(k)}(x)}{x}$$

$$= \lim_{x \to 0} \frac{1}{x} Q_{3k}\left(\frac{1}{x}\right) e^{-1/x^2} = \lim_{x \to 0} Q^*_{3k+1}\left(\frac{1}{x}\right) e^{-1/x^2} = 0,$$

where we use the fact that $Q^*_{3k+1}(x) = xQ_{3k}(x)$ is itself a polynomial, of degree $3k + 1$. Therefore, if $f^{(k)}(0)$ equals zero, so does $f^{(k+1)}(0)$. But $f'(0)$ equals zero, since

$$f'(0) = \lim_{x \to 0} \frac{f(x) - f(0)}{x} = \lim_{x \to 0} \frac{e^{-1/x^2}}{x} = 0,$$

and hence $f^{(n)}(0) = 0$ for all $n = 1, 2, \ldots$, by mathematical induction (see the Appendix, page A-1).

Thus the function f and *all* its derivatives (of arbitrarily high order) are zero at $x = 0$. This is because of the extreme "flatness" of the bottom of the curve $y = f(x)$ near $x = 0$ (see Figure 13). It follows that *every* Taylor polynomial $P_n(x)$ of f at $x = 0$ is identically zero, so that if $x \neq 0$, the percentage error of the approximation $f(x) \approx P_n(x)$ is always 100%, regardless of the value of n (see Problem 31, page 127). □

Proof of Theorem 17 (Optional)   Let

$$h(x) = f(x) - \sum_{k=0}^{n} \frac{f^{(k)}(a)}{k!}(x-a)^k, \qquad g(x) = (x-a)^{n+1}$$

(recall that $f^{(0)} \equiv f$). Then

$$h'(x) = f'(x) - f'(a) - f''(a)(x-a) - \cdots - \frac{f^{(n)}(a)}{(n-1)!}(x-a)^{n-1},$$

$$h''(x) = f''(x) - f''(a) - \cdots - \frac{f^{(n)}(a)}{(n-2)!}(x-a)^{n-2},$$

$$\vdots$$

$$h^{(n-1)}(x) = f^{(n-1)}(x) - f^{(n-1)}(a) - f^{(n)}(a)(x-a),$$
$$h^{(n)}(x) = f^{(n)}(x) - f^{(n)}(a),$$

after n consecutive differentiations of h, while

$$g'(x) = (n+1)(x-a)^n,$$
$$g''(x) = (n+1)n(x-a)^{n-1},$$

$$\vdots$$

$$g^{(n-1)}(x) = (n+1)n(n-1) \cdots 3(x-a)^2,$$
$$g^{(n)}(x) = (n+1)!(x-a),$$

so that in particular

$$h(a) = g(a) = 0, \quad h'(a) = g'(a) = 0, \ldots, \quad h^{(n)}(a) = g^{(n)}(a) = 0.$$

Applying Cauchy's mean value theorem to the functions h and g on the interval $[a, x]$ if $x > a$, or on $[x, a]$ if $x < a$, we get

$$\frac{h(x)}{g(x)} = \frac{h(x) - h(a)}{g(x) - g(a)} = \frac{h'(t_1)}{g'(t_1)},$$

where t_1 lies between a and x. Applying the same theorem again to the functions h' and g' on the interval $[a, t_1]$ or $[t_1, a]$, we find that

$$\frac{h(x)}{g(x)} = \frac{h'(t_1)}{g'(t_1)} = \frac{h'(t_1) - h'(a)}{g'(t_1) - g'(a)} = \frac{h''(t_2)}{g''(t_2)},$$

where t_2 lies between a and t_1, and hence between a and x. After a total of n such applications of Cauchy's mean value theorem, we finally obtain

$$\frac{h(x)}{g(x)} = \frac{h^{(n)}(t_n)}{g^{(n)}(t_n)} = \frac{f^{(n)}(t_n) - f^{(n)}(a)}{(n+1)!(t_n - a)},$$

where t_n lies between a and t_{n-1}, and hence between a and x. But by the mean value theorem applied to the function $f^{(n)}$ on the interval $[a, t_n]$ or $[t_n, a]$,

$$f^{(n)}(t_n) - f^{(n)}(a) = f^{(n+1)}(t)(t_n - a),$$

where t lies between a and t_n, and hence between a and x. Thus there is a point t between a and x such that

$$\frac{h(x)}{g(x)} = \frac{f^{(n+1)}(t)(t_n - a)}{(n+1)!(t_n - a)} = \frac{f^{(n+1)}(t)}{(n+1)!}.$$

It follows that

$$h(x) = \frac{f^{(n+1)}(t)}{(n+1)!} g(x) = \frac{f^{(n+1)}(t)}{(n+1)!} (x - a)^{n+1},$$

which is equivalent to formula (5), page 551. In general, the point t depends on a, x and n. ∎

Problems

Use the tangent line approximation (2) to estimate

1. $\sqrt{171}$
2. $\sqrt[3]{215}$
3. $1/2.01$
4. $\ln(0.98)$
5. $\tan 43°$
6. $\arctan(1.04)$

In each case give the answer to the number of decimal places known to be accurate after applying the error estimate (4).

Find the Taylor polynomial $P_n(x)$ and the remainder $R_n(x)$ for the given function, with the specified values of a and n.

7. $\sqrt{x}$, $a = 4$, $n = 2$
8. $1/x^2$, $a = -2$, $n = 3$
9. e^{-x}, $a = 1$, $n = 3$
10. xe^x, $a = 0$, $n = 4$
11. $\ln x$, $a = 2$, $n = 4$
13. $\cos x$, $a = 0$, $n = 6$
15. $\tan x$, $a = 0$, $n = 3$
17. $\sin^2 x$, $a = 0$, $n = 6$
19. $\sinh x$, $a = 1$, $n = 3$
12. $x \ln x$, $a = e$, $n = 3$
14. $\sin x$, $a = \pi/4$, $n = 3$
16. $x \cos x$, $a = \pi/2$, $n = 5$
18. $1/(1-x)$, $a = 2$, $n = 4$
20. $\ln(\cos x)$, $a = 0$, $n = 4$

21. Repeat the calculation in Problem 1, using Taylor's formula with $n = 2$ instead of the tangent line approximation, and giving the answer to the number of decimal places known to be accurate after estimating the remainder.
22. Do the same in Problem 2.
23. Do the same in Problem 3.

24. Repeat the calculation in Problem 4, using Taylor's formula with $n = 3$, and giving the answer to the number of decimal places known to be accurate after estimating the remainder.
25. Do the same in Problem 5.
26. Do the same in Problem 6.
27. Verify that the approximation $\sqrt{x} \approx 10 + \frac{1}{20}(x - 100)$ is accurate to two decimal places if $94 \leq x \leq 106$.
28. Let x be any point in the interval $[-1, 1]$. Show that the error of the approximation $\cos x \approx 1 - \frac{1}{2}x^2$ does not exceed $\frac{1}{24}$ in absolute value, while the error of the approximation $\sin x \approx x - \frac{1}{6}x^3$ does not exceed $\frac{1}{120}$ in absolute value.
29. Let $P_n(x)$ be the nth Taylor polynomial at $x = 0$ of a function f which is infinitely differentiable on the interval $[-a, a]$. Show that $P_n(x)$ contains only even powers of x if f is even on $[-a, a]$ and only odd powers of x if f is odd on $[-a, a]$.

Transform the given polynomial $Q(x)$ into a polynomial in the specified new variable.

30. $Q(x) = 1 + 3x + 5x^2 - 2x^3$, $x + 1$
31. $Q(x) = 4 - 3x^2 + 2x^4 - x^6$, $x - 1$
32. $Q(x) = 1 - x + x^2 - x^3 + x^4$, $x + 2$
33. $Q(x) = 1 + 2x^2 - 4x^3 + x^4$, $x - 5$
34. $Q(x) = x^3$, $x - \frac{1}{2}$
35. $Q(x) = x^4 + 1$, $x - 10$

Use Taylor's formula to evaluate the given limit.

36. $\lim_{x \to 0} \dfrac{e^x + e^{-x} - 2}{x^2}$

37. $\lim_{x \to 0} \dfrac{x - \ln(1 + x)}{x^2}$

38. $\lim_{x \to 0} \dfrac{\sin x - x \cos x}{x^3}$

39. $\lim_{x \to 0} \dfrac{\sin x - x + \frac{1}{6}x^3}{x^5}$

40. Prove the following generalization of the second derivative test for a local extremum (Theorem 9, page 171): If

$$f'(a) = f''(a) = \cdots = f^{(n-1)}(a) = 0 \quad (n \geq 2),$$

and if the nth derivative $f^{(n)}(a)$ is finite and different from zero, then f has a strict local minimum at a if n is even and $f^{(n)}(a) > 0$ and a strict local maximum at a if n is even and $f^{(n)}(a) < 0$, but no extremum at a if n is odd.

9.9 Taylor and Maclaurin Series

According to Taylor's formula (see page 551), if the function f has a finite derivative $f^{(n+1)}$ of order $n + 1$ at every point of an interval I containing the point a, then

$$f(x) = f(a) + f'(a)(x - a) + \frac{f''(a)}{2!}(x - a)^2 + \cdots + \frac{f^{(n)}(a)}{n!}(x - a)^n + R_n(x) \tag{1}$$

for all x in I, where the remainder $R_n(x)$ is given by

$$R_n(x) = \frac{f^{(n+1)}(t)}{(n+1)!}(x - a)^{n+1} \quad (t \text{ between } a \text{ and } x).$$

Suppose f is infinitely differentiable on I, so that f has derivatives of all orders on I. Then (1) holds for arbitrarily large n. This suggests that we investigate the infinite series

$$\sum_{n=0}^{\infty} \frac{f^{(n)}(a)}{n!}(x - a)^n$$

$$= f(a) + f'(a)(x - a) + \frac{f''(a)}{2!}(x - a)^2 + \cdots + \frac{f^{(n)}(a)}{n!}(x - a)^n + \cdots. \tag{2}$$

The series (2), which is a power series in the variable $x - a$, is called the *Taylor series (expansion) of f at $x = a$*, regardless of whether or not the series converges to f. This qualification is necessary, because there actually are cases in which the Taylor series of f fails to converge to f (see Example 6 below). However, the only situation of practical interest is where the Taylor series of f does converge to f, and we then say that "f is the sum of its own Taylor series."

Convergence of Taylor Series

You may suspect that whether or not f is the sum of its own Taylor series depends on the behavior of the remainder $R_n(x)$ in Taylor's formula (1). The following theorem shows that this is indeed true.

> **Theorem 18** *(Convergence criterion for a Taylor series). The Taylor series (2) converges to f on an interval I if and only if*
> $$\lim_{n \to \infty} R_n(x) = 0 \qquad (3)$$
> *for all x in I.*

Proof Formula (1) becomes
$$f(x) = P_n(x) + R_n(x)$$
in terms of the Taylor polynomials
$$P_n(x) = \sum_{k=0}^{n} \frac{f^{(k)}(a)}{k!}(x - a)^k \qquad (n = 0, 1, 2, \ldots),$$
and these polynomials are the partial sums of the Taylor series (2). Therefore (2) converges to f on I if and only if
$$f(x) = \lim_{n \to \infty} P_n(x) \qquad (x \text{ in } I),$$
or equivalently
$$\lim_{n \to \infty} [f(x) - P_n(x)] = 0 = \lim_{n \to \infty} R_n(x) \qquad (x \text{ in } I). \quad \blacksquare$$

Thus if the condition (3) is satisfied, we can write
$$f(x) = f(a) + f'(a)(x - a) + \frac{f''(a)}{2!}(x - a)^2 + \cdots + \frac{f^{(n)}(a)}{n!}(x - a)^n + \cdots \qquad (4)$$
in full confidence that the power series on the right actually converges to the function on the left. For $a = 0$ the Taylor series (4) reduces to
$$f(x) = f(0) + f'(0)x + \frac{f''(0)}{2!}x^2 + \cdots + \frac{f^{(n)}(0)}{n!}x^n + \cdots. \qquad (5)$$

A Taylor series of the special form (5) is often called a *Maclaurin series*, in honor of the Scottish mathematician Colin Maclaurin (1698–1746).

Example 1 Let $\sum c_n(x - a)^n$ be a power series with interval of convergence I and sum f. Show that $\sum c_n(x - a)^n$ is the *Taylor* series of f at $x = a$. (Thus the Taylor series of the sum function of a power series is just the power series itself.)

Solution Because of Theorem 14, page 540, we can differentiate the power series

$$f(x) = c_0 + c_1(x - a) + c_2(x - a)^2 + \cdots + c_n(x - a)^n + \cdots \quad (x \text{ in } I)$$

n times in succession. This gives

$$f^{(n)}(x) = n!c_n + (n+1)!c_{n+1}(x - a) + \frac{(n+2)!}{2!} c_{n+2}(x - a)^2 + \cdots \quad (x \text{ in } I),$$

which implies

$$c_n = \frac{f^{(n)}(a)}{n!} \quad (n = 0, 1, 2, \ldots)$$

after setting $x = a$ (note that $f^{(0)} \equiv f$, $0! = 1$). But then

$$f(x) = \sum_{n=0}^{\infty} c_n(x - a)^n = \sum_{n=0}^{\infty} \frac{f^{(n)}(a)}{n!} (x - a)^n,$$

that is, $\sum c_n(x - a)^n$ is the Taylor series of f at $x = a$. Naturally, this Taylor series converges to f at every point of I. □

Example 2 Find the Maclaurin series of e^x.

Solution If $f(x) = e^x$, then $f^{(n)}(x) = e^x$, $f^{(n)}(0) = 1$ for all $n = 0, 1, 2, \ldots$, and the Maclaurin series (5) becomes

$$e^x = 1 + x + \frac{x^2}{2!} + \cdots + \frac{x^n}{n!} + \cdots, \quad (6)$$

provided that the series on the right actually converges to e^x. To verify that this is so, we investigate the remainder

$$R_n(x) = \frac{f^{(n+1)}(t)}{(n+1)!} x^{n+1} = \frac{e^t}{(n+1)!} x^{n+1},$$

where t lies between 0 and x (bear in mind that t depends on n, as well as on x). Given any x, it is clear that

$$0 \leq |R_n(x)| \leq M \left| \frac{x^{n+1}}{(n+1)!} \right| \quad (7)$$

for every n, where M is the maximum of e^t on the interval $[0, x]$ if $x > 0$, or on $[x, 0]$ if $x < 0$, that is,

$$M = \begin{cases} e^x & \text{if } x > 0, \\ 1 & \text{if } x < 0. \end{cases}$$

Moreover,

$$\lim_{n \to \infty} \left| \frac{x^{n+1}}{(n+1)!} \right| = 0 \quad (8)$$

for any fixed x, since the power series with general term $x^n/n!$ is absolutely convergent by the ratio test (check this). Therefore, taking the limit as $n \to \infty$

in (7), we find that $|R_n(x)| \to 0$ as $n \to \infty$ or equivalently

$$\lim_{n \to \infty} R_n(x) = 0$$

for all x (note that $R_n(0) = 0$ for every n). Hence the series (6) converges to e^x on the whole interval $(-\infty, \infty)$. ∏

You will recall that the validity of (6) has already been established by another, entirely different method in Example 5, page 542.

The Maclaurin Series of sin x and cos x

Example 3 Find the Maclaurin series of sin x.

Solution If $f(x) = \sin x$, then

$$f(x) = \sin x, \qquad f(0) = 0,$$

$$f'(x) = \cos x = \sin\left(x + \frac{\pi}{2}\right), \qquad f'(0) = 1,$$

$$f''(x) = -\sin x = \sin\left(x + \frac{2\pi}{2}\right), \qquad f''(0) = 0,$$

$$f'''(x) = -\cos x = \sin\left(x + \frac{3\pi}{2}\right), \qquad f'''(0) = -1,$$

$$\vdots \qquad\qquad \vdots$$

$$f^{(n)}(x) = \sin\left(x + \frac{n\pi}{2}\right), \qquad f^{(n)}(0) = \sin\frac{n\pi}{2},$$

$$f^{(n+1)}(x) = \sin\left(x + \frac{(n+1)\pi}{2}\right),$$

and (5) becomes

$$\sin x = x - \frac{x^3}{3!} + \frac{x^5}{5!} - \frac{x^7}{7!} + \cdots . \qquad (9)$$

The remainder is now

$$R_n(x) = \frac{x^{n+1}}{(n+1)!} \sin\left(t + \frac{(n+1)\pi}{2}\right),$$

where t lies between 0 and x (here x is arbitrary but fixed). Since

$$\left|\sin\left(t + \frac{(n+1)\pi}{2}\right)\right| \leq 1$$

for arbitrary t and n, we have

$$0 \leq |R_n(x)| \leq \left|\frac{x^{n+1}}{(n+1)!}\right|,$$

and hence

$$\lim_{n \to \infty} R_n(x) = 0,$$

because of (8). Therefore (9) is the Maclaurin series of sin x on the whole interval $(-\infty, \infty)$. The series (9) can be written more concisely as

$$\sin x = \sum_{n=0}^{\infty} (-1)^n \frac{x^{2n+1}}{(2n+1)!}. \qquad \square$$

Example 4 Find the Maclaurin series of $\cos x$.

Solution We could give an argument like the one in the preceding example, but it is much simpler to differentiate the Maclaurin series for $\sin x$ term by term, with the help of Theorem 14, page 540. This immediately gives

$$\cos x = \frac{d}{dx} \sin x = \frac{d}{dx} \sum_{n=0}^{\infty} (-1)^n \frac{x^{2n+1}}{(2n+1)!} = \sum_{n=0}^{\infty} (-1)^n \frac{x^{2n}}{(2n)!},$$

that is,

$$\cos x = 1 - \frac{x^2}{2!} + \frac{x^4}{4!} - \frac{x^6}{6!} + \cdots. \qquad \square$$

Example 5 Find the Taylor series of $\sin x$ at $x = \pi/4$.

Solution This time we have

$$f(x) = \sin x, \qquad f\left(\frac{\pi}{4}\right) = \sin \frac{\pi}{4} = \frac{1}{\sqrt{2}}$$

$$f'(x) = \cos x = \sin\left(x + \frac{\pi}{2}\right), \qquad f'\left(\frac{\pi}{4}\right) = \sin \frac{3\pi}{4} = \frac{1}{\sqrt{2}}$$

$$f''(x) = -\sin x = \sin\left(x + \frac{2\pi}{2}\right), \qquad f''\left(\frac{\pi}{4}\right) = \sin \frac{5\pi}{4} = -\frac{1}{\sqrt{2}}$$

$$f'''(x) = -\cos x = \sin\left(x + \frac{3\pi}{2}\right), \qquad f'''\left(\frac{\pi}{4}\right) = \sin \frac{7\pi}{4} = -\frac{1}{\sqrt{2}}$$

$$\vdots \qquad \qquad \vdots$$

$$f^{(n)}(x) = \sin\left(x + \frac{n\pi}{2}\right), \qquad f^{(n)}\left(\frac{\pi}{4}\right) = \sin\left(\frac{n\pi}{2} + \frac{\pi}{4}\right),$$

$$f^{(n+1)}(x) = \sin\left(x + \frac{(n+1)\pi}{2}\right),$$

and (4) becomes

$$\sin x = \frac{1}{\sqrt{2}} \left[1 + \left(x - \frac{\pi}{4}\right) - \frac{1}{2!}\left(x - \frac{\pi}{4}\right)^2 - \frac{1}{3!}\left(x - \frac{\pi}{4}\right)^3 + \cdots \right],$$

or more concisely

$$\sin x = \frac{1}{\sqrt{2}} \sum_{n=0}^{\infty} \frac{(-1)^{[\![n/2]\!]}}{n!} \left(x - \frac{\pi}{4}\right)^n, \qquad (10)$$

where $[\![n/2]\!]$ is the integer part of $n/2$. Virtually the same remainder analysis as in Example 3 shows that

$$\lim_{n \to \infty} R_n(x) = 0$$

for all x. Hence the Taylor series (10) converges to $\sin x$ on the whole interval $(-\infty, \infty)$. $\square$

Example 6 Show that the function

$$f(x) = \begin{cases} e^{-1/x^2} & \text{if } x \neq 0, \\ 0 & \text{if } x = 0 \end{cases}$$

is not the sum of its own Taylor series at $x = 0$.

Solution Let $P_n(x)$ be the nth Taylor polynomial of f at $x = 0$. Then, as shown in Example 7, page 554, $P_n(x) \equiv 0$ for all $n = 0, 1, 2, \ldots$, so that

$$\lim_{n \to \infty} P_n(x) \equiv 0.$$

It follows that the sum of the Taylor series of f at $x = 0$ is not f, but the function which is identically equal to zero. □

It should be stressed that the direct use of formula (4) or (5) to find the Taylor or Maclaurin series of a given function often leads to calculations that are so formidable as to be infeasible. Thus you should always be on the lookout for ways to express a new Taylor series in terms of Taylor series which are already known. For example, to find the Maclaurin series of $x^4 e^x$, rather than calculate the derivatives of $x^4 e^x$, evaluate them at $x = 0$ and substitute the resulting values into formula (5), just multiply the known Maclaurin series of e^x by x^4, which immediately gives

$$x^4 e^x = x^4 \left(1 + x + \frac{x^2}{2!} + \cdots + \frac{x^n}{n!} + \cdots \right)$$
$$= x^4 + x^5 + \frac{x^6}{2!} + \cdots + \frac{x^{n+4}}{n!} + \cdots.$$

Similarly, to find the Taylor series of e^x at $x = 1$, you need only observe that

$$e^x = e^1 e^{x-1} = e \left[1 + (x-1) + \frac{(x-1)^2}{2!} + \cdots + \frac{(x-1)^n}{n!} + \cdots \right],$$

since it is legitimate to replace x by $x - 1$ in the Maclaurin series of e^x (why?).

If the Taylor series of a function f at $x = a$ converges to f, then it is precisely what has previously been called the power series expansion of f at $x = a$. This is an immediate consequence of the uniqueness property of power series discussed in Example 4, page 541. Thus in finding Taylor series, all the techniques of Section 9.7 are still at our disposal and should be used freely. With their help, it is often possible to find the Taylor series of a given function f indirectly, without having to calculate the derivatives of f or investigate the remainder $R_n(x)$. For instance, we have already found the Maclaurin series of arctan x in Example 8, page 544.

Problems

1. Show that

$$\sinh x = x + \frac{x^3}{3!} + \frac{x^5}{5!} + \frac{x^7}{7!} + \cdots = \sum_{n=0}^{\infty} \frac{x^{2n+1}}{(2n+1)!} \quad \text{(i)}$$

for all x. Start from formula (5), and investigate the behavior of the remainder $R_n(x)$ as $n \to \infty$.

2. Deduce the series

$$\cosh x = 1 + \frac{x^2}{2!} + \frac{x^4}{4!} + \frac{x^6}{6!} + \cdots = \sum_{n=0}^{\infty} \frac{x^{2n}}{(2n)!}, \quad \text{(i')}$$

valid for all x, directly from the series (i).

3. Show that

$$\ln(1+x) = x - \frac{x^2}{2} + \frac{x^3}{3} - \frac{x^4}{4} + \cdots$$
$$= \sum_{n=1}^{\infty} (-1)^{n-1} \frac{x^n}{n} \quad (0 \le x < 1). \quad \text{(ii)}$$

Start from formula (5), and investigate the behavior of the remainder $R_n(x)$ as $n \to \infty$. (The validity of this power series expansion of $\ln(1+x)$ on the larger interval $-1 < x < 1$ has already been established by another method in Example 7, page 544.)

4. By investigating the remainder $R_n(1)$, show that formula (ii) remains valid for $x = 1$, so that

$$\ln 2 = 1 - \frac{1}{2} + \frac{1}{3} - \frac{1}{4} + \cdots = \sum_{n=1}^{\infty} \frac{(-1)^{n-1}}{n}, \quad \text{(ii')}$$

as already shown on page 544 with the help of Abel's theorem.

Find the first five nonzero terms of the Maclaurin series of

5. $\sqrt{2} \cos\left(x - \frac{\pi}{4}\right)$ **6.** $\sin\left(x + \frac{\pi}{3}\right)$

7. $e^x \cos x$ **8.** $e^x \sin x$

Find the first three nonzero terms of the Maclaurin series of

9. $\ln(1 + e^x)$ **10.** $e^{\sin x}$

11. $e^{\cos x}$ **12.** $\ln(x + \sqrt{x^2 + 1})$

13. Find the first four terms of the Maclaurin series of $e^{1/(1-x)}$.

14. Use the formula $\cos^2 x = \frac{1}{2}(1 + \cos 2x)$ to find the Maclaurin series of the function $\cos^2 x$.

Find the Taylor series of $\sin x$ at

15. $\pi/6$ **16.** $-\pi/3$ **17.** $\pi/2$ **18.** π

Find the Taylor series of $\cos x$ at

19. $\pi/4$ **20.** $\pi/2$ **21.** $-\pi$ **22.** 2π

Find the Taylor series of

23. $\sqrt{x}$ at $x = 4$ **24.** $\sqrt[3]{x}$ at $x = -1$
25. $1/x^2$ at $x = 1$ **26.** e^x at $x = -2$
27. $e^{x/3}$ at $x = 2$ **28.** $\sin^2 x$ at $x = 0$

29. Use Maclaurin series to show that $\cosh x \le e^{ax^2}$ for all x if and only if $a \ge \frac{1}{2}$.

30. Find the Maclaurin series of the functions

$$S(x) = \int_0^x \sin t^2 \, dt, \quad C(x) = \int_0^x \cos t^2 \, dt,$$

known as the *Fresnel integrals*, which arise in the study of certain optical phenomena.

31. Find the Maclaurin series of the function

$$f(x) = \frac{1 - x + x^2}{1 + x + x^2}$$

by the "method of undetermined coefficients," that is, assuming that $f(x) = a_0 + a_1 x + \cdots + a_n x^n + \cdots$, choose the coefficients $a_0, a_1, \ldots, a_n, \ldots$ in such a way that

$$1 - x + x^2 \equiv (1 + x + x^2)(a_0 + a_1 x + \cdots + a_n x^n + \cdots).$$

(This process is equivalent to "long division" of the numerator by the denominator.) What is the value of $f^{(5)}(0)$?

Use the method of undetermined coefficients to find the first four terms of the Maclaurin series of

32. $\sec x$ **33.** $\tan x$

9.10 Newton's Method

Let the function f be differentiable (and hence continuous) on the interval $I = [a, b]$, and suppose the derivative f' is never zero on I. Suppose further that $f(a)f(b) < 0$, so that $f(a)$ and $f(b)$ have opposite signs. Then the equation $f(x) = 0$ has a solution or root r in (a, b), by the intermediate value theorem, and r is unique, since f is monotonic on I (why?). A method for approximating r to any desired accuracy, called the *bisection method*, was presented on page 97, but the method is of limited practical value because of the large number of operations required to achieve even modest accuracy. With the help of the extended mean value theorem (Taylor's theorem for $n = 1$), we now establish a much more powerful way of approximating r, called *Newton's method*. This method, also known as the *Newton–Raphson method*, generates a sequence $\{x_n\}$ of "successive approximations" to r, which in a great many cases converges very rapidly to r.

Theorem 19 *(Newton's method).* Let f be a function with a continuous second derivative f'' on an interval $I = [a, b]$, such that f' and f'' are nonzero on I and $f(a)f(b) < 0$. Let $\{x_n\}$ be the sequence defined by the recursion formula

$$x_{n+1} = x_n - \frac{f(x_n)}{f'(x_n)} \quad (n = 1, 2, \ldots), \tag{1}$$

where $x_1 = b$ if f' and f'' have the same sign, while $x_1 = a$ if f' and f'' have opposite signs. Then $\{x_n\}$ converges to r, the unique root of the equation $f(x) = 0$ in I, and moreover

$$|x_{n+1} - r| \leq \frac{M}{2m} |x_n - r|^2, \tag{2}$$

where m is the minimum of $|f'(x)|$ on I and M is the maximum of $|f''(x)|$ on I.

Geometric Meaning of Newton's Method

The proof of Theorem 19 is rather delicate, and hence is deferred to the end of the section. However, there is a simple geometric interpretation of Newton's method, which we illustrate in Figure 14 for the case where f' and f'' are both positive on $I = [a, b]$, so that f is increasing and concave upward on I with $f(a) < 0$ and $f(b) > 0$. Since f' and f'' have the same sign, the first term of the sequence x_n is chosen to be $x_1 = b$, the right endpoint of I. Let T_1 be the (left-hand) tangent to the curve $y = f(x)$ at the point $P_1 = (x_1, f(x_1))$. Then T_1 is the line

$$y = f'(x_1)(x - x_1) + f(x_1),$$

with x-intercept

$$x_2 = x_1 - \frac{f(x_1)}{f'(x_1)},$$

which is precisely formula (1) with $n = 1$. Similarly, if T_2 is the tangent to

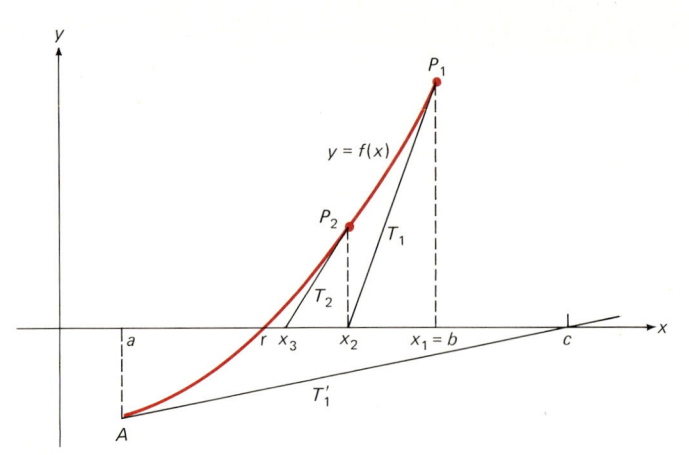

Geometric meaning of Newton's method

Figure 14

564   Chapter 9   Sequences and Series

the curve $y = f(x)$ at the point $P_2 = (x_2, f(x_2))$, then T_2 is the line

$$y = f'(x_2)(x - x_2) + f(x_2),$$

with x-intercept

$$x_3 = x_2 - \frac{f(x_2)}{f'(x_2)},$$

as given by formula (1) with $n = 2$, and so on for the desired number of steps. It is clear from the figure that under favorable conditions, the sequence of successive approximations $\{x_n\}$ will converge very rapidly to the x-intercept of the curve $y = f(x)$ itself, that is, to the root of the equation $f(x) = 0$. Algebraically, this is a consequence of formula (2), which shows that the absolute value of the error made at any step of the approximation process does not exceed the product of a fixed constant and the *square* of the error made at the preceding step. For example, suppose that $M/2m = 1$ and $|x_n - r| < 5 \times 10^{-5}$, so that the approximation $r \approx x_n$ is accurate to four decimal places. Then

$$|x_{n+1} - r| < |x_n - r|^2 < 25 \times 10^{-10} = 2.5 \times 10^{-9},$$

so that the next approximation $r \approx x_{n+1}$ is already accurate to eight decimal places!

Figure 14 also affords an example of how Newton's method can break down if we are not careful. Suppose that for the function f shown in the figure we choose the first term of the sequence $\{x_n\}$ to be $x_1 = a$ instead of $x_1 = b$, as required by Theorem 19. Then, since the tangent to $y = f(x)$ at $A = (a, f(a))$ is the line T'_1 whose x-intercept c lies *outside* the interval I on which f is defined, the approximation process comes to a halt after just one step. If other assumptions of the theorem are violated, it may well happen that the sequence $\{x_n\}$ generated by formula (1) is *divergent* (see Problems 20 and 21). Even in cases where the sequence $\{x_n\}$ is known to converge, it is of course desirable to choose the initial approximation x_1 as close to the root r as possible, by sketching the graph of f and guessing the value of r, or by using some other approximation method to make a preliminary estimate of r.

It follows from formula (1) that if $x_{n+1} = x_n$, then $f(x_n) = 0$, so that $r = x_n$. By the same token, if the first N decimal places of x_n and x_{n+1} coincide, it is customary to assume that the approximation $r \approx x_n$ is accurate to at least N decimal places, but to be on the safe side, one can do an error analysis based on the use of formula (2).

Remark The sequence

$$x_1, x_2 = g(x_1), x_3 = g(x_2), \ldots, x_{n+1} = g(x_n), \ldots$$

is said to be generated from its initial term x_1 by *iteration* with the function g. If x_1 and g satisfy suitable conditions, it can be shown that the sequence $\{x_n\}$ converges to a *fixed point* of g, that is, to a point r such that $g(r) = r$. Thus Newton's method corresponds to iteration with the function

$$g(x) = x - \frac{f(x)}{f'(x)}, \tag{3}$$

and Theorem 19 gives conditions guaranteeing the convergence of this iteration process. Notice that r is a fixed point of the iteration function (3) if and only if r is a root of the equation $f(x) = 0$.

Example 1 Use Newton's method to approximate $\sqrt{3}$ to an accuracy of eight decimal places.

Solution Since $(1.7)^2 = 2.89$ and $(1.8)^2 = 3.24$, we know that $\sqrt{3}$ lies between 1.7 and 1.8. If $f(x) = x^2 - 3$, then $f(r) = 0$ if and only if $r = \sqrt{3}$. Moreover $f'(x) = 2x$, and therefore, by formula (1),

$$x_{n+1} = x_n - \frac{f(x_n)}{f'(x_n)} = x_n - \frac{x_n^2 - 3}{2x_n} = \frac{x_n^2 + 3}{2x_n} = \frac{1}{2}\left(x_n + \frac{3}{x_n}\right).$$

Thus the $(n+1)$st approximation x_{n+1} is the average of the preceding approximation x_n and of 3 divided by x_n. Here the interval I is $[1.7, 1.8]$, and Theorem 19 calls for the initial approximation $x_1 = 1.8$, since $f'(x) = 2x$ and $f''(x) = 2$ are both positive on I. Using a calculator to compute the next few approximations to eight decimal places, we find that

$$x_1 = 1.8$$
$$x_2 = 1.73333333$$
$$x_3 = 1.73205128$$
$$x_4 = 1.73205081$$
$$x_5 = 1.73205081$$

Since x_4 and x_5 coincide to eight decimal places, we conclude that the approximation $r = \sqrt{3} \approx 1.73205081$ is accurate to eight decimal places, as confirmed in the next example. □

Example 2 In the preceding example how accurate are the approximations $\sqrt{3} \approx x_3$ and $\sqrt{3} \approx x_4$?

Solution Choosing $r = \sqrt{3}$ in formula (2), we have

$$|x_{n+1} - \sqrt{3}| \leq \frac{M}{2m}|x_n - \sqrt{3}|^2,$$

where m is the minimum of $|f'(x)| = 2|x|$ on $I = [1.7, 1.8]$ and M is the maximum of $|f''(x)| = 2$ on I. But $m = 2(1.7) = 3.4$, $M = 2$, and hence

$$|x_{n+1} - \sqrt{3}| \leq \frac{1}{3.4}|x_n - \sqrt{3}|^2 < \frac{3}{10}|x_n - \sqrt{3}|^2$$

($1/3.4 \approx 0.294$). Therefore

$$|x_2 - \sqrt{3}| < \frac{3}{10}|x_1 - \sqrt{3}|^2 = \frac{3}{10}|1.8 - \sqrt{3}|^2 < \frac{3}{10}\left(\frac{1}{10}\right)^2,$$

$$|x_3 - \sqrt{3}| < \frac{3}{10}|x_2 - \sqrt{3}|^2 < \left(\frac{3}{10}\right)^3\left(\frac{1}{10}\right)^4 = 3^3 \times 10^{-7} = 2.7 \times 10^{-6},$$

$$|x_4 - \sqrt{3}| < \frac{3}{10}|x_3 - \sqrt{3}|^2 < \left(\frac{3}{10}\right)^7\left(\frac{1}{10}\right)^8 = 3^7 \times 10^{-15} \approx 2.2 \times 10^{-12}.$$

Thus the approximation $\sqrt{3} \approx x_3$ is accurate to 5 decimal places, while the approximation $\sqrt{3} \approx x_4$ is accurate to 11 decimal places. Notice that this error estimate actually makes it unnecessary to calculate the next approximation $\sqrt{3} \approx x_5$, which turns out to be accurate to 23 decimal places. □

Example 3 Use Newton's method to solve the equation $x - e^x = 0$.

Solution It is apparent from Figure 15 that this equation has only one root r, equal to the abscissa of the point in which the line $y = x$ intersects the curve $y = e^{-x}$, and it looks as if $r \approx 0.5$ is a good initial approximation. Choosing $f(x) = x - e^{-x}$, $f'(x) = 1 + e^{-x}$ in formula (1), we get

$$x_{n+1} = x_n - \frac{x_n - e^{-x_n}}{1 + e^{-x_n}} = \frac{x_n e^{-x_n} + e^{-x_n}}{1 + e^{-x_n}} = \frac{x_n + 1}{e^{x_n} + 1}.$$

Calculating the first few approximations to six decimal places, we find that

$$x_1 = 0.5$$
$$x_2 = 0.566311$$
$$x_3 = 0.567143$$
$$x_4 = 0.567143$$

Thus we conclude that $r \approx 0.567143$ to six decimal places. If we take $x_1 = 0.6$ as our initial approximation, we obtain

$$x_1 = 0.6$$
$$x_2 = 0.566950$$
$$x_3 = 0.567143$$
$$x_4 = 0.567143$$

instead, which gives the same answer in the same number of steps. This does not contradict Theorem 19, which calls for choosing x_1 to be the left endpoint of the interval $I = [0.5, 0.6]$ since $f'(x) = 1 + e^{-x}$ and $f''(x) = -e^{-x}$ have opposite signs. After all, although the theorem does not *guarantee* convergence of the sequence $\{x_n\}$ if the other endpoint is chosen, it certainly does not say that $\{x_n\}$ will *fail* to converge, and if convergence does occur, there is no arguing with success! Notice that the sequence $\{x_n\}$ is monotonic if $x_1 = 0.5$, but not if $x_1 = 0.6$. How do you account for this? □

Proof of Theorem 19 (Optional) Choosing $x = r$ and $a = x_n$ in the extended mean value theorem (3), page 549, we have

$$0 = f(r) = f(x_n) + f'(x_n)(r - x_n) + \frac{f''(t_n)}{2}(r - x_n)^2,$$

where t_n lies between x_n and r. Therefore

$$x_n - \frac{f(x_n)}{f'(x_n)} = r + \frac{f''(t_n)}{2f'(x_n)}(r - x_n)^2,$$

or equivalently

$$x_{n+1} - r = \frac{f''(t_n)}{2f'(x_n)}(x_n - r)^2, \qquad (4)$$

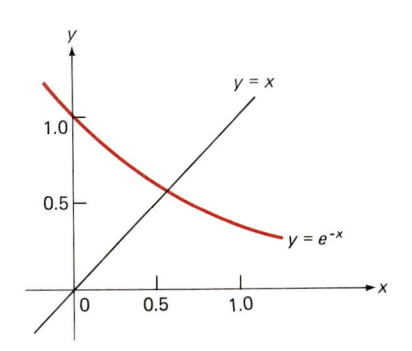

since

$$x_{n+1} = x_n - \frac{f(x_n)}{f'(x_n)}. \qquad (5)$$

It is easy to see that the sign of $f(x_1)$ is always the same as that of f''.† Hence it follows from (4) and (5) that $r < x_2 < x_1 = b$ if f' and f'' have the same sign, while $a = x_1 < x_2 < r$ if f' and f'' have opposite signs. Since x_1 and x_2 lie on the same side of r, $f(x_2)$ has the same sign as $f(x_1)$. Thus another application of formulas (4) and (5) shows that $r < x_3 < x_2 < x_1$ if f' and f'' have the same sign, while $x_1 < x_2 < x_3 < r$ if f' and f'' have opposite signs. In fact, repeated application of this argument shows that $r < x_{n+1} < x_n < \cdots < x_1$ for all n if f' and f'' have the same sign, while $x_1 < \cdots < x_n < x_{n+1} < r$ for all n if f' and f'' have opposite signs, so that in any event, $\{x_n\}$ is a bounded monotonic sequence (decreasing in the first case and increasing in the second). By Theorem 2, page 493, $\{x_n\}$ converges to a limit L. This limit is equal to r. In fact, taking the limit of both sides of (5) as $n \to \infty$, we get

$$L = L - \frac{f(L)}{f'(L)},$$

which implies $f(L) = 0$ and hence $L = r$. To complete the proof, we observe that the inequality (2) in the statement of the theorem is an immediate consequence of formula (4) and the meaning of the numbers m and M. ∎

---

† In detail, $f(a) < 0$, $f(b) > 0$ if $f' > 0$, while $f(a) > 0$, $f(b) < 0$ if $f' < 0$. If f' and f'' have the same sign, then either $f' > 0$, $f'' > 0$ or $f' < 0$, $f'' < 0$, and in both cases $f(x_1) = f(b)$ has the same sign as f''. On the other hand, if f' and f'' have opposite signs, then either $f' > 0$, $f'' < 0$ or $f' < 0$, $f'' > 0$, and in both cases $f(x_1) = f(a)$ again has the same sign as f''.

Problems

Use Newton's method to approximate the given quantity to an accuracy of four decimal places.

1. $\sqrt{2}$
2. $\sqrt[3]{11}$
3. $\sqrt[4]{75}$
4. $\sqrt[5]{1000}$
5. $\sqrt[6]{800}$

6. Let c be any positive number and k any integer greater than 1. Show that the sequence of successive approximations $\{x_n\}$ given by the recursion formula

$$x_{n+1} = \left(1 - \frac{1}{k}\right)x_n + \frac{1}{k}cx_n^{1-k} \qquad (n = 1, 2, \ldots)$$

converges to $\sqrt[k]{c}$ for every initial approximation $x_1 > 0$.

7. The quadratic equation $x^2 - x - 1 = 0$ has two roots $\frac{1}{2}(1 \pm \sqrt{5})$. For which values of the initial approximation x_1 does the sequence $\{x_n\}$ generated by Newton's method converge to the positive root? To the negative root? For which value of x_1 does the method fail?

8. Suppose Newton's method is used to approximate $\sqrt[3]{7}$, starting with the initial approximation $x_1 = 2$. Show that the approximation $\sqrt[3]{7} \approx x_4$ is accurate to nine decimal places.

9. In Example 1, page 96, the bisection method was used to find a crude approximation to the root r of the equation $2x^5 + 2x^2 + x - 3 = 0$ in the interval $(0, 1)$. Use Newton's method to approximate r to six decimal places.

10. Show graphically that the equation $e^{-x} - \sin x = 0$ has infinitely many positive roots. Then use Newton's method to approximate the two smallest roots to four decimal places.

11. Show that the equation $x^3 - 6x + 1 = 0$ has three distinct real roots, and approximate each root to four decimal places with the help of Newton's method.

Use Newton's method to solve the given equation to an accuracy of three decimal places.

12. $x^4 - 4x - 4 = 0 \quad (-1 < x < 0)$
13. $(x + 1)^2 x = 1$
14. $e^x - x^2 + 1 = 0$
15. $x + \ln x - 3 = 0$
16. $x^2 + \ln x - 2 = 0$
17. $4 \sin x - x = 0 \quad (x > 0)$
18. $x^2 - \cos x = 0 \quad (x < 0)$
19. $\tan x = x \quad (\pi/2 < x < 3\pi/2)$

568 Chapter 9 Sequences and Series

20. The function

$$f(x) = \begin{cases} \sqrt{x-r} & \text{if } x \geq r, \\ -\sqrt{x-r} & \text{if } x < r, \end{cases}$$

graphed in Figure 16, has a root at r and nowhere else. Show that if $x_1 \neq r$, the sequence $\{x_n\}$ generated by Newton's method diverges, oscillating back and forth between the values x_1 and $x_2 = 2r - x_1$. Interpret this graphically.

21. The function $f(x) = (x - r)^{1/3}$ has a root at r and nowhere else. Show that if $x_1 \neq r$, the sequence $\{x_n\}$ generated by Newton's method diverges, taking values of arbitrarily large absolute value. Interpret this graphically.

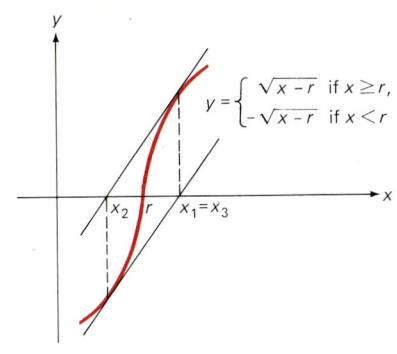

Figure 16

Key Terms and Topics
Infinite sequences
Recursion formulas
The limit of a sequence, convergent and divergent sequences
Bounded and unbounded sequences, monotonic sequences
Convergence of a bounded monotonic sequence
Infinite series, partial sums of a series
Convergent and divergent series, the sum of a series
The geometric series, the harmonic series, the p-series
Necessary condition for convergence
The remainder of a series
Convergence criterion for nonnegative series
Comparison tests, the integral test
Absolute vs. conditional convergence
Alternating series, the alternating series test
Rearrangement of series
The ratio and root tests
Power series, numerical series vs. series of functions
Interval of convergence and radius of convergence of a power series
The sum (function) of a power series
Term by term differentiation and integration of a power series
The binomial series
The extended mean value theorem
Taylor's theorem, Taylor's formula with remainder
Taylor polynomials
Taylor and Maclaurin series
Newton's method

Important Numerical Series

Geometric series $\sum_{n=0}^{\infty} ar^n$ (convergent to $a/(1 - r)$ if $|r| < 1$, divergent if $|r| \geq 1$)

Harmonic series $\sum_{n=1}^{\infty} \frac{1}{n}$ (divergent)

The p-series $\sum_{n=1}^{\infty} \frac{1}{n^p}$ (convergent if $p > 1$, divergent if $p \leq 1$)

Important Power Series

$$\frac{1}{1-x} = \sum_{n=0}^{\infty} x^n$$

$$e^x = \sum_{n=0}^{\infty} \frac{x^n}{n!}$$

$$\sin x = \sum_{n=0}^{\infty} (-1)^n \frac{x^{2n+1}}{(2n+1)!}$$

$$\cos x = \sum_{n=0}^{\infty} (-1)^n \frac{x^{2n}}{(2n)!}$$

$$\arctan x = \sum_{n=0}^{\infty} (-1)^n \frac{x^{2n+1}}{2n+1} \quad \text{(Gregory's series)}$$

$$(1+x)^r = 1 + \sum_{n=1}^{\infty} \frac{r(r-1)\cdots(r-n+1)}{n!} x^n \quad \text{(Binomial series)}$$

$$f(x) = \sum_{n=0}^{\infty} \frac{f^{(n)}(a)}{n!} (x-a)^n \quad \text{(Taylor series)}$$

Synopsis of Convergence Tests for Numerical Series†

If:	Then:
$\lim_{n \to \infty} a_n \neq 0$	$\sum a_n$ diverges (Theorem 3, page 502)
$a_n \geq 0,\ a_1 + a_2 + \cdots + a_n \leq C$	$\sum a_n$ converges (Theorem 4, page 508)
$a_n \geq 0,\ b_n \geq 0,\ a_n \leq b_n$	Convergence of $\sum b_n$ implies convergence of $\sum a_n$ Divergence of $\sum a_n$ implies divergence of $\sum b_n$ (Theorem 5, page 510)
$a_n > 0,\ b_n > 0,\ \lim_{n \to \infty} \frac{a_n}{b_n} = L$	Convergence of $\sum b_n$ implies convergence of $\sum a_n$ if $0 \leq L < \infty$ Divergence of $\sum b_n$ implies divergence of $\sum a_n$ if $0 < L \leq \infty$ (Theorem 6, page 511)
$a_n = f(n)$, f continuous, positive and decreasing on $[1, \infty)$	$\sum a_n$ converges if and only if $\int_1^{\infty} f(x)\,dx$ converges (Theorem 7, page 514)
$\sum \|a_n\|$ converges	$\sum a_n$ converges (Theorem 8, page 518)
$a_n > a_{n+1},\ \lim_{n \to \infty} a_n = 0$	$\sum (-1)^{n-1} a_n$ converges (Theorem 9, page 519)
$a_n \neq 0,\ \lim_{n \to \infty} \left\|\frac{a_{n+1}}{a_n}\right\| = L$	$\sum a_n$ converges absolutely if $0 \leq L < 1$ and diverges if $1 < L \leq \infty$ (Theorem 10, page 525)
$\lim_{n \to \infty} \sqrt[n]{\|a_n\|} = L$	$\sum a_n$ converges absolutely if $0 \leq L < 1$ and diverges if $1 < L \leq \infty$ (Theorem 11, page 527)

† $\sum$ is shorthand for $\sum_{n=1}^{\infty}$

Supplementary Problems

Find the limit (if it exists) of the sequence $\{a_n\}$, where a_n is the nth decimal place in the decimal representation of

1. $\frac{1}{6}$ 2. $\frac{1}{8}$ 3. $\frac{1}{7}$ 4. π

Find sequences $\{a_n\}$ and $\{b_n\}$, both converging to 0, such that

5. $\{a_n/b_n\}$ converges to 0
6. $\{a_n/b_n\}$ converges to 1
7. $\{a_n/b_n\}$ is divergent and bounded
8. $\{a_n/b_n\}$ is divergent and unbounded

9. Show that if $f(x)$ is a function defined on $(0, 1]$ such that $f(x) \to L$ as $x \to 0^+$, then $f(1/n) \to L$ as $n \to \infty$.

Evaluate the given limit.

10. $\lim_{n \to \infty} n \tan \frac{1}{n}$

11. $\lim_{n \to \infty} \cos\left(\frac{\ln n}{n}\right)$

12. $\lim_{n \to \infty} \frac{1}{\sqrt[n]{n!}}$

13. $\lim_{n \to \infty} (\sqrt{n^2 + n} - \sqrt{n^2 - 3n})$

14. $\lim_{n \to \infty} \frac{1 + 2 + \cdots + n}{n^2}$

15. $\lim_{n \to \infty} \frac{1 + a + a^2 + \cdots + a^n}{1 + b + b^2 + \cdots + b^n}$ ($|a| < 1$, $|b| < 1$)

16. $\lim_{n \to \infty} nc^n$ ($|c| < 1$)

17. $\lim_{n \to \infty} \arcsin \frac{2n - 1}{4n}$

18. $\lim_{n \to \infty} \arctan \sqrt[n]{n}$

19. $\lim_{n \to \infty} (\sqrt{2} \sqrt[4]{2} \cdots \sqrt[2^n]{2})$

20. Verify that

$$\lim_{n \to \infty} \left(\frac{1}{\sqrt{n^2 + 1}} + \frac{1}{\sqrt{n^2 + 2}} + \cdots + \frac{1}{\sqrt{n^2 + n}}\right) = 1,$$

while

$$\lim_{n \to \infty} \left(\frac{1}{\sqrt{n^2 + 1^2}} + \frac{1}{\sqrt{n^2 + 2^2}} + \cdots + \frac{1}{\sqrt{n^2 + n^2}}\right)$$
$$= \ln(1 + \sqrt{2}) \approx 0.88.$$

Hint. In the first case use the sandwich theorem for sequences given in Problem 54, page 497; in the second case interpret the left side as the limit of a Riemann sum for the function $1/\sqrt{1 + x^2}$ on $[0, 1]$.

21. Let c be any number such that $0 < c < 1$. Show that the sequence $\{a_n\}$ defined by the recursion formula

$$a_1 = c, \quad a_{n+1} = (2 - a_n)a_n \quad \text{if} \quad n \geq 2,$$

converges to 1.

22. Verify that the general term of the Fibonacci sequence $\{a_n\}$, defined by the recursion formula

$$a_1 = 1, \quad a_2 = 1, \quad a_n = a_{n-1} + a_{n-2} \quad \text{if} \quad n \geq 3$$

(as in Problem 35, page 496) is given by the explicit formula

$$a_n = \frac{1}{\sqrt{5}}\left[\left(\frac{1 + \sqrt{5}}{2}\right)^n - \left(\frac{1 - \sqrt{5}}{2}\right)^n\right].$$

23. Let c be any positive number. Show that the sequence

$$a_1 = \sqrt{c}, \quad a_2 = \sqrt{c + \sqrt{c}}, \quad a_3 = \sqrt{c + \sqrt{c + \sqrt{c}}}, \ldots,$$

defined by the recursion formula

$$a_1 = \sqrt{c}, \quad a_n = \sqrt{c + a_{n-1}} \quad \text{if} \quad n \geq 2,$$

converges to the limit

$$L = \frac{1 + \sqrt{1 + 4c}}{2}.$$

(Notice that $L = 2$ if $c = 2$.)

24. Show that the sequence $\{c_n\}$, defined by the recursion formula

$$c_1 = a, \quad c_2 = b, \quad c_n = \frac{1}{2}(c_{n-1} + c_{n-2}) \quad \text{if} \quad n \geq 3,$$

converges to the limit $\frac{1}{3}(a + 2b)$.
Hint. Notice that $c_n - c_{n-1} = -\frac{1}{2}(c_{n-1} - c_{n-2})$.

25. Let $\{a_n\}$ be a convergent sequence with limit L. Show that

$$\lim_{n \to \infty} \frac{a_1 + a_2 + \cdots + a_n}{n} = L. \qquad \text{(i)}$$

Does the validity of (i) imply that $\{a_n\}$ converges to L?

26. It can be shown that

$$\lim_{n \to \infty} \frac{n!}{(n/e)^n \sqrt{2\pi n}} = 1. \qquad \text{(ii)}$$

This result, known as *Stirling's formula*, leads to the approximation

$$n! \approx \sqrt{2\pi n}\, n^n e^{-n} \qquad \text{(ii')}$$

for n factorial. What is the percentage error of this approximation for $n = 5$? For $n = 10$?

Use formula (ii) to evaluate

27. $\lim_{n \to \infty} \dfrac{n}{\sqrt[n]{n!}}$

28. $\lim_{n \to \infty} \sqrt[n^2]{n!}$

29. $\lim_{n \to \infty} \dfrac{3^n n!}{n^n}$

30. $\lim_{n \to \infty} \dfrac{7^{n/2} n!}{n^n}$

31. $\lim_{n \to \infty} \dfrac{(n!)^2 2^{2n}}{(2n)! \sqrt{n}}$

32. $\lim_{n \to \infty} \left(\dfrac{1}{2} \dfrac{3}{4} \cdots \dfrac{2n-1}{2n} \right)$

Use formula (ii') to approximate

33. $\ln 40!$

34. $1 \cdot 3 \cdot 4 \cdots 49$

35. $\int_0^{\pi/2} \sin^{31} x \, dx$

36. $\int_0^\infty x^{50} e^{-x} \, dx$

37. $\int_0^\infty \dfrac{dx}{(x^2+1)^{11}}$

38. $\int_0^1 x^{15}(1-x)^{16} \, dx$

If the given series is convergent, find its sum. Otherwise identify it as divergent. In the case of a series specified by giving a few initial terms, assume that the law of formation suggested by these terms holds for all the terms of the series.

39. $\sum_{n=0}^\infty \left(\dfrac{\pi}{3} \right)^n$

40. $\sum_{n=0}^\infty \left(\dfrac{3}{\pi} \right)^n$

41. $\dfrac{1}{1 \cdot 4} + \dfrac{1}{4 \cdot 7} + \dfrac{1}{7 \cdot 10} + \cdots$

42. $\dfrac{3}{(1 \cdot 2)^2} + \dfrac{5}{(2 \cdot 3)^2} + \dfrac{7}{(3 \cdot 4)^2} + \cdots$

43. $\sum_{n=1}^\infty \dfrac{1}{(c+n)(c+n+1)}$

44. $\sum_{n=1}^\infty \dfrac{1}{(c+n)(c+n+1)(c+n+2)}$

45. $\sum_{n=1}^\infty (\sqrt{n+2} - 2\sqrt{n+1} + \sqrt{n})$

46. $\dfrac{1+2}{1-2} + \dfrac{1+2+4}{1-2+4} + \dfrac{1+2+4+8}{1-2+4-8} + \cdots$

47. The series $(\tfrac{3}{2} - 1) + (\tfrac{5}{4} - 1) + (\tfrac{9}{8} - 1) + \cdots$ is another way of writing the convergent geometric series $\tfrac{1}{2} + \tfrac{1}{4} + \tfrac{1}{8} + \cdots$. Show that the series obtained by removing the parentheses is divergent.

48. The *harmonic mean* of two positive numbers x and y is the number h such that

$$\dfrac{1}{h} = \dfrac{1}{2}\left(\dfrac{1}{x} + \dfrac{1}{y}\right),$$

that is, the reciprocal of h is the average (or arithmetic mean) of the reciprocals of x and y. Show that every term of the harmonic series

$$1 + \dfrac{1}{2} + \dfrac{1}{3} + \cdots + \dfrac{1}{n} + \cdots$$

except the first is the harmonic mean of the two adjacent terms.

49. Given two positive numbers x and y, let $g = \sqrt{xy}$ be their geometric mean, and let h be their harmonic mean, as defined in the preceding problem. Verify that $h < g$ unless $x = y$, in which case $h = g$. (Compare this with Problem 19, page 13.)

50. Let n be a positive integer. Show that $1/n$ is represented by a terminating decimal if and only if there are nonnegative integers p and q such that $n = 2^p 5^q$.

Find the rational number in lowest terms represented by the decimal

51. $0.1234\overline{5}$

52. $0.12\overline{345}$

53. $0.\overline{12345}$

54. Show that the decimal $0.12345678910111213\ldots$, obtained by writing all the positive integers in order after the decimal point, represents an irrational number.

Use any test to investigate the convergence of the given series. In the case of a series specified by giving a few initial terms, assume that the law of formation suggested by these terms holds for all the terms of the series. If the series has both positive and negative terms, distinguish between absolute and conditional convergence.

55. $\sum_{n=1}^\infty (-1)^{n-1} \dfrac{2^n n!}{n^n}$

56. $\sum_{n=1}^\infty \dfrac{(-1)^{n-1}}{(2n-1)^2}$

57. $1 + \dfrac{1}{2} - \dfrac{1}{3} + \dfrac{1}{4} + \dfrac{1}{5} - \dfrac{1}{6} + \cdots$

58. $\sum_{n=2}^\infty \dfrac{1}{\sqrt[n]{\ln n}}$

59. $1 + \dfrac{1}{2} - \dfrac{2}{3} + \dfrac{1}{4} + \dfrac{1}{5} - \dfrac{2}{6} + \cdots$

60. $\sum_{n=3}^\infty \dfrac{(-1)^n}{\ln(\ln n)}$

61. $\sum_{n=1}^\infty \dfrac{n}{(n+1)^n}$

62. $\dfrac{5}{2} + \dfrac{5 \cdot 8}{2 \cdot 6} + \dfrac{5 \cdot 8 \cdot 11}{2 \cdot 6 \cdot 10} + \cdots$

63. $100 - \dfrac{100 \cdot 101}{1 \cdot 3} + \dfrac{100 \cdot 101 \cdot 102}{1 \cdot 3 \cdot 5} - \cdots$

64. $\sum_{n=1}^\infty \dfrac{n^{n+(1/n)}}{[n+(1/n)]^n}$

65. $\sum_{n=1}^\infty \dfrac{\arctan n}{\tanh n}$

66. $\sum_{n=1}^\infty (-1)^n \sin n$

67. $\sum_{n=1}^\infty \dfrac{\pi^n n!}{n^n}$

572 Chapter 9 Sequences and Series

68. $\dfrac{1}{2} - \dfrac{3}{2^2} + \dfrac{1}{2^3} - \dfrac{3}{2^4} + \dfrac{1}{2^5} - \dfrac{3}{2^6} + \cdots$

69. $\displaystyle\sum_{n=2}^{\infty} (-1)^n \dfrac{1}{\ln(n!)}$

70. $\displaystyle\sum_{n=1}^{\infty} \left(1 + \dfrac{1}{2} + \cdots + \dfrac{1}{n}\right)^n$

71. Let d_n be the number of positive integers which are (exact) divisors of n. For example, $d_4 = 3$ since 4 has divisors 1, 2 and 4, $d_5 = 2$ since 5 has divisors 1 and 5, $d_6 = 4$ since 6 has divisors 1, 2, 3 and 6, and so on. Find the radius of convergence and interval of convergence of the power series $\sum d_n x^n = d_1 x + d_2 x^2 + \cdots + d_n x^n + \cdots$.

Expand the given function in a power series at $x = 0$.

72. $\dfrac{1-x}{1+x^2}$

73. $\dfrac{1}{1+x+x^2}$

74. $\dfrac{1}{(1-x)(1-x^2)}$

75. $\dfrac{x^2}{(1-x^3)^2}$

76. $(1-x^2)^{-3/2}$

77. $\dfrac{1}{1+x-2x^2}$

78. $\ln \dfrac{1+x}{1-x} + 2 \arctan x$

79. $[\ln(1-x)]^2$

80. $(\arctan x)^2$

81. Show that

$$\sum_{n=0}^{\infty} \dfrac{(-1)^n}{na+1} = 1 - \dfrac{1}{a+1} + \dfrac{1}{2a+1} - \dfrac{1}{3a+1} + \cdots$$

$$= \int_0^1 \dfrac{dx}{x^a+1} \quad (a>0). \qquad \text{(iii)}$$

Use formula (iii) to find the sum of the series

82. $1 - \dfrac{1}{4} + \dfrac{1}{7} - \dfrac{1}{10} + \cdots$

83. $1 - \dfrac{1}{5} + \dfrac{1}{9} - \dfrac{1}{13} + \cdots$

84. Starting from Machin's formula

$$\pi = 16 \arctan \dfrac{1}{5} - 4 \arctan \dfrac{1}{239}$$

(see Problem 62, page 297), use Gregory's series for $\arctan x$ to approximate π to eight decimal places.

85. Show that

$$\dfrac{1}{1-x-x^2} = \sum_{n=1}^{\infty} a_n x^{n-1}, \qquad \text{(iv)}$$

where $\{a_n\}$ is the Fibonacci sequence.

86. With the help of Problem 22, show that the radius of convergence of the series (iv) is $\tfrac{1}{2}(\sqrt{5}-1) \approx 0.618$.

87. Let a be any positive number. Verify that

$$\ln a = 2\left(b + \dfrac{b^3}{3} + \dfrac{b^5}{5} + \dfrac{b^7}{7} + \cdots\right),$$

where

$$b = \dfrac{a-1}{a+1},$$

and use this formula to approximate $\ln 3$ to four decimal places.

88. According to Einstein's special theory of relativity (see Problem 54, page 276), the total energy of a particle of mass m moving with velocity v is given by

$$E = \dfrac{mc^2}{\sqrt{1-(v/c)^2}},$$

where c is the velocity of light ($\approx 300{,}000$ km/sec). Show that if v is small compared to c, then

$$E = mc^2 + K,$$

to an excellent approximation, where $K = \tfrac{1}{2}mv^2$ is the Newtonian kinetic energy of the particle (see page 269). Notice that both formulas for E assign energy mc^2 to a particle of mass m at rest ($v=0$), thereby expressing the "equivalence of mass and energy."

89. Let $P(x)$ be the polynomial of degree 4 such that

$$P(2) = -1, \qquad P'(2) = 0, \qquad P''(2) = 2,$$
$$P'''(2) = -12, \qquad P^{(4)} = 24.$$

Find $P(-1)$, $P'(0)$ and $P''(1)$.

90. Estimate $\sin 55°$ by using Taylor's formula with $n=3$ and $a = 60° = \pi/3$, and show that the answer is accurate to five decimal places.

91. Let $P_n(x)$ be the nth Taylor polynomial of the function $f(x)$ at the point $x = a$, and let $R_n(x)$ be the corresponding remainder, so that $R_n(x) = f(x) - P_n(x)$. Show that $R_n(x)$ can be written in the *integral form*

$$R_n(x) = \dfrac{1}{n!} \int_a^x f^{(n+1)}(u)(x-u)^n\, du \qquad \text{(v)}$$

if $f^{(n+1)}(u)$ is continuous on the interval with endpoints a and x. Show that (v) implies

$$R_n(x) = \dfrac{f^{(n+1)}(t)}{(n+1)!}(x-a)^{n+1} \qquad (t \text{ between } a \text{ and } x),$$

the form of the remainder given on page 551.
Hint. Use the generalized mean value theorem for integrals (see Problem 27, page 275).

92. Use Taylor series to evaluate the fifth derivative of $x^2 \sqrt[4]{1+x}$ at $x = 0$

93. Use Taylor series to evaluate the tenth derivative of $x^6 e^x$ at $x = 0$

94. What is the Maclaurin series of a polynomial $P(x)$?

95. Find the first six terms of the Maclaurin series of e^{2x-x^2}.

Hint. Observe that
$$e^{2x-x^2} = (e^{2x})(e^{-x^2}).$$

Use power series to approximate the given integral to an accuracy of four decimal places.

96. $\int_0^1 \dfrac{\sin x}{x}\, dx$ **97.** $\int_0^1 \dfrac{1 - \cos x}{x}\, dx$

In each case define the integrand $f(x)$ at $x = 0$ by continuity, that is, set $f(0) = \lim\limits_{x \to 0} f(x)$.

Use Newton's method to solve the given equation to an accuracy of three decimal places.

98. $x \ln x = 1$ **99.** $xe^{x^2} = 1$
100. $x + \arctan x - 1 = 0$

10 Plane Analytic Geometry and Polar Coordinates

Analytic geometry, invented by René Descartes, is the subject in which algebraic methods are used to solve geometric problems, after introducing a suitable coordinate system. We now combine the techniques of analytic geometry and calculus to continue our study of curves in the plane. In Section 10.1 we begin by showing how the equation of a curve in rectangular coordinates changes when the coordinate system is shifted or rotated. Then in Sections 10.2–10.4 we make a detailed investigation of parabolas, ellipses and hyperbolas. These curves are of great importance in applied mathematics, and as explained in Section 10.5, each is a "conic section," that is, the intersection of a double right circular cone with a suitable cutting plane. Conic sections were a favorite topic of the ancient Greek geometers. In fact, Apollonius of Perga (255–170 B.C.) wrote a treatise on the subject, containing about 400 propositions, and it is to him that the terms parabola, ellipse and hyperbola are due. The conic sections all have second degree equations of the form $Ax^2 + Bxy + Cy^2 + Dx + Ey + F = 0$, and conversely the graph of any such equation is a conic section if allowance is made for certain exceptional cases described in Section 10.6.

In a cartesian coordinate system the position of a point P is specified by giving its rectangular coordinates. This amounts to indicating two lines, one horizontal and the other vertical, which intersect at P. There are also nonrectangular coordinate systems, and one of the most useful of these is the *polar coordinate system* studied in Sections 10.7–10.10. Here the position of a point P is specified by giving its polar coordinates, which amounts to indicating both a circle and a radius of the circle which intersect at P. As it happens, polar coordinates are particularly appropriate for writing the equations of conic sections.

10.1 Translation and Rotation of Axes

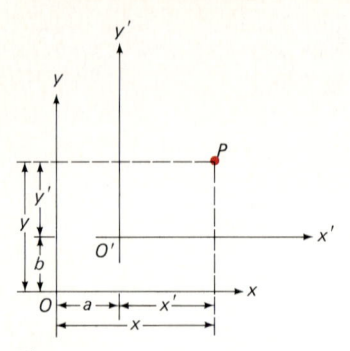

Figure 1

Figure 1 shows a rectangular coordinate system Oxy which has been *translated*, that is, shifted as a whole without rotation, until its origin O occupies a new position $O' = (a, b)$, thereby producing another rectangular coordinate system $O'x'y'$ in the same plane as Oxy. A fixed point P in the plane now has two pairs of coordinates, a pair (x, y) in the "old" xy-system and a pair (x', y') in the "new" $x'y'$-system. It is apparent from the figure that the relation between these coordinates is given by

$$x = x' + a, \qquad y = y' + b, \qquad (1)$$

or equivalently

$$x' = x - a, \qquad y' = y - b. \qquad (1')$$

Equations (1) and (1') are called the *translation equations*.

Example 1 Suppose the origin of the new $x'y'$-system is at the point $(a, b) = (3, -2)$ in the old xy-system. What are the new coordinates of the point P with old coordinates $x = -1$, $y = 4$?

Solution It follows from (1') with $a = 3$, $b = -2$ that

$$x' = x - 3, \qquad y' = y + 2.$$

Thus the new coordinates of the point P are

$$x' = -1 - 3 = -4, \qquad y' = 4 + 2 = 6,$$

as illustrated in Figure 2. □

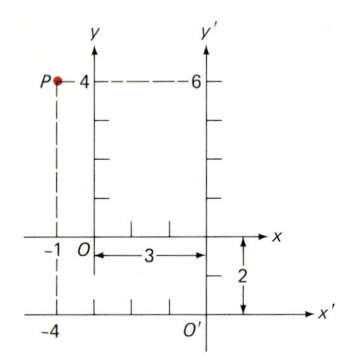

Figure 2

Example 2 By translating axes, find a system of coordinates x', y' in which the equation of the circle

$$(x + 2)^2 + (y - 3)^2 = 1$$

simplifies to

$$x'^2 + y'^2 = 1.$$

Solution The appropriate translation equations are clearly

$$x' = x + 2, \qquad y' = y - 3.$$

The new $x'y'$-system is the result of translating the old xy-system 2 units to the left and 3 units upward. This is to be expected, since $O'x'y'$ is the system in which the circle $(x + 2)^2 + (y - 3)^2 = 1$ of radius 1 centered at the point $(-2, 3)$ becomes the circle of radius 1 centered at the origin (see Figure 3). Observe also that the same translation carries the circle $x^2 + y^2 = 1$ (the dashed curve) into the circle $(x + 2)^2 + (y - 3)^2 = 1$ if the xy-system is held fixed. □

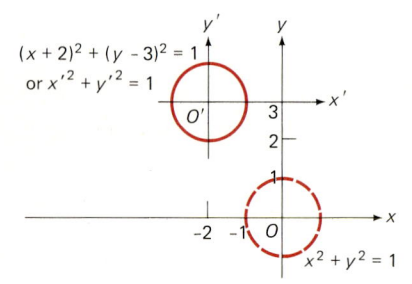

Figure 3

Rotation of Axes

Next we allow *rotations* of one coordinate system with respect to another. Figure 4(a) shows a rectangular coordinate system Oxy which has been rotated about its origin O through an angle θ in the counterclockwise direction, thereby producing another rectangular coordinate system $Ox'y'$ in the same plane as Oxy and with the same origin O. A fixed point P in the plane now has two pairs of coordinates, a pair (x, y) in the old xy-system and a

576   Chapter 10   Plane Analytic Geometry and Polar Coordinates

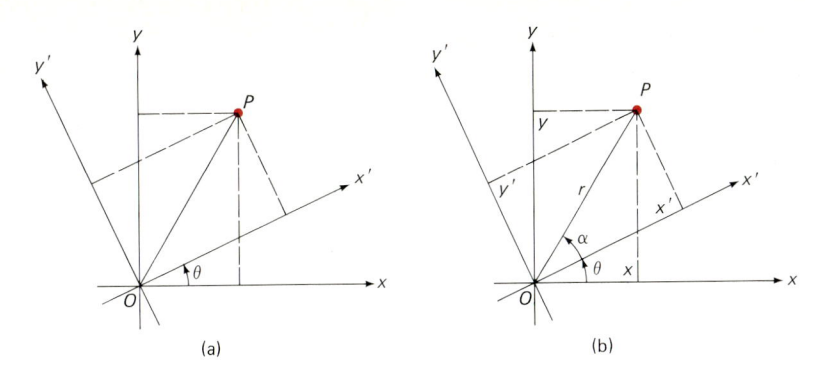

Figure 4

pair (x', y') in the new $x'y'$-system. To find the relation between these coordinates, let r be the length of the segment OP, and let α be the angle between OP and the x'-axis, measured in the counterclockwise direction. Then the angle between OP and the x-axis is $\theta + \alpha$, so that

$$x = r \cos(\theta + \alpha) = r \cos \theta \cos \alpha - r \sin \theta \sin \alpha,$$

$$y = r \sin(\theta + \alpha) = r \sin \theta \cos \alpha + r \cos \theta \sin \alpha.$$

But

$$x' = r \cos \alpha, \qquad y' = r \sin \alpha$$

[see Figure 4(b)], and hence the equations for x and y simplify to

$$x = x' \cos \theta - y' \sin \theta,$$
$$y = x' \sin \theta + y' \cos \theta. \tag{2}$$

The equations (2) express the old coordinates x and y in terms of the new coordinates x' and y'. The corresponding equations expressing x' and y' in terms of x and y are

$$x' = x \cos \theta + y \sin \theta,$$
$$y' = -x \sin \theta + y \cos \theta. \tag{2'}$$

We can obtain (2') by solving the pair of simultaneous equations (2) for x' and y', but it is much easier to recognize that if the system $Ox'y'$ makes the angle θ with the system Oxy, then Oxy makes the angle $-\theta$ with $Ox'y'$, so that the pairs (x, y) and (x', y') can be interchanged in (2), provided that θ is also changed to $-\theta$. As an exercise, verify that this does in fact transform (2) into (2') and vice versa. Equations (2) and (2') are called the *rotation equations*.

Example 3 Let P be the point $(1, 2)$, and suppose the xy-system is rotated through an angle of $30°$ in the counterclockwise direction, producing a new $x'y'$-system. What are the new coordinates of P?

Solution It follows from (2') with $\theta = 30°$ that

$$x' = x \cos 30° + y \sin 30° = \frac{\sqrt{3}}{2} x + \frac{1}{2} y,$$

$$y' = -x \sin 30° + y \cos 30° = -\frac{1}{2} x + \frac{\sqrt{3}}{2} y.$$

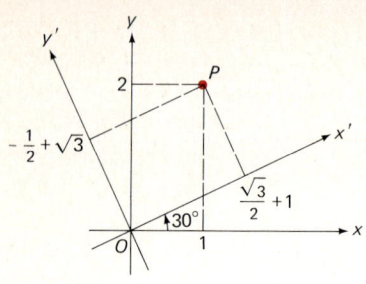

Figure 5

Substituting $x = 1$, $y = 2$ into these equations, we find that the new coordinates of the point P are

$$x' = \frac{\sqrt{3}}{2} + 1, \qquad y' = -\frac{1}{2} + \sqrt{3}$$

(see Figure 5). □

Example 4 Show that the equation

$$2xy = 1$$

transforms into the equation

$$x'^2 - y'^2 = 1$$

of the unit hyperbola (see page 353) if the $x'y'$-system is obtained by rotating the xy-system through an angle of 45° in the counterclockwise direction, as in Figure 6.

Solution Choosing $\theta = 45°$ in the rotation equations (2), we get

$$x = x' \cos 45° - y' \sin 45° = \frac{1}{\sqrt{2}}(x' - y'),$$

$$y = x' \sin 45° + y' \cos 45° = \frac{1}{\sqrt{2}}(x' + y').$$

Therefore

$$2xy = (x' - y')(x' + y') = x'^2 - y'^2,$$

so that $2xy = 1$ transforms into $x'^2 - y'^2 = 1$ in the $x'y'$-system. Observe also that the same 45° rotation carries the hyperbola $x^2 - y^2 = 1$ (the dashed graph in the figure) into the hyperbola $2xy = 1$ if the xy-system is held fixed. □

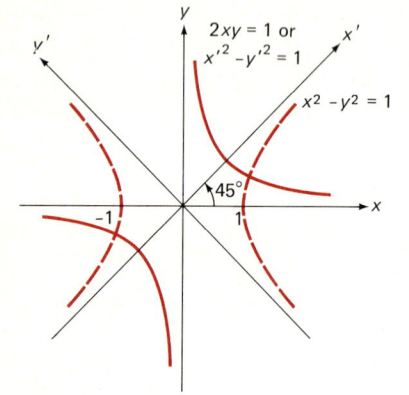

Figure 6

The Graph of F(x, y) = 0

Let $F(x, y)$ be any expression involving the two variables x and y. (Here, as on page 409, we anticipate the notation for functions of two variables.) Then by the graph of the equation

$$F(x, y) = 0, \qquad (3)$$

we mean the set of all points (x, y) in the xy-plane whose coordinates satisfy (3). For example, if $F(x, y) = x^2 + y^2 - 1$, the graph of (3) is the circle of unit radius centered at the origin, while if $F(x, y) = y - f(x)$, the graph of (3) is just the graph of the function $y = f(x)$. Let G be the graph of equation (3) and G' the graph of the equation

$$F(x - a, y - b) = 0. \qquad (3')$$

Then the point (x, y) satisfies (3) if and only if the point $(x + a, y + b)$ satisfies (3'). But $(x + a, y + b)$ is the result of subjecting (x, y) to a horizontal shift of $|a|$ units, to the right if $a > 0$ and to the left if $a < 0$, and a vertical shift of $|b|$ units, upward if $b > 0$ and downward if $b < 0$. Therefore G' is obtained by subjecting G to the same two shifts. For example, if the circle $x^2 + y^2 = 1$ is shifted 2 units to the left and 3 units upward, we get the circle $(x + 2)^2 + (y - 3)^2 = 1$, as already noted (see Figure 3).

578   Chapter 10 Plane Analytic Geometry and Polar Coordinates

Symmetry Tests

Next we consider possible symmetries of the graph G of an equation $F(x, y) = 0$. Suppose G is symmetric about the x-axis. This means that a point (a, b) belongs to G if and only if $(a, -b)$ also belongs to G, or equivalently that $F(a, b) = 0$ if and only if $F(a, -b) = 0$. In other words, G is *symmetric about the x-axis* if and only if the two equations $F(x, y) = 0$ and $F(x, -y) = 0$ have the same set of solutions. Similarly, G is *symmetric about the y-axis* if and only if $F(x, y) = 0$ and $F(-x, y) = 0$ have the same set of solutions, while G is *symmetric about the origin* if and only if $F(x, y) = 0$ and $F(-x, -y) = 0$ have the same set of solutions. In the same way, G is *symmetric about the line $y = x$* if and only if the equations $F(x, y) = 0$ and $F(y, x) = 0$ have the same set of solutions.

Example 5 The graph of the equation

$$(x - 1)^2 + (y - 1)^2 = 1 \qquad (4)$$

is not symmetric about either of the coordinate axes or the origin. In fact, replacing x by $-x$ in (4) gives the equation

$$(x + 1)^2 + (y - 1)^2 = 1,$$

replacing y by $-y$ gives

$$(x - 1)^2 + (y + 1)^2 = 1,$$

and replacing both x by $-x$ and y by $-y$ gives

$$(x + 1)^2 + (y + 1)^2 = 1,$$

where none of the last three equations has the same set of solutions as (4). For instance $x = 2$, $y = 1$ is a solution of equation (4), but not of the other equations. The fact that all four equations have different solution sets is immediately apparent from their graphs, shown in Figures 7(a)–7(d). The graphs are all circles of radius 1, but their centers are in different quadrants. However, the graph of (4) is symmetric about the line $y = x$, since interchanging x and y in (4) gives the equation

$$(y - 1)^2 + (x - 1)^2 = 1,$$

which is equivalent to (4), and therefore has the same set of solutions. This symmetry is to be expected, since (4) is the equation of a circle with its center on the line $y = x$.

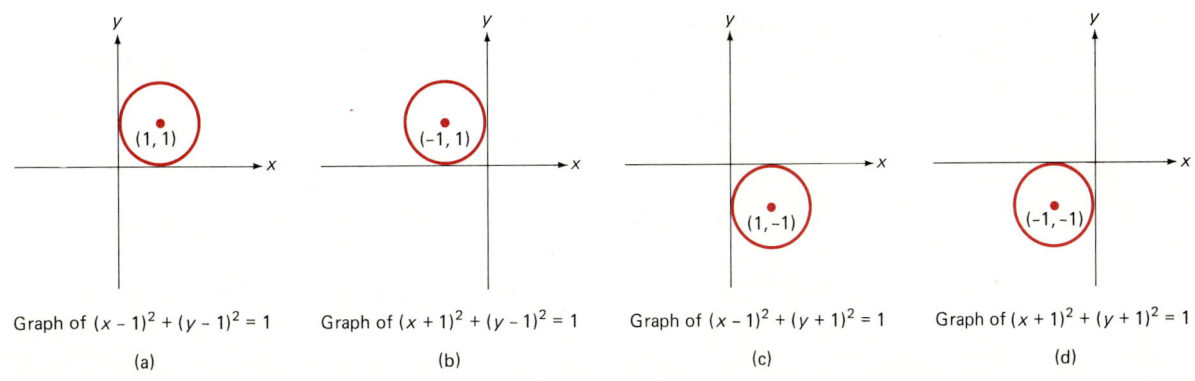

Graph of $(x - 1)^2 + (y - 1)^2 = 1$
(a)

Graph of $(x + 1)^2 + (y - 1)^2 = 1$
(b)

Graph of $(x - 1)^2 + (y + 1)^2 = 1$
(c)

Graph of $(x + 1)^2 + (y + 1)^2 = 1$
(d)

Figure 7

Section 10.1 Translation and Rotation of Axes

Distinct equations can have the same set of solutions. Thus the equations $x + y = 0$ and $x^2 + 2xy + y^2 = (x + y)^2 = 0$ have exactly the same set of solutions, namely all the points of the line $y = -x$. It should be noted that equations with different *sets* of solutions sometimes share certain solutions. For instance, the first two equations in Example 5 share the solution $x = 0$, $y = 1$, but no others. How do you explain this geometrically?

Two symmetries may imply a third. Thus a graph must be symmetric about the origin if it is symmetric about both coordinate axes, and it must be symmetric about one coordinate axis if it is symmetric about the other axis and the origin (verify these statements).

Problems

Suppose the xy-system is translated by shifting its origin to the given point. Write the corresponding translation equations.

1. $(0, 10)$ **2.** $(-5, 0)$ **3.** $(1, -2)$ **4.** $(-6, 4)$

Find the old coordinates of the new origin after making the given translation of axes.

5. $x' = x - 3$, $y' = y - 5$ **6.** $x' = x + 2$, $y' = y - 1$
7. $x' = x$, $y' = y + 1$ **8.** $x' = x + 5$, $y' = y$

9. Suppose the xy-system is translated 3 units to the right and 4 units downward, thereby producing a new $x'y'$-system. Let $A = (1, 3)$, $B = (-3, 0)$ and $C = (-1, 4)$ be three points in the new system. What are the old coordinates of A, B, and C?

Find the new coordinates of the points $A = (2, 1)$, $B = (-1, 3)$ and $C = (-2, 5)$ after shifting the origin of the xy-system to

10. The point A **11.** The point B **12.** The point C

Let G be the graph of the given equation. Find a translation of axes simplifying the equation, and then describe G.

13. $x^2 - 4x - 4y - 8 = 0$
14. $x^2 + y^2 + 10x - 12y + 12 = 0$
15. $xy - x + 2y - 3 = 0$
16. $3x^2 + 2y^2 - 6x + 8y + 5 = 0$

Suppose the xy-system is rotated about its origin through the given angle (in the counterclockwise direction). Write the corresponding rotation equations.

17. $90°$ **18.** $-30°$ **19.** $60°$ **20.** $135°$

Find the new coordinates of the points $A = (3, 1)$, $B = (-1, 5)$ and $C = (2, -3)$ after rotating the xy-plane about its origin through the given angle.

21. 3π **22.** $-90°$ **23.** $\arctan \frac{5}{12}$ **24.** $210°$

25. Suppose the xy-system is rotated about its origin through $60°$, thereby producing a new $x'y'$-system. Let $A = (2\sqrt{3}, -4)$, $B = (\sqrt{3}, 0)$ and $C = (0, -2\sqrt{3})$ be three points in the new system. What are the old coordinates of A, B and C?

Let G be the graph of the given equation. Simplify the equation by rotating the xy-system through $45°$ (in the counterclockwise direction), and then describe G.

26. $3x^2 - 2xy + 3y^2 - 8 = 0$
27. $x^2 + 6xy + y^2 - 2 = 0$
28. $x^2 + 2xy + y^2 + x - y = 0$

29. Suppose the ellipse $\frac{1}{2}x^2 + y^2 = 1$ is shifted 3 units to the left and 4 units upward. What is the equation of the new ellipse?

30. Suppose the hyperbola $x^2 - y^2 = 1$ is shifted 2 units to the right and 5 units downward. What is the equation of the new hyperbola?

Let G be the graph of the given equation. Which of the four symmetries discussed on page 579 does G have?

31. $x + y = 1$ **32.** $xy = 1$
33. $|x| + |y| = 1$ **34.** $(2 - x)x^2 - y^2 = 0$
35. $x^2y + y - 2x = 0$ **36.** $x^2 + 2y^2 = 1$

Characterize every function $y = f(x)$ whose graph is symmetric about

37. The x-axis   **38.** The y-axis
39. The origin   **40.** The line $y = x$

41. Find the new coordinates of the points $A = (5, 5)$, $B = (2, -1)$ and $C = (12, -6)$ after shifting the origin of the xy-system to the point B and rotating the axes through the angle $\arctan \frac{3}{4}$ about B.

580   Chapter 10   Plane Analytic Geometry and Polar Coordinates

10.2 Parabolas

Definition of a Parabola

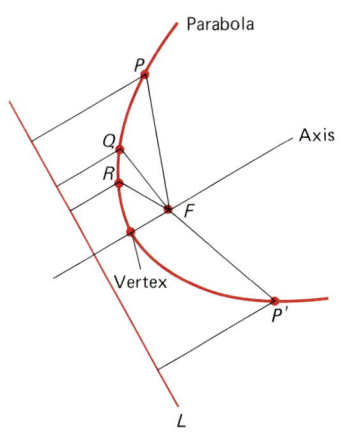

Figure 8

By definition, *a parabola is the set of all points in a plane which are equidistant from a fixed point F and a fixed line L not containing F*. Thus, for example, the point P in Figure 8 is equidistant from F and L, and the same is true of the points Q and R. The point F is called the *focus* and the line L the *directrix* of the parabola. It is clear that the parabola is symmetric about the line through F perpendicular to L (compare the positions of the points P and P' in the figure). This line is called the *axis of symmetry*, or simply the *axis*, of the parabola, and the unique point in which it intersects the parabola is called the *vertex* of the parabola. Notice that the vertex is halfway between the focus F and the directrix L.

To find an equation of the parabola, we introduce rectangular coordinates x, y in the plane of the parabola and place it in "standard position," with its vertex at the origin O and its axis of symmetry along the y-axis. Then the focus F is a point $(0, c)$ of the y-axis, and the directrix L is the line $y = -c$ (so that O is halfway between F and L). If $c > 0$, the parabola opens *upward* as in Figure 9(a), while if $c < 0$, it opens *downward* as in Figure 9(b). Let $P = (x, y)$ be any point of the parabola, and let $Q = (x, -c)$ be the foot of the perpendicular drawn from P to the directrix L. Then, by the defining property of the parabola,

$$|PF| = |PQ|,$$

which takes the form

$$\sqrt{x^2 + (y - c)^2} = \sqrt{(y + c)^2}$$

in terms of the coordinates of P and F. Squaring both sides of this equation, we get

$$x^2 + y^2 - 2cy + c^2 = y^2 + 2cy + c^2,$$

which simplifies to

$$x^2 = 4cy, \tag{1}$$

or

$$y = \frac{x^2}{4c}$$

if we write y as a function of x. Conversely, if the point $P = (x, y)$ is such that (1) holds, then by reversing the steps of this calculation, we find that $|PF| = |PQ|$. Therefore (1) is an equation of the parabola in the given position.

Another standard position of the parabola is to place it with its vertex at the origin O as before, but with its axis of symmetry along the x-axis instead of the y-axis. The focus F is now a point $(c, 0)$ of the x-axis, and the directrix L is the line $x = -c$. Virtually the same argument as before shows that an equation of the parabola in this position is

$$y^2 = 4cx, \tag{2}$$

or

$$x = \frac{y^2}{4c}$$

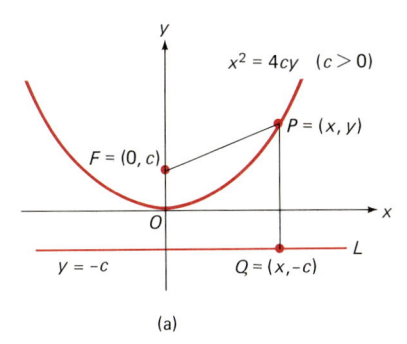

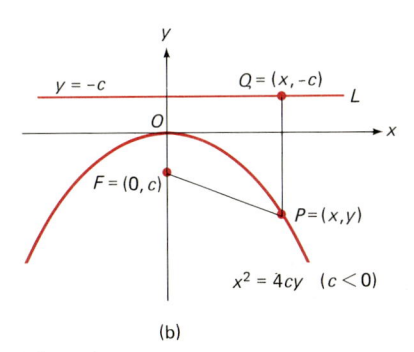

Figure 9

Section 10.2 Parabolas   **581**

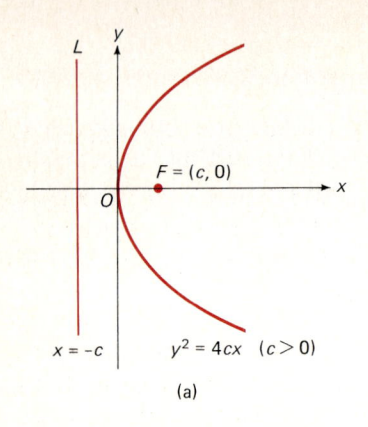

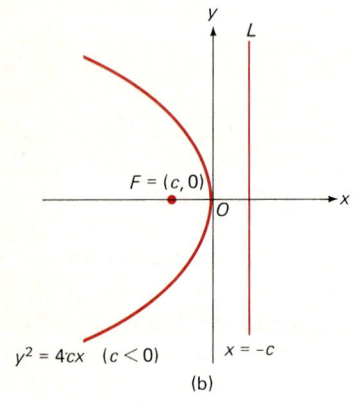

if we write x as a function of y. If $c > 0$, the parabola opens *to the right* as in Figure 10(a), while if $c < 0$, it opens *to the left* as in Figure 10(b). Observe that each of the equations (1) and (2) can be obtained from the other by interchanging the variables x and y.

Example 1 Find the graph of the equation $y^2 = 3x$.

Solution The equation is of the form (2) with $c = \frac{3}{4}$, and hence corresponds to a parabola with its vertex at the origin and the x-axis as its axis of symmetry (see Figure 11). The parabola opens to the right (since $c > 0$), its focus F is the point $(\frac{3}{4}, 0)$, and its directrix is the line $x = -\frac{3}{4}$. ☐

Equations in Standard Form

Each of the equations (1) and (2) represents a parabola with its vertex at the origin and a coordinate axis as its axis of symmetry. Suppose we subject each of these parabolas to a shift carrying its vertex into the point (a, b). Then equations (1) and (2) transform into

$$(x - a)^2 = 4c(y - b) \qquad (3)$$

and

$$(y - b)^2 = 4c(x - a) \qquad (4)$$

(recall equation (3'), page 578). In detail, (3) is an equation of the parabola with vertex (a, b), focus-to-vertex distance $|c|$, and the line $x = a$ as its axis, which opens upward if $c > 0$ and downward if $c < 0$, while (4) is an equation of the parabola with vertex (a, b), focus-to-vertex distance $|c|$, and the line $y = b$ as its axis, which opens to the right if $c > 0$ and to the left if $c < 0$. These equations are called the *equations of a parabola in standard form*. Naturally, (3) and (4) reduce to (1) and (2) in the special case where $a = b = 0$.

Example 2 Find the graph of the equation

$$x^2 - 2x + 8y + 9 = 0. \qquad (5)$$

Solution Completing the square, we find that (5) can be written as

$$(x - 1)^2 + 8(y + 1) = 0$$

582 Chapter 10 Plane Analytic Geometry and Polar Coordinates

or
$$(x - 1)^2 = -8(y + 1),$$

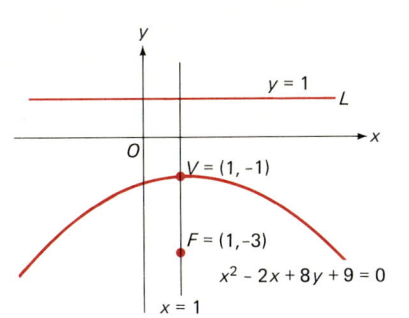

which is in the standard form (3) with $a = 1$, $b = -1$, $c = -2$. This is an equation of the parabola with vertex $V = (1, -1)$ and the line $x = 1$ as its axis (see Figure 12). The parabola opens downward (since $c < 0$), its focus is $F = (1, -3)$, and its directrix L is the line $y = 1$. To locate F and L, we have used the fact that the focus-to-directrix distance is $|c| = 2$, so that F is 2 units below the vertex V, while L is 2 units above V. □

Each of the equations (3) and (4) has the property of being quadratic in one of the two variables x and y and linear in the other, with no term containing the product xy. Conversely, any equation with this property can be written as either

$$x^2 + Ax + By + C = 0 \quad (B \neq 0) \tag{6}$$

or

$$y^2 + Ax + By + C = 0 \quad (A \neq 0) \tag{7}$$

by first dividing it by the coefficient of the quadratic term. But, just as in Example 2, we can bring each of the equations (6) and (7) into one of the standard forms (3) and (4) by completing the square, and hence the graph of each equation is a parabola with its axis parallel to one of the coordinate axes. In fact, completing the squares in (6) and (7), we get the equivalent equations

$$\left(x + \frac{A}{2}\right)^2 = -B\left(y + \frac{C}{B} - \frac{A^2}{4B}\right), \tag{6'}$$

and

$$\left(y + \frac{B}{2}\right)^2 = -A\left(x + \frac{C}{A} - \frac{B^2}{4A}\right), \tag{7'}$$

where (6') is of the form (3) and (7') is of the form (4). Thus a parabola has an equation of the form (6) or (7) if and only if its axis is parallel to one of the coordinate axes. In the next example we consider a parabola with an *oblique* axis.

Example 3 Find an equation of the parabola with the point $(1, 1)$ as its focus F and the line

$$x + y + 2 = 0 \tag{8}$$

as its directrix L.

Solution The point $P = (x, y)$ belongs to the parabola if and only if

$$\sqrt{(x - 1)^2 + (y - 1)^2} = \frac{|x + y + 2|}{\sqrt{2}},$$

where the left side is the distance between P and F, and the right side is the distance between P and the line (8), calculated with the help of Theorem 10, page 34. Squaring both sides of this equation, we obtain

$$(x - 1)^2 + (y - 1)^2 = \frac{1}{2}(x + y + 2)^2.$$

Section 10.2 Parabolas 583

It follows that
$$2(x^2 - 2x + 1 + y^2 - 2y + 1) = x^2 + 2xy + y^2 + 4(x+y) + 4$$
or
$$x^2 - 2xy + y^2 - 8x - 8y = 0, \qquad (9)$$
which can also be written as
$$(x-y)^2 = 8(x+y). \qquad (10)$$

Notice that equation (9) is *not* of the form (6) or (7), and in fact has a term containing the product xy. Figure 13 shows the parabola in question, which has the origin as its vertex and the *oblique* line $y = x$ as its axis. Algebraically, the symmetry of the parabola about the line $y = x$ corresponds to the fact that equation (9) or (10) remains the same if the variables x and y are interchanged. ☐

Example 4 Figure 13 suggests that the parabola (10) is the result of rotating the parabola
$$y^2 = 4\sqrt{2}x \qquad (11)$$
about its vertex through an angle of 45° in the counterclockwise direction. Verify this algebraically.

Solution Choosing $\theta = 45°$ in the rotation equations (2), page 577, we get
$$x = \frac{1}{\sqrt{2}}(x' - y'), \qquad y = \frac{1}{\sqrt{2}}(x' + y'),$$
as in Example 4, page 578. Substitution of these expressions into (10) gives
$$\left[\frac{1}{\sqrt{2}}(-2y')\right]^2 = \frac{8}{\sqrt{2}}(2x'),$$
or equivalently
$$y'^2 = 4\sqrt{2}x', \qquad (11')$$
which is the same as (11) except that x and y are primed. Thus equation (10) simplifies to (11') in an $x'y'$-system obtained by rotating the xy-system through an angle of 45° in the counterclockwise direction. Correspondingly, a counterclockwise rotation of the parabola (11) through 45° about the origin (its vertex) carries it into the parabola (10). ☐

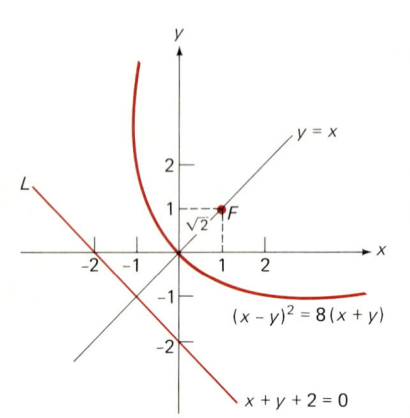

Figure 13

Reflection Property of the Parabola

An interesting property of the parabola, leading to a variety of practical applications, is its *reflection property*. Let T be the tangent line at any point $P = (x_0, y_0)$ of the parabola
$$y^2 = 4cx \qquad (c > 0), \qquad (12)$$
with focus $F = (c, 0)$, and let Q be the point in which T intersects the x-axis [see Figure 14(a)]. Differentiating (12) with respect to x, we get
$$2y\frac{dy}{dx} = 4c$$
or
$$\frac{dy}{dx} = \frac{2c}{y}.$$

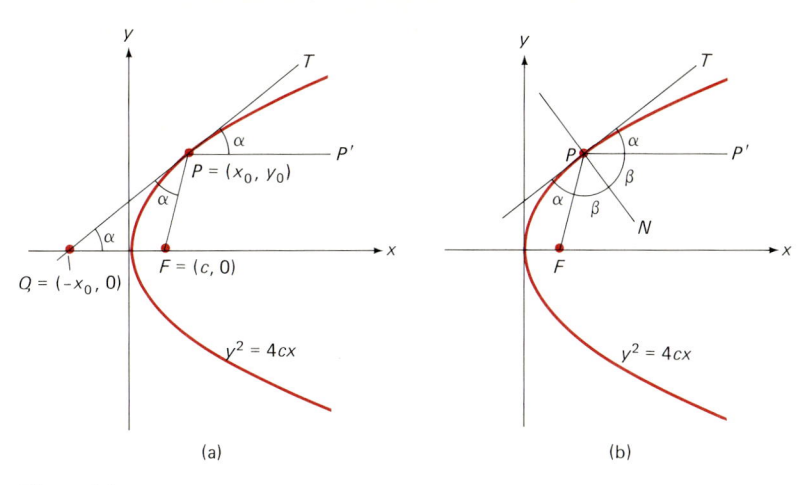

Figure 14

Hence if $y_0 \neq 0$, the tangent T is the line with equation

$$y - y_0 = \frac{2c}{y_0}(x - x_0),$$

or equivalently

$$yy_0 = 2c(x - x_0) + y_0^2 = 2c(x - x_0) + 4cx_0,$$

which simplifies to

$$yy_0 = 2c(x + x_0).$$

Therefore T has x-intercept $-x_0$, that is, $Q = (-x_0, 0)$ and

$$|FQ| = x_0 + c$$

(see the figure). But

$$|FP| = \sqrt{(x_0 - c)^2 + y_0^2} = \sqrt{(x_0 - c)^2 + 4cx_0} = \sqrt{(x_0 + c)^2} = x_0 + c,$$

so that $|FP| = |FQ|$. Hence the triangle FPQ is isosceles, with equal angles at P and Q.† It follows that the angle between T and PF is equal to the angle between T and the x-axis, which is in turn equal to the angle between T and PP', where PP' is the line through P parallel to the x-axis (see Figure 14(a), where this angle is denoted by α). Equivalently, if N is the normal line to the parabola at the point P, *the angle between N and PF is equal to the angle between N and PP'* (see Figure 14(b), where this angle is denoted by β).

Now let S be a parabolic reflector, that is, a reflector shaped like the surface (a paraboloid of revolution) generated by revolving a parabola about its axis of symmetry. Then it follows from the above considerations and the *law of reflection* established in Example 6, page 208, which is valid for a smooth curved surface as well as for a plane, that the rays emitted from a point source of light placed at the focus of S (the focus of the generating parabola) are converted into a parallel beam after reflection from S, as illustrated in Figure 15(a). For this reason, parabolic reflectors are used in

Figure 15

---

† Here we tacitly assume that the point of tangency P does not lie on the x-axis (the axis of the parabola); if it does, then the segment FP also lies on the x-axis and makes an angle of 90° with the tangent T which is now vertical (why?). We note in passing that a parabola intersects its axis at right angles.

Section 10.2 Parabolas 585

headlights and searchlights, radar antennas, and so on. By the same argument "in reverse," we see that a parallel beam of light rays incident on a parabolic reflector is brought to a focus at a single point on its axis of symmetry, as illustrated in Figure 15(b). This is why such great pains are taken to construct accurately parabolic mirrors for use in reflecting telescopes.

Parabolas play an important role in the solution of mechanical problems in the plane and in space. In particular, a projectile acted on only by gravity describes a parabolic trajectory (see Example 1, page 682), and the cable of a suspension bridge assumes the form of a parabola if the weight of the roadway is uniformly distributed (see Example 6, page 687).

Problems

Write the equation in standard form of the parabola which
1. Opens upward, with vertex (0, 0) and focus-to-directrix distance $\frac{1}{4}$
2. Opens downward, with vertex (0, 0) and focus-to-vertex distance 8
3. Opens to the left, with vertex (0, 0) and focus-to-vertex distance 5
4. Opens to the right, with vertex (0, 0) and focus-to-directrix distance $\frac{7}{2}$
5. Opens downward, with vertex (1, −2) and focus-to-vertex distance 9
6. Opens upward, with vertex (−3, 4) and focus-to-directrix distance $\frac{5}{2}$
7. Opens to the right, with vertex (7, 6) and focus-to-directrix distance $\frac{11}{4}$
8. Opens to the left, with vertex (−5, −10) and focus-to-vertex distance 3

Find an equation of the parabola with the given point and line as its focus and directrix.
9. (7, 2), $x - 5 = 0$
10. (4, 3), $y + 1 = 0$
11. (−8, 1), $y - 2 = 0$
12. (5, −3), $x - 4 = 0$
13. (2, −1), $x - y - 1 = 0$
14. (0, 0), $x + 2y - 3 = 0$

Graph the parabola with the given equation, and locate its vertex V, focus F and directrix L.
15. $y^2 = 8x$
16. $x^2 = 6y$
17. $x^2 = -4y$
18. $y^2 = -5x$
19. $(y + 1)^2 = -6(x + 2)$
20. $(x - \frac{1}{2})^2 = -8(y + 3)$
21. $(x + 4)^2 = 2(y - 3)$
22. $(y - 2)^2 = 10x$
23. $x^2 - 4x - 4y - 8 = 0$
24. $y^2 - 8x + 8y = 0$
25. $y^2 + 8x - 6y + 17 = 0$
26. $x^2 - 2x + 12y + 25 = 0$
27. $x^2 + 2xy + y^2 + 4x - 4y = 0$

28. A parabola with focus F and directrix L can be constructed as follows: To a drawing board attach a ruler with one edge along L, and place the short leg of an artist's triangle against the ruler. At the opposite vertex Q of the triangle fasten one end of a piece of string whose length is the same as that of the long leg of the triangle, and fasten the other end of the string at F. Slide the triangle along the ruler, holding the string taut with a pencil, as in Figure 16. Then the point P of the pencil traces out part of a parabola. Explain why this construction works.

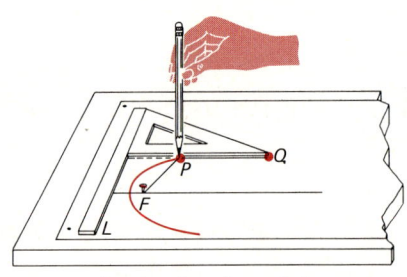

Construction of a parabola
Figure 16

29. A line segment with both endpoints on a parabola is called a *chord* of the parabola, and the chord through the focus F perpendicular to the axis and parallel to the directrix L is called the *latus rectum* (this is the chord AB in Figure 17). Show that the length of the latus rectum is twice the distance from F to L.

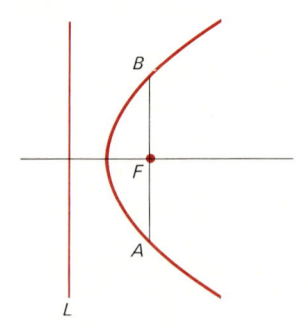

Figure 17

30. Show that the circle with the latus rectum of a parabola as its diameter is tangent to the directrix of the parabola.

31. Show that the curve with parametric equations
$$x = ct^2, \quad y = 2ct \quad (-\infty < t < \infty) \quad \text{(i)}$$
is a parabola. Discuss the orientation of the curve.

32. Show that the tangent to the parabola (i) at the point with parameter value t is the line $x - ty + ct^2 = 0$.

33. Find the locus of the midpoints of all chords of the parabola $y^2 = 4cx$ which have the vertex as one endpoint.

34. Consider the parabola $y^2 = 4cx$ $(c > 0)$, with vertex $V = (0, 0)$ and focus $F = (c, 0)$. Show that the point $A = (a, 0)$ on the axis of the parabola is closer to V than to any other point of the parabola if $0 < a \leq 2c$, but is closer to some point of the parabola other than V if $a > 2c$. (In particular, the focus F is closer to V than to any other point of the parabola.)

Find the points of intersection (if any) of the given line and parabola.

35. $3x - 2y + 6 = 0, \ y^2 = 6x$
36. $x + y - 3 = 0, \ x^2 = 4y$
37. $x + 6y + 6 = 0, \ x^2 = -18y$
38. $3x + 4y - 12 = 0, \ y^2 = -9x$

39. By making a suitable rotation of axes, show that the graph of the equation $\sqrt{x} + \sqrt{y} = 1$ is part of a parabola. Locate the vertex V, focus F and directrix L of this parabola.

40. A parabolic reflector has the cross section shown in Figure 18. How far from the vertex of the reflector should a light source be placed in order to produce an emergent beam of parallel rays?

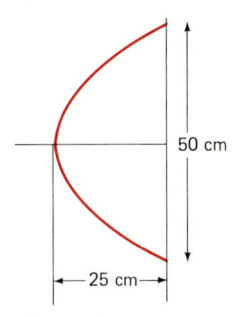

Figure 18

41. Find the two lines through the point (2, 9) tangent to the parabola $y^2 = 36x$.

42. Find the tangent to the parabola $x^2 = 16y$ perpendicular to the line $2x + 4y + 7 = 0$.

43. Find the tangent to the parabola $y^2 = 12x$ parallel to the line $3x - 2y + 30 = 0$. What is the distance between the tangent and the line?

10.3 Ellipses

Definition of an Ellipse

By definition, *an ellipse is the set of all points in a plane the sum of whose distances from two fixed points F_1 and F_2 is a constant.* Thus, for example, in Figure 19 the sum of the distances from the point P to the two points F_1 and F_2 is the same as the corresponding sum for the point Q or the point R. The points F_1 and F_2 are called the *foci* of the ellipse, and the midpoint of the segment F_1F_2 joining the foci is called the *center* of the ellipse.

This definition leads to the following simple construction of an ellipse. Suppose we put two tacks in a sheet of paper at the points F_1 and F_2, and place a loop of string around the tacks, as in Figure 20. (Naturally, the length

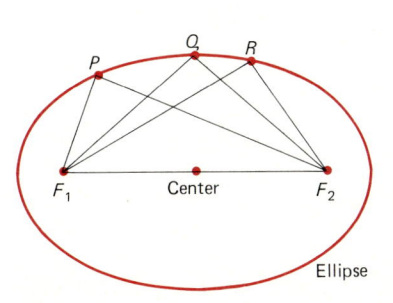

Figure 19

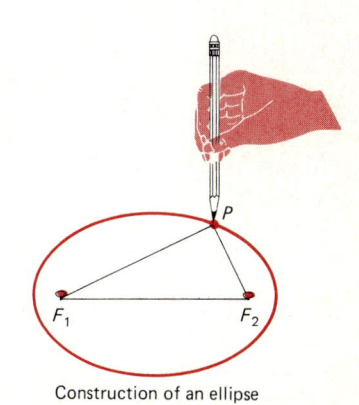

Figure 20   Construction of an ellipse

Section 10.3 Ellipses   **587**

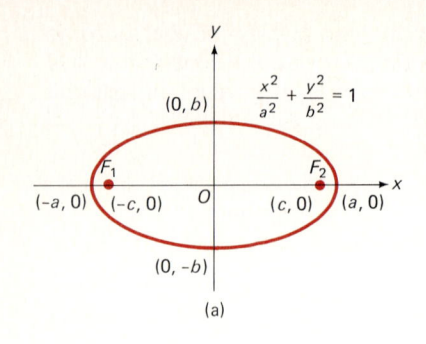

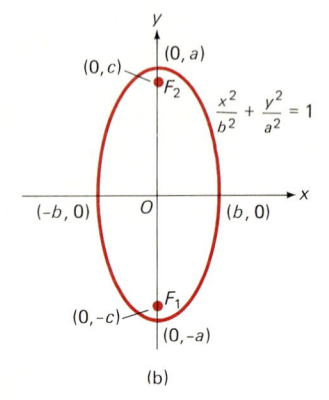

of the loop cannot be less than twice the distance between the tacks.) Then if we insert a pencil in the loop and move the pencil while holding it taut against the string, the point P of the pencil will trace out an ellipse with foci F_1 and F_2. This is because the sum $|PF_1| + |PF_2|$ is equal to the length of the loop of string minus the distance between the foci for every position of P. If the foci coincide, the ellipse reduces to a circle.

To find an equation of the ellipse, we introduce rectangular coordinates x and y in the plane of the ellipse and place it in "standard position," with its center at the origin O and its foci along the x-axis, as in Figure 21(a). The foci are then a pair of points $F_1 = (-c, 0)$ and $F_2 = (c, 0)$ a distance $2c$ apart ($c > 0$). Let $P = (x, y)$ be any point of the ellipse, and let $2a$ be the constant sum of the distances of P from the foci. Then $a > c$, except for the limiting case $a = c$ in which the ellipse "degenerates" to the line segment joining the foci. By the defining property of the ellipse,

$$|PF_1| + |PF_2| = 2a,$$

which takes the form

$$\sqrt{(x + c)^2 + y^2} + \sqrt{(x - c)^2 + y^2} = 2a$$

or

$$\sqrt{(x + c)^2 + y^2} = 2a - \sqrt{(x - c)^2 + y^2}$$

in terms of the coordinates of P, F_1 and F_2. Squaring both sides of this equation, we get

$$x^2 + 2cx + c^2 + y^2 = 4a^2 - 4a\sqrt{(x - c)^2 + y^2} + x^2 - 2cx + c^2 + y^2,$$

which simplifies to

$$a\sqrt{(x - c)^2 + y^2} = a^2 - cx.$$

Squaring both sides again, we find that

$$a^2(x^2 - 2cx + c^2 + y^2) = a^4 - 2a^2 cx + c^2 x^2,$$

and therefore

$$x^2(a^2 - c^2) + a^2 y^2 = a^2(a^2 - c^2). \tag{1}$$

Introducing a positive constant b such that

$$b^2 = a^2 - c^2,$$

or equivalently

$$a^2 = b^2 + c^2, \tag{2}$$

we can write (1) in the form

$$x^2 b^2 + a^2 y^2 = a^2 b^2,$$

which in turn becomes

$$\frac{x^2}{a^2} + \frac{y^2}{b^2} = 1, \tag{3}$$

588   Chapter 10   Plane Analytic Geometry and Polar Coordinates

after dividing both sides by a^2b^2. Conversely, if the point $P = (x, y)$ is such that (3) holds, then by carefully reversing the steps of this calculation, it is found that $|PF_1| + |PF_2| = 2a$ (the details are left as an exercise). Therefore (3) is an equation of the ellipse in standard position.

Major and Minor Axes

Since equation (3) is unaltered if we replace x by $-x$ and y by $-y$, both coordinate axes are axes of symmetry of the ellipse (3). More generally, every ellipse has two axes of symmetry, the line through the foci F_1 and F_2 and the perpendicular bisector of the segment F_1F_2. An ellipse cuts off two chords from its axes of symmetry. The longer chord is called the *major axis* of the ellipse, and the shorter chord is called the *minor axis*. The ellipse (3) has x-intercepts $(\pm a, 0)$ and y-intercepts $(0, \pm b)$, and since $a > b$, because of formula (2), the chord joining the points $(\pm a, 0)$ is longer than the chord joining the points $(0, \pm b)$. Hence the major axis of this ellipse is along the x-axis and the minor axis is along the y-axis [see Figure 21(a)]. The endpoints of the major axis, in this case the x-intercepts $(\pm a, 0)$, are called the *vertices* of the ellipse. The foci of the ellipse are at the points $(\pm c, 0)$, where

$$c = \sqrt{a^2 - b^2},$$

as we find by solving (2) for c.

Interchanging the variables x and y in equation (3), we get another ellipse

$$\frac{x^2}{b^2} + \frac{y^2}{a^2} = 1, \qquad (4)$$

which is also said to be in standard position. The ellipse (4) has x-intercepts $(\pm b, 0)$ and y-intercepts $(0, \pm a)$, where $a > b$. Thus the major axis is now along the y-axis, and its endpoints (the vertices) are $(0, \pm a)$, while the minor axis with endpoints $(\pm b, 0)$ is now along the x-axis [see Figure 21(b)]. Correspondingly, the foci of the ellipse (4) are the points $(0, \pm c)$, where c is again equal to $\sqrt{a^2 - b^2}$.

Formula (2) has a simple geometric interpretation. The points $(0, 0)$, $(0, b)$ and $(c, 0)$ in Figure 21(a), or the points $(0, 0)$, $(b, 0)$ and $(0, c)$ in Figure 21(b), lie at the vertices of a right triangle. This triangle has legs of lengths b and c, and its hypotenuse is of length a, since each endpoint of the minor axis of an ellipse is at distance a from the foci (why?). Therefore $a^2 = b^2 + c^2$, by the Pythagorean theorem. Since a is half the length of the major axis and b is half the length of the minor axis, the two lengths a and b are customarily called the *semimajor axis* and the *semiminor axis*.

Example 1 Find the graph of the equation $\dfrac{x^2}{9} + \dfrac{y^2}{4} = 1$.

Solution The equation is of the form (3) with $a = 3$, $b = 2$, and

$$c = \sqrt{a^2 - b^2} = \sqrt{3^2 - 2^2} = \sqrt{9 - 4} = \sqrt{5}.$$

Therefore its graph is an ellipse centered at the origin, which has a horizontal major axis with endpoints $(\pm 3, 0)$ a vertical minor axis with endpoints $(0, \pm 2)$, and foci $F_1 = (-\sqrt{5}, 0)$, $F_2 = (\sqrt{5}, 0)$ on the x-axis [see Figure 22(a)]. ☐

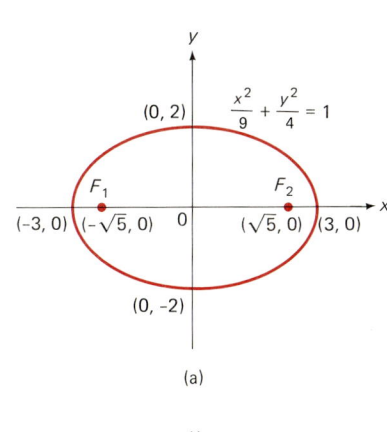

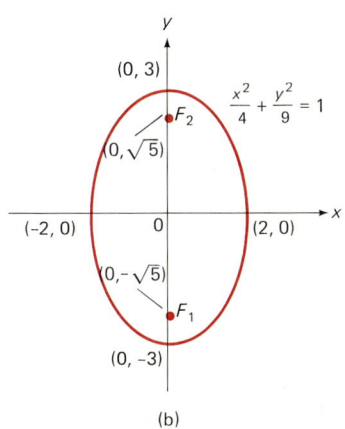

Figure 22

Section 10.3 Ellipses

Example 2 Find the graph of the equation $\dfrac{x^2}{4} + \dfrac{y^2}{9} = 1$.

Solution The equation is of the form (4) with the same values of a, b and c as in Example 1. Therefore its graph is an ellipse centered at the origin, which has a vertical major axis with endpoints $(0, \pm 3)$, a horizontal minor axis with endpoints $(\pm 2, 0)$, and foci $F_1 = (0, -\sqrt{5})$, $F_2 = (0, \sqrt{5})$ on the y-axis [see Figure 22(b)]. □

Equations in Standard Form

Equations (3) and (4) represent ellipses which are centered at the origin and have the coordinate axes as their axes of symmetry. The more general equations

$$\frac{(x - x_0)^2}{a^2} + \frac{(y - y_0)^2}{b^2} = 1 \tag{5}$$

and

$$\frac{(x - x_0)^2}{b^2} + \frac{(y - y_0)^2}{a^2} = 1 \tag{6}$$

represent ellipses which are of the same size and shape as the ellipses (3) and (4), but are centered at the point (x_0, y_0) and have their axes along lines parallel to the coordinate axes. In detail, (5) is an equation of the ellipse with center (x_0, y_0), major axis along the horizontal line $y = y_0$ and minor axis along the vertical line $x = x_0$, while (6) is an equation of the ellipse with center (x_0, y_0), major axis along the vertical line $x = x_0$ and minor axis along the horizontal line $y = y_0$. In both cases, the major axis is of length $2a$ and the minor axis is of length $2b$. These equations are called the *equations of an ellipse in standard form*. Naturally, (5) and (6) reduce to (3) and (4) in the special case where $x_0 = y_0 = 0$.

Example 3 Find the graph of the equation

$$x^2 + 4y^2 + 4x - 8y + 7 = 0. \tag{7}$$

Solution Completing the squares, we find that (7) can be written as

$$(x + 2)^2 + 4(y - 1)^2 = 1$$

or equivalently

$$(x + 2)^2 + \frac{(y - 1)^2}{(\frac{1}{2})^2} = 1,$$

which is in the standard form (5) with $x_0 = -2$, $y_0 = 1$, $a = 1$, $b = \frac{1}{2}$. The graph of this equation is an ellipse centered at the point $(-2, 1)$, with a horizontal major axis and a vertical minor axis (see Figure 23). The endpoints of the major and minor axes are $(-2 \pm 1, 1)$ and $(-2, 1 \pm \frac{1}{2})$, that is, $(-3, 1)$, $(-1, 1)$ and $(-2, \frac{1}{2})$, $(-2, \frac{3}{2})$. Moreover, since the distance from each focus to the center is $c = \sqrt{a^2 - b^2} = \sqrt{1 - \frac{1}{4}} = \frac{1}{2}\sqrt{3}$, the foci are $F_1 = (-2 - \frac{1}{2}\sqrt{3}, 1)$ and $F_2 = (-2 + \frac{1}{2}\sqrt{3}, 1)$. □

Each of the equations (5) and (6) has the following property: It is quadratic in both variables x and y, with no term containing the product xy, and moreover the coefficients of x^2 and y^2 have the same sign. Conversely,

$x^2 + 4y^2 + 4x - 8y + 7 = 0$

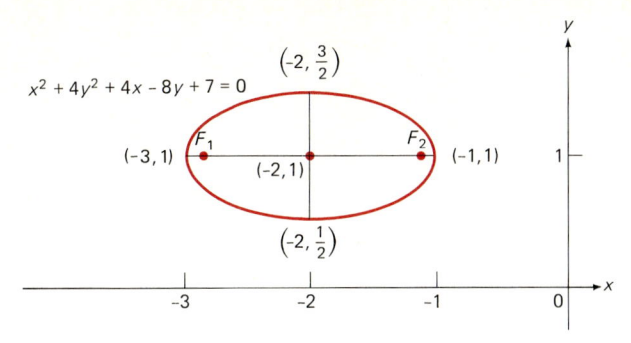

Figure 23

any equation which has this property can be written in the form

$$x^2 + ky^2 + Ax + By + C = 0 \qquad (k > 0) \tag{8}$$

after first dividing it by the coefficient of x^2. But, just as in Example 3, we can bring equation (8) into one of the standard forms (5) and (6) by completing the squares, and therefore, apart from certain exceptional cases, the graph of (8) is an ellipse with its axes parallel to the coordinate axes. In fact, after completing the squares we can write (8) as

$$\left(x + \frac{A}{2}\right)^2 + k\left(y + \frac{B}{2k}\right)^2 = D, \tag{8'}$$

where

$$D = \frac{A^2}{4} + \frac{B^2}{4k} - C.$$

If $D = 0$, the graph of (8'), and hence of (8), is the single point $(-A/2, -B/2k)$, while if $D < 0$, there are no points (x, y) whose coordinates satisfy (8), that is, the graph of (8) is "empty." These are the exceptional cases mentioned above. However, if $D > 0$, we can write (8') in the equivalent form

$$\frac{[x + (A/2)]^2}{(\sqrt{D})^2} + \frac{[y + (B/2k)]^2}{(\sqrt{D/k})^2} = 1,$$

which you will recognize as the equation of an ellipse with its center at the point $(-A/2, -B/2k)$. If $k > 1$, the ellipse has a horizontal major axis of length $2\sqrt{D}$ and a vertical minor axis of length $2\sqrt{D/k}$, while if $k < 1$, it has a vertical major axis of length $2\sqrt{D/k}$ and a horizontal minor axis of length $2\sqrt{D}$. If $k = 1$, the ellipse reduces to a circle of radius $\sqrt{D}$, and then this discussion reduces to the analysis of the equation $x^2 + y^2 + Ax + By + C = 0$, already given on pages 24–25. Thus, to summarize, an ellipse has an equation of the form (8) if and only if its axes are parallel to the coordinate axes.

Example 4 Find the area A enclosed by an ellipse with semimajor axis a and semiminor axis b.

Solution There is no loss of generality in assuming that the ellipse is the graph of the equation $(x^2/a^2) + (y^2/b^2) = 1$. Solving for y, we find that

$$y = \frac{b}{a}\sqrt{a^2 - x^2} \qquad (-a \leq x \leq a),$$

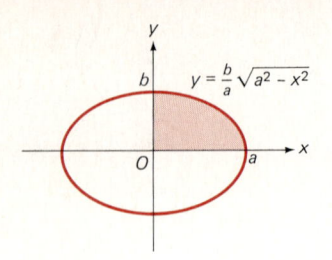

Figure 24

assuming that $y \geq 0$. By symmetry, A is four times the area between this curve and the x-axis in the first quadrant (the shaded area in Figure 24). Therefore

$$A = 4\frac{b}{a}\int_0^a \sqrt{a^2 - x^2}\, dx = 4\frac{b}{a}\left(\frac{1}{4}\pi a^2\right) = \pi ab,$$

with the help of Example 2, page 389. Notice that if $a = b$, A reduces to the area πa^2 enclosed by a circle of radius a.   □

Eccentricity of the Ellipse

The shape of an ellipse can be conveniently described by a single number between 0 and 1, namely

$$e = \frac{c}{a}$$

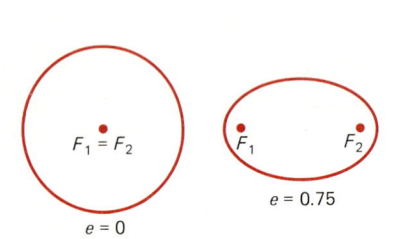

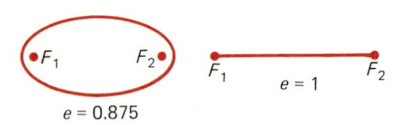

(not to be confused with the base of the natural logarithm). This number, called the *eccentricity*, is the ratio of the distance between the foci to the length of the major axis, and measures the degree of "ovalness" of the ellipse. If e is near 0, the ellipse is almost circular, while if e is near 1, the ellipse is very elongated and "cigar-shaped." We can allow the values $e = 0$ and $e = 1$ by regarding the circle of radius a and the line segment of length $2a$ as limiting cases of an ellipse (the foci coalesce if $e = 0$ and are a distance $2a$ apart if $e = 1$). The way the shape of an ellipse depends on its eccentricity e is illustrated in Figure 25, where the major axis has the same length in each case and the foci are F_1 and F_2.

Figure 25 Ellipses of various eccentricities

Example 5 What is the eccentricity of the ellipse in Figure 22(a)?

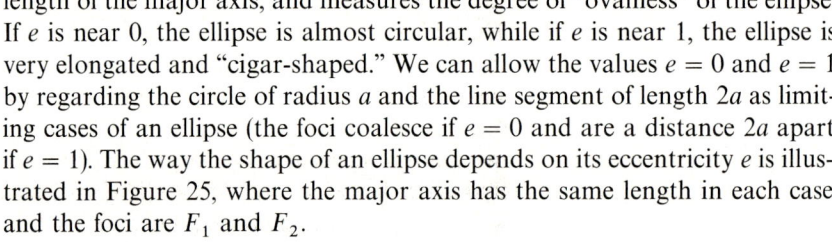

Solution Here $a = 3$ and $c = \sqrt{5}$, so that the eccentricity is

$$e = \frac{c}{a} = \frac{\sqrt{5}}{3} \approx 0.745.$$

The ellipse in Figure 22(b) has the same eccentricity.   □

Example 6 Find an equation of the ellipse with semimajor axis 4 and eccentricity $\frac{1}{2}$ if the center is at the origin and the major axis is horizontal.

Solution We are given that $a = 4$ and

$$e = \frac{c}{a} = \frac{\sqrt{a^2 - b^2}}{a} = \frac{\sqrt{16 - b^2}}{4} = \frac{1}{2}.$$

Therefore $b^2 = 12$ and the ellipse has an equation of the form

$$\frac{x^2}{16} + \frac{y^2}{12} = 1. \quad \square$$

Reflection Property of the Ellipse

The ellipse, like the parabola, has an interesting *reflection property*. Let T be the tangent line at any point $P = (x_0, y_0)$ of the ellipse

$$\frac{x^2}{a^2} + \frac{y^2}{b^2} = 1, \qquad (9)$$

with foci $F_1 = (-c, 0)$ and $F_2 = (c, 0)$, and let Q_1 and Q_2 be the points of intersection of T and the perpendiculars drawn from F_1 and F_2 to T [see

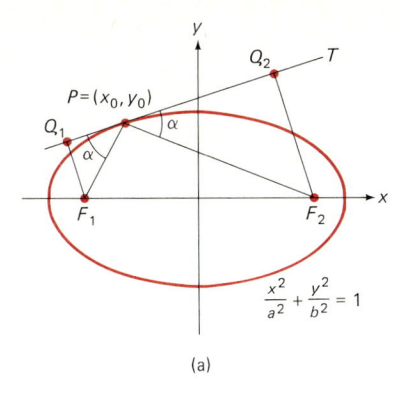

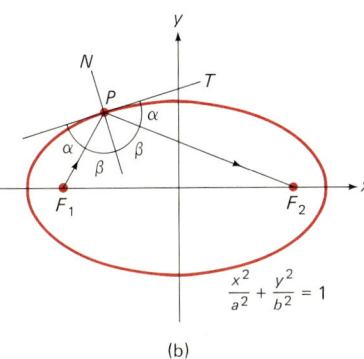

Figure 26(a)]. Differentiating (9) with respect to x, we get

$$\frac{2x}{a^2} + \frac{2y}{b^2}\frac{dy}{dx} = 0$$

or

$$\frac{dy}{dx} = -\frac{b^2 x}{a^2 y}.$$

Hence if $y_0 \neq 0$, the tangent T is the line with equation

$$y - y_0 = -\frac{b^2 x_0}{a^2 y_0}(x - x_0)$$

or

$$b^2 x_0 x + a^2 y_0 y - b^2 x_0^2 - a^2 y_0^2 = 0.$$

But

$$b^2 x_0^2 + a^2 y_0^2 = a^2 b^2, \tag{10}$$

since P lies on the ellipse, so that the equation of T reduces to

$$b^2 x_0 x + a^2 y_0 y - a^2 b^2 = 0, \tag{11}$$

or equivalently

$$\frac{x_0 x}{a^2} + \frac{y_0 y}{b^2} = 1. \tag{11'}$$

Next we establish the formula

$$\frac{|F_1 Q_1|}{|F_1 P|} = \frac{|F_2 Q_2|}{|F_2 P|}, \tag{12}$$

which shows that the right triangles $F_1 PQ_1$ and $F_2 PQ_2$ are similar.† In fact, with the help of (11) and Theorem 10, page 34, we can write (12) as

$$\frac{|-b^2 x_0 c - a^2 b^2|}{\sqrt{b^4 x_0^2 + a^4 y_0^2}\sqrt{(x_0 + c)^2 + y_0^2}} = \frac{|b^2 x_0 c - a^2 b^2|}{\sqrt{b^4 x_0^2 + a^4 y_0^2}\sqrt{(x_0 - c)^2 + y_0^2}},$$

or equivalently

$$\frac{(x_0^2 c^2 + a^4) + 2a^2 c x_0}{(x_0^2 + c^2 + y_0^2) + 2c x_0} = \frac{(x_0^2 c^2 + a^4) - 2a^2 c x_0}{(x_0^2 + c^2 + y_0^2) - 2c x_0},$$

which simplifies to $a^2(x_0^2 + c^2 + y_0^2) = x_0^2 c^2 + a^4$ or

$$(a^2 - c^2) x_0^2 + a^2 y_0^2 = a^2(a^2 - c^2). \tag{12'}$$

Since $a^2 - c^2 = b^2$, (12') is equivalent to (10), and this observation proves the validity of (12') and hence of (12). Since the triangles $F_1 PQ_1$ and $F_2 PQ_2$ are similar, the angle between T and PF_1 is equal to the angle between T and PF_2 (see Figure 26(a), where this angle is denoted by α). Equivalently, if N is the normal line to the ellipse at the point P, *the angle between N and PF_1 is equal to the angle between N and PF_2* (see Figure 26(b), where this angle is denoted by β).

---

† Here we tacitly assume that the point of tangency P does not lie on the major axis; if it does, then the segments PF_1 and PF_2 also lie on the major axis, and make the same angle of 90° with the tangent T which is now vertical (why?). We note in passing that an ellipse intersects both its axes at right angles, since T is horizontal if P lies on the minor axis and vertical if P lies on the major axis.

It follows from these considerations and the law of reflection established in Example 6, page 208, that if S is a reflector shaped like the surface generated by revolving an ellipse about its major axis, then the rays emitted by a point source of light or sound placed at one focus of S all converge at the other focus after reflection from S (for example, the ray leaving F_1 in Figure 26(b) is reflected into F_2). Rooms have been constructed which are shaped like the upper half of such a surface. In such a room, known as a "whispering gallery," a low conversation at one focus can be heard clearly at the other focus but not at intermediate points.

The ellipse plays a key role in problems of orbital motion. For example, the earth revolves around the sun in an elliptic orbit of small eccentricity with the sun at one focus. We will make a detailed study of orbital motion in Section 12.5.

Problems

Write the equation in standard form of the ellipse such that
1. The vertices are $(\pm 4, 0)$ and the foci are $(\pm 3, 0)$
2. The vertices are $(0, \pm 10)$ and the foci are $(0, \pm 5)$
3. The endpoints of the axes are $(2, 4), (10, 4), (6, 1), (6, 7)$
4. The center is $(-1, 1)$, one focus is $(-1, 3)$ and the semimajor axis is 5
5. The center is $(2, -2)$, one focus is $(2, 1)$ and the semiminor axis is 6
6. The center is $(-3, 2)$, one vertex is $(0, 2)$ and one focus is $(-5, 2)$
7. The origin is the center and the two points $(1, -1)$, $(1/\sqrt{2}, \sqrt{2})$ lie on the ellipse
8. The foci are $(1, \pm 3)$ and the eccentricity is $\frac{3}{5}$

9. Find all points of the ellipse

$$\frac{x^2}{25} + \frac{y^2}{4} = 1$$

whose abscissas have absolute value 3.

10. Show that the ellipse $(x^2/a^2) + (y^2/b^2) = 1$ is contained in the rectangular region $-a \leq x \leq a$, $-b \leq y \leq b$.

Graph the ellipse with the given equation, and locate the center, the foci and the endpoints of the major and minor axes.

11. $\dfrac{x^2}{9} + \dfrac{y^2}{25} = 1$

12. $\dfrac{x^2}{49} + \dfrac{y^2}{16} = 1$

13. $\dfrac{(x-2)^2}{36} + \dfrac{(y+4)^2}{9} = 1$

14. $\dfrac{(x+3)^2}{4} + \dfrac{(y-1)^2}{64} = 1$

15. $3x^2 + y^2 = 2$
16. $9x^2 + 25y^2 = 25$
17. $16x^2 + 25y^2 - 64x - 100y - 236 = 0$
18. $4x^2 + y^2 + 8x - 6y - 3 = 0$
19. $2x^2 + 2xy + 2y^2 - 1 = 0$
20. $9x^2 + 4y^2 + 36x = 0$

21. Find an equation of the ellipse with semimajor axis 1 and foci $F_1 = (-\frac{1}{2}, -\frac{1}{2})$, $F_2 = (\frac{1}{2}, \frac{1}{2})$.

22. A line segment with both endpoints on an ellipse is called a *chord* of the ellipse, and a chord through either focus F_1 or F_2 perpendicular to the major axis is called a *latus rectum* (this is the chord AB or CD in Figure 27). Show that the latus rectum is of length $2b^2/a$, where a is the semimajor axis and b is the semiminor axis of the ellipse.

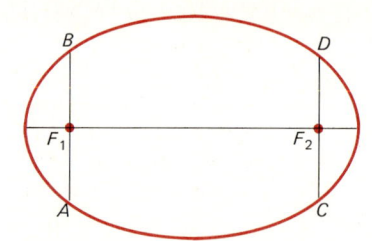

Figure 27

23. Find every point of the ellipse $x^2 + 5y^2 = 20$ for which the lines joining the point to the foci are perpendicular.

24. For what values of the constant c is the graph of the equation $x^2 + 2y^2 - 6x + 4y = c$ an ellipse? A single point? Empty?

25. Show that the curve with parametric equations

$$x = x_0 + a \cos t, \quad y = y_0 + b \sin t \quad (0 \le t \le 2\pi) \quad (i)$$

is an ellipse (assume that $a > 0, b > 0$).

Find the points of intersection (if any) of the given line and ellipse.

26. $x + 2y - 7 = 0, \ x^2 + 4y^2 = 25$
27. $3x + 10y - 25 = 0, \ 4x^2 + 25y^2 = 100$
28. $3x - 4y - 24 = 0, \ 9x^2 + 16y^2 = 144$
29. $x + y - 2 = 0, \ 2x^2 + y^2 = 3$

30. Show that as a ladder slides down a wall, any fixed point P of the ladder other than one of its endpoints traces out one fourth of an ellipse. (See Figure 28, where AB is the ladder, a is the distance between A and P, and b is the distance between P and B.)

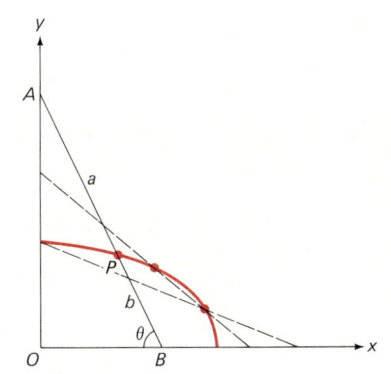

Figure 28

31. Let A be a variable point of the circle $x^2 + y^2 = a^2$ of radius a, let B be the point in which the radius OA of this circle intersects the concentric circle $x^2 + y^2 = b^2$ of radius b, and let P be the point in which the horizontal line through B intersects the vertical line through A, as in Figure 29, where θ is the angle between OA and the x-axis. Show that as A describes the circle $x^2 + y^2 = a^2$, the point P describes the ellipse $(x^2/a^2) + (y^2/b^2) = 1$.

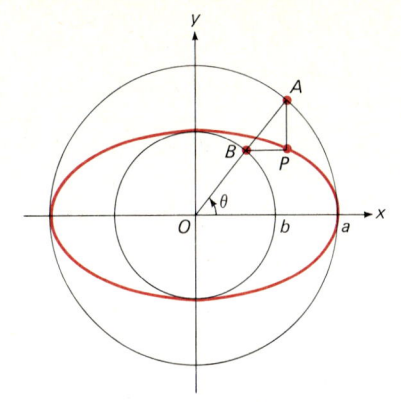

Figure 29

Find the eccentricity of the ellipse with the given semiaxes.

32. 13, 5
33. 17, 15
34. 25, 7
35. 53, 28

36. Find the area enclosed by the ellipse

$$\frac{(x-2)^2}{25} + \frac{(y+3)^2}{36} = 1.$$

37. Find the largest area of a rectangle inscribed in the ellipse $(x^2/a^2) + (y^2/b^2) = 1$ with two sides of the rectangle parallel to the major axis.

Among the tangents to the ellipse $\frac{1}{20}x^2 + \frac{1}{5}y^2 = 1$ find those which are

38. Perpendicular to the line $2x - 2y - 3 = 0$
39. Going through the point $(\frac{10}{3}, \frac{5}{3})$ outside the ellipse

40. Show that the product of the distances between the foci F_1, F_2 of an ellipse and any tangent to the ellipse is equal to the square of the semiminor axis.

41. The point nearest the sun in an orbit around the sun is called the *perihelion*, and the point farthest from the sun is called the *aphelion*. The earth moves in an elliptical orbit with the sun at one focus such that the distance between the earth and the sun is about 147.1 million kilometers at perihelion and about 152.1 million kilometers at aphelion. What is the approximate eccentricity of the earth's orbit?

10.4 Hyperbolas

Definition of a Hyperbola

By definition, *a hyperbola is the set of all points in a plane the difference of whose distances from two fixed points F_1 and F_2 is a constant*. (It is understood that the smaller of these distances is always subtracted from the larger,

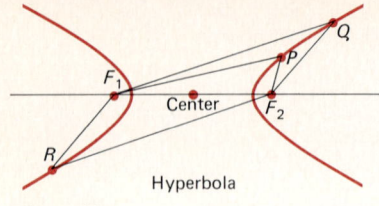

Figure 30

so that the constant is positive.) Thus, for example, in Figure 30 the difference between the distances from the point P to the points F_1 and F_2, called the *foci* of the hyperbola, is the same as the corresponding difference for the point Q or the point R. Notice that the hyperbola consists of two separate parts, called *branches*. On one branch (the right branch in the figure) every point of the hyperbola is closer to F_2 than to F_1, while on the other branch (the left branch in the figure) every point is closer to F_1 than to F_2. The midpoint of the segment F_1F_2 joining the foci is called the *center* of the hyperbola. Like the parabola (but unlike the ellipse), the hyperbola is an *unbounded* curve, that is, there are points of the curve which are arbitrarily far apart.

To find an equation of the hyperbola, we introduce rectangular coordinates x and y in the plane of the hyperbola and place it in "standard position," with its center at the origin O and its foci along the x-axis. The foci are then a pair of points $F_1 = (-c, 0)$ and $F_2 = (c, 0)$, a distance $2c$ apart ($c > 0$). Let $P = (x, y)$ be any point of the hyperbola, and let $2a$ be the constant value of the difference between the distances from P to the foci. By the defining property of the hyperbola,

$$|PF_1| - |PF_2| = 2a$$

if P is closer to F_2 than to F_1, or

$$|PF_2| - |PF_1| = 2a$$

if P is closer to F_1 than to F_2. It follows that

$$|PF_1| = \pm 2a + |PF_2|,$$

which takes the form

$$\sqrt{(x+c)^2 + y^2} = \pm 2a + \sqrt{(x-c)^2 + y^2}$$

in terms of the coordinates of P, F_1 and F_2. Squaring both sides of this equation, we get

$$x^2 + 2cx + c^2 + y^2 = 4a^2 \pm 4a\sqrt{(x-c)^2 + y^2} + x^2 - 2cx + c^2 + y^2,$$

which simplifies to

$$\pm a\sqrt{(x-c)^2 + y^2} = cx - a^2.$$

Squaring both sides again, we find that

$$a^2(x^2 - 2cx + c^2 + y^2) = c^2x^2 - 2a^2cx + a^4,$$

and therefore

$$x^2(c^2 - a^2) - a^2y^2 = a^2(c^2 - a^2). \tag{1}$$

To simplify (1) further, we observe that

$$|PF_1| < |F_1F_2| + |PF_2| = 2c + |PF_2|,$$
$$|PF_2| < |F_1F_2| + |PF_1| = 2c + |PF_1|,$$

since the length of one side of the triangle F_1PF_2 is less than the sum of the lengths of the other two sides. Therefore

$$2c > |PF_1| - |PF_2|, \quad 2c > |PF_2| - |PF_1|,$$

596   Chapter 10   Plane Analytic Geometry and Polar Coordinates

which implies
$$c > a,$$
since one of the differences on the right is equal to $2a$. Introducing a positive constant b such that
$$b^2 = c^2 - a^2, \tag{2}$$
we can write (1) in the form
$$x^2 b^2 - a^2 y^2 = a^2 b^2,$$
which in turn becomes
$$\frac{x^2}{a^2} - \frac{y^2}{b^2} = 1 \tag{3}$$
after dividing both sides by $a^2 b^2$. Conversely if the point $P = (x, y)$ is such that (3) holds, then by carefully reversing the steps in this calculation, it is found that $|PF_1| - |PF_2| = \pm 2a$ (the details are left as an exercise). Therefore (3) is an equation of the hyperbola in standard position.

Transverse and Conjugate Axes

Since equation (3) is unaltered if we replace x by $-x$ or y by $-y$, both coordinate axes are axes of symmetry of the hyperbola (3). More generally, every hyperbola has two axes of symmetry, the line through the foci F_1 and F_2 and the perpendicular bisector of the segment $F_1 F_2$. The hyperbola intersects the line through its foci in two points called *vertices*, and the segment joining the vertices is called the *transverse axis*. For the hyperbola (3) the vertices are the points $(\pm a, 0)$ and the transverse axis is horizontal [see Figure 31(a)]. The foci of the hyperbola are at the points $(\pm c, 0)$, where
$$c = \sqrt{a^2 + b^2},$$
as we find by solving (2) for c. Interchanging the variables x and y in equation (3), we get another hyperbola
$$\frac{y^2}{a^2} - \frac{x^2}{b^2} = 1, \tag{4}$$
which is also said to be in standard position. This hyperbola has vertices $(0, \pm a)$ and foci $(0, \pm c)$ on the y-axis, and correspondingly its transverse axis is vertical [see Figure 31(b)].

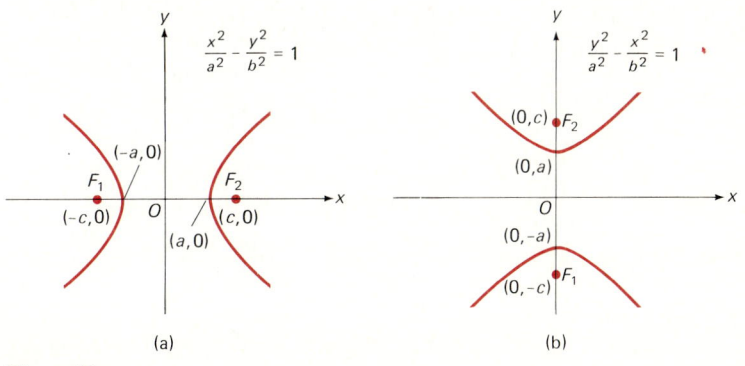

Figure 31

Section 10.4 Hyperbolas

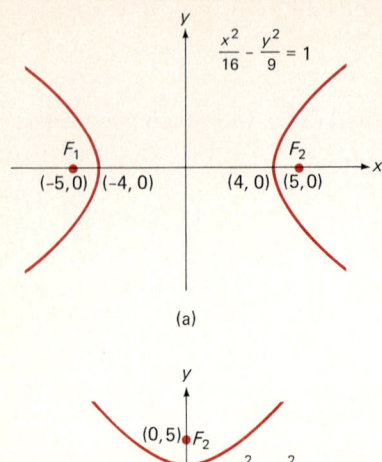

Example 1 Find the graph of the equation $\dfrac{x^2}{16} - \dfrac{y^2}{9} = 1$.

Solution The equation is of the form (3) with $a = 4$, $b = 3$, and
$$c = \sqrt{a^2 + b^2} = \sqrt{4^2 + 3^2} = \sqrt{16 + 9} = \sqrt{25} = 5.$$

Therefore its graph is the hyperbola with center at the origin, vertices $(\pm 4, 0)$ and foci $(\pm 5, 0)$ on the x-axis, and a horizontal transverse axis [see Figure 32(a)]. □

Example 2 Find the graph of the equation $\dfrac{y^2}{16} - \dfrac{x^2}{9} = 1$.

Solution The equation is of the form (4) with the same values of a, b and c as in Example 1. Therefore its graph is the hyperbola with center at the origin, vertices $(0, \pm 4)$ and foci $(0, \pm 5)$ on the y-axis, and a vertical transverse axis [see Figure 32(b)]. □

By the *conjugate axis* of a hyperbola we mean the line segment of length $2b$ which is perpendicular to the transverse axis and has the same midpoint, namely the center of the hyperbola. Thus in Example 1 the conjugate axis is the vertical segment joining the points $(0, -3)$ and $(0, 3)$, while in Example 2 it is the horizontal segment joining the points $(-3, 0)$ and $(3, 0)$. These axes are the analogues of the major and minor axes of an ellipse. However, the major axis of an ellipse is always longer than its minor axis ($a > b$), whereas there is no restriction on the relative sizes of the transverse and conjugate axes of a hyperbola, that is, we can have $a < b$, $a = b$ or $a > b$. This is because the condition $a^2 = b^2 + c^2$ for an ellipse is replaced by $c^2 = a^2 + b^2$ for a hyperbola, so that specifying a (in the defining property of the hyperbola) still leaves b free to take any value at all.

Figure 32

Asymptotes of a Hyperbola

Next we show that a hyperbola has two *asymptotes*. To this end, we solve the equation $(x^2/a^2) - (y^2/b^2) = 1$ for y, obtaining the functions

$$y = \frac{b}{a}\sqrt{x^2 - a^2} \qquad (5)$$

and

$$y = -\frac{b}{a}\sqrt{x^2 - a^2}. \qquad (5')$$

Each function is defined only if $x \leq -a$ or $x \geq a$, and in fact (5) represents the part of the hyperbola in the upper half-plane $y \geq 0$, while (5') represents the part in the lower half-plane $y \leq 0$. The line $y = (b/a)x$ is an oblique asymptote of the hyperbola (3). To see this, let $P = (x, y)$ be a variable point of the part of the hyperbola in the first quadrant. Then x and y are both positive, and the distance $d(x)$ between P and the point of the line $y = (b/a)x$ with the same x-coordinate is

$$d(x) = \frac{b}{a}(x - \sqrt{x^2 - a^2}) = \frac{b(x - \sqrt{x^2 - a^2})(x + \sqrt{x^2 - a^2})}{a} \cdot \frac{1}{x + \sqrt{x^2 - a^2}}$$

$$= \frac{ab}{x + \sqrt{x^2 - a^2}}.$$

But
$$\lim_{x \to \infty} d(x) = \lim_{x \to \infty} \frac{ab}{x + \sqrt{x^2 - a^2}} = 0,$$

and hence the graph of (5) gets closer and closer to the line $y = (b/a)x$ as $x \to \infty$, that is, the line is an asymptote of the part of the hyperbola in the first quadrant. A similar argument shows that the line $y = (b/a)x$ is an asymptote of the part of the hyperbola in the third quadrant. Then, since reflection in the x-axis carries the hyperbola (3) into itself and the line $y = (b/a)x$ into the line $y = -(b/a)x$, it follows that the line $y = -(b/a)x$ is an asymptote of the parts of the hyperbola in the second and fourth quadrants. All this is apparent from Figure 33.

To summarize, the hyperbola $(x^2/a^2) - (y^2/b^2) = 1$ has the lines

$$y = \pm \frac{b}{a} x$$

as asymptotes. As an exercise, show that the asymptotes of the hyperbola $(y^2/a^2) - (x^2/b^2) = 1$, with a vertical transverse axis, are the lines $y = \pm(a/b)x$.

There is a simple way of constructing the asymptotes of the hyperbola $(x^2/a^2) - (y^2/b^2) = 1$. Construct the *central rectangle*, that is, the rectangle $ABCD$ shown in Figure 33, with two sides on the lines $x = \pm a$ and two sides on the lines $y = \pm b$. Then the asymptotes are just the lines obtained by extending the diagonals of the rectangle. Observe that the semidiagonal OA of the rectangle is of length c, since $c = \sqrt{a^2 + b^2}$, so that the two points in which the circle of radius $|OA|$ intersects the x-axis are the foci $F_1 = (-c, 0)$ and $F_2 = (c, 0)$, as indicated in the figure which shows two arcs of this circle. Note also that the asymptotes are the graph of the equation

$$\frac{x^2}{a^2} - \frac{y^2}{b^2} = 0,$$

which differs from the equation

$$\frac{x^2}{a^2} - \frac{y^2}{b^2} = 1$$

of the hyperbola by having 0 instead of 1 as the right side.

Construction of the asymptotes and foci of a hyperbola

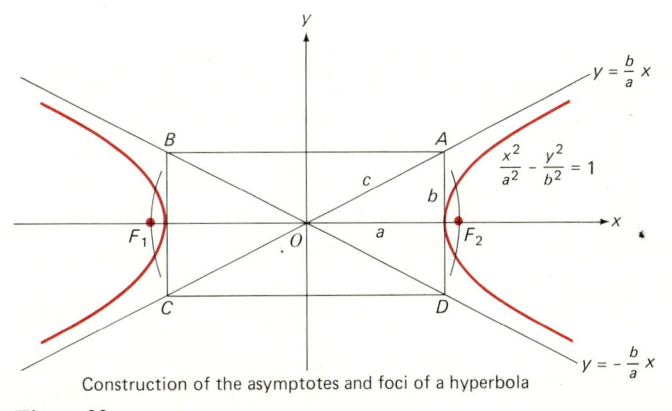

Figure 33

Section 10.4 Hyperbolas 599

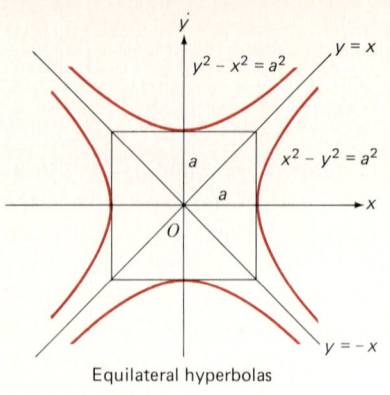

Figure 34 Equilateral hyperbolas

Example 3 If the transverse and conjugate axes have the same length, then $a = b$ in equation (3) or (4), and we get an *equilateral hyperbola*
$$x^2 - y^2 = a^2$$
or
$$y^2 - x^2 = a^2,$$
with asymptotes $y = \pm x$ (see Figure 34). In this case the asymptotes are perpendicular, and the central rectangle is a square. We have already encountered an equilateral hyperbola, namely the *unit hyperbola*
$$x^2 - y^2 = 1,$$
discussed on page 353 in connection with the hyperbolic functions.

Equations in Standard Form

Equations (3) and (4) represent hyperbolas which are centered at the origin and have the coordinate axes as their axes of symmetry. The more general equations
$$\frac{(x - x_0)^2}{a^2} - \frac{(y - y_0)^2}{b^2} = 1 \tag{6}$$
and
$$\frac{(y - y_0)^2}{a^2} - \frac{(x - x_0)^2}{b^2} = 1 \tag{7}$$
represent hyperbolas which are of the same size and shape as the hyperbolas (3) and (4), but are centered at the point (x_0, y_0) and have their axes along lines parallel to the coordinate axes. In detail, (6) is an equation of the hyperbola with center (x_0, y_0), transverse axis along the horizontal line $y = y_0$ and conjugate axis along the vertical line $x = x_0$, while (7) is an equation of the hyperbola with center (x_0, y_0), transverse axis along the vertical line $x = x_0$ and conjugate axis along the horizontal line $y = y_0$. In both cases, the transverse axis is of length $2a$ and the conjugate axis is of length $2b$. These equations are called the *equations of a hyperbola in standard form*. Naturally, (6) and (7) reduce to (3) and (4) in the special case where $x_0 = y_0 = 0$.

Example 4 Find the graph of the equation
$$x^2 - 4y^2 - 2x + 16y - 14 = 0. \tag{8}$$

Solution Completing the squares, we find that (8) can be written as
$$(x - 1)^2 - 4(y - 2)^2 + 1 = 0,$$
or equivalently
$$\frac{(y - 2)^2}{(\frac{1}{2})^2} - (x - 1)^2 = 1,$$
which is in the standard form (7) with $x_0 = 1$, $y_0 = 2$, $a = \frac{1}{2}$, $b = 1$. The graph of this equation is a hyperbola centered at the point $(1, 2)$, with a vertical transverse axis (see Figure 35). The endpoints of the transverse axis are $(1, 2 - \frac{1}{2})$ and $(1, 2 + \frac{1}{2})$, that is, $(1, \frac{3}{2})$ and $(1, \frac{5}{2})$. Also, since the distance from each focus to the center is $c = \sqrt{a^2 + b^2} = \sqrt{\frac{1}{4} + 1} = \frac{1}{2}\sqrt{5}$, the foci are $F_1 = (1, 2 - \frac{1}{2}\sqrt{5})$ and $F_2 = (1, 2 + \frac{1}{2}\sqrt{5})$. You can easily verify that the asymptotes are the lines $y = \frac{1}{2}x + \frac{3}{2}$ and $y = -\frac{1}{2}x + \frac{5}{2}$ of slope $\pm\frac{1}{2}$ intersecting in the point $(1, 2)$, as shown in the figure. □

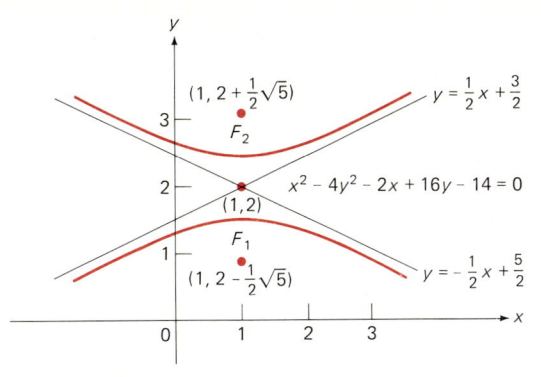

Figure 35

Each of the equations (6) and (7) has the following property: It is quadratic in both variables x and y, with no term containing the product xy, and moreover the coefficients of x^2 and y^2 have opposite signs (unlike the case of the ellipse discussed on pages 590–591, where they have the same sign). Conversely, any equation which has this property can be written in the form

$$x^2 - ky^2 + Ax + By + C = 0 \quad (k > 0) \tag{9}$$

after first dividing it by the coefficient of x^2. But, just as in Example 4, we can bring (9) into one of the standard forms (6) and (7) by completing the squares, and hence, apart from a certain exceptional case, the graph of (9) is a hyperbola with its axes parallel to the coordinate axes. In fact, after completing the squares we can write (9) as

$$\left(x + \frac{A}{2}\right)^2 - k\left(y - \frac{B}{2k}\right)^2 = D, \tag{9'}$$

where

$$D = \frac{A^2}{4} - \frac{B^2}{4k} - C.$$

If $D = 0$, the graph of (9'), and hence of (9), is the pair of lines

$$y - \frac{B}{2k} = \pm \frac{1}{\sqrt{k}}\left(x + \frac{A}{2}\right),$$

which intersect in the point $(-A/2, B/2k)$, and this is the exceptional case mentioned above. However, if $D \neq 0$, we can write (9') in one of the equivalent forms

$$\frac{[x + (A/2)]^2}{(\sqrt{D})^2} - \frac{[y - (B/2k)]^2}{(\sqrt{D/k})^2} = 1 \quad (D > 0)$$

or

$$\frac{[y - (B/2k)]^2}{(\sqrt{|D|/k})^2} - \frac{[x + (A/2)]^2}{(\sqrt{|D|})^2} = 1 \quad (D < 0),$$

depending on the sign of D, which you will recognize as the equation of a hyperbola with its center at the point $(-A/2, B/2k)$. If $D > 0$, the hyperbola has a horizontal transverse axis of length $2\sqrt{D}$ and a vertical conjugate axis

Section 10.4 Hyperbolas

of length $2\sqrt{D/k}$, while if $D < 0$, it has a vertical transverse axis of length $2\sqrt{|D|}$ and a horizontal conjugate axis of length $2\sqrt{|D|/k}$. Thus, to summarize, a hyperbola has an equation of the form (9) if and only if its axes are parallel to the coordinate axes.

There are of course hyperbolas whose equations have terms containing the product xy. For instance, the graph of the equation $2xy = 1$ is the hyperbola obtained by rotating the equilateral hyperbola $x^2 - y^2 = 1$ through 45° in the counterclockwise direction (see Example 4, page 578).

Eccentricity of the Hyperbola

Just as in the case of the ellipse, the shape of a hyperbola can be conveniently characterized by the number

$$e = \frac{c}{a},$$

called its *eccentricity*, but now e can only take values greater than 1, since $c > a$.

Example 5 What is the eccentricity of the hyperbola in Figure 32?

Solution Here $a = 4$ and $c = 5$, so that the eccentricity is

$$e = \frac{c}{a} = \frac{5}{4} = 1.25. \quad \square$$

Like the ellipse, the hyperbola plays an important part in the study of orbital motion. A number of comets are thought to have hyperbolic orbits, which means that they appear only once and then leave the solar system forever.† In 1911 Ernest Rutherford was able to deduce the existence of atomic nuclei by analyzing the behavior of alpha particles deflected into hyperbolic orbits by thin metallic foils. On page 353 we have already discussed the role of the hyperbola $x^2 - y^2 = 1$ in the definition of the hyperbolic functions. The use of hyperbolas in problems of range finding is illustrated in Problems 43 and 44.

† This is to be contrasted with the behavior of comets with elliptic orbits, which periodically return to perihelion (there have been 27 recorded appearances of Halley's comet). It is often difficult to determine whether the orbit of a comet is a hyperbola or a very elongated ellipse. Parabolic orbits are another possibility, at least in theory.

Problems

Write the equation in standard form of the hyperbola such that

1. The vertices are $(\pm 3, 0)$ and the foci are $(\pm 4, 0)$
2. The vertices are $(0, \pm 2)$ and the foci are $(0, \pm 5)$
3. The endpoints of the transverse axis are $(-7, 2)$, $(5, 2)$, and those of the conjugate axis are $(-1, 7)$, $(-1, -3)$
4. The center is $(3, 1)$, one focus is $(9, 1)$ and the transverse axis is of length 8
5. The center is $(-2, 0)$, one focus is $(-2, 4)$ and the conjugate axis is of length 4
6. The center is $(2, -1)$, one focus is $(-4, -1)$ and one vertex is $(7, -1)$
7. The origin is the center and the two points $(6, -1)$, $(8, \sqrt{8})$ lie on the hyperbola
8. The foci are $(\pm 10, 0)$ and the asymptotes are the lines $y = \pm \frac{4}{3}x$
9. The vertices are $(0, \pm 24)$ and the asymptotes are the lines $y = \pm \frac{12}{5}x$
10. The foci are $(\pm 7, 0)$ and the eccentricity is $\frac{7}{5}$

11. The origin is the center, the eccentricity is $\sqrt{2}$ and the point $(-5, 3)$ lies on the hyperbola
12. The origin is the center, the asymptotes are $y = \pm\frac{2}{3}x$ and the point $(\frac{9}{2}, -1)$ lies on the hyperbola

Graph the hyperbola with the given equation and locate the center, the foci, the endpoints of the transverse axis, and the asymptotes.

13. $\dfrac{x^2}{9} - \dfrac{y^2}{4} = 1$

14. $\dfrac{y^2}{36} - \dfrac{x^2}{25} = 1$

15. $\dfrac{(y-1)^2}{16} - \dfrac{(x+2)^2}{4} = 1$

16. $\dfrac{x^2}{64} - \dfrac{(y-3)^2}{16} = 1$

17. $3x^2 - y^2 = 4$
18. $2y^2 - 5x^2 = 10$
19. $2x^2 - y^2 + 4x - 2y + 17 = 0$
20. $9x^2 - 16y^2 - 90x + 32y + 65 = 0$
21. $81x^2 - 25y^2 - 324x - 50y - 1726 = 0$
22. $9x^2 - 4y^2 + 16y = 0$

23. Find the foci of the hyperbola $2xy = 1$ (see Figure 6, page 578), and deduce its equation from the defining property of the hyperbola.

24. A line segment with both endpoints on a hyperbola is called a *chord* of the hyperbola, and a chord through either focus F_1 or F_2 perpendicular to the transverse axis is called a *latus rectum* (this is the chord AB or CD in Figure 36). Show that the latus rectum is of length $2b^2/a$, where $2a$ is the length of the transverse axis and $2b$ is the length of the conjugate axis.

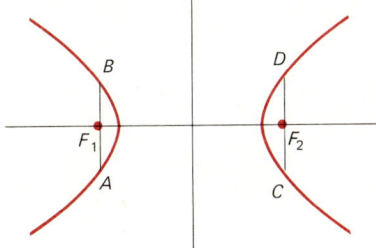

Figure 36

25. Find every point of the hyperbola $3y^2 - x^2 = 12$ for which the lines joining the point to the foci are perpendicular.
26. For what values of the constant c is the graph of the equation $x^2 - y^2 + 6x - 2y = c$ a hyperbola with a horizontal transverse axis? A hyperbola with a vertical transverse axis? A pair of intersecting lines?
27. Show that the curve with parametric equations

$$x = x_0 + a \cosh t, \quad y = y_0 + b \sinh t \quad (-\infty < t < \infty) \quad \text{(i)}$$

is the right branch of a hyperbola with a horizontal transverse axis if $a > 0$ and the left branch if $a < 0$. Write parametric equations for the branches of a hyperbola with a vertical transverse axis.

Find the points of intersection (if any) of the given line and hyperbola.
28. $2x - y - 10 = 0$, $x^2 - 4y^2 = 20$
29. $4x - 3y - 16 = 0$, $16x^2 - 25y^2 = 400$
30. $2x + y + 1 = 0$, $4x^2 - 9y^2 = 36$
31. $3x - y - 2 = 0$, $2x^2 - y^2 = 1$

32. Figure 37 shows a family of concentric circular arcs centered at F_1 and a family of concentric circles centered at F_2. The arcs and circles are labeled with their radii. Explain why the heavy dots in the figure lie on part of a hyperbola with foci F_1 and F_2.

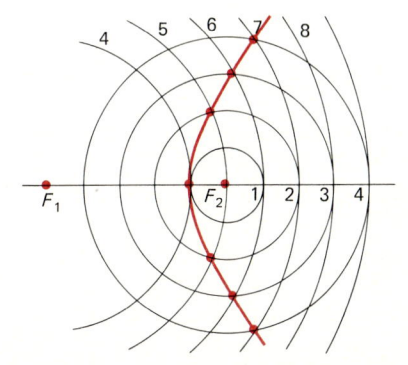

Figure 37

33. A hyperbola with foci F_1 and F_2 can be constructed as follows: To a drawing board attach one end of a ruler at F_1 in such a way that it is free to rotate about F_1. At the opposite end Q of the ruler fasten one end of a piece of string whose length s is less than the length r of the ruler, and fasten the other end of the string at F_2. Rotate the ruler about F_1, holding the string taut with a pencil, as in Figure 38. Then the point P of the pencil traces out part of a hyperbola.
Explain why this construction works.

Construction of a hyperbola

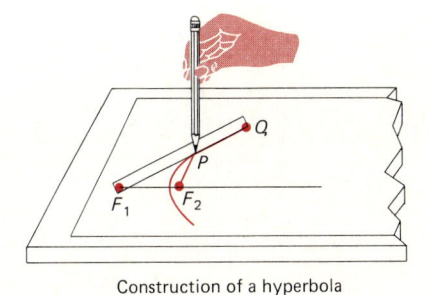

Figure 38

34. Each of the two hyperbolas

$$\dfrac{x^2}{a^2} - \dfrac{y^2}{b^2} = 1, \qquad \dfrac{x^2}{a^2} - \dfrac{y^2}{b^2} = -1$$

Section 10.4 Hyperbolas 603

is said to be *conjugate* to the other. Show that conjugate hyperbolas have the same asymptotes.

35. Find the eccentricity of a hyperbola whose transverse axis subtends an angle of 60° at either focus of its conjugate hyperbola.

36. Show that the distance from either focus of the hyperbola $(x^2/a^2) - (y^2/b^2) = 1$ to either asymptote is equal to b.

37. Find the hyperbola of eccentricity 2 whose foci coincide with those of the ellipse $9x^2 + 25y^2 = 225$.

38. Let T be the tangent line at any point $P = (x_0, y_0)$ of the hyperbola $(x^2/a^2) - (y^2/b^2) = 1$. Show that T is the line with equation

$$\frac{x_0 x}{a^2} - \frac{y_0 y}{b^2} = 1. \qquad \text{(ii)}$$

39. Find the two lines through the point $(-1, 7)$ tangent to the hyperbola $x^2 - y^2 = 16$.

40. Let T be the tangent to a hyperbola at any point P, and let A and B be the points in which T intersects the asymptotes of the hyperbola (see Figure 39, where the asymptotes intersect at O). Show that the point of tangency P is the midpoint of the segment AB, and show that the triangle OAB has the same area ab for every position of P.

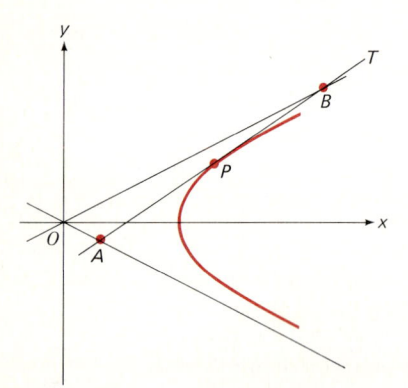

Figure 39

41. Let T be the tangent line at any point $P = (x_0, y_0)$ of the hyperbola $(x^2/a^2) - (y^2/b^2) = 1$. Show that T bisects the angle between the segments PF_1 and PF_2 joining P to the foci F_1 and F_2, as illustrated in Figure 40(a).

42. Show that a ray of light directed at one focus of a hyperbolic mirror is reflected toward the other focus, as illustrated in Figure 40(b).

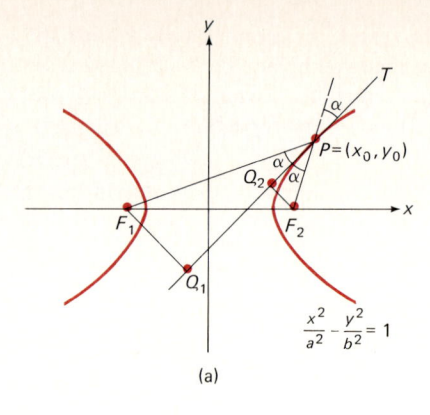

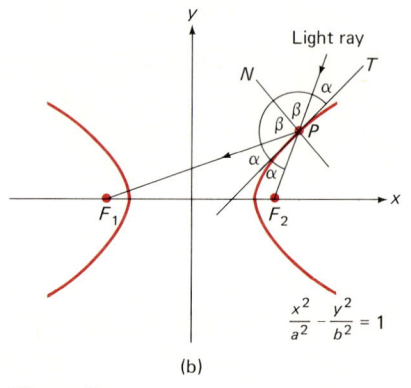

Figure 40

43. Three sentries are posted 1 km apart along a straight road leading to a tank depot. An explosion occurs while they are having a three-way telephone conversation, and this enables them to determine that the sentry nearest the depot heard the explosion 1 sec before the middle sentry and 3 sec before the sentry farthest from the depot. Find the two possible locations of the explosion, assuming that the speed of sound is $\frac{1}{3}$ km/sec.

44. Two radio beacons at positions A and B on a long straight coastline running north-south are synchronized to transmit simultaneous pulsed signals. The beacon at A is 540 km north of the one at B. Suppose the signal from A is received by a ship at sea 600 microseconds (that is, 600×10^{-6} sec) before the signal from B. How far offshore is the ship if it is due west of A? If it is due west of the coastal point 90 km south of A? (The speed of radio signals, equal to the speed of light, is about 300,000 km/sec, or equivalently 300 meters per microsecond.)

10.5 Conic Sections

The parabola, ellipse and hyperbola are all called *conic sections* for a reason to be explained presently. You will recall that the parabola was defined in terms of its "focus-directrix property," that is, as the set of all points equidistant from a fixed point called the focus and a fixed line called the directrix. As we now show, the ellipse and hyperbola also have a focus-directrix property, which involves the eccentricity e. This fact will allow us to give a unified treatment of all three conic sections, in which the eccentricity of the parabola is taken to be 1.

Consider the ellipse

$$\frac{x^2}{a^2} + \frac{y^2}{b^2} = 1, \qquad (1)$$

with foci $F_1 = (-c, 0)$, $F_2 = (c, 0)$, and eccentricity

$$e = \frac{c}{a} \qquad (0 < e < 1),$$

where

$$c^2 = a^2 - b^2.$$

The distances from a variable point $P = (x, y)$ of the ellipse to the foci are

$$|PF_1| = \sqrt{(x + c)^2 + y^2}$$

and

$$|PF_2| = \sqrt{(x - c)^2 + y^2}.$$

Therefore

$$|PF_1| = \sqrt{x^2 + 2cx + c^2 + b^2\left(1 - \frac{x^2}{a^2}\right)} = \sqrt{x^2\left(\frac{a^2 - b^2}{a^2}\right) + 2cx + c^2 + b^2}$$

$$= \sqrt{x^2 \frac{c^2}{a^2} + 2cx + a^2} = \sqrt{\left(a + \frac{c}{a}x\right)^2} = \sqrt{(a + ex)^2},$$

so that

$$|PF_1| = |a + ex|,$$

and similarly

$$|PF_2| = |a - ex|.$$

But $-a \le x \le a$, $0 < e < 1$, and therefore

$$|PF_1| = a + ex, \qquad |PF_2| = a - ex. \qquad (2)$$

Notice that

$$|PF_1| + |PF_2| = (a + ex) + (a - ex) = 2a,$$

as in the defining property of the ellipse given on page 588.

An analogous calculation for the hyperbola

$$\frac{x^2}{a^2} - \frac{y^2}{b^2} = 1, \qquad (3)$$

with foci $F_1 = (-c, 0)$, $F_2 = (c, 0)$, and eccentricity

$$e = \frac{c}{a} \qquad (e > 1),$$

Section 10.5 Conic Sections **605**

where
$$c^2 = a^2 + b^2,$$
gives
$$|PF_1| = \sqrt{x^2 + 2cx + c^2 + b^2\left(\frac{x^2}{a^2} - 1\right)} = \sqrt{x^2\left(\frac{a^2 + b^2}{a^2}\right) + 2cx + c^2 - b^2}$$
$$= \sqrt{x^2\frac{c^2}{a^2} + 2cx + a^2} = \sqrt{\left(a + \frac{c}{a}x\right)^2} = \sqrt{(a + ex)^2}.$$

Therefore
$$|PF_1| = |a + ex|,$$
as in the case of the ellipse, and similarly
$$|PF_2| = |a - ex|.$$
If P lies on the right branch of the hyperbola, then $x \geq a$ and
$$|PF_1| = a + ex, \qquad |PF_2| = -a + ex \tag{4}$$
(remember that $e > 1$), while if P lies on the left branch, then $x \leq -a$ and
$$|PF_1| = -a - ex, \qquad |PF_2| = a - ex. \tag{4'}$$
Notice that
$$|PF_1| - |PF_2| = 2a$$
in the first case and
$$|PF_2| - |PF_1| = 2a$$
in the second, just as on page 596.

Directrices of the Ellipse and Hyperbola

We now define the *directrices* of the ellipse (1) or the hyperbola (3) as the vertical lines L_1 and L_2 with equations $x = -a/e = -a^2/c$ and $x = a/e = a^2/c$, respectively (notice that L_1 and L_2 are a distance $h = 2a/e = 2a^2/c$ apart). The location of L_1 and L_2 relative to the ellipse (1) is shown in Figure 41(a); since $0 < e < 1$ for an ellipse, we have $a/e > a$, so that L_2 lies to the right of the vertex $(a, 0)$ while L_1 lies to the left of the vertex $(-a, 0)$. Similarly, Figure 41(b) shows the location of L_1 and L_2 relative to the hyperbola (3); since $e > 1$ for a hyperbola, this time we have $a/e < a$, so that L_2 lies to the left of the vertex $(a, 0)$ and to the right of the origin, while L_1 lies to the right of the vertex $(-a, 0)$ and to the left of the origin. Let Q_1 and Q_2 be the points in which the horizontal line through $P = (x, y)$ intersects L_1 and L_2. Then, as can be read off from the figures,
$$|PQ_1| = \frac{a}{e} + x, \qquad |PQ_2| = \frac{a}{e} - x \tag{5}$$
for the ellipse, while
$$|PQ_1| = \frac{a}{e} + x, \qquad |PQ_2| = -\frac{a}{e} + x \tag{6}$$

606 Chapter 10 Plane Analytic Geometry and Polar Coordinates

for the right branch of the hyperbola and

$$|PQ_1| = -\frac{a}{e} - x, \qquad |PQ_2| = \frac{a}{e} - x \qquad (6')$$

for the left branch. Comparing (2) with (5), (4) with (6), and (4') with (6'), we find in every case that

$$|PF_1| = e|PQ_1|, \qquad |PF_2| = e|PQ_2| \qquad (7)$$

(check the details).

It follows from (7) that *if P is a variable point of the ellipse* (1) *or hyperbola* (3), *then the distance from P to either focus is the product of the eccentricity e and the distance from P to the corresponding directrix.* Here the directrix corresponding to a given focus is always taken to be the directrix nearer the focus. The distance between each directrix and the corresponding focus is

$$d = \frac{a^2}{c} - c = \left(\frac{1}{e} - e\right)a \qquad (8)$$

for an ellipse and

$$d = c - \frac{a^2}{c} = \left(e - \frac{1}{e}\right)a \qquad (8')$$

for a hyperbola. (Each of the expressions (8) and (8') equals b^2/c, where b is half the length of the minor axis in the first case and half the length of the conjugate axis in the second case.) Similarly, the distance between each directrix and the corresponding vertex is

$$\frac{a^2}{c} - a = \left(\frac{1}{e} - 1\right)a$$

for an ellipse and

$$a - \frac{a^2}{c} = \left(1 - \frac{1}{e}\right)a$$

for a hyperbola. The directrices themselves are a distance $h = 2a/e = 2a^2/c$ apart, as noted above. These formulas can be used to find the directrices even if the ellipse or hyperbola does not have an equation of the form (1) or (3).

Thus we have just shown that every point P of an ellipse or hyperbola of eccentricity e satisfies the *focus-directrix property*

$$|PF| = e|PQ|, \qquad (9)$$

where F is either focus and Q is the foot of the perpendicular drawn from P to the directrix L nearer F. We now prove the converse, namely that every point $P = (x, y)$ satisfying (9) lies on an ellipse of eccentricity e if $0 < e < 1$ or on a hyperbola of eccentricity e if $e > 1$. To this end, let the focus be the point $(d, 0)$ and let the directrix L be the y-axis, as in Figure 42. Then

$$|PF| = \sqrt{(x-d)^2 + y^2}, \qquad |PQ| = |x|,$$

and (9) becomes

$$\sqrt{(x-d)^2 + y^2} = e|x|.$$

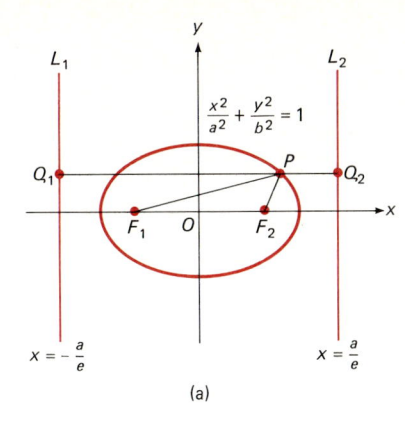

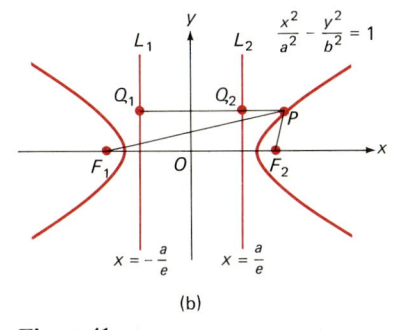

Figure 41

Focus-Directrix Property

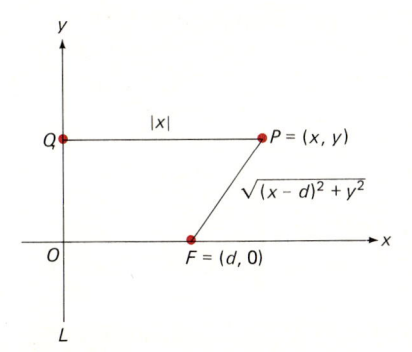

Figure 42

Section 10.5 Conic Sections 607

Therefore
$$(x - d)^2 + y^2 = e^2 x^2$$
or
$$(1 - e^2)x^2 - 2dx + y^2 + d^2 = 0,$$
where $1 - e^2 \neq 0$. Factoring out $1 - e^2$ and completing the square, we can write this equation as
$$(1 - e^2)\left(x - \frac{d}{1 - e^2}\right)^2 + y^2 = \frac{d^2}{1 - e^2} - d^2 = \frac{e^2 d^2}{1 - e^2},$$
which is equivalent to
$$\left(x - \frac{d}{1 - e^2}\right)^2 + \frac{y^2}{1 - e^2} = \left(\frac{ed}{1 - e^2}\right)^2.$$
The last equation can in turn be written in the form
$$\frac{\left(x - \dfrac{d}{1 - e^2}\right)^2}{\left(\dfrac{ed}{1 - e^2}\right)^2} + \frac{y^2}{\left(\dfrac{ed}{\sqrt{1 - e^2}}\right)^2} = 1 \tag{10}$$
if $0 < e < 1$, or in the form
$$\frac{\left(x + \dfrac{d}{e^2 - 1}\right)^2}{\left(\dfrac{ed}{e^2 - 1}\right)^2} - \frac{y^2}{\left(\dfrac{ed}{\sqrt{e^2 - 1}}\right)^2} = 1 \tag{10'}$$
if $e > 1$. But (10) is the equation of an ellipse, while (10') is the equation of a hyperbola. As an exercise, describe the ellipse (10) and the hyperbola (10'), and verify that each has eccentricity e.

The following theorem summarizes all these considerations, together with the observation that for $e = 1$ formula (9) becomes the defining property of the parabola (see page 581).

> **Theorem 1** *(Focus-directrix property of conic sections)*. *Given a positive number e, the distance from a variable point P to a fixed point F is equal to e times the distance from P to a fixed line L not containing F if and only if P lies on a conic section with focus F, directrix L and eccentricity e. The conic section is*
>
> (i) *An ellipse if $0 < e < 1$;*
> (ii) *A parabola if $e = 1$;*
> (iii) *A hyperbola if $e > 1$.*

The meaning of this theorem is illustrated in Figure 43, where the focus is the point $(1, 0)$, the directrix is the y-axis, and the eccentricity e takes the values 0.5, 0.75, 1, 1.5 and 2. The parabola is the graph of $y^2 = 2x - 1$, and the other curves are the graphs of (10) and (10') for $d = 1$ and the indicated values of e.

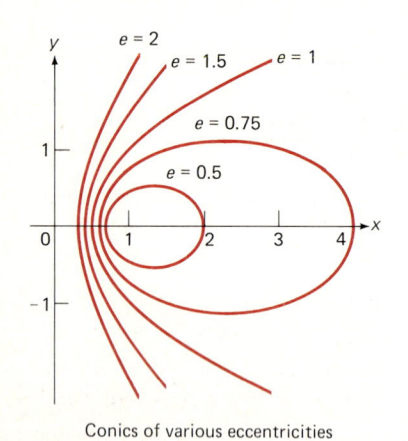

Conics of various eccentricities
Figure 43

608 Chapter 10 Plane Analytic Geometry and Polar Coordinates

Example 1 Find an equation of the ellipse with vertices $(\pm 6, 0)$ whose directrices are a distance $h = \frac{72}{5}$ apart. What is its eccentricity?

Solution The ellipse has an equation of standard form $(x^2/a^2) + (y^2/b^2) = 1$, where $2a$ is the length of the major axis and $2b$ the length of the minor axis. If $2c$ is the distance between the foci, then $2a^2/c$ is the distance between the directrices. We are given that

$$a = 6, \qquad h = \frac{2a^2}{c} = \frac{72}{5},$$

and consequently $c = 5$, $b^2 = a^2 - c^2 = 36 - 25 = 11$. Therefore an equation of the ellipse is

$$\frac{x^2}{36} + \frac{y^2}{11} = 1,$$

and its eccentricity is

$$e = \frac{c}{a} = \frac{5}{6}.$$

Figure 44 shows the ellipse, together with its foci $(\pm 5, 0)$ and directrices $x = \pm \frac{36}{5}$. □

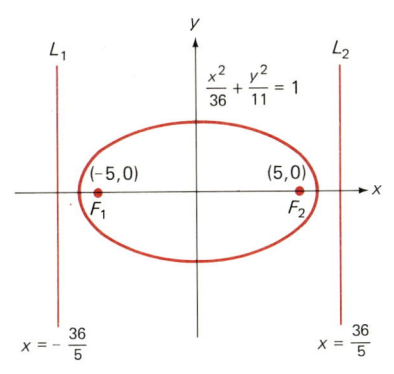

Figure 44

Example 2 Find an equation of the hyperbola with vertices $(0, \pm 2)$ and eccentricity $e = \frac{3}{2}$. Where are the directrices?

Solution The hyperbola in question has an equation of standard form $(y^2/a^2) - (x^2/b^2) = 1$, where $2a$ is the length of the transverse axis and $2b$ the length of the conjugate axis. If $2c$ is the distance between the foci, then c/a is the eccentricity. We are given that

$$a = 2, \qquad e = \frac{c}{a} = \frac{3}{2},$$

so that $c = 3$, $b^2 = c^2 - a^2 = 9 - 4 = 5$. Therefore an equation of the hyperbola is

$$\frac{y^2}{4} - \frac{x^2}{5} = 1,$$

and the distance between its directrices is

$$h = \frac{2a^2}{c} = \frac{2a}{e} = \frac{8}{3}.$$

Figure 45 shows the hyperbola, together with its foci $(0, \pm 3)$ and its directrices $y = \pm \frac{4}{3}$. □

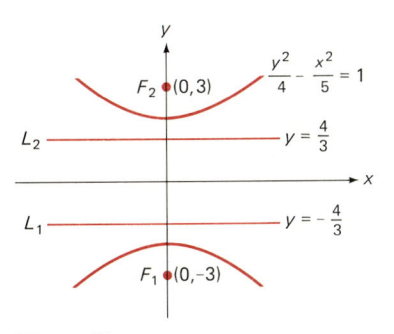

Figure 45

As promised, we now explain why the parabola, ellipse and hyperbola are all called *conic sections*, or simply *conics*. The name suggests that conics have something to do with a cone, and indeed each is the curve of intersection of a double right circular cone and a suitable cutting plane. Figure 46 shows such a cone. It is a surface consisting of two parts called *nappes*, which meet at a point V. This point is called the *vertex* of the cone, and every line

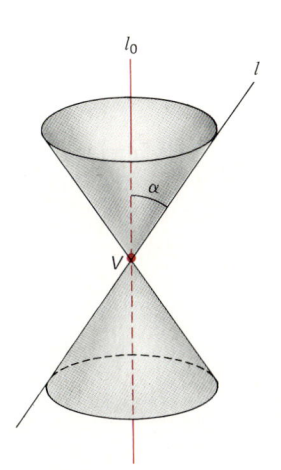

Right circular cone of two nappes
Figure 46

Section 10.5 Conic Sections   609

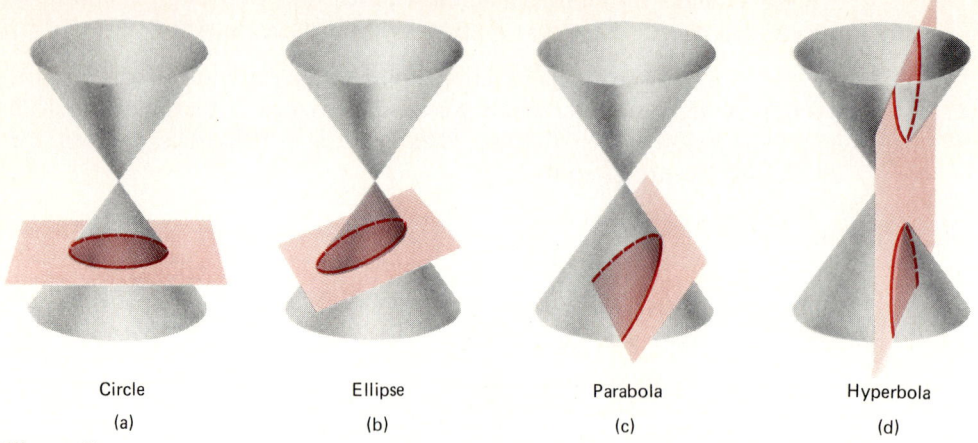

Figure 47

Circle (a)   Ellipse (b)   Parabola (c)   Hyperbola (d)

l lying on the cone and passing through V is called a *generator* of the cone and makes the same acute angle α with the axis l_0 of the cone (see the figure). In fact, any generator l sweeps out both nappes of the cone if it is revolved about l_0 while holding the angle α fixed. Let Π (capital Greek pi) be a plane which makes an angle β with the axis of the cone and does not go through the vertex of the cone. Then the curve, called a *conic section*, in which the plane Π intersects the cone is

(i) A circle if $\beta = 90°$, as in Figure 47(a);
(ii) An ellipse if $\alpha < \beta < 90°$, as in Figure 47(b);
(iii) A parabola if $\alpha = \beta$, as in Figure 47(c);
(iv) A hyperbola if $0 \le \beta < \alpha$, as in Figure 47(d).

Notice that it is only in case (iv) that the plane cuts both nappes of the cone. The proof of (i) is immediate (use congruent triangles). The proof of (ii)–(iv) requires more ingenuity, and the details will be given only for the ellipse. However, the same argument works with minor modifications for the parabola and the hyperbola.

Optional

Thus we now investigate case (ii) more closely, making the construction shown in Figure 48. Let E be the curve in which the plane Π intersects one nappe of the cone with vertex V, and inscribe in the cone a sphere S which is tangent to Π at a point F and to the cone along a circle C. Let P be any point of the curve E, and let A be the point in which the generator of the cone through P intersects C. Also let Σ be the plane of the circle C, let L be the line of intersection of the planes Σ and Π, and let Q be the point of intersection of L and the perpendicular drawn from P to L.

The segments PA and PF have the same length, that is,

$$|PA| = |PF|, \tag{11}$$

since they are tangents drawn from the same point P to the sphere S. Let B be the foot of the perpendicular dropped from P to the plane Σ, so that PB

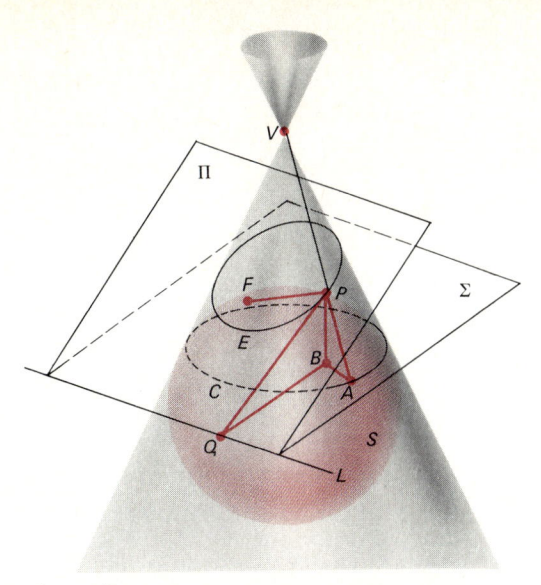

Figure 48

is parallel to the axis of the cone. Then

$$\frac{|PB|}{|PA|} = \cos \alpha, \qquad (12)$$

where α is the angle between any generator and the axis of the cone, while

$$\frac{|PB|}{|PQ|} = \cos \beta, \qquad (12')$$

where β is the angle between the plane Π and the axis (see Figure 49, an enlargement of part of Figure 48). Dividing (12') by (12), we get

$$\frac{|PA|}{|PQ|} = \frac{\cos \beta}{\cos \alpha},$$

or equivalently

$$|PF| = \frac{\cos \beta}{\cos \alpha} |PQ|, \qquad (13)$$

because of (11). But $\alpha < \beta < 90°$, and hence the ratio $\cos \beta / \cos \alpha$ is a number between 0 and 1. Denoting this number by e, we can write (13) in the form

$$|PF| = e|PQ|,$$

which you will recognize as the definition of an ellipse in terms of its focus-directrix property. Thus the curve E is the ellipse with focus F, directrix L and eccentricity e, and the proof of (ii) is complete.

There are also three "degenerate" conics, obtained when the cutting plane goes through the vertex of the cone. They are

(i) A pair of intersecting lines if the plane contains two generators, as in Figure 50(a);

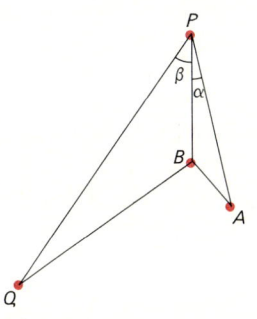

Figure 49

Section 10.5   Conic Sections   **611**

(a) (b) (c)

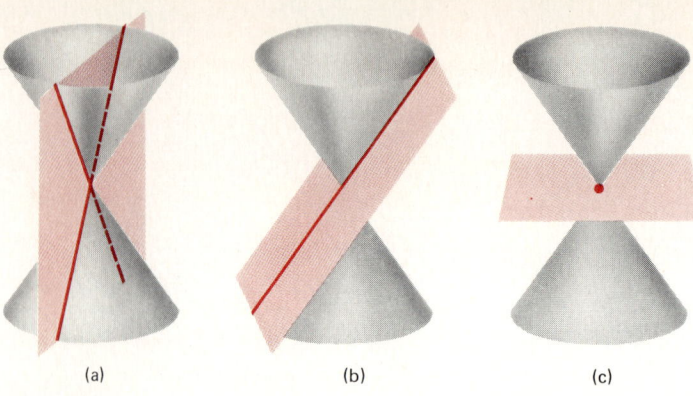

Figure 50

(ii) A single line if the plane contains only one generator, as in Figure 50(b);

(iii) A single point if the plane does not contain a generator, as in Figure 50(c).

We will find in the next section that conics, both nondegenerate and degenerate, arise naturally in studying the graph of the general second degree equation in two variables.

Problems

Let h be the distance between the directrices of an ellipse centered at the origin. Find an equation of the ellipse if

1. $h = 18$, the major axis is vertical and its length is 12
2. $h = 5$, the major axis is horizontal and the distance between the foci is 4
3. $h = 13$, the major axis is horizontal and the minor axis is of length 6
4. $h = 16$, the major axis is vertical and the eccentricity is $\frac{1}{2}$

Let h be the distance between the directrices of a hyperbola centered at the origin. Find an equation of the hyperbola if

5. $h = \frac{32}{5}$, the transverse axis is horizontal and the conjugate axis is of length 6
6. $h = 4$, the transverse axis is vertical and its length is 8
7. $h = \frac{8}{3}$, the transverse axis is vertical and the eccentricity is $\frac{3}{2}$
8. $h = \frac{16}{5}$, the transverse axis is horizontal and the asymptotes are $y = \pm\frac{3}{4}x$

Find the eccentricity and the directrices of

9. The ellipse $25x^2 + 9y^2 = 225$
10. The ellipse $x^2 + 4y^2 + 4x - 8y + 7 = 0$ (see Figure 23, page 591)
11. The ellipse $3x^2 - 2xy + 3y^2 - 2 = 0$ (see Problem 21, page 594)
12. The hyperbola $11x^2 - 25y^2 = 275$
13. The hyperbola $2xy = 1$ (see Figure 6, page 578)
14. The hyperbola $x^2 - 4y^2 - 2x + 16y - 14 = 0$ (see Figure 35, page 601)

15. Find the path traversed by a point $P = (x, y)$ which moves in such a way that its distance from the line $x = -4$ is always twice its distance from the point $(-1, 0)$.
16. Find the path traversed by a point $P = (x, y)$ which moves in such a way such that its distance from the point $(-8, 0)$ is always twice its distance from the line $x = -2$.
17. The point $(7, 0)$ lies on a hyperbola with the point $F = (-4, -2)$ as a focus and the line $x = -3$ as the directrix nearer F. Write the equation of the hyperbola in standard form.
18. The point $(-1, 5)$ lies on an ellipse with the point $F = (3, 2)$ as a focus and the line $x = \frac{21}{4}$ as the directrix nearer F. Write the equation of the ellipse in standard form.
19. Find an equation of the ellipse of eccentricity $\frac{1}{2}$ with the point $F = (2, 0)$ as a focus and the line $x + y - 1 = 0$ as the directrix nearer F.
20. Find an equation of the hyperbola of eccentricity 2 with the point $F = (-1, 1)$ as a focus and the line $3x - 4y + 2 = 0$ as the directrix nearer F.
21. The sphere S in Figure 48 lies below the cutting plane Π and is tangent to Π at a point F and to the cone along a circle C. One can also inscribe another sphere S' in the cone, which lies *above* Π and is tangent to Π at a point F' and to the cone along a circle C'. Then the same argument as used to show that F is one focus of the ellipse E in which Π intersects the cone shows that F' is the other focus. Use this construction to give a direct proof of the two-focus property of E, namely that $|PF| + |PF'| = 2a$. What is the geometric meaning of the constant $2a$ in terms of the cone and spheres?

612   Chapter 10   Plane Analytic Geometry and Polar Coordinates

10.6 Second Degree Curves (Optional)

By a *second degree equation in two variables x and y* we mean an equation of the form

$$Ax^2 + Bxy + Cy^2 + Dx + Ey + F = 0, \qquad (1)$$

where the coefficients A, B and C are not all zero (otherwise (1) reduces to a first degree or linear equation). The graph of any such equation is called a *second degree curve*, and can be found in the following way.

Were it not for the presence of the "cross product term" Bxy, equation (1) would be of a type with which we are already familiar (see below). Thus if $B \neq 0$, the first step in the analysis of (1) is to go over to a new coordinate system in which there is no cross product term. To this end, we rotate the xy-system about its origin O through an angle θ in the counterclockwise direction, thereby producing a new $x'y'$-system with the same origin. The old and new coordinates are related by the rotation equations

$$\begin{aligned} x &= x' \cos\theta - y' \sin\theta, \\ y &= x' \sin\theta + y' \cos\theta, \end{aligned} \qquad (2)$$

derived on page 577. Substituting (2) into (1), we get

$$\begin{aligned} &A(x'\cos\theta - y'\sin\theta)^2 + B(x'\cos\theta - y'\sin\theta)(x'\sin\theta + y'\cos\theta) \\ &+ C(x'\sin\theta + y'\cos\theta)^2 + D(x'\cos\theta - y'\sin\theta) \\ &+ E(x'\sin\theta + y'\cos\theta) + F = 0. \end{aligned}$$

After collecting terms, this rather formidable equation reduces to

$$A'x'^2 + B'x'y' + Cy'^2 + D'x' + E'y' + F' = 0, \qquad (1')$$

in terms of the new coefficients

$$\begin{aligned} A' &= A\cos^2\theta + B\cos\theta\sin\theta + C\sin^2\theta, \\ B' &= B(\cos^2\theta - \sin^2\theta) + 2(C - A)\sin\theta\cos\theta, \\ C' &= A\sin^2\theta - B\sin\theta\cos\theta + C\cos^2\theta, \\ D' &= D\cos\theta + E\sin\theta, \\ E' &= -D\sin\theta + E\cos\theta, \\ F' &= F. \end{aligned} \qquad (3)$$

Elimination of the Cross Product Term

We now choose a value of θ such that $B' = 0$, that is, such that

$$B' = B(\cos^2\theta - \sin^2\theta) + 2(C - A)\sin\theta\cos\theta = 0.$$

With the help of the double-angle formulas (see page 58), this equation can be written as

$$B\cos 2\theta + (C - A)\sin 2\theta = 0,$$

or equivalently

$$\cot 2\theta = \frac{A - C}{B} \qquad (4)$$

Section 10.6 Second Degree Curves 613

after dividing both sides by $B \sin 2\theta$. Inspection of the graph of the cotangent function (see Figure 24, page 63) shows that we can always find an angle θ in the interval $(0, \pi/2)$ satisfying (4), regardless of the value of $(A - C)/B$.

With this choice of θ, B' equals zero and equation (1') takes the form

$$A'x'^2 + C'y'^2 + D'x' + E'y' + F' = 0, \tag{5}$$

in which there is no cross product term. We are then back on familiar ground. If $A'C' = 0$, then either $A' \neq 0$, $C' = 0$ or $A' = 0$, $C' \neq 0$ (why is the case $A' = C' = 0$ impossible?). Therefore (5) can be written as either

$$x'^2 + ax' + by' + c = 0 \tag{6}$$

or

$$y'^2 + ax' + by' + c = 0, \tag{6'}$$

where

$$a = \frac{D'}{A'}, \quad b = \frac{E'}{A'}, \quad c = \frac{F'}{A'} \tag{7}$$

in the first case and

$$a = \frac{D'}{C'}, \quad b = \frac{E'}{C'}, \quad c = \frac{F'}{C'}$$

in the second. If $A'C' > 0$, we can write (5) in the form

$$x'^2 + ky'^2 + ax' + by' + c = 0, \tag{8}$$

where $k = C'/A' > 0$ and a, b, c are given by (7), while if $A'C' < 0$, (5) can be written as

$$x'^2 - ky'^2 + ax' + by' + c = 0, \tag{8'}$$

where this time $k = -C'/A' > 0$ and a, b, c are again given by (7). But (6), (6'), (8) and (8') are precisely the equations already investigated in detail on pages 583, 591 and 601, apart from slight differences in notation. In this way, any second degree equation can be transformed into an equation we already know how to handle.

If $b \neq 0$, the graph of equation (6) is a parabola with the y'-axis as its axis of symmetry, while if $a \neq 0$, the graph of (6') is a parabola with the x'-axis as its axis of symmetry. If these conditions are not satisfied, that is, if (6) reduces to

$$x'^2 + ax' + c = 0 \tag{9}$$

and (6') to

$$y'^2 + by' + c = 0, \tag{9'}$$

the graph will degenerate into a pair of parallel lines, a single line or the empty set (the set containing no points at all). In the case of (9), this is because the equation has two solutions

$$x' = \frac{-a \pm \sqrt{a^2 - 4c}}{2}$$

if $a^2 > 4c$, a single solution $x' = -a/2$ if $a^2 = 4c$, and no (real) solution if $a^2 < 4c$, and an analogous statement can be made about the solutions of (9').

As for equation (8), its graph is an ellipse (possibly a circle if $k = 1$), a single point or the empty set (see page 591), while the graph of (8′) is a hyperbola or a pair of intersecting lines (see page 601). If the graph is an ellipse or a hyperbola, its axes are parallel to the coordinate axes of the $x'y'$-plane.

Second Degree Curves as Conics

Taking all these possibilities into account, we see that every second degree curve must be an ellipse, circle, parabola or hyperbola, that is, one of the conics, or a pair of intersecting lines, a pair of parallel lines, a single line, a single point or the empty set. On page 612 we classified a pair of intersecting lines, a single line and a single point as degenerate conics, and showed that each of these "figures" is the intersection of a plane and a right circular cone of two nappes. We now decide to regard a pair of parallel lines and the empty set as degenerate conics too. The merit of this convention is that it allows us to assert that *every second degree curve is a conic, nondegenerate or degenerate*.

> **Remark** It is somewhat of a misnomer to call a pair of parallel lines or the empty set a degenerate *conic*, for it is geometrically apparent that the intersection of a plane and an infinite double cone cannot be a pair of parallel lines, nor can it be empty. Accordingly, it is customary to enlarge the definition of a conic to include intersections of a plane with a right circular *cylinder*, as well as with a right circular cone. We can then get a pair of parallel lines as in Figure 51(a), or the empty set (no intersection at all) as in Figure 51(b). We can also get a circle or a straight line (how?), or an ellipse as in Problem 17.

With the help of the identity

$$\cot 2\theta = \frac{\cos^2 \theta - \sin^2 \theta}{2 \sin \theta \cos \theta} = \frac{\cot^2 \theta - 1}{2 \cot \theta},$$

we can write (4) in the form

$$\frac{\cot^2 \theta - 1}{2 \cot \theta} = \frac{A - C}{B}. \tag{10}$$

(a)   (b)

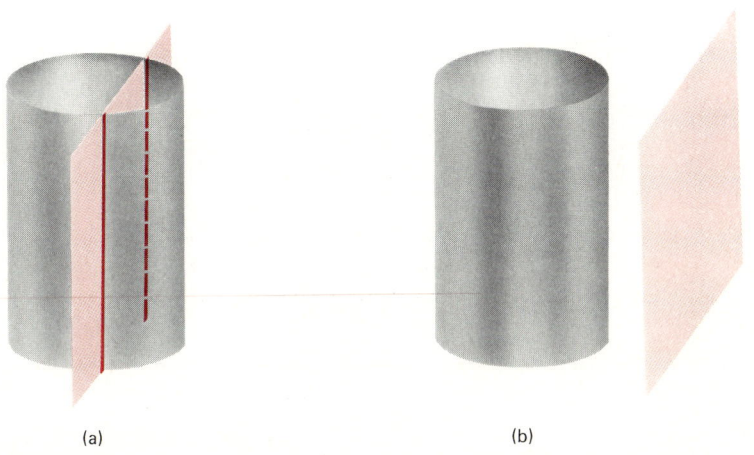

Figure 51

This is equivalent to the quadratic equation
$$B \cot^2 \theta - 2(A - C) \cot \theta - B = 0,$$
with two real solutions
$$\cot \theta = \frac{(A - C) \pm \sqrt{(A - C)^2 + B^2}}{B}.$$

One of these solutions is positive and the other is negative (why?). Choosing the positive solution, we can find the corresponding values of $\cos \theta$ and $\sin \theta$ needed to calculate the coefficients (3) by using the formulas
$$\cos \theta = \frac{\cot \theta}{\sqrt{1 + \cot^2 \theta}}, \qquad \sin \theta = \frac{1}{\sqrt{1 + \cot^2 \theta}}. \qquad (11)$$

(verify these formulas).

Example 1 Find the graph of the equation
$$5x^2 + 4xy + 2y^2 - 24x - 12y + 18 = 0. \qquad (12)$$

Solution Here $A = 5$, $B = 4$, $C = 2$, $D = -24$, $E = -12$, $F = 18$, and (10) becomes
$$\frac{\cot^2 \theta - 1}{2 \cot \theta} = \frac{3}{4}$$
or equivalently
$$2 \cot^2 \theta - 3 \cot \theta - 2 = 0,$$
with solutions
$$\cot \theta = \frac{3 \pm \sqrt{9 + 16}}{4} = \frac{3 \pm 5}{4}.$$
Choosing the plus sign to make $\cot \theta$ positive, we obtain
$$\cot \theta = \frac{8}{4} = 2.$$
Therefore, by (11) or simply by inspection of the right triangle in Figure 52,
$$\cos \theta = \frac{2}{\sqrt{1 + 4}} = \frac{2}{\sqrt{5}}, \qquad \sin \theta = \frac{1}{\sqrt{1 + 4}} = \frac{1}{\sqrt{5}},$$

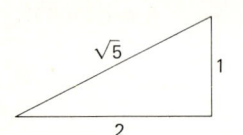

Figure 52

and for these values of $\cos \theta$ and $\sin \theta$ the coefficients (3) are equal to
$$A' = \frac{4}{5} A + \frac{2}{5} B + \frac{1}{5} C = 4 + \frac{8}{5} + \frac{2}{5} = 6,$$
$$B' = 0,$$
$$C' = \frac{1}{5} A - \frac{2}{5} B + \frac{4}{5} C = 1 - \frac{8}{5} + \frac{8}{5} = 1,$$
$$D' = \frac{2}{\sqrt{5}} D + \frac{1}{\sqrt{5}} E = -\frac{48}{\sqrt{5}} - \frac{12}{\sqrt{5}} = -\frac{60}{\sqrt{5}} = -12\sqrt{5},$$

$$E' = -\frac{1}{\sqrt{5}}D + \frac{2}{\sqrt{5}}E = \frac{24}{\sqrt{5}} - \frac{24}{\sqrt{5}} = 0,$$

$$F' = F = 18.$$

Hence a counterclockwise rotation of axes through the angle arccot $2 \approx 26.5°$ transforms equation (12) into

$$A'x'^2 + C'y'^2 + D'x' + E'y' + F' = 6x'^2 + y'^2 - 12\sqrt{5}x' + 18 = 0.$$

The last equation can be written as

$$6(x' - \sqrt{5})^2 + y'^2 = 12$$

after completing the square, which is equivalent to

$$\frac{(x' - \sqrt{5})^2}{2} + \frac{y'^2}{12} = 1$$

after dividing by 12. This is the equation in standard form of an ellipse with center $(\sqrt{5}, 0)$ in the $x'y'$-system, major axis along the line $x' = \sqrt{5}$ and minor axis along the x'-axis. The semimajor axis a is $\sqrt{12} = 2\sqrt{3}$, and the semiminor axis b is $\sqrt{2}$. Substituting $x' = \sqrt{5}$, $y' = 0$ into the rotation equations (2), we find that the coordinates of the center of the ellipse in the xy-system are

$$x = \sqrt{5}\cos\theta - 0\sin\theta = \sqrt{5}\frac{2}{\sqrt{5}} = 2,$$

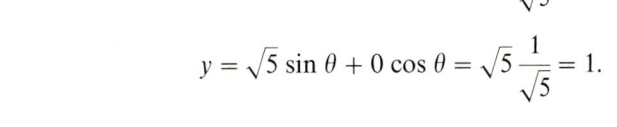

$$y = \sqrt{5}\sin\theta + 0\cos\theta = \sqrt{5}\frac{1}{\sqrt{5}} = 1.$$

Thus the graph of (12) is the ellipse shown in Figure 53. An interesting feature of this graph is considered in Problem 15. □

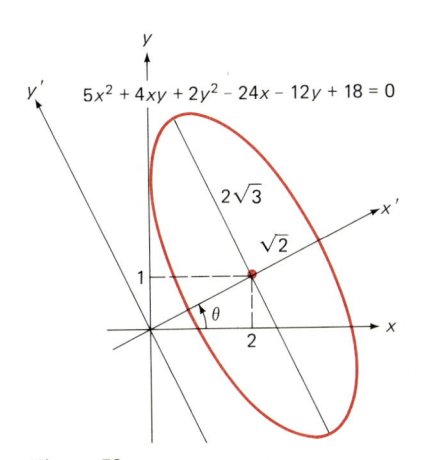

Figure 53

Example 2 Find the graph of the equation

$$9x^2 - 12xy + 4y^2 - 52x - 78y - 338 = 0. \tag{13}$$

Solution This time $A = 9$, $B = -12$, $C = 4$, $D = -52$, $E = -78$, $F = -338$, and (10) becomes

$$\frac{\cot^2\theta - 1}{2\cot\theta} = -\frac{5}{12}$$

or equivalently

$$6\cot^2\theta + 5\cot\theta - 6 = 0,$$

with solutions

$$\cot\theta = \frac{-5 \pm \sqrt{25 + 144}}{12} = \frac{-5 \pm 13}{12}.$$

Choosing the plus sign to make $\cot\theta$ positive, we obtain

$$\cot\theta = \frac{8}{12} = \frac{2}{3}.$$

Section 10.6 Second Degree Curves 617

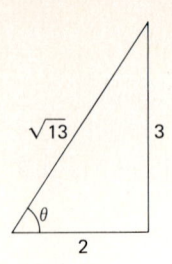

Figure 54

Therefore

$$\cos\theta = \frac{\frac{2}{3}}{\sqrt{1+\frac{4}{9}}} = \frac{2}{\sqrt{13}}, \qquad \sin\theta = \frac{1}{\sqrt{1+\frac{4}{9}}} = \frac{3}{\sqrt{13}},$$

by (11) or by inspection of the right triangle in Figure 54. For these values of $\cos\theta$ and $\sin\theta$ the coefficients (3) are equal to

$$A' = \frac{4}{13}A + \frac{6}{13}B + \frac{9}{13}C = \frac{36}{13} - \frac{72}{13} + \frac{36}{13} = 0,$$

$$B' = 0,$$

$$C' = \frac{9}{13}A - \frac{6}{13}B + \frac{4}{13}C = \frac{81}{13} + \frac{72}{13} + \frac{16}{13} = \frac{169}{13} = 13,$$

$$D' = \frac{2}{\sqrt{13}}D + \frac{3}{\sqrt{13}}E = \frac{2(-52)}{\sqrt{13}} + \frac{3(-78)}{\sqrt{13}} = -\frac{338}{\sqrt{13}} = -26\sqrt{13},$$

$$E' = -\frac{3}{\sqrt{13}}D + \frac{2}{\sqrt{13}}E = -\frac{3(-52)}{\sqrt{13}} + \frac{2(-78)}{\sqrt{13}} = 0,$$

$$F' = F = -338.$$

Hence a counterclockwise rotation of axes through the angle $\operatorname{arccot}\frac{2}{3} \approx 56.3°$ transforms equation (13) into

$$A'x'^2 + C'y'^2 + D'x' + E'y' + F' = 13y'^2 - 26\sqrt{13}\,x' - 338 = 0.$$

The last equation is equivalent to

$$y'^2 = 2\sqrt{13}(x' + \sqrt{13}),$$

which is the equation of a parabola with vertex $(-\sqrt{13}, 0)$ in the $x'y'$-system and the x'-axis as its axis of symmetry. Substituting $x' = -\sqrt{13}$, $y' = 0$ into the rotation equations (2), we find that the coordinates of the vertex of the parabola in the xy-system are

$$x = -\sqrt{13}\cos\theta - 0\sin\theta = -\sqrt{13}\,\frac{2}{\sqrt{13}} = -2,$$

$$y = -\sqrt{13}\sin\theta + 0\cos\theta = -\sqrt{13}\,\frac{3}{\sqrt{13}} = -3,$$

while the axis of symmetry is the line through the origin with inclination θ, that is, the line

$$y = x\tan\theta = \frac{x}{\cot\theta} = \frac{3}{2}x.$$

$9x^2 - 12xy + 4y^2 - 52x - 78y - 338 = 0$

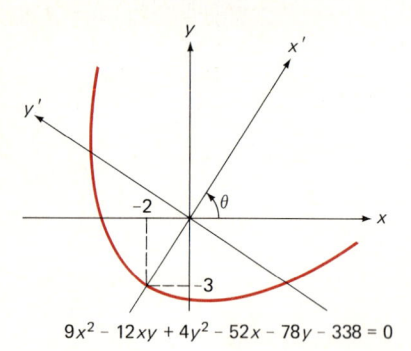

Figure 55

Thus the graph of (13) is the parabola shown in Figure 55. □

Suppose x^2 and y^2 have the same coefficient in the original second degree equation. Then $A = C$ and equation (10) becomes $\cot 2\theta = 0$, with solution $\theta = 45°$. Therefore in this case a $45°$ rotation of axes eliminates the cross product term. For example, a counterclockwise rotation of axes

618 Chapter 10 Plane Analytic Geometry and Polar Coordinates

through 45° transforms the equations

$$2xy - 1 = 0, \quad x^2 - 2xy + y^2 - 8x - 8y = 0$$

into the equations

$$x'^2 - y'^2 - 1 = 0, \quad y'^2 - 4\sqrt{2}x' = 0,$$

respectively, as you can check by substituting

$$x = x' \cos 45° - y' \sin 45° = \frac{1}{\sqrt{2}}(x' - y'),$$

$$y = x' \sin 45° + y' \cos 45° = \frac{1}{\sqrt{2}}(x' + y').$$

If these equations look familiar, it is because they have already appeared in two earlier examples involving 45° rotations, namely Example 4, page 578, and Examples 3–4, pages 583–584. You should now take another look at these examples from the standpoint of the theory developed in this section.

Finally we note that there is a simple test, given in Problem 19, for determining the nature of a second degree curve without carrying out explicit calculations like those in Examples 1 and 2.

Problems

After making a suitable rotation of axes, identify the conic which is the graph of the given equation. Sketch the conic if it is nondegenerate.

1. $3x^2 + 10xy + 3y^2 - 2x - 14y - 13 = 0$
2. $x^2 - 6xy + 9y^2 + 4x - 12y + 4 = 0$
3. $5x^2 - 2xy + 5y^2 - 4x + 20y + 20 = 0$
4. $19x^2 + 6xy + 11y^2 + 38x + 6y + 29 = 0$
5. $160x^2 - 56xy + 265y^2 - 2448x - 612y + 7956 = 0$
6. $16x^2 - 8xy + y^2 + 12x - 3y - 10 = 0$
7. $4x^2 + 24xy + 11y^2 + 16x + 38y + 15 = 0$
8. $5x^2 - 6xy + 5y^2 - 32 = 0$
9. $50x^2 - 8xy + 35y^2 + 100x - 8y + 67 = 0$
10. $5x^2 + 24xy - 5y^2 = 0$
11. $x^2 + 6xy + 9y^2 - 50x - 50y + 100 = 0$
12. $13x^2 + 10xy + 13y^2 + 16x - 16y - 272 = 0$
13. $25x^2 - 20xy + 4y^2 - 15x + 6y - 4 = 0$
14. $41x^2 + 24xy + 34y^2 + 34x - 112y + 129 = 0$

15. Prove that the ellipse in Example 1 is tangent to the y-axis, as shown in Figure 53. Find the point of tangency.
16. Show that equation (4) is satisfied by two distinct values of θ in the interval $(-\pi/2, \pi/2)$ which differ by $\pi/2$. Each choice leads to an equation of the form (5). How do the two equations differ?
17. By using the construction in Figure 56(a), show that the curve E in which a plane Π intersects a right circular cylinder is an ellipse if Π makes an acute angle with the axis of the cylinder. Also show that E is an ellipse by using the construction in Figure 56(b). (Here S and S' are spheres inscribed in the cylinder, which they intersect in the circles C and C'.)

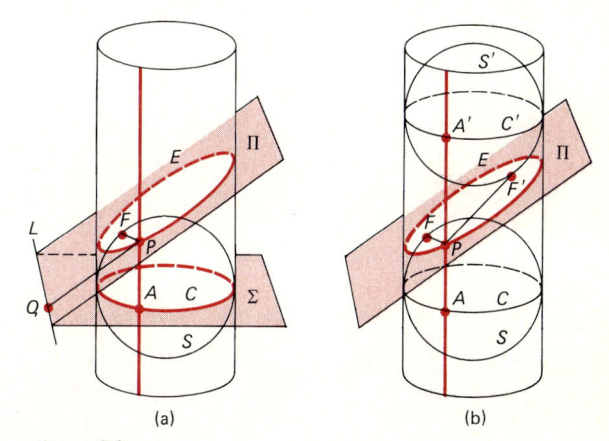

Figure 56

18. With the same notation as in (3), verify that

$$A' + C' = A + C, \quad B'^2 - 4A'C' = B^2 - 4AC$$

for every choice of the angle of rotation θ.

19. The quantity $\Delta = B^2 - 4AC$ is called the *discriminant* of the equation $Ax^2 + Bxy + Cy^2 + Dx + Ey + F = 0$. Thus the second formula in the preceding problem can be written

concisely as $\Delta' = \Delta$, and it says that the discriminant is *rotation invariant*, that is, does not change after a rotation of axes. Use this fact to show that the graph of a second degree equation is

(i) An ellipse (possibly a circle, a single point or the empty set) if $\Delta < 0$;

(ii) A parabola (possibly a pair of parallel lines, a single line or the empty set) if $\Delta = 0$;

(iii) A hyperbola (possibly a pair of intersecting lines) if $\Delta > 0$.

20. Test the validity of Problem 19 by applying it to Problems 1–14.

10.7 Polar Coordinates

Until now the position of a point in the plane has been specified by giving its rectangular or cartesian coordinates, defined in the way described on page 21. However, it is often convenient to specify the position of a point P in the plane in another way, by giving its "polar coordinates." These are defined as follows. Let l be a fixed ray or half-line, called the *polar axis*, emanating from a fixed point O, called the *origin* or *pole* (l is usually drawn horizontally and to the right, as in Figure 57). Let $r = |OP|$ be the distance between O and P, and let θ be the angle between l and the segment OP, measured from l to OP in the counterclockwise direction. Then the point P is said to have *polar coordinates* r and θ, and we denote P by the ordered pair (r, θ), writing $P = (r, \theta)$. For the purpose of drawing graphs, we can measure θ in either degrees or radians, but radian measure is usually called for in calculus problems. If $P = (r, \theta)$, we call r the *radial coordinate* and θ the *angular coordinate* of P.

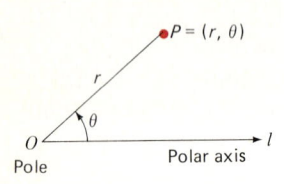

Figure 57

We will also allow r to take negative values. This is done by defining $P = (r, \theta)$, where $r < 0$, to be the reflection of the point $P' = (|r|, \theta)$ in the origin O. In other words, to find the point P with polar coordinates $r < 0$ and θ, instead of measuring off $|r|$ units along the ray making angle θ with the polar axis l, we measure off $|r|$ units along the oppositely directed ray. This is illustrated in Figure 58(a), where we plot the point $P = (-4, \pi/3)$, and also the point $P' = (4, \pi/3)$ of which P is the reflection in O. It is apparent from Figures 58(b) and 58(c) that $P = (-4, \pi/3)$ is also represented by the ordered pairs $(4, -2\pi/3)$ and $(4, 4\pi/3)$.

Given an ordered pair (r, θ), the point with polar coordinates r and θ is uniquely determined. On the other hand, as is already apparent from Figure 58, the polar coordinates of a given point P (unlike its rectangular coordinates) are not uniquely determined. In fact, if $P = (r, \theta)$ is any point other than the pole O, we also have

$$P = (r, \theta + 2n\pi) \qquad (n = \pm 1, \pm 2, \ldots)$$

and

$$P = (-r, \theta + (2n + 1)\pi) \qquad (n = 0, \pm 1, \pm 2, \ldots).$$

(This situation can be described by saying that the correspondence between ordered pairs of polar coordinates and points in the plane is *many-to-one*, rather than *one-to-one* as in the case of rectangular coordinates.) By convention, the pole O is represented by any ordered pair of the form $(0, \theta)$, regardless of the value of θ. This is simply because the radial coordinate of the pole is zero, and therefore its angular coordinate θ is indeterminate.

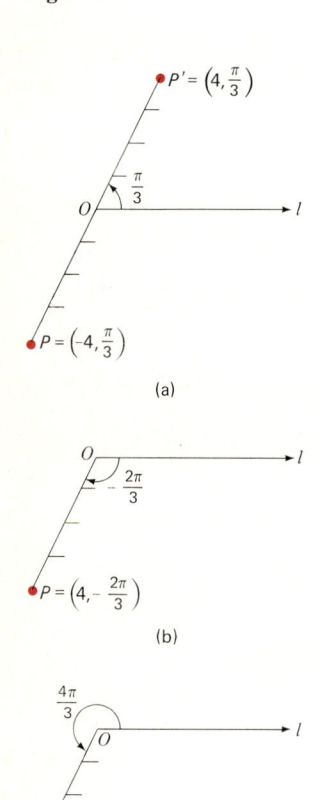

Figure 58

620 Chapter 10 Plane Analytic Geometry and Polar Coordinates

Relation Between Polar and Rectangular Coordinates

Polar coordinates and rectangular coordinates are often used simultaneously, by choosing the pole and polar axis to be the origin and the positive x-axis of a rectangular coordinate system. Then it is apparent from Figure 59 that the point with polar coordinates r and θ has rectangular coordinates

$$x = r \cos \theta, \qquad y = r \sin \theta. \tag{1}$$

You should verify that these formulas remain valid for negative r. If $r > 0$, it follows from (1) that

$$r = \sqrt{x^2 + y^2}, \qquad \tan \theta = \frac{y}{x} \quad (x \neq 0). \tag{2}$$

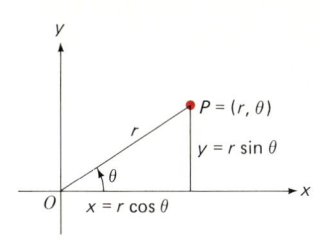

Figure 59

Example 1 If $P = (2, 3\pi/4)$ in polar coordinates, find the rectangular coordinates of P.

Solution It follows from (1) with $r = 2$ and $\theta = 3\pi/4$ that

$$x = 2 \cos \frac{3\pi}{4} = -\frac{2}{\sqrt{2}} = -\sqrt{2}, \qquad y = 2 \sin \frac{3\pi}{4} = \frac{2}{\sqrt{2}} = \sqrt{2}. \quad \square$$

Example 2 If $P = (-3, -4)$ in rectangular coordinates, find polar coordinates for P.

Solution It follows from (2) with $x = -3$ and $y = -4$ that

$$r = \sqrt{(-3)^2 + (-4)^2} = \sqrt{25} = 5$$

and also

$$\tan \theta = \frac{-4}{-3} = \frac{4}{3}.$$

To get negative values of x and y, we must choose a value of θ for which $\cos \theta$ and $\sin \theta$ are both negative. The angle $\arctan \frac{4}{3} \approx 53.1°$ does not satisfy this requirement, but the angle $\arctan \frac{4}{3} + 180° \approx 233.1°$ does. Alternatively, we can choose $\theta = \arctan \frac{4}{3}$ if the radial coordinate of P is taken to be -5, instead of 5. Thus $P = (5, \arctan \frac{4}{3} + \pi)$ and $P = (-5, \arctan \frac{4}{3})$ are two possible representations of P in polar coordinates. $\square$

Graphs of Polar Equations

By the *graph* of a function

$$r = f(\theta) \tag{3}$$

or more generally of an equation

$$F(r, \theta) = 0, \tag{4}$$

involving polar coordinates r and θ, we mean the set of all points with at least one pair of polar coordinates satisfying (3) or (4). For example, the point with polar coordinates $r = 1$ and $\theta = 1$ (angle in radians) belongs to the graph of the equation

$$r = \theta, \tag{5}$$

although the same point has polar coordinates $r = 1$ and $\theta = 2\pi + 1$ which do not satisfy (5). An equation like (3) or (4) is called a *polar equation*, and the graph of such an equation is called a *polar curve*.

Section 10.7 Polar Coordinates

Example 3 The graph of the equation $r = a$ ($a > 0$) is the circle of radius a centered at the pole O. The graph of the equation $\theta = \alpha$ (α arbitrary) is the ray emanating from O and making angle α with the polar axis l if $r \geq 0$, or the line through O making angle α with l if no condition is imposed on r (remember that r is allowed to be negative).

Example 4 Graph the function

$$r = 4 \sin \theta. \tag{6}$$

Solution Giving the angle θ various values whose sines are familiar, we construct the following table:

θ (radians)	0	$\dfrac{\pi}{6}$	$\dfrac{\pi}{4}$	$\dfrac{\pi}{3}$	$\dfrac{\pi}{2}$	$\dfrac{2\pi}{3}$	$\dfrac{3\pi}{4}$	$\dfrac{5\pi}{6}$	π
$\sin \theta$	0	$\dfrac{1}{2}$	$\dfrac{1}{\sqrt{2}}$	$\dfrac{\sqrt{3}}{2}$	1	$\dfrac{\sqrt{3}}{2}$	$\dfrac{1}{\sqrt{2}}$	$\dfrac{1}{2}$	0
$r = 4 \sin \theta$	0	2	$2\sqrt{2}$	$2\sqrt{3}$	4	$2\sqrt{3}$	$2\sqrt{2}$	2	0

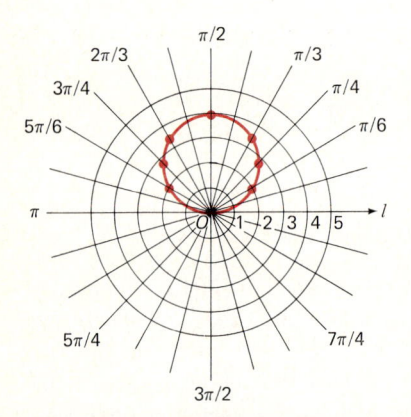

Plotting these points on polar coordinate graph paper (which shows rays through the pole O at various angles and concentric circles centered at O) and joining them by a smooth curve, we get the circle shown in Figure 60. The variable point (r, θ) traces out the right half of the circle as θ varies from 0 to $\pi/2$ and then the left half of the circle as θ varies from $\pi/2$ to π. Moreover, as you can easily verify, (r, θ) traces out the right semicircle again as θ varies from π to $3\pi/2$ and the left semicircle as θ varies from $3\pi/2$ to 2π (notice that r is negative when θ is between π and 2π). In fact, the circle is traced out exactly once as θ varies over *any* interval of length π (why?).

To verify that the curve in Figure 60 is indeed a circle, we introduce a rectangular coordinate system with its x-axis along the polar axis. Multiplying equation (6) by r, we get

$$r^2 = 4r \sin \theta,$$

which transforms into the cartesian equation

$$x^2 + y^2 = 4y \tag{6'}$$

after using formulas (1) and (2). But (6') is equivalent to the equation

$$x^2 + (y - 2)^2 = 4,$$

whose graph is the circle of radius 2 with its center at the point with rectangular coordinates $x = 0$ and $y = 2$, as in the figure. □

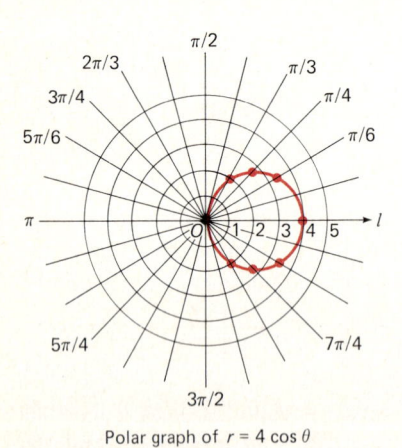

Example 5 Graph the function

$$r = 4 \cos \theta. \tag{7}$$

Solution A point (r, θ) belongs to the graph of the polar equation $F(r, \theta) = 0$ if and only if the point $(r, \theta - 90°)$ belongs to the graph of the polar equation $F(r, \theta + 90°) = 0$. Hence the graph of $F(r, \theta + 90°) = 0$ is obtained by rotating the graph of $F(r, \theta) = 0$ through 90° in the *clockwise* direction. Replacing θ by $\theta + 90°$ in the equation $r = 4 \sin \theta$, we get equation (7):

$$r = 4 \sin (\theta + 90°) = 4 \cos \theta.$$

Thus a 90° clockwise rotation of the circle $r = 4 \sin \theta$, shown in Figure 60,

Polar graph of $r = 4 \sin \theta$
Figure 60

Polar graph of $r = 4 \cos \theta$
Figure 61

carries it into the graph of the function (7). It follows that the graph of (7) is the circle of radius 2 with its center at the point with rectangular coordinates $x = 2$ and $y = 0$ (see Figure 61). What is the cartesian equation of this circle? □

Symmetry Tests

In graphing a polar equation $F(r, \theta) = 0$, always be on the lookout for possible symmetries of its graph. There are a number of tests for such symmetries. Suppose polar and rectangular coordinates are used simultaneously, with the pole as the common origin O and the polar axis along the x-axis. Let $P = (r, \theta)$ be any point other than the pole itself, let P', Q and P'' be the images of P under reflection in the x-axis, the origin and the y-axis, respectively, and let G be the graph of the equation $F(r, \theta) = 0$. Then it follows from the polar representations of the points P', Q and P'' given in Figures 62 and 63 that

(i) G is symmetric about the x-axis (the polar axis) if the set of solutions of $F(r, \theta) = 0$ is the same as that of $F(r, -\theta) = 0$ or $F(-r, \pi - \theta) = 0$;

(ii) G is symmetric about the origin O (the pole) if the set of solutions of $F(r, \theta) = 0$ is the same as that of $F(-r, \theta) = 0$ or $F(r, \pi + \theta) = 0$;

(iii) G is symmetric about the y-axis (the line through O perpendicular to the polar axis) if the set of solutions of $F(r, \theta) = 0$ is the same as that of $F(-r, -\theta) = 0$ or $F(r, \pi - \theta) = 0$.

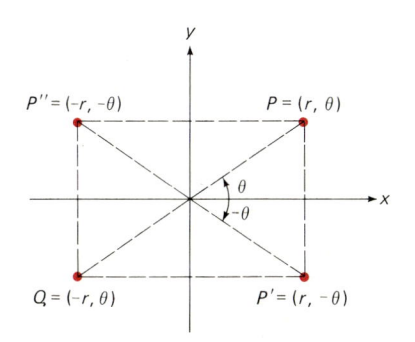

Figure 62

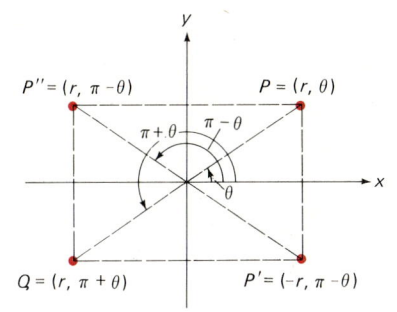

Figure 63

Example 6 The equation

$$F(r, \theta) = r - 4 \sin \theta = 0,$$

equivalent to (6), has the same set of solutions as the equation

$$F(-r, -\theta) = -r - 4 \sin (-\theta) = -r + 4 \sin \theta = 0$$

or

$$F(r, \pi - \theta) = r - 4 \sin (\pi - \theta) = r - 4 \sin \theta = 0.$$

According to (iii), this means that the graph of $F(r, \theta) = 0$ is symmetric about the y-axis, and in fact the circle in Figure 60 has this property. However, the circle is *not* symmetric about either the x-axis or the origin. Therefore, because of (i) and (ii), none of the equations

$$F(r, -\theta) = 0, \qquad F(-r, \pi - \theta) = 0, \qquad F(-r, \theta) = 0, \qquad F(r, \pi + \theta) = 0$$

can have the same set of solutions as the equation $F(r, \theta) = 0$, as you can verify directly.

Example 7 Graph the function

$$r = \sin 2\theta. \tag{8}$$

Solution First we look for symmetries. Writing (8) in the equivalent form

$$F(r, \theta) = r - \sin 2\theta = 0, \tag{8'}$$

we easily find that

$$F(-r, \pi - \theta) \equiv -F(r, \theta), \qquad F(r, \pi + \theta) \equiv F(r, \theta), \qquad F(-r, -\theta) \equiv -F(r, \theta)$$

Section 10.7 Polar Coordinates **623**

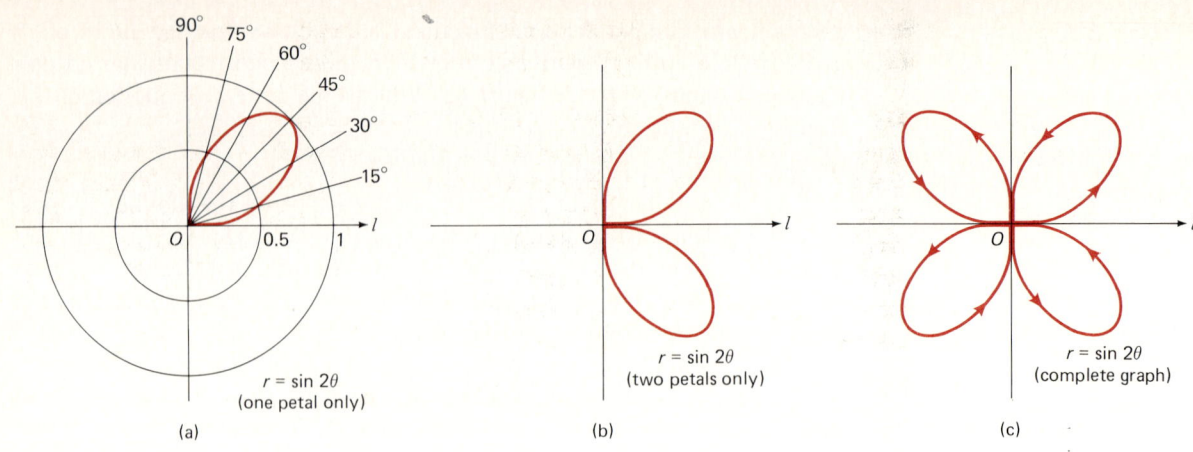

Figure 64

(a) $r = \sin 2\theta$ (one petal only)

(b) $r = \sin 2\theta$ (two petals only)

(c) $r = \sin 2\theta$ (complete graph)

(check the details). It follows from the tests (i)–(iii) that the graph of (8) is symmetric about both coordinate axes and the origin. Using the table

θ (degrees)	0°	15°	30°	45°	60°	75°	90°
$r = \sin 2\theta$	0	$\dfrac{1}{2}$	$\dfrac{\sqrt{3}}{2}$	1	$\dfrac{\sqrt{3}}{2}$	$\dfrac{1}{2}$	0

we plot a few points on the graph and connect them with a smooth curve. This gives the closed loop shown in Figure 64(a), which is only part of the graph. But the rest of the graph can now be found at once by exploiting the symmetry of the graph. In fact, reflecting the curve in Figure 64(a) in the x-axis, we get the curve in Figure 64(b), and then reflecting this curve in the y-axis, we get the flower-shaped curve in Figure 64(c), called a *four-petaled rose*, which is the complete graph of the function (8). (What other pairs of consecutive reflections generate the rose from the petal in part (a) of the figure?) As an exercise, show that the next three petals to be traced out as θ increases from 90° to 360° are the petals in the fourth, third and second quadrants *in that order*, as indicated by the arrowheads in Figure 64(c). ☐

Example 8 Graph the function

$$r = 2(1 - \cos\theta). \tag{9}$$

Solution Equation (9) does not change if θ is replaced by $-\theta$. Therefore, by test (i), the graph of (9) is symmetric about the polar axis l. Using this symmetry and the following table (note that $r \geq 0$ for all θ), we find that the graph of (9) is the heart-shaped curve shown in Figure 65, known as the *cardioid*, which has already put in an appearance in Problem 27, page 465.

Polar graph of $r = 2(1 - \cos\theta)$

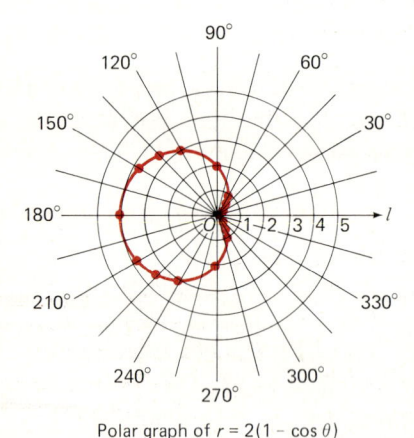

Figure 65

θ (degrees)	0°	30°	45°	60°	90°	120°	135°	150°	180°
$r = 2(1 - \cos\theta)$	0	$2 - \sqrt{3}$ ≈ 0.27	$2 - \sqrt{2}$ ≈ 0.59	1	2	3	$2 + \sqrt{2}$ ≈ 3.41	$2 + \sqrt{3}$ ≈ 3.73	4

Example 9 Find the points of intersection of the circle $r = 4\cos\theta$ considered in Example 5 and the cardioid $r = 2(1 - \cos\theta)$ considered in Example 8.

Solution In Figure 66(a) we graph both curves in the same system of polar coordinates. It is apparent from this figure, together with Figure 66(b) which shows a threefold enlargement of the circle and the cardioid near the origin, that the two curves have exactly three points of intersection, the origin or pole O and two points P_1 and P_2 which are symmetric about the polar axis l (so that each of these points is the image of the other under reflection in l). To find polar coordinates for P_1 and P_2, we observe that if $r = 4\cos\theta$ and $r = 2(1 - \cos\theta)$ hold simultaneously, then

$$4\cos\theta = 2(1 - \cos\theta), \tag{10}$$

or equivalently

$$\cos\theta = \frac{1}{3}, \tag{10'}$$

which implies $r = \frac{4}{3}$. Two values of θ satisfying (10') are $\theta = \pm\arccos\frac{1}{3} \approx \pm 70.5°$, and these correspond to points on opposite sides of the polar axis. Moreover, all the other solutions of (10') differ from $\theta = \pm\arccos\frac{1}{3}$ by integer multiples of 2π. Therefore we can conclude that

$$P_1 = (\tfrac{4}{3}, \arccos\tfrac{1}{3}), \qquad P_2 = (\tfrac{4}{3}, -\arccos\tfrac{1}{3}). \quad \square$$

In the preceding example, one of the points of intersection of the two polar curves $r = 4\cos\theta$ and $r = 2(1 - \cos\theta)$, namely the origin, was lost in solving equation (10). To see why, consider two variable points $P = (4\cos\theta, \theta)$ and $P' = (2(1 - \cos\theta), \theta)$, the first tracing out the circle $r = 4\cos\theta$ and the second tracing out the cardioid $r = 2(1 - \cos\theta)$ as the variable θ, which can be regarded as the time, increases from 0 to 2π. Then P arrives at the origin when $\theta = \pi/2$ or $\theta = 3\pi/2$, while P' starts at the origin and does not return until $\theta = 2\pi$. Hence the two points P and P' are never at the origin at the same time, although each describes a path going through the origin, so that solving (10) will not reveal that the origin is a point of intersection of the two curves.

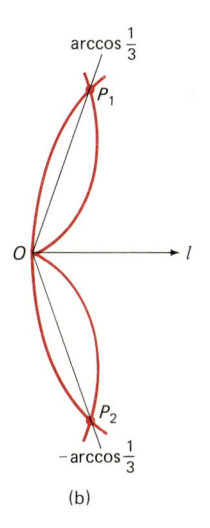

Figure 66

Intersection of Polar Curves

Thus, unlike the cartesian case, simultaneous solution of the equations of two polar curves $r = f(\theta)$ and $r = g(\theta)$ may not disclose all the points of intersection of the curves. This is because every point has infinitely many polar representations, each consisting of a different pair of polar coordinates, and there may well be points (like the origin in Example 9) with one representation satisfying the equation of one curve and another representation satisfying the equation of the other curve, but no representation satisfying the equations of *both* curves. To find points of intersection not revealed by solving the equation $f(\theta) = g(\theta)$, superimpose the two curves $r = f(\theta)$ and $r = g(\theta)$ in the same system of polar coordinates, just as was done in Example 9. In this regard, bear in mind that ink lines, unlike mathematical lines, have nonzero width, so that almost coincident intersections cannot be told apart if the scale of the drawing is too small. For example, in Figure 66(a) it would be hard to tell the three points of intersection O, P_1 and P_2 apart if the drawing were much smaller. There is a purely analytic method for finding

Section 10.7 Polar Coordinates 625

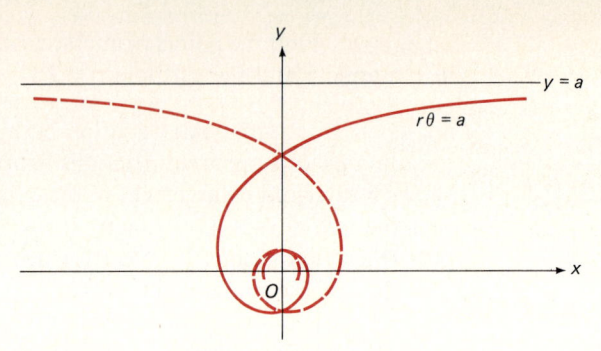

Figure 67

all points of intersection of two polar curves,† but its use can be avoided by making enlarged drawings [like the one in Figure 66(b)] whenever you suspect that points of intersection may lie too close together to be distinguished.

Example 10 Find the graph of the polar equation

$$r\theta = a \quad (a > 0). \tag{11}$$

Solution For nonnegative r the graph of (11) is the solid curve shown in Figure 67, called a *hyperbolic spiral* because of the resemblance between (11) and the equation $xy = a$, which represents a hyperbola in rectangular coordinates. The essential features of the graph are easily deduced from (11) by observing that

$$\lim_{\theta \to 0^+} r = \lim_{\theta \to 0^+} \frac{a}{\theta} = \infty,$$

while

$$\lim_{\theta \to \infty} r = \lim_{\theta \to \infty} \frac{a}{\theta} = 0.$$

It follows that as θ increases from some small positive value to ∞, a variable point $P = (r, \theta)$ on the graph of (11) "comes in from infinity" and winds around the origin or pole O in the counterclockwise direction, while r tends steadily to zero; in calculating r from (11), θ must be measured in radians (why?). The y-coordinate of P is

$$y = r \sin \theta = a \frac{\sin \theta}{\theta},$$

so that

$$\lim_{\theta \to 0^+} y = a \lim_{\theta \to 0^+} \frac{\sin \theta}{\theta} = a.$$

This, together with the fact that $r \to \infty$ as $\theta \to 0^+$, shows that the line $y = a$ is a horizontal asymptote of the spiral (see the figure).

To find the rest of the spiral, corresponding to negative values of r and θ, we reflect the solid curve in the y-axis, obtaining the dashed curve in the figure. This symmetry about the y-axis follows from the fact that $(-r)(-\theta) \equiv r\theta$. Why doesn't the origin O itself belong to the spiral? □

---

† For the record, the curves $r = f(\theta)$ and $r = g(\theta)$ intersect at a point $(f(\alpha), \alpha)$ other than the origin if and only if $f(\alpha) = g(\alpha + 2n\pi)$ or $f(\alpha) = -g(\alpha + (2n + 1)\pi)$ for some integer n, and the curves intersect at the origin if and only if $f(\alpha) = g(\beta) = 0$ for suitable α and β.

Problems

In all subsequent problems involving both polar and rectangular coordinates, the pole and polar axis of the polar coordinate system coincide with the origin and nonnegative x-axis of the rectangular coordinate system. All points are given in polar coordinates except in Problems 7–12.

Find the rectangular coordinates of the point with the given polar coordinates.
1. $(0, \pi^2)$
2. $(-8, 2\pi/3)$
3. $(12, -\pi/6)$
4. $(2, 5\pi/6)$
5. $(-10, -3\pi/2)$
6. (π, π)

Find *all* representations in polar coordinates (including those for which r is negative) of the point with the given rectangular coordinates.
7. $(-\sqrt{2}, -\sqrt{2})$
8. $(1, -\sqrt{3})$
9. $(2\sqrt{3}, 2)$
10. $(0, 4)$
11. $(-3, 0)$
12. $(-\pi, \pi)$

13. The points $A = (3, -4\pi/9)$ and $B = (5, 3\pi/14)$ are two vertices of a parallelogram $ABCD$ whose diagonals intersect at the pole. Find the other two vertices of the parallelogram.
14. Find the midpoint of the segment joining the points $(8, -2\pi/3)$ and $(6, \pi/3)$.
15. Show that the distance between the two points $P_1 = (r_1, \theta_1)$ and $P_2 = (r_2, \theta_2)$ is equal to

$$|P_1 P_2| = \sqrt{r_1^2 + r_2^2 - 2r_1 r_2 \cos(\theta_2 - \theta_1)}. \quad \text{(i)}$$

Use formula (i) to find the distance between the points
16. $(5, -\pi/12), (8, \pi/4)$
17. $(12, 3\pi/4), (-16, 5\pi/4)$

18. Find the area of each of the squares with the points $(12, -\pi/10)$ and $(3, \pi/15)$ as two adjacent vertices.
19. Find the area of the square with the points $(6, -105°)$ and $(4, 30°)$ as two opposite vertices.
20. Let O be the pole, and let $P_1 = (r_1, \theta_1)$, $P_2 = (r_2, \theta_2)$, where r_1, r_2 are positive and $0 \leq \theta_1 < \theta_2 \leq \pi$. Show that the area of the triangle $OP_1 P_2$ is equal to $\frac{1}{2} r_1 r_2 \sin(\theta_2 - \theta_1)$. Find the area of the triangle with vertices O, $P_1 = (5, \pi/3)$ and $P_2 = (10, 7\pi/12)$.
21. Show that the polar equation of the circle of radius a centered at (r_1, θ_1) is

$$r^2 - 2r_1 r \cos(\theta - \theta_1) + r_1^2 = a^2. \quad \text{(ii)}$$

22. Use formula (ii) to verify that the circles shown in Figures 60 and 61 have polar equations $r = 4 \sin \theta$ and $r = 4 \cos \theta$. Also verify these equations by using the fact that every triangle inscribed in a circle with a diameter as one side is a right triangle.

Find the polar equation of the circle with the given radius and center.
23. $3, (6, \pi/4)$
24. $4, (-4, \pi/3)$
25. $\sqrt{5}, (2, \pi/2)$

26. Let L be a straight line which does not go through the pole O. Show that the polar equation of L is

$$r \cos(\theta - \alpha) = p, \quad \text{(iii)}$$

where p is the length and α the inclination of the perpendicular drawn from O to L (see Figure 68). What is the cartesian equation equivalent to (iii)?

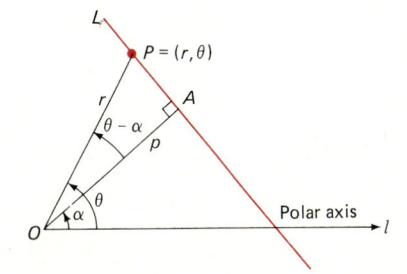

Figure 68

Write the given cartesian equation in polar coordinates.
27. $y = x$
28. $3x + 4y = 5$
29. $x^2 + y^2 = 9$
30. $x^2 - y^2 = 4$
31. $x^2 + y^2 - ax = 0$
32. $x^2 + y^2 + ay = 0$
33. $y^2 = 4x$
34. $2xy = 1$
35. $x^2 - y^2 = (x^2 + y^2)^2$
36. $x^4 = 9(x^2 + y^2)$

Write the given polar equation in rectangular coordinates, and identify its graph.
37. $\theta = \pi$
38. $r \sin \theta = 3$
39. $r \cos \theta = -2$
40. $\sin \theta = \frac{1}{2}$
41. $r \cos\left(\theta - \frac{\pi}{3}\right) = 6$
42. $r \sin\left(\theta + \frac{\pi}{4}\right) = \sqrt{2}$
43. $r = -\sin \theta$
44. $r = 8 \cos \theta$
45. $r^2 \sin 2\theta = 4$
46. $r = 2(\cos \theta - \sin \theta)$

The curve with polar equation

$$r = a \sin(n\theta + \phi) \quad (a > 0, \phi \text{ arbitrary}, n = 2, 3, \ldots)$$

is called a *rose* (see Example 7 for an analysis of the rose $r = \sin 2\theta$ corresponding to the choice $a = 1$, $\phi = 0$, $n = 2$). Graph the rose with equation
47. $r = 4 \sin 3\theta$
48. $r = 2 \cos 4\theta$
49. $r = \cos 6\theta$
50. $r = \sin(5\theta + \pi)$

Notice that the rose has n petals if n is odd and $2n$ petals if n is even.

The curve with polar equation

$$r = a + b \cos(\theta + \phi) \quad (a > 0, b > 0, \phi \text{ arbitrary})$$

is called a *limaçon*. If $a = b$, the limaçon reduces to a cardioid (see Example 8 for an analysis of the cardioid $r = 2(1 - \cos\theta)$ corresponding to the choice $a = b = 2$, $\phi = \pi$). Graph the limaçon with equation

51. $r = 3 + 2\cos\theta$
52. $r = 3 - \sin\theta$
53. $r = 1 + 2\cos\theta$
54. $r = 2 + 3\sqrt{2}\cos\left(\theta - \dfrac{\pi}{4}\right)$

Graph the given polar equation, making use of any symmetries.

55. $r^2 = 4\cos 2\theta$ (lemniscate)
56. $r^2 = -\sin 2\theta$ (lemniscate)
57. $r = \sin\theta \tan\theta$ (cissoid)
58. $r = \cot\theta$ (kappa curve)
59. $r = 1 + 2\sin(\theta/2)$ (nephroid)
60. $r = \sec\theta - 4\cos\theta$ (trisectrix)

61. $r = 2\theta$ (Archimedean spiral; general form $r = a\theta$)
62. $r = e^{\theta/10}$ or equivalently $\ln r = \theta/10$ (logarithmic spiral; general form $r = e^{a\theta}$)
63. $r^2 = \theta$ (parabolic spiral; general form $r^2 = a^2\theta$)
64. $r^2\theta = 1$ (lituus, the Latin word for "trumpet"; general form $r^2\theta = a^2$)

In Problems 61–64 use a solid curve for the part of the spiral with $r > 0$, $\theta > 0$, and a dashed curve for the rest of the spiral.

Find *all* points of intersection of the given pair of polar curves.

65. $\theta = \pi/4$, $r = \theta$
66. $r = 1 + \cos\theta$, $r = 1 - \cos\theta$
67. $r = 1 + \sin\theta$, $r = 1 - \cos\theta$
68. $r = 2\cos 3\theta$, $r = 1$
69. $r = \sin 2\theta$, $r = \cos\theta$
70. $r = \sin\theta$, $r = |\cos\theta|$
71. $r^2 = \sin 2\theta$, $r^2 = \cos 2\theta$
72. $r\theta = 2$, $r = 1$

10.8 Conics in Polar Coordinates

As shown by the following theorem, polar coordinates are particularly well suited for writing equations of nondegenerate conics.

Theorem 2 *(Polar equation of a conic).* The conic with eccentricity e and focus-to-directrix distance d has equation

$$r = \dfrac{ed}{1 - e\cos\theta} \qquad (1)$$

in polar coordinates if the pole O is at the focus and the polar axis is perpendicular to the directrix L in the direction pointing away from L.

Proof In the case of an ellipse or a hyperbola, there are two directrices and L is the one nearer the focus. The geometry of the situation is shown in Figure 69, where the directrix lies to the left of the focus. By Theorem 1, page 608, we have

$$\dfrac{|OP|}{|PQ|} = \dfrac{r}{|PQ|} = e. \qquad (2)$$

But

$$|PQ| = |AO| + r\cos\theta = d + r\cos\theta,$$

so that (2) becomes

$$r = e(d + r\cos\theta).$$

Solving for r, we get formula (1). ∎

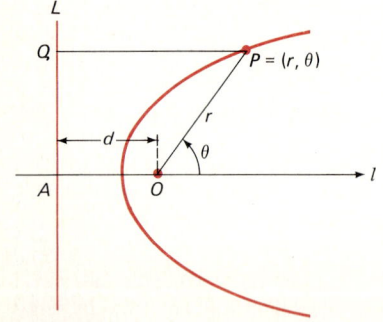

Figure 69

628   Chapter 10   Plane Analytic Geometry and Polar Coordinates

Example 1 Identify the conic with polar equation

$$r = \frac{9}{5 - 4\cos\theta}. \tag{3}$$

Solution Writing (3) in the form

$$r = \frac{\frac{9}{5}}{1 - \frac{4}{5}\cos\theta}, \tag{3'}$$

and then comparing (3') with (1), we find that

$$e = \frac{4}{5}, \quad ed = \frac{4}{5}d = \frac{9}{5}, \quad d = \frac{9}{4}.$$

Since $e < 1$, the conic is an ellipse, and in fact graphing equation (3), we get the ellipse E shown in Figure 70, with focus at the pole O, directrix L and focus-to-directrix distance $\frac{9}{4}$. Notice that the polar equation of L is $r\cos\theta = -\frac{9}{4}$ or equivalently $r = -\frac{9}{4}\sec\theta$.

To get a cartesian equation for the ellipse E, let the major axis be of length $2a$, and let the distance between the foci be $2c$. Then $e = c/a$ and

$$d = \frac{a^2}{c} - c,$$

by formula (8), page 607. Therefore

$$\frac{c}{a} = \frac{4}{5}, \quad \frac{a^2 - c^2}{c} = \frac{9}{4},$$

from which it follows that

$$a = 5, \quad c = 4, \quad b = \sqrt{a^2 - c^2} = \sqrt{9} = 3,$$

where $2b$ is the length of the minor axis. Thus E has the equation

$$\frac{x^2}{25} + \frac{y^2}{9} = 1$$

in a system of rectangular coordinates with its origin at the center of E and its x-axis along the major axis. In a system of rectangular coordinates with its origin at the left focus, the equation is

$$\frac{(x - 4)^2}{25} + \frac{y^2}{9} = 1$$

instead (why?). As an exercise, derive this equation directly from equation (3) by making the substitutions $r\cos\theta = x$ and $r^2 = x^2 + y^2$. □

Example 2 Write a polar equation for the hyperbola with cartesian equation

$$\frac{x^2}{144} - \frac{y^2}{25} = 1. \tag{4}$$

Solution Let $2a$ be the length of the transverse axis and $2b$ the length of the conjugate axis. Then

$$a = 12, \quad b = 5, \quad c = \sqrt{a^2 + b^2} = \sqrt{169} = 13,$$

where $2c$ is the distance between the foci F_1 and F_2, which lie on the x-axis

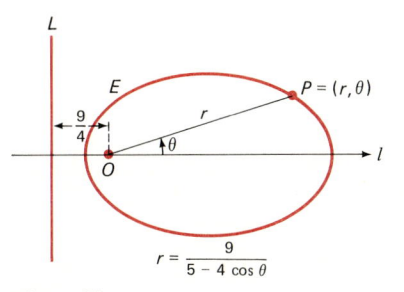

Section 10.8   Conics in Polar Coordinates   629

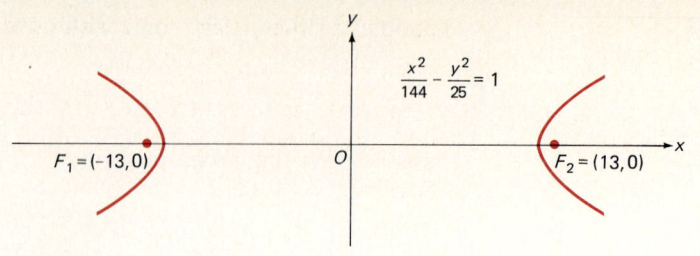

Figure 71

(see Figure 71). As before, the eccentricity is $e = c/a$, but now the distance between the focus and directrix

$$d = c - \frac{a^2}{c},$$

by formula (8'), page 607. Therefore

$$e = \frac{13}{12}, \quad d = 13 - \frac{144}{13} = \frac{25}{13},$$

so that the hyperbola has polar equation

$$r = \frac{ed}{1 - e\cos\theta} = \frac{\frac{25}{12}}{1 - \frac{13}{12}\cos\theta},$$

or equivalently,

$$r = \frac{25}{12 - 13\cos\theta}. \tag{5}$$

More exactly, the graph of (5) is the right branch of the hyperbola (4) if $\theta_0 < |\theta| \leq \pi$, where $\theta_0 = \arccos\frac{12}{13} \approx 22.6°$. If $0 \leq |\theta| < \theta_0$, then $r < 0$ and the graph of (5) is the left branch instead (the details are left as an exercise). In both cases the pole is at the right focus F_2, and the polar axis is along the positive x-axis. □

Variants of the Polar Equation of a Conic

Suppose we replace θ by $\theta - \alpha$ in equation (1), obtaining

$$r = \frac{ed}{1 - e\cos(\theta - \alpha)}. \tag{6}$$

Then the graph of (6) is a conic whose major axis, transverse axis or axis of symmetry (depending on whether the conic is an ellipse, hyperbola or parabola) lies along the line $\theta = \alpha$, rather than along the polar axis $\theta = 0$. In particular, the graphs of the equations

$$r = \frac{ed}{1 + e\cos\theta}, \quad r = \frac{ed}{1 - e\sin\theta}, \quad r = \frac{ed}{1 + e\sin\theta} \tag{6'}$$

are all conics, since these are the equations obtained from (6) by setting $\alpha = \pi, \pi/2, -\pi/2$ in turn. Specifically, the first of these equations describes a conic with directrix perpendicular to the polar axis, d units to the *right* of the focus, while the second and third equations describe conics with directrices

630   Chapter 10   Plane Analytic Geometry and Polar Coordinates

parallel to the polar axis, d units below the focus and d units above it, respectively.

Example 3 Write a polar equation for the parabola with focus-to-directrix distance 10 which has a vertical axis of symmetry and opens downward.

Solution Setting $e = 1$, $d = 10$ in the last of the equations (6'), we get

$$r = \frac{10}{1 + \sin \theta}.$$

The graph of this equation is the parabola shown in Figure 72. The vertex V of the parabola is the point $(5, \pi/2)$, while the directrix L is the line with polar equation $r \sin \theta = 10$ or equivalently $r = 10 \csc \theta$. ☐

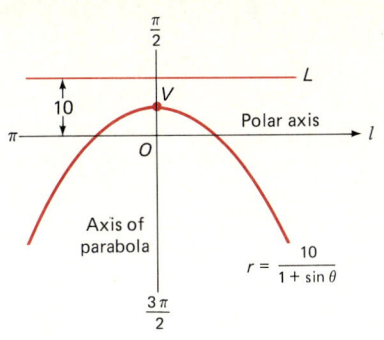

Figure 72

Problems

Identify and graph the conic with the given polar equation. Find the eccentricity e, locate the vertices, and write a polar equation for the directrix L corresponding to the focus at the pole O.

1. $r = \dfrac{5}{1 + \cos \theta}$
2. $r = \dfrac{4}{3 - 2 \cos \theta}$
3. $r = \dfrac{6}{2 - 3 \sin \theta}$
4. $r = \dfrac{12}{3 - 4 \cos \theta}$
5. $r = \dfrac{9}{4 + 3 \sin \theta}$
6. $r = 4 \csc^2 \dfrac{\theta}{2}$
7. $r = \dfrac{10}{1 + 2 \cos \theta}$
8. $r = \dfrac{8}{2 - \cos \theta + \sin \theta}$
9. $r = \dfrac{20}{2 - \sqrt{3} \cos \theta - \sin \theta}$

Write a polar equation for the conic with the given cartesian equation, placing the pole at a focus and the polar axis along the positive x-axis.

10. $\dfrac{x^2}{256} + \dfrac{y^2}{16} = 1$
11. $\dfrac{y^2}{64} - \dfrac{x^2}{36} = 1$
12. $4x^2 - y^2 = 1$
13. $y^2 = 6x$
14. $2xy = 1$
15. $3x^2 - 2xy + 3y^2 - 2 = 0$

Assuming that the pole is at a focus, write a polar equation for the conic satisfying the given conditions (e is the eccentricity).

16. Vertices at $(6, 0)$ and $(2, \pi)$.
17. Vertices at $(1, \pi/2)$ and $(5, \pi/2)$.
18. $e = 1$, vertex at $(2, \pi/4)$.
19. $e = \tfrac{2}{3}$, vertex at $(3, -\pi/2)$.
20. $e = \tfrac{3}{4}$, directrix $r = 2 \csc \theta$.
21. $e = \tfrac{4}{3}$, directrix $r = 6 \sec \theta$.
22. Let $r = ed/(1 - e \cos \theta)$ be the polar equation of a hyperbola. Then the denominator equals zero for $\theta = \theta_0 = \arccos (1/e)$. What is the connection between the angle θ_0 and the asymptotes of the hyperbola?
23. A comet moves in an elliptic orbit of high eccentricity ($e \approx 1$) with the sun at one focus. As the comet moves to perihelion (the point nearest the sun) from a point 50 million miles away, the line from the sun to the comet rotates through $45°$. Find the distance from the comet to the sun at perihelion.

10.9 The Tangent Line to a Polar Curve

We now consider the problem of finding the tangent line to a polar curve. In addition to the underlying polar coordinate system, we introduce rectangular coordinates with the pole O as origin and the polar axis as the positive x-axis. Let the curve be the graph of a differentiable function

$$r = f(\theta),$$

and let the tangent to the curve at the point P be the line of inclination

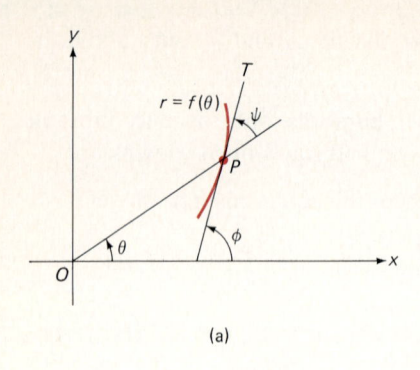

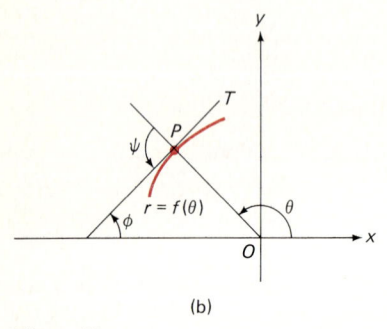

ϕ ($0 \le \phi < \pi$), as in Figure 73. Then the slope of T is

$$m = \tan \phi = \frac{dy}{dx}.$$

Besides the angle of inclination ϕ, we are also interested in the *radial-tangential angle* ψ (lowercase Greek psi). This is the angle between the extension of the radial line OP and the tangent T, measured from OP to T in the counterclockwise direction and chosen to lie in the interval $0 \le \psi < \pi$.

The connection between ϕ, ψ and the angular coordinate θ is given by the formula

$$\phi = \psi + \theta, \tag{1}$$

which can be read off from Figure 73(a). Here it is assumed that ϕ is an exterior angle of an acute triangle bounded by the x-axis, the tangent T and the radial line OP, with θ as an interior angle (thus in particular $0 < \theta < \phi$), but of course there are other possibilities. However, it is easy to see that (1) and the equivalent formula

$$\psi = \phi - \theta \tag{1'}$$

always hold to within an integer multiple of π [for example, $\phi = \psi + \theta - \pi$ in Figure 73(b)]. Therefore, since the tangent function is periodic with period π, we always have

$$\tan \psi = \tan(\phi - \theta) = \frac{\tan \phi - \tan \theta}{1 + \tan \phi \tan \theta}, \tag{2}$$

so that the same formula for $\tan \psi$ in terms of $\tan \phi$ and $\tan \theta$ is obtained in all cases.

Next we observe that the point $P = (r, \theta)$ has rectangular coordinates

$$x = r \cos \theta, \qquad y = r \sin \theta, \tag{3}$$

and also that

$$\tan \theta = \frac{y}{x} \qquad (x \ne 0). \tag{4}$$

Substituting $r = f(\theta)$ into (3), we obtain

$$x = x(\theta) = f(\theta) \cos \theta, \qquad y = y(\theta) = f(\theta) \sin \theta. \tag{5}$$

This is a parametric representation of the curve under discussion, with the angular coordinate θ as parameter. Therefore, by the considerations on pages 456–457,

$$m = \tan \phi = \frac{dy}{dx} = \frac{dy/d\theta}{dx/d\theta}. \tag{6}$$

Substitution of (4) and (6) into (2) gives

$$\tan \psi = \frac{\dfrac{dy}{dx} - \dfrac{y}{x}}{1 + \dfrac{dy}{dx}\dfrac{y}{x}} = \frac{\dfrac{dy/d\theta}{dx/d\theta} - \dfrac{y}{x}}{1 + \dfrac{dy/d\theta}{dx/d\theta}\dfrac{y}{x}} = \frac{x\dfrac{dy}{d\theta} - y\dfrac{dx}{d\theta}}{x\dfrac{dx}{d\theta} + y\dfrac{dy}{d\theta}}.$$

632   Chapter 10   Plane Analytic Geometry and Polar Coordinates

But the derivatives of the functions (5) are

$$\frac{dx}{d\theta} = f'(\theta)\cos\theta - f(\theta)\sin\theta, \qquad \frac{dy}{d\theta} = f'(\theta)\sin\theta + f(\theta)\cos\theta, \qquad (7)$$

and hence

$$\begin{aligned} x\frac{dy}{d\theta} - y\frac{dx}{d\theta} &= r\cos\theta\,[f'(\theta)\sin\theta + f(\theta)\cos\theta] \\ &\quad - r\sin\theta\,[f'(\theta)\cos\theta - f(\theta)\sin\theta] \\ &= (\cos^2\theta + \sin^2\theta)\,rf(\theta) = rf(\theta), \end{aligned}$$

while

$$\begin{aligned} x\frac{dx}{d\theta} + y\frac{dy}{d\theta} &= r\cos\theta\,[f'(\theta)\cos\theta - f(\theta)\sin\theta] \\ &\quad + r\sin\theta\,[f'(\theta)\sin\theta + f(\theta)\cos\theta] \\ &= (\cos^2\theta + \sin^2\theta)\,rf'(\theta) = rf'(\theta). \end{aligned}$$

It follows that

$$\tan\psi = \frac{rf(\theta)}{rf'(\theta)} = \frac{f(\theta)}{f'(\theta)},$$

that is,

$$\tan\psi = \frac{r}{dr/d\theta} = \frac{r}{r'}, \qquad (8)$$

provided that $dr/d\theta \neq 0$. The simplicity of this formula for $\tan\psi$ is striking if you compare it with the formula for the slope of T, namely

$$\tan\phi = \frac{f'(\theta)\sin\theta + f(\theta)\cos\theta}{f'(\theta)\cos\theta - f(\theta)\sin\theta},$$

or equivalently

$$\tan\phi = \frac{(dr/d\theta)\sin\theta + r\cos\theta}{(dr/d\theta)\cos\theta - r\sin\theta}, \qquad (9)$$

obtained by substituting (7) into (6).

Example 1 Find the angles ψ and ϕ for the cardioid $r = 1 + \sin\theta$ at each of the points $P = (1, 0)$ and $Q = (3/2, \pi/6)$.

Solution Here $dr/d\theta = \cos\theta$, and formula (8) becomes

$$\tan\psi = \frac{r}{dr/d\theta} = \frac{1 + \sin\theta}{\cos\theta}.$$

Therefore

$$\tan\psi\big|_{\theta=0} = 1, \qquad \tan\psi\big|_{\theta=\pi/6} = \frac{1 + \frac{1}{2}}{\frac{1}{2}\sqrt{3}} = \frac{2(\frac{3}{2})}{\sqrt{3}} = \sqrt{3},$$

which implies

$$\psi\big|_{\theta=0} = \frac{\pi}{4}, \qquad \psi\big|_{\theta=\pi/6} = \frac{\pi}{3}.$$

Section 10.9 The Tangent Line to a Polar Curve

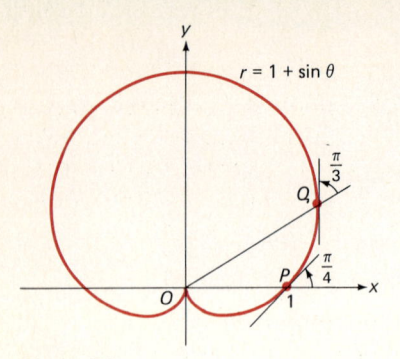

Figure 74

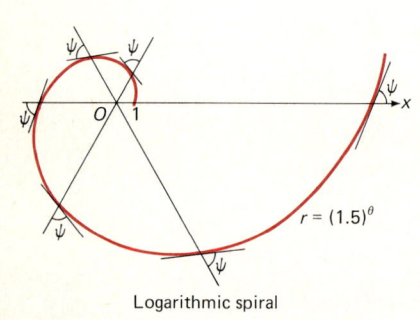

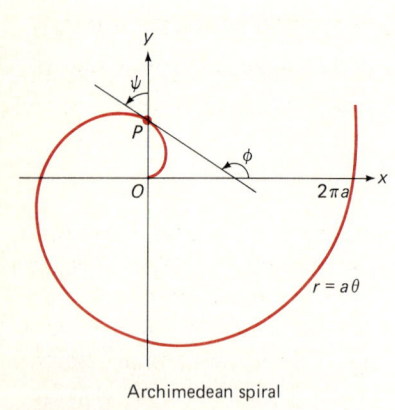

(see Figure 74). At P and Q we have $0 \leq \theta < \phi$, and hence $\phi = \psi + \theta$. It follows that

$$\phi|_{\theta=0} = \frac{\pi}{4} + 0 = \frac{\pi}{4}, \qquad \phi|_{\theta=\pi/6} = \frac{\pi}{3} + \frac{\pi}{6} = \frac{\pi}{2}.$$

Thus the tangent at P has slope 1, while the tangent at Q is vertical. It is easy to see that the tangent at P is the line $y = x - 1$, while the tangent at Q is the line $x = \frac{3}{4}\sqrt{3}$. □

Example 2 Show that the logarithmic spiral

$$r = e^{a\theta} \qquad (\theta \geq 0, a > 0)$$

is *equiangular*, in the sense that the angle ψ between the radial line and the tangent has the same value at every point of the spiral.

Solution Since $dr/d\theta = ae^{a\theta}$, we have

$$\tan \psi = \frac{r}{dr/d\theta} = \frac{e^{a\theta}}{ae^{a\theta}} = \frac{1}{a},$$

so that $\psi \equiv \text{arccot } a$. Figure 75 illustrates the equiangularity for the case $a = \ln \frac{3}{2}$, $r = (\frac{3}{2})^\theta$, $\psi \equiv \text{arccot}(\ln \frac{3}{2}) \approx 67.9°$. □

Example 3 Find the inclination ϕ of the tangent to the Archimedean spiral

$$r = a\theta \qquad (\theta \geq 0, a > 0)$$

at the point $P = (a\pi/2, \pi/2)$.

Solution This time $dr/d\theta = a$ and using formula (9) directly, we have

$$\tan \phi = \frac{a \sin \frac{\pi}{2} + \frac{a\pi}{2} \cos \frac{\pi}{2}}{a \cos \frac{\pi}{2} - \frac{a\pi}{2} \sin \frac{\pi}{2}} = -\frac{2}{\pi},$$

so that

$$\phi = \arctan\left(-\frac{2}{\pi}\right) + \pi \approx 147.5°$$

(why do we add π?). Notice that ϕ is independent of the choice of the constant a. Alternatively, by (8),

$$\tan \psi = \frac{a\pi/2}{a} = \frac{\pi}{2},$$

and hence

$$\psi = \arctan \frac{\pi}{2} \approx 57.5°, \qquad \phi = \psi + \theta = \arctan \frac{\pi}{2} + \frac{\pi}{2} \approx 147.5°$$

(see Figure 76). The equivalence of the two expressions for ϕ follows from formula (16), page 296. As an exercise, show that the tangent line at P has equation $4x + 2\pi y - a\pi^2 = 0$. □

Figure 75 Logarithmic spiral

Figure 76 Archimedean spiral

634   Chapter 10   Plane Analytic Geometry and Polar Coordinates

Problems

Find $\tan \psi$ first and then the angles ψ and ϕ for the given polar curve at the specified point P.

1. $r = 6 \cos \theta$, $P = (3, \pi/3)$
2. $r = 8 \sin \theta$, $P = (4, \pi/6)$
3. $r = \sin 2\theta$, $P = (1/\sqrt{2}, \pi/8)$
4. $r = \sqrt{2}\,|\cos \theta|$, $P = (1, 3\pi/4)$
5. $r = 3 \sec^2 \theta$, $P = (4, \pi/6)$
6. $r = 1 + 2 \sin(\theta/2)$, $P = (1, 4\pi)$
7. $r = 2 - \cos \theta$, $P = (2, 3\pi/2)$
8. $r = 3 \cot \theta$, $P = (\sqrt{3}, \pi/3)$
9. $r = \dfrac{5}{1 - \cos \theta}$, $P = (5, \pi/2)$
10. $r = \dfrac{3}{\sqrt{2} + \cos \theta}$, $P = (\sqrt{2}, \pi/4)$
11. $r = \theta^2/4$, $P = (\pi^2, 2\pi)$
12. $r = \pi/\theta$, $P = (\tfrac{1}{3}, 3\pi)$

13. Given a polar curve $r = f(\theta)$, suppose $f(\theta_0) = 0$. Show that the line $\theta = \theta_0$ is tangent to the curve at the pole.
14. Show that the cardioid $r = 2(1 - \cos \theta)$ is tangent to the polar axis at the pole (see Figure 65, page 624).
15. The three-petaled rose $r = \cos 3\theta$ has three distinct tangents at the pole. Find them.
16. At which points of the cardioid $r = 1 + \cos \theta$ is the tangent vertical?
17. At which points of the lemniscate $r^2 = \sin 2\theta$ is the tangent horizontal?
18. Use formula (8′) to show that the tangent to a circle is perpendicular to the radius drawn to the point of tangency.

As in Figure 17, page 122, the angle α ($0 \le \alpha < \pi$) between two intersecting curves C_1 and C_2 is defined as the angle between their tangents T_1 and T_2 at the point of intersection, measured from T_1 to T_2 in the counterclockwise direction.† Find the angle between the curves.

19. $r = 4 \cos \theta$ and $r = 2(1 - \cos \theta)$ at $(\tfrac{4}{3}, \arccos \tfrac{1}{3})$ (see Example 9, page 625)
20. $r = \sin 2\theta$ and $r = \cos \theta$ at $(\sqrt{3}/2, \pi/6)$
21. $r = 1/(1 - \sin \theta)$ and $r = 3/(1 + \sin \theta)$ at $(2, \pi/6)$ and $(2, 5\pi/6)$
22. $r^2 = \sin 2\theta$ and $r^2 = \cos 2\theta$ at $(1/\sqrt[4]{2}, \pi/8)$ and $(1/\sqrt[4]{2}, 9\pi/8)$

Hint. Work with the radial-tangential angles ψ_1 and ψ_2 corresponding to T_1 and T_2, observing that $\alpha = \psi_2 - \psi_1$ (see Figure 77).

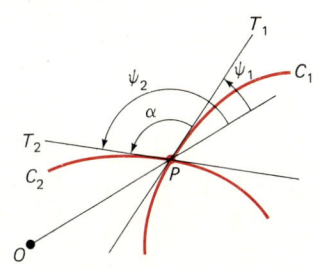

The angle between the curves C_1 and C_2 at P is $\alpha = \psi_2 - \psi_1$.

Figure 77

† Alternatively, the angle between C_1 and C_2 is often defined as the smaller of the two supplementary angles between T_1 and T_2, regardless of the direction of measurement. If β is this angle, then $\beta = \alpha$ if $0 \le \alpha \le \pi/2$, but $\beta = \pi - \alpha$ if $\pi/2 < \alpha < \pi$.

10.10 Area in Polar Coordinates; Length of a Polar Curve

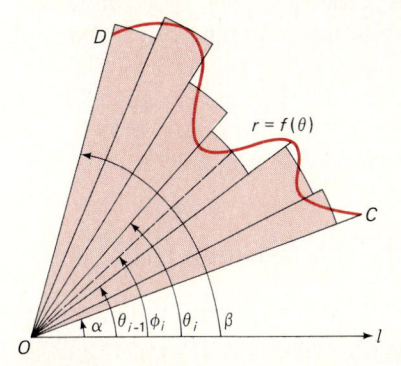

Figure 78

Consider the problem of finding the area A of the region OCD shown in Figure 78, bounded by the ray $\theta = \alpha$, the ray $\theta = \beta$ ($\beta > \alpha$) and the curve with polar equation

$$r = f(\theta) \qquad (\alpha \le \theta \le \beta),$$

where the function f is continuous and nonnegative. We can think of OCD as the generalization of a circular sector in which the curved side is no longer a circular arc. Since elementary geometry has nothing to say about such regions, we must find the proper *definition* of the area A before we can calculate it. Except for small details, the strategy is the same as that used to define the area under a curve with equation $y = f(x)$ in rectangular coordinates (see pages 229–231).

Thus we begin by dividing the interval $[\alpha, \beta]$ into a large number n of smaller subintervals, by introducing points of subdivision

$$\alpha = \theta_0, \theta_1, \theta_2, \ldots, \theta_{n-1}, \theta_n = \beta$$

satisfying the inequalities

$$\alpha = \theta_0 < \theta_1 < \theta_2 < \cdots < \theta_{n-1} < \theta_n = \beta.$$

Let

$$\Delta\theta_i = \theta_i - \theta_{i-1} \quad (i = 1, 2, \ldots, n), \tag{1}$$

and let

$$\mu = \max\{\Delta\theta_1, \Delta\theta_2, \ldots, \Delta\theta_n\}$$

be the maximum size of all the angles (1). Then the rays $\theta = \theta_0, \theta_1, \ldots, \theta_n$ divide the region OCD into n narrow pie-shaped slices. The function f is continuous, and hence its value changes only slightly on the subinterval $[\theta_{i-1}, \theta_i]$, at least if $\Delta\theta_i$ is small enough. Thus it is a good approximation to regard f as having the constant value $f(\phi_i)$ on $[\theta_{i-1}, \theta_i]$, where ϕ_i is an *arbitrary* point of $[\theta_{i-1}, \theta_i]$. Replacing $f(\theta)$ by $f(\phi_i)$ on each subinterval $[\theta_{i-1}, \theta_i]$ is equivalent to replacing the slices by the shaded circular sectors shown in the figure. The sum of the areas of these sectors is given by

$$\frac{1}{2}\sum_{i=1}^{n} [f(\phi_i)]^2 \Delta\theta_i \tag{2}$$

(recall formula (8), page 54). It is reasonable to regard (2) as a good approximation to the area A of the region OCD, where the approximation gets better and better as the number μ gets smaller and smaller. Motivated by this argument, we now *define* A as the limit

$$A = \lim_{\mu \to 0} \frac{1}{2} \sum_{i=1}^{n} [f(\phi_i)]^2 \Delta\theta_i,$$

which is just the integral

$$A = \frac{1}{2}\int_\alpha^\beta [f(\theta)]^2 \, d\theta = \frac{1}{2}\int_\alpha^\beta r^2 \, d\theta. \tag{3}$$

Here we use the fact that the sum in (2) is a Riemann sum for f^2 on $[\alpha, \beta]$, and the existence of the integral follows from the continuity of f^2, which is in turn implied by the continuity of f.

More generally, in the case of a region like the shaded region $CEFD$ in Figure 79, bounded by two rays $\theta = \alpha$, $\theta = \beta$ and the graphs of two continuous functions $r = f(\theta)$, $r = g(\theta)$, where $f(\theta) \geq g(\theta) \geq 0$, we can argue that

$$A = \text{Area of } CEFD = (\text{Area of } OEF) - (\text{Area of } OCD),$$

since the regions OEF and OCD share no points other than their common boundary CD. But

$$\text{Area of } OEF = \frac{1}{2}\int_\alpha^\beta [f(\theta)]^2 \, d\theta, \quad \text{Area of } OCD = \frac{1}{2}\int_\alpha^\beta [g(\theta)]^2 \, d\theta,$$

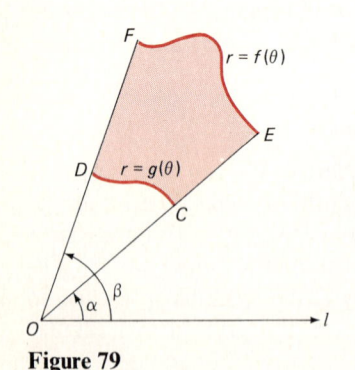

Figure 79

by two applications of formula (3), and therefore

$$A = \frac{1}{2}\int_\alpha^\beta [f(\theta)]^2\, d\theta - \frac{1}{2}\int_\alpha^\beta [g(\theta)]^2\, d\theta,$$

or equivalently

$$A = \frac{1}{2}\int_\alpha^\beta \{[f(\theta)]^2 - [g(\theta)]^2\}\, d\theta. \tag{3'}$$

Example 1 Find the area inside the cardioid $r = 1 + \cos\theta$, and also the area outside the cardioid and inside the circle $r = 3\cos\theta$.

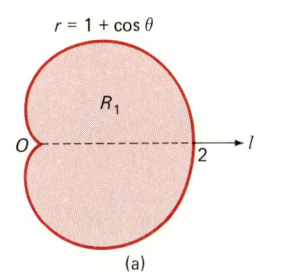

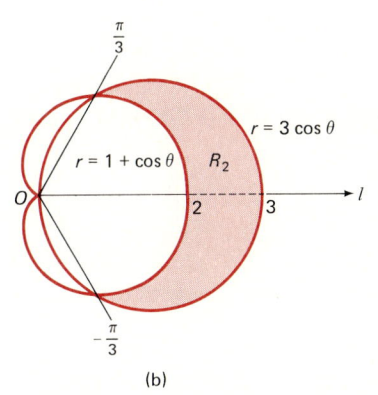

Solution For the region R_1 inside the cardioid [see Figure 80(a)] the limits of integration are $\alpha = -\pi$, $\beta = \pi$, and the rays $\theta = \alpha$, $\theta = \beta$ bounding the region both "shrink" to a single point, namely the pole O, since $r|_{\theta=-\pi} = r|_{\theta=\pi} = 0$. It follows from formula (3) that the area of R_1 is

$$A_1 = \frac{1}{2}\int_{-\pi}^\pi (1+\cos\theta)^2\, d\theta = \int_0^\pi (1 + 2\cos\theta + \cos^2\theta)\, d\theta$$

$$= \int_0^\pi \left(1 + 2\cos\theta + \frac{1+\cos 2\theta}{2}\right) d\theta$$

(at the second step we use the fact that R_1 is symmetric about the polar axis). Therefore

$$A_1 = \left[\theta + 2\sin\theta + \frac{\theta}{2} + \frac{\sin 2\theta}{4}\right]_0^\pi = \frac{3}{2}\pi.$$

As for the region R_2 outside the cardioid and inside the circle $r = 3\cos\theta$ (of radius $\frac{3}{2}$), the limits of integration are $\alpha = -\pi/3$, $\beta = \pi/3$ instead, since these are the angular coordinates of the points of intersection of the cardioid and the circle [see Figure 80(b)], and moreover $3\cos\theta \geq 1 + \cos\theta$ if $-\pi/3 \leq \theta \leq \pi/3$. Hence, by formula (3') with $f(\theta) = 3\cos\theta$, $g(\theta) = 1 + \cos\theta$, $\alpha = -\pi/3$, $\beta = \pi/3$, the area of R_2 is

$$A_2 = \frac{1}{2}\int_{-\pi/3}^{\pi/3} [(3\cos\theta)^2 - (1+\cos\theta)^2]\, d\theta = \int_0^{\pi/3} (8\cos^2\theta - 2\cos\theta - 1)\, d\theta$$

$$= \int_0^{\pi/3} \left(8 \cdot \frac{1+\cos 2\theta}{2} - 2\cos\theta - 1\right) d\theta = \int_0^{\pi/3} (4\cos 2\theta - 2\cos\theta + 3)\, d\theta$$

(again we use the fact that R_2 is symmetric about the polar axis). Therefore

$$A_2 = \left[2\sin 2\theta - 2\sin\theta + 3\theta\right]_0^{\pi/3} = 2\sin\frac{2\pi}{3} - 2\sin\frac{\pi}{3} + \pi = \pi,$$

which is $\frac{4}{9}$ of the area $(\frac{3}{2})^2\pi = \frac{9}{4}\pi$ bounded by the circle alone. ∎

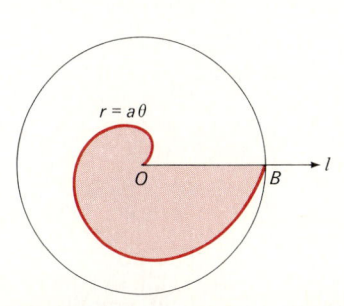

Example 2 Find the area A of the shaded region in Figure 81, bounded by the polar axis l and the first "turn" of the Archimedean spiral

$$r = a\theta \qquad (\theta \geq 0, a > 0),$$

that is, the part of the spiral corresponding to the interval $0 \leq \theta \leq 2\pi$.

Section 10.10 Area in Polar Coordinates; Length of a Polar Curve

Solution Here $\alpha = 0$, $\beta = 2\pi$, and

$$A = \frac{1}{2} \int_0^{2\pi} (a\theta)^2 \, d\theta = \frac{1}{6} a^2 \theta^3 \bigg|_0^{2\pi} = \frac{4}{3} \pi^3 a^2.$$

Notice that A is one third the area bounded by the indicated circle, of radius $|OB| = 2\pi a$. □

Example 3 Find the area A enclosed by the lemniscate

$$(x^2 + y^2)^2 = a^2(x^2 - y^2) \qquad (a > 0), \tag{4}$$

shown in Figure 82.

Solution First we transform to polar coordinates by setting $x = r \cos \theta$ and $y = r \sin \theta$ in (4). This gives $r^4 = a^2 r^2 (\cos^2 \theta - \sin^2 \theta)$ or

$$r^2 = a^2 \cos 2\theta$$

(justify the division by r^2). The lemniscate is symmetric about both the x-axis and the y-axis, and hence the total area A enclosed by the lemniscate is four times larger than the area of the shaded region in the figure, lying in the first quadrant between the rays $\theta = 0$ and $\theta = \pi/4$. Therefore

$$A = 4 \cdot \frac{1}{2} \int_0^{\pi/4} r^2 \, d\theta = 2a^2 \int_0^{\pi/4} \cos 2\theta \, d\theta = a^2 \sin 2\theta \bigg|_0^{\pi/4} = a^2. \quad \square$$

Figure 82 Lemniscate $r^2 = a^2 \cos 2\theta$

Length of a Polar Curve

We now turn to the problem of finding the length of a polar curve. Let C be the curve with polar equation

$$r = f(\theta) \qquad (\alpha \leq \theta \leq \beta).$$

Then C has the parametric representation

$$x = r \cos \theta = f(\theta) \cos \theta, \qquad y = r \sin \theta = f(\theta) \sin \theta \qquad (\alpha \leq \theta \leq \beta),$$

with the angular coordinate θ as parameter, and we can apply the theory of length already developed in Section 8.4. In fact, if f is continuously differentiable, then C is rectifiable, with length

$$L = \int_\alpha^\beta \sqrt{\left(\frac{dx}{d\theta}\right)^2 + \left(\frac{dy}{d\theta}\right)^2} \, d\theta$$

(see formula (2), p. 461). But

$$\left(\frac{dx}{d\theta}\right)^2 + \left(\frac{dy}{d\theta}\right)^2 = [f'(\theta) \cos \theta - f(\theta) \sin \theta]^2 + [f'(\theta) \sin \theta + f(\theta) \cos \theta]^2$$

$$= [f(\theta)]^2 + [f'(\theta)]^2,$$

so that

$$L = \int_\alpha^\beta \sqrt{[f(\theta)]^2 + [f'(\theta)]^2} \, d\theta,$$

or more concisely

$$L = \int_\alpha^\beta \sqrt{r^2 + \left(\frac{dr}{d\theta}\right)^2} \, d\theta. \tag{5}$$

Example 4 Find the perimeter L of the cardioid $r = 1 + \cos \theta$, graphed in Figure 80(a).

638 Chapter 10 Plane Analytic Geometry and Polar Coordinates

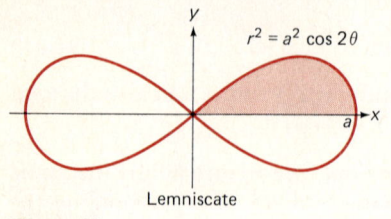

Solution Using formula (5) and the symmetry of the cardioid about the polar axis, we have

$$L = 2\int_0^\pi \sqrt{(1+\cos\theta)^2 + (-\sin\theta)^2}\, d\theta = 2\int_0^\pi \sqrt{2+2\cos\theta}\, d\theta$$

$$= 2\sqrt{2}\int_0^\pi \sqrt{1 + \cos^2\frac{\theta}{2} - \sin^2\frac{\theta}{2}}\, d\theta = 4\int_0^\pi \sqrt{\cos^2\frac{\theta}{2}}\, d\theta.$$

But $\cos(\theta/2) \geq 0$ if $0 \leq \theta \leq \pi$, and therefore

$$L = 4\int_0^\pi \cos\frac{\theta}{2}\, d\theta = 8\sin\frac{\theta}{2}\Big|_0^\pi = 8. \quad \square$$

Example 5 Find the length L of the first turn of the Archimedean spiral $r = e^{a\theta}$ considered in Example 2.

Solution Applying formula (5), we obtain

$$L = \int_0^{2\pi} \sqrt{a^2\theta^2 + a^2}\, d\theta = a\int_0^{2\pi} \sqrt{\theta^2 + 1}\, d\theta$$

$$= \frac{a}{2}\left[\theta\sqrt{\theta^2+1} + \ln(\theta + \sqrt{\theta^2+1})\right]_0^{2\pi}$$

$$= \frac{a}{2}\left[2\pi\sqrt{4\pi^2+1} + \ln(2\pi + \sqrt{4\pi^2+1})\right] \approx 21.3a,$$

with the help of formula (6'), page 391. $\quad \square$

Example 6 Find the total length of the logarithmic spiral

$$r = e^{-a\theta} \quad (0 \leq \theta < \infty, a > 0).$$

Solution As θ increases, the spiral winds around the pole O in the counterclockwise direction, as shown in Figure 83. The appropriate definition of the total length L of the spiral is given by the *improper* integral

$$L = \int_0^\infty \sqrt{r^2 + \left(\frac{dr}{d\theta}\right)^2}\, d\theta = \int_0^\infty \sqrt{(e^{-a\theta})^2 + (-ae^{-a\theta})^2}\, d\theta$$

$$= \sqrt{1+a^2}\int_0^\infty e^{-a\theta}\, d\theta = \sqrt{1+a^2}\lim_{u\to\infty}\int_0^u e^{-a\theta}\, d\theta$$

$$= \frac{\sqrt{1+a^2}}{a}\lim_{u\to\infty}(1 - e^{-au}) = \frac{\sqrt{1+a^2}}{a}. \quad \square$$

Example 7 Find the area A of the surface of revolution generated by revolving the lemniscate $r^2 = a^2\cos 2\theta$ about the x-axis (see Figure 82).

Solution By symmetry, A is twice the area of the surface of revolution generated by revolving the arc

$$r^2 = a^2\cos 2\theta \quad (0 \leq \theta \leq \pi/4, r \geq 0)$$

about the x-axis. Therefore, by formula (1), page 466, with t replaced by θ,

$$A = 4\pi\int_0^{\pi/4} y\sqrt{\left(\frac{dx}{d\theta}\right)^2 + \left(\frac{dy}{d\theta}\right)^2}\, d\theta = 4\pi\int_0^{\pi/4} r\sin\theta\sqrt{r^2 + \left(\frac{dr}{d\theta}\right)^2}\, d\theta.$$

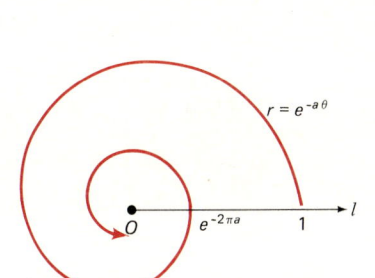

Differentiating $r^2 = a^2 \cos 2\theta$ with respect to θ, we get

$$2r \frac{dr}{d\theta} = -2a^2 \sin 2\theta,$$

so that

$$r\sqrt{r^2 + \left(\frac{dr}{d\theta}\right)^2} = \sqrt{r^4 + \left(r\frac{dr}{d\theta}\right)^2} = \sqrt{(a^2 \cos 2\theta)^2 + (-a^2 \sin 2\theta)^2}$$

$$= \sqrt{a^4 (\cos^2 2\theta + \sin^2 2\theta)} = a^2.$$

Therefore

$$A = 4\pi a^2 \int_0^{\pi/4} \sin \theta \, d\theta = -4\pi a^2 \cos \theta \Big|_0^{\pi/4} = 4\pi a^2 \left(1 - \frac{1}{\sqrt{2}}\right),$$

which is about 29% of the surface area $4\pi a^2$ of a sphere of radius a. □

Problems

Find the area of the region bounded by the given polar curve and the specified pair of rays.
1. $r = 3 \sec \theta, \theta = 0, \theta = \pi/4$
2. $r = -4 \csc \theta, \theta = 5\pi/4, \theta = 7\pi/4$
3. $r = 2 \tan \theta, \theta = \pi/6, \theta = \pi/3$
4. $r = 1/(1 + \theta), \theta = -\pi/4, \theta = \pi/4$
5. $r = \sqrt{2}/(1 + \cos \theta), \theta = \pi/3, \theta = \pi/2$
6. $r = 1/\theta, \theta = 2\pi/3, \theta = \pi$
7. $r = 2^\theta, \theta = 0, \theta = \pi/2$
8. $r = 5\theta^2, \theta = -\pi/2, \theta = 0$

Find the area inside the given polar curve.
9. $r = 8 \sin \theta$
10. $r = -\sqrt{3} \cos \theta$
11. $r = 10 |\cos \theta|$
12. $r = \cos \theta + \sin \theta$
13. $r = 3 - 3 \sin \theta$ (cardioid)
14. $r = 2 - \cos \theta$ (limaçon)
15. $r = 3 + 2 \cos \theta$ (limaçon)
16. $r^2 = 4 \sin 2\theta$ (lemniscate)
17. $r = \sqrt{5} \sec^3 (\theta/3)$ $(-\pi \le \theta \le \pi)$
18. $r = \sec \theta - 4 \cos \theta$ $(-\pi/3 \le \theta \le \pi/3)$

Find the total area enclosed by the given rose.
19. $r = \sin 2\theta$
20. $r = \cos 4\theta$
21. $r = \cos 3\theta$
22. $r = \sin 5\theta$
Hint. First calculate the area of a single petal.

In each case find the area inside the first polar curve and outside the second one.
23. $r = \cos \theta, r = \sin \theta$
24. $r = 4 \sin \theta, r = 2$
25. $r = 1, r = 1 + \cos \theta$
26. $r = 2 - \cos \theta, r = 3 \cos \theta$
27. $r = \sin 2\theta, r = \frac{1}{2}$
28. $r^2 = \sin 2\theta, r^2 = \cos 2\theta$

In each case find the area inside *both* polar curves.
29. $r = \cos \theta, r = \cos (\theta - (\pi/4))$
30. $r = 1 + \cos \theta, r = 1 - \cos \theta$
31. $r = \sin 2\theta, r = \cos \theta$
32. $r = 2 \cos 3\theta, r = 1$

33. Find the area enclosed by the inner loop of the limaçon $r = 1 + 2 \cos \theta$. Also find the area between the inner loop and the outer loop, that is, the area of the shaded region in Figure 84.

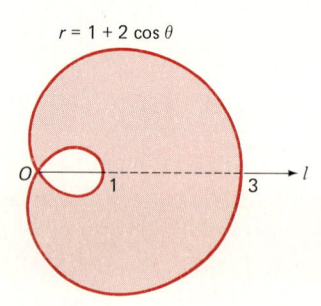

$r = 1 + 2 \cos \theta$

Figure 84

34. With as little effort as possible, show that the area bounded by the curve

$$r = \frac{p}{1 - e \cos \theta} \quad (0 < e < 1)$$

is equal to $\pi p^2 (1 - e^2)^{-3/2}$.
Hint. See Example 4, page 591.

35. The following calculation is clearly incorrect: The area enclosed by the lemniscate $r^2 = a^2 \cos 2\theta$ is

$$A = \frac{1}{2} \int_0^{2\pi} r^2 \, d\theta = \frac{a^2}{2} \int_0^{2\pi} \cos 2\theta \, d\theta = \frac{a^2}{4} \sin 2\theta \Big|_0^{2\pi} = 0$$

(see Example 3 for the correct answer). Where is the mistake?

36. Figure 85 shows a curve which has both a polar equation $r = r(\theta)$ on an interval $\alpha \leq \theta \leq \beta$ and a cartesian equation $y = y(x)$ on an interval $a \leq x \leq b$. There are now two expressions for the area A of the region OCD, namely

$$A = \frac{1}{2}\int_\alpha^\beta r^2 \, d\theta$$

and

$$A = \int_a^b y \, dx + (\text{Area of triangle } OaD)$$
$$- (\text{Area of triangle } ObC)$$

Verify that the same value of A is obtained in both cases, as required by a consistent theory of area in the plane.

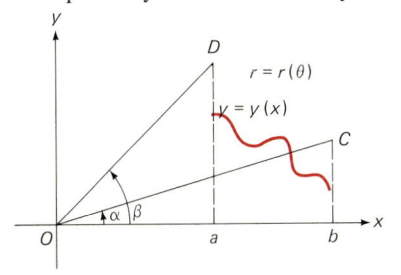

Figure 85

Find the length of the given polar curve (the constant a is positive wherever it appears).

+ 37. $r = a \sec \theta$ $(-\pi/4 \leq \theta \leq \pi/4)$
38. $r = a \csc \theta$ $(\pi/3 \leq \theta \leq 2\pi/3)$
+ 39. $r = 2a \sin \theta$ $(\pi \leq \theta \leq 2\pi)$
40. $r = \cos \theta + \sin \theta$ $(0 \leq \theta \leq \pi)$
+ 41. $r = e^\theta$ $(0 \leq \theta \leq 1)$
42. $r = \theta^2$ $(-\pi \leq \theta \leq \pi)$
43. $r = 1/\theta$ $(\frac{1}{2} \leq \theta \leq 2)$
44. $r = 2^\theta$ $(-\infty < \theta \leq 0)$
45. $r = 1/(1 + \cos \theta)$ $(-\pi/2 \leq \theta \leq \pi/2)$
46. $r = a \cos^2 (\theta/2)$ $(0 \leq \theta \leq 2\pi)$
47. $r = a \sin^3 (\theta/3)$ $(0 \leq \theta \leq 3\pi)$
48. $r = a \tanh (\theta/2)$ $(0 \leq \theta \leq 2\pi)$

49. Find the area of the surface of revolution generated by revolving the cardioid $r = a(1 + \cos \theta)$ about the x-axis.

50. Find the area of the surface of revolution generated by revolving the lemniscate $r^2 = a^2 \cos 2\theta$ about the y-axis (rather than about the x-axis, as in Example 7).

Key Terms and Topics
Translation and rotation of axes, the translation and rotation equations
Symmetry tests for the graph of $F(x, y) = 0$
Definition of a parabola
Focus, directrix, axis and vertex of a parabola
Equations of a parabola in standard form
Reflection property of the parabola
Definition of an ellipse
Foci, major axis, minor axis and vertices of an ellipse
Equations of an ellipse in standard form
Eccentricity of an ellipse
Reflection property of the ellipse
Definition of a hyperbola, branches of a hyperbola
Foci, transverse axis, conjugate axis, vertices and asymptotes of a hyperbola
Equations of a hyperbola in standard form
Eccentricity of a hyperbola
Directrices of the ellipse and hyperbola
The focus-directrix property of conic sections
Second degree curves as conic sections
Definition of polar coordinates
Relation between polar and rectangular coordinates
Graphs of polar equations, symmetry tests for the graph of $F(r, \theta) = 0$
Conics in polar coordinates
The tangent line to a polar curve, the radial-tangential angle
Area in polar coordinates, length of a polar curve

Supplementary Problems

1. A translation of axes carries the point $(3, -4)$ in the xy-system into a point of the x'-axis and the point $(2, 3)$ into a point of the y'-axis. What are the corresponding translation equations?

2. It is geometrically apparent that (a) the result of two consecutive translations of axes is itself a translation of axes, and (b) the distance between two points $P_1 = (x_1, y_1)$ and $P_2 = (x_2, y_2)$ in the plane is *translation invariant*, that is, does not change after a translation of axes. Verify (a) and (b) algebraically.

3. It is geometrically apparent that (a) the result of two consecutive rotations about the same point O is itself a rotation about O, and (b) the distance between two points $P_1 = (x_1, y_1)$ and $P_2 = (x_2, y_2)$ is *rotation invariant*, that is, does not change after a rotation of axes. Verify (a) and (b) algebraically.

4. An event E occurring at the position x (along a line) and time t can be specified by the ordered pair (x, t). Suppose the same event E is observed in two "reference frames," the xt-system and another $x't'$-system moving relative to the xt-system with constant velocity v, and let

$$\theta = \tanh^{-1}\frac{v}{c},$$

where c is the velocity of light (θ is called the *velocity parameter*). Then, according to Einstein's special theory of relativity (see Problem 54, page 276), the coordinates x' and t' are related to the coordinates x and t by the formulas

$$\begin{aligned} x' &= x \cosh \theta - ct \sinh \theta, \\ ct' &= -x \sinh \theta + ct \cosh \theta, \end{aligned} \quad \text{(i)}$$

called a *Lorentz transformation*. Show that the result of two consecutive Lorentz transformations is itself a Lorentz transformation.
Hint. Use formulas (6) and (7), page 351.

5. In relativity theory the *spacetime interval* between two events $E_1 = (x_1, t_1)$ and $E_2 = (x_2, t_2)$ is defined as

$$s = \sqrt{(x_1 - x_2)^2 - c^2(t_1 - t_2)^2},$$

where c is the velocity of light. Show that the quantity s is *Lorentz invariant*, that is, does not change as a result of a Lorentz transformation (i).

6. Let G be the graph of the equation $\sin(x + y) = 0$. Which of the four symmetries discussed on page 579 does G have? Describe G.

7. Find an equation of the parabolic arch of base b and height h shown in Figure 86. What is the area under the arch?

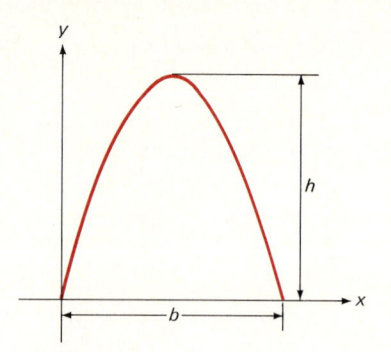

Figure 86

8. A chord of a parabola is called a *focal chord* if it goes through the focus F of the parabola. Show that the tangents to a parabola at the endpoints of a focal chord are perpendicular and intersect on the directrix L (see Figure 87, where PP' is a focal chord).
Hint. Use Problems 31 and 32, page 587.

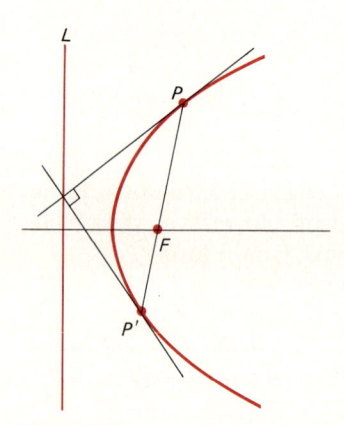

Figure 87

9. The normal N at an arbitrary point P of the parabola $y^2 = 4cx$ ($c > 0$), with focus $F = (c, 0)$, intersects the x-axis in a point B. Show that there are two positions of P for which the triangle FPB is equilateral, with side length $4c$. (See Figure 88, where L is the directrix of the parabola and Q is the foot of the perpendicular drawn from P to L.)

642 Chapter 10 Plane Analytic Geometry and Polar Coordinates

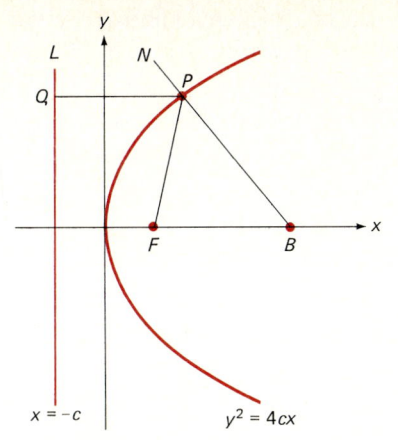

Figure 88

10. Let N, P and B be the same as in the preceding problem, and let A be the foot of the perpendicular drawn from P to the x-axis (see Figure 89). Show that the segment AB, called the *subnormal* of the parabola, is of constant length $2c$ for every position of the point P.

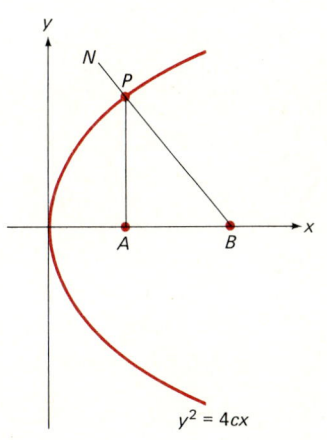

Figure 89

11. Show that the locus of the midpoints of all the chords of the parabola $y^2 = 4cx$ with the same nonzero inclination θ is a straight line parallel to the x-axis (see Figure 90).

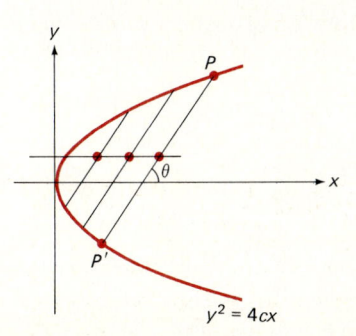

Figure 90

12. Show that the perpendicular drawn from the focus F of the parabola $y^2 = 4cx$ $(c > 0)$ to the tangent T at a point P of the parabola always intersects T in a point Q of the y-axis (see Figure 91). Also show that if $r = |FP|$ and $p = |FQ|$, then $p^2 = cr$ for every position of P.

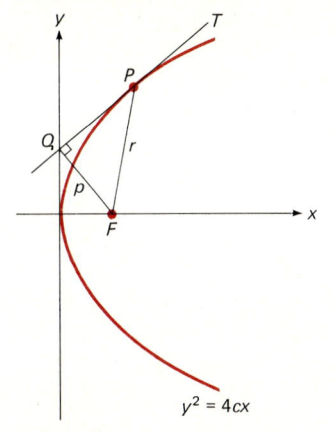

Figure 91

13. Let E be the ellipse with equation $7x^2 + 3y^2 = 55$. Which of the six points $P_1 = (2, -3)$, $P_2 = (-2, 2)$, $P_3 = (0, -5)$, $P_4 = (3, -2)$, $P_5 = (1, 4)$, $P_6 = (-3, 1)$ lie on E? Inside E? Outside E?

14. Find the area of the quadrilateral with two vertices at the foci of the ellipse $\frac{1}{20}x^2 + \frac{1}{4}y^2 = 1$ and the other two vertices at the endpoints of the minor axis.

15. An ellipse symmetric about both coordinate axes goes through the points $(6, 0)$ and $(4, \sqrt{5})$. Write an equation of the ellipse. What is its eccentricity?

16. Find the points in which the ellipse $x^2 + 4y^2 = 4$ intersects the circle going through the foci and the upper y-intercept of the ellipse.

17. Find the tangents to the ellipse $\frac{1}{30}x^2 + \frac{1}{24}y^2 = 1$ parallel to the line $6x - 3y + 11 = 0$. What is the distance between the tangents?

18. A chord of an ellipse which goes through the center of the ellipse is called a *diameter*. Show that the tangents to the ellipse $(x^2/a^2) + (y^2/b^2) = 1$ at the endpoints of a diameter are parallel (see Figure 92).

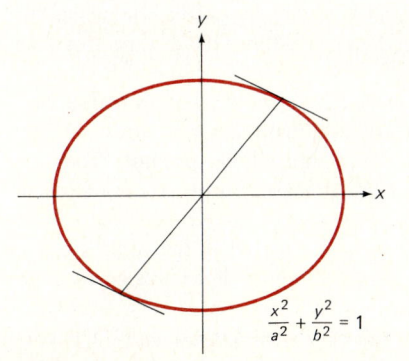

Figure 92

Supplementary Problems **643**

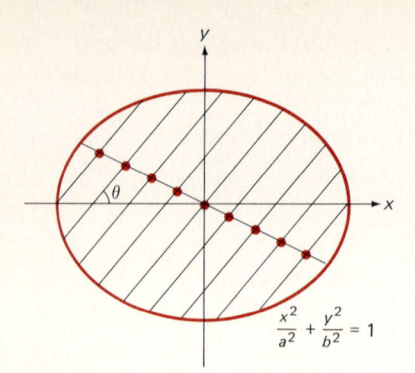

Figure 93

19. Show that the locus of the midpoints of all the chords of the ellipse $(x^2/a^2) + (y^2/b^2) = 1$ with the same inclination θ is a diameter of the ellipse (see Figure 93).

20. On the ellipse $\frac{1}{18}x^2 + \frac{1}{8}y^2 = 1$ find the point P nearest the line $2x - 3y + 25 = 0$. What is the distance between P and the line?

21. Show that the product of the distances from a point P of the hyperbola $(x^2/a^2) - (y^2/b^2) = 1$ with foci $(\pm c, 0)$ to its asymptotes has the same value for every position of P. What is this value?

22. Show that the curve with parametric equations

$$x = x_0 + a \sec t, \quad y = y_0 + b \tan t \quad (-\pi/2 < t < \pi/2) \quad \text{(ii)}$$

is the right branch of a hyperbola with a horizontal transverse axis if $a > 0$ and the left branch if $a < 0$. Write parametric equations for the branches of a hyperbola with a vertical transverse axis.

23. Find the tangents to the hyperbola $\frac{1}{20}x^2 - \frac{1}{5}y^2 = 1$ perpendicular to the line $4x + 3y - 7 = 0$.

24. Find the tangents to the hyperbola $\frac{1}{8}y^2 - \frac{1}{16}x^2 = 1$ parallel to the line $2x + 4y - 5 = 0$. What is the distance between the tangents?

25. Find the point of intersection of the tangent to the hyperbola $xy = 1$ going through the point $(2, \frac{1}{2})$ with the tangent going through the point $(\frac{1}{2}, 2)$.

26. Find the eccentricity of any ellipse for which the distance between the directrices is three times the distance between the foci.

27. Find the eccentricity of any hyperbola for which the distance between the foci is five times the distance between the directrices.

28. The distance between the directrices of an ellipse with its center at $(-3, 4)$ and a horizontal major axis is 36, and there is a point on the ellipse whose distances from the foci are 9 and 15. What is the equation of the ellipse in standard form?

29. Find the hyperbola whose foci lie at the vertices of the ellipse $\frac{1}{100}x^2 + \frac{1}{64}y^2 = 1$, and whose directrices go through the foci of the ellipse.

30. A chord of an ellipse is called a *focal chord* if it goes through a focus F of the ellipse. Show that the tangents to an ellipse at the endpoints of a focal chord intersect on the directrix L nearer the focus (see Figure 94, where PP' is a focal chord).

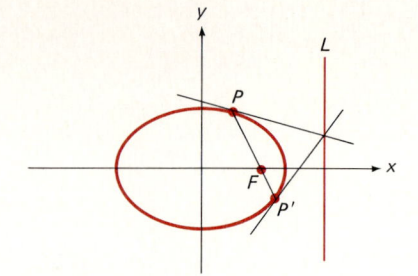

Figure 94

31. Show that the locus of all points from which a pair of perpendicular tangent lines can be drawn to the ellipse $(x^2/a^2) + (y^2/b^2) = 1$ is the circle $x^2 + y^2 = a^2 + b^2$, called the *director circle* of the ellipse. (See Figure 95, which also shows the directrices L_1 and L_2 of the ellipse.)

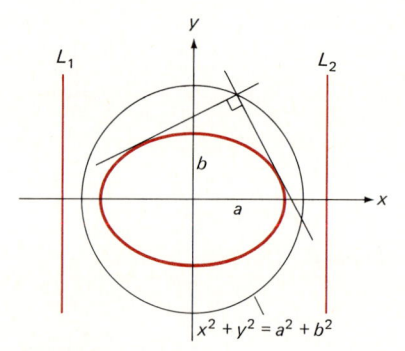

Figure 95

32. Find the radius of the director circle of an ellipse of eccentricity $\frac{4}{5}$ and semimajor axis 15.

33. According to Problem 30, the tangents at the endpoints of a focal chord of an ellipse intersect on a directrix. Show that unlike the case of the parabola (see Problem 8), these tangents are never perpendicular.

After making a suitable rotation of axes, identify the conic which is the graph of the given equation. Sketch the conic if it is nondegenerate.

34. $4xy + 3y^2 + 16x + 12y + 36 = 0$
35. $9x^2 + 30xy + 25y^2 - 42x - 70y + 49 = 0$
36. $7x^2 + 6xy - y^2 + 28x + 12y + 28 = 0$
37. $x^2 - 2xy + y^2 - 3x + y = 0$

38. Show that if $A + C$ and $4AC - B^2$ are both positive, then the graph of the equation $Ax^2 + Bxy + Cy^2 = 1$ is an ellipse enclosing an area equal to $2\pi/\sqrt{4AC - B^2}$.

644   Chapter 10   Plane Analytic Geometry and Polar Coordinates

39. Give a simple geometric description of the second degree curve with equation
$$(x - x_1)(x - x_2) + (y - y_1)(y - y_2) = 0.$$

40. Show that the graph of the equation
$$xy + Ax + By + C = 0$$
is either an equilateral hyperbola or a pair of perpendicular lines.

Let C_k be the curve with equation
$$\frac{x^2}{a^2 + k} + \frac{y^2}{b^2 + k} = 1, \qquad \text{(iii)}$$
where $a^2 > b^2$, $k > -a^2$, $k \neq -b^2$.

41. Show that C_k is an ellipse if $k > -b^2$ and a hyperbola if $-a^2 < k < -b^2$.
42. Show that all the C_k have the same foci (for this reason the curves C_k are said to be *confocal*).
43. Graph equation (iii) for $a^2 = 4$, $b^2 = 1$ and $k = -3, -2, 0, 1$, locating the common foci F_1 and F_2.
44. Show that every ellipse C_k ($k > -b^2$) is *orthogonal* to every hyperbola $C_{k'}$ ($-a^2 < k' < -b^2$), that is, show that the curves C_k and $C_{k'}$ intersect at right angles at all their points of intersection.

45. A circle has polar equation
$$r^2 - 3\sqrt{3}\, r \cos\theta - 3r \sin\theta - 7 = 0.$$
Find its radius and center (in polar coordinates).

46. Find the new polar coordinates of the points $A = (4, \pi/3)$, $B = (1, 2\pi/3)$ and $C = (5, \pi/4)$ after rotating the polar axis until it goes through A.

47. Find the midpoint of the segment joining the points with polar coordinates $(12, 4\pi/9)$ and $(12, -2\pi/9)$.

48. The curve with polar equation
$$r = a + b \sec\theta \qquad (a > 0, b > 0)$$
is called a *conchoid*. One of the three conchoids $r = 2 + \sec\theta$, $r = 2 + 2\sec\theta$, $r = 2 + 3\sec\theta$ has a "loop." Which one?

Give an example of a polar equation $F(r, \theta) = 0$ whose graph is symmetric about the y-axis even though its solution set does not coincide with that of the equation

49. $F(-r, -\theta) = 0$
50. $F(r, \pi - \theta) = 0$
51. $F(-r, -\theta) = 0$ or $F(r, \pi - \theta) = 0$

52. A point P moves in such a way that the product of its distances from two fixed points F_1 and F_2 has the constant value b^2 (thus $|PF_1||PF_2| = b^2$). Find both a cartesian equation and a polar equation for the path traversed by P, called a *Cassinian oval*, choosing $F_1 = (-a, 0)$, $F_2 = (a, 0)$ in rectangular coordinates or equivalently $F_1 = (a, \pi)$, $F_2 = (a, 0)$ in polar coordinates. Graph the Cassinian ovals for $b^2 = 8$, $a^2 = 6, 8, 10$, and describe the qualitative differences between these three cases.

53. A line through the focus F of a nondegenerate conic of eccentricity e and focus-to-directrix distance d intersects the conic in the points A and B. Show that the sum
$$\frac{1}{|AF|} + \frac{1}{|BF|}$$
has the same value for every position of the line. What is this value?

54. Use polar coordinates to give another proof of the reflection property of the parabola, established on page 585.

55. Show that the circles $r = a \sin\theta$ and $r = b \cos\theta$ are orthogonal (that is, intersect at right angles) at both points of intersection, regardless of the choice of the positive constants a and b.

56. The cardioids $r = a(1 - \cos\theta)$ and $r = b(1 + \cos\theta)$ intersect at the pole and at two points P_1 and P_2 which are symmetric about the polar axis (see Figure 96). Show that the cardioids are orthogonal at P_1 and P_2, regardless of the choice of the positive constants a and b. What is the angle between the cardioids at the pole?

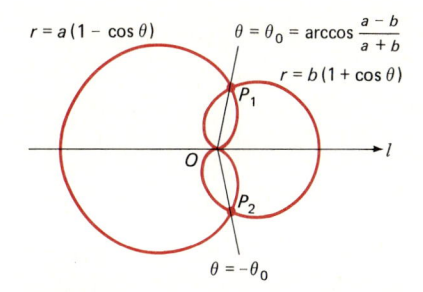

Figure 96

Let A_n be the area of one petal of the rose $r = a \sin n\theta$, and let A be the total area enclosed by the whole rose. Show that

57. $A_n = \dfrac{\pi a^2}{4n}$

58. $A = \dfrac{\pi a^2}{2}$ if n is even, while $A = \dfrac{\pi a^2}{4}$ if n is odd.

59. Find the area inside both cardioids $r = 1 + \sin\theta$ and $r = 1 - \cos\theta$.

60. Let A_n be the area bounded by the polar axis and the first n turns of the Archimedean spiral $r = a\theta$ ($\theta \geq 0$, $a > 0$). Show that $A_{n+1} - A_n = 8\pi^3 a^2 n$ ($n \geq 1$). Find A_1, A_2, A_3 and A_4.

61. Give an example of two simple closed polar curves enclosing different areas, each of which cuts off a segment of constant length $2a$ from every line through the pole.

Supplementary Problems 645

62. Use polar coordinates to show that the area enclosed by the loop of the *folium of Descartes*

$$x = \frac{3at}{t^3 + 1}, \quad y = \frac{3at^2}{t^3 + 1} \quad (-\infty < t < \infty)$$

(see Figure 41, page 459) is equal to $\frac{3}{2} a^2$.

Hint. The folium has cartesian equation $x^3 + y^3 = 3axy$.

63. If $0 < a < b$, the limaçon $r = a + b \cos \theta$ has two loops, an inner loop and an outer loop. Find the area between the two loops.

64. Show that the hyperbolic spiral $r\theta = 1$ $(1 \leq \theta < \infty)$ is not of finite length.

65. Find the length of the polar curve

$$\theta = \frac{1}{2}\left(r + \frac{1}{r}\right) \quad (1 \leq r \leq 3).$$

66. Find the length of the curve with "polar parametric equations"

$$r = t, \quad \theta = \ln t \quad (0 < t \leq 1).$$

Check the calculation by first expressing r as a function of θ.

11 Vectors in the Plane

In this chapter we show how the concepts of calculus can be extended from ordinary functions, mapping numbers into numbers, to *vector-valued* functions, mapping numbers into *vectors*. A vector is a quantity characterized not only by a number, called its *magnitude*, but also by a *direction*. The development of vector methods is largely due to the pioneering work of the great American chemist and mathematical physicist Josiah Willard Gibbs (1839–1903), who showed how their use greatly facilitates the solution of problems arising in applied science.

Our exposition of vector calculus proceeds in easy stages. In this chapter we confine our attention to the case of vectors in the plane, in which most of the essential features of the subject emerge. In the next chapter we go from two to three dimensions, and consider vectors in *space*. The final step of allowing our vector functions to have vector *arguments* as well as vector *values* will be taken later, after we have acquired the techniques of the calculus of functions of several variables.

11.1 The Vector Concept; Operations on Vectors

Scalars and Vectors

In science and technology it is important to distinguish between two kinds of quantities, scalars and vectors. By a *scalar* we mean a quantity which is completely specified by giving a single number (in appropriate units of measurement). Thus the pressure of a confined gas, the altitude of an airplane, and the temperature of an oven are all scalars. By a *vector*, on the other hand, we mean a quantity whose specification requires not only a number, called the *magnitude* of the vector, but also a *direction*. For example, the wind velocity at a weather station, the position of a target relative to a naval gun, and the force exerted on a moving electron by a magnetic field are all vectors.

Symbolically, vectors are denoted by boldface letters like

$$\mathbf{a}, \mathbf{b}, \mathbf{u}, \mathbf{v}, \ldots$$

(preferred in printing), or alternatively by putting little arrows over the corresponding lightface letters, as in

$$\vec{a}, \vec{b}, \vec{u}, \vec{v}, \ldots$$

(recommended in handwriting). As for scalars, they are denoted by ordinary lightface letters without arrows.

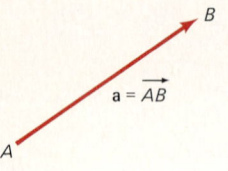

To represent a vector geometrically, we use a directed line segment or arrow, which points in the same direction as the vector and whose length is equal to the magnitude of the vector (note that the magnitude of a vector is inherently nonnegative). Every vector has an *initial point* and a *final point*, the latter marked by the arrowhead. For example, of the two endpoints of the vector **a** in Figure 1, A is the initial point and B is the final point; the directed line segment going from A to B in that order is indicated by $\overrightarrow{AB}$, and hence $\mathbf{a} = \overrightarrow{AB}$. The magnitude of a vector is indicated by the corresponding lightface letter (after all, the magnitude is a scalar), or by putting the vector inside the absolute value sign. Thus for the vector in Figure 1 we have

$$a = |\mathbf{a}| = |\overrightarrow{AB}|.$$

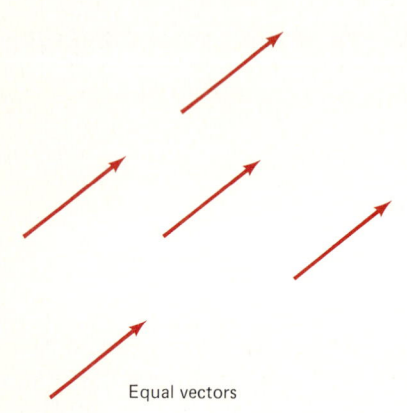

Two vectors are said to be *equal* if they are parallel (or collinear), point in the same direction, and have the same magnitude. Notice that *equality* of two vectors does not mean *identity* of the vectors, any more than equality of the fractions $\frac{1}{2}$ and $\frac{2}{4}$ means that they are identical.

Example 1 The five vectors shown in Figure 2, of which the top two are collinear, are all equal.

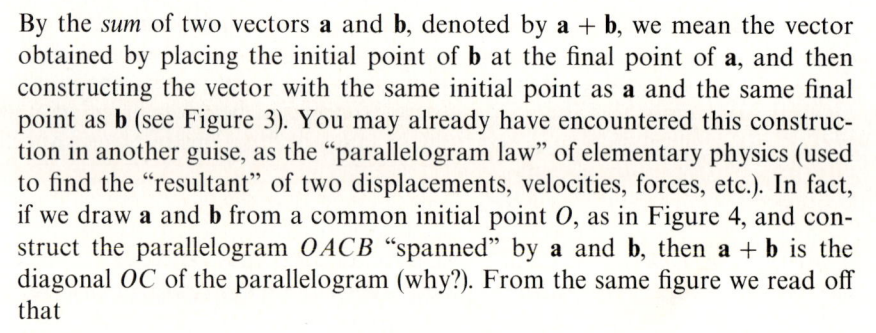

Addition of Vectors

By the *sum* of two vectors **a** and **b**, denoted by $\mathbf{a} + \mathbf{b}$, we mean the vector obtained by placing the initial point of **b** at the final point of **a**, and then constructing the vector with the same initial point as **a** and the same final point as **b** (see Figure 3). You may already have encountered this construction in another guise, as the "parallelogram law" of elementary physics (used to find the "resultant" of two displacements, velocities, forces, etc.). In fact, if we draw **a** and **b** from a common initial point O, as in Figure 4, and construct the parallelogram $OACB$ "spanned" by **a** and **b**, then $\mathbf{a} + \mathbf{b}$ is the diagonal OC of the parallelogram (why?). From the same figure we read off that

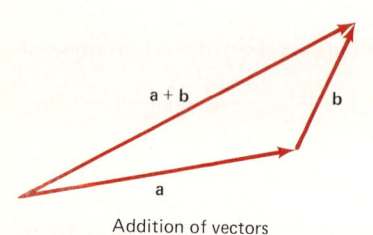

$$\mathbf{a} + \mathbf{b} = \overrightarrow{OA} + \overrightarrow{AC} = \overrightarrow{OC}$$

and

$$\mathbf{b} + \mathbf{a} = \overrightarrow{OB} + \overrightarrow{BC} = \overrightarrow{OC},$$

which together imply

$$\mathbf{a} + \mathbf{b} = \mathbf{b} + \mathbf{a}, \tag{1}$$

showing that vector addition is *commutative*.

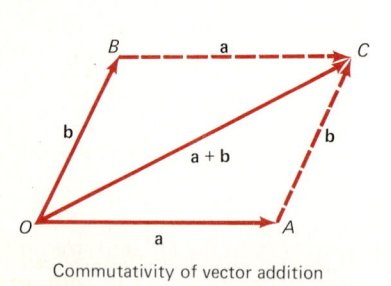

648   Chapter 11   Vectors in the Plane

Figure 5

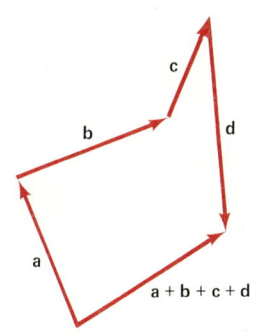

Figure 6 Associativity of vector addition

Figure 7

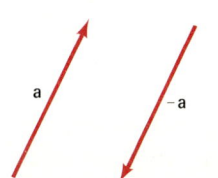

Figure 8

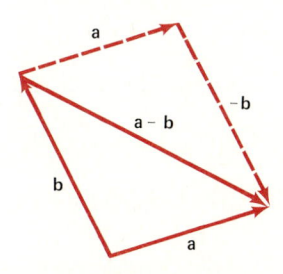

Figure 9 Subtraction of vectors

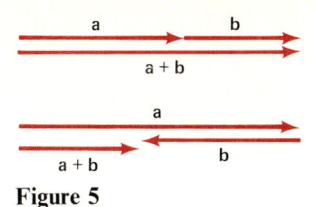

Remark The triangle in Figure 3 "collapses" if **a** and **b** are parallel. We then have one of the two cases illustrated in Figure 5.

The construction given in Figure 6 shows that vector addition is also *associative*, which means that

$$(\mathbf{a} + \mathbf{b}) + \mathbf{c} = \mathbf{a} + (\mathbf{b} + \mathbf{c}). \qquad (2)$$

It follows by repeated application of formulas (1) and (2) that the sum of any number of vectors is independent of the order of the terms or the way in which they are grouped together. In particular, this allows us to omit parentheses and brackets in writing sums of vectors. For example,

$$\mathbf{a} + \mathbf{b} + \mathbf{c} + \mathbf{d} = [(\mathbf{d} + \mathbf{b}) + \mathbf{c}] + \mathbf{a} = [\mathbf{c} + (\mathbf{a} + \mathbf{d})] + \mathbf{b},$$

and so on.

The fact that the length of one side of a triangle cannot exceed the sum of the lengths of the other two sides implies the *triangle inequality*

$$|\mathbf{a} + \mathbf{b}| \leq |\mathbf{a}| + |\mathbf{b}| \qquad (3)$$

for arbitrary vectors **a** and **b** (see Figure 3). Notice that the triangle inequality for scalars (Theorem 5, page 15) is the special case of (3) obtained when **a** and **b** are collinear, as in Figure 5.

Example 2 In Figure 7 the sum $\mathbf{a} + \mathbf{b} + \mathbf{c} + \mathbf{d}$ is the vector "closing" the polygonal path obtained by placing the initial point of **b** at the final point of **a**, the initial point of **c** at the final point of **b**, and the initial point of **d** at the final point of **c**.

By the *zero vector*, denoted by **0** (boldface zero), we mean the "vector" whose initial and final points coincide. Hence the zero vector has magnitude 0, but no well-defined direction. Clearly the vector **0** plays the same role in vector addition as the number 0 plays in ordinary (scalar) addition, that is

$$\mathbf{a} + \mathbf{0} = \mathbf{a}$$

for arbitrary **a**.

Given any vector **a**, the vector with the same magnitude as **a** and the opposite direction is called the *negative* of **a**, denoted by $-\mathbf{a}$ (see Figure 8). It follows from the definition of vector addition that

$$\mathbf{a} + (-\mathbf{a}) = \mathbf{0},$$

and since $\mathbf{0} + \mathbf{0} = \mathbf{0}$, we set $-\mathbf{0} = \mathbf{0}$. The *difference* $\mathbf{a} - \mathbf{b}$ is defined as the sum $\mathbf{a} + (-\mathbf{b})$. To construct $\mathbf{a} - \mathbf{b}$, we can place **a** and **b** so that their initial points coincide, and then draw the vector from the final point of **b** to the final point of **a**. Figure 9 shows why this construction works.

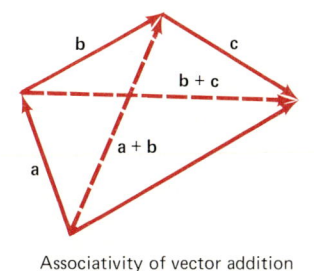

Section 11.1 The Vector Concept; Operations on Vectors

Scalar Multiples of a Vector

The product $p\mathbf{a}\ (=\mathbf{a}p)$ of a nonzero scalar p and a nonzero vector $\mathbf{a}$ is defined to be the vector whose magnitude is $|p|$ times that of $\mathbf{a}$ (so that $|p\mathbf{a}| = |p||\mathbf{a}|$, with the same direction as $\mathbf{a}$ if $p > 0$ and the opposite direction if $p < 0$ (see Figure 10). In particular, $(-1)\mathbf{a} = -\mathbf{a}$. If $p = 0$ or $\mathbf{a} = \mathbf{0}$, we set $p\mathbf{a} = \mathbf{0}$ by definition. Given two scalars p and q, we see at once that $p(q\mathbf{a}) = pq\mathbf{a}$, where $pq\mathbf{a}$ is the product of the scalar pq and the vector $\mathbf{a}$. Also, given arbitrary vectors $\mathbf{a}, \mathbf{b}$ and scalars p, q, we have the two *distributive laws*

$$(p + q)\mathbf{a} = p\mathbf{a} + q\mathbf{a}, \tag{4}$$

$$p(\mathbf{a} + \mathbf{b}) = p\mathbf{a} + p\mathbf{b}. \tag{5}$$

Figure 10

Figure 11 demonstrates the validity of (5), and the proof of (4) is left as an exercise. Division of a vector by a scalar is defined in the natural way, that is, by setting

$$\frac{\mathbf{a}}{p} = \frac{1}{p}\mathbf{a} \qquad (p \ne 0).$$

A vector of length 1 is called a *unit vector*. Given any nonzero vector $\mathbf{a}$, the vector

$$\frac{\mathbf{a}}{|\mathbf{a}|}$$

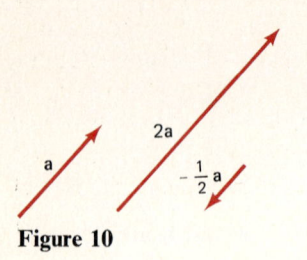

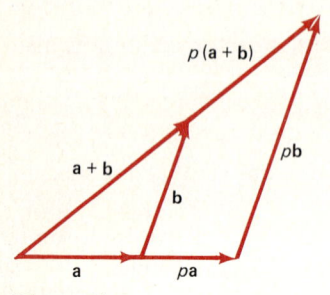

Figure 11

is a unit vector, since

$$\left|\frac{\mathbf{a}}{|\mathbf{a}|}\right| = \frac{1}{|\mathbf{a}|}|\mathbf{a}| = 1.$$

Components of a Vector

So far our discussion of vectors has been purely geometric and "coordinate-free." We now introduce a system of rectangular coordinates x and y in the plane, with origin O. Let $\mathbf{i}$ be a unit vector along the positive x-axis and $\mathbf{j}$ a unit vector along the positive y-axis, as shown in Figure 12(a). Then any vector $\mathbf{a}$ in the plane has a unique representation of the form

$$\mathbf{a} = \alpha_1 \mathbf{i} + \alpha_2 \mathbf{j}. \tag{6}$$

The scalars α_1 and α_2, called the *components* of $\mathbf{a}$, are determined as follows. Shift the vector $\mathbf{a}$, that is, move it parallel to itself without rotation, until its initial point coincides with the origin O. Then the final point of $\mathbf{a}$ will be a point $A = (\alpha_1, \alpha_2)$ of the xy-plane, and of course $\mathbf{a} = \overrightarrow{OA}$, since all shifted versions of a vector are equal to each other. But, as is apparent from the construction in Figure 12(b), the coordinates α_1 and α_2 of the point A are precisely the scalars called for in the representation or "expansion" (6). Moreover, the magnitude of $\mathbf{a}$ is just the distance from O to A, that is,

$$|\mathbf{a}| = \sqrt{\alpha_1^2 + \alpha_2^2}. \tag{7}$$

The figure also shows that if θ is the angle between the positive x-axis and the vector $\mathbf{a}$, then

$$\alpha_1 = |\mathbf{a}| \cos \theta, \qquad \alpha_2 = |\mathbf{a}| \sin \theta.$$

By a *basis* in the plane we mean any two fixed vectors $\mathbf{e}_1$ and $\mathbf{e}_2$, called *basis vectors*, such that an arbitrary vector $\mathbf{a}$ in the plane has a unique representation of the form

$$\mathbf{a} = \alpha_1 \mathbf{e}_1 + \alpha_2 \mathbf{e}_2, \tag{6'}$$

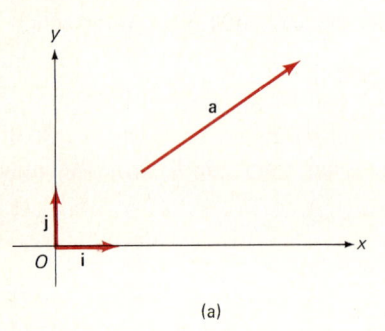

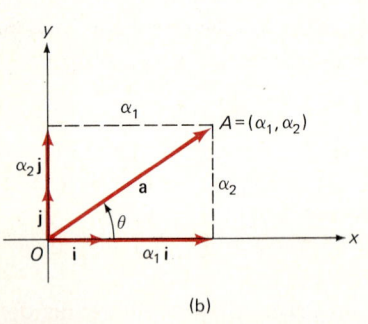

Figure 12

called the *expansion* of **a** with respect to $\mathbf{e}_1$ and $\mathbf{e}_2$. The scalars α_1 and α_2 are then called the *components* of **a** (with respect to $\mathbf{e}_1$ and $\mathbf{e}_2$). It can be shown that two nonzero vectors $\mathbf{e}_1$ and $\mathbf{e}_2$ form a basis in the plane if and only if they are *noncollinear*, that is, if and only if there is no line containing (or parallel to) both vectors. If the vectors $\mathbf{e}_1$ and $\mathbf{e}_2$ of a basis are perpendicular, the basis is said to be *orthogonal*, and if they are *unit* vectors as well as being perpendicular, the basis is said to be *orthonormal*. Thus the unit vectors **i** and **j** along the coordinate axes form an orthonormal basis. Notice that formula (7) is valid only for an orthonormal basis (why?).

Vectors as Ordered Pairs

Guided by the above considerations, we will henceforth employ ordered pairs to represent both points and vectors in the plane. Thus (α_1, α_2) can mean either the point with rectangular *coordinates* α_1 and α_2 or the vector $\alpha_1 \mathbf{i} + \alpha_2 \mathbf{j}$ with *components* α_1 and α_2 (with respect to the underlying orthonormal basis **i** and **j**). Since $\mathbf{i} = 1\mathbf{i} + 0\mathbf{j}$ and $\mathbf{j} = 0\mathbf{i} + 1\mathbf{j}$, the ordered pairs representing the basis vectors themselves are

$$\mathbf{i} = (1, 0), \qquad \mathbf{j} = (0, 1).$$

We can now interpret algebraic operations on vectors from the standpoint of ordered pairs. Let $\mathbf{a} = (\alpha_1, \alpha_2)$ and $\mathbf{b} = (\beta_1, \beta_2)$ be arbitrary vectors. Then

$$\mathbf{a} + \mathbf{b} = (\alpha_1 \mathbf{i} + \alpha_2 \mathbf{j}) + (\beta_1 \mathbf{i} + \beta_2 \mathbf{j}) = (\alpha_1 \mathbf{i} + \beta_1 \mathbf{i}) + (\alpha_2 \mathbf{j} + \beta_2 \mathbf{j}),$$

and therefore

$$\mathbf{a} + \mathbf{b} = (\alpha_1 + \beta_1)\mathbf{i} + (\alpha_2 + \beta_2)\mathbf{j}, \tag{8}$$

with the help of (4), or

$$(\alpha_1, \alpha_2) + (\beta_1, \beta_2) = (\alpha_1 + \beta_1, \alpha_2 + \beta_2). \tag{8'}$$

If p is any scalar, then $p\mathbf{a} = p(\alpha_1 \mathbf{i} + \alpha_2 \mathbf{j})$, and therefore

$$p\mathbf{a} = p\alpha_1 \mathbf{i} + p\alpha_2 \mathbf{j}, \tag{9}$$

with the help of (5), or

$$p(\alpha_1, \alpha_2) = (p\alpha_1, p\alpha_2). \tag{9'}$$

It follows from (8') that

$$(\alpha_1, \alpha_2) + (0, 0) = (\alpha_1 + 0, \alpha_2 + 0) = (\alpha_1, \alpha_2)$$

for every vector $\mathbf{a} = (\alpha_1, \alpha_2)$, so that $(0, 0)$ *is the ordered pair representing the zero vector* **0**. To get the ordered pair representing $-\mathbf{a}$, we set $p = -1$ in formula (9), obtaining $-\mathbf{a} = -\alpha_1 \mathbf{i} - \alpha_2 \mathbf{j}$, or equivalently

$$-\mathbf{a} = (-\alpha_1, -\alpha_2).$$

Moreover, $\mathbf{a} - \mathbf{b} = \mathbf{a} + (-\mathbf{b})$, so that

$$\mathbf{a} - \mathbf{b} = (\alpha_1, \alpha_2) + (-\beta_1, -\beta_2) = (\alpha_1 - \beta_1, \alpha_2 - \beta_2) \tag{10}$$

or

$$(\alpha_1, \alpha_2) - (\beta_1, \beta_2) = (\alpha_1 - \beta_1, \alpha_2 - \beta_2). \tag{10'}$$

Two vectors $\mathbf{a} = (\alpha_1, \alpha_2)$ and $\mathbf{b} = (\beta_1, \beta_2)$ are equal if and only if $\alpha_1 = \beta_1$ and $\alpha_2 = \beta_2$, that is, if and only if they are represented by the same ordered pair. This is apparent from the construction in Figure 12(b), since equal vectors are either coincident or shifted versions of each other. Thus if $\mathbf{b} = \mathbf{a}$, the vector $\mathbf{b}$ will coincide with $\overrightarrow{OA}$ after the initial point of $\mathbf{b}$ is shifted to the origin, so that $\mathbf{b}$ is represented by the same ordered pair $A = (\alpha_1, \alpha_2)$ as $\mathbf{a}$ itself.

Example 3 Let $\mathbf{a} = (5, -1)$, $\mathbf{b} = (1, 6)$ and $\mathbf{c} = (0, -2)$. Calculate the vector $\mathbf{a} - 2\mathbf{b} + 4\mathbf{c}$, and then find its magnitude.

Solution Making free use of the rules (8)–(10), we have

$$\mathbf{a} - 2\mathbf{b} + 4\mathbf{c} = (5, -1) - 2(1, 6) + 4(0, -2)$$
$$= (5 - 2 + 0, -1 - 12 - 8) = (3, -21),$$

or equivalently

$$\mathbf{a} - 2\mathbf{b} + 4\mathbf{c} = 3\mathbf{i} - 21\mathbf{j}.$$

By formula (7), the magnitude of this vector is

$$|3\mathbf{i} - 21\mathbf{j}| = \sqrt{3^2 + (-21)^2} = \sqrt{450} = 15\sqrt{2}. \quad \square$$

Example 4 Find the unit vector $\mathbf{u}$ with the same direction as the vector $12\mathbf{i} + 5\mathbf{j}$.

Solution The magnitude of $12\mathbf{i} + 5\mathbf{j}$ is

$$|12\mathbf{i} + 5\mathbf{j}| = \sqrt{12^2 + 5^2} = \sqrt{169} = 13,$$

and hence

$$\mathbf{u} = \frac{12\mathbf{i} + 5\mathbf{j}}{|12\mathbf{i} + 5\mathbf{j}|} = \frac{12}{13}\mathbf{i} + \frac{5}{13}\mathbf{j}. \quad \square$$

Example 5 Find the unit vector $\mathbf{u}$ making angle θ with the positive x-axis.

Solution If the initial point of $\mathbf{u}$ is placed at the origin O, its final point P has polar coordinates $|\mathbf{u}| = 1, \theta$ and rectangular coordinates $|\mathbf{u}| \cos \theta = \cos \theta$, $|\mathbf{u}| \sin \theta = \sin \theta$, as in Figure 13. Therefore

$$\mathbf{u} = (\cos \theta, \sin \theta) = (\cos \theta)\mathbf{i} + (\sin \theta)\mathbf{j}.$$

For the unit vector $\mathbf{u}$ in Example 4 we have $\theta = \arctan \frac{5}{12} \approx 22.6°$. $\quad \square$

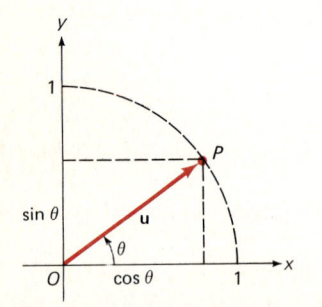

Figure 13

Example 6 Find the components of the vector $\overrightarrow{AB}$ with initial point $A = (\alpha_1, \alpha_2)$ and final point $B = (\beta_1, \beta_2)$.

Solution Drawing the *position vectors* of the points A and B, that is, the vectors joining the origin O to A and B, we find that $\overrightarrow{AB} = \overrightarrow{OB} - \overrightarrow{OA}$ as in Figure 14. Therefore

$$\overrightarrow{AB} = \overrightarrow{OB} - \overrightarrow{OA} = (\beta_1, \beta_2) - (\alpha_1, \alpha_2) = (\beta_1 - \alpha_1, \beta_2 - \alpha_2)$$

or $\overrightarrow{AB} = (\beta_1 - \alpha_1)\mathbf{i} + (\beta_2 - \alpha_2)\mathbf{j}$. For example, if $A = (-2, 3)$, $B = (4, -1)$, then $\overrightarrow{AB} = (4 - (-2), -1 - 3) = (6, -4) = 6\mathbf{i} - 4\mathbf{j}. \quad \square$

As illustrated in the next example, vectors are a powerful tool for proving geometric propositions.

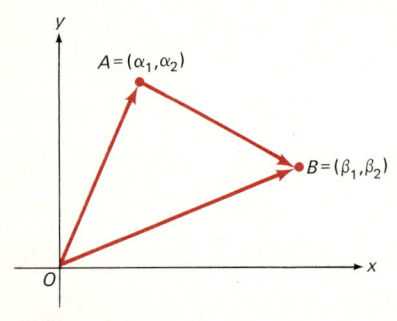

Figure 14

652   Chapter 11   Vectors in the Plane

Example 7 Show that the figure obtained by joining the midpoints of the adjacent sides of *any* quadrilateral $ABCD$ is a parallelogram.

Solution Let M, N, P and Q be the midpoints of the sides AB, BC, CD and DA (see Figure 15). Then

$$\overrightarrow{MN} - \overrightarrow{QP} = \overrightarrow{MN} + \overrightarrow{PQ} = (\overrightarrow{MB} + \overrightarrow{BN}) + (\overrightarrow{PD} + \overrightarrow{DQ})$$

$$= \frac{1}{2}\overrightarrow{AB} + \frac{1}{2}\overrightarrow{BC} + \frac{1}{2}\overrightarrow{CD} + \frac{1}{2}\overrightarrow{DA}$$

$$= \frac{1}{2}(\overrightarrow{AB} + \overrightarrow{BC} + \overrightarrow{CD} + \overrightarrow{DA}) = \frac{1}{2}\overrightarrow{AA} = \mathbf{0}.$$

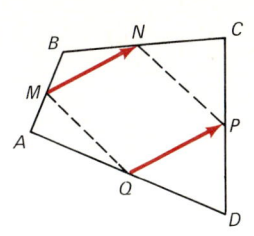

Figure 15

Therefore $\overrightarrow{MN} = \overrightarrow{QP}$, that is, two opposite sides of the quadrilateral $MNPQ$ are equal and parallel, so that $MNPQ$ is a parallelogram. The same proof works in three dimensions, so that the quadrilateral $ABCD$ doesn't even have to be a plane figure! □

Some Ideas from Linear Algebra

Finally we touch upon some ideas that play a key role in *linear algebra*, a standard subject in the undergraduate mathematics curriculum. By a *linear combination* of n vectors $\mathbf{a}_1, \mathbf{a}_2, \ldots, \mathbf{a}_n$ we mean an expression of the form

$$c_1\mathbf{a}_1 + c_2\mathbf{a}_2 + \cdots + c_n\mathbf{a}_n, \tag{11}$$

with scalar coefficients $c_1, c_2, \ldots, c_n$. Clearly (11) is itself a vector, and in fact we have already seen that an arbitrary vector in the plane is a linear combination of the unit vectors $\mathbf{i}$ and $\mathbf{j}$. The linear combination (11) is said to be *trivial* if all the coefficients $c_1, c_2, \ldots, c_n$ equal zero (in this case it equals the zero vector), and *nontrivial* if at least one of the coefficients is nonzero. The vectors $\mathbf{a}_1, \mathbf{a}_2, \ldots, \mathbf{a}_n$ are said to be *linearly dependent* if some nontrivial linear combination of $\mathbf{a}_1, \mathbf{a}_2, \ldots, \mathbf{a}_n$ equals zero, that is, if there is an expression of the form (11) with at least one nonzero coefficient which is equal to the zero vector; otherwise the vectors $\mathbf{a}_1, \mathbf{a}_2, \ldots, \mathbf{a}_n$ are said to be *linearly independent*. For example, the unit vectors $\mathbf{i}$ and $\mathbf{j}$ are linearly independent, since

$$c_1\mathbf{i} + c_2\mathbf{j} = c_1(1, 0) + c_2(0, 1) = (c_1, c_2) = \mathbf{0}$$

if and only if $c_1 = c_2 = 0$, while the vectors $\mathbf{a}_1 = (1, 2)$, $\mathbf{a}_2 = (3, 4)$, $\mathbf{a}_3 = (2, 3)$ are linearly dependent, since

$$\mathbf{a}_1 + \mathbf{a}_2 - 2\mathbf{a}_3 = (1, 2) + (3, 4) - 2(2, 3)$$
$$= (1 + 3 - 4, 2 + 4 - 6) = (0, 0) = \mathbf{0}.$$

More generally, it can be shown that *any three vectors in the plane are linearly dependent* and that *two vectors in the plane are linearly independent if and only if they are noncollinear*.

Problems

Sketch the given vectors in the same rectangular coordinate system, choosing the origin as the common initial point.

1. $-2\mathbf{i}, \mathbf{i} + \mathbf{j}, -\mathbf{i} + 2\mathbf{j}, 2\mathbf{i} - \mathbf{j}, -3\mathbf{j}$
2. $\frac{3}{2}\mathbf{j}, \mathbf{i} - \mathbf{j}, -2\mathbf{i} + \mathbf{j}, 2\mathbf{i} + 3\mathbf{j}, -\mathbf{i} - 2\mathbf{j}$

In each of the following problems find the vectors $\mathbf{a} + \mathbf{b}$, $\mathbf{a} - \mathbf{b}$, $2\mathbf{a} + 3\mathbf{b}$ and $3\mathbf{a} - 4\mathbf{b}$.

3. $\mathbf{a} = (1, -5)$, $\mathbf{b} = (3, 6)$
4. $\mathbf{a} = (2, 0)$, $\mathbf{b} = (0, -2)$

Section 11.1 The Vector Concept; Operations on Vectors

5. $\mathbf{a} = \mathbf{i} + \mathbf{j}, \mathbf{b} = \mathbf{i} - \mathbf{j}$
6. $\mathbf{a} = \mathbf{i}, \mathbf{b} = -\mathbf{i} - \mathbf{j}$
7. $\mathbf{a} = \mathbf{i} + 3\mathbf{j}, \mathbf{b} = -3\mathbf{i} + 2\mathbf{j}$
8. $\mathbf{a} = 7\mathbf{i} - 9\mathbf{j}, \mathbf{b} = 2\mathbf{i} + 5\mathbf{j}$
9. $\mathbf{a} = (1, 1), \mathbf{b} = (1, -1)$
10. $\mathbf{a} = (6, -5), \mathbf{b} = (-4, 3)$

Find the vector $\overrightarrow{AB}$ with the given endpoints.
11. $A = (-3, 0), B = (0, 6)$
12. $A = (3, 5), B = (4, 7)$
13. $A = (1, -4), B = (-1, -9)$
14. $A = (-5, 10), B = (0, 0)$
15. $A = (\frac{3}{2}, -\frac{1}{2}), B = (\frac{1}{2}, \frac{7}{2})$
16. $A = (-\sin\frac{1}{4}\pi, \cos\frac{1}{4}\pi), B = (\cos\frac{1}{4}\pi, \sin\frac{1}{4}\pi)$

17. What is the initial point of the vector $-7\mathbf{i} + 2\mathbf{j}$ if its final point is $(3, -11)$?
18. What is the final point of the vector $5\mathbf{i} - 6\mathbf{j}$ if its initial point is $(-4, 8)$?

Find the magnitude of the given vector.
19. $\frac{4}{5}\mathbf{i} - \frac{3}{5}\mathbf{j}$
20. $\alpha\mathbf{i}$
21. $-\beta\mathbf{j}$
22. $-7\mathbf{i} + 24\mathbf{j}$
23. $35\mathbf{i} + 12\mathbf{j}$
24. $-\sqrt{5}\mathbf{i} + \sqrt{11}\mathbf{j}$

25. Which of the vectors $3\mathbf{i} + 3\mathbf{j}$ and $4\mathbf{i} - \mathbf{j}$ has the larger magnitude?
26. Find the vector with half the magnitude of $16\mathbf{i} - 12\mathbf{j}$ and with the opposite direction.

Find the unit vector with the same direction as the given vector. What is the angle from the positive x-axis to this direction?
27. $\mathbf{i} + \mathbf{j}$
28. $\mathbf{i} - \mathbf{j}$
29. $-3\mathbf{i} + 2\mathbf{j}$
30. $-\mathbf{i} + 4\mathbf{j}$
31. $-6\mathbf{i} - 8\mathbf{j}$
32. $40\mathbf{i} + 9\mathbf{j}$

Find the unit vector making the given angle with the positive x-axis.
33. $5\pi/3$
34. $7\pi/6$
35. $3\pi/4$

Find the vector of magnitude 4 making the given angle with the positive x-axis.
36. $3\pi/2$
37. $\arctan\frac{2}{3}$
38. $-\pi/4$

39. Let $\mathbf{a} = (7, -1), \mathbf{b} = (13, 2), \mathbf{c} = (4, 5)$. Find the vector $\mathbf{x}$ such that $\mathbf{a} + \mathbf{b} + \mathbf{x} = 3\mathbf{c} - \mathbf{x}$.
40. Starting with formula (8'), give an algebraic proof of the commutativity and associativity of vector addition.
41. Let $\mathbf{a}$ and $\mathbf{b}$ be the position vectors of two points A and B with respect to any origin O. Show that the final point P of the vector $(1 - t)\mathbf{a} + t\mathbf{b}$ $(0 \leq t \leq 1)$ divides the segment AB in the ratio $t:(1 - t)$, and in particular that the midpoint of AB has the position vector $\frac{1}{2}(\mathbf{a} + \mathbf{b})$.

42. When does the triangle inequality (3) become an equality? (Assume that $\mathbf{a} \neq \mathbf{0}, \mathbf{b} \neq \mathbf{0}$.)
43. Verify the inequality $|\mathbf{a} - \mathbf{b}| \geq ||\mathbf{a}| - |\mathbf{b}||$ for arbitrary vectors $\mathbf{a}$ and $\mathbf{b}$.

Use vectors to prove that
44. The diagonals of a parallelogram bisect each other
45. The segment joining the midpoints of two sides of a triangle is half as long as the third side and parallel to it (see Figure 16)

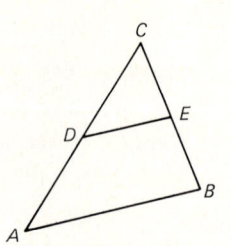

Figure 16

46. The medians of a triangle intersect in a common point which trisects each median (see Figure 17)
Hint. Use Problem 41.

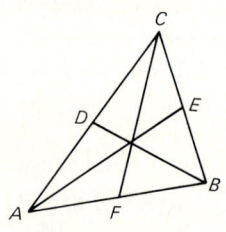

Figure 17

47. Show that the vectors $\mathbf{e}_1 = (1, 1)$ and $\mathbf{e}_2 = (1, 2)$ form a nonorthogonal basis. Expand the vector $\mathbf{a} = (-3, 5)$ with respect to $\mathbf{e}_1$ and $\mathbf{e}_2$.
48. Express each of the vectors $\mathbf{a} = (3, -2), \mathbf{b} = (-2, 1), \mathbf{c} = (7, -4)$ as a linear combination of the other two.
49. Find all vectors forming an orthogonal basis with the vector $2\mathbf{i} + \mathbf{j}$.
50. Let $\mathbf{a} = (3, -1), \mathbf{b} = (1, -2), \mathbf{c} = (-1, 7)$. Express the vector $\mathbf{a} + \mathbf{b} + \mathbf{c}$ as a linear combination of $\mathbf{a}$ and $\mathbf{b}$.
51. The captain of a fishing boat traveling due east at a speed of 10 knots finds that the wind seems to blow directly from north to south. When the speed of the boat is doubled, the wind seems to blow from the northeast instead. What is the true wind velocity?
52. A plane flies due north with a ground speed of 240 km/hr in a 60 km/hr wind blowing directly from east to west. Find the plane's airspeed and the direction of its velocity relative to the air.

11.2 The Dot Product

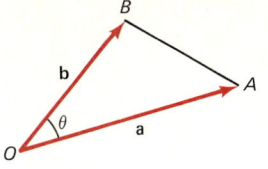

Figure 18 The angle between **a** and **b** is θ.

Figure 19

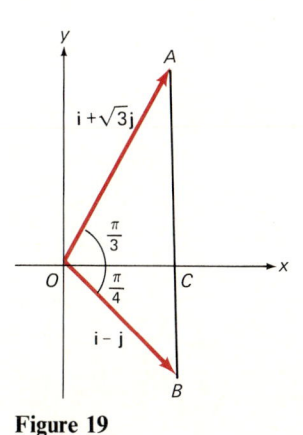

The concept of the "dot product" of two vectors arises both in the geometric problem of finding the component of one vector along another vector and in the physical problem of finding the work done by a force which does not act in the direction of motion of an object to which it is applied (both problems will be solved at the end of this section). To define the dot product of two vectors, we will need to know the angle between them. Suppose two nonzero vectors **a** and **b** are placed so that their initial points lie at the same point O, and let $\mathbf{a} = \overrightarrow{OA}$, $\mathbf{b} = \overrightarrow{OB}$ as in Figure 18. Then by the *angle between* **a** *and* **b** (in either order) we mean the angle θ at the vertex O of the triangle AOB. The triangle "collapses" if **a** and **b** are parallel, that is, if $\mathbf{a} = p\mathbf{b}$ where p is a scalar, and in this case we define $\theta = 0$ if $p > 0$ and $\theta = \pi$ if $p < 0$. Notice that θ always lies in the interval $0 \leq \theta \leq \pi$.

Example 1 Find the angle between the vectors $\mathbf{a} = \mathbf{i} + \sqrt{3}\mathbf{j}$ and $\mathbf{b} = \mathbf{i} - \mathbf{j}$.

Solution Examining Figure 19, we find that the angle AOC equals $\arctan \sqrt{3} = \pi/3$, while the angle BOC equals $\arctan 1 = \pi/4$. Hence the angle AOB between **a** and **b** is the sum

$$\frac{\pi}{3} + \frac{\pi}{4} = \frac{7\pi}{12} = 105°.$$

Definition of the Dot Product

We now define the *dot product* $\mathbf{a} \cdot \mathbf{b}$ of two vectors **a** and **b** by the formula

$$\mathbf{a} \cdot \mathbf{b} = |\mathbf{a}| |\mathbf{b}| \cos \theta, \tag{1}$$

where θ is the angle between **a** and **b**. If either $\mathbf{a} = \mathbf{0}$ or $\mathbf{b} = \mathbf{0}$, then θ is undefined, and we set $\mathbf{a} \cdot \mathbf{b} = 0$ by definition. The dot product, so called because of the dot appearing in the expression $\mathbf{a} \cdot \mathbf{b}$, is also called the *scalar product*, since $\mathbf{a} \cdot \mathbf{b}$ is a number, that is, a scalar.† It is an immediate consequence of the definition (1) that the dot product is commutative:

$$\mathbf{a} \cdot \mathbf{b} = \mathbf{b} \cdot \mathbf{a}.$$

The angle between a vector and itself is zero. Therefore

$$\mathbf{a} \cdot \mathbf{a} = |\mathbf{a}| |\mathbf{a}| \cos 0 = |\mathbf{a}|^2.$$

In particular, $\mathbf{a} \cdot \mathbf{a} = 0$ if and only if $|\mathbf{a}| = 0$, that is, if and only if $\mathbf{a} = \mathbf{0}$. The dot product $\mathbf{a} \cdot \mathbf{b}$ equals zero if and only if **a** is perpendicular to **b**, written $\mathbf{a} \perp \mathbf{b}$, where the zero vector is regarded as perpendicular to every vector. In fact, the formula

$$\mathbf{a} \cdot \mathbf{b} = |\mathbf{a}| |\mathbf{b}| \cos \theta = 0 \qquad (0 \leq \theta \leq \pi)$$

holds if and only if $\cos \theta = 0$ and hence $\theta = \pi/2$, or at least one of the vectors **a** and **b** is zero, so that $\mathbf{a} \perp \mathbf{b}$ in any event.

---

† In Section 12.3 we will introduce another kind of product of two vectors **a** and **b**, which is a vector rather than a scalar and is written $\mathbf{a} \times \mathbf{b}$ instead of $\mathbf{a} \cdot \mathbf{b}$.

Component Form of the Dot Product

Next we find an expression for the dot product $\mathbf{a} \cdot \mathbf{b}$ in terms of the components of $\mathbf{a}$ and $\mathbf{b}$ with respect to the basis vectors $\mathbf{i} = (1, 0)$ and $\mathbf{j} = (0, 1)$.

Theorem 1 *(Component form of the dot product).* If $\mathbf{a} = \alpha_1 \mathbf{i} + \alpha_2 \mathbf{j}$ and $\mathbf{b} = \beta_1 \mathbf{i} + \beta_2 \mathbf{j}$, or equivalently $\mathbf{a} = (\alpha_1, \alpha_2)$ and $\mathbf{b} = (\beta_1, \beta_2)$, then

$$\mathbf{a} \cdot \mathbf{b} = \alpha_1 \beta_1 + \alpha_2 \beta_2. \qquad (2)$$

Proof Let

$$\mathbf{c} = \mathbf{a} - \mathbf{b} = (\alpha_1 - \beta_1)\mathbf{i} + (\alpha_2 - \beta_2)\mathbf{j}.$$

Applying the law of cosines (Theorem 2, page 56) to the triangle in Figure 20, with sides formed by $\mathbf{a}$, $\mathbf{b}$ and $\mathbf{c}$, we get

$$|\mathbf{c}|^2 = |\mathbf{a}|^2 + |\mathbf{b}|^2 - 2|\mathbf{a}||\mathbf{b}| \cos \theta,$$

where θ is the angle between $\mathbf{a}$ and $\mathbf{b}$. Therefore

$$|\mathbf{c}|^2 = |\mathbf{a}|^2 + |\mathbf{b}|^2 - 2\mathbf{a} \cdot \mathbf{b},$$

so that

$$\mathbf{a} \cdot \mathbf{b} = \frac{1}{2}(|\mathbf{a}|^2 + |\mathbf{b}|^2 - |\mathbf{c}|^2). \qquad (3)$$

But

$$|\mathbf{a}|^2 = \alpha_1^2 + \alpha_2^2, \qquad |\mathbf{b}|^2 = \beta_1^2 + \beta_2^2,$$

$$|\mathbf{c}|^2 = (\alpha_1 - \beta_1)^2 + (\alpha_2 - \beta_2)^2 = \alpha_1^2 - 2\alpha_1\beta_1 + \beta_1^2 + \alpha_2^2 - 2\alpha_2\beta_2 + \beta_2^2,$$

and substituting these expressions for $|\mathbf{a}|^2$, $|\mathbf{b}|^2$ and $|\mathbf{c}|^2$ into (3), we immediately obtain (2). ∎

Corollary *If p and q are scalars, then*

$$(p\mathbf{a}) \cdot (q\mathbf{b}) = pq(\mathbf{a} \cdot \mathbf{b}) \qquad (4)$$

for arbitrary vectors $\mathbf{a}$ and $\mathbf{b}$. The dot product also satisfies the distributive laws

$$\mathbf{a} \cdot (\mathbf{b} + \mathbf{c}) = \mathbf{a} \cdot \mathbf{b} + \mathbf{a} \cdot \mathbf{c}, \qquad (5)$$

$$(\mathbf{a} + \mathbf{b}) \cdot \mathbf{c} = \mathbf{a} \cdot \mathbf{c} + \mathbf{b} \cdot \mathbf{c} \qquad (5')$$

for arbitrary vectors $\mathbf{a}$, $\mathbf{b}$ and $\mathbf{c}$.

Proof Let

$$\mathbf{a} = \alpha_1 \mathbf{i} + \alpha_2 \mathbf{j}, \qquad \mathbf{b} = \beta_1 \mathbf{i} + \beta_2 \mathbf{j}, \qquad \mathbf{c} = \gamma_1 \mathbf{i} + \gamma_2 \mathbf{j}$$

(γ is lowercase Greek gamma). Then

$$p\mathbf{a} = p\alpha_1 \mathbf{i} + p\alpha_2 \mathbf{j}, \qquad q\mathbf{b} = q\beta_1 \mathbf{i} + q\beta_2 \mathbf{j},$$

and

$$\mathbf{b} + \mathbf{c} = (\beta_1 + \gamma_1)\mathbf{i} + (\beta_2 + \gamma_2)\mathbf{j}.$$

Therefore

$$(p\mathbf{a}) \cdot (q\mathbf{b}) = (p\alpha_1)(q\beta_1) + (p\alpha_2)(q\beta_2) = pq(\alpha_1\beta_1 + \alpha_2\beta_2) = pq(\mathbf{a} \cdot \mathbf{b}),$$

which proves (4). Also

$$\mathbf{a} \cdot (\mathbf{b} + \mathbf{c}) = \alpha_1(\beta_1 + \gamma_1) + \alpha_2(\beta_2 + \gamma_2) = \alpha_1\beta_1 + \alpha_1\gamma_1 + \alpha_2\beta_2 + \alpha_2\gamma_2,$$

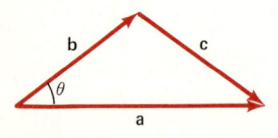

Figure 20

while
$$\mathbf{a} \cdot \mathbf{b} + \mathbf{a} \cdot \mathbf{c} = (\alpha_1 \beta_1 + \alpha_2 \beta_2) + (\alpha_1 \gamma_1 + \alpha_2 \gamma_2),$$
which proves (5), since the right sides of these two formulas are equal. Finally, to prove (5′), we use (5) and the fact that the dot product is commutative:
$$(\mathbf{a} + \mathbf{b}) \cdot \mathbf{c} = \mathbf{c} \cdot (\mathbf{a} + \mathbf{b}) = \mathbf{c} \cdot \mathbf{a} + \mathbf{c} \cdot \mathbf{b} = \mathbf{a} \cdot \mathbf{c} + \mathbf{b} \cdot \mathbf{c}. \quad \blacksquare$$

Example 2 By formula (2), the dot product of the vectors $\mathbf{a} = (2, 5)$ and $\mathbf{b} = (4, -1)$ is $\mathbf{a} \cdot \mathbf{b} = 2(4) + 5(-1) = 8 - 5 = 3$.

Example 3 Using both distributive laws (5) and (5′), we find that
$$(\mathbf{a} + \mathbf{b}) \cdot (\mathbf{c} + \mathbf{d}) = \mathbf{a} \cdot (\mathbf{c} + \mathbf{d}) + \mathbf{b} \cdot (\mathbf{c} + \mathbf{d}) = \mathbf{a} \cdot \mathbf{c} + \mathbf{a} \cdot \mathbf{d} + \mathbf{b} \cdot \mathbf{c} + \mathbf{b} \cdot \mathbf{d}$$
for arbitrary vectors $\mathbf{a}$, $\mathbf{b}$, $\mathbf{c}$ and $\mathbf{d}$. In particular,
$$|\mathbf{a} + \mathbf{b}|^2 = (\mathbf{a} + \mathbf{b}) \cdot (\mathbf{a} + \mathbf{b}) = \mathbf{a} \cdot \mathbf{a} + \mathbf{a} \cdot \mathbf{b} + \mathbf{b} \cdot \mathbf{a} + \mathbf{b} \cdot \mathbf{b}$$
$$= |\mathbf{a}|^2 + 2\mathbf{a} \cdot \mathbf{b} + |\mathbf{b}|^2.$$
Here $2\mathbf{a} \cdot \mathbf{b}$ means $2(\mathbf{a} \cdot \mathbf{b})$, and more generally, $p\mathbf{a} \cdot \mathbf{b}$ means $p(\mathbf{a} \cdot \mathbf{b})$. $\square$

For the unit vectors $\mathbf{i} = (1, 0)$ and $\mathbf{j} = (0, 1)$ we have
$$\mathbf{i} \cdot \mathbf{i} = 1(1) + 0(0) = 1, \qquad \mathbf{i} \cdot \mathbf{j} = 1(0) + 0(1) = 0, \qquad \mathbf{j} \cdot \mathbf{j} = 0(0) + 1(1) = 1,$$
or more concisely
$$\mathbf{i} \cdot \mathbf{i} = 1, \qquad \mathbf{i} \cdot \mathbf{j} = \mathbf{j} \cdot \mathbf{i} = 0, \qquad \mathbf{j} \cdot \mathbf{j} = 1.$$
These basic formulas should be memorized. Using them, we find that if $\mathbf{a} = \alpha_1 \mathbf{i} + \alpha_2 \mathbf{j}$, then
$$\mathbf{a} \cdot \mathbf{i} = (\alpha_1 \mathbf{i} + \alpha_2 \mathbf{j}) \cdot \mathbf{i} = \alpha_1 \mathbf{i} \cdot \mathbf{i} + \alpha_2 \mathbf{j} \cdot \mathbf{i} = \alpha_1,$$
and similarly
$$\mathbf{a} \cdot \mathbf{j} = (\alpha_1 \mathbf{i} + \alpha_2 \mathbf{j}) \cdot \mathbf{j} = \alpha_1 \mathbf{i} \cdot \mathbf{j} + \alpha_2 \mathbf{j} \cdot \mathbf{j} = \alpha_2.$$

Example 4 For what value of the parameter t are the vectors $\mathbf{a} = 3\mathbf{i} + \mathbf{j}$ and $\mathbf{b} = -2\mathbf{i} + t\mathbf{j}$ parallel? Perpendicular?

Solution The vectors $\mathbf{a}$ and $\mathbf{b}$ are parallel (or collinear) if and only if $\mathbf{a} = p\mathbf{b}$ for some scalar $p \neq 0$, that is, $3\mathbf{i} + \mathbf{j} = p(-2\mathbf{i} + t\mathbf{j})$. Taking components and solving the resulting system of equations $-2p = 3$, $pt = 1$, we find that $p = -\frac{3}{2}$, $t = 1/p = -\frac{2}{3}$. The vectors are perpendicular if and only if $\mathbf{a} \cdot \mathbf{b} = 0$, that is, $(3\mathbf{i} + \mathbf{j}) \cdot (-2\mathbf{i} + t\mathbf{j}) = -6 + t = 0$ or $t = 6$. $\square$

Let θ be the angle between two nonzero vectors $\mathbf{a} = \alpha_1 \mathbf{i} + \alpha_2 \mathbf{j}$ and $\mathbf{b} = \beta_1 \mathbf{i} + \beta_2 \mathbf{j}$. Then it follows from the definition (1) of the dot product that

$$\cos \theta = \frac{\mathbf{a} \cdot \mathbf{b}}{|\mathbf{a}| |\mathbf{b}|}. \tag{6}$$

But $|\cos \theta| \leq 1$, and hence
$$\left| \frac{\mathbf{a} \cdot \mathbf{b}}{|\mathbf{a}| |\mathbf{b}|} \right| \leq 1,$$
or equivalently
$$|\mathbf{a} \cdot \mathbf{b}| \leq |\mathbf{a}| |\mathbf{b}|. \tag{7}$$

Section 11.2 The Dot Product

In terms of the components of **a** and **b**, formulas (6) and (7) become

$$\cos \theta = \frac{\alpha_1 \beta_1 + \alpha_2 \beta_2}{\sqrt{\alpha_1^2 + \alpha_2^2}\sqrt{\beta_1^2 + \beta_2^2}} \tag{6'}$$

and

$$|\alpha_1 \beta_1 + \alpha_2 \beta_2| \leq \sqrt{\alpha_1^2 + \alpha_2^2}\sqrt{\beta_1^2 + \beta_2^2}. \tag{7'}$$

The inequality (7') is a special case of the *Cauchy–Schwarz inequality* (see Problem 47). Observe that equality holds in (7) if and only if $\cos \theta = \pm 1$, that is, if and only if $\theta = 0$ or $\theta = \pi$; in this case **a** and **b** are parallel, with the same direction if $\theta = 0$ and opposite directions if $\theta = \pi$. Also note that the angle θ is acute ($0 < \theta < \pi/2$) if $0 < \cos \theta < 1$ and obtuse ($\pi/2 < \theta < \pi$) if $-1 < \cos \theta < 0$.

Example 5 Find the angle θ between the vectors $\mathbf{a} = \mathbf{i} - \mathbf{j}$ and $\mathbf{b} = 2\mathbf{i} + \mathbf{j}$.

Solution Here we have

$$\mathbf{a} \cdot \mathbf{b} = 1(2) - 1(1) = 1, \qquad |\mathbf{a}| = \sqrt{1^2 + (-1)^2} = \sqrt{2},$$
$$|\mathbf{b}| = \sqrt{2^2 + 1^2} = \sqrt{5}.$$

Therefore, by (6),

$$\cos \theta = \frac{\mathbf{a} \cdot \mathbf{b}}{|\mathbf{a}||\mathbf{b}|} = \frac{1}{\sqrt{2}\sqrt{5}} = \frac{1}{\sqrt{10}},$$

which implies

$$\theta = \arccos \frac{1}{\sqrt{10}} \approx 71.6°$$

(see Figure 21). □

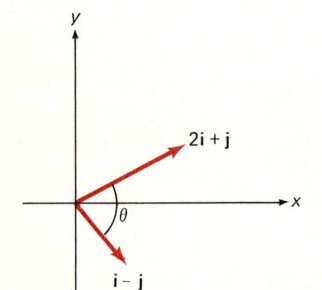

Figure 21

Projection of One Vector onto Another

Let **a** and **b** be two nonzero vectors. Then by the *component of* **a** *along* **b**, written $\text{comp}_\mathbf{b}\, \mathbf{a}$, we mean the *scalar* equal to the dot product $\mathbf{a} \cdot \mathbf{u}_\mathbf{b}$, where $\mathbf{u}_\mathbf{b}$ is a unit vector in the direction of **b**. Since

$$\mathbf{u}_\mathbf{b} = \frac{\mathbf{b}}{|\mathbf{b}|},$$

we have

$$\text{comp}_\mathbf{b}\, \mathbf{a} = \frac{\mathbf{a} \cdot \mathbf{b}}{|\mathbf{b}|},$$

or equivalently

$$\text{comp}_\mathbf{b}\, \mathbf{a} = |\mathbf{a}| \cos \theta,$$

where θ is the angle between **a** and **b**. By the *projection of* **a** *onto* **b**, denoted by $\text{proj}_\mathbf{b}\, \mathbf{a}$, we mean the *vector* of magnitude $\text{comp}_\mathbf{b}\, \mathbf{a}$ parallel to **b**, that is,

$$\text{proj}_\mathbf{b}\, \mathbf{a} = (\text{comp}_\mathbf{b}\, \mathbf{a})\mathbf{u}_\mathbf{b}.$$

In terms of the dot product we can write

$$\text{proj}_\mathbf{b}\, \mathbf{a} = \frac{\mathbf{a} \cdot \mathbf{b}}{|\mathbf{b}|}\mathbf{u}_\mathbf{b} = \frac{(\mathbf{a} \cdot \mathbf{b})\mathbf{b}}{|\mathbf{b}|^2}.$$

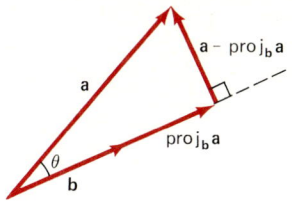

Resolution of a into components along b and orthogonal to b

Figure 22

The geometric meaning of $\text{proj}_b\,\mathbf{a}$ is illustrated in Figure 22. Notice that if the initial points of $\mathbf{a}$ and $\mathbf{b}$ coincide, the final point of $\text{proj}_b\,\mathbf{a}$ is the foot of the perpendicular dropped from the final point of $\mathbf{a}$ to the line containing $\mathbf{b}$. It follows that the vector $\mathbf{a} - \text{proj}_b\,\mathbf{a}$ is perpendicular to $\mathbf{b}$, as shown in the figure.

Thus $\mathbf{a}$ can be represented as the sum of the vector $\text{proj}_b\,\mathbf{a}$ parallel to $\mathbf{b}$ and the vector $\mathbf{a} - \text{proj}_b\,\mathbf{a}$ perpendicular to $\mathbf{b}$. These vectors are also known as the *vector component of* $\mathbf{a}$ *along* $\mathbf{b}$ (or *parallel to* $\mathbf{b}$) and the *vector component of* $\mathbf{a}$ *orthogonal to* $\mathbf{b}$ ("orthogonal" is a synonym for "perpendicular").

Example 6 Find $\text{comp}_b\,\mathbf{a}$ and $\text{proj}_b\,\mathbf{a}$ if $\mathbf{a} = 3\mathbf{i} + 2\mathbf{j}$ and $\mathbf{b} = 5\mathbf{i} - \mathbf{j}$. Represent $\mathbf{a}$ as the sum of a vector parallel to $\mathbf{b}$ and a vector orthogonal to $\mathbf{b}$.

Solution Since

$$\mathbf{a} \cdot \mathbf{b} = 3(5) + 2(-1) = 13, \qquad |\mathbf{b}| = \sqrt{5^2 + (-1)^2} = \sqrt{26},$$

we have

$$\text{comp}_b\,\mathbf{a} = \frac{\mathbf{a} \cdot \mathbf{b}}{|\mathbf{b}|} = \frac{13}{\sqrt{26}},$$

$$\text{proj}_b\,\mathbf{a} = (\text{comp}_b\,\mathbf{a})\mathbf{u}_b = \frac{13}{\sqrt{26}} \frac{5\mathbf{i} - \mathbf{j}}{\sqrt{26}} = \frac{5}{2}\mathbf{i} - \frac{1}{2}\mathbf{j}.$$

The vector

$$\mathbf{a} - \text{proj}_b\,\mathbf{a} = (3\mathbf{i} + 2\mathbf{j}) - \left(\frac{5}{2}\mathbf{i} - \frac{1}{2}\mathbf{j}\right) = \frac{1}{2}\mathbf{i} + \frac{5}{2}\mathbf{j}$$

is orthogonal to $\mathbf{a}$. Hence the representation of $\mathbf{a}$ as the sum of a vector parallel to $\mathbf{b}$ and a vector orthogonal to $\mathbf{b}$ is

$$\mathbf{a} = \left(\frac{5}{2}\mathbf{i} - \frac{1}{2}\mathbf{j}\right) + \left(\frac{1}{2}\mathbf{i} + \frac{5}{2}\mathbf{j}\right). \qquad \square$$

Work as a Dot Product

Suppose an object moves a distance d subject to a constant force F acting along the line of motion. Then, as we know from Example 5, page 270, the work done by the force is given by the formula

$$W = Fd. \tag{8}$$

To extend this formula to the case of a constant *vector* force $\mathbf{F}$, which in general does *not* act along the line of motion, we reason as follows. Let the displacement of the object be a vector $\mathbf{d}$, and let θ denote the angle between $\mathbf{F}$ and $\mathbf{d}$. Also let

$$\mathbf{F} = \mathbf{F}_{\parallel} + \mathbf{F}_{\perp},$$

where

$$\mathbf{F}_{\parallel} = \text{proj}_d\,\mathbf{F} = \frac{(\mathbf{F} \cdot \mathbf{d})\mathbf{d}}{|\mathbf{d}|^2}$$

is the vector component of $\mathbf{F}$ parallel to $\mathbf{d}$ and

$$\mathbf{F}_{\perp} = \mathbf{F} - \text{proj}_d\,\mathbf{F}$$

Section 11.2 The Dot Product 659

is the vector component of **F** orthogonal to **d** (see Figure 23). The component $\mathbf{F}_\perp$ can produce no motion along **d**, and hence from the standpoint of motion along **d** the force **F** can be replaced by $\mathbf{F}_\parallel$. Therefore all the work is done by $\mathbf{F}_\parallel$, so that

$$W = \begin{cases} |\mathbf{F}_\parallel||\mathbf{d}| & \text{if } 0 \le \theta \le \pi/2, \\ -|\mathbf{F}_\parallel||\mathbf{d}| & \text{if } \pi/2 < \theta \le \pi, \end{cases}$$

by formula (8). But

$$|\mathbf{F}_\parallel||\mathbf{d}| = \frac{|\mathbf{F} \cdot \mathbf{d}||\mathbf{d}|}{|\mathbf{d}|^2}|\mathbf{d}| = |\mathbf{F} \cdot \mathbf{d}|$$

$$= \begin{cases} \mathbf{F} \cdot \mathbf{d} & \text{if } 0 \le \theta \le \pi/2, \\ -\mathbf{F} \cdot \mathbf{d} & \text{if } \pi/2 < \theta \le \pi, \end{cases}$$

and hence

$$W = \mathbf{F} \cdot \mathbf{d}. \tag{8'}$$

*Thus the work is the dot product of the force **F** and the displacement **d**.* Here it is assumed that **F** is constant and the displacement rectilinear. Otherwise the proper definition of work requires the use of a new concept, the "line integral," introduced in Section 15.1.

Example 7 A wooden block weighing 4 lb is pushed up a frictionless inclined plane making an angle of 30° with the horizontal. How much work is done against gravity in moving the block 5 ft?

Solution As is Figure 24, we introduce unit vectors **i** and **j** in the horizontal and vertical directions. Then the unit vector in the direction of the displacement is $\mathbf{u} = \cos 30°\,\mathbf{i} + \sin 30°\,\mathbf{j}$. The force of gravity acts vertically downward and is of magnitude 4 lb, so that $\mathbf{F} = -4\mathbf{j}$, and the vector displacement corresponding to pushing the block 5 ft up the inclined plane is $\mathbf{d} = 5\mathbf{u}$. Therefore, in making this displacement, the work done *by* gravity is

$$\mathbf{F} \cdot \mathbf{d} = (-4\mathbf{j}) \cdot 5\mathbf{u} = -20\mathbf{j} \cdot (\cos 30°\,\mathbf{i} + \sin 30°\,\mathbf{j})$$
$$= -20 \sin 30° = -10 \text{ ft-lb},$$

and the work done *against* gravity is $-\mathbf{F} \cdot \mathbf{d} = 10$ ft-lb. □

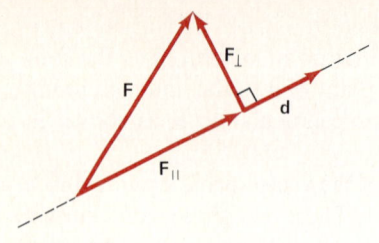

Figure 23

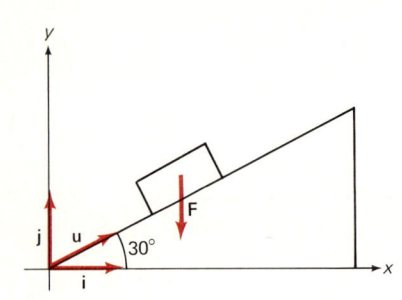

Figure 24

Problems

Find the dot product **a** · **b** of the given vectors, and also find the angle between the vectors.

1. $\mathbf{a} = (-2, 1)$, $\mathbf{b} = (3, 6)$
2. $\mathbf{a} = (1, 1)$, $\mathbf{b} = (-2, -2)$
3. $\mathbf{a} = (1, 1)$, $\mathbf{b} = (1, 0)$
4. $\mathbf{a} = (3, 4)$, $\mathbf{b} = (6, -8)$
5. $\mathbf{a} = -\mathbf{i} - \mathbf{j}$, $\mathbf{b} = 2\mathbf{i}$
6. $\mathbf{a} = 16\mathbf{i}$, $\mathbf{b} = -19\mathbf{j}$
7. $\mathbf{a} = 12\mathbf{i} + 5\mathbf{j}$, $\mathbf{b} = 4\mathbf{i} + 3\mathbf{j}$
8. $\mathbf{a} = 7\mathbf{i} - 24\mathbf{j}$, $\mathbf{b} = -3\mathbf{i} + \mathbf{j}$

Given that the angle between the vectors **a** and **b** is $2\pi/3$, and $|\mathbf{a}| = 3$, $|\mathbf{b}| = 4$, find

9. $\mathbf{a} \cdot \mathbf{b}$
10. $|\mathbf{a} - \mathbf{b}|^2$
11. $|3\mathbf{a} + 2\mathbf{b}|^2$
12. $(3\mathbf{a} - 2\mathbf{b}) \cdot (\mathbf{a} + 2\mathbf{b})$

13. Does $\mathbf{a} \cdot \mathbf{b} = \mathbf{a} \cdot \mathbf{c}$ where $\mathbf{a} \ne 0$ imply $\mathbf{b} = \mathbf{c}$? Explain your answer.
14. Show that $(\mathbf{a} \cdot \mathbf{b})\mathbf{c} - \mathbf{a}(\mathbf{b} \cdot \mathbf{c})$ is orthogonal to **b**.
15. Let **a**, **b** and **c** be unit vectors such that $\mathbf{a} + \mathbf{b} + \mathbf{c} = 0$. Find $\mathbf{a} \cdot \mathbf{b} + \mathbf{b} \cdot \mathbf{c} + \mathbf{c} \cdot \mathbf{a}$.
16. When does the formula $(\mathbf{a} \cdot \mathbf{b})\mathbf{c} = \mathbf{a}(\mathbf{b} \cdot \mathbf{c})$ hold for nonzero vectors **a**, **b** and **c**?
17. Find the angles of the triangle with vertices $A = (1, 0)$, $B = (1, 3)$ and $C = (2, 1)$.
18. Check Example 1 by using formula (6).
19. Show that the vector $\mathbf{v} = |\mathbf{b}|\mathbf{a} + |\mathbf{a}|\mathbf{b}$ bisects the angle between **a** and **b**.

20. Show that the vectors $\mathbf{v} = |\mathbf{b}|\mathbf{a} + |\mathbf{a}|\mathbf{b}$ and $\mathbf{w} = |\mathbf{a}|\mathbf{b} - |\mathbf{b}|\mathbf{a}$ are orthogonal.

Find the two unit vectors orthogonal to the given vector.
21. $4\mathbf{i} + 2\mathbf{j}$
22. $-\mathbf{i} + 3\mathbf{j}$
23. $6\mathbf{i} - 8\mathbf{j}$
24. $-5\mathbf{i} - 12\mathbf{j}$

25. Find the two unit vectors making an angle of $\pi/4$ with the vector $-4\mathbf{i} + \mathbf{j}$.

26. If $|\mathbf{a}| = 3$, $|\mathbf{b}| = 5$, for what values of t are the vectors $\mathbf{a} + t\mathbf{b}$ and $\mathbf{a} - t\mathbf{b}$ perpendicular?

27. For what value of t are the vectors $\mathbf{a} = 2\mathbf{i} + 3\mathbf{j}$ and $\mathbf{b} = 4t\mathbf{i} - 5\mathbf{j}$ parallel? Perpendicular?

28. For what value of t are the vectors $\mathbf{a} = -\mathbf{i} + 2\mathbf{j}$ and $\mathbf{b} = 3\mathbf{i} + 6t\mathbf{j}$ parallel? Perpendicular?

29. For what values of t is the angle between the vectors $\mathbf{a} = \mathbf{i} + t\mathbf{j}$ and $\mathbf{b} = -\mathbf{i} + \mathbf{j}$ equal to $\pi/3$?

Let $\mathbf{a}$ and $\mathbf{b}$ be nonzero vectors. Show that
30. $\mathbf{a}$ and $\mathbf{b}$ are perpendicular if $|\mathbf{a} + \mathbf{b}| = |\mathbf{a} - \mathbf{b}|$.
31. The angle between the vectors $\mathbf{a}$ and $\mathbf{b}$ is less than $\pi/2$ if $|\mathbf{a} + \mathbf{b}| > |\mathbf{a} - \mathbf{b}|$.
32. The angle between the vectors $\mathbf{a}$ and $\mathbf{b}$ is greater than $\pi/2$ if $|\mathbf{a} + \mathbf{b}| < |\mathbf{a} - \mathbf{b}|$.

33. Verify that $|\mathbf{a} + \mathbf{b}|^2 + |\mathbf{a} - \mathbf{b}|^2 = 2|\mathbf{a}|^2 + 2|\mathbf{b}|^2$ for arbitrary vectors $\mathbf{a}$ and $\mathbf{b}$.

34. Find $|\mathbf{a} + \mathbf{b}|$ if $|\mathbf{a}| = 11$, $|\mathbf{b}| = 23$, $|\mathbf{a} - \mathbf{b}| = 30$.

35. Show that the diagonals of a rhombus (a parallelogram with sides of equal length) are perpendicular.

36. Give an *algebraic* proof of the triangle inequality $|\mathbf{a} + \mathbf{b}| \leq |\mathbf{a}| + |\mathbf{b}|$.

Find $\text{proj}_\mathbf{b}\, \mathbf{a}$, the vector component of $\mathbf{a}$ along $\mathbf{b}$, for the given vectors $\mathbf{a}$ and $\mathbf{b}$, and also find the vector component of $\mathbf{a}$ orthogonal to $\mathbf{b}$.
37. $\mathbf{a} = (3, 2)$, $\mathbf{b} = (-1, -1)$
38. $\mathbf{a} = (5, 1)$, $\mathbf{b} = (-2, 10)$
39. $\mathbf{a} = (-4, 7)$, $\mathbf{b} = (3, 6)$
40. $\mathbf{a} = (8, 0)$, $\mathbf{b} = (4, 2)$
41. $\mathbf{a} = -\mathbf{i} + 2\mathbf{j}$, $\mathbf{b} = -3\mathbf{j}$
42. $\mathbf{a} = 7\mathbf{i} - \mathbf{j}$, $\mathbf{b} = \mathbf{i} + \mathbf{j}$
43. $\mathbf{a} = 2\mathbf{i} + 5\mathbf{j}$, $\mathbf{b} = 4\mathbf{i} - \mathbf{j}$
44. $\mathbf{a} = 6\mathbf{i}$, $\mathbf{b} = 2\mathbf{i} + 4\mathbf{j}$

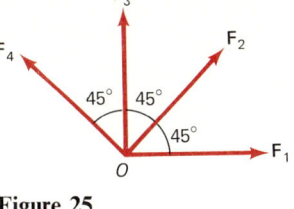

Figure 25

45. Each of the four forces shown in Figure 25 is of magnitude 10 lb and is applied to the same point O. Find the magnitude of the resultant force $\mathbf{F} = \mathbf{F}_1 + \mathbf{F}_2 + \mathbf{F}_3 + \mathbf{F}_4$.

46. Give a direct proof of the inequality (7'), without using vectors.

47. Prove the general *Cauchy–Schwarz inequality*
$$\left|\sum_{i=1}^{n} \alpha_i \beta_i\right| \leq \sqrt{\sum_{i=1}^{n} \alpha_i^2} \sqrt{\sum_{i=1}^{n} \beta_i^2},$$
valid for arbitrary real numbers α_i, β_i ($i = 1, 2, \ldots, n$).

48. A child drags a wagon 12 ft along level ground with a force of 50 lb making an angle of $60°$ with the horizontal. How much work is done by the force?

49. Find the work done by the force $\mathbf{F} = 2\mathbf{i} - 8\mathbf{j}$ on an object moving along the x-axis from the point $(-2, 0)$ to the point $(13, 0)$. Along the y-axis from the origin to the point $(0, -6)$. (Measure force in pounds and distance in feet.)

50. How much work is done against gravity in pushing a 10-kg mass a distance of 3 m up a frictionless inclined plane making an angle of $45°$ with the horizontal? (Take g, the acceleration due to gravity, to be 9.8 m/sec².)

51. An object moves along a straight line from the point $(-3, -2)$ to $(1, 11)$ under the combined action of two forces $\mathbf{F}_1 = 5\mathbf{i} - 4\mathbf{j}$ and $\mathbf{F}_2 = -7\mathbf{i} + 6\mathbf{j}$. Find the work done by the two forces acting together. (Measure force in dynes and distance in centimeters.)

52. Use vectors to prove that the altitudes of a triangle intersect in a common point.

53. Show that the vector (A, B) is perpendicular to the line $Ax + By + C = 0$.

54. Use vectors to give an alternative proof of Theorem 10, page 34, on the distance between a point $P_1 = (x_1, y_1)$ and a line $Ax + By + C = 0$.

11.3 Vector Functions; Velocity and the Unit Tangent Vector

In the language used so far, a function is a rule leading from one *scalar*, its argument, to another *scalar*, its value (where the value is uniquely determined by the argument). For example, if $f(t) = \cos t$, the function f takes the value $\frac{1}{2}$ when its argument t is equal to $\pi/3$. Having introduced quantities that are more general than scalars, namely vectors, it now seems natural to consider what happens if the value or argument of a function is allowed to be a *vector*. We begin by considering *vector-valued functions*, or briefly *vector functions*, that is, functions of a scalar argument whose values are vectors. For example, the vector function

$$\mathbf{f}(t) = (\cos t)\mathbf{i} + (\sin t)\mathbf{j}$$

takes the value $\frac{1}{2}\mathbf{i} + \frac{1}{2}\sqrt{3}\mathbf{j}$ when its argument t, a scalar, is equal to $\pi/3$. The more complicated case of a "vector field," that is, a vector-valued function of a *vector* argument, will be considered later (see Chapter 15).

The Limit of a Vector Function

We now develop calculus for vector functions, beginning with the concept of the *limit* of a vector function. Suppose there is a vector $\mathbf{L}$ such that

$$\lim_{t \to a} |\mathbf{f}(t) - \mathbf{L}| = 0, \tag{1}$$

where you will observe that $|\mathbf{f}(t) - \mathbf{L}|$ is a scalar, being the magnitude of a vector. Then we say that $\mathbf{f}(t)$ approaches the *limit* $\mathbf{L}$ as $t \to a$, and we write $\mathbf{f}(t) \to \mathbf{L}$ as $t \to a$ or

$$\lim_{t \to a} \mathbf{f}(t) = \mathbf{L}.$$

As you may have guessed, a vector function $\mathbf{f}$ approaches a limit $\mathbf{L}$ if and only if its components separately approach limits equal to the components of $\mathbf{L}$. The next theorem shows why.

Theorem 2 *(Limit of a vector function in terms of its components).* *The vector function*

$$\mathbf{f}(t) = f_1(t)\mathbf{i} + f_2(t)\mathbf{j},$$

which has components $f_1(t)$ and $f_2(t)$ with respect to the basis vectors $\mathbf{i}$ and $\mathbf{j}$, approaches the limit $\mathbf{L} = A\mathbf{i} + B\mathbf{j}$ if and only if its components approach those of $\mathbf{L}$, that is,

$$\lim_{t \to a} f_1(t) = A, \quad \lim_{t \to a} f_2(t) = B. \tag{2}$$

Proof (Optional) Suppose first that $\mathbf{f}(t) \to \mathbf{L}$ as $t \to a$, or equivalently $|\mathbf{f}(t) - \mathbf{L}| \to 0$ as $t \to a$. Clearly

$$|f_1(t) - A| = \sqrt{[f_1(t) - A]^2} \leq \sqrt{[f_1(t) - A]^2 + [f_2(t) - B]^2} = |\mathbf{f}(t) - \mathbf{L}|,$$
$$|f_2(t) - B| = \sqrt{[f_2(t) - B]^2} \leq \sqrt{[f_1(t) - A]^2 + [f_2(t) - B]^2} = |\mathbf{f}(t) - \mathbf{L}|,$$

where we use the fact that $f_1(t) - A$ and $f_2(t) - B$ are the components of $\mathbf{f}(t) - \mathbf{L}$. But the right sides of these inequalities approach 0 as $t \to a$, and therefore so do the left sides, which proves (2).

Conversely, suppose the formulas (2), or equivalently

$$\lim_{t \to a} |f_1(t) - A| = 0, \quad \lim_{t \to a} |f_2(t) - B| = 0, \tag{2'}$$

hold. By the triangle inequality for vectors (formula (3), page 649), we have

$$|\mathbf{f}(t) - \mathbf{L}| = |[f_1(t) - A]\mathbf{i} + [f_2(t) - B]\mathbf{j}| \leq |[f_1(t) - A]\mathbf{i}| + |[f_2(t) - B]\mathbf{j}|$$
$$= |f_1(t) - A||\mathbf{i}| + |f_2(t) - B||\mathbf{j}| = |f_1(t) - A| + |f_2(t) - B|.$$

But the right side of this inequality approaches 0 as $t \to a$, because of (2'), and hence so does the left side. Therefore $|\mathbf{f}(t) - \mathbf{L}| \to 0$ as $t \to a$, that is, $\mathbf{f}(t) \to \mathbf{L}$ as $t \to a$. ∎

Theorem 2 justifies the kind of calculation made in the following example, where the limit of a vector function is taken "component by component."

Example 1 If
$$\mathbf{f}(t) = (t^2 + 2)\mathbf{i} + (\arctan t)\mathbf{j},$$
then
$$\lim_{t \to 1} \mathbf{f}(t) = \left[\lim_{t \to 1} (t^2 + 2)\right]\mathbf{i} + \left[\lim_{t \to 1} \arctan t\right]\mathbf{j}$$
$$= (1^2 + 2)\mathbf{i} + (\arctan 1)\mathbf{j} = 3\mathbf{i} + \frac{\pi}{4}\mathbf{j}.$$

Differentiation of a Vector Function

Continuity and differentiability of vector functions are defined in the same way as for scalar functions. Thus a vector function $\mathbf{f}(t)$ is said to be *continuous at a* if
$$\lim_{t \to a} \mathbf{f}(t) = \mathbf{f}(a)$$
and *differentiable at a* if the limit
$$\lim_{t \to a} \frac{\mathbf{f}(t) - \mathbf{f}(a)}{t - a}, \tag{3}$$
or equivalently
$$\lim_{\Delta t \to 0} \frac{\mathbf{f}(a + \Delta t) - \mathbf{f}(a)}{\Delta t} \tag{3'}$$

($\Delta t = t - a$), exists and is finite. The limit (3) or (3') is called the *derivative* of $\mathbf{f}(t)$ at a, denoted by $\mathbf{f}'(a)$.

It follows from Theorem 2 that a vector function $\mathbf{f}(t) = f_1(t)\mathbf{i} + f_2(t)\mathbf{j}$ is continuous at a point a if and only if its components $f_1(t)$ and $f_2(t)$ are both continuous at a. Similarly, applying Theorem 2 to the components
$$\frac{f_1(t) - f_1(a)}{t - a}, \quad \frac{f_2(t) - f_2(a)}{t - a}$$
of the difference quotient in (3), we find that
$$\mathbf{f}'(a) = f'_1(a)\mathbf{i} + f'_2(a)\mathbf{j},$$
where the prime denotes differentiation with respect to t. Thus to find the derivative of a differentiable vector function $\mathbf{f}(t)$, we form the vector function whose components are the derivatives of the components of $\mathbf{f}(t)$. As always, continuity or differentiability *on an interval I* means continuity or differentiability at every point of I. Other notations for the derivative $\mathbf{f}'(t)$ are $\mathbf{D}_t \mathbf{f}(t)$ and $d\mathbf{f}(t)/dt$.

Example 2 Since e^t and $\ln t$ are both continuous and differentiable on $I = (0, \infty)$, the vector function
$$\mathbf{f}(t) = (e^t)\mathbf{i} + (\ln t)\mathbf{j}$$

is also continuous and differentiable on I, with derivative

$$\frac{d}{dt}\mathbf{f}(t) = \left(\frac{d}{dt}e^t\right)\mathbf{i} + \left(\frac{d}{dt}\ln t\right)\mathbf{j} = (e^t)\mathbf{i} + \left(\frac{1}{t}\right)\mathbf{j}.$$

Example 3 Let $c(t)$ be a scalar function and $\mathbf{f}(t)$ a vector function, both differentiable on the same interval I. Then

$$\frac{d}{dt}[c(t)\mathbf{f}(t)] = \frac{d}{dt}[c(t)f_1(t)\mathbf{i} + c(t)f_2(t)\mathbf{j}]$$

$$= \frac{dc(t)}{dt}f_1(t)\mathbf{i} + c(t)\frac{df_1(t)}{dt}\mathbf{i} + \frac{dc(t)}{dt}f_2(t)\mathbf{j} + c(t)\frac{df_2(t)}{dt}\mathbf{j},$$

and therefore

$$\frac{d}{dt}[c(t)\mathbf{f}(t)] = \frac{dc(t)}{dt}\mathbf{f}(t) + c(t)\frac{d\mathbf{f}(t)}{dt}.$$

In the special case where $c(t) \equiv c = $ constant, this formula simplifies to

$$\frac{d}{dt}[c\mathbf{f}(t)] = c\frac{d\mathbf{f}(t)}{dt}.$$

Example 4 Let $\mathbf{f}(t)$ and $\mathbf{g}(t)$ be two vector functions, both differentiable on the same interval I, with components $f_1(t), f_2(t)$ and $g_1(t), g_2(t)$, respectively. Then, omitting arguments for brevity, we find that the derivative of the dot product $\mathbf{f}(t) \cdot \mathbf{g}(t)$ is

$$\frac{d}{dt}(\mathbf{f} \cdot \mathbf{g}) = \frac{d}{dt}(f_1 g_1 + f_2 g_2) = \frac{df_1}{dt}g_1 + f_1\frac{dg_1}{dt} + \frac{df_2}{dt}g_2 + f_2\frac{dg_2}{dt}$$

$$= \left(\frac{df_1}{dt}\mathbf{i} + \frac{df_2}{dt}\mathbf{j}\right) \cdot (g_1\mathbf{i} + g_2\mathbf{j}) + (f_1\mathbf{i} + f_2\mathbf{j}) \cdot \left(\frac{dg_1}{dt}\mathbf{i} + \frac{dg_2}{dt}\mathbf{j}\right),$$

that is,

$$\frac{d}{dt}(\mathbf{f} \cdot \mathbf{g}) = \frac{d\mathbf{f}}{dt} \cdot \mathbf{g} + \mathbf{f} \cdot \frac{d\mathbf{g}}{dt}. \tag{4}$$

Notice the resemblance between this formula and the analogous rule for differentiating the product of two scalar functions. However, the products in the right side of (4) are not ordinary products, but *dot* products.

Example 5 Let $\mathbf{f}(t)$ be a vector function of constant magnitude, but of variable direction. Then

$$|\mathbf{f}(t)|^2 = \mathbf{f}(t) \cdot \mathbf{f}(t) = c,$$

where c is a scalar constant, and hence

$$\frac{d}{dt}|\mathbf{f}(t)|^2 = \frac{d}{dt}c = 0.$$

With the help of (4), this can be written as

$$\frac{d}{dt}|\mathbf{f}|^2 = \frac{d}{dt}(\mathbf{f} \cdot \mathbf{f}) = \frac{d\mathbf{f}}{dt} \cdot \mathbf{f} + \mathbf{f} \cdot \frac{d\mathbf{f}}{dt} = 2\mathbf{f} \cdot \frac{d\mathbf{f}}{dt} = 0$$

(we again omit arguments). Therefore

$$\mathbf{f} \cdot \frac{d\mathbf{f}}{dt} = 0,$$

so that $\mathbf{f}$ is orthogonal to $d\mathbf{f}/dt$. In other words, *a vector function of constant magnitude is always orthogonal to its own derivative.*

Integration of a Vector Function

The *definite integral* of a vector function $\mathbf{f}(t) = f_1(t)\mathbf{i} + f_2(t)\mathbf{j}$ is defined by the formula

$$\int_a^b \mathbf{f}(t)\, dt = \left(\int_a^b f_1(t)\, dt \right) \mathbf{i} + \left(\int_a^b f_2(t)\, dt \right) \mathbf{j}, \tag{5}$$

that is, by "componentwise" integration. If the function $\mathbf{f}(t)$ is continuous, so are its components, and hence both integrals on the right in (5) exist, which implies the existence of the integral on the left. The same formula (5) is obtained if we define the integral of $\mathbf{f}(t)$ as the limit of an appropriate vector Riemann sum (see Problem 34).

Example 6 If

$$\mathbf{f}(t) = (e^t)\mathbf{i} + (\ln t)\mathbf{j} \qquad (0 < t < \infty),$$

as in Example 2, then

$$\int_1^2 \mathbf{f}(t)\, dt = \left(\int_1^2 e^t\, dt \right) \mathbf{i} + \left(\int_1^2 \ln t\, dt \right) \mathbf{j}$$

$$= e^t \Big|_1^2 \mathbf{i} + (t \ln t - t) \Big|_1^2 \mathbf{j} = (e^2 - e)\mathbf{i} + (2 \ln 2 - 1)\mathbf{j}. \quad \square$$

By an *antiderivative* of a vector function $\mathbf{f}(t)$, defined on an interval I, we mean any other function $\mathbf{F}(t)$ defined on I such that

$$\frac{d\mathbf{F}(t)}{dt} = \mathbf{f}(t).$$

If $\mathbf{f}(t) = f_1(t)\mathbf{i} + f_2(t)\mathbf{j}$ and if $F_1(t)$, $F_2(t)$ are antiderivatives of $f_1(t)$, $f_2(t)$, respectively, then clearly $\mathbf{F}(t) = F_1(t)\mathbf{i} + F_2(t)\mathbf{j}$ is an antiderivative of $\mathbf{f}(t)$. It is an immediate consequence of the corresponding proposition for scalar functions (Theorem 4, page 249) that two antiderivatives of $\mathbf{f}(t)$ on the same interval I can differ only by a constant vector $\mathbf{C}$. The general antiderivative of $\mathbf{f}(t)$, namely $\mathbf{F}(t) + \mathbf{C}$, is called the *indefinite integral* of $\mathbf{f}(t)$ and is denoted by $\int \mathbf{f}(t)\, dt$, with no limits of integration. Componentwise indefinite integration of $\mathbf{f}(t)$ immediately gives

$$\int \mathbf{f}(t)\, dt = \left(\int f_1(t)\, dt \right) \mathbf{i} + \left(\int f_2(t)\, dt \right) \mathbf{j}.$$

The analogue for vector functions of the fundamental theorem of calculus is of course

$$\int_a^b \mathbf{f}(t)\, dt = \mathbf{F}(b) - \mathbf{F}(a) = \mathbf{F}(t) \Big|_a^b.$$

Example 7 The vector function $\mathbf{F}(t) = (\sin t)\mathbf{i} - (\cos t)\mathbf{j} + (3\mathbf{i} - 4\mathbf{j})$ is an antiderivative of the vector function $\mathbf{f}(t) = (\cos t)\mathbf{i} + (\sin t)\mathbf{j}$ on every interval.

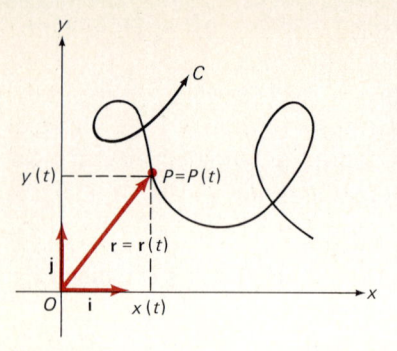

Figure 26

We now describe the use of vector functions to study motion in the plane. Let

$$\mathbf{r} = \mathbf{r}(t) = x(t)\mathbf{i} + y(t)\mathbf{j} \qquad (t \text{ in } I)$$

be a continuous vector function defined on an interval I, and suppose the initial point of $\mathbf{r} = \mathbf{r}(t)$ is placed at the origin O of the xy-plane, as in Figure 26. Then the final point of $\mathbf{r} = \mathbf{r}(t)$ is a variable point $P = P(t) = (x(t), y(t))$ of the plane, and $\mathbf{r} = \mathbf{r}(t)$ is called the *position vector* (or *radius vector*) of $P = P(t)$. (Here we use the symbol $\mathbf{r}$ instead of $\mathbf{f}$, because of its prevalence in physics and applied mathematics.) As t increases, $P = P(t)$ traces out a plane curve called the *graph* of $\mathbf{r} = \mathbf{r}(t)$, namely the curve C with parametric equations

$$x = x(t), \qquad y = y(t) \qquad (t \text{ in } I). \tag{6}$$

It is particularly suggestive to think of the parameter t as the time, and we will do so with the understanding that this is not the only possibility.

Velocity and Speed

If $\mathbf{r}(t)$ is differentiable, the derivative

$$\frac{d\mathbf{r}}{dt} = \frac{dx}{dt}\mathbf{i} + \frac{dy}{dt}\mathbf{j} = x'(t)\mathbf{i} + y'(t)\mathbf{j}$$

is the time rate of change of the position vector $\mathbf{r} = \mathbf{r}(t)$ specifying the position of the variable point $P = P(t)$. Because of its physical meaning, this quantity is known as the *velocity vector*, or simply the *velocity*, denoted by $\mathbf{v} = \mathbf{v}(t)$. Since $\mathbf{r}(t) = x(t)\mathbf{i} + y(t)\mathbf{j}$, we have

$$\mathbf{v}(t) = \frac{d\mathbf{r}(t)}{dt} = \frac{dx(t)}{dt}\mathbf{i} + \frac{dy(t)}{dt}\mathbf{j},$$

from which it is apparent that $\mathbf{v}(t)$ is the generalization to two dimensions of the one-dimensional velocity defined in Section 2.1.

By the *speed* of the variable point $P = P(t)$ we mean the magnitude of the velocity vector $\mathbf{v} = \mathbf{v}(t)$, that is, the scalar

$$v = |\mathbf{v}| = \left|\frac{d\mathbf{r}}{dt}\right| = \sqrt{\left(\frac{dx}{dt}\right)^2 + \left(\frac{dy}{dt}\right)^2}. \tag{7}$$

We will henceforth assume that

$$\left(\frac{dx}{dt}\right)^2 + \left(\frac{dy}{dt}\right)^2 \neq 0 \qquad (t \text{ in } I). \tag{8}$$

It follows that the speed, and hence the velocity, of the point P tracing out the curve C is never zero, so that the point never stops moving.

To interpret the velocity $\mathbf{v} = \mathbf{v}(t)$ geometrically, let $\phi = \phi(t)$ be the angle from the positive x-axis to $\mathbf{v}$; this means that the final point of the vector $\mathbf{v}$ would have polar coordinates $|\mathbf{v}|$ and ϕ if its initial point were placed at the origin. (Do not confuse ϕ with the angle between $\mathbf{v}$ and the unit vector $\mathbf{i}$, which unlike ϕ is restricted to the interval $[0, \pi]$.) Let a be any point of I, and suppose that $x'(a) \neq 0$. Then

$$\tan \phi = \frac{y'(a)}{x'(a)},$$

Chapter 11 Vectors in the Plane

since $\mathbf{v}(a) = x'(a)\mathbf{i} + y'(a)\mathbf{j}$. But as on page 456, this quotient is precisely the slope of the tangent to the curve C with parametric equations (6) at the point $P(a) = (x(a), y(a))$. Moreover, if $x'(a) = 0$, then $y'(a) \neq 0$, because of the condition (8), and then $\mathbf{v}(a) = y'(a)\mathbf{j}$, so that $\mathbf{v}(a)$ is vertical, while the tangent to the curve C at $P(a)$ is also vertical (show this by going back to the definition of the tangent as the limiting position of the secant line through the points $(x(a), y(a))$ and $(x(t), y(t))$ as $t \to a$). Thus, in any event, the velocity $\mathbf{v}(a)$ is always parallel to the tangent to the curve C at $P(a)$. Therefore if its initial point is placed at $P(a)$, the vector $\mathbf{v}(a)$ will always point along the tangent to C at $P(a)$.

Actually, we can be even more explicit about the direction of $\mathbf{v}(a)$. In fact, of the two opposite directions on the tangent line, $\mathbf{v}(a)$ *points in the direction in which* $P = P(t)$ *traverses* C *as t increases*. This is apparent from the construction in Figure 27(a) or 27(b), where in each case the direction of increasing t is indicated by an arrowhead on the curve C itself, and the difference quotient vector

$$\frac{\Delta \mathbf{r}}{\Delta t} = \frac{\mathbf{r}(a + \Delta t) - \mathbf{r}(a)}{\Delta t} \qquad (9)$$

rotates about $P(a)$ as $\Delta t \to 0$, eventually approaching $\mathbf{v}(a)$. It is assumed in both figures that Δt is positive, so that $P(a)$ comes *before* $P(a + \Delta t)$ along the curve C, which is traversed in the direction of increasing t. However, if Δt is negative, the limiting direction of the vector (9) as $\Delta t \to 0$ is the same as for positive Δt, even though $P(a)$ now comes *after* $P(a + \Delta t)$, because if Δt is negative, we can write (9) in the equivalent form

$$\frac{\Delta \mathbf{r}}{\Delta t} = \frac{\mathbf{r}(a + \Delta t) - \mathbf{r}(a)}{-|\Delta t|},$$

so that $\Delta \mathbf{r}/\Delta t$ still points along C in the direction of increasing t (why?).

> **Remark** If a is an endpoint of I, we take $x'(a)$ and $y'(a)$ to be the appropriate one-sided derivatives. Also we will relax the condition (8) to the extent of requiring it to hold only at interior points of I. This enables us to consider cases in which the initial or final velocity of a motion is zero.

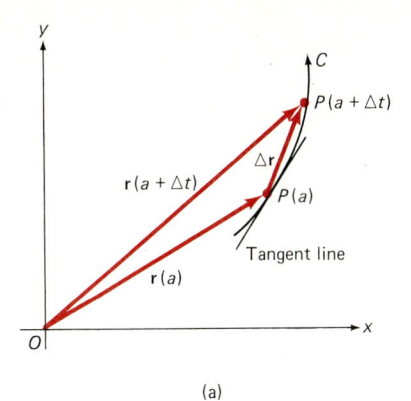

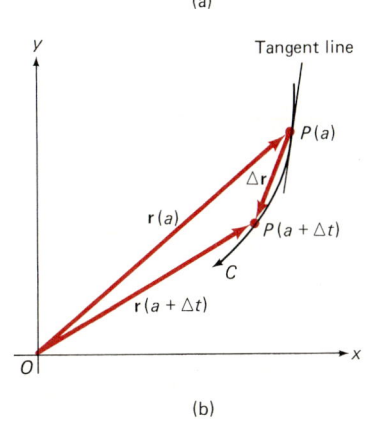

Figure 27

The Arc Length Function

So far, so good. But intuition demands that the speed of a point P moving along a curve C be equal to the time rate of change of the distance traveled by P along C, as measured from any fixed point of C (after all, this is what speed means in the special case where C is a straight line). Hence we now show that the speed v, defined as the magnitude $|\mathbf{v}|$ of the velocity vector $\mathbf{v}$, is equal to the derivative ds/dt, where s is arc length along C. To this end, let C be the graph of the vector function $\mathbf{r} = \mathbf{r}(t) = x(t)\mathbf{i} + y(t)\mathbf{j}$ as t varies over the interval I, and suppose the functions $x(t)$ and $y(t)$ are continuously differentiable on I. Choosing any fixed point a in I, we define the *arc length function*

$$s = s(t) \qquad (t \geq a)$$

as the length of the arc of C with fixed initial point $P(a) = (x(a), y(a))$ and variable final point $P(t) = (x(t), y(t))$. The geometric meaning of $s(t)$ is

Section 11.3 Vector Functions; Velocity and the Unit Tangent Vector illustrated in Figure 28. By formula (2), page 461,

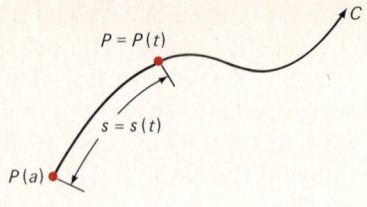

$$s = s(t) = \int_a^t \sqrt{\left(\frac{dx}{du}\right)^2 + \left(\frac{dy}{du}\right)^2} \, du, \tag{10}$$

where we pick u as the variable of integration to avoid confusion with the upper limit of integration t. Differentiating (10) with respect to t, we get

$$\frac{ds}{dt} = \sqrt{\left(\frac{dx}{dt}\right)^2 + \left(\frac{dy}{dt}\right)^2} \tag{10'}$$

Geometric meaning of the arc length function
Figure 28

(use Theorem 5, page 254), and comparison of this formula with (7) gives the desired result

$$v = |\mathbf{v}| = \frac{ds}{dt}. \tag{11}$$

Using vector functions, we can write (10) in the concise form

$$s(t) = \int_a^t |\mathbf{r}'(u)| \, du,$$

or equivalently

$$s(t) = \int_a^t |\mathbf{v}(u)| \, du,$$

from which the validity of formula (11) is immediately apparent.

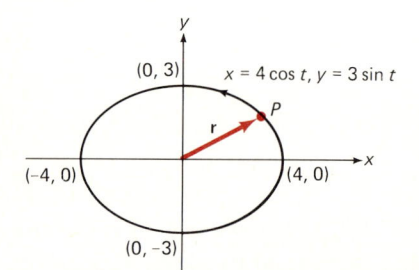

Example 8 Find the velocity and speed of the point P with position vector

$$\mathbf{r} = (4 \cos t)\mathbf{i} + (3 \sin t)\mathbf{j} \qquad (t \geq 0). \tag{12}$$

When and where does the speed take its maximum and minimum values?

Figure 29

Solution As t increases, P repeatedly traverses the ellipse with cartesian equation $\frac{1}{16}x^2 + \frac{1}{9}y^2 = 1$, starting from the point $(4, 0)$ at time $t = 0$ and moving in the counterclockwise direction (see Figure 29). Differentiating (12), we get

$$\mathbf{v} = \frac{d\mathbf{r}}{dt} = (-4 \sin t)\mathbf{i} + (3 \cos t)\mathbf{j}.$$

The speed is the magnitude of $\mathbf{v}$, equal to

$$v = |\mathbf{v}| = \sqrt{(-4 \sin t)^2 + (3 \cos t)^2} = \sqrt{16 \sin^2 t + 9 \cos^2 t}.$$

Since

$$v^2 = 16 \sin^2 t + 9 \cos^2 t = 7 \sin^2 t + 9(\sin^2 t + \cos^2 t) = 7 \sin^2 t + 9,$$

the maximum value of v^2 is 16, achieved when $\sin t = \pm 1$, $\cos t = 0$, and the minimum value of v^2 is 9, achieved when $\sin t = 0$, $\cos t = \pm 1$. Hence the maximum speed is 4, occurring when P is at the endpoints $(0, \pm 3)$ of the minor axis of the ellipse, and the minimum speed is 3, occurring when P is at the endpoints $(\pm 4, 0)$ of the major axis. □

668 Chapter 11 Vectors in the Plane

Definition of the Unit Tangent Vector

As before, we assume that the velocity vector $\mathbf{v} = d\mathbf{r}/dt$ is nonzero, with its initial point at $P = P(t)$, the final point of the position vector $\mathbf{r} = \mathbf{r}(t)$. Let C be the graph of $\mathbf{r} = \mathbf{r}(t)$. Then $\mathbf{v} = \mathbf{v}(t)$ is tangent to C at the point P and points in the direction of increasing t. Dividing the vector $\mathbf{v}$ by its magnitude, we get the unit vector

$$\mathbf{T} = \mathbf{T}(t) = \frac{\mathbf{v}(t)}{|\mathbf{v}(t)|}, \tag{13}$$

which like $\mathbf{v}$ itself is tangent to C at the point P and points in the direction of increasing t. Therefore $\mathbf{T}$ is called the *unit tangent vector* to the curve C at the point P. It follows from (13) that

$$\mathbf{T} = \frac{d\mathbf{r}/dt}{|d\mathbf{r}/dt|},$$

but we can derive an even simpler formula for $\mathbf{T}$, namely

$$\mathbf{T} = \frac{d\mathbf{r}}{ds}, \tag{14}$$

by going over to a description of C with the arc length s as parameter. This is done as follows. The function $s = s(t)$ is continuous and differentiable, and also increasing since the integrand in (10) is always positive (why?). Therefore $s(t)$ has a continuous and differentiable inverse function $t = t(s)$, with derivative

$$\frac{dt}{ds} = \frac{1}{ds/dt}$$

(see Theorem 4, page 286). But then with the help of (11),

$$\mathbf{T} = \frac{\mathbf{v}}{|\mathbf{v}|} = \frac{d\mathbf{r}/dt}{ds/dt} = \frac{d\mathbf{r}}{dt}\frac{dt}{ds} = \frac{d\mathbf{r}}{ds},$$

where the chain rule for vector functions (see Problem 21) is applied at the last step, and the proof of formula (14) is complete.

Thus *the unit tangent vector $\mathbf{T}$ is the derivative of the position vector $\mathbf{r}$ with respect to the arc length s.* In terms of the components x and y of the position vector $\mathbf{r}$, (14) takes the form

$$\mathbf{T} = \frac{dx}{ds}\mathbf{i} + \frac{dy}{ds}\mathbf{j}. \tag{14'}$$

In particular, this implies

$$\left(\frac{dx}{ds}\right)^2 + \left(\frac{dy}{ds}\right)^2 = 1,$$

since $\mathbf{T}$ is a unit vector.

In terms of the originally given function $\mathbf{r} = \mathbf{r}(t)$, the function being differentiated in formula (14) is the composite function $\mathbf{r} = \mathbf{r}(t(s))$, where $t = t(s)$ is the inverse of the arc length function $s = s(t)$. It is often difficult or impossible to calculate the function $s = s(t)$ explicitly, but this does not diminish the *conceptual* value of regarding the curve C as "parametrized" by the arc length s.

Section 11.3 Vector Functions; Velocity and the Unit Tangent Vector

Example 9 Find the unit tangent vector **T** for the graph of

$$\mathbf{r}(t) = (\cos t + t \sin t)\mathbf{j} + (\sin t - t \cos t)\mathbf{j} \qquad (t \geq 0),$$

both as a function **T**(t) of the parameter t and as a function **T**(s) of the arc length s measured from the point $\mathbf{r}(0) = \mathbf{i} = (1, 0)$.

Solution The derivative of **r**(t) is

$$\mathbf{r}'(t) = (-\sin t + \sin t + t \cos t)\mathbf{i} + (\cos t - \cos t + t \sin t)\mathbf{j}$$
$$= (t \cos t)\mathbf{i} + (t \sin t)\mathbf{j},$$

with magnitude

$$|\mathbf{r}'(t)| = \sqrt{(t \cos t)^2 + (t \sin t)^2} = t,$$

and therefore

$$\mathbf{T}(t) = \frac{\mathbf{r}'(t)}{|\mathbf{r}'(t)|} = (\cos t)\mathbf{i} + (\sin t)\mathbf{j}.$$

Moreover,

$$s = s(t) = \int_0^t |\mathbf{r}'(u)|\, du = \int_0^t u\, du = \frac{1}{2}t^2,$$

and hence $t = \sqrt{2s}$, so that

$$\mathbf{T}(s) = (\cos \sqrt{2s}\,)\mathbf{i} + (\sin \sqrt{2s}\,)\mathbf{j}.$$

The graph of the function **r**(t) is the spiral curve shown in Figure 30. □

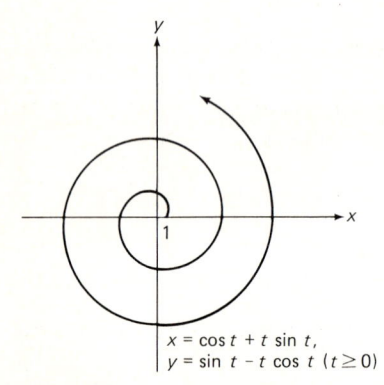

Figure 30
$x = \cos t + t \sin t,$
$y = \sin t - t \cos t \ (t \geq 0)$

Problems

Evaluate the given limit.

1. $\lim_{t \to 2} (t^2\mathbf{i} + t^3\mathbf{j})$

2. $\lim_{t \to 0} \left(\frac{\sin t}{t}\mathbf{i} - 2e^t\mathbf{j}\right)$

3. $\lim_{t \to 1} \left(\frac{1}{t^2 + 1}\mathbf{i} + \cos \frac{\pi t}{3}\mathbf{j}\right)$

4. $\lim_{t \to 0^+} \left(\frac{\arccos t}{5}\mathbf{i} + \sqrt{t}\mathbf{j}\right)$

5. $\lim_{t \to \infty} \left(\frac{2}{\ln t}\mathbf{i} + \frac{1-t}{1+t}\mathbf{j}\right)$

6. $\lim_{t \to -1} \left(\frac{t^2 - 1}{t - 1}\mathbf{i} + \frac{t^3 + 1}{t + 1}\mathbf{j}\right)$

7. Is the vector function $\mathbf{f}(t) = (\ln t)\mathbf{i} + (\ln (\ln t))\mathbf{j}$ continuous at $t = e$? At $t = 1$?

Show that if $\mathbf{f}(t) \to \mathbf{L}$ and $\mathbf{g}(t) \to \mathbf{M}$ as $t \to a$, then

8. $\lim_{t \to a} c\mathbf{f}(t) = c\mathbf{L}$ (c any scalar)

9. $\lim_{t \to a} [\mathbf{f}(t) + \mathbf{g}(t)] = \mathbf{L} + \mathbf{M}$

10. $\lim_{t \to a} [\mathbf{f}(t) \cdot \mathbf{g}(t)] = \mathbf{L} \cdot \mathbf{M}$

11. $\lim_{t \to a} |\mathbf{f}(t)| = |\mathbf{L}|$

Differentiate the given vector function.

12. $(2t^3 - 5)\mathbf{i} + (\sin 2t)\mathbf{j}$

13. $(t \ln t)\mathbf{i} + (t^2 e^t)\mathbf{j}$

14. $(\tan t)\mathbf{i} + (\ln (\ln t))\mathbf{j}$

15. $(\sec t)\mathbf{i} - (\sinh t)\mathbf{j}$

16. $(\arctan t)\mathbf{i} + (\cos (\sin t))\mathbf{j}$

17. $(e^t \sin t)\mathbf{i} + (e^{-t} \cos t)\mathbf{j}$

Show that

18. The derivative of a constant vector function is the zero vector

19. If $\mathbf{f}(t)$ is differentiable at a, then $\mathbf{f}(t)$ is continuous at a

20. If $\mathbf{f}(t)$ and $\mathbf{g}(t)$ are differentiable at a, with derivatives $\mathbf{f}'(a)$ and $\mathbf{g}'(a)$, then $\mathbf{f}(t) + \mathbf{g}(t)$ is differentiable at a, with derivative $\mathbf{f}'(a) + \mathbf{g}'(a)$

21. If $\mathbf{f}(t)$ is a differentiable function of t, and t is a differentiable function of u, then

$$\frac{d\mathbf{f}}{du} = \frac{d\mathbf{f}}{dt}\frac{dt}{du} \qquad \text{(chain rule for vector functions)}$$

provided that u is such that $t(u)$ lies in the domain of $\mathbf{f}(t)$

Find the antiderivative $\mathbf{F}(t)$ of the given vector function $\mathbf{f}(t)$ which satisfies the specified condition.

22. $\mathbf{f}(t) = 2t^2\mathbf{i} - t^3\mathbf{j}, \ \mathbf{F}(1) = 4\mathbf{j}$

23. $\mathbf{f}(t) = (\cos t)\mathbf{i} + (e^t)\mathbf{j}, \ \mathbf{F}(0) = 2\mathbf{i} + \mathbf{j}$

24. $\mathbf{f}(t) = (2/t)\mathbf{i} + (\ln t)\mathbf{j}, \ \mathbf{F}(e) = -\mathbf{i} + 3\mathbf{j}$

25. $\mathbf{f}(t) = \sec t[(\tan t)\mathbf{i} + (\sec t)\mathbf{j}], \ \mathbf{F}(\pi/3) = \mathbf{0}$

Evaluate the given integral of a vector function.

26. $\int_1^2 (3t\mathbf{i} - 4t^2\mathbf{j})\,dt$

27. $\int_4^9 \left(\frac{1}{\sqrt{t}}\mathbf{i} + \sqrt{t}\,\mathbf{j}\right)dt$

28. $\int [(t \ln t)\mathbf{i} + (\csc^2 t)\mathbf{j}]\,dt$

29. $\int_0^{10} |(\sin t)\mathbf{i} + (\cos t)\mathbf{j}|\,dt$

30. $\int_0^1 (2^t\mathbf{i} + 3^{-t}\mathbf{j})\,dt$

31. $\int [(\sin^2 t)\mathbf{i} + (\cos^2 t)\mathbf{j}]\,dt$

32. Let $\mathbf{f}(t)$ and $\mathbf{g}(t)$ be vector functions which are both continuous on $[a, b]$. Show that

$$\int_a^b c\mathbf{f}(t)\,dt = c\int_a^b \mathbf{f}(t)\,dt \qquad (c \text{ any scalar})$$

and

$$\int_a^b [\mathbf{f}(t) + \mathbf{g}(t)]\,dt = \int_a^b \mathbf{f}(t)\,dt + \int_a^b \mathbf{g}(t)\,dt.$$

33. Let $\mathbf{f}(t)$ be continuous on $[a, b]$, and let $\mathbf{c}$ be any constant vector. Show that

$$\int_a^b \mathbf{c} \cdot \mathbf{f}(t)\,dt = \mathbf{c} \cdot \int_a^b \mathbf{f}(t)\,dt.$$

34. Let $a = t_0 < t_1 < \cdots < t_{i-1} < t_i < \cdots < t_{n-1} < t_n = b$, $\Delta t_i = t_i - t_{i-1}$, $\mu = \max\{\Delta t_1, \Delta t_2, \ldots, \Delta t_n\}$, and let p_i be an arbitrary point in the subinterval $[t_{i-1}, t_i]$. Show that if $\mathbf{f}(t) = (f_1(t), f_2(t))$ is a continuous vector function on $[a, b]$

and if $\int_a^b \mathbf{f}(t)\,dt$ is defined as $\lim_{\mu \to 0} \sum_{i=1}^n \mathbf{f}(p_i)\Delta t_i$, then

$$\int_a^b \mathbf{f}(t)\,dt = \left(\int_a^b f_1(t)\,dt, \int_a^b f_2(t)\,dt\right).$$

Find the velocity $\mathbf{v}$ and speed $v = |\mathbf{v}|$ of the point $P = P(t)$ with the given position vector $\mathbf{r} = \mathbf{r}(t)$. Sketch the path traversed by P, indicating the direction of increasing t.

35. $\mathbf{r}(t) = 3t\mathbf{i} + 2\mathbf{j}$
36. $\mathbf{r}(t) = t\mathbf{i} - 4t\mathbf{j}$
37. $\mathbf{r}(t) = t^2\mathbf{i} + 2t\mathbf{j}$
38. $\mathbf{r}(t) = e^t\mathbf{i} + e^{2t}\mathbf{j}$
39. $\mathbf{r}(t) = (2 \sin t)\mathbf{i} - (\cos t)\mathbf{j}$
40. $\mathbf{r}(t) = (3 \cosh t)\mathbf{i} + (4 \sinh t)\mathbf{j}$

Find the unit tangent vector $\mathbf{T}$ to the graph of the given vector function $\mathbf{r} = \mathbf{r}(t)$, both as a function $\mathbf{T}(t)$ of the parameter t and as a function $\mathbf{T}(s)$ of the arc length s measured from the point with position vector $\mathbf{r}(0)$.

41. $\mathbf{r}(t) = 5t\mathbf{i} + \mathbf{j}$
42. $\mathbf{r}(t) = 6t\mathbf{i} - 8t\mathbf{j}$
43. $\mathbf{r}(t) = (\cos 3t)\mathbf{i} + (\sin 3t)\mathbf{j}$
44. $\mathbf{r}(t) = (4 \sin t)\mathbf{i} + (4 \cos t)\mathbf{j}$
45. $\mathbf{r}(t) = \frac{3}{2}t^2\mathbf{i} + t^3\mathbf{j} \quad (t > 0)$
46. $\mathbf{r}(t) = (e^t \cos t)\mathbf{i} + (e^t \sin t)\mathbf{j}$

47. Describe the path traversed by the point with the position vector

$$\mathbf{r}(t) = (4 \cos^3 t)\mathbf{i} + (4 \sin^3 t)\mathbf{j} \qquad (t \geq 0).$$

When and where does the speed v of the point P take its maximum and minimum values? Does P ever stop moving?

11.4 The Unit Normal Vector; Curvature and Acceleration

To continue our study of motion in the plane, let the position vector

$$\mathbf{r} = \mathbf{r}(t) = x(t)\mathbf{i} + y(t)\mathbf{j} \qquad (t \text{ in } I),$$

the velocity vector

$$\mathbf{v} = \mathbf{v}(t) = \mathbf{r}'(t) = x'(t)\mathbf{i} + y'(t)\mathbf{j},$$

the arc length function

$$s(t) = \int_a^t |\mathbf{v}(u)|\,du, \tag{1}$$

and the unit tangent vector

$$\mathbf{T} = \mathbf{T}(t) = \frac{\mathbf{r}'(t)}{|\mathbf{r}'(t)|} = \frac{d\mathbf{r}}{ds}$$

be the same as in the preceding section. The prime denotes differentiation with respect to the parameter t, which we regard as the time, and a is any fixed point of the interval I on which the vector function $\mathbf{r}(t)$ is defined. As before, we assume that $\mathbf{r}(t)$ is differentiable on I and satisfies the condition

$$|\mathbf{r}'(t)|^2 = [x'(t)]^2 + [y'(t)]^2 \neq 0$$

at every interior point of I (see the remark on page 667), but in addition we now assume that the *second* derivatives $x''(t)$ and $y''(t)$ exist and are continuous on I. Recall also that formula (1) implies

$$\frac{ds}{dt} = |\mathbf{v}(t)| = |\mathbf{r}'(t)|. \tag{1'}$$

As before, let $P = P(t) = (x(t), y(t))$ be the final point of the position vector $\mathbf{r} = \mathbf{r}(t)$, whose initial point is at the origin O of the xy-plane. Then as t increases, $P(t)$ traces out the graph of $\mathbf{r} = \mathbf{r}(t)$, that is, the plane curve C with parametric equations

$$x = x(t), \quad y = y(t) \quad (t \text{ in } I).$$

The initial point of the unit tangent vector $\mathbf{T} = \mathbf{T}(t)$ is at the point $P(t)$, and as t increases, the *direction* of $\mathbf{T}$ will in general vary from point to point along the curve C. Thus the derivative of $\mathbf{T}$ with respect to t is in general nonzero, despite the fact that the *magnitude* of $\mathbf{T}$ has the constant value 1. Since $|\mathbf{T}|^2 = \mathbf{T} \cdot \mathbf{T} = 1$, we have

$$\frac{d}{dt}(\mathbf{T} \cdot \mathbf{T}) = \frac{d\mathbf{T}}{dt} \cdot \mathbf{T} + \mathbf{T} \cdot \frac{d\mathbf{T}}{dt} = 2\mathbf{T} \cdot \frac{d\mathbf{T}}{dt} = 0,$$

and therefore

$$\mathbf{T} \cdot \frac{d\mathbf{T}}{dt} = 0 \tag{2}$$

(compare with Example 5, page 664). Hence the unit tangent vector $\mathbf{T}$ is always orthogonal to its derivative $d\mathbf{T}/dt$, which is in general not a unit vector, and as the direction of $\mathbf{T}$ changes, so does the direction of $d\mathbf{T}/dt$.

The geometric meaning of formula (2) is illustrated in Figure 31, where the arrowhead on the curve C indicates the direction in which C is traversed as t increases. The vector $d\mathbf{T}/dt$ always points toward the *concave side* of C, that is, the side to which the curve C turns as it is traced out by the moving point $P(t)$. To see why this is true, we make drawings like the ones in Figure 32, where $\mathbf{T}(t)$ is the unit tangent vector at the time t and $\mathbf{T}(t + \Delta t)$ is the unit tangent vector at a somewhat later time $t + \Delta t$. It is clear from the figures that the direction in which the curve C bends as it is traced out by $P(t)$ is the same as the direction of the difference vector $\mathbf{T}(t + \Delta t) - \mathbf{T}(t)$, which is in turn the same as the direction of the derivative $d\mathbf{T}/dt$ (justify the last statement).

Example 1 Find the vectors $\mathbf{T}$ and $d\mathbf{T}/dt$ for the graph of the vector function

$$\mathbf{r} = \mathbf{r}(t) = (\cos 2t)\mathbf{i} + (\sin 2t)\mathbf{j}. \tag{3}$$

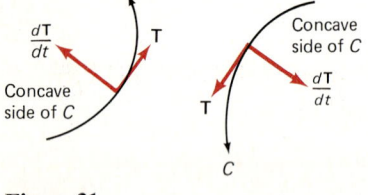

Figure 31

Figure 32

Solution The graph of (3) is the circle of radius 1 with its center at the origin, traversed in the counterclockwise direction as t increases. The unit tangent vector is

$$\mathbf{T} = \frac{\mathbf{r}'(t)}{|\mathbf{r}'(t)|} = \frac{(-2\sin 2t)\mathbf{i} + (2\cos 2t)\mathbf{j}}{\sqrt{(-2\sin 2t)^2 + (2\cos 2t)^2}}$$

$$= \frac{1}{2}[(-2\sin 2t)\mathbf{i} + (2\cos 2t)\mathbf{j}] = (-\sin 2t)\mathbf{i} + (\cos 2t)\mathbf{j}$$

[see Figure 33(a)], and its derivative is

$$\frac{d\mathbf{T}}{dt} = -2[(\cos 2t)\mathbf{i} + (\sin 2t)\mathbf{j}] = -2\mathbf{r}(t).$$

Since $-\mathbf{r}(t)$, the negative of the position vector, points toward the origin, $d\mathbf{T}/dt$ points toward the center of the circle C, and hence toward the concave side of C. This is shown in Figure 33(b), where $P = (\cos 2t, \sin 2t)$ is the common initial point of the vectors $\mathbf{T}$ and $d\mathbf{T}/dt$. □

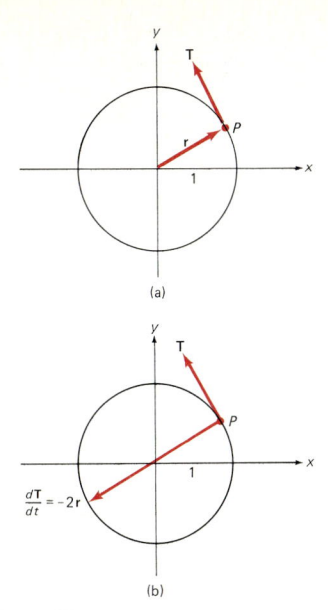

Figure 33

Definition of the Unit Normal Vector

Besides the unit tangent vector $\mathbf{T} = \mathbf{T}(t)$, we now introduce another unit vector, denoted by $\mathbf{N} = \mathbf{N}(t)$ and called the *unit normal vector* to the curve C at the point $P = P(t)$. This is the unit vector

$$\mathbf{N} = \frac{d\mathbf{T}/dt}{|d\mathbf{T}/dt|} \qquad \left(\frac{d\mathbf{T}}{dt} \neq \mathbf{0}\right)$$

in the direction of $d\mathbf{T}/dt = \mathbf{T}'(t)$, with its initial point at P. Since $d\mathbf{T}/dt$ is orthogonal to $\mathbf{T}$ and points toward the concave side of C, the same is true of the vector $\mathbf{N}$ (see Figure 34). If C is a straight line, the direction of $\mathbf{T}$ does not change, and then $d\mathbf{T}/dt \equiv \mathbf{0}$. In this case we take $\mathbf{N}$ to be the vector obtained by rotating $\mathbf{T}$ through 90° in the counterclockwise direction, as illustrated in Figure 35.

The unit tangent and normal vectors $\mathbf{T}$ and $\mathbf{N}$

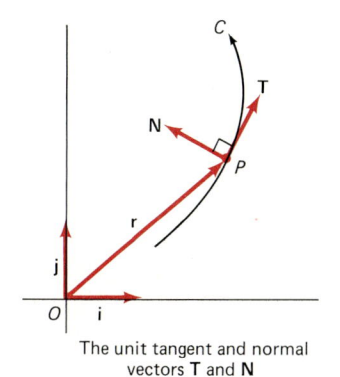

Figure 34

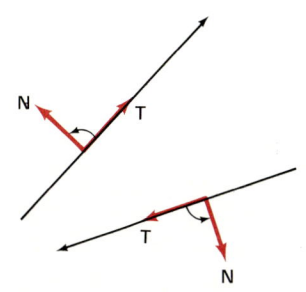

Figure 35

Example 2 Find the unit normal vector to the curve

$$x = 2t, \qquad y = t^2 \qquad (4)$$

at the origin O and also at the points $(-2, 1)$ and $(4, 4)$.

Solution Eliminating the parameter t from the equations (4), we find that the curve is the parabola with cartesian equation $y = \frac{1}{4}x^2$. The position

Section 11.4 The Unit Normal Vector; Curvature and Acceleration

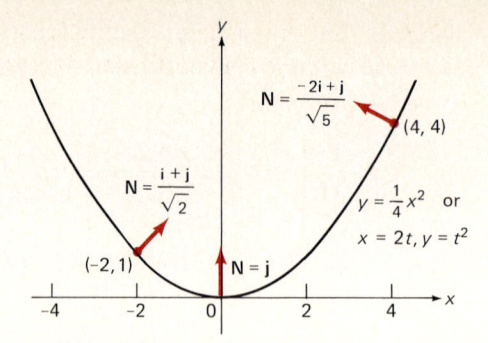

Figure 36

vector corresponding to (4) is $\mathbf{r}(t) = 2t\mathbf{i} + t^2\mathbf{j}$, with derivative $\mathbf{r}'(t) = 2\mathbf{i} + 2t\mathbf{j}$. Hence the unit tangent vector is

$$\mathbf{T}(t) = \frac{\mathbf{r}'(t)}{|\mathbf{r}'(t)|} = \frac{2\mathbf{i} + 2t\mathbf{j}}{\sqrt{4 + 4t^2}} = \frac{\mathbf{i} + t\mathbf{j}}{\sqrt{1 + t^2}}.$$

Differentiating $\mathbf{T}(t)$, we get the vector

$$\mathbf{T}'(t) = \frac{-t\mathbf{i} + \mathbf{j}}{(1 + t^2)^{3/2}}$$

with magnitude

$$|\mathbf{T}'(t)| = \frac{\sqrt{(-t)^2 + 1^2}}{(1 + t^2)^{3/2}} = \frac{\sqrt{1 + t^2}}{(1 + t^2)^{3/2}} = \frac{1}{1 + t^2}.$$

Thus the unit normal vector to the parabola at the point $P(t) = (2t, t^2)$ is

$$\mathbf{N}(t) = \frac{\mathbf{T}'(t)}{|\mathbf{T}'(t)|} = \frac{-t\mathbf{i} + \mathbf{j}}{\sqrt{1 + t^2}}.$$

The origin O corresponds to the parameter value $t = 0$, and hence the unit normal vector at O is $\mathbf{N}(0) = \mathbf{j}$. The points $(-2, 1)$ and $(4, 4)$ correspond to the parameter values $t = -1$ and $t = 2$. Therefore

$$\mathbf{N}(-1) = \frac{\mathbf{i} + \mathbf{j}}{\sqrt{2}}$$

is the unit normal vector at $(-2, 1)$, while

$$\mathbf{N}(2) = \frac{-2\mathbf{i} + \mathbf{j}}{\sqrt{5}}$$

is the unit normal vector at $(4, 4)$, as shown in Figure 36. ☐

The unit vectors $\mathbf{T}$ and $\mathbf{N}$ form an orthonormal basis (page 651), which unlike the fixed orthonormal basis $\mathbf{i} = (1, 0)$ and $\mathbf{j} = (0, 1)$, varies from point to point along the curve C, as shown in Figure 37. For this reason, the basis consisting of $\mathbf{T}$ and $\mathbf{N}$ is called the *local basis*. For example, the local basis for the circular motion considered in Example 1 is

$$\mathbf{T} = (-\sin 2t)\mathbf{i} + (\cos 2t)\mathbf{j}, \qquad \mathbf{N} = (-\cos 2t)\mathbf{i} + (-\sin 2t)\mathbf{j}$$

(check this statement).

The local basis

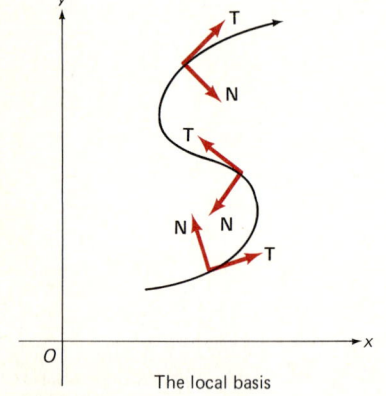

Figure 37

674 Chapter 11 Vectors in the Plane

Curvature

If the arc length s is chosen as the parameter, then the position vector $\mathbf{r} = \mathbf{r}(s)$, its final point $P = P(s)$, the unit tangent vector $\mathbf{T} = \mathbf{T}(s) = d\mathbf{r}/ds$, and the unit normal vector

$$\mathbf{N} = \mathbf{N}(s) = \frac{d\mathbf{T}/ds}{|d\mathbf{T}/ds|} \qquad \left(\frac{d\mathbf{T}}{ds} \neq \mathbf{0}\right) \qquad (5)$$

are all functions of s, as indicated by the notation. It follows from (5) that

$$\frac{d\mathbf{T}}{ds} = \left|\frac{d\mathbf{T}}{ds}\right| \mathbf{N}.$$

Let C be the graph of $\mathbf{r} = \mathbf{r}(s)$. Then the positive scalar $|d\mathbf{T}/ds|$ is called the *curvature* of C at P, denoted by κ (lowercase Greek kappa), and the last equation can be written in the form

$$\frac{d\mathbf{T}}{ds} = \kappa \mathbf{N}. \qquad (6)$$

The curvature κ measures the rate at which the curve C bends as it is traced out by the moving point $P = P(s)$. In fact, let $\phi = \phi(s)$ be the angle from the positive x-axis to $\mathbf{T} = \mathbf{T}(s)$, which means that the final point of $\mathbf{T}$ would have polar coordinates $|\mathbf{T}| = 1$ and ϕ if the initial point of $\mathbf{T}$ were placed at the origin. Then

$$\mathbf{T} = (\cos \phi)\mathbf{i} + (\sin \phi)\mathbf{j},$$

and

$$\frac{d\mathbf{T}}{ds} = \left(-\sin \phi \frac{d\phi}{ds}\right)\mathbf{i} + \left(\cos \phi \frac{d\phi}{ds}\right)\mathbf{j},$$

with the help of the chain rule. Therefore

$$\frac{d\mathbf{T}}{ds} = \frac{d\phi}{ds} \mathbf{T}_\perp,$$

where

$$\mathbf{T}_\perp = (-\sin \phi)\mathbf{i} + (\cos \phi)\mathbf{j}$$

is the unit vector obtained by rotating $\mathbf{T}$ through $90°$ in the counterclockwise direction, as can be seen by examining Figure 38. But $|\mathbf{T}_\perp| = 1$, and hence

$$\left|\frac{d\mathbf{T}}{ds}\right| = \left|\frac{d\phi}{ds}\right| |\mathbf{T}_\perp| = \left|\frac{d\phi}{ds}\right|,$$

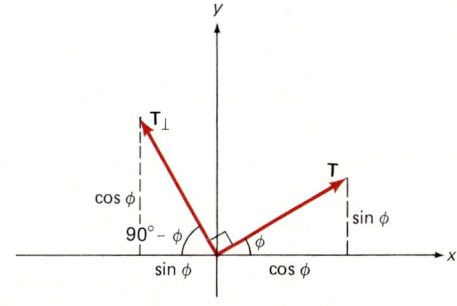

Figure 38

so that
$$\kappa = \left|\frac{d\mathbf{T}}{ds}\right| = \left|\frac{d\phi}{ds}\right|.$$

Thus the curvature κ is the absolute value of the rate of change of the angle ϕ with respect to arc length s. The larger $\kappa = \kappa(s)$, the faster the unit tangent vector $\mathbf{T} = \mathbf{T}(s)$ turns as it moves along C, that is, the more sharply the curve C bends. This is illustrated in Figure 39, where one curve has large curvature and bends sharply, while the other curve has small curvature and bends gradually.

Example 3 Show that every straight line has constant zero curvature $\kappa = 0$.

Solution For a straight line the angle ϕ is constant, as shown in Figure 40. Therefore $d\phi/ds = 0$ and
$$\kappa = \left|\frac{d\phi}{ds}\right| = 0. \quad \square$$

Large curvature

Small curvature

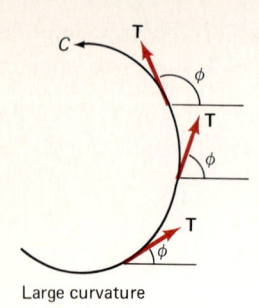

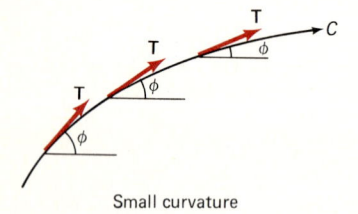

Figure 39

Constant zero curvature

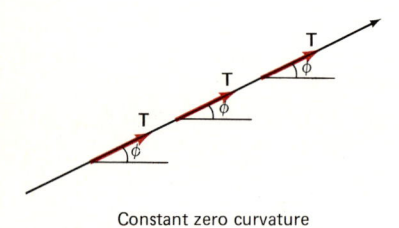

Figure 40

The curve C is usually given as the graph of a position vector function $\mathbf{r} = \mathbf{r}(t)$, where t is a parameter other than the arc length s. It is then an easy matter to express the curvature κ as a function of t. In fact,
$$\kappa = \left|\frac{d\mathbf{T}}{ds}\right| = \left|\frac{d\mathbf{T}}{dt}\frac{dt}{ds}\right| = \frac{|d\mathbf{T}/dt|}{ds/dt} = \frac{1}{v}\left|\frac{d\mathbf{T}}{dt}\right|,$$

where we use the chain rule in the first step and the formula for the derivative of an inverse function in the second step (note that $ds/dt > 0$). Equivalently, since $v = |\mathbf{r}'(t)|$,
$$\kappa = \kappa(t) = \frac{|\mathbf{T}'(t)|}{|\mathbf{r}'(t)|}. \tag{7}$$

Example 4 Find the curvature of the circle
$$x = a \cos t, \quad y = a \sin t \quad (0 \le t \le 2\pi)$$
of radius a.

Solution Here
$$\mathbf{r}(t) = (a \cos t)\mathbf{i} + (a \sin t)\mathbf{j}, \quad \mathbf{r}'(t) = a[(-\sin t)\mathbf{i} + (\cos t)\mathbf{j}],$$
so that
$$|\mathbf{r}'(t)| = a\sqrt{(-\sin t)^2 + \cos^2 t} = a$$
and
$$\mathbf{T}(t) = \frac{\mathbf{r}'(t)}{|\mathbf{r}'(t)|} = (-\sin t)\mathbf{i} + (\cos t)\mathbf{j}, \quad \mathbf{T}'(t) = (-\cos t)\mathbf{i} + (-\sin t)\mathbf{j}.$$

Therefore $|\mathbf{T}'(t)| = 1$ and
$$\kappa = \frac{|\mathbf{T}'(t)|}{|\mathbf{r}'(t)|} = \frac{1}{a},$$
that is, a circle of radius a has constant curvature $\kappa = 1/a$. Thus the smaller the radius a, the greater the rate of change of the direction of the unit tangent vector $\mathbf{T}$ with respect to the distance traveled along the circle. This is why

the smaller the turning radius of a car, the more rapidly it can change its direction of motion per unit of distance traveled. ☐

With a little effort, we can find a formula for the curvature κ in terms of the first and second derivatives of the components $x(t)$ and $y(t)$ of the position vector $\mathbf{r}(t)$. First we observe that

$$\mathbf{T} = \frac{\mathbf{r}'(t)}{|\mathbf{r}'(t)|} = \frac{x'\mathbf{i} + y'\mathbf{j}}{(x'^2 + y'^2)^{1/2}},$$

where the argument t of the derivatives x', y', x'' and y'' will be omitted for brevity. Using the quotient rule to differentiate $\mathbf{T}$, we obtain

$$\frac{d\mathbf{T}}{dt} = \frac{\frac{d(x'\mathbf{i} + y'\mathbf{j})}{dt}(x'^2 + y'^2)^{1/2} - (x'\mathbf{i} + y'\mathbf{j})\frac{d}{dt}(x'^2 + y'^2)^{1/2}}{x'^2 + y'^2}$$

$$= \frac{(x''\mathbf{i} + y''\mathbf{j})(x'^2 + y'^2) - (x'\mathbf{i} + y'\mathbf{j})(x'x'' + y'y'')}{(x'^2 + y'^2)^{3/2}}$$

$$= \frac{(x''y'^2 - x'y'y'')\mathbf{i} + (x'^2y'' - x'x''y')\mathbf{j}}{(x'^2 + y'^2)^{3/2}}$$

$$= \frac{(x'y'' - y'x'')(-y'\mathbf{i} + x'\mathbf{j})}{(x'^2 + y'^2)^{3/2}},$$

after some algebraic manipulation and cancellation of terms. Therefore

$$\left|\frac{d\mathbf{T}}{dt}\right| = \frac{|x'y'' - y'x''|}{(x'^2 + y'^2)^{3/2}}|-y'\mathbf{i} + x'\mathbf{j}| = \frac{|x'y'' - y'x''|}{x'^2 + y'^2},$$

since $|-y'\mathbf{i} + x'\mathbf{j}| = \sqrt{x'^2 + y'^2}$. Thus, finally,

$$\kappa = \frac{1}{|\mathbf{r}'(t)|}|\mathbf{T}'(t)| = \frac{1}{(x'^2 + y'^2)^{1/2}}\frac{|x'y'' - y'x''|}{x'^2 + y'^2}$$

that is,

$$\kappa = \frac{|x'y'' - y'x''|}{(x'^2 + y'^2)^{3/2}}. \tag{8}$$

Example 5 Find the curvature $\kappa = \kappa(t)$ of the parabola $x = 2t$, $y = t^2$, considered in Example 2 and shown in Figure 36.

Solution Here $x' = 2$, $x'' = 0$, $y' = 2t$, $y'' = 2$, and substituting these values into formula (8), we get

$$\kappa = \frac{4}{(4 + 4t^2)^{3/2}} = \frac{1}{2(1 + t^2)^{3/2}}. \tag{9}$$

It is apparent from (9) that the parabola has its maximum curvature $\kappa(0) = \frac{1}{2}$ at the origin. The parts of the parabola corresponding to large values of $|t|$ are very straight, and correspondingly (9) shows that they have very small curvature. For instance, if $|t| = 100$, the curvature is about $\frac{1}{2000}$. ☐

If C is the graph of a function $y = f(x)$, then C has the parametric representation $x = t$, $y = f(t)$. Therefore $x' = 1$, $x'' = 0$, so that formula (8)

Section 11.4 The Unit Normal Vector; Curvature and Acceleration **677** reduces to

$$\kappa = \frac{|y''|}{(1+y'^2)^{3/2}}, \qquad (8')$$

where $y' = dy/dx$ and $y'' = d^2y/dx^2$. For instance, if C is the straight line $y = mx + b$, then $y'' = 0$ and (8') implies that $\kappa = 0$, as we already know from Example 3.

The Radius of Curvature

Suppose κ is the curvature of a curve C at a point P. Then by the *radius of curvature* of C at P we mean the number

$$R = \frac{1}{\kappa},$$

that is, the reciprocal of the curvature. It follows from Example 4 that the radius of curvature of a circle is just its ordinary radius. If κ is tiny, R is huge. Therefore a very straight curve has a very large radius of curvature, and a straight line ($\kappa = 0$) can be regarded as having an infinite radius of curvature. By the same token, we set $R = 0$ at the points, if any, at which κ approaches infinity. If $\kappa \neq 0$, the curve C has a finite radius of curvature R at the point P. Then by the *circle of curvature* of C at P we mean the circle of radius R passing through P whose center Q lies on the concave side of C along the normal to C at P, as illustrated in Figure 41. This circle has the same tangent and radius of curvature as the curve C at P, and therefore fits C very closely at P. For this reason, the circle of curvature is also called the *osculating circle* ("to osculate" is fancy language for "to kiss").

Circle of curvature

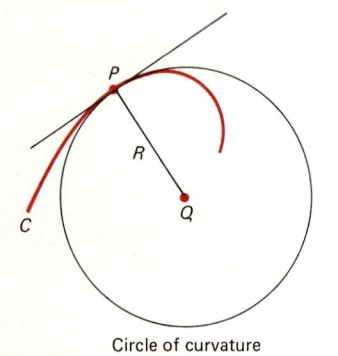

Figure 41

Example 6 Find the radius of curvature $R = R(t)$ of the ellipse

$$x = 2\cos t, \qquad y = \sin t \qquad (0 \le t \le 2\pi)$$

at the points (2, 0) and (0, 1).

Solution Notice that the ellipse has the cartesian equation $\frac{1}{4}x^2 + y^2 = 1$. Calculating the first and second derivatives of x and y, we obtain

$$x' = -2\sin t, \qquad y' = \cos t,$$
$$x'' = -2\cos t, \qquad y'' = -\sin t,$$

so that

$$x'y'' - y'x'' = 2\sin^2 t + 2\cos^2 t = 2,$$
$$x'^2 + y'^2 = 4\sin^2 t + \cos^2 t = 3\sin^2 t + 1.$$

Therefore, with the help of formula (8),

$$R(t) = \frac{1}{\kappa(t)} = \frac{(x'^2 + y'^2)^{3/2}}{|x'y'' - y'x''|} = \frac{1}{2}(3\sin^2 t + 1)^{3/2}.$$

The points (2, 0) and (0, 1) correspond to the parameter values $t = 0$ and $t = \pi/2$. Hence the radius of curvature of the ellipse is $R(0) = \frac{1}{2}$ at (2, 0) and $R(\pi/2) = \frac{1}{2}(4)^{3/2} = 4$ at (0, 1). The osculating circle at (2, 0) is the circle of radius $\frac{1}{2}$ centered at $(2 - \frac{1}{2}, 0) = (\frac{3}{2}, 0)$, and the osculating circle at (0, 1) is the circle of radius 4 centered at $(0, 1 - 4) = (0, -3)$. Figure 42 shows the ellipse together with both of these osculating circles. ∎

[Figure: Osculating circle at (2, 0); curve x = 2 cos t, y = sin t; Osculating circle at (0, 1)]

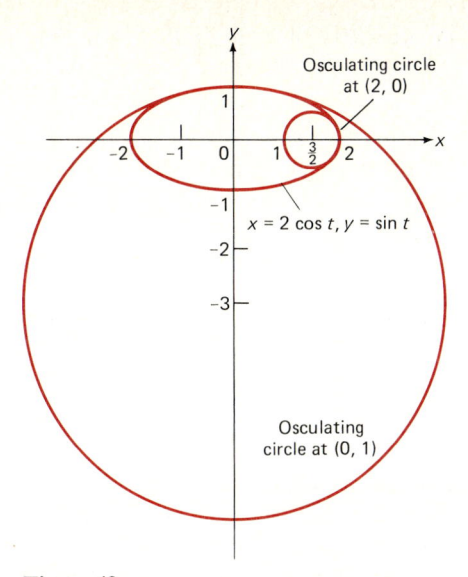

Figure 42

Acceleration and Its Components

Once again we let $P = P(t)$ be the final point of the position vector $\mathbf{r} = \mathbf{r}(t)$, and regard the parameter t as the time. Then, just as in the one-dimensional case, the *acceleration* $\mathbf{a}$ of the moving point P is defined as the time derivative of the velocity $\mathbf{v} = \mathbf{v}(t)$. Thus $\mathbf{a} = \mathbf{a}(t)$ is the vector function

$$\mathbf{a} = \frac{d\mathbf{v}}{dt}.$$

We now expand $\mathbf{a}$ with respect to the local basis consisting of the unit tangent and normal vectors $\mathbf{T}$ and $\mathbf{N}$. Since

$$\mathbf{T} = \frac{d\mathbf{r}/dt}{|d\mathbf{r}/dt|} = \frac{\mathbf{v}}{v},$$

we have

$$\mathbf{v} = v\mathbf{T}.$$

Therefore, by the product and chain rules,

$$\mathbf{a} = \frac{d\mathbf{v}}{dt} = \frac{d}{dt}(v\mathbf{T}) = \frac{dv}{dt}\mathbf{T} + v\frac{d\mathbf{T}}{dt} = \frac{dv}{dt}\mathbf{T} + v\frac{d\mathbf{T}}{ds}\frac{ds}{dt}.$$

But $ds/dt = v$ and $d\mathbf{T}/ds = \kappa\mathbf{N}$, where κ is the curvature. It follows that

$$\mathbf{a} = \frac{dv}{dt}\mathbf{T} + \kappa v^2 \mathbf{N}, \tag{10}$$

or equivalently

$$\mathbf{a} = \frac{dv}{dt}\mathbf{T} + \frac{v^2}{R}\mathbf{N}, \tag{10'}$$

where $R = 1/\kappa$ is the radius of curvature. The (scalar) components of $\mathbf{a}$ with respect to the local basis **T** and **N** are called the *tangential and normal components* of the acceleration, denoted by a_T and a_N. Thus

$$\mathbf{a} = a_T \mathbf{T} + a_N \mathbf{N},$$

where

$$a_T = \frac{dv}{dt}, \qquad a_N = \kappa v^2 = \frac{v^2}{R}. \tag{11}$$

Notice that (11) can also be written as

$$a_T = \frac{d^2 s}{dt^2}, \qquad a_N = \kappa \left(\frac{ds}{dt}\right)^2 = \frac{1}{R}\left(\frac{ds}{dt}\right)^2. \tag{11'}$$

The geometric meaning of the expansion $\mathbf{a} = a_T \mathbf{T} + a_N \mathbf{N}$ is shown in Figure 43.

Example 7 Find the acceleration of a particle traversing a circle of radius R with constant speed v.

Solution Since v is constant, we have $dv/dt \equiv 0$. Moreover, the radius of curvature of the circle is just its ordinary radius R. Therefore, by (10'),

$$\mathbf{a} = \frac{v^2}{R} \mathbf{N}.$$

Thus the tangential component of the acceleration is zero, or more informally, the acceleration "has no tangential component." In fact, the acceleration **a** is *centripetal*, meaning that it is directed toward the center of the circle, and its magnitude has the constant value v^2/R. □

Example 8 Find the tangential and normal components of the acceleration of a particle moving along the parabola $x = 2t$, $y = t^2$.

Solution Here

$$v = \sqrt{\left(\frac{dx}{dt}\right)^2 + \left(\frac{dy}{dt}\right)^2} = \sqrt{4 + 4t^2} = 2\sqrt{1 + t^2}$$

and

$$\kappa = \frac{1}{2(1 + t^2)^{3/2}},$$

as in Example 5. Therefore, by (11), the tangential and normal components of the acceleration a are

$$a_T = \frac{dv}{dt} = \frac{2t}{\sqrt{1 + t^2}}, \qquad a_N = \kappa v^2 = \frac{2}{\sqrt{1 + t^2}}.$$

For instance, $a_T = \sqrt{2}$ and $a_N = \sqrt{2}$ at the point $(2, 1)$ occupied by the particle at time $t = 1$, while $a_T = 4/\sqrt{5}$ and $a_N = 2/\sqrt{5}$ at the point $(4, 4)$ occupied at time $t = 2$. Actually, since

$$\mathbf{a} = \frac{d^2 \mathbf{r}}{dt^2} = \frac{d^2}{dt^2}(2t\mathbf{i} + t^2 \mathbf{j}) = 2\mathbf{j},$$

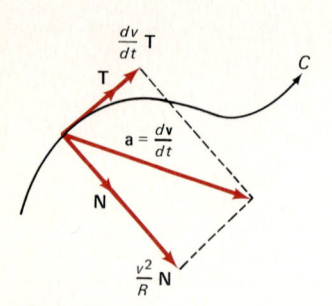

the acceleration is of constant magnitude 2 and is always directed vertically upward. Also note that the normal component of the acceleration can be found without calculating the curvature κ. In fact, since $|\mathbf{a}|^2 = a_T^2 + a_N^2$ (why?), we have

$$a_N^2 = |\mathbf{a}|^2 - a_T^2 = 4 - \frac{4t^2}{1+t^2} = \frac{4}{1+t^2},$$

and hence $a_N = 2/\sqrt{1+t^2}$. □

Problems

Find the unit tangent and normal vectors **T** and **N** to the graph of the given vector function $\mathbf{r} = \mathbf{r}(t)$.
1. $\mathbf{r}(t) = 2t\mathbf{i} - 5\mathbf{j}$
2. $\mathbf{r}(t) = 3t\mathbf{i} + t^3\mathbf{j}$ $(t > 0)$
3. $\mathbf{r}(t) = 2t^3\mathbf{i} + 3t^2\mathbf{j}$ $(t > 0)$
4. $\mathbf{r}(t) = (\cos t)\mathbf{i} + (2 \sin t)\mathbf{j}$
5. $\mathbf{r}(t) = (\sinh t)\mathbf{i} + (\cosh t)\mathbf{j}$
6. $\mathbf{r}(t) = e^{-t}\mathbf{i} + e^t\mathbf{j}$

Find the curvature κ of the given curve at the point corresponding to the specified value of the parameter t.
7. $x = 3t^2$, $y = 3t - t^3$, $t = 1$
8. $x = \frac{1}{3}t^3$, $y = t$, $t = -1$
9. $x = t \cos t$, $y = t \sin t$, $t = \sqrt{3}$
10. $x = \cos^3 t$, $y = \sin^3 t$, $t = \pi/4$
11. $x = e^t \cos t$, $y = e^t \sin t$, $t = 0$
12. $x = \cos t + t \sin t$, $y = \sin t - t \cos t$, $t = -2$

Find the curvature κ of the graph of the given function at the specified point.
13. $y = 4x - x^2$ at $(2, 4)$
14. $y = x^3 + 1$ at $(-1, 0)$
15. $y = 2/x$ at $(1, 2)$
16. $y = \ln x$ at $(1, 0)$
17. $y = xe^{-x}$ at $(1, 1/e)$
18. $y = e^{-x^2}$ at $(0, 1)$

19. What is the maximum curvature of the curve $y = e^x$, and where is it achieved?
20. Show that the radius of curvature of the curve $y = \cosh x$ at the point $P = (x, y)$ is equal to y^2, the square of the y-coordinate.
21. Suppose a circle of radius a rolls without slipping along a horizontal straight line. Then, as shown in Example 7, page 455, a fixed point P of the circumference of the circle traces out the cycloid $x = a(t - \sin t)$, $y = a(1 - \cos t)$. Find the radius of curvature R at an arbitrary point of the cycloid. Where is R equal to zero? Equal to a? What is the maximum value of R, and where is it achieved?
22. It was shown in Example 4 that the curvature of a circle of radius a is equal to $1/a$. Use the construction in Figure 44 to show this by direct calculation of $|d\phi/ds|$.

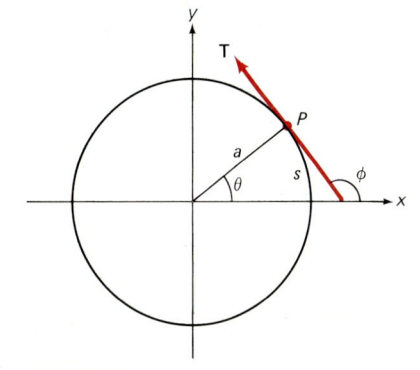

Figure 44

Write an equation for the osculating circle to the given curve at the point $(0, 1)$. In each case graph the curve and the circle.
23. $y = 1/(x^2 + 1)$
24. $y = \cos(x/\sqrt{2})$
25. $y = e^x$
26. $y = \sec x$

27. At which points of the parabola $x^2 = 8y$ is its radius of curvature equal to $\frac{125}{16}$?
28. Determine the radius of curvature of the ellipse $(x^2/a^2) + (y^2/b^2) = 1$ at the endpoints of its major and minor axes.
29. Let C be the graph of the equation $x^2 + xy + y^2 = 3$. What is the radius of curvature of C at the point $(1, 1)$?
30. Show that the graph of the function

$$y = 6x^5 - 15x^4 + 10x^3 \quad (0 < x < 1)$$

connects the two separate pieces of the discontinuous function

$$y = \begin{cases} 0 & \text{if } x \leq 0, \\ 1 & \text{if } x \geq 1 \end{cases}$$

in such a way that the resulting curve, shown in Figure 45, has both continuous slope and continuous curvature. Can you think of a real-life situation in which this mathematical problem might arise?

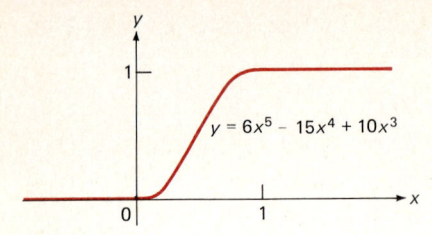

Figure 45

$y = 6x^5 - 15x^4 + 10x^3$

31–36. Find the tangential and normal components a_T and a_N of the acceleration of the point $P = P(t)$ with the same position vector $\mathbf{r} = \mathbf{r}(t)$ as in Problems 1–6.

37. Verify that the curvature $\kappa = \kappa(\theta)$ of the polar curve $r = r(\theta)$ is given by

$$\kappa = \frac{|r^2 + 2r'^2 - rr''|}{(r^2 + r'^2)^{3/2}}, \qquad (i)$$

where $r' = dr/d\theta$ and $r'' = d^2r/d\theta^2$.

Use formula (i) to find the curvature of the given polar curve at the point with the specified value of θ.

38. $r = 4\cos\theta$, $\theta = 10$
39. $r = 1 - \cos\theta$, $\theta = \pi$
40. $r = e^\theta$, $\theta = \ln 3$

41. Let $\mathbf{T}$ and $\mathbf{N}$ be the unit tangent and normal vectors at the same point of the graph of a position vector function $\mathbf{r} = x\mathbf{i} + y\mathbf{j}$, let ϕ be the angle from the positive x-axis to $\mathbf{T}$, and let $\mathbf{T}_\perp$ be the unit vector obtained by rotating $\mathbf{T}$ through $90°$ in the counterclockwise direction. Show that

$$\mathbf{N} = \begin{cases} \mathbf{T}_\perp & \text{if } d\phi/ds > 0, \\ -\mathbf{T}_\perp & \text{if } d\phi/ds < 0. \end{cases}$$

In particular, show that $\mathbf{N}$ is obtained from $\mathbf{T}$ by changing $\mathbf{i}$ to $\mathbf{j}$ and $\mathbf{j}$ to $-\mathbf{i}$ if $x'y'' - y'x'' > 0$, while $\mathbf{N}$ is obtained from $\mathbf{T}$ by changing $\mathbf{i}$ to $-\mathbf{j}$ and $\mathbf{j}$ to $\mathbf{i}$ if $x'y'' - y'x'' < 0$.

42–46. Use the preceding problem to simplify the calculations leading from $\mathbf{T}$ to $\mathbf{N}$ in Problems 2–6.

11.5 Applications to Mechanics

In this section we use vectors to solve a number of two-dimensional problems of Newtonian mechanics. Consider a particle of mass m acted on by a force $\mathbf{F}$. Then, according to *Newton's second law of motion*,

$$\mathbf{F} = m\mathbf{a}, \qquad (1)$$

where $\mathbf{a}$ is the particle's acceleration. This is the natural generalization of the one-dimensional version of Newton's law, studied in Section 4.7, but now the force $\mathbf{F}$ and the acceleration $\mathbf{a}$ are both *vectors*. Suppose $\mathbf{F}$ and $\mathbf{a}$ have components F_1, F_2 and a_1, a_2 with respect to a suitably chosen orthonormal basis $\mathbf{e}_1, \mathbf{e}_2$. Then the single vector equation (1) is equivalent to the *pair* of scalar equations

$$F_1 = ma_1, \qquad F_2 = ma_2, \qquad (1')$$

obtained by taking the components of both sides of (1).

Motion of a Projectile

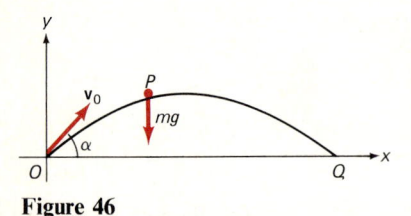

Example 1 A projectile is fired from a gun whose angle of elevation is α. What is the trajectory of the projectile if its initial speed is v_0? Ignore the earth's curvature and rotation, and also the effect of air resistance.

Solution We choose a rectangular coordinate system in the plane of the trajectory, that is, in the vertical plane containing the initial velocity vector $\mathbf{v}_0$, with the y-axis pointing vertically upward and the origin O at the position of the gun.† Then the vector $\mathbf{v}_0$, of magnitude v_0, makes angle α with the positive x-axis (see Figure 46, where the projectile is momentarily at P and eventually lands at Q), so that $\mathbf{v}_0 = (v_0 \cos\alpha)\mathbf{i} + (v_0 \sin\alpha)\mathbf{j}$ in terms of the

Figure 46

---

† The fact that the trajectory of the projectile lies in the fixed vertical plane containing the vector $\mathbf{v}_0$ is actually a consequence of Newton's law in three dimensions (see Problem 31, page 736).

orthonormal basis $\mathbf{i} = (1, 0)$ and $\mathbf{j} = (0, 1)$. The only force acting on the projectile is its weight, that is, the downward force

$$\mathbf{F} = -mg\mathbf{j},$$

where g is the acceleration due to gravity. Let $\mathbf{r}(t) = x(t)\mathbf{i} + y(t)\mathbf{j}$ be the position vector of the projectile relative to the gun at time t. Then Newton's law (1') gives a pair of second order differential equations

$$m\frac{d^2x}{dt^2} = 0, \qquad m\frac{d^2y}{dt^2} = -mg,$$

or equivalently

$$\frac{d^2x}{dt^2} = 0, \qquad \frac{d^2y}{dt^2} = -g, \tag{2}$$

after dividing through by the mass m, which plays no further role in the problem.

Integrating the differential equations (2), we obtain

$$\frac{dx}{dt} = A_1, \qquad \frac{dy}{dt} = -gt + B_1, \tag{3}$$

and another integration then gives

$$x = A_1 t + A_2, \qquad y = -\frac{1}{2}gt^2 + B_1 t + B_2. \tag{4}$$

To determine the constants of integration A_1, A_2, B_1 and B_2, we impose the initial conditions

$$x(0) = y(0) = 0, \qquad x'(0) = v_0 \cos \alpha, \qquad y'(0) = v_0 \sin \alpha \tag{5}$$

($x' = dx/dt$, $y' = dy/dt$). These equations express the fact that the projectile, which is initially at rest in the barrel of the gun, acquires the velocity $\mathbf{v}_0 = (v_0 \cos \alpha, v_0 \sin \alpha)$ at the time of firing $t = 0$. It follows from (3)–(5) that

$$A_1 = v_0 \cos \alpha, \qquad A_2 = 0, \qquad B_1 = v_0 \sin \alpha, \qquad B_2 = 0,$$

and substitution of these values into (4) then gives a pair of equations

$$x = (v_0 \cos \alpha)t, \qquad y = -\frac{1}{2}gt^2 + (v_0 \sin \alpha)t \tag{6}$$

for the trajectory of the projectile, with the time t as the parameter. Eliminating t from the parametric equations (6), we get the cartesian equation

$$y = x \tan \alpha - \frac{gx^2}{2v_0^2 \cos^2 \alpha}, \tag{6'}$$

whose graph is a parabola (see page 583), as anticipated in the figure. ☐

The motion of projectiles is investigated further in Problems 1–12. We now turn to problems involving circular motion.

Example 2 An artificial satellite describes a circular orbit around the earth at a constant altitude of 1000 miles.† Find its orbital speed v.

---

† The fact that circular orbits are possible if the satellite is inserted into orbit with the appropriate speed will be shown in Example 10, page 734.

Solution Here we choose our orthonormal basis to be the local basis consisting of the unit tangent and normal vectors **T** and **N** to the satellite's orbit, a circle which has the same center O as the earth itself; thus **N** points toward the center of the earth. The only force acting on the satellite is the earth's gravitational attraction. As in Example 8, page 271, this force is

$$\mathbf{F} = \frac{GMm}{R^2}\mathbf{N},$$

where G is the universal gravitational constant, M is the mass of the earth and m is the mass of the satellite. By formula (10'), page 679, the satellite's acceleration is

$$\mathbf{a} = \frac{dv}{dt}\mathbf{T} + \frac{v^2}{R}\mathbf{N},$$

where R is the radius (of curvature) of the circular orbit, so that Newton's second law $\mathbf{F} = m\mathbf{a}$ or $\mathbf{a} = \mathbf{F}/m$ becomes

$$\frac{dv}{dt}\mathbf{T} + \frac{v^2}{R}\mathbf{N} = \frac{GM}{R^2}\mathbf{N}$$

in the local basis. Taking tangential and normal components of both sides of this vector equation, we get two scalar equations

$$\frac{dv}{dt} = 0$$

and

$$v^2 = \frac{GM}{R}, \tag{7}$$

the first of which tells us that the satellite's speed is constant. Also, by formula (18), page 272,

$$G = \frac{gR_0^2}{M},$$

where R_0 is the earth's radius and g is the acceleration due to gravity. It follows from (7) and the last equation that

$$v = \sqrt{\frac{g}{R}}\, R_0, \tag{8}$$

Setting $g = 32$ ft/sec^2, $R_0 = 3960$ mi, and $R = R_0 + 1000 = 4960$ mi in formula (8), and noting that 1 mi $= 5280$ ft, we finally get

$$v = \sqrt{\frac{32}{5280(4960)}}\; 3960 \approx 4.4 \text{ mi/sec.} \qquad \square$$

Uniform Circular Motion

A particle P describing a circular orbit with constant speed, as in Example 2, is said to be in *uniform circular motion*. Suppose P has speed v and the circle is of radius R. Then P makes one complete revolution around the circle in the time

$$T = \frac{2\pi R}{v}, \tag{9}$$

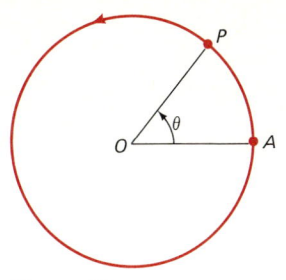

Figure 47

called the *period* of the motion. Let O be the center of the circle, and let θ be the angle from any fixed radius OA to the radius OP of the circle (see Figure 47). Then θ is a linear function of the time t, and θ increases by 2π during every period (assume that the motion is counterclockwise). Therefore

$$\theta = \frac{2\pi}{T} t + \theta_0,$$

where the constant θ_0 is the value of θ at $t = 0$, or equivalently

$$\theta = \omega t + \theta_0,$$

where the quantity

$$\omega = \frac{2\pi}{T}$$

is called the *angular speed* (ω is lowercase Greek omega). Thus

$$T = \frac{2\pi}{\omega},$$

and comparison of this formula with (9) shows that

$$v = R\omega.$$

It should also be noted that the normal component of the acceleration of a particle in uniform circular motion is

$$a_N = \frac{v^2}{R} = R\omega^2. \tag{10}$$

Example 3 A communications relay satellite describes a circular orbit of radius R in the plane of the equator, at a constant speed of v mi/sec. The choice of R is such that the satellite appears permanently suspended over a fixed point of the earth's surface. How high above the earth is the satellite?

Solution Since the satellite appears stationary in the sky, its motion is synchronous with the rotation of the earth, that is, the satellite revolves around the earth at exactly the same rate as the earth rotates about its axis (see Figure 48). In other words, the period of the orbital motion is $T = 1$ day $= 24(60)^2$ sec. Therefore, by formula (9),

$$24(60)^2 = \frac{2\pi R}{v},$$

which gives

$$R = \frac{12(3600)v}{\pi} = \frac{12(3600)R_0}{\pi}\sqrt{\frac{g}{R}},$$

after substituting for v from formula (8). Hence

$$R^{3/2} = \frac{12(3600)R_0 \sqrt{g}}{\pi},$$

and accordingly

$$R = \left[\frac{12(3600)(3960)}{\pi}\sqrt{\frac{32}{5280}}\right]^{2/3} \approx 26{,}190 \text{ mi.}$$

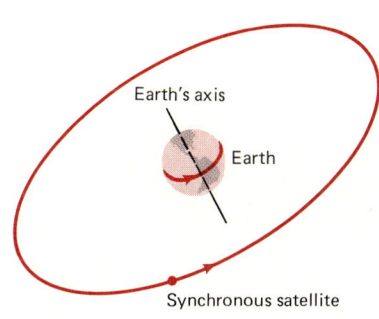

Figure 48

Section 11.5 Applications to Mechanics

Thus the height of the satellite above the earth is about 26,190 − 3960 = 22,230 mi. As an exercise, show that its speed is about 1.9 mi/sec. □

It follows from formulas (7) and (9) that the period T of a satellite in uniform circular motion around a center of gravitational attraction of mass M is

$$T = \frac{2\pi}{\sqrt{GM}} R^{3/2}, \tag{11}$$

or equivalently

$$T^2 = \frac{4\pi^2}{GM} R^3. \tag{11'}$$

Thus the square of the period of the satellite is proportional to the cube of the radius of its orbit. This is the form taken by *Kepler's third law* for the special case where the orbit of the satellite is circular (in general the orbit is an ellipse). A detailed discussion of Kepler's laws of planetary motion will be given in Section 12.5.

Example 4 The moon's orbit is nearly circular, and its period of revolution around the earth is about 27.3 days. What is the radius of the moon's orbit?

Solution By (11),

$$T_m = \frac{2\pi}{\sqrt{GM}} R_m^{3/2},$$

where T_m and R_m are the period of revolution of the moon and the radius of its orbit (G is the universal gravitational constant, and M is the mass of the earth). Similarly,

$$1 \text{ (day)} = \frac{2\pi}{\sqrt{GM}} R_s^{3/2},$$

where R_s is the radius of the orbit of a satellite in synchronous motion about the earth. Dividing the first of these equations by the second, we find that

$$T_m = \left(\frac{R_m}{R_s}\right)^{3/2},$$

so that

$$R_m = T_m^{2/3} R_s.$$

But $R_s \approx 26{,}190$ mi, as found in Example 3, and therefore

$$R_m \approx (27.3)^{2/3}(26{,}190) \approx 237{,}500 \text{ mi.} \quad \square$$

Example 5 To condition a test pilot to withstand the effects of large accelerations, he is whirled in a horizontal circle of radius 10 ft with the help of a giant centrifuge. At what angular speed ω does he experience an acceleration of $3g$?

Solution Setting the normal acceleration (10) equal to $3g$, we get $R\omega^2 = 3g$ or

$$\omega = \sqrt{\frac{3g}{R}},$$

which gives
$$\omega = \sqrt{9.6} \approx 3.1 \text{ rad/sec}$$
after setting $R = 10$ ft and $g = 32$ ft/sec^2. In more familiar units of measurement, $\omega = 60\sqrt{9.6}/2\pi \approx 29.6$ rpm (revolutions per minute). □

The Suspension Bridge

Finally we consider some problems of *statics*, in which a "system" of particles is in equilibrium, so that there is no motion at all.

Example 6 Find the shape of a suspension bridge loaded by a roadway of weight w per unit length. Neglect the weight of the cable itself compared to that of the roadway.

Solution Choose a rectangular coordinate system in the plane of the cable, with the y-axis pointing vertically upward and the origin O at the lowest point of the cable, as in Figure 49. (If there are two parallel cables, solve the same problem with each cable loaded by one half the weight of the roadway.) Let $P = (x, y)$ be a point of the cable to the right of O. Then the piece of cable OP is subject to three forces, the horizontal tension H pulling OP to the left at O, the tangential tension of magnitude T pulling OP to the right and upward at P, and the weight wx of x feet of roadway pulling OP vertically downward. Hence the total force acting on OP is

$$\mathbf{F} = (T \cos \theta - H)\mathbf{i} + (T \sin \theta - wx)\mathbf{j},$$

where the unit vectors $\mathbf{i}$ and $\mathbf{j}$ have their usual meaning and θ is the inclination of the tangent to the cable at P (see the figure). Equilibrium of OP requires that $\mathbf{F} = \mathbf{0}$, since otherwise Newton's second law of motion would lead to acceleration of OP.† Therefore

$$T \cos \theta = H, \quad T \sin \theta = wx, \tag{12}$$

and dividing the second of these equations by the first, we get

$$\tan \theta = \frac{dy}{dx} = \frac{w}{H} x,$$

---

† Here we are actually applying Newton's law to the *system* of particles making up OP; this can be justified by the argument given at the beginning of Section 14.3.

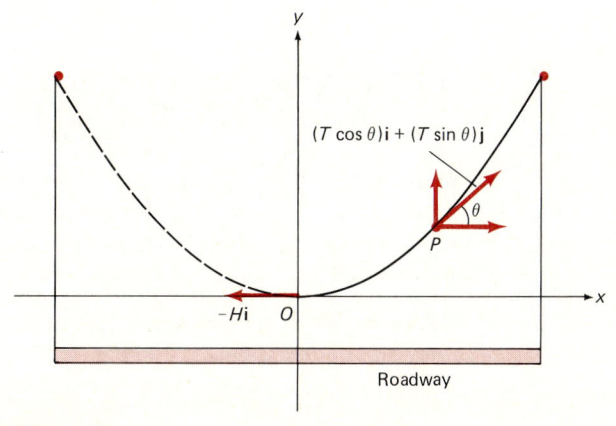

Figure 49

Section 11.5 Applications to Mechanics

where $y = y(x)$ is the equation of the cable in equilibrium. (Why is the same result obtained if P lies to the left of O?) Integration of this differential equation gives

$$y = \int \frac{w}{H} x\, dx + C = \frac{w}{2H} x^2 + C.$$

But $y(0) = 0$, since the origin has been chosen to lie on the curve $y = y(x)$. Hence the constant of integration C is zero and

$$y = \frac{w}{2H} x^2.$$

Thus the cable is shaped like a parabola. □

The Hanging Chain

In the absence of any roadway at all, we must take account of the weight of the cable itself. Then the curve $y = y(x)$ is no longer a parabola, as shown by the next example.

Example 7 A chain of weight w per unit length is suspended from two supports of equal height (imagine Figure 49 with the roadway removed). Find the shape of the hanging chain.

Solution As in Example 6, we let $P = (x, y)$ be a point of the chain and analyze the forces acting on the piece of chain OP, where the origin O is at the lowest point of the chain. Again OP is subject to a horizontal tension H pulling on OP at O and a tangential tension of magnitude T pulling on OP at P, but now the downward force is ws (not wx), where s is the length of the piece of chain OP. Thus instead of the equilibrium equations (12), we now have

$$T \cos \theta = H, \qquad T \sin \theta = ws, \tag{12'}$$

which differs from (12) only by the presence of s instead of x in the second equation. It follows from (12') that

$$\tan \theta = \frac{dy}{dx} = \frac{w}{H} s.$$

Differentiating this equation with respect to x, we get

$$\frac{d^2y}{dx^2} = \frac{w}{H} \frac{ds}{dx}. \tag{13}$$

But

$$\frac{ds}{dx} = \sqrt{1 + \left(\frac{dy}{dx}\right)^2}$$

(set $t = x$ in formula (10'), page 668). Therefore (13) becomes

$$\frac{d^2y}{dx^2} = \frac{w}{H} \sqrt{1 + \left(\frac{dy}{dx}\right)^2},$$

or

$$\frac{dp}{dx} = \frac{w}{H} \sqrt{1 + p^2} \tag{14}$$

in terms of the auxiliary variable $p = dy/dx$.

Separating variables in the differential equation (14) and integrating, we get

$$\int \frac{dp}{\sqrt{1+p^2}} = \int \frac{w}{H} dx + C_1,$$

where C_1 is a constant of integration. Therefore

$$\sinh^{-1} p = \frac{wx}{H} + C_1,$$

with the help of formula (2), page 357, or equivalently

$$p = \frac{dy}{dx} = \sinh\left(\frac{wx}{H} + C_1\right).$$

The slope $p = dy/dx$ of the curve $y = y(x)$ is zero at its lowest point (why?), so that $p|_{x=0} = 0$, which implies $C_1 = 0$. Hence

$$p = \frac{dy}{dx} = \sinh \frac{wx}{H},$$

and integrating again, we obtain

$$y = \int \sinh \frac{wx}{H} dx + C_2 = \frac{H}{w} \cosh \frac{wx}{H} + C_2, \tag{15}$$

where C_2 is another constant of integration. To determine C_2, we impose the condition $y(0) = 0$ (the origin lies on the curve), which implies

$$0 = \frac{H}{w} + C_2$$

or $C_2 = -H/w$. Therefore (15) becomes

$$y = \frac{H}{w}\left(\cosh \frac{wx}{H} - 1\right). \tag{16}$$

The graph of this equation, shown in Figure 50, is called a *catenary* (from the Latin word "catena" for "chain"). The horizontal line $y = -H/w$ is called the *directrix* of the catenary. It follows from (16) that if Y is the distance from the directrix to the point $P = (x, y)$ of the catenary, then

$$Y = \frac{H}{w} \cosh \frac{wx}{H}. \qquad \square \tag{16'}$$

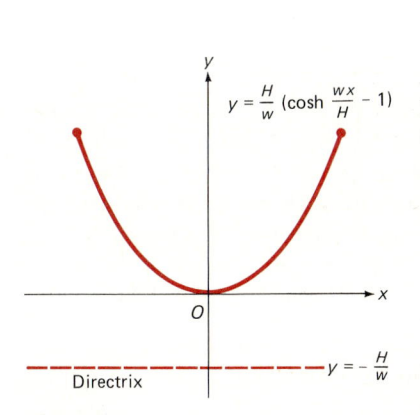

Figure 50

Problems

As in Example 1, a projectile is fired from a gun with angle of elevation α and initial speed ("muzzle velocity") v_0.

1. What is the total time of flight of the projectile from the gun to the target?

2. What is the maximum height of the projectile above ground, and at what time is it achieved?

3. Find the (horizontal) range of the gun, that is, the distance $|OQ|$ in Figure 46.

4. Show that the maximum range of the gun is v_0^2/g, achieved when the angle of elevation is 45°.

5. What muzzle velocity is required to give the gun a maximum range of 20 mi?

6. Show that any target whose distance from the gun is less than v_0^2/g (the maximum range) can be hit by projectiles fired at two different angles of elevation.

7. Find the vertex and directrix of the parabolic trajectory of the projectile.

8. Show that the height of the directrix is just the height to which the projectile would rise if fired vertically upward with muzzle velocity v_0 (and thus does not depend on the gun's angle of elevation).

9. A cannon has a maximum range of 10 mi. What is its range if fired at an elevation of 30°?

10. Find the two angles of elevation at which a howitzer of muzzle velocity 2000 ft/sec can hit a target 15 mi away.

11. A plane flying horizontally at an altitude of 200 ft with a constant speed of 300 mph drops a bomb on an enemy ammunition dump. The bomb is released when the direct line of sight from the plane to the dump makes a specified angle with the horizontal. What should this angle be to guarantee a direct hit?

12. A long fly ball rises to a height of 75 ft, and is caught at a distance of 400 ft from home plate. How long is the ball in the air? Find the angle α between the ball's trajectory and the horizontal at the instant the ball leaves the bat. What is the ball's initial speed v_0? (Neglect air resistance.)

Find the speed v in miles per second and the period T in minutes of a satellite in a circular orbit around the earth at the given altitude (height above the earth's surface).

13. 500 mi 14. 2500 mi 15. 5000 mi 16. 10,000 mi

17. Find the speed v in miles per second and the period T in minutes of a satellite in a circular orbit that just skims the surface of the earth. (Neglect air resistance.)

18. In Example 5 what angular speed is required to produce an acceleration of $5g$? Of $10g$?

19. In a device called the *conical pendulum*, an object (the "bob" of the pendulum) fastened to a cord of length L is whirled in a horizontal circle at a constant speed v in such a way that the cord sweeps out a right circular cone with a vertical axis (see Figure 51). Find v if the cord is 120 cm long and the angle α between the cord and the vertical is 30°. What is the tension T in the cord if the mass of the object is 50 grams? (Use $g = 980$ cm/sec^2.)

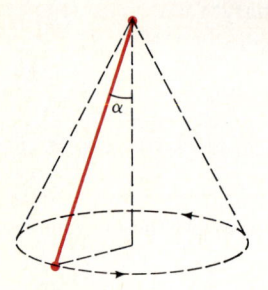

Figure 51

20. An object is fastened to a rope and whirled in a *vertical* circle of radius R. This circular motion (which is *not* uniform) can be established only if the speed of the object at the top of the circle is at least as large as a certain critical speed v_{cr}. Show that $v_{cr} = \sqrt{gR}$.

21. A water-filled bucket is swung at the end of a rope in a vertical circle of radius 80 cm. What angular speed is required to keep the water from spilling?

22. The earth's orbit is quite close to being a circle of radius 1.5×10^8 km. Estimate the mass of the sun, given that the universal gravitational constant G is about 6.67×10^{-20} km^3/kg sec.

23. Jupiter has 14 moons, of which four, the "Galilean satellites" Io, Europa, Ganymede and Callisto, were discovered by Galileo in 1610. Let T be the period and n the orbital radius of a Jovian moon, measured in units of Jupiter's radius. Then according to data obtained by Newton's contemporary, the astronomer royal John Flamsteed, the ratio n^3/T^2 has the same value of about 7.5×10^{-9} sec^{-2} for all four Galilean satellites, in dramatic confirmation of Kepler's third law. Use this fact and the value of G given in the preceding problem to show that the density ρ of Jupiter is about the same as that of water.

24. When a car rounds a curve, the centripetal acceleration is supplied by the frictional force exerted on the car's tires by the road. This force is proportional to the car's weight, with a constant of proportionality μ called the *coefficient of friction*. In the absence of friction, the car would "go off at a tangent," that is, it would skid. If $\mu = 0.5$, how fast can a car round a curve of radius of curvature 625 ft without skidding?

25. A car can be kept from skidding on a curved road by "banking" the curve, as shown in Figure 52. Then, even in the absence of appreciable friction between the road and the tires, the horizontal component of the normal force of reaction **F** exerted by the road on the car can supply the necessary centripetal acceleration, provided that the car's speed is not too high. The *rated speed* v_r of a banked curve is defined as the maximum speed at which a car can avoid skidding with no help at all from friction (imagine that the car runs onto a patch of wet ice). Show that $v_r = \sqrt{gR \tan \alpha}$, where R is the radius of curvature of the road and α is the banking angle.

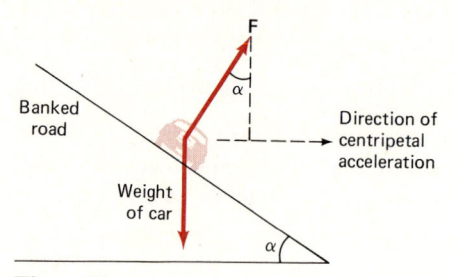

Figure 52

26. A stunt man rides a bicycle in a horizontal circle around the interior of a giant cylindrical barrel of radius R. This feat is possible only if there is friction between the barrel and the bicycle tires and the speed of the bicycle is at least

as large as a certain critical speed v_{cr}. Show that $v_{cr} = \sqrt{gR/\mu}$, where μ is the coefficient of friction. (In reality, the bicycle must also be tilted upward, but ignore this and regard the bicycle and its rider as a particle.)

27. The distance a between the points of support of a cable (or chain) is called its *span*, and the vertical distance d between the points of support and the lowest point of the cable is called its *sag* (see Figure 53). How are the span and sag related for the parabolic cable in Example 6? Show that the maximum tension in the cable is $\sqrt{H^2 + \frac{1}{4}w^2 a^2}$ at each point of support. What and where is the minimum tension?

28. The span of a two-cable suspension bridge is 200 ft, the sag of each cable is 50 ft, and the weight of the roadway is 400 tons. Assuming that the load is uniform, what is the tension in each cable at its midpoint? At each point of support?

29. Let s be the length of the catenary (16) between its lowest point and the point (x, y). Show that

$$s = \frac{H}{w} \sinh \frac{wx}{H}.$$

30. How are the span and sag related for the chain in Example 7? Show that the maximum tension in the chain is $H + wd$ at each point of support. What and where is the minimum tension?

31. Show that the shape of a tightly stretched catenary (large H/w) is close to being parabolic.

32. A heavy rope of length 40 m has a sag of 10 m. What is its span?

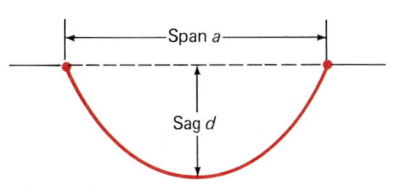

Figure 53

Key Terms and Topics
Scalars and vectors
Algebraic operations on vectors
Basis vectors, components of a vector
Orthogonal and orthonormal bases
Representation of vectors by ordered pairs
The dot product
Projection of one vector onto another
Work as a dot product
Vector functions
The limit of a vector function
Differentiation and integration of vector functions
Velocity and speed
The arc length function, arc length as a parameter
The unit tangent vector
The unit normal vector
Curvature, radius of curvature, circle of curvature
Acceleration, tangential and normal components of acceleration
Motion of a projectile
Uniform circular motion
The suspension bridge and the hanging cable

Supplementary Problems

Let $\mathbf{a} = \mathbf{i} + \mathbf{j}$, $\mathbf{b} = \mathbf{i} - \mathbf{j}$ and $\mathbf{c} = 5\mathbf{i} + 3\mathbf{j}$. Find

1. $-\mathbf{a} + 3\mathbf{b} - 2\mathbf{c}$

2. $\mathbf{a} - 2\mathbf{b} + \mathbf{c}$

3. $4\mathbf{a} + \mathbf{b} - \mathbf{c}$

4. $\dfrac{\mathbf{a} + \mathbf{b}}{|\mathbf{a} + \mathbf{b}|}$

5. $\dfrac{\mathbf{a} + \mathbf{b}}{|\mathbf{b} + \mathbf{c}|}$

6. $\dfrac{\mathbf{a} + \mathbf{b} + \mathbf{c}}{|\mathbf{a} - \mathbf{b} - \mathbf{c}|}$

7. What condition on $|\mathbf{a}|$ and $|\mathbf{b}|$ guarantees that $\mathbf{a} + \mathbf{b}$ bisects the angle between $\mathbf{a}$ and $\mathbf{b}$?

8. Let $OABCDE$ be a regular hexagon of side length 1. Express $\overrightarrow{OB}, \overrightarrow{BC}, \overrightarrow{EO}, \overrightarrow{OD}$ and $\overrightarrow{DA}$ as linear combinations of the unit vectors $\mathbf{u} = \overrightarrow{OA}$ and $\mathbf{v} = \overrightarrow{AB}$.

9. The vertices of a regular polygon are $P_1, P_2, \ldots, P_n$, and its center is O. Show that $\overrightarrow{OP_1} + \overrightarrow{OP_2} + \cdots + \overrightarrow{OP_n} = \mathbf{0}$.

10. A man can row a boat at a speed of 5 mph. He wants to cross a straight river 1 mile wide in which there flows a 3-mph current. In which direction should he row to cross the river as quickly as possible? How long will the trip take, and where will he land? In which direction should he row to reach the point directly opposite the river, and how long will this trip last?

11. Find $|\mathbf{a} - \mathbf{b}|$ if $|\mathbf{a}| = 13$, $|\mathbf{b}| = 19$, $|\mathbf{a} + \mathbf{b}| = 24$.

12. Show that $\mathbf{a} \cdot \mathbf{b} = \frac{1}{4}|\mathbf{a} + \mathbf{b}|^2 - \frac{1}{4}|\mathbf{a} - \mathbf{b}|^2$.

13. Find $|\mathbf{a} + \mathbf{b}|$ and $|\mathbf{a} - \mathbf{b}|$ if $|\mathbf{a}| = 5$, $|\mathbf{b}| = 8$, and the angle between $\mathbf{a}$ and $\mathbf{b}$ is $2\pi/3$.

14. Show that a quadrilateral whose diagonals bisect each other must be a parallelogram.

15. For what values of t are the vectors $\mathbf{a} = 2t\mathbf{i} + \mathbf{j}$ and $\mathbf{b} = \mathbf{i} + 2t\mathbf{j}$ parallel? Perpendicular?

16. Let $\mathbf{u}_1$ and $\mathbf{u}_2$ be two unit vectors making an angle of $\pi/3$ with each other. Find the lengths of the diagonals of the parallelogram spanned by the vectors $\mathbf{a} = 2\mathbf{u}_1 + \mathbf{u}_2$ and $\mathbf{b} = \mathbf{u}_1 - 2\mathbf{u}_2$.

17. What is the acute angle between the diagonals of a rectangle of length 5 and width 3?

18. What is the angle between the segments drawn from a vertex of a rectangle of length 6 and width 4 to the midpoints of the opposite sides?

19. A 2100-lb car is driven up a 30° ramp at a constant speed of 25 mph. What is the minimum power of the car's engine? Ignore all frictional effects.

20. Find the magnitude and direction of the resultant of the three forces shown in Figure 54, which are all applied at the same point O.

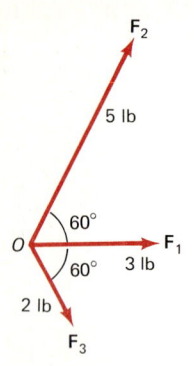

Figure 54

21. Calculate $\operatorname{proj}_{\overrightarrow{AB}} \overrightarrow{CD}$ and $\operatorname{proj}_{\overrightarrow{CD}} \overrightarrow{AB}$ for the points $A = (1, 2)$, $B = (2, 3)$, $C = (3, 4)$ and $D = (4, 1)$.

22. Use vectors to show that any angle inscribed in a semicircle is a right angle (see Figure 55).

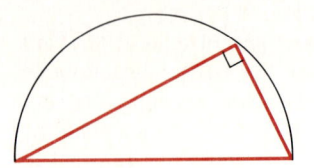

Figure 55

23. Let $\mathbf{r}$ be the position vector of a variable point in the plane, and let $\mathbf{a}$ be the position vector of a fixed point. Use the dot product to write a vector equation for the circle through the origin centered at the final point of $\mathbf{a}$.

24. Use vectors to show that the line joining the centers of two intersecting circles is perpendicular to the line joining the points of intersection.

Evaluate

25. $\lim\limits_{t \to 0} \left(\dfrac{1 - \cos t}{t} \mathbf{i} + \dfrac{t - \sin t}{t^3} \mathbf{j} \right)$

26. $\lim\limits_{t \to -\infty} \left(\dfrac{2}{\tanh t} \mathbf{i} + \dfrac{|t|}{t} \mathbf{j} \right)$

27. $\dfrac{d}{dt} [(\arcsin t)\mathbf{i} - (\arccos t)\mathbf{j}]$

28. $\dfrac{d}{dt} \left(\sqrt{t + 1}\, \mathbf{i} + \dfrac{2}{t + 1} \mathbf{j} \right)$

29. $\dfrac{d}{dt} [(t^2 e^t)\mathbf{i} + (\tanh^{-1} t)\mathbf{j}]$

30. $\displaystyle\int_0^1 \left(\dfrac{1}{1 + t^2} \mathbf{i} - \dfrac{1}{4 - t^2} \mathbf{j} \right) dt$

31. $\displaystyle\int_0^{\pi/3} [(\tan t)\mathbf{i} + (\sec t)\mathbf{j}]\, dt$

32. $\displaystyle\int [(t \sin t)\mathbf{i} + (te^{-t})\mathbf{j}]\, dt$

33. Does $\mathbf{r} \cdot (d\mathbf{r}/dt) \equiv 0$ imply that the vector function $\mathbf{r} = \mathbf{r}(t)$ is of constant magnitude? Explain your answer.

34. Find the solution of the vector differential equation $d\mathbf{r}/dt = c\mathbf{r}$ satisfying the initial condition $\mathbf{r}(0) = \mathbf{r}_0$. Here c is a constant scalar, and $\mathbf{r}_0$ is a constant vector.

Find the unit tangent and normal vectors $\mathbf{T}$ and $\mathbf{N}$ to the given curve at the point corresponding to the specified value of the parameter t.

35. $x = t^4 - 2t^2$, $y = t^3 + 1$, $t = -1$

36. $x = t + \cos t$, $y = \sin t$, $t = \pi/6$

37. $x = \ln(t + 1)$, $y = e^t$, $t = 0$

38. $x = \sec t$, $y = \tan t$, $t = \pi/4$

39. Find the curvature $\kappa = \kappa(x)$ of the curve $y = \ln(\sec x)$.
40. Show that the radius of curvature $R = R(\theta)$ of the lemniscate $r^2 = \cos 2\theta$ at any point other than the origin is inversely proportional to the radial coordinate r.

Write an equation for the osculating circle to the given curve at the point (1, 1). In each case graph the curve and the circle.

41. The parabola $y = x^2$ **42.** The hyperbola $xy = 1$

43. A ball thrown horizontally out of a window 64 ft high strikes the ground 100 ft from the side of the building. What is the initial speed of the ball?

44. A shell is fired from a howitzer with muzzle velocity 600 m/sec at an elevation of 60°. How high does the shell rise? How far from the howitzer does the shell land, and how long is it in the air? (Use $g = 9.8$ m/sec².)

45. A person traveling in an open car down a straight road fires a gun vertically upward. Where does the bullet land if the car's speed does not change? (Neglect air resistance.)

46. A hunter using a bow and arrow aims directly at an animal hanging from a branch of a tree. The arrow does not quite reach the branch, but instead hits the trunk of the tree somewhere below the branch. Show that the animal will get hit anyway if it makes the mistake of dropping from the branch at the instant the arrow is released from the bow.

47. At what two angles of elevation is a gun of maximum range d, located at the origin, able to hit a target located at the point $(\frac{1}{2}d, \frac{1}{4}d)$?

48. Show that the gun in the preceding problem can hit any target inside or on the parabola $x^2 + 2dy - d^2 = 0$, but no target outside this parabola, as illustrated in Figure 56.

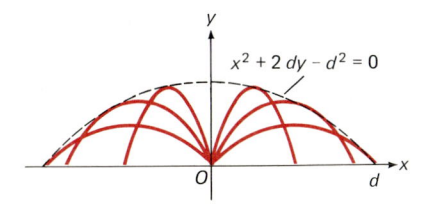

Figure 56

49. A pilot can momentarily withstand an acceleration of $8g$, but no more. What is the minimum radius of curvature with which the pilot can safely turn the plane upward at the end of a dive if its speed is 420 mph?

50. What is the rated speed (see Problem 25, page 690) of a circular roadway of radius 2250 ft which is banked at an angle of 30°?

51. A circular section of railroad track, of radius 1 mi, is banked to be safe at speeds up to 120 mph. The distance between the rails is 4 ft $8\frac{1}{2}$ in. (standard gauge). Find the "superelevation" h, that is, the height of the outer rail above the inner rail.

52. A cylindrical pail partly filled with water is rotated about its axis with constant angular speed ω. The water, initially at rest, eventually acquires the rotational motion of the pail. Show that the equilibrium shape of the surface of the water is a paraboloid of revolution. Find ω if the diameter of the pail is 1 ft and the water level at the center of the pail is 4 in. below the water level at the periphery.

Hint. Choose rectangular coordinates x and y as in Figure 57, and let $y = y(x)$ be the curve in which the xy-plane intersects the surface of the water. A water particle $P = (x, y)$ at the surface is acted on by a normal force of reaction $\mathbf{F}$ exerted on P by the rest of the fluid, and $\mathbf{F}$ must both balance the particle's weight and supply its centripetal acceleration. Show that this leads to an easily solved differential equation for $y = y(x)$.

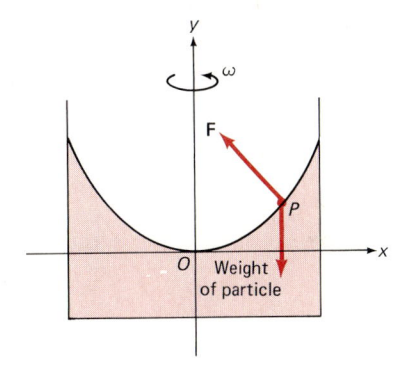

Figure 57

A heavy rope 105 ft long is suspended between two support towers a distance 100 ft apart.

53. What is the sag of the rope?

54. If the rope weighs 2 lb/ft, what is the minimum tension in the rope? The maximum tension?

Supplementary Problems   **693**

12 Vectors in Space and Solid Analytic Geometry

It is now time to make the move from the two-dimensional plane into three-dimensional space, and we do so in this and subsequent chapters. We begin by introducing rectangular coordinates in space and studying some simple surfaces. Then we extend the algebra of vectors from two to three dimensions. At first the considerations of the preceding chapter carry over with only minor modifications, but then in Section 12.3 a new concept, the *cross product* of two space vectors, enters the picture. Lines and planes in space are studied in Section 12.4, and in Section 12.5 we discuss three-dimensional vector functions, and present a modern version of Newton's derivation of Kepler's laws of planetary motion. We conclude our study of three-dimensional or "solid" analytic geometry by investigating the possibilities that can arise in graphing a second degree equation in three variables.

12.1 Rectangular Coordinates in Space

Rectangular coordinates in space are the natural extension of rectangular coordinates in the plane. Suppose we construct three mutually perpendicular lines Ox, Oy and Oz, known as the *coordinate axes*, each equipped with a *positive direction* (as indicated by the arrowheads in Figure 1), which intersect in a point O called the *origin*. The coordinates axes Ox, Oy and Oz are called the *x-axis*, the *y-axis* and the *z-axis*, respectively. These axes determine three mutually perpendicular *coordinate planes*, the *xy-plane*, containing the *x*- and *y*-axes, the *xz-plane* containing the *x*- and *z*-axes, and the *yz-plane* containing the *y*- and *z*-axes. The *yz*-plane is the plane of the paper in both parts of Figure 1, but in part (a) the *x*-axis points toward the reader, while in part (b) it points away from the reader. The coordinate system in part (a) is *right-handed*, meaning that if the fingers of the *right* hand are curled so that they go from the positive *x*-axis to the positive *y*-axis, the thumb then points along the positive *z*-axis, while the coordinate system in part (b) is *left-handed*,

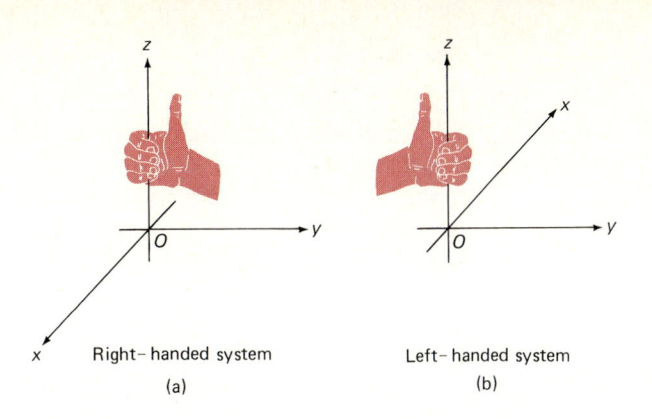

Right-handed system
(a)

Left-handed system
(b)

Figure 1

meaning that it has the same property with respect to the *left* hand. Notice that the two coordinate systems are reflections of each other in the yz-plane. From now on, we will deal exclusively with right-handed coordinate systems. In such a system a screwdriver whose blade is twisted through a 90° angle from the positive x-axis to the positive y-axis would cause an ordinary screw (with a right-handed thread) to advance along the positive z-axis.

Ordered Triples and Rectangular Coordinates

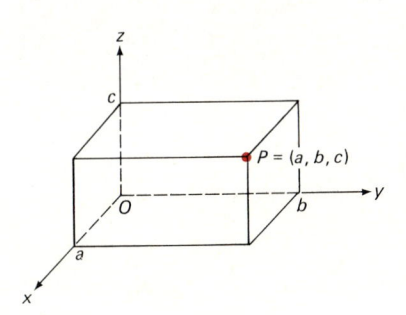

Now let (a, b, c) be an *ordered triple* of real numbers, where the notation indicates that a comes first, b second and c third. Choosing the same unit of length along all three coordinate axes, we plot a as a point of Ox (regarded as a number line), b as a point of Oy, and c as a point of Oz. We then construct the plane perpendicular to Ox at a, the plane perpendicular to Oy at b, and the plane perpendicular to Oz at c. As in Figure 2, these three planes intersect in a point P, which we regard as representing the ordered triple (a, b, c). The point P is said to have *rectangular* (or *cartesian*) *coordinates* a, b and c, or more exactly, x-coordinate a, y-coordinate b and z-coordinate c. By reversing this construction, that is, by drawing planes through P perpendicular to the coordinate axes (or alternatively, planes through P parallel to the coordinate planes), we can find the coordinates and hence the ordered triple corresponding to any given point P.

Figure 2

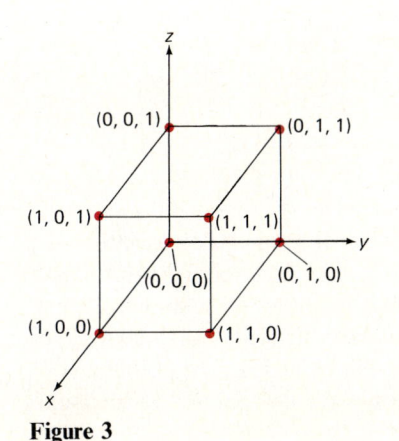

Thus there is a unique point in space corresponding to any given ordered triple, and conversely a unique ordered triple corresponding to any given point in space. Because of this one-to-one correspondence, we will usually make little or no distinction between ordered triples and the points representing them. In particular, $P = (a, b, c)$ means that P is the point with x-coordinate a, y-coordinate b and z-coordinate c. (Some authors write $P(a, b, c)$ for this point.) Notice that the origin O is the point $(0, 0, 0)$. Equality of two ordered triples (a, b, c) and (d, e, f) means of course that the two triples have the same first element, the same second element and the same third element, so that $a = d$, $b = e$ and $c = f$. Thus $(\sqrt{9}, 4, 0) = (3, 2^2, 0)$, but $(1, -1, 2) \neq (-1, 1, 2)$.

Example 1 Given four vertices $(0, 0, 0)$, $(1, 0, 0)$, $(0, 1, 0)$ and $(0, 0, 1)$ of a cube, find the other four.

Solution The answer is apparent from Figure 3. Notice that the eight vertices of the cube correspond to the eight distinct ordered triples with each

Figure 3

Section 12.1 Rectangular Coordinates in Space   **695**

element equal to either 0 or 1. All twelve edges of the cube have unit length, and all six faces have unit area. The cube itself has unit volume. ☐

A point (x, y, z) lies in the yz-plane if and only if $x = 0$, and on the z-axis if and only if $x = y = 0$. Alternatively, the yz-plane consists of all points of the form $(0, y, z)$, while the z-axis consists of all points of the form $(0, 0, z)$. As an exercise, give similar descriptions of the other coordinate planes and coordinate axes. The coordinate planes divide space into eight unbounded regions called *octants*. The *first* octant consists of all points (x, y, z) for which $x > 0, y > 0, z > 0$, but the other octants are not usually given names.

The Distance Formula

The distance between two points P_1 and P_2 in space is denoted by $|P_1P_2|$, just as in the case of points in the plane, and the following theorem is the three-dimensional generalization of Theorem 6, page 22.

Theorem 1 (*Distance between two points in space*). *The distance between two points $P_1 = (x_1, y_1, z_1)$ and $P_2 = (x_2, y_2, z_2)$ in space is given by the formula*

$$|P_1P_2| = \sqrt{(x_1 - x_2)^2 + (y_1 - y_2)^2 + (z_1 - z_2)^2}. \qquad (1)$$

Proof Through the points P_1 and P_2 we draw lines perpendicular to the xy-plane and planes perpendicular to the z-axis. Then P_1P_2 is the hypotenuse of the right triangle P_1QP_2 shown in Figure 4, where $Q = (x_2, y_2, z_1)$. It is clear that $|P_1Q| = |AB|$ and $|QP_2| = |CD|$, where A and B are the points (x_1, y_1) and (x_2, y_2) regarded as points in the xy-plane, while C and D are the points z_1 and z_2 regarded as points on the z-axis. Therefore, by the Pythagorean theorem and the distance formula in the plane and on the line, we have

$$|P_1P_2|^2 = |P_1Q|^2 + |QP_2|^2 = |AB|^2 + |CD|^2$$
$$= [(x_1 - x_2)^2 + (y_1 - y_2)^2] + (z_1 - z_2)^2,$$

which is equivalent to (1). ∎

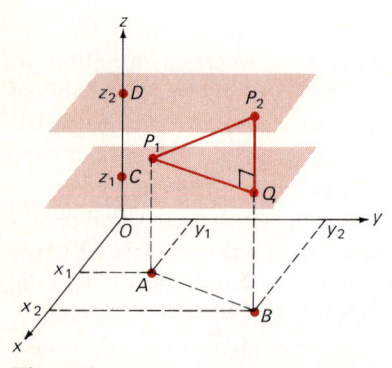

Figure 4

Figure 4 is drawn under the assumption that $x_1 < x_2$, $y_1 < y_2$ and $z_1 < z_2$, but it is easy to see that the same distance formula (1) is obtained if the direction of one (or more) of these inequalities is reversed.

Example 2 Find the distance between the points $P_1 = (3, 1, 9)$ and $P_2 = (-1, 4, -3)$.

Solution By formula (1),

$$|P_1P_2| = \sqrt{(3 + 1)^2 + (1 - 4)^2 + (9 + 3)^2}$$
$$= \sqrt{4^2 + 3^2 + 12^2} = \sqrt{169} = 13. \qquad \square$$

Graphs of Equations and Inequalities

By the *graph* of one or more equations or inequalities in three variables x, y and z we mean the set of points (x, y, z) in space whose coordinates satisfy the given equations or inequalities. Not all the variables need be present, and then the values of the missing variables are unrestricted. For instance, the graph of $x = a$ is the plane parallel to the yz-plane through the point $(a, 0, 0)$, while the graph of the equations $x = a$, $y = b$ is the line parallel to the z-axis through the point $(a, b, 0)$.

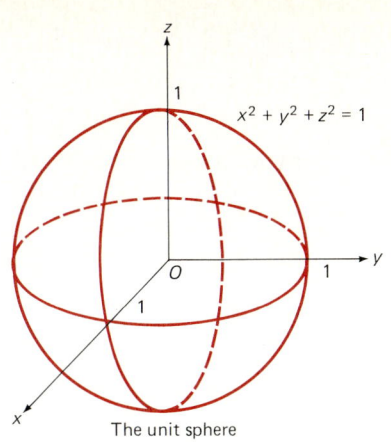

Example 3 Graph the equation

$$x^2 + y^2 + z^2 = 1. \qquad (2)$$

Solution Since $x^2 + y^2 + z^2$ is the square of the distance between the point (x, y, z) and the origin $O = (0, 0, 0)$, the point (x, y, z) belongs to the graph of (2) if and only if the distance between (x, y, z) and O is equal to 1. Thus the graph of (2) is the *unit sphere*, that is, the sphere of radius 1 with its center at O (see Figure 5). □

Example 4 Graph the inequality

$$x^2 + y^2 + z^2 < 1. \qquad (3)$$

Solution According to (3), the square of the distance between the point (x, y, z) and the origin O is less than 1, and hence the same is true of the distance itself. Thus the graph of (3) is the region inside the unit sphere (2); this region is called the *open unit ball*. The graph of the inequality

$$x^2 + y^2 + z^2 \le 1 \qquad (3')$$

is the *closed unit ball*, that is, the set consisting of the open unit ball together with its boundary (the unit sphere). □

Figure 5 The unit sphere

Spheres

Generalizing Example 3, we find that the coordinates of the point (x, y, z) satisfy the equation

$$(x - a)^2 + (y - b)^2 + (z - c)^2 = r^2 \qquad (r > 0) \qquad (4)$$

if and only if the square of the distance between (x, y, z) and the fixed point (a, b, c) equals r^2, or equivalently if and only if the distance between (x, y, z) and (a, b, c) equals r. Thus the coordinates of (x, y, z) satisfy (4) if and only if (x, y, z) lies on the sphere of radius r centered at (a, b, c). Notice that (4) reduces to equation (2) of the unit sphere if we choose $a = b = c = 0$ and $r = 1$.

Example 5 Graph the equation

$$x^2 + y^2 + z^2 + 4x - 6y + 2z - 11 = 0. \qquad (5)$$

Solution Completing the squares, we have

$$x^2 + 4x = (x + 2)^2 - 4, \quad y^2 - 6y = (y - 3)^2 - 9, \quad z^2 + 2z = (z + 1)^2 - 1,$$

and substituting these expressions into (5), we get

$$(x + 2)^2 + (y - 3)^2 + (z + 1)^2 = 25,$$

which you will recognize at once as an equation of the sphere of radius 5 centered as the point $(-2, 3, -1)$. Hence the original equation (5) also has this sphere as its graph. □

Cylinders

Let C be a plane curve, and let L be a line which is not parallel to (or in) the plane of C. Then the surface S made up of all lines through C parallel to L is called a *cylinder*. The curve C is called the *directrix* of the cylinder S, and the infinitely many lines parallel to L of which S is formed are called the *rulings* (or *generators*) of S.

For example, Figure 6 shows part of a cylinder S with a curve C in the xy-plane as its directrix and with the z-axis as the line L, so that the rulings are all parallel to the z-axis. Let $F(x, y)$ be any expression involving the two

Section 12.1 Rectangular Coordinates in Space

variables x and y, and suppose the curve C in the figure is the graph of the simultaneous equations

$$F(x, y) = 0, \qquad z = 0, \tag{6}$$

where the second equation $z = 0$ merely tells us that C is a curve in the xy-plane. Then the cylinder S, with directrix C and its rulings parallel to the z-axis, is the graph of the single equation

$$F(x, y) = 0, \tag{7}$$

obtained from (6) by omitting the second equation, that is, by allowing z to take unrestricted values and not just the value 0. In fact, given any point $P = (x, y, z)$, let $P_0 = (x, y, 0)$ be the point in which the line through P parallel to the z-axis intersects the xy-plane (see Figure 6). Then P lies on S if and only if P_0 lies on C, that is, if and only if the x- and y-coordinates of P_0, and hence of P, satisfy the first of the equations (6). But this means that S is the graph of equation (7), as asserted.

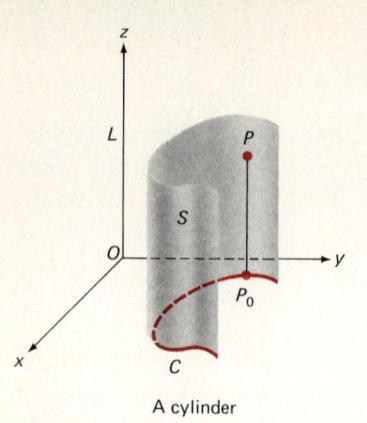

The fact that the coordinate z does not appear in equation (7) shows that the rulings of S are parallel to the z-axis. Similarly,

$$F(x, z) = 0$$

is the equation of a cylinder with rulings parallel to the y-axis, and

$$F(y, z) = 0$$

is the equation of a cylinder with rulings parallel to the x-axis. In each case the rulings are parallel to the axis labeled by the *missing coordinate*.

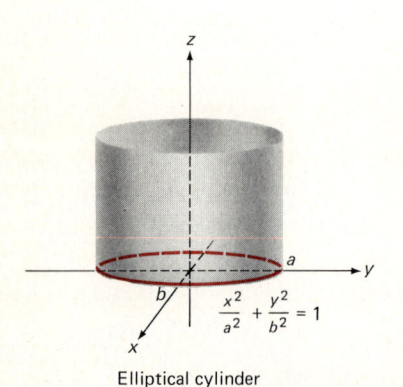

Example 6 The graph of the equation

$$\frac{x^2}{a^2} + \frac{y^2}{b^2} = 1 \tag{8}$$

is a cylinder with rulings parallel to the z-axis. The *trace* of this cylinder in the xy-plane, that is, its intersection with the xy-plane, is an ellipse (see Figure 7), namely the ellipse with the same equation (8) regarded as an equation in the coordinates of a variable point (x, y) of the xy-plane ("2-space") rather than as an equation in the coordinates of a variable point (x, y, z) of three-dimensional space ("3-space"). For this reason, the cylinder is called an *elliptic cylinder*. Every cross section of the cylinder by a plane parallel to the xy-plane is an ellipse congruent to its trace in the xy-plane. If $a = b = r$, the elliptic cylinder reduces to a right circular cylinder of radius r with the z-axis as its axis of symmetry.

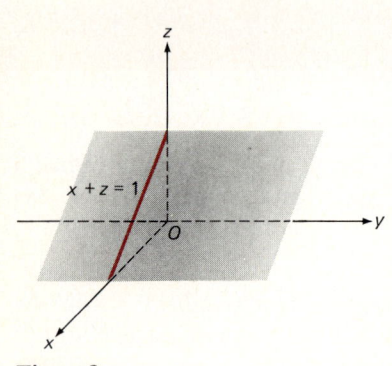

Example 7 In the xz-plane the graph of

$$x + z = 1 \tag{9}$$

is a line, but in 3-space the graph of the same linear equation (9) is the *plane* shown in Figure 8. Interpret this plane as a cylinder with its rulings parallel to the y-axis.

Figure 6 A cylinder

Figure 7 Elliptical cylinder

Figure 8

Surfaces of Revolution

The surface generated by revolving a plane curve about a line in its plane is called a *surface of revolution*. (We have already dealt with surfaces of revolution in Section 8.5, without attempting to find their equations in 3-space.)

698 Chapter 12 Vectors in Space and Solid Analytic Geometry

For example, let C be the curve in the yz-plane which is the graph of the equation

$$F(y, z) = 0 \quad (x = 0), \tag{10}$$

and suppose we revolve C around the z-axis, thereby generating the surface of revolution S shown in Figure 9. Then S is the graph of the equation

$$F(\sqrt{x^2 + y^2}, z) = 0, \tag{11}$$

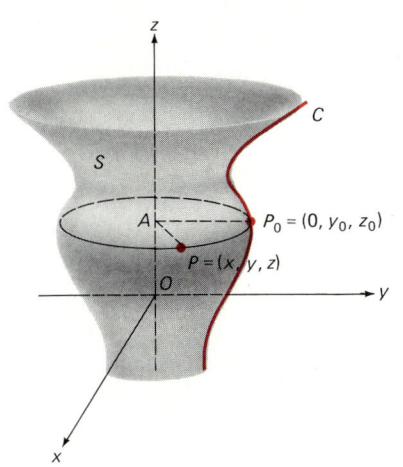

obtained from $F(y, z) = 0$ by replacing y by $\sqrt{x^2 + y^2}$ (here we temporarily assume that every point of C has a nonnegative y-coordinate). In fact, given any point $P = (x, y, z)$, let $P_0 = (0, y_0, z_0)$ be the point of intersection of the yz-plane with the circle through P parallel to the xy-plane with its center A on the z-axis. Then P lies on S if and only if P_0 lies on C, that is, if and only if $F(y_0, z_0) = 0$. It is apparent from the figure that $z_0 = z$ and

$$y_0 = |AP_0| = |AP| = \sqrt{x^2 + y^2}.$$

Therefore $F(y_0, z_0) = 0$ is equivalent to (11), and P lies on S if and only if (11) holds, that is, S is the graph of equation (11), as asserted.

At any point of the curve C with a negative y-coordinate, we have $y_0 = -\sqrt{x^2 + y^2}$ instead of $y_0 = \sqrt{x^2 + y^2}$. Thus, more generally, the surface S has an equation of the form

$$F(\pm\sqrt{x^2 + y^2}, z) = 0, \tag{11'}$$

where we choose the plus sign if $y_0 \geq 0$ and the minus sign if $y_0 < 0$.

If the same curve (10) in the yz-plane is revolved about the y-axis instead of the z-axis, a similar argument shows that the surface of revolution generated in this way has an equation of the form

$$F(y, \sqrt{x^2 + z^2}) = 0$$

(a minus sign may be needed before the radical in certain cases). The above formulas, together with the analogous results for curves in the xy- and xz-planes, are given in the following table. Make sure to verify the entries that have not already been discussed.

Curve	Axis of Revolution	Surface of Revolution
$F(y, z) = 0 \quad (x = 0)$	y-axis	$F(y, \sqrt{x^2 + z^2}) = 0$
	z-axis	$F(\sqrt{x^2 + y^2}, z) = 0$
$F(x, z) = 0 \quad (y = 0)$	x-axis	$F(x, \sqrt{y^2 + z^2}) = 0$
	z-axis	$F(\sqrt{x^2 + y^2}, z) = 0$
$F(x, y) = 0 \quad (z = 0)$	x-axis	$F(x, \sqrt{y^2 + z^2}) = 0$
	y-axis	$F(\sqrt{x^2 + z^2}, y) = 0$

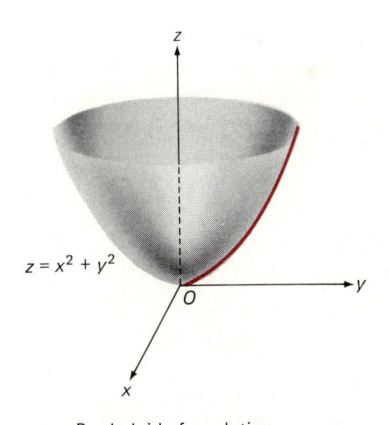

Example 8 Revolving the parabola

$$z = y^2 \quad (x = 0, y \geq 0)$$

in the first quadrant of the yz-plane about the z-axis (its axis of symmetry), we get the *paraboloid of revolution* shown in Figure 10, with equation

$$z = (\sqrt{x^2 + y^2})^2,$$

A surface of revolution
Figure 9

$z = x^2 + y^2$

Paraboloid of revolution
Figure 10

Section 12.1 Rectangular Coordinates in Space 699

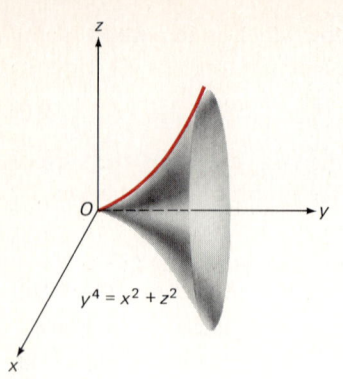

Figure 11

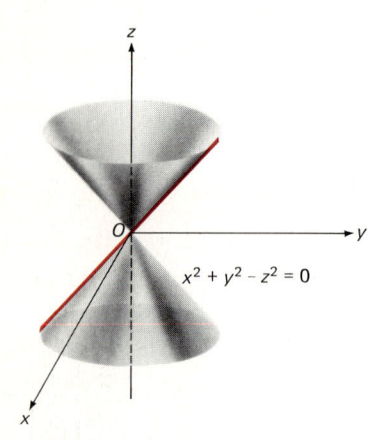

$y^4 = x^2 + z^2$

or equivalently

$$z = x^2 + y^2. \qquad (12)$$

The trace of this surface in the plane $z = c > 0$ is the circle of radius $\sqrt{c}$ with its center at the point $(0, 0, c)$, while its trace in the xz-plane is the parabola $z = x^2$, as we find by setting $y = 0$ in equation (12). Revolving the same parabola about the y-axis, we get the entirely different funnel-shaped surface shown in Figure 11, with equation

$$\sqrt{x^2 + z^2} = y^2,$$

or equivalently

$$y^4 = x^2 + z^2,$$

where $y \geq 0$.

Example 9 Revolving the line

$$y = z \qquad (x = 0)$$

in the yz-plane about the z-axis, we get the surface of revolution S shown in Figure 12, with equation

$$\pm\sqrt{x^2 + y^2} = z$$

(the minus sign is necessary when $y < 0$), or equivalently

$$x^2 + y^2 - z^2 = 0.$$

You will recognize S as a right circular cone of two nappes with its vertex at the origin, every generator of which makes an angle of 45° with the axis of the cone, namely the z-axis. The trace of S in the plane $z = c > 0$ is the circle of radius c (not $\sqrt{c}$ as in Example 8) with its center at $(0, 0, c)$. In keeping with the considerations of Section 10.5, show that the trace of S in every plane $x = c$ or $y = c$ is a hyperbola if $c \neq 0$ and a pair of intersecting lines if $c = 0$. What is the trace of S in the xy-plane?

$x^2 + y^2 - z^2 = 0$

Right circular cone of two nappes
Figure 12

Problems

1. Of the three rectangular coordinate systems shown in Figure 13, only two are right-handed. Which ones?

2. One of the coordinate systems in Figure 13 is left-handed. How can it be made right-handed?

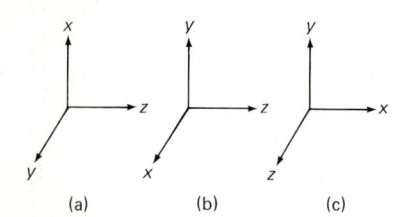

(a)   (b)   (c)
Figure 13

3. A rectangular parallelepiped (box) with faces parallel to the coordinate planes has the origin O and the point $(2, -3, 4)$ as two of its vertices. Graph the parallelepiped and find the other six vertices.

4. A rectangular parallelepiped with edges parallel to the coordinate axes has the points $(2, 1, -1)$ and $(-1, 3, 1)$ as two of its vertices. Graph the parallelepiped and find the other six vertices.

5. In what point does the line through $(6, -7, 9)$ parallel to the y-axis intersect the plane through $(8, -4, 5)$ parallel to the xz-plane?

Find the distance between the point (a, b, c) and

6. The x-axis   **7.** The y-axis   **8.** The z-axis

Find the distance between the given pair of points.

9. $(0, 0, 0), (12, -15, 16)$ **10.** $(0, 1, 0), (-4, 3, 4)$
11. $(4, -5, 3), (6, -2, -3)$ **12.** $(1, 1, 5), (10, 3, -1)$

13. $(8, 11, 9), (4, 12, 2)$
14. $(3, \pi, -14), (-9, \pi, -9)$
15. $(5, 0, -3), (0, 4, 0)$
16. $(1, -1, 1), (-1, 1, -1)$

17. Find the point on the y-axis equidistant from the points $(1, -3, 7)$ and $(5, 7, -5)$.
18. Show that the points $A = (3, -1, 6)$, $B = (-1, 7, -2)$ and $C = (1, -3, 2)$ are the vertices of a right triangle. Which side is the hypotenuse?
19. Which of the points $(-3, 0, 2)$, $(2, 1, 3)$, $(-1, 3, 1)$ and $(2, 2, -2)$ is closest to the origin?
20. Let M be the midpoint of the segment $P_1 P_2$ joining the points $P_1 = (x_1, y_1, z_1)$ and $P_2 = (x_2, y_2, z_2)$. Show that
$$M = \left(\frac{x_1 + x_2}{2}, \frac{y_1 + y_2}{2}, \frac{z_1 + z_2}{2}\right).$$

Hint. See Example 4, page 23.

Find the midpoint of the segment joining the given pair of points.

21. $(1, -7, 0), (-9, 11, 12)$
22. $(-4, 9, 2), (6, 3, 8)$
23. $(-1, 3, -5), (2, -4, 6)$
24. $(-5, 10, -20), (5, -10, 20)$

Two distinct points P and Q in space are said to be *symmetric about a point M* if M is the midpoint of the segment PQ, and *symmetric about a line L or a plane Π* if L or Π goes through the midpoint M of PQ and is perpendicular to PQ. Let P be the point $(5, -3, 2)$. Find the point symmetric to P about

25. The origin
26. The point $(3, 1, -2)$
27. The x-axis
28. The y-axis
29. The z-axis
30. The xy-plane
31. The xz-plane
32. The yz-plane

Find an equation of the sphere with the given radius and center.

33. $5, (0, 0, 0)$
34. $8, (-1, 1, -1)$
35. $\sqrt{11}, (3, -2, 4)$
36. $15, (10, -5, 10)$

Describe the graph of the given equation.

37. $x^2 + y^2 + z^2 + 6x - 4y + 10z - 83 = 0$
38. $x^2 + y^2 + z^2 - 2x + 2y + 2z - 6 = 0$
39. $x^2 + y^2 + z^2 + 10x + 12y - 8z + 77 = 0$
40. $2x^2 + 2y^2 + 2z^2 - 3y - 4z = 0$
41. $x^2 + y^2 + z^2 - 10y - 16z + 90 = 0$
42. $x^2 + y^2 + z^2 - 20x + 14y - 22z + 270 = 0$

43. Find an equation of the sphere which goes through the point $(4, -1, -1)$ and is tangent to all three coordinate planes.
44. Consider the second degree equation
$$x^2 + y^2 + z^2 + Ax + By + Cz + D = 0 \quad \text{(i)}$$
in which the coefficients of x^2, y^2 and z^2 are all equal to 1, and let
$$E = \frac{A^2}{4} + \frac{B^2}{4} + \frac{C^2}{4} - D.$$
Show that the graph of (i) is the sphere of radius $\sqrt{E}$ with its center at $(-A/2, -B/2, -C/2)$ if $E > 0$, the single point $(-A/2, -B/2, -C/2)$ if $E = 0$, and the empty set if $E < 0$.

Describe the graph in 3-space of the given equation or inequality.

45. $xy = 0$
46. $xyz = 0$
47. $x^2 + y^2 + z^2 > 1$
48. $1 < y^2 + z^2 < 4$

Describe and sketch the graph in 3-space of the given equation.

49. $z^2 - y^2 = 0$
50. $x^2 + z^2 = 2z$
51. $xy = -1$
52. $y^2 + 4z^2 = 4$
53. $y^2 - 8x = 0$
54. $x^2 - y^2 + z^2 = 0$
55. $2x^2 + 2y^2 + z^2 = 1$
56. $y^2 + z^2 - x = 0$

Find an equation of the surface of revolution generated by revolving the given curve about the specified axis.

57. $x = 4z^2$ $(y = 0)$, z-axis
58. Same curve, x-axis
59. $x^{2/3} + y^{2/3} = 1$ $(z = 0)$, x-axis
60. Same curve, y-axis
61. $y = \sqrt{z}$ $(x = 0)$, y-axis
62. Same curve, z-axis

12.2 From Vectors in the Plane to Vectors in Space

On pages 647–650 we defined vectors in the plane and various operations on them, namely addition, subtraction and multiplication by scalars. These definitions, and the rules governing the operations, carry over without change to the case of vectors in space, for the simple reason that they are purely geometric and "coordinate-free." From now on, by the word "vector" without further qualification, we will always mean a vector in ordinary three-dimensional space, rather than in some plane.

Components of a Vector

However, when coordinates are introduced, we can of course expect certain differences between the two-dimensional and three-dimensional cases. Specifically, suppose we introduce a system of rectangular coordinates x, y and z in space, with origin O. Let $\mathbf{i}$ be a unit vector along the positive x-axis, $\mathbf{j}$ a unit vector along the positive y-axis, and $\mathbf{k}$ a unit vector along the positive z-axis, as shown in Figure 14(a). Then any vector $\mathbf{a}$ (in space) has a unique representation of the form

$$\mathbf{a} = \alpha_1 \mathbf{i} + \alpha_2 \mathbf{j} + \alpha_3 \mathbf{k}. \tag{1}$$

The scalars α_1, α_2 and α_3, called the *components* of $\mathbf{a}$, are determined as follows. Shift the vector $\mathbf{a}$, that is, move it parallel to itself without rotation, until its initial point coincides with the origin O. Then the final point of $\mathbf{a}$ will be a point $A = (\alpha_1, \alpha_2, \alpha_3)$ of space, and of course $\mathbf{a} = \overrightarrow{OA}$, since all shifted versions of a vector are equal to each other. But, as is apparent from the construction in Figure 14(b), the coordinates α_1, α_2 and α_3 of the point A are precisely the scalars called for in the representation or "expansion" (1). Moreover, the magnitude of $\mathbf{a}$ is just the distance from O to A, that is,

$$|\mathbf{a}| = \sqrt{\alpha_1^2 + \alpha_2^2 + \alpha_3^2}. \tag{2}$$

By a *basis* in space we mean any three fixed vectors $\mathbf{e}_1$, $\mathbf{e}_2$ and $\mathbf{e}_3$, called *basis vectors*, such that an arbitrary vector $\mathbf{a}$ has a unique representation of the form

$$\mathbf{a} = \alpha_1 \mathbf{e}_1 + \alpha_2 \mathbf{e}_2 + \alpha_3 \mathbf{e}_3, \tag{1'}$$

called the *expansion* of $\mathbf{a}$ with respect to $\mathbf{e}_1$, $\mathbf{e}_2$ and $\mathbf{e}_3$. The scalars α_1, α_2 and α_3 are then called the *components* of $\mathbf{a}$ (with respect to $\mathbf{e}_1$, $\mathbf{e}_2$ and $\mathbf{e}_3$). It can be shown that three nonzero vectors $\mathbf{e}_1$, $\mathbf{e}_2$ and $\mathbf{e}_3$ form a basis in space if and only if they are *noncoplanar*, that is, if and only if there is no plane containing (or parallel to) all three vectors. If the vectors $\mathbf{e}_1$, $\mathbf{e}_2$ and $\mathbf{e}_3$ of a basis are mutually perpendicular, the basis is said to be *orthogonal*, and if they are *unit* vectors as well as being perpendicular, the basis is said to be *orthonormal*. Thus the unit vectors $\mathbf{i}$, $\mathbf{j}$ and $\mathbf{k}$ along the coordinate axes form an orthonormal basis. Notice that formula (2) is valid only for an orthonormal basis (why?).

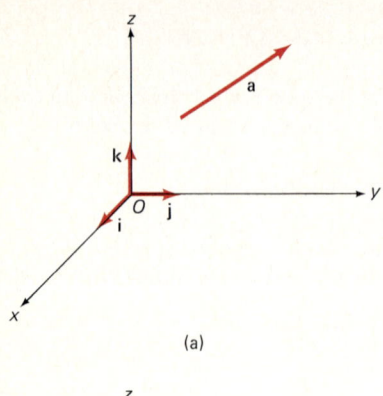

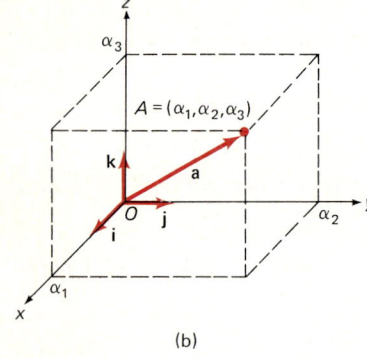

Figure 14

Vectors as Ordered Triples

Guided by the above considerations, we will henceforth employ ordered triples to represent both points and vectors in space. Thus $(\alpha_1, \alpha_2, \alpha_3)$ can mean either the point with rectangular *coordinates* α_1, α_2 and α_3 or the vector $\alpha_1 \mathbf{i} + \alpha_2 \mathbf{j} + \alpha_3 \mathbf{k}$ with *components* α_1, α_2 and α_3 (with respect to the underlying orthonormal basis $\mathbf{i}$, $\mathbf{j}$ and $\mathbf{k}$). Also, since $\mathbf{i} = 1\mathbf{i} + 0\mathbf{j} + 0\mathbf{k}$, $\mathbf{j} = 0\mathbf{i} + 1\mathbf{j} + 0\mathbf{k}$ and $\mathbf{k} = 0\mathbf{i} + 0\mathbf{j} + 1\mathbf{k}$, the ordered triples representing the basis vectors themselves are

$$\mathbf{i} = (1, 0, 0), \quad \mathbf{j} = (0, 1, 0), \quad \mathbf{k} = (0, 0, 1).$$

We can now interpret algebraic operations from the standpoint of ordered triples. Let $\mathbf{a} = (\alpha_1, \alpha_2, \alpha_3)$ and $\mathbf{b} = (\beta_1, \beta_2, \beta_3)$ be arbitrary vectors in space. Then

$$\mathbf{a} + \mathbf{b} = (\alpha_1 \mathbf{i} + \alpha_2 \mathbf{j} + \alpha_3 \mathbf{k}) + (\beta_1 \mathbf{i} + \beta_2 \mathbf{j} + \beta_3 \mathbf{k}),$$

and therefore

$$\mathbf{a} + \mathbf{b} = (\alpha_1 + \beta_1)\mathbf{i} + (\alpha_2 + \beta_2)\mathbf{j} + (\alpha_3 + \beta_3)\mathbf{k} \tag{3}$$

or
$$(\alpha_1, \alpha_2, \alpha_3) + (\beta_1, \beta_2, \beta_3) = (\alpha_1 + \beta_1, \alpha_2 + \beta_2, \alpha_3 + \beta_3). \tag{3'}$$

If p is any scalar, then $p\mathbf{a} = p(\alpha_1\mathbf{i} + \alpha_2\mathbf{j} + \alpha_3\mathbf{k})$, and therefore
$$p\mathbf{a} = p\alpha_1\mathbf{i} + p\alpha_2\mathbf{j} + p\alpha_3\mathbf{k} \tag{4}$$
or
$$p(\alpha_1, \alpha_2, \alpha_3) = (p\alpha_1, p\alpha_2, p\alpha_3). \tag{4'}$$

It follows from (3') that
$$(\alpha_1, \alpha_2, \alpha_3) + (0, 0, 0) = (\alpha_1 + 0, \alpha_2 + 0, \alpha_3 + 0) = (\alpha_1, \alpha_2, \alpha_3)$$
for every vector $\mathbf{a} = (\alpha_1, \alpha_2, \alpha_3)$, so that $(0, 0, 0)$ is the ordered triple representing the zero vector $\mathbf{0}$. To get the ordered triple representing $-\mathbf{a}$, we set $p = -1$ in formula (4), obtaining $-\mathbf{a} = -\alpha_1\mathbf{i} - \alpha_2\mathbf{j} - \alpha_3\mathbf{k}$, or equivalently
$$-\mathbf{a} = (-\alpha_1, -\alpha_2, -\alpha_3).$$

Moreover, $\mathbf{a} - \mathbf{b} = \mathbf{a} + (-\mathbf{b})$, so that
$$\mathbf{a} - \mathbf{b} = (\alpha_1, \alpha_2, \alpha_3) + (-\beta_1, -\beta_2, -\beta_3) = (\alpha_1 - \beta_1, \alpha_2 - \beta_2, \alpha_3 - \beta_3) \tag{5}$$
or
$$(\alpha_1, \alpha_2, \alpha_3) - (\beta_1, \beta_2, \beta_3) = (\alpha_1 - \beta_1, \alpha_2 - \beta_2, \alpha_3 - \beta_3). \tag{5'}$$

Also two vectors $\mathbf{a} = (\alpha_1, \alpha_2, \alpha_3)$ and $\mathbf{b} = (\beta_1, \beta_2, \beta_3)$ in space are equal if and only if $\alpha_1 = \beta_1, \alpha_2 = \beta_2$ and $\alpha_3 = \beta_3$, by the exact analogue of the argument given on pages 651–652 for the case of vectors in the plane.

Example 1 Let $\mathbf{a} = (2, -1, 3)$, $\mathbf{b} = (-1, 4, -2)$ and $\mathbf{c} = (1, 8, 7)$. Calculate the vector $2\mathbf{a} + \mathbf{b} - \mathbf{c}$, and then find its magnitude.

Solution Making free use of the rules (3)–(5), we have
$$2\mathbf{a} + \mathbf{b} - \mathbf{c} = 2(2, -1, 3) + (-1, 4, -2) - (1, 8, 7)$$
$$= (4 - 1 - 1, -2 + 4 - 8, 6 - 2 - 7) = (2, -6, -3),$$
or equivalently
$$2\mathbf{a} + \mathbf{b} - \mathbf{c} = 2\mathbf{i} - 6\mathbf{j} - 3\mathbf{k}.$$

By formula (2), the magnitude of this vector is
$$|2\mathbf{i} - 6\mathbf{j} - 3\mathbf{k}| = \sqrt{2^2 + (-6)^2 + (-3)^2} = \sqrt{49} = 7. \quad \square$$

Example 2 Find the unit vector $\mathbf{u}$ with the same direction as the vector $15\mathbf{i} - 12\mathbf{j} + 16\mathbf{k}$.

Solution The magnitude of $15\mathbf{i} - 12\mathbf{j} + 16\mathbf{k}$ is
$$|15\mathbf{i} - 12\mathbf{j} + 16\mathbf{k}| = \sqrt{15^2 + (-12)^2 + 16^2} = \sqrt{625} = 25,$$
and hence
$$\mathbf{u} = \frac{15\mathbf{i} - 12\mathbf{j} + 16\mathbf{k}}{|15\mathbf{i} - 12\mathbf{j} + 16\mathbf{k}|} = \frac{3}{5}\mathbf{i} - \frac{12}{25}\mathbf{j} + \frac{16}{25}\mathbf{k}. \quad \square$$

The Dot Product

As in the case of vectors in the plane, the *dot product* of two vectors $\mathbf{a}$ and $\mathbf{b}$ in space is defined by the formula
$$\mathbf{a} \cdot \mathbf{b} = |\mathbf{a}| |\mathbf{b}| \cos \theta, \tag{6}$$

where θ is the angle between **a** and **b** ($0 \le \theta \le \pi$). It follows from (6) that $\mathbf{a} \cdot \mathbf{b} = \mathbf{b} \cdot \mathbf{a}$ and $\mathbf{a} \cdot \mathbf{a} = |\mathbf{a}|^2$, and also that $\mathbf{a} \cdot \mathbf{b} = 0$ if and only if **a** and **b** are perpendicular (the zero vector is regarded as perpendicular to every vector). If $\mathbf{a} = (\alpha_1, \alpha_2, \alpha_3)$ and $\mathbf{b} = (\beta_1, \beta_2, \beta_3)$, then

$$\mathbf{a} \cdot \mathbf{b} = \alpha_1 \beta_1 + \alpha_2 \beta_2 + \alpha_3 \beta_3. \tag{7}$$

The proof is virtually the same as that of Theorem 1, page 656, and is left as an exercise. Using formula (7), you can easily verify that if p and q are scalars, then

$$(p\mathbf{a}) \cdot (q\mathbf{b}) = pq(\mathbf{a} \cdot \mathbf{b})$$

for arbitrary vectors **a** and **b**, while

$$\mathbf{a} \cdot (\mathbf{b} + \mathbf{c}) = \mathbf{a} \cdot \mathbf{b} + \mathbf{a} \cdot \mathbf{c},$$
$$(\mathbf{a} + \mathbf{b}) \cdot \mathbf{c} = \mathbf{a} \cdot \mathbf{c} + \mathbf{b} \cdot \mathbf{c}$$

for arbitrary vectors **a**, **b** and **c**, just as in the corollary on page 656.

Example 3 By formula (7), the dot product of the vectors $\mathbf{a} = (4, -3, 6)$ and $\mathbf{b} = (2, 5, -1)$ is $\mathbf{a} \cdot \mathbf{b} = 4(2) - 3(5) + 6(-1) = 8 - 15 - 6 = -13$.

The unit vectors $\mathbf{i} = (1, 0, 0)$, $\mathbf{j} = (0, 1, 0)$ and $\mathbf{k} = (0, 0, 1)$ satisfy the important formulas

$$\mathbf{i} \cdot \mathbf{i} = 1, \quad \mathbf{j} \cdot \mathbf{j} = 1, \quad \mathbf{k} \cdot \mathbf{k} = 1,$$
$$\mathbf{i} \cdot \mathbf{j} = \mathbf{j} \cdot \mathbf{i} = 0, \quad \mathbf{j} \cdot \mathbf{k} = \mathbf{k} \cdot \mathbf{j} = 0, \quad \mathbf{i} \cdot \mathbf{k} = \mathbf{k} \cdot \mathbf{i} = 0. \tag{8}$$

For example, $\mathbf{i} \cdot \mathbf{i} = 1(1) + 0(0) + 0(0) = 1$, $\mathbf{i} \cdot \mathbf{j} = 1(0) + 0(1) + 0(0) = 0$, and so on. The formulas (8) are also an immediate consequence of the definition (6) and the fact that **i**, **j** and **k** form an orthonormal basis. Notice that if $\mathbf{a} = \alpha_1 \mathbf{i} + \alpha_2 \mathbf{j} + \alpha_3 \mathbf{k}$, then

$$\mathbf{a} \cdot \mathbf{i} = (\alpha_1 \mathbf{i} + \alpha_2 \mathbf{j} + \alpha_3 \mathbf{k}) \cdot \mathbf{i} = \alpha_1 \mathbf{i} \cdot \mathbf{i} + \alpha_2 \mathbf{j} \cdot \mathbf{i} + \alpha_3 \mathbf{k} \cdot \mathbf{i} = \alpha_1,$$

with the help of (8), and similarly $\mathbf{a} \cdot \mathbf{j} = \alpha_2$ and $\mathbf{a} \cdot \mathbf{k} = \alpha_3$.

Let θ be the angle between two nonzero vectors **a** and **b**. Then it follows from (6) that

$$\cos \theta = \frac{\mathbf{a} \cdot \mathbf{b}}{|\mathbf{a}| |\mathbf{b}|}. \tag{9}$$

Example 4 Find the angle between the vectors $\mathbf{a} = \mathbf{i} - 2\mathbf{j} + 4\mathbf{k}$ and $\mathbf{b} = -4\mathbf{i} + \mathbf{j} - 2\mathbf{k}$.

Solution Here we have

$$\mathbf{a} \cdot \mathbf{b} = 1(-4) - 2(1) + 4(-2) = -14,$$
$$|\mathbf{a}| = \sqrt{1^2 + (-2)^2 + 4^2} = \sqrt{21}, \quad |\mathbf{b}| = \sqrt{(-4)^2 + 1^2 + (-2)^2} = \sqrt{21}.$$

Therefore

$$\cos \theta = \frac{\mathbf{a} \cdot \mathbf{b}}{|\mathbf{a}| |\mathbf{b}|} = \frac{-14}{\sqrt{21}\sqrt{21}} = -\frac{2}{3},$$

which implies

$$\theta = \arccos\left(-\frac{2}{3}\right) \approx 131.8°. \quad \square$$

Let **a** and **b** be two nonzero vectors in space. Then the *component of* **a** *along* **b**, denoted by comp$_b$ **a**, and the *projection of* **a** *onto* **b**, denoted by proj$_b$ **a**, are defined by the same formulas

$$\text{comp}_b\,\mathbf{a} = \frac{\mathbf{a}\cdot\mathbf{b}}{|\mathbf{b}|}, \qquad \text{proj}_b\,\mathbf{a} = \frac{(\mathbf{a}\cdot\mathbf{b})\mathbf{b}}{|\mathbf{b}|^2}$$

as for vectors in the plane (see page 658). Notice that comp$_b$ **a** is a scalar, while proj$_b$ **a** is a vector.

Example 5 Find comp$_b$ **a** and proj$_b$ **a** if $\mathbf{a} = 2\mathbf{i} - 3\mathbf{j} + \mathbf{k}$ and $\mathbf{b} = \mathbf{i} + 2\mathbf{j} - 2\mathbf{k}$. Represent **a** as the sum of a vector parallel to **b** and a vector orthogonal to **b** ("orthogonal" is a synonym for "perpendicular").

Solution Since

$$\mathbf{a}\cdot\mathbf{b} = 2(1) - 3(2) + 1(-2) = -6, \qquad |\mathbf{b}| = \sqrt{1^2 + 2^2 + (-2)^2} = \sqrt{9} = 3,$$

we have

$$\text{comp}_b\,\mathbf{a} = \frac{\mathbf{a}\cdot\mathbf{b}}{|\mathbf{b}|} = -\frac{6}{3} = -2,$$

$$\text{proj}_b\,\mathbf{a} = \frac{(\mathbf{a}\cdot\mathbf{b})\mathbf{b}}{|\mathbf{b}|^2} = \frac{-6(\mathbf{i} + 2\mathbf{j} - 2\mathbf{k})}{9} = -\frac{2}{3}\mathbf{i} - \frac{4}{3}\mathbf{j} + \frac{4}{3}\mathbf{k}.$$

The vector proj$_b$ **a** is parallel to **b**, and the vector

$$\mathbf{a} - \text{proj}_b\,\mathbf{a} = (2\mathbf{i} - 3\mathbf{j} + \mathbf{k}) - \left(-\frac{2}{3}\mathbf{i} - \frac{4}{3}\mathbf{j} + \frac{4}{3}\mathbf{k}\right) = \frac{8}{3}\mathbf{i} - \frac{5}{3}\mathbf{j} - \frac{1}{3}\mathbf{k}$$

is orthogonal to **b** (see Figure 22, page 659). Thus the representation of **a** as the sum of a vector parallel to **b** and a vector orthogonal to **b** is

$$\mathbf{a} = \left(-\frac{2}{3}\mathbf{i} - \frac{4}{3}\mathbf{j} + \frac{4}{3}\mathbf{k}\right) + \left(\frac{8}{3}\mathbf{i} - \frac{5}{3}\mathbf{j} - \frac{1}{3}\mathbf{k}\right). \qquad \square$$

Direction Angles and Direction Cosines

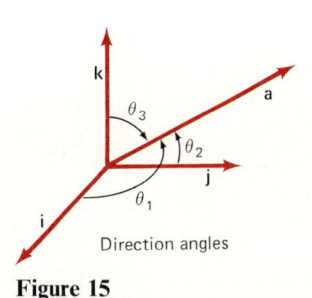

Given a nonzero vector $\mathbf{a} = \alpha_1\mathbf{i} + \alpha_2\mathbf{j} + \alpha_3\mathbf{k}$, let θ_1, θ_2 and θ_3 be the angles between **a** and the unit vectors **i**, **j** and **k** (see Figure 15). Then it follows from (9) that

$$\cos\theta_1 = \frac{\mathbf{a}\cdot\mathbf{i}}{|\mathbf{a}||\mathbf{i}|} = \frac{\alpha_1}{|\mathbf{a}|}, \qquad \cos\theta_2 = \frac{\mathbf{a}\cdot\mathbf{j}}{|\mathbf{a}||\mathbf{j}|} = \frac{\alpha_2}{|\mathbf{a}|}, \qquad \cos\theta_3 = \frac{\mathbf{a}\cdot\mathbf{k}}{|\mathbf{a}||\mathbf{k}|} = \frac{\alpha_3}{|\mathbf{a}|}, \tag{10}$$

or equivalently

$$\alpha_1 = |\mathbf{a}|\cos\theta_1, \qquad \alpha_2 = |\mathbf{a}|\cos\theta_2, \qquad \alpha_3 = |\mathbf{a}|\cos\theta_3. \tag{10'}$$

Direction angles

Figure 15

The angles θ_1, θ_2 and θ_3 are called the *direction angles* of the vector **a** (or of any directed line L with the same direction as **a**), while the numbers $\cos\theta_1$, $\cos\theta_2$ and $\cos\theta_3$ are called the *direction cosines* of **a** (or L). The direction cosines completely specify the direction of **a**, but say nothing about the magnitude of **a**. Substituting (10') into the formula $|\mathbf{a}|^2 = \alpha_1^2 + \alpha_2^2 + \alpha_3^2$ for the square of the magnitude of **a**, we find that

$$|\mathbf{a}|^2 = |\mathbf{a}|^2(\cos^2\theta_1 + \cos^2\theta_2 + \cos^2\theta_3).$$

Hence the direction cosines $\cos \theta_1$, $\cos \theta_2$ and $\cos \theta_3$ must satisfy the condition

$$\cos^2 \theta_1 + \cos^2 \theta_2 + \cos^2 \theta_3 = 1. \tag{11}$$

Example 6 Find the direction cosines and direction angles of the vector $\mathbf{a} = (4, -8, 1)$.

Solution Using (10) with $\alpha_1 = 4$, $\alpha_2 = -8$, $\alpha_3 = 1$ and

$$|\mathbf{a}| = \sqrt{\alpha_1^2 + \alpha_2^2 + \alpha_3^2} = \sqrt{4^2 + (-8)^2 + 1^2} = \sqrt{81} = 9,$$

we find that the vector $\mathbf{a}$ has direction cosines

$$\cos \theta_1 = \frac{4}{9}, \quad \cos \theta_2 = -\frac{8}{9}, \quad \cos \theta_3 = \frac{1}{9},$$

and corresponding direction angles

$$\theta_1 = \arccos \frac{4}{9} \approx 63.6°,$$

$$\theta_2 = \arccos \left(-\frac{8}{9}\right) \approx 152.7°,$$

$$\theta_3 = \arccos \frac{1}{9} \approx 83.6°. \quad \square$$

Example 7 Can a vector or directed line have direction angles $\theta_1 = 45°$, $\theta_2 = 135°$ and $\theta_3 = 60°$?

Solution No, since the condition (11) is not satisfied. In fact,

$$\cos \theta_1 = \frac{1}{\sqrt{2}}, \quad \cos \theta_2 = -\frac{1}{\sqrt{2}}, \quad \cos \theta_3 = \frac{1}{2},$$

and therefore

$$\cos^2 \theta_1 + \cos^2 \theta_2 + \cos^2 \theta_3 = \frac{1}{2} + \frac{1}{2} + \frac{1}{4} \neq 1. \quad \square$$

Finally, we mention that *linear dependence* and *linear independence* for vectors in space are defined in exactly the same way as for vectors in the plane (see page 653). It can be shown that *any four vectors in space are linearly dependent* and that *three vectors in space are linearly independent if and only if they are noncoplanar*. For example, the vectors $\mathbf{a}_1 = (0, 1, 1)$, $\mathbf{a}_2 = (1, 0, 1)$ and $\mathbf{a}_3 = (1, 1, 0)$ are noncoplanar (why?), and hence they are linearly independent. This can be checked by showing that $c_1 \mathbf{a}_1 + c_2 \mathbf{a}_2 + c_3 \mathbf{a}_3 = \mathbf{0}$ implies $c_1 = c_2 = c_3 = 0$.

Problems

In each case find the vectors $\mathbf{a} + 2\mathbf{b}$ and $2\mathbf{a} - 3\mathbf{b}$.
1. $\mathbf{a} = (1, -2, 4)$, $\mathbf{b} = (3, 2, -1)$
2. $\mathbf{a} = (4, 0, 7)$, $\mathbf{b} = (2, 5, 0)$
3. $\mathbf{a} = \mathbf{i} + \mathbf{j} + \mathbf{k}$, $\mathbf{b} = \mathbf{i} - \mathbf{j} - 2\mathbf{k}$
4. $\mathbf{a} = -\mathbf{j} + \mathbf{k}$, $\mathbf{b} = \mathbf{i} - \mathbf{k}$

Find the vector $\overrightarrow{AB}$ with the given endpoints.
5. $A = (-3, 2, 5)$, $B = (4, 1, -1)$
6. $A = (7, 0, 3)$, $B = (6, 2, 9)$
7. $A = (\frac{1}{3}, \frac{2}{3}, -\frac{1}{3})$, $B = (-\frac{2}{3}, \frac{4}{3}, \frac{1}{6})$
8. $A = (13, -11, 5)$, $B = (15, 17, -9)$

9. What is the final point of the vector $\mathbf{a} = 3\mathbf{i} - \mathbf{j} + 4\mathbf{k}$ if its initial point is $(1, 2, -3)$?

10. What is the initial point of the vector $\mathbf{a} = 2\mathbf{i} - 3\mathbf{j} - \mathbf{k}$ if its final point is $(-4, 0, 5)$?

Find the magnitude of the given vector.

11. $\mathbf{i} - \mathbf{j} + \mathbf{k}$
12. $2\mathbf{i} + \mathbf{j} - 2\mathbf{k}$
13. $6\mathbf{i} - 9\mathbf{j} + 2\mathbf{k}$
14. $-4\mathbf{i} + 12\mathbf{j} - 3\mathbf{k}$

Find the dot product $\mathbf{a} \cdot \mathbf{b}$ of the given vectors, and also find the angle between the vectors.

15. $\mathbf{a} = 4\mathbf{i} - 3\mathbf{j} - 2\mathbf{k}$, $\mathbf{b} = 3\mathbf{i} - 2\mathbf{j} + 4\mathbf{k}$
16. $\mathbf{a} = 12\mathbf{i} - 5\mathbf{k}$, $\mathbf{b} = 3\mathbf{j} + 4\mathbf{k}$
17. $\mathbf{a} = (1, 1, -1)$, $\mathbf{b} = (-1, 1, 1)$
18. $\mathbf{a} = (12, -15, 16)$, $\mathbf{b} = (2, 2, 1)$

19. Find the value of t for which the vectors $\mathbf{a} = t\mathbf{i} - 5\mathbf{j} + 2\mathbf{k}$ and $\mathbf{b} = \mathbf{i} + 4\mathbf{j} - t\mathbf{k}$ are orthogonal.

20. Find the values of s and t for which the vectors $\mathbf{a} = -2\mathbf{i} + 3\mathbf{j} + s\mathbf{k}$ and $\mathbf{b} = t\mathbf{i} - 6\mathbf{j} + 2\mathbf{k}$ are parallel.

Calculate $\text{proj}_\mathbf{b}\,\mathbf{a}$ for the given vectors $\mathbf{a}$ and $\mathbf{b}$.

21. $\mathbf{a} = 3\mathbf{i} + 2\mathbf{j} - \mathbf{k}$, $\mathbf{b} = 2\mathbf{i} + 3\mathbf{k}$
22. $\mathbf{a} = -2\mathbf{i} + \mathbf{j} + 3\mathbf{k}$, $\mathbf{b} = \mathbf{i} - 2\mathbf{j} + \mathbf{k}$
23. $\mathbf{a} = 4\mathbf{i} - \mathbf{j} + \mathbf{k}$, $\mathbf{b} = \mathbf{i} + 2\mathbf{j} - 2\mathbf{k}$
24. $\mathbf{a} = \mathbf{i} + \mathbf{j}$, $\mathbf{b} = \mathbf{j} + \mathbf{k}$

Can a vector have the given angles as direction angles?

25. $45°, 45°, 60°$
26. $90°, 150°, 60°$
27. $45°, 60°, 120°$

Can a vector make the given angles with two of the three positive coordinate axes?

28. $30°, 45°$
29. $150°, 30°$
30. $60°, 60°$

Find the direction cosines and direction angles of the given vector.

31. $2\mathbf{i} - 6\mathbf{j} + 9\mathbf{k}$
32. $12\mathbf{i} + 4\mathbf{j} - 3\mathbf{k}$
33. $6\mathbf{i} + 2\mathbf{j} + 3\mathbf{k}$
34. $-15\mathbf{i} + 16\mathbf{j} + 12\mathbf{k}$

35. A vector makes the same acute angle θ with all three positive coordinate axes. Find θ.

36. A cube has the vectors $2\mathbf{i}$, $2\mathbf{j}$ and $2\mathbf{k}$ as three of its edges. Find the angle between the vector joining the origin O to the center of the front face of the cube and the vector joining O to the center of the upper face.

37. A square card is marked with one of its diagonals, and also with two other lines which divide the card into three congruent rectangular strips. The card is then folded along the edges of the rectangular strips into a regular triangular prism. The folding causes the diagonal to become a polygonal path made up of three segments, one on each lateral face of the prism. Find the angle between two consecutive segments of this path.

38. Express the vector $\mathbf{b} = (4, 2, 3)$ as a linear combination of the vectors $\mathbf{a}_1 = (0, 1, 1)$, $\mathbf{a}_2 = (1, 0, 1)$ and $\mathbf{a}_3 = (1, 1, 0)$.

39. Give an example of a nonorthogonal basis in space.

12.3 The Cross Product

The concept of the "cross product" of two vectors arises both in the geometric problem of finding a vector perpendicular to two given vectors and in a variety of physical problems, including the behavior of a spinning top and the motion of a charged particle in a magnetic field. By the *cross product* $\mathbf{a} \times \mathbf{b}$ of two vectors $\mathbf{a}$ and $\mathbf{b}$ (in that order) we mean the vector of magnitude

$$|\mathbf{a} \times \mathbf{b}| = |\mathbf{a}||\mathbf{b}| \sin \theta, \tag{1}$$

where θ is the angle between $\mathbf{a}$ and $\mathbf{b}$ (defined as on page 655), such that $\mathbf{a} \times \mathbf{b}$ is perpendicular to the plane determined by $\mathbf{a}$ and $\mathbf{b}$. Also, of the two directions along the perpendicular we choose $\mathbf{a} \times \mathbf{b}$ to have the direction for which the vectors $\mathbf{a}$, $\mathbf{b}$ and $\mathbf{a} \times \mathbf{b}$ form a "right-handed system." This means that if the fingers of the right hand are curled so that they go from $\mathbf{a}$ to $\mathbf{b}$ through the angle θ, then the thumb points in the direction of $\mathbf{a} \times \mathbf{b}$, as illustrated in Figure 16. Alternatively, $\mathbf{a} \times \mathbf{b}$ points in the direction of advance of an ordinary screw (with a right-handed thread) driven by a screwdriver whose blade is turned from $\mathbf{a}$ to $\mathbf{b}$ through θ. If either $\mathbf{a} = \mathbf{0}$ or $\mathbf{b} = \mathbf{0}$, then θ is undefined, and we set $\mathbf{a} \times \mathbf{b} = \mathbf{0}$ by definition.

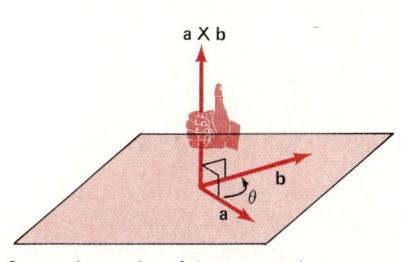

Geometric meaning of the cross product

Figure 16

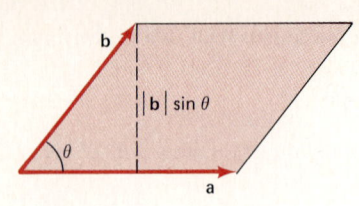

Figure 17

The cross product, so called because of the cross appearing in the expression **a** × **b**, is also called the *vector product*, since **a** × **b** is a *vector*. This is to be contrasted with the dot product **a** · **b**, which is a *scalar*. (Notice that the cross product can be defined only for vectors *in space*.) It is clear from formula (1) and Figure 17 that |**a** × **b**| is the area of the parallelogram spanned by the vectors **a** and **b**. In fact, |**a** × **b**| is the product of the base |**a**| and the altitude |**b**| sin θ of this parallelogram.

Properties of the Cross Product

The angle between a vector and itself is zero. Therefore
$$|\mathbf{a} \times \mathbf{a}| = |\mathbf{a}||\mathbf{a}| \sin 0 = 0,$$
so that
$$\mathbf{a} \times \mathbf{a} = \mathbf{0}. \tag{2}$$

The cross product **a** × **b** equals **0** if and only if **a** is parallel to **b**, written **a** ∥ **b**, meaning that **a** = p**b** where p is a scalar. (Choosing $p = 0$, we see that the zero vector is parallel to every vector.) In fact, **a** × **b** = **0** is equivalent to
$$|\mathbf{a} \times \mathbf{b}| = |\mathbf{a}||\mathbf{b}| \sin \theta = 0 \qquad (0 \leq \theta \leq \pi),$$
and this formula holds if and only if sin $\theta = 0$ and hence $\theta = 0$ or $\theta = \pi$, or at least one of the vectors **a** and **b** equals **0**, so that **a** ∥ **b** in any event.

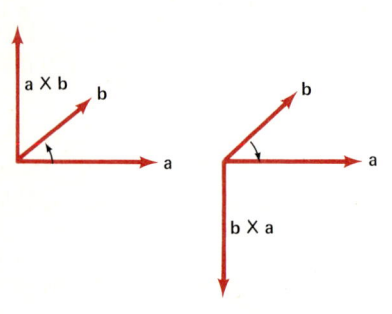

If the thumb of your right hand points in one direction when the fingers are curled from **a** to **b**, the thumb will point in the opposite direction when the fingers are curled from **b** to **a**. (Try this "thumbs up, thumbs down" experiment and see!) Thus **a** × **b** and **b** × **a** have opposite directions. But **a** × **b** and **b** × **a** have the same magnitude, as we find by comparing formula (1) with the formula
$$|\mathbf{b} \times \mathbf{a}| = |\mathbf{b}||\mathbf{a}| \sin \theta.$$
Therefore
$$\mathbf{b} \times \mathbf{a} = -(\mathbf{a} \times \mathbf{b}) \tag{3}$$

The cross product is anticommutative.

Figure 18

as in Figure 18, since vectors with the same magnitude and opposite directions are the negatives of each other. Formula (3) shows that the cross product is *not* commutative; in fact, it is *anticommutative*, which means that *changing the order of the factors* **a** *and* **b** *changes the sign of the product* **a** × **b**.

Example 1 Show that
$$\begin{array}{lll} \mathbf{i} \times \mathbf{i} = \mathbf{0}, & \mathbf{j} \times \mathbf{j} = \mathbf{0}, & \mathbf{k} \times \mathbf{k} = \mathbf{0}, \\ \mathbf{i} \times \mathbf{j} = \mathbf{k}, & \mathbf{j} \times \mathbf{k} = \mathbf{i}, & \mathbf{k} \times \mathbf{i} = \mathbf{j}, \\ \mathbf{j} \times \mathbf{i} = -\mathbf{k}, & \mathbf{k} \times \mathbf{j} = -\mathbf{i}, & \mathbf{i} \times \mathbf{k} = -\mathbf{j}, \end{array} \tag{4}$$

where **i**, **j** and **k** are the unit basis vectors of a right-handed system of rectangular coordinates.

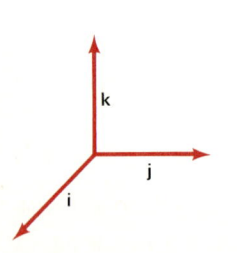

Figure 19

Solution The first three formulas are an immediate consequence of formula (2). To prove the middle three formulas, we observe that **i**, **j**, **k** form a right-handed system, and so do **j**, **k**, **i** and **k**, **i**, **j** (see Figure 19), while
$$|\mathbf{i} \times \mathbf{j}| = |\mathbf{i}||\mathbf{j}| \sin 90° = 1,$$

708 Chapter 12 Vectors in Space and Solid Analytic Geometry

and similarly $|\mathbf{j} \times \mathbf{k}| = |\mathbf{k} \times \mathbf{i}| = 1$. Since the cross product is anticommutative, the last three formulas follow at once from the middle three. ∎

Notice that any cross product formula involving all three vectors $\mathbf{i}, \mathbf{j}$ and $\mathbf{k}$ remains true if we replace $\mathbf{i}$ by $\mathbf{j}$, $\mathbf{j}$ by $\mathbf{k}$ and $\mathbf{k}$ by $\mathbf{i}$. Thus, to reproduce all of the important formulas (4), it is enough to remember the single formula $\mathbf{i} \times \mathbf{j} = \mathbf{k}$, together with the rules (2) and (3).

The following theorem is the analogue for cross products of the corollary on page 656.

Theorem 2 *(Further properties of the cross product).* *If p and q are scalars, then*

$$(p\mathbf{a}) \times (q\mathbf{b}) = pq(\mathbf{a} \times \mathbf{b}) \tag{5}$$

for arbitrary vectors $\mathbf{a}$ and $\mathbf{b}$. The cross product also satisfies the distributive laws

$$\mathbf{a} \times (\mathbf{b} + \mathbf{c}) = (\mathbf{a} \times \mathbf{b}) + (\mathbf{a} \times \mathbf{c}), \tag{6}$$

$$(\mathbf{a} + \mathbf{b}) \times \mathbf{c} = (\mathbf{a} \times \mathbf{c}) + (\mathbf{b} \times \mathbf{c}) \tag{6'}$$

for arbitrary vectors $\mathbf{a}$, $\mathbf{b}$ and $\mathbf{c}$.

The proof of this theorem is a bit tedious, and hence is deferred to the end of the section.

Example 2 Using Theorem 2 and the fact that the cross product is anticommutative, we find that

$$(\mathbf{a} + 2\mathbf{b}) \times (2\mathbf{a} - 3\mathbf{b}) = 2(\mathbf{a} \times \mathbf{a}) + 4(\mathbf{b} \times \mathbf{a}) - 3(\mathbf{a} \times \mathbf{b}) - 6(\mathbf{b} \times \mathbf{b})$$
$$= 0 - 4(\mathbf{a} \times \mathbf{b}) - 3(\mathbf{a} \times \mathbf{b}) - 0 = -7\mathbf{a} \times \mathbf{b},$$

where $p\mathbf{a} \times \mathbf{b}$ is shorthand for $p(\mathbf{a} \times \mathbf{b})$.

Example 3 Cross products are *nonassociative*, that is, $(\mathbf{a} \times \mathbf{b}) \times \mathbf{c}$ need not equal $\mathbf{a} \times (\mathbf{b} \times \mathbf{c})$; thus we cannot drop the parentheses and simply write $\mathbf{a} \times \mathbf{b} \times \mathbf{c}$. For example, by (4) and (6),

$$(\mathbf{i} \times \mathbf{j}) \times (\mathbf{i} + \mathbf{j}) = \mathbf{k} \times (\mathbf{i} + \mathbf{j}) = (\mathbf{k} \times \mathbf{i}) + (\mathbf{k} \times \mathbf{j}) = \mathbf{j} - \mathbf{i},$$

while

$$\mathbf{i} \times (\mathbf{j} \times (\mathbf{i} + \mathbf{j})) = \mathbf{i} \times [(\mathbf{j} \times \mathbf{i}) + (\mathbf{j} \times \mathbf{j})] = \mathbf{i} \times (-\mathbf{k}) = \mathbf{j},$$

so that

$$(\mathbf{i} \times \mathbf{j}) \times (\mathbf{i} + \mathbf{j}) \neq \mathbf{i} \times (\mathbf{j} \times (\mathbf{i} + \mathbf{j})).$$

Determinants

Before proceeding, we digress to introduce an algebraic concept which greatly simplifies the study of cross products. Symbols of the form

$$\begin{vmatrix} a_1 & a_2 \\ b_1 & b_2 \end{vmatrix}, \quad \begin{vmatrix} a_1 & a_2 & a_3 \\ b_1 & b_2 & b_3 \\ c_1 & c_2 & c_3 \end{vmatrix}, \tag{7}$$

where a square array of real numbers appears between vertical bars, are called *determinants* (in general there are n rows, each containing n numbers).

The number of rows or columns in a determinant is called its *order*, and a determinant of order n is also called an $n \times n$ ("n by n") *determinant*. Thus the first symbol in (7) is a 2×2 determinant, and the second symbol is a 3×3 determinant. Each of these symbols is a very concise way of writing a certain number. Specifically, the first symbol in (7) denotes the number

$$\begin{vmatrix} a_1 & a_2 \\ b_1 & b_2 \end{vmatrix} = a_1 b_2 - a_2 b_1, \tag{8}$$

while the second symbol denotes the number

$$\begin{vmatrix} a_1 & a_2 & a_3 \\ b_1 & b_2 & b_3 \\ c_1 & c_2 & c_3 \end{vmatrix} = a_1 \begin{vmatrix} b_2 & b_3 \\ c_2 & c_3 \end{vmatrix} - a_2 \begin{vmatrix} b_1 & b_3 \\ c_1 & c_3 \end{vmatrix} + a_3 \begin{vmatrix} b_1 & b_2 \\ c_1 & c_2 \end{vmatrix}$$

$$= a_1(b_2 c_3 - b_3 c_2) - a_2(b_1 c_3 - b_3 c_1) + a_3(b_1 c_2 - b_2 c_1). \tag{9}$$

Notice that in formula (9) the 2×2 determinant multiplying a_i ($i = 1, 2$ or 3) on the right is obtained from the 3×3 determinant on the left by deleting both the row and the column containing a_i, as indicated schematically by the following diagrams:

$$\begin{vmatrix} \not{a_1} & \not{a_2} & \not{a_3} \\ \not{b_1} & b_2 & b_3 \\ \not{c_1} & c_2 & c_3 \end{vmatrix}, \quad \begin{vmatrix} \not{a_1} & \not{a_2} & \not{a_3} \\ b_1 & \not{b_2} & b_3 \\ c_1 & \not{c_2} & c_3 \end{vmatrix}, \quad \begin{vmatrix} \not{a_1} & \not{a_2} & \not{a_3} \\ b_1 & b_2 & \not{b_3} \\ c_1 & c_2 & \not{c_3} \end{vmatrix}.$$

The fact that a_2 appears with a minus sign in the right side of (9) is not a typographical error, but an intrinsic part of the definition of a 3×3 determinant.

Example 4 By formula (8),

$$\begin{vmatrix} 5 & 3 \\ -1 & 4 \end{vmatrix} = 5(4) - 3(-1) = 20 + 3 = 23.$$

Example 5 By formula (9),

$$\begin{vmatrix} 2 & 3 & 8 \\ 4 & 7 & -2 \\ 1 & 5 & 9 \end{vmatrix} = 2 \begin{vmatrix} 7 & -2 \\ 5 & 9 \end{vmatrix} - 3 \begin{vmatrix} 4 & -2 \\ 1 & 9 \end{vmatrix} + 8 \begin{vmatrix} 4 & 7 \\ 1 & 5 \end{vmatrix}$$

$$= 2[7(9) + 2(5)] - 3[4(9) + 2(1)] + 8[4(5) - 7(1)]$$
$$= 2(73) - 3(38) + 8(13) = 136.$$

Component Form of the Cross Product

We now derive a formula for the cross product $\mathbf{a} \times \mathbf{b}$ in terms of the components of $\mathbf{a}$ and $\mathbf{b}$ with respect to the unit basis vectors $\mathbf{i}$, $\mathbf{j}$ and $\mathbf{k}$, and then show how $\mathbf{a} \times \mathbf{b}$ can be written as a determinant.

Theorem 3 *(Component form of the cross product)*. If

$$\mathbf{a} = \alpha_1 \mathbf{i} + \alpha_2 \mathbf{j} + \alpha_3 \mathbf{k} \quad \text{and} \quad \mathbf{b} = \beta_1 \mathbf{i} + \beta_2 \mathbf{j} + \beta_3 \mathbf{k},$$

then

$$\mathbf{a} \times \mathbf{b} = (\alpha_2 \beta_3 - \alpha_3 \beta_2)\mathbf{i} + (\alpha_3 \beta_1 - \alpha_1 \beta_3)\mathbf{j} + (\alpha_1 \beta_2 - \alpha_2 \beta_1)\mathbf{k}. \tag{10}$$

Proof With the help of Theorem 2 and the formulas (4) for the cross products of the vectors **i**, **j** and **k**, we have

$$\begin{aligned}\mathbf{a} \times \mathbf{b} &= (\alpha_1\mathbf{i} + \alpha_2\mathbf{j} + \alpha_3\mathbf{k}) \times (\beta_1\mathbf{i} + \beta_2\mathbf{j} + \beta_3\mathbf{k}) \\ &= \alpha_1\beta_1(\mathbf{i} \times \mathbf{i}) + \alpha_1\beta_2(\mathbf{i} \times \mathbf{j}) + \alpha_1\beta_3(\mathbf{i} \times \mathbf{k}) \\ &\quad + \alpha_2\beta_1(\mathbf{j} \times \mathbf{i}) + \alpha_2\beta_2(\mathbf{j} \times \mathbf{j}) + \alpha_2\beta_3(\mathbf{j} \times \mathbf{k}) \\ &\quad + \alpha_3\beta_1(\mathbf{k} \times \mathbf{i}) + \alpha_3\beta_2(\mathbf{k} \times \mathbf{j}) + \alpha_3\beta_3(\mathbf{k} \times \mathbf{k}) \\ &= \alpha_1\beta_1\mathbf{0} + \alpha_1\beta_2\mathbf{k} + \alpha_1\beta_3(-\mathbf{j}) + \alpha_2\beta_1(-\mathbf{k}) + \alpha_2\beta_2\mathbf{0} \\ &\quad + \alpha_2\beta_3\mathbf{i} + \alpha_3\beta_1\mathbf{j} + \alpha_3\beta_2(-\mathbf{i}) + \alpha_3\beta_3\mathbf{0} \\ &= \alpha_1\beta_2\mathbf{k} - \alpha_1\beta_3\mathbf{j} - \alpha_2\beta_1\mathbf{k} + \alpha_2\beta_3\mathbf{i} + \alpha_3\beta_1\mathbf{j} - \alpha_3\beta_2\mathbf{i},\end{aligned}$$

which is equivalent to (10). ∎

Inspection of formula (10) shows that it can also be written as

$$\mathbf{a} \times \mathbf{b} = \begin{vmatrix}\alpha_2 & \alpha_3 \\ \beta_2 & \beta_3\end{vmatrix}\mathbf{i} - \begin{vmatrix}\alpha_1 & \alpha_3 \\ \beta_1 & \beta_3\end{vmatrix}\mathbf{j} + \begin{vmatrix}\alpha_1 & \alpha_2 \\ \beta_1 & \beta_2\end{vmatrix}\mathbf{k} \tag{10'}$$

in terms of 2×2 determinants. Comparing the sum on the right with the analogous sum in formula (9), we find that

$$\mathbf{a} \times \mathbf{b} = \begin{vmatrix}\mathbf{i} & \mathbf{j} & \mathbf{k} \\ \alpha_1 & \alpha_2 & \alpha_3 \\ \beta_1 & \beta_2 & \beta_3\end{vmatrix}. \tag{11}$$

This 3×3 determinant is "symbolic" in the sense that its first row consists of vectors rather than numbers, but it is written with the understanding that it is just a very concise way of representing the sum in (10').

Example 6 Find the cross product $\mathbf{a} \times \mathbf{b}$ of the vectors $\mathbf{a} = \mathbf{i} + 3\mathbf{j} - \mathbf{k}$ and $\mathbf{b} = 2\mathbf{i} - \mathbf{j} + \mathbf{k}$.

Solution By formula (11),

$$\mathbf{a} \times \mathbf{b} = \begin{vmatrix}\mathbf{i} & \mathbf{j} & \mathbf{k} \\ 1 & 3 & -1 \\ 2 & -1 & 1\end{vmatrix} = \begin{vmatrix}3 & -1 \\ -1 & 1\end{vmatrix}\mathbf{i} - \begin{vmatrix}1 & -1 \\ 2 & 1\end{vmatrix}\mathbf{j} + \begin{vmatrix}1 & 3 \\ 2 & -1\end{vmatrix}\mathbf{k}$$

$$= (3 - 1)\mathbf{i} - (1 + 2)\mathbf{j} + (-1 - 6)\mathbf{k}$$
$$= 2\mathbf{i} - 3\mathbf{j} - 7\mathbf{k}. \quad \square$$

Example 7 Find the area A of the triangle PQR with vertices $P = (1, 2, 0)$, $Q = (3, 0, -3)$ and $R = (5, 2, 6)$.

Solution The area of the parallelogram spanned by the vectors $\overrightarrow{PQ}$ and $\overrightarrow{PR}$ is $|\overrightarrow{PQ} \times \overrightarrow{PR}|$, and this is twice the area A. Since

$$\overrightarrow{PQ} = OQ - OP = (3, 0, -3) - (1, 2, 0) = (2, -2, -3)$$
$$\overrightarrow{PR} = OR - OP = (5, 2, 6) - (1, 2, 0) = (4, 0, 6)$$

(where O is the origin), we have

$$\overrightarrow{PQ} \times \overrightarrow{PR} = \begin{vmatrix}\mathbf{i} & \mathbf{j} & \mathbf{k} \\ 2 & -2 & -3 \\ 4 & 0 & 6\end{vmatrix} = \begin{vmatrix}-2 & -3 \\ 0 & 6\end{vmatrix}\mathbf{i} - \begin{vmatrix}2 & -3 \\ 4 & 6\end{vmatrix}\mathbf{j} + \begin{vmatrix}2 & -2 \\ 4 & 0\end{vmatrix}\mathbf{k}$$

$$= (-12 + 0)\mathbf{i} - (12 + 12)\mathbf{j} + (0 + 8)\mathbf{k} = -12\mathbf{i} - 24\mathbf{j} + 8\mathbf{k}.$$

Therefore
$$2A = |-12\mathbf{i} - 24\mathbf{j} + 8\mathbf{k}| = |(-2)(6\mathbf{i} + 12\mathbf{j} - 4\mathbf{k})|,$$
so that
$$A = |6\mathbf{i} + 12\mathbf{j} - 4\mathbf{k}| = \sqrt{36 + 144 + 16} = \sqrt{196} = 14 \qquad \square$$

The Scalar Triple Product

Of the various products involving three or more vectors, the most important is the *scalar triple product* $\mathbf{a} \cdot (\mathbf{b} \times \mathbf{c})$, which is just the scalar or dot product of the vectors $\mathbf{a}$ and $\mathbf{b} \times \mathbf{c}$. To interpret the number $\mathbf{a} \cdot (\mathbf{b} \times \mathbf{c})$ geometrically, we write it in the form

$$\mathbf{a} \cdot (\mathbf{b} \times \mathbf{c}) = |\mathbf{b} \times \mathbf{c}| \frac{\mathbf{a} \cdot (\mathbf{b} \times \mathbf{c})}{|\mathbf{b} \times \mathbf{c}|} = |\mathbf{b} \times \mathbf{c}| \, \text{comp}_{\mathbf{b} \times \mathbf{c}} \, \mathbf{a}, \qquad (12)$$

assuming that $\mathbf{b} \times \mathbf{c} \neq \mathbf{0}$. Here $\text{comp}_{\mathbf{b} \times \mathbf{c}} \, \mathbf{a}$ is the component of $\mathbf{a}$ along $\mathbf{b} \times \mathbf{c}$. The absolute value of this component is the altitude h of the parallelepiped P spanned by the vectors $\mathbf{a}$, $\mathbf{b}$ and $\mathbf{c}$, as in Figure 20 where $\text{comp}_{\mathbf{b} \times \mathbf{c}} \, \mathbf{a} > 0$. Let V be the volume of P. Then, since $|\mathbf{b} \times \mathbf{c}|$ is the area of the base of P, it follows from (12) that

$$\mathbf{a} \cdot (\mathbf{b} \times \mathbf{c}) = \begin{cases} V & \text{if } \text{comp}_{\mathbf{b} \times \mathbf{c}} \, \mathbf{a} > 0, \\ -V & \text{if } \text{comp}_{\mathbf{b} \times \mathbf{c}} \, \mathbf{a} < 0, \end{cases}$$

so that in any event

$$V = |\mathbf{a} \cdot (\mathbf{b} \times \mathbf{c})|. \qquad (13)$$

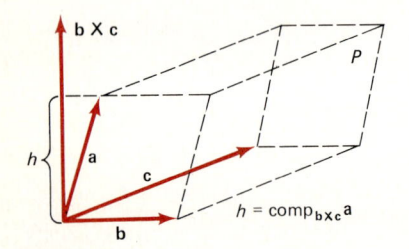

Figure 20

The component of $\mathbf{a}$ along $\mathbf{b} \times \mathbf{c}$ is zero if and only if $\mathbf{a}$ lies in the plane determined by $\mathbf{b}$ and $\mathbf{c}$. Therefore, by (12), *three nonzero vectors* $\mathbf{a}$, $\mathbf{b}$ *and* $\mathbf{c}$ *are coplanar if and only if* $\mathbf{a} \cdot (\mathbf{b} \times \mathbf{c}) = 0$. (Why is this true even if the case $\mathbf{b} \times \mathbf{c} = \mathbf{0}$ is allowed?)

To express the scalar triple product $\mathbf{a} \cdot (\mathbf{b} \times \mathbf{c})$ in terms of the components of $\mathbf{a}$, $\mathbf{b}$ and $\mathbf{c}$, let

$$\mathbf{a} = \alpha_1 \mathbf{i} + \alpha_2 \mathbf{j} + \alpha_3 \mathbf{k}, \qquad \mathbf{b} = \beta_1 \mathbf{i} + \beta_2 \mathbf{j} + \beta_3 \mathbf{k}, \qquad \mathbf{c} = \gamma_1 \mathbf{i} + \gamma_2 \mathbf{j} + \gamma_3 \mathbf{k}.$$

Then

$$\mathbf{b} \times \mathbf{c} = \begin{vmatrix} \mathbf{i} & \mathbf{j} & \mathbf{k} \\ \beta_1 & \beta_2 & \beta_3 \\ \gamma_1 & \gamma_2 & \gamma_3 \end{vmatrix},$$

by formula (11), so that

$$\mathbf{a} \cdot (\mathbf{b} \times \mathbf{c}) = (\alpha_1 \mathbf{i} + \alpha_2 \mathbf{j} + \alpha_3 \mathbf{k}) \cdot \left(\begin{vmatrix} \beta_2 & \beta_3 \\ \gamma_2 & \gamma_3 \end{vmatrix} \mathbf{i} - \begin{vmatrix} \beta_1 & \beta_3 \\ \gamma_1 & \gamma_3 \end{vmatrix} \mathbf{j} + \begin{vmatrix} \beta_1 & \beta_2 \\ \gamma_1 & \gamma_2 \end{vmatrix} \mathbf{k} \right)$$

$$= \alpha_1 \begin{vmatrix} \beta_2 & \beta_3 \\ \gamma_2 & \gamma_3 \end{vmatrix} - \alpha_2 \begin{vmatrix} \beta_1 & \beta_3 \\ \gamma_1 & \gamma_3 \end{vmatrix} + \alpha_3 \begin{vmatrix} \beta_1 & \beta_2 \\ \gamma_1 & \gamma_2 \end{vmatrix},$$

or equivalently

$$\mathbf{a} \cdot (\mathbf{b} \times \mathbf{c}) = \begin{vmatrix} \alpha_1 & \alpha_2 & \alpha_3 \\ \beta_1 & \beta_2 & \beta_3 \\ \gamma_1 & \gamma_2 & \gamma_3 \end{vmatrix}. \qquad (14)$$

Example 8 Are the vectors $\mathbf{a} = (2, 3, -1)$, $\mathbf{b} = (1, -1, 3)$ and $\mathbf{c} = (1, 9, -11)$ coplanar?

Solution Yes, since by (14) we have

$$\mathbf{a} \cdot (\mathbf{b} \times \mathbf{c}) = \begin{vmatrix} 2 & 3 & -1 \\ 1 & -1 & 3 \\ 1 & 9 & -11 \end{vmatrix} = 2 \begin{vmatrix} -1 & 3 \\ 9 & -11 \end{vmatrix} - 3 \begin{vmatrix} 1 & 3 \\ 1 & -11 \end{vmatrix} - \begin{vmatrix} 1 & -1 \\ 1 & 9 \end{vmatrix}$$

$$= 2(11 - 27) - 3(-11 - 3) - (9 + 1) = -32 + 42 - 10 = 0. \quad \square$$

Example 9 Find the volume V of the parallelepiped spanned by the vectors $\mathbf{i} + \mathbf{j}$, $\mathbf{j} + \mathbf{k}$ and $\mathbf{k} + \mathbf{i}$.

Solution Here

$$\mathbf{a} \cdot (\mathbf{b} \times \mathbf{c}) = \begin{vmatrix} 1 & 1 & 0 \\ 0 & 1 & 1 \\ 1 & 0 & 1 \end{vmatrix} = \begin{vmatrix} 1 & 1 \\ 0 & 1 \end{vmatrix} - \begin{vmatrix} 0 & 1 \\ 1 & 1 \end{vmatrix} = 1 - (-1) = 2,$$

and therefore $V = 2$, by (13). $\quad \square$

Example 10 Show that

$$\mathbf{a} \cdot (\mathbf{b} \times \mathbf{c}) = (\mathbf{a} \times \mathbf{b}) \cdot \mathbf{c} \tag{15}$$

for arbitrary vectors $\mathbf{a} = (\alpha_1, \alpha_2, \alpha_3)$, $\mathbf{b} = (\beta_1, \beta_2, \beta_3)$ and $\mathbf{c} = (\gamma_1, \gamma_2, \gamma_3)$.

Solution It follows from (10′) that

$$(\mathbf{a} \times \mathbf{b}) \cdot \mathbf{c} = \left(\begin{vmatrix} \alpha_2 & \alpha_3 \\ \beta_2 & \beta_3 \end{vmatrix} \mathbf{i} - \begin{vmatrix} \alpha_1 & \alpha_3 \\ \beta_1 & \beta_3 \end{vmatrix} \mathbf{j} + \begin{vmatrix} \alpha_1 & \alpha_2 \\ \beta_1 & \beta_2 \end{vmatrix} \mathbf{k} \right) \cdot (\gamma_1 \mathbf{i} + \gamma_2 \mathbf{j} + \gamma_3 \mathbf{k})$$

$$= \gamma_1 \begin{vmatrix} \alpha_2 & \alpha_3 \\ \beta_2 & \beta_3 \end{vmatrix} - \gamma_2 \begin{vmatrix} \alpha_1 & \alpha_3 \\ \beta_1 & \beta_3 \end{vmatrix} + \gamma_3 \begin{vmatrix} \alpha_1 & \alpha_2 \\ \beta_1 & \beta_2 \end{vmatrix} = \begin{vmatrix} \gamma_1 & \gamma_2 & \gamma_3 \\ \alpha_1 & \alpha_2 & \alpha_3 \\ \beta_1 & \beta_2 & \beta_3 \end{vmatrix}.$$

If you calculate both this determinant and the determinant (14), you will find that they have the same value, thereby proving formula (15). The algebraic details are left as an exercise. $\quad \square$

Actually, there is no ambiguity in writing $\mathbf{a} \cdot (\mathbf{b} \times \mathbf{c})$ and $(\mathbf{a} \times \mathbf{b}) \cdot \mathbf{c}$ as $\mathbf{a} \cdot \mathbf{b} \times \mathbf{c}$ and $\mathbf{a} \times \mathbf{b} \cdot \mathbf{c}$, since the expressions $(\mathbf{a} \cdot \mathbf{b}) \times \mathbf{c}$ and $\mathbf{a} \times (\mathbf{b} \cdot \mathbf{c})$ are both meaningless (why?). Thus formula (15) tells us that the dot and the cross can be interchanged in the expression $\mathbf{a} \cdot \mathbf{b} \times \mathbf{c}$ without changing its value.

Proof of Theorem 2 (Optional) It is an immediate consequence of the geometric meaning of the cross product that

$$(p\mathbf{a}) \times \mathbf{b} = p(\mathbf{a} \times \mathbf{b}) \tag{16}$$

for $p \neq 0$; the cases $p > 0$ and $p < 0$ have to be distinguished, as indicated in Figure 21, but there is no difficulty since $\sin(\pi - \theta) = \sin \theta$. Combining (16) with the companion formula

$$\mathbf{a} \times (q\mathbf{b}) = q(\mathbf{a} \times \mathbf{b}) \tag{16′}$$

for $q \neq 0$, we find at once that

$$(p\mathbf{a}) \times (q\mathbf{b}) = p[\mathbf{a} \times (q\mathbf{b})] = p[q(\mathbf{a} \times \mathbf{b})],$$

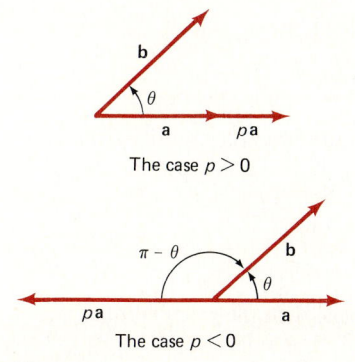

Figure 21

Section 12.3 The Cross Product

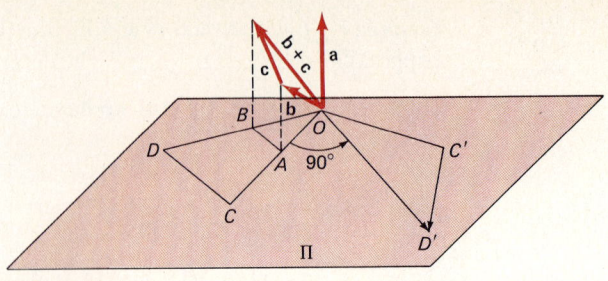

Figure 22

so that

$$(p\mathbf{a}) \times (q\mathbf{b}) = pq(\mathbf{a} \times \mathbf{b}),$$

and this formula clearly remains true if one or both of the scalars p and q is zero.

To prove the distributive laws

$$\mathbf{a} \times (\mathbf{b} + \mathbf{c}) = (\mathbf{a} \times \mathbf{b}) + (\mathbf{a} \times \mathbf{c}), \qquad (\mathbf{a} + \mathbf{b}) \times \mathbf{c} = (\mathbf{a} \times \mathbf{c}) + (\mathbf{b} \times \mathbf{c}), \quad (17)$$

let the vectors $\mathbf{a}$, $\mathbf{b}$ and $\mathbf{b} + \mathbf{c}$ have a common initial point O, and let Π be the plane through O perpendicular to $\mathbf{a}$, as in Figure 22. The perpendiculars drawn from the final points of $\mathbf{b}$ and $\mathbf{b} + \mathbf{c}$ to Π intersect Π in points A and B. Enlarging the triangle OAB by a factor of $|\mathbf{a}|$, we obtain the triangle OCD (the "enlargement is actually a reduction if $|\mathbf{a}| < 1$, and OCD coincides with OAB if $|\mathbf{a}| = 1$). We then rotate OCD through $90°$ about O in the plane Π; the rotation carries OCD into a congruent triangle $OC'D'$, and of the two possible directions of rotation we choose the one such that the vectors $\overrightarrow{OC}$, $\overrightarrow{OC'}$ and $\mathbf{a}$ form a right-handed system. It is apparent from the figure that

$$\overrightarrow{OD'} = \overrightarrow{OC'} + \overrightarrow{C'D'}. \tag{18}$$

But

$$\overrightarrow{OC'} = \mathbf{a} \times \mathbf{b}, \qquad \overrightarrow{OD'} = \mathbf{a} \times (\mathbf{b} + \mathbf{c}), \qquad \overrightarrow{C'D'} = \mathbf{a} \times \mathbf{c},$$

and substitution of these expressions into (18) gives the first of the formulas (17). To get the second formula, we use the first formula and the fact that the cross product is anticommutative:

$$(\mathbf{a} + \mathbf{b}) \times \mathbf{c} = -[\mathbf{c} \times (\mathbf{a} + \mathbf{b})] = -[(\mathbf{c} \times \mathbf{a}) + (\mathbf{c} \times \mathbf{b})]$$
$$= -(\mathbf{c} \times \mathbf{a}) - (\mathbf{c} \times \mathbf{b}) = (\mathbf{a} \times \mathbf{c}) + (\mathbf{b} \times \mathbf{c}). \blacksquare$$

Problems

Find the cross product $\mathbf{a} \times \mathbf{b}$ of the given vectors.
1. $\mathbf{a} = 3\mathbf{k}$, $\mathbf{b} = -2\mathbf{j}$
2. $\mathbf{a} = 2\mathbf{i} + \mathbf{j}$, $\mathbf{b} = \mathbf{j} - \mathbf{k}$
3. $\mathbf{a} = (0, 2, 1)$, $\mathbf{b} = (1, 0, 2)$
4. $\mathbf{a} = (10, 0, 5)$, $\mathbf{b} = (0, -2, 6)$
5. $\mathbf{a} = \mathbf{i} - \mathbf{j} - \mathbf{k}$, $\mathbf{b} = 3\mathbf{i} + 6\mathbf{j} + 2\mathbf{k}$
6. $\mathbf{a} = \mathbf{i} + 3\mathbf{j} + 2\mathbf{k}$, $\mathbf{b} = -\mathbf{i} + 4\mathbf{j} - 3\mathbf{k}$
7. $\mathbf{a} = (1, 3, 4)$, $\mathbf{b} = (2, 6, -3)$
8. $\mathbf{a} = (9, -7, 1)$, $\mathbf{b} = (8, 5, -2)$

Find the area of the parallelogram spanned by the given vectors.
9. $\mathbf{a} = \mathbf{i} + 2\mathbf{j} - \mathbf{k}$, $\mathbf{b} = -2\mathbf{i} + 3\mathbf{j} + \mathbf{k}$
10. $\mathbf{a} = 3\mathbf{i} - 4\mathbf{j} + 5\mathbf{k}$, $\mathbf{b} = -6\mathbf{k}$

Find the area of the triangle with the given vertices.
11. $P = (3, 4, 7)$, $Q = (0, 6, 1)$, $R = (5, -2, 4)$
12. $P = (-1, 4, 5)$, $Q = (1, 3, 7)$, $R = (2, 5, 6)$

13. Use the cross product to show that the area of the parallelogram spanned by the diagonals of a given parallelogram P is twice the area of P itself.
14. Show that $|a \times b|^2 + (a \cdot b)^2 = |a|^2|b|^2$ for arbitrary vectors a and b.

Find a unit vector perpendicular to both of the given vectors.
15. $a = i + j$, $b = j + k$
16. $a = 2i - k$, $b = i - 2j$
17. $a = (2, 0, -1)$, $b = (-2, 1, 0)$
18. $a = (3, 1, 2)$, $b = (-1, 3, -1)$

19. Does $a \times b = a \times c$ where $a \neq 0$ imply $b = c$? Explain your answer.
20. Show that $|a \times b| \leq |a||b|$. When does equality hold?

Evaluate the given determinant.

21. $\begin{vmatrix} -1 & 4 \\ -5 & 2 \end{vmatrix}$
22. $\begin{vmatrix} 5 & 2 \\ 7 & 3 \end{vmatrix}$
23. $\begin{vmatrix} x & 1 \\ x^2 & x \end{vmatrix}$
24. $\begin{vmatrix} 1 & 1 \\ x & y \end{vmatrix}$

25. $\begin{vmatrix} \cos \alpha & -\sin \alpha \\ \sin \alpha & \cos \alpha \end{vmatrix}$
26. $\begin{vmatrix} \sin \alpha & \cos \alpha \\ \sin \beta & \cos \beta \end{vmatrix}$

27. $\begin{vmatrix} 2 & 1 & 3 \\ 5 & 3 & 2 \\ 1 & 4 & 3 \end{vmatrix}$
28. $\begin{vmatrix} 3 & 4 & -5 \\ 8 & 7 & -2 \\ 2 & -1 & 8 \end{vmatrix}$

29. $\begin{vmatrix} 2 & 0 & 3 \\ 7 & 1 & 6 \\ 6 & 0 & 5 \end{vmatrix}$
30. $\begin{vmatrix} 4 & 5 & 11 \\ -2 & -3 & 1 \\ 3 & 8 & -2 \end{vmatrix}$

31. $\begin{vmatrix} 0 & a & 0 \\ b & c & d \\ 0 & e & 0 \end{vmatrix}$
32. $\begin{vmatrix} a & x & x \\ x & b & x \\ x & x & c \end{vmatrix}$
33. $\begin{vmatrix} 1 & 1 & 1 \\ x & y & z \\ x^2 & y^2 & z^2 \end{vmatrix}$

34. Show that $|a \cdot (b \times c)| \leq |a||b||c|$. When does equality hold?

Find the scalar triple product $a \cdot (b \times c)$ of the given vectors.
35. $a = (1, -1, 3)$, $b = (-2, 2, 1)$, $c = (3, -2, 5)$
36. $a = (1, 2, 5)$, $b = (1, -1, 3)$, $c = (3, -6, -1)$
37. $a = (-4, 2, 1)$, $b = (-5, 1, 2)$, $c = (-1, -1, 1)$
38. $a = (2, -1, 6)$, $b = (3, -5, 1)$, $c = (4, -7, 1)$

39. Show that $a \cdot (a \times b) = a \cdot (b \times a) = b \cdot (a \times a) = 0$. Use this to show that a 3×3 determinant equals zero if two of its rows are identical.
40. Regard as known the fact that a determinant changes its sign if any two of its rows are interchanged, and use this to prove that $a \cdot (b \times c) = b \cdot (c \times a) = c \cdot (a \times b)$.

Are the given vectors coplanar? Explain your answer.
41. $a = (-1, 2, 2)$, $b = (2, -3, 1)$, $c = (-4, 7, 3)$
42. $a = (3, -2, 1)$, $b = (2, 1, 2)$, $c = (3, -1, -4)$

43. Show that the four points $A = (1, 2, -1)$, $B = (0, 1, 5)$, $C = (-1, 2, 1)$ and $D = (2, 1, 3)$ are coplanar.

Find the volume of the parallelepiped spanned by the given vectors.
44. $a = i \times j$, $b = j \times k$, $c = k \times i$
45. $a = (1, 3, -1)$, $b = (-2, 1, 2)$, $c = (3, 5, -2)$
46. $a = (-4, 5, 0)$, $b = (6, 2, 5)$, $c = (2, 1, 7)$

47. Find the volume of the tetrahedron with vertices $A = (-1, 2, 1)$, $B = (5, 5, 4)$, $C = (2, 3, -1)$ and $D = (1, 4, 3)$.
48. A triple product of the form $a \times (b \times c)$ is called a *vector triple product*. Notice that $a \times (b \times c)$ is a vector, unlike $a \cdot (b \times c)$ which is a scalar. Verify by direct calculation that $a \times (b \times c) = (a \cdot c)b - (a \cdot b)c$. When written in the form

$$a \times (b \times c) = b(a \cdot c) - c(a \cdot b),$$

this is called the "*bac-cab* rule," and is a useful formula well worth memorizing.

Calculate the vector triple products $a \times (b \times c)$ and $(a \times b) \times c$ for the given vectors.
49. $a = (2, 1, 3)$, $b = (1, -2, 2)$, $c = (1, 1, 1)$
50. $a = (4, 0, 5)$, $b = (0, -1, 6)$, $c = (1, 2, 0)$
51. $a = i + j$, $b = j + k$, $c = i + k$

52. A charged particle of charge q moving with velocity v in a magnetic field B experiences a force $F = qv \times B$. Suppose the magnetic field is constant and perpendicular to the plane of the particle's motion, as in the device known as a *cyclotron*. Show that the particle describes a circular path of radius mv/qB, where $v = |v|$, $B = |B|$, and m is the mass of the particle. Show that this path is described with angular speed qB/m, known as the *cyclotron frequency*.

12.4 Lines and Planes in Space

Parametric Equations of Lines   A line L in space is uniquely determined by specifying any fixed point $P_1 = (x_1, y_1, z_1)$ on L and any nonzero vector $m = (a, b, c)$ parallel to L. Let $P = (x, y, z)$ be a variable point in space. Then the point P lies on L if and only if the vectors $\overrightarrow{P_1P}$ and $m = (a, b, c)$ are parallel, as shown in

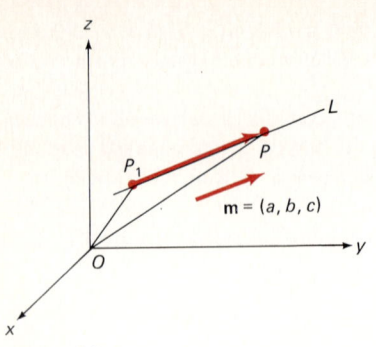

Figure 23, or equivalently

$$\overrightarrow{P_1P} = t\mathbf{m} = t(a, b, c),$$

where t is any scalar. But

$$\overrightarrow{P_1P} = \overrightarrow{OP} - \overrightarrow{OP_1} = (x - x_1, y - y_1, z - z_1),$$

where O is the origin, and therefore P lies on L if and only if

$$x - x_1 = at, \qquad y - y_1 = bt, \qquad z - z_1 = ct, \qquad (1)$$

or

$$x = x_1 + at, \qquad y = y_1 + bt, \qquad z = z_1 + ct. \qquad (2)$$

As t increases from $-\infty$ to ∞, the point with coordinates (2) traces out L. Thus the equations (2) are *parametric equations* of L, with t as the parameter.

Symmetric Equations of Lines

The numbers a, b and c are called *direction numbers* of the line L. If they are all nonzero, we can eliminate t from the equations (1). This gives the *symmetric equations*

$$\frac{x - x_1}{a} = \frac{y - y_1}{b} = \frac{z - z_1}{c} \qquad (3)$$

for a line in space. We can continue to use these equations even if one (or more) of the direction numbers is zero, with the understanding that in this case the corresponding numerator is also zero. For example, if $c = 0$, the third of the equations (2) tells us that $z = z_1$, and the symmetric equations then become

$$\frac{x - x_1}{a} = \frac{y - y_1}{b}, \qquad z = z_1.$$

Notice that if a, b and c are direction numbers for L, then so are pa, pb and pc, where p is any nonzero constant, since the vector $p\mathbf{m}$ is also parallel to L. In particular, the direction cosines of the vector $\mathbf{m} = (a, b, c)$ are direction numbers of the line L.

Example 1 Write parametric and symmetric equations for the line through the point $P_1 = (3, -1, 2)$ parallel to the vector $\mathbf{m} = (-2, 4, 5)$.

Solution Here $x_1 = 3$, $y_1 = -1$, $z_1 = 2$ and $a = -2$, $b = 4$, $c = 5$, so that the parametric equations (2) become

$$x = 3 - 2t, \quad y = -1 + 4t, \quad z = 2 + 5t \qquad (-\infty < t < \infty),$$

and the symmetric equations (3) become

$$\frac{x - 3}{-2} = \frac{y + 1}{4} = \frac{z - 2}{5}. \qquad \square$$

To find the line L going through two given points $P_1 = (x_1, y_1, z_1)$ and $P_2 = (x_2, y_2, z_2)$, we choose the vector $\mathbf{m} = (a, b, c)$ parallel to L to be the vector

$$\overrightarrow{P_1P_2} = (x_2 - x_1, y_2 - y_1, z_2 - z_1)$$

lying on L. Then, by (2) and (3), L has parametric equations

$$x = x_1 + (x_2 - x_1)t, \quad y = y_1 + (y_2 - y_1)t, \quad z = z_1 + (z_2 - z_1)t, \qquad (4)$$

where $-\infty < t < \infty$, and symmetric equations

$$\frac{x - x_1}{x_2 - x_1} = \frac{y - y_1}{y_2 - y_1} = \frac{z - z_1}{z_2 - z_1}. \tag{4'}$$

Example 2 Write parametric and symmetric equations for the line L through the points $P_1 = (3, 2, -1)$ and $P_2 = (4, -1, 1)$. Find the point of intersection of L with the yz-plane.

Solution Since $\overrightarrow{P_1P_2} = (1, -3, 2)$, it follows from (4) and (4') that L has parametric equations

$$x = 3 + t, \quad y = 2 - 3t, \quad z = -1 + 2t \quad (-\infty < t < \infty) \tag{5}$$

and symmetric equations

$$\frac{x - 3}{1} = \frac{y - 2}{-3} = \frac{z + 1}{2}. \tag{5'}$$

The point of intersection of L with the yz-plane is the point of the form $(0, y, z)$ whose coordinates satisfy (5'). Setting $x = 0$ in (5'), we get

$$\frac{y - 2}{-3} = \frac{z + 1}{2} = -3,$$

so that $y = 11$, $z = -7$. Thus the line L intersects the yz-plane in the point $(0, 11, -7)$. ☐

The following theorem affords a striking example of the power of vector methods in analytic geometry.

Theorem 4 *(Distance between a point and a line in space).* Given a line L in space and a point P not on L, let $\mathbf{m}$ be any vector parallel to L and let Q be any point on L. Then the distance d between P and L is given by

$$d = \frac{|\mathbf{m} \times \overrightarrow{QP}|}{|\mathbf{m}|}. \tag{6}$$

Proof Let θ be the angle between $\mathbf{m}$ and $\overrightarrow{QP}$ (see Figure 24). Then $0 \le \theta \le \pi$, and (6) follows at once by comparing the formulas

$$d = |\overrightarrow{QP}| \sin \theta$$

and

$$|\mathbf{m} \times \overrightarrow{QP}| = |\mathbf{m}| |\overrightarrow{QP}| \sin \theta. \quad \blacksquare$$

Example 3 Find the distance between the point $P = (4, 2, -2)$ and the line L with parametric equations

$$x = 3 - 2t, \quad y = 6t, \quad y = -1 + 9t \quad (-\infty < t < \infty).$$

Solution Setting $t = 0$ in these equations, we find that $Q = (3, 0, -1)$ is a point on L. Since $\mathbf{m} = (-2, 6, 9)$ and $\overrightarrow{QP} = (1, 2, -1)$, we have

$$\mathbf{m} \times \overrightarrow{QP} = \begin{vmatrix} \mathbf{i} & \mathbf{j} & \mathbf{k} \\ -2 & 6 & 9 \\ 1 & 2 & -1 \end{vmatrix} = \begin{vmatrix} 6 & 9 \\ 2 & -1 \end{vmatrix} \mathbf{i} - \begin{vmatrix} -2 & 9 \\ 1 & -1 \end{vmatrix} \mathbf{j} + \begin{vmatrix} -2 & 6 \\ 1 & 2 \end{vmatrix} \mathbf{k}$$

$$= -24\mathbf{i} + 7\mathbf{j} - 10\mathbf{k}.$$

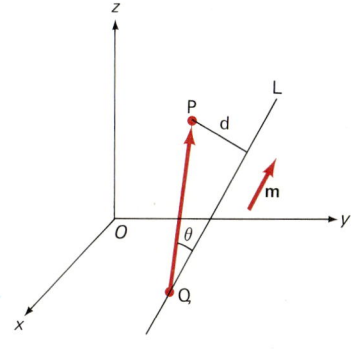

Figure 24

Therefore, by (6),

$$d = \frac{|\mathbf{m} \times \overrightarrow{QP}|}{|\mathbf{m}|} = \frac{|-24\mathbf{i} + 7\mathbf{j} - 10\mathbf{k}|}{|-2\mathbf{i} + 6\mathbf{j} + 9\mathbf{k}|} = \frac{\sqrt{(-24)^2 + 7^2 + (-10)^2}}{\sqrt{(-2)^2 + 6^2 + 9^2}}$$

$$= \frac{\sqrt{725}}{\sqrt{121}} = \frac{5}{11}\sqrt{29} \approx 2.45. \quad \square$$

Planes and Their Equations

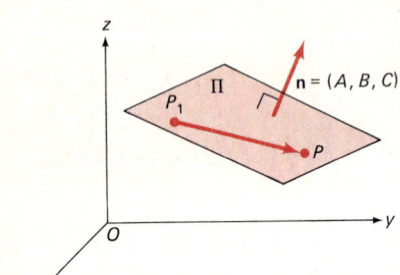

Next we consider planes in space. Just as a line L is determined by any point P_1 on L and any vector $\mathbf{m}$ *parallel* to L, a plane Π is determined by any point P_1 in Π and any nonzero vector $\mathbf{n} = (A, B, C)$ *perpendicular* to Π. Such a vector $\mathbf{n}$ is called a *normal* to the plane Π. Let $P = (x, y, z)$ be a variable point in space. Then Figure 25 shows that P lies in Π if and only if the vectors $\mathbf{n} = (A, B, C)$ and $\overrightarrow{P_1P}$ are perpendicular, or equivalently

$$\mathbf{n} \cdot \overrightarrow{P_1P} = 0.$$

But $\overrightarrow{P_1P} = (x - x_1, y - y_1, z - z_1)$, and therefore P lies in Π if and only if

$$A(x - x_1) + B(y - y_1) + C(z - z_1) = 0. \tag{7}$$

It follows from (7) that

$$Ax + By + Cz + D = 0, \tag{8}$$

where

$$D = -Ax_1 - By_1 - Cz_1. \tag{8'}$$

Conversely, the graph of any equation of the form (8), where A, B and C are not all zero, is a plane with normal $\mathbf{n} = (A, B, C)$. In fact, let x_1, y_1 and z_1 be any three numbers satisfying the condition (8'). (Why can such numbers always be found?) Then substitution of (8') into (8) gives an equation equivalent to (7), which is an equation of the plane through the point $P_1 = (x_1, y_1, z_1)$ perpendicular to the vector $\mathbf{n} = (A, B, C)$. Notice that the graph of (8) is a plane through the origin if $D = 0$, a plane parallel to the z-axis if $C = 0$, and a plane perpendicular to the z-axis if $A = B = 0$. You should examine other cases in which some of the numbers A, B, C and D are zero.

Example 4 Find an equation of the plane through the point $P_1 = (-2, 1, 3)$ perpendicular to the vector $\mathbf{n} = (4, 5, -1)$.

Solution Here $x_1 = -2$, $y_1 = 1$, $z_1 = 3$ and $A = 4$, $B = 5$, $C = -1$, so that equation (7) becomes

$$4(x + 2) + 5(y - 1) - (z - 3) = 0,$$

or equivalently

$$4x + 5y - z + 6 = 0. \quad \square$$

Example 5 Find the line of intersection L of the two planes

$$2x - 3y + 4z - 1 = 0 \quad \text{and} \quad x + 2y - z + 3 = 0. \tag{9}$$

Solution The first plane has the normal $\mathbf{n}_1 = (2, -3, 4)$ and the second plane has the normal $\mathbf{n}_2 = (1, 2, -1)$. Since the line L lies in both planes, it must be perpendicular to both $\mathbf{n}_1$ and $\mathbf{n}_2$. Thus L is parallel to the vector

$$\mathbf{n}_1 \times \mathbf{n}_2 = \begin{vmatrix} \mathbf{i} & \mathbf{j} & \mathbf{k} \\ 2 & -3 & 4 \\ 1 & 2 & -1 \end{vmatrix} = \begin{vmatrix} -3 & 4 \\ 2 & -1 \end{vmatrix}\mathbf{i} - \begin{vmatrix} 2 & 4 \\ 1 & -1 \end{vmatrix}\mathbf{j} + \begin{vmatrix} 2 & -3 \\ 1 & 2 \end{vmatrix}\mathbf{k}$$

$$= -5\mathbf{i} + 6\mathbf{j} + 7\mathbf{k}.$$

To find a point on L, we set $z = 0$ in both equations (9), and solve the resulting system of equations $2x - 3y - 1 = 0$, $x + 2y + 3 = 0$ for x and y, obtaining $x = -1$, $y = -1$. Hence the point $(-1, -1, 0)$ lies on L, and L has the symmetric equations

$$\frac{x+1}{-5} = \frac{y+1}{6} = \frac{z}{7}. \quad \square$$

It only takes two points to determine a line, but it takes *three* points to determine a plane.

Example 6 Write an equation for the plane Π going through the points $P_1 = (2, -1, 3)$, $P_2 = (1, 2, 2)$ and $P_3 = (-2, 1, 1)$.

Solution Since the points P_1, P_2 and P_3 lie in Π, so do the vectors $\overrightarrow{P_1P_2} = (-1, 3, -1)$ and $\overrightarrow{P_1P_3} = (-4, 2, -2)$. Hence the vector

$$\overrightarrow{P_1P_2} \times \overrightarrow{P_1P_3} = \begin{vmatrix} \mathbf{i} & \mathbf{j} & \mathbf{k} \\ -1 & 3 & -1 \\ -4 & 2 & -2 \end{vmatrix} = \begin{vmatrix} 3 & -1 \\ 2 & -2 \end{vmatrix}\mathbf{i} - \begin{vmatrix} -1 & -1 \\ -4 & 2 \end{vmatrix}\mathbf{j} + \begin{vmatrix} -1 & 3 \\ -4 & 2 \end{vmatrix}\mathbf{k}$$

$$= -4\mathbf{i} + 6\mathbf{j} + 10\mathbf{k},$$

which is perpendicular to both $\overrightarrow{P_1P_2}$ and $\overrightarrow{P_1P_3}$, is a normal to Π. Since Π is the plane through $P_1 = (2, -1, 3)$ with normal $\mathbf{n} = (-4, 6, 10)$, we find with the help of (7) that Π has an equation of the form

$$-4(x - 2) + 6(y + 1) + 10(z - 3) = 0,$$

or equivalently

$$2x - 3y - 5z + 8 = 0. \quad \square$$

Example 7 Find the angle between the planes $6x + 6y - 3z + 5 = 0$ and $x - 2y + 2z - 4 = 0$.

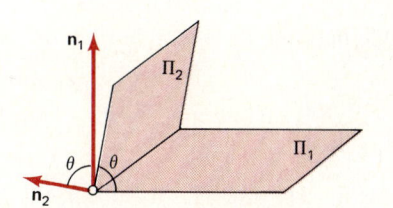

Solution The angle between two planes Π_1 and Π_2 is defined as the angle θ between their normals $\mathbf{n}_1$ and $\mathbf{n}_2$ if $0 \leq \theta \leq \pi/2$ (see Figure 26), or as $\pi - \theta$ if $\pi/2 < \theta \leq \pi$ (in this way the angle between two planes is always the smaller of the two possible choices). Here $\mathbf{n}_1 = (6, 6, -3)$ and $\mathbf{n}_2 = (1, -2, 2)$, so that

$$\cos\theta = \frac{\mathbf{n}_1 \cdot \mathbf{n}_2}{|\mathbf{n}_1||\mathbf{n}_2|} = \frac{6(1) + 6(-2) - 3(2)}{\sqrt{6^2 + 6^2 + (-3)^2}\sqrt{1^2 + (-2)^2 + 2^2}}$$

$$= \frac{-12}{\sqrt{81}\sqrt{9}} = \frac{-12}{9(3)} = -\frac{4}{9}.$$

The angle between the planes Π_1 and Π_2 is θ.
Figure 26

Since $\cos\theta$ is negative, θ is obtuse and the angle between the given planes is equal to

$$\pi - \theta = \pi - \arccos\left(-\frac{4}{9}\right) = \arccos\frac{4}{9} \approx 63.6°. \quad \square$$

Section 12.4   Lines and Planes in Space   719

Our next theorem is the analogue for planes of Theorem 10, page 34.

Theorem 5 (*Distance between a point and a plane*). *The distance d between the point $P_1 = (x_1, y_1, z_1)$ and the plane Π with equation*

$$Ax + By + Cz + D = 0$$

is given by

$$d = \frac{|Ax_1 + By_1 + Cz_1 + D|}{\sqrt{A^2 + B^2 + C^2}}. \tag{10}$$

Proof The vector $\mathbf{n} = (A, B, C)$ is normal to the plane Π. Let $Q = (a, b, c)$ be any point in Π, as in Figure 27. Then

$$d = |\text{comp}_\mathbf{n} \overrightarrow{QP_1}|,$$

where

$$\text{comp}_\mathbf{n} \overrightarrow{QP_1} = \frac{\mathbf{n} \cdot \overrightarrow{QP_1}}{|\mathbf{n}|}$$

is the component of $\overrightarrow{QP_1} = (x_1 - a, y_1 - b, z_1 - c)$ along $\mathbf{n}$. Therefore

$$d = \frac{|A(x_1 - a) + B(y_1 - b) + C(z_1 - c)|}{\sqrt{A^2 + B^2 + C^2}}$$

$$= \frac{|Ax_1 + By_1 + Cz_1 - (Aa + Bb + Cc)|}{\sqrt{A^2 + B^2 + C^2}}. \tag{11}$$

But $Aa + Bb + Cc + D = 0$, since $Q = (a, b, c)$ lies in Π, so that

$$D = -(Aa + Bb + Cc).$$

Substituting this expression for D into (11), we immediately get (10). ∎

Example 8 Find the distance between the point $(2, 6, -1)$ and the plane $3x - 4y + 12z - 22 = 0$.

Solution With the help of formula (10), we find that

$$d = \frac{|3(2) - 4(6) + 12(-1) - 22|}{\sqrt{3^2 + (-4)^2 + 12^2}} = \frac{|-52|}{\sqrt{169}} = \frac{52}{13} = 4. \quad \square$$

It is geometrically apparent that two planes are parallel if and only if their normals are parallel vectors. The angle between parallel planes is zero (why?).

Example 9 Verify that the planes $3x + 6y - 12z + 7 = 0$ and $x + 2y - 4z - 1 = 0$ are parallel, and find the distance d between them.

Solution The two planes, which we denote by Π_1 and Π_2, have normals $\mathbf{n}_1 = (3, 6, -12)$ and $\mathbf{n}_2 = (1, 2, -4)$. Since $\mathbf{n}_2 = \frac{1}{3}\mathbf{n}_1$, the vectors $\mathbf{n}_1$ and $\mathbf{n}_2$ are parallel, and hence so are the planes Π_1 and Π_2. Clearly d equals the distance between Π_1 and any point of Π_2, or between Π_2 and any point of Π_1. The point $(1, 0, 0)$ lies in Π_2, and therefore

$$d = \frac{|3(1) + 6(0) - 12(0) + 7|}{\sqrt{3^2 + 6^2 + (-12)^2}} = \frac{|3 + 7|}{3\sqrt{1^2 + 2^2 + (-4)^2}} = \frac{10}{3\sqrt{21}} \approx 0.73. \quad \square$$

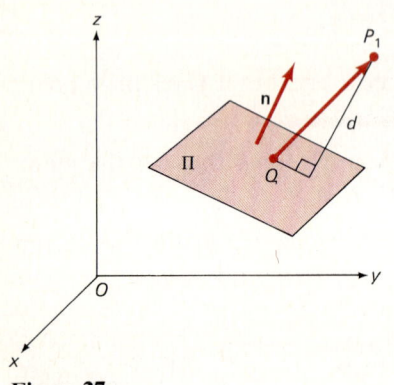

Figure 27

720 Chapter 12 Vectors in Space and Solid Analytic Geometry

Problems

Write parametric equations for the line through the point $(1, -2, 4)$ parallel to
1. The vector $\mathbf{m} = (2, 3, -1)$
2. The vector $\mathbf{m} = (5, 1, 0)$
3. The line $\dfrac{x-1}{-2} = \dfrac{x-2}{5} = \dfrac{z+1}{-3}$
4. The line $x = -1 + 3t,\ y = 3 - 2t,\ z = 2 + 5t$

Write symmetric equations for the line through the point $(-3, 2, 0)$ parallel to
5. The vector $\mathbf{m} = (9, -4, 3)$
6. The vector $\mathbf{m} = (-1, 1, -1)$
7. The line $x = 6t,\ y = 4,\ z = 10 - 5t$
8. The line $\dfrac{x-6}{-2} = \dfrac{y+4}{5} = \dfrac{z+7}{8}$

Write parametric and symmetric equations for the line through the given pair of points.
9. $(1, -2, 2),\ (3, 1, -1)$
10. $(0, 0, 1),\ (0, 2, -2)$
11. $(4, 1, 4),\ (-1, 5, 3)$
12. $(3, -6, 5),\ (10, 4, 8)$

Find the distance between the point P and the given line.
13. $P = (1, 3, 2),\ x = 3 - 2t,\ y = 1 + 2t,\ z = -2 + t$
14. $P = (4, -1, 2),\ x = 2 + 3t,\ y = -3 - 4t,\ z = 1 + 12t$
15. $P = (0, 1, 0),\ \dfrac{x+1}{-4} = \dfrac{y-1}{2} = \dfrac{z}{4}$
16. $P = (-6, 5, -7),\ \dfrac{x+7}{3} = \dfrac{y-4}{-2} = \dfrac{z+8}{6}$

Find an equation of the plane through
17. The origin with normal $\mathbf{n} = (4, 5, -3)$
18. The point $(2, 1, -1)$ with normal $\mathbf{n} = (1, -2, 3)$
19. The point $(3, -1, 5)$ perpendicular to the line through this point and the point $(6, -7, 9)$
20. The point $(3, 4, -5)$ parallel to both vectors $\mathbf{a} = (3, 1, -1)$ and $\mathbf{b} = (1, -2, 1)$
21. The points $(2, -1, 3)$ and $(3, 1, 2)$ parallel to the vector $\mathbf{a} = (3, -1, -4)$
22. The points $(3, -1, 2),\ (4, -1, -1)$ and $(2, 0, 2)$
23. The points $(1, -1, -2)$ and $(3, 1, 1)$ perpendicular to the plane $x - 2y - 3z - 5 = 0$
24. The point $(2, -1, 1)$ perpendicular to both planes $y = 0$ and $2x - z + 1 = 0$

25. Find the point in which the line
$$\dfrac{x-1}{2} = \dfrac{y+1}{-1} = \dfrac{z}{3}$$
intersects the plane $2x + 3y + z - 11 = 0$.

26. Let L be the line through the points $(6, -6, 5)$ and $(-12, 6, -1)$. Find the points in which L intersects the coordinate planes.

Find the line of intersection of the given pair of planes.
27. $x - 2y + 3z - 6 = 0,\ 3x + 2y - 5z - 10 = 0$
28. $4x + y + z = 0,\ 2x + 3y - 2z + 5 = 0$
29. $x - 2y + 3z + 2 = 0,\ 2x + y - 4z - 16 = 0$

30. Show that the line
$$\dfrac{x+5}{3} = \dfrac{y-1}{-2} = \dfrac{z+4}{1}$$
and the line of intersection of the planes $x + y - z + 3 = 0$ and $x - y - 5z = 0$ are parallel.

Find the angle between the given pair of planes.
31. $6x + 3y - 2z = 0,\ x + 2y + 2z = 0$
32. $x - \sqrt{2}y + z + 4 = 0,\ x + \sqrt{2}y - z - 6 = 0$
33. $3y - z + 1 = 0,\ 2x + z - 2 = 0$
34. $9x - 2y + 6z + 5 = 0,\ 4x + 2y - 4z + 1 = 0$

Find the distance between the point P and the given plane.
35. $P = (4, -1, 1),\ 16x - 12y + 15z + 9 = 0$
36. $P = (1, 6, -3),\ 6x - 2y - 9z + 12 = 0$
37. $P = (8, 3, -2),\ 12y - 5z - 27 = 0$

Find the distance between the given pair of parallel planes.
38. $x - 2y + 2z + 12 = 0,\ x - 2y + 2z - 6 = 0$
39. $6x + 18y - 9z - 21 = 0,\ 4x + 12y - 6z + 7 = 0$
40. $15x - 16y + 12z + 5 = 0,\ 30x - 32y + 24z - 5 = 0$

41. For what value of c is the plane $x + y + z = c$ tangent to the sphere $x^2 + y^2 + z^2 = 12$ in the first octant? Find the point of tangency.

42. Two lines in space are said to be *skew* if they are nonparallel and nonintersecting. Show that the lines L_1 and L_2 with symmetric equations
$$\dfrac{x-x_1}{a_1} = \dfrac{y-y_1}{b_1} = \dfrac{z-z_1}{c_1},\qquad \dfrac{x-x_2}{a_2} = \dfrac{y-y_2}{b_2} = \dfrac{z-z_2}{c_2} \tag{i}$$

and skew if and only if the determinant
$$\begin{vmatrix} x_2 - x_1 & y_2 - y_1 & z_2 - z_1 \\ a_1 & b_1 & c_1 \\ a_2 & b_2 & c_2 \end{vmatrix}$$
is nonzero.

Hint. After setting $P_1 = (x_1, y_1, z_1)$, $P_2 = (x_2, y_2, z_2)$, $\mathbf{m}_1 = (a_1, b_1, c_1)$ and $\mathbf{m}_2 = (a_2, b_2, c_2)$, observe that L_1 and L_2 are parallel or intersecting if and only if the vectors $\overrightarrow{P_1P_2}$, $\mathbf{m}_1$ and $\mathbf{m}_2$ are coplanar (as they are in Figure 28).

Section 12.4 Lines and Planes in Space **721**

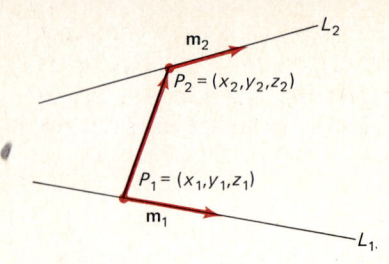

Figure 28

43. Show that if the lines (i) are skew, then the (shortest) distance between them is

$$d = \frac{|\overrightarrow{P_1P_2} \cdot (\mathbf{m}_1 \times \mathbf{m}_2)|}{|\mathbf{m}_1 \times \mathbf{m}_2|}, \quad \text{(ii)}$$

where P_1, P_2, $\mathbf{m}_1$ and $\mathbf{m}_2$ have the same meaning as in the preceding problem.

44. Show that the lines

$$\frac{x+4}{2} = \frac{y-4}{-1} = \frac{z+1}{2}, \quad \frac{x-1}{4} = \frac{y+2}{-3} = \frac{z-4}{5}$$

are skew. Use formula (ii) to find the distance between them.

45. Show that the lines

$$\frac{x+1}{2} = \frac{y-1}{3} = \frac{z-2}{-4}, \quad \frac{x-4}{1} = \frac{y-3}{-4} = \frac{z+5}{1}$$

intersect. Find the point of intersection.

46. The sphere $x^2 + y^2 + z^2 = 4z$ is illuminated by a beam of light parallel to the line $x = 0$, $y = z$. Find the shadow cast on the xy-plane (see Figure 29).

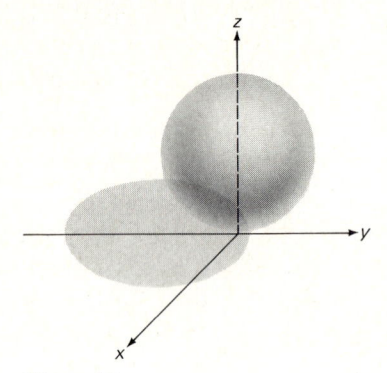

Figure 29

12.5 Space Curves and Orbital Motion

By a *curve in space* or a *space curve* we mean the graph of a *triple* of (parametric) equations

$$x = x(t), \quad y = y(t), \quad z = z(t), \quad (1)$$

that is, the set of all points (x, y, z) whose coordinates satisfy (1); here $x(t)$, $y(t)$ and $z(t)$ are three continuous functions with the same domain of definition, which we always take to be some interval I. We also assume that $x(t)$, $y(t)$ and $z(t)$ are not all constant functions, since otherwise the curve (1) would reduce to a single point. As the parameter t, which may be regarded as the time, varies over the interval I, the point $P = (x, y, z)$ takes various positions in space, and traces out the (parametric) curve (1). By an *arc* of the curve (1) we mean any curve with the same parametric equations, but where the domain of $x(t)$, $y(t)$ and $z(t)$ is some subinterval of I. All this is just the natural generalization to three dimensions of the definition of a plane curve given on page 451.

Example 1 The curve with parametric equations

$$x = x_1 + at, \quad y = y_1 + bt, \quad z = z_1 + ct \quad (-\infty < t < \infty)$$

is the straight line through the point (x_1, y_1, z_1) with direction numbers a, b and c.

Example 2 Let a and b be positive numbers. Then the curve C with parametric equations

$$x = a \cos t, \quad y = a \sin t, \quad z = bt \quad (0 \leq t < \infty) \quad (2)$$

is the corkscrew-shaped curve C shown in Figure 30, known as a *circular helix*. Since $x^2 + y^2 = a^2(\cos^2 t + \sin^2 t) = a^2$, the helix C lies on the surface

Circular helix

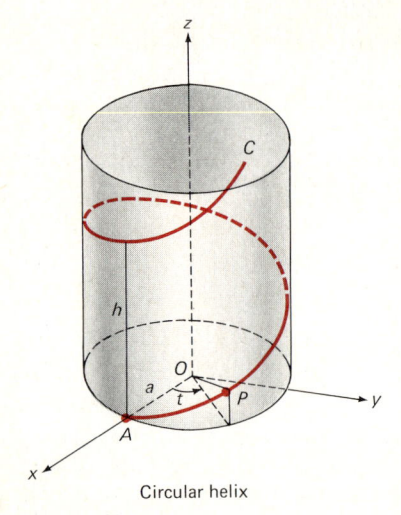

Figure 30

722   Chapter 12   Vectors in Space and Solid Analytic Geometry

of a right circular cylinder of radius a, with the z-axis as its axis of symmetry. As t increases, the helix winds around the cylinder, starting from the initial point $A = (a, 0, 0)$ and going around the cylinder once every time t increases by 2π. The vertical distance h between adjacent "turns" of C, equal to $2\pi b$, is called the *pitch* of the helix.

Remark The helix in the figure is *right-handed*, in the sense that it resembles the threads of a right-handed screw. If we drop the requirement that b be positive, we get a circle for $b = 0$ and a *left-handed* helix for $b < 0$ (check this).

Length of a Space Curve

The length of a space curve C is defined in the same way as the length of a plane curve, namely as the limiting length of a polygonal path inscribed in C, provided that this limit exists and is finite, in which case C is said to be *rectifiable*. An argument like that on page 461 shows that if $x(t)$, $y(t)$ and $z(t)$ are continuously differentiable on an interval $[a, b]$, then the curve

$$x = x(t), \quad y = y(t), \quad z = z(t) \qquad (a \le t \le b)$$

is rectifiable, with length L given by

$$L = \int_a^b \sqrt{\left(\frac{dx}{dt}\right)^2 + \left(\frac{dy}{dt}\right)^2 + \left(\frac{dz}{dt}\right)^2} \, dt \tag{3}$$

(it is assumed that C has no more than a finite number of self-intersections).

Example 3 Find the length L of one turn of the helix $x = 3 \cos t$, $y = 3 \sin t$, $z = 4t$.

Solution As t increases by 2π, the point $P = (3 \cos t, 3 \sin t, 4t)$ traces out one turn of the helix. Therefore, choosing 0 and 2π as the limits of integration in (3), we find that

$$L = \int_0^{2\pi} \sqrt{(-3 \sin t)^2 + (3 \cos t)^2 + 4^2} \, dt$$

$$= \int_0^{2\pi} \sqrt{9 + 16} \, dt = 5 \int_0^{2\pi} dt = 10\pi. \quad \square$$

Let C be the curve with parametric equations (1). Then, just as on pages 667–668, we can introduce the *arc length function*

$$s = s(t) = \int_a^t \sqrt{\left(\frac{dx}{du}\right)^2 + \left(\frac{dy}{du}\right)^2 + \left(\frac{dz}{du}\right)^2} \, du, \tag{4}$$

equal to the length of the arc of C with fixed initial point $P(a) = (x(a), y(a), z(a))$ and variable final point $P(t) = (x(t), y(t), z(t))$. Here we pick u as the variable of integration to avoid confusion with the upper limit of integration t. Differentiating (4) with respect to t, we get

$$\frac{ds}{dt} = \sqrt{\left(\frac{dx}{dt}\right)^2 + \left(\frac{dy}{dt}\right)^2 + \left(\frac{dz}{dt}\right)^2}. \tag{4'}$$

Section 12.5 Space Curves and Orbital Motion

Vector Functions in Space

Vector-valued functions (or simply vector functions) in space are defined in exactly the same way as in the plane, except that the vectors now have a z-component as well as x- and y-components. Apart from this minor difference, limits, derivatives and integrals of three-dimensional vector functions are calculated "component by component," just as in the two-dimensional case. For instance, the limit of the vector function

$$\mathbf{r}(t) = (2 \tan t)\mathbf{i} + (4t)\mathbf{j} + (\sec t)\mathbf{k},$$

as $t \to \pi/4$ is

$$\lim_{t \to \pi/4} \mathbf{r}(t) = \left(\lim_{t \to \pi/4} 2 \tan t\right)\mathbf{i} + \left(\lim_{t \to \pi/4} 4t\right)\mathbf{j} + \left(\lim_{t \to \pi/4} \sec t\right)\mathbf{k} = 2\mathbf{i} + \pi\mathbf{j} + \sqrt{2}\mathbf{k},$$

and its derivative is

$$\frac{d}{dt}\mathbf{r}(t) = \left(\frac{d}{dt} 2 \tan t\right)\mathbf{i} + \left(\frac{d}{dt} 4t\right)\mathbf{j} + \left(\frac{d}{dt} \sec t\right)\mathbf{k}$$

$$= (2 \sec^2 t)\mathbf{i} + 4\mathbf{j} + (\sec t \tan t)\mathbf{k}.$$

Example 4 Find the derivative of the cross product of two differentiable vector functions $\mathbf{r}_1 = \mathbf{r}_1(t)$ and $\mathbf{r}_2 = \mathbf{r}_2(t)$.

Solution Instead of differentiating components, we start from the definition of the derivative. Thus

$$\frac{d}{dt}(\mathbf{r}_1 \times \mathbf{r}_2) = \lim_{u \to t} \frac{\mathbf{r}_1(u) \times \mathbf{r}_2(u) - \mathbf{r}_1(t) \times \mathbf{r}_2(t)}{u - t}$$

$$= \lim_{u \to t} \frac{\mathbf{r}_1(u) \times \mathbf{r}_2(u) - \mathbf{r}_1(t) \times \mathbf{r}_2(u) + \mathbf{r}_1(t) \times \mathbf{r}_2(u) - \mathbf{r}_1(t) \times \mathbf{r}_2(t)}{u - t}$$

$$= \lim_{u \to t} \left(\frac{\mathbf{r}_1(u) - \mathbf{r}_1(t)}{u - t} \times \mathbf{r}_2(u) + \mathbf{r}_1(t) \times \frac{\mathbf{r}_2(u) - \mathbf{r}_2(t)}{u - t}\right)$$

$$= \left(\lim_{u \to t} \frac{\mathbf{r}_1(u) - \mathbf{r}_1(t)}{u - t}\right) \times \left(\lim_{u \to t} \mathbf{r}_2(u)\right) + \mathbf{r}_1(t) \times \left(\lim_{u \to t} \frac{\mathbf{r}_2(u) - \mathbf{r}_2(t)}{u - t}\right).$$

But

$$\lim_{u \to t} \frac{\mathbf{r}_1(u) - \mathbf{r}_1(t)}{u - t} = \frac{d\mathbf{r}_1}{dt}, \quad \lim_{u \to t} \mathbf{r}_2(u) = \mathbf{r}_2(t), \quad \lim_{u \to t} \frac{\mathbf{r}_2(u) - \mathbf{r}_2(t)}{u - t} = \frac{d\mathbf{r}_2}{dt},$$

and therefore

$$\frac{d}{dt}(\mathbf{r}_1 \times \mathbf{r}_2) = \left(\frac{d\mathbf{r}_1}{dt} \times \mathbf{r}_2\right) + \left(\mathbf{r}_1 \times \frac{d\mathbf{r}_2}{dt}\right). \tag{5}$$

There are many steps here, and make sure you understand why they all work. □

Notice the resemblance between formula (5) and the analogous rule for differentiating the product of two scalar functions. However, the products in the right side of (5) are not ordinary products, but *cross* products. The companion formula

$$\frac{d}{dt}(\mathbf{r}_1 \cdot \mathbf{r}_2) = \left(\frac{d\mathbf{r}_1}{dt} \cdot \mathbf{r}_2\right) + \left(\mathbf{r}_1 \cdot \frac{d\mathbf{r}_2}{dt}\right) \tag{5'}$$

for *dot* products can be established in virtually the same way.

Velocity and Speed

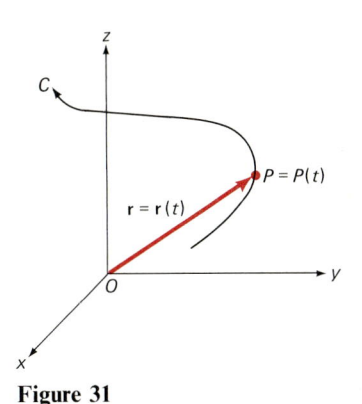

We now describe the use of vector functions to study motion in space. Let

$$\mathbf{r} = \mathbf{r}(t) = x(t)\mathbf{i} + y(t)\mathbf{j} + z(t)\mathbf{k} \qquad (t \text{ in } I)$$

be a differentiable vector function defined on an interval I, and suppose the initial point of $\mathbf{r} = \mathbf{r}(t)$ is placed at the origin O, as in Figure 31. Then the final point of $\mathbf{r} = \mathbf{r}(t)$ is a variable point $P = P(t) = (x(t), y(t), z(t))$ in space, and $\mathbf{r} = \mathbf{r}(t)$ is called the *position vector* of $P = P(t)$. As t increases, $P = P(t)$ traces out a space curve, called the *graph* of $\mathbf{r} = \mathbf{r}(t)$, namely the curve C with parametric equations

$$x = x(t), \quad y = y(t), \quad z = z(t) \qquad (t \text{ in } I),$$

and from now on we will think of the parameter t as the time. The derivative

$$\mathbf{v} = \mathbf{v}(t) = \frac{d\mathbf{r}}{dt} = \frac{dx}{dt}\mathbf{i} + \frac{dy}{dt}\mathbf{j} + \frac{dz}{dt}\mathbf{k}$$

Figure 31

is called the *velocity* of the moving point $P = P(t)$, and its magnitude

$$v = |\mathbf{v}| = \left|\frac{d\mathbf{r}}{dt}\right| = \sqrt{\left(\frac{dx}{dt}\right)^2 + \left(\frac{dy}{dt}\right)^2 + \left(\frac{dz}{dt}\right)^2}$$

is called the *speed*. Comparison of this formula with (4') gives

$$v = \frac{ds}{dt},$$

so that the speed of P is just the time rate of change of the distance traveled by P along C. All these ideas have already appeared in the last chapter, and we are merely reviewing them here, at the same time allowing $\mathbf{r}(t)$ and $\mathbf{v}(t)$ to have three components instead of only two.

The Unit Tangent Vector

We will henceforth assume that the velocity vector $\mathbf{v} = d\mathbf{r}/dt$ is nonzero, with its initial point at $P = P(t)$, the final point of the position vector $\mathbf{r} = \mathbf{r}(t)$. Let C be the graph of $\mathbf{r} = \mathbf{r}(t)$. Then, as on page 667, $\mathbf{v} = \mathbf{v}(t)$ is tangent to C at the point P, in the sense that $\mathbf{v}$ has the "limiting direction" as $\Delta t \to 0$ of the vector $\Delta \mathbf{r}/\Delta t$, where $\Delta \mathbf{r} = \mathbf{r}(t + \Delta t) - \mathbf{r}(t)$. The vector $\Delta \mathbf{r}$ is shown in Figure 32, where it is assumed that $\Delta t > 0$. Regardless of the sign of Δt, the vector $\Delta \mathbf{r}/\Delta t$, and hence the velocity $\mathbf{v}$, always points along C in the direction of increasing t (why?); in the figure this direction is indicated by the arrowhead on C. As in the two-dimensional case, the *unit tangent vector* $\mathbf{T}$ to the curve C at the point P is defined by

$$\mathbf{T} = \mathbf{T}(t) = \frac{\mathbf{v}(t)}{|\mathbf{v}(t)|} = \frac{d\mathbf{r}/dt}{|d\mathbf{r}/dt|},$$

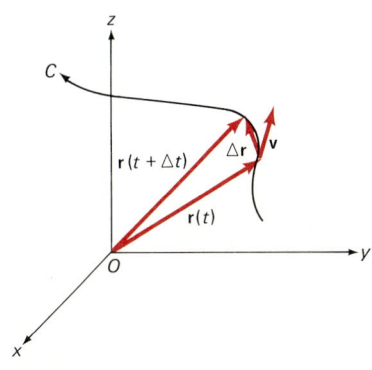

Figure 32

and like $\mathbf{v}$ itself is tangent to C at the point P and points in the direction of increasing t. The same argument as on page 669 shows that

$$\mathbf{T} = \frac{d\mathbf{r}}{ds}$$

if we go over to a description of C with the arc length s as parameter.

Example 5 Find the velocity $\mathbf{v}$, speed v and unit tangent vector $\mathbf{T}$ for the helix $x = 3\cos t$, $y = 3\sin t$, $z = 4t$. Show that $\mathbf{T}$ makes a constant angle with the z-axis.

Solution Here the position vector is

$$\mathbf{r}(t) = (3\cos t)\mathbf{i} + (3\sin t)\mathbf{j} + (4t)\mathbf{k}. \tag{6}$$

Therefore

$$\mathbf{v} = \frac{d\mathbf{r}}{dt} = (-3\sin t)\mathbf{i} + (3\cos t)\mathbf{j} + 4\mathbf{k},$$

and

$$v = |\mathbf{v}| = \sqrt{(-3\sin t)^2 + (3\cos t)^2 + 16} = \sqrt{9 + 16} = 5,$$

so that the speed has the constant value 5. The unit tangent vector is

$$\mathbf{T} = \frac{\mathbf{v}}{v} = \left(-\frac{3}{5}\sin t\right)\mathbf{i} + \left(\frac{3}{5}\cos t\right)\mathbf{j} + \frac{4}{5}\mathbf{k}.$$

If θ is the angle between $\mathbf{T}$ and the z-axis, then

$$\cos\theta = \frac{\mathbf{T}\cdot\mathbf{k}}{|\mathbf{T}||\mathbf{k}|} = \frac{4}{5},$$

so that θ has the constant value $\arccos\frac{4}{5} \approx 36.9°$. If we measure arc length along the helix from the point $\mathbf{r}(0) = 3\mathbf{i} = (3, 0, 0)$, then

$$s = \int_0^t v\,dt = 5t,$$

so that $t = s/5$ and

$$\mathbf{T} = \left(-\frac{3}{5}\sin\frac{s}{5}\right)\mathbf{i} + \left(\frac{3}{5}\cos\frac{s}{5}\right)\mathbf{j} + \frac{4}{5}\mathbf{k}.$$

Notice that the same expression for $\mathbf{T}$ is obtained if we substitute $t = s/5$ in formula (6) and then calculate the derivative $\mathbf{T} = d\mathbf{r}/ds$. □

The Unit Normal Vector

Next suppose the position vector function $\mathbf{r}(t)$ has a continuous second derivative on the interval I, and again let $\mathbf{T} = \mathbf{T}(t)$ be the unit tangent vector to the curve C at the point $P = P(t)$. Then, since $|\mathbf{T}|^2 = \mathbf{T}\cdot\mathbf{T} = 1$, we have

$$\frac{d}{dt}(\mathbf{T}\cdot\mathbf{T}) = \frac{d\mathbf{T}}{dt}\cdot\mathbf{T} + \mathbf{T}\cdot\frac{d\mathbf{T}}{dt} = 2\mathbf{T}\cdot\frac{d\mathbf{T}}{dt} = 0,$$

and therefore

$$\mathbf{T}\cdot\frac{d\mathbf{T}}{dt} = 0,$$

so that $d\mathbf{T}/dt$ is orthogonal to $\mathbf{T}$. Besides the unit tangent vector $\mathbf{T} = \mathbf{T}(t)$, we now introduce another vector, denoted by $\mathbf{N} = \mathbf{N}(t)$ and called the *unit normal vector* to the curve C at the point $P = P(t)$. This is the unit vector

$$\mathbf{N} = \frac{d\mathbf{T}/dt}{|d\mathbf{T}/dt|} \quad \left(\frac{d\mathbf{T}}{dt} \neq \mathbf{0}\right)$$

in the direction of $d\mathbf{T}/dt$, with its initial point at P. Since $d\mathbf{T}/dt$ is orthogonal to $\mathbf{T}$ and points in the direction in which C is bending, the same is true of the vector $\mathbf{N}$ (see Figure 33).

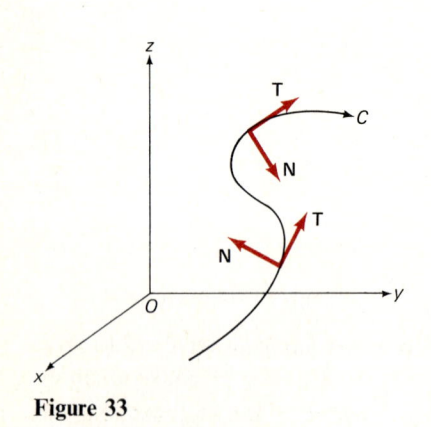

Figure 33

Remark The vector **N** is often called the *principal* unit normal vector, to emphasize that in space, as opposed to the plane, **N** is only one of infinitely many unit vectors orthogonal to **T** (see Figure 34).

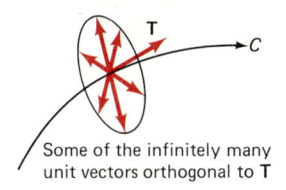

Figure 34

Curvature

If the arc length s is chosen as the parameter, the position vector $\mathbf{r} = \mathbf{r}(s)$, its final point $P = P(s)$, the unit tangent vector $\mathbf{T} = \mathbf{T}(s) = d\mathbf{r}/ds$, and the unit normal vector

$$\mathbf{N} = \mathbf{N}(s) = \frac{d\mathbf{T}/ds}{|d\mathbf{T}/ds|} \quad \left(\frac{d\mathbf{T}}{ds} \neq \mathbf{0}\right), \tag{7}$$

are all functions of s, as indicated by the notation. It follows from (7) that

$$\frac{d\mathbf{T}}{ds} = \left|\frac{d\mathbf{T}}{ds}\right| \mathbf{N}.$$

Let C be the graph of $\mathbf{r} = \mathbf{r}(s)$. Then the positive scalar $|d\mathbf{T}/ds|$ is called the *curvature* of C at P, denoted by κ, and the last equation can be written in the form

$$\frac{d\mathbf{T}}{ds} = \kappa \mathbf{N}. \tag{8}$$

Notice that the definitions of **T**, **N** and κ are essentially the same as for curves in the plane (see Section 11.4).

The curve C is usually given as the graph of a position vector function $\mathbf{r} = \mathbf{r}(t)$, where t is a parameter other than the arc length s, and we now use the cross product to find a concise expression for the curvature κ in terms of the first and second derivatives of $\mathbf{r}(t)$. First we observe that

$$\mathbf{v} = \frac{d\mathbf{r}}{dt} = v\mathbf{T},$$

since $\mathbf{T} = \mathbf{v}/v$ by definition, so that

$$\frac{d\mathbf{v}}{dt} = \frac{d^2\mathbf{r}}{dt^2} = \frac{dv}{dt}\mathbf{T} + v\frac{d\mathbf{T}}{dt} = \frac{dv}{dt}\mathbf{T} + v\frac{ds}{dt}\frac{d\mathbf{T}}{ds},$$

by the chain rule, or equivalently

$$\frac{d\mathbf{v}}{dt} = \frac{dv}{dt}\mathbf{T} + \kappa v^2 \mathbf{N}, \tag{9}$$

after substituting from (8) and using the fact that $v = ds/dt$. Forming the cross product of $\mathbf{v} = v\mathbf{T}$ and $d\mathbf{v}/dt$, we obtain

$$\mathbf{v} \times \frac{d\mathbf{v}}{dt} = v\frac{dv}{dt}\mathbf{T} \times \mathbf{T} + \kappa v^3 \mathbf{T} \times \mathbf{N} = \kappa v^3 \mathbf{T} \times \mathbf{N},$$

since $\mathbf{T} \times \mathbf{T} = \mathbf{0}$. But v is always positive, because of our assumption that

$\mathbf{v} \neq \mathbf{0}$, and $|\mathbf{T} \times \mathbf{N}| = 1$, since $\mathbf{T}$ and $\mathbf{N}$ are orthogonal unit vectors. Thus, taking the magnitude of $\mathbf{v} \times d\mathbf{v}/dt$ and solving for the curvature κ, we get

$$\kappa = \frac{\left|\mathbf{v} \times \dfrac{d\mathbf{v}}{dt}\right|}{v^3} \quad \text{or} \quad \kappa = \frac{\left|\dfrac{d\mathbf{r}}{dt} \times \dfrac{d^2\mathbf{r}}{dt^2}\right|}{\left|\dfrac{d\mathbf{r}}{dt}\right|^3},$$

which can be written more concisely as

$$\kappa = \frac{|\mathbf{r}' \times \mathbf{r}''|}{|\mathbf{r}'|^3}, \tag{10}$$

where the prime denotes differentiation with respect to t.

Acceleration and Its Components

As always, the time derivative $d\mathbf{v}/dt$ of the velocity $\mathbf{v}$ is called the *acceleration*, denoted by $\mathbf{a}$. Thus (9) can be written in the form

$$\mathbf{a} = \frac{dv}{dt}\mathbf{T} + \kappa v^2 \mathbf{N},$$

or equivalently

$$\mathbf{a} = a_T \mathbf{T} + a_N \mathbf{N},$$

where the scalars

$$a_T = \frac{dv}{dt}, \quad a_N = \kappa v^2 = \frac{v^2}{R} \tag{11}$$

are the *tangential and normal components* of the acceleration, and $R = 1/\kappa$ is the *radius of curvature*. These formulas are exactly the same as their counterparts for plane curves (see pages 679–680).

Example 6 Find the curvature κ of a plane curve C.

Solution There is no loss of generality in assuming that C lies in the xy-plane. Then $\mathbf{r} = x\mathbf{i} + y\mathbf{j}$, $\mathbf{r}' = x'\mathbf{i} + y'\mathbf{j}$ and $\mathbf{r}'' = x''\mathbf{i} + y''\mathbf{j}$, so that

$$\mathbf{r}' \times \mathbf{r}'' = \begin{vmatrix} \mathbf{i} & \mathbf{j} & \mathbf{k} \\ x' & y' & 0 \\ x'' & y'' & 0 \end{vmatrix} = (x'y'' - y'x'')\mathbf{k}.$$

Thus in this case (10) gives

$$\kappa = \frac{|\mathbf{r}' \times \mathbf{r}''|}{|\mathbf{r}'|^3} = \frac{|x'y'' - y'x''|}{(x'^2 + y'^2)^{3/2}},$$

as found by a different argument on page 677. $\square$

Example 7 Find the unit normal vector $\mathbf{N}$, the acceleration $\mathbf{a}$ and the curvature κ for the helix $x = 3\cos t$, $y = 3\sin t$, $z = 4t$.

Solution As already shown in Example 5, the velocity $\mathbf{v}$, speed v and unit tangent vector $\mathbf{T}$ are

$$\mathbf{v} = (-3\sin t)\mathbf{i} + (3\cos t)\mathbf{j} + 4\mathbf{k}, \quad v = |\mathbf{v}| \equiv 5,$$

$$\mathbf{T} = \left(-\frac{3}{5}\sin t\right)\mathbf{i} + \left(\frac{3}{5}\cos t\right)\mathbf{j} + \frac{4}{5}\mathbf{k}.$$

The derivative of **T** is

$$\frac{d\mathbf{T}}{dt} = -\frac{3}{5}[(\cos t)\mathbf{i} + (\sin t)\mathbf{j}],$$

with constant magnitude

$$\left|\frac{d\mathbf{T}}{dt}\right| = \frac{3}{5}\sqrt{(\cos t)^2 + (\sin t)^2} = \frac{3}{5}.$$

Hence the unit normal vector is

$$\mathbf{N} = \frac{d\mathbf{T}/dt}{|d\mathbf{T}/dt|} = -[(\cos t)\mathbf{i} + (\sin t)\mathbf{j}],$$

which always lies in a plane parallel to the xy-plane and points toward the z-axis (why?). Differentiating the velocity, we find that the acceleration is

$$\mathbf{a} = \frac{d\mathbf{v}}{dt} = -3[(\cos t)\mathbf{i} + (\sin t)]\mathbf{j} = 3\mathbf{N},$$

with tangential component 0 and normal component 3. Since the normal component of the acceleration is κv^2, the curvature has the constant value

$$\kappa = \frac{3}{v^2} = \frac{3}{25}. \qquad \square$$

Example 8 Find the curvature of the *twisted cubic* $x = t$, $y = \frac{1}{2}t^2$, $z = \frac{1}{3}t^3$. Also find the tangential and normal components of the acceleration.

Solution Here $x' = 1$, $y' = t$, $z' = t^2$, $x'' = 0$, $y'' = 1$ and $z'' = 2t$, so that

$$v = |\mathbf{r}'| = \sqrt{x'^2 + y'^2 + z'^2} = \sqrt{1 + t^2 + t^4},$$

$$\mathbf{r}' \times \mathbf{r}'' = \begin{vmatrix} \mathbf{i} & \mathbf{j} & \mathbf{k} \\ 1 & t & t^2 \\ 0 & 1 & 2t \end{vmatrix} = t^2\mathbf{i} - 2t\mathbf{j} + \mathbf{k},$$

and

$$|\mathbf{r}' \times \mathbf{r}''| = \sqrt{t^4 + 4t^2 + 1}.$$

Therefore, by formula (10),

$$\kappa = \frac{|\mathbf{r}' \times \mathbf{r}''|}{|\mathbf{r}'|^3} = \frac{\sqrt{t^4 + 4t^2 + 1}}{(t^4 + t^2 + 1)^{3/2}}.$$

For instance, the curvature at the origin, corresponding to the parameter value $t = 0$, is $\kappa = 1$, while the curvature at the point $(1, \frac{1}{2}, \frac{1}{3})$ corresponding to the parameter value $t = 1$, is

$$\kappa = \frac{\sqrt{6}}{3\sqrt{3}} = \frac{\sqrt{2}}{3} \approx 0.47.$$

Notice that $\kappa \to 0$ as $t \to \pm\infty$, which shows that the curve looks very much like a straight line for large values of $|t|$. By (11), the tangential and normal components of the acceleration are

$$a_T = \frac{dv}{dt} = \frac{d}{dt}\sqrt{t^4 + t^2 + 1} = \frac{2t^3 + t}{\sqrt{t^4 + t^2 + 1}}$$

and

$$a_N = \kappa v^2 = \sqrt{\frac{t^4 + 4t^2 + 1}{t^4 + t^2 + 1}}.$$

For instance, $a_T = 0$ and $a_N = 1$ at the origin, while $a_T = \sqrt{3}$ and $a_N = \sqrt{2}$ at the point $(1, \frac{1}{2}, \frac{1}{3})$. □

Kepler's Laws (Optional)

Finally, we use vectors to solve a basic problem of mechanics, namely the problem of determining the orbital motion of a body P of mass m, subject to the gravitational attraction of a body of much larger mass M. For example, P might be a planet revolving around the sun, or a satellite (real or artificial) revolving around the earth or some other planet. The solution to this problem was given, without benefit of vectors, by Newton in his celebrated treatise *Principia Mathematica* (1687). Choosing the origin O at the body of mass M, let $\mathbf{r} = \mathbf{r}(t)$ be the position vector of P at the time t. Also let $\mathbf{u} = \mathbf{u}(t)$ be a unit vector in the direction of $\mathbf{r} = \mathbf{r}(t)$, as in Figure 35, so that $\mathbf{u} = \mathbf{r}/r$ or $\mathbf{r} = r\mathbf{u}$, where $r = |\mathbf{r}|$. Then, by Newton's inverse square law of gravitation (already encountered on pages 271 and 684, the force acting on P is

$$\mathbf{F} = -\frac{GMm}{r^2}\mathbf{u},$$

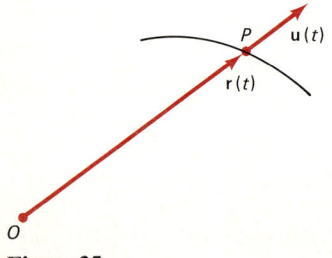

Figure 35

where G is the universal gravitational constant and the minus sign makes the force attractive.† Substituting this force into Newton's second law of motion

$$\mathbf{F} = m\mathbf{a} = m\frac{d^2\mathbf{r}}{dt^2},$$

where all vectors are now space vectors, we get

$$\frac{d^2\mathbf{r}}{dt^2} = \frac{d\mathbf{v}}{dt} = -C\frac{\mathbf{u}}{r^2} = -C\frac{\mathbf{r}}{r^3}, \qquad (12)$$

where $C = GM$ and $\mathbf{v} = d\mathbf{r}/dt$ is the velocity of P. Notice that equation (12) does not involve the smaller mass m.

The vector method can now be used to great advantage. Taking the cross product of (12) with $\mathbf{r}$, we obtain

$$\mathbf{r} \times \mathbf{r}'' = -\frac{C}{r^3}\mathbf{r} \times \mathbf{r} = \mathbf{0},$$

where the prime denotes differentiation with respect to t. But

$$\frac{d}{dt}(\mathbf{r} \times \mathbf{r}') = (\mathbf{r}' \times \mathbf{r}') + (\mathbf{r} \times \mathbf{r}'') = \mathbf{r} \times \mathbf{r}''$$

(see Example 4). Hence the last two equations together imply

$$\frac{d}{dt}(\mathbf{r} \times \mathbf{r}') = \frac{d}{dt}(\mathbf{r} \times \mathbf{v}) = \mathbf{0}.$$

† In treating celestial bodies as points, we rely on the fact, proved by Newton himself, that the gravitational attraction of a solid sphere is the same as if all its mass were concentrated at the center of the sphere (see Problem 35, page 896). A more detailed analysis of the problem shows that the position of the body of mass M can be regarded as fixed if M greatly exceeds m.

Integrating this vector differential equation, we get

$$\mathbf{r} \times \mathbf{v} = \mathbf{h}, \tag{13}$$

where $\mathbf{h}$ is a constant vector. It follows from (13) that the body P always lies in the plane through O perpendicular to $\mathbf{h}$. In fact, taking the dot product of (13) with $\mathbf{r}$, we find that $\mathbf{r} \cdot (\mathbf{r} \times \mathbf{v}) = \mathbf{r} \cdot \mathbf{h} = 0$ (why?), so that the projection of the position vector $\mathbf{r} = \mathbf{r}(t)$ along $\mathbf{h}$ is always zero. Observe that this conclusion is true for any *central force*, that is, for any force which always acts along the line connecting P to a fixed center O, and not just for the inverse square law of gravitational attraction.

The vector $\mathbf{h}$ can be written in the form

$$\mathbf{h} = \mathbf{r} \times \mathbf{v} = \mathbf{r} \times \mathbf{r}' = r\mathbf{u} \times \frac{d}{dt}(r\mathbf{u}) = r\mathbf{u} \times (r'\mathbf{u} + r\mathbf{u}') = r^2 \mathbf{u} \times \mathbf{u}'$$

(recall Example 3, page 664, which applies equally well to vectors in space). Therefore, by (12),

$$\frac{d\mathbf{v}}{dt} \times \mathbf{h} = -\frac{C}{r^2} \mathbf{u} \times \mathbf{h} = -C\mathbf{u} \times (\mathbf{u} \times \mathbf{u}'). \tag{14}$$

But

$$\mathbf{u} \times (\mathbf{u} \times \mathbf{u}') = \mathbf{u}(\mathbf{u} \cdot \mathbf{u}') - \mathbf{u}'(\mathbf{u} \cdot \mathbf{u}) = -\mathbf{u}',$$

with the help of the formula in Problem 48, page 715, and the fact that $\mathbf{u} \cdot \mathbf{u}' = 0$ (to see this, differentiate $|\mathbf{u}|^2 = \mathbf{u} \cdot \mathbf{u} = 1$). Moreover,

$$\frac{d}{dt}(\mathbf{v} \times \mathbf{h}) = \left(\frac{d\mathbf{v}}{dt} \times \mathbf{h}\right) + \left(\mathbf{v} \times \frac{d\mathbf{h}}{dt}\right) = \frac{d\mathbf{v}}{dt} \times \mathbf{h},$$

since $d\mathbf{h}/dt = \mathbf{0}$, so that (14) is equivalent to

$$\frac{d}{dt}(\mathbf{v} \times \mathbf{h}) = C\mathbf{u}'. \tag{15}$$

Integrating this equation, we get

$$\mathbf{v} \times \mathbf{h} = C\mathbf{u} + \mathbf{q}, \tag{16}$$

where $\mathbf{q}$ (like $\mathbf{h}$) is a constant vector.

Taking the dot product of (16) with $\mathbf{r}$, we obtain

$$(\mathbf{v} \times \mathbf{h}) \cdot \mathbf{r} = (C\mathbf{u} + \mathbf{q}) \cdot r\mathbf{u} = Cr + qr\cos\theta, \tag{17}$$

where $q = |\mathbf{q}| = $ constant and θ is the angle between the vectors $\mathbf{q}$ and $\mathbf{r}$. The left side of (17) equals

$$\mathbf{r} \cdot (\mathbf{v} \times \mathbf{h}) = (\mathbf{r} \times \mathbf{v}) \cdot \mathbf{h} = \mathbf{h} \cdot \mathbf{h} = h^2$$

(use Example 10, page 713), where $h = |\mathbf{h}| = $ constant. Thus (17) can be written in the form

$$h^2 = Cr + qr\cos\theta,$$

which implies

$$r = \frac{h^2/C}{1 + (q/C)\cos\theta}, \tag{18}$$

or equivalently

$$r = \frac{ed}{1 + e \cos \theta},\tag{19}$$

where

$$e = \frac{q}{C} = \frac{q}{GM}, \quad d = \frac{h^2}{q}.\tag{20}$$

Recalling the first of the formulas (6'), page 630, we recognize (19) as the equation in polar coordinates r and θ of a conic section with focus at the origin, focus-to-directrix distance d and eccentricity e. Thus the orbit of the body P is a conic. Specifically, the orbit is a circle if $e = 0$, an ellipse if $0 < e < 1$, a parabola if $e = 1$, and a hyperbola if $e > 1$, where the value of e can be determined in the way described in Example 10 below.

The case of particular interest in astronomy is where the orbit is elliptical. The body P is then "bound" to the center of attraction and does not "recede to infinity," as it does in the case of a parabolic or hyperbolic orbit. Thus, as applied to the solar system, we have just verified *Kepler's first law*, announced by Kepler in 1609: *The planets revolve around the sun in elliptical orbits, with the sun at one focus.* It should be stressed that the German astronomer Johannes Kepler (1571–1630) discovered this law, and the other two laws that bear his name, purely *empirically* by analyzing observational data amassed by his Danish predecessor Tycho Brahe (1546–1601). Newton's achievement in explaining these laws *theoretically*, and inventing the calculus in order to do so, still staggers the imagination.

Kepler's second law (1609) states that *the line joining the sun to each planet sweeps out equal areas in equal times*. This can be proved as follows. Let **i**, **j** and **k** be an orthonormal basis with **k** pointing in the direction of the constant vector **h**, and introduce polar coordinates r and θ in the plane of the orbit, with the polar axis along **i**. Then $\mathbf{r} = (r \cos \theta)\mathbf{i} + (r \sin \theta)\mathbf{j}$, and hence

$$\mathbf{h} = \mathbf{r} \times \mathbf{r}' = \begin{vmatrix} \mathbf{i} & \mathbf{j} & \mathbf{k} \\ r \cos \theta & r \sin \theta & 0 \\ r' \cos \theta - r\theta' \sin \theta & r' \sin \theta + r\theta' \cos \theta & 0 \end{vmatrix} = r^2 \theta' \mathbf{k}.$$

Since $\mathbf{h} = h\mathbf{k}$, it follows that

$$r^2 \frac{d\theta}{dt} = h = \text{constant}.$$

Let t_0 be a fixed time and t a variable time later than t_0. Then by the formula for area in polar coordinates (see page 636), the area A swept out by the radius vector $\mathbf{r} = \mathbf{r}(t)$ during the interval $[t_0, t]$, equal to the area of the shaded region in Figure 36, is given by

$$A = \frac{1}{2} \int_{t_0}^{t} r^2 \frac{d\theta}{d\tau} d\tau$$

(τ is lowercase Greek tau, used to avoid confusion with the variable upper limit of integration t). Therefore

$$\frac{dA}{dt} = \frac{1}{2} r^2 \frac{d\theta}{dt} = \frac{1}{2} h = \text{constant},\tag{21}$$

732 Chapter 12 Vectors in Space and Solid Analytic Geometry

*A is the area swept out by the position vector **r**(t) during the interval $[t_0, t]$.*

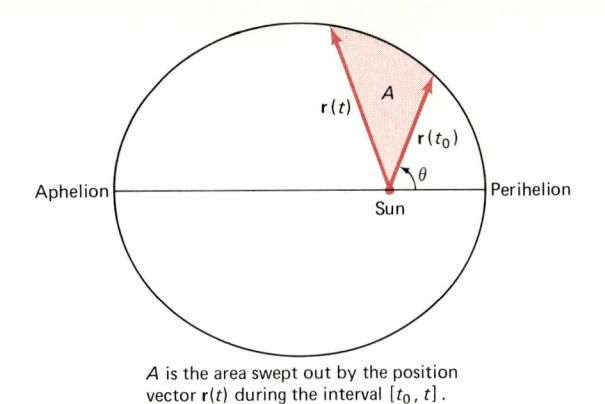

Figure 36

which is precisely Kepler's second law. Like the planarity of the orbit, the conclusion $dA/dt = $ constant is true for any *central force* (why?).

Kepler's third law, which he announced in 1619 after another 10 years of data analysis, states that *the square of the period of revolution of a planet is proportional to the cube of the semimajor axis of its elliptical orbit*. To show this, let the orbit of the planet be an ellipse with semimajor axis a and semiminor axis b. Then the area enclosed by the ellipse is πab (see Example 4, page 591), and the period of the planet is the time it takes to sweep out this area, namely

$$T = \frac{\pi ab}{dA/dt} = \frac{2\pi ab}{h}, \tag{22}$$

where formula (21) is used at the second step. Multiplying the expressions (20) for the eccentricity e and the focus-to-directrix distance d gives

$$ed = \frac{h^2}{GM},$$

and we also have

$$d = \left(\frac{1}{e} - e\right) a,$$

by formula (8), page 607, so that

$$ed = (1 - e^2)a = \left(1 - \frac{a^2 - b^2}{a^2}\right) a = \frac{b^2}{a}.$$

Equating these two expressions for the product ed, we obtain

$$h^2 = GM \frac{b^2}{a}.$$

Therefore $h = b\sqrt{GM/a}$, so that (22) becomes

$$T = \frac{2\pi ab}{b} \sqrt{\frac{a}{GM}} = \frac{2\pi}{\sqrt{GM}} a^{3/2} \tag{23}$$

(where G is the universal gravitational constant and M is the mass of the sun), or equivalently

$$T^2 = \frac{4\pi^2}{GM} a^3, \tag{23'}$$

and this is precisely Kepler's third law. By the same token, T^2 is proportional to a^3 for all satellites of a given planet, with a fixed constant of proportionality $4\pi^2/GM$, where M is now the mass of the planet. Kepler's third law for circular orbits was anticipated in formula (11'), page 686.

For elliptical orbits around the sun the point nearest the sun is called the *perihelion*, and the point farthest from the sun is called the *aphelion*, as illustrated in Figure 36. The analogous points for orbits around the earth are called the *perigee* and *apogee*. (The general terms for an arbitrary center of attraction are *pericenter* and *apocenter*.)

Example 9 Let $\mathbf{r}_0$ and $\mathbf{v}_0$ be the position vector and velocity of a planet P at perihelion. Show that the constants h and q are given by

$$h = r_0 v_0, \qquad q = r_0 v_0^2 - GM, \tag{24}$$

in terms of the *perihelion distance* $r_0 = |\mathbf{r}_0|$ and *perihelion speed* $v_0 = |\mathbf{v}_0|$.

Solution The vectors $\mathbf{r}_0$ and $\mathbf{v}_0$ are perpendicular, since the tangents to an ellipse at the endpoints of the major axis are perpendicular to the major axis. Therefore

$$h = |\mathbf{h}| = |\mathbf{r}_0 \times \mathbf{v}_0| = |\mathbf{r}_0| |\mathbf{v}_0| \sin \frac{\pi}{2} = r_0 v_0.$$

At perihelion the angular coordinate θ of P is 0, since this value of θ leads to the smallest value of the radial coordinate r in formula (18) or (19). Setting $\theta = 0$ in (18), we find that

$$r_0 = r|_{\theta=0} = \frac{h^2/C}{1 + (q/C)} = \frac{r_0^2 v_0^2}{C + q},$$

which implies

$$q = r_0 v_0^2 - C = r_0 v_0^2 - GM. \qquad \square$$

In the case of parabolic and hyperbolic orbits, there is a perihelion but no aphelion, since the body P approaches the center of attraction only once and then recedes to infinity. However, the formulas (24) are still applicable, since r_0 and v_0 are perihelion quantities.

Example 10 Let r_0 and v_0 be the perihelion (perigee) distance and speed of a body subject to the gravitational attraction of the sun (earth). Show that *the orbit of the body is a circle if $r_0 v_0^2 = GM$, an ellipse if $GM < r_0 v_0^2 < 2GM$, a parabola if $r_0 v_0^2 = 2GM$, and a hyperbola if $r_0 v_0^2 > 2GM$*.

Solution With the help of formulas (20) and (24), we find that the eccentricity of the orbit is

$$e = \frac{q}{GM} = \frac{r_0 v_0^2}{GM} - 1.$$

Problems

Find the length of the given space curve.
1. $x = 12 \sin t$, $y = 12 \cos t$, $z = 5t$ ($0 \le t \le 4\pi$)
2. $x = 1 - \cos t$, $y = t - \sin t$, $z = 4 \sin (t/2)$ ($0 \le t \le 2\pi$)
3. $x = t$, $y = t^2$, $z = \frac{2}{3}t^3$ ($0 \le t \le 6$)
4. $x = 2t$, $y = t^2$, $z = \ln t$ ($1 \le t \le 8$)
5. $x = \sqrt{2}\,t$, $y = e^t$, $z = e^{-t}$ ($0 \le t \le \ln 2$)

6. Show that the length of n turns of the helix $x = a \cos t$, $y = a \sin t$, $z = bt$ is $2n\pi\sqrt{a^2 + b^2}$.
7. Each helix of the double helix making up the DNA molecule is of diameter 20 Å and pitch 34 Å (the symbol Å denotes the angstrom unit = 10^{-8} cm). There are about 290,000,000 turns in each helix of human DNA. How long would such a helix be if it could be unraveled and stretched out to its full length?
8. Find the arc length function $s = s(t)$ for the space curve $x = e^t \sin t$, $y = e^t \cos t$, $z = e^t$. Choose the point $(0, 1, 1)$ corresponding to the parameter value $t = 0$ as the initial point.

Evaluate

9. $\lim\limits_{t \to 0} \left(\dfrac{1 - \cos t}{t^2} \mathbf{i} + \dfrac{\sin t}{t} \mathbf{j} + e^t \mathbf{k} \right)$

10. $\lim\limits_{t \to \pi} \left(\dfrac{\tan t}{t} \mathbf{i} + \dfrac{1 + \cos t}{t} \mathbf{j} + \dfrac{10t}{\pi} \mathbf{k} \right)$

11. $\dfrac{d}{dt}[(\arcsin t)\mathbf{i} - e^{t^2}\mathbf{j} + (\cosh t)\mathbf{k}]$

12. $\dfrac{d}{dt}[(-\cot t)\mathbf{i} + (\csc t)\mathbf{j} + (\sinh \sqrt{t})\mathbf{k}]$

13. $\displaystyle\int_0^1 \dfrac{\mathbf{i} + 2t\mathbf{j} + t^2\mathbf{k}}{t^2 + 1}\, dt$

14. $\displaystyle\int \left(te^t \mathbf{i} + \dfrac{t-1}{t+1}\mathbf{j} + \dfrac{\ln t}{t}\mathbf{k} \right) dt$

15. Show that

$$\dfrac{d}{dt}[\mathbf{r}_1 \cdot (\mathbf{r}_2 \times \mathbf{r}_3)]$$

$$= \dfrac{d\mathbf{r}_1}{dt} \cdot (\mathbf{r}_2 \times \mathbf{r}_3) + \mathbf{r}_1 \cdot \left(\dfrac{d\mathbf{r}_2}{dt} \times \mathbf{r}_3 \right) + \mathbf{r}_1 \cdot \left(\mathbf{r}_2 \times \dfrac{d\mathbf{r}_3}{dt} \right),$$

where $\mathbf{r}_1 = \mathbf{r}_1(t)$, $\mathbf{r}_2 = \mathbf{r}_2(t)$ and $\mathbf{r}_3 = \mathbf{r}_3(t)$ are differentiable vector functions.

16. Use the result of the previous problem to prove the following rule for differentiating a 3×3 determinant each of whose rows consists of differentiable scalar functions:

$$\dfrac{d}{dt} \begin{vmatrix} x_1 & y_1 & z_1 \\ x_2 & y_2 & z_2 \\ x_3 & y_3 & z_3 \end{vmatrix}$$

$$= \begin{vmatrix} x'_1 & y'_1 & z'_1 \\ x_2 & y_2 & z_2 \\ x_3 & y_3 & z_3 \end{vmatrix} + \begin{vmatrix} x_1 & y_1 & z_1 \\ x'_2 & y'_2 & z'_2 \\ x_3 & y_3 & z_3 \end{vmatrix} + \begin{vmatrix} x_1 & y_1 & z_1 \\ x_2 & y_2 & z_2 \\ x'_3 & y'_3 & z'_3 \end{vmatrix}.$$

(The prime denotes differentiation with respect to t.)

Find the velocity $\mathbf{v}$ and speed $v = |\mathbf{v}|$ of the point $P = P(t)$ with the given position vector $\mathbf{r} = \mathbf{r}(t)$, and also find the unit tangent and normal vectors $\mathbf{T}$ and $\mathbf{N}$ to the graph of $\mathbf{r} = \mathbf{r}(t)$.

17. $\mathbf{r}(t) = 2t\mathbf{i} + t\mathbf{j} - 2t\mathbf{k}$
18. $\mathbf{r}(t) = t\mathbf{i} + (1 - \cos t)\mathbf{j} + (\sin t)\mathbf{k}$
19. $\mathbf{r}(t) = 15t\mathbf{i} + (8 \cos t)\mathbf{j} + (8 \sin t)\mathbf{k}$
20. $\mathbf{r}(t) = (\sin t)\mathbf{i} + (\sin t)\mathbf{j} + (\sqrt{2} \cos t)\mathbf{k}$

Find the tangential and normal components a_T and a_N of the acceleration of the point $P = P(t)$ with the given position vector $\mathbf{r} = \mathbf{r}(t)$, and also find the curvature κ of the graph of $\mathbf{r} = \mathbf{r}(t)$.

21. $\mathbf{r}(t) = (t + \frac{1}{3}t^3)\mathbf{i} + (t - \frac{1}{3}t^3)\mathbf{j} + t^2\mathbf{k}$
22. $\mathbf{r}(t) = (\cos t)\mathbf{i} + (\sin t)\mathbf{j} + (\cosh t)\mathbf{k}$
23. $\mathbf{r}(t) = (a \cos \omega t)\mathbf{i} + (a \sin \omega t)\mathbf{j} + bt\mathbf{k}$ ($a > 0$)
24. $\mathbf{r}(t) = (\arctan t)\mathbf{i} + (t - \arctan t)\mathbf{j} + (1/\sqrt{2}) \ln (t^2 + 1)\mathbf{k}$

The plane containing the unit tangent and normal vectors $\mathbf{T}$ and $\mathbf{N}$ to a space curve C at a point P is called the *osculating plane* to C at P (because it fits C very closely at P). Find the osculating plane to
25. The curve in Problem 21 at the point for which $t = -1$
26. The curve in Problem 24 at the point for which $t = 1$

Let C be a space curve, and let $\mathbf{T}$ be the unit tangent vector to C at a point P. Then the line through P containing $\mathbf{T}$ is called the *tangent (line)* to C at P, and the plane through

P perpendicular to $\mathbf{T}$ is called the *normal plane* to C at P. Find the tangent line and normal plane to

27. The curve of intersection of the plane $x + y = 0$ and the parabolic cylinder $z = x^2$ at the point $(2, -2, 4)$
28. The curve of intersection of the cylinders $x^2 + y^2 = 10$ and $x^2 + z^2 = 10$ at the point $(3, 1, 1)$.
Hint. First represent the curves parametrically.

29. At which points is the tangent to the curve $x = t$, $y = t^2$, $z = t^3$ parallel to the plane $x + 2y + z - 1 = 0$?
30. Let the curve C be the graph of a continuous vector function

$$\mathbf{r} = \mathbf{r}(t) \quad (a \le t \le b)$$

in two or three dimensions with a nonzero derivative $\mathbf{r}'(t)$ at every interior point of $[a, b]$, and suppose that $\mathbf{r}(a) = \mathbf{r}(b)$, so that the endpoints of C coincide (such a curve is said to be *closed*, as on page 459). Show that there is a point of the curve C at which its tangent is perpendicular to any preassigned nonzero vector $\mathbf{m}$.
Hint. Use Rolle's theorem.

31. In Example 1, page 682, on the motion of a projectile, choose a z-axis perpendicular to the xy-plane (the vertical plane containing the initial velocity vector $\mathbf{v}_0$). Then Newton's second law gives $d^2\mathbf{r}/dt^2 = -g\mathbf{j}$, where g is the acceleration due to gravity. Integrate this vector differential equation subject to the initial conditions $\mathbf{r}(0) = \mathbf{0}$, $\mathbf{v}(0) = \mathbf{v}_0$, where $\mathbf{v} = d\mathbf{r}/dt$, and show that the projectile always remains in the xy-plane. By setting $g = 0$, prove *Newton's first law of motion*, which says that unless acted on by an external force, a body at rest ($\mathbf{v}_0 = \mathbf{0}$) remains at rest, and a body in motion ($\mathbf{v}_0 \ne \mathbf{0}$) continues to move with constant velocity $\mathbf{v}_0$ in a straight line.
32. By the *momentum* of a particle is meant the product $\mathbf{p} = m\mathbf{v}$ of its mass m and velocity $\mathbf{v} = d\mathbf{r}/dt$. In terms of the momentum, Newton's second law takes the form $d\mathbf{p}/dt = \mathbf{F}$, where $\mathbf{F}$ is the force acting on the particle. The cross product $\mathbf{L} = \mathbf{r} \times \mathbf{p}$ is called the *angular momentum* of the particle about the origin, that is, about the initial point of its position vector $\mathbf{r} = \mathbf{r}(t)$. Show that the angular momentum of a particle acted on by a central force is constant (or, as physicists say, is *conserved*). In terms of the angular momentum $\mathbf{L}$, what is the vector $\mathbf{h}$ defined by formula (13)? Describe the motion if $\mathbf{h} = \mathbf{0}$.
33. Show that the speed of a planet at a point P of its orbit is inversely proportional to the perpendicular distance from the sun to the tangent to the orbit at P. Where is the speed largest? Where is it smallest?
34. In Example 8, page 271, it was shown that a rocket fired vertically upward from the earth's surface with speed $v_0 = \sqrt{2gR}$, where g is the acceleration due to gravity and R is the radius of the earth (≈ 3960 mi), will escape the gravitational pull of the earth and never return. Show that the same is true if the rocket is fired in a direction parallel to the surface of the earth (neglect air resistance). How large is v_0 compared to the speed required to put the rocket into a circular orbit that just skims the earth's surface?
35. The altitude and speed of an earth satellite at perigee are 400 mi and 5 mi/sec. What is the eccentricity of the orbit? Find the altitude and speed of the satellite at apogee. What is its period of revolution? (Use the approximation $g \approx 32.15$ ft/sec^2 for the acceleration due to gravity at the earth's surface, instead of the cruder approximation $g \approx 32$ ft/sec^2 used until now.)
36. The maximum and minimum altitudes of an earth satellite are 840 mi and 360 mi. What are its maximum and minimum speeds? What is its period of revolution?
37. A space vehicle is moving in a circular orbit around the earth at an altitude of 120 mi, when its thruster rockets are activated, producing an acceleration of 450 mph/sec. How long should the rockets burn in order to give the vehicle the extra speed required to completely escape the earth's gravitational pull?
38. Show that when a planet is at either endpoint of the minor axis of its orbit, the speed of the planet is the same as the speed it would have in a circular orbit of radius equal to the semimajor axis of the orbit.
39. A geophysical satellite is put into a circular orbit at an altitude equal to one third of the earth's radius. The orbit passes over both poles. As the satellite passes over the north pole, its retro-rockets are activated, reducing its speed to a value causing it to land at the equator. What is the required reduction of speed? What reduction of speed would cause the satellite to land at the south pole instead?
40. Ground control wants to transfer an earth satellite moving in a circular orbit of radius r_1 to a larger circular orbit of radius r_2. This is done in the following way. By briefly activating its thruster rockets, the satellite is given an extra burst of speed Δv_1 at a point P_1 of the smaller orbit, causing the satellite to enter an elliptical *transfer orbit* with apogee P_2 at distance r_2 from the center of the earth (see Figure 37). Then at P_2 the thruster rockets are reactivated, and the satellite is given a second burst of speed Δv_2, causing

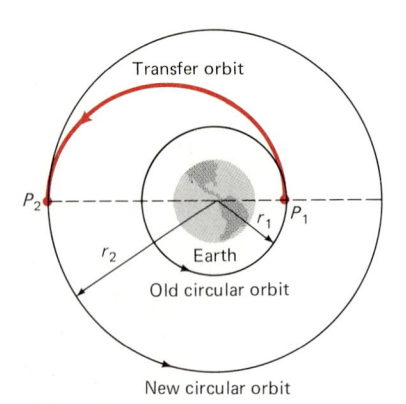

Figure 37

it to move in a circular orbit of radius r_2. Show that

$$\Delta v_1 = \left(\sqrt{\frac{2r_2}{r_1 + r_2}} - 1\right)\sqrt{\frac{g}{r_1}} R,$$

$$\Delta v_2 = \left(1 - \sqrt{\frac{2r_1}{r_1 + r_2}}\right)\sqrt{\frac{g}{r_2}} R.$$

Calculate Δv_1 and Δv_2 for $r_1 = 4500$ mi and $r_2 = 9000$ mi.

12.6 Quadric Surfaces

By a *second degree equation in three variables* x, y and z we mean an equation of the form

$$Ax^2 + By^2 + Cz^2 + Dxy + Exz + Fyz + Gx + Hy + Iz + J = 0, \quad (1)$$

where the coefficients A, B, C, D, E and F are not all zero (otherwise (1) reduces to a first degree equation). The graph of any such equation is called a *quadric surface*, or simply a *quadric*, and is the natural generalization of a conic in 2-space. As in the case of conics, there are a number of "degenerate" quadrics, which we list in the following table.

Degenerate Quadric	Sample Equation
Empty set	$x^2 + y^2 + z^2 + 1 = 0$
Single point	$x^2 + y^2 + z^2 = 0$
Single line	$x^2 + y^2 = 0$
Single plane	$(x - 1)^2 = 0$
Pair of parallel planes	$x^2 - 1 = 0$
Pair of intersecting planes	$x^2 - y^2 = 0$

If one of the three variables x, y and z is missing from equation (1), it reduces to the general second degree equation in the other two variables. For example, if z is missing, (1) takes the form

$$Ax^2 + By^2 + Dxy + Gx + Hy + J = 0.$$

But we already know from Section 10.6 that the graph of this equation is a conic in the xy-plane, and therefore, apart from certain degenerate cases, its graph is a *quadric cylinder*, specifically a parabolic, elliptic or hyperbolic cylinder. Such cylinders have already been discussed in Section 12.1, in addition to *quadric surfaces of revolution*, like the sphere and the paraboloid of revolution.

In the following examples we investigate quadrics which are neither cylinders nor surfaces of revolution, namely the general ellipsoid, two kinds of hyperboloids (single-sheeted and double-sheeted), the elliptic cone, and two kinds of paraboloids (elliptic and hyperbolic). These are essentially the only possibilities for the following reason: A detailed analysis of equation (1), which will not be given here, shows that we can always go over to a new system of rectangular coordinates x', y' and z' in which (1) transforms into an equation with no terms involving the products $x'y'$, $x'z'$ or $y'z'$, whose graph is either a degenerate quadric, a quadric cylinder, a quadric surface of revolution, or one of the six more general quadrics just listed. As in the two-dimensional case, the transformation from the old xyz-system to the new

$x'y'z'$-system involves both a rotation and a translation, but now the rotation and translation are both *spatial*.

Example 1 The graph of

$$\frac{x^2}{a^2} + \frac{y^2}{b^2} + \frac{z^2}{c^2} = 1 \qquad (2)$$

is the quadric surface called an *ellipsoid*, shown in Figure 38. There is no loss of generality in assuming that the numbers a, b and c are all positive (this assumption is also made in Examples 2–6). If two of these numbers are equal, the ellipsoid reduces to an ellipsoid of revolution or spheroid, while if $a = b = c$, it reduces to the sphere of radius a centered at the origin.

The ellipsoid (2) is symmetric about the origin, called the *center* of the ellipsoid, and about all three coordinate planes. Setting $y = z = 0$ in (2), we find that the ellipsoid intersects the x-axis in the points $(\pm a, 0, 0)$. It also intersects the y-axis in the points $(0, \pm b, 0)$ and the z-axis in the points $(0, 0, \pm c)$. Setting $z = 0$ in (2), we find that the trace of the ellipsoid in the xy-plane (that is, its intersection with the xy-plane) is the ellipse

$$\frac{x^2}{a^2} + \frac{y^2}{b^2} = 1 \qquad (z = 0).$$

Similarly, the trace of the ellipsoid in the xz-plane is the ellipse

$$\frac{x^2}{a^2} + \frac{z^2}{c^2} = 1 \qquad (y = 0),$$

and its trace in the yz-plane is the ellipse

$$\frac{y^2}{b^2} + \frac{z^2}{c^2} = 1 \qquad (x = 0).$$

Setting $z = k$ in (2), we get

$$\frac{x^2}{a^2} + \frac{y^2}{b^2} = 1 - \frac{k^2}{c^2} \qquad (z = k).$$

Hence the trace of the ellipsoid in the plane $z = k$ is an ellipse if $|k| < c$, a single point if $|k| = c$, and the empty set if $|k| > c$. The traces of the ellipsoid in planes parallel to the xz- and yz-planes behave in the same way.

Example 2 The graph of

$$\frac{x^2}{a^2} + \frac{y^2}{b^2} - \frac{z^2}{c^2} = 1 \qquad (3)$$

is the quadric called a *hyperboloid of one sheet*, shown in Figure 39. Like the ellipsoid, the hyperboloid (3) is symmetric about the origin, called the *center* of the hyperboloid, and about all three coordinate planes. It intersects the x-axis in the points $(\pm a, 0, 0)$ and the y-axis in the points $(0, \pm b, 0)$, but it does not intersect the z-axis at all, since the equation $-z^2/c^2 = 1$ obtained by setting $x = y = 0$ in (3) has no solutions. In fact, unlike the ellipsoid, the hyperboloid is an unbounded surface. This surface "opens out to infinity" in both directions along the z-axis, which is called the *axis of the hyperbola*.

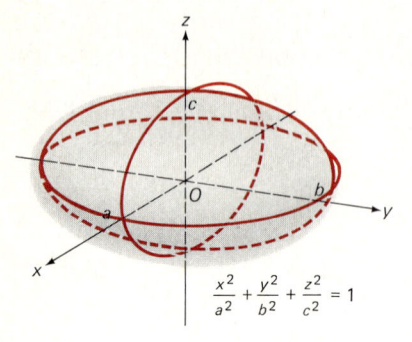

Ellipsoid

Figure 38

Hyperboloid of one sheet

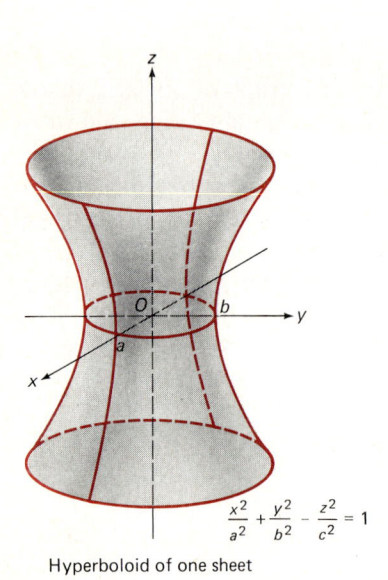

Figure 39

The trace of the hyperboloid (3) in the xy-plane is the ellipse

$$\frac{x^2}{a^2} + \frac{y^2}{b^2} = 1 \qquad (z = 0),$$

encircling the "throat" of the hyperboloid. More generally, the trace of the hyperboloid in the plane $z = k$ is the graph of

$$\frac{x^2}{a^2} + \frac{y^2}{b^2} = 1 + \frac{k^2}{c^2} \qquad (z = k),$$

which is an ellipse for every value of k. On the other hand, the trace of the hyperboloid in the xz-plane is the hyperbola

$$\frac{x^2}{a^2} - \frac{z^2}{c^2} = 1 \qquad (y = 0),$$

and its trace in the yz-plane is the hyperbola

$$\frac{y^2}{b^2} - \frac{z^2}{c^2} = 1 \qquad (x = 0).$$

If $a = b$, the graph of (3) reduces to the hyperboloid of revolution generated by revolving either of these hyperbolas about the z-axis. Note that equation (3) can be obtained from equation (2) of the ellipsoid by changing the sign of the term containing z^2. If the sign of the term containing x^2 or y^2 had been changed instead, the graph of the resulting equation would also be a hyperboloid of one sheet, but its axis would then be the x-axis or the y-axis.

The hyperboloid (3) consists of a single connected piece or "sheet," which is why it is called a hyperboloid *of one sheet*. The next example describes another kind of hyperboloid consisting of two disconnected sheets.

Example 3 The graph of

$$\frac{x^2}{a^2} + \frac{y^2}{b^2} - \frac{z^2}{c^2} = -1, \qquad (4)$$

or of the equivalent equation

$$-\frac{x^2}{a^2} - \frac{y^2}{b^2} + \frac{z^2}{c^2} = 1, \qquad (4')$$

is the quadric called a *hyperboloid of two sheets*, shown in Figure 40. This hyperboloid intersects the z-axis in the points $(0, 0, \pm c)$, but it does not intersect the x- and y-axes at all (why not?). Like the hyperboloid of one sheet, the hyperboloid (4) is an unbounded surface opening out to infinity along the z-axis, which is again called the *axis* of the hyperbola. Setting $z = k$ in (4), we get

$$\frac{x^2}{a^2} + \frac{y^2}{b^2} = \frac{k^2}{c^2} - 1 \qquad (z = k).$$

Hence the trace of the hyperboloid in the plane $z = k$ is an ellipse if $|k| > c$, a single point if $|k| = c$, and the empty set if $|k| < c$. On the other hand, the trace of the hyperboloid in the xz-plane is the hyperbola

$$\frac{z^2}{c^2} - \frac{x^2}{a^2} = 1 \qquad (y = 0),$$

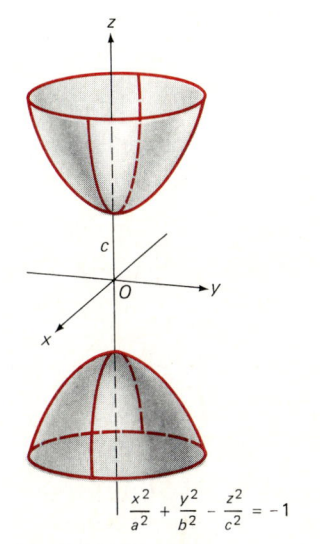

Hyperboloid of two sheets

Figure 40

and its trace in the yz-plane is the hyperbola

$$\frac{z^2}{c^2} - \frac{y^2}{b^2} = 1 \quad (x = 0).$$

If $a = b$, the graph of (4) reduces to the hyperboloid of revolution generated by revolving either of these hyperbolas about the z-axis. It is apparent from these considerations that the hyperboloid (4) consists of two disconnected pieces or sheets.

Example 4 Replacing the right side of equations (3) and (4) by zero, we get the equation

$$\frac{x^2}{a^2} + \frac{y^2}{b^2} - \frac{z^2}{c^2} = 0. \tag{5}$$

The graph of (5), called an *elliptic cone*, is shown in Figure 41. Setting $z = k$ in (5), we get

$$\frac{x^2}{a^2} + \frac{y^2}{b^2} = \frac{k^2}{c^2} \quad (z = k).$$

Hence the trace of the cone in the plane $z = k$ is the origin if $k = 0$ and an ellipse if $k \neq 0$, in particular the ellipse $(x^2/a^2) + (y^2/b^2) = 1$ if $k = \pm c$. As an exercise, show that the trace of the cone in the plane $x = k$ or $y = k$ is a pair of intersecting lines if $k = 0$ and a hyperbola if $k \neq 0$. If $a = b$, the elliptic cone reduces to the right circular cone

$$\frac{x^2}{a^2} + \frac{y^2}{a^2} - \frac{z^2}{c^2} = 0$$

(of two nappes), which is a surface of revolution. If in addition $a = c$, we get the particularly simple right circular cone

$$x^2 + y^2 - z^2 = 0,$$

considered in Example 9, page 700.

Example 5 The graph of

$$\frac{x^2}{a^2} + \frac{y^2}{b^2} = z \tag{6}$$

is the quadric surface called an *elliptic paraboloid*, shown in Figure 42. Setting $z = k$ in (6), we obtain

$$\frac{x^2}{ka^2} + \frac{y^2}{kb^2} = 1$$

if $k > 0$,

$$\frac{x^2}{a^2} + \frac{y^2}{b^2} = 0$$

if $k = 0$, and

$$\frac{x^2}{|k|a^2} + \frac{y^2}{|k|b^2} = -1$$

if $k < 0$. Thus the trace of the paraboloid in the plane $z = k$ is an ellipse if $k > 0$, the origin if $k = 0$, and the empty set if $k < 0$. Also, the trace of the

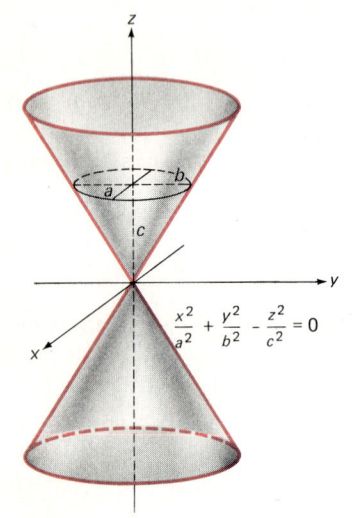

Figure 41

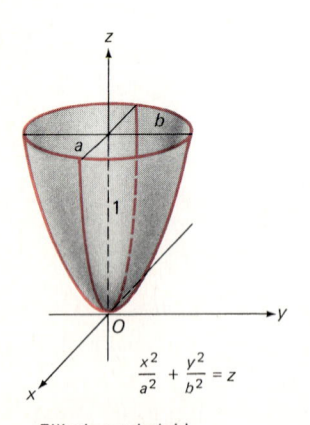

Figure 42 Elliptic paraboloid

paraboloid in the xz-plane is the parabola

$$\frac{x^2}{a^2} = z \quad (y = 0),$$

and its trace in the yz-plane is the parabola

$$\frac{y^2}{b^2} = z \quad (x = 0).$$

If $a = b$, the paraboloid reduces to the paraboloid of revolution generated by revolving either of these parabolas about the z-axis (the case $a = b = 1$ has already been considered in Example 8, page 699.)

Example 6 The graph of

$$\frac{x^2}{a^2} - \frac{y^2}{b^2} = z \tag{7}$$

is the quadric surface called a *hyperbolic paraboloid*, shown in Figure 43, which has a more complicated structure than the other quadric surfaces. Its trace in the xz-plane is the parabola

$$z = \frac{x^2}{a^2} \quad (y = 0)$$

which opens upward (*DOD'* in the figure), and its trace in the yz-plane is the parabola

$$z = -\frac{y^2}{b^2} \quad (x = 0)$$

which opens downward (*POQ* in the figure). The paraboloid (7) intersects the planes $x = \pm k$ in the parabolas

$$z = -\frac{y^2}{b^2} + \frac{k^2}{a^2} \quad (x = \pm k)$$

(*ADG* and *A'D'G'* in the figure), which open downward like *POQ* but have vertices a distance k^2/a^2 higher than the vertex O of *POQ*. Moreover, the

Hyperbolic paraboloid

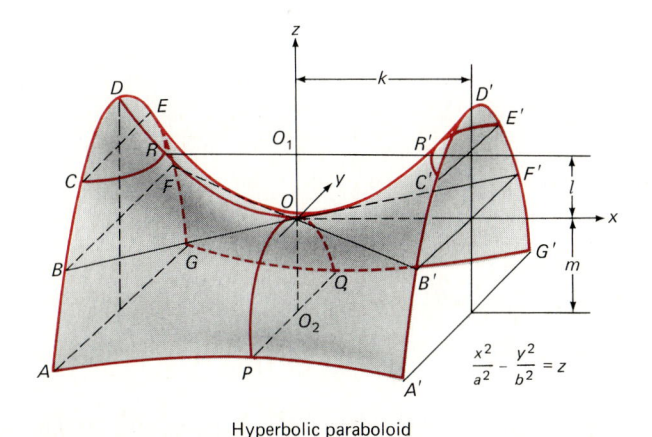

Figure 43

Section 12.6 Quadric Surfaces

trace of the paraboloid in the xy-plane is the pair of intersecting lines

$$\frac{x^2}{a^2} - \frac{y^2}{b^2} = 0 \quad (z = 0),$$

or equivalently

$$y = \pm\frac{b}{a}x \quad (z = 0)$$

(BOF' and FOB' in the figure). It also intersects the plane $z = l$ ($l > 0$) in the hyperbola

$$\frac{x^2}{la^2} - \frac{y^2}{lb^2} = 1 \quad (z = l)$$

with branches CRE, $C'R'E'$ and transverse axis RO_1R', and the plane $z = -m$ ($m > 0$) in the hyperbola

$$\frac{x^2}{ma^2} - \frac{y^2}{mb^2} = -1 \quad (z = -m)$$

with branches APA', GQG' and transverse axis PO_2Q.

Notice that the hyperbolic paraboloid is shaped like a saddle or a mountain pass near the origin O. The point O is called a *saddle point* or *minimax* of the surface. The aptness of the term "minimax" stems from the observation that although the height of the surface has neither a minimum nor a maximum at O, the lowest point of the parabola DOD' and the highest point of the parabola POQ both occur at O. □

Preliminary algebraic transformations may be called for in order to identify the graph of a second degree equation. For example, after completing a square, the equation

$$x^2 + y^2 - z^2 + 2xy = 0,$$

which has a cross product term $2xy$, becomes

$$(x + y)^2 - z^2 = 0$$

or equivalently

$$(x + y + z)(x + y - z) = 0,$$

and therefore its graph is the degenerate quadric surface consisting of the pair of intersecting planes $x + y + z = 0$ and $x + y - z = 0$. Similarly, the equation

$$36x^2 + 9y^2 + 4z^2 - 72x + 8z + 4 = 0$$

becomes

$$36(x - 1)^2 + 9y^2 + 4(z + 1)^2 = 36$$

after completing two squares, and then

$$(x - 1)^2 + \frac{y^2}{4} + \frac{(z + 1)^2}{9} = 1$$

after dividing by 36, so that its graph is an ellipsoid centered at the point $(1, 0, -1)$ instead of at the origin $(0, 0, 0)$.

Problems

Identify the quadric which is the graph of the given equation. Sketch the quadric if it is nondegenerate.
1. $2x^2 - \frac{1}{9}y^2 + z^2 = 1$
2. $\frac{1}{25}x^2 + \frac{1}{16}y^2 + \frac{1}{36}z^2 = 1$
3. $x^2 - y^2 - 2z^2 = 0$
4. $6x^2 + 2y^2 - 3z^2 + 6 = 0$
5. $x^2 + \frac{1}{4}y^2 + z = 1$
6. $x^2 - y^2 + z^2 + 2xz + 2y - 1 = 0$
7. $4x^2 + 2y^2 + z^2 - 8x + 4y + 7 = 0$
8. $z^2 - 4x^2 = y$
9. $2x^2 + y^2 + 3z^2 - 12z + 11 = 0$
10. $x^2 + 2y^2 - z^2 - 6x - 12y + 26 = 0$
11. $x^2 + z^2 - 2xz + 2x - 2z + 1 = 0$
12. $x^2 + 2y^2 + 3z^2 - 2x - 8y + 6z + 12 = 0$
13. $z = xy$
14. $z^2 = xy$
15. $x^2 = z - y$
16. $x^2 + y^2 + z^2 - xy - xz - yz = 0$
17. $x^2 - 2y^2 - z^2 = 1$
18. $x^2 + 6xy + 9y^2 - 4 = 0$

Hint. In Problems 13–15 make a preliminary rotation about the z-axis.

Let S be the solid which is enclosed by the ellipsoid $\frac{1}{4}x^2 + \frac{1}{25}y^2 + \frac{1}{9}z^2 = 1$. Find the area of the region in which S is intersected by

19. The plane $y = 3$
20. The plane $x = 1$

21. Find the points in which the line
$$\frac{x-4}{-6} = \frac{y-3}{3} = \frac{z+2}{4}$$
intersects the ellipsoid $\frac{1}{36}x^2 + \frac{1}{81}y^2 + \frac{1}{9}z^2 = 1$.

22. The elliptic paraboloid $\frac{1}{9}x^2 + \frac{1}{4}y^2 = 2z$ intersects the plane $2x - y - 2z - 10 = 0$ in a single point. Which one?
23. Find the locus of all points equidistant from the point $(0, 0, c)$ and the plane $z = -c$.
24. Use the method of cross sections (see Section 8.1) to show that the volume of the solid bounded by the elliptic paraboloid

$$\frac{x^2}{a^2} + \frac{y^2}{b^2} = h - z \qquad (h > 0)$$

and the xy-plane is equal to half the product of its base area and its height.
25. Use the method of cross sections to find the volume of the solid enclosed by the ellipsoid

$$\frac{x^2}{a^2} + \frac{y^2}{b^2} + \frac{z^2}{c^2} = 1.$$

26. Show that the elliptic cone

$$\frac{x^2}{a^2} + \frac{y^2}{b^2} - \frac{z^2}{c^2} = 0$$

is asymptotic to both the hyperboloid

$$\frac{x^2}{a^2} + \frac{y^2}{b^2} - \frac{z^2}{c^2} = 1$$

of one sheet and the hyperboloid

$$\frac{x^2}{a^2} + \frac{y^2}{b^2} - \frac{z^2}{c^2} = -1$$

of two sheets in the sense that both hyperboloids get arbitrarily close to the cone as $z \to \pm\infty$.

Key Terms and Topics
Rectangular coordinates in space
Distance between two points in space
Equations of cylinders and surfaces of revolution
Vectors in space and their representation as ordered triples
The cross product and its properties
Determinants
The scalar triple product
Parametric and symmetric equations of lines in space
Distance between a point and a line in space
Planes and their equations
The angle between two planes
Distance between a point and a plane
Space curves, vector functions in space
Orbital motion and Kepler's laws
Quadric surfaces

Supplementary Problems

1. Find the two points of the x-axis at distance 12 from the point $(-3, 4, 8)$.
2. Show that the points $A = (0, 1, 2)$, $B = (2, 0, 1)$ and $C = (1, 2, 0)$ are the vertices of an equilateral triangle. What is the side length of the triangle?
3. Show that the sphere $x^2 + y^2 + z^2 - 4x - 6y - 2z + 13 = 0$ is tangent to the xy-plane.
4. Find the surface of revolution generated by revolving the line $x = 0$, $y = c$ about the y-axis. About the z-axis.

Find an equation of the cylinder circumscribed about the sphere $x^2 + y^2 + z^2 = 2cx$ with its rulings parallel to

5. The x-axis   6. The y-axis   7. The z-axis

8. Let $(-3, -6, 2)$ and $(3, 4, -1)$ be two adjacent vertices of a parallelogram. Suppose the diagonals intersect in the point $(-1, 7, 4)$. Find the other two vertices.
9. Given three vertices $(3, -1, 2)$, $(1, 2, -4)$ and $(-1, 1, 2)$ of a parallelogram, find the other vertex. (There are three possibilities.)
10. A vector makes angles of $120°$ and $45°$ with the positive x- and z-axes. What angle does it make with the positive y-axis?
11. Find the vectors of magnitude 2 making angles of $60°$ and $120°$ with the positive x- and y-axes.
12. Find the angles of the triangle with vertices $A = (3, 2, -1)$, $B = (1, 1, 1)$ and $C = (5, 0, 0)$.
13. Show that if $\mathbf{a} + \mathbf{b} + \mathbf{c} = \mathbf{0}$, then $\mathbf{a} \times \mathbf{b} = \mathbf{b} \times \mathbf{c} = \mathbf{c} \times \mathbf{a}$. Use this to prove the law of sines (given in Problem 53, page 60).
14. Let $\mathbf{u}$ and $\mathbf{v}$ be unit vectors making an angle of $30°$ with each other. Find the area of the triangle spanned by the vectors $2\mathbf{u} + \mathbf{v}$ and $\mathbf{u} + 3\mathbf{v}$.
15. Let $\mathbf{u}$ and $\mathbf{v}$ be unit vectors making an angle of $45°$ with each other. Find the area of the parallelogram with the vectors $2\mathbf{u} - \mathbf{v}$ and $4\mathbf{u} + 2\mathbf{v}$ as its diagonals.
16. Show that the area of the triangle in the xy-plane with vertices $A = (x_1, y_1)$, $B = (x_2, y_2)$ and $C = (x_3, y_3)$ is equal to one half the absolute value of the determinant

$$\begin{vmatrix} 1 & 1 & 1 \\ x_1 & x_2 & x_3 \\ y_1 & y_2 & y_3 \end{vmatrix}.$$

Show that the points A, B and C are collinear if and only if this determinant equals zero.

Let $\mathbf{a} = (3, -2, 5)$, $\mathbf{b} = (-2, 2, 1)$ and $\mathbf{c} = (1, 4, -1)$. Find

17. $\mathbf{a} \cdot (\mathbf{b} \times \mathbf{c})$ 18. $\mathbf{a} \times (\mathbf{b} \times \mathbf{c})$
19. $(\mathbf{a} \times \mathbf{b}) \times \mathbf{c}$ 20. $(\mathbf{a} \times \mathbf{b}) \cdot (\mathbf{b} \times \mathbf{c})$

21. Find the volume of the parallelepiped spanned by the vectors $\mathbf{a} = (3, -7, 1)$, $\mathbf{b} = (-2, 0, 4)$ and $\mathbf{c} = (10, 6, 0)$.

22. Show that $\mathbf{a} \times (\mathbf{b} \times \mathbf{c}) + \mathbf{b} \times (\mathbf{c} \times \mathbf{a}) + \mathbf{c} \times (\mathbf{a} \times \mathbf{b}) = \mathbf{0}$ for any choice of the vectors $\mathbf{a}$, $\mathbf{b}$ and $\mathbf{c}$.
Hint. Use Problem 48, page 715.

23. Show that if the vectors $\mathbf{a}$, $\mathbf{b}$, $\mathbf{c}$ and $\mathbf{d}$ are coplanar, then $(\mathbf{a} \times \mathbf{b}) \times (\mathbf{c} \times \mathbf{d}) = \mathbf{0}$.

Let $\mathbf{a}$, $\mathbf{b}$, $\mathbf{c}$ and $\mathbf{d}$ be arbitrary vectors. Show that
24. $(\mathbf{a} \times \mathbf{b}) \cdot (\mathbf{c} \times \mathbf{d}) = (\mathbf{a} \cdot \mathbf{c})(\mathbf{b} \cdot \mathbf{d}) - (\mathbf{a} \cdot \mathbf{d})(\mathbf{b} \cdot \mathbf{c})$
25. $(\mathbf{a} \times \mathbf{b}) \times (\mathbf{c} \times \mathbf{d}) = [\mathbf{a} \cdot (\mathbf{b} \times \mathbf{d})]\mathbf{c} - [\mathbf{a} \cdot (\mathbf{b} \times \mathbf{c})]\mathbf{d}$
26. $[\mathbf{b} \cdot (\mathbf{c} \times \mathbf{d})]\mathbf{a} - [\mathbf{a} \cdot (\mathbf{c} \times \mathbf{d})]\mathbf{b} + [\mathbf{a} \cdot (\mathbf{b} \times \mathbf{d})]\mathbf{c} - [\mathbf{a} \cdot (\mathbf{b} \times \mathbf{c})]\mathbf{d} = \mathbf{0}$

27. Let $\mathbf{e}_1$, $\mathbf{e}_2$ and $\mathbf{e}_3$ be any three noncoplanar vectors, so that $\mathbf{e}_1$, $\mathbf{e}_2$ and $\mathbf{e}_3$ form a basis (see page 706). Show that the expansion of an arbitrary vector $\mathbf{a}$ with respect to this basis is $\mathbf{a} = \alpha_1 \mathbf{e}_1 + \alpha_2 \mathbf{e}_2 + \alpha_3 \mathbf{e}_3$, where

$$\alpha_1 = \frac{\mathbf{a} \cdot (\mathbf{e}_2 \times \mathbf{e}_3)}{\mathbf{e}_1 \cdot (\mathbf{e}_2 \times \mathbf{e}_3)}, \quad \alpha_2 = \frac{\mathbf{a} \cdot (\mathbf{e}_3 \times \mathbf{e}_1)}{\mathbf{e}_1 \cdot (\mathbf{e}_2 \times \mathbf{e}_3)}, \quad \alpha_3 = \frac{\mathbf{a} \cdot (\mathbf{e}_1 \times \mathbf{e}_2)}{\mathbf{e}_1 \cdot (\mathbf{e}_2 \times \mathbf{e}_3)}. \quad \text{(i)}$$

The vectors $\mathbf{e}_1 = (0, 1, 1)$, $\mathbf{e}_2 = (1, 0, 1)$ and $\mathbf{e}_3 = (1, 1, 0)$ are noncoplanar, and hence form a basis. Use the formulas (i) to expand the given vector $\mathbf{a}$ with respect to this basis.
28. $\mathbf{a} = (3, -6, 4)$ 29. $\mathbf{a} = (2, 1, 5)$
30. $\mathbf{a} = (-1, 2, -3)$

31. Find the line through the point $(5, -7, 6)$ parallel to the line

$$\frac{x+1}{-2} = \frac{y}{3} = \frac{z+4}{-1}.$$

32. Find parametric and symmetric equations for the line through the points $(0, -1, 2)$ and $(-2, 0, 4)$.
33. Find the distance between the point $(3, 5, 4)$ and the line $x = y = z$.
34. Show that the line $x = -4 + 3t$, $y = 2 - 4t$, $z = 6 + 4t$ is parallel to the plane $4x - 3y - 6z - 2 = 0$.
35. Find the line through the point $(-4, 2, 5)$ perpendicular to the plane $6x - 3y + 8z + 10 = 0$.
36. If a plane Π intersects the x-axis in a point $(a, 0, 0)$, the y-axis in a point $(0, b, 0)$ and the z-axis in a point $(0, 0, c)$, we call a the x-intercept, b the y-intercept and c the z-intercept of Π. Show that an equation of the plane with x-intercept a, y-intercept b and z-intercept c is

$$\frac{x}{a} + \frac{y}{b} + \frac{z}{c} = 1$$

if a, b and c are all nonzero.

Find an equation of the plane with the given intercepts.
37. $a = -1$, $b = 5$, $c = 2$
38. $a = 10$, $b = -20$, $c = 15$
39. No x-intercept, $b = 4$, $c = 8$
40. $a = 2$, $b = -3$, no z-intercept

Find the x-intercept a, y-intercept b and z-intercept c of the given plane.

41. $2x - 3y + 4z - 12 = 0$
42. $3x + 5y - 10z + 30 = 0$
43. $5x + 7y - 35 = 0$
44. $6y - 7z + 21 = 0$

45. For a certain value of a the planes $3x - 5y + az - 9 = 0$ and $x + 3y + 4z + 6 = 0$ are perpendicular. Find this value.

46. Determine the values of a and b for which the planes $2x + ay + 3z - 1 = 0$ and $bx - 10y - 6z + 5 = 0$ are parallel.

47. Find the plane through the point $(1, -9, 3)$ which is perpendicular to the line of intersection of the planes $x - 2y + z - 5 = 0$ and $x + y - z + 4 = 0$.

48. Find the angle between the plane $x + 2y - 2z - 1 = 0$ and each of the coordinate planes.

49. Find the distance between the point $(3, -4, 12)$ and the plane $x + y + z = 180$.

50. Show that the lines $x = y = z$ and

$$\frac{x+4}{3} = \frac{y-6}{-2} = \frac{z+10}{6}$$

intersect. Find the point of intersection.

51. Find the distance between the line $x = y = z$ and the line through the points $(3, -1, 2)$ and $(6, 1, 4)$.

52. Find the length of the space curve

$$x = \cos t, \quad y = \sin t, \quad z = \ln(\cos t) \quad (0 \le t \le \pi/4).$$

Evaluate

53. $\lim_{t \to 1} \left(\dfrac{\ln t}{t-1} \mathbf{i} - \dfrac{e^t - e}{t-1} \mathbf{j} + \dfrac{t+1}{t^2} \mathbf{k} \right)$

54. $\lim_{t \to 0} \left(\dfrac{\sqrt{1+t}-1}{t} \mathbf{i} + \dfrac{t-1}{t+1} \mathbf{j} - \dfrac{\sinh t}{t} \mathbf{k} \right)$

55. $\dfrac{d}{dt} \left(\dfrac{e^t}{t} \mathbf{i} + \dfrac{\ln t}{t} \mathbf{j} + \dfrac{\cos t}{t} \mathbf{k} \right)$

56. $\dfrac{d}{dt} [(\cosh t)\mathbf{i} + (\tanh t)\mathbf{j} + 10^t \mathbf{k}]$

57. $\displaystyle\int_{\pi/6}^{\pi/3} [(\cot t)\mathbf{i} - (\tan t)\mathbf{j} + 24\mathbf{k}] \, dt$

58. $\displaystyle\int \left(\dfrac{1}{t^2+4} \mathbf{i} + \dfrac{1}{t^2-1} \mathbf{j} + \dfrac{t^2-1}{t^2+1} \mathbf{k} \right) dt$

Find the unit tangent and normal vectors **T** and **N** to the given curve at the point P corresponding to the specified value of the parameter t. Also find the curvature κ at P.

59. $x = 1 - 6t$, $y = 3 + 2t$, $z = 4 - 3t$, $t = -1$
60. $x = 2t$, $y = t^2$, $z = \ln t$, $t = 1$
61. $x = \cosh t$, $y = \sinh t$, $z = t$, $t = 0$
62. $x = e^t \cos t$, $y = e^t \sin t$, $z = e^t$, $t = \pi/4$

63. Choosing the arc length s as the parameter, let C be the curve traced out by the point $P = P(s)$ with position vector $\mathbf{r} = \mathbf{r}(s)$, and let C have unit tangent and normal vectors **T** and **N** at P. Then the unit vector $\mathbf{B} = \mathbf{T} \times \mathbf{N}$ is called the *binormal* to C at P. Clearly $\mathbf{N} \times \mathbf{B} = \mathbf{T}$ and $\mathbf{B} \times \mathbf{T} = \mathbf{N}$. The vectors **T**, **N** and **B**, in that order, form a right-handed orthonormal basis, which varies with the position of P. This local basis is called the *moving trihedral* of C (see Figure 44). Assuming that $\mathbf{r} = \mathbf{r}(s)$ has a continuous third derivative, show that $d\mathbf{B}/ds$ is parallel to **N**. Thus

$$\frac{d\mathbf{B}}{ds} = -\tau \mathbf{N},$$

where the scalar τ is called the *torsion* of C at P; the choice of the minus sign is conventional, and leads to a positive value for the torsion of a right-handed helix (see Problem 65). Also show that

$$\frac{d\mathbf{N}}{ds} = \tau \mathbf{B} - \kappa \mathbf{T},$$

where κ is the curvature of C at P (we already know that $d\mathbf{T}/ds = \kappa \mathbf{N}$).

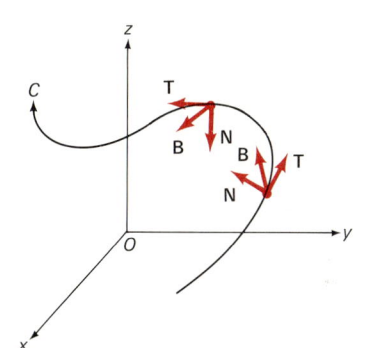

Figure 44

64. The curve C is usually given as the graph of a position vector function $\mathbf{r} = \mathbf{r}(t)$, where t is a parameter other than the arc length. Show that the torsion of C is given by the formula

$$\tau = \frac{\mathbf{r}' \cdot (\mathbf{r}'' \times \mathbf{r}''')}{|\mathbf{r}' \times \mathbf{r}''|^2}, \quad \text{(ii)}$$

where as usual the prime denotes differentiation with respect to t. Show that C is a plane curve if and only if its torsion is identically equal to zero.

65. Find the torsion τ of the general circular helix

$$x = a \cos \omega t, \quad y = a \sin \omega t, \quad z = bt \quad (a > 0).$$

66. Show that the curve $x = t^2$, $y = 1 - 3t$, $z = 4t - 2$ lies in a plane. Which one?

Let C be the twisted cubic $x = t$, $y = \frac{1}{2}t^2$, $z = \frac{1}{3}t^3$.

67. Find the maximum curvature of C.

68. Find the torsion τ and the vectors $\mathbf{T}$, $\mathbf{N}$ and $\mathbf{B}$ of the moving trihedral of C at the origin.

69. Find the torsion τ and the vectors $\mathbf{T}$, $\mathbf{N}$ and $\mathbf{B}$ at the point $(1, \frac{1}{2}, \frac{1}{3})$.

70. Find the maximum torsion of C.

71. What is the torsion of the curve $x = \cosh t$, $y = \sinh t$, $z = t$ at the point which corresponds to $t = \ln 2$?

72. Show that the curvature κ and torsion τ of the curve $x = 3t - t^3$, $y = 3t^2$, $z = 3t + t^3$ are equal at every point.

73. Show that the angular momentum of a projectile fired from a gun is not conserved if the initial point of the position vector of the projectile is at the gun. However, there is a point about which angular momentum is conserved. Where is it?

74. The altitude of an earth satellite at apogee is 660 mi, and the ratio of its maximum speed to its minimum speed is 1.1. What is the altitude of the satellite at perigee? What is its period of revolution? (Use $R = 3960$ mi for the radius of the earth and $g = 32.15$ ft/sec² for the acceleration due to gravity at the earth's surface.)

75. A space shuttle is coasting in a circular orbit 150 mi above the earth, when the pilot fires a brief burst from its thruster rockets, causing the shuttle to go 900 mph faster. What is the maximum altitude of the shuttle in its new orbit?

76. The perihelion distance of a planet is r_0, and the eccentricity of its orbit is e. Show that the radius of curvature of the orbit at the endpoints of the major axis is $(1 + e)r_0$.

77. The point $(3, -4, 7)$ lies on a right circular cone of two nappes with its axis along the z-axis and its vertex at the origin. Find an equation of the cone.

78. Find an equation of the prolate spheroid generated by revolving the ellipse

$$\frac{x^2}{a^2} + \frac{z^2}{b^2} = 1 \quad (a > b > 0)$$

about the x-axis. Of the oblate spheroid generated by revolving the same ellipse about the z-axis. (See Problem 46, page 485.)

79. Find an equation of the cone with vertex $(0, 0, 5)$ and generators tangent to the sphere $x^2 + y^2 + z^2 = 9$.

80. Let S be the set of all points in space the sum of whose distances from two fixed points is constant. Show that S is a prolate spheroid.

13 Partial Differentiation

Much of elementary calculus is concerned with scalar or vector quantities which are functions of a single variable, and hence are completely determined once we know just one number, the value of the independent variable. But many quantities of interest in mathematics and its applications are functions of two or more variables, that is, they are determined only when we specify two or more numbers, namely the values of these variables. There may be a great many variables in real-life situations. For example, the annual profit of a supermarket depends among other things on labor and maintenance costs, rental and delivery fees, and the quantities of various items sold.

The description of functions of several variables is relatively easy, at least in principle. The real challenge arises in constructing suitable multivariable generalizations of the derivative and integral. In this chapter we show how the ideas of *differential* calculus can be developed further until they are capable of dealing with functions of several variables. The goal of extending the concepts of *integral* calculus to functions of two and three variables will be accomplished in Chapters 14 and 15.

13.1 Functions of Several Variables

Definition of a Function of Two or Three Variables

Let D be a set of points (x, y) in the plane. Then by *a function f of two variables x and y* we mean a rule or procedure which assigns a unique real number, denoted by $f(x, y)$, to each point (x, y) in D. The set D is called the *domain* of f, the number $f(x, y)$ is called the *value* of f at (x, y), and the coordinates x and y of the point (x, y) are called the *independent variables* (or the *arguments* of f). A *function f of three variables x, y and z* is defined in the same way, except that D is now a set of points (x, y, z) in space and the value of f at (x, y, z) is denoted by $f(x, y, z)$. By "the function $f(x, y)$" we mean of course the function f whose value at (x, y) is $f(x, y)$, and this kind of abbreviated language is habitual. The set of all values taken by f at the points of D is called the *range* of f, and can be regarded as the set of all values of an extra *dependent* variable; for example, the range of $f(x, y, z)$ is the set of all values of the dependent variable $u = f(x, y, z)$ as (x, y, z) varies over the set D. Naturally, there is great freedom in the choice of the symbols used to denote the function itself and the independent and dependent variables. You will recognize all this as the immediate generalization of the corresponding ideas for functions of a single variable.

Functions of several variables arise in any situation where the values of two or more independent variables uniquely determine the value of another, dependent variable.

Example 1 Let A be the area of a rectangle of length l and width w. Then A is a function of l and w, and denoting this function by f, we have

$$A = f(l, w) = lw, \tag{1}$$

in accordance with a well-known formula of elementary geometry. Here l and w are the independent variables, which can take arbitrary positive values, and A is the dependent variable. But it is easy to make A one of the independent variables, and l or w the dependent variable. In fact, solving (1) for l, we get

$$l = g(A, w) = \frac{A}{w},$$

which expresses l as another function, denoted here by g, of the two variables A and w. Similarly, the formula

$$w = g(A, l) = \frac{A}{l}$$

expresses w as a function of the variables A and l. Why have we chosen to use the symbol g again for the function in this formula?

Example 2 Let

$$f(x, y, z) = \frac{x - y}{y - z}. \tag{2}$$

Then

$$f(1, 4, 0) = \frac{1 - 4}{4 - 0} = -\frac{3}{4}$$

and

$$f(2, 3, 4) = \frac{2 - 3}{3 - 4} = 1,$$

but $f(3, 1, 1)$ fails to exist, since calculation of this quantity immediately leads to a zero denominator. Whenever a function is given by an explicit formula like (2), with no information about the values of the independent variables, it is understood that the domain of the function is the *largest* set of points for which the formula makes sense; this set is called the *natural domain* of the function. Thus the natural domain of the function (2) is the set of all points (x, y, z) in space except those lying in the plane $y - z = 0$ parallel to the x-axis. As an exercise, show that the range of (2) is the whole real line.

Example 3 The natural domain of the function

$$f(x, y) = \arcsin x + \sqrt{xy} \qquad (3)$$

is the largest set of points (x, y) in the plane for which the function is defined. Since $\arcsin x$ is defined if and only if $-1 \leq x \leq 1$, while $\sqrt{xy}$ is defined if and only if $xy \geq 0$, the natural domain of (3) is the pair of "semi-infinite" shaded strips shown in Figure 1. The strips are closed, in the sense that they contain their boundaries.

Algebraic operations on functions of several variables are defined in the same way as for functions of a single variable. Thus

$$(f + g)(x, y) = f(x, y) + g(x, y),$$
$$(fg)(x, y, z) = f(x, y, z)g(x, y, z),$$

and so on. Composition of functions of several variables is illustrated in the following examples.

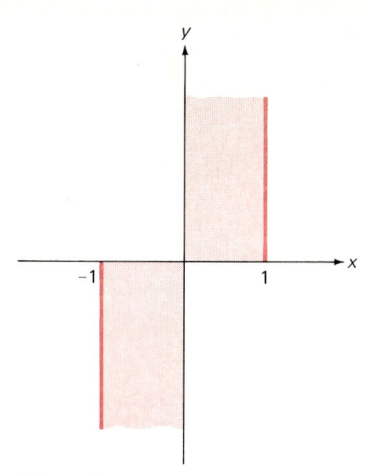

Figure 1

Example 4 Let

$$f(x, y) = \frac{1}{x^2 + y^2}, \qquad g(t) = t^2, \qquad h(t) = \sqrt{t}.$$

Then

$$f(g(t), h(t)) = \frac{1}{g^2(t) + h^2(t)} = \frac{1}{t^4 + t}$$

is a function of the single variable t, where we must require that $t > 0$ (why?).

Example 5 Let

$$f(u, v) = \frac{uv}{u + v}, \qquad g(x, y) = x + y, \qquad h(x, y) = x - y.$$

Then

$$f(g(x, y), h(x, y)) = \frac{g(x, y)h(x, y)}{g(x, y) + h(x, y)} = \frac{x^2 - y^2}{2x}$$

is a function of the two variables x and y.

> **Remark** You will recall that in addition to *numerical* functions of a single variable, whose values are real numbers, we have also considered *vector-valued* functions of a single variable (also called vector functions of a scalar argument). One can also consider vector functions of several variables, or equivalently vector functions of a variable point in the plane or in space. These are called *vector fields*, and will be investigated in Chapter 15.

n-Space and Functions of n Variables

While we're at it, we might as well go a step further and introduce (numerical) functions of more than three variables, defined on a set of points in a space of more than three dimensions. Although difficult to visualize, n-space or R^n is simply the set of all "points" $(x_1, x_2, \ldots, x_n)$, where $x_1, x_2, \ldots, x_n$ are real numbers designated as the *coordinates* of $(x_1, x_2, \ldots, x_n)$. The notation

Section 13.1 Functions of Several Variables

$(x_1, x_2, \ldots, x_n)$, sometimes called an "*ordered n-tuple*," tells us that the point $(x_1, x_2, \ldots, x_n)$ has x_1 as its first coordinate, x_2 as its second coordinate, and so on, with x_n as its nth coordinate. We are already familiar with the meaning of R^n for $n = 1$, 2 and 3. In fact, $R^1 = R$ is just the real line, R^2 is the plane (2-space), and R^3 is ordinary three-dimensional space (3-space). Let D be a subset of R^n. Then by a *function of n variables* $x_1, x_2, \ldots, x_n$ we mean a rule or procedure f which assigns a unique real number, denoted by $f(x_1, x_2, \ldots, x_n)$, to each point $(x_1, x_2, \ldots, x_n)$ in D. As always, D is called the *domain* of f, and the number $f(x_1, x_2, \ldots, x_n)$ is called the *value of f* at $(x_1, x_2, \ldots, x_n)$.

Example 6 Functions of n variables abound in statistics. For instance, given any n numbers $x_1, x_2, \ldots, x_n$ (called *sample values* in this context), their *average* or *mean* $\bar{x}$ is the function

$$\bar{x} = \frac{1}{n}(x_1 + x_2 + \cdots + x_n) = \frac{1}{n}\sum_{i=1}^{n} x_i,$$

and their *variance* s^2 is the function

$$s^2 = \frac{1}{n}\sum_{i=1}^{n}(x_i - \bar{x})^2. \quad \square$$

Given a function $z = f(x, y)$ of two variables x and y, by the *graph* of f we mean the graph of the equation

$$z = f(x, y),$$

that is, the set of all points (x, y, z) in R^3 whose (rectangular) coordinates satisfy this equation. However, we cannot graph a function of three or more variables, since there is "not enough room" in 3-space for the graph of an equation of the form $u = f(x_1, x_2, \ldots, x_n)$ with $n \geq 3$, which involves at least four variables $x_1, x_2, \ldots, x_n, u$.

Level Curves

By a *level curve* of a function $f(x, y)$ of two variables we mean the projection onto the xy-plane of the curve (or set) in which the graph of f intersects the horizontal plane $z = c$, where c is any constant in the range of f. Thus the level curve corresponding to c has the equation

$$f(x, y) = c, \tag{4}$$

regarded as a curve in the xy-plane. Notice that the intersection of the graph of f with the plane $z = c$ coincides with the level curve (4) if $c = 0$, or else lies directly above or below it at vertical distance $|c|$. By drawing a number of typical level curves of a function f, and labeling each curve with the appropriate value of c, we can get a good idea of the shape of the graph of f. For instance, in Figure 2 we have drawn a surface that looks like a mountain with two peaks, which we regard as the graph of some function $f(x, y)$, and the bottom part of the figure shows a few of the level curves $f(x, y) = c$ corresponding to this function. The level curve with $c = 5$ degenerates into a single point, and every level curve with $2 < c < 4$ consists of two disjoint loops (why?).

Example 7 The graph of the function

$$z = \sqrt{1 - x^2 - y^2} \tag{5}$$

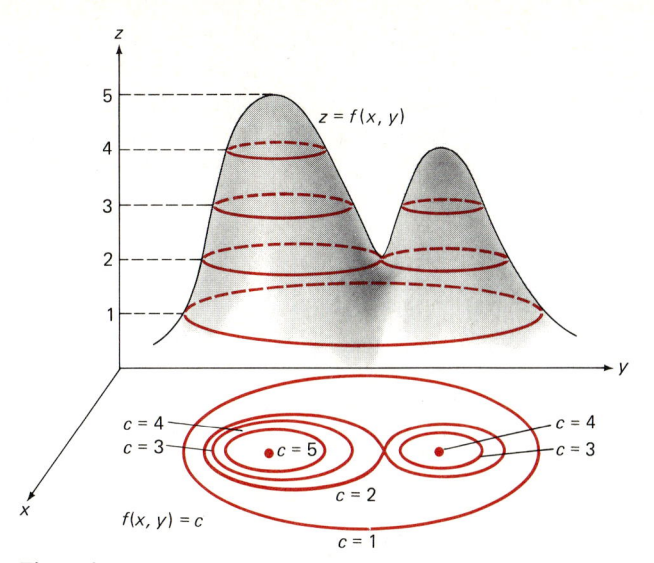

Figure 2

is the hemisphere shown in Figure 3(a) [the top half of the unit sphere $x^2 + y^2 + z^2 = 1$], while the graph of the somewhat different function

$$z = 1 - \sqrt{x^2 + y^2} \qquad (0 \le x^2 + y^2 \le 1) \tag{6}$$

is the right circular cone shown in Figure 4(a) [part of the bottom nappe of the cone $x^2 + y^2 + (z-1)^2 = 0$]. Both the hemisphere and the cone have circular level curves. In fact, substituting $z = c$ into (5) and (6), we get the equations

$$x^2 + y^2 = 1 - c^2 \tag{5'}$$

and

$$x^2 + y^2 = (1 - c)^2, \tag{6'}$$

both of whose graphs are circles for $0 \le c < 1$ and the single point $x = y = 0$ for $c = 1$. But a detailed comparison of the two sets of level curves, given in Figures 3(b) and 4(b), reveals an important difference in their behavior. In the case of the hemisphere the level curves with small c have radii near 1, since the climb up a hemispherical dome is initially very steep; also there

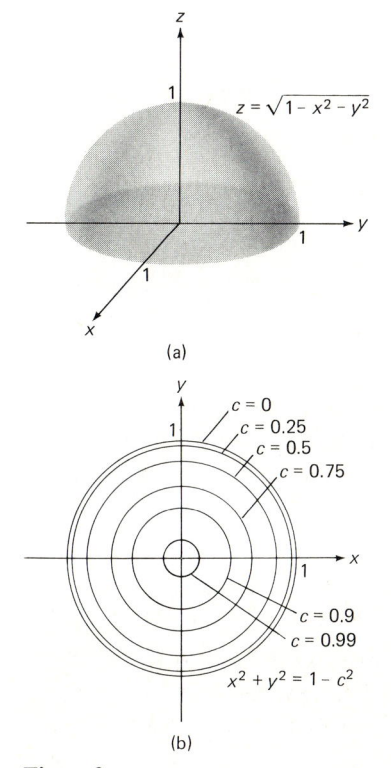

Figure 3

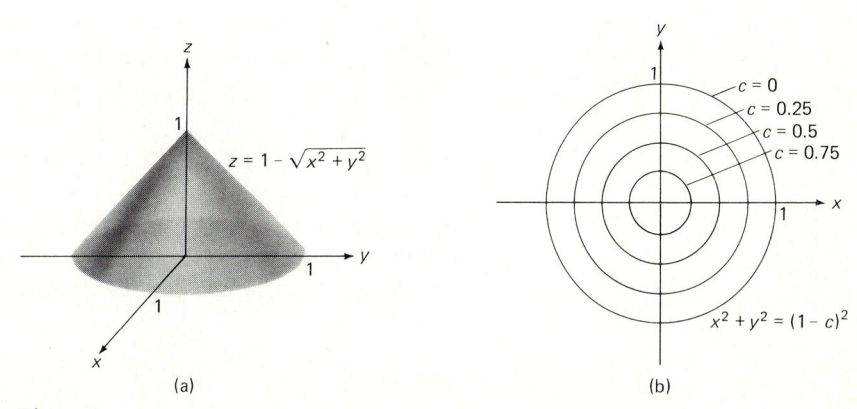

Figure 4

Section 13.1 Functions of Several Variables **751**

are level curves with values of c near 1 and radii that are still appreciably different from 0, since the relatively flat top of the dome is reached well before getting to the summit. On the other hand, the level curves of the cone are "evenly spaced" (equal increases in c lead to equal decreases in the radii of the curves). This is because a climb up the cone is always at the same angle of 45° with the horizontal, from the bottom of the cone to its vertex. □

Level curves are widely used in cartography (map-making) as a device for indicating elevation. They are also used in weather maps, where curves of constant pressure are called *isobars* and curves of constant temperature are called *isotherms*.

Level Surfaces

By a *level surface* of a function $f(x, y, z)$ of three variables we mean the graph of the equation

$$f(x, y, z) = c, \qquad (7)$$

where c is any constant in the range of f. Notice that equation (7) can be graphed in three-dimensional space, even though the function f itself cannot! This is because (7) is obtained from the equation $u = f(x, y, z)$ by giving the dependent variable u a *fixed* value.

Example 8 The set of level surfaces of the function

$$f(x, y, z) = x - 2y + 3z$$

is the set of parallel planes

$$x - 2y + 3z = c \qquad (-\infty < c < \infty),$$

or equivalently the set of all planes with $\mathbf{n} = (1, -2, 3)$ as a normal vector.

Computer Graphics

When properly programmed, the modern electronic computer is capable of generating marvelously detailed and accurate plots of the surfaces corresponding to graphs of functions of two variables. Four examples of such computer-generated figures are shown on the facing page. The curves on each surface are obtained by assigning x or y fixed values ranging from -1.5 to 1.5, and each surface is viewed from a position somewhat above the fourth quadrant of the xy-plane.

Problems

Find the value of the function $f(x, y) = \dfrac{2x - y}{x - 2y}$ at the point

1. $(0, 1)$ **2.** $(1, 0)$ **3.** $(-4, 5)$
4. $(11, -7)$ **5.** $(\sqrt{2}, \sqrt{2})$ **6.** $(\pi, -\pi)$

Find the value of the function $g(x, y) = \dfrac{\sin x}{\cos y}$ at the point

7. $(\pi/3, \pi/3)$ **8.** $(\pi/4, 3\pi/4)$
9. $(\pi, 2\pi)$ **10.** $(5\pi/6, -\pi/6)$
11. $(\arctan 2, \arctan 3)$ **12.** $(1, -1)$

Find the value of the function $f(x, y, z) = \ln xyz$ at the point

13. $(e, 1, 1)$ **14.** $(-1, 1, -1)$
15. (e, e^2, e^3) **16.** $(\sqrt{e}, \sqrt{e}, \sqrt{e})$

Find the value of the function $g(x, y, z) = 2^{xy/z}$ at the point

17. $(\sqrt{2}, \sqrt{3}, \sqrt{6})$ **18.** $(5, -1, 10)$
19. $(\log_2 \pi, 1, \tfrac{1}{2})$ **20.** $(9, 4, 12)$

Find the value of the function

$$f(x_1, x_2, \ldots, x_n) = \frac{x_1 + x_2 + \cdots + x_n}{x_1 x_2 \cdots x_n}$$

at the point

21. $(1, 1, \ldots, 1)$ **22.** $(a, a, \ldots, a)$
23. $(1/n, 1/n, \ldots, 1/n)$ **24.** $(1, 2, \ldots, n)$

$z = -10x^3y^2$

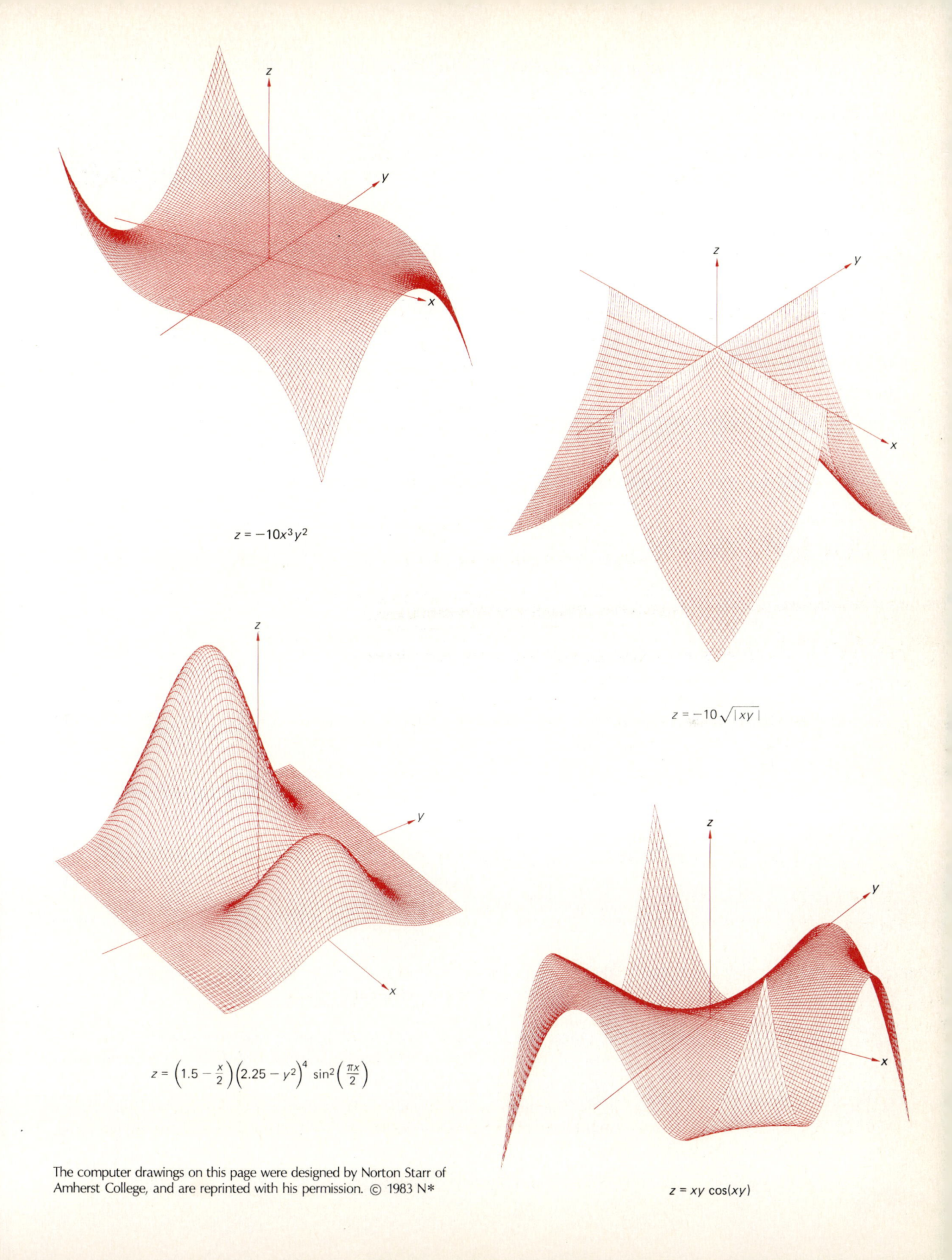

$z = -10\sqrt{|xy|}$

$z = \left(1.5 - \frac{x}{2}\right)\left(2.25 - y^2\right)^4 \sin^2\left(\frac{\pi x}{2}\right)$

$z = xy \cos(xy)$

The computer drawings on this page were designed by Norton Starr of Amherst College, and are reprinted with his permission. © 1983 N*

25. Let $f(x, y) = \ln(x/y)$, $g(t) = e^t$ and $h(t) = 1/t$. Find $f(g(t), h(t))$.
26. Find $f(x, y)$ if $f(x + y, x - y) = 2xy + y^2$.
27. Let $f(x, y, z) = x \cos yz$, $u(x, y) = 2 \sin(x/y)$, $v(x, y) = x^2 y$ and $w(x, y) = 1/xy^2$. Find $f(u(x, y), v(x, y), w(x, y))$.
28. Let $F(x, y, z) = e^{xyz}$, $f(t) = \ln t$, $g(t) = t^2$ and $h(t) = 1/t^3$. Find $F(f(t), g(t), h(t))$.
29. Find the function $f(s, s^3, \ldots, s^{2n-1})$, if it is given that $f(x_1, x_2, \ldots, x_n) = \ln(x_1 x_2 \cdots x_n)$.
30. Determine $\phi(u(1, -1, 1), v(1, -1, 1), w(1, -1, 1))$, if it is given that $\phi(u, v, w) = 2^u 3^v 4^w$, $u(x, y, z) = xyz$, $v(x, y, z) = x + y + z$ and $w(x, y, z) = xy + xz + yz$.

Describe the (natural) domain of the given function of several variables.

31. $z = \sqrt{9 - x^2 - y^2}$
32. $z = \dfrac{1}{\sqrt{4 - x^2 - y^2}}$
33. $z = x^2 - y^2$
34. $z = \dfrac{xy}{x - y}$
35. $z = \dfrac{1}{x} + \dfrac{1}{y}$
36. $z = \dfrac{1}{x^2 - y}$
37. $z = \ln(x + y)$
38. $z = \ln(xy)$
39. $z = y + \arccos x$
40. $u = \dfrac{1}{x} + \dfrac{1}{y} + \dfrac{1}{z}$
41. $u = \sqrt{x} + \sqrt{y} + \sqrt{z}$
42. $u = \sqrt{16 - x^2 - y^2 - z^2}$
43. $u = e^{xy/z}$
44. $u = \arcsin x + \arcsin y + \arcsin z$

Describe the graph of the given function of two variables.

45. $z = 2x + 3y$
46. $z = 1 - x^2 - y^2$
47. $z = -\sqrt{16 - x^2 - y^2}$
48. $z = \sqrt{x^2 + y^2}$
49. $z = \sqrt{x^2 + y^2 + 1}$
50. $z = \sqrt{x^2 + y^2 - 1}$

Sketch the level curves $f(x, y) = c$ of the given function of two variables for the specified values of c.

51. $f(x, y) = x - 2y$, $c = 0, \pm 1, \pm 2$
52. $f(x, y) = 1 - |x| - |y|$, $c = 0, \frac{1}{4}, \frac{1}{2}, \frac{3}{4}, 1$
53. $f(x, y) = x^2 + 4y^2$, $c = 0, 1, 2, 3, 4$
54. $f(x, y) = y/x^2$ ($f(0, 0) = 0$), $c = 0, \pm 1, \pm 2$
55. $f(x, y) = xy$, $c = 0, \pm 1, \pm 2$
56. $f(x, y) = y - \cos x$, $c = 0, \pm 1, \pm 2$

Describe the level surfaces $f(x, y, z) = c$ of the given function of three variables.

57. $f(x, y, z) = x + y + z$ $(-\infty < c < \infty)$
58. $f(x, y, z) = x^2 + y^2 + z^2$ $(0 \le c < \infty)$
59. $f(x, y, z) = x^2 + 2y^2 + 3z^2$ $(0 \le c < \infty)$
60. $f(x, y, z) = x^2 + y^2 - z^2$ $(-\infty < c < \infty)$

13.2 Limits and Continuity

Next we discuss limits and continuity for functions of several variables. We begin by extending the concept of a neighborhood to a space of two or more dimensions. Let $P_0 = (a, b)$ be a fixed point in the plane, let δ be a positive number, and let N be the set of all points $P = (x, y)$ such that

$$|P_0 P| < \delta, \tag{1}$$

where $|P_0 P|$ is the distance between P_0 and P. In terms of coordinates, N is the set of all points (x, y) such that

$$\sqrt{(x - a)^2 + (y - b)^2} < \delta, \tag{1'}$$

that is, the interior of the circle

$$(x - a)^2 + (y - b)^2 = \delta^2$$

of radius δ with its center at (a, b). A region of this type is called a *neighborhood* of the point $P_0 = (a, b)$. Clearly, N is the two-dimensional generalization of the one-dimensional neighborhood $\{x : |x - a| < \delta\}$.

The merit of the inequality (1), as opposed to (1'), is that it leaves the number n of independent variables unspecified. If $n = 3$, we write $P_0 =$

(a, b, c) and $P = (x, y, z)$, and then the set N defined by (1) is the interior of the *sphere*

$$(x - a)^2 + (y - b)^2 + (z - c)^2 = \delta^2$$

of radius δ with its center at (a, b, c). We again call N a neighborhood, this time of the point $P_0 = (a, b, c)$ in 3-space. More generally, let the distance between a fixed point $P_0 = (a_1, a_2, \ldots, a_n)$ and a variable point $P = (x_1, x_2, \ldots, x_n)$ in n-space be defined as

$$|P_0P| = \sqrt{(x_1 - a_1)^2 + (x_2 - a_2)^2 + \cdots + (x_n - a_n)^2} = \sqrt{\sum_{i=1}^{n} (x_i - a_i)^2},$$

where you will note that for $n = 1, 2$ and 3 this formula reduces to the formulas we already have for distance on the line, in the plane and in 3-space. Then by a neighborhood of P_0 we still mean the set of all points P satisfying the inequality (1), which can now be thought of as the interior of the "n-sphere"

$$\sum_{i=1}^{n} (x_i - a_i)^2 = \delta^2$$

of radius δ with its center at $(a_1, a_2, \ldots, a_n)$. Admittedly we can't draw a picture of an n-sphere if $n > 3$, but this does not diminish the conceptual value of such an idea.

By a *deleted neighborhood* of a point P_0 in n-space we mean any neighborhood of P_0 with the point P_0 itself left out. In other words, if N is the neighborhood of P_0 defined by the inequality (1), the corresponding deleted neighborhood of P_0 is the set defined by the double inequality $0 < |P_0P| < \delta$.

The Limit of a Function of Several Variables

To define the *limit* of a function of several variables, we need only make slight modifications in the definition of the limit of a function of a single variable, given on page 72. This is how it goes. Let $f(P)$ be a function of several variables defined in a deleted neighborhood of a point P_0; just as $P_0 = (a_1, a_2, \ldots, a_n)$ and $P = (x_1, x_2, \ldots, x_n)$ in terms of the coordinates, $f(P)$ is shorthand for $f(x_1, x_2, \ldots, x_n)$. Then $f(P)$ is said to approach a limit L as P approaches P_0 (or to have the limit L at P_0) if, given any $\varepsilon > 0$, we can find a $\delta > 0$ such that $|f(P) - L| < \varepsilon$ whenever $0 < |P_0P| < \delta$. This is expressed by writing

$$\lim_{P \to P_0} f(P) = L \tag{2}$$

or

$$f(P) \to L \quad \text{as} \quad P \to P_0. \tag{2'}$$

Also, to say that $f(P)$ *has a limit* at P_0 means that there is some number L such that $f(P) \to L$ as $P \to P_0$.

Let $f(P) = f(x, y)$ be a function of two variables, with a limit L at a point $P_0 = (a, b)$, and suppose $P = (x, y)$ approaches P_0 along some curve C passing through P_0. Then we get a "partial limit"

$$\lim_{P \to P_0,\ P \in C} f(P) \tag{3}$$

analogous to a one-sided limit in the case of a function of a single variable.

More exactly, if C has parametric equations
$$x = x(t), \quad y = y(t) \quad (\alpha \leq t \leq \beta),$$
where $x(t_0) = a$, $y(t_0) = b$ for some t_0 in (α, β), then (3) means
$$\lim_{t \to t_0} f(x(t), y(t)).$$

It is an immediate consequence of the definition of a limit that (3) exists and is equal to L, regardless of the choice of C. After all, the definition says nothing about *how* P approaches P_0. In particular, P can approach P_0 along the horizontal line $y = b$ or the vertical line $x = a$. This gives two limits

$$\lim_{x \to a} f(x, b) \tag{4}$$

and

$$\lim_{y \to b} f(a, y), \tag{4'}$$

each of which is the limit of a function of a single variable. These limits must both equal the limit (2), which we now indicate more explicitly as

$$\lim_{(x,y) \to (a,b)} f(x, y). \tag{5}$$

Thus if the limits (4) and (4') fail to exist, or exist and are unequal, we know that the limit (5) of the function $f(x, y)$ of two variables does not exist.

Example 1 The function

$$f(x, y) = \frac{x^2 - y^2}{x^2 + y^2} \quad (x^2 + y^2 \neq 0)$$

has no limit at the origin $(0, 0)$. In fact, letting (x, y) approach the origin first along the x-axis and then along the y-axis, we have

$$\lim_{x \to 0} f(x, 0) = \lim_{x \to 0} \frac{x^2}{x^2} = 1,$$

$$\lim_{y \to 0} f(0, y) = \lim_{y \to 0} \frac{-y^2}{y^2} = -1.$$

Since these two limits are unequal the limit

$$\lim_{(x,y) \to (0,0)} \frac{x^2 - y^2}{x^2 + y^2}$$

does not exist.

Example 2 If

$$f(x, y) = \frac{xy}{x^2 + y^2} \quad (x^2 + y^2 \neq 0),$$

then

$$\lim_{x \to 0} f(x, 0) = \lim_{x \to 0} 0 = 0,$$

$$\lim_{y \to 0} f(0, y) = \lim_{y \to 0} 0 = 0,$$

so that $f(x, y) \to 0$ as $P = (x, y)$ approaches the origin $(0, 0)$ along the x-axis, and also as P approaches $(0, 0)$ along the y-axis. Can we conclude from this that

$$\lim_{(x,y)\to(0,0)} \frac{xy}{x^2+y^2} \qquad (6)$$

exists and equals 0? No, we cannot! For suppose P is made to approach the origin along the line $y = mx$ of slope m. Then the corresponding partial limit is

$$\lim_{x\to 0} f(x, mx) = \lim_{x\to 0} \frac{mx^2}{x^2+m^2x^2} = \frac{m}{1+m^2},$$

whose value varies with m and in particular in nonzero if $m \ne 0$. Hence the limit (6) does not exist.

Continuity of a Function of Several Variables

Continuity for functions of several variables is defined in just the same way as for functions of a single variable. Thus if $f(P)$ is a function of several variables defined in a neighborhood of a point P_0, we say that $f(P)$ is *continuous at P_0* if $f(P)$ has a limit at P_0 and if this limit is equal to the value of $f(P)$ at P_0, so that

$$\lim_{P\to P_0} f(P) = f(P_0).$$

In ε, δ language $f(P)$ is said to be continuous at P_0 if given any $\varepsilon > 0$, there is a $\delta > 0$ such that $|f(P) - f(P_0)| < \varepsilon$ whenever $|P_0 P| < \delta$ (recall the related discussion on pages 72–73.)

Example 3 Suppose $f(x, y)$ is independent of y, so that $f(x, y) = g(x)$, and let $g(x)$ be continuous at a. Then $f(x, y)$ is continuous at (a, b) for arbitrary b. In fact, given any $\varepsilon > 0$, there is a $\delta > 0$ such that $|g(x) - g(a)| < \varepsilon$ whenever $|x - a| < \delta$. But then for the same ε and δ,

$$|P_0 P| = \sqrt{(x-a)^2 + (y-b)^2} < \delta$$

implies $|x - a| < \delta$, which in turn implies

$$|f(x, y) - f(a, b)| = |g(x) - g(a)| < \varepsilon,$$

showing that $f(x, y)$ is continuous at (a, b). Similarly, if $f(x, y)$ is independent of x, so that $f(x, y) = h(y)$, and if $h(y)$ is continuous at b, then $f(x, y)$ is continuous at (a, b) for arbitrary a.

Example 4 Suppose $f(x, y)$ and $g(x, y)$ are continuous at (a, b). Then so are $f(x, y) \pm g(x, y)$, $f(x, y)g(x, y)$ and $f(x, y)/g(x, y)$, provided that $g(a, b) \ne 0$ in the last case. This follows from the analogue of Theorem 7, page 83, for functions of several variables. Thus the function $(2x^3 - 3y)/(x^2 + xy^2 - 1)$ is continuous wherever it is defined. The same is true of the function $\tan(e^{xy})$, because of the following easily verified generalization of the corollary on page 88: If $g(x, y)$ is continuous at (a, b) and if $f(t)$ is continuous at $g(a, b)$, then $f(g(x, y))$ is continuous at (a, b). By similar reasoning, we can establish the continuity of such functions of three variables as $\sin(xy + xz + yz)$ and $\cosh xyz$.

Classification of Points and Sets in *n*-Space

The following terminology will be useful in studying functions of several variables. Let D be any set of points in *n*-space. Then a point P is called an *interior point* of D if there is *some* neighborhood of P containing only points in *D*. A point *P* is called a *boundary point* of *D* if *every* neighborhood of *P* contains both points in *D* and points not in *D*. These definitions are illustrated in Figure 5 for the case where *D* is a set of points in the plane. An interior point of *D* must be a point of *D* (why?), but a boundary point of *D* may or may not belong to *D*. A set *D* is said to be *open* if all its points are interior points, and *closed* if it contains all its boundary points. The set of all boundary points of *D* is called the *boundary* of *D*. Thus a closed set contains its boundary, but an open set does not.

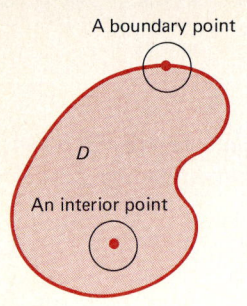

Figure 5

Example 5 The set of points in the *xy*-plane defined by the inequalities $0 < x < 1$, $0 < y < 1$ is open, the set defined by $0 \le x \le 1$, $0 \le y \le 1$ is closed, and the set defined by $0 \le x \le 1$, $0 < y < 1$ is neither open nor closed. All three sets have the same boundary, namely the square with vertices $(0, 0)$, $(0, 1)$, $(1, 0)$ and $(1, 1)$.

Continuity on a Set

A function f of several variables is said to be *continuous on a set D* in *n*-space (which need not be the domain of f) if f is continuous at every point of *D*, that is, if $f(P) \to f(P_0)$ as $P \to P_0$ for every point P_0 in *D*. Our previous definition of the limit of a function at a point works only for interior points of *D*, but a slight modification of the definition makes it work for boundary points as well. Specifically, let P_0 be a boundary point of *D*. Then $f(P)$ is said to have the limit *L* at P_0 if given any $\varepsilon > 0$, there is a $\delta > 0$ such that $|f(P) - L| < \varepsilon$ whenever *P* belongs to *D* and $0 < |P_0 P| < \delta$. In other words, we do not require that the inequality $|f(P) - L| < \varepsilon$ be satisfied for all points of *n*-space sufficiently near P_0, but only for all points of *the set D* sufficiently near P_0 (excluding the point P_0 itself). Notice that if P_0 is an interior point of *D*, this definition of a limit contains the earlier one (why?). We are now able to talk about continuity of f at a boundary point P_0 of *D*, provided of course that f is defined at P_0. The same approach was used on page 94 to define continuity of a function of one variable at the endpoints of an interval *I*.

Example 6 The function

$$f(x, y) = \begin{cases} 1 & \text{if } x^2 + y^2 \le 1, \\ 0 & \text{otherwise} \end{cases}$$

is clearly not continuous on the whole *xy*-plane, since $f(x, y)$ does not have a limit at any point of the unit circle $x^2 + y^2 = 1$ (why not?). But $f(x, y)$ *is* continuous on the closed unit disk defined by the inequality $x^2 + y^2 \le 1$, with the unit circle as its boundary. This follows at once from our modified definition of a limit and the fact that f has the constant value 1 on the disk *D*.

Problems

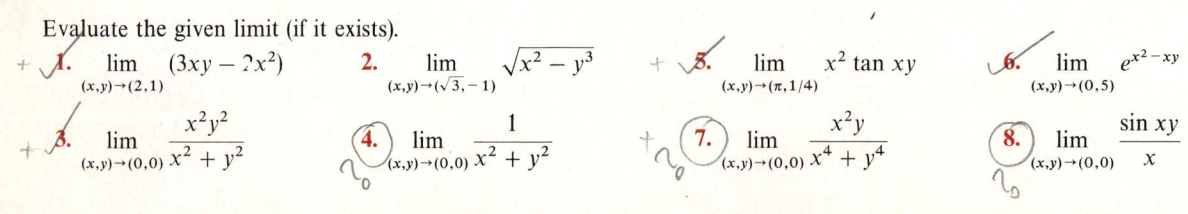

Evaluate the given limit (if it exists).

1. $\lim_{(x,y) \to (2,1)} (3xy - 2x^2)$
2. $\lim_{(x,y) \to (\sqrt{3}, -1)} \sqrt{x^2 - y^3}$
3. $\lim_{(x,y) \to (\pi, 1/4)} x^2 \tan xy$
4. $\lim_{(x,y) \to (0,5)} e^{x^2 - xy}$
5. $\lim_{(x,y) \to (0,0)} \dfrac{x^2 y^2}{x^2 + y^2}$
6. $\lim_{(x,y) \to (0,0)} \dfrac{1}{x^2 + y^2}$
7. $\lim_{(x,y) \to (0,0)} \dfrac{x^2 y}{x^4 + y^4}$
8. $\lim_{(x,y) \to (0,0)} \dfrac{\sin xy}{x}$

9. $\lim_{(x,y)\to(\infty,1)} \arctan \dfrac{x}{y}$

10. $\lim_{(x,y)\to(4,-\infty)} \tanh(x+y)$

11. $\lim_{(x,y,z)\to(0,0,0)} \dfrac{xyz}{x^2+y^2+z^2}$

12. $\lim_{(x,y,z)\to(1,-1,e)} \ln(xy^2 z^3)$

13. Show that the function

$$f(x,y) = \dfrac{xy^2}{x^2+y^4}$$

has no limit at the origin O, although it approaches the same limit along every straight line through O.

14. Give an example of a closed set in 2-space with no interior points.

Is the given function f continuous on the specified set D? Explain your answer.

15. $f(x,y) = \dfrac{1}{x^2-y^2}$, $D = \{(x,y): (x-1)^2 + y^2 < \tfrac{1}{2}\}$

16. $f(x,y) = \cot \pi xy$, $D = \{(x,y): 0 < x < 1, 0 < y \le 1\}$

17. $f(x,y) = \dfrac{1}{\sin \pi x} + \dfrac{1}{\sin \pi y}$,

$D = \{(x,y): (x-\tfrac{1}{2})^2 + (y-\tfrac{1}{2})^2 \le \tfrac{1}{4}\}$

18. $f(x,y) = \dfrac{1}{\sin^2 \pi x + \sin^2 \pi y}$,

same D as in Problem 17

19. $f(x,y,z) = \dfrac{x+y+z}{x^2+y^2+z^2-1}$,

$D = \{(x,y,z): x^2 + 4y^2 + 4z^2 \le 1\}$

20. $f(x,y,z) = \ln xyz$,

$D = \{(x,y,z): (x+1)^2 + (y+1)^2 + (z-1)^2 < 1\}$

13.3 Partial Derivatives

Let f be a function of n variables defined in a neighborhood of a point $(x_1, x_2, \ldots, x_n)$. Then by the *partial derivative of f with respect to x_i at* $(x_1, x_2, \ldots, x_n)$, denoted by the expression

$$\dfrac{\partial f(x_1, x_2, \ldots, x_n)}{\partial x_i}, \qquad (1)$$

we mean the limit

$$\lim_{\Delta x_i \to 0} \dfrac{f(x_1, \ldots, x_i + \Delta x_i, \ldots, x_n) - f(x_1, \ldots, x_i, \ldots, x_n)}{\Delta x_i},$$

where x_i is given an increment Δx_i, but all the other variables are held fixed, provided this limit exists and is finite. Clearly, we can consider n such partial derivatives, one for each of the n independent variables $x_1, x_2, \ldots, x_n$. A function $f(x,y)$ of two variables has two partial derivatives, namely

$$\dfrac{\partial f(x,y)}{\partial x} = \lim_{\Delta x \to 0} \dfrac{f(x+\Delta x, y) - f(x,y)}{\Delta x}$$

and

$$\dfrac{\partial f(x,y)}{\partial y} = \lim_{\Delta y \to 0} \dfrac{f(x, y+\Delta y) - f(x,y)}{\Delta y},$$

provided these limits exist and are finite.

Thinking of the symbol $\partial/\partial x_i$ as a single entity, whose effect is to form the partial derivative with respect to x_i of any function written after it, we can also write (1) as

$$\dfrac{\partial}{\partial x_i} f(x_1, x_2, \ldots, x_n). \qquad (1')$$

The symbol ∂ can still be pronounced "dee," even though we are now dealing with a "curved dee." Arguments are often omitted, and then the partial derivative (1) can be written concisely as $\partial f/\partial x_i$. To evaluate $\partial f/\partial x_i$, we need only treat all the independent variables except x_i as if they were constants. Thus no extra technique is required to calculate partial derivatives.

Example 1 Suppose that $f(x, y) = xe^{xy}$. Then, by the product rule,

$$\frac{\partial f(x, y)}{\partial x} = \frac{\partial}{\partial x}(xe^{xy}) = e^{xy} + xye^{xy},$$

calculated by regarding y as a constant and treating $\partial/\partial x$ like d/dx, and

$$\frac{\partial f(x, y)}{\partial y} = \frac{\partial}{\partial y}(xe^{xy}) = x^2 e^{xy},$$

calculated by regarding x as a constant and treating $\partial/\partial y$ like d/dy. These partial derivatives are defined at every point (x, y) of the plane.

Example 2 If

$$f(x, y, z) = \frac{x}{(x^2 + y^2 + z^2)^{1/2}},$$

then

$$\frac{\partial f}{\partial x} = \frac{1}{(x^2 + y^2 + z^2)^{1/2}} - \frac{1}{2}\frac{2x^2}{(x^2 + y^2 + z^2)^{3/2}} = \frac{y^2 + z^2}{(x^2 + y^2 + z^2)^{3/2}}$$

at every point (x, y, z) of space other than the origin $(0, 0, 0)$. You can easily verify that the other two partial derivatives of f are

$$\frac{\partial f}{\partial y} = -\frac{xy}{(x^2 + y^2 + z^2)^{3/2}}, \qquad \frac{\partial f}{\partial z} = -\frac{xz}{(x^2 + y^2 + z^2)^{3/2}}. \quad \square$$

There are other notations for partial derivatives. If a dependent variable $u = f(x_1, x_2, \ldots, x_n)$ is introduced, we can write $\partial u/\partial x_i$ instead of $\partial f/\partial x_i$. A prevalent notation is to write $\partial f/\partial x_i$ or $\partial u/\partial x_i$ as f_{x_i} or u_{x_i}, where the subscript x_i calls for (partial) differentiation with respect to x_i. For example, if

$$u = f(x, y, z) = xy^2 \sin yz,$$

then

$$\frac{\partial u}{\partial x} = f_x(x, y, z) = y^2 \sin yz,$$

$$\frac{\partial u}{\partial y} = f_y(x, y, z) = 2xy \sin yz + xy^2 z \cos yz,$$

$$\frac{\partial u}{\partial z} = f_z(x, y, z) = xy^3 \cos yz.$$

Example 3 According to the *ideal gas law*, the pressure p, volume V and absolute temperature T of one mole of a confined gas are related by the formula

$$pV = RT,$$

where R is a constant of proportionality known as the *universal gas constant*.

Since
$$V = \frac{RT}{p}, \qquad p = \frac{RT}{V}, \qquad T = \frac{pV}{R},$$
we have
$$\frac{\partial V}{\partial T} = \frac{R}{p}, \qquad \frac{\partial V}{\partial p} = -\frac{RT}{p^2},$$

$$\frac{\partial p}{\partial T} = \frac{R}{V}, \qquad \frac{\partial p}{\partial V} = -\frac{RT}{V^2},$$

$$\frac{\partial T}{\partial p} = \frac{V}{R}, \qquad \frac{\partial T}{\partial V} = \frac{p}{R}.$$

It follows that
$$\frac{\partial p}{\partial V} \frac{\partial V}{\partial T} \frac{\partial T}{\partial p} = \left(-\frac{RT}{V^2}\right)\left(\frac{R}{p}\right)\left(\frac{V}{R}\right) = -\frac{RT}{pV} = -1.$$

This formula should convince you that the expression on the left cannot be thought of as a product of three fractions!

In applied problems it is often helpful to think of a partial derivative as the rate of change of a dependent variable with respect to one independent variable, with the other independent variables held fixed. Thus in the preceding example we have calculated the rate of change of volume with respect to temperature at constant pressure ($\partial V/\partial T$) and the rate of change of temperature with respect to pressure at constant volume ($\partial T/\partial p$). We have also calculated four other rates of change (describe them).

It is easy to give a geometric interpretation of the partial derivatives of a function $z = f(x, y)$ of two variables. Let f be defined in a neighborhood of the point (a, b) of the xy-plane, and let $P = (a, b, f(a, b))$ be the corresponding point of the graph of f, which is some surface S in space. Then the plane $y = b$ intersects S in a curve C_x going through P, and the plane $x = a$ intersects S in another curve C_y which also goes through P (see Figure 6). Suppose the function f has partial derivatives $f_x(a, b)$ and $f_y(a, b)$ at the point (a, b). Then C_x has a tangent line at P, of slope $f_x(a, b)$, and C_y also has a tangent line at P, but its slope is $f_y(a, b)$ instead. These two tangent lines are denoted by T_x and T_y in the figure.

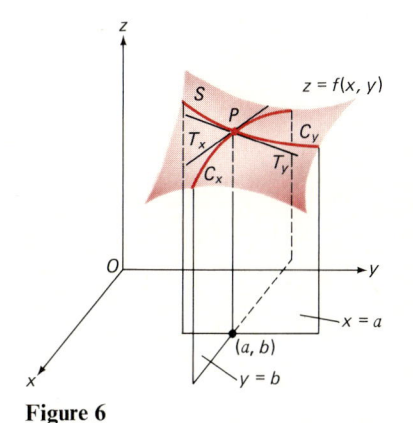

Figure 6

Partial Derivatives of Higher Order

The partial derivatives just introduced are *first* derivatives. Partial derivatives of higher order are defined in the natural way. For example, let $f(x, y)$ be a function of two variables which has (first) partial derivatives $\partial f/\partial x$ and $\partial f/\partial y$ at every point of some open set. Then f has four *second partial derivatives*, namely

$$\frac{\partial^2 f}{\partial x^2} = \frac{\partial}{\partial x}\left(\frac{\partial f}{\partial x}\right), \qquad \frac{\partial^2 f}{\partial y^2} = \frac{\partial}{\partial y}\left(\frac{\partial f}{\partial y}\right),$$

and the "mixed" derivatives

$$\frac{\partial^2 f}{\partial x\, \partial y} = \frac{\partial}{\partial x}\left(\frac{\partial f}{\partial y}\right), \qquad \frac{\partial^2 f}{\partial y\, \partial x} = \frac{\partial}{\partial y}\left(\frac{\partial f}{\partial x}\right),$$

which differ only in the order of differentiation (to calculate $\partial^2 f/\partial x\, \partial y$, we differentiate first with respect to y and then with respect to x, while $\partial^2 f/\partial y\, \partial x$ is calculated by carrying out the differentiations in the opposite order). Using the subscript notation, we can also write these second partial derivatives as

$$\frac{\partial^2 f}{\partial x^2} = f_{xx}, \qquad \frac{\partial^2 f}{\partial x\,\partial y} = f_{yx}, \qquad \frac{\partial^2 f}{\partial y\,\partial x} = f_{xy}, \qquad \frac{\partial^2 f}{\partial y^2} = f_{yy}.$$

Notice that in f_{yx} we differentiate first with respect to the first subscript y and then with respect to the second subscript x. This is the reverse of the order in which the variables x and y appear in the denominator of the expression $\partial^2 f/\partial x\,\partial y$, but it is the proper order since f_{yx} must mean $(f_y)_x$, the derivative with respect to x of f_y, which is in turn the partial derivative of f with respect to y.

Of course, all this is based on the assumption that the limits defining these second partial derivatives exist and are finite. An example of such a limit is

$$\frac{\partial^2 f(x, y)}{\partial x\,\partial y} = \lim_{\Delta x \to 0} \frac{f_y(x + \Delta x, y) - f_y(x, y)}{\Delta x}.$$

If there is a dependent variable $z = f(x, y)$, we can also write $\partial^2 z/\partial x^2$ or z_{xx} for $\partial^2 f/\partial x^2$, $\partial^2 z/\partial x\,\partial y$ or z_{yx} for $\partial^2 f/\partial x\,\partial y$, and so on. Partial derivatives of order greater than two, and higher order partial derivatives of functions of more than two variables are defined in the way you would expect. For example, if $u = f(x, y, z)$, then

$$\frac{\partial^3 u}{\partial x\,\partial y\,\partial z} = \frac{\partial}{\partial x}\left(\frac{\partial^2 u}{\partial y\,\partial z}\right) = (u_{zy})_x = u_{zyx},$$

$$\frac{\partial^4 u}{\partial y\,\partial x\,\partial z^2} = \frac{\partial}{\partial y}\left(\frac{\partial^3 u}{\partial x\,\partial z^2}\right) = (u_{zzx})_y = u_{zzxy},$$

and so on.

Example 4 If

$$f(x, y, z) = xy \ln z \qquad (z > 0),$$

then

$$\frac{\partial f}{\partial x} = y \ln z, \qquad \frac{\partial f}{\partial y} = x \ln z, \qquad \frac{\partial f}{\partial z} = \frac{xy}{z},$$

$$\frac{\partial^2 f}{\partial x^2} = \frac{\partial}{\partial x}(y \ln z) = 0,$$

$$\frac{\partial^2 f}{\partial y^2} = \frac{\partial}{\partial y}(x \ln z) = 0,$$

$$\frac{\partial^2 f}{\partial z^2} = \frac{\partial}{\partial z}\left(\frac{xy}{z}\right) = -\frac{xy}{z^2},$$

$$\frac{\partial^2 f}{\partial x\,\partial y} = \frac{\partial}{\partial x}(x \ln z) = \ln z, \qquad \frac{\partial^2 f}{\partial y\,\partial x} = \frac{\partial}{\partial y}(y \ln z) = \ln z,$$

$$\frac{\partial^2 f}{\partial x\,\partial z} = \frac{\partial}{\partial x}\left(\frac{xy}{z}\right) = \frac{y}{z}, \qquad \frac{\partial^2 f}{\partial z\,\partial x} = \frac{\partial}{\partial z}(y \ln z) = \frac{y}{z}.$$

$$\frac{\partial^2 f}{\partial y\,\partial z} = \frac{\partial}{\partial y}\left(\frac{xy}{z}\right) = \frac{x}{z}, \qquad \frac{\partial^2 f}{\partial z\,\partial y} = \frac{\partial}{\partial z}(x \ln z) = \frac{x}{z},$$

Observe that in the preceding example *the value of each mixed partial derivative is independent of the order of differentiation*. It is not hard to show that this is always true *if the mixed partial derivatives are continuous*, but the proof will not be given here.

Equations containing partial derivatives are called *partial differential equations*, as opposed to the ordinary differential equations that we have considered so far. An example of a partial differential equation is *Laplace's equation*

$$\frac{\partial^2 u}{\partial x^2} + \frac{\partial^2 u}{\partial y^2} = 0, \qquad (2)$$

which arises in a great variety of problems in both pure and applied mathematics. By the *order* of a partial differential equation we mean the order of the highest partial derivative appearing in the equation. Thus Laplace's equation is of the second order, and so is

$$\frac{\partial^2 u}{\partial x^2} = \frac{\partial u}{\partial t},$$

a special case of the partial differential equation governing one-dimensional diffusion or heat conduction. A solution of Laplace's equation (2) is called a *harmonic function*.

Example 5 The function $u = x^3 - 3xy^2$ is harmonic. In fact, we find at once that

$$\frac{\partial^2 u}{\partial x^2} + \frac{\partial^2 u}{\partial y^2} = \frac{\partial}{\partial x}\left(\frac{\partial u}{\partial x}\right) + \frac{\partial}{\partial y}\left(\frac{\partial u}{\partial y}\right)$$

$$= \frac{\partial}{\partial x}(3x^2 - 3y^2) + \frac{\partial}{\partial y}(-6xy) = 6x - 6x = 0.$$

As an exercise, show that the function $3x^2 y - y^3$ is also harmonic.

Problems

Find all first partial derivatives of the given function.

1. $f(x, y) = x^2 y^3 - x^3 y^4$
2. $f(x, y) = \dfrac{x + y}{x - y}$
3. $f(x, y) = \dfrac{xy}{x^2 + y^2}$
4. $f(x, y) = \sqrt{x^2 - y^2}$
5. $f(x, y) = \ln(x^3 - y^2)$
6. $f(x, y) = e^{-x/y}$
7. $f(x, y) = e^{\sin xy}$
8. $f(x, y) = \ln(x + \sqrt{x^2 + y^2})$
9. $f(s, t) = \arctan \dfrac{t}{s}$
10. $g(u, v) = \tan(u^2 - v)$
11. $h(x, y) = \ln(x + \ln y)$
12. $k(x, y) = \cos\dfrac{x}{y} \sin\dfrac{y}{x}$
13. $f(x, y) = x^y$
14. $f(x, y) = 10^{xy}$
15. $f(x, y, z) = xy + xz + yz$
16. $f(x, y, z) = \dfrac{x}{y} + \dfrac{y}{z} - \dfrac{z}{x}$
17. $f(x, y, z) = \ln xyz$
18. $f(x, y, z) = \sinh xyz$
19. $f(x, y, z) = x^{yz}$
20. $f(x, y, z) = e^{xy/z}$

If the mixed partial derivatives f_{xy} and f_{yx} are continuous, then $f_{xy} = f_{yx}$. Verify this by direct calculation for the given function.

21. $f(x, y) = x^2 y^4 - 2x^3 y^3 + 3x^6 y - 10$
22. $f(x, y) = x^2/y^2$
23. $f(x, y) = y^x$
24. $f(x, y, z) = e^{xyz}$

25. Let

$$f(x, y) = xy\frac{x^2 - y^2}{x^2 + y^2}$$

Section 13.3 Partial Derivatives 763

if $x^2 + y^2 \neq 0$, while $f(0, 0) = 0$. Show that $f_{xy} \neq f_{yx}$ at the origin. Reconcile this with the fact that the mixed partial derivatives are equal if they are continuous.

26. Let $x = r \cos \theta$ and $y = r \sin \theta$, as in the transformation from polar to rectangular coordinates. Find $\partial x/\partial r$, $\partial y/\partial r$, $\partial x/\partial \theta$ and $\partial y/\partial \theta$.

Find the second partial derivatives z_{xx}, $z_{xy} = z_{yx}$ and z_{yy} of the given function.

27. $z = x^4 y^2 - 3x^2 y^3 + 6xy$
28. $z = x^3 \sin y + y^3 \cos x$
29. $z = e^{x^2 y}$
30. $z = \ln(x^2 + y^2)$
31. $z = \sin^2(3x - 4y)$
32. $z = \arctan xy$

33. Find u_{xxy} if $u = 2^{xy}$.
34. Find v_{xyx} if $v = \cos(x + \sin y)$.
35. Find $\dfrac{\partial^{m+n} u}{\partial x^m \, \partial y^n}$ if $u = x^m y^n$.
36. Find $\dfrac{\partial^5 u}{\partial x^3 \, \partial y^2}$ if $u = \sin^2 x \cos^2 y$.
37. Find $\dfrac{\partial^3 u}{\partial x \, \partial y \, \partial z}$ if $u = \sin xy + \cos xz + \tan yz$.
38. Find $\dfrac{\partial^4 w}{\partial z^2 \, \partial x \, \partial y}$ if $w = \dfrac{x}{y+z}$.

39. How many distinct partial derivatives of order n does a function $f(x, y)$ of two variables have? Assume that each derivative is independent of the order of differentiation, an assumption which is justified if the mixed partial derivatives are continuous.

40. Evaluate
$$\frac{\partial^3}{\partial x \, \partial y \, \partial z} f(x) g(y) h(z),$$
where f, g and h are differentiable functions of a single variable.

The angle between the sides of a triangle is θ, and the sides have lengths x and y. Let A be the area of the triangle, and let z be the length of the third side.

41. Find the rate of change of A with respect to θ if x and y are held fixed.
42. Find the rate of change of A with respect to x if θ and y are held fixed.
43. Find the rate of change of z with respect to y if θ and x are held fixed.
44. Find the rate of change of z with respect to θ if x and y are held fixed.

45. The intersection of the plane $x = 1$ with the double-sheeted hyperboloid of revolution $x^2 + y^2 - z^2 = -1$ is a hyperbola. Write parametric equations for the tangent line to this hyperbola at the point $(1, 1, \sqrt{3})$.

46. The plane $y = -1$ intersects the elliptic paraboloid $2x^2 + y^2 - z = 0$ in a parabola. Write symmetric equations for the tangent line to this parabola at the point $(2, -1, 9)$.

In *complex analysis*, a branch of mathematics of great importance in applied science, a basic role is played by the system of first order partial differential equations

$$\frac{\partial u}{\partial x} = \frac{\partial v}{\partial y}, \qquad \frac{\partial u}{\partial y} = -\frac{\partial v}{\partial x},$$

known as the *Cauchy–Riemann equations*. Show that the given pair of functions u and v satisfy these equations.

47. $u = x^4 - 6x^2 y^2 + y^4$, $v = 4x^3 y - 4xy^3$
48. $u = e^x \cos y$, $v = e^x \sin y$
49. $u = \cos x \cosh y$, $v = -\sin x \sinh y$

50. Show that all six functions in Problems 47–49 are harmonic. Why is this to be expected?

51. Show that the function
$$u = \frac{1}{\sqrt{x^2 + y^2 + z^2}} \qquad (x^2 + y^2 + z^2 \neq 0)$$
is a solution of the three-dimensional version of Laplace's equation:
$$\frac{\partial^2 u}{\partial x^2} + \frac{\partial^2 u}{\partial y^2} + \frac{\partial^2 u}{\partial z^2} = 0.$$

52. Show that the function
$$u = \frac{1}{\sqrt{t}} e^{-x^2/4t} \qquad (t > 0)$$
is a solution of the partial differential equation $\partial u^2/\partial x^2 = \partial u/\partial t$. Find a simpler solution involving an exponential.

The second order partial differential equation
$$\frac{\partial^2 u}{\partial x^2} = \frac{1}{c^2} \frac{\partial^2 u}{\partial t^2},$$
where c is a positive constant, is called the *wave equation*. Show that the given function is a solution of the wave equation.

53. $u = x^2 + c^2 t^2$
54. $u = \sin x \sin ct$
55. $u = e^x \cosh ct$

56. Let f and g be arbitrary functions of a single variable with second derivatives f'' and g''. Show that the function $u = f(x + ct) + g(x - ct)$ is a solution of the wave equation. Express each of the functions in Problems 53–55 in this form.

764 Chapter 13 Partial Differentiation

13.4 Differentiability and Differentials

A function $f(x)$ of one variable is said to be differentiable at a point a if the derivative $f'(a)$ exists.† This says a lot about the behavior of f near a, and in particular it tells us that f has a tangent line approximation near a (see page 123). In the case of a function $f(x, y)$ of two variables, knowledge of the partial derivatives $f_x(a, b)$ and $f_y(a, b)$ at a point $P = (a, b)$ says relatively little about the behavior of f near P. In fact, these derivatives are determined entirely by the values of f on the lines through P parallel to the coordinate axes, and they remain the same if the values of f are altered anywhere else. In Figure 6, page 761, this means that the surface S can be deformed in any manner without changing the values of $f_x(a, b)$ and $f_y(a, b)$, just as long as the planes $y = b$ and $x = a$ intersect the new surface in the same curves C_x and C_y. Thus it would not be appropriate to define a differentiable function $f(x, y)$ as one for which the partial derivatives $f_x = \partial f/\partial x$ and $f_y = \partial f/\partial y$ merely exist. In particular, existence of the partial derivatives will not by itself lead to a generalization of the tangent line approximation, which is what we really want (in this regard see the discussion on page 782). So how *should* differentiability be defined for a function of two or more variables?

The answer to this question is suggested by a more careful examination of the meaning of differentiability for a function $f(x)$ of one variable. Consider the increment

$$\Delta f = f(a + \Delta x) - f(a), \tag{1}$$

which is the change in the value of f when x is changed from a to $a + \Delta x$. Then existence of $f'(a)$ means that

$$\lim_{\Delta x \to 0} \frac{\Delta f}{\Delta x} = f'(a),$$

or equivalently

$$\frac{\Delta f}{\Delta x} = f'(a) + \alpha(\Delta x),$$

where $\alpha(\Delta x)$ is a function of Δx such that

$$\lim_{\Delta x \to 0} \alpha(\Delta x) = 0 \tag{2}$$

(recall Corollary 2, page 81). Thus if $f'(a)$ exists, the increment (1) can be written as

$$\Delta f = f'(a)\, \Delta x + \alpha(\Delta x)\, \Delta x,$$

where $\alpha(\Delta x)$ satisfies (2). In other words, if $f(x)$ is differentiable at a, there is a constant A such that

$$\Delta f = A\, \Delta x + \alpha(\Delta x)\, \Delta x, \tag{3}$$

where $\alpha(\Delta x) \to 0$ as $\Delta x \to 0$. Of course $A = f'(a)$, but this fact is suppressed in (3), because we want a definition of differentiability which does not mention

---

† The word "exists" as applied to a derivative, ordinary or partial, will always mean "exists *and is finite.*"

derivatives! In fact, we will now define differentiability for a function $f(x, y)$ of two variables by generalizing formula (3) rather than using the partial derivatives f_x and f_y.

Definition of Differentiability

To this end, let $f(x, y)$ be a function of two variables defined in a neighborhood of a point (a, b), and consider the increment

$$\Delta f = f(a + \Delta x, b + \Delta y) - f(a, b), \qquad (4)$$

which is the change in the value of f when x is changed from a to $a + \Delta x$ and y is changed from b to $b + \Delta y$. Suppose that for every point $(a + \Delta x, b + \Delta y)$ in a deleted neighborhood of (a, b) we can represent Δf in the form

$$\Delta f = A \, \Delta x + B \, \Delta y + \alpha(\Delta x, \Delta y) \, \Delta x + \beta(\Delta x, \Delta y) \, \Delta y, \qquad (5)$$

analogous to (3), where A and B are constants, and $\alpha(\Delta x, \Delta y)$ and $\beta(\Delta x, \Delta y)$ are functions of Δx and Δy such that

$$\lim_{(\Delta x, \Delta y) \to (0,0)} \alpha(\Delta x, \Delta y) = 0, \qquad \lim_{(\Delta x, \Delta y) \to (0,0)} \beta(\Delta x, \Delta y) = 0. \qquad (6)$$

Then f is said to be *differentiable* at (a, b), and the expression $A \, \Delta x + B \, \Delta y$ is called the *(total) differential* of f at (a, b), denoted by df, so that

$$df = A \, \Delta x + B \, \Delta y. \qquad (7)$$

The fact that this is the proper definition of differentiability for a function f of two variables is confirmed by the following theorem, which shows that the properties of differentiable functions of two variables closely resemble those of differentiable functions of one variable.

> **Theorem 1** *(Implications of differentiability)*. *Let f be a function of two variables which is differentiable at (a, b). Then f is continuous at (a, b) and has partial derivatives f_x and f_y at (a, b).*

Proof It is an immediate consequence of formulas (4)–(6) that

$$\lim_{(\Delta x, \Delta y) \to (0,0)} \Delta f$$

$$= \lim_{(\Delta x, \Delta y) \to (0,0)} [f(a + \Delta x, b + \Delta y) - f(a, b)]$$

$$= \lim_{(\Delta x, \Delta y) \to (0,0)} [A \, \Delta x + B \, \Delta y + \alpha(\Delta x, \Delta y) \, \Delta x + \beta(\Delta x, \Delta y) \, \Delta y] = 0,$$

and this is the increment way of saying that f is continuous at (a, b). Setting $\Delta y = 0$ in (4) and (5), we get

$$\Delta f = f(a + \Delta x, b) - f(a, b) = A \, \Delta x + \alpha(\Delta x, 0) \, \Delta x,$$

and hence

$$f_x(a, b) = \lim_{\Delta x \to 0} \frac{f(a + \Delta x, b) - f(a, b)}{\Delta x} = \lim_{\Delta x \to 0} [A + \alpha(\Delta x, 0)] = A,$$

because of (6), that is, $f_x(a, b)$ exists and equals A. Similarly, $f_y(a, b)$ exists and equals B. ∎

Corollary *If f is differentiable at (a, b), then*

$$\Delta f = [f_x(a, b) + \alpha(\Delta x, \Delta y)] \, \Delta x + [f_y(a, b) + \beta(\Delta x, \Delta y)] \, \Delta y$$

for all sufficiently small $|\Delta x|$ and $|\Delta y|$, where $\alpha(\Delta x, \Delta y)$ and $\beta(\Delta x, \Delta y)$ both approach zero as $(\Delta x, \Delta y) \to (0, 0)$, and moreover

$$df = f_x(a, b) \Delta x + f_y(a, b) \Delta y. \tag{8}$$

Proof Substitute $A = f_x(a, b)$ and $B = f_y(a, b)$ into formulas (5) and (7). ∎

Example 1 If $f(x, y) = Ax + By$ and the point (a, b) is arbitrary, then

$$\begin{aligned}\Delta f &= f(a + \Delta x, b + \Delta y) - f(a, b) \\ &= A(a + \Delta x) + B(b + \Delta y) - (Aa + Bb) \\ &= A \Delta x + B \Delta y,\end{aligned}$$

which is of the form (5) with $\alpha(\Delta x, \Delta y) \equiv 0$ and $\beta(\Delta x, \Delta y) \equiv 0$. Therefore f is differentiable everywhere (that is, at every point of the plane), with differential $df = A \Delta x + B \Delta y$. Observe that $A = f_x$, $B = f_y$, as in the corollary. If $A = 1$ and $B = 0$, then $f(x, y) = x$ and $df = dx = \Delta x$, while if $A = 0$ and $B = 1$, then $f(x, y) = y$ and $df = dy = \Delta y$. Thus *the increments and the differentials of the independent variables are equal.*

Example 2 If $f(x, y) = 4x^2 - 3xy$ and (a, b) is arbitrary, then

$$\begin{aligned}\Delta f &= f(a + \Delta x, b + \Delta y) - f(a, b) \\ &= 4(a + \Delta x)^2 - 3(a + \Delta x)(b + \Delta y) - (4a^2 - 3ab) \\ &= 8a \Delta x + 4(\Delta x)^2 - 3b \Delta x - 3a \Delta y - 3 \Delta x \Delta y \\ &= (8a - 3b) \Delta x - 3a \Delta y + 4(\Delta x)^2 - 3 \Delta x \Delta y,\end{aligned}$$

which is of the form (5) with $A = 8a - 3b$, $B = -3a$, $\alpha = 4 \Delta x$ and $\beta = -3 \Delta x$. Therefore, since $\alpha \to 0$ and $\beta \to 0$ as $(\Delta x, \Delta y) \to (0, 0)$, f is differentiable everywhere; the differential is $df = (8a - 3b) \Delta x - 3a \Delta y$, where $A = f_x(a, b)$ and $B = f_y(a, b)$. Notice that in this case the choice of the functions α and β is not unique, and in fact we could just as well have chosen $\alpha = 4 \Delta x - 3 \Delta y$ and $\beta = 0$. □

Since $\Delta x = dx$ and $\Delta y = dy$, as shown in Example 1, we can write

$$df = f_x(a, b) \, dx + f_y(a, b) \, dy, \tag{8'}$$

instead of (8), which can in turn be written more concisely as

$$df = \frac{\partial f}{\partial x} dx + \frac{\partial f}{\partial y} dy. \tag{9}$$

This formula expresses the *total* differential df as a sum of the "partial differentials" $(\partial f/\partial x) \, dx$ and $(\partial f/\partial y) \, dy$. The generalization of formula (9) to the case of a function $f(x_1, x_2, \ldots, x_n)$ of n variables is

$$df = \frac{\partial f}{\partial x_1} dx_1 + \frac{\partial f}{\partial x_2} dx_2 + \cdots + \frac{\partial f}{\partial x_n} dx_n. \tag{9'}$$

To prove (9'), one first defines differentiability and the differential of a function of n variables, with the help of formulas that generalize (5)–(7) in the natural way. The details are left as an exercise.

Example 3 The total differential of the function $f(x, y, z) = xy \ln z$ considered in Example 4, page 762, is

$$df = \frac{\partial f}{\partial x} dx + \frac{\partial f}{\partial y} dy + \frac{\partial f}{\partial z} dz = (y \ln z) dx + (x \ln z) dy + \frac{xy}{z} dz.$$

The fact that f is actually differentiable at any point (x, y, z) with $z > 0$ can be proved directly, but it follows more easily from Theorem 2 below (as extended to functions of three variables).

Example 4 The function

$$f(x, y) = \begin{cases} 1 & \text{if } x > 0 \text{ and } y > 0, \\ 0 & \text{otherwise} \end{cases}$$

is clearly not continuous at the origin $O = (0, 0)$, since every neighborhood of O contains points where f takes the value 1 and also points where f takes the value 0. It follows from Theorem 1 that f is not differentiable at O. However, both partial derivatives f_x and f_y exist at O. In fact, f has the constant value 0 on both coordinate axes, and therefore

$$f_x(0, 0) = \lim_{\Delta x \to 0} \frac{f(\Delta x, 0) - f(0, 0)}{\Delta x} = \lim_{\Delta x \to 0} 0 = 0,$$

$$f_y(0, 0) = \lim_{\Delta y \to 0} \frac{f(0, \Delta y) - f(0, 0)}{\Delta y} = \lim_{\Delta y \to 0} 0 = 0.$$

The preceding example shows that a function of two variables may well fail to be differentiable even if its partial derivatives exist. On the other hand, as proved in the next theorem, a function with *continuous* partial derivatives must be differentiable.

Theorem 2 (*Continuity of partial derivatives implies differentiability*). *If f is a function of two variables with partial derivatives f_x and f_y in a neighborhood of (a, b), and if these derivatives are continuous at (a, b), then f is differentiable at (a, b).*

Proof (Optional) Let $(a + \Delta x, b + \Delta y)$ be any point in the given neighborhood, which we denote by N. Then

$$\Delta f = f(a + \Delta x, b + \Delta y) - f(a, b)$$
$$= [f(a + \Delta x, b + \Delta y) - f(a, b + \Delta y)] + [f(a, b + \Delta y) - f(a, b)]$$
$$= [g(a + \Delta x) - g(a)] + [h(b + \Delta y) - h(b)],$$

where we have introduced two auxiliary functions of a single variable

$$g(x) = f(x, b + \Delta y) \qquad (a \leq x \leq a + \Delta x)$$

(Δy is fixed in this part of the proof) and

$$h(y) = f(a, y) \qquad (b \leq y \leq b + \Delta y).$$

The functions g and h are both differentiable on their domains, which are line segments in the neighborhood N (see Figure 7). Hence we can apply the mean value theorem (see page 162) to both differences $g(a + \Delta x) - g(a)$

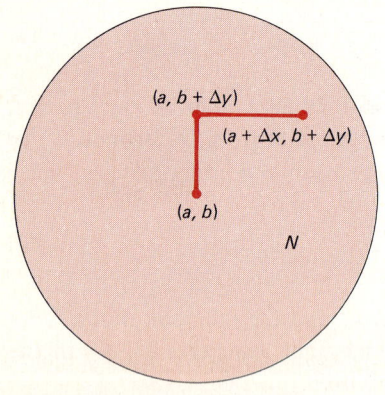

Figure 7

768   Chapter 13   Partial Differentiation

and $h(b + \Delta y) - h(b)$. As a result, we find that

$$\Delta f = \frac{dg(a + t\Delta x)}{dx} \Delta x + \frac{dh(b + u\Delta y)}{dy} \Delta y,$$

where t and u are numbers between 0 and 1; notice that $a + t\Delta x$ lies between a and $a + \Delta x$, and $b + u\Delta y$ lies between b and $b + \Delta y$, regardless of the signs of Δx and Δy, so that both points $(a + t\Delta x, b + \Delta y)$ and $(a, b + u\Delta y)$ lie in N. Moreover,

$$\frac{dg(a + t\Delta x)}{dx} = f_x(a + t\Delta x, b + \Delta y),$$

$$\frac{dh(b + u\Delta y)}{dy} = f_y(a, b + u\Delta y),$$

and therefore

$$\Delta f = f_x(a + t\Delta x, b + \Delta y)\Delta x + f_y(a, b + u\Delta y)\Delta y. \tag{10}$$

In terms of the functions

$$\alpha(\Delta x, \Delta y) = f_x(a + t\Delta x, b + \Delta y) - f_x(a, b),$$
$$\beta(\Delta x, \Delta y) = f_y(a, b + u\Delta y) - f_y(a, b),$$

we can write (10) in the form

$$\Delta f = f_x(a, b)\Delta x + f_y(a, b)\Delta y + \alpha(\Delta x, \Delta y)\Delta x + \beta(\Delta x, \Delta y)\Delta y. \tag{11}$$

But f_x and f_y are continuous at (a, b), by hypothesis, and $(\Delta x, \Delta y) \to (0, 0)$ implies $(a + t\Delta x, b + \Delta y) \to (a, b)$ and $(a, b + u\Delta y) \to (a, b)$. It follows that $\alpha(\Delta x, \Delta y) \to 0$ and $\beta(\Delta x, \Delta y) \to 0$ as $(\Delta x, \Delta y) \to (0, 0)$, and this together with formula (11) establishes the differentiability of f at (a, b). □

Approximation by Differentials

Just as in the case of functions of one variable, we can use differentials to make approximate calculations, relying on the fact that the approximation $\Delta f \approx df$ is good if the increments of the independent variables are small in absolute value.

Example 5 Estimate $\sqrt{(2.98)^2 + (4.03)^2}$.

Solution Let $f(x, y) = \sqrt{x^2 + y^2}$. Then

$$\sqrt{(2.98)^2 + (4.03)^2} = f(3 + \Delta x, 4 + \Delta y),$$

where $\Delta x = -0.02$ and $\Delta y = 0.03$. But

$$\Delta f = f(3 + \Delta x, 4 + \Delta y) - f(3, 4) \approx df = f_x(3, 4)\Delta x + f_y(3, 4)\Delta y,$$

or equivalently

$$f(3 + \Delta x, 4 + \Delta y) \approx f(3, 4) + f_x(3, 4)\Delta x + f_y(3, 4)\Delta y.$$

Here

$$f(3, 4) = \sqrt{3^2 + 4^2} = 5,$$

and

$$f_x = \frac{\partial f}{\partial x} = \frac{x}{\sqrt{x^2 + y^2}}, \quad f_y = \frac{\partial f}{\partial y} = \frac{y}{\sqrt{x^2 + y^2}},$$

so that

$$f_x(3, 4) = \frac{3}{\sqrt{3^2 + 4^2}} = \frac{3}{5}, \qquad f_y(3, 4) = \frac{4}{\sqrt{3^2 + 4^2}} = \frac{4}{5}.$$

Therefore

$$\sqrt{(2.98)^2 + (4.03)^2} \approx 5 + \frac{3}{5}(-0.02) + \frac{4}{5}(0.03) = 5.012,$$

as compared with the exact value of 5.012115 to six decimal places. □

Example 6 Estimate the volume of 1-mm thick sheet metal needed to make a closed rectangular box of length 48 cm, width 36 cm and height 30 cm (interior dimensions).

Solution Let x be the length, y the width and z the height of the box. Then the volume of the box is $f(x, y, z) = xyz$, and the volume of metal needed to make the box is

$$\Delta f = f(48.2, 36.2, 30.2) - f(48, 36, 30)$$

$$\approx df = \frac{\partial f}{\partial x}\Delta x + \frac{\partial f}{\partial y}\Delta y + \frac{\partial f}{\partial z}\Delta z = yz\,\Delta x + xz\,\Delta y + xy\,\Delta z,$$

where $x = 48$, $y = 36$, $z = 30$, and $\Delta x = \Delta y = \Delta z = 0.2$. Therefore

$$\Delta f \approx [36(30) + 48(30) + 48(36)](0.2) = 4248(0.2) = 849.6 \text{ cm}^3,$$

as compared with the exact value of $\Delta f = 48.2(36.2)(30.2) - 48(36)(30) = 854.168 \text{ cm}^3$. Although this approximation is not as good as the one in the preceding example, its *relative* error (see Problem 31, page 127) is only $4.568/854.168 \approx 0.005$.

Problems

1. Find the increment $\Delta f = f(a + \Delta x, y + \Delta y) - f(a, b)$ if $f(x, y) = \ln xy$, $(a, b) = (1, 1)$ and $\Delta x = -\frac{1}{2}$, $\Delta y = 3$.

2. Find the increment

$$\Delta f = f(a + \Delta x, b + \Delta y, c + \Delta z) - f(a, b, c)$$

if $f(x, y, z) = 2xy - 3xz + yz$, $(a, b, c) = (2, 1, 4)$ and $\Delta x = 3$, $\Delta y = -1$, $\Delta z = 2$.

Find the total differential of the given function.

3. $f(x, y) = x^2y - xy^3 + x^3y^2$

4. $f(x, y) = \dfrac{xy}{x - y}$

5. $f(x, y) = \tanh^{-1}\dfrac{x}{y}$

6. $f(x, y) = \text{arccot}\,\dfrac{y}{x}$

7. $g(s, t) = e^{st}$

8. $h(u, v) = \ln \sqrt{u^2 + v^4}$

9. $f(x, y, z) = 2^x 3^y 4^z$

10. $f(x, y, z) = x^{yz^2}$

11. $f(x, y, z) = \dfrac{x + y}{y + z}$

12. $g(s, t, u, v) = s^2 t^{-1} u^3 v^{-4}$

13. $f(x_1, x_2, \ldots, x_n) = x_1 x_2 \cdots x_n$

14. $f(x_1, x_2, \ldots, x_n) = 10^{x_1 + x_2 + \cdots + x_n}$

15. Use differentials to estimate the change in

$$f(x, y) = \arctan \frac{x}{y}$$

when (x, y) changes from $(1, 1)$ to $(0.96, 1.03)$.

16. Use differentials to estimate the change in

$$f(x, y, z) = \frac{x}{\sqrt{y^2 + z^2}}$$

when (x, y, z) changes from $(2, 3, 4)$ to $(2.01, 3.02, 3.97)$.

Use differentials to estimate

17. $(1.002)(2.003)^2(3.004)^3$

18. $\sqrt{(1.06)^2 + (1.97)^3}$

19. $(0.98)^{1.05}$

20. $\sin 31° \cos 58°$

21. $\ln(\sqrt{1.04} + \sqrt[3]{1.08} - 1)$

22. $\sqrt[4]{0.97}\sqrt[3]{(1.04)^2}$

23. Estimate the change in the length of either diagonal of a rectangle of length $x = 30$ cm and width $y = 16$ cm if x is increased by 3 mm and y is decreased by 1 mm.

24. Estimate the volume of 0.05-in. thick sheet metal needed to make a closed can of radius 4 in. and height 12 in. (interior dimensions).

25. Estimate the change in the volume of a conical frustum (see Problem 11, page 440) if its upper radius r_1 is increased from 8 to 8.3 cm, its base radius r_2 is decreased from 12 to 11.8 cm, and its height h is increased from 10 cm to 10.1 cm.

26. If two resistors, one of resistance R_1 and the other of resistance R_2, are connected in parallel as in Figure 8, their combined resistance R satisfies the formula

$$\frac{1}{R} = \frac{1}{R_1} + \frac{1}{R_2}.$$

Suppose that $R_1 = 200$ ohms and $R_2 = 300$ ohms, where these values may be in error by as much as 2 ohms. Estimate the maximum error in the calculated value of R.

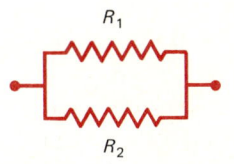

Figure 8

Prove that the given function is differentiable everywhere directly from the definition of differentiability, that is, find numbers A and B and functions α and β satisfying formulas (5) and (6) for every point (a, b) in the xy-plane.

27. $f(x, y) = x^2 + y^2$
28. $f(x, y) = (x + y)^2$
29. $f(x, y) = x^2 y - xy^2$
30. $f(x, y) = e^{xy}$

31. Show that the function $f(x, y) = \sqrt{x^2 + y^2}$ is not differentiable at the origin.

32. Show that the function

$$f(x, y) = \begin{cases} x & \text{if } |x| \leq |y|, \\ -x & \text{if } |x| > |y| \end{cases}$$

is continuous and has partial derivatives at the origin, but is not differentiable there. Show that this is consistent with Theorem 2.

33. Write the analogue of formula (5) for a function $f(x, y, z)$ of three variables. The expression for Δf is given in the statement of Problem 2.

34. Show that the function

$$f(x, y) = \begin{cases} (x^2 + y^2) \sin \dfrac{1}{\sqrt{x^2 + y^2}} & \text{if } x^2 + y^2 \neq 0, \\ 0 & \text{if } x = y = 0 \end{cases}$$

is differentiable at the origin, despite the fact that its partial derivatives are not continuous there. Thus the conditions for differentiability in Theorem 2 are sufficient, but not necessary. (Roughly speaking, differentiability of a function of several variables requires more than existence of the first partial derivatives, but less than their continuity.)

13.5 The Chain Rule and Implicit Differentiation

We now develop multivariable analogues of the chain rule

$$\frac{df}{dt} = \frac{df}{dx}\frac{dx}{dt},$$

which will allow us to differentiate composite functions of several variables. For simplicity we begin with a function of just two variables x and y, each a function of a single independent variable t. The more general case of a function with more than two arguments, each a function of several independent variables, will be considered afterwards.

> **Theorem 3** *(Chain rule for a function of two variables)*. Let $x = x(t)$ and $y = y(t)$ be functions of a single variable, both of which are differentiable at t, and let $f(x, y)$ be a function of two variables which is differentiable at $(x(t), y(t))$. Then the composite function $F(t)$, defined by $F(t) = f(x(t), y(t))$, is differentiable at t, and its derivative is given by
>
> $$F'(t) = f_x(x(t), y(t))x'(t) + f_y(x(t), y(t))y'(t), \tag{1}$$
>
> where the prime denotes differentiation with respect to t.

Proof (Optional) Notice that in this version of the chain rule, $F(t)$ is a function of a single variable, although $f(x, y)$ is a function of two variables. Since $f(x, y)$ is differentiable at $(x(t), y(t))$,

$$\Delta f = f(x + \Delta x, y + \Delta y) - f(x, y)$$
$$= [f_x(x, y) + \alpha(\Delta x, \Delta y)]\, \Delta x + [f_y(x, y) + \beta(\Delta x, \Delta y)]\, \Delta y,$$

by the corollary on page 767, where $\alpha(\Delta x, \Delta y)$ and $\beta(\Delta x, \Delta y)$ both approach zero as $(\Delta x, \Delta y) \to (0, 0)$. So far $\alpha(\Delta x, \Delta y)$ and $\beta(\Delta x, \Delta y)$ are defined only for $(\Delta x, \Delta y) \neq (0, 0)$, but we now enlarge the domain of these functions by setting $\alpha(0, 0) = 0$ and $\beta(0, 0) = 0$, thereby making both functions *continuous* at $(\Delta x, \Delta y) = (0, 0)$.

Next we divide Δf by the increment Δt of the independent variable, and take the limit as $\Delta t \to 0$. This gives

$$\lim_{\Delta t \to 0} \frac{f(x + \Delta x, y + \Delta y) - f(x, y)}{\Delta t}$$

$$= \lim_{\Delta t \to 0} [f_x(x, y) + \alpha(\Delta x, \Delta y)] \lim_{\Delta t \to 0} \frac{\Delta x}{\Delta t}$$

$$+ \lim_{\Delta t \to 0} [f_y(x, y) + \beta(\Delta x, \Delta y)] \lim_{\Delta t \to 0} \frac{\Delta y}{\Delta t}, \qquad (2)$$

where $\Delta x = x(t + \Delta t) - x(t)$ and $\Delta y = y(t + \Delta t) - y(t)$ are the increments of the dependent variables x and y. Moreover, $x(t)$ and $y(t)$ are differentiable at t, and hence continuous at t. Therefore $\Delta t \to 0$ implies $\Delta x \to 0$ and $\Delta y \to 0$, so that

$$\lim_{\Delta t \to 0} \alpha(\Delta x, \Delta y) = \lim_{(\Delta x, \Delta y) \to (0,0)} \alpha(\Delta x, \Delta y) = 0 \qquad (3)$$

and

$$\lim_{\Delta t \to 0} \beta(\Delta x, \Delta y) = \lim_{(\Delta x, \Delta y) \to (0,0)} \beta(\Delta x, \Delta y) = 0. \qquad (3')$$

The fact that $\alpha(0, 0) = 0$ and $\beta(0, 0) = 0$, and consequently that $\alpha(\Delta x, \Delta y)$ and $\beta(\Delta x, \Delta y)$ are continuous at $(\Delta x, \Delta y) = (0, 0)$, is crucial in drawing this conclusion, since we cannot guarantee that $(\Delta x, \Delta y) \neq (0, 0)$. (Remember that Δx and Δy are not arbitrary, but are determined by the value of Δt.) It follows from (2), (3) and (3') that

$$\lim_{\Delta t \to 0} \frac{f(x + \Delta x, y + \Delta y) - f(x, y)}{\Delta t} = f_x(x, y)x'(t) + f_y(x, y)y'(t),$$

or equivalently

$$\lim_{\Delta t \to 0} \frac{f(x(t + \Delta t), y(t + \Delta t)) - f(x(t), y(t))}{\Delta t}$$

$$= f_x(x(t), y(t))x'(t) + f_y(x(t), y(t))y'(t), \qquad (4)$$

since $x = x(t)$, $y = y(t)$ and $x + \Delta x = x(t + \Delta t)$, $y + \Delta y = y(t + \Delta t)$. The proof of the chain rule (1) is now complete, since the limit in (4) is just the derivative at t of the composite function $F(t) = f(x(t), y(t))$. ∎

Formula (1) can be written more concisely as

$$\frac{dF}{dt} = \frac{\partial f}{\partial x}\frac{dx}{dt} + \frac{\partial f}{\partial y}\frac{dy}{dt},$$

where arguments of functions are omitted. We can simplify this even further by writing

$$\frac{df}{dt} = \frac{\partial f}{\partial x}\frac{dx}{dt} + \frac{\partial f}{\partial y}\frac{dy}{dt}. \tag{5}$$

After all, since f is a function of two variables, the fact that we write an *ordinary* derivative df/dt on the left in (5) means that each argument of f is being thought of as a function of a single variable, namely t. With this understanding, we can do without the extra symbol F, which was introduced only to make the distinction between $f(x, y)$ and $f(x(t), y(t))$ more explicit.

Example 1 Find dw/dt if $w = f(x, y) = x^2 + xy$, $x = e^t$ and $y = \sin t$.

Solution Here we introduce another dependent variable w, in terms of which the chain rule (5) takes the form

$$\frac{dw}{dt} = \frac{\partial w}{\partial x}\frac{dx}{dt} + \frac{\partial w}{\partial y}\frac{dy}{dt}. \tag{5'}$$

Since $\partial w/\partial x = 2x + y$, $\partial w/\partial y = x$, $dx/dt = e^t$, $dy/dt = \cos t$, we find that

$$\frac{dw}{dt} = (2x + y)e^t + x \cos t = 2e^{2t} + e^t \sin t + e^t \cos t.$$

The same answer can be obtained without the help of (5') by first substituting $x = e^t$ and $y = \sin t$ into $w = x^2 + xy$, and then differentiating:

$$\frac{dw}{dt} = \frac{d}{dt}(e^{2t} + e^t \sin t) = 2e^{2t} + e^t \sin t + e^t \cos t. \quad \square$$

From Theorem 3 we can easily deduce a chain rule for the more general case where the arguments of $f(x, y)$ are functions of *several* variables. For example, suppose $x = x(t, u)$ and $y = y(t, u)$, where x and y are differentiable functions of two variables t and u. If u is held fixed, x and y reduce to functions of the single variable t, and we can apply the theorem at once, obtaining

$$\frac{\partial f}{\partial t} = \frac{\partial f}{\partial x}\frac{\partial x}{\partial t} + \frac{\partial f}{\partial y}\frac{\partial y}{\partial t}, \tag{6}$$

where all three ordinary derivatives in (5) now become partial derivatives. Similarly, if t is held fixed, x and y reduce to functions of the single variable u, and we obtain

$$\frac{\partial f}{\partial u} = \frac{\partial f}{\partial x}\frac{\partial x}{\partial u} + \frac{\partial f}{\partial y}\frac{\partial y}{\partial u}. \tag{6'}$$

Here again we might have introduced a composite function F, defined by $F(t, u) = f(x(t, u), y(t, u))$, but it is simpler to regard the two functions $f(x, y)$ and $f(x(t, u), y(t, u))$ as being the same, written in terms of different variables.

Section 13.5 The Chain Rule and Implicit Differentiation

Example 2 Find $\partial f/\partial t$ and $\partial f/\partial u$ if $f(x, y) = \ln(x^2 + y^2)$, $x = tu$ and $y = t/u$.

Solution Since

$$\frac{\partial f}{\partial x} = \frac{2x}{x^2 + y^2}, \quad \frac{\partial f}{\partial y} = \frac{2y}{x^2 + y^2},$$

$$\frac{\partial x}{\partial t} = u, \quad \frac{\partial x}{\partial u} = t, \quad \frac{\partial y}{\partial t} = \frac{1}{u}, \quad \frac{\partial y}{\partial u} = -\frac{t}{u^2},$$

it follows from (6) and (6′) that

$$\frac{\partial f}{\partial t} = \frac{2x}{x^2 + y^2} u + \frac{2y}{x^2 + y^2} \frac{1}{u} = \frac{2x^2}{t(x^2 + y^2)} + \frac{2y^2}{t(x^2 + y^2)} = \frac{2}{t},$$

$$\frac{\partial f}{\partial u} = \frac{2x}{x^2 + y^2} t + \frac{2y}{x^2 + y^2}\left(-\frac{t}{u^2}\right) = \frac{2}{t^2 u^2 + \frac{t^2}{u^2}}\left(t^2 u - \frac{t^2}{u^3}\right) = \frac{2(u^4 - 1)}{u(u^4 + 1)}.$$

As an exercise, calculate these derivatives without the help of (6) and (6′) by first substituting $x = tu$ and $y = t/u$ into $f(x, y) = \ln(x^2 + y^2)$. ☐

The natural generalization of formula (5) to the case of a function $f(x_1, x_2, \ldots, x_n)$ whose n arguments depend on a single variable t is given by

$$\frac{df}{dt} = \frac{\partial f}{\partial x_1}\frac{dx_1}{dt} + \frac{\partial f}{\partial x_2}\frac{dx_2}{dt} + \cdots + \frac{\partial f}{\partial x_n}\frac{dx_n}{dt} = \sum_{i=1}^{n} \frac{\partial f}{\partial x_i}\frac{dx_i}{dt}, \quad (7)$$

and is proved in much the same way (we omit the lengthy details). Similarly, the generalization of (6) and (6′) to the case of a function $f(x_1, x_2, \ldots, x_n)$ whose n arguments depend on m new independent variables $t_1, t_2, \ldots, t_m$ is given by

$$\frac{\partial f}{\partial t_j} = \frac{\partial f}{\partial x_1}\frac{\partial x_1}{\partial t_j} + \frac{\partial f}{\partial x_2}\frac{\partial x_2}{\partial t_j} + \cdots + \frac{\partial f}{\partial x_n}\frac{\partial x_n}{\partial t_j} = \sum_{i=1}^{n} \frac{\partial f}{\partial x_i}\frac{\partial x_i}{\partial t_j} \quad (j = 1, 2, \ldots, m).$$
(8)

The last two formulas are the "master" chain rules, of which all previous versions, including the formula

$$\frac{df}{dt} = \frac{df}{dx}\frac{dx}{dt},$$

are special cases. Notice the following common features of (7) and (8):

(i) The right side of the formula contains n terms, one for each "intermediate" variable $x_1, x_2, \ldots, x_n$.
(ii) Each of these terms is a product of two derivatives, with the intermediate variable appearing in the denominator of the first factor and in the numerator of the second factor.
(iii) In all n terms the numerator of the first factor and the denominator of the second factor correspond to the numerator and denominator of the derivative being calculated.

Example 3 Find dw/dt if $w = xyz$, $x = t^2$, $y = \ln t$ and $z = \sinh t$.

Solution By formula (7) with $n = 3$ and the dependent variable w written instead of f, we have

$$\frac{dw}{dt} = \frac{\partial w}{\partial x}\frac{dx}{dt} + \frac{\partial w}{\partial y}\frac{dy}{dt} + \frac{\partial w}{\partial z}\frac{dz}{dt}$$

$$= (yz)(2t) + (xz)\left(\frac{1}{t}\right) + (xy)(\cosh t)$$

$$= 2t \ln t \sinh t + t \sinh t + t^2 \ln t \cosh t. \qquad \square$$

The following example illustrates how the chain rule for functions of several variables can be used to solve problems involving related rates.

Example 4 How fast is the volume of a right circular cylinder changing if its radius r is 15 cm and increasing at the rate of 2 cm/sec, while its height h is 24 cm/sec and decreasing at the rate of 3 cm/sec? How fast is its surface area (including both ends) changing at the same instant?

Solution If V is the volume and A the surface area of the cylinder, then $V = \pi r^2 h$ and $A = 2\pi rh + 2\pi r^2$. Therefore

$$\frac{dV}{dt} = \frac{\partial V}{\partial r}\frac{dr}{dt} + \frac{\partial V}{\partial h}\frac{dh}{dt} = 2\pi rh \frac{dr}{dt} + \pi r^2 \frac{dh}{dt},$$

$$\frac{dA}{dt} = \frac{\partial A}{\partial r}\frac{dr}{dt} + \frac{\partial A}{\partial h}\frac{dh}{dt} = (2\pi h + 4\pi r)\frac{dr}{dt} + 2\pi r \frac{dh}{dt}.$$

Substituting the values $r = 15$, $h = 24$, $dr/dt = 2$ and $dh/dt = -3$ into these formulas, we find that at the given instant

$$\frac{dV}{dt} = (720\pi)(2) + (225\pi)(-3) = 765\pi \approx 2403.3 \text{ cm}^3/\text{sec},$$

$$\frac{dA}{dt} = (108\pi)(2) + (30\pi)(-3) = 126\pi \approx 395.8 \text{ cm}^2/\text{sec}. \qquad \square$$

Implicit Differentiation Reconsidered

Another application of the chain rule is to calculate derivatives of implicitly defined functions. For example, let $F(x, y)$ be a function of two variables, and suppose that

$$F(x, y) = 0 \qquad (9)$$

defines y as an implicit function of x. This means that there exists a function $y = y(x)$ such that

$$F(x, y(x)) = 0$$

for all x in some interval I. Then, assuming that the functions $F(x, y)$ and $y(x)$ are both differentiable, we can use the chain rule to differentiate (9) with respect to x:

$$\frac{dF(x, y)}{dx} = F_x(x, y)\frac{dx}{dx} + F_y(x, y)\frac{dy}{dx} = F_x(x, y) + F_y(x, y)\frac{dy}{dx} = 0.$$

Solving this equation for dy/dx, we get

$$y' = \frac{dy}{dx} = -\frac{F_x(x, y)}{F_y(x, y)}, \tag{10}$$

provided that $F_y(x, y) \neq 0$. This is just a more sophisticated version of the technique of implicit differentiation introduced in Section 2.7.

Example 5 The equation

$$x^2 - xy + y^3 = 1,$$

considered in Example 3, page 147, is of the form (9) if we set

$$F(x, y) = x^2 - xy + y^3 - 1.$$

Then

$$F_x(x, y) = 2x - y, \qquad F_y(x, y) = -x + 3y^2,$$

so that (10) becomes

$$y' = \frac{2x - y}{x - 3y^2},$$

which is precisely formula (7), p. 147.

The technique of implicit differentiation can also be used to calculate *partial* derivatives. Thus, given a function $F(x, y, z)$ of three variables, suppose the equation

$$F(x, y, z) = 0 \tag{11}$$

defines z as an implicit function of x and y, in the sense that there exists a function $z = f(x, y)$ such that

$$F(x, y, z(x, y)) = 0$$

on some open set D. Then, assuming that the functions $F(x, y, z)$ and $z(x, y)$ are both differentiable, we can use the chain rule to differentiate (11) with respect to x and y (dropping arguments for simplicity):

$$\frac{\partial F}{\partial x} = F_x \frac{\partial x}{\partial x} + F_y \frac{\partial y}{\partial x} + F_z \frac{\partial z}{\partial x} = F_x + F_z \frac{\partial z}{\partial x} = 0, \tag{12}$$

$$\frac{\partial F}{\partial y} = F_x \frac{\partial x}{\partial y} + F_y \frac{\partial y}{\partial y} + F_z \frac{\partial z}{\partial y} = F_y + F_z \frac{\partial z}{\partial y} = 0. \tag{12'}$$

Here we use the fact that

$$\frac{\partial x}{\partial x} = \frac{\partial y}{\partial y} = 1, \qquad \frac{\partial x}{\partial y} = \frac{\partial y}{\partial x} = 0.$$

Solving (12) and (12′) for $\partial z/\partial x$ and $\partial z/\partial y$, we obtain

$$z_x = \frac{\partial z}{\partial x} = -\frac{F_x(x, y, z)}{F_z(x, y, z)}, \qquad z_y = \frac{\partial z}{\partial y} = -\frac{F_y(x, y, z)}{F_z(x, y, z)}, \tag{13}$$

provided that $F_z(x, y, z) \neq 0$.

Example 6 Given that

$$e^{-xy} - 3z + e^z = 0,$$

find $\partial z/\partial x$ and $\partial z/\partial y$.

776   Chapter 13   Partial Differentiation

Solution This equation is of the form (11) with $F(x, y, z) = e^{-xy} - 3z + e^z$. Thus

$$F_x(x, y, z) = -ye^{-xy}, \quad F_y(x, y, z) = -xe^{-xy}, \quad F_z(x, y, z) = -3 + e^z,$$

and it follows from (13) that

$$\frac{\partial z}{\partial x} = \frac{ye^{-xy}}{e^z - 3}, \quad \frac{\partial z}{\partial y} = \frac{xe^{-xy}}{e^z - 3},$$

provided that $z \neq \ln 3$. Note that these expressions involve the function z itself, for which we have no explicit formula. □

The Implicit Function Theorem

To carry this subject further, we need conditions guaranteeing the existence and differentiability of a function defined implicitly. Such conditions are given in the *implicit function theorem*, proved in advanced calculus. Here's what the theorem says for the case of a function defined implicitly by an equation of the form $F(x, y) = 0$: Let (x_0, y_0) be a point such that $F(x_0, y_0) = 0$, and suppose $F(x, y)$ has continuous partial derivatives $F_x(x, y)$ and $F_y(x, y)$ in a neighborhood of (x_0, y_0). Suppose further that $F_y(x_0, y_0) \neq 0$. Then there is a unique implicit function $y = y(x)$ such that $y(x_0) = y_0$ and $F(x, y(x)) = 0$ for all x in some interval $I = (x_0 - \delta, x_0 + \delta)$, and moreover $y(x)$ is continuously differentiable on I with the derivative y' given by formula (10).

The roles of x and y can be interchanged in the implicit function theorem. Specifically, if $F(x_0, y_0) = 0$ and $F(x, y)$ has continuous partial derivatives in a neighborhood of (x_0, y_0), and if $F_x(x_0, y_0) \neq 0$, then there is a unique implicit function $x = x(y)$ such that $x(y_0) = x_0$ and $F(x(y), y) = 0$ for all y in some interval $J = (y_0 - \varepsilon, y_0 + \varepsilon)$, and in this case $x(y)$ is continuously differentiable on J with the derivative

$$\frac{dx}{dy} = -\frac{F_y(x, y)}{F_x(x, y)}. \tag{10'}$$

There is also an analogous implicit function theorem for a function of two variables defined implicitly by an equation of the form $F(x, y, z) = 0$. For example, suppose $F(x_0, y_0, z_0) = 0$ and $F(x, y, z)$ has continuous partial derivatives in a neighborhood of (x_0, y_0, z_0). Then, if $F_z(x_0, y_0, z_0) \neq 0$, there is a unique implicit function $z = z(x, y)$ such that $z(x_0, y_0) = z_0$ and $F(x, y, z(x, y)) = 0$ for all (x, y) in some neighborhood N of the point (x_0, y_0), and $z(x, y)$ has continuous partial derivatives z_x and z_y on N, given by the formulas (13).

Problems

Use the chain rule to find dw/dt if
1. $w = x^2 - xy + y^2, x = t^3, y = t^4$
2. $w = x^3 - xy^2, x = e^{-t}, y = \cos t$
3. $w = y/x, x = \ln t, y = \tan t$
4. $w = e^{xy} \ln(x + y), x = t^2, y = 2 - t^2$
5. $w = \sqrt{r + s}, r = \cos t, s = \sin t$
6. $w = u/(u - v), u = \cosh t, v = \sinh t$
7. $w = \sin(x + y - z), x = \sin t, y = e^{t^2}, z = \ln t$
8. $w = \sqrt{x^2 + y^2 + z^2}, x = 1, y = t, z = t^2$
9. $w = xy + xz + yz, x = e^t, y = e^{-t}, z = \ln t$
10. $w = \ln pqrs, p = \sin t, q = \cos t, r = e^t, s = e^{-t}$

In each case check the answer by expressing w explicitly as a function of t before differentiating.

Use the chain rule to find $\partial w/\partial t$ and $\partial w/\partial u$ if
11. $w = 1 - x^2 - y^2, x = t \cos u, y = t \sin u$
12. $w = x/y, x = e^t \cos u, y = e^t \sin u$

13. $w = e^{x/y}$, $x = tu$, $y = 1/tu$
14. $w = \tan xy$, $x = t^2 + u^2$, $y = t^2 - u^2$
15. $w = \arctan(x/y)$, $x = t^2 - u^2$, $y = 2tu$
16. $w = x/(x+y)$, $x = t\cosh u$, $y = t\sinh u$
17. $w = xyz$, $x = t + u$, $y = t - u$, $z = t^2 + u^2$
18. $w = \ln\sqrt{x^2 + y^2 + z^2}$, $x = t + u$, $y = t - u$, $z = 2\sqrt{tu}$

In each case check the answer by expressing w explicitly as a function of t and u before differentiating.

Use the chain rule to find $\partial u/\partial x$, $\partial u/\partial y$ and $\partial u/\partial z$ if
19. $u = \sin rs$, $r = x^2 + yz$, $s = x^2 - yz$
20. $u = \tanh^{-1}(r/s)$, $r = x\sin yz$, $s = x\cos yz$
21. $u = r^2 + s^2 + t^2$, $r = x + y + z$, $s = x - y + z$, $t = x - y - z$
22. $u = \ln(r + s + t)$, $r = xy$, $s = xz$, $t = yz$

In each case check the answer by expressing u explicitly as a function of x, y and z before differentiating.

23. Let $u = u(r, s, t)$, where $r = x - y$, $s = y - z$, $t = z - x$ and u is a differentiable function. Show that

$$\frac{\partial u}{\partial x} + \frac{\partial u}{\partial y} + \frac{\partial u}{\partial z} = 0.$$

24. Let $u = xf(x+y) + yg(x+y)$, where f and g are arbitrary functions of a single variable with second derivatives f'' and g''. Show that

$$\frac{\partial^2 u}{\partial x^2} - 2\frac{\partial^2 u}{\partial x\,\partial y} + \frac{\partial^2 u}{\partial y^2} = 0.$$

Let $x = r\cos\theta$ and $y = r\sin\theta$, as in the transformation from polar to rectangular coordinates, and let $u = u(x, y)$ be a function of two variables with continuous second partial derivatives. Show that

25. $\dfrac{\partial r}{\partial x} = \cos\theta$, $\dfrac{\partial \theta}{\partial x} = -\dfrac{\sin\theta}{r}$

26. $\dfrac{\partial r}{\partial y} = \sin\theta$, $\dfrac{\partial \theta}{\partial y} = \dfrac{\cos\theta}{r}$

27. $\left(\dfrac{\partial u}{\partial x}\right)^2 + \left(\dfrac{\partial u}{\partial y}\right)^2 = \left(\dfrac{\partial u}{\partial r}\right)^2 + \dfrac{1}{r^2}\left(\dfrac{\partial u}{\partial \theta}\right)^2$

28. $\dfrac{\partial^2 u}{\partial x^2} + \dfrac{\partial^2 u}{\partial y^2} = \dfrac{1}{r}\dfrac{\partial}{\partial r}\left(r\dfrac{\partial u}{\partial r}\right) + \dfrac{1}{r^2}\dfrac{\partial^2 u}{\partial \theta^2}$

29. A function $f(x, y)$ of two variables with domain D is said to be *homogeneous of degree n* if

$$f(tx, ty) = t^n f(x, y) \qquad (i)$$

for all (x, y) in D and $t > 0$; here n can be any real number, and it is assumed that if (x, y) belongs to D, then so does every point (tx, ty) with $t > 0$. For example, the function $f(x, y) = x^2 \ln(x/y) - xy$, defined on the first and third quadrants, is homogeneous of degree 2. Show that if $f(x, y)$ is homogeneous of degree n and differentiable at every point of its domain, then

$$xf_x(x, y) + yf_y(x, y) = nf(x, y), \qquad (ii)$$

a result known as *Euler's theorem on homogeneous functions*.

30. Verify that the function $f(x, y) = x^2 \ln(x/y) - xy$ satisfies Euler's theorem (ii) with $n = 2$.

31. Show that if a differentiable function $f(x, y)$ is homogeneous of degree n, then its partial derivatives $f_x(x, y)$ and $f_y(x, y)$ are homogeneous of degree $n - 1$.

Is the given function homogeneous, and if so what is its degree?

32. $f(x, y) = \sqrt{x} - \sqrt{y}$
33. $f(x, y) = 1/(x^2 + y^2)$
34. $f(x, y) = 1 + \sin(x/y)$
35. $f(x, y) = x^{5/3}y^{-2/3} + x^{2/3}y^{-5/3}$

If the function is homogeneous, verify that it satisfies Euler's theorem.

36. Use the chain rule to show that the formula

$$df = \frac{\partial f}{\partial x}dx + \frac{\partial f}{\partial y}dy$$

for the total differential of a function f of two variables x and y remains valid even if x and y are themselves functions of new independent variables t and u.

37. Two ships A and B approach a port, A sailing due west at 25 knots and B due south at 20 knots. At a certain time A is 3 (nautical) miles out of port, and B is 4 miles out of port. How fast is the distance between the ships decreasing at this time?

38. How fast is the volume of a right circular cone changing if its radius r is 18 cm and increasing at 1.5 cm/sec, while its height h is 30 cm and decreasing at 2 cm/sec?

39. Find the rate at which the pressure p of one mole of an ideal gas is increasing if its volume V is 150 cm^3 and decreasing at 5 cm^3/min, while its absolute temperature T is 300° and increasing at 2°/min. Use the ideal gas law $pV = RT$ (see Example 3, page 760) with $R = 82.07$ cm^3 atm/°K, measuring pressure in atmospheres (atm) and temperature in degrees Kelvin (°K).

40. The length l of a six-sided rectangular box is 15 in. and increasing at 3 in/sec, the width w is 10 in. and decreasing at 0.5 in/sec, and the height h is 8 in. and increasing at 2 in/sec. At what rate is the volume of the box changing? At what rate is the surface area of the box changing at the same instant?

Use implicit differentiation as in formula (10) to find $y' = dy/dx$ if x and y satisfy the given equation.

41. $x^4 + y^2 - 8x + 5y = 1$
42. $x^2y - xy^2 + 2y^3 = 4$
43. $y^6 - y - x^3 = 0$
44. $xe^{2y} - ye^{3x} = 0$
45. $x^2 + 2\sin xy + y^3 = 0$
46. $\cos(x - y) = xy$

47. $xy - \tan y = 0$
48. $x^y = y^x$
49. $xy^2 = e^y$
50. $\ln(x+y) = y$

Use implicit differentiation as in the formulas (13) to find $z_x = \partial z/\partial x$ and $z_y = \partial z/\partial y$ if x, y and z satisfy the given equation.

51. $x^2 + z^2 + xz - x^3y = 2$
52. $\sin xz + \cos yz = 0$
53. $e^z - xyz = 0$
54. $\ln(xy + xz + yz) = 0$

55. Let $x^2 + y^2 + z^2 = 49$. Calculate z_x and z_y at the point $(2, -3, 6)$ by both implicit and explicit differentiation.

56. Given that $2x^2 + y^2 + 3z^2 + xy - z - 9 = 0$, find $\partial^2 z/\partial x^2$, $\partial^2 z/\partial x \partial y$ and $\partial^2 z/\partial y^2$ at the point $(-2, 1, 1)$.

57. Find dz if $\sin^2 x + \sin^2 y + \sin^2 z = 1$.

58. Assuming that the equation $F(x, y, z) = 0$ defines each of the three variables x, y and z as an implicit function of the other two, show that

$$\frac{\partial x}{\partial y}\frac{\partial y}{\partial x} = 1, \qquad \frac{\partial x}{\partial y}\frac{\partial y}{\partial z}\frac{\partial z}{\partial x} = -1$$

at every point where the partial derivatives F_x, F_y and F_z are all nonzero. What has this to do with Example 3, page 760?

59. Prove the following two-dimensional version of the mean value theorem: Let $f(x, y)$ be differentiable in a neighborhood N of the point $A = (a, b)$, and suppose the point $B = (a + \Delta x, b + \Delta y)$ belongs to N. Then there is a point (α, β) of the line segment joining A and B (distinct from the endpoints of the segment) such that

$$\Delta f = f(a + \Delta x, b + \Delta y) - f(a, b) = f_x(\alpha, \beta)\Delta x + f_y(\alpha, \beta)\Delta y.$$
(iii)

60. Find the point (α, β) satisfying (iii) if $f(x, y) = x^2 + xy$, $(a, b) = (0, 0)$ and $(\Delta x, \Delta y) = (2, 3)$.

61. Show that if the function f is continuous on $[a, b]$ and if $a \le u(x) \le v(x) \le b$, then

$$\frac{d}{dx}\int_{u(x)}^{v(x)} f(t)\,dt = f(v(x))\frac{dv(x)}{dx} - f(u(x))\frac{du(x)}{dx},$$

provided that u and v are differentiable functions.

62. Evaluate $\dfrac{d}{dx}\displaystyle\int_{x^2}^{x^3} \ln t\,dt$ both directly and with the help of the preceding problem.

63. It is shown in Problem 51, page 823, that if the functions $f(x, y)$ and $\partial f(x, y)/\partial x$ are continuous on the rectangular region $\{(x, y)\colon a \le x \le b, c \le y \le d\}$, then

$$\frac{d}{dx}\int_c^d f(x, y)\,dy = \int_c^d \frac{\partial f(x, y)}{\partial x}\,dy \qquad (a \le x \le b).$$

Assuming the validity of this formula, show that

$$\frac{d}{dx}\int_{u(x)}^{v(x)} f(x, y)\,dy$$

$$= \int_{u(x)}^{v(x)} \frac{\partial f(x, y)}{\partial x}\,dy + f(x, v(x))\frac{dv}{dx} - f(x, u(x))\frac{du}{dx},$$

where u and v have the same meaning as in Problem 61.

64. Use Problem 63 to evaluate $\dfrac{d}{dx}\displaystyle\int_x^{x^2} \frac{\sin xy}{y}\,dy$. What prevents the direct calculation of this derivative?

65. Let $x = x(t, u)$ and $y = y(t, u)$ be functions of two variables which are both differentiable at (t, u), and let $f(x, y)$ be a function of two variables which is differentiable at $(x(t, u), y(t, u))$. Then, as shown in the discussion following Example 1, the composite function $f(x(t, u), y(t, u))$ has partial derivatives at (t, u), given by formulas (6) and (6'). Go a step further and show that $f(x(t, u), y(t, u))$ is actually *differentiable* at (t, u).

13.6 The Tangent Plane to a Surface

Next we show how to define the tangent plane to a surface S. Let S be the graph of the equation

$$F(x, y, z) = 0,$$

let $P_0 = (x_0, y_0, z_0)$ be any fixed point of S, and let the curve C, with parametric equations

$$x = x(t), \quad y = y(t), \quad z = z(t) \qquad (\alpha < t < \beta)$$

be any curve on S going through P_0, as illustrated in Figure 9. We will assume that C is *smooth*, which means that the functions $x(t)$, $y(t)$ and $z(t)$ are continuously differentiable on (α, β) and satisfy the condition

$$[x'(t)]^2 + [y'(t)]^2 + [z'(t)]^2 \neq 0,$$

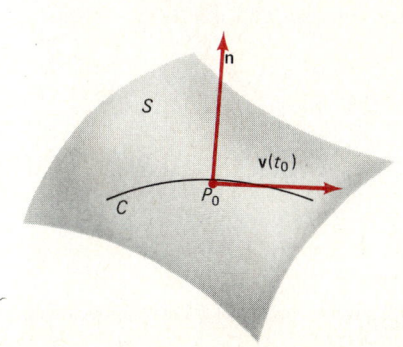

Figure 9

where as usual the prime denotes differentiation with respect to the argument t.

Let t_0 be the value of the parameter t corresponding to the point P_0, so that $P_0 = (x(t_0), y(t_0), z(t_0))$, and let $F(x, y, z)$ be differentiable at P_0, with partial derivatives that are not all zero simultaneously. Since C lies on S, we have

$$F(x(t), y(t), z(t)) = 0. \tag{1}$$

Using the chain rule to differentiate (1) with respect to t and afterwards setting $t = t_0$, we get

$$F_x(x_0, y_0, z_0)x'(t_0) + F_y(x_0, y_0, z_0)y'(t_0) + F_z(x_0, y_0, z_0)z'(t_0) = 0.$$

This formula says that the dot product of the vectors

$$x'(t_0)\mathbf{i} + y'(t_0)\mathbf{j} + z'(t_0)\mathbf{k} \tag{2}$$

and

$$\mathbf{n} = F_x(x_0, y_0, z_0)\mathbf{i} + F_y(x_0, y_0, z_0)\mathbf{j} + F_z(x_0, y_0, z_0)\mathbf{k} \tag{3}$$

is equal to zero, so that these two vectors are perpendicular. But (2) is just the velocity vector $\mathbf{v}(t_0)$ at P_0 of a variable point P moving along the curve C, and this vector is tangent to C at P_0, as we know from page 725. Thus $\mathbf{n}$ is perpendicular to $\mathbf{v}(t_0)$ for every (smooth) curve C on S through P_0.† It follows that the plane through P_0 with normal $\mathbf{n}$ contains the tangent line to *every* curve C on S that goes through P_0. This plane, which is the graph of the equation

$$F_x(x_0, y_0, z_0)(x - x_0) + F_y(x_0, y_0, z_0)(y - y_0) + F_z(x_0, y_0, z_0)(z - z_0) = 0, \tag{4}$$

is defined as the *tangent plane* to the surface S at the point P_0.

The Normal Line to a Surface

The vector $\mathbf{n}$ defined by (3) is said to be *normal* to the surface S at P_0. By the same token, the line through P_0 parallel to $\mathbf{n}$, that is, perpendicular to the tangent plane to S at P_0, is called the *normal line* to S at P_0. Thus the normal line to the surface S at P_0 is the line with symmetric equations

$$\frac{x - x_0}{F_x(x_0, y_0, z_0)} = \frac{y - y_0}{F_y(x_0, y_0, z_0)} = \frac{z - z_0}{F_z(x_0, y_0, z_0)}. \tag{5}$$

The geometric meaning of the tangent plane and normal line to a surface S is illustrated in Figure 10.

Example 1 Find the tangent plane and normal line to the ellipsoid

$$\frac{x^2}{a^2} + \frac{y^2}{b^2} + \frac{z^2}{c^2} = 1$$

at the point $P_0 = (x_0, y_0, z_0)$.

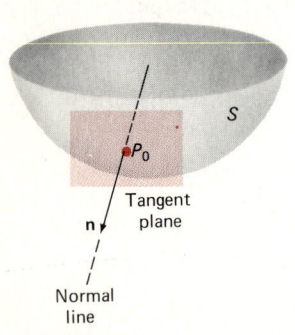

Figure 10

---

† It should be noted that both vectors $\mathbf{v}(t_0)$ and $\mathbf{n}$ are nonzero, because of the assumptions that we have made about the derivatives x', y', z' and F_x, F_y, F_z.

Solution Here

$$F(x, y, z) = \frac{x^2}{a^2} + \frac{y^2}{b^2} + \frac{z^2}{c^2} - 1$$

$$F_x(x, y, z) = \frac{2x}{a^2}, \qquad F_y(x, y, z) = \frac{2y}{b^2}, \qquad F_z(x, y, z) = \frac{2z}{c^2},$$

and hence, by (4), the tangent plane at P_0 has an equation of the form

$$\frac{2x_0}{a^2}(x - x_0) + \frac{2y_0}{b^2}(y - y_0) + \frac{2z_0}{c^2}(z - z_0) = 0$$

or

$$\frac{x_0 x}{a^2} + \frac{y_0 y}{b^2} + \frac{z_0 z}{c^2} = \frac{x_0^2}{a^2} + \frac{y_0^2}{b^2} + \frac{z_0^2}{c^2},$$

which simplifies to

$$\frac{x_0 x}{a^2} + \frac{y_0 y}{b^2} + \frac{z_0 z}{c^2} = 1,$$

since P_0 lies on the ellipsoid. This is the three-dimensional generalization of equation (11'), page 593. Moreover, by (5), the normal line to the ellipsoid at P_0 has symmetric equations

$$\frac{x - x_0}{x_0/a^2} = \frac{y - y_0}{y_0/b^2} = \frac{z - z_0}{z_0/c^2}. \quad \square$$

Suppose the surface S is the graph of a function

$$z = f(x, y),$$

which is differentiable at the point (x_0, y_0). This equation is of the form $F(x, y, z) = 0$, where

$$F(x, y, z) = f(x, y) - z,$$

so that

$$F_x(x, y, z) = f_x(x, y), \qquad F_y(x, y, z) = f_y(x, y), \qquad F_z(x, y, z) = -1.$$

Hence in the present case the equations (4) and (5) for the tangent plane and normal line at $P_0 = (x_0, y_0, z_0)$ reduce to

$$f_x(x_0, y_0)(x - x_0) + f_y(x_0, y_0)(y - y_0) - (z - z_0) = 0 \qquad (4')$$

and

$$\frac{x - x_0}{f_x(x_0, y_0)} = \frac{y - y_0}{f_y(x_0, y_0)} = \frac{z - z_0}{-1}, \qquad (5')$$

where $z_0 = f(x_0, y_0)$. Observe that equation (4') can also be written in the form

$$z = f(x_0, y_0) + f_x(x_0, y_0)(x - x_0) + f_y(x_0, y_0)(y - y_0), \qquad (6)$$

from which it is apparent that the graph of $z = f(x, y)$ cannot have a tangent plane parallel to the z-axis (why not?).

Example 2 Find the tangent plane and normal line to the surface $z = f(x, y) = xy$ at the point $(2, -2, -4)$.

Solution Since $f_x(x, y) = y$ and $f_y(x, y) = x$, we have $f_x(2, -2) = -2$ and $f_y(2, -2) = 2$. It follows from (4′) that the tangent plane at $(2, -2, -4)$ is $-2(x - 2) + 2(y + 2) - (z + 4) = 0$ or equivalently $2x - 2y + z - 4 = 0$, and from (5′) that the normal line at $(2, -2, -4)$ has symmetric equations

$$\frac{x - 2}{-2} = \frac{y + 2}{2} = \frac{z + 4}{-1}. \quad \square$$

The Tangent Plane Approximation

Having defined the tangent plane, we can now give a simple geometric interpretation of the meaning of differentiability for a function $z = f(x, y)$ of two variables. You will recall from the corollary on page 767 that if $f(x, y)$ is differentiable at (x_0, y_0), then

$$\Delta f = f(x_0 + \Delta x, y_0 + \Delta y) - f(x_0, y_0)$$
$$= f_x(x_0, y_0)\Delta x + f_y(x_0, y_0)\Delta y + \alpha(\Delta x, \Delta y)\Delta x + \beta(\Delta x, \Delta y)\Delta y,$$

where $\alpha(\Delta x, \Delta y)$ and $\beta(\Delta x, \Delta y)$ approach zero as $(\Delta x, \Delta y) \to (0, 0)$. Equivalently, writing $\Delta x = x - x_0$ and $\Delta y = y - y_0$, we have

$$f(x, y) = f(x_0, y_0) + f_x(x_0, y_0)(x - x_0) + f_y(x_0, y_0)(y - y_0)$$
$$+ \alpha^*(x, y)(x - x_0) + \beta^*(x, y)(y - y_0), \quad (7)$$

where $\alpha^*(x, y) = \alpha(x - x_0, y - y_0)$ and $\beta^*(x, y) = \beta(x - x_0, y - y_0)$ approach zero as $(x, y) \to (x_0, y_0)$. Comparing equations (6) and (7), we see that differentiability of $f(x, y)$ at (x_0, y_0) means that the graph of $f(x, y)$ has a tangent plane at (x_0, y_0) and is approximated by this tangent plane near (x_0, y_0) with an error $e(x, y) = \alpha^*(x, y)(x - x_0) + \beta^*(x, y)(y - y_0)$, where $e(x, y)$ satisfies the inequality

$$\left| \frac{e(x, y)}{\sqrt{(x - x_0)^2 + (y - y_0)^2}} \right| \leq |\alpha^*(x, y)| + |\beta^*(x, y)|$$

(why?), and hence approaches zero as $(x, y) \to (x_0, y_0)$ "faster" than the distance between the points (x, y) and (x_0, y_0). It can also be shown that the approximation of the graph of $f(x, y)$ near (x_0, y_0) by its tangent plane at (x_0, y_0) is better than by any other plane through (x_0, y_0). Notice the complete analogy between this *tangent plane approximation* and the corresponding tangent line approximation for a differentiable function of a single variable; in particular, take another look at Problems 29 and 30, page 127.

In Section 13.4 we used differentials to approximate increments of functions of n variables. This is tantamount to using the tangent plane approximation if $n = 2$ and its higher-dimensional generalization if $n > 2$.

Problems

Write equations for the tangent plane and normal line to the graph of the given equation at the specified point P.

1. $y^2 = 4x$, $P = (1, 2, 6)$
2. $z = 2x^2 + y^2$, $P = (1, -1, 3)$
3. $z = 2x^2 - 3y^2$, $P = (-2, 1, 5)$
4. $x^2 + y^2 - z^2 = -1$, $P = (2, -2, 3)$
5. $x^2 - 3xy + y^2 + xz - z^2 = 3$, $P = (1, -1, -1)$
6. $x^3 + y^3 + z^3 = 1$, $P = (-2, 1, 2)$
7. $xyz = a^3$, $P = (x_0, y_0, z_0)$
8. $z^2 = xy$, $P = (x_0, y_0, z_0)$
9. $z = \ln(x^2 + y^2)$, $P = (e, 0, 2)$
10. $Ax + By + Cz + D = 0$, $P = (x_0, y_0, z_0)$
11. $2^{x/z} + 2^{y/z} = 16$, $P = (3, 3, 1)$
12. $z = 2 \sin x \cos y$, $P = (\pi/4, \pi/4, 1)$

13. At what point of the cone $x^2 + y^2 - z^2 = 0$ are the tangent plane and normal line undefined? Explain your answer.

14. Write equations for the tangent plane and normal line to the quadric surface $Ax^2 + By^2 + Cz^2 = 1$ at the point $P_0 = (x_0, y_0, z_0)$, and show that these equations include the results of Example 1.

15. The ellipsoid $x^2 + 2y^2 + 3z^2 = 66$ has two tangent planes parallel to the plane $x + y + z = 1$. Find these planes and their points of tangency.

16. Show that the sum of the intercepts (the coordinates of the points of intersection with the coordinate axes) of any tangent plane to the surface $\sqrt{x} + \sqrt{y} + \sqrt{z} = \sqrt{a}$ $(a > 0)$ is equal to a.

17. Show that the sum of the squares of the intercepts of any tangent plane to the surface $x^{2/3} + y^{2/3} + z^{2/3} = a^{2/3}$ is equal to a^2. intercepts → ?

18. Show that every tetrahedron bounded by the coordinate planes and a tangent plane to the surface $xyz = a^3$ $(a > 0)$ has the same volume $\frac{9}{2}a^3$.

19. Show that the sphere $x^2 + y^2 + z^2 = 2$ and the hyperbolic cylinder $xy = 1$ are tangent to each other, that is, have the same tangent plane, at the point $(1, 1, 0)$. What is the common tangent plane?

20. Show that the surfaces $z = xy - y^2 + 8y - 5$ and $z = e^{2x+y+4}$ are tangent to each other at the point $(-3, 2, 1)$. What is the common tangent plane?

21. Find the plane through the point $(0, 0, 1)$ tangent to the surface $x^2 - y^2 + 3z = 0$ and parallel to the line
$$\frac{x}{2} = \frac{y}{1} = \frac{z}{-2}.$$

22. It is geometrically apparent that all the normal lines to the surface of revolution $z = f(\sqrt{x^2 + y^2})$ intersect the z-axis, which is the axis of revolution. Show this analytically.

Write equations analogous to (4') and (5') for the tangent plane and normal line at (p_0, y_0, z_0) to the graph of a function of the form

23. $x = g(y, z)$ 24. $y = h(x, z)$

25. The theory of this section is applicable to plane curves as well as to surfaces: Let the curve C be the graph of an equation $F(x, y) = 0$, and suppose $F(x, y)$ has continuous partial derivatives which are never zero simultaneously ($F_x^2 + F_y^2 \neq 0$). Show that the tangent line to C at $P_0 = (x_0, y_0)$ has an equation of the form

$$F_x(x_0, y_0)(x - x_0) + F_y(x_0, y_0)(y - y_0) = 0. \quad \text{(i)}$$

It then follows from Problem 53, page 661, that the vector $F_x(x_0, y_0)\mathbf{i} + F_y(x_0, y_0)\mathbf{j}$ is normal to this line and hence to the curve C at P_0.

26. Let the curve C be the same as in the preceding problem. Show that the normal line to C at $P_0 = (x_0, y_0)$ has an equation of the form

$$F_y(x_0, y_0)(x - x_0) - F_x(x_0, y_0)(y - y_0) = 0. \quad \text{(ii)}$$

Use formulas (i) and (ii) to find the tangent and normal lines to the graph of the given equation at the specified point P.

27. $x^3y - x^2y^2 + xy^3 = 6$, $P = (1, 2)$
28. $e^{x+y} - 2x^2 - xy = 0$, $P = (-1, 1)$
29. $\sin xy + \ln y = 0$, $P = (0, 1)$
30. $x^y = y^x$, $P = (1, 1)$

13.7 The Directional Derivative and Gradient

The partial derivatives of a function $f(P) = f(x, y, z)$ give the rates of change of f along the three coordinate axes. We now calculate the rate of change of f along an arbitrary line in space. Let f be defined in a neighborhood of a point $P_0 = (x_0, y_0, z_0)$, and let L be the directed line through P_0 with direction angles α, β and γ (see Figure 11). Then the direction of L is the same as that of the unit vector $\mathbf{u} = (\cos \alpha)\mathbf{i} + (\cos \beta)\mathbf{j} + (\cos \gamma)\mathbf{k}$, and L has parametric equations

$$x = x_0 + t \cos \alpha, \qquad y = y_0 + t \cos \beta, \qquad z = z_0 + t \cos \gamma. \quad (1)$$

By the *directional derivative* of f at P_0 in the direction of L or $\mathbf{u}$, denoted by

$$D_{\mathbf{u}}f(P_0) = D_{\mathbf{u}}f(x_0, y_0, z_0),$$

we mean the limit

$$\lim_{t \to 0} \frac{f(x_0 + t \cos \alpha, y_0 + t \cos \beta, y_0 + t \cos \gamma) - f(x_0, y_0, z_0)}{t}, \quad (2)$$

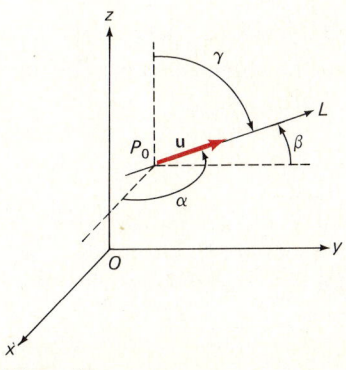

Figure 11

provided that this limit exists and is finite. Notice that if L is the x-axis, then $\alpha = 0$, $\beta = \gamma = \pi/2$, and (2) reduces to

$$\lim_{t \to 0} \frac{f(x_0 + t, y_0, z_0) - f(x_0, y_0, z_0)}{t},$$

which is just the partial derivative of f with respect to x at P_0. Similarly, if L is the y- or z-axis, (2) reduces to the partial derivative of f with respect to y or z at P_0.

The following theorem shows how to express the directional derivative of a function f in any direction in terms of the partial derivatives of f.

Theorem 4 *(Directional derivative in terms of partial derivatives).* If $f(P) = f(x, y, z)$ is differentiable at $P_0 = (x_0, y_0, z_0)$ and if the directed line L or the corresponding unit vector $\mathbf{u}$ has direction cosines $\cos \alpha$, $\cos \beta$ and $\cos \gamma$, then the directional derivative of f at P_0 in the direction of L or $\mathbf{u}$ is given by the formula

$$D_{\mathbf{u}} f(P_0) = \frac{\partial f(P_0)}{\partial x} \cos \alpha + \frac{\partial f(P_0)}{\partial y} \cos \beta + \frac{\partial f(P_0)}{\partial z} \cos \gamma. \tag{3}$$

Proof Let

$$F(t) = f(x_0 + t \cos \alpha, y_0 + t \cos \beta, y_0 + t \cos \gamma),$$

so that in particular $F(0) = f(x_0, y_0, z_0)$. Then $F(t)$ is differentiable at $t = 0$, and

$$D_{\mathbf{u}} f(P_0) = \lim_{t \to 0} \frac{F(t) - F(0)}{t} = \frac{dF(0)}{dt},$$

where the limit is a concise way of writing (2). But

$$\frac{dF}{dt} = \frac{\partial f}{\partial x} \frac{dx}{dt} + \frac{\partial f}{\partial y} \frac{dy}{dt} + \frac{\partial f}{\partial z} \frac{dz}{dt}$$

by the chain rule, and therefore

$$\frac{dF}{dt} = \frac{\partial f}{\partial x} \cos \alpha + \frac{\partial f}{\partial y} \cos \beta + \frac{\partial f}{\partial z} \cos \gamma,$$

after differentiating each of the formulas (1) with respect to t. Formula (3) is now obtained by evaluating dF/dt at $t = 0$ and the partial derivatives of f at the corresponding point $P = P_0$. ∎

The Gradient Vector

More insight into the structure of formula (3) can be obtained by introducing the vector

$$\operatorname{grad} f = \frac{\partial f}{\partial x} \mathbf{i} + \frac{\partial f}{\partial y} \mathbf{j} + \frac{\partial f}{\partial z} \mathbf{k},$$

called the *gradient* of f, with components equal to the partial derivatives of f. This vector is also denoted by ∇f, where the symbol ∇, an upside-down capital delta, is pronounced "del." The symbol ∇ (or grad) by itself stands for the operator

$$\nabla = \mathbf{i} \frac{\partial}{\partial x} + \mathbf{j} \frac{\partial}{\partial y} + \mathbf{k} \frac{\partial}{\partial z},$$

which has the effect of forming the gradient of any differentiable function of three variables written after it. Thus

$$\operatorname{grad} f = \nabla f = \left(\mathbf{i}\frac{\partial}{\partial x} + \mathbf{j}\frac{\partial}{\partial y} + \mathbf{k}\frac{\partial}{\partial z}\right)f = \frac{\partial f}{\partial x}\mathbf{i} + \frac{\partial f}{\partial y}\mathbf{j} + \frac{\partial f}{\partial z}\mathbf{k}.$$

We now recognize formula (3) as the dot product of the gradient of f and the unit vector $\mathbf{u} = (\cos \alpha)\mathbf{i} + (\cos \beta)\mathbf{j} + (\cos \gamma)\mathbf{k}$:

$$D_\mathbf{u} f(P_0) = \nabla f(P_0) \cdot \mathbf{u}. \tag{4}$$

If $\mathbf{a}$ is a nonzero vector which is not a unit vector, the directional derivative of f at P_0 in the direction of $\mathbf{a}$ is defined as the directional derivative in the direction of the unit vector $\mathbf{u}$ with the same direction as $\mathbf{a}$, namely as

$$D_\mathbf{u} f(P_0) = \nabla f(P_0) \cdot \frac{\mathbf{a}}{|\mathbf{a}|},$$

or equivalently

$$D_\mathbf{u} f(P_0) = \operatorname{comp}_\mathbf{a} \nabla f(P_0),$$

where $\operatorname{comp}_\mathbf{a} \nabla f(P_0)$ is the component of the gradient vector $\nabla f(P_0)$ along $\mathbf{a}$ (see page 705).

Example 1 Find the directional derivative of the function

$$f(P) = f(x, y, z) = x^2 + y^2 - z^2$$

at the point $P_0 = (1, 2, 3)$ in the direction of the vector $\mathbf{a} = -2\mathbf{i} + \mathbf{j} - 2\mathbf{k}$.

Solution Calculating the gradient of f, we get

$$\nabla f(P) = \frac{\partial f}{\partial x}\mathbf{i} + \frac{\partial f}{\partial y}\mathbf{j} + \frac{\partial f}{\partial z}\mathbf{k} = 2x\mathbf{i} + 2y\mathbf{j} - 2z\mathbf{k},$$

so that

$$\nabla f(P_0) = \nabla f(1, 2, 3) = 2\mathbf{i} + 4\mathbf{j} - 6\mathbf{k}.$$

Hence the directional derivative of f at P_0 in the direction of $\mathbf{a}$ is

$$\operatorname{comp}_\mathbf{a} \nabla f(P_0) = (2\mathbf{i} + 4\mathbf{j} - 6\mathbf{k}) \cdot \frac{-2\mathbf{i} + \mathbf{j} - 2\mathbf{k}}{|-2\mathbf{i} + \mathbf{j} - 2\mathbf{k}|}$$

$$= \frac{2(-2) + 4(1) - 6(-2)}{\sqrt{(-2)^2 + 1^2 + (-2)^2}} = \frac{12}{\sqrt{9}} = 4. \quad \square$$

Next we use (4) to derive a number of important properties of directional derivatives:

(i) If $\nabla f(P_0) = \mathbf{0}$, then $D_\mathbf{u} f(P_0) = \mathbf{0} \cdot \mathbf{u} = 0$ for every unit vector $\mathbf{u}$. Thus *the directional derivative of f at P_0 in every direction is zero if $\nabla f(P_0) = \mathbf{0}$.*
(ii) If $\nabla f(P_0) \neq \mathbf{0}$, then

$$D_\mathbf{u} f(P_0) = |\nabla f(P_0)||\mathbf{u}|\cos\theta = |\nabla f(P_0)|\cos\theta, \tag{5}$$

where θ ($0 \leq \theta \leq \pi$) is the angle between the vectors $\nabla f(P_0)$ and $\mathbf{u}$. But $\cos\theta$ takes its largest value, equal to 1, for $\theta = 0$. It follows that *if $\nabla f(P_0) \neq \mathbf{0}$, the directional derivative at P_0 with the largest value is the derivative in the direction of $\nabla f(P_0)$, and this largest value is $|\nabla f(P_0)|$, the magnitude of the gradient vector $\nabla f(P_0)$.*

(iii) Again let $\nabla f(P_0) \neq \mathbf{0}$. Then, since $\cos \theta$ takes its smallest value, equal to -1, for $\theta = \pi$, it follows from formula (5) that *if* $\nabla f(P_0) \neq \mathbf{0}$, *the directional derivative at P_0 with the smallest value is the derivative in the direction opposite to $\nabla f(P_0)$, and this smallest value is* $-|\nabla f(P_0)|$.

(iv) *If* $\nabla f(P_0) \neq \mathbf{0}$, *the directional derivative in any direction orthogonal to $\nabla f(P_0)$ is zero*, as we see at once by choosing $\theta = \pi/2$ in formula (5).

Thus, in particular, the direction of *maximum increase* of the function f at the point P_0 is the direction of the gradient vector $\nabla f(P_0)$, and the direction of *maximum decrease* of f is the opposite direction, that is, the direction of $-\nabla f(P_0)$.

Example 2 Find the largest and smallest values of the directional derivative of the function $f(P) = f(x, y, z) = x^2 + y^2 - z^2$ at the point $P_0 = (1, 2, 3)$.

Solution As in Example 1,
$$\nabla f(P_0) = 2\mathbf{i} + 4\mathbf{j} - 6\mathbf{k}.$$
Hence, by property (ii), the largest value of $D_{\mathbf{u}} f(P_0)$ is
$$|\nabla f(P_0)| = \sqrt{2^2 + 4^2 + (-6)^2} = \sqrt{56} = 2\sqrt{14}$$
in the direction of $\nabla f(P_0)$, and by property (iii), the smallest value of $D_{\mathbf{u}} f(P_0)$ is $-|\nabla f(P_0)| = -2\sqrt{14}$ in the direction opposite to $\nabla f(P_0)$. □

As on page 753, the graph of the equation
$$f(P) = f(x, y, z) = c, \tag{6}$$
where c is any constant in the range of f, is called a *level surface* of f. Let S be the graph of (6). Then the normal to S at the point $P_0 = (x_0, y_0, z_0)$ is the same as the normal at P_0 to the graph of $F(x, y, z) = 0$, where $F(x, y, z) = f(x, y, z) - c$. But clearly $F(x, y, z)$ and $f(x, y, z)$ have the same partial derivatives. Hence the vector $\mathbf{n}$ in formula (3), page 780, which is nonzero and normal to S at P_0, can be written in the form
$$\mathbf{n} = f_x(x_0, y_0, z_0)\mathbf{i} + f_y(x_0, y_0, z_0)\mathbf{j} + f_z(x_0, y_0, z_0)\mathbf{k},$$
or more concisely as
$$\mathbf{n} = \frac{\partial f(P_0)}{\partial x}\mathbf{i} + \frac{\partial f(P_0)}{\partial y}\mathbf{j} + \frac{\partial f(P_0)}{\partial z}\mathbf{k}.$$
But this is the vector we now denote by $\nabla f(P_0)$ and call the gradient of f at P_0. Therefore, *if $\nabla f(P_0) \neq \mathbf{0}$, then $\nabla f(P_0)$ is normal at P_0 to the level surface of f through P_0*, as illustrated schematically in Figure 12, which depicts three level surfaces of a function f and the gradient vector at a point P_0 of one of them.

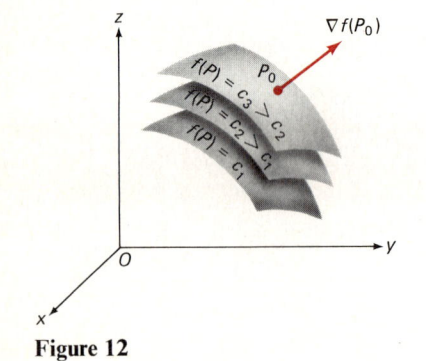

Figure 12

The Two-Dimensional Case

The preceding considerations have natural counterparts in two-space (R^2). Let $f(x, y)$ be a differentiable function of two variables. Then the gradient of f is defined by
$$\operatorname{grad} f = \nabla f = \frac{\partial f}{\partial x}\mathbf{i} + \frac{\partial f}{\partial y}\mathbf{j},$$

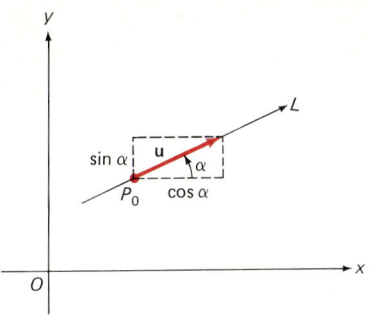

Figure 13

and the del operator simplifies to

$$\nabla = \mathbf{i}\frac{\partial}{\partial x} + \mathbf{j}\frac{\partial}{\partial y}.$$

Let L be a directed line in the xy-plane, making angle α with the positive x-axis. Then $\mathbf{u} = (\cos \alpha)\mathbf{i} + (\sin \alpha)\mathbf{j}$ is the unit vector in the direction of L (see Figure 13), and the directional derivative of f at $P_0 = (x_0, y_0)$ in the direction of L or $\mathbf{u}$ is given by

$$D_{\mathbf{u}} f(P_0) = \nabla f(P_0) \cdot \mathbf{u} = \frac{\partial f(P_0)}{\partial x} \cos \alpha + \frac{\partial f(P_0)}{\partial y} \sin \alpha. \tag{7}$$

Properties (i)–(iv) of the directional derivative remain valid for the derivative (7), and are proved in the same way as in the three-dimensional case. We now have level curves $f(x, y) = c$ instead of level surfaces, and ∇f is normal to the level curves if $\nabla f \neq \mathbf{0}$ (recall Problem 25, page 783). Moreover, ∇f points in the direction of maximum increase of f, and this is the direction of steepest ascent of the surface $z = f(x, y)$, which is the graph of f. For instance, in either Figure 3(b) or Figure 4(b), page 751, at every point (x, y) for which $0 < x^2 + y^2 < 1$ the gradient vector points toward the origin (check this), and a climber who wants to ascend the hemispherical hill in Figure 3(a) or the conical hill in Figure 4(a) as rapidly as possible should always head straight for the summit, which is the point of the hill directly above the origin of the xy-plane. By the same token, a skier wishing to descend either of these hills as rapidly as possible should always take a path heading in the direction of $-\nabla f$, that is, a path whose projection onto the xy-plane is a radius of the circle $x^2 + y^2 = 1$.

Example 3 Find the directional derivative of the function

$$f(P) = f(x, y) = xy^3 + x^2 y^2 - 2y$$

at the point $P_0 = (2, 1)$ in the direction from P_0 to the point $P_1 = (4, 0)$.

Solution The gradient is

$$\nabla f = \frac{\partial f}{\partial x}\mathbf{i} + \frac{\partial f}{\partial y}\mathbf{j} = (y^3 + 2xy^2)\mathbf{i} + (3xy^2 + 2x^2 y - 2)\mathbf{j}, \tag{8}$$

so that

$$\nabla f(P_0) = \nabla f(2, 1) = 5\mathbf{i} + 12\mathbf{j}.$$

Since $\overrightarrow{P_0 P_1} = 2\mathbf{i} - \mathbf{j}$, the direction from P_0 to P_1 is that of the unit vector

$$\mathbf{u} = \frac{2\mathbf{i} - \mathbf{j}}{\sqrt{5}},$$

and

$$D_{\mathbf{u}} f(P_0) = \nabla f(P_0) \cdot \mathbf{u} = (5\mathbf{i} + 12\mathbf{j}) \cdot \frac{2\mathbf{i} - \mathbf{j}}{\sqrt{5}} = \frac{10 - 12}{\sqrt{5}} = -\frac{2}{\sqrt{5}}. \quad \square$$

Example 4 Find the direction of steepest descent of the surface $z = f(x, y) = xy^3 + x^2y^2 - 2y$, starting from the point $(1, -2)$.

Solution It follows from formula (8) that $\nabla f(1, -2) = 6\mathbf{j}$, and this is the direction of steepest *ascent* at $(1, -2)$. Hence the direction of steepest *descent* at $(1, -2)$ is the direction of $-\nabla f(1, -2) = -6\mathbf{j}$, that is, due south. ☐

Example 5 Graph the level curve of the function $f(x, y) = \frac{1}{2}x^2 - y^2$ going through the point $P_0 = (3, 2)$, and draw the gradient vector at P_0.

Solution The level curve of f through P_0 is $f(x, y) = f(3, 2)$, which is the hyperbola $\frac{1}{2}x^2 - y^2 = \frac{1}{2}$ or $x^2 - 2y^2 = 1$ (see Figure 14). Since

$$\nabla f = \frac{\partial f}{\partial x}\mathbf{i} + \frac{\partial f}{\partial y}\mathbf{j} = x\mathbf{i} - 2y\mathbf{j},$$

the gradient at $P_0 = (3, 2)$ is $\nabla f(P_0) = 3\mathbf{i} - 4\mathbf{j}$, as shown in the figure. ☐

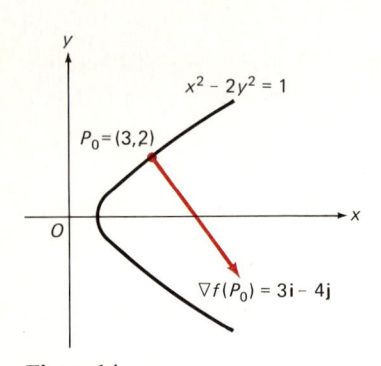

Figure 14

The gradient $\nabla f(P)$ is our first example of a vector-valued function whose argument is a point P in the plane or in space. Such a function is called a vector *field*. We can also regard vector fields as vector functions of vector arguments. In fact, if $\mathbf{r}$ is the position vector of a variable point P in the plane or in space, we can write $f(\mathbf{r})$ instead of $f(P)$ for a function of two or three variables and $\nabla f(\mathbf{r})$ instead of $\nabla f(P)$ for the gradient of the function. Thus the gradient can be thought of as a function mapping the variable vector $\mathbf{r}$ into another vector $\nabla f(\mathbf{r})$.

Example 6 Use the gradient to write an equation for the tangent plane to the surface $F(x, y, z) = 0$ at the point $P_0 = (x_0, y_0, z_0)$.

Solution Let $\mathbf{r}_0 = x_0\mathbf{i} + y_0\mathbf{j} + z_0\mathbf{k}$ be the position vector of P_0, and let $\mathbf{r} = x\mathbf{i} + y\mathbf{j} + z\mathbf{k}$ be the position vector of a variable point $P = (x, y, z)$. Then equation (4), page 780, for the tangent plane at P_0 is equivalent to the vector equation

$$\nabla F(\mathbf{r}_0) \cdot (\mathbf{r} - \mathbf{r}_0) = 0,$$

where it is assumed that $\nabla F(\mathbf{r}_0) \neq \mathbf{0}$. ☐

The study of vector fields is an important part of multivariable calculus, and will be pursued in Chapter 15.

Problems

Find the gradient ∇f of the given function f of two or three variables, and then use ∇f to calculate the directional derivative of f at the given point P in the direction of the specified vector $\mathbf{a}$.

1. $f(x, y) = 4x - 3y$, $P = (2, 1)$, $\mathbf{a} = 12\mathbf{i} - 5\mathbf{j}$
2. $f(x, y) = x^3 - 2xy + y^2 + 8$, $P = (-1, 4)$, $\mathbf{a} = \mathbf{i} + 2\mathbf{j}$
3. $f(x, y) = \sqrt{x^2 - y^2}$, $P = (25, 7)$, $\mathbf{a} = \mathbf{i} - \mathbf{j}$
4. $f(x, y) = \frac{x}{y} + \frac{y}{x}$, $P = (6, 6)$, $\mathbf{a} = -3\mathbf{i} - 4\mathbf{j}$
5. $f(x, y) = \sin(x + y)$, $P = (2, -2)$, $\mathbf{a} = 20\mathbf{i} - 21\mathbf{j}$
6. $f(x, y) = \ln(x^2 + y^2)$, $P = (1, 3)$, $\mathbf{a} = -\mathbf{i} + \mathbf{j}$
7. $f(x, y) = e^{x^2 y}$, $P = (-1, 0)$, $\mathbf{a} = 8\mathbf{i} + 15\mathbf{j}$
8. $f(x, y) = \arctan(x/y)$, $P = (3, 4)$, $\mathbf{a} = -6\mathbf{i} - 8\mathbf{j}$
9. $f(x, y, z) = xyz$, $P = (1, 2, -3)$, $\mathbf{a} = \mathbf{i} + 3\mathbf{j} + 5\mathbf{k}$
10. $f(x, y, z) = x^2 y^2 z^2$, $P = (-1, 3, 1)$, $\mathbf{a} = 2\mathbf{i} + \mathbf{j} - 2\mathbf{k}$
11. $f(x, y, z) = \sqrt{x^2 + y^2 + z^2}$, $P = (9, -6, 2)$, $\mathbf{a} = \mathbf{i} - 2\mathbf{j} + 2\mathbf{k}$
12. $f(x, y, z) = (x + y)/z$, $P = (6, -3, 3)$, $\mathbf{a} = \mathbf{i} + \mathbf{j} - \mathbf{k}$
13. $f(x, y, z) = z/xy$, $P = (-1, 1, 2)$, $\mathbf{a} = -10\mathbf{i} + 10\mathbf{j} - 5\mathbf{k}$

14. $f(x, y, z) = e^{x-y+2z}$, $P = (1, -3, -2)$, $\mathbf{a} = \mathbf{i} - 4\mathbf{j} + 7\mathbf{k}$
15. $f(x, y, z) = \cos(x - y + z)$, $P = (2, 1, -1)$, $\mathbf{a} = -\mathbf{i} + 2\mathbf{j} + 5\mathbf{k}$
16. $f(x, y, z) = \sinh xyz$, $P = (0, 2, 4)$ $\mathbf{a} = 12\mathbf{i} + 15\mathbf{j} - 16\mathbf{k}$

17. Let $f(x, y) = x^4 + 2xy + y^3$. Find the directional derivative of f at the point $(1, 2)$ in the northwest direction making an angle of $135°$ with the positive x-axis.
18. Find the directional derivative of $f(x, y) = \ln(e^x + e^y)$ at the origin in the northeast direction bisecting the first quadrant.
19. Find the directional derivative of $f(x, y) = (x + y)^2$ at the point $P_0 = (3, 2)$ in the direction from P_0 to $(6, 5)$.
20. Let $f(x, y) = x^2 y^2 - xy^3 + 2y + 1$. Find the directional derivative of f at the point $P_0 = (-2, 1)$ in the direction from P_0 to the origin.
21. Let $f(x, y, z) = \ln(x + y + z)$. Find the directional derivative of f at the point $(2, 3, -4)$ in the direction making equal acute angles with the positive coordinate axes.
22. Let $f(x, y, z) = xy^2 - xyz + z^3$. Find the directional derivative of f at the point $(1, 2, 1)$ in the direction with direction angles $60°$, $45°$ and $60°$.
23. Given that $f(x, y) = xy$, find a unit vector $\mathbf{u}$ such that $D_\mathbf{u} f(3, 4) = 0$.
24. Find the points at which the gradient of the function $f(x, y) = \ln(x^{-1} + y)$ is equal to $-\frac{16}{9}\mathbf{i} + \mathbf{j}$.

Graph the level curve of the given function going through the specified point P. Also draw the gradient vector at P.
25. $f(x, y) = -x + 2y$, $P = (2, 0)$
26. $f(x, y) = \frac{1}{4}x^2 + y^2$, $P = (1, \frac{1}{2}\sqrt{3})$
27. $f(x y) = x - \frac{1}{4}y^2$, $P = (1, -2)$
28. $f(x, y) = y^2 - x^2$, $P = (\frac{3}{4}, \frac{5}{4})$

29. What is the maximum rate of ascent of the surface $z = f(x, y) = x^y$ at the point $(e, 1)$, and in which direction does it occur?
30. What is the maximum rate of descent of the surface $z = f(x, y) = xy$ at the point $(40, -9)$, and in which direction does it occur?
31. What is the maximum rate of increase of the function $f(x, y, z) = xy^2 z^3$ at the point $(2, 1, -1)$, and in which direction does it occur?

32. Let f and g be differentiable functions of two or three variables. Show that the operator ∇ obeys the same rules

$$\nabla(f + g) = \nabla f + \nabla g, \qquad \nabla(cf) = c\nabla f \quad (c \text{ constant}),$$

$$\nabla(fg) = g\nabla f + f\nabla g, \qquad \nabla\left(\frac{f}{g}\right) = \frac{g\nabla f - f\nabla g}{g^2}$$

as the ordinary differentiation operator D.
33. Calculate $\nabla(f^p)$, where f is a differentiable function of two or three variables and p is any real number.
34. Let $\mathbf{r}(t)$ be differentiable at t, and let $f(\mathbf{r})$ be differentiable at $\mathbf{r}$, that is, at the point with position vector $\mathbf{r}$. Show that

$$\frac{d}{dt} f(\mathbf{r}(t)) = \nabla f(\mathbf{r}(t)) \cdot \frac{d\mathbf{r}(t)}{dt}$$

The elevation of a mountain above sea level is given by the formula $z = 1000 - 0.3x^2 - 0.2y^2$ meters, where the positive x-axis points east and the positive y-axis points north. A climber has arrived at the point $(5, -10, 972.5)$.
35. Will the climber ascend or descend, and at what rate, if she heads due west?
36. Will she ascend or descend, and at what rate, if she heads southeast?
37. In which direction should she walk in order to stay at the same elevation?

38. The temperature T at any point of a solid metal sphere with its center at the origin is given by the function $T = T_0 e^{-(x^2+y^2+z^2)}$. Where is the ball hottest? Show that at any point of the ball the temperature increases most rapidly in the direction pointing toward the origin.

Let $\mathbf{a}$ and $\mathbf{b}$ be constant vectors. Find the gradient of the given function $f(\mathbf{r})$ of the position vector $\mathbf{r} = x\mathbf{i} + y\mathbf{j} + z\mathbf{k}$.
39. $f(\mathbf{r}) = \mathbf{a} \cdot \mathbf{r}$
40. $f(\mathbf{r}) = (\mathbf{a} \cdot \mathbf{r})(\mathbf{b} \cdot \mathbf{r})$
41. $f(\mathbf{r}) = \mathbf{a} \cdot (\mathbf{b} \times \mathbf{r})$

42. Find the gradient of the function $f(\mathbf{r}) = |\mathbf{r} - \mathbf{r}_1|$, where $\mathbf{r} = x\mathbf{i} + y\mathbf{j}$ and $\mathbf{r}_1 = x_1\mathbf{i} + y_1\mathbf{j}$, and use the result to prove the reflection property of the ellipse, established without vectors in Section 10.3.
43. Let $f(x, y)$ be differentiable on the interior of a circle or a rectangle, and suppose $\nabla f(x, y) = \mathbf{0}$ at every point of D. Show that $f(x, y)$ is constant on D.

13.8 Extrema of Functions of Several Variables

Absolute vs. Local Extrema

We now consider *extrema* (or *extreme values*) of functions of several variables, concentrating on the case of functions of two variables, where we can draw graphs and make free use of geometric intuition. Absolute and local extrema of functions of two variables are defined in the same way as for functions of a single variable. Specifically, let $f(x, y)$ be a function of two variables defined on a set D of points in the xy-plane, and suppose there is a point (A, B) in D such that $f(A, B) \geq f(x, y)$ for all (x, y) in D. Then the number $f(A, B) = M$ is called the *maximum of f on D*. Similarly, if there is a point (a, b) in D such that $f(a, b) \leq f(x, y)$ for all (x, y) in D, the number $f(a, b) = m$ is called the *minimum of f on D*. These extrema are often called *absolute* extrema to distinguish them from the *local* extrema of f, which are defined as follows.

Suppose $f(a, b) \geq f(x, y)$ for all (x, y) sufficiently close to the point (a, b), that is, for all (x, y) in some neighborhood of (a, b). Then f is said to have a *local maximum at* (a, b), equal to the number $f(a, b)$, and the maximum is said to be *strict* if $f(a, b) > f(x, y)$, with $>$ instead of $\geq$, for all (x, y) in some *deleted* neighborhood of (a, b). Similarly, if $f(a, b) \leq f(x, y)$ for all (x, y) in some neighborhood of (a, b), then f is said to have a *local minimum at* (a, b), equal to $f(a, b)$, and the minimum is said to be *strict* if $f(a, b) < f(x, y)$, with $<$ instead of $\leq$, for all (x, y) in some deleted neighborhood of (a, b).

Example 1 Figure 15 shows a surface S which is the graph of a function $f(x, y)$ of two variables, whose domain is a closed set D in the xy-plane, bounded by a simple closed curve C which also serves as the "rim" of S. The surface S has two peaks, and each corresponds to a strict local maximum of f, one at (A, B) and the other at (A', B'). At first glance it might look as if f has a strict local minimum at (a, b), but closer inspection reveals that this is not true, and in fact f has a *saddle point* or *minimax* at (a, b) in the following sense (compare with Example 6, page 741): Let P be the point of S directly above (a, b), and let γ and γ' be the curves in which the planes $x = a$ and $y = b$ intersect S (both planes go through P). Then an ant crawling along γ will find P to be the *lowest* point of its trip, but an ant crawling along γ' instead will find P to be the *highest* point of its trip! In particular, in every neighborhood of (a, b) the function f takes values larger than $f(a, b)$ and values smaller than $f(a, b)$, so that f cannot have a local extremum at (a, b).

The absolute maximum of f on D is at (A', B'), and is equal to the height of the higher of the two peaks of the surface S. The absolute minimum of f on D is equal to 0, and is taken at every point of the curve C which is the boundary of the region D. Notice that f does *not* have local minima on C, simply because the definition of a local extremum involves comparison of the value of f at the given point (a, b) with the values of f at the points of a whole neighborhood of (a, b), and such a comparison is possible only if (a, b) is an *interior point of D*, for then there is some neighborhood of (a, b) containing only points of D. □

If f is a function with domain D, then an absolute extremum of f at an interior point of D is automatically a local extremum of f. (We have just seen that f cannot have a local extremum at a boundary point of D.) For example, suppose f has an absolute minimum on D at an interior point (a, b). Then $f(a, b) \leq f(x, y)$ for all (x, y) in D, and hence it is certainly true that $f(a, b) \leq f(x, y)$ for all (x, y) in any neighborhood of (a, b) small enough to consist entirely of points of D.

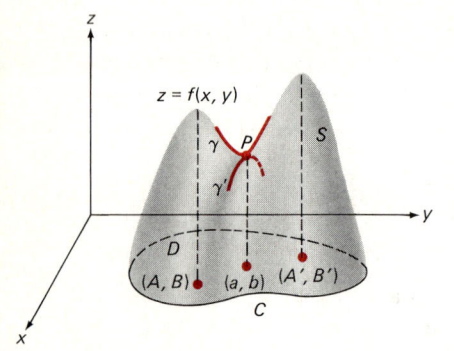

A set D of points in the xy-plane is said to be *bounded* if there is some circle $x^2 + y^2 = r^2$ large enough to enclose all the points of D. For example, the interior of every ellipse is bounded (regardless of the position of its center), but the first quadrant is not bounded. Here it is important to realize that the term "bounded" has nothing to do with the term "boundary." Thus, although unbounded, the first quadrant has a boundary, consisting of the nonnegative x- and y-axes. A function $f(x, y)$ of two variables is said to be *bounded on a set D* if there is a number $C > 0$ such that $|f(x, y)| \leq C$ for all (x, y) in D. For example, the function $\sin xy$ is bounded on the whole xy-plane, while the function $1/(x^2 + y^2)$ is unbounded on every set which has the origin as a boundary point (why?).

The Extreme Value Theorem

One of the key theorems of single variable calculus is the *extreme value theorem* (see page 99), which says that a function continuous on a bounded closed interval I is bounded on I and has both a maximum and a minimum on I. The analogous result is true for functions of two variables. As in the single variable case, we omit the proof, which is usually given in advanced calculus.

> **Theorem 5** *(Extreme value theorem for a function of two variables).* Let $f(x, y)$ be a function of two variables which is continuous on a bounded closed set D. Then f is bounded on D, and f has both a maximum M and a minimum m on D, that is, there are points (A, B) and (a, b) in D such that
> $$f(a, b) \leq f(x, y) \leq f(A, B)$$
> for all (x, y) in D.

Thus a function $f(x, y)$ continuous on a bounded closed set D always has absolute extrema. *These extrema will either be found among the local extrema of f (at certain interior points of D), or else they occur at boundary points of D.* Therefore, in searching for absolute extrema of f on D, we must first locate the local extrema of f, in order to compare their values with the values taken by f on the boundary of D. It will always be assumed that the function f under consideration is continuous, but we allow for the possibility that there are points at which f fails to be differentiable. Just as in the case of functions of one variable, there is a simple test involving differentiability that singles out the points at which f might have local extrema:

> **Theorem 6** *(Necessary condition for a local extremum of a function of two variables).* Let $f(x, y)$ be a function of two variables which has a local extremum at a point (a, b). Then either f is nondifferentiable at (a, b), or f is differentiable at (a, b) and $\nabla f(a, b) = \mathbf{0}$, that is, $f_x(a, b) = f_y(a, b) = 0$.

Proof Either f is nondifferentiable at (a, b), and then there is nothing more to prove, or f is differentiable at (a, b). In the latter case, the partial derivatives $f_x(a, b)$ and $f_y(a, b)$ both exist. But then both functions $f(x, b)$ and $f(a, y)$ of a single variable are differentiable, the first at $x = a$ and the second at $y = b$, with derivatives

$$\left.\frac{df(x, b)}{dx}\right|_{x=a} = f_x(a, b), \qquad \left.\frac{df(a, y)}{dy}\right|_{y=b} = f_y(a, b). \tag{1}$$

Since $f(x, y)$ has a local extremum at (a, b), $f(x, b)$ has a local extremum at a and $f(a, y)$ has a local extremum at b (why?). Hence the left sides of both

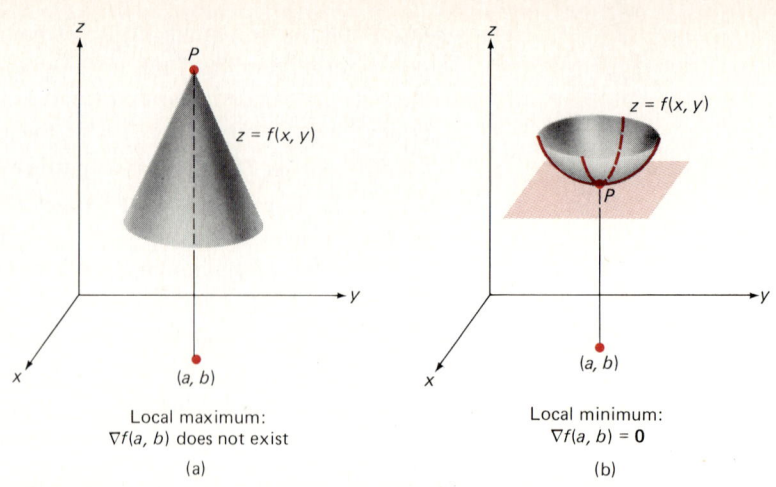

Figure 16

Local maximum:
$\nabla f(a, b)$ does not exist
(a)

Local minimum:
$\nabla f(a, b) = \mathbf{0}$
(b)

formulas (1) are zero, by Theorem 6, page 168 (the necessary condition for a local extremum of a function of a *single* variable), so that $f_x(a, b) = f_y(a, b) = 0$ or equivalently $\nabla f(a, b) = f_x(a, b)\mathbf{i} + f_y(a, b)\mathbf{j} = \mathbf{0}$. ∎

Interpreted geometrically, Theorem 6 says that if a function $f(x, y)$ has a local extremum at a point (a, b), then either the graph of f has no tangent plane at the point $P = (a, b, f(a, b))$, if $\nabla f(a, b)$ fails to exist, or the graph of f has a *horizontal* tangent plane at P, if $\nabla f(a, b) = \mathbf{0}$ (this follows from formula (6), page 781, with $x_0 = a$, $y_0 = b$). These two possibilities are illustrated in Figure 16(a) and 16(b), which show the graphs of two functions, each with a (strict) local extremum at (a, b).

Critical Points and Saddle Points

By a *critical point* of a function $f(x, y)$ of two variables we mean a point (a, b) at which f fails to be differentiable, so that the gradient $\nabla f(a, b)$ fails to exist, or at which $\nabla f(a, b)$ exists and is equal to the zero vector. According to Theorem 6, if f has a local extremum at (a, b), then (a, b) is a critical point of f. On the other hand, just as in the case of functions of a single variable, if (a, b) is a critical point of f, the function f may well fail to have a local extremum at (a, b). If $\nabla f(a, b) = \mathbf{0}$ and if every neighborhood of (a, b) contains points (x, y) such that $f(x, y) > f(a, b)$ and also other points such that $f(x, y) < f(a, b)$, we say that f has a *saddle point* (or *minimax*) at (a, b).

Example 2 Consider the function

$$f(x, y) = x^2 - y^2,$$

whose graph is the hyperbolic paraboloid shown in Figure 17. The partial derivatives

$$\frac{\partial f}{\partial x} = 2x, \qquad \frac{\partial f}{\partial y} = -2y$$

are both equal to zero at the origin $O = (0, 0)$, so that $\nabla f(0, 0) = \mathbf{0}$. Therefore O is a critical point of f. But f does not have a local extremum at O. In fact,

$$\frac{\partial^2 f}{\partial x^2} = 2, \qquad \frac{\partial^2 f}{\partial y^2} = -2,$$

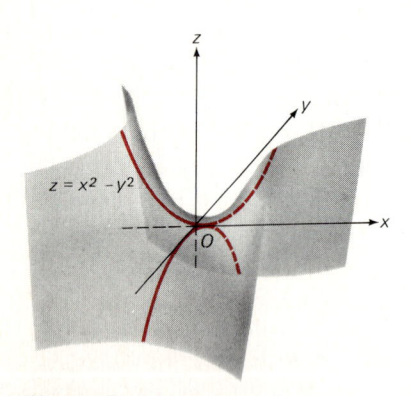

Figure 17

and hence, by the second derivative test for a function of a single variable (Theorem 9, page 171), the function $f(x, 0)$ has a local minimum at 0, while the function $f(0, y)$ has a local maximum at 0, and this clearly prevents $f(x, y)$ from having either a local maximum or a local minimum at $O = (0, 0)$. Actually, as is apparent from the figure, the origin O is a saddle point of f, since $f(0, 0) = 0$ and f takes both positive and negative values in every neighborhood of O.

Thus what we really want are conditions on a function f which *compel* f to have a local extremum at a given point (a, b). For the case of a function of two variables such conditions are given by the following two-dimensional generalization of the second derivative test, which we state without proof. For brevity we sometimes use the term "partials" as a synonym for "partial derivatives."

The Second Partials Test

Theorem 7 (*Second partials test for a local extremum of a function of two variables*). *Suppose $f(x, y)$ has continuous second partial derivatives in a neighborhood of a critical point (a, b), and let*

$$A = f_{xx}(a, b), \quad B = f_{xy}(a, b) = f_{yx}(a, b),$$
$$C = f_{yy}(a, b), \quad D = AC - B^2.$$

Then

(i) *f has a strict local maximum at (a, b) if $D > 0$ and $A < 0$;*
(ii) *f has a strict local minimum at (a, b) if $D > 0$ and $A > 0$;*
(iii) *f has a saddle point at (a, b) if $D < 0$.*

The expression D is called the *discriminant* of f at (a, b), and is equal to the determinant

$$\begin{vmatrix} f_{xx} & f_{xy} \\ f_{yx} & f_{yy} \end{vmatrix} = f_{xx}f_{yy} - f_{xy}^2$$

evaluated at (a, b). If $D = 0$, the second partials test is inconclusive. For example, each of the functions $f(x, y) = x^2 + y^4$ and $g(x, y) = x^2 + y^3$ has a single critical point at the origin O, and in both cases the discriminant is zero at O (check this). But f has a local minimum at O, since $f(0, 0) = 0$ and $f(x, y) > 0$ if $(x, y) \neq (0, 0)$, while g has no extremum at O, since $g(0, y) = y^3$ is increasing on the interval $-\infty < y < \infty$.

Example 3 Investigate the behavior of the function

$$f(x, y) = x^3 + y^3 - 3xy$$

at its critical points.

Solution Since f is differentiable everywhere, the coordinates of the critical points can be found by solving the simultaneous equations

$$\frac{\partial f}{\partial x} = 3x^2 - 3y = 0, \quad \frac{\partial f}{\partial y} = 3y^2 - 3x = 0.$$

The first equation implies $y = x^2$, which when substituted into the second equation gives $x^4 = x$, so that $x = 0$ or $x = 1$. Therefore f has exactly two critical points, namely (0, 0) and (1, 1). Since

$$A = \frac{\partial^2 f}{\partial x^2} = 6x, \qquad B = \frac{\partial^2 f}{\partial x\, \partial y} = -3, \qquad C = \frac{\partial^2 f}{\partial y^2} = 6y,$$

we have

$$A = 0, \qquad B = -3, \qquad C = 0, \qquad D = AC - B^2 = -9$$

at (0, 0), and

$$A = 6, \qquad B = -3, \qquad C = 6, \qquad D = AC - B^2 = 27$$

at (1, 1). It follows from Theorem 7 that f has a saddle point at (0, 0) and a strict local minimum at (1, 1). The minimum is equal to $f(1, 1) = -1$. □

Example 4 Find the local extrema of the function

$$f(x, y) = xy(3 - x - y). \tag{2}$$

Solution This time the coordinates of the critical points satisfy the equations

$$\frac{\partial f}{\partial x} = y(3 - 2x - y) = 0, \qquad \frac{\partial f}{\partial y} = x(3 - x - 2y) = 0.$$

The first equation tells us that $y = 0$ or $y = 3 - 2x$, and the second that $x = 0$ or $x = 3 - 2y$, from which it follows (check the details) that there are exactly four critical points, namely (0, 0), (3, 0), (0, 3) and (1, 1). Calculating the second partials, we obtain

$$A = \frac{\partial^2 f}{\partial x^2} = -2y, \qquad B = \frac{\partial^2 f}{\partial x\, \partial y} = 3 - 2x - 2y, \qquad C = \frac{\partial^2 f}{\partial y^2} = -2x,$$

and hence

$$D = AC - B^2 = 4xy - (3 - 2x - 2y)^2.$$

At each of the critical points (0, 0), (3, 0) and (0, 3) we find that $D = -9$. Thus, by the second partials test, f does not have a local extremum at any of these points, which are all saddle points. However, at (1, 1) we find that $A = -2$ and $D = 3$, so that f has a strict local maximum at (1, 1), equal to $f(1, 1) = 1$. □

Example 5 Find the *absolute* extrema of the function (2) on the closed region R between the nonnegative coordinate axes and the line $x + y = 4$ (see Figure 18).

Solution The function (2) is continuous, and the region R is bounded and closed. Therefore, by Theorem 5, f has absolute extrema on R. These extrema can occur only on the boundary of R, or at critical points of f which are interior points of R (explain). But of the four critical points of f found in the preceding example, only the point (1, 1) is an interior point of R. Thus the (absolute) maximum of f on R is the largest value taken by f at a boundary point of R or at (1, 1), while the minimum of f on R is the smallest value of f at a boundary point or at (1, 1).

We now compare these values. The boundary of R consists of three line segments, labeled OP, OQ and PQ in the figure. The function f is identically zero on OP and OQ, while on PQ, where $x + y = 4$, it can be

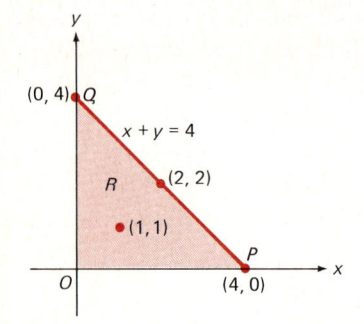

Figure 18

regarded as a function

$$g(x) = f(x, 4-x) = x(x-4) \qquad (0 \le x \le 4)$$

of the single variable x. Differentiating g with respect to x, we find that $g'(x) = 2x - 4$, so that the only critical point of g is at $x = 2$, where $g'(2) = 0$ and $g(2) = -4$. Also $g(0) = g(4) = 0$, so that the maximum of g on $0 \le x \le 4$ is $\max\{g(0), g(2), g(4)\} = \max\{0, -4, 0\} = 0$, while the minimum of g on $0 \le x \le 4$ is $\min\{g(0), g(2), g(4)\} = \min\{0, -4, 0\} = -4$. These are the maximum and minimum of $f(x, y)$ on the segment PQ, and in fact the maximum is achieved at the endpoints $P = (4, 0)$, $Q = (0, 4)$ and the minimum at the midpoint $(2, 2)$. Hence, if γ is the boundary of R, the maximum of f on γ is 0, taken on the segments OP and OQ, while the minimum of f on γ is -4, taken at the point $(2, 2)$. Also f takes the value 1 at the interior critical point $(1, 1)$. Thus, finally, the maximum of f on R is $\max\{0, 1\} = 1$, taken at the interior point $(1, 1)$, while the minimum of f on R is -4, taken at the boundary point $(2, 2)$. ☐

The following example is typical of the applied optimization problems that can be solved with the help of the techniques developed in this section.

Example 6 Find the least amount of plywood needed to construct a closed rectangular box of given volume V. Neglect the thickness of the wood, and assume that none is wasted.

Solution By the "amount" of wood we mean of course the total surface area S of the box. Let x be the length, y the width and z the height of the box (see Figure 19). Then

$$S = 2(xy + xz + yz), \qquad (3)$$

since the top and bottom of the box are of area xy and there are four sides, two of area xz and two of area yz. The problem is to minimize S subject to the condition that the volume

$$V = xyz \qquad (4)$$

of the box is constant. To incorporate this condition into the problem, we solve (4) for z, obtaining $z = V/xy$. Substitution of $z = V/xy$ into (3) then converts (3) into the function

$$S = S(x, y) = 2\left(xy + \frac{V}{y} + \frac{V}{x}\right) \qquad (3')$$

of two variables x and y, whose domain of definition is the first quadrant Q of the xy-plane (bear in mind that x and y are inherently positive).

Looking for critical points of S, we differentiate $(3')$ and set the first partials

$$\frac{\partial S}{\partial x} = 2\left(y - \frac{V}{x^2}\right), \qquad \frac{\partial S}{\partial y} = 2\left(x - \frac{V}{y^2}\right) = 0$$

equal to zero. The only solution in Q of the resulting simultaneous equations

$$y - \frac{V}{x^2} = 0, \qquad x - \frac{V}{y^2} = 0$$

is $x = V^{1/3}$, $y = V^{1/3}$, so that $(V^{1/3}, V^{1/3})$ is the only critical point of S in Q (check the calculations). At this point x, y and $z = V/xy$ are all equal to $V^{1/3}$,

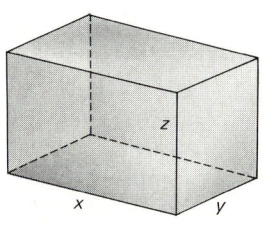

Figure 19

Section 13.8 Extrema of Functions of Several Variables

corresponding to a *cubical* box. But it is apparent from (3') that $S \to \infty$ if $x \to 0^+$ or $y \to 0^+$, and also if $x \to \infty$ or $y \to \infty$. Hence, by the two-dimensional analogue of Theorem 12, page 207, S has an absolute minimum on Q at $(V^{1/3}, V^{1/3})$, equal to $S(V^{1/3}, V^{1/3}) = 6V^{2/3}$. For instance, the least amount of wood needed to construct a box of volume 3375 cu. in. is $6(3375)^{2/3} = 1350$ sq. in., and the box should be a cube of edge length 15 in.

The function S has no absolute maximum on Q (why not?), but this does not contradict Theorem 5, since the set Q is neither bounded nor closed. Calculating the second partials of S, we find that

$$A = \frac{\partial^2 S}{\partial x^2} = \frac{4V}{x^3}, \qquad B = \frac{\partial^2 S}{\partial x \, \partial y} = 2, \qquad C = \frac{\partial^2 S}{\partial y^2} = \frac{4V}{y^3},$$

so that

$$A = 4, \qquad B = 2, \qquad C = 4, \qquad D = AC - B^2 = 12$$

at $P = (V^{1/3}, V^{1/3})$. Therefore Theorem 7 tells us that S has a local minimum at P, but we have already used another argument to show that S actually has an absolute minimum at P. □

The Case of More Than Two Variables

Finally we indicate how the results of this section can be extended to the case of functions of more than two variables. Absolute and local extrema of a function $f(x_1, \ldots, x_n)$ of n variables are defined in virtually the same way as on page 790, with $(x_1, \ldots, x_n)$ instead of (x, y) and $(a_1, \ldots, a_n)$ instead of (a, b). For example, if $f(a_1, \ldots, a_n) \geq f(x_1, \ldots, x_n)$ for all $(x_1, \ldots, x_n)$ in some neighborhood $\sqrt{(x_1 - a_1)^2 + \cdots + (x_n - a_n)^2} < \delta$ of $(a_1, \ldots, a_n)$, we say that f has a *local maximum* at $(a_1, \ldots, a_n)$, equal to $f(a_1, \ldots, a_n)$. Theorem 5 generalizes to functions of n variables with only slight modifications. Specifically, boundedness of a set D in n-space (R^n) means that there is some "n-sphere"

$$\sum_{i=1}^{n} x_i^2 = r^2$$

enclosing all the points of D, and as applied to a function $f(x_1, \ldots, x_n)$ of n variables and a bounded closed set D in R^n, the conclusion of Theorem 5 now reads "there are points $(A_1, \ldots, A_n)$ and $(a_1, \ldots, a_n)$ in D such that

$$f(a_1, \ldots, a_n) \leq f(x_1, \ldots, x_n) \leq f(A_1, \ldots, A_n)$$

for all $(x_1, \ldots, x_n)$ in D."

The generalization of Theorem 6 is also immediate, with essentially the same proof as for $n = 2$, and goes as follows: *Let $f(x_1, \ldots, x_n)$ be a function of n variables which has a local extremum at a point $(a_1, \ldots, a_n)$. Then either f is nondifferentiable at $(a_1, \ldots, a_n)$, or f is differentiable at $(a_1, \ldots, a_n)$ and all n first partial derivatives $f_{x_1}(a_1, \ldots, a_n), \ldots, f_{x_n}(a_1, \ldots, a_n)$ are equal to zero.* As in the two-dimensional case, the points of nondifferentiability of f or at which $f_{x_1} = \cdots = f_{x_n} = 0$ are called the *critical points* of f. For $n = 3$ it is customary to write $f(x, y, z)$ instead of $f(x_1, x_2, x_3)$ and (a, b, c) instead of (a_1, a_2, a_3). Thus, if $f(x, y, z)$ has a local extremum at (a, b, c), then (a, b, c) is a critical point of f, that is, either f is nondifferentiable at (a, b, c) or $\nabla f(a, b, c) = \mathbf{0}$. The generalization of Theorem 7 (the second partials test) is more complicated, and is investigated in advanced calculus.

Problems

Find all critical points of the given function. Then determine whether each critical point corresponds to a local maximum, a local minimum or a saddle point.
1. $f(x, y) = 2x^2 + (y - 1)^2$
2. $f(x, y) = (x + 1)^2 - 2y^2$
3. $f(x, y) = -x^2 + xy - y^2 + 6x - 9y + 15$
4. $f(x, y) = (x - y + 2)^2$
5. $f(x, y) = 2x^2 + 3xy + y^2 - 2x - 4y + 8$
6. $f(x, y) = 3x^2y + y^3 - 108y$
7. $f(x, y) = 4xy - x^4 - y^4$
8. $f(x, y) = xy + 20x^{-1} + 50y^{-1}$ $(x > 0, y > 0)$
9. $f(x, y) = x\sqrt{y} - x^2 + 9x - y$ $(y > 0)$
10. $f(x, y) = 8x^3 + y^3 - 12xy + 6$
11. $f(x, y) = (2x - x^2)(2y - y^2)$
12. $f(x, y) = x^2y^3(6 - x - y)$ $(x > 0, y > 0)$
13. $g(u, v) = u \sin v$
14. $h(u, v) = e^u \cos v$
15. $f(x, y) = \sin x + \sin y$ $(0 < x < \pi, 0 < y < \pi)$
16. $f(x, y) = \sin x + \sin y + \sin(x + y)$
    $(0 < x < \pi, 0 < y < \pi)$
17. $f(x, y) = (x^2 + 2x + y)e^y$
18. $f(x, y) = e^{xy}$
19. $f(x, y) = (x^2 + y^2)e^{-(x^2+y^2)}$
20. $f(x, y) = (x - 1)\ln xy$ $(xy > 0)$

Find the absolute extrema of the given function f on the specified region R.
21. $f(x, y) = x^2 - y^2$, $R = \{(x, y): x^2 + y^2 \le 4\}$
22. $f(x, y) = xy$, $R = \{(x, y): x^2 + y^2 \le 2\}$
23. $f(x, y) = 2xy + y^2 + 8x - 4y$,
    $R = \{(x, y): 0 \le x \le 2, 0 \le y \le 1\}$
24. $f(x, y) = (3x^2 + 2y^2)e^{-(x^2+y^2)}$,
    $R = \{(x, y): 1 \le x^2 + y^2 \le 4\}$

25. Find the largest value of the product of three positive numbers whose sum is 60.
26. Find the smallest value of the sum of three positive numbers whose product is 343.
27. Find the largest volume of a closed rectangular box of surface area 600 cm³.
28. Solve Example 6 for the case where the box has no top.
29. Find the parallelogram of largest area with a given perimeter p.
30. Let T be the triangle with vertices $(0, 0)$, $(1, 0)$ and $(0, 1)$. Find the point P inside T such that the sum of the squares of the distances from P to the vertices of T is the smallest.
31. In the ellipsoid $(x^2/a^2) + (y^2/b^2) + (z^2/c^2) = 1$ there is inscribed a rectangular box with its edges parallel to the coordinate axes. What is the largest volume of such a box?
32. Given a point $P = (a, b, c)$ in the first octant, find the plane through P which cuts off the tetrahedron of minimum volume from the first octant. What is the minimum volume?

33. Find the absolute extrema of the function
$$f(x, y, z) = (ax + by + cz)e^{-(x^2+y^2+z^2)}.$$

34. Given a function $f(x, y)$ of two variables with continuous partial derivatives, suppose that $f(x_0, y_0) = f_x(x_0, y_0) = f_y(x_0, y_0) = 0$ and
$$f_{xx}(x_0, y_0)f_{yy}(x_0, y_0) - f_{xy}^2(x_0, y_0) > 0.$$
Show that (x_0, y_0) is an *isolated point* of the graph of the equation $f(x, y) = 0$, which means that there is some neighborhood of (x_0, y_0) containing no point of the graph other than (x_0, y_0). Show that the origin is an isolated point of the graph of the equation $x^3 - y^3 + xy^2 - yx^2 + x^2 + y^2 = 0$, and describe the graph.
Hint. Use the second partials test.

35. The quantity which is conventionally used to measure the closeness between a line $y = f(x) = mx + b$ in the plane and n points $(x_1, y_1), (x_2, y_2), \ldots, (x_n, y_n)$ is the sum
$$S = \sum_{i=1}^{n} [f(x_i) - y_i]^2 = \sum_{i=1}^{n} (mx_i + b - y_i)^2,$$
which will be recognized as the sum of the *squares* of the deviations between the given values of y_i and the ordinates of the line $y = mx + b$ corresponding to the given values of x_i. This measure of closeness is better than the sum $\sum [f(x_i) - y_i]$ of the deviations themselves, in which positive and negative terms can cancel each other out, leading to the erroneous conclusion that the points lie near the line even when they actually do not. The line $y = mx + b$ minimizing S is called the *line of best fit* to the n points, or the *regression line*, to use the language of statistics. Show that this line has slope
$$m = \frac{\sum_{i=1}^{n} x_i y_i - n\bar{x}\bar{y}}{\sum_{i=1}^{n} x_i^2 - n\bar{x}^2} \quad (i)$$
and y-intercept
$$b = \bar{y} - m\bar{x}, \quad (ii)$$
where
$$\bar{x} = \frac{1}{n}\sum_{i=1}^{n} x_i, \quad \bar{y} = \frac{1}{n}\sum_{i=1}^{n} y_i$$
(recall Example 6, page 750).
Hint. Note that S is a function of m and b.

With the help of formulas (i) and (ii), find the regression line for the given set of points.
36. $(1, 0), (3, 2), (5, 4)$
37. $(-1, 0), (0, 1), (1, 1), (2, -1)$
38. $(0, 0), (1, 2), (2, 2), (3, 3), (4, 3)$

It is shown in advanced calculus that a function $f(x, y, z)$ of three variables with continuous second partial derivatives has a local minimum at a critical point (a, b, c) if all three quantities

$$A = f_{xx}, \quad D = \begin{vmatrix} f_{xx} & f_{xy} \\ f_{yx} & f_{yy} \end{vmatrix}, \quad E = \begin{vmatrix} f_{xx} & f_{xy} & f_{xz} \\ f_{yx} & f_{yy} & f_{yz} \\ f_{zx} & f_{zy} & f_{zz} \end{vmatrix}$$

are positive at (a, b, c), and a strict local maximum at (a, b, c) if $A < 0$, $D > 0$, $E < 0$ at (a, b, c). Use this test to find the local extrema of the given function.

39. $f(x, y, z) = xy + 3x - x^2 - y^2 - z^2$
40. $f(x, y, z) = x^2 + y^2 + 2z^2 + xy - 9x - 3y + 4z + 10$

Does the given function have a local extremum at the origin?

41. $f(x, y, z) = x^2 + y^2 + z^2 - xyz$
42. $f(x, y, z) = x^3 + y^3 + z^3 + xyz$

13.9 Lagrange Multipliers

You may have already noticed that in order to solve many of the optimization problems considered in the last section (and in Section 3.7), we had to find the maximum or minimum of a function of two or three variables, subject to certain *side conditions* or *constraints* imposed on the variables. Thus Example 1, page 204, asks for the largest area of a rectangular field enclosed by a fence 800 ft long, and Example 6, page 795, asks for the smallest amount of wood needed to construct a rectangular box of given volume V. Mathematically, the first of these problems consists in maximizing the function

$$f(x, y) = xy \tag{1}$$

subject to the constraint

$$2(x + y) = 800, \tag{1'}$$

while the second problem consists in minimizing the function

$$f(x, y, z) = 2(xy + xz + yz) \tag{2}$$

subject to the constraint

$$xyz = V. \tag{2'}$$

Our method of dealing with these problems was the same in both cases. First we solved the equation of constraint for one of the variables in terms of the other variable or variables, obtaining $y = 400 - x$ in the case of equation (1') and $z = V/xy$ in the case of equation (2'). Next we substituted these expressions back into (1) and (2), reducing (1) to a function

$$f(x, 400 - x) = 400x - x^2 \tag{3}$$

of a single variable x, and (2) to a function

$$f\left(x, y, \frac{V}{xy}\right) = 2\left(xy + \frac{V}{y} + \frac{V}{x}\right) \tag{4}$$

of just two variables x and y. We then used the theory of extrema to maximize (3) on the interval $0 \leq x \leq 400$ and to minimize (4) on the first quadrant. The details are given in the cited examples.

Constrained Extrema

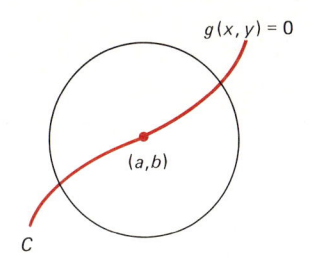

f(x, y) has a constrained local minimum at (a, b) if f(a, b) ≤ f(x, y) for all points (x, y) on the curve C inside some circle centered at (a, b).

Figure 20

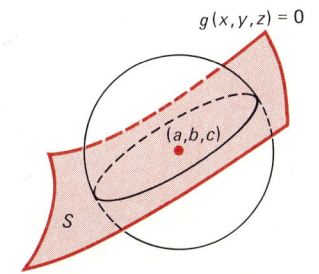

f(x, y, z) has a constrained local maximum at (a, b, c) if f(a, b, c) ≥ f(x, y, z) for all points (x, y, z) on the surface S inside some sphere centered at (a, b, c).

Figure 21

Lagrange Multiplier Rule: The Case of Two Variables

We now present an alternative way of finding *constrained extrema* (that is, extrema subject to side conditions or constraints), discovered by the great French mathematician Joseph Louis Lagrange (1736–1813), and known as the method of *Lagrange multipliers*. This method has two advantages. In the first place, it avoids the algebraic complications of solving the equation of constraint *explicitly* for one of the variables in terms of the others. Secondly, it treats all the variables of a problem "impartially," without singling out certain variables for special attention (like the variable x in the first of the above examples or the variables x and y in the second example).

Although the method of Lagrange multipliers can be used for functions of more than three variables, and for functions subject to more than one constraint, we will restrict ourselves to the two simplest cases, namely finding the extrema of a function $f(x, y)$ of two variables subject to a single constraint of the form $g(x, y) = 0$ and finding the extrema of a function $f(x, y, z)$ of three variables, subject to a single constraint of the form $g(x, y, z) = 0$.† In the first case we will assume that the graph of $g(x, y) = 0$ is some curve C in the plane, and in the second case that the graph of $g(x, y, z) = 0$ is some surface S in space. Then by a *constrained local extremum* of $f(x, y)$ or of $f(x, y, z)$ at a point (a, b) of C or at a point (a, b, c) of S we mean a local extremum in which the value of f at (a, b) or at (a, b, c) is compared not with *all* values of f at nearby points, but only with values of f at nearby points *that also lie on C or S*; here it is assumed that (a, b) is not an endpoint of the curve C, and that (a, b, c) does not belong to an "edge" of the surface S. What this means geometrically is illustrated in Figure 20 for the two-variable case and in Figure 21 for the three-variable case.

The following pair of closely related theorems governs the use of Lagrange multipliers in the two cases just discussed. In each theorem the Lagrange multiplier is a number traditionally denoted by λ (lowercase Greek lambda).

Theorem 8 *(Lagrange multiplier rule for functions of two variables).* Let $f(x, y)$ and $g(x, y)$ have continuous first partial derivatives on an open set containing the curve C which is the graph of the equation $g(x, y) = 0$. Let $\nabla g(x, y) \neq \mathbf{0}$ on C, and suppose $f(x, y)$ has a constrained local extremum at a point (a, b) of C. Then there is a number λ such that

$$\nabla f(a, b) = \lambda \nabla g(a, b), \qquad (5)$$

that is, the gradients of f and g are parallel at (a, b).

Proof (Optional) Since $\nabla g(x, y) \neq \mathbf{0}$ on C, either $g_x(x, y)$ or $g_y(x, y)$ is nonzero at every point of C, so that either $g_x(a, b) \neq 0$ or $g_y(a, b) \neq 0$. To be specific, suppose $g_y(a, b) \neq 0$. Then, by the implicit function theorem (page 777), the part of the curve C at and near (a, b) is the graph of a continuously differentiable function $y = y(x)$. Substituting $y = y(x)$ into $f(x, y)$ and assuming that $f(x, y)$ has a constrained local extremum at (a, b), we find that the differentiable function $F(x) = f(x, y(x))$ of a single variable must have a

---

† However, see Problems 31 and 32.

local extremum (in the ordinary sense) at $x = a$. Therefore $F'(a) = 0$, where the prime denotes differentiation with respect to x, so that

$$F'(a) = f_x(a, b) + f_y(a, b)y'(a) = 0,$$

after using the chain rule. Thus $f_x(a, b) = 0$ if $f_y(a, b) = 0$, while

$$y'(a) = -\frac{f_x(a, b)}{f_y(a, b)} \qquad (6)$$

if $f_y(a, b) \neq 0$. In the first case $\nabla f = f_x\mathbf{i} + f_y\mathbf{j} = \mathbf{0}$, and (5) holds with $\lambda = 0$. In the second case, differentiating the identity $g(x, y(x)) \equiv 0$, valid in a neighborhood of $x = a$, we get

$$g_x(a, b) + g_y(a, b)y'(a) = 0,$$

and hence

$$y'(a) = -\frac{g_x(a, b)}{g_y(a, b)}. \qquad (6')$$

Comparison of (6) and (6') then gives

$$f_x(a, b) = \frac{f_y(a, b)}{g_y(a, b)} g_x(a, b),$$

so that (5) again holds if we choose $\lambda = f_y(a, b)/g_y(a, b)$, a choice equivalent to setting $f_y(a, b)$ equal to $\lambda g_y(a, b)$. To complete the proof, we observe that an analogous argument works if it is assumed that $g_x(a, b) \neq 0$ instead of $g_y(a, b) \neq 0$ (check the details, reversing the roles of x and y). ∎

Lagrange Multiplier Rule: The Case of Three Variables

> **Theorem 9** *(Lagrange multiplier rule for functions of three variables)*. Let $f(x, y, z)$ and $g(x, y, z)$ have continuous first partial derivatives on an open set containing the surface S which is the graph of the equation $g(x, y, z) = 0$. Let $\nabla g(x, y, z) \neq \mathbf{0}$ on C, and suppose $f(x, y, z)$ has a constrained local extremum at a point (a, b, c) of S. Then there is a number λ such that
>
> $$\nabla f(a, b, c) = \lambda \nabla g(a, b, c), \qquad (7)$$
>
> that is, the gradients of f and g are parallel at (a, b, c).

Proof (Optional) We could give a proof like that of Theorem 8, but it is more instructive to proceed somewhat differently. As on page 779, let C be any smooth curve on S going through (a, b, c), with parametric equations

$$x = x(t), \quad y = y(t), \quad z = z(t) \qquad (\alpha < t < \beta),$$

and let t_0 be the value of the parameter t corresponding to the point (a, b, c). Suppose $f(x, y, z)$ has a constrained local extremum at (a, b, c). Then the differentiable function $F(t) = f(x(t), y(t), z(t))$ of a single variable must have a local extremum at $t = t_0$. Therefore $F'(t_0) = 0$, where the prime denotes differentiation with respect to t, so that

$$F'(t_0) = f_x(a, b, c)x'(t_0) + f_y(a, b, c)y'(t_0) + f_z(a, b, c)z'(t_0) = 0, \qquad (8)$$

after using the chain rule. Also

$$g_x(a, b, c)x'(t_0) + g_y(a, b, c)y'(t_0) + g_z(a, b, c)z'(t_0) = 0, \qquad (9)$$

800 Chapter 13 Partial Differentiation

as we find by differentiating the identity $g(x(t), y(t), z(t)) \equiv 0$. Let $\mathbf{r}(t) = x(t)\mathbf{i} + y(t)\mathbf{j} + z(t)\mathbf{k}$. Then $\mathbf{r}'(t_0) = x'(t_0)\mathbf{i} + y'(t_0)\mathbf{j} + z'(t_0)\mathbf{k}$, where $\mathbf{r}'(t_0) \neq \mathbf{0}$, since the curve C is smooth (see page 779). We can now write (8) and (9) as

$$\nabla f(a, b, c) \cdot \mathbf{r}'(t_0) = 0 \tag{8'}$$

and

$$\nabla g(a, b, c) \cdot \mathbf{r}'(t_0) = 0, \tag{9'}$$

in terms of dot products of the vector $\mathbf{r}'(t_0)$ with the gradients $\nabla f(a, b, c)$ and $\nabla g(a, b, c)$. Since $\mathbf{r}'(t_0)$ is tangent to the curve C at (a, b, c), it follows that $\nabla f(a, b, c)$ and $\nabla g(a, b, c)$ are both perpendicular to C at (a, b, c). But C is an *arbitrary* smooth curve on S going through (a, b, c), and therefore $\nabla f(a, b, c)$ and $\nabla g(a, b, c)$ must both lie along the normal line to S at (a, b, c). Hence the vectors $\nabla f(a, b, c)$ and $\nabla g(a, b, c)$ are parallel at (a, b, c), and formula (7) is proved. ■

The Lagrange multiplier rule (7) has a simple geometric interpretation. The gradient vector $\nabla f(a, b, c)$ is normal to the level surface $f(x, y, z) = f(a, b, c)$ at the point (a, b, c), and $\nabla g(a, b, c)$ is normal to the surface of constraint $g(x, y, z) = 0$ at (a, b, c); in fact, the surface of constraint is the level surface $g(x, y, z) = c$ corresponding to $c = 0$. Therefore formula (7) says that if $f(x, y, z)$ has a constrained local extremum at (a, b, c), then the level surface of f through (a, b, c) and the surface of constraint have a common normal and hence a common tangent plane at (a, b, c). Similarly, the two-dimensional Lagrange multiplier rule (5) says that if $f(x, y)$ has a constrained local extremum at the point (a, b), then the level curve of f through (a, b) and the curve of constraint $g(x, y) = 0$ have a common normal and hence a common tangent line at (a, b).

The points (a, b) or (a, b, c) on the curve C or surface S at which $\nabla f = \lambda \nabla g$ are only *candidates* for points where f has constrained local extrema, and f may well *fail* to have a constrained local extremum at such a point; in other words, the condition $\nabla f = \lambda \nabla g$ is necessary but not sufficient for a constrained local extremum. This fact is illustrated in Figure 22, which shows level curves of a function $f(x, y)$, together with a curve C which is the graph of a constraint $g(x, y) = 0$. Notice that there are three points M, N and P at which the curve C and the corresponding level curve of f have a common normal, and these are the only candidates for points at which f has constrained local extrema. Inspecting the figure, we find that although the function f has a constrained local maximum at M and a constrained local minimum at N, it does *not* have a constrained local extremum at P.

Thus additional analysis is required to prove that f actually has a constrained local extremum at a point of a curve C for which $\nabla f = \lambda \nabla g$, as well as to determine whether the extremum is a maximum or a minimum.† In this regard, the following facts should be kept in mind.

(i) If C is a bounded closed set and if f is continuous on C, then f has absolute extrema on C, by Theorem 5, page 791.

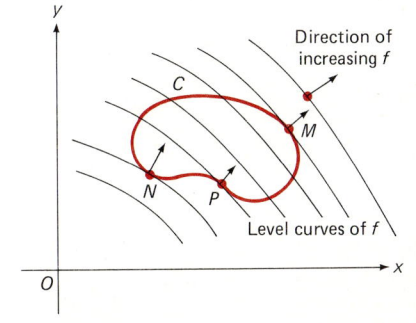

Figure 22

---

† Here for simplicity we talk about a curve C and f is assumed to be a function of two variables, but analogous comments apply to the case of a surface S and a function f of three variables.

Section 13.9 Lagrange Multipliers **801**

(ii) If f has absolute extrema on C and if f is differentiable on an open set containing C, these extrema are either constrained local extrema of f, in which case they can be discovered by the Lagrange multiplier rule $\nabla f = \lambda \nabla g$ (provided that $\nabla g \neq \mathbf{0}$), or else they occur at endpoints of C.

(iii) In many problems there are geometrical or physical considerations which allow us to conclude that f actually has an extremum at a point where $\nabla f = \lambda \nabla g$.

Remark If the constraint is of the form $g(x, y) = c$ or $g(x, y, z) = c$, where $c \neq 0$, we can still write the necessary condition for a constrained local extremum as $\nabla f = \lambda \nabla g$, since $\nabla(g - c) = \nabla g - \nabla c = \nabla g$. There are cases in which the Lagrange multiplier λ turns out to be zero. Then $\nabla f = \mathbf{0}$, but the statement that ∇f and ∇g are parallel at (a, b) or (a, b, c) remains true, since the zero vector is regarded as parallel to every vector. There is also the possibility that a constrained local extremum occurs at a point where $\nabla g = \mathbf{0}$, since this case is not covered by the Lagrange multiplier rule.

To illustrate the method of Lagrange multipliers, we first give alternative solutions of the two problems discussed at the beginning of the section. We then use Lagrange multipliers to solve some other typical problems involving constrained extrema.

Example 1 Find the rectangular field of largest area that can be enclosed by a fence 800 ft long.

Solution Let x be the width and y the length of the field. Then we want the absolute maximum of the area function $f(x, y) = xy$, subject to the constraint $g(x, y) = 2(x + y) = 800$. Here the variables x and y are inherently nonnegative, and in fact positive, since it would be absurd to choose $x = 0$ or $y = 0$ (this corresponds to folding the fence back on itself, and gives a field of zero area). Thus the domain of both functions f and g is the first quadrant Q of the xy-plane. Let C be the graph of the equation of constraint $g(x, y) = 800$ ($x > 0$, $y > 0$). As shown in Figure 23, C is the line segment which connects but does not include the points $A = (400, 0)$ and $B = (0, 400)$. The geometric meaning of the problem makes it clear that f has an absolute maximum on C, since the field is narrow with a tiny area if x or y is small and also if x or y is almost equal to 400. The gradients of f and g are

$$\nabla f = y\mathbf{i} + x\mathbf{j}, \qquad \nabla g = 2\mathbf{i} + 2\mathbf{j} \neq \mathbf{0},$$

and hence the Lagrange multiplier rule $\nabla f = \lambda \nabla g$ becomes $y\mathbf{i} + x\mathbf{j} = \lambda(2\mathbf{i} + 2\mathbf{j})$, or equivalently

$$y = 2\lambda, \qquad x = 2\lambda, \tag{10}$$

which implies $y = x$. Substituting $y = x$ into the constraint $2(x + y) = 800$, we find that $4x = 800$ or $x = 200$. Since there are no other points of C at which $\nabla f = \lambda \nabla g$ and since C does not have endpoints, we can be sure that $(200, 200)$ is the point at which f takes its absolute maximum on C. Thus the rectangular field of largest area enclosed by a fence 800 ft long is a *square* field 200 ft on a side and of area $(200)^2 = 40{,}000$ sq. ft, as shown by another method in Example 1, page 204. We arrived at this conclusion without

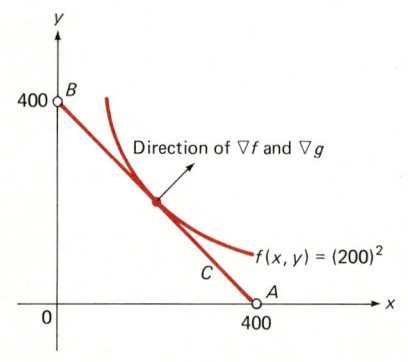

Figure 23

bothering to solve for the multiplier λ itself, which is of no direct interest. However, it follows at once from (10) that $\lambda = x/2 = y/2 = 100$. □

A somewhat different approach to the problem posed in Example 1 is to let the curve C include A and B as its endpoints, although this makes no practical sense from the standpoint of the fence builder. Then C is a bounded closed set, so that the continuous function f has both a maximum and a minimum on C. The maximum of f on C occurs at the point (200, 200) where the Lagrange multiplier rule $\nabla f = \lambda \nabla g$ holds, and the minimum of f on C is of course at the endpoints A and B, where f takes the value 0. Notice that the minimum of f on C cannot be discovered by the Lagrange multiplier rule, since as you can easily verify, there is no value of λ such that $\nabla f = \lambda \nabla g$ at A or B (here let the domain of f and g be the whole plane). Thus separate analysis of the behavior of f at the endpoints of a constraint curve C is called for in those cases in which C has endpoints.

Example 2 Find the least amount of plywood needed to construct a closed rectangular box of given volume V.

Solution Let x be the length, y the width and z the height of the box. Then we want the absolute minimum of the total surface area function $f(x, y, z) = 2(xy + xz + yz)$, subject to the constraint $g(x, y, z) = xyz = V$. Since the variables x, y and z are inherently positive (why?), the domain of both functions f and g is the first octant. Calculating the gradients of f and g, we get

$$\nabla f = 2(y + z)\mathbf{i} + 2(x + z)\mathbf{j} + 2(x + y)\mathbf{k}$$

and

$$\nabla g = yz\mathbf{i} + xz\mathbf{j} + xy\mathbf{k},$$

where it should be noted that $\nabla g \neq \mathbf{0}$ on the domain of g. Hence the Lagrange multiplier rule $\nabla f = \lambda \nabla g$ implies

$$2(y + z) = \lambda yz, \quad 2(x + z) = \lambda xz, \quad 2(x + y) = \lambda xy,$$

or equivalently

$$\frac{1}{z} + \frac{1}{y} = \frac{\lambda}{2}, \quad \frac{1}{z} + \frac{1}{x} = \frac{\lambda}{2}, \quad \frac{1}{y} + \frac{1}{x} = \frac{\lambda}{2} \tag{11}$$

(remember that $x > 0$, $y > 0$, $z > 0$). From (11) we immediately deduce that $x = y = z$, corresponding to a *cubical* box. This box must lead to a constrained absolute minimum of f, because there are noncubical boxes of volume V that use up an arbitrarily large amount of wood (show that $f(x, y, z) \to \infty$ as any of the variables x, y and z approaches 0 subject to the constraint $xyz = V$). For $x = y = z$ the constraint becomes $x^3 = V$, so that $x = V^{1/3}$. Thus the least amount of wood needed to construct a box of volume V is $f(V^{1/3}, V^{1/3}, V^{1/3}) = 6V^{2/3}$, as found by another method in Example 6, page 795, and the box should be a cube of edge length $V^{1/3}$. What is the value of λ for which $\nabla f = \lambda \nabla g$? □

Example 3 Investigate the extrema of the function $f(x, y) = x^2 y$ on the ellipse $2x^2 + y^2 = 3$.

Solution Here the equation of constraint is $g(x, y) = 2x^2 + y^2 = 3$, and the gradients of f and g are

$$\nabla f = 2xy\mathbf{i} + x^2\mathbf{j}, \quad \nabla g = 4x\mathbf{i} + 2y\mathbf{j},$$

where $\nabla g \neq \mathbf{0}$ on the ellipse. Thus the Lagrange multiplier rule $\nabla f = \lambda \nabla g$ gives
$$2xy = 4\lambda x, \qquad x^2 = 2\lambda y. \qquad (12)$$

The first equation tells us that either $x = 0$ or $\lambda = \frac{1}{2}y$. If $x = 0$, the constraint reduces to $y^2 = 3$ or $y = \pm\sqrt{3}$. If $\lambda = \frac{1}{2}y$, the second of the equations (12) becomes $x^2 = y^2$ or $y = \pm x$, and the constraint becomes $3x^2 = 3$ or $x = \pm 1$. Hence there are exactly *six* points of the ellipse $2x^2 + y^2 = 3$ at which $\nabla f = \lambda \nabla g$, namely $(0, \pm\sqrt{3})$, $(\pm 1, 1)$ and $(\pm 1, -1)$, as indicated in Figure 24. The ellipse, which we denote by C, is a bounded closed set, and f is continuous on C. Therefore f has absolute extrema on C. Since $f(0, \pm\sqrt{3}) = 0$, $f(\pm 1, 1) = 1$ and $f(\pm 1, -1) = -1$, the absolute maximum of f on C is 1, taken at $(\pm 1, 1)$, and the absolute minimum of f on C is -1, taken at $(\pm 1, -1)$. As for the points $(0, \pm\sqrt{3})$, f has a constrained local minimum at $(0, \sqrt{3})$ and a constrained local maximum at $(0, -\sqrt{3})$. This follows from the fact that f is zero if $x = 0$, but positive if $x \neq 0$, $y > 0$ and negative if $x \neq 0$, $y < 0$. It is easy to see that $\lambda = 0$ is the value of λ for which $\nabla f = \lambda \nabla g$ at the points $(0, \pm\sqrt{3})$.

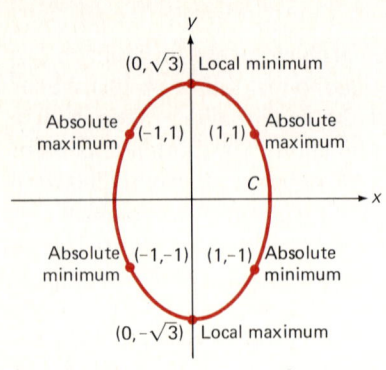

Further insight into this problem can be obtained by introducing parametric equations
$$x = \sqrt{\frac{3}{2}} \cos t, \qquad y = \sqrt{3} \sin t \qquad (0 \leq t \leq 2\pi)$$

for the ellipse C, where C is described once in the counterclockwise direction as t increases from 0 to 2π. Then the value of $f(x, y) = x^2 y$ at the point of C with parameter value t is given by the function
$$F(t) = \frac{3\sqrt{3}}{2} \cos^2 t \sin t.$$

The graph of $F(t)$ is shown in Figure 25, and is in complete accord with the behavior of $f(x, y)$ discovered by the method of Lagrange multipliers. ☐

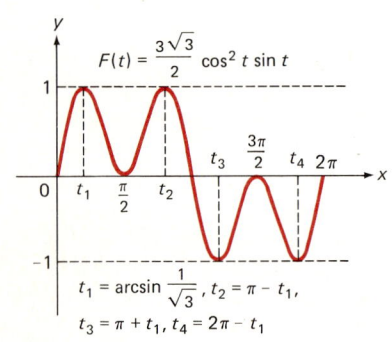

Example 4 Find the point of the paraboloid of revolution $z = \frac{1}{4}(x^2 + y^2) - 2$ which is closest to the point $(0, 1, 0)$.

Solution Since the square of the distance from the point $(0, 1, 0)$ to a variable point (x, y, z) of the paraboloid is $x^2 + (y-1)^2 + z^2$, we are looking for the absolute minimum of the function
$$f(x, y, z) = x^2 + (y-1)^2 + z^2,$$
subject to the constraint
$$g(x, y, z) = x^2 + y^2 - 4z = 8$$
(if the square of the distance is minimized, so is the distance itself). Taking gradients, we have
$$\nabla f = 2x\mathbf{i} + 2(y-1)\mathbf{j} + 2z\mathbf{k}, \qquad \nabla g = 2x\mathbf{i} + 2y\mathbf{j} - 4\mathbf{k} \neq \mathbf{0},$$
so that the Lagrange multiplier rule $\nabla f = \lambda \nabla g$ gives
$$2x = 2\lambda x, \qquad 2(y-1) = 2\lambda y, \qquad 2z = -4\lambda.$$

It follows from the first of these equations that $x = 0$ or $\lambda = 1$. But if $\lambda = 1$, the second equation becomes $-2 = 0$, which is absurd, and hence $x = 0$

(note that $y \neq 0$ for the same reason). Eliminating λ from the second and third equations, we get

$$z = -\frac{2(y-1)}{y}. \tag{13}$$

Substitution of $x = 0$ and formula (13) into the equation of constraint leads to

$$y^2 - 4z = y^2 + \frac{8(y-1)}{y} = 8,$$

which implies $y = 2$, and then (13) gives $z = -1$. Therefore $(0, 2, -1)$ is the only point satisfying the condition $\nabla f = \lambda \nabla g$. It is geometrically apparent that f has a minimum, but not a maximum, on the paraboloid. Hence the minimum must be at $(0, 2, -1)$, and this is the point of the paraboloid which is closest to the point $(0, 1, 0)$, a distance $\sqrt{f(0, 2, -1)} = \sqrt{2}$ away (see Figure 26). ☐

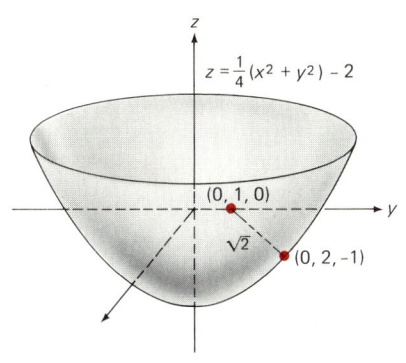

Figure 26

Problems

Use Lagrange multipliers to investigate the extrema of the function f subject to the given constraint.
1. $f(x, y) = 3x - 2y + 1$, $9x^2 + 4y^2 = 18$
2. $f(x, y) = xy$, $3x + 2y - 36 = 0$
3. $f(x, y) = x^2 - y^2$, $xy = 1$
4. $f(x, y) = x^2 + y^2$, $x^4 + y^4 = 2$
5. $f(x, y) = x^2 + 8y^2$, $\sqrt{x} + \sqrt{y} = 1$
6. $f(x, y) = x^2y^2$, $x^2 + y^2 = 1$
7. $f(x, y) = x^2 + y^2$, $x^2 - y^2 = 1$
8. $f(x, y) = x^2 + y^2$, $(x-1)^2 + (y+1)^2 = 8$
9. $f(x, y, z) = 2x - 3y + z - 1$, $x^2 + y^2 + z^2 = 14$
10. $f(x, y, z) = x^2 + y^2 + z^2$, $x - 2y + 2z = 6$
11. $f(x, y, z) = \frac{1}{8}x + \frac{1}{27}y + \frac{1}{64}z$, $xyz = 1$ (x, y, z positive)
12. $f(x, y, z) = x - y + z$, $4x^2 + y^2 + 9z^2 = 49$
13. $f(x, y, z) = x^2 + y^2 + z^2$, $2x^2 + y^2 - z^2 = 2$
14. $f(x, y, z) = xy + xz$, $x^2 + 2y^2 + 2z^2 = 4$

Use Lagrange multipliers to find the distance between
15. The point $(-1, 4)$ and the line $12x - 5y + 71 = 0$
16. The point $(1, 1, 1)$ and the plane $2x + 6y - 9z + 12 = 0$

17. Find the point of the parabola $(y - 1)^2 = 4x$ which is closest to the point $(2, 0)$.
18. Find the points of the hyperbola $x^2 - y^2 = 1$ which are closest to the point $(0, 4)$.
19. Find the point of the parabola $y = \frac{1}{4}x^2$ which is closest to the line $x - y - 4 = 0$.
20. Find the points of the ellipse $4x^2 + y^2 = 4$ which are farthest from and closest to the line $3x - 2y + 6 = 0$.
21. Determine the minimum of the function $f(x, y) = x^2 + (y + 1)^2$ subject to the constraint $g(x, y) = x^2 - y^3 = 0$. Is the Lagrange multiplier rule applicable to this problem? Explain your answer.

22. Use the Lagrange multiplier rule to locate the axes of the rotated ellipse $3x^2 - 2xy + 3y^2 - 2 = 0$. Sketch the ellipse.
Hint. Find the points of the ellipse which are farthest from and closest to the origin.

23. A tent is in the shape of a right circular cylinder of height H, surmounted by a cone of height h with the same radius r (see Figure 27). The volume of the tent is specified in advance. How should the dimensions of the tent be related in order to minimize the amount of canvas needed to construct the tent?

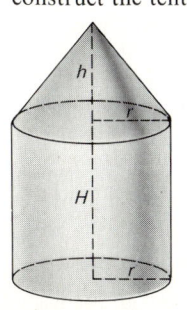

Figure 27

24. The wood used to make the sides of a closed rectangular box costs $1.00 per square foot, while the wood used to make the top and bottom costs $1.50 per square foot. The volume of the box is 12 cubic feet. What dimensions of the box minimize the cost of construction, and what is the minimum cost?

Use Lagrange multipliers to solve
25. Example 8, page 210
26. Problem 1, page 212
27. Problem 3, page 212
28. Problem 25, page 797
29. Problem 26, page 797
30. Problem 27, page 797

31. It can be shown that if $f(x, y, z)$ has a local extremum at (a, b, c) subject to *two* constraints $g(x, y, z) = 0$ and $h(x, y, z) = 0$, and if the gradients $\nabla g(a, b, c)$ and $\nabla h(a, b, c)$ are nonzero and nonparallel, then there are *two* numbers λ and μ, both called Lagrange multipliers, such that

$$\nabla f(a, b, c) = \lambda \nabla g(a, b, c) + \mu \nabla h(a, b, c)$$

(f, g and h are assumed to have continuous first partial derivatives). Use this generalization of Theorem 9 to find the extrema of the function $f(x, y, z) = 2x - y + 3z$ subject to the constraints $x^2 + y^2 = 5$ and $y + z = 4$.

32. Use the method of the preceding problem to find the distance between the origin and the line of intersection of the planes $x + 2y - z - 5 = 0$ and $x - y + z - 3 = 0$. Check the answer by using another method.

Key Terms and Topics
Functions of several variables
The space R^n of points $(x_1, x_2, \ldots, x_n)$
Level curves and surfaces
Limits and continuity of functions of n variables
Interior points and boundary points
Open and closed sets, continuity on a set
Partial derivatives of the first and higher order
Differentiability of a function of two or more variables
Implications of differentiability
Approximation by differentials
The chain rule for functions of several variables
Implicit differentiation and the implicit function theorem
The tangent plane and normal line to a surface
The tangent plane approximation
The directional derivative and gradient
Absolute and local extrema of functions of several variables
Critical points and saddle points
The second partials test
Constrained extrema and Lagrange multipliers

Supplementary Problems

1. Find a continuous function $f(x, y)$ defined on the whole xy-plane, whose values at the 16 points (m, n) with integer coordinates $m, n = 0, 1, 2, 3$ are given in the following table.

x \ y	0	1	2	3
0	1	3	5	7
1	-2	0	2	4
2	-5	-3	-1	1
3	-8	-6	-4	-2

2. Find the value of the function

$$f(x, y) = \frac{\arctan(x - y)}{\arctan(x + y)}$$

at the point $(\tfrac{1}{2}(\sqrt{3} + 1), \tfrac{1}{2}(\sqrt{3} - 1))$.

3. Express the area A of a triangle as a function of its side lengths x, y and z.

4. A right triangle with legs of lengths x and y is inscribed in a circle of radius R. Let $A = f(x, y)$ be the area of the triangle. What are the domain and range of f?

Describe the (natural) domain of the given function.

5. $z = \sqrt{x^2 - 1} + \sqrt{1 - y^2}$

6. $z = \sqrt{1 - x^2} + \sqrt{9 - y^2}$

7. $z = \sqrt{(x^2 + y^2 - 1)(4 - x^2 - y^2)}$

8. $z = \arcsin(y/x)$

9. $u = \dfrac{1}{\ln(1 - x^2 - y^2 - z^2)}$

10. $u = \sqrt{x + y + z - 1}$

11. Let
$$f(x, y) = \begin{cases} \sqrt{4x^2 + y^2} & \text{if } x < 0, \\ |y| & \text{if } x \geq 0. \end{cases}$$

Sketch the level curves $f(x, y) = c$ for $c = 0, 2, 4$.

12. Describe the level surfaces of the function $f(x, y, z) = 10^{2x + 3y - z}$.

Evaluate

13. $\lim_{(x,y) \to (0,0)} \dfrac{\sin(x^3 + y^3)}{x^2 + y^2}$

14. $\lim_{(x,y) \to (0,0)} \dfrac{\sqrt{x^2 + y^2 + 1} - 1}{x^2 + y^2}$

15. $\lim_{(x,y) \to (\infty, \infty)} \dfrac{2x + y}{x^2 + y^2}$

16. $\lim_{(x,y) \to (0,0)} xy \sin \dfrac{1}{x^2 + y^2}$

17. Give an example of a set which is both open and closed.

18. What is the boundary of the natural domain of the function $f(x, y) = \cot \pi(x^2 + y^2)$?

Find all first partial derivatives of the given function.

19. $f(x, y) = e^{x^2 + y^3}$
20. $f(x, y) = e^{(x + y^2)^2}$
21. $f(x, y) = x^{xy}$
22. $f(x, y) = (1 + xy)^x$
23. $f(x, y, z) = (x - y)(x - z)(y - z)$
24. $f(x, y, z) = \arctan(xy/z)$
25. $g(x, y, z) = x^{y^z}$
26. $h(x, y, z) = xy^2z^3 \cosh xyz$

27. Find f_{xyz}, f_{xxy} and f_{yyx} for the function in Problem 23.
28. Verify by direct calculation that $z_{xxyy} = z_{yxyx}$ if $z = \cos xy$.
29. The plane $x = 2$ intersects the surfaces $z = \frac{1}{3}(x^2 + y^2)$ and $z = \frac{1}{6}x^2 + y^2$ in a pair of curves. At what acute angle do these curves intersect?
30. Show that the function

$$f(x, y) = \begin{cases} \dfrac{xy}{x^2 + y^2} & \text{if } x^2 + y^2 \neq 0, \\ 0 & \text{if } x = y = 0 \end{cases}$$

has partial derivatives at every point of the plane, but is discontinuous at the origin.

31. Show that if $f(x, y)$ has *bounded* partial derivatives in a neighborhood of a point (a, b), then $f(x, y)$ is continuous at (a, b). Show that this is consistent with the preceding problem.

32. Show that the function

$$f(x, y) = \begin{cases} \dfrac{xy}{\sqrt{x^2 + y^2}} & \text{if } x^2 + y^2 \neq 0, \\ 0 & \text{if } x = y = 0 \end{cases}$$

is continuous and has partial derivatives at the origin, but is not differentiable there. Show that this is consistent with Theorem 2, page 768.

33. Find the total differential of
$$f(x, y) = \arctan \dfrac{x - y}{x + y}.$$

34. Two sides of a triangle are of lengths $x = 10$ cm and $y = 20$ cm, and the angle θ between them is $60°$. Estimate the change in the area of the triangle if x and y are both increased by 4 mm, while θ is decreased by $1°$. Also estimate the change in the length of the third side of the triangle.

35. Use differentials to estimate $(3.01/11.98)^{0.49}$.

36. Show that Laplace's equation $u_{xx} + u_{yy} = 0$ is satisfied by the function $u = \arctan(y/x)$, but not by the function $u = \sqrt{x^2 + y^2}$.

37. Let $u = f(x^2 + y^2)$, where f is an arbitrary differentiable function with derivative f'. Show that
$$y \dfrac{\partial u}{\partial x} - x \dfrac{\partial u}{\partial y} = 0.$$

Use the chain rule to find dw/dt if

38. $w = x_1 x_2 \cdots x_n$, $x_1 = x_2 = \cdots = x_n = t$
39. $w = \sqrt{x_1^2 + x_2^2 + \cdots + x_n^2}$, $x_1 = x_2 = \cdots = x_n = t > 0$
40. $w = x_1^2 + x_2^2 + \cdots + x_n^2$, $x_1 = t$, $x_2 = -t, \ldots, x_n = (-1)^{n-1} t$

41. Find $\partial w/\partial x$ and $\partial w/\partial y$ if $w = \ln(rstu)$ and $r = x - y$, $s = x + y$, $t = x^2 + y^2$, $u = x^4 + y^4$.

42. Prove the product and quotient rules for ordinary differentiation by applying the two-variable chain rule to the functions $u(x, y) = xy$ and $v(x, y) = x/y$, where $x = x(t)$ and $y = y(t)$.

43. Let $u = (1/x)[f(x + y) + g(x - y)]$, where f and g are arbitrary functions of a single variable with second derivatives f'' and g''. Show that
$$\dfrac{\partial}{\partial x}\left(x^2 \dfrac{\partial u}{\partial x}\right) = x^2 \dfrac{\partial^2 u}{\partial y^2}.$$

44. Give an example of a homogeneous function $f(x, y)$ of degree π (see Problem 29, page 778).

45. Show that if $f(x, y)$ is homogeneous of degree n and has continuous second partial derivatives, then
$$x^2 f_{xx}(x, y) + 2xy f_{xy}(x, y) + y^2 f_{yy}(x, y) = n(n - 1)f(x, y). \quad \text{(i)}$$

46. Show that the function $f(x, y) = x^2 y \arctan(x/y)$ is homogeneous of degree 3, and verify that it satisfies formula (i).

47. Can the area of a rectangle be increasing at the same time that the length of its diagonal is decreasing?

48. At a certain time t_0 the two equal sides of an isosceles triangle are 50 cm long and their length l is increasing at 2.5 cm/sec, while the angle θ between the sides is $45°$. How fast is θ decreasing at t_0 if it is known that the rate of change of the area of the triangle is exactly zero at t_0?

49. Let $u = u(x, y)$ and $v = v(x, y)$ satisfy the Cauchy–Riemann equations $\partial u/\partial x = \partial v/\partial y$, $\partial u/\partial y = -\partial v/\partial x$ (see Problems 47–49, page 764), and let $x = r\cos\theta$, $y = r\sin\theta$, as in the transformation from polar to rectangular coordinates. Find a pair of equations satisfied by $\partial u/\partial r$, $\partial u/\partial \theta$, $\partial v/\partial r$ and $\partial v/\partial \theta$.

50. Show that the function
$$u = (x_1^2 + x_2^2 + \cdots + x_n^2)^m \quad (m \neq 0, n > 2)$$
satisfies the n-dimensional version of Laplace's equation, namely
$$\frac{\partial^2 u}{\partial x_1^2} + \frac{\partial^2 u}{\partial x_2^2} + \cdots + \frac{\partial^2 u}{\partial x_n^2} = 0$$
if and only if $m = 1 - (n/2)$.

51. The ellipsoid $16x^2 + 9y^2 + z^2 = 144$ has eight tangent planes for which all three intercepts have the same absolute value. Find these planes.

52. The *angle between two surfaces* at a point of intersection P is defined as the angle between the tangent planes to the surfaces at P. Find the angle between the cylinder $x^2 + y^2 = 1$ and the sphere $(x - 1)^2 + y^2 + z^2 = 1$ at the point $(1, 0, 1)$. At the point $(\frac{1}{2}, \frac{1}{2}\sqrt{3}, 0)$.

Calculate $u_x = \partial u/\partial x$, $u_y = \partial u/\partial y$ and $u_z = \partial u/\partial z$ by implicit differentiation of the given equation.

53. $x^2 y^2 - 3xyzu + z^2 u^2 = 2$
54. $u^3 - u\cos xyz = 0$
55. $e^{xyu} = zu$
56. $\sinh(x + y + u) + \cosh(x + z - u) = 2$

57. Find the directional derivative of the function $f(x, y) = e^{\sin(x-y)}$ at the point $(1, 1)$ in the direction of the vector $\mathbf{a} = 8\mathbf{i} - 15\mathbf{j}$.

58. Find the directional derivative of the function
$$f(x, y, z) = \cosh(x + y + z)$$
at the point $(\ln 2, 1, -1)$ in the direction of the vector $\mathbf{a} = \mathbf{i} - 4\mathbf{j} + 8\mathbf{k}$.

59. Find the angle between the gradient vectors $\nabla f(3, 5)$ and $\nabla f(5, 13)$ if $f(x, y) = \arcsin(x/y)$.

60. Show that the directional derivative of the function $f(x, y) = x^2/y$ at any point of the ellipse $x^2 + 2y^2 = 1$ is zero in the direction of the normal to the ellipse.

61. What is the maximum rate of increase of the function $f(x, y, z) = xy + xz + yz$ at the point $(-3, 5, -1)$, and in which direction does it occur?

62. Show that the function
$$f(x, y) = \begin{cases} 0 & \text{if } y \leq 0 \text{ or } y \geq x^2, \\ 1 & \text{if } 0 < y < x^2 \end{cases}$$
has a directional derivative in every direction at the origin O, but is not differentiable at O.

Find all critical points of the given function. Then determine whether each critical point corresponds to a local maximum, a local minimum or a saddle point.

63. $f(x, y) = |x| + |y|$
64. $f(x, y) = x\sqrt{1 + y} + y\sqrt{1 + x}$ $(x > -1, y > -1)$
65. $f(x, y) = \sin x + \sin y + \cos(x + y)$
$(0 < x < \pi/2, 0 < y < \pi/2)$
66. $f(x, y) = (2x + y^2)e^x$
67. $f(x, y) = e^{x^2} - 3y^2$

68. Find the absolute extrema of the function $f(x, y) = \sin x + \sin y - \sin(x + y)$ on the triangular region $R = \{(x, y): x \geq 0, y \geq 0, x + y \leq 2\pi\}$.

69. Find the largest volume of a rectangular box whose diagonal is of length $\sqrt{3}a$.

70. Find the largest volume of a rectangular box, the sum of whose edge lengths is $12a$.

71. Find the triangle of largest area with perimeter $2c$.

72. Find the triangle of largest area that can be inscribed in a circle of radius r.

73. Show that if the function $f(x, y)$ or $f(x, y, z)$ has a local extremum subject to a constraint $g(x, y) = 0$ or $g(x, y, z) = 0$ at a point (a, b) or (a, b, c), then the function $F(x, y, \lambda) = f(x, y) - \lambda g(x, y)$ of three variables x, y and λ or the function $F(x, y, z, \lambda) = f(x, y, z) - \lambda g(x, y, z)$ of four variables x, y, z and λ has an *unconstrained* local extremum at (a, b, λ) or (a, b, c, λ) for some value of λ. This gives an alternative, but entirely equivalent, approach to the method of Lagrange multipliers.

74. Suppose a firm produces two commodities Q_1 and Q_2. Then the total cost to the firm of producing a quantity q_1 of Q_1 and a quantity q_2 of Q_2 is some function of q_1 and q_2, called the *cost function* and denoted by $C(q_1, q_2)$. There are now *two* marginal costs, $MC_1(q_1, q_2) = \partial C(q_1, q_2)/\partial q_1$, the marginal cost of the first commodity Q_1, and $MC_2(q_1, q_2) = \partial C(q_1, q_2)/\partial q_2$, the marginal cost of the second commodity Q_2. (This is just the natural extension of the considerations of Section 3.8 to the case of a two-commodity firm.) Let the firm's cost function be
$$C(q_1, q_2) = 3q_1^2 + 2q_1 q_2 + 3q_2^2.$$
Find the marginal costs $MC_1(q_1, q_2)$ and $MC_2(q_1, q_2)$. Suppose further that units of the commodities Q_1 and Q_2 are sold at prices $p_1 = \$480$ and $p_2 = \$720$, respectively. Write an expression for the firm's profit $P(q_1, q_2)$, which is now a function of two variables. What output levels of the two commodities maximize the profit, and what is the maximum profit?

75. In the preceding problem suppose profit is to be maximized subject to the constraint $q_1 + q_2 = 120$. For example, a furniture manufacturer may want to produce a total of 120 armchairs and sofas during a given period. What output levels maximize the profit in this case, and what is the maximum profit?

76. The amount of satisfaction or benefit to a consumer is called *utility* in economic theory. Let $U(q_1, q_2)$ be the utility to a consumer of acquiring a quantity q_1 of one commodity Q_1 and a quantity q_2 of another commodity Q_2, and suppose the consumer has an amount of money M to spend on the two commodities. Then the consumer's purchases are subject to the *budget constraint* $p_1 q_1 + p_2 q_2 = M$, where p_1 and p_2 are the prices of unit quantities of the commodities Q_1 and Q_2. With the help of a Lagrange multiplier λ, show that the consumer's utility is maximized at the point (q_1, q_2) for which

$$\frac{\partial U/\partial q_1}{p_1} = \frac{\partial U/\partial q_2}{p_2} = \lambda,$$

where the partial derivatives $\partial U/\partial q_1$ and $\partial U/\partial q_2$ are called the *marginal utilities* of the commodities Q_1 and Q_2, respectively. Thus the condition for utility maximization is that the ratio of marginal utility to price be the same for both commodities. By examining the way the maximum value of U depends on M, give an economic interpretation of the Lagrange multiplier λ. Find the values of q_1 and q_2 maximizing the utility, and also the corresponding value of λ, if $U(q_1, q_2) = q_1 q_2$, $p_1 = \$10$, $p_2 = \$5$ and $M = \$300$.

77. What is the largest volume of a closed rectangular box of surface area 16 sq. ft, the sum of whose edge lengths is 20 ft?

14 Multiple Integration

In this chapter we develop the integral calculus of functions of two and three variables. To avoid encroaching upon the territory of advanced calculus, where the same topics are treated from a mathematically more rigorous point of view, we will concentrate on concrete problems from geometry, physics and engineering. In fact, it was the need to solve such problems that led the founders of calculus to devise multiple integrals in the first place.

14.1 Double Integrals

In Chapter 4 we introduced the notion of the definite integral of a function of one variable. The analogous concept for a function of several variables is called a *multiple integral*. We begin with the case of the integral of a function of *two* variables, called a *double integral*. Here instead of a function $f(x)$ defined on a closed interval $[a, b]$, we consider a function $f(x, y)$ defined on a suitable *region of integration* R. But two-dimensional regions, unlike one-dimensional intervals, can be very complicated objects indeed. In fact, the boundary of a region R may be so irregular that it is impossible to assign R a well-defined area, which would make R unsuitable as a candidate for a region of integration. Thus from the outset we restrict ourselves to a special class of regions, said to be *normal* (in the sense of being free of "pathology"). These regions all have area, and moreover the class is large enough to include all regions encountered in practical applications of calculus. Here are the appropriate definitions.

The set of all points (x, y) satisfying the inequalities

$$a \leq x \leq b, \qquad g_1(x) \leq y \leq g_2(x), \tag{1}$$

where the functions g_1 and g_2 are continuous on the interval $[a, b]$, is called a *vertically simple region*, while the set of all points (x, y) satisfying the

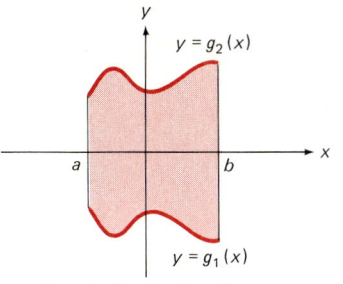

Figure 1 — A vertically simple region

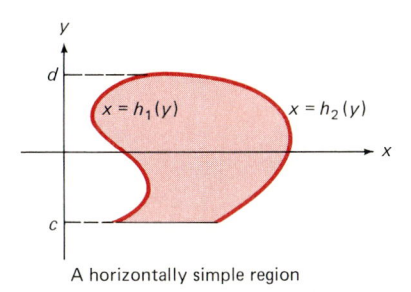

Figure 2 — A horizontally simple region

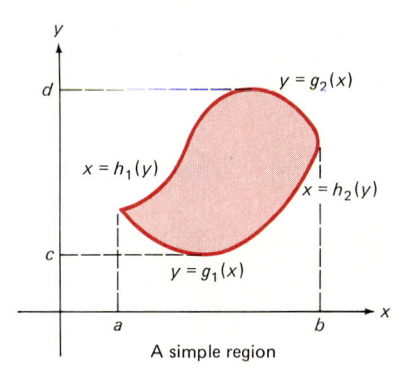

Figure 3 — A simple region

inequalities

$$c \leq y \leq d, \qquad h_1(y) \leq x \leq h_2(y), \tag{1'}$$

where the functions h_1 and h_2 are continuous on the interval $[c, d]$, is called a *horizontally simple region*. Thus the region in Figure 1 is vertically simple, while the region in Figure 2 is horizontally simple. By a *simple region*, without further qualification, we mean a region which is *both* vertically simple and horizontally simple. Such a region is shown in Figure 3. Finally, by a *normal region* we mean a bounded closed region which can be decomposed into a finite number of subregions, each vertically or horizontally simple (or both), by drawing suitable lines parallel to the coordinate axes (adjacent subregions share parts of their boundaries). Naturally, a region which is already vertically or horizontally simple is regarded as normal.

Example 1 The annular region $R = \{(x, y): 1 \leq x^2 + y^2 \leq 4\}$ in Figure 4 is neither horizontally nor vertically simple. However, R is normal. In fact, the x-axis divides R into two vertically simple subregions R_1 and R_2, as shown in Figure 5(a), while the y-axis divides R into two horizontally simple subregions R'_1 and R'_2, as shown in Figure 5(b). □

These regions all have well-defined and easily calculated areas. The area of the vertically simple region defined by the inequalities (1) is just the area between the upper curve $y = g_2(x)$ and the lower curve $y = g_1(x)$ from a to b, equal to

$$A = \int_a^b [g_2(x) - g_1(x)]\, dx, \tag{2}$$

as shown on page 242, while the area of the horizontally simple region defined by the inequalities (1') is the area between the right-hand curve $x = h_2(y)$ and the left-hand curve $x = h_1(y)$ from c to d, equal to

$$A = \int_c^d [h_2(y) - h_1(y)]\, dy, \tag{2'}$$

as shown on page 259. If a region is simple, its area is given by either of the formulas (2) and (2'), since both apply. Finally, the area of a normal region R is defined in the natural way as the sum of the areas of the horizontally or vertically simple subregions into which R can be decomposed by drawing lines parallel to the coordinate axes (we regard it as geometrically evident that the area of R does not depend on the particular manner in which this decomposition is made).

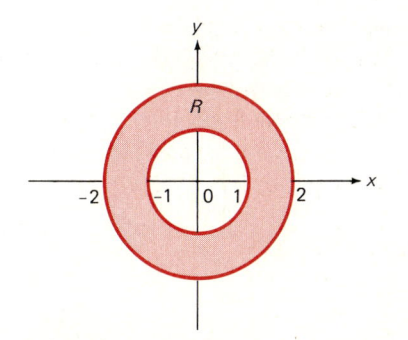

Figure 4

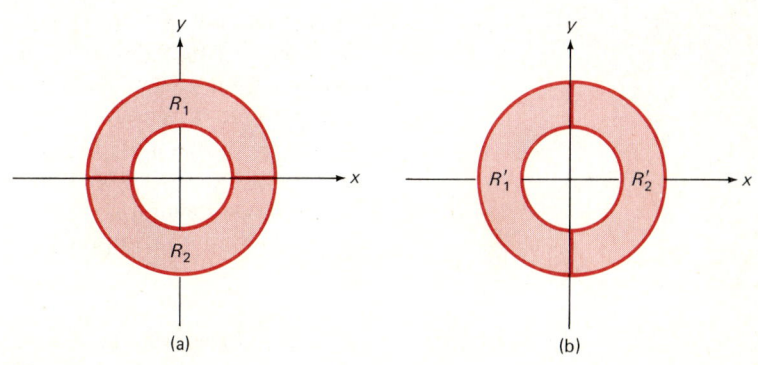

Figure 5
(a)   (b)

Section 14.1 Double Integrals **811**

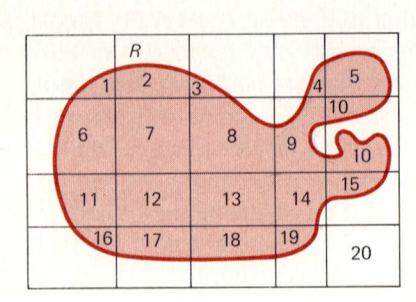

Figure 6

Not only does every normal region have a well-defined area, but if such a region R is partitioned by drawing horizontal and vertical lines parallel to the coordinate axes, the part of R lying in each cell of the partition (a rectangle) is also sure to have an area. This is illustrated in Figure 6, where a normal region R is drawn against a background of twenty rectangular cells (resulting from the intersection of five horizontal and six vertical lines), which partition R into nineteen nonempty subregions (observe that cell 20 contains no points of R). Subregions 1–8 and 11–19 are all simple, and in fact subregions 7, 12 and 13 are rectangular. Subregion 9 is horizontally simple, while subregion 10 consists of two disconnected pieces, one simple and the other normal, and its area is of course taken to be the sum of the areas of the two pieces. The figure is sufficiently general to exhibit all the possibilities (simple subregions, normal subregions and "subregions" consisting of several pieces), and in every case it is easy, at least in principle, to assign areas to the subregions.

Definition of the Double Integral

We are now ready to define the double integral of a function $f(x, y)$ of two variables over a normal region R. Let $Q = \{(x, y): a \leq x \leq b, c \leq y \leq d\}$ be any closed rectangular region with sides parallel to the coordinate axes which contains R (see Figure 7), and let the points x_j ($j = 0, 1, \ldots, J$) and y_k ($k = 0, 1, \ldots, K$) be partitions of the intervals $[a, b]$ and $[c, d]$, with mesh sizes μ_x and μ_y, respectively, as defined on page 230. This means that

$$a = x_0 < x_1 < x_2 < \cdots < x_{J-1} < x_J = b,$$
$$c = y_0 < y_1 < y_2 < \cdots < y_{K-1} < y_K = d,$$

and

$$\mu_x = \max\{x_1 - x_0, x_2 - x_1, \ldots, x_J - x_{J-1}\},$$
$$\mu_y = \max\{y_1 - y_0, y_2 - y_1, \ldots, y_K - y_{K-1}\}.$$

Then the two sets of lines

$$\begin{aligned} x &= a, \quad x = x_1, \quad x = x_2, \ldots, \quad x = x_{J-1}, \quad x = b, \\ y &= c, \quad y = y_1, \quad y = y_2, \ldots, \quad y = y_{K-1}, \quad y = d, \end{aligned} \tag{3}$$

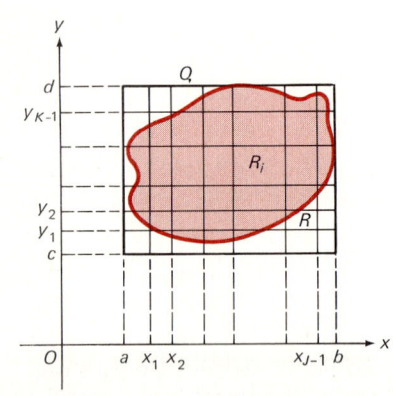

Figure 7

parallel to the coordinate axes, constitute a *partition* of the rectangle Q; these lines divide Q into JK subrectangles and the region R into n nonempty closed subregions $R_1, R_2, \ldots, R_n$, where $n \leq JK$ (typically $n < JK$), as illustrated in the figure. In general, some of the subregions will be nonrectangular, with boundaries consisting of parts of the lines (3) and parts of the boundary of R, but as already explained, since R is a normal region, every subregion R_i has a well-defined area ΔA_i. Given any function $f(x, y)$ of two variables, let (p_i, q_i) be an *arbitrary* point in R_i, and form the *Riemann sum*

$$S = \sum_{i=1}^{n} f(p_i, q_i) \Delta A_i.$$

Suppose S approaches a finite limit as the quantity

$$\mu = \max\{\mu_x, \mu_y\},$$

known as the *mesh size* of the partition, approaches zero, *regardless of the choice of the numbers x_j, y_k, p_i and q_i satisfying the stipulated conditions.* Then this limit is called the *(double) integral of f over R*, denoted by

$$\iint_R f(x, y)\, dA,$$

and the function f is said to be *integrable* on R, or over R. Thus

$$\iint_R f(x, y)\, dA = \lim_{\mu \to 0} \sum_{i=1}^{n} f(p_i, q_i) \Delta A_i, \qquad (4)$$

where f is called the *integrand* and R the *region of integration* of the integral on the left.

Example 2 If A is the area of R, then

$$A = \sum_{i=1}^{n} \Delta A_i$$

for every partition of the region R by lines parallel to the coordinate axes. Therefore, choosing $f(x, y) \equiv 1$ in formula (4), we get

$$\iint_R 1\, dA = \iint_R dA = \lim_{\mu \to 0} \sum_{i=1}^{n} \Delta A_i = \lim_{\mu \to 0} A = A.$$

It follows that

$$A = \iint_R dA. \qquad (5)$$

Although we have assumed that the boundary of the region of integration R is made up of graphs of continuous functions of x or y, we have made no assumption about the continuity of the integrand $f(x, y)$ of a double integral, and discontinuous functions may well be integrable. However, as you might expect, Theorem 1, page 232, on the integrability of continuous functions carries over to the case of double integrals.

Theorem 1 *(Continuity implies integrability on a region).* If the function $f(x, y)$ is continuous on a normal region R, then $f(x, y)$ is integrable on R.

We omit the proof, which belongs to the domain of advanced calculus.

Apart from technicalities pertaining to the nature of the region of integration R and the way it is partitioned, the definition of the double integral $\iint_R f(x, y)\, dA$ is virtually the same as that of the ordinary or "single" integral $\int_a^b f(x)\, dx$. Thus double integrals and single integrals obey a number of similar rules. We now list the most important of these rules, omitting proofs since they are analogous to the corresponding proofs for single integrals. In each case R is a normal region.

(i) If f is integrable on R and c is any constant, then cf is also integrable on R, and

$$\iint_R cf(x, y)\, dA = c \iint_R f(x, y)\, dA.$$

(ii) If f and g are both integrable on R, then the sum $f + g$ is also integrable on R, and

$$\iint_R [f(x, y) + g(x, y)] \, dA = \iint_R f(x, y) \, dA + \iint_R g(x, y) \, dA.$$

(iii) If f is integrable on R and $c \leq f(x, y) \leq C$, where c and C are constants, then

$$cA \leq \iint_R f(x, y) \, dA \leq CA,$$

where A is the area of R. In particular, if $f(x, y) \geq 0$, then

$$\iint_R f(x, y) \, dA \geq 0.$$

(iv) If f is continuous on R and if R is decomposed into two normal subregions R_1 and R_2 with no interior points in common (as illustrated in Figure 8), then

$$\iint_R f(x, y) \, dA = \iint_{R_1} f(x, y) \, dA + \iint_{R_2} f(x, y) \, dA.$$

This is the analogue of Theorem 2, page 239, and we assume continuity of f as an easy way to guarantee the existence of all three integrals.

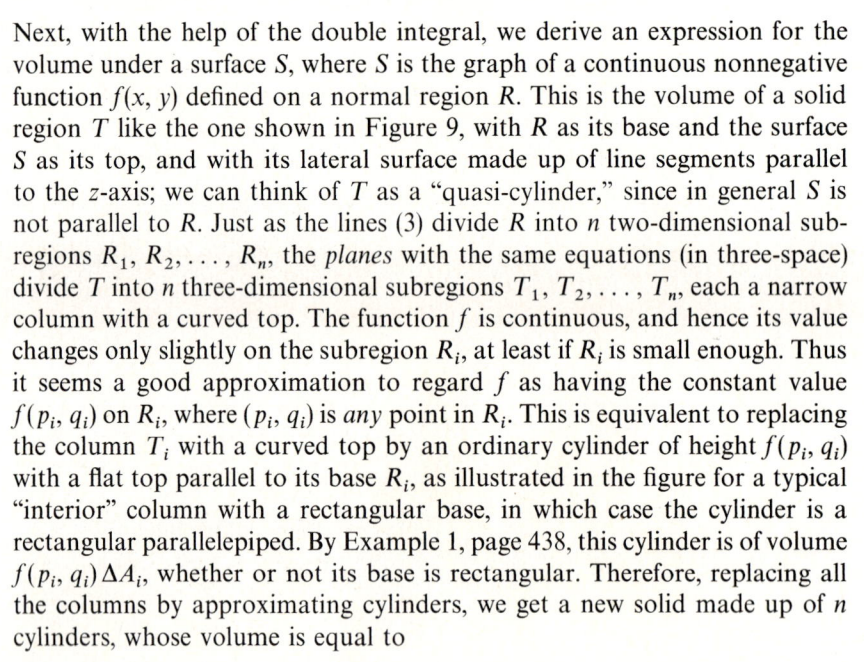

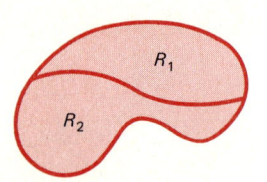

Figure 8

Volume Under a Surface as a Double Integral

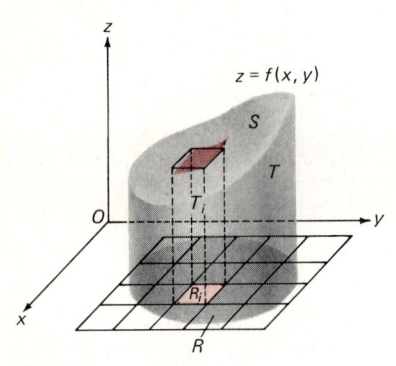

Next, with the help of the double integral, we derive an expression for the volume under a surface S, where S is the graph of a continuous nonnegative function $f(x, y)$ defined on a normal region R. This is the volume of a solid region T like the one shown in Figure 9, with R as its base and the surface S as its top, and with its lateral surface made up of line segments parallel to the z-axis; we can think of T as a "quasi-cylinder," since in general S is not parallel to R. Just as the lines (3) divide R into n two-dimensional subregions $R_1, R_2, \ldots, R_n$, the *planes* with the same equations (in three-space) divide T into n three-dimensional subregions $T_1, T_2, \ldots, T_n$, each a narrow column with a curved top. The function f is continuous, and hence its value changes only slightly on the subregion R_i, at least if R_i is small enough. Thus it seems a good approximation to regard f as having the constant value $f(p_i, q_i)$ on R_i, where (p_i, q_i) is *any* point in R_i. This is equivalent to replacing the column T_i with a curved top by an ordinary cylinder of height $f(p_i, q_i)$ with a flat top parallel to its base R_i, as illustrated in the figure for a typical "interior" column with a rectangular base, in which case the cylinder is a rectangular parallelepiped. By Example 1, page 438, this cylinder is of volume $f(p_i, q_i) \Delta A_i$, whether or not its base is rectangular. Therefore, replacing all the columns by approximating cylinders, we get a new solid made up of n cylinders, whose volume is equal to

$$\sum_{i=1}^{n} f(p_i, q_i) \Delta A_i,$$

the sum of the volumes of the n cylinders. It is reasonable to regard this sum as a good approximation to the volume V of the solid T, where the

Figure 9

approximation gets better and better as the size of each and every subregion R_i gets smaller and smaller, that is, as the mesh size μ of the partition of R approaches zero (notice that $\mu < \varepsilon$ implies that every $\Delta A_i < \varepsilon^2$). Motivated by this argument, we now *define* V as the limit

$$V = \lim_{\mu \to 0} \sum_{i=1}^{n} f(p_i, q_i) \Delta A_i,$$

that is,

$$V = \iint_R f(x, y)\, dA,$$

where the existence of the double integral follows from Theorem 1 and the assumed continuity of f.

More generally, let S_1 and S_2 be two surfaces whose projections onto the xy-plane are the same normal region R, and suppose S_1 and S_2 are the graphs of two continuous functions $f_1(x, y)$ and $f_2(x, y)$ satisfying the inequality $f_2(x, y) \geq f_1(x, y)$. Then the volume between S_1 and S_2 is given by

$$V = \iint_R [f_2(x, y) - f_1(x, y)]\, dA,$$

by the same kind of reasoning used in Section 4.3 to derive a formula for the area between two curves.

The direct evaluation of a double integral $\iint_R f(x, y)\, dA$ starting from its definition as the limiting value of a Riemann sum is quite unfeasible, except in the simplest cases (see Problem 23). Thus we must find another way of evaluating double integrals. To this end, we now introduce some integrals of a different type, which involve the consecutive evaluation of two single integrals.

Iterated Integrals

Let R be a vertically simple region defined by the inequalities $a \leq x \leq b$, $g_1(x) \leq y \leq g_2(x)$, where g_1 and g_2 are functions continuous on the interval $[a, b]$, and let $f(x, y)$ be a function of two variables defined on R. Then the integral

$$I_R = \int_a^b \left[\int_{g_1(x)}^{g_2(x)} f(x, y)\, dy \right] dx \qquad (6)$$

is called an *iterated integral* of f over R. To get I_R, we first integrate $f(x, y)$ with respect to y, *holding x fixed* both in the integrand $f(x, y)$ and in the limits of integration $g_1(x)$ and $g_2(x)$.† The result of the y-integration depends on x, and hence yields some function of x, say $i(x)$. We then integrate $i(x)$ with respect to x between the constant limits a and b, obtaining the number denoted by I_R in formula (6). Similarly, if R is a horizontally simple region, defined by the inequalities $c \leq y \leq d$, $h_1(y) \leq x \leq h_2(y)$, where h_1 and h_2 are functions continuous on the interval $[c, d]$, the integral

$$J_R = \int_c^d \left[\int_{h_1(y)}^{h_2(y)} f(x, y)\, dx \right] dy \qquad (6')$$

† This process constitutes a kind of "partial integration," analogous to the operation of partial differentiation (in calculating $\partial f(x, y)/\partial y$ we hold x fixed and differentiate with respect to y).

Section 14.1   Double Integrals   **815**

is also called an *iterated integral* of f over R. However, to get J_R, we first integrate $f(x, y)$ with respect to x, *holding y fixed* both in the integrand $f(x, y)$ and in the limits of integration $h_1(y)$, and $h_2(y)$, obtaining a function of y, say $j(y)$, which is then integrated with respect to y between the constant limits c and d. Notice that at each stage in the calculation of I_R and J_R, we are evaluating the integral of a function of a *single* variable, which allows us to make use of our basic tool for evaluating integrals, namely the fundamental theorem of calculus.

Remark In the above discussion we have tacitly assumed that the integrals (6) and (6′) *exist*. It can be shown that continuity of the functions f, g_1 and g_2 guarantees the continuity of the inner integral $i(x)$ in (6) and hence the existence of (6), while continuity of the functions f, h_1 and h_2 guarantees the continuity of the inner integral $j(y)$ in (6′) and hence the existence of (6′). You will recall that the continuity of g_1, g_2, h_1 and h_2 was assumed from the outset, in defining vertically and horizontally simple regions.

To keep the notation simple, it is customary to omit the inner brackets in formulas (6) and (7). Thus we will henceforth write the iterated integrals I_R and J_R simply as

$$I_R = \int_a^b \int_{g_1(x)}^{g_2(x)} f(x, y)\, dy\, dx, \qquad J_R = \int_c^d \int_{h_1(y)}^{h_2(y)} f(x, y)\, dx\, dy,$$

or alternatively as

$$I_R = \int_a^b dx \int_{g_1(x)}^{g_2(x)} f(x, y)\, dy, \qquad J_R = \int_c^d dy \int_{h_1(y)}^{h_2(y)} f(x, y)\, dx,$$

where the omission of the brackets is compensated by moving the differentials dx and dy apart, with dx appearing only behind the integral sign carrying the limits of x-integration and dy appearing only behind the integral sign carrying the limits of y-integration.

Example 3 Calculate both iterated integrals I_R and J_R of the function $f(x, y) = 3xy^2 - 2x^2 y$ over the rectangle R bounded by the lines $x = 0$, $x = 2$, $y = 1$ and $y = 2$.

Solution Both I_R and J_R exist, since f is continuous and R is clearly simple (see Figure 10). Calculating I_R, we get

$$I_R = \int_0^2 dx \int_1^2 (3xy^2 - 2x^2 y)\, dy = \int_0^2 \left[xy^3 - x^2 y^2 \right]_{y=1}^{2} dx$$

$$= \int_0^2 (7x - 3x^2)\, dx = \left[\frac{7}{2} x^2 - x^3 \right]_0^2 = 14 - 8 = 6,$$

where the fundamental theorem of calculus is used twice. Similarly, a calculation of J_R gives

$$J_R = \int_1^2 dy \int_0^2 (3xy^2 - 2x^2 y)\, dx = \int_1^2 \left[\frac{3}{2} x^2 y^2 - \frac{2}{3} x^3 y \right]_{x=0}^{2} dy$$

$$= \int_1^2 \left(6y^2 - \frac{16}{3} y \right) dy = \left[2y^3 - \frac{8}{3} y^2 \right]_1^2 = \frac{16}{3} - \left(-\frac{2}{3} \right) = 6.$$

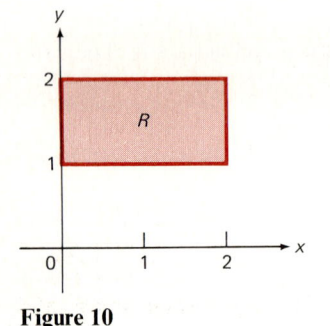

Figure 10

The fact that $I_R = J_R$ is of course not a coincidence, and we will see later that they are both equal to $\iint_R (3xy^2 - 2x^2y)\,dA$, the double integral of f over R. □

Example 4 Evaluate the iterated integral

$$I_R = \int_0^1 dx \int_0^{x^2} (x+y)\,dy,$$

and then reverse the order of integration.

Solution Since we are starting with an integral of the type I_R, it is immediately apparent that the region of integration R is vertically simple. In fact, R is the region defined by the inequalities $0 \le x \le 1$, $0 \le y \le x^2$, that is, the region shown in Figure 11, bounded by the x-axis, the line $x = 1$ and the parabola $y = x^2$. Evaluating I_R, we find that

$$I_R = \int_0^1 dx \int_0^{x^2} (x+y)\,dy = \int_0^1 \left[xy + \frac{1}{2}y^2\right]_{y=0}^{x^2} dx$$

$$= \int_0^1 \left(x^3 + \frac{1}{2}x^4\right) dx = \left[\frac{1}{4}x^4 + \frac{1}{10}x^5\right]_0^1 = \frac{1}{4} + \frac{1}{10} = \frac{7}{20}.$$

To reverse the order of integration, we must set up the other iterated integral J_R. This assumes that R is also horizontally simple. But if a horizontal line drawn from left to right intersects R, it enters R at a point of the parabola $y = x^2$ and leaves R at a point of the line $x = 1$. Thus, since $y = x^2$ is equivalent to $x = \sqrt{y}$ for $y \ge 0$, the region R can also be defined by the inequalities $0 \le y \le 1$, $\sqrt{y} \le x \le 1$, and hence is horizontally simple as well as vertically simple. Therefore we can reverse the order of integration, writing the other iterated integral

$$J_R = \int_0^1 dy \int_{\sqrt{y}}^1 (x+y)\,dx.$$

Evaluation of J_R gives

$$J_R = \int_0^1 \left[\frac{1}{2}x^2 + xy\right]_{\sqrt{y}}^1 dy = \int_0^1 \left(\frac{1}{2} + y - \frac{1}{2}y - y^{3/2}\right) dy$$

$$= \left[\frac{1}{2}y + \frac{1}{4}y^2 - \frac{2}{5}y^{5/2}\right]_0^1 = \frac{1}{2} + \frac{1}{4} - \frac{2}{5} = \frac{7}{20},$$

so that once again $I_R = J_R$. □

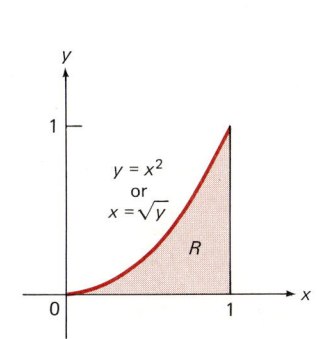

Figure 11

$y = x^2$ or $x = \sqrt{y}$

Evaluation of Double Integrals

We now show how to evaluate double integrals in terms of iterated integrals, and at the same time we explain why $I_R = J_R$ in the last two examples.

Theorem 2 *(Evaluation of a double integral over a vertically simple region).* If $f(x, y)$ is continuous on the vertically simple region R defined by the inequalities $a \le x \le b$, $g_1(x) \le y \le g_2(x)$, then

$$\iint_R f(x, y)\,dA = \int_a^b \int_{g_1(x)}^{g_2(x)} f(x, y)\,dy\,dx. \tag{7}$$

Proof As shown above, the double integral

$$V = \iint_R f(x, y)\, dA \tag{8}$$

is the volume V of the quasi-cylindrical solid T between the surface $z = f(x, y)$ and the region R (see Figure 12, where for simplicity we have chosen f to be nonnegative). Consider the iterated integral

$$I_R = \int_a^b \int_{g_1(x)}^{g_2(x)} f(x, y)\, dy\, dx. \tag{9}$$

It is apparent from the figure that if $x =$ constant, the inner integral

$$i(x) = \int_{g_1(x)}^{g_2(x)} f(x, y)\, dy$$

is just the area in the plane $x =$ constant under the *curve* $z = f(x, y)$ from $g_1(x)$ to $g_2(x)$. But this is in turn the area of the cross section cut from T by the plane $x =$ constant perpendicular to the x-axis. In other words, $i(x)$ is the continuous cross-sectional area function denoted by $A(x)$ in Section 8.1. Therefore, by formula (1), page 436, the volume of the solid T is also equal to

$$V = \int_a^b i(x)\, dx = I_R, \tag{10}$$

and comparing formulas (8) and (10), we find that

$$\iint_R f(x, y)\, dA = I_R,$$

which is equivalent to (7). ∎

As you might expect, there is an analogous theorem for double integrals over a *horizontally* simple region.

Theorem 2′ *(Evaluation of a double integral over a horizontally simple region).* If $f(x, y)$ is continuous on the horizontally simple region R defined by the inequalities $h_1(y) \le x \le h_2(y)$, $c \le y \le d$, then

$$\iint_R f(x, y)\, dA = \int_c^d \int_{h_1(y)}^{h_2(y)} f(x, y)\, dx\, dy. \tag{7′}$$

Proof The proof is virtually the same as that of Theorem 2. Let T be the quasi-cylindrical solid between the surface $z = f(x, y)$ and the region R (see Figure 13). Once again the volume of T is given by the double integral (8). Consider the iterated integral

$$J_R = \int_c^d \int_{h_1(y)}^{h_2(y)} f(x, y)\, dx\, dy. \tag{9′}$$

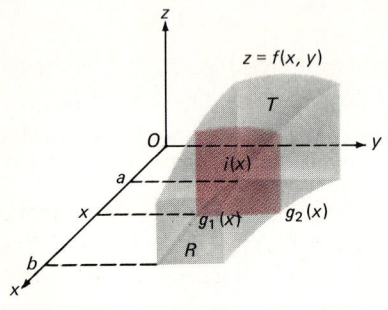

Figure 12

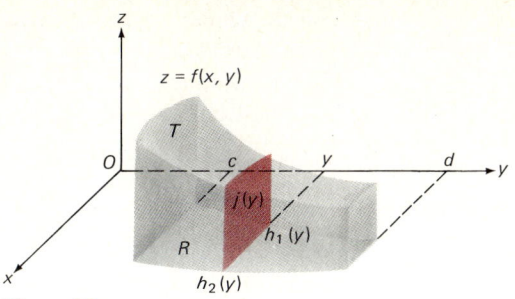

Figure 13

It is apparent from the figure that if $y = $ constant, the inner integral

$$j(y) = \int_{h_1(y)}^{h_2(y)} f(x, y)\, dx$$

is just the area in the plane $y = $ constant under the *curve* $z = f(x, y)$ from $h_1(y)$ to $h_2(y)$. But this is in turn the area of the cross section cut from T by the plane $y = $ constant perpendicular to the y-axis. Hence, by the method of cross sections, the volume of the solid T is also equal to

$$V = \int_c^d j(y)\, dy = J_R, \tag{10'}$$

and comparing formulas (8) and (10'), we find that

$$\iint_R f(x, y)\, dA = J_R,$$

which is equivalent to (7'). ∎

Juxtaposing Theorems 2 and 2', and assuming that the same region R appears on the left in both formulas (7) and (7'), we immediately obtain another basic result.

Theorem 3 (*Evaluation of a double integral over a simple region*). Let R be a simple region which is defined both by the inequalities $a \leq x \leq b$, $g_1(x) \leq y \leq g_2(x)$ and by the inequalities $c \leq y \leq d$, $h_1(y) \leq x \leq h_2(y)$, and let $f(x, y)$ be continuous on R. Then

$$\iint_R f(x, y)\, dA = I_R = J_R, \tag{11}$$

where I_R and J_R are the iterated integrals (9) and (9') of f over R. In particular

$$I_R = J_R,$$

so that the two iterated integrals of f over R are equal.

Remark In establishing Theorems 2 and 2', we have assumed rather naively that there is something called the "volume" of a solid, which can be defined in various equally satisfactory ways, leading to different expressions for the volume, which can then be equated without contradiction (the same assumption was made repeatedly in Chapter 8). A deeper investigation of the meaning of area and volume, given in advanced calculus, shows that this intuitive approach had not led us astray, and that formulas (7), (7′) and (11) are perfectly sound, provided that the integrand f and the region of integration R obey the stated conditions.

We have now completely solved the problem of practical evaluation of double integrals. Given a double integral $\iint_R f(x, y)\,dA$, where R is a normal region and f is continuous on R, we partition R (if necessary) into vertically or horizontally simple subregions $R_1, \ldots, R_n$ by drawing suitable lines parallel to the coordinate axes. This can be done, by the very definition of a normal region. Then by rule (iv), page 814,

$$\iint_R f(x, y)\,dA = \iint_{R_1} f(x, y)\,dA + \cdots + \iint_{R_n} f(x, y)\,dA.$$

But each of the integrals on the right can be replaced by an iterated integral, of the type I_R or J_R, and the evaluation of an iterated integral reduces to two consecutive integrations of functions of a *single* variable.

Example 5 Evaluate the double integral $\iint_R x^2\,dA$, where R is the region between the unit circle $x^2 + y^2 = 1$ and the ellipse $x^2 + 2y^2 = 1$.

Solution The x-axis divides R into two vertically simple regions R_1 and R_2, as shown in Figure 14. Hence, by rule (iv) and Theorem 2,

$$\iint_R x^2\,dA = \iint_{R_1} x^2\,dA + \iint_{R_2} x^2\,dA$$

$$= \int_{-1}^{1} dx \int_{\sqrt{(1-x^2)/2}}^{\sqrt{1-x^2}} x^2\,dy + \int_{-1}^{1} dx \int_{-\sqrt{1-x^2}}^{-\sqrt{(1-x^2)/2}} x^2\,dy$$

$$= 2\int_{-1}^{1} x^2\left(\sqrt{1-x^2} - \sqrt{\frac{1-x^2}{2}}\right) dx$$

$$= 2\left(1 - \frac{1}{\sqrt{2}}\right)\int_{-1}^{1} x^2\sqrt{1-x^2}\,dx$$

$$= 4\left(1 - \frac{1}{\sqrt{2}}\right)\int_{0}^{1} x^2\sqrt{1-x^2}\,dx,$$

where at the last step we use the evenness of the integrand. To evaluate the last integral, make the substitution $x = \sin t$, so that $dx = \cos t\,dt$ and

$$\int_0^1 x^2\sqrt{1-x^2}\,dx = \int_0^{\pi/2} \sin^2 t \cos^2 t\,dt = \frac{1}{4}\int_0^{\pi/2} \sin^2 2t\,dt$$

$$= \frac{1}{8}\int_0^{\pi/2}(1 - \cos 4t)\,dt = \frac{1}{8}\left[t - \frac{1}{4}\sin 4t\right]_0^{\pi/2} = \frac{\pi}{16}.$$

It follows that

$$\iint_R x^2\,dA = 4\left(1 - \frac{1}{\sqrt{2}}\right)\frac{\pi}{16} = \left(1 - \frac{1}{\sqrt{2}}\right)\frac{\pi}{4} \approx 0.23.$$

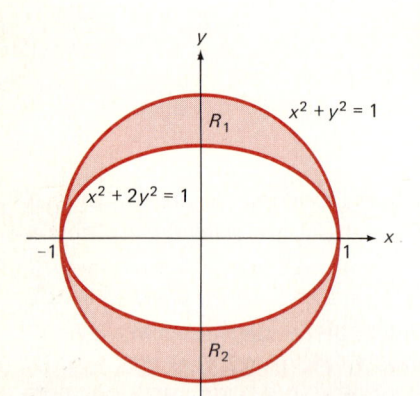

Figure 14

Geometrically, this is the volume under the parabolic cylinder $z = x^2$ and over the region R. □

Example 6 Evaluate the double integral $\iint_R e^{-x^2} dA$ over the region R bounded by the x-axis, the line $x = 1$ and the line $y = x$ (see Figure 15).

Solution The triangular region R is simple. Therefore, by Theorem 3,

$$\iint_R e^{-x^2} dA = \int_0^1 dx \int_0^x e^{-x^2} dy = \int_0^1 dy \int_y^1 e^{-x^2} dx$$

(explain all the limits of integration). If we try to evaluate the second iterated integral, we are blocked from the outset, since the function e^{-x^2} does not have an elementary antiderivative (see page 366). However, the first iterated integral can be easily evaluated:

$$\int_0^1 dx \int_0^x e^{-x^2} dy = \int_0^1 e^{-x^2} x\, dx = -\frac{1}{2} \int_0^1 d(e^{-x^2})$$

$$= -\frac{1}{2} e^{-x^2} \Big|_0^1 = \frac{1}{2}\left(1 - \frac{1}{e}\right) = \frac{e-1}{2e}.$$

Thus

$$\iint_R e^{-x^2} dA = \frac{e-1}{2e}. \quad □$$

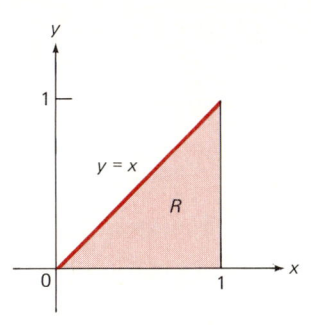

Figure 15

Calculation of Area and Volume

Since we already know how to calculate the area of a vertically or horizontally simple region as an area between two curves, and since we also know how to calculate volumes by the method of cross sections, the use of double integrals to calculate areas and volumes is more of a convenience than a necessity. But the convenience is *great*. The point is that once an area or volume has been expressed as a double integral, the rest of the calculation is usually routine. If the region of integration is not given explicitly, you may have to make a little drawing in order to find it, but a rough sketch is good enough if it reveals the essential features of the problem.

Example 7 Find the area A of the region R in Example 5.

Solution By formula (5), A is just the double integral $\iint_R dA$. Therefore, replacing the integrand x^2 by 1 in the calculations of Example 5, we get

$$A = \iint_R dA = 4\left(1 - \frac{1}{\sqrt{2}}\right) \int_0^1 \sqrt{1 - x^2}\, dx.$$

The integral on the right will be recognized as one fourth the area enclosed by the unit circle, namely π. It follows that

$$A = \left(1 - \frac{1}{\sqrt{2}}\right)\pi.$$

As an exercise, check this answer by subtracting the area enclosed by the ellipse $x^2 + 2y^2 = 1$ (see Example 4, page 591) from that enclosed by the unit circle. □

Example 8 Find the volume V under the plane $x + y - z = 0$ and over the region R bounded by the x-axis, the line $x = 1$ and the parabola $y = x^2$.

Solution The plane is the graph of the function $z = x + y$, so that

$$V = \iint_R (x + y)\, dA.$$

By Theorem 2, this double integral is equal to the iterated integral

$$I_R = \int_0^1 dx \int_0^{x^2} (x + y)\, dy.$$

But $I_R = \frac{7}{20}$, as already shown in Example 4, and therefore $V = \frac{7}{20}$. □

Finally we observe that it is sometimes convenient to write the double integral $\iint_R f(x, y)\, dA$ as

$$\iint_R f(x, y)\, dx\, dy.$$

There is no possibility of confusing this alternative form of the double integral with an iterated integral, since it contains the symbol $\iint_R$ in which the two integral signs do not carry separate limits of integration.

Problems

Evaluate the given iterated integral.

1. $\int_0^2 dx \int_0^1 (2x + y^2)\, dy$
2. $\int_1^5 dy \int_2^3 \frac{dx}{(x + y)^2}$
3. $\int_0^1 dy \int_0^1 \frac{y^2}{x^2 + 1}\, dx$
4. $\int_1^2 dx \int_{1/x}^x \frac{y^2}{x^2}\, dy$
5. $\int_2^8 dx \int_0^{\ln x} e^y\, dy$
6. $\int_0^1 dy \int_0^{\sqrt{1-y^2}} \sqrt{1 - x^2 - y^2}\, dx$
7. $\int_3^5 dy \int_y^{2y} \frac{x}{y}\, dx$
8. $\int_{-1}^1 dx \int_{3-x}^2 (x + 2y)\, dy$
9. $\int_0^1 dx \int_x^{x^2} (3x - y)\, dy$
10. $\int_0^\pi dy \int_0^{1+\cos y} x^2 \sin y\, dx$
11. $\int_0^1 dy \int_0^1 \frac{x}{(1 + x^2 + y^2)^{3/2}}\, dx$
12. $\int_0^{\pi/2} dx \int_{\sin x}^1 y^3\, dy$

Reverse the order of integration in the given iterated integral, after making a sketch of the region of integration R. (Assume that $f(x, y)$ is continuous.)

13. $\int_1^2 dx \int_x^{2x} f(x, y)\, dy$
14. $\int_0^1 dx \int_x^{\sqrt{2x - x^2}} f(x, y)\, dy$
15. $\int_0^2 dy \int_{-\sqrt{4-y^2}}^{\sqrt{4-y^2}} f(x, y)\, dx$
16. $\int_1^e dy \int_0^{\ln y} f(x, y)\, dx$

17. By reversing the order of integration, write the sum

$$\int_0^3 dx \int_0^x f(x, y)\, dy + \int_3^6 dx \int_0^{6-x} f(x, y)\, dy$$

as just one iterated integral.

18. Why can't the iterated integral

$$\int_0^1 dy \int_y^1 \sin \pi x^2\, dx$$

be evaluated as it stands? Evaluate it by reversing the order of integration.

Express the double integral $\iint_R f(x, y)\, dA$ in terms of iterated integrals if R is the region bounded by the given figure.

19. The triangle with vertices (1, 0), (2, 2), (0, 2)
20. The trapezoid with vertices (1, 1), (5, 1), (4, 4), (2, 4)

21. The parallelogram with vertices $(0, 0)$, $(2, -2)$, $(3, -1)$, $(1, 1)$
22. The polygon with vertices $(0, 0)$, $(2, 0)$, $(1, 1)$, $(2, 2)$, $(0, 2)$

23. Let R be the rectangular region defined by the inequalities $a \leq x \leq b$, $c \leq y \leq d$. Evaluate the double integral $\iint_R xy\,dA$ directly from the definition (4).

Hint. Partition R by horizontal and vertical lines, and form the Riemann sum based on the points at the centers of the resulting subrectangles.

24. Which of the double integrals

$$\iint_R (x^4 + 6x^2y^2 + y^4)\,dA, \qquad \iint_R (4x^3y + 4xy^3)\,dA$$

is larger?

Evaluate the given double integral.

25. $\iint_R \dfrac{x}{x^2 + 1}\,dA$, where R is the rectangular region bounded by the lines $x = 0$, $x = 1$, $y = 0$ and $y = 1$

26. $\iint_R xy^2 e^{xy}\,dA$, where R is the rectangular region bounded by the lines $x = 0$, $x = 1$, $y = 0$ and $y = 2$

27. $\iint_R \sqrt{4x^2 - y^2}\,dA$, where R is the triangular region bounded by the lines $x = 1$, $y = 0$ and $y = x$

28. $\iint_R y\,dA$, where R is the smaller of the two regions bounded by the circle $(x - 1)^2 + y^2 = 1$ and the line joining the points $(2, 0)$ and $(0, 2)$

29. $\iint_R \cos(x - y)\,dA$, where R is the triangular region bounded by the lines $x = 0$, $y = \pi$ and $y = x$

30. $\iint_R e^{y/x}\,dA$, where R is the region bounded by the lines $x = 1$, $y = 0$ and the parabola $y = x^2$

31. $\iint_R xy\,dA$, where R is the region bounded by the coordinate axes and the curve $\sqrt{x} + \sqrt{y} = 1$

32. $\iint_R x \ln y\,dA$, where R is the region bounded by the lines $x = 2$, $y = 1$ and the curve $xy = 1$

Use a double integral to find the area A of the region R bounded by the given curves.

33. $xy = 1$, $y = x$ and $x = e$
34. $y = \ln x$, $y = x - 1$ and $y = -2$

35. $y = \frac{1}{4}x^2 - 1$ and $y = 2 - x$
36. $x = y^2$ and $x = 4 - 3y^2$
37. $x^2 + 2y^2 = 1$ and $2x^2 + y^2 = 1$
38. $x^2 + y^2 = 4$ and $y^2 - 2x^2 = 1$ (R contains the origin)

Use a double integral to find the volume V of the specified solid region T.

39. T is bounded by the coordinate planes, the plane $z = x + 2y + 1$, and the planes $x = 1$ and $y = 2$
40. T is bounded by the coordinate planes and the plane $x + y + z = 3$
41. T lies in the first octant, and is bounded by the coordinate planes and the planes $x + 2y = 2$ and $x + 4y + 2z = 8$
42. T is bounded by the five planes $x = 0$, $z = 0$, $x + 3y = 6$, $2x + 3y = 12$ and $x + y + z = 6$
43. T is bounded by the elliptic paraboloid $z = \frac{1}{4}x^2 + \frac{1}{9}y^2$, the coordinate planes, and the planes $x = -2$ and $y = 3$.
44. T is bounded by the cone $z^2 = xy$ and the planes $x = 2$ and $y = 2$
45. T is bounded by the cone $z^2 = xy$ and the plane $x + y = 4$
46. T is bounded by the circular cylinder $x^2 + y^2 = 1$, the plane $z = 0$ and the plane $2x + 2y + 3z = 6$
47. T lies in the first octant and is bounded by the elliptic cylinder $4x^2 + z^2 = 1$, the plane $y = x$, and the planes $y = 0$ and $z = 0$
48. T lies in the first octant, and is bounded by the hyperbolic paraboloid $2z = xy$, the circular cylinder $x^2 + y^2 = 2x$ and the plane $z = 0$

49. Let R be a simple region. Use Theorem 3 to show that both formulas (2) and (2') give the same value for the area of R.

50. A set D is said to be (arcwise) connected if every pair of points P and Q in D can be joined by a continuous curve lying entirely in D, that is, by a curve with parametric equations $x = x(t)$, $y = y(t)$ ($0 \leq t \leq 1$), where $x(t)$ and $y(t)$ are continuous functions, $(x(t), y(t)) \in D$ for $0 \leq t \leq 1$, and $P = (x(0), y(0))$, $Q = (x(1), y(1))$. Show that if $f(x, y)$ is continuous on a connected normal region R of area A, then there is a point (a, b) in R such that

$$\iint_R f(x, y)\,dA = Af(a, b).$$

This is the mean value theorem for double integrals.
Hint. Apply the intermediate value theorem, page 96, to the function $F(t) = f(x(t), y(t))$.

51. Deduce from Theorem 3 that

$$\frac{d}{dx}\int_c^d f(x, y)\,dy = \int_c^d \frac{\partial f(x, y)}{\partial x}\,dy \qquad (a \leq x \leq b),$$

where $f(x, y)$ and $\partial f(x, y)/\partial x$ are continuous on the rectangular region $\{(x, y): a \leq x \leq b,\ c \leq y \leq d\}$. This result was anticipated in Problem 63, page 779.

Section 14.1 Double Integrals **823**

14.2 Triple Integrals

The considerations of the preceding section generalize at once to the case of *triple integrals*, that is, integrals over three-dimensional or solid regions. The big step has already been taken in going from ordinary single integrals to double integrals, and nothing essentially new occurs in the transition from two to three dimensions.

Let R_{xy} be a normal region in the xy-plane. Then the set of all points (x, y, z) such that

$$(x, y) \in R_{xy}, \qquad g_1(x, y) \leq z \leq g_2(x, y), \tag{1}$$

where the functions g_1 and g_2 are continuous on R_{xy} (and $\in$ denotes set membership), is called a *z-simple region* in space. Similarly, the set of all points (x, y, z) such that

$$(x, z) \in R_{xz}, \qquad h_1(x, z) \leq y \leq h_2(x, z), \tag{1'}$$

where h_1 and h_2 are continuous on a normal region R_{xz} in the xz-plane, is called a *y-simple region*, while the set of all points (x, y, z) such that

$$(y, z) \in R_{yz}, \qquad k_1(y, z) \leq x \leq k_2(y, z), \tag{1''}$$

where k_1 and k_2 are continuous on a normal region R_{yz} in the yz-plane, is called an *x-simple region*. Suppose that by drawing suitable planes parallel to the coordinate planes, we can decompose a solid region T into a finite number of subregions, each of which is simple with respect to at least one of the coordinates x, y and z. Then T is said to be *normal*. Naturally, a region which is already x-simple, y-simple or z-simple is regarded as normal.

Example 1 The solid region T shown in Figure 16 is x-simple, y-simple and z-simple. The figure also shows the regions R_{xy}, R_{xz} and R_{yz} corresponding to T, which are the projections of T onto the xy-, xz- and yz-planes.

These solid regions all have well-defined volumes. The volume of the z-simple region defined by (1) is just the volume between the upper surface $z = g_2(x, y)$ and the lower surface $z = g_1(x, y)$ projecting onto the region R_{xy} in the xy-plane, and this volume is equal to the double integral

$$V = \iint\limits_{R_{xy}} [g_2(x, y) - g_1(x, y)] \, dx \, dy,$$

by the argument on pages 814–815. Similarly, the volume of the y-simple and x-simple regions defined by (1') and (1'') are given by the double integrals

$$V = \iint\limits_{R_{xz}} [h_2(x, z) - h_1(x, z)] \, dx \, dz$$

and

$$V = \iint\limits_{R_{yz}} [k_2(y, z) - k_2(y, z)] \, dy \, dz.$$

If a solid region T is normal, its volume is defined as the sum of the volumes of the subregions, each simple with respect to at least one of the coordinates

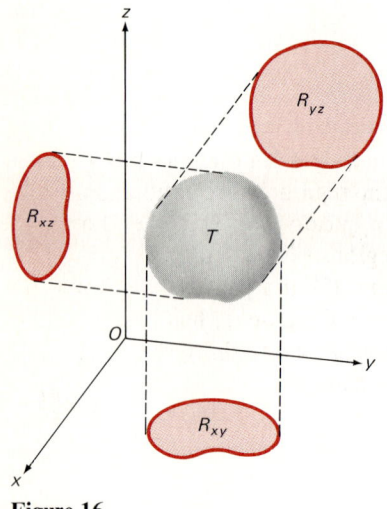

Figure 16

824   Chapter 14   Multiple Integration

x, y and z, into which T can be decomposed by drawing planes parallel to the coordinate planes. Just as in the case of normal plane regions, it can be shown that if a normal solid region T is partitioned by drawing planes parallel to the coordinate planes, the part of T lying in each cell of the partition (a rectangular box) is sure to have a volume.

Definition of the Triple Integral

We now define the triple integral of a function $f(x, y, z)$ of three variables over a normal three-dimensional region T. Let†

$$Q = \{(x, y, z): a \leq x \leq b, c \leq y \leq d, A \leq z \leq B\}$$

be any closed box (rectangular parallelepiped) with faces parallel to the coordinate planes which contains T (see Figure 17), and let the points x_j ($j = 0, 1, \ldots, J$), y_k ($k = 0, 1, \ldots, K$) and z_l ($l = 0, 1, \ldots, L$) be partitions of the intervals $[a, b]$, $[c, d]$ and $[A, B]$, with mesh sizes μ_x, μ_y and μ_z, respectively. This means that

$$a = x_0 < x_1 < x_2 < \cdots < x_{J-1} < x_J = b,$$
$$c = y_0 < y_1 < y_2 < \cdots < y_{K-1} < y_K = d,$$
$$A = z_0 < z_1 < z_2 < \cdots < z_{L-1} < z_L = B,$$

and

$$\mu_x = \max\{x_1 - x_0, x_2 - x_1, \ldots, x_J - x_{J-1}\},$$
$$\mu_y = \max\{y_1 - y_0, y_2 - y_1, \ldots, y_K - y_{K-1}\},$$
$$\mu_z = \max\{z_1 - z_0, z_2 - z_1, \ldots, z_L - z_{L-1}\}.$$

Then the three sets of planes

$$\begin{aligned} x = a, & \quad x = x_1, \quad x = x_2, \ldots, \quad x = x_{J-1}, \quad x = b, \\ y = c, & \quad y = y_1, \quad y = y_2, \ldots, \quad y = y_{K-1}, \quad y = d, \\ z = A, & \quad z = z_1, \quad z = z_2, \ldots, \quad z = z_{L-1}, \quad z = B, \end{aligned} \quad (2)$$

parallel to the coordinate planes, constitute a *partition* of the box Q; these planes divide Q into JKL subboxes and the region T into n nonempty subregions $T_1, T_2, \ldots, T_n$, where $n \leq JKL$ (typically $n < JKL$), as illustrated in the figure. In general, some of the subregions will be nonrectangular, with boundaries consisting of parts of the planes (2) and parts of the boundary of T, but since T is a normal region, every subregion T_i has a well-defined volume ΔV_i. Given any function $f(x, y, z)$ of three variables, let (p_i, q_i, r_i) be an *arbitrary* point in T_i, and form the *Riemann sum*

$$S = \sum_{i=1}^{n} f(p_i, q_i, r_i) \Delta V_i.$$

Suppose S approaches a finite limit as the quantity

$$\mu = \max\{\mu_x, \mu_y, \mu_z\},$$

known as the *mesh size* of the partition, approaches zero, *regardless of the choice of the numbers* x_j, y_k, z_l, p_i, q_i and r_i satisfying the stipulated conditions.

---

† It would be nice to write $e \leq z \leq f$ instead of $A \leq x \leq B$, but e suggests the base of the natural logarithms and f is already reserved for the integrand $f(x, y, z)$.

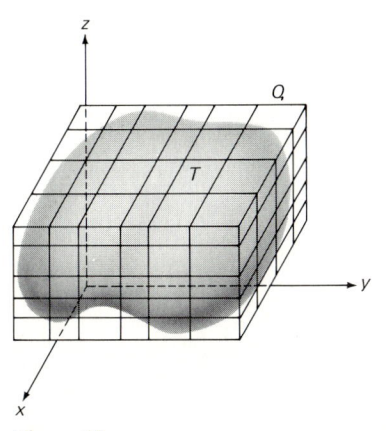

Figure 17

Then this limit is called the (*triple*) *integral of f over R*, denoted by

$$\iiint_T f(x, y, z)\, dV,$$

and the function f is said to be *integrable* on T, or over T. Thus

$$\iiint_T f(x, y, z)\, dV = \lim_{\mu \to 0} \sum_{i=1}^{n} f(p_i, q_i, r_i)\, \Delta V_i, \qquad (3)$$

where f is called the *integrand* and T the *region of integration* of the integral on the left.

Example 2 If V is the volume of T, then

$$V = \sum_{i=1}^{n} \Delta V_i$$

for every partition of the region by planes parallel to the coordinate planes. Therefore, choosing $f(x, y, z) \equiv 1$ in formula (3), we get

$$\iiint_T 1\, dV = \iiint_T dV = \lim_{\mu \to 0} \sum_{i=1}^{n} \Delta V_i = \lim_{\mu \to 0} V = V.$$

It follows that

$$V = \iiint_T dV. \qquad (4)$$

As you may have anticipated, Theorem 1, page 813, carries over to the case of triple integrals.

Theorem 4 *(Continuity implies integrability on a solid region). If the function $f(x, y, z)$ is continuous on a normal solid region T, then $f(x, y, z)$ is integrable on T.*

Triple integrals obey the same rules (i)–(iv), pages 813–814, as double integrals, with the appropriate slight changes:

(i) If f is integrable on T and c is any constant, then cf is also integrable on T, and

$$\iiint_T cf(x, y, z)\, dV = c \iiint_T f(x, y, z)\, dV.$$

(ii) If f and g are both integrable on T, then the sum $f + g$ is also integrable on T, and

$$\iiint_T [f(x, y, z) + g(x, y, z)]\, dV = \iiint_T f(x, y, z)\, dV + \iiint_T g(x, y, z)\, dV.$$

(iii) If f is integrable on T and $c \leq f(x, y, z) \leq C$, where c and C are constants, then

$$cV \leq \iiint_T f(x, y, z)\, dV \leq CV,$$

where V is the volume of T. In particular, if $f(x, y, z) \geq 0$, then

$$\iiint_T f(x, y, z)\, dV \geq 0.$$

(iv) If f is continuous on T and if T is decomposed into two normal subregions T_1 and T_2 with no interior points in common, then

$$\iiint_T f(x, y, z)\, dV = \iiint_{T_1} f(x, y, z)\, dV + \iiint_{T_2} f(x, y, z)\, dV.$$

Evaluation of Triple Integrals

To actually *evaluate* triple integrals, we need analogues of Theorems 2 and 2′ of the preceding section. Those theorems involve iterated integrals, where the first integration is between variable limits, which are functions of a single variable, and the second integration is between fixed limits; in fact, the fixed limits are the endpoints of the interval $[a, b]$ or $[c, d]$ obtained when the region of integration is projected onto the x-axis or y-axis. In the analogous theorems for triple integrals, the first integration is again between variable limits, which this time are functions of *two* variables, and the second integration consists of evaluating a *double* integral over the region obtained when the given three-dimensional region T is projected onto one of the coordinate planes.

Theorem 5 *(Evaluation of a triple integral over a z-simple region).* If $f(x, y, z)$ is continuous on the z-simple region T defined by $(x, y) \in R_{xy}$, $g_1(x, y) \leq z \leq g_2(x, y)$, where R_{xy} is a normal region in the xy-plane, then

$$\iiint_T f(x, y, z)\, dV = \iint_{R_{xy}} dA \int_{g_1(x,y)}^{g_2(x,y)} f(x, y, z)\, dz. \tag{5}$$

Theorem 5′ *(Evaluation of a triple integral over a y-simple region).* If $f(x, y, z)$ is continuous on the y-simple region T defined by $(x, z) \in R_{xz}$, $h_1(x, z) \leq y \leq h_2(x, z)$, where R_{xz} is a normal region in the xz-plane, then

$$\iiint_T f(x, y, z)\, dV = \iint_{R_{xz}} dA \int_{h_1(x,z)}^{h_2(x,z)} f(x, y, z)\, dy. \tag{5′}$$

Theorem 5″ *(Evaluation of a triple integral over an x-simple region).* If $f(x, y, z)$ is continuous on the x-simple region T defined by $(y, z) \in R_{yz}$, $k_1(y, z) \leq x \leq k_2(y, z)$, where R_{yz} is a normal region in the yz-plane, then

$$\iiint_T f(x, y, z)\, dV = \iint_{R_{yz}} dA \int_{k_1(y,z)}^{k_2(y,z)} f(x, y, z)\, dx. \tag{5″}$$

Make sure that you understand the differences and similarities between these theorems. We omit the proofs.

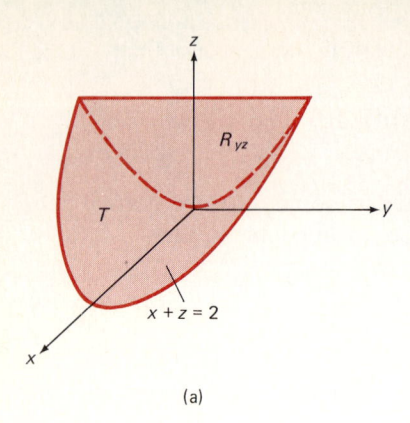

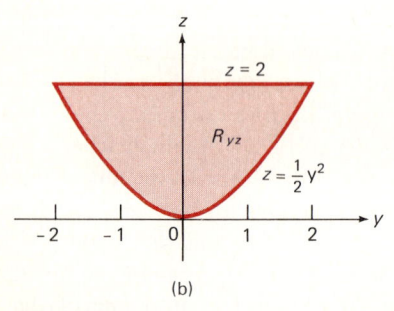

Figure 18

Suppose the region R_{xy} is simple, that is, both vertically and horizontally simple. Then the double integral in (5) can be evaluated in two different ways (recall Theorem 3, page 819). The same is true of the double integral in (5′) or (5″) if the region R_{xz} or R_{yz} is simple (with respect to both coordinates in its plane). The triple integral $\iiint_T f(x, y, z)\, dV$ is sometimes written in the form $\iiint_T f(x, y, z)\, dx\, dy\, dz$. What tells you that this expression is not an iterated integral?

Example 3 Use a triple integral to find the volume of the solid region T bounded by the parabolic cylinder $z = \frac{1}{2} y^2$, the plane $x = 0$ and the plane $x + z = 2$ [see Figure 18(a)].

Solution Since T is x-simple, y-simple and z-simple, we can find V by using any of the formulas (5), (5′) and (5″). Let's use (5″). The projection of T onto the yz-plane is the region R_{yz} shown in Figure 18(b), which is bounded by the parabola $z = \frac{1}{2} y^2$ and the line $z = 2$. Inspection of the figure shows that R_{yz} is vertically (and horizontally) simple, and is the set of points (y, z) such that $-2 \leq y \leq 2$, $\frac{1}{2} y^2 \leq z \leq 2$. Hence T is the set of points (x, y, z) such that $(y, z) \in R_{yz}$, $0 \leq x \leq 2 - z$, since x varies from the value 0 on the back face of T (the region R_{yz}) to the value $2 - z$ on the front face of T (part of the plane $x + z = 2$). Therefore, by formula (4) and formula (5″) with $f(x, y, z) \equiv 1$, $k_1(y, z) \equiv 0$ and $k_2(y, z) = 2 - z$,

$$V = \iiint_T dV = \iint_{R_{yz}} dA \int_0^{2-z} dx = \iint_{R_{yz}} (2 - z)\, dA$$

$$= \int_{-2}^2 dy \int_{y^2/2}^2 (2 - z)\, dz = \int_{-2}^2 \left[2z - \frac{1}{2} z^2 \right]_{y^2/2}^2 dy$$

$$= 2 \int_0^2 \left(2 - y^2 + \frac{1}{8} y^4 \right) dy,$$

where we use the evenness of the integrand. Thus, finally,

$$V = 2 \left[2y - \frac{1}{3} y^3 + \frac{1}{40} y^5 \right]_0^2 = 2 \left(4 - \frac{8}{3} + \frac{32}{40} \right) = \frac{64}{15}. \quad \square$$

Example 4 Use a triple integral to find the volume V of the solid region T bounded by the paraboloid of revolution $z = 1 + x^2 + y^2$ and the elliptic paraboloid $z = 4 - 2x^2 - 11y^2$ [see Figure 19(a)].

Solution Once again T is x-simple, y-simple and z-simple, so that V can be found by using any of the formulas (5), (5′) and (5″). This time let's use (5). Solving the equations of the paraboloids simultaneously, we find that $1 + x^2 + y^2 = 4 - 2x^2 - 11y^2$ or equivalently

$$x^2 + 4y^2 = 1. \tag{6}$$

The graph of equation (6) in space is an elliptic cylinder with its rulings parallel to the z-axis. Since the paraboloids intersect in a curve lying on this cylinder, the region R_{xy}, the projection of T onto the xy-plane, is bounded by the ellipse with the same equation, as shown in Figure 19(b). Examining the figure, we find that R_{xy} is horizontally (and vertically) simple, and is the set

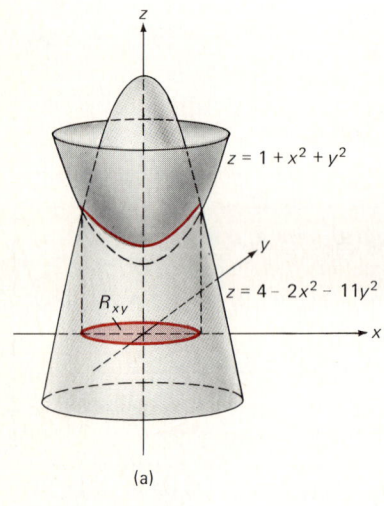

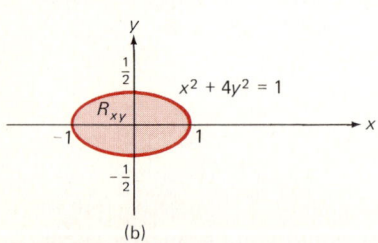

Figure 19

828 Chapter 14 Multiple Integration

of points (x, y) such that $-\frac{1}{2} \leq y \leq \frac{1}{2}$, $-\sqrt{1 - 4y^2} \leq x \leq \sqrt{1 - 4y^2}$. Therefore T is the set of points (x, y, z) such that $(x, y) \in R_{xy}$, $1 + x^2 + y^2 \leq z \leq 4 - 2x^2 - 11y^2$, since z varies from the value $1 + x^2 + y^2$ (depending on x and y) on the bottom paraboloid to the value $4 - 2x^2 - 11y^2$ on the top paraboloid. Hence by formula (5) with $f(x, y, z) \equiv 1$, $g_1(x, y) = 1 + x^2 + y^2$ and $g_2(x, y) = 4 - 2x^2 - 11y^2$,

$$V = \iiint_T dV = \iint_{R_{xy}} dA \int_{1+x^2+y^2}^{4-2x^2-11y^2} dz = \iint_{R_{xy}} (3 - 3x^2 - 12y^2) \, dA$$

$$= 3 \int_{-1/2}^{1/2} dy \int_{-\sqrt{1-4y^2}}^{\sqrt{1-4y^2}} (1 - x^2 - 4y^2) \, dx$$

$$= 3 \int_{-1/2}^{1/2} \left[(1 - 4y^2)x - \frac{1}{3}x^3 \right]_{x=-\sqrt{1-4y^2}}^{\sqrt{1-4y^2}} dy$$

$$= 4 \int_{-1/2}^{1/2} (1 - 4y^2)^{3/2} \, dy = 2 \int_{-\pi/2}^{\pi/2} \cos^4 t \, dt = 4 \int_0^{\pi/2} \cos^4 t \, dt,$$

after making the substitution $y = \frac{1}{2} \sin t$. Therefore, with the help of Problems 13 and 14, page 381, we finally obtain

$$V = 4 \left(\frac{1}{2} \frac{3}{4} \frac{\pi}{2} \right) = \frac{3\pi}{4}. \quad \square$$

Example 5 If the function $f(x, y, z)$ is a *product* $X(x)Y(y)Z(z)$ of three functions of a single variable, and if the region of integration T is a *box*, so that $T = \{(x, y, z): a \leq x \leq b, c \leq y \leq d, A \leq z \leq B\}$, then the triple integral of f over T simplifies to a product of three single integrals. In fact,

$$\iiint_T f(x, y, z) \, dV = \int_a^b dx \int_c^d dy \int_A^B X(x)Y(y)Z(z) \, dz$$

$$= \left(\int_a^b X(x) \, dx \right) \left(\int_c^d Y(y) \, dy \right) \left(\int_A^B Z(z) \, dz \right).$$

The details are left as an exercise. What is the two-dimensional version of this formula?

Calculation of Mass from Density

Next we discuss the use of triple integrals to determine the total mass of a solid object T from a knowledge of its *density function* $\rho(x, y, z)$. Let K be a cube of edge length d centered at a point (x, y, z) of T (see Figure 20), and let ΔV be the volume and Δm the mass of the part of T contained in K. Then, by definition,

$$\rho(x, y, z) = \lim_{d \to 0} \frac{\Delta m}{\Delta V}, \tag{7}$$

where $\rho(x, y, z) \geq 0$, since mass and density of mass are inherently nonnegative. Here we disregard the fact that macroscopic objects actually consist of huge numbers of submicroscopic particles and the empty space between them, and think of matter as being a "continuous medium." The reason why we are allowed to ignore the particulate nature of matter is that d can be

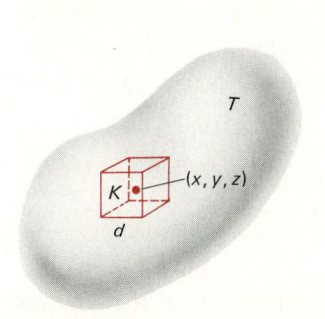

Figure 20

Section 14.2 Triple Integrals **829**

simultaneously very small compared to the size of macroscopic objects and very large compared to the size of atoms and molecules. Also it is physically evident that the same value of the density (7) would be obtained if instead of being a cube, K were a more general region containing (x, y, z) whose diameter (the maximum distance between points of K) is made to approach 0.

Now, as in the definition of the triple integral over T, let the region T (assumed to be normal) be divided into n subregions $T_1, T_2, \ldots, T_n$ by a set of planes parallel to the coordinate axes. Let ΔV_i and Δm_i be the volume and mass of T_i, and suppose T has a continuous density function $\rho(x, y, z)$. The function ρ changes its value only slightly in the subregion T_i, at least if T_i is small enough. Thus, since ρ is the limiting ratio of mass to volume, it seems a good approximation to write

$$\Delta m_i \approx \rho(p_i, q_i, r_i) \Delta V_i,$$

where (p_i, q_i, r_i) is *any* point in T_i. Moreover, if M is the total mass of T, then

$$M = \sum_{i=1}^{n} \Delta m_i,$$

so that M is approximated by

$$\sum_{i=1}^{n} \rho(p_i, q_i, r_i) \Delta V_i, \tag{8}$$

the sum of the approximate masses of the n subregions. It is reasonable to regard (8), which is a Riemann sum for ρ on T, as a good approximation to M, where the approximation gets better and better as the size of each and every subregion T_i gets smaller and smaller, that is, as the mesh size μ of the partition of T approaches zero. Hence we set M equal to the limit

$$M = \lim_{\mu \to 0} \sum_{i=1}^{n} \rho(p_i, q_i, r_i) \Delta V_i,$$

that is,

$$M = \iiint_T \rho(x, y, z)\, dV, \tag{9}$$

where the existence of the triple integral follows from Theorem 4 and the assumed continuity of ρ.

Example 6 Find the total mass M of the tetrahedron T in the first octant bounded by the plane $x + y + z = 1$ and the coordinate planes if its density function is

$$\rho(x, y, z) = \frac{16}{(1 + x + y + z)^3}.$$

Solution The projection of T onto the xy-plane is the triangle R in the first quadrant bounded by the line $x + y = 1$ and the positive coordinate axes (see Figure 21). Therefore, by formula (9) and Theorem 5,

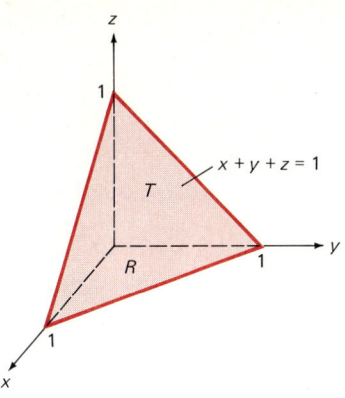

$$M = \iiint_T \frac{16}{(1+x+y+z)^3}\,dV = 16\iint_R dA \int_0^{1-x-y} \frac{dz}{(1+x+y+z)^3}$$

$$= 16\int_0^1 dx \int_0^{1-x} \left[-\frac{1}{2}\frac{1}{(1+x+y+z)^2}\right]_{z=0}^{1-x-y} dy$$

$$= 8\int_0^1 dx \int_0^{1-x}\left[\frac{1}{(1+x+y)^2} - \frac{1}{4}\right]dy$$

$$= 8\int_0^1 \left[-\frac{1}{1+x+y} - \frac{y}{4}\right]_{y=0}^{1-x} dx$$

$$= 8\int_0^1 \left(\frac{1}{x+1} + \frac{x-3}{4}\right)dx = 8\left[\ln(x+1) + \frac{1}{8}x^2 - \frac{3}{4}x\right]_0^1$$

$$= 8\ln 2 - 5 \approx 0.545. \quad \square$$

The Two-Dimensional Case

A thin plate of negligible thickness is called a *lamina*. We can use a *double* integral to find the total mass of a lamina, given its mass density $\rho(x, y)$. This is now a *surface* density, measured in units like grams per square centimeter, rather than a *volume* density, measured in units like grams per cubic centimeter. Virtually the same argument as used to derive formula (9) shows that the total mass M of the lamina is given by

$$M = \iint_R \rho(x, y)\,dA. \tag{9'}$$

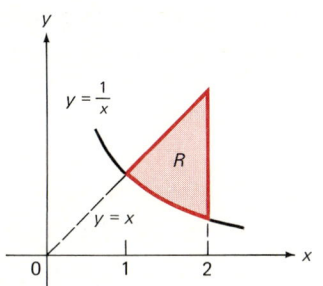

Example 7 A lamina is shaped like the region R bounded by the lines $x = 2$, $y = x$ and the hyperbola $xy = 1$, and its density function is

$$\rho(x, y) = \frac{x^2}{y^2}.$$

Find the total mass M of the lamina.

Solution The region R is vertically simple, and is defined by the inequalities $1 \le x \le 2$, $1/x \le y \le x$ (see Figure 22). Therefore, by formula (9'),

$$M = \iint_R \frac{x^2}{y^2}\,dA = \int_1^2 dx \int_{1/x}^x \frac{x^2}{y^2}\,dy = \int_1^2 \left[-\frac{x^2}{y}\right]_{y=1/x}^x dx$$

$$= \int_1^2 (x^3 - x)\,dx = \left[\frac{1}{4}x^4 - \frac{1}{2}x^2\right]_1^2 = \frac{9}{4}. \quad \square$$

Figure 22

Problems

Evaluate the given iterated integral.

1. $\int_0^3 \int_{-2}^0 \int_{-1}^1 dx\,dy\,dz$

2. $\int_0^1 \int_0^1 \int_0^1 (xy + xz + yz)\,dx\,dy\,dz$

3. $\int_0^1 dx \int_0^{x^2} dz \int_0^z (x+y+z)\,dy$

4. $\int_0^3 dy \int_0^y dx \int_1^e \frac{xy}{z}\,dz$

5. $\int_0^1 dx \int_x^1 dy \int_x^y xyz \, dz$

6. $\int_{-1}^1 dz \int_0^z dx \int_0^{x+z} x^2 y z^2 \, dy$

7. $\int_0^1 dy \int_0^1 dz \int_0^1 \dfrac{dx}{\sqrt{1+x+y+z}}$

8. $\int_0^\pi dz \int_0^{\pi/2} dy \int_1^2 x \cos y \sin z \, dx$

9. $\int_1^2 dz \int_0^{\ln z} dy \int_0^{\ln y} e^{x+y} \, dx$

10. $\int_0^1 dx \int_0^{\sqrt{1-x^2}} dy \int_0^{\sqrt{1-x^2-y^2}} \dfrac{dz}{\sqrt{1-x^2-y^2-z^2}}$

Evaluate the given triple integral.

11. $\iiint_T (x^2 + y^2 + z^2) \, dV$, where T is the box bounded by the planes $x = 1$, $x = 2$, $y = 0$, $y = 3$, $z = -1$ and $z = 1$

12. $\iiint_T xyz \, dV$, where T is the box bounded by the planes $x = 0$, $x = 2$, $y = -1$, $y = 3$, $z = 1$ and $z = 5$

13. $\iiint_T x \, dV$, where T is the prism in the first octant bounded by the coordinate planes and the planes $y = 5$ and $x + z = 2$

14. $\iiint_T y \, dV$, where T is the tetrahedron in the first octant bounded the plane $x + y + z = 1$ and the coordinate planes

15. $\iiint_T z^2 \, dV$, where T is the solid region bounded by the spheres $x^2 + y^2 + z^2 = 4$ and $x^2 + y^2 + z^2 = 4z$

16. $\iiint_T \dfrac{xy}{\sqrt{z}} \, dV$, where T is the solid region in the first octant bounded by the elliptic cone $z^2 = \tfrac{1}{4}x^2 + \tfrac{1}{9}y^2$ and the planes $x = 0$, $y = 0$ and $z = 1$

17. $\iiint_T \cos y \, dV$, where T is the solid region in the first octant bounded by the hyperbolic paraboloid $z = xy$, the plane $z = 0$ and the plane $x + y = \pi$.

18. $\iiint_T xe^{x+y+z} \, dV$, where T is the box bounded by the coordinate planes and the planes $x = 1$, $y = \ln 2$ and $z = \ln 3$

Use a triple integral to find the volume V of the specified solid region T.

19. T is bounded by the plane $6x + 2y + 3z = 12$ and the coordinate planes

20. T is bounded by the plane $z = 10 - 2x - 5y$ and the planes $x = 0$, $y = 1$, $y = x$ and $z = 0$

21. T is bounded by the paraboloids of revolution $z = x^2 + y^2$ and $2z = 1 - x^2 - y^2$

22. T is bounded by the paraboloid of revolution $2z = x^2 + y^2$ and the sphere $x^2 + y^2 + z^2 = 3$

23. T is bounded by the parabolic cylinders $z = 4 - x^2$ and $z = 2 + x^2$, and by the planes $y = -2$ and $y = 3$

24. T is bounded by the paraboloids $z = x^2 + y^2$ and $z = 2x^2 + y^2$, and by the planes $x = y$, $x = 3y$ and $y = 1$

Find the total mass of the cube T bounded by the planes $x = \pm 1$, $y = \pm 1$ and $z = \pm 1$ if the density function is

25. $\rho(x, y, z) = \cos(\pi x/2) \cos(\pi y/2) \cos(\pi z/2)$

26. $\rho(x, y, z) = x^2 + y^2 + z^2$

As in Example 6, let T be the tetrahedron in the first octant bounded by the plane $x + y + z = 1$ and the coordinate planes. Find the total mass of T if the density function is

27. $\rho(x, y, z) = x + y + z$

28. $\rho(x, y, z) = xyz$

29. Find the total mass of a lamina in the shape of the region between the parabolas $y = x^2$ and $x = y^2$ if its density function is $\rho(x, y) = xy$.

30. The density at a variable point P of a square lamina 1 ft on a side is proportional to the square of the distance from P to the center of the square (the point of intersection of the diagonals). Find the total mass of the lamina if the density is equal to 1 ounce per square inch at the corners of the square.

31. A solid T in the shape of a three-dimensional region R has electric charge density $\rho(x, y, z)$. Write a formula for Q, the total charge in T. Find Q if $\rho(x, y, z) = xy - 2yz$ and T is the box bounded by the planes $x = 0$, $x = 2$, $y = 1$, $y = 4$, $z = -1$ and $z = 2$. (Notice that unlike mass, electric charge can take negative values.)

32. The *mean value* of a function $f(x, y, z)$ over a three-dimensional region T of volume V is defined by

$$\dfrac{1}{V} \iiint_T f(x, y, z) \, dV$$

(this is the three-dimensional generalization of the expression

$$\dfrac{1}{b-a} \int_a^b f(x) \, dx$$

for the mean value of a function of one variable). Find the mean value of $f(x, y, z) = xy + xz + yz$ over the tetrahedron T bounded by the plane $x + y + z = 1$ and the coordinate planes. How do we know in advance that there is at least one point in T at which f takes its mean value over T? Find such a point.

14.3 The Center of Mass and Centroids

We begin by extending the ideas of Newtonian mechanics to a system of n particles $P_1, P_2, \ldots, P_n$ in space. Let $\mathbf{r}_i = \overrightarrow{OP_i}$ be the position vector of P_i with respect to a fixed origin O, and let m_i be the mass of P_i. Suppose P_i is acted on by an external force $\mathbf{F}_i$, that is, a force outside the n-particle system, and also by forces due to the other $n-1$ particles; for example, the particles exert gravitational forces on one another. Let $\mathbf{F}_{ij}$ ($j \neq i$) be the force exerted on the particle P_i by the particle P_j. Then, according to Newton's second law, the motion of P_i is governed by the vector differential equation

$$m_i \frac{d^2 \mathbf{r}_i}{dt^2} = \mathbf{F}_i + \sum_{j=1}^{n}{}' \mathbf{F}_{ij} \qquad (i = 1, 2, \ldots, n) \tag{1}$$

where t is the time and the symbol $\sum_{j=1}^{n}{}'$, with a prime, means that the sum is over all the subscripts j except i itself (the particle P_i does not exert a force on itself). For each particle P_i there is an equation of the form (1), and to get a differential equation governing the motion of the system of particles $P_1, P_2, \ldots, P_n$ as a whole, we add the equations (1) from $i = 1$ to n. This gives

$$\sum_{i=1}^{n} m_i \frac{d^2 \mathbf{r}_i}{dt^2} = \sum_{i=1}^{n} \mathbf{F}_i + \sum_{i,j=1}^{n}{}' \mathbf{F}_{ij}, \tag{2}$$

where the symbol $\sum_{i,j=1}^{n}{}'$, with a prime, means that the sum is over all pairs i, j such that $i \neq j$. Thus, for example,

$$\sum_{i,j=1}^{3}{}' \mathbf{F}_{ij} = \mathbf{F}_{12} + \mathbf{F}_{13} + \mathbf{F}_{21} + \mathbf{F}_{23} + \mathbf{F}_{31} + \mathbf{F}_{32}. \tag{3}$$

Equation (2) looks complicated, but it can be simplified at once by using *Newton's third law*, which states that "action equals reaction." Specifically, the force exerted on the jth particle by the ith particle is equal and opposite to the force exerted on the ith particle by the jth particle, that is,

$$\mathbf{F}_{ji} = -\mathbf{F}_{ij}.$$

It follows that the sum $\sum_{i,j=1}^{n}{}' \mathbf{F}_{ij}$ is equal to $\mathbf{0}$, since the terms cancel out in pairs; for example, rearranging terms in (3), we get

$$\sum_{i,j=1}^{3}{}' \mathbf{F}_{ij} = (\mathbf{F}_{12} + \mathbf{F}_{21}) + (\mathbf{F}_{13} + \mathbf{F}_{31}) + (\mathbf{F}_{23} + \mathbf{F}_{32})$$

$$= (\mathbf{F}_{12} - \mathbf{F}_{12}) + (\mathbf{F}_{13} - \mathbf{F}_{13}) + (\mathbf{F}_{23} - \mathbf{F}_{23}) = \mathbf{0}.$$

Thus equation (2) reduces to

$$\sum_{i=1}^{n} m_i \frac{d^2 \mathbf{r}_i}{dt^2} = \sum_{i=1}^{n} \mathbf{F}_i,$$

or equivalently

$$\frac{d^2}{dt^2} \sum_{i=1}^{n} m_i \mathbf{r}_i = \sum_{i=1}^{n} \mathbf{F}_i. \tag{4}$$

Suppose we introduce the vector

$$\bar{\mathbf{r}} = \frac{\sum_{i=1}^{n} m_i \mathbf{r}_i}{M}, \tag{5}$$

where

$$M = \sum_{i=1}^{n} m_i \tag{5'}$$

is the total mass of the system, namely the sum of the masses of all n particles. Then $\sum_{i=1}^{n} m_i \mathbf{r}_i = M\bar{\mathbf{r}}$, and equation (4) can be written in the concise form

$$M \frac{d^2 \bar{\mathbf{r}}}{dt^2} = \mathbf{F}, \tag{6}$$

where

$$\mathbf{F} = \sum_{i=1}^{n} \mathbf{F}_i$$

is the resultant of all the external forces acting on the separate particles P_1, $P_2, \ldots, P_n$. According to (6), regardless of the motion of the particles relative to one another, the system as a whole moves like a single particle of mass M and position vector $\bar{\mathbf{r}}$, acted on by the force $\mathbf{F}$. The point with position vector $\bar{\mathbf{r}}$ is called the *center of mass* of the system of n particles.

If the net external force $\mathbf{F}$ is zero, we can immediately integrate equation (6), obtaining first

$$\bar{\mathbf{v}} = \frac{d\bar{\mathbf{r}}}{dt} = \mathbf{c}_1$$

and then

$$\bar{\mathbf{r}} = \mathbf{c}_1 t + \mathbf{c}_2,$$

where $\mathbf{c}_1$ and $\mathbf{c}_2$ are vector constants of integration. If the center of mass has initial velocity zero, then $\bar{\mathbf{v}}|_{t=0} = \mathbf{0}$, so that $\mathbf{c}_1 = \mathbf{0}$ and $\bar{\mathbf{r}} = \mathbf{c}_2$, in which case the center of mass remains at rest. Even in this case, the center of mass remains an important concept. For example, it can be shown that a system of n particles attached to a thin sheet of rigid material of negligible weight will balance on a sharp vertical spike if the spike is placed directly under the center of mass of the system, but will otherwise tip over. By the same token, the balance point of a lamina in the shape of a plane region of mass density $\rho(x, y)$ is also its center of mass (as defined below). In engineering applications the external force acting on a structure (a bridge or building, say) is most often the force of gravity, and in this context the center of mass is usually called the *center of gravity*.

It is easy to express the coordinates of the center of mass of a system of particles in terms of the coordinates of the particles themselves. Introducing a system of rectangular coordinates x, y and z with O as the origin, let $\bar{\mathbf{r}} = (\bar{x}, \bar{y}, \bar{z})$ and

$$\mathbf{r}_i = (x_i, y_i, z_i) \qquad (i = 1, 2, \ldots, n).$$

Then, taking the components of both sides of (5) and using (5'), we find at once that

$$\bar{x} = \frac{\sum_{i=1}^{n} m_i x_i}{\sum_{i=1}^{n} m_i}, \qquad \bar{y} = \frac{\sum_{i=1}^{n} m_i y_i}{\sum_{i=1}^{n} m_i}, \qquad \bar{z} = \frac{\sum_{i=1}^{n} m_i z_i}{\sum_{i=1}^{n} m_i}. \qquad (7)$$

Example 1 A system of particles consists of three masses $m_1 = 3$, $m_2 = 4$ and $m_3 = 2$ at the points with position vectors $\mathbf{r}_1 = (2, -1, 3)$, $\mathbf{r}_2 = (5, 2, 4)$ and $\mathbf{r}_3 = (-2, 0, 1)$. Find its center of mass.

Solution Here $x_1 = 2$, $x_2 = 5$ and $x_3 = -2$, so that the first of the formulas (7) gives

$$\bar{x} = \frac{3(2) + 4(5) + 2(-2)}{3 + 4 + 2} = \frac{22}{9}.$$

Similarly,

$$\bar{y} = \frac{3(-1) + 4(2) + 2(0)}{3 + 4 + 2} = \frac{5}{9}$$

and

$$\bar{z} = \frac{3(3) + 4(4) + 2(1)}{3 + 4 + 2} = \frac{27}{9} = 3. \qquad \square$$

Moments of a System of Particles

The numerators of the expression (7) for the coordinates $\bar{x}$, $\bar{y}$ and $\bar{z}$ of the center of mass are called *moments* of the system S of particles $P_1, P_2, \ldots, P_n$. Specifically, since apart from their signs, the numbers x_i, y_i and z_i are the distances between P_i and the yz-, xz- and xy-planes, respectively, we call $\sum m_i x_i$ the *moment of S about the yz-plane*, $\sum m_i y_i$ the *moment of S about the xz-plane*, and $\sum m_i z_i$ the *moment of S about the xy-plane* (omitting limits of summation for brevity). If the particles all lie in a plane, taken to be the xy-plane, then apart from signs, x_i and y_i are the distances between P_i and the y- and x-axes, respectively, and we call $\sum m_i x_i$ the *moment of S about the y-axis* and $\sum m_i y_i$ the *moment of S about the x-axis*. Dividing these moments by the total mass $\sum m_i$ of the system S, we get back the coordinates of the center of mass ($\bar{x}$, $\bar{y}$ in the plane, $\bar{x}$, $\bar{y}$, $\bar{z}$ in space). The moments $\sum m_i x_i$, $\sum m_i y_i$ and $\sum m_i z_i$ are also called *first moments*, as opposed to the *second moments* or *moments of inertia*, to be introduced in Section 14.5.

A solid object can be regarded as a three-dimensional region T on which there is defined a continuous mass density $\rho(x, y, z)$, where $\rho(x, y, z) \equiv$ constant if the object is homogeneous. To find the center of mass of T, we partition T into a large number of subregions $T_1, T_2, \ldots, T_n$ of volumes $\Delta V_1, \Delta V_2, \ldots, \Delta V_n$, by drawing three sets of planes parallel to the coordinate planes, just as was done on page 825 in defining the triple integral. Let (x_i, y_i, z_i) be any point in T_i. Then T_i can be regarded as a particle of mass $\Delta m_i \approx \rho(x_i, y_i, z_i) \Delta V_i$ located at (x_i, y_i, z_i). Hence the center of mass of T is to a good approximation the same as that of the system of n particles T_1, $T_2, \ldots, T_n$. Let μ be the mesh size of the partition of T. Then the center of mass of T is defined as the limiting position of the center of mass of this system of particles as $\mu \to 0$. Therefore, with the help of (7), we find that the coordinates of the center of mass of T are given by the formulas

$$\bar{x} = \frac{\lim_{\mu \to 0} \sum_{i=1}^{n} x_i \Delta m_i}{\lim_{\mu \to 0} \sum_{i=1}^{n} \Delta m_i} = \frac{\lim_{\mu \to 0} \sum_{i=1}^{n} x_i \rho(x_i, y_i, z_i) \Delta V_i}{\lim_{\mu \to 0} \sum_{i=1}^{n} \rho(x_i, y_i, z_i) \Delta V_i},$$

$$\bar{y} = \frac{\lim_{\mu \to 0} \sum_{i=1}^{n} y_i \Delta m_i}{\lim_{\mu \to 0} \sum_{i=1}^{n} \Delta m_i} = \frac{\lim_{\mu \to 0} \sum_{i=1}^{n} y_i \rho(x_i, y_i, z_i) \Delta V_i}{\lim_{\mu \to 0} \sum_{i=1}^{n} \rho(x_i, y_i, z_i) \Delta V_i},$$

$$\bar{z} = \frac{\lim_{\mu \to 0} \sum_{i=1}^{n} z_i \Delta m_i}{\lim_{\mu \to 0} \sum_{i=1}^{n} \Delta m_i} = \frac{\lim_{\mu \to 0} \sum_{i=1}^{n} z_i \rho(x_i, y_i, z_i) \Delta V_i}{\lim_{\mu \to 0} \sum_{i=1}^{n} \rho(x_i, y_i, z_i) \Delta V_i}.$$

But each of the sums containing ρ is a Riemann sum, and approaches a triple integral over T as $\mu \to 0$ (the region T is assumed to be normal). In fact,

$$\lim_{\mu \to 0} \sum_{i=1}^{n} \rho(x_i, y_i, z_i) \Delta V_i = \iiint_{T} \rho(x, y, z) \, dV,$$

$$\lim_{\mu \to 0} \sum_{i=1}^{n} x_i \rho(x_i, y_i, z_i) \Delta V_i = \iiint_{T} x \rho(x, y, z) \, dV,$$

$$\lim_{\mu \to 0} \sum_{i=1}^{n} y_i \rho(x_i, y_i, z_i) \Delta V_i = \iiint_{T} y \rho(x, y, z) \, dV,$$

$$\lim_{\mu \to 0} \sum_{i=1}^{n} z_i \rho(x_i, y_i, z_i) \Delta V_i = \iiint_{T} z \rho(x, y, z) \, dV.$$

It follows that

$$\bar{x} = \frac{\iiint_{T} x\rho \, dV}{\iiint_{T} \rho \, dV}, \quad \bar{y} = \frac{\iiint_{T} y\rho \, dV}{\iiint_{T} \rho \, dV}, \quad \bar{z} = \frac{\iiint_{T} z\rho \, dV}{\iiint_{T} \rho \, dV}, \quad (8)$$

where for brevity we omit the arguments of the density function ρ. The integral $\iiint_{T} \rho \, dV$ in all three denominators is of course just the total mass M of the object (recall page 830). Thus we can write the formulas (8) more concisely as

$$\bar{x} = \frac{1}{M} \iiint_{T} x\rho \, dV, \quad \bar{y} = \frac{1}{M} \iiint_{T} y\rho \, dV, \quad \bar{z} = \frac{1}{M} \iiint_{T} z\rho \, dV. \quad (9)$$

Chapter 14 Multiple Integration

If the object is homogeneous, the density ρ has a constant value. Then the first of the formulas (8) reduces to

$$\bar{x} = \frac{\rho \iiint_T x\, dV}{\rho \iiint_T dV} = \frac{\iiint_T x\, dV}{\iiint_T dV},$$

and similarly

$$\bar{y} = \frac{\iiint_T y\, dV}{\iiint_T dV}, \quad \bar{z} = \frac{\iiint_T z\, dV}{\iiint_T dV}.$$

But $\iiint_T dV = V$, where V is the volume of T, and therefore

$$\bar{x} = \frac{1}{V}\iiint_T x\, dV, \quad \bar{y} = \frac{1}{V}\iiint_T y\, dV, \quad \bar{z} = \frac{1}{V}\iiint_T z\, dV \quad \text{(constant density)}.$$

(9′)

In this case the center of mass is called the *centroid*, of the solid region T, and is a purely geometric concept, quite independent of the physical idea of mass.

The Two-Dimensional Case

In the case of a thin plate or lamina, we have a plane region R and a two-dimensional density function $\rho(x, y)$, instead of a solid region T and a three-dimensional density function $\rho(x, y, z)$. Virtually the same arguments then show that the center of mass of the lamina has coordinates

$$\bar{x} = \frac{1}{M}\iint_R x\rho\, dA, \quad \bar{y} = \frac{1}{M}\iint_R y\rho\, dA, \tag{10}$$

where $M = \iint_R \rho\, dA$ is the total mass of the lamina. If the lamina is homogeneous, then $\rho(x, y) \equiv $ constant, and these formulas reduce to

$$\bar{x} = \frac{1}{A}\iint_R x\, dA, \quad \bar{y} = \frac{1}{A}\iint_A y\, dA \quad \text{(constant density)}, \tag{10′}$$

where A is the area of R. In this case the center of mass is again called the *centroid*, of the plane region R, and is a purely geometric concept.

Moments of a Continuous Mass Distribution

For a solid T or lamina R, as in the case of a system of particles, we again have various quantities known as *moments*, but they are now integrals instead of sums. Specifically, $M_{yz} = \iiint_T x\rho\, dV$ is the moment of T about the yz-plane, $M_{xz} = \iiint_T y\rho\, dV$ is the moment of T about the xz-plane, and $M_{xy} = \iiint_T z\rho\, dV$ is the moment of T about the xy-plane. Likewise $M_y = \iint_R x\rho\, dA$ is the moment of R about the y-axis, and $M_x = \iint_R y\rho \, dA$ is the moment of R about the x-axis. In terms of moments, formulas (9) and (10) become

$$\bar{x} = \frac{M_{yz}}{M}, \qquad \bar{y} = \frac{M_{xz}}{M}, \qquad \bar{z} = \frac{M_{xy}}{M}$$

and

$$\bar{x} = \frac{M_y}{M}, \qquad \bar{y} = \frac{M_x}{M},$$

respectively.

Example 2 A lamina is shaped like the circular sector R in the first quadrant bounded by the coordinate axes and an arc of the unit circle $x^2 + y^2 = 1$ (see Figure 23). Find the center of mass $(\bar{x}, \bar{y})$ of the lamina if its density function is $\rho(x, y) = x^2 y$.

Solution The total mass of the lamina is

$$M = \iint_R \rho \, dA = \iint_R x^2 y \, dA = \int_0^1 dx \int_0^{\sqrt{1-x^2}} x^2 y \, dy = \int_0^1 \left[\frac{1}{2} x^2 y^2\right]_{y=0}^{\sqrt{1-x^2}} dy$$

$$= \frac{1}{2} \int_0^1 x^2(1 - x^2) \, dx = \frac{1}{2}\left[\frac{1}{3} x^3 - \frac{1}{5} x^5\right]_0^1 = \frac{1}{15}.$$

The moment about the x-axis is

$$M_x = \iint_R y\rho \, dA = \iint_R x^2 y^2 \, dA = \int_0^1 dx \int_0^{\sqrt{1-x^2}} x^2 y^2 \, dy$$

$$= \int_0^1 \left[\frac{1}{3} x^2 y^3\right]_{y=0}^{\sqrt{1-x^2}} dx = \frac{1}{3} \int_0^1 x^2 (1-x^2)^{3/2} \, dx$$

$$= \frac{1}{3} \int_0^{\pi/2} \sin^2 t \cos^4 t \, dt = \frac{1}{3}\left[\frac{1}{16} t - \frac{1}{64} \sin 4t + \frac{1}{48} \sin^3 2t\right]_0^{\pi/2} = \frac{\pi}{96},$$

with the help of the substitution $x = \sin t$ and Example 3, page 383. The other moment, about the y-axis, is

$$M_y = \iint_R x\rho \, dA = \iint_R x^3 y \, dA = \int_0^1 dx \int_0^{\sqrt{1-x^2}} x^3 y \, dy = \int_0^1 \left[\frac{1}{2} x^3 y^2\right]_{y=0}^{\sqrt{1-x^2}} dx$$

$$= \frac{1}{2} \int_0^1 x^3(1 - x^2) \, dx = \frac{1}{2}\left[\frac{1}{4} x^4 - \frac{1}{6} x^6\right]_0^1 = \frac{1}{24}.$$

Consequently

$$\bar{x} = \frac{M_y}{M} = \frac{\frac{1}{24}}{\frac{1}{15}} = \frac{5}{8} = 0.625, \qquad \bar{y} = \frac{M_x}{M} = \frac{\frac{1}{96}\pi}{\frac{1}{15}} = \frac{5\pi}{32} \approx 0.491,$$

so that $(\frac{5}{8}, \frac{5}{32}\pi)$ is the center of mass of the lamina, and the lamina will balance on a vertical spike placed directly under this point. Note that $\bar{y} < \bar{x}$ even though the region R is symmetric about the line $y = x$. This is because there is more mass near the x-axis than near the y-axis (why?). ☐

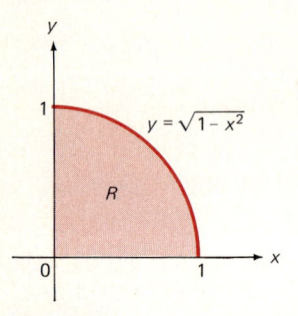

Figure 23

Example 3 Let R be the same region as in Example 2. Find the centroid of R.

Solution Since the density function is now constant, the symmetry of R about the line $y = x$ guarantees that $\bar{x} = \bar{y}$ (see Problem 4). By the first of the formulas (10'),

$$\bar{x} = \frac{1}{A} \iint_R x \, dA = \frac{4}{\pi} \iint_R x \, dA,$$

since A, the area of the region R, is one fourth the area π enclosed by the unit circle. Evaluating the double integral, we get

$$\iint_R x \, dA = \int_0^1 dx \int_0^{\sqrt{1-x^2}} x \, dy = \int_0^1 \sqrt{1-x^2} \, x \, dx = \left[-\frac{1}{3}(1-x^2)^{3/2} \right]_0^1 = \frac{1}{3}.$$

Therefore

$$\bar{x} = \bar{y} = \frac{4}{3\pi} \approx 0.424,$$

and the centroid is $(4/3\pi, 4/3\pi)$. □

Example 4 Find the center of mass of a solid right circular cone T of height h and base radius a if the density at each point of T is proportional to the distance between the point and the base of T.

Solution Choosing the z-axis along the axis of T with the base of T in the xy-plane, as in Figure 24, we have $\rho(x, y, z) = cz$, where c is a positive constant. Since every cross section of the cone T by a plane parallel to the xy-plane has its center of mass on the z-axis, the same must be true of T itself (explain further). Therefore $\bar{x} = \bar{y} = 0$, and we need only calculate

$$\bar{z} = \frac{\iiint_T z\rho \, dV}{\iiint_T \rho \, dV} = \frac{c \iiint_T z^2 \, dV}{c \iiint_T z \, dV} = \frac{\iiint_T z^2 \, dV}{\iiint_T z \, dV} \quad (11)$$

(since the constant c cancels out, its value has no effect on the answer). The easiest way to evaluate the triple integral is to use the method of cross sections. Specifically, let $A(z)$ be the area of the cross section of T at z, that is, the area of the circular disk of radius $r = r(z)$ in which the plane through the point $(0, 0, z)$ parallel to the xy-plane intersects T. By similar triangles,

$$\frac{r}{a} = \frac{h-z}{h},$$

so that

$$A(z) = \pi r^2 = \frac{\pi a^2}{h^2}(h-z)^2. \quad (12)$$

Then, just as the method of cross sections tells us that the volume of the cone T is equal to

$$V = \iiint_T dV = \int_0^h A(z) \, dz$$

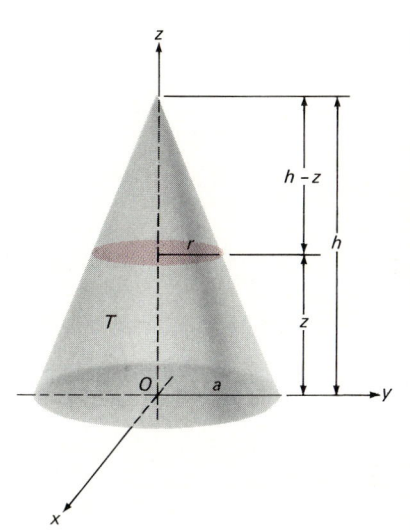

Figure 24

(check that this gives $V = \frac{1}{3}\pi a^2 h$), the same method tells us that the integral over T of a function $f(z)$ which depends only on z is equal to

$$\iiint_T f(z)\,dV = \int_0^h f(z)A(z)\,dz. \tag{13}$$

To see this more formally, observe that T is an x-simple region, and in fact

$$T = \{(x, y, z): (y, z) \in R_{yz}, k_1(y, z) \le x \le k_2(y, z)\},$$

where $R_{yz} = \{(y, z): 0 \le z \le h, g_1(z) \le y \le g_2(z)\}$; the functions g_1 and g_2 are continuous on $[0, h]$, while k_1 and k_2 are continuous on R_{yz} (we need not specify these functions, but $g_1 = -g_2$ and $k_1 = -k_2$ because of the symmetry). Therefore, by Theorem 5″, page 827,

$$\iiint_T f(z)\,dV = \iint_{R_{yz}} dA \int_{k_1(y,z)}^{k_2(y,z)} f(z)\,dx$$

$$= \int_0^h f(z)\left(\int_{g_1(z)}^{g_2(z)} [k_2(y, z) - k_1(y, z)]\,dy\right) dz. \tag{14}$$

But

$$\int_{g_1(z)}^{g_2(z)} [k_2(y, z) - k_1(y, z)]\,dy = A(z),$$

where $A(z)$ is the area of the cross section of T at z, and hence (14) reduces to (13), as anticipated.

Returning to the calculation of $\bar{z}$, we now deduce from (12) and (13) that

$$\iiint_T z\,dV = \frac{\pi a^2}{h^2}\int_0^h z(h-z)^2\,dz = \frac{\pi a^2}{h^2}\int_0^h (h^2 z - 2hz^2 + z^3)\,dz$$

$$= \frac{\pi a^2}{h^2}\left[\frac{1}{2}h^2 z^2 - \frac{2}{3}hz^3 + \frac{1}{4}z^4\right]_0^h = \frac{1}{12}\pi a^2 h^2,$$

$$\iiint_T z^2\,dV = \frac{\pi a^2}{h^2}\int_0^h z^2(h-z)^2\,dz = \frac{\pi a^2}{h^2}\int_0^h (h^2 z^2 - 2hz^3 + z^4)\,dz$$

$$= \frac{\pi a^2}{h^2}\left[\frac{1}{3}h^2 z^3 - \frac{1}{2}hz^4 + \frac{1}{5}z^5\right]_0^h = \frac{1}{30}\pi a^2 h^3,$$

and then (11) gives

$$\bar{z} = \frac{\frac{1}{30}\pi a^2 h^3}{\frac{1}{12}\pi a^2 h^2} = \frac{2}{5}h.$$

Thus $(0, 0, \frac{2}{5}h)$ is the center of mass of the cone T. ∎

Example 5 Let T be the same solid cone as in Example 4, but with constant density. Find the centroid of T.

Solution As before $\bar{x} = \bar{y} = 0$ by symmetry, but now

$$\bar{z} = \frac{1}{V} \iiint_T z \, dV,$$

by the third of the formulas (9′). The volume of T is $V = \frac{1}{3}\pi a^2 h$, and

$$\iiint_T z \, dV = \frac{1}{12} \pi a^2 h^2,$$

as just shown. Therefore

$$\bar{z} = \frac{\frac{1}{12}\pi a^2 h^2}{\frac{1}{3}\pi a^2 h} = \frac{1}{4} h,$$

and the centroid is $(0, 0, \frac{1}{4}h)$. Why is the centroid lower than the center of mass found in Example 4? □

Problems

1. A system of particles consists of four masses $m_1 = 6$, $m_2 = 1$, $m_3 = 2$ and $m_4 = 5$ with position vectors $\mathbf{r}_1 = (0, 3, 4)$, $\mathbf{r}_2 = (-1, 0, 6)$, $\mathbf{r}_3 = (2, -1, 1)$ and $\mathbf{r}_4 = (5, 8, 0)$. Find the moments M_{xy}, M_{xz} and M_{yz} of the system about the coordinate planes and also its center of mass $(\bar{x}, \bar{y}, \bar{z})$.

2. Let $\bar{\mathbf{r}}_1$ and $\bar{\mathbf{r}}_2$ be the position vectors of the centers of mass of two systems of particles S_1 and S_2. Show that the position vector of the system S obtained by combining S_1 and S_2 is $\bar{\mathbf{r}} = (M_1 \mathbf{r}_1 + M_2 \mathbf{r}_2)/M$, where M_1, M_2 and M are the total masses of the systems S_1, S_2 and S.

3. A child of weight 50 lb walks from one end to the other of a smooth plank of length 12 ft and weight 25 lb lying on a frozen pond. What happens to the plank? The ice is slippery, but there is traction between the child's shoes and the plank.

4. Show that if a solid region T is symmetric about a plane Π, then the centroid of T lies on Π. Show that if a plane region R is symmetric about a line L, then the centroid of R lies on L.

5. Suppose a plane region R of area A is decomposed into two subregions R_1 and R_2 of areas A_1 and A_2 with no interior points in common, as in Figure 8, page 814. Show that if the centroids of R, R_1 and R_2 are $(\bar{x}, \bar{y})$, $(\bar{x}_1, \bar{y}_1)$ and $(\bar{x}_2, \bar{y}_2)$, then

$$\bar{x} = \frac{1}{A}(A_1 \bar{x}_1 + A_2 \bar{x}_2), \quad \bar{y} = \frac{1}{A}(A_1 \bar{y}_1 + A_2 \bar{y}_2).$$

Show that if R is a lamina of variable density, the coordinates of the center of mass of the lamina are given by the same formulas with A, A_1 and A_2 replaced by M, M_1 and M_2, the masses of R, R_1 and R_2.

6. State and prove the analogue of Problem 5 for a solid region and for a solid object of variable density.

7. Let R be the plane region bounded by the lines $x = a$ and $x = b$, the x-axis and the curve $y = f(x)$ ($a \le x \le b$), where f is continuous and nonnegative on $[a, b]$. Show that the centroid of R is the point with coordinates

$$\bar{x} = \frac{1}{A} \int_a^b x f(x) \, dx, \quad \bar{y} = \frac{1}{2A} \int_a^b [f(x)]^2 \, dx, \quad \text{(i)}$$

where $A = \int_a^b f(x) \, dx$ is the area of R.

8. Solve Example 2 if the density function is $\rho(x, y) = x^2 + y^2$.

Find the centroid of the plane region R bounded by

9. The parabolas $y = x^2$ and $x = y^2$
10. The curve $y = 1 - x^3$ and the coordinate axes
11. The ellipse $(x^2/a^2) + (y^2/b^2) = 1$ and the positive coordinate axes
12. The part of the curve $y = \cos x$ from $x = -\pi/2$ to $x = \pi/2$ and the x-axis
13. The circles $x^2 + y^2 = 4$ and $(x - 1)^2 + y^2 = 1$
14. The curve $y = \sin x$ and the line $y = 2x/\pi$
15. The curve $\sqrt{x} + \sqrt{y} = 1$ and the coordinate axes
16. The curve $x^{2/3} + y^{2/3} = 1$ and the positive coordinate axes

17. Use an improper integral to find the centroid of the unbounded region R under the curve $y = e^x$ in the second quadrant.

18. Find the first moment of a circular disk of radius a about any of its tangent lines.

19. A cubical box of edge length 2 ft has no top. Where is its centroid?

20. Show that the force F on a submerged vertical plate R is given by $F = \delta A h$, where δ is the weight density of the fluid, A is the area of R, and h is the depth of the centroid of R below the surface of the fluid.
Hint. Recall the related discussion in Section 8.7.

Find the center of mass of the solid bounded by the coordinate planes and the planes $x = 1$, $y = 2$ and $y + z = 4$ if the density function is

21. $\rho(x, y, z) \equiv 16$
22. $\rho(x, y, z) = x$
23. $\rho(x, y, z) = xy$
24. $\rho(x, y, z) = xyz$

Find the centroid of the solid region T bounded by

25. The parabolic cylinder $z = \frac{1}{2}y^2$, the plane $x = 0$ and the plane $x + z = 2$ (see Example 3, page 828)
26. The plane $x + y + z = 1$ and the coordinate planes
27. The spheres $x^2 + y^2 + z^2 = 4$ and $x^2 + y^2 + z^2 = 4z$ (see Problem 15, page 832)
28. The paraboloid of revolution $2z = x^2 + y^2$ and the sphere $x^2 + y^2 + z^2 = 3$ (see Problem 22, page 832)
29. The ellipsoid $(x^2/a^2) + (y^2/b^2) + (z^2/c^2) = 1$ and the coordinate planes in the first octant
30. The hyperbolic paraboloid $z = xy$ and the planes $x = 2$ and $y = 3$ in the first octant

31. Find the center of mass of the solid T in Problem 27 if the density function is $\rho(x, y, z) = z$.

32. A tank in the shape of a solid region T of volume V is filled with fluid of density ρ. Show that the work W required to pump all the fluid to the top of the tank is the same as that required to lift an "equivalent particle" of mass ρV from the centroid of T to the highest point of T.
Hint. Recall Example 2, page 472.

14.4 More About Centroids; The Theorems of Pappus

In the preceding section we showed how to find the center of mass of a three-dimensional object or of a plane lamina. A related problem is that of finding the center of mass of a wire of variable density. Suppose the wire is in the shape of a plane curve C with parametric equations

$$x = x(t), \quad y = y(t) \quad (a \le t \le b),$$

where C has no more than a finite number of self-intersections. We assume that C is *smooth*, which means that the functions $x(t)$ and $y(t)$ have continuous derivatives $x'(t)$ and $y'(t)$ satisfying the condition $[x'(t)]^2 + [y'(t)]^2 \ne 0$. Then C is rectifiable, with length

$$L = \int_a^b \sqrt{[x'(t)]^2 + [y'(t)]^2} \, dt$$

(see page 461), and every arc of C is also rectifiable. Let γ be any arc of C containing the point (x, y) of C (see Figure 25), and let Δs be the length and Δm the mass of γ. Then, by definition,

$$\rho(x, y) = \lim_{\Delta s \to 0} \frac{\Delta m}{\Delta s}$$

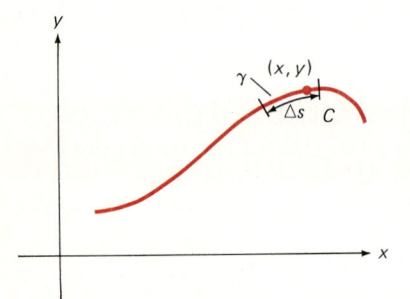

Figure 25

is the *density function* of C at (x, y), measured in units like grams per centimeter. Our aim is to determine the total mass of the wire C and its center of mass from a knowledge of the function $\rho(x, y)$, assumed to be continuous.

To this end, let the points t_i $(i = 0, 1, \ldots, n)$ be a partition of the interval $[a, b]$ of mesh size μ, so that $a = t_0 < t_1 < t_2 < \cdots < t_{n-1} < t_n = b$ and $\mu = \max \{\Delta t_1, \Delta t_2, \ldots, \Delta t_n\}$, where $\Delta t_i = t_i - t_{i-1}$ is the length of the ith subinterval $[t_{i-1}, t_i]$. Then the corresponding points

$$P_i = (x(t_i), y(t_i)) \quad (i = 0, 1, \ldots, n)$$

of the wire C divide it into n arcs

$$\gamma_i = \widehat{P_{i-1} P_i} \quad (i = 1, 2, \ldots, n)$$

(see Figure 26), where γ_i is of length

$$\Delta s_i = \int_{t_{i-1}}^{t_i} \sqrt{[x'(t)]^2 + [y'(t)]^2}\, dt.$$

By the mean value theorem for integrals (Theorem 3, page 245), there is a point τ_i in $[t_{i-1}, t_i]$ such that

$$\Delta s_i = \sqrt{[x'(\tau_i)]^2 + [y'(\tau_i)]^2}\, \Delta t_i.$$

Let

$$Q_i = (x_i, y_i) = (x(\tau_i), y(\tau_i)) \qquad (i = 1, 2, \ldots, n)$$

be the point corresponding to the parameter value τ_i. Then the arc γ_i can be regarded as a particle of mass

$$\Delta m_i \approx \rho(x(\tau_i), y(\tau_i))\, \Delta s_i = \rho(x(\tau_i), y(\tau_i))\sqrt{[x'(\tau_i)]^2 + [y'(\tau_i)]^2}\, \Delta t_i,$$

located at the point Q_i. Thus the center of mass of C is to a good approximation the same as that of the system of particles $\gamma_1, \gamma_2, \ldots, \gamma_n$. Therefore we define the center of mass $(\bar{x}, \bar{y})$ of C as the limiting position of the center of mass of this system of particles as $\mu \to 0$, that is,

$$\bar{x} = \frac{\lim\limits_{\mu \to 0} \sum\limits_{i=1}^{n} x_i\, \Delta m_i}{\lim\limits_{\mu \to 0} \sum\limits_{i=1}^{n} \Delta m_i} = \frac{\lim\limits_{\mu \to 0} \sum\limits_{i=1}^{n} x(\tau_i)\rho(x(\tau_i), y(\tau_i))\sqrt{[x'(\tau_i)]^2 + [y'(\tau_i)]^2}\, \Delta t_i}{\lim\limits_{\mu \to 0} \sum\limits_{i=1}^{n} \rho(x(\tau_i), y(\tau_i))\sqrt{[x'(\tau_i)]^2 + [y'(\tau_i)]^2}\, \Delta t_i},$$

$$\bar{y} = \frac{\lim\limits_{\mu \to 0} \sum\limits_{i=1}^{n} y_i\, \Delta m_i}{\lim\limits_{\mu \to 0} \sum\limits_{i=1}^{n} \Delta m_i} = \frac{\lim\limits_{\mu \to 0} \sum\limits_{i=1}^{n} y(\tau_i)\rho(x(\tau_i), y(\tau_i))\sqrt{[x'(\tau_i)]^2 + [y'(\tau_i)]^2}\, \Delta t_i}{\lim\limits_{\mu \to 0} \sum\limits_{i=1}^{n} \rho(x(\tau_i), y(\tau_i))\sqrt{[x'(\tau_i)]^2 + [y'(\tau_i)]^2}\, \Delta t_i}.$$

Each of the sums containing ρ is a Riemann sum approaching a definite integral as $\mu \to 0$. In fact,

$$\lim_{\mu \to 0} \sum_{i=1}^{n} \rho(x(\tau_i), y(\tau_i))\sqrt{[x'(\tau_i)]^2 + [y'(\tau_i)]^2}\, \Delta t_i$$
$$= \int_a^b \rho(x(t), y(t))\sqrt{[x'(t)]^2 + [y'(t)]^2}\, dt,$$

$$\lim_{\mu \to 0} \sum_{i=1}^{n} x(\tau_i)\rho(x(\tau_i), y(\tau_i))\sqrt{[x'(\tau_i)]^2 + [y'(\tau_i)]^2}\, \Delta t_i$$
$$= \int_a^b x(t)\rho(x(t), y(t))\sqrt{[x'(t)]^2 + [y'(t)]^2}\, dt,$$

$$\lim_{\mu \to 0} \sum_{i=1}^{n} y(\tau_i)\rho(x(\tau_i), y(\tau_i))\sqrt{[x'(\tau_i)]^2 + [y'(\tau_i)]^2}\, \Delta t_i$$
$$= \int_a^b y(t)\rho(x(t), y(t))\sqrt{[x'(t)]^2 + [y'(t)]^2}\, dt.$$

These three integrals are abbreviated by

$$\int_C \rho(x, y)\, ds, \qquad \int_C x\rho(x, y)\, ds, \qquad \int_C y\rho(x, y)\, ds,$$

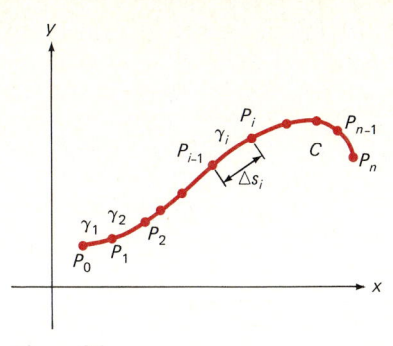

Figure 26

Section 14.4 More About Centroids; The Theorems of Pappus

and are traditionally called *line integrals*, or more exactly, line integrals along the curve C *with respect to the arc length s.*† To understand the notation, bear in mind that

$$ds = \sqrt{[x'(t)]^2 + [y'(t)]^2}\, dt$$

is the differential of the arc length function

$$s = s(t) = \int_a^t \sqrt{[x'(u)]^2 + [y'(u)]^2}\, du,$$

giving the length of the arc of C with fixed initial point $P(a) = (x(a), y(a))$ and variable final point $P(t) = (x(t), y(t))$. In terms of line integrals, the expressions for the coordinates $\bar{x}$ and $\bar{y}$ of the center of mass of a curved wire C with density function $\rho = \rho(x, y)$ take the concise form

$$\bar{x} = \frac{\int_C x\rho\, ds}{\int_C \rho\, ds}, \qquad \bar{y} = \frac{\int_C y\rho\, ds}{\int_C \rho\, ds}. \tag{1}$$

The integral $\int_C \rho\, ds$ in both denominators is of course just the total mass M of the wire. Hence we can write the formulas (1) more concisely as

$$\bar{x} = \frac{1}{M}\int_C x\rho\, ds, \qquad \bar{y} = \frac{1}{M}\int_C y\rho\, ds.$$

If the wire is homogeneous, the density ρ has a constant value. Then the formulas (1) simplify at once to

$$\bar{x} = \frac{\int_C x\, ds}{\int_C ds}, \qquad \bar{y} = \frac{\int_C y\, ds}{\int_C ds}.$$

But $\int_C ds = L$, where L is the length of C, and therefore

$$\bar{x} = \frac{1}{L}\int_C x\, ds, \qquad \bar{y} = \frac{1}{L}\int_C y\, ds \qquad \text{(constant density)}. \tag{2}$$

In this case the center of mass is called the *centroid*, and is a purely geometric concept. Thus we talk about the centroid of the *curve C*, which need not be thought of as a wire with mass.

Example 1 Let C be the curve

$$x = \cos t, \qquad y = \sin t \qquad (0 \leq t \leq \pi/2),$$

that is, the arc of the unit circle in the first quadrant. Find the center of mass of a wire shaped like C if its density function is $\rho(x, y) = x$.

Solution Here we have

$$ds = \sqrt{[x'(t)]^2 + [y'(t)]^2}\, dt = \sqrt{(-\sin t)^2 + (\cos t)^2}\, dt = dt,$$

---

† A better name would be "curve integrals," that is, integrals along curves. There is another kind of line integral, *with respect to the coordinates x and y*, to be introduced in Section 15.1.

844   Chapter 14   Multiple Integration

and hence, by (1),

$$\bar{x} = \frac{\int_C x^2\, ds}{\int_C x\, ds} = \frac{\int_0^{\pi/2} \cos^2 t\, dt}{\int_0^{\pi/2} \cos t\, dt}, \quad \bar{y} = \frac{\int_C yx\, ds}{\int_C x\, ds} = \frac{\int_0^{\pi/2} \sin t \cos t\, dt}{\int_0^{\pi/2} \cos t\, dt}.$$

Evaluating the integrals, we find that

$$\int_0^{\pi/2} \cos t\, dt = \sin t \Big|_0^{\pi/2} = 1,$$

$$\int_0^{\pi/2} \sin t \cos t\, dt = \frac{1}{2} \sin^2 t \Big|_0^{\pi/2} = \frac{1}{2},$$

$$\int_0^{\pi/2} \cos^2 t\, dt = \int_0^{\pi/2} \frac{1}{2}(1 + \cos 2t)\, dt = \left[\frac{1}{2} t + \frac{1}{4} \sin 2t\right]_0^{\pi/2} = \frac{\pi}{4},$$

and therefore

$$\bar{x} = \frac{\pi}{4}, \quad \bar{y} = \frac{1}{2}.$$

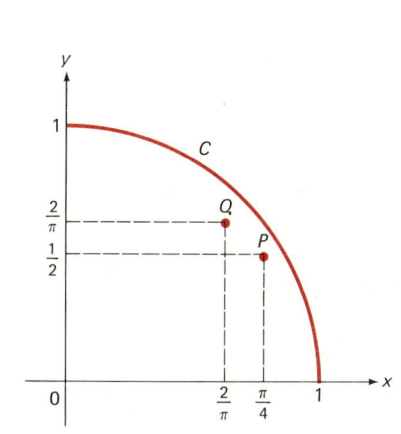

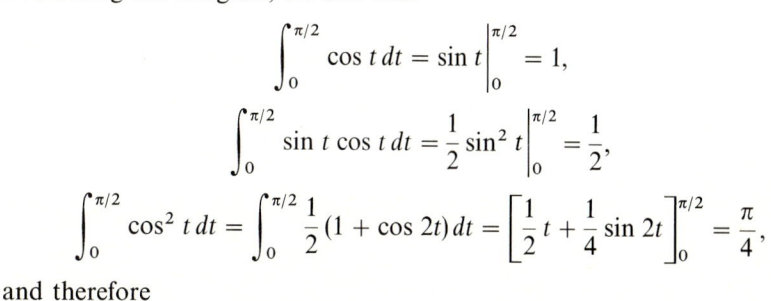

Thus the center of mass of the wire is the point $P = (\pi/4, 1/2)$ shown in Figure 27. Notice that the center of mass does not lie on the wire (it seldom does). Give a physical reason why $\bar{x} > \bar{y}$. ☐

Figure 27

Example 2 Let C be the same circular arc as in the preceding example. Find the centroid of C.

Solution The arc is of length $\pi/2$, and hence by (2),

$$\bar{x} = \frac{1}{L} \int_C x\, ds = \frac{2}{\pi} \int_0^{\pi/2} \cos t\, dt = \frac{2}{\pi} \sin t \Big|_0^{\pi/2} = \frac{2}{\pi},$$

$$\bar{y} = \frac{1}{L} \int_C y\, ds = \frac{2}{\pi} \int_0^{\pi/2} \sin t\, dt = -\frac{2}{\pi} \cos t \Big|_0^{\pi/2} = \frac{2}{\pi}.$$

Thus the centroid of the arc is the point $Q = (2/\pi, 2/\pi)$ shown in Figure 27. Why is the fact that $\bar{x} = \bar{y}$ to be expected? ☐

If C is a *space* curve with parametric equations

$$x = x(t), \quad y = y(t), \quad z = z(t) \quad (a \leq t \leq b),$$

formulas (1) and (2) continue to apply, with $\rho = \rho(x, y, z)$ and

$$ds = \sqrt{[x'(t)]^2 + [y'(t)]^2 + [z'(t)]^2}\, dt,$$

but now, as you can easily verify, there is an extra formula

$$\bar{z} = \frac{\int_C z\rho\, ds}{\int_C \rho\, ds} = \frac{1}{M} \int_C z\rho\, ds.$$

for the z-coordinate of the center of mass, which becomes

$$\bar{z} = \frac{1}{L}\int_C z\,ds$$

in the case of constant ρ.

Example 3 Let C be the curve

$$x = \cos t, \qquad y = \sin t, \qquad z = t \qquad (0 \le t \le \pi),$$

that is, half a turn of a circular helix of unit radius and pitch 2π (see Example 2, page 722). Find the centroid of C.

Solution Since

$$ds = \sqrt{[x'(t)]^2 + [y'(t)]^2 + [z'(t)]^2}\,dt = \sqrt{(-\sin t)^2 + (\cos t)^2 + 1}\,dt = \sqrt{2}\,dt,$$

the curve C is of length $L = \int_0^\pi \sqrt{2}\,dt = \sqrt{2}\pi$. Hence the coordinates of the centroid are

$$\bar{x} = \frac{1}{L}\int_0^\pi x\,ds = \frac{1}{\pi}\int_0^\pi \cos t\,dt = \frac{1}{\pi}\sin t\bigg|_0^\pi = 0,$$

$$\bar{y} = \frac{1}{L}\int_0^\pi y\,ds = \frac{1}{\pi}\int_0^\pi \sin t\,dt = -\frac{1}{\pi}\cos t\bigg|_0^\pi = \frac{2}{\pi},$$

$$\bar{z} = \frac{1}{L}\int_0^\pi z\,ds = \frac{1}{\pi}\int_0^\pi t\,dt = \frac{1}{2\pi}t^2\bigg|_0^\pi = \frac{\pi}{2}.$$

The values of $\bar{x}$ and $\bar{z}$ could have been predicted from the symmetry of C, without making any calculations (explain further). □

The Theorems of Pappus

Next, pursuing our study of centroids, we prove a pair of closely related theorems which are modern versions of results known since antiquity. In fact, they can be found in Book VII of a work written in about 320 A.D. by the Greek mathematician Pappus of Alexandria.

> **Theorem 6** (*Pappus' theorem for a solid of revolution*). Let T be the solid of revolution generated by revolving a plane region R of area A about an axis in its plane, and suppose the axis does not go through an interior point of R. Then V, the volume of T, is equal to the product of A and the distance traveled by the centroid of C.

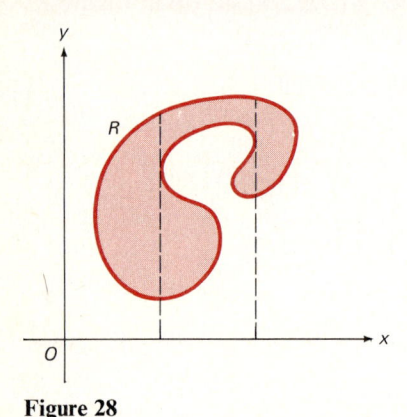

Proof Without loss of generality, we can assume that R lies in the upper half of the xy-plane with the x-axis as the axis of revolution. We also make the assumption (satisfied by all regions encountered in practical problems) that the region R is vertically simple, or else can be partitioned into a finite number of vertically simple regions by drawing lines parallel to the y-axis, as illustrated in Figure 28. Suppose R is vertically simple, so that $R = \{(x, y): a \le x \le b, g(x) \le y \le f(x)\}$, where the functions f and g are continuous and nonnegative on $[a, b]$. Then, according to formula (2), page 444, the solid T generated by revolving R about the x-axis is of volume

$$V = \pi \int_a^b \{[f(x)]^2 - [g(x)]^2\}\,dx. \tag{3}$$

Figure 28

But the y-coordinate of the centroid of R is

$$\bar{y} = \frac{1}{A} \iint_R y \, dA = \frac{1}{A} \int_a^b dx \int_{g(x)}^{f(x)} y \, dy$$

$$= \frac{1}{A} \int_a^b \left[\frac{1}{2} y^2 \right]_{g(x)}^{f(x)} dx = \frac{1}{2A} \int_a^b \{[f(x)]^2 - [g(x)]^2\} \, dx$$

(recall (10′), page 837), or equivalently

$$\int_a^b \{[f(x)]^2 - [g(x)]^2\} \, dx = 2A\bar{y}. \tag{4}$$

Substitution of (4) into (3) now gives

$$V = A(2\pi \bar{y}), \tag{5}$$

where $2\pi \bar{y}$ is the distance traveled by the centroid of R in making one revolution about the x-axis. This proves the theorem for the case where R is vertically simple.

If R is not vertically simple, we draw lines parallel to the y-axis partitioning R into vertically simple subregions $R_1, \ldots, R_n$ whose centroids have y-coordinates $\bar{y}_1, \ldots, \bar{y}_n$, respectively. Let A_i be the area of R_i, and let V_i be the volume of the solid of revolution generated by revolving R_i about the x-axis. Then, by formula (5), applied to each subregion R_i,

$$V = V_1 + \cdots + V_n = A_1(2\pi \bar{y}_1) + \cdots + A_n(2\pi \bar{y}_n).$$

But

$$\bar{y}_i = \frac{1}{A_i} \iint_{R_i} y \, dA \qquad (i = 1, \ldots, n),$$

and therefore, with the help of rule (iv), page 814,

$$V = V_1 + \cdots + V_n = 2\pi \iint_{R_1} y \, dA + \cdots + 2\pi \iint_{R_n} y \, dA$$

$$= 2\pi \iint_R y \, dA = 2\pi A \cdot \frac{1}{A} \iint_R y \, dA = A(2\pi \bar{y}),$$

so that formula (5) is valid for the whole region R. ■

Theorem 7 *(Pappus' theorem for a surface of revolution). Let S be the surface of revolution generated by revolving a simple plane curve C of length L about an axis in its plane, and suppose C does not cross the axis. Then A, the area of S, is equal to the product of L and the distance traveled by the centroid of C.*

Proof Again there is no loss of generality in assuming that C lies in the upper half of the xy-plane with the x-axis as the axis of revolution. Let C have parametric equations

$$x = x(t), \qquad y = y(t) \qquad (a \leq t \leq b),$$

where $x(t)$ and $y(t)$ are continuously differentiable on $[a, b]$. Then, by formula (1), page 466,

$$A = 2\pi \int_a^b y(t) \sqrt{[x'(t)]^2 + [y'(t)]^2}\, dt,$$

which can be written more concisely as

$$A = 2\pi \int_C y\, ds, \tag{6}$$

in terms of a line integral with respect to the arc length s. But, according to (2), the y-coordinate of the centroid of C is

$$\bar{y} = \frac{1}{L} \int_C y\, ds,$$

or equivalently

$$\int_C y\, ds = L\bar{y}. \tag{7}$$

Substitution of (7) into (6) now gives

$$A = L(2\pi \bar{y}),$$

where $2\pi \bar{y}$ is the distance traveled by the centroid of C. (Here it has been assumed that C is smooth, but the theorem remains valid if C is only "piecewise smooth," as defined on page 884.) ∎

The following examples illustrate the use of Pappus' theorems.

Example 4 Find the volume V of the torus generated by revolving a circular disk about a line in its plane that does not intersect the disk (see Figure 29).

Solution Let r be the radius of the disk, and let a be the perpendicular distance from the center of the disk to the axis. The centroid of the disk is clearly at its center, by symmetry, and the area of the disk is πr^2. Hence, by Theorem 6,

$$V = \pi r^2 (2\pi a) = 2\pi^2 r^2 a.$$

We have already used the method of washers to get the same answer in Example 4, page 446. □

Example 5 Find the centroid $(\bar{x}, \bar{y})$ of the semicircular region $x^2 + y^2 \leq r^2$, $y \geq 0$ of radius r.

Solution By symmetry, $\bar{x} = 0$. To find $\bar{y}$, we observe that the region is of area $\frac{1}{2}\pi r^2$, and revolving it about the x-axis generates a solid sphere of volume $V = \frac{4}{3}\pi r^3$. It follows from Theorem 6 that

$$V = \frac{4}{3}\pi r^3 = \frac{1}{2}\pi r^2 (2\pi \bar{y}),$$

which implies

$$\bar{y} = \frac{4r}{3\pi} \approx 0.42r.$$

The location of the centroid $(\bar{x}, \bar{y})$ is shown in Figure 30. □

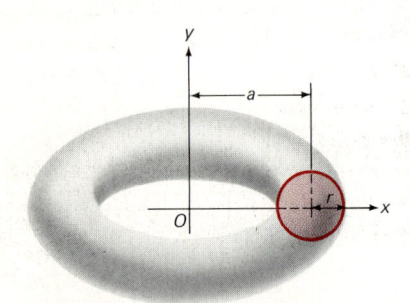

Figure 29

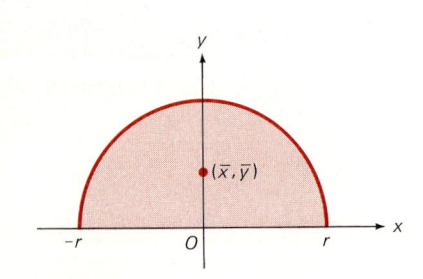

Figure 30

Example 6 Find the surface area A of the torus in Example 4.

Solution The centroid of the circle generating the surface is at its center, by symmetry, and the length (circumference) of the circle is $2\pi r$. Hence, by Theorem 7,

$$A = 2\pi r(2\pi a) = 4\pi^2 ra,$$

where a is the perpendicular distance from the center of the circle to the axis of revolution. The same answer was obtained by direct calculation in Example 2, page 469. □

Example 7 Find the centroid of the semicircular arc $x^2 + y^2 = r^2$, $y \geq 0$ of radius r.

Solution As in Example 5, $\bar{x} = 0$ by symmetry, but now we are dealing with a curve rather than a region. To find $\bar{y}$, we observe that revolving the semicircle of length πr about the x-axis generates a sphere of surface area $A = 4\pi r^2$. It follows from Theorem 7 that

$$A = 4\pi r^2 = \pi r(2\pi \bar{y}),$$

which implies

$$\bar{y} = \frac{2r}{\pi} \approx 0.64\, r.$$

The location of the centroid is shown in Figure 31. Why is the centroid of this arc higher than the centroid of the region in Figure 30 with the arc as part of its boundary? □

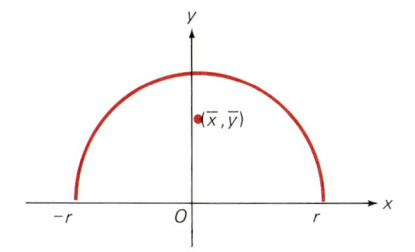

Figure 31

Problems

1. Find the centroid of the curve

$$x = \cos t, \quad y = \sin t \quad (0 \leq t \leq \theta \leq 2\pi),$$

a circular arc of unit radius and central angle θ.

2. Find the total mass of a wire shaped like the curve

$$y = \ln x \quad (\tfrac{1}{2} \leq x \leq 2)$$

if its density function is $\rho(x, y) = x^2$.

3. Find the total mass of a wire shaped like the curve

$$x = 2\cos t, \quad y = \sin t \quad (0 \leq t \leq \pi/2)$$

(an elliptic arc) if its density function is $\rho(x, y) = y$.

4. Find the centroid of the curve

$$y = \cosh x \quad (0 \leq x \leq \ln 2).$$

Find the total mass M and the center of mass $\bar{x}$ of a distribution of mass on the nonnegative x-axis with the given density function.

5. $\rho(x) = 1/(x^2 + 1)$ **6.** $\rho(x) = e^{-x}$
7. $\rho(x) = e^{-x^2}$ **8.** $\rho(x) = xe^{-x}$

Hint. The quantities M and $\bar{x}$ are defined in the natural way, as the improper integrals

$$M = \int_0^\infty \rho(x)\, dx, \quad \bar{x} = \frac{1}{M}\int_0^\infty x\rho(x)\, dx.$$

9. With the help of Example 5, page 464, find the centroid of the arc of the astroid $x^{2/3} + y^{2/3} = a^{2/3}$ ($a > 0$) in the first quadrant.

10. Find the total mass of a wire shaped like the curve

$$x = 3\cos t, \quad y = 3\sin t, \quad z = 4t \quad (0 \leq t \leq 2\pi)$$

(one turn of a circular helix) if its density function is $\rho(x, y, z) = x^2 + y^2 + z^2$.

11. Find the total mass of a wire shaped like the curve

$$x = t, \quad y = \tfrac{1}{2}t^2, \quad z = \tfrac{1}{3}t^3 \quad (0 \leq t \leq 1)$$

(a twisted cubic) if its density function is $\rho(x, y, z) = \sqrt{2y}$.

12. Use improper integrals to find the centroid of the curve

$$x = e^{-t}\cos t, \quad y = e^{-t}\sin t, \quad z = e^{-t} \quad (0 \leq t < \infty).$$

Use Theorem 6 to find the volume of the solid generated by revolving each of the following regions about the x-axis.

13. The region in Problem 9, page 841.

14–16. The regions in Problems 14–16, page 841.

Section 14.4 More About Centroids; The Theorems of Pappus

17. Find the volume of the solid generated by revolving a rectangular region of length 4 and width 3 about an axis in its plane which goes through a vertex P of the region and is perpendicular to the diagonal with endpoint P.

18. A square is revolved about an axis in its plane which intersects it in one of its vertices, but in no other points. For which position of the axis is the area of the resulting surface of revolution the largest?

19. In Problem 1, page 471, it was found that the area of the surface generated by revolving the curve

$$x = t^3, \qquad y = \tfrac{3}{2} t^2 \qquad (0 \le t \le 1)$$

about the x-axis is $A = \tfrac{6}{5}(\sqrt{2} + 1)\pi$. Use this and Theorem 7 to find $\bar{y}$, the ordinate of the centroid of the curve.

20. In Problem 15, page 471, it was found that the area of the surface generated by revolving the curve

$$y = \tfrac{1}{6}x^3 + \tfrac{1}{2}x^{-1} \qquad (1 \le x \le \sqrt{2})$$

about the x-axis is $A = \tfrac{47}{72}\pi$. Use this and Theorem 7 to find $\bar{y}$, the ordinate of the centroid of the curve.

21. Use Theorem 6 twice to find the centroid of the triangular region R bounded by the coordinate axes and the line

$$\frac{x}{a} + \frac{y}{b} = 1 \qquad (a > 0, b > 0).$$

22. The astroid $x^{2/3} + y^{2/3} = a^{2/3}$ $(a > 0)$ is revolved about an axis going through the points $(a, 0)$ and $(0, a)$. Use Pappus' theorems to find both the volume V of the resulting three-dimensional region and the area A of its surface.

14.5 Moments of Inertia

Suppose a rigid body T rotates with angular speed ω about an axis l, as shown in Figure 32. Then a point P in T moves with translational speed $v = r\omega$, where r is the distance from P to l (see page 685). Thus, if P is thought of as a particle of mass m, the kinetic energy of P is

$$\frac{1}{2} mv^2 = \frac{1}{2} mr^2\omega^2$$

(see page 269). More generally, let $P_1, P_2, \ldots, P_n$ be a system of n particles in T, with masses $m_1, m_2, \ldots, m_n$, respectively. Then the total kinetic energy K of the system, due to its rotation about l, is given by

$$K = \sum_{i=1}^{n} \frac{1}{2} m_i r_i^2 \omega^2,$$

where r_i is the distance from P_i to l. Therefore

$$K = \frac{1}{2} I_l \omega^2,$$

in terms of the quantity

$$I_l = \sum_{i=1}^{n} m_i r_i^2, \qquad (1)$$

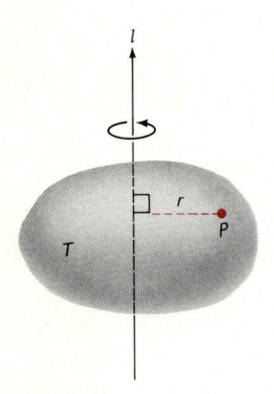

Figure 32

called the system's *moment of inertia about l*. This quantity is of basic importance in the investigation of dynamical problems involving rotation. If mass is measured in kilograms (kg) and distance in meters (m), the physical units of moments of inertia are kg-m². For example, the moment of inertia of the earth about its polar axis is about 8×10^{37} kg-m².

By now you should be able to anticipate how we will define the moment of inertia about l of the whole body T, regarded as a (normal) three-dimensional region on which there is defined a continuous mass density $\rho = \rho(P) = \rho(x, y, z)$. Let T be partitioned in the usual way into n subregions $T_1, T_2, \ldots, T_n$ (by drawing planes parallel to the coordinate planes), let μ be the mesh size of the partition, and let P_i be any point in T_i. Then T_i can be regarded as a particle of mass $\Delta m \approx \rho(P_i)\, \Delta V_i$ located at P_i, and if μ is small, the moment of inertia of T is to a good approximation the same as that of the system of n particles $T_1, T_2, \ldots, T_n$. Thus we define the moment of inertia I_l of T about l as the limit as $\mu \to 0$ of the moment of inertia of this system of particles about l, that is, guided by (1), we set

$$I_l = \lim_{\mu \to 0} \sum_{i=1}^{n} r^2(P_i)\, \Delta m_i = \lim_{\mu \to 0} \sum_{i=1}^{n} r^2(P_i)\rho(P_i)\, \Delta V_i,$$

where $r = r(P) = r(x, y, z)$ is the distance from the point P to the axis of revolution l. Since the sum on the right is a Riemann sum for the function $r^2(P)\rho(P)$ on T, it follows that

$$I_l = \iiint_T r^2(P)\rho(P)\, dV,$$

or more concisely

$$I_l = \iiint_T r^2 \rho\, dV. \tag{2}$$

In the case of a lamina R in the xy-plane, of density $\rho = \rho(x, y)$, the appropriate definition of I_l is of course

$$I_l = \iint_R r^2 \rho\, dA, \tag{2'}$$

with the triple integral in (2) replaced by a double integral. Here $r = r(x, y)$ is the distance from a variable point (x, y) of R to the axis l.

Making different choices of l, we get various moments of inertia. In the case of a solid T, we have $r = \sqrt{y^2 + z^2}$ if l is the x-axis, $r = \sqrt{x^2 + z^2}$ if l is the y-axis, and $r = \sqrt{x^2 + y^2}$ if l is the z-axis, so that

$$I_x = \iiint_T (y^2 + z^2)\rho\, dV,$$

$$I_y = \iiint_T (x^2 + z^2)\rho\, dV, \tag{3}$$

$$I_z = \iiint_T (x^2 + y^2)\rho\, dV.$$

In the case of a lamina R, we have $r = |y|$ if l is the x-axis, $r = |x|$ if l is the y-axis, and $r = \sqrt{x^2 + y^2}$ if l is the z-axis, so that

$$I_x = \iint_R y^2 \rho\, dA,$$

$$I_y = \iint_R x^2 \rho\, dA, \tag{4}$$

$$I_z = \iint_R (x^2 + y^2)\rho\, dA.$$

The Radius of Gyration

Since $\sqrt{x^2 + y^2}$ is the distance from P to the origin, as well as the distance from P to the z-axis, I_z is often denoted by I_O and called the *polar moment of inertia* (this term is used for laminas only). Notice that $I_O = I_z = I_x + I_y$.

As always,
$$M = \iint_R \rho(x, y)\, dA$$

is the total mass of the lamina R, and
$$M = \iiint_T \rho(x, y, z)\, dV$$

is the total mass of the solid T. In terms of M, we can write
$$I_x = Mk_x^2,$$

where
$$k_x = \sqrt{\frac{I_x}{M}}.$$

Thus the moment of inertia I_x of R or T about the x-axis is the same as that of a single "equivalent particle" of mass M, whose distance from the x-axis is k_x. The quantity k_x is called the *radius of gyration* of R or T about the x-axis. Similarly, the radii of gyration of a lamina or solid about the y- and z-axes are

$$k_y = \sqrt{\frac{I_y}{M}}, \quad k_z = \sqrt{\frac{I_z}{M}}.$$

In the case of a lamina, the last formula has the alternative version
$$k_O = \sqrt{\frac{I_O}{M}},$$

where $k_O = k_z$ is called the *polar radius of gyration*.

We can also talk about the moment of inertia or the radius of gyration of a plane or solid *region*, rather than of a plane lamina or solid object. By this we mean the corresponding moment of inertia or radius of gyration with a density function $\rho(x, y)$ or $\rho(x, y, z)$ which is identically equal to 1, in units of mass per unit area or volume.

Example 1 Find the moments of inertia and radii of gyration about the x- and y-axes of a plate of density $\rho(x, y) = y$, shaped like the region R bounded by the positive coordinate axes and the parabola $y^2 = 1 - x$ (see Figure 33).

Solution By the first two of the formulas (4),

$$I_x = \iint_R y^2 \rho\, dA = \int_0^1 dx \int_0^{\sqrt{1-x}} y^3\, dy = \frac{1}{4} \int_0^1 (1 - x)^2\, dx$$

$$= \frac{1}{4}\left[x - x^2 + \frac{1}{3} x^3 \right]_0^1 = \frac{1}{12},$$

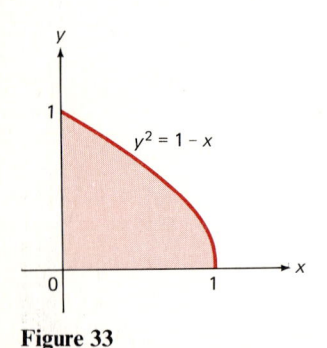

Figure 33

852 Chapter 14 Multiple Integration

$$I_y = \iint_R x^2 \rho \, dA = \int_0^1 dx \int_0^{\sqrt{1-x}} x^2 y \, dy = \frac{1}{2} \int_0^1 x^2(1-x) \, dx$$

$$= \frac{1}{2}\left[\frac{1}{3}x^3 - \frac{1}{4}x^4\right]_0^1 = \frac{1}{24}.$$

Since the mass of the plate is

$$M = \int_0^1 dx \int_0^{\sqrt{1-x}} y \, dy = \frac{1}{2}\int_0^1 (1-x) \, dx = \frac{1}{2}\left[x - \frac{1}{2}x^2\right]_0^1 = \frac{1}{4},$$

the corresponding radii of gyration are

$$k_x = \sqrt{\frac{I_x}{M}} = \frac{1}{\sqrt{3}}, \qquad k_y = \sqrt{\frac{I_y}{M}} = \frac{1}{\sqrt{6}}. \qquad \square$$

Example 2 Find the moments of inertia and radii of gyration about the x- and y-axes of the region R of area πab which is enclosed by the ellipse $(x^2/a^2) + (y^2/b^2) = 1$. Also find the polar moment of inertia and polar radius of gyration of R.

Solution Here $\rho(x, y) \equiv 1$, and using the symmetry of R, we have

$$I_x = \iint_R y^2 \, dA = 4\int_0^a dx \int_0^{b\sqrt{a^2-x^2}/a} y^2 \, dy = \frac{4b^3}{3a^3}\int_0^a (a^2 - x^2)^{3/2} \, dx,$$

$$I_y = \iint_R x^2 \, dA = 4\int_0^a dx \int_0^{b\sqrt{a^2-x^2}/a} x^2 \, dy = \frac{4b}{a}\int_0^a x^2 \sqrt{a^2 - x^2} \, dx.$$

Making the substitution $x = a \cos t$, we obtain

$$I_x = \frac{4ab^3}{3}\int_0^{\pi/2} \sin^4 t \, dt = \frac{4ab^3}{3}\frac{3\pi}{16} = \frac{\pi ab^3}{4},$$

with the help of Problem 13, page 381, and

$$I_y = 4a^3 b \int_0^{\pi/2} \cos^2 t \sin^2 t \, dt = 4a^3 b \left(\frac{\pi}{16}\right) = \frac{\pi a^3 b}{4}$$

(this integral was evaluated in the solution of Example 5, page 820). Since the mass of the region is $M = \pi ab$, the corresponding radii of gyration are

$$k_x = \sqrt{\frac{I_x}{M}} = \frac{b}{2}, \qquad k_y = \sqrt{\frac{I_y}{M}} = \frac{a}{2}.$$

To get the polar moment of inertia I_O, we observe that

$$I_O = I_z = I_x + I_y = \frac{\pi ab}{4}(a^2 + b^2).$$

Hence the polar radius of gyration is

$$k_O = \sqrt{\frac{I_O}{M}} = \frac{\sqrt{a^2 + b^2}}{2}.$$

If $a = b$, the ellipse reduces to the circle $x^2 + y^2 = a^2$ of radius a, and these formulas simplify to

$$I_x = I_y = \frac{\pi a^4}{4}, \qquad I_O = \frac{\pi a^4}{2}$$

and
$$k_x = k_y = \frac{a}{2}, \qquad k_O = \frac{a}{\sqrt{2}}. \qquad \square$$

Example 3 Find the moments of inertia and radii of gyration about the coordinate axes of the tetrahedron T in the first octant bounded by the coordinate planes and the plane $x + y + z = a$, where $a > 0$ (see Figure 34).

Solution Setting $\rho(x, y, z) \equiv 1$ in (3), we find that the moments of inertia are

$$I_x = \iiint_T (y^2 + z^2)\, dV, \quad I_y = \iiint_T (x^2 + z^2)\, dV, \quad I_z = \iiint_T (x^2 + y^2)\, dV,$$

where

$$\iiint_T x^2\, dV = \iiint_T y^2\, dV = \iiint_T z^2\, dV,$$

because of the symmetry of T. The cross section of T at z, that is, the figure in which the plane through the point $(0, 0, z)$ parallel to the xy-plane intersects T, is an isosceles right triangle with legs of length $a - z$ and area $A(z) = \frac{1}{2}(a - z)^2$. Therefore, by the method of cross sections,

$$\iiint_T z^2\, dV = \int_0^a z^2 A(z)\, dz = \frac{1}{2} \int_0^a z^2 (a - z)^2\, dz$$

$$= \frac{1}{2} \left[\frac{1}{3} a^2 z^3 - \frac{1}{2} a z^4 + \frac{1}{5} z^5 \right]_0^a = \frac{1}{60} a^5.$$

It follows that

$$I_x = I_y = I_z = \frac{1}{30} a^5.$$

The mass of the tetrahedron is

$$M = \frac{1}{2} \int_0^a (a - z)^2\, dz = \frac{1}{2} \left[a^2 z - a z^2 + \frac{1}{3} z^3 \right]_0^a = \frac{1}{6} a^3,$$

and hence the corresponding radii of gyration are

$$k_x = \sqrt{\frac{I_x}{M}} = \frac{a}{\sqrt{5}}, \quad k_y = \sqrt{\frac{I_y}{M}} = \frac{a}{\sqrt{5}}, \quad k_z = \sqrt{\frac{I_z}{M}} = \frac{a}{\sqrt{5}}. \qquad \square$$

Instead of introducing the radius of gyration, we can write any moment of inertia as an expression with the total mass of the lamina or solid as a factor. For instance, the polar moment of inertia of the ellipse in Example 2 can be written in the form

$$I_O = \frac{1}{4} M(a^2 + b^2).$$

Similarly, in Example 3 we can write the moments of inertia as

$$I_x = I_y = I_z = \frac{1}{5} M a^2.$$

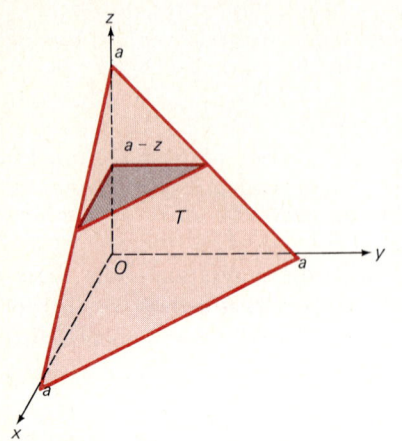

Problems

1. A system of particles consists of four masses $m_1 = 4$, $m_2 = 3$, $m_3 = 1$ and $m_4 = 6$ at the points $P_1 = (-3, 2, 0)$, $P_2 = (1, 5, -2)$, $P_3 = (0, 4, -1)$ and $P_4 = (3, 0, 1)$. Find the moments of inertia I_x, I_y, I_z and the radii of gyration k_x, k_y, k_z of the system.

2. Find the moments of inertia I_x and I_y of the triangular region R in the first quadrant bounded by the coordinate axes and the line $3x + 2y = 6$.

3. Find the moments of inertia I_x and I_y of the triangular region R with vertices $(1, 1)$, $(2, 1)$ and $(3, 3)$.

4. Let R be the region bounded by the lines $x = a$ and $x = b$, the x-axis and the curve

$$y = f(x) \quad (a \leq x \leq b),$$

where f is continuous and nonnegative on $[a, b]$. Show that the moments of inertia of R about the x- and y-axes are given by the formulas

$$I_x = \frac{1}{3} \int_a^b [f(x)]^3 \, dx, \quad I_y = \int_a^b x^2 f(x) \, dx. \quad \text{(i)}$$

5. Find the moments of inertia I_x and I_y of a rectangular lamina R bounded by the coordinate axes and the lines $x = 1$ and $y = 2$ if its density function is $\rho(x, y) = x^2 + y^2$.

Let R be a rectangular lamina of length a and width b. Find the moment of inertia of R about an axis perpendicular to R which goes through

6. The center of R

7. A vertex of R

8. Find the moment of inertia of the semicircular region $x^2 + y^2 \leq a^2$, $y \geq 0$ about its diameter.

9–12. Find the moments of inertia I_x and I_y of each of the regions in Problems 9–12, page 841.

13. Let l be the line through the origin with inclination ϕ, and let I_l be the moment of inertia about l of a plane region R. Show that

$$I_l = I_x \cos^2 \phi - 2J \sin \phi \cos \phi + I_y \sin^2 \phi$$

in terms of the moments of inertia I_x, I_y, and the quantity

$$J = \iint_R xy \, dA,$$

known as the *product of inertia*.

14. Show that the sum of the moments of inertia of a region R about two perpendicular axes in the plane of R through a fixed point O is the same for every orientation of the axes. *Hint.* Use the preceding problem.

15. Prove the *parallel axis theorem*, which says that if c is an axis through the center of mass of a solid or lamina of mass M, and if l is an axis parallel to c, then $I_l = I_c + Md^2$, where d is the distance between the axes.

16. Find the moment of inertia of a circular disk of radius a and unit density about any of its tangent lines. *Hint.* Use the parallel axis theorem.

Find the moments of inertia I_x, I_y and I_z of the rectangular parallelepiped T bounded by the coordinate planes and the planes $x = a$, $y = b$ and $z = c$ (a, b, c positive) if the density function is

17. $\rho(x, y, z) \equiv 1$ **18.** $\rho(x, y, z) = x$

19. $\rho(x, y, z) = xy$ **20.** $\rho(x, y, z) = xyz$

Write each answer in a form in which the total mass M appears as a factor.

21. Find the moment of inertia about the z-axis of the solid region T which is bounded by the hyperboloid of revolution $x^2 + y^2 - z^2 = 1$ and the planes $z = 0$ and $z = 1$.

22. Find the moments of inertia I_x, I_y and I_z of the solid of unit density bounded by the hyperbolic paraboloid $z = xy$ and the planes $x = 2$ and $y = 3$ in the first octant. Express each moment of inertia as a multiple of M, the total mass of T.

Let T be a solid right circular cone of unit density, with height h and base radius a. Find the moment of inertia of the cone T

23. About its axis

24. About a diameter of its base

14.6 Double Integrals in Polar Coordinates

It will be recalled from Section 14.1 that the double integral

$$\iint_R f(x, y) \, dx \, dy, \quad (1)$$

where R is a normal region and f is continuous on R, has been defined in

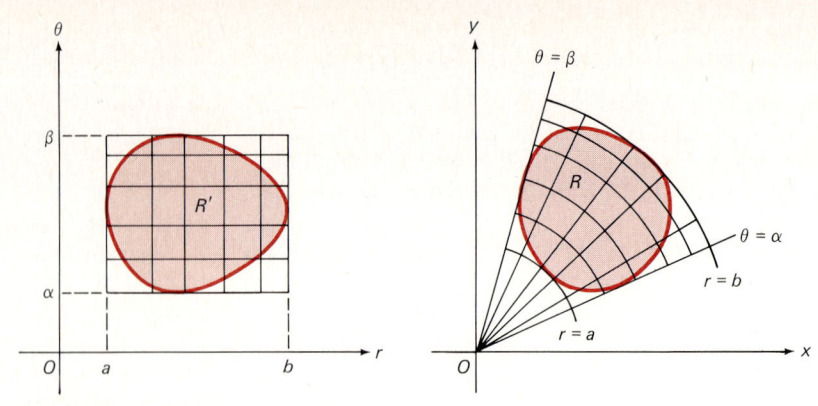

Figure 35

terms of an underlying system of rectangular coordinates x and y. But these may not be the best coordinates to employ in evaluating (1). For example, suppose we simultaneously introduce polar coordinates r and θ by choosing the pole and polar axis to be the origin and positive x-axis of the xy-plane. Then the description of R may be much simpler in polar coordinates than in rectangular coordinates, a fact which can be used to advantage after "transforming" (1) to polar coordinates. This is accomplished as follows.

Suppose we introduce another rectangular coordinate system, in another plane, with r as abscissa and θ as ordinate. Then, as shown schematically in Figure 35, R is the image under the coordinate transformation

$$x = r\cos\theta, \qquad y = r\sin\theta \tag{2}$$

of some region R' in the $r\theta$-plane, in the sense that every point (x, y) in R is related by the formulas (2) to a point (r, θ) in R'. We will assume that R' lies in the right half-plane $r \geq 0$ and in the horizontal strip $0 \leq \theta \leq 2\pi$. Then the correspondence between R' and R established by (2) is one-to-one except possibly on the boundary of R' (the points $(r, 0)$ and $(r, 2\pi)$ of the $r\theta$-plane are both mapped into the same point $(r, 0)$ of the xy-plane, and the line $r = 0$ is mapped as a whole into the origin of the xy-plane). Suppose R and R' are both normal regions, where R' is contained in the rectangle $a \leq r \leq b$, $\alpha \leq \theta \leq \beta$ in the $r\theta$-plane. Then R is contained in the "annular sector" or "polar rectangle" in the xy-plane defined by the same inequalities $a \leq r \leq b$, $\alpha \leq \theta \leq \beta$ (see the figure). We now have the following theorem, relating double integrals over R to double integrals over R'.

Theorem 8 (*Double integrals in polar coordinates*). *If $f(x, y)$ is continuous on R, then*

$$\iint_R f(x, y)\, dx\, dy = \iint_{R'} f(r\cos\theta, r\sin\theta)\, r\, dr\, d\theta. \tag{3}$$

Partial proof [†] Let the intervals $[a, b]$ and $[\alpha, \beta]$ be partitioned by the points $r_j\, (j = 0, 1, \ldots, J)$ and $\theta_k\, (k = 0, 1, \ldots, K)$, with mesh sizes μ_r

---

[†] A few tricky details, verging on advanced calculus, will not be explained in full. For example, the theorem remains valid if the transformation (2) fails to be one-to-one on the boundary of R', as is often the case.

856 Chapter 14 Multiple Integration

and μ_θ, respectively. This means that $a = r_0 < r_1 < r_2 < \cdots < r_{J-1} < r_J = b$, $\alpha = \theta_0 < \theta_1 < \theta_2 < \cdots < \theta_{K-1} < \theta_K = \beta$, and

$$\mu_r = \max \{r_1 - r_0, r_2 - r_1, \ldots, r_J - r_{J-1}\},$$
$$\mu_\theta = \max \{\theta_1 - \theta_0, \theta_2 - \theta_1, \ldots, \theta_K - \theta_{K-1}\}.$$

Then the lines

$$r = r_j \qquad (j = 0, 1, \ldots, J)$$

and

$$\theta = \theta_k \qquad (k = 0, 1, \ldots, K),$$

parallel to the coordinate axes of the $r\theta$-plane, divide R' into n closed subregions $R'_1, R'_2, \ldots, R'_n$, mostly rectangular, where $n \le JK$ (typically $n < JK$). The circles and rays in the xy-plane with the same equations $r = r_j$ and $\theta = \theta_k$ divide R into n closed subregions $R_1, R_2, \ldots, R_n$, mostly annular sectors, where R_i is the image of R'_i under the transformation (2) carrying polar coordinates into rectangular coordinates. Let

$$\mu = \max \{\mu_r, \mu_\theta\},$$

and let ΔA_i and $\Delta A'_i$ be the areas of R_i and R'_i (the fact that ΔA_i and $\Delta A'_i$ exist and are nonzero will be taken for granted). Also let $P'_i = (r'_i, \theta'_i)$ be an arbitrary point of R'_i (there is no connection between r'_i, θ'_i and the points of subdivision r_j and θ_k), and let $P_i = (p_i, q_i)$ be the image of P'_i under the transformation (2). Then

$$\iint_R f(x, y)\,dx\,dy = \lim_{\mu \to 0} \sum_{i=1}^n f(p_i, q_i)\,\Delta A_i$$
$$= \lim_{\mu \to 0} \sum_{i=1}^n f(r'_i \cos \theta'_i, r'_i \sin \theta'_i) \frac{\Delta A_i}{\Delta A'_i}\,\Delta A'_i. \tag{4}$$

Now suppose R'_i is the rectangular region

$$r^* \le r \le r^* + \Delta r, \qquad \theta^* \le \theta \le \theta^* + \Delta \theta$$

of area $\Delta A'_i = \Delta r\,\Delta\theta$. Then R_i is the annular sector shown in Figure 36, of area

$$\Delta A_i = \frac{1}{2}\left[(r^* + \Delta r)^2 - r^{*2}\right]\Delta\theta = \frac{1}{2}\left[2r^*\,\Delta r + (\Delta r)^2\right]\Delta\theta = \left(r^* + \frac{1}{2}\Delta r\right)\Delta r\,\Delta\theta,$$

since ΔA_i is the difference between the areas of two circular sectors with the same central angle $\Delta\theta$ and different radii r^* and $r^* + \Delta r$. Therefore

$$\Delta A_i = r_i^*\,\Delta A'_i, \tag{5}$$

where $r_i^* = r^* + \frac{1}{2}\Delta r$ is the average of the radii r^* and $r^* + \Delta r$ (the subscript on the quantity r_i^* shows that it goes with the region R_i). Clearly there are points in R'_i with abscissa r_i^*, and since $P'_i = (r'_i, \theta'_i)$ is an arbitrary point of R'_i, we can assume that $r'_i = r_i^*$. Then (5) takes the form $\Delta A_i = r'_i\,\Delta A'_i$, or equivalently

$$\frac{\Delta A_i}{\Delta A'_i} = r'_i. \tag{6}$$

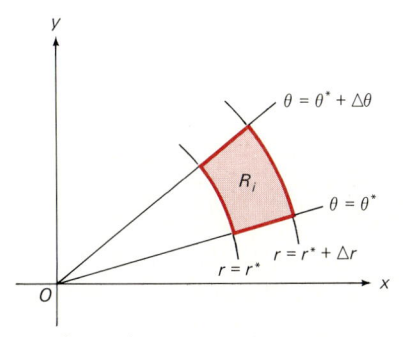

Figure 36
An annular sector or polar rectangle

Section 14.6 Double Integrals in Polar Coordinates **857**

If all the regions R'_i are rectangular, so that the corresponding regions R_i are annular sectors, we can substitute (6) into (4), obtaining

$$\iint_R f(x, y)\, dx\, dy = \lim_{\mu \to 0} \sum_{i=1}^{n} f(r'_i \cos \theta'_i, r'_i \sin \theta'_i) r'_i \Delta A'_i. \qquad (7)$$

But the sum on the right is a Riemann sum for the continuous function $f(r \cos \theta, r \sin \theta)r$ on the region R' of the $r\theta$-plane, in which r and θ are rectangular coordinates. Therefore, by Theorem 1, page 813, as $\mu \to 0$ this sum approaches the double integral $\iint_{R'} f(r \cos \theta, r \sin \theta)\, r\, dA$, also written as $\iint_{R'} f(r \cos \theta, r \sin \theta)\, r\, dr\, d\theta$, and formula (3) is proved. The fact that some of the regions R'_i are in general nonrectangular doesn't matter, since it can be shown that the sum in (7) approaches the same limit if we drop every term corresponding to a nonrectangular region. ∎

How to Remember Formula (3)

The nub of the proof of Theorem 8 consists in the observation that the region R_i in Figure 36 is almost rectangular if Δr and $\Delta \theta$ are both small, with straight sides of length Δr and curved sides of approximate length $r\, \Delta \theta$, where r is the radial coordinate of any point in R_i. Hence the area of R_i is approximately equal to $r\, \Delta r\, \Delta \theta$. This serves as a crude explanation of why $r\, dr\, d\theta$ appears in the right side of formula (3) when we take the limit as $\mu \to 0$. It is also a good way to remember the formula.

Naturally, to calculate the double integral

$$\iint_{R'} f(r \cos \theta, r \sin \theta)\, r\, dr\, d\theta,$$

we resort to iterated integrals in the usual way. For example, if R is the "radially simple" region in Figure 37, bounded by two rays $\theta = \alpha$, $\theta = \beta$ and the graphs of two functions $r = g_1(\theta)$, $r = g_2(\theta)$, where $g_1(\theta)$ and $g_2(\theta)$ are continuous on $[\alpha, \beta]$ and $g_2(\theta) \geq g_1(\theta) \geq 0$, then R' is a vertically simple region and

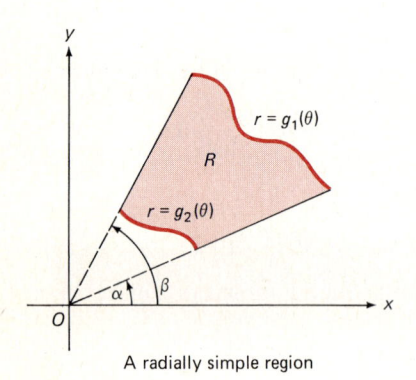

$$\iint_R f(x, y)\, dx\, dy = \iint_{R'} f(r \cos \theta, r \sin \theta)\, r\, dr\, d\theta$$

$$= \int_\alpha^\beta d\theta \int_{g_1(\theta)}^{g_2(\theta)} f(r \cos \theta, r \sin \theta)\, r\, dr.$$

Similarly, if R is the "angularly simple" region in Figure 38, bounded by two circles $r = a$, $r = b$ and the graphs of two functions $\theta = h_1(r)$, $\theta = h_2(r)$, where $h_1(r)$ and $h_2(r)$ are continuous on $[a, b]$ and $h_2(r) \geq h_1(r) \geq 0$, then R' is a horizontally simple region and

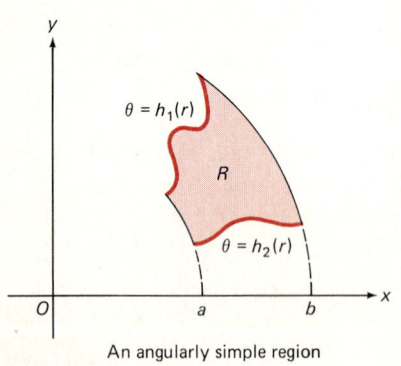

$$\iint_R f(x, y)\, dx\, dy = \iint_{R'} f(r \cos \theta, r \sin \theta)\, r\, dr\, d\theta$$

$$= \int_a^b dr \int_{h_1(r)}^{h_2(r)} f(r \cos \theta, r \sin \theta)\, r\, d\theta.$$

A radially simple region
Figure 37

An angularly simple region
Figure 38

Choosing $f \equiv 1$ in formula (3), we find that the area A of the radially simple region in Figure 37 is given by

$$A = \iint_{R'} r \, dr \, d\theta = \int_\alpha^\beta d\theta \int_{g_1(\theta)}^{g_2(\theta)} r \, dr = \int_\alpha^\beta \left[\frac{1}{2} r^2\right]_{g_1(\theta)}^{g_2(\theta)} d\theta$$

$$= \frac{1}{2} \int_\alpha^\beta \{[g_2(\theta)]^2 - [g_1(\theta)]^2\} \, d\theta.$$

This is in complete accord with the formula for A already derived in Section 10.10 (see formula (3'), page 637, where f and g are used instead of g_2 and g_1). As an exercise, show that the area of the angularly simple region in Figure 38 is

$$A = \int_a^b [h_2(r) - h_1(r)] r \, dr.$$

Example 1 Use polar coordinates to evaluate the integral

$$I = \iint_R \sqrt{x^2 + y^2} \, dx \, dy,$$

where R is the semicircular region $x^2 + y^2 \leq a^2$, $y \geq 0$ of radius a (see Figure 39).

Solution By Theorem 8,

$$I = \iint_{R'} \sqrt{(r \cos \theta)^2 + (r \sin \theta)^2} \, r \, dr \, d\theta = \iint_{R'} r^2 \, dr \, d\theta,$$

where R' is the rectangle $0 \leq r \leq a$, $0 \leq \theta \leq \pi$ in the $r\theta$-plane. Therefore

$$I = \int_0^\pi \int_0^a r^2 \, dr \, d\theta = \left(\int_0^\pi d\theta\right)\left(\int_0^a r^2 \, dr\right) = \frac{\pi a^3}{3}. \quad \square$$

Example 2 Evaluate the integral

$$I = \iint_R \ln(x^2 + y^2) \, dx \, dy,$$

where R is the annular region bounded by the circles $x^2 + y^2 = 4$ and $x^2 + y^2 = e^2$ (see Figure 40).

Solution By Theorem 8,

$$I = \iint_{R'} \ln(r^2 \cos^2 \theta + r^2 \sin^2 \theta) r \, dr \, d\theta = \iint_{R'} (\ln r^2) r \, dr \, d\theta,$$

where R' is the rectangle $2 \leq r \leq e$, $0 \leq \theta \leq 2\pi$ in the $r\theta$-plane. Therefore

$$I = \int_0^{2\pi} \int_2^e (\ln r^2) r \, dr \, d\theta = \left(\int_0^{2\pi} d\theta\right)\left(2 \int_2^e \ln r \, d(\tfrac{1}{2} r^2)\right)$$

$$= 4\pi \left[\frac{1}{2} r^2 \ln r - \frac{1}{4} r^2\right]_2^e = \pi(e^2 + 4 - 8 \ln 2),$$

with the help of integration by parts. $\quad \square$

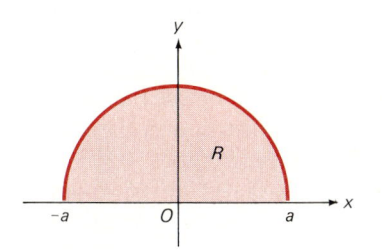

Figure 39

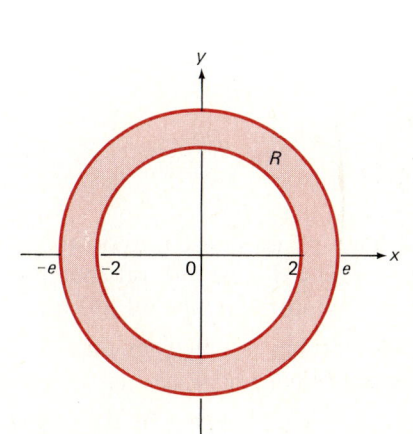

Figure 40

Example 3 Evaluate the improper integral

$$I = \int_0^\infty e^{-x^2}\,dx,$$

which was shown to be convergent in Example 9, page 430.

Solution The following trick, involving a *double* integral in polar coordinates, allows us to evaluate this seemingly intractable integral. First we observe that by definition $I = \lim_{a\to\infty} I_a$, where $I_a = \int_0^a e^{-x^2}\,dx$. Since any letter can be used as the dummy variable of integration, we have

$$I_a^2 = \left(\int_0^a e^{-x^2}\,dx\right)\left(\int_0^a e^{-x^2}\,dx\right)$$

$$= \left(\int_0^a e^{-x^2}\,dx\right)\left(\int_0^a e^{-y^2}\,dy\right) = \iint_R e^{-(x^2+y^2)}\,dx\,dy,$$

where R is the square region in the first quadrant defined by the inequalities $0 \le x \le a$, $0 \le y \le a$. But the integrand $e^{-(x^2+y^2)}$ is positive and the region R contains the quarter-disk R_1 defined by the inequalities $0 \le r \le a$, $0 \le \theta \le \pi/2$ in polar coordinates, and in turn is contained in the quarter-disk R_2 defined by the inequalities $0 \le r \le \sqrt{2}a$, $0 \le \theta \le \pi/2$ (see Figure 41). Therefore

$$\iint_{R_1} e^{-(x^2+y^2)}\,dx\,dy \le \iint_R e^{-(x^2+y^2)}\,dx\,dy = I_a^2 \le \iint_{R_2} e^{-(x^2+y^2)}\,dx\,dy,$$

where the inequalities are actually strict. Transforming to polar coordinates in which $x^2 + y^2 + r^2$, we find that

$$\iint_{R_1} e^{-(x^2+y^2)}\,dx\,dy = \int_0^{\pi/2} d\theta \int_0^a e^{-r^2} r\,dr = \frac{\pi}{2}\left[-\frac{1}{2}e^{-r^2}\right]_0^a = \frac{\pi}{4}(1 - e^{-a^2})$$

(here we go directly to an iterated integral), and similarly

$$\iint_{R_2} e^{-(x^2+y^2)}\,dx\,dy = \frac{\pi}{4}(1 - e^{-2a^2}),$$

so that

$$\frac{\pi}{4}(1 - e^{-a^2}) \le I_a^2 \le \frac{\pi}{4}(1 - e^{-2a^2}).$$

Hence I_a^2 lies between two expressions which approach the same limit $\pi/4$ as $a \to \infty$. Since $I_a^2 \to I^2$ as $a \to \infty$, it follows from Theorem 10, page 85 (the sandwich theorem), that $I^2 = \pi/4$. Thus, finally,

$$I = \int_0^\infty e^{-x^2}\,dx = \frac{\sqrt{\pi}}{2}. \quad \square$$

It is clear from the above examples that a transformation to polar coordinates is usually called for if the integrand of a double integral contains the expression $x^2 + y^2$ or its powers. Here is another example of this type.

Example 4 Find the volume V of the solid region T cut from the solid sphere $x^2 + y^2 + z^2 \le a^2$ by the circular cylinder $x^2 + y^2 = ax$ $(a > 0)$.

Solution The cylinder $x^2 + y^2 = ax$ intersects the xy-plane in the circle $(x - \tfrac{1}{2}a)^2 + y^2 = \tfrac{1}{4}a^2$ of radius $\tfrac{1}{2}a$ centered at the point $(\tfrac{1}{2}a, 0)$. As in Figure 42, let R be the semicircular region in the first quadrant of the xy-

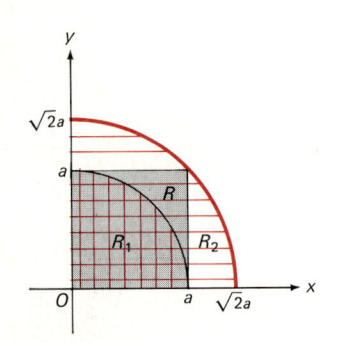

Figure 41

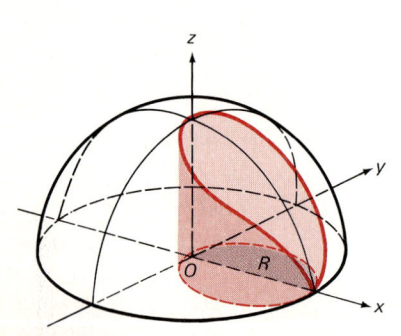

Figure 42

plane bounded by this circle and the positive x-axis. Then, by symmetry, V is four times the volume between the radially simple region R and the graph of $z = \sqrt{a^2 - x^2 - y^2}$ (the upper surface of the sphere), that is,

$$V = 4 \iint_R \sqrt{a^2 - x^2 - y^2}\, dx\, dy.$$

The semicircular part of the boundary of R has the polar equation $r = a\cos\theta$ ($0 \le \theta \le \pi/2$). Therefore, transforming to polar coordinates, we obtain

$$V = 4\int_0^{\pi/2} d\theta \int_0^{a\cos\theta} \sqrt{a^2 - r^2}\, r\, dr = 4\int_0^{\pi/2} \left[-\frac{1}{3}(a^2 - r^2)^{3/2}\right]_0^{a\cos\theta} d\theta$$

$$= \frac{4}{3} a^3 \int_0^{\pi/2} (1 - \sin^3\theta)\, d\theta = \frac{4}{3} a^3 \left(\frac{\pi}{2} - \int_0^{\pi/2} (1 - \cos^2\theta)\sin\theta\, d\theta\right)$$

$$= \frac{4}{3} a^3 \left(\frac{\pi}{2} - \left[\frac{1}{3}\cos^3\theta - \cos\theta\right]_0^{\pi/2}\right) = \frac{4}{3} a^3 \left(\frac{\pi}{2} - \frac{2}{3}\right),$$

where once again we go over directly to an iterated integral, omitting the step

$$\iint_R \sqrt{a^2 - x^2 - y^2}\, dx\, dy = \iint_{R'} \sqrt{a^2 - r^2}\, r\, dr\, d\theta$$

involving the vertically simple region $R' = \{(r, \theta): 0 \le \theta \le \pi/2,\ 0 \le r \le a\cos\theta\}$.

The volume V can also be found by evaluating the integral $\iint_R \sqrt{a^2 - x^2 - y^2}\, dx\, dy$ directly in rectangular coordinates, but the calculation turns out to be quite formidable. It is the simplicity of the polar equation $r = a\cos\theta$ for the semicircular part of the boundary of R, as opposed to the cartesian equation $y = \sqrt{ax - x^2}$, that makes the calculation of V so straightforward in polar coordinates. □

Problems

Evaluate the given double integral by transforming it from rectangular to polar coordinates.

1. $\iint_R \sqrt{x^2 + y^2}\, dx\, dy$, where R is the disk $x^2 + y^2 \le 9$

2. $\iint_R (x^2 + y^2)\, dx\, dy$, where R is the disk $x^2 + y^2 \le 4x$

3. $\iint_R \dfrac{dx\, dy}{x^2 + y^2 + 1}$, where R is the semicircular region $x^2 + y^2 \le 1,\ y \ge 0$

4. $\iint_R e^{-(x^2+y^2)}\, dx\, dy$, where R is the disk $x^2 + y^2 \le \ln 2$

5. $\iint_R \dfrac{dx\, dy}{\sqrt{x^2 + y^2}}$, where R is the annular region $1 \le x^2 + y^2 \le 4$

6. $\iint_R \dfrac{\sin\sqrt{x^2 + y^2}}{\sqrt{x^2 + y^2}}\, dx\, dy$, where R is the annular region $\pi^2/16 \le x^2 + y^2 \le \pi^2/9$

7. $\iint_R \cos\sqrt{x^2 + y^2}\, dx\, dy$, where R is the annular region $\pi^2 \le x^2 + y^2 \le 4\pi^2$

8. $\iint_R (1 - 3x + 2y)\, dx\, dy$, where R is the semicircular region $x^2 + y^2 \le 1,\ y \le 0$

9. $\iint_R \ln(1 + x^2 + y^2)\, dx\, dy$, where R is the disk $x^2 + y^2 \le e^2 - 1$

10. $\iint_R \arctan(y/x)\, dx\, dy$, where R is the annular sector $1 \le x^2 + y^2 \le 4,\ x/\sqrt{3} \le y \le \sqrt{3} x$

Transform the given iterated integral from rectangular coordinates to polar coordinates. (Assume that $f(x, y)$ is continuous.)

11. $\int_0^1 dy \int_{-y}^{y} f(x, y) \, dx$ 12. $\int_0^2 dx \int_0^{\sqrt{3}x} f(x^2 + y^2) \, dy$

13. $\int_0^1 dx \int_0^1 f(\sqrt{x^2 + y^2}) \, dy$ 14. $\int_1^4 dy \int_0^{\sqrt{4y - y^2}} f(x, y) \, dx$

Use a double integral in polar coordinates to find the area
15. Inside the cardioid $r = a(1 - \cos \theta)$ and outside the circle $r = a$
16. Between the cardioid $r = a(1 + \cos \theta)$ and the circle $r = a \cos \theta$
17. Bounded by the circle $r = \frac{1}{2}$ and the curves $\theta = \sin r$ and $\theta = \cos r$
18. Bounded by the circle $x^2 + y^2 = x$, the circle $x^2 + y^2 = 2x$, the x-axis and line $y = x$
19. Bounded by the circle $r = 2 \sin \theta$ and the ray $\theta = \phi$, inside the wedge $0 \leq \theta \leq \phi \leq 2\pi$
20. Between the Archimedean spirals $r = \theta$ and $r = 2\theta$ for $0 \leq \theta \leq 4\pi$

21. A circular sector of radius a and central angle 2ϕ is bisected by the x-axis (see Figure 43). Find its centroid $(\bar{x}, \bar{y})$.
22. The density at a variable point P of a washer of inner radius 1 cm and outer radius 3 cm is inversely proportional to the distance from P to the center of the washer. Find the total mass of the washer if the density at its outer edge is 2 g/cm^2.
23. Find the centroid $(\bar{x}, \bar{y})$ of the region enclosed by the cardioid $r = a(1 + \cos \theta)$.
24. Find the centroid $(\bar{x}, \bar{y})$ of the region enclosed by the petal of the rose $r = \sin 2\theta$ in the first quadrant (see Figure 64, page 624).
25. Find the polar moment of inertia of the region enclosed by the lemniscate $r^2 = a^2 \cos 2\theta$ (see Figure 82, page 638).
26. Find the moments of inertia about the x- and y-axes of the region R enclosed by the cardioid $r = a(1 + \cos \theta)$. Also find the polar moment of inertia of R.

Use a double integral in polar coordinates to find the volume V of the specified solid region T.
27. T is cut from a solid sphere of radius R by a circular cylinder of radius $a < R$ with its axis through the center of the sphere
28. T is bounded by the paraboloid $az = x^2 + y^2$ and the plane $z = a > 0$
29. T is bounded by the sphere $x^2 + y^2 + z^2 = 4$ and the paraboloid $x^2 + y^2 = 4(1 - z)$
30. T is bounded by the xy-plane, the cylinder $x^2 + y^2 = x$ and the cone $z = \sqrt{x^2 + y^2}$
31. T is bounded by the paraboloids $x^2 + y^2 - z = 0$ and $x^2 + y^2 + 2z = 1$
32. T is bounded by the cone $x^2 + y^2 - z^2 = 0$ and the sphere $x^2 + y^2 + z^2 = 9$, and lies outside the cone (of two nappes)

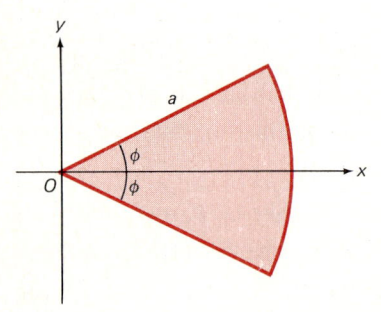

Figure 43

14.7 Triple Integrals in Cylindrical Coordinates

Next we introduce a system of coordinates in space which is the natural extension of the polar coordinate system in the plane. Suppose a point P in space has rectangular coordinates x, y and z. Then P also has *cylindrical coordinates* r, θ and z, obtained by retaining z and replacing x and y by the polar coordinates r and θ of the point Q which is the projection of P onto the xy-plane (see Figure 44). If a point has cylindrical coordinates r, θ and z as well as rectangular coordinates x, y and z, we write $P = (r, \theta, z)$ as well as $P = (x, y, z)$. As in the preceding section, we impose the conditions $r \geq 0$ and $0 \leq \theta \leq 2\pi$, but z is free to take any value in the interval $(-\infty, \infty)$. It is apparent from the figure that the point with cylindrical coordinates r, θ and z has rectangular coordinates

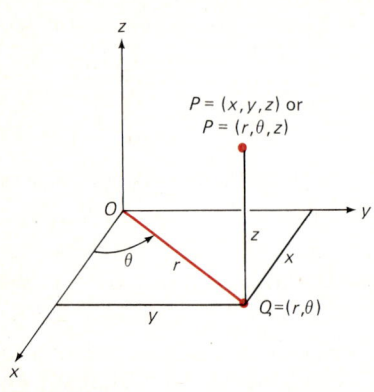

Figure 44

$$x = r \cos \theta, \quad y = r \sin \theta, \quad z = z, \quad (1)$$

and these formulas in turn imply

$$r = \sqrt{x^2 + y^2}, \qquad \tan\theta = \frac{y}{x} \quad (x \neq 0), \qquad z = z. \tag{2}$$

Example 1 According to (1), the point with cylindrical coordinates $r = \sqrt{2}$, $\theta = \pi/4$, $z = -1$ has rectangular coordinates

$$x = \sqrt{2}\cos\frac{\pi}{4} = 1, \qquad y = \sqrt{2}\sin\frac{\pi}{4} = 1, \qquad z = -1,$$

while according to (2), the point with rectangular coordinates $x = \sqrt{3}$, $y = 1$, $z = 2$ has cylindrical coordinates

$$r = \sqrt{(\sqrt{3})^2 + 1} = \sqrt{4} = 2, \qquad \theta = \arctan\frac{1}{\sqrt{3}} = \frac{\pi}{6}, \qquad z = 2. \quad \square$$

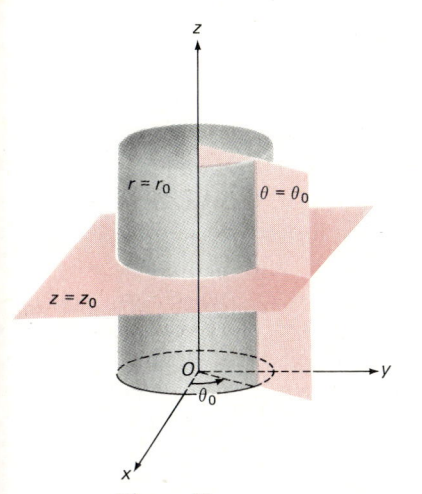

In cylindrical coordinates the graph of the equation $r = r_0 > 0$ is a circular cylinder of radius r_0 with the z-axis as its axis of symmetry, and the graph of $r = 0$ is just the z-axis itself. Similarly, the graph of $\theta = \theta_0$ is the half-plane with the z-axis as its edge, making angle θ_0 with the xz-plane, and the graph of $z = z_0$ is the plane parallel to the xy-plane going through the point $(0, 0, z_0)$ in rectangular coordinates. Observe that the three surfaces $r = r_0$, $\theta = \theta_0$ and $z = z_0$ are mutually orthogonal for every choice of $r_0 \neq 0$, θ_0 and z_0, as illustrated in Figure 45.

Figure 45

Example 2 Describe the graph of the equation

$$z = a\theta \qquad (a > 0) \tag{3}$$

in cylindrical coordinates.

Solution As θ increases so does z, but the value of r is not specified. Therefore r is free to take all nonnegative values for any given pair of values of θ and z. Hence the graph of (3) is an unbounded surface, called a *helicoid*, part of which is shown in Figure 46. The surface is shaped like a spiral ramp, and is generated by a ray with one end attached to the z-axis. In fact, as the end of the ray moves up the z-axis, the ray itself rotates about the z-axis and sweeps out the helicoid. $\quad\square$

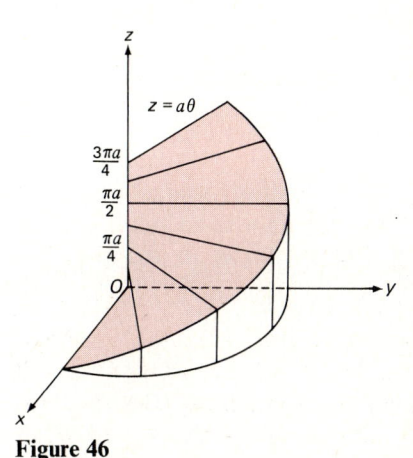

In evaluating a triple integral

$$\iiint_T f(x, y, z)\,dx\,dy\,dz$$

Figure 46

over a normal three-dimensional region T, it is often advantageous to transform the integral to cylindrical coordinates. This is called for in cases where the region T exhibits *cylindrical symmetry*, that is, symmetry about some axis which can be chosen as the z-axis. Suppose that in addition to ordinary xyz-space where every point has both rectangular coordinates x, y and z and cylindrical coordinates r, θ, and z, we introduce another "$r\theta z$-space" in which the coordinates r, θ and z serve as *rectangular* coordinates. Let T'' be the region in $r\theta z$-space which has T as its image under the coordinate

Section 14.7 Triple Integrals in Cylindrical Coordinates

transformation (1). Then we have the following analogue of Theorem 8, page 856, proved in virtually the same way.

> **Theorem 9** (*Triple integrals in cylindrical coordinates*). If $f(x, y, z)$ is continuous on T, then
> $$\iiint_T f(x, y, z)\, dx\, dy\, dz = \iiint_{T'} f(r\cos\theta, r\sin\theta, z)\, r\, dr\, d\theta\, dz. \qquad (4)$$

Partial proof Suppose the region T' is contained in a box $a \le r \le b$, $\alpha \le \theta \le \beta$, $A \le z \le B$ in $r\theta z$-space. Let the points r_j ($j = 0, 1, \ldots, J$) and θ_k ($k = 0, 1, \ldots, K$) be partitions of the intervals $[a, b]$ and $[\alpha, \beta]$, with mesh sizes μ_r and μ_θ, and let z_l ($l = 0, 1, \ldots, L$) be a partition of $[A, B]$ with mesh size μ_z. Then the planes $r = r_j$, $\theta = \theta_k$ and $z = z_l$, parallel to the coordinate planes of $r\theta z$-space, divide T' into n closed subregions $T'_1, T'_2, \ldots, T'_n$, mostly boxes, where $n \le JKL$ (typically $n < JKL$). The cylinders, half-planes and planes in xyz-space with the same equations $r = r_j$, $\theta = \theta_k$ and $z = z_l$ divide T into n closed subregions $T_1, T_2, \ldots, T_n$, mostly "cylindrical wedges" of a type to be described in a moment, where T_i is the image of T'_i under the coordinate transformation (1) carrying cylindrical coordinates into rectangular coordinates. Let

$$\mu = \max\{\mu_r, \mu_\theta, \mu_z\},$$

and let ΔV_i and $\Delta V'_i$ be the volumes of T_i and T'_i. Also let $P'_i = (r'_i, \theta'_i, z'_i)$ be an arbitrary point of T'_i, and let $P_i = (p_i, q_i, s_i)$ be the image of P'_i under the transformation (1). Then

$$\iiint_T f(x, y, z)\, dx\, dy\, dz = \lim_{\mu \to 0} \sum_{i=1}^n f(p_i, q_i, s_i)\, \Delta V_i$$

$$= \lim_{\mu \to 0} \sum_{i=1}^n f(r'_i \cos\theta'_i, r'_i \sin\theta'_i, z'_i)\, \frac{\Delta V_i}{\Delta V'_i}\, \Delta V'_i. \qquad (5)$$

Now suppose T'_i is the box or rectangular parallelepiped

$$r^* \le r \le r^* + \Delta r, \qquad \theta^* \le \theta \le \theta^* + \Delta\theta, \qquad z^* \le z \le z^* + \Delta z$$

of volume $\Delta V'_i = \Delta r\, \Delta\theta\, \Delta z$. Then T_i is the "cylindrical wedge" shown in Figure 47. Actually T_i is a right cylinder of height Δz with an annular sector of area $r_i^* \Delta r\, \Delta\theta$ as its base, where $r_i^* = r^* + \frac{1}{2}\Delta r$ as in the proof of Theorem 8. Hence the volume of T_i is

$$\Delta V_i = r_i^*\, \Delta r\, \Delta\theta\, \Delta z = r_i^*\, \Delta V'_i \qquad (6)$$

(recall Example 1, page 438). Clearly there are points in T'_i with r_i^* as the r-coordinate, and since $P'_i = (r'_i, \theta'_i, z'_i)$ is an arbitrary point of T'_i, we can assume that $r'_i = r_i^*$. Then (6) takes the form $\Delta V_i = r'_i\, \Delta V'_i$, or equivalently

$$\frac{\Delta V_i}{\Delta V'_i} = r'_i. \qquad (7)$$

If all the regions T'_i are boxes, so that the corresponding regions T_i are

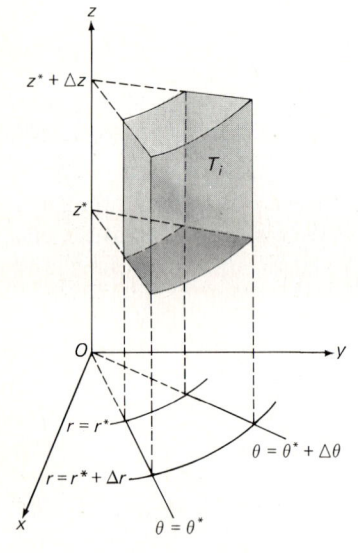

Figure 47

A cylindrical wedge

864   Chapter 14   Multiple Integration

cylindrical wedges, we can substitute (7) into (5), obtaining

$$\iiint_T f(x, y, z)\, dx\, dy\, dz = \lim_{\mu \to 0} \sum_{i=1}^{n} f(r'_i \cos \theta'_i, r'_i \sin \theta'_i, r'_i)\, r'_i \Delta V'_i$$

$$= \iiint_{T'} f(r \cos \theta, r \sin \theta, z)\, r\, dr\, d\theta\, dz,$$

and formula (4) is proved. In general, some of the regions T'_i are nonrectangular, but it can be shown that this does not affect the validity of (4). ∎

How to Remember Formula (4)

Another look at Figure 47 shows that the region T_i is almost a rectangular parallelepiped if Δr and $\Delta \theta$ are both small, with four straight edges of length Δr and four of length Δz, and four curved edges of approximate length $r \Delta \theta$, where r is the radial coordinate of any point in T_i. Hence the volume of T_i is approximately equal to $r \Delta r \Delta \theta \Delta z$. This explains why $r\, dr\, d\theta\, dz$ appears in the right side of formula (4) after taking the limit as $\mu \to 0$, and also serves as a way to remember the formula.

To evaluate the triple integral

$$\iiint_{T'} f(r \cos \theta, r \sin \theta, z)\, r\, dr\, d\theta\, dz$$

we resort to iterated integrals in the familiar fashion, as illustrated by the following examples.

Example 3 Use cylindrical coordinates to find the centroid $(\bar{x}, \bar{y}, \bar{z})$ of a solid hemisphere T of radius a.

Solution Let T be bounded above by the sphere $x^2 + y^2 + z^2 = a^2$ and below by the plane $z = 0$, as in Figure 48. Then $\bar{x} = \bar{y} = 0$ by symmetry, and

$$\bar{z} = \frac{1}{V} \iiint_T z\, dx\, dy\, dz,$$

where $V = \frac{2}{3}\pi a^3$ is the volume of the hemisphere. Using Theorem 9 to transform to cylindrical coordinates, we have

$$\iiint_T z\, dx\, dy\, dz = \iiint_{T'} zr\, dr\, d\theta\, dz,$$

where T' is the region in $r\theta z$-space bounded by the planes $r = 0$, $\theta = 0$, $\theta = 2\pi$, $z = 0$ and the cylinder $r^2 + z^2 = a^2$ (remember that $x^2 + y^2 = r^2$ in both polar and cylindrical coordinates). Going over to an iterated integral, we get

$$\iiint_{T'} zr\, dr\, d\theta\, dz = \int_0^{2\pi} d\theta \int_0^a dr \int_0^{\sqrt{a^2 - r^2}} rz\, dz = \int_0^{2\pi} d\theta \int_0^a r \left[\frac{1}{2} z^2\right]_0^{\sqrt{a^2 - r^2}} dr$$

$$= \pi \int_0^a r(a^2 - r^2)\, dr = \pi \left[\frac{1}{2} a^2 r^2 - \frac{1}{4} r^4\right]_0^a = \frac{1}{4} \pi a^4,$$

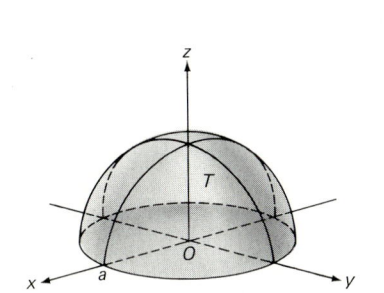

Figure 48

Section 14.7 Triple Integrals in Cylindrical Coordinates **865**

and therefore
$$\bar{z} = \frac{\frac{1}{4}\pi a^4}{\frac{2}{3}\pi a^3} = \frac{3}{8}a.$$

Thus the centroid of T is $(0, 0, \frac{3}{8}a)$. □

Example 4 Let T be a homogeneous solid circular cylinder of radius a and height h. Find the moment of inertia of T about its axis.

Solution Let T be bounded by the cylindrical surface $x^2 + y^2 = a^2$ and the planes $z = 0$ and $z = h$, as in Figure 49, and let ρ be the constant density of T. Then, going over to an iterated integral, we find that the moment of inertia of the cylinder T about its axis (the z-axis) is equal to

$$I_z = \iiint_T (x^2 + y^2)\rho\, dV = \int_0^h dz \int_0^{2\pi} d\theta \int_0^a (\rho r^2) r\, dr$$

$$= \rho \left(\int_0^h dz\right)\left(\int_0^{2\pi} d\theta\right)\left(\int_0^a r^3\, dr\right) = 2\pi\rho h \left[\frac{1}{4}r^4\right]_0^a = \frac{1}{2}\rho\pi a^4 h.$$

Since the total mass of T is $M = \rho\pi a^2 h$, we can write I_z in the concise form

$$I_z = \frac{1}{2}Ma^2. \quad \square$$

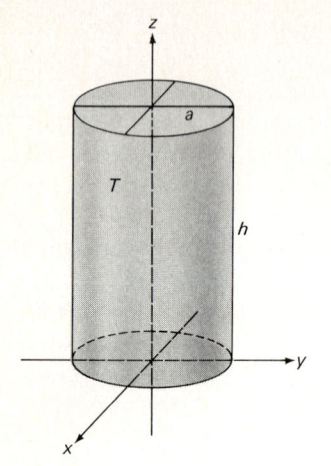

Figure 49

Example 5 Find the volume V of the solid region T bounded above by the sphere $x^2 + y^2 + z^2 = a^2$ of radius a and below by the cone

$$z = \sqrt{x^2 + y^2}\cot\phi \quad (0 < \phi < \pi),$$

whose generators make angle ϕ with the z-axis.

Solution For small values of the angle ϕ the region T is shaped like a filled ice cream cone, as in Figure 50. In cylindrical coordinates the sphere and the cone have equations $r^2 + z^2 = a^2$ and $z = r\cot\phi$, and it is apparent from the figure that the two surfaces intersect for $r = a\sin\phi$. Therefore, with the help of Theorem 9,

$$V = \iiint_T dV = \int_0^{2\pi} d\theta \int_0^{a\sin\phi} dr \int_{r\cot\phi}^{\sqrt{a^2-r^2}} r\, dz$$

$$= \int_0^{2\pi} d\theta \int_0^{a\sin\phi} (\sqrt{a^2 - r^2} - r\cot\phi) r\, dr$$

$$= 2\pi\left[-\frac{1}{3}(a^2 - r^2)^{3/2} - \frac{1}{3}r^3\cot\phi\right]_0^{a\sin\phi}$$

$$= \frac{2}{3}\pi a^3(1 - \cos^3\phi - \sin^3\phi\cot\phi)$$

$$= \frac{2}{3}\pi a^3[1 - \cos^3\phi - (1 - \cos^2\phi)\cos\phi],$$

so that finally

$$V = \frac{2}{3}\pi a^3(1 - \cos\phi). \tag{8}$$

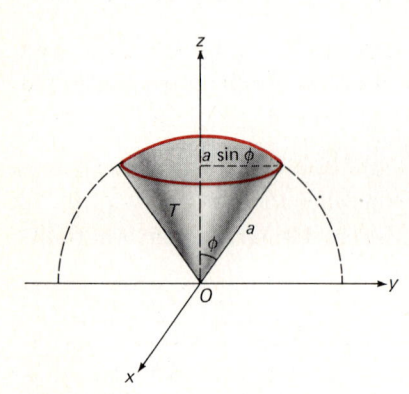

Figure 50

866 Chapter 14 Multiple Integration

In particular, it follows from (8) that $V = \frac{2}{3}\pi a^3$ if $\phi = \pi/2$, as is to be expected (why?). □

Problems

Find the rectangular coordinates of the point with the given cylindrical coordinates.

1. $r = \sqrt{3}, \theta = 5\pi/6, z = 0$
2. $r = 1, \theta = 1, z = 1$
3. $r = 10, \theta = \pi, z = 2\pi$
4. $r = 0, \theta = 9\pi/10, z = -5$

Find the cylindrical coordinates of the point with the given rectangular coordinates.

5. $x = -4, y = 4, z = -2$
6. $x = 3, y = \sqrt{3}, z = 13$
7. $x = 1, y = -\sqrt{3}, z = -6$
8. $x = -5, y = -12, z = 0$

Transform the given equation from rectangular to cylindrical coordinates.

9. $x^2 - y^2 = z^2$
10. $x + y + z = 8$
11. $x^2 + y^2 - z^2 = 1$
12. $z = 2xy$

Evaluate the given triple integral by transforming it from rectangular to cylindrical coordinates.

13. $\iiint_T (x^2 + y^2)\,dx\,dy\,dz$, where T is the solid region bounded by the cylinder $x^2 + y^2 = 2x$ and the planes $z = 0$ and $z = 2$

14. $\iiint_T xz\,dx\,dy\,dz$, where T is the solid region in the first octant bounded by the paraboloid $z = x^2 + y^2$ and the planes $x = 0, y = 0$ and $z = 4$

15. $\iiint_T (x^2 + y^2 + z^2)^2\,dx\,dy\,dz$, where T is the solid region bounded by the cylinder $x^2 + y^2 = 3$ and the planes $z = \pm 1$

16. $\iiint_T x^2 y^2\,dx\,dy\,dz$, where T is the solid region bounded by the cone $z = \sqrt{x^2 + y^2}$ and the plane $z = a > 0$

17. $\iiint_T y^2 z\,dx\,dy\,dz$, where T is the solid region bounded above by the sphere $x^2 + y^2 + z^2 = 4$ and below by the cone $z = \sqrt{x^2 + y^2}$

18. $\iiint_T (x^3 + y^3)\,dx\,dy\,dz$, where T is the solid region bounded by the cylinder $x^2 + y^2 = y$, the paraboloid $z = x^2 + y^2$ and the xy-plane

Use cylindrical coordinates to find the volume V of the specified solid region T

19. T is bounded by the xy-plane, the cylinder $x^2 + y^2 = a^2$ of radius a and the first turn of the helicoid $z = a\theta$ ($0 \leq \theta \leq 2\pi$)

20. T is bounded by the xy-plane, the cylinder $x^2 + y^2 = 4$ and the plane $x + y + z = 6$

21. T is bounded by the circular cylinder $x^2 + y^2 = 2x$ and the parabolic cylinder $z^2 = 4x$

22. T is bounded by the single-sheeted hyperboloid of revolution $x^2 + y^2 - z^2 = 1$ and the sphere $x^2 + y^2 + z^2 = 3$, and contains the origin

23. With the help of cylindrical coordinates, find the centroid of the solid region T considered in Example 4, page 860.

24. Use cylindrical coordinates to give alternative solutions of Examples 4 and 5, pages 839 and 840.

25. Find the moment of inertia of a homogeneous solid circular cylinder of mass M, radius a and height h about a diameter of its base. Use the parallel axis theorem (see Problem 15, page 855) to find the moment of inertia of the cylinder about an axis through its center of mass perpendicular to its axis of symmetry.

26. Find the moment of inertia of a homogeneous hollow circular cylinder of mass M, inner radius a_1 and outer radius a_2 about its axis of symmetry. Also find the moment of inertia of a thin-walled circular pipe of mass M and radius a about its axis of symmetry.

14.8 Triple Integrals in Spherical Coordinates

Just as cylindrical coordinates are particularly well suited to problems involving objects which exhibit cylindrical symmetry, the coordinates that we now introduce are usually the best to use in problems involving objects, like balls and spherical shells, which exhibit *spherical symmetry*, that is, symmetry about a point in space which can be chosen as the origin. Given a

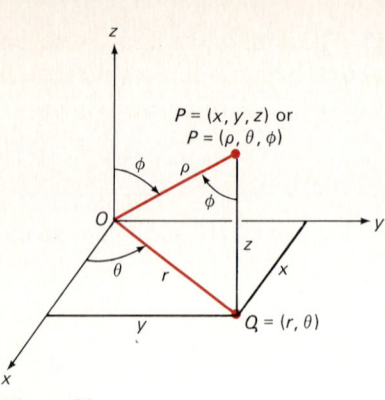

Figure 51

point P with rectangular coordinates x, y and z, let $\rho = |OP|$ be the distance between the origin O and P, and let ϕ be the angle measured (downward) from the positive z-axis to OP (see Figure 51). Also let θ be the same angle as in the case of cylindrical coordinates, namely the angle from the positive x-axis to OQ, where Q is the projection of P onto the xy-plane (θ is measured in the counterclockwise direction as seen by an observer looking down on the xy-plane from the side of positive z). Then the point P is said to have *spherical coordinates* ρ, θ and ϕ, and we write $P = (\rho, \theta, \phi)$ as well as $P = (x, y, z)$. The radial coordinate ρ can take any value in the interval $[0, \infty)$, but we impose the conditions $0 \leq \theta \leq 2\pi$ and $0 \leq \phi \leq \pi$ on the angular coordinates θ and ϕ. Guided by their geographical meaning, we call θ the *longitude* and ϕ the *colatitude* (the complement of the latitude). Since the symbol ρ has now been preempted by the radial coordinate, in problems involving spherically symmetric distributions of mass, make sure to use another symbol, say δ, for the density function.

Examining the figure, we find that the point with spherical coordinates ρ, θ and ϕ has cylindrical coordinates r, θ and z, where

$$r = \rho \sin \phi, \qquad \theta = \theta, \qquad z = \rho \cos \phi,$$

and hence has rectangular coordinates

$$x = r \cos \theta = \rho \sin \phi \cos \theta, \qquad y = r \sin \theta = \rho \sin \phi \sin \theta, \qquad z = \rho \cos \phi. \tag{1}$$

As you can easily verify, it follows from (1) that

$$\rho = \sqrt{x^2 + y^2 + z^2}, \qquad \tan \theta = \frac{y}{x} \quad (x \neq 0),$$

$$\cos \phi = \frac{z}{\sqrt{x^2 + y^2 + z^2}} \qquad (x^2 + y^2 + z^2 \neq 0). \tag{2}$$

Example 1 According to (1), the point with spherical coordinates $\rho = \sqrt{2}$, $\theta = \pi/3$, $\phi = \pi/4$ has rectangular coordinates

$$x = \sqrt{2} \sin \frac{\pi}{4} \cos \frac{\pi}{3} = \frac{1}{2}, \qquad y = \sqrt{2} \sin \frac{\pi}{4} \sin \frac{\pi}{3} = \frac{\sqrt{3}}{2},$$

$$z = \sqrt{2} \cos \frac{\pi}{4} = 1,$$

while according to (2), the point with rectangular coordinates $x = 1$, $y = -1$, $z = \sqrt{2}$ has spherical coordinates

$$\rho = \sqrt{1^2 + (-1)^2 + (\sqrt{2})^2} = \sqrt{4} = 2, \qquad \theta = \arctan(-1) + 2\pi = \frac{7\pi}{4},$$

$$\phi = \arccos \frac{\sqrt{2}}{2} = \frac{\pi}{4}$$

(why do we add 2π in the second formula?). □

In spherical coordinates the graph of the equation $\rho = \rho_0 > 0$ is a sphere of radius ρ_0 centered at the origin, and the graph of $\rho = 0$ is just the origin itself. Similarly, the graph of $\theta = \theta_0$ is the half-plane with the z-axis as its edge making angle θ_0 with the xz-plane, and the graph of $\phi = \phi_0$ is

868 Chapter 14 Multiple Integration

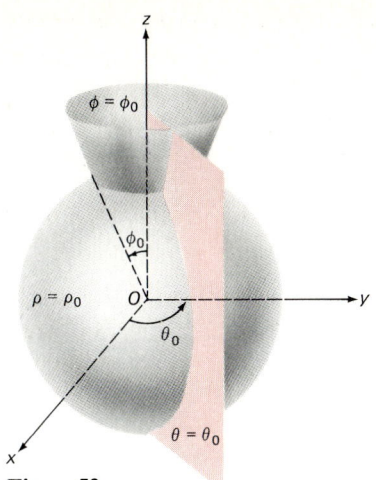

one nappe of a right circular cone with vertex at the origin and generators making angle ϕ_0 with the positive z-axis (see Figure 52). Notice that the graph of $\phi = \phi_0$ reduces to the positive z-axis if $\phi_0 = 0$, the xy-plane if $\phi_0 = \pi/2$, and the negative z-axis if $\phi_0 = \pi$.

Example 2 Write an equation in spherical coordinates for the sphere S with cartesian equation

$$x^2 + y^2 + z^2 = 2az \qquad (a > 0). \tag{3}$$

Solution Since (3) is equivalent to $x^2 + y^2 + (z - a)^2 = a^2$, S is a sphere of radius a centered at the point $(0, 0, a)$ of the z-axis. Substituting from (1), we find that (3) transforms into $\rho^2 = 2a\rho \cos \phi$ or equivalently

$$\rho = 2a \cos \phi$$

(justify the division by ρ). ▫

In evaluating a triple integral

$$\iiint_T f(x, y, z)\,dx\,dy\,dz$$

over a normal three-dimensional region T, a transformation to spherical coordinates is usually called for in cases where the region T exhibits spherical symmetry. Suppose that in addition to ordinary xyz-space, where every point has both rectangular coordinates x, y and z and spherical coordinates r, θ and ϕ, we introduce another "$\rho\theta\phi$-space," in which the coordinates ρ, θ and ϕ serve as *rectangular* coordinates. Let T' be the region in $\rho\theta\phi$-space which has T as its image under the coordinate transformation (1). Then we have the following theorem, analogous to Theorem 9, page 864, for cylindrical coordinates.

Theorem 10 *(Triple integrals in spherical coordinates).* If $f(x, y, z)$ is continuous on T, then

$$\iiint_T f(x, y, z)\,dx\,dy\,dz$$

$$= \iiint_{T'} f(\rho \sin \phi \cos \theta, \rho \sin \phi \sin \theta, \rho \cos \phi)\rho^2 \sin \phi\,d\rho\,d\theta\,d\phi. \tag{4}$$

Partial proof Suppose the region T' is contained in a box $a \leq \rho \leq b$, $\alpha \leq \theta \leq \beta$, $A \leq \phi \leq B$ in $\rho\theta\phi$-space. Let the points ρ_j ($j = 0, 1, \ldots, J$), θ_k ($k = 0, 1, \ldots, K$) and ϕ_l ($l = 0, 1, \ldots, L$) be partitions of the intervals $[a, b]$, $[\alpha, \beta]$ and $[A, B]$, with mesh sizes μ_ρ, μ_θ and μ_ϕ, respectively. Then the planes $\rho = \rho_j$, $\theta = \theta_k$ and $\phi = \phi_l$, parallel to the coordinate planes of $\rho\theta\phi$-space, divide T' into n closed subregions $T'_1, T'_2, \ldots, T'_n$, mostly boxes, where $n \leq JKL$ (typically $n < JKL$). The spheres, half-planes and cones in xyz-space with the same equations $\rho = \rho_j$, $\theta = \theta_k$ and $\phi = \phi_l$ divide T into n closed subregions $T_1, T_2, \ldots, T_n$, where T_i is the image of T'_i under the transformation (1). Let

$$\mu = \max\{\mu_\rho, \mu_\theta, \mu_\phi\},$$

and let ΔV_i and $\Delta V'_i$ be the volumes of T_i and T'_i. Also let $P'_i = (\rho'_i, \theta'_i, \phi'_i)$ be an arbitrary point of T'_i, and let $P_i = (p_i, q_i, s_i)$ be the image of P'_i under (1). Then

$$\iiint_T f(x, y, z)\, dx\, dy\, dz$$

$$= \lim_{\mu \to 0} \sum_{i=1}^{n} f(p_i, q_i, s_i)\, \Delta V_i$$

$$= \lim_{\mu \to 0} \sum_{i=1}^{n} f(\rho'_i \sin \phi'_i \cos \theta'_i, \rho'_i \sin \phi'_i \sin \theta'_i, \rho'_i \cos \phi'_i) \frac{\Delta V_i}{\Delta V'_i} \Delta V'_i. \quad (5)$$

Now suppose T'_i is the box

$$\rho^* \le \rho \le \rho^* + \Delta\rho, \qquad \theta^* \le \theta \le \theta^* + \Delta\theta, \qquad \phi^* \le \phi \le \phi^* + \Delta\phi$$

of volume $\Delta V'_i = \Delta\rho\, \Delta\theta\, \Delta\phi$. Then T_i is the "spherical wedge" shown in Figure 53, bounded by the two spheres $\rho = \rho^*$ and $\rho = \rho^* + \Delta\rho$, the two half-planes $\theta = \theta^*$ and $\theta = \theta^* + \Delta\theta$, and the two cones $\phi = \phi^*$ and $\phi = \phi^* + \Delta\phi$. Therefore, with the help of Example 5, page 866, we find that

$$\Delta V_i = \frac{2\pi}{3} \frac{\Delta\theta}{2\pi} [(\rho^* + \Delta\rho)^3 - \rho^{*3}][1 - \cos(\phi^* + \Delta\phi) - (1 - \cos \phi^*)]$$

$$= -\frac{\Delta\theta}{3}[(\rho^* + \Delta\rho)^3 - \rho^{*3}][\cos(\phi^* + \Delta\phi) - \cos \phi^*]. \quad (6)$$

But, by two applications of the mean value theorem,

$$(\rho^* + \Delta\rho)^3 - \rho^{*3} = 3\rho_i^{*2}\,\Delta\rho, \qquad \cos(\phi^* + \Delta\phi) - \cos \phi^* = -\sin \phi_i^*\, \Delta\phi, \quad (7)$$

where $\rho^* < \rho_i^* < \rho^* + \Delta\rho$ and $\phi^* < \phi_i^* < \phi^* + \Delta\phi$ (the subscripts on the quantities ρ_i^* and ϕ_i^* show that they go with the region T_i). Substituting (7) into (6), we get

$$\Delta V_i = \rho_i^{*2} \sin \phi_i^*\, \Delta\rho\, \Delta\theta\, \Delta\phi = \rho_i^{*2} \sin \phi_i^*\, \Delta V'_i. \quad (8)$$

Clearly there are points in T'_i with ρ-coordinate ρ_i^* and ϕ-coordinate ϕ_i^*, and since $P'_i = (\rho'_i, \theta'_i, \phi'_i)$ is an arbitrary point of T'_i, we can assume that $\rho'_i = \rho_i^*$ and $\phi'_i = \phi_i^*$. Then (8) takes the form $\Delta V_i = \rho_i'^2 \sin \phi'_i \Delta V'_i$, or equivalently

$$\frac{\Delta V_i}{\Delta V'_i} = \rho_i'^2 \sin \phi'_i. \quad (9)$$

If all the regions T'_i are boxes, we can substitute (9) into (5), obtaining

$$\iiint_T f(x, y, z)\, dx\, dy\, dz$$

$$= \lim_{\mu \to 0} \sum_{i=1}^{n} f(\rho'_i \sin \phi'_i \cos \theta'_i, \rho'_i \sin \phi'_i \sin \theta'_i, \rho'_i \cos \phi'_i) \rho_i'^2 \sin \phi'_i\, \Delta V'_i$$

$$= \iiint_{T'} f(\rho \sin \phi \cos \theta, \rho \sin \phi \sin \theta, \rho \cos \phi)\, \rho^2 \sin \phi\, d\rho\, d\theta\, d\phi,$$

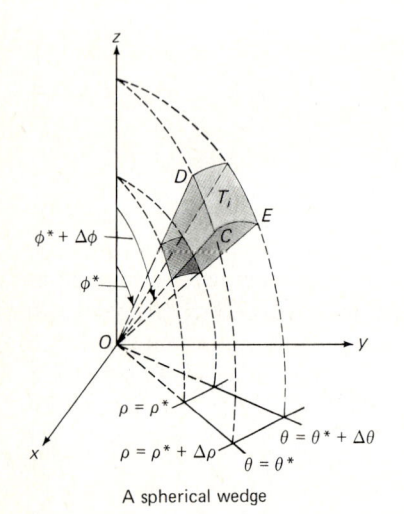

Figure 53 A spherical wedge

and formula (4) is proved. In general, some of the regions T'_i are nonrectangular, but it can be shown that this does not affect the validity of (4). ■

How to Remember Formula (4)

Inspection of Figure 53 shows that the region T_i is almost a rectangular parallelepiped if $\Delta\rho$, $\Delta\theta$ and $\Delta\phi$ are small, with four straight edges of length $\Delta\rho$, four curved edges like CD of approximate length $\rho\,\Delta\phi$, and four curved edges like CE of approximate length $\rho \sin\phi\,\Delta\theta$, where ρ and ϕ are the radial coordinate and colatitude of any point in T_i. Hence the volume of T_i is approximately equal to $\rho^2 \sin\phi\,\Delta\rho\,\Delta\theta\,\Delta\phi$. This explains the presence of the expression $\rho^2 \sin\phi\,d\rho\,d\theta\,d\phi$ in the triple integral over T', and makes it easy to remember formula (4).

Needless to say, the triple integral

$$\iiint_{T'} f(\rho \sin\phi \cos\theta, \rho \sin\phi \sin\theta, \rho \cos\phi)\rho^2 \sin\phi\,d\rho\,d\theta\,d\phi$$

is evaluated by using iterated integrals, as illustrated in the following examples.

Example 3 Use spherical coordinates to find the centroid $(\bar{x}, \bar{y}, \bar{z})$ of a solid hemisphere T of radius a.

Solution This problem has already been solved with the help of cylindrical coordinates in Example 3, page 865, but the solution is even easier in spherical coordinates. Let T be bounded above by the sphere $x^2 + y^2 + z^2 = a^2$ and below by the plane $z = 0$. Again $\bar{x} = \bar{y} = 0$ by symmetry, and

$$\bar{z} = \frac{1}{V} \iiint_T z\,dx\,dy\,dz,$$

where $V = \frac{2}{3}\pi a^3$ is the volume of the hemisphere. Using Theorem 10 to transform to spherical coordinates, we have

$$\iiint_T z\,dx\,dy\,dz = \iiint_{T'} (\rho \cos\phi)\rho^2 \sin\phi\,d\rho\,d\theta\,d\phi,$$

where T' is the box in $\rho\theta\phi$-space bounded by the planes $\rho = 0$, $\rho = a$, $\theta = 0$, $\theta = 2\pi$, $\phi = 0$ and $\phi = \pi/2$. Going over to an iterated integral, we get

$$\iiint_{T'} \rho^3 \cos\phi \sin\phi\,d\rho\,d\theta\,d\phi = \int_0^{2\pi} d\theta \int_0^{\pi/2} d\phi \int_0^a \rho^3 \cos\phi \sin\phi\,d\rho$$

$$= \left(\int_0^{2\pi} d\theta\right)\left(\int_0^{\pi/2} \sin\phi \cos\phi\,d\phi\right)\left(\int_0^a \rho^3\,d\rho\right)$$

$$= 2\pi \left[\frac{1}{2}\sin^2\phi\right]_0^{\pi/2}\left[\frac{1}{4}\rho^4\right]_0^a = \frac{1}{4}\pi a^4,$$

and hence

$$\bar{z} = \frac{\frac{1}{4}\pi a^4}{\frac{2}{3}\pi a^3} = \frac{3}{8}a,$$

as before. Thus the centroid of T is $(0, 0, \frac{3}{8}a)$. □

Example 4. Use spherical coordinates to find the moment of inertia I_l of a homogeneous solid sphere T of radius a about any axis l through its center.

Solution There is no loss of generality in choosing the center of T at the origin of a system of rectangular coordinates x, y and z, with l as the x-axis. Since the sphere T is homogeneous, its density δ is constant. Therefore

$$I_l = I_x = \delta \iiint_T (y^2 + z^2) \, dV$$

$$= 2\delta \iiint_T z^2 \, dV = \frac{2\delta}{3} \iiint_T (x^2 + y^2 + z^2) \, dV,$$

where we use the fact that

$$\iiint_T x^2 \, dV = \iiint_T y^2 \, dV = \iiint_T z^2 \, dV,$$

because of the symmetry. Transforming to spherical coordinates (without explicitly introducing the region T'), we have

$$I_l = \frac{2\delta}{3} \int_0^{2\pi} d\theta \int_0^{\pi} d\phi \int_0^a (\rho^2) \rho^2 \sin\phi \, d\rho$$

$$= \frac{2\delta}{3} \left(\int_0^{2\pi} d\theta \right) \left(\int_0^{\pi} \sin\phi \, d\phi \right) \left(\int_0^a \rho^4 \, d\rho \right)$$

$$= \frac{4\pi\delta}{3} \left[-\cos\phi \right]_0^{\pi} \left[\frac{\rho^5}{5} \right]_0^a = \frac{8\pi\delta}{15} a^5.$$

Let M be the total mass of T. Then

$$M = \frac{4\pi\delta}{3} a^3,$$

so that we can write I_l in the concise form

$$I_l = \frac{2}{5} Ma^2. \quad \square$$

Example 5 Let S be the sphere with equation $x^2 + y^2 + z^2 = 2az$ ($a > 0$), that is, the sphere of radius a centered at the point $(0, 0, a)$ of the z-axis. Show that the cone C with equation $x^2 + y^2 - z^2 = 0$ divides the interior of S into two pieces, one of which is three times larger than the other.

Solution It was found in Example 2 that S has the equation

$$\rho = 2a \cos\phi$$

in spherical coordinates. As Figure 54 shows in cross section, the upper nappe of C divides the interior of S into two pieces T_1 and T_2, and the generators of C

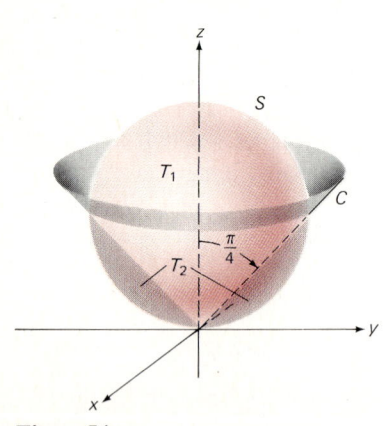

Figure 54

make angle $\pi/4$ with the z-axis. Hence the volume of T_1 is

$$V_1 = \int_0^{2\pi} d\theta \int_0^{\pi/4} d\phi \int_0^{2a\cos\phi} \rho^2 \sin\phi \, d\rho$$

$$= \int_0^{2\pi} d\theta \int_0^{\pi/4} \left[\frac{1}{3}\rho^3\right]_0^{2a\cos\phi} \sin\phi \, d\phi$$

$$= \frac{8a^3}{3}\left(\int_0^{2\pi} d\theta\right)\left(\int_0^{\pi/4} \cos^3\phi \sin\phi \, d\phi\right) = \frac{16\pi a^3}{3}\left[-\frac{1}{4}\cos^4\phi\right]_0^{\pi/4}$$

$$= \frac{4\pi a^3}{3}\left[1 - \left(\frac{1}{\sqrt{2}}\right)^4\right] = \pi a^3,$$

while the volume of T_2 is

$$V_2 = \frac{4}{3}\pi a^3 - V_1 = \frac{1}{3}\pi a^3.$$

Therefore $V_1 = 3V_2$, so that the volume of T_1 is three times larger than the volume of T_2. □

In Sections 14.6–14.8 we have shown how the evaluation of certain double and triple integrals can be simplified by transforming from rectangular to polar, cylindrical and spherical coordinates. It turns out that Theorems 8–10 are all consequences of a general treatment of change of variables in multiple integrals, given at the end of Section 15.4.

Problems

Find the rectangular coordinates of the point with the given spherical coordinates.
1. $\rho = 1, \theta = \pi, \phi = \pi/2$
2. $\rho = \sqrt{6}, \theta = 3\pi/4, \phi = \pi/3$
3. $\rho = 4, \theta = 5\pi/6, \phi = \pi/6$
4. $\rho = 2, \theta = 0, \phi = 3\pi/4$

Find the spherical coordinates of the point with the given rectangular coordinates.
5. $x = -1, y = 1, z = 1$
6. $x = 0, y = 1, z = -\sqrt{3}$
7. $x = 4, y = -12, z = 3$
8. $x = 6, y = 2, z = 9$

Transform the given equation from rectangular to spherical coordinates.
9. $x^2 + y^2 = 4$
10. $x^2 + y^2 + z^2 = 2z$
11. $x^2 + y^2 = z^2$
12. $x^2 - y^2 = z^2$

Evaluate the given triple integral by transforming it from rectangular to spherical coordinates.
13. $\iiint_T \sqrt{x^2 + y^2 + z^2} \, dx \, dy \, dz$, where T is the ball $x^2 + y^2 + z^2 \leq a^2 \quad (a > 0)$

14. $\iiint_T (x^2 + y^2) \, dx \, dy \, dz$, where T is the solid region bounded above by the unit sphere $x^2 + y^2 + z^2 = 1$ and below by the cone $z = -\sqrt{\frac{1}{3}(x^2 + y^2)}$

15. $\iiint_T yz \, dx \, dy \, dz$, where T is the part of the spherical shell $1 \leq x^2 + y^2 + z^2 \leq 4$ in the first octant

16. $\iiint_T \frac{dx \, dy \, dz}{x^2 + y^2 + z^2}$, where T is the bounded solid region between the cylinder $x^2 + y^2 = 4$ and the nappes of the cone $x^2 + y^2 - z^2 = 0$

17. $\iiint_T \frac{dx \, dy \, dz}{\sqrt{x^2 + y^2 + (z-2)^2}}$, where T is the unit ball $x^2 + y^2 + z^2 \leq 1$

18. $\iiint_T \frac{dx \, dy \, dz}{(x^2 + y^2 + z^2)^{3/2}}$, where T is the spherical shell $1 \leq x^2 + y^2 + z^2 \leq e$

19. $\iiint_T (x^2 + y^2 + z^2)\, dx\, dy\, dz$, where T is the ball $x^2 + y^2 + z^2 \leq z$

20. $\iiint_T \sqrt{1 + (x^2 + y^2 + z^2)^{3/2}}\, dx\, dy\, dz$, where T is the ball $x^2 + y^2 + z^2 \leq 3^{2/3}$

Use spherical coordinates to find the volume V of the specified solid region T.

21. T is cut out of the sphere $x^2 + y^2 + z^2 = 4z - 3$ by the cone $z^2 = 4(x^2 + y^2)$ and lies inside the cone
22. T lies inside the nappes of the cone $x^2 + y^2 - z^2 = 0$ between the spheres $x^2 + y^2 + z^2 = 1$ and $x^2 + y^2 + z^2 = 16$
23. T is enclosed by the surface $(x^2 + y^2 + z^2)^2 = 8z$
24. T is enclosed by the surface $(x^2 + y^2 + z^2)^3 = xyz$

25. The density at a variable point P of a spherical shell of inner radius 10 cm and outer radius 12 cm is inversely proportional to the distance from P to the center of the shell. Find the total mass of the shell if the density at its inner surface is 3 g/cm^3.

26. Find the centroid of the part T of the solid sphere $x^2 + y^2 + z^2 \leq a^2$ $(a > 0)$ in the first octant.

27. Find the centroid of the solid region T bounded above by the sphere $x^2 + y^2 + z^2 = 4$ and below by the cone $z = \sqrt{\frac{1}{3}(x^2 + y^2)}$.

28. Find the moment of inertia of a homogeneous hollow ball of mass M, inner radius a_1 and outer radius a_2 about any axis l through its center. Also find the moment of inertia of a thin-walled spherical shell of mass M and radius a about any such axis.

Key Terms and Topics

The double integral over a plane region
Area of a region and volume under a surface as double integrals
Iterated integrals and their use in evaluating double integrals
The triple integral over a solid region
Volume of a solid region as a triple integral
Evaluation of triple integrals in terms of iterated integrals
Calculation of the mass of a plane or solid region from its density
Center of mass and moments of a system of particles
Center of mass and moments of a lamina or solid
Centroid of a plane or solid region
Center of mass of a wire of variable density, centroid of a curve
Pappus' theorem for a solid of revolution
Pappus' theorem for a surface of revolution
Moments of inertia and radii of gyration
Double integrals in polar coordinates
Triple integrals in cylindrical and spherical coordinates

Supplementary Problems

Evaluate the given iterated integral.

1. $\int_0^3 dy \int_0^1 xy\, dx$
2. $\int_0^1 dx \int_{-1}^0 2^{x+y}\, dy$
3. $\int_1^e dx \int_{1/x}^x \ln x\, dy$
4. $\int_0^{\pi/2} dy \int_0^y \sin(x + y)\, dx$
5. $\int_0^1 dy \int_0^{y^2} e^{x/y}\, dx$
6. $\int_0^{\pi/2} d\theta \int_0^{2 \sin \theta} r^2 \cos \theta\, dr$

7. Evaluate
$$\iint_R \frac{y}{x^2 + y^2}\, dA,$$
where R is the region bounded by the line $y = x$ and the parabola $y = x^2$.

8. Let R be the region bounded by the x-axis and the curve
$$x = a(t - \sin t), \quad y = a(1 - \cos t) \quad (0 \leq t \leq 2\pi, a > 0)$$
(one arch of a cycloid). Show that $\iint_R y\, dA = \frac{5}{2}\pi a^3$. Find the centroid of R.

9. Find the volume of the solid region bounded by the elliptic cylinder $(x^2/a^2) + (y^2/b^2) = 1$, the xy-plane and any plane $z = \alpha x + \beta y + h$ lying above the trace of the cylinder in the xy-plane.

10. Find the volume of the solid region bounded by the paraboloid of revolution $z = x^2 + y^2$, the parabolic cylinder $y = x^2$, the xy-plane, and the plane $y = 1$.

11. Find the volume of the solid region cut from the circular cylinder $x^2 + y^2 = 2x$ by the paraboloid of revolution $y^2 + z^2 = 4x$.

12. Use a double integral to show that if $f(x)$ and $g(x)$ are continuous on $[a, b]$, then

$$\left| \int_a^b f(x)g(x)\,dx \right| \leq \sqrt{\int_a^b f^2(x)\,dx} \sqrt{\int_a^b g^2(x)\,dx}. \quad \text{(i)}$$

This is the *Cauchy–Schwarz inequality* for integrals (compare with the Cauchy–Schwarz inequality for sums, given in Problem 47, page 661.
Hint. Consider the double integral

$$\iint_R [f(x)g(y) - f(y)g(x)]^2\,dx\,dy,$$

where R is the square bounded by the lines $x = a$, $x = b$, $y = a$ and $y = b$.

13. Show that if the function $f(x)$ is positive and continuous on $[a, b]$, then

$$\sqrt{\int_a^b f(x)\,dx} \sqrt{\int_a^b \frac{dx}{f(x)}} \geq b - a. \quad \text{(ii)}$$

14. Verify the inequality (i) for the functions $f(x) = e^x$ and $g(x) = x$ on the interval $[0, 1]$. Verify the inequality (ii) for the function $f(x) = x$ on the interval $[1, 2]$.

15. Evaluate $\iiint_T \sin x \cos y \tan z\,dV$, where T is the box bounded by the planes $x = 0$, $x = \pi/2$, $y = 0$, $y = \pi/6$, $z = 0$ and $z = \pi/3$.

16. As shown on page 442, the volume V of the solid of revolution T generated by revolving the graph of a continuous nonnegative function $y = f(x)$ ($a \leq x \leq b$) about the x-axis is given by the formula $V = \pi \int_a^b [f(x)]^2\,dx$. Give an alternative derivation of this formula, based on the use of a triple integral.

17. The density at a variable point P of a square lamina 6 cm on a side is proportional to the square of the distance from P to one vertex of the square. Find the total mass of the lamina if the density is 3 g/cm² at the center of the square.

18. Find the total mass of the box bounded by the coordinate planes and the planes $x = 1$, $y = 3$ and $z = 5$ if the density function is $\rho(x, y, z) = x + y + z$.

19. Find the total mass of the object in the first octant bounded by the sphere $x^2 + y^2 + z^2 = 9$, the coordinate planes and the plane $3x + 2y = 6$ if the density function is $\rho(x, y, z) = 12z$.

20. Show that the center of mass of a system of three noncollinear particles A, B and C of equal mass lies at the point of intersection of the medians of the triangle ABC.

21. Find the centroid of the region R bounded by the parabolas $y^2 = 4 + 2x$ and $y^2 = 4 - 4x$.

22. Find the centroid of the region R bounded by the curve $y = \cos x$ and the lines $x = \pi/2$ and $y = 1$.

23. Find the centroid of the region R enclosed by the graph of the equation $y^2 = x^2(1 - x^2)$ ($x \geq 0$).

24. Find the center of mass of the tetrahedron T bounded by the coordinate planes and the plane $x + y + z = 1$ if the density of T at a variable point P is proportional to the distance from P to the xz-plane.

25. Find the centroid of the solid region T bounded by the planes $x = 1$, $x = 3$, $y = 0$, $z = 0$ and $y + 2z = 2$.

26. Find the centroid of the solid region T bounded by the elliptic paraboloid $z = 2x^2 + y^2$ and the plane $z = 4$.

27. Find the rectangular coordinates of the centroid of the plane curve $r = 2(1 - \cos \theta)$ ($0 \leq \theta \leq \pi$), which is the upper half of the cardioid in Figure 65, page 624.

28. Find the centroid of the curve in Problem 8 (one arch of a cycloid).

29. Use Pappus' theorems to find both the volume V and the surface area A of the three-dimensional region generated by revolving a semicircular region of radius r about the tangent line parallel to its diameter.

30. The ellipse $(x^2/a^2) + (y^2/b^2) = 1$ is revolved about the line $y = 3b$. What is the volume V of the resulting solid of revolution?

31. Find the moment of inertia I_l of the region bounded by the parabola $y^2 = x$ and the line $x = 1$ about the line l with equation $y = 1$.

32. Find the polar moment of inertia I_O of the region bounded by the curve $\sqrt{x} + \sqrt{y} = 1$ and the coordinate axes.

Find the moment of inertia I_z about the z-axis of the solid region T bounded by

33. The spheres $x^2 + y^2 + z^2 = 4$ and $x^2 + y^2 + z^2 = 4z$ (see Problem 15, page 832)

34. The paraboloid of revolution $2z = x^2 + y^2$ and the sphere $x^2 + y^2 + z^2 = 3$ (see Problem 22, page 832)

35. Use polar coordinates to evaluate

$$\iint_R \sqrt{\frac{1 - x^2 - y^2}{1 + x^2 + y^2}}\,dx\,dy,$$

where R is the unit disk $x^2 + y^2 \leq 1$.

36. Use polar coordinates to give an alternative solution of Example 6, page 821.

Evaluate the given iterated integral by transforming it to polar coordinates.

37. $\int_0^1 dx \int_0^{\sqrt{1-x^2}} x^2\sqrt{x^2 + y^2}\,dy$

38. $\int_0^1 dx \int_0^x \frac{x^2}{\sqrt{x^2 + y^2}}\,dx$

Supplementary Problems **875**

39. Use polar coordinates to find the area A of the region R enclosed by the curve $(x^2 + y^2)^2 = 2x^3$.

40. Find the centroid $(\bar{x}, \bar{y})$ of the region R in the preceding problem.

41. Find the moment of inertia of a circular disk of radius a about an axis l perpendicular to the disk at a point of its circumference.

42. Use cylindrical coordinates to evaluate the iterated integral

$$\int_0^2 dx \int_0^{\sqrt{2x-x^2}} dy \int_0^3 z\sqrt{x^2 + y^2} \, dz.$$

43. Use cylindrical coordinates to find the volume of the solid region between the paraboloid $z = x^2 + y^2$ and the cone $z = \sqrt{x^2 + y^2}$.

44. Show that the length of the space curve with parametric equations

$$r = r(t), \quad \theta = \theta(t), \quad z = z(t) \quad (a \le t \le b)$$

in cylindrical coordinates is

$$L = \int_a^b \sqrt{r'^2 + r^2\theta'^2 + z'^2} \, dt,$$

where the prime denotes differentiation with respect to t. After writing parametric equations for the circular helix $x = 3\cos t$, $y = 3\sin t$, $z = 4t$ in cylindrical coordinates, use this formula to find the length of one turn of the helix (compare with Example 3, page 723).

Find the cylindrical coordinates of the point with the given spherical coordinates.

45. $\rho = 2, \theta = \pi/3, \phi = \pi/2$

46. $\rho = \sqrt{2}, \theta = \pi, \phi = 3\pi/4$

47. $\rho = \sqrt{3}, \theta = \pi/4, \phi = \pi/3$

Find the spherical coordinates of the point with the given cylindrical coordinates.

48. $r = \sqrt{3}, \theta = 5\pi/6, z = -1$

49. $r = 2, \theta = \pi/6, z = 2$

50. $r = 3, \theta = \sqrt{\pi}, z = 4$

51. Evaluate the iterated integral

$$\int_0^1 dx \int_0^{\sqrt{1-x^2}} dy \int_0^{\sqrt{1-x^2-y^2}} (x^2 + y^2) \, dz$$

by transforming it to spherical coordinates.

52. The density at a variable point P of the solid hemisphere $x^2 + y^2 + z^2 \le 2ay$, $z \ge 0$ of radius a is equal to the distance from P to the origin. Find the total mass of the hemisphere.

53. Find the moment of inertia of a solid sphere of mass M and radius a about any of its tangent lines.

54. Find the mean value of the function $\sqrt{x^2 + y^2 + z^2}$ over the ball $x^2 + y^2 + z^2 \le a^2$ of radius a (see Problem 32, page 832). This number can be regarded as the "average distance" between the points of the ball and its center.

15 Line and Surface Integrals

In this chapter we resume the study of *line integrals* initiated in Section 14.4, and we introduce line integrals of a new kind, suitable for calculating the work done on a particle moving along a curve under the action of a variable force. We also generalize the concept of a double integral, by allowing the region of integration to be a curved surface. This leads to the concept of a *surface integral*, explored in Section 15.3.

Armed with line and surface integrals, we are able to establish a number of multivariable generalizations of the fundamental theorem of calculus, namely, *Green's theorem*, the *divergence theorem* and *Stokes' theorem*, presented in Section 15.4–15.6. These results are of great importance in applied science, and were in fact discovered while investigating such physical phenomena as electricity, and magnetism, gravitation, and fluid flow.

A collateral theme is that of *vector fields*, that is, vector functions of a *vector* argument, or equivalently, vector functions of a variable point in the plane or in space. From a formal standpoint, a vector field is a rule or procedure assigning one and only one vector to every point of a given set of points in two or three dimensions, but our approach will emphasize the concrete physical meaning of vector fields.

15.1 Line Integrals

Line Integrals with Respect to Arc Length

One kind of line integral, of the type

$$\int_C f(x, y)\, ds \qquad (1)$$

in the plane, or

$$\int_C f(x, y, z)\, ds \qquad (1')$$

in space, has already been introduced in Section 14.4 as a tool for calculating the center of mass of a curved wire C of variable density. In this application the integrand in (1) or (1') takes the form ρ, $x\rho$, $y\rho$ or $z\rho$, where $\rho = \rho(x, y)$ or $\rho = \rho(x, y, z)$ is the mass density of the wire, but the considerations leading to (1) and (1') apply equally well to any function f continuous on a smooth curve C in the plane or in space, with no more than a finite number of self-intersections. Thus

$$\int_C f(x, y)\, ds = \int_a^b f(x(t), y(t))\, \sqrt{[x'(t)]^2 + [y'(t)]^2}\, dt \qquad (2)$$

(see page 841) if C is a plane curve with parametric equations

$$x = x(t), \qquad y = y(t) \qquad (a \le t \le b),$$

and

$$\int_C f(x, y, z)\, ds = \int_a^b f(x(t), y(t), z(t))\, \sqrt{[x'(t)]^2 + [y'(t)]^2 + [z'(t)]^2}\, dt \quad (2')$$

(see page 843) if C is a space curve with parametric equations

$$x = x(t), \qquad y = y(t), \qquad z = z(t) \qquad (a \le t \le b).$$

Notice that (2) is obtained from (2') when the function $z(t)$ is suppressed.

The integrals (1) and (1') are called *line integrals of f with respect to the arc length s*. For positive f each integral can be interpreted as the total mass of a curved wire of variable density f.

Example 1 Evaluate

$$\int_C y e^{-x}\, ds,$$

where C is the plane curve

$$x = \ln(t^2 + 1), \qquad y = 2 \arctan t - t \qquad (0 \le t \le \sqrt{3}).$$

Solution Using (2), we have

$$\int_C y e^{-x}\, ds = \int_0^{\sqrt{3}} \frac{2 \arctan t - t}{t^2 + 1}\, \sqrt{\left(\frac{2t}{t^2 + 1}\right)^2 + \left(\frac{2}{t^2 + 1} - 1\right)^2}\, dt$$

$$= \int_0^{\sqrt{3}} \frac{2 \arctan t - t}{t^2 + 1}\, dt$$

$$= 2 \int_0^{\sqrt{3}} \arctan t\, d(\arctan t) - \int_0^{\sqrt{3}} \frac{t}{t^2 + 1}\, dt$$

$$= \left[(\arctan t)^2\right]_0^{\sqrt{3}} - \frac{1}{2}\left[\ln(t^2 + 1)\right]_0^{\sqrt{3}} = \frac{\pi^2}{9} - \ln 2. \qquad \square$$

Example 2 Evaluate

$$\int_C xyz\, ds,$$

where C is the space curve

$$x = t, \quad y = \frac{4}{3}t^{3/2}, \quad z = t^2 \quad (0 \leq t \leq 1).$$

Solution With the help of (2'),

$$\int_C xyz\, ds = \frac{4}{3}\int_0^1 t^{9/2}\sqrt{1+4t+4t^2}\, dt = \frac{4}{3}\int_0^1 t^{9/2}(1+2t)\, dt$$

$$= \frac{4}{3}\left[\frac{2}{11}t^{11/2} + \frac{4}{13}t^{13/2}\right]_0^1 = \frac{280}{429}. \quad \square$$

Work as a Line Integral

Another kind of line integral arises in extending the concept of the work done by a variable force from the one-dimensional case to the case of two or three dimensions. Consider a particle of mass m, moving in the plane or in space, whose path C is the graph of the position vector function

$$\mathbf{r} = \mathbf{r}(t) \quad (a \leq t \leq b),$$

where we think of the parameter t as the time. It is assumed that C has no more than a finite number of self-intersections and is *smooth*, which means that $\mathbf{r}(t)$ has a continuous nonzero derivative $d\mathbf{r}/dt$ on $[a, b]$. (However, as in the remark on page 667), we allow for the possibility that $d\mathbf{r}/dt = \mathbf{0}$ at the endpoints $t = a$ and $t = b$.) The particle's velocity at time t is

$$\mathbf{v} = \mathbf{v}(t) = \frac{d\mathbf{r}(t)}{dt}.$$

Suppose the particle is acted on by a force $\mathbf{F} = \mathbf{F}(\mathbf{r})$, where $\mathbf{F}(\mathbf{r})$ is a vector function of the position vector $\mathbf{r}$, or equivalently a function $\mathbf{F}(x, y)$ or $\mathbf{F}(x, y, z)$ of the coordinates of the point with position vector $\mathbf{r}$; such a function is called a *vector field*, in this case a force field. Thus in the plane

$$\mathbf{F} = P\mathbf{i} + Q\mathbf{j},$$

where $P = P(x, y)$ and $Q = Q(x, y)$ are the components of $\mathbf{F}$, both scalar functions of two variables, while in space

$$\mathbf{F} = P\mathbf{i} + Q\mathbf{j} + R\mathbf{k},$$

where the components $P = P(x, y, z)$, $Q = Q(x, y, z)$ and $R = R(x, y, z)$ of $\mathbf{F}$ are now scalar functions of three variables. We will assume that $\mathbf{F}(\mathbf{r})$ is continuous on some set D, meaning that the components of $\mathbf{F}(\mathbf{r})$ are continuous on D.

According to Newton's second law of motion,

$$m\frac{d\mathbf{v}}{dt} = \mathbf{F}. \tag{3}$$

Taking the dot product of (3) with $\mathbf{v} = d\mathbf{r}/dt$, we find that

$$m\frac{d\mathbf{v}}{dt} \cdot \mathbf{v} = \mathbf{F} \cdot \frac{d\mathbf{r}}{dt}$$

or equivalently

$$\frac{d}{dt}\left(\frac{1}{2}mv^2\right) = \mathbf{F} \cdot \frac{d\mathbf{r}}{dt} \tag{4}$$

in terms of the particle's speed $v = v(t) = |\mathbf{v}(t)|$, since

$$\frac{d}{dt}(v^2) = \frac{d}{dt}(\mathbf{v} \cdot \mathbf{v}) = 2\mathbf{v} \cdot \frac{d\mathbf{v}}{dt}.$$

Integrating (4) with respect to t from a to b, we obtain

$$\left[\frac{1}{2}mv^2\right]_a^b = \int_a^b \mathbf{F} \cdot \frac{d\mathbf{r}}{dt} dt,$$

that is,

$$\frac{1}{2}mv_B^2 - \frac{1}{2}mv_A^2 = \int_a^b \mathbf{F} \cdot \frac{d\mathbf{r}}{dt} dt, \tag{5}$$

where $v_A = v(a)$ and $v_B = v(b)$.

The quantity

$$K = \frac{1}{2}mv^2$$

is called the *kinetic energy* of the particle, as on page 269. Thus equation (5) says that as a result of the action of the force $\mathbf{F}$, the particle's kinetic energy changes by an amount

$$W = \int_a^b \mathbf{F} \cdot \frac{d\mathbf{r}}{dt} dt, \tag{6}$$

called the *work* done by the force on the particle as the particle traverses the path C from the initial point A of the path, with position vector $\mathbf{r}(a)$, to the final point B, with position vector $\mathbf{r}(b)$.

We now have line integrals of two kinds, the integrals (1) and (1′) which can both be written in the form $\int_C f \, ds$, where $f = f(\mathbf{r})$ is a scalar function (of a vector argument), and the integral on the right in (5) and (6), which can be written in the form

$$\int_C \mathbf{F} \cdot d\mathbf{r}, \tag{7}$$

and involves a continuous vector field $\mathbf{F} = \mathbf{F}(\mathbf{r})$, which in general does not have to be a force. The integral (7), which is shorthand for

$$\int_a^b \mathbf{F}(\mathbf{r}(t)) \cdot \frac{d\mathbf{r}(t)}{dt} dt, \tag{8}$$

is called the *line integral of* $\mathbf{F}$ *along* C, but the integration is now with respect to the position vector $\mathbf{r}$, rather than the arc length s as in the case of $\int_C f \, ds$. The differential $d\mathbf{r}$ of the vector function $\mathbf{r} = \mathbf{r}(t)$ is defined in the same way as for a scalar function. Thus

$$d\mathbf{r} = \frac{d\mathbf{r}}{dt} dt,$$

so that

$$d\mathbf{r} = d(x\mathbf{i} + y\mathbf{j}) = \left(\frac{dx}{dt}\mathbf{i} + \frac{dy}{dt}\mathbf{j}\right) dt = \left(\frac{dx}{dt} dt\right)\mathbf{i} + \left(\frac{dy}{dt} dt\right)\mathbf{j} = dx\,\mathbf{i} + dy\,\mathbf{j}$$

in the plane, and similarly $d\mathbf{r} = dx\,\mathbf{i} + dy\,\mathbf{j} + dz\,\mathbf{k}$ in space. In terms of the line integral (7), formulas (5) and (6) become

$$\frac{1}{2}mv_B^2 - \frac{1}{2}mv_A^2 = \int_C \mathbf{F} \cdot d\mathbf{r}$$

and

$$W = \int_C \mathbf{F} \cdot d\mathbf{r}.$$

It is easily verified that all the expressions for the work W given in earlier sections are special cases of the last formula.

Line Integrals with Respect to the Coordinates

Substituting the expressions $\mathbf{F} = P\mathbf{i} + Q\mathbf{j}$ and $d\mathbf{r} = dx\,\mathbf{i} + dy\,\mathbf{j}$ into (7), we get

$$\int_C \mathbf{F} \cdot d\mathbf{r} = \int_C P\,dx + Q\,dy \qquad (9)$$

in the plane, since $(P\mathbf{i} + Q\mathbf{j}) \cdot (dx\,\mathbf{i} + dy\,\mathbf{j}) = P\,dx + Q\,dy$, and similarly

$$\int_C \mathbf{F} \cdot d\mathbf{r} = \int_C P\,dx + Q\,dy + R\,dz \qquad (9')$$

in space. The right sides of (9) and (9') are also called line integrals along C. It would perhaps be less ambiguous to write these integrals as

$$\int_C (P\,dx + Q\,dy), \qquad \int_C (P\,dx + Q\,dy + R\,dz),$$

but this use of parentheses around the "differential forms" $P\,dx + Q\,dy$ and $P\,dx + Q\,dy + R\,dz$ is not customary.

To evaluate (9) and (9'), we write out (8) in full, obtaining

$$\int_C P\,dx + Q\,dy = \int_a^b P(x(t), y(t))\frac{dx}{dt}\,dt + \int_a^b Q(x(t), y(t))\frac{dy}{dt}\,dt \qquad (10)$$

in the plane, and

$$\int_C P\,dx + Q\,dy + R\,dz = \int_a^b P(x(t), y(t), z(t))\frac{dx}{dt}\,dt$$
$$+ \int_a^b Q(x(t), y(t), z(t))\frac{dy}{dt}\,dt + \int_a^b R(x(t), y(t), z(t))\frac{dz}{dt}\,dt$$
$$(10')$$

in space. The separate terms in these sums are denoted more concisely by

$$\int_C P\,dx, \qquad \int_C Q\,dy, \qquad \int_C R\,dz,$$

and are again called line integrals, this time *with respect to the coordinates* x, y and z. Thus in abbreviated notation (10) and (10') become

$$\int_C P\,dx + Q\,dy = \int_C P\,dx + \int_C Q\,dy$$

and

$$\int_C P\,dx + Q\,dy + R\,dz = \int_C P\,dx + \int_C Q\,dy + \int_C R\,dz.$$

In the case where C is the graph of a continuous function

$$y = f(x) \quad (a \le x \le b),$$

we can choose x as the parameter in formula (10), which then becomes

$$\int_C P\,dx + Q\,dy = \int_a^b P(x, f(x))\,dx + \int_a^b Q(x, f(x))f'(x)\,dx. \quad (11)$$

Similarly, if C is the graph of a continuous function of the form

$$x = g(y) \quad (a \le y \le b),$$

then

$$\int_C P\,dx + Q\,dy = \int_a^b P(g(y), y)g'(y)\,dy + \int_a^b Q(g(y), y)\,dy. \quad (11')$$

Let $s = s(t)$ be the arc length function of C, that is, the length of the arc of the curve C from the initial point of C to the point with position vector $\mathbf{r}(t)$. Then in addition to being the graph of $\mathbf{r} = \mathbf{r}(t)$ ($a \le t \le b$), C is also the graph of $\mathbf{r} = \mathbf{r}(s)$ ($0 \le s \le L$), where L is the length of C and $\mathbf{r}(s)$ is the composite function $\mathbf{r}(t(s))$ involving the inverse $t = t(s)$ of the function $s = s(t)$. Therefore

$$\int_C \mathbf{F} \cdot d\mathbf{r} = \int_a^b \mathbf{F} \cdot \frac{d\mathbf{r}}{dt}\,dt = \int_0^L \mathbf{F} \cdot \frac{d\mathbf{r}}{dt}\frac{dt}{ds}\,ds = \int_0^L \mathbf{F} \cdot \frac{d\mathbf{r}}{ds}\,ds, \quad (12)$$

with the help of the chain rule. But, as we know from pages 669 and 725, $d\mathbf{r}/ds = \mathbf{T}$, where $\mathbf{T}$ is the unit tangent vector along C. Thus the line integral $\int_C \mathbf{F} \cdot d\mathbf{r}$ has the alternative representation

$$\int_C \mathbf{F} \cdot \mathbf{T}\,ds,$$

from which it is apparent that the work done by a force $\mathbf{F}$ on a particle moving along C is entirely due to its tangential component, namely its component along the vector $\mathbf{T}$.

The same argument as just used to establish formula (12) shows that *the value of the line integral $\int_C \mathbf{F} \cdot d\mathbf{r}$ is independent of the parametric representation of C, provided that the orientation of C is preserved*. In fact, if the oriented curve C is the graph of $\mathbf{r} = \mathbf{r}(t)$, then C is also the graph of $\mathbf{r} = \mathbf{r}(t(\tau))$, where $t = t(\tau)$ is any continuously differentiable function such that $t(\alpha) = a$, $t(\beta) = b$ and $dt/d\tau$ is positive on $[\alpha, \beta]$. But then

$$\int_C \mathbf{F} \cdot d\mathbf{r} = \int_a^b \mathbf{F} \cdot \frac{d\mathbf{r}}{dt}\,dt = \int_\alpha^\beta \mathbf{F} \cdot \frac{d\mathbf{r}}{dt}\frac{dt}{d\tau}\,d\tau = \int_\alpha^\beta \mathbf{F} \cdot \frac{d\mathbf{r}}{d\tau}\,d\tau,$$

which proves the italicized statement.

The situation is different if the orientation of C is reversed. Let $-C$ be the curve obtained by reversing the orientation of C, as illustrated in Figure 1. Then $-C$ is the graph of $\mathbf{r} = \mathbf{r}(t(\tau))$, where $t = t(\tau)$ is any continuously differentiable function such that $t(\alpha) = b$, $t(\beta) = a$ and $dt/d\tau$ is *negative* on

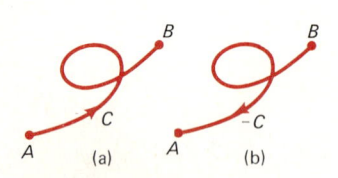

Figure 1

882 Chapter 15 Line and Surface Integrals

$[\alpha, \beta]$. Therefore

$$\int_{-C} \mathbf{F} \cdot d\mathbf{r} = \int_{\alpha}^{\beta} \mathbf{F} \cdot \frac{d\mathbf{r}}{d\tau} d\tau = \int_{b}^{a} \mathbf{F} \cdot \frac{d\mathbf{r}}{d\tau} \frac{d\tau}{dt} dt = \int_{b}^{a} \mathbf{F} \cdot \frac{d\mathbf{r}}{dt} dt = -\int_{a}^{b} \mathbf{F} \cdot \frac{d\mathbf{r}}{dt} dt,$$

so that

$$\int_{-C} \mathbf{F} \cdot d\mathbf{r} = -\int_{C} \mathbf{F} \cdot d\mathbf{r}. \tag{13}$$

If C has initial point A and final point B, then $-C$ has initial point B and final point A (see the figure), and we can write (13) in the alternative form

$$\int_{BA} \mathbf{F} \cdot d\mathbf{r} = -\int_{AB} \mathbf{F} \cdot d\mathbf{r}, \tag{13'}$$

with the understanding that AB denotes an arc starting at A and ending at B, and BA denotes the same arc traversed in the opposite direction. As an exercise, show that the value of a line integral $\int_C f(\mathbf{r}) \, ds$, of the type (1) or (1'), does not change if the orientation of the curve C is reversed.

The following examples illustrate the evaluation of line integrals of the type (7)–(11).

Example 3 Evaluate

$$I = \int_C 3y \, dx + x \, dy,$$

where C is the semicircle $x = \cos t$, $y = \sin t$ ($0 \le t \le \pi$).

Solution Using formula (10), we find that

$$I = \int_0^{\pi} 3 \sin t (-\sin t) \, dt + \int_0^{\pi} \cos t (\cos t) \, dt$$

$$= -3 \int_0^{\pi} \sin^2 t \, dt + \int_0^{\pi} \cos^2 t \, dt = -\frac{3\pi}{2} + \frac{\pi}{2} = -\pi. \quad \square$$

Example 4 Evaluate

$$I = \int_C yz \, dx + xz \, dy + xy \, dz,$$

where C is the twisted cubic $x = t$, $y = t^2$, $z = t^3$ ($0 \le t \le 1$).

Solution This time we use formula (10') and obtain

$$I = \int_0^1 t^5 \, dt + \int_0^1 t^4(2t) \, dt + \int_0^1 t^3(3t^2) \, dt = \int_0^1 6t^5 \, dt = 1. \quad \square$$

The next example shows that in general the values of a line integral along different curves with the same endpoints are different.

Example 5 Evaluate the line integral

$$I = \int_C xy \, dx + (y - x) \, dy \tag{14}$$

along each of the following curves joining the points $(0, 0)$ and $(1, 1)$:

(a) The line $y = x$;
(b) The parabola $y = x^2$;

Section 15.1 Line Integrals 883

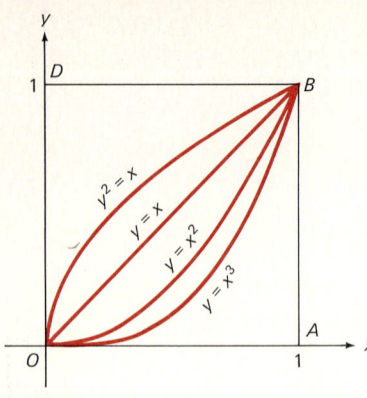

Figure 2

(c) The parabola $y^2 = x$;
(d) The curve $y = x^3$.

Solution The four curves are shown in Figure 2. Using formula (11) and making some easy calculations, we obtain

(a) $I = \int_0^1 [x^2 + (x - x)]\,dx = \dfrac{1}{3}$;

(b) $I = \int_0^1 [x^3 + 2(x^2 - x)x]\,dx = \dfrac{1}{12}$;

(c) $I = \int_0^1 \left[x^{3/2} + \dfrac{1}{2}(x^{1/2} - x)x^{-1/2}\right] dx = \dfrac{17}{30}$;

(d) $I = \int_0^1 [x^4 + 3(x^3 - x)x^2]\,dx = -\dfrac{1}{20}$. □

Piecewise Smooth Curves

Let C be a curve formed by joining a finite number of smooth curves C_1, $C_2, \ldots, C_n$ end to end, as in Figure 3. Such curves are said to be *piecewise smooth*, and can have corners at the "junction points." The line integral $\int_C \mathbf{F} \cdot d\mathbf{r}$ along a piecewise smooth curve C is defined by the formula

$$\int_C \mathbf{F} \cdot d\mathbf{r} = \int_{C_1} \mathbf{F} \cdot d\mathbf{r} + \int_{C_2} \mathbf{F} \cdot d\mathbf{r} + \cdots + \int_{C_n} \mathbf{F} \cdot d\mathbf{r}. \tag{15}$$

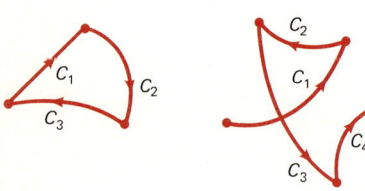

Piecewise smooth curves

Figure 3

Of course, as is easily verified, this formula holds automatically for a smooth curve C divided into n arcs $C_1, C_2, \ldots, C_n$.

Example 6 Evaluate the line integral (14) along the polygonal paths OAB and ODB, where $O = (0, 0)$, $A = (1, 0)$, $B = (1, 1)$ and $D = (0, 1)$ as in Figure 2.

Solution The paths OAB and ODB are piecewise smooth. In fact, OAB consists of the segments OA and AB, while ODB consists of the segments OD and DB. These line segments have the parametric representations

$$\begin{aligned}
OA: \quad & x = t, \quad y = 0 \quad (0 \le t \le 1), \\
AB: \quad & x = 1, \quad y = t \quad (0 \le t \le 1), \\
OD: \quad & x = 0, \quad y = t \quad (0 \le t \le 1), \\
DB: \quad & x = t, \quad y = 1 \quad (0 \le t \le 1),
\end{aligned}$$

respectively. Therefore, with the help of (15),

$$\int_{OAB} xy\,dx + (y - x)\,dy = \int_{OA} xy\,dx + (y - x)\,dy + \int_{AB} xy\,dx + (y - x)\,dy$$

$$= \int_0^1 0\,dt + \int_0^1 (t - 1)\,dt = -\dfrac{1}{2},$$

$$\int_{ODB} xy\,dx + (y - x)\,dy = \int_{OD} xy\,dx + (y - x)\,dy + \int_{DB} xy\,dx + (y - x)\,dy$$

$$= \int_0^1 t\,dt + \int_0^1 t\,dt = 1$$

884 Chapter 15 Line and Surface Integrals

(notice that $dx = 0$ on AB and OD, while $dy = 0$ on OA and DB). As an exercise, use formulas (11) and (11′) to calculate the same integrals. □

Example 7 Evaluate the line integral

$$I = \int_C y^2\,dx + 2xy\,dy$$

along the same curves as in Example 5.

Solution Straightforward calculations give

(a) $I = \int_0^1 (x^2 + 2x^2)\,dx = 1;$

(b) $I = \int_0^1 (x^4 + 4x^4)\,dx = 1;$

(c) $I = \int_0^1 (x + x)\,dx = 1;$

(d) $I = \int_0^1 (x^6 + 6x^6)\,dx = 1.$ □

Notice that unlike Example 5, the values of I in Example 7 are all the same. We will show why this is true in Example 1 of the next section.

Problems

Evaluate the given line integral of the type $\int_C f(\mathbf{r})\,ds$.

1. $\int_C (x + y)\,ds$, where C is the line segment from $(0, 0)$ to $(3, 4)$

2. $\int_C (x - y)\,ds$, where C is the line segment from $(1, 2)$ to $(5, -2)$

3. $\int_C \dfrac{ds}{x - y}$, where C is the line segment from $(0, -3)$ to $(6, 0)$

4. $\int_C xy\,ds$, where C is the parabolic arc $x = 2t$, $y = t^2$ $(0 \le t \le 1)$

5. $\int_C x^2 y\,ds$, where C is the rectangle with vertices $(0, 0)$, $(2, 0)$, $(2, 3)$ and $(0, 3)$ traversed once in the counterclockwise direction

6. $\int_C \dfrac{y}{\sqrt{x}}\,ds$, where C is the curve $x = 3t^2$, $y = 2t^3$ $(1 \le t \le 2)$

7. $\int_C (x^2 + y^2)^2\,ds$, where C is the circle $x = 2\cos t$, $y = 2\sin t$ $(0 \le t \le 2\pi)$

8. $\int_C \dfrac{z}{\sqrt{x^2 + y^2}}\,ds$, where C is the helical arc $x = 3\cos t$, $y = 3\sin t$, $z = 4t$ $(0 \le t \le 2\pi)$

9. $\int_C xyz\,ds$, where C is the circular arc $x = \cos t$, $y = \sin t$, $z = 1$ $(0 \le t \le \pi/2)$

10. $\int_C (2\sqrt{x^2 + y^2} - z)\,ds$, where C is the curve $x = t\cos t$, $y = t\sin t$, $z = t$ $(0 \le t \le \sqrt{2})$

11. $\int_C (x + y + z)\,ds$, where C is the intersection of the plane $y = x$ and the sphere $x^2 + y^2 + z^2 = 4$ in the first octant, traversed from $(\sqrt{2}, \sqrt{2}, 0)$ to $(0, 0, 2)$

12. Show that the line integral $\int_C f(x, y)\,ds$ along the curve C with polar equation $r = r(\theta)$ $(\alpha \le \theta \le \beta)$ is equal to

$$\int_\alpha^\beta f(r\cos\theta, r\sin\theta)\sqrt{r^2 + \left(\dfrac{dr}{d\theta}\right)^2}\,d\theta.$$

13. Use the preceding problem to evaluate $\int_C (x - y)\,ds$, where C is the circle $x^2 + y^2 = 2x$ traversed once in the counterclockwise direction.

14. Suppose a moving particle with velocity $\mathbf{v}$ is acted on by a force $\mathbf{F}$ which is always perpendicular to $\mathbf{v}$. Show that the work done by $\mathbf{F}$ on the particle is zero.

15. Find the work done by the force $\mathbf{F} = 3x^2\mathbf{i} + xy\mathbf{j}$ acting on a particle moving along the parabolic arc $y = 4x^2$ from $(0, 0)$ to $(1, 4)$.

16. Find the work done by the force $\mathbf{F} = (x + y)\mathbf{i} - xy^2\mathbf{j}$ acting on a particle moving once in the counterclockwise direction around the square C bounded by the lines $x = \pm 1$ and $y = \pm 1$.

17. A 150-lb repairman carrying a 25-lb bag of cement goes up a helical stairway surrounding a cylindrical silo of radius 10 ft. The height of the silo is 60 ft, and the stairway makes exactly three complete turns around the silo. How much work is done by the repairman against gravity in ascending the stairway?

18. In the preceding problem how much work is done if there is a hole in the bag, causing 12 lb of cement to be lost at a steady rate during the ascent?

Evaluate the given line integral of the type $\int_C P\,dx + Q\,dy$ or $\int_C P\,dx + Q\,dy + R\,dz$.

19. $\int_C x\,dy$, where C is the line segment from $(0, 0)$ to $(2, -3)$

20. $\int_C y\,dx$, where C is the line segment from $(2, 6)$ to $(7, 1)$

21. $\int_C x\,dy - y\,dx$, where C is the line segment from $(a, 0)$ to $(0, b)$

22. $\int_C \sin x\,dy - \cos y\,dx$, where C is the line segment from $(0, 0)$ to $(\pi/3, \pi/6)$

23. $\int_C (x^2 + y^2)\,dx + (x^2 - y^2)\,dy$, where C is the circular arc $x = \cos t$, $y = \sin t$ $(0 \le t \le \pi/2)$

24. $\int_C \dfrac{(x - y)\,dx + (x + y)\,dy}{x^2 + y^2}$, where C is the semicircle $x = 2\cos t$, $y = 2\sin t$ $(0 \le t \le \pi)$

25. $\int_C (x + y)\,dx + (x - y)\,dy$, where C is the ellipse $x = \sin t$, $y = 2\cos t$ $(0 \le t \le 2\pi)$

26. $\int_C \dfrac{dx + dy}{|x| + |y|}$, where C is the square with vertices $(1, 0)$, $(0, 1)$, $(-1, 0)$ and $(0, -1)$, traversed once in the counterclockwise direction.

27. $\int_C \dfrac{x\,dy - y\,dx}{x^2 + xy + y^2}$, where C is the circle $x = a\cos t$, $y = a\sin t$ $(0 \le t \le 2\pi, a > 0)$

28. $\int_C xy\,dx + x^2\,dy$, where C is the closed curve bounded by the parabolas $y = x^2$ and $y^2 = x$, traversed once in the counterclockwise direction.

29. $\int_C x\,dx + (x - y)\,dy + (x + y + z)\,dz$, where C is the line segment from $(1, 0, -1)$ to $(2, 3, 4)$

30. $\int_C (y - z)\,dx + (z - x)\,dy + (x - y)\,dz$, where C is the same helical arc as in Problem 8

31. $\int_C e^x\,dx + e^y\,dy + e^z\,dz$, where C is the same twisted cubic as in Example 4

Evaluate $\int_C (x^2 + y)\,dx + (y^2 - x)\,dy$ along the specified path from the point $(0, 0)$ to the point $(1, 2)$.

32. The line segment from $(0, 0)$ to $(1, 2)$
33. The polygonal path from $(0, 0)$ to $(1, 0)$ and then from $(1, 0)$ to $(1, 2)$
34. The parabolic arc $y = 2x^2$ $(0 \le x \le 1)$

15.2 Path Independence and Gradient Fields

Domains and Paths

Let D be a set of points in the plane or in space, and suppose that for every pair of points A and B in D there is a curve C with initial point A and final point B which is entirely contained in D. Then D is said to be (*arcwise*) *connected*. An *open* connected set is called a *domain* (not to be confused with the unrelated concept of the domain of a function). For example, Figure 4 shows a two-dimensional domain D with two of its points A and B joined by a curve C entirely contained in D. Note that since the domain D is open, it does not contain its boundary.

By a *simple closed curve* we mean the graph of a continuous vector function $\mathbf{r} = \mathbf{r}(t)$ $(a \le t \le b)$ in the plane or in space with coincident endpoints, so that $\mathbf{r}(a) = \mathbf{r}(b)$, but with no other self-intersections (recall page

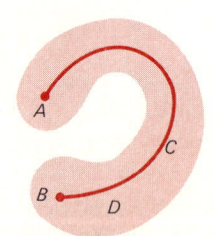

Figure 4

459). A two-dimensional domain D is said to be *simply connected* if given any simple closed curve C entirely contained in D, the region consisting of C and its interior is also entirely contained in D, but otherwise the domain is said to be *multiply connected*. Thus the domain in Figure 4 is simply connected, since the interior of every simple closed curve lying in D consists only of points that belong to D; this is simply because there are no holes in D. On the other hand, the domain E shown in Figure 5, which has a hole in it, is multiply connected, since the interior of any simple closed curve in E surrounding the hole contains points that do not belong to E.

Given two points A and B, by a *path* from A to B we mean a piecewise smooth curve C with initial point A and final point B (and no more than a finite number of self-intersections). In general, the value of a line integral $\int_C \mathbf{F} \cdot d\mathbf{r}$ along a path C from point A to point B depends not only on the points A and B, but also on the particular path from A to B. For instance, this is true in Example 5, page 883. However, there are many cases in which a line integral is *path independent*, in the sense that its value depends only on the endpoints of C, but not on the path C itself. As shown by the following theorem, a necessary and sufficient condition for path independence of a line integral $\int_C \mathbf{F} \cdot d\mathbf{r}$ on a domain D is that the integrand $\mathbf{F}$ be a *gradient field*, meaning that there is a differentiable scalar function $U = U(x, y)$ defined on D with $\mathbf{F}$ as its gradient, that is, $\mathbf{F} = \text{grad } U = \nabla U$. The theorem is given for a two-dimensional field $\mathbf{F}$, but is easily extended to the case where $\mathbf{F}$ is three-dimensional.

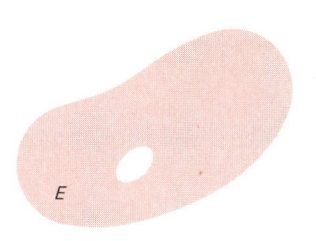

Figure 5

Theorem 1 *(Path independent line integrals and gradient fields)*. Let the vector field $\mathbf{F} = \mathbf{F}(x, y) = P(x, y)\mathbf{i} + Q(x, y)\mathbf{j}$ be continuous on a domain D. Then the line integral $\int_C \mathbf{F} \cdot d\mathbf{r} = \int_C P\, dx + Q\, dy$ is path independent on D if and only if there is a function $U = U(x, y)$ defined on D such that $\mathbf{F} = \text{grad } U$ or equivalently

$$P = \frac{\partial U}{\partial x}, \qquad Q = \frac{\partial U}{\partial y}. \tag{1}$$

Proof Suppose first that $\mathbf{F} = \text{grad } U$, and let C be any smooth curve in D going from A to B, with parametric equations

$$x = x(t), \qquad y = y(t) \qquad (a \le t \le b).$$

It follows from (1) that

$$\int_C \mathbf{F} \cdot d\mathbf{r} = \int_C P\, dx + Q\, dy = \int_a^b P(x(t), y(t)) x'(t)\, dt + \int_a^b Q(x(t), y(t))\, y'(t)\, dt$$

$$= \int_a^b [U_x(x(t), y(t))\, x'(t) + U_y(x(t), y(t))\, y'(t)]\, dt,$$

where $U_x = \partial U/\partial x$ and $U_y = \partial U/\partial y$, and the prime denotes differentiation with respect to t. Applying the chain rule, we obtain

$$\int_C \mathbf{F} \cdot d\mathbf{r} = \int_a^b \frac{d}{dt}[U(x(t), y(t))]\, dt = U(x(b), y(b)) - U(x(a), y(a)),$$

Section 15.2 Path Independence and Gradient Fields

with the help of the fundamental theorem of calculus. This can be written concisely as

$$\int_C \mathbf{F} \cdot d\mathbf{r} = U(B) - U(A), \tag{2}$$

where $U(A)$ and $U(B)$ are the values of U at $A = (x(a), y(a))$ and $B = (x(b), y(b))$. If C is only piecewise smooth, the same result is obtained after dividing C into smooth arcs (the details are left as an exercise). Since the value of the expression $U(B) - U(A)$ is path independent, the same is true of the line integral $\int_C \mathbf{F} \cdot d\mathbf{r}$ itself.

Conversely, suppose $\int_C \mathbf{F} \cdot d\mathbf{r}$ is path independent. Let (x_0, y_0) be a fixed point of D and (x, y) a variable point of D, and let $U = U(x, y)$ be the scalar function defined by the line integral

$$U = \int_C P\,dx + Q\,dy, \tag{3}$$

where C is an *arbitrary* path from (x_0, y_0) to (x, y). Notice that this definition makes sense only because of the path independence of the integral (3). We now hold y fixed and give x an increment Δx, choosing $|\Delta x|$ small enough to keep the horizontal line segment α from (x, y) to $(x + \Delta x, y)$ inside D, as shown in Figure 6 (this is always possible, since D is an open set). Then

$$U(x + \Delta x, y) = \int_{C+\alpha} P\,dx + Q\,dy = \int_C P\,dx + Q\,dy + \int_\alpha P\,dx + Q\,dy,$$

where $C + \alpha$ is the path obtained by joining the initial point of α to the final point of C. Using (3) and the fact that $dy = 0$ along α, we obtain

$$U(x + \Delta x, y) - U(x, y) = \int_\alpha P\,dx + Q\,dy = \int_\alpha P\,dx.$$

But

$$\int_\alpha P\,dx = \int_x^{x+\Delta x} P(t, y)\,dt \qquad (y \text{ fixed}),$$

and hence by the mean value theorem for ordinary integrals,

$$U(x + \Delta x, y) - U(x, y) = P(u, y)\,\Delta x \qquad (x \le u \le x + \Delta x),$$

where $u \to x$ as $\Delta x \to 0$. Therefore

$$\frac{\partial U}{\partial x} = \lim_{\Delta x \to 0} \frac{U(x + \Delta x, y) - U(x, y)}{\Delta x} = \lim_{\Delta x \to 0} P(u, y) = P(x, y),$$

by the continuity of P, and a similar argument involving a *vertical* line segment drawn from (x, y) to $(x, y + \Delta y)$ shows that

$$\frac{\partial U}{\partial y} = Q(x, y).$$

Thus $\mathbf{F} = \mathrm{grad}\,U$, where U is the function defined by the path independent line integral (3). ∎

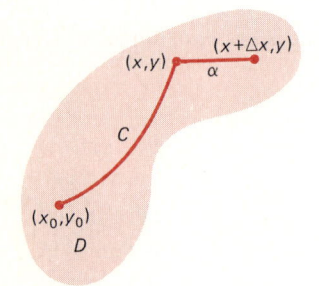

Figure 6

888   Chapter 15   Line and Surface Integrals

The Fundamental Theorem of Line Integrals

A path independent line integral $\int_C \mathbf{F} \cdot d\mathbf{r}$ depends only on the endpoints A and B of the path C, and hence is sometimes written in the form $\int_A^B \mathbf{F} \cdot d\mathbf{r}$ (this notation is of course inappropriate if the line integral depends on the path). Thus formulas (2) and (3) can also be written as

$$\int_A^B \mathbf{F} \cdot d\mathbf{r} = U(B) - U(A) \qquad (4)$$

and

$$U = \int_{(x_0, y_0)}^{(x,y)} P\, dx + Q\, dy.$$

Formula (4), or its equivalent

$$\int_A^B \nabla U \cdot d\mathbf{r} = U(B) - U(A)$$

($\nabla U = \text{grad } U$) is a multivariable analogue of the fundamental theorem of calculus, and for this reason is sometimes called the *fundamental theorem of line integrals*.

Remark Theorem 1 remains true in three dimensions if we set

$$\mathbf{F} = \mathbf{F}(x, y, z) = P(x, y, z)\mathbf{i} + Q(x, y, z)\mathbf{j} + R(x, y, z)\mathbf{k},$$

and

$$\int_C \mathbf{F} \cdot d\mathbf{r} = \int_C P\, dx + Q\, dy + R\, dz.$$

Naturally, instead of (1) we now have

$$P = \frac{\partial U}{\partial x}, \qquad Q = \frac{\partial U}{\partial y}, \qquad R = \frac{\partial U}{\partial z},$$

with an extra formula.

Example 1 In Example 7, page 885, we found that the line integral

$$I = \int_C y^2\, dx + 2xy\, dy$$

has the same value 1 along four different paths from $(0, 0)$ to $(1, 1)$. Using Theorem 1, we can now assert that $I = 1$ for *every* path C from $(0, 0)$ to $(1, 1)$, since $I = \int_C \mathbf{F} \cdot d\mathbf{r}$, where $\mathbf{F} = y^2 \mathbf{i} + 2xy \mathbf{j}$ is a gradient field (on the whole xy-plane). In fact, a moment's thought shows that $\mathbf{F}$ is the gradient of the function $U = U(x, y) = xy^2$, and formula (2) then gives

$$\int_C \mathbf{F} \cdot d\mathbf{r} = U(1, 1) - U(0, 0) = 1 - 0 = 1.$$

It is usually not so easy to recognize a gradient field, and later we will establish a test based on the components of $\mathbf{F}$ enabling us to decide whether or not there is a function U such that $\mathbf{F} = \text{grad } U$. If the function U exists, it can then be found by calculating the line integral (3).

Example 2 In Example 5, page 883, it was found that the line integral

$$I = \int_C xy\, dx + (y - x)\, dy$$

is *not* path independent, by showing that it takes different values for different paths connecting the same two points. The fact that I fails to be path independent can be deduced from Theorem 1 without calculating any line integrals at all! For suppose I were path independent. Then, by Theorem 1, there would be a function U such that

$$\frac{\partial U}{\partial x} = xy, \qquad \frac{\partial U}{\partial y} = y - x.$$

But this is impossible, since then

$$\frac{\partial^2 U}{\partial x\, \partial y} = -1, \qquad \frac{\partial^2 U}{\partial y\, \partial x} = x,$$

contradicting the known fact that the mixed second partial derivatives $\partial^2 U/\partial x\, \partial y$ and $\partial^2 U/\partial y\, \partial x$ must be identically equal if they are continuous (see page 763). □

Suppose that the line integral of $\mathbf{F} = \mathbf{F}(\mathbf{r})$ is path independent on a domain D, or equivalently that $\mathbf{F}$ is a gradient field on D. Let C be any *closed* path in D, that is, any path whose endpoints A and B coincide. Then

$$\int_C \mathbf{F} \cdot d\mathbf{r} = 0, \qquad (5)$$

as we find at once by setting $A = B$ in formula (2). Conversely, if (5) holds for every closed path in D, then the line integral of $\mathbf{F}$ is path independent on D. To see this, let C_1 and C_2 be two paths from A to B, as in Figure 7(a). Then the path C obtained by joining the initial point of $-C_2$ to the final point of C_1, as in Figure 7(b), is closed. Therefore

$$\int_{C_1} \mathbf{F} \cdot d\mathbf{r} + \int_{-C_2} \mathbf{F} \cdot d\mathbf{r} = \int_C \mathbf{F} \cdot d\mathbf{r} = 0,$$

by hypothesis. It follows that

$$\int_{C_1} \mathbf{F} \cdot d\mathbf{r} = -\int_{-C_2} \mathbf{F} \cdot d\mathbf{r},$$

or equivalently

$$\int_{C_1} \mathbf{F} \cdot d\mathbf{r} = \int_{C_2} \mathbf{F} \cdot d\mathbf{r},$$

with the help of formula (13), page 883.

Remark Actually, it can be shown that the assumption that (5) holds for every *simple* closed path C in the domain D is enough to guarantee that the line integral of $\mathbf{F}$ on D is path independent (see Problem 37).

Circulation

If C is a closed path, the quantity $\int_C \mathbf{F} \cdot d\mathbf{r}$ is called the *circulation of* $\mathbf{F}$ *around* C, and is often denoted by

$$\oint_C \mathbf{F} \cdot d\mathbf{r}.$$

In using the special symbol $\oint$ it is customary to assume that the path C is simple, with no self-intersections except for its coincident endpoints. It

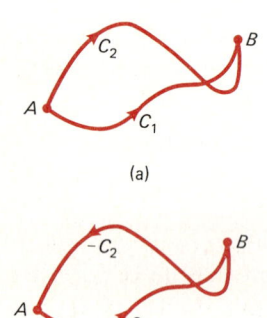

Figure 7

follows from Theorem 1 and the above considerations that the circulation of a gradient field around any closed path is zero.

Example 3 Calculate the circulation of the field

$$\mathbf{F} = \frac{-y\mathbf{i} + x\mathbf{j}}{x^2 + y^2} \qquad (x^2 + y^2 \neq 0)$$

around any circle C centered at the origin, traversed once in the counterclockwise direction.

Solution The field $\mathbf{F}$ is defined on the multiply connected domain D consisting of all the points of the xy-plane except the origin. If the circle C is of radius a, it has the parametric representation $x = a \cos t$, $y = a \sin t$ $(0 \le t \le 2\pi)$. Therefore

$$\oint_C \mathbf{F} \cdot d\mathbf{r} = \oint_C \frac{-y\,dx + x\,dy}{x^2 + y^2}$$

$$= \int_0^{2\pi} \frac{(-a \sin t)(-a \sin t) + (a \cos t)(a \cos t)}{(a \cos t)^2 + (a \sin t)^2}\,dt$$

$$= \int_0^{2\pi} dt = 2\pi.$$

Since the circulation around C is nonzero, we conclude that $\mathbf{F}$ is not a gradient field on D. □

Potential Energy and Conservation of Energy

There is an important physical interpretation of the function U figuring in Theorem 1, or rather of its negative. Think of the vector field $\mathbf{F}$ as a variable force acting on a particle of mass m. Then, as shown on page 881,

$$\frac{1}{2} mv_B^2 - \frac{1}{2} mv_A^2 = \int_C \mathbf{F} \cdot d\mathbf{r}, \tag{6}$$

where v_A is the speed of the particle at the point A, and v_B is its speed at the point B. The right side of (6) is the work W done by the force $\mathbf{F}$ on the particle as it moves along the path C from A to B, and the left side is the corresponding change in the particle's kinetic energy.

Now suppose $\mathbf{F}$ is a gradient field, so that $\mathbf{F} = \mathbf{F}(\mathbf{r}) = \nabla U(\mathbf{r})$. Then, by Theorem 1,

$$\int_C \mathbf{F} \cdot d\mathbf{r} = \int_A^B \mathbf{F} \cdot d\mathbf{r} = U(B) - U(A), \tag{7}$$

and combining (6) and (7), we get

$$\frac{1}{2} mv_A^2 - U(A) = \frac{1}{2} mv_B^2 - U(B). \tag{8}$$

In order to change the minus signs in (8) to plus signs, we introduce a new function $V(\mathbf{r}) = -U(\mathbf{r})$, called the *potential energy* of the particle or the *potential* of the field $\mathbf{F}$, so that $\mathbf{F} = -\nabla V(\mathbf{r})$. (Since the gradient of a constant is zero, both U and V are determined only to within an arbitrary additive constant.) Equation (8) then becomes

$$\frac{1}{2}mv_A^2 + V(A) = \frac{1}{2}mv_B^2 + V(B), \tag{8'}$$

and says that as a particle moves in a gradient field, the sum of its kinetic energy $\frac{1}{2}mv^2$ and potential energy V remains the same, or in the language of physics, is "conserved." This fact, which plays a key role in Newtonian mechanics, is called the *law of conservation of energy*, and the common value of the sums in (8') is called the *total energy* of the particle. By the same token, gradient fields, which lead to conservation of energy, are also known as *conservative fields*. In terms of the potential V, equation (7) takes the form

$$\int_A^B \mathbf{F} \cdot d\mathbf{r} = V(A) - V(B), \tag{7'}$$

and says that the amount of work W done by a conservative field $\mathbf{F}$ on a particle moving from A to B does not depend on the path joining A to B, and is equal to the decrease in V, or the "potential drop," in going from A to B.

A vector field is said to be *continuously differentiable* if the first partial derivatives of its components exist and are continuous. The next theorem describes certain relations which must be satisfied by the components of a continuously differentiable gradient field.

Theorem 2 *(Conditions on the components of a gradient field).* Suppose $\mathbf{F} = P\mathbf{i} + Q\mathbf{j} + R\mathbf{k}$ is a continuously differentiable gradient field on a domain D, with components $P = P(x, y, z)$, $Q = Q(x, y, z)$ and $R = R(x, y, z)$. Then

$$\frac{\partial Q}{\partial x} = \frac{\partial P}{\partial y}, \quad \frac{\partial R}{\partial y} = \frac{\partial Q}{\partial z}, \quad \frac{\partial P}{\partial z} = \frac{\partial R}{\partial x} \tag{9}$$

at every point of D.

Proof Since $\mathbf{F}$ is a gradient field, there exists a function $U = U(x, y, z)$ defined on D such that

$$P = \frac{\partial U}{\partial x}, \quad Q = \frac{\partial U}{\partial y}, \quad R = \frac{\partial U}{\partial z}.$$

Therefore

$$\frac{\partial Q}{\partial x} = \frac{\partial^2 U}{\partial x\, \partial y}, \quad \frac{\partial P}{\partial y} = \frac{\partial^2 U}{\partial y\, \partial x},$$

$$\frac{\partial R}{\partial y} = \frac{\partial^2 U}{\partial y\, \partial z}, \quad \frac{\partial Q}{\partial z} = \frac{\partial^2 U}{\partial z\, \partial y},$$

$$\frac{\partial P}{\partial z} = \frac{\partial^2 U}{\partial z\, \partial x}, \quad \frac{\partial R}{\partial x} = \frac{\partial^2 U}{\partial x\, \partial z},$$

where the assumed continuity of each first partial derivative on the left implies that of the second partial derivative on the right. But then

$$\frac{\partial^2 U}{\partial x\, \partial y} = \frac{\partial^2 U}{\partial y\, \partial x}, \quad \frac{\partial^2 U}{\partial y\, \partial z} = \frac{\partial^2 U}{\partial z\, \partial y}, \quad \frac{\partial^2 U}{\partial z\, \partial x} = \frac{\partial^2 U}{\partial x\, \partial z},$$

by the comment following Example 4, page 762, and these formulas are equivalent to (9). ∎

In the case of a two-dimensional gradient field $\mathbf{F} = P\mathbf{i} + Q\mathbf{j}$, the functions P and Q depend only on x and y, and there is no z-component R. The equations (9) then reduce to the single condition

$$\frac{\partial Q}{\partial x} = \frac{\partial P}{\partial y}. \tag{9'}$$

Thus failure of the conditions (9), or of the condition (9') in the two-dimensional case, tells us that $\mathbf{F}$ is *not* a gradient field. This immediately suggests the following question: Does the validity of (9) or (9') on a domain D *guarantee* that $\mathbf{F}$ is a gradient field on D? Surprisingly, the answer is in the negative, as shown by the next example.

Example 4 Let the field $\mathbf{F} = P\mathbf{i} + Q\mathbf{j}$ be the same as in Example 3. Then

$$P = \frac{-y}{x^2 + y^2}, \qquad Q = \frac{x}{x^2 + y^2} \qquad (x^2 + y^2 \neq 0),$$

and $\mathbf{F}$ is defined on the *multiply* connected domain D consisting of all the points of the xy-plane except the origin. Calculating the partial derivatives $\partial Q/\partial x$ and $\partial P/\partial y$, we easily find that

$$\frac{\partial Q}{\partial x} = \frac{y^2 - x^2}{(x^2 + y^2)^2} = \frac{\partial P}{\partial y},$$

so that the condition (9') holds at every point of D. Nevertheless, $\mathbf{F}$ is not a gradient field on D, as already shown in Example 3.

Test for a Gradient Field

Since the domain D in Example 4 is multiply connected, there still remains the possibility that the condition (9') guarantees that $\mathbf{F} = P\mathbf{i} + Q\mathbf{j}$ is a gradient field if D is *simply* connected. This is indeed true, as shown later (see Corollary 2, page 911). In the following theorem we prove the result only for the case of a *rectangular* domain, which is of course a simply connected domain of a particularly elementary type. The theorem also has a three-dimensional version, which you will find in Problem 20.

Theorem 3 *(Test for a gradient field on a rectangular domain)*. Suppose the components $P = P(x, y)$ and $Q = Q(x, y)$ of a continuously differentiable vector field $\mathbf{F} = P\mathbf{i} + Q\mathbf{j}$ satisfy the condition

$$\frac{\partial Q}{\partial x} = \frac{\partial P}{\partial y}$$

at every point of a rectangular domain $D = \{(x, y): a < x < b, c < y < d\}$. Then $\mathbf{F}$ is a gradient field on D, and in fact $\mathbf{F}$ is the gradient of the function

$$U = U(x, y) = \int_{x_0}^{x} P(t, y_0)\, dt + \int_{y_0}^{y} Q(x, t)\, dt, \tag{10}$$

where (x_0, y_0) is any fixed point of D and (x, y) is a variable point of D.

Proof Suppose $\mathbf{F} = P\mathbf{i} + Q\mathbf{j}$ is known to be a gradient field on D. Then the line integral of $\mathbf{F}$ is path independent on D, and as in the proof of Theorem 1, $\mathbf{F}$ is the gradient of the function

$$U = \int_C P\,dx + Q\,dy,$$

where C is any path in D from a fixed point (x_0, y_0) to a variable point (x, y) of D. We choose the path C consisting of the horizontal line segment α drawn from (x_0, y_0) to (x, y_0), followed by the vertical line segment β drawn from (x, y_0) to (x, y). As shown in Figure 8, this path always stays in the rectangular domain D, regardless of the choice of (x_0, y_0) and (x, y). Therefore

$$U = \int_C P\,dx + Q\,dy = \int_\alpha P\,dx + Q\,dy + \int_\beta P\,dx + Q\,dy = \int_\alpha P\,dx + \int_\beta Q\,dy,$$

since $dy = 0$ on α and $dx = 0$ on β, and this immediately gives formula (10), where we use the same dummy variable of integration t in both integrals on the right, which are ordinary definite integrals (x is held fixed in the second integrand).

Formula (10) was derived under the assumption that $\mathbf{F}$ is a gradient field, which is actually what we are trying to prove! But if it turns out that the gradient of U is equal to $\mathbf{F}$, we will have succeeded in proving the theorem anyway. Thus we now calculate the partial derivatives of U. The first integral on the right in (10) is independent of y, but both integrals depend on x. Thus we first differentiate (10) with respect to y, obtaining

$$\frac{\partial U(x, y)}{\partial y} = \frac{\partial}{\partial y} \int_{y_0}^{y} Q(x, t)\,dt = Q(x, y) \tag{11}$$

(recall Theorem 5, page 254). Next, using Problem 51, page 823, we differentiate (10) with respect to x. The result is

$$\frac{\partial U(x, y)}{\partial x} = \frac{\partial}{\partial x} \int_{x_0}^{x} P(t, y_0)\,dt + \frac{\partial}{\partial x} \int_{y_0}^{y} Q(x, t)\,dt = P(x, y_0) + \int_{y_0}^{y} \frac{\partial Q(x, t)}{\partial x}\,dt.$$

But

$$\int_{y_0}^{y} \frac{\partial Q(x, t)}{\partial x}\,dt = \int_{y_0}^{y} \frac{\partial P(x, t)}{\partial t}\,dt = P(x, y) - P(x, y_0),$$

because of the condition $\partial Q/\partial x = \partial P/\partial y$ with y changed to the dummy variable t. Combining the last two equations, we find that

$$\frac{\partial U(x, y)}{\partial x} = P(x, y_0) + P(x, y) - P(x, y_0) = P(x, y), \tag{11'}$$

which together with (11), shows that

$$P = \frac{\partial U}{\partial x}, \qquad Q = \frac{\partial U}{\partial y},$$

or equivalently $\mathbf{F} = \operatorname{grad} U$. ∎

Example 5 Let

$$P(x, y) = \frac{1}{y}, \qquad Q(x, y) = -\frac{x}{y^2} \qquad (y \neq 0), \tag{12}$$

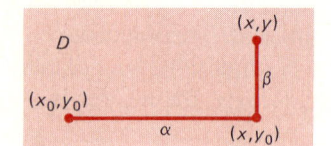

894 Chapter 15 Line and Surface Integrals

and let $D = \{(x, y): a < x < b, c < y < d\}$ be any rectangular domain containing no points of the x-axis (the line $y = 0$). Then, since

$$\frac{\partial Q}{\partial x} = -\frac{1}{y^2} = \frac{\partial P}{\partial y},$$

it follows from Theorem 3 that

$$\mathbf{F} = P\mathbf{i} + Q\mathbf{j} = \frac{y\mathbf{i} - x\mathbf{j}}{y^2}$$

is a gradient field on D. Therefore $\mathbf{F}$ is a gradient field on the upper half-plane $y > 0$, which can be regarded as the unbounded rectangular domain $D = \{(x, y): -\infty < x < \infty, 0 < y < \infty\}$, and also on the lower half-plane $y < 0$.

You may have already recognized $\mathbf{F}$ as the gradient of the function $U(x, y) = x/y$, but if not, you can construct $U(x, y)$ by using formula (10). In fact, substituting (12) into (10) and choosing $x_0 = 0$ and $y_0 = 1$, we get

$$U(x, y) = \int_0^x dt - \int_1^y \frac{x}{t^2} dt = \frac{x}{y}.$$

The effect of choosing different values of x_0 and y_0 is merely to introduce a constant of integration. For instance, the choice $x_0 = y_0 = 1$ gives

$$U(x, y) = \int_1^x dt - \int_1^y \frac{x}{t^2} dt = \frac{x}{y} - 1.$$

Thus the general form of $U(x, y)$ is

$$U(x, y) = \frac{x}{y} + C,$$

where C is an arbitrary constant. Make sure to check that $\mathbf{F} = \text{grad } U$ by direct calculation. ☐

The function $U(x, y)$ stands in the same relation to the vector field $\mathbf{F}(x, y) = P(x, y)\mathbf{i} + Q(x, y)\mathbf{j}$ as an antiderivative $U(x)$ to a function $F(x)$ of a single variable, and in particular $U(x, y)$ is defined only to within an arbitrary additive constant. The key difference is that whereas *every* function $F(x)$ continuous on an interval has an antiderivative, only certain fields $\mathbf{F}(x, y)$ continuously differentiable on a rectangular domain are "derivable" from a function $U(x, y)$, namely those that satisfy the "integrability condition" $\partial Q/\partial x = \partial P/\partial y$.

Problems

Is the vector field $\mathbf{F} = P\mathbf{i} + Q\mathbf{j}$ with the given components $P = P(x, y)$ and $Q = Q(x, y)$ a gradient field? If so, find a function U and a domain D on which $\mathbf{F} = \text{grad } U$.

1. $P = e^x \sin y, Q = e^x \cos y$
2. $P = x \ln y, Q = -x/y$
3. $P = x^2 + y^2, Q = x^3 + 2xy$
4. $P = x + \ln y, Q = (x/y) + \sin y$
5. $P = 4x^3 y^3 - 3y^2, Q = 3x^4 y^2 - 6xy$
6. $P = x^2 + xy^3, Q = x^2 y^2 - 2y$
7. $P = xe^y, Q = ye^x$
8. $P = x \sin 2y, Q = x^2 \cos 2y$
9. $P = y - (\sin^2 y)/x^2, Q = x + (\sin 2y)/x$
10. $P = 2x \cos^2 y, Q = 2y - x^2 \sin 2y$

Verify that the given force field $\mathbf{F} = P\mathbf{i} + Q\mathbf{j}$ is conservative, and then use formula (7) to calculate the work W done by $\mathbf{F}$ on a particle moving from point A to point B.

11. $\mathbf{F} = y\mathbf{i} + x\mathbf{j}$, $A = (0, 1)$, $B = (-3, 4)$
12. $\mathbf{F} = x\mathbf{i} + y\mathbf{j}$, $A = (-1, 1)$, $B = (2, 3)$
13. $\mathbf{F} = (x + y)(\mathbf{i} + \mathbf{j})$, $A = (-13, 0)$, $B = (5, 12)$
14. $\mathbf{F} = (2x + 3y)\mathbf{i} + (3x - 2y)\mathbf{j}$, $A = (-4, 6)$, $B = (1, -2)$
15. $\mathbf{F} = [(x + y + 1)e^x - e^y]\mathbf{i} + [e^x - (x + y + 1)e^y]\mathbf{j}$, $A = (0, 0)$, $B = (\ln 3, \ln 2)$
16. $\mathbf{F} = (\sinh x + \cosh y)\mathbf{i} + (x \sinh y + 2)\mathbf{j}$, $A = (0, \ln 4)$, $B = (\ln 2, 0)$

17. Calculate the circulation of the field
$$\mathbf{F} = \frac{-y\mathbf{i} + x\mathbf{j}}{x^2 + 4y^2} \quad (x^2 + y^2 \neq 0)$$
around any circle C centered at the origin, traversed once in the counterclockwise direction.

18. Show that Theorem 3 remains valid if formula (10) is replaced by
$$U = U(x, y) = \int_{x_0}^{x} P(t, y)dt + \int_{y_0}^{y} Q(x_0, t) dt.$$

19. The force of gravitational attraction exerted on a particle Q of unit mass at the point with position vector $\mathbf{R}$ by a particle P of mass M at the point with position vector $\mathbf{R}_1$ is $\mathbf{F} = -(GM/r^2)\mathbf{u}$, where G is the universal gravitational constant, $\mathbf{r} = \mathbf{R} - \mathbf{R}_1$, $r = |\mathbf{r}|$, and $\mathbf{u} = \mathbf{r}/r$ is the unit vector from P_1 to Q. Show that $\mathbf{F} = \operatorname{grad} U$ in terms of the gravitational potential $U = GM/r$ (physicists tend to call $V = -U$ the gravitational potential instead). More generally, show that if Q is attracted by n particles $P_1, \ldots, P_n$ of masses $M_1, \ldots, M_n$ at the points with position vectors $\mathbf{R}_1, \ldots, \mathbf{R}_n$, then the resultant force $\mathbf{F}$ acting on Q is again $\mathbf{F} = \operatorname{grad} U$, where now
$$U = \sum_{i=1}^{n} \frac{GM_i}{r_i} \quad (r_i = |\mathbf{R} - \mathbf{R}_i|).$$

20. Prove the following three-dimensional version of Theorem 3: Suppose the components $P = P(x, y, z)$, $Q = Q(x, y, z)$ and $R = R(x, y, z)$ of a continuously differentiable vector field $\mathbf{F} = P\mathbf{i} + Q\mathbf{j} + R\mathbf{k}$ satisfy the conditions
$$\frac{\partial Q}{\partial x} = \frac{\partial P}{\partial y}, \quad \frac{\partial R}{\partial y} = \frac{\partial Q}{\partial z}, \quad \frac{\partial P}{\partial z} = \frac{\partial R}{\partial x}$$
at every point of a rectangular domain
$$D = \{(x, y, z): a < x < b, c < y < d, A < z < B\}.$$
Then $\mathbf{F}$ is a gradient field on D, and in fact $\mathbf{F}$ is the gradient of the function
$$U = U(x, y, z)$$
$$= \int_{x_0}^{x} P(t, y_0, z_0) dt + \int_{y_0}^{y} Q(x, t, z_0) dt + \int_{z_0}^{z} R(x, y, t) dt, \quad \text{(i)}$$
where (x_0, y_0, z_0) is any fixed point of D and (x, y, z) is a variable point of D.

Is the vector field $\mathbf{F} = P\mathbf{i} + Q\mathbf{j} + R\mathbf{k}$ with the given components $P = P(x, y, z)$, $Q = Q(x, y, z)$ and $R = R(x, y, z)$ a gradient field? If so, find a function U and a domain D on which $F = \operatorname{grad} U$. (See Problem 20, and use formula (i) if necessary.)

21. $P = y + z$, $Q = x + z$, $R = x + y$
22. $P = y - z$, $Q = x - z$, $R = x - y$
23. $P = yz$, $Q = xz$, $R = xy$
24. $P = xyz$, $Q = \frac{1}{2}x^2z$, $R = \frac{1}{2}x^2y$
25. $P = 2xy$, $Q = \ln(1 + x^2)$, $R = \ln(1 + z^2)$
26. $P = yz \cos xy$, $Q = xz \cos xy$, $R = \sin xy$
27. $P = \ln y - \cos 2z$, $Q = (x/y) + z$, $R = y + 2x \sin 2z$
28. $P = Q = R = 1/(x + y + z)$

29. If there is a differentiable function $U = U(x, y)$ such that $dU = P\,dx + Q\,dy$, the differential form $P\,dx + Q\,dy$ is said to be an *exact differential*. Show that $P\,dx + Q\,dy$ is an exact differential if and only if $\mathbf{F} = P\mathbf{i} + Q\mathbf{j}$ is a gradient field.

30. Show that $(y + xy^2)\,dx - x\,dy$ is not an exact differential, but becomes one if multiplied by an appropriate "integrating factor" $f(x, y)$. Find $f(x, y)$ by inspection.

31. A roller coaster car is raised to a height of 121 ft and then released. What is the maximum speed the car can attain at the lowest point of the track? Neglect air resistance and friction, and use an argument involving the conversion of potential energy into kinetic energy. Does the answer depend on the weight of the car and its passengers? Take g, the acceleration due to gravity, to be 32 ft/sec^2.

32. Two children stand on the edge of a roof h ft above the ground. The first child throws a stone downward with speed v_0, and at exactly the same moment the second child throws another stone upward with speed v_0. Show that the stones hit the ground with the same speed $v_1 = \sqrt{v_0^2 + 2gh}$, but at different times t_1 and t_2. Find t_1, t_2 and $\Delta t = t_2 - t_1$.

33. Find the potential energy V of a stretched spring if the elastic restoring force is $F = -ks$, where k is the spring constant and s is the elongation of the spring beyond its natural length (recall page 270). Take V to be zero when $s = 0$.

34. A spider hangs from the ceiling by a single strand of web. Suppose the spider's weight doubles the natural length of the strand, stretching it from L to $2L$. In climbing back to the ceiling, the spider does work W, which is different from the work W_0 that it would do if the strand were inelastic, of length $2L$. Show that $W = \frac{3}{4}W_0$.

35. Let T be the solid sphere $x^2 + y^2 + z^2 \leq a^2$ of radius a, and let Q be a particle of unit mass at the point $(0, 0, Z)$ outside T. Suppose T has constant density δ, and partition T into "spherical wedges" T_i ($i = 1, \ldots, n$) of the kind shown in Figure 53, page 870. Then regarding each T_i as a particle P_i of mass M_i, where M_i is the product of δ and the volume of T_i, applying the result of Problem 19 with r_i the distance from any point of T_i to Q, and finally taking the limit as the maximum size of all the T_i approaches zero, we find

that Q is attracted by T with a force $\mathbf{F} = \operatorname{grad} U$, where the gravitational potential U is given by the integral

$$U = \iiint_T \frac{G\delta}{\sqrt{x^2 + y^2 + (z-Z)^2}} \, dV.$$

Calculate U, and show that the solid sphere T attracts P with the same force as if all its mass were concentrated at its center. Show that this is true not only for constant density δ, but also for any spherically symmetric density function $\delta = \delta(\sqrt{x^2 + y^2 + z^2})$.

36. Show that the net gravitational force exerted by a homogeneous spherical shell on a particle inside it is zero.

37. Figure 9 shows two paths C_1 and C_2 from A to B, where C_1 and C_2 intersect twice (at points other than A and B), and C_1 has three self-intersections. Suppose it is known only that $\int_C \mathbf{F} \cdot d\mathbf{r} = 0$ for every *simple* closed path C (in some underlying domain). By introducing another path C_3 from A to B, which does not intersect C_1 or C_2, show that $\int_{C_1} \mathbf{F} \cdot d\mathbf{r} = \int_{C_2} \mathbf{F} \cdot d\mathbf{r}$.

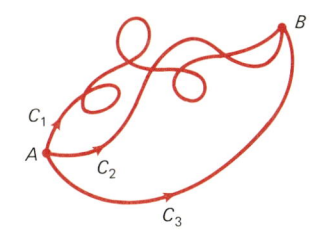

Figure 9

15.3 Surface Area and Surface Integrals

Surface Area

We now turn our attention to *surface integrals*, that is, integrals over surfaces which are in general curved. We begin with the problem of determining the area of the graph of a continuously differentiable function of two variables. Let the surface S be the graph of a function

$$z = f(x, y) \qquad ((x, y) \in R),$$

where R is a normal region in the xy-plane, as defined on page 811, and f is continuously differentiable (meaning that the partial derivatives $\partial f/\partial x$ and $\partial f/\partial y$ exist and are continuous). As in the definition of a double integral over R, given in Section 14.1, we partition R into n subregions $R_1, R_2, \ldots, R_n$ by drawing the lines

$$x = x_j \quad (j = 0, 1, \ldots, J), \qquad y = y_k \quad (k = 0, 1, \ldots, K), \qquad (1)$$

where

$$a = x_0 < x_1 < \cdots < x_{J-1} < x_J = b, \qquad c = y_0 < y_1 < \cdots < y_{K-1} < y_K = d,$$

and the rectangular region $\{(x, y): a \leq x \leq b, c \leq y \leq d\}$ contains the region R. The *planes* with the same equations (1) are parallel to the yz- and xz-planes, and divide S into n "surface elements" $S_1, S_2, \ldots, S_n$, like the element S_i shown in Figure 10 for the case $f(x, y) \geq 0$.

Next we choose an arbitrary point (p_i, q_i) in each subregion R_i, and find the point P_i on S_i with (p_i, q_i) as its projection onto the xy-plane. We then draw the tangent plane to S at each such point P_i. This gives n tangent planes which intersect the planes (1) in n plane regions $\Pi_1, \Pi_2, \ldots, \Pi_n$, adhering to S like loosely fitting shingles on a roof; such a region Π_i is shown in the figure for a typical "interior" surface element S_i, whose projection onto the xy-plane is rectangular. Let $\Delta\sigma_i$ be the area of Π_i (the symbol σ is lowercase Greek sigma). Then it seems as if the sum $\sum_{i=1}^{n} \Delta\sigma_i$ is a good approximation to what is meant intuitively by the area A_S of the surface S, where the approximation gets better and better as the size of each and every surface element S_i gets smaller and smaller, that is, as the mesh size

$$\mu = \max\{x_1 - x_0, \ldots, x_J - x_{J-1}, y_1 - y_0, \ldots, y_K - y_{K-1}\}$$

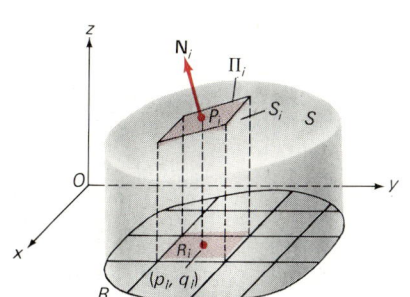

Figure 10

of the partition of R approaches zero. These considerations lead us to *define* A_S as the limit

$$A_S = \lim_{\mu \to 0} \sum_{i=1}^{n} \Delta\sigma_i. \tag{2}$$

Now let **i**, **j** and **k** be unit vectors pointing along the positive x-, y- and z-axes, and let $\mathbf{N}_i$ be any *upper* normal to Π_i (and hence to S_i or S) at P_i, that is, any nonzero vector perpendicular to Π_i which points *upward*, making an acute angle γ_i with the positive z-axis (and hence with **k**). To bring (2) into the form of a double integral over R, we observe that if ΔA_i is the area of the subregion R_i, then

$$\Delta\sigma_i = \Delta A_i \sec \gamma_i, \tag{3}$$

at least for a "shingle" Π_i with straight edges.† To see this, consider Figure 11, showing a typical straight-edged shingle Π_i, shaped like a parallelogram spanned by two vectors **a** and **b**. The area $\Delta\sigma_i$ of the shingle is the magnitude of the cross product $\mathbf{a} \times \mathbf{b}$, while the cross product $\mathbf{a} \times \mathbf{b}$ itself is a normal $\mathbf{N}_i$ to Π_i (for simplicity, we attach $\mathbf{N}_i$ to a corner of Π_i). Suppose the rectangular subregion R_i lying under Π_i in the xy-plane has side lengths Δx and Δy. Then $\Delta A_i = \Delta x \Delta y$ and it is clear from the figure that

$$\mathbf{a} = \overrightarrow{CC'} + \overrightarrow{C'D'} + \overrightarrow{D'D} = \Delta x\,\mathbf{i} + (\overrightarrow{CC'} + \overrightarrow{D'D}) = \Delta x\,\mathbf{i} + r\mathbf{k},$$
$$\mathbf{b} = \overrightarrow{CC'} + \overrightarrow{C'E'} + \overrightarrow{E'E} = \Delta y\,\mathbf{j} + (\overrightarrow{CC'} + \overrightarrow{E'E}) = \Delta y\,\mathbf{j} + s\mathbf{k},$$

where r and s are suitable scalars. Therefore

$$\mathbf{a} \times \mathbf{b} = (\Delta x\,\mathbf{i} + r\mathbf{k}) \times (\Delta y\,\mathbf{j} + s\mathbf{k}) = \Delta x\,\Delta y\,\mathbf{k} - r\,\Delta y\,\mathbf{i} - s\,\Delta x\,\mathbf{j},$$

so that

$$(\mathbf{a} \times \mathbf{b}) \cdot \mathbf{k} = \Delta x\,\Delta y$$

and hence

$$\Delta\sigma_i \cos \gamma_i = \Delta x\,\Delta y = \Delta A_i,$$

which is equivalent to (3).

It follows from (2) and (3) that

$$A_S = \lim_{\mu \to 0} \sum_{i=1}^{n} \Delta A_i \sec \gamma_i. \tag{4}$$

Choosing $F(x, y, z) = z - f(x, y)$ in formula (3), page 780, and observing that S is the graph of the equation $F(x, y, z) = 0$, we find that

$$\mathbf{N}_i = -f_x(p_i, q_i)\mathbf{i} - f_y(p_i, q_i)\mathbf{j} + \mathbf{k}, \tag{5}$$

is an upper normal to Π_i, at (p_i, q_i). Therefore

$$\cos \gamma_i = \frac{\mathbf{N}_i \cdot \mathbf{k}}{|\mathbf{N}_i||\mathbf{k}|} = \frac{1}{\sqrt{[f_x(p_i, q_i)]^2 + [f_y(p_i, q_i)]^2 + 1}},$$

or equivalently

$$\sec \gamma_i = \sqrt{[f_x(p_i, q_i)]^2 + [f_y(p_i, q_i)]^2 + 1} \tag{6}$$

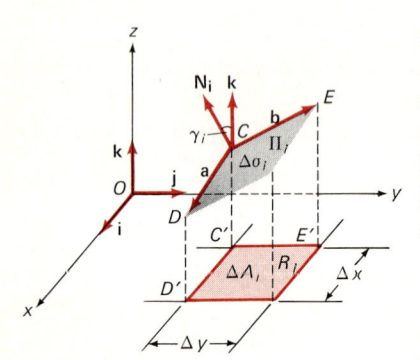

Figure 11

† It can be shown that the sum (2) approaches the same limit A_S even if we drop every term $\Delta\sigma_i$ corresponding to a shingle Π_i with curved edges.

Substituting (6) into (4), we finally obtain

$$A_S = \lim_{\mu \to 0} \sum_{i=1}^{n} \sqrt{[f_x(p_i, q_i)]^2 + [f_y(p_i, q_i)]^2 + 1} \, \Delta A_i.$$

You will recognize this limit as the double integral over R of the function

$$\sqrt{[f_x(x, y)]^2 + [f_y(x, y)]^2 + 1}.$$

Therefore A_S is given by the integral

$$A_S = \iint_R \sqrt{[f_x(x, y)]^2 + [f_y(x, y)]^2 + 1} \, dA, \tag{7}$$

whose existence is guaranteed by Theorem 1, page 813, and the continuity of the partial derivatives f_x and f_y on R, which implies the continuity of the integrand on R. Formula (7) can be written more concisely as

$$A_S = \iint_R \sqrt{\left(\frac{\partial z}{\partial x}\right)^2 + \left(\frac{\partial z}{\partial y}\right)^2 + 1} \, dA. \tag{7'}$$

Of course, there are similar formulas for the case where the surface S is the graph of a continuously differentiable function defined on a region in a coordinate plane other than the xy-plane. To get these formulas, we replace the integrand of (7') by

$$\sqrt{\left(\frac{\partial y}{\partial x}\right)^2 + \left(\frac{\partial y}{\partial z}\right)^2 + 1}$$

if R is a region in the xz-plane, and by

$$\sqrt{\left(\frac{\partial x}{\partial y}\right)^2 + \left(\frac{\partial x}{\partial z}\right)^2 + 1}$$

if R is a region in the yz-plane.

Example 1 Let S be the part of the parabolic cylinder $z = \frac{1}{2}y^2$ which lies over the rectangular region R in the xy-plane bounded by the y-axis, the line $x = 3$ and the lines $y = \pm 1$ (see Figure 12). Find the area A_S of the surface S.

Solution By symmetry, A_S is twice the area of the part of S lying over the first quadrant of the xy-plane. Therefore, applying formula (7'), we get

$$A_S = 2 \iint_R \sqrt{\left(\frac{\partial z}{\partial x}\right)^2 + \left(\frac{\partial z}{\partial y}\right)^2 + 1} \, dA = 2 \int_0^3 dx \int_0^1 \sqrt{y^2 + 1} \, dy$$

$$= 6 \left[\frac{1}{2} y \sqrt{y^2 + 1} + \frac{1}{2} \ln(y + \sqrt{y^2 + 1}) \right]_0^1$$

$$= 3[\sqrt{2} + \ln(1 + \sqrt{2})] \approx 6.89,$$

with the help of formula (6'), page 391. □

Application of formula (7) or (7') often leads to a convergent improper integral. In such cases the surface area A_S is taken to be the value of the improper integral.

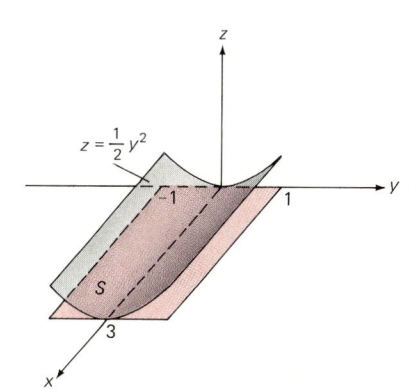

Figure 12

Example 2 Find the area A_S of the surface S which is the graph of the function $z = \sqrt{a^2 - x^2 - y^2}$ $(a > 0)$.

Solution Since S is a hemisphere of radius a, we already know that $A_S = \frac{1}{2}(4\pi a^2) = 2\pi a^2$. Hence this example is given only to illustrate the method. Although the function $z = \sqrt{a^2 - x^2 - y^2}$ is continuous on its domain of definition, namely the closed disk $R = \{(x, y): x^2 + y^2 \leq a^2\}$, it is not continuously differentiable on R, since the partial derivatives

$$\frac{\partial z}{\partial x} = -\frac{x}{\sqrt{a^2 - x^2 - y^2}}, \quad \frac{\partial z}{\partial y} = -\frac{y}{\sqrt{a^2 - x^2 - y^2}}$$

do not exist on the circle $x^2 + y^2 = a^2$ which is the boundary of R. However, $\sqrt{a^2 - x^2 - y^2}$ is continuously differentiable on every closed disk

$$R_u = \{(x, y): x^2 + y^2 \leq u^2\} \quad (0 < u < a)$$

of smaller radius u, and thus it is appropriate to define the area of S by the improper double integral

$$A_S = \iint_R \sqrt{\left(\frac{\partial z}{\partial x}\right)^2 + \left(\frac{\partial z}{\partial y}\right)^2 + 1} \, dA$$

$$= \iint_R \sqrt{\frac{x^2}{a^2 - x^2 - y^2} + \frac{y^2}{a^2 - x^2 - y^2} + 1} \, dA$$

$$= \iint_R \frac{a}{\sqrt{a^2 - x^2 - y^2}} \, dA = \lim_{u \to a^-} \iint_{R_u} \frac{a}{\sqrt{a^2 - x^2 - y^2}} \, dA.$$

Transforming to polar coordinates, because of the circular symmetry, we find that

$$\iint_{R_u} \frac{a}{\sqrt{a^2 - x^2 - y^2}} \, dA = a \int_0^{2\pi} d\theta \int_0^u \frac{r}{\sqrt{a^2 - r^2}} \, dr = 2\pi a \left[-\sqrt{a^2 - r^2} \right]_0^u$$

$$= 2\pi a (a - \sqrt{a^2 - u^2}),$$

and hence

$$A_S = 2\pi a \lim_{u \to a^-} (a - \sqrt{a^2 - u^2}) = 2\pi a^2. \quad \square$$

Surface Integrals

Next we define the integral of a continuous function over a curved surface S. Once again, let S be the graph of a continuously differentiable function $z = f(x, y)$ defined on a normal region R in the xy-plane and let the points of subdivision x_j, y_k, the mesh size μ, the subregions R_i, and the "elementary areas" ΔA_i and $\Delta \sigma_i$ have the same meaning as before. Let $g(x, y, z)$ be a function of three variables which is continuous on the surface S; this means that the composite function $g(x, y, f(x, y))$ is continuous on the region R which is the projection of S onto the xy-plane. Choosing an arbitrary point (p_i, q_i) in each subregion R_i, we form the sum

$$\sum_{i=1}^n g(p_i, q_i, f(p_i, q_i)) \, \Delta \sigma_i.$$

Suppose this sum approaches a finite limit as the mesh size μ approaches zero, regardless of the choice of the numbers x_j, y_k, p_i and q_i satisfying the stipulated conditions. Then the limit is called the (*surface*) *integral of g over S*, denoted by

$$\iint_S g(x, y, z)\, d\sigma,$$

and the function g is said to be *integrable* on S, or over S. Thus

$$\iint_S g(x, y, z)\, d\sigma = \lim_{\mu \to 0} \sum_{i=1}^{n} g(p_i, q_i, f(p_i, q_i))\, \Delta\sigma_i, \tag{8}$$

where g is called the *integrand* of the integral on the left.

Example 3 Choosing $g(x, y, z) \equiv 1$ in formula (8), we get

$$\iint_S 1\, d\sigma = \lim_{\mu \to 0} \sum_{i=1}^{n} \Delta\sigma_i = A_S,$$

where A_S is the area of the surface S. It follows that

$$A_S = \iint_S d\sigma. \quad \square$$

To evaluate a surface integral $\iint_S g(x, y, z)\, d\sigma$ over S, we parallel the argument used to derive formula (7) for the area of S. Thus we replace S by R and $\Delta\sigma_i$ by $\Delta A_i \sec \gamma_i$, where γ_i is the angle between the positive z-axis and any upper normal to S at $P_i = (p_i, q_i, f(p_i, q_i))$. As a result, with the help of (6) we obtain

$$\iint_S g(x, y, z)\, d\sigma = \lim_{\mu \to 0} \sum_{i=1}^{n} g(p_i, q_i, f(p_i, q_i))\, \Delta A_i \sec \gamma_i$$

$$= \lim_{\mu \to 0} \sum_{i=1}^{n} g(p_i, q_i, f(p_i, q_i)) \sqrt{[f_x(p_i, q_i)]^2 + [f_y(p_i, q_i)]^2 + 1}\, \Delta A_i,$$

where the limit on the right will be recognized as the double integral

$$\iint_R g(x, y, f(x, y)) \sqrt{[f_x(x, y)]^2 + [f_y(x, y)]^2 + 1}\, dA.$$

whose existence is guaranteed by the assumed continuity of the functions g, f_x and f_y on R. Accordingly,

$$\iint_S g(x, y, z)\, d\sigma = \iint_R g(x, y, f(x, y)) \sqrt{[f_x(x, y)]^2 + [f_y(x, y)]^2 + 1}\, dA, \tag{9}$$

or more concisely

$$\iint_S g(x, y, z)\, d\sigma = \iint_R g(x, y, z) \sqrt{\left(\frac{\partial z}{\partial x}\right)^2 + \left(\frac{\partial z}{\partial y}\right)^2 + 1}\, dA, \tag{9'}$$

where $z = f(x, y)$ in the right side. Of course, there are formulas analogous to (9) and (9′) for the case where the surface S is the graph of a continuously differentiable function defined on a region in the xz-plane or in the yz-plane.

Example 4 Evaluate the surface integral

$$\iint_S \ln z \, d\sigma,$$

where S is the part of the cone $z = \sqrt{x^2 + y^2}$ between the planes $z = 1$ and $z = 2$ (see Figure 13).

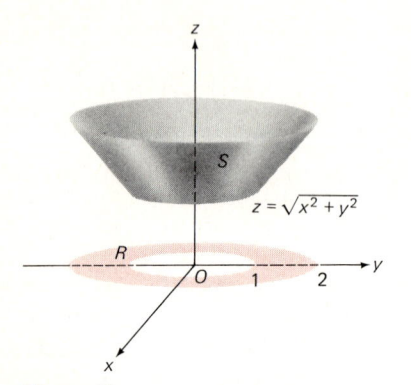

Solution The projection of S onto the xy-plane is the annular region $R = \{(x, y): 1 \leq x^2 + y^2 \leq 4\}$, and

$$\frac{\partial z}{\partial x} = \frac{x}{\sqrt{x^2 + y^2}}, \qquad \frac{\partial z}{\partial y} = \frac{y}{\sqrt{x^2 + y^2}}.$$

It follows from (9′) that

$$\iint_S \ln z \, d\sigma = \iint_R \ln \sqrt{x^2 + y^2} \sqrt{\frac{x^2}{x^2 + y^2} + \frac{y^2}{x^2 + y^2} + 1} \, dA$$

$$= \sqrt{2} \iint_R \ln \sqrt{x^2 + y^2} \, dA = \sqrt{2} \int_0^{2\pi} d\theta \int_1^2 r \ln r \, dr,$$

after transforming to polar coordinates. Therefore

$$\iint_S \ln z \, d\sigma = 2\pi\sqrt{2} \left[\frac{1}{2} r^2 \ln r - \frac{1}{4} r^2 \right]_1^2 = \frac{\pi}{\sqrt{2}} (8 \ln 2 - 3) \approx 5.65,$$

with the help of integration by parts. □

Example 5 Evaluate the surface integral

$$\iint_S xz \, d\sigma,$$

where S is the part of the plane $2x + 3y = 6$ in the first octant between the planes $z = 0$ and $z = 1$ (see Figure 14).

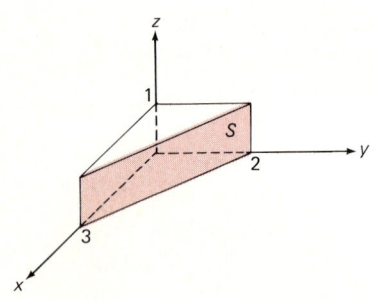

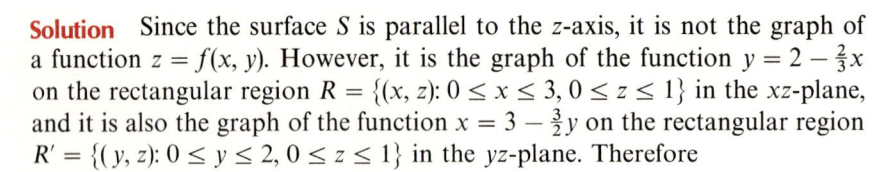

Solution Since the surface S is parallel to the z-axis, it is not the graph of a function $z = f(x, y)$. However, it is the graph of the function $y = 2 - \frac{2}{3}x$ on the rectangular region $R = \{(x, z): 0 \leq x \leq 3, 0 \leq z \leq 1\}$ in the xz-plane, and it is also the graph of the function $x = 3 - \frac{3}{2}y$ on the rectangular region $R' = \{(y, z): 0 \leq y \leq 2, 0 \leq z \leq 1\}$ in the yz-plane. Therefore

$$\iint_S xz \, d\sigma = \iint_R xz \sqrt{\left(\frac{\partial y}{\partial x}\right)^2 + \left(\frac{\partial y}{\partial z}\right)^2 + 1} \, dx \, dz = \frac{\sqrt{13}}{3} \int_0^3 dx \int_0^1 xz \, dz$$

$$= \frac{\sqrt{13}}{3} \left[\frac{1}{2} x^2\right]_0^3 \left[\frac{1}{2} z^2\right]_0^1 = \frac{\sqrt{13}}{3} \left(\frac{9}{2}\right)\left(\frac{1}{2}\right) = \frac{3\sqrt{13}}{4},$$

or alternatively

$$\iint_S xz\,d\sigma = \iint_{R'} \left(3 - \frac{3}{2}y\right)z\,\sqrt{\left(\frac{\partial x}{\partial y}\right)^2 + \left(\frac{\partial x}{\partial z}\right)^2 + 1}\,dy\,dz$$

$$= \frac{\sqrt{13}}{2}\int_0^2 dy \int_0^1 \left(3 - \frac{3}{2}y\right)z\,dz$$

$$= \frac{3\sqrt{13}}{2}\left[y - \frac{1}{4}y^2\right]_0^2 \left[\frac{1}{2}z^2\right]_0^1 = \frac{3\sqrt{13}}{2}\left(\frac{1}{2}\right) = \frac{3\sqrt{13}}{4}. \quad \square$$

Just as the total mass of a wire in the shape of a curve C is given by the line integral $\int_C \rho(x, y)\,ds$, where $\rho(x, y)$ is the density at the point (x, y) of C, measured in units like grams per centimeter, the total mass of a thin material sheet in the shape of a surface S is given by the surface integral $\iint_S \rho(x, y, z)\,d\sigma$, where $\rho(x, y, z)$ is the density at the point (x, y, z) of S, measured in units like grams per square centimeter. A detailed explanation is unnecessary, since it has already been given in Chapter 14 for the case of solids, plane laminas and wires of variable density. By the same token, we can also define the center of mass and various moments of inertia for a material surface of variable or constant density (see Problems 21–23).

We often want to calculate a surface integral over a surface S composed of a finite number of subsurfaces $S_1, S_2, \ldots, S_n$, each the graph of a continuously differentiable function, which share part or all of their boundary curves, but have no other points in common. The integral of a continuous function g over such a surface S is defined in the natural way, as the sum of the integrals of g over the subsurfaces $S_1, S_2, \ldots, S_n$. Specifically,

$$\iint_S g\,d\sigma = \iint_{S_1} g\,d\sigma + \iint_{S_2} g\,d\sigma + \cdots + \iint_{S_n} g\,d\sigma, \tag{10}$$

where for brevity we write just g instead of $g(x, y, z)$.

Example 6 Evaluate the surface integral

$$\iint_S xy^2z^3\,d\sigma,$$

where S is the cube in the first octant bounded by the coordinate planes and the planes $x = 1$, $y = 1$ and $z = 1$.

Solution Since $xy^2z^3 = 0$ on the coordinate planes, we need only consider the integrals over three faces of the cube, namely the front face S_1, the right face S_2 and the upper face S_3 (see Figure 15). Therefore, by (10),

$$\iint_S xy^2z^3\,d\sigma = \iint_{S_1} xy^2z^3\,d\sigma + \iint_{S_2} xy^2z^3\,d\sigma + \iint_{S_3} xy^2z^3\,d\sigma.$$

Each face is a place surface parallel to a coordinate plane, and hence $d\sigma = dA$

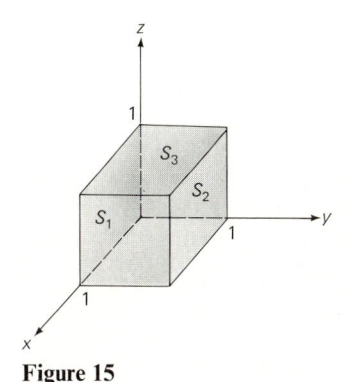

Figure 15

(for instance, the square root in (9′) is equal to 1 if $z =$ constant). Therefore, since $x = 1$ on S_1, $y = 1$ on S_2 and $z = 1$ on S_3, we have

$$\iint_S xy^2z^3\,d\sigma = \iint_{R_1} y^2z^3\,dy\,dz + \iint_{R_2} xz^3\,dx\,dz + \iint_{R_3} xy^2\,dx\,dy,$$

where $R_1 = \{(y, z): 0 \le y \le 1, 0 \le z \le 1\}$, $R_2 = \{(x, z): 0 \le x \le 1, 0 \le z \le 1\}$ and $R_3 = \{(x, y): 0 \le x \le 1, 0 \le y \le 1\}$. It follows that

$$\iint_S xy^2z^3\,d\sigma = \left(\int_0^1 y^2\,dy\right)\left(\int_0^1 z^3\,dz\right) + \left(\int_0^1 x\,dx\right)\left(\int_0^1 z^3\,dz\right)$$

$$+ \left(\int_0^1 x\,dx\right)\left(\int_0^1 y^2\,dy\right)$$

$$= \frac{1}{3}\left(\frac{1}{4}\right) + \frac{1}{2}\left(\frac{1}{4}\right) + \frac{1}{2}\left(\frac{1}{3}\right) = \frac{3}{8}. \quad \square$$

Unit Normal Vectors to a Surface

The set of *unit* normal vectors to a surface S constitutes a vector field on S. For example, suppose S is the graph of a continuously differentiable function

$$z = f(x, y) \quad ((x, y) \in R),$$

where R is a region in the xy-plane. Then the same argument as the one leading to formula (5) shows that

$$\mathbf{n} = \frac{-f_x(x, y)\mathbf{i} - f_y(x, y)\mathbf{j} + \mathbf{k}}{\sqrt{[f_x(x, y)]^2 + [f_y(x, y)]^2 + 1}}$$

is a unit normal vector to S at the point $P = (x, y, f(x, y))$, and so is the oppositely directed vector

$$-\mathbf{n} = \frac{f_x(x, y)\mathbf{i} + f_y(x, y)\mathbf{j} - \mathbf{k}}{\sqrt{[f_x(x, y)]^2 + [f_y(x, y)]^2 + 1}}.$$

In fact, $\mathbf{n}$ is the *upper* unit normal to S at P, which points upward from S, and $-\mathbf{n}$ is the *lower* unit normal to S at P, which points downward from S (bear in mind that $0 < \mathbf{n} \cdot \mathbf{k} \le 1$); as illustrated in Figure 16, there are only two unit normals to S at P. Let $\mathbf{n}'$ be the upper unit normal at another point P' of S (see the figure). Then, because of the continuity of the partial derivatives f_x and f_y, the angle between $\mathbf{n}$ and $\mathbf{n}'$ can be made as small as we please by choosing P' sufficiently close to P. This fact is expressed by saying that the unit normal $\mathbf{n}$ *varies continuously* on S, or that there is a *continuous unit normal* $\mathbf{n}$ on S. If $\mathbf{n}$ is a continuous unit normal on S, then clearly so is the oppositely directed unit normal $-\mathbf{n}$. A surface with a continuous unit normal is said to be *smooth*.

By the same token, we can define two continuous unit normals on a surface which is the graph of a continuously differentiable function of the form $y = g(x, z)$ or $x = h(y, z)$. For example, the two unit normals at the point $P = (x, g(x, z), z)$ of the graph of $y = g(x, z)$ are

$$\pm \frac{-g_x(x, z)\mathbf{i} - g_z(x, z)\mathbf{k} + \mathbf{j}}{\sqrt{[g_x(x, z)]^2 + [g_z(x, z)]^2 + 1}}.$$

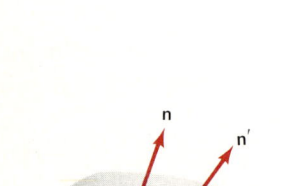

Figure 16

Figure 17

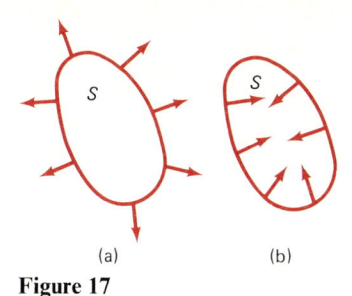

We can also define two fields of unit normal vectors on a "closed" surface S, like a sphere, ellipsoid or cube, which separates space into two regions, a bounded region said to be *inside* S and an unbounded region said to be *outside* S. Then one field of unit normals consists of the *outer* unit normals to S, pointing outward from S as in Figure 17(a), and the other field consists of the *inner* unit normals, pointing inward from S as in Figure 17(b). On a cube there is an abrupt change of direction when a unit normal goes from one face to another. Hence a cube is not a smooth surface, but it is *piecewise smooth*, meaning that it is made up of a finite number of smooth surfaces joined along their boundaries (in the case of a cube the smooth surfaces are its six faces).

> **Remark** All the surfaces under discussion are *two-sided*. For example, the graph of a function $z = f(x, y)$ has an upper side and a lower side, a sphere has an outside and an inside, and so on. Although we will be concerned exclusively with two-sided surfaces, you should be aware of the existence of *one-sided* surfaces (a classical example of such a surface is given in Problem 24).

Flux of a Vector Field Across a Surface

Next we introduce a special kind of surface integral, of basic importance in applied mathematics, which will play a key role in the last two sections of this chapter. Let S be a smooth (two-sided) surface on which we have chosen a continuous unit normal $\mathbf{n} = \mathbf{n}(\mathbf{r})$, where $\mathbf{r}$ is the position vector of a variable point of S, and let $\mathbf{F} = \mathbf{F}(\mathbf{r})$ be a vector field continuous on S. Then by the *flux of* $\mathbf{F}$ *across* S we mean the surface integral

$$\iint_S \mathbf{F} \cdot \mathbf{n}\, d\sigma, \tag{11}$$

involving the dot product $\mathbf{F} \cdot \mathbf{n}$. Reversing the direction of $\mathbf{n}$, that is, choosing the other continuous normal $-\mathbf{n}$ on S, changes the sign of the flux, since

$$\iint_S (\mathbf{F} \cdot -\mathbf{n})\, d\sigma = \iint_S -(\mathbf{F} \cdot \mathbf{n})\, d\sigma = -\iint_S \mathbf{F} \cdot \mathbf{n}\, d\sigma.$$

To give a physical interpretation of the flux integral (11), imagine that the surface S is submerged in a fluid moving with velocity $\mathbf{F} = \mathbf{F}(\mathbf{r})$, which varies from point to point but is time-independent (a "steady-state flow" in the language of fluid mechanics). Then the flux of $\mathbf{F}$ across S is just the net volume of fluid flowing across S per unit time, from the side determined by $-\mathbf{n}$ to the side determined by $\mathbf{n}$. To see this, we reason as follows, making intuitive use of differentials. In a brief time interval of duration dt, the volume of fluid flowing through a little piece of S, which we denote by dS, is equal to the volume of the (generally oblique) cylinder with base dS and altitude $F_n\, dt$, where F_n is the component of $\mathbf{F}$ along $\mathbf{n}$, as illustrated in Figure 18. This volume is just $F_n\, dt\, d\sigma$, where $d\sigma$ is the area of dS (recall Example 1, page 438). Here we assume that the angle between $\mathbf{F}$ and $\mathbf{n}$ does not exceed $90°$; otherwise, the cylinder will lie on the other side of S and the volume of fluid flowing through dS will turn out to be negative (of absolute value equal to the volume of the cylinder), since the flow is now from the side determined by $\mathbf{n}$ to the side determined by $-\mathbf{n}$. Hence the volume of fluid moving

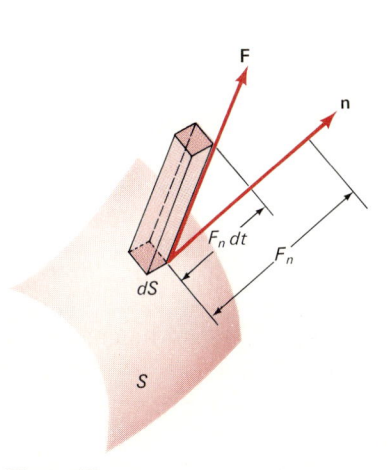

Figure 18

across dS per unit time is

$$\frac{F_n \, dt \, d\sigma}{dt} = F_n \, d\sigma = \mathbf{F} \cdot \mathbf{n} \, d\sigma,$$

and summing the contributions from all the little pieces of surface dS making up S, we get the flux integral (11). If you find the last step too cursory, partition S into little subsurfaces $S_1, S_2, \ldots, S_n$, let $\mathbf{r}_i$ be the position vector of an arbitrary point of S_i, and form the Riemann sum $\sum_{i=1}^{n} [\mathbf{F}(\mathbf{r}_i) \cdot \mathbf{n}(\mathbf{r}_i)] \Delta \sigma_i$, where $\Delta \sigma_i$ is the area of S_i. Then the flux integral (11) is the limit of this sum as μ, the maximum size of the subsurfaces, approaches zero. (The "size" of S_i is in turn the maximum distance between points of S_i.)

Example 7 Evaluate the flux of the field $\mathbf{F} = x^2 \mathbf{i} + y^2 \mathbf{j} + z \mathbf{k}$ across the surface S which is the graph of

$$z = 1 - x^2 - y^2 \qquad (z \geq 0),$$

choosing $\mathbf{n}$ to be the upper unit normal to S.

Solution The surface S is the "parabolic cap" shown in Figure 19. Since

$$\mathbf{N} = -\frac{\partial z}{\partial x} \mathbf{i} - \frac{\partial z}{\partial y} \mathbf{j} + \mathbf{k}$$

is an upper normal to S, the vector $\mathbf{n} = \mathbf{N}/|\mathbf{N}|$ is the upper *unit* normal to S (note that $\mathbf{n} \cdot \mathbf{k} > 0$). But $\mathbf{N} = 2x\mathbf{i} + 2y\mathbf{j} + \mathbf{k}$, and

$$|\mathbf{N}| = \sqrt{\left(\frac{\partial z}{\partial x}\right)^2 + \left(\frac{\partial z}{\partial y}\right)^2 + 1}.$$

Therefore, by formula (9'),

$$\iint_S \mathbf{F} \cdot \mathbf{n} \, d\sigma = \iint_R (\mathbf{F} \cdot \mathbf{n}) |\mathbf{N}| \, dA = \iint_R \left(\mathbf{F} \cdot \frac{\mathbf{N}}{|\mathbf{N}|}\right) |\mathbf{N}| \, dA$$

$$= \iint_R \mathbf{F} \cdot \mathbf{N} \, dA = \iint_R [x^2(2x) + y^2(2y) + z] \, dA$$

$$= \iint_R [1 - (x^2 + y^2) + 2(x^3 + y^3)] \, dx \, dy,$$

where R is the unit disk $x^2 + y^2 \leq 1$ in the xy-plane. Transforming to polar coordinates, we find that

$$\iint_S \mathbf{F} \cdot \mathbf{n} \, d\sigma = \int_0^{2\pi} d\theta \int_0^1 [1 - r^2 + 2r^3(\cos^3 \theta + \sin^3 \theta)] r \, dr$$

$$= \int_0^{2\pi} \left[\frac{1}{2} r^2 - \frac{1}{4} r^4 + \frac{2}{5} r^5(\cos^3 \theta + \sin^3 \theta)\right]_{r=0}^1 d\theta$$

$$= \int_0^{2\pi} \left[\frac{1}{4} + \frac{2}{5}(\cos^3 \theta + \sin^3 \theta)\right] d\theta = \frac{2\pi}{4} = \frac{\pi}{2},$$

since $\int_0^{2\pi} \cos^3 \theta \, d\theta = \int_0^{2\pi} \sin^3 \theta \, d\theta = 0$ (why?). □

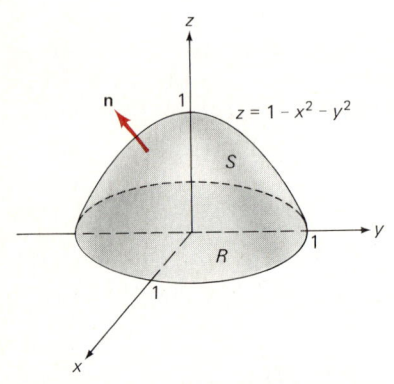

Figure 19

More generally, if S is a smooth surface which is the graph of a function $z = f(x, y)$ defined on a region R_{xy} in the xy-plane, then, by exactly the same method used to solve Example 7, we find that the flux of a continuous vector field $\mathbf{F} = F_1\mathbf{i} + F_2\mathbf{j} + F_3\mathbf{k}$ across S is given by

$$\iint_S \mathbf{F} \cdot \mathbf{n}\, d\sigma = \iint_{R_{xy}} \left(-F_1 \frac{\partial z}{\partial x} - F_2 \frac{\partial z}{\partial y} + F_3 \right) dx\, dy \qquad (\mathbf{n} \cdot \mathbf{k} > 0). \qquad (12)$$

Two other formulas for the flux, analogous to (12), are given in Problem 31.

Example 8 Find the flux of the field $\mathbf{F} = xy^2\mathbf{i} - yz\mathbf{j} + x^2z^2\mathbf{k}$ out of the cube S in the first octant bounded by the coordinate planes and the planes $x = 1$, $y = 1$ and $z = 1$ (the same surface was considered in Example 6).

Solution Naturally, the flux of $\mathbf{F}$ out of the cube S is defined as the sum of the fluxes of $\mathbf{F}$ across the six faces of S, from the inside of S to the outside of S. The following table lists these faces $S_1, \ldots, S_6$, together with the corresponding outer unit normals and the values of the dot product $\mathbf{F} \cdot \mathbf{n}$ (refer to Figure 15):

Face	Outer Unit Normal $\mathbf{n}$	$\mathbf{F} \cdot \mathbf{n}$
$S_1: x = 1$	$\mathbf{i}$	$xy^2 = y^2$
$S_2: y = 1$	$\mathbf{j}$	$-yz = -z$
$S_3: z = 1$	$\mathbf{k}$	$x^2z^2 = x^2$
$S_4: x = 0$	$-\mathbf{i}$	0
$S_5: y = 0$	$-\mathbf{j}$	0
$S_6: z = 0$	$-\mathbf{k}$	0

Since $\mathbf{F} \cdot \mathbf{n} = 0$ on S_4, S_5 and S_6, we need only consider the fluxes across three faces of the cube, namely the front face S_1, the right face S_2 and the upper face S_3. Therefore

$$\iint_S \mathbf{F} \cdot \mathbf{n}\, d\sigma = \iint_{S_1} \mathbf{F} \cdot \mathbf{n}\, d\sigma + \iint_{S_2} \mathbf{F} \cdot \mathbf{n}\, d\sigma + \iint_{S_3} \mathbf{F} \cdot \mathbf{n}\, d\sigma$$

$$= \iint_{R_1} y^2\, dy\, dz + \iint_{R_2} (-z)\, dx\, dz + \iint_{R_3} x^2\, dx\, dy,$$

where $R_1 = \{(y, z): 0 \leq y \leq 1, 0 \leq z \leq 1\}$, $R_2 = \{(x, z): 0 \leq x \leq 1, 0 \leq z \leq 1\}$ and $R_3 = \{(x, y): 0 \leq x \leq 1, 0 \leq y \leq 1\}$, as in Example 6; in transforming from an integral over S_i to an integral over R_i, we can replace $d\sigma$ by the product of coordinate differentials corresponding to R_i, since S_i is parallel to R_i. Thus, finally, the net flux of $\mathbf{F}$ out of the cube S is

$$\iint_S \mathbf{F} \cdot \mathbf{n}\, d\sigma = \left(\int_0^1 y^2\, dy \right)\left(\int_0^1 dz \right) - \left(\int_0^1 dx \right)\left(\int_0^1 z\, dz \right)$$

$$+ \left(\int_0^1 x^2\, dx \right)\left(\int_0^1 dy \right)$$

$$= \frac{1}{3} - \frac{1}{2} + \frac{1}{3} = \frac{1}{6}. \qquad \square$$

Problems

Find the area A_S of the specified surface S.

1. S is the part of the plane $3x + 2y + 6z = 12$ in the first octant
2. S is the part of the parabolic cylinder $z = x^2$ cut off by the planes $x = \sqrt{2}$, $y = x$ and $y = 2x$
3. S is the part of the cone $z^2 = 2xy$ lying between the planes $x = 0$, $y = 0$, $x = a$ and $y = b$ $(a > 0, b > 0)$
4. S is the part of the cone $z^2 = 2xy$ lying inside the sphere $x^2 + y^2 + z^2 = a^2$ of radius a
5. S is the part of the sphere $x^2 + y^2 + z^2 = a^2$ cut off by the circular cylinder $x^2 + y^2 = ax$ $(a > 0)$
6. S is the part of the circular cylinder $x^2 + y^2 = ax$ $(a > 0)$ inside the sphere $x^2 + y^2 + z^2 = a^2$
7. S is the parabolic cap $x = 12 - y^2 - z^2$ $(x \geq 0)$
8. S is the part of the sphere $x^2 + y^2 + z^2 = 4$ cut off by the elliptic cylinder $x^2 + \frac{1}{4}y^2 = 1$
9. S is the part of the cylinder $x^2 + z^2 = a^2$ inside the cylinder $x^2 + y^2 = a^2$ of the same radius a
10. S is the part of the cone $x^2 - y^2 - z^2 = 0$ in the first octant cut off by the plane $y + z = 4$
11. S is the part of the cone $z = \sqrt{x^2 + y^2}$ between the xy-plane and the cylinder $x^2 + y^2 = 2y$
12. S is the part of the parabolic cylinder $z^2 = 4x$ cut off by the parabolic cylinder $y^2 = 4x$ and the plane $x = 3$

Evaluate the given surface integral.

13. $\iint_S xyz \, d\sigma$, where S is the part of the plane $x + y + z = 1$ in the first octant

14. $\iint_S z^2 \, d\sigma$, where S is the sphere $x^2 + y^2 + z^2 = 16$

15. $\iint_S x^2 \, d\sigma$, where S is the part of the cone $x = \sqrt{y^2 + z^2}$ between the planes $x = 0$ and $x = 1$

16. $\iint_S z \, d\sigma$, where S is the part of the sphere $x^2 + y^2 + z^2 = 9$ in the first octant

17. $\iint_S x^2 y^2 \, d\sigma$, where S is the hemisphere $z = \sqrt{1 - x^2 - y^2}$

18. $\iint_S y \, d\sigma$, where S is the parabolic cap $y = 2 - x^2 - z^2$ $(y \geq 0)$

19. $\iint_S (x + y + z) \, d\sigma$, where S is the same cubical surface as in Example 6

20. $\iint_S (x^2y^2 + x^2z^2 + y^2z^2) \, d\sigma$, where S is the part of the cone $z = \sqrt{x^2 + y^2}$ cut off by the cylinder $x^2 + y^2 = 2x$

Let S be the part of the cone $z = \sqrt{x^2 + y^2}$ between the planes $z = 1$ and $z = 4$, and suppose the density at every point P of S is equal to the distance from P to the plane $z = 0$. Find

21. The total mass and center of mass of S
22. The moment of inertia of S about the z-axis

23. Find the centroid of S, where S is the part of the sphere $x^2 + y^2 + z^2 = 9$ in the first octant.
24. Suppose a rectangular piece of paper is given a half twist and the short ends are then pasted together. The resulting surface, known as a *Möbius strip*, is shown in Figure 20. Why is it appropriate to describe this surface as one-sided? *Hint.* Consider a bug crawling along the middle of the strip, starting from the point P, say.

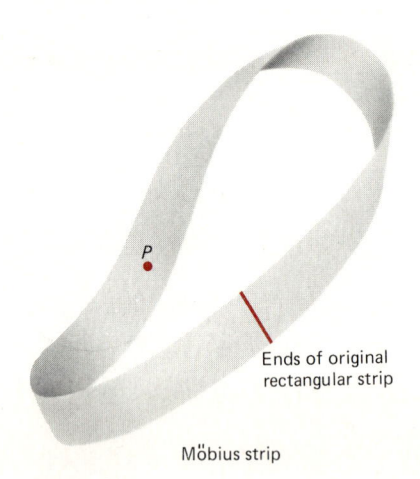

Figure 20

Calculate the flux $\iint_S \mathbf{F} \cdot \mathbf{n} \, d\sigma$, where $\mathbf{n}$ is the upper unit normal to S, for the given field $\mathbf{F}$ and surface S.

25. $\mathbf{F} = x\mathbf{i} + y\mathbf{j} + z\mathbf{k}$, S is the hemisphere $z = \sqrt{9 - x^2 - y^2}$
26. $\mathbf{F} = xz\mathbf{i} + yz\mathbf{j} + z^2\mathbf{k}$, S is the same parabolic cap as in Example 7
27. $\mathbf{F} = \mathbf{i} - y^2\mathbf{j} - z\mathbf{k}$, S is the part of the hyperbolic paraboloid $z = xy$ above the rectangular region $0 \leq x \leq 3$, $0 \leq y \leq 2$
28. $\mathbf{F} = -x\mathbf{i} - y\mathbf{j} + 3\mathbf{k}$, S is the part of the sphere $x^2 + y^2 + z^2 = 4$ in the first octant

29. Find the flux of the field $\mathbf{F} = x\mathbf{i} + y\mathbf{j} + z\mathbf{k}$ out of the closed surface S bounded by the cylinder $x^2 + y^2 = 1$ and the planes $z = 0$ and $z = 3$

30. Find the flux of the field $\mathbf{F} = x^2 y \mathbf{i} + xy^2 \mathbf{j} + xyz \mathbf{k}$ out of the same cube as in Example 8

31. Let $\mathbf{F} = F_1 \mathbf{i} + F_2 \mathbf{j} + F_3 \mathbf{k}$ be a continuous vector field on a smooth surface S which is the graph of a function $y = g(x, z)$, defined on a region R_{xz} in the xz-plane. Show that

$$\iint_S \mathbf{F} \cdot \mathbf{n} \, d\sigma = \iint_{R_{xz}} \left(-F_1 \frac{\partial y}{\partial x} + F_2 - F_3 \frac{\partial y}{\partial z} \right) dx \, dz.$$

Also show that if S is the graph of a function $x = h(y, z)$ defined on a region R_{yz} in the yz-plane, then

$$\iint_S \mathbf{F} \cdot \mathbf{n} \, d\sigma = \iint_{R_{yz}} \left(F_1 - F_2 \frac{\partial x}{\partial y} - F_3 \frac{\partial x}{\partial z} \right) dy \, dz.$$

Here $\mathbf{n}$ is a continuous unit normal on S satisfying the condition $\mathbf{n} \cdot \mathbf{j} > 0$ in the first case and $\mathbf{n} \cdot \mathbf{i} > 0$ in the second case.

32. Let S be a smooth surface which is simultaneously the graph of three functions $z = f(x, y)$, $y = g(x, z)$ and $x = h(y, z)$ defined on regions R_{xy}, R_{xz} and R_{yz} in the xy-, xz- and yz-planes, and let $\mathbf{F} = F_1 \mathbf{i} + F_2 \mathbf{j} + F_3 \mathbf{k}$ be a continuous vector function on S. Show that if S has a continuous unit normal $\mathbf{n}$ making an acute angle with all three positive coordinate axes, then

$$\iint_S \mathbf{F} \cdot \mathbf{n} \, d\sigma = \iint_{R_{yz}} F_1 \, dy \, dz + \iint_{R_{xz}} F_2 \, dx \, dz + \iint_{R_{xy}} F_3 \, dx \, dy. \quad \text{(i)}$$

33. With the help of formula (i), calculate $\iint_S \mathbf{F} \cdot \mathbf{n} \, d\sigma$, where

$$\mathbf{F} = yz \mathbf{i} + xz \mathbf{j} + xy \mathbf{k}, \quad S \text{ is the part of the plane}$$

$$x + y + z = a \quad (a > 0)$$

in the first octant, and $\mathbf{n}$ is the upper unit normal to S. Then use formula (12) to check the answer.

15.4 Green's Theorem; Change of Variables in Multiple Integrals

The following theorem, due to the self-taught English mathematician George Green (1793–1841), reveals a deep connection between a line integral along the boundary of a plane region and a related double integral over the region itself.

> **Theorem 4** *(Green's theorem).* Let R be a region in the xy-plane consisting of a piecewise smooth simple closed curve C and its interior, and let the functions $P = P(x, y)$ and $Q = Q(x, y)$ be continuously differentiable on R. Then
>
> $$\int_C P \, dx + Q \, dy = \iint_R \left(\frac{\partial Q}{\partial x} - \frac{\partial P}{\partial y} \right) dA, \quad (1)$$
>
> where C is traversed in the counterclockwise direction.

Partial proof The full proof of Green's theorem is an advanced calculus topic, and hence we will prove the theorem only for the special case where R is a simple region bounded by a piecewise smooth curve C (however, see Problem 18). Since the region R is simple, it is both vertically and horizontally simple, so that in particular

$$R = \{(x, y): a \leq x \leq b, g_1(x) \leq y \leq g_2(x)\},$$

where as just assumed, g_1 and g_2 have piecewise smooth graphs. This is illustrated in Figure 21(a), where C is the closed curve $KLMNK$, that is, the curve made up of the arcs KL, LM, MN and NK traversed in that order, and we allow for the possibility that one or both of the segments LM and NK may reduce to a single point. By Theorem 2, page 817, applied to the

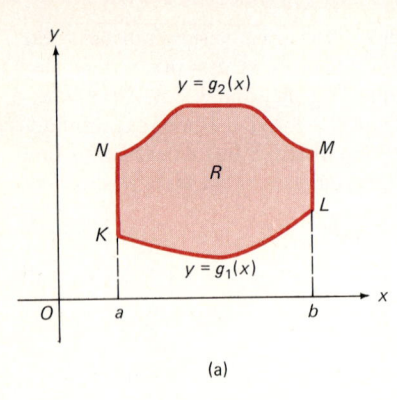

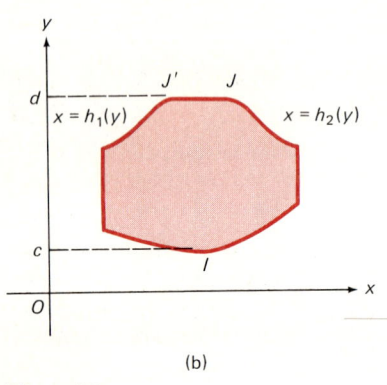

function $-\partial P/\partial y$,

$$\iint_R -\frac{\partial P}{\partial y} dA = -\int_a^b dx \int_{g_1(x)}^{g_2(x)} \frac{\partial P}{\partial y} dy = -\int_a^b [P(x, g_2(x)) - P(x, g_1(x))] dx$$

$$= \int_a^b P(x, g_1(x)) dx + \int_b^a P(x, g_2(x)) dx = \int_{KL} P\, dx + \int_{MN} P\, dx$$

(see formula (11), page 882). But

$$\int_{LM} P\, dx = \int_{NK} P\, dx = 0,$$

since $dx = 0$ on LM and NK, and therefore

$$\int_C P\, dx = \int_{KL} P\, dx + \int_{LM} P\, dx + \int_{MN} P\, dx + \int_{NK} P\, dx$$

$$= \int_{KL} P\, dx + \int_{MN} P\, dx,$$

so that

$$\int_C P\, dx = \iint_R -\frac{\partial P}{\partial y} dA. \qquad (2)$$

Similarly, representing R in the form

$$R = \{(x, y): c \leq y \leq d, h_1(y) \leq x \leq h_2(y)\},$$

where the functions h_1 and h_2 have piecewise smooth graphs on $[c, d]$, we find that

$$\iint_R \frac{\partial Q}{\partial x} dA = \int_c^d dy \int_{h_1(y)}^{h_2(y)} \frac{\partial Q}{\partial x} dx = \int_c^d [Q(h_2(y), y) - Q(h_1(y), y)] dy$$

$$= \int_c^d Q(h_2(y), y) dy + \int_d^c Q(h_1(y), y) dy = \int_{IJ} Q\, dy + \int_{J'I} Q\, dy$$

(see Figure 21(b), where R is the same region as in Figure 21(a), labeled differently). But

$$\int_{JJ'} Q\, dy = 0,$$

since $dy = 0$ on JJ', and therefore

$$\int_C Q\, dy = \int_{IJ} Q\, dy + \int_{JJ'} Q\, dy + \int_{J'I} Q\, dy = \int_{IJ} Q\, dy + \int_{J'I} Q\, dy,$$

so that

$$\int_C Q\, dy = \iint_R \frac{\partial Q}{\partial x} dA. \qquad (2')$$

Adding (2) and (2'), we finally get the desired formula (1). ∎

Corollary 1 *Let R and C be the same as in Green's theorem. Then the area of R is given by any of the formulas*

$$A = -\int_C y\,dx, \qquad A = \int_C x\,dy, \qquad A = \frac{1}{2}\int_C -y\,dx + x\,dy, \qquad (3)$$

where C is traversed in the counterclockwise direction.

Proof The choice $P = -y$, $Q = 0$ converts formula (1) into $\int_C -y\,dx = \iint_R dA$, while the choice $P = 0$, $Q = x$ converts it into $\int_C x\,dy = \iint_R dA$. But $\iint_R dA = A$, which proves the first two of the formulas (3). To get the third formula, add the first two and solve for A, or else choose $P = -\frac{1}{2}y$, $Q = \frac{1}{2}x$ in (1). ∎

Corollary 2 *Suppose the components $P = P(x, y)$ and $Q = Q(x, y)$ of a continuously differentiable vector field $\mathbf{F} = P\mathbf{i} + Q\mathbf{j}$ satisfy the condition $\partial Q/\partial x = \partial P/\partial y$ at every point of a simply connected domain D. Then $\mathbf{F}$ is path independent on D, or equivalently $\mathbf{F}$ is a gradient field on D.*

Proof Let C be any simple closed path contained in D, and suppose $\partial Q/\partial x = \partial P/\partial y$ at every point of D. Since D is simply connected, the interior of C is also contained in D (recall page 887). Therefore $\partial Q/\partial x = \partial P/\partial y$ on the region R consisting of C and its interior. But then, by Green's theorem,

$$\int_C P\,dx + Q\,dy = \iint_R \left(\frac{\partial Q}{\partial x} - \frac{\partial P}{\partial y}\right) dA = \int_R 0\,dA = 0.$$

Therefore $\int_C P\,dx + Q\,dy = 0$ for every simple closed path C contained in D. It follows from the remark on page 890 that the line integral of $\mathbf{F}$ is path independent on D, or equivalently that $\mathbf{F}$ is a gradient field on D (recall Theorem 1, page 887). ∎

Example 1 Use Green's theorem to evaluate the line integral

$$\int_C (x^2 + y^2)\,dx + (x + 2y)^2\,dy,$$

where C is the triangle with vertices $(0, 0)$, $(1, 1)$ and $(0, 2)$, traversed in the counterclockwise direction (see Figure 22).

Solution By formula (1) with $P = x^2 + y^2$, $Q = (x + 2y)^2$ and R the shaded region in the figure,

$$\int_C (x^2 + y^2)\,dx + (x + 2y)^2\,dy = \iint_R \left[\frac{\partial}{\partial x}(x + 2y)^2 - \frac{\partial}{\partial y}(x^2 + y^2)\right] dA$$

$$= \iint_R [2(x + 2y) - 2y]\,dA = 2\iint_R (x + y)\,dA = 2\int_0^1 dx \int_x^{2-x} (x + y)\,dy$$

$$= 2\int_0^1 \left[xy + \frac{1}{2}y^2\right]_{y=x}^{2-x} dx = 2\int_0^1 \left[x(2 - x) + \frac{1}{2}(2 - x)^2 - x^2 - \frac{1}{2}x^2\right] dx$$

$$= 4\int_0^1 (1 - x^2)\,dx = 4\left[x - \frac{1}{3}x^3\right]_0^1 = \frac{8}{3}.$$

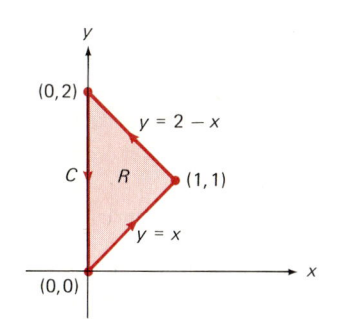

Figure 22

Check the answer by direct evaluation of the line integral, which you will discover to be a harder calculation. □

Example 2 Use Corollary 1 to find the area A enclosed by an ellipse with semimajor axis a and semiminor axis b.

Solution There is no loss of generality in assuming that the ellipse has the parametric representation $x = a \cos t$, $y = b \sin t$ ($0 \le t \le 2\pi$). Hence, by the third of the formulas (3),

$$A = \frac{1}{2}\int_C -y\,dx + x\,dy = \frac{1}{2}\int_0^{2\pi}[(-b\sin t)(-a\sin t) + (a\cos t)(b\cos t)]\,dt$$

$$= \frac{1}{2}ab\int_0^{2\pi}(\sin^2 t + \cos^2 t)\,dt = \frac{1}{2}ab\int_0^{2\pi}dt = \pi ab,$$

as we already know from Example 4, page 591. □

The Case of Two Boundary Curves

There is a version of Green's theorem involving the region R between *two* piecewise smooth simple closed curves C_1 and C_2. Suppose C_2 lies inside C_1, as in Figure 23(a), and let us agree that the *positive* direction of traversing the curve C_1 or C_2 is the direction such that R is always to the *left* of an observer moving along the curve. Thus the positive direction for the outer curve C_1 is counterclockwise, but the positive direction for the inner curve C_2 is *clockwise*, as indicated by the arrowheads attached to the curves. Suppose we draw two line segments, called "crosscuts," joining C_1 and C_2. Then the region R between C_1 and C_2 is divided into two subregions R_1 and R_2, as in Figure 23(b), each of which has just one simple closed curve as its boundary. Specifically, referring to Figure 23(c), we see that the boundary of R_1 is the contour consisting of the arc Γ_1, the segment KL, the arc γ_1, and the segment MN, traversed in that order with the indicated directions, while the boundary of R_2 is the contour consisting of the arc Γ_2, the segment NM, the arc γ_2, and the segment LK.

Now let the functions $P = P(x, y)$ and $Q = Q(x, y)$ be continuously differentiable on R, and hence on the subregions R_1 and R_2. Applying Green's theorem to R_1 and R_2 separately, we find that

$$\iint_R \left(\frac{\partial Q}{\partial x} - \frac{\partial P}{\partial y}\right) dA = \iint_{R_1}\left(\frac{\partial Q}{\partial x} - \frac{\partial P}{\partial y}\right) dA + \iint_{R_2}\left(\frac{\partial Q}{\partial x} - \frac{\partial P}{\partial y}\right) dA$$

$$= \left(\int_{\Gamma_1} P\,dx + Q\,dy + \int_{KL} \cdots + \int_{\gamma_1} \cdots + \int_{MN} \cdots\right)$$

$$+ \left(\int_{\Gamma_2} P\,dx + Q\,dy + \int_{NM} \cdots + \int_{\gamma_2} \cdots + \int_{LK} \cdots\right),$$

where each occurrence of the dots $\cdots$ stands for the expression $P\,dx + Q\,dy$. But $\int_{LK}\cdots = -\int_{KL}\cdots$ and $\int_{MN}\cdots = -\int_{NM}\cdots$, as in formula (13′), page 883. Hence the integrals along the crosscuts cancel each other out, and we

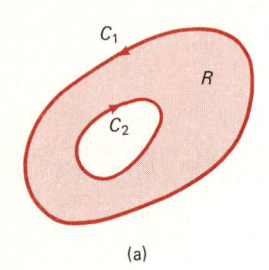

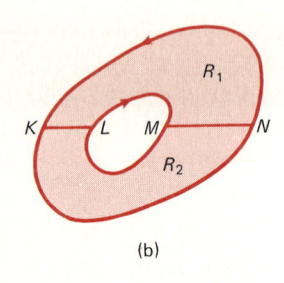

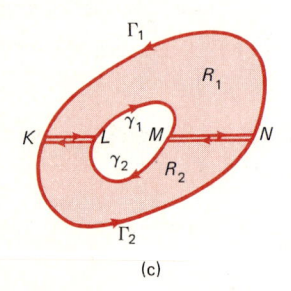

Figure 23

are left with

$$\iint_R \left(\frac{\partial Q}{\partial x} - \frac{\partial P}{\partial y}\right) dA = \left(\int_{\Gamma_1} P\,dx + Q\,dy + \int_{\Gamma_2} \cdots\right)$$

$$+ \left(\int_{\gamma_1} P\,dx + Q\,dy + \int_{\gamma_2} \cdots\right)$$

$$= \int_{C_1} P\,dx + Q\,dy + \int_{C_2} P\,dx + Q\,dy$$

(recall formula (15), page 884). Thus Green's theorem holds for the region R between the curves C_1 and C_2, provided that the integral $\int_C P\,dx + Q\,dy$ in formula (1) is interpreted as the sum of the line integrals $\int_{C_1} P\,dx + Q\,dy$ and $\int_{C_2} P\,dx + Q\,dy$, with the understanding that each of the curves C_1 and C_2 is traversed in the appropriate positive direction.

Example 3 Let C be any piecewise smooth simple closed curve enclosing the origin, traversed once in the counterclockwise direction. Show that

$$\int_C \frac{-y\,dx + x\,dy}{x^2 + y^2} = 2\pi. \tag{4}$$

Solution The functions $P = -y/(x^2 + y^2)$ and $Q = x/(x^2 + y^2)$ are defined and continuously differentiable everywhere except at the origin, and

$$\frac{\partial Q}{\partial x} = \frac{y^2 - x^2}{(x^2 + y^2)^2} = \frac{\partial P}{\partial y}$$

if $(x, y) \neq (0, 0)$. Let γ be a circle centered at the origin and traversed once in the *clockwise* direction, which is small enough to be enclosed by C (see Figure 24). Then P and Q are continuously differentiable and satisfy the condition $\partial Q/\partial x = \partial P/\partial y$ on the region R between C and γ. Therefore

$$\int_C P\,dx + Q\,dy + \int_\gamma P\,dx + Q\,dy = \iint_R \left(\frac{\partial Q}{\partial x} - \frac{\partial P}{\partial y}\right) dA = \iint_R 0\,dA = 0,$$

by the version of Green's theorem for two boundary curves. But then

$$\int_C P\,dx + Q\,dy = -\int_\gamma P\,dx + Q\,dy = \int_{-\gamma} P\,dx + Q\,dy,$$

where $-\gamma$ is traversed in the *counterclockwise* direction. This proves formula (4), since $\int_{-\gamma} P\,dx + Q\,dy = 2\pi$ by Example 3, page 891. As an exercise, use Corollary 2 to show that the right side of (4) becomes 0 instead of 2π if the origin lies outside C. □

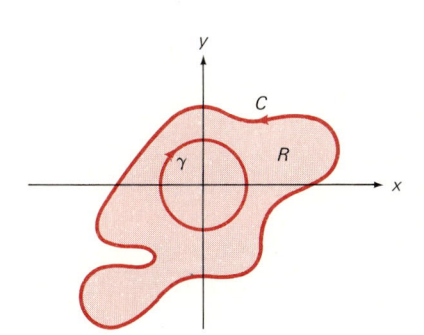

Figure 24

Coordinate Transformations and Jacobians

Next we use Green's theorem to establish a formula for changing variables in a double integral. An important example of such a formula has already been given in Theorem 8, page 856, where it was shown that if $f(x, y)$ is continuous on a region R, and if R is the image of a region R' in the $r\theta$-plane under the transformation

$$x = r\cos\theta, \quad y = r\sin\theta,$$

leading from polar coordinates r and θ to rectangular coordinates x and y, then

$$\iint_R f(x, y)\, dx\, dy = \iint_{R'} f(r \cos \theta, r \sin \theta)\, r\, dr\, d\theta. \tag{5}$$

We now seek the analogue of formula (5) for a general coordinate transformation of the form

$$x = x(u, v), \qquad y = y(u, v), \tag{6}$$

leading from the coordinates labeled u and v to rectangular coordinates x and y. It is assumed that the transformation (6) is defined on a region R' in the uv-plane consisting of a piecewise smooth simple closed curve C' and its interior, and that the image of R' under the transformation is a region R in the xy-plane consisting of another piecewise smooth simple closed curve C and its interior, as indicated schematically in Figure 25. In addition, we assume that as (u, v) traverses the curve C' once in the positive direction, the image (x, y) of the point (u, v) traverses the curve C once in either the positive or negative direction. We also make some further assumptions about the transformation (6):

(i) The functions $x(u, v)$ and $y(u, v)$ are continuous and have continuous first and second partial derivatives on R';

(ii) The determinant

$$\frac{\partial(x, y)}{\partial(u, v)} = \begin{vmatrix} \dfrac{\partial x}{\partial u} & \dfrac{\partial x}{\partial v} \\ \dfrac{\partial y}{\partial u} & \dfrac{\partial y}{\partial v} \end{vmatrix} = \frac{\partial x}{\partial u}\frac{\partial y}{\partial v} - \frac{\partial y}{\partial u}\frac{\partial x}{\partial v},$$

called the *Jacobian* of the transformation (6) and denoted concisely by $\partial(x, y)/\partial(u, v)$, is nonzero on R';

(iii) The transformation (6) is *one-to-one*, that is, not only is there a unique point (x, y) in R corresponding to any given point in R', but there is also a unique point (u, v) in R' corresponding to any given point (x, y) in R, and the transformation from R back to R' is denoted by

$$u = u(x, y), \qquad v = v(x, y). \tag{6'}$$

Under these conditions, it can be shown that the functions $u(x, y)$ and $v(x, y)$ are continuous and have continuous first and second partial deriva-

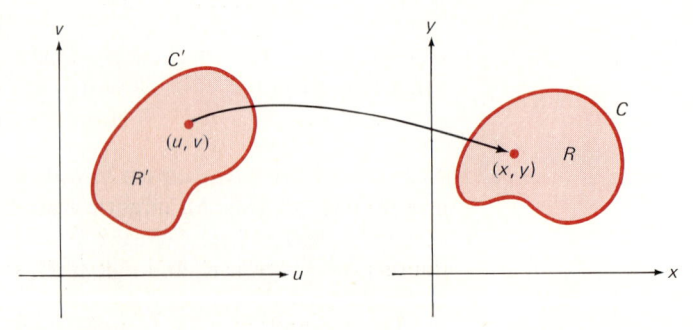

Figure 25

tives on R, and that the Jacobian

$$\frac{\partial(u, v)}{\partial(x, y)} = \begin{vmatrix} \dfrac{\partial u}{\partial x} & \dfrac{\partial u}{\partial y} \\ \dfrac{\partial v}{\partial x} & \dfrac{\partial v}{\partial y} \end{vmatrix} = \frac{\partial u}{\partial x}\frac{\partial v}{\partial y} - \frac{\partial v}{\partial x}\frac{\partial u}{\partial y}$$

of the inverse transformation (6′) is nonzero on R and satisfies the formula

$$\frac{\partial(x, y)}{\partial(u, v)} \frac{\partial(u, v)}{\partial(x, y)} = 1, \tag{7}$$

(see Problem 26). In particular, (7) implies that the Jacobians $\partial(u, v)/\partial(x, y)$ and $\partial(x, y)/\partial(u, v)$ have the same sign on their respective regions R and R'.

Change of Variables in Double Integrals

Making all of the above assumptions about the transformations (6) and (6′), the regions R and R', and the curves C and C', we now present a theorem which generalizes formula (5). The proof of the theorem is a straightforward, but rather technical application of Green's theorem, and hence is deferred to the end of the section.

Theorem 5 *(Change of variables in double integrals)*. If $f(x, y)$ is continuous on R, then

$$\iint_R f(x, y)\, dx\, dy = \iint_{R'} f(x(u, v), y(u, v)) \left| \frac{\partial(x, y)}{\partial(u, v)} \right| du\, dv. \tag{8}$$

It is shown in advanced calculus that Theorem 5 remains true under weaker hypotheses, in particular if the transformation (6) fails to be one-to-one on the boundary of the region R'.

Example 4 Show that formula (5) is a special case of formula (8).

Solution Let the variables u and v be the polar coordinates r and θ. Then the transformation (6) is

$$x = r \cos \theta, \qquad y = r \sin \theta,$$

where $r \geq 0$ (see page 856), and the Jacobian $\partial(x, y)/\partial(u, v)$ becomes

$$\frac{\partial(x, y)}{\partial(r, \theta)} = \begin{vmatrix} \dfrac{\partial x}{\partial r} & \dfrac{\partial x}{\partial \theta} \\ \dfrac{\partial y}{\partial r} & \dfrac{\partial y}{\partial \theta} \end{vmatrix} = \begin{vmatrix} \cos \theta & -r \sin \theta \\ \sin \theta & r \cos \theta \end{vmatrix} = r(\cos^2 \theta + \sin^2 \theta) = r.$$

Hence for this choice of u and v, formula (8) reduces to formula (5).

Example 5 Evaluate the double integral $\iint_R xy\, dx\, dy$, where R is the region shown in Figure 26(a), whose boundary C is the parallelogram with sides along the lines $x + 3y = -1$, $x + 3y = 3$, $x - y = 1$ and $x - y = 2$.

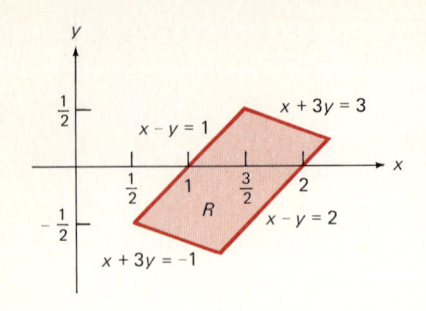

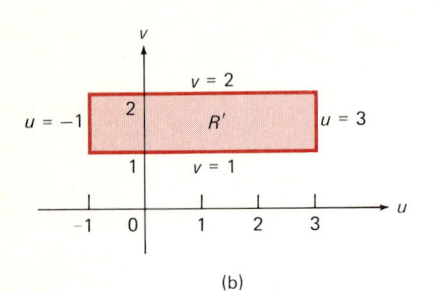

Figure 26

(a)

(b)

Solution The expressions $x + 3y$ and $x - y$ appear in the equations of the lines bounding R, and this suggests that we introduce the new variables

$$u = x + 3y, \qquad v = x - y, \tag{9}$$

since the equations of the lines then simplify to $u = -1$, $u = 3$, $v = 1$ and $v = 2$. Thus the image of R under the transformation (9) is the region R' in the uv-plane, whose boundary C' is the rectangle with sides along the lines $u = -1$, $u = 3$, $v = 1$ and $v = 2$, as shown in Figure 26(b). Solving the system of equations (9) for x and y in terms of u and v, we obtain

$$x = \frac{1}{4}(u + 3v), \qquad y = \frac{1}{4}(u - v), \tag{9'}$$

so that

$$\frac{\partial(x, y)}{\partial(u, v)} = \begin{vmatrix} \dfrac{\partial x}{\partial u} & \dfrac{\partial x}{\partial v} \\ \dfrac{\partial y}{\partial u} & \dfrac{\partial y}{\partial v} \end{vmatrix} = \begin{vmatrix} \dfrac{1}{4} & \dfrac{3}{4} \\ \dfrac{1}{4} & -\dfrac{1}{4} \end{vmatrix} = -\frac{4}{16} = -\frac{1}{4}. \tag{10}$$

Therefore, by formula (8),

$$\iint_R xy\, dx\, dy = \iint_{R'} \frac{1}{16}(u + 3v)(u - v) \left|-\frac{1}{4}\right| du\, dv$$

$$= \frac{1}{64} \int_{-1}^{3} du \int_{1}^{2} (u^2 + 2uv - 3v^2)\, dv$$

$$= \frac{1}{64} \int_{-1}^{3} \left[u^2 v + uv^2 - v^3 \right]_{v=1}^{2} du$$

$$= \frac{1}{64} \int_{-1}^{3} (u^2 + 3u - 7)\, du = \frac{1}{64} \left[\frac{1}{3}u^3 + \frac{3}{2}u^2 - 7u \right]_{-1}^{3} = -\frac{5}{48}.$$

As you can easily verify by examining the correspondence between the vertices of C and C', if either contour C or C' is traversed in one direction (clockwise or counterclockwise), the other is traversed in the opposite direction. This might be expected, since the Jacobian (10) is negative. Notice that (10) could also have been found by first calculating $\partial(u, v)/\partial(x, y)$ and then applying formula (7). As an exercise, evaluate $\iint_R xy\, dx\, dy$ directly, and see how *very* much easier it is to use the present method. □

Example 6 Evaluate the double integral $\iint_R (x^2/y^4)\, dx\, dy$, where R is the region bounded by the hyperbolas $xy = 2$, $xy = 4$ and the parabolas $y^2 = x$, $y^2 = 3x$ (see Figure 27).

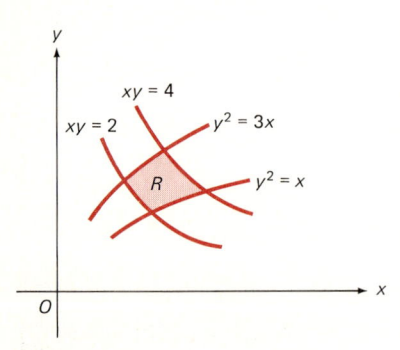

Solution Writing

$$xy = u, \qquad y^2 = vx,$$

or equivalently

$$u = xy, \qquad v = \frac{y^2}{x},$$

Figure 27

916 Chapter 15 Line and Surface Integrals

we see that the curves forming the boundary of R correspond to the values $u = 2$, $u = 4$, $v = 1$ and $v = 3$. Also

$$\frac{\partial(u, v)}{\partial(x, y)} = \begin{vmatrix} \frac{\partial u}{\partial x} & \frac{\partial u}{\partial y} \\ \frac{\partial v}{\partial x} & \frac{\partial v}{\partial y} \end{vmatrix} = \begin{vmatrix} y & x \\ -\frac{y^2}{x^2} & \frac{2y}{x} \end{vmatrix} = \frac{3y^2}{x},$$

and then (7) gives

$$\frac{\partial(x, y)}{\partial(u, v)} = \frac{1}{\frac{\partial(u, v)}{\partial(x, y)}} = \frac{x}{3y^2} = \frac{1}{3v}.$$

Choosing $f(x, y) = x^2/y^4 = 1/v^2$ in formula (8) and noting that R' is the square region $\{(u, v): 2 \leq u \leq 4, 1 \leq v \leq 3\}$ in the uv-plane, we find that

$$\iint_R \frac{x^2}{y^4}\, dx\, dy = \iint_{R'} \frac{du\, dv}{3v^3} = \frac{1}{3}\left(\int_2^4 du\right)\left(\int_1^3 \frac{dv}{v^3}\right) = \frac{2}{3}\left[-\frac{1}{2v^2}\right]_1^3 = \frac{8}{27}. \quad \square$$

Change of Variables in Triple Integrals

The above considerations can be extended to three dimensions. Let

$$x = x(u, v, w), \qquad y = y(u, v, w), \qquad z = z(u, v, w) \quad (11)$$

be a three-dimensional transformation mapping a solid region T' in uvw-space onto a solid region T in xyz-space, and suppose (11) satisfies conditions analogous to those imposed on the two-dimensional transformation (6). In particular, it is assumed that the 3×3 determinant

$$\begin{vmatrix} \frac{\partial x}{\partial u} & \frac{\partial x}{\partial v} & \frac{\partial x}{\partial w} \\ \frac{\partial y}{\partial u} & \frac{\partial y}{\partial v} & \frac{\partial y}{\partial w} \\ \frac{\partial z}{\partial u} & \frac{\partial z}{\partial v} & \frac{\partial z}{\partial w} \end{vmatrix},$$

called the *Jacobian* of the transformation (11) and denoted concisely by $\partial(x, y, z)/\partial(u, v, w)$, is nonzero on T'. Then we have the following analogue of Theorem 5, which we state without proof.

Theorem 6 *(Change of variables in triple integrals).* If $f(x, y, z)$ is continuous on T, then

$$\iiint_T f(x, y, z)\, dx\, dy\, dz$$

$$= \iiint_{T'} f(x(u, v, w), y(u, v, w), z(u, v, w)) \left|\frac{\partial(x, y, z)}{\partial(u, v, w)}\right| du\, dv\, dw. \quad (12)$$

Example 7 Find the volume V of the oblique parallelepiped T bounded by the six planes $x + y + 2z = \pm 3$, $x - 2y + z = \pm 2$ and $4x + y + z = \pm 6$.

Section 15.4 Green's Theorem; Change of Variables in Multiple Integrals

Solution Introducing new variables
$$u = x + y + 2z, \quad v = x - 2y + z, \quad w = 4x + y + z,$$
where $-3 \le u \le 3$, $-2 \le v \le 2$ and $-6 \le w \le 6$, we find that

$$\frac{\partial(u, v, w)}{\partial(x, y, z)} = \begin{vmatrix} 1 & 1 & 2 \\ 1 & -2 & 1 \\ 4 & 1 & 1 \end{vmatrix} = \begin{vmatrix} -2 & 1 \\ 1 & 1 \end{vmatrix} - \begin{vmatrix} 1 & 1 \\ 4 & 1 \end{vmatrix} + 2\begin{vmatrix} 1 & -2 \\ 4 & 1 \end{vmatrix}$$

$$= -3 + 3 + 2(9) = 18.$$

But
$$\frac{\partial(x, y, z)}{\partial(u, v, w)} \frac{\partial(u, v, w)}{\partial(x, y, z)} = 1,$$

by the analogue of formula (7), and hence
$$\frac{\partial(x, y, z)}{\partial(u, v, w)} = \frac{1}{18}.$$

Therefore, by formula (12) with $f(x, y, z) \equiv 1$,

$$V = \iiint_T dx\, dy\, dz = \frac{1}{18}\left(\int_{-3}^{3} du\right)\left(\int_{-2}^{2} dv\right)\left(\int_{-6}^{6} dw\right)$$

$$= \frac{1}{18}(6)(4)(12) = 16. \quad \square$$

Proof of Theorem 5 (Optional) Let
$$F(x, y) = \int_{x_0}^{x} f(t, y)\, dt,$$

where x_0 is fixed, so that $\partial F(x, y)/\partial x = f(x, y)$. By Green's theorem in the xy-plane, with $P = 0$ and $Q = F$,

$$\iint_R f(x, y)\, dx\, dy = \iint_R \frac{\partial F(x, y)}{\partial x}\, dx\, dy = \int_C F(x, y)\, dy, \tag{13}$$

where C is traversed once in the positive (counterclockwise) direction. Let
$$u = u(t), \quad v = v(t) \quad (a \le t \le b)$$
be parametric equations for the curve C' (the boundary of the region R'), traversed once in the positive direction. Then C has parametric equations
$$x = x(u(t), v(t)), \quad y = y(u(t), v(t)) \quad (a \le t \le b),$$
and as t varies from a to b, the point (x, y) traverses C once in the positive or negative direction (we will decide which at the end of the proof). The right side of (13) can be written as
$$\int_C F(x, y)\, dy = \int_a^b G(t)\frac{dy}{dt}\, dt$$

918   Chapter 15   Line and Surface Integrals

in terms of the composite function $G(t) = F(x(u(t), v(t)), y(u(t), v(t)))$. But
$$\frac{dy}{dt} = \frac{\partial y}{\partial u}\frac{du}{dt} + \frac{\partial y}{\partial v}\frac{dv}{dt},$$
by the chain rule, and hence
$$\int_C F(x, y)\,dy = \int_a^b G(t)\left(\frac{\partial y}{\partial u}\frac{du}{dt} + \frac{\partial y}{\partial v}\frac{dv}{dt}\right)dt$$
$$= \int_{C'} H(u, v)\frac{\partial y}{\partial u}\,du + H(u, v)\frac{\partial y}{\partial v}\,dv \qquad (14)$$

in terms of the composite function $H(u, v) = F(x(u, v), y(u, v))$. You will recognize the right side of (14) as a line integral in the uv-plane, of the form
$$\int_{C'} P\,du + Q\,dv,$$
where
$$P = H\frac{\partial y}{\partial u}, \qquad Q = H\frac{\partial y}{\partial v}.$$

To continue the proof, we now evaluate $\int_{C'} P\,du + Q\,dv$ by applying Green's theorem again, this time in the uv-plane. In detail,
$$\int_{C'} P\,du + Q\,dv = \iint_{R'} \left(\frac{\partial Q}{\partial u} - \frac{\partial P}{\partial v}\right) du\,dv$$
$$= \iint_{R'} \left[\frac{\partial}{\partial u}\left(H\frac{\partial y}{\partial v}\right) - \frac{\partial}{\partial v}\left(H\frac{\partial y}{\partial u}\right)\right] du\,dv$$
$$= \iint_{R'} \left(\frac{\partial H}{\partial u}\frac{\partial y}{\partial v} + H\frac{\partial^2 y}{\partial u\,\partial v} - \frac{\partial H}{\partial v}\frac{\partial y}{\partial u} - H\frac{\partial^2 y}{\partial v\,\partial u}\right) du\,dv$$
$$= \iint_{R'} \left(\frac{\partial H}{\partial u}\frac{\partial y}{\partial v} - \frac{\partial H}{\partial v}\frac{\partial y}{\partial u}\right) du\,dv, \qquad (15)$$

where at the last step we use the assumption that $y(u, v)$ has continuous second partial derivatives. But
$$\frac{\partial H}{\partial u} = \frac{\partial F}{\partial x}\frac{\partial x}{\partial u} + \frac{\partial F}{\partial y}\frac{\partial y}{\partial u}, \qquad \frac{\partial H}{\partial v} = \frac{\partial F}{\partial x}\frac{\partial x}{\partial v} + \frac{\partial F}{\partial y}\frac{\partial y}{\partial v}, \qquad (16)$$
by the chain rule and the definition of H. Thus, combining (13)–(16), we get
$$\iint_R f(x, y)\,dx\,dy = \pm\iint_{R'} \left[\left(\frac{\partial F}{\partial x}\frac{\partial x}{\partial u} + \frac{\partial F}{\partial y}\frac{\partial y}{\partial u}\right)\frac{\partial y}{\partial v} - \left(\frac{\partial F}{\partial x}\frac{\partial x}{\partial v} + \frac{\partial F}{\partial y}\frac{\partial y}{\partial v}\right)\frac{\partial y}{\partial u}\right] du\,dv$$
$$= \pm\iint_{R'} \frac{\partial F}{\partial x}\left(\frac{\partial x}{\partial u}\frac{\partial y}{\partial v} - \frac{\partial y}{\partial u}\frac{\partial x}{\partial v}\right) du\,dv$$
$$= \pm\iint_{R'} f(x(u, v), y(u, v))\frac{\partial(x, y)}{\partial(u, v)}\,du\,dv. \qquad (17)$$

Here the plus sign is chosen if the curve C is traversed in the positive direction as t increases from a to b, but if C is traversed in the *negative* direction, the minus sign must be chosen instead (why?).

Finally, suppose that $f(x, y) \equiv 1$. Then the left side of (17) is just the area of R, which is inherently positive, and this tells us to choose the plus sign if the Jacobian $J = \partial(x, y)/\partial(u, v)$ is positive and the minus sign if J is negative. In other words, we can drop the symbol $\pm$ in (17) if we replace the Jacobian J by its absolute value $|J| = |\partial(x, y)/\partial(u, v)|$. Formula (17) then becomes

$$\iint_R f(x, y)\, dx\, dy = \iint_{R'} f(x(u, v), y(u, v)) \left| \frac{\partial(x, y)}{\partial(u, v)} \right| du\, dv,$$

and the proof of Theorem 5 is complete. ■

Problems

Use Green's theorem to evaluate the given line integral. (In each case the curve C is traversed in the counterclockwise direction.)

1. $\int_C (y^4 - x)\, dx + (y - x^2)\, dy$, where C is the square with vertices $(0, 0), (1, 0) (1, 1)$ and $(0, 1)$

2. $\int_C (y + \ln x)\, dx + (x^3 + e^{-y})\, dy$, where C is the triangle with vertices $(1, 0), (2, 0)$ and $(2, 1)$

3. $\int_C (3x - 2y)\, dx + (2x + 3y)\, dy$, where C is the boundary of the region between the parabolas $y = x^2$ and $y^2 = x$

4. $\int_C \sqrt{x^2 + y^2}\, dx + [xy^2 + y \ln (x + \sqrt{x^2 + y^2})]\, dy$, where C is the rectangle with vertices $(1, 0), (4, 0), (4, 5)$ and $(1, 5)$

5. $\int_C -xy^2\, dx + xy^2\, dy$, where C is the circle $x^2 + y^2 = 4$

6. $\int_C (2x + y)\, dx + 2x\, dy$, where C is the ellipse $4x^2 + 9y^2 = 36$

7. $\int_C e^x \cos y\, dx - e^x \sin y\, dy$, where C is the ellipse $2x^2 + y^2 = 1$

8. $\int_C \frac{xy^2}{1 + x^2}\, dx + y \ln (1 + x^2)\, dy$, where C is the circle $x^2 + y^2 + 2y = 0$

9. $\int_C -y \sec^2 x\, dx + \tan x\, dy$, where C is the triangle with vertices $(0, 0), (\pi/4, 0)$ and $(\pi/4, \pi/2)$

10. $\int_C -xy\, dx + (y^2 + 16)\, dy$, where C consists of the top half of the ellipse $4x^2 + y^2 = 1$ and the line segment joining the points $(\pm \frac{1}{2}, 0)$

Use a line integral to find the area A of the given region R.

11. R is the region in the first quadrant enclosed by the hyperbola $xy = 4$ and the lines $y = x$ and $y = 4x$

12. R is the region in the first quadrant enclosed by the hyperbola $8xy = 1$ and the parabolas $y = x^2$ and $y^2 = x$, with the point $(1, 1)$ on its boundary

13. R is the region enclosed by the astroid

$$x = a \cos^3 t, \quad y = a \sin^3 t \quad (0 \leq t \leq 2\pi, a > 0)$$

(see Figure 45, page 464)

14. R is the region enclosed by the cardioid

$$x = 2a \cos t - a \cos 2t, \quad y = 2a \sin t - a \sin 2t$$
$$(0 \leq t \leq 2\pi, a > 0)$$

(see Figure 48, page 465)

15. Let R be a region in the xy-plane consisting of a piecewise smooth simple closed curve C and its interior. Show that the centroid of R has coordinates

$$\bar{x} = \frac{1}{2A} \int_C x^2\, dy, \quad \bar{y} = -\frac{1}{2A} \int_C y^2\, dx, \quad \text{(i)}$$

where A is the area of R.

16. Use the formulas (i) to find the centroid of the region in Problem 11.

17. Give an example of a path independent line integral $\int_C P\, dx + Q\, dy$ on a *multiply* connected domain D. (Notice that Corollary 2 does not exclude this possibility!)

18. Show that Green's theorem holds for any bounded closed region R which can be decomposed into a finite number of simple subregions by drawing suitable lines parallel to the coordinate axes.

920   Chapter 15   Line and Surface Integrals

Find the Jacobian $\partial(x, y)/\partial(u, v)$ of the given transformation.
19. $x = au + bv$, $y = cu + dv$
20. $x = u^2 - v^2$, $y = uv$
21. $x = \frac{1}{2} \ln(u^2 + v^2)$, $y = \arctan(v/u)$
22. $x = u^2 v$, $y = uv^2$
23. $x = e^u \cos v$, $y = e^u \sin v$
24. $x = ue^{-v}$, $y = ve^u$

25. Verify that formula (7) is satisfied by the transformation in Problem 23 and its inverse.
26. Verify that
$$\begin{vmatrix} a & b \\ c & d \end{vmatrix} \begin{vmatrix} A & B \\ C & D \end{vmatrix} = \begin{vmatrix} aA + bC & aB + bD \\ cA + dC & cB + dD \end{vmatrix},$$
and then use this rule for multiplying 2×2 determinants and several applications of the chain rule to establish formula (7).

Find the Jacobian $\partial(x, y, z)/\partial(u, v, w)$ of the given transformation.
27. $x = au$, $y = \alpha u + bv$, $z = \beta u + \gamma v + cw$
28. $x = vw$, $y = uw$, $z = uv$
29. $x = u + v + w$, $y = u^2 + v^2 + w^2$, $z = uv + uw + vw$
30. $x = e^u \cos v \sin w$, $y = e^u \cos v \cos w$, $z = e^u \sin v$

Use Theorem 5 to evaluate the given double integral, by making a suitable change of variables.

31. $\iint_R x^2 \, dx \, dy$, where R is the region bounded by the lines $2x + y = 1$, $2x + y = 3$, $x - 2y = -1$ and $x - 2y = 2$

32. $\iint_R \left(1 + \frac{y}{x}\right) dx \, dy$, where R is the region bounded by the lines $y = 0$, $y = 3x$, $x + y = 1$ and $x + y = 2$

33. $\iint_R \frac{1}{y} dx \, dy$, where R is the region bounded by the curves $y^3 = x^2$, $y^3 = 4x^2$ and the lines $y = x$, $y = 2x$

34. $\iint_R x^2 y^2 \, dx \, dy$, where R is the region bounded by the hyperbolas $xy = 2$, $xy = 3$ and the lines $y = x$, $y = 5x$

35. $\iint_R \frac{x^2 \sin xy}{y} dx \, dy$, where R is the region bounded by the parabolas $x^2 = \pi y/2$, $x^2 = \pi y$, $y^2 = x/2$ and $y^2 = x$

36. $\iint_R \cos \frac{x - y}{x + y} dx \, dy$, where R is the region bounded by the x-axis, the line $y = x$ and the line $x + y = \pi/2$

37. Evaluate the triple integral $\iiint_T yz \, dx \, dy \, dz$, where T is the region bounded by the six planes $x + y + z = \pm 2$, $x - y + z = \pm 3$ and $x + y - z = \pm 1$.
38. Show that Theorem 9, page 864, on triple integrals in cylindrical coordinates, is a special case of Theorem 6 on change of variables in triple integrals.
39. Show that Theorem 10, page 869, on triple integrals in spherical coordinates, is also a special case of Theorem 6.
40. First determine the area enclosed by the ellipse $(x^2/a^2) + (y^2/b^2) = 1$ by making the transformation $x = au$, $y = bv$. Then determine the volume enclosed by the ellipsoid $(x^2/a^2) + (y^2/b^2) + (z^2/c^2) = 1$ by making an analogous transformation.

15.5 The Divergence Theorem

We begin by writing Green's theorem

$$\int_C P \, dx + Q \, dy = \iint_R \left(\frac{\partial Q}{\partial x} - \frac{\partial P}{\partial y}\right) dA \tag{1}$$

in vector form. Here C is a piecewise smooth simple closed curve, traversed once in the counterclockwise direction, and $P = P(x, y)$ and $Q = Q(x, y)$ are functions which are continuously differentiable on the bounded plane region R consisting of the curve C and its interior. Introducing the two-dimensional vector field $\mathbf{F} = P\mathbf{i} + Q\mathbf{j}$, we recognize the left side of (1) as the line integral $\int_C \mathbf{F} \cdot d\mathbf{r}$ (recall formula (9), page 881). Thus (1) takes the form

$$\int_C \mathbf{F} \cdot d\mathbf{r} = \iint_R \left(\frac{\partial Q}{\partial x} - \frac{\partial P}{\partial y}\right) dA,$$

or equivalently

$$\int_C \mathbf{F} \cdot d\mathbf{r} = \iint_R \left(\frac{\partial F_2}{\partial x} - \frac{\partial F_1}{\partial y} \right) dA \tag{2}$$

if we write $\mathbf{F} = F_1\mathbf{i} + F_2\mathbf{j}$ instead of $\mathbf{F} = P\mathbf{i} + Q\mathbf{j}$. An alternative way of writing (2) is

$$\int_C \mathbf{F} \cdot \mathbf{T}\, ds = \iint_R \left(\frac{\partial F_2}{\partial x} - \frac{\partial F_1}{\partial y} \right) dA, \tag{2'}$$

where $\mathbf{T}$ is the unit tangent vector to C (recall the discussion following formula (12), page 882.

If $\mathbf{F}$ is interpreted as a *force*, formula (2) tells us that the work done by $\mathbf{F}$ on a particle traversing C is equal to the double integral in the right side of (2). For instance, it follows from Example 1, page 911, that the work done by the force $\mathbf{F} = (x^2 + y^2)\mathbf{i} + (x + 2y)^2\mathbf{j}$ on a particle traversing the triangle with vertices (0, 0), (1, 1) and (0, 2) in the counterclockwise direction is equal to $\frac{8}{3}$. Notice that if $\mathbf{F}$ is a conservative field, then $\partial F_2/\partial x = \partial F_1/\partial y$ and the force "does no work" as the particle traverses C.

We now consider another, equally interesting, vector form of Green's theorem. Let $\mathbf{n}$ be the *outer* unit normal to the curve C. Then $\mathbf{n} = \mathbf{T} \times \mathbf{k}$, as is apparent from Figure 28. Since $\mathbf{T} = d\mathbf{r}/ds$, it follows that

$$\mathbf{n} = \left(\frac{dx}{ds}\mathbf{i} + \frac{dy}{ds}\mathbf{j} \right) \times \mathbf{k} = \frac{dy}{ds}\mathbf{i} - \frac{dx}{ds}\mathbf{j}.$$

Figure 28

The line integral $\int_C \mathbf{F} \cdot \mathbf{n}\, ds$, which differs from the left side of (2') by having $\mathbf{n}$ instead of $\mathbf{T}$, can also be expressed as a double integral over the region R bounded by C. In fact,

$$\int_C \mathbf{F} \cdot \mathbf{n}\, ds = \int_C (F_1\mathbf{i} + F_2\mathbf{j}) \cdot \left(\frac{dy}{ds}\mathbf{i} - \frac{dx}{ds}\mathbf{j} \right) ds = \int_C -F_2\, dx + F_1\, dy,$$

and hence, by Green's theorem (1) with $P = -F_2$ and $Q = F_1$, we find at once that

$$\int_C \mathbf{F} \cdot \mathbf{n}\, ds = \iint_R \left(\frac{\partial F_1}{\partial x} + \frac{\partial F_2}{\partial y} \right) dA. \tag{3}$$

The left side of (3) is the exact analogue of the flux integral $\iint_S \mathbf{F} \cdot \mathbf{n}\, d\sigma$, where S is a surface and $\mathbf{F}$ is a three-dimensional vector field (see page 905). Correspondingly, $\int_C \mathbf{F} \cdot \mathbf{n}\, ds$ is called the *flux of* $\mathbf{F}$ *across the curve* C. To interpret $\int_C \mathbf{F} \cdot \mathbf{n}\, ds$ physically, imagine that the xy-plane is covered by a thin layer of fluid of negligible thickness moving with velocity $\mathbf{F} = \mathbf{F}(x, y)$. Then the flux of $\mathbf{F}$ across C is just the net area of fluid lowing across C per unit time, from the inside to the outside of C. This can be seen by making slight modifications in the argument given on pages 905–906, leading to the physical interpretation of the flux $\iint_S \mathbf{F} \cdot \mathbf{n}\, d\sigma$ across a surface S of a spatial flow with velocity $\mathbf{F} = \mathbf{F}(x, y, z)$. In particular, the surface S and the oblique cylinder

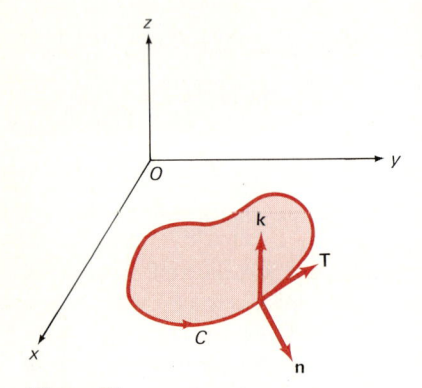

922   Chapter 15   Line and Surface Integrals

in Figure 18 are replaced by the curve C and a parallelogram with one side an arc of C and an adjacent side along $\mathbf{F} = \mathbf{F}(x, y)$.

Example 1 Find the flux of the field $\mathbf{F} = (x^2 + y^2)\mathbf{i} + (x + 2y)^2\mathbf{j}$ across the triangle C with vertices $(0, 0)$, $(1, 1)$ and $(0, 2)$, shown in Figure 22, page 911.

Solution By formula (3) with $F_1 = x^2 + y^2$ and $F_2 = (x + 2y)^2$, we have

$$\int_C \mathbf{F} \cdot \mathbf{n}\, ds = \int_0^1 dx \int_x^{2-x} \left[\frac{\partial}{\partial x}(x^2 + y^2) + \frac{\partial}{\partial y}(x + 2y)^2 \right] dy$$

$$= \int_0^1 dx \int_x^{2-x} (6x + 8y)\, dy = 2\int_0^1 \left[3xy + 2y^2 \right]_{y=x}^{2-x} dx$$

$$= 2\int_0^1 (8 - 2x - 6x^2)\, dx = 2\left[8x - x^2 - 2x^3 \right]_0^1 = 10. \quad \square$$

Divergence of a Vector Field

The integrand of the double integral over R in formula (3) is called the *two-dimensional divergence* of the field $\mathbf{F} = F_1\mathbf{i} + F_2\mathbf{j}$, and is written

$$\operatorname{div} \mathbf{F} = \frac{\partial F_1}{\partial x} + \frac{\partial F_2}{\partial y}.$$

In terms of this quantity, we can write (3) in the form

$$\int_C \mathbf{F} \cdot \mathbf{n}\, ds = \iint_R \operatorname{div} \mathbf{F}\, dA, \tag{4}$$

which is called the *two-dimensional divergence theorem*. According to (4), the flux of the field $\mathbf{F}$ across the simple closed curve C is equal to the integral of its divergence over the region R bounded by C.

More generally, the divergence of a *three-dimensional* vector field $\mathbf{F} = F_1\mathbf{i} + F_2\mathbf{j} + F_3\mathbf{k}$ is defined as

$$\operatorname{div} \mathbf{F} = \frac{\partial F_1}{\partial x} + \frac{\partial F_2}{\partial y} + \frac{\partial F_3}{\partial z}. \tag{5}$$

Notice that the operation of taking the divergence converts a vector field $\mathbf{F}$ into a scalar field. This is in contrast to the operation of taking the gradient, which converts a scalar field f into a vector field:

$$\operatorname{grad} f = \nabla f = \frac{\partial f}{\partial x}\mathbf{i} + \frac{\partial f}{\partial y}\mathbf{j} + \frac{\partial f}{\partial z}\mathbf{k}.$$

Observe also that div $\mathbf{F}$ is obtained if we take the formal dot product of the del operator

$$\nabla = \mathbf{i}\frac{\partial}{\partial x} + \mathbf{j}\frac{\partial}{\partial y} + \mathbf{k}\frac{\partial}{\partial z}$$

with $\mathbf{F}$, since

$$\nabla \cdot \mathbf{F} = \left(\mathbf{i}\frac{\partial}{\partial x} + \mathbf{j}\frac{\partial}{\partial y} + \mathbf{k}\frac{\partial}{\partial z} \right) \cdot (F_1\mathbf{i} + F_2\mathbf{j} + F_3\mathbf{k}) = \frac{\partial F_1}{\partial x} + \frac{\partial F_2}{\partial y} + \frac{\partial F_3}{\partial z}.$$

Example 2 Calculate div $(f\mathbf{F})$, where f is a differentiable scalar function and $\mathbf{F} = F_1\mathbf{i} + F_2\mathbf{j} + F_3\mathbf{k}$ is a differentiable vector field.

Solution By the definition (5) and the product rule for differentiation, we have

$$\text{div}(f\mathbf{F}) = \frac{\partial(fF_1)}{\partial x} + \frac{\partial(fF_2)}{\partial y} + \frac{\partial(fF_3)}{\partial z}$$

$$= \frac{\partial f}{\partial x}F_1 + f\frac{\partial F_1}{\partial x} + \frac{\partial f}{\partial y}F_2 + f\frac{\partial F_2}{\partial y} + \frac{\partial f}{\partial z}F_3 + f\frac{\partial F_3}{\partial z}$$

$$= \left(\frac{\partial f}{\partial x}F_1 + \frac{\partial f}{\partial y}F_2 + \frac{\partial f}{\partial z}F_3\right) + f\left(\frac{\partial F_1}{\partial x} + \frac{\partial F_2}{\partial y} + \frac{\partial F_3}{\partial z}\right)$$

$$= (\text{grad } f) \cdot \mathbf{F} + f \text{ div } \mathbf{F},$$

or more concisely

$$\nabla \cdot (f\mathbf{F}) = \nabla f \cdot \mathbf{F} + f\nabla \cdot \mathbf{F}. \quad \square$$

Next we show that (4) is just the two-dimensional version of a more general three-dimensional divergence theorem.

> **Theorem 7** (*Divergence theorem*).[†] *Let T be the solid region consisting of a piecewise smooth closed surface S and its interior, let* **n** *be the outer unit normal to S, and let* $\mathbf{F} = F_1\mathbf{i} + F_2\mathbf{j} + F_3\mathbf{k}$ *be a continuously differentiable vector field on T. Then*
>
> $$\iint_S \mathbf{F} \cdot \mathbf{n} \, d\sigma = \iiint_T \text{div } \mathbf{F} \, dV, \tag{6}$$
>
> *that is, the flux of* **F** *across S equals the integral of div* **F** *over the region T bounded by S.*

Partial proof We will prove the theorem only for the case where T is x-simple, y-simple and z-simple, as defined on page 824; the full proof is given in advanced calculus. Writing formula (6) in more detail as

$$\iint_S (F_1\mathbf{i} + F_2\mathbf{j} + F_3\mathbf{k}) \cdot \mathbf{n} \, d\sigma = \iiint_T \left(\frac{\partial F_1}{\partial x} + \frac{\partial F_2}{\partial y} + \frac{\partial F_3}{\partial z}\right) dV, \tag{6'}$$

we see that the theorem will be proved if we succeed in showing that

$$\iint_S (F_1\mathbf{i}) \cdot \mathbf{n} \, d\sigma = \iiint_T \frac{\partial F_1}{\partial x} \, dV, \tag{7}$$

$$\iint_S (F_2\mathbf{j}) \cdot \mathbf{n} \, d\sigma = \iiint_T \frac{\partial F_2}{\partial y} \, dV, \tag{8}$$

$$\iint_S (F_3\mathbf{k}) \cdot \mathbf{n} \, d\sigma = \iiint_T \frac{\partial F_3}{\partial z} \, dV, \tag{9}$$

since (6′) is the sum of the formulas (7)–(9). All three of these formulas are proved in the same way, and hence it is enough to prove just one of them, say (9).

[†] Also known as *Gauss' theorem* after Carl Friedrich Gauss (1777–1855), a leading contender for the title of the world's greatest mathematician.

924 Chapter 15 Line anol Surface Integrals

To this end, we first observe that since the region T is z-simple and S is piecewise smooth, T is a set of the form

$$\{(x, y, z) : (x, y) \in R_{xy}, g_1(x, y) \leq z \leq g_2(x, y)\},$$

where R_{xy} is the projection of T onto the xy-plane, and the functions g_1 and g_2 have piecewise smooth graphs. Thus T is shaped like the solid in Figure 29, made up of a lower surface S_1, an upper surface S_2, and a lateral surface S_3 generated by line segments parallel to the z-axis. In some cases S_3 may be absent, for example in the case of a sphere, but in any event S_3 makes no contribution to the integral $\iint_S (F_3 \mathbf{k}) \cdot \mathbf{n} \, d\sigma$, since on S_3 the vector $\mathbf{n}$ is parallel to the xy-plane and hence $\mathbf{k} \cdot \mathbf{n} = 0$. Thus, to prove (9), we need only show that

$$\iint_{S_1} (F_3 \mathbf{k}) \cdot \mathbf{n} \, d\sigma + \iint_{S_2} (F_3 \mathbf{k}) \cdot \mathbf{n} \, d\sigma = \iiint_T \frac{\partial F_3}{\partial z} \, dV. \tag{9'}$$

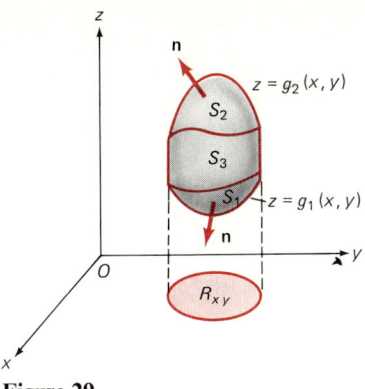

Figure 29

The proof of (9') is easily accomplished by using already established results. In the first place, it follows from Theorem 5, page 827, that

$$\iiint_T \frac{\partial F_3}{\partial z} \, dV = \iint_{R_{xy}} dA \int_{g_1(x,y)}^{g_2(x,y)} \frac{\partial F_3(x, y, z)}{\partial z} \, dz$$

$$= \iint_{R_{xy}} [F_3(x, y, g_2(x, y)) - F_3(x, y, g_1(x, y))] \, dA. \tag{10}$$

Secondly, by formula (12), page 907, with $S = S_2$, $\mathbf{F} = F_3 \mathbf{k}$, $F_1 = 0$, $F_2 = 0$ and $dx \, dy = dA$,

$$\iint_{S_2} (F_3 \mathbf{k}) \cdot \mathbf{n} \, d\sigma = \iint_{R_{xy}} F_3(x, y, g_2(x, y)) \, dA. \tag{11}$$

To get the corresponding formula for S_1, we must replace $\mathbf{n}$ by $-\mathbf{n}$, since the outer unit normal points downward on S_1, rather than upward as on S_2. This gives

$$\iint_{S_1} (F_3 \mathbf{k}) \cdot (-\mathbf{n}) \, d\sigma = \iint_{R_{xy}} F_3(x, y, g_1(x, y)) \, dA,$$

or equivalently

$$\iint_{S_1} (F_3 \mathbf{k}) \cdot \mathbf{n} \, d\sigma = -\iint_{R_{xy}} F_3(x, y, g_1(x, y)) \, dA. \tag{11'}$$

Adding (11) and (11'), we find that

$$\iint_{S_1} (F_3 \mathbf{k}) \cdot \mathbf{n} \, d\sigma + \iint_{S_2} (F_3 \mathbf{k}) \cdot \mathbf{n} \, d\sigma$$

$$= \iint_{R_{xy}} [F_3(x, y, g_2(x, y)) - F_3(x, y, g_1(x, y))] \, dA,$$

and then comparison of this formula with (10) establishes the validity of (9′) and hence of (9). As already noted, the rest of the proof consists in establishing formulas (7) and (8) by analogous arguments. ∎

Example 3 The flux of the field $\mathbf{F} = xy^2\mathbf{i} - yz\mathbf{j} + x^2z^2\mathbf{k}$ out of the cubical surface S bounded by the coordinate planes and the planes $x = 1$, $y = 1$ and $z = 1$ was found by direct calculation in Example 8, page 907. Use the divergence theorem to calculate the same quantity.

Solution The divergence of $\mathbf{F}$ is

$$\text{div } \mathbf{F} = \frac{\partial(xy^2)}{\partial x} + \frac{\partial(-yz)}{\partial y} + \frac{\partial(x^2z^2)}{\partial z} = y^2 - z + 2x^2z,$$

and hence, by the divergence theorem,

$$\iint_S \mathbf{F} \cdot \mathbf{n}\, d\sigma = \iiint_T \text{div } \mathbf{F}\, dV = \iiint_T (y^2 - z + 2x^2z)\, dV,$$

where T is the solid cube bounded by S. Therefore

$$\iint_S \mathbf{F} \cdot \mathbf{n}\, d\sigma = \int_0^1 dx \int_0^1 dy \int_0^1 (y^2 - z + 2x^2z)\, dz$$

$$= \int_0^1 dx \int_0^1 \left[y^2z - \frac{1}{2}z^2 + x^2z^2 \right]_{z=0}^1 dy$$

$$= \int_0^1 dx \int_0^1 \left(y^2 - \frac{1}{2} + x^2 \right) dy = \int_0^1 \left[\frac{1}{3}y^3 - \frac{1}{2}y + x^2y \right]_{y=0}^1 dx$$

$$= \int_0^1 \left(-\frac{1}{6} + x^2 \right) dx = \left[-\frac{1}{6}x + \frac{1}{3}x^3 \right]_0^1 = \frac{1}{6},$$

in accordance with the cited example. Notice how the divergence theorem eliminates the need to separately examine the behavior of $\mathbf{F} \cdot \mathbf{n}$ on each of the six faces of S! □

Example 4 Find the flux of the field $\mathbf{F} = x^3\mathbf{i} + y^3\mathbf{j} + z^3\mathbf{k}$ out of the sphere $x^2 + y^2 + z^2 = a^2$ of radius a.

Solution Denoting the sphere by S and the ball $x^2 + y^2 + z^2 \leq a^2$ by T, we apply the divergence theorem. Since

$$\text{div } \mathbf{F} = \frac{\partial(x^3)}{\partial x} + \frac{\partial(y^3)}{\partial y} + \frac{\partial(z^3)}{\partial z} = 3(x^2 + y^2 + z^2),$$

we find after transforming to spherical coordinates that

$$\iint_S \mathbf{F} \cdot \mathbf{n}\, d\sigma = \iiint_T \text{div } \mathbf{F}\, dV = \int_0^{2\pi} d\theta \int_0^{\pi} d\phi \int_0^a (3\rho^2)\rho^2 \sin\phi\, d\rho$$

$$= 6\pi \left(\int_0^{\pi} \sin\phi\, d\phi \right) \left(\int_0^a \rho^4\, d\rho \right) = \frac{12}{5}\pi a^5. \quad □$$

Example 5 Find the flux of the field $\mathbf{F} = yz\mathbf{i} + xz\mathbf{j} + xy\mathbf{k}$ out of any surface S satisfying the conditions of Theorem 7.

926 Chapter 15 Line and Surface Integrals

Solution Here

$$\text{div } \mathbf{F} = \frac{\partial(yz)}{\partial x} + \frac{\partial(xz)}{\partial y} + \frac{\partial(xy)}{\partial z} = 0,$$

so that if T is the region enclosed by S, we can immediately conclude that

$$\iint_S \mathbf{F} \cdot \mathbf{n}\, d\sigma = \iiint_T \text{div } \mathbf{F}\, dV = \iiint_T 0\, dV = 0. \quad \square$$

A field with zero divergence, like the field $\mathbf{F}$ in the preceding example, is said to be *solenoidal*.

Example 6 Find the flux of the field $\mathbf{F} = \mathbf{r} = x\mathbf{i} + y\mathbf{j} + z\mathbf{k}$ out of any surface S satisfying the conditions of Theorem 7.

Solution This time

$$\iint_S \mathbf{F} \cdot \mathbf{n}\, d\sigma = \iint_S \mathbf{r} \cdot \mathbf{n}\, d\sigma = \iiint_T \text{div } \mathbf{F}\, dV$$

$$= \iiint_T \left(\frac{\partial x}{\partial x} + \frac{\partial y}{\partial y} + \frac{\partial z}{\partial z}\right) dV = 3 \iiint_T dV = 3V,$$

where V is the volume of the region T enclosed by S. Solving for V, we get the interesting formula

$$V = \frac{1}{3} \iint_S \mathbf{r} \cdot \mathbf{n}\, d\sigma, \tag{12}$$

expressing the volume of a solid in terms of the flux of the vector $\mathbf{r}$ across its surface. $\quad \square$

Physical Meaning of the Divergence

We now comment on the physical meaning of the divergence. Let $\mathbf{v} = \mathbf{v}(\mathbf{r})$ be the steady-state velocity field of a moving incompressible fluid, where $\mathbf{v}$ is assumed to be continuously differentiable. Suppose div $\mathbf{v}(\mathbf{a}) > 0$, so that the divergence of $\mathbf{v}$ is positive at the point P with position vector $\mathbf{a}$. Let S_ε be the sphere of radius ε centered at P, with outer unit normal $\mathbf{n}$, and let T_ε be the ball enclosed by S_ε (see Figure 30). Then, since div $\mathbf{v}$ is continuous (why?), div $\mathbf{v}$ is positive on T_ε for some sufficiently small value of ε. Therefore, by the divergence theorem,

$$\iint_{S_\varepsilon} \mathbf{v} \cdot \mathbf{n}\, d\sigma = \iiint_{T_\varepsilon} \text{div } \mathbf{v}\, dV \tag{13}$$

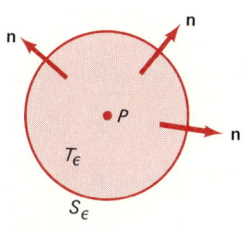

Figure 30

is also positive, so that there is a net outward flow of fluid across S_ε. In this case the flow is said to have a *source* at P. Similarly, if div $\mathbf{v}(\mathbf{a}) < 0$, then $\int_{S_\varepsilon} \mathbf{v} \cdot \mathbf{n}\, d\sigma < 0$, so that there is a net *inward* flow of fluid across S_ε and the flow is said to have a *sink* at P. For example, imagine a tub being filled with water at such a rate that the tap is below the surface of the water, but the water level remains constant because the drain is open. Then the flow is steady-state, with a source at the tap and a sink at the drain. In the absence of a source or sink at P, we have div $\mathbf{v}(\mathbf{a}) = 0$, since any other accumulation or loss of fluid inside S_ε is prevented by the incompressibility of the fluid.

Another interesting conclusion can be drawn from formula (13). Let V_ε be the volume of T_ε. Then (13) is equivalent to

$$\frac{1}{V_\varepsilon} \iiint_{T_\varepsilon} \text{div } \mathbf{v} \, dV = \frac{1}{V_\varepsilon} \iint_{S_\varepsilon} \mathbf{v} \cdot \mathbf{n} \, d\sigma,$$

where the expression on the left is the mean value or average of $\mathbf{v}$ over the region T_ε (recall Problem 32, page 832). But as $\varepsilon \to 0$, this average approaches div $\mathbf{v}(\mathbf{a})$, the value of the divergence of $\mathbf{v}$ at P (why?). Therefore

$$\text{div } \mathbf{v}(\mathbf{a}) = \lim_{\varepsilon \to 0} \frac{1}{V_\varepsilon} \iint_{S_\varepsilon} \mathbf{v} \cdot \mathbf{n} \, d\sigma, \tag{14}$$

which gives a *coordinate-free* interpretation of the divergence of $\mathbf{v}$ at P as the *flux of $\mathbf{v}$ per unit volume* at P. In particular, if div $\mathbf{v}(\mathbf{a})$ is positive, there is a net "divergence" or outward flow of fluid out of the sphere S_ε.

Problems

Find the divergence of the given two-dimensional field $\mathbf{F}$.

1. $\mathbf{F} = x\mathbf{i} + y\mathbf{j}$
2. $\mathbf{F} = \dfrac{-y}{x^2 + y^2}\mathbf{i} + \dfrac{x}{x^2 + y^2}\mathbf{j}$
3. $\mathbf{F} = (\cos^2 xy)\mathbf{i} + (\sin^2 xy)\mathbf{j}$
4. $\mathbf{F} = (\cosh^2 xy)\mathbf{i} + (\sinh^2 xy)\mathbf{j}$
5. $\mathbf{F} = \ln(\sqrt{x^2 + y^2})\mathbf{i} + \arctan(x/y)\mathbf{j}$
6. $\mathbf{F} = (x \ln |x|)\mathbf{i} + (y \ln |y|)\mathbf{j}$

Find the divergence of the given three-dimensional field $\mathbf{F}$.

7. $\mathbf{F} = 2x\mathbf{i} + 3y\mathbf{j} - z\mathbf{k}$
8. $\mathbf{F} = x^2\mathbf{i} + y^2\mathbf{j} + z^2\mathbf{k}$
9. $\mathbf{F} = xyz(\mathbf{i} + \mathbf{j} + \mathbf{k})$
10. $\mathbf{F} = x^2yz\mathbf{i} + xy^2z\mathbf{j} + xyz^2\mathbf{k}$
11. $\mathbf{F} = \text{grad } (e^{x+y+z})$
12. $\mathbf{F} = (\sin xy)\mathbf{i} + (\cos xz)\mathbf{j} + (\sinh yz)\mathbf{k}$

13. Calculate div $(a\mathbf{F} + b\mathbf{G})$, where a and b are scalar constants.

14. The divergence of the gradient of a scalar field f is called the *Laplacian* of f, denoted by $\nabla^2 f$. Thus

$$\nabla^2 f = \text{div } (\text{grad } f) = \text{div } (\nabla f).$$

The equation $\nabla^2 f = 0$ is called *Laplace's equation*, and solutions of this equation are called *harmonic functions* (see page 763). Show that

$$\nabla^2 f = \frac{\partial^2 f}{\partial x^2} + \frac{\partial^2 f}{\partial y^2}$$

in two dimensions, while

$$\nabla^2 f = \frac{\partial^2 f}{\partial x^2} + \frac{\partial^2 f}{\partial y^2} + \frac{\partial^2 f}{\partial z^2}$$

in three dimensions.

15. Show that the function $f(x, y) = \ln \sqrt{x^2 + y^2}$ is harmonic everywhere except at the origin.
16. Show that the function $f(x, y, z) = \ln \sqrt{x^2 + y^2 + z^2}$ is harmonic nowhere. What does the result of Problem 51, page 764, say in the language of harmonic functions?
17. Is the function $f(x, y, z) = e^x(\sin y + \cos z)$ harmonic? Explain your answer.
18. Show that div $(f \nabla f) = |\nabla f|^2 + f \nabla^2 f$.
19. Verify formula (12) for the sphere $x^2 + y^2 + z^2 = a^2$ of radius a.
20. Find the two-dimensional analogue of formula (12), and verify it for the circle $x^2 + y^2 = a^2$ of radius a.

Use the two-dimensional divergence theorem to calculate

21. The flux of the field $\mathbf{F} = 3x\mathbf{i} - 2\mathbf{j}$ out of the circle $x^2 + y^2 = 2$

22. The flux of the field $\mathbf{F} = x^2\mathbf{i} + y^2\mathbf{j}$ out of the square with vertices $(0, \pm 1)$ and $(2, \pm 1)$

23. Let $\partial f/\partial n$ denote the derivative of a function f of two or three variables in the direction of the outer unit normal $\mathbf{n}$ to a simple closed path C or closed surface S (so that $\partial f/\partial n = \nabla f \cdot \mathbf{n}$), and suppose f has continuous second partial derivatives on C or S. Show that

$$\iint_R \nabla^2 f \, dA = \int_C \frac{\partial f}{\partial n} \, ds, \qquad \text{(i)}$$

where R is the plane region consisting of the path C and its interior, or

$$\iiint_T \nabla^2 f \, dV = \iint_S \frac{\partial f}{\partial n} \, d\sigma, \qquad \text{(i')}$$

where T is the solid region consisting of the surface S and its interior.

24. Verify formula (i') for the function $f(x, y, z) = x^2 y^2 z^2$ and the cube $T = \{(x, y, z): 0 \le x \le 1, 0 \le y \le 1, 0 \le z \le 1\}$.

With the help of the divergence theorem, calculate the flux of the given field $\mathbf{F}$ out of the unit sphere $x^2 + y^2 + z^2 = 1$.

25. $\mathbf{F} = 4x\mathbf{i} - 2y\mathbf{j} + \mathbf{k}$

26. $\mathbf{F} = a\mathbf{i} + b\mathbf{j} + c\mathbf{k}$ $(a, b, c \text{ constants})$
27. $\mathbf{F} = -y^2\mathbf{i} + x^2\mathbf{j} + z^3\mathbf{k}$
28. $\mathbf{F} = 2x^3\mathbf{i} + y^3\mathbf{j} - z^2\mathbf{k}$

By calculating both sides of equation (6), verify the divergence theorem for the given solid region T and field $\mathbf{F}$.

29. $T = \{(x, y, z): x + y + z \le a, x \ge 0, y \ge 0, z \ge 0\}$,
$\mathbf{F} = x\mathbf{i} + y\mathbf{j} + z\mathbf{k}$
30. $T = \{(x, y, z): 0 \le x \le a, 0 \le y \le b, 0 \le z \le c\}$,
$\mathbf{F} = x^3\mathbf{i} + y^3\mathbf{j} + z^3\mathbf{k}$
31. $T = \{(x, y, z): x^2 + y^2 \le 2, 0 \le z \le 1\}$,
$\mathbf{F} = 3x\mathbf{i} - y^2\mathbf{j} + z^2\mathbf{k}$

32. Let $f = f(x, y, z)$ and $g = g(x, y, z)$ be two scalar functions with continuous second derivatives, and let S, $\mathbf{n}$ and T be the same as in the divergence theorem. Prove Green's first identity

$$\iiint_T (f\nabla^2 g + \nabla f \cdot \nabla g) \, dV = \iint_S f \nabla g \cdot \mathbf{n} \, d\sigma,$$

and then use it to prove Green's second identity

$$\iiint_T (f\nabla^2 g - g\nabla^2 f) \, dV = \iint_S (f \nabla g - g \nabla f) \cdot \mathbf{n} \, d\sigma.$$

15.6 Stokes' Theorem

Curl of a Vector Field

You will recall from the last section that the divergence div $\mathbf{F}$ of a vector field $\mathbf{F} = F_1\mathbf{i} + F_2\mathbf{j} + F_3\mathbf{k}$ is formed by taking the *dot* product of the operator ∇ with $\mathbf{F}$. We now consider the *curl* of $\mathbf{F}$, which is formed by taking the *cross* product of ∇ with $\mathbf{F}$. Thus

$$\text{curl } \mathbf{F} = \nabla \times \mathbf{F} = \left(\mathbf{i}\frac{\partial}{\partial x} + \mathbf{j}\frac{\partial}{\partial y} + \mathbf{k}\frac{\partial}{\partial z}\right) \times (F_1\mathbf{i} + F_2\mathbf{j} + F_3\mathbf{k})$$

$$= \left(\frac{\partial F_3}{\partial y} - \frac{\partial F_2}{\partial z}\right)\mathbf{i} + \left(\frac{\partial F_1}{\partial z} - \frac{\partial F_3}{\partial x}\right)\mathbf{j} + \left(\frac{\partial F_2}{\partial x} - \frac{\partial F_1}{\partial y}\right)\mathbf{k}, \quad (1)$$

which can be written more concisely as

$$\text{curl } \mathbf{F} = \begin{vmatrix} \mathbf{i} & \mathbf{j} & \mathbf{k} \\ \dfrac{\partial}{\partial x} & \dfrac{\partial}{\partial y} & \dfrac{\partial}{\partial z} \\ F_1 & F_2 & F_3 \end{vmatrix}. \quad (1')$$

This is not an ordinary numerical determinant, since its first row consists of vectors, its second row of partial differential operators, and its third row of scalar functions, but it is a convenient way of remembering the right side of formula (1). Notice that the operation of taking the curl converts a vector field into another vector field.

Example 1 Calculate curl (grad f), where f is a scalar field with continuous first and second partial derivatives.

Solution Since

$$\text{grad } f = \frac{\partial f}{\partial x}\mathbf{i} + \frac{\partial f}{\partial y}\mathbf{j} + \frac{\partial f}{\partial z}\mathbf{k},$$

it follows from formula (1) that

$$\text{curl (grad } f) = \left(\frac{\partial^2 f}{\partial y\, \partial z} - \frac{\partial^2 f}{\partial z\, \partial y}\right)\mathbf{i} + \left(\frac{\partial^2 f}{\partial z\, \partial x} - \frac{\partial^2 f}{\partial x\, \partial z}\right)\mathbf{j} + \left(\frac{\partial^2 f}{\partial x\, \partial y} - \frac{\partial^2 f}{\partial y\, \partial x}\right)\mathbf{k}.$$

But the expression in each pair of parentheses on the right is zero (why?), and therefore

$$\text{curl (grad } f) = \nabla \times \nabla f = \mathbf{0}. \quad \square$$

A field with zero curl, like every continuously differentiable gradient field, is said to be *irrotational* (for a reason given below). Having introduced the field curl **F**, we are now able to state *Stokes' theorem*. This is the three-dimensional generalization of Green's theorem in the form

$$\int_C \mathbf{F} \cdot d\mathbf{r} = \iint_R \left(\frac{\partial F_2}{\partial x} - \frac{\partial F_1}{\partial y}\right) dA, \tag{2}$$

just as the divergence theorem is the three-dimensional generalization of Green's theorem in the form

$$\int_C \mathbf{F} \cdot \mathbf{n}\, ds = \iint_R \left(\frac{\partial F_1}{\partial x} + \frac{\partial F_2}{\partial y}\right) dA$$

(recall the discussion at the beginning of Section 15.5). In (2) the simple closed plane curve C bounding the region R is traversed in the counterclockwise direction, which is the positive direction, as defined on page 912, in the sense that R is always to the left of an observer moving along C in this direction. More generally, if a smooth surface S equipped with a field of unit normals **n** is bounded by a simple closed *space* curve, the positive direction of traversing C, said to be "induced by **n**," is defined as the direction such that S is always to the left of an observer moving along C with his head pointing in the direction of **n**. If S is not smooth, but only piecewise smooth, so that S consists of a finite number m of smooth subsurfaces $S_1, \ldots, S_m$ joined along parts of their boundary curves $C_1, \ldots, C_m$, the assignment of the unit normal **n** to the subsurfaces $S_1, \ldots, S_m$ is such that the induced orientations of $C_1, \ldots, C_m$ have the following feature: Each arc of $C_1, \ldots, C_m$ shared by two adjacent subsurfaces is traversed twice, once in each direction, and the remaining arcs "fit together" to form the curve C bounding the surface S (the curve C is regarded as part of S). This is illustrated in Figure 31 for the case of three subsurfaces S_1, S_2 and S_3. Notice that if **n** is a unit normal to S, then $-\mathbf{n}$ is another choice of the unit normal (and the only other).

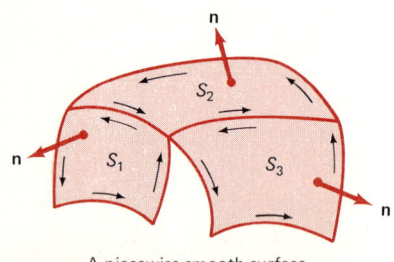

A piecewise smooth surface
Figure 31

930 Chapter 15 Line and Surface Integrals

Theorem 8 *(Stokes' theorem).*† *Let S be a piecewise smooth surface bounded by a piecewise smooth simple closed curve C, and choose a unit normal* **n** *to S. Let C be traversed in the positive direction (the direction induced by* **n**), *and let* $\mathbf{F} = F_1\mathbf{i} + F_2\mathbf{j} + F_3\mathbf{k}$ *be a continuously differentiable vector field on S. Then*

$$\int_C \mathbf{F} \cdot d\mathbf{r} = \iint_S (\text{curl } \mathbf{F}) \cdot \mathbf{n} \, d\sigma, \tag{3}$$

that is, the circulation of **F** *around C equals the flux of curl* **F** *across S.*

Partial proof (Optional) The full proof of Stokes' theorem is an advanced calculus topic, and hence we will prove the theorem only for the special case, illustrated in Figure 32, where S is the graph of a function $z = z(x, y)$ with continuous second partial derivatives, defined on a simple region R_{xy} in the xy-plane, whose boundary is a piecewise smooth simple closed curve C_{xy}. The region R_{xy} and its boundary curve C_{xy} are of course the projections onto the xy-plane of the surface S and its boundary curve C. It will also be assumed that **n** is the *upper* unit normal to S.

The left side of (3) is just

$$\int_C \mathbf{F} \cdot d\mathbf{r} = \int_C F_1 \, dx + F_2 \, dy + F_3 \, dz. \tag{4}$$

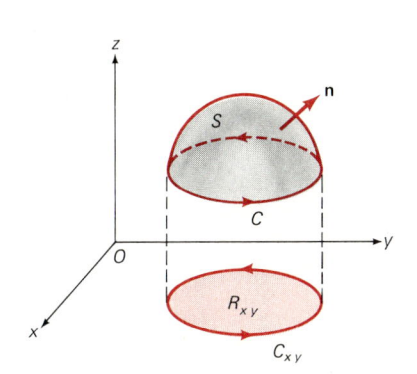

Figure 32

Using formula (12), page 907, with **F** replaced by curl **F**, we find that the right side of (3) is

$$\iint_S (\text{curl } \mathbf{F}) \cdot \mathbf{n} \, d\sigma = \iint_{R_{xy}} \left[-(\text{curl } \mathbf{F})_1 \frac{\partial z}{\partial x} - (\text{curl } \mathbf{F})_2 \frac{\partial z}{\partial y} + (\text{curl } \mathbf{F})_3 \right] dx \, dy$$

$$= \iint_{R_{xy}} \left[\left(\frac{\partial F_2}{\partial z} - \frac{\partial F_3}{\partial y} \right) \frac{\partial z}{\partial x} + \left(\frac{\partial F_3}{\partial x} - \frac{\partial F_1}{\partial z} \right) \frac{\partial z}{\partial y} \right.$$

$$\left. + \left(\frac{\partial F_2}{\partial x} - \frac{\partial F_1}{\partial y} \right) \right] dx \, dy. \tag{5}$$

Comparing (4) and (5), we see that formula (3) will be proved if we succeed in showing that

$$\int_C F_1 \, dx = -\iint_{R_{xy}} \left(\frac{\partial F_1}{\partial z} \frac{\partial z}{\partial y} + \frac{\partial F_1}{\partial y} \right) dx \, dy, \tag{6}$$

$$\int_C F_2 \, dy = \iint_{R_{xy}} \left(\frac{\partial F_2}{\partial z} \frac{\partial z}{\partial x} + \frac{\partial F_2}{\partial x} \right) dx \, dy, \tag{7}$$

$$\int_C F_3 \, dz = \iint_{R_{xy}} \left(\frac{\partial F_3}{\partial x} \frac{\partial z}{\partial y} - \frac{\partial F_3}{\partial y} \frac{\partial z}{\partial x} \right) dx \, dy, \tag{8}$$

since (3) is just the sum of the formulas (6)–(8).

---

† Named after Sir George Gabriel Stokes, professor of mathematics at Cambridge University, who investigated the theorem after it was suggested to him in 1850 by the eminent physicist William Thomson, better known as Lord Kelvin.

Section 15.6 Stokes' Theorem   **931**

To verify these formulas, let the curve C_{xy} traversed in the positive direction have parametric equations $x = x(t)$, $y = y(t)$ ($a \leq t \leq b$). Then C, the boundary curve of the surface S, has parametric equations

$$x = x(t), \qquad y = y(t), \qquad z = z(x(t), y(t)) \qquad (a \leq t \leq b),$$

and will also be traversed in the positive direction. Therefore

$$\int_C F_1 \, dx = \int_a^b F_1(x(t), y(t), z(x(t), y(t))) \frac{dx}{dt} \, dt$$

$$= \int_a^b G(x(t), y(t)) \frac{dx}{dt} \, dt = \int_{C_{xy}} G \, dx, \tag{9}$$

in terms of the composite function $G(x, y) = F_1(x, y, z(x, y))$. By Green's theorem (1), page 909, with $P = G$, $Q = 0$, $C = C_{xy}$, $R = R_{xy}$ and $dA = dx\,dy$,

$$\int_{C_{xy}} G \, dx = -\iint_{R_{xy}} \frac{\partial G}{\partial y} \, dx\,dy = -\iint_{R_{xy}} \left(\frac{\partial F_1}{\partial y} + \frac{\partial F_1}{\partial z} \frac{\partial z}{\partial y} \right) dx\,dy, \tag{9'}$$

after applying the chain rule. Comparing (9) and (9'), we get (6), and (7) is proved in virtually the same way (give the details).

To complete the proof, we now verify formula (8). Here

$$\int_C F_3 \, dz = \int_a^b F_3(x(t), y(t), z(x(t), y(t))) \frac{dz(x(t), y(t))}{dt} \, dt$$

$$= \int_a^b H(x(t), y(t)) \left(\frac{\partial z}{\partial x} \frac{dx}{dt} + \frac{\partial z}{\partial y} \frac{dy}{dt} \right) dt$$

$$= \int_{C_{xy}} H \frac{\partial z}{\partial x} \, dx + H \frac{\partial z}{\partial y} \, dy, \tag{10}$$

in terms of the composite function $H(x, y) = F_3(x, y, z(x, y))$. By Green's theorem again, we have

$$\int_{C_{xy}} H \frac{\partial z}{\partial x} \, dx + H \frac{\partial z}{\partial y} \, dy = \iint_{R_{xy}} \left[\frac{\partial}{\partial x} \left(H \frac{\partial z}{\partial y} \right) - \frac{\partial}{\partial y} \left(H \frac{\partial z}{\partial x} \right) \right] dx\,dy$$

$$= \iint_{R_{xy}} \left(\frac{\partial H}{\partial x} \frac{\partial z}{\partial y} + H \frac{\partial^2 z}{\partial x \partial y} - \frac{\partial H}{\partial y} \frac{\partial z}{\partial x} - H \frac{\partial^2 z}{\partial y \partial x} \right) dx\,dy$$

$$= \iint_{R_{xy}} \left(\frac{\partial H}{\partial x} \frac{\partial z}{\partial y} - \frac{\partial H}{\partial y} \frac{\partial z}{\partial x} \right) dx\,dy$$

$$= \iint_{R_{xy}} \left[\left(\frac{\partial F_3}{\partial x} + \frac{\partial F_3}{\partial z} \frac{\partial z}{\partial x} \right) \frac{\partial z}{\partial y} \right.$$

$$\left. - \left(\frac{\partial F_3}{\partial y} + \frac{\partial F_3}{\partial z} \frac{\partial z}{\partial y} \right) \frac{\partial z}{\partial x} \right] dx\,dy$$

$$= \iint_{R_{xy}} \left(\frac{\partial F_3}{\partial x} \frac{\partial z}{\partial y} - \frac{\partial F_3}{\partial y} \frac{\partial z}{\partial x} \right) dx\,dy, \tag{10'}$$

after further application of the chain rule (where did we use the assumption that $z(x, y)$ has continuous second partials?). Comparing (10) and (10'), we obtain (8), thereby completing the proof of Stokes' theorem for this special case. ∎

To recover Green's theorem in the vector form (2) from Stokes' theorem (3), choose the surface S to be a region R in the xy-plane, with boundary curve C and fixed upper unit normal equal to the basis vector $\mathbf{k}$, and let $\mathbf{F} = F_1\mathbf{i} + F_2\mathbf{j}$ be a continuously differentiable field in the plane. Then the positive direction of C is counterclockwise, the curl of $\mathbf{F}$ simplifies to

$$\text{curl } \mathbf{F} = \left(\frac{\partial F_2}{\partial x} - \frac{\partial F_1}{\partial y}\right)\mathbf{k},$$

and $d\sigma = dA$ (why?). Therefore Stokes' theorem reduces at once to Green's theorem:

$$\int_C \mathbf{F} \cdot d\mathbf{r} = \iint_R (\text{curl } \mathbf{F}) \cdot \mathbf{k} \, d\sigma = \iint_R \left(\frac{\partial F_2}{\partial x} - \frac{\partial F_1}{\partial y}\right) dA.$$

Example 2 Verify Stokes' theorem for the field $\mathbf{F} = -2z\mathbf{i} + 3x\mathbf{j} + 4y\mathbf{k}$, where S is the parabolic cap $z = 1 - x^2 - y^2$ ($z \geq 0$) with upper unit normal $\mathbf{n}$, whose boundary C is the unit circle $x^2 + y^2 = 1$ in the xy-plane (see Figure 19, page 906).

Solution Since

$$\text{curl } \mathbf{F} = \begin{vmatrix} \mathbf{i} & \mathbf{j} & \mathbf{k} \\ \frac{\partial}{\partial x} & \frac{\partial}{\partial y} & \frac{\partial}{\partial z} \\ -2z & 3x & 4y \end{vmatrix} = 4\mathbf{i} - 2\mathbf{j} + 3\mathbf{k},$$

it follows from formula (5) that

$$\iint_S (\text{curl } \mathbf{F}) \cdot \mathbf{n} \, d\sigma = \iint_R [(-4)(-2x) + 2(-2y) + 3] \, dx \, dy$$

$$= \iint_R (8x - 4y + 3) \, dx \, dy,$$

where R is the unit disk $x^2 + y^2 \leq 1$. Therefore, transforming to polar coordinates, we have

$$\iint_S (\text{curl } \mathbf{F}) \cdot \mathbf{n} \, d\sigma = \int_0^{2\pi} d\theta \int_0^1 (8r \cos \theta - 4r \sin \theta + 3) r \, dr$$

$$= \int_0^{2\pi} \left[\frac{8}{3} r^3 \cos \theta - \frac{4}{3} r^3 \sin \theta + \frac{3}{2} r^2\right]_{r=0}^1 d\theta$$

$$= \int_0^{2\pi} \left(\frac{8}{3} \cos \theta - \frac{4}{3} \sin \theta + \frac{3}{2}\right) d\theta = 3\pi.$$

As for the line integral $\int_C \mathbf{F} \cdot d\mathbf{r}$, since C is the unit circle $x^2 + y^2 = 1$ in the xy-plane, we have

$$\int_C \mathbf{F} \cdot d\mathbf{r} = \int_C -2z\,dx + 3x\,dy + 4y\,dz = \int_C 3x\,dy$$

($z = dz = 0$). Therefore, by Corollary 1 to Green's theorem (see page 911),

$$\int_C \mathbf{F} \cdot d\mathbf{r} = 3(\text{Area of } R) = 3\pi,$$

and therefore $\int_C \mathbf{F} \cdot d\mathbf{r}$ is equal to $\iint_S (\text{curl } \mathbf{F}) \cdot \mathbf{n}\,d\sigma$, in accordance with Stokes' theorem. □

Example 3 Verify Stokes' theorem for the field $\mathbf{F} = \mathbf{i} + x^3 y^2 \mathbf{j} + z\mathbf{k}$, where S is the spheroid $4x^2 + y^2 + z^2 = 4$ ($z \geq 0$), and $\mathbf{n}$ is the upper unit normal to S.

Solution The fact that $\mathbf{n}$ is horizontal along the rim of S, namely the ellipse $4x^2 + y^2 = 4$ in the xy-plane, leads to a slight modification of our partial proof of Stokes' theorem, in which we replace S by S_ε, the graph of

$$4x^2 + y^2 + z^2 = 4 \quad (z \geq \varepsilon > 0),$$

repeat the calculations, and then let $\varepsilon \to 0$, but the validity of the theorem is not affected. Since

$$\text{curl } \mathbf{F} = \begin{vmatrix} \mathbf{i} & \mathbf{j} & \mathbf{k} \\ \dfrac{\partial}{\partial x} & \dfrac{\partial}{\partial y} & \dfrac{\partial}{\partial z} \\ 1 & x^3 y^2 & z \end{vmatrix} = 3x^2 y^2 \mathbf{k},$$

it follows from formula (5) that

$$\iint_S (\text{curl } \mathbf{F}) \cdot \mathbf{n}\,d\sigma = \iint_R 3x^2 y^2\,dx\,dy = 6 \int_{-1}^{1} dx \int_0^{2\sqrt{1-x^2}} x^2 y^2\,dy$$

$$= 6 \int_{-1}^{1} \left[\frac{1}{3} x^2 y^3 \right]_{y=0}^{2\sqrt{1-x^2}} dx = 16 \int_{-1}^{1} x^2 (1 - x^2)^{3/2}\,dx,$$

where R is the elliptical region $4x^2 + y^2 \leq 4$ in the xy-plane. Making the substitution $x = \sin t$, we find that

$$\iint_S (\text{curl } \mathbf{F}) \cdot \mathbf{n}\,d\sigma = 32 \int_0^{\pi/2} \cos^4 t \sin^2 t\,dt. \quad (11)$$

As for the line integral $\int_C \mathbf{F} \cdot d\mathbf{r}$, since C is the ellipse $4x^2 + y^2 = 4$ in the xy-plane with parametric equations $x = \cos t$, $y = 2\sin t$, $z = 0$ ($0 \leq t \leq 2\pi$), we have

$$\int_C \mathbf{F} \cdot d\mathbf{r} = \int_C dx + x^3 y^2\,dy + z\,dz = \int_0^{2\pi} (-\sin t + 8\cos^4 t \sin^2 t)\,dt$$

$$= 8 \int_0^{2\pi} \cos^4 t \sin^2 t\,dt = 32 \int_0^{\pi/2} \cos^4 t \sin^2 t\,dt \quad (11')$$

(justify the last step). Comparing (11) and (11′), we find that Stokes' theorem has been confirmed in this case, without bothering to evaluate the integral $\int_0^{\pi/2} \cos^4 t \sin^2 t\, dt$, which happens to be $\pi/32$, as can be shown with the help of Example 3, page 383. □

Let $\mathbf{F} = F_1 \mathbf{i} + F_2 \mathbf{j} + F_3 \mathbf{k}$ be a continuously differentiable vector field. We already know from Theorem 2, page 892, in which F_1, F_2 and F_3 are written as P, Q and R instead, that if $\mathbf{F}$ is a gradient field, then

$$\frac{\partial F_2}{\partial x} = \frac{\partial F_1}{\partial y}, \qquad \frac{\partial F_3}{\partial y} = \frac{\partial F_2}{\partial z}, \qquad \frac{\partial F_1}{\partial z} = \frac{\partial F_3}{\partial x},$$

which is equivalent to curl $\mathbf{F} = \mathbf{0}$. Conversely, suppose curl $\mathbf{F} = \mathbf{0}$ at every point of a three-dimensional domain D, so that $\mathbf{F}$ is irrotational on D, and suppose D is such that every simple closed path C entirely contained in D is the boundary of a piecewise smooth surface S entirely contained in D; such a domain is said to be *simply connected*.† Then curl $\mathbf{F} = \mathbf{0}$ on S, and hence, by Stokes' theorem,

$$\int_C \mathbf{F} \cdot d\mathbf{r} = \iint_S (\text{curl } \mathbf{F}) \cdot \mathbf{n}\, d\sigma = \iint_S \mathbf{0} \cdot \mathbf{n}\, d\sigma = 0.$$

Therefore $\int_C \mathbf{F} \cdot d\mathbf{r} = 0$ for every simple closed path contained in D. It follows from the remark on page 890 which applies to both two- and three-dimensional fields, that the line integral of $\mathbf{F}$ is path independent on D, or equivalently that $\mathbf{F}$ is a gradient field on D. This is the three-dimensional analogue of Corollary 2 to Green's theorem, page 911.

Physical Meaning of the Curl

To clarify the physical meaning of the curl, imagine that a region in space is filled with a fluid rotating with constant angular speed ω (see p. 685) about an axis l, and choose the origin O at the center of the circle described by a rotating fluid particle P, as in Figure 33. The scalar ω does not tell us the *direction* of rotation, and hence we introduce the *vector* $\boldsymbol{\omega}$ of magnitude ω which lies along l and points in the direction such that the rotation appears counterclockwise to an observer at the tip of $\boldsymbol{\omega}$. In terms of this vector, called the *angular velocity*, the ordinary (translational) velocity of the particle P is given by the cross product

$$\mathbf{v} = \boldsymbol{\omega} \times \mathbf{r}, \tag{12}$$

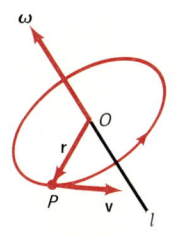

Figure 33

where $\mathbf{r} = x\mathbf{i} + y\mathbf{j} + z\mathbf{k}$ is the position vector of P. In fact, $\mathbf{v}$ is perpendicular to both $\boldsymbol{\omega}$ and $\mathbf{r}$, as is apparent from the figure, and the vectors $\mathbf{v}$, $\boldsymbol{\omega}$ and $\mathbf{r}$ form a right-handed system (to see this, place the initial point of $\mathbf{v}$ at O). Thus $\mathbf{v}$ and $\boldsymbol{\omega} \times \mathbf{r}$ have the same direction. But $\mathbf{v}$ and $\boldsymbol{\omega} \times \mathbf{r}$ also have the same magnitude, since $\boldsymbol{\omega}$ and $\mathbf{r}$ are perpendicular, so that $|\boldsymbol{\omega} \times \mathbf{r}| = \omega |\mathbf{r}|$, and

---

† This definition is essentially the same as that of a simply connected two-dimensional domain, given on page 887 (think of the plane region consisting of a simple closed curve C and its interior as a surface with C as its boundary). However, a simply connected domain in space can have holes; for instance, the domain between two concentric spheres is simply connected. As an example of a three-dimensional domain which is *multiply connected* (that is, not simply connected), consider the torus.

we already know from page 685 that $|\mathbf{v}| = \omega|\mathbf{r}|$. This proves (12), and the formula remains valid if O is chosen to be any other point of the axis l (why?).

We now calculate the curl of the velocity field (12). Let

$$\boldsymbol{\omega} = \omega_1 \mathbf{i} + \omega_2 \mathbf{j} + \omega_3 \mathbf{k}.$$

Then

$$\mathbf{v} = \boldsymbol{\omega} \times \mathbf{r} = \begin{vmatrix} \mathbf{i} & \mathbf{j} & \mathbf{k} \\ \omega_1 & \omega_2 & \omega_3 \\ x & y & z \end{vmatrix}$$

$$= (\omega_2 z - \omega_3 y)\mathbf{i} + (\omega_3 x - \omega_1 z)\mathbf{j} + (\omega_1 y - \omega_2 x)\mathbf{k},$$

so that

$$\operatorname{curl} \mathbf{v} = \begin{vmatrix} \mathbf{i} & \mathbf{j} & \mathbf{k} \\ \dfrac{\partial}{\partial x} & \dfrac{\partial}{\partial y} & \dfrac{\partial}{\partial z} \\ \omega_2 z - \omega_3 y & \omega_3 x - \omega_1 z & \omega_1 y - \omega_2 x \end{vmatrix} = 2\omega_1 \mathbf{i} + 2\omega_2 \mathbf{j} + 2\omega_3 \mathbf{k} = 2\boldsymbol{\omega}.$$

Thus the curl of the velocity field of a rotating fluid is, apart from a factor of 2, just the angular velocity of the fluid. This shows that curl **v** is a measure of the "rotational tendency" of a velocity field **v**, and explains why a field with zero curl is said to be "irrotational."

There is another interesting way to interpret the curl. Let S_ε be a disk of radius ε immersed in a steady-state velocity field $\mathbf{v} = \mathbf{v}(\mathbf{r})$, and let the center of S_ε be at the point P with radius vector $\mathbf{a}$. Then, by Stokes' theorem,

$$\iint_{S_\varepsilon} (\operatorname{curl} \mathbf{v}) \cdot \mathbf{n}\, d\sigma = \int_{C_\varepsilon} \mathbf{v} \cdot d\mathbf{r}, \qquad (13)$$

where S_ε has the unit normal **n** and C_ε is the circular boundary of S_ε, traversed in the positive direction relative to **n** (see Figure 34). Let A_ε be the area of S_ε. Then (13) is equivalent to

$$\frac{1}{A_\varepsilon} \iint_{S_\varepsilon} (\operatorname{curl} \mathbf{v}) \cdot \mathbf{n}\, d\sigma = \frac{1}{A_\varepsilon} \int_{C_\varepsilon} \mathbf{v} \cdot d\mathbf{r},$$

where $\int_{C_\varepsilon} \mathbf{v} \cdot d\mathbf{r}$ is the circulation of **v** around C_ε (recall page 890), and the expression on the left is the mean value or average of $(\operatorname{curl} \mathbf{v}) \cdot \mathbf{n}$ over the disk S_ε. But as $\varepsilon \to 0$, this average approaches $[\operatorname{curl} \mathbf{v}(\mathbf{a})] \cdot \mathbf{n}$, the value of the **n**-component of the curl of **v** at P. Therefore

$$[\operatorname{curl} \mathbf{v}(\mathbf{a})] \cdot \mathbf{n} = \lim_{\varepsilon \to 0} \frac{1}{A_\varepsilon} \int_{C_\varepsilon} \mathbf{v} \cdot d\mathbf{r},$$

which gives a *coordinate-free* interpretation of the **n**-component of the curl of **v** at P as the *circulation of* **v** *per unit area* in the plane through P

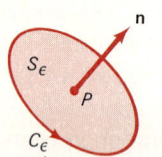

Figure 34

936   Chapter 15   Line and Surface Integrals

orthogonal to **n**. This formula should be compared with formula (14), page 928, for the divergence of **v** at P.

Problems

Find the curl of the given vector field F.
1. $\mathbf{F} = x\mathbf{i} + y\mathbf{j} + z\mathbf{k}$
2. $\mathbf{F} = -y\mathbf{i} + x\mathbf{j}$
3. $\mathbf{F} = (y - z)\mathbf{i} + (z - x)\mathbf{j} + (x - y)\mathbf{k}$
4. $\mathbf{F} = (x - y)\mathbf{i} + (y - z)\mathbf{j} + (x - z)\mathbf{k}$
5. $\mathbf{F} = (y\mathbf{i} - x\mathbf{j})/(x^2 + y^2)$
6. $\mathbf{F} = x^2\mathbf{i} + y^2\mathbf{j} + z^2\mathbf{k}$
7. $\mathbf{F} = x\mathbf{i} + xy\mathbf{j} + xyz\mathbf{k}$
8. $\mathbf{F} = z^3\mathbf{i} + y^3\mathbf{j} + x^3\mathbf{k}$
9. $\mathbf{F} = ye^z\mathbf{i} + xe^z\mathbf{j} + xy\mathbf{k}$
10. $\mathbf{F} = (\sin y)\mathbf{i} + (\cos x)\mathbf{j} + (\sin xy)\mathbf{k}$

11. Assuming that **F** has continuous second partial derivatives, calculate div (curl **F**).

12. Let f be a differentiable scalar field and **F** a differentiable vector field. Show that

$$\text{curl}(f\mathbf{F}) = \text{grad } f \times \mathbf{F} + f \text{ curl } \mathbf{F} = \nabla f \times \mathbf{F} + f\nabla \times \mathbf{F}$$

(note the analogy with Example 2, page 923).

Let **F** and **G** be differentiable vector functions. Show that
13. $\text{div}(\mathbf{F} \times \mathbf{G}) = \mathbf{G} \cdot \text{curl } \mathbf{F} - \mathbf{F} \cdot \text{curl } \mathbf{G}$
14. $\text{curl}(a\mathbf{F} + b\mathbf{G}) = a \text{ curl } \mathbf{F} + b \text{ curl } \mathbf{G}$ (a and b scalar constants)

Verify Stokes' theorem for the given vector field **F** and surface S. In each case choose **n** to be the upper unit normal.
15. $\mathbf{F} = z\mathbf{i} + x\mathbf{j} - y\mathbf{k}$, S is the graph of $z = x$ over the rectangle $0 \le x \le 1$, $0 \le y \le 2$
16. $\mathbf{F} = (x^2 + x)\mathbf{i} + 3xy\mathbf{j} + (2xz + z^2)\mathbf{k}$, S is the hemisphere $x^2 + y^2 + z^2 = 4$ ($z \ge 0$)
17. $\mathbf{F} = e^x\mathbf{i} + e^y\mathbf{j} + e^z\mathbf{k}$, S is the same parabolic cap as in Example 2
18. $\mathbf{F} = -y\mathbf{i} + x^2\mathbf{j} + z\mathbf{k}$, S is the elliptical region in which the plane $x + y + z = 2$ intersects the cylinder $x^2 + y^2 = 1$

Use Stokes' theorem to evaluate the integral $\int_C \mathbf{F} \cdot d\mathbf{r}$ if $\mathbf{F} = -y\mathbf{i} + x\mathbf{j} - z\mathbf{k}$ and C is the given curve.
19. C is the triangle with vertices $(0, 0, 3)$, $(1, 0, 3)$ and $(2, 1, 3)$ traversed in that order
20. C is the circle in which the plane $y = z$ intersects the sphere $x^2 + y^2 + z^2 = 1$, traversed in the counterclockwise direction as seen from the positive z-axis

Let f and g be scalar fields in space, with continuous second partial derivatives, and let S, C and **n** be the same as in Stokes' theorem. Show that

21. $\int_C f \nabla g \cdot d\mathbf{r} = \iint_S (\nabla f \times \nabla g) \cdot \mathbf{n}\, d\sigma$

22. $\int_C (f\nabla g + g\nabla f) \cdot d\mathbf{r} = 0$

23. Let C be a piecewise smooth simple closed curve which is the common boundary of two piecewise smooth surfaces S_1 and S_2 with unit normals $\mathbf{n}_1$ and $\mathbf{n}_2$, and let

$$I_1 = \iint_{S_1} (\text{curl } \mathbf{F}) \cdot \mathbf{n}_1\, d\sigma, \quad I_2 = \iint_{S_2} (\text{curl } \mathbf{F}) \cdot \mathbf{n}_2\, d\sigma$$

where **F** is a continuously differentiable vector field on both S_1 and S_2. Show that $I_1 = I_2$ if the direction of traversing C induced by both $\mathbf{n}_1$ and $\mathbf{n}_2$ is the same, but otherwise $I_1 = -I_2$.

24. Suppose the surface S in Stokes' theorem is *closed*, so that there is no boundary curve C. Show that in this case

$$\iint_S (\text{curl } \mathbf{F}) \cdot \mathbf{n}\, d\sigma = 0.$$

25. Let $\mathbf{F} = -y^2\mathbf{i} - xz\mathbf{j} + x^2\mathbf{k}$, let S be the cubical surface shown in Figure 15, page 903, but with the front face S_1 removed, and let **n** be the outer unit normal to S. Find the flux of curl **F** across S by direct calculation, by evaluating the circulation of **F** around the boundary of S, and also by using Problem 23 to express the flux as an integral over the missing face S_1.

26. Assuming that F has continuous second partial derivatives, show that

$$\text{curl}(\text{curl } \mathbf{F}) = -\nabla^2 \mathbf{F} + \text{grad}(\text{div } \mathbf{F}) \quad \text{(i)}$$

(interpret the expression $\nabla^2 \mathbf{F} = \nabla^2(F_1\mathbf{i} + F_2\mathbf{j} + F_3\mathbf{k})$ as $\nabla^2 F_1\mathbf{i} + \nabla^2 F_2\mathbf{j} + \nabla^2 F_3\mathbf{k}$).

27. Calculate curl (curl **F**) for $\mathbf{F} = x^2y\mathbf{i} + 2xz\mathbf{j} - 3yz\mathbf{k}$ both directly and by using formula (i).

28. Let T, S and **n** be the same as in the divergence theorem (see page 924), and let **G** be a continuously differentiable vector field on T. Show that

$$\iint_S \mathbf{n} \times \mathbf{G}\, d\sigma = \iiint_T \text{curl } \mathbf{G}\, dV.$$

Hint. Set $\mathbf{F} = \mathbf{G} \times \mathbf{c}$ in the divergence theorem, where **c** is any constant vector. Interpret surface and volume integrals of *vector* fields componentwise; for example

$$\iint_S \mathbf{F}\, d\sigma = \iint_S (F_1\mathbf{i} + F_2\mathbf{j} + F_3\mathbf{k})\, d\sigma$$

$$= \left(\iint_S F_1\, d\sigma\right)\mathbf{i} + \left(\iint_S F_2\, d\sigma\right)\mathbf{j} + \left(\iint_S F_3\, d\sigma\right)\mathbf{k}.$$

Key Terms and Topics

Line integrals with respect to the arc length
Line integrals with respect to the coordinates
Piecewise smooth curves, simply and multiply connected domains
Path independent line integrals, gradient fields
Potential energy, conservation of energy, conservative fields
Characterization of gradient fields
Surface area and surface integrals
Unit normals to a surface, piecewise smooth surfaces
Flux of a vector field across a surface
Green's theorem and its uses
Coordinate transformations and Jacobians
Change of variables in multiple integrals
Vector forms of Green's theorem
Divergence and curl of a vector field
The divergence theorem and Stokes' theorem

Supplementary Problems

Evaluate the given line integral of the type $\int_C f(\mathbf{r})\, ds$.

1. $\int_C xy\, ds$, where C is the elliptical arc $x = 3\cos t$, $y = 2\sin t$ $(0 \le t \le \pi/2)$

2. $\int_C \sqrt{x^2 + y^2}\, ds$, where C is the circle $x^2 + y^2 = 2y$ traversed once in the counterclockwise direction

3. $\int_C x\sqrt{x^2 - y^2}\, ds$, where C is the right loop of the lemniscate $r^2 = 4\cos 2\theta$ $(-\pi/4 \le \theta \le \pi/4)$

4. $\int_C \dfrac{ds}{x^2 + y^2 + z^2}$, where C is the helical arc $x = \cos t$, $y = \sin t$, $z = t$ $(0 \le t \le 2\pi)$

5. A particle moving along the circle $x^2 + y^2 = R^2$ of radius R in the counterclockwise direction is acted on by a constant force of magnitude F directed along the positive x-axis. Find the work done by the force on the particle as it describes the arc of the circle in the first and second quadrants. As it makes one complete revolution around the circle, starting at any point.

6. The magnitude of a force is inversely proportional to the distance between its point of application and the xy-plane, with constant of proportionality a. Suppose the force is always directed away from the origin. Find the work done by the force on a particle moving from the point $A = (1, 2, 2)$ to the point $B = (2, 4, 4)$ along the line segment AB.

7. Evaluate $\int_C \mathbf{F} \cdot d\mathbf{r}$, where $\mathbf{F} = x^2\mathbf{i} + y^2\mathbf{j} + z^2\mathbf{k}$ and C is the line segment joining the origin to the point (a, a, a).

8. Evaluate the line integral

$$\int_C \frac{x^2\, dy - y^2\, dx}{x^{5/3} + y^{5/3}},$$

where C is the astroid $x = \cos^3 t$, $y = \sin^3 t$ $(0 \le t \le 2\pi)$, traversed once in the counterclockwise direction.

9. Evaluate the line integral $\int_C y^2\, dx + z^2\, dy + x^2\, dz$, where C is the curve $x = t$, $y = t^2$, $z = t^3$ $(0 \le t \le 1)$.

10. Let $f(x)$ be continuously differentiable on $(-\infty, \infty)$. Show that the vector fields $\mathbf{F}(x, y) = f(x^2 + y^2)(x\mathbf{i} + y\mathbf{j})$ and $\mathbf{G}(x, y) = f(xy)(y\mathbf{i} + x\mathbf{j})$ are both conservative.

Is the vector field $\mathbf{F} = P\mathbf{i} + Q\mathbf{j}$ with the given components $P = P(x, y)$ and $Q = Q(x, y)$ a gradient field? If so, find a function U and a domain D on which $\mathbf{F} = \operatorname{grad} U$.

11. $P = (x^2 + xy + y^2)x$, $Q = (x^2 - xy + y^2)y$

12. $P = \cosh x + \cosh y$, $Q = x \sinh y$

13. $P = e^{x+y} - \sin(x - y)$, $Q = e^{x+y} + \sin(x - y)$

14. $P = \dfrac{1}{x} + \dfrac{1}{y}$, $Q = \dfrac{1}{y} - \dfrac{x}{y^2}$

Evaluate the given line integral, after verifying that it is path independent.

15. $\displaystyle\int_{(0,0)}^{(1,-1)} \arctan x\, dx + \arctan y\, dy$

16. $\displaystyle\int_{(-1, \pi/3)}^{(\pi/2, 1)} (\sin xy)(y\, dx + x\, dy)$

17. $\displaystyle\int_{(-1,1,-1)}^{(1,-1,\sqrt{3})} \frac{yz\, dx + xz\, dy + xy\, dz}{1 + x^2 y^2 z^2}$

18. $\displaystyle\int_{(-1,2,-2)}^{(3,-4,12)} \frac{x\,dx + y\,dy + z\,dz}{\sqrt{x^2 + y^2 + z^2}}$

Find the area of the specified surface S.

19. S is the part of the plane $3x + 4y + 12z = 1$ inside the cylinder $y^2 + z^2 = 1$

20. S is the part of the cone $x^2 - y^2 - z^2 = 0$ cut off by the hyperbolic cylinder $x^2 - y^2 = 1$ and the planes $y = \pm\sqrt{2}$

21. S is the part of the cone $x^2 + y^2 - z^2 = 0$ cut off by the cylinder $x^2 + z^2 = a^2$ of radius a

22. As shown on page 470, the area A of the surface of revolution S generated by revolving the graph of a continuous nonnegative function $y = f(x)$ $(a \le x \le b)$ about the x-axis is given by the formula

$$A = 2\pi \int_a^b y\sqrt{1 + \left(\frac{dy}{dx}\right)^2}\,dx.$$

Give an alternative derivation of this formula, based on the use of a surface integral.

23. Evaluate the surface integral

$$\iint_S \frac{d\sigma}{\sqrt{x^2 + y^2}},$$

where S is the part of the hyperbolic paraboloid $z = xy$ cut off by the cylinder $x^2 + y^2 = 1$.

24. Let S be the lateral surface of a conical frustum of unit density, upper radius r_1, base radius r_2 and height h (see Problem 11, page 440). Find the moment of inertia of S about its axis of symmetry.

25. The velocity field of a fluid is $\mathbf{v} = y\mathbf{i} + \mathbf{j} + z\mathbf{k}$ in meters per second. How many cubic meters per second of fluid are crossing the triangular surface S with vertices $(1, 0, 0)$, $(0, 2, 0)$ and $(0, 0, 3)$ in the upward direction?

26. Find the flux of the field $\mathbf{F} = x\mathbf{i} + y\mathbf{j} + z\mathbf{k}$ out of the closed surface S bounded by the cone $z = \sqrt{x^2 + y^2}$ and the plane $z = 1$.

27. Let C be the square contour with vertices $(\pm 1, \pm 1)$, traversed once in the counterclockwise direction. Use Green's theorem to evaluate the line integral

$$\int_C (ye^{xy} - y^2 \sin x)\,dx + (xe^{xy} + 2y\cos x)\,dy.$$

28. Let C be any piecewise smooth simple closed curve. What is the geometric meaning of the line integral $\int_C (2xy - y)\,dx + x^2\,dy$?

29. Let the functions $f = f(x, y)$ and $g = g(x, y)$ be continuously differentiable on a region R consisting of a piecewise smooth simple closed curve C and its interior. Show that

$$\int_C fg\,dx + fg\,dy = \iint_R \left[g\left(\frac{\partial f}{\partial x} - \frac{\partial f}{\partial y}\right) + f\left(\frac{\partial g}{\partial x} - \frac{\partial g}{\partial y}\right)\right]dA.$$

30. Use Green's theorem to show that the area enclosed by the simple closed polygonal path $P_1P_2P_3 \ldots P_nP_1$ with vertices

$$P_i = (x_i, y_i) \quad (i = 1, 2, \ldots, n)$$

is equal to one half the absolute value of the sum

$$(x_1y_2 - x_2y_1) + (x_2y_3 - x_3y_2) + \cdots + (x_ny_1 - x_1y_n).$$

In particular, verify Problem 16, page 744, on the area of the triangle with vertices (x_1, y_1), (x_2, y_2) and (x_3, y_3).

Hint. First show that $\int_L x\,dy - y\,dx = x_1y_2 - x_2y_1$ if L is the line segment from (x_1, y_1) to (x_2, y_2).

31. Use Problem 30 to find the area enclosed by the quadrilateral with vertices $(1, -1)$, $(3, -2)$, $(5, 1)$ and $(2, 6)$.

32. Evaluate the double integral

$$\iint_R \sqrt{\sqrt{x} + \sqrt{y}}\,dx\,dy,$$

where R is the region in the first quadrant bounded by the coordinate axes and the parabolic arc $\sqrt{x} + \sqrt{y} = 1$.

Hint. Make the transformation $x = u\cos^4 v$, $y = u\sin^4 v$.

Find the divergence of the given three-dimensional vector field, where $\mathbf{a}$ and $\mathbf{b}$ are constant vectors and $\mathbf{r} = x\mathbf{i} + y\mathbf{j} + z\mathbf{k}$.

33. $\mathbf{a} \times \mathbf{r}$
34. $(\mathbf{a} \cdot \mathbf{r})\mathbf{b}$
35. $(\mathbf{a} \cdot \mathbf{r})\mathbf{r}$
36. $(\mathbf{a} \times \mathbf{r}) \times \mathbf{b}$
37. $(\mathbf{a} \times \mathbf{r}) \times \mathbf{r}$
38. $(\mathbf{a} \cdot \mathbf{r})(\mathbf{b} \times \mathbf{r})$

39. For what choice of the constant c is the field $\mathbf{F} = (3x - y)\mathbf{i} + (2y - x^2)\mathbf{j} + (xy - cz)\mathbf{k}$ solenoidal?

40. Let T, S and $\mathbf{n}$ be the same as in the divergence theorem, and let A be the area of the surface S. Show that

$$\iiint_T \operatorname{div}\mathbf{n}\,dV = A.$$

41. Show that the divergence theorem remains true for the solid region T between two closed surfaces S' and S'' if the surface integral $\iint_S \mathbf{F} \cdot \mathbf{n}\,d\sigma$ is replaced by a sum of surface integrals over S' and S'' with the unit normal $\mathbf{n}$ always pointing away from T.

42. Let $\mathbf{F} = (q/r^2)\mathbf{u}$ be an inverse square law field, due to a source of attraction or repulsion (gravitational or electrostatic, say) at the origin; here q is a positive or negative constant, $\mathbf{r} = x\mathbf{i} + y\mathbf{j} + z\mathbf{k}$, $r = |\mathbf{r}|$, and $\mathbf{u} = \mathbf{r}/r$. With the help of the preceding problem, show that if S is any piecewise smooth closed surface surrounding the origin, with outer unit normal $\mathbf{n}$, then

$$\iint_S \mathbf{F} \cdot \mathbf{n}\,d\sigma = 4\pi q,$$

a result known as *Gauss' law*.

Hint. Use Problem 41 after showing that $\operatorname{div}\mathbf{F} = 0$ if $r \ne 0$.

Supplementary Problems **939**

43–48. Find the curl of each vector field in Problems 33–38. *Hint.* In Problems 46–48 use the "*bac-cab*" rule (see Problem 48, page 715).

49. Let f be a scalar field with continuous first and second partial derivatives. Show that the vector field $f\nabla f$ is irrotational.

50. Let T, S and $\mathbf{n}$ be the same as in the divergence theorem (see page 924), and let f be a continuously differentiable scalar field on T. Show that

$$\iint_S f\mathbf{n}\, d\sigma = \iiint_T \operatorname{grad} f\, dV.$$

Hint. Set $\mathbf{F} = f\mathbf{c}$ in the divergence theorem, where $\mathbf{c}$ is any constant vector.

51. Let S be a piecewise smooth closed surface with outer unit normal $\mathbf{n}$. Show that $\iint_S \mathbf{n}\, d\sigma = \mathbf{0}$.

52. Electromagnetic phenomena are governed by a system of first order partial differential equations, called *Maxwell's equations*, involving the *electric field* $\mathbf{E} = \mathbf{E}(x, y, z, t)$ and the *magnetic field* $\mathbf{H} = \mathbf{H}(x, y, z, t)$, which both depend on the time t as well as on the spatial coordinates x, y and z. In empty space Maxwell's equations take the form

$$\operatorname{div} \mathbf{E} = 0, \qquad \operatorname{curl} \mathbf{E} = -\frac{1}{c}\frac{\partial \mathbf{H}}{\partial t},$$

$$\operatorname{div} \mathbf{H} = 0, \qquad \operatorname{curl} \mathbf{H} = \frac{1}{c}\frac{\partial \mathbf{E}}{\partial t},$$

where c is the velocity of light. Show that each component of $\mathbf{E}$ and $\mathbf{H}$ satisfies the *wave equation*

$$\nabla^2 u = \frac{\partial^2 u}{\partial x^2} + \frac{\partial^2 u}{\partial y^2} + \frac{\partial^2 u}{\partial z^2} = \frac{1}{c^2}\frac{\partial^2 u}{\partial t^2},$$

from which Maxwell deduced the existence of electromagnetic waves traveling with the velocity of light. Radio waves, X-rays and light itself are all forms of electromagnetic radiation.

Hint. With the help of formula (i), page 937, show that Maxwell's equations imply

$$\nabla^2 \mathbf{E} = \frac{1}{c^2}\frac{\partial^2 \mathbf{E}}{\partial t^2}, \qquad \nabla^2 \mathbf{H} = \frac{1}{c^2}\frac{\partial^2 \mathbf{H}}{\partial t^2}.$$

16 Elementary Differential Equations

Differential equations have already put in an appearance in Sections 2.6 and 4.6, and differential equations of a special type, known as *separable* equations, were used extensively in Chapter 6 to solve a variety of applied problems. In this brief chapter we pursue the subject of differential equations a bit further. The subject is vast, and in the space that remains we can do no more than touch upon a few elementary topics of particular importance, confining ourselves to *ordinary* differential equations, that is, equations involving one or more derivatives of a function $y = y(x)$ of a single independent variable x. Of course, there are also *partial* differential equations, containing partial derivatives of a function $y = y(x_1, \ldots, x_n)$ of several independent variables $x_1, \ldots, x_n$, but the systematic investigation of such equations lies well beyond the scope of this course.

By the *order* of a differential equation we mean the order of the highest derivative of the unknown function y appearing in the equation. A differential equation of order n is said to be *linear* if it can be written in the form

$$a_0(x)y + a_1(x)y' + \cdots + a_n(x)y^{(n)} = b(x),$$

where $y', y'', \ldots, y^{(n)}$ are the first n derivatives of y, and $a_0(x), a_1(x), \ldots, a_n(x)$, $b(x)$ are functions of the independent variable x only (some or all of which may be constant), but otherwise the equation is said to be *nonlinear*. Thus

$$\frac{d^2y}{dx^2} + x^2 \frac{dy}{dx} - xy = e^x$$

is a linear differential equation of the second order, but

$$\frac{dy}{dx} + y^2 = 0, \qquad y\frac{dy}{dx} + \sin x = 0, \qquad x\frac{dy}{dx} + \sin y = 0$$

are all nonlinear differential equations of the first order. Most of the equations in Section 16.1 are nonlinear, but the rest of the chapter is primarily concerned with first and second order linear equations and their applications.

16.1 Exact Equations and Integrating Factors

Let $U = U(x, y)$ be a continuously differentiable function of two variables, and consider the equation

$$U(x, y) = C, \tag{1}$$

where C is an arbitrary constant. Assuming that (1) defines y implicitly as a differentiable function of x, we find with the help of the chain rule that

$$\frac{dU}{dx} = \frac{\partial U}{\partial x}\frac{dx}{dx} + \frac{\partial U}{\partial y}\frac{dy}{dx} = 0.$$

It follows that

$$P + Q\frac{dy}{dx} = 0, \tag{2}$$

where the functions $P = P(x, y)$ and $Q = Q(x, y)$ are the partial derivatives of U:

$$P = \frac{\partial U}{\partial x}, \quad Q = \frac{\partial U}{\partial y}. \tag{3}$$

Thus the *general solution* of the first order differential equation (2) is given by (1). This means that for every choice of the constant C, equation (1) defines an implicit function $y = y(x)$ which satisfies equation (2). There is also the possibility that (2) has *singular solutions*, that is, solutions which do not correspond to any choice of the constant C; for instance, a singular solution occurs in Example 4 below. Note that in terms of differentials, equation (2) takes the form

$$P\, dx + Q\, dy = 0. \tag{2'}$$

Exact Equations

Conversely, given a differential equation of the form (2), where $P = P(x, y)$ and $Q = Q(x, y)$ are continuously differentiable functions, suppose there exists a function $U = U(x, y)$ satisfying the conditions (3). Then (2) is said to be an *exact differential equation*, and by the same token

$$dU = \frac{\partial U}{\partial x}\, dx + \frac{\partial U}{\partial y}\, dy = P\, dx + Q\, dy$$

is called an *exact differential* (this last concept was anticipated in Problem 29, page 896). We already know from Chapter 15 that $P\, dx + Q\, dy$ is an exact differential on a simply connected domain D, or equivalently that $\mathbf{F} = P\mathbf{i} + Q\mathbf{j}$ is a gradient field on D, if and only if the condition

$$\frac{\partial Q}{\partial x} = \frac{\partial P}{\partial y} \tag{4}$$

is satisfied at every point of D. In fact, (3) implies (4), since then

$$\frac{\partial Q}{\partial x} = \frac{\partial^2 U}{\partial x\, \partial y} = \frac{\partial^2 U}{\partial y\, \partial x} = \frac{\partial P}{\partial y},$$

while (4) implies (3) as shown for a rectangular domain in Theorem 3, page 893, and for a general simply connected domain in Corollary 2, page 911.

Example 1 In Section 6.6 we introduced *separable* differential equations, namely equations of the form

$$y' = \frac{dy}{dx} = \frac{f(x)}{g(y)} \quad (g(y) \neq 0), \tag{5}$$

involving an unknown function $y = y(x)$ and two given functions $f(x)$ and $g(y)$. *Every separable equation is exact.* To see this, suppose $f(x)$ and $g(y)$ are continuously differentiable, and set

$$P = f(x), \quad Q = -g(y).$$

Then (5) can be written in the form (2), and the exactness condition is automatically satisfied, since $\partial Q/\partial x = 0 = \partial P/\partial y$.

Example 2 Consider the differential equation

$$(2x + y) + (x + y)\frac{dy}{dx} = 0. \tag{6}$$

Although not separable (why not?), this equation is exact. In fact, here $P = 2x + y$ and $Q = x + y$, so that $\partial Q/\partial x = 1 = \partial P/\partial y$. Equation (6) can be solved in various ways. One way is to write (6) in the equivalent form

$$(y\,dx + x\,dy) + 2x\,dx + y\,dy = 0, \tag{6'}$$

which will immediately be recognized as

$$d(xy) + d(x^2) + d(\tfrac{1}{2}y^2) = 0.$$

Thus (6') becomes

$$d(xy + x^2 + \tfrac{1}{2}y^2) = 0,$$

which leads at once to the general solution

$$xy + x^2 + \tfrac{1}{2}y^2 = C, \tag{7}$$

where C is an arbitrary constant. Thus we have easily solved equation (6) *by inspection*.

Another way of solving (6) is to write

$$P = 2x + y = \frac{\partial U}{\partial x}, \quad Q = x + y = \frac{\partial U}{\partial y}, \tag{8}$$

where the fact that $\partial Q/\partial x = \partial P/\partial y$ tells us that a function U satisfying this system of equations must exist. Integrating the first of the equations (8) with respect to x, with y held fixed, we get

$$U = x^2 + xy + f(y), \tag{9}$$

where $f(y)$ is a differentiable function of the variable y only. It follows that

$$\frac{\partial U}{\partial y} = x + f'(y)$$

in terms of the derivative $f'(y)$. Substituting this expression for $\partial U/\partial y$ into the second of the equations (8), we get $x + y = x + f'(y)$ or $f'(y) = y$, so that $f(y) = \tfrac{1}{2}y^2$ to within an additive constant. For this choice of $f(y)$, equation (9) becomes

$$U = x^2 + xy + \tfrac{1}{2}y^2,$$

and equating U to C, we obtain formula (7) again.

A third way of solving (6) is to use the formula

$$U = \int_{x_0}^{x} P(t, y_0)\,dt + \int_{y_0}^{y} Q(x, t)\,dt,$$

established on page 894. Choosing $x_0 = y_0 = 0$, we find that

$$U = \int_0^x 2t\,dt + \int_0^y (x+t)\,dt = \left[t^2\right]_0^x + \left[xt + \frac{1}{2}t^2\right]_{t=0}^{y} = x^2 + xy + \frac{1}{2}y^2$$

to within an arbitrary additive constant, which once again leads to formula (7) for the general solution.

Whenever you solve a differential equation, you should always check the answer by showing that direct differentiation of the alleged solution gives back the original differential equation. For instance, (implicit) differentiation of the general solution (7) in the above example leads at once to the differential equation (6).

Example 3 Find the particular solution of the exact equation (6) satisfying the initial condition $y(0) = 1$.

Solution Setting $x = 0$ and $y = 1$ in formula (7), we immediately get $C = \frac{1}{2}$, and then (7) becomes $xy + x^2 + \frac{1}{2}y^2 = \frac{1}{2}$, or equivalently

$$y^2 + 2xy + (2x^2 - 1) = 0. \tag{10}$$

It is easy to solve (10) explicitly for y. In fact, by the formula for the solution of a quadratic equation,

$$y = \frac{-2x \pm \sqrt{4x^2 - 4(2x^2 - 1)}}{2} = -x + \sqrt{1 - x^2},$$

since the plus sign must be chosen if y is to satisfy the initial condition $y(0) = 1$. Hence the desired particular solution is

$$y = -x + \sqrt{1 - x^2} \qquad (-1 < x < 1), \tag{11}$$

where the fact that (11) satisfies the original equation (6) should be checked by direct differentiation. The condition $-1 < x < 1$ guarantees both the existence and the differentiability of y. $\square$

The graph of a solution of a differential equation is called an *integral curve*. In Figure 1 we show not only the integral curve (11) of equation (6), but also, with the help of Problem 13, the integral curves corresponding to the particular solutions which satisfy the initial conditions

$$y(0) = y_0 \qquad (y_0 = -1, 2, -2).$$

Notice how the particular solutions satisfying the initial conditions $y(0) = 2$ and $y(0) = -2$ are defined on a larger interval $(-2 < x < 2)$ than those satisfying the conditions $y(0) = 1$ and $y(0) = -1$.

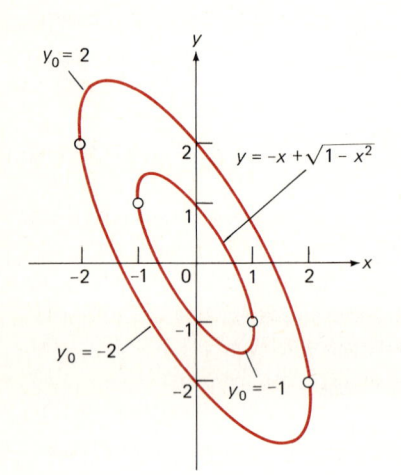

Figure 1

Integrating Factors

It is often feasible to convert a *nonexact* differential equation of the form

$$P + Q\frac{dy}{dx} = 0 \tag{12}$$

944   Chapter 16   Elementary Differential Equations

into an *exact* equation

$$\mu\left(P + Q\frac{dy}{dx}\right) = 0 \tag{12'}$$

by multiplying it by a suitable function $\mu = \mu(x, y)$, called an *integrating factor*. In fact, if μ is never zero or if μ is zero only at points (x, y) where P and Q are both zero, then every solution of (12') will also be a solution of the original equation (12).

Example 4 Solve the differential equation

$$y + xy^2 - x\frac{dy}{dx} = 0. \tag{13}$$

or equivalently

$$(y + xy^2)\,dx - x\,dy = 0. \tag{13'}$$

Solution Here $P = y + xy^2$, $Q = -x$, $\partial P/\partial y = 1 + 2xy$ and $\partial Q/\partial x = -1$, so that $\partial Q/\partial x \neq \partial P/\partial y$. Therefore (13) is not an exact equation. However, a shrewd guess shows that $\mu = 1/y^2$ is an integrating factor for (13). In fact, multiplying (13') by $1/y^2$, we get

$$\left(\frac{1}{y} + x\right)dx - \frac{x}{y^2}\,dy = \frac{y\,dx - x\,dy}{y^2} + x\,dx = d\left(\frac{x}{y}\right) + d\left(\frac{1}{2}x^2\right),$$

by inspection. The general solution of this differential equation is clearly

$$\frac{x}{y} + \frac{1}{2}x^2 = C,$$

where C is an arbitrary constant. Solving for y, we find that

$$y = \frac{2x}{2C - x^2}, \tag{14}$$

which, as you can easily check, satisfies the original nonexact equation (13). Notice that in dividing by y^2, we ran the risk of losing the solution $y \equiv 0$, and indeed it was lost, since $y \equiv 0$ is clearly a solution of (13). This is a *singular* solution, since it cannot be obtained from the "general" solution (14) for any choice of the constant C. ☐

Equation (13) is nonlinear (why?). It can be shown that a *linear* differential equation has no singular solutions. Hence this complication will not be encountered in the remaining sections.

Problems

First verify that the given differential equation of the form $P\,dx + Q\,dy = 0$ is exact, and then find its general solution.

1. $(x + y)\,dx + (x - y)\,dy = 0$
2. $(y^2 - x^2)\,dx + 2xy\,dy = 0$
3. $\left(4 - \frac{y^2}{x^2}\right)dx + \frac{2y}{x}\,dy = 0$
4. $\cos x \sin y\,dx + \sin x \cos y\,dy = 0$
5. $y^2 \cos xy\,dx + (\sin xy + xy \cos xy)\,dy = 0$
6. $y \cosh y\,dx + (x \cosh y + xy \sinh y)\,dy = 0$
7. $[(x + y)e^x - e^y]\,dx + [e^x - (x + y)e^y]\,dy = 0$
8. $e^x y\,dx + (e^x - 2)\,dy = 0$
9. $(2xye^{x^2} - \frac{1}{2}y^2)\,dx + (e^{x^2} - xy)\,dy = 0$
10. $(1 + e^{x/y})\,dx + e^{x/y}\left(1 - \frac{x}{y}\right)dy = 0$

11. $\left(3x^2 \tan y - \dfrac{2y^3}{x^3}\right) dx + \left(x^3 \sec^2 y + 2y + \dfrac{3y^2}{x^2}\right) dy = 0$

12. $\left(\dfrac{xy}{\sqrt{x^2+1}} + 3x^2 y - \dfrac{y}{x}\right) dx + (\sqrt{x^2+1} + x^3 - \ln|x|) dy = 0$

13. Show that the particular solution of equation (6) satisfying the initial condition $y(0) = y_0 \neq 0$ is

$$y = \begin{cases} -x + \sqrt{y_0^2 - x^2} & \text{if } y_0 > 0, \\ -x - \sqrt{y_0^2 - x^2} & \text{if } y_0 < 0. \end{cases}$$

Is there a particular solution satisfying the initial condition $y(0) = 0$? Show that every integral curve of (6) is the arc of an ellipse with its major axis along the line $x + y = 0$, as suggested by Figure 1.

14. Show that if $\mu = \mu(x, y)$ is an integrating factor of the nonexact ordinary differential equation $P\, dx + Q\, dy = 0$, then μ satisfies the partial differential equation

$$P \dfrac{\partial \mu}{\partial y} - Q \dfrac{\partial \mu}{\partial x} = \mu \left(\dfrac{\partial Q}{\partial x} - \dfrac{\partial P}{\partial y} \right), \qquad \text{(i)}$$

which reduces to

$$\dfrac{1}{\mu} \dfrac{d\mu}{dx} = \dfrac{1}{Q}\left(\dfrac{\partial P}{\partial y} - \dfrac{\partial Q}{\partial x} \right) \qquad \text{(ii)}$$

if μ is a function of x only, and to

$$\dfrac{1}{\mu} \dfrac{d\mu}{dy} = \dfrac{1}{P}\left(\dfrac{\partial Q}{\partial x} - \dfrac{\partial P}{\partial y} \right) \qquad \text{(ii')}$$

if μ is a function of y only; here of course it is assumed that the right side of (ii) or (ii') turns out to be a function of x or y only. Solving equation (i) in its full generality is unfeasible, but it is often easy to solve (ii) or (ii'). For example, show how the integrating factor in Example 4 can be found by solving (ii').

Determine an integrating factor μ for the given nonexact differential equation, and then find its general solution.

15. $(3xy + 2)\, dx + x^2\, dy = 0$
16. $y\, dx - x\, dy = 0$
17. $y\, dx - 3x\, dy = 0$
18. $2x \tan y\, dx + x^2\, dy = 0$

19. $(e^x - y^2)\, dx + 2y\, dy = 0$
20. $\cos x\, dx + (e^{-y} + \sin x)\, dy = 0$
21. $(x^2 + y^2 + y)\, dx - x\, dy = 0$
22. $(x^2 \sqrt{x^2+1} + y^2)\, dx + 2xy \ln|x|\, dy = 0$
23. $xy^2\, dx + (x^2 y - x)\, dy = 0$
24. $(y^2 \cos x + y \ln|y|)\, dx + (x + y \sin x)\, dy = 0$

A first order differential equation of the form $y' = dy/dx = f(x, y)$ is said to be *homogeneous* if there is a function g of a single variable such that $f(x, y) = g(y/x)$. For example, the equation

$$y' = \dfrac{2x^2 - y^2}{xy} = \dfrac{2 - (y/x)^2}{y/x} \qquad (xy \neq 0) \qquad \text{(iii)}$$

is homogeneous, with $g(u) = (2 - u^2)/u$. To solve a homogeneous equation, make the substitution $y = xu$. Then $y' = u + xu'$, so that $y' = g(u)$ takes the form

$$x \dfrac{du}{dx} + u = g(u),$$

or equivalently

$$\dfrac{dx}{x} + \dfrac{du}{u - g(u)} = 0$$

in which the variables x and u are separated. This equation has the general solution

$$\ln|x| + \int \dfrac{du}{u - g(u)} = C. \qquad \text{(iv)}$$

The general solution of the original homogeneous equation can then be found by substituting $u = y/x$ in (iv).

25. Use the method just described to find the general solution of (iii).
26. Find the particular solution of (iii) satisfying the condition $y(1) = 2$
27. Find the particular solution of (iii) satisfying the condition $y(2) = 1$

Find the general solution of the given homogeneous equation.

28. $x^2 y' = x^2 - xy + y^2$
29. $xyy' = x^2 + y^2$
30. $xy' = y - 2\sqrt{xy}$

16.2 First Order Linear Equations

An equation of the form

$$y' + py = q, \qquad (1)$$

where $p = p(x)$ and $q = q(x)$ are continuous functions of the independent variable x, is called a first order *linear* differential equation. To solve (1), we let $P = \int p(x)\,dx$ be any fixed antiderivative of p, and then multiply (1) by e^P. This gives

$$y'e^P + pe^P y = qe^P,$$

and since $(e^P)' = P'e^P = pe^P$, the expression on the left is recognizable as the derivative with respect to x of the product ye^P. Therefore

$$\frac{d}{dx}(ye^P) = qe^P,$$

which implies

$$ye^P = \int qe^P\,dx + C,$$

or equivalently

$$y = e^{-P}\left(\int qe^P\,dx + C\right) = e^{-P}\int qe^P\,dx + Ce^{-P}. \tag{2}$$

This formula, involving an arbitrary constant C, is the general solution of equation (1). When written out in full, (2) becomes

$$y = e^{-\int p(x)\,dx}\int q(x)e^{\int p(x)\,dx}\,dx + Ce^{-\int p(x)\,dx}. \tag{2'}$$

In keeping with the terminology introduced on page 945, the function $\mu = e^P$ is called an *integrating factor* for equation (1). We found μ by "inspection," that is, by astute guesswork, but μ can also be found more mechanically. In fact, suppose we multiply (1) by μ and require that the left side of the resulting equation

$$\mu y' + \mu p y = \mu q$$

be of the form $(\mu y)'$, so that it can be integrated without further ado. Then the condition $\mu y' + \mu p y = (\mu y)'$ leads at once to the separable equation $\mu' = p\mu$ for μ, which has $\mu = e^P$ as one of its solutions.

Example 1 Find the general solution of

$$y' + ay = bx, \tag{3}$$

where a and b are nonzero constants.

Solution Equation (3) is of the form (1) with $p = a$ and $q = bx$, so that

$$P = \int p\,dx = ax, \qquad \int qe^P\,dx = b\int xe^{ax}\,dx.$$

Integrating by parts, we find that

$$\int xe^{ax}\,dx = \int xd\left(\frac{e^{ax}}{a}\right) = \frac{xe^{ax}}{a} - \frac{1}{a}\int e^{ax}\,dx = \frac{xe^{ax}}{a} - \frac{e^{ax}}{a^2}.$$

Therefore, by (2), the general solution of (3) is

$$y = be^{-ax}\left(\frac{xe^{ax}}{a} - \frac{e^{ax}}{a^2}\right) + Ce^{-ax} = \frac{bx}{a} - \frac{b}{a^2} + Ce^{-ax}. \quad \square$$

Example 2 Find the particular solution of

$$xy' - y = x^4 \tag{4}$$

satisfying the initial condition $y(1) = 3$.

Solution Dividing (4) by x, we get

$$y' - \frac{1}{x}y = x^3, \tag{4'}$$

which is of the form (1) with $p = -1/x$ and $q = x^3$. Therefore

$$P = \int p\,dx = -\int \frac{dx}{x} = -\ln x, \quad e^P = \frac{1}{x}, \quad \int qe^P\,dx = \int x^2\,dx = \frac{1}{3}x^3,$$

and it follows from (2) that the general solution of (4') and hence of (4) is

$$y = \frac{1}{3}x^4 + Cx. \tag{5}$$

Although $p = -1/x$ is not defined at $x = 0$, this solution satisfies (4) on the whole real line (why?). To find the particular solution satisfying the condition $y(1) = 3$, we set $x = 1$ and $y = 3$ in (5), obtaining $C = \frac{8}{3}$. Thus the desired particular solution is

$$y = \frac{1}{3}x^4 + \frac{8}{3}x. \quad \square$$

Example 3 A switch is suddenly closed, applying a source of alternating voltage $V = V_0 \cos \omega t$ to the electric circuit shown in Figure 2, consisting of a resistor, of resistance R ohms, connected in series to an inductor, of inductance L henries. Find the resulting current $i = i(t)$ in the circuit.

Solution With these units i will be in amperes. The constant V_0 is the *peak voltage*, that is, the maximum value of V, and ω is the *angular frequency*, equal to $2\pi f$, where f is the frequency in cycles per second (see page 961). In Example 3, page 345, we considered the dc (direct current) case, in which V is constant, and found that the current i satisfies the linear differential equation

$$Ri + L\frac{di}{dt} = V,$$

or equivalently

$$\frac{di}{dt} + \frac{R}{L}i = \frac{V}{L}, \tag{6}$$

and we then solved (6) by separating variables. The same equation (6) holds in the present ac (alternating current) case, but V is now a function of t, which prevents (6) from being separable. However, we can still deal with (6) by using the method just developed.

To this end, we observe that (6) is of the form (1), with t instead of x, i instead of y, and

$$p = \frac{R}{L}, \quad q = \frac{V}{L} = \frac{V_0}{L}\cos \omega t,$$

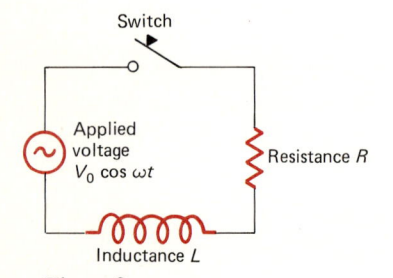

948 Chapter 16 Elementary Differential Equations

so that

$$P = \int p\, dt = \frac{Rt}{L}, \qquad \int qe^P\, dt = \frac{V_0}{L}\int e^{Rt/L} \cos \omega t\, dt.$$

By formula (9), page 378,

$$\int e^{Rt/L} \cos \omega t\, dt = e^{Rt/L}\, \frac{(R/L)\cos \omega t + \omega \sin \omega t}{(R/L)^2 + \omega^2},$$

and hence

$$\int qe^P\, dt = V_0 e^{Rt/L}\, \frac{R\cos \omega t + \omega L \sin \omega t}{R^2 + \omega^2 L^2}.$$

Thus, by formula (2), the general solution of (6) with $V = V_0 \cos \omega t$ is

$$i = \frac{V_0}{R^2 + \omega^2 L^2}(R\cos \omega t + \omega L \sin \omega t) + Ce^{-Rt/L}. \tag{7}$$

To determine the constant C, we apply the initial condition $i(0) = 0$ (there is no current in the circuit until the switch is closed at time $t = 0$). Setting $t = 0$ and $i = 0$ in (7), we find that

$$C = -\frac{V_0 R}{R^2 + \omega^2 L^2},$$

and then (7) becomes

$$i = \frac{V_0}{R^2 + \omega^2 L^2}\left[(R\cos \omega t + \omega L \sin \omega t) - Re^{-Rt/L}\right]. \tag{8}$$

Notice that this formula contains the solution to Example 3, page 345, as a special case, as you can easily verify by choosing $\omega = 0$ and dropping the subscript on V_0.

According to (8), i is the difference between two terms, a *steady-state* alternating current

$$i_{ac} = \frac{V_0}{R^2 + \omega^2 L^2}(R\cos \omega t + \omega L \sin \omega t) \tag{9}$$

and a *transient* current

$$i_{tr} = \frac{V_0 R}{R^2 + \omega^2 L^2}\, e^{-Rt/L},$$

which dies out rapidly in the same way as in the dc case (see page 346). Formula (9) looks complicated, but it can be written in the much simpler form

$$i_{ac} = \frac{V_0}{Z}\cos(\omega t - \phi), \tag{9'}$$

in terms of the *impedance*

$$Z = \sqrt{R^2 + \omega^2 L^2}$$

and the angle

$$\phi = \arctan \frac{\omega L}{R}.$$

(the details are left as an exercise). Thus i_{ac} has the same frequency as the applied voltage V, but "lags" V by ϕ radians, while the *peak current* i_0, that is, the maximum value of i_{ac}, is related to the peak voltage V_0 by the formula

$$i_0 = \frac{V_0}{Z},$$

involving the impedance Z. ☐

Problems

Find the general solution of the given first order linear equation.
1. $y' + ay = b$
2. $y' + ay = e^{bx}$
3. $y' - xy = x$
4. $y' - 2xy = xe^{-x^2}$
5. $y' + 2y = x^2 + x$
6. $y' + y = \sin x$
7. $y' - \frac{1}{x} y = x$
8. $y' + \frac{2}{x} y = x^2$
9. $y' + \frac{1}{x} y = x \cos x$
10. $y' + \frac{n}{x} y = \frac{1}{x^n}$
11. $y' - \frac{2y}{x+1} = (x+1)^2 e^x$
12. $xy' = y - 1$
13. $x(x-1)y' + y = x + 1$
14. $(x^2 + 1)y' - 4xy = (x^2 + 1)^2$
15. $(x^2 + 1)y' - y = \arctan x$
16. $y' + y \cos x = \sin^2 x \cos x$

Solve the given initial value problem.
17. $y' + 3y = 4$, $y(0) = 2$
18. $y' - y = e^{2x}$, $y(1) = 0$
19. $y' - 2y = \cos x$, $y(\pi/2) = 0$
20. $y' - \frac{1}{x} y = x \sin x$, $y(\pi) = 1$
21. $y' \cos x + y \sin x = 1$, $y(0) = -1$
22. $\sqrt{1-x^2} y' + y = 2$, $y(\tfrac{1}{2}) = 0$
23. $y' - \frac{y}{x \ln x} = \ln x$, $y(e) = 1$
24. $y' + \frac{xy}{1+x^2} = x$, $y(0) = 1$

25. Show that the first term in the formula

$$y = e^{-P} \int q e^{P} \, dx + C e^{-P} \qquad (P = \int p \, dx)$$

for the general solution of the linear equation $y' + py = q$ is a particular solution of this equation, while the second term is the general solution of the equation $y' + py = 0$, obtained by setting q equal to zero.

26. The nonlinear differential equation

$$y' + py = q y^n \qquad (n \neq 0, 1), \qquad \text{(i)}$$

where $p = p(x)$ and $q = q(x)$ are continuous functions, is known as *Bernoulli's equation*. Show that multiplication of equation (i) by $(1-n)y^{-n}$ transforms it into the linear differential equation

$$u' + (1-n)pu = (1-n)q \qquad \text{(i')}$$

in the new variable $u = y^{1-n}$.

Use the method of the preceding problem to solve the given Bernoulli equation.
27. $y' + xy = xy^2$
28. $xy' - y = y^3 \ln x$
29. $y' - y = xy^{-3}$

30. A switch is suddenly closed, applying a source of alternating voltage $V = V_0 \cos \omega t$ to an electric circuit consisting of a resistor, of resistance R ohms, connected in series to a capacitor, of capacitance C farads. Find the resulting current $i = i(t)$ in the circuit. Show that the steady-state alternating current is

$$i_{ac} = \frac{V_0}{Z} \cos(\omega t + \phi),$$

where

$$Z = \sqrt{\frac{1}{\omega^2 C^2} + R^2}, \qquad \phi = \arctan \frac{1}{\omega RC}.$$

Thus i_{ac} has the same frequency as the applied voltage V, but "leads" V by ϕ radians.
Hint. Recall Problem 23, page 349.

31. Solve the nonlinear differential equation $e^y(y' + 1) = x$ by reducing it to a linear equation in e^y.

32. Show that the linear equation $y' + py = q$ can be solved by separation of variables if the functions $p = p(x)$ and $q = q(x)$ are both constant.

16.3 Second Order Linear Equations with Constant Coefficients

We now turn to *second* order differential equations, confining ourselves to the study of equations of the particularly simple form

$$y'' + ay' + by = f(x). \tag{1}$$

where y' and y'' are the first and second derivatives of the unknown function $y = y(x)$, a and b are real constants, and $f(x)$ is a given continuous function of the independent variable x. A differential equation of this type is called a *second order linear equation with constant coefficients*. If $f(x)$ is not identically zero, equation (1) is said to be *nonhomogeneous*, but if $f(x) \equiv 0$, it simplifies to the *homogeneous* equation

$$y'' + ay' + by = 0. \tag{2}$$

(Here the term "homogeneous" has an entirely different meaning than in Problems 25–30, page 946.)

To solve the nonhomogeneous equation (1), we first make a detailed study of the associated homogeneous equation (2). The following theorem explains why.

> **Theorem 1** *(General solution of the nonhomogeneous equation).* The general solution of the nonhomogeneous equation (1) is equal to the sum of any particular solution of (1) and the general solution of the homogeneous equation (2).

Proof Let y_1 be an arbitrary solution and y_2 a fixed solution of the given nonhomogeneous equation (1). Then $y_1'' + ay_1' + by_1 = f(x)$ and $y_2'' + ay_2' + by_2 = f(x)$, and subtracting the second of these equations from the first, we get

$$(y_1'' + ay_1' + by_1) - (y_2'' + ay_2' + by_2) = f(x) - f(x) = 0,$$

or equivalently

$$(y_1 - y_2)'' + a(y_1 - y_2)' + b(y_1 - y_2) = 0.$$

Therefore $y_1 - y_2$ satisfies the homogeneous equation (2), that is, $y_1 = y_2 + u$ where u is *some* solution of (2). To complete the proof, we must still show that the sum of y_2 and an *arbitrary* solution u of (2) is a solution of (1). But this follows at once from the observation that $y_2'' + ay_2' + by_2 = f(x)$ and $u'' + au' + bu = 0$ together imply

$$(y_2 + u)'' + a(y_2 + u)' + b(y_2 + u) = (y_2'' + ay_2' + by_2) + (u'' + au' + bu)$$
$$= f(x) + 0 = f(x). \blacksquare$$

We begin our investigation of the homogeneous equation (2) by observing that *if y_1 and y_2 are any two solutions of (2), then so is every linear combination of y_1 and y_2*, that is, every expression of the form $C_1 y_1 + C_2 y_2$, where C_1 and C_2 are arbitrary constants. This is an immediate consequence of the fact that if $y_1'' + ay_1' + by_1 = 0$ and $y_2'' + ay_2' + by_2 = 0$, then

$$(C_1 y_1 + C_2 y_2)'' + a(C_1 y_1 + C_2 y_2)' + b(C_1 y_1 + C_2 y_2)$$
$$= C_1 y_1'' + C_2 y_2'' + aC_1 y_1' + aC_2 y_2' + bC_1 y_1 + bC_2 y_2$$
$$= C_1(y_1'' + ay_1' + by_1) + C_2(y_2'' + ay_2' + by_2)$$
$$= C_1(0) + C_2(0) = 0.$$

It is now natural to ask whether, conversely, *every* solution of (2) is a linear combination of two given solutions y_1 and y_2, so that $y = C_1 y_1 + C_2 y_2$ is the general solution of (2). As shown by the following example, this is not true if the choice of the solutions y_1 and y_2 is unrestricted.

Example 1 Consider the homogeneous linear equation

$$y'' - y = 0. \tag{3}$$

Clearly $y_1 \equiv 0$ is a solution of (3), and so is $y_2 = e^x$, since $(e^x)'' = (e^x)' = e^x$. But $y = C_1 y_1 + C_2 y_2 = C_2 e^x$ cannot be the general solution of (3), since (3) also has the solution $y = e^{-x}$ (check this), which is not a constant multiple of e^x. Failure of $y = C_1 y_1 + C_2 y_2$ to be the general solution of (3) can also occur even if y_1 and y_2 are both nonzero. For instance, if we choose $y_1 = 2e^x$, which is another solution of (3), we find that $y = C_1 y_1 + C_2 y_2 = (2C_1 + C_2)e^x = ke^x$, where $k = 2C_1 + C_2$, but this is not the general solution of (3) either, for the same reason as before, namely there is no constant k such that $e^{-x} = ke^x$.

The above example shows that we cannot claim that $y = C_1 y_1 + C_2 y_2$ is the general solution of the homogeneous equation (2) if either of the functions y_1 and y_2 is identically zero, or if either is a constant multiple of the other. However, according to the next theorem, which we cite without proof, $y = C_1 y_1 + C_2 y_2$ is indeed the general solution of (2) if we exclude these possibilities.

Theorem 2 *(General solution of the homogeneous equation). Let y_1 and y_2 be any two solutions of the homogeneous equation (2) such that neither is identically zero and neither is a constant multiple of the other (it will be shown below that such solutions always exist). Then the general solution of (2) is*

$$y = C_1 y_1 + C_2 y_2,$$

where C_1 and C_2 are arbitrary constants.

Two solutions y_1 and y_2 satisfying the conditions of Theorem 2 are said to form a *fundamental set* of solutions of equation (2). However, the choice of the set is not unique.

Example 2 It was found in Example 1 that the functions e^x and e^{-x} are both solutions of equation (3). Neither function is identically zero, and neither is a constant multiple of the other. Thus $y_1 = e^x$ and $y_2 = e^{-x}$ form a fundamental set of solutions of (3), and correspondingly (3) has the general solution $y = C_1 e^x + C_2 e^{-x}$. As you can easily verify, the functions $y_1 = \cosh x$ and $y_2 = \sinh x$ form another set of fundamental solutions of (3).

The Existence and Uniqueness Theorem

The reason for omitting the proof of Theorem 2 is that it rests on the following *existence and uniqueness theorem*, whose proof is beyond the scope of this course, although its meaning is perfectly clear. *Given any three real numbers x_0, y_0 and y'_0, there is one and only one solution $y = y(x)$ of the homogeneous equation (2) satisfying the initial conditions*

$$y(x_0) = y_0, \quad y'(x_0) = y'_0.$$

We will make tacit use of this result in solving initial value problems for equation (2).

Example 3 Find the general solution of the homogeneous equation

$$y'' + \omega^2 y = 0 \quad (\omega > 0). \tag{4}$$

Also find the particular solution of (4) satisfying the initial conditions

$$y(0) = 1, \quad y'(0) = -1. \tag{4'}$$

Solution Any function which when multiplied by ω^2 is the negative of its own second derivative is a solution of (4). Two such functions are $\cos \omega x$ and $\sin \omega x$, since $(\cos \omega x)'' = (-\omega \sin \omega x)' = -\omega^2 \cos \omega x$ and $(\sin \omega x)'' = (\omega \cos \omega x)' = -\omega^2 \sin \omega x$. Moreover, neither of these functions is identically zero, and neither is a constant multiple of the other (why not?). Therefore, by Theorem 2, the general solution of (4) is

$$y = C_1 \cos \omega x + C_2 \sin \omega x, \tag{5}$$

where C_1 and C_2 are arbitrary constants. Differentiation of (5) gives

$$y' = -\omega C_1 \sin \omega x + \omega C_2 \cos \omega x. \tag{5'}$$

To find the solution of (4) satisfying the conditions (4'), we set $x = 0$, $y = 1$ in (5) and $x = 0$, $y' = -1$ in (5'), and immediately obtain $C_1 = 1$, $C_2 = -1/\omega$. Choosing these values of C_1 and C_2 in (5), we get the desired particular solution

$$y = \cos \omega x - \frac{1}{\omega} \sin \omega x. \quad \square$$

The next example illustrates the use of Theorem 1 to find the general solution of a *nonhomogeneous* linear equation.

Example 4 Solve the equation

$$y'' - y = 2 - 3x. \tag{6}$$

Solution According to Theorem 1, the general solution of (6) is the sum of any particular solution of (6) and the general solution of the associated homogeneous equation $y'' - y = 0$. We already know from Example 2 that the general solution of $y'' - y = 0$ is $y = C_1 e^x + C_2 e^{-x}$, and a glance at (6) reveals that it has the particular solution $y = 3x - 2$ (note that $y'' = 0$ for this choice of y). Hence the general solution of (6) is

$$y = C_1 e^x + C_2 e^{-x} + 3x - 2. \quad \square$$

Analysis of the Homogeneous Equation

We now attack the problem of solving the general homogeneous linear equation (2), with the help of Theorem 2. To this end, we substitute the exponential $y = e^{rx}$ into (2), in the hope of finding values of the constant r for

Section 16.3 Second Order Linear Equations with Constant Coefficients 953 which equation (2) is satisfied. If $y = e^{rx}$, then $y' = re^{rx}$ and $y'' = r^2 e^{rx}$, so that this substitution converts $y'' + ay' + by = 0$ into

$$r^2 e^{rx} + are^{rx} + be^{rx} = (r^2 + ar + b)e^{rx} = 0. \tag{7}$$

But e^{rx} is never zero, and hence (7) holds if and only if r is a solution of the quadratic equation

$$r^2 + ar + b = 0, \tag{8}$$

known as the *characteristic equation* of (2).

Completing the square in (8), we get

$$\left(r + \frac{a}{2}\right)^2 - \frac{a^2}{4} + b = 0,$$

or equivalently

$$\left(r + \frac{a}{2}\right)^2 = \frac{a^2 - 4b}{4},$$

from which it follows that (8) has two distinct roots

$$r_1 = -\frac{a}{2} + \frac{\sqrt{a^2 - 4b}}{2}, \qquad r_2 = -\frac{a}{2} - \frac{\sqrt{a^2 - 4b}}{2}$$

if $a^2 - 4b > 0$, only one (double) root

$$r_1 = r_2 = -\frac{a}{2} \tag{9}$$

if $a^2 - 4b = 0$, and no real roots if $a^2 - 4b^2 < 0$. Each of these three cases leads to a fundamental set of solutions of equation (2), but the nature of the solutions is different in each case.

If $a^2 - 4b^2 > 0$, then $e^{r_1 x}$ and $e^{r_2 x}$ already form a fundamental set of solutions, since neither of these functions is identically zero and neither is a constant multiple of the other; in fact, an equality of the form $e^{r_1 x} = ke^{r_2 x}$ (k constant) can hold only at the single point $x = (\ln k)/(r_1 - r_2)$. Hence in this case Theorem 2 tells us that the general solution of the homogeneous equation (2) is

$$y = C_1 e^{r_1 x} + C_2 e^{r_2 x}.$$

Example 5 Find the general solution of

$$y'' + y' - 6y = 0. \tag{10}$$

Solution Here the characteristic equation is

$$r^2 + r - 6 = (r - 2)(r + 3) = 0,$$

with distinct roots $r_1 = 2$ and $r_2 = -3$. Hence the general solution of (10) is

$$y = C_1 e^{2x} + C_2 e^{-3x},$$

where C_1 and C_2 are arbitrary constants. □

If $a^2 - 4b = 0$, then $y_1 = e^{r_1 x} = e^{-ax/2}$ is one solution of the homogeneous equation $y'' + ay' + by = 0$, but we still need another solution y_2 forming a fundamental set with y_1. To find it, we resort to the following simple trick. Let $y = ue^{-ax/2}$, where $u = u(x)$ is a new unknown function.

Then

$$y' = u'e^{-ax/2} - \frac{a}{2}ue^{-ax/2},$$

$$y'' = u''e^{-ax/2} - au'e^{-ax/2} + \frac{a^2}{4}ue^{-ax/2},$$

so that $y'' + ay' + by = 0$ becomes

$$\left(u'' - au' + \frac{a^2}{4}u\right)e^{-ax/2} + a\left(u' - \frac{a}{2}u\right)e^{-ax/2} + bue^{-ax/2}$$

$$= \left(u'' + bu - \frac{a^2}{4}u\right)e^{-ax/2} = \left(u'' + \frac{4b - a^2}{4}u\right)e^{-ax/2} = u''e^{-ax/2} = 0, \tag{11}$$

where we use the fact that $4b - a^2 = 0$. Since $e^{-ax/2}$ is never zero, it follows that $u'' = 0$, which implies

$$u' = A, \qquad u = Ax + B,$$

where A and B are constants. Since we only need one solution u, we make the simple choice $A = 1$ and $B = 0$, which gives $u = x$. We then find that

$$y_2 = ue^{-ax/2} = xe^{-ax/2}$$

is another solution of the homogeneous equation. Neither of the solutions $y_1 = e^{-ax/2}$ and $y_2 = xe^{-ax/2}$ is identically zero, and neither is a constant multiple of the other; in fact, an equality of the form $e^{-ax/2} = kxe^{-ax/2}$ (k constant) can hold only at the single point $x = 1/k$. Thus in this case the general solution of (2) is

$$y = C_1 e^{-ax/2} + C_2 xe^{-ax/2}.$$

Example 6 Find the general solution of

$$y'' - 4y' + 4y = 0. \tag{12}$$

Solution This time the characteristic equation is

$$r^2 - 4r + 4 = (r - 2)^2 = 0,$$

with only one (double) root $r_1 = r_2 = 2$. It follows from (9) that this is also the value of the constant $-a/2$. Hence the general solution of (12) is

$$y = C_1 e^{2x} + C_2 xe^{2x}. \quad \square$$

If $a^2 - 4b < 0$, the characteristic equation (8) has no real roots, so that there are no solutions of the form e^{rx}. Nevertheless, a fundamental set of solutions of (2) can easily be found. To this end, we make the same substitution $y = ue^{-ax/2}$ as in the case $a^2 - 4b = 0$. Consulting (11), we find that the equation $y'' + ay' + by = 0$ again implies

$$\left(u'' + \frac{4b - a^2}{4}u\right)e^{-ax/2} = 0, \tag{13}$$

but now the coefficient of u is not zero, but a positive number. In fact, let

$$\omega = \frac{\sqrt{4b - a^2}}{2} > 0.$$

Then (13) is equivalent to
$$u'' + \omega^2 u = 0,$$
since $e^{-ax/2}$ is never zero, and as we know from Example 3, the general solution of this equation is
$$u = C_1 \cos \omega x + C_2 \sin \omega x.$$
Hence the general solution of (2), equal to $y = ue^{-ax/2}$, is in turn
$$y = e^{-ax/2}(C_1 \cos \omega x + C_2 \sin \omega x).$$

Example 7 Find the general solution of
$$y'' + 8y' + 25y = 0. \tag{14}$$

Solution Here $a = 8$ and $b = 25$, so that $a^2 - 4b = 64 - 100 = -36 < 0$ and $\omega = \frac{1}{2}\sqrt{36} = 3$. The general solution of (14) is therefore
$$y = e^{-4x}(C_1 \cos 3x + C_2 \sin 3x). \quad \square$$

The Method of Undetermined Coefficients

Finally, we return to the problem of solving the nonhomogeneous equation (1), of the form $y'' + ay' + by = f(x)$. You will recall from Theorem 1 that the general solution of (1) is the sum of any particular solution of (1) and the general solution of the associated homogeneous equation $y'' + ay' + by = 0$, often called the *reduced* equation of (1). But we have just shown how to find the general solution of the reduced equation. Thus what we now need is any particular solution of the nonhomogeneous equation itself (often called the *complete* equation). To find such a solution is in general a complicated problem, but fortunately there is an effective technique, called the *method of undetermined coefficients*, which works if $f(x)$ is either of the commonly encountered functions

$$P_n(x)e^{\alpha x} \cos \beta x \tag{15}$$

and

$$P_n(x)e^{\alpha x} \sin \beta x, \tag{15'}$$

involving a given polynomial

$$P_n(x) = c_0 + c_1 x + \cdots + c_n x^n$$

of degree n. In fact, let

$$A_n(x) = a_0 + a_1 x + \cdots + a_n x^n$$

and

$$B_n(x) = b_0 + b_1 x + \cdots + b_n x^n$$

be two other polynomials of the same degree n, but with undetermined (that is, as yet unknown) coefficients $a_0, a_1, \ldots, a_n$ and $b_0, b_1, \ldots, b_n$, and let $f(x)$ be of the form (15) or (15'). Then it can be shown that the substitution

$$y = [A_n(x)e^{\alpha x} \cos \beta x + B_n(x)e^{\alpha x} \sin \beta x] x^k$$
$$= \sum_{i=1}^{n} (a_i x^{i+k} e^{\alpha x} \cos \beta x + b_i x^{i+k} e^{\alpha x} \sin \beta x) \tag{16}$$

956 Chapter 16 Elementary Differential Equations

transforms the nonhomogeneous equation $y'' + ay' + by = f(x)$ into an equality which can be satisfied identically by making a suitable choice of the coefficients $a_0, a_1, \ldots, a_n$ and $b_0, b_1, \ldots, b_n$. Here x^k ($k = 0, 1, 2$) is the smallest power of x required to ensure that none of the terms $a_i x^{i+k} e^{\alpha x} \cos \beta x$ and $b_i x^{i+k} e^{\alpha x} \sin \beta x$ in the right side of (16) is a solution of the reduced equation $y'' + ay' + by = 0$. The substitution (16) can be replaced by just

$$y = A_n(x) x^k \tag{17}$$

if $f(x) = P_n(x)$, corresponding to $\alpha = \beta = 0$ in (16), and by just

$$y = A_n(x) x^k e^{\alpha x} \tag{17'}$$

if $f(x) = P_n(x) e^{\alpha x}$, corresponding to $\beta = 0$ in (16). On the other hand, if $f(x) = P_n(x) \cos \beta x$ or $f(x) = P_n(x) \sin \beta x$, corresponding to $\alpha = 0$ in (15) or (15'), the full expression

$$y = [A_n(x) \cos \beta x + B_n(x) \sin \beta x] x^k \tag{18}$$

is needed.

Example 8 Find the general solution of

$$y'' - y' - 2y = x^2 + x - 4. \tag{19}$$

Solution The right side of (19) is of the form (15) with $n = 2$ and $\alpha = \beta = 0$. Since none of the functions $y = a_i x^i$ ($i = 0, 1, 2$) is a solution of the reduced equation $y'' - y' - 2y = 0$, we make the substitution

$$y = Ax^2 + Bx + C, \tag{19'}$$

corresponding to formula (17) with $n = 2$ and $k = 0$; for simplicity, we denote the undetermined coefficients by the consecutive letters A, B and C. Then

$$y' = 2Ax + B, \qquad y'' = 2A,$$

and the substitution (19') transforms (19) into

$$2A - (2Ax + B) - 2(Ax^2 + Bx + C) = x^2 + x - 4,$$

which is an identity in x if and only if identical powers of x on both sides of the equation have the same coefficients. This requirement leads to the "triangular" system of linear algebraic equations

$$-2A = 1,$$
$$-2A - 2B = 1,$$
$$2A - B - 2C = -4,$$

with immediate solution $A = -\frac{1}{2}$, $B = 0$, $C = \frac{3}{2}$ (solve for A, B and C in succession). For these coefficients, (19') becomes

$$y = -\frac{1}{2} x^2 + \frac{3}{2},$$

which is a particular solution of (19), as is immediately verifiable. The characteristic equation of the reduced equation $y'' - y' - 2y = 0$ is $r^2 - r - 2 = (r - 2)(r + 1) = 0$, with distinct roots $r_1 = 2$ and $r_2 = -1$. Hence the general solution of the reduced equation is $y = C_1 e^{2x} + C_2 e^{-x}$, and it follows from Theorem 1 that the general solution of equation (19) is

$$y = C_1 e^{2x} + C_2 e^{-x} - \frac{1}{2}x^2 + \frac{3}{2}. \quad \square$$

Example 9 Find the general solution of

$$y'' - y' = x^2 + x - 4. \tag{20}$$

Solution Notice that the left side of (20) differs from that of (19) by the absence of the term $-2y$. As a result, one of the functions $y = a_i x^i$ ($i=0, 1, 2$), namely $y = a_0 x^0 = a_0$, is a solution of the reduced equation $y'' - y' = 0$. Hence instead of the substitution (19′), we must now make the substitution

$$y = Ax^3 + Bx^2 + Cx, \tag{20′}$$

corresponding to formula (17) with $n = 2$ and $k = 1$. Then

$$y' = 3Ax^2 + 2Bx + C, \qquad y'' = 6Ax + 2B,$$

and the substitution (20′) transforms (20) into

$$(6Ax + 2B) - (3Ax^2 + 2Bx + C) = x^2 + x - 4.$$

Equating coefficients of identical powers of x on both sides of this equation, we get the system

$$-3A = 1,$$
$$6A - 2B = 1,$$
$$2B - C = -4,$$

with solution $A = -\frac{1}{3}$, $B = -\frac{3}{2}$, $C = 1$. For these coefficients, (20′) becomes

$$y = -\frac{1}{3}x^3 - \frac{3}{2}x^2 + x.$$

The characteristic equation of the reduced equation $y'' - y' = 0$ is $r^2 - r = r(r - 1) = 0$, with distinct roots $r_1 = 0$ and $r_2 = 1$. Hence the general solution of the reduced equation is $y = C_1 + C_2 e^x$, and by Theorem 1, the general solution of equation (20) is

$$y = C_1 + C_2 e^x - \frac{1}{3}x^3 - \frac{3}{2}x^2 + x. \quad \square$$

Example 10 Find the general solution of

$$y'' - 2y' + y = 2e^x + \cos x. \tag{21}$$

Solution If y_1 is a particular solution of

$$y'' - 2y' + y = 2e^x \tag{22}$$

and y_2 is a particular solution of

$$y'' - 2y' + y = \cos x, \tag{22′}$$

then clearly $y_1 + y_2$ is a particular solution of (21). Both equations (22) and (22′) have the same reduced equation $y'' - 2y' + y = 0$ and characteristic equation $r^2 - 2r + 1 = (r - 1)^2$, with one (double) root $r_1 = r_2 = 1$. Hence the general solution of the reduced equation is $y = C_1 e^x + C_2 x e^x$, by the

958   Chapter 16   Elementary Differential Equations

argument following Example 5. Thus both e^{2x} and xe^{2x} are solutions of the reduced equation. Therefore, to solve (22), we make the substitution

$$y = y_1 = Ax^2 e^x,$$

corresponding to formula (17') with $n = 0$, $\alpha = 1$ and $k = 2$. Since

$$y_1' = (2x + x^2)Ae^x, \qquad y_1'' = (2 + 4x + x^2)Ae^x,$$

this substitution transforms (22) into

$$[(2 + 4x + x^2) - 2(2x + x^2) + x^2]Ae^x = 2Ae^x = 2e^x,$$

which implies $A = 1$. Thus $y_1 = x^2 e^x$ is a particular solution of (22), as you can easily verify.

As for equation (22'), since neither $\sin x$ nor $\cos x$ is a solution of the reduced equation, we make the substitution

$$y = y_2 = A \cos x + B \sin x,$$

corresponding to formula (18) with $n = 0$, $\beta = 1$ and $k = 0$. Since

$$y_2' = -A \sin x + B \cos x, \qquad y_2'' = -A \cos x - B \sin x,$$

this substitution transforms (22') into

$$(-A \cos x - B \sin x) - 2(-A \sin x + B \cos x) + (A \cos x + B \sin x)$$
$$= 2A \sin x - 2B \cos x = \cos x,$$

which implies $A = 0$, $B = -\frac{1}{2}$. Thus $y_2 = -\frac{1}{2} \sin x$ is a particular solution of (22'), as is easily checked. Adding the particular solutions y_1 and y_2 of equations (22) and (22'), we get a particular solution

$$y_1 + y_2 = x^2 e^x - \frac{1}{2} \sin x$$

of the original equation (21). But the general solution of the reduced equation $y'' - 2y' + y = 0$ is $y = C_1 e^x + C_2 x e^x$, and hence the general solution of (21) is

$$y = C_1 e^x + C_2 x e^x + x^2 e^x - \frac{1}{2} \sin x. \quad \square$$

Remark In cases where the method of undetermined coefficients does not apply, try using the method of *variation of parameters*, described in the preamble to Problems 19–24, page 969.

Problems

Find the general solution of the given homogeneous linear equation.

1. $y'' - 2y' - y = 0$
2. $y'' - 6y' + 9y = 0$
3. $y'' - 4y' + 13y = 0$
4. $y'' + 4y' + 29y = 0$
5. $y'' + 10y' + 25y = 0$
6. $5y'' + 3y' = 0$
7. $\frac{1}{4}y'' - 2y' + 4y = 0$
8. $y'' + 9y' + 14y = 0$
9. $4y'' - 8y' + 5y = 0$
10. $y'' - 2\sqrt{2}y' + 2y = 0$
11. $3y'' + 7y' + 2y = 0$
12. $y'' - 12y' + 52y = 0$

Find the particular solution $y = y(x)$ of the given homogeneous linear equation satisfying the specified conditions; some are initial conditions, others are boundary conditions (see page 265).

13. $y'' + y' - 2y = 0$, $y(0) = 1$, $y(\ln 2) = 0$
14. $y'' + y' = 0$, $y(1) = 0$, $y'(1) = 3$
15. $y'' - 2y' + 2y = 0$, $y(0) = 0$, $y'(0) = 2$
16. $y'' - 2y' + y = 0$, $y(0) = 0$, $y(1) = 1$

17. $7y'' + 19y' + 13y = 0$, $y(0) = y'(0) = 0$
18. $y'' + y = 0$, $y'(0) = 1$, $y'(\pi/4) = 0$

Find a particular solution of the nonhomogeneous linear equation $y'' + y = f(x)$, where $f(x)$ is the given function.

19. $(x + 1)^3$
20. $\cos x + 4x - 5$
21. $\cos 3x$
22. $\sinh x$
23. $e^x + \sin x$
24. $\sin x \sin 2x$
25. $x \sin x$
26. $e^x \cos x$

Find the general solution of the given nonhomogeneous linear equation.

27. $y'' + 4y' = 3x^2 - x + 2$
28. $y'' + 3y' - 4y = e^{-4x}$

29. $y'' + 2y' + 2y = e^x \sin x$
30. $y'' + 2y' + 2y = e^{-x} \cos x$
31. $y'' - 2y' + y = xe^x$
32. $y'' + 3y' = xe^{-3x}$
33. $y'' - y' = x^4$
34. $y'' + 2y' + y = 2\cosh x$
35. $y'' - 2y' - 8y = e^{6x}$
36. $y'' - 5y' + 6y = e^{3x} + e^{2x}$

37. Solve the initial value problem $y'' - 2y' = e^{2x} + x^2 - 1$, $y(0) = -1$, $y'(0) = 1$.

38. Solve the boundary value problem $y'' + y' = x$, $y(0) = 1$, $y(1) = 0$.

16.4 Simple Harmonic Motion; Damped and Forced Oscillations

Let P be a particle whose position is specified by a single coordinate $y = y(t)$ which is a function of the time t. Then P is said to execute *simple harmonic motion* if the coordinate y satisfies a second order linear differential equation of the form

$$\frac{d^2y}{dt^2} + \omega^2 y = 0 \qquad (\omega > 0). \tag{1}$$

As we know from Example 3, page 953 (after changing the independent variable from x to t), the general solution of this equation is

$$y = C_1 \cos \omega t + C_2 \sin \omega t, \tag{2}$$

where C_1 and C_2 are arbitrary constants.

To clarify the nature of the motion of P, we write (2) in another way. Let A and ϕ (with $A > 0$) be the polar coordinates of the point (C_2, C_1), regarded as a point in a rectangular coordinate system. Then

$$A \cos \phi = C_2, \qquad A \sin \phi = C_1, \tag{3}$$

so that $C_1^2 + C_2^2 = A^2(\sin^2 \phi + \cos^2 \phi) = A^2$, or equivalently

$$A = \sqrt{C_1^2 + C_2^2}. \tag{4}$$

Also, combining formulas (2) and (3), we obtain

$$y = A \cos \omega t \sin \phi + A \sin \omega t \cos \phi,$$

which simplifies to

$$y = A \sin (\omega t + \phi). \tag{5}$$

Alternatively, choosing A and ϕ as the polar coordinates of the point $(C_1, -C_2)$, we have $A \cos \phi = C_1$ and $A \sin \phi = -C_2$, where A is again given by (4), but now, as you can easily verify,

$$y = A \cos (\omega t + \phi), \tag{5'}$$

instead of (5). Note that the value of the angle ϕ in (5') is $\pi/2$ less than its value in (5).

It is apparent from both (5) and (5′) that a particle executing simple harmonic motion oscillates back and forth along the y-axis, and first repeats its motion after a time

$$T = \frac{2\pi}{\omega},$$

called the *period* of the oscillations. Observe that T is the fundamental period of both functions $\sin(\omega t + \phi)$ and $\cos(\omega t + \phi)$. The oscillations described by (5) or (5′) are said to be *sinusoidal* or *harmonic oscillations*. The reciprocal of T, namely

$$f = \frac{1}{T} = \frac{\omega}{2\pi},$$

is called the *frequency*, and if time is measured in seconds, f is measured in *cycles per second* (cps); the term "cycle" refers to one complete back and forth motion of the particle. One cycle per second is also called one *hertz* (Hz), after the German experimental physicist Heinrich Hertz (1857–1894), who was the first to produce radio waves. The quantity ω is called the *angular frequency*, and is measured in radians per second. The quantity A is called the *amplitude*, and is the largest value of $|y|$. In fact, the particle P oscillates back and forth between the two extreme positions $y = A$ and $y = -A$. The argument $\omega t + \phi$ of both functions $\sin(\omega t + \phi)$ and $\cos(\omega t + \phi)$ is called the *phase*, and the angle ϕ, equal to the value of the phase at time $t = 0$, is called the *initial phase*. In terms of the frequency f, we can write (5) as

$$y = \sin(2\pi f t + \phi) \tag{6}$$

and (5′) as

$$y = \cos(2\pi f t + \phi). \tag{6′}$$

There is an interesting and important connection between simple harmonic motion and uniform circular motion (see page 684). Suppose a point Q rotates with angular speed ω in the counterclockwise direction around the circle $x^2 + y^2 = A^2$ of radius A centered at the origin O. Then the angle between the radius OQ and the positive x-axis at the time t is equal to $\omega t + \phi$, where ϕ is the angle between OQ and the positive x-axis at time $t = 0$ (see Figure 3). Let P be the projection of Q onto the y-axis. Then as Q goes around the circle $x^2 + y^2 = A^2$, the point P oscillates back and forth along the y-axis, executing the simple harmonic motion (5), with amplitude A, angular frequency ω, phase $\omega t + \phi$ and initial phase ϕ. Also, as Q goes around the circle, the projection of Q onto the x-axis executes the simple harmonic motion (5′) with y replaced by x.

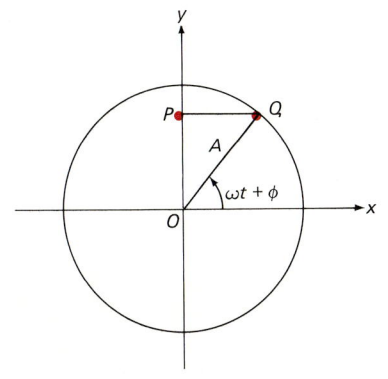

Figure 3

Example 1 Describe the simple harmonic motion

$$y = 3 \cos 4\pi t + 4 \sin 4\pi t. \tag{7}$$

Solution Equation (7) is of the form (2) with $C_1 = 3$ and $C_2 = 4$. Hence by (4),

$$A = \sqrt{3^2 + 4^2} = \sqrt{25} = 5,$$

Section 16.4 Simple Harmonic Motion; Damped and Forced Oscillations and (7) can be written as

$$y = 5 \sin(2\pi ft + \phi) = 5 \sin\left(\frac{2\pi t}{T} + \phi\right), \tag{7'}$$

where $f = 2$ cps, $T = 1/f = 0.5$ sec and

$$\phi = \arctan \frac{3}{4} \approx 0.6435 \text{ rad} \approx 36.87°$$

(divide the second of the equations (3) by the first and solve for ϕ). The graph of the function (7') is shown in Figure 4, where you will observe that the value of y at $t = 0$ is $5 \sin \phi = C_1 = 3$. □

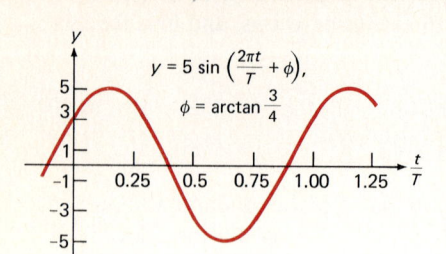

Figure 4

The Harmonic Oscillator (Undamped Case)

A mechanical system executing simple harmonic motion is called a *harmonic oscillator*. Such a system is described in the next example.

Example 2 A ball of mass m is attached to the lower end of a spring, whose upper end is attached to a rigid horizontal support. Suppose that at time $t = 0$ the ball is given an initial displacement y_0 and velocity v_0. Determine the subsequent motion of the ball, assuming that the ball (regarded as a particle) is subject to no forces other than its weight and the tension in the spring.

Solution According to *Hooke's law* (see page 270), the ball is acted on by an elastic restoring force $F = -ks$, where k is a positive constant, called the *stiffness* or *spring constant*, and s is the difference between the length of the stretched spring and its unstretched or natural length l. The ball is also acted on by its weight mg, which stretches the spring to the equilibrium length $l + s_0$, where $mg = ks_0$ (why?). As in Figure 5, let the y-axis point vertically downward with the origin ($y = 0$) at the equilibrium position of the ball. Then the total force acting on the ball in the displaced position with coordinate $y = y(t)$ is

$$F = mg - ks = mg - k(s_0 + y) = -ky,$$

and hence by Newton's second law of motion,

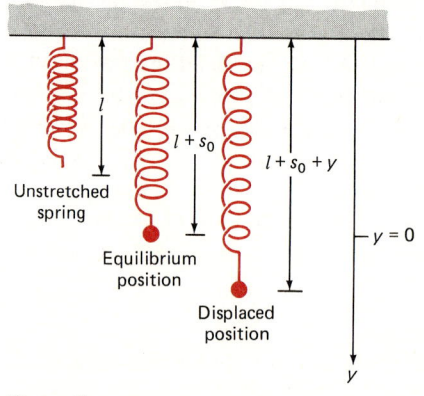

$$my'' = -ky, \tag{8}$$

Figure 5

962 Chapter 16 Elementary Differential Equations

where the prime denotes differentiation with respect to t, or equivalently

$$y'' + \omega^2 y = 0$$

in terms of the positive constant $\omega = \sqrt{k/m}$. But this is the differential equation leading to simple harmonic motion. Thus the position of the ball at time $t \geq 0$ is

$$y = C_1 \cos \omega t + C_2 \sin \omega t. \tag{9}$$

To determine the constants C_1 and C_2, we impose the initial conditions $y(0) = y_0$ and $y'(0) = v_0$. Setting $t = 0$, $y = y_0$ in formula (9) and $t = 0$, $y' = v_0$ in the formula

$$y' = -\omega C_1 \sin \omega t + \omega C_2 \cos \omega t, \tag{9'}$$

obtained by differentiating (9) with respect to t, we find that $C_1 = y_0$ and $C_2 = v_0/\omega$. Therefore (9) can be written in the form

$$y = A \sin(\omega t + \phi), \tag{10}$$

where the amplitude A and initial phase ϕ are given by

$$A = \sqrt{C_1^2 + C_2^2} = \sqrt{y_0^2 + \frac{v_0^2}{\omega^2}} \qquad \phi = \arctan \frac{C_1}{C_2} = \frac{\omega y_0}{v_0}.$$

Observe that the smaller the mass of the ball and the stiffer the spring, the greater the frequency

$$f = \frac{\omega}{2\pi} = \frac{1}{2\pi}\sqrt{\frac{k}{m}}$$

of the oscillations (10). Also note that $y = y_0 \cos \omega t$ if the ball is at rest ($v_0 = 0$) when released. □

The Harmonic Oscillator (Damped Case)

The oscillations described in the preceding example are *undamped*, that is, they go on indefinitely with constant amplitude. This is an idealization, since in reality the amplitude of the oscillations diminishes with time, due to intrinsic forces of resistance, like friction in the spring or air resistance. In fact, in some cases a resistive force is deliberately introduced to "damp out" oscillations, or even to prevent the occurrence of any oscillations in the first place.

For instance, Figure 6 shows a modification of the mechanical system in Example 2, in which a piston immersed in a cylinder filled with viscous fluid is now attached to the bottom of the ball. The effect of this damping mechanism, called a *dashpot*, is to subject the ball to an additional force of resistance $F_R = -bv$, where b is a positive constant called the *damping coefficient* and $v = dy/dt = y'$ is the velocity of the ball. Newton's second law now gives

$$my'' = -by' - ky \tag{11}$$

instead of (8), or equivalently

$$y'' + 2\lambda y' + \omega^2 y = 0, \tag{12}$$

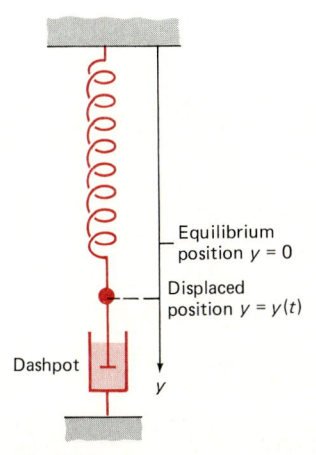

Figure 6

where

$$2\lambda = \frac{b}{m}, \qquad \omega^2 = \frac{k}{m}.$$

A mechanical system obeying the differential equation (12) is called a *damped harmonic oscillator*.

The characteristic equation of (12) is

$$r^2 + 2\lambda r + \omega^2 = 0,$$

or equivalently

$$(r + \lambda)^2 = \lambda^2 - \omega^2,$$

with two roots $r = -\lambda \pm \sqrt{\lambda^2 - \omega^2}$ if $\lambda > \omega$, one root $r = -\lambda$ if $\lambda = \omega$, and no real roots if $\lambda < \omega$. Let

$$\alpha = \sqrt{\lambda^2 - \omega^2} \quad \text{if} \quad \lambda > \omega, \qquad \beta = \sqrt{\omega^2 - \lambda^2} \quad \text{if} \quad \lambda < \omega.$$

Then, by the method described in Section 16.3, the solution of equation (12) is

$$y = e^{-\lambda t}(C_1 e^{\alpha t} + C_2 e^{-\alpha t}) \tag{13}$$

if $\lambda > \omega$ (the *overdamped case*),

$$y = e^{-\lambda t}(C_1 + C_2 t) \tag{14}$$

if $\lambda = \omega$ (the *critically damped case*), and

$$y = e^{-\lambda t}(C_1 \cos \beta t + C_2 \sin \beta t) \tag{15}$$

if $\lambda < \omega$ (the *underdamped case*). Assuming that the ball is initially at rest at the position $y = y_0$, from which it is released with zero velocity, we have the initial conditions

$$y(0) = y_0, \qquad y'(0) = 0,$$

which we now use to determine the constants C_1 and C_2 in each of the formulas (13), (14) and (15).

In the overdamped case ($\lambda > \omega$), we differentiate (13), obtaining

$$y' = -\lambda e^{-\lambda t}(C_1 e^{\alpha t} + C_2 e^{-\alpha t}) + \alpha e^{-\lambda t}(C_1 e^{\alpha t} - C_2 e^{-\alpha t}).$$

Setting $t = 0$, $y = y_0$ in (13) and $t = 0$, $y' = 0$ in the last formula, we get the system of equations

$$C_1 + C_2 = y_0, \qquad (\alpha - \lambda)C_1 - (\alpha + \lambda)C_2 = 0,$$

with solution

$$C_1 = \frac{\alpha + \lambda}{2\alpha} y_0, \qquad C_2 = \frac{\alpha - \lambda}{2\alpha} y_0,$$

so that (13) takes the form

$$y = y_0 e^{-\lambda t}\left(\frac{\alpha + \lambda}{2\alpha} e^{\alpha t} + \frac{\alpha - \lambda}{2\alpha} e^{-\alpha t}\right)$$

$$= y_0 e^{-\lambda t}\left(\cosh \alpha t + \frac{\lambda}{\alpha} \sinh \alpha t\right). \tag{16}$$

In the critically damped case ($\lambda = \omega$), differentiation of (14) gives
$$y' = -\lambda e^{-\lambda t}(C_1 + C_2 t) + C_2 e^{-\lambda t}.$$
Setting $t = 0$, $y' = 0$ here and $t = 0$, $y = y_0$ in (14), we find at once that
$$C_1 = y_0, \qquad -\lambda C_1 + C_2 = 0,$$
or equivalently $C_1 = y_0$, $C_2 = \lambda y_0$, so that (14) becomes
$$y = y_0 e^{-\lambda t}(1 + \lambda t). \tag{17}$$
In the underdamped case ($\lambda < \omega$), the derivative of (15) is
$$y' = -\lambda e^{-\lambda t}(C_1 \cos \beta t + C_2 \sin \beta t) + \beta e^{-\lambda t}(-C_1 \sin \beta t + C_2 \cos \beta t).$$
Thus, setting $t = 0$, $y' = 0$ here and $t = 0$, $y = y_0$ in (15), we immediately obtain
$$C_1 = y_0, \qquad -\lambda C_1 + \beta C_2 = 0,$$
or equivalently $C_1 = y_0$, $C_2 = \lambda y_0/\beta$, so that (15) becomes
$$y = y_0 e^{-\lambda t} \left(\cos \beta t + \frac{\lambda}{\beta} \sin \beta t \right). \tag{18}$$
As you can easily verify, (18) can be written in the alternative form
$$y = y_1 e^{-\lambda t} \sin (\beta t + \phi), \tag{18'}$$
where $y_1 = y_0 \sqrt{1 + (\lambda/\beta)^2}$ and $\phi = \arctan (\beta/\lambda)$.

It is apparent from (18') that the motion of the ball in the *underdamped* case is oscillatory, but as $t \to \infty$, the oscillations are "damped out" due to the presence of the exponential factor $e^{-\lambda t}$, at a rate depending on the size of λ (if λ is small, the oscillations can persist for a long time, but otherwise they are rapidly attenuated). The frequency of these "exponentially damped" oscillations is $\beta/2\pi$, and is smaller than the frequency $\omega/2\pi$ of the undamped oscillations that occur in the absence of resistive forces ($\lambda = 0$); actually, $\beta/2\pi$ should be called the "quasi-frequency" and $2\pi/\beta$ the "quasi-period," since the oscillations are not truly periodic. In the *overdamped* and *critically damped* cases, the motion is nonoscillatory. In fact, analysis of the functions (16) and (17) shows that they both decrease steadily to zero as $t \to \infty$, and never cross the t-axis. Figure 7 shows the graphs of (16)–(18) for $y_0 = 2$ and $\omega = 6$, with $\lambda = 10$ in the overdamped case, so that $\alpha = \sqrt{\lambda^2 - \omega^2} = 8$, and $\lambda = 1$ in the underdamped case, so that $\beta = \sqrt{\omega^2 - \lambda^2} = \sqrt{35} \approx 5.92$. Notice that the motion "dies out" faster in the critically damped case than in the overdamped case.

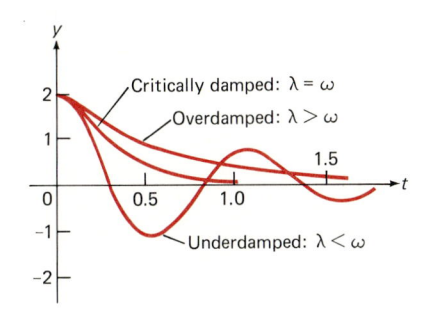

Figure 7

Forced Oscillations

Finally we consider the case of *forced oscillations*, in which the oscillating system is not free, but is subject to the continuing influence of an externally applied force, which we take to be $F_0 \cos \omega_0 t$ (a harmonic oscillation of frequency $\omega_0/2\pi$ and amplitude F_0). Let this force be applied to the ball in Figure 6, and for simplicity remove the dashpot, so that there is no resistive force. Then Newton's second law gives
$$my'' = -ky + F_0 \cos \omega_0 t$$

instead of (8), or equivalently

$$y'' + \omega^2 y = f_0 \cos \omega_0 t, \qquad (19)$$

where $\omega = \sqrt{k/m}$ and $f_0 = F_0/m$, and it is assumed that $\omega_0 \neq \omega$. Solving (19) by the method of Section 16.3, we find that

$$y = C_1 \cos \omega t + C_2 \sin \omega t + \frac{f_0}{\omega^2 - \omega_0^2} \cos \omega_0 t. \qquad (20)$$

The first two terms on the right are the general solution of the reduced equation $y'' + \omega^2 y = 0$, while the last term is a particular solution of the complete equation (19), as you can immediately verify. Suppose the ball is initially undisplaced and at rest. Then $y(0) = y'(0) = 0$, and the constants in (20) are easily found to be $C_1 = -f_0/(\omega^2 - \omega_0^2)$ and $C_2 = 0$, so that (20) becomes

$$y = \frac{f_0}{\omega^2 - \omega_0^2} (\cos \omega_0 t - \cos \omega t),$$

which can be written in the form

$$y = \frac{2f_0}{\omega^2 - \omega_0^2} \sin \left(\tfrac{1}{2}(\omega + \omega_0)t\right) \sin \left(\tfrac{1}{2}(\omega - \omega_0)t\right), \qquad (21)$$

with the help of the formula in Problem 62, page 61.

Suppose $|\omega - \omega_0|$ is small compared to $\omega + \omega_0$. Then (21) is a rapidly varying oscillation proportional to $\sin \left(\tfrac{1}{2}(\omega + \omega_0)t\right)$, "modulated" by the slowly varying oscillation $\sin \left(\tfrac{1}{2}(\omega - \omega_0)t\right)$. Thus the oscillation (21) is rapid, but its "envelope" undergoes slow periodic variations, known as *beats*. The sound of beats in the audible frequency range is experienced as dissonance; to confirm this, go to a piano and simultaneously strike two adjacent keys. Figure 8 illustrates the phenomenon of beats for the motion $y = 5 \sin 10t \sin 2t$, obtained from (21) by choosing $\omega = 12$, $\omega_0 = 8$ and $2f_0/(\omega^2 - \omega_0^2) = 5$.

If $\omega_0 = \omega$, so that the frequency of the "forcing function" $F = F_0 \cos \omega_0 t$ coincides with the natural frequency of the free system, we get a different solution of the differential equation (19), leading to an important phenomenon known as *resonance* (see Problem 21).

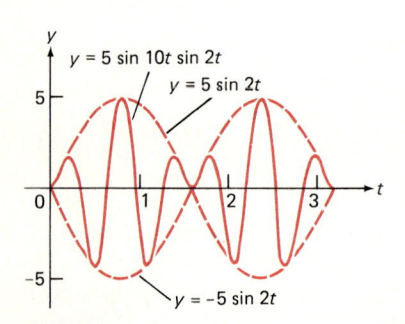

Figure 8

Problems

1. Show that every oscillation of the form (5′) can be written as an oscillation of the form (5) with the same amplitude and frequency.

2. Show that every harmonic oscillation of negative frequency can be written as a harmonic oscillation of positive frequency. Thus there is no loss of generality in assuming that $\omega > 0$.

3. Find the approximate length, along the groove, of one period of a sinusoidal oscillation of frequency 440 cps (A below middle C) cut into the outermost groove of a 12-inch phonograph record turning at the rate of $33\tfrac{1}{3}$ rpm (revolutions per minute).

4. Show that the sum of two harmonic oscillations of a given frequency is another harmonic oscillation of the same frequency.

5. Show that the speed of a particle executing simple harmonic motion is zero at the times when the particle's displacement from equilibrium is largest.

6. Show that the speed of a particle executing simple harmonic motion is largest at the times when the particle passes through its equilibrium position.

7. A particle executes simple harmonic motion with period $T = \pi$ sec, initial position $y(0) = -8$ cm and initial velocity $y'(0) = 12$ cm/sec. Find the amplitude A, angular frequency ω and initial phase ϕ of the motion. At what time t_1 does the particle first pass through the equilibrium position?

8. A spring is stretched 1.5 in. by a 3-lb weight. Suppose the weight is pulled 6 in. below the equilibrium position, and then released with zero initial velocity. Find the position y and velocity y' of the weight one quarter of a second later. (Choose $g = 32$ ft/sec^2 as the acceleration due to gravity.)

9. Was it really necessary to specify the weight of the object in the preceding problem? Explain your answer.

10. The period of oscillation of an object hung from a spring is 1 sec. Suppose the object is removed from the spring. How much shorter is the spring after it comes to rest?

Idealize the earth as a homogeneous spherical ball of radius $R = 3960$ miles, and suppose a hole has been drilled through the earth along a diameter. Then it can be shown (with the help of Problems 35 and 36 of Section 15.2) that an object inside the hole experiences an attractive force proportional to its distance from the center of the earth. Let an object be dropped from rest into the hole.

11. Find the time T it takes the object to return to its initial position at the surface of the earth.

12. Find the speed v with which the object passes through the center of the earth.

13. Show that T and v are the period and speed of a satellite in a circular orbit that just skims the earth.

14. Let K be the kinetic energy and V the potential energy of an undamped harmonic oscillator. Show by direct calculation that the total energy $E = K + V$ of the oscillator is constant.

15. A spring is stretched 4 in. by an 8-lb weight, attached to a dashpot with damping coefficient $b = 5$ lb sec/ft. Suppose the weight is pulled down 6 in. below its equilibrium position and then gently released. Determine the subsequent motion of the weight.

16. A spring is stretched 6 in. by a 4-lb weight, attached to a dashpot whose damping coefficient is b lb sec/ft. For what values of b is the motion overdamped? Critically damped? Underdamped?

17. Let A_n ($n = 0, 1, 2, \ldots$) be the consecutive maxima of the damped harmonic oscillations (18). Then the quantity $\delta = \ln(A_n/A_{n+1})$ is called the *logarithmic decrement*. Express δ in terms of the constant λ and the quasi-period $T = 2\pi/\beta$ of the damped oscillations.

18. Show that the displacement y of the critically damped oscillations (17) is less than that of the overdamped oscillations (16) for all $t > 0$.

19. Figure 9 shows a familiar mechanical device, known as the (*simple*) *pendulum*, in which a bob of mass m attached to a cord of length L swings in a vertical plane. Let $\theta = \theta(t)$ be the angular displacement of the bob from its rest position, and suppose the cord is weightless. Show that θ satisfies the differential equation

$$\frac{d^2\theta}{dt^2} + \frac{g}{L}\sin\theta = 0.$$

Describe the motion of the pendulum if θ is small enough to justify replacing $\sin\theta$ by θ. This "linearizing" approximation is equivalent to dropping all but the first term in the power series $\sin\theta = \theta - \frac{1}{6}\theta^3 + \cdots$. What is the period T of the pendulum in this case?

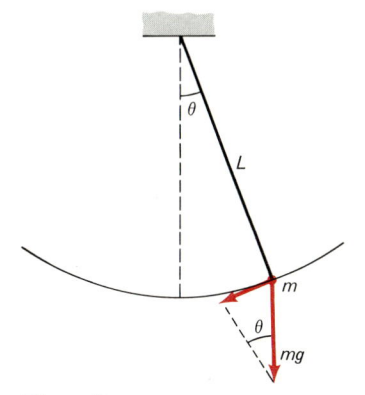

Figure 9

20. An electric circuit consists of an inductor, of inductance L henries, connected in series to a resistor, of resistance R ohms, and a capacitor, of capacitance C farads (see Figure 10). Let $q = q(t)$ and $i = i(t)$ be the charge on the capacitor and the current in the circuit at time t, and suppose that initially there is a charge on the capacitor, but no current in the circuit. Show that q and i are oscillatory if and only if $R < 2\sqrt{L/C}$.

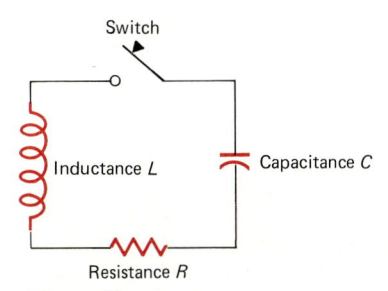

Figure 10

21. Suppose that $\omega_0 = \omega$ in equation (19), so that the frequency ω_0 of the forcing function coincides with the natural frequency of the free system. Solve (19) and show that the resulting oscillations get larger and larger until the spring eventually breaks. This phenomenon is known as *resonance*, and can lead to the collapse of mechanical structures. It is to avoid resonance that soldiers are ordered to break step when marching across small bridges.

22. Solve equation (19) for forced oscillations in the case where damping is present, so that (19) has an extra term and takes the form
$$y'' + 2\lambda y' + \omega^2 y = f_0 \cos \omega_0 t.$$
Assuming that $\lambda^2 < \tfrac{1}{2}\omega^2$, find the *resonant frequency*, that is, the frequency for which the resulting steady-state oscillations are the largest. Find the amplitude of the oscillations if $\omega_0 = \omega$.

Key Terms and Topics
Linear vs. nonlinear differential equations
Exact equations, condition for exactness
Integral curves
Integrating factors
First order linear equations
Homogeneous second order linear equations (with constant coefficients)
Nonhomogeneous second order linear equations
Existence and uniqueness theorem
Fundamental set of solutions
The characteristic equation
The method of undetermined coefficients
Simple harmonic motion, harmonic oscillations
Period, frequency, amplitude and phase
The undamped and damped harmonic oscillator
Overdamping, critical damping, underdamping
Forced oscillations

Supplementary Problems

Let F be a family of curves in the xy-plane. Then a curve orthogonal to every member of F is called an *orthogonal trajectory* of F. It is easy to see that if F consists of the integral curves of the differential equation

$$y' = f(x, y), \qquad \text{(i)}$$

where $f(x, y) \neq 0$, then every orthogonal trajectory of F is an integral curve of the differential equation

$$y' = -\frac{1}{f(x, y)}, \qquad \text{(ii)}$$

since the product of the slope of an integral curve of (i) with the slope of an integral curve of (ii) is -1 at any point of intersection of the two curves. For example, let F consist of all hyperbolas $x^2 - y^2 = C$, where C is an arbitrary constant. Then every member of F is an integral curve of the equation $y' = x/y$ (differentiate $x^2 - y^2 = C$ implicitly and then solve for y'), and therefore every orthogonal trajectory of F is an integral curve of the equation $y' = -y/x$, or equivalently $y\,dx + x\,dy = 0$, with general solution $xy = k$, where k is another arbitrary constant. Thus every hyperbola $xy = k$ is an orthogonal trajectory of the family of hyperbolas $x^2 - y^2 = C$, and by the same token, every hyperbola $x^2 - y^2 = C$ is an orthogonal trajectory of the family of hyperbolas $xy = k$, as illustrated in Figure 11.

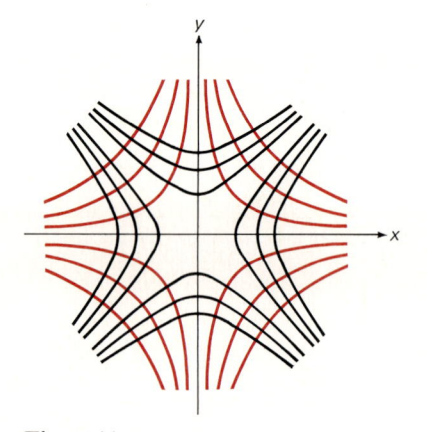

Figure 11

Use this method to find the orthogonal trajectories of the given family of curves.

1. The lines $y = Cx$
2. The parabolas $y^2 = Cx$
3. The circles $x^2 + y^2 = Cx$
4. The ellipses $x^2 + 2y^2 = C$
5. The hyperbolas $x^2 - 2y^2 = C$

6. Solve the linear differential equation $y' + py = q$ by substituting $y = uv$, where u and v are two unknown functions, choosing v to make the coefficient of u equal zero, and then integrating the equation so obtained.

7. It is often possible to solve a differential equation by interchanging the roles of the variables x and y. For example, the differential equation

$$\frac{dy}{dx}(\ln y - x) = y, \qquad \text{(iii)}$$

which is nonlinear in y, is linear in x, as can be seen by writing it in the equivalent form

$$\frac{dx}{dy} + \frac{x}{y} = \frac{\ln y}{y}. \qquad \text{(iii')}$$

Find the general solution of (iii'), and hence of (iii).

Use the method of the preceding problem to solve the given equation.

8. $(y - x)y' = 1$ 9. $(x - y^2)y' = y$
10. $(e^y + 2x)y' = 1$

Consider the second order differential equation

$$y'' = f(x, y'), \qquad \text{(iv)}$$

in which the right side depends on the independent variable x and the derivative y', but not on the unknown function y itself. Then (iv) reduces at once to the first order differential equation

$$p' = f(x, p) \qquad \text{(iv')}$$

in the new variable

$$p = y'.$$

Note that the last equation is actually a differential equation $y' = p$ of a very simple type. Thus, if $p = p(x, C_1)$ is the general solution of (iv') in explicit form, involving an arbitrary constant C_1, the general solution of the original equation (iv) is $y = \int p(x, C_1)\,dx + C_2$, where C_2 is another arbitrary constant. For example, if $y'' - y' = x$, then $p' - p = x$, and the general solution of this linear equation in p is easily found to be $p = -x - 1 + C_1 e^x$. It follows that $y = \int p\,dx = -\frac{1}{2}x^2 - x + C_1 e^x + C_2$.

Use the method just described to solve the given differential equation.

11. $xy'' + y' = 0$ 12. $y'' - y' = e^x$
13. $y'' = \sqrt{1 + y'^2}$ 14. $x^2 y'' + xy' = 1$

Consider the second order differential equation

$$y'' = f(y, y'), \qquad \text{(v)}$$

in which the right side depends on the unknown function y and its derivative y', but not on the independent variable x, and let $p = y' = dy/dx$ as in the preceding set of problems. Then

$$y'' = \frac{dp}{dy}\frac{dy}{dx} = \frac{dp}{dy}y' = \frac{dp}{dy}p,$$

so that (v) reduces to the first order differential equation

$$p\frac{dp}{dy} = f(y, p). \qquad \text{(v')}$$

Thus, if $p = p(y, C_1)$ is the general solution of (v') in explicit form, involving an arbitrary constant C_1, the general solution of the equation

$$\frac{dy}{dx} = p(y, C_1)$$

is in turn the general solution of (v). For example, if $yy'' - y'^2 = 0$, then $yp(dp/dy) - p^2 = 0$, or $y\,dp - p\,dy = 0$ after dividing by p. Multiplying the last equation by the integrating factor $1/y^2$ gives $(y\,dp - p\,dy)/y^2 = d(p/y) = 0$, so that $p/y = C_1$ or $dy/dx = p = C_1 y$. The general solution of the last equation (which is separable), and hence of the original equation $yy'' - y'^2 = 0$, is $y = C_2 e^{C_1 x}$. Note that the solution $y \equiv$ constant, lost in dividing by p, is recovered by setting $C_1 = 0$.

Use the method just described to solve the given differential equation.

15. $yy'' + y'^2 = 0$ 16. $yy'' - y'^2 = 1$
17. $y''(1 + y) = y'(1 + y')$ 18. $\sqrt{y}y'' = y'$

The following method of solving the second order nonhomogeneous linear equation $y'' + ay' + by = f(x)$ is called the method of *variation of parameters*. Let $y_1 = y_1(x)$ and $y_2 = y_2(x)$ be a fundamental set of solutions of the reduced equation $y'' + ay' + by = 0$, and let $u = u(x)$ and $v = v(x)$ be two unknown functions. Substitute $y = uy_1 + vy_2$ into the nonhomogeneous equation, and impose the condition $u'y_1 + v'y_2 = 0$ on the derivatives u' and v'. Then, as is easily verified, $(uy_1 + vy_2)'' + a(uy_1 + vy_2)' + b(uy_1 + vy_2) = f(x)$ simplifies to $u'y_1' + v'y_2' = f(x)$. The result is a system of two equations

$$\begin{aligned} u'y_1 + v'y_2 &= 0, \\ u'y_1' + v'y_2' &= f(x), \end{aligned} \qquad \text{(vi)}$$

and it can be shown that this system can always be solved for u' and v'. Evaluation of the integrals $u = \int u'\,dx$ and $v = \int v'\,dx$ (in cases where it is feasible) then gives the functions u and v, and a corresponding particular solution $y = uy_1 + vy_2$ of the nonhomogeneous equation. For instance, if $y'' - 2y' + y = 2e^x$, then $y_1 = e^x$ and $y_2 = xe^x$ are a fundamental set of solutions of the reduced equation $y'' - 2y' + y = 0$, as shown in Example 10, page 958, and the system (vi) becomes

$$\begin{aligned} u'e^x + v'xe^x &= 0, \\ u'e^x + v'(e^x + xe^x) &= 2e^x, \end{aligned}$$

with solution $u' = -2x$, $v' = 2$. Therefore $u = \int -2x\,dx = -x^2$, $v = \int 2\,dx = 2x$, so that $y = -x^2 e^x + 2x^2 e^x = x^2 e^x$ is a particular solution of the nonhomogeneous equation $y'' - 2y' + y = 2e^x$, as already found in the cited example.

Use the method just described to solve the given differential equation.

19. $y'' + y = \tan x$
20. $y'' + y = \sec x$
21. $y'' - y = 2\sin^2 x$
22. $y'' - 2y' + y = e^x/x$
23. $y'' - 2y' + y = e^x \ln x$
24. $y'' + 3y' + 2y = \cos(e^x)$

Find the homogeneous linear equation $y'' + ay' + by = 0$ with the given fundamental set of solutions.

25. $e^{-4x}\cos 5x, e^{-4x}\sin 5x$
26. e^{-10x}, xe^{-10x}
27. $e^{3x}, \sinh 3x$
28. e^{-99x}, e^{100x}

Solve the given system of first order differential equations, involving two unknown functions $x = x(t)$ and $y = y(t)$ of an independent variable t. (First find a second order differential equation satisfied by x by eliminating the other function y.)

29. $\dfrac{dx}{dt} = y, \dfrac{dy}{dt} = -x$

30. $\dfrac{dx}{dt} = x + y, \dfrac{dy}{dt} = x - y$

31. $\dfrac{dx}{dt} = x + y, \dfrac{dy}{dt} = t + x + y$

32. $\dfrac{dx}{dt} = 2x + y, \dfrac{dy}{dt} = x + 2y$

33. A cylindrical buoy floats partly submerged in a lake. Suppose the buoy is pushed down slightly deeper and then gently released. Show that the buoy executes simple harmonic motion in the vertical direction (neglect any resistive forces). Find the period T of the oscillations if the buoy sinks to a depth of 2 ft below the water's surface when in equilibrium.

34. Let E be the total energy of a damped harmonic oscillator. Show that E decreases at the rate bv^2, where v is the velocity of the oscillator and b is its damping coefficient.

It is often possible to solve a differential equation by assuming that the solution is a convergent power series

$$y = y(x) = \sum_{n=0}^{\infty} a_n x^n. \qquad \text{(vii)}$$

For example, consider the problem of solving the differential equation $y'' = 2xy' + 4y$. Differentiating (vii) twice, with the help of Theorem 14, page 540, we obtain the power series

$$y' = \sum_{n=1}^{\infty} na_n x^{n-1}, \quad y'' = \sum_{n=2}^{\infty} n(n-1)x^{n-2}, \quad \text{(vii')}$$

with the same radius of convergence as (vii). Substitution of (vii) and (vii') into $y'' = 2xy' + 4y$ gives

$$\sum_{n=2}^{\infty} n(n-1)a_n x^{n-2} = 2x \sum_{n=1}^{\infty} na_n x^{n-1} + 4 \sum_{n=0}^{\infty} a_n x^n,$$

$$(n = 1, 2, \ldots)$$

or equivalently

$$\sum_{n=0}^{\infty} (n+2)(n+1)a_{n+2}x^n = 2\sum_{n=1}^{\infty} na_n x^n + 4\sum_{n=0}^{\infty} a_n x^n.$$

If this equation is to be an identity, then identical powers of x must have the same coefficients (recall Example 4, page 541). Therefore $2a_2 = 4a_0$ or $a_2 = 2a_0$, and

$$(n+2)(n+1)a_{n+2} = (2n+4)a_n \quad (n = 1, 2, \ldots).$$

Hence the coefficients of the series (vii) satisfy the recursion formula

$$a_{n+2} = \frac{2}{n+1} a_n \quad (n = 0, 1, 2, \ldots). \qquad \text{(viii)}$$

The coefficients a_0 and a_1 can be given arbitrary values, but once these coefficients have been chosen, the remaining coefficients $a_2, a_3, \ldots$ are determined by formula (viii). In fact, it is easily verified that (viii) implies

$$a_{2k} = \frac{2^k}{(2k-1)(2k-3)\cdots 3 \cdot 1} a_0, \quad a_{2k+1} = \frac{1}{k!} a_1$$

$$(k = 1, 2, \ldots).$$

Substituting these values of the coefficients into (vii), we finally find that the general solution of the given differential equation $y'' = 2xy' + 4y$ is $y = a_0 y_0(x) + a_1 y_1(x)$, where the functions $y_0 = y_0(x)$ and $y_1 = y_1(x)$ have the power series expansions

$$y_0 = 1 + \sum_{k=1}^{\infty} \frac{2^k x^{2k}}{(2k-1)(2k-3)\cdots 3 \cdot 1}, \quad y_1 = \sum_{k=0}^{\infty} \frac{x^{2k+1}}{k!}.$$

It follows from the ratio test that both of these series converge for all x (actually y_1 is readily identified as the function xe^{x^2}).

Use the method just described to solve the given differential equation.

35. $y'' = 2xy' - 4y$
36. $y'' + xy' + y = 0$
37. $y'' = xy$
38. $y'' = x^2 y$

Appendix

Mathematical Induction

In mathematics one often encounters statements or formulas involving an arbitrary positive integer n. For example, consider the formula

$$\sum_{k=1}^{n} (2k-1) = 1 + 3 + \cdots + (2n-1) = n^2, \tag{1}$$

which says that the sum of the first n odd numbers $1, 3, \ldots, 2n-1$ equals n^2. A little experimentation ($1 + 3 = 4 = 2^2$, $1 + 3 + 5 = 9 = 3^2, \ldots$) certainly makes the formula plausible, but what we need is a *proof*. To prove a formula like (1), we can use the following important technique, known as the principle of *mathematical induction*. Suppose the formula (or statement) is known to be true for $n = 1$, and suppose that by *assuming* its truth for $n = m$, where m is an arbitrary positive integer, we can *deduce* its truth for the next integer, that is, for $n = m + 1$. Then the formula is true for all $n = 1, 2, \ldots$ In fact, from the validity of the formula for $n = 1$ we deduce its validity for $n = 1 + 1 = 2$, then from the validity of the formula for $n = 2$ (as just established), we deduce its validity for $n = 2 + 1 = 3$, and so on indefinitely. Repeating this argument as often as necessary, we can establish the validity of the formula for every positive integer n, *no matter how large*, and hence for all $n = 1, 2, \ldots$

Thus, to prove (1), we first note that the formula is certainly true for $n = 1$, since it then reduces to the trivial statement that $1 = 1^2$. Suppose (1) holds for $n = m$, so that

$$\sum_{k=1}^{m} (2k-1) = 1 + 3 + \cdots + (2m-1) = m^2.$$

Then adding $2m + 1$ to both sides of this equation (where $2m + 1$ is the next odd number after $2m - 1$), we get

$$\sum_{k=1}^{m+1} (2k-1) = 1 + 3 + \cdots + (2m-1) + (2m+1) = m^2 + 2m + 1,$$

that is,
$$\sum_{k=1}^{m+1}(2k-1) = 1 + 3 + \cdots + (2m+1) = (m+1)^2.$$

But this is just the form taken by (1) if $n = m + 1$ (note that $2(m + 1) - 1 = 2m + 1$). Therefore, by the principle of mathematical induction, formula (1) is true for all $n = 1, 2, \ldots$

The truth of the given statement (or formula) for $n = 1$ is only needed "to get the induction started." This condition can be relaxed. For example, suppose the statement is known to be true for $n = 8$, and suppose its truth for $n = m$ ($m \geq 8$) implies its truth for $n = m + 1$. Then the statement is true for all n starting from 8, that is, for all $n = 8, 9, \ldots$ Far from being a wild example, this is actually the situation in Problem 19.

Example 1 Use mathematical induction to show that
$$\sum_{k=1}^{n} k = 1 + 2 + \cdots + n = \frac{1}{2}n(n+1) \tag{2}$$
for all $n = 1, 2, \ldots$

Solution Formula (2) is true for $n = 1$, since it then reduces to $1 = \frac{1}{2} \cdot 1 \cdot 2$. Suppose (2) is true for $n = m$, so that
$$\sum_{k=1}^{m} k = 1 + 2 + \cdots + m = \frac{1}{2}m(m+1).$$
Then adding $m + 1$ to both sides of this equation, we get
$$\sum_{k=1}^{m+1} k = 1 + 2 + \cdots + m + (m+1)$$
$$= \frac{1}{2}m(m+1) + (m+1) = (m+1)\left(\frac{1}{2}m + 1\right) = \frac{1}{2}(m+1)(m+2),$$
which is precisely the result of substituting $n = m + 1$ into (2). Therefore, by mathematical induction, formula (2) holds for all $n = 1, 2, \ldots$ □

Example 2 Use mathematical induction to give another proof of the key differentiation formula
$$\frac{d}{dx}x^n = nx^{n-1}, \tag{3}$$
valid for every positive integer n.

Solution Formula (3) holds for $n = 1$, since it then says that $D_x x = 1$, which is true. Suppose (3) holds for $n = m$, so that
$$\frac{d}{dx}x^m = mx^{m-1}.$$
Then
$$\frac{d}{dx}x^{m+1} = \frac{d}{dx}(x^m \cdot x) = \frac{dx^m}{dx}x + x^m\frac{dx}{dx} = mx^{m-1}x + x^m \cdot 1 = mx^m + x^m,$$
with the help of the product rule for differentiation. Therefore
$$\frac{d}{dx}x^{m+1} = (m+1)x^m,$$

which is just the form taken by (3) for $n = m + 1$. The validity of (3) for every positive integer n now follows by mathematical induction. □

Example 3 Use mathematical induction to prove the binomial theorem

$$(a + b)^n = a^n + \binom{n}{1} a^{n-1} b + \binom{n}{2} a^{n-2} b^2 + \cdots + \binom{n}{n-1} a b^{n-1} + b^n, \quad (4)$$

(see Example 5, page 227), where

$$\binom{n}{k} = \frac{n!}{k!(n-k)!} = \frac{n(n-1) \cdots (n-k+1)}{k!}. \quad (5)$$

Solution Formula (4) holds for $n = 1$, since in this case it simply says that $(a + b)^1 = a + b$. Suppose (4) is true for $n = m$. Then

$$(a + b)^m = a^m + \binom{m}{1} a^{m-1} b + \binom{m}{2} a^{m-2} b^2 + \cdots + \binom{m}{m-1} a b^{m-1} + b^m.$$

Multiplying both sides of this equation by $a + b$, we get

$$(a + b)^{m+1} = (a + b) \left[a^m + \binom{m}{1} a^{m-1} b + \binom{m}{2} a^{m-2} b^2 \right.$$

$$\left. + \cdots + \binom{m}{m-1} a b^{m-1} + b^m \right]$$

$$= a^{m+1} + \left[1 + \binom{m}{1} \right] a^m b + \left[\binom{m}{1} + \binom{m}{2} \right] a^{m-1} b^2$$

$$+ \cdots + \left[\binom{m}{m-1} + 1 \right] a b^m + b^{m+1},$$

or equivalently

$$(a + b)^{m+1} = a^{m+1} + \left[\binom{m}{0} + \binom{m}{1} \right] a^m b + \left[\binom{m}{1} + \binom{m}{2} \right] a^{m-1} b^2$$

$$+ \cdots + \left[\binom{m}{m-1} + \binom{m}{m} \right] a b^m + b^{m+1}, \quad (6)$$

since

$$\binom{m}{0} = \binom{m}{m} = 1.$$

But

$$\binom{m}{k} + \binom{m}{k+1} = \frac{m(m-1) \cdots (m-k+1)}{k!} + \frac{m(m-1) \cdots (m-k)}{(k+1)!},$$

with the help of (5), and hence, since $(k+1)! = (k+1)k!$,

$$\binom{m}{k} + \binom{m}{k+1} = \frac{m(m-1) \cdots (m-k+1)}{k!} \left(1 + \frac{m-k}{k+1} \right)$$

$$= \frac{(m+1)m(m-1) \cdots (m-k+1)}{(k+1)!} = \binom{m+1}{k+1}.$$

Mathematical Induction **A-3**

Using this formula to simplify equation (6), we find that

$$(a+b)^{m+1} = a^{m+1} + \binom{m+1}{1}a^m b + \binom{m+1}{2}a^{m-1}b^2$$
$$+ \cdots + \binom{m+1}{m}ab^m + b^{m+1},$$

which is precisely the form taken by equation (4) for $n = m + 1$. Therefore, by mathematical induction, the binomial theorem (4) holds for every positive integer n. ☐

Problems

1. What is the sum of the first 1000 positive integers?

Let n be any positive integer. Use mathematical induction to show that

2. $1^2 + 2^2 + \cdots + n^2 = \dfrac{1}{6}n(n+1)(2n+1)$

3. $1^3 + 2^3 + \cdots + n^3 = \dfrac{1}{4}n^2(n+1)^2 = (1 + 2 + \cdots + n)^2$

4. $1 \cdot 2 + 2 \cdot 3 + \cdots + n(n+1) = \dfrac{1}{3}n(n+1)(n+2)$

5. $\dfrac{1}{1 \cdot 2} + \dfrac{1}{2 \cdot 3} + \cdots + \dfrac{1}{n(n+1)} = \dfrac{n}{n+1}$

6. $\dfrac{1}{1 \cdot 3} + \dfrac{1}{3 \cdot 5} + \cdots + \dfrac{1}{(2n-1)(2n+1)} = \dfrac{n}{2n+1}$

7. $\left(1 - \dfrac{1}{4}\right)\left(1 - \dfrac{1}{9}\right) \cdots \left(1 - \dfrac{1}{n^2}\right) = \dfrac{n+1}{2n}$ $(n > 1)$

8. $1 \cdot 1! + 2 \cdot 2! + \cdots + n \cdot n! = (n+1)! - 1$

9. $1^2 - 2^2 + 3^2 - 4^2 + \cdots + (-1)^{n-1}n^2 = \dfrac{1}{2}(-1)^{n-1}n(n+1)$

10. $\dfrac{1}{1 \cdot 4} + \dfrac{1}{4 \cdot 7} + \cdots + \dfrac{1}{(3n-2)(3n+1)} = \dfrac{n}{3n+1}$

11. $\dfrac{1^2}{1 \cdot 3} + \dfrac{2^2}{3 \cdot 5} + \cdots + \dfrac{n^2}{(2n-1)(2n+1)} = \dfrac{n(n+1)}{2(2n+1)}$

12. $1^4 + 2^4 + \cdots + n^4 = \dfrac{1}{30}n(n+1)(6n^3 + 9n^2 + n - 1)$

13. Use mathematical induction to show that

$$1 + x + x^2 + \cdots + x^n = \dfrac{x^{n+1} - 1}{x - 1} \quad (x \neq 1) \quad \text{(i)}$$

for all $n = 1, 2, \ldots$

14. An advisor to a certain king was asked what he would like as a reward for interpreting one of the king's dreams. He asked for a chessboard with one grain of rice on the first square, twice as much rice on the second square as on the first, twice as much on the third square as on the second, and so on. Why did the king have his advisor executed for insolence?

15. Arriving at a stadium filled with a capacity crowd of 50,000 spectators, a baseball fan learns from a ticket taker that the manager of the opposing team has just been fired. A minute later the fan tells this news to two other fans. Then a minute later each of these fans tells the news to two other fans who haven't heard the news yet, and so on, until the whole crowd has heard the news. How long does this take?

16. By differentiating formula (i), show that

$$1 + 2x + 3x^2 + \cdots + nx^{n-1} = \dfrac{nx^{n+1} - (n+1)x^n + 1}{(x-1)^2} \quad \text{(ii)}$$

for $x \neq 1$ and all $n = 1, 2, \ldots$

17. Use formula (ii) to show that $99x^{100} - 100x^{99} + 1$ has $x^2 - 2x + 1$ as a factor.

18. Show that the sum of the cubes of three consecutive positive integers is exactly divisible by 9.

19. Show that every integer greater than 7 can be written as a sum of threes and fives exclusively (thus $8 = 5 + 3$, $9 = 3 + 3 + 3$, $10 = 5 + 5$, $11 = 3 + 3 + 5$, and so on).

20. Show that n straight lines, no two of which are parallel and no three of which have a point in common, divide the plane into $\frac{1}{2}(n^2 + n + 2)$ regions.

Areas and Volumes

Parallelogram
Area = bh

Triangle
Area = $\frac{1}{2}bh$

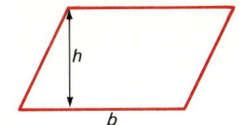

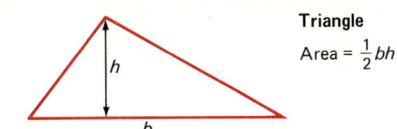

Circular disk
Area = πr^2
Circumference = $2\pi r$

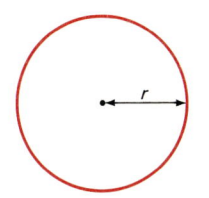

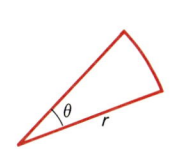

Circular sector
Area = $\frac{1}{2}r^2\theta$
Arc length = $r\theta$
(θ in radians)

Rectangular parallelepiped
Volume = abc
Surface area = $2(ab + ac + bc)$

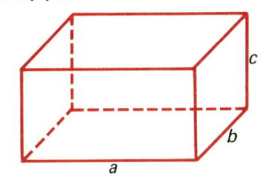

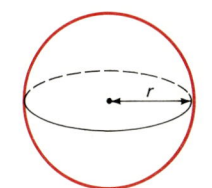

Sphere
Volume = $\frac{4}{3}\pi r^3$
Surface area = $4\pi r^2$

Right circular cylinder
Volume = $\pi r^2 h$
Lateral surface area = $2\pi rh$

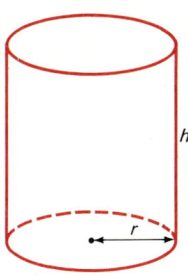

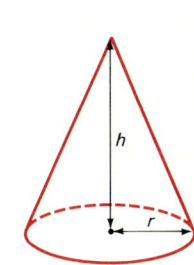

Right circular cone
Volume = $\frac{1}{3}\pi r^2 h$
Lateral surface area = $\pi r \sqrt{r^2 + h^2}$

The Greek Alphabet

α	A	alpha	ι	I	iota	ρ	P	rho
β	B	beta	κ	K	kappa	σ	Σ	sigma
γ	Γ	gamma	λ	Λ	lambda	τ	T	tau
δ	Δ	delta	μ	M	mu	υ	Υ	upsilon
ε	E	epsilon	ν	N	nu	ϕ	Φ	phi
ζ	Z	zeta	ξ	Ξ	xi	χ	X	chi
η	H	eta	o	O	omicron	ψ	Ψ	psi
θ	Θ	theta	π	Π	pi	ω	Ω	omega

Numerical Tables

Table 1. Trigonometric Functions, Degree Measure

Angle	sin	tan	cot	cos	—
0.0°	0.0000	0.0000	—	1.0000	**90.0°**
0.5°	0.0087	0.0087	114.59	1.0000	89.5°
1.0°	0.0175	0.0175	57.290	0.9998	89.0°
1.5°	0.0262	0.0262	38.188	0.9997	88.5°
2.0°	0.0349	0.0349	28.636	0.9994	88.0°
2.5°	0.0436	0.0437	22.904	0.9990	**87.5°**
3.0°	0.0523	0.0524	19.081	0.9986	87.0°
3.5°	0.0610	0.0612	16.350	0.9981	86.5°
4.0°	0.0698	0.0699	14.301	0.9976	86.0°
4.5°	0.0785	0.0787	12.706	0.9969	85.5°
5.0°	0.0872	0.0875	11.430	0.9962	**85.0°**
5.5°	0.0958	0.0963	10.385	0.9954	84.5°
6.0°	0.1045	0.1051	9.5144	0.9945	84.0°
6.5°	0.1132	0.1139	8.7769	0.9936	83.5°
7.0°	0.1219	0.1228	8.1443	0.9925	83.0°
7.5°	0.1305	0.1317	7.5958	0.9914	**82.5°**
8.0°	0.1392	0.1405	7.1154	0.9903	82.0°
8.5°	0.1478	0.1495	6.6912	0.9890	81.5°
9.0°	0.1564	0.1584	6.3138	0.9877	81.0°
9.5°	0.1650	0.1673	5.9758	0.9863	80.5°
10.0°	0.1736	0.1763	5.6713	0.9848	**80.0°**
10.5°	0.1822	0.1853	5.3955	0.9833	79.5°
11.0°	0.1908	0.1944	5.1446	0.9816	79.0°
11.5°	0.1994	0.2035	4.9152	0.9799	78.5°
12.0°	0.2079	0.2126	4.7046	0.9781	78.0°
12.5°	0.2164	0.2217	4.5107	0.9763	**77.5°**
13.0°	0.2250	0.2309	4.3315	0.9744	77.0°
13.5°	0.2334	0.2401	4.1653	0.9724	76.5°
14.0°	0.2419	0.2493	4.0108	0.9703	76.0°
14.5°	0.2504	0.2586	3.8667	0.9681	75.5°
15.0°	0.2588	0.2679	3.7321	0.9659	**75.0°**
15.5°	0.2672	0.2773	3.6059	0.9636	74.5°
16.0°	0.2756	0.2867	3.4874	0.9613	74.0°
16.5°	0.2840	0.2962	3.3759	0.9588	73.5°
17.0°	0.2924	0.3057	3.2709	0.9563	73.0°
17.5°	0.3007	0.3153	3.1716	0.9537	**72.5°**
18.0°	0.3090	0.3249	3.0777	0.9511	72.0°
18.5°	0.3173	0.3346	2.9887	0.9483	71.5°
19.0°	0.3256	0.3443	2.9042	0.9455	71.0°
19.5°	0.3338	0.3541	2.8239	0.9426	70.5°
20.0°	0.3420	0.3640	2.7475	0.9397	**70.0°**
20.5°	0.3502	0.3739	2.6746	0.9367	69.5°
21.0°	0.3584	0.3839	2.6051	0.9336	69.0°
21.5°	0.3665	0.3939	2.5386	0.9304	68.5°
22.0°	0.3746	0.4040	2.4751	0.9272	68.0°
22.5°	0.3827	0.4142	2.4142	0.9239	**67.5°**
—	cos	cot	tan	sin	Angle

Angle	sin	tan	cot	cos	—
22.5°	0.3827	0.4142	2.4142	0.9239	**67.5°**
23.0°	0.3907	0.4245	2.3559	0.9205	67.0°
23.5°	0.3987	0.4348	2.2998	0.9171	66.5°
24.0°	0.4067	0.4452	2.2460	0.9135	66.0°
24.5°	0.4147	0.4557	2.1943	0.9100	65.5°
25.0°	0.4226	0.4663	2.1445	0.9063	**65.0°**
25.5°	0.4305	0.4770	2.0965	0.9026	64.5°
26.0°	0.4384	0.4877	2.0503	0.8988	64.0°
26.5°	0.4462	0.4986	2.0057	0.8949	63.5°
27.0°	0.4540	0.5095	1.9626	0.8910	63.0°
27.5°	0.4617	0.5206	1.9210	0.8870	**62.5°**
28.0°	0.4695	0.5317	1.8807	0.8829	62.0°
28.5°	0.4772	0.5430	1.8418	0.8788	61.5°
29.0°	0.4848	0.5543	1.8040	0.8746	61.0°
29.5°	0.4924	0.5658	1.7675	0.8704	60.5°
30.0°	0.5000	0.5774	1.7321	0.8660	**60.0°**
30.5°	0.5075	0.5890	1.6977	0.8616	59.5°
31.0°	0.5150	0.6009	1.6643	0.8572	59.0°
31.5°	0.5225	0.6128	1.6319	0.8526	58.5°
32.0°	0.5299	0.6249	1.6003	0.8480	58.0°
32.5°	0.5373	0.6371	1.5697	0.8434	**57.5°**
33.0°	0.5446	0.6494	1.5399	0.8387	57.0°
33.5°	0.5519	0.6619	1.5108	0.8339	56.5°
34.0°	0.5592	0.6745	1.4826	0.8290	56.0°
34.5°	0.5664	0.6873	1.4550	0.8241	55.5°
35.0°	0.5736	0.7002	1.4281	0.8192	**55.0°**
35.5°	0.5807	0.7133	1.4019	0.8141	54.5°
36.0°	0.5878	0.7265	1.3764	0.8090	54.0°
36.5°	0.5948	0.7400	1.3514	0.8039	53.5°
37.0°	0.6018	0.7536	1.3270	0.7986	53.0°
37.5°	0.6088	0.7673	1.3032	0.7934	**52.5°**
38.0°	0.6157	0.7813	1.2799	0.7880	52.0°
38.5°	0.6225	0.7954	1.2572	0.7826	51.5°
39.0°	0.6293	0.8098	1.2349	0.7771	51.0°
39.5°	0.6361	0.8243	1.2131	0.7716	50.5°
40.0°	0.6428	0.8391	1.1918	0.7660	**50.0°**
40.5°	0.6494	0.8541	1.1708	0.7604	49.5°
41.0°	0.6561	0.8693	1.1504	0.7547	49.0°
41.5°	0.6626	0.8847	1.1303	0.7490	48.5°
42.0°	0.6691	0.9004	1.1106	0.7431	48.0°
42.5°	0.6756	0.9163	1.0913	0.7373	**47.5°**
43.0°	0.6820	0.9325	1.0724	0.7314	47.0°
43.5°	0.6884	0.9490	1.0538	0.7254	46.5°
44.0°	0.6947	0.9657	1.0355	0.7193	46.0°
44.5°	0.7009	0.9827	1.0176	0.7133	45.5°
45.0°	0.7071	1.0000	1.0000	0.7071	**45.0°**
—	cos	cot	tan	sin	Angle

Numerical Tables A-7

Table 2. Trigonometric Functions, Radian Measure

Radians	sin	cos	tan	Radians	sin	cos	tan
0.00	0.0000	1.0000	0.0000	0.40	0.3894	0.9211	0.4228
0.01	0.0100	1.0000	0.0100	0.41	0.3986	0.9171	0.4346
0.02	0.0200	0.9998	0.0200	0.42	0.4078	0.9131	0.4466
0.03	0.0300	0.9996	0.0300	0.43	0.4169	0.9090	0.4586
0.04	0.0400	0.9992	0.0400	0.44	0.4259	0.9048	0.4708
0.05	0.0500	0.9988	0.0500	0.45	0.4350	0.9004	0.4831
0.06	0.0600	0.9982	0.0601	0.46	0.4439	0.8961	0.4954
0.07	0.0699	0.9976	0.0701	0.47	0.4529	0.8916	0.5080
0.08	0.0799	0.9968	0.0802	0.48	0.4618	0.8870	0.5206
0.09	0.0899	0.9960	0.0902	0.49	0.4706	0.8823	0.5334
0.10	0.0998	0.9950	0.1003	0.50	0.4794	0.8776	0.5463
0.11	0.1098	0.9940	0.1104	0.51	0.4882	0.8727	0.5594
0.12	0.1197	0.9928	0.1206	0.52	0.4969	0.8678	0.5726
0.13	0.1296	0.9916	0.1307	0.53	0.5055	0.8628	0.5859
0.14	0.1395	0.9902	0.1409	0.54	0.5141	0.8577	0.5994
0.15	0.1494	0.9888	0.1511	0.55	0.5227	0.8525	0.6131
0.16	0.1593	0.9872	0.1614	0.56	0.5312	0.8473	0.6269
0.17	0.1692	0.9856	0.1717	0.57	0.5396	0.8419	0.6410
0.18	0.1790	0.9838	0.1820	0.58	0.5480	0.8365	0.6552
0.19	0.1889	0.9820	0.1923	0.59	0.5564	0.8309	0.6696
0.20	0.1987	0.9801	0.2027	0.60	0.5646	0.8253	0.6841
0.21	0.2085	0.9780	0.2131	0.61	0.5729	0.8196	0.6989
0.22	0.2182	0.9759	0.2236	0.62	0.5810	0.8139	0.7139
0.23	0.2280	0.9737	0.2341	0.63	0.5891	0.8080	0.7291
0.24	0.2377	0.9713	0.2447	0.64	0.5972	0.8021	0.7445
0.25	0.2474	0.9689	0.2553	0.65	0.6052	0.7961	0.7602
0.26	0.2571	0.9664	0.2660	0.66	0.6131	0.7900	0.7761
0.27	0.2667	0.9638	0.2768	0.67	0.6210	0.7838	0.7923
0.28	0.2764	0.9611	0.2876	0.68	0.6288	0.7776	0.8087
0.29	0.2860	0.9582	0.2984	0.69	0.6365	0.7712	0.8253
0.30	0.2955	0.9553	0.3093	0.70	0.6442	0.7648	0.8423
0.31	0.3051	0.9523	0.3203	0.71	0.6518	0.7584	0.8595
0.32	0.3146	0.9492	0.3314	0.72	0.6594	0.7518	0.8771
0.33	0.3240	0.9460	0.3425	0.73	0.6669	0.7452	0.8949
0.34	0.3335	0.9428	0.3537	0.74	0.6743	0.7385	0.9131
0.35	0.3429	0.9394	0.3650	0.75	0.6816	0.7317	0.9316
0.36	0.3523	0.9359	0.3764	0.76	0.6889	0.7248	0.9505
0.37	0.3616	0.9323	0.3879	0.77	0.6961	0.7179	0.9697
0.38	0.3709	0.9287	0.3994	0.78	0.7033	0.7109	0.9893
0.39	0.3802	0.9249	0.4111	0.79	0.7104	0.7038	1.009

Table 2. Trigonometric Functions, Radian Measure (*cont.*)

Radians	sin	cos	tan
0.80	0.7174	0.6967	1.030
0.81	0.7243	0.6895	1.050
0.82	0.7311	0.6822	1.072
0.83	0.7379	0.6749	1.093
0.84	0.7446	0.6675	1.116
0.85	0.7513	0.6600	1.138
0.86	0.7578	0.6524	1.162
0.87	0.7643	0.6448	1.185
0.88	0.7707	0.6372	1.210
0.89	0.7771	0.6294	1.235
0.90	0.7833	0.6216	1.260
0.91	0.7895	0.6137	1.286
0.92	0.7956	0.6058	1.313
0.93	0.8016	0.5978	1.341
0.94	0.8076	0.5898	1.369
0.95	0.8134	0.5817	1.398
0.96	0.8192	0.5735	1.428
0.97	0.8249	0.5653	1.459
0.98	0.8305	0.5570	1.491
0.99	0.8360	0.5487	1.524
1.00	0.8415	0.5403	1.557
1.01	0.8468	0.5319	1.592
1.02	0.8521	0.5234	1.628
1.03	0.8573	0.5148	1.665
1.04	0.8624	0.5062	1.704
1.05	0.8674	0.4976	1.743
1.06	0.8724	0.4889	1.784
1.07	0.8772	0.4801	1.827
1.08	0.8820	0.4713	1.871
1.09	0.8866	0.4625	1.917
1.10	0.8912	0.4536	1.965
1.11	0.8957	0.4447	2.014
1.12	0.9001	0.4357	2.066
1.13	0.9044	0.4267	2.120
1.14	0.9086	0.4176	2.176
1.15	0.9128	0.4085	2.234
1.16	0.9168	0.3993	2.296
1.17	0.9208	0.3902	2.360
1.18	0.9246	0.3809	2.427
1.19	0.9284	0.3717	2.498

Radians	sin	cos	tan
1.20	0.9320	0.3624	2.572
1.21	0.9356	0.3530	2.650
1.22	0.9391	0.3436	2.733
1.23	0.9425	0.3342	2.820
1.24	0.9458	0.3248	2.912
1.25	0.9490	0.3153	3.010
1.26	0.9521	0.3058	3.113
1.27	0.9551	0.2963	3.224
1.28	0.9580	0.2867	3.341
1.29	0.9608	0.2771	3.467
1.30	0.9636	0.2675	3.602
1.31	0.9662	0.2579	3.747
1.32	0.9687	0.2482	3.903
1.33	0.9711	0.2385	4.072
1.34	0.9735	0.2288	4.256
1.35	0.9757	0.2190	4.455
1.36	0.9779	0.2092	4.673
1.37	0.9799	0.1994	4.913
1.38	0.9819	0.1896	5.177
1.39	0.9837	0.1798	5.471
1.40	0.9854	0.1700	5.798
1.41	0.9871	0.1601	6.165
1.42	0.9887	0.1502	6.581
1.43	0.9901	0.1403	7.055
1.44	0.9915	0.1304	7.602
1.45	0.9927	0.1205	8.238
1.46	0.9939	0.1106	8.989
1.47	0.9949	0.1006	9.887
1.48	0.9959	0.0907	10.98
1.49	0.9967	0.0807	12.35
1.50	0.9975	0.0707	14.10
1.51	0.9982	0.0608	16.43
1.52	0.9987	0.0508	19.67
1.53	0.9992	0.0408	24.50
1.54	0.9995	0.0308	32.46
1.55	0.9998	0.0208	48.08
1.56	0.9999	0.0108	92.62
1.57	1.0000	0.0008	1256.

Table 3. Natural Logarithms

	0.00	0.01	0.02	0.03	0.04	0.05	0.06	0.07	0.08	0.09
1.0	0.0000	0.0100	0.0198	0.0296	0.0392	0.0488	0.0583	0.0677	0.0770	0.0862
1.1	0.0953	0.1044	0.1133	0.1222	0.1310	0.1398	0.1484	0.1570	0.1655	0.1740
1.2	0.1823	0.1906	0.1989	0.2070	0.2151	0.2231	0.2311	0.2390	0.2469	0.2546
1.3	0.2624	0.2700	0.2776	0.2852	0.2927	0.3001	0.3075	0.3148	0.3221	0.3293
1.4	0.3365	0.3436	0.3507	0.3577	0.3646	0.3716	0.3784	0.3853	0.3920	0.3988
1.5	0.4055	0.4121	0.4187	0.4253	0.4318	0.4383	0.4447	0.4511	0.4574	0.4637
1.6	0.4700	0.4762	0.4824	0.4886	0.4947	0.5008	0.5068	0.5128	0.5188	0.5247
1.7	0.5306	0.5365	0.5423	0.5481	0.5539	0.5596	0.5653	0.5710	0.5766	0.5822
1.8	0.5878	0.5933	0.5988	0.6043	0.6098	0.6152	0.6206	0.6259	0.6313	0.6366
1.9	0.6419	0.6471	0.6523	0.6575	0.6627	0.6678	0.6729	0.6780	0.6831	0.6881
2.0	0.6931	0.6981	0.7031	0.7080	0.7130	0.7178	0.7227	0.7275	0.7324	0.7372
2.1	0.7419	0.7467	0.7514	0.7561	0.7608	0.7655	0.7701	0.7747	0.7793	0.7839
2.2	0.7885	0.7930	0.7975	0.8020	0.8065	0.8109	0.8154	0.8198	0.8242	0.8286
2.3	0.8329	0.8372	0.8416	0.8459	0.8502	0.8544	0.8587	0.8629	0.8671	0.8713
2.4	0.8755	0.8796	0.8838	0.8879	0.8920	0.8961	0.9002	0.9042	0.9083	0.9123
2.5	0.9163	0.9203	0.9243	0.9282	0.9322	0.9361	0.9400	0.9439	0.9478	0.9517
2.6	0.9555	0.9594	0.9632	0.9670	0.9708	0.9746	0.9783	0.9821	0.9858	0.9895
2.7	0.9933	0.9969	1.0006	1.0043	1.0080	1.0116	1.0152	1.0188	1.0225	1.0260
2.8	1.0296	1.0332	1.0367	1.0403	1.0438	1.0473	1.0508	1.0543	1.0578	1.0613
2.9	1.0647	1.0682	1.0716	1.0750	1.0784	1.0818	1.0852	1.0886	1.0919	1.0953
3.0	1.0986	1.1019	1.1053	1.1086	1.1119	1.1151	1.1184	1.1217	1.1249	1.1282
3.1	1.1314	1.1346	1.1378	1.1410	1.1442	1.1474	1.1506	1.1537	1.1569	1.1600
3.2	1.1632	1.1663	1.1694	1.1725	1.1756	1.1787	1.1817	1.1848	1.1878	1.1909
3.3	1.1939	1.1970	1.2000	1.2030	1.2060	1.2090	1.2119	1.2149	1.2179	1.2208
3.4	1.2238	1.2267	1.2296	1.2326	1.2355	1.2384	1.2413	1.2442	1.2470	1.2499
3.5	1.2528	1.2556	1.2585	1.2613	1.2641	1.2669	1.2698	1.2726	1.2754	1.2782
3.6	1.2809	1.2837	1.2865	1.2892	1.2920	1.2947	1.2975	1.3002	1.3029	1.3056
3.7	1.3083	1.3110	1.3137	1.3164	1.3191	1.3218	1.3244	1.3271	1.3297	1.3324
3.8	1.3350	1.3376	1.3403	1.3429	1.3455	1.3481	1.3507	1.3533	1.3558	1.3584
3.9	1.3610	1.3635	1.3661	1.3686	1.3712	1.3737	1.3762	1.3788	1.3813	1.3838
4.0	1.3863	1.3888	1.3913	1.3938	1.3962	1.3987	1.4012	1.4036	1.4061	1.4085
4.1	1.4110	1.4134	1.4159	1.4183	1.4207	1.4231	1.4255	1.4279	1.4303	1.4327
4.2	1.4351	1.4375	1.4398	1.4422	1.4446	1.4469	1.4493	1.4516	1.4540	1.4563
4.3	1.4586	1.4609	1.4633	1.4656	1.4679	1.4702	1.4725	1.4748	1.4770	1.4793
4.4	1.4816	1.4839	1.4861	1.4884	1.4907	1.4929	1.4952	1.4974	1.4996	1.5019
4.5	1.5041	1.5063	1.5085	1.5107	1.5129	1.5151	1.5173	1.5195	1.5217	1.5239
4.6	1.5261	1.5282	1.5304	1.5326	1.5347	1.5369	1.5390	1.5412	1.5433	1.5454
4.7	1.5476	1.5497	1.5518	1.5539	1.5560	1.5581	1.5602	1.5623	1.5644	1.5665
4.8	1.5686	1.5707	1.5728	1.5748	1.5769	1.5790	1.5810	1.5831	1.5851	1.5872
4.9	1.5892	1.5913	1.5933	1.5953	1.5974	1.5994	1.6014	1.6034	1.6054	1.6074
5.0	1.6094	1.6114	1.6134	1.6154	1.6174	1.6194	1.6214	1.6233	1.6253	1.6273
5.1	1.6292	1.6312	1.6332	1.6351	1.6371	1.6390	1.6409	1.6429	1.6448	1.6467
5.2	1.6487	1.6506	1.6525	1.6544	1.6563	1.6582	1.6601	1.6620	1.6639	1.6658
5.3	1.6677	1.6696	1.6715	1.6734	1.6752	1.6771	1.6790	1.6808	1.6827	1.6845
5.4	1.6864	1.6882	1.6901	1.6919	1.6938	1.6956	1.6974	1.6993	1.7011	1.7029

$$\ln(N \cdot 10^m) = \ln N + m \ln 10, \quad \ln 10 = 2.3026$$

Table 3. Natural Logarithms (*cont.*)

	0.00	0.01	0.02	0.03	0.04	0.05	0.06	0.07	0.08	0.09
5.5	1.7047	1.7066	1.7084	1.7102	1.7120	1.7138	1.7156	1.7174	1.7192	1.7210
5.6	1.7228	1.7246	1.7263	1.7281	1.7299	1.7317	1.7334	1.7352	1.7370	1.7387
5.7	1.7405	1.7422	1.7440	1.7457	1.7475	1.7492	1.7509	1.7527	1.7544	1.7561
5.8	1.7579	1.7596	1.7613	1.7630	1.7647	1.7664	1.7682	1.7699	1.7716	1.7733
5.9	1.7750	1.7766	1.7783	1.7800	1.7817	1.7834	1.7851	1.7867	1.7884	1.7901
6.0	1.7918	1.7934	1.7951	1.7967	1.7984	1.8001	1.8017	1.8034	1.8050	1.8066
6.1	1.8083	1.8099	1.8116	1.8132	1.8148	1.8165	1.8181	1.8197	1.8213	1.8229
6.2	1.8245	1.8262	1.8278	1.8294	1.8310	1.8326	1.8342	1.8358	1.8374	1.8390
6.3	1.8406	1.8421	1.8437	1.8453	1.8469	1.8485	1.8500	1.8516	1.8532	1.8547
6.4	1.8563	1.8579	1.8594	1.8610	1.8625	1.8641	1.8656	1.8672	1.8687	1.8703
6.5	1.8718	1.8733	1.8749	1.8764	1.8779	1.8795	1.8810	1.8825	1.8840	1.8856
6.6	1.8871	1.8886	1.8901	1.8916	1.8931	1.8946	1.8961	1.8976	1.8991	1.9006
6.7	1.9021	1.9036	1.9051	1.9066	1.9081	1.9095	1.9110	1.9125	1.9140	1.9155
6.8	1.9169	1.9184	1.9199	1.9213	1.9228	1.9242	1.9257	1.9272	1.9286	1.9301
6.9	1.9315	1.9330	1.9344	1.9359	1.9373	1.9387	1.9402	1.9416	1.9430	1.9445
7.0	1.9459	1.9473	1.9488	1.9502	1.9516	1.9530	1.9544	1.9559	1.9573	1.9587
7.1	1.9601	1.9615	1.9629	1.9643	1.9657	1.9671	1.9685	1.9699	1.9713	1.9727
7.2	1.9741	1.9755	1.9769	1.9782	1.9796	1.9810	1.9824	1.9838	1.9851	1.9865
7.3	1.9879	1.9892	1.9906	1.9920	1.9933	1.9947	1.9961	1.9974	1.9988	2.0001
7.4	2.0015	2.0028	2.0042	2.0055	2.0069	2.0082	2.0096	2.0109	2.0122	2.0136
7.5	2.0149	2.0162	2.0176	2.0189	2.0202	2.0215	2.0229	2.0242	2.0255	2.0268
7.6	2.0282	2.0295	2.0308	2.0321	2.0334	2.0347	2.0360	2.0373	2.0386	2.0399
7.7	2.0412	2.0425	2.0438	2.0451	2.0464	2.0477	2.0490	2.0503	2.0516	2.0528
7.8	2.0541	2.0554	2.0567	2.0580	2.0592	2.0605	2.0618	2.0631	2.0643	2.0656
7.9	2.0669	2.0681	2.0694	2.0707	2.0719	2.0732	2.0744	2.0757	2.0769	2.0782
8.0	2.0794	2.0807	2.0819	2.0832	2.0844	2.0857	2.0869	2.0882	2.0894	2.0906
8.1	2.0919	2.0931	2.0943	2.0956	2.0968	2.0980	2.0992	2.1005	2.1017	2.1029
8.2	2.1041	2.1054	2.1066	2.1078	2.1090	2.1102	2.1114	2.1126	2.1138	2.1150
8.3	2.1163	2.1175	2.1187	2.1190	2.1211	2.1223	2.1235	2.1247	2.1258	2.1270
8.4	2.1282	2.1294	2.1306	2.1318	2.1330	2.1342	2.1353	2.1365	2.1377	2.1389
8.5	2.1401	2.1412	2.1424	2.1436	2.1448	2.1459	2.1471	2.1483	2.1494	2.1506
8.6	2.1518	2.1529	2.1541	2.1552	2.1564	2.1576	2.1587	2.1599	2.1610	2.1622
8.7	2.1633	2.1645	2.1656	2.1668	2.1679	2.1691	2.1702	2.1713	2.1725	2.1736
8.8	2.1748	2.1759	2.1770	2.1782	2.1793	2.1804	2.1815	2.1827	2.1838	2.1849
8.9	2.1861	2.1872	2.1883	2.1894	2.1905	2.1917	2.1928	2.1939	2.1950	2.1961
9.0	2.1972	2.1983	2.1994	2.2006	2.2017	2.2028	2.2039	2.2050	2.2061	2.2072
9.1	2.2083	2.2094	2.2105	2.2116	2.2127	2.2138	2.2148	2.2159	2.2170	2.2181
9.2	2.2192	2.2203	2.2214	2.2225	2.2235	2.2246	2.2257	2.2268	2.2279	2.2289
9.3	2.2300	2.2311	2.2322	2.2332	2.2343	2.2354	2.2364	2.2375	2.2386	2.2396
9.4	2.2407	2.2418	2.2428	2.2439	2.2450	2.2460	2.2471	2.2481	2.2492	2.2502
9.5	2.2513	2.2523	2.2534	2.2544	2.2555	2.2565	2.2576	2.2586	2.2597	2.2607
9.6	2.2618	2.2628	2.2638	2.2649	2.2659	2.2670	2.2680	2.2690	2.2701	2.2711
9.7	2.2721	2.2732	2.2742	2.2752	2.2762	2.2773	2.2783	2.2793	2.2803	2.2814
9.8	2.2824	2.2834	2.2844	2.2854	2.2865	2.2875	2.2885	2.2895	2.2905	2.2915
9.9	2.2925	2.2935	2.2946	2.2956	2.2966	2.2976	2.2986	2.2996	2.3006	2.3016

Table 4. Exponential and Hyperbolic Functions

x	e^x	e^{-x}	sinh x	cosh x	tanh x
0.00	1.0000	1.0000	0.0000	1.0000	0.0000
0.01	1.0101	0.9900	0.0100	1.0001	0.0100
0.02	1.0202	0.9802	0.0200	1.0002	0.0200
0.03	1.0305	0.9704	0.0300	1.0005	0.0300
0.04	1.0408	0.9608	0.0400	1.0008	0.0400
0.05	1.0513	0.9512	0.0500	1.0013	0.0500
0.06	1.0618	0.9418	0.0600	1.0018	0.0599
0.07	1.0725	0.9324	0.0701	1.0025	0.0699
0.08	1.0833	0.9231	0.0801	1.0032	0.0798
0.09	1.0942	0.9139	0.0901	1.0041	0.0898
0.10	1.1052	0.9048	0.1002	1.0050	0.0997
0.11	1.1163	0.8958	0.1102	1.0061	0.1096
0.12	1.1275	0.8869	0.1203	1.0072	0.1194
0.13	1.1388	0.8781	0.1304	1.0085	0.1293
0.14	1.1503	0.9694	0.1405	1.0098	0.1391
0.15	1.1618	0.8607	0.1506	1.0113	0.1489
0.16	1.1735	0.8521	0.1607	1.0128	0.1586
0.17	1.1853	0.8437	0.1708	1.0145	0.1684
0.18	1.1972	0.8353	0.1810	1.0162	0.1781
0.19	1.2092	0.8270	0.1911	1.0181	0.1877
0.20	1.2214	0.8187	0.2013	1.0201	0.1974
0.21	1.2337	0.8106	0.2115	1.0221	0.2070
0.22	1.2461	0.8025	0.2218	1.0243	0.2165
0.23	1.2586	0.7945	0.2320	1.0266	0.2260
0.24	1.2712	0.7866	0.2423	1.0289	0.2355
0.25	1.2840	0.7788	0.2526	1.0314	0.2449
0.26	1.2969	0.7711	0.2629	1.0340	0.2543
0.27	1.3100	0.7634	0.2733	1.0367	0.2636
0.28	1.3231	0.7558	0.2837	1.0395	0.2729
0.29	1.3364	0.7483	0.2941	1.0423	0.2821
0.30	1.3499	0.7408	0.3045	1.0453	0.2913
0.31	1.3634	0.7334	0.3150	1.0484	0.3004
0.32	1.3771	0.7261	0.3255	1.0516	0.3095
0.33	1.3910	0.7189	0.3360	1.0549	0.3185
0.34	1.4049	0.7118	0.3466	1.0584	0.3275
0.35	1.4191	0.7047	0.3572	1.0619	0.3364
0.36	1.4333	0.6977	0.3678	1.0655	0.3452
0.37	1.4477	0.6907	0.3785	1.0692	0.3540
0.38	1.4623	0.6839	0.3892	1.0731	0.3627
0.39	1.4770	0.6771	0.4000	1.0770	0.3714
0.40	1.4918	0.6703	0.4108	1.0811	0.3799
0.41	1.5068	0.6637	0.4216	1.0852	0.3885
0.42	1.5220	0.6570	0.4325	1.0895	0.3969
0.43	1.5373	0.6505	0.4434	1.0939	0.4053
0.44	1.5527	0.6440	0.4543	1.0984	0.4136

Table 4. Exponential and Hyperbolic Functions (*cont.*)

x	e^x	e^{-x}	sinh x	cosh x	tanh x
0.45	1.5683	0.6376	0.4653	1.1030	0.4219
0.46	1.5841	0.6313	0.4764	1.1077	0.4301
0.47	1.6000	0.6250	0.4875	1.1125	0.4382
0.48	1.6161	0.6188	0.4986	1.1174	0.4462
0.49	1.6323	0.6126	0.5098	1.1225	0.4542
0.50	1.6487	0.6065	0.5211	1.1276	0.4621
0.51	1.6653	0.6005	0.5324	1.1329	0.4699
0.52	1.6820	0.5945	0.5438	1.1383	0.4777
0.53	1.6989	0.5886	0.5552	1.1438	0.4854
0.54	1.7160	0.5827	0.5666	1.1494	0.4930
0.55	1.7333	0.5769	0.5782	1.1551	0.5005
0.56	1.7507	0.5712	0.5897	1.1609	0.5080
0.57	1.7683	0.5655	0.6014	1.1669	0.5154
0.58	1.7860	0.5599	0.6131	1.1730	0.5227
0.59	1.8040	0.5543	0.6248	1.1792	0.5299
0.60	1.8221	0.5488	0.6367	1.1855	0.5370
0.61	1.8044	0.5434	0.6485	1.1919	0.5441
0.62	1.8589	0.5379	0.6605	1.1984	0.5511
0.63	1.8776	0.5326	0.6725	1.2051	0.5581
0.64	1.8965	0.5273	0.6846	1.2119	0.5649
0.65	1.9155	0.5220	0.6967	1.2188	0.5717
0.66	1.9348	0.5169	0.7090	1.2258	0.5784
0.67	1.9542	0.5117	0.7213	1.2330	0.5850
0.68	1.9739	0.5066	0.7336	1.2402	0.5915
0.69	1.9937	0.5016	0.7461	1.2476	0.5980
0.70	2.0138	0.4966	0.7586	1.2552	0.6044
0.71	2.0340	0.4916	0.7712	1.2628	0.6107
0.72	2.0544	0.4868	0.7838	1.2706	0.6169
0.73	2.0751	0.4819	0.7966	1.2785	0.6231
0.74	2.0959	0.4771	0.8094	1.2865	0.6291
0.75	2.1170	0.4724	0.8223	1.2947	0.6351
0.76	2.1383	0.4677	0.8353	1.3030	0.6411
0.77	2.1598	0.4630	0.8484	1.3114	0.6469
0.78	2.1815	0.4584	0.8615	1.3199	0.6527
0.79	2.2034	0.4538	0.8748	1.3286	0.6584
0.80	2.2255	0.4493	0.8881	1.3374	0.6640
0.81	2.2479	0.4449	0.9015	1.3464	0.6696
0.82	2.2705	0.4404	0.9150	1.3555	0.6751
0.83	2.2933	0.4360	0.9286	1.3647	0.6805
0.84	2.3164	0.4317	0.9423	1.3740	0.6858
0.85	2.3396	0.4274	0.9561	1.3835	0.6911
0.86	2.3632	0.4232	0.9700	1.3932	0.6963
0.87	2.3869	0.4190	0.9840	1.4029	0.7014
0.88	2.4109	0.4148	0.9981	1.4128	0.7064
0.89	2.4351	0.4107	1.0122	1.4229	0.7114

Table 4. Exponential and Hyperbolic Functions (*cont.*)

x	e^x	e^{-x}	sinh x	cosh x	tanh x
0.90	2.4596	0.4066	1.0265	1.4331	0.7163
0.91	2.4843	0.4025	1.0409	1.4434	0.7211
0.92	2.5093	0.3985	1.0554	1.4539	0.7259
0.93	2.5345	0.3946	1.0700	1.4645	0.7306
0.94	2.5600	0.3906	1.0847	1.4753	0.7352
0.95	2.5857	0.3867	1.0995	1.4862	0.7398
0.96	2.6117	0.3829	1.1144	1.4973	0.7443
0.97	2.6379	0.3791	1.1294	1.5085	0.7487
0.98	2.6645	0.3753	1.1446	1.5199	0.7531
0.99	2.6912	0.3716	1.1598	1.5314	0.7574
1.00	2.7183	0.3679	1.1752	1.5431	0.7616
1.05	2.8577	0.3499	1.2539	1.6038	0.7818
1.10	3.0042	0.3329	1.3356	1.6685	0.8005
1.15	3.1582	0.3166	1.4208	1.7374	0.8178
1.20	3.3201	0.3012	1.5085	1.8107	0.8337
1.25	3.4903	0.2865	1.6019	1.8884	0.8483
1.30	3.6693	0.2725	1.6984	1.9709	0.8617
1.35	3.8574	0.2592	1.7991	2.0583	0.8741
1.40	4.0552	0.2466	1.9043	2.1509	0.8854
1.45	4.2631	0.2346	2.0143	2.2488	0.8957
1.50	4.4817	0.2231	2.1293	2.3524	0.9051
1.55	4.7115	0.2122	2.2496	2.4619	0.9138
1.60	4.9530	0.2019	2.3756	2.5775	0.9217
1.65	5.2070	0.1920	2.5075	2.6995	0.9289
1.70	5.4739	0.1827	2.6456	2.8283	0.9354
1.75	5.7546	0.1738	2.7904	2.9642	0.9414
1.80	6.0496	0.1653	2.9422	3.1075	0.9468
1.85	6.3598	0.1572	3.1013	3.2585	0.9517
1.90	6.6859	0.1496	3.2682	3.4177	0.9562
1.95	7.0287	0.1423	3.4432	3.5855	0.9603
2.00	7.3891	0.1353	3.6269	3.7622	0.9640
2.05	7.7679	0.1287	3.8196	3.9483	0.9674
2.10	8.1662	0.1225	4.0219	4.1443	0.9705
2.15	8.5849	0.1165	4.2342	4.3507	0.9732
2.20	9.0250	0.1108	4.4571	4.5679	0.9757
2.25	9.4877	0.1054	4.6912	4.7966	0.9780
2.30	9.9742	0.1003	4.9370	5.0372	0.9801
2.35	10.486	0.0954	5.1951	5.2905	0.9820
2.40	11.023	0.0907	5.4662	5.5569	0.9837
2.45	11.588	0.0863	5.7510	5.8373	0.9852
2.50	12.182	0.0821	6.0502	6.1323	0.9866
2.55	12.807	0.0781	6.3645	6.4426	0.9879
2.60	13.464	0.0743	6.6947	6.7690	0.9890
2.65	14.154	0.0707	7.0417	7.1123	0.9901
2.70	14.880	0.0672	7.4063	7.4735	0.9910

Table 4. Exponential and Hyperbolic Functions (cont.)

x	e^x	e^{-x}	sinh x	cosh x	tanh x
2.75	15.643	0.0639	7.7894	7.8533	0.9919
2.80	16.445	0.0608	8.1919	8.2527	0.9926
2.85	17.288	0.0578	8.6150	8.6728	0.9933
2.90	18.174	0.0550	9.0596	9.1146	0.9940
2.95	19.106	0.0523	9.5268	9.5791	0.9945
3.00	20.086	0.0498	10.018	10.068	0.9951
3.05	21.115	0.0474	10.534	10.581	0.9955
3.10	22.198	0.0450	11.076	11.122	0.9959
3.15	23.336	0.0429	11.647	11.689	0.9963
3.20	24.533	0.0408	12.246	12.287	0.9967
3.25	25.790	0.0388	12.876	12.915	0.9970
3.30	27.113	0.0369	13.538	13.575	0.9973
3.35	28.503	0.0351	14.234	14.269	0.9975
3.40	29.964	0.0334	14.965	14.999	0.9978
3.45	31.500	0.0317	15.734	15.766	0.9980
3.50	33.115	0.0302	16.543	16.573	0.9982
3.55	34.813	0.0287	17.392	17.421	0.9983
3.60	36.598	0.0273	18.286	18.313	0.9985
3.65	38.475	0.0260	19.224	19.250	0.9986
3.70	40.447	0.0247	20.211	20.236	0.9988
3.75	42.521	0.0235	21.249	21.272	0.9989
3.80	44.701	0.0224	22.339	22.362	0.9990
3.85	46.993	0.0213	23.486	23.507	0.9991
3.90	49.402	0.0202	24.691	24.711	0.9992
3.95	51.935	0.0193	25.958	25.977	0.9993
4.00	54.598	0.0183	27.290	27.308	0.9993
4.10	60.340	0.0166	30.162	30.178	0.9995
4.20	66.686	0.0150	33.336	33.351	0.9996
4.30	73.700	0.0136	36.843	36.857	0.9996
4.40	81.451	0.0123	40.719	40.732	0.9997
4.50	90.017	0.0111	45.003	45.014	0.9998
4.60	99.484	0.0101	49.737	49.747	0.9998
4.70	109.95	0.0091	54.969	54.978	0.9998
4.80	121.51	0.0082	60.751	60.759	0.9999
4.90	134.29	0.0074	67.141	67.149	0.9999
5.00	148.41	0.0067	74.203	74.210	0.9999
5.20	181.27	0.0055	90.633	90.639	0.9999
5.40	221.41	0.0045	110.70	110.71	1.0000
5.60	270.43	0.0037	135.21	135.22	1.0000
5.80	330.30	0.0030	165.15	165.15	1.0000
6.00	403.43	0.0025	201.71	201.72	1.0000
7.00	1096.6	0.0009	548.32	548.32	1.0000
8.00	2981.0	0.0003	1490.5	1490.5	1.0000
9.00	8103.1	0.0001	4051.5	4051.5	1.0000
10.00	22026.	0.00005	11013.	11013.	1.0000

Answers and Hints to Odd-Numbered Problems

Chapter 0

Section 0.1 (page 7)
1. $\{0\}$ **3.** $\{6\}$ **5.** $\{0, 1, -1\}$ **7.** True **9.** False **11.** False **13.** False **15.** True **17.** 0.25 **19.** 0.125
21. 0.008 **23.** $0.\overline{3}$ **25.** $0.\overline{1}$ **27.** $-0.0\overline{45}$ **29.** $\frac{3}{7}$ **31.** $\frac{27}{121}$
33. Let the numbers be $r = m/n$, $r' = m'/n'$, and calculate $r + r'$, rr'. **35.** $\sqrt{2}$, $1 - \sqrt{2}$ **37.** 16

Section 0.2 (page 13)
1. If $b - a > 0$, then $-a - (-b) > 0$. **3.** If $b - a > 0$ and $d - c > 0$, then $(b - a) + (d - c) > 0$.
5. Use Theorem 2 repeatedly, and also Example 6. **7.** $1/\sqrt{2}$ **9.** 25^4 **11.** $\sqrt{7} - \sqrt{3}$ **13.** $\frac{10}{3}$ **15.** $\frac{7}{19}$ **17.** $a = b = 0$
19. $(x - y)^2 > 0$ unless $x = y$. **21.** $\frac{5}{12} \leq \frac{1}{a} + \frac{1}{b} \leq \frac{3}{4}$ **23.** $x \geq -1$ **25.** $x < -2$ or $x > 0$ **27.** $-3 < x < -2$ or $x > -1$
29. $-1 < x < 1$ **31.** -1 **33.** 2 **35.** $a^4 b^4 c^4$ **37.** 2^{-14} **39.** 2^{10} **41.** $4a^3$ **43.** $3|a|$ **45.** $\sqrt{3} - 1$ **47.** $3 + 2\sqrt{2}$
49. $1/25a^2$ **51.** 2 **53.** $\frac{125}{27}$ **55.** Add p or q to both sides of the inequality $p < q$.

Section 0.3 (page 20)
1. 9 **3.** $\sqrt{2} - 1$ **5.** 1 **7.** $\sqrt{11} - \pi$ **9.** $13 < a + b < 19$ **11.** $15 < a^2 - b^2 < 119$ **13.** $|x| < |x + 2|$
15. $|x - 1| = 2|x + 1|$ **17.** $x = 0$ or $x = -\frac{8}{3}$ **19.** $x > \frac{1}{2}$ **21.** $-2 < x < 2$ **23.** $x = 2$ **25.** $x = \frac{2}{3}$ or $x = -2$
27. $x = \frac{3}{2}$ or $x = -\frac{5}{2}$ **29.** $x \leq -3$ or $x \geq 1$ **31.** $x < -2$ or $x > 2$ **33.** $x < 1$ **35.** $x \neq \pm 1$
37. $x < -1$ or $-1 < x \leq 0$ **39.** $\frac{9}{2}$ **41.** -4 **43.** $(-2, 3]$ **45.** $-1 \leq x < 1$ **47.** $(5, 13)$ **49.** $3 < x < \infty$
51. $[-4, -2)$ **53.** $-2 \leq x \leq -1$ **55.** $(-\infty, -5]$ **57.** $-\infty < x < 5$ **59.** $[1, 5]$ **61.** $(-3, -1)$ **63.** $|x - 1| < 2$
65. $0 < |x + 1| < 1$

Section 0.4 (page 26)

1.

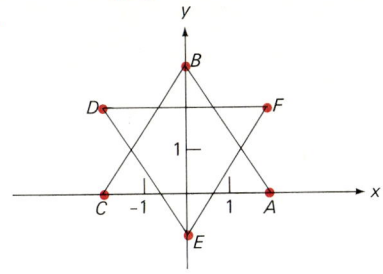

A six-pointed star

3. $(2, -1)$ **5.** $4\sqrt{2}$ **7.** 1 **9.** $2\sqrt{2}$ **11.** $2\sqrt{5}$ **13.** 5 **15.** $\sqrt{2\pi^2 + 4}$ **17.** Yes **19.** No **21.** $(5, 4)$ **23.** $(0, 0)$
25. $|AB| = |BC| = |CD| = |DA| = \sqrt{17}$; moreover, $|AC| = |BD| \,(= \sqrt{34})$, so $ABCD$ is a square.

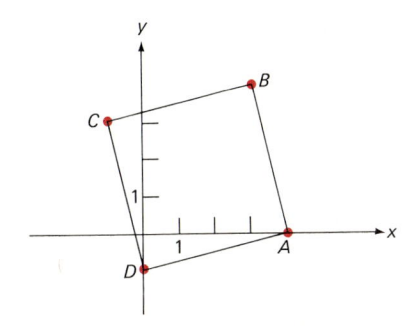

27. $(x + 1)^2 + (y - 1)^2 = 1$ **29.** $(x - 4)^2 + (y + 5)^2 = 9$ **31.** The circle of radius 8 with center $(-2, 0)$
33. The single point $(-5, 2)$ **35.** The circle of radius 5 with center $(1, -2)$ **37.** $(x - 1)^2 + (y - 4)^2 = 8$
39. $(x - 2)^2 + (y - 1)^2 = 1$ **41.** On   **43.** Inside

45.

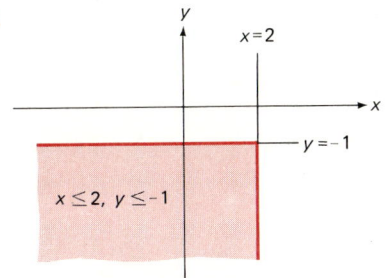

$x \leq 2, y \leq -1$

Boundary lines included

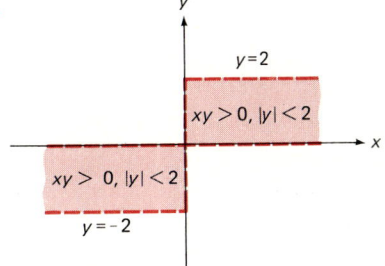

47.

$xy > 0, |y| < 2$

$xy > 0, |y| < 2$

Boundary lines excluded

49. Note that $ax^2 + bx + c = a\left[\left(x + \dfrac{b}{2a}\right)^2 + \dfrac{c}{a} - \left(\dfrac{b}{2a}\right)^2\right]$.

Section 0.5 (page 35)

1. 0.364   **3.** 1.192   **5.** -0.839 **7.** $\frac{5}{2}$ **9.** 0   **11.** Undefined   **13.** $90°$ **15.** $150°$ **17.** Set $b = 0$ in (5).
19. $y = -x + 1$ **21.** $y = -2x - 4$ **23.** $y = \frac{2}{3}x + 3$ **25.** $y = -2$ **27.** $y = -7x - 3$ **29.** $m = 5, a = -\frac{3}{5}, b = 3$
31. $m = -\frac{5}{2}, a = -\frac{2}{5}, b = -1$ **33.** $m = 0$, no x-intercept, $b = 2$ **35.** $2x - 4y + 1 = 0$ **37.** $x + 2y - 11 = 0$
39. $2x - y + 2 = 0$ **41.** $x - 8y - 4 = 0$ **43.** $x + 25y - 5 = 0$ **45.** $b = -\sqrt{3}a$ **47.** $(\frac{1}{4}, \frac{3}{4})$
49. No point of intersection   **51.** $x + y = 0$ **53.** $x + 9y - 19 = 0$ **55.** $x + 3y = 0$ **57.** $2x + y - 2 = 0$
59. $x - y = 0$ **61.** $6x + 8y - 17 = 0$ **63.** 3   **65.** $2\sqrt{5}$ **67.** $\frac{5}{2}$ **69.** $\frac{1}{2}$

Supplementary Problems (page 36)

1. $A \cup B = \{0, 1, 2, 3, 4\}$, $A \cap B = \{1, 2, 3\}$ **3.** $A \cup B = \{x: x \geq 1 \text{ or } x < -1\}$, $A \cap B = \{x: x > 1\}$
5. $\{3\}$ **7.** A **9.** $2^{20}, 4^{10}$ **11.** $\frac{311}{1000}$ **13.** Start from $(a - 1)^2 \geq 0$
15. Calculate max $\{a, b\}$ and min $\{a, b\}$ for both cases $a \geq b$ and $a < b$. **17.** It moves from a to b.
19. Let $x = \frac{1}{2}\frac{3}{4}\frac{5}{6}\cdots\frac{99}{100}$, $y = \frac{2}{3}\frac{4}{5}\frac{6}{7}\cdots\frac{100}{101}$. Then $x < y$, and hence $x^2 < xy = \frac{1}{101}$.
21. $M = a$, $m = a^n$ **23.** $M = m = 0$ if $a = 0$; $M = m = 1$ if $a = 1$; $M = 1$, $m = -1$ if $a = -1$.
25. $M = a^n$, $m = a^{n-1}$ if n is even; $M = a^{n-1}$, $m = a^n$ if n is odd. **27.** $[-1, 4)$ **29.** $(-2, 1]$
31. The circle of radius $\frac{1}{2}$ with center $(-\frac{1}{2}, 0)$ **33.** The circle of radius 1 with center $(-3, 2)$

35.

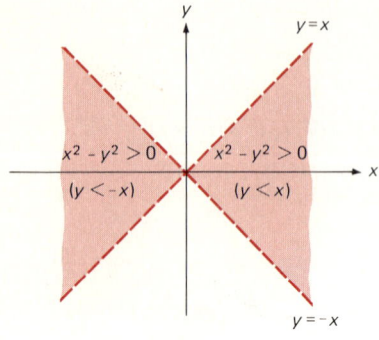

Boundary lines excluded

37. $(3, 3), (15, 15)$ **39.** 21 **41.** 20 **43.** $x + y - 3 = 0$; $(\frac{1}{2}, \frac{5}{2})$ **45.** $y = -10x$ **47.** $x = 0, y = 0$
49. $4x + 8y + 5 = 0$, $8x - 4y - 3 = 0$

51.

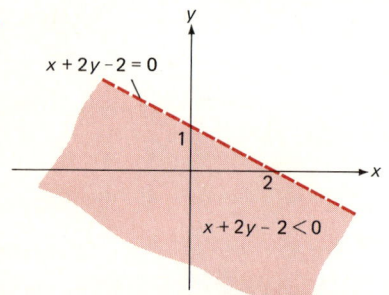

Boundary line excluded

53.

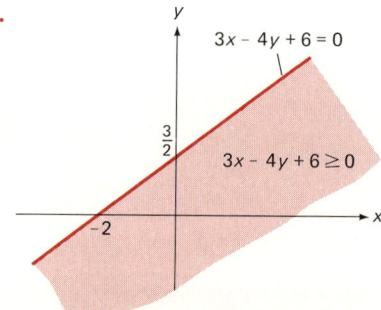

Boundary line included

55. $q_d = q_{max}\left(1 - \frac{p}{p_{max}}\right)$ **57.** $p_{eq} = \frac{a - c}{d - b}$, $q_{eq} = \frac{ad - bc}{d - b}$ **59.** $p_{eq} = 110$, $q_{eq} = 120$ **61.** $p_{eq} = 100$, $q_{eq} = 1800$

Chapter 1

Section 1.1 (page 43)
1. 5 **3.** 9 **5.** 33 **7.** -24 **9.** 0 **11.** 0 **13.** Undefined **15.** -1 **17.** 1 **19.** 0 **21.** $-\frac{1}{19}$ **23.** $3 - 2\sqrt{2}$
25. 1 **27.** 0.1 **29.** Undefined **31.** 15 **33.** $10 - \sqrt{5}$ **35.** $3 - 5\pi + 2\pi^2$ **37.** Yes **39.** Yes. No **41.** $f(5)$
43. $2 + a$ **45.** All $x \neq 0$ **47.** $0 < x < \infty$ **49.** All x **51.** $-4 \leq x \leq 4$ **53.** $1 \leq x < 2$ **55.** $y = \dfrac{1}{\sqrt{x(1 - x)}}$
57. $y = \sqrt{\dfrac{1 - x}{x}}$

Section 1.2 (page 51)

1. $\frac{7}{3}$ 3. Undefined 5. 25 7. No 9. Yes 11. 26 13. 3 15. 0 17. x $(x \neq -1)$ 19. $1/x$ $(x \neq 0, -1)$
21. $(\sqrt{x} - 1)^3$ 23. $(x - 1)^9$ 25. $\sqrt{x^3 - 1}$ 27. $f(x) = 2x - 2$ 29. $f(x) = -5x + 6$

31.

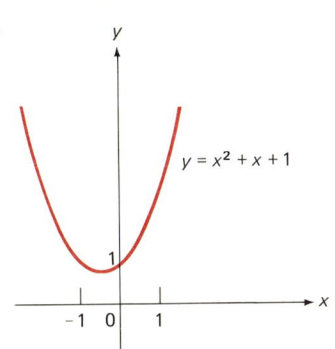

[Graph: $y = x^2 + x + 1$]

33.

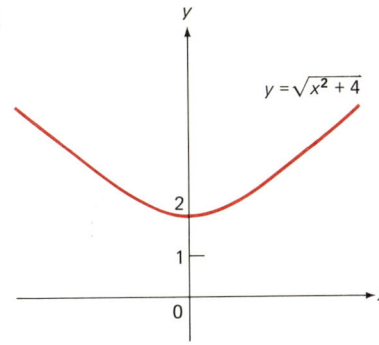

[Graph: $y = \sqrt{x^2 + 4}$]

35.

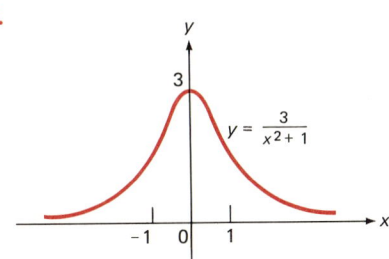

[Graph: $y = \dfrac{3}{x^2 + 1}$]

37.

[Graph: $y = |x+1| + |x-1|$, with asymptotes $y = -2x$ and $y = 2x$, and $y = 2$]

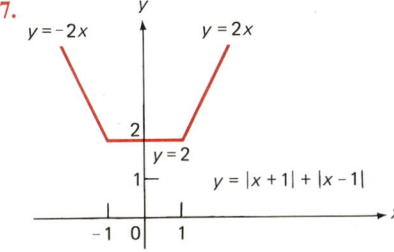

39. Even 41. Neither 43. Odd 45. Even 47. Odd
49. Examine the effect of changing x to $-x$ in each term of the sum   51. See Prob. 49.
53. Increasing on $[1, \infty)$, decreasing on $(-\infty, -1]$, constant on $[-1, 1]$ 55. $y = x^2 - 9x + 27$ 57. $y = x^2 + 5x$
59. A leftward shift of $\frac{1}{2}$ and a downward shift of $\frac{3}{4}$ 61. See Prob. 1, p. 13.

Section 1.3 (page 60)

1. $\pi/12$ 3. $25\pi/3$ 5. $-11\pi/9$ 7. $12°$ 9. $-15°$ 11. $\approx 565.5°$ 13. $L = \pi/2$, $A = \pi/2$ 15. $L = 10\pi/3$, $A = 25\pi/3$
17. $L = 5\pi/18$, $A = 125\pi/18$ 19. $c = \sqrt{34 - 15\sqrt{3}}$ 21. $b = 3$ 23. $a = 5(\sqrt{7} - \sqrt{2})$ 25. -1 27. $2/\sqrt{3}$ 29. 2
31. $2/\sqrt{3}$ 33. 2 35. $1/\sqrt{3}$ 37. $-\frac{1}{2}$ 39. $-2/\sqrt{3}$ 41. $\sqrt{2}$ 43. 1 45. $1/\sqrt{3}$ 47. 2 49. Zero 51. Negative
53. Draw the altitude of the triangle ABC. 55. $\frac{1}{4}\sqrt{3}\, s^2$ 57. Use formulas (13), (14) and (14'). 59. Use formula (14').
61. Use the same trick as in the derivation of (15). 63. Use formula (14). 65. Use formulas (14) and (14'). 67. $45°$
69. $135°$ 71. To 6 decimal places

Section 1.4 (page 65)

1. First show that $\sin n\pi = 0$, $\cos n\pi = (-1)^n$. 3. Note that $f(x - p) \equiv f((x - p) + p)$.
5. Use the oddness of $\sin x$ and the evenness of $\cos x$. 7. Neither 9. Even 11. $x = 2n\pi$, n any integer
13. $x = (2n - \frac{1}{2})\pi$, n any integer 15. $x = (2n + \frac{1}{6})\pi$ or $x = (2n + \frac{5}{6})\pi$, n any integer 17. $x = (n + \frac{1}{4})\pi$, n any integer
19. $x = (2n - \frac{1}{6})\pi$ or $x = (2n - \frac{5}{6})\pi$, n any integer 21. $x = (n + \frac{1}{2})\pi$, n any integer
23. $[(2n - \frac{1}{2})\pi, (2n + \frac{1}{2})\pi]$, n any integer 25. $[(2n - 1)\pi, 2n\pi]$, n any integer 27. 2π 29. π

31.

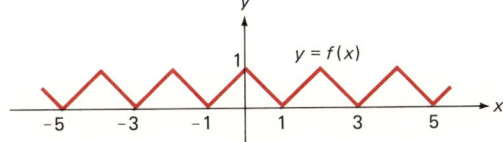

[Graph: $y = f(x)$, triangular wave]

Section 1.5 (page 71)
1. 6 **3.** 0 **5.** 3

7. The limit is 2. Note that $\dfrac{x-1}{\sqrt{x}-1} = \sqrt{x}+1$ if $x \neq 1$.

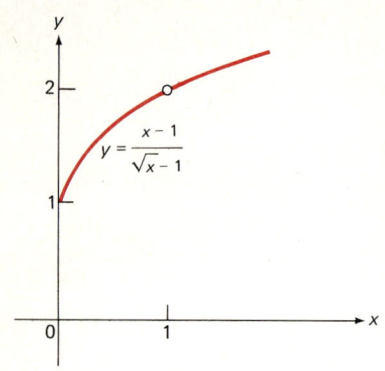

9. 0 **11.** 1 **13.** 1 **15.** 0 **17.** No limit **19.** No limit **21.** 4 **23.** 48 **25.** 0 **27.** $\sin 1 \approx 0.84$ **29.** -1
31. 1 **33.** 2 **35.** 0 **37.** Only at $x = 2$

Section 1.6 (page 79)
1. $L = 5a$; $\delta = \varepsilon/5$ **3.** $L = -10$; $\delta = 0.05$ **5.** $L = 4$; $\delta = 0.2$, say. **7.** $L = -1$; $\delta = 0.3$, say.
9. As $x \to a$, $g(x) \to L$ and $g(x) \to g(a)$. **11.** When f has no limit at a **13.** Yes **15.** 0 **17.** 0 **19.** Does not exist
21. 0 **23.** 0 **25.** Use inequality (5), p. 16. **27.** Let $f(x) = |x|/x$, say.
29. Apply rule (ii) to the function $g(x) - f(x)$. **31.** Use the triangle inequality.

Section 1.7 (page 87)
1. 16 **3.** -3 **5.** 48 **7.** $\frac{1}{4}$ **9.** 6 **11.** 10 **13.** $L = 2$; $\delta = \min\{\varepsilon, 2\}$ **15.** None **17.** None **19.** $x = \pm 2$
21. $x = 1, 2, 3$ **23.** $\cos 2 \approx -0.42$ **25.** -1 **27.** 0 **29.** $\frac{3}{4}$ **31.** $\sqrt{3}$

Section 1.8 (page 90)
1. $\sin 1$ **3.** 0 **5.** $\tan \frac{1}{2}$ **7.** 0 **9.** $\frac{1}{2}$ **11.** 2 **13.** $4\sqrt{5}$ **15.** $\tan 2$ **17.** π **19.** $\frac{2}{3}$ **21.** $\frac{1}{3}$ **23.** 2
25. $-\frac{1}{2}$ **27.** 1 **29.** 0 **31.** 2 **33.** $\pi/2$

35. Let $a = 0$, $f(x) = x \sin \dfrac{1}{x}$ $(x \neq 0)$, $g(x) = \begin{cases} x \text{ if } x \neq 0, \\ 1 \text{ if } x = 0. \end{cases}$ The assertion is true if $f(x) \neq L$ in some deleted neighborhood of a.

Section 1.9 (page 95)
1. -4 **3.** -1 **5.** 2 **7.** Does not exist **9.** 0 **11.** -1 **13.** 1 **15.** 0 **17.** 1 **19.** $\lim\limits_{x \to 0^+} f(x) = 1$, $\lim\limits_{x \to 0^-} f(x) = -1$
21. $(0, 1)$, $[0, 1)$ **23.** J, at the point a **25.** Discontinuities at $x = 0, \pm 1, \pm 2, \ldots$ No. Yes **27.** $a = -1$, $b = 1$

Section 1.10 (page 101)
1. f is discontinuous at $x = 0$. **3.** 2
5. If $f(x) = \cos \pi x - x$, then $f(0) = 1$, $f(\frac{1}{2}) = -\frac{1}{2}$, and f is decreasing on $(0, \frac{1}{2})$; $r = \frac{13}{32}$ to within $\frac{1}{32}$.
7. Apply Theorem 13 to the function $g(x) = f(x) - x$.
9. $M = 1$ at $x = 3$, $m = -3$ at $x = 1$; $J = [-3, 1]$ **11.** $M = 2$ at $x = 1$, $m = 0$ at $x = -1$; $J = [0, 2]$
13. M and m do not exist; $J = (0, \infty)$ **15.** $M = 1$ at $x = 0$, m does not exist; $J = (0, 1]$
17. $M = 0$ at $x = 1$, $m = -2$ at $x = 9$; $J = [-2, 0]$ **19.** $M = 1$ at $x = 0, \pi$; $m = -1$ at $x = \pi/2$; $J = [-1, 1]$
21. The maximum and minimum of $-f$ are $-m$ and $-M$.

Supplementary Problems (page 102)
1. $-\frac{1}{2}$ **3.** -8 **5.** Undefined **7.** $x = -1, 3$ **9.** $x = 1$ **11.** Yes **13.** 2 **15.** $\frac{1}{2}$ **17.** $\frac{1}{16}$ **19.** $f(f(x)) = f(x)$
21. $g(f(x)) = g(x)$ **23.** $f(x) = x$ **25.** $f(x) = \dfrac{|x+1| - |x-1|}{2}$ **27.** Odd **29.** Even **31.** $f(x) \equiv 0$
33. $y = -5x$, $y = \frac{1}{5}x$ **35.** $\pi/2$ **37.** Yes. No **39.** $f(0) = 0$ **41.** $f(0) = 1$ **43.** $n - 1$ **45.** -1 **47.** No

49. $f \circ g$ has jump discontinuities at $x = 0, \pm 1$; $g \circ f$ has no discontinuities. **51.** 3 **53.** $-\frac{1}{4}$ **55.** 1024 **57.** $\frac{1}{2}$ **59.** $\frac{3}{2}$

61. $M(x) = \dfrac{f(x) + g(x)}{2} + \dfrac{|f(x) - g(x)|}{2}$, $m(x) = \dfrac{f(x) + g(x)}{2} - \dfrac{|f(x) - g(x)|}{2}$

63. If $a < b$, let $n > 0$ be an integer such that $n(b - a) > 1$. Then $a < \dfrac{p}{n} < b$, $a < \dfrac{q}{\sqrt{2}n} < b$ for suitable integers p and q.

65. Limit 0 at $x = 0$, but no limit elsewhere **67.** No limit anywhere
69. If $f(x) = ax + b \sin x - c$, then $f(n\pi) = an\pi - c$ if n is an integer.

Chapter 2

Section 2.1 (page 114)
1. $v_{av}(1) = 215$ ft/sec, $v_{av}(0.1) = 210.5$ ft/sec, $v_{av}(0.01) = 210.05$ ft/sec, $v_{av}(0.001) = 210.005$ ft/sec; $v = 210$ ft/sec at $t = 20$
3. $\Delta y = 0.61$, $\Delta y/\Delta x = 6.1$ **5.** $\Delta y = 0.0016$, $\Delta y/\Delta x = -0.008$ **7.** $\Delta y = -\frac{3}{4}$, $\Delta y/\Delta x = \frac{1}{4}$ **9.** $\Delta y = -1$, $\Delta y/\Delta x = \frac{1}{7}$
11. $\Delta y = -1/\sqrt{2}$, $\Delta y/\Delta x = -2\sqrt{2}/\pi$ **13.** $3x^2 + 2x + 1$ **15.** $-3t^2 + 18t + 5$ **17.** $3x^2$ **19.** $3x^2 + 8x - 4$
21. $4x^3 + 6x$ **23.** $9x^8 + 6x^5 + 3x^2$ **25.** $100x^{99} - 100x^{49} + 25$ **27.** $4s^3 - 3s^2 + 6s$ **29.** $35u^6 - 42u^5 + 45u^4$
31. $4x^3$ **33.** The derivative of x^3 at -2, equal to 12 **35.** When $t = 3$ sec; 144 ft; -32 ft/sec^2 **37.** 128 ft; 240 ft
39. 9 amp. No. 6 sec **41.** Write $uf(x) - xf(u) = uf(x) - xf(x) + xf(x) - xf(u)$.

43. $D_x \dfrac{1}{x} = \lim\limits_{\Delta x \to 0} \dfrac{1}{\Delta x}\left(\dfrac{1}{x + \Delta x} - \dfrac{1}{x}\right) = -\dfrac{1}{x} \lim\limits_{\Delta x \to 0} \dfrac{1}{x + \Delta x} = -\dfrac{1}{x^2}$ $(x \neq 0)$

45. $D_x \dfrac{1}{x^2 + 1} = \lim\limits_{u \to x} \dfrac{1}{u - x}\left(\dfrac{1}{u^2 + 1} - \dfrac{1}{x^2 + 1}\right) = -\dfrac{1}{x^2 + 1} \lim\limits_{u \to x} \dfrac{u + x}{u^2 + 1} = -\dfrac{2x}{(x^2 + 1)^2}$

Section 2.2 (page 121)
1. $y = 2$ **3.** $y = x - \frac{1}{3}$ **5.** $y = \frac{1}{4}x + 1$ **7.** $y = -x + \frac{1}{2}\pi$ **9.** $y = 4x - 4$, $y = 8x - 16$ **11.** $y = \pm 4x - 3$ **13.** Yes
15. No **17.** Yes **19.** $(\frac{1}{2}, \frac{17}{4})$ **21.** $f'_+(0) = 2$, $f'_-(0) = -2$ **23.** $f'_+(0)$ and $f'_-(0)$ do not exist. **25.** $m = 2a$, $b = -a^2$
27. Only for $n = 1$ **29.** $x = 0$ **31.** $y = -\frac{1}{8}x - \frac{23}{4}$ **33.** $y = -\frac{1}{2}x + \frac{3}{2}$ **35.** $\alpha = 0°$ at $(0, 0)$, $\alpha \approx 8.1°$ at $(1, 1)$
37. $\alpha = 90°$ at $(\pm 1, \frac{1}{2})$ **39.** $c = \frac{1}{3}$

Section 2.3 (page 127)
1. $7\,dx$ **3.** $(5x^4 + 18x^2 + 1)\,dx$ **5.** $(30u^5 + \sin u)\,du$ **7.** $-(\sin x + \cos x)\,dx$ **9.** $df = 0.002$, $\Delta f = 0.002$
11. $df = 0.03$, $\Delta f = 0.0297$ to 4 decimal places **13.** $df = -0.025$, $\Delta f = -0.02516$ to 5 decimal places
15. $(3.1)^2 + (3.1)^3 + (3.1)^4 \approx 117 + 141(0.1) = 131.1$, exact value $= 131.7531$

17. $\sqrt{49.6} \approx 7 + \dfrac{0.6}{2\sqrt{49}} = 7.04286$, exact value $= 7.04273$, both to 5 decimal places

19. $\sin 28° \approx \dfrac{1}{2} - \dfrac{\sqrt{3}}{2}\dfrac{\pi}{90} = 0.46977$, exact value $= 0.46947$, both to 5 decimal places

21. Use the tangent line approximation at the origin. **23.** 10%; 6% **25.** 64.5 ft **27.** 2.7 mi

29. $\lim\limits_{\Delta x \to 0} e(\Delta x) = f(a) - f(a) - f'(a) \cdot 0$, $\lim\limits_{\Delta x \to 0} \dfrac{e(\Delta x)}{\Delta x} = f'(a) - f'(a)$

31. Approximate the curve $y = x^2$ by its tangent line at the origin.

Section 2.4 (page 133)
1. $-6x^{-3} + 15x^{-4}$ **3.** $-\dfrac{1}{x^2} - \dfrac{4}{x^3} - \dfrac{9}{x^4}$ **5.** $2x - 4x^3$ **7.** $-2x + 3x^2 - 5x^4$ **9.** $3x^2$ **11.** $-3x^2 - 1 - 4x^{-3}$
13. $10x^{-6} - 10x^{-11}$ **15.** $2x + 3x^2 + 4x^3 + 5x^4 + 6x^5 + 7x^6 + 9x^8$ **17.** $-\dfrac{1}{(1 + x)^2}$ **19.** $\dfrac{2x}{(1 + x^2)^2}$ **21.** $\dfrac{6x^2}{(x^3 + 1)^2}$
23. $\dfrac{2 - 4x}{(1 - x + x^2)^2}$ **25.** $\dfrac{2t^3 - 5t^2 - 4t - 3}{(t - 2)^2}$ **27.** $-\dfrac{1}{\sqrt{u}(1 + \sqrt{u})^2}$ **29.** $\dfrac{v \cos v - \sin v}{v^2}$ **31.** $\dfrac{\cot w}{2\sqrt{w}} - \sqrt{w} \csc^2 w$
33. $2 \csc x (1 - x \cot x)$ **35.** $\sin x \tan x + x \sin x + x \tan x \sec x$ **37.** $-\dfrac{\sin x}{(1 - \cos x)^2}$ **39.** $\dfrac{2 \sec^2 x}{(1 - \tan x)^2}$

41. $\csc x - 2\csc^3 x$ **43.** $(-2x^{-3} - x^{-2})\,dx$ **45.** $\dfrac{dx}{(1+x)^2}$ **47.** $\left(\dfrac{\sin x}{2\sqrt{x}} + \sqrt{x}\cos x\right)dx$ **49.** $\dfrac{\cos x + x\sin x}{\cos^2 x}\,dx$

51. Factor the numerator and denominator.

Section 2.5 (page 139)

1. $4(x-1)^3$ **3.** $8(2x+3)^3$ **5.** $-36x(1-3x^2)^5$ **7.** $(x^2-1)^2(x-1)(7x+1)$ **9.** $2(x+2)(x+3)^2(3x^2+11x+9)$

11. $-\dfrac{(1-x)(5-x)}{(1+x)^4}$ **13.** $\dfrac{2(x^2+1)(x^2-2x-1)}{(x-1)^3}$ **15.** $\dfrac{20(1+x)^9}{(1-x)^{11}}$ **17.** $-\dfrac{1}{(1+x)\sqrt{1-x^2}}$ **19.** $\tfrac{3}{2}x^{1/2} - \tfrac{4}{3}x^{1/3}$

21. $0.99(x^{-0.01} - x^{-1.99})$ **23.** $\dfrac{1}{3\sqrt[3]{t^2}} + \dfrac{1}{4\sqrt[4]{t^3}} + \dfrac{1}{5\sqrt[5]{t^4}}$ **25.** $\dfrac{2}{3\sqrt[3]{v}} + \dfrac{3}{2\sqrt{v^3}}$ **27.** $\dfrac{1}{6\sqrt[3]{x^2}\sqrt{1+\sqrt[3]{x}}}$

29. $2\sin x \cos x + 2\tan x \sec^2 x$ **31.** $6(1+\sin 2x)^2 \cos 2x$ **33.** $-6\sin 2x + 6\sec 3x \tan 3x$ **35.** $\sin\dfrac{1}{x} - \dfrac{1}{x}\cos\dfrac{1}{x}$

37. $\dfrac{\cos u}{2\sqrt{1+\sin u}}$ **39.** $-3w^2 \sin w^3$ **41.** $\sin(\cos x)\sin x$ **43.** $\cos(\sin(\sin x))\cos(\sin x)\cos x$ **45.** $12x^3(x^4-2)^2\,dx$

47. $(\tfrac{3}{5}x^{-2/5} + \tfrac{5}{3}x^{2/3})\,dx$ **49.** $-\dfrac{\sin\sqrt{x}}{2\sqrt{x}}\,dx$ **51.** $2u\sec^2 u^2 \, du$ **53.** $\tfrac{2}{3}(\sec w)^{2/3}\tan w\,dw$ **55.** $-100(2)^{49}$

57. $D_x g^{-1} = -g^{-2} D_x g$ **59.** Differentiate the identity $f(x+p) \equiv f(x)$, where p is the period of f.   **61.** $x - 12y + 17 = 0$
63. $x = 0$ **65.** Use Example 8, p. 78; $\Delta y = 0$ if $x = 1/n\pi$, where n is any nonzero integer.

Section 2.6 (page 144)

1. $y'' = 450x^8 + 200x^3$, $y''' = 3600x^7 + 600x^2$ **3.** $y'' = -\dfrac{2}{(1+x)^3}$, $y''' = \dfrac{6}{(1+x)^4}$

5. $y'' = 2\cos x - x\sin x$, $y''' = -3\sin x - x\cos x$ **7.** $y'' = 2\cos x^2 - 4x^2 \sin x^2$, $y''' = -12x\sin x^2 - 8x^3 \cos x^2$

9. $y'' = \sec x \tan^2 x + \sec^3 x$, $y''' = \sec x \tan^3 x + 5\sec^3 x \tan x$ **11.** 2880 **13.** $\dfrac{x^2}{x-1} = x + 1 + \dfrac{1}{x-1}$

15. $\dfrac{1}{x(1-x)} = \dfrac{1}{x} - \dfrac{1}{x-1}$ **17.** $a = 3,\ b = -3,\ c = 1$ **19.** $f'(x) = \begin{cases} 2x\sin\dfrac{1}{x} - \cos\dfrac{1}{x} & \text{if } x \neq 0 \\ 0 & \text{if } x = 0 \end{cases}$ **21.** Calculate y''.

Section 2.7 (page 149)

1. $y' = -\dfrac{9x}{4y}$ **3.** $y' = -\dfrac{y^3}{x^3}$ **5.** $y' = \dfrac{x^2 y^3 - 1}{1 - x^3 y^2}$ **7.** $y' = x^{-1/3} y^{5/3}$ **9.** $y' = \dfrac{2xy}{x^2+y^2}$ **11.** $y' = \dfrac{\cos x}{\sin y}$

13. $y' = -\dfrac{1 + y\sin xy}{x\sin xy}$ **15.** $\dfrac{d}{dy}(x^2+y^2) = \dfrac{d}{dy}r^2$, so that $2x\dfrac{dx}{dy} + 2y = 0$, $\dfrac{dx}{dy} = -\dfrac{y}{x}$.

17. $y' = -\dfrac{\sqrt{y}}{\sqrt{x}} = -\dfrac{4}{\sqrt{x}} + 1$, $y'|_{x=4} = -1$

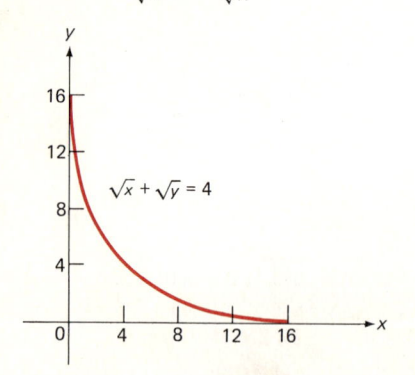

$\sqrt{x} + \sqrt{y} = 4$

19. $y = 2$ **21.** $y'|_{x=0} = \frac{1}{2}$, $y''|_{x=0, y=1} = -\frac{3}{4}$, $y''|_{x=0, y=-1} = \frac{3}{4}$, $y'''|_{x=0} = 0$
23. $y'|_{x=-1, y=0} = 2$, $y'|_{x=-1, y=-1} = -1$, $y''|_{x=-1, y=0} = -6$, $y''|_{x=-1, y=-1} = 6$, $y'''|_{x=-1, y=0} = 54$, $y'''|_{x=-1, y=-1} = -54$
25. $(2/\sqrt{3}, 1/\sqrt{3}), (-2/\sqrt{3}, -1/\sqrt{3})$ **27.** $(1, 1), (-1, -1)$ **29.** $x = y = 0$ is the only solution of $x^4 + y^4 = x^2 y^2$.

Section 2.8 (page 154)

1. x increases faster if $x < 4$; y increases faster if $x > 4$; x and y increase at the same rate if $x = 4$.
3. Increasing, at 40 cm²/sec   **5.** 6 ft; $5/4\pi \approx 0.4$ in/sec   **7.** $90/\sqrt{13} \approx 24.96$ knots   **9.** 1.25 ft/sec; $\frac{3}{20}$ rad/sec $\approx 8.59°$/sec
11. $5/\pi \approx 1.59$ in/min; $(5/\pi)2^{-4/3} \approx 0.63$ in/min; $96\pi/5 \approx 60.3$ min ≈ 1 hr. No   **13.** 300 km/hr   **15.** $\frac{2}{9}$ rad/sec $\approx 12.73°$/sec
17. $1.5\sqrt{3} \approx 2.6$ ft/sec   **19.** $-\frac{675}{256} \approx -2.64$ in/sec²; $-\frac{27}{20} = -1.35$ in/sec²   **21.** $\sqrt{3}v$

Supplementary Problems (page 157)

1. 356 ft (see Prob. 35, p. 115)   **3.** $a - \frac{b}{x^2}$ **5.** $3x^2 - 2ax - 2bx + ab$ **7.** $\frac{2a}{(x+a)^2}$ **9.** $-\frac{a+b}{(x-b)^2}$ **11.** $\frac{2(a-b)x}{(x^2-b^2)}$

13. $\frac{x}{\sqrt{x^2+a^2}}$ **15.** $\frac{a}{(a-x)\sqrt{a^2-x^2}}$ **17.** $-(2ax+b)\sin(ax^2+bx+c)$ **19.** The parabola $y = x^2$ except at $P = (0, 0)$

21. $y = -x - 2$ **23.** $y = x$ **25.** $y = -\frac{1}{2}x + 2$ **27.** $y = 4x - \frac{31}{4}$ **29.** $y = \frac{x}{\sqrt{2}} + \frac{1}{\sqrt{2}}\left(2 + \frac{\pi}{4}\right)$ **31.** $\approx 18.435°$

33. Right-hand tangent $y = x$ and left-hand tangent $y = 2 - x$ at $(1, 1)$, right-hand tangent $y = 2 - x$ and left-hand tangent $y = 2 + x$ at $(0, 2)$, right-hand tangent $y = 2 + x$ and left-hand tangent $y = -x$ at $(-1, 0)$

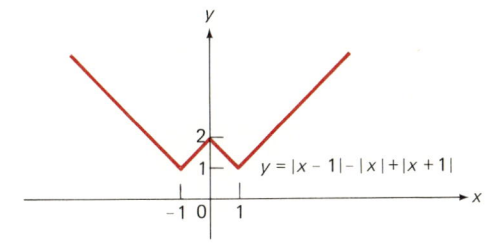

$y = |x - 1| - |x| + |x + 1|$

35. Yes   **37.** $df = 0.2$, $\Delta f = \frac{2}{9} = 0.222$ to 3 decimal places
39. $df = 0.181$, $\Delta f = 0.203$, both to 3 decimal places
41. $(2.9)^{-2} \approx \frac{1}{9} + \frac{2}{27}(0.1) = 0.11852$, exact value $= 0.11891$, both to 5 decimal places
43. $(8.6)^{2/3} \approx 4 + \frac{1}{3}(0.6) = 4.2$, exact value $= 4.1976$ to 4 decimal places
45. $\tan 63° \approx \sqrt{3} + 4(\pi/60) = 1.941$, exact value $= 1.963$, both to 3 decimal places
47. $\sec 1° \approx \sec 0 = 1$, exact value $= 1.00015$ to 5 decimal places
49. $\sqrt[4]{17} \approx 2 + \frac{1}{32} = 2.03125$, exact value $= 2.03054$ to 5 decimal places
51. $(7.8)^{-1/3} \approx \frac{1}{2} - \frac{1}{48}(-0.2) = 0.50417$, exact value $= 0.50424$, both to 5 decimal places
53. $f(4.001) \approx f(4) + f'(4)(0.001) = 0.008$ **55.** $y = \frac{1}{4}(x + 1)^4$ **57.** $y = \frac{1}{9}(x^3 + 1)^3$ **59.** $y = \frac{1}{6}x^3$ **61.** $y = \frac{1}{20}x^5$
63. $y = \frac{1}{24}x^4$ **65.** $-\frac{1}{(\sqrt{1+x^2})^3}$ **67.** $\frac{\cos(\tan\sqrt{x}) \sec^2\sqrt{x}}{4\sqrt{x}\sqrt{\sin(\tan\sqrt{x})}}$ **69.** $-2x \cos(\cos(\sin x^2)) \sin(\sin x^2) \cos x^2$

71. $f'g' = c$ implies $(f'g')' = 0$.   **73.** $\frac{7}{340}$ rad/sec $\approx 1.18°$/sec; $\frac{1}{40}$ rad/sec $\approx 1.43°$/sec; $-\frac{1}{20}$ rad/sec $\approx -2.865°$/sec
75. The tangent to the curve $y = \sin x$ at the point $(a, \sin a)$ is the line $y = (x - a) \cos a + \sin a$.

Chapter 3

Section 3.1 (page 165)

1. $c = \frac{3}{2}$ **3.** $c = 0$ **5.** $c = \frac{6}{5}$ **7.** $c = 5\pi/4$

9. $f'(x) = -\frac{2}{3}x^{-1/3}$ if $x \neq 0$, $f'(0)$ does not exist

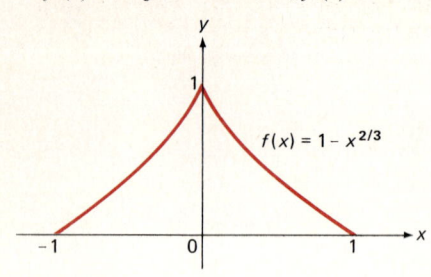

11. The derivative of $x^3 + x^2 + x + 1$ is never zero. **13.** The derivative of $x^5 - 5x + 1$ is zero at only two points.
15. $c = \sqrt{3}$ **17.** $c = 2\sqrt{3} - 1$ **19.** $c = (\frac{4}{3})^{3/2}$ **21.** $c \approx 0.88$ ($\cos c = 2/\pi$) **23.** Apply the mean value theorem.
25. Let $f(x) = [\![x]\!]$ for $n < x < n + 1$, but leave f undefined at $x = n$ (n any integer). **27.** $c = \frac{1}{3}(1 \pm \sqrt{7})$ **29.** $c = \frac{4}{3}$

Section 3.2 (page 173)

1. Local (and absolute) minimum $y = -1$ at $x = -1$, increasing on $[-1, \infty)$, decreasing on $(-\infty, -1]$

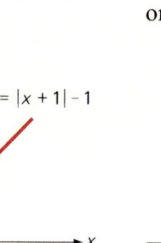

3. Local maximum $y = 2$ at $x = 1$, local minimum $y = -2$ at $x = -1$, increasing on $[-1, 1]$, decreasing on $(-\infty, -1]$ and $[1, \infty)$

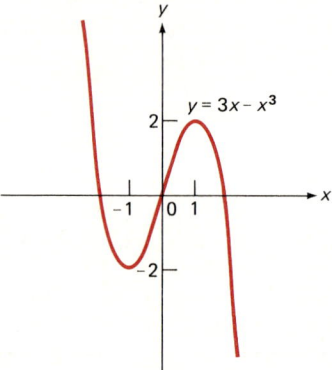

5. Local maximum $y = 4$ at $x = 0$, local (and absolute) minimum $y = 0$ at $x = \pm\sqrt{2}$, increasing on $[-\sqrt{2}, 0]$ and $[\sqrt{2}, \infty)$, decreasing on $(-\infty, -\sqrt{2}]$ and $[0, \sqrt{2}]$

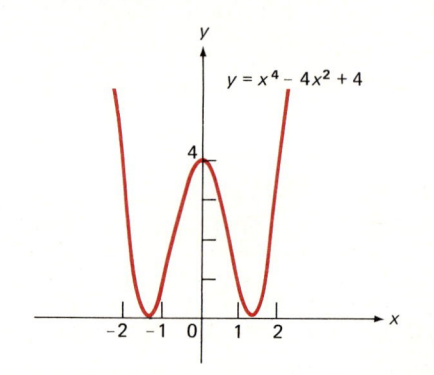

7. Local (and absolute) maximum $y = \frac{32}{7}$ at $x = -\frac{1}{2}$, increasing on $(-\infty, -\frac{1}{2}]$, decreasing on $[-\frac{1}{2}, \infty)$

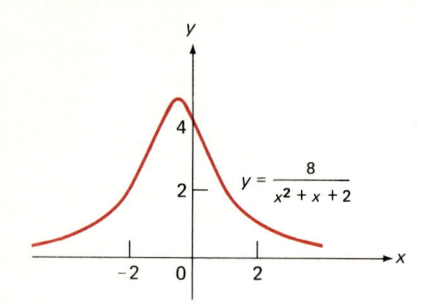

9. Local maximum $y = -4$ at $x = -2$, local minimum $y = 0$ at $x = 0$, increasing on $(-\infty, -2]$ and $[0, \infty)$, decreasing on $[-2, -1)$ and $(-1, 0]$

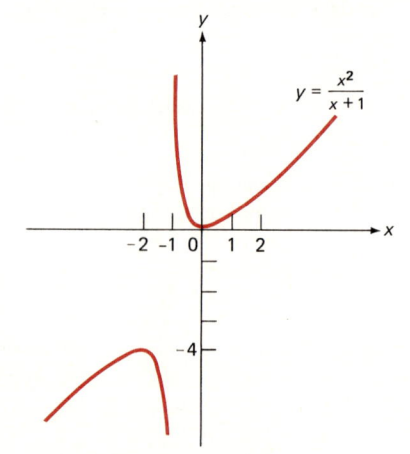

11. Local maxima $y = \sqrt{2}$ at $x = (2n + \frac{1}{4})\pi$, local minima $y = -\sqrt{2}$ at $x = (2n + \frac{5}{4})\pi$, increasing on $[(2n - \frac{3}{4})\pi, (2n + \frac{1}{4})\pi]$, decreasing on $[(2n + \frac{1}{4})\pi, (2n + \frac{5}{4})\pi]$, n any integer. Note that $\cos x + \sin x = \sqrt{2} \cos(x - \frac{1}{4}\pi)$.

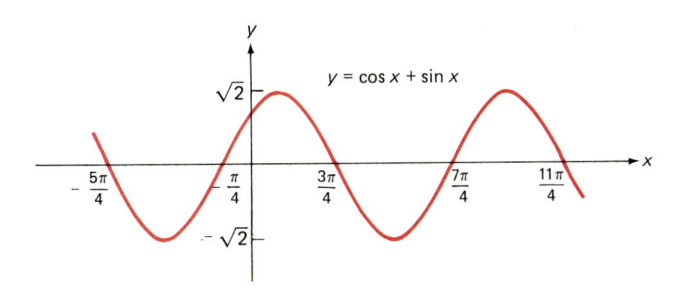

13. Maximum $M = 66$ at $x = 10$; minimum $m = 2$ at $x = 2$
15. Maximum $M = 132$ at $x = -10$; minimum $m = 0$ at $x = 1, 2$
17. Maximum $M = 100.01$ at $x = 0.01, 100$; minimum $m = 2$ at $x = 1$
19. Maximum $M = 1$ at $x = -1, 0$; minimum $m = \frac{1}{2}\sqrt{3}$ at $x = -\frac{1}{2}$
21. Maximum $M = 3$ at $x = \pi/5, 5\pi/6$; minimum $m = 2$ at $x = 0, \pi/2, \pi$
23. Maximum $M = 1$ at $x = \sqrt{\pi/2}$; minimum $m = \sin 4 \approx -0.76$ at $x = 2$
25. If f is even, then $f(-x) \equiv f(x)$. Only Prob. 5   27. $c = -\frac{1}{2}$

Section 3.3 (page 181)

1. $x = n\pi$, n any integer   3. $x = (n + \frac{1}{2})\pi$, n any integer   5. None   7. None
9. Compare the behavior of $f(x) = x^4$ and $g(x) = x^5$.
11. No extrema, inflection point at $x = 0$, increasing on $(-\infty, \infty)$, concave upward on $[0, \infty)$, concave downward on $(-\infty, 0]$

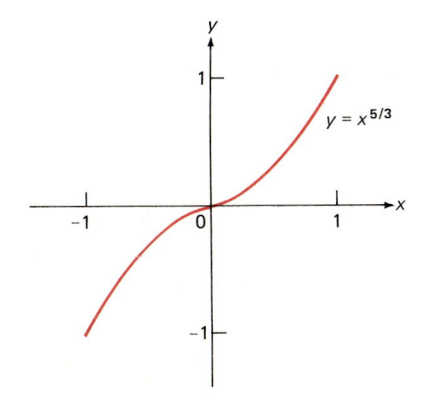

13. Local maximum $y = 4$ at $x = 0$, local minimum $y = 0$ at $x = 2$, inflection point at $x = 1$, increasing on $(-\infty, 0]$ and $[2, \infty)$, decreasing on $[0, 2]$, concave upward on $[1, \infty)$, concave downward on $(-\infty, 1]$

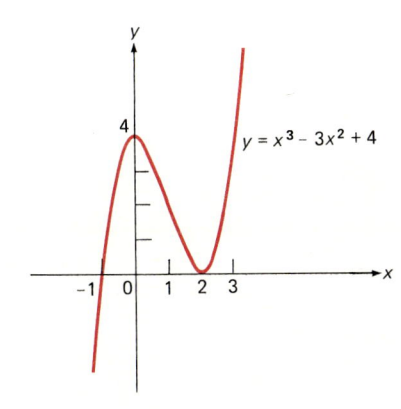

15. Local (and absolute) maximum $y = 2$ at $x = \pm 1$, local minimum $y = 0$ at $x = 0$, inflection points at $x = \pm 1/\sqrt{3}$, increasing on $(-\infty, -1]$ and $[0, 1]$, decreasing on $[-1, 0]$ and $[1, \infty)$, concave upward on $[-1/\sqrt{3}, 1/\sqrt{3}]$, concave downward on $(-\infty, -1/\sqrt{3}]$ and $[1/\sqrt{3}, \infty)$

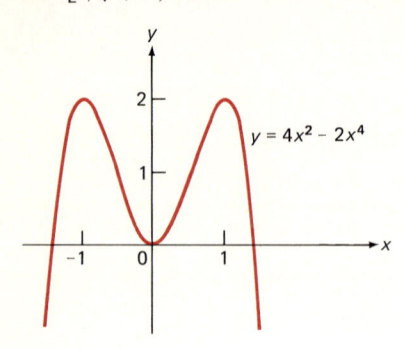

17. Local (and absolute) minimum $y = 0$ at $x = 0$, inflection points at $x = \pm 1$, increasing on $[0, \infty)$, decreasing on $(-\infty, 0]$, concave upward on $[-1, 1]$, concave downward on $(-\infty, -1]$ and $[1, \infty)$

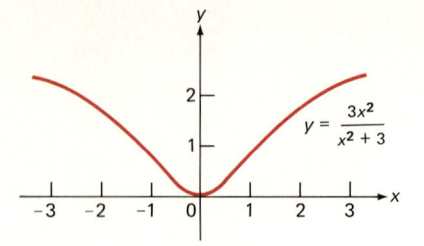

19. Apply the concavity test. Yes **21.** Apply rules (i) and (ii). **23.** $a = -\frac{3}{2}, b = \frac{9}{2}$
25. Apply the monotonicity test to g.

Section 3.4 (page 184)
1. ∞ **3.** $-\infty$ **5.** ∞ **7.** ∞ **9.** $-\infty$ **11.** 0 **13.** ∞ **15.** ∞ **17.** 0 **19.** 2 **21.** $-\frac{1}{2}$
23. $3 < x < \frac{3000}{999}$; $\frac{3000}{1001} < x < 3$ **25.** -1 **27.** $-\infty$ **29.** ∞ **31.** Modify the proof of Theorem 3, p. 77. **33.** 0
35. 0 **37.** 0 **39.** Does not exist **41.** $\frac{1}{2}$ **43.** -2 **45.** $\frac{1}{2}$ **47.** 1

Section 3.5 (page 195)
1. $f(x) = \dfrac{1}{(x-1)(x-2)\cdots(x-n)}$

3. Horizontal asymptote $y = \frac{1}{2}$, vertical asymptote $x = -2$

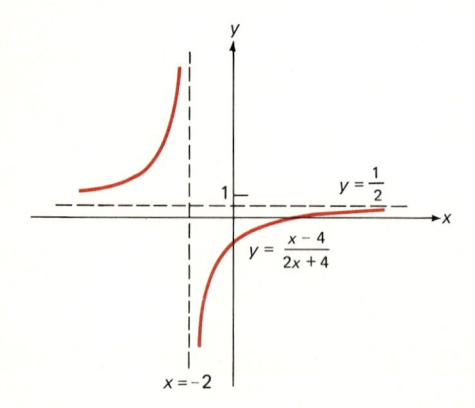

5. Horizontal asymptote $y = 1$, vertical asymptotes $x = \pm 2$

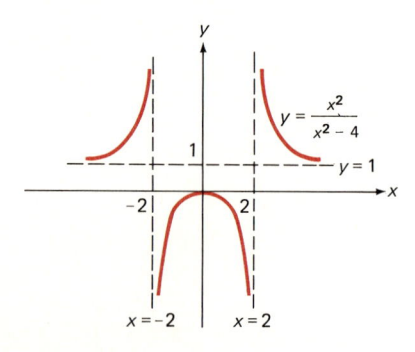

7. Horizontal asymptote $y = -1$

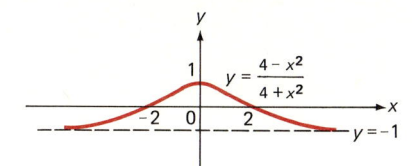

9. Horizontal asymptote $y = 1$, vertical asymptote $x = 1$

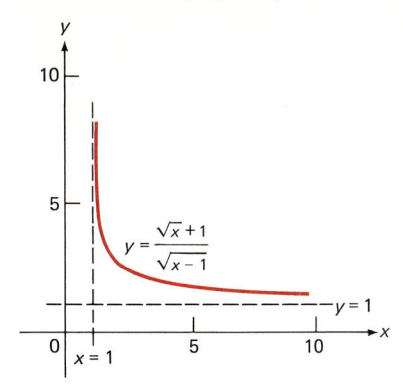

11. The x-axis and the line $x = -2$ **13.** Apply Theorem 10, p. 34.
15. Vertical asymptote $x = 0$, oblique asymptote $y = -x$ **17.** Oblique asymptotes $y = \pm 2x - 2$ **19.** No asymptotes

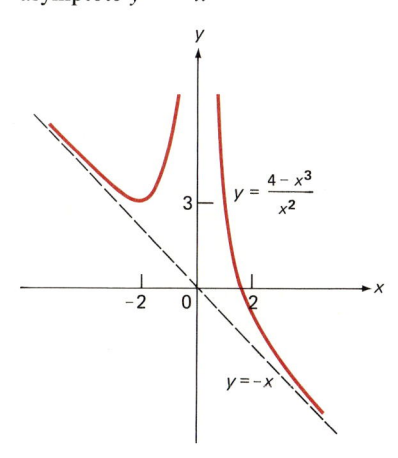

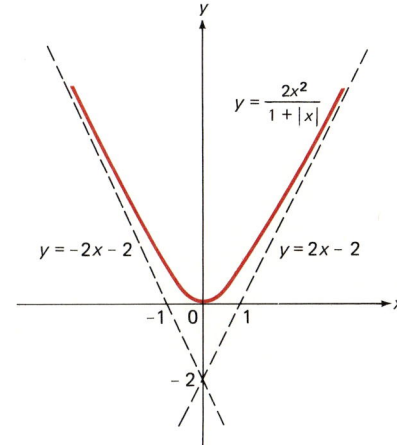

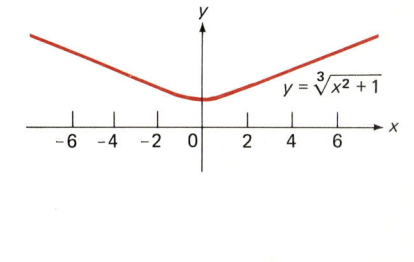

21. $\lim\limits_{x \to \pm\infty} \dfrac{f(x)}{x} = 1$, but $\lim\limits_{x \to \pm\infty} [f(x) - x]$ does not exist. **23.** T is the line $x = 1$; N is the x-axis. No
25. T is the line $x = 3$; N is the line $y = -1$. Yes **27.** At the points $((n + \tfrac{1}{2})\pi, 0)$, n any integer. No

Section 3.6 (page 203)

1. $0/0$; $-\tfrac{1}{11}$ **3.** $0/0$; $-\tfrac{10}{3}$ **5.** $0/0$; -2 **7.** $0 \cdot \infty$; 1 **9.** $0/0$; 4 **11.** $0 \cdot \infty$; $\tfrac{1}{2}$ **13.** $0/0$; $-\infty$ **15.** $\infty - \infty$; 0
17. $0/0$; $\tfrac{1}{4}$ **19.** Let $f(x) = F(x) - F(a)$, $g(x) = x - a$ in the modification of Theorem 11 for $x \to a$.   **21.** $-\tfrac{2}{3}$
23. Let $F(x) = \begin{cases} x^2 \sin \dfrac{1}{x} & \text{if } x \neq 0, \\ 0 & \text{if } x = 0. \end{cases}$ Then F is differentiable everywhere and $F'(0) = 0$, but $\lim\limits_{x \to 0} F'(x)$ does not exist.

Section 3.7 (page 212)

1. $80{,}000$ ft²   **3.** The square of side $\sqrt{A}$ **5.** The maximum is $c^2/4$, but there is no minimum.
7. $10\sqrt{2}$ in. long, $5\sqrt{2}$ in. wide   **9.** $2R^2$ **11.** $\dfrac{b}{4}\sqrt{s^2 - \dfrac{b^2}{4}}$ **13.** $\dfrac{3\sqrt{3}\,s^2}{4}$; $2s$
15. The distance from P to BC should be $\tfrac{3}{4}$ the diameter of the circle.
17. $(27\pi)^{-1/2} S^{3/2}$; $3\pi^{1/3} V^{2/3}$. If $V_{\max}(S)$ is the maximum volume of a cup of surface area S, while $S_{\min}(V)$ is the minimum surface area of a cup of volume V, then $V = V_{\max}(S_{\min}(V))$, $S = S_{\min}(V_{\max}(S))$.

19. The cylinder of radius $p/3$, height $p/6$ and volume $\pi p^3/54$.   21. $\theta = \sqrt{\frac{8}{3}\pi}$ rad $\approx 293.94°$ 23. $\frac{1}{2}\sqrt{3}\pi R^3$ 25. $\frac{4}{27}$
27. 25 mi   29. 1 hr, 33 min later; 8 mi   31. 2 mi   33. $5 - \frac{4}{\sqrt{3}} \approx 2.69$ mi   35. The line $\frac{x}{2a} + \frac{y}{2b} = 1$; $2ab$
37. $L^2/4(4+\pi)$. Make the side of the square equal to the diameter of the circle.   39. $L = (a^{2/3} + b^{2/3})^{3/2}$
41. The shortest path joining the points P_1 and P'_2 is the line segment $P_1 P'_2$.

Section 3.8 (page 221)
1. $MC(q) = b$, $AC(q) = (a/q) + b$. No
3. $MC(q) = 20 + 0.2q$, $AC(q) = (490/q) + 20 + 0.1q$. The minimum average cost is $AC(70) = MC(70) = \$34$.
$\Delta C(q) = C(q+1) - C(q) = MC(q) + 0.1$, but $\$0.1 = 10¢$ is negligible compared to either $\Delta C(q)$ or $MC(q)$.
5. $AVC(q) = 18 - 0.06q + 0.001q^2$. The minimum average variable cost is $AVC(30) = \$17.10$.
$MC(q) = 18 - 0.12q + 0.003q^2$, $MC(30) = \$17.10$; $D_q AVC(q) = 0$ implies $C'(q)q - C(q) + C(0) = 0$.
7. $C(0) > 0$ implies $a > 0$, $MC(0) > 0$ implies $b > 0$, $MC(q)$ has a minimum at $q_0 = -c/3d$ if $d > 0$,
$q_0 > 0$ implies $c < 0$, $MC(q_0) > 0$ implies $c^2 < 3bd$, $MC'(q_0) = 0$ and $MC''(q_0) \neq 0$ imply an inflection point of C at q_0.
9. If $R(p)$ is the revenue at price p, then $R(p) = pq = p(600 - 0.4p)$.   11. 4000   13. 35¢; 45¢; \$10 a day
15. $p = 160$; $e_D > 1$ if $p > 160$, $e_D < 1$ if $p < 160$.   17. $p = 30$; $e_D > 1$ if $p > 30$, $e_D < 1$ if $p < 30$.
19. $MR(q_0) = MC(q_0)$ is only a necessary condition for profit maximization. To get a sufficient condition, use the second derivative test.
21. Calculate e_{zx}, where $z = xf(x)$.   23. First show that $R'(p) = q(1 - e_D)$.   25. \$52,780 at $q = 125$
27. \$50,280 at $q = 200$ 29. Tax revenue \$12,500, tax rate \$5 a gallon, profit \$4250, output level 2500
31. In the absence of taxes, profit is maximized at $q = 5000$.

Supplementary Problems (page 223)
1. Let $b = a + \Delta x$ in Theorem 2, p. 162.   3. $t = 10 - 3\sqrt{10}$ 5. $t \approx 0.57$ ($\cos t = \sin 1 \approx 0.84$)
7. Consider the formula $f'(c) - kg'(c) = 0$, figuring in the proof of the theorem.   9. $c = \frac{1}{24}(1 + \sqrt{577})$
11. Apply Rolle's theorem n times in succession.   13. $c = -3\pi/4$, $\pi/4$ or $5\pi/4$ 15. $c = 0$ 17. $a = 1$, $b = 0$
19. $2^{1-r} > 1$ if $0 < r < 1$, $2^{1-r} \leq 1$ if $r \geq 1$ 21. $f(-\frac{1}{3}) = \frac{2}{3}$

23. Local (and absolute) minimum $y = -4$ at $x = 1$, no inflection points, increasing on $[1, \infty)$, decreasing on $[0, 1]$, concave upward on $(0, \infty)$, no asymptotes, one-sided vertical tangent at the origin

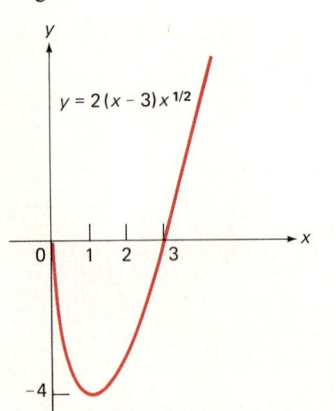

25. Local maximum $y = \frac{3}{5}(\frac{2}{5})^{2/3} \approx 0.33$ at $x = \frac{2}{5}$, local minimum $y = 0$ at $x = 0$, inflection point at $x = -\frac{1}{5}$, increasing on $[0, \frac{2}{5}]$, decreasing on $(-\infty, 0]$ and $[\frac{2}{5}, \infty)$, concave upward on $(-\infty, -\frac{1}{5}]$, concave downward on $[-\frac{1}{5}, 0)$ and $(0, \infty)$, no asymptotes, vertical tangent and cusp at the origin

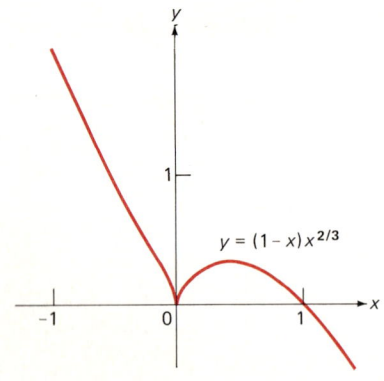

27. Local (and absolute) maximum $y = \frac{3}{4}\sqrt{3} \approx 1.30$ at $x = \frac{1}{2}$, absolute minimum $y = 0$ at $x = 0$ and $x = 2$, inflection point at $x = \frac{1}{2}(1 + \sqrt{3}) \approx 1.37$, increasing on $[0, \frac{1}{2}]$, decreasing on $[\frac{1}{2}, 2]$, concave upward on $[\frac{1}{2}(1+\sqrt{3}), 2)$, concave downward on $(0, \frac{1}{2}(1+\sqrt{3})]$, no asymptotes, one-sided vertical tangent at the origin

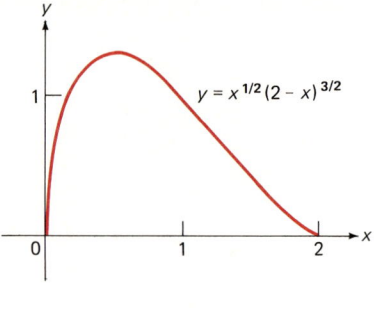

29. 5^{-5} 31. $\frac{1}{4}$ 33. $\frac{1}{2}$ 35. 1   37. $1/n$ 39. $n(n+1)/2$ 41. $(m-n)/2$ 43. $p^2/16$ 45. $\frac{8}{27}$ 47. $\frac{2}{3}$ 49. $ab/4$
51. 8192 in$^3 \approx 4.74$ ft^3 53. $16{,}384/\sqrt{3}$ in$^3 \approx 5.47$ ft^3
55. In each case the girth is proportional to the independent variable x, and the cross-sectional area is proportional to x^2.
57. θ is largest for $|PQ| = \sqrt{h(s+h)}$, where Q is the point in which the prolongation of the segment intersects L.   59. 125 ft

Chapter 4

Section 4.1 (page 228)
1. 135 **3.** −182 **5.** 700 **7.** $\frac{517}{280}$ **9.** 0 **11.** Expand the sum and make all possible cancellations. **13.** $\frac{7}{16}$
15. $\sum_{i=1}^{11}(3i-1)$ **17.** $\sum_{i=1}^{8}\frac{i+2}{i}$ **19.** True **21.** False **23.** 924 **25.** 3876 **27.** 465 **29.** $41+29\sqrt{2}$
31. $a^7 - 7a^6b + 21a^5b^2 - 35a^4b^3 + 35a^3b^4 - 21a^2b^5 + 7ab^6 - b^7$ **33.** Expand $(1+1)^n$.
35. $D_x f(x)g(x) = (D_s + D_t)f(s)g(t)|_{s=t=x}$ **37.** $-5040\sin x - 2880x\cos x + 540x^2\sin x + 40x^3\cos x - x^4\sin x$

Section 4.2 (page 237)
1. See Theorem 1. **3.** $A = \frac{1}{2}ab$ **5.** $\frac{1}{10}$ **7.** $\frac{11}{3}$ **9.** $\frac{3}{4}$ **11.** 0 **13.** $\frac{1}{12}$ **15.** $\frac{7}{2}$ **17.** 204

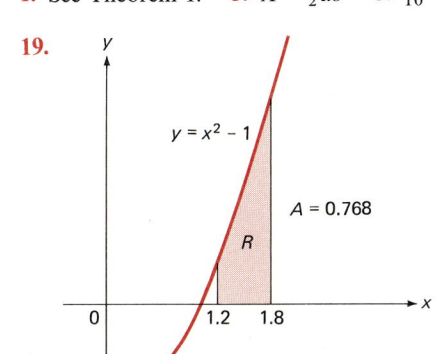

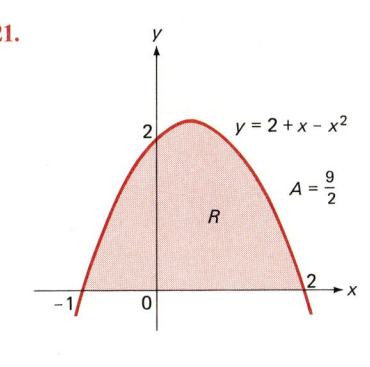

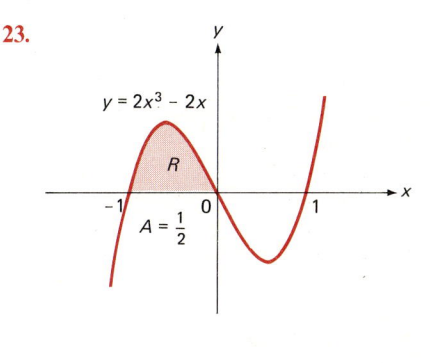

19.

21.

23.

25. The smallest integer not less than $\frac{b-a}{\mu}$ **27.** $\frac{1}{10}$. There is no largest value.
29. Estimate the Riemann sum. **31.** Use Prob. 63, p. 104.
33. The Riemann sum of an unbounded function can be made arbitrarily large (in absolute value).
35. Start from $-|f(x)| \le f(x) \le |f(x)|$, and use rule (v).

Section 4.3 (page 247)
1. If $a > b$, write $\int_a^b x^n\, dx = -\int_b^a x^n\, dx$. **3.** $-\frac{2}{3}$ **5.** 4 **7.** $\frac{2}{11}$ **9.** $\frac{55}{2}$ **11.** 20

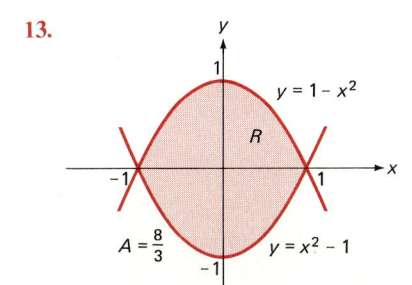

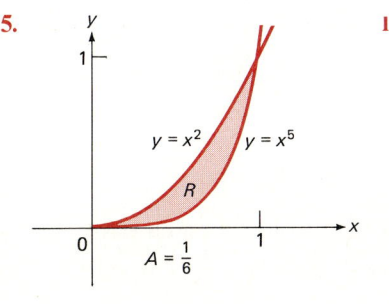

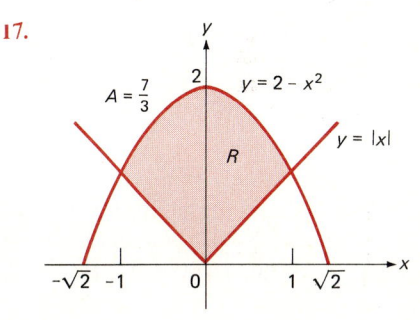

13.

15.

17.

19. $A = \frac{253}{12}$ **21.** The midpoint of $[a, b]$ **23.** $-\frac{19}{3}$ **25.** $\frac{5}{6}$ **27.** $c = 4$ **29.** $c = 2/\sqrt{3}$ **31.** $c = 0$ or $\pm\sqrt{2}$
33. Apply Example 8 to the function $f - g$. **35.** The first **37.** The second
39. Use Example 8, noting that if f is nonconstant, then f takes values between the maximum and minimum of f on $[a, b]$.

Section 4.4 (page 255)
1. $\frac{1}{3}x^3 + \frac{1}{2}x^2 + 2x + C$ **3.** $\frac{1}{50}x^{50} - \frac{1}{5}x^{25} + 2x^{10} - 10x + C$ **5.** $4x^{1/4} + 3x^{-1/3} + C$ **7.** $-2\cos x - 3\sin x + C$
9. $t - \frac{1}{5}t^5 + C$ **11.** $8v - \frac{27}{4}v^4 + C$ **13.** Use the chain rule **15.** f is a polynomial of degree less than n.
17. $\frac{1}{5}x^5 - x^3 + x^2 - 4x + C$ **19.** $\frac{1}{4}x^4 - \frac{1}{2}x^2 - x^{-1} + \frac{1}{3}\cos 3x + C$ **21.** $u - 3u^2 + \frac{11}{3}u^3 - \frac{3}{2}u^4 + C$

23. $\frac{1}{2}x - \frac{1}{4}\sin 2x + C$ **25.** $-\cot x - x + C$ **27.** $\sec x + C$ **29.** $2\sin x + C$ **31.** $\sin 2u - u + C$
33. $\frac{1}{4}w^4 + \frac{1}{3}w^3 + \frac{1}{2}w^2 + w + C$ **35.** $\frac{3}{2}x - \cos x - \frac{1}{4}\sin 2x + C$ **37.** 0   **39.** $f(b)$ **41.** $(1 + \sin t)^{25}$
43. Use Prob. 35, p. 238, after observing that if $F(x) = \int_a^x f(t)\,dt$, then $\dfrac{F(x + \Delta x) - F(x)}{\Delta x} - f(x) = \dfrac{1}{\Delta x}\int_x^{x+\Delta x} [f(t) - f(x)]\,dt$.
Then apply Theorem 2, p. 162, to $F(x)$.

Section 4.5 (page 260)

1. First show that $\int_{-a}^{0} f(x)\,dx = \int_0^a f(-x)\,dx$.   **3.** $-\frac{10}{3}$ **5.** $-\frac{5}{12}$ **7.** $\frac{235}{24}$ **9.** $-\frac{100}{3}$ **11.** 6   **13.** $\frac{116}{15}$ **15.** $4 - \pi$
17. π **19.** $\sqrt{3}$ **21.** $4/\sqrt{3}$ **23.** 0   **25.** $\sqrt{2} - 1$

27.

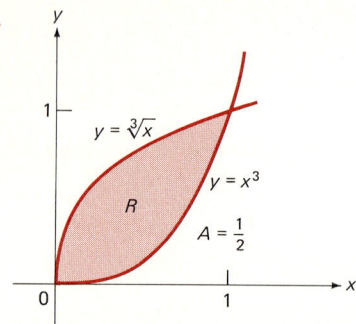

29.

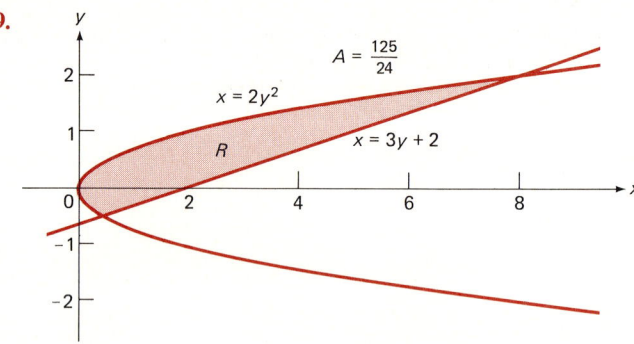

31.

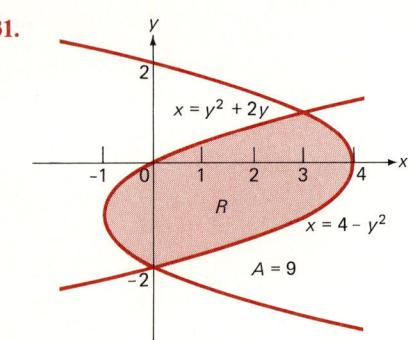

33.

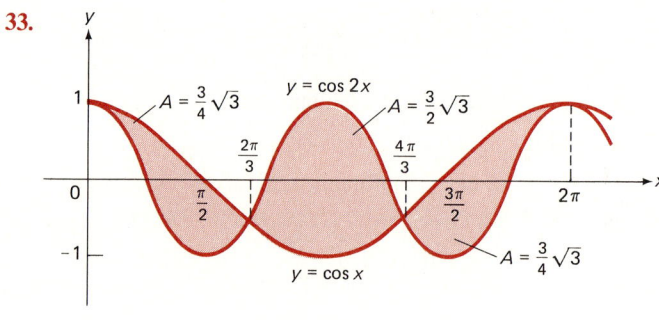

35. $\frac{1}{3}$ **37.** 0   **39.** $\dfrac{4}{\pi}$ **41.** $F(x_i) - F(x_{i-1}) = F'(p_i)(x_i - x_{i-1}) = f(p_i)\Delta x_i$, where $x_{i-1} < p_i < x_i$.

Section 4.6 (page 266)

1. Use the fundamental theorem of calculus.   **3.** 2000 cm, $\frac{50}{3}$ cm/sec   **5.** 2475 ft, 56.25 mph
7. $v_{av}(T) \to v(T)$ as $T \to \infty$; $v_{av}(10\text{ min}) \approx 73.77$ mph, $v(10\text{ min}) \approx 74.98$ mph   **9.** $y = x^2 - 3$ **11.** $y = \frac{1}{3}x^3 - \frac{1}{2}x^2 - 4$
13. $y = x^3 + \sin x + 2 - \pi^3$ **15.** $y = \frac{1}{12}x^4 + \frac{1}{6}x^3 + \frac{1}{6}x - \frac{5}{12}$ **17.** $y = \frac{4}{3}x^{3/2} - 4x + \frac{22}{3}$ **19.** $y = \frac{3}{2}x^{-1} + \frac{9}{4}x - \frac{25}{4}$
21. $y = -\dfrac{\cos x}{x}$

Section 4.7 (page 272)

1. $s = \dfrac{F}{2m}t^2$ **3.** $s = \dfrac{k}{6m}t^3 + v_0 t$ **5.** The slower particle has 4 times the mass of the faster one.
7. $\dfrac{25}{8m}$ **9.** $h = \dfrac{v_0^2}{2g}$. Doubling v_0 quadruples h. If $h = 100$ ft, then $v_0 = 80$ ft/sec.   **11.** Yes, 2.375 sec after A begins to fall.
13. Choose $s_1 = R + h$ in Example 8. Let $h \to \infty$ in formula (i).   **15.** ≈ 85 mi; ≈ 364 mi   **17.** ≈ 15 ft
19. ≈ 6500 km/sec   **21.** Set $R = R_0$ and $v_0 = c$ in formula (17).   **23.** 9 mm

25. There will be no contraction if the escape velocity of a galaxy from a sphere of radius R is less than HR. $\rho_c \approx 4.5 \times 10^{-18}$ kg/km³

27. Integrate $m\dfrac{dv}{dt} = F(t)$.   **29.** Integrate $mv\dfrac{dv}{dt} = Fv$.

Supplementary Problems (page 275)

1. 55   **3.** $-\frac{109}{85}$ **5.** $\frac{9}{100}$ **7.** The common domain of F and G is not an interval.
9. $x^3 + \dfrac{1}{x} - \dfrac{1}{x^2} + C$ **11.** $\dfrac{\sin(\pi x + \sqrt{2})}{\pi} + C$ **13.** $-\cot u - \tan u + C$ **15.** $\frac{7}{3}$ **17.** 0   **19.** $2 - \dfrac{\pi}{2}$
21. Use Prob. 1, p. 260.   **23.** $\frac{12}{5}$ **25.** $\frac{1}{2} + \frac{1}{3}\pi^2$
27. If M and m are the maximum and minimum of f on $[a, b]$, and if g is nonnegative, then $mg(x) \le f(x)g(x) \le Mg(x)$. Now parallel the proof of Theorem 3, p. 245.
29. $c = -\frac{3}{8}$ **31.** $c = \pi/6$

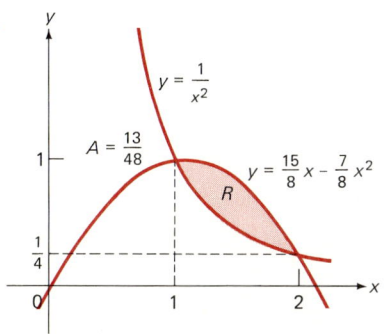

33.

35.

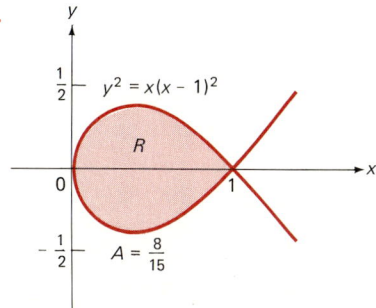

37. $\frac{1}{2}a$ **39.** $y = \frac{1}{6}x^6 - \frac{1}{4}x^4 + 4$ **41.** $y = \sec x + 6$ **43.** $y = \frac{4}{15}x^{5/2} + \frac{1}{3}x + \frac{7}{5}$ **45.** $y = \frac{9}{4}x^{7/3} - \frac{5}{4}x - 1$
47. $C(q) = q^3 - 45q^2 + 1200q + 7200$ dollars; \$7056; \$1076   **49.** $8.25t$ hp; 61.875 hp   **51.** ≈ 133.2 ft
53. Show that $\dfrac{d}{dr}(r^3 v) = \dfrac{r^3 g}{k}$, where r is the radius of the raindrop and v is its velocity.   **55.** 0
57. $3\pi - 5 - \sqrt{2}$ **59.** $\frac{101}{3}$ **61.** 35   **63.** $\frac{1}{2}(n-1)$

Chapter 5

Section 5.1 (page 285)

1. $a \ne 0$; $f^{-1}(x) = \dfrac{x-b}{a}$; $a \ne \pm 1$; $\left(\dfrac{b}{1-a}, \dfrac{b}{1-a}\right)$
3. It follows from $y = f(x)$, $x = f^{-1}(y)$, $f(-x) = -f(x)$, that $f^{-1}(-y) = -f^{-1}(y)$.   **5.** Yes   **7.** No   **9.** No   **11.** Yes
13. Yes   **15.** No   **17.** Yes   **19.** No   **21.** $[(n-\frac{1}{2})\pi, (n+\frac{1}{2})\pi]$, n any integer   **23.** $x = -y$ **25.** $x = 1/y$
27. $x = \dfrac{y}{1-y}$ **29.** $x = \sqrt[3]{y+2}$ **31.** $x = 4 - \sqrt{16-y^2}$ **33.** $f^{-1}(x) = \frac{1}{3}(1-x)$ **35.** $f^{-1}(x) = \sqrt{\dfrac{1}{x} - 1}$
37. $f^{-1}(x) = 3 + (2-x)^2$ **39.** $f^{-1}(x) = 1 - \sqrt{x-4}$
41. Reflection in the line $y = x$ leaves the graph of $y = f(x)$ unchanged.
43. If $y = \dfrac{1-x}{1+x}$, then $x = \dfrac{1-y}{1+y}$.   **45.** If $y = \dfrac{3x+5}{4x-3}$, then $x = \dfrac{3y+5}{4y-3}$.   **47.** x^n is one-to-one on $(-\infty, \infty)$ if n is odd.
49. Use Theorem 3.

Section 5.2 (page 289)

1. Use the chain rule to differentiate $f(x) = y$ or $f(f^{-1}(y)) = y$ with respect to y.   **3.** $\pm\frac{1}{13}$ (why two values?)   **5.** $\frac{1}{16}$
7. $\frac{1}{5}$ **9.** $\pm\frac{4}{3}$ (why two values?)   **11.** 1   **13.** $\frac{1}{4}$ **15.** $x + 4y - 9 = 0$ **17.** $x - 13y - 16 = 0$ **19.** $12x + 5y - 169 = 0$

21. $f'(x) > 0$ for all x; $(f^{-1})'(a) = 1/\sqrt{2}$, $(f^{-1})'(b) = 1$; $(f^{-1})'(0) = 1/\sqrt{2 + \sin^{11} 1} \approx 0.68$ 23. $8/\sqrt{65}$
25. Use the chain rule.   27. $\frac{4}{3}$ 29. 0   31. $\frac{1}{6}$

Section 5.3 (page 296)
1. $\pi/6$ 3. $\pi/4$ 5. $-\pi/6$ 7. $5\pi/6$ 9. $-\pi/4$ 11. $2\pi/3$ 13. $-\pi/2$ 15. $2\sqrt{2}/3$ 17. $3\pi/4$ 19. $\pi - \arcsin x$
21. $\dfrac{1}{\sqrt{x^2 + 1}}$ 23. $2x\sqrt{1 - x^2}$ 25. $1 - 2x^2$ 27. $\dfrac{1}{|2x + 1|\sqrt{x^2 + x}}$ 29. $-\dfrac{2}{1 + t^2}$ 31. $\dfrac{1}{\sqrt{1 - t^2}}$ 33. $\dfrac{1}{\sqrt{1 + 2u - u^2}}$
35. $\dfrac{1}{3 + v^2}$ 37. $\dfrac{3}{5 + 4 \cos v}$ 39. Use formulas (8) and (16).   41. $\dfrac{d}{dx}|x| = \dfrac{|x|}{x}$ if $x \neq 0$ 43. $\dfrac{1}{11} \arctan \dfrac{x}{11} + C$
45. $\dfrac{1}{2} \arcsin \dfrac{t}{2} + C$ 47. $\dfrac{1}{12} \text{arcsec} \dfrac{11|v|}{12} + C$ 49. $\dfrac{\pi}{6}$ 51. $\sqrt{18.75} \approx 4.33$ ft

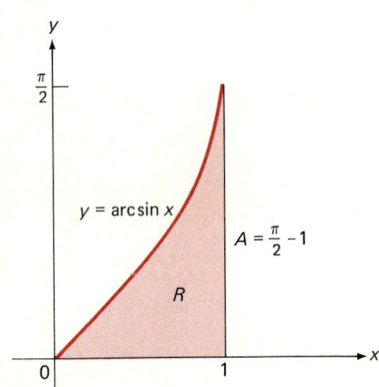

53.

55.

57. To get (ii), let $\theta = \arcsin x$, $\phi = \arcsin y$, and calculate $\sin(\theta + \phi)$. To get (iii), let $\theta = \arctan x$, $\phi = \arctan y$, and calculate $\tan(\theta + \phi)$.   59. Let $x = \frac{1}{2}$, $y = \frac{1}{3}$ in formula (iii).   61. Use formula (iii) twice.

Supplementary Problems (page 298)
1. If x, x' are distinct points of $[-a, 0]$, then $f(-x) \neq f(-x')$. No   3. No   5. Yes   7. $f^{-1}(s) = -\frac{1}{2} + \frac{1}{2}\sqrt{4s - 3}$
9. $h^{-1}(u) = (u^5 - 1)^{1/3}$ 11. Let f be the same as in Prob. 64, p. 104.
13. If $y = (a^{2/3} - x^{2/3})^{3/2}$ $(0 \leq x \leq a)$, then $x = (a^{2/3} - y^{2/3})^{3/2}$ $(0 \leq y \leq a)$.
15. If $y = \sqrt[3]{27 - x^3}$, then $x = \sqrt[3]{27 - y^3}$.   17. $\frac{1}{12}$ 19. $\frac{1}{2}$ 21. -22 23. $\frac{1}{8}$
25. The left side of (i) is the sum of the areas of the regions R_1 and R_2. When $b = f(a)$ 27. $5\pi/6$ 29. $\pi/4$ 31. $-\pi/6$
33. $\sqrt{5}$ 35. $\pi/2$ 37. $\sqrt{13}/3$ 39. $(\sqrt{8} - \sqrt{3})/6$ 41. $\frac{1}{21}$ 43. $\sqrt{a^2 - x^2}$ 45. $\dfrac{1}{2\sqrt{1 - x^2}}$ 47. $\pi/42$ 49. $\pi/60$
51. 1   53. $\frac{1}{6}$ 55.

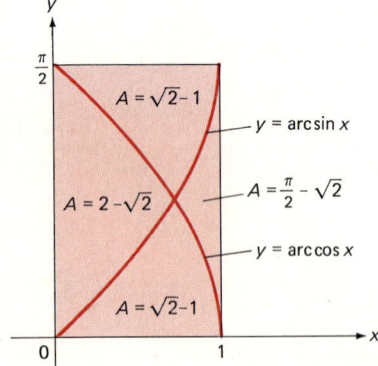

57. $f^{-1}(x) = \cos \dfrac{x}{4}$

Chapter 6

Section 6.1 (page 306)

1. No **3.** $4 \leq x \leq 6$ **5.** $1/e \leq x \leq e$ **7.** $x \geq 3$ **9.** $\ln 2 + \frac{1}{3}\ln 3 - \frac{2}{3}\ln 5$ **11.** $-2\ln 2 - 3\ln 5$

13. $3\ln 2 + 2\ln 3 + 4\ln 5$ **15.** $\dfrac{1}{b-a}\ln\dfrac{b}{a}$ **17.** $\dfrac{3x^2 - 2}{x^3 - 2x + 5}$ **19.** $2x\ln x + x$ **21.** $\dfrac{1 + x^2(1 - 2\ln x)}{x(x^2 + 1)^2}$

23. $2\sin(\ln x)$ **25.** $\dfrac{2}{1 - t^2}$ **27.** $\dfrac{1}{\sqrt{1 + t^2}}$ **29.** $\dfrac{274}{x^6} - \dfrac{120}{x^6}\ln x$ **31.** See Example 1.

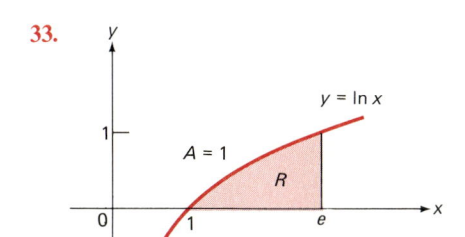

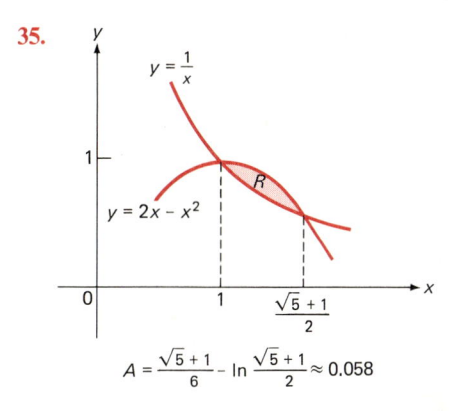

33.

35. $A = \dfrac{\sqrt{5}+1}{6} - \ln\dfrac{\sqrt{5}+1}{2} \approx 0.058$

37. Local (and absolute) minimum $y = -1/e$ at $x = 1/e$, no inflection points, increasing on $[1/e, \infty)$, decreasing on $(0, 1/e]$, concave upward on $(0, \infty)$, no asymptotes

39. Local (and absolute) minimum $y = 0$ at $x = 0$, inflection points at $x = \pm 1$, increasing on $[0, \infty)$, decreasing on $(-\infty, 0]$, concave upward on $[-1, 1]$, concave downward on $(-\infty, -1]$ and $[1, \infty)$, no asymptotes

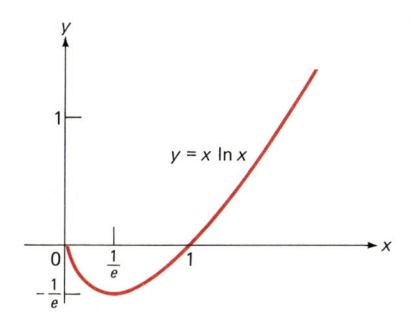

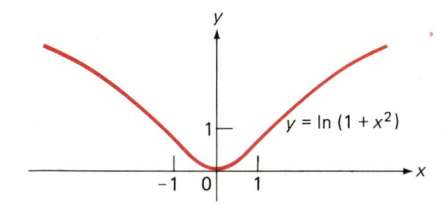

41. $\ln b - \ln a = \dfrac{b - a}{c}$ $(a < c < b)$ **43.** $x(2x^2 - 1)^{3/4}(x^3 + 1)^{4/3}\left(\dfrac{3}{2x^2 - 1} + \dfrac{4x}{x^3 + 1}\right)$

45. $\dfrac{1}{3}\sqrt[3]{\dfrac{x(x^2 + 1)}{(x - 1)^2}}\left(\dfrac{1}{x} + \dfrac{2x}{x^2 + 1} - \dfrac{2}{x - 1}\right)$ **47.** $\dfrac{(1 + \sin x)^5}{(1 - \cos x)^6}\left(\dfrac{5\cos x}{1 + \sin x} - \dfrac{6\sin x}{1 - \cos x}\right)$ **49.** 1 **51.** 0 **53.** $\frac{1}{2}$

Section 6.2 (page 311)

1. $\frac{1}{15}\ln|3x + 1| + C$ **3.** $-\frac{1}{7}\ln|7s - 11| + C$ **5.** $-\frac{1}{5}\ln 6$ **7.** $\frac{1}{4}x - \frac{11}{16}\ln|4x + 3| + C$ **9.** $-6t - 31\ln|t - 5| + C$

11. $6 - 5\ln 3$ **13.** $\frac{1}{3}\ln|x^3 - 1| + C$ **15.** $\ln|\ln x| + C$ **17.** 0 **19.** $\displaystyle\int \cot x\, dx = \int \dfrac{(\sin x)'}{\sin x}\, dx$ **21.** $\dfrac{1}{6}\ln\left|\dfrac{x - 5}{x + 1}\right| + C$

23. $\dfrac{1}{30}\ln\left|\dfrac{3x - 5}{3x + 5}\right| + C$ **25.** $-\frac{4}{5}\ln 2$

Section 6.3 (page 318)

1. $ce^x = e^k e^x$, where $k = \ln c$. **3.** $-6e^{-6x}$ **5.** $(2+x)xe^x$ **7.** $2xe^{x^2}$ **9.** $-\dfrac{e^{1/x}}{x^2}$ **11.** $\dfrac{e^{\sqrt{x}}}{2\sqrt{x}}$

13. $e^{\tan x} \sec^2 x$ **15.** $10^x(1 + x \ln 10)$ **17.** $2x \cdot 5^{x^2} \ln 5$ **19.** $2\exp_2(4^x) \cdot 4^x(\ln 2)^2$ **21.** $\dfrac{10^x \ln 2 + \ln 5}{5^x}$

23. $e^{-u}(1-u)$ **25.** $2e^{x^2}(4x^4 + 12x^2 + 3)$ **27.** $e^x\left(\ln x + \dfrac{5}{x} - \dfrac{10}{x^2} + \dfrac{20}{x^3} - \dfrac{30}{x^4} + \dfrac{24}{x^5}\right)$

29. $\dfrac{e^{x^2+2x}}{2^x \ln x}\left(2x + 2 - \ln 2 - \dfrac{1}{x \ln x}\right)$ **31.** If $y' \equiv y$, then $(ye^{-x})' \equiv 0$.

33.

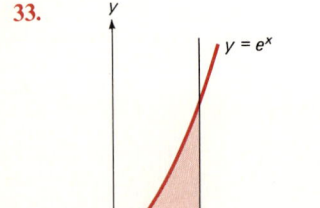

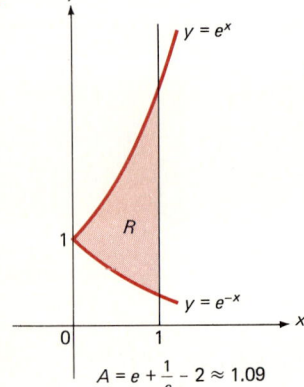

$y = e^x$, $y = e^{-x}$, R

$A = e + \dfrac{1}{e} - 2 \approx 1.09$

35.

$y = xe^{1-x}$, $y = 4x^2 - 3x$, R

$A = e - \dfrac{11}{6} \approx 0.885$

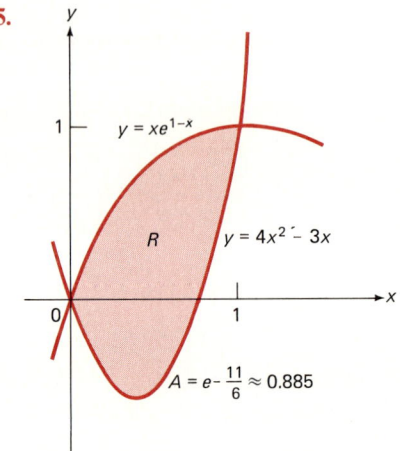

37. Local (and absolute) maximum $y = 1$ at $x = 0$, inflection points at $x = \pm 1$, increasing on $(-\infty, 0]$, decreasing on $[0, \infty)$, concave upward on $(-\infty, -1]$ and $[1, \infty)$, concave downward on $[-1, 1]$, horizontal asymptote $y = 0$

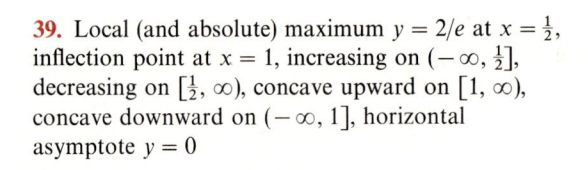

$y = e^{-x^2/2}$

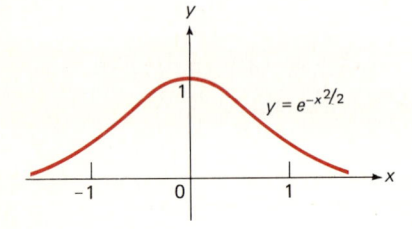

39. Local (and absolute) maximum $y = 2/e$ at $x = \frac{1}{2}$, inflection point at $x = 1$, increasing on $(-\infty, \frac{1}{2}]$, decreasing on $[\frac{1}{2}, \infty)$, concave upward on $[1, \infty)$, concave downward on $(-\infty, 1]$, horizontal asymptote $y = 0$

$y = 4xe^{-2x}$

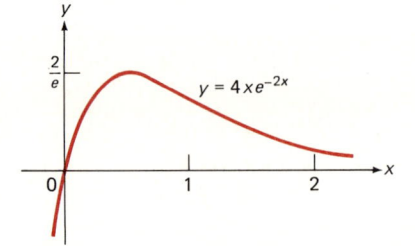

41. If h is continuous and nonzero at c, then h is nonzero in a neighborhood of c. But this is impossible, since $h(x) = 0$ if x is rational.

43. $y = ex$ **45.** The second **47.** The first **49.** $y' = 2ae^{2x} + 3be^{3x}$, $y'' = 4ae^{2x} + 9be^{3x}$ **51.** $\dfrac{xa^x}{\ln a} - \dfrac{a^x}{(\ln a)^2} + C$

53. $\dfrac{8}{3 \ln 3}$ **55.** $\dfrac{624}{5 \ln 5}$ **57.** $\dfrac{4e-1}{2\ln 2 + 1}$ **59.** $a - b$ **61.** 2 **63.** $\frac{1}{2}$ **65.** $\ln 3$ **67.** $\frac{2}{3}$

Section 6.4 (page 322)

1. 10 **3.** -4 **5.** $-\frac{1}{2}$ **7.** 2 **9.** $1 - \log_{10} 2$ **11.** $1 + 2\log_\pi |x|$ **13.** $\log_2 3$ **15.** 1 **17.** 0
19. $1 < a < b$ implies $0 < \ln a < \ln b$, while $0 < a < b < 1$ implies $\ln a < \ln b < 0$. No **21.** $y = 3(\frac{1}{9})^x$; $\log_4 y = \frac{1}{2}x - \frac{3}{2}$

23. $x \geq 1$ if $a > 1$, $0 < x \leq 1$ if $0 < a < 1$ **25.** $2 < x < 3$ **27.** $1 \leq x \leq 100$ **29.** $\log_{10} ex$ **31.** $\dfrac{1}{2x}$ **33.** $\dfrac{1}{x} 5^{\log_7 x} \log_7 5$
35. $x \log_2 \dfrac{x}{e} + C$ **37.** $\tfrac{2}{3} \log_3 2$ **39.** $(\log_{10} e)^2$ **41.** $\dfrac{1}{e} \log_2 e$ **43.** $\log_{10} a \approx \log_{10} b + k$
45. A mouse is about 3.5 orders of magnitude lighter than a man. **47.** $10 \log_{10} \dfrac{10I}{I_0} = 10 + \log_{10} \dfrac{I}{I_0}$. About 26%
49. ≈ 47 dB; ≈ 97 dB **51.** 100 times. About 10 million times **53.** ≈ 1.9 times **55.** ≈ 0.92 bit

Section 6.5 (page 332)

1. $x^{a-1}(a \ln x + 1)$ **3.** $x^{a-1}a^x(a + x \ln a)$ **5.** $x^{1/x}x^{-2}(1 - \ln x)$ **7.** $(\ln x)^x \left[\ln (\ln x) + \dfrac{1}{\ln x} \right]$
9. $x^{\tan x} \left(\sec^2 x \ln x + \dfrac{\tan x}{x} \right)$ **11.** $(\sin x)^{\cos x}[\cos x \cot x - \sin x \ln (\sin x)]$ **13.** $e^{x^x} x^x (\ln x + 1)$
15. $x^{4^x} 4^x \left(\ln 4 \ln x + \dfrac{1}{x} \right)$ **17.** See Prob. 1. **19.** $\tfrac{2}{3} x^{3/2} \ln x - \tfrac{4}{9} x^{3/2} + C$ **21.** $y = \tfrac{1}{2} x^\pi$; $\log_3 y = \sqrt{2} \log_3 x + 2$
23. Apply L'Hospital's rule repeatedly. **25.** $0/0$; e/π **27.** ∞/∞; ∞ **29.** 0^0; 1 **31.** 0^0; 2 **33.** 1^∞; e^π **35.** 1^∞; e^2
37. ∞^0; 1 **39.** 1^∞; e^{-1} **41.** ∞^0; e **43.** \$1485.95; \$1491.82 **45.** ≈ 7.3 yr; ≈ 12.1 yr **47.** $100/r^2$ **49.** 8.16%
51. $\approx 6.72\%$

Section 6.6 (page 339)

1. $y = \dfrac{2}{x}$ **3.** $y = \sqrt{\dfrac{5}{x} - 1}$ **5.** $y = 2\left(\dfrac{x}{x+1}\right)^2 - \dfrac{1}{2}$ **7.** $y = \dfrac{x+1}{x-1}$ **9.** $y = 2e^{3x}$
11. S grows exponentially at the rate of 5% per minute. **13.** Use formula (15). **15.** 7.7 billion **17.** ≈ 164.5 min
19. 75% type A, 25% type B; 13 times its original size **21.** 5.2% **23.** ≈ 131.6 yr **25.** $C = C_0 e^{-kt}$. About 155.6 hr
27. No **29.** 2380 yr **31.** $m_A = m_0 e^{-at}$, $m_B = m_0 \dfrac{a}{b-a}(e^{-at} - e^{-bt})$ **33.** ≈ 124.5 yr old; $\tfrac{4}{9} m_0$, about 51 yr from now

Section 6.7 (page 347)

1. ≈ 7673 **3.** 40 days **5.** ≈ 0.759 g **7.** ≈ 30.7 min **9.** If $N_0 > N_1$, then $1 + \left(\dfrac{N_1}{N_0} - 1\right)e^{-rt}$ is an increasing function.
11. Change r to $-r$ and s to $-s$ in (3) and (6). If $N_0 > N_1$, then $N \to \infty$ as $t \to \dfrac{1}{r} \ln \dfrac{N_0}{N_0 - N_1}$.
13. $N = \left(N_0 - \dfrac{s}{r}\right)e^{rt} + \dfrac{s}{r}$; $s < rN_0$; $s = rN_0$; $s > rN_0$ **15.** $T = T_1 + (T_0 - T_1)e^{-kt}$ **17.** 8°; $\approx 23.2°$ **19.** ≈ 61.3 sec
21. ≈ 4.7 days. ≈ 11.5 days. Yes, after about 23 days. No **23.** $q = CV(1 - e^{-t/RC})$; $i = \dfrac{V}{R} e^{-t/RC}$
25. $y = \dfrac{y_0}{y_0 + (1 - y_0)e^{-kt}} \to 1$ as $t \to \infty$. $T = \dfrac{1}{k} \ln \dfrac{1 - y_0}{y_0}$. The model assumes that infectors stay infectious forever.

Section 6.8 (page 356)

1. e^x **3.** $\dfrac{x^2 + 1}{2x}$ **5.** $\dfrac{x - 1}{2\sqrt{x}}$ **7.** Use formulas (5) and (8). **9.** Use formula (5). **11.** Use formulas (6) and (7).
13. $\cosh c = \sqrt{2}$, $\tanh c = -\dfrac{1}{\sqrt{2}}$, $\coth c = -\sqrt{2}$, $\mathrm{sech}\, c = \dfrac{1}{\sqrt{2}}$, $\mathrm{csch}\, c = -1$ **15.** $2 \sinh 2x$ **17.** $\dfrac{\sinh 2x}{\sqrt{\cosh 2x}}$
19. $-\tanh x$ **21.** $-\dfrac{\mathrm{csch}\,\sqrt{x}\,\coth\sqrt{x}}{2\sqrt{x}}$ **23.** $\dfrac{\mathrm{sech}^2 (\ln x)}{x}$ **25.** $-e^{\coth x} \mathrm{csch}^2 x$ **27.** $y'' - c^2 y = 0$; $y'' + c^2 y = 0$ **29.** No
31. Start from $\mathrm{sech}\, x = 1/\cosh x$; $\mathrm{sech}\, x$ is concave upward on $(-\infty, -\ln(1 + \sqrt{2})]$ and $[\ln(1 + \sqrt{2}), \infty)$, and concave downward on $[-\ln(1 + \sqrt{2}), \ln(1 + \sqrt{2})]$, with inflection points at $x = \pm \ln(1 + \sqrt{2})$.
33. $\tfrac{1}{4} \sinh 2x + \tfrac{1}{2} x + C$ **35.** $\ln |\sinh x| + C$ **37.** $\ln (x^2 + 1) + C$

Section 6.9 (page 360)

1. $\sinh^{-1} x + \dfrac{x}{\sqrt{x^2+1}}$ 3. $-\dfrac{\sin x}{\sqrt{\cos^2 x + 1}}$ 5. $\dfrac{1}{x[1-(\ln x)^2]}$ 7. $\dfrac{d}{dx} \ln(x + \sqrt{x^2+1}) = \dfrac{1}{x+\sqrt{x^2+1}}\left(1 + \dfrac{x}{\sqrt{x^2+1}}\right)$

9. $\sqrt{3}$, taken at $x = \tanh^{-1}(-\tfrac{1}{2})$ 11. $x = \cosh y$ implies $(e^y)^2 - 2xe^y + 1 = 0$, and hence $e^y = x + \sqrt{x^2 - 1}$.

13. $x = \coth y$ implies $x = \dfrac{e^y + e^{-y}}{e^y - e^{-y}}$, and hence $e^{2y} = \dfrac{x+1}{x-1}$.

15. $x = \operatorname{csch} y$ implies $x(e^y)^2 - 2e^y - x = 0$, and hence $e^y = \dfrac{1 \pm \sqrt{1+x^2}}{x}$ ($x \ne 0$), where the plus sign must be chosen if $x > 0$ and the minus sign if $x < 0$, so that $e^y = \dfrac{1}{x} + \dfrac{\sqrt{1+x^2}}{|x|}$.

17. $-\dfrac{\sin x}{\sqrt{\cos^2 x - 1}}$ 19. $\dfrac{1}{2\sqrt{x}(1-x)}$ 21. $-\dfrac{1}{x|\ln x|\sqrt{1+(\ln x)^2}}$ 23. 1 25. $\dfrac{1}{3}\ln\dfrac{3+\sqrt{8}}{2+\sqrt{3}}$ 27. $\dfrac{1}{6}\ln\dfrac{3}{7}$ 29. $\ln\dfrac{2+2\sqrt{5}}{1+\sqrt{17}}$

31. $\sinh^{-1} x$ and $\tanh^{-1} x$ at $x = 0$, $\operatorname{sech}^{-1} x$ at $x = \dfrac{1}{\sqrt{2}}$

Supplementary Problems (page 362)

1. $x = 6$ 3. $x = 32$ 5. Odd 7. $\tfrac{1}{2} \ln 2$ 9. $\ln 3$ 11. $e^{\cosh x} \sinh x$ 13. $2^{3^x} 3^x \ln 2 \ln 3$

15. $x^{\sinh x}\left(\cosh x \ln x + \dfrac{\sinh x}{x}\right)$ 17. Maximum $y = 10^{10} e^{-9}$ at $x = 9$, minimum $y = 0$ at $x = -1$

19. Maximum $y = (\log_2 e)^e$ at $x = e \log_2 e$ 21. Yes, at $x = 1/\sqrt{2}$ 23. No 25. Yes, at $x = 0$ and $x = 8$ 27. 2

29. $-\infty$ 31. π 33. e^{-1} 35. $y = x + \sqrt{x^2 + 1}$ 37. $y = \ln\dfrac{1}{2-e^x}$ 39. $y = 4\sqrt{\dfrac{x-1}{x+1}}$ 41. $y = Cx^n$

43. $\$106{,}186.10$ 45. $\approx 9.40\%$ 47. $\$3097.91$ 49. ≈ 27.8 min 51. $C = C_1(1 - e^{-kt})$

53. Separate variables in (i) and integrate. 55. Separate variables in $\dfrac{dm}{dt} = k_0 e^{-rt} m$ and integrate.

57. ≈ 5.9 billion years 59. Solve (iii) subject to the initial condition $N|_{t=0} = N_0$; $R_{cr} \approx 8.5$ cm, $m_{cr} \approx 48$ kg

61. Show that the solution of the initial value problem $m\dfrac{dv}{dt} = mg - bv^2$, $v|_{t=0} = 0$ ($b > 0$) is

$v = v_1 \dfrac{1 - e^{-2kv_1 t}}{1 + e^{-2kv_1 t}} = v_1 \tanh kv_1 t$, where $k = b/m$ and $v_1 = \sqrt{g/k}$, so that $v \to v_1$ as $t \to \infty$; $v_1 = \sqrt{mg/b}$; $s = \dfrac{1}{k}\ln(\cosh kv_1 t)$

63. Show that the solution of the initial value problem $m\dfrac{dv}{dt} = mg - bv$, $v|_{t=0} = 0$ ($b > 0$) is

$v = v_1(1 - e^{-kt})$, where $k = b/m$ and $v_1 = g/k$, so that $v \to v_1$ as $t \to \infty$; $v_1 = mg/b$; $s = v_1 t - \dfrac{v_1}{k}(1 - e^{-kt})$

65. $\cosh x - \cosh y = \cosh\left(\dfrac{x+y}{2} + \dfrac{x-y}{2}\right) - \cosh\left(\dfrac{x+y}{2} - \dfrac{x-y}{2}\right)$

67. Express $\sinh x$, $\sinh y$, $\sinh(x \pm y)$ in terms of exponentials. 69. Use formulas (6) and (8), pp. 351–2. 71. 1 73. $-\tfrac{1}{2}$

75. $\tfrac{1}{6}$ 77. 1 79. e 81. $(x^2 - 4x) - 2xy^3 + y^6 = 0$ 83. $\dfrac{1}{5}\ln(5x + \sqrt{25x^2 - 16}) + C$ 85. $\dfrac{1}{3}\ln\dfrac{|x|}{3 + \sqrt{9+49x^2}} + C$

Chapter 7

Section 7.1 (page 372)

1. $-\tfrac{1}{20}(1-2x)^{10} + C$ 3. $\tfrac{2}{15}(4+5x)^{3/2} + C$ 5. $\tfrac{2}{27}(3x-1)^{3/2} + \tfrac{2}{9}(3x-1)^{1/2} + C$ 7. $\tfrac{3}{7}(1-x)^{7/3} - \tfrac{3}{4}(1-x)^{4/3} + C$

9. $\tfrac{4}{3}(2+\sqrt{x})^{3/2} + C$ 11. $2 \operatorname{arcsec}\sqrt{x} + C$ 13. $\sin x - \tfrac{1}{3}\sin^3 x + C$ 15. $-\tfrac{1}{5}\cos^5 x + \tfrac{2}{3}\cos^3 x - \cos x + C$

17. $-\sqrt{1 + 2\cos x} + C$ 19. $\tfrac{1}{2}\sec x^2 + C$ 21. $\tfrac{1}{2}e^{x^2} + C$ 23. $-\dfrac{1}{\ln x} + C$ 25. $\tfrac{1}{3}\ln|e^{3x} - 1| + C$ 27. $\arctan(e^x) + C$

29. $\dfrac{\sinh^{-1}(2^x)}{\ln 2} + C = \dfrac{\ln(2^x + \sqrt{4^x + 1})}{\ln 2} + C$ **31.** $\tfrac{1}{2}\cosh 2\sqrt{x} + C$ **33.** $\ln\left|\cos\dfrac{1}{x}\right| + C$

35. Let $f(x) = \csc x + \cot x$ in formula (4). **37.** Show that $D_x \tanh^{-1}(\sin x) = \sec x$, $D_x \tanh^{-1}(\cos x) = -\csc x$.

39. Let $x = -u$ in $\int_{-a}^{0} f(x)\,dx$. **41.** $\dfrac{32}{3}$ **43.** $7 + 2\ln 2$ **45.** $\dfrac{\pi}{6}$ **47.** $\dfrac{8}{15}$ **49.** $\dfrac{\ln 5}{\ln 3}$ **51.** $\ln(3 + 2\sqrt{2})$

53. $\int_a^{a+p} f(x)\,dx = \int_a^0 f(x)\,dx + \int_0^p f(x)\,dx + \int_p^{a+p} f(x)\,dx$ **55.** See Prob. 39.

Section 7.2 (page 378)

1. $x \sin x + \cos x + C$ **3.** $\tfrac{1}{3}x^3 \ln x - \tfrac{1}{9}x^3 + C$ **5.** $x \arcsin x + \sqrt{1-x^2} + C$ **7.** $-\tfrac{1}{5}e^{-x}(\sin 2x + 2\cos 2x) + C$
9. $x(\ln x)^2 - 2x \ln x + 2x + C$ **11.** $\tfrac{2}{3}x^{3/2}\ln x - \tfrac{4}{9}x^{3/2} + C$ **13.** $\tfrac{2}{3}\sin x \sin 2x + \tfrac{1}{3}\cos x \cos 2x + C$
15. $(x^3 - 6x)\sin x + (3x^2 - 6)\cos x + C$ **17.** $x \tanh x - \ln(\cosh x) + C$ **19.** $\tfrac{1}{3}x(2x+3)^{3/2} - \tfrac{1}{15}(2x+3)^{5/2} + C$
21. $-\tfrac{1}{2}\csc x \cot x - \tfrac{1}{2}\ln|\csc x + \cot x| + C$ **23.** $\tfrac{1}{2}x[\sin(\ln x) - \cos(\ln x)] + C$ **25.** $1 - \dfrac{2}{e}$ **27.** -2π **29.** $\pi^3 - 6\pi$
31. $6 - 2e$ **33.** π

35.

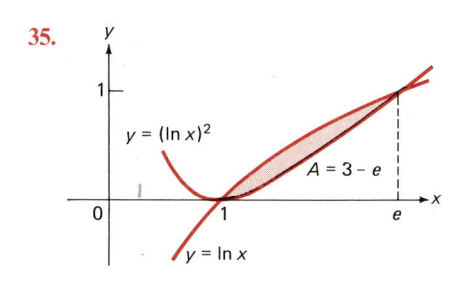

$y = (\ln x)^2$, $y = \ln x$, $A = 3 - e$

37. Integrate by parts n times. **39.** Let $u = 1 - x$; $\dfrac{m!\,n!}{(m+n+1)!}$

Section 7.3 (page 381)

1. Integrate by parts with $u = x^n$, $dv = e^x\,dx$. **3.** Integrate by parts with $u = x^n$, $dv = \sin x\,dx$.
5. $\tfrac{1}{4}\cos^3 x \sin x + \tfrac{3}{8}\cos x \sin x + \tfrac{3}{8}x + C$ **7.** $(x^5 - 5x^4 + 20x^3 - 60x^2 + 120x - 120)e^x + C$
9. $(-x^4 + 12x^2 - 24)\cos x + (4x^3 - 24x)\sin x + C$ **11.** $\dfrac{1}{6}\dfrac{x}{(x^2+1)^3} + \dfrac{5}{24}\dfrac{x}{(x^2+1)^2} + \dfrac{5}{16}\dfrac{x}{x^2+1} + \dfrac{5}{16}\arctan x + C$
13. If $I_n = \int_0^{\pi/2} \sin^n x\,dx$, then $I_0 = \dfrac{\pi}{2}$, $I_1 = 1$, and formula (1) implies $I_n = \dfrac{n-1}{n}I_{n-2}$. **15.** $\dfrac{63\pi}{512}$ **17.** $\dfrac{1}{120}$ **19.** $\dfrac{4096}{6435}$
21. Integrate by parts with $u = (\ln x)^n$, $dv = x^a\,dx$. **23.** $\tfrac{2}{3}x^{3/2}(\ln x)^3 - \tfrac{4}{3}x^{3/2}(\ln x)^2 + \tfrac{16}{9}x^{3/2}\ln x - \tfrac{32}{27}x^{3/2} + C$
25. $9(\ln 3)^2 - 6\ln 3 + 2 - \tfrac{5}{27}e^3$

Section 7.4 (page 387)

1. $\tfrac{1}{8}x - \tfrac{1}{32}\sin 4x + C$ **3.** $-\tfrac{1}{12}\cos^3 4x + \tfrac{1}{20}\cos^5 4x + C$
5. $-\tfrac{2}{3}(\cos x)^{3/2} + \tfrac{6}{7}(\cos x)^{7/2} - \tfrac{6}{11}(\cos x)^{11/2} + \tfrac{2}{15}(\cos x)^{15/2} + C$ **7.** $-\tfrac{1}{3}\csc^3 x + \csc x + C$ **9.** $\cos x + \sec x + C$
11. $3\pi/256$ **13.** $\tfrac{1}{6}\sec^3 2x - \tfrac{1}{2}\sec 2x + C$ **15.** $\tfrac{1}{5}\tan^5 x + \tfrac{1}{3}\tan^3 x + C$ **17.** $-\tfrac{1}{5}\cot^5 x - \tfrac{1}{3}\cot^3 x + C$
19. $\tfrac{1}{9}\tan^3 3x - \tfrac{1}{3}\tan 3x + x + C$ **21.** $-\tfrac{1}{2}\cot^2 x - \ln|\sin x| + C$ **23.** $\tfrac{4}{3}$ **25.** $-\tfrac{1}{16}\cos 8x - \tfrac{1}{4}\cos 2x + C$
27. $\tfrac{1}{6}\sin 3x - \tfrac{1}{18}\sin 9x + C$ **29.** $\tfrac{1}{2}\sin x + \tfrac{3}{2}\sin(x/3) + C$ **31.** $-\tfrac{6}{7}$ **33.** $\tfrac{1}{6}\sin 6a + \tfrac{1}{2}\sin 2a$

Section 7.5 (page 396)

1. $\tfrac{1}{2}x\sqrt{1-9x^2} + \tfrac{1}{6}\arcsin 3x + C$ **3.** $\tfrac{1}{2}x\sqrt{16x^2-1} - \tfrac{1}{8}\ln|4x + \sqrt{16x^2-1}| + C$ **5.** $\ln(x + \sqrt{x^2+64}) + C$
7. $-\dfrac{1}{6}x\sqrt{1-3x^2} + \dfrac{1}{6\sqrt{3}}\arcsin\sqrt{3}x + C$ **9.** $\dfrac{1}{4}\ln\dfrac{\sqrt{x^2+4}-2}{\sqrt{x^2+4}+2} + C$ **11.** $\dfrac{\sqrt{x^2-2}}{2x} + C$
13. $\tfrac{1}{3}(16-x^2)^{3/2} - 16(16-x^2)^{1/2} + C$ **15.** $\tfrac{1}{8}(2x^3 - 5x)\sqrt{x^2-1} + \tfrac{3}{8}\ln|x + \sqrt{x^2-1}| + C$

Answers and Hints to Odd-Numbered Problems

17. $-\dfrac{\sqrt{x^2-9}}{x}+\dfrac{1}{2}\ln\left|\dfrac{x+\sqrt{x^2-9}}{x-\sqrt{x^2-9}}\right|+C$ 19. $\dfrac{1}{8}\dfrac{\sqrt{x^2-4}}{x^2}+\dfrac{1}{16}\arccos\dfrac{2}{|x|}+C$

21. $\dfrac{1}{2}(x-1)\sqrt{2x-x^2}+\dfrac{1}{2}\arcsin(x-1)+C$ 23. $\ln\left|x+\dfrac{3}{2}+\sqrt{x^2+3x-4}\right|+C$ 25. $\dfrac{3}{4}\sqrt{3}+\dfrac{3}{8}\ln(2+\sqrt{3})$

27. $2(\sqrt{5}-\sqrt{2})+\ln\dfrac{1+\sqrt{2}}{2+\sqrt{5}}$ 29. $\dfrac{2}{9}\sqrt{2}-\dfrac{1}{18}\sqrt{5}$ 31. $\ln\dfrac{3+\sqrt{8}}{2+\sqrt{3}}$ 33. $\dfrac{\pi}{6}$ 35. $A=\pi-2a\sqrt{1-a^2}-2\arcsin a$

37. $A=\dfrac{1}{2}\theta$

Section 7.6 (page 406)

1. $x(x-1)(x^2+1)$ 3. $(x^2+1)(x^2+x+1)$ 5. $(x-1)(x+1)(x+9)(x-10)$

7. $\dfrac{1}{(x+a)(x+b)}=\dfrac{1}{b-a}\left(\dfrac{1}{x+a}-\dfrac{1}{x+b}\right)$ 9. $2x-\dfrac{5}{4}\ln|4x+1|+C$

11. $\dfrac{1}{2}\ln\left|\dfrac{3x+4}{x+2}\right|+C$ 13. $\dfrac{2}{5}\ln|x+2|+\dfrac{3}{5}\ln|x-3|+C$ 15. $x-6\ln|x+2|-\dfrac{9}{x+2}+C$

17. $\dfrac{1}{3}\ln|x+1|-\dfrac{1}{6}\ln(x^2-x+1)+\dfrac{1}{\sqrt{3}}\arctan\dfrac{2x-1}{\sqrt{3}}+C$ 19. $\ln\left|\dfrac{x^2}{x+1}\right|+\dfrac{6}{x+1}+C$

21. $-\dfrac{1}{x}-\dfrac{3}{2}\arctan x-\dfrac{x}{2(x^2+1)}+C$ 23. $\dfrac{x+1}{x^2+1}+C$ 25. $\ln|2x-1|-6\ln|2x-3|+5\ln|2x-5|+C$

27. $\dfrac{1}{4}\ln\dfrac{x^2+1}{x^2+3}+\arctan x-\dfrac{1}{\sqrt{3}}\arctan\dfrac{x}{\sqrt{3}}+C$ 29. $-\dfrac{1}{96}(x-1)^{-96}-\dfrac{3}{97}(x-1)^{-97}-\dfrac{3}{98}(x-1)^{-98}-\dfrac{1}{99}(x-1)^{-99}+C$

31. $\dfrac{1}{8}\ln\left|\dfrac{x-2}{x+2}\right|+\dfrac{1}{4}\arctan\dfrac{x}{2}+C$ 33. $\dfrac{2x^6-3x^2}{4(x^4-1)}+\dfrac{3}{8}\ln\left|\dfrac{x^2-1}{x^2+1}\right|+C$ 35. $12\ln 2+7\ln 3-10\ln 5$ 37. $\dfrac{29}{48}+\dfrac{5\pi}{32}$

Section 7.7 (page 412)

1. $\dfrac{4}{3}x^{3/4}-2\sqrt{x}+4\sqrt[4]{x}-4\ln(\sqrt[4]{x}+1)+C$ 3. $\dfrac{6}{5}x^{5/6}-3\sqrt[3]{x}+\ln\dfrac{\sqrt{x}+1}{(\sqrt[6]{x}+1)^3}+2\sqrt{3}\arctan\dfrac{2\sqrt[6]{x}-1}{\sqrt{3}}+C$

5. $\dfrac{6}{7}x^{7/6}-\dfrac{6}{5}x^{5/6}-\dfrac{3}{2}x^{2/3}+2\sqrt{x}+3\sqrt[3]{x}-6\sqrt[6]{x}-3\ln(\sqrt[3]{x}+1)+6\arctan\sqrt[6]{x}+C$

7. $\ln(x+\sqrt{x^2-1})+\sqrt{x^2-1}+C$ $\left(\text{let }u^2=\dfrac{x+1}{x-1}\right)$ 9. $\dfrac{4}{3}(1+\sqrt{x})^{3/2}-4\sqrt{1+\sqrt{x}}+C$ 11. $\dfrac{1}{2}\ln\left|\dfrac{e^x+1}{e^x-1}\right|+C$

13. $\dfrac{2}{\sqrt{3}}\arctan\left(\dfrac{1}{\sqrt{3}}\tan\dfrac{x}{2}\right)+C$ 15. $\dfrac{1}{\sqrt{2}}\ln\left|\tan\left(\dfrac{x}{2}+\dfrac{\pi}{8}\right)\right|+C$ 17. $-\dfrac{1}{4}\sin^2 x-\dfrac{1}{4}\sin x+\dfrac{3}{8}\ln|2\sin x-1|+C$

19. $\ln\left|\sin\dfrac{x}{2}\right|-\dfrac{x}{2}+C$ 21. $\dfrac{1}{6}\ln\left|\dfrac{(\cos x+1)(\cos x-2)^2}{(\cos x-1)^3}\right|+C$

23. $\dfrac{1}{a^2+b^2}(ax+b\ln|a\cos x+b\sin x|)+C$ 25. $\dfrac{16u^3}{(u^2+3)(u^2+1)^3}=\dfrac{6}{u^2+3}-\dfrac{6}{u^2+1}+\dfrac{12}{(u^2+1)^2}-\dfrac{8}{(u^2+1)^3}$

27. $2-\dfrac{\pi}{2}$ 29. $\dfrac{\pi^2}{4}$ 31. $\sqrt{2}\arctan\dfrac{1}{\sqrt{2}}$

Section 7.8 (page 423)

1. 3.613 (midpoint), 3.734 (trapezoidal) 3. 0.949 (midpoint), 0.946 (trapezoidal) 5. 3.056 (midpoint), 3.065 (trapezoidal)
7. $T_{2n}=\dfrac{1}{2}(T_n+M_n)$ 9. $n\geq 29;\ n\geq 41$ 11. $E_M(n)=E_T(n)=0$ if n is even; $E_M(n)=1/n^2,\ E_T(n)=-1/n^2$ if n is odd.
13. Verify (i) for $P(x)=x^3$. 15. $\dfrac{32}{3}$ 17. 0.957 19. 1.3506 21. 0.5890
23. $f^{(4)}(x)=4(4x^4-12x^2+3)e^{-x^2}=4[4(x^2-\dfrac{3}{2})^2-6]e^{-x^2}$. No

Section 7.9 (page 432)

1. $\dfrac{1}{2}$ 3. π 5. Divergent 7. 12 9. $-\dfrac{75}{4}$ 11. $\ln 3$ 13. $\pi/6$ 15. 1 17. Divergent 19. -1 21. 2 23. $\dfrac{1}{5}$
25. Divergent 27. $\dfrac{1}{2}$ 29. $\pi^2/8$ 31. Consider the four cases $p\geq 0$, $-1<p<0$, $p=-1$ and $p<-1$.
33. Use Probs. 31 and 32. 35. $a/(a^2+b^2)$

37.

(figure: region between $y = x^{-1/2}$ and $y = \frac{1}{2}x^{-1/3}$, $A = \frac{5}{4}$)

39.

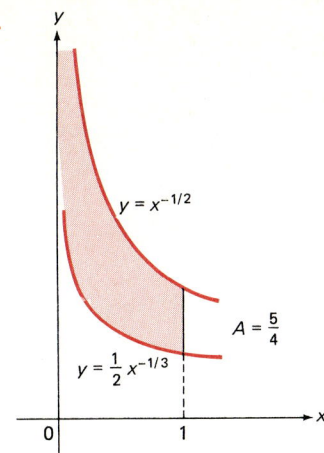

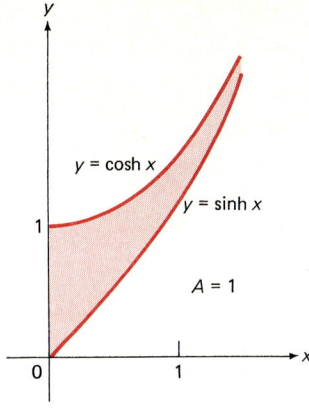

(figure: region between $y = \cosh x$ and $y = \sinh x$, $A = 1$)

41. Divergent **43.** Convergent **45.** Convergent **47.** Divergent **49.** Let $x = t^2$. **51.** $x = \ln 2$ **53.** $x = 1$
55. $f(x) = x$

Supplementary Problems (page 433)

1. $\frac{1}{3}(x^3 - 1)e^{x^3} + C$ **3.** $-\frac{1}{6(x-1)^6} - \frac{2}{7(x-1)^7} - \frac{1}{8(x-1)^8} + C$ **5.** $x + \frac{1}{6}\ln\frac{x^2 - x + 1}{x^2 + 2x + 1} - \frac{1}{\sqrt{3}}\arctan\frac{2x - 1}{\sqrt{3}} + C$

7. $\frac{1}{4}x(x^2 + 1)^{3/2} + \frac{3}{8}x\sqrt{x^2 + 1} + \frac{3}{8}\ln(x + \sqrt{x^2 + 1}) + C$

9. $\frac{1}{4}\ln\frac{\sqrt[4]{x^4 + 1} + x}{\sqrt[4]{x^4 + 1} - x} - \frac{1}{2}\arctan\frac{\sqrt[4]{x^4 + 1}}{x} + C$ $\left(\text{let } u = \frac{\sqrt[4]{x^4 + 1}}{x}\right)$

11. $\ln\left|\tan\frac{x}{2}\right| + \frac{1}{\sqrt{2}}\ln\left|\frac{\sqrt{2}\cos x + 1}{\sqrt{2}\cos x - 1}\right| + C$ (let $u = \cos x$)

13. $\frac{1}{6}\ln\left|\frac{\sqrt{x^6 + x^3} + x^3}{\sqrt{x^6 + x^3} - x^3}\right| + \frac{1}{3}\sqrt{x^6 + x^3} + C$ $\left(\text{let } u = \sqrt{\frac{x^3 + 1}{x^3}}\right)$ **15.** $\frac{2}{3}(y + 1)^{3/2} - y + C$

17. $x\tanh^{-1} x + \frac{1}{2}\ln(1 - x^2) + C$ **19.** $\frac{1}{6}x(1 - x^2)^{5/2} + \frac{5}{24}x(1 - x^2)^{3/2} + \frac{15}{48}x(1 - x^2)^{1/2} + \frac{15}{48}\arcsin x + C$

21. $\arctan(\sinh x) + C = 2\arctan(e^x) + C = \arcsin(\tanh x) + C$ **23.** $\frac{1}{4}(\arcsin x)^2 + \frac{1}{2}x\sqrt{1 - x^2}\arcsin x - \frac{1}{4}x^2 + C$

25. $\frac{1}{2}x^2 - x\coth x + \ln|\sinh x| + C$ **27.** $-\frac{1}{2}xe^x\cos x + \frac{1}{2}xe^x\sin x + \frac{1}{2}e^x\cos x + C$

29. $2\ln\frac{x^2 + 1}{x^2 + x + 1} + \frac{2}{\sqrt{3}}\arctan\frac{2x + 1}{\sqrt{3}} + C$ (see Prob. 3, p. 406)   **31.** $\arctan(\tan^2 x) + C$

33. $\frac{1}{2}x\sqrt{x^2 + 2} - \ln(x + \sqrt{x^2 + 2}) + C$ **35.** $-\frac{\cos x}{5}\left(\frac{8}{3\sin x} + \frac{4}{3\sin^3 x} + \frac{1}{\sin^5 x}\right) + C$ **37.** $2(\tan y - \sec y) - y + C$

39. $2x\sqrt{x}\sin\sqrt{x} + 6x\cos\sqrt{x} - 12\sqrt{x}\sin\sqrt{x} - 12\cos\sqrt{x} + C$ **41.** $\frac{3}{16}\ln\left|\frac{x + 1}{x - 1}\right| - \frac{3x^2 + 3x - 2}{8(x - 1)(x + 1)^2}$

43. $\frac{1}{2}\sin x\cosh x - \frac{1}{2}\cos x\sinh x + C$ **45.** $2\sqrt{1 + x}\arcsin x + 4\sqrt{1 - x} + C$

47. $-\frac{1}{60}\ln|x - 1| + \frac{1}{176}\ln|x + 1| - \frac{73}{1520}\ln|x + 9| + \frac{37}{627}\ln|x - 10| + C$ (see Prob. 5, p. 406)

49. $-\frac{\cos x}{x\sin x + \cos x} + C$ **51.** $x - 4\sqrt{x + 1} + 4\ln(\sqrt{x + 1} + 1) + C$

53. $\tan x - \sqrt{2}\arctan\left(\frac{\tan x}{\sqrt{2}}\right) + C$ **55.** $\sin x + \frac{3}{4}\ln\frac{1 - \sin x}{1 + \sin x} + \frac{\sin x}{2\cos^2 x} + C$

57. $\frac{1}{4\sqrt{2}}\ln\frac{x^2 + \sqrt{2}x + 1}{x^2 - \sqrt{2}x + 1} + \frac{1}{2\sqrt{2}}\arctan(\sqrt{2}x + 1) + \frac{1}{2\sqrt{2}}\arctan(\sqrt{2}x - 1) + C$

59. $\dfrac{1}{\sqrt{3}} \arctan \dfrac{2x+1}{\sqrt{3}} + \dfrac{1}{\sqrt{3}} \arctan \dfrac{2x-1}{\sqrt{3}} + C$ **61.** $e^{-s} + \dfrac{1}{2} \ln \left| \dfrac{e^s - 1}{e^s + 1} \right| + C$ **63.** $\dfrac{1}{t} - \dfrac{1}{3t^3} + \dfrac{1}{2} \ln(t^2 + 1) + \arctan t + C$
65. $\sqrt{\tan z} + C$ **67.** $\dfrac{1}{\sin 2} \ln \left| \dfrac{\sin(x-1)}{\sin(x+1)} \right| + C$ **69.** $\tfrac{1}{5} y^5 \ln y - \tfrac{1}{25} y^5 + C$ **71.** $\dfrac{1}{2\sqrt{3}} \ln \left| \dfrac{\sqrt{3} + \tan v}{\sqrt{3} - \tan v} \right| + C$
73. $\tfrac{1}{2} \operatorname{sech} x \tanh x + \tfrac{1}{2} \arctan(\sinh x) + C$ **75.** $\dfrac{1}{4} \ln \left| \tan \dfrac{t}{2} \right| + \dfrac{1}{8} \tan^2 \dfrac{t}{2} + C$
77. $-\dfrac{1}{2} \ln(2 + \sin x + \cos x) + \dfrac{x}{2} - \sqrt{2} \arctan \left(\dfrac{\tan(x/2) + 1}{\sqrt{2}} \right) + C$
79. Expand $P(x)/Q(x)$ in partial fractions, determine the coefficients, and integrate.
81. $\tfrac{1}{2} \ln |x - 1| - 3 \ln |x + 2| + \tfrac{7}{2} \ln |x + 3| + C$
83. Let $x = \tan u$, and use the identity $\tan u + 1 = \dfrac{\sqrt{2}}{\cos u} \left(\sin \dfrac{\pi}{4} + u \right)$. **85.** 1.206 (midpoint), 1.170 (trapezoidal)
87. $n \geq 10$ **89.** Divergent **91.** $\pi/2\sqrt{2}$ **93.** $\ln(1 + \sqrt{2})$ **95.** Divergent **97.** Divergent **99.** $\tfrac{1}{3}$
101. Let $t = \ln x$. **103.** See Probs. 13–14, p. 381 **105.** 720 **107.** $-\tfrac{2}{27}$ **109.** Integrate by parts.

Chapter 8

Section 8.1 (page 440)

1. $\tfrac{28}{15}$ **3.** 2 **5.** 144 **7.** $36\sqrt{3}$ **9.** $\pi r^2 h$ **11.** $\tfrac{1}{3} \pi h(r_1^2 + r_1 r_2 + r_2^2)$ **13.** $\tfrac{8}{15} \sqrt{3}$ **15.** 2 **17.** $\tfrac{1}{9} \pi h^3$
19. If the x-axis is perpendicular to the plane of R, with the origin at P, then $A(x) = (A_0/h^2) x^2$. **21.** $\tfrac{1}{6} abc$
23. $\pi/8$ **25.** Use Cavalieri's principle. **27.** $\tfrac{1}{6} \pi h(3r_1^2 + 3r_2^2 + h^2)$ **29.** 3. Infinite

Section 8.2 (page 450)

1. $\dfrac{33\pi}{5}$ **3.** $\dfrac{6\pi}{5}$ **5.** $\dfrac{74\pi}{3}$ **7.** $\sqrt{3}\pi$ **9.** $\left(\dfrac{\sinh 2}{2} - 1 \right) \pi$ **11.** $2 \left(1 - \dfrac{2}{e} \right) \pi$ **13.** 3π **15.** π **17.** Use formula (8).
19. Use formula (2'). **21.** 8π **23.** $\dfrac{32\pi}{5}$ **25.** $\dfrac{256\pi}{15}$ **27.** $\dfrac{8\pi}{3}$ **29.** $\dfrac{32\pi}{15}, \dfrac{8\pi}{15}$ **31.** 3π **33.** $\dfrac{\pi}{30\sqrt{2}}$

Section 8.3 (page 458)

1.

3.

5.

7.

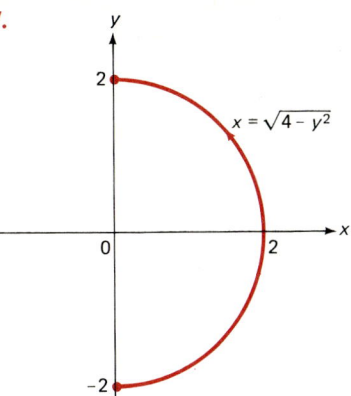

9.

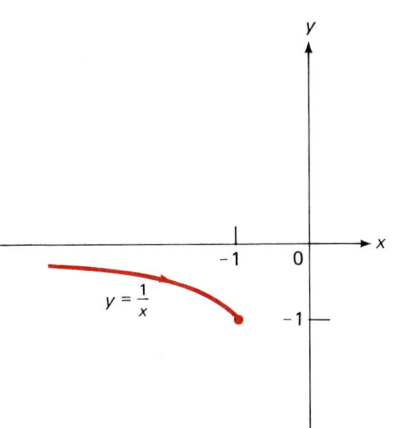

11.

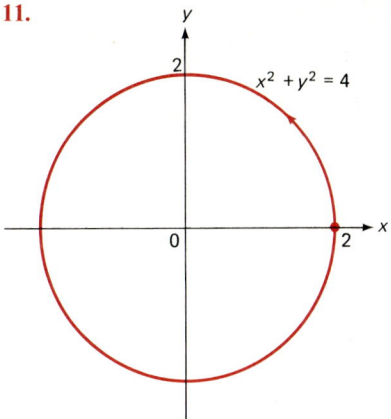

13.

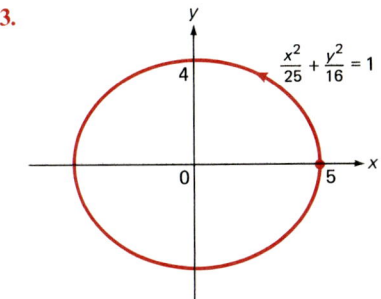

15.

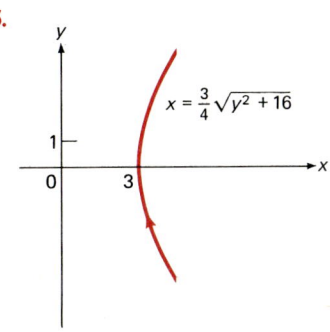

17. $x = t, y = 4 - t$ $(-\infty < t < \infty)$ **19.** $x = 2 + t, y = -3t$ $(-\infty < t < \infty)$
21. $x = -3 + 7t, y = 2 + 5t$ $(-\infty < t < \infty)$ **23.** $x = -2 + 5\cos t, y = 3 + 5\sin t$ $(0 \le t \le 2\pi)$
25. $x = 2\cos t, y = -3\sin t$ $(0 \le t \le 2\pi)$ **27.** $x = -3\cosh t, y = 4\sinh t$ $(-\infty < t < \infty)$
29. $x = \cos^3 t, y = \sin^3 t$ $(0 \le t \le \pi/2)$ **31.** $\frac{1}{4}$ **33.** $1/3e^3$ **35.** $-\frac{3}{2}$ **37.** $x = a\sec t, y = b\tan t$ $(-\pi/2 < t < \pi/2)$
39. $y = 0, x - 3y = 0$ **41.** $10x - 3y - 8 = 0$ **43.** $\frac{9}{16}$ **45.** 2 **47.** $(1, -1), (-3, 1)$ **49.** $(2, 0)$
51. Compare the behavior of $x = t^4, y = \sin^2 t$ and $x = t^2, y = \sin^2 t$ at the origin.
53. $\frac{dy}{dx} = t$ **55.** $y - 2 = 0, 6x - y - 4 = 0, 24x - y - 22 = 0$

Section 8.4 (page 464)

1. $4(2^{3/2} - 1)$ **3.** $\sqrt{2} - \frac{1}{2}\sqrt{5} + \ln\frac{2 + \sqrt{5}}{1 + \sqrt{2}}$ **5.** $\frac{1}{2}\sqrt{5} + \frac{1}{4}\ln(2 + \sqrt{5})$ **7.** 8 **9.** $\sqrt{2}(e^{-a} - e^{-b})$
11. The segment has parametric equations $x = x_1 + (x_2 - x_1)t, y = y_1 + (y_2 - y_1)t$ $(0 \le t \le 1)$. **13.** $\frac{14}{3}$ **15.** $\frac{1022}{27}$
17. $1 + \frac{1}{2}\ln\frac{3}{2}$ **19.** $\ln 3$ **21.** $\sqrt{2}(e - e^{-1})$
23. Choosing $x_0 = 0, x_1 = \frac{1}{n}, x_2 = \frac{1}{n-1}, \ldots, x_{n-1} = \frac{1}{2}, x_n = 1$, show that the polygonal path $P_0P_1P_2 \ldots P_{n-1}P_n$ inscribed in C with vertices $P_i = (x_i, f(x_i))$ is of length greater than $\sqrt{n}$.
25. If u and v are the angles shown in Figure 47, then $at = bu$ and $v = t + u - \frac{\pi}{2}$. **27.** $16a$ **29.** $8(a + b); 8(a - b)$
31. $L = 4\sqrt{2}\int_0^{\pi/2}\sqrt{1 - \frac{1}{2}\sin^2 x}\,dx; L \approx 7.64$

Section 8.5 (page 471)

1. $\frac{6}{5}(\sqrt{2} + 1)\pi$ **3.** $\sqrt{2}\pi$ **5.** $\frac{2}{5}\sqrt{2}(e^\pi - 2)\pi$ **7.** $\frac{12}{5}\pi a^2$ **9.** $\frac{1}{2}a$ **11.** $[\frac{1}{2}\sqrt{10} + \frac{1}{6}\ln(3 + \sqrt{10})]\pi$ **13.** $\frac{20}{3}\sqrt{2}\pi$ **15.** $\frac{47}{72}\pi$
17. $\frac{1}{16}(e^4 - 9)\pi$ **19.** $2\pi Rh$ **21.** $\pi(4 - \pi)/\sqrt{2}$

Section 8.6 (page 476)

1. 13,720 joules **3.** 59,375 ft-lb **5.** ≈28.4 hr **7.** $W = \int_{h-\delta}^{h} \rho g A(s) s \, ds$ **9.** $(\frac{1}{4}r + \frac{2}{3}h)\pi\rho gr^3$
11. ≈3016 ft-lb; ≈6032 ft-lb **13.** ≈2.33 × 10¹² joules, about 300,000 man-years of work
15. ≈3468 ft-lb; ≈10,629 ft-lb **17.** Show that $\dfrac{dp}{dh} = -\dfrac{g}{k}p$.

Section 8.7 (page 481)

1. 422.5 tons **3.** δh. No. See the remark, p. 478. **5.** $\delta a^3/\sqrt{2}$ **7.** $\frac{1}{3}(a+2b)\delta h^2$ **9.** $\frac{8}{15}\delta ah^2$ **11.** $\frac{1}{2}\delta ab^2$
13. $(2 + \cos\theta)\pi\delta$ **15.** $\dfrac{1}{2}\left(\dfrac{1-\alpha}{\alpha}\right)^2 wh$ **17.** $\sqrt[3]{\alpha}H$ **19.** $\left(\dfrac{3}{4}\sqrt[3]{\alpha} - 1 + \dfrac{1}{4\alpha}\right)wH$

Supplementary Problems (page 483)

1. $\sqrt[3]{\frac{1}{2}}H$ **3.** $\frac{16}{3}ab^2$ **5.** $\frac{2}{3}\pi ab^2$ **7.** $\frac{1}{4}\pi$ **9.** $2e(2e-1)\pi$ **11.** $\dfrac{4\pi}{e^2+1}$ **13.** π **15.** 24π **17.** $\frac{1}{4}\pi^2$ **19.** 2π
21. $\left(\dfrac{\pi^2}{4} - \pi + 2\right)\pi$ **23.** $\left(2 - \dfrac{\pi}{4}\right)\pi$ **25.** $t = (2n+1)\pi$, n any integer

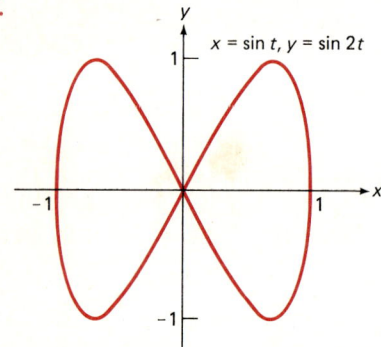

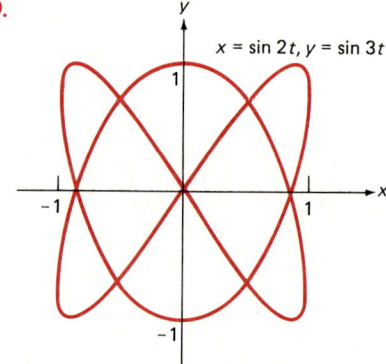

27. $x = \sin t, y = \sin 2t$

29. $x = \sin 2t, y = \sin 3t$

31. See Suppl. Prob. 7, p. 222. **33.** $f'_\pm(0) = \mp\infty$; $Q = (t, 0)$ **35.** $x^2(x+9a) - 27ay^2 = 0$; $x \pm \sqrt{3}y = 0$ **37.** 2π
39. 24 **41.** $\ln\dfrac{\sqrt{2}+1}{\sqrt{2}-1}$ **43.** $18\pi^2$ **45.** $\frac{64}{55}\pi$ **47.** $V = \frac{4}{3}\pi ab^2$, $V' = \frac{4}{3}\pi a^2 b$ **49.** $4\sqrt{3}\pi a^2$; $2\pi a^3$ **51.** $\left(\dfrac{2r}{3} + \dfrac{\pi h}{2}\right)\rho g L r^2$
53. $\pi\delta h(r_2^2 - r_1^2)$ **55.** $5\pi\delta/12$

Chapter 9

Section 9.1 (page 495)

1. $\frac{1}{2}, 1, \frac{5}{4}, \frac{7}{5}, \frac{3}{2}, \frac{11}{7}$; $L = 2$ **3.** $\frac{3}{4}, \frac{7}{9}, \frac{13}{16}, \frac{21}{25}, \frac{31}{36}, \frac{43}{49}$; $L = 1$ **5.** $1, 2, \frac{1}{3}, 4, \frac{1}{5}, 6$; L does not exist.
7. $1, \frac{5}{9}, \frac{1}{3}, \frac{17}{81}, \frac{11}{81}, \frac{65}{729}$; $L = 0$ **9.** $0, \frac{1}{2}, 0, \frac{1}{4}, 0, \frac{1}{6}$; $L = 0$ **11.** 1.4, 1.41, 1.414, 1.4142, 1.41421, 1.414213; $L = \sqrt{2}$
13. $\frac{3}{2}, \frac{1}{2}, \frac{5}{6}, \frac{3}{4}, \frac{7}{10}, \frac{5}{6}$; L does not exist **15.** $\frac{1}{2}, -\frac{1}{2}, -1, -\frac{1}{2}, \frac{1}{2}, 1$; L does not exist
17. $\frac{1}{2}, \frac{4}{9}, \frac{27}{64}, \frac{256}{625}, \frac{3125}{7776}, \frac{46656}{117469}$; $L = 1/e$ **19.** $0, -\ln 2, -\ln 3, -\ln 4, -\ln 5, -\ln 6$; $L = -\infty$ **21.** $a_n = n^2 - 1$; $L = \infty$
23. $a_n = (-1)^n \dfrac{2n-1}{2n+1}$; L does not exist **25.** $a_n = \dfrac{1}{n(2n-1)}$; $L = 0$ **27.** $a_n = 10 - 5n$; $L = -\infty$ **29.** $n \geq 20$
31. 0, 2, 8, 26, 80, 242 **33.** $\frac{1}{2}, 3, \frac{4}{3}, \frac{7}{4}, \frac{11}{7}, \frac{18}{11}$ **35.** 1, 1, 2, 3, 5, 8, 13, 21, 34, 55 **37.** $s_n = a_1 + a_2 + \cdots + a_n$; $s_n = n^2$
39. $\{2^{-n}\}$ **41.** $\{(-1)^n(n-1)/n\}$ **43.** See Theorem 3, p. 77. **45.** Only the first three
47. 0 if $|r| < 1$, $\frac{1}{2}$ if $r = 1$, 1 if $|r| > 1$ **49.** 0 **51.** $a_n = (-1)^n$, $b_n = (-1)^{n+1}$ **53.** Use L'Hospital's rule to evaluate $\lim\limits_{x\to\infty} \dfrac{x}{c^x}$.
55. See Theorem 11, p. 87. **57.** a_3 **59.** a_{14} **61.** $A_n = \dfrac{\pi s^2}{8} \dfrac{n(n+1)}{(n+\sqrt{3}-1)^2}$

Section 9.2 (page 506)
1. 3, 6, 9, 12, 15   3. $1, \frac{3}{2}, \frac{5}{3}, \frac{41}{24}, \frac{103}{60}$ 5. 0, ln 2, ln 6, ln 24, ln 120   7. $0, \frac{1}{2}, \frac{3}{4}, \frac{7}{8}, \frac{15}{16}$ 9. $a_1 = s_1, a_n = s_n - s_{n-1}$ if $n \geq 2$
11. $\frac{1}{2} + \frac{1}{4} + \cdots + \frac{1}{2^n} + \cdots$ 13. $\frac{1}{2}$ 15. Divergent   17. $\frac{24}{11}$ 19. $\frac{3}{4}$ 21. Divergent   23. 3   25. 1   27. $\frac{1}{4}$
29. $R_n = \frac{1}{e^n}$ 31. $R_n = \frac{1}{n+1} + \frac{1}{n+2} + \frac{1}{n+3}$ 33. No   35. $1 + \frac{1}{2} + \frac{1}{4} + \cdots = 2$ 37. ≈ 205.25 ft; 225 ft   39. 187.5 ft
41. About 5 min, 27 sec after 1 P.M.   43. $\frac{1}{1 \cdot 2} + \frac{1}{2 \cdot 3} + \cdots + \frac{1}{n(n+1)} + \cdots$
45. $C = C_0 e^{-kt} + C_0 e^{-k(t-T)} + \cdots + C_0 e^{-k(t-(n-1)T)}$ during the interval $(n-1)T < t \leq nT$.

Section 9.3 (page 515)
1. $\frac{1}{2}$ 3. $\frac{215}{999}$ 5. $\frac{5501}{1375}$ 7. $\frac{1}{21}$ 9. $\frac{893}{175}$ 11. Divergent   13. Convergent   15. Divergent   17. Convergent
19. Convergent   21. Divergent   23. Convergent   25. Convergent   27. Divergent   29. Divergent   31. Convergent
33. Convergent   35. Divergent   37. Convergent   39. Convergent   41. Divergent   43. Convergent
45. Convergent   47. Divergent   49. Convergent   51. Divergent   53. $a_n < 1$ for all sufficiently large n
55. Use Theorem 6 with $b_n = \frac{1}{n}$; $1 + \frac{1}{2^2} + \frac{1}{3^2} + \frac{1}{4} + \frac{1}{5^2} + \frac{1}{6^2} + \frac{1}{7^2} + \frac{1}{8^2} + \frac{1}{9} + \frac{1}{10^2} + \cdots$ 57. No
59. If $a_n = \frac{1}{n^p}$, then $\sum_{k=0}^{\infty} 2^k a_{2^k} = \sum_{k=0}^{\infty} (2^{1-p})^k$.
61. If $p > 0$, use L'Hospital's rule to show that $(\ln n)^p < n$ for all sufficiently large n.

Section 9.4 (page 524)
1. Conditionally convergent   3. Absolutely convergent   5. Divergent   7. Divergent   9. Conditionally convergent
11. Absolutely convergent   13. Conditionally convergent   15. Absolutely convergent   17. Divergent
19. Absolutely convergent   21. Conditionally convergent
23. Absolutely convergent if $0 < a < 1$, conditionally convergent if $a = 1$, divergent if $a > 1$
25. R_{99} is negative, $|R_{99}| < \frac{1}{100}$ 27. R_6 is positive, $R_6 < 2.1 \times 10^{-9}$ 29. R_8 is negative, $|R_8| < 0.093$
31. 3   33. 44   35. $1 + \frac{1}{\sqrt{3}} - \frac{1}{\sqrt{2}} + \frac{1}{\sqrt{5}} + \frac{1}{\sqrt{7}} - \frac{1}{\sqrt{4}} + \cdots$ 37. $1 + \frac{1}{3} + \frac{1}{5} + \frac{1}{7} - \frac{1}{2} + \frac{1}{9} + \frac{1}{11} + \frac{1}{13} + \frac{1}{15} - \frac{1}{4} + \cdots$
39. $\sum_{i=1}^{n} a_i b_i = a_n B_n + \sum_{i=1}^{n-1} (a_i - a_{i+1}) B_i$
41. The partial sums of the series $1 + 2 - 3 + 1 + 2 - 3 + \cdots$ can only take the values 1, 3, 0.

Section 9.5 (page 529)
1. Convergent   3. Divergent   5. Absolutely convergent   7. Convergent   9. Divergent   11. Convergent
13. Convergent   15. Divergent   17. Convergent   19. Divergent   21. Conditionally convergent
23. Absolutely convergent   25. Use Theorem 3, p. 502.   27. Use Prob. 25.   29. $\frac{1}{2} + 1 + \frac{1}{8} + \frac{1}{4} + \frac{1}{32} + \frac{1}{16} + \cdots$
31. Use Prob. 29.

Section 9.6 (page 536)
1. $0, [\pi, \pi]$ 3. $1, [-1, 1)$ 5. $\infty, (-\infty, \infty)$ 7. $1, (-1, 1)$ 9. $1/e, (-1/e, 1/e)$ 11. $\infty, (-\infty, \infty)$
13. $2/\sqrt{3}, (-2/\sqrt{3}, 2/\sqrt{3})$ 15. $0, [-e, -e]$ 17. $R = \max\{a, b\}, (-R, R)$ 19. $\frac{1}{3}, [-\frac{1}{3}, \frac{1}{3})$
21. $\frac{1}{2}, (-\frac{1}{2}, \frac{1}{2})$ 23. $1, [3, 5]$ 25. $3, (-3, 3)$ 27. $1, (-1, 1)$ if $0 \leq c \leq 1$; $1/c, (-1/c, 1/c)$ if $c > 1$ 29. $1, [-2, 0)$
31. Use the ratio or root test.   33. $4R$ 35. R^p 37. $\frac{1}{4}, (-\frac{1}{4}, \frac{1}{4})$

Section 9.7 (page 547)
1. $\sum_{n=0}^{\infty} \frac{x^n}{3^{n+1}}$ $(R = 3)$ 3. $\sum_{n=1}^{\infty} (-1)^{n-1} n x^{n-1}$ $(R = 1)$ 5. $\sum_{n=0}^{\infty} (-1)^n 2^n x^{n+1}$ $(R = \frac{1}{2})$ 7. $\sum_{n=1}^{\infty} n x^{2n-1}$ $(R = 1)$
9. $\sum_{n=0}^{\infty} \left(1 - \frac{1}{2^{n+1}}\right) x^n$ $(R = 1)$ 11. $\sum_{n=0}^{\infty} (-1)^{n+1}(x-3)^n$ $(2 < x < 4)$ 13. $\sum_{n=0}^{\infty} \frac{(-1)^n}{n!} x^{n+2}$ $(R = \infty)$
15. $\sum_{n=0}^{\infty} \frac{x^{2n+1}}{(2n+1)!}$ $(R = \infty)$ 17. $\sum_{n=0}^{\infty} \frac{(\ln a)^n}{n!} x^n$ $(R = \infty)$ 19. $2 \sum_{n=0}^{\infty} \frac{x^{2n+1}}{2n+1}$ $(R = 1)$ 21. $-2 \sum_{n=1}^{\infty} \frac{x^n}{n} \cos \frac{n\pi}{3}$ $(R = 1)$
23. $1 + \frac{1}{3} x - \frac{1}{9} x^2 + \frac{5}{81} x^3 - \frac{10}{243} x^4$ 25. $8 - 12(x/4) + 3(x/4)^2 + \frac{1}{2}(x/4)^3 + \frac{3}{16}(x/4)^4$ 27. $1 + 10x^2 + 55x^4 + 220x^6 + 715x^8$

Answers and Hints to Odd-Numbered Problems   **A-43**

29. $\sum_{n=1}^{\infty} (-1)^{n-1} \dfrac{(x+1)^{2n}}{n}$ ($|x+1| < 1$)   **31.** $\sqrt{1.04} \approx 1 + \tfrac{1}{50} - \tfrac{1}{5000} = 1.01980$ **33.** $\sqrt[4]{79} \approx 3(1 - \tfrac{1}{162} - \tfrac{1}{17496}) \approx 2.98131$

35. $\sqrt[6]{65} \approx 2(1 + \tfrac{1}{384} - \tfrac{5}{294912}) \approx 2.00517$ **37.** Start from $\arcsin x = \int_0^x \dfrac{dt}{\sqrt{1-t^2}}$, $\sinh^{-1} x = \int_0^x \dfrac{dt}{\sqrt{1+t^2}}$.   **39.** Let $x = \dfrac{1}{\sqrt{3}}$.

41. See Prob. 19.   **43.** $\dfrac{1-2x}{(1+x)^2}$ **45.** $\sum_{n=1}^{\infty} \dfrac{x^{2n-1}}{(2n-1)n!}$ **47.** $\sum_{n=1}^{\infty} (-1)^{n-1} \dfrac{x^n}{n^2}$ **49.** $\int_0^{1/2} \sqrt{1+x^4}\, dx \approx \tfrac{1}{2} + \tfrac{1}{320} \approx 0.5031$

51. $\int_0^1 e^{x^2} dx \approx 1 + \tfrac{1}{3} + \tfrac{1}{10} + \tfrac{1}{42} + \tfrac{1}{216} + \tfrac{1}{1320} + \tfrac{1}{9360} \approx 1.4626$ **53.** $g(x) = \sum_{n=0}^{\infty} s_n x^n$, where $s_n = a_0 + a_1 + \cdots + a_n$.

Section 9.8 (page 556)

1. $\sqrt{171} \approx 13 + \dfrac{1}{13} \approx 13.077$ **3.** $\dfrac{1}{2.01} \approx \dfrac{1}{2} - \dfrac{0.01}{4} = 0.4975$ **5.** $\tan 43° \approx 1 - \dfrac{\pi}{45} \approx 0.93$

7. $P_2(x) = 2 + \tfrac{1}{4}(x-4) - \tfrac{1}{64}(x-4)^2$, $R_2(x) = \dfrac{t^{-5/2}}{16}(x-4)^3$ (t between 4 and $x > 0$)

9. $P_3(x) = e^{-1}[1 - (x-1) + \tfrac{1}{2}(x-1)^2 - \tfrac{1}{6}(x-1)^3]$, $R_3(x) = \dfrac{e^{-t}}{24}(x-1)^4$ (t between 1 and x)

11. $P_4(x) = \ln 2 + \tfrac{1}{2}(x-2) - \tfrac{1}{8}(x-2)^2 + \tfrac{1}{24}(x-2)^3 - \tfrac{1}{64}(x-2)^4$, $R_4(x) = \dfrac{(x-2)^5}{5t^5}$ (t between 2 and $x > 0$)

13. $P_6(x) = 1 - \tfrac{1}{2}x^2 + \tfrac{1}{24}x^4 - \tfrac{1}{720}x^6$, $R_6(x) = \dfrac{\sin t}{5040} x^7$ (t between 0 and x)

15. $P_3(x) = x + \tfrac{1}{3}x^3$, $R_3(x) = (\tfrac{2}{3}\tan t + \tfrac{5}{3}\tan^3 t + \tan^5 t)x^4$ (t between 0 and x, $|x| < \pi/2$).

17. $P_6(x) = x^2 - \tfrac{1}{3}x^4 + \tfrac{2}{45}x^6$, $R_6(x) = -\dfrac{4\sin t}{315} x^7$ (t between 0 and x)

19. $P_3(x) = \sinh 1 + (x-1)\cosh 1 + \tfrac{1}{2}(x-1)^2 \sinh 1 + \tfrac{1}{6}(x-1)^3 \cosh 1$, $R_3(x) = \dfrac{\sinh t}{24}(x-1)^6$ (t between 1 and x)

21. $\sqrt{171} \approx 13 + \dfrac{1}{13} - \dfrac{1}{2 \cdot 13^3} \approx 13.07670$ **23.** $\dfrac{1}{2.01} \approx \dfrac{1}{2} - \dfrac{0.01}{4} + \dfrac{(0.01)^2}{8} \approx 0.497513$

25. $\tan 43° \approx 1 - \dfrac{\pi}{45} + 2\left(\dfrac{\pi}{90}\right)^2 - \dfrac{8}{3}\left(\dfrac{\pi}{90}\right)^3 \approx 0.93251$

27. $\sqrt{x} = \sqrt{100} + \dfrac{1}{2\sqrt{100}}(x-100) - \dfrac{1}{8}\dfrac{(x-100)^2}{t^{3/2}}$ (t between 100 and x)

29. $f^{(k)}(x) = (-1)^k f^{(k)}(-x)$ if f is even, while $f^{(k)}(x) = (-1)^{k+1} f^{(k)}(-x)$ if f is odd.
31. $2 - 4(x-1) - 6(x-1)^2 - 12(x-1)^3 - 13(x-1)^4 - 6(x-1)^5 + (x-1)^6$
33. $176 + 220(x-5) + 92(x-5)^2 + 16(x-5)^3 + (x-5)^4$
35. $10001 + 4000(x-10) + 600(x-10)^2 + 40(x-10)^3 + (x-10)^4$ **37.** $\tfrac{1}{2}$ **39.** $\tfrac{1}{120}$

Section 9.9 (page 562)

1. If $f(x) = \sinh x$, then $f(0) = f''(0) = f^{(4)}(0) = \cdots = 0$, $f'(0) = f'''(0) = f^{(5)}(0) = \cdots = 1$.
3. If $f(x) = \ln(1+x)$, then $f(0) = 0$, $f'(0) = 1$, $f''(0) = -1$, $f'''(0) = 2!, \ldots, f^{(n)}(0) = (-1)^{n-1}(n-1)!$
5. $1 + x - \tfrac{1}{2}x^2 - \tfrac{1}{6}x^3 + \tfrac{1}{24}x^4$ **7.** $1 + x - \tfrac{1}{3}x^3 - \tfrac{1}{6}x^4 - \tfrac{1}{30}x^5$ **9.** $\ln 2 + \tfrac{1}{2}x + \tfrac{1}{8}x^2$ **11.** $e(1 - \tfrac{1}{2}x^2 + \tfrac{1}{6}x^4)$

13. $e(1 + x + \tfrac{3}{2}x^2 + \tfrac{13}{6}x^3)$.   **15.** $\dfrac{1}{2} + \dfrac{\sqrt{3}}{2}\left(x - \dfrac{\pi}{6}\right) - \dfrac{1}{4}\left(x - \dfrac{\pi}{6}\right)^2 - \dfrac{\sqrt{3}}{12}\left(x - \dfrac{\pi}{6}\right)^3 + \cdots$

17. $1 - \dfrac{1}{2!}\left(x - \dfrac{\pi}{2}\right)^2 + \dfrac{1}{4!}\left(x - \dfrac{\pi}{2}\right)^4 - \dfrac{1}{6!}\left(x - \dfrac{\pi}{2}\right)^6 + \cdots$ **19.** $\dfrac{1}{\sqrt{2}}\left[1 - \left(x - \dfrac{\pi}{4}\right) - \dfrac{1}{2}\left(x - \dfrac{\pi}{4}\right)^2 + \dfrac{1}{6}\left(x - \dfrac{\pi}{4}\right)^3 + \cdots\right]$

21. $-1 + \dfrac{1}{2!}(x+\pi)^2 - \dfrac{1}{4!}(x+\pi)^4 + \dfrac{1}{6!}(x+\pi)^6 + \cdots$

23. $2\left[1 + \dfrac{x-4}{2^3} - \dfrac{(x-4)^2}{2^6 \cdot 2!} + \dfrac{1 \cdot 3(x-4)^3}{2^9 \cdot 3!} - \dfrac{1 \cdot 3 \cdot 5(x-4)^4}{2^{12} \cdot 4!} + \cdots\right]$ **25.** $1 - 2(x-1) + 3(x-1)^2 - 4(x-1)^3 + \cdots$

27. $e^{2/3}\left[1 + \dfrac{x-2}{3} + \dfrac{(x-2)^2}{3^2 \cdot 2!} + \dfrac{(x-2)^3}{3^3 \cdot 3!} + \cdots\right]$ **29.** $\sum_{n=0}^{\infty} \dfrac{x^{2n}}{(2n)!} \leq \sum_{n=0}^{\infty} \dfrac{x^{2n}}{2^n n!}$. Also $\lim_{x \to 0} \dfrac{e^{ax^2} - \cosh x}{x^2} = a - \dfrac{1}{2}$.

31. $f(x) = 1 - 2x + 2x^2 - 2x^4 + 2x^5 - \cdots$; 240   **33.** $1 + \frac{1}{3}x^3 + \frac{2}{15}x^5 + \frac{17}{315}x^7$

Section 9.10 (page 568)
1. 1.4142   **3.** 2.9428   **5.** 3.0468   **7.** $x_1 > \frac{1}{2}$; $x_1 < \frac{1}{2}$; $x_1 = \frac{1}{2}$ **9.** 0.830484   **11.** $-2.5289, 0.1674, 2.3615$ **13.** 0.466
15. 2.208   **17.** 2.475   **19.** 4.493   (write $\tan x = x$ as $\sin x - x \cos x = 0$)   **21.** $x_{n+1} = x_n - 3(x_n - r)$

Supplementary Problems (page 571)
1. 6   **3.** Does not exist   **5.** $a_n = \frac{1}{n^2}$, $b_n = \frac{1}{n}$ **7.** $a_n = \frac{1}{n}$, $b_n = \frac{(-1)^n}{n}$ **9.** $x \to 0^+$ implies $\frac{1}{x} \to \infty$.   **11.** 1   **13.** 2
15. $\frac{1-b}{1-a}$ **17.** $\frac{\pi}{6}$ **19.** 2   **21.** Use mathematical induction (see the Appendix) to show that $0 < a_n < 1$ for all n.
23. Show that $a_n < a_{n+1}$, $a_n < \sqrt{c} + 1$ for all n.   **25.** First let $L = 0$. No   **27.** e **29.** ∞ **31.** $\sqrt{\pi}$ **33.** 110.32
35. 0.22   **37.** 0.28   **39.** Divergent   **41.** $\frac{1}{3}$ **43.** $1/(c+1)$ **45.** $1 - \sqrt{2}$ **47.** See Theorem 3, p. 502.
49. See Prob. 19, p. 13.   **51.** $\frac{11111}{90000}$ **53.** $\frac{4115}{33333}$ **55.** Absolutely convergent   **57.** Divergent
59. Conditionally convergent   **61.** Convergent   **63.** Absolutely convergent   **65.** Divergent   **67.** Divergent
69. Conditionally convergent   **71.** $1, (-1, 1)$ **73.** $\frac{2}{\sqrt{3}} \sum_{n=0}^{\infty} [\sin \frac{2}{3}(n+1)\pi] x^n$ $(|x| < 1)$
75. $\sum_{n=1}^{\infty} nx^{3n-1}$ $(|x| < 1)$ **77.** $\frac{1}{3} \sum_{n=0}^{\infty} [1 - (-2)^{n+1}] x^n$ $(|x| < \frac{1}{2})$ **79.** $2 \sum_{n=1}^{\infty} \left(1 + \frac{1}{2} + \cdots + \frac{1}{n}\right) \frac{x^{n+1}}{n+1}$ $(|x| < 1)$
81. Integrate the series $1 - x^a + x^{2a} - x^{3a} + \cdots$.   **83.** $\frac{1}{2\sqrt{2}} \left[\ln(\sqrt{2} + 1) + \frac{\pi}{2} \right]$
85. Multiply both sides of (iv) by $1 - x - x^2$.   **87.** See Prob. 19, p. 547.   **89.** $P(-1) = 143$, $P'(0) = -60$, $P''(1) = 26$
91. Integrate by parts repeatedly, starting from $f(x) - f(a) = \int_a^x f'(u)\, du = -\int_a^x f'(u)\, d(x-u)$.   **93.** $\frac{10!}{4!}$
95. $1 + 2x + x^2 - \frac{2}{3}x^3 - \frac{5}{6}x^4 - \frac{1}{15}x^5$ **97.** 0.2398   **99.** 0.653

Chapter 10

Section 10.1 (page 580)
1. $x' = x$, $y' = y - 10$ **3.** $x' = x - 1$, $y' = y + 2$ **5.** $(3, 5)$ **7.** $(0, -1)$ **9.** $A = (4, -1)$, $B = (0, -4)$, $C = (2, 0)$
11. $A = (3, -2)$, $B = (0, 0)$, $C = (-1, 2)$ **13.** $x' = x - 2$, $y' = y + 3$; a parabola   **15.** $x' = x + 2$, $y' = y - 1$; a hyperbola
17. $x' = y$, $y' = -x$ **19.** $x' = \frac{1}{2}x + \frac{1}{2}\sqrt{3}y$, $y' = -\frac{1}{2}\sqrt{3}x + \frac{1}{2}y$ **21.** $A = (-3, -1)$, $B = (1, -5)$, $C = (-2, 3)$
23. $A = (\frac{41}{13}, -\frac{3}{13})$, $B = (1, 5)$, $C = (\frac{9}{13}, -\frac{46}{13})$ **25.** $A = (3\sqrt{3}, 1)$, $B = (\frac{1}{2}\sqrt{3}, \frac{3}{2})$, $C = (3, -\sqrt{3})$
27. The hyperbola $x^2 - \frac{1}{2}y^2 = 1$ rotated through $45°$ **29.** $\frac{1}{2}(x+3)^2 + (y-4)^2 = 1$ **31.** Symmetry about the line $y = x$
33. All four   **35.** Symmetry about the origin   **37.** f is identically zero.
39. f is odd.   **41.** $A = (6, 3)$, $B = (0, 0)$, $C = (5, -10)$

Section 10.2 (page 586)
1. $x^2 = \frac{1}{2}y$ **3.** $y^2 = -20x$ **5.** $(x-1)^2 = -36(y+2)$ **7.** $(y-6)^2 = \frac{11}{2}(x-7)$ **9.** $y^2 - 4x - 4y + 28 = 0$
11. $x^2 + 16x + 2y + 61 = 0$ **13.** $x^2 + 2xy + y^2 - 6x + 2y + 9 = 0$

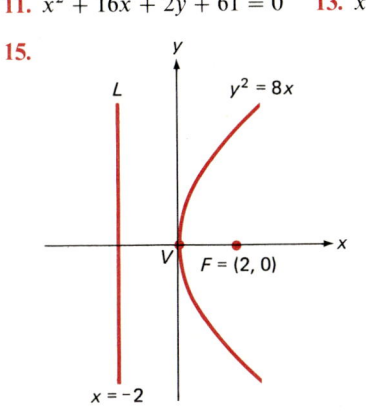

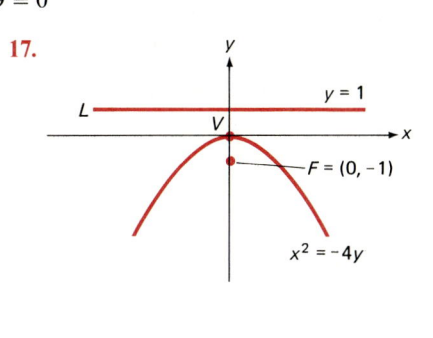

15.

17.

19.

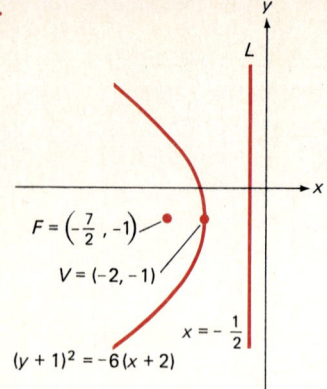

(Figure: parabola $(y+1)^2 = -6(x+2)$, $F = (-\frac{7}{2}, -1)$, $V = (-2, -1)$, directrix $x = -\frac{1}{2}$)

21.

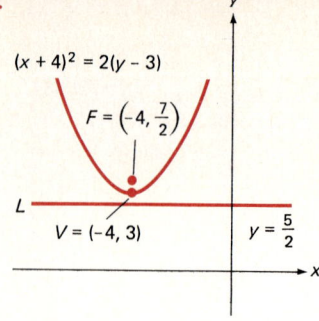

(Figure: parabola $(x+4)^2 = 2(y-3)$, $F = (-4, \frac{7}{2})$, $V = (-4, 3)$, directrix $y = \frac{5}{2}$)

23.

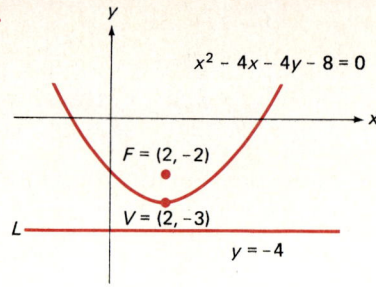

(Figure: parabola $x^2 - 4x - 4y - 8 = 0$, $F = (2, -2)$, $V = (2, -3)$, directrix $y = -4$)

25.

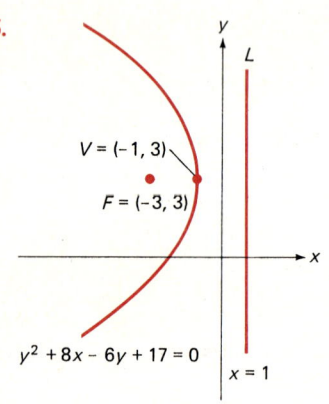

(Figure: parabola $y^2 + 8x - 6y + 17 = 0$, $V = (-1, 3)$, $F = (-3, 3)$, directrix $x = 1$)

27.

(Figure: parabola $x^2 + 2xy + y^2 + 4x - 4y = 0$, $F = (-\frac{1}{2}, \frac{1}{2})$, directrix $x - y - 1 = 0$)

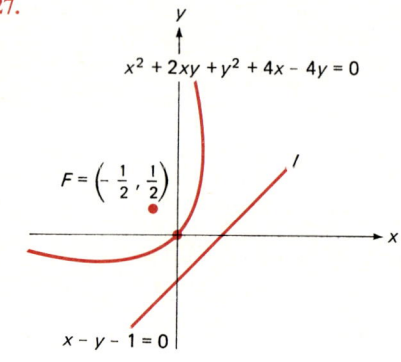

29. The line $x = c$ intersects the parabola $y^2 = 4cx$ in the points $(c, \pm 2c)$. **31.** $y^2 = 4c(ct^2) = 4cx$
33. The parabola $y^2 = 2cx$ **35.** There are none. **37.** $(6, -2), (-3, -\frac{1}{2})$

39.

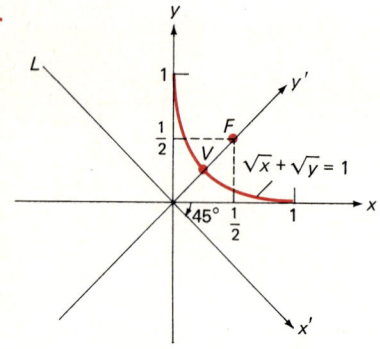

(Figure: curve $\sqrt{x} + \sqrt{y} = 1$ with rotated axes at 45°)

41. $3x - y + 3 = 0$, $3x - 2y + 12 = 0$ **43.** $3x - 2y + 4 = 0$; $2\sqrt{13}$

Section 10.3 (page 594)

1. $\dfrac{x^2}{16} + \dfrac{y^2}{7} = 1$ **3.** $\dfrac{(x-6)^2}{16} + \dfrac{(y-4)^2}{9} = 1$ **5.** $\dfrac{(x-2)^2}{36} + \dfrac{(y+2)^2}{45} = 1$ **7.** $\dfrac{2x^2}{3} + \dfrac{y^2}{3} = 1$ **9.** $(\pm 3, \frac{8}{5}), (\pm 3, -\frac{8}{5})$

11.

(0, 5), (0, 4), (-3, 0), (3, 0), (0, -4), (0, -5)

$$\frac{x^2}{9} + \frac{y^2}{25} = 1$$

13.

(2, -1), (-4, -4), (2, -4), (8, -4), $(2 - \sqrt{27}, -4)$, (2, -7), $(2 + \sqrt{27}, -4)$

$$\frac{(x-2)^2}{36} + \frac{(y+4)^2}{9} = 1$$

15.

$\left(0, \sqrt{\frac{4}{3}}\right)$, $(0, \sqrt{2})$, $3x^2 + y^2 = 2$, $\left(-\sqrt{\frac{2}{3}}, 0\right)$, $\left(\sqrt{\frac{2}{3}}, 0\right)$, $\left(0, -\sqrt{\frac{4}{3}}\right)$, $(0, -\sqrt{2})$

17.

(2, 6), (-3, 2), (-1, 2), (2, 2), (5, 2), (7, 2), (2, -2)

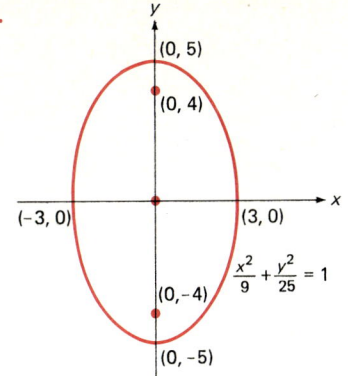

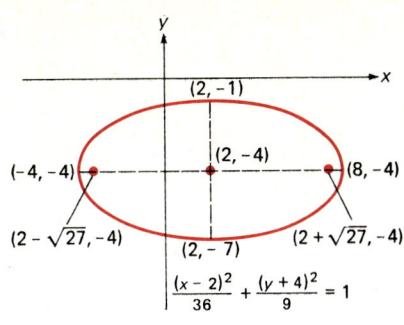

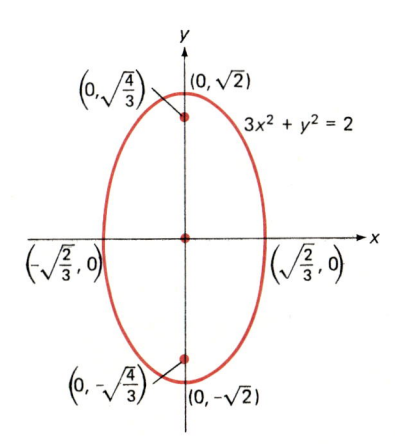

$16x^2 + 25y^2 - 64x - 100y - 236 = 0$

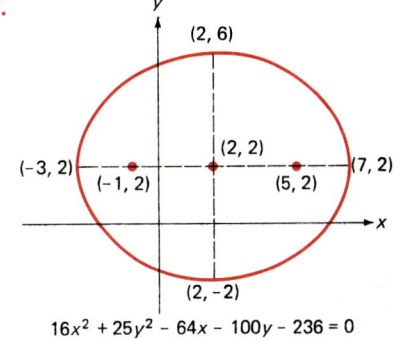

19.

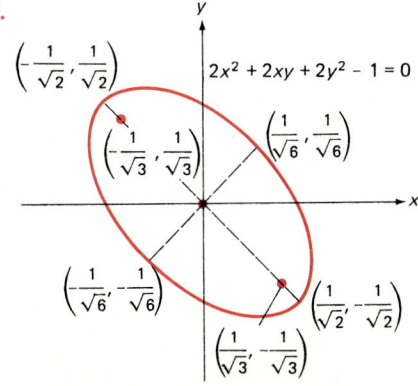

$\left(-\frac{1}{\sqrt{2}}, \frac{1}{\sqrt{2}}\right)$, $2x^2 + 2xy + 2y^2 - 1 = 0$, $\left(-\frac{1}{\sqrt{3}}, \frac{1}{\sqrt{3}}\right)$, $\left(\frac{1}{\sqrt{6}}, \frac{1}{\sqrt{6}}\right)$, $\left(-\frac{1}{\sqrt{6}}, -\frac{1}{\sqrt{6}}\right)$, $\left(\frac{1}{\sqrt{3}}, -\frac{1}{\sqrt{3}}\right)$, $\left(\frac{1}{\sqrt{2}}, -\frac{1}{\sqrt{2}}\right)$

21. $3x^2 - 2xy + 3y^2 - 2 = 0$ **23.** $(\sqrt{15}, \pm 1), (-\sqrt{15}, \pm 1)$ **25.** See Example 5, p. 453.   **27.** $(3, \frac{8}{5})$
29. $(1, 1), (\frac{1}{3}, \frac{5}{3})$ **31.** P has coordinates $x = a \cos \theta, y = b \sin \theta$.   **33.** $\frac{8}{17}$ **35.** $\frac{45}{53}$ **37.** $2ab$
39. $x + 4y - 10 = 0, x + y - 5 = 0$ **41.** 0.0167

Section 10.4 (page 602)

1. $\frac{x^2}{9} - \frac{y^2}{7} = 1$ **3.** $\frac{(x+1)^2}{36} - \frac{(y-2)^2}{25} = 1$ **5.** $\frac{y^2}{12} - \frac{(x+2)^2}{4} = 1$ **7.** $\frac{x^2}{32} - \frac{y^2}{8} = 1$ **9.** $\frac{y^2}{576} - \frac{x^2}{100} = 1$
11. $x^2 - y^2 = 16$

13.

[Graph: hyperbola $\frac{x^2}{9} - \frac{y^2}{4} = 1$ with vertices $(\pm 3, 0)$, foci $(\pm\sqrt{13}, 0)$, asymptotes $y = \pm\frac{2}{3}x$]

15.

[Graph: hyperbola $\frac{(y-1)^2}{16} - \frac{(x+2)^2}{4} = 1$ with center $(-2, 1)$, vertices $(-2, 5)$ and $(-2, -3)$, foci $(-2, 1 \pm \sqrt{20})$, asymptotes $y = 2x+5$ and $y = -2x - 3$]

17.

[Graph: hyperbola $3x^2 - y^2 = 4$ with vertices $\left(\pm\frac{2}{\sqrt{3}}, 0\right)$, foci $\left(\pm\frac{4}{\sqrt{3}}, 0\right)$, asymptotes $y = \pm\sqrt{3}\,x$]

19.

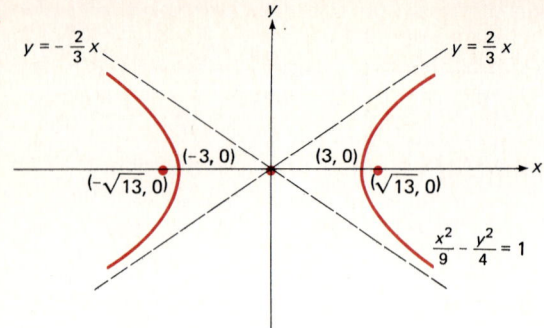

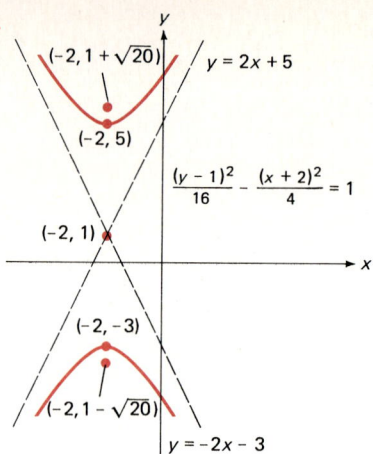

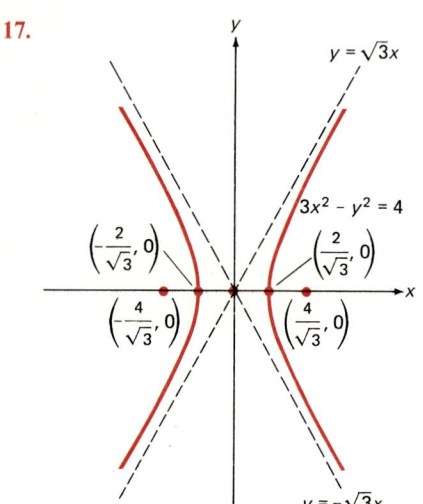

[Graph: hyperbola $2x^2 - y^2 + 4x - 2y + 17 = 0$ with center $(-1, 1)$, vertices $(-1, 3)$ and $(-1, -1)$; points $(-1, \sqrt{24})$, $(-1, -\sqrt{24})$; asymptotes $y = \sqrt{2}x + \sqrt{2} - 1$ and $y = -\sqrt{2}x - \sqrt{2} - 1$]

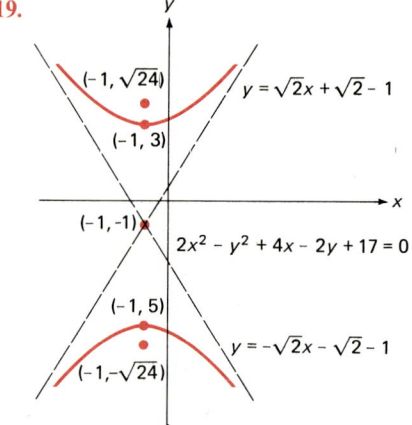

21.

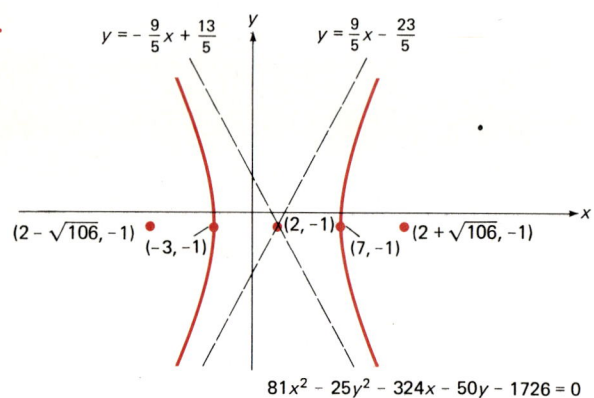

[Graph: hyperbola $81x^2 - 25y^2 - 324x - 50y - 1726 = 0$ with center $(2, -1)$, vertices $(-3, -1)$ and $(7, -1)$, foci $(2 \pm \sqrt{106}, -1)$, asymptotes $y = -\frac{9}{5}x + \frac{13}{5}$ and $y = \frac{9}{5}x - \frac{23}{5}$]

23. $F_1 = (-1, -1)$, $F_2 = (1, 1)$, $|PF_1| - |PF_2| = \pm 2$ **25.** $(3, \pm\sqrt{7})$, $(-3, \pm\sqrt{7})$
27. See Example 6, p. 454.   **29.** $\left(\frac{25}{4}, 3\right)$ **31.** $(1, 1)$, $\left(\frac{5}{7}, \frac{1}{7}\right)$ **33.** $|F_1P| - |F_2P| = r - s$ **35.** $\sqrt{3}$ **37.** $\frac{x^2}{4} - \frac{y^2}{12} = 1$
39. $5x - 3y - 16 = 0$, $13x - 5y + 48 = 0$ **41.** Show that the right triangles F_1PQ_1 and F_2PQ_2 are similar.
43. $\frac{1}{6}$ km closer to the depot than the nearest sentry and $\sqrt{\frac{10}{3}}$ km from the road on either side

Section 10.5 (page 612)

1. $\dfrac{x^2}{20} + \dfrac{y^2}{36} = 1$ **3.** $\dfrac{x^2}{13} + \dfrac{y^2}{9} = 1$ or $\dfrac{4x^2}{117} + \dfrac{y^2}{9} = 1$ **5.** $\dfrac{x^2}{16} - \dfrac{y^2}{9} = 1$ **7.** $\dfrac{y^2}{4} - \dfrac{x^2}{5} = 1$ **9.** $\dfrac{4}{5}$, $y = \pm\dfrac{25}{4}$

11. $\dfrac{1}{\sqrt{2}}$, $y = -x \pm 2$ **13.** $\sqrt{2}$, $y = -x \pm 1$ **15.** The ellipse $\dfrac{x^2}{4} + \dfrac{y^2}{3} = 1$ **17.** $\dfrac{(x-1)^2}{20} - \dfrac{(y+2)^2}{5} = 1$

19. $7x^2 - 2xy + 7y^2 - 30x + 2y + 31 = 0$ **21.** $2a$ is the distance from C to C' along a generator of the cone.

Section 10.6 (page 619)

1. Hyperbola, $\theta = 45°$, transformed equation
$\left(x' - \dfrac{1}{\sqrt{2}}\right)^2 - \dfrac{1}{4}\left(y' + \dfrac{3}{\sqrt{2}}\right)^2 = 1$

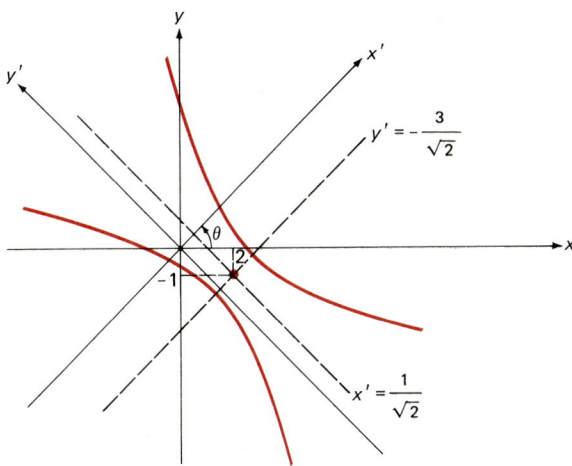

3. The single point $(0, -2)$

5. Ellipse, $\theta = \operatorname{arccot} 4 \approx 14.0°$, transformed equation
$\dfrac{1}{16}(x' - 2\sqrt{17})^2 + \dfrac{1}{9}y'^2 = 1$

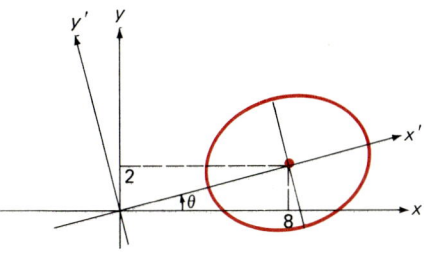

$3x^2 + 10xy + 3y^2 - 2x - 14y - 13 = 0$

$160x^2 - 56xy + 265y^2 - 2448x - 612y + 7956 = 0$

7. The lines $2x + y + 3 = 0$, $2x + 11y + 5 = 0$ intersecting in the point $\left(-\dfrac{7}{5}, -\dfrac{1}{5}\right)$ **9.** The empty set

11. Parabola, $\theta = \operatorname{arccot} \dfrac{1}{3} \approx 71.6°$, transformed equation $(x' - \sqrt{10})^2 = -\sqrt{10}\,y'$

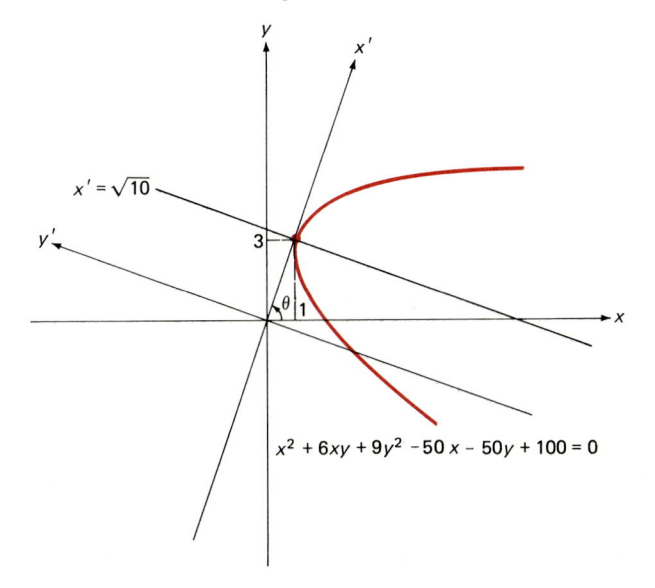

$x^2 + 6xy + 9y^2 - 50x - 50y + 100 = 0$

13. The parallel lines $5x - 2y - 4 = 0$, $5x - 2y + 1 = 0$ **15.** $(0, 3)$ **17.** $|PA| = |PF|$, $|PA'| = |PF'|$

19. $\Delta = B^2 - 4AC = -4A'C'$ if $B' = 0$.

Answers and Hints to Odd-Numbered Problems   **A-49**

Section 10.7 (page 627)

1. $(0, 0)$ **3.** $(6\sqrt{3}, -6)$ **5.** $(0, -10)$ **7.** $\left(2, \frac{5\pi}{4} + 2n\pi\right), \left(-2, \frac{\pi}{4} + 2n\pi\right)$, n any integer

9. $\left(4, \frac{\pi}{6} + 2n\pi\right), \left(-4, \frac{7\pi}{6} + 2n\pi\right)$, n any integer **11.** $(3, (2n+1)\pi), (-3, 2n\pi)$, n any integer

13. $C = (3, 5\pi/9), D = (5, 17\pi/14)$ **15.** Use the law of cosines. **17.** 20 **19.** $26 + 12\sqrt{2}$ **21.** See Prob. 15.

23. $r^2 - 12r \cos\left(\theta - \frac{\pi}{4}\right) + 27 = 0$ **25.** $r^2 - 4r \sin\theta - 1 = 0$ **27.** $\theta = \frac{\pi}{4} + n\pi$, n any integer **29.** $r = 3$

31. $r = a \cos\theta$ **33.** $r = 4 \cot\theta \csc\theta$ **35.** $r^2 = \cos 2\theta$ **37.** $y = 0$, the x-axis **39.** $x = -2$, a vertical line

41. $x + \sqrt{3}y - 12 = 0$, an oblique line **43.** $x^2 + y^2 + y = 0$, the circle of radius $\frac{1}{2}$ centered at $(0, -\frac{1}{2})$

45. $xy = 2$, an equilateral hyperbola

47. $r = 4 \sin 3\theta$, Three-petaled rose

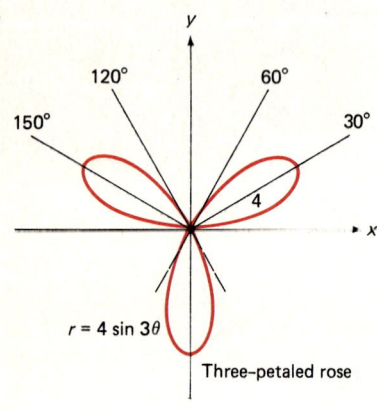

49. $r = \cos 6\theta$, Twelve-petaled rose

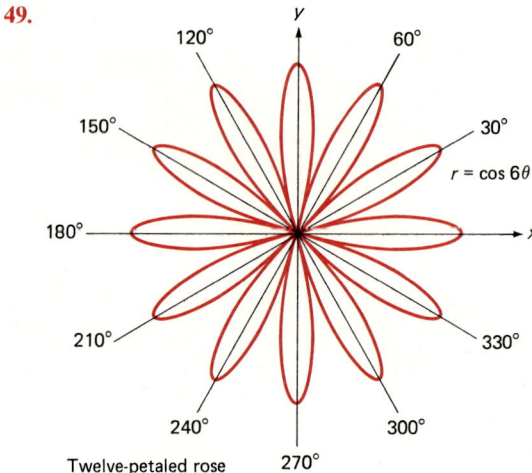

51. $r = 3 + 2\cos\theta$, Limaçon

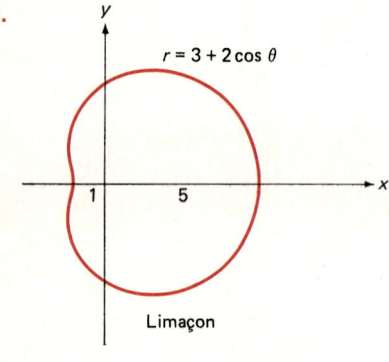

53. $r = 1 + 2\cos\theta$, Limaçon

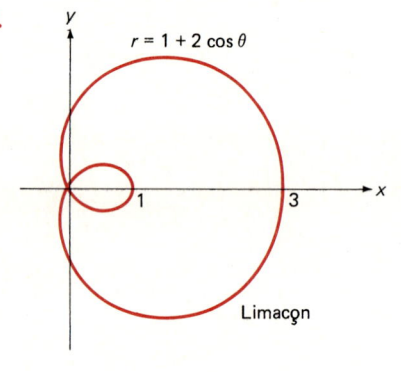

55. $r^2 = 4\cos 2\theta$, Lemniscate

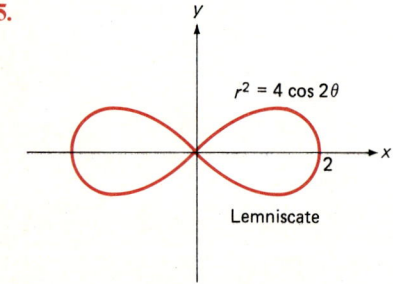

57.

Cissoid
$r = \sin\theta \tan\theta$, $x=1$

59.

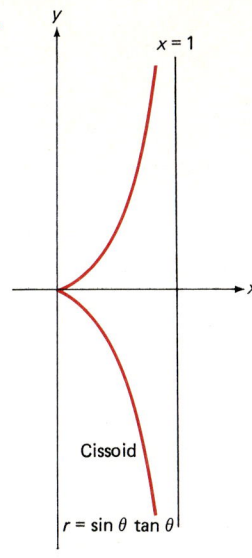

Nephroid
$r = 1 + 2\sin\dfrac{\theta}{2}$

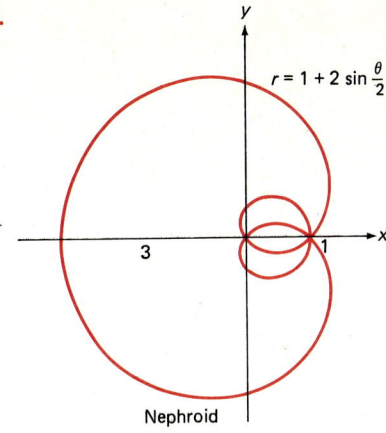

61.

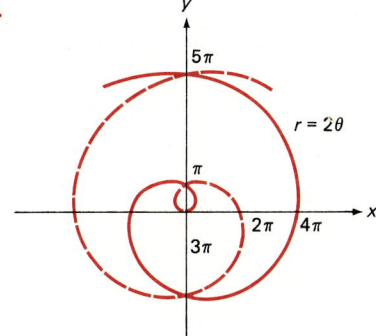

Archimedean spiral
$r = 2\theta$

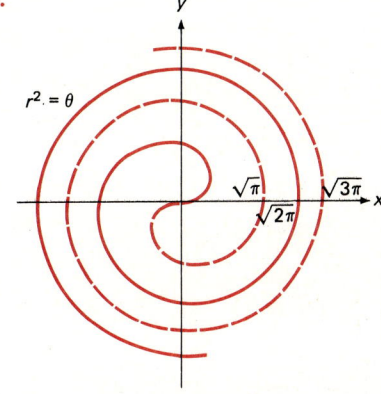

63.

Parabolic spiral
$r^2 = \theta$

65. $\left(\dfrac{\pi}{4} + n\pi, \dfrac{\pi}{4} + n\pi\right)$, n any integer   **67.** Origin, $\left(1 + \dfrac{1}{\sqrt{2}}, \dfrac{3\pi}{4}\right), \left(1 - \dfrac{1}{\sqrt{2}}, -\dfrac{\pi}{4}\right)$

69. Origin, $\left(\dfrac{\sqrt{3}}{2}, \dfrac{\pi}{6}\right), \left(-\dfrac{\sqrt{3}}{2}, \dfrac{5\pi}{6}\right)$; the last of these points can also be represented as $\left(\dfrac{\sqrt{3}}{2}, -\dfrac{\pi}{6}\right)$

71. Origin, $\left(\dfrac{1}{\sqrt[4]{2}}, \dfrac{\pi}{8}\right), \left(\dfrac{1}{\sqrt[4]{2}}, \dfrac{9\pi}{8}\right)$

Section 10.8 (page 631)

1. Parabola, $e = 1$, vertex $(\frac{5}{2}, 0)$

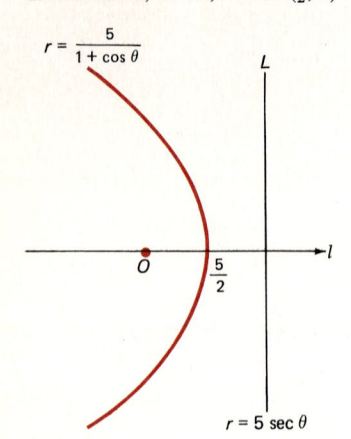

$r = \dfrac{5}{1 + \cos \theta}$

$r = 5 \sec \theta$

3. Hyperbola, $e = \frac{3}{2}$, vertices $(\frac{6}{5}, \frac{3}{2}\pi)$, $(6, \frac{3}{2}\pi)$

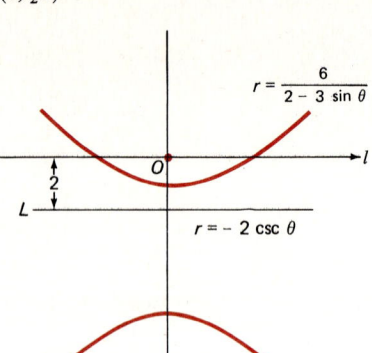

$r = \dfrac{6}{2 - 3 \sin \theta}$

$r = -2 \csc \theta$

5. Ellipse, $e = \frac{3}{4}$, vertices $(\frac{9}{7}, \frac{1}{2}\pi)$, $(9, \frac{3}{2}\pi)$

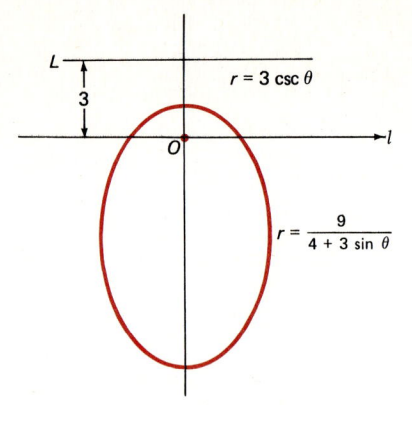

$r = 3 \csc \theta$

$r = \dfrac{9}{4 + 3 \sin \theta}$

7. Hyperbola, $e = 2$, vertices $(\frac{10}{3}, 0)$, $(10, 0)$

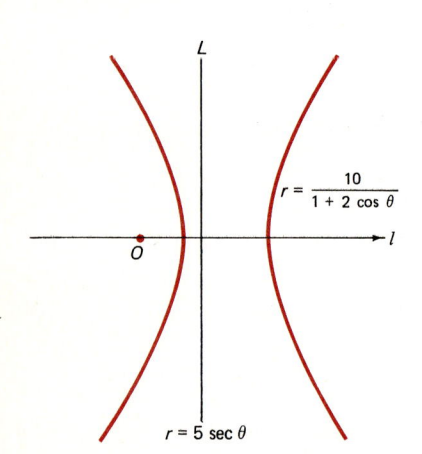

$r = \dfrac{10}{1 + 2 \cos \theta}$

$r = 5 \sec \theta$

9. Parabola, $e = 1$, vertex $(5, \frac{7}{6}\pi)$

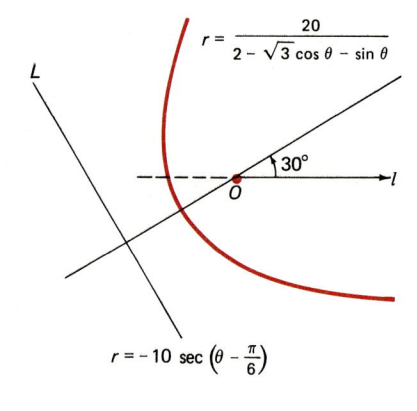

$r = \dfrac{20}{2 - \sqrt{3} \cos \theta - \sin \theta}$

$r = -10 \sec\left(\theta - \dfrac{\pi}{6}\right)$

11. $r = \dfrac{18}{4 - 5 \sin \theta}$ **13.** $r = \dfrac{3}{1 - \cos \theta}$ **15.** $r = \dfrac{1}{4 - \sqrt{6}(\cos \theta + \sin \theta)}$ **17.** $r = \dfrac{5}{2 + 3 \sin \theta}$

19. $r = \dfrac{15}{3 - 2 \sin \theta}$ or $r = \dfrac{3}{3 + 2 \sin \theta}$ **21.** $r = \dfrac{24}{3 + 4 \cos \theta}$ **23.** ≈ 7.3 million miles

Section 10.9 (page 635)

1. $\tan \psi = -\dfrac{1}{\sqrt{3}}$, $\psi = \dfrac{5\pi}{6}$, $\phi = \dfrac{\pi}{6}$ **3.** $\tan \psi = \dfrac{1}{2}$, $\psi = \arctan \dfrac{1}{2} \approx 26.6°$, $\phi = \arctan \dfrac{1}{2} + \dfrac{\pi}{8} \approx 49.1°$

5. $\tan \psi = \dfrac{\sqrt{3}}{2}$, $\psi = \arctan \dfrac{\sqrt{3}}{2} \approx 40.9°$, $\phi = \arctan \dfrac{\sqrt{3}}{2} + \dfrac{\pi}{6} \approx 70.9°$

7. $\tan \psi = -2$, $\psi = \pi - \arctan 2 \approx 116.6°$, $\phi = \dfrac{\pi}{2} - \arctan 2 \approx 26.6°$

9. $\tan \psi = -1$, $\psi = 3\pi/4$, $\phi = \pi/4$ **11.** $\tan \psi = \pi$, $\psi = \phi = \arctan \pi \approx 72.3°$

13. Show that $\tan \psi|_{\theta = \theta_0} = 0$ if $f(\theta_0) = 0$, $f'(\theta_0) \neq 0$; use another argument if $f(\theta_0) = f'(\theta_0) = 0$. **15.** $\theta = \pi/6, \pi/2, 5\pi/6$

17. Pole, $\left(\sqrt[4]{\dfrac{3}{4}}, \dfrac{\pi}{3}\right)$, $\left(\sqrt[4]{\dfrac{3}{4}}, \dfrac{4\pi}{3}\right)$. There is also a vertical tangent at the pole. **19.** $\dfrac{\pi}{2} - \dfrac{1}{2} \arccos \dfrac{1}{3} \approx 54.7°$ **21.** $\dfrac{\pi}{2}$

A-52   Answers and Hints to Odd-Numbered Problems

Section 10.10 (page 640)

1. $\dfrac{9}{2}$ **3.** $\dfrac{4\sqrt{3}-\pi}{3}$ **5.** $\dfrac{2}{3}-\dfrac{5\sqrt{3}}{27}$ **7.** $\dfrac{2^\pi - 1}{4\ln 2}$ **9.** 16π **11.** 50π **13.** $\dfrac{27\pi}{2}$ **15.** 11π **17.** $72\sqrt{3}$ **19.** $\dfrac{\pi}{2}$ **21.** $\dfrac{\pi}{4}$
23. $\dfrac{\pi}{8}+\dfrac{1}{4}$ **25.** $2-\dfrac{\pi}{4}$ **27.** $\dfrac{\pi}{6}+\dfrac{\sqrt{3}}{4}$ **29.** $\dfrac{3\pi}{16}-\dfrac{\sqrt{2}}{8}$ **31.** $\dfrac{\pi}{4}-\dfrac{3\sqrt{3}}{16}$ **33.** $\pi - \dfrac{3\sqrt{3}}{2}; \pi + 3\sqrt{3}$
35. The equation $r^2 = a^2\cos 2\theta$ has no solutions for $\pi/4 < \theta < 3\pi/4$ or $5\pi/4 < \theta < 7\pi/4$. **37.** $2a$ **39.** $2\pi a$
41. $\sqrt{2}(e-1)$ **43.** $\dfrac{\sqrt{5}}{2}+\ln\dfrac{3+\sqrt{5}}{2}$ **45.** $\sqrt{2}+\ln(\sqrt{2}+1)$ **47.** $\dfrac{3\pi a}{2}$ **49.** $\dfrac{32\pi a^2}{5}$

Supplementary Problems (page 642)

1. $x' = x - 2$, $y' = y + 4$
3. Show that $(x'_1 - x'_2)^2 + (y'_1 - y'_2)^2 = (x_1 - x_2)^2 + (y_1 - y_2)^2$, where $P_1 = (x'_1, y'_1)$, $P_2 = (x'_2, y'_2)$ in the new system.
5. Show that $(x'_1 - x'_2)^2 - c^2(t'_1 - t'_2)^2 = (x_1 - x_2)^2 - c^2(t_1 - t_2)^2$, where $E_1 = (x'_1, t'_1)$, $E_2 = (x'_2, t'_2)$ in the new system.
7. $y = h - \dfrac{4h}{b^2}\left(x - \dfrac{b}{2}\right)^2; \dfrac{2}{3}bh$ **9.** The positions are $P = (3c, \pm 2\sqrt{3}c)$.
11. If the chords have nonzero slope m, their midpoints lie on the line $y = 2c/m$.
13. P_1, P_5; only P_2; P_3, P_4, P_6 **15.** $(x^2/36) + (y^2/9) = 1; \sqrt{3}/2$ **17.** $2x - y - 12 = 0, 2x - y + 12 = 0; 24/\sqrt{5}$
19. If the chords have nonzero slope m, their midpoints lie on the line $y = -b^2 x/a^2 m$. **21.** $a^2 b^2/c^2$
23. $3x - 4y - 10 = 0, 3x - 4y + 10 = 0$ **25.** $(\tfrac{4}{5}, \tfrac{4}{5})$ **27.** $\sqrt{5}$ **29.** $(x^2/60) - (y^2/40) = 1$
31. The line $y = mx + \sqrt{m^2 a^2 + b^2}$ is tangent to the ellipse.
33. The director circle lies between the directrices (see Figure 95). **35.** The single line $3x + 5y - 7 = 0$
37. Parabola, $\theta = 45°$, transformed equation $\left(y' + \dfrac{1}{\sqrt{2}}\right)^2 = \dfrac{1}{\sqrt{2}}\left(x' + \dfrac{1}{\sqrt{2}}\right)$

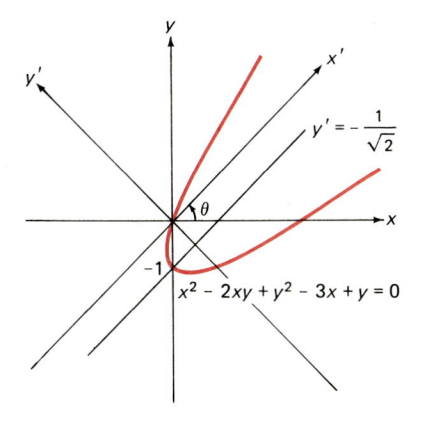

39. The circle with (x_1, y_1) and (x_2, y_2) as endpoints of a diameter
41. If $-a^2 < k < -b^2$, write (iii) in the form $\dfrac{x^2}{a^2 + k} - \dfrac{y^2}{-b^2 - k} = 1$.
43.

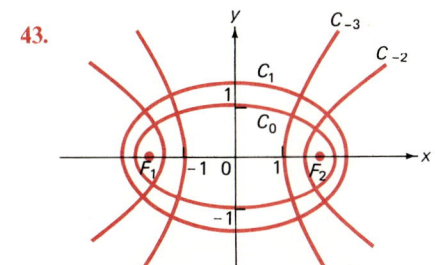

45. 4, $(3, \pi/6)$ **47.** $(6, \pi/9)$ **49.** $r = 1 + \sin\theta$ **51.** $r = \theta^2 + (3\pi - \theta)^2$ **53.** $2/ed$
55. Use Prob. 13, p. 635, at the pole, and calculate radial-tangential angles at the other point of intersection.
57. $A_n = \dfrac{1}{2}\displaystyle\int_0^{\pi/n} r^2\,d\theta$ **59.** $\dfrac{3\pi}{2} - 2\sqrt{2}$ **61.** $r = a, r = a(1 + \cos\theta)$ **63.** $\left(a^2 + \dfrac{1}{2}b^2\right)\left(\pi - 2\arccos\dfrac{a}{b}\right) + 3a\sqrt{b^2 - a^2}$
65. $2 + \tfrac{1}{2}\ln 3$

Chapter 11

Section 11.1 (page 653)

1.

3. $(4, 1), (-2, -11), (11, 8), (-9, -39)$ **5.** $2\mathbf{i}, 2\mathbf{j}, 5\mathbf{i} - \mathbf{j}, -\mathbf{i} + 7\mathbf{j}$ **7.** $-2\mathbf{i} + 5\mathbf{j}, 4\mathbf{i} + \mathbf{j}, -7\mathbf{i} + 12\mathbf{j}, 15\mathbf{i} + \mathbf{j}$
9. $(2, 0), (0, 2), (5, -1), (-1, 7)$ **11.** $(3, 6)$ **13.** $(-2, -5)$ **15.** $(-1, 4)$ **17.** $(10, -13)$ **19.** 1
21. $|\boldsymbol{\beta}|$ **23.** 37 **25.** $3\mathbf{i} + 3\mathbf{j}$ **27.** $(\mathbf{i} + \mathbf{j})/\sqrt{2}, \pi/4$ **29.** $(-3\mathbf{i} + 2\mathbf{j})/\sqrt{13}, \arctan(-\tfrac{2}{3}) + \pi \approx 146.3°$
31. $-(3\mathbf{i} + 4\mathbf{j})/5, \arctan\tfrac{4}{3} + \pi \approx 233.1°$ **33.** $(\mathbf{i} - \sqrt{3}\mathbf{j})/2$ **35.** $(-\mathbf{i} + \mathbf{j})/\sqrt{2}$ **37.** $(12\mathbf{i} + 8\mathbf{j})/\sqrt{13}$ **39.** $\mathbf{x} = (-4, 7)$
41. If $\overrightarrow{AP} = t\overrightarrow{AB}$, then P divides AB in the ratio $t:(1 - t)$. **43.** See formula (5), p. 16.
45. In Figure 16 show that $\overrightarrow{DE} = \tfrac{1}{2}\overrightarrow{AB}$. **47.** $\mathbf{e}_1$ and $\mathbf{e}_2$ are nonzero, noncollinear and nonperpendicular; $\mathbf{a} = -11\mathbf{e}_1 + 8\mathbf{e}_2$
49. $p(-\mathbf{i} + 2\mathbf{j})$ for arbitrary $p \neq 0$ **51.** $10\sqrt{2}$ knots from the northwest

Section 11.2 (page 660)

1. $0, \pi/2$ **3.** $1, \pi/4$ **5.** $-2, 3\pi/4$ **7.** $63, \arccos\tfrac{63}{65} \approx 14.25°$ **9.** -6 **11.** 73 **13.** No **15.** $-\tfrac{3}{2}$
17. $45°$ at A, $\arccos(2/\sqrt{5}) \approx 26.6°$ at B, $\pi - \arccos(1/\sqrt{10}) \approx 108.4°$ at C **19.** Show that $(\mathbf{v}\cdot\mathbf{a})/|\mathbf{a}| = (\mathbf{v}\cdot\mathbf{b})/|\mathbf{b}|$.
21. $\pm(\mathbf{i} - 2\mathbf{j})/\sqrt{5}$ **23.** $\pm(4\mathbf{i} + 3\mathbf{j})/5$ **25.** $(-3\mathbf{i} + 5\mathbf{j})/\sqrt{34}, -(5\mathbf{i} + 3\mathbf{j})/\sqrt{34}$ **27.** $-\tfrac{5}{6}; \tfrac{15}{8}$ **29.** $2 \pm \sqrt{3}$
31. Start from $(\mathbf{a} + \mathbf{b})\cdot(\mathbf{a} + \mathbf{b}) > (\mathbf{a} - \mathbf{b})\cdot(\mathbf{a} - \mathbf{b})$. **33.** $|\mathbf{a} + \mathbf{b}|^2 + |\mathbf{a} - \mathbf{b}|^2 = (\mathbf{a} + \mathbf{b})\cdot(\mathbf{a} + \mathbf{b}) + (\mathbf{a} - \mathbf{b})\cdot(\mathbf{a} - \mathbf{b})$
35. $(\mathbf{a} + \mathbf{b})\cdot(\mathbf{a} - \mathbf{b}) = |\mathbf{a}|^2 - |\mathbf{b}|^2$ **37.** $(\tfrac{5}{2}, \tfrac{5}{2}), (\tfrac{1}{2}, -\tfrac{1}{2})$ **39.** $(2, 4), (-6, 3)$ **41.** $2\mathbf{j}, -\mathbf{i}$ **43.** $(12\mathbf{i} - 3\mathbf{j})/17, (22\mathbf{i} + 88\mathbf{j})/17$
45. $10\sqrt{4 + 2\sqrt{2}} \approx 26.13$ lb **47.** $\displaystyle\sum_{i=1}^{n}(\alpha_i x + \beta_i)^2 \equiv Ax^2 + Bx + C \geq 0$ for all x; this imposes a condition on A, B and C.
49. 30 ft-lb; 48 ft-lb **51.** 18 ergs **53.** If (x_1, y_1) and (x_2, y_2) are two points on the line, then $A(x_1 - x_2) + B(y_1 - y_2) = 0$.

Section 11.3 (page 670)

1. $4\mathbf{i} + 8\mathbf{j}$ **3.** $\tfrac{1}{2}\mathbf{i} + \tfrac{1}{2}\mathbf{j}$ **5.** $-\mathbf{j}$ **7.** Yes. No **9.** Take components. **11.** Use Prob. 43, p. 654.
13. $(\ln t + 1)\mathbf{i} + (2t + t^2)e^t\mathbf{j}$ **15.** $(\sec t \tan t)\mathbf{i} - (\cosh t)\mathbf{j}$ **17.** $(\sin t + \cos t)(e^t\mathbf{i} - e^{-t}\mathbf{j})$ **19.** See Theorem 2, p. 120.
21. See Theorem 6, p. 134. **23.** $\mathbf{F}(t) = (\sin t)\mathbf{i} + e^t\mathbf{j} + 2\mathbf{i}$ **25.** $\mathbf{F}(t) = (\sec t)\mathbf{i} + (\tan t)\mathbf{j} - 2\mathbf{i} - \sqrt{3}\mathbf{j}$ **27.** $2\mathbf{i} + \tfrac{38}{3}\mathbf{j}$ **29.** 10
31. $\tfrac{1}{4}(2t - \sin 2t)\mathbf{i} + \tfrac{1}{4}(2t + \sin 2t)\mathbf{j} + \mathbf{C}$ **33.** If $\mathbf{c} = (c_1, c_2)$ and $\mathbf{f}(t) = (f_1(t), f_2(t))$, then $\mathbf{c}\cdot\mathbf{f}(t) = c_1 f_1(t) + c_2 f_2(t)$.

35. $\mathbf{v} = 3\mathbf{i}, v = |\mathbf{v}| = 3$

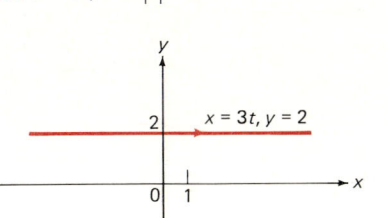

[Graph: $x = 3t, y = 2$, horizontal line at $y=2$]

37. $\mathbf{v} = 2t\mathbf{i} + 2\mathbf{j}, v = |\mathbf{v}| = 2\sqrt{t^2 + 1}$

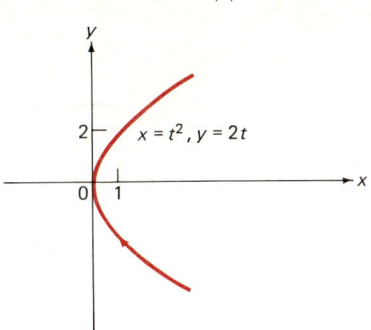

[Graph: $x = t^2, y = 2t$, sideways parabola]

39. $\mathbf{v} = (2\cos t)\mathbf{i} + (\sin t)\mathbf{j}$, $v = |\mathbf{v}| = \sqrt{3\cos^2 t + 1}$

[Graph: $x = 2\sin t, y = -\cos t$, ellipse]

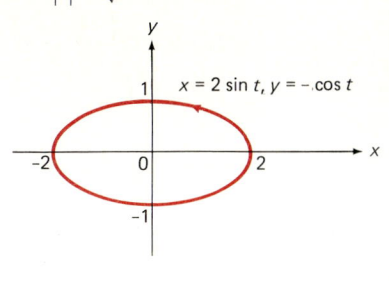

41. $\mathbf{T}(t) = \mathbf{T}(s) = \mathbf{i}$ **43.** $\mathbf{T}(t) = (-\sin 3t)\mathbf{i} + (\cos 3t)\mathbf{j}$, $\mathbf{T}(s) = (-\sin s)\mathbf{i} + (\cos s)\mathbf{j}$

45. $\mathbf{T}(t) = \dfrac{\mathbf{i} + t\mathbf{j}}{\sqrt{1 + t^2}}$, $\mathbf{T}(s) = \dfrac{\mathbf{i} + \sqrt{(s+1)^{2/3} - 1}\,\mathbf{j}}{(s+1)^{1/3}}$

47. See Figure 45, p. 464. Maximum 6 when $t = \pi/4, 3\pi/4, 5\pi/4, 7\pi/4, \ldots$ at $(\sqrt{2}, \sqrt{2}), (-\sqrt{2}, \sqrt{2}), (-\sqrt{2}, -\sqrt{2})$ and $(\sqrt{2}, -\sqrt{2})$; minimum 0 when $t = 0, \pi/2, \pi, 3\pi/2, \ldots$ at $(4, 0), (0, 4), (-4, 0)$ and $(0, -4)$. Yes, it does.

Section 11.4 (page 681)

1. $\mathbf{T} = \mathbf{i}, \mathbf{N} = \mathbf{j}$ **3.** $\mathbf{T} = \dfrac{t\mathbf{i} + \mathbf{j}}{\sqrt{t^2 + 1}}, \mathbf{N} = \dfrac{\mathbf{i} - t\mathbf{j}}{\sqrt{t^2 + 1}}$ **5.** $\mathbf{T} = \dfrac{(\cosh t)\mathbf{i} + (\sinh t)\mathbf{j}}{\sqrt{\cosh 2t}}, \mathbf{N} = \dfrac{(-\sinh t)\mathbf{i} + (\cosh t)\mathbf{j}}{\sqrt{\cosh 2t}}$

7. $\dfrac{1}{6}$ **9.** $\dfrac{5}{8}$ **11.** $\dfrac{1}{\sqrt{2}}$ **13.** 2 **15.** $\dfrac{4}{5^{3/2}}$ **17.** $\dfrac{1}{e}$ **19.** $\dfrac{2}{3\sqrt{3}}$ at $\left(-\dfrac{1}{2}\ln 2, \dfrac{1}{\sqrt{2}}\right)$

21. $R = a\sqrt{8(1 - \cos t)}$. At the cusps. At the points with ordinate $\tfrac{1}{8}a$. Maximum $4a$ at the points with ordinate $2a$

23.

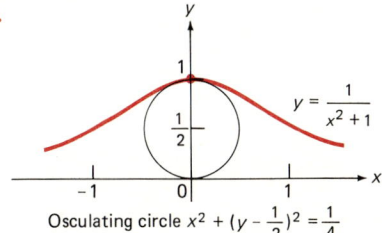

[Graph: $y = \dfrac{1}{x^2+1}$ with osculating circle $x^2 + (y - \tfrac{1}{2})^2 = \tfrac{1}{4}$]

25.

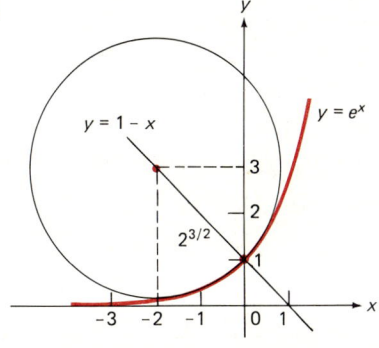

[Graph: $y = e^x$ with line $y = 1-x$ and osculating circle $(x+2)^2 + (y-3)^2 = 8$]

27. $(\pm 3, \tfrac{9}{8})$ **29.** $3\sqrt{2}$ **31.** $a_T = 0, a_N = 0$ **33.** $a_T = \dfrac{6(2t^2 + 1)}{\sqrt{t^2 + 1}}, a_N = \dfrac{6t}{\sqrt{t^2 + 1}}$

35. $a_T = \dfrac{\sinh 2t}{\sqrt{\cosh 2t}}, a_N = \sqrt{\cosh 2t}$ **37.** Calculate κ with $x = r\cos\theta, y = r\sin\theta, t = \theta$ **39.** $\tfrac{3}{4}$

41. Reverse the directions of the two curves in Figure 31. **43.** Change $\mathbf{i}$ to $-\mathbf{j}$ and $\mathbf{j}$ to $\mathbf{i}$. **45.** Change $\mathbf{i}$ to $\mathbf{j}$ and $\mathbf{j}$ to $-\mathbf{i}$.

Section 11.5 (page 689)

1. $(2v_0/g)\sin\alpha$ **3.** $(v_0^2/g)\sin 2\alpha$ **5.** ≈ 1838 ft/sec **7.** Vertex $((v_0^2/2g)\sin 2\alpha, (v_0^2/2g)\sin^2\alpha)$, directrix $y = v_0^2/2g$
9. ≈ 8.66 mi **11.** $\arctan(\sqrt{2}/11) \approx 7.3°$ **13.** $v \approx 4.6$ mi/sec, $T \approx 101$ min **15.** $v \approx 3.3$ mi/sec, $T \approx 288$ min
17. $v \approx 4.9$ mi/sec, $T \approx 85$ min **19.** $v \approx 184.25$ cm/sec, $T \approx 56{,}580$ dynes **21.** 3.5 rad/sec **23.** $\rho \approx 1060$ kg/m³

25. $F \cos \alpha = mg$, $F \sin \alpha = mv_r^2/R$, where $F = |\mathbf{F}|$ and m is the car's mass.
27. $d = (w/8H)a^2$. Square and add the equations (12). The minimum tension is H at the midpoint of the cable.
29. Use formula (5), p. 462. **31.** Expand $\cosh(wx/H)$ in Maclaurin series.

Supplementary Problems (page 691)

1. $-8\mathbf{i} - 10\mathbf{j}$ **3.** 0 **5.** $\mathbf{i}/\sqrt{10}$ **7.** $|\mathbf{a}| = |\mathbf{b}|$ **9.** Rotate about O through the angle $2\pi/n$. **11.** 22
13. $|\mathbf{a} + \mathbf{b}| = 7$, $|\mathbf{a} - \mathbf{b}| = \sqrt{129}$ **15.** $\pm\frac{1}{2}; 0$ **17.** $\arccos\frac{8}{17} \approx 61.9°$ **19.** 70 hp **21.** $(-1, -1), (-\frac{1}{5}, \frac{3}{5})$
23. $\mathbf{r} \cdot (\mathbf{r} - 2\mathbf{a}) = 0$ **25.** $\frac{1}{6}\mathbf{j}$ **27.** $\dfrac{\mathbf{i} + \mathbf{j}}{\sqrt{1 - t^2}}$ **29.** $(2te^t + t^2 e^t)\mathbf{i} + \dfrac{\mathbf{j}}{1 - t^2}$
31. $(\ln 2)\mathbf{i} + \ln(2 + \sqrt{3})\mathbf{j}$ **33.** Yes **35.** $\mathbf{T} = \mathbf{j}, \mathbf{N} = \mathbf{i}$ **37.** $\mathbf{T} = (\mathbf{i} + \mathbf{j})/\sqrt{2}$, $\mathbf{N} = (-\mathbf{i} + \mathbf{j})/2$ **39.** $|\cos x|$

41.

Osculating circle $(x + 4)^2 + (y - \frac{7}{2})^2 = \frac{125}{4}$

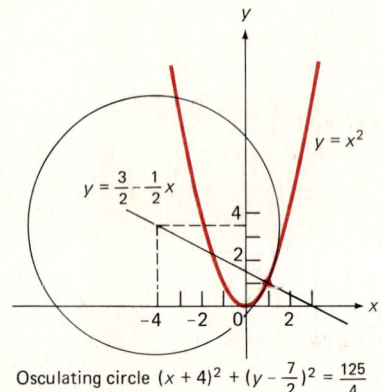

43. 50 ft/sec **45.** Back in the car **47.** 45°, $\arctan 3 \approx 71.6°$ **49.** ≈ 1482 ft **51.** ≈ 10.2 in. **53.** ≈ 13.93 ft

Chapter 12

Section 12.1 (page 700)

1. Those in parts (a) and (c) **3.** The other vertices are $(2, 0, 0), (2, -3, 0), (0, -3, 0), (0, 0, 4), (2, 0, 4)$ and $(0, -3, 4)$.

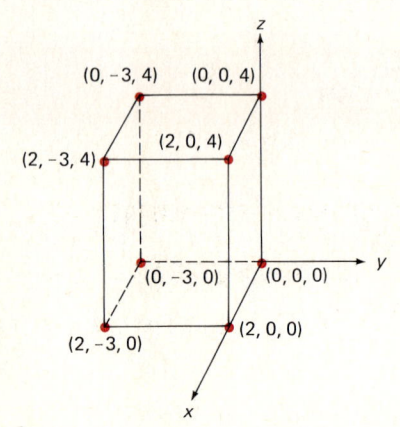

5. $(6, -4, 9)$ **7.** $\sqrt{a^2 + c^2}$ **9.** 25 **11.** 7 **13.** $\sqrt{66}$ **15.** $5\sqrt{2}$ **17.** $(0, 2, 0)$ **19.** $(-1, 3, 1)$ **21.** $(-4, 2, 6)$
23. $(\frac{1}{2}, -\frac{1}{2}, \frac{1}{2})$ **25.** $(-5, 3, -2)$ **27.** $(5, 3, -2)$ **29.** $(-5, 3, 2)$ **31.** $(5, 3, 2)$ **33.** $x^2 + y^2 + z^2 = 25$
35. $(x - 3)^2 + (y + 2)^2 + (z - 4)^2 = 11$ **37.** The sphere of radius 11 with center $(-3, 2, -5)$
39. The single point $(-5, -6, 4)$ **41.** The empty set **43.** $(x - 3)^2 + (y + 3)^2 + (z + 3)^2 = 9$
45. The yz- and xz-planes **47.** The exterior of the unit sphere

49.

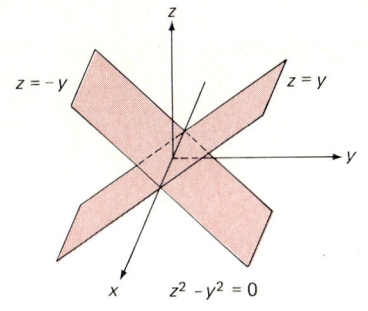

Pair of intersecting perpendicular planes
$z^2 - y^2 = 0$, $z = -y$, $z = y$

51.

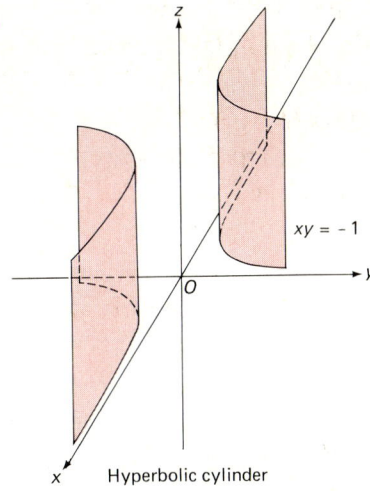

Hyperbolic cylinder
$xy = -1$

53.

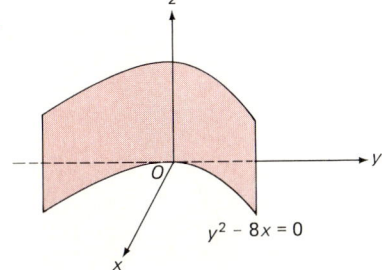

Parabolic cylinder
$y^2 - 8x = 0$

55.

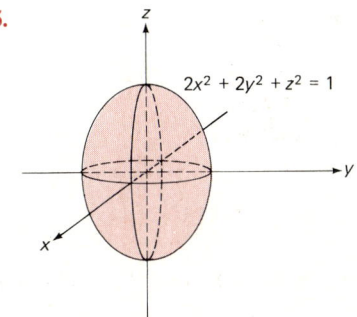

Ellipsoid of revolution
$2x^2 + 2y^2 + z^2 = 1$

57. $x^2 + y^2 = 16z^4$ **59.** $x^{2/3} + (y^2 + z^2)^{1/3} = 1$ **61.** $y^4 = x^2 + z^2$

Section 12.2 (page 706)

1. (7, 2, 2), (−7, −10, 11)   **3.** $3\mathbf{i} - \mathbf{j} - 3\mathbf{k}$, $-\mathbf{i} + 5\mathbf{j} + 8\mathbf{k}$ **5.** (7, −1, 6)   **7.** $(-1, \frac{2}{3}, \frac{1}{2})$ **9.** (4, 1, 1)   **11.** $\sqrt{3}$ **13.** 11
15. 10, arccos $\frac{10}{29} \approx 69.8°$ **17.** −1, arccos $(-\frac{1}{3}) \approx 109.5°$ **19.** −20   **21.** $\frac{6}{13}\mathbf{i} + \frac{9}{13}\mathbf{k}$ **23.** 0   **25.** No   **27.** Yes
29. No   **31.** $\frac{2}{11}, -\frac{6}{11}, \frac{9}{11}$, and arccos $\frac{2}{11} \approx 79.5°$, arccos $(-\frac{6}{11}) \approx 123.1°$, arccos $\frac{9}{11} \approx 35.1°$
33. $\frac{6}{7}, \frac{2}{7}, \frac{3}{7}$, and arccos $\frac{6}{7} \approx 31.0°$, arccos $\frac{2}{7} \approx 73.4°$, arccos $\frac{3}{7} \approx 64.6°$ **35.** arccos $(1/\sqrt{3}) = 54.7°$
37. arccos $(-\frac{1}{4}) \approx 104.5°$ **39.** The vectors $\mathbf{a}_1, \mathbf{a}_2, \mathbf{a}_3$ in Prob. 38

Section 12.3 (page 714)

1. $6\mathbf{i}$ **3.** (4, −1, −2)   **5.** $4\mathbf{i} - 5\mathbf{j} + 9\mathbf{k}$ **7.** (−33, 11, 0)   **9.** $5\sqrt{3}$ **11.** $\frac{49}{2}$ **13.** $|(\mathbf{a} + \mathbf{b}) \times (\mathbf{a} - \mathbf{b})| = 2|\mathbf{a} \times \mathbf{b}|$
15. $\pm(\mathbf{i} - \mathbf{j} + \mathbf{k})/\sqrt{3}$ **17.** $\pm\frac{1}{3}(\mathbf{i} + 2\mathbf{j} + 2\mathbf{k})$ **19.** No   **21.** 18   **23.** 0   **25.** 1   **27.** 40   **29.** −8   **31.** 0
33. $(x - y)(y - z)(z - x)$ **35.** −7   **37.** 0   **39.** $\mathbf{a} \times \mathbf{b}$ is orthogonal to $\mathbf{a}$.   **41.** Yes   **43.** $\overrightarrow{AB} \cdot (\overrightarrow{AC} \times \overrightarrow{AD}) = 0$ **45.** 7
47. 3 (see Prob. 19, p. 441)   **49.** $\mathbf{a} \times (\mathbf{b} \times \mathbf{c}) = -18\mathbf{j} + 6\mathbf{k}$, $(\mathbf{a} \times \mathbf{b}) \times \mathbf{c} = 4\mathbf{i} - 13\mathbf{j} + 9\mathbf{k}$
51. $\mathbf{a} \times (\mathbf{b} \times \mathbf{c}) = -\mathbf{i} + \mathbf{j}$, $(\mathbf{a} \times \mathbf{b}) \times \mathbf{c} = -\mathbf{i} + \mathbf{k}$

Section 12.4 (page 721)

1. $x = 1 + 2t$, $y = -2 + 3t$, $z = 4 - t$ **3.** $x = 1 - 2t$, $y = -2 + 5t$, $z = 4 - 3t$ **5.** $\dfrac{x + 3}{9} = \dfrac{y - 2}{-4} = \dfrac{z}{3}$

7. $\dfrac{x + 3}{6} = \dfrac{z}{-5}$, $y = 2$ **9.** $x = 1 + 2t$, $y = -2 + 3t$, $z = 2 - 3t$, $\dfrac{x - 1}{2} = \dfrac{y + 2}{3} = \dfrac{z - 2}{-3}$

11. $x = 4 - 5t$, $y = 1 + 4t$, $z = 4 - t$, $\dfrac{x - 4}{-5} = \dfrac{y - 1}{4} = \dfrac{z - 4}{-1}$ **13.** $2\sqrt{2}$ **15.** $\frac{1}{3}\sqrt{5}$ **17.** $4x + 5y - 3z = 0$

19. $3x - 6y + 4z - 35 = 0$ **21.** $9x - y + 7z - 40 = 0$ **23.** $3y - 2z - 1 = 0$ **25.** $(7, -4, 9)$ **27.** $\dfrac{x-4}{2} = \dfrac{y+1}{7} = \dfrac{z}{4}$

29. $\dfrac{x-4}{1} = \dfrac{y}{2} = \dfrac{z+2}{1}$ **31.** $\arccos \tfrac{8}{21} \approx 67.6°$ **33.** $\pi - \arccos(-1/5\sqrt{2}) \approx 81.9°$ **35.** 4   **37.** $\tfrac{1}{13}$ **39.** $\tfrac{3}{2}$

41. 6; (2, 2, 2)   **43.** Find parallel planes containing L_1 and L_2.   **45.** $(3, 7, -6)$

Section 12.5 (page 735)

1. 52π **3.** 150   **5.** $\tfrac{3}{2}$ **7.** About 2 meters   **9.** $\tfrac{1}{2}\mathbf{i} + \mathbf{j} + \mathbf{k}$ **11.** $\dfrac{\mathbf{i}}{\sqrt{1-t^2}} - 2te^{t^2}\mathbf{j} + (\sinh t)\mathbf{k}$

13. $\dfrac{\pi}{4}\mathbf{i} + (\ln 2)\mathbf{j} + \left(1 - \dfrac{\pi}{4}\right)\mathbf{k}$ **15.** Use formulas (5) and (5′).

17. $\mathbf{v} = 2\mathbf{i} + \mathbf{j} - 2\mathbf{k}$, $v = 3$, $\mathbf{T} = \tfrac{2}{3}\mathbf{i} + \tfrac{1}{3}\mathbf{j} - \tfrac{2}{3}\mathbf{k}$, $\mathbf{N}$ not defined
19. $\mathbf{v} = 15\mathbf{i} - (8 \sin t)\mathbf{j} + (8 \cos t)\mathbf{k}$, $v = 17$, $\mathbf{T} = \tfrac{15}{17}\mathbf{i} - (\tfrac{8}{17} \sin t)\mathbf{j} + (\tfrac{8}{17} \cos t)\mathbf{k}$, $\mathbf{N} = -[(\cos t)\mathbf{j} + (\sin t)\mathbf{k}]$

21. $a_T = 2\sqrt{2}\, t$, $a_N = 2$, $\kappa = \dfrac{1}{(1+t^2)^2}$ **23.** $a_T = 0$, $a_N = a\omega^2$, $\kappa = \dfrac{a\omega^2}{a^2\omega^2 + b^2}$ **25.** $3x + 3z + 1 = 0$

27. $x - y + 4z - 20 = 0$ **29.** $(-1, 1, 1)$, $(-\tfrac{1}{3}, \tfrac{1}{9}, -\tfrac{1}{27})$ **31.** $\mathbf{r} = -\tfrac{1}{2}gt^2\mathbf{j} + \mathbf{v}_0 t$, $\mathbf{r} \cdot \mathbf{k} = 0$
33. Use $\mathbf{r} \times \mathbf{v} = \mathbf{h}$. At perihelion. At aphelion.   **35.** ≈ 0.1415; ≈ 1838 mi, ≈ 3.76 mi/sec; ≈ 7360 sec
37. ≈ 16 sec   **39.** ≈ 0.57 mi/sec; ≈ 0.32 mi/sec

Section 12.6 (page 743)

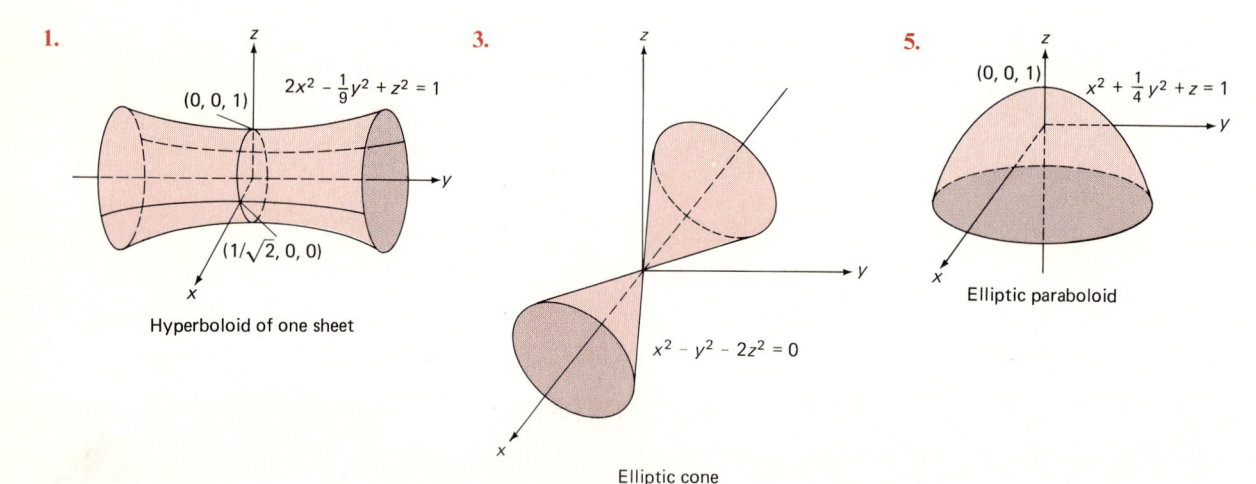

1.

Hyperboloid of one sheet
$2x^2 - \tfrac{1}{9}y^2 + z^2 = 1$
$(0, 0, 1)$
$(1/\sqrt{2}, 0, 0)$

3.

Elliptic cone
$x^2 - y^2 - 2z^2 = 0$

5.

Elliptic paraboloid
$x^2 + \tfrac{1}{4}y^2 + z = 1$
$(0, 0, 1)$

7. The empty set

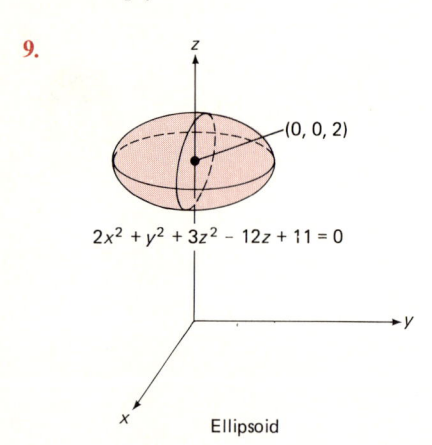

9.

Ellipsoid
$2x^2 + y^2 + 3z^2 - 12z + 11 = 0$
$(0, 0, 2)$

11. The plane $x - z + 1 = 0$

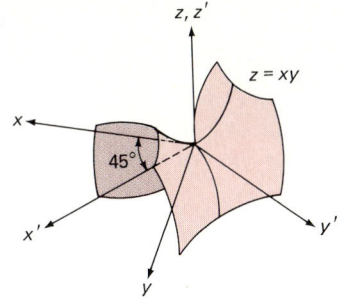

13. Hyperbolic paraboloid, $z = xy$, $45°$

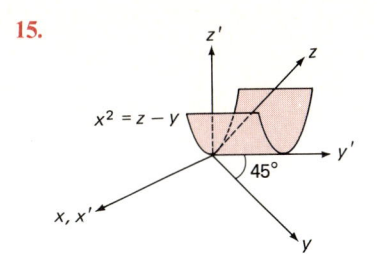

15. Parabolic cylinder, $x^2 = z - y$, $45°$

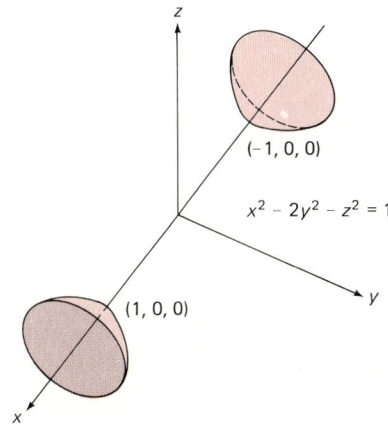

17. Hyperboloid of two sheets, $x^2 - 2y^2 - z^2 = 1$, $(-1, 0, 0)$, $(1, 0, 0)$

19. $96\pi/25$ **21.** $(4, 3, -2), (-2, 6, 2)$ **23.** The elliptic paraboloid $x^2 + y^2 = 4cz$ **25.** $\frac{4}{3}\pi abc$

Supplementary Problems (page 744)

1. $(5, 0, 0), (-11, 0, 0)$ **3.** First complete the squares. **5.** $y^2 + z^2 = c^2$ **7.** $x^2 + y^2 = 2cx$
9. $(5, 0, -4), (1, -2, 8), (-3, 4, -4)$ **11.** $\mathbf{i} - \mathbf{j} \pm \sqrt{2}\mathbf{k}$ **13.** Calculate $\mathbf{a} \times (\mathbf{a} + \mathbf{b} + \mathbf{c})$ and $\mathbf{b} \times (\mathbf{a} + \mathbf{b} + \mathbf{c})$. **15.** $2\sqrt{2}$
17. -66 **19.** $5\mathbf{i} - 10\mathbf{j} - 35\mathbf{k}$ **21.** 364 **23.** $\mathbf{a} \times \mathbf{b}$ and $\mathbf{c} \times \mathbf{d}$ are parallel. **25.** Use Prob. 48, p. 715.
27. Choose $\mathbf{b} = \mathbf{e}_1, \mathbf{c} = \mathbf{e}_2, \mathbf{d} = \mathbf{e}_3$ in Prob. 26. **29.** $\mathbf{a} = 2\mathbf{e}_1 + 3\mathbf{e}_2 - \mathbf{e}_3$ **31.** $\dfrac{x-5}{-2} = \dfrac{y+7}{3} = \dfrac{z-6}{-1}$ **33.** $\sqrt{2}$
35. $\dfrac{x+4}{6} = \dfrac{y-2}{-3} = \dfrac{z-5}{8}$ **37.** $10x - 2y - 5z + 10 = 0$ **39.** $2y + z - 8 = 0$ **41.** $a = 6, b = -4, c = 3$
43. $a = 7, b = 5$, no z-intercept **45.** 3 **47.** $x + 2y + 3z + 8 = 0$ **49.** 13 **51.** $3/\sqrt{2}$ **53.** $\mathbf{i} - e\mathbf{j} + 2\mathbf{k}$
55. $\dfrac{1}{t^2}[e^t(t-1)\mathbf{i} + (1 - \ln t)\mathbf{j} - (t \sin t + \cos t)\mathbf{k}]$ **57.** $\frac{1}{2}(\ln 3)\mathbf{i} - \frac{1}{2}(\ln 3)\mathbf{j} + 4\pi\mathbf{k}$
59. $\mathbf{T} = -\frac{6}{7}\mathbf{i} + \frac{2}{7}\mathbf{j} - \frac{3}{7}\mathbf{k}$, $\mathbf{N}$ is not defined, $\kappa = 0$ **61.** $\mathbf{T} = (\mathbf{j} + \mathbf{k})/\sqrt{2}$, $\mathbf{N} = \mathbf{i}$, $\kappa = \frac{1}{2}$
63. Differentiate $\mathbf{B} \cdot \mathbf{B} = 1$; $\dfrac{d\mathbf{N}}{ds} = \dfrac{d}{ds}(\mathbf{B} \times \mathbf{T}) = \left(\dfrac{d\mathbf{B}}{ds} \times \mathbf{T}\right) + \left(\mathbf{B} \times \dfrac{d\mathbf{T}}{ds}\right)$ **65.** $\dfrac{b\omega}{a^2\omega^2 + b^2}$
67. ≈ 1.017 for $t \approx \pm 0.271$ **69.** $\tau = \frac{1}{3}$, $\mathbf{T} = (\mathbf{i} + \mathbf{j} + \mathbf{k})/\sqrt{3}$, $\mathbf{N} = (-\mathbf{i} + \mathbf{k})/\sqrt{2}$, $\mathbf{B} = (\mathbf{i} - 2\mathbf{j} + \mathbf{k})/\sqrt{6}$
71. $\frac{8}{25}$ **73.** See Prob. 31, p. 736. The center of the earth **75.** ≈ 1129 mi **77.** $49x^2 + 49y^2 - 25z^2 = 0$
79. $16x^2 + 16y^2 - 9z^2 + 90z - 225 = 0$

Chapter 13

Section 13.1 (page 752)

1. $\frac{1}{2}$ **3.** $\frac{13}{14}$ **5.** -1 **7.** $\sqrt{3}$ **9.** 0 **11.** $2\sqrt{2}$ **13.** 1 **15.** 6 **17.** 2 **19.** π^2 **21.** n **23.** n^n
25. $t + \ln t$ $(t > 0)$ **27.** $\sin(2x/y)$ **29.** $n^2 \ln s$ **31.** All points inside and on the circle $x^2 + y^2 = 9$
33. The whole plane **35.** All points not on the coordinate axes **37.** All points to the right of the line $x + y = 0$
39. All points between and on the lines $x = \pm 1$ **41.** The first octant (see p. 696) **43.** All points not in the xy-plane
45. The plane $2x + 3y - z = 0$ through the origin **47.** The bottom half of the sphere $x^2 + y^2 + z^2 = 16$
49. The upper sheet of the double-sheeted hyperboloid of revolution $x^2 + y^2 - z^2 = -1$

51.

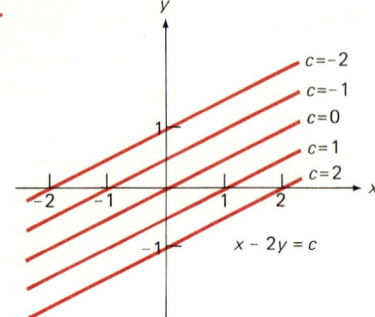

53.

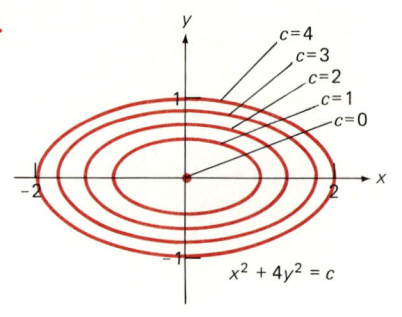

55.

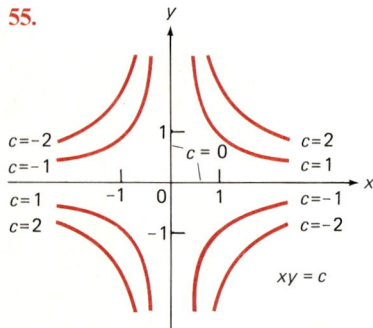

57. Parallel planes **59.** Concentric ellipsoids

Section 13.2 (page 758)

1. -2 **3.** 0 **5.** π^2 **7.** Does not exist **9.** $\pi/2$ **11.** 0 **13.** Let $x = y^2$ **15.** Yes **17.** No **19.** No

Section 13.3 (page 763)

1. $f_x = 2xy^3 - 3x^2y^4$, $f_y = 3x^2y^2 - 4x^3y^3$ **3.** $f_x = \dfrac{y^3 - x^2y}{(x^2 + y^2)^2}$, $f_y = \dfrac{x^3 - xy^2}{(x^2 + y^2)^2}$ **5.** $f_x = \dfrac{3x^2}{x^3 - y^2}$, $f_y = -\dfrac{2y}{x^3 - y^2}$

7. $f_x = ye^{\sin xy} \cos xy$, $f_y = xe^{\sin xy} \cos xy$ **9.** $f_s = -\dfrac{t}{s^2 + t^2}$, $f_t = \dfrac{s}{s^2 + t^2}$ **11.** $h_x = \dfrac{1}{x + \ln y}$, $h_y = \dfrac{1}{y(x + \ln y)}$

13. $f_x = yx^{y-1}$, $f_y = x^y \ln x$ **15.** $f_x = y + z$, $f_y = x + z$, $f_z = x + y$ **17.** $f_x = 1/x$, $f_y = 1/y$, $f_z = 1/z$
19. $f_x = yzx^{yz-1}$, $f_y = zx^{yz} \ln x$, $f_z = yx^{yz} \ln x$ **21.** $f_{xy} = f_{yx} = 8xy^3 - 18x^2y^2 + 18x^5$ **23.** $f_{xy} = f_{yx} = xy^{x-1} \ln y + y^{x-1}$
25. $f_{yx}(0, 0) = 1$, but $f_{xy}(0, 0) = -1$; f_{xy} is not continuous at $(0, 0)$.
27. $z_{xx} = 12x^2y^2 - 6y^3$, $z_{xy} = 8x^3y - 18xy^2 + 6$, $z_{yy} = 2x^4 - 18x^3y$
29. $z_{xx} = (2y + 4x^2y^2)e^{x^2y}$, $z_{xy} = (2x + 2x^3y)e^{x^2y}$, $z_{yy} = x^4e^{x^2y}$
31. $z_{xx} = 18\cos(6x - 8y)$, $z_{xy} = -24\cos(6x - 8y)$, $z_{yy} = 32\cos(6x - 8y)$ **33.** $y\,2^{xy}(\ln 2)^2(2 + xy \ln 2)$ **35.** $m!n!$ **37.** 0

39. $n + 1$ **41.** $\dfrac{\partial A}{\partial \theta} = \dfrac{1}{2}xy \cos \theta$ **43.** $\dfrac{\partial z}{\partial y} = \dfrac{y - x\cos \theta}{\sqrt{x^2 + y^2 - 2xy \cos \theta}}$ **45.** $x = 1$, $y = 1 + \sqrt{3}t$, $z = \sqrt{3} + t$

47. $u_x = v_y = 4x^3 - 12xy^2$, $u_y = -v_x = -12x^2y + 4y^3$ **49.** $u_x = v_y = -\sin x \cosh y$, $u_y = -v_x = \cos x \sinh y$

A-60 Answers and Hints to Odd-Numbered Problems

51. $u_{xx} = \dfrac{2x^2 - y^2 - z^2}{(x^2 + y^2 + z^2)^{5/2}}$, $u_{yy} = \dfrac{2y^2 - x^2 - z^2}{(x^2 + y^2 + z^2)^{5/2}}$, $u_{zz} = \dfrac{2z^2 - x^2 - y^2}{(x^2 + y^2 + z^2)^{5/2}}$ 53. $u_{xx} = 2$, $u_{tt} = 2c^2$
55. $u_{xx} = e^x \cosh ct$, $u_{tt} = c^2 e^x \cosh ct$

Section 13.4 (page 770)

1. $\Delta f = \ln 2$ 3. $df = (2xy - y^3 + 3x^2 y^2)\,dx + (x^2 - 3xy^2 + 2x^3 y)\,dy$ 5. $df = \dfrac{y}{y^2 - x^2}\,dx - \dfrac{x}{y^2 - x^2}\,dy$

7. $dg = ts^{t-1} e^{st}\,ds + s^t e^{st} \ln s\,dt$ 9. $df = 2^x 3^y 4^z (\ln 2\,dx + \ln 3\,dy + \ln 4\,dz)$ 11. $df = \dfrac{(y+z)\,dx + (z-x)\,dy - (x+y)\,dz}{(y+z)^2}$

13. $df = x_2 x_3 \cdots x_n\,dx_1 + x_1 x_3 \cdots x_n\,dx_2 + \cdots + x_1 x_2 \cdots x_{n-1}\,dx_n$ 15. -0.035 17. 108.972 19. 0.98 21. 0.0467
23. 0.2184 25. $16.8\pi\text{ cm}^3$ 27. $\Delta f = A\,\Delta x + B\,\Delta y + \alpha\,\Delta x + \beta\,\Delta y$, where $A = 2a$, $B = 2b$, $\alpha = \Delta x$, $\beta = \Delta y$
29. $\Delta f = A\,\Delta x + B\,\Delta y + \alpha\,\Delta x + \beta\,\Delta y$, where $A = 2ab - b^2$, $B = a^2 - 2ab$, $\alpha = b\,\Delta x + 2a\,\Delta y + \Delta x\,\Delta y$,
$\beta = -(a\,\Delta y + 2b\,\Delta x + \Delta x\,\Delta y)$ and there are other choices of α and β.
31. The partial derivatives do not exist at the origin.
33. $\Delta f = A\,\Delta x + B\,\Delta y + C\,\Delta z + \alpha\,\Delta x + \beta\,\Delta y + \gamma\,\Delta z$, where α, β and γ are functions of Δx, Δy and Δz that approach 0 as $(\Delta x, \Delta y, \Delta z) \to (0, 0, 0)$.

Section 13.5 (page 777)

1. $6t^5 - 7t^6 + 8t^7$ 3. $\dfrac{\sec^2 t}{\ln t} - \dfrac{\tan t}{t(\ln t)^2}$ 5. $\dfrac{-\sin t + \cos t}{2\sqrt{\cos t + \sin t}}$ 7. $(\cos t + 2te^{t^2} - t^{-1}) \cos(\sin t + e^{t^2} - \ln t)$

9. $2 \sinh t \ln t + 2t^{-1} \cosh t$ 11. $\partial w/\partial t = -2t$, $\partial w/\partial u = 0$ 13. $\partial w/\partial t = 2tu^2 e^{t^2 u^2}$, $\partial w/\partial u = 2t^2 u e^{t^2 u^2}$
15. $\partial w/\partial t = 2u/(t^2 + u^2)$, $\partial w/\partial u = -2t/(t^2 + u^2)$ 17. $\partial w/\partial t = 4t^3$, $\partial w/\partial u = -4u^3$
19. $\partial u/\partial x = 4x^3 \cos(x^4 - y^2 z^2)$, $\partial u/\partial y = -2yz^2 \cos(x^4 - y^2 z^2)$, $\partial u/\partial z = -2y^2 z \cos(x^4 - y^2 z^2)$
21. $\partial u/\partial x = 6x - 2y + 2z$, $\partial u/\partial y = 6y - 2x + 2z$, $\partial u/\partial z = 6z + 2x + 2y$ 23. $\dfrac{\partial u}{\partial x} = \dfrac{\partial u}{\partial r} - \dfrac{\partial u}{\partial t}, \dfrac{\partial u}{\partial y} = -\dfrac{\partial u}{\partial r} + \dfrac{\partial u}{\partial s}, \dfrac{\partial u}{\partial z} = -\dfrac{\partial u}{\partial s} + \dfrac{\partial u}{\partial t}$

25. Solve $1 = \dfrac{\partial x}{\partial x} = \dfrac{\partial x}{\partial r}\dfrac{\partial r}{\partial x} + \dfrac{\partial x}{\partial \theta}\dfrac{\partial \theta}{\partial x}$ and $0 = \dfrac{\partial y}{\partial x} = \dfrac{\partial y}{\partial r}\dfrac{\partial r}{\partial x} + \dfrac{\partial y}{\partial \theta}\dfrac{\partial \theta}{\partial x}$ for $\dfrac{\partial r}{\partial x}$ and $\dfrac{\partial \theta}{\partial x}$.

27. $\dfrac{\partial u}{\partial x} = \dfrac{\partial u}{\partial r} \cos \theta - \dfrac{\partial u}{\partial \theta} \dfrac{\sin \theta}{r}, \dfrac{\partial u}{\partial y} = \dfrac{\partial u}{\partial r} \sin \theta + \dfrac{\partial u}{\partial \theta} \dfrac{\cos \theta}{r}$ 29. Differentiate (i) with respect to t, and then set $t = 1$.

31. Differentiate (i) with respect to x, and also with respect to y.   33. Yes, of degree -2 35. No   37. 31 knots

39. ≈ 6.57 atm/min   41. $y' = \dfrac{8 - 4x^3}{2y + 5}$ 43. $y' = \dfrac{3x^2}{6y^5 - 1}$ 45. $y' = -\dfrac{2x + 2y \cos xy}{3y^2 + 2x \cos xy}$ 47. $y' = \dfrac{y}{\sec^2 y - x}$

49. $y' = \dfrac{y^2}{e^y - 2xy}$ 51. $z_x = \dfrac{3x^2 y - 2x - z}{x + 2z}$, $z_y = \dfrac{x^3}{x + 2z}$ 53. $z_x = \dfrac{yz}{e^z - xy}$, $z_y = \dfrac{xz}{e^z - xy}$ 55. $z_x = -\tfrac{1}{3}$, $z_y = \tfrac{1}{2}$

57. $dz = -\dfrac{\sin 2x\,dx + \sin 2y\,dy}{\sin 2z}$ 59. Apply the ordinary mean value theorem to $F(t) = f(a + t\,\Delta x, b + t\,\Delta y)$.

61. Let $F(u, v) = \displaystyle\int_u^v f(t)\,dt$, where $u = u(x)$, $v = v(x)$, and use the chain rule to calculate $\dfrac{dF}{dx}$.

63. Let $F(x, u, v) = \displaystyle\int_u^v f(x, y)\,dy$, where $u = u(x)$, $v = v(x)$, and use the chain rule to calculate $\dfrac{dF}{dx}$.

65. In the increment $\Delta f = (f_x + \alpha)\,\Delta x + (f_y + \beta)\,\Delta y$ substitute $\Delta x = (x_t + \gamma)\,\Delta t + (x_u + \delta)\,\Delta u$, $\Delta y = (y_t + \varepsilon)\,\Delta t + (y_u + \theta)\,\Delta u$, where γ, δ, ε and θ all approach 0 as $(\Delta t, \Delta u) \to (0, 0)$.

Section 13.6 (page 782)

1. Plane $x - y + 1 = 0$, line $\dfrac{x-1}{1} = \dfrac{y-2}{-1}$, $z = 6$ 3. Plane $8x + 6y + z + 5 = 0$, line $\dfrac{x+2}{8} = \dfrac{y-1}{6} = \dfrac{z-5}{1}$

5. Plane $4x - 5y + 3z - 6 = 0$, line $\dfrac{x-1}{4} = \dfrac{y+1}{-5} = \dfrac{z+1}{3}$ 7. Plane $\dfrac{x}{x_0} + \dfrac{y}{y_0} + \dfrac{z}{z_0} = 3$, line $\dfrac{x - x_0}{y_0 z_0} = \dfrac{y - y_0}{x_0 z_0} = \dfrac{z - z_0}{x_0 y_0}$

9. Plane $2x - ez = 0$, line $\dfrac{x - e}{2/e} = \dfrac{z - 2}{-1}$, $y = 0$ 11. Plane $x + y - 12z + 6 = 0$, line $\dfrac{x-3}{1} = \dfrac{y-3}{1} = \dfrac{z-1}{-12}$

13. $(0, 0, 0)$ 15. $x + y + z - 11 = 0$ at $(6, 3, 2)$, $x + y + z + 11 = 0$ at $(-6, -3, -2)$

17. The intercepts are $a^{2/3}x_0^{1/3}$, $a^{2/3}y_0^{1/3}$, $a^{2/3}z_0^{1/3}$ **19.** $x + y - 2 = 0$ **21.** $4x - 2y + 3z - 3 = 0$
23. Plane $-(x - x_0) + g_y(y_0, z_0)(y - y_0) + g_z(y_0, z_0)(z - z_0) = 0$, line $\dfrac{x - x_0}{-1} = \dfrac{y - y_0}{g_y(y_0, z_0)} = \dfrac{z - z_0}{g_z(y_0, z_0)}$
25. Use formulas (10) and (10') of Sec. 13.5. **27.** Tangent $2x + 3y - 8 = 0$, normal $3x - 2y + 1 = 0$
29. Tangent $x + y - 1 = 0$, normal $x - y + 1 = 0$

Section 13.7 (page 788)

1. $4\mathbf{i} - 3\mathbf{j}, \frac{63}{13}$ **3.** $(25\mathbf{i} - 7\mathbf{j})/24, 4/3\sqrt{2}$ **5.** $\mathbf{i} + \mathbf{j}, -\frac{1}{29}$ **7.** $2xye^{x^2y}\mathbf{i} + x^2 e^{x^2y}\mathbf{j}, \frac{15}{17}$ **9.** $-6\mathbf{i} - 3\mathbf{j} + 2\mathbf{k}, -\sqrt{\frac{5}{7}}$
11. $\frac{9}{11}\mathbf{i} - \frac{6}{11}\mathbf{j} + \frac{2}{11}\mathbf{k}, \frac{25}{33}$ **13.** $-2\mathbf{i} + 2\mathbf{j} - \mathbf{k}, 3$ **15.** $0, 0$ **17.** $3\sqrt{2}$ **19.** $10\sqrt{2}$ **21.** $\sqrt{3}$ **23.** $\mathbf{u} = \pm(\frac{3}{5}\mathbf{i} - \frac{4}{5}\mathbf{j})$

25.

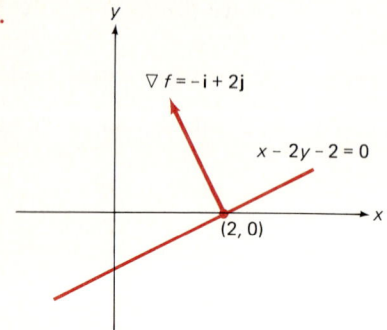

27.

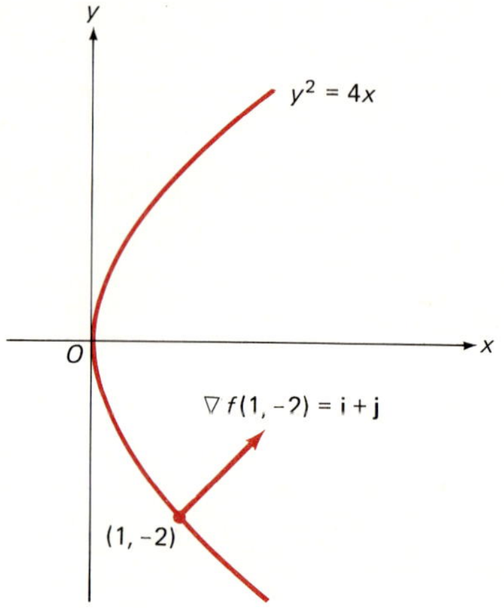

29. $\sqrt{1 + e^2}$ in the direction of $\mathbf{i} + e\mathbf{j}$ **31.** $\sqrt{53}$ in the direction of $-\mathbf{i} - 4\mathbf{j} + 6\mathbf{k}$ **33.** $pf^{p-1}\nabla f$
35. Ascend at the rate of 3 meters per meter **37.** N 53.13° E or S 53.13° W **39.** a **41.** $\mathbf{a} \times \mathbf{b}$ **43.** See Prob. 59, p. 779.

Section 13.8 (page 797)

1. Minimum 0 at (0, 1) **3.** Maximum 36 at (1, −4) **5.** Saddle point at (8, −10)
7. Maxima equal to 2 at (1, 1) and (−1, −1), saddle point at (0, 0) **9.** Maximum 27 at (6, 9)
11. Maximum 1 at (1, 1), saddle points at (0, 0), (2, 0), (0, 2) and (2, 2) **13.** Saddle point at $(0, n\pi)$, n any integer
15. Maximum 1 at $(\pi/2, \pi/2)$ **17.** Minimum -1 at $(-1, 0)$
19. Minimum 0 at (0, 0), maxima equal to e^{-1} at every point of the circle $x^2 + y^2 = 1$
21. Maximum 4 at $(\pm 2, 0)$, minimum -4 at $(0, \pm 2)$ **23.** Maximum 17 at (2, 1), minimum -3 at (0, 1)
25. 8000 **27.** 1000 cm^3 **29.** The square of side length $p/4$ **31.** $8abc/3\sqrt{3}$
33. Maximum $d/2\sqrt{e}$ at $(a/d, b/d, c/d)$, minimum $-d/2\sqrt{e}$ at $(-a/d, -b/d, -c/d)$, where $d = \sqrt{2(a^2 + b^2 + c^2)}$
35. Solve the simultaneous equations $\partial S/\partial m = 0$, $\partial S/\partial b = 0$. **37.** $y = -0.3x + 0.4$ **39.** Maximum 3 at (2, 1, 0) **41.** Yes

Section 13.9 (page 805)

1. Maximum 7 at $(1, -\frac{3}{2})$, minimum -5 at $(-1, \frac{3}{2})$ **3.** No extrema **5.** Maximum 8 at (0, 1), minimum $\frac{8}{27}$ at $(\frac{4}{9}, \frac{1}{9})$
7. Minimum 1 at $(\pm 1, 0)$ **9.** Maximum 13 at (2, −3, 1), minimum -15 at $(-2, 3, -1)$ **11.** Minimum $\frac{1}{8}$ at $(\frac{1}{3}, \frac{9}{8}, \frac{8}{3})$
13. Minimum 1 at $(\pm 1, 0, 0)$, saddle points at $(0, \pm\sqrt{2}, 0)$ **15.** 3 **17.** $(1, -1)$ **19.** $(2, 1)$ **21.** 1. No
23. $r = \sqrt{5}h/2$, $H = h/2$ **25.** Set $\nabla f = \lambda \nabla g$, where $f(x, y) = (x - 2)^2 + y^2$, $g(x, y) = \sqrt{x} - y = 0$, and also examine $f(0, 0)$.
27. Set $\nabla f = \lambda \nabla g$, where $f(x, y) = 2(x + y)$, $g(x, y) = xy = A$.
29. Set $\nabla f = \lambda \nabla g$, $f(x, y, z) = x + y + z$, $g(x, y, z) = xyz = 343$. **31.** Maximum 22 at (1, −2, 6), minimum 2 at (−1, 2, 2)

Supplementary Problems (page 806)

1. $f(x, y) = 2x - 3y + 1$ **3.** $A = \frac{1}{4}\sqrt{(x + y + z)(x + y - z)(x - y + z)(y + z - x)}$

5.

Boundary lines included

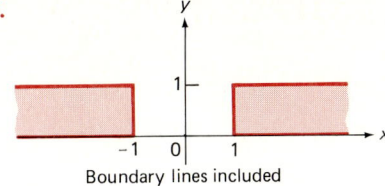

7.

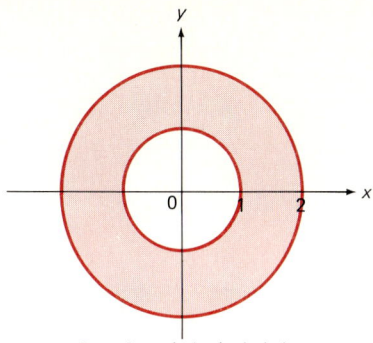

Boundary circles included

9. All points inside the unit sphere $x^2 + y^2 + z^2 < 1$ except the origin

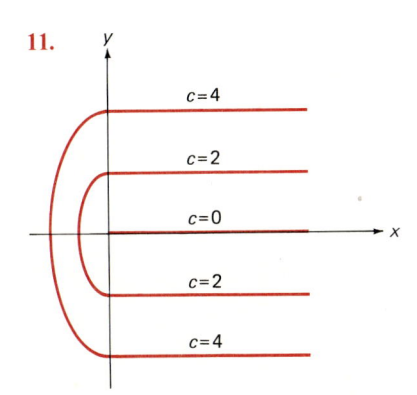

11.

13. 0 **15.** 0 **17.** R^n (n-space) **19.** $f_x = 2xe^{x^2+y^3}$, $f_y = 3y^2e^{x^2+y^3}$ **21.** $f_x = x^{xy}x^{y-1}(y \ln x + 1)$, $f_y = x^{xy}x^y(\ln x)^2$
23. $f_x = (2x - y - z)(y - z)$, $f_y = (x - 2y + z)(x - z)$, $f_z = (2z - x - y)(x - y)$
25. $g_x = y^z x^{y^z-1}$, $g_y = zy^{z-1}x^{y^z} \ln x$, $g_z = y^z x^{y^z} \ln x \ln y$ **27.** $f_{xyz} = 0$, $f_{xxy} = 2$, $f_{yyz} = -2$ **29.** $\arctan \frac{4}{7} \approx 29.7°$
31. Apply the mean value theorem twice to the right side of the inequality $|f(x, y) - f(a, b)| \le$
$|f(x, y) - f(a, y)| + |f(a, y) - f(a, b)|$. In Prob. 30, f_x is unbounded in every neighborhood of the origin.

33. $df = \dfrac{y\,dx - x\,dy}{x^2 + y^2}$ **35.** 0.50818 **37.** $\dfrac{\partial u}{\partial x} = 2xf'(x^2 + y^2)$, $\dfrac{\partial u}{\partial y} = 2yf'(x^2 + y^2)$

39. $\sqrt{n}$ **41.** $\dfrac{\partial w}{\partial x} = \dfrac{8x^7}{x^8 - y^8}$, $\dfrac{\partial w}{\partial y} = -\dfrac{8y^7}{x^8 - y^8}$ **43.** $\dfrac{\partial}{\partial x}(x^2 u_x) = x[f''(x + y) + g''(x - y)]$

45. Differentiate $f(tx, ty) = t^n f(x, y)$ twice with respect to t, and then set $t = 1$. **47.** Yes

49. $\dfrac{\partial u}{\partial r} = \dfrac{1}{r}\dfrac{\partial v}{\partial \theta}$, $\dfrac{\partial u}{\partial \theta} = -r\dfrac{\partial v}{\partial r}$ **51.** $\pm x \pm y \pm z = 13$ **53.** $u_x = \dfrac{3yzu - 2xy^2}{2z^2u - 3xyz}$, $u_y = \dfrac{3xzu - 2x^2y}{2z^2u - 3xyz}$, $u_z = \dfrac{3xyu - 2zu^2}{2z^2u - 3xyz}$

55. $u_x = \dfrac{yue^{xyu}}{z - xye^{xyu}}$, $u_y = \dfrac{xue^{xyu}}{z - xye^{xyu}}$, $u_z = \dfrac{-u}{z - xye^{xyu}}$ **57.** $\frac{23}{17}$ **59.** $\arccos\dfrac{40}{\sqrt{17}\sqrt{97}} \approx 9.9°$

61. 6 in the direction of $2\mathbf{i} - 2\mathbf{j} + \mathbf{k}$
63. Minimum 0 at (0, 0). All other points of the coordinate axes are critical points, but none corresponds to an extremum or a saddle point.
65. Maximum $\frac{3}{2}$ at $(\frac{1}{6}\pi, \frac{1}{6}\pi)$ **67.** Saddle point at (0, 0) **69.** a^3 **71.** The equilateral triangle of side length $\frac{2}{3}c$
73. If F has an unconstrained local extremum at a point P, all the first partials of F at P are zero. In particular, $\partial F/\partial \lambda = 0$ is just the equation of constraint.
75. $q_1 = 30$, $q_2 = 90$, $P = \$46{,}800$ **77.** $\frac{112}{27} \approx 4.15$ sq. ft

Answers and Hints to Odd-Numbered Problems   **A-63**

Chapter 14

Section 14.1 (page 822)

1. $\dfrac{14}{3}$ **3.** $\dfrac{\pi}{12}$ **5.** 24 **7.** 12 **9.** $-\dfrac{11}{60}$ **11.** $\ln\dfrac{2+\sqrt{2}}{1+\sqrt{3}}$

13. $\displaystyle\int_1^2 dy \int_1^y f(x,y)\,dx + \int_2^4 dy \int_{y/2}^2 f(x,y)\,dx$ **15.** $\displaystyle\int_{-2}^2 dx \int_0^{\sqrt{4-x^2}} f(x,y)\,dy$

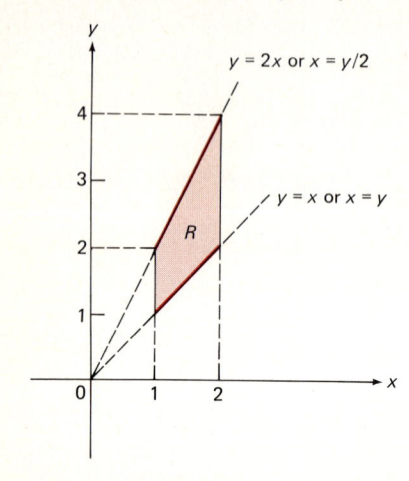

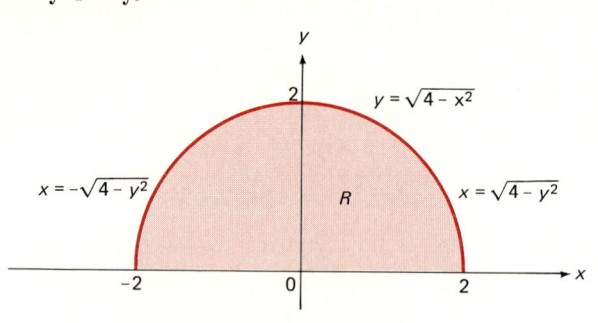

17. $\displaystyle\int_0^3 dy \int_y^{6-y} f(x,y)\,dx$ **19.** $\displaystyle\int_0^1 dx \int_{2-2x}^2 f(x,y)\,dy + \int_1^2 dx \int_{2x-2}^2 f(x,y)\,dy$ or $\displaystyle\int_0^2 dy \int_{1-(y/2)}^{1+(y/2)} f(x,y)\,dx$

21. $\displaystyle\int_0^1 dx \int_{-x}^x f(x,y)\,dy + \int_1^2 dx \int_{-x}^{2-x} f(x,y)\,dy + \int_2^3 dx \int_{x-4}^{2-x} f(x,y)\,dy$

or $\displaystyle\int_{-2}^{-1} dy \int_{-y}^{y+4} f(x,y)\,dx + \int_{-1}^{0} dy \int_{-y}^{2-y} f(x,y)\,dx + \int_0^1 dy \int_y^{2-y} f(x,y)\,dy$

23. $\dfrac{1}{4}(b^2-a^2)(d^2-c^2)$ **25.** $\dfrac{1}{2}\ln 2$ **27.** $\dfrac{1}{2\sqrt{3}}+\dfrac{\pi}{9}$ **29.** 2 **31.** $\dfrac{1}{280}$ **33.** $\dfrac{3}{2}$ **35.** $\dfrac{64}{3}$

37. $\sqrt{2}\arcsin\dfrac{2\sqrt{2}}{3}$ (use Prob. 57, formula (ii), p. 297) **39.** 7 **41.** $\dfrac{23}{3}$ **43.** 4 **45.** $\dfrac{16\pi}{3}$ **47.** $\dfrac{1}{12}$ **49.** Choose $f(x,y) \equiv 1$.

51. Let $F(x) = \displaystyle\int_c^d f(x,y)\,dy$, $G(x) = \displaystyle\int_c^d \dfrac{\partial f(x,y)}{\partial x}\,dy$, and reverse the order of integration in the right side of

$\displaystyle\int_a^u G(x)\,dx = \int_a^u dx \int_c^d \dfrac{\partial f(x,y)}{\partial x}\,dy$, thereby showing that F is an antiderivative of G.

Section 14.2 (page 831)

1. 12 **3.** $\dfrac{13}{84}$ **5.** $\dfrac{1}{48}$ **7.** $\dfrac{8}{15}(31+12\sqrt{2}-27\sqrt{3})$ **9.** $2\ln 2 - \dfrac{7}{4}$ **11.** 34 **13.** $\dfrac{20}{3}$ **15.** $\dfrac{59\pi}{15}$ **17.** $6-\dfrac{\pi^2}{2}$ **19.** 8

21. $\dfrac{\pi}{12}$ **23.** $\dfrac{40}{3}$ **25.** $\dfrac{64}{\pi^3}$ **27.** $\dfrac{1}{8}$ **29.** $\dfrac{1}{12}$ **31.** $Q = \displaystyle\iiint_T \rho(x,y,z)\,dV;\ Q=0$

Section 14.3 (page 841)

1. $M_{xy}=32,\ M_{xz}=56,\ M_{yz}=28,\ (\bar{x},\bar{y},\bar{z})=(2,4,\tfrac{16}{7})$ **3.** The plank moves 8 ft in the direction opposite to the child's motion.
5. Use rule (iv), p. 814. **7.** Write $\iint_R x\,dA$ and $\iint_R y\,dA$ as iterated integrals, and do the inner integrations. **9.** $(\tfrac{9}{20},\tfrac{9}{20})$
11. $(4a/3\pi,\ 4b/3\pi)$ **13.** $(-\tfrac{1}{3},0)$ **15.** $(\tfrac{1}{5},\tfrac{1}{5})$ **17.** $(-1,\tfrac{1}{4})$ **19.** 9.6 in. above the center of the bottom of the box
21. $(\tfrac{1}{2},\tfrac{8}{9},\tfrac{14}{9})$ **23.** $(\tfrac{2}{3},\tfrac{5}{4},\tfrac{11}{8})$ **25.** $(\tfrac{4}{7},0,\tfrac{6}{7})$ **27.** $(0,0,1)$ **29.** $(\tfrac{3}{8}a,\tfrac{3}{8}b,\tfrac{3}{8}c)$ **31.** $(0,0,\tfrac{59}{50})$

Section 14.4 (page 849)

1. $\left(\dfrac{\sin\theta}{\theta}, \dfrac{1-\cos\theta}{2}\right)$ 3. $\dfrac{1}{2}+\dfrac{2\pi}{3\sqrt{3}}$ 5. $M=\dfrac{\pi}{2}$, but $\bar{x}$ does not exist. 7. $M=\dfrac{\sqrt{\pi}}{2}$ (see Example 9, p. 430), $\bar{x}=\dfrac{1}{\sqrt{\pi}}$

9. $\left(\dfrac{2a}{5},\dfrac{2a}{5}\right)$ 11. $\dfrac{3\sqrt{3}-1}{8}+\dfrac{3}{16}\ln\dfrac{3+2\sqrt{3}}{3}$ 13. $\dfrac{3\pi}{10}$ 15. $\dfrac{\pi}{15}$ 17. 60π 19. $\dfrac{15+9\sqrt{2}}{35}$ 21. $\left(\dfrac{a}{3},\dfrac{b}{3}\right)$

Section 14.5 (page 855)

1. $I_x=126$, $I_y=112$, $I_z=200$, $k_x=3$, $k_y=2\sqrt{2}$, $k_z=10/\sqrt{7}$ 3. $I_x=3$, $I_y=\dfrac{25}{6}$ 5. $I_x=\dfrac{328}{45}$, $I_y=\dfrac{58}{45}$ 7. $\dfrac{1}{3}ab(a^2+b^2)$

9. $I_x=I_y=\dfrac{3}{35}$ 11. $I_x=\dfrac{1}{16}\pi ab^3$, $I_y=\dfrac{1}{16}\pi a^3 b$ 13. The distance between l and (x,y) is $\dfrac{|x\tan\phi-y|}{\sqrt{1+\tan^2\phi}}$.

15. Choose the origin at the center of mass. 17. $I_x=\dfrac{1}{3}M(b^2+c^2)$, $I_y=\dfrac{1}{3}M(a^2+c^2)$, $I_z=\dfrac{1}{3}M(a^2+b^2)$

19. $I_x=\dfrac{1}{2}M(b^2+\dfrac{2}{3}c^2)$, $I_y=\dfrac{1}{2}M(a^2+\dfrac{2}{3}c^2)$, $I_z=\dfrac{1}{2}M(a^2+b^2)$ 21. $I_z=\dfrac{14}{15}\pi$ 23. $I_z=\dfrac{1}{10}\pi a^4 h$

Section 14.6 (page 861)

1. 18π 3. $\dfrac{\pi}{2}\ln 2$ 5. 2π 7. 4π 9. $\pi(e^2+1)$ 11. $\displaystyle\int_{\pi/4}^{3\pi/4}d\theta\int_0^{\csc\theta}f(r\cos\theta,r\sin\theta)\,r\,dr$

13. $\displaystyle\int_0^{\pi/4}d\theta\int_0^{\sec\theta}f(r)\,r\,dr+\int_{\pi/4}^{\pi/2}d\theta\int_0^{\csc\theta}f(r)\,r\,dr$ 15. $\left(2+\dfrac{\pi}{4}\right)a^2$ 17. $\dfrac{3}{2}\cos\dfrac{1}{2}-\dfrac{1}{2}\sin\dfrac{1}{2}-1$ 19. $\phi-\dfrac{1}{2}\sin 2\phi$

21. $\left(\dfrac{2a\sin\phi}{3\phi},0\right)$ 23. $\left(\dfrac{5a}{6},0\right)$ 25. $\dfrac{\pi a^4}{8}$ 27. $\dfrac{4\pi}{3}[R^3-(R^2-a^2)^{3/2}]$ 29. $\dfrac{10\pi}{3}$ 31. $\dfrac{\pi}{12}$

Section 14.7 (page 867)

1. $x=-\dfrac{3}{2}$, $y=\dfrac{\sqrt{3}}{2}$, $z=0$ 3. $x=-10$, $y=0$, $z=2\pi$ 5. $r=4\sqrt{2}$, $\theta=\dfrac{3\pi}{4}$, $z=-2$ 7. $r=2$, $\theta=\dfrac{5\pi}{3}$, $z=-6$

9. $r^2\cos 2\theta=z^2$ 11. $r^2-z^2=1$ 13. 3π 15. $\dfrac{126\pi}{5}$ 17. $\dfrac{2\pi}{3}$ 19. $\pi^2 a^3$ 21. $\dfrac{128\sqrt{2}}{15}$ 23. $\left(\dfrac{12a}{5(3\pi-4)},0,0\right)$

25. $M(\dfrac{1}{4}a^2+\dfrac{1}{3}h^2)$; $M(\dfrac{1}{4}a^2+\dfrac{1}{12}h^2)$

Section 14.8 (page 873)

1. $x=-1$, $y=0$, $z=0$ 3. $x=-\sqrt{3}$, $y=1$, $z=2\sqrt{3}$ 5. $\rho=\sqrt{3}$, $\theta=\dfrac{3\pi}{4}$, $\phi=\arccos\dfrac{1}{\sqrt{3}}\approx 54.7°$

7. $\rho=13$, $\theta=\arctan(-3)+2\pi\approx 288.4°$, $\phi=\arccos\dfrac{3}{13}\approx 76.7°$ 9. $\rho=2\csc\phi$ $(0<\phi<\pi)$ 11. $\phi=\dfrac{\pi}{4},\dfrac{3\pi}{4}$

13. πa^4 15. $\dfrac{31}{15}$ 17. $\dfrac{2\pi}{3}$ 19. $\dfrac{\pi}{15}$ 21. $\dfrac{92\pi}{75}$ 23. $\dfrac{8\pi}{3}$ 25. 2.64π kg 27. $(0,0,\dfrac{9}{8})$

Supplementary Problems (page 874)

1. $\dfrac{9}{4}$ 3. $\dfrac{1}{4}(e^2-1)$ 5. $\dfrac{1}{2}$ 7. $1-\dfrac{1}{4}\pi$ 9. πabh 11. $\dfrac{16}{3}+2\pi$ 13. Use formula (i). 15. $\dfrac{1}{2}\ln 2$ 17. 72 g 19. 123

21. $(-\dfrac{2}{5},0)$ 23. $(\dfrac{3}{16}\pi,0)$ 25. $(2,\dfrac{2}{3},\dfrac{1}{3})$ 27. $(-\dfrac{8}{5},\dfrac{8}{5})$ 29. $V=\left(1-\dfrac{4}{3\pi}\right)\pi^2 r^3$, $A=2\left(1-\dfrac{4}{3\pi}\right)\pi^2 r^2$ 31. $\dfrac{8}{5}$ 33. $\dfrac{53\pi}{15}$

35. $\pi\left(\dfrac{\pi}{2}-1\right)$ 37. $\dfrac{\pi}{20}$ 39. $\dfrac{5\pi}{8}$ 41. $\dfrac{3\pi a^4}{2}$ 43. $\dfrac{\pi}{6}$ 45. $r=2$, $\theta=\dfrac{\pi}{3}$, $z=0$ 47. $r=\dfrac{3}{2}$, $\theta=\dfrac{\pi}{4}$, $z=\dfrac{\sqrt{3}}{2}$

49. $\rho=2\sqrt{2}$, $\theta=\dfrac{\pi}{6}$, $\phi=\dfrac{\pi}{4}$ 51. $\dfrac{\pi}{15}$ 53. $\dfrac{7}{5}Ma^2$

Chapter 15

Section 15.1 (page 885)

1. $\frac{35}{2}$ **3.** $\sqrt{5}\ln 2$ **5.** 26 **7.** 64π **9.** $\frac{1}{2}$ **11.** $4(\sqrt{2}+1)$ **13.** 2π **15.** $\frac{37}{5}$ **17.** 10,500 ft-lb **19.** -3 **21.** ab
23. $-\frac{2}{3}$ **25.** 0 **27.** $4\pi/\sqrt{3}$ **29.** $\frac{43}{2}$ **31.** $3(e-1)$ **33.** 1

Section 15.2 (page 895)

1. Yes. $U = e^x \sin y$, D the whole plane **3.** No **5.** Yes. $U = x^4 y^3 - 3xy^2$, D the whole plane **7.** No
9. Yes. $U = xy - \dfrac{\cos 2y}{2x} + \dfrac{1}{2x}$, D any rectangular domain containing no points of the y-axis **11.** -12 **13.** 60 **15.** $\ln 6$
17. π **19.** If $\mathbf{R} = (x, y, z)$ and $\mathbf{R}_1 = (x_1, y_1, z_1)$, then $\dfrac{\partial r}{\partial x} = \dfrac{x - x_1}{r}$, $\dfrac{\partial r}{\partial y} = \dfrac{y - y_1}{r}$, $\dfrac{\partial r}{\partial z} = \dfrac{z - z_1}{r}$.
21. Yes. $U = xy + xz + yz$, D all of space **23.** Yes. $U = xyz$, D all of space **25.** No
27. Yes. $U = x \ln y + yz - x \cos 2z$, D the half-space $y > 0$ **29.** See formula (9), p. 767. **31.** 60 mph. No **33.** $V = \frac{1}{2}ks^2$
35. Transform to spherical coordinates.
37. C_3 and C_2 form the boundary of a simple closed path, and so do C_3 and the path obtained by deleting the loops from C_1.

Section 15.3 (page 908)

1. 14 **3.** $\frac{4}{3}(a+b)\sqrt{2ab}$ **5.** $2a^2(\pi - 2)$ **7.** 57π **9.** $8a^2$ **11.** $\sqrt{2\pi}$ **13.** $\sqrt{3}/120$ **15.** $\pi/\sqrt{2}$ **17.** $2\pi/15$ **19.** 9
21. $42\sqrt{2}\pi$, $(0, 0, \frac{85}{28})$ **23.** $(\frac{3}{2}, \frac{3}{2}, \frac{3}{2})$ **25.** 54π **27.** -3 **29.** 9π
31. If S is the graph of $y = g(x, z)$, then $-\dfrac{\partial y}{\partial x}\mathbf{i} + \mathbf{j} - \dfrac{\partial y}{\partial z}\mathbf{k}$ is a normal to S, while if S is the graph of $x = h(y, z)$, then $\mathbf{i} - \dfrac{\partial x}{\partial y}\mathbf{j} - \dfrac{\partial x}{\partial z}\mathbf{k}$ is a normal to S. **33.** $\frac{1}{8}a^4$

Section 15.4 (page 920)

1. -2 **3.** $\frac{4}{3}$ **5.** 4π **7.** 0 **9.** $\pi - 2\ln 2$ **11.** $4\ln 2$ **13.** $\frac{3}{8}\pi a^2$ **15.** Use Green's theorem.
17. $P = x/(x^2 + y^2)$, $Q = y/(x^2 + y^2)$, D the whole plane with the origin deleted **19.** $ad - bc$ **21.** $1/(u^2 + v^2)$ **23.** e^{2u}
25. See Prob. 21. **27.** abc **29.** 0 **31.** $\frac{134}{125}$ **33.** $\frac{7}{8}$ **35.** $1 - \frac{1}{3}\sqrt{2}$ **37.** 4 **39.** $|\partial(x, y, z)/\partial(\rho, \theta, \phi)| = \rho^2 \sin \phi$

Section 15.5 (page 928)

1. 2 **3.** $(x - y)\sin 2xy$ **5.** 0 **7.** 4 **9.** $yz + xz + xy$ **11.** $3e^{x+y+z}$ **13.** a div $\mathbf{F} + b$ div $\mathbf{G}$
15. $\nabla^2 f = \dfrac{\partial}{\partial x}\left(\dfrac{x}{x^2 + y^2}\right) + \dfrac{\partial}{\partial y}\left(\dfrac{y}{x^2 + y^2}\right)$ **17.** Yes **19.** $V = \frac{1}{3}aA$ **21.** 6π **23.** Choose $\mathbf{F} = \nabla f$ in (4) or (6). **25.** $8\pi/3$
27. $4\pi/5$ **29.** Both sides equal $\frac{1}{2}a^3$. **31.** Both sides equal 8π.

Section 15.6 (page 937)

1. 0 **3.** $-2(\mathbf{i} + \mathbf{j} + \mathbf{k})$ **5.** 0 **7.** $xz\mathbf{i} - yz\mathbf{j} + y\mathbf{k}$ **9.** $(1 - e^z)(x\mathbf{i} - y\mathbf{j})$ **11.** 0 **13.** $\mathbf{G} \cdot$ curl $\mathbf{F} - \mathbf{F} \cdot$ curl $\mathbf{G}$
15. Both sides equal 4. **17.** Both sides equal 0. **19.** 1 **21.** Use Prob. 12. **23.** Use Stokes' theorem. **25.** -1
27. $(2x - 3)\mathbf{j}$

Supplementary Problems (page 938)

1. $\frac{38}{5}$ **3.** $\frac{16}{3}\sqrt{2}$ **5.** $2FR$; 0 **7.** a^3 **9.** $\frac{21}{20}$ **11.** No **13.** Yes. $U = e^{x+y} + \cos(x - y)$ **15.** $\frac{1}{2}\pi - \ln 2$ **17.** $-\frac{7}{12}\pi$
19. $\frac{13}{3}\pi$ **21.** $2\pi a^2$ **23.** $[\sqrt{2} + \ln(1 + \sqrt{2})]\pi$ **25.** 4.5 m³/sec **27.** 0 **29.** Use Green's theorem. **31.** 17 **33.** 0
35. $4\mathbf{a} \cdot \mathbf{r}$ **37.** $2\mathbf{a} \cdot \mathbf{r}$ **39.** $c = 5$ **41.** Partition T into two subregions, each bounded by a single closed subsurface.
43. $2\mathbf{a}$ **45.** $\mathbf{a} \times \mathbf{r}$ **47.** $3\mathbf{a} \times \mathbf{r}$ **49.** Use Prob. 12, p. 937. **51.** Use Prob. 50.

Chapter 16

Section 16.1 (page 945)

1. $\partial Q/\partial x = 1 = \partial P/\partial y$, $\frac{1}{2}(x^2 - y^2) + xy = C$ **3.** $\partial Q/\partial x = -2y/x^2 = \partial P/\partial y$, $4x + (y^2/x) = C$
5. $\partial Q/\partial x = 2y \cos xy - xy^2 \sin xy = \partial P/\partial y$, $y \sin xy = C$ **7.** $\partial Q/\partial x = e^x - e^y = \partial P/\partial y$, $(x + y - 1)(e^x - e^y) = C$
9. $\partial Q/\partial x = 2xe^{x^2} - y = \partial P/\partial y$, $ye^{x^2} - \frac{1}{2}xy^2 = C$ **11.** $\partial Q/\partial x = 3x^2 \sec^2 y - 6(y^2/x^3) = \partial P/\partial y$, $x^3 \tan y + y^2 + (y^3/x^2) = C$

A-66 Answers and Hints to Odd-Numbered Problems

13. Set $x = 0$, $y = y_0$ in (7). No. Rotate the coordinate system. 15. $x^3y + x^2 = C$ 17. $y^3/x = C$ 19. $x + y^2e^{-x} = C$
21. $x + \arctan(x/y) = C$ 23. $xy - \ln|y| = C$ ($y \equiv 0$ is also a solution) 25. $x^2|y^2 - x^2| = C > 0$
27. $y = \sqrt{x^2 - (12/x^2)}$ ($x > 0$) 29. $\ln x^2 - (y/x)^2 = C$

Section 16.2 (page 950)

1. $y = \dfrac{b}{a} + Ce^{-ax}$ 3. $y = Ce^{x^2/2} - 1$ 5. $y = \tfrac{1}{2}x^2 + Ce^{-2x}$ 7. $y = x^2 + Cx$ 9. $y = x \sin x + 2 \cos x - 2\dfrac{\sin x}{x} + \dfrac{C}{x}$

11. $y = (x + 1)^2(e^x + C)$ 13. $y = \dfrac{x \ln x - 1 + Cx}{x - 1}$ 15. $y = Ce^{\arctan x} - (\arctan x + 1)$ 17. $y = \tfrac{4}{3} + \tfrac{2}{3}e^{-3x}$

19. $y = -\tfrac{2}{5}\cos x + \tfrac{1}{5}\sin x - \tfrac{1}{5}e^{2x-\pi}$ 21. $y = \sin x - \cos x$ 23. $y = x \ln x + (1-e)\ln x$

25. The equation $y' + py = 0$ is separable. 27. $y = \dfrac{1}{1 + Ce^{x^2/2}}$ ($y \equiv 0$ is also a solution) 29. $y = \sqrt[4]{Ce^{4x} - x - \tfrac{1}{4}}$

31. $y = \ln(x - 1 + Ce^{-x})$

Section 16.3 (page 959)

1. $y = C_1 e^{(1+\sqrt{2})x} + C_2 e^{(1-\sqrt{2})x}$ 3. $y = e^{2x}(C_1 \cos 3x + C_2 \sin 3x)$ 5. $y = C_1 e^{-5x} + C_2 xe^{-5x}$ 7. $y = C_1 e^{4x} + C_2 xe^{4x}$
9. $y = e^x[C_1 \cos(x/2) + C_2 \sin(x/2)]$ 11. $y = C_1 e^{-2x} + C_2 e^{-x/3}$ 13. $y = \tfrac{8}{7}e^{-2x} - \tfrac{1}{7}e^x$ 15. $y = 2e^x \sin x$
17. $y \equiv 0$ (no calculations needed!) 19. $y = x^3 + 3x^2 - 3x - 5$ 21. $y = -\tfrac{1}{8}\cos 3x$ 23. $y = \tfrac{1}{2}e^x - \tfrac{1}{2}x \cos x$
25. $y = -\tfrac{1}{4}x^2 \cos x + \tfrac{1}{4}x \sin x$ 27. $y = C_1 + C_2 e^{-4x} + \tfrac{1}{4}x^3 - \tfrac{5}{16}x^2 + \tfrac{21}{32}x$
29. $y = e^{-x}(C_1 \cos x + C_2 \sin x) + \tfrac{1}{8}e^x(\sin x - \cos x)$ 31. $y = C_1 e^x + C_2 xe^x + \tfrac{1}{6}x^3 e^x$
33. $y = C_1 + C_2 e^x - \tfrac{1}{5}x^5 - x^4 - 4x^3 - 12x^2 - 24x$ 35. $y = C_1 e^{4x} + C_2 e^{-2x} + \tfrac{1}{16}e^{6x}$
37. $y = -\tfrac{9}{8} + \tfrac{1}{8}e^{2x} + \tfrac{1}{2}xe^{2x} - \tfrac{1}{6}x^3 - \tfrac{1}{4}x^2 + \tfrac{1}{4}x$

Section 16.4 (page 966)

1. See formula (12), p. 57. 3. 0.0476 in. 5. If $|\sin(\omega t + \phi)| = 1$, then $\cos(\omega t + \phi) = 0$.
7. $A = 10$ cm, $\omega = 2$ rad/sec, $\phi = -\arctan \tfrac{4}{3}$; $t_1 \approx 0.464$ sec 9. No 11. ≈ 85 min
13. See Prob. 17, p. 690. 15. $y = 6e^{-10t}(3e^{2t} - 2e^{-2t})$ in. 17. $\delta = \lambda T$
19. The tangential force on the bob is $F = -mg \sin \theta$; $\theta = C \sin(\omega t + \phi)$, where $\omega = \sqrt{g/L}$; $T = 2\pi\sqrt{L/g}$
21. $y = C_1 \cos \omega t + C_2 \sin \omega t + (f_0 t/2\omega)\sin \omega t$, where $f_0 t/2\omega \to \infty$ as $t \to \infty$.

Supplementary Problems (page 968)

1. $x^2 + y^2 = k$ 3. $x^2 + y^2 = ky$ 5. $x^2 y = k$ 7. $x = \ln y - 1 + (C/y)$ 9. $x = Cy - y^2$ 11. $y = C_1 \ln|x| + C_2$
13. $y = \cosh(x + C_1) + C_2$ 15. $y = \pm\sqrt{C_1 x + C_2}$ 17. $y = (1/C_1) - 1 + C_2 e^{C_1 x}$ ($y \equiv -1$ is also a solution)
19. $y = C_1 \cos x + C_2 \sin x - (\ln|\sec x + \tan x|)\cos x$ 21. $y = C_1 e^x + C_2 e^{-x} - \tfrac{2}{5}\sin^2 x - \tfrac{4}{5}$
23. $y = C_1 e^x + C_2 xe^x + \tfrac{1}{2}x^2 e^x \ln x - \tfrac{3}{4}x^2 e^x$ 25. $y'' + 8y' + 41y = 0$ 27. $y'' - 9y = 0$
29. $x = C_1 \cos t + C_2 \sin t$, $y = -C_1 \sin t + C_2 \cos t$ 31. $x = C_1 + C_2 e^{2t} - \tfrac{1}{4}(t^2 + t)$, $y = -C_1 + C_2 e^{2t} + \tfrac{1}{4}(t^2 - t - 1)$
33. Use Archimedes' principle, p. 480. $T \approx 1.57$ sec

35. $y = a_0 y_0 + a_1 y_1$, where $y_0 = 1 - 2x^2$ and $y_1 = x + \displaystyle\sum_{k=1}^{\infty} \dfrac{(4k-6)(4k-10)\cdots(2)(-2)}{(2k+1)!} x^{2k+1}$

37. $y = a_0 y_0 + a_1 y_1$, where $y_0 = 1 + \displaystyle\sum_{k=1}^{\infty} \dfrac{x^{3k}}{(3k)(3k-1)(3k-3)(3k-4)\cdots 3 \cdot 2}$ and

$y_1 = x + \displaystyle\sum_{k=1}^{\infty} \dfrac{x^{3k+1}}{(3k+1)(3k)(3k-2)(3k-3)\cdots 4 \cdot 3}$.

Appendix (page A-4)

1. 500,500 3. $\tfrac{1}{4}m^2(m+1)^2 + (m+1)^3 = \tfrac{1}{4}(m+1)^2[m^2 + 4(m+1)] = \tfrac{1}{4}(m+1)^2(m+2)^2$

5. $\dfrac{m}{m+1} + \dfrac{1}{(m+1)(m+2)} = \dfrac{m^2 + 2m + 1}{(m+1)(m+2)} = \dfrac{m+1}{m+2}$ 7. $\dfrac{m+1}{2m}\left[1 - \dfrac{1}{(m+1)^2}\right] = \dfrac{m+1}{2m}\dfrac{m(m+2)}{(m+1)^2} = \dfrac{m+2}{2(m+1)}$

9. $\tfrac{1}{2}(-1)^{m-1}m(m+1) + (-1)^m(m+1)^2 = \tfrac{1}{2}(-1)^m(m^2 + 3m + 2) = \tfrac{1}{2}(-1)^m(m+1)(m+2)$

11. $\dfrac{m(m+1)}{2(2m+1)} + \dfrac{(m+1)^2}{(2m+1)(2m+3)} = \dfrac{m+1}{2(2m+1)}\dfrac{2m^2 + 5m + 2}{2m+3} = \dfrac{(m+1)(m+2)}{2(2m+3)}$ 13. $\dfrac{x^{m+1} - 1}{x - 1} + x^{m+1} = \dfrac{x^{m+2} - 1}{x - 1}$

15. 15 min 17. Let $n = 99$. 19. Replace a five by 2 threes or 3 threes by 2 fives.

Index

References to problems are indicated by numbers in italic type.

A

Abel, Niels (1802-1829), 544
Abel's theorem on power series, 544
Abscissa, 21
Absolute error, *127*
Absolute extrema, 167, 790, 796
   test for, 167
Absolute value:
   definition of, 14
   properties of, 14-15
Absolute value function, 47
Absolutely convergent series, 517
   convergence of, 518
Absorption coefficient, *348*
Absorption of light, *348*
Acceleration, 107, 143, 266, 679-681, 728
   centripetal, 680
   due to gravity (g), 108, 268, 272
   normal component of, 680, 728
   tangential component of, 680, 728
Accumulation of litter, *349*
Accuracy to n decimal places, 542-543
Achilles and the tortoise, *506*, *507*
Addition laws for sine and cosine, 57
Adiabatic expansion, 475
Air resistance, *365*
Algebraic function, *365*
Algebraic sum, 83
Alternating current (ac), 948

Alternating harmonic series, 517
Alternating series, 518-523
   error in truncating, 519
Alternating series test, 519
Amount of information, *324*
Amplitude, 961
Analytic geometry:
   plane, 575-646
   solid, 694-476
Anchor ring (*see* Torus)
Angle:
   between two curves, *122*, *635*
   between two lines, 59
   between two planes, 719
   between two surfaces, *808*
   between two vectors, 655
   of incidence, 209, 210
   of inclination (*see* Inclination)
   of reflection, 209
   of refraction, 210
Angstrom unit, *735*
Angular coordinate, 620
Angular frequency, 948, 961
Angular momentum, *736*
   conservation of, *736*
Angular speed, 685
Angular velocity, 935
   and curl, 936
Angular sector (polar rectangle), 856, 857
Angularly simple region, 858
Annulus, 444
Antiderivative, 248
   general, 249
   of a vector function, 665

Aortic valve, *477*
Aphelion, *595*, 734
Apocenter, 734
Apogee, 734
Appolonius of Perga (255-170 B.C.), 575
Approximate integration, 413-424
   by the midpoint rule, 413-416
   by Simpson's rule, 418-423
   by the trapezoidal rule, 416-418
Arc length function, 667-668, 723
Arc of a curve, 451, 722
Archimedean spiral, *628*, *634*, *637*, *639*, *645*
Archimedes (287-212 B.C.), xvi
Archimedes' principle, 480
Area:
   between two curves, 241-244, 258-260
   as a double integral, 813
   as a line integral, 911
   of a normal region, 811
   of a surface, 897-900
      of revolution, 466-471, 639-640, *939*
   of an unbounded region, 426, 428, 429
   in polar coordinates, 635-638, 859, 862
      compared with area in rectangular coordinates, *641*
   under a curve, 229-231
      as an integral, 231
Areas and volumes, table of, A-5
Argument (of a function), 41, 747

Arithmetic mean, 13, *277*
Associative laws, 6
Astroid, 464
   length of, 464
Asymptotes:
   horizontal, 190
   of a hyperbola, 598-599
      construction of, 599
   oblique, 193, *196*
   vertical, 192
Atomic bomb, *364*
Average:
   of a function, 244
   of *n* numbers, *277*, 750
   of two numbers, *13*
Average cost, 215
   minimum, 216
Average density, 108
Average revenue, 218
Average velocity, 106, *166*, *266*
Axis:
   of an ellipse, 589
      major, 589
      minor, 589
   of a hyperbola, 597, 598
      conjugate, 598
      transverse, 597
   of a parabola, 581
   of revolution, 442

B

bac-cab rule, *715*
Balloon problems, 150-151, *154*
Banked curve, *690*
   rated speed of, *690*
Barometric equation, *477*
Basis:
   local, 674
   orthogonal, 651, 702
   orthonormal, 651, 702
   in the plane, 650
   in space, 702
Basis vectors, 650, 702
Beats, 966
Bernoulli, John (1667-1748), *196n*
Bernoulli's equation, *950*
Big bang, *273*
Big crunch, *273*
Binary fission (of bacteria), 337, *340*
Binomial coefficients, 227
   calculation of, 228
Binomial series, 546-547
Binomial theorem, 111, 227, A-3
Binormal, *745*
Birth rate, 337, 343
Bisection method, 97, 179
Bisectors of angles between two lines, 37

Bit (unit of information), *324*
Black box, 45
Black hole, *273*
Blood pressure, *477*, *508*
Bouncing ball problem, *507*
Boundary (of a set), 758
Boundary conditions, 265
Boundary point, 758
Boundary value problem, 265
Bounded function, 62
   near infinity, *190*
   on an interval, 62
   on a set in the plane, 791
Bounded (or finite) intervals, 18-19
Bounded sequence, 491
Bounded set, 790
Boyle's law, 475, *477*
Brachistochrone, 456
Brahe, Tycho (1546-1601), 732
Branches, 63
   of a hyperbola, 353, 454
Break-even point, *221*
Budget constraint, *809*
Bullet fired into resistive medium, 346 347, *348*
Buoyancy, 480-481
Business and economic problems, 214-222

C

Callisto, *690*
Canopus, *323*
Cardioid(s), *465*, *624*, *625*, *628*, 633-634, *637*, 638-639, *640*, *641*
   orthogonal, *645*
Carrying capacity, *348*
Cartesian coordinates (*see* Rectangular coordinates)
Cartography, *752*
Cassinian oval, *645*
Catenary, 689, *691*
   directrix of, 689
Cathode ray oscilloscope, *483*
Cauchy, Augustin Louis (1789-1857), 164, 543
Cauchy-Riemann equations, *764*, *808*
Cauchy-Schwarz inequality, 658, *661*
   for integrals, *875*
Cauchy's condensation test, *516*
Cauchy's mean value theorem, 164-165, 166
   variant of, *223*
Cauchy's theorem on multiplication of power series, 543
Cavalieri's principle:
   for area, 247

   for volume, *441*
Center of gravity, 834
Center of mass:
   of a lamina, 837
   of a solid, 835-837
   of a system of particles, 834-835
   of a wire, 842-844, 878
Central force, 731, 733
Central rectangle, 599
Centripetal acceleration, 680
Centroid:
   of a curve, 844
   of a plane region, 837
   of a solid region, 837
Chain reaction, *364*
Chain rule, 133-140
   for elasticities, 222
   for a function of *n* variables, 774
   for a function of two variables, 771-774
   informal treatment of, 135-136
   for vector functions, 669, *670*
Characteristic equation, 954
Chord:
   of an ellipse, *594*
   of a hyperbola, *603*
   of a parabola, *586*
Chromatic scale, *14*
Circle of curvature, 678
Circles, 24-25
   parametric equations of, 453-454
   polar equations of, 622-623, *627*
Circular arc, length of, 54
Circular cone:
   oblique, *440*
   right (*see* Right circular cone)
Circular functions, 353
Circular sector, area of, 54
Circulation, 890
Cissoid, *628*
Clipped function, *104*
Closed interval, 18
Closed set, 758
Closure of a set under arithmetical operations, 3
Coefficient of friction, *690*, *691*
Coefficient of linear expansion, 156
Coefficient of volume expansion, 156
Coextensive equations, 453
Coin tossing, *324*
Colatitude, 868
Common logarithm, 301, 303, 320
Commutative laws, 6
Comparison test:
   for improper integrals, 429, 430
   for infinite series, 510
Complete equation, 956
Completing the square, 24
Complex analysis, *764*
Complex numbers, 12

Component(s) of a vector:
  with respect to a basis, 650, 651, 702
  along another vector, 658, 705
Composite function, 45
  continuity of, 88
  derivative of, 133
  limit of, 87
Composition, 45
  associativity of, 47
  noncommutativity of, 46
Compound interest, 330-331, 332
Computer graphics, 752, 753
Computer science, 203
Concave function, 174
Concave side (of a curve), 672
Concavity test, 175, 179
Conchoid, *645*
Conditionally convergent series, 518
Confocal ellipses and hyperbolas, *645*
Conic sections, 575, 604-612
  degenerate, 611-612, 615
  focus-directrix property of, 608
  geometric meaning of, 609-611
  in polar coordinates, 628-631
Conical pendulum, *690*
Conics (*see* Conic sections)
Conjugate axis, 598
Conjugate hyperbolas, *603-604*
Conservation of energy, 270, 892
Conservative field, 892
Constant of integration, 250
Constrained extrema, 799
Constraints, 798, 799
Continuity:
  of combinations of functions, 83, 757
  of a composite function, 88, 757
  definition of, 69, 72-73, 757
  implied by differentiability, 120, 766
  failure of, reasons for, 69-70
  of a function of several variables, 757
  of $f(x) = c$, 73-74
  of $f(x) = x$, 73
  of $f(x) = |x|$, 74
  integrability as a consequence of, 232, 813, 826
  on an interval, 94-95
  from the left, 94
  of the nth root function, 88, 284
  of polynomials, 70, 84
  of rational functions, 84
  from the right, 94
  of a set in n-space, 758
  of trigonometric functions, 70, 84
  uniform, 232, 461
  of a vector function, 663
Continuous differentiability, 460
Continuous function(s), 69, 72-73, 757
  algebraic operations on, 83-85, 757

  existence of antiderivative of, 254
  extrema of:
    on a bounded closed interval, 99
    on a bounded closed set, 791
    image of an interval under, 100
    integrability of, 232, 813, 826
    properties of, 96-101
Continuous one-to-one function, 284
  continuity of inverse of, 284-285
  monotonicity of, 284
Continuously compounded interest, 331
Continuously differentiable function, 460
Convergent improper integral, 425, 426, 427, 428, 431, 498
Convergent sequence, 490
  boundedness of, 492
Convergent series, 498
  necessary condition for, 502
Coordinate axes, 21, 694
Coordinate planes, 694
Coordinate system:
  left-handed, 694
  nonrectangular, 575
  polar, 575
  rectangular, 21, 575, 694
  right-handed, 694
Coordinate transformation, 914, 917
  inverse, 914
  Jacobian of, 914, 917
  one-to-one, 914
Coordinates:
  cylindrical, 862
  on the line, 16
  missing from an equation, 698
  in n-space, 749
  in the plane, 21
  polar, 620
  rectangular, 21, 453, 695
  in space, 695
  spherical, 868
Corners, 119, 195
  vs. cusps, 195
Correspondence:
  many-to-one, 620
  one-to-one, 6, 21, 620
Cosecant function, 52
  derivative of, 132
  fundamental period of, 64
  graph of, 64
Cosine function, 52
  derivative of, 112
  evenness of, 55
  fundamental period of, 62
  graph of, 61
  integral of, 251
  Maclaurin series of, 561
Cost:
  average, 215
  elasticity of, *224*

  marginal, 215
  total, 214
Cost function, 214, *808*
  cubic, *221*
Cotangent function, 52
  branches of, 63
  derivative of, 132
  fundamental period of, 63
  graph of, 63
Critical density, *274*
Critical mass, *364*
Critical point:
  of a derivative, 177
  of a function of n variables, 796
  of a function of one variable, 169, 195
  of a function of two variables, 792
Critical radius, *364*
Critical speed, *690, 691*
Critically damped oscillations, 964, 965
Cross product, 694, 707
  anticommutativity of, 708
  component form of, 710
  derivative of, 724
  as a determinant, 711
  distributive laws for, 709, 714
  nonassociativity of, 708
Cross sections (of a solid), 436
Crosscuts, 912
Curl, 929
  physical meaning of, 935-937
Current, 114
  steady-state, 346, 949
  transient, 346, 949
Curvature:
  circle of, 678
  in the plane, 675-678, 728
  in polar coordinates, *682*
  radius of, 678, 728
  in space, 727-728
Curve(s) in parametric form, 450-459
  arc of, 451
  closed, 459
  curvature of (*see* Curvature)
  definition of, 451
  endpoints of, 459
  final point of, 459
  initial point of, 459
  nonrectifiable, 460, *463*
  orientation of, 452, 882
    reversal of, 882
  piecewise smooth, 884
  polygonal path inscribed in, 460
  rectifiable, 460
    length of, 460-461
  simple, 459
  simple closed, 459, 886
    exterior of, 459
    interior of, 459
  slope of, 456

Curve(s) in parametric form (*cont.*):
  smooth, 779, 842, 879
  in space (*see* Space curve)
  tangent line to, 456-457
  unit normal vector to, 673-674
  unit tangent vector to, 669-670
Curve sketching, 179
  summary of techniques for, 179
Cusps, 194-195
  vs. corners, 195
Cycles per second, 961
Cycloid, 455-456, 462, 469
  as brachistochrone, 456
  as tautochrone, 456
Cyclotron, 715
Cyclotron frequency, 715
Cylinder, 697-698
  directrix of, 697
  generators of, 697
  quadric, 737
Cylindrical coordinates, 862
  in relation to rectangular coordinates, 862-863
  triple integrals in, 862-867
Cylindrical symmetry, 863
Cylindrical wedge, 864

D

Damped oscillations, 964-965
Damping coefficient, 963
Dashpot, 963
Death rate, 337, 343
Decay constant, 338
Deceleration, 107
Decibels, 323
Decimals, 5-6
  and infinite series, 500, 508-510
  nonrepeating, 5
  repeating, 5, 509
  terminating, 5
Decreasing function, 49
Decreasing sequence, 493
  strictly, 493
Definite integral, 229
  additivity of, on adjacent intervals, 239
  of a constant multiple of a function, 235
  definition of, 231, 232
  integrand of, 231
  of a linear combination of functions, 236
  of a nonnegative function, 236
  of a sum of functions, 235-236
  of a vector function, 665, *671*
Degree measure, 54
  vs. radian measure, 54
Del operator, 784, 786, 923, 929
  rules obeyed by, 789

Deleted neighborhood, 19
  in *n*-space, 755
Deltoid, *484*
Demand:
  elastic, 220
  elasticity of, 220
  inelastic, 220
Demand function (or curve):
  general, *222*
  linear, *38*, 217
Density, 108, 829, 831
  average, 108
  calculation of mass from, 829-831
  charge, *832*
  critical, *274*
  exact, 108
  surface vs. volume, 831
  of a wire, 842
Dependent variable, 39, 748
Derivative(s), 66, 105
  of a^x, 317
  of a composite function, 133
  of a constant, 110
  of a constant multiple of a function, 110
  of $\cos x$, 112
  definition of, 108-109
  directional (*see* Directional derivative)
  of e^x, 314
  as a function, 109
  higher, 140-142
  of hyperbolic functions, 355
  infinite, 194
  of an inverse function, 286-290
  of inverse hyperbolic functions, 357, *359*, *360*, *361*
  of inverse trigonometric functions, 295
  left-hand, 119
  of $\ln |f(x)|$, 306
  of $\ln |x|$, 302
  of $\log_a x$, 321
  of $\log_x a$, 322
  logarithmic, 306
  *n*th (of order *n*), 141
  one-sided, 119
  partial (*see* Partial derivatives)
  of a polynomial, 142
  of a product, 128
  of a quotient, 131
  of rational powers, 138-139, 148
  of a reciprocal, 129
  right-hand, 119
  second, 140
  of $\sin x$, 112
  third, 141
  of trigonometric functions, 131-132
  of a vector function, 663
    of constant magnitude, 665
  of x^a, 326

  of x^n, 111, 124, A-2
  of $\sqrt{x}$, 112
  zeroth, 141
Descartes, René (1596-1650), xvi, 453, 575
Determinants, 709-713
Diameter of an ellipse, *643*, *644*
Difference of sets, 36
Difference quotient, 109, 128
Differentiability, 109, 195
  of functions of several variables, 765-769
    conditions for, 768-769
    geometric meaning of, 782
    implications of, 766-767
  on an interval, 140
  of a vector function, 663
Differentiable function:
  of one variable, 109, 765
  of functions of several variables, 766, 767
Differential(s), 123-127
  approximation by, 125, 769-770
  exact, *896*, *942*
  of functions of several variables, 766, 767
  of the independent variable(s), 124, 767
  partial, 767
  total, 766, 767
  of the variable of integration, 367
Differential calculus, 225, 747
Differential equation(s), 143
  exact, *942*, *945*
  first order, 262-264, 942-950
  general solution of, 263, 264
  homogeneous, *946*
  integral curves of, 944
  linear (*see* Linear differential equation)
  nonlinear, *941*
  order of, 143, *941*
  ordinary vs. partial, 763, *941*
  particular solutions of, 263, 264
  power series solutions of, *970*
  separable, 333, *941*, *943*
  with separated variables, 333
  singular solution of, *942*, *945*
  of the form $y'' = f(x, y')$, *969*
  of the form $y'' = f(y, y')$, *969*
Differentiation, 109
  implicit, 145
  logarithmic, 306
  partial, 747-809
  of power series, 540
  rules for, 110-112
  of a vector function, 663-665
Direct current (dc), 948
Direction angles, 705
Direction cosines, 705, 716
Direction numbers, 716

Directional derivative:
  definition of, 783, 785
  in terms of the gradient, 785
  in terms of partial derivatives, 784
  properties of, 785-786
Director circle, *644*
Directrices, distance between, 606, 607
Directrix:
  of a catenary, 689
  of a cylinder, 697
  of an ellipse, 606
  of a hyperbola, 606
  of a parabola, 581
Directrix-focus distance:
  of an ellipse, 607
  of a hyperbola, 607
Directrix-vertex distance:
  of an ellipse, 607
  of a hyperbola, 607
Dirichlet's test, *525*
Discontinuity, 69
  nonremovable, *79*
  removable, *79*
Discontinuous function, 69
Discriminant, *619*, 793
Disjoint sets, 2
Distance:
  between parallel lines, 35
  between a point and a line, 35
    in space, 717
  between a point and a plane, 720
  between skew lines, *722*
  between two points, 16, 22
    in polar coordinates, 627
    rotation invariance of, *642*
    in space, 696
    translation invariance of, *642*
Distance formula:
  on the line, 16
  in the plane, 22
  in space, 696
Distillery problem, *222*
Distinct sets, 2
Distributive laws, 6, 649, 656
Divergence, 923
  physical meaning of, 927-928
Divergence theorem, 877, 923, 924
Divergent improper integral, 425, 426, 427, 428, 431
Divergent sequence, 490
Divergent series, 498
  *n*th term test for, 502
DNA molecule, *735*
Domain (of definition) of a function, 39, *43*, 747
  natural, 41, 748
Domain (open connected set), 886
  multiply connected, 887, 935*n*
  simply connected, 887, 935
Domination of one series by another, 510

Dot product, 655, 703, 708
  commutativity of, 655, 704
  component form of, 656, 704
  derivative of, 664, 724
  distributive laws for, 656, 704
  work as, 659-660
Double helix, *735*
Double integral, 810
  change of variables in, 913-917, 918-920
  definition of, 812-813
  evaluation of, 817-820
  mean value theorem for, *823*
  in polar coordinates, 855-862, 914, 915
  rules obeyed by, 813-814
Double-angle formulas, 58, *140*
Doubling time, 337
Drug dosage problem, *508*
Dummy index, 226, 498
Dummy variable, 231
Dyne, 276, 474*n*

E

e (base of the natural logarithm), 304-305
  as the limit of a sequence, 329, 491
  as the sum of a series, 542
E. coli, 337, 344
Eccentricity:
  of an ellipse, *485*, 592
  of a hyperbola, 602
  of a parabola, 605
Economies of scale, 216
Einstein, Albert (1871-1955), 273, 276, 573, 642
Elastic restoring force, 271, 962
Elasticity (in business and economics), 219-221
  of cost, *224*
  definition of, 220, *222*
  of demand, 220
Electric charge, 114
  current as rate of change of, 114
Electric circuits, 345-346, *349*, 948-950, *967*
Electric field, *940*
Electromagnetic radiation, *940*
Elementary function, 366
Elements (of a set), 1
  infinitely many, 63*n*
Elimination constant (of a drug), 340, *508*
Ellipse, 453, 455, 575
  area enclosed by, 591-592
  center of, 587
  chord(s) of, 594, 643, 644
  construction of, 587-588
  definition of, 587
  diameter of, 643, 644

director circle of, *644*
directrices of, 606
eccentricity of, 592
equations of, 588-589, 590-591
  parametric, 453, *595*
  in standard form, 590
focal chord of, *644*
foci of, 587
latus rectum of, *594*
major axis of, 589
minor axis of, 589
reflection property of, 592-594, *789*
semimajor axis of, 590
semiminor axis of, 590
tangent to, 593
vertices of, 589
Ellipsoid, 737, 738
  center of, 738
  normal line to, 781
  of revolution (or spheroid), *485*, 738
  tangent plane to, 781
Elliptic cone, 737, 740
Elliptic cylinder, 698, 737
Elliptic integral, *466*
Elliptic paraboloid, 737, 740-741
Empty set, 2
Endpoints:
  of a curve, 459
  of an interval, 18
Energy:
  conservation of, 270, 892
  kinetic, 269, *274*, 573, 880
  potential, 891
  total, 892
Entropy (function), *324*
  graph of, *324*
Epicycloid, *465*
Epsilon-delta method, 72
  geometric meaning of, 75
Equal temperament, *14*
Equations in standard form:
  of an ellipse, 590-591
  of a hyperbola, 600-602
  of a parabola, 582-583
Equilateral hyperbola, 600
Equilibrium demand, *38*
Equilibrium price, *38*
Equilibrium supply, *38*
Equivalence of mass and energy, 573
Erg, 276, 474*n*
Error of an approximation, 81
  absolute, *127*
  relative, *127*, 770
  round-off vs. truncation, 543
Error function, 423
Escape velocity:
  for the earth, 272
  for the moon, 273
  for a neutron star, 273
  for the sun, 273
  for a white dwarf, 273

Euler, Leonhard (1707-1783), 40
Euler's constant, 502
Euler's theorem on homogeneous
 functions, *778*
Europa, *690*
Evaporating mothball, *154, 348*
Even function, 48, 179
Even number, 3
Exact density, 108
Exact differential, *896, 942*
Exact differential equations, 942-944
 general solution of, 942
Expansion of the universe, *273*
Exponential (function), 300, 312
 approximation of, by Taylor
  polynomials, 552
 concavity of, 314
 definition of, 312
 derivative of, 314
 graph of, 312, 314
 integral of, 314
 inverse of, 312
 limiting behavior of, 312, 314, *332*
 Maclaurin series of, 559
 monotonicity of, 312
 power series of, 542
 of a sum, 312-313
 x-axis as asymptote of, 312, *318*
Exponential to the base a:
 definition of, 315
 derivative of, 317
 graph of, 317
 integral of, 317
 limiting behavior of, 317
 properties of, 315-316
Exponential growth and decay, 335-339
Exponentially damped oscillations, 965
Exponentially decreasing function, 335
Exponentially increasing function, 335
Exponents, laws of, 11, 13, 316, 325
Extended mean value theorem, 548-550
Exterior (of a simple closed curve), 459
Extreme value theorem, 99
 for a function of two variables, 791
 for functions approaching
  infinity, 207
Extreme values (extrema), 96, 789
 absolute, 167, 790
 constrained, 799
 local, 167, 790, 798

F

Factor theorem, 103
Falling stone problem, 42, 107,
 267-268, 270
Fermat's principle, 208, 209
Fibonacci, sequence, *496, 571, 573*
Final point:
 of a curve, 459
 of a vector, 648

Finite (or bounded) intervals, 18-19
Finite set, 63*n*
First derivative test, 170, 179
First quadrant, 22
Fixed point, *101*, 565
Flamsteed, John (1646-1719), *690*
Fluid pressure, 477-482
Flux of a vector field:
 across a curve, 922-923
 across a surface, 905-907
Focus:
 of an ellipse, 587
 of a hyperbola, 596
 of a parabola, 581
Focus-directrix property:
 of an ellipse, 607-608, 611
 of a hyperbola, 607-608
 of a parabola, 604-605
Folium of Descartes, *459, 646*
Force:
 absence of, 267
 buoyancy, 480-481
 central, 731, 733
 constant, 270
 elastic restoring, 271, 962
 of gravitational attraction, 271, 684,
  730, *896*
  by a solid sphere, *897*
  by a spherical shell, *897*
 on a particle, 143, 266
 resistive, *365*, 963
 on a submerged plate, 477, 478-480,
  842
 weight as a, 267, 683
Forced oscillations, 965-966
Four-petaled rose, 624
Fourth quadrant, 22
Fractional linear transformation, *298*
Free neutrons, *364*
Frequency, *14*
 cyclotron, *715*
 of harmonic oscillations, 961
 resonant, *968*
Fresnel integrals, *563*
Frustum:
 of a general solid, 444
 of a right circular cone, *440*
Function(s):
 algebraic, *365*
 algebraic operations on, 44, 749
 antiderivative of, 248
 argument(s) of, 41, 747
 asymptotes of (*see* Asymptotes)
 average of, 244
 bounded, 62, 791
 composite (*see* Composite function)
 concave downward, 174
 concave upward, 174
 constant, 42, 49
 continuous (*see* Continuous function)
 continuously differentiable, 460
 critical point of, 169, 792, 796

decreasing, 49
defined on a set, 40
definite integral of (*see* Definite
 integral)
definition of, 39
derivative of (*see* Derivative)
differentiable, 109, 766
 on an interval, 140
differential of, 123, 766
directional derivative of, 783
discontinuous, 69
domain (of definition) of, 39, *43*,
 747
elasticity of, 220
elementary, 366
equality of, 44
even, 48, 179
exponentially decreasing, 335
exponentially increasing, 335
graph of, 46, 750
greatest integer, 94
harmonic, 763, *928*
homogeneous, *778*
identically equal, 44
identity, 281
implicit, 145, 775-777
increasing, 49
increment of, 109, 123
indefinite integral of (*see* Indefinite
 integral)
infinitely differentiable, 540
as an input-output device, 45
integrable, 231, 813, 826, 901
inverse function (*see* Inverse
 function)
jump of, *95*
limit of, 66, 755
linear, 47
as a mapping, 45
maximum of, 97, 790
mean value of, 244, *832*
minimum of, 97, 790
monotonic, 169
of n variables, 750-752
natural domain of, 41
nonintegrable, *238*
notation for, 40
odd, 48, 179
one-to-one, 278-279
parity of, 49
partial derivatives of (*see* Partial
 derivatives)
periodic, 62
piecewise continuous, *276-277*
piecewise linear, 47
polynomial, 70
range of, 39, 43, 747
real, 42
saddle point of, 792
as a set of ordered pairs, 43
of several variables, 747-754
 limits and continuity of, 754-759

Function(s) (cont.):
  sum of, 44
  of three variables, 747
    integral of, 826
    level surfaces of, 752
  transcendental, *365*
  trigonometric (see Trigonometric functions)
  of two variables, 747
    integral of, 813
    level curves of, 750
  unbounded, 62
  value of, 40, *43*, 747
  vector-valued (see Vector function)
  with zero derivative, 164
Fundamental period, 62
Fundamental set of solutions, 952
Fundamental theorem of calculus, 225, 256-257, *261*, 816
  generalizations of, 877
  for a vector function, 665
Fundamental theorem of line integrals, 889

G

Gabriel's horn, *471*
Galilean satellites, *690*
Galileo Galilei (1564-1642), *690*
Gamma function, *435*
Ganymede, *690*
Gauss, Carl Friedrich (1777-1855), 924*n*
Gauss' law, *939*
Gauss' theorem, 924*n*
General cone, *441*
General power function (x^a):
  definition of, 324, 326
  derivative of, 326
  graph of, 326
  integral of, 326
  limiting behavior of, 326
  properties of, 325
General solution, 263, 264, 334
General term:
  of a sequence, 488
  of a series, 498
General theory of relativity, *273*
Generators:
  of a cone, 610
  of a cylinder, 697
Geometric mean, *13, 572*
Geometric series, 499
  convergence behavior of, 499-500
  ratio of, 499
  remainder of, 505
  sum of convergent, 500
Gibbs, Josiah Willard (1839-1903), 647
Gompertz growth law, *364*

Gradient field, 887
  conditions on components of, 892
  and path independence, 887
  test for, 893, *896*, 911, 935
Gradient (vector), 784, 786
Graph:
  of an equation, 23, 696
  of a function, 46
    in polar coordinates, 621
    shifting of, 50-51
    vertical line property of, 46
  of a function of two variables, 750
  of $F(r, \theta) = 0$, 621
    symmetries of, 623
  of $F(x, y) = 0$, 578-580
    shifting of, 578
    symmetries of, 579, 580
  of an inequality, 23, 696
  of a one-to-one function, 280
    horizontal line property of, 280
Gravitational potential, *896, 897*
Gravitational radius, *273*
Great Pyramid of Cheops, *476*
Greatest integer function, 94, 95
Greek alphabet, A-5
Green, George (1793-1841), 909
Green's first identity, *929*
Green's second identity, *929*
Green's theorem, 877, 909-921
  with two boundary curves, 912
  vector forms of, 922, 930
Gregory, James (1638-1675), 545
Gregory's series, 545, *573*
Growth of a bacterial culture, 337-338, 344-345
Growth of money in a bank, 330-331, 336, 337
Growth rate, 336
  relative (or fractional) vs. absolute, 336
Growth of tumors, *364*
Gudermannian, *365*

H

Half-angle substitution, 410
  universality of, 412
Half-life, 339
Half-open interval, 18
Halley's comet, 602*n*
Hanging chain, 349*n*, 688-689
Harmonic function, 763, *928*
Harmonic mean, *572*
Harmonic oscillations, 961
  amplitude of, 961
  frequency of, 961
    angular, 961
  period of, 961
Harmonic oscillator, 962
  damped, 964
Harmonic series, 501

alternating, 517
divergence of, 502
rate of, 502
Heart, model of, *477*
Heine, Heinrich Eduard (1821-1881), 72
Helicoid, 863
Helix, 722-723, 725-726
  double, *735*
  left-handed, 723
  length of, 723
  right-handed, 723
  pitch of, 723
Hertz (one cycle per second), 961
Hertz, Heinrich Rudolf (1857-1894), 961
Homogeneous differential equation, *946*
Homogeneous function of degree n, *778, 807*
Hooke's law, 270, 962
Horizontal asymptote, 190
Horizontal line property, 280, 451
Horizontally simple region, 811
  double integral over, 818
Horsepower, *274*, 474
Hubble's constant, *273*
Hydrostatic paradox, 478
Hyperbola, 454-455, 575
  asymptotes of, 598-599
  branches of, 353, 454, 596
  center of, 596
  central rectangle of, 599
  chord of, *603*
  conjugate axis of, 598
  construction of, *603*
  definition of, 595
  directrices of, 606
  eccentricity of, 602
  equations of, 596-597, 600-602
    parametric, 454, *603*, 644
    in standard form, 600
  equilateral, 600
  foci of, 596
  latus rectum of, *603*
  reflection property of, *604*
  tangent to, *604*
  transverse axis of, 597
  unit, 353, 600
  vertices of, 597
Hyperbolic argument, *365*
Hyperbolic cosecant, 354
  derivative of, 355
  graph of, *357*
Hyperbolic cosine, 349
  concavity of, 351
  derivative of, 350
  graph of, 350
  integral of, 350
  limiting behavior of, 350
Hyperbolic cotangent, 353
  derivative of, 355
  graph of, *356*

Index I-7

Hyperbolic cylinder, 737
Hyperbolic functions, 300, 349-357, 600, 602
   definitions of, 349, 353-354
   derivatives of, 355
   graphs of, 350, 354, *356*, *357*
   inverse, 300, 357-361
Hyperbolic identities, 351-352
Hyperbolic paraboloid, 737, 741-742, 792
Hyperbolic secant, 353
   derivative of, 355
   graph of, *356*
Hyperbolic sector, 353, *397*
Hyperbolic sine, 349
   concavity of, 351
   derivative of, 350
   graph of, 350
   integral of, 350
   limiting behavior of, 351
Hyperbolic spiral, 626, *646*
Hyperbolic substitutions, 391, 393, 396
Hyperbolic tangent, 353
   concavity of, 356
   derivative of, 354
   graph of, 354
   integral of, 354
   limiting behavior of, 354
Hyperboloid of one sheet, 737, 738-739
   axis of, 738
   center of, 738
Hyperboloid of revolution, 739, 740
Hyperboloid of two sheets, 737, 739-740
   axis of, 739
Hypocycloid, *465*
   of four cusps, *466*
   of three cusps, *484*

I

Ideal gas law, 760, *778*
Identical equality, 44
Identity, 44
   of polynomials, 401
   of power series, 541
Identity function, 281
Image:
   of a number under a function, 45, 100
   of a set under a function, 100
Impedance, 949
Implicit differentiation, 145, 775-777
Implicit function, 145, 775-777
Implicit function theorem, 777
Improper integral(s), 366, 424-433
   comparison test for, 429, 430
   convergent, 425, 426, 427, 428, 431
   divergent, 425, 426, 427, 428, 431
   of mixed type, 431

   with an unbounded integrand, 427-429
   over an unbounded interval, 425-427
Impulse, *274*
Inclination, 28
   in relation to slope, 28-29
Income stream, *433*
Incompressible fluid, 927
Increasing function, 49
Increasing sequence, 493
   strictly, 493
Increment:
   of the dependent variable, 109
   of a function, 109, 123, 765, 766
   of the independent variable(s), 109, 124, 767
Increment notation, 27, 106, 109
Indefinite integral, 249
   of a constant multiple of a function, 251
   definition of, 249
   of a function of $ax + b$, 252
   integrand of, 250
   of a linear combination of functions, 252
   of a sum of functions, 251-252
   of a vector function, 665
Independent variable, 39, 747
Indeterminacy (*see* Indeterminate form)
Indeterminate form:
   0/0, 4, 89-90, 106, 109, 187-188, 196-199
   $0 \cdot \infty$, 187, 188, 328
   ∞/∞, 187, 188, 199-200
   $\infty - \infty$, 187, 188-189, 201
   0^0, 187*n*, 328
   ∞^0, 187*n*, 328, 329
   1^∞, 187*n*, 328, 329-330
Index of summation, 226
   changes of, 505
Inequalities:
   addition rule for, 8
   combination of, 8
   multiplication rule for, 8
   reciprocals rule for, 9
   reversal of, 9
   strict, 11
   transitivity of, 9
Inequality:
   greater than, 7
   greater than or equal to, 10
   less than, 7
   less than or equal to, 10
Infinite derivatives, 194
Infinite (or unbounded) intervals, 19-20
Infinite limits, 182
   vs. limits at infinity, 182
   operations on, 185-187
Infinite sequence (*see* Sequence)
Infinite series (*see* Series)
Infinite set, 63*n*

Infinitely differentiable function, 540
Infinity, 19, 182
   minus, 19, 182
Inflection point, 175
   necessary condition for, 177, 179
   second derivative test for, 178, 179
   third derivative test for, 178, 179
Inflectional tangent, 176
Information theory, *324*
Initial condition(s), 262, 263, 267, 268, 334
Initial point:
   of a curve, 459
   of a vector, 648
Initial value problem, 265
Input-output device, 45
Integer part, 94
Integer powers (of a number), 11
Integers, 2
   negative, 2
   positive, 2
Integrability of continuous functions, 232, 813, 826
Integrable function, 231, 813, 826
   boundedness of, *238*
   on a surface, 901
Integral(s), 225
   definite (*see* Definite integral)
   double, 810-823
   improper (*see* Improper integral)
   indefinite (*see* Indefinite integral)
   iterated, 815-816, 827
   leading to logarithms, 307-312
   not evaluable in closed form, 366
   Riemann, 231
   surface, 897-909
   triple, 824-832
   with a variable upper limit, 254
      differentiation of, 254
Integral calculus, 225, 747
Integral curve, 944
Integral sign, 231
Integral test, 514
Integrand, 231, 250, 813, 826, 901
Integrating factor, *896*, *945*, *946*, 947
Integration:
   approximate, 413-424
   by change of variables, 367
   constant of, 250
   interval of, 231
   lower limit of, 231
   methods of, 366-435
   multiple, 810-876
   operation of, 231, 232, 250
   by parts, 366, 373-379
   of power series, 540
   of rational functions, 397-407
   of rational functions in sin x and cos x, 409-412
   by rationalizing substitutions, 407-413

Integration (*cont.*):
  by reduction formulas, 378-382
  rules for definite, 235-237
  rules for indefinite, 251-252
  by substitution, 252, 366, 367-373
  by trigonometric and hyperbolic substitutions, 388-397
  upper limit of, 231
  variable of, 231, 250
  of a vector function, 665-666
  of the velocity function, 261-262
Integration of products:
  of powers of cot x and csc x, 386-387
  of powers of sin x and cos x, 382-384
  of powers of tan x and sec x, 384-386
  of sines and cosines with different arguments, 387
Intercepts of a line, 29-30
Interest rate, 330, 336
  effective vs. nominal, *332*
Interior (of a simple closed curve), 459
Interior point, 18, 757, 790
Intermediate value theorem, 96
Intersection of sets, *36*
Interval(s):
  closed, 18
  connectedness of, 20
  continuity on, 94
  continuous mapping of, 100
  of convergence, 531-532, 534
  endpoints of, 18
  finite (or bounded), 18-19
  half-open, 18
  infinite (or unbounded), 19-20
  of integration, 231
  length of, 18
  open, 18
  partition of, 230
Interval mapping theorem, 100
Inverse cosecant, 293-294
Inverse cosine, 292
Inverse cotangent, 293
Inverse function(s), 278-299
  continuity of, 284-285
  derivative of, 286-287
  graph of, 281-282
Inverse hyperbolic cosecant, *361*
Inverse hyperbolic cosine, *360*
Inverse hyperbolic cotangent, *360*
Inverse hyperbolic functions, 300, 357-361
  derivatives of, 357, 359, *360*, *361*
  graphs of, 357, 359, *360*, *361*
Inverse hyperbolic secant, *360*
Inverse hyperbolic sine, 357
  in terms of the logarithm, 358
Inverse hyperbolic tangent, 359
  in terms of the logarithm, 359

Inverse secant, 294-295
Inverse sine, 290-291
Inverse square law, 271, 730, 731
Inverse tangent, 292-293
  power series of, 545
Inverse trigonometric functions, 278, 290-297
  derivatives of, 295
  graphs of, 290, 292-294
Io, *690*
Irrational numbers, 2
  decimal representation of, 5-6
Irrationality of $\sqrt{2}$, 4-5
Irreducible quadratic polynomials, 398
Irrotational field, 930, 936
Isobars, 752
Isolated point, *797*
Isothermal expansion, 475
Isotherms, 752
Iterated integrals, 815-816, 827
  equality of, 817, 819
Iteration, 565
Iterative methods, 487

J

Jacobian, 914, 917
  of inverse transformation, 915
Jordan curve theorem, 459
Joule, 474*n*
Jump, *95*
Jump discontinuity, *95*
Jupiter:
  density of, *690*
  moons of, *690*
  oblateness of, *485*

K

Kappa curve, *628*
Kelvin, Lord (William Thomson, 1824-1907), 931*n*
Kepler, Johannes (1571-1630), 694, 732
Kepler's first law, 732
Kepler's laws of planetary motion, 732-734
Kepler's second law, 732-733
Kepler's third law, 686, *690*, 733-734
Keyboard insruments, *14*
Keynes, John Maynard (1883-1940), *507*
Kinetic energy, 269, *274*, *573*, 880
  conservation of, 270

L

Lagrange, Joseph Louis (1736 -1813), 799

Lagrange multiplier method, 798-806, *808*
Lagrange multiplier rule:
  for functions of three variables, 800-801
  for functions of two variables, 799-800
  geometric interpretation of, 801
  with two constraints, *806*
Lamina, 831
  balance point of, 834
  center of mass of, 837
  density of, 831
    calculation of mass from, 831
  moments of, 837-838
  moments of inertia of, 851-852
Laplace's equation:
  in n dimensions, *808*
  in three dimensions, *764*, *928*
  in two dimensions, *763*, *928*
Laplacian, *928*
Latus rectum:
  of an ellipse, *594*
  of a hyperbola, *603*
  of a parabola, *586*
Law of cosines, 56, 656
Law of reflection, 209, 585, 594
Law of refraction, 210
Law of sines, *60*
Laws of exponents, 11, 13, 316, 325
Least squares estimate, *224*
Least upper bound, 494
Left ventricle, *477*
Left-hand derivative, 119
Left-hand limit, 92
Left-hand tangent, 119
Left-handed coordinate system, 694
Leibniz, Gottfried Wilhelm von (1646-1716), xvi, 72, 124, 231, 519
Leibniz notation, 124
Leibniz's rule, *229*
Lemniscate, *628*, *635*, *638*, *639*, *640*, *641*
Length:
  of an interval, 18
  of a plane curve, 459-466
    as an integral, 461
    in polar coordinates, 638-639
  of a space curve, 723
    in cylindrical coordinates, *876*
    as an integral, 723
Level curves, 750-752, 787
Level surfaces, 752, 786
L'Hospital, Marquis de (1661-1704), 196*n*
L'Hospital's rule, 159, 196-203
  for 0/0, 196-197
  for ∞/∞, 199-200
  simplified version of, *203*
Light year, *273*

Index **I-9**

Limaçon, *628, 640, 646*
Limit(s):
  algebraic operations on, 80-83
  of a composite function, 87
  of cos x as $x \to 0$, 69
  of a difference of functions, 80
  existence of, 68
    conditions for, 93
  finite, 182
  formal definition of, 72
  of a function, 66
    of n variables, 755
    of two variables, 755-756
  geometric interpretation of, 75
  infinite, 182
    operations on, 185-187
  of an infinite sequence, 489
  at infinity, 182
    operations on, 184-185
  informal definition of, 68
  left-hand, 92
  one-sided, 91
    vs. ordinary, 92
  of a product of functions, 81
  of a quotient of functions, 82
  right-hand, 92
  rules obeyed by, 76-77, 492
  of $(\sin x)/x$ as $x \to 0$, 67, 85-86
  of a sum of functions, 80
  uniqueness of, *80*
  of a vector function, 662-663
  of $x \sin(1/x)$ as $x \to 0$, 78, 87
Limit comparison test, 511
Limiting velocity (*see* Terminal velocity)
Limits of integration, 231
Limits of summation, 225
Line (*see* Straight line)
Line of best fit, *797*
Line integral(s), 660, 877
  with respect to arc length, 844, 878
  with respect to coordinates, 844*n*, 881
  along a curve, 880, 881
  fundamental theorem of, 889
  independence of parametric representation, 882
  path independent, 887
    condition for, 887, 911
    and gradient fields, 887
  with respect to position vector, 880
  in terms of unit tangent vector, 882
Linear algebra, 653
Linear differential equation:
  definition of, 941
  first order, 946-950
    general solution of, 947
  second order (*see* Second order linear differential equation)
Linear equation, 30
  homogeneous, 402*n*

in n variables, 402*n*
Linear functions, 47
Lissajous figures, *483*
Litmus paper, *323*
Lituus, *628*
Local basis, 674
Local extremum (extrema):
  vs. absolute extrema, 167, 790
  constrained, 799
  definition of, 167, 790, 796
  first derivative test for, 170, 179
  of a function of three variables, *798*
  of a function of two variables, 790
    second partials test for, 793
  necessary condition for, 168, 179, 791, 796
    geometric interpretation of, 168, 792
  second derivative test for, 171, 179
  sufficient conditions for, 169
Local maximum, 166, 790, 796, *798*
  strict, 166, 790
Local minimum, 166, 790, *798*
  strict, 166, 790
Logarithm to the base *a*:
  definition of, 320
  derivative of, 321
  graph of, 320
  limiting behavior of, 320
Logarithm to the base *e* (*see* Natural logarithm)
Logarithmic decrement, *967*
Logarithmic derivative, 306
  growth rate as, 336
  integration of, 308-309
Logarithmic differentiation, 306
Logarithmic spiral, *628, 634*
  equiangular property of, 634
  length of, 639
Logistic growth, 341-345
  law of, 342
Longest beam problem, *214*
Longitude, 868
Lorentz invariance, *642*
Lorentz transformation, *642*
Loudness, *323*

M

Machin, John (1680-1751), *297*
Machin's formula, *297, 573*
Maclaurin, Colin (1698-1746), *558*
Maclaurin series, 558
  of cos x, 561
  of cosh x, *562*
  of e^x, 559
  of sin x, 560-561
  of sinh x, *562*
Magnetic field, *940*
Magnitude (of a vector), 647, 650, 702

Major axis, 589
Manhattan Island, 126
Mapping by a function, 45
Mapping diagram, 279
Marginal cost, 215, 808
Marginal profit, 219
Marginal propensity:
  to consume, *507*
  to save, *507*
Marginal revenue, 218
Marginality (in business and economics), 214-219
Mass:
  of a lamina, 831
  of a material surface, 903
  of a solid object, 829-831
  of a wire, 844
Mass function, 108
Mathematical induction, 228, 555, A-1
Maximum:
  of a function, 97, 790
  local, 166, 790, 796, *798*
  of a set of numbers, 11
  of two functions, *104*
Maximum decrease, direction of, 786
Maximum increase, direction of, 786
Maxwell's equations, *940*
Mean value theorem, 159-166
  Cauchy's, 164-165, *166*
  extended, 549
  geometric interpretation of, 160
  kinematic interpretation of, *166*
  proof of, 162
  two-dimensional version of, *779*
Mean value theorem for integrals, 245-247
  generalized, *275, 573*
  geometric interpretation of, 245-246
  proof of, 245
Members (of a set), 1
Mesh size, 230, 232, 812, 825
Method of disks, 442-444
Method of shells, 446-450
Method of undetermined coefficients, 563, 956
Method of washers, 444-446
Midpoint formula:
  in the plane, 23
  in space, *701*
Midpoint rule (for approximate integration), 413-416
  error of, 414-415
Milky Way, *273*
Minimax (*see* Saddle point)
Minimum:
  of a function, 97, 790
  local, 166, 790, *798*
  of a set of numbers, 11
  of two functions, *104*
Minor axis, 589
Mitral valve, *477*

Möbius strip, *908*
Moments:
  of a lamina, 837-838
  of a solid, 837
  of a system of particles, 835
Moment(s) of inertia:
  of the earth, 850
  of a lamina, 851
  polar, 852
  of a solid, 851
  of a system of particles, 835, 850
Momentum, 274, 276, *736*
  angular, *736*
Monopolistic firm, 220
Monotonic function, 169, 282
  inverse of, 282-283
Monotonic sequence, 493
  convergence of bounded, 493
Monotonicity test, 169, 179
Motion:
  of alpha particles, 602
  of a charged particle in a magnetic field, *715*
  of comets, 602, *631*
  of the earth, 594, *595*
  in a hole through center of the earth, *967*
  on a line (rectilinear), 105-108, 261-262, 266-274
  of the moon, 686
  orbital, 594, 602, 730-735, *736-737*
  in the plane, 682-691
  of a planet, 730-735
  of a projectile, 586, 682-683, 689-690, *693*
  of a rocket, 271-272, *273*, *736*
  of a rotating pail of water, *693*
  of a satellite, 683-684, 730-735, *736-737*, *746*
    in synchronous orbit, 685-686
  simple harmonic, 960
  in space, 725-730
  uniform circular, 684-685
Mount Palomar Observatory, *323*
Moving trihedral, *745*
Multiple integral, 810
  change of variables in, 873, 915-920
Multiple reflection problem, *214*
Multiplier:
  in Keynesian economics, *507*
  Lagrange, 799, *806*
Multiplier effect, *507*
Multiply connected domain, 887, 935*n*
Mutations in a bacterial culture, *363*
Muzzle velocity, 274, 689, *693*

N

Napkin ring, 445
Napoleon brandy, *332*
Nappes, 609
National income, *507*
Natural logarithm, 255, 300
  concavity of, 304
  definition of, 301
  derivative of, 301
  geometric interpretation of, 301
  graph of, 304
  inverse of, 312
  limiting behavior of, 304, 305, 327-328
  monotonicity of, 301
  of a product, 302-303, 370
  of a rational power, 303
  y-axis as asymptote of, 304, *307*
Necessary condition:
  for convergence of a series, 502
  for an inflection point, 177, 179
  for a local extremum, 168, 179, 791, 796
Negative number, 7
Neighborhood, 19
  deleted, 19, 755
  in 2-space, 754
  in 3-space, 755
  in *n*-space, 755
Nephroid, *628*
Neutron star, *273*
New York Stock Exchange, *42*
Newton, Sir Isaac (1642-1727), xvi, 72, 266, *690*, 694, 730, 732
Newton (unit of force), 474*n*
Newtonian mechanics:
  one-dimensional, 266-274
  two-dimensional, 682-691
Newton-Raphson method (*see* Newton's method)
Newton's first law of motion, 267, *736*
Newton's law of cooling, *348*
Newton's law of gravitation, 271, 730
Newton's method, 487, 563-569
  geometric interpretation of, 564-565
  proof of, 567-568
Newton's second law of motion, 143, 266-267
  integration of, 267-269
  in terms of momentum, 276
  applied to a system of particles, 833-834
  vector form of, 682, 730
Newton's third law of motion, 833
n factorial, 142, 226, 488, *571*
Nonhomogeneous rod problem, 108, *115*
Nonnegative number, 8
Nonrectifiable curve, *465*
Norm (*see* Mesh size)
Normal component of acceleration, 680, 728
Normal line:
  to a curve, 120, *783*
  to a surface, 780, 781
Normal region, 810, 811
  area of, 811
  in space, 824
  volume of, 824-825, 826
Normal (vector) to a plane, 718
n-space (R^n), 749
  points in, 749
n-sphere, 755, 796
*n*th power, 11
*n*th root, 12
*n*th term:
  of a sequence, 488
  of a series, 498
*n*th term test for divergence, 502
Nuclear fission, *364*
Number(s):
  complex, 12
  decimal representation of, 5-6
  even, 3
  irrational, 2, 4-5
  negative, 7
  nonnegative, 8
  odd, 3
  positive, 7
  powers of, 11, 12-13
  rational (*see* Rational number)
  real (*see* Real number)
  roots of, 11-12
Number line, 2
  origin of, 2
  positive direction of, 3
Numerical integration (*see* Approximate integration)
Nutrient uptake, *363*

O

Oblate spheroid, *485*, *746*
Oblique asymptote, 193, *196*
Octants, 696
Odd function, 48, 179
Odd number, 3
Ohm's law, 345
One-sided derivatives, 119
One-sided limits, 91-94
One-to-one correspondence:
  between decimals and real numbers, 6
  between ordered pairs and points in the plane, 21
  between ordered triples and points in space, 695
One-to-one function, 278-279
  continuous, 284
  graph of, 280
  inverse of, 279
  as a set of ordered pairs, *286*
Open interval, 18
Open set, 758

Index I-11

Optimization problems, 159, 203-214
  step-by-step procedure for solving, 211
Ordered *n*-tuple, 750
Ordered pairs, 21, 651
  equality of, 21
  vectors as, 651
Ordered triples, 695
  equality of, 695
  vectors as, 702
Orders of magnitude, *323*
Ordinate, 21
Orientation of a curve, 452
  reversal of, 882-883
Origin, 2, 21, 694, 695
  (or pole) in polar coordinates, 620
  symmetry about, 48, 179, 579, 623
Orthogonal basis, 651, 702
Orthogonal curves, *122*, *645*
Orthogonal trajectories, *968*
Orthonormal basis, 651, 702
Oscillations:
  critically damped, 964, 965
  damped, 964-965
  exponentially damped, 965
  forced, 965-966
  harmonic (*see* Harmonic oscillations)
  overdamped, 964
  undamped, 963
  underdamped, 964, 965
Oscillatory phenomena, 387, 960-967
Osculating circle, 678
Osculating plane, *735*
Out of phase, 63
Overdamped oscillations, 964
Overhead, 216

P

Pappus of Alexandria (ca. 300 A.D.), 846
Pappus' theorem:
  for a solid of revolution, 846
  for a surface of revolution, 847
Parabola, 48, 575
  axis (of symmetry) of, 581
  chord(s) of, *586*, *643*
  construction of, *586*
  definition of, 581
  directrix of, 581
  equations of, 581-584
    parametric, *587*
    in standard form, 582
  focal chord of, *642*
  focus of, 581
  latus rectum of, *586*
  reflection property of, 584-586, *645*
  subnormal of, *643*
  symmetry of, 48, 581
  tangent to, 116, 117, *121*, 585
  vertex of, 581
Parabolic cylinder, 737
Parabolic reflector, 585-586, *587*
Parabolic spiral, *628*
Paraboloid:
  elliptic, 737, 740-741
  hyperbolic, 737, 741-742, 792
  of revolution, 444, 471, 585, *693*, 699, 737, 741
Parallel axis theorem, *855*
Parallel lines, 32
  distance between, 35
Parallelogram law, 648
Parameter, 451
Parametric curve (*see* Curve in parametric form)
Parity (of a function), 49
  change of, under differentiation, *140*
Partial derivatives, 759
  equality of mixed, 763
  geometric interpretation of, 761
  of higher order, 761-763
  of implicit functions, 776
Partial differential equation, 763
  for diffusion or heat conduction, 763
  order of, 763
Partial fractions, 400
Partial sum, 498
Particle, 105
Particular solution, 263, 264, 334
Partition:
  of a box, 825
  of an interval, 230
  of a rectangle, 812
Parts, integration by, 366, 373-379
Pascal's principle, 477
Path, 887
Peak current, 950
Peak voltage, 948
Pendulum:
  conical, *690*
  simple, *967*
Pericenter, 734
Perigee, 734
Perihelion, *595*, 602*n*, *631*, 734
  distance at, 734
  speed at, 734
Period, 62, 961
  fundamental, 62
  of harmonic oscillations, 961
  of uniform circular motion, 685
Periodic function, 62
Periodicity, 62
  preservation of, under differentiation, *140*
Perpendicular lines, 32-33
Perpendicularity condition, 32
pH (of a solution), *323*
Phase, 961
  initial, 961
Phonograph record problems, *61*, *966*
Photons, *273*
Pi (π), 5
  approximation of, *61*
  decimal places of, *297*
  as the sum of a series, 545, *573*
  Wallis' formula for, *497*
Piecewise continuous functions, 276-277
  integration of, 276
Piecewise linear function, 47, 167
Piecewise smooth curve, 884
Piecewise smooth surface, 905, 930
Planes, 698, 718-720
  angle between two, 719
  equations of, 718-719
  parallel, 720
Plutonium, *340*
Point-slope equation of a line, 30
Points of subdivision, 229
  equally spaced, 413
Polar axis, 620
  symmetry about, 623
Polar coordinates, 575, 620-641
  area in, 635-638, 859, *862*
  definition of, 620
  double integrals in, 855-862
  graph paper for, 622
  length in, 638-639
  in relation to rectangular coordinates, 621
Polar curve(s), 621
  angle between, *635*
  intersection of, 625-626
  length of, 638-639
  tangent line to, 631-635
Polar equation(s), 621
  of a conic, 628-631
  graph of, 621
Polar parametric equations, *646*
Polygonal path, 460
  inscribed in a curve, 460, 723
  vertices of, 460
Polynomial(s), 70
  coefficients of, 70
  continuity of, 70, 84
  degree of, 70
  derivative of, 113
  factorization of, 399
  identity of, 401
  irreducible quadratic, 398
  *n*th derivative of, 142
  in two variables, 409
Population drain, *348*
Population explosion, 341, *348*
Population growth:
  of endangered species, *348*
  exponential, 336-338
    doubling time of, 337
    with emigration, *348*
  logistic, 341-345

Population size:
 critical, *348*
 limiting (or stable), 343
Position function, 105, 261, 262
Position vector, 652, 666, 725
 graph of, 666, 725
Positive direction, 3, 694
Positive number, 7
Potential, 891
 gravitational, *896*, *897*
Potential drop, 892
Potential energy, 891
 of a stretched spring, *896*
Power (rate of change of work), *274*, *474*
Power series, 487, 525-548
 Abel's theorem on, 543
 of arctan x, 545
 basic convergence property of, 531
 Cauchy's theorem on, 543
 coefficients of, 529
 used to solve differential equations, *970*
 differentiation of, 540
 of e^x, 542
 identity of, 541
 integration of, 540
 interval of convergence of, 531-532, 534
 of $\ln(1 + x)$, 544
 multiplication of, 543
 radius of convergence of, 532, 534
 sum (function) of, 537, 538
Precalculus, 1-38
Present (or discounted) value, 331
Pressure:
 exerted by an expanding gas, 474
 on a submerged plate, 477
Principal (initial amount), 330
Principal unit normal vector, 727
Principia Mathematica, 730
Prism, 441*n*
Prismoid, *441*
 bases of, *441*
Prismoidal formula, *424*, *441*
Probability theory, 423
Product of inertia, *855*
Product rule, 128
Profit:
 marginal, 219
 total, 218
Projectile, motion of, 586, 682-683, 689-690, *693*
Projection of a vector (along another vector), 658, 705
Prolate spheroid, *485*, *746*
Proper subset, 1
p-series, 511, 515, 521
Pursuit curve, *484*
Pyramid, 438, *476*
 volume of, 438-439

Pythagorean theorem, 4, 22, 23, 56

Q

Quadrants, 22
Quadratic equation, solution of, *26*
Quadric cylinders, 737
Quadric surfaces (or quadrics), 737-743
 degenerate, 737
 of revolution, 737
Quasi-cylinder, 814
Quasi-frequency, 965
Quasi-period, 965, *967*
Quotient rule, 131

R

Rabbit colony, growth of, *496*
Radar antennas, 586
Radial coordinate, 620
Radially simple region, 858
Radial-tangential angle, 632
Radian measure, 53
 vs. degree measure, 54
Radio beacons, *604*
Radioactive decay, 338-339
 half-life of, 339
Radioactive fallout, 339
Radioactive tracers, *340*
Radiocarbon dating, *340*
Radium in the earth's crust, *340*
Radius of convergence, 532, 534
 of differentiated power series, 539
 of integrated power series, 539
Radius of curvature, 678, 728
Radius of gyration, 852
 polar, 852
Radius vector (*see* Position vector)
Rainy Day Co., operation of, 215-219, *220*, *222*
Range (of a function), 39, *43*, 748
Range finding, 602, *604*
Rate of change, 105, 107, 225
 of charge with respect to time, 114
 of distance with respect to time, 107
 of mass with respect to distance, 108
 of y-coordinate with respect to x-coordinate, 117
Rate law (of a chemical reaction), *364*
Rated speed, 690
Ratio test, 525
Rational functions, 83
 continuity of, 84
 improper, 397
 integration of, 311, 397-407
 partial fraction expansion of, 400-406
 proper, 397
 in $\sin x$ and $\cos x$, 409
 integration of, 409-412

 in two variables, 409
Rational numbers, 2, 3-4
 decimal representation of, 5-6
 in lowest terms, 3
Rational powers, 12-13
 logarithm of, 303
Rationalizing substitution, 407
Real line, 5
 coordinates on, 16
 as 1-space, 750
Real number(s), 2
 arithmetical operations on, 6
 decimal representation of, 5-6
 negative of, 6
 reciprocal of, 6
Real number system, 5
 completeness of, 96, 494*n*
Rearrangement of series, 523
Rectangular coordinates:
 in the plane, 21-26, 453
 in space, 694-701
Rectifiable curve, 460
Recursion formulas, 489
Reduced equation, 956
Reduction formulas, 378-382
Reflecting telescopes, 586
Reflection by a plane mirror, 208-209
Reflection property:
 of an ellipse, 592-594, *789*
 of a hyperbola, *604*
 of a parabola, 584-586, *645*
RC circuit, *349*, *950*
Refraction by a plane interface, 209-210
Region:
 angularly simple, 858
 area of, 229-231, 241-244, 258-260, 811
 bounded (or finite), 426
 centroid of, 837
 horizontally simple, 811
 of integration, 810, 813, 826
 normal, 810, 811, 824
 radially simple, 858
 simple, 811
 solid (*see* Solid region)
 unbounded (or infinite), 426
 area of, 426, 428, 429
 vertically simple, 810
Regression line, 797
Regular tetrahedron, *441*
Related rates problems, 150-156
 step-by-step procedure for solving, 154
Relative error, 127, 770
Relativistic effects, *276*
Relativity:
 general theory of, *273*
 special theory of, *276*, *573*, *642*
Remainder:
 of a series, 505

Index **I-13**

Remainder (*cont.*):
  of a series (*cont.*):
    after *n* terms, 505
    of Taylor's formula, 551, 558
      integral form of, *573*
Residual concentration, 508
Resolving an indeterminacy, 89, 187
Resonance, 966, *967*
Resonant frequency, *968*
Revenue:
  average, 218
  marginal, 218
  total, 217-218
Riemann, Georg Friedrich Bernhard (1826-1866), 231
Riemann sum, 231, 235, 236, 812, 825, 858
Right circular cone, 609, 700, 740
  generators of, 610
  nappes of, 609
  vertex of, 609
  volume of, 438, 448-449
Right-hand derivative, 119
Right-hand limit, 92
Right-hand tangent, 119
Right-handed coordinate system, 694
Rise-to-run ratio, 27
RL circuit, 345-346, *348*
  steady-state current in, 346, 949
  time constant of, 346, 949
  transient current in, 346
RLC circuit, *967*
Rolle, Michel (1652-1719), 160*n*
Rolle's theorem, 160-162, 164, 165, 168
  generalized, *223*
  geometric meaning of, 160
Root (of an equation), 96
Root test, 527
Rose:
  area enclosed by, *645*
  four-petaled, 624
  general form of, *627*, *640*
  petals of, *627*
  three-petaled, *635*
Rotation of axes, 576-578, 613
Rotation equations, 577
Rotation invariance, *619*, *642*
Round-off error, 416, 418, 422, 423, 543
Rulings (*see* Generators)
Rutherford, Ernest (1871-1937), 602

S

Saddle point, 742, 790
  of a function of two variables, 792
  test for, 793
Sag (of a cable or chain), *691*

Sample values, 750
  average (or mean) of, 750
  variance of, 750
Sandwich theorem, 85
  for sequences, *497*
Satellite motion, 683-686, 730-735, *736-737*, 746
Scalar product (*see* Dot product)
Scalar triple product, 712
Scalars vs. vectors, 647
Secant function, 52
  derivative of, 132
  fundamental period of, 64
  graph of, 64
  integral of, 369
Secant line, 116
Second degree curves, 613-620
  as conics, 615-618
Second degree equation in *x* and *y*, 575, 613
  cross product term of, 613
    elimination of, 613-614
  discriminant of, *619*
    rotation invariance of, *619*
Second degree equation in *x*, *y* and *z*, 737
Second derivative test:
  for an inflection point, 178, 179
  for a local extremum, 171, 179
    generalization of, *557*, 793
Second order linear differential equation (with constant coefficients), 951-960
  existence and uniqueness theorem for, 953
  homogeneous, 951
    characteristic equation of, 954
    fundamental set of solutions of, 952
    general solution of, 952, 953-956
  nonhomogeneous, 951
    general solution of, 951, 956-959
Second partials test, 793
Second quadrant, 22
Semimajor axis, 590
Semiminor axis, 590
Separable differential equation, 333-335, 941
  exactness of, 943
  general solution of, 334
  particular solution of, 334
Separation of variables, 333
Sequence(s), 329, 487-497
  algebraic operations on, 492
  constant, 493
  convergent, 490
  decreasing, 493
    strictly, 493
  definition of, 488
  divergent, 490

  to ∞, 490
  to $-\infty$, 490
  Fibonacci, *496*
  general (or *n*th) term of, 488
  generation of, by iteration, 565
  graphic representation of, 489
  increasing, 493
    strictly, 493
  law of formation of, 488
  limit of, 489-490
  limit rules for, 492
  monotonic, 493
  recursively defined, 489
  terms of, 488
  upper bound of, 494
    least, 494
Series, 487, 497-563
  absolutely convergent, 517
  algebraic operations on, 503
  alternating, 518-523
    test for convergence of, 519
  binomial, 546-547
  Cauchy's condensation test for, *516*
  conditionally convergent, 518
  convergent, 498
  Dirichlet's test for, *525*
  divergent, 498
  general (or *n*th) term of, 498
  geometric, 499
  harmonic, 501
  and improper integrals (analogy between), 498, 513
  integral test for, 514-515
  Maclaurin, 558
  nonnegative, 508-516
    comparison test for, 510
    convergence test for, 508
    limit comparison test for, 511
  numerical, 530, 537
  partial sum of, 498
  power (*see* Power series)
  ratio test for, 525
  rearrangement of, 523
  remainder of, 505
  root test for, 527
  sum of, 498
  summation (summing) of, 498
  Taylor, 558
  terms of, 498
Set(s):
  boundary of, 758
  boundary point of, 758
  bounded, 790
  closed, 758
  (arcwise) connected, *823*, 886
  difference of, *36*
  disjoint, 2
  distinct, 2
  elements of, 1
  empty, 2

Set(s) (cont.):
  equality of, 2
  finite, 63n
  infinite, 63n
  interior point of, 757
  intersection of, 36
  members of, 1
  open, 758
  subset of, 1
    proper, 1
  union of, 36
Shannon, Claude E., *324*
Sheet(s) of a hyperboloid, 739
Shifting graphs, 50-51
Shortest ladder problem, *224*
Side conditions (*see* Constraints)
Sigma notation, 225-229, 498
Simple closed curve, 459, 886
Simple harmonic motion, 960
  in relation to uniform circular motion, 961
Simple region, 811
  double integral over, 819
Simply connected domain, 887, 935
Simpson's rule, 418-423, 545
  error of, 420-421
Sine function, 52
  derivative of, 112
  fundamental period of, 62
  graph of, 61
  integral of, 251
  Maclaurin series of, 560-561
  oddness of, 55
Singular solution, 942, 945
Sink, 927
Sinusoidal oscillations (*see* Harmonic oscillations)
Sirius, *323*
Skew lines, *721*
  condition for, *721*
  distance between, *722*
Sliding ladder problem, 151-152, *155-156*
Slope:
  of a curve, 117
  of a line, 26
  in relation to inclination, 28-29
Slope-intercept equation of a line, 30
Slug, 268n
Smooth curve, 779, 842, 879
  piecewise, 884
Smooth surface, 904
  piecewise, 905, 930
Snell's law, 210
Solar system, 732
Solenoidal field, 927
Solid (object):
  center of mass of, 835-837
  density of, 829-830
    calculation of mass from, 830

moments of, 837
moments of inertia of, 850-851
Solid region, 824
  centroid of, 837
  normal, 824
  volume of, 436-450, 814-815, 824-825, 826
  x-simple, 824
  y-simple, 824
  z-simple, 824
Solid of revolution, 442
  Pappus' theorem for, 846
  volume of, 442-450, *875*
Sound:
  intensity of, *323*
  loudness of, *323*
  speed of, *276*, *323*
Source, 927
Space curve(s), 722-730
  arc of, 722
  binormal of, *745*
  closed, *736*
  curvature of, 727-728
    radius of, 728
  moving trihedral of, *745*
  normal plane to, *736*
  osculating plane to, *735*
  rectifiable, 723
    length of, 723
  tangent line to, *736*
  torsion of, *745*
  unit normal vector to, 726
  unit tangent vector to, 725
Space shuttle, *746*
Spacetime interval, *642*
  Lorentz invariance of, *642*
Span (of a cable or chain), *691*
Special theory of relativity, *276*, *573*, *642*
Specific gravity, 481
Speed, 107, 666, 725
  angular, 685
  critical, *690*, *691*
  as derivative of arc length, 667-668, 725
  rated, *690*
Speed (velocity) of light, *273*, *276*, *323*, *604*, *642*, *940*
Speed of sound, *276*, *323*
Spheres, 697
Spherical coordinates, 868
  in relation to rectangular coordinates, 868
  triple integrals in, 867-874
Spherical segment, 442
Spherical symmetry, 867
Spherical wedge, 870
Spherical zone, *471*
Spheroid, *485*, 738
  oblate, *485*, *746*

prolate, *485*, *746*
Spiral:
  Archimedean, *628*, 634, 637, 639, *645*
  hyperbolic, 626, *646*
  logarithmic, *628*, 634, 639
  parabolic, *628*
Spread of a disease, *349*
Spread of a rumor, *349*
Spring constant, 270, 962
Spring problems, 270-271, *273*, 962-966, *967*
Square root, 11
Starr, Norton, 753
Statistics, 750, *797*
Steady-state current, 346, 949
Steady-state flow, 905, 928
Stellar magnitudes, *323*
Stiffness (*see* Spring constant)
Stirling's formula, *571*
Stokes, Sir George Gabriel (1819-1903), 931n
Stokes' theorem, 877, 930, 931
Straight line(s), 26-36
  angle between two, 59
  inclination of, 28
  intercepts of, 29-30
  oblique, 29
  parallel, 32
  parametric equations of, 453
  perpendicular, 32-33
  point-slope equation of, 30
  polar equation of, *627*
  slope of, 26
  slope-intercept equation of, 30
  in space (*see* Straight line in space)
  two-intercept equation of, 31
  two-point equation of, 31
  x-intercept of, 29
  y-intercept of, 30
Straight line(s) in space:
  direction numbers of, 716
  parametric equations of, 715-716
  skew, *721*
  symmetric equations, 716
  two-point equations of, 716-717
Strontium-90, 339
Subnormal, *643*
Subsequence, 495
Subset, 1
  proper, 1
Substitution, integration by, 252, 366, 367-373, 388-397, 407-413
Successive approximations, 563, 565
Sum:
  algebraic, 83
  of an infinite series, 498
  partial, 498
  of a power series, 537, 538
    continuity of, 540

Index I-15

Sum (*cont.*):
  of a power series (*cont.*):
    infinite differentiability of, 540
    Taylor series of, 559
  Riemann, 231, 235, 236, 812, 825, 858
  telescoping, *228*, *233*, *234*
Summation, 225
  index of, 226
  lower limit of, 225
  by parts, *229*
  upper limit of, 225
Summation sign, 225
Superelevation, *693*
Supply curve, *38*
Surface:
  function integrable on, 901
  one-sided, 905, *908*
  piecewise smooth, 905, 930
  smooth, 904
  two-sided, 905
Surface area, 466-471, 897-900
  as a double integral, 899
  as a surface integral, 901
Surface integral, 877, 897, 901
  definition of, 900-901
  evaluation of, 901-907
Surface of revolution, 466, 698-700
  area of, 466-471, 639-640, *939*
  as an integral, 468
  Pappus' theorem for, 847
Suspension bridge, 586, 687-688
Symmetric equations (of a line in space), 716
Symmetry (of a graph):
  about the line $y = x$, 579
  about the origin, 48, 179, 579, 623
  about the polar axis, 623
  about the pole, 623
  about the x-axis, 579, 623
  about the y-axis, 48, 179, 579, 623
Symmetry of two points:
  about a line, *701*
  about a plane, *701*
  about a point, *701*
Symmetry tests:
  for the graph of $F(r, \theta) = 0$, 623
  for the graph of $F(x, y) = 0$, 579

T

Tangent function, 52
  branches of, 63
  derivative of, 132
  fundamental period of, 63
  graph of, 63
  integral of, 309
Tangent line:
  to a circle, 116
  to the curve $F(x, y) = 0$, 783
  to the curve $y = f(x)$, 115-117
  to an ellipse, 593
  to a hyperbola, *604*
  inflectional, 176
  left-hand, 120
  to a parabola, 116, 117, *121*, 585
  to a parametric curve, 456-457
  to a polar curve, 631-635
  right-hand, 119
  vertical, 176, 194, 457
Tangent line approximation, 123, 159
  accuracy of, 126, 550
  as best linear approximation, *127*
  error of, *127*, 549, 550
  geometric meaning of, 123
Tangent plane:
  in terms of the gradient, 788
  to the surface $F(x, y, z) = 0$, 780
  to the surface $z = f(x, y)$, 781
Tangent plane approximation, 782
  error of, 782
Tangential component of acceleration, 680, 728
Tautochrone, 456
Tax revenue, *222*
Taylor, Brook (1685-1731), 548
Taylor polynomials, 551-552, 558, *573*
Taylor series, 487, 557-563
  convergence criterion for, 558
Taylor's formula, 551, 557
  applications of, 553-554
  remainder of, 551, 558
Taylor's theorem, 551
  proof of, 556-557
Telescoping sum, *228*, *233*, *234*
Tetrahedron, *441*, 830-831, 854
  regular, *441*
Terminal velocity, 365
Third derivative test (for an inflection point), 178, 179
Third quadrant, 22
Three-space (R^3), 750
Threshold of hearing, *323*
Threshold of pain, *323*
Torsion, *745*
Torus, 446
  surface area of, 469, 849
  volume of, 446, 848
Total energy, 892
  conservation of, 892
Tractrix, *484*
  equitangential property of, *484*
Transcendental function, *365*
Transfer orbit, *736*
Transient current, 346, 949
Transitivity (of inequalities), 9
Translation of axes, 576
Translation equations, 576
Translation invariance, *642*
Transverse axis, 597
Trapezoidal rule, 416-418
  error of, 417

Triangle inequality, 15, 649
  for integrals, *238*
  for vectors, 649
Trigonometric functions, 52-65
  continuity of, 70, 84
  definitions of, 52
  derivatives of, 131-132
  graphs of, 61, 63, 64
  inverse, 278, 290-297
Trigonometric inequalities, 64-65
Trigonometric integrals, 382-388
Trigonometric substitutions, 388-390, 392-393, 394-395
Triple integral, 824
  change of variables in, 917-918
  in cylindrical coordinates, 862-867, *921*
  definition of, 825-826
  evaluation of, 827
  rules obeyed by, 826-827
  in spherical coordinates, 867-874, *921*
Trisectrix, *628*
Truncation error, 543
Tschirnhausen's cubic, *484*
Twisted cubic, 729, *746*
Two-intercept equation of a line, 31
Two-point equation of a line, 31
Two-space (R^2), 750

U

Unbounded function, 62
  on an interval, 62
Unbounded sequence, 491
  divergence of, 492
Undamped oscillations, 963
Underdamped oscillations, 964, 965
Uniform circular motion, 684-685
  angular speed of, 685
  period of, 685
  in relation to simple harmonic motion, 961
Union of sets, *36*
Unit ball:
  closed, 697
  open, 697
Unit circle, 24
Unit hyperbola, 353, 600
  branches of, 353
Unit normal vector, 673-674, 726
  principal, 727
Unit normal vectors to a surface:
  continuity of, 904
  lower, 904
  upper, 904
Unit sphere, 697
Unit tangent vector, 669-670, 725
Unit vector, 650
Universal gas constant, 760

Universal gravitational constant, 271, *273*, 684, *690*, 730, 734, *896*
Upper bound, 494
  least, 494
Upper normal:
  to a plane, 898
  to a surface, 898
Uranium, isotopes of, *364*
U.S. Steel, 42
Utility, *809*
  marginal, *809*
  maximization of, *809*

V

Value (of a function), 40, *43*, 747
Variable:
  dependent, 39, 748
  independent, 39, 747
  of integration, 231, 250
    differential of, 367
  real, 42
Variable cost, *221*
  average, *221*
Variance, 750
Variation of parameters, 959, *969*
Vector(s):
  addition of, 648, 651, 702-703
    associativity of, 649
    commutativity of, 648
  angle between, 655
  basis, 650, 702
  component of (along another vector), 658, 705
  components of, 650, 651, 702
  cross product of (*see* Cross product)
  difference of, 649, 651, 703
  direction of, 647
  direction angles of, 705
  direction cosines of, 705
  dot product of (*see* Dot product)
  equality of, 648, 651-652, 703
  expansion of, 651, 702
  final point of, 648
  gradient, 784, 786
  i and **j**, 650, 651
    dot products of, 657
  i, j and **k**, 702
    cross products of, 708
    dot products of, 704
  initial point of, 648
  linear combination of, 653
    nontrivial, 653
    trivial, 653
  linearly dependent, 653, 706
  linearly independent, 653, 706
  magnitude of, 647, 650, 702
  negative of, 649, 651, 703
  as ordered pairs, 651
  as ordered triples, 702
  position, 652, 666, 725
  projection of (onto another vector), 658, 705
  scalar multiples of, 649-650, 651, 703
    distributive laws for, 649
  scalar triple product of, 712
  in space, 701-715
  subtraction of, 649, 651, 703
  sum of, 648, 651
  unit, 650, 702
  unit normal, 673-674, 726-727
  unit tangent, 669-670, 725
  vector triple product of, *715*
  velocity, 666, 671, 725
  zero, 649, 651, 703
Vector component:
  along another vector, 658
  orthogonal to another vector, 658
Vector field(s), 662, 749, 788, 877, 879, 880
  circulation of, 890
  conservative, 892
  continuity of, 879
  continuously differentiable, 892
  curl of, 929
  divergence of, 923
  flux of, 905-907, 922-923
  irrotational, 930
  potential of, 891
  solenoidal, 927
Vector function(s), 647, 661-665, 724
  antiderivative of, 665
  chain rule for, 669, *670*
  continuity of, 663
  definite integral of, 665, *671*
  derivative of, 663
  differentiability of, 663
  differentiation of, 663-665
  fundamental theorem of calculus for, 665
  indefinite integral of, 665
  integration of, 665
  limit of, 662-663
  of several variables, 749
  in space, 724
Vector product (*see* Cross product)
Vector triple product, *715*
  bac-cab rule for, *715*
Velocity:
  average, 106, *166*
  instantaneous, 106, *166*, 261, 266
    integration of, 261-262
  in the plane, 666
  in space, 725
Velocity field, 905, 922, 927
  of a rotating fluid, 935-936
Velocity parameter, *642*
Vertex:
  of an ellipse, 589
  of a hyperbola, 597
  of a parabola, 581
Vertical asymptote, 192
Vertical line property, 46, 451
Vertical tangent, 176, 194, 457
Vertically simple region, 810
  double integral over, 817
Volume:
  between two surfaces, 815
  of a conical frustum, *440*
  of a general cone, *441*
  by the method of cross sections, 436-442
  of a napkin ring, 445
  of a normal region in space, 824-825
    as a triple integral, 826
  of a parallelepiped, 712
  of a pyramid, 438-439
  of a right circular cone, 438, 448-449
  of a solid of revolution, 442-450, *875*
    by the method of disks, 442-444
    by the method of shells, 446-450
    by the method of washers, 444-446
  of a solid sphere, 443
  of a spherical segment, *442*
  of a tetrahedron, *441*
  of a torus, 446, 848
  under a surface, 814-815
    as a double integral, 815

W

Wallis, John (1616-1703), *497*
Wallis' formula for π, *497*
Water resistance, *348*
Watt, 474
Wave equation, *764*, 940
Weather maps, 752
Weierstrass, Karl Theodor Wilhelm (1815-1897), 72
Weight, 267
  vs. mass, 268*n*
Whispering gallery, 594
White dwarf, *273*
Width of a region between two curves, *247*
Wire:
  center of mass of, 844
  density of, 842
  mass of, 844
Witch of Agnesi, *484*
Work, 269, *274*, 471-477, 659-660, 877, 879-881
  done in compressing a gas, 475-476
  done by a constant force, 270, 474
  as a dot product, 659-660
  done by earth's gravitational pull, 271
  done by an expanding gas, 474-476
  as a line integral, 877, 879-881
  done in pulling up a rope, 472
  done in pumping a fluid, 472-474, 842

Index **I-17**

X

x-axis, 21
   negative, 22
   positive, 22
   symmetry about, 579, 623
x-coordinate, 21, 695
x-intercept:
   of a curve, 179
   of a line, 29
   of a plane, *744*
x-simple region, 824
   triple integral over, 827
xy-plane, 21, 694
xz-plane, 694

Y

y-axis, 21
   negative, 22
   positive, 22
   symmetry about, 48, 179, 579, 623
y-coordinate, 21, 695
y-intercept:
   of a line, 30
   of a plane, *744*
Young's inequality, *298*
y-simple region, 824
   triple integral over, 827
yz-plane, 694

Z

z-coordinate, 695
Zeno's paradox, *507*
Zero, 2
   exclusion of division by, 4
Zero factorial, 227
Zero vector, 649, 651, 703
Zeroth term, 226, 499, 530
z-intercept (of a plane), *744*
z-simple region, 824
   triple integral over, 827

57. $\int \arcsec x \, dx = x \arcsec x - \ln|x + \sqrt{x^2 - 1}| + C$ 58. $\int \arccsc x \, dx = x \arccsc x + \ln|x + \sqrt{x^2 - 1}| + C$

59. $\int x \arcsin x \, dx = \frac{1}{4}(2x^2 - 1) \arcsin x + \frac{x}{4}\sqrt{1 - x^2} + C$

60. $\int x \arctan x \, dx = \frac{1}{2}(x^2 + 1) \arctan x - \frac{x}{2} + C$

Logarithms and Exponentials

61. $\int \ln x \, dx = x \ln x - x + C$

62. $\int x^n \ln x \, dx = \frac{x^{n+1}}{n+1}\left(\ln x - \frac{1}{n+1}\right) + C$

63. $\int \frac{dx}{x \ln x} = \ln|\ln x| + C$

64. $\int (\ln x)^n \, dx = x(\ln x)^n - n \int (\ln x)^{n-1} \, dx$

65. $\int x e^x \, dx = x e^x - x + C$ 66. $\int x^n e^x \, dx = x^n e^x - n \int x^{n-1} e^x \, dx$

67. $\int e^{ax} \cos bx \, dx = e^{ax} \frac{a \cos bx + b \sin bx}{a^2 + b^2} + C$ 68. $\int e^{ax} \sin bx \, dx = e^{ax} \frac{a \sin bx - b \cos bx}{a^2 + b^2} + C$

Hyperbolic Functions

69. $\int \tanh x \, dx = \ln(\cosh x) + C$ 70. $\int \coth x \, dx = \ln|\sinh x| + C$

71. $\int \sech x \, dx = 2 \arctan(e^x) + C$ 72. $\int \csch x \, dx = \ln\left|\tanh \frac{x}{2}\right| + C$

73. $\int \sinh^2 x \, dx = \frac{1}{4} \sinh 2x - \frac{x}{2} + C$ 74. $\int \cosh^2 x \, dx = \frac{1}{4} \sinh 2x + \frac{x}{2} + C$

75. $\int \tanh^2 x \, dx = x - \tanh x + C$ 76. $\int \coth^2 x \, dx = x - \coth x + C$

77. $\int e^{ax} \cosh bx \, dx = e^{ax} \frac{a \cosh bx - b \sinh bx}{a^2 - b^2} + C$ 78. $\int e^{ax} \sinh bx \, dx = e^{ax} \frac{a \sinh bx - b \cosh bx}{a^2 - b^2} + C$

Irrational Functions

79. $\int x^2 \sqrt{a^2 - x^2} \, dx = \frac{x}{8}(2x^2 - a^2)\sqrt{a^2 - x^2} + \frac{a^4}{8} \arcsin \frac{x}{a} + C$

80. $\int x^2 \sqrt{x^2 \pm a^2} \, dx = \frac{x}{8}(2x^2 \pm a^2)\sqrt{x^2 \pm a^2} - \frac{a^4}{8} \ln|x + \sqrt{x^2 \pm a^2}| + C$

81. $\int \frac{\sqrt{a^2 \pm x^2}}{x} \, dx = \sqrt{a^2 \pm x^2} - a \ln\left|\frac{a + \sqrt{a^2 \pm x^2}}{x}\right| + C$

35. $\int \tan^3 x \, dx = \frac{1}{2} \tan^2 x + \ln |\cos x| + C$

36. $\int \cot^3 x \, dx = -\frac{1}{2} \cot^2 x - \ln |\sin x| + C$

37. $\int \sec^3 x \, dx = \frac{1}{2} \sec x \tan x + \frac{1}{2} \ln |\sec x + \tan x| + C$

38. $\int \csc^3 x \, dx = -\frac{1}{2} \csc x \cot x + \frac{1}{2} \ln |\csc x - \cot x| + C$

39. $\int \sin ax \sin bx \, dx = \frac{\sin (a-b)x}{2(a-b)} - \frac{\sin (a+b)x}{2(a+b)} + C$

40. $\int \cos ax \cos bx \, dx = \frac{\sin (a-b)x}{2(a-b)} + \frac{\sin (a+b)x}{2(a+b)} + C$

41. $\int \sin ax \cos bx \, dx = -\frac{\cos (a-b)x}{2(a-b)} - \frac{\cos (a+b)x}{2(a+b)} + C$

42. $\int \sin^n x \, dx = -\frac{1}{n} \sin^{n-1} x \cos x + \frac{n-1}{n} \int \sin^{n-2} x \, dx$

43. $\int \cos^n x \, dx = \frac{1}{n} \cos^{n-1} x \sin x + \frac{n-1}{n} \int \cos^{n-2} x \, dx$

44. $\int \tan^n x \, dx = \frac{1}{n-1} \tan^{n-1} x - \int \tan^{n-2} x \, dx$

45. $\int \cot^n x \, dx = -\frac{1}{n-1} \cot^{n-1} x - \int \cot^{n-2} x \, dx$

46. $\int \sec^n x \, dx = \frac{1}{n-1} \sec^{n-2} x \tan x + \frac{n-2}{n-1} \int \sec^{n-2} x \, dx$

47. $\int \csc^n x \, dx = -\frac{1}{n-1} \csc^{n-2} x \cot x + \frac{n-2}{n-1} \int \csc^{n-2} x \, dx$

48a. $\int \sin^m x \cos^n x \, dx = -\frac{\sin^{m-1} x \cos^{n+1} x}{m+n} + \frac{m-1}{m+n} \int \sin^{m-2} x \cos^n x \, dx$

48b. $\int \sin^m x \cos^n x \, dx = \frac{\sin^{m+1} x \cos^{n-1} x}{m+n} + \frac{n-1}{m+n} \int \sin^m x \cos^{n-2} x \, dx$

49. $\int x \sin x \, dx = \sin x - x \cos x + C$

50. $\int x \cos x \, dx = \cos x + x \sin x + C$

51. $\int x^n \sin x \, dx = -x^n \cos x + n \int x^{n-1} \cos x \, dx$

52. $\int x^n \cos x \, dx = x^n \sin x - n \int x^{n-1} \sin x \, dx$

Inverse Trigonometric Functions

53. $\int \arcsin x \, dx = x \arcsin x + \sqrt{1 - x^2} + C$

54. $\int \arccos x \, dx = x \arccos x - \sqrt{1 - x^2} + C$

55. $\int \arctan x \, dx = x \arctan x - \frac{1}{2} \ln (x^2 + 1) + C$

56. $\int \text{arccot } x \, dx = x \, \text{arccot } x + \frac{1}{2} \ln (x^2 + 1) + C$